W9-AQF-197

INDEX OF TABLES

UNIVERSITY PHYSICS

SEVENTH EDITION

The end-of-chapter Problem Sets were revised for the *Seventh Edition* by A. Lewis Ford, Texas A&M University.

Contributions to the Problem Sets were made by Lawrence B. Coleman, University of California, Davis; James L. Monroe, Pennsylvania State University, The Beaver Campus; Terry F. O'Dwyer, Nassau Community College.

Craig Watkins, Massachusetts Institute of Technology, provided assistance to both Professor Young and to Professor Ford in the development of the manuscript.

UNIVERSITY PHYSICS

SEVENTH EDITION

Francis W. Sears

Late Professor Emeritus
Dartmouth College

Mark W. Zemansky

Late Professor Emeritus
City College of the City University of New York

Hugh D. Young

Professor of Physics
Carnegie-Mellon University

ADDISON-WESLEY PUBLISHING COMPANY
Reading, Massachusetts ▪ Menlo Park, California
Don Mills, Ontario ▪ Wokingham, England ▪ Amsterdam
Sydney ▪ Singapore ▪ Tokyo ▪ Madrid ▪ Bogotá
Santiago ▪ San Juan

This book is in the
Addison-Wesley Series in Physics

Sponsoring Editor: Bruce Spatz, Debra Hunter, Steve Mautner
Developmental Editor: David M. Chelton
Production Supervisor: Marion E. Howe
Copy Editor: Jacqueline M. Dormitzer
Text Designer: Catherine L. Dorin
Layout Artist: Lorraine Hodsdon
Illustrators: Oxford Illustrators, Ltd.
Art Consultant: Loretta Bailey
Manufacturing Supervisor: Ann DeLacey
Cover: Marshall Henrichs

Photo credits: page 1, NASA. 3, Lockheed Missiles and Space Company. 25, Education
Development Center. 49 and 71, Dr. Harold Edgerton, M.I.T., Cambridge MA. 90, National
Center for Atmospheric Research/National Science Foundation. 119, AP/Wide World Photos.
143, Dr. Harold Edgerton, M.I.T., Cambridge MA. 145, Matson Navigation Co. 182,
Education Development Center. 210, R. V. Willstrop/Anglo-Australian Telescope Board. 249,
Golden Gate Bridge, Highway, and Transportation District. 263, Dr. Harold Edgerton,
M.I.T., Cambridge MA. 289, U.S. Geological Survey. 291, M. S. Paterson, Australian National
University. 306, Department of Aeronautics, Imperial College of Science and Technology.
340, N.Y. State Department of Commerce. 357, Nancy Rodger/Exploratorium. 374, Lockheed
Corp. 389, Barry L. Runk/Grant Heilman Photography. 403, Pacific Gas & Electric. 424, Kurt
Rogers, San Francisco Examiner. 451, Lawrence Berkeley Laboratory. 473, M. P. Moller, Inc.
475, Fundamental Photographs. 496, The Exploratorium. 512, Steinway & Sons. 527, AT&T
Bell Laboratories. 529, Lockyer Collection. 546, Education Development Center. 575,
Wellcome Institute for the History of Medicine. 604, Chip Clark. 625, American Institute of
Physics. 653 and 681, AT&T Bell Laboratories. 683, Fundamental Photographs. 714, Pacific
Gas & Electric. 742, Chip Clark. 767, Varian Associates, Inc. 788, Pacific Gas & Electric. 813,
NASA. 835, Bausch & Lomb Optical Co. 837, Corning Glass Works. 868, Chip Clark. 887,
Palomar Observatory. 953, Lawrence Livermore National Laboratory. 955, U.S. Dept. of
Energy. 980, IBM Corp. 1007, Manfred Kage/Peter Arnold, Inc. 1036, Chip Clark. 1060,
Stanford Linear Accelerator.

Library of Congress Cataloging-in-Publication Data

Sears, Francis Weston, 1898–
    University physics.

    Includes index.
    1. Physics.   I. Zemansky, Mark Waldo, 1900–
II. Young, Hugh D.   III. Title.
QC21.2.S36 1986   530   85–28801
ISBN 0-201-06681-5

ABCDEFGHIJ-MU-8987

PREFACE

In this new edition of *University Physics*, we have tried to preserve those qualities and features that users of previous editions have found useful. Yet this is the most comprehensive revision in the long, successful history of the book. Physics courses and physics students have changed substantially in recent years, and new editions must keep pace with these changes.

Our basic goals have not changed. Our objective is to provide a broad, rigorous introduction to physics at the beginning college level. This book is appropriate for students of science and engineering who are taking an introductory calculus course concurrently. We place primary emphasis on physical principles and the development of problem-solving ability, rather than on historical background or specialized applications. The complete text may be taught in an intensive two- or three-semester course, and the book is also adaptable to a wide variety of shorter courses. It is available as a single volume or as two volumes. Volume I includes mechanics, heat, and mechanical waves, and Volume II includes electricity and magnetism, optics, and atomic and nuclear physics.

Here are some of the most important new features in this edition:

Table of Contents. After careful consideration and consultation with many users of our book, we have reorganized the chapters on mechanics. We now conform to the usual order in introductory courses, beginning with kinematics and dynamics and treating statics later as a special case of dynamics. Getting students into the study of motion immediately helps to build motivation for the study of physics, and it also helps to tie the physics course in with the often-concurrent calculus course.

Problems. The end-of-chapter problem collections have been extensively revised and augmented. We now group each collection into three categories: *Exercises*, single-concept problems that are keyed to specific sections of the text; *Problems*, usually requiring two or more nontrivial steps for their solution; and *Challenge Problems*, intended to challenge the strongest students. The number of problems has grown by 12%; the total number is now approximately 1800, of which over 25% are new. We have also added to the lists of thought-provoking questions at the ends of the chapters, about 700 questions in all. The revision of the problem collections was carried out by Professor A.

Lewis Ford (Texas A. & M. University), with the assistance of Mr. Craig Watkins (Massachusetts Institute of Technology). Additional problems were contributed by Professors Lawrence B. Coleman (University of California, Davis), James L. Monroe (Pennsylvania State University, The Beaver Campus), and Terry F. O'Dwyer (Nassau Community College).

Problem-Solving Strategies. The remark heard most often in the freshman Physics classroom is: "I understood the material, but I couldn't do the problems!" To respond to this universal cry for help, we have included in each chapter one or more sections called *Problem-Solving Strategy,* where we list suggestions for developing a methodical and systematic approach to solving problems. Our most important objective in this book is to help students learn to apply physical principles to a wide variety of problems, and these new strategy sections should be a substantial help.

Chapter Summaries. Each chapter concludes with a list of *Key terms,* which have been highlighted in boldface type in the text, and a *summary,* in prose and equations, of the most important principles presented in the chapter. These will be a useful aid for the student, especially in identifying and emphasizing the concepts and relationships that are of central importance.

Chapter Introductions and Perspectives. Each chapter now begins with an introductory paragraph summarizing briefly the content of the chapter and relating it to what has come before. In addition, a *Prospectus* and seven *Perspectives* are included at intervals throughout the book. Their object is to enhance continuity by looking both backward and forward to show how various areas of physics are interrelated and to exhibit as clearly as possible the beauty and fundamental unity of all branches of physics.

Study Notes. Brief notes inserted in the margins act as references for the reader and identify key ideas and concepts in the text. They help the student locate quickly a particular discussion or example, and they provide capsule summaries of paragraphs and sections of text, useful for re-study and review of troublesome areas.

Mathematical Level. The level of mathematical sophistication has not changed substantially, but we have increased somewhat the use of unit vectors and calculus in worked-out examples. There are also more problems providing opportunities to use unit vectors and calculus.

Changes in subject matter. Many topics have been added or treated in greater depth than in previous editions. A partial list includes:

> estimates and orders of magnitude
> gravitational field
> automotive power
> damped and forced oscillations
> Maxwell–Boltzmann distribution
> sign conventions for Kirchhoff's rules
> Maxwell's equations
> circular apertures and resolving power
> relativistic Doppler effect
> Planck radiation law
> superconductivity
> band theory of solids

The elasticity chapter has been rewritten so that each type of stress is introduced along with its corresponding strain. The treatment of fluid mechanics has been condensed and reduced from two chapters to one. The material on acoustic phenomena has been reduced, as has the treatment of magnetic materials. The discussion of polarization of light has been shortened and incorporated into the chapter on Nature and Propagation of Light. The material on refracting surfaces has been reorganized so that the thin-lens equation can be presented earlier.

Units and Notation. We have moved closer to 100% SI units. English units are retained in a few examples and problems in the first half of the text, but SI units are used exclusively in the second half. We use the joule as the standard unit of energy of all forms, including heat. In examples, units are always carried through all stages of numerical calculations. As usual, boldface symbols are used for vector quantities, and in addition boldface +, −, and = signs are used to remind the student at every opportunity of the crucial distinctions between operations with vectors and those with numbers.

Supplements. A textbook should stand on its own feet. Yet some students benefit from supplementary materials designed to be used with the text. With this thought in mind, we offer a *Study Guide* and a *Solutions Manual.* The *Study Guide,* prepared by Professors James R. Gaines and William F. Palmer, includes for each chapter & statement of objectives, a review of central concepts, problem-solving hints, additional worked-out examples, and a short quiz. The *Solutions Guide,* prepared by Professor A. Lewis Ford, includes completely worked-out solutions for about one third of the problems in the book, drawn from odd-numbered problems only. Answers to all odd-numbered problems are listed at the end of the text, and a booklet containing answers to all even-numbered problems can be obtained by instructors from the publisher.

Reviewers. Many of the changes in this edition are a direct result of recommendations from colleagues who have used earlier editions with their students. The views and suggestions collected through reviews, written questionnaires, telephone surveys, and discussion groups have been invaluable. In addition to those named in connection with the problem revisions, I gratefully acknowledge the very helpful and valuable contributions of the following reviewers and discussion-group participants:

Alex Azima, Lansing Community College
Dilip Balamore, Nassau Community College
Arun Bansil, Northeastern University
Albert Bartlett, University of Colorado
Lev I. Berger, San Diego State University
James Brooks, Boston University
Nicholas E. Brown, California Polytechnic University–San Luis Obispo
Hans Courant, University of Minnesota
Gayl Cook, University of Colorado
Bruce A. Craver, University of Dayton
Steve Detweiler, University of Florida
Lewis Ford, Texas A & M University
Walter S. Gray, University of Michigan
Graham D. Gutsche, U. S. Naval Academy–Annapolis
Michael J. Harrison, Michigan State University
Howard Hayden, University of Connecticut
Lorella Jones, University of Illinois
Jean P. Krisch, University of Michigan
Alfred Leitner, Rensselaer Polytechnic University
David Markowitz, University of Connecticut
Joseph L. McCauley, University of Houston
T. K. McCubbin, Jr., Pennsylvania State University
Thomas Meyer, Texas A & M University
Herbert Muether, S.U.N.Y.–Stony Brook
Jack Munsee, California State University–Long Beach

Lorenzo Narducci, Drexel University
Van E. Neie, Purdue University
David A. Nordling, U. S. Naval Academy–Annapolis
W. F. Parks, University of Missouri
Arnold Perlmutter, University of Miami
John S. Risley, North Carolina State University
Richard Roth, Eastern Michigan University
Rajarshi Roy, Georgia Institute of Technology
Russell A. Roy, Santa Fe Community College
Stan Shepherd, Pennsylvania State University
Malcolm Smith, University of Lowell
James Stith, U. S. Military Academy–West Point
Conley Stutz, Bradley University
G. David Toot, Alfred University
George Williams, University of Utah
John Williams, Auburn University
D. H. Ziebell, Manatee Community College
George O. Zimmerman, Boston University

With the departure of Professors Sears and Zemansky from this life, I have assumed sole responsibility for the book. I feel a little like a violinmaker who is asked to take a Stradivarius apart and repair it. It is an honor to be asked to do it, but it is also an awesome responsibility. I have taken great care to remain true to the original spirit of this text, while making it as useful for today's students as the first edition was for its users a few decades ago.

Acknowledgments. A special debt of gratitude is owed to the author's colleagues at Carnegie-Mellon, especially Professors Robert Kraemer, Bruce Sherwood, and Helmut Vogel, for many stimulating discussions about physics pedagogy, and to Professor Kraemer for major contributions to the high-energy physics material. An equally important debt of a different kind is owed to Dr. Michael Schur for his support and encouragement when they were most needed. Finally and most important, I offer my gratitude to my wife Alice and our children Gretchen and Rebecca for their love, support, and emotional sustenance during a difficult period in my life. May all men be blessed with love such as theirs.

As always, I welcome communications from students and professors, especially when they concern errors or deficiencies that may be found in this edition. I have written the best book I know how to write; I hope it will help you to teach and learn physics. In turn, you can help me by letting me know what still needs to be improved!

Pittsburgh, Pennsylvania   H. D. Y.
November 1986

CONTENTS

28

CURRENT, RESISTANCE, AND ELECTROMOTIVE FORCE 625

29

DIRECT-CURRENT CIRCUITS 653

PART SIX

ELECTRODYNAMICS 681

30

MAGNETIC FIELDS AND MAGNETIC FORCES 683

31

SOURCES OF MAGNETIC FIELD

32

ELECTROMAGNETIC INDUCTION

33

INDUCTANCE

34

ALTERNATING CURRENTS

MECHANICS—FUNDAMENTALS

PROSPECTUS

The study of physics is an adventure. It is challenging, sometimes frustrating, occasionally painful, and often richly rewarding and satisfying. It appeals to the emotions and the aesthetic sense as well as to the intellect. The achievements of such scientific giants as Galileo, Newton, Maxwell, and Einstein form the foundation for our present understanding of the physical world. You can share the excitement of discovery that they experienced when you learn the value of physics in solving practical problems and in gaining insight into everyday phenomena, and its significance as an achievement of the human intellect in its quest for understanding of the world we all live in.

We begin our study of physics with the subject of *mechanics:* the study of motion and its causes. This is a natural starting point; everyday experience offers abundant examples of mechanical principles, more than for any other area of physics. In the opening chapter we introduce several elements of the language of physics, including units, calculational techniques, and vector algebra. In the following chapters we develop detailed language for describing motion. We begin with motion of a single particle, which is a body with no size and no shape; we represent such a body as a geometric point. The simplest case is motion of a point along a straight line. We then progress to examples of motion in a plane, including motion in parabolic and circular paths. Next we consider the relationships between motion and the forces that are always associated with it; whenever a particle speeds up, slows down, or changes the direction of its motion, there is always an associated force. The relationships of force and motion are summarized neatly in Newton's three laws of motion.

In studying these laws, we also study the interplay of theory and experiment; every physical theory must be grounded in experimental observations of phenomena in the physical world. As we learn how to apply these laws to a variety of practical problems, we develop systematic problem-solving procedures that help up set up problems and carry out solutions efficiently and accurately. We also begin to appreciate the role of idealized models: approximate representations of physical situations that are simplifed to facilitate analysis and calculation. Equally important, we begin to develop a sense of the beauty of physics as we learn how the essential relationships in the entire area of mechanics are wrapped up in a wonderfully neat and compact package that we label "Newton's laws of motion."

1
UNITS, PHYSICAL QUANTITIES, AND VECTORS

WE BEGIN OUR STUDY OF PHYSICS WITH SEVERAL IMPORTANT MATTERS of language. The first of these is the concept of *units*. To describe physical phenomena in a quantitative way, we need to use numbers; we usually describe a quantity as a multiple of some standard unit of that quantity. In this chapter we study various unit systems, the standard notations used with them, and procedures for converting quantities from one set of units to another. Closely related to these concepts are the ideas of *precision* of a measurement and its number of *significant figures*. We also study examples of problems where no precise calculations are possible but where rough *estimates* can be interesting and useful. Finally, we study several aspects of *vector algebra*. Vectors are mathematical entities that we use to describe physical quantities having directions in space as well as numerical magnitudes. As such, vectors are an essential part of the language of all areas of physics; it is important to learn this language thoroughly, and the other items we mentioned, at the beginning.

1–1 INTRODUCTION

Physics is an experimental science. Everything we know about the physical world and about the principles that govern its behavior has been learned through experiment, that is, through observations of the phenomena of nature. The ultimate test of any physical theory is its agreement with experimental observations. These observations usually involve measurements; thus physics is inherently a science of *experiment* and *measurement*. Figure 1–1 shows two well-known experimental facilities.

To learn the laws of nature, we must *observe* nature.

Any number used to describe a physical phenomenon quantitatively is called a **physical quantity.** A physical quantity is defined in one of two ways: We may specify a procedure for *measuring* the quantity, or we may describe a way to *calculate* the quantity from other quantities that we can measure. For example, in the first case we might use a ruler to measure a distance or a stopwatch to measure a time interval. In the second case we might define the average speed of a moving object as the distance traveled (measured with a ruler) divided by the time of travel (measured with a stopwatch).

3

(a)

(b)

1–1 Two research laboratories. (a) The Cathedral of Pisa (Italy), with the Baptistry in the foreground, the Cathedral behind it, and the famous Leaning Tower at the far right. According to legend, Galileo studied the motion of freely falling bodies by dropping them from the tower. He is also alleged to have gained insights into pendulum motion by observing the swinging of the hanging chandeliers in the Cathedral. The nearly parabolic dome of the Baptistry produces interesting acoustical effects. (Art Resource) (b) The Hubble Space Telescope. When this 2.4-m reflecting telescope is placed in orbit 500 km above the surface of the earth, it will permit observation of celestial objects 100,000,000 times fainter in brightness than the faintest objects visible with the best earth-based telescope. (Courtesy Lockheed Missiles and Space Company)

Operational definitions: describing how to measure a quantity

A definition that gives a procedure for measuring the defined quantity is called an **operational definition.** Some quantities, such as mass, length, and time, are so fundamental that they can be defined only with operational definitions. Later we will encounter operational definitions of other fundamental quantities such as temperature (Chapter 14) and electric current (Chapter 31).

1–2 STANDARDS AND UNITS

Units: standards for describing magnitudes of physical quantities

When we measure a quantity, we always compare it with some reference standard. When we say a rope is 30 meters long, we mean that it is 30 times as long as an object, such as a meter stick, that is defined to be one meter long. Such a standard is called a **unit** of the quantity. Thus the meter is a unit of distance, and the second is a unit of time.

To make precise measurements we need definitions of the units of measurement that do not change and that can be duplicated by observers in various locations. When the metric system was established in 1791 by the Paris Academy of Sciences, the **meter** was originally defined as one ten-millionth of the distance from the equator to the North Pole, and the **second** as the time for a pendulum one meter long to swing from one side to the other.

The International System of units (SI)

These definitions were cumbersome and hard to duplicate precisely; in more recent years they have been replaced by more refined definitions. Since 1889 the definitions of the basic units have been established by an international organization, the General Conference on Weights and Measures. The system of units defined by this organization is based on the metric system, and since 1960 it has been known officially as the **International System,** or SI (the abbreviation for the French equivalent, Système International).

Until 1960 the unit of time was based on a certain fraction of the mean solar day, the average time interval between successive arrivals of the sun at its highest point in the sky. The present standard, adopted in 1967, is an atomic one, based on the two lowest energy states of the cesium atom. These two states have slightly different energies, depending on whether the spin of the outermost electron is parallel or antiparallel to the nuclear spin. Electromagnetic radiation (microwaves) of precisely the proper frequency causes transitions from one state to the other. We now define one second as the time required for 9,192,631,770 cycles of this radiation. Figure 1–2 shows the instrument currently used to establish this standard.

Definition of the second, the unit of time

In 1960 the meter was defined by an atomic standard, in terms of the wavelength of the orange-red light emitted by atoms of krypton (^{86}Kr) in a glow discharge tube; one meter was defined as 1,650,763.73 of these wavelengths. In November 1983 the standard was changed again in a more radical way. The new definition of the meter is the distance light travels in 1/299,792,458 second. This has the effect of defining the speed of light to be precisely 299,792,458 m·s^{-1}; we then define the meter to be consistent with this number and with the definition of the second given above. The reason for this change is that at present we can measure the speed of light and intervals of time much more precisely than distances.

Definition of the meter, the unit of length

The standard of *mass* is the mass of a particular cylinder of platinum-iridium alloy. Its mass is defined to be one **kilogram,** and it is kept at the International Bureau of Weights and Measures at Sèvres, near Paris. An atomic standard of mass has not yet been adopted because at present we cannot measure masses on an atomic scale with as much precision as on a macroscopic scale.

Definition of the kilogram, the unit of mass

Once the fundamental units are defined, it is easy to introduce larger and smaller units for the same physical quantities. In all versions of the metric system, including SI, these other units are always related to the fundamental units by multiples of 10 or 1/10. Thus one kilometer (1 km) is 1000 meters,

1–2 NBS-6 is the latest of six generations of primary atomic frequency standards developed by the National Bureau of Standards (NBS). Consisting of a 6-m cesium beam tube, NBS-6 achieves an accuracy of better than one part in 10^{13}, and when operated as a clock, can keep time to within 3 millionths of a second per year. (Courtesy National Bureau of Standards.)

TABLE 1–1 PREFIXES FOR POWERS OF TEN

Power of ten	10^{-18}	10^{-15}	10^{-12}	10^{-9}	10^{-6}	10^{-3}	10^{-2}	10^{3}	10^{6}	10^{9}	10^{12}	10^{15}	10^{18}
Prefix	atto-	femto-	pico-	nano-	micro-	milli-	centi-	kilo-	mega-	giga-	tera-	peta-	exa-
Abbreviation	a	f	p	n	μ	m	c	k	M	G	T	P	E

How to use unit prefixes

one centimeter (1 cm) is 1/100 meter, and so on. We usually express the multiplicative factors in exponential notation; thus $1000 = 10^3$, $1/1000 = 10^{-3}$, and so on. The names of the additional units are always derived by adding a **prefix** to the name of the fundamental unit. For example, the prefix "kilo-", abbreviated k, always means a unit larger by a factor of 1000; thus

$$1 \text{ kilometer} = 1 \text{ km} = 10^3 \text{ meters} = 10^3 \text{ m},$$
$$1 \text{ kilogram} = 1 \text{ kg} = 10^3 \text{ grams} = 10^3 \text{ g},$$
$$1 \text{ kilowatt} = 1 \text{ kW} = 10^3 \text{ watts} = 10^3 \text{ W}.$$

Table 1–1 lists the standard SI prefixes with their meanings and abbreviations. We note that most of these are multiples of 10^3.

When pronouncing unit names with prefixes, we always accent the first syllable; some examples are KIL-o-gram, KIL-o-meter, CEN-ti-meter, and MIC-ro-meter.

Here are several examples of the use of multiples of 10 and their prefixes. Some additional time units are also included.

$$1 \text{ nanometer} = 1 \text{ nm} = 10^{-9} \text{ m (a few times the size of an atom)}$$
$$1 \text{ micrometer} = 1 \text{ } \mu\text{m} = 10^{-6} \text{ m (size of some bacteria and cells)}$$
$$1 \text{ millimeter} = 1 \text{ mm} = 10^{-3} \text{ m (point of a ballpoint pen)}$$
$$1 \text{ centimeter} = 1 \text{ cm} = 10^{-2} \text{ m (diameter of your little finger)}$$
$$1 \text{ kilometer} = 1 \text{ km} = 10^{3} \text{ m (a 10-minute walk)}$$
$$1 \text{ microgram} = 1 \text{ } \mu\text{g} = 10^{-9} \text{ kg}$$
$$1 \text{ milligram} = 1 \text{ mg} = 10^{-6} \text{ kg}$$
$$1 \text{ gram} = 1 \text{ g} = 10^{-3} \text{ kg (mass of a paper clip)}$$
$$1 \text{ nanosecond} = 1 \text{ ns} = 10^{-9} \text{ s (time for light to travel 0.3 m)}$$
$$1 \text{ microsecond} = 1 \text{ } \mu\text{s} = 10^{-6} \text{ s}$$
$$1 \text{ millisecond} = 1 \text{ ms} = 10^{-3} \text{ s (time for sound to travel 0.35 m)}$$
$$1 \text{ minute} = 1 \text{ min} = 60 \text{ s}$$
$$1 \text{ hour} = 1 \text{ hr} = 3600 \text{ s}$$
$$1 \text{ day} = 1 \text{ da} = 86,400 \text{ s}$$

The British system of units

Finally, we should mention the British system of units. These units are used only in the United States and a few other countries, and they are rapidly being replaced by SI in the latter. British units are now officially defined in terms of SI units, as follows:

Length: 1 inch = 2.54 cm (exactly).
Force: 1 pound = 4.448221615260 newtons (exactly).

The fundamental British unit of time is the second, defined the same way as in SI. In physics, British units are used only in mechanics and thermodynamics; there is no British system of electrical units. In this book we use SI units for all examples and problems, but occasionally in the early chapters we give approximate equivalents in British units. A few of the exercises also use British units.

1–3 UNIT CONSISTENCY AND CONVERSIONS

We often use equations to express relations among physical quantities that are represented by algebraic symbols. An algebraic symbol always denotes both a number and a unit. For example, d might represent a distance of 10 m, t a time of 5 s, and v (for velocity) a speed of 2 m/s or 2 m·s^{-1}. (In this book we usually use negative exponents with units to avoid use of the fraction bar.)

An equation must always be **dimensionally consistent;** this means that two terms may be added or equated only if they have the same units. For example, if a body moving with constant speed v travels a distance d in a time t, these quantities are related by the equation

$$d = vt. \qquad (1-1)$$

Unit consistency: You can't add apples and artichokes.

If d is measured in meters, then the product vt must also be expressed in meters. Using the numbers above as an example, we may write

$$10 \text{ m} = (2 \text{ m·s}^{-1})(5 \text{ s}).$$

Because the unit s^{-1} or 1/s cancels the unit s on the right side, the product vt is indeed expressed in meters, as it must be. In calculations, units are always treated just like algebraic symbols with respect to multiplication and division.

When a problem requires calculations using numbers with units, the numbers should always be written with the correct units, and the units should be carried through the calculation as in the example above. This provides a very useful check for calculations. If at some stage in a calculation you find that an equation or an expression has inconsistent units, you know you have made an error somewhere. In this book we will always carry units through all calculations, and we strongly urge you to follow this practice when you solve problems.

PROBLEM-SOLVING STRATEGY: *Unit conversions*

Units are multiplied and divided just like ordinary algebraic symbols. This fact provides a convenient procedure for converting a quantity from one set of units to another. The key to the procedure is the fact that we can use equality to represent the same physical quantity when we express it in two different units. For example, to say that 1 min = 60 s does not mean that the number 1 is equal to the number 60; it means that 1 min represents the same physical time interval as 60 s. Thus we may multiply a quantity by 1 min and then divide it by 60 s, or multiply by the quantity (1 min/60 s), without changing its physical meaning. To find the number of seconds in 3 min, we write

$$3 \text{ min} = (3 \text{ min})\left(\frac{60 \text{ s}}{1 \text{ min}}\right) = 180 \text{ s}.$$

This procedure is sometimes called the **factor-label method.**

EXAMPLE 1–1 American women in the age group 19 to 22 years have an average height of 5 ft, 4 in. What is this height in centimeters? In meters?

SOLUTION We first express the height in inches:

$$5 \text{ ft} = \left(\frac{12 \text{ in.}}{1 \text{ ft}}\right) 5 \text{ ft} = 60 \text{ in.}$$

5 ft, 4 in. = 5 ft + 4 in. = 60 in. + 4 in. = 64 in.

Then

$$64 \text{ in.} = \left(\frac{2.54 \text{ cm}}{1 \text{ in.}}\right) 64 \text{ in.} = 163 \text{ cm}$$

(The product has been rounded to the nearest centimeter.) Finally,

$$163 \text{ cm} = \left(\frac{1 \text{ m}}{100 \text{ cm}}\right) 163 \text{ cm} = 1.63 \text{ m}.$$

EXAMPLE 1–2   A woman drives a car in Germany at 50 km·hr^{-1} (50 kilometers per hour). Express this speed in meters per second.

SOLUTION

$$50 \text{ km·hr}^{-1} = (50 \text{ km·hr}^{-1})\left(\frac{1000 \text{ m}}{1 \text{ km}}\right)\left(\frac{1 \text{ hr}}{3600 \text{ s}}\right) = 13.89 \text{ m·s}^{-1}.$$

1–4 PRECISION AND SIGNIFICANT FIGURES

Precision: How exact is that number?

Measurements always have uncertainties. When we measure a distance with an ordinary ruler, it is usually reliable only to the nearest millimeter, while a precision micrometer caliper can measure distances dependably to 0.01 mm or even less. We often indicate the precision of a number by writing the number, the symbol $\pm$, and a second number indicating the maximum likely uncertainty. If the diameter of a steel rod is given as 56.47 $\pm$ 0.02 mm, this means that the true value is unlikely to be less than 56.45 mm or greater than 56.49 mm.

Percent uncertainty

We can also express precision in terms of the maximum likely fractional or percent uncertainty. A resistor labeled "47 ohms, 10%" probably has a true resistance differing from 47 ohms by no more than 10% of 47 ohms, or about 5 ohms; that is, the resistance is between about 42 and 52 ohms. In the steel rod example above, the fractional uncertainty is (0.02 mm)/(56.47 mm), or about 0.00035; the percent uncertainty is (0.00035)(100%), or about 0.035%.

When we use numbers having uncertainties or errors to compute other numbers, the computed numbers are also uncertain. It is especially important to understand this when comparing a number obtained from measurements with a value obtained from a theoretical prediction. Suppose you want to verify the value of π, the ratio of the circumference to the diameter of a circle. The true value of this ratio, to ten digits, is 3.141592654. To make your own calculation, you draw a large circle and measure its diameter and circumference to the nearest millimeter, obtaining the values 135 mm and 424 mm. You punch these into your calculator and obtain the quotient 3.140740741. Does this agree with the true value or not?

Significant figures: an indication of precision

To answer this question we must first recognize that at least the last six digits in your answer are meaningless because they imply greater precision than is possible with your measurements. The number of meaningful digits in a number is called the number of **significant figures;** usually a numerical result has no more significant figures than the numbers from which it is computed. Thus your value of π has only three significant figures and should be stated simply as 3.14 or possibly as 3.141 (rounded to four figures). Within the limit of three significant figures, your value does agree with the true value.

In the examples and problems in this book, we usually assume that the numerical values we give are precise to three or at most four significant figures, and thus your answers should show at most four significant figures. You may do the arithmetic with a calculator having a display with five to ten digits. But you should not give a ten-digit answer for a calculation using numbers with three significant figures. To do so is not only unnecessary but it is also genuinely wrong because it misrepresents the precision of the results. Always round your answer to keep only the correct number of significant figures, or in doubtful cases one more at most. Thus in Example 1–2 the result should have been stated as 13.9, or 14 m·s^{-1}. Of course, significant figures provide only a crude representation of the reliability of a number. The fractional uncertainty of 104 is not appreciably less than that of 96, despite the difference in number of significant figures. When a better representation of uncertainty is needed, more sophisticated statistical methods are used.

In calculations with very large or very small numbers, we can show significant figures much more easily by using powers-of-ten notation, sometimes called **scientific notation.** The distance from the earth to the sun is about 149,000,000,000 m, but to write the number in this form gives no indication of the number of significant figures. Certainly not all 12 are significant! Instead, we move the decimal point 11 places to the left (corresponding to dividing by 10^{11}) and multiply by 10^{11}. That is,

$$149{,}000{,}000{,}000 \text{ m} = 1.49 \times 10^{11} \text{ m}.$$

Scientific notation: using powers of ten to represent very large and very small numbers

In this form it is clear that the number of significant figures is three. In scientific notation the usual practice is to express the quantity as a number between 1 and 10 multiplied by the appropriate power of ten.

We can use the same technique when we multiply or divide very large or very small numbers. For example, the energy E corresponding to the mass m of an electron is given by the equation

$$E = mc^2, \tag{1-2}$$

where c is the speed of light. The appropriate numbers are $m = 9.11 \times 10^{-31}$ kg and $c = 3.00 \times 10^8$ m·s^{-1}. We find

$$
\begin{aligned}
E &= (9.11 \times 10^{-31} \text{ kg})(3.00 \times 10^8 \text{ m·s}^{-1})^2 \\
&= (9.11)(3.00)^2(10^{-31})(10^8)^2 \text{ kg·m}^2\text{·s}^{-2} \\
&= (82.0)(10^{[-31+(2\times8)]}) \text{ kg·m}^2\text{·s}^{-2} \\
&= 8.20 \times 10^{-14} \text{ kg·m}^2\text{·s}^{-2}.
\end{aligned}
$$

Most calculators use scientific notation and do this addition of exponents automatically for you; but you should be able to do such calculations by hand when necessary. Incidentally, the value used for c has three significant figures even though two of them are zeros. To greater accuracy, $c = 2.997925 \times 10^8$ m·s^{-1}; thus it would *not* be correct to write $c = 3.000 \times 10^8$ m·s^{-1}.

1–5 ESTIMATES AND ORDERS OF MAGNITUDE

We have discussed the importance of knowing the precision of numbers that represent physical quantities. But there are also situations where even a very crude estimate of a quantity may provide useful information. We may know how to calculate a certain quantity if we are given the necessary input data, but

Estimates: A good guess is better than no information at all.

those data may be unavailable or may have to be guessed at. Or the calculation may be too complicated to carry out exactly. In any case, the result of such a calculation is also a guess, but in some situations a guess is useful, even if it is uncertain by a factor or two or ten or more. Such calculations are often called **order-of-magnitude** calculations or estimates.

EXAMPLE 1–3   You are writing an international espionage novel in which the hero escapes across the border with a billion dollars' worth of gold in a suitcase. Is this possible? Would it fit? Would it be too heavy to carry?

SOLUTION   Gold sells for around $400 an ounce. On a particular day it may be $200 or $600, but never mind. An ounce is about 30 grams. Actually, an ordinary (avoirdupois) ounce is 28.35 g; an ounce of gold is a troy ounce, which is 9.45% more. Again, never mind. Ten dollars' worth of gold has a mass somewhere around one gram, so a billion (10^9) dollars' worth is a hundred million (10^8) grams, or a hundred thousand (10^5) kilograms. This corresponds to a weight in British units of around 200,000 lb, or a hundred tons. Whether the precise number is 50 tons or 200 does not matter; either way, our hero is not about to carry it across the border in a suitcase.

We can also estimate the *volume* of this gold. If its density were the same as that of water (1 g·cm^{-3}), the volume would be 10^8 cm^3, or 100 m^3. But gold is a heavy metal; we might guess its density as ten times that of water. It is actually 19.3 times as dense as water. But guessing ten, we find a volume of 10 m^3. Visualize ten cubical stacks of gold bricks, each one meter on a side, and ask whether it would fit in a suitcase!

Exercises 1–15 through 1–23 at the end of this chapter are of the estimating or "order-of-magnitude" variety. Some are silly, and most require much guesswork for the needed input data. Do not try to look up a lot of data; make the best guesses you can. Even when they are off by a factor of ten, the results can be useful and interesting.

1–6 VECTORS AND VECTOR ADDITION

Some physical quantities, such as time, temperature, mass, density, and electric charge, can be described completely by a single number with a unit. Many other quantities, however, have a *directional* quality that cannot be described by a single number. A familiar example is velocity. To describe the motion of a body we must say not only how fast it is moving but also in what direction. Force is another example. When we push or pull on a body, we exert a force on it. To describe a force we need to describe the direction in which it acts, as well as its magnitude, or "how hard" the force pushes or pulls.

Scalar and vector quantities: Does the quantity have direction or only a "how much" number?

A physical quantity that is described by a single number is called a **scalar quantity;** a quantity having both magnitude (the "how much" or "how big" part) and direction is called a **vector quantity.** Calculations with scalar quantities use the operations of ordinary arithmetic, but calculations with vector quantities are somewhat different. Vector quantities play an essential role in all areas of physics, and so we turn now to a discussion of their nature and the operation of vector addition.

We begin with a vector quantity called **displacement.** When a particle, which we represent as a point, moves from one location in space to another, it undergoes a displacement. In Fig. 1–3a we represent the change of position from point P_1 to point P_2 by the directed line segment P_1P_2, with an arrowhead at P_2 to represent the direction of motion. Displacement is a vector quantity because we must state not only how far the particle moves but also in what direction. A displacement of 3 km north is not the same as a displacement of 3 km southeast.

We usually represent a displacement by a single letter, such as A in Fig. 1–3a. In this book we always print vector symbols in boldface type as a reminder that vector quantities have different properties from scalar quantities. In handwriting, vector symbols are usually underlined or written with an arrow above, as shown in Fig. 1–3c, to indicate that they represent vector quantities.

The vector from point P_3 to point P_4 in Fig. 1–3b has the same length and direction as the one from P_1 to P_2. These two displacements are equal, even though they start at different points. By definition, two vector quantities are equal if they have the same magnitude (length) and direction, no matter where they are located in space. The vector B, however, is not equal to A because its direction is opposite to that of A. We define the *negative* of a vector as a vector having the same magnitude as, but the opposite direction to, the original vector. The negative of vector quantity A is denoted as $-A$, and we use a boldface "minus" to emphasize the vector nature of the quantities. Thus the relation between A and B may be written as $A = -B$ or $B = -A$. The vectors A and B are *antiparallel.* Note that a boldface "equals" sign is also used to emphasize that equality of two vector quantities is not the same relationship as equality of scalar quantities.

Displacement is always a straight-line segment, directed from the starting point to the endpoint, even though the path of the particle may be curved. Thus in Fig. 1–4, when the particle moves along the curved path shown from P_1 to P_2, the displacement is still the vector A shown. Also, when it continues on to P_3 and then returns to P_1, the displacement for the entire trip is zero.

We represent the **magnitude** of a vector quantity (its length, in the case of a displacement vector) by the same letter used for the vector, but in light italic type rather than boldface italic. An alternative notation is the vector symbol with vertical bars on both sides. Thus

$$\text{(Magnitude of } A) = A = |A|. \tag{1–3}$$

By definition, the magnitude of a vector quantity is a scalar quantity (a single number) and is always positive. We also note that a vector quantity can never be equal to a scalar one because they are different kinds of quantities. The expression $A = 6$ m is just as wrong as 2 oranges = 3 apples or 6 lb = 7 km!

Now suppose a particle undergoes a displacement A, followed by a second displacement B, as shown in Fig. 1–5a. The final result is the same as though it had started at the same initial point and undergone a single displacement C, as shown. We call displacement C the **vector sum** of displacements A and B; the relationship is expressed symbolically as

$$C = A + B. \tag{1–4}$$

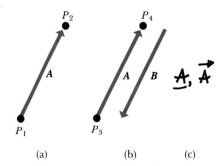

1–3 (a) Vector A is the displacement from point P_1 to point P_2. (b) The displacement from P_3 to P_4 is equal to that from P_1 to P_2, but displacement B is the negative of displacement A.

Equality of vector quantities

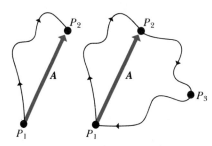

1–4 A displacement is always a straight line segment, directed from the starting point to the endpoint, even if the actual path is curved. When a point ends at the same place it started, the displacement is zero.

Magnitude of a vector quantity

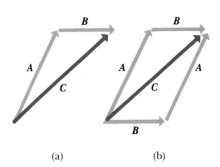

1–5 (a) Vector C is the vector sum of vectors A and B. (b) The order in vector addition is immaterial.

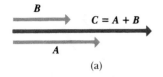

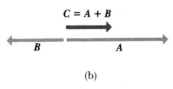

1–6 Vector sum of (a) two parallel vectors, and of (b) two antiparallel vectors.

Resultant: another name for vector sum

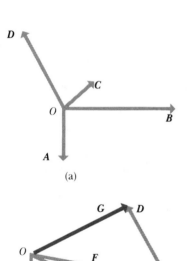

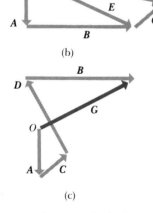

1–7 Polygon method of vector addition.

The boldface "plus" sign emphasizes that adding two vector quantities requires a geometrical process and is not the same operation as adding two scalar quantities, such as $2 + 3 = 5$.

If we make the displacements A and B in the reverse order, as in Fig. 1–5b, with B first and A second, the result is the same, as the figure shows. Thus

$$C = B + A \quad \text{and} \quad A + B = B + A. \tag{1–5}$$

This shows that vector addition obeys the *commutative law;* The order of the terms in the sum does not matter.

Figure 1–5b also suggests an alternative graphical representation of the vector sum: When vectors A and B are both drawn from a common point, vector C is the diagonal of a parallelogram constructed with A and B as two adjacent sides.

The vector sum is often called the **resultant;** thus the vector sum C of vectors A and B can be called the resultant displacement.

Figure 1–6 shows a special case in which two vectors A and B are parallel, as in (a), or antiparallel, as in (b). When they are parallel, the magnitude of the vector sum equals the *sum* of the magnitudes of A and B; when they are antiparallel, it equals the *difference* of their magnitudes. The vectors in Fig. 1–6 have been displaced slightly sidewise to show them more clearly, but they actually lie along the same geometric line.

When more than two vectors are to be added, we may first find the vector sum of any two, add this vectorially to the third, and so on. This process is shown in Fig. 1–7; part (a) shows four vectors A, B, C, and D. In Fig. 1–7b, vectors A and B are first added, giving a vector sum E; vectors E and C are then added by the same process, to obtain the vector sum F; finally F and D are added, to obtain the vector sum

$$G = A + B + C + D.$$

We do not need to draw vectors E and F; all we need do is draw the given vectors in succession, with the tail of each at the head of the one preceding it, and complete the polygon by a vector G from the *tail* of the first to the *head* of the last vector. The order makes no difference, as shown in Fig. 1–7c; we invite you to try other orders.

Diagrams for addition of displacement vectors do not need to be drawn actual size. It is often convenient to use a scale similar to those used for maps, where the distance on the diagram is proportional to the actual distance, such as 1 cm for 5 km. When we work with other vector quantities whose units are not distance units, we *must* use a scale. For example, in a diagram for force vectors we might use a scale in which a vector 1 cm long represents a force of magnitude 5 N. (The newton, abbreviated N, is the SI unit of force.) A 20-N force would then be represented by a vector 4 cm long with the appropriate direction.

A vector quantity such as a displacement can be multiplied by a scalar quantity (an ordinary number). The displacement $2A$ is a displacement (vector quantity) in the same direction as the vector A but twice as long. The scalar quantity used to multiply a vector may be a physical quantity having units. For example, you may be familiar with the relationship $F = ma$; force F (a vector quantity) is equal to the product of mass m (a scalar quantity) and acceleration *a* (a vector quantity). The magnitude of the force is equal to the mass multiplied by the magnitude of the acceleration, and the unit of the magnitude of force is the product of the unit of mass and that of the magnitude of acceleration.

We have already mentioned the special case of multiplication by -1: $(-1)A = -A$ is by definition a vector having the same magnitude as A but the opposite direction. This provides the basis for defining vector subtraction. We define the difference $A - B$ of the two vectors A and B to be the vector sum of A and $-B$:

$$A - B = A + (-B). \tag{1–6}$$

The boldface **+**, **−**, and **=** signs remind us of the vector nature of these operations.

Subtraction of vector quantities

1–7 COMPONENTS OF VECTORS

Addition and subtraction of vectors are usually carried out by the use of **components.** To define components we use a rectangular (cartesian) coordinate-axis system as in Fig. 1–8. We can represent any vector lying in the *xy*-plane as the sum of a vector parallel to the *x*-axis and a vector parallel to the *y*-axis. These two vectors are labeled A_x and A_y in the figure; they are called the component vectors of vector A. The relation is expressed formally as

$$A = A_x + A_y. \tag{1–7}$$

By definition, each component vector lies along a coordinate-axis direction. Thus only a single number is needed to describe each one. For example, the number A_x is, apart from a possible negative sign, the magnitude of the component vector A_x. We further specify that A_x is positive when A_x points in the positive axis direction and negative when it points in the opposite direction. The two numbers A_x and A_y are called simply the components of A.

If we know the magnitude A of the vector A and its direction, given by angle θ in Fig. 1–8, we can calculate the components. From the definitions of the trigonometric functions,

$$\frac{A_x}{A} = \cos\theta \qquad \text{and} \qquad \frac{A_y}{A} = \sin\theta;$$
$$A_x = A\cos\theta \qquad \text{and} \qquad A_y = A\sin\theta. \tag{1–8}$$

In Fig. 1–9, the component B_x is negative, because its direction is opposite to that of the positive *x*-axis. This is consistent with Eqs. (1–8); the cosine of an angle in the second quadrant is negative. The component B_y is positive, but both C_x and C_y are negative.

We can use either the magnitude and direction or the *x*- and *y*-components of a vector quantity to describe it completely. Equations (1–8) show how to obtain the components if the magnitude and direction are given. Or if we are given the components, we can find the magnitude and direction. Applying the Pythagorean theorem to Fig. 1–8, we find

$$A = \sqrt{A_x{}^2 + A_y{}^2}. \tag{1–9}$$

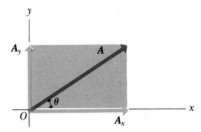

1–8 Vectors A_x and A_y are the rectangular components of A in the directions of the *x*- and *y*-axes.

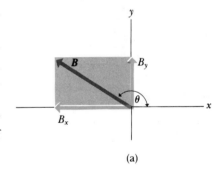

(a)

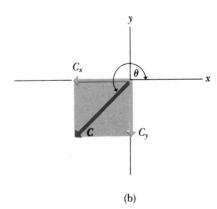

(b)

1–9 Components of a vector may be positive or negative numbers.

Also, from the definition of the tangent of an angle,

$$\tan \theta = \frac{A_y}{A_x} \quad \text{and} \quad \theta = \arctan \frac{A_y}{A_x}. \quad (1\text{--}10)$$

There is one slight complication in using Eq. (1–10) to find θ. Suppose $A_x = 2$ m and $A_y = -2$ m; then $\tan \theta = -1$. But there are two angles having tangents of -1, namely 135° and 315° (or $-45°$). To decide which is correct we must look at the individual components; because A_x is positive and A_y is negative, the angle must be in the fourth quadrant; thus $\theta = 315°$ (or $-45°$) is the correct value. Most pocket calculators give arctan $(-1) = -45°$; in this case that is correct, but if instead we have $A_x = -2$ m and $A_y = 2$ m, then the correct angle is 135°. Thus you should always draw a sketch to check which of the two possibilities is the correct one.

Components provide an efficient means of calculating the vector sum of several vectors. In principle, such sums can always be carried out by using diagrams such as those in Figs. 1–5 and 1–7; but to do this we must either draw and measure a scale diagram (which is hard to do accurately) or carry out a trigonometric solution of oblique triangles (which can be very complicated). The component method, by contrast, requires only right triangles and simple computations, and it can be carried out with great accuracy.

Here is the basic idea of the component method. Figure 1–10 shows two vectors A and B and their vector sum (resultant) C, along with x- and y-components of all three vectors. You can see from the diagram that the x-component C_x of the vector sum is simply the sum $(A_x + B_x)$ of the x-components of the vectors being added. The same is true for the y-components.

$$C_x = A_x + B_x,$$
$$C_y = A_y + B_y. \quad (1\text{--}11)$$

Thus once we know the components of A and B, perhaps by using Eqs. (1–8), we can compute the components of the vector sum. Then if the magnitude and direction of C are needed, we can obtain them from Eqs. (1–9) and (1–10).

This procedure for finding the sum of two vectors can easily be extended to any number. Let R be the vector sum of $A, B, C, D, E, \ldots$. Then

$$R_x = A_x + B_x + C_x + D_x + E_x + \cdots,$$
$$R_y = A_y + B_y + C_y + D_y + E_y + \cdots. \quad (1\text{--}12)$$

When we have to add a large number of vectors, the component method is often the only reasonable method.

Using components to calculate vector sums

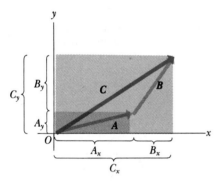

1–10 C_x is the x-component of the vector sum C of vectors A and B, and is equal to the sum of the x-components of A and B. The y-components are similarly related.

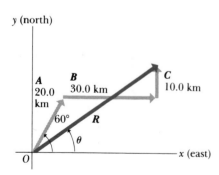

1–11 Three successive displacements $A, B,$ and $C,$ and the resultant or vector sum displacement $R = A + B + C.$

EXAMPLE 1–4 The pilot of a private plane flies 20.0 km in a direction 60° north of east, then 30.0 km straight east, then 10.0 km straight north. How far and in what direction is the plane from the starting point?

SOLUTION The situation is shown in Fig. 1–11. We have chosen the x-axis as east and the y-axis as north, the usual choice for maps. Let A be the first displacement, B the second, C the third, and R the vector sum or resultant displacement. We estimate from the diagram that R is about 50 km, at an angle of about 30°. We can later check this estimate against our calculated results.

The components of A are

$$A_x = (20.0 \text{ km})(\cos 60°) = 10.0 \text{ km},$$
$$A_y = (20.0 \text{ km})(\sin 60°) = 17.3 \text{ km}.$$

The components of all the displacements and the calculations can be arranged systematically, as in Table 1–2.

TABLE 1–2

Distance	Angle	x-component	y-component
A = 20.0 km	60°	10.0 km	17.3 km
B = 30.0 km	0°	30.0 km	0
C = 10.0 km	90°	0	10.0 km
		$R_x = 40.0$ km	$R_y = 27.3$ km

$$R = \sqrt{(40.0 \text{ km})^2 + (27.3 \text{ km})^2} = 48.4 \text{ km},$$

$$\theta = \arctan \frac{27.3 \text{ km}}{40.0 \text{ km}} = 34.3°.$$

Alternatively, we can find θ first, as above, then use Eqs. (1–8) to find R:

$$R = \frac{R_x}{\cos \theta} = \frac{40.0 \text{ km}}{\cos 34.3°} = 48.4 \text{ km},$$

or

$$R = \frac{R_y}{\sin \theta} = \frac{27.3 \text{ km}}{\sin 34.3°} = 48.4 \text{ km}.$$

This discussion has been confined to vectors lying in the xy-plane, but we can easily generalize it to vectors having any direction in space. We introduce a z-axis perpendicular to the xy-plane; then in general a vector A has components A_x, A_y, and A_z in the three coordinate directions. The magnitude A is given by

$$A = \sqrt{A_x^2 + A_y^2 + A_z^2}. \tag{1–13}$$

1–8 UNIT VECTORS

A **unit vector** is a vector having a magnitude of unity, with no units. Its only purpose is to describe a direction in space. Unit vectors provide a convenient notation in many expressions involving components of vectors.

Unit vectors: a handy way to describe directions in space

In an xy-coordinate system we can define a unit vector i that points in the direction of the positive x-axis, and a unit vector j in the direction of the positive y-axis. Then we can express the relationships between component vectors and components, described at the beginning of Section 1–7, as follows:

$$A_x = A_x i, \qquad A_y = A_y j. \tag{1–14}$$

Similarly, we write vector A in terms of its components as

$$A = A_x i + A_y j. \tag{1–15}$$

Equations (1–14) and (1–15) are *vector* equations; each term, such as $A_x i$, is a vector quantity, and the boldface **=** and **+** signs denote vector equality and addition.

When two vectors A and B are represented in terms of their components, we can express the vector sum using unit vectors, as follows:

$$A = A_x i + A_y j, \qquad B = B_x i + B_y j,$$
$$C = A + B$$
$$= (A_x i + A_y j) + (B_x i + B_y j)$$
$$= (A_x + B_x)i + (A_y + B_y)j$$
$$= C_x i + C_y j. \tag{1–16}$$

Equation (1–16) restates the content of Eqs. (1–11) in the form of a single vector equation rather than two ordinary equations.

If the vectors do not all lie in the xy-plane, then a third component is needed. We introduce a third unit vector k in the z-axis direction. The generalized forms of Eqs. (1–15) and (1–16) are

$$A = A_x i + A_y j + A_z k, \tag{1–17}$$
$$C = (A_x + B_x)i + (A_y + B_y)j + (A_z + B_z)k$$
$$= C_x i + C_y j + C_z k. \tag{1–18}$$

1–9 PRODUCTS OF VECTORS

Many physical relationships can be expressed concisely by the use of *products* of vectors. Because vectors are not ordinary numbers, ordinary multiplication is not directly applicable to vectors. Just as addition of vectors is different from addition of scalars, so it is with multiplication. In fact, there are two different kinds of vector products. The first, called the scalar product, yields a result that is a scalar quantity, while the second, the vector product, yields another vector.

Scalar product: one of the ways to multiply vectors

Here is the definition of the **scalar product** of two vectors A and B. We draw the two vectors from a common point, as in Fig. 1–12a. The angle between their directions is θ, as shown. We define the scalar product, denoted by $A \cdot B$, as

$$A \cdot B = AB \cos \theta = |A||B| \cos \theta \tag{1–19}$$

Because of this notation, the scalar product is also called the *dot product*. It is a scalar quantity, not a vector, and it may be positive or negative. When θ is between zero and 90°, the scalar product is positive; when θ is between 90° and 180°, it is negative; and when $\theta = 90°$, $A \cdot B = 0$. *The scalar product of two perpendicular vectors is always zero.*

The scalar product of two vectors is a scalar, not a vector.

The scalar product obeys the *commutative* law of multiplication; the order of the two vectors doesn't matter. For any two vectors A and B, $A \cdot B = B \cdot A$. This property follows directly from the definition.

1–12 (a) Two vectors drawn from a common starting point to define their scalar product. (b) $B \cos \theta$ is the component of B in the direction of A, and $A \cdot B$ is the product of this component with the magnitude of A.

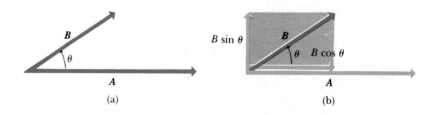
(a)   (b)

We can represent vector B in terms of a component parallel to A and a component perpendicular to A, as shown in Fig. 1–12b; the component parallel to A is ($B \cos \theta$). Thus, from Eq. (1–19), $A \cdot B$ is equal to the component of B parallel to A, multiplied by the magnitude of A. Alternatively, it is also the component of A parallel to B, multiplied by the magnitude of B.

If we know the components of A and B, we can calculate their scalar product. The easiest procedure is to use the unit vector representation introduced in Section 1–8. First,

Calculating the scalar product from the components of the vectors

$$A \cdot B = (A_x i + A_y j + A_z k) \cdot (B_x i + B_y j + B_z k). \qquad (1\text{–}20)$$

We expand the product of the two sets of parentheses on the right, obtaining nine terms in all, as follows:

$$\begin{aligned}
A \cdot B = {} & A_x i \cdot B_x i + A_x i \cdot B_y j + A_x i \cdot B_z k \\
& + A_y j \cdot B_x i + A_y j \cdot B_y j + A_y j \cdot B_z k \\
& + A_z k \cdot B_x i + A_z k \cdot B_y j + A_z k \cdot B_z k.
\end{aligned} \qquad (1\text{–}21)$$

Each of these terms contains the scalar product of two vectors that are either parallel or perpendicular. For example, in $A_x i \cdot B_x i$, the two vectors are parallel, the angle between them is zero, its cosine is unity, and the scalar product is simply the product $A_x B_x$ of the magnitudes. But in $A_x i \cdot B_y j$, the two vectors are perpendicular and the scalar product is zero. Thus six of the nine terms are zero, and the three that survive give simply

$$A \cdot B = A_x B_x + A_y B_y + A_z B_z. \qquad (1\text{–}22)$$

EXAMPLE 1–5 Find the angle between the two vectors

$$A = 2i + 3j + 4k, \qquad B = i - 2j + 3k.$$

SOLUTION We have

$$\begin{array}{ll}
A_x = 2 & B_x = 1 \\
A_y = 3 & B_y = -2 \\
A_z = 4 & B_z = 3
\end{array}$$

The scalar product is given by either Eq. (1–19) or (1–22). Equating these two and rearranging, we obtain

$$\cos \theta = \frac{A_x B_x + A_y B_y + A_z B_z}{AB}. \qquad (1\text{–}23)$$

In our example,

$$A_x B_x + A_y B_y + A_z B_z = (2)(1) + (3)(-2) + (4)(3) = 8,$$

$$A = \sqrt{2^2 + 3^2 + 4^2} = \sqrt{29}, \qquad B = \sqrt{1^2 + (-2)^2 + 3^2} = \sqrt{14},$$

$$\cos \theta = \frac{8}{\sqrt{29}\sqrt{14}} = 0.397,$$

and

$$\theta = 66.6°.$$

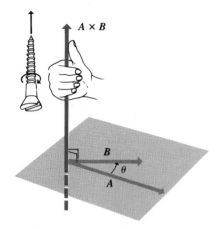

1–13 Vectors A and B lie in a plane; the vector product $A \times B$ is perpendicular to this plane, in a direction determined by the "right-hand rule."

Our first physical application of the scalar product will come in Chapter 7 with the concept of work. When a constant force F is applied to a body that undergoes a displacement d, the word W (a scalar quantity) done by the force is given by

$$W = F \cdot d. \qquad (1\text{--}24)$$

The **vector product** of two vectors A and B is denoted by $A \times B$. To define the vector product (also called the *cross product* because of this notation) we again draw A and B from a common point. The two vectors then lie in a plane. We define the vector product as a vector quantity with a direction perpendicular to this plane (i.e., perpendicular to both A and B) and a magnitude given by $AB \sin \theta$. That is, if $C = A \times B$, then

$$C = AB \sin \theta. \qquad (1\text{--}25)$$

We measure the angle θ from A toward B and take it always to be between 0 and 180°. Thus C in Eq. (1–25) is always positive, as a vector magnitude must be. We note also that when A and B are parallel or antiparallel, $\theta = 0$ or 180° and $C = 0$. That is, *the vector product of two parallel or antiparallel vectors is always zero.*

There are always *two* directions perpendicular to a given plane. To distinguish between these, we imagine rotating vector A about the perpendicular line until it is aligned with B (choosing the smaller of the two possible angles). We then curl the fingers of the right hand around this perpendicular line so that the fingertips point in the direction of rotation; the thumb then gives the direction of the vector product. This rule is shown in Fig. 1–13. Alternatively, the direction of the vector product is the direction a right-hand-thread screw advances if turned in the sense from A toward B, as shown in the figure.

Similarly, we determine the direction of the vector product $B \times A$ by rotating B into A in Fig. 1–13. This yields a result *opposite* to that for $A \times B$. The vector product is not commutative! In fact, for any two vectors A and B,

$$A \times B = -B \times A. \qquad (1\text{--}26)$$

If we know the components of A and B, we can calculate the components of the vector product by using a procedure similar to that for the scalar product. We expand the expression

$$
\begin{aligned}
A \times B &= (A_x i + A_y j + A_z k) \times (B_x i + B_y j + B_z k), \\
&= A_x i \times B_x i + A_x i \times B_y j + A_x i \times B_z k \\
&\quad + A_y j \times B_x i + A_y j \times B_y j + A_y j \times B_z k \\
&\quad + A_z k \times B_x i + A_z k \times B_y j + A_z k \times B_z k.
\end{aligned} \qquad (1\text{--}27)
$$

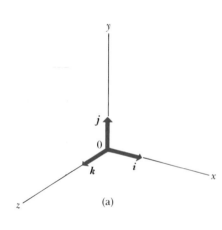

The individual terms may also be rewritten as $A_x i \times B_y j = A_x B_y i \times j$, and so on. Each term in which the same unit vector appears twice, such as $i \times i$, is zero because it is a product of two parallel vectors. To evaluate the others we refer to the axis system of Fig. 1–14a. We find, for example, $i \times j = k$, and $j \times i = -k$. Thus $A_x i \times B_y j = A_x B_y k$, and so on. We obtain finally

$$A \times B = (A_y B_z - A_z B_y) i + (A_z B_x - A_x B_z) j + A_x B_y - A_y B_x) k. \qquad (1\text{--}28)$$

If $C = A \times B$, the components of C are given by

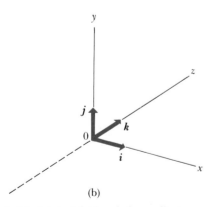

1–14 (a) A right-handed coordinate system, in which $i \times j = k, j \times k = i,$ and $k \times i = j.$ (b) A left-handed coordinate system, in which $i \times j = -k,$ and so on. Usually only right-handed systems are used.

$$
\begin{aligned}
C_x &= A_y B_z - A_z B_y, \\
C_y &= A_z B_x - A_x B_z, \\
C_z &= A_x B_y - A_y B_x.
\end{aligned} \qquad (1\text{--}29)
$$

Our first physical application of the vector product will come in Chapter 9 with the definitions of torque and angular momentum. We will also use it extensively in the chapters on magnetic fields.

Here is an interesting quirk of the vector product. In Fig. 1–14a, suppose we reverse the direction of the z-axis, giving the axis system shown in Fig. 1–14b. Then, as you may verify, the definition of the vector product gives $i \times j = -k$ instead of $i \times j = k$. If two axis directions are reversed, we obtain again $i \times j = k$, and if all three are reversed, $i \times j = -k$. Thus there are two kinds of coordinate-axis systems, differing in the signs of the products of unit vectors. In using the vector product, we must specify which kind we are using, to avoid ambiguity.

An axis systems in which $i \times j = k$ is called a **right-handed system.** The usual practice is to use *only* right-handed systems; we will follow that practice throughout this book.

Coordinate axis systems, like people, are either right-handed or left-handed.

EXAMPLE 1–6 Vector A has magnitude 6 units and is in the direction of the $+x$-axis; vector B has magnitude 4 units and lies in the xy-plane, making an angle of 30° with the $+x$-axis and an angle of 60° with the $+y$-axis. Find the vector product $A \times B$.

SOLUTION From Eq. (1–25), the magnitude of the vector product is

$$AB \sin \theta = (6)(4)(\sin 30°) = 12.$$

From the right-hand rule, the direction of $A \times B$ is that of the $+z$-axis. Alternatively, we may write the components of A and B and use Eqs. (1–29):

$$A_x = 6, \quad A_y = 0, \quad A_z = 0,$$
$$B_x = 4 \cos 30° = 2\sqrt{3}, \quad B_y = 4 \cos 60° = 2, \quad B_z = 0.$$

If $C = A \times B$, then

$$C_x = (0)(0) - (0)(2) = 0,$$
$$C_y = (0)(2\sqrt{3}) - (6)(0) = 0.$$
$$C_z = (6)(2) - (0)(2\sqrt{3}) = 12.$$

The vector product C has only a z-component, and it lies along the z-axis. The magnitude agrees with the above result.

SUMMARY

The fundamental physical quantities of mechanics are mass, length, and time; the corresponding SI units are the kilogram, the meter, and the second, respectively. Other units are related to these by multiples of powers of ten and are identified by adding a prefix to the basic unit. Derived units for other physical quantities are products or quotients of the basic units.

Equations must be dimensionally consistent; two terms can be added or equated only when they have the same units. Treating units like algebraic symbols makes it possible to check this consistency and to convert units easily from one system to another.

The uncertainty of a number can be indicated by the number of significant figures or by an expressed uncertainty. The result of a calculation never has more significant figures than the input data. Even when only crude esti-

KEY TERMS
physical quantity
operational definition
unit
second
kilogram
meter
International System
prefix
dimensional consistency
factor-label method
significant figures

scientific notation
order-of-magnitude estimate
scalar quantity
vector quantity
displacement
magnitude
vector sum
resultant
component
unit vector
scalar product
vector product
right-handed system

mates are available for the magnitudes of the input data, useful estimates (called order-of-magnitude estimates) can often be obtained.

Scalar quantities are numbers and are combined by the usual rules of arithmetic. Vector quantities have direction as well as magnitude and are combined according to the rules of vector addition. Vector addition can be carried out by using components of vectors. If A_x and A_y are the components of vector **A**, and B_x and B_y the components of vector **B**, the components of the vector sum **C** = **A** + **B** are given by

$$C_x = A_x + B_x,$$
$$C_y = A_y + B_y. \tag{1-11}$$

Unit vectors describe directions in space. A unit vector has a magnitude of unity, with no units. Unit vectors aligned with the coordinate axes of a rectangular coordinate system are especially useful.

There are two kinds of products of vectors: the scalar product (or dot product) and the vector product (or cross product). The scalar product **C** = **A** · **B** of two vectors **A** and **B** is a scalar quantity, defined as

$$\mathbf{A} \cdot \mathbf{B} = AB \cos \theta = |\mathbf{A}||\mathbf{B}| \cos \theta. \tag{1-19}$$

The scalar product can also be expressed in terms of the components of the vectors:

$$\mathbf{A} \cdot \mathbf{B} = A_x B_x + A_y B_y + A_z B_z. \tag{1-22}$$

The scalar product is commutative; for any two vectors **A** and **B**, **A** · **B** = **B** · **A**. The scalar product of two perpendicular vectors is zero.

The vector product **C** = **A** × **B** of two vectors **A** and **B** is another vector **C**, with magnitude given by

$$C = AB \sin \theta. \tag{1-25}$$

Its direction is perpendicular to the plane of the two vectors, as given by the right-hand rule. The components of the vector product can be expressed in terms of the components of the two vectors being multiplied, as follows:

$$C_x = A_y B_z - A_z B_y,$$
$$C_y = A_z B_x - A_x B_z,$$
$$C_z = A_x B_y - A_y B_x. \tag{1-29}$$

The vector product is not commutative; the order of the factors must not be interchanged. For any two vectors **A** and **B**, **A** × **B** = −**B** × **A**. The vector product of two parallel or antiparallel vectors is zero.

QUESTIONS

1–1 What are the units of the number π?

1–2 The rate of climb of a mountain trail was described in the guidebook as 150 meters per kilometer. How can this be expressed as a number with no units?

1–3 Suppose you are asked to compute the cosine of 3 meters. Is this possible?

1–4 Hydrologists describe the rate of volume flow of rivers in "second-feet." Is this unit technically correct? If not, what would be a correct unit?

1–5 A highway contractor stated that in building a bridge deck he had poured 200 yards of concrete. What do you think he meant?

1–6 Does a vector having zero length have a direction?

1–7 What is your weight in newtons?

1–8 What is your height in centimeters?

1–9 What physical phenomena (other than a pendulum or cesium clock) could be used to define a time standard?

1–10 Could some atomic quantity be used for a definition of a unit of mass? What advantages or disadvantages would this have compared to the 1-kilogram platinum cylinder kept at Sèvres?

1–11 How could you measure the thickness of a sheet of paper with an ordinary ruler?

1–12 Can you find two vectors with different lengths that have a vector sum of zero? What length restrictions are required for three vectors to have a vector sum of zero?

1–13 What is the displacement when a bicyclist travels from the north side of a circular race track of radius 500 meters to the south side? When she makes one complete circle around the track?

1–14 What are the units of volume? Suppose a student tells you a cylinder of radius r and height h has volume given by $\pi r^3 h$. Explain why this cannot be right.

1–15 An angle (measured in radians) is a number with no units, since it is a ratio of two lengths. Think of other geometrical or physical quantities that are unitless.

1–16 Can you find a vector quantity that has components different from zero but a magnitude of zero?

1–17 One sometimes speaks of the "direction of time," evolving from past to future. Does this mean that time is a vector quantity?

1–18 Is the scalar product of two vectors commutative? Explain.

1–19 What is the scalar product of a vector with itself? The vector product?

EXERCISES

Section 1–2 Standards and Units

Section 1–3 Unit Consistency and Conversions

1–1 Starting with the definition 1.00 in. = 2.54 cm, compute the number of kilometers in 1 mi.

1–2 The density of water is 1 g·cm^{-3}. What is this value in kilograms per cubic meter?

1–3 The Concorde is the fastest airliner used for commercial service and can cruise at 1450 mi·hr^{-1} (about two times the speed of sound, or in other words Mach 2).

a) What is the cruise speed of the Concorde in mi·s^{-1}?

b) What is the cruise speed of the Concorde in m·s^{-1}?

1–4 Compute the number of seconds in a day (24 hr), and in a year (365 da).

1–5 The piston displacement of a certain automobile engine is given as 2.0 liters (L). Using only the facts that 1.0 L = 1000 cm^3 and 1.0 in. = 2.54 cm, express this volume in cubic inches.

1–6 If one Deutschmark (the West German unit of currency) is worth 40 cents and gasoline costs 1.30 Deutschmarks per liter, what is its cost in dollars per gallon? Use the conversion factors in Appendix E. How does your answer compare with the cost of gasoline in the United States?

1–7 The gasoline consumption of a small car is 17.0 km·L^{-1}. How many miles per gallon is this? Use conversion factors in Appendix E.

1–8 The speed limit on a highway in Lower Slobbovia was given as 150,000 furlongs per fortnight. How many miles per hour is this? (One furlong is 1/8 mile, and a fortnight is 14 days. A furlong originally referred to the length of a plowed furrow.)

1–9 One standard of frequency is the radio waves the hydrogen maser has as its output. These radio waves have a frequency of 1,420,405,751.786 hertz. (A hertz is just a special name for cycles per second.) A clock controlled by a hydrogen maser is off by only 1 s in 100,000 years. For the following questions use only three significant figures. (The large number of significant figures for the frequency was given to illustrate the remarkable precision to which it has been measured.)

a) What is the time for one cycle of the radio wave?

b) How many cycles occur in 1 hour?

c) How many cycles would have occurred during the age of the universe, which is estimated to be 10 billion years?

d) By how many seconds would a hydrogen maser clock be off during the lifetime of the earth, which is estimated to be 4600 million years?

Section 1–4 Precision and Significant Figures

1–10 What is the percent error in each of the following approximations to π?

a) 22/7   b) 355/113

1–11 What is the fractional error in the approximate statement 1 yr = $\pi \times 10^7$ s? (Assume that a year is 365 days.)

1–12 Estimate the percent error in measuring

a) a distance of about 50 cm with a meter stick;

b) a mass of about 1 g with a chemical balance;

c) a time interval of about 4 min with a stopwatch.

1–13 The mass of the earth is 5.98×10^{24} kg, and its radius is 6.38×10^6 m. Compute the density of the earth, using powers-of-ten notation and the correct number of significant figures. (The density of an object is defined as its mass divided by its volume. The formula for the volume of a sphere is given in Appendix B.)

1–14 An angle is given, to one significant figure, as 5°, meaning that its value is between 4.5° and 5.5°. Find the corresponding range of possible values of the cosine of the angle. Is this a case where there are more significant figures in the result than in the input data?

Section 1–5 Estimates and Orders of Magnitude

1–15 A box of typewriter paper is 11″ × 17″ × 9″; it is marked "10 M." Does that mean it contains 10 thousand sheets or 10 million?

1–16 What total volume of air does a person breathe in a lifetime? How does that compare with the volume of the Houston Astrodome?

1–17 Could the water-supply needs of Los Angeles be met by hauling water in by truck? By railroad?

1–18 How many kernels of corn does it take to fill a 1-L soft-drink bottle?

1–19 How many hairs do you have on your head?

1–20 How many times does a human heart beat during a lifetime? How many gallons of blood does it pump?

1–21 How much would it cost to paper the continental United States with dollar bills?

1–22 How many dollar bills would have to be stacked up to reach the moon? Would that be cheaper than building and launching a spacecraft?

1–23 How many cars can pass through a two-lane tunnel through a mountain in 1 hour?

Section 1–6 Vectors and Vector Addition

Section 1–7 Components of Vectors

1–24 A bug starts at the center of a 12-in. phonograph record and crawls along a straight radial line to the edge. While this is happening, the record turns through an angle of 45°. Draw a sketch of the situation and describe the magnitude and direction of the bug's final displacement from its starting point.

1–25 Find the magnitude and direction of the vector represented by each of the following pairs of components:

a) $A_x = 3.0$ cm, $A_y = -4.0$ cm;

b) $A_x = -5.0$ m, $A_y = -12.0$ m;

c) $A_x = -2.0$ km, $A_y = 3.0$ km.

1–26 Vector A has components $A_x = 2.0$ cm, $A_y = 3.0$ cm, and vector B has components $B_x = 4.0$ cm, $B_y = -2.0$ cm. Find

a) the components of the vector sum $A + B$;

b) the magnitude and direction of $A + B$;

c) the components of the vector difference $A - B$;

d) the magnitude and direction of $A - B$.

1–27 A disoriented physics professor drives 5.0 km east, then 4.0 km south, then 2.0 km west. Find the magnitude and direction of the resultant displacement.

1–28 A postal employee drives a delivery truck 1 mi north, then 2 mi east, then 5 mi northwest. Determine the resultant displacement.

1–29 Find the magnitude and direction of

a) the vector sum $A + B$;

b) the vector difference $A - B$

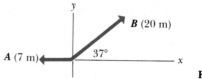

FIGURE 1–15

for the vectors A and B given in Fig. 1–15.

1–30 Vector A is 2.0 cm long and is 60° above the x-axis in the first quadrant. Vector B is 2.0 cm long and is 60° below the x-axis in the fourth quadrant. Find

a) the vector sum $A + B$;

b) the vector differences $A - B$ and $B - A$.

1–31 Vector M, of magnitude 5.0 cm, is at 36.9° counterclockwise from the +x-axis. It is added to vector N, and the resultant is a vector of magnitude 5.0 cm, at 53.1° counterclockwise from the +x-axis. Find

a) the components of N;

b) the magnitude and direction of N.

Section 1–8 Unit Vectors

1–32 Write each of the two vectors in Fig. 1–15 in terms of the unit vectors i and j.

1–33 Given two vectors $A = 2i + 3j$ and $B = i - 2j$, do the following:

a) Find the magnitude of each vector.

b) Write an expression for the vector sum, using unit vectors.

c) Find the magnitude and direction of the vector sum.

d) Write an expression for the vector difference $A - B$, using unit vectors.

e) Find the magnitude and direction of the vector difference $A - B$.

Section 1–9 Products of Vectors

1–34 Find the scalar product $A \cdot B$ of the two vectors in Fig. 1–15.

1–35 Write out a multiplication table for the scalar products of all possible pairs of unit vectors, such as $i \cdot i = ?$, $i \cdot j = ?$, and so on.

1–36 Find the scalar product of the two vectors given in Exercise 1–33.

1–37 Find the angle between the two vectors $A = -2i + 5j$ and $B = 3i - j$.

1–38

a) Find the magnitude and direction of the vector product $A \times B$ of the two vectors in Fig. 1–15.

b) Find the magnitude and direction of the vector product $B \times A$ of the two vectors in Fig. 1–15.

1–39 Write out a multiplication table for the vector products of all possible pairs of unit vectors, such as $i \times i = ?$, $i \times j = ?$, and so on, using a right-handed coordinate system.

1–40 Find the vector product of the two vectors given in Exercise 1–33. What is the magnitude of this vector product?

1–41

a) Find the scalar product and the vector product for the two vectors $A = 2i - 5j$ and $B = 5i + 2j$.

b) Find the scalar product and the vector product for the two vectors $A = 2i - 5j$ and $B = 4i - 10j$.

1–42 Which of the following are legitimate mathematical operations?

a) $A \cdot (B - C)$ b) $(A - B) \times C$

c) $A \cdot (B \times C)$ d) $A \times (B \times C)$

e) $A \times (B \cdot C)$

1–43

a) Show that Eq. (1–29) can be written in terms of a determinant as follows:

$$C = \begin{vmatrix} i & j & k \\ A_x & A_y & A_z \\ B_x & B_y & B_z \end{vmatrix}.$$

b) Show that

$$(A \times B) \cdot C = \begin{vmatrix} A_x & A_y & A_z \\ B_x & B_y & B_z \\ C_x & C_y & C_z \end{vmatrix}.$$

PROBLEMS

1–44 Physicists, mathematicians, and others often deal with large numbers. The number 10^{100} has been given the whimsical name of a *googol* by mathematicians. (See Edward Kasner and J. R. Newman in Vol. 3 of *The World of Mathematics,* ed. by J. R. Newman. New York: Simon and Schuster.) Let us compare some large numbers in physics with the googol. (*Note.* The following problem requires numerical values that can be found in the Appendixes of the book, with which you should become familiar.)

a) Approximately how many atoms make up the earth? For simplicity, take the average atomic mass of the atoms to be 14 g·mole^{-1}. Avogadro's number gives the number of atoms in a mole.

b) Approximately how many neutrons are in a neutron star? Neutron stars are made up of neutrons and have approximately twice the mass of the sun.

c) In one theory of the origin of the universe, the universe at a very early time had a density (mass divided by volume) of 10^{15} g·cm^{-3}, while its radius was approximately the present distance of the earth to the sun. Assuming $\frac{1}{3}$ of the particles were protons, $\frac{1}{3}$ of the particles were neutrons, and the remaining $\frac{1}{3}$ were electrons, how many particles then made up the universe?

1–45 A spelunker is surveying a cave. He follows a passage 100 m straight east, then 50 m in a direction 30° west of north, then 150 m at 45° west of south. After a fourth unmeasured displacement he finds himself back where he started. Determine the fourth displacement (magnitude and direction).

1–46 A sailboat sails 2.0 km east, then 4.0 km southeast, then an additional distance in an unknown direction. Its final position is 6.0 km directly east of the starting point. Find the magnitude and direction of the third leg of the journey.

1–47 Two points P_1 and P_2 are described by their x- and y-coordinates, (x_1, y_1) and (x_2, y_2), respectively. Show that the components of the displacement A from P_1 to P_2 are

$A_x = x_2 - x_1$ and $A_y = y_2 - y_1$. Also derive expressions for the magnitude and direction of this displacement.

1–48

a) Find the scalar product $A \cdot B$ of the two vectors A and B in Exercise 1–30.

b) Find the magnitude and direction of the vector product $A \times B$ of the two vectors A and B in Exercise 1–30.

1–49 When two vectors A and B are drawn from a common point, the angle between them is θ.

a) Show that the magnitude of their vector sum is given by

$$\sqrt{A^2 + B^2 + 2AB \cos \theta}.$$

b) If A and B have the same magnitude, under what circumstances will their vector sum have the same magnitude as A or B?

c) Derive a result analogous to that in (a) for the magnitude of the vector difference $A - B$.

d) If A and B have the same magnitude, under what circumstance will $A - B$ have this same magnitude?

1–50 Given two vectors $A = -i + 2j - 5k$ and $B = 2i + 3j - 2k$, obtain the following:

a) Find the magnitude of each vector.

b) Write an expression for the vector sum, using unit vectors.

c) Find the magnitude of the vector sum.

d) Write the expression for the vector difference $A - B$, using unit vectors.

e) Find the magnitude of the vector difference $A - B$. Is this the same as the magnitude of $B - A$? Explain.

1–51 Find the scalar product of the two vectors given in Problem 1–50.

1–52 Find the vector product of the two vectors given in Problem 1–50. What is the magnitude of this vector product?

1–53 Find the angle between the two vectors $A = 3i + 4j + 5k$ and $B = 3i + 4j - 5k$.

CHALLENGE PROBLEMS

1–54 Obtain a *unit vector* perpendicular to the two vectors given in Problem 1–50.

1–55 Prove that for any three vectors A, B, and C,

$$A \cdot (B \times C) = (A \times B) \cdot C.$$

1–56

a) Prove that for any three vectors A, B, and C,

$$A \times (B \times C) = B(A \cdot C) - C(A \cdot B).$$

b) Consider the two repeated vector products $A \times (B \times C)$ and $(A \times B) \times C$. Are these two products equal in either magnitude *or* direction? Prove your answer.

1–57 The dot and cross product will appear later in many physical applications; for now, consider the following applications.

a) Let A and B be vectors along the sides of a parallelogram, where the sides have lengths A and B.

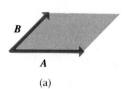
(a)

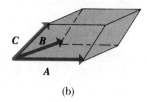
(b)

FIGURE 1–16

(See Fig. 1–16a.) Show that C is the area of the parallelogram, where $C = A \times B$.

b) The volume of a parallelepiped with sides of length A, B, and C can be expressed in an elegant way as a product of three vectors A, B, and C, where A is along the edge of length A, and so on. (See Fig. 1–16b.) Express the volume of the parallelepiped as the magnitude of the appropriate product of the three vectors A, B, and C.

2

MOTION ALONG
A STRAIGHT LINE

OUR STUDY OF PHYSICS BEGINS IN THE AREA OF *MECHANICS*, WHICH DEALS
with the relations among force, matter, and motion. In this chapter and the
next we discuss mathematical methods for describing motion, and in subsequent chapters we begin to explore the relationship of motion to force. The
part of mechanics concerned with describing motion is called **kinematics.**

Motion is a continuous change of position. We can think of a moving body
as a particle when it is small and when there is no rotation or change of shape.
We say that the particle is a **model** for a moving body; it provides a simplified,
idealized description of the position and motion of the body.

The simplest case is motion of a particle along a straight line, and we will
always take that line to be a **coordinate axis.** Later we will consider more
general motions in space, but these can always be represented by means of
their projections onto three coordinate axes. Displacement is in general a vector quantity, as discussed in Chapter 1, but we first consider situations in
which only one component of displacement is different from zero. Then the
particle moves along one coordinate axis, and its position is described by a
single **coordinate.**

Coordinates: describing the position of a
point

2–1 AVERAGE VELOCITY

Let us consider a hockey player skating the puck down the center line of the
ice toward the opposition's net. The puck moves along a straight line, which
we will use as the *x*-axis of our coordinate system, as shown in Fig. 2–1a. The
puck's distance from the origin O at center court is given by the coordinate x,
which varies with time. At time t_1 the puck is at point P, with coordinate x_1,
and at time t_2 it is at point Q, where its coordinate is x_2. The displacement
during the time interval from t_1 to t_2 is the vector from P to Q: the *x*-compo-
nent of this vector is $(x_2 - x_1)$, and all other components are zero.

It is convenient to represent the quantity $(x_2 - x_1)$, the *change* in x, by
means of a notation using the Greek letter Δ (capital delta) to designate a
change in any quantity. Thus we write

$$\Delta x = x_2 - x_1 \qquad (2\text{--}1)$$

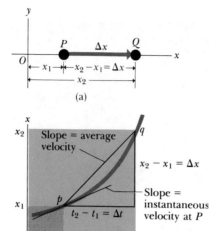

(a)

(b)

2–1 (a) Hockey puck moving on the x-axis. (b) Coordinate–time graph of the motion. The average velocity between t_1 and t_2 equals the slope of the line pq. The instantaneous velocity at P equals the slope of the tangent at p.

in which Δx is not a product but is a single symbol representing the change in the quantity. Similarly, we denote the time interval from t_1 to t_2 as $\Delta t = t_2 - t_1$.

The **average velocity** of the puck is defined as the ratio of the displacement Δx to the time interval Δt. We represent average velocity by the letter v with a subscript "av" to signify average value. Thus

Average velocity: distance divided by time

$$v_{\mathrm{av}} = \frac{x_2 - x_1}{t_2 - t_1} = \frac{\Delta x}{\Delta t}. \qquad (2\text{--}2)$$

Strictly speaking, average velocity is a vector quantity, and this defines the x-component of the average velocity. However, in this chapter all vectors have *only* x-components, so we do not need to distinguish between a vector and its x-component. In later chapters we will return to this distinction in two-dimensional motion.

In Fig. 2–1b the coordinate x is graphed as a function of time. The curve in this figure does not represent the path of the puck in space; as Fig. 2–1a shows, the path is a straight line. Rather, the graph is a representation of the change of position with time. The points on the coordinate–time graph corresponding to points P and Q are labeled p and q.

In Fig. 2–1b, the average velocity v_{av} of the puck is represented by the *slope* of the line pq, that is, the ratio of vertical to horizontal intervals of the triangle. Thus v_{av} is the ratio of $x_2 - x_1$ or Δx to $t_2 - t_1$ or Δt.

We have not specified whether the speed of the hockey puck is or is not constant during the time interval $\Delta t = t_2 - t_1$. It may have started from rest, reached a maximum speed, and then slowed down. But that does not matter: To calculate the average velocity we need only the total displacement $\Delta x = x_2 - x_1$ and the total time interval $\Delta t = t_2 - t_1$.

2–2 INSTANTANEOUS VELOCITY

Instantaneous velocity: What do we mean by "velocity at a point"?

Even when the velocity of a moving particle varies, we can still define a velocity at any one specific instant of time or at one specific point in the path. Such a velocity is called **instantaneous velocity,** and it needs to be defined carefully.

Suppose we want to find the instantaneous velocity of the hockey puck in Fig. 2–1 at the point P. We associate the average velocity between points P and Q with the entire displacement Δx and with the entire time interval Δt. But now we can imagine the second point Q as taken closer and closer to the first point P, and we can compute the average velocity over these shorter and shorter displacements and time intervals. We can then define the instantaneous velocity at P as the limiting value approached by this series of average velocities for shorter and shorter intervals as Q becomes closer and closer to P.

In the notation of calculus, the limit of $\Delta x / \Delta t$ as Δt approaches zero is written dx/dt and is called the **derivative** of x with respect to t. Thus instantaneous velocity is defined as

The derivative: calculus as the language of physics

$$v = \lim_{\Delta t \to 0} \frac{\Delta x}{\Delta t} = \frac{dx}{dt}. \qquad (2\text{–}3)$$

Since Δt is assumed positive, v has the same algebraic sign as Δx. Hence if the positive x-axis points to the right, a positive velocity indicates motion toward the right.

In more general motion, velocity must be treated as a vector quantity. In that case Eq. (2–3) becomes the x-component of the instantaneous velocity. However, in one-dimensional motion along a straight line, it is customary to call v the instantaneous velocity. When we use the term *velocity*, we always mean instantaneous rather than average velocity, unless we state otherwise.

As point Q approaches point P in Fig. 2–1a, point q approaches point p in Fig. 2–1b. In the limit, the slope of the line pq equals the slope of the tangent to the curve at point p, obtained by dividing a vertical interval (with distance units) by a horizontal interval (with time units). *The instantaneous velocity at any point of a coordinate–time graph therefore equals the slope of the tangent to the graph at that point.* If the tangent slopes upward to the right, its slope is positive, the velocity is positive, and the motion is toward the right. If the tangent slopes downward to the right, the velocity is negative. At a point where the tangent is horizontal, its slope is zero and the velocity is zero.

If distance is expressed in meters and time in seconds, velocity is expressed in meters per second (m·s^{-1}). Other common units of velocity are feet per second (ft·s^{-1}), centimeters per second (cm·s^{-1}), miles per hour (mi·hr^{-1}), and knots (1 knot = 1 nautical mile or 6080 ft per hour).

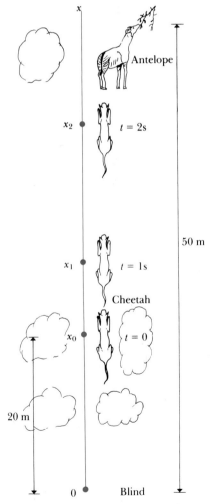

2–2 A cheetah attacking an antelope from ambush.

EXAMPLE 2–1 A cheetah is crouched in ambush 20 m to the north of an observer's blind (see Fig. 2–2). At time $t = 0$, the cheetah charges an antelope in a clearing 50 m north of the observer. The observer estimates that during the first 2 s of the attack, the cheetah's motion is described by the equation $x = a + bt^2$, where $a = 20$ m and $b = 5 \text{ m·s}^{-2}$.

a) Find the displacement of the cheetah in the time interval between $t_1 = 1$ s and $t_2 = 2$ s.

At time $t_1 = 1$ s, the cheetah's position is

$$x_1 = 20 \text{ m} + (5 \text{ m·s}^{-2})(1 \text{ s})^2 = 25 \text{ m}.$$

At time $t_2 = 2$ s,

$$x_2 = 20 \text{ m} + (5 \text{ m·s}^{-2})(2 \text{ s})^2 = 40 \text{ m}.$$

The displacement is therefore

$$x_2 - x_1 = 40 \text{ m} - 25 \text{ m} = 15 \text{ m}.$$

b) Find the average velocity in this time interval:

$$v_{av} = \frac{x_2 - x_1}{t_2 - t_1} = \frac{15 \text{ m}}{1 \text{ s}} = 15 \text{ m·s}^{-1}.$$

c) Find the instantaneous velocity at time $t_1 = 1$ s.
   The position at time $t = 1$ s $+ \Delta t$ is

$$x = 20 \text{ m} + (5 \text{ m·s}^{-2})(1 \text{ s} + \Delta t)^2$$
$$= 25 \text{ m} + (10 \text{ m·s}^{-1})\Delta t + (5 \text{ m·s}^{-2})(\Delta t)^2.$$

The displacement during the interval Δt is

$$\Delta x = 25 \text{ m} + (10 \text{ m·s}^{-1})\Delta t + (5 \text{ m·s}^{-2})(\Delta t)^2 - 25 \text{ m}$$
$$= (10 \text{ m·s}^{-1})\Delta t + (5 \text{ m·s}^{-2})(\Delta t)^2.$$

The average velocity during Δt is

$$v_{av} = \frac{\Delta x}{\Delta t} = 10 \text{ m·s}^{-1} + (5 \text{ m·s}^{-2}) \, \Delta t.$$

For the instantaneous velocity at $t = 1$ s, we let Δt approach zero in the expression for v_{av}: $v = 10 \text{ m·s}^{-1}$. This corresponds to the slope of the tangent at point p in Fig. 2–1b.

d) Derive a general expression for the instantaneous velocity as a function of time, and from it find v at $t = 1$ s and $t = 2$ s.
   We take the derivative of $x = a + bt^2$, obtaining

$$v = \frac{dx}{dt} = 2bt = (10 \text{ m·s}^{-1})t.$$

At time $t = 1$ s, $v = 10 \text{ m·s}^{-1}$, in agreement with part (c). At time $t = 2$ s, $v = 20 \text{ m·s}^{-1}$.

Speed: a somewhat ambiguous term

The term **speed** has two different meanings. It may mean the *magnitude* of the instantaneous velocity. For example, when two cars travel at 50 km·hr^{-1}, one north and the other south, both have speeds of 50 km·hr^{-1}. In a different sense, referring to an *average* quantity, the speed of a body is the total length of path, divided by the elapsed time. Thus if a car travels 90 km in 3 hr, its average speed is 30 km·hr^{-1}, even if the trip starts and ends at the same point. In the latter case the average velocity would be zero, because the total displacement is zero.

2–3 AVERAGE AND INSTANTANEOUS ACCELERATION

When the velocity of a moving body changes with time, we say that the body has an *acceleration*. Just as velocity is a quantitative description of the rate of change of position with time, so acceleration is a quantitative description of the rate of change of velocity with time.

Average acceleration: change in velocity divided by time

Considering again the motion of a particle along the x-axis, suppose that at time t_1 the particle is at point P and has velocity v_1, and that at a later time t_2 it is at point Q and has velocity v_2.

The **average acceleration** a_{av} of the particle as it moves from P to Q is defined as the ratio of the change in velocity to the elapsed time.

$$a_{av} = \frac{v_2 - v_1}{t_2 - t_1} = \frac{\Delta v}{\Delta t}. \tag{2-4}$$

Again, strictly speaking, v_1 and v_2 are values of the x-component of instantaneous velocity, and Eq. (2–4) defines the x-component of average acceleration.

EXAMPLE 2–2 An astronaut has left the space shuttle on a tether line to test a new personal maneuvering device. Her on-board partner measures her velocity before and after certain maneuvers, as follows:

a) $v_1 = 0.8 \text{ m·s}^{-1}$, $v_2 = 1.2 \text{ m·s}^{-1}$;

b) $v_1 = 1.6 \text{ m·s}^{-1}$, $v_2 = 1.2 \text{ m·s}^{-1}$;

c) $v_1 = -0.4 \text{ m·s}^{-1}$, $v_2 = -1.0 \text{ m·s}^{-1}$;

d) $v_1 = -1.6 \text{ m·s}^{-1}$, $v_2 = -0.8 \text{ m·s}^{-1}$.

If $t_1 = 2$ s and $t_2 = 4$ s, find the average acceleration for each set of data.

SOLUTION

a) $a_{av} = \dfrac{1.2 \text{ m·s}^{-1} - 0.8 \text{ m·s}^{-1}}{4 \text{ s} - 2 \text{ s}} = 0.2 \text{ m·s}^{-2}$.

b) $a_{av} = \dfrac{1.2 \text{ m·s}^{-1} - 1.6 \text{ m·s}^{-1}}{4 \text{ s} - 2 \text{ s}} = -0.2 \text{ m·s}^{-2}$.

c) $a_{av} = \dfrac{-1.0 \text{ m·s}^{-1} - (-0.4 \text{ m·s}^{-1})}{4 \text{ s} - 2 \text{ s}} = -0.3 \text{ m·s}^{-2}$.

d) $a_{av} = \dfrac{-0.8 \text{ m·s}^{-1} - (-1.6 \text{ m·s}^{-1})}{4 \text{ s} - 2 \text{ s}} = +0.4 \text{ m·s}^{-2}$.

When the acceleration has the *same* direction as the initial velocity, the body goes faster; when it is in the opposite direction, the particle slows down. When the body moves in the negative direction with increasing speed (c), the acceleration is negative; when it moves in the negative direction with decreasing speed (d), the acceleration is positive.

We can now define **instantaneous acceleration,** following the same procedure used for instantaneous velocity. Consider this situation: A sports car driver has just entered the final straightaway at the Grand Prix. He reaches point P at time t_1, moving with velocity v_1, and crosses the finish line at point Q at time t_2 with velocity v_2, as shown in Fig. 2–3a. Figure 2–3b is a graph of instantaneous velocity v plotted as a function of time, points p and q corresponding to positions P and Q in Fig. 2–3a. The average acceleration is represented by the slope of the line pq, computed by using the appropriate scales and units of the graph.

The instantaneous acceleration of a body—that is, its acceleration at some one instant of time or at some one point of its path—is defined in the same way as instantaneous velocity. We take the second point Q in Fig. 2–3a to be closer and closer to the first point P, so the average acceleration is computed over shorter and shorter intervals of time. The instantaneous acceleration at the first point is the limit approached by the average acceleration when the

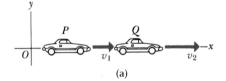

(a)

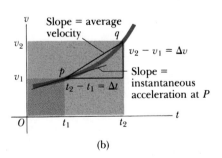

(b)

2–3 (a) Car moving on the x-axis. (b) Velocity–time graph of the motion. The average acceleration between t_1 and t_2 equals the slope of the line pq. The instantaneous acceleration at P equals the slope of the tangent at p.

Instantaneous acceleration: instantaneous rate of change of velocity second point is taken closer and closer to the first:

$$a = \lim_{\Delta t \to 0} \frac{\Delta v}{\Delta t} = \frac{dv}{dt}. \tag{2-5}$$

Instantaneous acceleration plays an essential role in the laws of mechanics, while average acceleration is less frequently used. From now on when we use the term *acceleration,* we always mean instantaneous acceleration.

As point Q in Fig. 2–3a approaches point P, point q in the graph (Fig. 2–3b) approaches point p, and the slope of the line pq approaches the slope of the line tangent to the curve at point p. Thus *the instantaneous acceleration at any point on the graph equals the slope of the line tangent to the curve at that point.* As with velocity, we have to use appropriate scales and units for the graph.

The acceleration $a = dv/dt$ can be expressed in various ways. Since $v = dx/dt$, it follows that

$$a = \frac{dv}{dt} = \frac{d}{dt}\left(\frac{dx}{dt}\right) = \frac{d^2x}{dt^2}. \tag{2-6}$$

Acceleration as a second derivative

The acceleration is therefore the *second* derivative of the coordinate with respect to time.

If we express velocity in meters per second and time in seconds, then acceleration is in meters per second, per second ($\text{m·s}^{-1}\text{·s}^{-1}$). This is usually written as m·s^{-2}, and is read "meters per second squared." Other common units of acceleration are feet per second squared (ft·s^{-2}) and centimeters per second squared (cm·s^{-2}).

EXAMPLE 2–3   Suppose the velocity of the car in Fig. 2–3 is given by the equation

$$v = m + nt^2,$$

where $m = 10 \text{ m·s}^{-1}$ and $n = 2 \text{ m·s}^{-3}$.

a)   Find the change in velocity of the car in the time interval between $t_1 = 2$ s and $t_2 = 5$ s.

At time $t_1 = 2$ s,

$$v_1 = 10 \text{ m·s}^{-1} + (2 \text{ m·s}^{-3})(2 \text{ s})^2$$
$$= 18 \text{ m·s}^{-1}.$$

At time $t_2 = 5$ s,

$$v_2 = 10 \text{ m·s}^{-1} + (2 \text{ m·s}^{-3})(5 \text{ s})^2$$
$$= 60 \text{ m·s}^{-1}.$$

The change in velocity is therefore

$$v_2 - v_1 = 60 \text{ m·s}^{-1} - 18 \text{ m·s}^{-1}$$
$$= 42 \text{ m·s}^{-1}.$$

b)   Find the average acceleration in this time interval:

$$a_{av} = \frac{v_2 - v_1}{t_2 - t_1} = \frac{42 \text{ m·s}^{-1}}{3 \text{ s}} = 14 \text{ m·s}^{-2}.$$

This corresponds to the slope of the line pq in Fig. 2–3b.

c) Find the instantaneous acceleration at time $t_1 = 2$ s.

At time $t = 2$ s $+ \Delta t$,

$$v = 10 \text{ m·s}^{-1} + (2 \text{ m·s}^{-3}) (2 \text{ s} + \Delta t)^2$$
$$= 18 \text{ m·s}^{-1} + (8 \text{ m·s}^{-2}) \Delta t + (2 \text{ m·s}^{-3}) (\Delta t)^2.$$

The change in velocity during Δt is

$$\Delta v = 18 \text{m·s}^{-1} + (8 \text{ m·s}^{-2}) \Delta t + (2 \text{ m·s}^{-3})(\Delta t)^2 - 18 \text{ m·s}^{-1}$$
$$= (8 \text{ m·s}^{-2}) \Delta t + (2 \text{ m·s}^{-3}) (\Delta t)^2.$$

The average acceleration during Δt is

$$a_{\text{av}} = \frac{\Delta v}{\Delta t} = 8 \text{ m·s}^{-2} + (2 \text{ m·s}^{-3}) \Delta t.$$

The instantaneous acceleration at $t = 2$ s, obtained by letting Δt approach zero, is $a = 8 \text{ m·s}^{-2}$. This corresponds to the slope of the tangent at point p in Fig. 2–3b.

d) Derive an expression for the instantaneous acceleration at any time, and use it to find the acceleration at $t = 2$ s and at $t = 5$ s.

Starting with $v = 10 \text{m·s}^{-1} + (2 \text{ m·s}^{-1})t^2$ and using Eq. (2–6), we find

$$a = \frac{dv}{dt}$$

$$= \frac{d}{dt}[10 \text{ m·s}^{-1} + (2 \text{ m·s}^{-3})t^2] = 2(2 \text{ m·s}^{-3})t.$$

When $t = 2$ s,

$$a = (2) (2 \text{m·s}^{-3}) (2 \text{ s}) = 8 \text{ m·s}^{-2}.$$

When $t = 5$ s,

$$a = (2) (2 \text{ m·s}^{-3}) (5 \text{ s}) = 20 \text{ m·s}^{-2}.$$

Note that neither of these values is equal to the average acceleration found in (b); the instantaneous acceleration varies with time. Automotive engineers sometimes call the time rate of change of acceleration the "jerk."

A few remarks about the *sign* of acceleration may be helpful. When the acceleration and velocity of a body have the same sign, the body is speeding up. If both are positive, the body moves in the positive direction with increasing speed. If both are negative, the body moves in the negative direction with a velocity that becomes more and more negative with time, and again the body's speed increases.

What does the sign of acceleration mean?

When v and a have opposite signs, the body is slowing down. If v is positive and a negative, the body moves in the positive direction with decreasing speed. If v is negative and a positive, the body moves in the negative direction with a velocity that is becoming less negative, and again the body slows down.

The term *deceleration* is sometimes used, either for a negative value of a or for a decrease in speed. Because of the ambiguity, it is best to avoid this term. We do not use it in this book; instead we recommend careful attention to the interpretation of the algebraic sign of a in relation to that of v.

Deceleration is a bad word.

Constant acceleration: an important
special case.

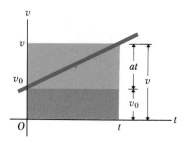

2–4 Velocity–time graph for straight-
line motion with constant acceleration.

2–4 MOTION WITH CONSTANT ACCELERATION

The simplest accelerated motion is straight-line motion with *constant* acceleration, when the velocity changes at the same rate throughout the motion. The
graph of velocity as a function of time is then a straight line, as in Fig. 2–4; the
velocity increases by equal amounts in equal time intervals. Hence in Eq. (2–4)
we can replace the average acceleration a_{av} by the constant (instantaneous)
acceleration a. We then have

$$a = \frac{v_2 - v_1}{t_2 - t_1}. \tag{2–7}$$

Now let $t_1 = 0$ and let t_2 be any arbitrary later time t. Let v_0 represent the
velocity when $t = 0$ (called the *initial* velocity), and let v be the velocity at the
later time t. Then Eq. (2–7) becomes

$$a = \frac{v - v_0}{t - 0}, \qquad \text{or} \qquad v = v_0 + at. \tag{2–8}$$

How to find the velocity at any time

We can interpret this equation as follows: The acceleration a is the constant rate of change of velocity, or the change per unit time. The term at is the
product of the change in velocity per unit time, a, and the time interval t.
Therefore it equals the *total* change in velocity. The velocity v at any time t
then equals the initial velocity v_0 (at the time $t = 0$) plus the change in velocity
at. Graphically, we can consider the ordinate v at time t in Fig. 2–4 as the sum
of two segments: one with length v_0 equal to the initial velocity, the other with
length at equal to the change in velocity during time t.

Here is an alternative route to Eq. (2–8). We know that v is some function
of t. If the derivative of that function with respect to t is the constant a, what
must the function itself be? (Such a function might be called an *antiderivative*,
but the more usual term is *indefinite integral*.) One possibility is the function at;
its derivative with respect to t is the constant a. But the derivative of the
function $at + C$, where C is any constant, is also equal to a, because the derivative of any constant is zero. Thus we conclude that the function for v must
have the form

$$v = at + C. \tag{2–9}$$

Now this function must also satisfy the additional requirement that at time
$t = 0$ it yields the value v_0. Substituting $t = 0$ and equating the result to v_0, we
find

$$v_0 = a(0) + C, \qquad \text{or } C = v_0. \tag{2–10}$$

Putting all this together, we again obtain

$$v = v_0 + at.$$

How to find the position at any time

We may use a similar procedure to find an expression for the position x of
the particle as a function of time, if the position at time $t = 0$ is denoted by x_0.
We know that $v = dx/dt$; what function of t must x be, in order for its derivative
with respect to t to equal $v_0 + at$? That is, what is the antiderivative of $v_0 + at$?
The function

$$x = v_0 t + \tfrac{1}{2}at^2 + D, \tag{2–11}$$

where D is any constant, satisfies this requirement; this may be checked easily
by taking the derivative of Eq. (2–11).

We also know that at time $t = 0$, Eq. (2–11) must give the value x_0. Substituting the value $t = 0$ and equating the result to x_0, we find $x_0 = D$. Combining

this with Eq. (2–11), we obtain

$$x = x_0 + v_0 t + \tfrac{1}{2}at^2. \tag{2–12}$$

This equation states that if at time $t = 0$ a particle is at position x_0 and has velocity v_0, its new position x at any later time t is the sum of three terms: its initial position x_0, plus the distance $v_0 t$ it would move if its velocity were constant, plus an additional distance $at^2/2$ caused by the changing velocity.

We may also combine Eqs. (2–8) and (2–12) to obtain a relation for x, v, and a that does not contain t. We first solve Eq. (2–8) for t and then substitute the resulting expression into Eq. (2–12) and simplify:

Getting rid of the time variable: relation of position to velocity

$$t = \frac{v - v_0}{a},$$

$$x = x_0 + v_0\left(\frac{v - v_0}{a}\right) + \frac{1}{2}a\left(\frac{v - v_0}{a}\right)^2.$$

Transfer the term x_0 to the left side and multiply through by $2a$:

$$2a(x - x_0) = 2v_0 v - 2v_0{}^2 + v^2 - 2v_0 v + v_0{}^2,$$

or, finally,

$$v^2 = v_0{}^2 + 2a(x - x_0). \tag{2–13}$$

Equations (2–8), (2–12), and (2–13) are the *equations of motion with constant acceleration*. Any kinematic problem involving motion of a particle on a straight line with constant acceleration can be solved by using these equations.

The curve in Fig. 2–5 is a graph of the coordinate x as a function of time for motion with constant acceleration, for a case where v_0 and a are positive. That is, it is a graph of Eq. (2–12); the curve is a *parabola*. The slope of the tangent at $t = 0$ equals the initial velocity v_0, and the slope of the tangent at any time t equals the velocity v at that time. The slope continuously increases with t, and measurements would show that the *rate* of increase with t is constant.

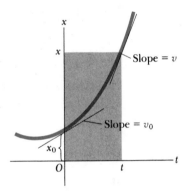

2–5 Coordinate–time graph for motion with constant acceleration.

PROBLEM-SOLVING STRATEGY: *Motion with constant acceleration*

1. It is essential to decide at the beginning of a problem where the origin of coordinates is and which axis direction is positive. The choices are usually made on the basis of convenience; often it is easiest to place the particle at the origin at time $t = 0$. A diagram showing these choices is always helpful.

2. Once you have chosen the positive axis direction, the positive directions for velocity and acceleration are also determined. It would be inconsistent to define x as positive to the right of the origin and velocities toward the left as positive.

3. In constant-acceleration problems it is always useful to make a list of unknown quantities such as x, x_0, v, v_0, and a. In general, some of these will be known, some unknown. When possible, select an equation from Eqs. (2–8), (2–12), and (2–13) that contains only one of the unknowns; solve for the unknown, then substitute the known values and compute the value of the unknown.

4. It often helps to restate the problem in prose first, then translate that description into symbols and equations. *When* does the particle arrive at a certain point? (I.e., at what value of t?) *Where* is the particle when its velocity has a certain value? (I.e., what is the value of x when v has the specified value?) In the following example we ask: "Where is the motorcyclist when his velocity is 5 m·s⁻¹?" Translated into symbols, this becomes: "What is the value of x when $v = 5$ m·s⁻¹?"

5. Take a hard look at your results to see whether they make sense. Are they within the general range of magnitudes you expected?

How fast can a motorcyclist get out of
town?

EXAMPLE 2–4 A motorcyclist heading east through a small Iowa town accelerates as he passes the signpost at $x = 0$ marking the city limits. His acceleration is constant, $a = 4$ m·s^{-2}. At time $t = 0$ he is at $x = 5$ m and has velocity $v = 3$ m·s^{-1}.

a)   Find his position and velocity at time $t = 2$ s.

b)   Where is he when his velocity is 5 m·s^{-1}?

SOLUTION   We take the signpost as the origin of coordinates, and the positive x-axis points east. At the initial time $t = 0$, the initial position is $x_0 = 5$ m, and the initial velocity is $v_0 = 3$ m·s^{-1}. We want to know the position and velocity (i.e., the values of x and v) at the later time $t = 2$ s. With reference to Eqs. (2–8), (2–12), and (2–13), we have

$$x_0 = 5 \text{ m}, \qquad v_0 = 3 \text{ m·s}^{-1}, \qquad a = 4 \text{ m·s}^{-2}.$$

a)   From Eq. (2–12), which gives position x as a function of time t,

$$\begin{aligned} x &= x_0 + v_0 t + \tfrac{1}{2}at^2 \\ &= 5 \text{ m} + (3 \text{ m·s}^{-1})(2 \text{ s}) + \tfrac{1}{2}(4 \text{ m·s}^{-2})(2 \text{ s})^2 \\ &= 19 \text{ m}. \end{aligned}$$

From Eq. (2–8), which gives velocity v as a function of time t,

$$\begin{aligned} v &= v_0 + at \\ &= 3 \text{ m·s}^{-1} + (4 \text{ m·s}^{-2})(2 \text{ s}) \\ &= 11 \text{ m·s}^{-1}. \end{aligned}$$

b)   From Eq. (2–13)

$$v^2 = v_0{}^2 + 2a(x - x_0),$$
$$(5 \text{ m·s}^{-1})^2 = (3 \text{ m·s}^{-1})^2 + 2(4 \text{m·s}^{-2})(x - 5 \text{ m}),$$
$$x = 7 \text{ m}.$$

Alternatively, we may use Eq. (2–8) to find first the *time* when $v = 5$ m·s^{-1}:

$$5 \text{ m·s}^{-1} = 3 \text{ m·s}^{-1} + (4 \text{ m·s}^{-2})(t),$$
$$t = \tfrac{1}{2} \text{ s}.$$

Then use Eq. (2–12),

$$\begin{aligned} x &= 5 \text{ m} + (3 \text{ m·s}^{-1})(\tfrac{1}{2} \text{ s}) + \tfrac{1}{2}(4 \text{ m·s}^{-2})(\tfrac{1}{2} \text{ s})^2 \\ &= 7 \text{ m}. \end{aligned}$$

2–5 VELOCITY AND COORDINATE BY INTEGRATION

In deriving the equations of Section 2–4 we assumed that the acceleration is constant; when it is not, we cannot use them. When a varies with time, we can still use the relation $v = dx/dt$ to find the velocity v as a function of time if the position x is a given function of time. Similarly, we can use $a = dv/dt$ to find the acceleration a as a function of time if the velocity v is a given function of time.

Integration: Reversing the process of
differentiation

We can also reverse this process. Suppose v is known as a function of time; how can we find x as a function of time? To answer this question, we first consider a graphical approach. Figure 2–6 shows a velocity-versus-time curve for a situation where the acceleration (the slope of the curve) is not constant but increases with time. Considering the motion during the interval between times t_1 and t_2, we divide this total interval into many smaller intervals, calling a typical one Δt. Let the velocity during that interval be v. Of course, the velocity changes during Δt, but if the interval is very small, the change will also be very small. This displacement during that interval, neglecting the variation of v, is given by

$$\Delta x = v \, \Delta t.$$

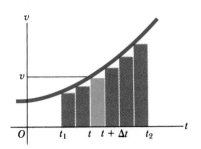

2–6 The area under a velocity–time graph equals the displacement.

This corresponds graphically to the area of the shaded strip with height v and width Δt, that is, the area under the curve corresponding to the interval Δt. Since the total displacement in any interval (say, t_1 to t_2) is the sum of the displacements in the small subintervals, the total displacement is given graphically by the *total* area under the curve between the vertical lines t_1 and t_2. In the limit, when all the Δt's become very small and their number very large, this is simply the **integral** of v (which is in general a function of t) from t_1 to t_2. Thus x_1 is the position at time t_1 and x_2 the position at time t_2:

$$x_2 - x_1 = \int_{x_1}^{x_2} dx = \int_{t_1}^{t_2} v \, dt. \tag{2–14}$$

A similar analysis with the acceleration-versus-time curve, where a is in general a function of t, shows that if v_1 is the velocity at time t_1 and v_2 the velocity at time t_2, the change in velocity Δv during a small time interval Δt is approximately equal to $a \, \Delta t$, and the total change in velocity $(v_2 - v_1)$ during the interval $t_2 - t_1$ is given by

$$v_2 - v_1 = \int_{v_1}^{v_2} dv = \int_{t_1}^{t_2} a \, dt. \tag{2–15}$$

EXAMPLE 2–5 An automobile travels along a straight highway. Its acceleration is given as a function of time by

$$a = 2.0 \text{ m·s}^{-2} - (0.1 \text{ m·s}^{-3})t.$$

The position at time $t = 0$ is given by $x_0 = 0$, and the velocity at time $t = 0$ is $v_0 = 10 \text{ m·s}^{-1}$.

a) Derive expressions for the velocity and position as functions of time.

b) At what time is the velocity of the car greatest?

c) What is the maximum velocity?

SOLUTION

a) We let t_1 be the initial time $t = 0$ and t_2 be any later time t. Similarly, v_1 is the initial velocity (at time $t = 0$), $v_0 = 10 \text{ m·s}^{-1}$. We use Eq. (2–15) to find the velocity v at time t:

$$v - 10 \text{ m·s}^{-1} = \int_0^t [2.0 \text{ m·s}^{-2} - (0.1 \text{ m·s}^{-3})t] \, dt$$
$$= (2.0 \text{ m·s}^{-2})t - \tfrac{1}{2}(0.1 \text{ m·s}^{-2})t^2.$$

Then we use Eq. (2–14) to find x as a function of t, taking x_1 as the initial position $x_0 = 0$:

$$x - 0 = \int_0^t [10 \text{ m·s}^{-1} + (2.0 \text{ m·s}^{-2})t - \tfrac{1}{2}(0.1 \text{ m·s}^{-2})t^2] \, dt$$
$$= (10 \text{ m·s}^{-1})t + \tfrac{1}{2}(2.0 \text{ m·s}^{-1})t^2 - \tfrac{1}{6}(0.1 \text{ m·s}^{-2})t^3.$$

b) The maximum value of v occurs at the time when v stops increasing and begins to decrease. At this instant, $dv/dt = 0$. Taking the derivative of the expression above for v and setting it equal to zero, we obtain

$$0 = (2.0 \text{ m·s}^{-2}) - (0.1 \text{ m·s}^{-3})t,$$
$$t = \frac{2.0 \text{ m·s}^{-2}}{0.1 \text{ m·s}^{-3}} = 20 \text{ s}.$$

This result can also be obtained immediately from the expression for a by noting that a is positive between $t = 0$ and $t = 20$ s, and negative after that. It is zero at $t = 20$ s, corresponding to the fact that $dv/dt = 0$ at this time. Thus the automobile speeds up until $t = 20$ s and then begins to slow down again.

c) The maximum velocity is obtained by substituting $t = 20$ s (the time when the maximum velocity is attained) into the general velocity equation:

$$v_{\max} = (2.0 \text{ m·s}^{-2})(20 \text{ s}) - \tfrac{1}{2}(0.1 \text{ m·s}^{-3})(20 \text{ s})^2 + 10 \text{ m·s}^{-1}$$
$$= 30 \text{ m·s}^{-1}$$

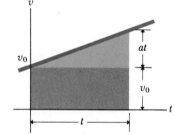

2–7 The area under a velocity–time graph equals the displacement $x - x_0$.

We can also use Eq. (2–14) for an alternative derivation of Eq. (2–12), the constant-acceleration case. Using Eq. (2–8) in Eq. (2–14) and carrying out the integration, we obtain:

$$x - x_0 = \int_0^t (v_0 + at) \, dt$$
$$= v_0 t + \tfrac{1}{2}at^2.$$

Finally, we can use the graphical approach that led us to Eq. (2–14) for still another derivation of Eq. (2–12). The graph of Eq. (2–8) is shown in Fig. 2–7. The total displacement $x - x_0$ from time $t = 0$ to any later time t is represented by the area under the graph. This can be obtained by finding the rectangular and triangular areas and adding. Carrying out this program, we find

$$x - x_0 = v_0 t + \tfrac{1}{2}(t)(at)$$
$$= v_0 t + \tfrac{1}{2}at^2.$$

Thus the various methods we have described are all consistent, and they lead to the same results in the special case of constant acceleration.

2–6 FREELY FALLING BODIES

Free fall: Do heavy bodies fall faster than light ones?

The most familiar example of motion with (nearly) constant acceleration is that of a body falling toward the earth. The motion of a falling body under the influence of the earth's gravity alone is often called **free fall**; it has held the attention of philosophers and scientists since ancient times. Aristotle thought (erroneously) that heavy objects fall faster than light objects, in proportion to their weight. Galileo argued that the motion of a falling body should be nearly independent of its weight and should have constant acceleration. According to legend, Galileo tested his theory by dropping cannonballs and bullets from the Leaning Tower of Pisa, although there is no reference to such experiments in his own writings.

In more recent times the motion of falling bodies has been studied with great precision. When air resistance can be neglected, we find that all bodies at a particular location fall with the same acceleration, regardless of their size or weight. If the distance of fall is small compared to the radius of the earth, the acceleration is constant throughout the fall. In the following discussion we neglect the effects of air resistance and the decrease of acceleration with increasing altitude. We call this idealized motion *free fall*, although it includes rising as well as falling motion.

The constant acceleration of a freely falling body is called the **acceleration due to gravity,** the *acceleration of gravity,* or (most accurately) the *acceleration of free fall,* and its magnitude is denoted by the letter g. At or near the earth's surface, the value of g is approximately 9.8 m·s^{-2}, 980 cm·s^{-2}, or 32 ft·s^{-2}. Later we will discuss the variation of g with latitude and elevation and the effect of the earth's rotation. On the surface of the moon, the acceleration of gravity is caused by the attractive force of the moon rather than the earth, and there $g = 1.67 \text{ m·s}^{-2}$. Near the surface of the sun, $g = 274 \text{ m·s}^{-2}$!

In the following examples we use the constant-acceleration equations developed in Section 2–4. We suggest you review the problem-solving strategies discussed in that section before you study these examples.

A body falls with constant acceleration, independent of its mass (almost).

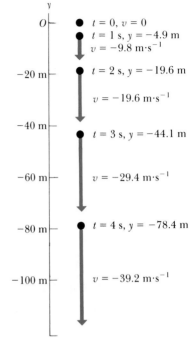

2–8 Position and velocity of a freely falling body.

EXAMPLE 2–6 A bullet is dropped from the Leaning Tower of Pisa; it starts from rest and falls freely. Compute its position and velocity after 1, 2, 3, and 4 s. Take the origin O at the elevation of the starting point, the y-axis as vertical, and the upward direction as positive.

SOLUTION The initial coordinate y_0 and the initial velocity v_0 are both zero. The acceleration is downward, in the negative y-direction, so $a = -g = -9.8 \text{ m·s}^{-2}$.
From Eqs. (2–12) and (2–8),

$$y = v_0 t + \tfrac{1}{2}at^2 = 0 - \tfrac{1}{2}gt^2 = (-4.9 \text{ m·s}^{-2})t^2,$$

$$v = v_0 + at = 0 - gt = (-9.8 \text{m·s}^{-2})t.$$

When $t = 1$ s, $y = (-4.9 \text{ m·s}^{-2})(1 \text{ s})^2 = -4.9$ m, and $v = (-9.8 \text{ m·s}^{-2})(1 \text{ s}) = -9.8 \text{ m·s}^{-1}$. The body is therefore 4.9 m below the origin (y is negative) and has a downward velocity (v is negative) of magnitude 9.8 m·s^{-1}.

The position and velocity at 2, 3, and 4 s are found in the same way. The results are shown in Fig. 2–8: check the numerical values for yourself.

EXAMPLE 2–7 Suppose you throw a ball vertically upward from the cornice of a tall building, and it leaves your hand with an upward speed of 15 m·s^{-1}. On its way back down it just misses the cornice. In Fig. 2–9, the downward path is displaced a little to the right from its actual position, for clarity. Find (a) the position and velocity of the ball 1 s and 4 s after it leaves your hand; (b) the velocity when the ball is 5 m above the cornice; (c) the maximum height reached and the time at which it is reached. Take the origin at the elevation at which the ball leaves your hand, the y-axis vertical and positive upward.

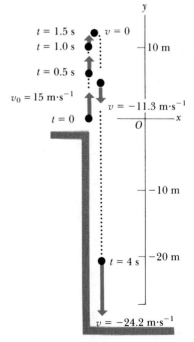

2–9 Position and velocity of a body thrown vertically upward.

SOLUTION The initial position y_0 is zero. The initial velocity v_0 is $+15$ m·s^{-1}, and the acceleration is $a = -9.8$ m·s^{-2}. The velocity at any time is

$$v = v_0 + at = 15 \text{ m·s}^{-1} + (-9.8 \text{ m·s}^{-2})t. \qquad (2\text{--}16)$$

The position at any time is

$$y = v_0t + \tfrac{1}{2}at^2 = (15 \text{ m·s}^{-1})t + \tfrac{1}{2}(-9.8 \text{ m·s}^{-2})t^2. \qquad (2\text{--}17)$$

The velocity at any position is

$$v^2 = v_0{}^2 + 2ay = (15 \text{ m·s}^{-1})^2 + 2(-9.8 \text{ m·s}^{-2})y. \qquad (2\text{--}18)$$

a) When $t = 1$ s, Eqs. (2–16) and (2–17) give

$$y = +10.1 \text{ m}, \qquad v = +5.2 \text{ m·s}^{-1}.$$

The ball is 10.1 m above the origin (y is positive) and it has an upward velocity (v is positive) of 5.2 m·s^{-1} (less than the initial velocity, as expected).

When $t = 4$ s, again from Eqs. (2–16) and (2–17),

$$y = -18.4 \text{ m}, \qquad v = -24.2 \text{ m·s}^{-1}.$$

The ball has passed its highest point and is 18.4 m *below* the origin (y is negative). It has a *downward* velocity (v is negative) of magnitude 24.4 m·s^{-1}. Note that it is not necessary to find the highest point reached or the time at which it was reached. The equations of motion give the position and velocity at *any* time, whether the ball is on the way up or the way down.

b) When the ball is 5 m above the origin,

$$y = +5 \text{ m}$$

and, from Eq. (2–18),

$$v^2 = 127 \text{ m}^2\text{·s}^{-2}, \qquad v = \pm 11.3 \text{ m·s}^{-1}.$$

The ball passes this point *twice*, once on the way up and again on the way down. The velocity on the way up is $+11.3$ m·s^{-1}, and on the way down it is -11.3 m·s^{-1}.

c) At the highest point, $v = 0$. Hence, from Eq. (2–18),

$$0 = (15 \text{ m·s}^{-1})^2 - (19.6 \text{ m·s}^{-2})y$$

and

$$y = 11.5 \text{ m}.$$

The time can now be found from Eq. (2–16), setting $v = 0$:

$$0 = 15 \text{ m·s}^{-1} + (-9.8 \text{ m·s}^{-2})t.$$

$$t = 1.53 \text{ s}.$$

Alternatively, to find the maximum height we may ask first *when* the maximum height is reached. That is, at what t is $v = 0$? As just shown, $v = 0$ when $t = 1.53$ s; substituting this value of t back into Eq. (2–17), we find

$$y = (15 \text{ m·s}^{-1})(1.53 \text{ s}) + \tfrac{1}{2}(-9.8 \text{ m·s}^{-2})(1.53 \text{ s})^2$$
$$= 11.5 \text{ m}.$$

Although at the highest point the velocity is instantaneously zero, the *accelera-*

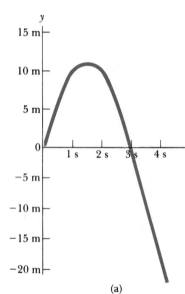

(a)

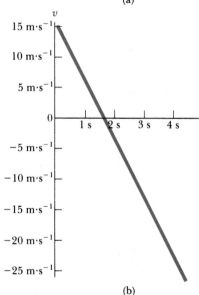

(b)

2–10 Graphs of y and v as functions of t for Example 2–7.

tion at this point is still -9.8 m·s^{-2}. The ball stops for an instant, but its velocity is continuously changing; the acceleration is constant throughout.

Figure 2–10 shows graphs of position and velocity as functions of time for this problem.

Figure 2–11 is a multiflash photograph of a falling golf ball, taken with a stroboscopic light source. This light source produces a series of intense flashes, each so short (a few millionths of a second) that there is no blur in the image of even a rapidly moving body. The time interval between flashes can be adjusted; the camera shutter is left open during the entire motion, and as each flash occurs, the position of the ball at that instant is recorded on the film.

The equally spaced light flashes subdivide the motion into equal time intervals Δt, so the velocity of the ball between any two flashes is directly proportional to the separation of its corresponding images in the photograph. If the velocity were constant, the images would be equally spaced. The increasing separation of the images during the fall shows that the velocity is continually increasing; the ball accelerates. By comparing two successive displacements of the ball, we can find the *change* in velocity in the corresponding time interval. Careful measurement shows that this change is the same in each time interval, indicating that the motion is one of *constant* acceleration.

Strobe photography: a useful tool to study motion

2–7 RELATIVE VELOCITY

The velocity of a body, like its position, is always described with reference to a coordinate system or **frame of reference.** Often this system is taken to be fixed in some other body, but this second body may also be in motion relative to a third, and so on. Thus one frame of reference may be moving with respect to another. When we speak of "the velocity of a car," we usually mean its velocity relative to the earth. But the earth is in motion relative to the sun, the sun is in motion relative to some other star, and so on. We use the term **relative velocity** to describe such relationships.

Suppose a long train of flatcars is moving to the right along a straight, level track, as in Fig. 2–12. Such trains are used in the Alps to transport automobiles through long tunnels. Suppose a daredevil is driving a car to the right along the flatcars. In Fig. 2–12, v_{FE} represents the velocity of the flatcars F relative to the earth E, and v_{AF} the velocity of the automobile A relative to the flatcars. In any time interval, the total displacement of the automobile relative to the earth is the sum of its displacement relative to the flatcars and their displacement relative to the earth. Thus the automobile's velocity relative to the earth, v_{AE}, is equal to the *sum* of the relative velocities v_{AF} and v_{FE}:

$$v_{AE} = v_{AF} + v_{FE}. \qquad (2\text{--}19)$$

Thus if the flatcars are traveling relative to the earth at 30 km·hr^{-1} ($= v_{FE}$), and the automobile is traveling relative to the flatcars at 40 km·hr^{-1} ($= v_{AF}$), the velocity of the automobile relative to the earth (v_{AE}) is 70 km·hr^{-1}.

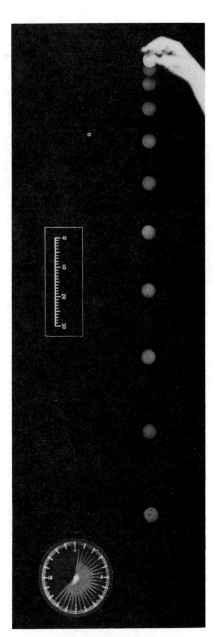

2–11 Multiflash photograph (retouched) of freely falling golf ball.

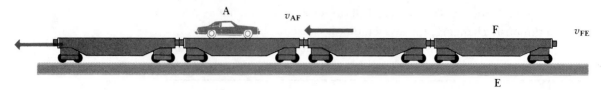

2–12 Vector v_{FE} is the velocity of the flatcars F relative to the earth E, and v_{AF} is the velocity of automobile A relative to the flatcars.

In straight-line motion the velocity v_{AE} is the *algebraic* sum of v_{AF} and v_{FE}. Thus if the automobile were traveling to the *left* with a velocity of 40 km·hr^{-1} relative to the flatcars, $v_{AF} = -40$ km·hr^{-1} and the velocity of the automobile relative to the earth would be -10 km·hr^{-1}; that is, it would be traveling to the left, relative to the earth.

PROBLEM-SOLVING STRATEGY: *Relative velocity*

Note the order of the double subscripts on the velocities: v_{AB} always means "velocity of A relative to B." In writing equations such as Eq. (2–19), make sure that the first subscript on the left side of the equation is the same as the first subscript in the *first* term on the right, and the second on the left is the same as the second in the *last* term on the right. Also, adjacent subscripts in adjacent terms on the right must match. Thus

$$v_{AE} = v_{AB} + v_{BC} + v_{CD} + v_{DE}.$$

Catching a speeding driver: an example of relative velocity

EXAMPLE 2–8   A man is driving a car at 65 km·hr^{-1} on a straight, level road where the speed limit is 40 km·hr^{-1}. He is spotted by a motorcycle officer, who accelerates in pursuit. By the time the driver sees the motorcycle's flashing blue light, the motorcycle is traveling at 80 km·hr^{-1}. What is the motorcycle's velocity relative to the car?

SOLUTION   Let the car be C, the motorcycle M, and the earth E. Then

$$v_{CE} = 65 \text{ km·hr}^{-1} \quad \text{and} \quad v_{ME} = 80 \text{ km·hr}^{-1},$$

and we wish to find v_{MC}.

From the rule for combining velocities,

$$v_{ME} = v_{MC} + v_{CE}.$$

Thus

$$80 \text{ km·hr}^{-1} = v_{MC} + 65 \text{ km·hr}^{-1},$$

$$v_{MC} = 15 \text{ km·hr}^{-1},$$

and the officer is overtaking the driver at 15 km·hr^{-1}.

EXAMPLE 2–9   How would the relative velocities be altered if the motorcycle were ahead of the car?

SOLUTION   Not at all. The relative *positions* of the bodies do not matter. The velocity of M relative to C is still $+15$ km·hr^{-1}, but he is now pulling ahead of C at this rate.

In more general motion, where the velocities do not all lie along a straight line, vector addition must be used to combine velocities. We will return to these more general problems at the end of Chapter 3.

SUMMARY

When a particle moves along a straight line, we describe its position with respect to an origin O by means of a coordinate such as x.

The particle's average velocity during a time interval Δt is defined as

$$v_{av} = \frac{\Delta x}{\Delta t}.$$ (2-2)

The instantaneous velocity at any time t is defined as

$$v = \lim_{\Delta t \to 0} \frac{\Delta x}{\Delta t} = \frac{dx}{dt}.$$ (2-3)

The average acceleration during a time interval Δt is defined as

$$a_{av} = \frac{\Delta v}{\Delta t}.$$ (2-4)

The instantaneous acceleration at any time t is defined as

$$a = \lim_{\Delta t \to 0} \frac{\Delta v}{\Delta t} = \frac{dv}{dt}.$$ (2-5)

Positive and negative directions for velocity are determined by the choice of positive direction for the coordinate.

When the acceleration is constant, the position x and velocity v at any time t are related to the acceleration a, the initial position x_0, and the initial velocity v_0 (both at time $t = 0$) by the following equations:

$$x = x_0 + v_0 t + \tfrac{1}{2}at^2,$$ (2-12)

$$v = v_0 + at,$$ (2-8)

$$v^2 - v_0{}^2 = 2a(x - x_0).$$ (2-13)

When the acceleration is not constant but is a known function of time, we can find the velocity and coordinate as functions of time by integrating the acceleration function.

When a body A moves relative to a body F, and F moves relative to the earth E or other frame of reference, we denote the velocity of A relative to F by v_{AF}, the velocity of F relative to E by v_{FE}, and the velocity of A relative to E by v_{AE}. These velocities are related by

$$v_{AE} = v_{AF} + v_{FE}.$$ (2-19)

KEY TERMS

kinematics
model
coordinate axis
coordinate
average velocity
instantaneous velocity
derivative
speed
average acceleration
instantaneous acceleration
integral
free fall
acceleration due to gravity
frame of reference
relative velocity

QUESTIONS

2-1 When a particle moves in a circular path, how many coordinates are required to describe a position, assuming that the radius of the circle is given?

2-2 What is meant by the statement that space is three-dimensional?

2-3 Does the speedometer of a car measure speed, or velocity?

2-4 Some European countries have highway speed limits of 100 km·hr^{-1}. What is the equivalent number of miles per hour?

2–5 A student claims that a speed of 60 mi·hr^{-1} is the same as 88 ft·s^{-1}. Is this relationship exact, or only approximate?

2–6 In a given time interval, is the total displacement of a particle equal to the product of the average velocity and the time interval, even when the velocity is not constant?

2–7 Under what conditions is average velocity equal to instantaneous velocity?

2–8 When one flies in an airplane at night in smooth air, there is no sensation of motion even though the plane may be moving at 800 km·hr^{-1} (500 mi·hr^{-1}). Why is this?

2–9 An automobile is traveling north. Can it have a velocity toward the north and at the same time have an acceleration toward the south? Under what circumstances?

2–10 A ball is thrown straight up in the air. What is its acceleration at the instant it reaches it highest point?

2–11 Is the acceleration of a car greater when the accelerator is pushed to the floor, or when the brake pedal is pushed hard?

2–12 Under constant acceleration, the average velocity of a particle is half the sum of its initial and final velocities. Is this still true if the acceleration is *not* constant?

2–13 A baseball is thrown straight up in the air. Is the acceleration greater while it is being thrown, or after it is thrown?

2–14 How could you measure the acceleration of an automobile by using only instruments located within the automobile?

2–15 If the initial position and initial velocity of a vehicle are known, and a record is kept of the acceleration at each instant, can its position after a certain time be computed from these data? Explain how this might be done.

EXERCISES

Section 2–1 Average Velocity

2–1 In 1954 Roger Bannister became the first human to run a mile in less than 4 min. Suppose that a runner on a straight track covers a distance of 1 mi in exactly 4 min. What was the magnitude of his average velocity in

a) mi·hr^{-1} b) ft·s^{-1}

2–2 A hiker travels in a straight line for 40 min with an average velocity of 1.6 m·s^{-1}. After the 40 min what distance has he covered?

2–3 A mouse moves along a straight line; its distance from the origin at any instant is given by the equation $x = at + bt^2$, where $a = 8$ cm·s^{-1} and $b = -3$ cm·s^{-2}. Find the average velocity of the mouse in the interval from $t = 0$ to $t = 1$ s, and in the interval from $t = 0$ to $t = 4$ s.

Section 2–2 Instantaneous Velocity

2–4 A physics professor leaves his house and walks along the sidewalk toward campus. After 5 min he realizes it is raining and returns home. His distance from his house as a function of time is shown in Fig. 2–13. At which of the labeled points is his velocity

a) zero?   b) constant and positive?

c) constant and negative?   d) increasing in magnitude?

e) decreasing in magnitude?

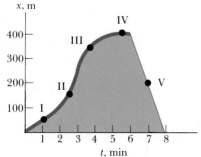

FIGURE 2–13

2–5 A hobbyist is testing a new model rocket engine by using it to propel a cart along a model railroad track. He determines that its motion along the x-axis is described by the equation $x = bt^2$, where $b = 10$ cm·s^{-2}. Compute the instantaneous velocity of the cart at time $t = 3$ s.

2–6 A car is initially stopped at a traffic light. It then travels along a straight road such that its distance from the light is given by $x = bt + ct^2$, where $b = 4.0$ m·s^{-1} and $c = 0.5$ m·s^{-2}.

a) Calculate the average velocity of the car for the time interval $t = 0$ to $t = 10$ s.

b) Calculate the instantaneous velocity of the car at
i) $t = 0$;
ii) $t = 5$ s;
iii) $t = 10$ s.

Section 2–3 Average and Instantaneous Acceleration

2–7 A test driver at Incredible Motors, Inc., is testing a new model car whose speedometer is calibrated to read m·s^{-1} rather than mi·hr^{-1}. The following series of speedometer readings was obtained during a test run:

Time (s)	0	2	4	6	8	10	12	14	16
Velocity (m·s^{-1})	0	0	2	5	10	15	20	22	22

a) Compute the average acceleration during each 2-s interval. Is the acceleration constant? Is it constant during any part of the time?

b) Make a velocity–time graph of the data above, using scales of 1 cm = 1 s horizontally, and 1 cm = 2 m·s^{-1} vertically. Draw a smooth curve through the plotted points. By measuring the slope of your curve, find the instantaneous acceleration at $t = 8$ s, 13 s, and 15 s.

2–8 An astronaut has left Spacelab V to test a new space scooter for use in constructing Space Habitat I. His partner measures the following velocity changes, each taking place in a 10-s interval. What are the magnitude, the algebraic sign, and the direction of the average acceleration in each interval?

a) At the beginning of the interval the astronaut is moving toward the right along the x-axis at 5 m·s^{-1}, and at the end of the interval he is moving toward the right at 20 m·s^{-1}.

b) At the beginning he is moving toward the right at 20 m·s^{-1}, and at the end he is moving toward the right at 5 m·s^{-1}.

c) At the beginning he is moving toward the left at 5 m·s^{-1}, and at the end he is moving toward the left at 20 m·s^{-1}.

d) At the beginning he is moving toward the left at 20 m·s^{-1}, and at the end he is moving toward the left at 5 m·s^{-1}.

e) At the beginning he is moving toward the right at 20 m·s^{-1}, and at the end he is moving toward the left at 20 m·s^{-1}.

f) At the beginning he is moving toward the left at 20 m·s^{-1}, and at the end he is moving toward the right at 20 m·s^{-1}.

g) In which of these cases is the scooter speeding up? In which cases is it slowing down?

2–9 Figure 2–14 is a graph of the coordinate of a spider crawling along the x-axis. Sketch the graphs of its velocity and acceleration as functions of time.

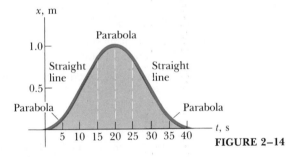

FIGURE 2–14

2–10 An automobile has a velocity as a function of time given by $v(t) = \alpha + \beta t^2$, where $\alpha = 5.0$ m·s^{-1} and $\beta = 0.3$ m·s^{-3}. Calculate

a) the average acceleration for the time interval $t = 0$ to $t = 5$ s;

b) the instantaneous acceleration for
   i) $t = 0$,
   ii) $t = 5$ s.

Section 2–4 Motion with Constant Acceleration

2–11 The makers of a certain automobile advertise that it will accelerate from 15 to 50 mi·hr^{-1} in 13 s. Compute

a) the acceleration in ft·s^{-2};

b) the distance the car travels in this time, assuming the acceleration to be constant.

2–12 An airplane travels 500 m down the runway before taking off. If it starts from rest, moves with constant

acceleration, and becomes airborne in 20 s, with what velocity in m·s^{-1} does it take off?

2–13 A car moving with constant acceleration covers the distance between two points 60 m apart in 6 s. Its velocity as it passes the second point is 15 m·s^{-1}.

a) What is its velocity at the first point?

b) What is the acceleration?

2–14 The fastest time for a 440 yd race from a standing start is 7.08 s according to the 1984 *Guinness Book of World Records*. This was done on a 1200 cc motorcycle in 1980. Assuming constant acceleration:

a) What was the acceleration of the cycle (in ft·s^{-2})?

b) What was the final velocity of the cycle (in ft·s^{-1})?

c) With this acceleration how long would it take to go from 0 to 55 mph?

2–15 The graph in Fig. 2–15 shows the velocity of a motorcycle police officer plotted as a function of time.

a) Find the instantaneous acceleration at $t = 3$ s, at $t = 7$ s, and at $t = 11$ s.

b) How far does the officer go in the first 5 s? The first 9 s? The first 13 s?

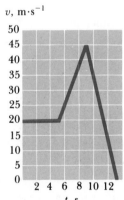

FIGURE 2–15

2–16 Figure 2–16 is a graph of the acceleration of a model railroad locomotive moving on the x-axis. Sketch the graphs of its velocity and coordinate as functions of time, if $x = v = 0$ when $t = 0$.

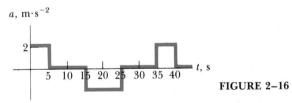

FIGURE 2–16

2–17 The "reaction" time of the average automobile driver is about 0.7 s. (The reaction time is the interval between the perception of a signal to stop and the application of the brakes.) If an automobile can slow down with an acceleration -16 ft·s^{-2}, compute the total distance covered in coming to a stop after a signal is observed

a) from an initial velocity of 15 mi·hr^{-1} (in a school zone);

b) from an initial velocity of 55 mi·hr^{-1}.

2–18 A subway train starts from rest at a station and accelerates at a rate of 2 m·s^{-2} for 10 s. It runs at constant speed for 30 s, and slows down at -4 m·s^{-2} until it stops at the next station. Find the *total* distance covered.

2–19 A hypothetical spaceship takes a straight-line path from the earth to the moon, a distance of about 400,000 km. Suppose it accelerates at 10 m·s^{-2} for the first 10 min of the trip, then travels at constant speed until the last 10 min, when it accelerates at -10 m·s^{-2}, just coming to rest as it reaches the moon.

a) What is the maximum speed attained?

b) What fraction of the total distance is traveled at constant speed?

c) What total time is required for the trip?

Section 2–5 Velocity and Coordinate by Integration

2–20 The motion of a particle along a straight line is described by the function

$$x = (6 \text{ m}) + (5 \text{ m·s}^{-2})t^2 - (1 \text{ m·s}^{-4})t^4.$$

Assume that t is positive.

a) Find the position, velocity, and acceleration at time $t = 2$ s.

b) During what time interval is the velocity positive?

c) During what time interval is x positive?

d) What is the maximum positive velocity attained by the particle?

2–21 The acceleration of a motorcycle is given by $a = (1.2 \text{ m·s}^{-3})t - (0.12 \text{ m·s}^{-4})t^2$. It is at rest at the origin at time $t = 0$.

a) Find its position and velocity as functions of time.

b) Calculate the maximum velocity it attains.

2–22 The acceleration of a bus is given by $a = (2 \text{ m·s}^{-3})t$.

a) If the bus's velocity at time $t = 1$ s is 5 m·s^{-1}, what is its velocity at time $t = 2$ s?

b) If the bus's position at time $t = 1$ s is 6 m, what is its position at time $t = 2$ s?

Section 2–6 Freely Falling Bodies

2–23

a) With what velocity must a ball be thrown vertically upward in order to rise to a height of 20 m?

b) How long will it be in the air?

2–24 A brick is dropped from the roof of a building. The brick strikes the ground in 4 s.

a) How tall, in meters, is the building?

b) What is the velocity of the brick just before it reaches the ground?

2–25 A ball is thrown vertically downward from the top of a building, leaving the thrower's hand with a velocity of 10 m·s^{-1}.

a) What will be its velocity after falling for 2 s?

b) How far will it fall in 2 s?

c) What will be its velocity after falling 10 m?

d) If it moved a distance of 1 m while in the thrower's hand, find its acceleration while in his hand (assuming it is constant).

e) If the point where the ball leaves the thrower's hand is 40 m above the ground, in how many seconds will the ball strike the ground?

f) What will the velocity of the ball be just before it strikes the ground?

2–26 A hot-air balloonist, rising vertically with a velocity of 5 m·s^{-1}, releases a sandbag at an instant when the balloon is 20 m above the ground.

a) Compute the position and velocity of the sandbag at the following times after its release: $\frac{1}{4}$ s, $\frac{1}{2}$ s, 1 s, 2 s.

b) How many seconds after its release will the bag strike the ground?

c) With what velocity will it strike?

2–27 A ball is thrown nearly vertically upward from a point near the cornice of a tall building. It just misses the cornice on the way down, and passes a point 160 ft below its starting point 5 s after it leaves the thrower's hand.

a) What was the initial velocity of the ball?

b) How high did it rise above its starting point?

c) What were the magnitude and direction of its velocity at the highest point?

d) What were the magnitude and direction of its acceleration at the highest point?

e) What was the magnitude of its velocity as it passed a point 64 ft below the starting point?

2–28 A ball is thrown vertically upward from the ground with a velocity of 30 m·s^{-1}.

a) How long will it take to rise to its highest point?

b) How high does the ball rise?

c) How long after being thrown will the ball have a velocity of 10 m·s^{-1} upward?

d) Of 10 m·s^{-1} downward?

e) When is the displacement of the ball zero?

f) When is the magnitude of the ball's velocity equal to half its initial velocity?

g) When is the magnitude of the ball's displacement equal to half the greatest height to which it rises?

h) What are the magnitude and direction of the acceleration while the ball is moving upward?

i) While moving downward?

j) When at the highest point?

2–29 The rocket-driven sled Sonic Wind No. 2, used for investigating the physiological effects of large accelerations, runs on a straight, level track 3500 ft long. Starting from rest, it can reach a speed of 1000 mi·hr^{-1} in 1.8 s.

a) Compute the acceleration in ft·s^{-2}, assuming it to be constant.

b) What is the ratio of this acceleration to that of a free-falling body, g?

c) What is the distance covered in 1.8 s?

d) A magazine article states that at the end of a certain run the speed of the sled was decreased from 632 mi·hr^{-1} to zero in 1.4 s, and that during this time its passenger was subjected to more than 40 times the pull of gravity (that is, the acceleration was greater than $40\,g$). Are these figures consistent?

2–30 Suppose the acceleration of gravity were only 1.0 m·s^{-2}, instead of 9.8 m·s^{-2}.

a) Estimate the height to which you could jump vertically from a standing start.

b) How high could you throw a baseball?

c) Estimate the maximum height of a window from which you would care to jump to a concrete sidewalk below. (Each story of an average building is about 4 m high.)

d) With what speed, in kilometers per hour, would you strike the sidewalk?

e) How many seconds would be required for you to reach the sidewalk?

Section 2–7 Relative Velocity

2–31 A "moving sidewalk" in an airport terminal building moves 1 m·s^{-1} and is 150 m long. If a man steps on at one end and walks 2 m·s^{-1} relative to the moving sidewalk, how much time does he require to reach the opposite end if he walks

a) in the same direction the sidewalk is moving?

b) in the opposite direction?

2–32 Two piers A and B are located on a river, 1610 m apart. Two men must make round trips from pier A to pier B and return. One man is to row a boat at a velocity of 6.4 km·hr^{-1} relative to the water, and the other man is to walk on the shore at a velocity of 6.4 km·hr^{-1}. The velocity of the river is 3.2 km·hr^{-1} in the direction from A to B. How long does it take each man to make the round trip?

PROBLEMS

2–33 A motorcycle's position is described by $x = A + Bt + Ct^3$, where A, B, and C are numerical constants.

a) Calculate the velocity of the cycle, as a function of time.

b) From your answer to (a), calculate the acceleration of the motorcycle as a function of time.

2–34 A jet-propelled motorcycle starts from rest, moves in a straight line with constant acceleration, and covers a distance of 64 m in 4 s.

a) What is the final velocity?

b) How much time was required to cover half the total distance?

c) What is the distance covered in one-half the total time?

d) What is the velocity when half the total distance has been covered?

e) What is the velocity after one-half the total time?

f) When will the instantaneous velocity equal the average velocity for the 0-to-4 s time interval?

2–35 A sled starts from rest at the top of a hill and slides down with constant acceleration. At some later time it is 32.0 m from the top; two seconds after that it is 50.0 m from the top, two seconds later 72.0 m from the top, and two seconds later 98.0 m.

a) What is the average velocity of the sled during each of the 2-s intervals after passing the 32.0-m point?

b) What is the acceleration of the sled?

c) What was the velocity of the sled when it passed the 32.0-m point?

d) How long did it take to go from the top to the 32.0-m point?

e) How far did the sled go during the first second after passing the 32.0-m point?

f) How long does it take the sled to go from the 32.0-m point to the midpoint between the 32.0-m and the 50.0-m points?

g) What is the velocity of the sled as it passes the midpoint in part (f)?

2–36 The engineer of a passenger train traveling at 30 m·s^{-1} sights a freight train whose caboose is 200 m ahead on the same track. The freight train is traveling in the same direction as the passenger train with a velocity of 10 m·s^{-1}. The engineer of the passenger train immediately applies the brakes, causing a constant acceleration of $-1\ \text{m·s}^{-2}$, while the freight train continues with constant speed.

a) Will there be a collision?

b) If so, where will it take place?

2–37 At the instant the traffic light turns green, an automobile that has been waiting at an intersection starts ahead with a constant acceleration of 2 m·s^{-2}. At the same instant a truck, traveling with a constant velocity of 10 m·s^{-1}, overtakes and passes the automobile.

a) How far beyond its starting point will the automobile overtake the truck?

b) How fast will it be traveling?

2–38 An automobile and a truck start from rest at the same instant, with the automobile initially at some distance behind the truck. The truck has a constant acceleration of 2 m·s^{-2} and the automobile an acceleration of 3 m·s^{-2}. The automobile overtakes the truck after the truck has moved 75 m.

a) How long does it take the automobile to overtake the truck?

b) How far was the automobile behind the truck initially?

c) What is the velocity of each when they are abreast?

2–39 A delivery truck's velocity is given by

$$v(t) = (4 \text{ m·s}^{-2})t + (3 \text{ m·s}^{-4})t^3.$$

It is at $x = 0$ when $t = 0$. Calculate its position and acceleration as functions of time.

2–40 An object's velocity is measured to be

$$v(t) = 4 \text{ m·s}^{-1} - (5 \text{ m·s}^{-3})t^2.$$

At $t = 0$ the object is at $x = 0$.

a) Calculate the object's position and acceleration as functions of time.

b) What is the object's maximum *positive* displacement from the origin?

2–41 The position of a particle is given by $x = A \sin \omega t$, where A and ω are constants.

a) Find the velocity of the particle as a function of time.

b) Find the acceleration as a function of time.

c) Find the velocity as a function of x. At what point is the magnitude of the velocity greatest? Least?

d) Find the acceleration as a function of x. At what point is the magnitude of the acceleration greatest? Least?

e) What is the maximum distance of the particle from the origin ($x = 0$)?

f) What is the maximum velocity?

g) What is the maximum acceleration?

h) Sketch graphs of x, v, and a as functions of time.

2–42 A student determined to test the law of gravity for himself walks off a skyscraper 300 m high, stopwatch in hand, and starts his free fall (zero initial velocity). Five seconds later, Superman arrives at the scene and dives off the roof to save the student.

a) What must Superman's initial velocity be for him to catch the student just before the ground is reached? (Assume that Superman's acceleration is that of any free-falling body.)

b) If the height of the skyscraper is less than some minimum value, even Superman cannot save the student. What is this minimum height?

2–43 A football is kicked vertically upward from the ground, and a student gazing out of the window sees it moving upward past her at 5 m·s^{-1}. The window is 10 m above the ground.

a) How high does the ball go above ground?

b) How long does it take to go from a height of 10 m to its highest point?

c) Find its velocity and acceleration $\frac{1}{2}$ s after it leaves the ground.

2–44 A flower pot falls off a window sill and past the window below. It takes 0.25 s to pass this window, which is 3 m high. How far is the top of the window below the window sill above?

2–45 A ball is released from rest at the top of an inclined plane 18 m long and reaches the bottom 3 s later. At the same instant that the first ball is released, a second ball is projected upward along the plane from its bottom with a certain initial velocity. The second ball is to travel part way up the plane, stop, and return to the bottom so that it arrives simultaneously with the first ball. Both balls have the same constant acceleration.

a) Find the acceleration of the balls.

b) What must be the initial velocity of the second ball?

c) How far up the plane will the second ball travel?

2–46 A stone is dropped from the top of a tall cliff, and 1 s later a second stone is thrown vertically downward with a velocity of 20 m·s^{-1}. How far below the top of the cliff will the second stone overtake the first?

2–47 A marble is dropped, and then 1 s later and from a point 5.0 m lower a second marble is dropped. When will the two marbles be 15.0 m apart?

2–48 A juggler performs in a room whose ceiling is 3 m above the level of his hands. He throws a ball vertically upward so that it just reaches the ceiling.

a) With what initial velocity does he throw the ball?

b) How much time is required for the ball to reach the ceiling?

He throws the second ball upward with the same initial velocity, at the instant that the first ball is at the ceiling.

c) How long after the second ball is thrown do the two balls pass each other?

d) When the balls pass each other, how far are they above the juggler's hands?

2–49 The driver of a car wishes to pass a truck that is traveling at a constant speed of 20 m·s^{-1} (about 45 mi·hr^{-1}). The car initially is also traveling at 20 m·s^{-1}. The car's maximum acceleration in this speed range is 0.5 m·s^{-2}. Initially the vehicles are separated by 25 m, and the car pulls back into the truck's lane after it is 25 m ahead of the truck. The car is 5 m long and the truck 20 m.

a) How much time is required for the car to pass the truck?

b) What distance does the car travel during this time?

c) What is the final speed of the car, assuming its acceleration is constant while passing the truck?

2–50 Two cars, A and B, travel in a straight line. The distance of A from the starting point is given as a function of time by $x_A = (4 \text{ m·s}^{-1})t + (1 \text{ m·s}^{-2})t^2$; the distance of B from the starting point is $x_B = (2 \text{ m·s}^{-2})t^2 + (2 \text{ m·s}^{-3})t^3$.

a) Which car is ahead just after they leave the starting point?

b) At what values of t are the cars at the same point?

c) At what values of t is the velocity of B relative to A zero?

d) At what values of t is the distance from A to B neither increasing nor decreasing?

CHALLENGE PROBLEMS

2–51 As implied by Eqs. 2–3 and 2–5, the definition of the derivative df/dt of the function $f(t)$ is

$$\frac{df}{dt} = \lim_{\Delta t \to 0} \frac{\Delta f}{\Delta t},$$

where $\Delta f = f(t + \Delta t) - f(t)$. Use this definition to prove the following:

a) $(d/dt)t^n = nt^{n-1}$. (*Hint:* Use the binomial theorem, which may be found in Appendix B.)

b) $(d/dt)[af(t)] = a(df/dt)$.

c) $(d/dt)[f(t) + g(t)] = df/dt + dg/dt$.

2–52 An alert hiker sees a boulder fall from the top of a distant cliff, and notes that it takes 2 s for the boulder to fall the last third of the way to the ground.

a) What is the height of the cliff, in meters?

b) If in part (a) you get two roots to a quadratic equation and use one for your answer, what does the other root represent?

2–53 A student is running to catch the campus shuttle bus, which is stopped at the bus stop. The student is running at a constant velocity of 6 m·s^{-1}; she cannot run any faster. When the student is still 80 m from the bus, it starts to pull away. The bus moves with a constant acceleration of 0.2 m·s^{-2}.

a) For how much time and how far will the student have to run before she overtakes the bus?

b) When she reaches the bus, how fast will the bus be traveling?

c) Sketch a graph showing $x(t)$ for both the student and the bus. Take $x = 0$ as the initial position of the student.

d) The equations you used in (a) to find the time have a second solution, corresponding to a later time for which the student and bus will again be at the same place if they continue their specified motions. Explain the significance of this second solution. How fast will the bus be traveling at this point?

e) If the student's constant velocity is 4 m·s^{-1}, will she catch the bus?

f) What is the *minimum* speed the student must have just to catch up with the bus? How long and how far will she have to run in that case?

2–54 A ball is thrown straight up from the edge of the roof of a building. A second ball is dropped from the roof 2 s later.

a) If the height of the building is 40 m, what must be the initial velocity of the first ball if both are to hit the ground at the same time?

Consider the same situation as above, but now let the initial velocity v_0 of the first ball be given and treat the height h of the building as an unknown.

b) What must the height of the building be for both balls to reach the ground at the same time for each of the

following values of v_0:

  i) 13.0 m·s^{-1};

  ii) 19.5 m·s^{-1}?

c) If v_0 is greater than some value v_{max}, a value of h does not exist that allows both balls to hit the ground at the same time. Solve for v_{max}. The value v_{max} has a simple physical interpretation; what is it?

d) If v_0 is less than some value v_{min}, a value of h does not exist that allows both balls to hit the ground at the same time. Solve for v_{min}. The value v_{min} also has a simple physical interpretation; what is it?

2–55

a) Suppose cars are lined up bumper to bumper at a red light. Assume each car is 4.6 m in length. When the light turns green, the first car accelerates at 1.22 m·s^{-2}. When this car is 6 m in front of the second car, the second car begins to accelerate at 1.22 m·s^{-2}. When the second car is 6 m in front of the third car, it begins to accelerate at 1.22 m·s^{-2}. This continues in similar fashion for the fourth, fifth, etc., car. If the light stays green for a minute and a half, how many cars have made it to the beginning of the intersection before the light turns red again?

b) Now assume instead that the cars stopped at the red light are spaced such that there is 9 m between each pair of cars. If, when the light turns green, each car immediately accelerates at 1.22 m·s^{-2} for 12 s and then goes at constant velocity, how many cars make it to the beginning of the intersection before the light turns red again? Assume as in (a) that the light is green for 90 s and that each car is 4.6 m long. Compare your answer to that of (a); you may find the results surprising.

2–56 The acceleration of an object suspended from a spring and oscillating vertically is $a = -Ky$, where K is a constant and y is the coordinate measured from the equilibrium position. Suppose that an object moving in this way is given an initial velocity v_0 at the coordinate y_0. Find the expression for the velocity v of the object as a function of its coordinate y. (*Hint:* Use the expression $a = v(dv/dy)$)

2–57 The motion of an object falling from rest in a resisting medium is described by the equation

$$\frac{dv}{dt} = A - Bv,$$

where A and B are constants. In terms of A and B, find

a) the initial acceleration;

b) the velocity at which the acceleration becomes zero (the terminal velocity).

c) Show that the velocity at any given t is given by

$$v = \frac{A}{B}(1 - e^{-Bt}).$$

2–58 After the engine of a moving motorboat is cut off, the boat has an acceleration in the opposite direction to its velocity and directly proportional to the square of its velocity. That is, $dv/dt = -kv^2$, where k is constant.

a) Show that the magnitude v of the velocity at a time t after the engine is cut off is given by

$$\frac{1}{v} = \frac{1}{v_0} + kt.$$

b) Show that the distance x traveled in a time t is

$$x = \frac{1}{k} \ln (v_0 kt + 1).$$

c) Show that the velocity after traveling a distance x is

$$v = v_0 e^{-kx}.$$

As a numerical example, suppose the engine is cut off when the velocity is 6 m·s^{-1}, and that the velocity decreases to 3 m·s^{-1} in a time of 15 s.

d) Find the numerical value of the constant k, and the unit in which it is expressed.

e) Find the acceleration at the instant the engine is cut off.

f) Calculate x, v, and a at 10-s intervals for the first 60 s after the engine is cut off. Sketch graphs of x, v, and a as functions of t, for the first 60 s of the motion.

3
MOTION IN A PLANE

IN CHAPTER 2 WE STUDIED MOTION ALONG A STRAIGHT LINE. NOW WE broaden our discussion to include motion along a path that lies in a plane. A few familiar examples are the flight of a thrown or batted baseball, a projectile shot from a gun, a ball whirled in a circle at the end of a cord, the motion of the moon or of a satellite around the earth, and the motions of the planets around the sun. To describe these motions we need to generalize the kinematic language introduced in Chapter 2. Velocity and acceleration are *vector* quantities, and so we need to use the methods of vector algebra introduced in Chapter 1.

The content of this chapter, like that of the preceding one, is classified as *kinematics;* we are concerned only with describing motion, not with relating motion to its causes. The language introduced here, however, will be an essential tool in later chapters when we use Newton's laws of motion to study the relation between force and motion.

3–1 AVERAGE AND INSTANTANEOUS VELOCITY

To begin our study of motion along a curved path in a plane, we consider the situation of Fig. 3–1a. A particle moves along a curved path in the *xy*-plane; points P and Q represent the positions of the particle at two different times. We can describe the position of the particle at P by means of the displacement vector r from the origin 0 to P; we call r the **position vector** of the particle. The components of r are the coordinates x and y, and we may write

$$r = x\boldsymbol{i} + y\boldsymbol{j}, \qquad (3\text{--}1)$$

where $\boldsymbol{i}$ and $\boldsymbol{j}$ are the unit vectors introduced in Section 1–8.

As the particle moves from P to Q, its displacement is the change Δr in the position vector r, as shown. The *x*- and *y*-components of Δr are the quantities Δx and Δy, as shown.

Let Δt be the time interval during which the particle moves from P to Q. The **average velocity** v_{av} during this interval is defined to be a vector quantity

How to describe the position of a point in a plane

Average velocity: displacement divided by time

49

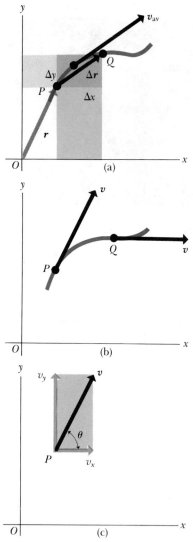

3–1 (a) The displacement Δr between points P and Q has components Δx and Δy. The average velocity v_{av} has the same direction as Δr. (b) The instantaneous velocity v at each point is always tangent to the path at that point. (c) Components of the instantaneous velocity at P are shown.

equal to the displacement divided by the time interval:

$$v_{av} = \frac{\Delta r}{\Delta t}. \tag{3–2}$$

That is, the average velocity is a vector quantity having the same direction as Δr and a magnitude equal to the magnitude of Δr divided by Δt. The magnitude of Δr is always the straight-line distance from P to Q, regardless of the actual shape of the path taken by the particle. Thus the average velocity would be the same for any path that would take the particle from P to Q in the time interval Δt.

The **instantaneous velocity** v at point P is defined in magnitude and direction as the *limit approached by the average velocity* when point Q is taken closer and closer to point P (as $\Delta t \to 0$).

$$\text{Instantaneous velocity } v = \lim_{\Delta t \to 0} \frac{\Delta r}{\Delta t} = \frac{dr}{dt}. \tag{3–3}$$

As point Q approaches point P, the direction of the vector Δr approaches that of the tangent to the path at P, so that the instantaneous velocity vector at any point is tangent to the path at that point. The instantaneous velocities at points P and Q are shown in Fig. 3–1b. Just as in Chapter 2, we define average velocity with reference to a finite time interval, while instantaneous velocity describes the motion at a specific point and at a specific instant of time.

It follows from Eq. (3–1) that the components $(v_x)_{av}$ and $(v_y)_{av}$ of average velocity are the corresponding components of displacement, divided by Δt. That is,

$$(v_x)_{av} = \frac{\Delta x}{\Delta t}, \qquad (v_y)_{av} = \frac{\Delta y}{\Delta t}.$$

Similarly, the instantaneous rates of change of x and y, denoted by v_x and v_y, are equal respectively to the x- and y-components of the instantaneous velocity v, and so we may write

$$v_x = \lim_{\Delta t \to 0} \frac{\Delta x}{\Delta t} = \frac{dx}{dt}, \qquad v_y = \lim_{\Delta t \to 0} \frac{\Delta y}{\Delta t} = \frac{dy}{dt}. \tag{3–4}$$

The magnitude of the instantaneous velocity is given by

$$|v| = \sqrt{v_x{}^2 + v_y{}^2},$$

and the angle θ in Fig. 3–1c by

$$\tan \theta = \frac{v_y}{v_x}.$$

In terms of unit vectors,

$$v = \frac{dr}{dt} = \frac{d}{dt}(xi + yj) = \frac{dx}{dt}i + \frac{dy}{dt}j. \tag{3–5}$$

If we know x and y as functions of time, we can compute the derivatives in Eqs. (3–4) and (3–5) and determine the velocity at any time. In many of the problems of this chapter our objective will be to obtain x and y as functions of time; these functions then provide a complete description of the motion.

Thus we may represent velocity, a vector quantity, in terms of its components or in terms of its magnitude and direction, just as with other vector quantities such as displacement and force. The *direction* of the instantaneous velocity of a particle at any point is *always* tangent to the path at that point.

3–2 AVERAGE AND INSTANTANEOUS ACCELERATION

In Fig. 3–2a the vectors v_1 and v_2 represent the instantaneous velocities, at points P and Q, of a particle moving in a curved path. The velocity v_2 may differ in both magnitude and direction from the velocity v_1.

The **average acceleration** a_{av} of the particle as it moves from P to Q is defined as the *vector change in velocity*, Δv, divided by the time interval Δt:

$$\text{Average acceleration} = a_{av} = \frac{\Delta v}{\Delta t}. \qquad (3\text{–}6)$$

Average acceleration is a vector quantity, in the same direction as the vector Δv.

The vector change in velocity, Δv, means the vector difference $v_2 - v_1$:

$$\Delta v = v_2 = v_1,$$

or

$$v_2 = v_1 + \Delta v.$$

Thus v_2 is the vector sum of the original velocity v_1 and the change Δv. This relationship is shown in Fig. 3–2b. The average acceleration vector, $a_{av} = \Delta v/\Delta t$, is shown in Fig. 3–2a.

The **instantaneous acceleration** a at point P is defined in magnitude and direction as the limit approached by the average acceleration when point Q approaches point P and Δv and Δt both approach zero:

$$\text{Instantaneous acceleration} = a = \lim_{\Delta t \to 0} \frac{\Delta v}{\Delta t} = \frac{dv}{dt}. \qquad (3\text{–}7)$$

Just as with velocity, average acceleration refers to a finite time interval during which the velocity changes, while instantaneous acceleration is the rate of change of velocity at a specific point and at a specific instant of time. The instantaneous acceleration vector at point P is shown in Fig. 3–2c. Note that it does not have the same direction as the velocity vector; in general there is no reason it should. Reference to the construction in Fig. 3–2c shows that the acceleration vector must always lie on the *concave* side of the curved path.

3–3 COMPONENTS OF ACCELERATION

We often represent the acceleration of a particle in terms of the components of this vector quantity. Figure 3–3 again shows the motion of a particle as described in a rectangular coordinate system. The x- and y-components, a_x and a_y, of the instantaneous acceleration a of the particle are

$$a_x = \lim_{\Delta t \to 0} \frac{\Delta v_x}{\Delta t} = \frac{dv_x}{dt}, \qquad a_y = \lim_{\Delta t \to 0} \frac{\Delta v_y}{\Delta t} = \frac{dv_y}{dt}. \qquad (3\text{–}8)$$

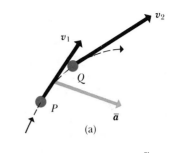

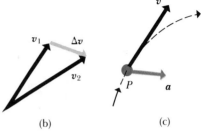

3–2 (a) The vector $a_{av} = \Delta v/\Delta t$ represents the average acceleration between P and Q. (b) Construction for obtaining $\Delta v = v - v_1$. (c) Instantaneous acceleration a at point P. Vector v is tangent to the path; vector a points toward the concave side of the path.

Instantaneous acceleration: acceleration at a point

Acceleration is a vector quantity; it has components.

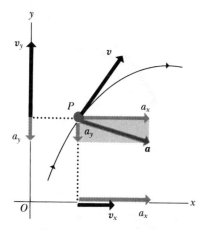

3–3 The acceleration a is resolved into its components a_x and a_y.

In terms of unit vectors,

$$a = \frac{dv_x}{dt}i + \frac{dv_y}{dt}j. \qquad (3\text{–}9)$$

By using Eqs. (3–4) and (3–5), we may also express the acceleration as

$$a_x = \frac{d^2x}{dt^2}, \qquad a_y = \frac{d^2y}{dt^2}, \qquad (3\text{–}10)$$

or

$$a = \frac{d^2x}{dt^2}i + \frac{d^2y}{dt^2}j. \qquad (3\text{–}11)$$

If we know the components a_x and a_y, we can find the magnitude and direction of the acceleration a, just as with velocity.

EXAMPLE 3–1 The coordinates of a particle moving in the xy-plane are given as functions of time by

$$x = 1 \text{ m} + (2 \text{ m·s}^{-2})t^2,$$
$$y = (2 \text{ m·s}^{-1})t + (1 \text{ m·s}^{-3})t^3.$$

Find the particle's position, velocity, and acceleration at time $t = 2$ s.

SOLUTION At time $t = 2$ s,

$$x = 1 \text{ m} + (2 \text{ m·s}^{-2})(2 \text{ s})^2 = 9 \text{ m},$$
$$y = (2 \text{ m·s}^{-1})(2 \text{ s}) + (1 \text{ m·s}^{-3})(2 \text{ s})^3 = 12 \text{ m},$$

or

$$r = (9 \text{ m})i + (12 \text{ m})j.$$

The particle's distance from the origin at this time is

$$r = \sqrt{x^2 + y^2} = \sqrt{(9 \text{ m})^2 + (12 \text{ m})^2} = 15 \text{ m}.$$

The velocity at any given time t is given by

$$v_x = \frac{dx}{dt} = 2(2 \text{ m·s}^{-2})t, \qquad v_y = \frac{dy}{dt} = (2 \text{ m·s}^{-1}) + 3(1 \text{ m·s}^{-3})t^2.$$

The velocity at time $t = 2$ s is given by

$$v_x = 2(2 \text{ m·s}^{-2})(2 \text{ s}) = 8 \text{ m·s}^{-1},$$
$$v_y = (2 \text{ m·s}^{-1}) + 3(1 \text{ m·s}^{-3})(2 \text{ s})^2 = 14 \text{ m·s}^{-1},$$

or

$$v = (8 \text{ m·s}^{-1})i + (14 \text{ m·s}^{-1})j.$$

The magnitude of the velocity at this time is

$$v = \sqrt{v_x{}^2 + v_y{}^2} = \sqrt{(8 \text{ m·s}^{-1})^2 + (14 \text{ m·s}^{-1})^2} = 16.1 \text{ m·s}^{-1},$$

and its direction with respect to the positive x-axis is

$$\theta = \arctan \frac{14 \text{ m·s}^{-1}}{8 \text{ m·s}^{-1}} = 60.3°.$$

The acceleration at any time is given by

$$a_x = \frac{dv_x}{dt} = \frac{d^2x}{dt^2} = 2(2 \text{ m·s}^{-2}), \qquad a_y = \frac{dv_y}{dt} = \frac{d^2y}{dt^2} = 6(1 \text{ m·s}^{-3})t.$$

The acceleration at time $t = 2$ s is

$$a_x = 4 \text{ m·s}^{-2}, \qquad a_y = 6(1 \text{ m·s}^{-3})(2 \text{ s}) = 12 \text{ m·s}^{-2},$$

or

$$\boldsymbol{a} = (4 \text{ m·s}^{-2})\boldsymbol{i} + (12 \text{ m·s}^{-2})\boldsymbol{j}.$$

The magnitude of the acceleration at this time is

$$a = \sqrt{a_x{}^2 + a_y{}^2} = \sqrt{(4 \text{ m·s}^{-2})^2 + (12 \text{ m·s}^{-2})^2} = 12.6 \text{ m·s}^{-2},$$

and its direction with respect to the positive x-axis is

$$\theta = \arctan \frac{a_y}{a_x} = \arctan \frac{12 \text{ m·s}^{-2}}{4 \text{ m·s}^{-2}} = 71.6°.$$

Note that the direction of the acceleration is different from that of the velocity, and that it is *not* tangent to the particle's path.

The acceleration of a particle moving in a curved path can also be represented in terms of rectangular components $a_\perp$ and $a_\parallel$, in directions **normal** (perpendicular) and **tangential** (parallel) to the path, as shown in Fig. 3–4a. Unlike the rectangular components referred to a set of fixed axes, the normal and tangential components do not have fixed directions in space. They do, however, have a direct physical significance. The parallel component $a_\parallel$ corresponds to a change in the *magnitude* of the velocity vector $\boldsymbol{v}$, while the normal component $a_\perp$ is associated with a change in the *direction* of the velocity.

Figure 3–4 shows two special cases. In Fig. 3–4b, the acceleration is *parallel* to the velocity $\boldsymbol{v}_1$. Then because $\boldsymbol{a}$ gives the rate of change of velocity, the change in $\boldsymbol{v}$ during a small time interval Δt is a vector $\Delta \boldsymbol{v}$ having the same direction as $\boldsymbol{a}$ and hence the same direction as $\boldsymbol{v}_1$. Thus the velocity $\boldsymbol{v}_2$ at the end of Δt, given by $\boldsymbol{v}_2 = \boldsymbol{v}_1 + \Delta \boldsymbol{v}$, is a vector having the same direction as $\boldsymbol{v}_1$ but somewhat greater magnitude.

In Fig. 3–4c, the acceleration is *perpendicular* to the velocity. In an interval Δt, the change $\Delta \boldsymbol{v}$ is a vector perpendicular to $\boldsymbol{v}_1$, as shown. Again $\boldsymbol{v}_2 = \boldsymbol{v}_1 + \Delta \boldsymbol{v}$, but in this case $\boldsymbol{v}_1$ and $\boldsymbol{v}_2$ have different directions. As the time interval Δt approaches zero, the angle θ in the figure also approaches zero, $\Delta \boldsymbol{v}$ becomes perpendicular to both $\boldsymbol{v}_1$ and $\boldsymbol{v}_2$, and $\boldsymbol{v}_1$ and $\boldsymbol{v}_2$ have the same magnitude.

Thus when $\boldsymbol{a}$ is parallel to $\boldsymbol{v}$, its effect is to change the magnitude of $\boldsymbol{v}$ but not its direction; when $\boldsymbol{a}$ is perpendicular to $\boldsymbol{v}$, its effect is to change the direction of $\boldsymbol{v}$ but not its magnitude. In general $\boldsymbol{a}$ may have components both parallel and perpendicular to $\boldsymbol{v}$, but the above statements are still valid for the individual components. In particular, when a particle travels along a curved path with constant speed, its acceleration is not zero, even though the magnitude of $\boldsymbol{v}$ does not change. In this case the acceleration is always perpendicular to $\boldsymbol{v}$ at each point. When a particle moves in a circle with constant speed, the acceleration is at each instant directed toward the center of the circle. We will consider this special case in detail in Section 3–5.

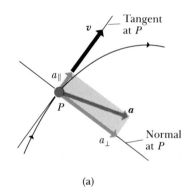

(a)

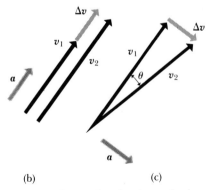

(b)   (c)

3–4 (a) The acceleration is resolved into a component $a_\parallel$ parallel or *tangent* to the path and a component $a_\perp$ perpendicular or *normal* to the path. (b) When $\boldsymbol{a}$ is parallel to $\boldsymbol{v}$, the magnitude of $\boldsymbol{v}$ increases but its direction does not change. (c) When $\boldsymbol{a}$ is perpendicular to $\boldsymbol{v}$, the direction of $\boldsymbol{v}$ changes but its magnitude is constant.

3–4 PROJECTILE MOTION

Projectile motion: What do a kicked football and a fireworks rocket have in common?

A **projectile** is any body that is given an initial velocity and then follows a path determined by the effect of the gravitational acceleration and by air resistance. A batted baseball, a thrown football, an object dropped from an airplane, and a bullet shot from a rifle are all examples of projectiles. The path followed by a projectile is called its *trajectory*. The motion of a freely falling body, discussed in Chapter 2, is a special case of projectile motion; in that case the trajectory is a vertical straight line.

To simplify the analysis we will consider only trajectories that have a range short enough that the acceleration of gravity may be considered constant in both magnitude and direction. We will also omit effects associated with air resistance and the rotation of the earth. These simplifications form the basis of an idealized *model* of the physical problem; we neglect minor details in order to focus attention on the most important features of the problem.

For our analysis of projectile motion we will use a set of rectangular coordinate axes, taking the *x*-axis horizontal and the *y*-axis vertically upward.

After the diver's feet leave the board, his center of mass follows a parabolic path. (Dr. Harold Edgerton, M.I.T., Cambridge, Massachusetts)

There is no acceleration in the horizontal direction, and the acceleration in the vertical direction is the acceleration due to gravity, which we studied in Section 2–6. Thus the components of a are

$$a_x = 0, \qquad a_y = -g. \qquad (3\text{--}12)$$

The negative sign for a_y arises because we have chosen the positive y-axis to be upward, while the acceleration due to gravity is downward.

Thus the horizontal component of acceleration is zero and the vertical component is downward and equal to that of a freely falling body. Since zero acceleration means constant velocity, the motion can be described as a combination of *horizontal motion with constant velocity* and *vertical motion with constant acceleration*.

The key to analysis of projectile motion is the fact that we can express all the needed vector relationships in terms of separate equations for the x- and y-components of these vector quantities. Each component of velocity is the rate of change of the corresponding coordinate, and each component of acceleration is the rate of change of the corresponding velocity component. In this sense the x and y motions are independent and may be analyzed separately. The actual motion is then the superposition of these separate motions.

Suppose that at time $t = 0$ our particle is at the point (x_0, y_0) and has velocity components v_{0x} and v_{0y}. The components of acceleration are $a_x = 0$, $a_y = -g$. The time variation of each coordinate is an example of motion with constant acceleration, and we can use Eqs. (2–8) and (2–12) directly. Substituting v_{0x} for v_0 and 0 for a, we find for x

$$v_x = v_{0x}, \qquad (3\text{--}13)$$
$$x = x_0 + v_{0x}t. \qquad (3\text{--}14)$$

Similarly, substituting v_{0y} for v_0 and $-g$ for a,

$$v_y = v_{0y} - gt, \qquad (3\text{--}15)$$
$$y = y_0 + v_{0y}t - \tfrac{1}{2}gt^2. \qquad (3\text{--}16)$$

The content of Eqs. (3–13) through (3–16) can also be represented by the vector equations

$$\boldsymbol{v} = \boldsymbol{v}_0 - gt\boldsymbol{j}, \qquad (3\text{--}17)$$
$$\boldsymbol{r} = \boldsymbol{r}_0 + \boldsymbol{v}_0 t - \tfrac{1}{2}gt^2\boldsymbol{j}, \qquad (3\text{--}18)$$

where $\boldsymbol{r}_0$ is the position vector at time $t = 0$.

Usually it is convenient to take the initial position as the origin; in this case, $x_0 = y_0 = 0$ or $\boldsymbol{r}_0 = \boldsymbol{0}$. This might be, for example, the position of a ball at the instant it leaves the thrower's hand, or the position of a bullet at the instant it leaves the gun barrel.

Figure 3–5 shows the path of a projectile that passes through the origin at time $t = 0$. The position, velocity, and velocity components of the projectile are shown at a series of times separated by equal intervals. As the figure shows, v_x does not change, but v_y changes by equal amounts in successive intervals, corresponding to constant y-acceleration.

The initial velocity v_0 may be represented by its magnitude v_0 (the initial speed) and the angle θ_0 it makes with the positive x-axis. In terms of these quantities, the *components* v_{0x} and v_{0y} of initial velocity are

$$v_{0x} = v_0 \cos \theta_0,$$
$$v_{0y} = v_0 \sin \theta_0. \qquad (3\text{--}19)$$

The x- and y-motions can be analyzed separately.

Equations of projectile motion: all you ever wanted to know about projectiles

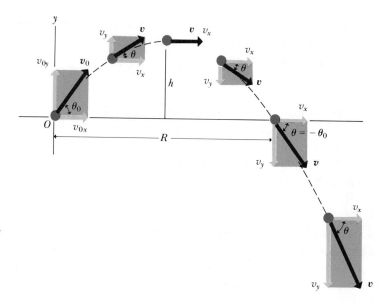

3–5 Trajectory of a body projected with an initial velocity v_0 at an angle of departure θ_0. *The distance R is the* horizontal range, and h is the maximum height.

Using these relations in Eqs. (3–13) through (3–16) and setting $x_0 = y_0 = 0$, we obtain

$$x = (v_0 \cos \theta_0)t, \tag{3-20}$$
$$y = (v_0 \sin \theta_0)t - \tfrac{1}{2}gt^2, \tag{3-21}$$
$$v_x = v_0 \cos \theta_0, \tag{3-22}$$
$$v_y = v_0 \sin \theta_0 - gt. \tag{3-23}$$

These equations describe the position and velocity of the projectile in Fig. 3–5 at any time t.

We can obtain a variety of additional information from Eqs. (3–20) through (3–23). For example, the distance r of the projectile from the origin at any time (the magnitude of the position vector r) is given by

$$r = \sqrt{x^2 + y^2}. \tag{3-24}$$

The projectile's speed (the magnitude of its resultant velocity) is

$$v = \sqrt{v_x^2 + v_y^2}. \tag{3-25}$$

The *direction* of the velocity, in terms of the angle θ it makes with the positive x-axis, is given by

$$\tan \theta = \frac{v_y}{v_x}. \tag{3-26}$$

The velocity vector v is tangent to the trajectory at each point.

Equations (3–20) and (3–21) give the position of the particle in terms of the parameter t. We can also obtain an equation for the shape of the trajectory, in terms of x and y, by eliminating t. We find $t = x/v_0 \cos \theta_0$ and

$$y = (\tan \theta_0)x - \frac{g}{2v_0^2 \cos^2 \theta_0}x^2. \tag{3-27}$$

The quantities v_0, $\tan \theta_0$, $\cos \theta_0$, and g are constants, so the equation has the form

$$y = ax - bx^2,$$

where a and b are constants. This is the equation of a *parabola*.

Figure 3–6 shows parabolic trajectories of a bouncing golf ball.

3–6 Stroboscopic photograph of a bouncing golf ball, showing parabolic trajectories after each bounce. Successive images are separated by equal time intervals, as in Fig. 3–5. Each peak in the trajectories is lower than the preceding one because of energy loss during the "bounce" or collision with the horizontal surface. (Dr. Harold Edgerton, M.I.T., Cambridge, Massachusetts.)

PROBLEM-SOLVING STRATEGY: *Projectile problems*

The same strategies used in Section 2–4 for solving problems with constant acceleration along a straight line are also useful here.

1. Define your coordinate system. Make a sketch showing your axes; label the positive direction for each and show the location of the origin.

2. Make lists of known and unknown quantities. In some problems the components (or magnitude and direction) of initial velocity will be given, and you can use Eqs. (3–20) and (3–21) to find the coordinates and velocity components at some later time. In other problems you may know two points on the trajectory and be asked to find the initial velocity. Be sure you know which quantities are given and which are to be found.

3. It often helps to state the problem in prose and then translate into symbols. *When* does the particle arrive at a certain point (i.e., at what value of t)? *Where* is the particle when its velocity has a certain value (i.e., what are the values of x and y when v_x or v_y has the specified value)? And so on.

4. At the highest point in a trajectory, $v_y = 0$. So the question "When does the projectile reach its highest point?" translates into "What is the value of t when $v_y = 0$?" Similarly, if $y_0 = 0$, then "When does the projectile return to its initial elevation?" translates into "What is the value of t when $y = 0$?" And so on.

EXAMPLE 3–2   A motorcycle stunt rider rides off the edge of a cliff with a horizontal velocity of magnitude 5 m·s^{-1}. Find the rider's position and velocity after $\frac{1}{4}$ s (see Fig. 3–7).

An adventurous motorcyclist rides off a cliff.

SOLUTION   The coordinate system is shown in Fig. 3–7. The initial angle θ_0 is zero, so $v_{0x} = 5$ m·s^{-1} and $v_{0y} = 0$. The horizontal velocity component equals the initial velocity and is constant.

Where is the motorcycle at $t = \frac{1}{4}$ s? The x- and y-coordinates, when $t = \frac{1}{4}$ s, are

$$x = v_x t = (5 \text{ m·s}^{-1})(\tfrac{1}{4} \text{ s}) = 1.25 \text{ m},$$
$$y = -\tfrac{1}{2}gt^2 = -\tfrac{1}{2}(9.8 \text{ m·s}^{-2})(\tfrac{1}{4} \text{ s})^2 = -0.306 \text{ m}.$$

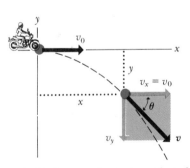

The distance from the origin at this time is

$$r = \sqrt{x^2 + y^2} = \sqrt{(1.25 \text{ m})^2 + (-0.306 \text{ m})^2} = 1.29 \text{ m}.$$

3–7 Trajectory of a body projected horizontally.

What is the velocity at time $t = \frac{1}{4}$ s? The components of velocity at time $t = \frac{1}{4}$ s are

$$v_x = v_0 = 5 \text{ m·s}^{-1},$$
$$v_y = -gt = (-9.8 \text{ m·s}^{-2})(\tfrac{1}{4} \text{ s}) = -2.45 \text{ m·s}^{-1}.$$

The resultant velocity has magnitude

$$v = \sqrt{v_x{}^2 + v_y{}^2} = 5.57 \text{ m·s}^{-1}.$$

The angle θ is

$$\theta = \arctan \frac{v_y}{v_x} = \arctan \frac{-2.45 \text{ m·s}^{-1}}{5.00 \text{ m·s}^{-1}} = -26.1°.$$

That is, at this time the velocity is 26.1° *below* the horizontal.

EXAMPLE 3–3   In Fig. 3–5, suppose the projectile is a baseball hit by Superman with an initial speed $v_0 = 50$ m·s^{-1} at an initial angle $\theta_0 = 53.1°$. The ball is probably struck a meter or so above ground level; we neglect this distance and assume it

How far can Superman hit a major-league fastball?

starts at ground level, where $y = 0$. Then

$$v_{0x} = v_0 \cos \theta_0 = (50 \text{ m·s}^{-1})(0.60) = 30 \text{ m·s}^{-1},$$
$$v_{0y} = v_0 \sin \theta_0 = (50 \text{ m·s}^{-1})(0.80) = 40 \text{ m·s}^{-1}.$$

a) Find the position of the ball, and the magnitude and direction of its velocity, when $t = 2.0$ s.

Using the coordinate system shown in Fig. 3–5, we want to find x, y, v_x, and v_y at time $t = 2.0$ s. From Eqs. (3–20) through (3–23),

$$x = (30 \text{ m·s}^{-1})(2.0 \text{ s}) = 60 \text{ m},$$
$$y = (40 \text{ m·s}^{-1})(2.0 \text{ s}) - \tfrac{1}{2}(9.8 \text{ m·s}^{-2})(2.0 \text{ s})^2 = 60.4 \text{ m},$$
$$v_x = 30 \text{ m·s}^{-1},$$
$$v_y = 40 \text{ m·s}^{-1} - (9.8 \text{ m·s}^{-2})(2.0 \text{ s}) = 20.4 \text{ m·s}^{-1},$$
$$v = \sqrt{v_x^2 + v_y^2} = 36.3 \text{ m·s}^{-1},$$
$$\theta = \arctan \frac{20.4 \text{ m·s}^{-1}}{30 \text{ m·s}^{-1}} = \arctan 0.680 = 34.2°.$$

b) Find the time at which the ball reaches the highest point of its flight, and find the height of this point.

At the highest point, the vertical velocity v_y is zero. When does this happen? If t_1 is the time at which this point is reached,

$$v_y = 0 = 40 \text{ m·s}^{-1} - (9.8 \text{ m·s}^{-2})t_1, \qquad t_1 = 4.08 \text{ s}.$$

The height h of the point is the value of y when $t = 4.08$ s:

$$h = (40 \text{ m·s}^{-1})(4.08 \text{ s}) - \tfrac{1}{2}(9.8 \text{ m·s}^{-2})(4.08 \text{ s})^2 = 81.6 \text{ m}.$$

c) Find the *horizontal range R*, that is, the horizontal distance from the starting point to the point where the ball returns to earth, where $y = 0$.

When does the ball return to earth? Let t_2 be the time when y becomes zero. Then

$$y = 0 = (40 \text{ m·s}^{-1})t_2 - \tfrac{1}{2}(9.8 \text{ m·s}^{-2})t_2^2.$$

This is a quadratic equation for t_2; it has two roots,

$$t_2 = 0 \qquad \text{and} \qquad t_2 = 8.16 \text{ s},$$

corresponding to the two times at which $y = 0$. The value $t_2 = 0$ is, of course, the time the ball left the ground; $t_2 = 8.16$ s is the time of its return. Note that this is just twice the time to reach the highest point. The time of descent therefore equals the time of rise. (This is *always* true if the starting point and end point are at the same elevation; can you prove this?)

The horizontal range R is the value of x when the ball returns to the ground, that is, when $t = 8.16$ s:

$$R = v_x t_2 = (30 \text{ m·s}^{-1})(8.16 \text{ s}) = 245 \text{ m}.$$

Thus, the ball is a home run. The vertical component of velocity at this point is

$$v_y = 40 \text{ m·s}^{-1} - (9.8 \text{ m·s}^{-2})(8.16 \text{ s}) = -40 \text{ m·s}^{-1}.$$

That is, the vertical velocity has the same magnitude as the initial vertical velocity, but the opposite direction. Since v_x is constant, the angle *below* the horizontal at this point equals the initial angle θ_0.

d) If the ball did not hit the ground, it would continue to travel on below its original level. The playing field might be located atop a flat-topped hill that drops off steeply on one side. Then negative values of y, corresponding to times greater than 8.16 s, are possible. We challenge the reader to compute the position and velocity at a time 10 s after the start, corresponding to the last position shown in Fig. 3–5. The results are

$$x = 300 \text{ m}, \qquad y = -90 \text{ m},$$
$$v_x = 30 \text{ m·s}^{-1}, \qquad v_y = -58 \text{ m·s}^{-1}.$$

EXAMPLE 3–4 In Fig. 3–8, a boy shoots an arrow from ground level at an apple hanging in a tree. At the same instant he releases the arrow, the apple falls from the tree and drops straight down, starting from rest. Show that the arrow's path curves just enough for it to hit the apple, regardless of its initial velocity.

A very talented archer hits the apple.

SOLUTION We have to prove that the apple and the arrow both arrive at the same time at some point directly below the apple's initial location. The initial elevation of the apple is $x \tan \theta_0$, and in time t it falls a distance $\frac{1}{2}gt^2$. Its elevation at the instant of collision is therefore

$$y = x \tan \theta_0 - \tfrac{1}{2}gt^2.$$

In this same time the arrow travels the distance x with constant x-component of velocity $v_0 \cos \theta_0$, so $x = v_0 \cos \theta_0 t$. Solving this for t and substituting in the preceding equation, we obtain

$$y = x \tan \theta_0 - \frac{1}{2}g\left(\frac{x}{v_0 \cos \theta_0}\right)^2.$$

But this is the same expression as Eq. (3–27) for the path of the arrow. Thus at the instant the arrow reaches the line along which the apple is falling, the two heights are the same, and the two meet at this point.

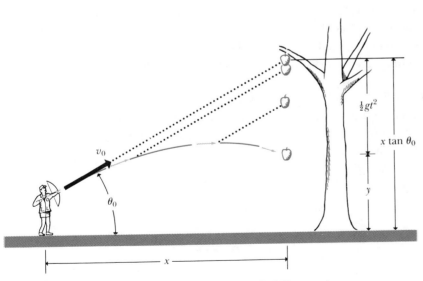

3–8 Trajectory of an arrow shot directly at a freely falling apple.

EXAMPLE 3–5 For a projectile launched with speed v_0 at initial angle θ_0, derive general expressions for the maximum height h and range R shown in Fig. 3–5. For a given v_0, what value of θ_0 gives maximum range?

How to find the range and maximum height of a projectile

SOLUTION We follow the same pattern as in Example 3–3, part (b). First, when does the projectile reach its maximum height? At this point $v_y = 0$, so the time t_1

at maximum height is given by

$$v_y = v_0 \sin \theta_0 - gt_1 = 0, \qquad t_1 = \frac{v_0 \sin \theta_0}{g}.$$

Next, what is the value of y at this time? From Eq. (3–21),

$$h = v_0 \sin\theta_0 \left(\frac{v_0 \sin \theta_0}{g} \right) - \frac{1}{2} g \left(\frac{v_0 \sin \theta_0}{g} \right)^2$$

$$= \frac{v_0^2 \sin^2 \theta_0}{2g}. \tag{3–28}$$

As expected, the maximum value of h occurs when the projectile is launched straight up, for then $\theta_0 = 90°$, $\sin \theta_0 = 1$, and $h = v_0^2/2g$. If it is launched horizontally, $\theta_0 = 0$ and the maximum height is zero!

To find the range, we first find the time t_2 when the projectile returns to the ground. At that time $y = 0$ and, from Eq. (3–21),

$$t_2(v_0 \sin \theta_0 - \tfrac{1}{2}gt_2) = 0.$$

The two roots of this quadratic equation for t_2 are $t_2 = 0$ and $t_2 = 2v_0 \sin \theta_0/g$. The first is obviously the time the projectile *leaves* the ground; the range R is the value of x at the second time. From Eq. (3–20),

$$R = (v_0 \cos \theta_0)\left(\frac{2v_0 \sin \theta_0}{g} \right).$$

According to a familiar trigonometric identity, $2 \sin \theta_0 \cos \theta_0 = \sin 2\theta_0$, so

$$R = \frac{v_0^2 \sin 2\theta_0}{g}. \tag{3–29}$$

The maximum value of $\sin 2\theta_0$, namely unity, occurs when $2\theta_0 = 90°$, or $\theta_0 = 45°$, and this angle gives the maximum range for a given initial speed. Finally, Eqs. (3–28) and (3–29) can be used only when the initial and final values of y are equal. Be careful; there are many end-of-chapter problems where they are *not* applicable.

How to be a Super Bowl quarterback

EXAMPLE 3–6 In some problems we want to know what the departure angle θ_0 should be for a given v_0 to produce a certain range R. Suppose a football player wants to throw a football at 20 m·s^{-1} to a receiver 30 m away. At what angle should he throw it?

SOLUTION From Eq. (3–29),

$$\theta_0 = \frac{1}{2} \arcsin \frac{Rg}{v_0^2}$$

$$= \frac{1}{2} \arcsin \frac{(30 \text{ m})(9.80 \text{ m·s}^{-2})}{(20 \text{ m·s}^{-1})^2}$$

$$= \frac{1}{2} \arcsin 0.735.$$

There are *two* values of θ_0 between 0 and 90° satisfying this equation: arcsin 0.735 = 47.3° or 132.7°, giving $\theta_0 = 23.7°$ or 66.3°. Both these angles give the same range; the time of flight and the maximum height are greater for the higher-angle trajectory. Incidentally, the sum of these two values of θ_0 is exactly 90°. This is not a coincidence. Can you prove this?

Figure 3–9 is copied from a multiflash photograph of the trajectory of a ball; x- and y-axes and the initial velocity vector have been added. The horizontal distances between consecutive positions are all equal, showing that the horizontal velocity component is constant. The vertical distances first decrease and then increase, indicating that the vertical motion is accelerated.

Figure 3–10 is made from a composite photograph of three trajectories of a ball projected from a spring gun with angles of 30°, 45°, and 60°. The initial speed v_0 is approximately the same in all three cases. The horizontal ranges are nearly the same for the 30° and 60° angles, and the range for 45° is greater than either.

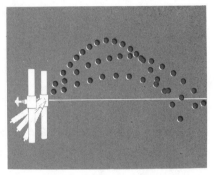

3–9 Trajectory of a body projected at an angle with the horizontal. (Reproduction of a multiflash photograph.)

3–5 CIRCULAR MOTION

We discussed components of acceleration in Section 3–3. When a particle moves along a curved path, it must have a component of acceleration perpendicular (normal) to the path, even if its speed is constant. When the path is a *circle*, there is a simple relation between the normal component of acceleration, the speed of the particle, and the radius of the circle. We now derive this relation for the special case when a particle moves in a circle with *constant speed*. This motion is called **uniform circular motion.** Note that this case is different from that of Section 3–4 because here the acceleration is *not* constant.

Figure 3–11 shows a particle moving in a circular path of radius R with center at O. The vector change in velocity, Δv, is shown in Fig. 3–11b. The particle moves from P to Q in a time Δt.

The triangles OPQ and opq in Fig. 3–11 are similar, since both are isosceles triangles and the angles labeled $\Delta\theta$ are the same. Hence

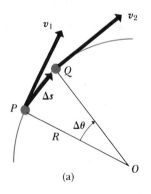

3–10 An angle of departure of 45° gives the maximum horizontal range. (Reproduction of a multiflash photograph.)

$$\frac{\Delta v}{v_1} = \frac{\Delta s}{R} \qquad \text{or} \qquad \Delta v = \frac{v_1}{R}\Delta s.$$

The magnitude of the average normal acceleration $(a_\perp)_{\text{av}}$ during Δt is therefore

$$(a_\perp)_{\text{av}} = \frac{\Delta v}{\Delta t} = \frac{v_1}{R}\frac{\Delta s}{\Delta t}.$$

The *instantaneous* acceleration $a_\perp$ at point P is the limit of this expression, as point Q is taken closer and closer to point P:

$$a_\perp = \lim_{\Delta t \to 0} \frac{v_1}{R}\frac{\Delta s}{\Delta t} = \frac{v_1}{R}\lim_{\Delta t \to 0}\frac{\Delta s}{\Delta t}.$$

However, the limit of $\Delta s/\Delta t$ is the speed v_1 at point P, and since P can be any point of the path, we can drop the subscript from v_1 and let v represent the speed at any point. Then

$$a_\perp = \frac{v^2}{R}. \tag{3–30}$$

The magnitude of the instantaneous normal acceleration is equal to the square of the speed divided by the radius of the circle. The direction is perpendicular to v and inward along the radius. Because the acceleration is always directed toward the center of the circle, it is sometimes called **centripetal acceleration;** the word *centripetal* is derived from two Greek words meaning "seeking the center."

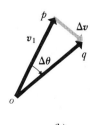

3–11 Construction for finding change in velocity, Δv, of a particle moving in a circle.

A particle moving in a circle with constant speed *must* have $a_\perp = v^2/R$. When a particle travels with a constant speed of 4 m·s^{-1} in a circle of radius 2 m, its acceleration has magnitude

$$a = \frac{(4 \text{ m·s}^{-1})^2}{2 \text{ m}} = 8 \text{ m·s}^{-2}.$$

Figure 3–12 shows the directions of the velocity and acceleration vectors at several points for a particle moving with uniform circular motion.

The magnitude of the acceleration can also be expressed in terms of the **period** τ of the motion, the time for one revolution. If a particle travels once around the circle, a distance of $2\pi R$, in a time τ, its speed v is given by

$$v = \frac{2\pi R}{\tau}. \tag{3–31}$$

Thus Eq. (3–30) can also be written as

$$a_\perp = \frac{4\pi^2 R}{\tau^2}. \tag{3–32}$$

We have assumed that the particle's speed is constant. If the speed varies, Eq. (3–30) still gives the normal component of acceleration, but in that case there is also a *tangential* component of acceleration. From the discussion at the end of Section 3–3, we see that the tangential component of acceleration is equal to the rate of change of speed:

$$a_\parallel = \lim_{\Delta t \to 0} \frac{\Delta v_\parallel}{\Delta t} = \frac{dv_\parallel}{dt}. \tag{3–33}$$

If the speed is constant, there is no tangential component of acceleration and the acceleration is purely normal, resulting from the continuous change in direction of the velocity.

The period is the time for one revolution.

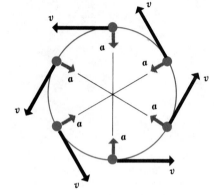

3–12 Velocity and acceleration vectors of a particle in uniform circular motion.

EXAMPLE 3–7 A car traveling at a constant speed of 20 m·s^{-1} rounds a curve of radius 100 m. What is its acceleration?

SOLUTION The magnitude of the acceleration is given by Eq. (3–30):

$$a_\perp = \frac{v^2}{R} = \frac{(20 \text{ m·s}^{-1})^2}{100 \text{ m}} = 4.0 \text{ m·s}^{-2}.$$

The direction of $\boldsymbol{a}$ at each instant is perpendicular to the velocity and directed toward the center of the circle.

EXAMPLE 3–8 In a carnival ride, the passengers travel in a circle of radius 5.0 m, making one complete circle in 4.0 s. What is the acceleration?

SOLUTION The speed is the circumference of the circle divided by the period τ (the time for one revolution):

$$v = \frac{2\pi R}{\tau} = \frac{2\pi(5.0 \text{ m})}{4.0 \text{ s}} = 7.85 \text{ m·s}^{-1}.$$

The centripetal acceleration is

$$a_\perp = \frac{v^2}{R} = \frac{(7.85 \text{ m·s}^{-1})^2}{5.0 \text{ m}} = 12.3 \text{ m·s}^{-2}.$$

Or, from Eq. (3–32),

$$a_\perp = \frac{4\pi^2(5.0 \text{ m})}{(4.0 \text{ s})^2} = 12.3 \text{ m·s}^{-2}.$$

As in the preceding example, the direction of $\boldsymbol{a}$ is always toward the center of the circle. The magnitude of $\boldsymbol{a}$ is greater than g, the acceleration due to gravity, so this is quite a wild ride!

3–6 RELATIVE VELOCITY

In Section 2–7 we introduced the concept of **relative velocity** for motion along a straight line. We can easily extend this concept to include motion in a plane or in space. Suppose that in Section 2–7 (Fig. 2–12) the automobile is not traveling in the same direction as the train, but perpendicular to this direction, across the flatcars. In any time interval the displacement of the automobile relative to the earth is the vector sum of its displacement relative to the flatcars and their displacement relative to the earth. Thus the velocity v_{AE} of the automobile relative to the earth is the vector sum of its velocity v_{AF} relative to the flatcars and their velocity v_{FE} relative to the earth. That is, Eq. (2–19) is a special case of the more general *vector equation:*

$$v_{AE} = v_{AF} + v_{FE}. \tag{3–34}$$

Thus if the automobile is moving across the flatcars at 40 km·hr^{-1}, and they are moving relative to the earth at 30 km·hr^{-1}, the appropriate vector diagram is shown in Fig. 3–13. The automobile's velocity relative to the earth is 50 km·hr^{-1} at an angle of 53° to the direction of the train's motion.

We can extend Eq. (3–34) to include any number of relative velocities. For example, if a bug B crawls along the floor of the automobile with a velocity v_{BA} relative to the automobile, the bug's velocity relative to the earth is the vector sum of its velocity relative to the automobile and that of the velocity of the automobile relative to the earth:

$$v_{BE} = v_{BA} + v_{AE}.$$

Combining this with Eq. (3–34), we find

$$v_{BE} = v_{BA} + v_{AF} + v_{FE}. \tag{3–35}$$

This equation illustrates the general rule for combining relative velocities.

Relative velocity again: What you see depends on how you're moving. Vector addition of velocities.

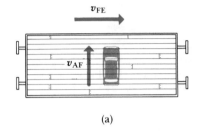

(a)

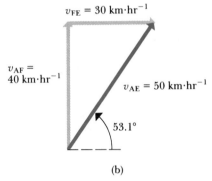

(b)

3–13 (a) An automobile being driven across a moving railroad flatcar. (b) Velocity vector diagram showing how the automobile's velocity v_{AE} relative to the earth is related to its velocity v_{AF} relative to the flatcar and the flatcar's velocity v_{FE} relative to the earth.

PROBLEM-SOLVING STRATEGY: Relative velocity

The same strategy introduced in Section 2–7 is also useful here.

1. Write each velocity with a double subscript, *in the proper order,* as "velocity of (first subscript) relative to (second subscript)."

2. In adding relative velocities, make sure that the first subscript of each velocity is the same as the last subscript of the preceding velocity.

3. The *first* subscript of the *first* velocity in the sum is the same as the *first* subscript of the velocity representing the sum, and the *last* subscript of the *last* velocity in the sum is the same as the *last* subscript of the velocity representing the sum. This sounds complicated, but referring to Eq. (3–35) will help you see how it works.

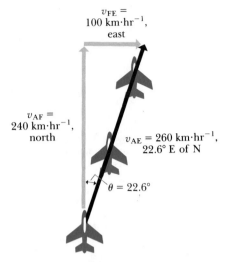

3–14 Vector diagram for aircraft flying north, wind blowing east, and resultant velocity vector of the plane.

Any of the relative velocities in an equation such as Eq. (3–34) can be transferred from one side of the equation to the other, with the sign reversed. Thus Eq. (3–34) can be written

$$v_{AF} = v_{AE} - v_{FE}.$$

The velocity of the automobile relative to the flatcar equals the *vector difference* between the velocities of automobile and flatcar, each relative to the earth.

One more point should be noted. The velocity of body A relative to body B, v_{AB}, is the negative of the velocity of B relative to A, v_{BA}:

$$v_{AB} = -v_{BA}.$$

That is, v_{AB} is equal in magnitude and opposite in direction to v_{BA}. If the automobile is traveling to the *left* at 40 mi·hr⁻¹, relative to the flatcars, the flatcars are traveling to the *right* at 40 mi·hr⁻¹, relative to the automobile.

EXAMPLE 3–9 The compass of an aircraft indicates that it is headed due north, and its airspeed indicator shows that it is moving through the air at 240 km·hr⁻¹. If there is a wind of 100 km·hr⁻¹ from west to east, what is the velocity of the aircraft relative to the earth?

SOLUTION Let subscript A refer to the aircraft, and subscript F to the moving air (which now corresponds to the flatcar in Fig. 2–12). Subscript E refers to the earth. We have given

$$v_{AF} = 240 \text{ km·hr}^{-1}, \text{ due north},$$
$$v_{FE} = 100 \text{ km·hr}^{-1}, \text{ due east},$$

and we wish to find the magnitude and direction of v_{AE}:

$$v_{AE} = v_{AF} + v_{FE}.$$

The three relative velocities and their relationship are shown in Fig. 3–14. It follows from this diagram that

$$v_{AE} = 260 \text{ km·hr}^{-1}, \qquad \theta = \arctan \frac{100 \text{ km·hr}^{-1}}{240 \text{ km·hr}^{-1}} = 22.6° \text{ E of N}.$$

EXAMPLE 3–10 In Example 3–9, what direction should the pilot head in order to travel due north? What will then be the plane's velocity relative to the earth? (The magnitude of the airspeed and the wind velocity are the same as in the preceding example.)

SOLUTION We now have given

$$v_{AF} = 240 \text{ km·hr}^{-1}, \text{ direction unknown},$$
$$v_{FE} = 100 \text{ km·hr}^{-1}, \text{ due east}.$$

We want to find v_{AE}; its magnitude is unknown, but we know that its direction is due north. (Note that both this and the preceding example require us to determine two unknown quantities. In Example 3–9 these were the magnitude and direction of v_{AE}; in this example the unknowns are the direction of v_{AF} and the magnitude of v_{AE}.)

The three relative velocities must still satisfy the vector equation

$$v_{AE} = v_{AF} + v_{FE}.$$

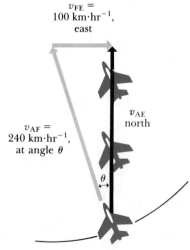

3–15 The resultant vector v_{AF} shows the direction in which the pilot should point the plane in order to have the plane travel due north.

The appropriate vector diagram is shown in Fig. 3–15. We find

$$v_{AE} = \sqrt{(240 \text{ km·hr}^{-1})^2 - (100 \text{ km·hr}^{-1})^2} = 218 \text{ km·hr}^{-1},$$

$$\theta = \arcsin \frac{100 \text{ km·hr}^{-1}}{240 \text{ km·hr}^{-1}} = 24.6°.$$

That is, the pilot should head 24.6° W of N, and his ground speed is then 218 km·hr^{-1}.

SUMMARY

The position vector r of a point P in a plane is the displacement vector from the origin to P. Its components are the coordinates x and y.

The average velocity v_{av} during the time interval Δt is the displacement Δr (the change in the position vector r) divided by Δt, that is, $v_{av} = \Delta r / \Delta t$. Its components are $(v_x)_{av} = \Delta x / \Delta t$ and $(v_y)_{av} = \Delta y / \Delta t$.

Instantaneous velocity is $v = dr/dt$; its components are $v_x = dx/dt$ and $v_y = dy/dt$.

The average acceleration a_{av} during the time interval Δt is the velocity change Δv divided by Δt; that is, $a_{av} = \Delta v / \Delta t$; its components are $(a_x)_{av} = \Delta v_x / \Delta t$ and $(a_y)_{av} = \Delta v_y / \Delta t$.

Instantaneous acceleration is $a = dv/dt$. Its components are $a_x = dv_x/dt$ and $a_y = dv_y/dt$.

Acceleration can also be represented in terms of its components parallel and normal to the path, $a_{\parallel} = dv/dt$ and $a_{\perp}$, respectively.

In projectile motion, $a_x = 0$ and $a_y = -g$. The coordinates and velocity components are given as functions of time by

$$x = (v_0 \cos \theta_0)t, \quad y = (v_0 \sin \theta_0)t - \tfrac{1}{2}gt^2, \quad v_x = v_0 \cos \theta_0, \quad v_y = v_0 \sin \theta_0 - gt.$$

The shape of the path in projectile motion is always a parabola.

When a particle moves in a circular path of radius R with speed v, it has an acceleration with magnitude $a_{\perp} = v^2/R$, directed always toward the center of the circle and perpendicular to v. The period τ of a circular motion is the time for one revolution. If the speed is constant, then $v = 2\pi R/\tau$ and $a_{\perp} = 4\pi^2 R/\tau^2$. If the speed is not constant, there is also a component of a parallel to the path, equal to $a_{\parallel} = dv/dt$.

When a body A moves relative to a body F, and F moves relative to the earth E or other reference frame, we denote the velocity of A relative to F by v_{AF}, the velocity of F relative to E by v_{FE}, and the velocity of A relative to E by v_{AE}. These velocities are related by

$$v_{AE} = v_{AF} + v_{FE}.$$

KEY TERMS

position vector
average velocity
instantaneous velocity
average acceleration
instantaneous acceleration
normal component
tangential component
projectile
uniform circular motion
centripetal acceleration
period
relative velocity

QUESTIONS

3–1 A simple pendulum (a mass swinging at the end of a string) swings back and forth in a circular arc. What is the direction of its acceleration at the ends of the swing? At the midpoint?

3–2 A football is thrown in a parabolic path. Is there any point at which the acceleration is parallel to the velocity? Perpendicular to the velocity?

3–3 When a rifle is fired at a distant target, the barrel is not lined up exactly on the target. Why not? Does the angle of correction depend on the distance of the target?

3–4 The acceleration of a falling body is measured in an elevator traveling upward at a constant speed of 9.8 m·s^{-1}. What result is obtained?

3–5 One can play catch with a softball in an airplane in level flight just as though the plane were at rest. Is this still possible when the plane is making a turn?

3–6 A package is dropped out of an airplane in level flight. If air resistance could be neglected, how would the motion of the package look to the pilot? To an observer on the ground?

3–7 No matter what the initial velocity, the motion of a projectile (neglecting air resistance) is *always* confined to a single plane. Why?

3–8 If a jumper can give himself the same initial speed regardless of the direction he jumps (forward or straight up), how is his maximum vertical jump (high jump) related to his maximum horizontal jump (broad jump)?

3–9 A manned space-flight projectile is launched in a parabolic trajectory. A man in the capsule feels weightless. Why? In what sense is he weightless?

3–10 The maximum range of a projectile occurs when it is aimed at 45°, if air resistance is neglected. Is this still true if air resistance is included? If not, is the optimum angle greater or less than 45°? Why?

3–11 A passenger in a car rounding a sharp curve is thrown toward the outside of the curve. What force throws him in this direction?

3–12 In uniform circular motion, what is the *average* velocity during one revolution? The average acceleration?

3–13 In uniform circular motion the acceleration is perpendicular to the velocity at every instant, even though both change continuously in direction. Is there any other motion having this property, or is uniform circular motion unique?

3–14 If an artificial earth satellite is in an orbit with a period of exactly one day, how does its motion look to an observer on the rotating earth? (Such an orbit is said to be *geosynchronous;* most communications satellites are placed in such orbits.)

3–15 Raindrops hitting the side windows of a car in motion often leave diagonal streaks. Why? What about diagonal streaks on the windshield? Is the explanation the same or different?

3–16 In a rainstorm with a strong wind, what determines the best position in which to hold an umbrella?

EXERCISES

Section 3–3 Components of Acceleration

3–1 The coordinates of a particle moving in the xy-plane are given as functions of time by $x = 2 \text{ m} - (4 \text{ m·s}^{-1})t$ and $y = (3 \text{ m·s}^{-2})t^2$.

a) Calculate the velocity and acceleration vectors of the particle as functions of time.

b) Calculate the magnitude and direction of the particle's velocity and acceleration at $t = 3$ s.

3–2 A particle moves in the xy-plane with acceleration

$$\boldsymbol{a} = (2 \text{ m·s}^{-4})t^2\boldsymbol{i} + (4 \text{ m·s}^{-3})t\boldsymbol{j}.$$

a) Assuming that the particle is at rest at the origin at time $t = 0$, derive expressions for the velocity and position vectors as functions of time.

b) Sketch the path of the particle.

c) Find the magnitude and direction of the velocity at $t = 3$ s.

Section 3–4 Projectile Motion

3–3 A golf ball is hit horizontally from a tee at the edge of a cliff. Its x- and y-coordinates are given as functions of time by

$$x = (20 \text{ m·s}^{-1})t, \qquad y = -(4.9 \text{ m·s}^{-2})t^2.$$

a) Compute the x- and y-coordinates at times $t = 0$ s, 1 s, 2 s, 3 s, and 4 s. Plot these positions on graph paper and sketch the trajectory.

b) Determine the ball's initial velocity vector and its acceleration vector.

c) Find the x- and y-components of the velocity at time $t = 2$ s. Plot the velocity vector at the appropriate point on the trajectory obtained in (a). Is the velocity tangent to the trajectory?

3–4 A ball rolls off the edge of a table top 1 m above the floor and strikes the floor at a point 1.5 m horizontally from the edge of the table.

a) Find the time of flight.

b) Find the initial velocity.

c) Find the magnitude and direction of the velocity of the ball just before it strikes the floor. Draw a diagram to scale.

3–5 A book slides off a horizontal table top with a speed of 4 m·s^{-1}. It is observed to strike the floor in 0.5 s. Find

a) the height of the table top above the floor;

b) the horizontal distance from the edge of the table to the point where the book strikes the floor;

c) the horizontal and vertical components of its velocity when it reaches the floor.

3–6 An airplane flying horizontally at a speed of 100 m·s^{-1} drops a box at an elevation of 2000 m.

a) How much time is required for the box to reach the earth?

b) How far does it travel horizontally while falling?

c) Find the horizontal and vertical components of its velocity when it strikes the earth.

d) Where is the airplane when the box strikes the earth?

3–7 A marksman fires a .22-caliber rifle horizontally at a target; the bullet has a muzzle velocity of 900 ft·s^{-1}. How much does the bullet drop in flight if the target is

a) 50 yd away?

b) 150 yd away?

3–8 A baseball is thrown with an initial upward velocity component of 20 m·s^{-1} and a horizontal velocity component of 25 m·s^{-1}.

a) Find the horizontal and vertical components of the displacement and velocity of the ball after 1 s, 2 s, 3 s, and 4 s.

b) How much time is required to reach the highest point of the trajectory?

c) How high is this point?

d) How much time (after being thrown) is required for the ball to return to its original level? How does this compare with the time calculated in part (b)?

e) How far has it traveled horizontally during this time? Show your results in a neat sketch large enough to show all features clearly.

3–9 A baseball is thrown at an angle of 53° above the horizontal with an initial speed of 40 m·s^{-1}.

a) What is the maximum height above the thrower's hand reached by the baseball?

b) How long does it take to reach the maximum height?

c) At what *two* times is the baseball at a height of 25 m above the point from which it was thrown?

d) Calculate the horizontal and vertical components of the baseball's velocity at each of the two times calculated in (c).

e) What are the magnitude and direction of the baseball's velocity when it returns to the level from which it was thrown?

3–10 A batted baseball leaves the bat at an angle of 30° above the horizontal and is caught by an outfielder 400 ft from the plate. Assume that the height of the point where it was struck by the bat equals the height of the point where it was caught.

a) What was the initial speed of the ball?

b) How high did it rise above the point where it struck the bat?

3–11 A man stands on the roof of a building that is 50 m tall and throws a rock with a velocity of 60 m·s^{-1} at an angle of 37° above the horizontal. Calculate:

a) the maximum height, above the roof, reached by the rock;

b) the magnitude of the resultant velocity of the rock just before it strikes the ground;

c) the horizontal distance from the base of the building to the point where the rock strikes the ground.

3–12 Suppose the departure angle θ_0 in Fig. 3–8 is 14° and the distance x is 5 m. Where will the arrow and apple meet

if the initial speed of the arrow is

a) 20 m·s^{-1}?

b) 12 m·s^{-1}?

Sketch both trajectories.

c) What will happen if the initial speed of the arrow is 8 m·s^{-1}?

Section 3–5 Circular Motion

3–13 The earth has a radius of 6.38×10^6 m and turns around on its axis once in 24 hr. What is the radial acceleration of an object at the equator of the earth, in units of m·s^{-2}?

3–14 The radius of the earth's orbit around the sun (assumed circular) is 1.49×10^{11} m, and the earth travels around this orbit in 365 days.

a) What is the magnitude of the orbital velocity of the earth, in meters per second?

b) What is the radial acceleration of the earth toward the sun, in meters per second squared?

3–15 A Ferris wheel of radius 12 m is turning about a horizontal axis through its center, such that the linear speed of a passenger on the rim is constant and equal to 9 m·s^{-1}.

a) What are the magnitude and direction of the acceleration of the passenger as he passes through the lowest point in his circular motion?

b) How long does it take the Ferris wheel to make one revolution?

3–16 A model of a helicopter rotor has four blades, each 2 m in length from the central shaft to the blade tip, and is rotated in a wind tunnel at 1500 rev·min^{-1}.

a) What is the linear speed of the blade tip, in meters per second?

b) What is the radial acceleration of the blade tip, expressed as a multiple of the acceleration of gravity, g?

Section 3–6 Relative Velocity

3–17 An airplane pilot wishes to fly due north. A wind of 96 km·hr^{-1} (about 60 mi·hr^{-1}) is blowing toward the west.

a) If the flying speed of the plane (its speed in still air) is 290 km·hr^{-1} (about 180 mi·hr^{-1}), in what direction should the pilot head?

b) What is the speed of the plane over the ground? Illustrate with a vector diagram.

3–18 A passenger on a ship traveling due east with a speed of 18 knots observes that the stream of smoke from the ship's funnels makes an angle of 20° with the ship's wake. The wind is blowing from south to north. Assume that the smoke acquires a velocity (with respect to the earth) equal to the velocity of the wind, as soon as it leaves the funnels. Find the magnitude of the velocity of the wind, in knots. (A knot is a unit of speed used by sailors; 1 knot = 1.852 km·hr^{-1}.)

3–19 A river flows due north with a velocity of 2 m·s^{-1}. A man rows a boat across the river; his velocity relative to the water is 3 m·s^{-1} due east; the river is 1000 m wide.

a) What is his velocity relative to the earth?

b) How much time is required to cross the river?

c) How far north of his starting point will he reach the opposite bank?

3–20

a) In what direction should the rowboat in Exercise 3–19 be headed in order to reach a point on the opposite bank directly east from the starting point?

b) What will be the velocity of the boat relative to the earth?

c) How much time is required to cross the river?

PROBLEMS

3–21 The coordinates of a particle moving in the xy-plane are given as functions of time by

$$x = (2 \text{ m·s}^{-1})t, \qquad y = 19 \text{ m} - (2 \text{ m·s}^{-2})t^2.$$

a) What is the particle's distance from the origin at time $t = 2$ s?

b) What is the particle's velocity (magnitude and direction) at time $t = 2$ s?

c) What is the particle's acceleration (magnitude and direction) at time $t = 2$ s?

d) At what times is the particle's velocity perpendicular to its acceleration?

e) At what times is the particle's velocity perpendicular to its position vector? What are the locations of the particle at these times?

f) What is the particle's minimum distance from the origin? At what time does this minimum occur?

g) Sketch the path of the particle.

3–22 A faulty model rocket moves in the xy-plane with an acceleration whose components in a coordinate system in which the positive y-direction is vertically upward are given by $a_x = (3 \text{ m·s}^{-4})t^2$ and $a_y = 10 \text{ m·s}^{-2} - (2 \text{ m·s}^{-3})t$. At $t = 0$ the rocket is at the origin and has an initial velocity $v_0 = 2 \text{ m·s}^{-1}i + 6 \text{ m·s}^{-1}j$.

a) Calculate the velocity and position vectors as functions of time.

b) What is the maximum height reached by the rocket?

c) What is the horizontal displacement of the rocket when it returns to $y = 0$?

3–23 A bird flies in the xy-plane with a velocity vector given by

$$v = (2 \text{ m·s}^{-1} - 3 \text{ m·s}^{-3}t^2)i + (5 \text{ m·s}^{-2}t)j,$$

where the positive y-direction is vertically upward. At $t = 0$ the bird is at the origin.

a) Calculate the position and acceleration vectors of the bird as functions of time.

b) What is the bird's altitude (y-coordinate) as it flies over $x = 0$ for the first time after $t = 0$?

3–24 A player kicks a football at an angle of 37° with the horizontal and with an initial speed of 15 m·s^{-1}. A second player standing at a distance of 30 m from the first in the direction of the kick starts running to meet the ball at the instant it is kicked. How fast must he run in order to catch the ball just before it hits the ground?

3–25 According to the *Guiness Book of World Records* the longest home run ever measured was hit by Roy "Dizzy" Carlyle in a minor league game and traveled 618 ft before landing on the ground outside the ballpark.

a) Assuming the ball's initial velocity was 45° above the horizontal and neglecting air resistance, what would the initial speed of the ball need to be to produce such a home run if hit at a point 3.0 ft above ground level? Assume that the ground is perfectly flat.

b) How far above a fence 10 ft in height would the ball be if the fence were 380 ft from home plate?

3–26 A projectile shot at an angle of 60° above the horizontal strikes a building 30 m away at a point 15 m above the point of projection.

a) Find the initial velocity of the projectile, the velocity with which it is projected.

b) Find the magnitude and direction of the velocity of the projectile just before it strikes the building.

3–27 An airplane diving at an angle of 36.9° with the horizontal drops a bag of sand from an altitude of 800 m. The bag is observed to strike the ground 5 s after its release.

a) What is the speed of the plane?

b) How far does the bag travel horizontally during its fall?

c) What are the horizontal and vertical components of its velocity just before it strikes the ground?

3–28 A snowball rolls off a roof that slopes downward at an angle of 40°, as shown in Fig. 3–16. The edge of the roof is

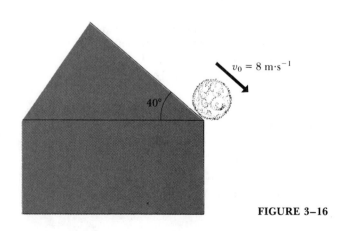

$v_0 = 8 \text{ m·s}^{-1}$

40°

FIGURE 3–16

12 m above the ground, and the snowball has a speed of 8 m·s^{-1} as it rolls off the roof.

a) How far from the edge of the house will the snowball strike the ground, if it does not strike anything else while falling?

b) A man 2 m tall is standing 4 m from the edge of the house. Will he be hit by the snowball?

3–29 Prove that a projectile launched at angle θ_0 has the same range as one launched with the same speed at angle $(90° - \theta_0)$.

3–30 A rock is thrown from the roof of a building with a velocity v_0 at an angle of θ relative to the horizontal. The building has height h. Calculate the magnitude of the resultant velocity of the rock just before it strikes the ground, and show that this velocity is independent of θ.

3–31 A particle moves in the xy-plane; its coordinates are given as functions of time by

$$x = R \cos \omega t, \qquad y = R \sin \omega t,$$

where R and ω are constants.

a) Show that the particle's distance from the origin is constant and equal to R, i.e., that its path is a circle of radius R.

b) Show that at every point the particle's velocity is perpendicular to its position vector.

c) Show that the particle's acceleration is always opposite in direction to its position vector and has magnitude $\omega^2 R$.

d) Show that the magnitude of the particle's velocity is constant and equal to ωR.

e) Combine the results of parts (c) and (d) to show that the particle's acceleration has constant magnitude v^2/R.

3–32 In an action-adventure film the hero is supposed to throw a grenade from his car, which is going 80.5 km·hr^{-1}, to his enemy's car, which is going 133 km·hr^{-1}. The enemy's car is 14.6 m in front of the hero's when he lets go of the grenade. If the hero throws the grenade so its initial velocity relative to the hero is at an angle of 45° with the horizontal, what should be the magnitude of the velocity? The cars are both traveling in the same direction on a level road. Find the magnitude of the velocity both relative to the hero and relative to the earth.

3–33 An airplane pilot sets a compass course due west and maintains an air speed of 240 km·hr^{-1}. After flying for $\frac{1}{2}$ hr, he finds himself over a town that is 150 km west and 40 km south of his starting point.

a) Find the wind velocity, in magnitude and direction.

b) If the wind velocity were 120 km·hr^{-1} due south, in what direction should the pilot set his course in order to travel due west? Take the same air speed of 240 km·hr^{-1}.

3–34 When a train's speed is 10 m·s^{-1} eastward, raindrops that are falling vertically with respect to the earth make traces that are inclined 30° to the vertical on the windows of the train.

a) What is the horizontal component of a drop's velocity with respect to the earth? With respect to the train?

b) What is the magnitude of the velocity of the raindrop with respect to the earth? With respect to the train?

3–35 A motorboat is observed to travel 16 km·hr^{-1} relative to the earth in the direction 37° north of east. If the velocity of the boat due to the wind is 3.2 km·hr^{-1} eastward and that due to the current is 6.4 km·hr^{-1} southward, what are the magnitude and direction of the velocity of the boat due to its own power?

CHALLENGE PROBLEMS

3–36 A man is riding on a flatcar traveling with a constant speed of 9.1 m·s^{-1} (Fig. 3–17). He wishes to throw a ball though a stationary hoop 4.9 m above the height of his hands in such a manner that the ball will move horizontally as it passes through the hoop. He throws the ball with a speed of 12.2 m·s^{-1} with respect to himself.

a) What must be the vertical component of the initial velocity of the ball?

b) How many seconds after he releases the ball will it pass through the hoop?

c) At what horizontal distance in front of the hoop must he release the ball?

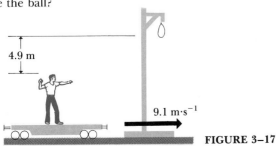

4.9 m

9.1 m·s^{-1}

FIGURE 3–17

3–37 A physics professor did daredevil stunts in his spare time. His last stunt was to attempt to jump across a river on a motorcycle (Fig. 3–18). The take-off ramp was inclined at 53°, the river was 40 m wide, and the far bank was 15 m

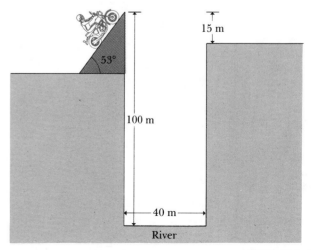

15 m

53°

100 m

40 m

River

FIGURE 3–18

lower than the top of the ramp. The river itself was 100 m below ramp. What should his velocity have been at the top of the ramp to have just made it to the edge of the far bank?

3–38 For a basketball free throw, at what angle and with what initial velocity should you shoot a basketball so that the ball goes through the basket at the smallest angle of entry (smallest angle relative to the horizontal) possible without touching the rim? For a free throw you stand at the foul line 13.8 ft from the center of the basket which is 10.0 ft above the floor. Assume that the ball leaves your hands at a point 5.0 ft above the floor. The basketball is 30 in. in circumference and the basket is 18 in. in diameter according to the official rules. (For more on the physics involved in basketball, see the article by P. J. Brancazio in *Amer. Jour. of Phys.*, Vol. 49 (1981), pp. 356–365.)

3–39 A projectile is given an initial velocity v_0 at an angle ϕ above the surface of an incline, which is in turn inclined at an angle θ above the horizontal; see Fig. 3–19.

a) Calculate the distance, measured along the incline, from the launch point to where the object strikes the incline. Your answer will be in terms of v_0, g, θ, and ϕ.

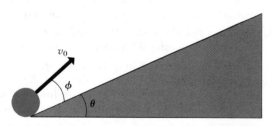

FIGURE 3–19

b) What angle ϕ gives the maximum range, measured along the incline?

(*Note*. You may be interested in the three different methods of solution presented by I. R. Lapidus in *Amer. Jour. of Phys.*, Vol. 51 (1983), pp. 806 and 847. See also H. A. Buckmaster in *Amer. Jour. of Phys.*, Vol. 53 (1985), pp. 638–641, for a thorough study of this and some similar problems.)

3–40 Refer to the previous problem.

a) An archer on ground of constant upward slope of 30° aims at a target 50 m farther up the incline. The arrow in the bow and the bull's-eye at the center of the target are each 1.5 m above the ground. Let the initial velocity of the arrow just after it leaves the bow be 28 m·s^{-1}. At what angle above the *horizontal* should the archer aim to hit the bull's-eye? If there are two such angles, calculate the smaller of the two. You may have to solve the equation for the angle by iteration, that is, by trial and error. How does the angle compare to that required when the ground is level, with zero slope?

b) Repeat the above, for ground of constant *downward* slope of 30°.

3–41 An object is traveling in a circle of radius $R = 2$ m with a constant speed of $v = 5$ m·s^{-1}. Let v_1 be the velocity vector at time t_1, and v_2 be the velocity vector at time t_2. Consider $\Delta v = v_2 - v_1$ and $\Delta t = t_2 - t_1$. Recall that $a_{av} = \Delta v / \Delta t$. For $\Delta t = 0.5$ s, 0.1 s, and 0.05 s, calculate the magnitude (to four significant figures) and direction (relative to v_1) of the average acceleration a_{av}. Compare your results to the general expression for the instantaneous acceleration a for uniform circular motion that is derived in the text.

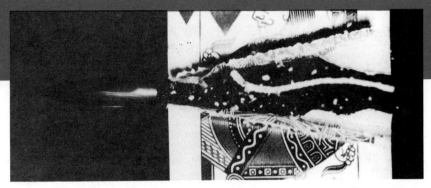

4

NEWTON'S LAWS OF MOTION

IN THE TWO PRECEDING CHAPTERS WE STUDIED *KINEMATICS,* THAT IS, THE methods for *describing* motion. With this chapter we begin a study of more general problems involving the relation of motion to its causes; these problems form the area called **dynamics.** In addition to the kinematic quantities introduced in Chapters 2 and 3, we need two additional concepts: *force* and *mass. Statics* is a division of dynamics dealing with special cases where the acceleration is zero, that is, with bodies in *equilibrium.*

Dynamics: what makes it move

All of dynamics is based on three principles called *Newton's laws of motion.* The first law states that when the vector sum of forces on a body is zero, the acceleration of the body is also zero. The second law relates force to acceleration when the vector sum of forces is *not* zero. And the third law relates the pairs of forces that interacting bodies exert on each other.

Newton's laws: the cornerstone of mechanics

Newton's laws are *empirical* laws, deduced from experiment; they cannot be derived from anything more fundamental. They were clearly stated for the first time by Sir Isaac Newton (1642–1727) and were published in 1686 in his *Philosophiae Naturalis Principia Mathematica* (The Mathematical Principles of Natural Science). Many other scientists before Newton contributed to the foundations of mechanics, especially Galileo Galilei (1564–1642), who died the same year Newton was born. Indeed Newton himself said, "If I have been able to see a little farther than other men, it is because I have stood on the shoulders of giants."

4–1 FORCE

Force is a central concept in all of physics. A force on a body resulting from direct contact with another body is called a *contact force.* For example, when we push or pull on a body, we exert a force on it. A stretched spring exerts forces on the bodies attached to its ends; compressed air exerts a force on the wall of its container; and a locomotive exerts a force on the train it is pulling or pushing. Viewed on an atomic scale, contact forces result principally from the electrical attractions and repulsions of the electrons and nuclei in the atoms of the materials.

Force: a measure of interaction

71

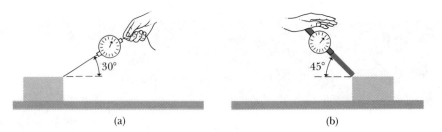

4–1 Force may be exerted on the box by either (a) pulling it, or (b) pushing it.

(a)   (b)

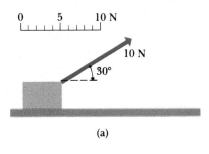

(a)

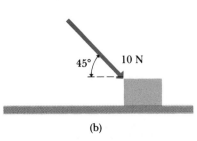

(b)

4–2 Force diagram for the forces acting on the box in Fig. 4–1.

Measuring force

Forces can be combined by vector addition.

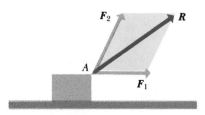

4–3 A force represented by the vector **R**, equal to the vector sum of F_1 and F_2, produces the same effect as the forces F_1 and F_2 acting simultaneously.

In contrast, gravitational forces, as well as electric and magnetic forces, can act through empty space; bodies do not have to be in contact for these forces to act. The force of gravitational attraction exerted on a body by the earth is called the *weight* of the body.

As we mentioned in Section 1–6, force is a *vector* quantity. This means that to describe a force, we need to describe the *direction* in which it acts as well as its *magnitude*, the quantity that describes "how much" or "how hard" the force pushes. In Section 4–4 we will define units of force in terms of the units of mass, length, and time. As mentioned in Section 1–6, the SI unit of force is the *newton*, abbreviated N. In the cgs version of the metric system (discussed in Section 4–4) the unit of force is the *dyne*, equal to 10^{-5} N, and in the British system the unit of force is the *pound*.

A familiar instrument for measuring forces is the spring balance. It consists of a coil spring, enclosed in a case for protection, with a pointer attached to one end. When forces are applied to the ends of the spring, it stretches; the amount of stretch depends on the force. We can calibrate such an instrument by using a number of identical bodies, each having a weight of exactly 1 N. Then when two, three, or more of these are suspended simultaneously from the balance, the total force stretching the spring is 2 N, 3 N, and so on, and the corresponding positions of the pointer can be labeled 2 N, 3 N, and so on. Then we can use the instrument to measure the magnitude of an unknown force.

Suppose we slide a box along the floor by pulling it with a string or pushing it with a stick, as in Fig. 4–1. The interaction of the box with the other bodies that push or pull it is described in terms of the *forces* they exert on the box. Thus our point of view is that the motion of the box is caused by the *forces* these bodies exert on it. The forces in the two cases can be represented as in Fig. 4–2. The labels indicate the magnitude and direction of the force; the length of the arrow, to some chosen scale, also shows the magnitude.

What happens when *two* forces, represented by the vectors F_1 and F_2 in Fig. 4–3, are applied simultaneously at the same point A of a body? Is it possible to produce the same effect on the body by applying a *single* force at A? If so, what should be its magnitude and direction? These questions can be answered only by experiment. Investigation shows that a single force equal to the *vector sum* $F_1 + F_2$ of the original forces produces the same effect as the two forces together. This single force is often called the **resultant** of the two forces. Hence the mathematical process of *vector addition* of two force vectors corresponds to the physical operation of finding the *resultant of two forces,* simultaneously applied at a given point. The same statement can be extended to combining any number of forces.

The fact that forces can be combined by vector addition is of the utmost importance, as we will see in the following chapters. Furthermore, this fact also allows us to represent a force by means of *components*, as we have done with displacements in Section 1–7. In Fig. 4–4a, force F acts on a body at point O. The components of F in the directions Ox and Oy are F_x and F_y; we find that simultaneous application of the forces F_x and F_y, as in Fig. 4–4b, is equivalent in all respects to the effect of the original force. *Any force can be replaced by its components, acting at the same point.*

As a numerical example, let

$$F = 10.0 \text{ N}, \qquad \theta = 30°.$$

Then,

$$F_x = F \cos \theta = (10.0 \text{ N})(0.866) = 8.66 \text{ N},$$
$$F_y = F \sin \theta = (10.0 \text{ N})(0.500) = 5.00 \text{ N},$$

and the effect of the original 10.0-N force is equivalent to the simultaneous application of a horizontal force, to the right, with magnitude 8.66 N, and a lifting force with magnitude 5.00 N.

The axes we use to obtain components of a vector need not be vertical and horizontal. For example, Fig. 4–5 shows a block being pulled up an inclined plane by a force F, represented by its components F_x and F_y, parallel and perpendicular to the sloping surface of the plane.

We will often need to find the vector sum of several forces acting on a body, which again may be called their *resultant*. The Greek letter Σ (sigma) is often used in a shorthand notation for this sum. If the forces are labeled F_1, F_2, F_3, and so on, and their resultant is R, then the operation

$$R = F_1 + F_2 + F_3 + \cdots$$

is often abbreviated

$$R = \Sigma F, \qquad (4–1)$$

where ΣF is read "sum of the forces." (The Greek letter Σ, equivalent to the Roman S, is an abbreviation for "sum.") This means, of course, the *vector* sum. In terms of components, we may write

$$R_x = \Sigma F_x, \qquad R_y = \Sigma F_y, \qquad (4–2)$$

in which ΣF_x is read "the sum of the x-components of the forces." We also note that a boldface Σ is used in Eq. (4–1) as a reminder that the sum is a vector sum, and a lightface Σ is used in Eqs. (4–2) because components are ordinary numbers.

Finally, we can combine these components to form the resultant R. Its magnitude is

$$R = \sqrt{R_x{}^2 + R_y{}^2},$$

since R_x and R_y are perpendicular to each other.

The angle α between R and the x-axis can now be found from any one of its trigonometric functions. For example, $\tan \alpha = R_y/R_x$. As with the individual vectors, the components R_x and R_y may be positive or negative, and the angle α may be in any of the four quadrants.

Forces can be represented by components.

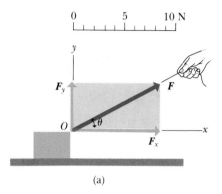

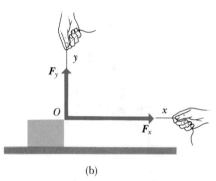

(a)

(b)

4–4 The inclined force F may be replaced by its rectangular components F_x and F_y. $F_x = F \cos \theta$, $F_y = F \sin \theta$.

Resultant force: the vector sum of forces

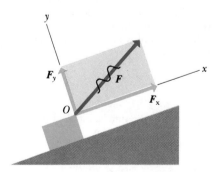

4–5 F_x and F_y are the rectangular components of F, parallel and perpendicular to the sloping surface of the inclined plane.

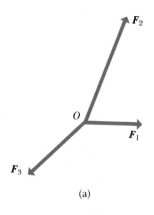

(a)

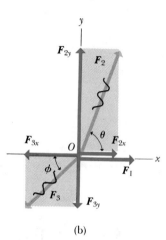

(b)

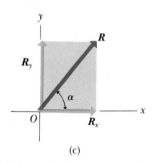

(c)

4–6 Vector $\boldsymbol{R}$, the resultant of $\boldsymbol{F}_1$, $\boldsymbol{F}_2$, and $\boldsymbol{F}_3$, is obtained by the method of rectangular resolution. The rectangular components of $\boldsymbol{R}$ are $R_x = \Sigma F_x$ and $R_y = \Sigma F_y$.

Equilibrium: the simplest state of motion

EXAMPLE 4–1   In Fig. 4–6a, three forces $\boldsymbol{F}_1$, $\boldsymbol{F}_2$, $\boldsymbol{F}_3$ lying in the *xy*-plane act at point *O*. Let $F_1 = 120$ N, $F_2 = 200$ N, $F_3 = 150$ N, $\theta = 60°$, and $\phi = 45°$. Find the *x*- and *y*-components of the resultant or vector sum $\boldsymbol{R}$, and find its magnitude and direction.

SOLUTION   The computations can be arranged systematically as in Table 4–1.

TABLE 4–1

Force	Angle	x-component	y-component
$F_1 = 120$ N	0	+120 N	0
$F_2 = 200$ N	60°	+100 N	−173 N
$F_3 = 150$ N	45°	−106 N	−106 N
		$R_x = \Sigma F_x = +114$ N	$R_y = \Sigma F_y = +67$ N

$$R = \sqrt{(114 \text{ N})^2 + (67 \text{ N})^2} = 132 \text{ N},$$
$$\alpha = \arctan \frac{67 \text{ N}}{114 \text{ N}} = \arctan 0.588 = 30.4°.$$

4–2 EQUILIBRIUM AND NEWTON'S FIRST LAW

We have mentioned that one possible effect of a force is to alter the state of motion of the body on which it acts. In general this motion can consist of motion of the body as a whole, called *translational* motion, and of *rotational* motion of the body about its center. A familiar example is a thrown football, which spins as it moves through the air. In contrast, the motion of a *particle*, represented as a single point, is described in terms of translational motion only. In this chapter and the next we assume the body can be represented as a point, so we need not be concerned yet with the possibility of rotation.

When several forces act on a body at the same time, their effects can compensate or cancel one another. When this happens, Newton's first law predicts that there is *no* change in the motion, and we say that the body is in **equilibrium.** This means that the body either remains at rest or moves in a straight line with constant velocity.

In Fig. 4–7 a small body such as a hockey puck rests on a horizontal surface having negligible friction, such as an air-hockey table or a slab of wet ice. If the body is initially at rest and a single force $\boldsymbol{F}_1$ acts on it, as in Fig. 4–7a, the body immediately starts to move. If it is in motion at the start, the effect of the force is to change the motion, either in direction or speed, or both. In either case, the body is *not* in equilibrium.

Now suppose we apply a second force $\boldsymbol{F}_2$, as in Fig. 4–7b, equal in magnitude to $\boldsymbol{F}_1$ but opposite in direction. Experiments show that the body is then in equilibrium; if it is initially at rest, it remains at rest, and if it is initially moving, it continues to move in the same direction with constant speed. In this case the two forces are negatives of each other, $\boldsymbol{F}_2 = -\boldsymbol{F}_1$, and so their vector sum is zero:

$$\boldsymbol{R} = \boldsymbol{F}_1 + \boldsymbol{F}_2 = 0.$$

For brevity, we speak of two forces as being "equal and opposite," meaning that their magnitudes are equal and that one is the negative of the other.

We can generalize this discussion to a body with any number of forces acting on it. When a body is in equilibrium, the vector sum, or resultant, of all the forces acting on it must be zero. Each component of the resultant must therefore be zero. Hence, for a body in equilibrium,

$$R = \Sigma F = 0,$$

or

$$\Sigma F_x = 0, \qquad \Sigma F_y = 0. \tag{4–3}$$

When Eqs. (4–3) are satisfied, the body is in equilibrium, provided that it can be represented as a point or that all the forces acting on it are applied at the same point.

The condition for equilibrium represented by Eqs. (4–3) is called *Newton's first law of motion*. Newton did not state his first law in exactly these words. His original statement (translated from the Latin in which the *Principia* was written) reads:

> Every body continues in its state of rest, or of uniform motion in a straight line, unless it is compelled to change that state by forces impressed on it.

Newton's first law of motion is not as self-evident as it may seem. This law asserts that in the absence of any applied force a body either remains at rest or moves uniformly in a straight line. It follows that *once a body has been set in motion, no force is needed to keep it moving.*

This assertion may seem to be contradicted by everyday experience. Suppose you exert a force with your hand to push a book along a horizontal table top. After the book has left your hand, and you are no longer exerting a force on it, it does *not* continue to move indefinitely, but slows down and eventually comes to rest. To keep it moving uniformly you have to continue to push it. But the reason this force is needed is because a frictional force is exerted on the sliding book by the table top, in a direction *opposite* to the book's motion. The smoother the surfaces of the book and table, the smaller the frictional force and the smaller the force needed to keep the book moving. The first law asserts that if the frictional force could be eliminated completely, no forward force at all would be required to keep the book moving once it is set in motion. The law also states that if the *resultant* force on the book is zero, as it is when the frictional force is balanced by an equal forward force, the book also continues to move uniformly. In other words, to keep the book moving uniformly, *zero resultant force is equivalent to no force at all.*

4–3 MASS AND NEWTON'S SECOND LAW

We know from experience that a body at rest never starts to move by itself; some other body has to apply a push or pull to it. Similarly, when a body is already in motion, a force is required to slow it down or stop it. To make a moving body deviate from straight-line motion, we must apply a sideways force. All these processes (speeding up, slowing down, or changing direction) involve a *change* in either the magnitude or the direction of the velocity. In each case the body has an *acceleration*, and a force must act on it to cause this acceleration.

Let us consider several fundamental experiments. A small body, which we may model as a particle, rests on a level, frictionless surface and moves to the

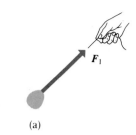

(a)

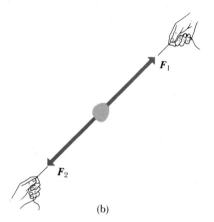

(b)

4–7 A small body acted on by two forces is in equilibrium if the forces are equal in magnitude and opposite in direction.

A moving body keeps moving unless some force makes it stop.

No force at all is equivalent to a set of forces with a vector sum of zero.

Bodies resist *changes* in their motion.

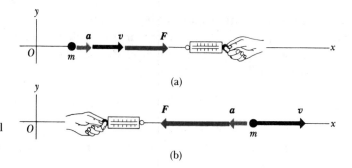

4–8 The acceleration a is proportional to the force F and is in the same direction as the force.

right along the x-axis of a reference system, as in Fig. 4–8a. We apply a constant horizontal force F to the body. This force might be supplied by a spring balance, as described in Section 4–1, with the spring stretched a constant amount. We find that during the time the force is acting, the velocity of the body increases at a constant rate; that is, the body moves with *constant acceleration*. If the magnitude of the force is changed, the acceleration changes in the same proportion. Doubling the force doubles the acceleration, halving the force halves the acceleration, and so on. When the force is removed, the acceleration becomes zero: The body then moves with constant velocity, as discussed in Section 4–2.

Before the time of Galileo and Newton, it was generally believed that a force was necessary just to keep a body moving, even on a level, frictionless surface or in outer space. Galileo and Newton realized that *no* net force is necessary to keep a body moving, once it has been set in motion, and that the effect of a force is not to *maintain* the velocity of a body, but to *change* its velocity. The *rate of change* of velocity for a given body is directly proportional to the force acting on it.

In Fig. 4–8b, the velocity of the body is also toward the right, but the *force* is toward the left. Under these conditions the body moves more and more slowly to the right until it stops. It then reverses its direction of motion and begins to move more and more rapidly toward the left. During this entire process the direction of the acceleration is toward the *left*, in the same direction as the force F. Hence the *magnitude* of the acceleration is proportional to that of the force, and the *direction* of the acceleration is the same as that of the force, regardless of the direction of the velocity.

Mass: the ratio of force to acceleration

To say that the acceleration of a body is directly proportional to the force exerted on it is to say that the *ratio* of the force to the acceleration is a constant, regardless of the magnitude of the force. This ratio is called the **mass** m of the body. Thus

$$m = \frac{F}{a},$$

or

$$F = ma. \qquad (4\text{–}4)$$

We can think of the mass of a body as the *force per unit of acceleration*. For example, if the acceleration of a certain body is $5 \; \text{m·s}^{-2}$ when the force is $20 \; \text{N}$, the mass of the body is

$$m = \frac{20 \; \text{N}}{5 \; \text{m·s}^{-2}} = 4 \; \text{N·m}^{-1}\text{·s}^2,$$

and a force of $4 \; \text{N}$ must be exerted on the body for each m·s^{-2} of acceleration.

This relationship can also be used to compare masses quantitatively. Suppose we apply a certain force F to a body having mass m_1 and observe an acceleration of a_1. We then apply the *same* force to another body having mass m_2, observing an acceleration a_2. Then, according to Eq. (4–4),

$$m_1 a_1 = m_2 a_2,$$

or

Comparing masses by comparing accelerations

$$\frac{m_2}{m_1} = \frac{a_1}{a_2}. \tag{4–5}$$

We can use this relation to compare any mass with a standard mass. If m_1 is a standard mass and m_2 an unknown mass, we can apply the same force to each and measure the accelerations; the ratio of the masses is the inverse of the ratio of the accelerations. When a large force is needed to give a body a certain acceleration (i.e., speed it up, slow it down, or deviate it if it is in motion), the mass of the body is large; if only a small force is needed for the same acceleration, the mass is small. Thus the mass of a body is a quantitative measure of the property described in everyday language as *inertia*.

Mass: a quantitative description of inertia

To identify another important property of mass, we measure the masses of two bodies, using the procedure just described, and then fasten them together and measure the mass of the composite body. If m_1 and m_2 are the individual masses, the mass of the composite body is always found to be $m_1 + m_2$. This very important result shows that mass is an *additive* quantity, and that it is directly correlated with quantity of matter. Indeed, the concept of mass is one way to give the term *quantity of matter* a precise meaning.

Mass is an *additive* property of matter.

In the discussion above, the particle moves along a straight line (the x-axis), and the force also lies along this direction. This is of course a special case. More generally, the force may also have a component in the y-direction, and the particle's motion need not be confined to a straight line. Furthermore, more than one force may act on the particle. Thus this formulation needs to be generalized to include motion in a plane or in space and the possibility of several forces acting simultaneously.

Experiments show that *when several forces act on a particle at the same time, the acceleration is the same as would be produced by a single force equal to the vector sum of these forces.* This sum is usually most conveniently handled by using the method of components. When several forces act on a particle moving along the x-axis,

$$\sum F_x = ma. \tag{4–6}$$

When a particle moves in a plane, with position described by coordinates (x,y), the velocity is a vector quantity with components v_x and v_y equal to the time rates of change of x and y, respectively, and the acceleration is a vector quantity with components a_x and a_y equal to the rates of change of v_x and v_y, respectively. Then a more general formulation of the relation of force to acceleration is

$$\sum F_x = ma_x, \qquad \sum F_y = ma_y. \tag{4–7}$$

This pair of equations is equivalent to the single vector equation

$$\sum \boldsymbol{F} = m\boldsymbol{a}, \tag{4–8}$$

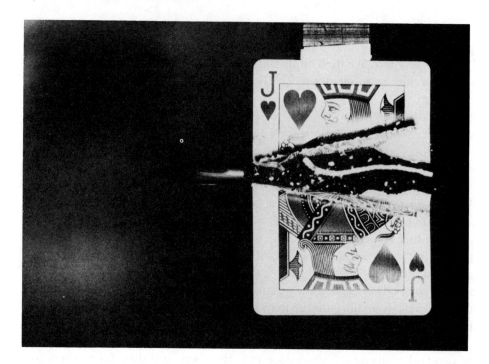

A playing card is cut in two by a .30-caliber bullet striking it edge-on. Although the card accelerates under the action of the force applied by the bullet, it does not have time to move any appreciable distance during the time of the bullet's travel through it (roughly 1/10,000 s). The actual exposure time of the photograph was less than a millionth of a second. (Dr. Harold Edgerton, M.I.T., Cambridge, Massachusetts.)

Newton's second law: the most important relation in mechanics

where we write the left-hand side explicitly as $\Sigma \boldsymbol{F}$ to emphasize that the acceleration is determined by the resultant of all the forces acting on the particle. If the particle moves in three dimensions, then of course Eqs. (4–7) include a third equation for the z-components, $\Sigma F_z = ma_z$.

Equation (4–8), or the equivalent pair of equations (4–7), is the mathematical statement of Newton's *second law of motion*. The acceleration of a body (the rate of change of its velocity) is equal to the resultant (vector sum) of all forces acting on the particle, divided by its mass, and has the same direction as the resultant force.

In this chapter we consider only straight-line motion, and we usually take the line of motion to be the x-axis. In these problems v_y and a_y are always zero. Individual forces may have y-components, but the *sum* of the y-components will always be zero. In Chapter 5 we consider more general motion in which v_y, a_y, and ΣF_y need not be zero.

4–4 SYSTEMS OF UNITS

In the preceding section we did not discuss the *units* to be used to measure force, mass, and acceleration. From the equation $\Sigma \boldsymbol{F} = m\boldsymbol{a}$ it is clear that a force of one unit gives a mass of one unit an acceleration of one unit, no matter what system we use. Thus the three units cannot be chosen independently; if we choose units for mass and acceleration, the unit of force is determined.

Several systems of units are in common use in the United States; one is the meter-kilogram-second (mks) system, which forms the basis of the Système International (SI) introduced in Chapter 1. Other systems are the British or foot-pound-second system and the now-obsolete centimeter-gram-second (cgs) version of the metric system. This book uses principally SI units, not only because this system is convenient and widely used in scientific work, but also because there seems little doubt that SI will eventually be adopted worldwide for commerce and industry as well as scientific work. The United States and Great Britain are among the few major countries that do not use the metric system exclusively, and Great Britain is now in the process of converting.

In SI units, the meter, the kilogram, and the second are defined as described in Section 1–2. The unit of force in this system is then defined as *the magnitude of force that gives an acceleration of 1 m·s^{-2} to a body of mass one kilogram.* This force is called one **newton** (1 N). Thus in SI units

Definitions of units of force

$$F \text{ (N)} = m \text{ (kg)} \times a(\text{m·s}^{-2}).$$

That is,

$$1 \text{ N} = 1 \text{ kg·m·s}^{-2}.$$

This is a very useful relationship to remember; we will use it often in the next several chapters.

In the *centimeter-gram-second* (cgs) system, the unit of mass is one gram (1 g), equal to 1/1000 kg, and the unit of acceleration is 1 cm·s^{-2}. The unit of force is called one *dyne;* one dyne gives one gram an acceleration of 1 cm·s^{-2}. Since 1 kg = 10^3 g and 1 m·s^{-2} = 10^2 cm·s^{-2}, it follows that 1 N = 10^5 dyn. In the cgs system,

$$F \text{ (dyn)} = m \text{ (g)} \times a(\text{cm·s}^{-2}).$$

In the SI and cgs systems, we first selected units of mass and acceleration, and defined the unit of force in terms of these. In the British system, we first select a unit of force (1 lb) and a unit of acceleration (1 ft·s^{-2}), and then define the unit of mass as *the mass of a body whose acceleration is* 1 ft·s^{-2} *when the resultant force on the body is* 1 lb. This unit of mass is called one *slug.* In the British system,

$$F \text{ (lb)} = m \text{ (slugs)} \times a \text{ (ft·s}^{-2}).$$

The pound is used in everyday life as a unit of quantity of matter (e.g., a pound of butter), but properly speaking it is a unit of *force* or *weight*. Thus a pound of butter is an amount having a weight of 1 lb. A useful fact in converting between mks and British units is that an object with a *mass* of 1 kg has a *weight* of about 2.2 lb (more precisely, 2.2046 lb).

Confusion between mass and weight: We don't always say what we mean.

The units of force, mass, and acceleration in the three systems are summarized in Table 4–2.

TABLE 4–2

System of units	Force	Mass	Acceleration
SI	newton (N)	kilogram (kg)	m·s^{-2}
cgs	dyne (dyn)	gram (g)	cm·s^{-2}
Engineering	pound (lb)	slug	ft·s^{-2}

We emphasize again that because of the definitions given above, there is an equivalence between force units and those of mass times acceleration. For example, 1 N = 1 kg·m·s^{-2}; in checking equations for consistency of units it is always appropriate to replace "N" wherever it appears by "kg·m·s^{-2}." Similarly, 1 lb can be replaced by 1 slug·ft·s^{-2}; this conversion can be used in unit checks.

4–5 MASS AND WEIGHT

In the introductory discussion of force in Section 4–1 we mentioned that the **weight** of a body on earth, a familiar kind of force, is the result of the gravitational interaction of the body with the earth. We will study gravitational interactions in detail in Chapter 6, but some preliminary analysis is appropriate at this point. The terms *mass* and *weight* are often misused and interchanged in everyday conversation, and it is absolutely essential for us to keep clearly in mind the distinctions between these two physical quantities.

Relation between mass and weight

Mass, on the one hand, characterizes the *inertial* properties of a body. The greater the mass, the greater the force needed to produce a given acceleration; this meaning is reflected in Newton's second law, $\Sigma F = ma.$ Weight, on the other hand, is a *force*, exerted on a body by the earth or some other large body. Of course, everyday experience shows us that bodies having large mass also have large weight; a cart loaded with bricks is hard to get rolling because of its large mass, and it is also hard to lift off the ground because of its large weight. Thus we are led to ask what the relationship is between mass and weight.

Weight and free fall: Newton's apple tree

The answer to this question, according to legend, came to Newton as he sat under an apple tree watching the apples fall. A falling body has an acceleration, and according to Newton's second law this requires a force. If a 1-kg body falls with an acceleration of 9.80 m·s^{-2}, the force required to cause this acceleration is

$$F = ma = (1\ \text{kg})(9.8\ \text{m·s}^{-2}) = 9.8\ \text{kg·m·s}^{-2}$$
$$= 9.8\ \text{N}.$$

But the force that makes the body accelerate downward is its *weight.* Hence any body having a mass of 1 kg *must* have a weight of 9.8 N. We can generalize this result: Denoting the weight of the body by w, we say that a body having a mass m must have a weight w given by

$$w = mg. \tag{4–9}$$

Variation of g with location

The value of g varies somewhat from point to point on the earth's surface. Part of this variation results from the fact that the earth's density is not uniform because of local deposits of ore, oil, or other materials of anomalous density. In addition, the earth is not perfectly spherical but is flattened at the poles and is slightly egg shaped. There are also complications associated with the rotation of the earth. For now we ignore these, but we will consider some aspects of these effects in Chapter 6.

Weight depends on location.

As mentioned above, the weight of a standard kilogram at a point where $g = 9.80$ m·s^{-2} is $w = 9.80$ N. At a second point where $g = 9.78$ m·s^{-2}, the weight is $w = 9.78$ N. Thus, unlike the mass of a body, which is constant, the weight of a body varies from one location to another. If we take a standard kilogram to the moon, where the acceleration of free fall is 1.67 m·s^{-2}, its weight is 1.67 N but its mass is still 1 kg.

Because acceleration and force are both vector quantities, Eq. (4–9) may be written as a vector equation:

$$\mathbf{w} = m\mathbf{g}. \tag{4–10}$$

It is important to understand that the weight of a body, as given by Eq. (4–10), acts on the body all the time, whether it is actually in free fall or not. When a 1-kg body hangs suspended from a string, it is in equilibrium, and its acceleration is zero. But its weight is still acting on it and is given by Eq. (4–10). In this case there is also an upward force on the body provided by the string; the *vector sum* of the forces is zero, and the body is in equilibrium.

The mass of a body can be measured in several different ways. One way is to use the relationship $m = F/a$. We apply a known force to the body, measure its acceleration, and compute the mass as the ratio of force to acceleration. This method, or some variation of it, is often used to measure the masses of atomic and subatomic particles.

Another method is to use a comparison technique, finding by trial some other body whose mass is equal to that of the given body but is already known. Often it is easier to compare *weights* than masses. The weight of a body equals the product of its mass and the acceleration of gravity, so if the weights of two bodies are equal at a given location, their masses are also equal. Balances can be used to determine with great precision (up to 1 part in 10^6) when the weights of two bodies are equal, and hence when their masses are equal.

Measuring masses by comparing weights

Thus mass plays two rather different roles in mechanics. On the one hand, the gravitational force acting on a body is proportional to its mass, so mass is *the property of matter that causes bodies to exert attractive gravitational forces on each other*. We may call this property *gravitational mass*. On the other hand, Newton's second law tells us that the force (which need not be gravitational) required to cause an acceleration of a body is proportional to its mass. This inertial property of the body may be called its *inertial mass*.

It is not obvious that the *gravitational* mass of a particle has to be the same as its *inertial* mass, but very precise experiments have established (within about 1 part in 10^{12}) that in fact the two are the same. That is, if we have to push twice as hard on body A as on body B to cause a given acceleration, then the weight of A at a given location is *precisely* twice that of body B at the same location. Thus inertial and gravitational masses really are identical, and we do not have to distinguish between them. This equivalence is also the fundamental reason why the acceleration of free fall is independent of mass.

Inertial and gravitational masses are identical.

Finally, we remark that the SI units for mass and weight are frequently misused in everyday speech. Expressions such as "This box weighs 6 kg" are nearly universal. What is meant, of course, is that the *mass* of the box (probably determined indirectly by *weighing*) is 6 kg. This usage is so common that there is probably no hope of straightening things out, but it is essential to recognize that the term *weight* is often used when *mass* is meant.

4–6 NEWTON'S THIRD LAW

Any individual force on a body is only one aspect of a mutual interaction between *two* bodies; I cannot push on you unless you push back on me at the same time. Experiment shows that *whenever one body exerts a force on another, the second body exerts simultaneously on the first a force that is equal in magnitude and opposite in direction to the force of the first on the second.* Thus it is impossible to have a single isolated force. A force on a particular body must be exerted by some other body or bodies, and it in turn exerts forces back on the other body or bodies. This property of forces is Newton's *third law of motion.* In his words,

Newton's third law: Action and reaction are equal.

> To every action there is always opposed an equal reaction; or, the mutual actions of two bodies upon each other are always equal, and directed to contrary parts.

An astronaut outside the cabin of the Space Shuttle Challenger uses a jet of nitrogen to maneuver. As the nitrogen escapes through the jet orifice, it exerts a reaction force on its container. This force causes the acceleration of the astronaut needed for him to maneuver in space. (Courtesy of NASA.)

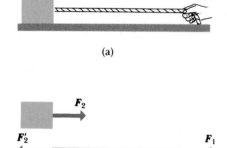

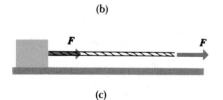

4–9 (a) A man pulls on a rope attached to a block. The forces that the rope exerts on the block and on the man are shown. (b) Separate diagrams showing the forces acting on the block, on the rope, and on the man. (c) If the rope is in equilibrium so that $F'_2 = F_1$, the rope can be considered to transmit a force from the man to the block, and vice versa.

The two forces involved in a mutual interaction between two bodies are often called an **action** and a **reaction.** This does not imply any basic difference in their nature or that one is cause and the other result. Either force may be considered the action and the other the reaction.

As an example, suppose that a man pulls on one end of a rope attached to a box, as in Fig. 4–9. The box may or may not be in equilibrium. The resulting action–reaction pairs of horizontal forces are shown in the figure. (The lines of action of all the forces lie along the rope; the force vectors have been offset from this line to show them more clearly.) Vector F_1 represents the force exerted *on* the rope *by* the man. Its reaction is the equal and opposite force F'_1 exerted *on* the man *by* the rope. Vector F_2 represents the force exerted on the box by the rope. The reaction to it is the equal and opposite force F'_2, exerted on the rope by the box:

$$F'_1 = -F_1, \qquad F'_2 = -F_2. \qquad (4-11)$$

It is very important to realize that the forces F_1 and F'_2, although they are opposite in direction and have the same line of action, are *not* an action–reaction pair. These forces act on the *same* body (the rope), whereas an action and its reaction must act on *different* bodies. Furthermore, the forces F_1 and F'_2 are not necessarily equal in magnitude. If the box and rope are moving to the right with increasing speed, the rope is not in equilibrium because it is accelerating. In that case F_1 is greater in magnitude than F'_2. Only in the special case when the rope remains at rest or moves with constant speed are the forces F_1 and F'_2 equal in magnitude—but this is an example of Newton's *first* law, not his *third*. Even when the speed of the rope is changing, however, the action–reaction forces F_1 and F'_1 are equal in magnitude to each other, and the action–reaction forces F_2 and F'_2 are equal in magnitude to each other, although then F_1 is not equal to F'_2.

In the special case when the rope is in equilibrium, and when no forces act on it except those at its end, F_2' equals $-F_1$ because of Newton's *first* law. Since F_2 *always* equals $-F_2'$ by Newton's *third* law, then in this special case F_2 also equals F_1, and the force exerted on the box by the rope is equal to the force exerted on the rope by the man. The rope can therefore be considered to "transmit" to the box, without change, the force exerted on it by the man, as in Fig. 4–9c. This point of view is often useful, but we must remember that it applies only under the restricted conditions as stated.

Tension in a rope: Forces act at both ends.

A body such as the rope in Fig. 4–9, which is subjected to pulls at its ends, is said to be in **tension.** The tension at any point equals the force exerted at that point. Thus in Fig. 4–9b the tension at the right-hand end of the rope equals the magnitude of F_1 (or of F_1'), and the tension at the left-hand end equals the magnitude of F_2 (or of F_2'). If the rope is in equilibrium and if no forces act except at its ends, the tension is the same at both ends. If, for example, the magnitudes of F_1 and F_2 are each 50 N, the tension in the rope is 50 N (*not* 100 N).

One statement made above needs emphasis: The two forces in an action–reaction pair can *never* act on the same body. Remembering this general principle can often help clear up confusion regarding action–reaction pairs.

4–7 APPLICATIONS OF NEWTON'S LAWS

The examples in this section illustrate applications of Newton's laws and the usefulness of the following strategy.

Free-body diagrams: an indispensable problem-solving aid

PROBLEM-SOLVING STRATEGY: *Newton's laws of motion*

1. Always define your coordinate system; a diagram showing the location of the origin and positive axis directions is always helpful. It is often convenient to take the positive direction to be the direction of the acceleration, if this is known.

2. Be consistent with signs. If the positive end of the x-axis is defined as positive, then velocities, accelerations, and forces to the right are also positive.

3. In applying Newton's laws, always decide on a specific body to which the laws will be applied. Draw a diagram showing all the forces (magnitudes and directions) acting on the body, but *do not* include forces that the body exerts on any other body. Such a

diagram is called a **free-body diagram.** The acceleration of the body is determined by the forces acting *on it*, not by the forces it exerts on something else.

4. Identify the known and unknown quantities, and give each unknown an algebraic symbol.

5. Always check for unit consistency; use the conversion $1 \text{ N} = 1 \text{ kg} \cdot \text{m} \cdot \text{s}^{-2}$ when appropriate.

6. We will elaborate this strategy further in the following chapters in the context of more complex problems. It is important to use it consistently from the start, however, to develop good habits in the systematic analysis of problems.

EXAMPLE 4–2   A constant horizontal force with magnitude 2 N is applied to a box of mass 4 kg resting on a level, frictionless surface. What is the acceleration of the box?

SOLUTION   We take the $+x$-axis in the direction of the horizontal force. The free-body diagram (Fig. 4–10) includes this force and the two vertical forces: the weight of the box and the upward supporting force exerted on it by the surface. The acceleration is given by Newton's second law, Eq. (4–5). There is only one

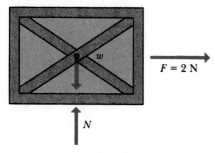

4–10 Free-body diagram for Example 4–2, showing side view of box.

horizontal force, and we have

$$a = \frac{F}{m} = \frac{2\ N}{4\ kg} = 0.5\ N \cdot kg^{-1} = 0.5\ (kg \cdot m \cdot s^{-2}) \cdot kg^{-1}$$
$$= 0.5\ m \cdot s^{-2}.$$

The force is constant, so the acceleration is also constant. Hence if we know the initial position and velocity of the box, we can find the position and velocity at any later time from the equations of motion with constant acceleration.

EXAMPLE 4–3   A student shoves a lab notebook of mass 0.2 kg toward the right along a level lab bench. As the book leaves contact with the student's hand, it has an initial velocity of 0.4 m·s⁻¹. As it slides, it slows down because of the horizontal friction force exerted on it by the bench top. It slides a distance of 1.0 m before coming to rest. What are the magnitude and direction of the friction force F acting on it?

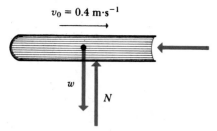

4–11 Free-body diagram for Example 4–3, showing side view of lab notebook.

SOLUTION   Suppose the book slides along the x-axis in the +x-direction, starting at the point x_0 with the given initial velocity. A free-body diagram is shown in Fig. 4–11. We assume that the friction force is constant. The acceleration is then constant also. From the equations of motion with constant acceleration, we have

$$v^2 = v_0{}^2 + 2ax,\qquad 0 = (0.4\ m \cdot s^{-1})^2 + (2a)(1.0\ m);$$
$$a = -0.08\ m \cdot s^{-2}.$$

The negative sign means that the acceleration is toward the *left* (although the velocity is toward the right). The friction force on the book is

$$F = ma = (0.2\ kg)(-0.08\ m \cdot s^{-2})$$
$$= -0.016\ N,$$

and is toward the left also. (A force of equal magnitude, but directed toward the right, is exerted *on* the table *by* the sliding book.)

EXAMPLE 4–4   A machine part having a mass of 5 kg slides along a rail aligned with the x-axis. Its position is given as a function of time by

$$x = (18\ m \cdot s^{-2})t^2 - (3\ m \cdot s^{-3})t^3.$$

Find the force acting on the body, as a function of time. What is the force at time $t = 5$ s? For what times is the force positive? Negative? Zero?

SOLUTION   The free-body diagram is just like that of Example 4–2. We first find the acceleration a by taking the second derivative of x:

$$a = d^2x/dt^2 = 36\ m \cdot s^{-2} - (18\ m \cdot s^{-3})t.$$

Then, from Newton's second law,

$$F = ma = (5\ kg)[36\ m \cdot s^{-2} - (18\ m \cdot s^{-3})t]$$
$$= [180 - (90\ s^{-1})t]\ N.$$

At time $t = 5$ s, the force is $180\ N - (90\ s^{-1})(5\ s)\ N = -270\ N$. The force is zero when $36\ m \cdot s^{-2} - (18\ m \cdot s^{-3})t = 0$, that is, when $t = 2$ s. When $t < 2$ s, F is positive, and when $t > 2$ s, F is negative.

EXAMPLE 4–5 A flowerpot having a mass of 10 kg is suspended by a chain from the ceiling. What is its weight? What force (magnitude and direction) does the chain exert on it? What is the tension in the chain? Assume the weight of the chain itself is negligible.

SOLUTION Free-body diagrams for the flowerpot and the chain are shown in Fig. 4–12. The weight of the pot is given by Eq. (4–9):

$$w = (10 \text{ kg})(9.8 \text{ m·s}^{-2}) = 98 \text{ N}.$$

The pot is in equilibrium, so the vector sum of forces acting on it must be zero. Thus the chain must pull *up* with a force T of magnitude 98 N. By Newton's third law, the pot pulls *down* on the chain with a force of 98 N. Because the chain is also in equilibrium, there must be an upward force of 98 N exerted on the chain at its top end to make the vector sum of forces on the chain equal zero.

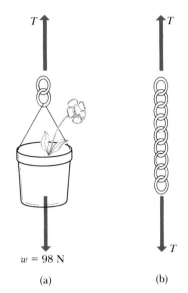

4–12 Free-body diagrams for Example 4–5. (a) Forces acting on the flowerpot; (b) forces acting on the chain. Each body is in equilibrium, and the sum of the forces on each must be zero. The weight of the chain is assumed to be negligible.

4–8 INERTIAL FRAMES OF REFERENCE

A set of coordinate axes attached to (or moving with) some specified body or bodies is called a **frame of reference** (or a *reference system*). We always describe the motion of a given body with reference to, or relative to, some other body. Its motion relative to one body may be very different from that relative to another. Thus a passenger in an airplane making its takeoff run is at rest relative to the airplane but is moving faster and faster relative to the earth.

Newton's first law defines by implication a special class of frames of reference. Suppose we consider a frame of reference attached to the airplane mentioned above. During the takeoff run, while the plane is going faster and faster, a passenger feels the back of his seat pushing him forward, although he remains at rest relative to the airplane. Therefore Newton's first law does not describe the situation correctly; a forward force does act on the passenger, but nevertheless (relative to the airplane) he remains at rest.

But now suppose that the passenger is standing in the aisle on roller skates. He then starts to move *backward,* relative to the plane, when the takeoff run begins, even though no backward force acts on him. Again Newton's first law is not obeyed.

Thus Newton's first law is obeyed in some frames of reference and not in others. We now define an **inertial frame of reference** as a frame of reference in which Newton's first law is obeyed. Relative to an inertial frame, a body does remain at rest or move uniformly in a straight line when no force (or no resultant force) acts on it.

An airplane gaining speed during takeoff is evidently *not* an inertial frame. For many purposes a reference system attached to the earth can be considered an inertial system, although it is not precisely so because of effects related to the earth's rotation and other motions. But if a frame of reference *is* inertial, then a second frame moving uniformly (i.e., with constant speed in a straight line, without rotation) relative to it is also inertial, since any body in uniform motion as seen by an observer in the first frame will also appear to an observer in the second frame to be in uniform motion. The speed and direction appear different to the two observers, but both observe that Newton's first law is obeyed.

Inertial frames of reference: the coordinate systems where Newton's laws work

There is no single, unique inertial frame of reference. There are infinitely many, but the motion of any one relative to any other is uniform in the sense noted above. It follows that the concepts of "absolute rest" and "absolute motion" have no physical meaning. An airplane in steady flight over the earth is just as suitable an inertial frame of reference as the earth itself.

In most of this book we assume that the earth is an adequate approximation to an inertial frame of reference and that errors resulting from its being not precisely inertial are negligible. When we observe a body at rest relative to the earth begin to move, or when a moving body speeds up, slows down, or changes direction, we know that a force must be acting on it, or perhaps several forces with a vector sum or resultant that is different from zero. The quantitative relationship between the body's acceleration and the vector sum of the forces is Newton's second law. Like the first law, it is obeyed only in inertial frames of reference.

SUMMARY

KEY TERMS

dynamics

force

resultant

equilibrium

mass

newton

weight

action and reaction

tension

free-body diagram

frame of reference

inertial frame of reference

Force is a quantitative measure of the mechanical interaction between two bodies. It is a *vector* quantity, having a magnitude and a direction. When several forces act on a body at the same time, the effect is the same as when a single force, equal to the *vector sum* or *resultant* of the forces, acts on the body.

Newton's first law states that when no force acts on a body, or when the vector sum of all forces acting on it is zero, the body is in *equilibrium.* If the body is initially at rest, it remains at rest; if initially in motion, it continues to move with constant velocity.

The inertial properties of a body are characterized by its *mass.* The acceleration of a body under the action of a given force has a magnitude that is directly proportional to the magnitude of the force and inversely proportional to the mass of the body. This relationship is Newton's second law, $\Sigma F = ma.$

Units of force are defined in terms of the fundamental units of mass, length, and time. In SI the unit of force is 1 newton (1 N), equal to 1 kg·m·s^{-2}. In the cgs system it is 1 dyne (1 dyn), equal to 1 g·cm·s^{-2}.

The weight of a body is the gravitational force exerted on it by the earth (or whatever other body exerts the gravitational force). Weight is a force and is therefore a vector quantity. The magnitude of the weight of a body at any specific location is equal to the product of its mass and the magnitude of the acceleration of free fall (acceleration due to gravity) at that location; that is, $w = mg.$ The weight of a body depends on its location, but the mass is independent of location.

Newton's third law states that when two bodies interact, they exert forces on each other simultaneously; these forces are equal in magnitude and opposite in direction. This principle is often stated as "action equals reaction." The two forces in an action-reaction pair always act on two *different* bodies; they never act on the same body.

Newton's laws of motion are valid only in *inertial frames of reference.* If one frame of reference is inertial, any other frame moving relative to it with constant velocity is also inertial, but a frame that accelerates relative to the first is *not* inertial. The earth is approximately an inertial frame, but not precisely so because of its rotation about its axis and orbital motion around the sun.

QUESTIONS

4–1 When a heavy weight is lifted by a string that is barely strong enough, the weight can be lifted by a steady pull, but if the string is jerked it will break. Why?

4–2 When a car accelerates, starting from rest, where is the force applied to the car to cause its acceleration? By what other body is the force exerted?

4–3 When a car stops suddenly, the passengers are thrown forward, away from their seats. What force causes this motion?

4–4 For medical reasons it is important for astronauts in outer space to determine their body mass at regular intervals. Devise a scheme for measuring body mass in a zero-gravity environment.

4–5 In SI units, suppose the fundamental units had been chosen to be force, length, and time instead of mass, length, and time. What would be the units of mass, in terms of the fundamental units?

4–6 When a bullet is fired from a gun, what is the origin of the force that accelerates the bullet?

4–7 Why can a person dive into water from a height of 10 m without injury, while a person who jumps off the roof of a 10-m building and lands on a concrete street is likely to be seriously injured?

4–8 A passenger in a bus notices that a ball that has been at rest in the aisle suddenly starts to move toward the rear of the bus. Think of two different possible explanations, and devise a way to decide which is correct.

4–9 Two bodies on the two sides of an equal-arm balance exactly balance each other. If the balance is placed in an elevator and given an upward acceleration, do they still balance?

4–10 If a man in an elevator drops his briefcase but it does not fall to the floor, what can he conclude about the elevator's motion?

4–11 Suppose you are sealed inside a box on the planet Vulcan. After some time you notice that objects do not fall to the floor when released. How can you determine whether you and the box are in free fall or whether the Vulcans have somehow found a way to turn off the force of gravity?

4–12 Automotive engineers, in discussing the vertical motion of an automobile driving over a bump, call the time derivative of the acceleration the "jerk." Why is this a useful quantity in characterizing the riding qualities of an automobile?

EXERCISES

Section 4–1 Force

4–1 A raccoon pushes a tin box along the floor as in Fig. 4–1b with a force of 40 N that makes an angle of 30° with the horizontal. Find the horizontal and vertical components of the force.

4–2 A man is dragging a trunk up the loading ramp of a mover's truck. The ramp has a slope angle of 20°, and the man pulls upward with force F whose direction makes an angle of 30° with the ramp.

a) How large a force F is necessary for the component F_x parallel to the plane to be 16 N?

b) How large will the component F_y then be?

4–3 Two men pull horizontally on ropes attached to a post; the angle between the ropes is 45°. If man A exerts a force of 68 lb and man B a force of 50 lb, find the magnitude of the resultant force and the angle it makes with A's pull.

4–4 Two forces, F_1 and F_2, act at a point. The magnitude of F_1 is 8 N and its direction is 60° above the x-axis in the first quadrant. The magnitude of F_2 is 5 N and its direction is 53° below the x-axis in the fourth quadrant.

a) What are the horizontal and vertical components of the resultant force?

b) What is the magnitude of the resultant?

c) What is the magnitude of the vector difference $F_1 - F_2$?

Section 4–5 Mass and Weight

4–5

a) What is the mass of a book that weighs 1 N at a point where $g = 9.80$ m·s^{-2}?

b) At the same point, what is the weight of a dog whose mass is 8 kg?

4–6

a) What is the mass of a radio that weighs 8 lb at a point where $g = 32$ ft·s^{-2}?

b) At the same point, what is the weight of a television set whose mass is 0.8 slug?

4–7 At the surface of Mars the acceleration due to gravity is $g = 3.7$ m·s^{-2}. A watermelon weighs 52 N at the surface of the earth. What would be its mass and weight on the surface of Mars?

Section 4–6 Newton's Third Law

4–8 Imagine that you are holding a book weighing 4 N at rest on the palm of your hand. Complete the following sentences:

a) A downward force of magnitude 4 N is exerted on the book by _____.

b) An upward force of magnitude _____ is exerted on _____ by the hand.

c) Is the upward force (b) the reaction to the downward force (a)?

d) The reaction to force (a) is a force of magnitude _____, exerted on _____ by _____. Its direction is _____.

e) The reaction to force (b) is a force of magnitude _____, exerted on _____ by _____. Its direction is _____.

f) That the forces (a) and (b) are equal and opposite is an example of Newton's _____ law.

g) That forces (b) and (e) are equal and opposite is an example of Newton's _____ law.

Suppose now that you exert an upward force of magnitude 5 N on the book.

h) Does the book remain in equilibrium?

i) Is the force exerted on the book by the hand equal and opposite to the force exerted on the book by the earth?

j) Is the force exerted on the book by the earth equal and opposite to the force exerted on the earth by the book?

k) Is the force exerted on the book by the hand equal and opposite to the force exerted on the hand by the book?

Finally, suppose that you snatch your hand away while the book is moving upward.

l) How many forces then act on the book?

m) Is the book in equilibrium?

n) What balances the downward force exerted on the book by the earth?

4–9 A bottle is given a push along a table top and slides off the edge of the table. Neglect air resistance.

a) What forces are exerted on it while it is falling from the table to the floor?

b) What is the reaction to each force; that is, on what body and by what body is the reaction exerted?

Section 4–7 Applications of Newton's Laws

4–10 A crate of mass 1.2 slugs initially at rest on a frictionless horizontal plane is acted on by a horizontal force of 36 lb.

a) What acceleration is produced?

b) How far will the crate travel in 10 s?

c) What will be its velocity at the end of 10 s?

4–11 A constant horizontal force of 40 N acts on a block of ice on a smooth horizontal plane. The block starts from rest and is observed to move 100 m in 5 s.

a) What is the mass of the block of ice?

b) If the force ceases to act at the end of 5 s, how far will the block move in the next 5 s?

4–12 A .22 rifle bullet, traveling at 360 m·s^{-1}, strikes a block of soft wood, which it penetrates to a depth of 0.1 m. The mass of the bullet is 1.8 g. Assume a constant retarding force.

a) How much time was required for the bullet to stop?

b) What was the accelerating force, in newtons?

4–13 A hockey puck of mass 0.5 kg is at rest at the origin $x = 0$ on a frictionless horizontal surface. At time $t = 0$ a force of 0.001 N is applied to the puck parallel to the x-axis, and 5 s later this force is removed

a) What are the position and velocity of the puck at $t = 5$ s?

b) If the same force is again applied at $t = 15$ s, what are the position and velocity of the puck at $t = 20$ s?

4–14 An electron (mass = 9.11×10^{-31} kg) leaves the cathode of a vacuum tube with zero initial velocity and travels in a straight line to the anode, which is 1 cm away. It reaches the anode with a velocity of 6.00×10^6 m·s^{-1}. If the accelerating force is constant, compute

a) the acceleration;

b) the time to reach the anode;

c) the accelerating force, in newtons.

(The gravitational force on the electron may be neglected.)

4–15 An object of mass m moves along the x-axis. Its position as a function of time is given by $x(t) = At + Bt^3$, where A and B are constants. Calculate the resultant force on the object, as a function of time.

PROBLEMS

4–16 The three forces shown in Fig. 4–13 act on an object located at the origin.

a) Find the x- and y-components of each of the three forces.

b) Find the resultant of these forces.

c) Find the magnitude and direction of a fourth force that must be added to make the resultant force zero. Indicate the fourth force by a diagram.

4–17 Two forces F_1 and F_2 act on an object such that the resultant force R has a magnitude equal to that of F_1 and makes an angle of 90° with F_1. Let $F_1 = R = 10$ N. Find the magnitude of the second force, and its direction (relative to F_1).

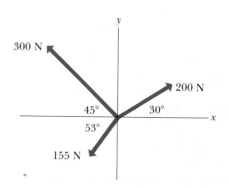

FIGURE 4–13

4–18 The resultant of four forces is 1000 N in the direction 30° west of north. Three of the forces are 400 N, 60° north of east; 200 N, south; and 400 N, 53° west of south. Find the magnitude and direction of the fourth force.

4–19 Two men and a boy want to push a crate in the direction marked x in Fig. 4–14. The two men push with forces F_1 and F_2, whose magnitudes and directions are indicated in the figure. Find the magnitude and direction of the *smallest* force that the boy should exert.

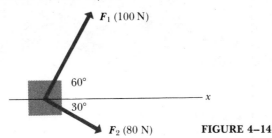

FIGURE 4–14

4–20 A short commuter train consists of a locomotive and two cars. The mass of the locomotive is 6000 kg and that of each car 2000 kg. The train pulls away from a station with an acceleration of 0.5 m·s^{-2}.

a) Find the tension in the coupler joining the locomotive to the first car, and in the coupler joining the two cars.

b) What total horizontal force must the locomotive wheels exert on the track?

4–21 A loaded elevator with very worn cables has a total mass of 2000 kg, and the cables can withstand a maximum tension of 24,000 N.

a) What is the maximum upward acceleration for the elevator if the cables are not to break?

b) What is the answer to (a) if the elevator is taken to the moon, where $g = 1.67 \text{ m·s}^{-2}$?

4–22 An 80-kg man steps off a platform 2.0 m above the ground. He keeps his legs straight as he falls, but at the moment his feet touch the ground his knees begin to bend, and his torso moves an additional 0.6 m before coming to rest.

a) What is his speed at the instant his feet touch the ground?

b) What is the acceleration of his torso as he slows down, assuming the acceleration is constant?

c) What force do his feet exert on the ground while he slows down? Express this force in newtons and also as a multiple of his weight.

4–23 An object of mass 2 kg moves in the x–y plane such that its x- and y-coordinates vary in time according to

$$x(t) = 2 \text{ m} - (5 \text{ m·s}^{-3})t^3, \qquad y(t) = (16 \text{ m·s}^{-2})t^2.$$

a) Calculate the x- and y-components of the resultant force on the object, as functions of time.

b) What are the magnitude and direction of the resultant force at $t = 3$ s?

4–24 An object of mass m initially at rest is acted on by a force $F = k_1 i + k_2 t^2 j$, where k_1 and k_2 are constants. Calculate the velocity $v(t)$ of the object as a function of time.

CHALLENGE PROBLEM

4–25 An object of mass m is at rest at the origin at time $t = 0$. A force $F(t)$ is then applied that has components

$$F_x(t) = k_1 + k_2 y, \qquad F_y(t) = k_3 t,$$

where k_1, k_2, and k_3 are constants. Calculate the position $r(t)$ and velocity $v(t)$ vectors as functions of time.

5

APPLICATIONS OF NEWTON'S LAWS—I

NEWTON'S LAWS OF MOTION FORM THE FOUNDATION FOR ALL OF CLASSICAL mechanics, including statics (the study of bodies in equilibrium) and dynamics (the general relationships between force and motion). In this chapter we encounter no new fundamental principles; instead, we concentrate on learning to apply Newton's laws to problems of fundamental and practical importance. We begin with a discussion of the various classes of forces found in nature. Then we study in detail the contact forces between two bodies, including friction forces. Next we consider examples of bodies in equilibrium, using Newton's first law, the concept of an inertial frame of reference, and the vector language introduced in Chapter 1. Finally, we study several examples of Newton's second law, analyzing in detail the relationships between force and motion for bodies that are *not* in equilibrium. We find that much of the same problem-solving strategy can be used for both statics and dynamics problems.

5–1 FORCES IN NATURE

Our present-day understanding of the physical world includes four distinct classes of forces. Two are familiar in everyday experience. The other two are concerned with interactions of fundamental particles. We cannot observe the latter with the unaided senses; studying them requires sophisticated and elaborate experiments.

Forces found in nature: two familiar kinds

Of the two familiar classes of force, **gravitational interactions** were the first to be studied in detail. The *weight* of a body results from the earth's gravitational attraction acting on it. The sun's gravitational force on the earth is responsible for making the earth move in a nearly circular orbit instead of following a straight-line path as it would do if there were no force. Indeed, one of Newton's great achievements was to recognize that both the motions of the planets around the sun and the free fall of objects on earth are manifestations of gravitational forces. We will study gravitational interactions in greater detail in Chapter 6, and will analyze their vital role in the motions of planets and satellites.

The second familiar class of forces, **electromagnetic interactions,** includes electric and magnetic forces. When you run a comb through your hair, you can then use the comb to pick up bits of paper or fluff; this interaction is the result of electric charge on the comb. We encounter magnetic forces in interactions between magnets or between a magnet and a piece of iron. These may seem to fall in a different category, but detailed study shows that magnetic interactions are actually the result of electric charges in motion. An electromagnet causes magnetic interactions as a result of an electric current in a coil of wire. We will study electric and magnetic interactions in detail in the second half of this book.

These two familiar kinds of interactions differ enormously in their strength. For example, the electrical repulsion between two protons at a given

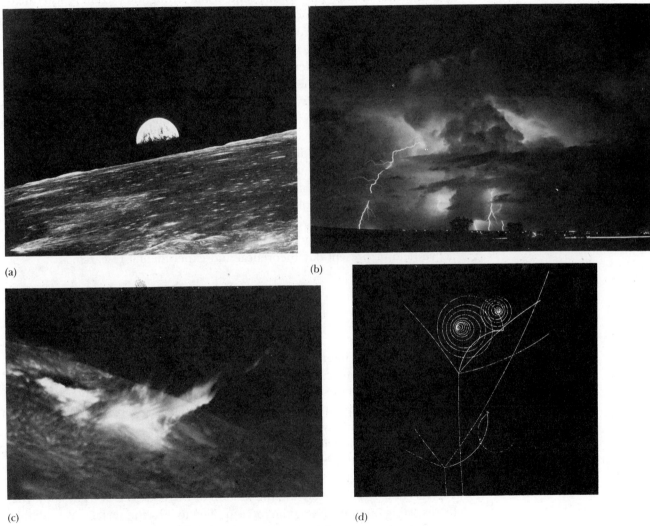

(a)   (b)

(c)   (d)

Photographs of phenomena associated with the four classes of forces occurring in nature. (a) The orbital motions of the earth around the sun and of the moon around the earth result from *gravitational* forces. (NASA.) (b) Lightning is a flow of electric charge resulting from the electrical interactions of charged particles, one aspect of the *electromagnetic* interaction. (National Center for Atmospheric Research/National Science Foundation.) (c) A solar flare, the result of a thermonuclear reaction associated with *strong* interactions between nuclear particles. (Sacramento Peak Observatory Association of Universities for Research in Astronomy, Inc.) (d) An experiment to detect the presence of neutrinos, which have only *weak* interactions. (Lawrence Berkeley Laboratory.)

distance apart is stronger than the gravitational attraction by a factor of the order of 10^{35}. Thus gravitational forces play no significant role in determining the microscopic structure of atoms, molecules, and materials. But in bodies of astronomical size, protons and electrons are ordinarily present in nearly equal numbers; their charges are opposite and nearly cancel out. Gravitational interactions are the dominant influence in the motion of planets and also in the internal structure of stars.

The strong interaction holds nuclei together.

The other two classes of interactions are less familiar. One, the **strong interaction,** is responsible for holding the nuclei of atoms together. Nuclei contain electrically neutral and positively charged particles. The charged particles repel each other; a nucleus could not be stable it it were not for the presence of an attractive force of a different kind that counteracts the repulsive electrical interactions. In this context the strong interaction is also called the *nuclear force.* It has shorter range than electrical interactions, but within its range it is much stronger.

The weak interaction is responsible for the decay of some unstable particles.

Finally, there are the **weak interactions;** they play no direct role in the behavior of ordinary matter but are of vital importance in interactions among fundamental particles. The weak interaction is responsible for the emission of electrons (beta particles) from radioactive nuclei. In a beta-emitting nucleus a neutron is converted into a proton, an electron, and a third particle called a neutrino. The weak interaction is also responsible for the decay of many unstable particles produced in high-energy collisions of fundamental particles.

Grand unification: Are the four kinds of forces really all the same?

Since about 1970, attempts have been made to understand all four classes of interactions on the basis of a single unified theory called a *grand unified theory.* Such theories are still very speculative; as yet none has been completely successful. In 1983, however, evidence was found that supports a unified theory of the electromagnetic and weak interactions. The entire area is a very active field of present-day theoretical and experimental research.

5–2 CONTACT FORCES AND FRICTION

Describing the interaction of bodies in contact

When two bodies interact by direct contact (touching) of their surfaces, the interaction forces are called **contact forces.** We have already remarked that on a microscopic level these forces arise mainly from the electrical interactions of the electrons and positively charged nuclei of the atoms in the bodies. But in this chapter we simply describe the behavior of the contact forces without going into their electrical basis.

Normal and frictional components of the contact force

To discuss the nature of the contact force, let us first think about a body sliding across a surface. It might be, for example, the lab notebook in Example 4–3 (Section 4–7). In studying the motion of the body, we need to know the forces acting on it, including the force applied to it by the surface. We call this force the contact force; the force acting on the body can always be represented in terms of a component perpendicular to the surface and a component parallel to the surface. We call the perpendicular component the **normal force,** denoted by $\mathcal{n}$. (*Normal* is a synonym for *perpendicular.*) The component parallel to the surface is the **friction force,** denoted by $\mathcal{f}$.

The friction force $\mathcal{f}$ is always the component of the contact force *parallel* to the surface, and the normal force $\mathcal{n}$ is always the component *normal* (perpendicular) to the surface. We use special script symbols for these quantities to emphasize their special role in representing the contact force. In addition, use of the script letter $\mathcal{n}$ prevents our confusing this symbol for normal force with the abbreviation for newton, N. The normal force is not necessarily equal in magnitude to the weight of the body, because there may be additional components of force normal to the surface. Nevertheless, $\mathcal{n}$ always denotes the normal component of contact force exerted by the surface on the body. By definition, $\mathcal{n}$ and $\mathcal{f}$ are always perpendicular.

If the surface is perfectly smooth and perfectly lubricated so the body can slide without any resistance, then there is *no* friction force. In the real world this is an unattainable idealization, but we sometimes talk about an idealized "frictionless surface," where the friction force is negligibly small. In such cases the contact force has *only* a normal component and is perpendicular to the surface.

The general behavior of friction forces is rather complex, but some aspects are quite familiar. The friction force on each interacting body is opposite in direction to the motion of that body relative to the other. Thus when a book slides from left to right along a table top, the friction force on it acts to the left. According to Newton's third law, the book exerts on the table a force of equal magnitude, directed to the right. It tries to drag the table along with it.

The *magnitude* of the friction force depends on the magnitude of the normal force, usually increasing with the normal force. It is harder to push a full oil barrel across a floor than to push the same barrel when it is empty. Often the friction force $\mathcal{f}_k$ is found to be approximately *proportional* to the normal force; in such cases we represent the relation by the equation

$$\mathcal{f}_k = \mu_k \mathcal{n}, \qquad (5\text{--}1)$$

where μ_k is a constant called the **coefficient of kinetic friction.** The adjective *kinetic* and the corresponding subscript k refer to the fact that the two surfaces are moving relative to each other. Note that because the friction force and the normal force are always perpendicular, Eq. (5–1) is not a vector equation but a relation between the *magnitudes* of the two forces.

The coefficient of kinetic friction depends on the nature of the materials and the textures of the surfaces. For teflon on steel, μ_k is about 0.04; for rubber tires on dry concrete, μ_k can be greater than unity. The more slippery the surface, the smaller the coefficient of friction. Table 5–1 shows a few representative values of μ_k. Friction forces can also depend on the relative *velocity* of the interacting surfaces; we will ignore that complication here in order to concentrate on the simplest cases.

The frictional force is proportional to the normal force (sometimes).

TABLE 5–1 Coefficients of Friction

Materials	Static, μ_s	Kinetic, μ_k
Steel on steel	0.74	0.57
Aluminum on steel	0.61	0.47
Copper on steel	0.53	0.36
Brass on steel	0.51	0.44
Zinc on cast iron	0.85	0.21
Copper on cast iron	1.05	0.29
Glass on glass	0.94	0.40
Copper on glass	0.68	0.53
Teflon on Teflon	0.04	0.04
Teflon on steel	0.04	0.04
Rubber on concrete (dry)	1.0	0.80
Rubber on concrete (wet)	0.30	0.25

Friction forces may also act when there is *no* relative motion. For example, when we push horizontally on a heavy packing case resting on the floor, the case may not move because the floor exerts an equal and opposite friction force on the case. This is called a **static friction force** $\mathcal{F}_s$, exerted on the box by the surface.

Figure 5–1 shows a box at rest on a horizontal surface. Its weight w and the upward normal force n exerted on it by the surface have equal magnitude, and it is in equilibrium. Now we attach a rope to the box as in Fig. 5–1b and gradually increase the tension T in the rope. At first the box remains at rest because the force of static friction $\mathcal{F}_s$ is equal in magnitude and opposite in direction to the force T. As T increases further, a critical value is reached at which the box suddenly "breaks loose" and starts to slide. In other words, there is a certain *maximum* value that the force of static friction $\mathcal{F}_s$ can have. When that value is not sufficient to prevent motion, the box starts to move. Figure 5–1c is the force diagram when T is just below its critical value. If T exceeds this value, the box is no longer in equilibrium.

For a given pair of surfaces, the magnitude of the maximum value of $\mathcal{F}_s$ depends on the magnitude of the normal force n; when n increases, $\mathcal{F}_s$ also increases, and a greater force is needed to make the box start to slide. In some cases the maximum value of $\mathcal{F}_s$ is approximately *proportional* to n; we call the proportionally factor μ_s the **coefficient of static friction.** The actual force of static friction can have any magnitude between zero (when there is no other force parallel to the surface) and a maximum value given by $\mu_s n$. Thus,

$$\mathcal{F}_s \leq \mu_s n. \tag{5–2}$$

The equality sign holds only when the applied force T, parallel to the surface, has such a value that motion is about to start (Fig. 5–1c). When T is less than this value (Fig. 5–1b), the inequality sign holds and the magnitude of the friction force must be computed from the conditions of equilibrium.

As soon as sliding begins, the friction force usually (but not always) *decreases.* Thus for any given pair of surfaces, the coefficient of kinetic friction is usually less than the coefficient of static friction. A few representative values of both μ_s and μ_k are given in Table 5–1. This table lists values only for solids. Liquids and gases also show frictional effects, but the simple equation $\mathcal{F} = \mu n$ does not hold. The friction force between two surfaces sliding over each other with a layer of liquid or gas between them is determined by a property of the fluid called *viscosity.* Gases have much smaller viscosities than

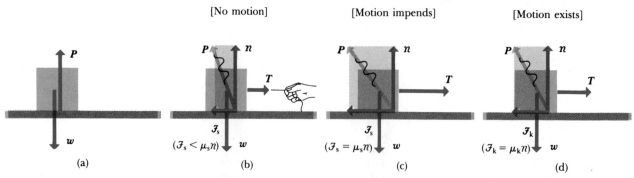

5–1 The magnitude of the friction force $\mathcal{F}_s$ is less than or equal to $\mu_s n$ when there is no relative motion and is equal to $\mu_k n$ when there is motion.

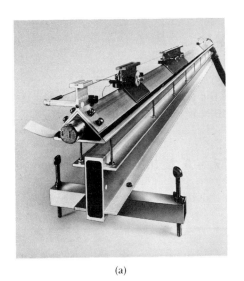

(a)   (b)

5–2 (a) The Ealing–Stull linear air track. Inverted Y-shaped sliders ride on a layer of air streaming through many fine holes in the inverted V-shaped surface. (b) The Ealing–Daw two-dimensional air table. Plastic pucks slide on a cushion of air issuing from more than a thousand minute holes in the tabletop. (Courtesy of the Ealing Corporation.)

liquids at normal temperatures, and therefore friction can be made very small if a body can be made to slide on a layer of gas.

This is the principle of the hovercraft, a type of vehicle containing powerful blowers that maintain a cushion of air between the vehicle and the ground so it literally floats on a layer of air. A familiar laboratory example of this principle is the linear air track, shown in Fig. 5–2a. The frictional force is velocity dependent, but at typical speeds the effective coefficient of friction is of the order of 0.001.

A similar device is the frictionless air table shown in Fig. 5–2b. The pucks are supported by an array of small air jets about 2 cm apart. Two-dimensional collisions can be demonstrated on this table. The same principle is used in "air-hockey" games.

EXAMPLE 5–1   Suppose the body shown in Fig. 5–1 is a 500-N crate full of home exercise equipment that a delivery company has just unloaded in your driveway. You find that to get it moving toward your garage you have to push with a horizontal force of magnitude 200 N; but once it is moving, you can keep it moving with constant speed by pushing with 100-N force. What are the coefficients of static and kinetic friction?

The role of friction in moving heavy boxes

SOLUTION   When you are getting the crate moving, the appropriate force diagram is Fig. 5–1c; from this and the above data, we have

$$\Sigma F_y = n - w = n - 500\text{ N} = 0, \qquad n = 500\text{ N},$$
$$\Sigma F_x = T - \mathcal{F}_s = 200\text{ N} - \mathcal{F}_s = 0, \qquad \mathcal{F}_s = 200\text{ N}.$$
$$\mathcal{F}_s = \mu_s n \qquad \text{(motion impends)}.$$

Hence

$$\mu_s = \frac{\mathcal{F}_s}{n} = \frac{200\text{ N}}{500\text{ N}} = 0.40.$$

After the crate is moving, the forces are as in Fig. 5–1d, and we have

$$\Sigma F_y = n - w = n - 500\text{ N} = 0, \qquad n = 500\text{ N},$$
$$\Sigma F_x = T - \mathcal{F}_k = 100\text{ N} - \mathcal{F}_k = 0, \qquad \mathcal{F}_k = 100\text{ N}.$$
$$\mathcal{F}_k = \mu_k n \qquad \text{(motion exists)}.$$

Hence

$$\mu_k = \frac{\mathcal{J}_k}{n} = \frac{100 \text{ N}}{500 \text{ N}} = 0.20.$$

EXAMPLE 5–2 In Example 5–1, what is the friction force if the crate is at rest on the surface and a horizontal force of 50 N is exerted on it?

SOLUTION We have

$$\Sigma F_x = T - \mathcal{J}_s = 50 \text{ N} - \mathcal{J}_s = 0 \qquad \text{(first law)}.$$

$$\mathcal{J}_s = 50 \text{ N}.$$

Note that in this case $\mathcal{J}_s < \mu_s n$

Rolling friction: why the wheel is a useful invention

Ever since the invention of the wheel, we have known that it is a lot easier to move a heavy load across a horizontal surface by using a cart with wheels than it is to drag the same load across the surface. How much easier? We can define a coefficient of rolling friction μ_r for a wheeled vehicle as the ratio of the horizontal force needed to make it move with constant speed, to the upward normal force exerted by the surface (equal in this case to the weight of the vehicle and load). Transportation engineers call this ratio the *tractive resistance.* Typical values of tractive resistance are 0.002 to 0.003 for steel wheels on steel rails, and 0.01 to 0.02 for rubber tires on concrete. These values show why railroad trains are in general much more fuel-efficient than highway trucks.

EXAMPLE 5–3 A 1200-kg car weighs about 12,000 N (about 2700 lb). If the tractive resistance is $\mu_r = 0.01$, what horizontal force must be provided to make the car move with constant speed?

SOLUTION From the definition of tractive resistance, a horizontal force of $(0.01)(12,000 \text{ N}) = 120$ N (about 50 lb) is required.

The actual behavior of tractive resistance forces is a lot more complicated than this. An important factor is speed: Air-resistance forces usually increase proportionally to the square of the speed. The numerical values cited above are for relatively slow speeds.

5–3 EQUILIBRIUM OF A PARTICLE

Equilibrium: The forces on a body must add up to zero.

In this section we consider several problems involving the **equilibrium** of particles. It is surprising how many situations of interest and importance in engineering, in the life and earth sciences, and in everyday life involve equilibrium of particles. The essential physical principle is Newton's first law: When a particle is at rest in an inertial frame of reference, the vector sum of all the forces acting on it must be zero. In symbols,

$$\Sigma F = 0. \tag{5–3}$$

We will usually use this in component form:

$$\sum F_x = 0,$$
$$\sum F_y = 0. \qquad\qquad (5\text{--}4)$$

In the analysis of equilibrium problems, we strongly recommend strict adherence to the following strategy.

PROBLEM-SOLVING STRATEGY: *Equilibrium of a particle*

1. Make a simple sketch of the apparatus or structure, showing dimensions and angles.

2. Choose some object (a knot in a rope, for example) as the particle in equilibrium. Draw a separate diagram of this object and show with arrows (use a colored pencil or pen) *all* of the forces exerted *on* it by other bodies. This is called the *force diagram* or **free-body diagram.** When a system is composed of several particles, you may need to draw a separate free-body diagram for each one. Do *not* show, in the free-body diagram of a chosen particle, any of the forces exerted *by* it on other bodies.

3. Draw a set of coordinate axes and resolve into components all forces acting on the particle. Cross out lightly those forces that have been resolved. Often one particular choice of axes can simplify the problem considerably. For example, when friction is present, it is usually simplest to take the axes in the directions of the normal and frictional forces, even when these are not vertical and horizontal.

4. Set the algebraic sum of all *x*-forces (or force components) equal to zero, and the algebraic sum of all *y*-forces (or components) equal to zero. This provides two independent equations, which can be solved simultaneously for two unknown quantities; these may be forces, angles, distances, and so on.

A remark about notation is in order. We have consistently denoted vector quantities by boldface italic symbols, and we will continue this usage throughout this book. However, it is sometimes convenient in diagrams to label a vector by using a lightface italic symbol to represent the *magnitude* of the quantity, and to describe its direction (if that is not obvious in the diagram) separately, perhaps by use of an angle. When both light and bold symbols appear on a diagram, you can avoid confusion by recalling that a bold symbol always represents the vector quantity itself, while a light symbol represents either the magnitude of a vector quantity or a component.

In many of the problems we will encounter, one of the forces is the *weight* of a body, that is, the force of gravitational attraction that the earth exerts on the body. In such a case, the body attracts the earth with a force equal in magnitude and opposite in direction to the earth's force on the body. Thus if a body weighs 10 N, the earth pulls *down* on it with a force of 10 N, and the body pulls *up* on the earth with a force of 10 N. This is an example of Newton's third law and an action–reaction pair.

How to label forces on free-body diagrams

The weight of a body

EXAMPLE 5–4 A gymnast has just begun climbing up a rope hanging from a gymnasium ceiling, as in Fig. 5–3a. She stops, suspended from the lower end of the rope by her hands. Her weight is 600 N, and the weight of the rope is 100 N. Analyze the forces on the gymnast and on the rope.

SOLUTION Figure 5–3b is a free-body diagram for the gymnast. The forces acting on her are her weight w_1 and the upward force T_1 exerted on her by the rope. If

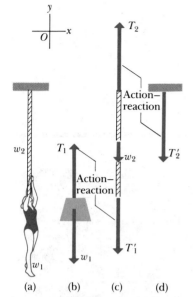

5–3 (a) Gymnast hanging at rest from vertical rope. (b) The gymnast is isolated and all forces acting on her are shown. (c) Forces on the rope. (d) Downward force on the ceiling. Lines connect action–reaction pairs.

we take the x-axis horizontal and the y-axis vertical, there are no x-components of force. The y-components are those associated with the forces T_1 and w_1. Force T_1 acts in the positive y-direction, and its y-component is just the magnitude T_1, a positive (scalar) quantity. But w_1 acts in the negative y-direction, and its y-component is the *negative* of the magnitude w_1. Thus the algebraic sum of y-components is $T_1 - w_1$. Then from the equilibrium condition, $\Sigma F_y = 0$, we have

$$\Sigma F_y = T_1 - w_1 = 0,$$
$$T_1 = w_1.$$

The forces w_1 and T_1 are *not* an action–reaction pair, although they are equal in magnitude, opposite in direction, and have the same line of action. The weight w_1 is a force of attraction exerted on the body by the earth. Its reaction is an equal and opposite force of attraction exerted on the earth by the body. This reaction is one of the set of forces acting *on* the earth, and therefore it does not appear in the free-body diagram of the suspended body. The reaction to the force T_1 is a downward force, T_1', of equal magnitude:

$$T_1 = T_1' \qquad \text{(third law)}.$$

The force T_1' is shown in part (c), which is the free-body diagram for the rope. The other forces on the rope are its own weight w_2 and the upward force T_2 exerted on its upper end by the ceiling. The y-component of T_2 is positive, while those of w_2 and T_2' are negative. Since the rope is also in equilibrium,

$$\Sigma F_y = T_2 - w_2 - T_1' = 0,$$
$$T_2 = w_2 + T_1' \qquad \text{(first law)}.$$

The reaction to T_2 is the downward force T_2' in part (d), exerted on the ceiling by the rope:

$$T_2 = T_2' \qquad \text{(third law)}.$$

With the numbers given, the gymnast's weight w_1 is 600 N, and the weight of the rope w_2 is 100 N. Then

$$T_1 = w_1 = 600 \text{ N},$$
$$T_1' = T_1 = 600 \text{ N},$$
$$T_2 = w_2 + T_1' = 100 \text{ N} + 600 \text{ N} = 700 \text{ N},$$
$$T_2' = T_2 = 700 \text{ N}.$$

EXAMPLE 5–5 In Fig. 5–4a, a hanging lamp of weight w hangs from a cord, which is knotted at point O to two other cords, one fastened to the ceiling, the other to the wall. We wish to find the tensions in these three cords, assuming the weights of the cords to be negligible.

SOLUTION To use the conditions of equilibrium to determine unknown forces (in this case, the tensions in the cords), we must consider some body that is in equilibrium and on which the desired forces act. The suspended lamp is one such body. As shown in the preceding examples, the tension T_1 in the vertical cord supporting the lamp is equal in magnitude to the weight of the lamp. The other cords do not exert forces on the lamp (because they are not attached directly to it), but they do act on the knot at O. Hence we consider the *knot* as a particle in equilibrium, considering the weight of the knot itself as negligible.

Free-body diagrams for the lamp and the knot are shown in Fig. 5–4b, where T_1, T_2, and T_3 are the *magnitudes* of the forces shown; the directions of these forces are indicated by the vectors on the diagram. An *xy*-coordinate axis system is also shown, and the force of magnitude T_3 has been resolved into its *x*- and *y*-components.

Considering the lamp first, we note that there are no *x*-components of force. The *y*-component of force exerted by the cord is in the positive *y*-direction and is just T_1. The *y*-component of the weight is, however, in the negative *y*-direction and is $-w$. The equilibrium condition for the lamp, that the algebraic sum of *y*-components of force must be zero, is

$$T_1 + (-w) = 0.$$

Hence,

$$T_1 = w.$$

The tension in the vertical cord equals the weight of the lamp, as already noted.

We now consider the knot at *O*. Both *x*- and *y*-components of force are acting on it, so there are two separate equilibrium conditions. The algebraic sum of the *x*-components must be zero, and the algebraic sum of *y*-components must separately be zero. (Note that *x*- and *y*-components are *never* added together in a single equation.) We find

$$\Sigma F_x = 0: \quad T_3 \cos 60° - T_2 = 0.$$
$$\Sigma F_y = 0: \quad T_3 \sin 60° - T_1 = 0.$$

Because $T_1 = w$, the second equation can be rewritten as

$$T_3 = \frac{T_1}{\sin 60°} = \frac{w}{\sin 60°} = 1.155\,w.$$

This result can now be used in the first equation:

$$T_2 = T_3 \cos 60° = (1.155w)\cos 60° = 0.577w.$$

Thus all three tensions can be expressed as multiples of the weight *w* of the lamp, which is assumed to be known. To summarize,

$$T_1 = w,$$
$$T_2 = 0.577w,$$
$$T_3 = 1.155w.$$

If the lamp's weight is $w = 50.0$ N, then

$$T_1 = 50.0 \text{ N},$$
$$T_2 = (0.577)(50.0 \text{ N}) = 28.9 \text{ N},$$
$$T_3 = (1.155)(50.0 \text{ N}) = 57.7 \text{ N}.$$

We note that T_3 is greater than the weight of the lamp. If this seems strange, note that T_3 must be large enough so that its vertical component is equal to *w* in magnitude; thus T_3 itself must have somewhat *larger* magnitude than *w*.

EXAMPLE 5–6 A repairman of weight w_1 is trying to find a leak in an icy, perfectly frictionless roof with slope angle *θ*. He is secured by a rope passing over the top and down the vertical side of the building. The other end of the rope is held by his helper (weight w_2) on the ground. The helper finds that he must put his full weight on the rope to hold the man on the roof. If we know the weight w_1 of the repairman, what is the weight w_2 of the helper?

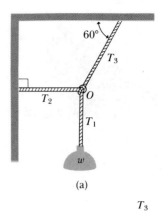

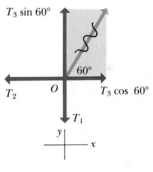

5–4 (a) A lamp of weight *w* is suspended from a cord knotted at *O* to two other cords. (b) Free-body diagrams for the lamp and for the knot, showing the components of force acting on each.

Fixing a slippery roof: how physics can help

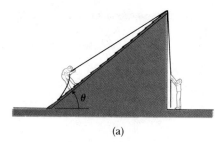

(a)

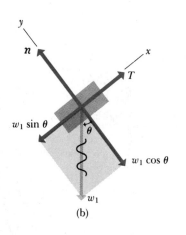

(b)

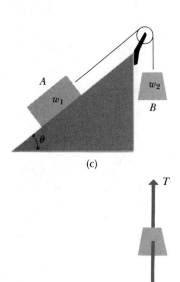

(c)

T

w_2

(d)

5–5 Forces on a body in equilibrium on a frictionless inclined plane. (a) The physical situation; (b) free-body diagram for body A; (c) idealized model of situation; (d) free-body diagram for body B.

Moving heavy boxes again, this time with a rope

SOLUTION We represent the situation by the idealized model shown in Fig. 5–5, where A is the repairman and B the helper. Free-body diagrams for the two bodies are shown in parts (b) and (d) of the figure. Here, as in preceding examples, we label and refer to forces by using light italic symbols, which in equations represent the *magnitudes* of these forces. The forces on body B are its weight w_2 and the force T exerted on it by the cord. Because it is in equilibrium,

$$T = w_2.$$

Body A is acted on by its weight w_1, the force T exerted on it by the rope, and the force n exerted on it by the plane. We use the same symbol T for the force exerted on each body because, as discussed above, these forces are equivalent to an action–reaction pair and have the same magnitude. The pulley, which corresponds to the frictionless top edge of the roof, changes the directions of the forces exerted by the rope but not their magnitudes; and the two forces at the ends of the rope have the same magnitude even though their directions are different. Because there is no friction, the force n can have no component along the plane that would resist the motion of the body on the plane; hence it must be perpendicular, or *normal,* to the surface of the plane.

It is simplest to choose x- and y-axes parallel and perpendicular to the surface of the plane, because then only the weight w_1 needs to be resolved into components. The conditions of equilibrium for body A give

$$\Sigma F_x = T - w_1 \sin \theta = 0,$$
$$\Sigma F_y = n - w_1 \cos \theta = 0.$$

Thus if $w_1 = 800$ N and $\theta = 30°$, we find

$$w_2 = T = w_1 \sin \theta = (800 \text{ N})(0.500) = 400 \text{ N},$$

and

$$n = w_1 \cos \theta = (800 \text{ N})(0.866) = 693 \text{ N}.$$

Note carefully that, *if there is no friction,* the same weight w_2 of 400 N is required for equilibrium whether the system remains at rest or moves with constant speed in *either* direction. This is not the case when friction is present.

EXAMPLE 5–7 In Example 5–1 (Section 5–2) suppose you try to move the crate by tying a rope around it and pulling upward on the rope at an angle of 30° to the horizontal. What tension in the rope is required to make the crate move with constant speed?

SOLUTION Figure 5–6 is a free-body diagram showing the forces on the crate. We note that the normal force n is *not* equal in magnitude to the weight of the crate because the force exerted by the rope has an additional vertical component. From the equilibrium conditions,

$$\Sigma F_x = T \cos 30° - 0.2n = 0,$$
$$\Sigma F_y = T \sin 30° + n - 500 \text{ N} = 0.$$

These are two simultaneous equations for the two unknown quantities T and n. To solve them we can eliminate one unknown and solve for the other, as follows. Rearrange the second equation to the form

$$n = 500 \text{ N} - T \sin 30°.$$

Substitute this expression for n back into the first equation, obtaining

$$T \cos 30° - 0.2(500 \text{ N} - T \sin 30°) = 0.$$

Finally, solve this equation for T, then substitute the result back into either of the original equations to obtain n. The results are

$$T = 103 \text{ N}, \qquad n = 448 \text{ N}.$$

EXAMPLE 5-8 A toboggan loaded with vacationing students slides down a long, snow-covered slope having a coefficient of kinetic friction μ_k. The slope has just the right angle to make the toboggan slide with constant speed. Find the slope angle.

SOLUTION We represent the situation by a block sliding down a plane inclined at an angle θ, as shown in Fig. 5-7. The forces on the block (toboggan) are its weight w and the normal and frictional components of the force exerted on it by the plane. We take axes perpendicular and parallel to the surface of the plane and represent the weight in terms of its components in these two directions, as shown. The equilibrium conditions in component form are then

$$\Sigma F_x = w \sin \theta - \mathcal{F}_k = w \sin \theta - \mu_k n = 0,$$
$$\Sigma F_y = n - w \cos \theta = 0.$$

Hence

$$\mu_k n = w \sin \theta, \qquad n = w \cos \theta.$$

Dividing the first of these equations by the second, we find

$$\mu_k = \frac{\sin \theta}{\cos \theta} = \tan \theta.$$

It follows that a body, regardless of its weight, slides down an inclined plane with constant speed if the tangent of the slope angle of the plane equals the coefficient of kinetic friction. Measurement of this angle then provides a simple experimental method for determining the coefficient of kinetic friction.

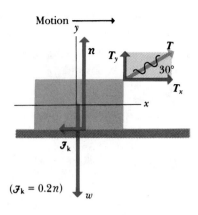

5-6 Forces on a box being dragged to the right on a level surface at constant speed.

A not-too-fast toboggan ride down a not-too-steep slope

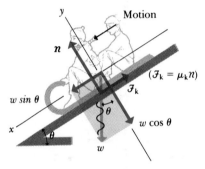

5-7 Forces on a toboggan sliding down a hill (with friction) at constant speed.

5-4 APPLICATIONS OF NEWTON'S SECOND LAW

We now discuss several **dynamics** problems, showing applications of Newton's second law to systems that are *not* in equilibrium. In this case,

$$\Sigma \mathbf{F} = m\mathbf{a}. \qquad (5-5)$$

We will usually use this relation in component form:

$$\Sigma F_x = ma_x,$$
$$\Sigma F_y = ma_y. \qquad (5-6)$$

The problem-solving strategies appropriate for these problems are very similar to those used for equilibrium problems in Section 5-3. We urge you to study these carefully, watch for their applications in the example problems, and use them when you solve end-of-chapter problems.

Dynamics: the relation of force to acceleration

PROBLEM-SOLVING STRATEGY: *Newton's second law*

1. Select a body for analysis. Newton's second law will be applied to this body.

2. Draw a free-body diagram. Be sure to include all the forces acting *on* the chosen body, but be equally careful *not* to include any force exerted *by* the body on some other body. Some of the forces may be unknown; label them with algebraic symbols. Usually one of the forces will be the body's weight. Often it is useful to label this immediately as *mg* rather than *w*. If a numerical value of mass is given, the corresponding numerical value of weight can be computed.

3. Show your coordinate axes explicitly in the free-body diagram, and then determine components of forces with reference to these axes. When a force is represented in terms of its components, cross out the original force so as not to include it twice. When the direction of the acceleration is known in advance, it is

usually best to take its direction as that of the +*x*-axis. When there are two or more bodies, it is often useful to use a separate axis system for each body; there is no reason you have to use the same axis system for all the bodies.

4. If more than one body is involved, steps 1–3 must be carried out for each body. In addition there may be *geometrical* relationships between the motions of two or more bodies; express these in algebraic form.

5. Write the Newton's-second-law equations for each body, usually Eqs. (4–3), and solve to find the unknown quantities.

6. Check special cases or extreme values of quantities, where possible, and compare the results for these particular cases with your intuitive expectations. Ask: "Does this result make sense?"

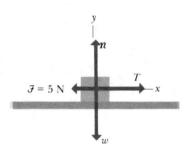

5–8 Force diagram for block in Example 5–9.

EXAMPLE 5–9 A 10.0-kg block of ice rests on a horizontal surface. What constant horizontal force *T* is required to give it a velocity of 4.0 m·s^{-1} in 2.0 s? Assume that the block starts from rest, that the friction force between the block and the surface is constant and equal to 5.0 N, and that all forces act at the center of the block (see Fig. 5–8).

SOLUTION The mass of the block is given. The *y*-component of its acceleration is zero. The *x*-component of acceleration can be found from the data on the velocity acquired in a given time. Since the forces are constant, the *x* acceleration is constant, and, from the equations of motion with constant acceleration,

$$a_x = \frac{v - v_0}{t} = \frac{4 \text{ m·s}^{-1} - 0}{2 \text{ s}} = 2 \text{ m·s}^{-2}.$$

The resultant of the *x*-forces is

$$\sum F_x = T - \mathcal{F}.$$

and that of the *y*-forces is

$$\sum F_y = n - w.$$

(Note that the *magnitude* of the friction force, $\mathcal{F} = 5.0$ N, is a positive quantity, but the *component* of this force in the *x*-direction is negative, equal to $-\mathcal{F}$ or -5.0 N. Similarly, the magnitude of the weight is the positive quantity *w*, but the *y*-component of this force is $-w$.) Hence, from Newton's second law in component form,

$$T - \mathcal{F} = ma_x, \qquad n - w = ma_y = 0.$$

From the second equation, we find that

$$n = w = mg = (10.0 \text{ kg})(9.80 \text{ m·s}^{-2}) = 98.0 \text{ N},$$

and from the first,

$$T = \mathcal{F} + ma_x = 5.0 \text{ N} + (10.0 \text{ kg})(2.0 \text{ m·s}^{-2}) = 25 \text{ N}.$$

EXAMPLE 5–10 An elevator and its load, shown in Fig. 5–9, have a total mass of 800 kg. Find the tension T in the supporting cable when the elevator, originally moving downward at 10 m·s^{-1}, is brought to rest with constant acceleration in a distance of 25 m.

Is the elevator cable strong enough?

SOLUTION The weight of the elevator is

$$w = mg = (800 \text{ kg})(9.8 \text{ m·s}^{-2}) = 7840 \text{ N}.$$

From the equations of motion with constant acceleration,

$$v^2 = v_0{}^2 + 2a(y - y_0), \qquad a = \frac{v^2 - v_0{}^2}{2(y - y_0)}.$$

The initial velocity v_0 is -10 m·s^{-1}; the final velocity v is zero. If we take the origin at the point where the acceleration begins, then $y_0 = 0$ and $y = -25$ m. Hence

$$a = \frac{0 - (-10 \text{ m·s}^{-1})^2}{2(-25 \text{ m})} = 2 \text{ m·s}^{-2}.$$

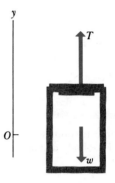

5–9 The resultant vertical force is $T - w$.

The acceleration is therefore positive (upward). From the free-body diagram (Fig. 5–9) the resultant force is

$$\Sigma F = T - w = T - 7840 \text{ N}.$$

Since $\Sigma F = ma$,

$$T - 7840 \text{ N} = (800 \text{ kg})(2 \text{ m·s}^{-2}) = 1600 \text{ N},$$

$$T = 9440 \text{ N}.$$

The tension must be *greater* than the weight by 1600 N to cause the upward acceleration while the elevator is stopping.

EXAMPLE 5–11 In Example 5–10, with what force do the feet of a passenger press downward on the floor, if the passenger's mass is 80.0 kg?

SOLUTION We first find the force the floor exerts *on* the passenger's feet, which is the reaction force to the force required. The forces on the passenger are his weight $(80.0 \text{ kg})(9.80 \text{ m·s}^{-2}) = 784$ N and the upward force F caused by the floor. Thus Newton's second law applied to the passenger, whose acceleration is the same as the elevator's, gives

$$F - 784 \text{ N} = (80.0 \text{ kg})(2.00 \text{ m·s}^{-2}) = 160 \text{ N},$$

and

$$F = 944 \text{ N}.$$

Thus the passenger pushes down on the floor with a force of 944 N while the elevator is stopping, and he *feels* a greater strain in his legs and feet than while standing at rest. If he stands on a bathroom scale calibrated in newtons, what will the scale read?

EXAMPLE 5–12 A child slides down a frictionless playground slide inclined at an angle θ with the horizontal (Fig. 5–10a). What is the child's acceleration?

Does a heavy child slide faster than a light child?

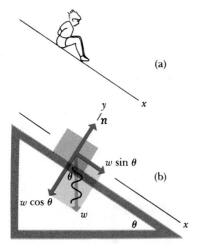

5–10 (a) A body on a frictionless inclined plane. (b) Free-body diagram.

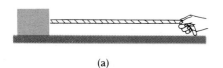

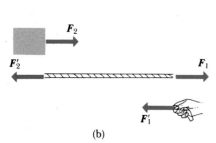

5–11 (a) A man pulls on a rope attached to a block. (b) Forces acting on the block, the rope, and the man's hand.

SOLUTION The only forces acting on the child are the weight w and the normal force n exerted by the plane (Fig. 5–10b).

Take axes parallel and perpendicular to the surface of the plane and resolve the weight into x- and y-components. Then

$$\Sigma F_y = n - w \cos \theta,$$
$$\Sigma F_x = w \sin \theta.$$

But we know that $a_y = 0$, so from the equation $\Sigma F_y = ma_y$, we find that $n = w \cos \theta$. From the equation $\Sigma F_x = ma_x$, we have

$$w \sin \theta = ma_x,$$

and since $w = mg$,

$$a_x = g \sin \theta.$$

The mass does not appear in the final result, which means that any body, regardless of its mass, will slide down a frictionless inclined plane with an acceleration of $g \sin \theta$. In particular, when $\theta = 0$, $a_x = 0$, and when the plane is vertical, $\theta = 90°$ and $a_x = g$, as we should expect.

EXAMPLE 5–13 Consider the situation of Fig. 5–11 (which is the same figure as Fig. 4–9). Let the mass of the block be 4 kg and that of the rope be 0.5 kg. If the force F_1 is 9 N, what are the forces F_1', F_2, and F_2'? The surface on which the block moves is level and frictionless.

SOLUTION We know from Newton's third law that $F_1 = F_1'$ and that $F_2 = F_2'$. Hence $F_1' = 9$ N. The force F_2 could be computed by applying Newton's second law to the block, if its acceleration were known, or the force F_2' could be computed by applying this law to the rope if its acceleration were known. The acceleration is not given, but it can be found by considering the block and rope together as a single system. We know there is no vertical motion, so the vertical forces need not be considered. Since there is no friction, the resultant force acting *on* the system is the force F_1. (The forces F_2 and F_2' are not forces acting on the system but rather are *internal* forces exerted by one part of the system on another. The force F_1' does not act on the system, but *on the man*.) Then, from Newton's second law,

$$\Sigma F = ma,$$
$$9 \text{ N} = (4 \text{ kg} + 0.5 \text{ kg})a,$$
$$a = 2 \text{ m·s}^{-2}.$$

We can now apply Newton's second law to the block:

$$\Sigma F = ma,$$
$$F_2 = (4 \text{ kg})(2 \text{ m·s}^{-2}) = 8 \text{ N}.$$

If we consider the rope alone, the resultant force on it is

$$\Sigma F = F_1 - F_2' = 9 \text{ N} - F_2',$$

and from the second law,

$$9 \text{ N} - F_2' = (0.5 \text{ kg})(2 \text{ m·s}^{-2}) = 1 \text{ N},$$
$$F_2' = 8 \text{ N}.$$

We find that F_2 and F_2' are equal in magnitude. This result agrees with Newton's third law, which we used tacitly when we omitted the forces F_2 and F_2' in considering the system as a whole. Note, however, that the forces F_1 and F'_2 are *not* equal and opposite (the rope is not in equilibrium) and that these forces are *not* an action–reaction pair.

EXAMPLE 5–14 In Fig. 5–12, a block of mass m_1 and weight $w_1 = m_1g$ moves on a level, frictionless surface. It is connected by a light, flexible cord passing over a small, frictionless pulley to a second hanging block of mass m_2 and weight $w_2 = m_2g$. What is the acceleration of the system, and what is the tension in the cord connecting the two blocks?

How helpful would a frictionless pulley be?

SOLUTION The diagram shows the forces acting on each block. Because there is no friction in the pulley and the cord is considered massless, the tension is the same throughout the string. Thus the forces applied to the blocks by the ends of the cord have equal magnitude; we have used the same symbol T for each. For the block on the surface,

$$\Sigma F_x = T = m_1a_x,$$
$$\Sigma F_y = n - m_1g = m_1a_y = 0.$$

Applying Newton's second law to the hanging block, we obtain

$$\Sigma F_y = m_2g - T = m_2a_y.$$

Note that there is no necessity to use the same coordinate axes for the two bodies. It is convenient to take the $+y$ direction as downward for m_2, so both bodies move in positive axis directions.

Now assuming that the string does not stretch, the speeds of the two bodies at any instant are equal, and so their accelerations have the same magnitude. That is, a_x for m_1 equals a_y for m_2. We can thus denote this common magnitude simply as a. Then the two equations containing accelerations are

$$T = m_1a.$$
$$m_2g - T = m_2a.$$

These are two simultaneous equations for the unknowns T and a. Adding the two equations eliminates T, giving

$$m_2g = m_1a + m_2a = (m_1 + m_2)a$$

and

$$a = \frac{m_2}{m_1 + m_2}g.$$

Substituting this back into the first equation yields

$$T = \frac{m_1m_2}{m_1 + m_2}g.$$

We see that the tension T is not equal to the weight m_2g of mass m_2, but is *less* by a factor of $m_1/(m_1 + m_2)$. Further thought shows that T *cannot* be equal to m_2g; if it were, m_2 would be in equilibrium, and it is not.

As an example of step 6 in the problem strategy, we can check some special cases. If $m_1 = 0$, we would expect that m_2 would fall freely and there would be no tension in the string. The equations do give $T = 0$ and $a = g$ for this case. Also, if $m_2 = 0$, we expect no tension and no acceleration; for this case the equations give $T = 0$ and $a = 0$. Thus in these two cases the general analytical results agree with intuitive expectations.

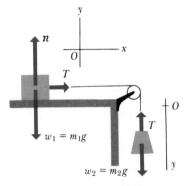

5–12 Force diagram for block on a frictionless horizontal surface and for the hanging block.

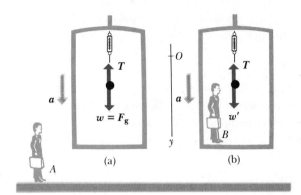

5–13 (a) To observer A, the body has a downward acceleration a, and he writes $w - T = ma$. (b) To observer B, the acceleration of the body is zero. He writes $w' = T$.

Can you weigh things in an accelerating elevator?

EXAMPLE 5–15   A body of mass m is suspended from a spring balance attached to the roof of an elevator, as shown in Fig. 5–13. What is the reading of the balance if the elevator has an acceleration a relative to the earth? Consider the earth's surface to be an inertial reference system.

SOLUTION   The forces on the body are its weight w (the gravitational force F_g exerted on it by the earth) and the upward force T exerted on it by the balance. The reading on the scale of the balance is T. The body is at rest relative to the elevator and hence has an acceleration a relative to the earth. (We take the downward direction as positive.) The resultant force on the body is $w - T$, so, from Newton's second law,

$$w - T = ma, \qquad T = w - ma = m(g - a).$$

By Newton's third law, the body pulls down on the balance with a force equal and opposite to T, or equal to $(w - ma)$, so the balance reading T equals $(w - ma)$. The same analysis can be applied to a person standing on a bathroom scale in an accelerating elevator, and the result is the same; in that case T is the scale reading.

If the same body were suspended in equilibrium from a balance attached to the earth, the balance reading would equal the weight w. To an observer riding in the accelerating elevator, the body *appears* (in this noninertial frame of reference) to be acted on by a downward force w' equal in magnitude to the balance reading, as in Fig. 5–13b. This apparent force w' can be called the *apparent weight* of the body. Then

$$w' = w - ma = m(g - a). \qquad (5\text{--}7)$$

In a noninertial frame of reference, apparent weight is different from true weight.

If the elevator is at rest, or moving vertically (either up or down) with constant velocity, $a = 0$ and the apparent weight equals the true weight. If the acceleration is *downward*, as in Fig. 5–13, so that a is positive, the apparent weight is less than the true weight: the body appears "lighter." If the acceleration is *upward*, a is negative, the apparent weight is greater than the true weight, and the body appears "heavier." These effects can be felt by taking a few steps in an elevator that is starting to move up or down after a stop. If the elevator falls freely, $a = g$, and since the true weight w also equals mg, the apparent weight w' is zero and the body appears "weightless." It is in this sense that an astronaut orbiting the earth in a space capsule is said to be "weightless."

The effect of this condition is exactly the same as though the body were in outer space with no gravitational force at all. The physiological effect of prolonged weightlessness is an interesting medical problem that is being actively explored. Gravity plays a role in blood distribution in the body, and one reaction to weightlessness is a decrease in volume of blood through increased excretion of water. In some cases, astronauts returning to earth have experienced temporary impairment of their sense of balance and a greater tendency toward motion sickness.

EXAMPLE 5–16 Figure 5–14a represents a simple *accelerometer*. A small body is fastened at one end of a light rod pivoted freely at point P. When the system has an acceleration a toward the right, the rod makes an angle θ with the vertical. In a practical instrument, some form of damping must be provided to keep the rod from swinging when the acceleration changes. For example, the rod might hang in a tank of oil. You can make a primitive version of this device by tying a thread to the ceiling light in an automobile, tying a key or nearly any other small object to the other end, and using a protractor. The problem: Given m and θ, what is a?

How to make an accelerometer

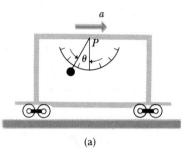

(a)

SOLUTION As shown in the free-body diagram, Fig. 5–14b, two forces act on the body: its weight $w = mg$ and the tension T in the rod. (We neglect the weight of the rod.) The resultant horizontal force is

$$\Sigma F_x = T \sin \theta,$$

and the resultant vertical force is

$$\Sigma F_y = T \cos \theta - mg.$$

The x-acceleration is the acceleration a of the system, and the y-acceleration is zero. Hence

$$T \sin \theta = ma, \qquad T \cos \theta = mg.$$

When the first equation is divided by the second, we get

$$a = g \tan \theta,$$

and the acceleration a is proportional to the tangent of the angle θ. When $\theta = 0$, the acceleration is zero; when $\theta = 45°$, $a = g$, and so on. We note that θ can never be 90° because that would require an infinite acceleration.

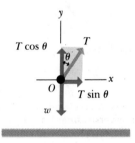

(b)

5–14 (a) A simple accelerometer. (b) The forces on the body are w and T.

EXAMPLE 5–17 Sometimes a frictional force opposing the motion of a body is not constant but increases with the body's speed. Two familiar examples are air resistance on a falling body and the viscous resisting force on a body falling through a liquid. Suppose we drop a penny into a deep pond. For small speeds the resisting force f of the liquid is approximately proportional to the penny's speed v:

$$f = kv, \tag{5-8}$$

Free fall with air resistance

where k is a proportionality constant that depends on the dimensions of the penny and the properties of the fluid. Taking the positive direction to be downward and neglecting any force associated with buoyancy in the fluid, we find that the net vertical component of force is $mg - kv$. Newton's second law gives

$$mg - kv = ma.$$

When the penny first starts to move, $v = 0$, the resisting force is zero, and the initial acceleration is $a = g$. As its speed increases, the resisting force also increases, until finally it equals the weight in magnitude. At this time $mg - kv = 0$, the acceleration becomes 0, and there is no further increase in speed. The final speed, also

Terminal speed

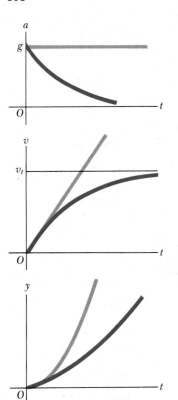

5–15 Graphs of acceleration, velocity, and position versus time for a body falling in a viscous fluid, shown as solid color curves. The light color curves show the corresponding relations if there is *no* viscous friction.

called the *terminal* speed, is given by this equation. Solving for v, we find

$$v_{\mathrm{t}} = \frac{mg}{k}. \tag{5-9}$$

Figure 5–15 shows how the acceleration, velocity, and position vary with time.

To find the relation between speed and time during the interval before the terminal speed is reached, we go back to Newton's second law, which we rewrite as

$$m \frac{dv}{dt} = mg - kv.$$

After rearranging terms and replacing mg/k by v_{t}, we find

$$\frac{dv}{v - v_{\mathrm{t}}} = -\frac{k}{m}\, dt.$$

We now integrate both sides, noting that $v = 0$ when $t = 0$:

$$\int_0^v \frac{dv}{v - v_{\mathrm{t}}} = -\frac{k}{m} \int_0^t dt,$$

which integrates to

$$\ln \frac{v_{\mathrm{t}} - v}{v_{\mathrm{t}}} = -\frac{k}{m}\, t, \qquad \text{or} \qquad 1 - \frac{v}{v_{\mathrm{t}}} = e^{-(k/m)t},$$

and finally

$$v = v_{\mathrm{t}}(1 - e^{-(k/m)t}). \tag{5-10}$$

From this result we can derive expressions for the acceleration and position as functions of time. The derivations are left as an exercise. The results are

$$a = ge^{-(k/m)t},$$

$$y = v_{\mathrm{t}}[t - \frac{m}{k}(1 - e^{-(k/m)t})]. \tag{5-11}$$

Now look again at Fig. 5–15, which shows graphs of these three relations.

In high-speed motion through air, the resisting force is approximately proportional to v^2 rather than to v; it is then called *air drag* or simply *drag*. Falling raindrops, airplanes, and cars moving at high speed all experience air drag. In this case Eq. (5–8) is replaced by $f = Kv^2$. We challenge the reader to show that the terminal speed is then given by

$$v_{\mathrm{t}} = \sqrt{\frac{mg}{K}}, \tag{5-12}$$

and that the units of the constant K are $\mathrm{N \cdot s^2 \cdot m^{-2}}$ or $\mathrm{kg \cdot m^{-1}}$.

The physics of sky-diving

EXAMPLE 5–18 For a human body falling through air, the numerical value of the constant K in Eq. (5–12) is found to be about $0.25\ \mathrm{kg \cdot m^{-1}}$. If an 80-kg sky diver jumps out of an airplane and reaches terminal velocity before he opens his parachute, the terminal velocity is, from Eq. (5–12),

$$v_t = \sqrt{\frac{(80 \text{ kg})(9.8 \text{ m·s}^{-2})}{0.25 \text{ kg·m}^{-1}}}$$
$$= 56 \text{ m·s}^{-1}.$$

SUMMARY

Four classes of forces are found in nature. Two—the gravitational and electromagnetic interactions—are familiar in everyday life; the other two—the strong interaction and the weak interaction—are seen directly only in experiments with fundamental particle interactions.

The contact force between two bodies can always be represented in terms of a normal component n perpendicular to the interaction surface and a frictional component $\mathcal{F}$ parallel to the surface. When sliding occurs between the surfaces, the friction force $\mathcal{F}_k$ is approximately proportional to the normal force, and the proportionality constant is μ_k, the coefficient of kinetic friction:

$$\mathcal{F}_k = \mu_k n. \tag{5–1}$$

When there is no relative motion, the maximum possible friction force is approximately proportional to the normal force, and the proportionality constant is μ_s, the coefficient of static friction:

$$\mathcal{F}_s \leq \mu_s n. \tag{5–2}$$

The actual static friction force may be anything from zero to this maximum value, and it is determined by the details of the specific problem at hand. Usually μ_k is less than μ_s for a given pair of surfaces.

When a particle is in equilibrium in an inertial frame of reference, the vector sum of all forces acting on it must be zero:

$$\Sigma F = 0. \tag{5–3}$$

The vector sum is often calculated most easily by using components:

$$\Sigma F_x = 0,$$
$$\Sigma F_y = 0. \tag{5–4}$$

Free-body diagrams are useful in identifying the forces acting on the body being considered. Newton's third law is also frequently needed in equilibrium problems. The two forces in an action–reaction pair never act on the same body.

When the vector sum of forces on a body is not zero, the body has an acceleration determined by Newton's second law:

$$\Sigma F = ma. \tag{5–5}$$

In problems this is usually used in component form:

$$\Sigma F_x = ma_x,$$
$$\Sigma F_y = ma_y. \tag{5–6}$$

If you apply the problem-solving strategies outlined in this chapter, you will be rewarded with an improved percentage of correct problem solutions!

KEY TERMS

gravitational interactions
electromagnetic interactions
strong interaction
weak interactions
contact forces
normal force
friction force
coefficient of kinetic friction
static friction force
coefficient of static friction
equilibrium
free-body diagram

QUESTIONS

5–1 Can a body be in equilibrium when only one force acts on it?

5–2 A helium balloon hovers in midair, neither ascending nor descending. Is it in equilibrium? What forces act on it?

5–3 If the two ends of a rope in equilibrium are pulled with forces of equal magnitude and opposite direction, why is the total tension in the rope not zero?

5–4 A horse is hitched to a wagon. Since the wagon pulls back on the horse just as hard as the horse pulls on the wagon, why doesn't the wagon remain in equilibrium, no matter how hard the horse pulls?

5–5 A clothesline is hung between two poles, and then a shirt is hung near the center. No matter how tightly the line is stretched, it always sags a little at the center. Explain why.

5–6 A man sits in a chair that is suspended from a rope. The rope passes over a pulley suspended from the ceiling, and the man holds the other end of the rope in his hands. What is the tension in the rope, and what force does the chair exert on the man?

5–7 How can pushing *down* on a bicycle pedal make the bicycle move *forward*?

5–8 A car is driven up a steep hill at constant speed. Discuss all the forces acting on the car; in particular, what pushes it up the hill?

5–9 Can the coefficient of friction ever be greater than unity? If so, give an example; if not, explain why not.

5–10 A block rests on an inclined plane with enough friction to prevent sliding down. To start the block moving, is it easier to push it up the plane, down the plane, or sideways? Why?

5–11 In pushing a box up a ramp, is it better to push horizontally or to push parallel to the ramp?

5–12 In stopping a car on an icy road, is it better to push the brake pedal hard enough to "lock" the wheels and make them slide, or to push gently so the wheels continue to roll? Why?

5–13 When one stands with bare feet in a wet bathtub, the grip feels fairly secure, and yet a catastrophic slip is quite possible. Discuss this situation with respect to the two coefficients of friction.

5–14 The horrible squeak made by a piece of chalk held at the wrong angle against a blackboard results from alternate sticking and slipping of the chalk against the blackboard. Interpret this phenomenon in terms of the two coefficients of friction. Can you think of other examples of "slip-stick" behavior?

EXERCISES

Section 5–2 Contact Forces and Friction

5–1

a) A large rock rests on a rough horizontal surface. A bulldozer pushes on the rock with a horizontal force T that is slowly increased, started from zero. Draw a graph with T along the x-axis and the friction force $\mathcal{F}$ along the y-axis, starting at $T = 0$ and showing the region of no motion, the point where motion impends, and the region where motion exists.

b) A block of weight w rests on a rough horizontal plank. The slope angle of the plank θ is gradually increased until the block starts to slip. Draw two graphs, both with θ along the x-axis. In one graph show the ratio of the normal force to the weight, n/w as a function of θ. In the second graph show the ratio of the friction force to the weight, $\mathcal{F}/w$. Indicate the region of no motion, the point where motion impends, and the region where motion exists.

5–2 A box of bananas weighing 20 N rests on a horizontal surface. The coefficient of static friction between the block and the surface is 0.40, and the coefficient of sliding friction is 0.20.

a) How large is the friction force exerted on the box?

b) How great will the friction force be if a horizontal force of 5 N is exerted on the box?

c) What is the minimum force that will start the box in motion?

d) What is the minimum force that will keep the box in motion once it has been started?

e) If the horizontal force is 10 N, how great is the friction force?

5–3 A box of mass 0.6 slug is moving with a constant velocity of 15 ft·s^{-1} on a horizontal surface. The coefficient of sliding friction between the box and the surface is 0.20.

a) What horizontal force is required to maintain the motion?

b) If the force is removed, how soon will the box come to rest?

5–4 A hockey puck leaves a player's stick with a velocity of 10 m·s^{-1} and slides 40 m before coming to rest. Find the coefficient of friction between the puck and the ice.

5–5

a) If the coefficient of friction between tires and dry pavement is 0.8, what is the shortest distance in which an automobile can be stopped by locking the brakes when traveling at 27 m·s^{-1} (about 60 mi·hr^{-1})?

b) On wet pavement the coefficient of friction may be only 0.25. How fast should you drive on wet pavement in order to be able to stop in the same distance as in part (a)?

(*Note.* Locking the brakes is *not* the safest way to stop.)

5–6 Find the ratio of the stopping distance of an automobile on wet concrete with the wheels locked ($\mu_k = 0.25$) to the stopping distance of the same automobile stopped by tractive friction only (where the tractive resistance is 0.015). Assume the same initial velocity for both cases.

5–7 Air pressure greatly affects the tractive resistance of bicycle tires. To study this effect, two tires are set rolling with the same initial speed of 5 m·s^{-1} along a long, straight road, and the distance each travels before its speed is reduced by half is measured. One tire is inflated to a pressure of 40 psi and the other is at 105 psi. The low-pressure tire goes 45.6 m, and the high-pressure tire goes 213 m. What is the tractive resistance for each? Assume the net horizontal force is due to tractive friction only.

Section 5–3 Equilibrium of a Particle

5–8 In each of the situations in Fig. 5–16 the objects suspended from the rope have weight w. The pulleys are frictionless. Calculate in each case the tension T in the rope in terms of the weight w.

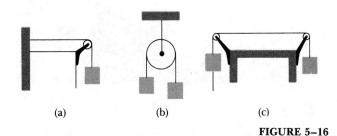

(a)   (b)   (c)

FIGURE 5–16

5–9 Two 10-N weights are suspended at opposite ends of a rope that passes over a light, frictionless pulley. The pulley is attached to a chain that goes to the ceiling.

a) What is the tension in the rope?

b) What is the tension in the chain?

5–10 A picture frame hung against a wall is suspended by two wires attached to its upper corners. If the two wires make the same angle with the vertical, what must this angle be for the tension in each wire to equal the weight of the frame?

5–11 Figure 5–17 illustrates a mountaineering technique called a Tyrolean traverse. A rope is stretched tightly

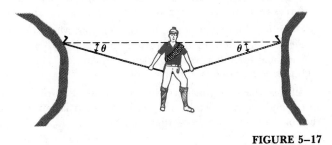

FIGURE 5–17

between two points, and the climber slides across the rope. The climber's weight is 800 N, and the breaking strength of the rope (typically nylon, 11 mm diameter) is 20,000 N.

a) If the angle θ is 15°, find the tension in the rope.

b) What is the smallest value the angle θ can have if the rope is not to break?

5–12 Find the tension in each cord in Fig. 5–18 if the weight of the suspended object is w.

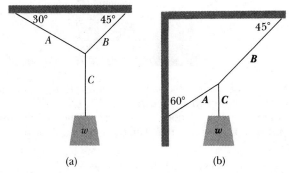

(a)   (b)

FIGURE 5–18

5–13 In Fig. 5–19 the tension in the diagonal string is 20 N.

a) Find the magnitudes of the horizontal forces F_1 and F_2 that must be applied to hold the system in the position shown.

b) What is the weight of the suspended block?

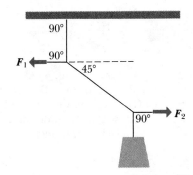

FIGURE 5–19

5–14 A tether ball leans against the post it is attached to, as shown in Fig. 5–20. If the string the ball is attached to is 1.8 m long, the ball has a radius of 0.20 m, and the ball has a mass of 0.20 kg, what are the tension in the rope and the force the pole exerts on the ball? Assume there is so little friction between the ball and the pole that it can be neglected.

FIGURE 5–20

5–15 Two blocks, each of weight w, are held in place on a frictionless incline as shown in Fig. 5–21. In terms of w and the angle θ of the incline, calculate the tension in

a) the rope connecting the blocks;

b) the rope that connects block A to the wall.

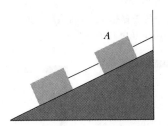

FIGURE 5–21

5–16 A man is pushing a piano weighing 315 lb at constant velocity up a ramp that is inclined at 37° above the horizontal. Neglect friction. If the force applied by the man is parallel to the incline, calculate the magnitude of this force.

5–17 Two crates connected by a rope lie on a horizontal surface as shown in Fig. 5–22. Crate A has weight w_A, and crate B weight w_B. The coefficient of kinetic friction between the crates and the surface is μ_k. The crates are pulled to the right at constant velocity by a horizontal force P. In terms of w_A, w_B, and μ_k calculate

a) the magnitude of the force P;

b) the tension in the rope connecting the blocks.

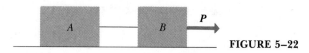

FIGURE 5–22

5–18 Consider the system shown in Fig. 5–23. The coefficient of kinetic friction between block A, of weight w_A, and the table top is μ_k. Calculate the weight w_B of the hanging block required if this block is to descend at constant speed once it has been set into motion.

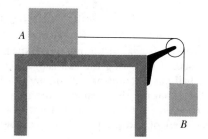

FIGURE 5–23

5–19 A safe weighing 2000 N is to be lowered at constant speed down skids 4 m long from a truck 2 m high.

a) If the coefficient of sliding friction between safe and skids is 0.30, will the safe need to be pulled down or held back?

b) How great a force parallel to the skids is needed?

5–20

a) If a force of 86 N parallel to the surface of a 20° inclined plane will push a 120-N block up the plane at constant

speed, what force parallel to the plane will push it down at constant speed?

b) What is the coefficient of sliding friction?

5–21 A large crate of weight w rests on a horizontal floor. The coefficients of friction between the crate and the floor are μ_s and μ_k. A man pushes on the crate with a force P that is directed at an angle θ below the horizontal.

a) What magnitude of force P is required to keep the crate moving at constant velocity?

b) If μ_s is larger than some critical value, the man cannot start the crate moving no matter how hard he pushes. Calculate this critical value of μ_s.

5–22 Two blocks A and B, are placed as in Fig. 5–24 and connected by ropes to block C. Both A and B weigh 20 N each, and the coefficient of sliding friction between each block and the surface is 0.5. Block C descends with constant velocity.

a) Draw two separate force diagrams showing the forces acting on A and on B.

b) Find the tension in the rope connecting blocks A and B.

c) What is the weight of block C?

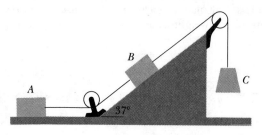

FIGURE 5–24

Section 5–4 Applications of Newton's Second Law

5–23 The first two steps in the solution of Newton's-second-law problems are to select an object for analysis and then to draw free-body diagrams for that object. Do exactly that for the following situations:

a) a mass M sliding down a frictionless inclined plane of angle θ;

b) a mass M sliding up a frictionless inclined plane of angle θ;

c) a mass M sliding up an inclined plane of angle θ with kinetic friction present;

d) masses M and m sliding down an inclined plane of angle θ with friction present, as shown in Fig. 5–25a. Here draw free-body diagrams for both m and M. Identify the forces that are action–reaction pairs.

e) Draw free-body diagrams for masses m and M shown in Fig. 5–25b. Identify all action–reaction pairs. There is a frictional force between all surfaces in contact. The pulley is frictionless and massless.

In all cases be sure you have the correct direction of the forces and are completely clear on what object is causing the force in your free-body diagram.

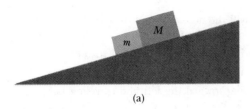

(a)

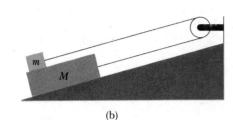

(b)   **FIGURE 5–25**

5–24 An elevator with mass 2000 kg rises with an acceleration of 1 m·s^{-2}. What is the tension in the supporting cable?

5–25 A 4-kg bucket of water is accelerated upward by a cord whose breaking strength is 60 N. Find the maximum upward acceleration that can be given to the bucket without breaking the cord.

5–26 A large fish hangs from a spring balance supported from the roof of an elevator.

a) If the elevator has an upward acceleration of 2.45 m·s^{-2} and the balance reads 50 N, what is the true weight of the fish?

b) Under what circumstances will the balance read 30 N?

c) What will the balance read if the elevator cable breaks?

5–27 A train (an engine plus four cars) is accelerating at 1.5 ft·s^{-2}. If each car has a mass of 2700 slugs and if each car has negligible frictional forces acting on it, what is

a) the force of the engine on the first car?

b) the force of the first car on the second car?

c) the force of the second car on the third car?

d) the force of the fourth car on the third car?

e) What would be these same four forces if the train were slowing down with an acceleration of −1.5 ft·s^{-2}?

When solving, note the importance of selecting the correct set of cars as your object.

5–28 A transport plane is to take off from a level landing field with two gliders in tow, one behind the other. Each glider has a mass of 1200 kg, and the friction force or drag on each may be assumed constant and equal to 2000 N. The tension in the towrope between the transport plane and the first glider is not to exceed 10,000 N.

a) If a velocity of 40 m·s^{-1} is required for takeoff, what minimum length of runway is needed?

b) What is the tension in the towrope between the two gliders while they are accelerating for the takeoff?

5–29 A 5-kg block slides down a plane inclined at 30° to the horizontal. Find the acceleration of the block

a) if the plane is frictionless;

b) if the coefficient of kinetic friction is 0.2.

5–30 A physics student playing with an air hockey table (just a frictionless surface) finds that if she gives the puck a velocity of 15.0 ft·s^{-1} along the length (5.5 ft) of the table at one end, by the time it has reached the other end the puck has drifted 1.1 in. to the right but still has a velocity along the length of 15.0 ft·s^{-1}. She concludes correctly that the table is not level, and correctly calculates its inclination from the above information. What does she get for the angle of inclination?

5–31 A 600-N man stands on a bathroom scale in an elevator. As the elevator starts moving, the scale reads 800 N.

a) Find the acceleration of the elevator (magnitude and direction).

b) What is the acceleration if the scale reads 450 N?

c) If the scale reads zero, should the man worry? Explain.

5–32 A packing crate rests on an inclined plane that makes an angle θ with the horizontal. The coefficient of sliding friction is 0.50, and the coefficient of static friction is 0.75.

a) As the angle θ is increased, find the minimum angle at which the crate starts to slip.

b) At this angle, find the acceleration (in m·s^{-2}) once the crate has begun to move.

c) How much time is required for the crate to slip 8 m along the inclined plane?

5–33 A block of mass 6 kg resting on a horizontal surface is connected by a cord passing over a light, frictionless pulley to a hanging block of mass 4 kg. The coefficient of kinetic friction between the block and the horizontal surface is 0.5. After the blocks are released find

a) the acceleration of each block;

b) the tension in the cord.

5–34 A box of mass m is dragged across a rough, level floor having a coefficient of kinetic friction μ_k by a rope that is pulled upward at an angle θ to the horizontal with a force of magnitude F. In terms of m, μ_k, θ, F, and g, obtain expressions for

a) the normal force;

b) the frictional force;

c) the acceleration of the box;

d) the magnitude of force required to move the box with constant speed.

5–35 Two 0.2-kg blocks hang at the ends of a light, flexible cord passing over a small, frictionless pulley, as in Fig. 5–26. A 0.1-kg block is placed on the block on the right, and then removed after 2 s.

a) How far will each block move in the first second after the 0.1-kg block is removed?

b) What was the tension in the cord before the 0.1-kg block was removed? After it was removed?

c) What was the tension in the cord supporting the pulley before the 0.1-kg block was removed? Neglect the weight of the pulley.

0.2 kg

0.2 kg   **FIGURE 5–26**

5–36 Which way will the accelerometer in Fig. 5–14 deflect under the following conditions?

a) The cart is moving toward the right and traveling faster.

b) The cart is moving toward the right and traveling slower.

c) The cart is moving toward the left and traveling faster.

d) The cart is moving toward the left and traveling slower.

e) The cart is at rest on a sloping surface.

f) The cart is given an upward velocity on a frictionless inclined plane. It first moves up, then stops, and then moves down. What is the deflection in each stage of the motion?

PROBLEMS

5–37 In Fig. 5–27, a man lifts a weight w by pulling down on a rope with a force F. The upper pulley is attached to the ceiling by a chain, and the lower pulley is attached to the weight by another chain. In terms of w, find the tension in each chain and the force F if the weight is lifted at constant speed. Assume the weights of the rope, pulleys, and chains to be negligible.

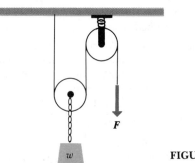

F

w

FIGURE 5–27

5–38 Figure 5–28 shows a technique called rappelling, used by mountaineers for descending vertical rock faces. The

15°

FIGURE 5–28

climber sits in a rope seat, and the rappel rope passes through a friction device attached to this seat. Suppose the rock is perfectly smooth (no friction), and that the climber's feet push horizontally onto the rock. If the climber's weight is 800 N, find the tension in the rope and the force his feet exert on the rock face.

5–39 A refrigerator of weight w is being pushed up an incline at constant velocity by a man who applies a force **P**. The surface of the incline is at an angle θ above the horizontal. *Neglect friction.* If the force **P** is *horizontal*, calculate its magnitude in terms of w and θ.

5–40 Most problems in this book do not give the mass of ropes, cords, or cables, and therefore we do not consider their mass or weight. In some problems the cords are said to be massless. It is understood that the ropes, cords, and cables, compared to other objects in the problem, have so little mass that their mass can be neglected. But if the rope is the *only* object in the problem, then clearly its mass cannot be neglected. For example, suppose we have a clothesline attached to two poles as in Fig. 5–29. The clothesline has a mass M, and each end makes an angle θ with the horizontal. What is

a) the tension at the ends of the clothesline?

b) the tension at the lowest point?

The curve of the clothesline, or of any flexible cable hanging under its own weight, is called a catenary curve. For a more advanced treatment of this curve, see J. L. Synge and B. A. Griffith, *Principles of Mechanics* (New York: McGraw-Hill, 1959), pp. 94–97.

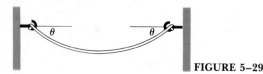

θ θ **FIGURE 5–29**

5–41 If the coefficient of static friction between a table and a rope is μ_s, what fraction of the rope can hang over the edge of a table without the rope sliding?

5–42 A dictionary is pulled to the right at constant velocity by a 10-N force acting 30° above the horizontal. The coefficient of sliding friction between the dictionary and the surface is 0.5. What is the weight of the dictionary?

5–43

a) Block A in Fig. 5–30 weighs 100 N. The coefficient of static friction between the block and the surface on which it rests is 0.30. The weight w is 20 N and the system is in equilibrium. Find the friction force exerted on block A.

b) Find the maximum weight w for which the system will remain in equilibrium.

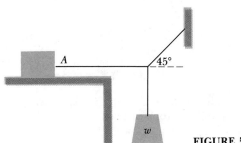

FIGURE 5–30

5–44 A block weighing 100 N is placed on an inclined plane of slope angle 30° and is connected to a second hanging block of weight w by a cord passing over a small, frictionless pulley, as in Fig. 5–31. The coefficient of static friction is 0.40, and the coefficient of sliding friction is 0.30.

a) Find the weight w for which the 100-N block moves up the plane at constant speed, once it has been set in motion.

b) Find the weight w for which it moves down the plane at constant speed.

c) For what range of values of w will the block remain at rest, if it is released from rest?

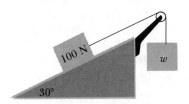

FIGURE 5–31

5–45 A window washer pushes his scrub brush up a vertical window by applying a force P as shown in Fig. 5–32. The brush weighs 6 N, and the coefficient of kinetic friction is $\mu_k = 0.4$. Calculate

a) the magnitude of the force P;

b) the normal force exerted by the window on the brush.

FIGURE 5–32

5–46 Block A in Fig. 5–33 weighs 4 N and block B weighs 8 N. The coefficient of sliding friction between all surfaces is 0.25. Find the force P necessary to drag block B to the left at constant speed

a) if A rests on B and moves with it;

b) if A is held at rest.

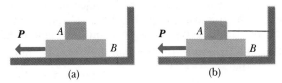

(a)   (b)

FIGURE 5–33

5–47 Block A, of weight w, slides down an inclined plane S of slope angle 37° at a constant velocity while plank B, also of weight w, rests on top of A. The plank is attached by a cord to the top of the plane (Fig. 5–34).

a) Draw a diagram of all the forces acting on block A.

b) If the coefficient of friction is the same between surfaces A and B and between S and A, determine its value.

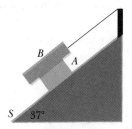

FIGURE 5–34

5–48 A man attempts to push a box of books of weight w up a ramp inclined at an angle θ above the horizontal. The coefficients of friction between the ramp and the box are μ_s and μ_k. The force P applied by the man is *horizontal*.

a) If μ_s is greater than some critical value, the man cannot start the box moving up the ramp no matter how hard he pushes. Calculate this critical value of μ_s.

b) Assume that μ_s is less than this critical value. What magnitude of force must the man apply to keep the box moving up the ramp at constant velocity?

5–49 A 40-kg packing case is initially at rest on the floor of a truck. The coefficient of static friction between the case and the truck floor is 0.30, and the coefficient of sliding friction is 0.20. Find the magnitude and direction of the friction force acting on the case

a) when the truck accelerates at 2.5 m·s^{-2};

b) when it accelerates at -3.0 m·s^{-2}.

5–50 A 20-kg box rests on the flat floor of a truck. The coefficients of friction between box and floor are $\mu_s = 0.15$ and $\mu_k = 0.10$. The truck stops at a stop sign and then starts to move, with an acceleration of 2.0 m·s^{-2}. If the box is 5.0 m from the rear of the truck when it starts, how much time elapses before it falls off the rear of the truck? How far does the truck travel in this time?

5–51 A truck traveling down a 10° slope is slowing down with an acceleration of magnitude 2.0 m·s^{-2}. A box weighing 900 N is resting on the back of the truck.

a) Will the box slide forward on the truck bed if the coefficient of static friction between the box and truck bed is 0.30?

b) If your answer in (a) is yes, what will be the acceleration of the box relative to the truck bed if the coefficient of kinetic friction is 0.22?

5–52 A book whose mass is 2 kg is projected up a long 30° incline with an initial velocity of 22 m·s⁻¹. The coefficient of kinetic friction between the book and the plane is 0.3.

a) Find the friction force acting on the book as it moves up the plane.

b) How much time does it take the book to move up the plane?

c) How *far* does the book move up the plane?

d) How much time does it take the book to slide from its position in part (c) back to its starting point?

e) With what velocity does it arrive at this point?

f) If the mass of the book had been 5 kg instead of 2 kg, would the answers in the preceding parts be different?

5–53 Block A in Fig. 5–35 has a mass of 2 kg and block B 20 kg. The coefficient of kinetic friction between B and the horizontal surface is 0.1.

a) What is the mass of block C if the acceleration of B is 2 m·s⁻² toward the right?

b) What is the tension in each cord when B has the acceleration stated above?

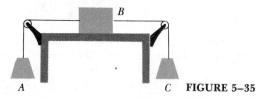

A   *C*   **FIGURE 5–35**

5–54 Two blocks connected by a cord passing over a small, frictionless pulley rest on frictionless planes, as shown in Fig. 5–36.

a) Which way will the system move?

b) What is the acceleration of the blocks?

c) What is the tension in the cord?

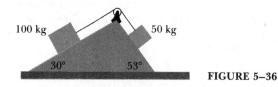

FIGURE 5–36

5–55 In terms of m_1, m_2, and g, find the accelerations of both blocks in Fig. 5–37. There is no friction anywhere in the system.

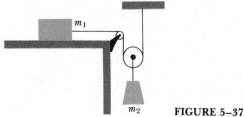

FIGURE 5–37

5–56 The masses of objects A and B in Fig. 5–38 are 20 kg and 10 kg, respectively. They are initially at rest on the floor and are connected by a weightless string passing over a weightless and frictionless pulley. An upward force **F** is applied to the pulley. Find the accelerations a_1 of object A and a_2 of object B when **F** is

a) 98 N,   b) 196 N,   c) 394 N,   d) 788 N.

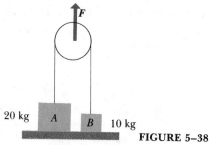

FIGURE 5–38

5–57 The two blocks in Fig. 5–39 are connected by a heavy, uniform rope of mass 4 kg. An upward force of 200 N is applied as shown.

a) What is the acceleration of the system?

b) What is the tension at the top of the heavy rope?

c) What is the tension at the midpoint of the rope?

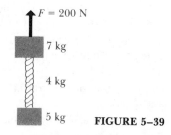

FIGURE 5–39

5–58 Two blocks with masses of 4 kg and 8 kg, respectively, are connected by a string and slide down a 30° inclined plane, as in Fig. 5–40. The coefficient of sliding friction between the 4-kg block and the plane is 0.25; between the 8-kg block and the plane it is 0.50.

a) Calculate the acceleration of each block.

b) Calculate the tension in the string.

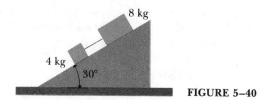

FIGURE 5–40

5–59 Two objects with masses of 5 kg and 2 kg, respectively, hang 1 m above the floor from the ends of a cord 3 m long passing over a frictionless pulley. Both objects start from rest. Find the maximum height reached by the 2-kg object.

5–60 A man of mass 80 kg stands on a platform of mass 40 kg. He pulls on a rope that is fastened to the platform and runs over a pulley on the ceiling. With what force does he have to pull in order to give himself and the platform an upward acceleration of 1 m·s⁻²?

5–61 What acceleration must the cart in Fig. 5–41 have in order that the block A will not fall? The coefficient of static friction between the block and the cart is μ_s. How would the behavior of the block be described by an observer on the cart?

FIGURE 5–41

5–62 A 20-kg monkey has a firm hold on a light rope that passes over a frictionless pulley and is attached to a 20-kg bunch of bananas, as shown in Fig. 5–42. The monkey looks upward, sees the bananas, and starts to climb the rope to get at the bananas.

a) As he climbs, do the bananas move up, down, or remain at rest?

b) As he climbs, does the distance between him and the bananas decrease, increase, or remain constant?

c) The monkey releases his hold on the rope. What about the distance between him and the bananas while he is falling?

d) Before he reaches the ground, he grabs the rope to stop his fall. What do the bananas do?

FIGURE 5–42

CHALLENGE PROBLEMS

5–63 A box of weight w is pulled at constant speed along a level floor by a force $\boldsymbol{P}$ that is at an angle ϕ above the horizontal. The coefficient of kinetic friction between the floor and box is μ_k.

a) In terms of ϕ, μ_k, and w, calculate P.

b) For $w = 400$ N and $\mu_k = 0.4$, calculate P for ϕ ranging from 0° to 90° in increments of 10°. Sketch a graph of P versus ϕ.

c) From the general expression in (a), calculate the value of ϕ for which the P required to maintain constant speed is a minimum. For the special case of $w = 400$ N and $\mu_k = 0.4$, evaluate this optimal ϕ and compare your result to the graph you constructed in part (b).

(*Note.* At the value of ϕ where the function $P(\phi)$ has a minimum, $dP(\phi)/d\phi = 0$.)

5–64 A rock of mass $m = 5$ kg falls from rest in a viscous medium. The rock is acted on by a net constant downward force of 20 N and by a viscous retarding force proportional to its speed and equal to $(5 \text{ N·s·m}^{-1})v$, where v is the speed in meters per second. (See Example 5–17.)

a) Find the initial acceleration, a_0.

b) Find the acceleration when the speed is 3 m·s^{-1}.

c) Find the speed when the acceleration equals $0.1a_0$.

d) Find the terminal velocity, v_t.

e) Find the coordinate, velocity, and acceleration 2 s after the start of the motion.

f) Find the time required to reach a speed $0.9v_t$.

g) Construct a graph of v versus t, for the first 3 s of the motion.

5–65 The assumption that the coefficient of kinetic friction is independent of speed when an object is sliding is true only over a suitably small range of speeds. For example, in some cases for a speed of 130 km·hr^{-1} (about 80 mi·hr^{-1}), μ_k may be only $\frac{1}{2}$ what it is for speeds of 8 to 16 km·hr^{-1}. For a car sliding with its brakes locked, take $\mu_k = 1.0$ at 8 km·hr^{-1} and $\mu_k = 0.5$ at 130 km·hr^{-1}, and assume that μ_k varies linearly with speed. Then how far would a car slide after the brakes are applied at 130 km·hr^{-1} before coming to rest? (For a still more realistic characterization of the friction forces involved in sliding, see J. D. Edwards, Jr., *Amer. Jour. Phys.*, Vol. 48 (1980), pp. 253–254.)

5–66 A marble falls from rest through a medium that exerts a resisting force that varies directly with the square of the velocity ($R = -Kv^2$).

a) Draw a diagram showing the direction of motion and indicate with the aid of vectors all of the forces acting on the marble.

b) Apply Newton's second law and infer from the resulting equation the general properties of the motion.

c) Show that the marble acquires a terminal velocity that is as given in Eq. (5–8).

d) Derive the equation for the velocity at any time.

$\Bigg($*Note.*

$$\int \frac{dx}{a^2 - x^2} = \frac{1}{\text{arctanh}(x/a)},$$

where the hyperbolic tangent is defined by

$$\tanh(x) = \frac{e^x - e^{-x}}{e^x + e^{-x}} = \frac{e^{2x} - 1}{e^{2x} + 1}.\Bigg)$$

5–67 A particle of mass m, originally at rest, is subjected to a force whose direction is constant but whose magnitude varies with time according to the relation

$$F = F_0\left[1 - \left(\frac{t - T}{T}\right)^2\right],$$

where F_0 and T are constants. The force acts only for the time interval $2T$.

a) Make a rough graph of F versus t.

b) Prove that the speed v of the particle after time $2T$ has elapsed is equal to $4F_0T/3m$.

c) Choose numbers for v, T, and m that might be appropriate to a batted baseball, and calculate the force F_0. Judge whether the answer is sensible.

5–68

a) A wedge of mass M rests on a frictionless horizontal table top. A block of mass m is placed on the wedge, as shown in Fig. 5–43a. There is no friction between the block and the wedge. The system is released from rest.

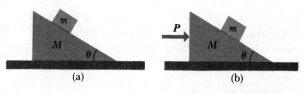

FIGURE 5–43

Calculate

 i) the acceleration of the wedge;

 ii) the horizontal and vertical components of the acceleration of the block.

Do your answers reduce to the correct results when M is very large?

b) The wedge and block are as in part (a). Now a horizontal force P is applied to the wedge, as in Fig. 5–43b. What magnitude must P have if the block is to remain at constant height above the table top?

6
APPLICATIONS OF NEWTON'S LAWS—II

IN CHAPTER 5 WE STUDIED SEVERAL SIMPLE DYNAMICS PROBLEMS SHOWING
applications of Newton's laws. In this chapter we continue our study of dynamics, concentrating especially on *circular motion.* Examples of motion of a
particle in a circular path turn up in a wide variety of situations, such as
racetracks, children's swings, and the motions of planets and satellites; thus it
is worthwhile to study this special class of motions in detail. We also return to
gravitational interactions for a detailed study of Newton's law of gravitation;
this law plays a central role in planet and satellite motion and also provides a
fuller understanding of the concept of *weight.* We suggest you review the discussion of the kinematics of circular motion in Section 3–5 to help you understand what is to come in this chapter.

6–1 FORCE IN CIRCULAR MOTION

When a particle moves in a circular path of radius R with constant speed v, the
motion is called **uniform circular motion.** In Section 3–5 we showed that in
this motion the acceleration of the particle is always directed toward the center
of the circle, perpendicular to the instantaneous velocity. This is called **centripetal acceleration.** Its magnitude, as derived in Section 3–5, is

$$a_\perp = \frac{v^2}{R}. \tag{6–1}$$

Uniform circular motion is accelerated
motion.

The symbol $\perp$ is a reminder that at each point the acceleration is perpendicular to the instantaneous velocity. The **period** τ is the time for one revolution. It is given by

$$\tau = \frac{2\pi R}{v}. \tag{6–2}$$

In terms of the period, the centripetal acceleration can also be expressed as

$$a_\perp = \frac{4\pi^2 R}{\tau^2}. \tag{6–3}$$

Circular motion, like all other motion of a particle, is governed by Newton's second law. The particle's acceleration toward the center of the circle

must be caused by a *force* also directed radially toward the center. The magnitude of the radial acceleration is given by v^2/R, so the magnitude of the net radial force F on a particle of mass m must be

$$F = ma_\perp = m\frac{v^2}{R}. \tag{6-4}$$

A familiar example of such a radial force occurs when you tie an object to a string and whirl it in a circle. The string must constantly pull in toward the center; if it breaks, then the inward force no longer acts, and the object flies off along a tangent to the circle.

More generally, several forces may be acting on a body in uniform circular motion. In this case the *vector sum* of all forces on the body must have the magnitude given by Eq. (6-4) and must be directed toward the center of the circle.

The force in Eq. (6-4) is sometimes called *centripetal force*. This usage is unfortunate because it implies that this force is somehow different from ordinary forces, or that the fact of circular motion somehow generates an additional force. Neither of these assumptions is correct. *Centripetal* refers to the *effect* of the force—specifically, to the fact that it causes circular motion in which the *direction* of the velocity changes but not its *magnitude*. In the equation $\Sigma \mathbf{F} = m\mathbf{a}$, the sum of forces must include only the real physical forces: pushes or pulls exerted by strings, rods, or other agencies. The quantity mv^2/R does *not* appear in $\Sigma \mathbf{F}$ but belongs on the $m\mathbf{a}$ side of the equation.

What force causes centripetal acceleration?

PROBLEM-SOLVING STRATEGY: *Circular motion*

The strategies for dynamics problems outlined at the beginning of Section 5–4 are equally applicable here, and we suggest you reread them before beginning your study of the examples below.

A serious peril in circular-motion problems is the temptation to regard mv^2/R as a *force*, as though the body's circular motion somehow generates an extra force in addition to the real physical forces exerted by

strings, contact with other bodies, gravitational interactions, or whatever. Resist this temptation! Draw the free-body diagram with *green* arrows for the forces and a *red* arrow for the acceleration. Then in the $\Sigma \mathbf{F} = m\mathbf{a}$ equations, write the force terms in green and the mv^2/R term in red. This may sound silly, but if it helps, do it!

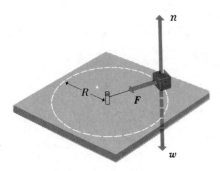

6–1 Body revolving uniformly in a circle on a horizontal frictionless surface.

EXAMPLE 6–1 A small plastic box having mass 0.200 kg revolves uniformly in a circle on a frictionless horizontal surface such as an air-hockey table. It is attached by a cord 0.200 m long to a pin set in the surface. If the body makes two complete revolutions per second, find the force F exerted on it by the cord. (See Fig. 6–1.)

SOLUTION The circumference of the circle is

$$2\pi(0.200 \text{ m}) = 1.26 \text{ m},$$

so the speed is $v = 2.51 \text{ m·s}^{-1}$. The magnitude of the centripetal acceleration is

$$a = \frac{v^2}{R} = \frac{(2.51 \text{ m·s}^{-1})^2}{0.200 \text{ m}} = 31.6 \text{ m·s}^{-2}.$$

The body has no vertical acceleration, so the forces $\mathcal{n}$ and $\boldsymbol{w}$ are equal and opposite. Thus the force $\boldsymbol{F}$ is the resultant force, and the tension F in the cord is

$$F = ma = (0.200 \text{ kg})(31.6 \text{ m·s}^{-2})$$
$$= 6.32 \text{ N}.$$

EXAMPLE 6–2   A child's indoor swing consists of a rope of length L anchored to the ceiling, with a seat at the lower end. The total mass of child and seat is m. They swing in a horizontal circle with constant speed v, as shown in Fig. 6–2; as they swing around, the rope makes a constant angle θ with the vertical. Assuming the time τ for one revolution (i.e., the period) is known, find the tension T in the rope and the angle θ.

Dynamics of a child's swing

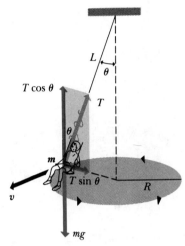

6–2 A child's indoor swing.

SOLUTION   First, note that the center of the circular path is at the center of the shaded area, *not* at the top end of the rope. A free-body diagram for the system is shown in the figure. The forces on the system in the position shown are the weight mg and the tension T in the rope. The tension has a horizontal component $T \sin \theta$ and a vertical component $T \cos \theta$. (Note that the force diagram in Fig. 6–2 is exactly like that in Fig. 5–14b. The only difference is that in this case the acceleration a is the *radial* acceleration, v^2/R.) The system has no vertical acceleration, so the sum of the vertical components of force is zero. The horizontal force must equal the mass m times the radial (centripetal) acceleration. Thus the $\Sigma \boldsymbol{F} = m\boldsymbol{a}$ equations are

$$T \sin \theta = m \frac{v^2}{R}, \qquad T \cos \theta = mg.$$

When the first of these equations is divided by the second, the result is

$$\tan \theta = \frac{v^2}{Rg}. \tag{6–5}$$

Also, the radius R of the circle is given by $R = L \sin \theta$, and the speed is the circumference $2\pi L \sin \theta$ divided by the period τ:

$$v = \frac{2\pi L \sin \theta}{\tau}.$$

When these relations are used to eliminate R and v from Eq. (6–5), we obtain

$$\cos \theta = \frac{g\tau^2}{4\pi^2 L}, \tag{6–6}$$

or

$$\tau = 2\pi \sqrt{L(\cos \theta)/g}. \tag{6–7}$$

Once θ is determined, the tension T can be found from $T = mg/\cos \theta$. For a given length L, $\cos \theta$ decreases as the time is made shorter, and the angle θ increases. The angle never becomes 90°, however, since this requires that $\tau = 0$ or $v = \infty$. Equation (6–7) is similar in form to the expression for the time of swing of a simple pendulum, which will be derived in Chapter 11. Because of this similarity, and because the moving rope traces out a cone, the present system is called a *conical pendulum.*

You may feel tempted to add to the forces in Fig. 6–2 an extra outward force, to "keep the body out there" at angle θ, or to "keep it in equilibrium." Perhaps you have heard the term *centrifugal force; centrifugal* means "fleeing from the center."

Don't make up imaginary forces that aren't there.

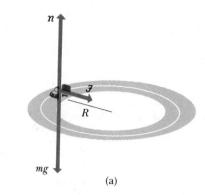

(a)

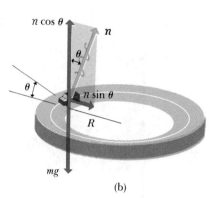

(b)

6–3 (a) Forces on a vehicle rounding a curve on a level track. (b) Forces when the track is banked.

Friction keeps the car on the road.

The temptation to add an extra force must be resisted with all possible strength. First, the body does not stay "out there"; it is in constant motion around its circular path. Its velocity is constantly changing in direction, and the body is *not* in equilibrium. Second, if there were an additional outward ("centrifugal") force to balance the inward force, there would then be *no* resultant inward force to cause the circular motion. In that case the body would move in a straight line, not a circle. What is sometimes called centrifugal force is, at least in an inertial frame of reference, not a force at all but a reflection of Newton's first law. In the absence of force, a body moves in a straight line with constant speed, and for circular motion to occur there must be a net *inward* force.

EXAMPLE 6–3   The driver of a sports car is rounding a flat, unbanked curve of radius R. If the coefficient of friction between tires and road is μ_s, what is the maximum speed v at which he can take the curve without sliding?

SOLUTION   Figure 6–3a is a free-body diagram for the situation. The acceleration v^2/R toward the center of the curve must be caused by the friction force $\mathcal{F}$, and there is no vertical acceleration. Thus we have

$$\mathcal{F} = m\frac{v^2}{R}, \qquad n = mg.$$

The maximum friction force available is $\mathcal{F} = \mu_s n = \mu_s mg$. Combining these relations to eliminate $\mathcal{F}$, we find

$$\mu_s mg = m\frac{v^2}{R},$$

or

$$v = \sqrt{\mu_s gR}.$$

If $\mu_s = 0.50$ and $R = 25$ m, then

$$v = \sqrt{(0.50)(9.8\ \mathrm{m\cdot s^{-2}})(25\ \mathrm{m})} = 11.1\ \mathrm{m\cdot s^{-1}},$$

or about 25 mi·hr^{-1}. Why did we use the coefficient of *static* friction for a moving car?

Banking a curve to help the car make the turn

EXAMPLE 6–4   An engineer proposes to rebuild the curve in Example 6–3, banking it so that at a certain speed v no friction at all is needed for the car to make the curve. At what angle θ should it be banked?

SOLUTION   The free-body diagram is shown in Fig. 6–3b. The normal force is no longer vertical but is perpendicular to the roadway, at an angle θ with the vertical. Thus it has a vertical component $n\cos\theta$ and a horizontal component $n\sin\theta$, as shown. The horizontal component of n must now cause the acceleration v^2/R, and again there is no vertical acceleration. Thus

$$n\sin\theta = \frac{mv^2}{R}, \qquad n\cos\theta = mg.$$

Dividing the first of these equations by the second, we find

$$\tan\theta = \frac{v^2}{Rg}. \tag{6–8}$$

If $R = 25$ m and $v = 11.1$ m·s^{-1}, as in Example 6–3, then

$$\theta = \arctan \frac{(11.1 \text{ m·s}^{-1})^2}{(25 \text{ m})(9.8 \text{ m·s}^{-2})} = 26.6°.$$

Equation (6–8) shows that the tangent of the banking angle is proportional to the square of the speed and inversely proportional to the radius. For a given radius, no one angle is correct for all speeds. Hence in the design of highways and railroads, curves are banked for the *average speed* of the traffic over them. The same considerations apply to the correct banking angle of a plane when it makes a turn in level flight. We also note that the banking angle is given by the same expression as that for the angle of a conical pendulum, Example 6–2, Eq. (6–5). In fact, the free-body diagrams of Figs. 6–2 and 6–3b are identical.

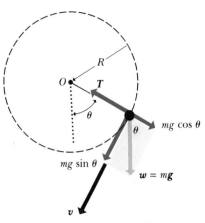

6–4 Forces on a body whirling in a vertical circle with center at O.

6–2 MOTION IN A VERTICAL CIRCLE

Figure 6–4 shows a small ball attached to a string of length R and whirling in a *vertical* circle about a fixed point O; the other end of the string is attached to this point. The motion is circular, but it is *not* uniform (unlike the examples in Section 6–1), because the speed increases on the way down and decreases on the way up. The normal component of acceleration is still given by $a_\perp = v^2/R$, as discussed above, but v changes from point to point along the path. Because the magnitude of the velocity is changing, there must also be a component of acceleration $a_\parallel$ *parallel* or *tangent* to the path, as discussed in Section 3–3.

The forces on the body at any point are its weight $w = mg$ and the tension T in the cord. We resolve the weight into a normal component, of magnitude $mg \cos \theta$, and a tangential component of magnitude $mg \sin \theta$, as in Fig. 6–4. The resultant tangential and normal forces are then

$$F_\parallel = mg \sin \theta \quad \text{and} \quad F_\perp = T - mg \cos \theta.$$

The tangential acceleration, from Newton's second law, is

$$a_\parallel = \frac{F_\parallel}{m} = g \sin \theta;$$

this is the same as the acceleration of a body sliding on a frictionless inclined plane with slope angle θ. The normal (radial) acceleration $a_\perp = v^2/R$ is

$$a_\perp = \frac{F_\perp}{m} = \frac{T - mg \cos \theta}{m} = \frac{v^2}{R};$$

To find the tension in the cord, we solve this for T:

$$T = m \left(\frac{v^2}{R} + g \cos \theta \right). \tag{6–9}$$

At the lowest point of the path, $\theta = 0$, $\sin \theta = 0$, and $\cos \theta = 1$. Hence at this point $F_\parallel = 0$, $a_\parallel = 0$, and the acceleration is purely radial (upward). The magnitude of the tension, from Eq. (6–9), is

$$T = m \left(\frac{v^2}{R} + g \right).$$

As the passengers in this amusement-park ride travel in a nearly circular path, they experience accelerations toward the center of the circle. (Tidal Wave, Great America, California)

At the highest point, $\theta = 180°$, $\sin \theta = 0$, $\cos \theta = -1$, and the acceleration is again purely radial (downward). The tension is

$$T = m\left(\frac{v^2}{R} - g\right). \tag{6-10}$$

With motion of this sort, it is a familiar fact that when the speed at the highest point is less than some critical value v_c, the cord becomes slack and the path is no longer circular. To find this critical speed, we set $T = 0$ in Eq. (6-10):

$$0 = m\left(\frac{v_c^2}{R} - g\right), \qquad v_c = \sqrt{Rg}.$$

If $R = 1$ m, then

$$v_c = \sqrt{(1 \text{ m})(9.8 \text{ m·s}^{-2})} = 3.13 \text{ m·s}^{-1}.$$

EXAMPLE 6-5 In Fig. 6-5, a small body of mass $m = 0.10$ kg swings in a vertical circle at the end of a cord of length $R = 1.0$ m. If its speed is 2.0 m·s^{-1} when the cord makes an angle $\theta = 30°$ with the vertical, find

a) the radial and tangential components of its acceleration at this instant;

b) the magnitude and direction of the resultant acceleration;

c) the tension T in the cord.

SOLUTION

a) The radial component of acceleration is

$$a_\perp = \frac{v^2}{R} = \frac{(2.0 \text{ m·s}^{-1})^2}{1.0 \text{ m}} = 4.0 \text{ m·s}^{-2}.$$

The tangential component of acceleration, due to the tangential force $mg \sin \theta$, is

$$a_\| = g \sin \theta = (9.8 \text{ m·s}^{-2})(0.50) = 4.9 \text{ m·s}^{-2}.$$

b) The magnitude of the resultant acceleration (see Fig. 6-5b) is

$$a = \sqrt{a_\|^2 + a_\perp^2} = 6.33 \text{ m·s}^{-2}.$$

The angle ϕ is

$$\phi = \arctan \frac{a_\|}{a_\perp} = 50.8°.$$

c) The tension in the cord is given by $F_\perp = ma_\perp$: $T - mg \cos \theta = mv^2/R$, so

$$T = m\left(\frac{v^2}{R} + g \cos \theta\right) = 1.25 \text{ N}.$$

Note that the magnitude of the tangential acceleration is not constant but is proportional to the sine of the angle θ. Hence the equations of motion with constant acceleration *cannot* be used to find the speed at other points of the path. We will show in the next chapter, however, how the speed at any point can be found from *energy* considerations.

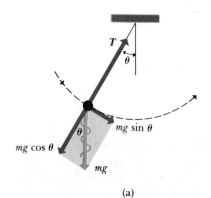

(a)

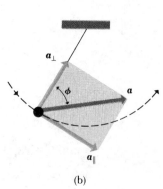

(b)

6-5 (a) Forces on a body swinging in a vertical circle. (b) The radial and tangential components of acceleration are combined to obtain the resultant acceleration a.

6–3 NEWTON'S LAW OF GRAVITATION

We have already encountered gravitational interactions several times. The **weight** of a body, the force that attracts it toward the earth (or whatever celestial body it is near), is an example of gravitational attraction. In this section we study this phenomenon in more detail.

Newton discovered the **law of gravitation** during his study of the motions of planets around the sun, and he published it in 1686. It may be stated as follows:

A general formulation of gravitational interactions

> Every particle of matter in the universe attracts every other particle with a force that is directly proportional to the product of the masses of the particles and inversely proportional to the square of the distance between them.

Thus

$$F_g = G \frac{m_1 m_2}{r^2}, \qquad (6\text{--}11)$$

where F_g is the magnitude of the gravitational force on either particle, m_1 and m_2 are their masses, r is the distance between them, and G is a fundamental physical constant called the **gravitational constant.** The numerical value of G depends on the system of units used.

The directions of the gravitational forces on the interacting bodies are along the straight line joining them. These forces form an action–reaction pair. Even when the masses of the particles are different, the interaction forces have equal magnitude.

Newton's law of gravitation describes the interaction forces between two *particles.* But these forces also obey the **principle of superposition:** If two masses each exert forces on a third, the *total* force on the third mass is the vector sum of the individual forces of the first two. It can also be shown that *the gravitational force exerted on or by any homogeneous sphere is the same as though the entire mass of the sphere were concentrated in a point at its center.* Thus if the earth were a homogeneous sphere, of mass m_E, the force exerted by it on a small body of mass m, at a distance r from its center, would be

Superposition: how gravitational forces combine

$$F_g = G \frac{m m_E}{r^2}, \qquad (6\text{--}12)$$

provided that the body lies outside the earth, i.e., that r is greater than the radius of the earth. A force of the same magnitude would be exerted *on* the earth by the whole body. We will learn how to prove these statements in Chapter 25 in connection with analogous problems with electric charge. (See also Challenge Problem 6–42.)

At points *inside* the earth, these statements need to be modified. If we could drill a hole to the center of the earth and measure the force of gravity on a body at various distances from the center, the force would be found to *decrease* as the center is approached, rather than increasing as $1/r^2$. Qualitatively, it is easy to see why this should be so: As the body enters the interior of the earth (or other spherical body), some of the earth's mass is on the side of the body opposite from the center of the earth and pulls the body in the opposite direction. Exactly at the center of the earth, the gravitational force on the body is zero!

The Cavendish balance: a way to measure very small forces

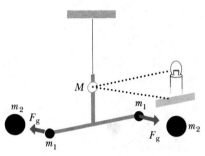

6–6 Principle of the Cavendish balance.

The magnitude of the gravitational constant G can be found experimentally by measuring the force of gravitational attraction between two bodies of known masses m_1 and m_2, at a known separation. For bodies of moderate size the force is extremely small, but it can be measured with an instrument invented by the Rev. John Mitchell and first used for this purpose by Sir Henry Cavendish in 1798. The same type of instrument was also used by Coulomb for studying forces of electrical and magnetic attraction and repulsion.

The Cavendish balance consists of a light, rigid T-shaped rod (Fig. 6–6) supported by a fine vertical fiber such as a quartz thread or a thin metallic ribbon. Two small spheres, each of mass m_1, are mounted at the ends of the horizontal portion of the T, and a small mirror M, fastened to the vertical portion, reflects a beam of light onto a scale. To use the balance, two large spheres, each of mass m_2, are brought up to the positions shown. The forces of gravitational attraction between the large and small spheres twist the system through a small angle, thereby moving the reflected light beam along the scale.

By use of an extremely thin fiber, the deflection of the mirror may be made sufficiently large so that the gravitational force can be measured quite accurately. The gravitational constant obtained in this way is found to be (in SI units)

$$G = 6.670 \times 10^{-11} \text{ N·m}^2\text{·kg}^{-2}.$$

Gravitational forces on small bodies are tiny!

EXAMPLE 6–6 The mass m_1 of one of the small spheres of a Cavendish balance is 0.00100 kg; the mass m_2 of one of the large spheres is 0.500 kg; and the center-to-center distance between the spheres is 0.0500 m. Find the gravitational force on each sphere.

SOLUTION The force is

$$F_g = 6.67 \times 10^{-11} \text{ N·m}^2\text{·kg}^{-2} \frac{(0.00100 \text{ kg})(0.500 \text{ kg})}{(0.0500 \text{ m})^2}$$
$$= 1.33 \times 10^{-11} \text{ N},$$

or about one hundred-billionth of a newton!

EXAMPLE 6–7 Suppose the spheres in Example 6–6 are placed 0.0500 m from each other at a point in space far removed from all other bodies. What is the acceleration of each, relative to an inertial system?

SOLUTION The acceleration a_1 of the smaller sphere is

$$a_1 = \frac{F_g}{m_1} = \frac{1.33 \times 10^{-11} \text{ N}}{1.0 \times 10^{-3} \text{ kg}} = 1.33 \times 10^{-8} \text{ m·s}^{-2}.$$

The acceleration a_2 of the larger sphere is

$$a_2 = \frac{F_g}{m_2} = \frac{1.33 \times 10^{-11} \text{ N}}{0.5 \text{ kg}} = 2.67 \times 10^{-11} \text{ m·s}^{-2}.$$

In this case, the accelerations are *not* constant because the gravitational force increases as the spheres approach each other.

EXAMPLE 6–8 Some of the masses from Example 6–6 are arranged as shown in Fig. 6–7. Find the magnitude and direction of the total force exerted on the small sphere by both large ones.

SOLUTION We use the principle of superposition: The total force on the small sphere is the vector sum of the two forces due to the two individual large spheres. These forces have different directions, so we have to compute a vector sum. We first find the magnitudes of the forces. The force F_1 on the small mass due to the upper large one is

$$F_1 = \frac{(6.67 \times 10^{-11}\ \text{N·m}^2\text{·kg}^{-2})(0.5\ \text{kg})(0.001\ \text{kg})}{(0.2\ \text{m})^2 + (0.2\ \text{m})^2}$$

$$= 4.17 \times 10^{-13}\ \text{N},$$

and the force F_2 due to the lower large mass is

$$F_2 = \frac{(6.67 \times 10^{-11}\ \text{N·m}^2\text{·kg}^{-2})(0.5\ \text{kg})(0.001\ \text{kg})}{(0.2\ \text{m})^2}$$

$$= 8.34 \times 10^{-13}\ \text{N}.$$

The x- and y-components of these forces are

$$F_{1x} = (4.17 \times 10^{-13}\ \text{N}) \cos 45° = 2.95 \times 10^{-13}\ \text{N},$$
$$F_{1y} = (4.17 \times 10^{-13}\ \text{N}) \sin 45° = 2.95 \times 10^{-13}\ \text{N},$$
$$F_{2x} = 8.34 \times 10^{-13}\ \text{N},$$
$$F_{2y} = 0.$$

Or, in terms of unit vectors,

$$\boldsymbol{F}_1 = (2.95 \times 10^{-13}\ \text{N})\boldsymbol{i} + (2.95 \times 10^{-13}\ \text{N})\boldsymbol{j},$$
$$\boldsymbol{F}_2 = (8.34 \times 10^{-13}\ \text{N})\boldsymbol{i}.$$

Thus the total force on the small mass is the vector sum

$$\boldsymbol{F} = [(2.95 + 8.34) \times 10^{-13}\ \text{N}]\boldsymbol{i} + (2.95 \times 10^{-13}\ \text{N})\boldsymbol{j},$$

with components

$$F_x = 11.29 \times 10^{-13}\ \text{N},$$
$$F_y = 2.95 \times 10^{-13}\ \text{N}.$$

The magnitude of this force is

$$F = \sqrt{(11.29 \times 10^{-13}\ \text{N})^2 + (2.95 \times 10^{-13}\ \text{N})^2}$$
$$= 11.7 \times 10^{-13}\ \text{N},$$

and its direction is

$$\theta = \arctan \frac{2.95}{11.29} = 14.6°.$$

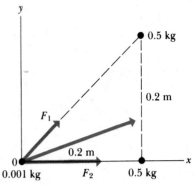

6–7 The total gravitational force on the 0.001-kg mass is the vector sum of the forces exerted by the two 0.5-kg masses.

We can now define the *weight* of a body more generally as *the resultant gravitational force exerted on the body by all other bodies in the universe.* When the body is near the surface of the earth, the earth's attraction is much greater than that of any other body; in that case we can ignore all other gravitational forces and consider the weight as due entirely to the earth's attraction. Similarly, at the surface of the moon, or of another planet, the weight of a body

A more general definition of *weight*

results almost entirely from the gravitational attraction of the moon or the planet. Thus if the earth were a homogeneous sphere of radius R and mass m_E, the weight w of a small body of mass m at or near its surface would be

$$w = F_g = \frac{Gmm_E}{R^2}. \tag{6-13}$$

Equation (6–13) shows that the weight of a given body decreases inversely with the square of its distance from the earth's center; at a radial distance of two earth radii, for example, it has decreased to one quarter of its value at the earth's surface.

The *apparent* weight of a body at the surface of the earth differs slightly in magnitude and direction from the earth's force of gravitational attraction because the rotation of the earth about its axis makes it not precisely an inertial frame of reference. For the present, we ignore this effect and assume that the earth *is* an inertial system. Then when a body is allowed to fall freely, the force accelerating it is its weight w, and the acceleration produced by this force is the acceleration due to gravity, g. The general relation

$$\Sigma F = ma$$

therefore becomes, for the special case of a freely falling body,

$$w = mg. \tag{6-14}$$

Since, according to Eq. (6–13)

$$w = mg = G\frac{mm_E}{R^2},$$

it follows that

$$g = \frac{Gm_E}{R^2}, \tag{6-15}$$

showing that the acceleration due to gravity is the same for *all* bodies (since m cancels out) and is very nearly constant (since G and m_E are constants and R varies only slightly from point to point on the earth).

Determining the mass of the earth

When Eq. (6–15) is solved for m_E, we find

$$m_E = \frac{R^2 g}{G}, \tag{6-16}$$

where R is the earth's radius. All of the quantities on the right are known, so the mass of the earth, m_E, can be calculated. Taking $R = 6370$ km $= 6.37 \times 10^6$ m, and $g = 9.80$ m·s^{-2}, we find

$$m_E = 5.96 \times 10^{24} \text{ kg.}$$

The volume of the earth is

$$V = \frac{4}{3}\pi R^3 = 1.08 \times 10^{21} \text{ m}^3.$$

Density of the earth

The mass of a body divided by its volume is known as its average *density*. (The density of water is 1 g·cm^{-3} = 1000 kg·m^{-3}.) The average density of the earth is therefore

$$\frac{m_E}{V} = 5520 \text{ kg·m}^{-3} = 5.52 \text{ g·cm}^{-3}.$$

The density of most rock near the earth's surface, such as granites and gneisses, is about 3 g·cm^{-3} = 3000 kg·m^{-3}, so the interior of the earth must have much higher density. Some basaltic rock found on the surface has a density of about 5 g·cm^{-3}.

6–4 GRAVITATIONAL FIELD

Newton's law of gravitation can be restated in a useful way with the concept of **gravitational field.** Instead of calculating the interaction forces between two masses by using Eq. (6–11) directly, we consider a two-stage process. First, we think of one mass as creating a change at each point in the space surrounding it. This change is called a *gravitational field.* To describe it, we associate a vector quantity g with each point in space. The significance of g is that a particle of mass m at a point where the gravitational field is g experiences a gravitational *force* given by $F = mg$. To measure a gravitational field we may place a small mass, called a *test mass*, at the point. If the mass experiences a gravitational force, then there is a gravitational field at that point.

In Example 6–8, instead of calculating directly the forces on the 0.001-kg body by the large bodies, we may take the point of view that the large bodies create a *gravitational field* at the location of the small body, and that the field exerts a force on any mass located at the point.

Because force is a vector quantity, gravitational field is also a vector quantity. We define the gravitational field g at a point as the quotient of the force F experienced by a test mass m and the mass. Thus

$$g = \frac{F}{m}. \qquad (6-17)$$

To say it another way: At a point where the gravitational field is g, the force F on a mass m is given by

$$F = mg. \qquad (6-18)$$

The gravitational field due to a point mass m, at a distance r away from it, has magnitude

$$g = \frac{Gm}{r^2}, \qquad (6-19)$$

and is directed away from the point mass. Gravitational fields also obey the principle of superposition: The *total* gravitational field at a point, due to several point masses, is the *vector sum* of the gravitational fields of the separate masses. Gravitational field is a very useful concept; when we know the field, we can calculate quickly the gravitational force on any body.

The gravitational field in the vicinity of any collection of masses varies from one point to another in the region. Thus it is not a single vector quantity but rather a whole collection of vector quantities, one vector associated with each point in space. Indeed, this is what is meant by the general term *vector field*. Another familiar example of a vector field is the velocity in a flowing fluid: different parts of the fluid have different velocities, and so we speak of the *velocity field* in the fluid. During our study of electricity and magnetism, we will discuss in detail the properties of the electric and magnetic fields; both of these are vector fields.

Gravitational field: a useful reformulation of gravitational interactions

Vector field: a vector quantity associated with each point in a region of space

We assert that when a particle of mass m is acted on *only* by gravitational force, its acceleration at each instant is equal in magnitude and direction to the gravitational field g at its instantaneous location. Can you prove this?

EXAMPLE 6–9 In Example 6–8, find the magnitude of the gravitational field at the location of the 0.001-kg mass.

SOLUTION From Eq. (6–17), the gravitational field at this point has magnitude

$$g = \frac{11.7 \times 10^{-13} \text{ N}}{0.001 \text{ kg}} = 11.7 \times 10^{-10} \text{ m·s}^{-2}.$$

The direction of the field is the same as that of the force on the small body.

EXAMPLE 6–10 A ring-shaped body of radius a has total mass M. Find the gravitational field at point P, at a distance x from the center, along the line perpendicular to the plane of the ring, through its center.

SOLUTION The situation is shown in Fig. 6–8. We consider first a small segment Δs of the ring; we call the mass of this segment ΔM. At point P, this segment produces a gravitational field of magnitude

$$g = \frac{G\,\Delta M}{r^2} = \frac{G\,\Delta M}{x^2 + a^2}.$$

The component of this field along the x-axis is given by

$$g_x = -g \cos \theta = -\frac{G\,\Delta M}{x^2 + a^2} \frac{x}{\sqrt{x^2 + a^2}} = -\frac{G\,\Delta Mx}{(x^2 + a^2)^{3/2}}. \tag{6–20}$$

Now x is the same for every segment around the ring, so to find the *total* x-component of gravitational field from all the mass elements ΔM, we simply replace ΔM by the *total* mass M. The result is

$$g_x = -\frac{GMx}{(x^2 + a^2)^{3/2}}. \tag{6–21}$$

The y-component of gravitational field at point P is zero because the y-components of field produced by two segments at opposite sides of the ring cancel each other. Similarly, the contributions of all such pairs added over the entire ring sum to zero.

Equation (6–21) shows that at the center of the ring ($x = 0$) the total gravitational field is zero. We should expect this. Mass elements on opposite sides pull in opposite directions, and their fields cancel. Also, when x is much larger than a, Eq. (6–21) becomes approximately equal to GM/x^2, corresponding to the fact that at distances much greater than the dimensions of the ring, it appears as a point mass.

6–8 Gravitational field of a ring of mass M and radius a.

Calculations with gravitational fields

6–5 SATELLITE MOTION

In our discussion of projectile motion in Section 3–4, we assumed that the acceleration of the projectile was constant in magnitude and direction. Because this acceleration is caused by the projectile's weight w, an equivalent assumption is that the weight is the same at all points in the trajectory, corresponding to motion in a uniform gravitational field **g.** These conditions are approximately satisfied if the projectile remains near the earth's surface and if the length of flight is small compared to the earth's radius. The trajectory is then a *parabola* (provided that air resistance can be neglected).

When the size of the trajectory becomes comparable to the radius of the earth, we can no longer ignore the variation of weight with position. At any point outside the earth, the gravitational force on the projectile is directed toward the center of the earth, and its magnitude is inversely proportional to the square of the distance from the center; thus the force is *not* constant in either magnitude or direction. It turns out that under such an attractive "inverse-square" force, the projectile's path may be a circle, an ellipse, a parabola, or a hyperbola. Circular paths will be of special interest to us, and they are also the easiest paths to analyze.

To gain intuitive understanding of satellite motion, let us imagine a tall tower, as in Fig. 6–9a. We launch a projectile from point *A* in the direction *AB* tangent to the earth's surface. If the initial speed is rather small, the trajectory will be like that labeled (1) in the figure; this is a portion of an ellipse with the

Orbits of satellites and planets

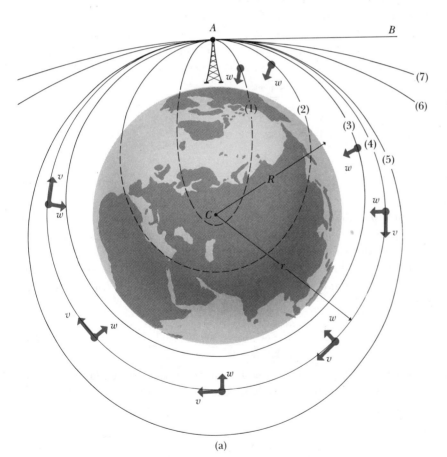

(a)

(b)

6–9 (a) Trajectories of a body projected from point *A* in the direction *AB* with different initial velocities. Orbits (1) and (2) would be completed as shown if the earth were a point mass at *C*. (b) A communications satellite before launch. The main body is 2.16 m in diameter and 2.82 m in height. Solar-energy collector panels on the sides and a parabolic antenna at the end can be seen. (Courtesy Hughes Aircraft Co.)

Halley's comet moves in an elliptical path around the sun under the action of the sun's gravitational attraction. A complete trip around the very flat ellipse requires about 76 years; the comet was in its closest proximity to the sun and the planets of the solar system during 1985 and 1986. (AP/ Wide World Photos)

earth's center C at one focus. If the trajectory is so short that changes in the magnitude and direction of w can be neglected, the ellipse is approximately the same shape as a parabola.

Trajectories (2) to (7) illustrate the effect of increasing the initial velocity. Trajectory (2) is again a portion of an ellipse. Trajectory (3), which just misses the earth, is a *complete* ellipse, and the projectile has become an earth satellite. Its velocity when it returns to point A is the same as its initial velocity, and if there is no retarding force (such as air resistance), it repeats its motion indefinitely. The earth's rotation will have moved the tower to a different point by the time the satellite returns to point A, but this does not affect the orbit.

Trajectory (4) is a special case in which the orbit is a circle. Trajectory (5) is again an ellipse, (6) is a parabola, and (7) is a hyperbola. Trajectories (6) and (7) are not closed orbits; for these the projectile never returns to its starting point, but "escapes" permanently from the earth's gravitational field.

We will consider only the simplest case, *circular* orbits. Many manmade satellites have nearly circular orbits, and the orbits of the planets of our solar system around the sun are also nearly circular. We have learned that a particle in uniform circular motion with speed v and radius r experiences a centripetal acceleration $a_\perp = v^2/r$ directed always toward the center of the circle. The *force* that provides this acceleration is the gravitational attraction of the earth (mass m_E) for the satellite (mass m). If the radius of the circular orbit, measured from the *center* of the earth, is r, then the gravitational force is given by $F = Gm_Em/r^2$, as discussed in Section 6–3. Using Newton's second law, we equate this to the product of the satellite's mass m and its acceleration v^2/r:

$$\frac{Gm_Em}{r^2} = \frac{mv^2}{r}.$$

When this equation is solved for v, the result is

$$v = \sqrt{\frac{Gm_E}{r}}. \tag{6–22}$$

This relation shows that the orbit radius r and the satellite's speed v cannot be chosen independently but must be related by Eq. (6–22).

We can also derive a relation between the orbit radius r and the *period* τ, the time for one revolution. As pointed out in Section 3–5, the speed v is the distance $2\pi r$ traveled in one revolution, divided by the time τ for one revolution. Thus

$$v = \frac{2\pi r}{\tau} = 2\pi rf, \tag{6–23}$$

The period of a satellite depends on its orbit radius but not on its mass.

where f is the *frequency* (number of revolutions per unit time), equal to $1/\tau$. We can obtain an expression for τ by solving this equation for τ and combining it with Eq. (6–22):

$$\tau = \frac{2\pi r}{v} = 2\pi r\sqrt{\frac{r}{Gm_E}} = \frac{2\pi r^{3/2}}{\sqrt{Gm_E}}. \tag{6–24}$$

Equations (6–22) and (6–24) show that larger orbits correspond to longer periods and slower speeds.

EXAMPLE 6–11 Suppose you want to place the communications satellite shown in Fig. 6–9b into a circular orbit 300 km above the earth's surface. What must be its speed, its period, and its radial acceleration? The earth's radius is 6380 km = 6.38×10^6 m, and its mass is 5.98×10^{24} kg. (See Appendix F.)

SOLUTION The radius of the orbit is

$$r = 6380 \text{ km} + 300 \text{ km} = 6680 \text{ km} = 6.68 \times 10^6 \text{ m}.$$

From Eq. (6–22),

$$v = \sqrt{\frac{(6.67 \times 10^{-11} \text{ N·m}^2\text{·kg}^{-2})(5.98 \times 10^{24} \text{ kg})}{6.68 \times 10^6 \text{ m}}}$$
$$= 7728 \text{ m·s}^{-1}.$$

From Eq. (6–24),

$$\tau = \frac{2\pi(6.68 \times 10^6 \text{ m})}{7728 \text{ m·s}^{-1}} = 5431 \text{ s} = 90.5 \text{ min}.$$

The radial acceleration is

$$a_\perp = \frac{v^2}{r} = \frac{(7728 \text{ m·s}^{-1})^2}{6.68 \times 10^6 \text{ m}} = 8.94 \text{ m·s}^{-2}.$$

This is somewhat less than the value of the free-fall acceleration g at the earth's surface and is, in fact, equal to the value of g at a height of 300 km above the earth's surface.

Our discussion has centered around the motion of manmade earth satellites, but it should be clear that these principles are also applicable to the circular motion of *any* body under its gravitational attraction to a stationary body. Other familiar examples are the motion of our moon and that of the moons of Jupiter and the motions of the planets around the sun in nearly circular orbits. The rings of Saturn, shown in Fig. 6–10, are composed of small particles in circular orbits around the planet.

Historically the study of planetary motion played a pivotal role in the development of physics. Johannes Kepler (1571–1630) spent several painstaking years analyzing the motions of the planets, basing his work on remarkably precise measurements made by the Danish astronomer Tycho Brahe (1546–1601) *without the aid of a telescope.* (The telescope was invented in 1608.) Kepler discovered that the orbits of the planets are (nearly circular) ellipses, and that the period of a planet in its orbit is proportional to the three-halves power of the orbit radius, as Eq. (6–24) shows.

It remained for Newton (1642–1727) to show, with his laws of motion and law of gravitation, that this behavior of the planets could be understood on the basis of the very same physical principles he had developed to analyze *terrestrial* motion. According to legend the law of gravitation came to Newton as he watched an apple fall out of a tree and asked himself what force accelerated it toward the earth. But to arrive at the $1/r^2$ form of the law of gravitation he had to do something much subtler: He had to take Brahe's observations of planetary motions, as systematized by Kepler, and show that Kepler's rules demanded a $1/r^2$ force law.

6–10 The rings of Saturn, photographed from Voyager 1 on November 1, 1980, from a range of 700,000 km. Individual particles move in circular orbits, with periods given by Eq. (6–24), using Saturn's mass. The outer portions of the rings take more time for one revolution than the inner portions. (Courtesy of Jet Propulsion Laboratory/NASA.)

The Newtonian synthesis: getting it all together

From our historical perspective three hundred years later, there is absolutely no doubt that this **Newtonian synthesis,** as it has come to be called, is one of the greatest achievements in the entire history of science, certainly comparable in significance to the development of quantum mechanics, the theory of relativity, and the understanding DNA in our own century.

6–6 EFFECT OF THE EARTH'S ROTATION ON g

On a rotating earth, apparent weight is different from real weight.

Because the earth rotates on its axis, it is not precisely an inertial frame of reference. For this reason the apparent weight of a body on earth is not precisely equal to the earth's gravitational attraction w_0, which we define as the **true weight** of the body. Figure 6–11 is a cutaway view of the earth, showing three observers, each holding a body of mass m hanging from a spring balance. Each balance applies a certain tension force T to the body hanging from it, and the reading on each balance is the magnitude T of this force. Each observer *thinks* that his scale reading equals the weight of the body. Also, each observer, if he is unaware of the earth's rotation, thinks the body on his spring balance is in equilibrium; if so, he thinks, the tension T must be opposed by an equal and opposite force w, which we define to be the **apparent weight.** As we will see, the apparent weight is different at different points on the earth.

If we assume that the earth is spherically symmetric, then the true weight w_0 has magnitude Gmm_E/R^2, where m_E and R are the mass and radius of the earth. This value is the same for all points on the earth's surface. If the center of the earth can be taken as the origin of an inertial coordinate system, then the body at the north pole really *is* in equilibrium in an inertial system, and the reading on that observer's spring balance is equal to w_0. But the body at the equator is moving in a circle of radius R with speed v, and there must be a net inward force equal to the mass times the centripetal acceleration:

$$w_0 - T = \frac{mv^2}{R}. \tag{6–25}$$

Thus the magnitude of the apparent weight (equal to T at this location) is

$$w = w_0 - \frac{mv^2}{R}. \tag{6–26}$$

We also note that if the earth were not rotating, the body when released would have a free-fall acceleration g_0 given by $g_0 = w_0/m$, but that its actual acceleration relative to the observer at the equator is given by $g = w/m$. Dividing Eq. (6–26) by m and using these relations, we find

$$g = g_0 - \frac{v^2}{R}. \tag{6–27}$$

To evaluate v^2/R, we note that in one day (86,400 s) a point on the equator moves a distance equal to the earth's circumference, $2\pi R = 2\pi(6.38 \times 10^6 \text{ m})$. Thus

$$v = \frac{2\pi(6.38 \times 10^6 \text{ m})}{86,400 \text{ s}} = 464 \text{ m·s}^{-1},$$

and

$$\frac{v^2}{R} = \frac{(464 \text{ m·s}^{-1})^2}{6.38 \times 10^6 \text{ m}} = 0.0337 \text{ m·s}^{-2}.$$

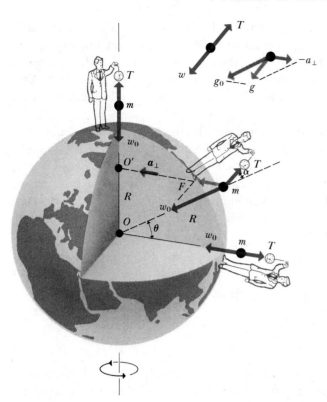

6–11 The resultant of the forces T and w_0 is equal to m times the centripetal acceleration $a_\perp = v^2/r$.

So for a spherically symmetric earth, the acceleration due to gravity should be about 0.03 m·s^{-2} less at the equator than at the poles.

At locations intermediate between the equator and the poles, we can write a vector equation corresponding to Eq. (6–26). From Fig. 6–11 we see that the appropriate equation is

$$\boldsymbol{w} = \boldsymbol{w}_0 - m\boldsymbol{a}_\perp = m\boldsymbol{g}_0 - m\boldsymbol{a}_\perp. \tag{6–28}$$

The difference in the magnitudes of g and g_0 is intermediate between zero and 0.0337 m·s^{-2}, and the *direction* of the apparent weight differs from the radial direction by an angle α, as shown.

Table 6–1 gives the values of g at several locations, showing variations with latitude of approximately this magnitude. There are also small additional variations due to the lack of perfect spherical symmetry of the earth, local variations in density, and differences in elevation.

In this discussion we have ignored the *orbital* motion of the earth around the sun; this motion makes the center of the earth not precisely an inertial

The acceleration of free fall on earth depends on location.

TABLE 6–1 Variations of g with Latitude and Elevation

Station	North latitude	Elevation, m	g, m·s^{-2}	g, ft·s^{-2}
Canal Zone	9°	0	9.78243	32.0944
Jamaica	18°	0	9.78591	32.1059
Bermuda	32°	0	9.79806	32.1548
Denver	40°	1638	9.79609	32.1393
Cambridge, Mass.	42°	0	9.80398	32.1652
Pittsburgh, Pa.	40.5°	235	9.80118	32.1561
Greenland	70°	0	9.82534	32.2353

Weightlessness: Is a body in an orbiting satellite *really* weightless?

origin. We can estimate the magnitude of this error by using the fact that in one year (about 3.15×10^7 s) the earth travels once around a nearly circular orbit of radius 1.49×10^{11} m. We challenge you to show that the centripetal acceleration of the orbital motion is about 0.006 m·s^{-2}. This is only 20% of the acceleration due to the earth's rotation about its axis, so it is a less important effect.

Our discussion of apparent weight can also be applied to the phenomenon of "weightlessness" in satellites. Bodies in an orbiting satellite are *not* weightless; the earth's gravitational attraction continues to act on them just as though they were at rest relative to the earth. But a space vehicle in orbit has an acceleration $\mathbf{a}_\perp$ toward the earth's center equal to the value of the acceleration of gravity g_0 at its orbit radius. The apparent weight w is given as before by

$$w = w_0 - m\mathbf{a}_\perp = mg_0 - m\mathbf{a}_\perp.$$

But in this case,

$$g_0 = \mathbf{a}_\perp,$$

so

$$w = 0.$$

It is in this sense that an astronaut or other body in the spacecraft is said to be "weightless" or in a state of "zero g." (See Fig. 6–12.) The earth's gravity acts on both the spacecraft and the bodies in it, giving each the same acceleration. Thus a body released inside the spacecraft does not fall relative to it, and it appears to be weightless. The situation is similar to that of the freely falling elevator in Example 5–15 (Section 5–4), except that the spacecraft also has a large constant tangential speed along its orbit.

6–12 Astronauts conducting experiments in zero-gravity environment. (Courtesy of NASA, Johnson Space Center.)

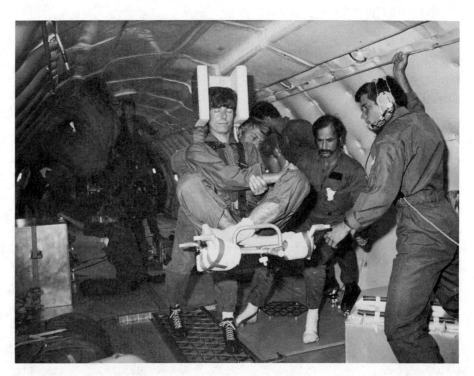

SUMMARY

A body in uniform circular motion with speed v and radius R has an acceleration $a_\perp$ directed toward the center of the circle, with magnitude

$$a_\perp = v^2/R. \qquad (6\text{--}1)$$

The period τ is the time for one revolution:

$$\tau = 2\pi R/v. \qquad (6\text{--}2)$$

According to Newton's second law, this acceleration requires a resultant force on the body, directed toward the center, with magnitude mv^2/R. Care must be taken not to regard mv^2/R as an additional force; it is the ma side of $\Sigma \mathbf{F} = m\mathbf{a}$.

A particle moving in a *vertical* circle has a centripetal acceleration $a_\perp = v^2/R$, and in addition a tangential component of acceleration $a_\parallel$ determined by the tangential components of force.

Newton's law of gravitation states that two bodies having masses m_1 and m_2, a distance r apart, attract each other with forces of magnitudes

$$F = \frac{Gm_1m_2}{r^2}. \qquad (6\text{--}11)$$

These forces form an action–reaction pair and obey Newton's third law. When two or more bodies exert gravitational forces on a particular body, the total gravitational force on it is the vector sum of the forces exerted by the individual bodies.

The weight of a body is defined as the total gravitational force exerted on it by all other bodies in the universe. Near the surface of the earth, where this is essentially equal to the earth's gravitational force alone, w is given by

$$w = \frac{Gmm_\mathrm{E}}{R^2}, \qquad (6\text{--}12)$$

and the acceleration due to gravity is given by

$$g = \frac{Gm_\mathrm{E}}{R^2}. \qquad (6\text{--}15)$$

The gravitational field $\mathbf{g}$ at a point in space is defined as the gravitational force exerted on a mass m at that point, divided by m. The gravitational field caused by a mass m at a distance r from it has magnitude

$$g = \frac{Gm}{r^2}. \qquad (6\text{--}19)$$

The total gravitational field at any point due to several masses is the vector sum of gravitational fields due to the individual masses.

When a satellite moves in a circular orbit, the centripetal acceleration is provided by the gravitational attraction of the earth. The speed and period of a satellite in an orbit of radius r are given by

$$v = \sqrt{\frac{Gm_\mathrm{E}}{r}}, \qquad (6\text{--}22)$$

$$\tau = \frac{2\pi r^{3/2}}{\sqrt{Gm_\mathrm{E}}}. \qquad (6\text{--}24)$$

KEY TERMS

uniform circular motion
centripetal acceleration
period
weight
Newton's law of gravitation
gravitational constant
principle of superposition
gravitational field
Newtonian synthesis
true weight
apparent weight

Because of the earth's rotation, the apparent weight of a body on earth differs from its true weight by about 0.3% at the equator, and the free-fall acceleration g differs from the value it would have without rotation by the same amount.

QUESTIONS

6–1 "It's not the fall that hurts you; it's the sudden stop at the bottom." Translate this saying into the language of Newton's laws of motion.

6–2 If action and reaction are always equal in magnitude and opposite in direction, why don't they always cancel each other and leave no net force for acceleration of the body?

6–3 Scales for weighing objects can be classified as those that use springs and those using standard weights to balance unknown weights. Which group would have greater accuracy when used in an accelerating elevator? When used on the moon? Does it matter whether you are trying to determine mass or weight?

6–4 Because of air resistance, two objects of unequal mass do *not* fall at precisely the same rate. If two bodies of identical shape but unequal mass are dropped simultaneously from the same height, which one reaches the ground first?

6–5 In discussing the Cavendish balance, a student claimed that the reason the small pivoted masses move toward the larger stationary masses rather than the larger toward the smaller is that the larger masses exert a stronger gravitational pull than the smaller. Please comment.

6–6 Since the earth is constantly attracted toward the sun by the gravitational interaction, why does it not fall into the sun and burn up?

6–7 A student wrote: "The only reason an apple falls downward to meet the earth instead of the earth falling upward to meet the apple is that the earth is much more massive and so exerts a much greater pull." Please comment.

6–8 Cavendish described his measurement of the gravitational constant G as "weighing the earth." This is an apt description not of the experiment itself but of the significance of the result. Why?

6–9 In uninformed discussions of satellites, one hears questions such as: "What keeps the satellite moving in its orbit?" and "What keeps the satellite up?" How do you answer these questions? Are your answers applicable to the moon?

6–10 A certain centrifuge is claimed to operate at $100,000\,g$. What does this mean?

EXERCISES

Section 6–1 Forces in Circular Motion

6–1 A stone of mass 1 kg is attached to one end of a string 1 m long, of breaking strength 500 N, and is whirled in a horizontal circle on a frictionless table top. The other end of the string is kept fixed. Find the maximum velocity the stone can attain without breaking the string.

6–2 A flat (unbanked) curve on a highway has a radius of 240 m, and a car rounds the curve at a speed of $22\ \mathrm{m\cdot s^{-1}}$. What must be the minimum coefficient of friction to prevent sliding?

6–3 A highway curve of radius 1200 ft is to be banked so that a car traveling $50\ \mathrm{mi\cdot hr^{-1}}$ will not skid sideways even in the absence of friction. At what angle should it be banked?

6–4 A coin placed on a record of diameter 12 in. will revolve with the record when it is brought up to a speed of $33\frac{1}{3}\ \mathrm{rev\cdot min^{-1}}$, provided that the coin is not more than 4 in. from the axis.

a) What is the coefficient of static friction between the coin and the record?

b) How far from the axis can the coin be placed, without slipping, if the turntable rotates at $45\ \mathrm{rev\cdot min^{-1}}$?

6–5 One of the problems for humans living in outer space is the fact that they would be weightless. One way around this problem is to design space stations as shown in Fig. 6–13 and have them spin about their center at a constant rate. This would create a sort of artificial gravity. If the diameter of the space station were 1000 m, how many revolutions per minute would be needed for the artificial gravity acceleration to be $9.8\ \mathrm{m\cdot s^{-2}}$?

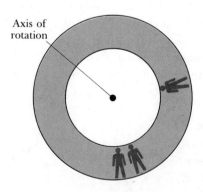

Axis of rotation

FIGURE 6–13

6–6 The "Giant Swing" at a county fair consists of a vertical central shaft with a number of horizontal arms attached at its upper end. Each arm supports a seat suspended from a cable 5 m long, with the upper end of the cable fastened to the arm at a point 4 m from the central shaft.

a) Find the time of one revolution of the swing if the cable supporting a seat makes an angle of 30° with the vertical.

b) Does the angle depend on the weight of the passenger for a given rate of revolution?

Section 6–2 Motion in a Vertical Circle

6–7 A cord is tied to a pail of water, and the pail is swung in a vertical circle of radius 1 m. What must be the minimum velocity of the pail at the highest point of the circle if no water is to spill from the pail?

6–8 A bowling ball weighing 16 lb is attached to the ceiling by a 15-ft rope. The ball is pulled to one side and released; it then swings back and forth as a pendulum. As it swings through the vertical, its velocity is 12 ft·s^{-1}.

a) What is the centripetal acceleration, in magnitude and direction, at this instant?

b) What is the tension in the rope at this instant?

6–9 A stunt pilot who has been diving vertically at a velocity of 180 m·s^{-1} pulls out of the dive by changing his course to a circle in a vertical plane.

a) What is the minimum radius of the circle for the acceleration at the lowest point not to exceed "7g"?

b) What is the apparent weight of a 80-kg pilot at the lowest point of the pullout?

Section 6–3 Newton's Law of Gravitation

6–10 A communications satellite of mass 200 kg is in a circular orbit of radius 40,000 km measured from the center of the earth. What is the gravitational force on the satellite, and what fraction is this of its weight at the surface of the earth?

6–11 What is the ratio of the gravitational pull of the sun on the moon to that of the earth on the moon? Use the data in Appendix F. What is the significance of the result to the motion of the moon?

6–12 The mass of the moon is about one eighty-first, and its radius one fourth, that of the earth. What is the acceleration due to gravity on the surface of the moon?

6–13 In an experiment using the Cavendish balance to measure the gravitational constant G, it is found that a mass of 0.8 kg attracts another sphere of mass 0.004 kg with a force of 13×10^{-11} N when the distance between the centers of the spheres if 0.04 m. The acceleration of gravity at the earth's surface is 9.80 m·s^{-2}, and the radius of the earth is 6400 km. Compute the mass of the earth from these data.

6–14 A 0.1-kg point mass is located on the line between a 5-kg and a 10-kg point mass. It is 4 m to the right of the 5-kg mass and 6 m to the left of the 10-kg mass. What are the magnitude and direction of the force on the 0.1-kg mass?

Section 6–4 Gravitational Field

6–15 What is the magnitude of the gravitational field 3 m from a 5-kg point mass?

6–16 What are the magnitude and direction of the gravitational force on a 0.1-kg point mass placed at a point where the gravitational field has components $g_x = 4$ m·s^{-2} and $g_y = -8$ m·s^{-2}?

6–17 The gravitational force F on a 0.01-kg test mass is found to be $F = (-0.1 \text{ N})i + (0.4 \text{ N})j$ at a certain point. What are the components of the gravitational field vector at that point?

6–18 What are the magnitude and direction of the gravitational field at a point midway between two point masses, one of mass m_1 and the other of mass m_2 where $m_2 > m_1$? The two point masses are a distance d apart.

Section 6–5 Satellite Motion

6–19 To launch a satellite in a circular orbit 1000 km above the surface of the earth, what orbital speed must be imparted to the satellite?

6–20 What is the period of revolution of a manmade satellite of mass m that is orbiting the earth in a circular path of radius 8000 km (about 1620 km above the surface of the earth)?

6–21 An earth satellite rotates in a circular orbit of radius 7000 km (about 620 km above the earth's surface) with an orbital speed of 7550 m·s^{-1}.

a) Find the time of one revolution.

b) Find the acceleration of gravity at the orbit.

Section 6–6 Effect of the Earth's Rotation on g

6–22 The weight of a man as determined by a spring balance at the equator is 800 N. By how much does this differ from the true force of gravitational attraction at the same point?

6–23 The acceleration due to gravity at the north pole of Jupiter is approximately 25 m·s^{-2}. Jupiter has mass 1.9×10^{27} kg, radius 7.1×10^7 m, and makes one revolution about its axis in about 10 hr.

a) What is the gravitational force on a 2-kg object at the north pole of Jupiter?

b) What is the apparent weight of this same object at the equator of Jupiter?

PROBLEMS

6–24 You are riding in a bus. As the bus rounds a flat curve of radius 60 m, a horseshoe of mass 0.5 kg suspended from the ceiling of the bus by a 2-m-long string is found to be in equilibrium when the string makes an angle of 37° with respect to the vertical. What is the speed v of the bus?

6–25 Consider a roadway banked as in Example 6–4, where there is a coefficient of static friction of 0.35 and a coefficient of kinetic friction of 0.25 between the tires and the roadway. The radius of the curve is $R = 25$ m.

a) If the banking angle is $\theta = 25°$, what is the *maximum* velocity the automobile can have before sliding *up* the banking?

b) What is the *minimum* velocity the automobile can have before sliding *down* the banking?

6–26 A curve of 200-m radius on a level road is banked at the correct angle for a velocity of 15 m·s^{-1}. If an automobile rounds this curve at 30 m·s^{-1}, what is the minimum coefficient of friction between tires and road so that the automobile will not skid? Assume all forces to act at the center of gravity.

6–27 In the ride "Spindletop" at the Six Flags Over Texas amusement park, people stand against the inner wall of a hollow vertical cylinder of radius 3 m. The cylinder starts to rotate, and when it reaches a constant rotation rate of 0.5 rev·s^{-1}, the floor on which people are standing drops about 0.5 m. The people remain pinned against the wall.

a) Draw a force diagram for a person in this ride, after the floor has dropped.

b) What minimum coefficient of static friction is required if the person in the ride is not to slide downward to the new position of the floor?

c) Does your answer in (b) depend on the mass of the passenger?

(*Note.* When the ride is over, the cylinder is slowly brought to rest; as it slows down, people slide down the walls to the floor.)

6–28 The 4-kg block in Fig. 6–14 is attached to a vertical rod by means of two strings. When the system rotates about the axis of the rod, the strings are extended as shown in the diagram.

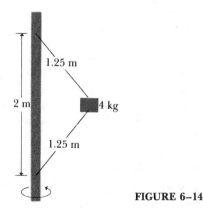

FIGURE 6–14

a) How many revolutions per minute must the system make for the tension in the upper string to be 60 N?

b) What is then the tension in the lower string?

6–29 A bead can slide without friction on a circular hoop of radius 0.1 m in a vertical plane. The hoop rotates at a constant rate of 2 rev·s^{-1} about a vertical diameter, as in Fig. 6–15.

a) Find the angle θ at which the bead is in vertical equilibrium. (Of course, it has a radial acceleration toward the axis.)

b) Is it possible for the bead to "ride" at the same elevation as the center of the hoop?

c) What will happen if the hoop rotates at 1 rev·s^{-1}?

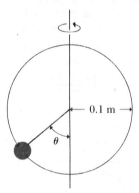

FIGURE 6–15

6–30 A physics major is working to pay his college tuition by performing in a traveling carnival. He rides a motorcycle inside a hollow steel sphere of radius 15 m. The surface of the sphere is full of holes so the audience can see in. After gaining sufficient speed, he travels in a vertical circle. The physics major has mass 60 kg, and his motorcycle has mass 40 kg.

a) What minimum velocity must he have at the top of the circle if the tires of the motorcycle are not to lose contact with the steel sphere?

b) At the bottom of the circle his velocity is twice the value calculated in (a). What then is the magnitude of the normal force exerted on the motorcycle by the steel sphere at this point?

6–31 The radius of a Ferris wheel is 5 m, and it makes one revolution in 10 s.

a) Find the difference between the apparent weight of a passenger at the highest and lowest points, expressed as a fraction of his weight. (That is, find the difference between the upward force exerted on the passenger by the seat at these two points.)

b) What would be the time for one revolution if his apparent weight at the highest point were zero?

c) What would then be his apparent weight at the lowest point?

d) What would happen, at this rate of revolution, if his seat belt broke at the highest point, and if he did not hang onto his seat?

6–32 A ball is held at rest at position A in Fig. 6–16 by two light cords. The horizontal cord is cut, and the ball starts swinging as a pendulum. Point B is the farthest to the right that the ball goes as it swings back and forth. What is the ratio of the tension in the supporting cord in position B to what it was at A before the horizontal cord was cut?

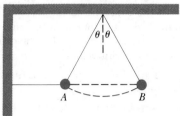

FIGURE 6–16

6–33 A spaceship travels from Earth directly toward the sun. At what distance from the center of the earth do the gravitational forces of the sun and the earth on the ship exactly cancel? Use the data in Appendix F.

6–34 Two spheres, each of mass 6.4 kg, are fixed at points A and B (Fig. 6–17). Find the magnitude and direction of the initial acceleration of a sphere of mass 0.010 kg if released from rest at point P and acted on only by forces of gravitational attraction of the spheres at A and B.

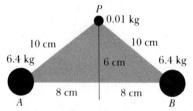

FIGURE 6–17

6–35 Three point masses are fixed at positions shown in Fig. 6–18.

a) Calculate the x- and y-components of the gravitational field at point P, which is at the origin of coordinates, due to these three point masses.

b) What would be the magnitude and direction of the force on a 0.01-kg mass placed at P?

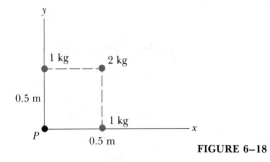

FIGURE 6–18

6–36 Two point masses are located at (x,y) coordinates as follows: 4 kg at (1 m, 0) and 3 kg at (0, −0.5 m).

a) What are the components of the gravitational field at the origin, due to these two point masses?

b) What will be the magnitude and direction of the gravitational force on a 0.01-kg test mass placed at the origin?

6–37 The asteroid Toro was discovered in 1964. Its radius is about 5 km.

a) Assuming the density (mass per unit volume) of Toro is the same as that of the earth, find its total mass, and find the acceleration due to gravity at its surface.

b) Suppose an object is to be placed in a circular orbit around Toro, with a radius just slightly larger than the asteroid's radius. What is the speed of the object? Could you launch yourself into orbit around Toro by running?

6–38 Several satellites are moving in a circle in the earth's equatorial plane. They are at such a height above the earth's surface that they always remain above the same point. Find the height of these satellites. (Such an orbit is said to be geosynchronous.)

6–39 What would be the length of a day (that is, the time required for one revolution of the earth on its axis) if the rate of revolution of the earth were such that $g = 0$ at the equator?

CHALLENGE PROBLEMS

6–40 Mass M is distributed uniformly along a line of length $2L$. Calculate the gravitational field components perpendicular and parallel to the line at a point that is at a distance a from the line, on its perpendicular bisector. Does your result reduce to the correct expression as a becomes very large?

6–41 Mass M is distributed uniformly over a disk of radius a. Find the gravitational force (magnitude and direction) between this disk-shaped mass and a point mass m located a distance x above the center of the disk. Does your result reduce to the correct expression as x becomes very large? (*Hint:* Divide the disk into infinitesimally thin concentric rings, use the expression derived in Section 6–4 for the

gravitational field due to each ring, and integrate to find the total field.)

6–42 Mass M is distributed uniformly over a thin spherical shell of radius R. Calculate the gravitational field at a distance r from the center of the shell for

a) $r > R$;   b) $r < R$.

(*Hint:* Divide the spherical shell into infinitesimally thin concentric rings, use the expression derived in Section 6–4 for the gravitational field due to each ring, and integrate to find the total field.)

6–43 A small block of mass m is placed inside an inverted cone that is rotating about a vertical axis such that the time for one revolution of the cone is τ (see Fig. 6–19). The walls of the cone make an angle θ with the vertical. The coefficient of static friction between the block and the cone is μ_s. If the block is to remain at a constant height h above the apex of the cone, what are the maximum and minimum values of τ?

6–44 A small block of mass m on a frictionless horizontal table is attached to a spring of unstretched length l_0. The other end of the spring is looped around a vertical rod at the center of the table, so that the block and spring can rotate freely about the rod. The spring has negligible mass and force constant k, defined by $F_{\text{spring}} = -k(l - l_0)$ where l is the length of the stretched spring. The block is set into uniform circular motion with velocity of magnitude v.

a) Calculate the length l of the spring as a function of the speed v of the block.

b) Express your result in (a) as the length l of the spring in terms of the frequency f of revolution of the block, where f is the number of revolutions it makes each second. Make a sketch of l as a function of f.

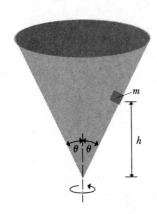

FIGURE 6–19

PART TWO

MECHANICS—FURTHER DEVELOPMENTS

PERSPECTIVE

In the opening chapters of this book we developed mathematical language for describing the position and motion of a particle, using the concepts of displacement, velocity, and acceleration. We used the particle as a simplified *model* to represent any body small enough to be considered as a point, or whose shape is irrelevant in the problem at hand. With the aid of two additional concepts, *force* and *mass*, we have also been able to state general principles relating the motion of a particle to the forces acting on it. We learned how to determine the motion of a particle under the action of given forces and how to determine the forces when the motion is given. In particular, we studied in detail the motion of a body in a uniform gravitational field, including free fall and parabolic trajectories, and we applied Newton's laws to the motion of a variety of other simple mechanical systems. With the aid of one additional principle, the *law of gravitation*, we can also analyze in detail the motions of planets around the sun and of satellites around the earth.

However, there are many problems where it is not feasible to apply Newton's laws directly, either because the motion is too complicated to describe in detail or because the forces are not known. An example is a collision between two bodies, where large, variable, and unknown forces act on the bodies during their interaction. To discuss this and similar situations, we need additional principles that will enable us to relate the motion of a system at one time to its motion at a later time, without having to know all the details of what happened in between. The concepts of *energy* and *momentum*, which we study next, enable us to formulate such principles and the related *conservation laws*. These concepts have the added significance that their relevance extends to other areas of physics, such as thermodynamics, electromagnetism, and optics;

thus they provide unifying threads that help to integrate the whole fabric of physics.

Next we broaden the scope of our mechanical principles to include bodies that cannot be represented adequately as particles. In particular, we consider in detail the rotational motion of an extended body. We introduce the idealized model of a *rigid body:* a body that does not deform or change shape when forces are applied. We develop kinematic language for describing this motion and the dynamical principles that relate angular motion to the forces acting on the body. These principles also allow us to deal with a broader range of equilibrium problems than we could handle using only particle models. After this we undertake a detailed study of *periodic motion*, a class of motions in which a physical system repeats the same motion over and over again in a definite cycle. Periodic motion has far-reaching applications that go beyond mechanics and extend into nearly all other areas of physics. The study of periodic motion prepares us for later work in analyzing mechanical waves, electrical circuits, optics, and many other areas of physics. Once again, a concept developed in a mechanical context serves as a unifying theme, tying together several areas of physics.

In the opening chapters, we relied heavily on experimental evidence as the basis for the general principles we call Newton's laws. Our approach there was largely *inductive*, generalizing from specific observations to general principles. In the next few chapters our process is more *deductive*, deriving additional results from Newton's laws. Thus these chapters have a somewhat different flavor from that of the opening chapters. Taken together, these first eleven chapters show the vital interplay between inductive and deductive reasoning in the development of physical science.

7
WORK AND ENERGY

ENERGY IS ONE OF THE MOST IMPORTANT UNIFYING CONCEPTS IN ALL OF physical science. Its importance stems from the principle of **conservation of energy,** which states that in any isolated system the total energy of all forms is constant. In this chapter we concentrate on mechanical energy—energy associated with the motion and position of mechanical systems. We begin by introducing the concepts of work and kinetic energy and the relationship between them, and we use these in a variety of problems. Potential energy provides a convenient way to calculate work associated with certain kinds of forces. Finally, we introduce the term *power* to characterize the rate of doing work or transferring energy in or out of a system.

7–1 CONSERVATION OF ENERGY

The concept of **energy** appears throughout every area of physics, and yet it is difficult to define in a general way just what energy is. Energy plays a central role in one of a class of fundamental laws of nature called **conservation laws,** and looking at this role is as good a way as any to approach the question of what energy is. The various conservation laws of physics, including conservation of energy, momentum, angular momentum, electric charge, and others, play vital roles in every area of physics and help to unify the subject.

A conservation law always concerns a transformation or an interaction that occurs within some physical system or between a system and its surroundings. The state or condition of the system at any time is described by physical quantities such as position and velocity. These quantities may change during the transformation or interaction, but one or more quantities may remain constant, or be *conserved.* A familiar example is conservation of *mass* in chemical reactions. A large amount of experimental evidence has established that the total mass of the reactants in a chemical reaction is always equal to the total mass of all the products of the reaction. That is, the total mass is always the same after the reaction occurs as before. This generalization is called the principle of *conservation of mass,* and it is obeyed in all chemical reactions.

Conservation laws play a vital role in all of physical science.

Conservation of mass: Matter cannot be created or destroyed.

145

Kinetic energy is energy associated with mass in motion.

Something similar happens in collisions between bodies. For a body of mass m moving with speed v, we can define a quantity $\frac{1}{2}mv^2$ associated with the motion and called the *kinetic energy* of the body. When two highly elastic or "springy" bodies (such as two hard steel ball bearings) collide, we find that the individual speeds change, but the total kinetic energy (the sum of the $\frac{1}{2}mv^2$ quantities for all the colliding bodies) is the same after the collision as before. We say that kinetic energy is *conserved* in such collisions. This result does not tell us what kinetic energy *is*, but only that it is useful in representing a conservation principle in certain kinds of interactions.

In contrast, when two soft, deformable bodies, such as two balls of putty or chewing gum, collide, experiment shows that kinetic energy is *not* conserved. However, something else happens: The bodies become *warmer*. Furthermore, it turns out that there is a definite relationship between the temperature rise of the material and the loss of kinetic energy. We can define a new quantity, *internal energy*, that increases with temperature in a definite way, so that the *sum* of kinetic energy and internal energy is conserved in these collisions.

Conservation of energy can be generalized to include energy associated with every kind of physical phenomenon.

The significant discovery here is that it is *possible* to extend the principle of conservation of energy to a broader class of phenomena by defining a new form of energy. This is precisely how the principle has developed. In any situation where it seems that the total energy in all known forms is *not* conserved, it has been found possible to define a new form of energy so that the total energy, including the new form, *is* conserved. Thus we find energy associated with heat, with elastic deformations, with electric and magnetic fields, and, in relativity theory, even with mass itself. Conservation of energy has the status, along with a small number of partners, of a *universal* conservation principle: No exception to its validity has ever been found.

In this chapter we are concerned primarily with *mechanical* energy, that is, with energy associated with motion, position, and deformation of material bodies. In some interactions mechanical energy is conserved, and in others there is a conversion from mechanical energy to other forms of energy, or the reverse, so that mechanical energy by itself is not conserved. In Chapters 15 and 18 we will study in greater detail the relation of mechanical energy to heat, and in later chapters still other forms of energy will be considered.

7–2 WORK

A force acting on a body does work on the body only when the body moves.

In everyday life *work* is any activity that requires muscular or mental exertion. Physicists use the term in a much more specific sense, involving a *force* acting on a body while the body undergoes a *displacement*. When a body moves a distance s along a straight line while a constant force of magnitude F, directed along the line, acts on it, the **work** W done by the force is defined as

$$W = Fs. \tag{7-1}$$

In this chapter we will develop a very useful relationship between work and quantities that describe the *motion* of the body.

The force need not have the same direction as the displacement. In Fig. 7–1, the force F, assumed constant, makes an angle θ with the displacement. The work W done by this force when its point of application undergoes a displacement s is defined as the product of the magnitude of the displacement and the *component* of the force in the direction of the displacement.

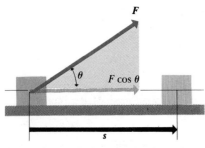

7–1 The work done by the force F during a displacement s is $(F \cos \theta)s$.

The component of F in the direction of s is $F \cos \theta$. Then

$$W = (F \cos \theta)s. \tag{7–2}$$

An alternative interpretation of Eq. (7–2) is that $s \cos \theta$ is the component of displacement in the direction of F. Thus the work is also the component of displacement in the direction of F multiplied by the magnitude of F.

Comparing Eq. (7–2) with the definition of the *scalar product* of two vectors, Eq. (1–19), we see that work can also be expressed compactly as

$$W = F \cdot s. \tag{7–3}$$

Although it is calculated from two vector quantities, work itself is a *scalar* quantity. A 5-N force toward the east acting on a body that is displaced 6 m to the east does exactly the same amount of work as a 5-N force toward the north acting on a body displaced 6 m to the north. Work is an *algebraic* quantity: It can be positive or negative. When the component of the force is in the *same* direction as the displacement, the work W is *positive*. When it is *opposite* to the displacement, the work is *negative*. If the force is at *right angles* to the displacement, it has no component in the direction of the displacement, and the work is *zero*.

Thus, when a body is lifted, the work done by the lifting force is positive; when a spring is stretched, the work done by the stretching force is positive; when a gas is compressed in a cylinder, the work done by the compressing force is positive. On the other hand, the work done by the *gravitational* force on a body being lifted is negative, since the (downward) gravitational force is opposite to the (upward) displacement. It is considered "hard work" to hold a heavy object stationary at arm's length, but no work is done in the technical sense because there is no motion! Even if you walk along a level floor while carrying the object, no work would be done, because the (vertical) supporting force has no component in the direction of the horizontal motion. Similarly, when a body slides along a surface, the work done by the normal force acting on the body is zero, and when a body moves in a circle, the work done by the centripetal force on the body is also zero.

The unit of work in any system is the unit of force multiplied by the unit of distance. In SI units the unit of force is the newton and the unit of distance is the meter; thus in this system the unit of work is 1 *newton meter* (1 N·m). This combination of units appears so frequently in mechanics that it is given a special name, the **joule** (abbreviated J).

$$1 \text{ joule} = (1 \text{ newton})(1 \text{ meter}) \qquad \text{or} \qquad 1 \text{ J} = 1 \text{ N·m}.$$

In the cgs system, the unit of work is 1 dyne-centimeter, also called 1 *erg*. Because 1 dyn = 10^{-5} N and 1 cm = 10^{-2} m,

$$1 \text{ erg} = 10^{-7} \text{ J}.$$

In the British system the unit of work is 1 *foot-pound* (1 ft-lb); no special name is given this unit. The following conversions are useful:

$$1 \text{ J} = 0.7376 \text{ ft·lb}, \qquad 1 \text{ ft·lb} = 1.356 \text{ J}.$$

When several external forces act on a body, it is useful to consider the work done by each separate force. Each of these may be computed from the definition of work in Eq. (7–2). Then, since work is a scalar quantity, the total work is the algebraic sum of the individual works.

Work is a scalar quantity, even though force and displacement are vectors.

When force and displacement are perpendicular, the force does no work.

The joule is the SI unit of work and of energy of all kinds.

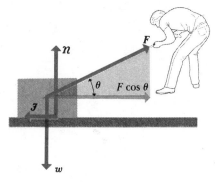

7–2 A box on a rough horizontal surface moving to the right under the action of a force **F** inclined at an angle θ.

When several forces act on a moving body, the total work can be calculated in several ways.

EXAMPLE 7–1   Figure 7–2 shows a box being dragged along a horizontal surface by a worker applying a constant force **F** that makes a constant angle θ with the direction of motion. The other forces on the box are its weight w, the upward normal force $\mathcal{n}$ exerted by the surface, and the friction force $\mathcal{F}$. What is the work done by each force when the box moves a distance s along the surface to the right?

SOLUTION   The component of **F** in the direction of motion is $F \cos \theta$. The work done by the force **F** is therefore

$$W_F = (F \cos \theta)s.$$

The forces w and $\mathcal{n}$ are both at right angles to the displacement. Hence

$$W_w = 0, \qquad W_n = 0.$$

The friction force $\mathcal{F}$ is opposite to the displacement, so the work done by the friction force is

$$W_{\mathcal{F}} = -\mathcal{F}s.$$

Since work is a scalar quantity, the total work ΣW done by all forces on the box is the *algebraic* sum (not the vector sum) of the individual works:

$$\Sigma W = W_F + W_w + W_n + W_{\mathcal{F}}$$
$$= (F \cos \theta)s + 0 + 0 - \mathcal{F}s$$
$$= (F \cos \theta - \mathcal{F})s.$$

But $(F \cos \theta - \mathcal{F})$ is the *resultant* force on the box. Hence *the total work done by all forces is equal to the work done by the resultant force.*

Suppose that $F = 50$ N, $\mathcal{F} = 15$ N, $\theta = 36.9°$, and $s = 20$ m. Then

$$W_F = (F \cos \theta)s$$
$$= (50 \text{ N})(0.800)(20 \text{ m}) = 800 \text{ N·m},$$

$$W_{\mathcal{F}} = -\mathcal{F}s$$
$$= (-15 \text{ N})(20 \text{ m}) = -300 \text{ N·m},$$

$$\Sigma W = W_F + W_{\mathcal{F}} = 500 \text{ N·m}.$$

As a check, the total work may be expressed as

$$\Sigma W = (F \cos \theta - \mathcal{F})s$$
$$= (40 \text{ N} - 15 \text{ N})(20 \text{ m}) = 500 \text{ N·m}.$$

When several forces act on a body, there are always two equivalent ways to calculate the total work. We may calculate the work done by each force separately and take the algebraic sum of these works, or we may compute the vector sum or resultant of the forces and compute the work done by the resultant.

7–3 WORK DONE BY A VARYING FORCE

In the preceding section we defined the work done by a *constant* force. Often, however, work is done by a force that varies in magnitude or direction during the displacement of the body on which it acts. Thus when a spring is stretched slowly, the force required to stretch it increases steadily as the spring elongates; when a body is projected vertically upward, the gravitational force exerted on it by the earth decreases inversely with the square of its distance from the earth's center.

Suppose a particle moves along a line under the action of a force directed along the line but varying with the particle's position. In Fig. 7–3a the force magnitude is shown as a function of the particle's coordinate x. To find the work done by this force, we divide the displacement into short segments Δx_1, Δx_2, and so on, as in Fig. 7–3b. We approximate the varying force by one that is constant within each segment. The force then has approximately the value F_1 in segment Δx_1, F_2 in segment Δx_2, and so on. The work done in the first segment is then $F_1\,\Delta x_1$, that in the second is $F_2\,\Delta x_2$, and so on. The *total* work is

$$W = F_1\,\Delta x_1 + F_2\,\Delta x_2 + \cdots.$$

As the number of segments becomes very large and the size of each very small, this sum becomes (in the limit) the *integral* of F from x_1 to x_2:

$$W = \int_{x_1}^{x_2} F\,dx. \tag{7–4}$$

Note that $F_1\,\Delta x_1$ represents the *area* of the first vertical strip in Fig. 7–3b, and so on, and that the integral in Eq. (7–4) represents the area under the curve in Fig. 7–3a.

If the force also varies in *direction* during the displacement, then F in Eq. (7–4) must be replaced by the *component* of force in the direction of displacement. This is given by $F \cos\theta$, where θ is the angle between F and the x-axis at each point. The angle θ may vary from point to point. Then we have

$$W = \int_{x_1}^{x_2} F \cos\theta\,dx. \tag{7–5}$$

As an example of this method, let us compute the work done when a spring is stretched. To keep a spring stretched at an elongation x beyond its unstretched length, we must apply a force F at each end, as shown in Fig. 7–4. If the elongation is not too great, F is directly proportional to x:

$$F = kx, \tag{7–6}$$

where k is a constant called the **force constant,** or the *stiffness,* of the spring. This direct proportion between force and elongation, for elongations that are not too great, was discovered by Robert Hooke in 1678 and is known as **Hooke's law.** We will discuss it more fully in Chapter 12.

Suppose that equal and opposite forces are applied to the ends of the spring, and that the forces are gradually increased from zero. One end of the spring is held stationary; no work is done by the force at this end, but work is

When the force acting on a body varies during the displacement, the work can be expressed as an integral.

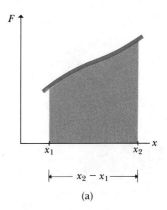

(a)

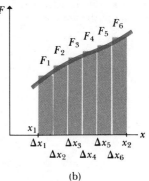

(b)

7–3 (a) Curve showing how F varies with x. (b) If the area is partitioned into small rectangles, their sum approximates the total work done during the displacement; the greater the number of rectangles used, the closer is the approximation.

Hooke's law: The force required to stretch a spring is proportional to the amount of stretch.

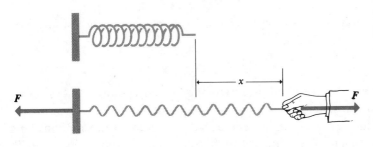

7–4 The force to stretch a spring is proportional to its elongation: $F = kx$.

The work required to stretch a spring is proportional to the square of the amount of stretch.

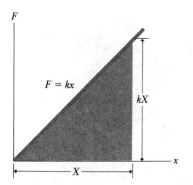

7–5 The work done in stretching a spring is equal to the area of the shaded triangle.

Bathroom scales often use springs to measure weight.

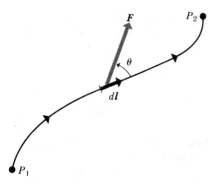

7–6 A particle moves along a curved path from point P_1 to P_2, acted on by a force **F** that varies in magnitude and direction. During an infinitesimal displacement d**l** (a small segment of the path), the work dW done by the force is given by $dW = \boldsymbol{F} \cdot d\boldsymbol{l}$.

done by the varying force $\boldsymbol{F}$ at the moving end. In Fig. 7–5, F is plotted vertically, and the displacement x of the moving end, representing the elongation of the spring, is plotted horizontally.

The work required to stretch the spring from $x = 0$ (no elongation) to $x = X$ is

$$W = \int_0^X F\,dx = \int_0^X kx\,dx = \tfrac{1}{2}kX^2. \tag{7-7}$$

This result can also be obtained graphically: The area of the shaded triangle in Fig. 7–5, representing the total work, is equal to half the product of base and altitude, or

$$W = \tfrac{1}{2}(X)(kX) = \tfrac{1}{2}kX^2,$$

in agreement with the above result.

The work is therefore proportional to the *square* of the final elongation X. When the elongation is doubled, the work increases by a factor of four. Another view of this relationship is that the total work W is equal to the *average* force $\tfrac{1}{2}kX$ multiplied by the total displacement X.

The spring also exerts a force on the hand, which moves during the stretching process, and we may ask what work the spring does on the hand. The displacement of the hand is the same as that of the moving end of the spring, but the force on it is the negative of the force on the spring because the two forces form an action–reaction pair. Thus the work done on the hand is the negative of the work done on the spring, namely, $-\tfrac{1}{2}kX^2$. In work calculations, it is always essential to specify on which body the work is being done!

In some cases, such as a spring with spaces between the coils, Hooke's law holds for compression as well as stretching. In this case, the work done on the spring in compressing it a distance X is also given by Eq. (7–7).

EXAMPLE 7–2   A woman weighing 600 N steps on a bathroom scale containing a heavy spring. The spring is compressed 1.0 cm under her weight. Find the force constant of the spring and the total work done on it during the compression.

SOLUTION   From Eq. (7–6),

$$k = \frac{F}{x} = \frac{600 \text{ N}}{0.01 \text{ m}} = 60,000 \text{ N} \cdot \text{m}^{-1}.$$

Then from Eq. (7–7),

$$W = \tfrac{1}{2}kX^2 = \tfrac{1}{2}(60,000 \text{ N} \cdot \text{m}^{-1})(0.01 \text{ m})^2 = 3 \text{ N} \cdot \text{m} = 3 \text{ J}.$$

The work done on a spring in changing its elongation or compression from x_1 to x_2 is

$$W = \tfrac{1}{2}kx_2{}^2 - \tfrac{1}{2}kx_1{}^2,$$

provided that $x = 0$ corresponds to zero elongation or compression, as above. (Can you prove this?)

The definition of work can be generalized further to include motion along a *curved* path. In Fig. 7–6, suppose the particle moves from point P_1 to P_2

along the curve. We imagine dividing the portion of the curve between these points into many infinitesimal vector displacements, and we call a typical one of these dl. Each dl is tangent to the path at its position. Let F be the force at a typical point along the path, and θ the angle between F and dl at this point. Then the small element of work dW done on the particle during the displacement dl may be written as

$$dW = F \cos \theta \, dl = F_\| \, dl = \boldsymbol{F} \cdot d\boldsymbol{l},$$

where as before $F_\| = F \cos \theta$ is the component of F in the direction of dl. The total work done on the particle by F as the particle moves from P_1 to P_2 is then represented symbolically as

$$W = \int_{P_1}^{P_2} F \cos \theta \, dl = \int_{P_1}^{P_2} F_\| \, dl = \int_{P_1}^{P_2} \boldsymbol{F} \cdot d\boldsymbol{l}. \qquad (7\text{–}8)$$

This integral is called a *line integral*. Actually evaluating Eq. (7–8) in a specific problem requires that we have some sort of detailed description of the path and of the variation of F along the path.

When a particle moves along a curved path, the work done on it can be expressed as a line integral.

EXAMPLE 7–3 A child of weight w sits on a swing of length R, as shown in Fig. 7–7. A *variable* horizontal force P, which starts at zero and gradually increases, is used to pull the child very slowly (so that equilibrium exists at all times) until the ropes make an angle θ_0 with the vertical. Calculate the work of the force P.

Work done while pushing a child on a swing: an example of motion along a curved path

SOLUTION The body is in equilibrium, so the sum of the horizontal forces equals zero:

$$P = T \sin \theta.$$

The sum of the vertical forces is also zero:

$$w = T \cos \theta.$$

Dividing these two equations, we find

$$P = w \tan \theta.$$

The point of application of P swings through the arc s. Since $s = R \theta$, $dl = R \, d\theta$, and

$$
\begin{aligned}
W &= \int \boldsymbol{P} \cdot d\boldsymbol{l} = \int P \cos \theta \, dl \\
&= \int_0^{\theta_0} w \tan \theta \cos \theta \, R \, d\theta \\
&= wR \int_0^{\theta_0} \sin \theta \, d\theta = wR \, (1 - \cos \theta_0). \qquad (7\text{–}9)
\end{aligned}
$$

If $\theta_0 = 0$, there is no displacement; in that case $\cos \theta_0 = 1$ and $W = 0$, as expected. If $\theta_0 = 90°$, then $\cos \theta_0 = 0$ and $W = wR$. In that case the work is the same as though the body had been lifted straight up a distance R by a force equal to its weight w. In fact, as you should verify, the quantity $R(1 - \cos \theta_0)$ is the increase in the height of the body during the displacement; so for any value of θ_0, the work done by force P is the change in height multiplied by the weight. We will prove this result more generally in Section 7–5.

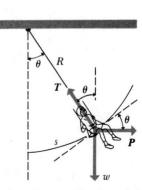

7–7 A variable horizontal force P acts on a body while the displacement varies from zero to s.

7–4 WORK AND KINETIC ENERGY

The work done on a body by a force is related very directly to the resulting change in the body's motion. To develop this relationship, we consider first a body of mass m moving along a straight line under the action of a constant resultant force of magnitude F directed along the line. The body's acceleration is given by Newton's second law, $F = ma$. Suppose the speed increases from v_1 to v_2 while the body undergoes a displacement $s = x_2 - x_1$. Then, from the analysis of motion with constant acceleration, and from Eq. (2–13), we have

$$v_2{}^2 = v_1{}^2 + 2as,$$
$$a = \frac{v_2{}^2 - v_1{}^2}{2s}.$$

Hence

$$F = m\,\frac{v_2{}^2 - v_1{}^2}{2s},$$

and

$$Fs = \tfrac{1}{2}mv_2{}^2 - \tfrac{1}{2}mv_1{}^2. \tag{7–10}$$

The product Fs is the work W done by the resultant force F. The quantity $\tfrac{1}{2}mv^2$, one-half the product of the mass of the body and the square of its speed, is called its **kinetic energy,** K:

$$K = \tfrac{1}{2}mv^2. \tag{7–11}$$

The work done on a body equals its change in kinetic energy.

The first term on the right side of Eq. (7–10) is the final kinetic energy of the body, $K_2 = \tfrac{1}{2}mv_2{}^2$, and the second term is the initial kinetic energy, $K_1 = \tfrac{1}{2}mv_1{}^2$. The difference between these terms is the *change* in kinetic energy, and we have the important result that *the work done by the resultant external force on a body is equal to the change in kinetic energy of the body:*

$$W = K_2 - K_1 = \Delta K. \tag{7–12}$$

Kinetic energy is a scalar quantity, even though the particle's velocity is a vector quantity.

Kinetic energy, like work, is a *scalar* quantity. The kinetic energy of a moving body depends only on its speed (the *magnitude* of its velocity) but not on the *direction* in which it is moving. The *change* in kinetic energy depends only on the work $W = Fs$ and not on the individual values of F and s. That is, the force F could have been large and the displacement s small, or the reverse. If the mass m and the speeds v_1 and v_2 are known, the work done by the resultant force can be found without any knowledge of the force F and the displacement s.

If the work W is *positive*, the final kinetic energy is greater than the initial kinetic energy and the kinetic energy *increases*. If the work is *negative*, the kinetic energy *decreases*. In the special case in which the work is *zero*, the kinetic energy remains *constant*.

In Eq. (7–12), W is the work done by the *resultant* force, that is, the *vector sum* of all the forces acting on the body. Alternatively, one may calculate the work done by each separate force. W is then the *algebraic sum* of all these quantities of work. Example 7–1 (Section 7–2) illustrates these two alternatives.

Although we derived Eq. (7–12) for the special case of a constant resultant force, it is true even when the force varies in an arbitrary way. The work done on a particle by any resultant force equals the change in kinetic energy of the body. To prove this statement, we divide the total displacement x into a large number of small segments Δx, just as in the calculation of work done by a varying force. The change of kinetic energy in segment Δx_1 is equal to the work $F_1 \Delta x_1$, and so on; the total kinetic energy change is the sum of the changes in the individual segments and is thus equal to the total work done.

In SI units, m is measured in kilograms and v in meters per second. From Eq. (7–12), kinetic energy must have the same units as work. To verify this we recall that $1\ N = 1\ kg \cdot m \cdot s^{-2}$. Hence

The units of work and kinetic energy are the same.

$$1\ J = 1\ N \cdot m = 1\ kg \cdot m \cdot s^{-2} \cdot m = 1\ kg \cdot m^2 \cdot s^{-2}.$$

The joule is thus the SI unit of *both* work and kinetic energy and, as we will see later, of all kinds of energy. Similarly, in the cgs system, with m in grams and v in centimeters per second, we have

$$1\ erg = 1\ dyn \cdot cm = 1\ g \cdot cm \cdot s^{-2} \cdot cm = 1\ g \cdot cm^2 \cdot s^{-2}.$$

In the British system.

$$1\ ft \cdot lb = 1\ ft \cdot slug \cdot ft \cdot s^{-2} = 1\ slug \cdot ft^2 \cdot s^{-2}.$$

PROBLEM-SOLVING STRATEGY: *Work and kinetic energy*

1. Make a list of all the forces acting on the body, and calculate the work done by each force. In some cases one or more forces may be unknown; represent the unknowns by algebraic symbols. Be sure to check signs: When a force has a component in the *same* direction as the displacement, its work is positive; when it has a component in the direction opposite to the displacement, its work is negative.

2. Add the works done by the separate forces to find the total work. Again be careful with signs. Sometimes, though not often, it may be easier to calculate the resultant (vector sum) of the forces first, and then find the work done by the resultant force.

3. List the initial and final kinetic energies K_1 and K_2. If a quantity, such as v_1 or v_2 is unknown, leave it in terms of the corresponding algebraic symbol.

4. Use the relationship $W = K_2 - K_1$; insert the results from the above steps and solve for whatever unknown is required.

EXAMPLE 7–4 Consider again the box in Fig. 7–2 and the numbers in Example 7–1. We found that the total work done by the forces was $500\ N \cdot m = 500\ J$. Hence the kinetic energy of the box must increase by 500 J. Suppose the initial speed v_1 is $4\ m \cdot s^{-1}$ and the mass of the box is 10 kg. What is the final speed?

Pushing a box with increasing speed: an application of the work-energy relationship

SOLUTION The initial kinetic energy is

$$K_1 = \tfrac{1}{2}mv_1{}^2 = \tfrac{1}{2}(10\ kg)(4\ m \cdot s^{-1})^2 = 80\ J.$$

The final kinetic energy is given by Eq. (7–12):

$$K_2 = K_1 + W = 80\ J + 500\ J = 580\ J.$$

Thus

$$K_2 = \tfrac{1}{2}(10\ kg)v_2{}^2 = 580\ J$$

and

$$v_2 = 10.8\ m \cdot s^{-1}.$$

To verify this we may find the acceleration from $F = ma$ and then use the equations of motion with constant acceleration to find v_2:

$$a = \frac{F}{m} = \frac{40 \text{ N} - 15 \text{ N}}{10 \text{ kg}} = 2.5 \text{ m·s}^{-2}.$$

Then

$$v_2{}^2 = v_1{}^2 + 2as = (4 \text{ m·s}^{-1})^2 + 2(2.5 \text{ m·s}^{-2})(20 \text{ m})$$
$$= 116 \text{ m}^2\text{·s}^{-2},$$
$$v_2 = 10.8 \text{ m·s}^{-1}.$$

This is the same result we obtained with the work-energy approach, but there we avoided the intermediate step of finding the acceleration. Several examples and problems in this chapter can be done without using energy considerations but are easier when energy methods are used. Example 7–4 illustrates this point.

7–5 GRAVITATIONAL POTENTIAL ENERGY

When a gravitational force acts on a body while the body undergoes a vertical displacement, the force does work on the body. This work can be expressed conveniently in terms of the initial and final *positions* of the body. In Fig. 7–8a, a body having mass m and weight $w = mg$ moves vertically from a height y_1 above some reference level to a height y_2. The positive direction for y is upward. We assume that the body remains close enough to the earth's surface so that g and w can be considered constant. In the figure, P represents the result of all other forces on the body. The direction of w is opposite to the upward displacement, and the work done by this force is

$$W_{\text{grav}} = Fs = -w(y_2 - y_1) = -(mgy_2 - mgy_1). \tag{7–13}$$

Gravitational potential energy: a convenient way to calculate the work done by a gravitational force

You should verify that this expression also has the correct sign when the body moves *downward* (so y_2 is less than y_1) and also that it gives the correct resultant when y_1, or y_2, or both, are negative, corresponding to positions *below* the reference plane.

Thus we can express W_{grav} in terms of the values of the quantity mgy at the beginning and end of the displacement. This quantity, the product of the

7–8 Work done by the gravitational force w during the motion of an object from one point in a gravitational field to another. (a) A vertical displacement from height y_1 to y_2. (b) A displacement along a curved path. (c) The work done by the gravitational force w depends only on the vertical component of displacement Δy.

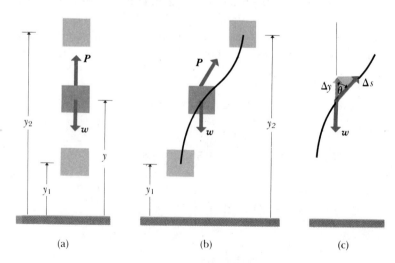

weight mg and the height y above the reference level (the origin of coordinates), is called the **gravitational potential energy,** U:

$$U = mgy. \qquad (7\text{--}14)$$

Later, when we also consider other kinds of potential energy, we will sometimes denote it as $U_{\text{grav}} = mgy$. The initial value of gravitational potential energy is $U_1 = mgy_1$ and the final value is $U_2 = mgy_2$. We can then express the work W_{grav} done by the gravitational force during the displacement from y_1 to y_2 as

$$W_{\text{grav}} = U_1 - U_2 = -\Delta U. \qquad (7\text{--}15)$$

Thus when the body moves *downward*, y decreases, the gravitational force does *positive* work, and the potential energy *decreases*. When the body moves *upward*, the work done by the gravitational force is *negative* and the potential energy *increases*.

The gravitational potential energy of a body depends on its position.

Note that if we shift the origin for y, then y_1 and y_2 change, but the difference $(y_1 - y_2)$ does not. Similarly, U_1 and U_2 change, but the difference $(U_1 - U_2)$ is the same as before. The choice of origin is arbitrary; the physically significant quantity is not the value of U at a particular point, but only the *difference* in U between two points.

Now let W_{other} represent the total work done by $\boldsymbol{P}$, that is, by all forces other than the gravitational force. The total work done by *all* forces is then $W = W_{\text{grav}} + W_{\text{other}}$. Since the total work equals the change in kinetic energy,

$$W_{\text{other}} + W_{\text{grav}} = K_2 - K_1 = \Delta K,$$
$$W_{\text{other}} - (mgy_2 - mgy_1) = (\tfrac{1}{2}mv_2{}^2 - \tfrac{1}{2}mv_1{}^2). \qquad (7\text{--}16)$$

The quantities $\tfrac{1}{2}mv_2{}^2$ and $\tfrac{1}{2}mv_1{}^2$ depend only on the final and initial *speeds;* the quantities mgy_2 and mgy_1 depend only on the initial and final *elevations*. Let us therefore rearrange this equation, transferring the quantities mgy_2 and mgy_1 from the "work" side of the equation to the "energy" side:

The work-energy theorem can be expressed in terms of gravitational potential energy and other work.

$$W_{\text{other}} = (\tfrac{1}{2}mv_2{}^2 - \tfrac{1}{2}mv_1{}^2) + (mgy_2 - mgy_1)$$
$$= \Delta K + \Delta U. \qquad (7\text{--}17)$$

The left side of Eq. (7–17) contains only the work done by the force $\boldsymbol{P}$. The terms on the right depend only on the final and initial states of the body (its speed and position). The first expression in parentheses on the right of Eq. (7–17) is the change in kinetic energy of the body, and the second is the change in its gravitational potential energy.

The sum of kinetic and potential energies is called the **total mechanical energy,** $E = K + U$. Equation (7–17) can also be written

Total mechanical energy is the sum of kinetic and potential energies.

$$W_{\text{other}} = (\tfrac{1}{2}mv_2{}^2 + mgy_2) - (\tfrac{1}{2}mv_1{}^2 + mgy_1)$$
$$= (K_2 + U_2) - (K_1 + U_1) = E_2 - E_1 = \Delta E. \qquad (7\text{--}18)$$

Hence *the work done by all forces acting on the body,* **with the exception of the gravitational force,** *equals the change in the total mechanical energy of the body.* If the work W_{other} is positive, the mechanical energy increases. If W_{other} is negative, the mechanical energy decreases.

In the special case where the *only* force on the body is the gravitational force, the work W_{other} is zero. Equation (7–18) can then be written as $E_1 = E_2$, or

$$K_1 + U_1 = K_2 + U_2,$$

or

$$\tfrac{1}{2}mv_1{}^2 + mgy_1 = \tfrac{1}{2}mv_2{}^2 + mgy_2. \qquad (7\text{--}19)$$

Total mechanical energy is constant (conserved) when the only work on the body is the work done by the gravitational force.

Under these conditions *the total mechanical energy is constant;* that is, it is *conserved.* This is a particular case of the principle of **conservation of mechanical energy.**

PROBLEM-SOLVING STRATEGY: Conservation of mechanical energy I

1. Decide what the initial and final states of the system are; use the subscript 1 for the initial state, 2 for the final state.

2. Define your coordinate system, particularly the level at which $y = 0$. This will be used to compute potential energies. Equation (7–14) assumes that the positive direction for y is upward; we suggest you use this choice consistently.

3. Make a list of the initial and final kinetic and potential energies, that is, K_1, K_2, U_1, and U_2. In general, some of these will be known, some unknown.

Use algebraic symbols for any unknown coordinates or velocities.

4. Identify all nongravitational forces that do work. A free-body diagram is often helpful. Calculate the work done by all these forces. If some of the needed quantities are unknown, represent them by algebraic symbols.

5. Use Eq. (7–16) or (7–17) to relate these quantities. If there is no nongravitational work, Eq. (7–19) may be used. Then solve to find whatever unknown quantity is required.

Throwing a ball in the air: a problem made simpler by use of conservation of energy

EXAMPLE 7–5 A man holds a ball of mass $m = 0.2$ kg at rest in his hand. He then throws the ball vertically upward. In this process, his hand moves up 0.5 m before the ball leaves his hand with an upward velocity of 20 m·s^{-1}. Discuss the motion of the ball from the work-energy standpoint, assuming $g = 10$ m·s^{-2}.

SOLUTION First, consider the throwing process. Take the reference level ($y = 0$) at the initial position of the ball. Then $K_1 = 0$, $U_1 = 0$. Take point 2 at the point where the ball leaves the thrower's hand. Then

$$U_2 = mgy_2 = (0.2 \text{ kg})(10 \text{ m·s}^{-2})(0.5 \text{ m}) = 1.0 \text{ J},$$
$$K_2 = \tfrac{1}{2}mv_2{}^2 = \tfrac{1}{2}(0.2 \text{ kg})(20 \text{ m·s}^{-1})^2 = 40 \text{ J}.$$

Let P represent the upward force exerted on the ball by the man in the throwing process. The work W_{other} is then the work done by this force and is equal to the sum of the changes in kinetic and potential energy of the ball. The kinetic energy of the ball increases by 40 J and its potential energy by 1 J. The work W_{other} done by the upward force P is therefore 41 J.

If the force P is constant, the work done by this force is given by

$$W_{\text{other}} = P(y_2 - y_1),$$

and the force P is then

$$P = \frac{W_{\text{other}}}{y_2 - y_1} = \frac{41 \text{ J}}{0.5 \text{ m}} = 82 \text{ N}.$$

However, the *work* done by the force P is 41 J whether the force is constant or not.

Now consider the flight of the ball after it leaves the thrower's hand. In the absence of air resistance, the only force of the ball is then its weight $w = m\mathbf{g}$. Hence the total mechanical energy of the ball remains constant. The calculations are simpler if we take a new reference level at the point where the ball leaves the thrower's hand. Calling this point 1, we have

$$K_1 = 40 \text{ J}, \qquad U_1 = 0,$$
$$K_1 + U_1 = K_2 + U_2 = 40 \text{ J} + 0,$$

and the total mechanical energy at *any* point in the path is 40 J.

Suppose we want to find the speed of the ball at a height of 15 m above the reference level. Its potential energy at this elevation is

$$U_2 = mgy = (0.2 \text{ kg})(10 \text{ m·s}^{-2})(15 \text{ m}) = 30 \text{ J}.$$

Therefore the kinetic energy at this point is $K_2 = 10$ J, and the speed v_2 at this point is given by

$$\tfrac{1}{2}mv^2 = K_2, \qquad v = \pm \sqrt{2K_2/m} = \pm 10 \text{ m·s}^{-1}.$$

The significance of the $\pm$ sign is that the ball passes this point *twice*, once on the way up and again on the way down. Its *potential* energy at this point is the same whether it is moving up or down. Hence its kinetic energy is the same and its speed is the same. The algebraic sign of the velocity is $+$ when the ball is moving up and $-$ when it is moving down.

Next, let us find the highest point the ball reaches. At this point $v = 0$ and $K = 0$. Therefore at this point $U = 40$ J, and the maximum height h of the ball above the thrower's hand is given by

$$mgh = (0.2 \text{ kg})(10 \text{ m·s}^{-2})h = 40 \text{ J}$$

and

$$h = 20 \text{ m}.$$

Finally, suppose we are asked to find the ball's speed at a point 30 m above the reference level. The potential energy at this point would be 60 J. But the *total* energy is only 40 J, so the ball can never reach a height of 30 m.

What happens when a body travels from an initial elevation y_1 to a final elevation y_2 along a slanted or curved path, as shown in Fig. 7–8b? We assert that the work done by the gravitational force during this displacement is the same as when the body travels straight up, as in Fig. 7–8a. To prove this we divide the path into a large number of small segments Δs; a typical one is shown in Fig. 7–8c. The work done during this displacement is the component of displacement in the direction of the force, multiplied by the magnitude of the force. As shown in Fig. 7–8c, the vertical component of displacement has magnitude $\Delta s \cos \theta = \Delta y$, but its direction is opposite to that of the force w. Thus the work done by w is

$$-mg \, \Delta s \cos \theta = -mg \, \Delta y.$$

This is the same as though the body had been displaced straight up a distance Δy. Similarly, the *total* work done by the gravitational force depends only on the *total* vertical displacement $(y_2 - y_1)$. It is given by $-mg(y_2 - y_1)$ and is independent of any horizontal motion that may occur.

In a uniform gravitational field, the change of potential energy of a body depends only on the change in elevation.

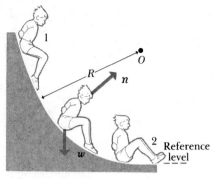

7–9 A child sliding down a frictionless curved slide.

A child on a playground slide: another problem simplified by energy considerations

EXAMPLE 7–6   A child slides down a curved playground slide that is one quadrant of a circle of radius R, as in Fig. 7–9. If he starts from rest and there is no friction, find his speed at the bottom of the track. (The motion of this child is exactly the same as that of a child on a swing of length R, with the other end held at point O.)

SOLUTION   We cannot use the equations of motion with constant acceleration because the acceleration decreases during the motion. (The slope angle of the slide becomes smaller and smaller as the body descends.) If there is no friction, however, the only force on the child other than his weight is the normal force n exerted on him by the slide. The work done by this force is zero because at each point it is perpendicular to the small element of displacement near that point. Thus $W_{other} = 0$ and mechanical energy is conserved. Take point 1 at the starting point and point 2 at the bottom of the slide. Take the reference level at point 2. Then $y_1 = R$, $y_2 = 0$, and

$$K_2 + U_2 = K_1 + U_1.$$
$$\tfrac{1}{2}mv_2^2 + 0 = 0 + mgR,$$
$$v_2 = \pm\sqrt{2gR}.$$

The speed is therefore the same as if the child had fallen *vertically* through a height R. (What is now the significance of the $\pm$ sign?)
   As a numerical example, let $R = 3.00$ m. Then

$$v = \pm\sqrt{2(9.80 \text{ m·s}^{-2})(3.00 \text{ m})} = \pm 7.67 \text{ m·s}^{-2}.$$

EXAMPLE 7–7   Suppose a child of mass 25.0 kg slides down a slide of radius $R = 3.00$ m, like that in Fig. 7–9, but his speed at the bottom is only 3.00 m·s^{-1}. What work was done by the frictional force acting on the child?

SOLUTION   In this case, $W_{other} = W_{\mathscr{F}}$, and

$$\begin{aligned} W_{\mathscr{F}} &= (\tfrac{1}{2}mv_2^2 - \tfrac{1}{2}mv_1^2) + (mgy_2 - mgy_1) \\ &= \tfrac{1}{2}(25.0 \text{ kg})(3.00 \text{ m·s}^{-1})^2 - 0 + 0 - (25.0 \text{ kg})(9.80 \text{ m·s}^{-2})(3.00 \text{ m}) \\ &= 112 \text{ J} - 735 \text{ J} = -623 \text{ J}. \end{aligned}$$

The frictional work was therefore -623 J, and the total mechanical energy decreased by 623 J. The mechanical energy of a body is *not* conserved when friction forces act on it.

EXAMPLE 7–8   In the absence of air resistance, the only force on a ball after it is thrown is its weight, and the mechanical energy of the ball is constant. Figure 7–10 shown two trajectories of a ball with the same initial speed (hence the same total energy) but with different angles of departure. At all points at the same elevation the potential energy is the same; hence the kinetic energy is the same and the speed is the same.

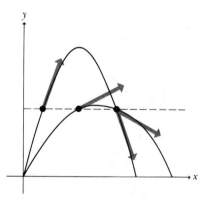

7–10 For the same initial speed, the speed is the same at all points at the same elevation.

EXAMPLE 7–9   A child of weight w sits on a swing of length l, as shown in Fig. 7–11. A *variable* horizontal force P that starts at zero and gradually increases is used to pull the child very slowly (so the kinetic energy is negligibly small) until the swing makes an angle θ with the vertical. Calculate the work done by the force P.

Child on a swing again: an easy way to calculate work

SOLUTION   The sum of the works W_{other} done by all the forces other than the gravitational force must equal the change of total energy, that is, the change of kinetic energy plus the change of gravitational potential energy. Hence

$$W_{\text{other}} = W_P + W_T = \Delta K + \Delta U = \Delta E.$$

Since T is perpendicular to the path at its point of application, $W_T = 0$; and since the swing was pulled very slowly at all times, the change of kinetic energy is also zero. Hence

$$W_P = \Delta U = w \, \Delta y,$$

where Δy is the distance that the child has been raised. From Fig. 7–11, $\Delta y = l(1 - \cos \theta)$. Therefore

$$W_P = wl(1 - \cos \theta).$$

Note that this result agrees with that obtained in Example 7–3, where W_P was calculated directly.

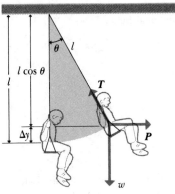

7–11 $\Delta y = l(1 - \cos \theta)$.

Thus far in this section we have assumed that the gravitational force on a body is constant in magnitude and direction. But we learned in Section 6–3 that the earth's gravitational force on a body of mass m is given more generally by

Potential energy in a nonuniform gravitational field, obtained by integration

$$F_g = \frac{Gmm_E}{r^2}, \tag{7-20}$$

where m_E is the mass of the earth and r is the distance of the body from the earth's center. When the change in r is sufficiently large, we may no longer consider the gravitational force constant; it decreases as $1/r^2$. To find the corresponding generalization of gravitational potential energy, we must compute the work W_{grav} done by the gravitational force when r increases from r_1 to r_2. This is given by

$$W_{\text{grav}} = \int_{r_1}^{r_2} F_r \, dr,$$

where F_r is the *component* of the gravitational force in the direction of increasing r, that is, the direction outward from the center of the earth. Because this force actually points in the direction of *decreasing* r, F_r differs from Eq. (7–20) by a minus sign. That is, the magnitude of the gravitational force is positive, but its component along the direction of increasing r (the radial direction) is negative.

Thus W_{grav} is given by

$$W_{\text{grav}} = -Gmm_E \int_{r_1}^{r_2} \frac{dr}{r^2} = \frac{Gmm_E}{r_2} - \frac{Gmm_E}{r_1}. \tag{7-21}$$

The quantity $-Gmm_E/r$ is therefore the general expression for the gravitational potential energy of a body attracted by the earth:

$$U(\text{gravitational}) = -G\frac{mm_E}{r}. \tag{7-22}$$

If the gravitational force is the only force on the body, then W_{grav} is equal to the change in kinetic energy from r_1 to r_2:

$$Gmm_E\left(\frac{1}{r_2} - \frac{1}{r_1}\right) = \frac{1}{2}mv_2^2 - \frac{1}{2}mv_1^2$$

$$\frac{1}{2}mv_1^2 - \frac{Gmm_E}{r_1} = \frac{1}{2}mv_2^2 - \frac{Gmm_E}{r_2}. \tag{7-23}$$

The *total mechanical energy* of the body, the sum of its kinetic energy and potential energy, is

$$E = K + U = \frac{1}{2}mv^2 - G\frac{mm_E}{r}. \tag{7-24}$$

If the only force on the body is the gravitational force, the total mechanical energy remains constant, or is *conserved*.

EXAMPLE 7–10 In Jules Verne's story "From the Earth to the Moon" (written in 1865) three men were shot to the moon in a shell fired from a giant cannon sunk in the earth in Florida. What muzzle velocity would be needed (a) to raise a total mass m to a height above the earth equal to the earth's radius; (b) to escape from the earth completely? To simplify the calculation, neglect the gravitational pull of the moon.

SOLUTION

a) In Eq. (7–23), let m be the total projectile mass and let v_1 be the initial velocity. Then

$$r_1 = R, \qquad r_2 = 2R, \qquad v_2 = 0,$$

and

$$\frac{1}{2}mv_1^2 - G\frac{mm_E}{R} = 0 - G\frac{mm_E}{2R},$$

or

$$v_1^2 = \frac{Gm_E}{R}.$$

$$v_1 = \sqrt{\frac{(6.67 \times 10^{-11}\ \text{N·m}^2\text{·kg}^{-2})(5.98 \times 10^{24}\ \text{kg})}{6.38 \times 10^6\ \text{m}}}$$

$$= 7920\ \text{m·s}^{-1}\ (= 17{,}715\ \text{mi·hr}^{-1}).$$

b) When v_1 is the escape velocity, $r_1 = R$, $r_2 = \infty$, $v_2 = 0$. Then

$$\frac{1}{2}mv_1^2 - G\frac{mm_E}{R} = 0 \qquad \text{or} \qquad v_1^2 = \frac{2Gm_E}{R}.$$

$$v_1 = \sqrt{\frac{2(6.67 \times 10^{-11}\ \text{N·m}^2\text{·kg}^{-2})(5.98 \times 10^{24}\ \text{kg})}{6.38 \times 10^6\ \text{m}}}$$

$$= 1.12 \times 10^4\ \text{m·s}^{-1}\ (= 25{,}050\ \text{mi·hr}^{-1}).$$

Note that the speed of an earth satellite in a circular orbit of radius just slightly greater than R is, from Eq. (6–22), the same as the result of part (a), and that the escape velocity is larger than this by exactly a factor of $\sqrt{2}$.

It may seem strange that the general expression for gravitational potential energy, Eq. (7–22), should contain a minus sign. The reason lies in the choice of reference level at which the potential energy is considered zero. If we set $U = 0$ in Eq. (7–22) and solve for r, we find $r = \infty$. That is, *the gravitational potential energy of a body is zero when the body is at an infinite distance from the earth.*

Since the potential energy decreases as the body approaches the earth, it must be negative at any finite distance from the earth. The *change* in potential energy of a body as it moves from one point to another is the same, whatever the choice of reference level, and it is only changes in potential energy that are significant.

Finally, note that Eq. (7–21) can be rewritten as

$$W_{grav} = Gmm_E \left(\frac{r_1 - r_2}{r_1 r_2} \right).$$

If the particle stays close to the earth, then in the denominator we may replace r_1 and r_2 by R, the earth's radius, obtaining

$$W_{grav} = Gmm_E \left(\frac{r_1 - r_2}{R^2} \right).$$

But according to Eq. (6–15), $g = Gm_E/R^2$, so we finally obtain

$$W_{grav} = mg(r_1 - r_2),$$

which agrees with Eq. (7–13) with the y's replaced by r's. Thus Eq. (7–13) may be considered a special case of the more general Eq. (7–21).

The relationship between the gravitational force on a body, given by Eq. (7–20), and its gravitational potential energy, given by Eq. (7–22), can be expressed in a different way. When the distance r changes by a small amount dr, the work done by the gravitational force is $dW = F_r\, dr$, and the corresponding change in potential energy dU can be represented as $dU = (dU/dr)\, dr$. Because of the general relation of potential energy to work, we also know that $dW = -dU$. Thus we have the general relation

Relation between potential energy and force: Force is the negative of the derivative of potential energy.

$$F_r = -\frac{dU}{dr}. \qquad (7-25)$$

For the gravitational force of the earth on a body of mass m, we have

$$F_r = -\frac{Gmm_E}{r^2},$$

$$\frac{dU}{dr} = \frac{Gmm_E}{r^2},$$

which is consistent with Eq. (7–25). We will find analogous relationships when we study electric fields and their associated potential energies in Chapters 25 and 26.

7–6 ELASTIC POTENTIAL ENERGY

The concept of potential energy is useful in calculations of work done by *elastic* forces, such as the spring discussed in Section 7–3. Figure 7–12 shows a body of mass m on a level surface. One end of a spring is attached to the body, and the other end is held stationary. We take our origin of coordinates ($x = 0$) as the position of the body when the spring is neither stretched nor compressed. We now apply a force P, causing the body to accelerate. As soon as the spring begins to stretch, it begins to exert a force F on the body; we may call this an *elastic force*. If the force P is removed, the elastic force pulls the body back

Elastic potential energy: how to calculate work involved in stretching a spring

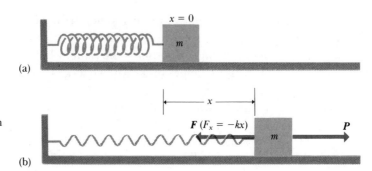

7–12 When an applied force P produces an extension x of a spring, an elastic restoring force F acts on the mass m. The x-component of F is $F_x = -kx$.

toward the original position, where the spring is unstretched. Hence F may be called a *restoring force*.

In Section 7–3 we spoke of the work done on a spring. Now, in the light of the relation between work done on a body and its change in kinetic energy, we shift our attention to the work that the spring does on the body of mass m; this is the negative of the work done on the spring. Thus in an elongation from $x = 0$ to a final value x, the work done *on* the mass *by* the spring is $-\frac{1}{2}kx^2$. Similarly, in a displacement from an initial elongation x_1 to a final elongation x_2, the elastic restoring force does an amount of work W_{el} given by

$$W_{el} = -\tfrac{1}{2}kx_2^2 - (-\tfrac{1}{2}kx_1^2).$$

The quantity $\frac{1}{2}kx^2$, one-half the product of the force constant and the square of the displacement from the unstretched position, is called the **elastic potential energy** of the system, denoted by U. That is,

$$U = \tfrac{1}{2}kx^2. \tag{7–26}$$

Elastic potential energy is proportional to the square of the amount of stretch or compression.

We use the symbol U for any form of potential energy. When there is more than one kind of potential energy in a problem, we will use $U_{el} = \frac{1}{2}kx^2$ for elastic potential energy. Note that $U = 0$ when $x = 0$. We could add any arbitrary constant to Eq. (7–26) if we chose, because the difference between the values of U at any two points would not change.

Thus we can express the work done on the body by the elastic force in terms of a change in potential energy, just as we did for the work done by a gravitational force:

$$W_{el} = \tfrac{1}{2}kx_1^2 - \tfrac{1}{2}kx_2^2 = U_1 - U_2. \tag{7–27}$$

When x increases, the work done on the mass by the elastic force is negative and the potential energy increases; when x decreases, the elastic force does positive work and the potential energy decreases. If the spring can be compressed as well as stretched, then x is negative when it is compressed. In this case U is still positive, and Eq. (7–26) is still valid.

As before, we let W_{other} be the work done by forces other than the elastic force, in this case the work done by the force P. Setting the total work equal to the change in kinetic energy of the body, we have

$$W_{other} + W_{el} = \Delta K,$$
$$W_{other} - (\tfrac{1}{2}kx_2^2 - \tfrac{1}{2}kx_1^2) = (\tfrac{1}{2}mv_2^2 - \tfrac{1}{2}mv_1^2).$$

The quantities $\frac{1}{2}kx_2^2$ and $\frac{1}{2}kx_1^2$ depend only on the initial and final positions of the body. When we transfer them from the "work" side of the equa-

Conservation of energy with elastic forces

tion to the "energy" side, we obtain:

$$W_{\text{other}} = (\tfrac{1}{2}mv_2{}^2 - \tfrac{1}{2}mv_1{}^2) + (\tfrac{1}{2}kx_2{}^2 - \tfrac{1}{2}kx_1{}^2). \qquad (7\text{–}28)$$

Hence the work W_{other} of the force **P** equals the sum of the change in the kinetic energy of the body and the change in its elastic potential energy.

Equation (7–28) can also be written

$$\begin{aligned} W_{\text{other}} &= (\tfrac{1}{2}mv_2{}^2 + \tfrac{1}{2}kx_2{}^2) - (\tfrac{1}{2}mv_1{}^2 + \tfrac{1}{2}kx_1{}^2) \\ &= (K_2 + U_2) - (K_1 + U_1) \\ &= E_2 - E_1 = \Delta E. \qquad (7\text{–}29) \end{aligned}$$

The sum of the kinetic and potential energies of the body is its total mechanical energy; and *the work done by all forces acting on the body,* **with the exception of the elastic force,** *equals the change in the total mechanical energy of the body.*

If the work W_{other} is positive, the mechanical energy increases. If W_{other} is negative, it decreases. In the special case where $W_{\text{other}} = 0$, the mechanical energy is constant, or is conserved. In that special case we can rewrite Eq. (7–29) as

$$\tfrac{1}{2}mv_1{}^2 + \tfrac{1}{2}kx_1{}^2 = \tfrac{1}{2}mv_2{}^2 + \tfrac{1}{2}kx_2{}^2. \qquad (7\text{–}30)$$

We can easily generalize these results to cases where we have *both* gravitational and elastic forces and their associated potential energies. We may still use the relationship

$$W_{\text{other}} = (K_2 + U_2) - (K_1 + U_1) = E_2 - E_1, \qquad (7\text{–}31)$$

where now U_1 and U_2 are the initial and final values of the *total* potential energy, including both gravitational and elastic. If the gravitational and elastic forces are the *only* forces that do work on the body, then $W_{\text{other}} = 0$; in that special case we have

$$K_1 + U_1 = K_2 + U_2, \qquad (7\text{–}32)$$

where again U_1 and U_2 are the total potential energies.

PROBLEM-SOLVING STRATEGY: *Conservation of energy II*

The strategy outlined in Section 7–5 is equally useful here. In the list of kinetic and potential energies in item 3, include both gravitational and elastic potential energies where appropriate. Keep in mind that when gravitational, elastic, and other forces are present, the work done by the gravitational and elastic forces is accounted for by the respective potential energies, and that the work of the other forces, W_{other}, has to be included separately. In some cases, though, there may be "other" forces that do no work. This was the case in Example 7–6 (Section 7–5): The normal force exerted by the slide on the child does no work.

EXAMPLE 7–11 In Fig. 7–12, let the force constant k of the spring be $24.0\ \text{N} \cdot \text{m}^{-1}$, and let the mass of the body be 4.00 kg. The body is initially at rest, and the spring is initially stretched 0.500 m. Then the body is released and moves

Motion of a body pulled by a spring: a problem where energy methods are easier than Newton's laws

back toward its equilibrium position. What is its speed when the spring is stretched 0.300 m?

SOLUTION The speed can be found most easily by using energy considerations. The spring force is the only force that does work on the body. Thus $W_{\text{other}} = 0$, and we may use Eq. (7–32). The various energy quantities can be listed as follows:

$$K_1 = \tfrac{1}{2}(4.00 \text{ kg})(0)^2 = 0,$$
$$U_1 = \tfrac{1}{2}(24.0 \text{ N·m}^{-1})(0.500 \text{ m})^2 = 3.00 \text{ J},$$
$$K_2 = \tfrac{1}{2}(4.00 \text{ kg})(v_2)^2,$$
$$U_2 = \tfrac{1}{2}(24.0 \text{ N·m}^{-1})(0.300 \text{ m})^2 = 1.08 \text{ J}.$$

Then from Eq. (7–32),

$$0 + 3.00 \text{ J} = (2.00 \text{ kg})v_2{}^2 + 1.08 \text{ J},$$

from which we obtain $v_2 = \pm 0.98 \text{ m·s}^{-1}$. (What is the physical significance of the $\pm$ sign?)

Alternatively, we may use Eq. (7–30). We have $v_1 = 0, x_1 = 0.5 \text{ m}, x_2 = 0.3 \text{ m}$, and v_2 is to be found. From Eq. (7–30),

$$\tfrac{1}{2}(4.00 \text{ kg})(0)^2 + \tfrac{1}{2}(24.0 \text{ N·m}^{-1})(0.500 \text{ m})^2$$
$$= \tfrac{1}{2}(4.00 \text{ kg})(v_2)^2 + \tfrac{1}{2}(24.0 \text{ N·m}^{-1})(0.300 \text{ m})^2.$$

Evaluating the left side, we find that the total mechanical energy is 3 J, and solving for v_2 again yields the result

$$v_2 = \pm 0.98 \text{ m·s}^{-1}.$$

Note that this problem *cannot* be done by using the equations of motion with constant acceleration because the spring force varies with position. The energy method, by contrast, offers a simple and elegant solution.

EXAMPLE 7–12 For the system of Example 7–11, suppose the body is initially at rest and the spring is initially unstretched. Then a constant force P of magnitude 10.0 N is applied to the body. There is no friction. What is the speed of the body when it has moved 0.500 m?

The equations of motion with constant acceleration cannot be used with spring problems because the force isn't constant.

SOLUTION The equations of motion with constant acceleration cannot be used, since the resultant force on the body varies as the spring is stretched. However, the speed can be found from energy considerations:

$$W_{\text{other}} = \Delta K + \Delta U,$$
$$(10.0 \text{ N})(0.500 \text{ m}) = \tfrac{1}{2}(4.00 \text{ kg})v_2{}^2 - 0$$
$$+ \tfrac{1}{2}(24.0 \text{ N·m}^{-1})(0.500 \text{ m})^2 - 0,$$
$$v_2 = 1.00 \text{ m·s}^{-1}.$$

EXAMPLE 7–13 In Example 7–12, suppose the force P is removed when the body has moved 0.500 m. How much *farther* does the body move before coming to rest?

SOLUTION The elastic force is now the only force, and total mechanical energy is conserved. The initial kinetic energy is $\tfrac{1}{2}mv^2 = 2 \text{ J}$, and the potential energy is $\tfrac{1}{2}kx^2 = 3 \text{ J}$. The total energy is therefore 5 J (equal to the work of the force P). When the body comes to rest, its kinetic energy is zero and its potential energy is therefore 5 J. Hence

$$\tfrac{1}{2}kx_{max}^{2} = \tfrac{1}{2}(24.0 \text{ N·m}^{-1})x_{max}^{2} = 5.00 \text{ J},$$
$$x_{max} = 0.645 \text{ m}.$$

EXAMPLE 7-14 A brick of mass m, initially at rest, is dropped from a height h onto a spring whose force constant is k. Find the maximum distance y that the spring will be compressed (see Fig. 7-13).

Dropping a brick onto a spring-mounted platform: using elastic and gravitational potential energies together

SOLUTION The gravitational and elastic forces are the only forces acting on the brick. Their work can be represented in terms of potential energies, and the total energy $K + U$ is conserved. At the instant of release, the brick is at rest; hence $v_1 = 0$ and $K_1 = 0$. At the instant when maximum compression occurs, the brick is again at rest, so $K_2 = 0$. Hence the total potential energy is the same at the beginning and at the end. The gravitational potential energy *decreases* by an amount $mg(h + y)$, and the elastic potential energy *increases* by an amount $\tfrac{1}{2}ky^2$. Equating these two quantities, we find

$$mg(h + y) = \tfrac{1}{2}ky^{2}, \quad \text{or} \quad y^{2} - \frac{2mg}{k}y - \frac{2mgh}{k} = 0.$$

Therefore, from the quadratic formula,

$$y = \frac{1}{2}\left[\frac{2mg}{k} \pm \sqrt{\left(\frac{2mg}{k}\right)^{2} + \frac{8mgh}{k}}\right].$$

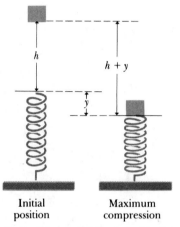

The positive root is the desired result; the negative root corresponds to the height to which the brick and spring would rebound if they were fastened together after contact.

Initial position   Maximum compression

7-13 The total fall of the block is $h + y$.

7-7 CONSERVATIVE AND DISSIPATIVE FORCES

We have seen that when a body acted on by a gravitational force moves from one position to another, the work done by the gravitational force is independent of the body's path. It can be represented as the difference between the initial and final values of a function called the gravitational potential energy, $W = U_1 - U_2$. If the gravitational force is the *only* force acting on the body, the total mechanical energy (the sum of kinetic and potential energies) is constant, or conserved. For this reason the gravitational force is called a **conservative force.** When a body ascends in the earth's gravitational force, the work done by this force is negative, the kinetic energy decreases, and the potential energy increases. When it descends to its original level, the gravitational force does positive work and the kinetic energy increases to its original value. Thus the kinetic energy lost during the ascent is completely recovered during descent.

When kinetic energy decreases because a force does negative work, can it be recovered? Maybe!

A similar situation occurs when a body is attached to a spring and moved from one position to another, changing the extension or compression of the spring. When the extension increases, the spring does negative work, the kinetic energy decreases, and the potential energy increases. When the body returns to its original position, the spring does positive work and the kinetic energy returns to its original value. In each case the work done by the spring force can again be represented as $W = U_1 - U_2$.

The work done by a conservative force is reversible; when the direction of motion reverses, so does the work.

In both these situations, the work is *reversible:* On the return trip the work is always exactly the negative of that on the first part of the trip. Thus the work done by a conservative force always has these properties:

1. It is independent of the path of the body and depends only on the starting point and endpoint.
2. It is equal to the difference between the initial and final values of a *potential energy* function.
3. It is completely reversible.
4. When the starting point and endpoint are the same—that is, the path forms a closed loop—the total work is zero.

Work done by a friction force is not reversible, and the force is not conservative.

For comparison, we may consider the friction force acting on a body sliding on a stationary surface. There is no potential-energy function here. When the body slides in one direction and then back to its original position, the total work done on it by the frictional force is *not* zero. The reason is that when the direction of motion reverses, so does the friction force. Thus it does negative work in both directions. There is no such change in the direction of a gravitational force. When a body slides across a surface with decreasing speed (and therefore decreasing kinetic energy), there is no way to recover this lost kinetic energy. When friction forces act in such situations, the total mechanical energy is *not* conserved. Such a force is therefore called a *nonconservative force* or a **dissipative force.** We can then describe the energy relations in terms of additional kinds of energy, and a more general energy-conservation principle includes these additional energies. For example, when a body slides on a rough surface, the surface becomes hotter, and energy is associated with this change in the condition of the material.

EXAMPLE 7–15 In Example 7–7 (Section 7–5) a body is acted on by a dissipative friction force. The initial mechanical energy of the body is its initial potential energy of 735 J. Its final mechanical energy is its final kinetic energy of 112 J. The work W_f done by the friction force is −623 J. As the body slides down the track, the surfaces become warmer, and the same temperature changes could have been produced by adding 623 J of heat to the bodies. Thus their *internal* energy increases by 623 J; the sum of this energy and the final mechanical energy equals the initial mechanical energy, and the total energy of the system is conserved.

7–8 INTERNAL WORK AND ENERGY

We have discussed mechanical energy in the context of bodies that can be represented as particles. For more complex systems that have to be represented in terms of many particles, new subtleties in energy considerations appear. We cannot discuss these in detail, but here are three examples.

First consider a man standing on frictionless roller skates on a level surface, facing a rigid wall, as shown in Fig. 7–14. He pushes against the wall, setting himself in motion backward (to the right). The forces acting on him are his weight w, the upward normal forces n_1 and n_2 exerted by the ground on his skates, and the horizontal force P exerted on him by the wall. There is no

7–14 External forces acting on a man who is pushing against a wall. The work done by these forces is zero.

vertical displacement, so w, n_1, and n_2 do no work. The force P is the horizontal force that gives the system its acceleration to the right, but the point where that force is applied (i.e., the man's hands) does not move, so the force P also does no work. Where does the man's kinetic energy come from?

The difficulty is that representing the man as a single point is not an adequate model. For the motion to occur as we have described it, different parts of the man's body must have different motions. One part of the body may exert forces and do work on another part; therefore there may be changes in the *total* kinetic energy of this composite system, even though no work is done by forces applied by bodies outside the system. This would not be possible with a system that can be represented as a single point. In Chapter 8 we will consider the motion of a collection of moving particles interacting with each other. We will discover that the total kinetic energy of such a system can change even when no work is done on any part of the system by anything outside it.

Some systems cannot be represented by a single point. One part of a system can do work on another part of the system and change the kinetic energy even when no work is done by external forces.

As a second example, we consider a body sliding across a rough horizontal surface. The friction force on the body acts in the direction opposite to its motion relative to the surface, so it does negative work on the body. If that is the only horizontal force on the body, the body's kinetic energy decreases. At the same time, the body exerts an equal and opposite reaction force on the surface. But because the surface does not move, that force does no work. The work done on the surface is *not* the negative of the work done on the body. Hence there is a decrease in the body's kinetic energy but no corresponding increase in kinetic energy of the surface. So where does the energy go?

We have already pointed out in Section 7–7 that in this situation, mechanical energy is not conserved. The surfaces become warmer; the corresponding energy transferred to the materials is called *internal energy*. In later chapters we will study internal energy and its relation to temperature changes, heat, and work in considerable detail. This is the heart of the area of physics called *thermodynamics*.

Internal energy and energy associated with heat: a preview of things to come

As a final example, suppose we drop a box onto a moving conveyor belt. If the box is dropped from a very small initial height, it is essentially at rest the instant it contacts the belt. It slips at first, but eventually acquires the same velocity as the belt. In doing so, it acquires kinetic energy, and some force has to act on the box, doing positive work on it, to give it this kinetic energy. What is the force?

Clearly, the answer has to be "the friction force exerted on the box by the belt"; there is no other horizontal force on the box. So here is a case where a friction force on a moving body does *positive* work on the body and *increases* its kinetic energy. The same thing happens when a truck starts from rest and accelerates. If there is a box sitting in the truck, the force that gives the box its forward acceleration and adds to its kinetic energy is again the friction force exerted on the box, this time by the moving floor of the truck. Although we are accustomed to think of moving bodies slowing down and stopping as a result of friction forces, in these two examples friction forces do positive work and increase the kinetic energy of the bodies on which they act.

Box on a conveyer belt: a case where friction does positive work

Still, we are left with the nagging feeling that somehow friction forces ought to dissipate or waste mechanical energy. After all, don't we oil bearings and sliding surfaces to decrease friction and make them more efficient? True enough! Going back to the conveyor belt, let us consider the work done *on the belt* by the friction force exerted on the belt by the box. This is negative,

7–15 Stroboscopic photograph of a pole-vaulter. The athlete's initial kinetic energy is partly converted to elastic potential energy in the flexed pole, then to gravitational potential energy as he rises and clears the bar, then back to kinetic energy as he drops on the other side. The elastic and gravitational forces are conservative. (Dr. Harold Edgerton, M.I.T., Cambridge, Massachusetts.)

because the force on the belt has opposite direction to its motion. By Newton's third law, the two forces have equal magnitude. Furthermore, during the sliding of the box the belt moves farther than the box; thus the negative work done on the belt has greater magnitude than the positive work done on the box. The *total* work done by these two forces is negative, and in this energy-exchange process there is indeed a net loss of mechanical energy. As mentioned above, there is a corresponding increase of internal energy of the materials, showing itself as a rise in temperature.

Despite these complications, it remains true that for any system that can be adequately represented as a moving *point* mass, the total change in kinetic energy in any process is *always* equal to the total work done by all the forces acting on the system. Only when a system consists of several interacting masses do complications develop. A dramatic example of these complications is shown in Fig. 7–15.

7–9 POWER

Power: a description of how quickly work is done

Time considerations are not involved in the definition of work. When you lift a body weighing 100 N through a vertical distance of 0.5 m, you do 50 J of work, whether it takes you 1 second, 1 hour, or 1 year to do it. There are many situations, though, where it is important for us to know how quickly work is done, that is, the *rate* of doing work. The time rate at which work is done or energy transferred is called **power.**

When a quantity of work ΔW is done during a time interval Δt, the **average power** P_{av} is defined as

$$\text{Average power} = \frac{\text{work done}}{\text{time interval}}, \qquad P_{av} = \frac{\Delta W}{\Delta t}.$$

If the rate at which work is done is not constant, this ratio may vary; in this case we may define an *instantaneous power* P as the limit of this quotient as $\Delta t \rightarrow 0$:

$$P = \lim_{\Delta t \to 0} \frac{\Delta W}{\Delta t} = \frac{dW}{dt}. \qquad (7\text{--}33)$$

The SI unit of power, 1 joule per second (1 J·s^{-1}), is called 1 *watt* (1 W). The kilowatt (1 kW = 10^3 W) and the megawatt (1 MW = 10^6 W) are also commonly used. The cgs power unit is 1 erg per second (1 erg·s^{-1}).

In the British system, where work is expressed in foot-pounds and time in seconds, the unit of power is 1 foot-pound per second. A larger unit called the *horsepower* (hp) is also used:

$$\begin{aligned} 1 \text{ hp} &= 550 \text{ ft·lb·s}^{-1} \\ &= 33{,}000 \text{ ft·lb·min}^{-1}. \end{aligned}$$

That is, a 1-hp motor running at full load does 33,000 ft·lb of work every minute.

The watt is a familiar unit of *electrical* power; a 100-W light bulb converts electrical energy into light and heat at the rate of 100 joules per second. But there is nothing inherently electrical about the watt; the electric power consumption of a light bulb could be expressed in horsepower, and many automobile manufacturers now rate their engines in kilowatts rather than horsepower.

The watt, the kilowatt, and the horsepower: all units of power

From the relations between the newton, pound, meter, and foot, we can show that

$$1 \text{ hp} = 746 \text{ W} = 0.746 \text{ kW},$$

or about $\frac{3}{4}$ of a kilowatt, a useful figure to remember.

Because power is energy or work per unit time, the units of power may be used to define new units of work or energy. The kilowatt-hour (kWh) is commonly used as a unit of electrical energy. One kilowatt-hour is the work done in 1 hour by an agent working at a constant power of 1 kilowatt (10^3 J·s^{-1}). Such an agent does 1000 J of work each second; the work done in 1 hr is $3600 \times 1000 = 3{,}600{,}000$ J:

$$1 \text{ kWh} = 3.6 \times 10^6 \text{ J} = 3.6 \text{ MJ}.$$

The kilowatt-hour is a unit of *work* or *energy*, not power.

The kilowatt-hour: a familiar unit of energy, not power

Although energy is an abstract physical quantity, it nevertheless has a monetary value. A newton of force or a meter per second of velocity is not a thing that is bought and sold as such, but a kilowatt-hour of energy is a quantity offered for sale at a definite market rate. In the form of electrical energy, a kilowatt-hour can be purchased at a price varying from a few tenths of a cent to around 10 cents, depending on the locality and the quantity purchased.

When a force acts on a moving body, doing work on it, the corresponding power can be expressed in terms of the force and the velocity. Suppose a force $\boldsymbol{F}$ acts on a body while it undergoes a displacement of magnitude Δs. If $F_\parallel$ is the component of $\boldsymbol{F}$ tangent to the path, then work is given by $\Delta W = F_\parallel \Delta s$, and the average power is

$$P_{\text{av}} = \frac{\Delta W}{\Delta t} = F_\parallel \frac{\Delta s}{\Delta t} = F_\parallel v_{\text{av}}.$$

In the limit as $\Delta t \rightarrow 0$ we obtain the instantaneous power:

$$P = F_{\parallel}v, \tag{7-34}$$

where v is the magnitude of the instantaneous velocity. We can also express this relation in terms of the scalar product:

$$P = \boldsymbol{F} \cdot \boldsymbol{v}. \tag{7-35}$$

Power output of a jet engine.

EXAMPLE 7–16 A jet airplane engine develops a thrust (a forward force on the plane) of 15,000 N (roughly 3000 lb). When the plane is flying at 300 m·s^{-1} (roughly 600 mph), what horsepower does the engine develop?

SOLUTION

$$P = Fv = (1.50 \times 10^4 \text{ N})(300 \text{ m·s}^{-1}) = 4.50 \times 10^6 \text{ W}$$

$$= (4.50 \times 10^6 \text{ W})\left(\frac{1 \text{ hp}}{746 \text{ W}}\right) = 6030 \text{ hp}.$$

EXAMPLE 7–17 As part of a charity fund-raising drive, a Chicago marathon runner of mass 50.0 kg runs up the stairs to the top of the Sears tower, the tallest building in the United States (443 m), in 15.0 minutes. What is her average power output, in watts? In kilowatts? In horsepower?

SOLUTION The total work is

$$W = mgh = (50.0 \text{ kg})(9.80 \text{ m·s}^{-2})(443 \text{ m}) = 2.17 \times 10^5 \text{ J}.$$

The time is 15 min = 900 s, so the average power is

$$P_{\text{av}} = \frac{2.17 \times 10^5 \text{ J}}{900 \text{ s}} = 241 \text{ W} = 0.241 \text{ kW} = 0.323 \text{ hp}.$$

Alternatively, the average vertical component of velocity is (443 m)/(900 s) = 0.492 m·s^{-1}, so the average power is

$$P_{\text{av}} = Fv_{\text{av}} = (mg)v_{\text{av}}$$
$$= (50.0 \text{ kg})(9.80 \text{ m·s}^{-2})(0.492 \text{ m·s}^{-1}) = 241 \text{ W}.$$

7–10 AUTOMOTIVE POWER

Energy relations in a gasoline-powered automobile provide a familiar and interesting example of some of the energy and power concepts in this chapter. First, burning 1 liter of gasoline provides about 3.5×10^7 J of energy. Not all of this is converted to *mechanical* energy; the laws of thermodynamics, which we will encounter in Chapters 18 and 19, impose fundamental limitations on the efficiency of converting heat to mechanical energy or work. In a typical car engine, two-thirds of the heat from gasoline combustion is wasted in the cooling system and the exhaust. Another 20% or so is converted to mechanical energy but is lost in friction in the drive train or is used by accessories such as electric-power generators, air-conditioners, and power steering. This leaves about 15% of the energy to propel the car.

Rolling friction: It's easier to roll things than to drag them.

Most of this energy is used to overcome rolling friction and air resistance. Each of these can be described in terms of a force that resists the motion of the car. We described rolling friction in Section 5–2 in terms of an effective coefficient of rolling friction called *tractive resistance* μ_r. A typical value of μ_r for rubber tires on hard pavement is 0.015. For a 1000-kg compact car weighing about 10,000 N, the resisting force of rolling friction is thus

$$F_{rol} = (0.015)(10,000 \text{ N}) = 150 \text{ N}.$$

This force is nearly independent of car speed.

The air resistance force increases rapidly with speed, and in fact the relationship can be expressed approximately by the equation

$$F_{air} = \tfrac{1}{2}CA\rho v^2, \tag{7–36}$$

The drag coefficient: a way to calculate air resistance of a moving car

where A is the silhouette area of the car (seen from the front), ρ is the density of air (about 1.2 kg·m^{-3} at ordinary temperatures), v is the car's speed, and C is a dimensionless constant, called the *drag coefficient*, that depends on the shape of the moving body. A typical value of C for a car designed with reasonable attention to streamlining is 0.5. Assuming $A = 2 \text{ m}^2$ for our car, we find an air-resistance force of

$$F_{air} = \tfrac{1}{2}(0.5)(2 \text{ m}^2)(1.2 \text{ kg·m}^{-3})v^2$$
$$= (0.6 \text{ N·s}^2\text{·m}^{-2})v^2.$$

Thus in a residential speed zone, where $v = 10$ m·s^{-1} (about 22 mi·hr^{-1}), the air-resistance force is about

$$F_{air} = (0.6 \text{ N·s}^2\text{·m}^{-2})(10 \text{ m·s}^{-1})^2 = 60 \text{ N}.$$

At a moderate speed of 15 m·s^{-1} (34 mi·hr^{-1}), F_{air} is 135 N, and at a highway speed of 30 m·s^{-1} (67 mi·hr^{-1}), it is 540 N. Thus at slow speeds air resistance is less important than rolling friction. At moderate speeds they are comparable, and at highway speeds air resistance is the dominant effect.

What does this mean in terms of the *power* needed from the engine? In constant-speed driving, the forward force supplied by the driven wheels must just balance the sum of these two resisting forces, and the power is this force multiplied by the velocity. Thus for our hypothetical car, the power needed for constant speed v is

$$P = (F_{rol} + F_{air})v$$
$$= [150 \text{ N} + (0.6 \text{ N·s}^2\text{·m}^{-2})v^2]v.$$

You can do the arithmetic yourself. For the three speeds mentioned above, you will find the following results:

v (m·s^{-1})	F_{rol} (N)	F_{air} (N)	F_{tot} (N)	P (kW)	P (hp)
10	150	60	210	2.10	2.81
15	150	135	285	4.28	5.73
30	150	540	690	20.70	27.70

Now, how is all this related to fuel consumption? Let's look at the 15 m·s^{-1} case. The power required is 4.28 kW = 4280 J·s^{-1}. In one hour (3600 s) the total energy required is

$$(4280 \text{ J·s}^{-1})(3600 \text{ s}) = 1.54 \times 10^7 \text{ J}.$$

During that hour the car travels a distance of

$$(15 \text{ m·s}^{-1})(3600 \text{ s}) = 5.4 \times 10^4 \text{ m} = 54 \text{ km}.$$

If the entire 3.5×10^7 J of energy obtained by burning one liter of gasoline were available to propel the car, the amount consumed in an hour would be

$$(1.54 \times 10^7 \text{ J})/(3.5 \times 10^7 \text{ J·liter}^{-1}) = 0.44 \text{ liter}.$$

But because only 15% of the energy is actually available for propulsion, as explained above, the total amount consumed is $(0.44 \text{ liter})/0.15 = 2.93$ liter. That amount of gasoline gets us 54 km, so the rate of consumption per unit distance is $(54 \text{ km})/(2.93 \text{ liter}) = 18.4 \text{ km·liter}^{-1}$. You may want to work out the conversion factor between kilometers per liter and miles per gallon; this comes out to about 43.3 miles per gallon.

The power required for a steady 15 m·s^{-1} is 4.28 kW, but the power required for acceleration and hill climbing may be much greater. Suppose the car accelerates from zero to 30 m·s^{-1} in 20 s. The final kinetic energy is

$$K = \tfrac{1}{2}mv^2 = \tfrac{1}{2}(1000 \text{ kg})(30 \text{ m·s}^{-1})^2 = 4.5 \times 10^5 \text{ J}.$$

The average power required is

$$P_{av} = (4.5 \times 10^5 \text{ J})/(20 \text{ s}) = 2.25 \times 10^4 \text{ W} = 22.5 \text{ kW} = 30.2 \text{ hp}.$$

Thus this relatively rapid acceleration requires five times as much power as cruising at a steady, moderate speed.

What about hill climbing? A 5% grade, which is about the maximum found on most interstate highways, rises 5 meters for every 100 meters of horizontal distance. A car driving at 15 m·s^{-1} up a 5% grade is gaining elevation at the rate of $(0.05)(15 \text{ m·s}^{-1}) = 0.75$ m·s^{-1}. To lift a car weighing 10,000 N at this rate requires a power of

$$P = Fv = (10,000 \text{ N})(0.75 \text{ m·s}^{-1})$$
$$= 7500 \text{ J·s}^{-1} = 7.50 \text{ kW} = 10.0 \text{ hp}.$$

The *total* power required is this amount plus the 4.28 kW needed to maintain 15 m·s^{-1} on a level road, that is,

$$P_{tot} = 7.50 \text{ kW} + 4.28 \text{ kW} = 11.78 \text{ kW}.$$

Finally, let us compare these energy and power quantities with some purely thermal considerations. This is a little premature; we will study the relation of heat to mechanical energy in detail in later chapters, but we can anticipate that discussion here. We will learn that to heat one kilogram of water from 0°C to 100°C requires an energy input to the water of 4.18×10^5 J. Thus the 3.5×10^7 J obtained from one liter of gasoline is enough to heat $(3.5 \times 10^7)/(4.18 \times 10^5) = 83.7$ kg of water from freezing to boiling. That is not much water, only about 23 gallons. So the amount of energy needed to heat 23 gallons of water from freezing to boiling is enough to push our car over 18 km! Does that surprise you?

SUMMARY

KEY TERMS

conservation of energy

energy

The concept of energy plays a central role in one of the universal conservation laws of physics. Mechanical energy includes kinetic energy, associated with motion, and potential energy, associated with position.

When a force F acts on a particle that undergoes a displacement s, the work W done by the force is defined as

$$W = Fs \cos \theta = F \cdot s,$$

where θ is the angle between the directions of F and s. The unit of work in SI units is 1 newton·meter = 1 joule (1 N·m = 1 J). Work is a scalar quantity; it has an algebraic sign (positive or negative) but no direction in space.

When the force varies during the displacement, the work done by the force is given by

$$W = \int_{x_1}^{x_2} F \, dx \qquad (7\text{--}4)$$

if the force has the same direction as the displacement, or

$$W = \int_{x_1}^{x_2} F \cos \theta \, dx \qquad (7\text{--}5)$$

if it makes an angle θ with the displacement. The angle θ may vary during the displacement.

The kinetic energy K of a particle having mass m and speed v is given by $K = \frac{1}{2}mv^2$. Kinetic energy is a scalar quantity; it has a magnitude (always positive) but no direction in space. When forces act on a particle while it undergoes a displacement, the total work W_{tot} done on it by all the forces equals the change in the particle's kinetic energy:

$$W_{\text{tot}} = K_2 - K_1.$$

The work done on a particle in a uniform gravitational field can be represented in terms of a potential energy $U = mgy$, as

$$W_{\text{grav}} = U_1 - U_2 = mgy_1 - mgy_2. \qquad (7\text{--}15)$$

More generally, the potential energy of a particle of mass m in the gravitational field of a particle of mass m_E is

$$U = -\frac{Gmm_E}{r}. \qquad (7\text{--}22)$$

If gravitational and other forces act on a body, the work W_{other} done by the forces other than gravitational potential energy:

$$W_{\text{other}} = K_2 - K_1 + U_2 - U_1 = E_2 - E_1,$$

where $E = K + U$ is the total energy (kinetic plus potential).

The work done by a stretched or compressed spring that exerts a force $F = -kx$ on a particle, where x is the amount of stretch or compression, can be represented in terms of a potential-energy function $U = \frac{1}{2}kx^2$:

$$W_{\text{el}} = U_1 - U_2.$$

If other forces also act on the body, then the work of the other forces equals the total change in kinetic energy plus elastic potential energy:

$$W_{\text{other}} = K_2 - K_1 + U_2 - U_1 = E_2 - E_1$$
$$= (\tfrac{1}{2}mv_2^2 + \tfrac{1}{2}kx_2^2) - (\tfrac{1}{2}mv_1^2 + \tfrac{1}{2}kx_1^2). \qquad (7\text{--}31)$$

conservation laws
work
joule
force constant
Hooke's law
kinetic energy
gravitational potential energy
total mechanical energy
conservation of mechanical energy
elastic potential energy
conservative force
dissipative force
power
average power

In the special cases where no forces other than elastic or gravitational do work on the body, the total energy $E = K + U$ is conserved or constant, and

$$K_1 + U_1 = K_2 + U_2,$$

where U may in general include both gravitational and elastic potential energies.

A conservative force is one for which the work–kinetic energy relation is completely reversible. The work of a conservative force can always be represented in terms of a potential energy, but the work of a nonconservative force cannot.

Power is time rate of doing work. If an amount of work ΔW is done in a time Δt, the average power P_{av} is defined as

$$P_{av} = \frac{\Delta W}{\Delta t},$$

and the instantaneous power P is defined as

$$P = \frac{dW}{dt}. \tag{7–33}$$

When a force $\boldsymbol{F}$ acts on a particle moving with velocity $\boldsymbol{v}$, the instantaneous power, or rate at which the force does work, is

$$P = \boldsymbol{F} \cdot \boldsymbol{v}.$$

Power, like work and kinetic energy, is a scalar quantity.

QUESTIONS

7–1 An elevator is hoisted by its cables at constant speed. Is the total work done on the elevator positive, negative, or zero?

7–2 A rope tied to a body is pulled, causing the body to accelerate. But according to Newton's third law the body pulls back on the rope with an equal and opposite force. Is the total work done then zero? If so, how can the body's kinetic energy change?

7–3 In Fig. 7–10, the projectile has the same initial kinetic energy in each case. Why does it not then rise to the same maximum height in each case?

7–4 Are there any cases where a frictional force can *increase* the mechanical energy of a system? If so, give examples.

7–5 An automobile jack is used to lift a heavy weight by exerting a force much smaller in magnitude than the weight. Does this mean that less work is done than if the weight had been lifted directly?

7–6 A compressed spring is clamped in its compressed position and is then dissolved in acid. What becomes of its potential energy?

7–7 A child standing on a playground swing can increase the amplitude of his motion by "pumping up," pulling back on the swing ropes at the appropriate points during the motion. Where does the added energy come from?

7–8 In a siphon, water is lifted above its original level during its flow from one container to another. Where does it get the needed potential energy?

7–9 A man bounces on a trampoline, going a little higher with each bounce. Explain how he increases his total mechanical energy.

7–10 Is it possible for the second hill on a roller-coaster track to be higher than the first? What would happen if it were higher?

7–11 Does the kinetic energy of a car change more when it speeds up from 10 to 15 $m \cdot s^{-1}$ or from 15 to 20 $m \cdot s^{-1}$?

7–12 A car accelerates from an initial speed to a greater final speed while the engine develops constant power. Is the acceleration greater at the beginning of this process or at the end?

7–13 Time yourself while running up a flight of steps, and compute your maximum power, in horsepower. Are you stronger than a horse?

7–14 When a constant force is applied to a body moving with constant acceleration, is the power of the force constant? If not, how would the force have to vary with speed for the power to be constant?

7–15 An advertisement for a power saw states: "This power tool uses the energy of the motor to stop the blade rotation within 5 seconds after the switch is turned off." Is this an accurate statement? Please explain.

EXERCISES

Section 7-2 Work

7-1 A block is pushed 2 m along a horizontal table top by a horizontal force of 2 N. The opposing force of friction is 0.4 N.

a) How much work is done by the 2-N force?

b) What is the work done by the friction force?

7-2 A 40-kg crate is pushed a distance of 5 m along a level floor at a constant speed by a force that is at an angle of 30° below the horizontal. The coefficient of kinetic friction between the crate and floor is 0.25.

a) How much work is done by this force?

b) How much work is done by friction?

c) How much work is done by the normal force? By gravity?

7-3 A 22-lb block is pulled up a frictionless plane inclined at 30° to the horizontal, by a force **P** of 18 lb acting parallel to the plane. If the block travels 15 ft along the incline, calculate

a) the work done by the force **P**,

b) the work done by the gravity force,

c) the work done by the normal force,

d) the total work done on the block.

7-4 A 5-kg package slides 2 m down a surface that is inclined at 53° below the horizontal. The coefficient of kinetic friction between the package and the surface is $\mu_k = 0.4$. Calculate

a) the work done by friction on the package,

b) the work done by gravity,

c) the work done by the normal force,

d) the total work done on the package.

Section 7-3 Work Done by a Varying Force

7-5 A force of 100 N is observed to stretch a certain spring a distance of 0.4 m.

a) What force is required to stretch the spring 0.1 m? To compress the spring 0.2 m?

b) How much work must be done to stretch the spring 0.1 m beyond its unstretched length? How much work to compress the spring 0.2 m from its equilibrium position?

7-6 A force **F** that is parallel to the x-axis acts on an object. The magnitude of the force varies with the x-coordinate of the object as shown in Fig. 7-16. Calculate the work done by the force **F** when the object moves from $x = 0$ to $x = 7$ m.

7-7 An object is attracted toward the origin with a force given by $F = -(6\text{N}\cdot\text{m}^{-3})x^3$, where F is in newtons and x is in meters.

a) What is the force F when the object is at point a, 1 m from the origin?

b) At point b, 2 m from the origin?

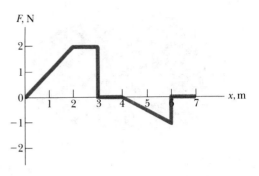

FIGURE 7-16

c) How much work is done by the force F when the object moves from a to b? Is this work positive or negative?

7-8 An object is attracted toward the origin with a force given by $F = -k/x^2$. Calculate the work done by the force F when the object moves in the x-direction from x_1 to x_2. If $x_2 > x_1$, is the work done by F positive or negative?

Section 7-4 Work and Kinetic Energy

7-9 Compute the kinetic energy, in joules, of a 2-g rifle bullet traveling at 500 m·s⁻¹.

7-10 An object of mass 2 slugs is initially at rest on a frictionless horizontal surface. It is then pulled 4 ft by a horizontal force of magnitude 25 lb. Use the work-energy relation (Eq. 7-12) to find its final speed.

7-11 An object of mass 2 kg moves in a straight line on a frictionless horizontal surface. It has an initial speed of 10 m·s⁻¹ and then is pulled 4 m by a force of magnitude 25 N and in the direction of the initial velocity. What is the object's final speed? Use the work-energy relation.

7-12 A block of mass 8 kg moves in a straight line on a frictionless horizontal surface. At one point in its path its speed is 4 m·s⁻¹, and after it has traveled 3 m its speed is 5 m·s⁻¹ in the same direction. Use the work-energy relation to find the force acting on the block, assuming that this force is constant.

7-13 A car is traveling on a level road with velocity v_0 at the instant the brakes are applied.

a) Use the work-energy relation (Eq. 7-12) to calculate the minimum stopping distance of the car in terms of v_0, the coefficient of friction μ between the tires and the road, and the acceleration g due to gravity.

b) It is observed that the car stops in a distance of 140 ft if $v_0 = 40$ mi·hr⁻¹. What is the stopping distance if $v_0 = 60$ mi·hr⁻¹? Assume that μ and hence the friction force remain the same.

7-14 A small block of mass 0.05 kg is attached to a cord passing through a hole in a frictionless horizontal surface, as in Fig. 7-17. The block is originally revolving at a distance of 0.2 m from the hole with a speed of 0.6 m·s⁻¹. The cord is then pulled from below, shortening the radius of the circle in which the block revolves to 0.1 m. At this new distance the speed of the block is observed to be 1.2 m·s⁻¹. How much work was done by the person who pulled the cord?

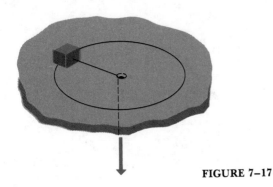

FIGURE 7–17

7–15 A proton of mass 1.67×10^{-27} kg is propelled at an initial speed of $2 \times 10^4 \mathrm{m \cdot s^{-1}}$ directly toward a gold nucleus that is 4 m away. The proton is repelled by the gold nucleus with a force of magnitude $F = (1.82 \times 10^{-26} \, \mathrm{N \cdot m^2})/x^2$, where x is the separation between the two objects. Assume that the gold nucleus remains at rest.

a) What is the speed of the proton when it is 1.0×10^{-7} m from the gold nucleus?

b) How close to the gold nucleus does the proton get? (Calculate the position of the proton for which the work done by the repulsive force has reduced the velocity of the proton momentarily to zero.)

7–16 An object of mass 5 kg is initially at rest on a frictionless horizontal surface. A force F is then applied to the object, and as a result the object moves along the x-axis such that its position as a function of time is given by $x(t) = (5\mathrm{m \cdot s^{-2}})t^2 + (2\mathrm{m \cdot s^{-3}})t^3$, where t is in seconds and x is in meters.

a) Calculate the velocity of the object when $t = 4$ s.

b) Calculate the work done by the force F during the first 4 s of the motion. (*Hint:* Use Eq. 7–12.)

Section 7–5 Gravitational Potential Energy

7–17 What is the potential energy of an 800-kg elevator at the top of the Empire State Building, 380 m above street level? Assume the potential energy at street level to be zero.

7–18 A 5-kg block is lifted vertically at a constant velocity of 4 m·s^{-1} through a height of 12 m.

a) How great a force is required?

b) How much work is done? What becomes of this work?

7–19 A man of mass 80 kg sits on a platform suspended from a movable pulley as shown in Fig. 7–18 and raises himself at constant speed by a rope passing over a fixed pulley. The platform and the pulleys have negligible mass. Assuming no friction losses, find

a) the force he must exert;

b) the increase in his energy when he raises himself 1 m.

Answer part (b) by calculating his increase in potential energy, and also by computing the product of the force on the rope and the length of the rope passing through his hands.

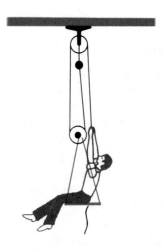

FIGURE 7–18

7–20 A barrel of mass 120 kg is suspended by a rope 10 m long.

a) What horizontal force is necessary to hold the barrel in a position displaced sideways 2 m from the vertical?

b) How much work is done in moving it to this position?

7–21 A ball is thrown from the roof of a 90-ft-tall building with an initial velocity of 60 ft·s^{-1} at an angle of 37° above the horizontal.

a) What is the speed of the ball just before it strikes the ground? Use energy methods.

b) What would be the answer in (a) if the initial velocity was at an angle of 37° *below* the horizontal?

7–22 A small sphere of mass m is fastened to a weightless string of length 0.5 m to form a pendulum. The pendulum is swinging so as to make a maximum angle of 60° with the vertical. What is the speed of the sphere when it passes through the vertical position?

7–23 An object of mass 0.1 kg is released from rest at point A, which is at the top edge of a hemispherical bowl of radius $R = 0.4$ m (Fig. 7–19). When it reaches point B at the bottom of the bowl, the object is observed to have speed $v = 1.8$ m·s^{-1}. Calculate the *work* done by friction on the block when it moves from A to B. (*Note.* The friction force is not constant. You can easily solve for the friction *work* but not for the friction *force*.)

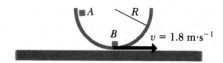

FIGURE 7–19

7–24 A 5-kg block is pushed up a frictionless plane inclined at 30° to the horizontal. It is pushed 2.0 m along the incline by a constant 100 N force parallel to the plane. If its speed at the bottom is 2 m·s^{-1}, what is its speed at the top? Use energy methods.

7–25 A 12-kg block is pushed 20 m up the sloping surface of a plane inclined at an angle of 37° to the horizontal, by a constant force F of 120 N acting parallel to the plane. The coefficient of kinetic friction between the block and plane is 0.25.

a) What is the work of the force F?

b) What is the work done by the friction force?

c) Compute the increase in potential energy of the block.

d) Use your answers to (a), (b), and (c) to calculate the increase in kinetic energy of the block.

e) Use $\Sigma F = ma$ to calculate the acceleration of the block. Assuming that the block is initially at rest, use the acceleration to calculate the block's velocity after traveling 20 m. Compute from this the increase in kinetic energy of the block, and compare to the answer you got in (d).

7–26 Use the results of Example 7–10 to calculate the escape velocity for an object

a) from the surface of the moon,

b) from the surface of Saturn.

c) Why is the escape velocity for an object independent of the object's mass?

7–27 The asteroid Toro, discovered in 1964, has a radius of about 5 km and a mass of about 2×10^{15} kg. Calculate the escape velocity for an object at the surface of Toro. Could a person *jump* off Toro, i.e., jump with a velocity greater than the escape velocity?

7–28 An object of mass m is shot vertically upward from the surface of the earth. If the object's initial velocity is 6.0×10^3 m·s^{-1}, to what height above the surface of the earth will it rise?

Section 7–6 Elastic Potential Energy

7–29 A force of 1200 N stretches a certain spring a distance of 0.1 m.

a) What is the potential energy of the spring when it is stretched 0.1 m? Compressed 0.05 m?

b) What is the potential energy of the spring when a 60-kg mass hangs vertically from it?

7–30 The spring of a spring gun has a force constant of $k = 500$ N·m^{-1}. The spring is compressed 0.05 m and a ball of mass 0.01 kg is placed in the barrel against the compressed spring. The spring is then released and the ball is propelled out the barrel.

a) Compute the speed with which the ball leaves the gun, if friction forces are negligible.

b) Determine the speed of the ball as it leaves the end of the barrel if a constant resisting force of 10 N acts on the ball while it travels the 0.05 m along the barrel. Note that in this case the maximum speed does not occur at the end of the barrel.

7–31 A 2-kg block is dropped from a height of 0.4 m onto a spring whose force constant k is 1960 N·m^{-1}. Find the maximum distance the spring will be compressed.

7–32 A block of mass 5 kg is moving at 8 m·s^{-1} along a frictionless horizontal surface toward a spring of force constant $k = 500$ N·m^{-1} that is attached to a wall (Fig. 7–20). Find the maximum distance the spring will be compressed. The spring has negligible mass.

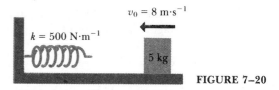

FIGURE 7–20

7–33 A block of mass 0.5 kg is pressed against a vertical spring of force constant $k = 500$ N·m^{-1} such that the spring is compressed 0.2 m. When the block is released, how high will the block rise from this position? (The block and the spring are *not* attached. The spring has negligible mass.)

7–34 A 2-kg block is pushed against a spring of force constant $k = 400$ N·m^{-1}, compressing it 0.3 m. The block is then released and moves along a frictionless horizontal surface and then up a frictionless incline of slope 37° (Fig. 7–21).

a) What is the speed of the block as it slides along the horizontal surface after having left the spring?

b) How far does the object travel up the incline before starting to slide back down?

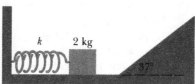

FIGURE 7–21

7–35 A block of mass 1 kg is forced against a horizontal spring of negligible mass, compressing the spring a distance $x_1 = 0.2$ m. When released, the block moves on a horizontal table top a distance $x_2 = 1.0$ m before coming to rest. The spring constant k is 100 N·m^{-1} (Fig. 7–22). What is the coefficient of kinetic friction, μ_k, between the block and the table?

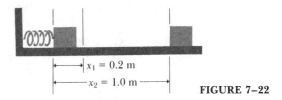

FIGURE 7–22

Section 7–9 Power

7–36 At 5 cents per kWh, what does it cost to operate a 10-hp motor for 8 hr?

7–37 A tandem (two-person) bicycle team must overcome a force of 34.0 lb to maintain a speed of 30.0 ft·s^{-1}. Find the power required per rider, assuming they each contribute equally. Express your answer in horsepower.

7–38 The total consumption of electrical energy in the United States is of the order of 10^{19} joules per year.

a) What is the average rate of electrical energy consumption, in watts?

b) If the population of the United States is 200 million, what is the average rate of electrical energy consumption per person?

c) The sun transfers energy to the earth by radiation, at a rate of approximately 1.4 kW per square meter of surface. If this energy could be collected and converted to electrical energy with 100% efficiency, how great an area (in square kilometers) would be required to collect the electrical energy used by the United States?

7–39 A man whose mass is 70 kg walks up to the third floor of a building. This is a vertical height of 12 m above the street level.

a) By how much has he increased his potential energy?

b) If he climbs the stairs in 20 s, what is his rate of working, in watts?

7–40 The hammer of a pile driver has a weight of 1100 lb and must be lifted a vertical distance of 6 ft in 3 s. What horsepower engine is required?

7–41 The Grand Coulee Dam is 1270 m long and 170 m high. The electrical power output from generators at its base is approximately 2000 megawatts. How much water in $m^3 \cdot s^{-1}$ must be flowing over the dam to produce this amount of power? (The density, or mass per unit volume, of water is 1.0×10^3 $kg \cdot m^{-3}$.)

7–42 A ski tow is to be operated on a 37° slope 300 m long. The rope is to move at 12 $km \cdot hr^{-1}$, and power must be provided for 80 riders at one time, with an average mass

per rider of 70 kg. Estimate the power required to operate the tow.

7–43 The engine of a motor boat delivers 30 kW to the propeller while the boat is moving at 10 $m \cdot s^{-1}$. What would be the tension in the towline if the boat were being towed at the same speed?

7–44 The engine of an automobile develops 15 kW (20 hp) when the automobile is traveling at 50 $km \cdot hr^{-1}$.

a) What is the resisting force acting on the automobile, in newtons?

b) If the resisting force is proportional to the velocity, what power will drive the car at 25 $km \cdot hr^{-1}$? At 100 $km \cdot hr^{-1}$? Give your answers in kilowatts and in horsepower.

Section 7–10 Automotive Power

7–45 For a touring bicyclist, the drag coefficient is 1.0, the frontal area is 0.463 m^2, and the tractive resistance is 0.0045. The rider weighs 712 N and his bike 111 N.

a) To maintain a speed of 14 $m \cdot s^{-1}$ (about 31 $mi \cdot hr^{-1}$) on a level road, what must the rider's power output be?

b) For racing, the same rider uses a different bike, one with tractive resistance 0.003 and weight 89 N. He also crouches down, reducing his drag coefficient to 0.88 and reducing his frontal area to 0.366 m^2. What then must his power output be to maintain a speed of 14 $m \cdot s^{-1}$?

c) For the situation in part (b), what power output is required to maintain a speed of 7 $m \cdot s^{-1}$? Note the great drop in power requirement when the speed is only halved. (See "The Aerodynamics of Human-Powered Land Vehicles," in *Scientific American*, December 1983. The article discusses aerodynamic speed limitations for a wide variety of human-powered vehicles.)

PROBLEMS

7–46 A projectile of mass m is fired from a gun with a muzzle velocity of v_0 at an angle of θ above the horizontal. Neglecting air resistance and using energy methods, find the maximum height attained by the projectile.

7–47 The system in Fig. 7–23 is released from rest with the 12-kg block 3 m above the floor. Use the principle of conservation of energy to find the speed with which the

block strikes the floor. Neglect friction and the inertia of the pulley.

7–48 A small object of mass m slides without friction around the loop-the-loop apparatus shown in Fig. 7–24. It starts from rest at point A at a height h above the bottom of the loop.

a) What is the minimum value of h (in terms of R) such that the object moves around the loop without falling off at the top (point B)?

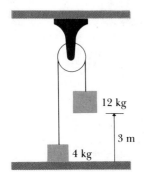

12 kg

3 m

4 kg

FIGURE 7–23

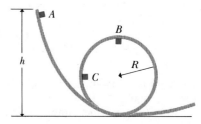

A

B

h

C   *R*

FIGURE 7–24

b) If $h = 3R$, compute the velocity, radial acceleration, and tangential acceleration of the object at point C, which is at the end of a horizontal diameter.

7–49 A skier starts at the top of a very large frictionless spherical snowball, with a very small initial velocity, and skis straight down the side. At what point does he lose contact with the snowball and fly off at a tangent? That is, at the instant he loses contact with the snowball, what angle does a radial line from the center to the skier make with the vertical?

7–50 In an ore-processing plant a conveyor belt carrying small pieces of the refined metal moves at a constant velocity v_0. The conveyor belt goes around a cylinder of radius R as shown in Fig. 7–25. At some angle θ the pieces of metal no longer contact the conveyor belt. In order to design the next phase of the plant the engineer needs to know this angle θ. Find θ in terms of g, R, and v_0.

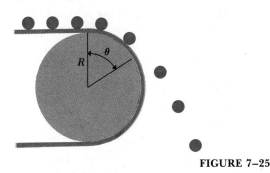

FIGURE 7–25

7–51 A ball of mass 0.5 kg is tied to a string of length 1.0 m, and the other end of the string is tied to a rigid support. The ball is held straight out horizontally from the point of support, with the string pulled taut, and is then released.

a) What is the speed of the ball at the lowest point of its motion?

b) What is the tension in the string at this point?

7–52 A ball is tied to a cord and set in rotation in a vertical circle. Prove that the tension in the cord at the lowest point exceeds that at the highest point by six times the weight of the ball.

7–53 A block of mass 2 kg is released from rest at point A on a track that is one quadrant of a circle of radius 1 m (Fig. 7–26). It slides down the track and reaches point B with a speed of 4 m·s^{-1}. From point B it slides on a level surface a distance of 3 m to point C, where it comes to rest.

a) What was the coefficient of sliding friction on the horizontal surface?

b) How much work was done by friction as the block slid down the circular arc from A to B?

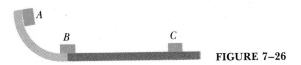

FIGURE 7–26

7–54 Consider the system shown in Fig. 7–27. The coefficient of kinetic friction between the 8 kg block and the table top is $\mu_k = 0.3$. Neglect the mass of the rope and of the pulley, and assume that the pulley is frictionless. Use

FIGURE 7–27

energy methods to calculate the speed of the 6-kg block after it has descended 5 m, starting from rest.

7–55 A pump is required to lift 800 kg (about 200 gal) of water per minute from a well 10 m deep and eject it with a speed of 20 m·s^{-1}.

a) How much work is done per minute in lifting the water?

b) How much in giving it kinetic energy?

c) What power engine is needed?

7–56 An automobile of mass m accelerates, starting from rest, while the engine supplies constant power P.

a) Show that the speed is given as a function of time by $v = (2Pt/m)^{1/2}$.

b) Show that the displacement is given as a function of time by $x - x_0 = (8P/9m)^{1/2}t^{3/2}$.

7–57 The force exerted by a gas in a cylinder on a piston whose area is A is given by $F = pA$, where p is the force per unit area, or *pressure*. The work W in a displacement of the piston from x_1 to x_2 is

$$W = \int_{x_1}^{x_2} F \, dx = \int_{x_1}^{x_2} pA \, dx = \int_{V_1}^{V_2} p \, dV,$$

where dV is the accompanying infinitesimal change of volume of the gas.

a) During an expansion of a gas at constant temperature (isothermal), the pressure depends on the volume according to the relation

$$p = \frac{nRT}{V},$$

where n and R are constants and T is the constant temperature. Calculate the work done by the force F exerted by the gas when the gas expands isothermally from volume V_1 to volume V_2.

b) During an expansion of a gas at constant entropy (adiabatic), the pressure depends on the volume according to the relation

$$p = \frac{k}{V^\gamma},$$

where k and γ are constants. Calculate the work done by the gas when it expands adiabatically from V_1 to V_2.

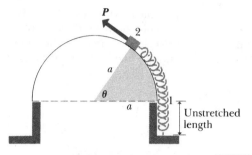

Unstretched length

FIGURE 7–28

7–58 A variable force P is maintained tangent to a frictionless surface of radius a, as shown in Fig. 7–28. By slowly varying the force, a block of weight w is moved and the spring to which it is attached is stretched from position 1 to position 2. Calculate the work done by the force P.

7–59 An object has several forces acting on it. One of these forces is $F = 3xy\mathbf{i}$, a force in the x-direction whose magnitude depends on the position of the object. If x and y are in meters, then F is in newtons. Calculate the work done by this force for each of the following displacements of the object:

a) The object starts at the point $x = 0$, $y = 4$ m and moves parallel to the x-axis to the point $x = 2$ m, $y = 4$ m.

b) The object starts at this point $x = 2$ m, $y = 0$ and moves in the y-direction to the point $x = 2$ m, $y = 4$ m.

c) The object starts at the origin and moves on the line $y = 2x$ to the point $x = 2$ m, $y = 4$ m.

7–60 An object has several forces acting on it. One of these forces is $F = -2xy^2\mathbf{j}$, a force in the negative y-direction whose magnitude depends on the position of the object. For x and y in meters, F is newtons. Consider the displacement of the object from the origin to the point $x = 2$ m, $y = 2$ m.

a) Calculate the work done by F if this displacement is along the straight line $y = x$ that connects these two points.

b) Calculate the work done by F if this displacement is carried out by the object first moving out along the x-axis to the point $x = 2$ m, $y = 0$, and then moving parallel to the y-axis to $x = 2$ m, $y = 2$ m.

c) Compare the work done by F along these two paths. Is the force F conservative or nonconservative?

7–61 There are two equations from which a change in the gravitational potential energy U of a mass m can be calculated. One is $U = mgy$ (Eq. 7–14). The other is $U = -G(mm_E/r)$ (Eq. 7–22). As shown in Section 7–5, the first equation is correct only if g is a constant over the change in height Δy. The second is always correct. Actually g is never exactly constant over any change in height, but the variation may be so small that we ignore it. Consider the difference in U between a mass at the earth's surface and a distance h above it by using both equations, and find for what height h Eq. (7–14) is in error by 1%. Express this h as a fraction of the earth's radius, and also obtain a numerical value for it.

7–62 An object is thrown straight up from the surface of the asteroid Toro (see Exercise 7–27) with a velocity of 15 m·s^{-1}, which is greater than the escape velocity. What is the speed of the object when it is far from Toro? Neglect the gravitational forces due to all other astronomical objects.

7–63 An object of mass m is dropped from a height h above the earth's surface. This height is not necessarily small compared with the radius R_E of the earth. If air resistance is neglected, derive an expression for the speed v of the object when it reaches the surface of the earth. Your expression should involve h, R_E, and g.

7–64 A shaft is drilled from the surface to the center of the earth. The gravitational force on an object of mass m that is inside the earth at a distance r from the center has magnitude $F = mgr/R_E$, where R_E is the radius of the earth, and points toward the center of the earth. If an object is released in the shaft at the surface of the earth, what speed will it have when it reaches the center of the earth? (See Challenge Problem 6–42 and Problem 11–34.)

7–65 A certain spring is found *not* to obey Hooke's law but rather exerts a restoring force $F(x) = -80x - 15x^2$ if it is stretched or compressed a distance x. The units of the numerical factors are such that if x is in meters, then F will be in newtons.

a) Calculate the potential energy function $U(x)$ for this spring. Let $U = 0$ when $x = 0$.

b) An object of mass 2 kg is attached to this spring, pulled a distance 1.0 m to the right on a frictionless horizontal surface, and released. What is the velocity of the object when it is 0.5 m to the right of the $x = 0$ equilibrium position?

7–66 An 80 kg man jumps from a height of 2 m onto a platform mounted on springs. As the springs compress, the platform is pushed down a maximum distance of 0.2 m below its initial position, and then it rebounds. The platform and springs have negligible mass.

a) What is the man's speed at the instant the platform is depressed 0.1 m?

b) If the man had just stepped gently onto the platform, how far would it have been pushed down?

7–67 A block of mass 5 kg is placed against a compressed spring at the bottom of an incline of slope 37° (point A). When the spring is released, it projects the block up the incline. At point B, a distance of 6 m up the incline from A, the block is observed to have a velocity of 4 m·s^{-1}, directed up the incline. The coefficient of kinetic friction between the block and incline is $\mu_k = 0.5$. Calculate the amount of potential energy that was initially stored in the spring.

7–68 If an object is attached to a vertical spring and slowly lowered to its equilibrium position, it is found to stretch the spring by an amount d. If the same object is attached to the unstretched spring and then allowed to fall from rest, through what maximum distance does it stretch the spring? (*Hint:* Calculate the force constant of the spring in terms of the distance d and the mass m of the object.)

CHALLENGE PROBLEMS

7–69 A 0.5-kg block *attached* to a spring of length 0.6 m and force constant $k = 40$ N·m^{-1} is at rest at point A on a frictionless horizontal surface (Fig. 7–29). A constant horizontal force $P = 20$ N is applied to the block and moves the block to the right along the surface.

a) What is the velocity of the block when it reaches point B, which is 0.25 m to the right of point A?

b) When the block reaches point B, the force P is suddenly removed. In the subsequent motion, how close will the block get to the wall?

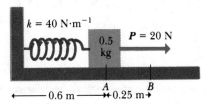

FIGURE 7–29

7–70 A 2-kg block is released on an incline 5 m from a long spring of force constant $k = 80$ N·m^{-1} that is attached at the bottom of the incline. The incline makes an angle of 53° with the horizontal, and the coefficients of friction between the block and the incline are $\mu_s = 0.4$ and $\mu_k = 0.2$.

a) What is the velocity of the block just before it reaches the spring?

b) What will be the maximum compression of the spring?

c) The spring will propel the block back up the incline. How close will the block get to its initial position?

7–71 A particle of mass m moves with a one-dimensional potential-energy function

$$U(x) = \frac{\alpha}{x^2} - \frac{\beta}{x}.$$

The particle is released from rest at $x_0 = \alpha/\beta$.

a) Show that $U(x)$ can be written as

$$U(x) = \frac{\alpha}{x_0^2}\left[\left(\frac{x_0}{x}\right)^2 - \left(\frac{x_0}{x}\right)\right].$$

Sketch $U(x)$. Calculate $U(x_0)$ and thereby locate the point x_0 on the sketch.

b) Calculate $v(x)$, the velocity of the particle as a function of position. Draw a sketch of v as a function of x, and give a qualitative description of the motion.

c) For what value of x is the velocity of the particle a maximum, and what is the value of that maximum velocity?

d) What is the force on the particle at the point where the velocity is maximum?

e) Now let the particle be released instead at $x_1 = 3(\alpha/\beta)$. Locate the point x_1 on the sketch of $U(x)$. Calculate $v(x)$ and give a qualitative description of the motion.

f) In each case, when the particle is released at $x = x_0$ and at $x = x_1$, what are the maximum and minimum values of x reached during the motion?

8

IMPULSE AND MOMENTUM

IN THE PRECEDING CHAPTER WE STUDIED THE CONCEPT OF ENERGY AND THE principle of conservation of energy, one of the great conservation principles of physical science. In the course of this study, we developed the concepts of work and mechanical energy from Newton's laws of motion. In this chapter we develop an equally important and fundamental conservation law, conservation of *momentum*. We introduce two new concepts, impulse and momentum; we show how they also arise from Newton's laws and how they are related. This information is especially useful in situations such as impacts and collisions, where very large forces act for short times and cause sudden changes in motion of a body. The concept of momentum and the principle of conservation of momentum appear in many other areas of physics as well as in mechanics.

8–1 IMPULSE AND MOMENTUM

To introduce the concepts of impulse and momentum, let us first consider a particle of mass m moving along a straight line. For the moment we assume that the force F on this particle is constant and directed along the line of motion. If the particle's velocity at some initial time $t = 0$ is v_0, the velocity v at a later time t is given by

$$v = v_0 + at,$$

where the constant acceleration a is determined from the relation $F = ma$. When we multiply the above equation by m and replace ma by F, the result is

$$mv = mv_0 + Ft,$$

or

$$mv - mv_0 = Ft. \tag{8–1}$$

The right side of this equation, the product of the force and the time during which it acts, is called the **impulse** of the force, denoted by J. More generally, when a constant force F acts during a time interval from t_1 to t_2, the impulse of the force is defined to be

$$\text{Impulse} = J = F(t_2 - t_1). \tag{8-2}$$

The left side of Eq. (8–1) contains the product of mass and velocity of the particle at two different times. This product is also given a special name: **momentum.** It is also sometimes called *linear momentum* to distinguish it from a similar quantity called *angular momentum,* to be discussed later. We use the symbol p for momentum:

$$\text{Momentum} = p = mv. \tag{8-3}$$

In terms of these newly defined quantities, the meaning of Eq. (8–1) is that the impulse of the force from time zero to time t is equal to the change of momentum during that interval. There is nothing special about these two times; we may also say that if the particle's velocity at time t_1 is v_1, and its velocity at time t_2 is v_2, then

$$F(t_2 - t_1) = mv_2 - mv_1. \tag{8-4}$$

Momentum depends on the motion of the particle; impulse depends on the force acting on it. Thus Eq. (8–4) is a relation between force and motion. Indeed, it is closely related to Newton's second law, which can be restated in the form

$$F = ma = m\,\frac{dv}{dt} = \frac{d(mv)}{dt}$$
$$= \frac{dp}{dt}.$$

Straight-line motion is, of course, a special case. Strictly speaking, the quantities F, v, and a should have been called the *components* along the straight line of the vector quantities $\boldsymbol{F}$, $\boldsymbol{v}$, and $\boldsymbol{a}$, respectively. The definitions of momentum and impulse and the relationship expressed by Eq. (8–4) are easily generalized for motion not confined to a straight line. We define impulse and momentum to be *vector quantities*, as follows:

$$\text{Impulse} = \boldsymbol{J} = \boldsymbol{F}(t_2 - t_1), \tag{8-5}$$

$$\text{Momentum} = \boldsymbol{p} = m\boldsymbol{v}. \tag{8-6}$$

The corresponding generalization of Eq. (8–4) is the vector equation

$$\boldsymbol{F}(t_2 - t_1) = m\boldsymbol{v}_2 - m\boldsymbol{v}_1. \tag{8-7}$$

This relationship between impulse and momentum change has the same form as the relation between work and change of kinetic energy developed in Chapter 7. Both are relations between force and change of motion. There are important differences, however. First, impulse is a product of a force and a *time* interval, while work is a product of force and a *distance* and depends on the angle between force and displacement. Furthermore, both force and velocity are *vector* quantities, so impulse and momentum are also vector quantities, unlike work and kinetic energy, which are scalars. Even in straight-line motion, where only one component of a vector is involved, force and velocity may have components along this line that are either positive or negative.

EXAMPLE 8–1   A particle of mass 2 kg moves along the x-axis with an initial velocity of 3 m·s^{-1}. A force $F = -6$ N (i.e., in the negative x-direction) is applied for a period of 3 s. Find the final velocity.

SOLUTION   From Eq. (8–4),

$$(-6 \text{ N})(3 \text{ s}) = (2 \text{ kg})v_2 - (2 \text{ kg})(3 \text{ m·s}^{-1}),$$

or

$$v_2 = -6 \text{ m·s}^{-1}.$$

The particle's final velocity is in the negative x-direction.

Units of impulse and momentum

The unit of impulse, in any system, equals the product of the units of force and time in that system. Thus, in SI units, the unit is one newton-second (1 N·s); in the cgs system it is one dyne-second (1 dyn·s); and in the British system it is one pound-second (1 lb·s).

The unit of momentum in SI units is one kilogram meter per second (1 kg·m·s^{-1}); in the cgs system it is one gram centimeter per second (1 g·cm·s^{-1}), and in the British system it is 1 slug foot per second (1 slug·ft·s^{-1}). Because

$$1 \text{ kg·m·s}^{-1} = (1 \text{ kg·m·s}^{-2})\text{s} = 1 \text{ N·s},$$

the units of momentum and impulse are equivalent.

We have defined impulse only for a constant force, but the concept can be generalized to include forces that vary with time; both the magnitude and the direction of the force may vary. Consider a particle of mass m moving in the xy-plane, or in space, acted on by a varying resultant force $\boldsymbol{F}$. Newton's second law states that

$$\boldsymbol{F} = m\boldsymbol{a} = m\frac{d\boldsymbol{v}}{dt},$$

or

$$\boldsymbol{F} \, dt = m \, d\boldsymbol{v}.$$

When the force varies with time, the impulse can be expressed as an integral.

If $\boldsymbol{v}_1$ is the particle's velocity when $t = t_1$, and $\boldsymbol{v}_2$ is its velocity when $t = t_2$, then

$$\int_{t_1}^{t_2} \boldsymbol{F} \, dt = \int_{\boldsymbol{v}_1}^{\boldsymbol{v}_2} m \, d\boldsymbol{v}. \tag{8–8}$$

The integral on the left is the impulse $\boldsymbol{J}$ of the force $\boldsymbol{F}$ in the time interval $t_2 - t_1$, and is a vector quantity:

$$\text{Impulse} = \boldsymbol{J} = \int_{t_1}^{t_2} \boldsymbol{F} \, dt.$$

This integral can be evaluated, of course, only when the force is known as a function of time. The integral on the right, however, is just m multiplied by the total change in $\boldsymbol{v}$:

$$\int_{\boldsymbol{v}_1}^{\boldsymbol{v}_2} m \, d\boldsymbol{v} = m \int_{\boldsymbol{v}_1}^{\boldsymbol{v}_2} d\boldsymbol{v} = m(\boldsymbol{v}_2 - \boldsymbol{v}_1).$$

Thus Eq. (8–8) may be written as

$$\int_{t_1}^{t_2} \mathbf{F} \, dt = m\mathbf{v}_2 - m\mathbf{v}_1. \qquad (8\text{–}9)$$

In the special case when $\mathbf{F}$ is constant, this reduces to Eq. (8–7). When $\mathbf{F}$ varies with time, we can define an *average* force $\mathbf{F}_{av}$ such that the product $\mathbf{F}_{av}(t_2 - t_1)$ represents the same area under the curve in Fig. 8–1. This can be done even when the force varies in direction, so the statement

$$\mathbf{F}_{av}(t_2 - t_1) = \mathbf{J} = m\mathbf{v}_2 - m\mathbf{v}_1 \qquad (8\text{–}10)$$

is perfectly general, whether the actual force $\mathbf{F}$ is constant or not.

Equation (8–9) is a vector equation; it is equivalent (for forces and velocities in the xy-plane) to the two scalar equations

$$\int_{t_1}^{t_2} F_x \, dt = mv_{2x} - mv_{1x},$$
$$\int_{t_1}^{t_2} F_y \, dt = mv_{2y} - mv_{1y}. \qquad (8\text{–}11)$$

The impulse of any force component, or any force whose direction is constant, can be represented graphically by plotting the force vertically and time horizontally, as in Fig. 8–1. The *area* under the curve, between vertical lines at t_1 and t_2, is equal to the impulse of the force in this time interval.

When the component of the impulse of a force, in a given direction, is *positive*, the corresponding component of momentum of the body on which the impulse acts *increases* algebraically; when the component of impulse is negative, the momentum component decreases. If the impulse is zero, there is no change in momentum.

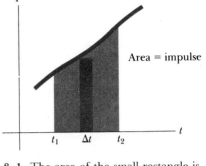

8–1 The area of the small rectangle is approximately equal to the total change in momentum in the interval Δt.

Impulse and momentum can be expressed in terms of their components.

EXAMPLE 8–2 Consider the changes in momentum produced by the following forces:

a) A hockey puck on a frictionless horizontal ice surface moves along the x-axis; it is acted on for 2 s by a constant force of 10 N toward the right.

b) The puck is acted on for 2 s by a constant force of 10 N toward the right and then for 2 s by a constant force of 20 N toward the left.

c) The puck is acted on for 2 s by a constant force of 10 N toward the right, then for 1 s by a constant force of 20 N toward the left. The forces for each case are shown graphically in Fig. 8–2.

SOLUTION

a) The impulse of the force is $(+10 \text{ N})(2 \text{ s}) = +20$ N·s. Hence the momentum of *any* body on which this force acts increases by 20 kg·m·s^{-1}. This change is the same whatever the mass of the body and whatever the magnitude and direction of its initial velocity.

Suppose the mass of the puck is 2 kg and that it is initially at rest. Its *final* momentum then equals its *change* in momentum, and its final *velocity* is 10 m·s^{-1} toward the right. You may verify this by computing the acceleration.

Hitting a hockey puck: analysis in terms of impulse and momentum

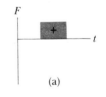

(a)

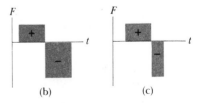

(b)   (c)

8–2 Graphic representation of the forces in Example 8–2.

If the puck had been moving initially toward the *right* at 5 m·s^{-1}, its initial momentum would have been 10 kg·m·s^{-1} toward the right, its final momentum 30 kg·m·s^{-1}, and its final velocity 15 m·s^{-1} toward the right.

If the puck has been moving initially toward the *left* at 5 m·s^{-1}, its initial momentum would have been −10 kg·m·s^{-1}, its final momentum +10 kg·m·s^{-1}, and its final velocity 5 m·s^{-1} toward the right. That is, the constant force of 10 N toward the right would first have brought the body to rest and then given it a velocity in the direction opposite to its initial velocity.

b) The impulse of this force is

$$(+10 \text{ N})(2 \text{ s}) - (20 \text{ N})(2 \text{ s}) = -20 \text{ N·s}.$$

The momentum of *any* body on which this impulse acts is changed by −20 kg·m·s^{-1}. We invite you to check various possibilities.

c) The impulse of this force is

$$(+10 \text{ N})(2 \text{ s}) - (20 \text{ N})(1 \text{ s}) = 0.$$

Hence the momentum of any body on which it acts is not changed. Of course, the momentum of the body is increased during the first 2 s, but it is *decreased* by an equal amount in the next second. As an exercise, describe the motion of a body of mass 2 kg, moving initially to the left at 5 m·s^{-1}, and acted on by this force. It helps to construct a graph of velocity versus time.

A ball bouncing off a wall: using impulse and momentum to find the impact force

EXAMPLE 8–3 A ball of mass 0.40 kg is thrown against a brick wall. When it strikes the wall, the ball is moving horizontally to the left at 30 m·s^{-1}, and it rebounds horizontally to the right at 20 m·s^{-1}. Find the impulse of the force exerted on the ball by the wall. If the ball is in contact with the wall for 0.010 s, find the average force on the ball during the impact.

SOLUTION The initial momentum of the ball is

$$(0.40 \text{ kg})(-30 \text{ m·s}^{-1}) = -12 \text{ kg·m·s}^{-1}.$$

The final momentum is +8.0 kg·m·s^{-1}. The *change* in momentum is

$$mv_2 - mv_1 = 8.0 \text{ kg·m·s}^{-1} - (-12 \text{ kg·m·s}^{-1})$$
$$= 20 \text{ kg·m·s}^{-1} = J.$$

Hence the impulse of the force exerted on the ball was 20 kg·m·s^{-1} = 20 N·s. Since the impulse is *positive*, the force must be toward the right.

The general nature of the force–time graph is shown by one of the curves in Fig. 8–3. The force is zero before impact, rises to a maximum, and decreases to zero when the ball leaves the wall. If the ball is relatively rigid, like a baseball, the time of collision is small and the maximum force is large, as in curve (a). If the ball is more yielding, like a tennis ball, the collision time is larger and the maximum force is less, as in curve (b). In any event, the *area* under the curve represents the impulse $J = 20$ N·s.

If the collision time is 0.01 s, then from Eq. (8–10),

$$F_{av}(0.01 \text{ s}) = 20 \text{ N·s}, \qquad F_{av} = 2000 \text{ N}.$$

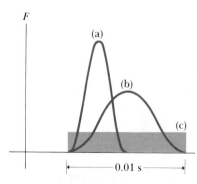

8–3 Force–time graph for Example 8–3.

This average force is represented by the horizontal line (c) in Fig. 8–3. The actual force might vary with time as shown by curve (b) in the figure.

Figure 8–4 is a stroboscopic photograph showing the impact of a tennis ball and racket during a serve.

EXAMPLE 8–4 Suppose the ball in Example 8–3 is a soccer ball with mass 0.4 kg. Initially it is moving to the left at 30 m·s^{-1}, but then it is kicked and given a velocity at 45° upward and to the right, with magnitude 30 m·s^{-1}. Find the impulse of the force and the average force, assuming a collision time of 0.010 s.

SOLUTION The velocities are not along the same line, and we must consider the vector nature of momentum and impulse explicitly. Taking the *x*-axis horizontal to the right and the *y*-axis vertically upward, we find the following velocity components:

$$v_{1x} = -30 \text{ m·s}^{-1}, \qquad v_{1y} = 0,$$

$$v_{2x} = v_{2y} = (0.707)(30 \text{ m·s}^{-1}) = 21.2 \text{ m·s}^{-1}.$$

Alternatively,

$$\boldsymbol{v}_1 = (-30 \text{ m·s}^{-1})\boldsymbol{i},$$

$$\boldsymbol{v}_2 = (21.2 \text{ m·s}^{-1})\boldsymbol{i} + (21.2 \text{ m·s}^{-1})\boldsymbol{j}.$$

The *x*-component of impulse is equal to the *x*-component of momentum change, and similarly for the *y*-components:

$$\begin{aligned} J_x &= m(v_{2x} - v_{1x}) \\ &= (0.4 \text{ kg})[21.2 \text{ m·s}^{-1} - (-30 \text{ m·s}^{-1})] \\ &= 20.5 \text{ kg·m·s}^{-1}, \end{aligned}$$

$$\begin{aligned} J_y &= m(v_{2y} - v_{1y}) \\ &= (0.4 \text{ kg})(21.2 \text{ m·s}^{-1} - 0) \\ &= 8.48 \text{ kg·m·s}^{-1}. \end{aligned}$$

Or:

$$\boldsymbol{J} = (20.5 \text{ kg·m·s}^{-1})\boldsymbol{i} + (8.48 \text{ kg·m·s}^{-1})\boldsymbol{j}.$$

The components of average force on the ball are

$$F_x = J_x/\Delta t = 2050 \text{ N}, \qquad F_y = J_y/\Delta t = 848 \text{ N},$$

and

$$\boldsymbol{F}_{av} = (2050 \text{ N})\boldsymbol{i} + (848 \text{ N})\boldsymbol{j}.$$

The magnitude and direction of the average force are

$$F_{av} = \sqrt{(2050 \text{ N})^2 + (848 \text{ N})^2} = 2218 \text{ N},$$

$$\theta = \arctan \frac{848 \text{ N}}{2050 \text{ N}} = 22.5°.$$

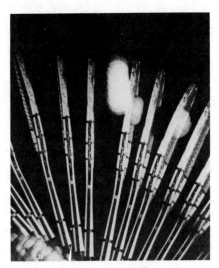

8–4 Multiflash stroboscopic photograph of a tennis racket hitting a ball during a serve. The exposure rate was 300 pictures per second. The ball is in contact with the racket for approximately 0.01 s. The ball flattens noticeably on both sides, and the frame of the racket bends and vibrates during and after the impact. (Dr. Harold Edgerton, M.I.T., Cambridge, Massachusetts.)

8–2 CONSERVATION OF MOMENTUM

The concept of momentum is most useful in situations involving several interacting bodies. To explore this topic, we consider first a system consisting of two bodies that interact with each other but not with anything else. Each body exerts a force on the other, so the momentum of each body changes. According to Newton's third law, the forces the bodies exert on each other are always equal in magnitude and opposite in direction. Thus the *impulses* given to the two bodies in any time interval are also equal and opposite, and therefore the *momentum changes* of the two bodies are equal and opposite.

To state this principle more compactly, we define the **total momentum** of the system, **P,** as the vector sum of momenta of the bodies in the system:

The total momentum of a system is the vector sum of the momenta of its parts.

$$P = mv_1 + mv_2 + \cdots = p_1 + p_2 + \cdots. \qquad (8\text{--}12)$$

If the change in momentum of one body is exactly the negative of that of the other, then the change in the *total* momentum must be zero. Thus *when two bodies interact only with each other, their total momentum is constant.*

A force that one part of a system exerts on another is called an **internal force,** and a force exerted on a part of the system by some agency outside the system is an **external force.** When no external forces act on a system, we call it an **isolated system.** Thus we may state the principle of **conservation of momentum** as follows:

Internal and external forces: Is a force on one part of a system due to another part or to something outside the system?

The total momentum of an isolated system is constant, or conserved.

When no external forces act, the total momentum is constant.

More generally, if there are external forces but the vector sum of all external forces is zero, the total momentum is constant. The *internal* forces can change the momentum of an individual body within the system, but they cannot change the *total* momentum of the system.

The principle of conservation of momentum is one of the most fundamental and important principles in mechanics. In some respects it is more general than the principle of conservation of mechanical energy. For example, it is valid even if the internal forces are not conservative, while mechanical energy is conserved only when they *are* conservative. In this chapter we will see situations in which both momentum and energy are conserved, and others in which only momentum is conserved.

PROBLEM-SOLVING STRATEGY: *Conservation of momentum*

Momentum is a vector quantity, and the rules of vector addition *must* be used in computing the total momentum of a system. Usually the use of components offers the simplest method.

1. Define your coordinate system. Make a sketch showing the coordinate axes, including the positive direction for each. Often the analysis is simplified if you choose the *x*-axis to have the same direction as one of the initial velocities.

2. Draw a sketch, including the coordinate axes and vectors representing all known velocity vectors. Label the vectors with magnitudes, angles, components, or whatever information is given, and give algebraic symbols to all unknown magnitudes, angles, or components.

3. Compute the *x*- and *y*-components of momentum of each body, both before and after the collision, impact, or other interaction, by using the relations $p_x = mv_x$ and $p_y = mv_y$. Even when all the velocities lie along a line (such as the *x*-axis), the components of velocity along this line can be positive or negative, and careful attention to signs is essential.

4. Write an equation equating the total *initial* *x*-component of momentum to the total *final* *x*-component of momentum. Write another equation for the *y*-components. These two equations express conservation of momentum in component form. Some of the components will be expressed in terms of symbols representing unknown quantities.

5. Solve these equations to determine whatever results are required. In some problems you will have to convert from the *x*- and *y*-components of a velocity to its magnitude and direction, or the reverse.

6. Remember that the *x*- and *y*-components of velocity or momentum are *never* added together in the same equation.

7. In some problems energy considerations give additional relationships among the various velocities, as we will see later in this chapter.

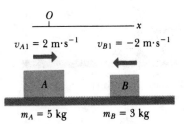

8-5 Momentum diagram for Example 8-5.

EXAMPLE 8-5 Figure 8-5 shows two gliders moving toward each other on a linear air track. The resultant vertical force on each body is zero; the resultant force on each glider is the force exerted on it by the other glider, and the total momentum of the system is constant in magnitude and direction.

Suppose that, after the gliders collide, B moves away with a final velocity of $+2 \text{ m·s}^{-1}$. What is the final velocity of A?

SOLUTION Let the final velocity of A be v_{A2}. Then write an equation showing the equality of the total momentum before and after the collision:

$$(0.5 \text{ kg})(2 \text{ m·s}^{-1}) + (0.3 \text{ kg})(-2 \text{ m·s}^{-1}) = (0.5 \text{ kg})v_{A2} + (0.3 \text{ kg})(+2 \text{ m·s}^{-1}).$$

Solving this equation for v_{A2}, we find $v_{A2} = -0.4 \text{ m·s}^{-1}$.

Collision of two gliders on an air track: Which one moves away faster?

EXAMPLE 8-6 Figure 8-6 shows two chunks of ice sliding on a frictionless frozen pond. Chunk A, having mass $m_A = 5$ kg, moves with initial velocity $v_{A1} = 2 \text{ m·s}^{-1}$ parallel to the x-axis. It collides with chunk B, which has mass $m_B = 3$ kg and is initially at rest. After the collision, the velocity of A is found to be $v_{A2} = 1 \text{ m·s}^{-1}$ in a direction making an angle $\theta = 30°$ with the initial direction. What is the final velocity of B?

Colliding ice chunks on a skating pond: use of components in a conservation-of-momentum calculation

SOLUTION The total momentum of the system is the same before and after the collision. The velocities are not all along a single line, and we have to recognize the *vector* nature of momentum. The simplest procedure is to find the components of each momentum in the x- and y-directions. Then momentum conservation requires that the sum of the x-components before the collision must equal the sum after the collision, and similarly for the y-components. Just as with force equilibrium problems, we write a separate equation for each component. For the x-components, we have

$$(5 \text{ kg})(2 \text{ m·s}^{-1}) + (3 \text{ kg})(0) = (5 \text{ kg})(1 \text{ m·s}^{-1})(\cos 30°) + (3 \text{ kg})v_{B2x}.$$

From this we find $v_{B2x} = 1.9 \text{ m·s}^{-1}$.

Similarly, conservation of the y-component of total momentum gives

$$(5 \text{ kg})(0) + (3 \text{ kg})(0) = (5 \text{ kg})(1 \text{ m·s}^{-1})(\sin 30°) + (3 \text{ kg})v_{B2y},$$

from which $v_{B2y} = -0.83 \text{ m·s}^{-1}$.

The magnitude of v_{B2} is

$$v_{B2} = \sqrt{(1.9 \text{ m·s}^{-1})^2 + (-0.83 \text{ m·s}^{-1})^2}$$
$$= 2.06 \text{ m·s}^{-1},$$

and the angle of its direction from the positive x-axis is

$$\phi = \arctan \frac{-0.83 \text{ m·s}^{-1}}{1.9 \text{ m·s}^{-1}}$$
$$= -24°.$$

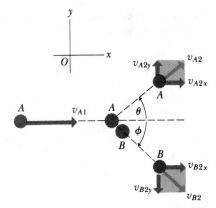

8-6 Velocity vector diagram for Example 8-6.

8-3 COLLISIONS

Momentum considerations alone are not always sufficient to predict all the details of the outcome of a collision. For example, suppose we are given the masses m_A and m_B and initial velocities v_{A1} and v_{B1} of two colliding bodies, and we want to find their velocities v_{A2} and v_{B2} after the collision. If the collision is

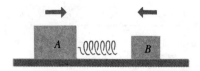

8–7 Body A (with spring) and body B approach each other on a frictionless surface.

In an elastic collision, the total kinetic energy is conserved, and so is the total momentum.

In an inelastic collision, the total momentum is conserved, but the total kinetic energy is not.

When two colliding bodies stick together, the collision is inelastic.

"head-on," as in Example 8–5, all the velocities are along the same line. Momentum conservation gives the equation

$$m_A v_{A1} + m_B v_{B1} = m_A v_{A2} + m_B v_{B2} \qquad (8\text{--}13)$$

for the two unknowns v_{A2} and v_{B2}. This equation by itself does not provide enough information to determine two unknowns. If the collision is of the more general type of Example 8–6, then we have two equations, one each for the x- and y-components of momentum; but then we have four unknowns, the x- and y-components of each final velocity. Again we need more information to determine these unknowns.

If the interaction forces between the bodies are *conservative*, the total *kinetic energy* of the system is the same after the collision as before, and this provides an additional relation among the velocities. Such a collision is called an **elastic collision,** or sometimes *completely elastic* or *perfectly elastic collision*. A collision between two glass balls or two hard steel balls or two ivory billiard balls is almost completely elastic. As a model for elastic collisions, we may consider the situation shown in Fig. 8–7. When the bodies collide, the spring is momentarily compressed and some of the original kinetic energy is momentarily converted to elastic potential energy. The spring then expands, and when the bodies separate, this potential energy is reconverted to kinetic energy.

A collision in which the total kinetic energy after the collision is *less* than before is called an **inelastic collision.** In one kind of inelastic collision, the colliding bodies stick together and move *as a unit* after the collision. In Fig. 8–7, we could replace the spring with a ball of putty or chewing gum that squashes and sticks the two bodies together. A gum ball striking a window shade, a bullet embedding itself in a block of wood, or two cars colliding and bending their fenders are examples of inelastic collisions.

8–4 INELASTIC COLLISIONS

In this section we consider inelastic collisions of the particular type where the two bodies stick together after the collision. Their two final velocities are then equal:

$$\boldsymbol{v_{A2} = v_{B2} = v_2.}$$

When this is combined with the principle of conservation of momentum, we obtain

$$m_A v_{A1} + m_B v_{B1} = (m_A + m_B) v_2, \qquad (8\text{--}14)$$

and the final velocity can be computed if the initial velocities and the masses are known.

The total kinetic energy of a system after an inelastic collision is always less than before the collision. Suppose, for example, a body having mass m_A and initial velocity v_1 collides inelastically with a body of mass m_B initially at rest. After the collision the two bodies have a common velocity v_2 given by Eq. (8–14):

$$v_2 = \frac{m_A}{m_A + m_B} v_1. \qquad (8\text{--}15)$$

The kinetic energies K_1 and K_2 before and after the collision, respectively, are

$$K_1 = \tfrac{1}{2} m_A v_1{}^2,$$

$$K_2 = \tfrac{1}{2}(m_A + m_B)\left(\frac{m_A}{m_A + m_B}\right)^2 v_1{}^2,$$

so the ratio of final to initial kinetic energy is

$$\frac{K_2}{K_1} = \frac{m_A}{m_A + m_B}. \qquad (8\text{–}16)$$

The right side is always less than unity because the denominator of the fraction is always greater than the numerator. Thus in such a collision the final kinetic energy is always less than the initial value. Even when the initial velocity of m_B is not zero, it is not difficult to prove that the kinetic energy after a collision where the bodies stick together is always less than before.

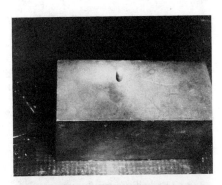

EXAMPLE 8–7 Suppose that in the collision in Fig. 8–5 the two bodies stick together after the collision; the masses and initial velocities are as shown. From momentum conservation,

$$(5 \text{ kg})(2 \text{ m·s}^{-1}) + (3 \text{ kg})(-2 \text{ m·s}^{-1}) = (5 \text{ kg} + 3 \text{ kg})v_2,$$

and

$$v_2 = 0.5 \text{ m·s}^{-1}.$$

Since v_2 is positive, the system moves to the right after the collision. The kinetic energy of body A before the collision is

$$\tfrac{1}{2}m_A v_{A1}{}^2 = \tfrac{1}{2}(5 \text{ kg})(2 \text{ m·s}^{-1})^2 = 10 \text{ J},$$

and that of body B is

$$\tfrac{1}{2}m_B v_{B1}{}^2 = \tfrac{1}{2}(3 \text{ kg})(-2 \text{ m·s}^{-1})^2 = 6 \text{ J}.$$

The total kinetic energy before collision is therefore 16 J. Note that the kinetic energy of body B is positive, although its velocity v_{B1} and its momentum mv_{B1} are both negative.

The kinetic energy after the collision is

$$\tfrac{1}{2}(m_A + m_B)v_2{}^2 = \tfrac{1}{2}(5 \text{ kg} + 3 \text{ kg})(0.5 \text{ m·s}^{-1})^2 = 1 \text{ J}.$$

Hence, far from remaining constant, the final kinetic energy is only $\frac{1}{16}$ of the original, and $\frac{15}{16}$ is "lost" in the collision.

The energy is not really lost, of course; it is converted from mechanical energy to various other forms. If there is a ball of putty between the masses, it squashes irreversibly and becomes warmer. If the masses couple together like two freight cars, the energy goes into elastic waves that are eventually dissipated. If there is a spring between the bodies that is compressed when they are locked together, then the energy is stored as potential energy of the spring. In all of these cases the *total* energy of the system is conserved, although the *kinetic* energy is not. In an isolated system, however, momentum is *always* conserved, whether the collision is elastic or not.

EXAMPLE 8–8 Figure 8–8 shows a *ballistic pendulum*, a device for measuring the speed of a bullet. The bullet is allowed to make a completely inelastic collision with a body of much greater mass. The momentum of the system immediately after the collision equals the original momentum of the bullet, but since the *velocity* is very much smaller, it can be determined more easily. Although the ballistic pendulum has now been superseded by other devices, it is still an important laboratory experiment for illustrating the concepts of momentum and energy.

The impact of this bullet on a hard surface is an inelastic collision; most of the bullet's kinetic energy is lost, and it becomes appreciably hotter as its kinetic energy is converted to internal energy. (Dr. Harold Edgerton, M.I.T., Cambridge, Massachusetts.)

In an inelastic collision, the kinetic energy is always less after the collision than before.

Ballistic pendulum: using conservation of momentum to measure the speed of a bullet

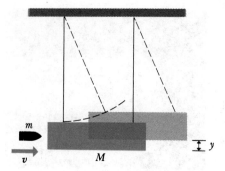

8-8 The ballistic pendulum.

In Fig. 8–8, the pendulum, consisting perhaps of a large wooden block of mass M, hangs vertically by two cords. A bullet of mass m, traveling with a velocity v, strikes the pendulum and remains embedded in it. If the collision time is very small compared with the time of swing of the pendulum, the supporting cords remain practically vertical during the collision. Hence no external horizontal forces act on the system during the collision, and the horizontal momentum is conserved. Then if V represents the velocity of bullet and block immediately after the collision,

$$mv = (m + M)V, \qquad v = \frac{m + M}{m}V.$$

The kinetic energy of the system immediately after the collision is $K = \frac{1}{2}(m + M)V^2$.

The pendulum now swings to the right and upward until its kinetic energy is converted to gravitational potential energy. Hence

$$\tfrac{1}{2}(m + M)V^2 = (m + M)gy,$$

$$V = \sqrt{2gy},$$

and

$$v = \frac{m + M}{m}(\sqrt{2gy}).$$

By measuring m, M, and y, we can compute the original velocity v of the bullet.

Suppose $m = 5.00 \text{ g} = 0.00500 \text{ kg}$, $M = 2.00 \text{ kg}$, and $h = 3.00 \text{ cm} = 0.0300 \text{ m}$. Working backwards, we find the velocity V of the block just after impact:

$$V = \sqrt{2(9.80 \text{ m·s}^{-2})(0.0300 \text{ m})} = 0.767 \text{ m·s}^{-1}.$$

Then we use momentum conservation to find the bullet's velocity v just before impact:

$$(0.00500 \text{ kg})v = (2.00 \text{ kg} + 0.00500 \text{ kg})(0.767 \text{ m·s}^{-1}),$$

$$v = 307 \text{ m·s}^{-1}.$$

Note that the kinetic energy just before impact is $\frac{1}{2}(0.00500 \text{ kg})(307 \text{ m·s}^{-1})^2 = 236 \text{ J}$, while just after impact it is $\frac{1}{2}(2.005 \text{ kg})(0.767 \text{ m·s}^{-1})^2 = 0.589 \text{ J}$. Nearly all the kinetic energy disappears as the wood splinters and the bullet becomes hotter.

8–5 ELASTIC COLLISIONS

Next let us consider an *elastic* collision between two bodies A and B. We first discuss a head-on collision in which all the initial and final velocities lie along the same line. Later in this section we will consider more general collisions.

Conservation of kinetic energy in an elastic collision

Both kinetic energy and momentum are conserved. From conservation of kinetic energy we have

$$\tfrac{1}{2}m_A v_{A1}{}^2 + \tfrac{1}{2}m_B v_{B1}{}^2 = \tfrac{1}{2}m_A v_{A2}{}^2 + \tfrac{1}{2}m_B v_{B2}{}^2,$$

and from conservation of momentum,

$$m_A v_{A1} + m_B v_{B1} = m_A v_{A2} + m_B v_{B2}.$$

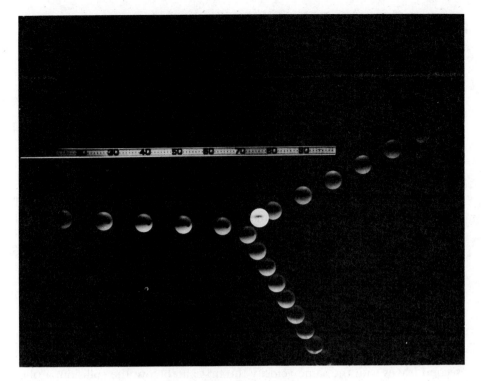

In this multiflash photograph, a billiard ball approaching from the left collides elastically with a second ball having equal mass, initially at rest. Both momentum and kinetic energy are conserved, and the final velocities of the two balls are perpendicular. The speeds can be determined from the distances between successive photographic images (representing equal time intervals), and the conservation principles can be verified directly from this photograph. (*PSSC Physics*, 2e, 1965; D.C. Heath and Company, Inc., with Education Development Center, Inc., Newton, Mass.)

If the masses and initial velocities are known, these are two independent equations that can be solved simultaneously to find the two final velocities. The general solution of these equations is somewhat involved, so we will concentrate on a few particular cases, such as having one body at rest before the collision.

Suppose mass m_B is initially at rest. We can then simplify the velocity notation by letting v be the initial velocity of A, and v_A and v_B the final velocities of A and B, respectively. Then the kinetic energy and momentum conservation equations are, respectively,

$$\tfrac{1}{2}m_A v^2 = \tfrac{1}{2}m_A v_A^2 + \tfrac{1}{2}m_B v_B^2, \tag{8-17}$$

and

$$m_A v = m_A v_A + m_B v_B. \tag{8-18}$$

Assuming the masses and v are known, we may solve for v_A and v_B. The simplest approach is somewhat indirect and uncovers an additional interesting feature of elastic collisions.

We first rearrange Eqs. (8–17) and (8–18) as follows:

$$m_B v_B^2 = m_A(v^2 - v_A^2) = m_A(v - v_A)(v + v_A), \tag{8-19}$$

and

$$m_B v_B = m_A(v - v_A). \tag{8-20}$$

We now divide Eq. (8–19) by Eq. (8–20) to obtain

$$v_B = v + v_A. \tag{8-21}$$

We now substitute this back into Eq. (8–20) to eliminate v_B and then solve for v_A:

$$m_B(v + v_A) = m_A(v - v_A),$$

$$v_A = \frac{m_A - m_B}{m_A + m_B} v. \tag{8–22}$$

Finally, we substitute this result back into Eq. (8–21) to obtain

$$v_B = \frac{2m_A}{m_A + m_B} v. \tag{8–23}$$

A somewhat tedious calculation, but there is no substitute for persistence!

Now we can interpret the results. Suppose first that A is a ping-pong ball and B is a bowling ball. Then we expect A to bounce off with a velocity nearly equal to the original value but in the opposite direction, and we expect B's velocity to be much smaller. Indeed, when m_A is much smaller than m_B, the fraction in Eq. (8–22) is approximately equal to (-1), and the fraction in Eq. (8–23) is much smaller than unity. We challenge you to invent a similar check for the opposite case where A is the bowling ball and B the ping-pong ball.

Another interesting case is that of *equal* masses. If $m_A = m_B$, then Eqs. (8–22) and (8–23) give $v_A = 0$ and $v_B = v$. That is, the first body stops and the second leaves with the same velocity the first had before the collision, a phenomenon familiar to pool players.

Now comes the dividend. Equation (8–21) may be rewritten as

$$v = v_B - v_A. \tag{8–24}$$

In an elastic collision, the relative velocity has the same magnitude before and after: an alternative definition of an elastic collision.

Now $v_B - v_A$ is just the velocity of B relative to A after the collision, and v, apart from sign, is the velocity of B relative to A before the collision. Hence *the relative velocity has the same magnitude before and after the collision.* Although we have proved this only for one special case, it turns out to be a general property of *all* elastic collisions, even when both bodies are moving initially and the velocities do not all lie along the same line. The constancy of the magnitude of relative velocity provides an alternative definition of an elastic collision; in any collision where this condition is satisfied, the total kinetic energy is also conserved.

EXAMPLE 8–9 Suppose the collision shown in Fig. 8–5 is completely elastic. What are the velocities of A and B after the collision?

SOLUTION From the principle of conservation of momentum,

$$(5 \text{ kg})(2 \text{ m·s}^{-1}) + (3 \text{ kg})(-2 \text{ m·s}^{-1}) = (5 \text{ kg})v_{A2} + (3 \text{ kg})v_{B2},$$

$$5v_{A2} + 3v_{B2} = 4 \text{ m·s}^{-1}.$$

Because the collision is completely elastic,

$$v_{B2} - v_{A2} = -(v_{B1} - v_{A1})$$
$$= -(-2 \text{ m·s}^{-1} - 2 \text{ m·s}^{-1}) = 4 \text{ m·s}^{-1}.$$

Solving these equations simultaneously, we obtain

$$v_{A2} = -1 \text{ m·s}^{-1}, \qquad v_{B2} = 3 \text{ m·s}^{-1}.$$

Both bodies therefore reverse their directions of motion, A traveling to the left at 1 m·s^{-1} and B to the right at 3 m·s^{-1}.

The total kinetic energy after the collision is

$$\tfrac{1}{2}(5 \text{ kg})(-1 \text{ m·s}^{-1})^2 + \tfrac{1}{2}(3 \text{ kg})(3 \text{ m·s}^{-1})^2 = 16 \text{ J},$$

which equals the total kinetic energy before the collision.

If an elastic collision is not head-on, the velocities do not all lie on a single line. Each final velocity then has two unknown components, and we have four unknowns in all. Conservation of energy and of the x- and y-components of momentum gives only three equations, so additional information is needed. This may take the form of the direction of one of the final velocities, perhaps obtained from geometrical considerations, or of one of the final velocity magnitudes.

EXAMPLE 8–10   In Fig. 8–6 suppose the 5-kg mass m_A has an initial velocity of 4 m·s^{-1} in the positive direction and a final velocity of 2 m·s^{-1} in an unknown direction. Find the final speed v_{B2} of the 3-kg mass, and the angles θ and ϕ in the figure, assuming the collision is perfectly elastic.

An elastic collision: using momentum components to find unknown directions

SOLUTION   Because the collision is elastic, the total final kinetic energy is the same as the initial kinetic energy. Thus

$$\tfrac{1}{2}(5 \text{ kg})(4 \text{ m·s}^{-1})^2 = \tfrac{1}{2}(5 \text{ kg})(2 \text{ m·s}^{-1})^2 + \tfrac{1}{2}(3 \text{ kg})v_{B2}{}^2,$$

from which

$$v_{B2} = 4.47 \text{ m·s}^{-1}.$$

Conservation of the x- and y-components of total momentum gives, respectively,

$$(5 \text{ kg})(4 \text{ m·s}^{-1}) = (5 \text{ kg})(2 \text{ m·s}^{-1}) \cos\theta + (3 \text{ kg})(4.47 \text{ m·s}^{-1}) \cos\phi,$$

$$0 = (5 \text{ kg})(2 \text{ m·s}^{-1}) \sin\theta - (3 \text{ kg})(4.47 \text{ m·s}^{-1}) \sin\phi.$$

These are two simultaneous equations for θ and ϕ; the simplest solution is to eliminate ϕ as follows. We solve the first equation for $\cos\phi$ and the second for $\sin\phi$; we then square each equation and add; since $\sin^2\phi + \cos^2\phi = 1$, this eliminates ϕ and leaves an equation that may be solved for $\cos\theta$ and hence for θ. This value may then be substituted back into either of the two equations and the result solved for ϕ. The details are left as an exercise; the results are

$$\theta = 36.9°, \qquad \phi = 26.6°.$$

The examples in this section and the preceding one show that collisions can be classified according to energy considerations. A collision in which kinetic energy is conserved is called an *elastic collision*. A collision in which the total kinetic energy decreases is called an *inelastic* or *completely inelastic collision*. We have discussed inelastic collisions where the two bodies stick together after the collision. There are also inelastic collisions where some kinetic energy is

Can the kinetic energy ever be greater after the collision than before?

lost, but not enough for the two bodies to reach a common final velocity. In some cases the final kinetic energy is *greater* than the initial value. This occurs, for example, when an explosive between the two bodies is detonated during the collision, blowing them apart with great speed. Such a collision might be called a *superelastic collision,* although this term is not in common use. A special case of a superelastic collision is **recoil,** discussed in the next section.

Finally, momentum conservation can be applied even to systems that are not really isolated. Some external forces may be acting on the colliding bodies; but if the internal forces during the collision are much larger than the external, then the external forces during the actual collision can be neglected. This is certainly the case, for example, when two cars collide at an icy intersection.

8–6 RECOIL

Recoil: a case where the total kinetic energy increases

Figure 8–9 shows two toy cars A and B, with a compressed spring between them. We hold both cars, and then release them without giving them any initial motion. The spring begins to expand, pushing the cars apart. As it does so, it exerts equal and opposite forces on the cars until it has expanded back to its original uncompressed length. It then drops to the surface, while the cars roll on. The initial momentum of the system was zero, and if frictional forces can be neglected, the resultant external force on the system is zero. The total momentum of the system therefore remains constant and equal to zero during this process. Then if v_A and v_B are the velocities acquired by A and B, we have

$$m_A v_A + m_B v_B = 0, \qquad \frac{v_A}{v_B} = -\frac{m_B}{m_A}. \qquad (8\text{–}25)$$

The velocities are of opposite sign, and their magnitudes are inversely proportional to the corresponding masses.

The original kinetic energy of the system is also zero. The final kinetic energy is $K = \frac{1}{2}m_A v_A{}^2 + \frac{1}{2}m_B v_B{}^2$. The source of this energy is the original elastic potential energy of the spring. The ratio of kinetic energies is

$$\frac{\frac{1}{2}m_A v_A{}^2}{\frac{1}{2}m_B v_B{}^2} = \frac{m_A}{m_B}\left(\frac{v_A}{v_B}\right)^2 = \frac{m_B}{m_A}. \qquad (8\text{–}26)$$

Thus although the final momenta are equal in magnitude, the final kinetic energies are inversely proportional to the corresponding masses; the body of smaller mass receives the larger share of the original potential energy. The reason is that the change in *momentum* of a body equals the *impulse* of the force acting on it, while the change in *kinetic energy* equals the *work* of the force. The forces on the two bodies are equal in magnitude and act for equal *times,* so they produce equal and opposite changes in momentum. The points of application of the forces, however, *do not* move through equal *distances* (except when $m_A = m_B$), since the acceleration, velocity, and displacement of the smaller body are greater than those of the larger. Hence more *work* is done on the body of smaller mass.

8–9 Conservation of momentum in recoil.

We can apply these considerations to the firing of a rifle. The initial momentum of the system is zero. When the rifle is fired, the bullet and the powder gases acquire a forward momentum, and the rifle (together with the person holding it) acquires a rearward momentum of the same magnitude. Because of the relatively large mass of the rifle compared with that of the bullet and powder charge, the velocity and kinetic energy of the rifle are much *smaller* than those of the bullet and powder gases.

Firing a rifle: an application of momentum conservation

EXAMPLE 8–11 A hunter holds a 3.00-kg rifle loosely in his hands and fires a bullet of mass 5.00 g with a muzzle velocity of 300 m·s^{-1}. What is the recoil velocity of the rifle? What is the final kinetic energy of the bullet? Of the rifle?

SOLUTION The total momentum is zero before and after firing. Thus

$$0 = (0.00500 \text{ kg})(300 \text{ m·s}^{-1}) + (3.00 \text{ kg})v,$$

$$v = -0.500 \text{ m·s}^{-1}.$$

The negative sign means that the recoil is in the direction opposite to that of the bullet.

The kinetic energy of the bullet is

$$K_B = \tfrac{1}{2}(0.00500 \text{ kg})(300 \text{ m·s}^{-1})^2 = 225 \text{ J},$$

and the kinetic energy of the rifle is

$$K_R = \tfrac{1}{2}(3.00 \text{ kg})(0.500 \text{ m·s}^{-1})^2 = 0.375 \text{ J}.$$

The bullet acquires much more kinetic energy than the rifle because the interaction force on it acts over a much longer distance, and hence does more work, than on the rifle. Finally, we note that

$$\frac{K_B}{K_R} = \frac{225 \text{ J}}{0.375 \text{ J}} = 600 = \frac{3.00 \text{ kg}}{0.00500 \text{ kg}}.$$

in agreement with Eq. (8–26).

8–7 CENTER OF MASS

The principle of conservation of momentum for an isolated system can be restated in a useful way by referring to the concept of **center of mass.** Suppose we have a collection of particles having masses m_1, m_2, and so on. Let the coordinates of m_1 be (x_1, y_1), those of m_2 be (x_2, y_2), and so on. The center of mass of the system is defined as the point having coordinates (X, Y) given by

$$X = \frac{m_1x_1 + m_2x_2 + m_3x_3 + \cdots}{m_1 + m_2 + m_3 + \cdots},$$

$$Y = \frac{m_1y_1 + m_2y_2 + m_3y_3 + \cdots}{m_1 + m_2 + m_3 + \cdots}. \tag{8–27}$$

Thus the center of mass represents a weighted average position of the particles.

Center of mass: the average position of a set of particles

The center of the mass of this wrench is marked with an X. The total external force acting on the wrench is zero. As it spins on a smooth horizontal surface, the center of mass moves in a straight line with constant velocity. (*PSSC Physics*, 2e, 1965; D. C. Heath and Company, Inc., with Education Development Center, Inc., Newton, Mass.)

The x- and y-components of velocity of the center of mass are dX/dt and dY/dt. Taking the time derivatives of Eqs. (8–27), we see that the *velocity* of the center of mass is given by

$$V = \frac{m_1 v_1 + m_2 v_2 + m_3 v_3 + \cdots}{m_1 + m_2 + m_3 + \cdots}. \tag{8–28}$$

We denote the *total* mass $m_1 + m_2 + \cdots$ by M; we can then rewrite Eq. (8–28) as

$$MV = m_1 v_1 + m_2 v_2 + m_3 v_3 + \cdots = P. \tag{8–29}$$

The right side is simply the total momentum P of the system; hence we have proved that *the total momentum is equal to the total mass times the velocity of the center of mass*. It follows that for an isolated system, where the total momentum is constant, the velocity of the center of mass is also constant.

Proceeding one additional step, we note that the *rates of change* of the various velocities, that is, the accelerations, are related in the same way:

$$MA = m_1 a_1 + m a_2 + m_3 a_3 + \cdots. \tag{8–30}$$

The total momentum of a system depends on the velocity of its center of mass.

Now $m_1 a_1$ is equal to the vector sum of forces on the first particle, and so on, and the right side of Eq. (8–30) is equal to the vector sum of *all* the forces on all the particles. Just as in Section 8–2, we may classify each force as internal or external; because of Newton's third law, the internal forces all cancel in pairs. That is, $\Sigma F = \Sigma F_{\text{ext}} + \Sigma F_{\text{int}}$, and $\Sigma F_{\text{int}} = 0$, from Newton's third law. Thus what survives on the right side is the sum of the *external* forces only, and we have

$$\Sigma F_{\text{ext}} = MA. \tag{8–31}$$

Newton's second law: The center of mass moves as though all the mass and force were concentrated at that point.

That is, *when a body or a collection of particles is acted on by external forces, the center of mass moves just as though all the mass were concentrated at that point and it were acted on by a resultant force equal to the sum of the external forces on the system*. For example, suppose we mark the center of mass of a hammer, which will be at some point partway down the handle. We throw the hammer spinning through the air. The motion appears complicated, but the point representing the position of the center of mass follows a parabolic path just as though all the mass were concentrated at the center of mass. Momentum is not conserved because there is an external force (the hammer's weight), but the effect of the force is to cause the parabolic motion of the center of mass. Or suppose a shell traveling in a parabolic trajectory explodes in flight, splitting into two fragments of equal mass. The fragments follow new parabolic paths, but the center of mass continues on the original trajectory just as though all the mass were still concentrated at that point. This phenomenon is shown in Fig. 8–10. A

8–10 A shell explodes in flight. The fragments follow individual parabolic paths, but the center of mass continues on the same trajectory as the shell's path before exploding.

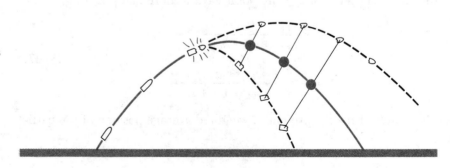

Fourth-of-July skyrocket exploding in air is a more spectacular example of the same idea.

This property of the center of mass is important in the analysis of the motion of rigid bodies, where the motion is described as a combination of motion of the center of mass and rotational motion about an axis through the center of mass. We return to this topic in Chapter 9.

Finally, we note that $M\mathbf{A} = M\,d\mathbf{V}/dt = d\mathbf{P}/dt$, so the rate of change of total momentum of the system equals the sum of the external forces:

$$\Sigma \mathbf{F}_{ext} = \frac{d\mathbf{P}}{dt}. \tag{8–32}$$

Thus when $\Sigma\mathbf{F} = 0$, $\mathbf{A} = 0$, $\mathbf{V}$ is constant, and $\mathbf{P}$ is constant; conservation of momentum corresponds to constant velocity of the center of mass and to zero total external force.

In principle the position of the center of mass of a body can always be calculated by regarding it as a collection of particles and using Eqs. (8–27). In practice these calculations can become quite complex. Usually we will deal with symmetric bodies in which the center of mass lies at the geometric center. We will discuss other calculation techniques in Chapter 10 in connection with the related concept of *center of gravity*.

EXAMPLE 8–12 A 2-kg body and a 3-kg body are moving along the *x*-axis. At a particular instant the 2-kg body is 1.0 m from the origin and has a velocity of $3\ \text{m·s}^{-1}$, and the 3-kg body is 2 m from the origin and has a velocity of $-1\ \text{m·s}^{-1}$. Find the position and velocity of the center of mass, and also find the total momentum.

SOLUTION From Eq. (8–27),

$$X = \frac{(2\ \text{kg})(1\ \text{m}) + (3\ \text{kg})(2\ \text{m})}{2\ \text{kg} + 3\ \text{kg}} = 1.6\ \text{m},$$

and from Eq. (8–28),

$$V = \frac{(2\ \text{kg})(3\ \text{m·s}^{-1}) + (3\ \text{kg})(-1\ \text{m·s}^{-1})}{2\ \text{kg} + 3\ \text{kg}} = 0.60\ \text{m·s}^{-1}.$$

Strictly speaking, this is the *x*-component of velocity of the center of mass. The total momentum (actually, its *x*-component) is

$$P = (2\ \text{kg})(3\ \text{m·s}^{-1}) + (3\ \text{kg})(-1\ \text{m·s}^{-1}) = 3\ \text{kg·m·s}^{-1}.$$

Alternatively, from Eq. (8–29),

$$P = (5\ \text{kg})(0.6\ \text{m·s}^{-1}) = 3\ \text{kg·m·s}^{-1}.$$

8–8 ROCKET PROPULSION

We can use momentum and impulse considerations as a basis for analyzing rocket propulsion. A rocket is propelled by rearward ejection of part of its mass. The forward force on the rocket is the reaction to the backward force on the ejected material, and as more material is ejected, the mass of the rocket

In a rocket, burned fuel is ejected backward; the reaction force pushes the rocket forward.

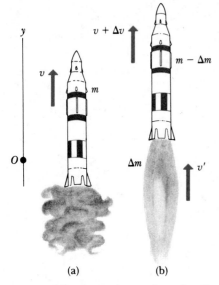

8–11 (a) Rocket at time t after takeoff, with mass m and upward velocity v. Its momentum is mv. (b) At time $t + \Delta t$, the mass of the rocket (and unburned fuel) is $m - \Delta m$, its velocity is $v + \Delta v$, and its momentum is $(m - \Delta m)$ $(v + \Delta v)$. The ejected gas has momentum $\Delta m(v - v_r)$.

The mass of a rocket decreases during flight.

Calculating the thrust of a rocket: an application of momentum and impulse

decreases. As a simple example, consider a rocket fired vertically upward in a uniform gravitational field; neglect air resistance.

Figure 8–11a represents the rocket at a time t after takeoff, when its mass is m and its upward velocity is v. The total momentum at this instant is thus mv. In a short time interval dt, a mass dm of gas is ejected from the rocket. Let v_r represent the downward speed of this gas *relative to the rocket*. The velocity v' of the gas relative to the earth is then

$$v' = v - v_r,$$

and its momentum is

$$dm\, v' = dm(v - v_r).$$

At the end of the time interval dt, the mass of rocket and unburned fuel has decreased to $m - dm$, and its velocity has increased to $v + dv$. Its momentum is therefore

$$(m - dm)(v + dv).$$

Thus the *total* momentum at time $t + dt$ is

$$(m - dm)(v + dv) + dm(v - v_r).$$

Figure 8–11b represents rocket and ejected gas at this time.

We now use the impulse–momentum relation. The product of the resultant external force F on a system and the time interval dt during which it acts is equal to the change in momentum of this system. If air resistance is neglected, the external force F on the rocket is its weight, $-mg$. (We take the upward direction as positive.) The change in momentum, in time dt, is the difference between the momentum of the system at the end and at the beginning of the time interval. Hence

$$-mg\, dt = (m - dm)(v + dv) + dm(v - v_r) - mv$$

$$= m\, dv - dm\, v_r - dm\, dv.$$

The term $dm\, dv$ may be dropped because it is a product of two small quantities and thus is much smaller than the other terms. Dropping this term, dividing by dt, and rearranging, we obtain

$$m\left(\frac{dv}{dt}\right) = v_r\left(\frac{dm}{dt}\right) - mg. \qquad (8\text{--}33)$$

Now dv/dt is the acceleration of the rocket, so the left side of this equation (mass times acceleration) equals the resultant force on the rocket. The first term on the right equals the upward thrust on the rocket, and the resultant force equals the difference between this thrust and the weight of the rocket mg. We see that the upward thrust is proportional both to the relative velocity, v_r, of the ejected gas and to the mass of gas ejected per unit time, dm/dt.

The acceleration of the rocket is

$$\frac{dv}{dt} = \frac{v_r}{m}\left(\frac{dm}{dt}\right) - g. \qquad (8\text{--}34)$$

As the rocket rises, the value of g decreases, according to Newton's law of gravitation. (In "outer space," far from all other bodies, g becomes negligibly small.) If the values of v_r and dm/dt remain approximately constant while the fuel is being consumed and m continually decreases, the acceleration *increases* until all the fuel is burned.

Equation (8–34) can be integrated to find a relation between the velocity at any time and the remaining mass. From Eq. (8–34),

$$dv = v_\mathrm{r} \frac{dm}{m} - g\,dt.$$

Now dm is a positive quantity, representing the mass ejected in time dt, so the change in mass of the rocket in that time is $-dm$. Thus in computing the total mass change *in the rocket* we must change the sign of the term containing dm.

Let the mass and velocity at time $t = 0$ be m_0 and v_0, respectively; then

$$\int_{v_0}^{v} dv = -\int_{m_0}^{m} v_\mathrm{r} \frac{dm}{m} - \int_0^t g\,dt,$$

and

$$v - v_0 = -v_\mathrm{r} \ln \frac{m}{m_0} - gt,$$

$$v = v_0 + v_\mathrm{r} \ln \frac{m_0}{m} - gt. \tag{8–35}$$

EXAMPLE 8–13 In the first second of its flight, a rocket ejects $\frac{1}{60}$ of its mass with a relative velocity of 2400 m·s^{-1}. What is the acceleration of the rocket?

SOLUTION We have $dm = m/60$, $dt = 1$ s. From Eq. (8–34),

$$\frac{dv}{dt} = \frac{2400 \text{ m·s}^{-1}}{(60)(1 \text{ s})} - 9.8 \text{ m·s}^{-2}$$
$$= 30.2 \text{ m·s}^{-2}.$$

EXAMPLE 8–14 Suppose the ratio of initial mass m_0 to final mass m for the rocket in Example 8–13 is 4, and that the fuel is consumed in a time $t = 60$ s. Find the rocket's velocity at the end of this time.

SOLUTION From Eq. (8–35),

$$v = (2400 \text{ m·s}^{-1})(\ln 4) - (9.8 \text{ m·s}^{-2})(60 \text{ s})$$
$$= 2740 \text{ m·s}^{-1}.$$

At the start of the flight, when the velocity of the rocket is zero, the ejected gases are moving downward, relative to the earth, with a velocity equal to the relative velocity v_r. When the velocity of the rocket has increased to v_r, the ejected gases have a velocity zero relative to the earth. When the rocket velocity becomes greater than v_r, the velocity of the ejected gases is in the same direction as that of the rocket. Thus the velocity acquired by the rocket can be greater (and is often much greater) than the relative velocity v_r. In the example above, where the final velocity of the rocket was 2740 m·s^{-1} and the relative velocity was 2400 m·s^{-1}, the last portion of the ejected fuel had an upward velocity (relative to the earth) of $(2740 - 2400)$ m·s^{-1} = 340 m·s^{-1}.

8–12 Launch of a space shuttle, a dramatic example of rocket propulsion. Successful launch and landing of the Space Shuttle are the first steps toward feasibility of a manned space station in orbit around the earth. Despite a catastrophic set-back, research and progress continue. (Courtesy NASA, Kennedy Space Center.)

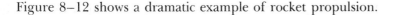

Figure 8–12 shows a dramatic example of rocket propulsion.

SUMMARY

The impulse J of a force F acting for a time interval from t_1 to t_2 is defined for a constant force as

$$J = F(t_2 - t_1) \qquad (8\text{–}5)$$

and for a force that varies with time as

$$J = \int_{t_1}^{t_2} F \, dt. \qquad (8\text{–}8)$$

The momentum p of a particle of mass m moving with velocity v is defined as

$$p = mv. \qquad (8\text{–}6)$$

In any time interval, the total impulse of all forces acting on a body equals the change in its momentum:

$$J = mv_2 - mv_1 = p_2 - p_1. \qquad (8\text{–}9)$$

When the force is not constant, the impulse still equals the average force multiplied by the time interval:

$$F_{av}(t_2 - t_1) = J = m(v_2 - v_1). \qquad (8\text{–}10)$$

The total momentum P of a system is the vector sum of momenta of all bodies in the system:

$$P = p_1 + p_2 + \cdots = m_1 v_1 + m_2 v_2 + \cdots. \qquad (8\text{–}12)$$

The total momentum of an isolated system is constant, or conserved. When momenta are represented in terms of their components, each component of momentum is separately conserved.

Collisions can be classified according to energy relations and final velocities:

Elastic collision: Energy is conserved, and the initial and final relative velocities have the same magnitude.

Inelastic collision: The final energy is less than the initial; the two bodies may or may not have the same final velocity.

Superelastic collision: The final energy is greater than the initial.

In recoil phenomena the total momentum is often zero both before and after the interaction.

The coordinates of the center of mass of a system of particles are defined as

$$X = \frac{m_1 x_1 + m_2 x_2 + m_3 x_3 + \cdots}{m_1 + m_2 + m_3 + \cdots},$$

$$Y = \frac{m_1 y_1 + m_2 y_2 + m_3 y_3 + \cdots}{m_1 + m_2 + m_3 + \cdots} \qquad (8\text{–}27)$$

The total momentum of the system equals the total mass M multiplied by the velocity V of the center of mass:

$$MV = m_1 v_1 + m_2 v_2 + \cdots = P. \qquad (8\text{–}29)$$

In rocket propulsion the mass of the rocket changes as the fuel is burned. Analysis of momentum relations must include the momentum carried away by the fuel as well as the momentum of the rocket itself.

QUESTIONS

8–1 Suppose you catch a baseball, and then someone invites you to catch a bullet with the same momentum or with the same kinetic energy. Which would you choose?

8–2 In splitting logs with a hammer and wedge, is a heavy hammer more effective than a lighter hammer? Why?

8–3 When a large, heavy truck collides with a passenger car, the occupants of the car are much more likely to be hurt than the truck driver. Why?

8–4 A glass dropped on the floor is more likely to break if the floor is concrete than if it is wood. Why?

8–5 "It ain't the fall that hurts you; it's the sudden stop at the bottom." Discuss.

8–6 When a person fires a rifle or shotgun, it is advisable to hold the butt firmly against the shoulder rather than a bit away from it, to minimize the impact on the shoulder. Why?

8–7 When a catcher in a baseball game catches a fast ball, he does not hold his arms rigid, but relaxes them so the mitt moves several inches while the ball is being caught. Why is this important?

8–8 When rain falls from the sky, what becomes of its momentum as it hits the ground? Is your answer also valid for Newton's famous apple?

8–9 A man stands in the middle of a perfectly smooth, frictionless frozen lake. He can set himself in motion by throwing things, but suppose he has nothing to throw. Can he propel himself to shore *without* throwing anything?

8–10 A machine gun is fired at a steel plate. Is the average force on the plate from the bullet impact greater if the bullets bounce off, or if they are squashed and stick to the plate?

8–11 How do Mexican jumping beans work? Do they violate conservation of momentum? Of energy?

8–12 Early critics of Robert Goddard, a pioneer in the use of rocket propulsion, claimed that rocket engines could not be used in outer space where there is no air for the rocket to push against. How would you answer such criticism? Would the criticism be valid for a jet engine in an ordinary airplane?

8–13 In a zero-gravity environment, can a rocket-propelled spaceship ever attain a speed greater than the relative speed with which the burnt fuel is exhausted?

EXERCISES

Section 8–1 Impulse and Momentum

8–1

a) What is the momentum of a 10,000-kg truck whose speed is 20 m·s^{-1}? What speed must a 5000-kg truck attain in order to have

b) the same momentum?

c) the same kinetic energy?

8–2 A bullet having a mass of 0.05 kg, moving with a speed of 400 m·s^{-1}, penetrates a distance of 0.1 m into a wooden block firmly attached to the earth. Assume the force that stops it is constant. Compute

a) the acceleration of the bullet,

b) the accelerating force,

c) the time of the acceleration,

d) the impulse of the force.

Compare the answer to part (d) with the initial momentum of the bullet.

8–3 A block of ice of mass 2 kg is moving on a frictionless horizontal surface. At $t = 0$ the block is moving to the right with a velocity of 6 m·s^{-1}. Calculate the velocity of the block after each of the following forces has been applied for 5 s:

a) a force of 5 N, directed to the right;

b) a force of 7 N, directed to the left.

8–4 A baseball has a mass of about 0.2 kg.

a) If the velocity of a pitched ball is 30 m·s^{-1}, and after being batted it is 50 m·s^{-1} in the opposite direction, find the change in momentum of the ball and the impulse applied to it by the bat.

b) If the ball remains in contact with the bat for 0.002 s, find the average force applied by the bat.

8–5 A golf ball of mass 0.10 kg initially at rest is given a speed of 50 m·s^{-1} when it is struck by a club. If the club and ball are in contact for 0.002 s, what average force acted on the ball? Is the effect of the ball's weight during the time of contact significant?

8–6 A baseball of mass 0.25 kg is struck by a bat. Just before impact, the ball is traveling horizontally at 40 m·s^{-1}, and it leaves the bat at an angle of 30° above horizontal with a speed of 60 m·s^{-1}. If the ball and bat were in contact for 0.005 s, find the horizontal and vertical components of the average force on the ball.

8–7 A force of magnitude $F(t) = A + Bt^2$ and directed to the right is applied to an object of mass m starting at $t = 0$ and continuing until $t = t_2$.

a) What is the impulse J of the force?

b) If the object was initially at rest, what is its velocity at time t_2?

Section 8-2 Conservation of Momentum

8-8 On a frictionless horizontal surface block A, of mass 3 kg, is moving toward block B, of mass 5 kg, which is initially at rest. After the collision block A has a velocity of 1.2 m·s^{-1} to the left and block B has a velocity of 7.9 m·s^{-1} to the right.

a) What was the velocity of block A before the collision?

b) Calculate the decrease in kinetic energy that occurs during the collision.

8-9 Ice hockey star Wayne Gretzky is skating at 13 m·s^{-1} toward a defender, who is in turn skating at 4 m·s^{-1} toward Gretzky. Gretzky's weight is 756 N, that of the defender is 900 N. Immediately after the collision Gretzky is moving at 8 m·s^{-1} in his original direction. Neglect external horizontal forces applied by the ice to the skaters during the collision.

a) What is the velocity of the defender immediately after the collision?

b) Calculate the decrease in total kinetic energy of the two players.

8-10 A railroad handcar is moving along straight, frictionless tracks. In each of the following cases, the car initially has a total mass (car and contents) of 200 kg and is traveling with a velocity of 4 m·s^{-1}. Find the *final velocity* of the car in each of the three cases.

a) A 20-kg mass is thrown sideways out of the car with a velocity of 2 m·s^{-1} relative to the car.

b) A 20-kg mass is thrown backwards out of the car with a velocity of 4 m·s^{-1} relative to the *initial* motion of the car.

c) A 20-kg mass is thrown into the car with a velocity of 6 m·s^{-1} relative to the ground and opposite in direction to the velocity of the car.

8-11 A hockey puck B rests on a smooth ice surface and is struck by a second puck A, which was originally traveling at 30 m·s^{-1} and which is deflected 30° from its original direction (Fig. 8–13). Puck B acquires a velocity at 45° with the original velocity of A.

a) Compute the speed of each puck after the collision.

b) What fraction of the original kinetic energy of puck A is "lost"?

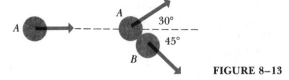

FIGURE 8-13

8-12 A block of mass 0.60 kg is initially at rest. It is struck by a second block of mass 0.40 kg initially moving with a velocity of 0.125 m·s^{-1} toward the right along the x-axis. After the collision the 0.40-kg block has a velocity of 0.10 m·s^{-1} at an angle of 37° above the x-axis in the first quadrant. Both blocks move on a frictionless horizontal plane.

a) What are the magnitude and direction of the velocity of the 0.60-kg block after the collision?

b) What is the loss of kinetic energy during the collision?

Section 8-4 Inelastic Collisions

8-13

a) An empty freight car of mass 10,000 kg rolls at 2 m·s^{-1} along a level track and collides with a loaded car of mass 20,000 kg, standing at rest with brakes released. If the cars couple together, find their speed after the collision.

b) Find the decrease in kinetic energy as a result of the collision.

c) With what speed should the loaded car be rolling toward the empty car for both to be brought to rest by the collision?

8-14 On a frictionless table, a 3-kg block moving 4 m·s^{-1} to the right collides with an 8-kg block moving 1.5 m·s^{-1} to the left.

a) If the two blocks stick together, what is the final velocity?

b) How much mechanical energy is dissipated in the collision?

8-15 A 2000-kg automobile going eastward on Chestnut Street at 60 km·hr^{-1} collides with a 4000-kg truck that is going southward *across* Chestnut Street at 20 km·hr^{-1}. If they become coupled on collision, what are the magnitude and direction of their velocity immediately after colliding?

8-16 A rifle bullet of mass 10 g is fired with a velocity of 800 m·s^{-1} into a ballistic pendulum of mass 5 kg, suspended from a cord 1 m long. Compute

a) the vertical height through which the pendulum rises,

b) the initial kinetic energy of the bullet,

c) the kinetic energy of the bullet and pendulum immediately after the bullet becomes embedded in the pendulum.

8-17 A 5-g bullet is fired horizontally into a 3-kg wooden block resting on a horizontal surface. The coefficient of sliding friction between block and surface is 0.20. The bullet remains embedded in the block, which is observed to slide 0.25 m along the surface before stopping. What was the velocity of the bullet?

Section 8-5 Elastic Collisions

8-18 Two blocks of mass 0.30 kg and 0.20 kg are moving toward each other along a frictionless horizontal surface with velocities of 0.50 m·s^{-1} and 1.00 m·s^{-1}, respectively.

a) If the blocks collide and stick together, find their final velocity.

b) Find the loss of kinetic energy during the collision.

c) Find the final velocity of each block if the collision is completely elastic, so the blocks do not stick together.

8-19 Supply the details of the calculation of θ and ϕ in Example 8–10.

8-20 A small steel ball moving with speed v_0 in the positive x-direction makes a perfectly elastic noncentral collision with an incident ball originally at rest. After impact, the first ball moves with speed v_1 in the first quadrant at an angle θ_1 with the x-axis and the second with speed v_2 in the fourth quadrant at an angle θ_2 with the x-axis.

a) Write the equations expressing conservation of linear momentum in the x-direction and in the y-direction.

b) Square these equations and add them.

c) At this point, introduce the fact that the collision is perfectly elastic.

d) Prove that $\theta_1 + \theta_2 = \pi/2$.

8–21 In Exercise 8–11, suppose the collision is perfectly elastic, and A is deflected 30° from its initial direction. Find the final speed of each puck and the direction of B's velocity.

Section 8–6 Recoil

8–22 An 80-kg man standing on ice throws a 0.2-kg ball horizontally with a speed of 30 m·s⁻¹. With what speed and in what direction will the man begin to move?

8–23 Block A in Fig. 8–14 has a mass of 1 kg, and block B has a mass of 2 kg. The blocks are forced together, compressing a spring S between them, and the system is released from rest on a level, frictionless surface. The spring is not fastened to either block and drops to the surface after it has expanded. Block B acquires a speed of 0.5 m·s⁻¹.

a) What is the final velocity acquired by block A?

b) How much potential energy was stored in the compressed spring?

FIGURE 8–14

8–24 An open-topped freight car of mass 10,000 kg is coasting without friction along a level track. It is raining very hard, with the rain falling vertically. The car is originally empty and moving with a velocity of 1 m·s⁻¹. What is the velocity of the car after it has traveled long enough to collect 1000 kg of rain water?

8–25 One of James Bond's adversaries is standing on a frozen lake. He throws his steel-lined hat with a velocity of 20 m·s⁻¹ at 37° above the horizontal, hoping to hit James. If his mass is 160 kg and that of his hat is 9 kg, what is his horizontal recoil velocity?

Section 8–7 Center of Mass

8–26 A 1000-kg automobile is moving along a straight highway at 10 m·s⁻¹. Another car, with mass 2000 kg and speed 20 m·s⁻¹ is 30 m ahead of the first.

a) Find the position of the center of mass of the two automobiles.

b) Find the total momentum from the above data.

c) Find the velocity of the center of mass.

d) Find the total momentum, using the velocity of the center of mass. Compare your result with that of part (b).

8–27 Find the position of the center of mass of the earth–moon system. Use the data in Appendix F.

8–28 Three particles have the following masses and coordinates: (1) 2 kg, (3 m, 2 m); (2) 3 kg, (1 m, −4 m); (3) 4 kg, (−3 m, 5 m). Find the coordinates of the center of mass of the system.

Section 8–8 Rocket Propulsion

8–29 A small rocket burns 0.05 kg of fuel per second, ejecting it as a gas with a velocity relative to the rocket of 5000 m·s⁻¹.

a) What force does this gas exert on the rocket?

b) Would the rocket operate in free space?

c) If so, how would you steer it? Could you brake it?

8–30 If a single-stage rocket, fired vertically from rest at the earth's surface, burns its fuel in a time of 30 s, and the relative velocity $v_r = 3000$ m·s⁻¹, what must be the mass ratio m_0/m for a final velocity v of 8 km·s⁻¹ (about equal to the orbital velocity of an earth satellite)?

8–31 Obviously rockets can be made to go very fast, but what is a reasonable top speed? Assume that a rocket is fired from rest at a space station where there is no gravitational force.

a) If the rocket ejects gas at a relative velocity of 2400 m·s⁻¹ and you want the rocket's velocity eventually to be $0.001c$, where c is the velocity of light, what fraction of the initial mass of the rocket and fuel is *not* fuel?

b) What is this fraction if the final velocity is to be 3000 m·s⁻¹?

PROBLEMS

8–32 A steel ball of mass 0.5 kg is dropped from a height of 4 m onto a horizontal steel slab. The collision is elastic, and the ball rebounds to its original height.

a) Calculate the impulse delivered to the ball during impact.

b) If the ball is in contact with the slab for 0.002 s, find the average force on the ball during impact.

8–33 A bullet emerges from the muzzle of a gun with a velocity of 300 m·s⁻¹. The resultant force on the bullet,

while it is in the gun barrel, is given by

$$F = 400 \text{ N} - \frac{(4 \times 10^5 \text{ N·s}^{-1})t}{3}.$$

a) Construct a graph of F versus t.

b) Compute the time required for the bullet to travel the length of the barrel, assuming the force becomes zero just at the end of the barrel.

c) Find the impulse of the force.

d) Find the mass of the bullet.

8–34 A tennis ball weighing 0.5 N has $v_1 = (22 \text{ m·s}^{-1})i - (4 \text{ m·s}^{-1})j$ before being struck by a racket. The racket applies a force $F = -(600 \text{ N})i + (80 \text{ N})j$ that we will assume to be constant during the 0.004 s that the racket and ball are in contact.

a) What are the x- and y-components of the impulse of the force applied to the ball?

b) What are the x- and y-components of the final velocity of the ball?

8–35 A force $F = (20 \text{ N·s}^{-2}t^2)i - (10 \text{ N} - 30 \text{ N·s}^{-1}t)j$ is applied to an object of mass 2 kg. If the object was originally at rest, what is its velocity after the force has acted for 0.5 s?

8–36 Objects A, of mass 2 kg; B, of mass 3 kg; and C, of mass 5 kg, are each approaching the origin as shown in Fig. 8–15. The initial velocities of A and B are given in the figure. All three objects arrive at the origin at the same time. What must be the x- and y-components of the initial velocity of C if all three objects are to end up at rest after the collision?

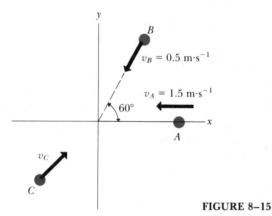

FIGURE 8–15

8–37 A station wagon traveling west collides with a pickup truck traveling north. They stick together as a result of the collision. The station wagon has mass 1400 kg and the pickup truck has mass 900 kg. A police officer estimates from the skid marks that immediately after the collision the combined object was traveling at 9 m·s⁻¹ in a direction 37° west of north. Calculate the velocities of the station wagon and pickup just before the collision.

8–38 A bullet of mass 5 g is shot *through* a 1-kg wood block suspended on a string 2 m long. The center of gravity of the block is observed to rise a distance of 0.50 cm. Find the speed of the bullet as it emerges from the block if the initial speed is 300 m·s⁻¹.

8–39 A bullet of mass 2 g, traveling in a horizontal direction with a velocity of 500 m·s⁻¹, is fired into a wooden block of mass 1 kg, initially at rest on a level surface. The bullet passes through the block and emerges with its velocity reduced to 100 m·s⁻¹. The block slides a distance of 0.20 m along the surface from its initial position.

a) What was the coefficient of sliding friction between block and surface?

b) What was the decrease in kinetic energy of the bullet?

c) What was the kinetic energy of the block at the instant after the bullet passed through it?

8–40 A rifle bullet of mass 0.01 kg strikes and embeds itself in a block of mass 0.99 kg, which rests on a frictionless horizontal surface and is attached to a coil spring, as shown in Fig. 8–16. The impact compresses the spring 10 cm. Calibration of the spring shows that a force of 1.0 N is required to compress the spring 1 cm.

a) Find the maximum potential energy of the spring.

b) Find the velocity of the block just after impact.

c) What was the initial velocity of the bullet?

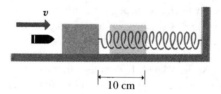

|← 10 cm →| **FIGURE 8–16**

8–41 A frame of mass 0.20 kg, when suspended from a certain coil spring, is found to stretch the spring 0.10 m. A lump of putty of mass 0.20 kg is dropped from rest onto the frame from a height of 0.30 m (Fig. 8–17). Find the maximum distance the frame moves downward.

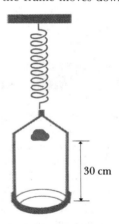

30 cm

FIGURE 8–17

8–42

a) Prove that when a moving object makes a perfectly inelastic collision with a second object of equal mass, initially at rest, one-half the original kinetic energy is "lost."

b) Prove that when a very heavy particle makes a perfectly elastic head-on collision with a very light particle that is at rest, the light one goes off with twice the velocity of the heavy one.

8–43 Two railroad cars roll along and couple with a third car, which is initially at rest. These three roll along and couple to a fourth. This process continues until the speed of the final collection of railroad cars is one-tenth the speed of the initial two railroad cars. Ignoring friction, how many cars are in the final collection of cars? All the cars are identical.

8–44 A block of mass 0.20 kg, sliding with a velocity of 0.12 m·s^{-1} on a smooth, level surface, makes a perfectly elastic head-on collision with a block of mass m, initially at rest. After the collision, the velocity of the 0.20 kg block is 0.04 m·s^{-1} in the same direction as its initial velocity. Find

a) the mass m;   **b)** its velocity after the collision.

8–45 A stone whose mass is 0.10 kg rests on a frictionless horizontal surface. A bullet of mass 2.5 g, traveling horizontally at 400 m·s^{-1}, strikes the stone and rebounds horizontally at right angles to its original direction with a speed of 300 m·s^{-1}.

a) Compute the magnitude and direction of the velocity of the stone after it is struck.

b) Is the collision perfectly elastic?

8–46 A neutron of mass m collides elastically with a nucleus of mass M, which is initially at rest. Show that if the neutron's initial kinetic energy is K_0, the maximum kinetic energy that it can *lose* during the collision is

$$4mMK_0/(M + m)^2.$$

(*Hint:* The maximum energy loss occurs in a head-on collision.)

8–47 A neutron of mass 1.67×10^{-27} kg, moving with a velocity of $2.0 \times 10^4 \text{ m·s}^{-1}$, makes a head-on collision with a boron nucleus of mass 17.0×10^{-27} kg, originally at rest.

a) If the collision is completely inelastic, what is the final kinetic energy of the system, expressed as a fraction of the original kinetic energy?

b) If the collision is perfectly elastic, what fraction of its original kinetic energy does the neutron transfer to the boron nucleus?

8–48 A man and a woman are sitting in a sleigh that is at rest on frictionless ice. The weight of the man is 800 N, the weight of the woman is 600 N, and that of the sleigh is 1200 N. The people suddenly see a poisonous spider on the floor of the sleigh and jump out. The man jumps to the left with a velocity of 5 m·s^{-1} at 30° above the horizontal, and the woman to the right at 9 m·s^{-1} at 37° above the horizontal. Calculate the horizontal velocity (magnitude and direction) that the sleigh has after they jump out.

8–49 Fission, the process that supplies energy in nuclear power plants, occurs when a heavy nucleus is split into two medium-sized nuclei. One such reaction would occur if a neutron colliding with a ^{235}U nucleus split that nucleus into a ^{141}Ba nucleus and a ^{92}Kr nucleus. In this reaction two neutrons also would be split off from the original ^{235}U. Before the collision we have the arrangement in Fig. 8–18a. After the collision we have the Ba nucleus moving in the $+z$-direction and the Kr nucleus in the $-z$-direction. The three neutrons are moving in the xy-plane as shown in Fig. 8–18b. If the incoming neutron has an initial velocity of $3.0 \times 10^6 \text{ m·s}^{-1}$ and a final velocity of $1.5 \times 10^6 \text{ m·s}^{-1}$ in the directions shown, what are the velocities of the other two neutrons and what can you say about the velocities of the Ba and Kr nuclei? (The mass of the Ba nucleus is approximately 2.3×10^{-25} kg and that of Kr is about 1.5×10^{-25} kg.)

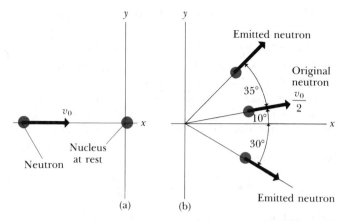

FIGURE 8–18

8–50 A uniform steel rod 1 m in length is bent in a 90° angle at its midpoint. Determine the position of its center of mass (*Hint:* The mass of each side of the angle may be assumed to be concentrated at its center.)

8–51 Two asteroids with masses m_A and m_B are moving with velocities v_A and v_B with respect to an astronomer in a space vehicle.

a) Show that the total kinetic energy as measured by the astronomer is

$$K = \tfrac{1}{2}MV^2 + \tfrac{1}{2}(m_A v_A'^2 + m_B v_B'^2),$$

with V and M defined as in Section 8–7, $v_A' = v_A - V$, and $v_B' = v_B - V$. In this expression the total kinetic energy of the two asteroids is the energy associated with their center of mass plus that associated with the internal motion relative to the center of mass.

b) If the asteroids collide, what is the *minimum* possible kinetic energy they can have after the collision, as measured by the astronomer?

8–52 A man of mass 80 kg stands up in a 30-kg canoe of length 5 m. He walks from a point 1 m from one end to a point 1 m from the other end. If resistance to motion of the canoe in the water can be neglected, how far does the canoe move during this process?

8–53 This problem illustrates the advantage of using a multistage rather than a single-stage rocket. Suppose that the first stage of a two-stage rocket has a total mass of 12,000 kg, of which 9000 kg is fuel. The total mass of the second stage is 1000 kg, of which 750 kg is fuel. Assume that the relative velocity v_r of ejected material is constant, and neglect any effect of gravity. (The latter effect is small during the firing period if the rate of fuel consumption is large.)

a) Suppose that the entire fuel supply carried by the two-stage rocket was utilized in a single-stage rocket of the same total mass of 13,000 kg. What would be the velocity of the rocket, starting from rest, when its fuel was exhausted?

b) What is the velocity when the fuel of the first stage is exhausted, if the first stage carries the second stage with it to this point? This velocity then becomes the initial velocity of the second stage. At this point the second stage separates from the first stage.

c) What is the final velocity of the second stage?

d) What value of v_r would be required to give the second stage of the rocket a velocity of 8 km·s^{-1}?

8–54 In the rocket-propulsion problem the mass is variable. Another such problem is a raindrop falling through a cloud of small water droplets. Some of these droplets adhere to the raindrop, thereby *increasing* its mass as it falls. The force on the raindrop is

$$F_{ext} = \frac{dp}{dt} = m\frac{dv}{dt} + v\frac{dm}{dt}.$$

Suppose the mass of the raindrop depends on how far it has fallen. Then $m = kx$, where k is a constant, and $dm/dt = kv$. Since $F_{ext} = mg$, this gives

$$mg = m\frac{dv}{dt} + v(kv).$$

Or, dividing by k,

$$xg = x\frac{dv}{dt} + v^2.$$

This is a differential equation that has a solution of the form $v = at$. Take the initial velocity of the raindrop to be zero.

a) Using the proposed solution for v, find the acceleration a.

b) Find the distance the raindrop has fallen in $t = 0.4$ s.

c) Given that $k = 2.0$ g·m^{-1}, find the mass of the raindrop at $t = 0.4$ s.

For many more intriguing aspects of this problem, see K. S. Krane, *Amer. Jour. Phys.*, Vol. 49, pp. 113–117 (1981).

CHALLENGE PROBLEMS

8–55 This problem illustrates the usefulness of the concept of center of mass. Suppose one-third of a rope of length l is hanging down over the edge of a frictionless table. The rope has a linear density (mass per unit length) λ, and the end already on the table is held by a person. How much work is done by the person when he pulls on the rope to raise the rest of the rope slowly onto the table? Do the problem in two ways.

a) Find the force that the person must exert to raise the rope, and from this the work done. Note this is a variable force because at different times different amounts of rope are hanging over the edge.

b) Suppose the segment of the rope initially hanging over the edge has all of its mass concentrated at the center of mass. Find the work necessary to raise this to table height. You will probably find this approach simpler than that of part (a). How do the answers compare?

8–56 A 20-kg projectile is fired at an angle of 60° above the horizontal and with a muzzle velocity of 250 m·s^{-1}. At the highest point of its trajectory the projectile explodes into two fragments of equal mass, one of which falls vertically with zero initial speed.

a) How far from the point of firing does the other fragment strike if the terrain is level?

b) How much energy was released during the explosion?

8–57 Block A of mass m_A is moving on a frictionless horizontal surface in the +x-direction with velocity v_{A1} and makes an elastic collision with block B of mass m_B that is initially at rest.

a) Calculate the velocity of the center of mass (cm) of the two-block system before the collision.

b) Consider a coordinate system whose origin is at the cm and moves with it. Is this an inertial reference frame?

c) What are the initial velocities u_{A1} and u_{B1} of the two blocks in this cm reference frame, and what is the total momentum in this frame?

d) Use conservation of momentum and energy, applied in the cm frame, to relate the final momentum of each block to its initial momentum, and hence the final velocity of each block to its initial velocity. Your results should show that a one-dimensional elastic collision has a very simple description in cm coordinates.

e) Let $m_A = 2$ kg, $m_B = 4$ kg, and $v_{A1} = 2$ m·s^{-1}. Find the cm velocities u_{A1} and u_{B1}, apply the simple result found in (d), and transform back to velocities in a stationary frame to find the final velocities of the blocks. Does your result agree with Eqs. (8–22) and (8–23)?

8–58 A jet of liquid of cross-sectional area A and density ρ moves with speed v_J in the positive x-direction and impinges against a perfectly smooth blade B, which deflects the stream at right angles but does not slow it down, as shown in Fig. 8–19.

a) If the blade is *stationary*, prove that the rate of arrival of mass at the blade is $dm/dt = \rho A v_J$.

b) If the impulse–momentum theorem is applied to a small mass dm, prove that the x-component of the force acting on this mass for the time interval dt is given by

$$F_x = -\frac{dm}{dt}v_J.$$

c) Prove that the *steady* force exerted *on* the blade in the x-direction is

$$F_x = \rho A v_J^2.$$

Consider now that the blade moves to the right with a speed v_B ($v_B < v_J$). The stream is deflected at right angles to

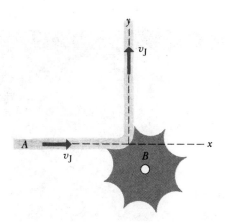

FIGURE 8-19

where dm/dt is the rate of change of the mass of the rocket.

b) Suppose that the rate of ejection of mass by the rocket is constant; that is, the rate of decrease of mass is $dm/dt = -km_0$, where k is a positive constant and m_0 is the initial mass. What is the numerical value of k and in what units is it expressed for the rocket in Exercise 8–30?

c) Show that the mass at any time t is given by the equation

$$m = m_0(1 - kt).$$

d) Show that the acceleration at any time is equal to

$$\frac{v_r k}{1 - kt} - g.$$

e) Find the initial acceleration of the rocket in Exercise 8–30, in terms of the acceleration of gravity, g.

f) Find the acceleration 15 s after the motion starts.

g) Sketch the acceleration–time graph.

the *moving* blade. Derive the equations for

d) the rate of arrival of mass at the moving blade,

e) the force F_x on the blade,

f) the power delivered to the blade.

8–59

a) Show that the acceleration of a rocket fired vertically upward is given by

$$a = -\frac{v_r}{m}\frac{dm}{dt} - g,$$

ROTATIONAL MOTION

THE MOTIONS OF REAL-WORLD BODIES CAN BE VERY COMPLEX. A BODY CAN have rotational as well as translational motions, and it can deform as the forces acting on it stretch, twist, and squeeze it. Our discussion of motion thus far has centered around a pointlike particle, an idealized model that is adequate when rotation and deformation can be ignored. Now we go on to a more sophisticated idealized model, a body that has finite size and definite shape and that can have rotational as well as translational motion. We continue to neglect deformations and assume that the body has a perfectly definite and unchanging shape and size. This idealized model is called a **rigid body,** and the study of rotational motion of a rigid body is the principal subject of this chapter. We introduce language for *describing* rotational motion and then develop the dynamic principles relating the forces on the body to its motion. During this development we introduce several new physical quantities, including torque, moment of inertia, and angular momentum. As we will see, several aspects of rotational motion have direct analogs in translational motion. We postpone until Chapter 12 a detailed study of the deformations of real-world bodies when forces act on them.

A rigid body: an idealized body that doesn't change shape at all

9–1 ANGULAR VELOCITY

Let us first consider a rigid body that rotates about a stationary axis; it might be a motor shaft, a chunk of roast beef on a barbeque skewer, or possibly a merry-go-round. In Fig. 9–1, a rigid body rotates about a stationary line passing through point O, perpendicular to the plane of the diagram. Line OP is fixed in the body and rotates with it. The angle between this line and the horizontal line in the figure is θ. Once we know the position of the axis of rotation, θ describes the position of the body completely. Thus θ serves as a *coordinate* to describe the rotational position of the body.

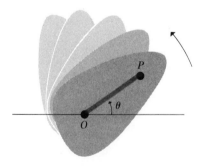

9–1 Body rotating about a fixed axis through point O.

An angle can be measured in degrees or radians; in rotational motion, radians are more convenient.

It is convenient to measure the angle θ in **radians.** As shown in Fig. 9–2a, one radian (1 rad) is the angle subtended at the center of a circle by an arc of length equal to the radius of the circle. The circumference is 2π times the radius, so there are 2π or about 6.283 radians in one complete revolution or

360°. Hence

$$1 \text{ rad} = \frac{360°}{2\pi} = 57.3°,$$

$$360° = 2\pi \text{ rad} = 6.28 \text{ rad},$$

$$180° = \pi \text{ rad} = 3.14 \text{ rad},$$

$$90° = \pi/2 \text{ rad} = 1.57 \text{ rad},$$

$$60° = \pi/3 \text{ rad} = 1.05 \text{ rad},$$

$$45° = \pi/4 \text{ rad} = 0.79 \text{ rad, and so on.}$$

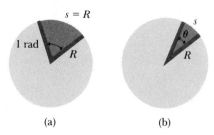

9-2 An angle θ in radians is defined as the ratio of the arc s to the radius R.

Figure 9–2b shows an arbitrary angle θ subtended by an arc of length s of the circumference of a circle of radius R; θ (in radians) is equal to s divided by R:

$$\theta = \frac{s}{R}, \qquad s = R\theta. \tag{9–1}$$

For example, if $\theta = 2\pi$ rad, then s is the circumference of the circle.

An angle in radians is the quotient of a length and a length, so it is a pure number, without units. If $s = 1.5$ m and $R = 1$ m, the angle is usually described as $\theta = 1.5$ rad, but it would be equally correct to say simply $\theta = 1.5$.

Rotational *motion* of a body can be described in terms of the rate of change of θ. In Fig. 9–3, a reference line OP in a rotating body makes an angle θ_1 with the reference line Ox, at a time t_1. At a later time t_2 the angle has changed to θ_2. We define the **average angular velocity** of the body, ω_{av}, in the time interval $\Delta t = t_2 - t_1$, as the ratio of the angular displacement $\theta_2 - \theta_1$, or $\Delta\theta$, to Δt:

Angular velocity: describing how fast a body is rotating

$$\omega_{av} = \frac{\Delta\theta}{\Delta t}.$$

The **instantaneous angular velocity** ω is defined as the limit of this ratio as Δt approaches zero, that is, the derivative of θ with respect to t:

$$\omega = \lim_{\Delta t \to 0} \frac{\Delta\theta}{\Delta t} = \frac{d\theta}{dt}. \tag{9–2}$$

Because the body is rigid, *all* lines in it rotate through the same angle in the same time, and the angular velocity is characteristic of the body as a whole. If the angle θ is in radians, the unit of angular velocity is one radian per second (1 rad·s^{-1} or simply 1 s^{-1}). Other units, such as the revolution per minute (rev·min^{-1}), are in common use. We note that $1 \text{ rev·s}^{-1} = 2\pi \text{ rad·s}^{-1}$, and 1 rev·min$^{-1} = 1$ rpm $= (2\pi/60)$rad·s^{-1}.

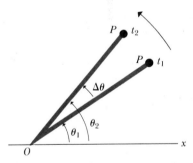

9-3 Angular displacement $\Delta\theta$ of a rotating body.

9-2 ANGULAR ACCELERATION

When the angular velocity of a body changes, it has an angular acceleration. If ω_1 and ω_2 are the instantaneous angular velocities at times t_1 and t_2, we define the **average angular acceleration** α_{av} as

Angular acceleration: describing the rate of change of rotational motion

$$\alpha_{av} = \frac{\omega_2 - \omega_1}{t_2 - t_1} = \frac{\Delta\omega}{\Delta t},$$

and the **instantaneous angular acceleration** α as the limit of this ratio as

$\Delta t \to 0$:

$$\alpha = \lim_{\Delta t \to 0} \frac{\Delta \omega}{\Delta t} = \frac{d\omega}{dt}. \tag{9–3}$$

The unit of angular acceleration is 1 rad·s^{-2} or 1 s^{-2}. Angular velocity and angular acceleration are exactly analogous to linear velocity and acceleration. In each case, velocity is the time derivative of position, and acceleration is the time derivative of velocity.

Because $\omega = d\theta/dt$, the angular acceleration can also be written:

$$\alpha = \frac{d}{dt}\frac{d\theta}{dt} = \frac{d^2\theta}{dt^2}. \tag{9–4}$$

Note that we are using Greek letters for angular kinematic quantities: θ for angular position, ω for angular velocity, and α for angular acceleration. These are analogous to x for position, v for velocity, and a for acceleration, respectively, in straight-line motion.

Notation: using Greek letters for angular quantities

9–3 ROTATION WITH CONSTANT ANGULAR ACCELERATION

Rotation with constant angular acceleration is completely analogous to straight-line motion with constant acceleration.

In Chapter 2 we found that analyzing straight-line motion is particularly simple when the acceleration is *constant*. This is also true of rotational motion; when the angular acceleration is constant, it is easy to derive equations for angular velocity and angular position as functions of time by integration. We begin with

$$\frac{d\omega}{dt} = \alpha = \text{constant}.$$

We integrate this with respect to t:

$$\int d\omega = \int \alpha \, dt, \qquad \omega = \alpha t + C_1,$$

where C_1 is an integration constant. If ω_0 is the angular velocity when $t = 0$, C_1 is equal to ω_0 and

$$\omega = \omega_0 + \alpha t. \tag{9–5}$$

Also, $\omega = d\theta/dt$; integrating again, we find

$$\int d\theta = \int \omega_0 \, dt + \int \alpha t \, dt, \qquad \theta = \omega_0 t + \frac{1}{2}\alpha t^2 + C_2.$$

The integration constant C_2 is the value of θ when $t = 0$ (the initial position), which we denote as θ_0. Thus

$$\theta = \theta_0 + \omega_0 t + \tfrac{1}{2}\alpha t^2. \tag{9–6}$$

We can also derive an equation relating ω and θ, by the same procedure we used to derive Eq. (2–13): Solve Eq. (9–5) for t, substitute the result into Eq. (9–6) to eliminate t, and simplify the resulting equation. We leave the details for you to work out; the final result is

$$\omega^2 = \omega_0{}^2 + 2\alpha(\theta - \theta_0). \tag{9–7}$$

Table 9–1 shows the similarity between Eqs. (9–5), (9–6), and (9–7) for motion with constant angular acceleration and the equations for motion with constant linear acceleration.

TABLE 9–1

Motion with constant linear acceleration	Motion with constant angular acceleration
$a = $ constant	$\alpha = $ constant
$v = v_0 + at$	$\omega = \omega_0 + \alpha t$
$x = x_0 + v_0 t + \frac{1}{2}at^2$	$\theta = \theta_0 + \omega_0 t + \frac{1}{2}\alpha t^2$
$v^2 = v_0^2 + 2a(x - x_0)$	$\omega^2 = \omega_0^2 + 2\alpha(\theta - \theta_0)$

EXAMPLE 9–1   The angular velocity of a bicycle wheel is 4.0 rad·s^{-1} at time $t = 0$, and its angular acceleration is constant and equal to 2.0 rad·s^{-2}. A spoke OP on the wheel is horizontal at time $t = 0$.

a)   What angle does this spoke make with the horizontal at time $t = 3.0$ s?

b)   What is the wheel's angular velocity at this time?

SOLUTION   We can use Eqs. (9–5) and (9–6) to find θ and ω at any time, in terms of the given initial conditions.

Motion of a speeding-up bicycle wheel: an example of rotational kinematics

a)   The angle θ is given as a function of time by

$$\theta = \theta_0 + \omega_0 t + \tfrac{1}{2}\alpha t^2.$$

At time $t = 3.0$ s,

$$\theta = 0 + (4.0 \text{ rad·s}^{-1})(3.0 \text{ s}) + \tfrac{1}{2}(2.0 \text{ rad·s}^{-2})(3.0 \text{ s})^2$$

$$= 21 \text{ rad} = \frac{21}{2\pi} \text{ rev} = 3.34 \text{ rev}.$$

The body turns through three complete revolutions plus an additional 0.34 rev or (0.34 rev) $(2\pi$ rad·rev$^{-1}) = 2.15$ rad $= 123°$. The line OP thus turns through 123° and makes an angle of 57° with the horizontal.

b)   In general, $\omega = \omega_0 + \alpha t$. At time $t = 3.0$ s,

$$\omega = 4.0 \text{ rad·s}^{-1} + (2.0 \text{ rad·s}^{-2})(3.0 \text{ s}) = 10 \text{ rad·s}^{-1}.$$

Alternatively, from Eq. (9–7),

$$\omega^2 = \omega_0^2 + 2\alpha(\theta - \theta_0) = (4 \text{ rad·s}^{-1})^2 + 2(2 \text{ rad·s}^{-2})(21 \text{ rad})$$

$$= 100 \text{ rad}^2\text{·s}^{-2},$$

$$\omega = 10 \text{ rad·s}^{-1}.$$

9–4 RELATION BETWEEN ANGULAR AND LINEAR VELOCITY AND ACCELERATION

When a rigid body rotates about a stationary axis, every particle in the body moves in a circle lying in a plane perpendicular to this axis, with the center of the circle on the axis. In Section 3–5 we worked out a relation for the acceleration of a particle moving in a circular path, in terms of its speed and the radius; this relation is still valid when the particle is part of a rotating rigid body.

A spiral galaxy. As material moves away from the rapidly rotating center of the galaxy, its angular momentum is approximately constant. Thus the angular velocity must decrease as the distance from the center increases, forming the spiral pattern. Can you deduce which way the galaxy is rotating? (Photo by R. V. Willstrop; © 1975 Anglo-Australian Telescope Board.)

The speed of a particle in a rotating rigid body depends on its distance from the axis and on the angular velocity.

The speed of a particle in a rigid body is directly proportional to the body's angular velocity. In Fig. 9–4, point P is a distance r away from the axis of rotation, and it moves in a circle of radius r. When the angle θ increases by a small amount $\Delta\theta$ in a time interval Δt, the particle moves through an arc length $\Delta s = r\,\Delta\theta$. If $\Delta\theta$ is very small, the arc is nearly a straight line, and the average speed of the particle is given by

$$v_{\text{av}} = \frac{\Delta s}{\Delta t} = r\frac{\Delta\theta}{\Delta t}. \tag{9–8}$$

In the limit, as $\Delta t \to 0$, this becomes

$$v = r\frac{d\theta}{dt} = r\omega. \tag{9–9}$$

The *direction* of the particle's velocity is tangent to its circular path at each point.

If the angular velocity changes by $\Delta\omega$, the particle's speed changes by an amount Δv given by

$$\Delta v = r\,\Delta\omega.$$

This corresponds to a component of acceleration $a_\parallel$ tangent to the circle. If these changes take place in a small time interval Δt, then

$$\frac{\Delta v}{\Delta t} = r\frac{\Delta\omega}{\Delta t}.$$

The acceleration of a particle in a rotating rigid body can be expressed in terms of tangential and radial components.

and, in the limit, as $\Delta t \to 0$,

$$a_\parallel = r\frac{d\omega}{dt} = r\alpha, \tag{9–10}$$

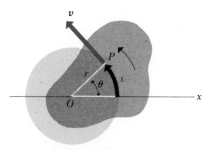

9–4 The distance s moved through by point P equals $r\theta$.

where $a_\parallel$ is the **tangential component of acceleration** of a point at a distance r from the axis.

The **radial component of acceleration** of the point, as worked out in Section 3–5, is $a_\perp = v^2/r$. We can also express this in terms of ω by using Eq. (9–9):

$$a_\perp = \frac{v^2}{r} = \omega^2 r. \qquad (9\text{–}11)$$

This is true at each instant *even when ω and v are not constant.*

The tangential and radial components of acceleration of a point P in a rotating body are shown in Fig. 9–5. Their vector sum is the acceleration $\boldsymbol{a}$, as shown.

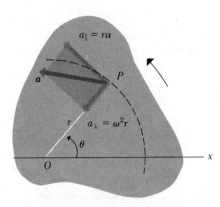

9–5 Nonuniform rotation about a fixed axis through point O. The tangential component of acceleration of point P equals $r\alpha$; the radial component equals $\omega^2 r$.

EXAMPLE 9–2   A discus thrower turns with angular acceleration $\alpha = 50$ rad·s^{-2}. What are the radial and tangential components of acceleration of the discus at the instant when the angular velocity is 10 rad·s^{-1}? The thrower's arm is 0.80 m long.

SOLUTION   The acceleration components are given by Eqs. (9–10) and (9–11):

$$a_\perp = \omega^2 r = (10 \text{ s}^{-1})^2 (0.80 \text{ m}) = 80 \text{ m·s}^{-2},$$

$$a_\parallel = r\alpha = (0.80 \text{ m})(50 \text{ s}^{-2}) = 40 \text{ m·s}^{-2}.$$

The magnitude of the acceleration vector is

$$a = \sqrt{a_\perp{}^2 + a_\parallel{}^2} = 89 \text{ m·s}^{-2},$$

or about nine times the acceleration due to gravity. Note that in using Eqs. (9–10) and (9–11) we *must* express the angular quantities in radians. Also note that in unit cancellations, we may drop out "radians" when it is convenient because an angle expressed in radians is a unitless number.

9–5 KINETIC ENERGY OF ROTATION

Because a rotating rigid body consists of particles in motion, it has kinetic energy. It turns out that we can express the kinetic energy simply in terms of the body's angular velocity and a quantity called the *moment of inertia* of the body. To develop this relationship, we use Eq. (9–9) to find the speed v of a particle in a rigid body that rotates with angular velocity ω about a stationary axis. That is, $v = r\omega$, where r is the particle's distance from the axis. If the particle has mass m, its kinetic energy is

$$\tfrac{1}{2}mv^2 = \tfrac{1}{2}mr^2\omega^2.$$

The *total* kinetic energy of the body is the *sum* of the kinetic energies of all particles in the body,

$$K = \tfrac{1}{2}m_1 r_1{}^2\omega^2 + \tfrac{1}{2}m_2 r_2{}^2\omega^2 + \cdots$$
$$= \Sigma \tfrac{1}{2}mr^2\omega^2.$$

Because ω is the same for all particles in a rigid body, we can rewrite this as

$$K = \tfrac{1}{2}(m_1 r_1{}^2 + m_2 r_2{}^2 + \cdots)\omega^2$$
$$= \tfrac{1}{2}[\Sigma mr^2]\omega^2.$$

The kinetic energy of a rotating rigid body can be expressed in terms of angular velocity and moment of inertia.

To obtain the sum Σmr^2, we subdivide the body (in our imagination) into a large number of particles, multiply the mass of each particle by the square of its distance from the axis, and add these products for all particles. The result is called the **moment of inertia** I of the body, about the axis of rotation:

$$I = \Sigma mr^2. \qquad (9\text{--}12)$$

In SI units the unit of moment of inertia is 1 kilogram-meter^2 (1 kg·m^2). We can now express the rotational kinetic energy of a rigid body as

$$K = \tfrac{1}{2}I\omega^2. \qquad (9\text{--}13)$$

Moment of inertia is the rotational analog of mass.

The form of this expression is analogous to that for translational kinetic energy:

$$K = \tfrac{1}{2}mv^2.$$

That is, for rotation about a stationary axis, moment of inertia I is analogous to mass m, and angular velocity ω is analogous to velocity v.

Calculating the moment of inertia of a set of particles by adding their separate moments of inertia

EXAMPLE 9–3 An engineer is designing a one-piece machine part consisting of three heavy connectors linked by light molded struts, as in Fig. 9–6. The connectors can be considered as massive particles connected by massless rods.

What is the moment of inertia of this machine part

a) about an axis through point A, perpendicular to the plane of the diagram?

b) about an axis coinciding with rod BC?

c) If the body rotates about an axis through A perpendicular to the plane of the diagram, with angular velocity $\omega = 4.0$ rad·s^{-1}, what is the rotational kinetic energy?

SOLUTION

a) The particle at point A lies on the axis. Its distance *from* the axis is zero, and it contributes nothing to the moment of inertia. Therefore, from Eq. (9–12),

$$\begin{aligned} I = \Sigma mr^2 &= (0.10 \text{ kg})(0.50 \text{ m})^2 + (0.20 \text{ kg})(0.40 \text{ m})^2 \\ &= 0.057 \text{ kg·m}^2 \end{aligned}$$

b) The particles at B and C both lie on the axis, so neither contributes to the moment of inertia; only A contributes, and we have

$$I = \Sigma mr^2 = (0.30 \text{ kg})(0.40 \text{ m})^2 = 0.048 \text{ kg·m}^2.$$

This illustrates the important fact that the moment of inertia of a body, unlike its mass, is *not* a unique property of the body; it depends on the position of the axis about which it is computed.

c) From Eq. (9–13),

$$K = \tfrac{1}{2}I\omega^2 = \tfrac{1}{2}(0.057 \text{ kg·m}^2)(4.0 \text{ rad·s}^{-1})^2 = 0.456 \text{ J}.$$

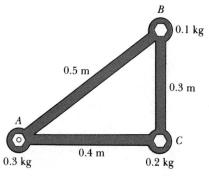

9–6 A strangely shaped machine part.

In Example 9–3 the body could be represented as several point masses, and the sum in Eq. (9–12) could be evaluated directly. When the body is a *continuous* distribution of matter, such as a solid cylinder or plate, this sum must be evaluated by integration. Several examples of calculations of mo-

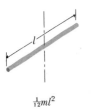

$\frac{1}{12}ml^2$

(a) Slender rod,
axis through
center

$\frac{1}{12}m(a^2 + b^2)$

(b) Rectangular plate,
axis through
center

$\frac{1}{2}m\,(R_1{}^2 + R_2{}^2)$

(c) Hollow cylinder

$\frac{1}{2}mR^2$

(d) Solid cylinder

mR^2

(e) Thin-walled hollow
cylinder

$\frac{2}{5}mR^2$

(f) Solid sphere

9–7 Moments of inertia. For each body, the axis is shown as a broken line.

ments of inertia are given in Section 9–6; meanwhile, Fig. 9–7 gives moments of inertia for several simple shapes.

You may be tempted to try to compute the moment of inertia of a body by assuming that all the mass is concentrated at the center of mass, and then multiplying the total mass by the square of the distance from the center of mass to the axis. Resist that temptation; it doesn't work! For example, when a uniform, thin rod of length L and mass M is pivoted about an axis through one end, perpendicular to the rod, the moment of inertia is $I = ML^2/3$. If we took the mass as concentrated at the center, a distance $L/2$ from the axis, we would obtain the *incorrect* result $I = M(L/2)^2 = ML^2/4$.

Now that we have learned how to calculate the kinetic energy of a rotating rigid body, we can apply the energy principles of Chapter 7 to rotational motion. The following examples illustrate this technique.

PROBLEM-SOLVING STRATEGY: *Rotational energy*

1. We suggest you review the strategy outlined in Section 7–5; it is equally useful here. The only new idea is that the kinetic energy K is expressed in terms of the moment of inertia I and angular velocity ω of the body instead of its mass M and speed v. We can then use work–energy relations and conservation of energy, where appropriate, to find relations involving position and motion of a rotating body.

2. The kinematic relations of Section 9–4, especially Eqs. (9–9) and (9–10), are often useful, especially when a rotating cylindrical body functions as any sort of pulley. Example 9–4 illustrates this point.

EXAMPLE 9–4 A light, flexible rope is wrapped several times around a solid cylinder of mass 50 kg and diameter 0.12 m, which rotates on frictionless bearings about a stationary horizontal axis. The free end of the rope is pulled with a constant force of magnitude 9.0 N for a distance of 2.0 m. If the cylinder is initially at rest, find its final angular velocity and the final speed of the rope.

An example of the work–energy theorem with rotational kinetic energy

SOLUTION Because no energy is lost in friction, the final kinetic energy $\frac{1}{2}I\omega^2$ of the cylinder is equal to the work Fd done by the force, which is (9.0 N) (2.0 m) = 18 J. From Fig. 9–7, the moment of inertia is

$$I = \tfrac{1}{2}MR^2 = \tfrac{1}{2}(50 \text{ kg})(0.060 \text{ m})^2 = 0.090 \text{ kg·m}^2.$$

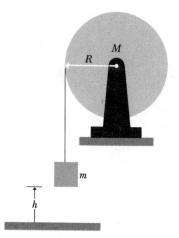

9–8 As the cylinder rotates, the rope unwinds and mass m drops.

An example of conservation of energy with rotational kinetic energy

The work–energy relation then gives

$$\tfrac{1}{2}(0.090 \text{ kg·m}^2)\omega^2 = 18 \text{ J},$$

$$\omega = 20 \text{ rad·s}^{-1}.$$

The final speed of the rope is equal to the tangential speed of the cylinder, which is given by Eq. (9–9):

$$v = r\omega = (0.060 \text{ m})(20 \text{ rad·s}^{-1}) = 1.2 \text{ m·s}^{-1}.$$

EXAMPLE 9–5 We wrap a light, flexible rope around a solid cylinder of mass M and radius R, which rotates with no friction about a stationary horizontal axis, as in Fig. 9–8. We tie the free end of the rope to a mass m and release the mass with no initial velocity, at a distance h above the floor. Find its speed and the angular velocity of the cylinder just as mass m strikes the floor.

SOLUTION Initially the system has no kinetic energy ($K_1 = 0$), but it has potential energy $U_1 = mgh$. Just as mass m strikes the floor, the potential energy is zero ($U_2 = 0$), but both this mass and the cylinder have kinetic energy. The total kinetic energy K_2 is

$$K_2 = \tfrac{1}{2}mv^2 + \tfrac{1}{2}I\omega^2. \qquad (9–14)$$

Now, according to Fig. 9–7, the moment of inertia of the cylinder is $I = \tfrac{1}{2}MR^2$. Furthermore, v and ω are related by $v = R\omega$, since the speed of mass m must be equal to the tangential speed at the outer surface of the cylinder. Using these relations and the energy relation $K_1 + U_1 = K_2 + U_2$, we obtain

$$mgh = \tfrac{1}{2}mv^2 + \tfrac{1}{2}(\tfrac{1}{2}MR^2)\left(\frac{v}{R}\right)^2 = \tfrac{1}{2}(m + \tfrac{1}{2}M)v^2,$$

$$v = \sqrt{\frac{2gh}{1 + M/2m}}.$$

When M is much larger than m, v is very small, as might be expected. When M is much smaller than m, v is nearly equal to the speed of a body in free fall with initial height h, namely, $\sqrt{2gh}$.

9–6 MOMENT-OF-INERTIA CALCULATIONS

Although the moment of inertia of a solid body is defined in principle by Eq. (9–12), this equation can be applied directly only in cases where the body consists of a few point masses, as was the case in Example 9–3. When the body consists of a *continuous* distribution of matter, we can express the sum in terms of an integral.

Imagine dividing the entire volume of the body into small volume elements dV so that all points in a particular element are very nearly the same distance from the axis of rotation; we call this distance r, as before. Let dm be the mass in a volume element dV. The moment of inertia may then be expressed as

$$I = \int r^2 \, dm. \qquad (9–15)$$

Density ρ is mass per unit volume, $\rho = dm/dV$, so we may also write

Finding the moment of inertia of a solid body by integration

$$I = \int r^2 \, \rho \, dV.$$

If the body is homogeneous (uniform in density), then ρ may be taken outside the integral:

$$I = \rho \int r^2 \, dV. \qquad (9\text{–}16)$$

In using this equation, we express the volume element dV in terms of the differentials of the integration variables, usually the coordinates of the volume element. The element dV must always be chosen so that all points within it are at very nearly the same distance from the axis of rotation. For regularly shaped bodies this integration can often be carried out quite easily. Three examples are given below.

EXAMPLE 9–6 *Uniform, slender rod; axis perpendicular to length.* Figure 9–9 shows a uniform, slender rod of mass M and length l. We wish to compute its moment of inertia about an axis through O, at an arbitrary distance h from one end. Using Eq. (9–15), we choose as an element of mass a short section having length dx at a distance x from point O. The ratio of the mass dm of this element to the total mass M is equal to the ratio of its length dx to the total length l. Thus

$$\frac{dm}{M} = \frac{dx}{l}.$$

We solve this for dm, substitute into Eq. (9–15), and add the appropriate integration limits on x, to obtain

$$I_0 = \int x^2 \, dm = \frac{M}{l} \int_{-h}^{l-h} x^2 \, dx$$

$$= \frac{M}{l} \frac{x^3}{3} \bigg]_{-h}^{l-h} = \frac{1}{3} M(l^2 - 3lh + 3h^2).$$

From this general expression we can find the moment of inertia about an axis through any point on the rod. For example, if the axis is at the left end, $h = 0$ and

$$I = \frac{1}{3} M l^2. \qquad (9\text{–}17)$$

If the axis is at the right end, $h = l$ and

$$I = \frac{1}{3} M l^2,$$

as would be expected. If the axis passes through the center,

$$h = \frac{l}{2} \quad \text{and} \quad I = \frac{1}{12} M l^2, \qquad (9\text{–}18)$$

as shown also in Fig. 9–7.

EXAMPLE 9–7 *Hollow or solid cylinder; axis of symmetry.* Figure 9–10 shows a hollow cylinder of length l and inner and outer radii R_1 and R_2. We choose as the most convenient volume element a thin cylindrical shell of radius r, thickness dr, and length l. The volume of this shell is very nearly equal to that of a flat sheet of thickness dr, length l, and width $2\pi r$ (the circumference of the shell). Then

$$dm = \rho \, dV = 2\pi \rho l r \, dr.$$

Dividing a thin rod into short segments to compute its moment of inertia

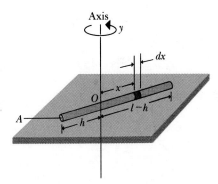

9–9 Moment of inertia of a thin rod. The mass element is a segment of length dx.

Dividing a cylinder into cylindrical shells to compute its moment of inertia

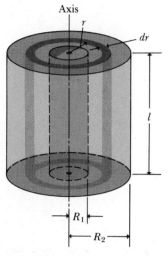

9–10 Moment of inertia of a hollow cylinder. The mass element is a cylindrical shell of radius r and thickness dr.

The moment of inertia is given by

$$
\begin{aligned}
I &= \rho \int r^2 \, dV \\
&= 2\pi\rho l \int_{R_1}^{R_2} r^3 \, dr \\
&= \frac{\pi\rho l}{2} (R_2{}^4 - R_1{}^4) \\
&= \frac{\pi\rho l}{2} (R_2{}^2 - R_1{}^2)(R_2{}^2 + R_1{}^2).
\end{aligned}
\tag{9–19}
$$

It is usually more convenient to express the moment of inertia in terms of the total mass M of the body, which is its density multiplied by the total volume. The volume is given by

$$\pi l (R_2{}^2 - R_1{}^2).$$

Hence

$$M = \pi l \rho (R_2{}^2 - R_1{}^2),$$

and the moment of inertia is

$$I = \tfrac{1}{2} M (R_1{}^2 + R_2{}^2), \tag{9–20}$$

as shown also in Fig. 9–7.

If the cylinder is solid, $R_1 = 0$; letting the outer radius be R, we find that the moment of inertia of a solid cylinder of radius R is

$$I = \tfrac{1}{2} M R^2. \tag{9–21}$$

If the cylinder is very thin walled (like a stovepipe), R_1 and R_2 are very nearly equal; if R represents this common radius,

$$I = M R^2.$$

Note that the moment of inertia of a cylinder about an axis coinciding with its axis of symmetry does not depend on the length l. Two hollow cylinders of the same inner and outer radii, one of wood and one of brass, but having the same mass M, have equal moments of inertia even though the length of the wood cylinder is much greater. Moment of inertia depends only on the *radial* distribution of mass, not on its distribution along the axis. Thus Eq. (9–20) holds also for a very short cylinder, such as a washer, and Eq. (9–21) for a thin disk.

EXAMPLE 9–8 *Uniform sphere of radius R; axis through center.* Divide the sphere into thin disks, as indicated in Fig. 9–11. The radius r of the disk shown is

$$r = \sqrt{R^2 - x^2}.$$

Its volume is

$$dV = \pi r^2 \, dx = \pi (R^2 - x^2) \, dx;$$

and its mass is

$$dm = \rho \, dV.$$

Hence from Eq. (9–21) its moment of inertia is

$$dI = \frac{\pi\rho}{2} (R^2 - x^2)^2 \, dx.$$

Slicing a sphere into disks to compute its moment of inertia

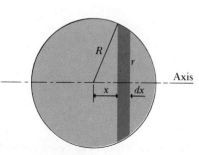

9–11 Moment of inertia of a sphere. The mass element is a disk of thickness dx.

Integrating this expression from 0 to R gives the moment of inertia of the right hemisphere; from symmetry, the total I for the entire sphere is just twice this:

$$I = (2)\frac{\pi\rho}{2} \int_0^R (R^2 - x^2)^2 \, dx.$$

Carrying out the integration, we obtain

$$I = \frac{8\pi\rho}{15} R^5.$$

The mass M of the sphere is

$$M = \rho V = \frac{4\pi\rho R^3}{3}.$$

Hence

$$I = \tfrac{2}{5} M R^2.$$

9–7 PARALLEL-AXIS THEOREM

Here is a theorem that is often useful in finding moments of inertia with respect to various axes. If the moment of inertia I_{cm} of a body about an axis through its center of mass is known, then the moment of inertia I_P about any other axis parallel to the original one but displaced from it by a distance d is easily obtained by means of a relation called the **parallel-axis theorem,** which states that

$$I_P = I_{cm} + Md^2. \tag{9–22}$$

To prove this theorem we consider the body shown in Fig. 9–12. The origin of coordinates has been chosen to coincide with the center of mass. We wish to compute the moment of inertia about an axis through point P, perpendicular to the plane of the figure. Point P has coordinates (a, b), and its distance from the origin is d. We note that $d^2 = a^2 + b^2$.

How to relate the moments of inertia of a body about two different axes

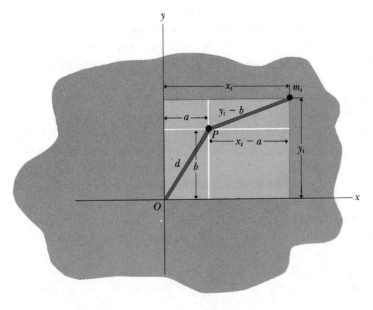

9–12 The mass element m_i has coordinates (x_i, y_i) with respect to the center of mass at O, and coordinates $(x_i - a, y_i - b)$ with respect to point P.

Let m_i be a typical mass element, with coordinates (x_i, y_i). Then the moment of inertia about an axis through O perpendicular to the figure is

$$I_{cm} = \Sigma m_i(x_i^2 + y_i^2),$$

and the moment of inertia about the axis through P is

$$I_P = \Sigma m_i\left[(x_i - a)^2 + (y_i - b)^2\right].$$

We expand the squared terms and regroup, obtaining

$$I_P = \Sigma m_i(x_i^2 + y_i^2) - 2a\Sigma m_i x_i - 2b\Sigma m_i y_i + (a^2 + b^2)\Sigma m_i.$$

The first sum is I_{cm}. The second and third sums are zero because they represent the x- and y-coordinates of the center of mass, which are zero because we have taken the origin to be the center of mass. The final term is d^2 multiplied by the total mass, so the theorem is proved.

EXAMPLE 9–9   The moment of inertia of a thin rod about an axis through its midpoint, perpendicular to the rod, is $I = ML^2/12$ (as shown in Fig. 9–7). Find the moment of inertia about an axis perpendicular to the rod at one end.

SOLUTION   We have $I_{cm} = ML^2/12$ and $d = L/2$; Eq. (9–22) yields

$$I_P = \frac{ML^2}{12} + M\left(\frac{L^2}{4}\right) = \frac{ML^2}{3}.$$

EXAMPLE 9–10   Find the moment of inertia of a thin, uniform disk about an axis perpendicular to its plane at the edge.

SOLUTION   From Fig. 9–7 we have $I_{cm} = MR^2/2$, and in this case $d = R$. Thus

$$I_P = \frac{MR^2}{2} + MR^2 = \frac{3MR^2}{2}.$$

Torque is the tendency of a force to cause a rotation.

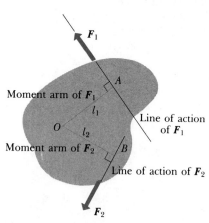

9–13 The moment or torque of a force about an axis is the product of the force and its moment arm.

9–8 TORQUE

In studying dynamics of a particle, we made extensive use of Newton's second law, which relates the acceleration of a particle to the forces acting on it. Now we need to develop an analogous relation between the *angular acceleration* of a rotating rigid body and the forces acting on it. This relation includes a new concept, *torque*.

A torque is always associated with a force. Qualitatively speaking, torque is the tendency of a force to cause a rotation of the body on which it acts. This tendency depends on the magnitude and direction of the force, and also on the location of the point where it acts. For example, it is easier to push a door open by pushing near the doorknob side than near the hinge side.

Torque is always defined with reference to a specific *axis* of rotation. In the problems in this chapter, the axis will usually be an actual axis of rotation about which the body can turn. In Chapter 10, when we study equilibrium of a rigid body, we will see that it is often useful to compute torques with respect to several different axes.

To define torque quantitatively, we consider in Fig. 9–13 a body that can rotate about an axis perpendicular to the plane of the figure, through point O.

Forces F_1 and F_2 act on the body; both forces act along lines that lie in a plane perpendicular to the axis. The tendency of force F_1 in Fig. 9–13 to cause a rotation about the axis through O depends on both the magnitude F_1 of the force and the perpendicular distance l_1 between the line of action of the force and the axis. If $l_1 = 0$, there is *no* tendency to cause rotation. The role of distance l_1 is analogous to that of a wrench handle; we can turn a tight bolt more easily by using a long-handled wrench than with a short-handled one. The distance l_1 is called the **moment arm** of force F_1 about the axis through O, and the product F_1l_1 is called the **torque,** or **moment,** of the force about point O, denoted by the Greek letter Γ (capital "gamma"). The terms *torque* and *moment* are synonymous; we will usually use *torque,* but *moment arm* is the more usual term for the distance l_1:

$$\Gamma = Fl. \tag{9–23}$$

The moment arm of F_1 is the perpendicular distance OA or l_1, and the moment arm of F_2 is the perpendicular distance OB or l_2.

Force F_1 tends to cause *counterclockwise* rotation about the axis, while F_2 tends to cause *clockwise* rotation. To distinguish between these directions of rotation, we will usually use the convention that *counterclockwise torques are positive and clockwise torques are negative.* Hence the torque Γ_1 of the force F_1 about the axis through O is

$$\Gamma_1 = +F_1l_1,$$

and the torque Γ_2 of F_2 is

$$\Gamma_2 = -F_2l_2.$$

When the line along which a force acts passes through the axis of rotation, the moment arm for that force is zero and its torque with respect to that axis is zero. In SI units, where the unit of force is the newton and the unit of length the meter, the unit of torque is the newton-meter.

In problems involving torques in this and the following chapter, we use the symbol

to indicate the choice of positive direction of rotation.

Often one of the important forces acting on a body is its *weight.* This force is not concentrated at a point but is distributed over the entire body. Nevertheless, it is always possible to calculate the corresponding torque by assuming all the weight to be concentrated at the center of mass of the body. We postpone proof of this statement until Chapter 10, but meanwhile we can use it in some of the problems in this chapter.

Figure 9–14 shows that there are several alternative ways to calculate torque. We can find the moment arm l and use $\Gamma = Fl$. Or we can determine the angle θ and use $\Gamma = rF \sin \theta$. Finally, we can represent F in terms of components F_1 parallel to r and F_2 perpendicular to r, as shown by the light-colored vectors. Then $F_2 = F \sin \theta$, and $\Gamma = F_2r = rF \sin \theta$. The component F_1 has no torque with respect to O because its moment arm with respect to that point is zero. Whenever the line of action of a force goes through the point we are using to calculate torques, the torque of that force is zero.

In more advanced work, where rotations about axes in various directions have to be considered, we generalize the definition of torque as follows. When

The torque of a force depends on the point where the force is applied, with reference to the axis of rotation.

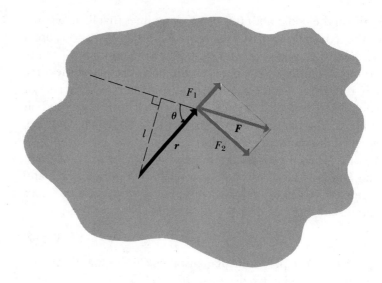

9–14 The vector torque $\boldsymbol{\Gamma}$ of force $\boldsymbol{F}$ with respect to point O is defined as $\boldsymbol{\Gamma} = \boldsymbol{r} \times \boldsymbol{F}$. In this example $\boldsymbol{\Gamma}$ points into the plane of the figure.

The torque of a force can be expressed as a vector product.

a force $\boldsymbol{F}$ acts at a point having a position vector $\boldsymbol{r}$ with respect to an origin O, as in Fig. 9–14, the torque $\boldsymbol{\Gamma}$ of the force with respect to O is defined to be the vector quantity

$$\boldsymbol{\Gamma} = \boldsymbol{r} \times \boldsymbol{F}. \qquad (9\text{--}24)$$

Recalling the definition of the vector product from Section 1–9, we note that the magnitude of $\boldsymbol{\Gamma}$ is $rF \sin \theta$. From the figure we see that $r \sin \theta = l$, so the magnitude of $\boldsymbol{\Gamma}$ is equal to Fl, in agreement with our previous definition of torque. The *direction* of $\boldsymbol{\Gamma}$ is perpendicular to the plane of the figure, in this case *into* the plane, as determined by the right-hand rule. If the sense of rotation were opposite, the direction of $\boldsymbol{\Gamma}$ would be *out of* the plane. This would be the case, for instance, if $\boldsymbol{F}$ were applied at the bottom edge of the body.

9–9 TORQUE AND ANGULAR ACCELERATION

We are now ready to consider the *dynamics* of rotational motion of a rigid body about a stationary axis. As we will see, the angular acceleration of a rotating body is directly proportional to the sum of the torques with respect to the axis of rotation. The proportionality factor is the moment of inertia.

To develop this relationship, we again imagine the body as made up of a large number of particles. A typical particle has mass m and is at distance r from the axis. We represent the *total force* acting on this particle in terms of a component $F_\perp$ that acts along the radial direction and a component $F_\parallel$ that is tangent to the circle of radius r in which the particle moves during rotation. Applying Newton's second law to this particle gives

$$F_\parallel = ma_\parallel. \qquad (9\text{--}25)$$

Now $a_\parallel$ may be expressed in terms of the angular acceleration α, according to Eq. (9–10): $a_\parallel = r\alpha$. Using this relation and multiplying both sides of Eq. (9–25) by r, we obtain

$$F_\parallel r = mr^2\alpha. \qquad (9\text{--}26)$$

Note that $F_{\parallel}r$ is just the torque of the force, and that mr^2 is the moment of inertia of the particle. (The component $F_{\perp}$ acts along a line passing through the axis and so has no torque with respect to the axis.) Thus Eq. (9–26) may be rewritten as

$$\Gamma = I\alpha.$$

We can write an equation like this for every particle in the body and add all these equations. The left side of the resulting equation is the sum of all the torques acting on all the particles, and the right side is the total moment of inertia multiplied by the angular acceleration, which is the same for every particle. Thus for the entire body

$$\Sigma\Gamma = I\alpha. \tag{9–27}$$

Finally, the sum $\Sigma\Gamma$ includes only the torques of the *external* forces, that is, the forces exerted on the body by agencies outside it. Torques corresponding to internal forces that the particles of the body exert on each other cancel out in pairs. The two forces that a given pair of particles exert on each other are equal and opposite, and if they act along the line joining the particles, their moment arms with respect to any axis are equal. Hence the torques of these two forces add to zero. Similarly, all other internal torques cancel out in pairs. (The requirement that the interaction forces for each pair of particles are not only equal and opposite but also act along the same line is called the *strong form* of Newton's third law; in previous developments we have needed only the fact that the forces are equal and opposite.)

Equation (9–27) is the rotational analog of Newton's second law, **$\Sigma F = ma$.** It provides the basis for relating the rotational motion of a rigid body to the forces acting on it.

Torque and angular acceleration: the rotational analog of Newton's second law

---

PROBLEM-SOLVING STRATEGY: *Rotational dynamics*

The strategy we recommend for problems in rotational dynamics is very similar to that used in Section 5–4 for applications of Newton's laws:

1. Select a body for analysis. $\Sigma F = ma$ or $\Sigma\Gamma = I\alpha$ will be applied to this body.

2. Draw a free-body diagram. Be sure to include all the forces acting *on* the chosen body, but be equally careful *not* to include any force exerted *by* the body on some other body. Some of the forces may be unknown; label them with algebraic symbols. One of the forces may be the body's weight; it is often useful to label it immediately as mg rather than w. If a numerical value of mass is given, the corresponding numerical value of weight may be computed.

3. Choose coordinate axes for each body and also indicate a positive sense of rotation for each rotating body. If you know the direction of α in advance, it is usually easiest to pick that as the positive sense of rotation. When appropriate, determine components of

force with reference to the chosen axes. When a force is represented in terms of its components, cross out the original force so as not to include it twice.

4. If more than one body is involved, carry out steps 1–3 for each body. Some problems will include one or more bodies having translational motion and one or more others having rotational motion. There may also be geometrical relations between the motions of two or more bodies. Express these in algebraic form, usually as relations between two accelerations or an acceleration and an angular acceleration.

5. Write the appropriate dynamical equations, cited in step 1, and solve them to find the unknown quantities. Often this involves solving a set of simultaneous equations.

6. Check special cases or extreme values of quantities, where possible, and compare the results for these particular cases with your intuitive expectations. Ask: "Does this result make sense?"

Spinning a cylinder with a rope: a simple example of rotational dynamics

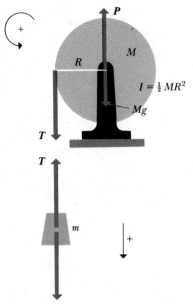

9–15 Free-body diagrams for Example 9–12.

Torque and angular acceleration in a two-body problem

The tension in the string does not equal the weight of the hanging body!

EXAMPLE 9–11 A rope is wrapped several times around a uniform solid cylinder of radius 0.1 m and mass 50 kg, pivoted so it can rotate about its axis. What is the angular acceleration when the rope is pulled with a force of 20 N?

SOLUTION The torque is $\Gamma = (20 \text{ N})(0.1 \text{ m}) = 2.0 \text{ N·m}$, and the angular acceleration is

$$\alpha = \frac{\Gamma}{I} = \frac{2.0 \text{ N·m}}{\frac{1}{2}(50 \text{ kg})(0.1 \text{ m})^2} = 8 \text{ rad·s}^{-2}.$$

EXAMPLE 9–12 In Example 9–5 (Section 9–5), find the acceleration of mass m and the angular acceleration of the cylinder.

SOLUTION We treat the two bodies separately. Figure 9–15 shows the forces acting on each body. We take the positive sense of rotation for the cylinder to be counterclockwise and the positive direction for m to be downward. Applying Newton's second law to m yields the relation

$$mg - T = ma.$$

Applying Eq. (9–27) to the cylinder gives

$$RT = I\alpha = \frac{1}{2}MR^2 \alpha.$$

Now the velocity of the mass and string at any instant is equal to the tangential velocity of the point on the cylinder where the string is tangent. Thus the acceleration a of mass m must equal the tangential acceleration of a point on the surface of the cylinder, which, according to Eq. (9–10), is given by $a_{\parallel} = R\alpha$. We replace $(R\alpha)$ with a in the cylinder equation, divide by R, and substitute the resulting expression for T into the equation for m, obtaining

$$mg - \frac{1}{2}Ma = ma, \qquad a = \frac{mg}{m + M/2} = \frac{g}{1 + M/2m}.$$

Note that the tension in the rope is *not* equal to the weight mg of mass m; if it were, m could not accelerate. From the above relations,

$$T = mg - ma = \frac{mg}{1 + 2m/M}.$$

When M is much larger than m, the tension is nearly equal to mg, and the acceleration is correspondingly much less than g. When M is zero, $T = 0$ and $a = g$; the mass then falls freely.

If mass m starts from rest at a height h above the floor, its velocity v when it strikes the ground is given by $v^2 - v_0^2 = 2ah$. In this case $v_0 = 0$ and

$$v = \sqrt{2ah} = \sqrt{\frac{2gh}{1 + M/2m}},$$

in agreement with the result obtained from energy considerations in Section 9–5.

EXAMPLE 9–13 In Fig. 9–16a, mass m_1 slides without friction on the horizontal surface, the pulley is in the form of a thin cylindrical shell of mass M and radius R, and the string turns the pulley without slipping. Find the acceleration of each mass, the angular acceleration of the pulley, and the tension in each part of the string.

SOLUTION Figure 9–16b shows free-body diagrams for the three bodies involved, and also shows a choice of positive directions for the various coordinates. Note that the two tensions T_1 and T_2 *cannot* be equal; if they were, the pulley could not have an angular acceleration. Hence to label the tension in both parts of the string as simply T would be a serious error.

The equations of motion for masses m_1 and m_2 are

$$T_1 = m_1 a_1 \qquad (9\text{–}28)$$

and

$$m_2 g - T_2 = m_2 a_2. \qquad (9\text{–}29)$$

The unknown normal force N_2 acting on the axis of the pulley has no torque with respect to the axis of rotation, and the equation of motion of the pulley is

$$T_2 R - T_1 R = I\alpha = (MR^2)\alpha. \qquad (9\text{–}30)$$

Assuming the string does not stretch or slip, we have the additional *kinematic* relations

$$a_1 = a_2 = R\alpha. \qquad (9\text{–}31)$$

(The accelerations of m_1 and m_2 have different directions but the same magnitude.)

Equation (9–31) can be used to eliminate a_2 and α from Eqs. (9–28) through (9–30). The result is the following set of equations for the three unknowns T_1, T_2, and a_1:

$$T_1 = m_1 a_1,$$

$$m_2 g - T_2 = m_2 a_1,$$

$$T_2 - T_1 = M a_1.$$

These may be solved simultaneously; the simplest procedure is to add the three equations to eliminate T_1 and T_2, and then solve for a_1. The result is

$$a_1 = \frac{m_2 g}{m_1 + m_2 + M}.$$

This result may then be substituted back into Eqs. (9–28) and (9–29) to find the tensions. The results are

$$T_1 = \frac{m_1 m_2 g}{m_1 + m_2 + M}, \qquad T_2 = \frac{(m_1 + M)m_2 g}{m_1 + m_2 + M}.$$

Note that if either m_1 or M is much larger than m_2, the accelerations are very small and T_2 is approximately $m_2 g$, while if m_2 is much larger than either m_1 or M, the acceleration is approximately g, as should be expected.

Three interacting masses: linear and rotational dynamics

(a)

(b)

9–16 (a) System for Example 9–13. (b) Free-body diagrams.

9–10 WORK AND POWER IN ROTATIONAL MOTION

A force applied to a rotating body does *work* on the body. In problems involving energy changes and power it is important to know how to calculate this work, which may be expressed in terms of the torque of the force and the angular displacement.

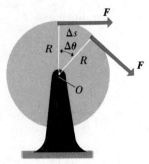

9–17 A force applied to a rotating body does work on the body.

How to calculate the power transmitted by a rotating motor shaft

Suppose a force F acts as shown in Fig. 9–17 at the rim of a pivoted wheel of radius R while the wheel rotates through a small angle $d\theta$. If this angle is small enough, the force may be regarded as constant during the correspondingly small time interval. By definition, the work done by the force F is

$$dW = F \, ds.$$

But $ds = R \, d\theta$, so that

$$dW = FR \, d\theta.$$

Now FR is the *torque*, Γ, due to the force F, so we have finally

$$dW = \Gamma \, d\theta. \tag{9–32}$$

If the torque is constant while the angle changes by a finite amount from θ_1 to θ_2,

$$W = \Gamma(\theta_2 - \theta_1) = \Gamma \, \Delta\theta. \tag{9–33}$$

That is, *the work done by a constant torque equals the product of the torque and the angular displacement.* If Γ is expressed in newton·meters, the work is in joules.

The force in Fig. 9–17 has no component along the *radial* direction. Such a component, if it existed, would do no work, since the displacement of the point of application has no radial component. Similarly, a radial component of force would make no contribution to the torque. Hence Eqs. (9–32) and (9–33) are still correct even when F does have a radial component.

When both sides of Eq. (9–32) are divided by the time interval dt during which the displacement occurs, we obtain

$$\frac{dW}{dt} = \Gamma\left(\frac{d\theta}{dt}\right).$$

But dW/dt is the rate of doing work, or the *power*, and $d\theta/dt$ is the angular velocity. Hence

$$P = \Gamma\omega. \tag{9–34}$$

That is, *the instantaneous power developed by an agent exerting a torque equals the product of the torque and the instantaneous angular velocity.* This is the analog of $P = Fv$ for linear motion.

EXAMPLE 9–14 The driveshaft of an automobile rotates at 3600 rpm and transmits 80 hp from the engine to the rear wheels. Compute the torque developed by the engine.

SOLUTION We first find ω (in rad·s^{-1}) and then use Eq. (9–34):

$$\omega = \frac{(3600 \text{ rev·min}^{-1})(2\pi \text{ rad·rev}^{-1})}{(60 \text{ s·min}^{-1})} = 120 \, \pi \text{ rad·s}^{-1},$$

$$80 \text{ hp} = (80 \text{ hp})(746 \text{ W·hp}^{-1}) = 59{,}700 \text{ W},$$

$$\Gamma = \frac{P}{\omega} = \frac{59{,}700 \text{ W}}{120\pi \text{ rad·s}^{-1}} = 158 \text{ N·m}.$$

EXAMPLE 9–15 Suppose an electric motor exerts a constant torque of $\Gamma = 10$ N·m on a grindstone mounted on its shaft; the moment of inertia of the grindstone is $I = 2$ kg·m^2. The system starts from rest. Find the work done by the motor in 8 s and the kinetic energy at the end of this time. What was the average power of the motor?

SOLUTION From $\Gamma = I\alpha$, the angular acceleration is 5 s^{-2}. The angular velocity after 8 s is

$$\omega = \alpha t = (5 \text{ s}^{-1})(8 \text{ s}) = 40 \text{ s}^{-1}.$$

The kinetic energy at this time is

A motor-driven grindstone: The work done by the rotating motor increases the grindstone's kinetic energy.

$$K = \tfrac{1}{2}I\omega^2 = \tfrac{1}{2}(2 \text{ kg·m}^2)(40 \text{ s}^{-1})^2 = 1600 \text{ J}.$$

The total angle through which the system turns in 8 s is

$$\theta = \tfrac{1}{2}\alpha t^2 = \tfrac{1}{2}(5 \text{ s}^{-1})(8 \text{ s})^2 = 160 \text{ rad},$$

and the total work done by the torque is

$$W = \Gamma\theta = (160 \text{ rad})(10 \text{ N·m}) = 1600 \text{ J}.$$

This equals the total kinetic energy, as of course it must.

The average power is the total work divided by the time interval:

$$P_{av} = \frac{1600 \text{ J}}{8 \text{ s}} = 200 \text{ J·s}^{-1} = 200 \text{ W}.$$

The instantaneous power, given by $P = \Gamma\omega$, is not constant, since ω increases continuously. But we can compute the total work by taking the time integral of P, as follows:

$$W = \int P\, dt = \int \Gamma\omega\, dt = \int \Gamma(\alpha t)\, dt$$
$$= \int_0^{8 \text{ s}} (10 \text{ N·m})(5 \text{ s}^{-2})t\, dt$$
$$= 1600 \text{ J},$$

as we found previously. The instantaneous power increases from zero at the start to $(10 \text{ N·m})(40 \text{ s}^{-1}) = 400 \text{ W}$ at time $t = 8 \text{ s}$. The angular velocity and the power increase uniformly with time, so the *average* power is just half this maximum value, or 200 W.

9–11 ROTATION ABOUT A MOVING AXIS

Our analysis of the dynamics of rotational motion of a rigid body (Section 9–9) can be extended to some cases where the axis of rotation moves, that is, where both translational and rotational motion occur at once. Familiar examples of such motion include a ball rolling down a hill or a yo-yo unwinding at the end of a string. The key to this more general analysis, which we shall not derive in detail, is that $\Sigma\Gamma = I\alpha$ remains valid when the axis of rotation moves, *if the axis passes through the center of mass of the body and does not change its direction.*

A rigid body can have translational and rotational motion at the same time.

Another useful relationship, which we also state without proof, is an expression for the total kinetic energy of a rigid body having both translational and rotational motion. For a body of mass M, moving with a center-of-mass velocity V and rotating with angular velocity ω about an axis through the center of mass, the *total kinetic energy* of the body is

$$K = \tfrac{1}{2}MV^2 + \tfrac{1}{2}I_c\omega^2,$$

where I_c is the moment of inertia about the axis through the center of mass.

PROBLEM-SOLVING STRATEGY: Rotation about a moving axis

The strategy outlined in Section 9–9 is equally useful here. There is one new wrinkle: When a body undergoes translational and rotational motion at the same time, we need two separate equations of motion *for the same body*. One of these is based on $\Sigma F = ma$ for the translational motion of the center of mass. The discussion of Section 8–7 shows that the acceleration of the center of mass is the same as that of a point mass equal to the total mass of the body, acted on by all the forces on the actual body. The other equation of motion is based on $\Sigma \Gamma = I\alpha$ for the rotational motion about the axis through the center of mass. In addition, there is often a kinematic relation between the two motions, such as a wheel that rolls without slipping or a string that unwinds from a pulley while turning it. Such relations are needed to relate the various accelerations.

Analyzing a yo-yo: dynamics of combined translational and rotational motion

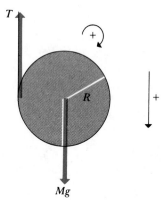

9–18 A cylinder rotates and drops as string unwinds.

EXAMPLE 9–16 A string is wrapped several times around a solid cylinder. The end of the string is held stationary while the cylinder is released with no initial motion. Find the downward acceleration of the cylinder and the tension in the string.

SOLUTION Figure 9–18 shows a free-body diagram for the situation, including the choice of positive coordinate directions. The equation for the translational motion of the center of mass is

$$Mg - T = Ma, \tag{9–35}$$

and the equation for rotational motion about the axis through the center of mass is

$$TR = I\alpha = \tfrac{1}{2}(MR^2)\alpha. \tag{9–36}$$

In addition, if the string unwinds without slipping, we have the kinematic relation

$$a = R\alpha. \tag{9–37}$$

This may be used to eliminate α from Eq. (9–36), and then Eqs. (9–35) and (9–36) may be solved simultaneously for T and a. The results are

$$a = \tfrac{2}{3}g, \qquad T = \tfrac{1}{3}Mg.$$

A ball rolling down a hill: another example of combined translational and rotational dynamics.

EXAMPLE 9–17 A solid bowling ball rolls without slipping down a ramp inclined at angle θ to the horizontal. What is the acceleration of its center?

SOLUTION Figure 9–19 shows a free-body diagram, with positive coordinate directions indicated. The equations of motion for translational and rotational motion, respectively, are

$$mg \sin \theta - \mathcal{F} = ma,$$
$$\mathcal{F}R = I\alpha = (\tfrac{2}{5}mR^2)\alpha.$$

If the ball rolls without slipping, then $a = R\alpha$. We use this to eliminate α and then solve for a and $\mathcal{F}$ to obtain

$$a = \tfrac{5}{7}g \sin \theta, \qquad \mathcal{F} = \tfrac{2}{7}mg \sin \theta.$$

Note that the acceleration is just $\tfrac{5}{7}$ as large as it would be if the ball could *slide* without friction down the slope. Also, the friction force is essential to prevent slipping and thus to cause the angular acceleration of the ball. An expression for the minimum coefficient of friction can be obtained by noting that the normal force is $\mathcal{n} = mg \cos \theta$. To prevent slipping, the coefficient of (static) friction must be at least as great as

$$\mu_s = \frac{\mathcal{F}}{\mathcal{n}} = \frac{\frac{2}{7} mg \sin \theta}{mg \cos \theta} = \frac{2}{7} \tan \theta.$$

If the plane is tilted only slightly, θ is small and only a small value of μ_s is needed to prevent slipping; but as the angle increases, the required value of μ_s increases, as we might expect intuitively.

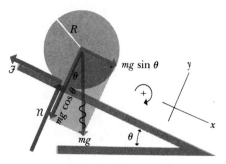

9–19 Free-body diagram for a ball rolling down a plane.

9–12 ANGULAR MOMENTUM AND ANGULAR IMPULSE

The concept of **angular momentum** plays a role in rotational motion that is closely analogous to that of momentum in particle motion, and there is a corresponding conservation principle. To introduce this concept, we begin by defining angular momentum of a particle.

Figure 9–20a shows a particle with mass m moving in the plane of the figure with velocity v and momentum mv. We define the angular momentum L of the particle, about an axis that passes through O and is perpendicular to the plane of the figure, as the product of the magnitude of its momentum and the perpendicular distance from the axis to its instantaneous line of motion:

$$\text{Angular momentum} = L = mvr. \qquad (9\text{–}38)$$

This definition is analogous to the definition of torque of a force.

Now we define the *total* angular momentum of a body of finite size as the sum of the angular momenta of the particles of the body. For a rotating rigid body, this can be expressed easily in terms of the body's moment of inertia and angular velocity. Figure 9–20b shows a rigid body rotating about an axis through O. The speed v of a small element of the body is related to the angular velocity ω of the body by $v = \omega r$. The angular momentum of the element is therefore

$$L = mvr = \omega mr^2,$$

and the total angular momentum of the body is

$$\Sigma \omega mr^2 = \omega \Sigma mr^2.$$

But Σmr^2 is the moment of inertia of the body about its axis of rotation. Hence the angular momentum can be written as $I\omega$. Again denoting angular momentum by the symbol L, we have

$$L = I\omega. \qquad (9\text{–}39)$$

This is analogous to the definition of linear momentum mv for a particle.

We can use the same sign convention for angular momentum as for angular velocity. I is always positive, so the sign of L is the same as that of ω. In Fig. 9–20b, where the positive direction for ω is defined to be clockwise, L is positive for clockwise rotations, negative for counterclockwise.

When a constant torque Γ acts on a body having moment of inertia I, for a time interval from t_1 to t_2, the angular velocity changes from ω_1 to ω_2, accord-

Angular momentum of a particle depends on its momentum and its position with reference to the axis of rotation.

The angular momentum of a rigid body can be expressed in terms of its moment of inertia and angular velocity.

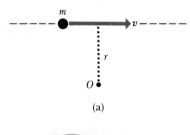

(a)

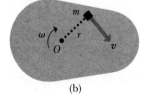

(b)

9–20 Angular momentum.

ing to the relation

$$\Gamma = I\alpha = I\left(\frac{\omega_2 - \omega_1}{t_2 - t_1}\right).$$

Rearranging this equation, we obtain

$$\Gamma(t_2 - t_1) = I\omega_2 - I\omega_1 = L_2 - L_1 = \Delta L. \tag{9-40}$$

The product of the torque and the time interval during which it acts is called the **angular impulse** of the torque; we denote this quantity by J_θ.

$$\text{Angular impulse} = J_\theta = \Gamma(t_2 - t_1). \tag{9-41}$$

This quantity is the rotational analog of the impulse of a force, defined in Section 7–1. Equation (9–40) states that *the angular impulse acting on a body equals the change of angular momentum of the body about the same axis.*

For a torque that varies with time, we generalize the definition of angular impulse to

$$J_\theta = \int_{t_1}^{t_2} \Gamma\, dt. \tag{9-42}$$

Thus the general relationship between angular impulse and angular momentum is

$$J_\theta = I\omega_2 - I\omega_1 = L_2 - L_1, \tag{9-43}$$

with J_θ given by Eq. (9–41) if Γ is constant, and by Eq. (9–42) otherwise. A torque that is large but acts only during a short time interval is called an *impulsive torque.*

The basic dynamic relation for rigid-body rotation, given by Eq. (9–27), may be restated in terms of angular momentum. Taking the time derivative of Eq. (9–39), we find

$$\frac{dL}{dt} = I\frac{d\omega}{dt} = I\alpha.$$

Combining this with Eq. (9–27) yields the relation

$$\Sigma\Gamma = \frac{dL}{dt}. \tag{9-44}$$

In more complex rotational motion, where the direction of the axis of rotation may change, we generalize the definition of angular momentum of a particle to a vector quantity L defined as

$$\boldsymbol{L} = \boldsymbol{r} \times m\boldsymbol{v}, \tag{9-45}$$

where $\boldsymbol{r}$ is the position vector of the particle with respect to the origin O. Comparing this definition with Fig. 9–20a, we see that in this figure L is a vector perpendicular to the plane of the figure, pointing *into* the plane, with magnitude mvr, in agreement with Eq. (9–38). The basic dynamic relationship for rigid body motion can then be stated as

$$\boldsymbol{\Sigma\Gamma} = \frac{d\boldsymbol{L}}{dt}, \tag{9-46}$$

with the vector torque $\boldsymbol{\Gamma}$ defined as in Eq. (9–24). In Section 9–14 we will explore the usefulness of this formulation.

9–13 CONSERVATION OF ANGULAR MOMENTUM

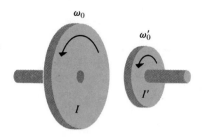

Here is an example of the usefulness of the concepts of angular impulse and angular momentum. Figure 9–21 shows two disks with moments of inertia I and I'; initially they are rotating with constant angular velocities ω_0 and ω_0', respectively. The disks are then pushed together by a force directed along the axis, which does not exert any torque on either disk. After a short time the disks reach a common final angular velocity ω.

During this time the larger disk exerts a torque Γ' on the smaller, and the smaller exerts a torque Γ on the larger. Both Γ and Γ' vary during the contact, and both become zero after the common final angular velocity is reached. At each instant the two torques are equal in magnitude and opposite in direction, because of Newton's third law. To see why this must be so, consider a point of contact between the two bodies. The two forces that the bodies exert on each other at this common point have equal magnitude and opposite direction, according to Newton's third law. Because they act at the same point, they also have the same moment arm with respect to any axis. Thus the torque of one force is the negative of the torque of the other. This is true of all the pairs of forces at all the other contact points. Thus at any instant $\Gamma = -\Gamma'$, and the total impulses on the two bodies are related by $J_\theta = -J_\theta'$.

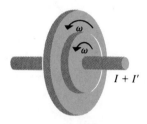

Now, according to Eq. (9–43), the impulse on each disk equals the change of angular momentum of that disk:

$$J_\theta = I\omega - I\omega_0,$$

$$J_\theta' = I'\omega - I'\omega_0'.$$

9–21 An impulsive torque acts when two rotating disks engage.

When no external torques act on a system, its total angular momentum is constant.

But because $J_\theta = -J_\theta'$, these changes are equal and opposite:

$$I\omega - I\omega_0 = -(I'\omega - I'\omega_0')$$

or

$$I\omega_0 + I'\omega_0' = (I + I')\omega. \qquad (9\text{–}47)$$

The left side of Eq. (9–47) is the total angular momentum of the system before contact, and the right side is the total angular momentum after contact. We have therefore derived the important conclusion that the total angular momentum of the whole system is unaltered. When both disks are regarded as one system, then the torques Γ and Γ' are *internal* torques; as the disks are pushed together, no *external* torque acts. *When the resultant external torque on a system is zero, the angular momentum of the system remains constant;* hence any internal interaction between the parts of a system cannot alter its total angular momentum. This is the principle of **conservation of angular momentum,** and it ranks with the principles of conservation of linear momentum and conservation of energy as one of the most fundamental of physical laws.

A circus acrobat, a diver, or a skater performing a pirouette on the toe of one skate, all take advantage of this principle. Suppose an acrobat has just left a swing as in Fig. 9–22, with arms and legs extended and with a counterclockwise angular momentum. When he pulls his arms and legs in, his moment of inertia I becomes much smaller. His angular momentum $I\omega$ remains constant and I decreases, so his angular velocity ω increases. (Note that the change of ω cannot be determined from $\Sigma\Gamma = I\alpha$ because I is not constant.)

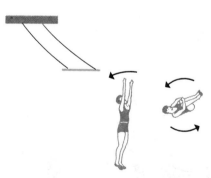

9–22 Conservation of angular momentum.

Acrobats, divers, and skaters: examples of conservation of angular momentum

EXAMPLE 9–18   In the situation of Fig. 9–21, suppose the large disk has mass 2 kg, radius 0.2 m, and initial angular velocity 50 rad·s^{-1}, and the small disk has mass 4 kg, radius 0.1 m, and initial angular velocity 200 rad·s^{-1} (about 1900 rpm). Find the common final angular velocity after the disks are pushed into contact. Is kinetic energy conserved during this process?

SOLUTION   The moments of inertia of the two disks are

$$I = \tfrac{1}{2}(2 \text{ kg})(0.2 \text{ m})^2 = 0.04 \text{ kg·m}^2,$$

and

$$I' = \tfrac{1}{2}(4 \text{ kg})(0.1 \text{ m})^2 = 0.02 \text{ kg·m}^2.$$

From conservation of angular momentum, we have

$$(0.04 \text{ kg·m}^2)(50 \text{ rad·s}^{-1}) + (0.02 \text{ kg·m}^2)(200 \text{ rad·s}^{-1})$$
$$= (0.04 \text{ kg·m}^2 + 0.02 \text{ kg·m}^2)\omega,$$
$$\omega = 100 \text{ rad·s}^{-1}.$$

The initial kinetic energy is

$$K_0 = \tfrac{1}{2}(0.04 \text{ kg·m}^2)(50 \text{ rad·s}^{-1})^2 + \tfrac{1}{2}(0.02 \text{ kg·m}^2)(200 \text{ rad·s}^{-1})^2$$
$$= 450 \text{ J}.$$

The final kinetic energy is

$$K = \tfrac{1}{2}(0.04 \text{ kg·m}^2 + 0.02 \text{ kg·m}^2)(100 \text{ rad·s}^{-1})^2 = 300 \text{ J}.$$

One-third of the kinetic energy was lost during this "angular collision," the rotational analog of an inelastic collision. We should not expect kinetic energy to be conserved, even though the resultant external force and torque are zero, since nonconservative (frictional) internal forces act during the contact.

EXAMPLE 9–19   In Fig. 9–23, a man stands at the center of a turntable, holding his arms extended horizontally, with a 5-kg mass in each hand. He is set rotating about a vertical axis with an angular velocity of one revolution in 2 s. Find his new angular velocity if he drops his hands to his sides. The man's moment of inertia may be assumed constant and equal to 6 kg·m^2. The original distance of the weights from the axis is 1 m, and their final distance is 0.2 m.

SOLUTION   If friction in the turntable is neglected, no external torques act about the vertical axis, and the angular momentum about this axis is constant. That is,

$$I_i\omega_i = I_f\omega_f,$$

where I_i and ω_i are the initial moment of inertia and angular velocity, and I_f and ω_f are the final values of these quantities. In each case, $I = I_{man} + I_{weights}$.

$$I_i = 6 \text{ kg·m}^2 + 2(5 \text{ kg})(1.0 \text{ m})^2 = 16 \text{ kg·m}^2,$$

$$I_f = 6 \text{ kg·m}^2 + 2(5 \text{ kg})(0.2 \text{ m})^2 = 6.4 \text{ kg·m}^2,$$

$$\omega_i = 2\pi(\tfrac{1}{2})\text{rad·s}^{-1}$$

$$\omega_f = \omega_i \frac{I_i}{I_f} = \pi \text{ rad·s}^{-1}\frac{16 \text{ kg·m}^2}{6.4 \text{ kg·m}^2}$$

$$= 2.5\pi \text{ rad·s}^{-1} = 1.25 \text{ rev·s}^{-1}.$$

9–23 Conservation of angular momentum about a fixed axis.

That is, the angular velocity is more than doubled.

The initial kinetic energy is

$$K_0 = \tfrac{1}{2}(16 \text{ kg·m}^2)(\pi \text{ rad·s}^{-1})^2 = 79 \text{ J}.$$

The final kinetic energy is

$$K = \tfrac{1}{2}(6.4 \text{ kg·m}^2)(2.5\pi \text{ rad·s}^{-1})^2 = 197 \text{ J}.$$

Where did the extra energy come from?

EXAMPLE 9–20   A door 1.0 m wide, having a mass of 15 kg, is hinged at one side so it can rotate without friction about a vertical axis. A bullet having mass 10 g and speed 400 m·s^{-1} is fired into the door, in a direction perpendicular to the plane of the door, and embeds itself at the exact center of the door. Find the angular velocity of the door just after the bullet embeds itself. Is kinetic energy conserved?

Firing a bullet into a door: a dangerous way to open a door, but a good way to illustrate conservation of angular momentum

SOLUTION   There is no external torque about the axis defined by the hinges, so angular momentum about this axis is conserved. The initial angular momentum of the bullet is given by Eq. (9–38):

$$L = mvr = (0.01 \text{ kg})(400 \text{ m·s}^{-1})(0.5 \text{ m}) = 2.0 \text{ kg·m}^2\text{·s}^{-1}.$$

This is equal to the final angular momentum $I\omega$, where $I = I_{\text{door}} + I_{\text{bullet}}$ is the total moment of inertia of door and bullet. To find I for the door alone, note that dimensions parallel to the axis are irrelevant, and that its moment of inertia is the same as that of a rod 1 m long, pivoted at one end. As discussed in Example 9–9 (Section 9–7), this is given by

$$I_{\text{door}} = \frac{ML^2}{3} = \frac{(15 \text{ kg})(1.0 \text{ m})^2}{3} = 5.0 \text{ kg·m}^2.$$

The moment of inertia of the bullet is

$$I_{\text{bullet}} = mr^2 = (0.01 \text{ kg})(0.5 \text{ m})^2 = 0.0025 \text{ kg·m}^2.$$

Conservation of angular momentum requires that $mvr = I\omega$, or

$$2.0 \text{ kg·m}^2\text{·s}^{-1} = (5.0 \text{ kg·m}^2 + 0.0025 \text{ kg·m}^2)\omega,$$

$$\omega = 0.4 \text{ rad·s}^{-1}.$$

The collision of bullet and door is inelastic, so we do not expect energy to be conserved. To check, we calculate initial and final kinetic energies:

$$K_i = \tfrac{1}{2}mv^2 = \tfrac{1}{2}(0.01 \text{ kg})(400 \text{ m·s}^{-1})^2 = 800 \text{ J};$$

$$K_f = \tfrac{1}{2}I\omega^2 = \tfrac{1}{2}(5.0025 \text{ kg·m}^2)(0.4 \text{ rad·s}^{-1})^2 = 0.4 \text{ J}.$$

The final kinetic energy is only $\frac{1}{2000}$ of the initial value.

9–14 VECTOR REPRESENTATION OF ANGULAR QUANTITIES

A rotational quantity associated with an axis of rotation, such as angular velocity, angular momentum, or torque, can be represented by a *vector* lying along that axis. Thus we can define angular velocity as a vector quantity $\boldsymbol{\omega}$ having a magnitude equal to the number of radians through which the body turns per unit time, and a direction along the axis of rotation.

A vector lying along a given line can have either of two opposite directions on that line. In defining the direction of the angular momentum vector, we use the same right-hand rule used for the vector product in Section 1–9. This rule is shown in Fig. 9–24; the direction of $\boldsymbol{\omega}$ is the direction in which a screw with a right-hand thread would advance if the screw turned with the body. Alternatively, wrap the fingers of your right hand around the axis, with your fingers pointing in the direction of rotation; your thumb then points in the direction of the vector $\boldsymbol{\omega}$.

We mentioned in Section 9–12 the vector definition of angular momentum of a particle. Using this definition, we can prove that for a body rotating about an axis of symmetry, the total angular momentum L is given by

$$L = I\boldsymbol{\omega}, \tag{9-48}$$

where I is the moment of inertia about the axis of rotation. Deriving this equation would be beyond our scope, but note that it is valid *only* when the axis of rotation is a symmetry axis of the body, such as a motor shaft or the axle of a turning wheel. When a body rotates about an axis that is *not* a symmetry axis, such as a disk with an axis through its center but not perpendicular to its plane, the angular-momentum vector does not have the same direction as the angular velocity. As a result, a body rotating with constant angular velocity *does not* have constant angular momentum. This makes the body tend to wobble, and torques must be supplied by the bearings supporting the body to prevent this. A general analysis of rigid body motion is a problem of considerable complexity.

The direction of the torque vector, as defined by Eq. (9–24), can be understood as follows. A torque tends to cause a rotation about a certain axis, namely, the axis about which the body would begin to rotate if it were initially at rest with only the torque under consideration acting on it. As mentioned in Section 9–12, the generalized relation between torque and angular momentum is

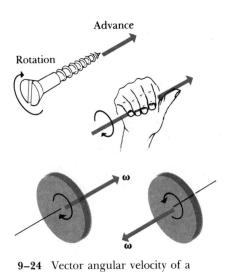

9–24 Vector angular velocity of a rotating body.

$$\Sigma\boldsymbol{\Gamma} = \frac{d\boldsymbol{L}}{dt}. \tag{9-49}$$

We have not needed this generalized equation thus far in this chapter, because when the axis of rotation always keeps the same direction, only one component of angular velocity is different from zero. Such rotational motion is analogous to motion of a particle along a straight line. But Eq. (9–49) also includes the possibility that $\boldsymbol{\Gamma}$ and $\boldsymbol{L}$ may have different directions; in that case, $\boldsymbol{L}$ and $d\boldsymbol{L}/dt$ have different directions, and the direction of the axis of rotation may change. In such situations it is essential to consider the vector nature of the various angular quantities.

We cannot discuss the general formulation of the dynamics of rotation in detail here, but here is an example of its application. Figure 9–25 shows a familiar toy gyroscope. We set the flywheel spinning by wrapping a string around its shaft and pulling. When the shaft is supported at only one end, as shown in the figure, one possible motion is a steady circular motion of the axis in a horizontal plane. This is an interesting phenomenon, and quite unexpected if you have not seen it before. Intuition suggests that the free end of the axis should simply drop if it is not supported. Vector angular momentum considerations provide the key to understanding this behavior.

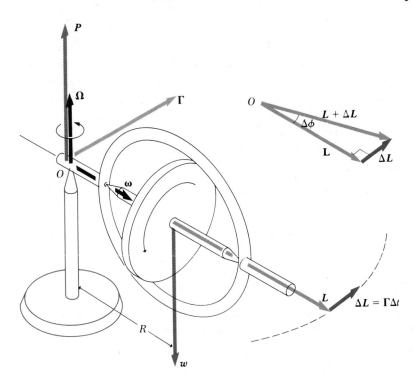

9–25 Vector $\Delta\mathbf{L}$ is the change in angular momentum produced in time Δt by the moment $\boldsymbol{\Gamma}$ of the force w. Vectors $\Delta\mathbf{L}$ and $\boldsymbol{\Gamma}$ are in the same direction.

The forces acting on the gyroscope are its weight $\mathbf{w}$, acting downward at the center of mass, and the upward force $\mathbf{P}$ at the pivot point O. These forces acting on a body *at rest* would tend to cause a rotation about a horizontal axis perpendicular to the gyroscope's axis; the associated torque $\boldsymbol{\Gamma}$ is in the direction shown. However, the body is *not* at rest, and the effect of this torque is not to *initiate* a rotational motion but to *change* a motion that already exists. At the instant shown, the body has a definite, nonzero angular momentum described by the vector $\mathbf{L}$, and the *change* $\Delta\mathbf{L}$ in angular momentum in a short time interval following this is given by Eq. (9–49). Thus $\Delta\mathbf{L}$ has the same direction as $\boldsymbol{\Gamma}$. After time Δt, the angular momentum is $\mathbf{L} + \Delta\mathbf{L}$ and, as the vector diagram shows, this means that the gyroscope axis has turned through a small angle $\Delta\phi$ given by $\Delta\phi = |\Delta\mathbf{L}|/|\mathbf{L}|$. Thus the motion of the axis is consistent with the torque–angular-momentum relationship. This motion of the axis is called **precession.**

The rate at which the axis moves, $\Delta\phi/\Delta t$, is called the *precession angular velocity.* Denoting this quantity by Ω, we find

$$\Omega = \frac{\Delta\phi}{\Delta t} = \frac{|\Delta L|/|L|}{\Delta t} = \frac{\Gamma}{L} = \frac{wR}{I\omega}. \tag{9–50}$$

Thus the precession angular velocity Ω is *inversely* proportional to the angular velocity ω of spin about the axis. A rapidly spinning gyroscope precesses slowly; and as it slows down, the precession angular velocity *increases*!

We may now return to the question of why the unsupported end of the axis does not fall, the reason is that the upward force $\mathbf{P}$ exerted on the gyroscope by the pivot is just equal in magnitude to its weight w. The resultant vertical *force* is zero, and the vertical acceleration of the center of gravity is zero. However, the resultant *torque* of these forces is *not* zero, and the angular momentum changes.

Precession of a gyroscope: obeying the laws of motion by defying common sense

If the top were not rotating, it would have no angular momentum L initially. Its angular momentum ΔL after a time Δt would be that acquired from the torque acting on it and would have the same direction as the torque. In other words, the top would rotate about an axis through O in the direction of the vector $\boldsymbol{\Gamma}$. But if the top is originally rotating, the *change* in its angular momentum produced by the torque adds vectorially to the large angular momentum it already has. Since ΔL is horizontal and perpendicular to L, the result is a motion of precession, with both the angular momentum and the axis remaining horizontal.

This example gives us a glimpse of the richness of the dynamics of rotational motion, which can involve some very complex phenomena. In the next chapter we consider a *simpler* set of problems involving the *equilibrium* of rigid bodies.

SUMMARY

KEY TERMS

rigid body

radians

average angular velocity

instantaneous angular velocity

average angular acceleration

instantaneous angular acceleration

tangential component of acceleration

radial component of acceleration

moment of inertia

parallel-axis theorem

moment arm

torque

moment

angular momentum

angular impulse

conservation of angular momentum

precession

When a rigid body rotates about a stationary axis, its position is described by an angular coordinate θ. The angular velocity ω is defined as the derivative of the angular coordinate.

$$\omega = \frac{d\theta}{dt}, \tag{9-2}$$

and the angular acceleration α is defined as the derivative of the angular velocity, or the second derivative of the angular coordinate:

$$\alpha = \frac{d\omega}{dt} = \frac{d^2\theta}{dt^2}. \tag{9-3}$$

When a body rotates with constant angular acceleration, the angular position, velocity, and acceleration are related by

$$\omega = \omega_0 + \alpha t, \tag{9-5}$$

$$\theta = \theta_0 + \omega_0 t + \tfrac{1}{2}\alpha t^2, \tag{9-6}$$

$$\omega^2 = \omega_0{}^2 + 2\alpha(\theta - \theta_0), \tag{9-7}$$

where θ_0 and ω_0 are the initial values of angular position and angular velocity, respectively.

A particle in a rotating rigid body has a speed v given by

$$v = r\omega \tag{9-9}$$

and an acceleration $\boldsymbol{a}$ with tangential component

$$a_\parallel = r\alpha \tag{9-10}$$

and radial component

$$a_\perp = \frac{v^2}{r} = \omega^2 r. \tag{9-11}$$

The moment of inertia I of a body with respect to a given axis is defined as

$$I = \Sigma m r^2, \tag{9-12}$$

where r is the distance of mass m from the axis of rotation. The kinetic energy of a rotating rigid body is given by

$$K = \tfrac{1}{2}I\omega^2. \qquad (9\text{--}13)$$

When the moment of inertia I_{cm} of a body about an axis through the center of mass is known, the moment of inertia I_P about a parallel axis at a distance d from the first axis is given by

$$I_P = I_{\text{cm}} + Md^2. \qquad (9\text{--}22)$$

When a force $\boldsymbol{F}$ acts on a body, the torque Γ of that force with respect to a point O is given by

$$\Gamma = Fl. \qquad (9\text{--}23)$$

A sign convention is needed, such as "counterclockwise positive, clockwise negative" or the reverse. A generalized definition of torque as a vector quantity $\boldsymbol{\Gamma}$ is

$$\boldsymbol{\Gamma} = \boldsymbol{r} \times \boldsymbol{F}, \qquad (9\text{--}24)$$

where $\boldsymbol{r}$ is the position vector of the point at which the force acts.

The angular acceleration α of a rigid body rotating about a stationary axis is related to the total torque $\Sigma\Gamma$ and the moment of inertia I of the body by

$$\Sigma\Gamma = I\alpha. \qquad (9\text{--}27)$$

This relation is also valid for a moving axis, provided the axis passes through the center of mass and does not change direction.

When a torque Γ acts on a rigid body that undergoes an angular displacement $d\theta$, the work dW done by the torque is given by

$$dW = \Gamma \, d\theta. \qquad (9\text{--}32)$$

When the body rotates with angular velocity ω, the power P (rate of doing work) is given by

$$P = \Gamma\omega. \qquad (9\text{--}34)$$

The angular momentum L with respect to a point O of a particle of mass m moving with speed v is given by

$$L = mvr, \qquad (9\text{--}38)$$

where r is the perpendicular distance from the line of motion to point O. When a rigid body having moment of inertia I rotates with angular velocity ω about a stationary axis, its angular momentum with respect to that axis is given by

$$L = I\omega. \qquad (9\text{--}39)$$

A generalized definition of angular momentum of a particle with respect to a point O, as a vector quantity $\boldsymbol{L}$, is

$$\boldsymbol{L} = \boldsymbol{r} \times m\boldsymbol{v}, \qquad (9\text{--}45)$$

where $\boldsymbol{r}$ is the position vector of the particle with respect to O. The angular impulse J_θ of a torque acting on a body during a time interval from t_1 to t_2 is given by

$$J_\theta = \Gamma(t_2 - t_1) \qquad (9\text{--}41)$$

if the torque is constant, and in general by

$$J_\theta = \int_{t_1}^{t_2} \Gamma \, dt \qquad (9\text{–}42)$$

for any torque, constant or not.

The relation between angular impulse and angular momentum is

$$J_\theta = I\omega_2 - I\omega_1 = L_2 - L_1. \qquad (9\text{–}43)$$

In terms of vector torque $\mathbf{\Gamma}$ and angular momentum $\mathbf{L}$, the basic dynamic relation for rotational motion can be restated as

$$\Sigma\mathbf{\Gamma} = \frac{d\mathbf{L}}{dt}. \qquad (9\text{–}46)$$

If a system consists of bodies that interact with each other but not with anything else, or if the total torque associated with the external forces is zero, the total angular momentum of the system is constant, or conserved.

QUESTIONS

9–1 What is the difference between tangential and radial acceleration, for a point on a rotating body?

9–2 A flywheel rotates with constant angular velocity. Does a point on its rim have a tangential acceleration? A radial acceleration? Are these accelerations constant? In magnitude? In direction?

9–3 A flywheel rotates with constant angular acceleration. Does a point on its rim have a tangential acceleration? A radial acceleration? Are these accelerations constant? In magnitude? In direction?

9–4 Can a single force applied to a body change both its translational and rotational motion?

9–5 How might you determine experimentally the moment of inertia of an irregularly shaped body?

9–6 Can you think of a body that has the same moment of inertia for all possible axes? For all axes passing through a certain point? What point?

9–7 A cylindrical body has mass M and radius R. Is it ever possible for its moment of inertia to be greater than MR^2?

9–8 In order to maximize the moment of inertia of a flywheel while minimizing its weight, what shape should it have?

9–9 In tightening cylinder-head bolts in an automobile engine, the critical quantity is the *torque* applied to the bolts. Why is this more important than the actual *force* applied to the wrench handle?

9–10 The flywheel of an automobile engine is included to increase the moment of inertia of the engine crankshaft. Why is this desirable?

9–11 A solid ball and a hollow ball with the same mass and radius roll down a slope. Which one reaches the bottom first?

9–12 A solid ball, a solid cylinder, and a hollow cylinder roll down a slope. Which reaches the bottom first? Last? Does it matter whether the radii are the same?

9–13 When an electrical motor is turned on, it takes longer to come up to final speed if there is a grinding wheel attached to the shaft. Why?

9–14 Experienced cooks can tell whether an egg is raw or hard-boiled by rolling it down a slope (taking care to catch it at the bottom). How is this possible?

9–15 Consider the idea of an automobile powered by energy stored in a rotating flywheel, which can be "recharged" by using an electric motor. What advantages and disadvantages would such a scheme have, compared to more conventional drive mechanisms? Could as much energy be stored as in a tank of gasoline? What factors would limit the maximum energy storage?

9–16 An electric grinder coasts for a minute or more after the power is turned off, while an electric drill coasts for only a few seconds. Why is there a difference?

9–17 Part of the kinetic energy of a moving automobile is in rotational motion of its wheels. When the brakes are applied hard on an icy street, the wheels "lock" and the car starts to slide. What becomes of the rotational kinetic energy?

EXERCISES

Section 9–1 Angular Velocity

9–1

a) What angle in radians is subtended by an arc 3 m in length, on the circumference of a circle whose radius is 2 m?

b) What angle in radians is subtended by an arc of length 78.54 cm on the circumference of a circle of diameter 100 cm? What is this angle in degrees?

c) The angle between two radii of a circle of radius 2.00 m is 0.60 rad. What length of arc is intercepted on the circumference of the circle by the two radii?

9–2 Compute the angular velocity in $rad \cdot s^{-1}$, of the crankshaft of an automobile engine that is rotating at 4800 $rev \cdot min^{-1}$.

9–3 A merry-go-round is being pushed by a child. The angle the merry-go-round has turned through varies with time according to $\theta(t) = (2 \ rad \cdot s^{-1})t + (0.05 \ rad \cdot s^{-3})t^3$.

a) Calculate the angular velocity of the merry-go-round as a function of time.

b) What is the initial value of the angular velocity?

c) Calculate the instantaneous value of the angular velocity ω at $t = 5$ s and the average angular velocity ω_{av} for the time interval $t = 0$ to $t = 5$ s.

Section 9–2 Angular Acceleration

9–4 A rigid object rotates with angular velocity that is given by $\omega(t) = 4 \ rad \cdot s^{-1} - (8 \ rad \cdot s^{-3})t^2$.

a) Calculate the angular acceleration as a function of time.

b) Calculate the instantaneous angular acceleration α at $t = 2$ s, and the average angular acceleration α_{av} for the time interval $t = 0$ to $t = 2$ s.

9–5 The angle θ through which a bicycle wheel turns is given by $\theta(t) = a + bt^2 + ct^3$, where a, b, and c are constants such that for t in seconds θ will be in radians. Calculate the angular acceleration of the wheel as a function of time.

Section 9–3 Rotation with Constant Angular Acceleration

9–6 A circular saw blade 0.6 m in diameter starts from rest and accelerates with constant angular acceleration to an angular velocity of 100 $rad \cdot s^{-1}$ in 20 s. Find the angular acceleration and the angle through which the blade has turned.

9–7 An electric motor is turned off, and its angular velocity decreases uniformly from 1000 $rev \cdot min^{-1}$ to 400 $rev \cdot min^{-1}$ in 5 s.

a) Find the angular acceleration and the number of revolutions made by the motor in the 5-s interval.

b) How many more seconds are required for the motor to come to rest?

9–8 A flywheel requires 3 s to rotate through 234 rad. Its angular velocity at the end of this time is 108 $rad \cdot s^{-1}$. Find

a) the angular velocity at the beginning of the 3-s interval;

b) the constant angular acceleration.

9–9 A flywheel whose angular acceleration is constant and equal to 2 $rad \cdot s^{-2}$ rotates through an angle of 100 rad in 5 s. How long had it been in motion at the beginning of the 5-s interval if it started from rest?

9–10 A bicycle wheel of radius 0.33 m turns with angular acceleration $\alpha(t) = 1.2 \ rad \cdot s^{-2} - (0.4 \ rad \cdot s^{-3})t$. It is at rest at $t = 0$.

a) Calculate the angular velocity and angular displacement as functions of time.

b) Calculate the maximum positive angular velocity and maximum positive angular displacement of the wheel.

Section 9–4 Relation between Angular and Linear Velocity and Acceleration

9–11

a) A cylinder 0.15 m in diameter rotates in a lathe at 750 $rev \cdot min^{-1}$. What is the tangential velocity of the surface of the cylinder?

b) The proper tangential velocity for machining cast iron is about 0.60 $m \cdot s^{-1}$. At how many $rev \cdot min^{-1}$ should a piece of stock 0.05 m in diameter be rotated in a lathe?

9–12 Find the required angular velocity of an ultracentrifuge, in rpm, for the radial acceleration of a point 1 cm from the axis to equal 300,000 g (i.e., 300,000 times the acceleration of gravity).

9–13 A wheel rotates with a constant angular velocity of 10 $rad \cdot s^{-1}$.

a) Compute the radial acceleration of a point 0.5 m from the axis, from the relation $a_\perp = \omega^2 r$.

b) Find the tangential velocity of the point, and compute its radial acceleration from the relation $a_\perp = v^2/r$.

9–14 An electric fan blade 3.0 ft in diameter is rotating about a fixed axis with an initial angular velocity of 2 $rev \cdot s^{-1}$. The angular acceleration is 3 $rev \cdot s^{-2}$.

a) Compute the angular velocity after 1 s.

b) Through how many revolutions has the blade turned in this time interval?

c) What is the tangential velocity of a point on the tip of the blade at $t = 1$ s?

d) What is the resultant acceleration of a point on the tip of the blade at $t = 1$ s?

9–15 A flywheel of radius 0.30 m starts from rest and accelerates with constant angular acceleration of 0.50 $rad \cdot s^{-2}$. Compute the tangential acceleration, the radial acceleration, and the resultant acceleration of a point on its rim

a) at the start;

b) after it has turned through 120°;

c) after it has turned through 240°.

Section 9-5 Kinetic Energy of Rotation

9-16 Small blocks, each of mass m, are clamped at the ends and at the center of a light rod of length L. Compute the moment of inertia of the system about an axis perpendicular to the rod and passing through a point one-quarter of the length from one end. Neglect the moment of inertia of the light rod.

9-17 Four small spheres, each of mass 3 kg, are arranged in a square 0.5 m on a side and are connected by light rods. Find the moment of inertia of the system about an axis

a) through the center of the square, perpendicular to its plane;

b) bisecting two opposite sides of the square.

9-18 Find the moment of inertia of a rod 4 cm in diameter and 2 m long, of mass 8 kg, for the following axes. Use the formulas of Fig. 9-7.

a) About an axis perpendicular to the rod and passing through its center.

b) About a longitudinal axis passing through the center of the rod.

9-19 A wagon wheel is constructed as in Fig. 9-26. The radius of the wheel is 0.3 m and the rim has mass 1.0 kg. Each of the four spokes, which lie along diameters, has a mass of 0.4 kg. What is the moment of inertia of the wheel about an axis through its center and perpendicular to the plane of the wheel? (Use the formulas given in Fig. 9-7.)

FIGURE 9-26

9-20 A grinding wheel 0.2 m in diameter, of mass 3 kg, is rotating at 3600 rev·min^{-1} about an axis through its center.

a) What is its kinetic energy?

b) How far would it have to drop in free fall to acquire the same kinetic energy?

9-21 The flywheel of a gasoline engine is required to give up 300 J of kinetic energy while its angular velocity decreases from 600 rev·min^{-1} to 540 rev·min^{-1}. What moment of inertia is required?

9-22 A light, flexible rope is wrapped several times around a solid cylinder of weight 40 lb and radius 0.75 ft, which rotates without friction about a fixed horizontal axis. The free end of the rope is pulled with a constant force P for a distance of 15 ft. What must P be for the final speed of the end of the rope to be 12.0 ft·s^{-1}?

Section 9-6 Moment-of-Inertia Calculations

Section 9-7 Parallel-Axis Theorem

9-23 A thin, rectangular sheet of steel is 0.3 m by 0.4 m and has mass 24 kg. Find the moment of inertia about an axis

a) through the center, parallel to the long sides;

b) through the center, parallel to the short sides;

c) through the center, perpendicular to the plane.

9-24 The four objects shown in Fig. 9-27 have equal masses m. Object A is a solid cylinder of radius R. Object B is a hollow, thin cylinder of radius R. Object C is a solid square whose length of side $= 2R$. Object D is the same size as C, but hollow (i.e., made up of four thin sticks). The objects have axes of rotation perpendicular to the page and through the center of gravity of each object.

a) Which object has the smallest moment of inertia?

b) Which object has the largest moment of inertia?

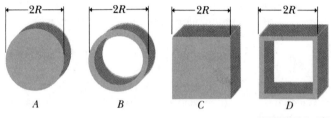

FIGURE 9-27

9-25 Use the parallel-axis theorem to calculate the moment of inertia of a square sheet of metal of side length a and mass M for an axis perpendicular to the sheet and passing through one corner.

9-26

a) Use the integration methods of Section 9-6 to calculate the moment of inertia of a uniform, slender rod of mass M and length l for an axis a distance $l/4$ from one end and perpendicular to the rod.

b) Calculate the same moment of inertia as in (a), but by using Fig. 9-7 and the parallel-axis theorem.

Section 9-8 Torque

9-27 Calculate the torque (magnitude and direction) about point O due to the force F in each of the situations sketched in Fig. 9-28. In each case the object to which the force is applied has length 4 ft, and the force $F = 20$ lb.

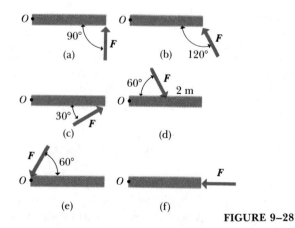

FIGURE 9-28

9–28 Calculate the resultant torque about point O for the two forces applied as in Fig. 9–29.

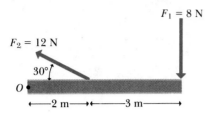

FIGURE 9–29

9–29 A force F is given by $F = (2\ \text{N})i - (3\ \text{N})j$, and the vector r from the axis to point of application of the force is $r = (-0.3\ \text{m})i + (0.5\ \text{m})j$. Calculate the vector torque produced by this force.

Section 9–9 Torque and Angular Acceleration

9–30 A cord is wrapped around the rim of a flywheel 0.5 m in radius, and a steady pull of 50 N is exerted on the cord. The wheel is mounted in frictionless bearings on a horizontal shaft through its center. The moment of inertia of the wheel is 4 kg·m². Compute the angular acceleration of the wheel.

9–31 A grindstone 1.0 m in diameter, of mass 50 kg, is rotating at 900 rev·min⁻¹. A tool is pressed against the rim with a normal force of 200 N, and the grindstone comes to rest in 10 s. Find the coefficient of friction between the tool and the grindstone. Neglect friction in the bearings.

9–32 A thin rod of length l and mass M is pivoted about a vertical axis at one end. A constant force F is applied to the other end, causing the rod to rotate. The force is maintained perpendicular to the rod. Calculate the angular acceleration α of the rod.

9–33 A bucket of water of mass 20 kg is suspended by a rope wrapped around a windlass in the form of a solid cylinder 0.2 m in diameter, also of mass 20 kg. The cylinder is pivoted on a frictionless axle through its center. The bucket is released from rest at the top of a well and falls 20 m to the water. Neglect the weight of the rope.

a) What is the tension in the rope while the bucket is falling?

b) With what velocity does the bucket strike the water?

c) What was the time of fall?

d) While the bucket is falling, what is the force exerted on the cylinder by the axle?

Section 9–10 Work and Power in Rotational Motion

9–34 What is the power output in horsepower of an electric motor turning at 3600 rpm and developing a torque of 2.5 ft·lb?

9–35 A grindstone in the form of a solid cylinder has a radius of 0.5 m and a mass of 50 kg.

a) What torque will bring it from rest to an angular velocity of 300 rev·min⁻¹ in 10 s?

b) Use Eq. (9–33) to calculate the work done by the torque.

c) What is the grindstone's kinetic energy when it is rotating at 300 rev·min⁻¹? Compare your answer to the result in (b).

9–36 The flywheel of a motor has a mass of 300 kg and a moment of inertia of 675 kg·m². The motor develops a constant torque of 2000 N·m, and the flywheel starts from rest.

a) What is the angular acceleration of the flywheel?

b) What will be its angular velocity after making four revolutions?

c) How much work is done by the motor during the first four revolutions?

9–37

a) Compute the torque developed by an airplane engine whose output is 1.5×10^6 W at an angular velocity of 2400 rev·min⁻¹.

b) If a drum of negligible mass, 0.5 m in diameter, were attached to the motor shaft, and the power output of the motor were used to raise a weight hanging from a rope wrapped around the drum, how large a weight could be lifted?

c) With what velocity would it rise?

Section 9–11 Rotation about a Moving Axis

9–38 A solid cylinder of mass 4 kg rolls, without slipping, down a 30° slope. Find the acceleration, the frictional force, and the minimum coefficient of friction needed to prevent slipping.

9–39 A string is wrapped several times around the rim of a small hoop. The hoop has radius 0.08 m and mass 1.2 kg. If the free end of the string is held in place and the hoop is released from rest, calculate

a) the tension in the string while the hoop is descending;

b) the time it takes the hoop to descend 0.5 m;

c) the angular velocity of the rotating hoop after it has descended 0.5 m.

Section 9–12 Angular Momentum and Angular Impulse

9–40 Calculate the angular momentum of a uniform sphere of radius 0.20 m and mass 4 kg if it is rotating about an axis along a diameter at 6 rad·s⁻¹.

9–41 What is the angular momentum of the hour hand on a clock, about an axis through the center of the clock face, if the clock hand has a length of 25.0 cm and a mass of 20.0 g? Take the hour hand to be a slender rod rotating about one end.

9–42 A solid wood door 1.0 m wide and 2.0 m high is hinged along one side and has a total mass of 50 kg. Initially open and at rest, the door is struck at its center with a hammer. During the blow an average force of 2000 N acts for 0.01 s. Find the angular velocity of the door after the impact.

9–43 A rock of mass 2 kg is thrown with velocity $v = 12$ m·s^{-1}. When it is at point P in Fig. 9–30, what is its angular momentum relative to point O? Assume that the rock travels in a straight line.

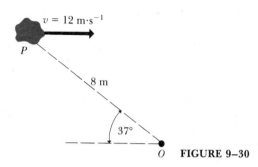

$v = 12$ m·s^{-1}

P

8 m

37°

O **FIGURE 9–30**

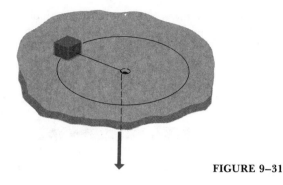

FIGURE 9–31

9–44 A man of mass 70 kg is standing on the rim of a large disk that is rotating at 0.5 rev·s^{-1} about an axis through its center. The disk has mass 120 kg and radius 4 m. Calculate the total angular momentum of the man-plus-disk system.

Section 9–13 Conservation of Angular Momentum

9–45 On an old-fashioned rotating piano stool, a man sits holding a pair of dumbbells at a distance of 0.6 m from the axis of rotation of the stool. He is given an angular velocity of 5 rad·s^{-1}, after which he pulls the dumbbells in until they are only 0.2 m distant from the axis. The man's moment of inertia about the axis of rotation is 5 kg·m^2 and may be considered constant. Each dumbbell has a mass of 5 kg and may be considered a point mass. Neglect friction.

a) What is the initial angular momentum of the system?

b) What is the angular velocity of the system after the dumbbells are pulled in toward the axis?

c) Compute the kinetic energy of the system before and after the dumbbells are pulled in. Account for the difference, if any.

9–46 The outstretched arms of a figure skater preparing for a spin can be considered a slender rod pivoting about an axis through its center. When her arms are brought in and wrapped around her body to execute the spin, they can be considered a thin-walled hollow cylinder. If her original angular velocity is 1 rev·s^{-1}, what is her final angular velocity? Her arms have a combined mass of 8 kg. When outstretched, they span 1.8 m; when wrapped, they form a cylinder of radius 25 cm. The moment of inertia of the remainder of her body is constant and equal to 3 kg·m^2.

9–47 A puck on a frictionless air-hockey table has a mass of 0.05 kg and is attached to a cord passing through a hole in the table surface, as in Fig. 9–31. The puck is originally revolving at a distance of 0.2 m from the hole, with an angular velocity of 3 rad·s^{-1}. The cord is then pulled from below, shortening the radius of the circle in which the puck revolves to 0.1 m. The puck may be considered a point mass.

a) What is the new angular velocity?

b) Find the change in kinetic energy of the puck.

c) How much work was done by the person who pulled the cord?

9–48 A turntable rotates about a fixed vertical axis, making one revolution in 10 s. The moment of inertia of the turntable about this axis is 1200 kg·m^2. A man of mass 80 kg, initially standing at the center of the turntable, runs out along a radius. What is the angular velocity of the turntable when the man is 2 m from the center?

9–49 In models of stellar evolution, one scenario is for stars to undergo sudden gravitational collapse in which the radius of the star decreases drastically while most of the star's mass is retained. Consider the hypothetical collapse of our sun. Data are given in Appendix F. Also, the present period for rotation on its axis for the sun is about 25 days. If the sun undergoes gravitational collapse to a new radius of 32 km (about 20 mi) while keeping its present mass, calculate its new revolution rate in rev·s^{-1}. (This illustrates one possible model for the formation of pulsars.)

9–50 A large wooden turntable of radius 2.0 m and total mass 120 kg is rotating about a vertical axis through its center, with an angular velocity of 3.0 rad·s^{-1}. From a very small height a sandbag of mass 100 kg is dropped vertically onto the turntable, at a point near the outer edge.

a) Find the angular velocity of the turntable after the sandbag is dropped.

b) Compute the kinetic energy of the system before and after the sandbag is dropped.

Why are these kinetic energies not equal?

9–51 The door in Exercise 9–42 is struck at its center by a handful of sticky mud of mass 0.5 kg, traveling 10 m·s^{-1} just before impact. Find the final angular velocity of the door. Is the moment of inertia of the mud significant?

Section 9–14 Vector Representation of Angular Quantities

9–52 The mass of the rotor of a toy gyroscope is 0.150 kg, and its moment of inertia about its axis is 1.5×10^{-4} kg·m^2. The mass of the frame is 0.030 kg. The gyroscope is supported on a single pivot, as in Fig. 9–32, with its center of gravity a horizontal distance of 4 cm from the pivot. The gyroscope is precessing in a horizontal plane at the rate of one revolution in 6 s.

a) Find the upward force exerted by the pivot.

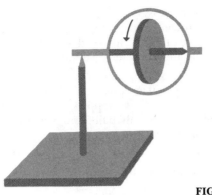

FIGURE 9–32

b) Find the angular velocity with which the rotor is spinning about its axis, expressed in rev·min^{-1}.

c) Copy the diagram, and show by vectors the angular momentum of the rotor and the torque acting on it.

9–53 The stabilizing gyroscope of a ship is a solid disk of mass 50,000 kg; its radius is 2 m, and it rotates about a vertical axis with an angular velocity of 900 rev·min^{-1}.

a) How long a time is required to bring it up to speed, starting from rest, with a constant power input of 7.46×10^4 W?

b) Find the torque needed to cause the axis to precess in a vertical fore-and-aft plane at the rate of 1°·s^{-1}.

PROBLEMS

9–54

a) Prove that when a body starts from rest and rotates about a fixed axis with constant angular acceleration, the radial acceleration of a point in the body is directly proportional to its angular displacement.

b) Through what angle has the body turned at the instant when the resultant acceleration of a point makes an angle of 60° with the radial direction?

9–55 The moment of inertia of a sphere of uniform density is $0.400MR^2$ (Fig. 9–7). Recent satellite observations show the earth's moment of inertia to be $0.3308MR^2$. The earth has a mantle and a core. The average density of the core is approximately 11.0×10^3 kg·m^{-3}; the mantle's is 5.0×10^3 kg·m^{-3}. The mantle occurs from the earth's surface to a depth of 3600 km, and the remainder is core. For this two-part model of the earth, what is the earth's moment of inertia? Express your answer in terms of M and R as above.

9–56 A solid uniform disk of mass m and radius R is pivoted about a horizontal axis through its center, and a small object of mass m is attached to the rim of the disk. If the disk is released from rest with the small object at the end of a horizontal radius, find the angular velocity when the small object is at the bottom.

9–57 The flywheel of a punch press has a moment of inertia of 25 kg·m^2 and runs at 300 rev·min^{-1}. The flywheel supplies all the energy needed in a quick punching operation.

a) Find the speed in rev·min^{-1} to which the flywheel will be reduced by a sudden punching operation requiring 4000 J of work.

b) What must be the constant power supply to the flywheel (in watts) to bring it back to its initial speed in a time of 5 s?

9–58 A magazine article described a passenger bus in Zurich, Switzerland, that derived its motive power from the energy stored in a large flywheel. The wheel was brought up to speed periodically, when the bus stopped at a station, by an electric motor, which could then be attached to the electric power lines. The flywheel was a solid cylinder of mass 1000 kg and diameter 1.8 m; its top speed was 3000 rev·min^{-1}.

a) At this speed, what was the kinetic energy of the flywheel?

b) If the average power required to operate the bus was 1.86×10^4 W, how long could it operate between stops?

9–59 A meter stick of mass 0.4 kg is pivoted about one end so it can rotate without friction about a horizontal axis. The meter stick is held in a horizontal position and released. As it swings through the vertical, calculate

a) the angular velocity of the stick;

b) the linear velocity of the end of the stick opposite the axis.

9–60 Consider the system sketched in Fig. 9–33. The pulley has radius 0.2 m and moment of inertia. 0.32 kg·m^2. The rope does not slip on the pulley rim. Use energy methods to calculate the velocity of the 4-kg block just before it strikes the floor.

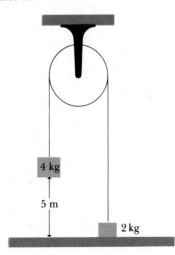

FIGURE 9–33

9-61 Consider the system sketched in Fig. 9-34. The pulley has radius R and moment of inertia I. The rope does not slip over the pulley. The coefficient of friction between block A and the table top is μ_k. The system is released from rest, and block B descends. Use energy methods to calculate the velocity of block B as a function of the distance d that it has descended.

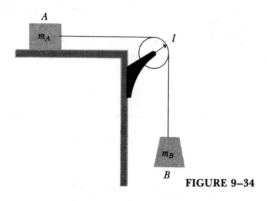

FIGURE 9-34

9-62 A uniform, thin rod is bent into a square of side length a. If the total mass is M, find the moment of inertia about an axis through the center, perpendicular to the plane of the square. (*Hint:* Use the parallel-axis theorem.)

9-63 A constant torque of 20 N·m is exerted on a pivoted wheel for 10 s, during which time the angular velocity of the wheel increases from zero to 100 rev·min^{-1}. The external torque is then removed, and the wheel is brought to rest by friction in its bearings in 100 s. Compute

a) the moment of inertia of the wheel;

b) the friction torque;

c) the total number of revolutions made by the wheel.

9-64 A 60-kg grindstone is 1 m in diameter and has a moment of inertia of 3.75 kg·m^2. A tool is pressed down on the rim with a normal force of 50 N. The coefficient of sliding friction between the tool and the stone is 0.6, and there is a constant friction torque of 5 N·m between the axle of the stone and its bearings

a) How much force must be applied normally at the end of a crank handle 0.5 m long to bring the stone from rest to 120 rev·min^{-1} in 9 s?

b) After attaining a speed of 120 rev·min^{-1}, what must the normal force at the end of the handle become to maintain a constant speed of 120 rev·min^{-1}?

c) How long will it take the grindstone to come from 120 rev·min^{-1} to rest if it is acted on by the axle friction alone?

9-65 Dirk the Dragonslayer is exploring a castle. He is spotted by a dragon who chases him down a hallway. Dirk runs into a room and attempts to swing the heavy door shut before the dragon gets him. The door is initially perpendicular to the wall, so it must be turned through 90°

to close. The door is 3 m tall and 1 m wide and weighs 600 N. The friction at the hinges can be neglected. If Dirk applies a force of 180 N at the edge of the door and perpendicular to it, how long will it take him to close the door?

9-66 A flywheel 1.0 m in diameter is pivoted on a horizontal axis. A rope is wrapped around the outside of the flywheel, and a steady pull of 50 N is exerted on the rope. Ten meters of rope are unwound in 4 s.

a) What was the angular acceleration of the flywheel?

b) What is its final angular velocity?

c) What is its final kinetic energy?

d) What is its moment of inertia?

9-67 A 5-kg block rests on a frictionless horizontal surface. A cord attached to the block passes over a pulley, whose diameter is 0.2 m, to a hanging block also of mass 5 kg. The system is released from rest, and the blocks are observed to move 4 m in 2 s.

a) What was the tension in each part of the cord?

b) What was the moment of inertia of the pulley?

9-68 Figure 9-35 represents an Atwood's machine. Find the linear accelerations of blocks A and B, the angular acceleration of the wheel C, and the tension in each side of the cord

a) if the surface of the wheel is frictionless;

b) if there is no slipping between the cord and the surface of the wheel.

In each case let the masses of the blocks A and B be 4 kg and 2 kg, respectively; the moment of inertia of the wheel about its axis, 0.2 kg·m^2; and the radius of the wheel, 0.1 m.

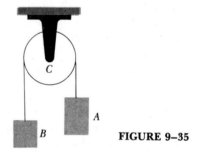

FIGURE 9-35

9-69 A block of mass $m = 5$ kg slides down a surface inclined 37° to the horizontal, as shown in Fig. 9-36. The coefficient of sliding friction is 0.25. A string attached to the

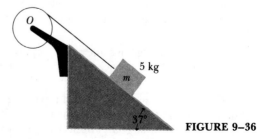

FIGURE 9-36

block is wrapped around a flywheel on a fixed axis at O. The flywheel has a mass $M = 20$ kg, an outer radius $R = 0.2$ m, and a moment of inertia with respect to the axis of 0.2 kg·m^2.

a) What is the acceleration of the block down the plane?

b) What is the tension in the string?

9–70 A yo-yo is made from two uniform disks, each of mass m and radius R, connected by a light axle of radius b. A string is wound several times around the axle and then held stationary while the yo-yo is released from rest, dropping as the string unwinds. Find the acceleration of the yo-yo and the tension in the string.

9–71 A lawn roller in the form of a hollow cylinder of mass M is pulled horizontally with a constant force F applied by a handle attached to the axle. If it rolls without slipping, find the acceleration and the frictional force.

9–72 A uniform rod of mass 0.03 kg and length 0.2 m rotates in a horizontal plane about a fixed axis through its center. Two small objects, each of mass 0.02 kg, are mounted so that they can slide along the rod. They are initially held by catches at positions 0.05 m on each side of the center of the rod, and the system is rotating at 15 rev·min^{-1}. Without otherwise changing the system, the catches are released and the masses slide outward along the rod and fly off at the ends.

a) What is the angular velocity of the system at the instant when the small masses reach the ends of the rod?

b) What is the angular velocity of the rod after the small masses leave it?

9–73 A small block of mass 4 kg is attached to a cord passing through a hole in a frictionless horizontal surface. The block is originally revolving in a circle of radius 0.5 m about the hole, with a tangential velocity of 4 m·s^{-1}. The cord is then pulled slowly from below, shortening the radius of the circle in which the block revolves. The breaking strength of the cord is 600 N. What will be the radius of the circle when the cord breaks?

9–74 Figure 9–37 shows part of a "fly-ball" governor, a speed-controlling device used in old-fashioned steam engines. Each of the steel balls A and B has a mass of 0.50 kg and is rotating about the vertical axis with an angular velocity of 4 rad·s^{-1} at a distance of 0.15 m from

the axis. Collar C is now forced down until the balls are at a distance of 0.05 m from the axis. How much work must be done to move the collar down?

9–75 Disks A and B are mounted on a shaft SS and may be connected or disconnected by a clutch C, as in Fig. 9–38. The moment of inertia of disk A is one-half that of disk B. With the clutch disconnected, A is brought up to an angular velocity ω_0. The accelerating torque is then removed from A, and A is coupled to disk B by the clutch. (Bearing friction may be neglected.) It is found that 2000 J of heat are developed in the clutch when the connection is made. What was the original kinetic energy of disk A?

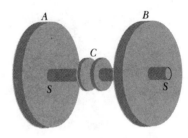

FIGURE 9–38

9–76 A man of mass 60 kg runs around the edge of a horizontal turntable that is mounted on a frictionless vertical axis through its center. The velocity of the man, relative to the earth, is 1 m·s^{-1}. The turntable is rotating in the opposite direction with an angular velocity of 0.2 rad·s^{-1}, relative to the earth. The radius of the turntable is 2 m, and its moment of inertia about the axis of rotation is 400 kg·m^2. Find the final angular velocity of the system if the man comes to rest, relative to the turntable.

9–77 The moment of inertia of the front wheel of a bicycle is 0.4 kg·m^2, its radius is 0.5 m, and the forward speed of the bicycle is 5 m·s^{-1}. With what angular velocity must the front wheel be turned about a vertical axis to counteract the capsizing torque due to a mass of 60 kg, 0.02 m horizontally to the right or left of the line of contact of wheels and ground? (Bicycle riders: Compare your own experience and see if your answer seems reasonable.)

9–78 A demonstration gyroscope wheel is constructed by removing the tire from a bicycle wheel 0.70 m in diameter, wrapping lead wire around the rim, and taping it in place. The shaft projects 0.2 m at each side of the wheel, and a man holds the ends of the shaft in his hands. The mass of the system is 5 kg; its entire mass may be assumed to be located at its rim. The shaft is horizontal, and the wheel is spinning about the shaft at 5 rev·s^{-1}. Find the magnitude and direction of the force each hand exerts on the shaft

a) when the shaft is at rest;

b) when the shaft is rotating in a horizontal plane about its center at 0.04 rev·s^{-1};

c) when the shaft is rotating in a horizontal plane about its center at 0.20 rev·s^{-1}.

d) At what rate must the shaft rotate in order to be supported at one end only?

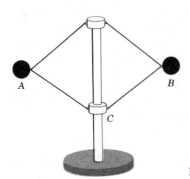

FIGURE 9–37

CHALLENGE PROBLEMS

9–79 A cylinder of radius R and mass M has density that increases linearly with distance r from the cylinder axis, $\rho = \alpha r$. Calculate the moment of inertia of the cylinder about its axis of symmetry, in terms of M and R. Compare the result to that for a cylinder of uniform density. Is the proper qualitative relation obtained?

9–80 Calculate the moment of inertia of a solid cone of uniform density about an axis through the center of the cone, as shown in Fig. 9–39. The cone has mass M and altitude h. The radius of its circular base is R.

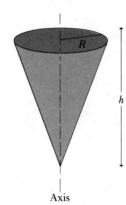

Axis   **FIGURE 9–39**

9–81 A block of mass m is revolving with linear velocity v_1 in a circle of radius r_1 on a frictionless horizontal surface, as shown in Fig. 9–31. The string is slowly pulled from below until the radius of the circle in which the object is revolving is reduced to r_2.

a) Calculate the tension T in the string as a function of r, the distance of the block from the hole. Your answer will be in terms of the initial velocity v_1 and the radius r_1.

b) Use $W = \int_{r_1}^{r_2} \boldsymbol{T}(r) \cdot d\boldsymbol{r}$ to calculate the work done by $\boldsymbol{T}$ when r changes from r_1 to r_2.

c) Compare the results of (b) to the change in the kinetic energy of the block.

9–82 In a binary star system, one star has mass m_A and velocity $\boldsymbol{v}_A$ with respect to an observer, and is located at a position $\boldsymbol{r}_A$. The other star has mass m_B, velocity $\boldsymbol{v}_B$, and position $\boldsymbol{r}_B$. Let $\boldsymbol{R}$ and $\boldsymbol{V}$ be the center of mass position and velocity vectors. Define *relative* positions and velocities by $\boldsymbol{R}_A = \boldsymbol{r}_A - \boldsymbol{R}$ and $\boldsymbol{V}_A = \boldsymbol{v}_A - \boldsymbol{V}$, and similarly for star B.

a) Show that $\boldsymbol{L} = M\boldsymbol{R} \times \boldsymbol{V} + (m_A\boldsymbol{R}_A \times \boldsymbol{V}_A + m_B\boldsymbol{R}_B \times \boldsymbol{V}_B)$. Hence $\boldsymbol{L}$ consists of two parts: $M\boldsymbol{R} \times \boldsymbol{V}$, the angular momentum *of* the center of mass, and $(m_A\boldsymbol{R}_A \times \boldsymbol{V}_A + m_B\boldsymbol{R}_B \times \boldsymbol{V}_B)$, the angular momentum *about* the center of mass.

b) If the only forces acting on the stars are their mutual gravitational attractions, show that $d\boldsymbol{L}/dt = \boldsymbol{0}$. (This result is in fact valid for any number of objects, including planets, asteroids, natural and artificial satellites, comets, and even dust.)

9–83 When an object is rolling without slipping, the rolling friction force is much less than the friction force when the object is sliding; a silver dollar will roll on its edge much farther than it will slide on its flat side. (See Section 5–2.) When an object is rolling without slipping on a horizontal surface, one can therefore take the friction force to be zero so that a and α are approximately zero and v and ω are approximately constant. Rolling without slipping means $v = r\omega$ and $a = r\alpha$. If an object is set into motion on a surface *without* these equalities, sliding (kinetic) friction will act on the object as it slips until rolling without slipping is established.

A solid cylinder of mass M and radius R, rotating with angular velocity ω_0 about an axis through its center, is set on a horizontal surface for which the kinetic friction coefficient is μ_k.

a) Draw a free-body force diagram for the cylinder on the surface. Think carefully about the direction of the kinetic friction force on the cylinder. Calculate the accelerations a of the center of mass (cm) and α of rotation about the cm.

b) The cylinder is initially slipping completely, as initially $\omega = \omega_0$ but $v = 0$. Rolling without slipping sets in when $v = R\omega$. Calculate the *distance* the cylinder rolls before slipping stops.

c) Calculate the work done by the friction force on the cylinder as it moves from where it was set down to where it begins to roll without slipping.

9–84 A uniform sphere is released from rest at the top of an inclined plane of slope θ, as in Example 9–17. The top of the incline is a height h above the horizontal surface on which the incline rests. The coefficient of static friction is *less* than that required to maintain rolling without slipping. The coefficient of sliding friction is $\mu_k = \frac{1}{7} \tan \theta$, and this coefficient determines the friction force.

a) Calculate the linear acceleration a of the center of mass (cm) and the angular acceleration α for rotation about the cm of the sphere.

b) How long does it take the sphere to reach the bottom of the incline, and what are the translational velocity v of the cm and the angular velocity ω of rotation about the cm when it gets there?

c) If the coefficient of kinetic friction between the horizontal surface and the sphere is the same ($\frac{1}{7} \tan \theta$) as between the sphere and the incline, how far will the sphere roll along the horizontal surface before rolling without slipping sets in? What is the (approximately constant) final translational velocity of the sphere as it rolls without slipping along the horizontal surface?

d) Calculate the total work done by friction on the sphere from when the sphere was first released to when it starts to roll without slipping.

10
EQUILIBRIUM OF A RIGID BODY

IN CHAPTER 5 WE DISCUSSED EQUILIBRIUM OF BODIES THAT COULD BE represented as *particles*. Such bodies are in equilibrium whenever the sum of the forces acting on them is zero. Now that we have learned the basic principles of rotational dynamics, we can return to the subject of equilibrium to study the more general problem of equilibrium of a rigid body. We will find that there is a second condition for equilibrium: The sum of the torques about any axis must also be zero for the body not to have any tendency to rotate. Often one of the forces acting on a rigid body is its weight; we show how to compute the torque due to the weight of a body, using the concept of center of mass from Section 8–7 and the related concept of center of gravity, which is introduced in this chapter.

10–1 THE SECOND CONDITION FOR EQUILIBRIUM

We learned in Section 5–3 that for a particle acted on by several forces to be in equilibrium, the vector sum of the forces acting on the particle must be zero, $\Sigma F = 0$. This is often called the **first condition for equilibrium.** In components,

$$\Sigma F_x = 0, \qquad \Sigma F_y = 0. \tag{10–1}$$

For a rigid body there is a **second condition for equilibrium:**

*The sum of the torques of all forces acting on the
body, with respect to some specified axis, must be zero.*

This condition is needed so the body has no tendency to *rotate*. The body does not actually have to be pivoted about the axis chosen. If a rigid body is to be in equilibrium, it must not have any tendency to rotate about *any* axis; thus the sum of torques must be zero, *no matter what axis is chosen.*

A body in equilibrium has no tendency to rotate about any axis.

$$\Sigma \Gamma = 0 \quad \text{about any axis.} \tag{10–2}$$

Although the choice of axis is arbitrary, we must use the *same* axis to calculate all the torques. An important element of problem-solving strategy is to pick the axis or axes so as to simplify the calculations as much as possible.

The second condition for equilibrium is based on the dynamics of rotational motion in exactly the same way that the first condition is based on Newton's first law. For a rigid body at rest to remain at rest, it must have no angular acceleration about any axis, so the sum of torques with respect to every axis must be zero. Note that the term *equilibrium* is used here in a more restricted sense than for particles. A particle in uniform motion (constant velocity) is in equilibrium, but for rigid bodies we use the word *equilibrium* to mean a body actually *at rest* in an inertial frame of reference. Uniform rotational motion with $\Sigma\Gamma = 0$ is *not* considered to be an equilibrium state because the individual particles of the body have accelerations.

10–2 CENTER OF GRAVITY

Center of gravity: how to find the torque due to the weight of a body

In many equilibrium problems, one of the forces acting on the body is its *weight*. We need to be able to calculate the torque of this force with respect to any axis. This is not a completely trivial problem, because the weight does not act at a single point but is distributed over the entire body. However, we can always calculate the torque due to the body's weight by assuming that the entire force of gravity (weight) is concentrated at the **center of mass** of the body, which in this context is also called the **center of gravity.** We used this result without proof in rigid-body dynamics problems in Section 9–9, and now we return for a detailed proof.

Throughout this discussion the body is assumed to be in a *uniform* gravitational field, so the acceleration due to gravity g has the same magnitude and direction at every point in the body. Every particle in the body experiences a gravitational force, and the total weight of the body is the vector sum (resultant) of a large number of parallel forces.

Let us consider the gravitational torque on a flat body of arbitrary shape in the xy-plane, as shown in Fig. 10–1a. We imagine the body as subdivided into a large number of small particles of weights w_1, w_2, and so on, with coordinates (x_1, y_1), (x_2, y_2), and so on. The total weight W of the body is

$$W = w_1 + w_2 + \cdots = \Sigma w.$$

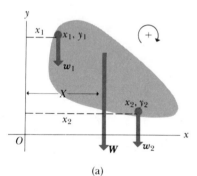

Each particle's weight also contributes to the total torque acting on the body. Computing torques about point O, we see that the torque associated with particle 1 is w_1x_1, that for particle 2 it is w_2x_2, and so on, and that the total torque is

$$w_1x_1 + w_2x_2 + \cdots = \Sigma wx.$$

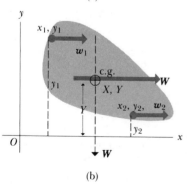

10–1 The body's weight W is the resultant of a large number of parallel forces. The line of action of W always passes through the center of gravity.

For convenience, we are taking *clockwise* torques to be positive, the opposite of the convention we usually used in Chapter 9.

We now ask: Along what line must the total weight W act if the total torque is to equal the expression given above? Suppose W acts along a line at a distance X to the right of the origin; then its torque is WX. Hence it must be true that $w_1x_1 + w_2x_2 + \cdots = WX$, or

$$X = \frac{w_1x_1 + w_2x_2 + \cdots}{W} = \frac{\Sigma wx}{W} = \frac{\Sigma wx}{\Sigma w} \qquad (10\text{–}3)$$

That is, in computing the torque due to the weight of a body, we must consider the weight to act at a point whose x-coordinate X is given by Eq. (10–3).

Now we rotate the object and the reference axes 90° clockwise (or equivalently, we rotate the gravitational forces 90° counterclockwise), as in Fig. 10–1b. The total weight W is unaltered, but now it must act along a line a distance Y above the origin, such that $WY = \Sigma wy$, or

$$Y = \frac{w_1 y_1 + w_2 y_2 + \cdots}{w_1 + w_2 + \cdots} = \frac{\Sigma wy}{\Sigma w} = \frac{\Sigma wy}{W}. \qquad (10\text{–}4)$$

The point of intersection of the lines of action of W in the two parts of Fig. 10–1 has coordinates (X, Y) and is called the *center of gravity* of the body. By considering some arbitrary orientation of the object, we can show that the line of action of W *always* passes through the center of gravity, and thus that the torque of W can always be obtained correctly by taking W to act at the center of gravity.

The center of gravity is identical to the center of mass in a uniform gravitational field. Otherwise center of gravity is meaningless.

If we divide numerator and denominator in Eqs. (10–3) and (10–4) by g and use the fact that $w_1 = m_1 g$, and so on, the right sides of these equations become identical to the equations defining the center of mass in Section 8–7, Eqs. (8–27). Thus *the center of gravity of any body is identical to its center of mass.* Note, however, that the definition of center of gravity makes sense only in a uniform gravitational field, where g has the same value everywhere; otherwise the w's would depend on position. The center of *mass,* conversely, is defined independently of any gravitational effect.

The role of the center of gravity in calculating gravitational torques can be established in a more elegant, though perhaps more acrobatic, way by using the more general vector definition of torque given by Eq. (9–24), $\Gamma = r \times F$. We let the vector g represent the acceleration due to gravity (magnitude and direction). Then the weight w_1 of a particle in the body having mass m_1 is given by $w_1 = m_1 g$. If r_1 is the position vector of this particle with respect to an arbitrary origin O, the torque Γ_1 of this force with respect to O is

An elegant vector derivation of the relation of gravitational torque to center of gravity

$$\Gamma_1 = r_1 \times m_1 g,$$

and the total torque due to the gravitational forces on all the particles is

$$\begin{aligned} \Gamma &= r_1 \times m_1 g + r_2 \times m_2 g + \cdots \\ &= (m_1 r_1 + m_2 r_2 + \cdots) \times g. \end{aligned} \qquad (10\text{–}5)$$

When we multiply and divide this by the total mass

$$M = m_1 + m_2 + \cdots,$$

we obtain

$$\Gamma = \frac{m_1 r_1 + m_2 r_2 + \cdots}{M} \times M g. \qquad (10\text{–}6)$$

The fraction is just the position vector R of the center of mass, with components X and Y as given by Eq. (8–27), and Mg is the total weight $W = mg$ of the body. Thus

$$\Gamma = R \times M g = R \times W, \qquad (10\text{–}7)$$

and the total torque is the same as though the total weight were acting at the position R of the center of mass (center of gravity).

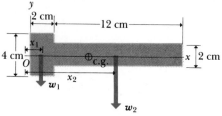

10-2 Machine part for Example 10–1.

If the center of gravity of each of a number of bodies has been determined, the coordinates of the center of gravity of the combination can be computed from Eqs. (10–3) and (10–4), letting w_1, w_2, and so on, be the weights of the bodies and (x_1, y_1), (x_2, y_2), and so on, be the coordinates of the center of gravity of each.

Symmetry considerations are often useful in finding the position of the center of gravity. The center of gravity of a homogeneous sphere, cube, circular disk, or rectangular plate is at its geometric center. The center of gravity of a right circular cylinder or cone is on the axis of symmetry, and so on.

Determining the center of gravity of a composite machine part

EXAMPLE 10–1 Locate the center of gravity of the machine part in Fig. 10–2. The part consists of a disk 4 cm in diameter and 2 cm long, attached to a rod 2 cm in diameter and 12 cm long. Both pieces have the same uniform density.

SOLUTION By symmetry, the center of gravity lies along the axis and the center of gravity of each part is midway between its ends. The volume of the disk is 8π cm^3 and that of the rod is 12π cm^3. Since the weights of the two parts are proportional to their volumes,

$$\frac{w(\text{disk})}{w(\text{rod})} = \frac{w_1}{w_2} = \frac{8\pi}{12\pi} = \frac{2}{3}.$$

Take the origin O at the left face of the disk, on the axis. Then

$$x_1 = 1 \text{ cm}, \qquad x_2 = 8 \text{ cm},$$

and

$$X = \frac{w_1(1 \text{ cm}) + \frac{3}{2}w_1(8 \text{ cm})}{w_1 + \frac{3}{2}w_1} = 5.2 \text{ cm}.$$

The center of gravity is on the axis, 5.2 cm to the right of O.

The center of gravity (center of mass) has several other important properties. First, when a body is suspended in a gravitational field from a single point, the center of gravity always hangs directly below the point of suspension; otherwise the body could not be in rotational equilibrium. This fact can be used for an experimental determination of the location of the center of gravity of an irregular body. Figure 10–3 shows two examples of locating the center of gravity of a body by finding its balance point.

Second, a force applied to a body at its center of gravity does not tend to cause the body to rotate, although a force applied at any other point tends to cause both rotational and translational motion. Third, as discussed in Section 8–7, the total momentum of a body equals the product of its total mass and its center-of-mass velocity. The acceleration of the center of mass is the same as that of a point mass equal to the total mass of the body, acted on by the resultant external force.

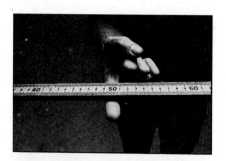

10-3 Locating the center of gravity of a body by balancing. The meter stick balances at its midpoint, but the hammer has a heavy head at one end. (Nancy Rodgers, Exploratorium.)

10–3 EXAMPLES

As we have seen, the principles in rigid-body equilibrium are few and simple; the vector sum of the forces on the body must be zero, and the sum of torques about any point must be zero. The hard part is *applying* these principles to

A view of the Golden Gate Bridge (in San Francisco Bay) during its construction. Every element of the bridge had to be in equilibrium during each step in the construction. (Golden Gate Bridge, Highway and Transportation District.)

specific problems. As always, careful and systematic problem-solving methodology always pays off! The following strategy is very similar to that outlined in Section 5–3 for equilibrium of a particle.

PROBLEM-SOLVING STRATEGY *Equilibrium of a rigid body*

1. Make a sketch of the physical situation, including dimensions.

2. Choose some appropriate body as the body in equilibrium, and draw a free-body diagram showing the forces acting *on* this body and no others. *Do not* include forces exerted *by* this body on other bodies.

3. Draw coordinate axes and specify a positive sense of rotation for torques. Represent forces in terms of their components with respect to the axes you have chosen.

4. Write equations expressing the equilibrium conditions. Remember that $\Sigma F_x = 0$, $\Sigma F_y = 0$, and $\Sigma \Gamma = 0$ are always separate equations; *never* add x- and y-components in a single equation.

5. In choosing an axis to compute torques, note that if the line of action of a force goes through a particular axis, the torque of the force with respect to that axis is zero. Thus unknown forces or components can sometimes be eliminated from the torque equation by a clever choice of axis location. Also remember that when a force is represented in terms of its components, the torque of that force can be obtained by computing separately the torques of the components, each with its appropriate moment arm and sign, and adding the results. This is often easier than determining the moment arm of the original force.

6. Depending on the number of unknowns, you may need to compute torques with respect to two or more axes to obtain as many equations as unknowns. Often there are several equally good sets of force-and-torque equations in a particular problem. When you have as many equations as unknowns, the equations can be solved simultaneously.

EXAMPLE 10–2 A hanging flower basket B having weight w_2 is hung out over the edge of a balcony railing on a horizontal beam that rests on the balcony railing and is counterbalanced at its other end by a body of weight w_1, as shown in Fig. 10–4a. Find the weight w_1 needed to balance the basket, and the total upward force P exerted on the beam at point O. (This illustrates the principle of *cantilever* construction, by which decks, stairways, and the like can be built over empty space, supported entirely from one side.)

Decorating a balcony: an example of torque calculations and rigid-body equilibrium

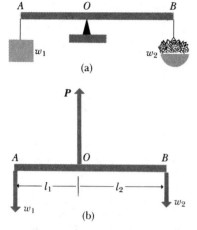

(a)

(b)

10-4 A flower basket cantilevered out over a balcony railing.

A ladder leaning against a wall: using equilibrium conditions to find forces on a body in equilibrium

SOLUTION Figure 10–4b is a free-body diagram for the beam. The first condition for equilibrium gives

$$\Sigma F_y = P - w_1 - w_2 = 0.$$

If we take torques about an axis through O, perpendicular to the diagram, the second condition for equilibrium gives

$$\Sigma \Gamma_0 = w_1 l_1 - w_2 l_2 = 0.$$

Let $l_1 = 1.2$ m, $l_2 = 1.6$ m, and $w_2 = 15$ N. Then, from the equations above,

$$w_1 = 20 \text{ N}, \qquad P = 35 \text{ N}.$$

To illustrate the point that the total torque about *any* axis is zero, let us compute torques about an axis through point A:

$$\begin{aligned}
\Sigma \Gamma_A &= P l_1 - w_2(l_1 + l_2) \\
&= (35 \text{ N})(1.2 \text{ m}) - (15 \text{ N})(2.8 \text{ m}) = 0
\end{aligned}$$

The axis about which torques are computed need not lie on the rod. To verify this, calculate the total torque about a point 0.4 m to the left of A and 0.4 m above it.

EXAMPLE 10–3 The ladder in Fig. 10–5a is 10 m long and has a weight of 400 N that we can consider as concentrated at its center. It leans in equilibrium against a frictionless vertical wall and makes an angle of 53.1° with the horizontal (conveniently forming a 3–4–5 right triangle). Find the magnitudes and directions of the forces F_1 and F_2.

SOLUTION Figure 10–5b shows a free-body diagram. If the wall is frictionless, F_1 is horizontal. The direction of F_2 is unknown; except in special cases, its direction does *not* lie along the ladder. Instead of considering its magnitude and direction as unknowns, it is simpler to resolve the force F_2 into (unknown) x- and y-components and solve for these. The magnitude and direction of F_2 may then be computed. The first condition of equilibrium therefore provides the equations

$$\left. \begin{aligned}
\Sigma F_x &= F_{2x} - F_1 = 0, \\
\Sigma F_y &= F_{2y} - 400 \text{ N} = 0.
\end{aligned} \right\} \quad \text{(First condition)}$$

10-5 (a) A ladder in equilibrium leaning against a frictionless wall. (b) Free-body diagram for the ladder. The force at the base is represented in terms of its x- and y-components.

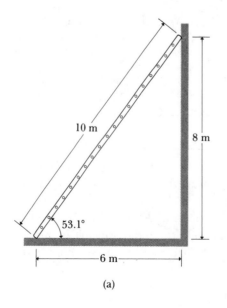

(a)

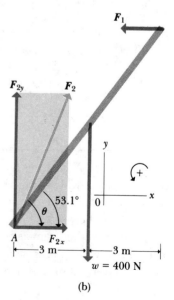

(b)

In writing the second condition, we may compute torques about an axis through any point. The resulting equation is simplest if we choose a point through which the lines of action of two or more forces pass, since these forces then do not appear in the equation. Let us therefore take torques about an axis through point A.

$$\Sigma \Gamma_A = F_1(8 \text{ m}) - (400 \text{ N})(3 \text{ m}) = 0. \quad \text{(Second condition)}$$

From the second equation, $F_{2y} = 400$ N, and from the third,

$$F_1 = \frac{1200 \text{ N·m}}{8 \text{ m}} = 150 \text{ N}.$$

Then from the first equation,

$$F_{2x} = 150 \text{ N}.$$

Hence

$$F_2 = \sqrt{(400 \text{ N})^2 + (150 \text{ N})^2} = 427 \text{ N},$$

$$\theta = \arctan \frac{400 \text{ N}}{150 \text{ N}} = 69.4°.$$

EXAMPLE 10–4 Figure 10–6a shows a human arm lifting a dumbbell, and Fig. 10–6b is a free-body diagram for the forearm, showing the forces involved. The forearm is in equilibrium under the action of the weight w of the dumbbell, the tension T in the tendon connected to the biceps muscle, and the forces exerted on the forearm by the upper arm, at the elbow joint. For clarity, the tendon force has been displaced away from the elbow farther than its actual position. The weight w and the angle θ are given; we want to find the tendon tension and the two components of force at the elbow (three unknown scalar quantities in all).

The human arm: forces in joints and tendons under equilibrium conditions

SOLUTION First we represent the tendon force in terms of its components T_x and T_y, using the given angle θ and the unknown magnitude T:

$$T_x = T \cos \theta, \qquad T_y = T \sin \theta.$$

We also represent the force at the elbow in terms of its components E_x and E_y. Next we note that if we take torques about the elbow joint, the resulting torque equation does not contain E_x, E_y, or T_x, because the lines of action of all these forces pass through this point. The torque equation is then simply

$$lw - dT_y = 0.$$

From this we immediately find

$$T_y = \frac{lw}{d} \qquad \text{and} \qquad T = \frac{lw}{d \sin \theta}.$$

To find E_x and E_y, we could now use the first conditions for equilibrium, $\Sigma F_x = 0$ and $\Sigma F_y = 0$. Instead, for added practice in using torques, we take torques about the point A where the tendon is attached:

$$(l - d)w + dE_y = 0 \qquad \text{and} \qquad E_y = -\frac{(l - d)w}{d}.$$

The negative sign of this result shows that our initial guess for the direction of E_y was wrong; it is actually vertically *downward*.

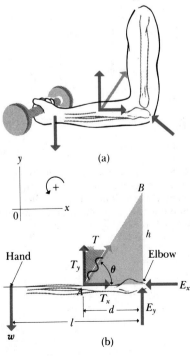

10–6 Force diagrams for human forearm.

Finally, we take torques about point B in the figure:

$$lw - hE_x = 0 \quad \text{and} \quad E_x = \frac{lw}{h}.$$

We note that each stage of this calculation is simplified by choosing the point for calculating torques so as to eliminate one or more of the unknown quantities. In the last step, the force T has no torque about point B; thus when the torques of T_x and T_y are computed separately, they must add to zero. We invite you to verify this statement in detail.

As a specific example, suppose $w = 50$ N, $d = 0.10$ m, $l = 0.50$ m, and $\theta = 80°$. Then $\tan \theta = h/d$, and we find

$$h = d \tan \theta = (0.10 \text{ m})(5.67) = 0.567 \text{ m}.$$

From the previous general results, we find

$$T = \frac{(0.5 \text{ m})(50 \text{ N})}{(0.1 \text{ m})(0.985)} = 254 \text{ N}.$$

$$E_y = -\frac{(0.5 \text{ m} - 0.1 \text{ m})(50 \text{ N})}{0.1 \text{ m}} = -200 \text{ N},$$

$$E_x = \frac{(0.5 \text{ m})(50 \text{ N})}{0.567 \text{ m}} = 44.1 \text{ N}.$$

The magnitude of the force at the elbow is

$$E = \sqrt{E_x{}^2 + E_y{}^2} = 205 \text{ N}.$$

As mentioned above, we have not explicitly used the first condition for equilibrium, that the vector sum of the forces be zero. As a check, you should verify that this condition is satisfied by the results.

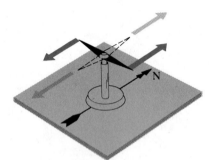

10–7 Forces on the poles of a compass needle.

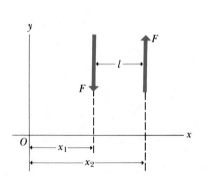

10–8 Two equal and opposite forces having different lines of action are called a *couple*. The torque of the couple is the same about all points and is equal to Fl.

The torque of a couple is the same for every axis.

10–4 COUPLES

Sometimes the forces on a body include two forces of equal magnitude and opposite direction, with lines of action that are parallel but do not coincide. Such a pair of forces is called a **couple.** A common example is a compass needle in the earth's magnetic field, as shown in Fig. 10–7. The north and south poles of the needle are acted on by equal forces, one toward the north and the other toward the south. Except when the needle points in the N–S direction, the two forces do not have the same line of action.

Figure 10–8 shows a couple consisting of two forces, each of magnitude F, separated by a perpendicular distance l. The resultant R of the forces is

$$R = F - F = 0.$$

The fact that the resultant force is zero means that a couple has no effect in producing translation (as a whole) of the body on which it acts. The only effect of a couple is to produce rotation.

The resultant torque of the couple in Fig. 10–8 about an arbitrary point O is

$$\begin{aligned} \Sigma\Gamma_0 &= x_2 F - x_1 F \\ &= (x_1 + l)F - x_1 F \\ &= lF. \end{aligned} \tag{10–8}$$

The distances x_1 and x_2 do not appear in the result, and we conclude that *the torque of a couple is the same about all points in the plane of the forces; it is equal to the product of the magnitude of either force and the perpendicular distance between their lines of action.*

A plumber tightening a pipe fitting uses the principle of the couple. He applies a force to the end of the wrench handle and holds the pipe close to the fitting with the free hand, exerting an opposite force. Thus there is no net force sideways on the pipe, which would put a strain on the opposite end, but only a torque tending to screw the two parts together.

How plumbers use couples to avoid breaking pipes

SUMMARY

For a rigid body to be in equilibrium, two conditions must be satisfied: The vector sum of forces must be zero, and the sum of torques about any axis must be zero. The force condition is usually expressed most conveniently in terms of components; the equations expressing the equilibrium conditions are then

$$\Sigma F_x = 0, \qquad \Sigma F_y = 0, \tag{10-1}$$

and

$$\Sigma \Gamma_A = 0, \tag{10-2}$$

where A represents an arbitrarily chosen axis.

The torque due to the weight of a body can be obtained by assuming the entire weight to be concentrated at the center of mass, also called in this context the center of gravity. The coordinates of the center of gravity are given by

$$X = \frac{w_1 x_1 + w_2 x_2 + \cdots}{W} = \frac{\Sigma wx}{W} = \frac{\Sigma wx}{\Sigma w} \tag{10-3}$$

and

$$Y = \frac{w_1 y_1 + w_2 y_2 + \cdots}{w_1 + w_2 + \cdots} = \frac{\Sigma wy}{\Sigma w} = \frac{\Sigma wy}{W}. \tag{10-4}$$

The torque of a force can be computed by finding the torque of each component, using its appropriate moment arm and sign, and adding these.

A couple consists of two forces of equal magnitude F and opposite direction, acting on a body. If the lines of action of the forces are separated by a distance l, the total torque due to the couple is given (in magnitude) by

$$\Gamma = lF \tag{10-8}$$

and is the same for all axes.

KEY TERMS

first condition for equilibrium
second condition for equilibrium
center of mass
center of gravity
couple

QUESTIONS

10-1 Mechanics sometimes extend the length of a wrench handle by slipping a section of pipe over the handle. Why is this a dangerous procedure?

10-2 Manuals for car-engine repair always specify the *torque* (moment) to be applied when tightening the cylinder-head bolts. Why is torque specified, rather than the *force* to be applied to the wrench handle?

10-3 Car tires are sometimes "balanced" on a machine that pivots the tire and wheel about the center, by laying weights around the wheel rim until it does not tip from the horizontal plane. Discuss this procedure in terms of the center of gravity.

10-4 Is it possible for a solid body to have no matter at its center of gravity? Consider, for example, a disk with a hole in the center, an empty bottle, and a hollow cylinder.

10-5 A man climbs a tall, old stepladder that has a tendency to sway. He feels much more unstable when standing near the top than when near the bottom. Why?

10–6 What determines whether a body in equilibrium is stable or unstable? As an example, discuss a cone resting on its base on a horizontal floor; balanced on its point; and lying on its side.

10–7 When a heavy load is placed in the back end of a truck, behind the rear wheels, the effect is to *raise* the front end of the truck. Why?

10–8 When a tall, heavy object, such as a refrigerator, is pushed across a rough floor, what determines whether it slides or tips?

10–9 Why is a tapered water glass with a narrow base easier to tip over than one with straight sides? Does it matter whether the glass is full or empty?

10–10 People sometimes prop open a door by wedging some object in the space between the hinged side of the door and the frame. Explain why this often results in the hinge screws being ripped out.

10–11 Discuss the action of a claw hammer in pulling out nails, in terms of torques.

10–12 In trying to lift an object that is much too heavy to be lifted directly, one often speaks of the importance of "leverage." Discuss this term in relation to moments, and explain how it is related to the concept of leverage in trading stock options and other securities.

10–13 Why is it easier to hold a 10-kg body in your hand at your side than to hold it with your arm extended horizontally?

10–14 In pioneer days when a Conestoga wagon was stuck in the mud, a man would grasp a wheel spoke and try to turn the wheel, rather than simply pushing the wagon. Why?

10–15 Does the center of gravity of a solid body always lie within the body? If not, give a counterexample.

EXERCISES

Section 10–2 Center of Gravity

10–1 A ball of mass 1 kg is attached by a light rod 0.4 m in length to a second ball of mass 3 kg. Where is the center of gravity of this system?

10–2 In Exercise 10–1, suppose the rod is uniform and has mass 2 kg. Where is the system's center of gravity?

10–3 Three small objects of equal mass are located in the xy-plane, at points having coordinates (0.1 m, 0), (0, 0.1 m), (0.1 m, 0.1 m). Determine the coordinates of the center of gravity.

Section 10–3 Examples

10–4 Two men are carrying a uniform ladder that is 20 ft long and weighs 200 lb. If one man can lift a maximum of 80 lb and lifts at one end, at what point should the other man lift?

10–5 A heavy electric motor is to be carried by two men by placing it on a light board 2.0 m long. To lift the board with the motor on it, one man must lift at one end with a force of 600 N and the other at the opposite end with a force of 400 N. What is the weight of the motor, and where is it located?

10–6 In Exercise 10–5, suppose the board is not light but weighs 200 N. The two men each exert the same force as before. What is the weight of the motor in this case, and where is it located?

10–7 A uniform plank 15 m long, weighing 400 N, rests symmetrically on two supports 8 m apart, as shown in Fig. 10–9. A boy weighing 640 N starts at point A and walks toward the right.

a) Construct in the same diagram two graphs showing the upward forces F_A and F_B exerted on the plank at points A and B, as functions of the coordinate x of the boy. Let 1 cm = 100 N vertically, and 1 cm = 1.0 m horizontally.

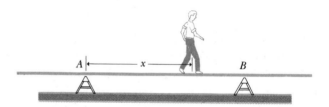

FIGURE 10–9

b) Find from your diagram how far beyond point B the boy can walk before the plank tips.

c) How far from the right end of the plank should support B be placed so that the boy can walk just to the end of the plank without causing it to tip?

10–8 The strut in Fig. 10–10 weighs 200 N, and its center of gravity is at its center. Find

a) the tension in the cable;

b) the horizontal and vertical components of the force exerted on the strut at the wall.

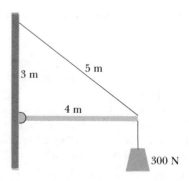

FIGURE 10–10

10–9 Find the tension T in each cable, and the magnitude and direction of the force exerted on the strut by the pivot, in each of the arrangements in Fig. 10–11. Let the weight of the suspended object in each case be w. Neglect the weight of the strut.

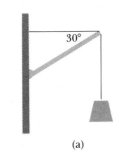

(a)

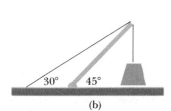

(b)   **FIGURE 10–11**

10–10 The boom in Fig. 10–12 is uniform and weighs 2500 N.

a) Find the tension in the guy wire, and the horizontal and vertical components of the force exerted on the boom at its lower end.

b) Does the line of action of the force exerted on the boom at its lower end lie along the boom?

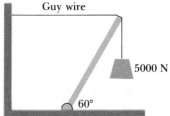

FIGURE 10–12

10–11 A door 1.0 m wide and 2.5 m high weighs 200 N and is supported by two hinges, one 0.5 m from the top and the other 0.5 m from the bottom. Each hinge supports half the total weight of the door. Assuming the door's center of gravity is at its center, find the components of force exerted on the door by each hinge.

Section 10–4 Couples

10–12 Two equal parallel forces of magnitude $F_1 = F_2 = 8$ N are applied to a rod as shown in Fig. 10–13.

a) What should be the distance l between the forces if they are to provide a net torque of 12 N·m about the left-hand end of the rod?

b) Is this torque clockwise or counterclockwise?

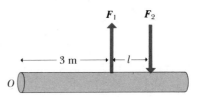

FIGURE 10–13

10–13 Two forces of magnitude $F_1 = F_2 = 3$ N are applied to a rod as shown in Fig. 10–14.

a) Calculate the net torque about point O due to these two forces by calculating the torque due to each separate force.

b) Calculate the net torque about point P due to these two forces by calculating the torque due to each separate force.

c) Compare your results to Eq. (10–8).

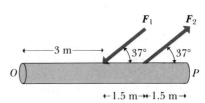

FIGURE 10–14

PROBLEMS

10–14 A certain automobile has a wheelbase (distance between front and rear axles) of 3.0 m. If 60% of the weight rests on the front wheels, how far behind the front wheels is the center of gravity?

10–15 A horizontal boom 4 m long is hinged to a vertical wall at one end, and a 500-N object hangs from its other end. The boom is supported by a guy wire from its outer end to a point on the wall directly above the boom. The weight of the boom can be neglected.

a) If the tension in this wire is not to exceed 1000 N, what is the minimum height above the boom at which it may be fastened to the wall?

b) By how many newtons would the tension be increased if the wire were fastened 0.5 m below this point, the boom remaining horizontal?

10–16 A station wagon weighing 1.96×10^4 N has a wheelbase of 3.0 m. Ordinarily 10,780 N rests on the front wheels and 8820 N on the rear wheels. A box weighing 1960 N is now placed on the tailgate, 1.0 m behind the rear axle. How much total weight now rests on the front wheels? On the rear wheels?

10–17 End A of the bar AB in Fig. 10–15 rests on a frictionless horizontal surface, while end B is hinged. A horizontal force P of 60 N is exerted on end A. Neglect the weight of the bar. What are the horizontal and vertical components of the force exerted by the bar on the hinge at B?

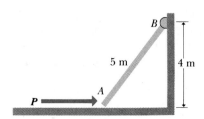

FIGURE 10–15

10–18 A single additional force is to be applied to the bar in Fig. 10–16 to maintain it in equilibrium in the position shown. The weight of the bar can be neglected.

a) What are the horizontal and vertical components of the required force?

b) What is the angle the force must make with the bar?

c) What is the magnitude of the required force?

d) Where should the force be applied?

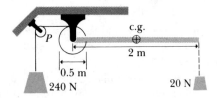

FIGURE 10–16

10–19 A circular disk 0.5 m in diameter, pivoted about a horizontal axis through its center, has a cord wrapped around its rim. The cord passes over a frictionless pulley P and is attached to an object of weight 240 N. A uniform rod 2 m long is fastened to the disk, with one end at the center of the disk. The apparatus is in equilibrium, with the rod horizontal, as shown in Fig. 10–17.

a) What is the weight of the rod?

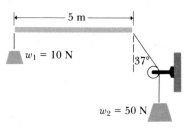

FIGURE 10–17

b) What is the new equilibrium direction of the rod when a second object weighing 20 N is suspended from the other end of the rod, as shown by the broken line? That is, what is then the angle the rod makes with the horizontal?

10–20 The objects in Fig. 10–18 are constructed of uniform wire bent into the shapes shown. Find the position of the center of gravity of each.

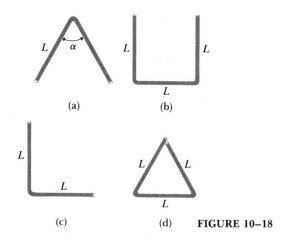

(c)   (d)   **FIGURE 10–18**

10–21 A playground equipment company is designing a new swing with the dimensions shown in Fig. 10–19. For safety the company wants to allow for someone as heavy as 1200 N (about 270 lb) sitting on each of the two seats. Adding an additional safety factor of 30%, they consider the downward force at the top of each end of the swing frame to be 1560 N. The two 3.0-m side pieces each weigh 120 N and are joined at the top by a frictionless hinge.

a) What vertical force is exerted by the ground on each side piece?

b) What is the tension in the horizontal rod, whose weight is negligible?

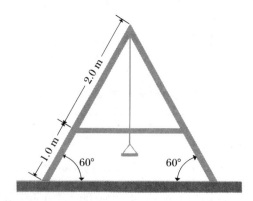

FIGURE 10–19

10–22 A uniform ladder 10 m long rests against a frictionless vertical wall with its lower end 6 m from the wall. The ladder weighs 400 N. The coefficient of static friction between the foot of the ladder and the ground is 0.40. A man weighing 800 N climbs slowly up the ladder.

a) What is the maximum frictional force that the ground can exert on the ladder at its lower end?

b) What is the actual frictional force when the man has climbed 3 m along the ladder?

c) How far along the ladder can the man climb before the ladder starts to slip?

10–23 A gate 4 m long and 2 m high weighs 400 N. Its center of gravity is at its center, and it is hinged at A and B. To relieve the strain on the top hinge, a wire CD is connected as shown in Fig. 10–20. The tension in CD is increased until the horizontal force at hinge A is zero.

a) What is the tension in the wire CD?

b) What is the magnitude of the horizontal component of the force at hinge B?

c) What is the combined vertical force exerted by hinges A and B?

a) In a diagram drawn to scale, show the line of action of the resultant normal force exerted on the block by the plank when the angle $\theta = 15°$.

b) If end B of the plank is slowly raised, will the block start to slide down the plank before it tips over? Find the angle θ at which it starts to slide, or at which it tips.

c) What would be the answer to part (b) if the coefficient of static friction were 0.60?

10–26 A rectangular block 0.25 m wide and 0.5 m high is dragged to the right along a level surface at constant speed by a horizontal force P, as shown in Fig. 10–22. The coefficient of sliding friction is 0.40; the block weighs 25 N; and its center of gravity is at its center.

a) Find the magnitude of the force P.

b) Find the position of the line of action of the normal force n exerted on the block by the surface, if the height $h = 0.125$ m.

c) Find the value of h at which the block just starts to tip.

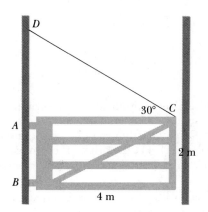

FIGURE 10–20

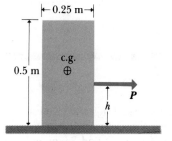

FIGURE 10–22

10–24 A roller whose diameter is 1.0 m weighs 360 N. What is the minimum horizontal force necessary to pull the roller over a brick 0.1 m high when the force is applied

a) at the center?

b) at the top of the roller?

10–25 A uniform rectangular block, 0.5 m high and 0.25 m wide, rests on a plank AB, as in Fig. 10–21. The coefficient of static friction between block and plank is 0.40.

10–27 A garage door is mounted on an overhead rail, as in Fig. 10–23. The wheels at A and B have rusted so that they do not roll, but rather slide along the track. The coefficient of sliding friction is 0.5. The distance between the wheels is 2 m, and each is 0.5 m in from the vertical sides of the door. The door is symmetrical and weighs 800 N. It is pushed to the left at constant velocity by a horizontal force P.

a) If the distance h is 1.5 m, what is the vertical component of the force exerted on each wheel by the track?

b) Find the maximum value h can have without causing one wheel to leave the track.

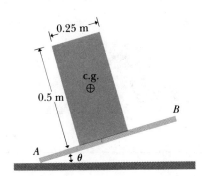

FIGURE 10–21

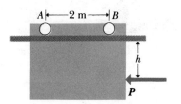

FIGURE 10–23

CHALLENGE PROBLEMS

10-28 One end of a post weighing 500 N rests on a rough horizontal surface with $\mu_s = 0.3$. The upper end is held by a rope fastened to the surface and making an angle of 37° with the post, as in Fig. 10-24. A horizontal force F is exerted on the post as shown.

a) If the force F is applied at the midpoint of the post, what is the largest value it can have without causing the post to slip?

b) How large can the force be, without causing the post to slip, if its point of application is $\frac{7}{10}$ of the way from the ground to the top of the post?

c) Show that if the point of application of the force is too high, the post cannot be made to slip, no matter how great the force. Find the critical height for the point of application.

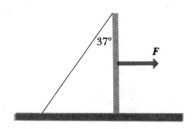

FIGURE 10-24

10-29 A bookcase weighing 1300 N rests on a horizontal surface for which the coefficient of static friction is $\mu_s = 0.3$. The bookcase is 1.8 m tall and 2.0 m wide, and its center of gravity is at its geometrical center. The bookcase rests on four short legs that are each 0.1 m in from the edge of the bookcase. A man pulls on a rope attached to an upper corner of the bookcase with a force P that makes an angle θ with the bookcase, as shown in Fig. 10-25.

a) If $\theta = 90°$, so P is horizontal, show that as P is increased from zero the bookcase will start to slide before it tips, and calculate the magnitude of P that will start the bookcase sliding.

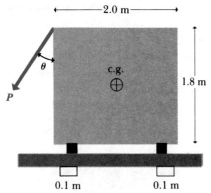

FIGURE 10-25

b) If $\theta = 0°$, so P is vertical, show that the bookcase will tip over rather than slide, and calculate the magnitude of P that will cause the bookcase to start to tip.

c) Calculate as a function of θ the magnitude of P that will cause the bookcase to start to slide and the magnitude that will cause it to start to tip. What is the smallest value θ can have for the bookcase to start to slide before it starts to tip?

10-30 In Section 10-2 the calculation of center of gravity was done by considering objects composed of a *finite* number of point masses or that by symmetry could be represented by a finite number of point masses. For a solid object whose mass distribution does not allow for a simple determination of the center of gravity by symmetry, the sums of Eqs. (10-3) and (10-4) must be generalized to integrals:

$$X = \frac{1}{W} \int xg \, dm$$

and

$$Y = \frac{1}{W} \int yg \, dm,$$

where x and y are the coordinates of the small piece of the object that has mass dm. The integration is over the whole of the object. Consider a thin rod of length L, mass M, and cross-sectional area A. Let the origin of coordinates be at the left-hand end of the rod and the positive x-axis lie along the rod.

a) If the density $\rho = m/V$ of the object is *constant*, perform the integration described above to show that the x-coordinate of the center of gravity of the rod is at its geometrical center.

b) If the density of the object varies linearly with x, $\rho = \alpha x$ where α is a constant, calculate the x-coordinate of the rod's center of gravity.

10-31 Use the methods of Challenge Problem 10-30 to calculate the x- and y-coordinates of the center of gravity of a semicircular metal plate of uniform density ρ and thickness t. Let the radius of the plate be a. The mass of the plate is thus $M = \frac{1}{2}\rho\pi a^2 t$. Use the coordinate system indicated in Fig. 10-26.

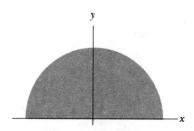

FIGURE 10-26

11

PERIODIC MOTION

IN THIS CHAPTER WE CONSIDER A CLASS OF PROBLEMS IN WHICH A BODY undergoes motion that is repetitive or cyclic. In these problems the body has a stable equilibrium position. When it is displaced from equilibrium and released, it moves back toward equilibrium, overshoots, and goes into a back-and-forth motion on both sides of the equilibrium position. A familiar example is the pendulum, a mass suspended from a string. In the equilibrium position it hangs straight down; when we displace and release it, it does not simply return to the straight-down position but instead swings back and forth in a regular, repetitive manner.

Such a motion is called a **periodic motion,** or an **oscillation.** Many other familiar examples come to mind: the balance wheel of a watch, the vibrating strings in a musical instrument, the action of a door buzzer. The molecules in a solid material vibrate around their equilibrium positions. In many types of *wave motion* in a material, the motions of particles of the material are periodic. Thus a study of periodic motion gives us an important foundation for further study in many different areas of physics.

11–1 BASIC CONCEPTS

One of the simplest systems that can undergo periodic motion is a mass attached to a spring, as shown in Fig. 11–1. For example, the body could be a glider on a linear air track, as described in Section 5–2. The body is attached to one end of a spring, and the other end of the spring is held stationary.

Let x be the displacement of the body from its equilibrium position. When $x = 0$, the spring is neither stretched nor compressed. When the body is displaced to the right, x is positive and the spring stretches. The x-component of force F_x that the spring exerts on the body is toward the left (the negative x-direction), toward the equilibrium position, and F_x is negative. When the body is displaced to the left, x is negative and the spring is compressed. The force on the body is toward the right (the positive x-direction), again toward

Restoring force: a force that always pulls a body toward its equilibrium position

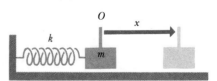

11–1 Model for periodic motion. If the spring obeys Hooke's law, the motion is simple harmonic motion.

Hooke's law: a restoring force
proportional to the displacement from
equilibrium

equilibrium, and F_x is positive. Thus the sign of F_x is always opposite to the sign of x itself. We call such a force a *restoring force*.

In Section 7–3 we noted that for some springs the force is *directly proportional* to the deformation, at least for small deformations. This proportionality is called Hooke's law. Thus in Fig. 11–1, if the spring obeys Hooke's law, we may represent the relationship of F_x to x as

$$F_x = -kx, \tag{11–1}$$

where k is the **force constant** for the spring. This relation is valid for both positive and negative x; in both cases F_x and x have opposite signs.

Suppose we displace the body to the right a distance A and release it, with no initial velocity. The spring exerts a force toward the equilibrium position, and the body accelerates in this direction. The acceleration is not constant, because the force decreases as the body approaches the equilibrium position.

When the body reaches $x = 0$, the force and acceleration have decreased to zero, but the velocity that the body has acquired causes it to "overshoot" the equilibrium position and continue to move to the left. The force then reverses direction, and the body's speed starts to decrease. The body comes to rest at some point to the left of O and starts back toward the equilibrium position.

Simple harmonic motion is the simplest
periodic motion to analyze.

In the next section we will show that the motion is confined to a range $x = \pm A$ on both sides of the equilibrium position, and that each complete back-and-forth trip takes the same amount of time. If there were no loss of mechanical energy due to friction, the motion would continue forever. This specific motion, under the influence of a restoring force proportional to displacement and without any friction, is called **simple harmonic motion,** abbreviated **SHM.**

Here are some terms we will use frequently in this chapter:

A complete vibration, or **cycle,** is one round trip, say from A to $-A$ and back to A, or from O to A to O to $-A$ to O. Note that motion from one side to the other (say A to $-A$) is a half-cycle, not a full cycle.

A language lesson: the terms used to
describe periodic motion

The **period** of the motion, denoted by τ, is the time required for one cycle.

The **frequency,** f, is the number of cycles per unit time. The frequency is the reciprocal of the period: $f = 1/\tau$. The SI unit of frequency, one cycle per second, is called 1 *hertz* (1 Hz = 1 s^{-1}).

The **amplitude,** A, is the maximum displacement from equilibrium, that is, the maximum value of $|x|$. The total overall range of the motion is therefore $2A$.   .

Simple harmonic motion is the simplest of all periodic motions to analyze. In more complex examples the force may depend on displacement in a more complicated way; but only when it is *directly proportional* to displacement do we use the expression "simple harmonic motion." However, many more complex periodic motions are *approximately* simple harmonic, if the displacements are small enough. Thus simple harmonic motion is a *model* that serves for an approximate representation of many periodic motions, and as such it is worth analyzing in detail.

11–2 EQUATIONS OF SIMPLE HARMONIC MOTION

In this section we will obtain expressions for position, velocity, and acceleration as functions of time, for a body undergoing simple harmonic motion. Caution: Do not try to use the equations for motion with *constant* acceleration from Chapter 2! In simple harmonic motion, the acceleration is *not* constant.

Figure 11–2 shows the vibrating body of Fig. 11–1 at a distance x from its equilibrium position O. The mass is m; the body moves along a straight line (the x-axis); and the force component F_x along that line is the elastic restoring force: $F_x = -kx$.

From Newton's second law,

$$F_x = -kx = ma_x,$$

or

$$a_x = -\frac{k}{m}x. \tag{11–2}$$

Thus *at each instant the acceleration is proportional to the negative of the displacement.* When x has its maximum positive value A, the acceleration has its maximum negative value $-kA/m$; at the instants when the body passes the equilibrium position, its acceleration is zero. Of course, its velocity is *not* zero at these times.

We can use conservation of energy to derive expressions for the body's position x and velocity v as functions of time in simple harmonic motion. Because the elastic restoring force is a *conservative* force, we can represent the work done by this force in terms of a potential energy $U = \frac{1}{2}kx^2$, just as in Section 7–6. The kinetic energy is $K = \frac{1}{2}mv^2$, and according to the principle of conservation of energy, the total energy $E = K + U$ is constant. That is,

$$E = \tfrac{1}{2}mv^2 + \tfrac{1}{2}kx^2 = \text{constant.} \tag{11–3}$$

The total energy E is also directly related to the amplitude A of the motion. When the body reaches its maximum displacement $\pm A$, it stops and turns back toward equilibrium. At this instant $v = 0$, so the body has no kinetic energy. The (constant) total energy is equal to the potential energy at this point, which is $\frac{1}{2}kA^2 = E$. Thus

$$\tfrac{1}{2}mv^2 + \tfrac{1}{2}kx^2 = \tfrac{1}{2}kA^2,$$

or

$$v = \pm\sqrt{\frac{k}{m}}\,\sqrt{A^2 - x^2}. \tag{11–4}$$

We can use this relation to obtain the velocity (apart from a sign) for any given position.

The significance of Eq. (11–3) is shown by the graph in Fig. 11–3, where energy is plotted vertically and the coordinate x horizontally. The curve represents the potential energy, $U = \frac{1}{2}kx^2$; this curve is a parabola. The horizontal line at height E represents the constant total energy of the body. We see that the body's motion is restricted to values of x lying between the points where the horizontal line intersects the parabola. If x were outside this range, the potential energy would exceed the total energy, which is impossible.

If we draw a vertical line at any value of x within the permitted range, the length of the segment between the x-axis and the parabola represents the

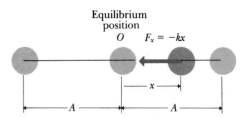

11–2 Coordinate system and free-body diagram for a body in simple harmonic motion.

In simple harmonic motion, the acceleration is proportional to the negative of the displacement.

In simple harmonic motion, energy is conserved; this gives a simple relation between position and speed of the vibrating body.

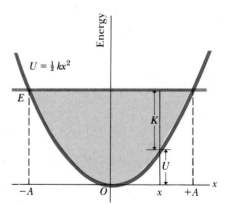

11–3 Relation between total energy E, potential energy U, and kinetic energy K, for a body oscillating with SHM.

The maximum speed and maximum displacement are related to the total energy in a simple way.

potential energy U at that value of x. The length of the segment between the parabola and the horizontal line at height E represents the corresponding kinetic energy K. At the endpoints the energy is all potential, and at the midpoint it is all kinetic. The speed has its maximum value v_{max} at the midpoint:

$$\frac{1}{2}mv_{max}^2 = E, \qquad v_{max} = \sqrt{\frac{2E}{m}}. \tag{11-5}$$

The fact that both the maximum potential energy and the maximum kinetic energy are equal to the total energy E (and thus to each other) may be used to relate the maximum velocity to the amplitude:

$$E = \tfrac{1}{2}kA^2 = \tfrac{1}{2}mv_{max}^2, \qquad v_{max} = \sqrt{\frac{k}{m}}A. \tag{11-6}$$

The position-velocity relation given by Eq. (11–4) is very useful, but it does not tell us where the body is at any given *time*. To have a complete description of the motion we need to know the position, the velocity, and the acceleration as functions of time. We can obtain these relations either by integrating the energy equation, Eq. (11–4), or by working directly with Newton's second law, Eq. (11–2). Let us take the energy approach first.

In Eq. (11–4) we replace v by dx/dt, temporarily disregarding the $\pm$ sign:

$$\frac{dx}{dt} = \sqrt{\frac{k}{m}} \sqrt{A^2 - x^2}. \tag{11-7}$$

Integrating the energy equation to find position as a function of time

The next step is called *separation of variables*. We write all factors containing x on one side, and all those containing t on the other, so that each side can be integrated:

$$\int \frac{dx}{\sqrt{A^2 - x^2}} = \sqrt{\frac{k}{m}} \int dt.$$

Carrying out the integration, we find

$$\arcsin \frac{x}{A} = \sqrt{\frac{k}{m}}t + C,$$

where C is an integration constant. Solving for x, we obtain

$$x = A \sin\left(\sqrt{\frac{k}{m}}t + C\right). \tag{11-8}$$

The quantity in parentheses plays the role of an angle (measured always in radians). The fact that the sine function is *periodic* shows that x is a periodic function of time, as we expected, and specifically that it is a *sinusoidal* function.

The period and frequency of simple harmonic motion don't depend on amplitude; this is one of the most important properties of SHM.

The *period* of motion, denoted by τ, is the time for one complete cycle, or oscillation. The sine function repeats itself whenever the quantity in parentheses increases by 2π. Thus if we start at time $t = 0$, the time τ at which one cycle has been completed is given by

$$\sqrt{\frac{k}{m}}\,\tau = 2\pi,$$

or

$$\tau = 2\pi \sqrt{\frac{m}{k}}. \tag{11-9}$$

Thus the period of the motion is determined by the mass m and the force constant k. It *does not* depend on the amplitude or on the total energy. For given values of m and k, the time of one complete oscillation is the same whether the amplitude is large or small. A periodic motion in which the period is independent of amplitude is said to be **isochronous.**

The frequency f, which is the number of complete oscillations per unit time, is the reciprocal of the period τ:

$$f = \frac{1}{\tau} = \frac{1}{2\pi}\sqrt{\frac{k}{m}}. \qquad (11-10)$$

The basic unit for f is "cycles per second," or simply s^{-1}. In SI units, one cycle per second is given the special name 1 *hertz* (1 Hz), in honor of Heinrich Hertz, one of the pioneers in investigating electromagnetic waves during the latter part of the nineteenth century. Thus

$$1 \text{ hertz} = 1 \text{ Hz} = 1 \text{ cycle} \cdot s^{-1}.$$

We can simplify many of the relationships in simple harmonic motion by using the **angular frequency** ω, defined as $\omega = 2\pi f$ and expressed in radians per second. It follows from either of the two preceding equations that

$$\omega = \sqrt{\frac{k}{m}}, \qquad (11-11)$$

and Eqs. (11–8) and (11–4) can be written more compactly as

$$x = A \sin (\omega t + C), \qquad (11-12)$$
$$v = \pm\omega\sqrt{A^2 - x^2}. \qquad (11-13)$$

Using ω rather than f saves us from having to write factors of 2π. By convention, the hertz is *not* used as a unit of angular frequency, which is usually given the unit $rad \cdot s^{-1}$, or simply s^{-1}.

In SI units, m is expressed in kilograms and k in newtons per meter. The frequency f is then in vibrations per second, or hertz, and the period τ is expressed in seconds per vibration.

The general form of Eqs. (11–9) and (11–10) should not be surprising. When m is large, we expect the motion to be slow and ponderous, corresponding to small f and large τ. A large value of k means a very stiff spring, corresponding to large f and small τ. These equations *do not* contain the amplitude A of the motion. Suppose we give a certain spring-mass system an initial displacement, release it, and measure its frequency. Then we stop it, give it a *different* displacement, and release it. We find that the two frequencies are the same! To be sure, the maximum displacement, the maximum speed, and the maximum acceleration are all different in the two cases, but *not* the frequency. Indeed, the most important characteristic of simple harmonic motion is that *the frequency does not depend on the amplitude of the motion.* This is why a tuning fork vibrates with the same frequency, regardless of amplitude. If it were not for simple harmonic motion, it would be impossible to play most musical instruments in tune.

In Eq. (11–8) the sine function can never be greater than $+1$ or less than -1, so x can never be greater than A or less than $-A$; this confirms again that A is the *amplitude* of the motion. The integration constant C determines the position of the body at time $t = 0$. If $C = 0$, then at time $t = 0$, $x = A \sin 0 = 0$, and the body starts at the origin. But if $C = \pi/2$, then at time $t = 0$, $x = A \sin \pi/2 = A$, and the body starts at its maximum positive displacement. In that

Angular frequency is usually more convenient than is ordinary frequency in SHM problems.

The differential equation for SHM; its solutions are the possible motions.

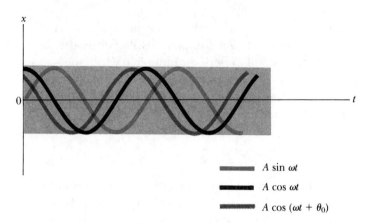

x

0   *t*

━━━ $A \sin \omega t$

━━━ $A \cos \omega t$

━━━ $A \cos (\omega t + \theta_0)$

11–4 Graphs of Eqs. (11–16), showing phase relationships.

case we can use the identity $\sin (\theta + \pi/2) = \cos \theta$ to rewrite Eq. (11–12) as

$$x = A \cos \omega t. \qquad (11\text{–}14)$$

Before exploring further the meaning of Eqs. (11–12) and (11–14), we return to an alternative derivation starting with Newton's second law as expressed by Eq. (11–2). This will give us additional insight into the meaning of these relations. Replacing a_x by d^2x/dt^2 and using $\omega^2 = k/m$, we find

$$\frac{d^2x}{dt^2} = -\omega^2 x. \qquad (11\text{–}15)$$

This is a *differential equation;* it does not give x as a function of t, but says that x must be a function such that when its second derivative is calculated, the result is a negative constant $(-\omega^2)$ multiplied by the original function. What functions have this property? Sines and cosines immediately come to mind. All the following functions have the required property:

$$x = A \sin \omega t, \qquad (11\text{–}16a)$$
$$x = A \cos \omega t, \qquad (11\text{–}16b)$$
$$x = A \cos (\omega t + \theta_0). \qquad (11\text{–}16c)$$

We invite you to verify that each of Eqs. (11–16) is a *solution* of Eq. (11–15). To do this, calculate the second derivative of each function, substitute it in the left side of Eq. (11–15), substitute the function itself into the right side, and verify that the left and right sides are indeed equal. This in turn shows that each function represents a motion that is consistent with Newton's second law and is thus a physically possible motion.

The three functions in Eqs. (11–16) are plotted as functions of time in Fig. 11–4. We see that the essential difference among them is the position of the body at the instant of time we choose to call $t = 0$.

The function that describes position as a function of time depends on the initial conditions.

Equations (11–16) will form the basis of our further description of simple harmonic motion. For given values of A and θ_0, they differ in the position of the particle at time $t = 0$, that is, in the particular point in the cycle at which $t = 0$. If the body is given an initial displacement A at time $t = 0$ and released with no initial velocity, the motion is described by Eq. (11–16b). If the body is given an initial velocity v_0 at the equilibrium position $(x = 0)$ at time $t = 0$, the appropriate equation is Eq. (11–16a). In that case, reference to Eq. (11–6) shows that v_0 and A are related by

$$v_0 = \omega A,$$

since the velocity at $x = 0$ is the maximum velocity.

When the body is given *both* an initial displacement x_0 and an initial velocity v_0 at time $t = 0$, we must use Eq. (11–16c). If x_0 and v_0 are given, we can determine A and θ_0. Angle θ_0 is called a **phase angle,** and its purpose is to describe the point in the cycle at which $t = 0$. To determine A and θ_0, we first note that when $t = 0$ is substituted into Eq. (11–16c), the result must equal x_0; thus

$$x_0 = A \cos \theta_0. \qquad (11\text{--}17)$$

Also, the velocity v at any time t is given by

$$v = \frac{dx}{dt} = -\omega A \sin(\omega t + \theta_0). \qquad (11\text{--}18)$$

Again substituting $t = 0$ and equating the result to v_0, we find

$$v_0 = -\omega A \sin \theta_0. \qquad (11\text{--}19)$$

This and Eq. (11–17) may be solved for A and θ_0. To eliminate A, we divide Eq. (11–19) by Eq. (11–17), obtaining

$$\tan \theta_0 = -\frac{v_0}{\omega x_0}. \qquad (11\text{--}20)$$

To eliminate θ_0, we divide Eq. (11–19) by ω, square it, square Eq. (11–17), and add the two, using the identity $\sin^2 \theta_0 + \cos^2 \theta_0 = 1$, to obtain

$$A^2 = x_0{}^2 + \frac{v_0{}^2}{\omega^2}. \qquad (11\text{--}21)$$

The amplitude is not equal to the initial displacement. This is reasonable: If at time $t = 0$ the particle has an initial displacement x_0 in the positive direction and also a positive velocity v_0 in that direction, then it will move *farther* in that direction before returning; hence A must be greater than x_0. The frequency and period relations, Eqs. (11–9) and (11–10), are unchanged.

The phase angle describes what part of the cycle the body is in at the initial time zero.

PROBLEM-SOLVING STRATEGY: Simple harmonic motion

1. It is important to distinguish between quantities that represent basic physical properties of the system and quantities that describe a particular motion that occurs when the system is set in motion in a specific way. The physical properties include the mass m, the force constant k, and the quantities derived from these, including the period τ, the frequency f, and the angular frequency $\omega = 2\pi f$. In some problems m or k, or both, can be determined from other information given about the system. Quantities that describe a particular motion include the amplitude A, the maximum velocity v_{max}, the phase angle θ_0, and any quantity representing the position, the velocity, or the acceleration at a particular time.

2. If the problem involves a relation among position velocity, and acceleration without reference to time, it is usually easiest to use Eqs. (11–2) or (11–4) than to use the general expressions for position as a function of time given by Eqs. (11–16).

3. When detailed information about positions, velocities, and accelerations at various times is required, then Eqs. (11–16) must be used. If the body has an initial velocity v_0 but no initial displacement ($x_0 = 0$), we use Eq. (11–16a), with $v_0 = \omega A$. If $v_0 = 0$—that is, the body has an initial displacement x_0 but no initial velocity—we use Eq. (11–16b) with $A = x_0$. If the initial position x_0 and initial velocity v_0 are both different from zero, we use Eq. (11–16c). The amplitude A and the phase angle θ_0 are related to x_0 and v_0 by Eqs. (11–20) and (11–21).

4. The energy equation, Eq. (11–3), together with the relation $E = \frac{1}{2}kA^2$, sometimes provides a convenient alternative for relations between velocity and position, especially when energy quantities are also required.

EXAMPLE 11–1: A spring is mounted as in Fig. 11–1. By attaching a spring balance to the free end and pulling sideways, we determine that the force is proportional to the displacement and that a force of 4 N causes a displacement of 0.02 m. We attach a 2-kg body to the end, pull it aside a distance of 0.04 m, and release it.

a) Find the force constant of the spring:

$$k = \frac{F}{x} = \frac{4 \text{ N}}{0.02 \text{ m}} = 200 \text{ N} \cdot \text{m}^{-1}.$$

b) Find the period and frequency of vibration:

$$\tau = 2\pi\sqrt{\frac{m}{k}} = 2\pi\sqrt{\frac{2 \text{ kg}}{200 \text{ N} \cdot \text{m}^{-1}}} = \frac{\pi}{5}\text{s} = 0.628 \text{ s};$$

$$f = \frac{1}{\tau} = \frac{5}{\pi}\text{s}^{-1} = 1.59 \text{ s}^{-1} = 1.59 \text{ Hz};$$

$$\omega = 2\pi f = \sqrt{\frac{k}{m}} = 10 \text{ s}^{-1}.$$

c) Compute the maximum velocity attained by the vibrating body.
   The maximum velocity occurs at the equilibrium position, where $x = 0$. For any x,

$$v = \pm\omega\sqrt{A^2 - x^2},$$

so when $x = 0$,

$$v = v_{max} = \pm\omega A = \pm(10 \text{ s}^{-1})(0.04 \text{ m}) = \pm 0.4 \text{ m} \cdot \text{s}^{-1}.$$

Also,

$$A = 0.04 \text{ m},$$
$$v_{max} = \pm(10 \text{ s}^{-1})(0.04 \text{ m}) = \pm 0.4 \text{ m} \cdot \text{s}^{-1}.$$

d) Compute the maximum acceleration.
   From Eq. (11–2) or (11–15),

$$a_x = -\frac{k}{m}x = -\omega^2 x.$$

The maximum acceleration occurs at the ends of the path, where $x = \pm A$. Therefore

$$a_{max} = \mp\omega^2 A = \mp(10 \text{ s}^{-1})^2(0.04 \text{ m}) = \mp 4.0 \text{ m} \cdot \text{s}^{-2}.$$

e) Compute the velocity and acceleration when the body has moved halfway in to the center from its initial position.
   At this point, from Eq. (11–4),

$$x = \frac{A}{2} = 0.02 \text{ m},$$

$$v = -(10 \text{ s}^{-1})\sqrt{(0.04 \text{ m})^2 - (0.02 \text{ m})^2}$$

$$= -\left(\frac{2\sqrt{3}}{10}\right) \text{m} \cdot \text{s}^{-1} = -0.346 \text{ m} \cdot \text{s}^{-1},$$

$$a_x = -\omega^2 x = -(10 \text{ s}^{-1})^2(0.02 \text{ m}) = -2.0 \text{ m} \cdot \text{s}^{-2}.$$

f) How much time is required for the body to move halfway in to the center from its initial position?

The position at any time is given by $x = A \cos \omega t$. From this,

$$A/2 = A \cos (10 \text{ s}^{-1})t,$$

$$\cos (10 \text{ s}^{-1})t = \frac{1}{2},$$

$$(10 \text{ s}^{-1})t = \arccos \frac{1}{2} = \frac{\pi}{3},$$

$$t = \frac{\pi}{30} \text{ s.}$$

EXAMPLE 11-2: The system in Example 11-1 is given an initial displacement of 0.05 m and an initial velocity of 2 m·s^{-1}. Find the amplitude, the phase angle, and the total energy of the motion, and write an equation for the position as a function of time.

SOLUTION From Eq. (11-21),

$$A = \sqrt{x_0{}^2 + (v_0/\omega)^2}$$

$$= \sqrt{(0.05 \text{ m})^2 + (2 \text{ m·s}^{-1}/10 \text{ s}^{-1})^2}$$

$$= 0.206 \text{ m.}$$

From Eq. (11-20),

$$\theta_0 = \arctan \frac{-v_0}{\omega x_0}$$

$$= \arctan \frac{-2 \text{ m·s}^{-1}}{(10 \text{ s}^{-1})(0.05 \text{ m})}$$

$$= -76.0°$$

$$= -1.33 \text{ rad.}$$

From Eq. (11-3) and the following discussion,

$$E = \tfrac{1}{2}kA^2 = \tfrac{1}{2}(200 \text{ N·m}^{-1})(0.206 \text{ m})^2$$

$$= 4.24 \text{ J.}$$

Alternatively, from the initial conditions,

$$E = \tfrac{1}{2}mv_0{}^2 + \tfrac{1}{2}kx_0{}^2$$

$$= \tfrac{1}{2}(2 \text{ kg})(2 \text{ m·s}^{-1})^2 + \tfrac{1}{2}(200 \text{ N·m}^{-1})(0.05 \text{ m})^2$$

$$= 4.24 \text{ J.}$$

The position is given by Eq. (11-16c):

$$x = (0.206 \text{ m}) \cos [(10 \text{ s}^{-1})t - 1.33 \text{ rad}].$$

The same system as before, but more complicated initial conditions

Suppose we turn the system of Fig. 11-1 by 90°, so the mass hangs vertically from the spring, in a uniform gravitational field. The motion does not change in any essential way. In Fig. 11-5a a body of mass m hangs in equilibrium from a spring with force constant k. In this position the spring is

A body hanging from a spring. The equilibrium point is shifted, but all the essential characteristics of SHM are the same as before.

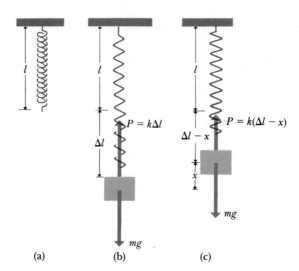

11–5 The restoring force on a body suspended by a spring is proportional to the coordinate measured from the equilibrium position.

(a)   (b)   (c)

stretched an amount Δl just great enough so that the spring's upward vertical force $k\,\Delta l$ on the body balances its weight mg. In that case,

$$k\,\Delta l = mg.$$

When the body is at a distance x *above* its equilibrium position, as in Fig. 11–5c, the extension of the spring is $\Delta l - x$. The upward force it exerts on the body is then $k(\Delta l - x)$, and the resultant force F on the body is

$$F = k(\Delta l - x) - mg = -kx,$$

that is, a net downward force of magnitude kx. Similarly, when the body is *below* the equilibrium position, there is a net upward force proportional to x. If the body is set in vertical motion, it oscillates with SHM, with the same angular frequency as though it were horizontal, $\omega = (k/m)^{1/2}$.

EXAMPLE 11–3: A body of mass 5 kg is suspended by a spring, which stretches 0.1 m when the body is attached. The body is then displaced downward an additional 0.05 m and released. Find the amplitude, the period, and the frequency of the resulting simple harmonic motion.

SOLUTION: Since the initial position is 0.05 m from equilibrium and there is no initial velocity, $A = 0.05$ m. To find the period we first find the force constant of the spring. The spring is stretched 0.1 m by a force of $(5\text{ kg})(9.8\text{ m}\cdot\text{s}^{-2})$, so

$$k = \frac{mg}{\Delta l} = \frac{(5\text{ kg})(9.8\text{ m}\cdot\text{s}^{-2})}{0.1\text{ m}} = 490\text{ N}\cdot\text{m}^{-1},$$

$$\tau = 2\pi\sqrt{\frac{m}{k}} = 2\pi\sqrt{\frac{5\text{ kg}}{490\text{ N}\cdot\text{m}^{-1}}} = 0.635\text{ s},$$

$$f = \frac{1}{\tau} = 1.57\text{ Hz}.$$

11–3 CIRCLE OF REFERENCE

We can gain additional insight into simple harmonic motion through a geometric representation called the **circle of reference.** This representation makes use of a close relationship between SHM and uniform circular motion, which we studied in Section 3–5. The basic idea is shown in Fig. 11–6. Point Q moves counterclockwise around a circle with a radius A that is equal to the amplitude of the actual simple harmonic motion, with constant angular velocity ω (measured in rad·s^{-1}). Thus ω is the rate of change of the angle θ; $\omega = d\theta/dt$.

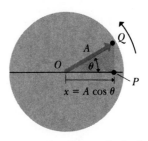

11–6 Coordinate of a body in simple harmonic motion.

The vector from O to Q is the position vector of point Q relative to O. This vector has constant magnitude A and at time t is at an angle θ, measured counterclockwise from the positive x-axis. As Q moves, this vector rotates counterclockwise with constant angular velocity $\omega = d\theta/dt$. The horizontal component of this vector represents the actual motion of the body under study. Such a rotating vector is called a **phasor.** This representation is also useful in many other areas of physics where we encounter quantities that vary sinusoidally with time, including ac-circuit analysis (Chapter 34) and interference phenomena in optics (Chapter 39).

Uniform circular motion of a particle is closely related to SHM.

In Fig. 11–6, point P lies on the horizontal diameter of the circle, directly below Q. We call Q the *reference point,* the circle the *reference circle,* and P the *projection* of Q onto the diameter. The location of P is that of a *shadow* of Q on the x-axis, cast by a light beam parallel to the y-axis. As Q revolves, P moves back and forth along the diameter, staying always directly below (or above) Q. We are about to show that the motion of P is *simple harmonic motion.*

The displacement of P from the origin O at any time t is the distance OP or x. From Fig. 11–6 we see that

$$x = A \cos \theta.$$

If point Q is at the extreme right end of the diameter at time $t = 0$, then $\theta = 0$ when $t = 0$, and the time variation of θ is given by

$$\theta = \omega t.$$

Hence

$$x = A \cos \omega t. \tag{11–22}$$

Now ω, the angular velocity of Q in radians per second, is related to f, the number of complete revolutions of Q per second, by

$$\omega = 2\pi f,$$

since there are 2π radians in one complete revolution. Furthermore, the point P makes one complete back-and-forth vibration for each revolution of Q. Hence f is also the number of vibrations per second, or the *frequency* of vibration of point P. Thus Eq. (11–22) may also be written

$$x = A \cos 2\pi f t. \tag{11–23}$$

We can find the instantaneous velocity of P with the aid of Fig. 11–7. The reference point Q moves with a tangential velocity given by Eq. (9–9):

$$v_{\parallel} = \omega A = 2\pi f A.$$

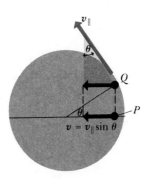

11–7 Velocity in simple harmonic motion.

Since point P is always directly below or above the reference point, the velocity of P at each instant must equal the x-component of the velocity of Q. That is, from Fig. 11–7,

$$v = -v_\parallel \sin \theta = -\omega A \sin \theta;$$

$$v = -\omega A \sin \omega t.$$

(11–24)

The minus sign is needed because the direction of the velocity is toward the left. When Q is below the horizontal diameter, the velocity of P is toward the right; but since $\sin \theta$ is negative at such points, the minus sign is still needed. Equation (11–24) gives the velocity of point P at any time.

We can also find the acceleration of point P by making use again of the fact that since P is always directly below or above Q, its acceleration must equal the x-component of the acceleration of Q. Because point Q moves in a circular path with constant angular velocity ω, it has at each instant an acceleration toward the center given by Eq. (9–11):

$$a_\perp = \frac{v_\parallel^2}{A} = \omega^2 A.$$

From Fig. 11–8, the x-component of this acceleration is

$$a_x = -a_\perp \cos \theta;$$

$$a_x = -\omega^2 A \cos \omega t.$$

(11–25)

The minus sign is needed because the acceleration is toward the left. When Q is to the left of the center, the acceleration of P is toward the right; but since $\cos \theta$ is negative at such points, the minus sign is still required. Equation (11–25) gives the acceleration of P at any time.

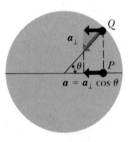

11–8 Acceleration in simple harmonic motion.

Now comes the crucial step in showing that the motion of P is simple harmonic. We combine Eqs. (11–22) and (11–25), obtaining

$$a = -\omega^2 x.$$

(11–26)

Because ω is constant, the acceleration a at each instant equals a negative constant times the displacement x at that instant. But this is just the essential feature of simple harmonic motion, as given by Eq. (11–2): Force and acceleration are proportional to displacement from equilibrium. Hence the motion of P is indeed simple harmonic!

In order to make Eqs. (11–2) and (11–26) agree precisely, we must choose an angular velocity ω for the reference point Q such that

$$\omega^2 = \frac{k}{m}.$$

Thus the angular *velocity* of point Q is identical with the angular *frequency* of motion defined by Eq. (11–11).

Throughout this discussion we have assumed that the initial position of the particle (at time $t = 0$) is its maximum positive displacement A, but this is not an essential restriction. Different initial positions correspond to different initial positions of the reference point Q. For example, if at time $t = 0$ the phasor OQ makes an angle θ_0 with the positive x-axis, then the angle θ at time t is given not by $\theta = \omega t$ as before, but by

$$\theta = \theta_0 + \omega t.$$

(11–27)

The only change in the discussion is to replace (ωt) in Eqs. (11–22), (11–24), and (11–25) by $(\omega t + \theta_0)$. These equations then become

$$x = A \cos (\omega t + \theta_0),$$
$$v = -\omega A \sin (\omega t + \theta_0),$$
$$a_x = -\omega^2 A \cos (\omega t + \theta_0) = -\omega^2 x.$$

(11–28)

The initial position x_0 and initial velocity v_0 (at time $t = 0$) are then given by

$$x_0 = A \cos \theta_0, \qquad v_0 = -\omega A \sin \theta_0,$$

in agreement with Eqs. (11–17) and (11–19).

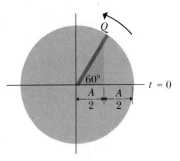

11–9 Phasor diagram for Example 11–4.

EXAMPLE 11–4: In Example 11–1, use the circle of reference to find the time for the displacement from $x = A$ to $x = A/2$, requested in part (f).

SOLUTION While the body moves halfway in, the reference point revolves through an angle of 60° (Fig. 11–9). Since the reference point moves with constant angular velocity and in this example makes one complete revolution during one period $(\tau = \pi/5$ s), the time to rotate through 60° is

$$\left(\frac{1}{6}\right)\left(\frac{\pi}{5} \text{ s}\right) = \frac{\pi}{30} \text{ s} = 0.105 \text{ s}.$$

11–4 ANGULAR SIMPLE HARMONIC MOTION

Simple harmonic motion has a direct *rotational* analog. Suppose a body is pivoted so it can rotate about an axis and is acted on by a torque that is directly proportional to its angular displacement from some equilibrium position. Such a torque might be supplied by a coil spring coaxial with the rotation axis, for example. The resulting rotational motion is directly analogous to linear SHM, and the corresponding equations can be written down immediately from our previous analogies between linear and angular quantities. Caution: In the following discussion, be careful not to confuse the rotational motion of a body in angular SHM with the motion of point Q in the circle-of-reference representation of SHM.

A restoring torque Γ proportional to angular displacement θ is expressed by

$$\Gamma = -k'\theta,$$

(11–29)

where k' is a proportionality constant called the *torque constant*. The equation of motion, from $\Gamma = I\alpha$, is

$$-k'\theta = I \frac{d^2\theta}{dt^2}, \qquad \text{or} \qquad \frac{d^2\theta}{dt^2} = -\frac{k'}{I}\theta = -\omega^2 \theta.$$

This has the same form as Eq. (11–15); the moment of inertia of the pivoted body corresponds to the mass of a body in linear motion. Hence the formula for the period of angular harmonic motion is

$$\tau = 2\pi \sqrt{\frac{I}{k'}},$$

(11–30)

When torque is proportional to angular displacement, the motion is angular simple harmonic motion.

where k' is the constant in Eq. (11–29). The angular frequency is now given by

$$\omega = \sqrt{\frac{k'}{I}}.\qquad(11-31)$$

We need to be cautious in our choice of symbols; we must not use ω for the angular velocity of the body (a quantity that varies with time) because it has already been used for the angular frequency of the motion (a constant for any given system). A natural symbol for the body's angular velocity is Ω. Then the appropriate formulas are obtained by replacing x everywhere by θ, v by Ω, and a_x by α.

The balance wheel of a watch or mechanical clock is an example of angular harmonic motion. If the hairspring behaves according to Eq. (11–29), the motion is *isochronous:* The period is constant even though the amplitude decreases somewhat as the mainspring unwinds.

11–5 THE SIMPLE PENDULUM

The simple pendulum: a mass on a string. The motion is not precisely simple harmonic.

A **simple pendulum** is an idealized model consisting of a point mass suspended by a weightless, unstretchable string in a uniform gravitational field. When the mass is pulled to one side of its straight-down equilibrium position and released, it oscillates about the equilibrium position. We can now analyze the motion of this system, asking in particular whether it is simple harmonic.

As Fig. 11–10 shows, the path is not a straight line but the arc of a circle of radius L, equal to the length of the string. We use as our coordinate the distance x measured along the arc. If the motion is simple harmonic, the restoring force must be directly proportional to x or, since $x = L\theta$, to θ. Is it?

In Fig. 11–10 the forces on the mass are represented in terms of tangential and radial components; the restoring force F is

$$F = -mg \sin \theta \qquad(11-32)$$

The restoring force is therefore proportional *not* to θ but to $\sin \theta$, so the motion is *not* simple harmonic. However, *if the angle θ is small*, $\sin \theta$ is very nearly equal to θ. For example, when $\theta = 0.1$ rad (about 6°), $\sin \theta = 0.0998$, a difference of only 0.2%. With this approximation, Eq. (11–32) becomes

$$F = -mg\theta = -mg\frac{x}{L},$$

or

$$F = -\frac{mg}{L}x.\qquad(11-33)$$

The restoring force is then proportional to the coordinate *for small displacements*, and the constant mg/L represents the force constant k. The angular frequency of a simple pendulum when its amplitude is small is therefore

$$\omega = \sqrt{\frac{k}{m}} = \sqrt{\frac{mg/L}{m}} = \sqrt{\frac{g}{L}}.\qquad(11-34)$$

The corresponding frequency and period relations are

$$f = \frac{\omega}{2\pi} = \frac{1}{2\pi}\sqrt{\frac{g}{L}},\qquad(11-35)$$

$$\tau = \frac{2\pi}{\omega} = \frac{1}{f} = 2\pi\sqrt{\frac{L}{g}}.\qquad(11-36)$$

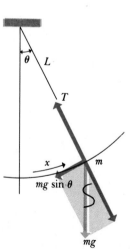

11–10 Forces on the bob of a simple pendulum.

Note that these expressions do not contain the *mass* of the particle; this is because the restoring force, a component of the particle's weight, is proportional to m. Thus the mass appears on both sides of $F = ma$ and may be canceled out. For small oscillations the period of a pendulum for a given value of g is determined entirely by its length.

An elegant argument invented by Galileo four hundred years ago also leads to this conclusion. Make a simple pendulum, said Galileo, measure its period, and then split it down the middle, string and all. Splitting it should not change the motion, so each half must swing with the same period as the original! Thus the period should not depend on the mass.

The form of the dependence on L and g in Eqs. (11–34) through (11–36) is generally what we should expect. It is a familiar fact that long pendulums have longer periods than shorter ones. Increasing g would increase the restoring force, which would cause the frequency to increase and the period to decrease.

We emphasize again that the motion of a pendulum is only *approximately* simple harmonic; when the amplitude is not small, the departures from simple harmonic motion can be substantial. But how small is "small"? The period can be expressed by an infinite series; when the maximum angular displacement is Θ, the period τ is given by

$$\tau = 2\pi\sqrt{\frac{L}{g}}\left(1 + \frac{1^2}{2^2}\sin^2\frac{\Theta}{2} + \frac{1^2 \cdot 3^2}{2^2 \cdot 4^2}\sin^4\frac{\Theta}{2} + \cdots\right). \quad (11\text{--}37)$$

We can compute the period to any desired degree of precision by taking enough terms in the series. When $\Theta = 15°$ (on either side of the central position), the true period differs from that given by the approximate Eq. (11–36) by less than 0.5%.

The usefulness of the pendulum as a timekeeper is based on the fact that the motion is *very nearly* isochronous (period independent of amplitude). So as a pendulum clock runs down and the amplitude of the swings becomes slightly smaller, the clock still keeps very nearly correct time.

The simple pendulum is also a precise and convenient method for measuring the acceleration of gravity g, since L and τ may readily be measured. Such measurements are often used in the field of geophysics. Local deposits of ore or oil affect the local value of g because their density differs from that of their surroundings. Precise measurements of this quantity over an area being prospected often furnish valuable information regarding the nature of underlying deposits.

11–6 THE PHYSICAL PENDULUM

A **physical pendulum** is any *real* pendulum, as contrasted with the idealized model of the *simple* pendulum, where all the mass is concentrated at a point. For small oscillations, analyzing the motion of a real pendulum is almost as easy as for a simple pendulum. Figure 11–11 shows a body of irregular shape pivoted so it can turn without friction about an axis through point O. In the equilibrium position the center of gravity is directly below the pivot; in the position shown in the figure, the body is displaced from equilibrium by an angle θ, which serves as a coordinate for the system. The distance from O to the center of gravity is h, the moment of inertia of the body about the axis through O is I, and the total mass is m. When the body is displaced as shown,

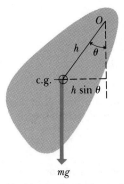

11–11 A physical pendulum.

Motion of a physical pendulum is
approximately simple harmonic motion,
for small amplitudes.

the weight mg causes a restoring torque

$$\Gamma = -(mg)(h \sin \theta). \tag{11-38}$$

When released, the body oscillates about its equilibrium position. As with the simple pendulum, the motion is not simple harmonic, since the torque Γ is proportional not to θ but to $\sin \theta$. However, if θ is small, we can again approximate $\sin \theta$ by θ, and the motion is approximately harmonic. With this approximation,

$$\Gamma \approx -(mgh)\theta,$$

and the effective torque constant is

$$k' = -\frac{\Gamma}{\theta} = mgh.$$

The angular frequency is

$$\omega = \sqrt{\frac{k'}{I}} = \sqrt{\frac{mgh}{I}}, \tag{11-39}$$

and the period is

$$\tau = 2\pi\sqrt{\frac{I}{k'}} = 2\pi\sqrt{\frac{I}{mgh}}. \tag{11-40}$$

EXAMPLE 11-5: Let the body in Fig. 11-11 be a meterstick pivoted at one end, and let L be the total length (1 m). Find the period of oscillation.

SOLUTION: We worked out the moment of inertia in Example 9-9 (Section 9-7): $I = \frac{1}{3}mL^2$. The distance h from the center of gravity to the point of suspension is $h = L/2$. From Eq. (11-40), we obtain

$$I = \frac{1}{3}mL^2, \quad h = \frac{L}{2}, \quad g = 9.8 \text{ m·s}^{-2},$$

$$\tau = 2\pi\sqrt{\frac{\frac{1}{3}mL^2}{mgL/2}} = 2\pi\sqrt{\frac{2L}{3g}}$$

$$= 2\pi\sqrt{(2/3)(1 \text{ m})/9.8 \text{ m·s}^{-2}} = 1.64 \text{ s}.$$

EXAMPLE 11-6: How can the period of a physical pendulum be used to determine its moment of inertia?

SOLUTION: Equation (11-40) may be solved for the moment of inertia I, giving

$$I = \frac{\tau^2 mgh}{4\pi^2}.$$

11-12 The moment of inertia of this connecting rod can be determined by measuring its period of oscillation.

0.2 m

c.g.

Measuring moment of inertia by
measuring the period of a physical
pendulum

The quantities on the right of the equation can all be measured directly. Hence the moment of inertia of a body of any complex shape may be found by suspending the body as a physical pendulum and measuring its period of vibration. We can find the center of gravity by balancing. Since τ, m, g, and h are known, we can compute I. For example, Fig. 11-12 shows a connecting rod pivoted about a horizontal knife edge. Suppose that when the rod is set into oscillation, it is found to make 100 complete vibrations in 120 s, so $\tau = 120 \text{ s}/100 = 1.2 \text{ s}$. Therefore,

$$I = \frac{(1.2 \text{ s})^2(2 \text{ kg})(9.8 \text{ m·s}^{-2})(0.2 \text{ m})}{4\pi^2} = 0.143 \text{ kg·m}^2.$$

11-7 DAMPED OSCILLATIONS

In the idealized oscillating systems we have discussed so far, there is no friction. Thus the systems are *conservative;* the total mechanical energy is constant, and a system set into motion continues oscillating forever with no decrease in amplitude.

Real-world systems always have some friction, however, and oscillations do die out with time unless some means is provided for replacing the mechanical energy lost to friction. A pendulum clock continues to run because the potential energy stored in the spring is used to replace the mechanical energy lost due to friction in the pendulum and the gears. But when the spring "runs down" and no more energy is available, the pendulum swings decrease in amplitude and stop.

The decrease in amplitude caused by dissipative forces is called **damping,** and the corresponding motion is called **damped oscillation.** The simplest case to analyze in detail is that of a frictional damping force directly proportional to the *velocity* of the oscillating body. This behavior occurs in friction involving viscous fluid flow, such as sliding between oil-lubricated surfaces, shock absorbers, and many other systems of practical importance. We then have an additional force on the body due to friction, $F_x = -bv$, where $v = dx/dt$ is the velocity and b a constant that describes the strength of the damping force. (Why is the negative sign needed?) The total force on the body is then

$$F_x = -kx - bv, \qquad (11-41)$$

and the Newton's-second-law formulation of Eq. (11-2) becomes

$$-kx - bv = ma_x,$$

or

$$-kx - b\frac{dx}{dt} = m\frac{d^2x}{dt^2}. \qquad (11-42)$$

Finding solutions of this equation is a straightforward problem in differential equations, but we will not go into the details here. If the damping force is relatively small and the body is given an initial displacement A, the motion is described by

$$x = Ae^{-(b/2m)t}\cos \omega't, \qquad (11-43)$$

where the frequency of oscillation ω' is given by

$$\omega' = \sqrt{\frac{k}{m} - \frac{b^2}{4m^2}}. \qquad (11-44)$$

This motion differs from that of the undamped case in two ways. First, the amplitude is not constant but decreases with time because of the exponential factor $e^{-(b/2m)t}$. The larger the value of b, the more quickly the amplitude decreases. Second, the angular frequency of oscillation is no longer equal to $(k/m)^{1/2}$ but is somewhat smaller. Figure 11-13 shows graphs of Eq. (11-43) for two different values of the constant b.

Note that in Eq. (11-44) the frequency becomes zero when b becomes so large that

$$\frac{k}{m} - \frac{b^2}{4m^2} = 0, \qquad \text{or} \qquad b = \sqrt{4km}. \qquad (11-45)$$

When friction is present, oscillations die out rather than repeating themselves forever.

A damping force proportional to velocity is easy to analyze.

Damping decreases the frequency of a harmonic oscillator.

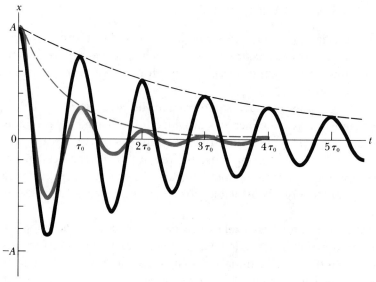

11–13 Graphs of Eq. (11–43), showing damped harmonic motion. The period when there is no damping ($b = 0$) is τ_0. The black curve shows the motion when $b = 0.1\sqrt{km}$, and the color curve is for $b = 0.4\sqrt{km}$. The broken lines show the exponential factor $Ae^{-(b/2m)t}$ for each case. The amplitude decreases more rapidly for the larger value of b. Close inspection of the points where the curves cross the t-axis also shows that the period increases slightly with increasing b. The critical-damping condition is $b = 2\sqrt{km}$.

If the damping force is too large, the system doesn't oscillate at all.

When b exceeds this value, the system no longer oscillates but returns to its equilibrium position without oscillation when it is displaced and released. When Eq. (11–45) is satisfied, the condition is called **critical damping.** The nonoscillating motion that occurs when b is even larger is called **overdamping.** In these cases the solutions of Eq. (11–42) are decreasing exponential functions without any sinusoidal factors. When b is less than the critical value, the situation is called **underdamping.** In all cases, both overdamping and underdamping, the mechanical energy of the system continuously decreases, approaching zero after a sufficiently long time.

Damping can be desirable or undesirable, depending on the situation.

The suspension system of an automobile is a familiar example of damped oscillations. The shock absorbers provide a velocity-dependent damping force so that when the car goes over a bump, it does not continue bouncing forever. For optimal passenger comfort, the system should be critically damped or perhaps slightly underdamped. As the shocks get old and worn, the value of b decreases and the bouncing persists longer. Not only is this nauseating, but it is bad for steering because the front wheels have less positive contact with the ground. Thus damping is an advantage in this system. Conversely, in a system such as a clock or an electrical oscillating system of the type found in radio transmitters, it is usually desirable to minimize damping.

11–8 FORCED OSCILLATIONS

A force that varies periodically with time can make a system vibrate with the same frequency as the force.

There are many practical situations where we would like to maintain oscillations of constant amplitude in a damped oscillating system. A familiar example is a child sitting on a swing. We set the system into motion by pulling the child back from the straight-down equilibrium position and releasing him. If that is all we do, the system oscillates with decreasing amplitude and eventually comes to rest. But by giving the system a little push once each cycle, we can maintain a nearly constant amplitude. More generally, we can maintain a constant-amplitude oscillation in a damped harmonic oscillator by applying an oscillating force, that is, a force that varies with time in a periodic or cyclic way. We call this additional force a **driving force.**

Furthermore, the frequency of variation of the force need not be the same as the natural oscillation frequency of the system. If we apply a periodically varying driving force to the mass of the harmonic oscillator system of Fig. 11–1, the mass undergoes a periodic motion *with the same frequency as that of the driving force.* We call this motion a **forced oscillation,** or a *driven oscillation;* it is different from the motion that occurs when the system is simply set into motion and then left alone to oscillate with a natural frequency determined by m, k, and b.

When the frequency of the driving force is *equal* to the natural frequency of the system, we would expect the amplitude of the resulting oscillation to be larger than when the two are very different, and this expectation is borne out by more detailed analysis and experiment. The easiest case to analyze is that of a *sinusoidally* varying force, say $F_x = F_{max} \sin \omega_d t$, where ω_d is not necessarily equal to the natural frequency ω' of the system given by Eq. (11–44). If we vary the frequency ω_d of the driving force, the amplitude of the resulting forced oscillation varies in an interesting way, as shown in Fig. 11–14. When there is very little damping, the amplitude goes through a sharp peak as the driving frequency passes through the natural oscillation frequency. For increased damping, the peak becomes broader and smaller in height and shifts toward lower frequencies.

The fact that there is an amplitude peak at driving frequencies close to the natural frequency of the system is called **resonance.** Physics is full of examples of resonance; building up the oscillations of a child on a swing is one. A vibrating rattle in a car that occurs only at a certain engine speed is another familiar example. You have probably heard of the dangers of a band marching

The amplitude of a forced oscillation is greatest when the driving frequency is close to the natural frequency of the system.

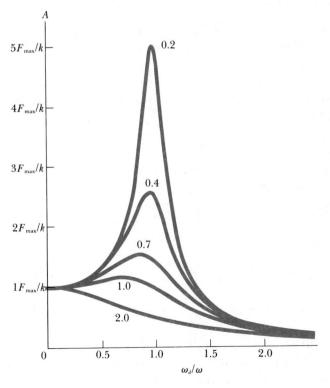

11–14 Graph of the amplitude A of forced oscillation of a damped harmonic oscillator, as a function of the frequency ω_d of the driving force, plotted on the horizontal axis as the ratio of ω_d to the angular frequency $\omega = \sqrt{k/m}$ of an undamped oscillator. Each curve is labeled with the value of the dimensionless quantity $b/\sqrt{km}$, which characterizes the amount of damping. The highest curve has $b = 0.2 \sqrt{km}$, the next has $b = 0.4 \sqrt{km}$, and so on. As b increases, the peak becomes broader and less sharp and shifts toward lower frequencies. When b is as large as $2\sqrt{km}$, the peak disappears completely.

across a bridge; if the frequency of their steps is close to a natural vibration frequency of the bridge, dangerously large oscillations can build up. A tuned circuit in a radio or television receiver responds strongly to waves having frequencies near its resonant frequency, and this is used to select a particular station and reject the others. We will study resonance in electric circuits in detail in Chapter 34.

SUMMARY

When a body of mass m is acted on by a force $F_x = -kx$ that is directly proportional to the body's displacement from its equilibrium position ($x = 0$), the resulting motion is called simple harmonic motion. The amplitude A is the maximum displacement from equilibrium. The period τ is the time for one complete cycle; it is given by

$$\tau = 2\pi\sqrt{m/k}. \tag{11-9}$$

The frequency f is the number of cycles per unit time and is the reciprocal of the period:

$$f = \frac{1}{\tau} = \frac{1}{2\pi}\sqrt{\frac{k}{m}}. \tag{11-10}$$

The angular frequency ω is given by

$$\omega = 2\pi f = \sqrt{\frac{k}{m}}. \tag{11-11}$$

In simple harmonic motion the acceleration at any instant is proportional to the negative of the displacement.

Conservation of energy leads to the following relation among the position and velocity at any time and the amplitude and total energy:

$$E = \tfrac{1}{2}kA^2 = \tfrac{1}{2}mv^2 + \tfrac{1}{2}kx^2 = \text{constant.} \tag{11-3}$$

If the body is given an initial displacement A and released with no initial velocity, the position is given as a function of time by

$$x = A\cos\omega t. \tag{11-16b}$$

If it is given an initial velocity v_0 and no initial displacement, the position is given as a function of time by

$$x = A\sin\omega t, \tag{11-16a}$$

with A given by $A = v_0/\omega$. If the body is given an initial displacement x_0 and an initial velocity v_0, the position is given by

$$x = A\cos(\omega t + \theta_0), \tag{11-16c}$$

with A and θ_0 given by

$$A^2 = x_0^2 + \frac{v_0^2}{\omega^2}, \tag{11-21}$$

$$\theta_0 = \arctan\left(-\frac{v_0}{\omega x_0}\right). \tag{11-20}$$

The circle-of-reference construction uses a rotating vector, called a phasor, having a length equal to the amplitude of the motion. Its projection on the horizontal axis represents the actual motion of the body.

Angular simple harmonic motion occurs when a pivoted body experiences a torque $\Gamma = -k'\theta$ proportional to its angular displacement θ from an equilibrium position. For a body having moment of inertia I, the angular frequency is given by

$$\omega = \sqrt{\frac{k'}{I}}. \qquad (11\text{-}31)$$

A simple pendulum consists of a point mass m at the end of a string of length L. Its motion is approximately simple harmonic for sufficiently small amplitude; the period, the frequency, and the angular frequency are then given by

$$\tau = 2\pi\sqrt{\frac{L}{g}}, \qquad (11\text{-}36)$$

$$f = \frac{1}{2\pi}\sqrt{\frac{g}{L}}, \qquad (11\text{-}35)$$

$$\omega = \sqrt{\frac{g}{L}}. \qquad (11\text{-}34)$$

These quantities are independent of m.

A physical pendulum is a body suspended from an axis of rotation a distance h from its center of gravity. If the moment of inertia about the axis of rotation is I, the angular frequency and period are given by

$$\omega = \sqrt{\frac{mgh}{I}}, \qquad (11\text{-}39)$$

$$\tau = 2\pi\sqrt{\frac{I}{mgh}}. \qquad (11\text{-}40)$$

When a damping force $F = -bv$ proportional to velocity is added to a simple harmonic oscillator, the motion is described as a damped oscillation:

$$x = Ae^{-(b/2m)t}\cos\omega't, \qquad (11\text{-}43)$$

provided that $b^2 < 4km$. This condition is called underdamping. When $b^2 = 4km$, the system is critically damped and no longer oscillates. When b is still larger, the system is overdamped.

When a sinusoidally varying driving force is added to a damped harmonic oscillator, the resulting motion is called a forced oscillation. Its amplitude reaches a peak at driving frequencies close to the natural oscillation frequency of the system. This behavior is called resonance.

QUESTIONS

11-1 Think of several examples in everyday life of motion that is at least approximately simple harmonic. In what respects does each differ from SHM?

11-2 Does a tuning fork or similar tuning instrument undergo simple harmonic motion? Why is this a crucial question to musicians?

11-3 If a spring is cut in half, what is the force constant of each half? How would the frequency of SHM using a half-spring differ from that using the same mass and the entire spring?

11-4 The analysis of SHM in this chapter neglected the mass of the spring. How does the spring mass change the characteristics of the motion?

11-5 The system shown in Fig. 11-5 is mounted in an elevator, which accelerates upward with constant acceleration. Does the period increase, decrease, or remain the same?

11-6 A highly elastic "superball" bouncing on a hard floor has a motion that is approximately periodic. In what ways is the motion similar to SHM? In what ways is it different?

11–7 How could one determine the force constants of a car's springs by bouncing each end up and down?

11–8 Do the pistons in an automobile engine undergo simple harmonic motion?

11–9 Why is the "springiness" of a diving board adjusted for different dives and different weights of divers? How is the adjustment made?

11–10 In Fig. 11–1, suppose the stationary end of the spring is connected instead to another mass, equal to the original mass and free to slide along the same line. Could such a system undergo SHM? How would the period compare with that of the original system?

11–11 For the mass-spring system of Fig. 11–1, is there any point during the motion at which the mass is in equilibrium?

11–12 In any periodic motion, unavoidable friction always causes the amplitude to decrease with time. Does friction also affect the *period* of the motion? Give a qualitative argument to support your answer.

11–13 If a pendulum clock is taken to a mountain top, does it gain or lose time, assuming it is correct at a lower elevation?

11–14 When the amplitude of a simple pendulum increases, should its period increase or decrease? Give a qualitative argument; do not rely on Eq. (11–37). Is your argument also valid for a physical pendulum?

11–15 A pendulum is mounted in an elevator that accelerates upward with constant acceleration. Does the period increase, decrease, or remain the same?

11–16 At what point in the motion of a simple pendulum is the string tension greatest? Least?

11–17 A child on a swing can increase his amplitude by "pumping up." Where does the extra energy come from? (The answer is not "from the child.")

11–18 Could a standard of time be based on the period of a certain standard pendulum? What advantages and disadvantages would such a standard have, compared to the actual present-day standard discussed in Section 1–2?

EXERCISES

Section 11–1 Basic Concepts

11–1 A vibrating object goes through five complete vibrations in 1 s. Find the angular frequency and the period of the motion.

11–2 In Fig. 11–1 the mass is displaced 0.12 m from its equilibrium position and released with no initial velocity. After 2.0 s its displacement is found to be 0.12 m on the opposite side, and it has passed the equilibrium position once during this interval. Find

a) the amplitude,   b) the period,

c) the frequency,   d) the angular frequency.

Section 11–2 Equations of Simple Harmonic Motion

11–3 A harmonic oscillator has a mass of 4 kg and a spring with force constant 100 N·m^{-1}. Find the period, the frequency, and the angular frequency.

11–4 A harmonic oscillator is made with a block of mass 0.5 kg and a spring of unknown force constant. It is found to have a period of 0.20 s. Find the force constant of the spring.

11–5 A block of unknown mass is attached to a spring of force constant 200 N·m^{-1}, in the arrangement shown in Fig. 11–1. It is found to vibrate with a frequency of 3 Hz. Find the period, the angular frequency, and the mass.

11–6 A tuning fork labeled 512 Hz has the tip of each of its two prongs vibrating through a maximum displacement of 0.80 mm.

a) What is the maximum speed of the tip of a prong?

b) What is the maximum acceleration of the tip of a prong?

c) If the maximum displacement were cut in half, what would then be the answers to (a) and (b)?

11–7 An object is vibrating with simple harmonic motion of amplitude 15 cm and frequency 4 Hz. Compute

a) the maximum values of the acceleration and velocity,

b) the acceleration and velocity when the coordinate is 9 cm,

c) the time required to move from the equilibrium position to a point 12 cm distant from it.

11–8 An object of mass 4 kg is attached to a spring of force constant $k = 100$ N·m^{-1}. The object is given an initial velocity of $v_0 = 12$ m·s^{-1} and an initial displacement of $x_0 = 0$. Find the amplitude, the phase angle, and the total energy of the motion, and write an equation for the position as a function of time.

11–9 Repeat Exercise 11–8, but assume the object is given an initial velocity of $v_0 = -6$ m·s^{-1} and an initial displacement of $x_0 = +0.2$ m.

11–10 An object of mass 0.25 kg is acted on by an elastic restoring force of force constant $k = 25$ N·m^{-1}.

a) Construct the graph of elastic potential energy U as a function of displacement x, over a range of x from -0.3 m to $+0.3$ m. Let 1 cm = 0.1 J vertically, and 1 cm = 0.05 m horizontally.

The object is set into oscillation with an initial potential energy of 0.6 J and an initial kinetic energy of 0.2 J. Answer the following questions by reference to the graph:

b) What is the amplitude of oscillation?

c) What is the potential energy when the displacement is one-half the amplitude?

d) At what displacement are the kinetic and potential energies equal?

e) What is the speed of the object at the midpoint of its path (that is, at $x = 0$)?

Find

f) the period τ,   g) the frequency f,

h) the angular frequency ω.

i) What is the initial phase angle θ_0 if the initial velocity v_0 is negative?

11–11 A block of mass 2 kg is suspended from a spring of negligible mass and is found to stretch the spring 0.20 m.

a) What is the force constant of the spring?

b) What is the period of oscillation of the block if pulled down and released?

c) What would be the period of a block of mass 4 kg hanging from the same spring?

11–12 The scale of a spring balance reading from zero to 180 N is 9 cm long. A fish suspended from the balance is observed to oscillate vertically at 1.5 Hz. What is the mass of the fish? Neglect the mass of the spring.

11–13 A block of mass 5 kg hangs from a spring and oscillates with a period of 0.5 s. How much will the spring shorten when the block is removed?

11–14

a) A block suspended from a spring vibrates with simple harmonic motion. At an instant when the displacement of the block is equal to one-half the amplitude, what fraction of the total energy of the system is kinetic and what fraction is potential? Assume $U = 0$ at equilibrium.

b) When the block is in equilibrium, the length of the spring is an amount s greater than in the unstretched state. Prove that $\tau = 2\pi\sqrt{s/g}$.

11–15 A body of mass 4 kg is attached to a coil spring and oscillates vertically in simple harmonic motion. The amplitude is 0.5 m, and at the highest point of the motion the spring has its natural unstretched length. Calculate the elastic potential energy of the spring, the kinetic energy of the body, its gravitational potential relative to the lowest point of the motion, and the sum of these three energies, when the body is

a) at its lowest point,

b) at its equilibrium position,

c) at its highest point.

Section 11–3 Circle of Reference

11–16 An object is undergoing simple harmonic motion with period $\tau = 0.4$ s. Use the circle of reference to calculate the time it takes the object to go from $x = 0$ to $x = A/4$.

11–17 An object is undergoing simple harmonic motion with period $(\pi/2)$s and amplitude $A = 0.2$ m. At $t = 0$ the object is at $x = 0$. How far is the object from the equilibrium position when $t = (\pi/10)$ s?

Section 11–4 Angular Simple Harmonic Motion

11–18 The balance wheel of a watch vibrates with an angular amplitude of π rad and a period of 0.5 s.

a) Find its maximum angular velocity.

b) Find its angular velocity when its displacement is one-half its amplitude.

c) Find its angular acceleration when its displacement is 45°.

11–19 A certain alarm clock ticks four times each second, each tick representing half a period. The balance wheel consists of a thin rim of radius 1.5 cm, connected to the balance staff by thin spokes of negligible mass. The total mass of the balance wheel is 0.8 g.

a) What is the moment of inertia of the balance wheel?

b) What is the torque constant of the hairspring?

Section 11–5 The Simple Pendulum

11–20 A simple pendulum 4 m long swings with an amplitude of 0.2 m.

a) Compute the linear velocity v of the pendulum at its lowest point.

b) Compute its linear acceleration a at the end of its path.

11–21 Find the length of a simple pendulum whose period is exactly 1 s at a point where $g = 9.80$ m·s^{-2}.

11–22 A certain simple pendulum has a period on earth of 2.0 s. What is its period on the surface of the moon, where $g = 1.7$ m·s^{-2}?

Section 11–6 The Physical Pendulum

11–23 A thin, uniform rod of length L and mass m is pivoted about a perpendicular axis through the rod at a distance $L/4$ from one end.

a) Find the moment of inertia about this axis.

b) Find the period of oscillation of the rod.

11–24 A monkey wrench is pivoted at one end and allowed to swing as a physical pendulum. The period is 0.9 s, and the pivot is 0.20 m from the center of gravity.

a) What is the ratio of moment of inertia to mass for the wrench, about an axis through the pivot?

b) If the wrench was initially displaced 0.1 rad from its equilibrium position, what is the angular velocity of the wrench as it passes through the equilibrium position?

Section 11–7 Damped Oscillations

11–25 A mass 0.4 kg is moving on the end of a spring of force constant $k = 300$ N·m^{-1} and is acted on by a damping force $F_x = -bv$.

a) If the constant b has the value 5 kg·s^{-1}, what is the frequency of oscillation of the mass?

b) For what value of the constant b will the motion be critically damped?

11–26 A mass 0.2 kg moving on the end of a spring of force constant $k = 250$ N·m^{-1} has an initial displacement of 0.3 m. There is a damping force $F_x = -bv$ acting on the mass. It is observed that the amplitude of the motion has decreased to 0.1 m in 5 s. Calculate the magnitude of the damping constant b.

PROBLEMS

11–27 The motion of the piston of an automobile engine is approximately simple harmonic.

a) If the stroke of an engine (twice the amplitude) is 0.10 m, and the engine runs at 3600 rev·min^{-1}, compute the acceleration of the piston at the end of its stroke.

b) If the piston has a mass of 0.5 kg, what resultant force must be exerted on it at this point?

c) What is the velocity of the piston, in meters per second, at the midpoint of its stroke?

11–28 Four passengers whose combined mass is 300 kg are observed to compress the springs of an automobile by 5 cm when they enter the automobile. If the total load supported by the springs is 900 kg, find the period of vibration of the loaded automobile.

11–29 A small block is executing simple harmonic motion in a horizontal plane with an amplitude of 0.10 m. At a point 0.06 m away from equilibrium, the velocity is 0.24 m·s^{-1}.

a) What is the period?

b) What is the displacement when the velocity is ±0.12 m·s^{-1}?

c) If a small object placed on the oscillating block is just on the verge of slipping at the endpoint of the path, what is the coefficient of static friction?

11–30 An object of mass 0.010 kg moves with simple harmonic motion of amplitude 0.24 m and period 4 s. The coordinate is +0.24 m when $t = 0$. Compute

a) the position of the object when $t = 0.5$ s,

b) the magnitude and direction of the force acting on the object when $t = 0.5$ s,

c) the minimum time required for the object to move from its initial position to the point where $x = -0.12$ m,

d) the velocity of the object when $x = -0.12$ m.

11–31 A rubber raft bobs up and down, executing simple harmonic motion due to the waves on a lake. The amplitude of the motion is 2.0 ft, and the period is 5.0 s. A stable dock is next to the raft and is at a level equal to the highest level of the raft. People wish to step off the raft onto the dock but can do so comfortably only if the level of the raft is within 1.0 ft of the dock level. How much time do the people have to get off comfortably during each period of the simple harmonic motion?

11–32 A force of 30 N stretches a vertical spring 0.15 m.

a) What mass must be suspended from the spring so that the system will oscillate with a period of $(\pi/4)$ s?

b) If the amplitude of the motion is 0.05 m, where is the object and in what direction is it moving $(\pi/12)$ s after it has passed the equilibrium position, moving downward?

c) What force does the spring exert on the object when it is 0.03 m below the equilibrium position, moving upward?

11–33 An object of mass 0.100 kg hangs from a long spiral spring. When pulled down 0.10 m below its equilibrium position and released, it vibrates with a period of 2 s.

a) What is its velocity as it passes through the equilibrium position?

b) What is its acceleration when it is 0.05 m above the equilibrium position?

c) When it is moving upward, how much time is required for it to move from a point 0.05 m below its equilibrium position to a point 0.05 m above it?

d) How much will the spring shorten if the object is removed?

11–34 A very interesting, though impractical, example of simple harmonic motion occurs when considering the motion of a particle dropped down a hole that extends from one side of the earth, through its center, to the other side. Prove that the motion is simple harmonic and find the period. (*Note.* Use the following property of the gravitational force: The force on an object of mass m due to a spherical mass M of radius R is toward the center of the sphere and has magnitude GmM'/r^2, where r is the distance from the center of M out to the location of point mass m and M' is the mass of that part of M inside a sphere of radius r. Hence if $r > R$ then $M' = M$. But if for example, $r = \frac{1}{2}R$ then $M' = \frac{1}{8}M$ because only $\frac{1}{8}$ the volume and hence $\frac{1}{8}$ the mass of M is inside the sphere of radius r. [This property of the gravitational force follows directly from what was shown in Challenge Problem 6--42.])

11–35 A block of mass 0.50 kg sits on top of a block of mass 5.0 kg. The larger block is attached to a horizontal spring that has a force constant of 15 N·m^{-1}. What is the largest amplitude the 5.0 kg mass can have for the smaller mass to remain at rest relative to the larger block? The coefficient of static friction between the two blocks is 0.20. There is no friction between the larger block and the floor.

11–36 The general equation of simple harmonic motion,

$$x = A \cos (\omega t + \theta_0),$$

can be written in the equivalent form

$$x = B \sin \omega t + C \cos \omega t.$$

a) Find the expressions for the amplitudes B and C in terms of the amplitude A and the initial phase angle θ_0.

b) Interpret these expressions in terms of a phasor diagram.

11–37 To measure g in an unorthodox manner, a student places a ball bearing on the concave side of a lens, as shown in Fig. 11–15. She attaches the lens to a simple harmonic oscillator (actually a small stereo speaker) whose amplitude is A and whose frequency f can be varied. She can measure both A and f with a strobe light.

a) If the ball bearing has a mass m, find the normal force exerted by the lens on the ball bearing as a function of time. Your result should be in terms of A, f, m, g, and a phase angle θ_0.

b) The frequency is slowly increased. When it reaches a value f_b, the ball is heard to bounce. What is g in terms of A and f_b?

FIGURE 11–15

11–38 A block of mass m_1 attached to a horizontal spring of force constant k is moving with simple harmonic motion of amplitude A. At the instant it passes through its equilibrium position, a lump of putty of mass m_2 is dropped vertically onto the block from a very small height and sticks to it.

a) Find the new period and amplitude.

b) Was there a loss of mechanical energy? If so, where did it go? Calculate the ratio between the final and the initial mechanical energy.

c) Would the answers be the same if the putty had been dropped on the block when it was at one end of its path?

11–39 A solid disk of radius $R = 12$ cm oscillates as a physical pendulum about an axis perpendicular to the plane of the disk at a distance r from its center. (See Fig. 11–16.)

a) Calculate the period of oscillation (for small amplitudes) for the following values of r: 0, $R/4$, $R/2$, $3R/4$, R, and $3R/2$. (*Hint:* Use the parallel-axis theorem, Section 9–7.)

b) Let τ_0 represent the period when $r = R$, and τ the period at any other value of r. Construct a graph of the dimensionless ratio τ/τ_0 as a function of the dimensionless ratio r/R. (Note that the graph then describes the behavior of *any* solid disk, whatever its radius.)

c) Find the value of r/R that minimizes the period, and calculate the minimum value of the period.

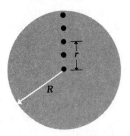

FIGURE 11–16

11–40 Show that $x(t)$ as given in Eq. (11–43) is a solution of Newton's second law for damped oscillations (Eq. 11–42), if ω' is as defined in Eq. (11–44).

11–41 It is desired to construct a pendulum of period 4 s.

a) What is the length of a *simple* pendulum having this period?

b) Suppose the pendulum must be mounted in a case not over 0.50 m high. Can you devise a pendulum, having a period of 4 s, that will satisfy this requirement?

CHALLENGE PROBLEMS

11–42 Two springs with the same unstretched length but different force constants k_1 and k_2 are attached to a block of mass m on a level, frictionless surface. Calculate the effective force constant in each of the three cases (a), (b), and (c) depicted in Fig. 11–17.

d) A body of mass m, suspended from a spring with a force constant k, vibrates with a frequency f_1. When the spring is cut in half and the same body is suspended from one of the halves, the frequency is f_2. What is the ratio f_2/f_1?

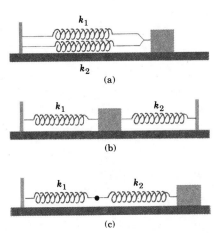

FIGURE 11–17

11–43 Two springs, each of unstretched length 0.2 m, but having different force constants k_1 and k_2, are attached to opposite ends of a block of mass m on a level, frictionless surface. The outer ends of the springs are now attached to two pins P_1 and P_2, 0.10 m from the original positions of the ends of the springs. Let

$$k_1 = 1 \text{ N·m}^{-1}, \qquad k_2 = 3 \text{ N·m}^{-1}, \qquad m = 0.1 \text{ kg}.$$

(See Fig. 11–18.)

a) Find the length of each spring when the block is in its new equilibrium position, after the springs have been attached to the pins.

b) Find the period of vibration of the block if it is slightly displaced from its new equilibrium position and released.

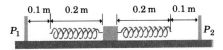

FIGURE 11–18

11–44 All the previous problems in this chapter have assumed that the springs had negligible mass. But of course no spring is completely massless. To find the effect of the spring's mass, consider a spring of mass M, equilibrium length L_0, and spring constant k. When stretched or compressed to a length L, the potential energy is $\frac{1}{2}kx^2$, where $x = L - L_0$.

a) Consider a spring as described above that has one end fixed and the other end moving with speed v. Assume that the speed of points along the length of the spring varies linearly with distance l from the fixed end. Assume also that the mass M of the spring is distributed uniformly along the length of the spring. Calculate the kinetic energy of the spring in terms of M and v. (*Hint:* Divide the spring into pieces of length dl; find the speed of each piece in terms of l, v, and L; find the mass of each piece in terms of dl, M, and L; and integrate from 0 to L. The result is *not* $\frac{1}{2}Mv^2$, since not all of the spring moves with the same speed.)

b) Take the time derivative of the conservation of energy equation, Eq. (11–3), for a mass m moving on the end of a *massless* spring. By comparing your results to Eq. (11–15) that defines ω, show that the angular frequency of oscillation is $\omega = \sqrt{k/m}$.

c) Apply the procedure of part (b) to obtain the angular frequency of oscillation ω of the spring considered in part (a). If the *effective mass M'* of the spring is defined by $\omega = \sqrt{k/M'}$, what is M' in terms of M?

11–45

a) What is the change $\Delta\tau$ in the period of a simple pendulum when the acceleration of gravity g changes by Δg? (*Hint:* The new period $\tau + \Delta\tau$ is obtained by substituting $g + \Delta g$ for g:

$$\tau + \Delta\tau = 2\pi\sqrt{\frac{L}{g + \Delta g}}.$$

To obtain an approximate expression, expand the factor $[g + \Delta g]^{-1/2}$, using the binomial theorem [Appendix B]

and keeping only the first two terms:

$$[g + \Delta g]^{-1/2} = g^{-1/2} - \tfrac{1}{2}g^{-3/2}\,\Delta g + \cdots.$$

The other terms contain higher powers of Δg and are very small if Δg is small.)

b) Find the *fractional* change in period $\Delta\tau/\tau$ in terms of the fractional change $\Delta g/g$.

c) A pendulum clock, which keeps correct time at a point where $g = 9.8000\ \mathrm{m \cdot s^{-2}}$, is found to lose 10 s each day at a higher elevation. Use the result of part (a) or (b) to find approximately the value of g at this new location.

11–46 An experimenter pivots a thin, uniform rod of length l about a perpendicular axis through the rod and measures its period as a physical pendulum. The rod is then inverted, and by trial and error he finds two other points for which the period is the same as for the first point. One is the point symmetrically opposite the first, of course, but he finds a second such point as well.

a) Show that in this case the period depends only on the distance L between the first point and the second nonsymmetric point and is given by $\tau = 2\pi(L/g)^{1/2}$.

b) For this to happen, what is the *least* distance the first point could have been from the center?

11–47 A meterstick hangs from a horizontal axis at one end and oscillates as a physical pendulum. A body of small dimensions, and of mass equal to that of the meterstick, can be clamped to the stick at a distance d below the axis. Let τ represent the period of the system with the body attached, and τ_0 the period of the meterstick alone.

a) Find the ratio τ/τ_0. Evaluate your expression for d ranging from 0 to 1.0 m in steps of 0.1 m, and sketch a graph of τ/τ_0 versus d.

b) Is there any value of d for which $\tau = \tau_0$? If so, find it and explain why the period is unchanged when d has this value.

PART THREE

MECHANICAL AND THERMAL PROPERTIES OF MATTER

PERSPECTIVE

In the last several chapters we introduced the concepts of momentum and energy; we have seen how these concepts and the associated conservation principles can be used to obtain needed information in problems where the forces and motion of a particle or of a system are too complicated to describe in full detail. We have also applied Newton's laws of motion to the analysis of rotational motion of a *rigid body,* an idealized model to describe a body that does not deform in any way when forces are applied. This analysis involved two new dynamic concepts: *torque* and *angular momentum.* The concept of torque is also useful in the analysis of equilibrium of rigid bodies. Angular momentum and its associated conservation principle extend further our ability to analyze systems that include rotating rigid bodies. Finally, we studied a variety of examples of *periodic motion* in mechanical systems. The importance of periodic motion is not by any means limited to macroscopic mechanical vibrations. We will see later that this discussion lays the foundation for understanding molecular vibrations and molecular spectra, heat capacities of molecules and solids, alternating currents in electric circuits, and many other areas of physics that have great practical as well as fundamental significance.

With this background in the fundamentals of mechanics, we are now ready for an extensive discussion of mechanical and thermal properties of matter. Moving beyond the rigid-body model of a solid body, we consider two kinds of generalizations. The first includes *elastic deformations* of solids and fluids that occur when forces are applied; the second concerns specifically the equilibrium and motion of *fluids,* substances having no definite shape. Included in this discussion are such diverse topics as buoyancy, surface tension, energy relations in fluid flow, and turbulent flow. Then we undertake an extensive discussion of *thermal phenomena* and their associated energy relations. The relation of heat to mechanical energy is vital to the operation of such familiar systems as automobile engines, refrigerators, and electric-power plants, so important practical applications of these principles are always close at hand.

We begin by defining *temperature* and *heat,* and then we study quantitative relationships involving transfer of energy in the form of heat from one body to another. We return to describing the behavior of matter, this time in situations where the temperature, volume, and pressure of a substance may all change.

The heart of this section, though, is the study of the first and second laws of *thermodynamics.* The first law expresses quantitatively the relation of heat to mechanical energy, using the concept of *internal energy* of a substance or a system. The second law is a formulation of fundamental restrictions on processes in which conversion of heat to mechanical energy occurs, including the maximum theoretical efficiency of engines and refrigerators. Another new concept, *entropy,* helps us formulate this second principle concisely and shows its relation to the basic one-way character of natural processes, which tend to proceed always toward states of greater randomness or disorder.

Finally, we look at some aspects of the relationship between the *macroscopic* properties of matter and its *microscopic* structure. We find that many properties of gases, such as the relation of pressure, volume, and temperature, and heat capacities, can be derived from a microscopic model. Although the principles of thermodynamics are developed first without reference to a molecular model, the study of their relation to molecular processes gives us added insight. The relationship between macroscopic properties of a material and its microscopic structure permits understanding of mechanical, thermal, and other properties on the basis of molecular structure. Hence we are sometimes able to design materials having specific desired properties. This ability is a vital part of contemporary technology.

12

ELASTICITY

CHAPTERS 9 AND 10 WERE CONCERNED WITH MOTION AND EQUILIBRIUM OF rigid bodies. The rigid body is an idealized *model* to represent a body that has a definite size and shape and that does not stretch, squeeze, or twist when forces act on it. Of course, real materials always do deform to some extent when forces act on them. In this chapter we consider how the forces and deformations are related. The concepts of stress, strain, and elastic modulus can be used to describe the elastic properties of materials in a way that does not depend on the dimensions of a particular specimen. For sufficiently small forces, the deformation of a material is proportional to the force, but with larger forces the behavior is more complex. For sufficiently large forces, materials can deform irreversibly or break, and we can use the concept of stress to characterize the *strength* of a material. The study of the elastic properties of materials is of tremendous practical importance in the design of buildings, automobiles, and many of the necessities of present-day life.

12–1 TENSILE STRESS AND STRAIN

The simplest elastic behavior to understand is the stretching of a bar, rod, or wire when its ends are pulled. Figure 12–1a shows a bar with uniform cross-sectional area A, subjected to equal and opposite forces F pulling at its ends. We say that the bar is in **tension.** Tensions in ropes and strings were discussed in earlier chapters; the concept is the same here. Imagine a cross section through the bar perpendicular to its length, as shown by the broken line. Every portion of the bar is in equilibrium, so the portion to the right of the section must be pulling on the portion to the left with a force F, and vice versa. If the forces at the ends are applied uniformly over the end surfaces, then the forces at every other cross section are also distributed uniformly over the section, as shown by the short arrows in Fig. 12–1b. We define the **stress** at this section as the ratio of the force F to the cross-sectional area A:

$$\text{Stress} = \frac{F}{A}. \tag{12–1}$$

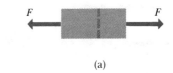

(a)

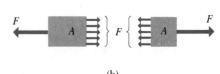

(b)

12–1 (a) A bar in tension. (b) The stress at a perpendicular section equals F/A.

When you pull on something, it stretches: tensile stress and strain.

291

This stress is called a **tensile stress** because each portion exerts tension on the other.

The SI unit of stress is the newton per square meter ($N \cdot m^{-2}$). This unit is also given a special name, the *pascal* (abbreviated Pa):

$$\text{One pascal} = 1\ \text{Pa} = 1\ N \cdot m^{-2}.$$

Other units of stress are the dyne per square centimeter ($dyn \cdot cm^{-2}$) and the pound per square foot ($lb \cdot ft^{-2}$). In the British system the hybrid unit, the pound per square inch ($lb \cdot in^{-2}$, or psi), is commonly used. The units of stress are the same as those of *pressure*, which we will encounter frequently in later chapters. The pascal is a fairly small unit; air pressure in automobile tires is typically of the order of 2×10^5 Pa, and steel cables are commonly used with tensile stresses of the order of 10^8 Pa.

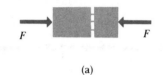

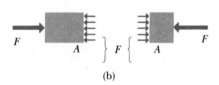

(a)

(b)

12–2 A bar in compression.

EXAMPLE 12–1 A human biceps (upper-arm muscle) may exert a force of the order of 600 N on the bones to which it is attached. If the muscle has a cross-sectional area at its center of 50 cm^2 = 0.005 m^2, and the tendon attaching its lower end to the bones below the elbow joint has a cross section of 0.5 cm^2 = 5×10^{-5} m^2, find the tensile stress in each of these cross sections.

SOLUTION In each case the stress is the force per unit area. For the muscle,

$$\text{Tensile stress} = \frac{600\ \text{N}}{0.005\ \text{m}^2} = 120{,}000\ N \cdot m^{-2} = 1.2 \times 10^5\ \text{Pa}.$$

For the tendon,

$$\text{Tensile stress} = \frac{600\ \text{N}}{5 \times 10^{-5}\ \text{m}^2} = 1.2 \times 10^7\ \text{Pa}.$$

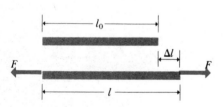

12–3 The longitudinal strain is defined as $\Delta l / l_0$.

When the forces acting on the ends of a bar are pushes rather than pulls, as in Fig. 12–2a, we say that the bar is in **compression.** The stress on the cross section shown by a broken line is now a **compressive stress:** Each portion pushes rather than pulls on the other.

The fractional change of length (stretch) of a body subjected to a tensile stress is called the **tensile strain.** Figure 12–3 shows a bar of natural length l_0 that stretches to a length $l = l_0 + \Delta l$ when equal and opposite forces F are applied to its ends. The elongation Δl does not occur only at the ends; every part of the bar stretches in the same proportion as does the bar as a whole. The tensile strain is defined as the ratio of the elongation Δl to the original length l_0:

$$\text{Tensile strain} = \frac{l - l_0}{l_0} = \frac{\Delta l}{l_0}. \qquad (12\text{–}2)$$

Strain is a ratio of two lengths, and the two lengths are always measured in the same units. Thus strain is a pure (dimensionless) number with no units. The **compressive strain** of a bar in compression is defined in the same way as tensile strain; it is the ratio of the decrease in length to the original length.

TABLE 12-1 Approximate Elastic Constants

Material	Young's Modulus, Y		Shear Modulus, S		Bulk Modulus, B		Poisson's Ratio, σ
	Pa	lb·in^{-2}	Pa	lb·in^{-2}	Pa	lb·in^{-2}	
Aluminum	0.70×10^{11}	10×10^{6}	0.30×10^{11}	3.4×10^{6}	0.70×10^{11}	10×10^{6}	0.16
Brass	0.91×10^{11}	13×10^{6}	0.36×10^{11}	5.1×10^{6}	0.61×10^{11}	8.5×10^{6}	0.26
Copper	1.1×10^{11}	16×10^{6}	0.42×10^{11}	6.0×10^{6}	1.4×10^{11}	20×10^{6}	0.32
Glass	0.55×10^{11}	7.8×10^{6}	0.23×10^{11}	3.3×10^{6}	0.37×10^{11}	5.2×10^{6}	0.19
Iron	1.9×10^{11}	26×10^{6}	0.70×10^{11}	10×10^{6}	1.0×10^{11}	14×10^{6}	0.27
Lead	0.16×10^{11}	2.3×10^{6}	0.056×10^{11}	0.8×10^{6}	0.077×10^{11}	1.1×10^{6}	0.43
Nickel	2.1×10^{11}	30×10^{6}	0.77×10^{11}	11×10^{6}	2.6×10^{11}	34×10^{6}	0.36
Steel	2.0×10^{11}	29×10^{6}	0.84×10^{11}	12×10^{6}	1.6×10^{11}	23×10^{6}	0.19
Tungsten	3.6×10^{11}	51×10^{6}	1.5×10^{11}	21×10^{6}	2.0×10^{11}	29×10^{6}	0.20

The tensile or compressive *strain* clearly depends on the tensile or compressive *stress:* The harder you pull on something, the more it stretches. Robert Hooke (1635–1703), a contemporary of Newton, discovered that when the forces are not too large, this relation can be represented approximately as a direct proportion; stress is proportional to strain, and the ratio of stress to strain is then constant. This proportionality is called **Hooke's law.**

The quotient of any stress and the corresponding strain is called an **elastic modulus.** For the particular case of tensile or compressive stress and strain, it is called **Young's modulus,** denoted by Y:

> An elastic modulus describes a basic property of a material, not just a particular piece of the material.

$$Y = \frac{\text{tensile stress}}{\text{tensile strain}} = \frac{\text{compressive stress}}{\text{compressive strain}},$$

$$Y = \frac{F/A}{\Delta l/l_0} = \frac{l_0}{A}\frac{F}{\Delta l}. \tag{12–3}$$

Since a strain is a pure number, the units of Young's modulus are the same as those of stress, namely, force per unit area. Some typical values are listed in Table 12–1.

When a material stretches under tensile stress, the dimensions *perpendicular* to the direction of stress become *smaller* by an amount proportional to the fractional change in length. When you stretch a wire or a rubber band, it gets a little thinner as well as longer. If w_0 is the original width and Δw is the change in width, then

> When a piece of material is stretched, it gets thinner as well as longer.

$$\frac{\Delta w}{w_0} = -\sigma \frac{\Delta l}{l_0}, \tag{12–4}$$

where σ is a dimensionless constant, different for different materials, called **Poisson's ratio.** For many common materials, σ has a value between 0.1 and 0.3. Similarly, a material under compressive stress "bulges" at the sides, and again the fractional change in width is given by Eq. (12–4).

Experiments have shown that for many materials the ratio of compressive strain to compressive stress is the same as the ratio of tensile strain to tensile stress. Hence Young's modulus describes the behavior of a material in both tension and compression.

EXAMPLE 12–2 In a small elevator, a load of 500 kg hanging from a steel cable of length 3 m and cross section 0.20 cm² was found to stretch the cable 0.4 cm above its no-load length. What were the stress, the strain, and the value of Young's modulus for the steel in the cable?

SOLUTION We use the definitions of stress, strain, and Young's modulus as given by Eqs. (12–1), (12–2), and (12–3):

$$\text{Stress} = \frac{F}{A} = \frac{(500 \text{ kg})(9.8 \text{ m·s}^{-2})}{2.0 \times 10^{-5} \text{ m}^2} = 2.45 \times 10^8 \text{ Pa};$$

$$\text{Strain} = \frac{\Delta l}{l_0} = \frac{0.004 \text{ m}}{3 \text{ m}} = 0.00133;$$

$$Y = \frac{\text{stress}}{\text{strain}} = \frac{2.45 \times 10^8 \text{ Pa}}{0.00133} = 1.84 \times 10^{11} \text{ Pa}.$$

Young's modulus is related to the force constant of a stretched wire or cable.

Young's modulus characterizes the elastic properties of a material under tension or compression in a way that is independent of the size or shape of the particular specimen. It does not indicate directly how a particular rod, cable, or spring made of the material will distort under given forces. We may solve Eq. (12–3) for F, obtaining

$$F = \frac{YA}{l_0} \Delta l.$$

Now suppose we represent the quantity YA/l_0 by a single letter k, which we call the **force constant;** we also call the elongation x instead of Δl. Then we have

$$F = kx.$$

We have already encountered this relationship in Sections 7–3 and 7–6 in connection with work and potential energy for a spring. It appeared again in our discussion of simple harmonic motion in Chapter 11. The elongation of a body in tension is *directly proportional* to the stretching force, and the shortening of a body in compression is directly proportional to the compressing force. Hooke's law was originally stated in this form; it was reformulated much later, by other physicists, in terms of stress and strain.

When a helical or coil spring is stretched or compressed, the stress and strain in the wire are nearly pure *shear* (to be discussed in Section 12–3), but the elongation or compression is still proportional to the stretching or compressing force, within certain limits of maximum force. That is, the equation $F = kx$ is still valid. Of course, a coil spring can be compressed only if there are spaces between the coils!

The units of the force constant are newtons per meter, dynes per centimeter, or pounds per foot. The reciprocal of the force constant—that is, the ratio of elongation or compression to force—is called the *compliance* of the spring.

12–2 BULK STRESS AND STRAIN

We have discussed tensile and compressive stresses and strains. A different stress–strain situation occurs when a solid or fluid material is subjected to a uniform pressure over its whole surface. The resulting deformation can be described in terms of the volume change of the material. A familiar example is the compression of a gas under pressure. The term *fluid* means a substance that can *flow,* so the term applies to both liquids and gases.

A fluid at rest cannot have shear stress.

A force transmitted through a cross section of a fluid at rest must always be perpendicular to that section; if we tried to exert a force parallel to a section, the fluid would slip sideways to counteract the effort. In the language to be introduced in Section 12–3, there can be no *shear stress* in a fluid at rest. Similarly, when a solid is immersed in a fluid and both are at rest, the forces that the fluid exerts on the surface of the solid are always perpendicular to the surface at each point.

Figure 12–4 shows a fluid in a cylinder with a piston. We apply a downward force to the piston, and this is transmitted throughout the fluid. The triangle is a side view of a wedge-shaped portion of the fluid. We will show that the force per unit area (pressure) is the same on all surfaces of this wedge of fluid. If we neglect the weight of the fluid, the only forces on this wedge are those exerted on its imaginary surfaces by the surrounding fluid, and each force must be perpendicular to the corresponding surface. Let F_x, F_y, and F represent the forces on the three faces, as shown. Since the fluid is in equilibrium, it follows that

$$F \sin \theta = F_x, \qquad F \cos \theta = F_y.$$

Also,

$$A \sin \theta = A_x, \qquad A \cos \theta = A_y.$$

Dividing the upper equations by the lower, we find

$$\frac{F}{A} = \frac{F_x}{A_x} = \frac{F_y}{A_y}.$$

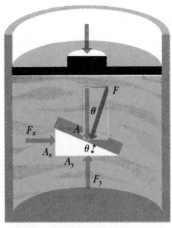

12–4 A fluid under hydrostatic pressure. The force on a surface in any direction is normal to the surface.

Hence the force per unit area is the *same* on all these surfaces. It does not depend on their orientation, and it is always a compression. The force per unit area on any of these surfaces is called the **pressure** p in the fluid:

Pressure is force per unit area.

$$p = \frac{F}{A}, \qquad F = pA. \tag{12–5}$$

The fact that pressure applied to the surface of a fluid is transmitted unchanged to all parts of the fluid is called Pascal's law.

Pressure has the same units as stress. Commonly used units include 1 Pa (= 1 N·m^{-2}), 1 dyn·cm^{-2}, 1 lb·ft^{-2}, and 1 lb·in^{-2}. Also in common use is the *atmosphere,* abbreviated atm. One atmosphere is defined to be the average pressure of the earth's atmosphere at sea level:

One atmosphere = 1 atm = 1.013×10^5 Pa = 14.7 lb·in^{-2}.

Pressure is a scalar quantity, not a vector quantity; no direction can be assigned to it. The force against any area within (or at the boundary surface of) a fluid at rest and under pressure is perpendicular to the area, regardless of the orientation of the area. This is what is meant by the statement that the pressure in a fluid is the same in all directions.

The stress within a solid material is a pressure, if the force per unit area is the same at *all* points of the surface, and if the force at each point is normal to the surface and directed inward. This is *not* the case in Fig. 12–2, where forces

If weight can be neglected, the pressure in a fluid at rest is the same everywhere in the fluid.

are applied at the ends of the bar only, but it is automatically the case if a solid is immersed in a fluid under pressure.

Pressure is the *stress* in a volume deformation. The corresponding strain is **volume strain;** it is defined as the fractional change in volume, the ratio of the volume change ΔV to the original volume V_0:

$$\text{Volume strain} = \frac{\Delta V}{V_0}. \qquad (12\text{–}6)$$

Volume strain is change in volume, compared to original volume.

Like tensile or compressive strain, volume strain is a pure number, without units.

Experimental evidence shows that for sufficiently small pressures, the volume strain is proportional to the stress (pressure). The corresponding elastic modulus (ratio of stress to strain) is called the **bulk modulus,** denoted by B. The general definition of the bulk modulus is the (negative) ratio of an infinitesimal pressure change dp to the volume strain dV/V_0 (fractional change in volume) that it causes:

$$B = -\frac{dp}{dV/V_0} = -V_0\frac{dp}{dV}. \qquad (12\text{–}7)$$

The minus sign is included in the definition of B because an *increase* in pressure always causes a *decrease* in volume. That is, if dp is positive, dV is negative. By including a minus sign in its definition, we make B itself a positive quantity.

Bulk modulus might be called incompressibility: The larger the bulk modulus, the less compressible the material is.

If the pressure change is not too great, the ratio dp/dV is constant, the bulk modulus is constant, and we can replace dp and dV in Eq. (12–7) by finite changes Δp and ΔV in pressure and volume. The bulk modulus of a *gas,* however, changes markedly with pressure, and the general definition of B must be used for gases.

The reciprocal of the bulk modulus is called the **compressibility** k. From Eq. (12–7),

$$k = \frac{1}{B} = -\frac{dV/V_0}{dp} = -\frac{1}{V_0}\frac{dV}{dp}. \qquad (12\text{–}8)$$

The compressibility of a material thus equals the *fractional decrease in volume,* $-dV/V_0$, *for a small increase dp in pressure.*

Table 12–1 includes values of the bulk modulus. Its units are the same as those of pressure, and the units of compressibility are those of *reciprocal pressure.* The compressibility of water, from Table 12–2, is 46.4×10^{-6} atm^{-1}. This means that the volume decreases by 46.4-millionths of the original volume for each atmosphere increase in pressure.

TABLE 12–2   Compressibilities of Liquids

Liquid	Compressibility, k		
	Pa^{-1}	(lb·in^{-2})$^{-1}$	atm^{-1}
Carbon disulfide	93×10^{-11}	64×10^{-7}	94×10^{-6}
Ethyl alcohol	110×10^{-11}	76×10^{-7}	111×10^{-6}
Glycerine	21×10^{-11}	14×10^{-7}	21×10^{-6}
Mercury	3.7×10^{-11}	2.6×10^{-7}	3.8×10^{-6}
Water	45.8×10^{-11}	31.6×10^{-7}	46.4×10^{-6}

EXAMPLE 12-3 The volume of oil contained in a certain hydraulic press is $0.2 \text{ m}^3 = 200$ liters. Find the decrease in volume of the oil when subjected to a pressure increase of 2.04×10^7 Pa. The compressibility of the oil is $20 \times 10^{-6} \text{ atm}^{-1}$.

SOLUTION Because we are given the compressibility in atm^{-1}, we first convert the pressure to atmospheres:

$$2.04 \times 10^7 \text{ Pa} = 201 \text{ atm}.$$

From Eq. (12-8),

$$\Delta V = -kV\,\Delta p = -(20 \times 10^{-6} \text{ atm}^{-1})(0.2 \text{ m}^3)(201 \text{ atm})$$
$$= -8.04 \times 10^{-4} \text{ m}^3 = -0.804 \text{ L}.$$

This represents a substantial compression of the oil under the action of a very large pressure, nearly 3000 pounds per square inch.

Oil in a hydraulic press is often under great pressure, and the compression can be quite appreciable.

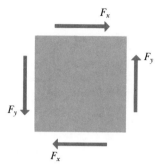

12-5 A body under shear stress.

12-3 SHEAR STRESS AND STRAIN

A third kind of stress is shown in Fig. 12-5. This stress is called **shear stress,** and we define it as the force tangent to a material surface, divided by the area on which the force acts. Shear stress, like the other two types of stress, is a force per unit area.

A shear deformation is shown in Fig. 12-6. The lightly shaded outline *abcd* represents an unstressed block of material, and the lines *a'b'c'd'* represent the block under stress. The centers of the stressed and unstressed block coincide in part (a). In part (b) the deformation is the same as in (a), but the edges *ad* and *a'd'* coincide. Under shear stress the lengths of the faces remain very nearly constant; all dimensions parallel to the diagonal *ac* increase in length, and those parallel to the diagonal *bd* decrease in length. This type of strain is called a **shear strain;** it is defined as the ratio of the displacement *x* of corner *b* to the transverse dimension *h*:

$$\text{Shear strain} = \frac{x}{h} = \tan\phi. \qquad (12\text{-}9)$$

In practice, x is nearly always much smaller than h, $\tan\phi$ is very nearly equal to ϕ, and the strain is simply the angle ϕ (measured in radians, of course). Like all strains, shear strain is a pure number with no units because it is a ratio of two lengths.

Once again, we find experimentally that if the forces are not too large, the shear strain is proportional to the shear stress. The corresponding elastic modulus (ratio of shear stress to shear strain) is called the **shear modulus,** denoted by S:

$$S = \frac{\text{shear stress}}{\text{shear strain}}$$
$$= \frac{F/A}{x/h} = \frac{h}{A}\frac{F}{x} = \frac{F/A}{\phi}, \qquad (12\text{-}10)$$

with x and h defined as in Fig. 12-6.

Shear stress results when forces are applied tangent to the surfaces of a solid material.

Shear strain is distortion of shapes and angles without volume change.

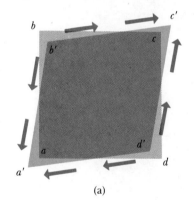

(a)

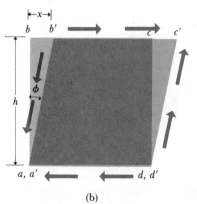

(b)

12-6 Change in shape of a block in shear. The shear strain is defined as x/h.

For most materials, the shear modulus is one-third to one-half as large as Young's modulus. The shear modulus is also called the *modulus of rigidity,* or the *torsion modulus.* Representative values of shear modulus are given in Table 12–1. The shear modulus has significance only for *solid* materials. A liquid or gas flows freely under the action of a shear stress, and a fluid at rest cannot sustain such a stress.

EXAMPLE 12–4   Suppose the object in Fig. 12–6 is a brass plate 1.0 m square and 0.5 cm thick. How large a force F must be exerted on each of its edges if the displacement x in Fig. 12–6b is 0.02 cm? The shear modulus of brass is 0.36×10^{11} Pa.

SOLUTION   The shear stress on each edge is

$$\text{Shear stress} = \frac{F}{A} = \frac{F}{(1.0 \text{ m})(0.005\text{m})} = (200 \text{ m}^{-2})F.$$

The shear strain is

$$\text{Shear strain} = \frac{x}{h} = \frac{2 \times 10^{-4} \text{ m}}{1.0 \text{ m}} = 2.0 \times 10^{-4}.$$

$$\text{Shear modulus } S = \frac{\text{stress}}{\text{strain}} = 0.36 \times 10^{11} \text{ Pa} = \frac{(200 \text{ m}^{-2})F}{2.0 \times 10^{-4}},$$

$$F = 3.6 \times 10^4 \text{ N.}$$

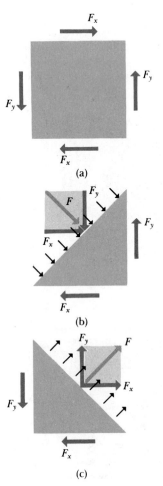

12–7 (a) A body in shear. The stress on one diagonal, part (b), is a pure compression; that on the other, part (c), is a pure tension.

Although the three elastic moduli and Poisson's ratio have been discussed separately, these four quantities are not completely independent. For materials having no distinction between various directions (i.e., *isotropic* materials), only two of these are really independent. For example, the bulk and shear moduli may be expressed in terms of Young's modulus and Poisson's ratio:

$$B = \frac{Y}{3(1 - 2\sigma)}, \qquad S = \frac{Y}{2(1 + \sigma)}. \tag{12–11}$$

For materials having directional properties, such as wood (which has a grain direction) or a single crystal of a material, these relations do not hold, and the elastic behavior is more complex.

Another aspect of the relations among the elastic constants is that different cross sections in a body have different states of stress. As an example, consider the body under shear stress in Fig. 12–7, acted on by the pairs of forces F_x and F_y distributed over its surfaces. The block is in equilibrium, and every portion of it must also be in equilibrium. Thus if we consider the triangular section shown in Fig. 12–7b, the distributed forces over the diagonal face must have a resultant F whose components are equal in magnitude to F_x and F_y. Thus the stress at the diagonal face is a pure *compression,* even though the stresses at the right and bottom faces are shear stresses. Similarly, the diagonal face shown in Fig. 12–7c is in pure *tension.*

The same thing happens with the stretched bar we used in Section 12–1 to introduce tensile stress. A cross section perpendicular to the length of the bar

TABLE 12–3   Stresses and Strains

Type of Stress	Stress	Strains	Elastic Modulus	Name of Modulus
Tension or compression	$\dfrac{F_\perp}{A}$	$\dfrac{\Delta l}{l_0}$	$Y = \dfrac{F_\perp/A}{\Delta l/l_0}$	Young's modulus
Hydrostatic pressure	$p\left(= \dfrac{F_\perp}{A}\right)$	$\dfrac{\Delta V}{V_0}$	$B = -\dfrac{p}{\Delta V/V_0}$	Bulk modulus
Shear	$\dfrac{F_\parallel}{A}$	$\tan\phi \approx \phi$	$S = \dfrac{F_\parallel/A}{\phi}$	Shear modulus

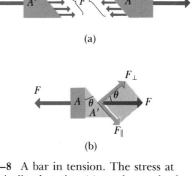

12–8 A bar in tension. The stress at an inclined section (a) can be resolved into (b) a normal stress $F_\perp/A'$, and a tangential or shear stress $F_\parallel/A'$.

has a purely tensile stress, as shown in Fig. 12–1. But if we take a cross section at an angle, as in Fig. 12–8a, the stress at this face can be represented as having both tensile and shear components, as shown in Fig. 12–8b. The force acting at a particular cross section has a definite direction and magnitude and can be represented by means of its components, but these depend also on the orientation of the section. To describe completely the state of stress in a material, we must describe three mutually perpendicular cross-sectional orientations, perhaps using three unit vectors, and then describe the three components of force (per unit area) at each cross section. The resulting set of nine numbers is called the *stress tensor* and is an example of a class of physical quantities called *tensors*.

The various types of stress, strain, and elastic moduli are summarized in Table 12–3.

12–4 ELASTICITY AND PLASTICITY

We are now ready to examine the limitations of Hooke's law. Suppose we plot a graph of stress as a function of the corresponding strain. If Hooke's law is obeyed, stress is directly proportional to strain and the graph is a straight line. Real materials show several types of departures from this idealized behavior.

Figure 12–9 shows a typical stress–strain graph for a metal such as copper or soft iron. The stress in this case is a simple tensile stress, and the strain is shown as the percent elongation. The first portion of the curve, up to a strain of less than 1%, is a straight line, indicating Hooke's-law behavior with stress directly proportional to strain. This straight-line portion ends at point *a*; the stress at this point is called the **proportional limit.**

From *a* to *b*, stress and strain are no longer proportional, but if the load is removed at any point between *O* and *b*, the curve is retraced and the material returns to its original length. In the entire region *Ob* the material is said to be *elastic* or to show *elastic behavior*. Point *b*, the end of this region, is called the **yield point,** and the corresponding stress is called the **elastic limit.** Up to this point the forces exerted by the material are *conservative*. When the load is removed, the material returns to its original shape, and the energy put into the material in causing the deformation is recovered. The deformation is said to be *reversible*.

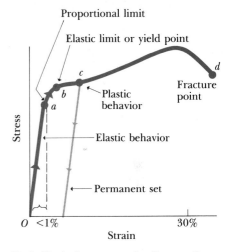

12–9 Typical stress–strain diagram for a ductile metal under tension.

When the stress exceeds the proportional limit, stress and strain are no longer proportional.

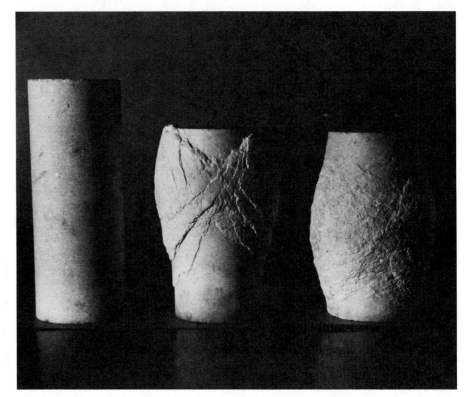

A marble cylinder is deformed by compressive stresses applied at its ends, with different confining pressures on the side surface. (a) The original undeformed shape; (b) 20% compressive strain with a confining pressure of 270 atm; (c) 20% strain with a confining pressure of 445 atm. The sample is brittle at the lower confining pressure but becomes ductile at higher confining pressure. (Photo by M. S. Paterson, Australian National University.)

When the stress exceeds the elastic limit, the material does not snap back all the way when the stress is removed.

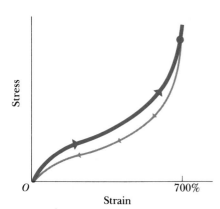

12–10 Typical stress–strain diagram for vulcanized rubber, showing elastic hysteresis.

If the stress is increased further, the strain increases rapidly, but when the load is removed at some point beyond *b*, say *c*, the material does not come back to its original length but traverses the thin line in Fig. 12–9. The length at zero stress is now *greater* than the original length, and the material is said to have a *permanent set*. Further increase of load beyond *c* produces a large increase in strain (even if the stress decreases) until a point *d* is reached at which *fracture* takes place. From *b* to *d*, the material is said to undergo *plastic flow*, or *plastic deformation*. A plastic deformation is *irreversible;* when the stress is removed, the material does not return to its original state. If a large amount of plastic deformation takes place between the elastic limit and the fracture point, the metal is said to be *ductile;* but if fracture occurs soon after the elastic limit is passed, the metal is said to be *brittle*. A soft iron wire that can have considerable permanent stretch without breaking is ductile, while a spring-steel wire that breaks soon after its elastic limit is reached is brittle.

Figure 12–10 shows a stress–strain curve for a typical sample of vulcanized rubber that has been stretched to over seven times its original length. During *no* portion of this curve is the stress proportional to the strain! The substance, however, is elastic, in the sense that when the load is removed, the rubber returns to its original length. On decreasing the load, the stress–strain curve is *not* retraced but follows the thin curve of Fig. 12–10.

The lack of coincidence of the curves for increasing and decreasing stress is known as *elastic hysteresis*. When the stress–strain relation has this behavior, the associated forces are *not* conservative, since the work done by the material in returning to its original shape is *less* than the work required to deform it. It can be shown that the area bounded by the two curves—that is, the area of the *hysteresis loop*—is proportional to the energy dissipated within the elastic material.

Some types of rubber have large elastic hysteresis, and these materials are very useful as vibration absorbers. If a block of such material is placed between a piece of vibrating machinery and the floor, elastic hysteresis takes place during each cycle of vibration. Mechanical energy is converted to a form known as internal energy, causing a rise in temperature of the material. As a result, only a small amount of energy of vibration is transmitted to the floor.

The stress required to cause actual fracture of a material is called the **breaking stress,** or the *ultimate strength.* Two materials, such as two steels, may have very similar elastic constants but vastly different breaking stresses. Table 12–4 gives a few typical values of breaking stress for several materials in tension.

When the stress exceeds the breaking stress, the material cracks or breaks.

TABLE 12–4 Breaking Stresses of Materials

Material	Breaking Stress (Pa or $N \cdot m^{-2}$)
Aluminum	2.2×10^8
Brass	4.7×10^8
Glass	10×10^8
Iron	3.0×10^8
Phosphor bronze	5.6×10^8
Steel	11.0×10^8

SUMMARY

Tensile stress is tensile force per unit area, F/A. Tensile strain is fractional change in length, $\Delta l/l_0$. Young's modulus Y is the ratio of tensile stress to tensile strain:

$$Y = \frac{F/A}{\Delta l/l_0}. \qquad (12\text{–}3)$$

Compressive stress and strain are defined the same way as tensile stress and strain; for many materials, Young's modulus has the same value for both tensile and compressive stresses and strains. Poisson's ratio σ is the negative of the ratio of the fractional change of width $\Delta w/w_0$ of a bar under tension or compression and the fractional change of length $\Delta l/l_0$:

$$\frac{\Delta w}{w_0} = -\sigma \frac{\Delta l}{l_0}. \qquad (12\text{–}4)$$

The force constant k is the ratio of force F to elongation or compression x for a specific specimen of material or a specific spring: $F = kx$. The reciprocal of the force constant is called compliance.

The bulk modulus B is the negative of the ratio of pressure change dp (bulk stress) to fractional volume change dV/V_0:

$$B = -\frac{dp}{dV/V_0}. \qquad (12\text{–}7)$$

Compressibility k is the reciprocal of bulk modulus: $k = 1/B$.

Shear stress is force per unit area F/A for a force applied parallel to a surface. Shear strain is the angle ϕ shown in Fig. 12–6. The shear modulus S is the ratio of shear stress to shear strain:

$$S = \frac{F/A}{\phi}. \qquad (12\text{–}10)$$

The proportional limit is the maximum stress for which stress and strain are proportional, that is, for which Hooke's law is valid. The elastic limit is the stress beyond which irreversible deformation occurs. The breaking stress, or ultimate strength, is the stress at which the material breaks.

KEY TERMS
tension
stress
tensile stress
tensile strain
compression
compressive stress
compressive strain
Hooke's law
elastic modulus
Young's modulus
Poisson's ratio
force constant
pressure
volume strain
bulk modulus
compressibility
shear stress
shear strain
shear modulus
proportional limit
yield point
elastic limit
breaking stress

QUESTIONS

12–1 When a wire is bent back and forth, it becomes hot. Why?

12–2 When a wire is stretched, the cross section decreases somewhat from the undeformed value. How does this affect the definition of tensile stress? Should the original or the decreased value be used?

12–3 Is the work required to stretch a metal rod proportional to the amount of stretch? Explain.

12–4 Why is concrete with steel reinforcing rods embedded in it stronger than plain concrete?

12–5 Is the shear modulus of steel greater or less than that of jelly? By roughly what factor?

12–6 Is the bulk modulus for steel greater or less than that of air? By roughly what factor?

12–7 Looking at the molecular structure of matter, discuss why gases are generally more compressible than liquids and solids.

12–8 Climbing ropes used by mountaineers are usually made of nylon. Would a steel cable of equal strength be just as good? What advantages and disadvantages would it have, compared to nylon?

12–9 How could you measure the force constants of the springs in an automobile?

12–10 In a nylon mountaineering rope, is a lot of mechanical hysteresis desirable or undesirable?

12–11 A spring scale for measuring weight has a spring and a scale that indicates how much the spring stretches under a given weight. Such scales are often illegal in commerce. (Cf. the inscription "no springs, honest weight" found on some grocery-store scales.) Why? (What property of a metal spring could affect its accuracy?)

12–12 Coil springs found in automobile suspension systems are sometimes designed *not* to obey Hooke's law. How can a spring be made so as to achieve this result? Why is it desirable?

12–13 Compare the mechanical properties of a steel cable, made by twisting many thin wires together, with those of a solid steel wire of the same diameter. What advantages does each have?

12–14 Electric power lines are sometimes made by using wires with steel core and copper jacket, or strands of copper and steel twisted together. Why?

12–15 A spring is compressed, clamped in its compressed position, and then dissolved in acid. What becomes of the elastic potential energy?

12–16 When rubber mounting blocks are used to absorb machine vibrations through mechanical hysteresis, as discussed in Section 12–4, what becomes of the energy associated with the vibrations?

EXERCISES

Section 12–1 Tensile Stress and Strain

12–1 A certain metal rod 4 m long and 0.5 cm^2 in cross section is found to stretch 0.2 cm under a tension of 12,000 N. What is Young's modulus for this metal?

12–2 A nylon rope used by mountaineers elongates 1.5 m under the weight of an 80-kg climber.

a) If the rope is 50 m in length and 9 mm in diameter, what is Young's modulus for this material?

b) If Poisson's ratio for nylon is 0.2, find the change in diameter under this stress.

12–3 A steel bar 0.2 cm square and 5 m long is stretched with a force of 400 N at each end. Find the stress, the strain, the total elongation, and the fractional change in thickness of the bar. Use the data for steel given in Table 12–1.

12–4 Two round rods, one of steel, the other of brass, are joined end to end. Each rod is 0.5 m long and 2 cm in diameter. The combination is subjected to tensile forces of 5000 N.

a) What is the strain in each rod?

b) What is the elongation of each rod?

c) What is the change in diameter of each rod?

Use the data for steel given in Table 12–1.

12–5 A 5-kg mass hangs on a vertical steel wire 0.5 m long and 0.004 cm^2 in cross section. Hanging from the bottom of this mass is a similar steel wire that supports a 10-kg mass. Compute the

a) longitudinal strain   b) elongation

of each wire. Use the data for steel given in Table 12–1.

12–6 A copper wire of length 2 m has a diameter of 3 mm. What is the force constant k for this wire?

12–7 A circular steel wire 2 m long is to stretch no more than 0.2 cm when a tensile force of 300 N is applied to each end of the wire. What minimum diameter of wire is required?

12–8 A mass of 100 kg suspended from a wire whose unstretched length l_0 is 4 m is found to stretch the wire by 0.004 m. The cross-sectional area of the wire, which can be assumed constant, is 0.1 cm^2.

a) If the load is pulled down a small additional distance and released, find the frequency at which it will vibrate.

b) Compute Young's modulus for the wire.

Section 12–2 Bulk Stress and Strain

12–9 A specimen of oil having an initial volume of 1000 cm^3 is subjected to a pressure increase of 12×10^5 Pa, and the volume is found to decrease by 0.3 cm^3. What is the bulk modulus of the material? The compressibility?

12–10 In the Challenger Deep of the Marianas Trench, the depth of sea water is 10.9 km, and the pressure is 1.10×10^8 Pa (about 1.09×10^3 atm).

a) If a cubic meter of water is taken from the surface to this depth, what is the change in its volume? (Normal atmospheric pressure is about 1.0×10^5 Pa.)

b) What is the density of sea water at this depth? (At the surface, sea water has a density of 1.03×10^3 kg·m^{-3}.)

Section 12–3 Shear Stress and Strain

12–11 Two strips of metal are riveted together at their ends by four rivets, each of diameter 0.5 cm. What is the maximum tension that can be exerted by the riveted strip if the shearing stress on each rivet is not to exceed 6×10^8 Pa? Assume each rivet to carry one-quarter of the load.

12–12 In Fig. 12–5, suppose the object is a square steel plate, 10 cm on a side and 1 cm thick. Find the magnitude of force required on each of the four sides to cause a shear strain of 0.01.

Section 12–4 Elasticity and Plasticity

12–13 The elastic limit of a steel elevator cable is 2.75×10^8 N·m^{-2}. Find the maximum upward acceleration that can be given a 900-kg elevator when supported by a cable whose cross section is 3 cm^2, if the stress is not to exceed $\frac{1}{4}$ of the elastic limit.

12–14 A steel wire has the following properties:

Length = 5 m

Cross section = 0.05 cm^2

Young's modulus = 1.8×10^{11} Pa

Shear modulus = 0.6×10^{11} Pa

Proportional limit = 3.6×10^8 Pa

Breaking stress = 7.2×10^8 Pa

The wire is fastened at its upper end and hangs vertically.

a) How great a load can be supported without exceeding the proportional limit?

b) How much will the wire stretch under this load?

c) What is the maximum load that can be supported?

PROBLEMS

12–15 A 15-kg mass, fastened to the end of a steel wire of unstretched length 0.5 m, is whirled in a vertical circle with an angular velocity of 2 rev·s^{-1} at the bottom of the circle. The cross section of the wire is 0.02 cm^2. Calculate the elongation of the wire when the weight is at the lowest point of the path.

12–16 A copper wire 4 m long and 1.0 mm in diameter was given the test below. A load of 20 N was originally hung from the wire to keep it taut. The position of the lower end of the wire was read on a scale:

Added load, N	Scale reading, cm
0	3.02
10	3.07
20	3.12
30	3.17
40	3.22
50	3.27
60	3.32
70	4.27

a) Make a graph of these values, plotting the increase in length horizontally and the added load vertically.

b) Calculate the value of Young's modulus.

c) What was the stress at the proportional limit?

12–17 A rod 1.05 m long, whose weight is negligible, is supported at its ends by wires A and B of equal length, as shown in Fig. 12–11. The cross section of A is 1 mm^2; that

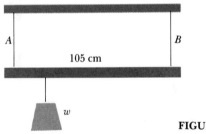

FIGURE 12–11

of B, 2 mm^2. Young's modulus for wire A is 2.4×10^{11} Pa; and for B, 1.6×10^{11} Pa. At what point along the bar should a weight w be suspended in order to produce

a) equal stresses in A and B?

b) equal strains in A and B?

12–18 One area where compressive strength is of everyday importance is in our bones. Young's modulus for bone is about 1.0×10^{10} N·m^{-2}. Bone can take only about a 1.0% change in its length before fracturing.

a) What is the maximum force that can be applied to a bone whose minimum cross-sectional area is 3.0 cm^2, which is approximately the cross-sectional area of a tibia at its narrowest point in the human leg?

b) Estimate from what maximum height a 70-kg person (one weighing about 150 lb) could jump and not fracture the tibia. Take the time between when the person first touches the floor and when he has stopped to be 0.02 s.

12–19 A student has a sling-shot made of a rubber band whose unstretched shape is shown by the shorter band in Fig. 12–12. The student pulls back on the band so it is then in the longer position. To do this the student pulls straight back with a force of 20.0 N. What is Young's modulus for the rubber bands if their cross-sectional area is 10.0 mm²?

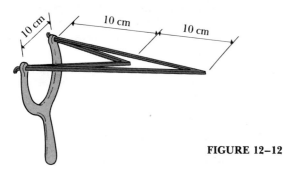

FIGURE 12–12

12–20 An amusement park ride consists of seats attached to cables as shown in Fig. 12–13. Each steel cable has a length of 20.0 m and cross-sectional area of 7.00 cm².

a) What is the amount the cable is stretched when the ride is at rest? Assume the seats plus two people seated in them have a total weight of 4000 N.

b) The ride, when turned on, has a maximum angular velocity of 0.90 rad·s⁻¹. How much will the cable then be stretched?

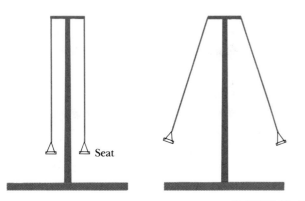

Seat

FIGURE 12–13

12–21 A copper rod of length 2 m and cross-sectional area 2.0 cm² is fastened end to end to a steel rod of length L and cross-sectional area 1.0 cm². The compound rod is subjected to equal and opposite pulls of magnitude 3×10^4 N at its ends.

a) Find the length L of the steel rod if the elongations of the two rods are equal.

b) What is the stress in each rod?

c) What is the strain in each rod?

12–22 A moonshiner produces pure ethanol (ethyl alcohol), late at night, and stores it in a stainless steel tank in the form of a cylinder 0.20 m in diameter with a tight-fitting piston at the top. The total volume of the tank is 200 L (0.2 m³). In an attempt to squeeze a little more into the tank, he piles lead bricks on the piston, so the total mass of bricks and piston is 120 kg. What additional volume of ethanol can he squeeze into the tank?

12–23 A bar of cross section A is subjected to equal and opposite tensile forces F at its ends. Consider a plane through the bar making an angle θ with a plane at right angles to the bar (Fig. 12–14).

a) What is the tensile (normal) stress at this plane, in terms of F, A, and θ?

b) What is the shear (tangential) stress at the plane, in terms of F, A, and θ?

c) For what value of θ is the tensile stress a maximum?

d) For what value of θ is the shear stress a maximum?

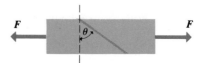

FIGURE 12–14

12–24 In Fig. 12–5, suppose the object is a square aluminum plate 0.2 m on a side. When forces are applied to the four edges, of equal magnitude 1.0×10^6 N each, we want the resulting shear strain to be no greater than 0.01. What minimum thickness of plate is required?

CHALLENGE PROBLEMS

12–25 The compressibility of sodium is to be measured by observing the displacement of the piston in Fig. 12–4 when a force is applied. The sodium is immersed in an oil that fills the cylinder below the piston. Assume that the piston and walls of the cylinder are perfectly rigid and that there is no friction and no oil leak. Compute the compressibility of the sodium in terms of the applied force F, the piston displacement x, the piston area A, the initial volume of oil V_0, the initial volume of sodium v_0, and the compressibility of oil k_0.

12–26 The equation of state (the equation relating pressure, volume, and temperature) for an ideal gas is $pV = nRT$, where n and R are constants.

a) Show that if the gas is compressed while the temperature T is held constant, the bulk modulus is equal to the pressure.

b) When an ideal gas is compressed without the transfer of any heat in or out of it, the pressure and volume are related by $pV^\gamma = $ constant, where γ is a constant having different values for different gases. Show that in this case the bulk modulus is given by $B = \gamma p$.

12–27 A 5-kg mass is hung from a vertical steel wire 2 m long and 0.004 cm^2 in cross section. The wire is securely fastened to the ceiling. Calculate

a) the amount the wire is stretched by the hanging mass;

b) the external force P needed to pull the mass very slowly downward 0.05 cm from its equilibrium position;

c) the work done by gravity when the mass moves downward 0.05 cm;

d) the work done by the force P;

e) the work done by the force the wire exerts on the mass;

f) the change in the elastic potential energy (the potential energy associated with the tensile stress in the wire) when the mass moves downward 0.05 cm.

13
FLUID MECHANICS

A fluid has no definite shape.

A fluid at rest is an equilibrium situation.

IN THIS CHAPTER WE RESUME OUR STUDY OF THE MECHANICAL PROPERTIES OF real materials. Chapter 12 involved the *elastic* properties of materials, and now we study the mechanical properties of fluids. The behavior of fluids is of great current interest both in pure science and in industry. A *fluid* is any substance that can flow; we use the term for both liquids and gases. In most circumstances a gas is easily compressed, while liquids are nearly incompressible. In this chapter we neglect the small volume changes that occur in a liquid under pressure.

Fluid *statics* is the study of fluids at rest, and fluid *dynamics* is the study of fluids in motion. We begin with equilibrium situations, including the concepts of density, pressure, buoyancy, and surface tension. The conditions for equilibrium are based on Newton's first law. Fluid dynamics is much more complex—indeed, it is one of the most complex branches of mechanics. Fortunately, many important situations can be represented by idealized models that are simple enough to permit detailed analysis. Even so, we will barely scratch the surface of this broad and interesting topic.

13–1 DENSITY

The **density** of a material is defined as its mass per unit volume. A homogeneous material has the same density throughout. The SI unit of density is one kilogram per cubic meter (1 $kg \cdot m^{-3}$). The cgs unit, one gram per cubic centimeter (1 $g \cdot cm^{-3}$), is also widely used. We use the Greek letter ρ (rho) for density. If a mass m of material has volume V, the density ρ is given by

$$\rho = \frac{m}{V}, \qquad m = \rho V. \qquad (13-1)$$

Densities of several common solids and liquids at ordinary temperatures are listed in Table 13–1. The conversion factor

$$1 \, g \cdot cm^{-3} = 1000 \, kg \cdot m^{-3}$$

is useful with this table. The densest material found on earth is the metal

TABLE 13–1 Densities

Material	Density $g \cdot cm^{-3}$
Aluminum	2.7
Brass	8.6
Copper	8.9
Gold	19.3
Ice	0.92
Iron	7.8
Lead	11.3
Platinum	21.4
Silver	10.5
Steel	7.8
Mercury	13.6
Ethyl alcohol	0.81
Benzene	0.90
Glycerin	1.26
Water	1.00
Sea water	1.03

osmium (22.5 g·cm^{-3}). The density of air is about 0.0012 g·cm^{-3}, but the density of white-dwarf stars is of the order of 10^6 g·cm^{-3}, and neutron stars 10^{15} g·cm^{-3}!

The **specific gravity** of a material is the ratio of its density to that of water; it is a pure (unitless) number. "Specific gravity" is a poor term, since it has nothing to do with gravity; "relative density" would be preferable.

Density measurements are an important analytical technique. For example, we can determine the charge condition of a storage battery by measuring the density of its electrolyte, a sulfuric acid solution. As the battery discharges, the sulfuric acid (H_2SO_4) combines with lead in the battery plates to form insoluble lead sulfate ($PbSO_4$), decreasing the concentration of the solution. The density decreases from about 1.30 g·cm^{-3} for a fully charged battery to 1.15 g·cm^{-3} for a discharged battery. Similarly, permanent-type antifreeze is usually a solution of ethylene glycol (density 1.12 g·cm^{-3}) in water, with small quantities of additives to retard corrosion. The glycol concentration, which determines the freezing point of the solution, can be found from a simple density measurement. Both these measurements are performed routinely in service stations with the aid of a hydrometer, which measures density by observation of the level at which a calibrated body floats in a sample of the solution. The hydrometer is discussed in Section 13–3.

13–2 PRESSURE IN A FLUID

When we introduced the concept of fluid pressure in Section 12–2, we neglected the *weight* of the fluid and assumed that the pressure was the same everywhere in the fluid. This is not really true, of course. Atmospheric pressure is greater at sea level than on high mountains, and the pressure of water in a lake or in the ocean increases with increasing depth below the surface. Thus we need to refine our concept of pressure. We define the **pressure** p at a point in a fluid as the ratio of the normal force dF on a small area dA around that point, to the area:

$$p = \frac{dF}{dA}, \qquad dF = p \, dA. \tag{13–2}$$

If the pressure is the same at all points of a finite plane surface of area A, these equations reduce to Eq. (12–5):

$$p = \frac{F}{A}, \qquad F = pA.$$

We can now derive a general relation between the pressure p at any point in a fluid in a gravitational field and the elevation of the point y. If the fluid is in equilibrium, every volume element is in equilibrium. Consider an element in the form of a thin slab, shown in Fig. 13–1, with thickness dy and face area A. If ρ is the density of the fluid (assumed to be uniform throughout), the mass of the volume element is $\rho A \, dy$ and its weight dw is $\rho g A \, dy$. The force exerted on the element by the surrounding fluid is everywhere normal to its surface. By symmetry, the resultant horizontal force on its vertical sides is zero. The upward force on its lower face is pA, and the downward force on its upper face is $(p + dp)A$. Since it is in equilibrium,

$$\Sigma F_y = 0, \qquad pA - (p + dp)A - \rho g A \, dy = 0,$$

Specific gravity compares the density of a material to the density of water.

Some practical uses for density measurements

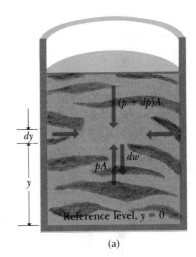

(a)

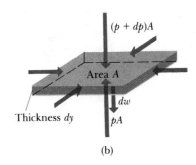

(b)

13–1 Forces on an element of fluid in equilibrium.

Because of the weight of a fluid, pressure varies with depth.

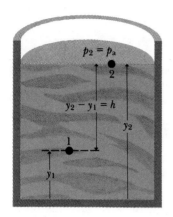

13–2 The pressure at a depth h in a liquid is greater than the surface pressure p_a by $\rho g h$.

Pressure applied to the surface of a fluid is transmitted throughout the fluid.

How a hydraulic jack works: lifting a heavy weight with a small force

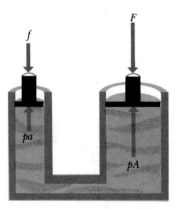

13–3 Principle of the hydraulic jack, an application of Pascal's law.

Gauge pressure: the difference between inside and outside pressures

and therefore

$$\frac{dp}{dy} = -\rho g. \tag{13–3}$$

Since ρ and g are both positive quantities, it follows that a positive dy (an increase of elevation) is accompanied by a negative dp (decrease of pressure). If p_1 and p_2 are the pressures at elevations y_1 and y_2 above some reference level, and ρ and g are constant, then

$$p_2 - p_1 = -\rho g(y_2 - y_1) \tag{13–4}$$

Let us apply this equation to a liquid in an open container, as shown in Fig. 13–2. Take point 1 at any level and let p represent the pressure at this point. Take point 2 at the surface of the liquid, where the pressure is atmospheric pressure, p_a. Then

$$p_a - p = -\rho g(y_2 - y_1),$$
$$p = p_a + \rho g h. \tag{13–5}$$

The pressure depends only on depth; the *shape* of the container does not matter. It also follows from Eq. (13–5) that if the pressure p_a is increased in any way, say by inserting a piston on the top surface and pressing down on it, the pressure p at any depth must increase by exactly the same amount. This fact was recognized in 1653 by the French scientist Blaise Pascal (1623–1662) and is called **Pascal's law:** *Pressure applied to an enclosed fluid is transmitted undiminished to every portion of the fluid and the walls of the containing vessel.*

The hydraulic jack shown schematically in Fig. 13–3 illustrates Pascal's law. A piston of small cross-sectional area a is used to exert a small force f directly on a liquid such as oil. The pressure $p = f/a$ is transmitted through the connecting pipe to a larger piston of area A. Since the pressure is the same in both cylinders,

$$p = \frac{f}{a} = \frac{F}{A} \quad \text{and} \quad F = \frac{A}{a}f.$$

Thus the hydraulic jack is a force-multiplying device with a multiplication factor equal to the ratio of the areas of the two pistons. Barber chairs, dentist chairs, car lifts and jacks, and hydraulic brakes all use this principle.

In deriving Eq. (13–4) we have assumed that the density ρ of the fluid is constant; this is a reasonable assumption for liquids, which are relatively incompressible, but it is *not* realistic for gases. In calculating the variation with altitude of the pressure in the earth's atmosphere, we must include this variation in density. We will carry out this calculation in Section 17–2. However, the density of gases is usually very much less than that of liquids, so over *small* changes of elevation (say, a few meters) the change in pressure in a gas is usually negligibly small.

In many practical situations the significant quantity is the *difference* between the pressure in a container and atmospheric pressure. For example, if the pressure inside a tire is just equal to atmospheric pressure, the tire is flat. When we say the pressure in a car tire is 32 pounds (actually 32 lb·in^{-2}), we mean it is *greater* than atmospheric pressure (14.7 lb·in^{-2}) by this amount. The *total* pressure in the tire is then 46.7 lb·in^{-2}. The excess pressure above atmospheric pressure is usually called **gauge pressure,** and the total pressure is called **absolute pressure.** Engineers use the abbreviations psig and psia for "pounds per square inch gauge" and "pounds per square inch absolute." Normal atmospheric pressure at sea level is

$$p_a = 1.013 \times 10^5 \text{ Pa} = 14.7 \text{ lb·in}^2 = 1 \text{ atm.}$$

EXAMPLE 13–1 A household hot-water heating system has an expansion tank in the attic, 12 m above the boiler. If the tank is open to the atmosphere, what is the gauge pressure in the boiler? The absolute pressure?

In a hot-water heating system, how much pressure must the pipes withstand?

SOLUTION From Eq. (13–5), the absolute pressure is

$$
\begin{aligned}
p &= p_a + \rho gh \\
&= (1.01 \times 10^5 \text{ Pa}) + (1000 \text{ kg·m}^{-3})(9.8 \text{ m·s}^{-2})(12 \text{ m}) \\
&= 2.19 \times 10^5 \text{ Pa} = 2.16 \text{ atm} = 31.8 \text{ lb·in}^{-2}
\end{aligned}
$$

The gauge pressure is

$$
\begin{aligned}
p - p_a &= (2.19 - 1.01) \times 10^5 \text{ Pa} = 1.18 \times 10^5 \text{ Pa} \\
&= 1.16 \text{ atm} = 17.1 \text{ lb·in}^{-2}.
\end{aligned}
$$

If the furnace has a pressure gauge, it is always calibrated to read gauge rather than absolute pressure.

We can also estimate the variation in atmospheric pressure over this height. The density of air at sea level and a temperature of 20°C is about 1.2 kg·m^{-3}; that is, about 0.12% of the density of water. If this density were constant over the 12-m height, then the atmospheric pressure at furnace level would be greater than that in the attic by

$$\rho gh = (1.2 \text{ kg·m}^{-3})(9.8 \text{ m·s}^{-2})(12 \text{ m}) = 141 \text{ Pa} = 0.0014 \text{ atm,}$$

or a very small fraction of the pressure calculated above. Thus the variation in air pressure over this height is negligible.

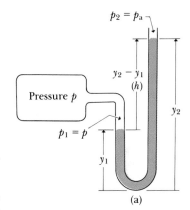

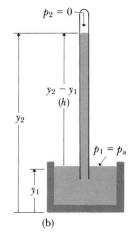

13–4 (a) The open-tube manometer. (b) The barometer.

The simplest pressure gauge is the open-tube manometer, shown in Fig. 13–4a. The U-shaped tube contains a liquid; one end of the tube is connected to the container where the pressure is to be measured, and the other end is open to the atmosphere at pressure p_a. The pressure at the bottom of the left column is $p + \rho gy_1$, and the pressure at the bottom of the right column (the same point) is $p_a + \rho gy_2$, where ρ is the density of the liquid in the manometer. Since these pressures must be equal,

$$p + \rho gy_1 = p_a + \rho gy_2,$$

and

$$p - p_a = \rho g(y_2 - y_1) = \rho gh. \tag{13–6}$$

A simple pressure gauge using a liquid in a tube

The pressure p is the *absolute pressure;* the difference $p - p_a$ between absolute and atmospheric pressure is the *gauge pressure.* Thus the gauge pressure is proportional to the difference in height of the liquid columns.

The **mercury barometer** is a long glass tube, closed at one end, that has been filled with mercury and then inverted in a dish of mercury, as shown in Fig. 13–4b. The space above the mercury column contains only mercury

Mercury barometers are used to measure atmospheric pressure for weather forecasting.

vapor; the pressure of the vapor at room temperature is negligibly small. From Eq. (13–5),

$$p_a = \rho g(y_2 - y_1) = \rho gh. \qquad (13-7)$$

Several different units are used to describe pressure.

Thus the mercury barometer reads atmospheric pressure directly from the height of the mercury column. The SI unit of pressure is *one pascal* (1 Pa), equal to one newton per square meter (1 N·m^{-2}). Two related units are the *bar,* defined as 10^5 Pa, and the *millibar,* defined as 10^{-3} bar or 10^2 Pa. Atmospheric pressures are of the order of 1000 millibars; the National Weather Service normally uses this unit. A pressure of 1.013×10^5 Pa = 1.013 bar is called one **atmosphere.** Because mercury manometers and barometers are commonly used laboratory instruments, pressures are also sometimes described in terms of the height of the corresponding mercury column, as so many "inches of mercury" or "millimeters of mercury" (abbreviated mm Hg). The pressure due to a column of mercury one millimeter high has even been given a special name: one *Torr* (after Torricelli, inventor of the mercury barometer). Such units depend on the density of mercury, which varies with temperature, and on the value of g, which varies with location. For these and other reasons, they are gradually passing out of common use in favor of the pascal.

What do blood-pressure readings mean?

One common type of blood-pressure gauge, called a *sphygmomanometer,* includes a manometer similar to that shown in Fig. 13–4a. Blood-pressure readings, such as 130/80, refer to the maximum and minimum gauge pressures, measured in millimeters of mercury or Torrs. Because of height differences, the hydrostatic pressure varies at different points in the body. The standard reference point is the upper arm, level with the heart. Pressure is also affected by the viscous nature of blood flow, by valves throughout the vascular system, and by the body's changing the diameters of blood vessels to regulate pressure.

Measuring atmospheric pressure with a mercury barometer

EXAMPLE 13–2   Compute the atmospheric pressure on a day when the height of mercury in a barometer is 76.0 cm.

SOLUTION   The height of the mercury column depends on ρ and g as well as on the atmospheric pressure. As mentioned, ρ varies with temperature, and g varies with latitude and elevation above sea level. If we assume $g = 9.8$ m·s^{-2} and $\rho = 13.6 \times 10^3$ kg·m^{-3},

$$p_a = \rho gh = (13.6 \times 10^3 \text{ kg·m}^{-3})(9.8 \text{ m·s}^{-2})(0.76 \text{ m})$$
$$= 101{,}300 \text{ N·m}^{-2} = 1.013 \times 10^5 \text{ Pa}.$$

In British units,

$$0.76 \text{ m} = 30 \text{ in.} = 2.5 \text{ ft},$$
$$\rho g = 850 \text{ lb·ft}^{-3},$$
$$p_a = 2120 \text{ lb·ft}^{-2} = 14.7 \text{ lb·in}^{-2.}$$

A pressure gauge with a dial and no moving liquid

Another type of pressure gauge, often more convenient than a liquid manometer, is the Bourdon pressure gauge. It consists of a flattened brass tube closed at one end and bent into a circular or spiral shape. The closed end of the tube is connected by a gear and pinion to a pointer that moves over a scale. The open end is connected to the container where the pressure is to be measured. When pressure increases within the flattened tube, it straightens slightly, just as a bent rubber hose straightens when the water is turned on. The resulting motion of the closed end is transmitted to the pointer.

13–3 BUOYANCY

Buoyancy is a familiar phenomenon. A body immersed in water seems to have less weight than when immersed in air, and a body with average density less than that of the fluid in which it is immersed can *float* in that fluid. Examples are the human body in water or a helium-filled balloon in air.

Archimedes' principle states: *When a body is immersed in a fluid, the fluid exerts an upward force on the body equal to the weight of the fluid that is displaced by the body.* To prove this principle, we consider an arbitrary portion of fluid at rest. In Fig. 13–5, the irregular outline is the surface bounding this portion of fluid. The arrows represent the forces exerted by the surrounding fluid on small elements of the boundary surface.

Since the entire fluid is at rest, the x-component of the resultant of these surface forces is zero. The y-component of their resultant, F_y, must be equal in magnitude to the weight mg of the fluid inside the surface, and its line of action must pass through the center of gravity of this fluid.

Now suppose that the fluid inside the surface is removed and is replaced by a solid body having exactly the same shape. The pressure at every point is exactly the same as before. Therefore the force exerted on the body by the surrounding fluid is unaltered and is equal in magnitude to the weight mg of fluid displaced. This upward force is called the buoyant force on the solid body. The line of action of this force again passes through the center of gravity of the displaced fluid.

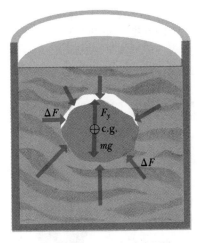

13–5 Archimedes' principle. The buoyant force F_y equals the weight of the displaced fluid.

Buoyancy: Why do bodies float?

Ice is less dense than liquid water (left), so ice cubes float in water. Solid benzene is more dense than liquid benzene, so benzene cubes sink in liquid benzene. (Photograph by Richard Byrnes.)

Balloons and submarines depend on buoyancy for their operation.

A simple density-measuring instrument using buoyancy

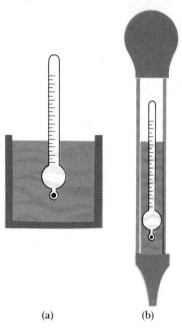

(a)   (b)

13–6 (a) A simple hydrometer.
(b) Hydrometer used as a tester for battery acid or antifreeze.

When a balloon floats in air, its weight must be the same as the weight of an amount of air having the same volume as the balloon. That is, the average density of a floating balloon must be the same as that of the surrounding air. Similarly, when a submerged submarine is in equilibrium, its average density must be equal to that of the surrounding water.

A body whose average density is less than that of a liquid can float partially submerged at the free upper surface of the liquid. A familiar example is the hydrometer, shown in Fig. 13–6. The instrument sinks in the fluid until the weight of the fluid it displaces is exactly equal to its own weight. In a fluid of greater density, less fluid is displaced for the same weight, so less of the instrument is immersed; that is, in denser fluids the hydrometer floats *higher*. Similarly, when you swim in sea water (density 1.03 g·cm^{-3}), your body floats higher than in fresh water.

The hydrometer is weighted at its bottom end so the upright position is stable, and a scale in the top stem permits direct density readings. Figure 13–6b shows a hydrometer commonly used to measure density of battery acid or antifreeze. The bottom of the tube is immersed in the liquid; the bulb is then squeezed to expel air and then released (like a giant medicine dropper). The resulting pressure difference causes the liquid to rise into the outer tube, and the hydrometer floats in this sample of the liquid.

EXAMPLE 13–3   A brass block with mass 0.5 kg and density 8.0×10^3 kg·m^{-3} is suspended from a string. What is the tension in the string if the block is in air? If it is completely immersed in water?

SOLUTION   If the very small buoyant force of air can be neglected, the tension when the block is in air is just equal to the weight of the block:

$$mg = (0.5 \text{ kg})(9.8 \text{ m·s}^{-2}) = 4.9 \text{ N}.$$

When immersed in water, the block experiences an upward buoyant force equal to the weight of water displaced. To find this, we first find the volume of the block:

$$V = \frac{m}{\rho} = \frac{0.5 \text{ kg}}{8.0 \times 10^3 \text{ kg·m}^{-3}} = 6.25 \times 10^{-5} \text{ m}^3.$$

The weight of this volume of water is

$$
\begin{aligned}
w = mg &= \rho V g \\
&= (1.0 \times 10^3 \text{ kg·m}^{-3})(6.25 \times 10^{-5} \text{ m}^3)(9.8 \text{ m·s}^{-2}) \\
&= 0.612 \text{ N}.
\end{aligned}
$$

The tension in the string is the actual weight less the upward buoyant force, or

$$4.9 \text{ N} - 0.612 \text{ N} = 4.29 \text{ N}.$$

A shortcut leading to the same result is as follows: The density of water is less than that of brass by a factor

$$\frac{1.0 \times 10^3 \text{ kg·m}^{-3}}{8.0 \times 10^3 \text{ kg·m}^{-3}} = \frac{1}{8}.$$

Therefore the weight of an amount of water with volume equal to that of the brass will be less than the weight of the brass by the same factor. Thus the weight of the displaced water is $\frac{1}{8}(4.9 \text{ N}) = 0.612 \text{ N}$.

EXAMPLE 13–4 Suppose you place the water container in Example 13–3 on a scale. How does the scale reading change when the brass is immersed in the water?

SOLUTION Taking the water and the brass together as a system, we note that its total weight does not change when the brass is immersed. Thus the sum of the two supporting forces, the string tension and the upward force of the scale on the container, must be the same in both cases. But we found that the string tension decreases by 0.612 N when the brass is immersed, so the scale reading must *increase* by 0.612 N. Qualitatively, we can see that the scale force *must* increase because when the brass is immersed, the water level in the container rises, increasing the pressure of water on the bottom of the container.

13–4 SURFACE TENSION

There are many interesting phenomena associated with the *boundary surface* that exists between a liquid and some other substance. A liquid flowing slowly from the tip of a medicine dropper emerges not as a continuous stream but as a succession of drops. A sewing needle, if placed carefully on a water surface, makes a small depression in the surface and rests there without sinking, even though its density may be as much as ten times that of water. Some insects can walk on the surface of water, their feet making indentations in the surface but not penetrating it. When a clean glass tube of small diameter is dipped into water, the water rises in the tube; but if the tube is dipped in mercury, the mercury is depressed.

How can insects walk on water? Physics or magic?

In all these phenomena the surface behaves as though it were under *tension*. For any line on the surface, the portions of surface on the two sides of the line exert pulls on it. The situation of Fig. 13–7 shows this effect. A wire ring has a loop of thread attached to it as shown. When the ring and thread are dipped in a soap solution and removed, a thin film of liquid is formed in which the thread "floats" freely, as shown in part (a). If we then puncture the film inside the loop of thread, the thread springs out into a circular shape as in part (b); the surfaces of the liquid pull radially outward on it, as shown by the arrows. These same forces were acting even before the film was punctured,

The surface of a liquid acts as though it were under tension.

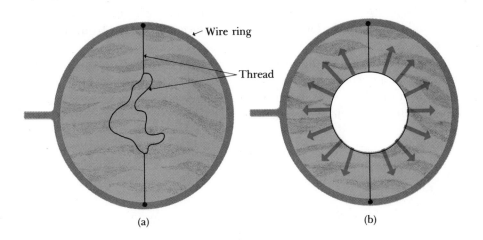

(a)   (b)

13–7 A wire ring with a flexible loop of thread, dipped in a soap solution, (a) before and (b) after puncturing the surface films inside the loop.

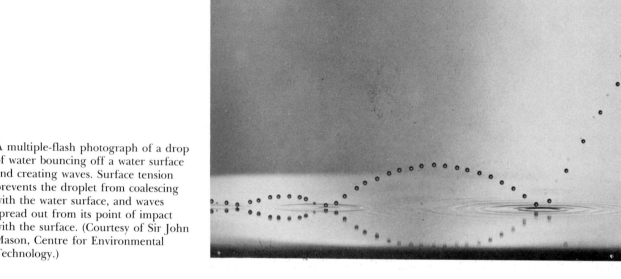

A multiple-flash photograph of a drop of water bouncing off a water surface and creating waves. Surface tension prevents the droplet from coalescing with the water surface, and waves spread out from its point of impact with the surface. (Courtesy of Sir John Mason, Centre for Environmental Technology.)

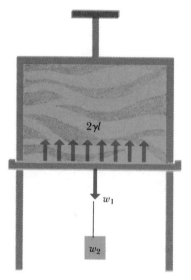

13–8 The horizontal slide wire is in equilibrium under the action of the upward surface force $2\gamma l$ and downward pull $w_1 + w_2$.

Surface tension is surface force per unit length.

but then there was film on *both* sides of the thread, and the net force on a section of thread was zero.

Another simple apparatus for demonstrating surface tension is shown in Fig. 13–8. A piece of wire is bent into a U shape and a second piece of wire is used as a slider. When the apparatus is dipped in a soap solution and removed, the slider (if its weight w_1 is not too great) is quickly pulled up to the top of the U. The slider may be held in equilibrium by adding a second weight w_2. Surprisingly, the same total force $F = w_1 + w_2$ will hold the slider at rest in *any* position, regardless of the area of the liquid film, provided the film remains at constant temperature. That is, the force does not increase as the surface is stretched farther. This is very different from the elastic behavior of a sheet of rubber, for which the force required would be greater as the sheet was stretched. Figure 13–9 shows a similar situation.

Although a soap film like that in Fig. 13–8 is very thin, its thickness is still enormous compared with the size of a molecule. Hence we can consider a soap film as made up chiefly of bulk liquid, bounded by two surface layers a few molecules thick. When the slider in Fig. 13–8 is pulled down and the area of the film is increased, molecules formerly in the main body of the liquid move into the surface layers. That is, these layers are not "stretched" as a rubber sheet would be; more surface is created by molecules moving from the bulk liquid.

Let l be the length of the wire slider. Since the film has two surfaces, the total length along which the surface force acts on the slider is $2l$. The **surface tension** γ in the film is defined as *the ratio of the surface force F to the length d* (perpendicular to the force) *along which the force acts:*

$$\gamma = \frac{F}{d}. \tag{13–8}$$

Hence, in this case, $d = 2l$ and

$$\gamma = \frac{F}{2l}.$$

TABLE 13-2   Experimental Values of Surface Tension

Liquid in Contact with Air	T, °C	Surface tension, dyn·cm^{-1}
Benzene	20	28.9
Carbon tetrachloride	20	26.8
Ethyl alcohol	20	22.3
Glycerin	20	63.1
Mercury	20	465.0
Olive oil	20	32.0
Soap solution	20	25.0
Water	0	75.6
Water	20	72.8
Water	60	66.2
Water	100	58.9
Oxygen	−193	15.7
Neon	−247	5.15
Helium	−269	0.12

Note that although the term *tension* has previously been used in reference to a force, surface tension is a *force per unit length*. The SI unit of surface tension is the newton per meter (1 N·m^{-1}). This unit is not in common use; the usual unit is the cgs unit, the dyne per centimeter (1 dyn·cm^{-1}). The conversion factor is

$$1 \text{ N·m}^{-1} = 1000 \text{ dyn·cm}^{-1}.$$

Some typical values of surface tension are shown in Table 13–2. Surface tension usually decreases as temperature increases; Table 13–2 shows this behavior for water.

A surface under tension contracts until it has the minimum area consistent with any fixed boundaries that may be present and with any pressure differences that may be present on opposite sides of the surface. For example, a drop of liquid under no external forces, or in free fall in vacuum, is always spherical in shape because the sphere has smaller surface area for a given volume than has any other geometric shape. Figure 13–10 is a beautiful example of formation of spherical droplets in a very complex phenomenon, the impact of a drop on a rigid surface.

13-9 A paper clip "floating" on water. The clip is supported not by buoyant forces (Section 13–3) but by surface tension of the water. (Nancy Rodgers, Exploratorium.)

13-10 A drop of milk splashes on a hard surface. (Dr. Harold Edgerton, M.I.T., Cambridge, Massachusetts.)

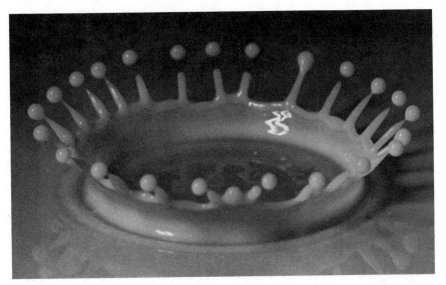

Surface tension causes increased pressure inside a bubble or liquid drop.

Surface tension causes a pressure difference between the inside and outside of a soap bubble or a liquid drop. A soap bubble consists of two spherical surface films very close together, with a thin layer of liquid between. Surface tension causes the films to tend to contract; but as the bubble contracts, it compresses the inside air, increasing the interior pressure to a point that prevents further contraction. We can derive a relation between this pressure and the radius of the bubble.

We assume first that there is no external pressure. Then we consider the equilibrium state of one-half the soap bubble, as shown in Fig. 13–11. At the surface where this half joins the upper half, two forces act: the upward force of surface tension and the downward force due to the pressure of air in the upper half. If the radius of the spherical surface is R, the circumference of the circle along which the surface tension acts is $2\pi R$. (Assuming the thickness of the bubble is very small, we neglect the difference between inner and outer radii.) The total force of surface tension for each surface (inner and outer) is $\gamma(2\pi R)$, so the total force of surface tension is $(2\gamma)(2\pi R)$. The force due to air pressure is the pressure p times the area πR^2 of the circle over which it acts. For the resultant of these forces to be zero, we must have

$$(2\gamma)(2\pi R) = p(\pi R^2),$$

The pressure increase inside a bubble or liquid drop is inversely proportional to its radius.

or

$$p = \frac{4\gamma}{R}. \tag{13–9}$$

Now the pressure outside the bubble is in general *not* zero. But from the derivation we can see that Eq. (13–9) gives the *difference* between inside and outside pressure. If the outside pressure is p_a, then

$$p - p_a = \frac{4\gamma}{R} \quad \text{(soap bubble).} \tag{13–10}$$

For a liquid drop, which has only *one* surface film, the difference between pressure of the liquid and that of the outside air is half that for a soap bubble:

$$p - p_a = \frac{2\gamma}{R} \quad \text{(liquid drop).} \tag{13–11}$$

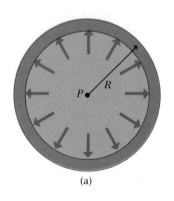

(a)

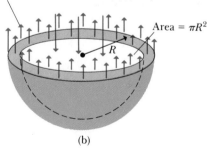

Surface tension force $(2\gamma)(2\pi R)$

Area $= \pi R^2$

(b)

13–11 (a) Cross section of a soap bubble, showing the two surfaces, the thin layer of liquid, and inside air. (b) Equilibrium of half a soap bubble. The surface-tension force exerted by the other half is $(2\gamma)(2\pi R)$, and the force exerted by the air inside the bubble is the pressure p times the area πR^2.

EXAMPLE 13–5 Calculate the excess pressure inside a drop of mercury of diameter 4 mm at 20°C.

SOLUTION From Table 13–2,

$$\gamma = 465 \text{ dyn·cm}^{-1} = 465 \times 10^{-3} \text{ N·m}^{-1}.$$

From Eq. (13–11),

$$p - p_a = \frac{2\gamma}{R} = \frac{(2)(465 \times 10^{-3} \text{ N·m}^{-1})}{0.002 \text{ m}}$$

$$= 465 \text{ N·m}^{-2}$$

$$\cong 0.005 \text{ atm.}$$

Thus far we have talked about surface films at the boundary between a liquid and a gas. Additional surface-tension effects occur when such a film meets a solid surface, such as the wall of a container, as shown in Fig. 13–12.

In general, the liquid-gas surface curves up or down as it approaches the solid surface, and the angle θ at which it meets the surface is called the **contact angle.** When the surface curves up, as with water and glass, θ is less than 90°; when it curves down, as with mercury and glass, θ is greater than 90°. The first case is characteristic of a liquid that tends to "wet" or adhere to the solid surface; the second, of a nonwetting liquid.

An important surface-tension phenomenon is the elevation or depression of the liquid in a narrow tube, as shown in Fig. 13–13. We call this effect **capillarity;** this term stems from the description of such tubes as *capillary,* which means "hairlike." For a liquid such as water, which wets the tube, the contact angle is less than 90° and the liquid rises until it reaches an equilibrium height y. This height is determined by the requirement that the total surface-tension force around the line where the liquid-gas surface meets the solid should just balance the extra weight of the liquid in the tube. The curved liquid surface is called a *meniscus.* Figure 13–13b shows the situation with a nonwetting liquid such as mercury. In this case the contact angle is greater than 90°. Again a meniscus forms, but the surface is depressed, pulled down by the surface-tension forces.

Capillarity is responsible for the rise of ink in blotting paper, the rise of fuel in the wick of a cigarette lighter, and many other everyday phenomena. It is also an essential part of many life processes. A familiar example is the rising of water (actually a dilute aqueous solution) from the roots of a plant to its foliage, due partly to capillarity and partly to osmotic pressure developed in the roots. In the higher animals, including humans, blood is pumped through the arteries and veins, but capillarity is still important in the smallest blood vessels, which indeed are called capillaries.

A related phenomenon is that of *negative pressure.* Ordinarily the stress in a liquid is compressive in nature, but in some circumstances liquids can sustain actual *tensile* stresses. Imagine a cylindrical tube, closed at one end and having a tight-fitting piston at the other. Suppose the tube is completely filled with a liquid that wets both the inner surface of the tube and the piston face; that is, the molecules of liquid adhere to the surfaces. If we pull the piston out, we expect that the liquid will pull away from the piston, leaving an empty space. If the surfaces are very clean and the liquid very pure, however, an actual tensile stress, accompanied by an increase in volume, can be observed. The liquid behaves as though we were stretching it; adhesive forces prevent it from pulling away from the walls of the container.

With water, tensile stresses as large as 300 atm have been observed. In the laboratory this situation is highly unstable; a liquid under tension tends to break up into many small droplets. In tall trees, however, negative pressures are a regular occurrence. Negative pressure is thought to be an important mechanism for transport of water and nutrients from the roots to the leaves in the small xylem tubes (diameter of the order of 0.1 mm) in the growing layers of the tree.

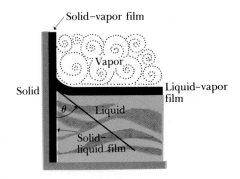

13–12 Surface films exist at the solid–vapor boundary as well as at the liquid–vapor boundary.

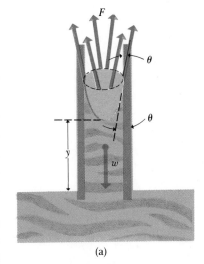

(a)

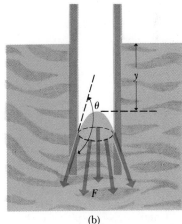

(b)

13–13 Surface-tension forces on a liquid in a capillary tube. The liquid (a) rises if $\theta < 90°$ and (b) is depressed if $\theta > 90°$.

13–5 FLUID FLOW

We are now ready to consider *motion* of a fluid. This motion can be extremely complex, as shown by the flow of a river in flood or the swirling flames of a campfire. Despite this, some situations can be represented by relatively simple idealized models. We begin by introducing one of these models, along with

An ideal fluid has constant density and no internal friction.

13–14 A flow tube bounded by flow lines.

Flow lines describe the paths of fluid particles.

A bundle of flow lines forms a flow tube.

several new terms that are useful in our discussion. An **ideal fluid** is a fluid that is *incompressible* and has no internal friction or viscosity. The assumption of incompressibility is usually a good approximation for liquids, and we may also treat a gas as incompressible whenever the pressure differences associated with the flow are not too great. Internal friction in a fluid causes shear stresses when two adjacent layers of fluid move relative to each other, as when the fluid flows inside a tube or around an obstacle. In some cases, these shear forces are negligible compared with forces arising from gravitation and pressure differences.

The path of an individual element of a moving fluid is called a **flow line.** In general, the velocity of a fluid element changes in both magnitude and direction as it moves along its line of flow. If every element passing through a given point follows the same flow line as that of preceding elements, the flow is said to be *steady* or *stationary*. In steady flow, the fluid velocity at each point of space remains constant in time, although the velocity of a particular particle of the fluid may change as it moves from one point to another.

A **streamline** is a curve whose tangent, at any point, is in the direction of the fluid velocity at that point. In steady flow, the streamlines coincide with the flow lines.

If we construct all the streamlines passing through the edge of an imaginary element of area, such as the area A in Fig. 13–14, these lines enclose a tube called a **flow tube.** From the definition of a streamline, no fluid can cross the side walls of a flow tube; in steady flow there can be no mixing of the fluids in different flow tubes.

Figure 13–15 shows fluid flow from left to right around a number of obstacles and in a channel of varying cross section. The photographs were made by using an apparatus in which alternate streams of clear and colored water flow around opaque obstacles between two closely spaced glass plates. Each obstacle is surrounded by a flow tube. The tube splits into two portions at a *stagnation point* (a point where the velocity is zero) on the upstream (left) side of the obstacle. These portions rejoin at a second stagnation point on the downstream (right) side. The velocity at the stagnation points is zero. The

Streamlines in symmetric laminar flow of water past an airfoil. This photograph was made by injecting dye into the water upstream from the airfoil. The left and right edges are stagnation points, where the flow velocity is zero. The Reynolds number is 7000, based on the left-to-right chord length. (ONERA photograph, Werlé 1974.)

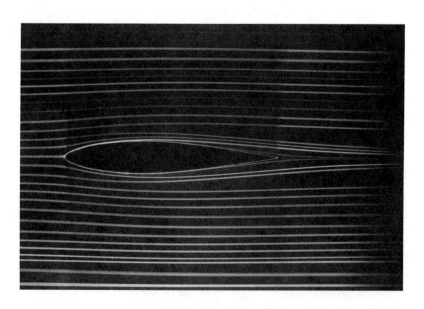

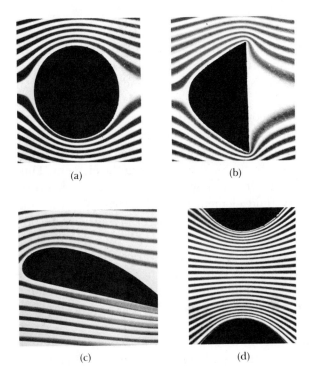

(a)   (b)

(c)   (d)

13–15 (a), (b), (c): Streamline flow around obstacles of various shapes. (d) Flow in a channel of varying cross-sectional area.

cross sections of all flow tubes decrease at a constriction and increase again when the channel widens.

The patterns of Fig. 13–15 are typical of **laminar flow,** in which adjacent layers of fluid slide smoothly past each other. At sufficiently high flow rates, or when boundary surfaces cause abrupt changes in velocity, the flow becomes irregular and much more complex; this is called **turbulent flow.** In turbulent flow *there is no steady-state pattern,* since the flow pattern continuously changes.

We can use the principle of conservation of mass to derive an important relationship in fluid flow called the **continuity equation.** For an incompressible fluid in steady flow, the derivation takes the following form. Figure 13–16 shows a portion of a flow tube, between two fixed cross sections of area A_1 and A_2. Let v_1 and v_2 be the speeds at these sections. No fluid flows across the side wall of the tube because the fluid velocity is tangent to the wall at every point on the wall. The volume of fluid that flows into the tube across A_1 in a time interval Δt is the fluid contained in the short cylindrical element of base A_1 and height $v_1 \Delta t$, that is, $A_1 v_1 \Delta t$. If the density of the fluid is ρ, the *mass* flowing in is $\rho A_1 v_1 \Delta t$. Similarly, the mass that flows out across A_2 in the same time is $\rho A_2 v_2 \Delta t$. The volume between A_1 and A_2 is constant, and since the flow is steady, the mass flowing out equals that flowing in. Hence

$$\rho A_1 v_1 \Delta t = \rho A_2 v_2 \Delta t,$$

or

$$A_1 v_1 = A_2 v_2. \tag{13–12}$$

That is, the product Av is constant along any given flow tube. It follows that when the cross section of a flow tube decreases, as in the constriction in Fig. 13–15d, the velocity increases. We can observe this by introducing small particles into the fluid and watching their motion.

Sometimes there is a smooth, steady flow pattern; sometimes not.

Conservation of mass in a fluid: What goes in must come out.

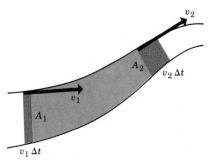

13–16 Flow into and out of a portion of a tube of flow.

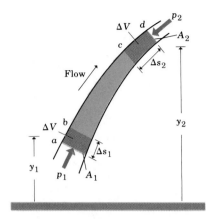

13–17 The net work done on the shaded element equals the increase in its kinetic and potential energy.

Bernoulli's equation: the work–energy theorem for a moving fluid

Work done on a fluid element by pressure differences

Change in kinetic energy for a fluid element

Change in gravitational potential energy for a fluid element

13–6 BERNOULLI'S EQUATION

We can now use the concepts introduced in Section 13–5, along with the energy principle, to derive an important relationship called **Bernoulli's equation**. This equation relates the pressure, flow velocity, and height for flow of an ideal fluid.

First we note that when an incompressible fluid flows along a horizontal flow tube of varying cross section, its velocity must change; Eq. (13–12) shows this. Each element of fluid must have an acceleration. The force that causes this acceleration has to be applied by the fluid surrounding the element. Thus the pressure must be different in different regions; if it were the same everywhere, the net force on every fluid element would be zero. When a fluid element speeds up, it must move from a region of larger pressure to a region of smaller pressure in order to experience a net forward force that makes it speed up. Thus when the cross section of a flow tube varies, the pressure *must* vary along the tube, even when there is no difference of elevation. If the elevation also changes, there is an additional pressure difference. Bernoulli's equation relates the pressure difference between two points in a flow tube to changes in both velocity and elevation.

To derive the Bernoulli equation, we apply the work–energy theorem to the fluid in a section of a flow tube. In Fig. 13–17, we consider the fluid that at some initial time lies between the two cross sections a and c. In a time interval Δt, the fluid initially at a moves to b, a distance $\Delta s_1 = v_1 \, \Delta t$, where v_1 is the speed at this end of the tube. In the same time interval the fluid initially at c moves to d, a distance $\Delta s_2 = v_2 \, \Delta t$. The cross-sectional areas at the two ends are A_1 and A_2, as shown. Because of the continuity relation, the volume of fluid ΔV passing any cross section in time Δt is $\Delta V = A_1 \, \Delta s_1 = A_2 \, \Delta s_2$.

Let us compute the *work* done on this fluid during Δt. The force on the cross section at a is $p_1 A_1$, and the force at c is $p_2 A_2$, where p_1 and p_2 are the pressures at the two ends. The net work done on the element during this displacement is therefore

$$W = p_1 A_1 \, \Delta s_1 - p_2 A_2 \, \Delta s_2 = (p_1 - p_2) \, \Delta V \qquad (13\text{–}13)$$

The negative sign in the second term arises because the force at c is opposite in direction to the displacement.

We now equate this work to the total change in energy, kinetic and potential, of the element. The kinetic energy of the fluid between sections b and c does not change. At the beginning of Δt, the fluid between a and b has volume $A_1 \, \Delta s_1$, mass $\rho A_1 \, \Delta s_1$, and kinetic energy $\frac{1}{2}\rho(A_1 \, \Delta s_1)v_1^2$. Similarly, at the end of Δt, the fluid between c and d has kinetic energy $\frac{1}{2}\rho(A_2 \, \Delta s_2)v_2^2$. Thus the net change in kinetic energy is

$$\Delta K = \tfrac{1}{2}\rho \, \Delta V(v_2^2 - v_1^2). \qquad (13\text{–}14)$$

The change in potential energy is obtained similarly. The potential energy of the mass entering at a in time Δt is $\Delta m \, gy_1 = \rho \, \Delta V \, gy_1$, and that of the mass leaving at c is $\Delta m \, gy_2 = \rho \, \Delta V \, gy_2$. The net change in potential energy is

$$\Delta U = \rho \, \Delta V \, g(y_2 - y_1). \qquad (13\text{–}15)$$

Combining Eqs. (13–13), (13–14), and (13–15) in the work–energy theorem $W = \Delta K + \Delta U$, we obtain

$$(p_1 - p_2) \, \Delta V = \tfrac{1}{2}\rho \, \Delta V \, (v_2^2 - v_1^2) + \rho \, \Delta V \, g(y_2 - y_1)$$

or

$$p_1 - p_2 = \tfrac{1}{2}\rho(v_2{}^2 - v_1{}^2) + \rho g(y_2 - y_1). \qquad (13\text{--}16)$$

In this form, Bernoulli's equation represents the equality of the work per unit volume of fluid $(p_1 - p_2)$ to the sum of the changes in kinetic and potential energies per unit volume that occur during the flow. Or we may interpret Eq. (13–16) in terms of pressures. The second term on the right is the pressure difference arising from the weight of the fluid and the difference in elevation of the two ends of the fluid element. The first term on the right is the additional pressure difference associated with the change of velocity of the fluid.

Equation (13–16) can also be written

$$p_1 + \rho g y_1 + \tfrac{1}{2}\rho v_1{}^2 = p_2 + \rho g y_2 + \tfrac{1}{2}\rho v_2{}^2, \qquad (13\text{--}17)$$

and since the subscripts 1 and 2 refer to *any* two points along the tube of flow, Bernoulli's equation may also be written

$$p + \rho g y + \tfrac{1}{2}\rho v^2 = \text{constant}. \qquad (13\text{--}18)$$

Note carefully that p is the *absolute* (not gauge) pressure, and that a consistent set of units must be used. In SI units, pressure is expressed in Pa, density in $\mathrm{kg \cdot m^{-3}}$, and velocity in $\mathrm{m \cdot s^{-1}}$.

PROBLEM-SOLVING STRATEGY: *Bernoulli's equation*

Bernoulli's equation is derived from the work–energy relationship, so it isn't surprising that much of the problem-solving strategy suggested in Sections 7–5 and 7–6 is equally applicable here. In particular:

1. Always begin by identifying clearly points 1 and 2, with reference to Bernoulli's equation.

2. Make lists of the known and unknown quantities in Eq. (13–17). The variables are p_1, p_2, v_1, v_2, y_1, and y_2, and the constants are ρ and g. What is given? What do you need to determine?

3. In some problems you will need to use the continuity equation, Eq. (13–12), to get a relation between the two velocities in terms of cross-sectional areas of pipes or containers. Or perhaps you will know both velocities and need to determine one of the areas.

4. The volume flow rate dV/dt across any area A is given by $dV/dt = Av$, and the corresponding mass flow rate dm/dt is $dm/dt = \rho\, Av$. These relations sometimes come in handy.

5. As you read the following example and the applications discussed in Section 13–7, watch for applications of this strategy.

EXAMPLE 13–6   Water enters a house through a pipe 2.0 cm in inside diameter, at an absolute pressure of 4×10^5 Pa (about 4 atm). The pipe leading to the second-floor bathroom 5 m above is 1.0 cm in diameter. When the flow velocity at the inlet pipe is 4 m·s⁻¹, find the flow velocity and pressure in the bathroom.

Variation in water pressure in a house: an application of Bernoulli's equation

SOLUTION   Let point 1 be at the inlet pipe and point 2 at the bathroom. The velocity v_2 at the bathroom is obtained from the continuity equation:

$$v_2 = \frac{A_1}{A_2}v_1 = \frac{\pi(1.0 \text{ cm})^2}{\pi(0.5 \text{ cm})^2}(4 \text{ m·s}^{-1}) = 16 \text{ m·s}^{-1}.$$

We take $y_1 = 0$ (at the inlet) and $y_2 = 5$ m (at the bathroom). We are given p_1 and v_1; we can find p_2 from Bernoulli's equation:

$$p_2 = p_1 - \tfrac{1}{2}\rho(v_2{}^2 - v_1{}^2) - \rho g(y_2 - y_1)$$
$$= 4 \times 10^5 \text{ Pa} - \tfrac{1}{2}(1.0 \times 10^3 \text{ kg·m}^{-3})(256 \text{ m}^2\text{·s}^{-2} - 16 \text{ m}^2\text{·s}^{-2})$$
$$- (1.0 \times 10^3 \text{ kg·m}^{-3})(9.8 \text{ m·s}^{-2})(5 \text{ m})$$
$$= 2.3 \times 10^5 \text{ Pa} \cong 2.3 \text{ atm.}$$

Note that when the water is turned off, the second term on the right vanishes and the pressure rises to 3.5×10^5 Pa.

13–7 APPLICATIONS OF BERNOULLI'S EQUATION

Here are several applications of Bernoulli's equation to situations of practical interest.

1. *Fluid statics.* Just as equilibrium of a particle is a special case of $\mathbf{F} = m\mathbf{a}$, the equations of fluid statics are special cases of Bernoulli's equation when the velocity is zero everywhere. Thus when v_1 and v_2 are zero, Eq. (13–16) reduces to

$$p_1 - p_2 = \rho g(y_2 - y_1), \qquad (13\text{–}19)$$

which is the same as Eq. (13–4).

How fast does fluid run out of a container?

2. *Speed of efflux: Torricelli's theorem.* Figure 13–18 shows a tank with cross-sectional area A_1, filled to a depth h with a liquid of density ρ. The space above the top of the liquid contains air at pressure p, and the liquid flows out of an orifice of area A_2. Let us consider the entire volume of moving fluid as a single flow tube, and let v_1 and v_2 be the speeds at points 1 and 2. The quantity v_2 is called the *speed of efflux*. The pressure at point 2 is atmospheric, p_a. Applying Bernoulli's equation to points 1 and 2, and taking the bottom of the tank as our reference level, we find

$$p + \tfrac{1}{2}\rho v_1{}^2 + \rho gh = p_a + \tfrac{1}{2}\rho v_2{}^2,$$

or

$$v_2{}^2 = v_1{}^2 + 2\frac{p - p_a}{\rho} + 2gh. \qquad (13\text{–}20)$$

If the tank is open to the atmosphere, then

$$p = p_a \qquad \text{and} \qquad p - p_a = 0.$$

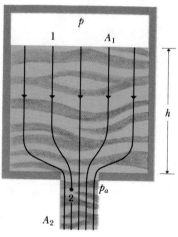

13–18 Flow of a liquid out of an orifice.

Also, in Fig. 13–18, if A_2 is much smaller than A_1, then $v_1{}^2$ is very much less than $v_2{}^2$ and can be neglected. Equation (13–20) then becomes

$$v_2 = \sqrt{2gh}. \qquad (13\text{–}21)$$

That is, *the speed of efflux is the same as the speed a body would acquire in falling freely through a height h.* This is *Torricelli's theorem.* It is not restricted to an opening in the bottom of a vessel, but applies also to a hole in a side wall at a depth h below the surface.

Rocket propulsion: What pushes a rocket forward?

3. *Rocket propulsion.* A flow of fluid out of an opening in a container causes a *reaction force* on the remainder of the system. This is the principle of rocket propulsion. Returning to Fig. 13–18, let us suppose that the vessel contains

only a gas, having density ρ and pressure p. Again we assume that A_2 is much smaller than A_1, so that the v_1^2 term in Bernoulli's can be neglected. Suppose also that p is so large that the term $2gh$ is negligible compared to the pressure term. The speed of efflux is then given by

$$v_2 = \sqrt{2(p - p_a)/\rho}. \qquad (13-22)$$

We can now derive an expression for the reaction force, or *thrust*, acting on the system. We drop the subscripts, letting A be the orifice area and v the speed of efflux. Then the volume of fluid flowing out in a time interval Δt is $Av\,\Delta t$, the corresponding mass is $\rho Av\,\Delta t$, and the momentum (mass × velocity) is $\rho Av^2\,\Delta t$. The escaping fluid acquires this momentum during the time Δt, so its *rate of change* of momentum is ρAv^2. (This analysis uses the same principles we used in Section 8–8.) From Newton's second law, this equals the force acting on the escaping fluid, and from Newton's third law, an equal and opposite reaction force acts on the remainder of the system. Using Eq. (13–22) for the speed of efflux, we can write the reaction force as

$$F = \rho Av^2 = \rho A \frac{2(p - p_a)}{\rho},$$

or

$$F = 2A(p - p_a). \qquad (13-23)$$

Thus, while the *speed* of efflux is inversely proportional to the fluid density, the *thrust* is independent of the density and depends only on the area of the orifice and the gauge pressure $p - p_a$.

4. The *Venturi tube*, shown in Fig. 13–19, consists of a constriction or throat inserted in a pipeline, with inlet and outlet tapers to avoid turbulence. Bernoulli's equation, applied to the wide and constricted portions of the pipe, respectively (with $y_1 = y_2$), becomes

$$p_1 + \tfrac{1}{2}\rho v_1^2 = p_2 + \tfrac{1}{2}\rho v_2^2.$$

From the equation of continuity, the speed v_2 is greater than the speed v_1, and hence the pressure p_2 in the throat is *less* than the pressure p_1. Thus a net force to the right accelerates the fluid as it enters the throat, and a net force to the left slows it as it leaves. The pressures p_1 and p_2 can be measured by attaching vertical side tubes, as shown in the diagram. When we know these pressures and the areas A_1 and A_2, we can compute the velocities and the mass rate of flow. When used for this purpose, the device is called a *Venturi meter*.

The effect of reduced pressure at a constriction has many practical applications. In the carburetor of a gasoline engine, gasoline is sprayed from a nozzle into a low-pressure region produced in a Venturi throat. The *aspirator pump* forces water through a Venturi throat. Air is pushed into the low-pressure water rushing through the constricted portion by the higher pressure outside.

5. *Measurement of pressure in a moving fluid.* The pressure p in a fluid flowing in an enclosed channel can be measured with an open-tube manometer, as shown in Fig. 13–20. A *probe* is inserted in the stream. The probe should be small enough so that the flow is not appreciably disturbed and should be shaped so as to avoid turbulence. The difference h in height of the liquid in the arms of the manometer is proportional to the difference between atmos-

Momentum relations for rocket propulsion

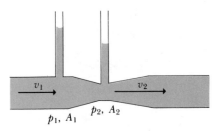

13–19 The Venturi tube.

In a Venturi tube, flow speed increases and pressure drops.

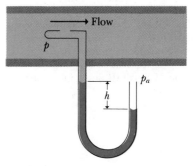

13–20 Pressure gauge for measuring the static pressure p in a fluid flowing in an enclosed channel.

pheric pressure p_a and the fluid pressure p. That is,

$$p_a - p = \rho_m gh,$$
$$p = p_a - \rho_m gh, \qquad (13\text{–}24)$$

where ρ_m is the density of the manometer liquid. A variation of this setup is used as an airspeed indicator in airplanes.

6. *Lift on an aircraft wing.* Figure 13–21 shows flow lines around a cross section of an aircraft wing or an airfoil. The orientation of the wing relative to the flow direction causes the flow lines to crowd together above the wing. This corresponds to increased flow velocity in this region, just as in the throat of a Venturi. Hence the region above the wing has increased velocity and reduced pressure, while below the wing the pressure remains nearly atmospheric. Because the upward force on the underside of the wing is greater than the downward force on the top side, there is a net upward force, or *lift*.

We can also understand this phenomenon on the basis of Newton's laws. In Fig. 13–21, the air flowing past the wing has a net *downward* change in the vertical component of its momentum, corresponding to the downward force the wing exerts on it. The reaction force *on* the wing is thus *upward*, in agreement with our discussion above. As the angle of the wing relative to the air flow increases, turbulent flow occurs in a larger and larger region above the wing. When this happens, the pressure drop is no longer as great as that predicted by Bernoulli's equation. The lift on the wing decreases, and in extreme cases the airplane stalls.

7. *The curved flight of a spinning ball.* This effect is somewhat more subtle than the lift on an airplane wing, but the same basic principles are involved. Figure 13–22a represents a stationary ball in a blast of air moving from right to left. The motion of the airstream around and past the ball is the same as though the ball were moving through still air from left to right. Because of the large velocities ordinarily involved, a region of turbulent flow occurs behind the ball, as shown.

When the ball is spinning, as in Fig. 13–22b, the viscosity of air causes layers of air near the ball's surface to be pulled around in the direction of spin. The velocity of air relative to the ball's surface becomes greater at the top of the ball than at the bottom. The region of turbulence becomes asymmetric, with turbulence occurring farther forward on the top side than on the bottom. This asymmetry causes a pressure difference. The average pressure at the top of the ball becomes greater than that at the bottom. The corresponding net downward force deflects the ball as shown. In a baseball curve pitch, the ball spins at an angle, and the actual deflection is sideways. In that case, Fig. 13–22 shows a *top* view of the situation. Tennis balls with spin behave in much the same way.

A similar effect occurs with golf balls, which always have "backspin" from impact with the slanted face of the golf club. The resulting pressure difference between top and bottom of the ball causes a "lift" force that keeps the ball in the air considerably longer than would be possible without spin. A well-hit drive appears from the tee to "float" or even curve *upward* during the initial portion of its flight, and this is a real effect, not an illusion. The dimples on the ball play an essential role: An undimpled ball has much shorter trajectory than a dimpled one given the same initial velocity and spin. Some manufacturers even claim that the *shape* of the dimples is significant; one states that polygonal

How does an airplane stay in the air?

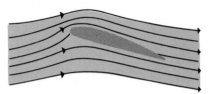

13–21 Flow lines around an airfoil.

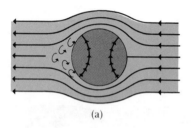

(a)

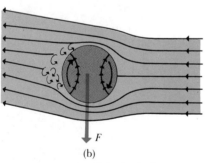

(b)

13–22 (a) Air flow past a stationary ball, showing symmetric region of turbulence behind ball. (b) Air flow past a spinning ball, showing asymmetric region of turbulence and deflection of airstream. The net force on the ball is in the direction shown. Motion of the air from right to left relative to the ball corresponds to motion of a ball through still air from left to right.

Spin is important for golf balls as well as baseballs.

dimples are superior to round ones! Figure 13–23 shows the backspin of a golf ball just after being struck by a club.

13–8 VISCOSITY

Another important characteristic of fluids is **viscosity,** which is internal friction of a fluid. Because of viscosity, a force is needed whenever one layer of a fluid slides past another, or when one surface slides past another with a layer of fluid between the surfaces. Viscosity plays a vital role in the flow of fluids in pipes, the flow of blood, the lubrication of engine parts, and many other areas of practical importance. Both liquids and gases have viscosity, but liquids are much more viscous than gases.

The simplest example of flow of a viscous fluid is flow between two parallel plates, as shown in Fig. 13–24. The bottom plate is stationary, while the top plate moves with constant speed v. We find that the fluid in contact with each surface has the same speed as that surface; thus at the top surface the fluid has speed v, while the fluid adjacent to the bottom surface is at rest. The speeds of intermediate layers of fluid increase uniformly from one surface to the other, as shown by the arrows.

Flow of this type is called *laminar.* (A lamina is a thin sheet.) In laminar flow, the layers of liquid slide over one another, and a portion of the liquid, which at some instant has the shape *abcd*, will a moment later take the shape *abc'd'* and will become more and more distorted as the motion continues. That is, the liquid is in a state of continually increasing shear strain.

To maintain the motion, we must apply a constant force to the right on the upper, moving plate, and hence indirectly on the upper liquid surface. This force tends to drag the fluid, and the lower plate as well, to the right. Therefore we must apply an equal force toward the left on the lower plate to hold it stationary. These forces are labeled F in Fig. 13–24. If A is the area of the fluid over which the forces are applied (i.e., the surface area of the plates), the ratio F/A is the shear stress exerted on the fluid.

When we apply a shear stress to a *solid*, the effect is to produce a certain displacement of the solid, such as *dd'*. The shear strain is defined as the ratio of this displacement to the transverse dimension l, and within the elastic limit the shear stress is proportional to the shear strain. With a *fluid*, however, the shear strain increases without limit as long as the stress is applied. Experiments show that the stress depends not on the shear strain but on its *rate of change*. The strain in Fig. 13–24 (at the instant when the volume of fluid has the shape *abc'd'*) is *dd'/ad*, or *dd'/l*. Since l is constant, the *rate of change* of strain equals $1/l$ times the rate of change of *dd'*. But the rate of change of *dd'* is simply the *speed* of point d'; that is, the speed v of the moving wall. Hence

$$\text{Rate of change of shear strain} = \frac{v}{l}. \qquad (13\text{–}25)$$

The rate of change of shear strain is also referred to simply as the *strain rate*.

The *coefficient of viscosity* of the fluid, or simply its *viscosity* η, is defined as the ratio of the shear stress, F/A, to the rate of change of shear strain:

$$\eta = \frac{\text{shear stress}}{\text{rate of change of shear strain}} = \frac{F/A}{v/l},$$

13–23 Stroboscopic photograph of a golf ball being struck by a club. The picture was taken at 1000 flashes per second; the ball rotates once in four pictures, corresponding to an angular velocity of 250 revolutions per second. (Dr. Harold Edgerton, M.I.T., Cambridge, Massachusetts.)

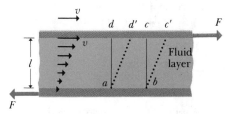

13–24 Laminar flow of a viscous fluid.

Forces are needed to make layers of a viscous fluid slide past each other.

Shear stress in a fluid depends on the rate of change of shear strain.

As a drop of milk falls freely through air, the forces of gravity, surface tension, and air resistance combine to deform and flatten the drop. Note that the drop does not have the classical teardrop shape often drawn by artists. (Dr. Harold Edgerton, M.I.T., Cambridge, Massachusetts.)

Viscosities of liquids increase with decreasing temperature: one reason it's harder to start your car in the winter

Flow of a viscous fluid in a pipe: The fluid sticks to the walls.

TABLE 13–3 Typical Values of Viscosity

Temperature, C°	Viscosity of Castor Oil, Poise	Viscosity of Water, Centipoise	Viscosity of Air, Micropoise
0	53.00	1.792	171
20	9.86	1.005	181
40	2.31	0.656	190
60	0.80	0.469	200
80	0.30	0.357	209
100	0.17	0.284	218

or

$$F = \eta A \frac{v}{l}. \tag{13–26}$$

For a liquid that flows readily, such as water or kerosene, the shear stress needed for a given strain rate is relatively small. For a liquid such as molasses or glycerine, a greater shear stress is necessary for the same strain rate, and the viscosity is correspondingly greater. Viscosities of all fluids are markedly dependent on temperature, increasing for gases and decreasing for liquids as the temperature increases—hence the expression "as slow as molasses in January." An important goal in the design of oils for engine lubrication is to *reduce* the temperature variation of viscosity as much as possible.

From Eq. (13–26), the unit of viscosity is that of force times distance, divided by area times velocity. In SI units it is

$$1 \text{ N·m·m}^{-2}(\text{m·s}^{-1})^{-1} = 1 \text{ N·s·m}^{-2}.$$

The corresponding cgs unit, 1 dyn·s·cm^{-2}, is the only viscosity unit in common use and is called 1 *poise* in honor of the French scientist Poiseuille. Thus

$$1 \text{ poise} = 1 \text{ dyn·s·cm}^{-2} = 10^{-1} \text{ N·s·m}^{-2}.$$

Small viscosities are expressed in *centipoises* ($1 \text{ cp} = 10^{-2}$ poise) or *micropoises* ($1 \text{ μp} = 10^{-6}$ poise). A few typical values are given in Table 13–3.

Not all fluids behave according to the direct proportionality of force and velocity predicted by Eq. (13–26). An interesting exception is blood, for which velocity increases more rapidly than force. This behavior results from the fact that blood is not a homogeneous fluid but rather a suspension of solid particles in a liquid. As the strain rate increases, these particles deform and become preferentially oriented to facilitate flow. The fluids that lubricate human joints show similar behavior. Some paints cling to the brush but flow when brushed on. A fluid for which Eq. (13–26) does hold is called a **Newtonian fluid;** it is a useful though approximate model for the behavior of many pure substances. Fluids that are suspensions or dispersions are often non-Newtonian in their viscous behavior.

When a viscous fluid flows in a stationary tube or pipe, the flow velocity is different at different points of a cross section. The outermost layer of fluid clings to the walls of the tube, and its velocity is zero. The walls exert a backward drag on this layer, which in turn drags backward on the next layer beyond it, and so on. If the velocity is not too great, the flow is *laminar*, with a velocity that is greatest at the center and decreases to zero at the walls. The flow is like that of a number of telescoping tubes sliding relative to one another, the central tube advancing most rapidly and the outer tube remaining at rest.

We can now derive an equation for the variation of velocity with radius for a cylindrical pipe of inner radius R. We start by considering the flow of a cylindrical element of fluid coaxial with the pipe, of radius r and length L, as shown in Fig. 13–25a. The force on the left end is $p_1\pi r^2$, and that on the right end $p_2\pi r^2$, as shown. The net force is thus

$$F = (p_1 - p_2)\pi r^2.$$

Since the element does not accelerate, this force must just balance the viscous retarding force at the surface of this element. The force is given by Eq. (13–26), but since the velocity does not vary uniformly with distance from the center, we must replace v/l in this expression with $-dv/dr$, where dv is the small change of velocity when we go from distance r to $r + dr$ from the axis. The area over which the viscous force acts is $A = 2\pi rL$. Thus the viscous force is

$$F = \eta 2\pi rL\frac{dv}{dr}.$$

Equating this to the net force due to pressure on the ends and rearranging, we find that

$$\frac{dv}{dr} = -\frac{(p_1 - p_2)r}{2\eta L}.$$

This relation shows that the velocity changes more and more rapidly as we go from the center ($r = 0$) to the pipe wall ($r = R$). The negative sign must be introduced because v decreases as r increases. Integrating, we find

$$-\int_v^0 dv = \frac{p_1 - p_2}{2\eta L}\int_r^R r\,dr,$$

and

$$v = \frac{p_1 - p_2}{4\eta L}(R^2 - r^2). \tag{13–27}$$

The velocity decreases from a maximum value $(p_1 - p_2)R^2/4\eta L$ at the center to zero at the wall. Thus the maximum velocity is proportional to the pressure change per unit length $(p_1 - p_2)/L$, called the *pressure gradient*. The curve in Fig. 13–25b is a graph of Eq. (13–27) with the v-axis horizontal and the r-axis vertical.

We can use Eq. (13–27) to find the *total* rate of fluid flow through the pipe. Note that the velocity at each point is proportional to the pressure gradient $(p_1 - p_2)/L$, so the total flow rate must also be proportional to this quantity. Now consider the thin-walled fluid element in Fig. 13–25c. The volume of fluid dV crossing the ends of this element in a time dt is $v\,dA\,dt$, where v is the velocity at the radius r and dA is the shaded area, equal to $2\pi r\,dr$. Taking the expression for v from Eq. (13–27), we get

$$dV = \frac{p_1 - p_2}{4\eta L}(R^2 - r^2)2\pi r\,dr\,dt.$$

The volume flowing across the entire cross section is obtained by integrating

Flow velocity in a cylindrical pipe depends on the distance from the axis.

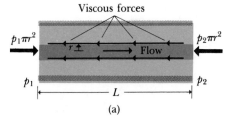

(a)

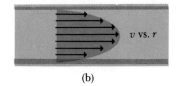

(b)

(c)

13–25 (a) Forces on a cylindrical element of a viscous fluid. (b) Velocity distribution for viscous flow.

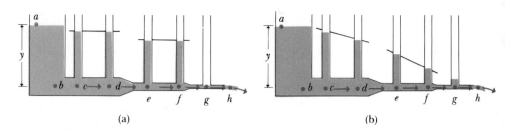

13–26 Pressures along a horizontal tube in which is flowing (a) an ideal fluid, (b) a viscous fluid.

over all elements between $r = 0$ and $r = R$:

$$dV = \frac{\pi(p_1 - p_2)}{2\eta L} \int_0^R (R^2 - r^2)r \, dr \, dt$$

$$= \frac{\pi}{8} \frac{R^4}{\eta} \frac{p_1 - p_2}{L} dt.$$

Volume flow rate of a viscous fluid in a pipe is proportional to the fourth power of the radius!

The total volume of flow per unit time, dV/dt, is given by

$$\frac{dV}{dt} = \frac{\pi}{8} \frac{R^4}{\eta} \frac{p_1 - p_2}{L}. \qquad (13\text{–}28)$$

This relation was first derived by Poiseuille and is called **Poiseuille's law.** It shows that the volume rate of flow is inversely proportional to viscosity, as might be expected. The volume flow rate is also proportional to the pressure gradient along the pipe, and it varies as the *fourth power* of the radius. For example, if the radius is halved, the flow rate is reduced by a factor of 16. This relation is familiar to physicians in connection with the selection of needles for hypodermic syringes. Needle size is much more important than thumb pressure in determining the flow rate from the needle; doubling the needle diameter has the same effect as increasing the thumb force sixteenfold. Similarly, blood flow in arteries and veins can be controlled over a wide range by relatively small changes in diameter, an important temperature-control mechanism in warm-blooded animals.

The difference between the flow of an ideal nonviscous fluid and one having viscosity is shown in Fig. 13–26, where fluid is flowing along a horizontal tube of varying cross section. The height of the fluid in each small vertical tube is proportional to the gauge pressure at the corresponding point in the horizontal tube.

Some examples of the differences in behavior between viscous and nonviscous liquids

In part (a), the fluid has no viscosity. The pressure at b is very nearly the static pressure ρgy, since the velocity is small in the large tank. The pressure at c is less than at b because the fluid must accelerate between these points. The pressures at c and d are equal, however, since the velocity and elevation at these points are the same. Further pressure drops occur between d and e, and between f and g. The pressure at g is atmospheric, and the gauge pressure at this point is zero.

Figure 13–26b shows the effect of viscosity. Again, the pressure at b is nearly the static pressure ρgy. The pressure drops from b to c, due now in part to viscous effects, and there is a further drop from c to d. The pressure gradient in this part of the tube is represented by the slope of the dotted line. The drop from d to e results in part from acceleration and in part from viscosity. The pressure gradient between e and f is greater than between c and d because of the smaller radius in this portion. Finally, the pressure at g is somewhat above atmospheric, since there is now a pressure gradient between this point and the end of the tube.

One more useful relation in viscous fluid flow is the expression for the force F exerted on a sphere of radius r moving with speed v through a fluid with viscosity η. This relation is called **Stokes's law** and is as follows:

$$F = 6\pi\eta r v. \qquad (13\text{–}29)$$

We have encountered this kind of velocity-proportional force before, in Example 5–17 (Section 5–4).

A sphere falling in a viscous fluid reaches a *terminal velocity* v_T at which the viscous retarding force plus the buoyant force equals the weight of the sphere. Let ρ be the density of the sphere and ρ' the density of the fluid. The weight of the sphere is then $\frac{4}{3}\pi r^3\rho g$, and the buoyant force is $\frac{4}{3}\pi r^3\rho' g$; when the terminal velocity is reached, the total force is zero, and

$$\tfrac{4}{3}\pi r^3\rho' g + 6\pi\eta\, r v_T = \tfrac{4}{3}\pi r^3\rho g,$$

or

$$v_T = \frac{2}{9}\frac{r^2 g}{\eta}(\rho - \rho'). \qquad (13\text{–}30)$$

When the terminal velocity of a sphere of known radius and density is measured, the viscosity of the fluid in which it is falling can be found from Eq. (13–30). Conversely, if the viscosity is known, the radius of the sphere can be determined by measuring the terminal velocity. This method was used by Millikan to determine the radius of very small electrically charged oil drops by observing their free fall in air. He used these drops to measure the charge of an individual electron, as we will discuss further in Section 26–6.

The Stokes's-law force is what keeps clouds in the air. The terminal velocity for a water droplet of radius 10^{-5} m is of the order of 1 mm·s^{-1}. Similarly, raindrops have terminal velocities of a few meters per second; otherwise they would smash everything in their path.

13–9 TURBULENCE

When the speed of a fluid flowing in a pipe exceeds a certain critical value, which depends on the properties of the fluid and the diameter of the pipe, the nature of the flow becomes extremely complicated. Within a thin layer adjacent to the pipe walls, called the *boundary layer,* the flow is still laminar. Beyond the boundary layer, the motion is highly irregular. Random local circular currents called *vortices* develop within the fluid, with a large increase in the resistance to flow. Flow of this sort is called *turbulent.*

Experiments show that a combination of four factors determines whether the flow of a fluid through a pipe is laminar or turbulent. This combination is known as the **Reynolds number,** N_R, and is defined as

$$N_R = \frac{\rho v D}{\eta}, \qquad (13\text{–}31)$$

where ρ is the density of the fluid, v is its average forward velocity, η its viscosity, and D the diameter of the pipe. (The average velocity of the fluid is defined as the uniform velocity over the entire cross section of the pipe that would result in the same volume rate of flow.) The Reynolds number, $\rho v D/\eta$, is a *dimensionless* quantity and has the same numerical value in any consistent system of units. For example, for water at 20°C flowing in a pipe of diameter

Stokes's law: The force on a sphere moving through a viscous fluid is proportional to its speed, its radius, and the viscosity.

Using Stokes's law in a measurement of the electron charge

Turbulence: when orderly flow turns to chaos

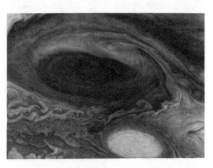

Photograph of the Great Red Spot of Jupiter, sent back by the Voyager spacecraft. This feature, first observed in the seventeenth century, is a large-scale atmospheric disturbance similar to a cyclone. (Courtesy of NASA.)

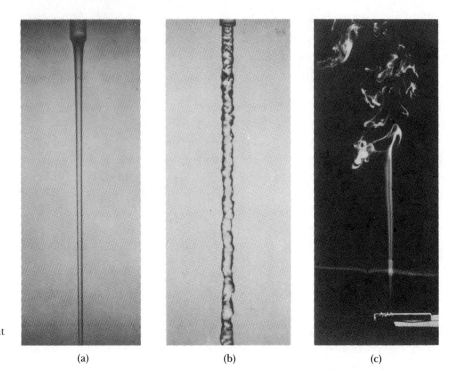

13-27 (a) Laminar flow. (b) Turbulent flow. (c) First laminar, then turbulent flow.

(a)   (b)   (c)

1 cm with an average velocity of 10 cm·s^{-1}, the Reynolds number is

$$N_R = \frac{\rho v D}{\eta} = \frac{(1 \text{ g·cm}^{-3})(10 \text{ cm·s}^{-1})(1 \text{ cm})}{0.01 \text{ dyn·s·cm}^{-2}} = 1000.$$

The Reynolds number: predicting when turbulence will occur

Had the four quantities been expressed originally in SI units, the same value of 1000 would have been obtained.

A variety of experiments have shown that for fluid flow in a cylindrical pipe, the flow is laminar when the Reynolds number is less than about 2000. When it is above about 3000 the flow is turbulent. In the transition region between 2000 and 3000 the flow is unstable and may change from one type to the other. Thus for water at 20°C flowing in a pipe 1 cm in diameter, the flow is laminar when

$$\frac{\rho v D}{\eta} \le 2000,$$

or when

$$v \le \frac{(2000)(0.01 \text{ dyn·s·cm}^{-2})}{(1 \text{ g·cm}^{-3})(1 \text{ cm})} = 20 \text{ cm·s}^{-1}.$$

Above about 30 cm·s^{-1} the flow is turbulent. If air at the same temperature were flowing at 30 cm·s^{-1} in the same pipe, the Reynolds number would be

13-28 An example of the onset of turbulence produced by a cylindrical obstacle. Upstream flow is laminar, but downstream flow shows the formation of eddies or vortices. The alternating series of vortices is called a Karman vortex street. The pattern is made visible by injecting smoke filaments into the air upstream from the obstacle. (Courtesy Department of Aeronautics, Imperial College of Science and Technology.)

$$N_R = \frac{(0.0013 \text{ g·cm}^{-3})(30 \text{ cm·s}^{-1})(1 \text{ cm})}{181 \times 10^{-6} \text{ dyn·s·cm}^{-2}} = 215.$$

Since this is much less than 3000, the flow would be laminar; it would not become turbulent unless the velocity were at least 420 cm·s^{-1}.

The distinction between laminar and turbulent flow is shown in the photographs of Fig. 13-27. In (a) and (b) the fluid is water; in (c), air and smoke particles. Figure 13-28 shows the onset of turbulence in the flow of a liquid past a cylindrical obstacle. The study of turbulent flow, and of chaotic behavior in general, is an area of great current interest in theoretical physics.

SUMMARY

Density is mass per unit volume. If a mass m of material has volume V, its density ρ is given by

$$\rho = m/V. \qquad (13\text{--}1)$$

Specific gravity is the ratio of the density of a material to the density of water. It is a dimensionless number.

Pressure is force per unit area. Pascal's law: Pressure applied to the surface of an enclosed fluid is transmitted undiminished to every portion of the fluid. Absolute pressure is the total pressure in a fluid; gauge pressure is the difference between absolute pressure and atmospheric pressure.

The SI unit of pressure is the pascal (Pa). $1\ \text{Pa} = 1\ \text{N}\cdot\text{m}^{-2}$. Other units of pressure are:

$1\ \text{bar} = 10^5\ \text{Pa}$

$1\ \text{millibar} = 10^{-3}\ \text{bar} = 10^2\ \text{Pa}.$

$1\ \text{Torr} = 1\ \text{millimeter of mercury (mmHg)}$

$1\ \text{atmosphere} = 1.013 \times 10^5\ \text{Pa} = 760\ \text{Torr} = 1.013\ \text{bar}$

Archimedes' principle: When a body is immersed in a fluid, the fluid exerts an upward force on the body equal to the weight of the fluid that is displaced by the body.

A boundary surface between a liquid and a gas behaves like a surface under tension; the force per unit length across a line on such a surface is called the surface tension, denoted by γ. Surface tension is responsible for pressure differences between the interior and exterior of drops and bubbles, and for the phenomenon of capillarity.

An ideal fluid is an incompressible fluid with no viscosity. A flow line is the path followed by an individual fluid particle; a streamline is a curve tangent at each point in the fluid to the velocity vector at that point. A flow tube is a tube bounded at its sides by flow lines. A stagnation point is a point where the flow velocity is zero. In laminar flow, layers of fluid slide smoothly past each other, while in turbulent flow there is great disorder and a constantly changing flow pattern.

Conservation of mass in an incompressible fluid is expressed by the equation of continuity: For two cross sections A_1 and A_2 in a flow tube, the velocities v_1 and v_2 are related by

$$A_1 v_1 = A_2 v_2. \qquad (13\text{--}12)$$

Bernoulli's equation relates the pressure, flow speed, and elevation at different points in an ideal fluid; if subscripts 1 and 2 refer to any two points in the fluid,

$$p_1 + \rho g y_1 + \tfrac{1}{2}\rho v_1{}^2 = p_2 + \rho g y_2 + \tfrac{1}{2}\rho v_2{}^2. \qquad (13\text{--}17)$$

The viscosity of a fluid characterizes its resistance to shear strain. A fluid in which the viscous force is proportional to strain rate is called a Newtonian fluid. When such a fluid flows in a cylindrical pipe of inner radius R, the flow speed v at a distance r from the axis is given by

$$v = \frac{p_1 - p_2}{4\eta L}(R^2 - r^2), \qquad (13\text{--}27)$$

KEY TERMS

density
specific gravity
pressure
Pascal's law
gauge pressure
absolute pressure
mercury barometer
atmosphere
buoyancy
Archimedes' principle
surface tension
contact angle
capillarity
ideal fluid
flow line
streamline
flow tube
laminar flow
turbulent flow
continuity equation
Bernoulli's equation
viscosity
Newtonian fluid
Poiseuille's law
Stokes's law
Reynolds number

where L is the length of pipe, p_1 and p_2 are the pressures at the two ends, and η is the viscosity. The total volume rate in this situation is given by Poiseuille's equation:

$$\frac{dV}{dt} = \frac{\pi}{8} \frac{R^4}{\eta} \frac{p_1 - p_2}{L}. \tag{13-28}$$

A sphere of radius r moving with speed v through a fluid having viscosity η experiences a resisting force given by Stokes's law:

$$F = 6\pi\eta r v. \tag{13-29}$$

The Reynolds number for fluid flow in a pipe is defined as

$$N_R = \rho v D / \eta. \tag{13-31}$$

The transition from laminar to turbulent flow occurs at a value of N_R in the range of 2000 to 3000.

QUESTIONS

13–1 A rubber hose is attached to a funnel, and the free end is bent around to point upward. When water is poured into the funnel, it rises in the hose to the same level as in the funnel, despite the fact that the funnel has a lot more water in it than the hose. Why?

13–2 Is the pressure inside a siphon tube greater or less than atmospheric pressure? Does this suggest a limitation on the maximum height over which a siphon can be used?

13–3 How can one instrument function as both a barometer and an altimeter?

13–4 In describing the size of a large ship, one uses such expressions as "it displaces 20,000 tons." What does this mean? Can the weight of the ship be obtained from this information?

13–5 A popular, though expensive, sport is hot-air ballooning, in which a large nylon balloon is filled with air heated by a gas burner at the bottom. Why must the air be heated? How does the balloonist control ascent and descent?

13–6 A rigid lighter-than-air balloon filled with helium cannot continue to rise indefinitely. Why not? What determines the maximum height it can attain?

13–7 Does the buoyant force on an object immersed in water change if the object and the container of water are placed in an elevator that accelerates upward?

13–8 If a person can barely float in sea water, will he or she float more easily in fresh water, or be unable to float at all? Why?

13–9 The purity of gold can be tested by weighing it in air and in water. How? Why is it not feasible to make a fake gold brick by gold-plating some cheaper material?

13–10 A possible new technology made feasible by the "zero-gravity" environment of an orbiting space station is the making of metal foam. Why is it easier to make metal foam under zero-g conditions than on earth?

13–11 Why can't a skin diver obtain an air supply at any desired depth by breathing through a "snorkel," a tube connected to a face mask and having its upper end above the water surface?

13–12 Suppose the door of a room makes an airtight but frictionless fit in its frame. Do you think you could open the door if the air pressure on one side were standard atmospheric pressure and that on the other side differed from standard by 1%?

13–13 Is the continuity relation, Eq. (13–12), valid for compressible fluids? If not, is there a similar relation that *is* valid?

13–14 If the velocity at each point in space in steady-state fluid flow is constant, how can a fluid particle accelerate?

13–15 Does the "lift" of an airplane wing depend on altitude?

13–16 How does a baseball pitcher give the ball the spin that makes it curve? Can he make it curve in either direction? Does it matter whether he is right-handed or left-handed? What is a spitball? Why is it illegal?

13–17 In a store-window vacuum cleaner display, a table-tennis ball is suspended in midair in a jet of air blown from the outlet hose of a tank-type vacuum cleaner. The ball bounces around a little but always returns to the center of the jet, even if it is tilted. How does this behavior illustrate the Bernoulli relation?

13–18 Why do jet airplanes usually fly at altitudes of about 30,000 ft, though it takes a lot of fuel to climb that high?

13–19 What causes the sharp hammering sound sometimes heard from a water pipe when an open faucet is suddenly turned off?

13–20 When a smooth-flowing stream of water comes out of a faucet, it narrows as it falls, and if it falls far enough it eventually breaks up into drops. Why does it narrow? Why does it break up?

13–21 A tornado consists of a rapidly whirling air vortex. Why is the pressure always much lower in the center than at the outside? How does this condition account for the destructive power of a tornado?

13–22 When paddling a canoe, one can attain a certain critical speed with relatively little effort, and then a much greater effort is required to make the canoe go even a little faster. Why?

13–23 Why does air escaping from the open end of a small pipe make a lot more noise than air escaping at the same volume rate from a large pipe?

EXERCISES

Section 13–1 Density

13–1 A rectangular block of unidentified material has dimensions $5 \times 15 \times 30$ cm and a mass of 2.0 kg.

a) What is the density?

b) Will the material float in water?

13–2 A lead cube has a total mass of 50 kg. What is the length of a side?

Section 13–2 Pressure in a Fluid

13–3 A diving bell is to be designed to withstand the pressure of sea water at a depth of 600 m.

a) What is the gauge pressure at this depth?

b) What is the total force on a circular glass window 15 cm in diameter, if the pressure inside the dividing bell equals the pressure at the surface of the water?

13–4 What gauge pressure must a pump produce in order to pump water from the bottom of Grand Canyon (730-m elevation) to Indian Gardens (1370 m)? Express your results in pascals and in atmospheres.

13–5 A barrel contains a 0.15-m layer of oil floating on water that is 0.25 m deep. The density of oil is 600 kg·m⁻³.

a) What is the gauge pressure at the oil–water interface?

b) What is the gauge pressure at the bottom of the barrel?

13–6 The piston of a hydraulic automobile lift is 0.30 m in diameter. What gauge pressure, in pascals, is required to lift a car of mass 1500 kg? Also express this pressure in atmospheres.

13–7 The liquid in the open-tube manometer in Fig. 13–4a is mercury, and $y_1 = 3$ cm, $y_2 = 8$ cm. Atmospheric pressure is 970 millibars.

a) What is the absolute pressure at the bottom of the U-tube?

b) What is the absolute pressure in the open tube, at a depth of 5 cm below the free surface?

c) What is the absolute pressure of the gas in the tank?

d) What is the gauge pressure of the gas, in pascals?

e) What is the gauge pressure of the gas, in "cm of mercury"?

f) What is the gauge pressure in "cm of water"?

Section 13–3 Buoyancy

13–8 A solid brass statue has a weight in air of 125 N.

a) What is its volume?

b) What is its apparent weight when immersed in water?

13–9 A certain object weighs 12 N in air and 10 N when immersed in water. Find its total volume and its density.

13–10 When an iceberg floats in sea water, what fraction of its volume is submerged? (The ice, being of glacial origin, is fresh water.)

13–11 A slab of ice floats on a freshwater lake. What minimum volume must the slab have in order for an 80-kg man to be able to stand on it without getting his feet wet?

13–12 A large hollow plastic sphere is held below the surface of a freshwater lake by a cable anchored to the bottom of the lake. The sphere has a volume of 0.3 m³, and the tension in the cable is observed to be 600 N.

a) Calculate the buoyant force exerted by the water on the sphere.

b) What is the mass of the sphere?

c) The cable breaks, and the sphere rises to the surface of the lake. When it comes to rest again, what fraction of its volume will be submerged?

13–13 A cubical block of wood 10 cm on a side floats at the interface between oil and water as in Fig. 13–29, with its lower surface 2 cm below the interface. The density of the oil is 600 kg·m⁻³.

a) What is the gauge pressure at the upper face of the block?

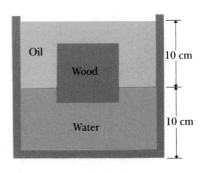

FIGURE 13–29

b) What is the gauge pressure at the lower face of the block?

c) What is the mass of the block?

Section 13–4 Surface Tension

13–14 Find the excess pressure

a) inside a water drop of radius 2 mm;

b) inside one of radius 0.01 mm, typical of water droplets in fog.

Assume $T = 20°C$.

13–15 Find the gauge pressure, in pascals, in a soap bubble 5 cm in diameter. The surface tension is 25×10^{-3} N·m^{-1}.

Section 13–5 Fluid Flow

13–16 Water is flowing in a pipe of varying cross section. At point 1 the cross-sectional area of the pipe is 0.08 m^2 and the fluid velocity is 4 m·s^{-1}.

a) What is the fluid velocity at points in the pipe where the cross-sectional area is

   i) 0.06 m^2?   ii) 0.112 m^2?

b) Calculate the volume of water discharged from the open end of the pipe in 1 s.

13–17 Water is flowing in a circular pipe of varying cross section.

a) At one place in the pipe the radius of the pipe is 0.2 m. What must the water velocity be at this point if the volume flow rate in the pipe is 0.8 m^3·s^{-1}?

b) At a second point in the pipe the water velocity is 3.8 m·s^{-1}. What is the radius of the pipe at this point?

Section 13–7 Applications of Bernoulli's Equation

13–18 A circular hole 2 cm in diameter is cut in the side of a large standpipe, 10 m below the water level in the standpipe. The top of the standpipe is open to the air. Find

a) the velocity of efflux;

b) the volume discharged per unit time.

Neglect the contraction of the streamlines after they emerge from the hole.

13–19 A sealed tank containing sea water to a height of 2 m also contains air above the water at a gauge pressure of 40 atm. Water flows out from a hole at the bottom. The cross-sectional area of the hole is 10 cm^2.

a) Calculate the efflux velocity of the water.

b) Calculate the reaction force on the tank exerted by the water in the emergent stream.

13–20 At a certain point in a horizontal pipeline, the velocity is 2 m·s^{-1} and the gauge pressure is 1.0×10^4 Pa above atmospheric. Find the gauge pressure at a second point in the line if the cross-sectional area at the second point is one-half that at the first. The liquid in the pipe is water.

13–21 Water flowing in a horizontal pipe discharges at the rate of 0.004 m^3·s^{-1}. At a point in the pipe where the cross section is 0.001 m^2, the absolute pressure is 1.2×10^5 Pa. What must be the cross section of a constriction in the pipe such that the pressure there is reduced to 1.0×10^5 Pa?

13–22 At a certain point in a pipeline the velocity is 4 m·s^{-1} and the gauge pressure is 3.0×10^5 Pa. Find the gauge pressure at a second point in the line 20 m lower than the first, if the cross section at the second point is one-half that at the first. The liquid in the pipe is water.

13–23 What gauge pressure is required in the city mains for a stream from a fire hose connected to the mains to reach a vertical height of 20 m?

13–24 An airplane of mass 6000 kg has a wing area of 60 m^2. If the pressure on the lower wing surface is 0.60×10^5 Pa during level flight at an elevation of 4000 m, what is the pressure on the upper wing surface?

13–25 Assume that air is streaming horizontally past a small airplane's wings such that the velocity is 40 m·s^{-1} over the top surface and 30 m·s^{-1} past the bottom surface. If the plane has a mass of 300 kg and a wing area of 5 m^2, what is the net force (including the effects of gravity) on the airplane? The density of air is 1.3 kg·m^{-3}.

Section 13–8 Viscosity

13–26 Water at 20°C flows through a pipe of radius 1.0 cm. If the flow velocity at the center is 0.10 m·s^{-1}, find the pressure drop along a 2-m section of pipe due to viscosity.

13–27 Water at 20°C is flowing in a pipe of radius 20 cm. If the fluid velocity in the center of the pipe is 3 m·s^{-1}, what is the fluid velocity

a) 10 cm from the center of the pipe (i.e., halfway between the center and the walls)?

b) at the walls of the pipe?

13–28 A viscous liquid flows through a tube with laminar flow, as in Fig. 13–25b. Prove that the volume rate of flow is the same as if the velocity were uniform at all points of a cross section and equal to half the velocity at the axis.

13–29 Water at 20°C is flowing in a horizontal pipe that is 15 m long. A pump maintains a gauge pressure of 800 Pa at a large tank at one end of the pipe. The other end of the pipe is open to the air.

a) If the pipe has a diameter of 8 cm, what is the volume flow rate?

b) What gauge pressure must the pump provide to achieve the same volume flow rate for a pipe of diameter 4 cm?

c) For the pipe in part (a), what does the volume flow rate become if the water is at a temperature of 60°C?

13–30 A copper sphere of mass 0.20 g is observed to fall with a terminal velocity of 0.08 m·s^{-1} in an unknown liquid. If the density of copper is 8900 kg·m^{-3} and that of the liquid is 2800 kg·m^{-3}, what is the viscosity of the liquid?

13–31 What velocity must a gold sphere of radius 4 mm have in water at 20°C for the viscous drag force to be one-fourth the weight of the sphere?

Section 13–9 Turbulence

13–32 Water at 20°C flows with a speed of 0.35 m·s^{-1} through a pipe of diameter 3 mm.

a) What is the Reynolds number?

b) What is the nature of the flow?

13–33 Water at a temperature of 80°C is flowing at a speed of 2 m·s^{-1} through a tube of diameter 4 mm.

a) What is the Reynolds number?

b) What is the nature of the flow?

PROBLEMS

13–34 A swimming pool measures 25 × 8 × 3 m deep. Compute the force exerted by the water against

a) the bottom, and

b) either end, which has width 8 m. (*Hint:* Calculate the force on a thin horizontal strip at a depth h, and integrate this over the end of the pool.)

Do not include the force due to air pressure.

13–35. The upper edge of a gate in a dam runs along the water surface. The gate is 2 m high and 3 m wide and is hinged along a horizontal line through its center. Calculate the torque about the hinge arising from the force due to the water. (*Hint:* Use a procedure similar to that used in Problem 13–34: Calculate the torque of a thin horizontal strip at a depth h and integrate this over the gate.)

13–36 A U-shaped tube open to the air at both ends contains some mercury. A quantity of water is carefully poured into the left arm of the tube until the vertical height of the water column is 15 cm, as shown in Fig. 13–30.

a) What is the gauge pressure at the water–mercury interface?

b) Calculate the vertical distance h from the top of the mercury in the right-hand arm of the tube to the top of the water in the left-hand arm.

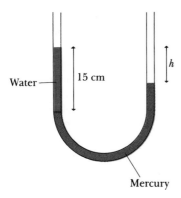

Water — 15 cm

h

Mercury

FIGURE 13–30

13–37 The deepest point known in any of the earth's oceans is in the Marianas Trench and is 10.92 km deep.

a) Assuming water to be incompressible, what is the pressure at this depth? Use the density of sea water.

b) The actual pressure is 1.17×10^8 Pa. Your calculated value of the pressure will be less because the density actually varies with depth. Take your value for the pressure and, using the compressibility of water, find its

density at the bottom of the Marianas Trench. What is the percent change in the density of the water?

13–38 According to advertising claims, a certain small car will float in water.

a) If the mass of the car is 1000 kg and its interior volume 4.0 m^3, what fraction of the car is immersed when it floats? The buoyancy of steel and other materials may be neglected.

b) As water gradually leaks in and displaces the air in the car, what fraction of the interior volume is filled with water when the car sinks?

13–39 The densities of air, helium, and hydrogen (at standard conditions) are, respectively, 1.29 kg·m^{-3}, 0.178 kg·m^{-3}, and 0.0899 kg·m^{-3}.

a) What is the volume in cubic meters displaced by a hydrogen-filled dirigible that has a total "lift" of 10,000 kg? (The "lift" is the total mass of the dirigible that can be supported by the buoyant force, in addition to the mass of the gas with which it is filled.)

b) What would be the "lift" if helium were used instead of hydrogen? In view of your answer, why is helium the gas actually used? (*Hint:* Remember the *Hindenberg.*)

13–40 A hydrometer consists of a spherical bulb and a cylindrical stem of cross section 0.4 cm^2. The total volume of bulb and stem is 13.2 cm^3. When immersed in water, the hydrometer floats with 8 cm of the stem above the water surface. In alcohol, 1 cm of the stem is above the surface. Find the density of alcohol. (*Note.* This illustrates the accuracy of such a hydrometer. Relatively small density differences give rise to relatively large differences in hydrometer readings.)

13–41 A barge is 20 m wide and has a longitudinal cross section as shown in Fig. 13–31. (The dimension not shown in the figure is 20 m.) If the barge is made out of 7.5-cm-thick steel plate on each of its four sides and its bottom, what mass of coal can the barge carry without sinking? Is there enough room in the barge to hold this amount of coal? (The density of coal is about 1.5 g·cm^{-3}.)

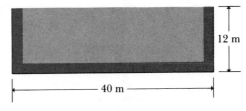

12 m

40 m

FIGURE 13–31

13–42 Consider an object of height h, mass M, and uniform cross section A floating upright in a liquid of density ρ.

a) Calculate the vertical distance from the surface of the liquid to the bottom of the floating object, at equilibrium.

b) A downward force F is applied to the top of the object. At the new equilibrium position, how much farther below the surface of the liquid is the bottom of the object than it was in part (a)? (Assume that F is small enough for some of the object to remain above the surface of the liquid.)

c) Your result in (b) shows that if F is suddenly removed, the object will oscillate up and down in simple harmonic motion. Calculate the period of this motion, in terms of the density ρ of the liquid and the mass M and cross section A of the object.

d) A 1500-kg cylindrical can buoy floats vertically in salt water (specific gravity = 1.03). The diameter of the buoy is 1.0 m. Calculate the additional distance the buoy will sink when a 100-kg man stands on top.

e) Calculate the period of the resulting vertical simple harmonic motion when the man dives off.

f) Calculate the period of oscillation of an ice cube 3 cm on a side floating in water of density $1 \times 10^3 \text{ kg·m}^{-3}$, if it is pushed down and released.

13–43 A piece of wood is 0.5 m long, 0.2 m wide, and 0.02 m thick. Its specific gravity is 0.6. What volume of lead must be fastened underneath to sink the wood in calm water so that its top is just even with the water level?

13–44 When a life preserver of volume 0.03 m³ is immersed in sea water (specific gravity 1.03), it will just support an 80-kg man (specific gravity 1.2) with $\frac{2}{10}$ of his volume above water. What is the density of the material composing the life preserver?

13–45 A block of balsa wood placed in one scale pan of an equal-arm balance is found to be exactly balanced by a 0.100-kg brass mass in the other scale pan. Find the true mass of the balsa wood, if its specific gravity is 0.15. Neglect the buoyancy of the brass in air. (The density of air is 1.29 kg·m⁻³.)

13–46 Block A in Fig. 13–32 hangs by a cord from spring balance D and is submerged in a liquid C contained in beaker B. The mass of the beaker is 1 kg; the mass of the

liquid is 1.5 kg. Balance D reads 2.5 kg and balance E reads 7.5 kg. The volume of block A is 0.003 m³.

a) What is the mass per unit volume of the liquid?

b) What will each balance read if block A is pulled up out of the liquid?

13–47 A piece of gold-aluminum alloy weighs 45 N. When suspended from a spring balance and submerged in water, the balance reads 36 N. What is the weight of gold in the alloy if the specific gravity of gold is 19.3 and the specific gravity of aluminum is 2.5?

13–48 A cubical block of wood 0.10 m on a side and of density 500 kg·m⁻³ floats in a jar of water. Oil of density 800 kg·m⁻³ is poured on the water until the top of the oil layer is 0.04 m below the top of the block.

a) How deep is the oil layer?

b) What is the gauge pressure at the lower face of the block?

13–49 A cubical block of wood 0.30 m on a side is weighted so that its center of gravity is at the point shown in Fig. 13–33a, and it floats in water with one-half its volume submerged. Compute the restoring torque when the block is "heeled" at an angle of 45° as in Fig. 13–33b.

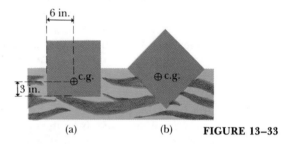

(a)   (b)   **FIGURE 13–33**

13–50 A glass tube of inside diameter 1 mm is dipped vertically into a container of mercury, with its lower end 1 cm below the mercury surface. What must be the gauge pressure of air in the tube to blow a hemispherical bubble at its lower end?

13–51 A cylindrical vessel, open at the top, is 0.20 m high and 0.10 m in diameter. A circular hole whose cross-sectional area is 1 cm² is cut in the center of the bottom of the vessel. Water flows into the vessel from a tube above it at the rate of 1.40×10^{-4} m³·s⁻¹. How high will the water in the vessel rise?

13–52 Water stands at a depth H in a large open tank whose side walls are vertical (Fig. 13–34). A hole is made in one of the walls at a depth h below the water surface.

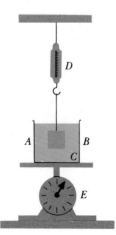

FIGURE 13–32

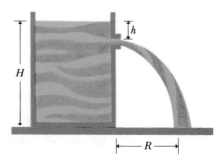

FIGURE 13–34

a) At what distance R from the foot of the wall does the emerging stream of water strike the floor?

b) Let $H = 12$ m and $h = 3$ m. At what height below the water surface could a second hole be cut so that the stream emerging from it would have the same range as for the first hole?

13–53 Water flows through a horizontal pipe of cross-sectional area 10 cm². At one section the cross-sectional area is reduced to 5 cm². The pressure difference between the two sections is 300 Pa. How many cubic meters of water will flow out of the pipe in 1 min?

13–54 Water flows steadily from a reservoir, as in Fig. 13–35. The elevation of point 1 is 10 m; of points 2 and 3, 1 m. The cross section at point 2 is 0.04 m²; at point 3, 0.02 m². The area of the reservoir is very large compared with the cross sections of the pipe. Assuming Bernoulli's equation applies, compute

a) the gauge pressure at point 2;

b) the discharge rate in cubic meters per second.

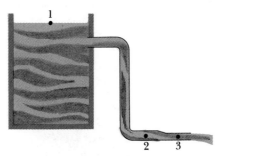

FIGURE 13–35

13–55 Modern airplane design calls for a "lift" of about 1000 N per square meter of wing area. Assume that air flows past the wing of an aircraft with streamline flow. If the velocity of flow past the lower wing surface is 100 m·s⁻¹, what is the required velocity over the upper surface to give a "lift" of 1000 N·m⁻²? The density of air is 1.29 kg·m⁻³.

13–56 The section of pipe shown in Fig. 13–36 has a cross section of 40 cm² at the wider portions and 10 cm² at the constriction. Water is flowing in the pipe, and the discharge from the pipe is 3.0×10^{-3} m³·s⁻¹.

a) Find the velocities at the wide and the narrow portions.

b) Find the pressure difference between these portions.

c) Find the difference in height between the mercury columns in the U-shaped tube.

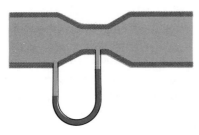

FIGURE 13–36

13–57 Two very large open tanks, A and F (Fig. 13–37), both contain the same liquid. A horizontal pipe BCD, having a constriction at C, leads out of the bottom of tank A, and a vertical pipe E opens into the constriction at C and dips into the liquid in tank F. Assume streamline flow and no viscosity. If the cross section at C is one-half that at D, and if D is a distance h_1 below the level of the liquid in A, to what height h_2 will liquid rise in pipe E? Express your answer in terms of h_1. Neglect changes in atmospheric pressure with elevation.

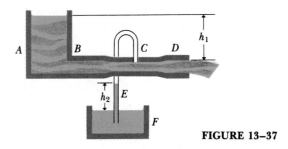

FIGURE 13–37

13–58 Water stands at a depth of 1 m in an enclosed tank whose side walls are vertical. The space above the water surface contains air at a gauge pressure of 8×10^5 Pa. The tank rests on a platform 2 m above the floor. A hole of cross-sectional area 1 cm² is made in one of the side walls just above the bottom of the tank.

a) Where does the stream of water from the hole strike the floor?

b) What is the vertical force exerted on the floor by the stream?

c) What is the horizontal force exerted on the tank?

Assume that the water level and pressure in the tank remain constant, and neglect any effect of viscosity.

13–59 Figure 13–27a shows a liquid flowing from a vertical pipe. Notice that the vertical stream of the liquid has a very definite shape as it flows from the pipe.

a) We will now get the equation for this shape. Assume that the liquid is in free fall once it leaves the pipe and then find an equation for the velocity of the liquid as a function of the distance it has fallen. Combining this with the equation of continuity, find an expression for the radius of the stream of liquid.

b) If water flows out of a vertical pipe at 1.0 m·s⁻¹, how far below the outlet will the radius be one half the original radius of the stream?

13–60 The densities of steel and of glycerine are 8500 kg·m⁻³ and 1320 kg·m⁻³, respectively. The viscosity of glycerine is 8.3 poise.

a) With what velocity is a steel ball 1 mm in radius falling in a tank of glycerine at an instant when its acceleration is one half that of a freely falling body?

b) What is the terminal velocity of the ball?

13–61

a) With what terminal velocity will an air bubble 1 mm in diameter rise in a liquid of viscosity 1.5 poise and density 900 kg·m⁻³?

b) What is the terminal velocity of the same bubble in water at 20°C?

13-62 Oil having a viscosity of 3.0 poise and a density of 900 kg·m^{-3} is to be pumped from one large open tank to another through 1 km of smooth steel pipe 0.15 m in diameter. The line discharges into the air at a point 30 m above the level of the oil in the supply tank.

a) What gauge pressure, in Pa and in atmospheres, must the pump exert in order to maintain a flow of 0.05 m^3·s^{-1}?

b) What is the power consumed by the pump?

13-63 The tank at the left in Fig. 13-26a has a very large cross section and is open to the atmosphere. The depth $y = 0.4$ m. The cross sections of the horizontal tubes leading out of the tank are, respectively, 1 cm^2, 0.5 cm^2, and 0.2 cm^2. The liquid is ideal, having zero viscosity.

a) What is the volume rate of flow out of the tank?

b) What is the velocity in each portion of the horizontal tube?

c) What are heights of the liquid in the vertical side tubes?

Suppose that the liquid in Fig. 13-26b has a viscosity of 0.5 poise and a density of 800 kg·m^{-3}, and that the depth of liquid in the large tank is such that the volume rate of flow is the same as in part (a) above. The distance between the side tubes at c and d, and between those at e and f, is 0.20 m. The cross sections of the horizontal tubes are the same in both diagrams.

d) What is the difference in level between the tops of the liquid columns in tubes c and d?

e) In tubes e and f.

f) What is the flow velocity on the axis of each part of the horizontal tube?

g) Is it reasonable to suppose that the flow of this second liquid is laminar?

h) Would the flow be laminar if the liquid were water at 20°C, with the same volume flow rate?

CHALLENGE PROBLEMS

13-64 The following is quoted from a letter. How would you reply?

> It is the practice of carpenters hereabouts, when laying out and leveling up the foundations of relatively long buildings, to use a garden hose filled with water, into the ends of the hose being thrust glass tubes 10 to 12 inches long.
>
> The theory is that water, seeking a common level, will be of the same height in both the tubes and thus effect a level. Now the question rises as to what happens if a bubble of air is left in the hose. Our greybeards contend the air will not affect the reading from one end to the other. Others say that it will cause important inaccuracies.

Can you give a relatively simple answer to this question, together with an explanation? Figure 13-38 gives a rough sketch of the situation that caused the dispute.

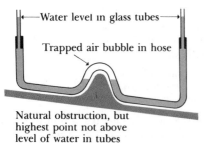

← Water level in glass tubes →

Trapped air bubble in hose

Natural obstruction, but highest point not above level of water in tubes

FIGURE 13-38

13-65 Suppose a piece of styrofoam, $\rho = 1.0 \times 10^2$ kg·m^{-3}, is held completely submerged in water, as shown in Fig. 13-39.

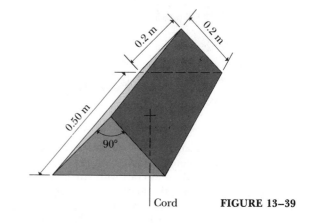

0.2 m
0.2 m
0.50 m
90°
Cord   **FIGURE 13-39**

a) What is the tension in the cord? Find this by using Archimedes' principle.

b) Use $p = p_a + \rho g h$ to calculate directly the force on the two sloped sides and the bottom of the styrofoam; then show that the sum of these forces is the buoyant force.

13-66 In designing a large sea aquarium, the engineer needs to know the force on the circular windows that are planned as observation windows. Each window's center is 4.0 m below the water's surface, and each window has a diameter of 1.0 m. What is the force exerted on each window? This is a marine aquarium, so the density of sea water should be used. (Refer to Table 13-1.)

13-67 A U-shaped tube of length l (see Fig. 13-40) contains a liquid. What is the difference in height between the liquid columns in the vertical arms

a) if the tube has an acceleration a toward the right?

b) if the tube is mounted on a horizontal turntable rotating with an angular velocity ω, with one of the vertical arms on the axis of rotation?

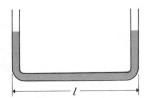

FIGURE 13-40

c) Explain why the difference in height does not depend on the density of the liquid or on the cross-sectional area of the tube. Would it be the same if the vertical tubes did not have equal cross sections? Would it be the same if the horizontal portion were tapered from one end to the other?

13-68 A rock of mass $m = 5$ kg is suspended from the roof of an elevator by a light cord. The rock is totally immersed in a bucket of water that sits on the floor of the elevator.

a) When the elevator is at rest, the tension in the cord is observed to be 36 N. Calculate the volume of the rock.

b) Derive an expression for the tension in the cord when the elevator is accelerating *upward* with an acceleration a. Calculate the tension when $a = 6$ m·s^{-2} upward.

c) Derive an expression for the tension in the cord when the elevator is accelerating *downward* with an acceleration a. Calculate the tension when $a = 6$ m·s^{-2} downward.

d) What is the tension when the elevator is in free fall, with a downward acceleration equal to g?

13-69

a) Derive the expression for the height of capillary rise in the space between two parallel plates dipping in a liquid of density ρ. Assume a contact angle θ. Let the separation between the plates be x.

b) Two glass plates, parallel to each other and separated by 0.5 mm, are dipped in water at 20°C. To what height will the water rise between them? Assume zero angle of contact.

13-70 A large tank of diameter D and open to the air contains water to a height H. A small hole of diameter d ($d \ll D$) is made at the base of the tank. Neglecting any effects of viscosity, calculate the time it takes for the tank to drain completely.

13-71 A siphon, as depicted in Fig. 13-41, is a convenient device for removing liquids from containers. To establish the flow the tube must initially be filled with fluid. Let the

fluid have density ρ. Assume that the cross-sectional area of the tube is the same at all points along the tube.

a) If the lower end of the siphon is at a distance h below the surface of the liquid in the container, what is the velocity of the fluid as it flows out the lower end of the siphon? (Assume that the container has a very large diameter, and neglect any effects of viscosity.)

b) A curious feature of a siphon is that the fluid initially flows "uphill." What is the greatest height H that the high point of the tube can have for the siphon flow still to occur?

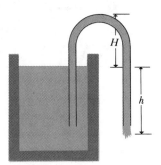

FIGURE 13-41

13-72 For an object falling in air, Stokes's law is obeyed only for small velocities. The air drag on an object falling through the earth's atmosphere is more accurately given by $\frac{1}{2}CA\rho v^2$, where A is the area of the object perpendicular to the motion, ρ is the density of the air, and C is the drag coefficient. (Refer to Section 7-10.) C is determined experimentally; for a spherical object having the velocities of this problem, it is approximately 0.5. (*Note.* For a graph of the values of C over a wide range of values of velocity, see J. E. A. John and W. Haberman, *Introduction to Fluid Mechanics* [Englewood Cliffs, N.J.: Prentice-Hall], p. 198.

A ball bearing of radius 0.5 cm and mass 5.0 g is dropped from the top of the Sears Tower in Chicago, which is 443 m tall.

a) Calculate the velocity the ball bearing would have at the ground if air resistance were negligible.

b) Taking air resistance into account, what is the terminal velocity of the ball bearing?

c) What is its actual velocity just before reaching the ground? Take the density of air to be 1.29 kg·m^{-3}. Note that $\int \tanh x \, dx = \ln (\cosh x)$. (Refer to Challenge Problem 5-66.)

14

TEMPERATURE AND EXPANSION

WITH THIS CHAPTER WE BEGIN A STUDY OF A CLASS OF PHENOMENA ASSO-ciated with *heat*. For this study we need to introduce a new fundamental physical quantity called *temperature*, in addition to the three fundamental quantities (mass, length, and time) used in our study of mechanical systems in preceding chapters. We discuss the concept of thermal equilibrium and various temperature-measuring instruments—that is, *thermometers*—along with the various temperature scales in common use. Finally, we consider the expansion and contraction of materials caused by temperature changes.

This chapter serves as an introduction to the next several chapters, where we study the broad area known as *thermodynamics*. Such diverse topics as expansion joints in bridges, the efficiency of an automobile engine, solar energy, and the design of better refrigerators are all directly related to this subject area.

14-1 TEMPERATURE AND THERMAL EQUILIBRIUM

Temperature: using numbers to describe hot and cold

The concept of **temperature** originates with qualitative ideas of "hot" and "cold" based on our sense of touch. A body that feels hot to the touch is said to have a higher temperature than the same body when it feels cold. More generally, many properties of matter depend on temperature. The length of a metal rod increases when the rod becomes hotter. The pressure of gas in a container increases when the gas becomes hotter; a steam boiler may explode if it becomes too hot. The electrical properties of materials change with temperature; most metals are poorer electrical conductors when hot than when cold. When a material is extremely hot, it glows "red-hot" or "white-hot," with a color that depends on its temperature.

All of this behavior can be understood in greater depth on the basis of the molecular structure of materials and the dependence of molecular configuration and motion on temperature. In Chapter 20 we will consider the relationship between temperature and molecular motion in more detail. It is important to understand, however, that temperature and heat are inherently *macroscopic* concepts, and that they can and should be defined independently of any detailed consideration of molecular motion.

To define temperature *quantitatively*, we must devise a scheme to assign numbers to various degrees of hotness or coldness. To do this, we can make use of some measurable property of a system that varies with its hotness or coldness. A simple example is a liquid such as mercury or ethanol in a bulb attached to a thin tube, as in Fig. 14–1a. When the system becomes hotter, the liquid rises in the tube and the value of L increases. Another simple system is a quantity of gas in a constant-volume container, as shown in Fig. 14–1b. The pressure p, measured by the gauge, increases or decreases as the gas becomes hotter or colder. A third example is the electrical resistance R of a conducting wire, which also varies when the wire becomes hotter or colder.

In each of these examples a significant quantity that changes with the state of hotness or coldness of the system, such as the length L, the pressure p, or the resistance R, can be used to assign a number to a system to describe how hot it is. Thus each device can be used as a **thermometer.**

In each of the above examples, the quantity measured indicates the temperature *of the thermometer.* How can such a thermometer be used to measure the temperature of any other body? The answer to this question leads us to the concept of *thermal equilibrium* and an important related principle. Suppose we want to measure the temperature of a cup of hot coffee, using a mercury thermometer initially at room temperature. We immerse the thermometer in the coffee; the thermometer becomes hotter, and the coffee cools off a little. After we observe that no further change is taking place as a result of this interaction, we read the thermometer. This *equilibrium* condition, in which no further change in the parts of the system occurs as a result of their interaction with each other, is called a state of **thermal equilibrium.** But how do we know that at thermal equilibrium the temperature of the thermometer is the same as that of the coffee? To establish this, we need some additional experiments.

First we consider the interaction of two systems. Let system A be the tube-and-liquid system of Fig. 14–1a and system B the container of gas and pressure gauge of Fig. 14–1b. When the two systems are brought into contact, in general the values of L and p change as the systems move toward thermal equilibrium. Initially one system is hotter than the other, and each system changes the state of the other.

However, if the systems are separated by an insulating material, or **insulator,** such as wood, plastic, foam, or fiberglass, they influence each other more slowly. An ideal insulator is defined as a material that permits *no* interactions at all between the two systems. Such an ideal insulator *prevents* the attainment of thermal equilibrium if the two systems are not initially in thermal equilibrium. This, incidentally, is exactly what the insulating material in a camping cooler does. We do not want the cold food inside to warm up and attain thermal equilibrium with the hot summer air outside; the insulation prevents this, although not forever.

Now consider three systems, A, B, and C, as shown in Fig. 14–2. We do not need to describe in detail what they are, but suppose that initially they are *not* in thermal equilibrium. Perhaps one is very hot, one very cold, and the third lukewarm. We surround the three systems with an insulating box so they cannot interact with anything except each other. We also separate systems A and B by an insulating wall, the gray slab in Fig. 14–2a, but we let system C interact

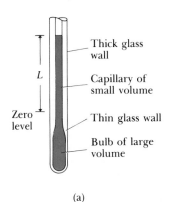

(a)

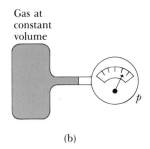

(b)

14–1 (a) A system whose state is specified by the value of the length L. (b) A system whose state is given by the value of the pressure p.

Two bodies in thermal equilibrium have the same temperature: a fundamental truth.

An insulator is a material that prevents or slows down thermal interaction.

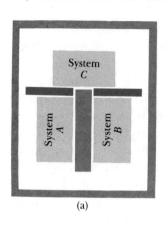

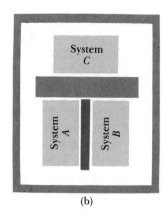

(a)   (b)

14–2 The zeroth law of thermodynamics. (a) If *A* and *B* are each in thermal equilibrium with *C*, then (b) *A* and *B* are in thermal equilibrium with each other. Thick gray layers represent insulating walls; thinner colored layers represent conducting walls.

A conductor is a material that permits thermal interaction.

The zeroth law: a strange name for an essential principle

with both *A* and *B*, as indicated symbolically by a thin colored slab, a thermal **conductor.** (A conducting material is one that permits thermal interactions of the sort we are discussing between the systems it separates.) We wait until no further change is occurring as a result of this interaction, that is, until thermal equilibrium is attained. We then remove *C* from *A* and *B* by inserting an insulating wall, as in Fig. 14–2b, and we also remove the insulating wall between *A* and *B*, replacing it by a conducting wall. What happens?

Experiment shows that *nothing* happens; the states of *A* and *B* do not change! Thus the experiment shows that if *C* is initially in thermal equilibrium with both *A* and *B*, then when *A* and *B* are placed in contact they are also in thermal equilibrium with each other. This outcome may have seemed obvious, but we must verify our intuitive expectations experimentally.

We may state these experimental facts concisely as follows: *Two systems in thermal equilibrium with a third system are in thermal equilibrium with each other.* This principle is called the **zeroth law of thermodynamics.** This odd-sounding name refers to the fact that the basic principles of thermodynamics are called the first, second, and third laws of thermodynamics. The principle just discussed is fundamental to all of these, although it was not clearly recognized until later. Thus it deserves to be called the zeroth law.

The property of any system that determines whether or not two systems are in thermal equilibrium is their *temperature. Two systems in thermal equilibrium must have the same temperature.* As we have seen, the temperature of a system may be represented by a number. Establishing a temperature scale, such as the Fahrenheit or Celsius scale, will be discussed in Section 14–3; this is simply a specific scheme for assigning numbers to temperatures. Once this is done, the condition for thermal equilibrium of two systems is that they have the same temperature. When the temperatures of two systems are different, they *cannot* be in thermal equilibrium.

14–2 THERMOMETERS

One form of thermometer commonly used in everyday life is the tube-and-liquid type shown in Fig. 14–1a. The liquid is usually mercury or colored ethanol. If we label the liquid level at the freezing temperature of water "zero" and the level at the boiling temperature "100," and if we divide the distance between these two points into 100 equal intervals, we obtain the familiar Celsius scale, discussed in detail in the next section.

Another type of thermometer in common use contains a metallic strip consisting of two dissimilar metals welded together, as shown in Fig. 14–3a. Because one metal expands more than the other with temperature increases, the strip bends with temperature changes. In practical thermometers this strip is usually made in the form of a spiral, with the outer end anchored to the thermometer case and the inner end attached to a pointer, as shown in Fig. 14–3b. The pointer rotates in response to temperature changes, and a circular scale is provided.

Various other systems serve as thermometers in situations having special requirements, such as great precision, speed of attaining equilibrium with the body being measured, or extreme temperature ranges.

In a *resistance thermometer* we measure the changing electrical resistance of a coil of fine wire. Resistance can be measured with great precision, and so the resistance thermometer is one of the most precise instruments for the measurement of temperature. At extremely low temperatures, a small carbon cylinder or a small piece of germanium crystal is used instead of a coil of wire.

To measure very high temperatures, an *optical pyrometer* can be used. It measures the brightness of light emitted by a red-hot or white-hot substance. As shown in Fig. 14–4, it consists of a telescope T with a small electric light bulb L mounted in the tube. The bulb is connected to a variable power supply P, and the brightness of the bulb is varied until it matches that of the substance being observed. From previous calibration at known temperatures, the power-supply adjustment can be marked to read temperature directly. No part of the instrument touches the hot substance, so the optical pyrometer can be used at temperatures that would destroy most other thermometers.

Among all temperature-dependent properties, or **thermometric properties,** the pressure of a gas kept at constant volume is outstanding in its sensitivity, accuracy of measurement, and reproducibility. The constant-volume gas

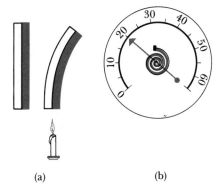

14–3 (a) A bimetallic strip bends when heated because one metal expands more than the other. (b) A bimetallic strip, usually in the form of a spiral, may be used as a thermometer.

Some thermometers depend on electrical resistance.

Thermometers that measure very high temperatures by using the emitted light from a hot body

The gas thermometer: a cumbersome, high-precision instrument with fundamental significance

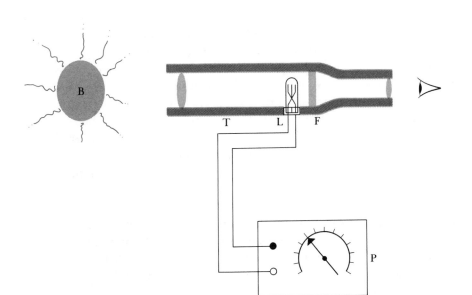

14–4 An optical pyrometer. The telescope T is pointed toward the hot body B, and the power supply P is adjusted until the brightness of the filament in the lamp L is the same as that of the hot body. The filter F reduces the overall brightness level. The power supply can be calibrated to read temperature directly.

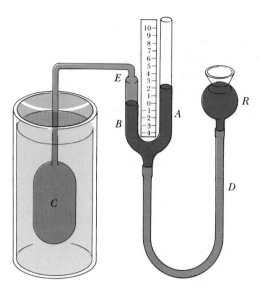

14–5 The constant-volume gas thermometer.

thermometer is shown schematically in Fig. 14–5. The gas, usually helium, is contained in bulb C, and its pressure is measured by the open-tube mercury manometer. As the temperature of the gas increases, the gas expands, forcing the mercury down in tube B and up in tube A. Tubes A and B are connected through a rubber tube D to a mercury reservoir R. By raising R, we can bring the mercury level in B back to a reference mark E. The gas is thus kept at constant volume. The pressure is determined by the difference in mercury levels in tubes A and B.

Gas thermometers are used mainly in bureaus of standards and in research laboratories. They are large, bulky, and slow in coming to thermal equilibrium.

14–3 THE CELSIUS AND FAHRENHEIT SCALES

Two familiar temperature scales and how they are related

The **Celsius temperature scale** (formerly called the *centigrade* scale in English-speaking countries) is defined so that the freezing temperature of pure water at normal atmospheric pressure is zero or 0°C (read "zero degrees Celsius") and the boiling temperature of pure water at normal atmospheric pressure is one hundred or 100°C. We interpolate between or extrapolate beyond these reference temperatures by using one of the thermometers described in Section 14–2. A temperature corresponding to a condition colder than freezing water is a negative number. The Celsius scale is used, both in everyday life and in science and industry, everywhere in the world except for a few English-speaking countries, and all of these except the United States are in the process of converting to Celsius.

The temperature scale used in everyday life in the United States is the **Fahrenheit scale.** It places the freezing temperature of water at 32°F (thirty-two degrees Fahrenheit) and the boiling temperature of water at 212°F, both at normal atmospheric pressure. Thus there are 180 degrees between freezing and boiling, compared to 100 on the Celsius scale, and one Fahrenheit degree represents only $\frac{100}{180}$ or $\frac{5}{9}$ as great a temperature change as one Celsius degree.

To convert temperatures from Celsius to Fahrenheit, we note that a Celsius temperature T_C is the number of Celsius degrees above freezing; to obtain the number of Fahrenheit degrees above freezing we must multiply this by $\frac{9}{5}$. But freezing on the Fahrenheit scale is at 32°F, so to obtain the actual Fahrenheit temperature we must first multiply the Celsius value by $\frac{9}{5}$ and then add 32°. Symbolically,

$$T_F = \tfrac{9}{5} T_C + 32°. \qquad (14\text{–}1)$$

To convert Fahrenheit to Celsius, we solve this equation for T_C, obtaining

$$T_C = \tfrac{5}{9}(T_F - 32°). \qquad (14\text{–}2)$$

That is, we first subtract 32° to obtain the number of Fahrenheit degrees above freezing, and then multiply by $\frac{5}{9}$ to obtain the number of Celsius degrees above freezing, that is, the Celsius temperature.

We do not recommend memorizing Eqs. (14–1) and (14–2) to convert between Celsius and Fahrenheit temperatures. Instead, try to understand the reasoning that led to them well enough so you can derive them on the spot when you need them.

We often need to distinguish between an actual temperature on a certain scale and a temperature *interval*, representing a *difference* in the temperature of two bodies or a *change* of temperature of a body. It is customary to represent a temperature *interval* of 10° as 10 C° (ten Celsius degrees) and an actual temperature of 20° as 20°C (twenty degrees Celsius). Thus a beaker of water heated from 20°C to 30°C undergoes a temperature change of 10 C°. This notation will be followed throughout this book.

A notation for distinguishing between temperatures and temperature intervals

A fundamental problem in defining temperature scales is that when two thermometers, such as a liquid-in-tube system and a resistance thermometer, are calibrated so that they agree at 0°C and 100°C, they *do not* necessarily agree at intermediate temperatures. Different thermometric properties such as length and resistance do not necessarily change with temperature in precisely the same way. Thus a temperature scale defined in this way always depends somewhat on the specific properties of the material used in the thermometer.

The closest approach to consistency is obtained by using a constant-volume gas thermometer with the lowest practical pressures. Low-pressure gas thermometers using various gases are found to agree very closely, and such thermometers are used for high-precision standards. But we are still at the mercy of properties of specific materials. We might well wish for a temperature scale that is *completely* independent of material properties. It is, in fact, possible to define such a scale; we will return to this fundamental problem in Chapter 19, after developing the necessary thermodynamic principles.

Thermometers aren't always consistent with each other.

The bimetallic strip shown in Fig. 14–3a is also used in most thermostats for controlling heating systems. One end of the bimetallic strip is attached to an electrical contact set, which makes or breaks contact as the bimetallic strip bends or straightens in response to temperature changes. This electrical contact turns the furnace on or off so as to maintain a constant temperature. This is a rudimentary example of a *feedback control system;* a physical quantity is measured, and if it does not have the desired value, a system is activated to correct the situation.

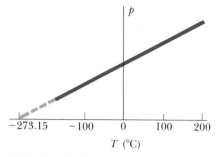

14–6 Graph of pressure versus temperature for a constant-volume gas thermometer. A straight-line extrapolation predicts that the pressure would become zero at $-273.15°C$ if that temperature could be reached and the proportionality of pressure to temperature held exactly.

14–4 THE KELVIN AND RANKINE SCALES

Calibration of a constant-volume gas thermometer might well make use of a graph similar to Fig. 14–6, showing the relation between gas pressure and temperature. Such a graph suggests that there is a hypothetical temperature at which the gas pressure would become zero. Surprisingly, this temperature turns out to be the same for a wide variety of gases: namely, $-273.15°C$. It is not possible actually to observe this zero-pressure condition, for several reasons. One is that most gases liquefy and solidify at very low temperatures, and the proportionality of pressure to temperature no longer holds.

The discovery that the extrapolated zero-pressure temperature is the same for many gases raises the possibility that there may be a more fundamental way to define the zero point on a temperature scale than the freezing point of water. This is in fact the basis of the **Kelvin temperature scale,** named after Lord Kelvin (1824–1907). The degrees are the same size as on the Celsius scale, but the zero is shifted so that $0 \text{ K} = -273.15°C$ and $273.15 \text{ K} = 0°C$; that is,

$$T_K = T_C + 273.15°. \tag{14–3}$$

Thus ordinary room temperature, 20°C, becomes 20 + 273.15, or about 293 K.

In SI nomenclature, "degree" is not used with the Kelvin scale; the temperature mentioned above is read "293 kelvins," not "degrees Kelvin." Also, Kelvin is capitalized when it refers to the temperature scale, but the *unit* of temperature is the *kelvin.* This is not capitalized but is abbreviated K. (The phrase "degrees Kelvin," although officially obsolete, is still fairly common.)

The Kelvin scale has been defined with reference to the Celsius scale, which has two fixed points, the normal freezing and boiling temperatures of water. However, it may also be defined with reference to a gas thermometer (at very low pressure) by means of a single reference temperature. We define the ratio of any two temperatures T_1 and T_2 on the Kelvin scale as the ratio of the corresponding gas-thermometer pressures p_1 and p_2:

$$\frac{T_2}{T_1} = \frac{p_2}{p_1}. \tag{14–4}$$

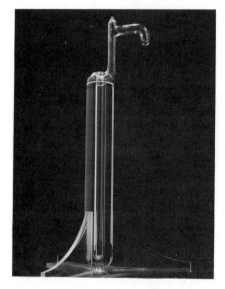

14–7 A cell used to establish the triple-point temperature of water. The cylindrical pyrex container is nearly filled with water of the highest possible purity and then permanently sealed. Partial freezing of the sealed water leads to the triple-point solid-liquid-vapor equilibrium condition and establishes the triple-point temperature of 0.0100°C or 273.1600 K. Calibrations with a single triple-point cell are precise to within 0.0001 K, and triple-point cells agree with each other within 0.0002 K. (Courtesy National Bureau of Standards.)

To complete the definition, we need only specify the Kelvin temperature of a single specific state. For reasons of precision and reproducibility, the state chosen is not a freezing or boiling point but the *triple point* of water. This is the unique condition under which solid water, liquid, and vapor can all coexist together. This occurs at a temperature of 0.01°C and a pressure of 610 Pa (about 0.006 atm). The triple-point temperature of water is assigned the value 273.16 K. A cell used to establish the triple-point temperature is shown in Fig. 14–7. Thus, with reference to Eq. (14–4), if p_3 is the pressure in a gas thermometer at the triple point and p is the pressure at some other temperature T, then T is given on the Kelvin scale by

$$T = T_3 \frac{p}{p_3}, \tag{14–5}$$

with T_3 *defined* to be 273.16 K.

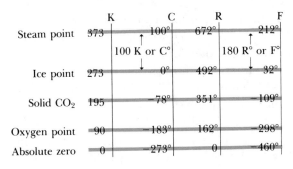

	K	C	R	F
Steam point	373	100°	672°	212°
		↑ 100 K or C°		↑ 180 R° or F°
Ice point	273	0°	492°	32°
Solid CO_2	195	−78°	351°	−109°
Oxygen point	90	−183°	162°	−298°
Absolute zero	0	−273°	0	−460°

14–8 Relations among Kelvin, Celsius, Rankine, and Fahrenheit temperature scales. Temperatures have been rounded off to the nearest degree.

EXAMPLE 14–1 Suppose a constant-volume gas thermometer has a pressure of 1.50×10^4 Pa at the triple point of water and a pressure of 1.95×10^4 Pa at some unknown temperature T. What is T?

SOLUTION From Eq. (14–5),

$$T = (273\ \text{K})\frac{1.95 \times 10^4\ \text{Pa}}{1.50 \times 10^4\ \text{Pa}} = 355\ \text{K}$$

$$= 82°\text{C}.$$

The **Rankine temperature scale,** named after William Rankine (1820–1872), has the same relation to the Fahrenheit scale as Kelvin has to Celsius. The zero point on the Kelvin scale is −273.15°C, which on the Fahrenheit scale is $(\tfrac{9}{5})(-273.15°) + 32° = -459.7°$F. Thus we define the Rankine scale so that 0°R = −459.7°F and 0°F = 459.7°R. The size of a degree is the same on the Rankine and Fahrenheit scales, and

$$T_R = T_F + 459.7°.$$

The relationships of the four temperature scales discussed above are shown graphically in Fig. 14–8.

The Kelvin and Rankine scales are called **absolute temperature scales,** and their zero point is called **absolute zero.** To define completely what is meant by absolute zero, we need to use thermodynamic principles that are developed in the next several chapters. We will return to the concept of absolute zero in Chapter 19.

What is absolute zero?

14–5 THERMAL EXPANSION

Most solid materials expand when heated. Suppose a rod of material has a length L_0 at some initial temperature T_0, and that, when the temperature increases by an amount ΔT, the length increases by ΔL. Experiment shows that if ΔT is not too large, ΔL is *directly proportional* to ΔT. But ΔL is also proportional to L_0; if two rods of the same material have the same temperature change, but one rod is initially twice as long as the other, then the *change* in its length is also twice as great. Introducing a proportionality constant α (which is different for different materials), we may summarize this relation as follows:

Thermal expansion: When things get hotter, they get bigger.

$$\Delta L = \alpha L_0\ \Delta T. \qquad (14\text{–}6)$$

The horizontal black lines on the roadway of the George Washington suspension bridge in New York City are expansion joints; these are needed to accommodate changes in the lengths of the sections that result from thermal expansion. (Courtesy of New York State Department of Commerce.)

The constant α, which characterizes the thermal expansion properties of a particular material, is called the *temperature coefficient of linear expansion* or, more briefly, the **coefficient of linear expansion.** The units of α are K^{-1} or $(C°)^{-1}$. For materials having no preferential directions, every linear dimension changes according to Eq. (14–6). Thus L could equally well represent the thickness of the rod, the side length of a square sheet, or the diameter of a hole in the material. There are some exceptional cases. Wood, for example, has different expansion properties along the grain and across the grain; and single crystals of material can have different properties along different crystal-axis directions. We exclude these exceptional cases from the present discussion.

We emphasize that the direct proportionality expressed by Eq. (14–6) is not exact, but that it is *approximately* correct for sufficiently small temperature changes. For any temperature, we can define a coefficient of thermal expansion by the equation

$$\alpha = \frac{1}{L}\frac{dL}{dT}. \qquad (14-7)$$

It is found that α for a given material varies somewhat with the initial temperature T_0 and the size of the temperature interval. Because Eq. (14–6) is at best an approximation, we will ignore this variation. Average values of α for several materials are listed in Table 14–1. Within the precision of these values, we do not need to worry about whether T_0 is 0°C or 20°C or some other temperature.

PROBLEM-SOLVING STRATEGY: Linear expansion

1. Identify which of the quantities in Eq. (14–6) are known and which are unknown. Often you will be given two temperatures and will have to compute ΔT. Or you may be given an initial temperature T_0 and may have to find a final temperature corresponding to a given length change. In this case, find ΔT first; then the final temperature is $T_0 + \Delta T$.

2. Unit consistency is crucial, as always. L_0 and ΔL must have the same units, and if you use a value of α in $(C°)^{-1}$, then ΔT must be in Celsius degrees $(C°)$.

3. Remember that sizes of holes in a material expand with temperature just the same way as any other linear dimension.

4. Points 1 and 2 are applicable also to volume expansion, which we will discuss following this first example.

Correcting a measuring tape for temperature variations

EXAMPLE 14–2   A carpenter uses a steel measuring tape 5 m long calibrated at a temperature of 20°C. What is its length on a hot summer day when the temperature is 35°C?

SOLUTION   From Eq. (14–6),

$$\Delta L = [5\ m][1.2 \times 10^{-5}(C°)^{-1}][35°C - 20°C]$$
$$= 0.9 \times 10^{-3}\ m = 0.9\ mm.$$

Thus the length at 35°C is 5.0009 m.

Increasing temperature usually causes increases in *volume*, for both solid and liquid materials. Experiments show that if the temperature change ΔT is not too great, the increase in volume ΔV is approximately *proportional* to the temperature change. The volume change is also proportional to the initial volume V_0, just as with linear expansion. The relationship can be expressed as follows:

$$\Delta V = \beta V_0 \, \Delta T. \qquad (14\text{–}8)$$

The constant β, which characterizes the volume expansion properties of a particular material, is called the *temperature coefficient of volume expansion*, or the **coefficient of volume expansion.** The units of β are K^{-1} or $(C°)^{-1}$. Like the coefficient of linear expansion, β varies somewhat with temperature, and Eq. (14–8) should be regarded as an approximate relation, valid for sufficiently small temperature changes. For many substances, β decreases as the temperature is lowered, approaching zero as the Kelvin temperature approaches zero. Also, metals with high melting temperatures usually have small volume-expansion coefficients.

Some values of β in the neighborhood of room temperature are listed in Table 14–2. Note that the values for liquids are much larger than those for solids.

Fluids usually expand when they get hotter.

TABLE 14–1   Coefficient of Linear Expansion

Material	α, $(C°)^{-1}$
Aluminum	2.4×10^{-5}
Brass	2.0×10^{-5}
Copper	1.7×10^{-5}
Glass	$0.4\text{–}0.9 \times 10^{-5}$
Steel	1.2×10^{-5}
Invar	0.09×10^{-5}
Quartz (fused)	0.04×10^{-5}

TABLE 14–2   Coefficient of Volume Expansion

Solids	β, $(C°)^{-1}$	Liquids	β, $(C°)^{-1}$
Aluminum	7.2×10^{-5}	Ethanol	75×10^{-5}
Brass	6.0×10^{-5}	Carbon disulfide	115×10^{-5}
Copper	5.1×10^{-5}	Glycerin	49×10^{-5}
Glass	$1.2\text{–}2.7 \times 10^{-5}$	Mercury	18×10^{-5}
Steel	3.6×10^{-5}		
Invar	0.27×10^{-5}		
Quartz (fused)	0.12×10^{-5}		

If there is a hole in a solid body, the volume of the hole increases when the body expands, just as if the hole were a solid of the same material as the body. This remains true even if the hole is so large that the surrounding body is only a thin shell. Thus the volume enclosed by a thin-walled bottle or thermometer bulb increases just as would a solid body of glass of the same size.

Holes in solid bodies expand too.

EXAMPLE 14–3   A glass flask of volume 200 cm^3 is just filled with mercury at 20°C. How much mercury will overflow when the temperature of the system is raised to 100°C? The coefficient of volume expansion of the glass is $1.2 \times 10^{-5} (C°)^{-1}$.

SOLUTION   The increase in the volume of the flask is

$$\Delta V = (1.2 \times 10^{-5} (C°)^{-1})(200 \text{ cm}^3)(100° - 20°) = 0.192 \text{ cm}^3.$$

The increase in the volume of the mercury is

$$\Delta V = (18 \times 10^{-5} (C°)^{-1})(200 \text{ cm}^3)(100° - 20°) = 2.88 \text{ cm}^3.$$

The volume of mercury overflowing is therefore

$$2.88 \text{ cm}^3 - 0.19 \text{ cm}^3 = 2.69 \text{ cm}^3.$$

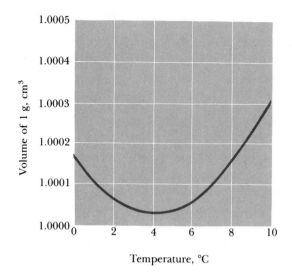

14–9 Volume of 1 gram of water, in the temperature range from 0°C to 10°C. If Eq. (14–8) were obeyed, the curve would be a straight line.

Water sometimes shrinks when it gets warmer. Why is this important?

Water, in the temperature range from 0°C to 4°C, *decreases* in volume with increasing temperature, contrary to the behavior of most substances. That is, between 0°C and 4°C the coefficient of expansion of water is *negative*. Above 4°C, water expands when heated. Since the volume of a given mass of water is smaller at 4°C than at any other temperature, the density of water is maximum at 4°C. Water also expands upon freezing, unlike most materials.

This anomalous behavior of water has an important effect on plant and animal life in lakes. When a lake cools, the cooled water at the surface flows to the bottom because of its greater density. But when the temperature reaches 4°C, this flow ceases and the water near the surface remains colder (and less dense) than that at the bottom. As the surface freezes, the ice floats because it is less dense than water. The water at the bottom remains at 4°C until nearly the entire lake is frozen. If water behaved like most substances, contracting continuously on cooling and freezing, lakes would freeze from the bottom up. Circulation due to density differences would continuously carry warmer water to the surface for efficient cooling, and lakes would freeze solid much more easily. This would destroy all plant and animal life that can withstand cold water but not freezing.

The anomalous expansion of water in the temperature range 0°C to 10°C is shown in Fig. 14–9. Table 14–3 covers a wider range of temperatures.

The volume-expansion coefficient for a solid material is related to the linear-expansion coefficient. To obtain the relation, consider a solid body in the form of a rectangular block with dimensions L_1, L_2, and L_3. Then the volume is

$$V_0 = L_1 L_2 L_3.$$

According to Eq. (14–6), when the temperature increases by ΔT, each linear dimension changes, and the new volume is

$$V_0 + \Delta V = [L_1(1 + \alpha\,\Delta T)][L_2(1 + \alpha\,\Delta T)][L_3(1 + \alpha\,\Delta T)]$$
$$= L_1 L_2 L_3 (1 + \alpha\,\Delta T)^3$$
$$= V_0 (1 + \alpha\,\Delta T)^3$$
$$= V_0[1 + 3\alpha\,\Delta T + 3\alpha^2(\Delta T)^2 + \alpha^3(\Delta T)^3].$$

TABLE 14–3   Volume and Density of Water

T, °C	Volume of 1 g, cm³	Density, g·cm⁻³
0	1.0002	0.9998
4	1.0000	1.0000
10	1.0003	0.9997
20	1.0018	0.9982
50	1.0121	0.9881
75	1.0258	0.9749
100	1.0434	0.9584

If ΔT is small, the terms containing $(\Delta T)^2$ and $(\Delta T)^3$ are very small and may be neglected. Dropping these and subtracting V_0 from both sides, we find

$$\Delta V = (3\alpha)V_0 \, \Delta T.$$

Comparing this with Eq. (14–8), we obtain

$$\beta = 3\alpha. \tag{14–9}$$

We invite you to check the validity of this relation for some of the materials listed in Tables 14–1 and 14–2.

Linear- and volume-expansion coefficients are related in a very simple way.

14–6 THERMAL STRESSES

If the ends of a rod are rigidly held so as to prevent expansion or contraction and the temperature of the rod is changed, tensile or compressive stresses, called **thermal stresses,** are set up in the rod. These stresses may become large enough to stress the rod beyond its elastic limit or even beyond its breaking strength. Hence in the design of any structure that is subject to changes in temperature, provision must be made for expansion. Long steam pipes have expansion joints or U-shaped sections of pipe. In bridges, one end may be rigidly fastened to its abutment while the other rests on rollers.

Thermal stress: What happens when a material wants to expand and you don't let it?

We can compute the thermal stress set up in a rod that is not free to expand or contract. We simply compute the amount the rod would contract if not held, and then compute the stress needed to stretch it back to its original length. Suppose that a rod of length L_0 and cross-sectional area A has its ends rigidly fastened while the temperature is reduced by an amount ΔT, causing a tensile stress. The fractional change in length if the rod were free to contract would be

$$\frac{\Delta L}{L_0} = \alpha \, \Delta T. \tag{14–10}$$

Both ΔL and ΔT are negative. Since the rod is *not* free to contract, the tension must increase enough to produce an equal and opposite fractional change in length. But from the definition of Young's modulus, Eq. (12–3),

$$Y = \frac{F/A}{\Delta L/L_0}, \qquad \frac{\Delta L}{L_0} = \frac{F}{AY}. \tag{14–11}$$

The tensile force F is determined by the requirement that the *total* fractional change in length, thermal expansion plus elastic strain, must be zero:

$$\alpha \, \Delta T + \frac{F}{AY} = 0,$$

$$F = -AY\alpha \, \Delta T. \tag{14–12}$$

Since ΔT represents a decrease in temperature, it is negative, so F is positive. The tensile *stress* F/A in the rod is

$$\frac{F}{A} = -Y\alpha \, \Delta T. \tag{14–13}$$

If, instead, ΔT represents an *increase* in temperature, then F and F/A become negative, corresponding to *compressive* force and stress, respectively.

A photograph of the surface of Mars, showing flaking of rock as a result of thermal stresses created by alternating extremes of hot and cold. (Courtesy of NASA.)

Thermal stress: how to break a thick-walled glass container

Thermal stresses can also be induced by nonuniform temperature. Even if a solid body at uniform temperature has no internal stresses, stress may be induced by nonuniform expansion due to temperature differences. Have you ever broken a thick glass container such as a vase by pouring very hot water into it? Thermal stress caused by temperature differences exceeded the breaking stress of the material, and it cracked. Heat-resistant glasses such as Pyrex have exceptionally low expansion coefficients and usually also have high strength to permit thin-wall construction to minimize temperature difference.

Similar phenomena occur with volume expansion. If you fill a bottle completely with water, tightly cap it and then heat it, it will break because the thermal expansion coefficient for water is greater than that for glass. If a material is enclosed in a very rigid container so that its volume cannot change, then a rise in temperature ΔT is accompanied by an increase in pressure Δp. An analysis similar to that leading to Eq. (14–13) shows that the pressure increase is given by

$$\Delta p = B\beta\, \Delta T, \tag{14–14}$$

where B is the bulk modulus for the material.

KEY TERMS

temperature

thermometer

thermal equilibrium

insulator

conductor

zeroth law of thermodynamics

thermometric properties

Celsius scale

Fahrenheit scale

Kelvin scale

Rankine scale

absolute temperature scales

absolute zero

coefficient of linear expansion

coefficient of volume expansion

thermal stresses

SUMMARY

A thermometer measures temperature. Any thermometric property can be used to make a thermometer. Two bodies in thermal equilibrium must have the same temperature. A conducting material between two bodies permits interaction leading to thermal equilibrium; an insulating material prevents or impedes this interaction.

The Celsius temperature scale is based on the freezing (0°C) and boiling (100°C) temperatures of water, with one hundred degrees (100 C°) between the two. The Fahrenheit scale is also based on the freezing (32°F) and boiling (212°F) temperatures, but with 180 F° between the two. Celsius and Fahrenheit temperatures are related by

$$T_{\mathrm{F}} = \tfrac{9}{5}T_{\mathrm{C}} + 32°, \tag{14–1}$$

$$T_{\mathrm{C}} = \tfrac{5}{9}(T_{\mathrm{F}} - 32°). \tag{14–2}$$

The Kelvin scale has its zero at the extrapolated zero-pressure temperature for a gas thermometer, which is $-273.15°C$. Thus $0\text{ K} = -273.15°C$, and

$$T_K = T_C + 273.15. \tag{14-3}$$

The Rankine scale is related similarly to the Fahrenheit scale; $0°\text{ R} = -459.7°F$, and

$$T_R = T_F + 459.7°. $$

In the gas-thermometer scale, the ratio of two temperatures is defined to be equal to the ratio of the two corresponding gas-thermometer pressures:

$$\frac{T_2}{T_1} = \frac{p_2}{p_1}. \tag{14-4}$$

The definition of this scale is completed by defining the triple-point temperature of water $(0.01°C)$ to be 273.16 K.

The change ΔL of any linear dimension L_0 of a solid body with a temperature change ΔT is given approximately by

$$\Delta L = \alpha L_0\,\Delta T, \tag{14-6}$$

where α is the coefficient of linear expansion.

The change ΔV in the volume V_0 of any solid or liquid material with a temperature change ΔT is given approximately by

$$\Delta V = \beta V_0\,\Delta T, \tag{14-8}$$

where β is the coefficient of volume expansion. For solid materials,

$$\beta = 3\alpha. \tag{14-9}$$

When a material is cooled and held so it cannot contract, the tensile stress F/A is given by

$$F/A = -Y\alpha\,\Delta T, \tag{14-13}$$

where Y is Young's modulus, α is the coefficient of linear expansion, and ΔT is the temperature change. This expression also gives the compressive stress accompanying a rise in temperature if expansion is prevented.

The pressure Δp resulting from thermal stress when a solid or liquid material is heated by an amount ΔT and prevented from expanding is given by

$$\Delta p = B\beta\,\Delta T, \tag{14-14}$$

where B is the bulk modulus and β the volume-expansion coefficient for the material.

QUESTIONS

14–1 Does it make sense to say that one body is twice as hot as another?

14–2 A student claimed that thermometers are useless because a thermometer always registers *its own* temperature. How would you respond?

14–3 What other properties of matter, in addition to those mentioned in the text, might be used as thermometric properties? How could they be used to make a thermometer?

14–4 A thermometer is laid out in direct sunlight. Does it measure the temperature of the air, or of the sun, or what?

14–5 Thermometers sometimes contain red or blue liquid, which is often ethanol. What advantages and disadvantages does this have compared with mercury?

14–6 Could a thermometer similar to that shown in Fig. 14–1a be made by using water as the liquid? What difficulties would this thermometer present?

14–7 What is the temperature of vacuum?

14–8 Is there any particular reason for constructing a temperature scale with higher numbers corresponding to hotter bodies, rather than the reverse?

14–9 If a brass pin is a little too large to fit in a hole in a steel block, should you heat the pin and cool the block, or the reverse?

14–10 When a block with a hole in it is heated, why doesn't the material around the hole expand into the hole and make it smaller?

14–11 Many automobile engines have cast-iron cylinders and aluminum pistons. What kinds of problems could occur if the engine gets too hot?

14–12 When a hot-water faucet is turned on, the flow often decreases gradually before it settles down. This is an annoyance in the shower. Why does it happen?

14–13 Two bodies made of the same material have the same external dimensions and appearance, but one is solid and the other is hollow. When they are heated, is the overall volume expansion the same or different?

14–14 A thermostat for controlling household heating systems often contains a bimetallic element consisting of two strips of different metals, welded together face to face. When the temperature changes, this composite strip bends one way or the other. Why?

14–15 Why is it sometimes possible to loosen caps on screw-top bottles by dipping the cap briefly in hot water?

14–16 The rate at which a pendulum clock runs depends on the length of the pendulum. Would a pendulum clock gain time in hot weather and lose in cold, or the reverse? Could one design a pendulum, perhaps using two different metals, that would *not* change length with temperature?

14–17 When a rod is cooled but prevented from contracting, as in Section 14–6, thermal tension develops. Under these circumstances, does the *thickness* of the rod change? If so, how would the change be calculated?

EXERCISES

Section 14–3 The Celsius and Fahrenheit Scales

14–1

a) If you feel sick in France and are told you have a temperature of 40°C, should you be concerned?

b) What is normal body temperature on the Celsius scale?

c) At what temperature do the Fahrenheit and Celsius scales coincide?

14–2 When the United States finally converts officially to metric units, the Celsius temperature scale will replace the Fahrenheit scale for everyday use. As a familiarization exercise, find the Celsius temperatures corresponding to

a) a cool room (68°F);

b) a hot summer day (95°F);

c) a cold winter day (5°F).

Section 14–4 The Kelvin and Rankine Scales

14–3 The normal boiling point of liquid oxygen is −182.97°C. What is this temperature on the Kelvin and Rankine scales?

14–4 The ratio of the pressure of a gas at the melting point of lead and at the triple point of water, when the gas is kept at constant volume, is found to be 2.19816. What is the Kelvin temperature of the melting point of lead?

14–5 In a rather primitive experiment with a constant-volume gas thermometer, the pressure at the triple point of water was found to be 4.0×10^4 Pa, and the pressure at the normal boiling point 5.4×10^4 Pa. According to these data, what is the temperature of absolute zero on the Celsius scale?

Section 14–5 Thermal Expansion

14–6 A building with a steel framework is 200 m tall when the temperature is 0°C. How much taller is the building on a hot summer day when the temperature is 30°C?

14–7 A steel bridge is built in the summer when the temperature is 25°C. At the time of construction its length is 80.00 m. What is the length of the bridge on a cold winter day when the temperature is −10°C?

14–8 The pendulum shaft of a clock is made of aluminum. What is the fractional change in length of the shaft when it is cooled from 25°C to 10°C?

14–9 A metal rod is observed to be 80.000 cm long at 20°C and 80.024 cm long at 40°C. From these data calculate the average coefficient of linear expansion of the rod for this temperature range.

14–10 To ensure a tight fit, the aluminum rivets used in airplane construction are made slightly larger than the rivet holes and cooled by "dry ice" (solid CO_2) before being driven. If the diameter of a hole is 0.6000 cm, what should be the diameter of a rivet at 20°C if its diameter is to equal that of the hole when the rivet is cooled to −78°C, the temperature of dry ice? Assume the expansion coefficient remains constant at the value given in Table 14–1.

14–11 A machinist bores a hole 2.500 cm in diameter in a brass plate at a temperature of 20°C. What is the diameter of the hole when the temperature of the plate is increased to 200°C? Assume the expansion coefficient remains constant.

14–12 An area is measured on the surface of a solid body. If the area is A_0 at some initial temperature and then changes by ΔA when the temperature changes by ΔT, show that

$$\Delta A = (2\alpha)A_0 \, \Delta T.$$

14–13 A brass cylinder is initially at 20°C. At what temperature will its volume be 0.2% larger?

14–14 Use the data of Table 14–3 to find the average coefficient of volume expansion of water in the temperature range

a) 0°C to 4°C;

b) 75°C to 100°C.

14–15 An underground tank with a capacity of 500 L (0.5 m³) is filled with ethanol that has an initial temperature of 25°C. After the ethanol has cooled off to the temperature of the tank and ground, which is 10°C, how much air space will there be above the ethanol in the tank?

Section 14–6 Thermal Stresses

14–16 The cross section of a steel rod is 10 cm². What force is needed to prevent it from contracting while cooling from 520°C to 20°C?

14–17

a) A wire that is 3 m long at 20°C is found to increase in length by 1.5 cm when heated to 520°C. Compute its average coefficient of linear expansion.

b) Find the stress in the wire if it is stretched taut at 520°C and cooled to 20°C without being allowed to contract. Young's modulus for the wire is 2×10^{11} Pa.

14–18 Steel railroad rails 18 m long are laid on a winter day when the temperature is 3°C.

a) How much space must be left between adjacent rails if they are to just touch on a summer day when the temperature is 39°C?

b) If the rails were originally laid in contact, what would be the stress in them on a summer day when the temperature is 39°C?

14–19 What hydrostatic pressure is necessary to prevent a copper block from expanding when its temperature is increased from 20°C to 39°C?

14–20

a) A block of steel at an initial pressure of 1 atm and a temperature of 20°C is kept at constant volume. If the temperature is raised to 32°C, what is the final pressure?

b) If the block is maintained at constant volume by rigid walls that can withstand a maximum pressure of 1200 atm (1.22×10^8 Pa), what is the highest temperature to which the system may be raised? Assume B and β remain practically constant at the values 1.6×10^{11} Pa and 3.6×10^{-5}(C°)$^{-1}$, respectively.

PROBLEMS

14–21 An aluminum cube 0.10 m on a side is heated from 10°C to 30°C.

a) What is the change in its volume?

b) In its density?

14–22 A surveyor's 30-m steel tape is correct at a temperature of 20°C. The distance between two points, as measured by this tape on a day when the temperature is 35°C, is 25.970 m. What is the true distance between the points?

14–23 A steel ring of 3.000 in. inside diameter at 20°C is to be heated and slipped over a brass shaft measuring 3.002 in. in diameter at 20°C.

a) To what temperature should the ring be heated?

b) If the ring and shaft together are cooled by some means such as liquid air, at what temperature will the ring just slip off the shaft?

14–24 Suppose that a steel hoop could be constructed around the earth's equator, just fitting it at a temperature of 20°C. What would be the thickness of the space between the hoop and the earth if the temperature of the hoop were increased by 1 C°?

14–25 A metal rod 30.0 cm long expands by 0.075 cm when its temperature is raised from 0°C to 100°C. A rod of a different metal and of the same length expands by 0.045 cm for the same rise in temperature. A third rod, also 30.0 cm long, is made up of pieces of each of the above metals placed end to end, and expands 0.065 cm between 0°C and 100°C. Find the length of each portion of the composite bar.

14–26 At a temperature of 20°C, the volume of a certain glass flask, up to a reference mark on the stem of the flask, is exactly 100 cm³. The flask is filled to this point with a liquid whose coefficient of volume expansion is 40×10^{-5}(C°)$^{-1}$, with both flask and liquid at 20°C. The coefficient of volume expansion of the glass is 2×10^{-5}(C°)$^{-1}$. The cross section of the stem is 6 mm² and can be considered constant. How far will the liquid rise or fall in the stem when the temperature is raised to 40°C?

14–27 A glass flask whose volume is exactly 1000 cm³ at 0°C is completely filled with mercury at this temperature. When flask and mercury are heated to 100°C, 15.2 cm³ of mercury overflow. If the coefficient of volume expansion of mercury is 18×10^{-5} per Celsius degree, compute the coefficient of volume expansion of the glass.

14–28 A steel rod of length 0.40 m and a copper rod of length 0.36 m, both of the same diameter, are placed end to end between rigid supports, with no initial stress in the rods. The temperature of the rods is now raised by 50 C°. What is the stress in each rod?

14–29 A heavy brass bar has projections at its ends, as in Fig. 14–10. Two fine steel wires fastened between the projections are just taut (zero tension) when the whole system is at 0°C. What is the tensile stress in the steel wires

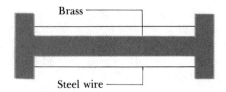

Brass

Steel wire

FIGURE 14–10

when the temperature of the system is raised to 300°C? Make any simplifying assumptions you think are justified, but state what they are.

14–30 Prove that if a body under hydrostatic pressure is raised in temperature but not allowed to expand, the increase in pressure is

$$\Delta p = B\beta\, \Delta T,$$

where the bulk modulus B and the average coefficient of volume expansion β are both assumed positive and constant.

14–31 Table 14–3 lists the density of water, and the volume of 1 g, at atmospheric pressure. A steel bomb is filled with water at 10°C and atmospheric pressure, and the system is heated to 75°C. What is then the pressure in the bomb? Assume the bomb to be sufficiently rigid so that its volume is not affected by the increased pressure.

14–32 A liquid is enclosed in a metal cylinder provided with a piston of the same metal. The system is originally at atmospheric pressure and at a temperature of 80°C. The piston is forced down until the pressure on the liquid is increased by 100 atm (1.013×10^7 Pa), and is then clamped in this position. Find the new temperature at which the pressure of the liquid is again 1 atm (1.013×10^5 Pa). Assume that the cylinder is sufficiently strong so that its volume is not altered by changes in pressure but only by changes in temperature.

Compressibility of liquid:
$k = 50 \times 10^{-6}$ atm^{-1}.

Coefficient of volume expansion of liquid:
$\beta = 53 \times 10^{-5}$(C°)$^{-1}$.

Coefficient of volume expansion of metal:
$\beta = 3.0 \times 10^{-5}$(C°)$^{-1}$.

CHALLENGE PROBLEMS

14–33 A clock whose pendulum makes one vibration in 2 s is correct at 25°C. The pendulum shaft is of steel and its mass may be neglected compared with that of the bob.

a) What is the fractional change in length of the shaft when it is cooled to 15°C?

b) How many seconds per day will the clock gain or lose at 15°C? (The calculation technique outlined for Challenge Problem 11–45 may be used.)

c) How closely must the temperature be controlled if the clock is not to gain or lose more than 1 s a day? Does the answer depend on the period of the pendulum?

14–34

a) The pressure p, volume V, number of moles n, and Kelvin temperature T of an ideal gas are related by the equation $pV = nRT$. Prove that the coefficient of volume expansion is equal to the reciprocal of the Kelvin temperature, if the expansion occurs at constant pressure.

b) Compare the coefficients of volume expansion of copper and air at a temperature of 20°C. Assume that air may be treated as an ideal gas and that the pressure remains constant.

14–35

a) For any material, density ρ, mass m, and volume V are related by $\rho = m/V$. Prove that

$$\beta = -\frac{1}{\rho}\frac{d\rho}{dT}.$$

b) The density of rock salt in units of g·cm^{-3} between -193°C and -13°C is given by the empirical formula

$$\rho = 2.1680(1 - 11.2 \times 10^{-5}T - 0.5 \times 10^{-7}T^2),$$

with T measured on the Celsius scale. Calculate β at -100°C.

14–36 In going from Eq. (14–7) to (14–6), we make two assumptions: that ΔL is small compared to L_0 and that $\alpha(T)$ can be taken to be constant over the temperature range in question. If ΔT is sufficiently large, or if a high degree of accuracy is required, both these approximations can break down.

a) Eq. (14–7) gives $dL/L = \alpha(T)\, dT$. Assuming $\alpha(T)$ to be constant, integrate this expression from an initial temperature T_0, where the length is L_0, to a final temperature T, where the length is L, and thereby obtain an equation from which L can be calculated.

b) Consider a copper rod 80.000 cm long at 0°C. Calculate its length at 800°C both from the equation you derived in (a) and also from Eq. (14–6). In each case assume that α remains constant at the value given in Table 14–1. What percentage error does Eq. (14–6) make in the calculation of ΔL?

c) If $\alpha(T)$ is given by $\alpha(T) = A + BT + CT^2$, where A, B, and C are constants, derive an expression analogous to Eq. (14–6) for ΔL, assuming ΔL is small.

15

QUANTITY OF HEAT

IN THE PRECEDING CHAPTER WE DISCUSSED THE CONCEPT OF TEMPERATURE in connection with thermal equilibrium; two bodies in thermal equilibrium must have the same temperature. When two bodies that are *not* initially in thermal equilibrium are placed in contact, their temperatures change until they reach thermal equilibrium. The study of the interaction that takes place during the approach to thermal equilibrium leads us to the concept of *heat,* the subject of this chapter. We define what we mean by quantity of heat, and we introduce units to measure quantity of heat. Then we apply this concept to a study of the quantities of heat involved in temperature changes and changes of phase of materials.

15–1 HEAT TRANSFER

Suppose we have two systems, *A* and *B*, with *A* initially at higher temperature than *B*. We place them in contact; when they have reached thermal equilibrium, *A*'s temperature has decreased and *B*'s has increased. It is natural to speculate that during this process *A* loses something and that this "something" flows into *B*. Indeed, the interaction that takes place while the temperatures are changing is called a **heat transfer** or a *heat flow* from *A* to *B*.

The process of heat transfer was formerly thought to be a flow of an invisible, weightless fluid called *caloric*. But during the eighteenth and nineteenth centuries the relationship of heat to mechanical energy and work gradually emerged. Count Rumford (1753–1814) studied the heat evolved during the drilling of cannon barrels, especially with dull drills. Sir James Joule (1818–1889) discovered that water can be heated by vigorous stirring with a paddle wheel and that there is a correlation between the temperature rise and the work needed to turn the wheel. These and many other similar experimental studies established that heat flow is really *energy transfer* and that there is an equivalence between heat and work. The relation of heat to work is at the core of the branch of physics called *thermodynamics,* the subject of the next several chapters. The concept of caloric did not include this equivalence, and it has been discarded.

The decline and fall of the caloric theory

Heat flow is energy transfer caused by temperature differences.

In many systems, energy can be transferred by both heat and work.

How you describe energy transfer depends on what system you're talking about: the importance of defining your system carefully.

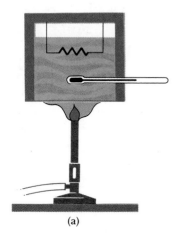

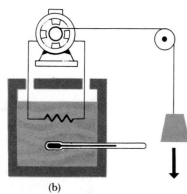

15–1 The same temperature change of the same system may be accomplished by either (a) a heat flow or (b) the performance of work.

Heat transfer is energy transfer that takes place solely because of a temperature difference. For example, water is converted to steam in a steam boiler by contact with a metal pipe or container that is kept at a high temperature by a hot flame from burning coal or gas. Water in the form of steam has greater ability to do *work* (by pushing against a turbine blade or the piston in a steam engine) than in the form of cold liquid water. Therefore the water must have received *energy* in a process involving a heat transfer from the hot flame (through the container) to the cooler water.

Of course, energy transfer can also occur without heat transfer. In an air compressor, a moving piston pushes against a mass of air, doing work on the air that it compresses. In this compressed state, the gas is capable of doing more work than before and hence has acquired energy.

Finally, if the air and the piston in the air compressor are at different temperatures, heat flow between the piston and the air can occur. This process involves two kinds of energy transfer simultaneously: transfer of *heat* and performance of *work*. Such processes are of great importance in many practical devices, such as internal-combustion engines. We will study them in detail in Chapters 18 and 19.

In many cases where a particular change of state of a quantity of material involves energy exchange, the change can be brought about in various ways. In Fig. 15–1a, the temperature of a quantity of water is raised by a gas flame. In Fig. 15–1b, the same change of state is accomplished by a falling weight that turns a generator and powers an electric heating coil in the water. We can regard the change as caused either by heat transfer or by the performance of mechanical work, depending on whether or not we consider the generator and weight as part of our system. Thus in describing various energy-transfer processes, we must always be careful to specify what is and is not included in the system under discussion.

This example shows again that there must be an equivalence between heat and mechanical work, since the same change of state of a system can be produced by *either* heat flow *or* work. A detailed study of this equivalence leads to the *first law of thermodynamics,* which we will study in Chapter 18.

15–2 QUANTITY OF HEAT

We will use the term *heat* only in reference to *transfer* or *flow.* A heat transfer is an energy transfer brought about solely by a temperature difference. The amount of energy transferred may be described as a quantity of heat, but we must be careful *not* to take the view that a certain body *contains* a certain quantity of heat. Such a statement has no meaning. This is a somewhat subtle point, and a thorough discussion of it leads to the concept of *internal energy,* to be introduced in Chapter 18.

We can define a unit of quantity of heat with reference to temperature change of any particular material. The *calorie* (abbreviated cal) was originally defined in the eighteenth century as the amount of heat required to raise the temperature of one gram of water by one Celsius degree (one kelvin). It was found later that this is ambiguous because more heat is required to raise the temperature from, say 90° to 91° than from 20° to 21°. In one of several refinements of the definition, the calorie was defined to be the amount of heat required to heat one gram of water from 14.5°C to 15.5°C. This leads to the

15–2 Sugar cubes obtained in a restaurant in Germany. A rough translation is "This package has 22 Calories (i.e., kilocalories), equal to 92 kilojoules" and "Get into the swing with sugar." By law, foods marketed in Germany must show the energy content in joules; the equivalent in calories is optional.

"15-degree calorie." Several alternative and slightly different definitions are unfortunately also in current use, so the ambiguity persists even today.

A corresponding unit defined in terms of Fahrenheit degrees and British units is the *British thermal unit,* or Btu. By definition, 1 Btu is the quantity of heat required to raise the temperature of one pound of water from 63°F to 64°F. A third unit in common use, especially in measuring food energy, is the *kilocalorie* (kcal), equal to 1000 cal. The relations among these three units are

$$1 \text{ Btu} = 252 \text{ cal} = 0.252 \text{ kcal}.$$

The food-value calorie is really 1000 calories.

The unit of quantity of heat is fundamentally a unit of energy; thus there must be a relation between the above units and the familiar mechanical-energy units such as the joule. It has been found experimentally that

$$1 \text{ cal} = 4.186 \text{ joules} = 4.186 \text{ J},$$
$$1 \text{ kcal} = 4186 \text{ J},$$
$$1 \text{ Btu} = 778 \text{ ft·lb} = 252 \text{ cal} = 1055 \text{ J}.$$

Figure 15–2 shows an example of this equivalence.

The International Committee on Weights and Measures no longer recognizes the calorie as a fundamental unit but recommends instead that the joule be used for quantity of heat as well as for all other forms of energy. Although the calorie is a convenient unit for heat-transfer problems involving water, it is awkward when both heat and other forms of energy are involved. Eventually the joule will probably become the universal unit of energy; we use it in most of the examples and problems of this and the following chapters.

Using the joule for all forms of energy, including heat

15–3 HEAT CAPACITY

We use the symbol Q for quantity of heat. When the quantity is associated with a temperature change ΔT, we usually call it ΔQ, and for an infinitesimal temperature change dT, we use dQ. The quantity of heat ΔQ required to increase the temperature of a mass m of a certain material by an amount ΔT is found to be approximately proportional to ΔT. It is also proportional to m; twice as much heat is needed to heat two cups of water to make tea than to heat only one cup to the same temperature. The quantity of heat needed also depends

How to calculate the heat needed for a certain temperature change

on the nature of the material; to raise the temperature of one gram of water 1 C° requires over five times as much heat as the same temperature increase for the same mass of aluminum.

Specific heat capacity: Different materials have different relationships of heat to temperature change.

Thus the relationship among all these quantities can be expressed as

$$\Delta Q = mc\,\Delta T, \tag{15-1}$$

where c is a constant, different for different materials, called the **specific heat capacity** for the material. Strictly speaking, it is not precisely constant but depends somewhat on temperature. A more precise definition, obtained by rearranging Eq. (15-1) and replacing ΔQ by dQ and ΔT by dT, is

$$c = \frac{1}{m}\frac{dQ}{dT}. \tag{15-2}$$

With this definition, c for a given material depends somewhat on the temperature at which the derivative dQ/dT is evaluated; in solving problems we will usually ignore this variation. The specific heat capacity of water is approximately

$$4.19\ \text{J·g}^{-1}\text{·(C°)}^{-1}, \quad 4190\ \text{J·kg}^{-1}\text{·(C°)}^{-1},$$
$$1\ \text{cal·g}^{-1}\text{·(C°)}^{-1}, \quad \text{or} \quad 1\ \text{Btu·lb}^{-1}\text{·(F°)}^{-1}.$$

When you have a fever, how much extra heat does your body produce to warm itself up?

EXAMPLE 15-1   During a bout with the flu, an 80-kg man ran a fever of 2 C° above normal, that is, a body temperature of 39°C or 102.2°F. To raise his temperature by that amount, his body had to produce extra heat by means of chemical reactions in his cells. Assuming the human body is mostly water, how much heat was needed? Express the result in joules and calories.

SOLUTION   From Eq. (15-1),

$$\Delta Q = (80\ \text{kg})(4190\ \text{J·kg}^{-1}\text{·(C°)}^{-1})(2\ \text{C°})$$
$$= 6.70 \times 10^5\ \text{J},$$

or

$$\Delta Q = (80{,}000\ \text{g})(1\ \text{cal·g}^{-1}\text{·(C°)}^{-1})(2\ \text{C°})$$
$$= 1.60 \times 10^5\ \text{cal}$$
$$= 160\ \text{kcal (160 food-value calories)}.$$

Describing a quantity of material by the number of moles instead of the number of kilograms

It is often convenient to describe a quantity of substance by use of the number of *moles* rather than the *mass* of material. One mole (1 mol) of any substance is a quantity of matter such that its mass in grams is numerically equal to the *molecular mass M*. (This quantity is often called *molecular weight*, but *molecular mass* is preferable because the quantity depends on the mass of a molecule, not its weight.) To calculate the number of moles n, we divide the mass m in grams by the molecular mass M; thus $n = m/M$. Replacing the mass m in Eq. (15-1) by the product nM, we find

$$Mc = \frac{\Delta Q}{n\,\Delta T}.$$

The product Mc is called the **molar heat capacity** and is represented by the symbol C. Hence, by definition

$$C = Mc = \frac{\Delta Q}{n \, \Delta T}, \qquad (15\text{–}3)$$

$$\Delta Q = nC \, \Delta T. \qquad (15\text{–}4)$$

The molecular mass of water is $M = 18 \text{ g·mol}^{-1}$; its molar heat capacity is approximately

$$C = 75.3 \text{ J·mol}^{-1}\cdot(\text{C}°)^{-1} \quad \text{or} \quad 18 \text{ cal·mol}^{-1}\cdot(\text{C}°)^{-1}.$$

The quantity defined by Eq. (15–2) is sometimes called simply *specific heat*, and the molar heat capacity defined by Eq. (15–3) is often called the *molar specific heat*. However, we will use the terms *specific heat capacity* and *molar heat capacity* in this book. Representative values of specific and molar heat capacity are given in Table 15–1.

From Eq. (15–1), the total quantity of heat Q that must be supplied to a body of mass m to change its temperature from T_1 to T_2 is

$$Q = mc(T_2 - T_1), \qquad (15\text{–}5)$$

assuming c may be considered constant through this temperature interval. If T_2 is less than T_1, Q is negative, indicating transfer of heat *out of* the body rather than *into* it.

When the specific heat capacity of a material varies appreciably with temperature, the value used in Eq. (15–5) should be the *mean* specific heat capacity over the interval $(T_2 - T_1)$. For small temperature changes, this variation may often be ignored.

The specific or molar heat capacity of a substance is not the only physical property involving quantity of heat. Heat conductivity, heat of fusion, heat of vaporization, and heat of combustion are a few other examples of *thermal properties* of matter. We will discuss some of these in Section 15–5. The field of physics and physical chemistry concerned with the measurement of thermal properties is called **calorimetry.**

TABLE 15–1 Mean Specific and Molar Heat Capacities (Constant Pressure, Temperature Range 0°C to 100°C)

Metal	Specific (c) $J\cdot kg^{-1}\cdot(C°)^{-1}$	Specific (c) $cal\cdot g^{-1}\cdot(C°)^{-1}$	$M,$ $g\cdot mol^{-1}$	Molar (C) $J\cdot mol^{-1}\cdot(C°)^{-1}$
Aluminum	910	0.217	27.0	24.6
Beryllium	1970	0.471	9.01	17.7
Copper	390	0.093	63.5	24.8
Ethanol	2428	0.58	46.0	112.0
Ethylene glycol	2386	0.57	62.0	148.0
Ice (−25°C to 0°C)	2000	0.48	18.0	36.5
Iron	470	0.112	55.9	26.3
Lead	130	0.031	207.0	26.9
Marble ($CaCO_3$)	879	0.21	100.0	87.9
Mercury	138	0.033	201.0	27.7
Salt	879	0.21	58.5	51.4
Silver	234	0.056	108.0	25.3
Water	4190	1.00	18.0	75.4

Do you know the difference between temperature and heat?

Finally, we remark that it is absolutely essential to distinguish carefully between the two physical quantities *temperature* and *heat*. Temperature depends on the physical state of a material and is a quantitative description of its hotness or coldness; heat is energy in transit from one body to another. We can change the temperature of a body by adding heat to it or taking heat away, but the temperature can also be changed in other ways that do not involve heat transfer.

15–4 MEASUREMENT OF HEAT CAPACITY

An experimental setup to measure heat capacities.

To measure a heat capacity we need to add a measured quantity of heat to a measured quantity of a material and observe the resulting temperature change. For maximum precision, the measurements are often made electrically. In a typical laboratory procedure, we supply heat input by passing a current through a heater wire wound around the specimen. By measuring the voltage, current, and time interval, we can determine the total energy (heat) input Q. We measure the temperature change ΔT by using a resistance thermometer or thermocouple embedded in the specimen, and we measure the mass m by weighing. We can then use Eq. (15–1) to determine the specific heat capacity c. This sounds simple, but great experimental skill is needed to avoid or compensate for unwanted heat transfer between the sample and its surroundings.

Figure 15–3 shows the results of specific heat capacity measurements for water. The quantity of heat needed to raise the temperature of 1 g of water from 14.5°C to 15.5°C is 4.186 J, and this is defined to be one 15° calorie. Two other calories are (unfortunately) frequently used. The *international table calorie* (IT cal) is defined to be precisely

$$1 \text{ IT cal} = \frac{1 \text{ W·hr}}{860} = \frac{3600 \text{ J}}{860} = 4.186 \text{ J}.$$

The terrible ambiguity of the calorie, and how to avoid it

It is almost identical with the 15° calorie. The *thermochemical calorie*, however, is defined as 4.1840 J. From Fig. 15-3 we see that this corresponds to about a 17° calorie. This variety of definitions of the calorie is an additional argument in favor of eliminating the calorie completely and using the joule as the fundamental unit of quantity of heat as well as of all other forms of energy. There are also several different definitions of the Btu, differing by as much as 0.5%.

15–3 Specific heat capacity of water as a function of temperature.

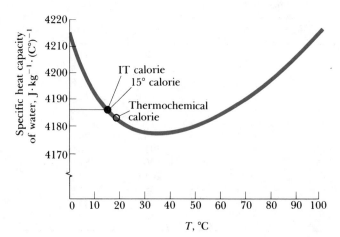

The heat capacity of a material also depends on the conditions imposed during the heat transfer. The two conditions of greatest practical usefulness are for the system to be kept at *constant pressure* or *constant volume;* the corresponding specific heat capacities are denoted as c_p and c_v, respectively, and the molar heat capacities as C_p and C_v, respectively. Laboratory measurements are most often made under constant-pressure conditions. The two heat capacities differ because if the system can expand during the heating, additional energy exchange occurs through the performance of *work* by the system on its surroundings. If the volume is held constant, the system does no work. For gases the difference between c_p and c_v is substantial. We will study heat capacities of gases in detail in Section 18–7.

When measuring heat capacities, do you keep the volume or the pressure constant?

A few values of heat capacities of metals and some familiar compounds are listed in Table 15–1 (Section 15–3). All the specific heat capacities are less than that of water, and they generally decrease with increasing molecular mass. The last column is of particular interest; it shows that the molar heat capacities for all metals except those with very small molecular mass are about the same, about $25 \text{ J·mol}^{-1}·(\text{C}°)^{-1}$. This correlation is called the **law of Dulong and Petit,** after its discoverers. Although only an approximate rule, it contains the germ of a very important idea.

Molar heat capacities of some materials obey a surprisingly simple rule.

The number of molecules in one mole is the same for all substances. Thus about the same amount of heat is required *per molecule* to raise the temperature of each of these metals by a given amount, even though the *mass* of a molecule of lead (for example) is nearly ten times as great as that of a molecule of aluminum. The heat required for a given temperature increase depends only on *how many* molecules the sample contains, and not on the mass of an individual molecule. Peter Debye (1884–1966) showed that the behavior of the molar heat capacity of *nonmetals* over the entire temperature range can be accounted for by the vibrational motion of the molecules in the crystal lattice. We will study the molecular basis of heat capacities in greater detail in Chapter 20.

A molecular viewpoint on heat capacities

15–5 PHASE CHANGES

The term **phase,** as we use it here, refers to a specific state of matter, such as a solid, liquid, or gaseous state. For example, the chemical compound H_2O exists in the *solid phase* as ice, in the *liquid phase* as water, and in the *gaseous phase* as steam. All substances that do not decompose at high temperatures can exist in any of these phases under proper conditions of temperature and pressure. A transition from one phase to another is called a **phase change** or phase transition. For any given pressure, a phase change takes place at a definite temperature. Usually a phase change is accompanied by absorption or liberation of heat and a change of volume and density.

Phase changes: melting, boiling, and all that

A familiar example of a phase change is the melting of ice. When heat is added to ice at 0°C and normal atmospheric pressure, the temperature of the ice *does not* increase; instead some of it melts to form liquid water. If the heat is added slowly, so that thermal equilibrium is maintained between the ice and liquid water, then the temperature remains at 0°C as the heat is added, until all the ice is melted. Thus the effect of adding heat to this system is not to raise its temperature but to change its *phase* from solid to liquid.

Phase changes require heat exchange.

Experiments show that to change 1 kg of ice at 0°C to 1 kg of liquid water at 0°C requires the addition of 3.34×10^5 J of heat. This quantity of heat is called the **heat of fusion** of water. The term *latent* heat of fusion is sometimes used; this is somewhat redundant and will not be used here, but we will use the symbol L, with a subscript, for quantities of heat associated with phase transitions. Thus the heat of fusion L_F of water is

$$L_F = 3.34 \times 10^5 \text{ J·kg}^{-1}.$$

In other units the value is

$$L_F = 79.7 \text{ cal·g}^{-1}$$
$$= 143 \text{ Btu·lb}^{-1}.$$

More generally, to melt a mass m of material requires the addition of a quantity of heat Q given by

$$Q = mL_F. \tag{15–6}$$

The heat of fusion is, of course, different for different materials, and it also varies somewhat with pressure.

Phase equilibrium: when two phases can exist together

The phase change described above is *reversible*. The conversion of liquid water to ice at 0°C requires the *removal* of heat; the amount of heat is again given by Eq. (15–6), but in this case it is considered negative because it is removed rather than added. We also note that at a given pressure it is possible for liquid water and ice to coexist only at one very specific temperature, which we call the *melting temperature*. This coexistence of two phases is called **phase equilibrium.**

This entire discussion can be repeated for boiling, a phase transition between liquid and gaseous phases. The corresponding heat of the phase transition is called the **heat of vaporization L_V**; both L_V and the boiling temperature of a material depend on pressure. Water boils at a lower temperature in Denver than in Pittsburgh because the average atmospheric pressure in Denver is less due to its higher elevation. The heat of vaporization is somewhat greater at this lower temperature. At normal atmospheric pressure, the heat of vaporization of water is

$$L_V = 2.26 \times 10^6 \text{ J·kg}^{-1}$$
$$= 539 \text{ cal·g}^{-1}$$
$$= 970 \text{ Btu·lb}^{-1}.$$

15–4 The metal gallium is one of the few *elements* that melt in the vicinity of room temperature; its melting temperature is 29.8°C, its heat of fusion is 8.04×10^4 J·kg^{-1}, or 19.2 cal·g^{-1}. A crystal of gallium is shown melting in a person's hand. (Photo by Chip Clark.)

It is interesting to note that over five times as much heat is required to boil a quantity of water at 100°C as to raise its temperature from 0° to 100°C.

Table 15–2 lists heats of fusion and vaporization for several materials. Also given are "normal" melting and boiling temperatures, that is, the melting and boiling temperatures at normal atmospheric pressure. Very few *elements* have melting temperatures in the vicinity of ordinary room temperatures; one of the few is the metal gallium, as shown in Fig. 15–4.

Sublimation: a direct solid-to-vapor transition

Under some conditions of temperature and pressure, a substance can change directly from the solid to the gaseous phase without passing through the liquid phase. The transfer from solid to vapor is called **sublimation,** and the solid is said to *sublime.* "Dry ice" (solid carbon dioxide) sublimes at atmospheric pressure. Liquid carbon dioxide cannot exist at a pressure lower than about 5×10^5 Pa (about 5 atm). Heat is absorbed in the process of sublimation

TABLE 15–2 Heats of Fusion and Vaporization

Substance	Normal Melting Point		Heat of Fusion, L_F, $J \cdot kg^{-1}$	Normal Boiling Point		Heat of Vaporization, L_V, $J \cdot kg^{-1}$
	K	°C		K	°C	
Helium	3.5	−269.65	5.23×10^3	4.216	−268.93	20.9×10^3
Hydrogen	13.84	−259.31	58.6×10^3	20.26	−252.89	452×10^3
Nitrogen	63.18	−209.97	25.5×10^3	77.34	−195.81	201×10^3
Oxygen	54.36	−218.79	13.8×10^3	90.18	−182.97	213×10^3
Ethyl alcohol	159	−114	104.2×10^3	351	78	854×10^3
Mercury	234	−39	11.8×10^3	630	357	272×10^3
Water	273.15	0.00	334×10^3	373.15	100.00	2256×10^3
Sulfur	392	119	38.1×10^3	717.75	444.60	326×10^3
Lead	600.5	327.3	24.5×10^3	2023	1750	871×10^3
Antimony	903.65	630.50	165×10^3	1713	1440	561×10^3
Silver	1233.95	960.80	88.3×10^3	2466	2193	2336×10^3
Gold	1336.15	1063.00	64.5×10^3	2933	2660	1578×10^3
Copper	1356	1083	134×10^3	1460	1187	5069×10^3

and liberated in the reverse process. The quantity of heat absorbed per unit mass is called the **heat of sublimation.** In freeze-drying of food, the food is first frozen, and then water is removed by sublimation. The heat of sublimation is often supplied by infrared or microwave radiation.

Under some conditions a material can be cooled below the normal phase-change temperature without a phase change occurring. The resulting state is unstable and is described as *supercooled.* Very pure water can be cooled several degrees below the normal freezing point under ideal conditions; when a small ice crystal is dropped in or the water is agitated, it crystallizes very quickly. Similarly, supercooled water vapor condenses quickly into fog droplets when a disturbance, such as dust particles or ionizing radiation, is introduced. This phenomenon is used in the *cloud chamber,* where charged particles such as protons and electrons induce condensation of supercooled vapor along their path, thus making the path visible. The same principle is involved in "seeding" clouds, which often contain supercooled water vapor, to cause condensation and rain.

Similarly, the temperature of a liquid can sometimes be raised above its normal boiling temperature, and the liquid is then said to be *superheated.* Again, any small disturbance—such as agitation, a dust particle, or the passage of a charged particle through the material—causes local boiling with bubble formation. This phenomenon is used in the bubble chamber, to be discussed in Chapter 44, to show the tracks of high-energy particles.

As an illustration of phase changes, suppose we take crushed ice from a freezer at −25°C, place it in a container with a thermometer, and surround it with a heating coil that supplies heat at a constant rate. We insulate the system from its surroundings so no other heat enters it, and we observe the rise in temperature with time. The result is shown in Fig. 15–5, a graph of temperature as a function of time. We find that the temperature of the ice increases steadily, as shown by the portion of the graph from *a* to *b*, until it has risen to 0°C. In this temperature range the specific heat capacity of ice is approximately

$$2.0 \times 10^3 \ J \cdot kg^{-1} \cdot (C°)^{-1} \quad \text{or} \quad 0.48 \ cal \cdot g^{-1} \cdot (C°)^{-1}.$$

Superheating and supercooling: postponing the inevitable

An experiment showing energy relations in phase transitions

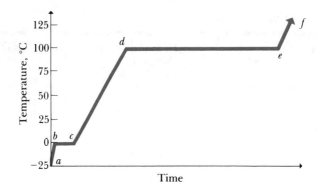

15–5 The temperature remains constant during each change of phase, provided the pressure remains constant.

As soon as this temperature is reached, the ice begins to *melt*, a *change of phase*, from the solid phase to the liquid phase. The thermometer, however, shows *no increase in temperature*, and although heat is being supplied at the same rate as before, the temperature remains at 0°C until all the ice is melted (point *c*).

As soon as the last of the ice has melted, the temperature begins to rise again at a uniform rate (from *c* to *d*), although this rate is *slower* than for ice. When a temperature of 100°C is reached (point *d*), bubbles of steam (gaseous water or water vapor) start to escape from the liquid surface; that is, the water begins to *boil*. The temperature remains constant at 100°C until all the water has boiled away. Another change of phase has taken place, from the liquid phase to the gaseous phase.

If all the water vapor had been trapped and not allowed to diffuse away (a very large container would be needed), the heating process could be continued as from *e* to *f*. The gas would now be called "superheated steam."

An essential point in this discussion is that when heat is added slowly (to maintain thermal equilibrium) to a substance that can exist in different phases, *either* the temperature rises *or* some of the substance undergoes a phase change, but *never* both at the same time. Once the temperature for a phase change (e.g., the melting or boiling temperature) has been reached, no further temperature change occurs until *all* the substance has undergone the phase change.

Although we have used water as an example, the same type of curve as in Fig. 15–5 is obtained for many other substances. Some, of course, decompose before reaching a melting or boiling point, and others, such as glass or tar, do not change phase at a definite temperature but become gradually softer as their temperature is raised. Crystalline substances, such as ice, or a metal melt at a definite temperature. Glass and tar are amorphous solids with no definite crystal structure, and they behave like liquids of very high viscosity.

When heat is removed from a gas, its temperature falls; at the same temperature at which it boiled, it returns to the liquid phase, or *condenses*. In so doing it gives up to its surroundings the same quantity of heat that was required to vaporize it. The heat so given up, per unit mass, is equal to the heat of vaporization. Similarly, a liquid returns to the solid phase, or *freezes*, when cooled to the temperature at which it melted, and it gives up heat equal to the heat of fusion.

Steam-heating systems use a boiling-condensing process to transfer heat from the furnace to the radiators. Each kilogram of water that is turned to steam in the furnace absorbs 2.26×10^6 J (the heat of vaporization of water)

Amorphous solids: When is a solid not really solid?

Steam heating systems: how to get the heat from the furnace to the radiators

from the furnace, and gives up this same amount when it condenses in the radiators. (This figure is correct if the steam pressure is 1 atm. It is slightly smaller at higher pressures.) Thus a steam-heating system does not need to circulate as much water as a hot-water heating system. If water leaves a hot-water furnace at 60°C and returns at 40°C, dropping 20°C, about 27 kg of water must circulate to carry the same amount of heat as is transferred in the form of heat of vaporization by 1 kg of steam.

The temperature-control mechanisms of many warm-blooded animals operate on a similar principle. As sweat (chiefly water) evaporates from the surface of the body, it removes heat from the body as heat of vaporization. At normal human body temperature (37°C) the heat of vaporization of water is 2.41×10^6 J·kg^{-1}, so each gram of sweat that evaporates carries away 2410 J of heat with it. Evaporative cooling makes it possible for humans to maintain normal body temperature in hot, dry desert climates where the air temperature may reach 55°C (about 130°F). Evaporation keeps the skin temperature as much as 20 C° cooler than the surrounding air. Adequate water intake is essential under these conditions; a normal person may perspire several liters per day, and unless this lost water is replaced regularly, dehydration, heat stroke, and death result. Old-time desert rats (such as the author) state that in the desert any canteen that holds less than a gallon is to be regarded as only a toy!

Temperature control in warm-blooded animals: how to stay cool in the desert

The same principle is applied in evaporative cooling systems. Hot, dry air is pushed through wet filters. As some of the water evaporates, it takes its heat of vaporization from the air, which thus becomes cooler by as much as 20 C° in hot, dry climates. Evaporative cooling is also used to condense and recirculate "used" steam in coal-fired or nuclear-powered electric power plants.

Finally, we mention that definite quantities of heat are also involved in chemical reactions. A familiar example is combustion; complete combustion of one gram of gasoline produces about 46,000 J or about 11,000 cal, and the **heat of combustion** of gasoline is

Heat of combustion: an example of heat associated with a chemical reaction

$$46,000 \text{ J·g}^{-1} \quad \text{or} \quad 46 \times 10^6 \text{ J·kg}^{-1}.$$

Energy values of foods are defined similarly; the unit of food energy, although called a calorie, is really a kilocalorie, equal to 1000 cal or 4186 J. When we say a gram of peanut butter "contains" 12 calories, we mean that, when it reacts with oxygen, with the help of enzymes, to convert the carbon and hydrogen completely to CO_2 and H_2O, the total energy liberated as heat is 12 kcal, 12,000 cal, or 50,200 J. Not all this energy is directly useful for mechanical work; we will take up the matter of *efficiency* of utilization of energy in Chapter 19.

Heat equivalents of food energy, and commonly used units

15–6 EXAMPLES

The basic principle in heat calculations is very simple: When heat flow occurs between two bodies, the amount of heat lost by one body must equal the amount gained by the other. Heat is energy in transit, so this is really just conservation of energy. We take each quantity of heat *added to* a body as *positive,* and each quantity *leaving* a body as *negative.* Then in general when several bodies interact, the *algebraic sum* of the quantities of heat transferred to each body must be zero. This is the basic principle of calorimetry, in some respects the simplest of all physical theories.

Calorimetry problems involve a simple conservation principle.

PROBLEM-SOLVING STRATEGY: Calorimetry problems

1. To avoid confusion about algebraic signs when calculating quantities of heat, use Eqs. (15–1) and (15–6) consistently for each body, noting that each Q (or ΔQ) is positive when heat enters a body and negative when it leaves. Then the algebraic sum of all the Q's must be zero.

2. Often you will need to find an unknown temperature. Represent it by an algebraic symbol such as T. Then if a body has an initial temperature of 20°C and an unknown final temperature T, the temperature change for the body is $\Delta T = T_{\text{final}} - T_{\text{initial}} = T - 20°C$ (*not* $20°C - T$!). And so on.

3. In problems where a phase change takes place, as when ice melts, you may not know in advance whether *all* the material undergoes a phase change or only part of it. You can always assume one or the other, and if the resulting calculation gives an absurd result (such as a final temperature higher or lower than *any* of the initial temperatures), you know the initial assumption was wrong. Back up and try again!

How much does your coffee cool when you pour it into a cold cup?

EXAMPLE 15–2 An ethnic restaurant down by the docks serves coffee in copper mugs. A waiter fills a cup of mass 0.1 kg, initially at 20°C, with 0.2 kg of coffee initially at 70°C. What is the final temperature after the coffee and the cup attain thermal equilibrium? (Assume coffee has the same specific heat capacity as water.)

SOLUTION Let the final temperature be T. The (negative) heat added to the coffee is

$$Q_{\text{coffee}} = mc_{\text{water}}\,\Delta T$$
$$= (0.2\text{ kg})(4190\text{ J·kg}^{-1}\text{·(C°)}^{-1})(T - 70°C).$$

Similarly, the heat gained by the copper cup is

$$Q_{\text{copper}} = mc_{\text{copper}}\,\Delta T$$
$$= (0.1\text{ kg})(390\text{ J·kg}^{-1}\text{·(C°)}^{-1})(T - 20°C).$$

We equate the sum of these two quantities of heat to zero, obtaining an algebraic equation for T: $Q_{\text{coffee}} + Q_{\text{copper}} = 0$, or

$$(0.2\text{ kg})(4190\text{ J·kg}^{-1}\text{·(C°)}^{-1})(T - 70°C)$$
$$+ (0.1\text{ kg})(390\text{ J·kg}^{-1}\text{·(C°)}^{-1})(T - 20°C) = 0.$$

Solution of this equation gives $T = 67.8°C$. The final temperature is much closer to the initial temperature of the coffee than that of the cup because of the much larger specific heat capacity of water. We can also find the quantities of heat by substituting this value for T back into the original equations. We leave it for you to show that $Q_{\text{coffee}} = -1860$ J, and $Q_{\text{copper}} = +1860$ J; Q_{coffee} is negative, as expected.

Cooling a soft drink: How much ice do you need?

EXAMPLE 15–3 A physics student wants to cool 0.25 kg of Omni-Cola (mostly water), which is initially at 20°C, by adding ice initially at −20°C. How much ice should be added so the final temperature will be 0°C with all the ice melted, if the heat capacity of the container can be neglected?

SOLUTION The (negative) heat added to the water is

$$Q = mc_{\text{water}}\,\Delta T$$
$$= (0.25\text{ kg})(4190\text{ J·kg}^{-1}\text{·(C°)}^{-1})(0°C - 20°C)$$
$$= -20{,}900\text{ J}.$$

The specific heat capacity of ice is approximately $2000 \text{ J·kg}^{-1}·(\text{C}°)^{-1}$. Let the mass of ice be m; then the heat needed to heat it from $-20°C$ to $0°C$ is

$$Q = mc_{\text{ice}} \, \Delta T$$
$$= m(2000 \text{ J·kg}^{-1}·(\text{C}°)^{-1})(0°C - (-20°C))$$
$$= m(40{,}000 \text{ J·kg}^{-1}).$$

The additional heat needed to melt the ice is the heat of fusion times the mass:

$$Q = mL_F = m(334{,}000 \text{ J·kg}^{-1}).$$

The sum of these three quantities must equal zero:

$$-20{,}900 \text{ J} + m(40{,}000 \text{ J·kg}^{-1}) + m(334{,}000 \text{ J·kg}^{-1}) = 0,$$

from which $m = 0.056 \text{ kg} = 56 \text{ g}$, or roughly two medium-sized ice cubes.

EXAMPLE 15–4 A certain gasoline camping lantern emits as much light as a 25-watt electric light bulb. Assuming that the efficiency of converting heat into light is the same for the lantern and the bulb (which is not actually correct), how much gasoline does the lantern burn in 10 hours?

Energy relations in a gasoline lantern

SOLUTION The rate of energy conversion in the lantern is 25 W, so in 10 hours (36,000 s) the total energy needed is

$$(25 \text{ J·s}^{-1})(36{,}000 \text{ s}) = 0.9 \times 10^6 \text{ J}.$$

As mentioned in Section 15–6, combustion of one gram of gasoline produces 46,000 J, so the mass of gasoline required is

$$\frac{0.9 \times 10^6 \text{ J}}{46{,}000 \text{ J·g}^{-1}} = 19.6 \text{ g}.$$

Actual lanterns are much less efficient than this, and typically require on the order of 300 to 400 g of gasoline (roughly one pint) to operate for ten hours.

SUMMARY

Heat is energy in transit; the term is properly used only in reference to transfer of energy from one body to another as a result of a temperature difference. The quantity of heat ΔQ required to raise the temperature of a mass m of material by a small amount ΔT is

$$\Delta Q = mc \, \Delta T, \tag{15–1}$$

where c is the specific heat capacity of the material. If c can be considered constant over a finite temperature range T_1 to T_2, then the amount of heat Q needed to cause this temperature change is given by

$$Q = mc(T_2 - T_1) = mc \, \Delta T. \tag{15–5}$$

When the quantity of material is represented in terms of the number of moles n, the corresponding relation is

$$\Delta Q = nC \, \Delta T, \tag{15–4}$$

where C is the molar heat capacity; $C = Mc$, where M is the molecular mass. The number of moles n and the mass m of material are related by $m = nM$.

KEY TERMS
heat transfer
specific heat capacity
molar heat capacity
calorimetry
law of Dulong and Petit
phase
phase change
heat of fusion
phase equilibrium
heat of vaporization
sublimation
heat of sublimation
heat of combustion

The molar heat capacities of most metals at sufficiently high temperatures approach the value $25 \, \text{J} \cdot \text{mol}^{-1} \cdot (\text{C}°)^{-1}$; this is the law of Dulong and Petit.

A phase change that occurs at constant temperature usually requires the addition or removal of heat. To change a mass m of solid material to liquid at the same temperature requires the addition of a quantity of heat Q given by

$$Q = mL_F, \qquad\qquad (15\text{–}6)$$

where L_F is called the heat of fusion. When the material changes back to the solid state, an equal quantity of heat must be removed. The corresponding quantities associated with boiling and sublimation are called the heat of vaporization and heat of sublimation, respectively.

When heat is added to a body, the corresponding Q is positive; when it is removed, Q is negative. The basic principle of calorimetry, stemming from conservation of energy, is that in a system whose parts interact by heat exchange the algebraic sum of the Q's for all parts of the system must be zero.

QUESTIONS

15–1 Suppose you have a thermos bottle half full of cold coffee. Can you warm it up to drinking temperature by shaking it? Is this possible *in principle?* Is it feasible in practice? Are you adding heat to the coffee?

15–2 When the oil in an automatic transmission is churned up by the turbine blades, it becomes hot, and usually an oil-cooling system is required. Is the engine adding heat to the oil?

15–3 A student asserted that a suitable unit for specific heat capacity was $1 \, \text{m}^2 \cdot \text{s}^{-2} \cdot (\text{C}°)^{-1}$. Is this correct?

15–4 The specific heat capacity of water has the same numerical value when expressed in $\text{cal} \cdot \text{g}^{-1} \cdot (\text{C}°)^{-1}$ as when expressed in $\text{Btu} \cdot \text{lb}^{-1} \cdot (\text{F}°)^{-1}$. Is this a coincidence? Does the same relation hold for heat capacities of other materials?

15–5 A student claimed that when two bodies not initially in thermal equilibrium are placed in contact, the temperature rise of the cooler body must always equal the temperature drop of the warmer. Do you agree? Is there a principle of conservation of temperature, or something like that?

15–6 In choosing a fluid to circulate inside a gasoline engine to cool it (such as water or antifreeze), should you choose a material with a large or a small specific heat capacity? Why? What other considerations are important?

15–7 Any heat of a phase transition has a numerical value that is $\frac{5}{9}$ as great when expressed in $\text{cal} \cdot \text{g}^{-1}$ as when expressed in $\text{Btu} \cdot \text{lb}^{-1}$. Why is the conversion so simple?

15–8 Why do you think the heat of vaporization for water is so much larger than the heat of fusion?

15–9 Some household air conditioners used in dry climates cool air by blowing it through a water-soaked filter, evaporating some of the water. How does this work? Would such a system work well in a high-humidity climate?

15–10 Why does food cook faster in a pressure cooker than in boiling water?

15–11 How does the human body maintain a temperature of $37.0°\text{C}$ ($98.6°\text{F}$) in the desert where the temperature is $50°\text{C}$ ($122°\text{F}$)?

15–12 Desert travelers sometimes keep water in a canvas bag. Some water seeps through the bag and evaporates. How does this cool the water inside?

15–13 When water is placed in ice-cube trays in a freezer, why does the water not freeze all at once when the temperature has reached $0°\text{C}$? In fact, it freezes first in a layer adjacent to the sides of the tray. Why?

15–14 When an automobile engine overheats and the radiator water begins to boil, the car can still be driven some distance before catastrophic engine damage occurs. Why? What determines the onset of really disastrous overheating?

15–15 Why do automobile manufacturers recommend that antifreeze (typically a 50% solution of ethylene glycol in water) be kept in the engine in summer as well as in winter?

15–16 When you step out of the shower you feel cold, but as soon as you are dry you feel warmer, even though the room temperature is the same. Why?

15–17 Suppose the heat of fusion of ice were only $10 \, \text{J} \cdot \text{g}^{-1}$ instead of $334 \, \text{J} \cdot \text{g}^{-1}$. Would this change the way we make iced tea? Martinis? Lemonade?

15–18 Why is the climate of regions adjacent to large bodies of water usually more moderate than that of regions far from large bodies of water?

EXERCISES

Section 15–2 Quantity of Heat

15–1 A taxi driver drives an automobile of mass 1500 kg at a speed of 5 m·s^{-1}. How many joules of heat are generated in the brake mechanism when the automobile is brought to rest?

15–2 A crate of mass 50 kg slides down a ramp inclined at 53° below the horizontal. The ramp is 6 m long. If the crate was at rest at the top of the incline and has a velocity of 6 m·s^{-1} at the bottom, how much heat is generated by friction? Express your answer in joules, calories, and Btu.

Section 15–3 Heat Capacity

15–3

a) How much heat is required to raise the temperature of 0.20 kg of water from 20°C to 30°C?

b) If this amount of heat is added to an equal mass of mercury initially at 20°C, what is the final temperature?

c) If this amount of heat is added to an equal volume of mercury initially at 20°C, what is the final temperature?

15–4 A student uses a 100-watt electric immersion heater to heat 0.2 kg of water from 20°C to 100°C to make tea.

a) How much heat must be added to the water?

b) How much time is required?

15–5 A copper tea kettle of mass 2.0 kg, containing 4.0 kg of water, is placed on a stove. How much heat must be added to raise the temperature from 20°C to 80°C?

15–6 A copper cup of mass 0.20 kg contains 0.40 kg of water. The water is heated by a friction device that dissipates mechanical energy, and it is observed that the temperature of the system rises at the rate of $3 \text{ C}°\text{·min}^{-1}$. Neglect heat losses to the surroundings. What power in watts is being dissipated in the water?

15–7 A technician measures the specific heat capacity of an unidentified liquid by immersing an electrical resistor in it. Electrical energy is dissipated for 100 s at a constant rate of 50 W. The mass of the liquid is 0.530 kg, and its temperature increases from 17.64°C to 20.77°C. Find the mean specific heat capacity of the liquid in this temperature range.

Section 15–5 Phase Changes

15–8 An ice-cube tray contains 0.8 kg of water at 20°C. How much heat must be removed to cool the water to 0°C and freeze it?

15–9 How much heat is required to convert 1 g of ice at −10°C to steam at 100°C? Express you answer in joules, calories, and Btu.

15–10 An open vessel contains 0.50 kg of ice at −20°C. The mass of the container can be neglected. Heat is supplied to the vessel at the constant rate of 420 J·min^{-1} for 500 min.

a) After how many minutes does the ice *start* to melt?

b) After how many minutes does the temperature start to rise above 0°C?

c) Plot a curve showing the elapsed time as abscissa and the temperature as ordinate.

15–11 An automobile engine whose output is $3.0 \times 10^4 \text{W}$ (about 40 hp) uses 4.5 gal of gasoline per hour. The heat of combustion is 12×10^7 J per gallon. What is the efficiency of the engine? That is, what fraction of the heat of combustion is converted to mechanical work?

15–12 The nominal food-energy value of butter is about 6.0 kcal·g^{-1}. If all this energy could be converted completely to mechanical energy, how much butter would be required to power an 80-kg mountaineer on his journey from Lupine Meadows (elevation 2070 m) to the summit of Grand Teton (4196 m)?

15–13 What must be the initial velocity of a lead bullet at a temperature of 25°C so that the heat developed when it is brought to rest will be just sufficient to melt it?

15–14 Evaporation of sweat is an important mechanism for temperature control in warm-blooded animals. What mass of water must evaporate from the surface of an 80-kg human body to cool it 1C°? The specific heat capacity of the human body is approximately $1.0 \text{ cal·g}^{-1}\text{·(C°)}^{-1}$, and the heat of vaporization of water at body temperature (37°C) is 577 cal·g^{-1}.

15–15 The capacity of air conditioners is sometimes expressed in "tons," the number of tons of ice that can be frozen from water at 0°C in 24 hr by the unit. Express the capacity of a one-ton air conditioner in watts and in Btu·hr^{-1}.

Section 15–6 Examples

15–16 In a physics lab experiment, a student immersed 100 copper pennies (mass 3.0 g each) in boiling water. After they reached thermal equilibrium, she fished them out and dropped them into 0.2 kg of water at 20°C. What was the final temperature?

15–17 An aluminum can of mass 0.500 kg contains 0.118 kg of water at a temperature of 20°C. A 0.200-kg block of iron at 75°C is dropped into the can. Find the final temperature, assuming no heat loss to the surroundings.

15–18 A 0.050-kg sample of unknown material, at a temperature of 100°C, is dropped into a calorimeter containing 0.200 kg of water initially at 20°C. The calorimeter is of copper, and its mass is 0.100 kg. The final temperature of the calorimeter is 22°C. Compute the specific heat capacity of the sample.

15–19 A 2-kg iron block is taken from a furnace where its temperature was 650°C and placed on a large block of ice at 0°C. Assuming that all the heat given up by the iron is used to melt the ice, how much ice is melted?

15–20 A copper calorimeter of mass 0.100 kg contains 0.150 kg of water and 0.008 kg of ice in thermal equilibrium at atmospheric pressure. If 0.500 kg of lead at a temperature of 200°C are dropped into the calorimeter, what is the final temperature, assuming no heat is lost to the surroundings?

15–21 A beaker of very small mass contains 0.500 kg of water at a temperature of 80°C. How many grams of ice at a temperature of −20°C must be dropped into the water so that the final temperature of the system will be 50°C?

15–22 A thirsty farmer cools a 1-L bottle of soft drink (mostly water) by pouring the contents into a large copper mug of mass 0.278 kg and adding 0.050 kg of ice initially at −16°C. If soft drink and mug are initially at 20°C, what is the final temperature of the system, assuming no heat losses?

15–23 A vessel whose walls are thermally insulated contains 2.10 kg of water and 0.200 kg of ice, all at a temperature of 0°C. The outlet of a tube leading from a boiler in which water is boiling at atmospheric pressure is inserted into the water. How many grams of steam must condense to raise the temperature of the system to 20°C? Neglect the heat capacity of the container.

PROBLEMS

15–24 An artificial satellite, constructed of aluminum, encircles the earth at a speed of 9000 m·s⁻¹.

a) Find the ratio of its kinetic energy to the energy required to raise its temperature by 600 C°. (The melting point of aluminum is 660°C.) Assume a constant specific heat capacity of 910 J·kg⁻¹·(C°)⁻¹.

b) Discuss the bearing of your answer on the problem of the reentry of a satellite into the earth's atmosphere.

15–25 A piece of ice at 0°C falls from rest into a lake whose temperature is 0°C, and one-half of 1% of the ice melts. Compute the minimum height from which the ice falls.

15–26 The windlass is a rotating drum or cylinder over which a rope or cord slides in order to provide a great amplification of the rope's tension while keeping both ends free. See Fig. 15–6. Since the added tension in the rope is due to friction, the windlass generates heat.

a) If the difference in tension between the two parts of the rope is 100 N, and the windlass has a diameter of 10 cm and turns once in 1 s, find the rate at which heat is being generated. Why does the number of turns not matter?

b) If the windlass is made of iron and has a mass of 5 kg, at what rate does its temperature rise? Assume that the temperature in the windlass is uniform.

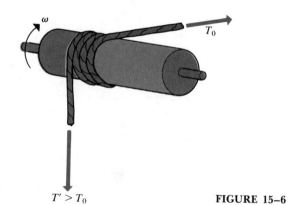

$T' > T_0$ **FIGURE 15–6**

15–27 At very low temperatures, the molar heat capacity of rock salt varies with temperature according to "Debye's T^3 law"; thus

$$C = k\frac{T^3}{\Theta^3},$$

where $k = 1940 \text{ J·mol}^{-1}\cdot K^{-1}$ and $\Theta = 281$ K.

a) How much heat is required to raise the temperature of 2 mol of rock salt from 10 K to 50 K? (*Hint:* Use Eq. (15–2) for dQ and integrate.)

b) What is the mean molar heat capacity in this range?

c) What is the true molar heat capacity at 50 K?

15–28 The molar heat capacity at constant pressure of a certain substance varies with temperature according to the empirical equation

$$C_p = 27.2 \text{ J·mol}^{-1}\cdot K^{-1} + (4 \times 10^{-3} \text{ J·mol}^{-1}\cdot K^{-2})T,$$

where T is in kelvins. How much heat is necessary to change the temperature of 10 mol of this substance from 27°C to 527°C? (*Hint:* Use Eq. (15–2) for dQ and integrate.)

15–29

a) A home owner in a cold climate has a coal-burning furnace that burns 9000 kg (about 10 tons) of coal during a winter. The heat of combustion of coal is $2.50 \times 10^7 \text{ J·kg}^{-1}$. If stack losses are 15% (the amount of heat energy lost up the chimney), how many joules were actually used to heat the house?

b) The home owner proposes to install a solar heating system, heating large tanks of water by solar radiation during the summer and using the stored energy for heating during the winter. Find the required dimensions of the storage tank, assuming it to be a cube, to store a quantity of energy equal to that computed in part (a). Assume that the water is raised to 49°C (120°F) in the summer and cooled to 27°C (80°F) in the winter.

15–30 In a household hot-water heating system, water is delivered to the radiators at 60°C (140°F) and leaves at 38°C (100°F). The system is to be replaced by a steam system in which steam at atmospheric pressure condenses in the radiators, the condensed steam leaving the radiators at 82°C (180°F). How many kilograms of steam will supply the same heat as was supplied by 1 kg of hot water in the first installation?

15–31 A "solar house" has storage facilities for 4.2×10^9 J (about 4 million Btu). Compare the space requirements for this storage on the assumption

a) that the heat is stored in water heated from a minimum temperature of 27°C (80°F) to a maximum of 49°C (120°F);

b) that the heat is stored in Glauber salt (Na$_2$SO$_4$·10 H$_2$O) heated in the same temperature range.

Properties of Glauber Salt	
Specific heat capacity	
Solid	1930 J·kg^{-1}·(C°)$^{-1}$
Liquid	2850 J·kg^{-1}·(C°)$^{-1}$
Specific gravity	1.6
Melting point	32°C
Heat of fusion	2.42 × 10^5 J·kg^{-1}

15–32 A calorimeter contains 0.100 kg of water at 0°C. A 1-kg copper cylinder and a 1-kg lead cylinder, both at 100°C, are placed in the calorimeter. Find the final temperature if there is no loss of heat to the surroundings. Neglect the heat capacity of the calorimeter itself.

15–33 An ice cube whose mass is 0.050 kg is taken from a refrigerator where its temperature was −10°C and dropped into a glass of water at 0°C. If no heat is gained or lost from outside, how much water will freeze onto the cube?

15–34 A tube leads from a flask in which water is boiling under atmospheric pressure to a calorimeter. The mass of the calorimeter is 0.150 kg, its specific heat capacity is 420·J·kg^{-1}·(C°)$^{-1}$, and it contains originally 0.340 kg of water at 15°C. Steam is allowed to condense in the calorimeter until its temperature increases to 71°C, after which the total mass of calorimeter and contents is found to be 0.525 kg. Compute the heat of condensation of steam from these data.

15–35 A copper calorimeter can of mass 0.322 kg contains 0.050 kg of ice. The system is initially at 0°C. If 0.012 kg of steam at 100°C and 1 atm pressure are admitted into the calorimeter, what will be the final temperature of the calorimeter and its contents?

CHALLENGE PROBLEMS

15–36 An electric heater is to provide a continuous supply of hot water. One trial design is shown in Fig. 15–7. Water is flowing at the rate of 0.300 kg·min^{-1}, the inlet thermometer registers 15°C, the voltmeter reads 120 V, and the ammeter reads 10 A (corresponding to a power input of [120 V][10 A] = 1200 W).

a) When a steady state is finally reached, what will be the reading of the outlet thermometer?

b) Why is it unnecessary to take into account the heat capacity mc of the apparatus itself?

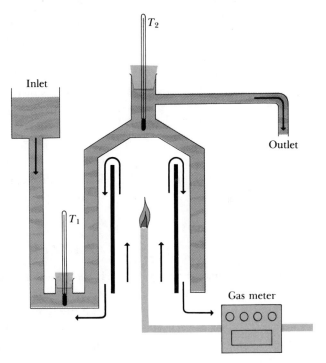

FIGURE 15–8

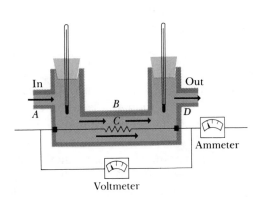

FIGURE 15–7

15–37 An engineer has designed a continuous-flow calorimeter to measure the heat of combustion of a gaseous fuel. A sketch of his design is shown in Fig. 15–8. Water is supplied at the rate of 5.67 kg·min^{-1} and natural gas at 5.67 × 10^{-4} m^3·min^{-1}. In the steady state, the inlet and outlet thermometers register 15.6°C and 24.4°C, respectively. What is the heat of combustion of natural gas in J·m^{-3}? Why should the gas flow be made as small as possible?

15–38 An aluminum rod of cross-sectional area 0.04 cm^2 and length 80.00 cm at a temperature of 140°C is laid alongside a copper rod of cross-sectional area 0.02 cm^2 and length 79.94 cm at temperature T. The two rods are laid alongside each other so they are in thermal contact. No heat is lost to the surroundings, and after they have come to thermal equilibrium they are observed to be of the same length. Calculate the original temperature T of the copper rod and the final temperature of the rods after they come to equilibrium.

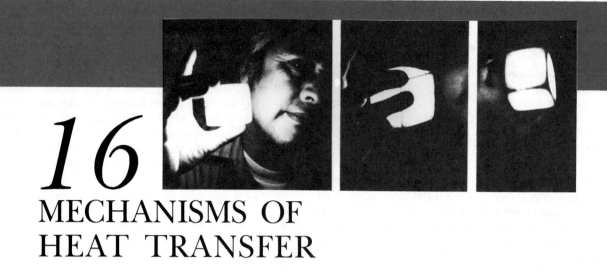

16
MECHANISMS OF HEAT TRANSFER

WE DISCUSSED CONDUCTION OF HEAT QUALITATIVELY IN CHAPTER 15 IN connection with the transfer of heat between two bodies at different temperatures. We did not need to be concerned there with the *rate* of transfer of heat from one body to another. In many situations, however, we need to know how quickly or slowly heat is transferred, and how the rate of heat transfer depends on the properties of the system. In this chapter we study the three mechanisms of heat transfer: conduction, convection, and radiation. Conduction occurs within a body and between two bodies in actual contact with each other. Convection involves motion of mass from one region of space to another. Radiation is heat transfer by electromagnetic radiation, with no need for matter to be present in the space between bodies.

16–1 CONDUCTION

Conduction: energy transfer through a material

If we place one end of a metal rod in a flame and hold the other end, the end we are holding gets hotter and hotter, even though it is not in direct contact with the flame. Heat reaches the cooler end by **conduction** through the material of the rod. Microscopically, the molecules at the hot end increase the energy of their vibrations as the temperature increases. They then interact with their more slowly moving neighbors farther from the flame; they share some of their energy with these neighbors, who in turn pass it on to those still farther from the flame. Thus energy associated with thermal motion is passed along from one molecule to the next, from the hotter end to the cooler, while each individual molecule remains at its original position.

Metals are good conductors.

Most metals are good conductors of electricity and also good conductors of heat. The ability of a metal to conduct an electric current is due to the fact that some electrons in the material have become detached from their parent atoms. These "free" electrons also provide an effective mechanism for heat transfer from the hotter to the cooler portions of the metal. The best electrical conductors (silver, copper, aluminum, and gold) are also the best thermal conductors.

Conduction of heat takes place in a body only when different parts of the body are at different temperatures, and the direction of heat flow is always from points of higher temperature to points of lower temperature. Figure 16–1 shows a rod of material with cross-sectional area A and length L. Let the left end of the rod be kept at a temperature T_2, and the right end at a lower temperature T_1. The direction of the flow of heat is then from left to right through the rod. We assume that the sides of the rod are covered by an insulating material, so no heat transfer occurs at the sides. This is, of course, an idealized model; all real materials, even the best thermal insulators, conduct heat to some extent.

Experiments show that the *rate* of flow of heat through the rod is proportional to the area A, proportional to the temperature difference $(T_2 - T_1)$, and inversely proportional to the length L. We can express these proportions as an equation by introducing a constant k whose numerical value depends on the material of the rod. The quantity k is called the **thermal conductivity** of the material:

$$\frac{dQ}{dt} = H = \frac{kA(T_2 - T_1)}{L}, \tag{16–1}$$

where H is the quantity of heat flowing through the rod per unit time, also called the **heat current**. The quantity $(T_2 - T_1)/L$ represents the temperature difference per unit length and is called the **temperature gradient**.

Equation (16–1) may also be used to compute the rate of heat flow through a slab or through *any* homogeneous body having a uniform cross section perpendicular to the direction of flow, provided that the flow has attained steady-state conditions and the ends are kept at constant temperatures.

The units of heat flow or heat current are energy per unit time; thus the SI unit of heat current is the joule per second, or watt. Other units such as the calorie per second or Btu per second are also sometimes used. The unit of k is obtained by solving Eq. (16–1) for k:

$$k = \frac{HL}{A(T_2 - T_1)}. \tag{16–2}$$

This shows that in SI units, the unit of k is

$$\frac{(1 \text{ J·s}^{-1})(1 \text{ m})}{(1 \text{ m})^2(1 \text{ C}°)} = 1 \text{ J·(s·m·C}°)^{-1}.$$

Values of thermal conductivity are sometimes tabulated by using cgs units, with the calorie as the energy unit. The unit of k then is $1 \text{ cal·(s·cm·C}°)^{-1}$. The conversion is

$$1 \text{ cal·(s·cm·C}°)^{-1} = 419 \text{ J·(s·m·C}°)^{-1}.$$

Some numerical values of k, at temperatures near room temperature, are given in Table 16–1. The properties of materials used commercially as heat insulators are sometimes expressed in a mixed system in which the unit of heat current is 1 Btu·hr^{-1}, the unit of area is 1 ft^2, and the unit of temperature gradient is one Fahrenheit degree per inch $(1 \text{ F}°\text{·in.}^{-1})$! What better argument could there be for adoption of the metric system?

Heat conduction in a material requires temperature differences between different regions.

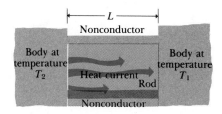

16–1 Steady-state heat flow in a uniform rod.

Thermal conductivity describes how well a material conducts heat.

TABLE 16–1   Thermal Conductivities (k)

	$J \cdot s^{-1} \cdot m^{-1} \cdot (C°)^{-1}$	$cal \cdot s^{-1} \cdot cm^{-1} \cdot (C°)^{-1}$
Metals		
Aluminum	205.0	0.49
Brass	109.0	0.26
Copper	385.0	0.92
Lead	34.7	0.083
Mercury	8.3	0.020
Silver	406.0	0.97
Steel	50.2	0.12
Various solids		
(Representative values)		
Insulating brick	0.15	0.00035
Red brick	0.6	0.0015
Concrete	0.8	0.002
Cork	0.04	0.0001
Felt	0.04	0.0001
Fiberglass	0.04	0.0001
Glass	0.8	0.002
Ice	1.6	0.004
Rock wool	0.04	0.0001
Styrofoam	0.01	0.00002
Wood	0.12–0.04	0.0003–0.0001
Gases		
Air	0.024	0.000057
Argon	0.016	0.000039
Helium	0.14	0.00033
Hydrogen	0.14	0.00033
Oxygen	0.023	0.000056

Equation (16–1) shows that the larger the thermal conductivity k, the larger the heat current, if other factors are equal. A good heat conductor has a large value of k, and a good insulator has a small k. A perfect insulator would have $k = 0$. This is an unattainable idealization, but Table 16–1 shows that the metals as a group have much greater thermal conductivities than the nonmetals, and that thermal conductivities of gases are extremely small. Figure 16–2 shows a "space-age" ceramic material with very unusual thermal properties.

Thermal resistance is a convenient way to describe the insulating properties of building insulation materials.

For thermal insulation in buildings, engineers use the concept of **thermal resistance,** denoted by R. The thermal resistance R of a slab of material with thickness L is defined to be

$$R = \frac{L}{k}. \tag{16–3}$$

Using this concept, we may rewrite Eq. (16–1) as

$$H = \frac{A(T_2 - T_1)}{R}. \tag{16–4}$$

The heat current H is *inversely* proportional to R; hence the term thermal *resistance*. In the units used for commercial insulating materials, H is expressed in $Btu \cdot hr^{-1}$, A in ft^2, and $(T_2 - T_1)$ in F°. The units of R are then $ft^2 \cdot F° \cdot hr \cdot Btu^{-1}$. Values of R are quoted without units; thus a six-inch-thick layer of fiberglass has an R value of 19, a two-inch slab of polyurethane foam a value of 12, and so on. Doubling the thickness doubles the R-value, of course.

16–2 This protective tile, developed for use on the space shuttle *Columbia*, has extraordinary thermal properties. The extremely small thermal conductivity and small heat capacity of the material make it possible to hold the tile by its edges, even though it is hot enough to emit the light for these photographs. (Courtesy Lockheed Corp.)

Common practice in new construction in severe northern climates is to specify R values of around 30 for exterior walls and ceilings. When the insulating material is in layers, such as a plastered wall, fiberglass insulation, and wood exterior siding, the R values are additive.

PROBLEM-SOLVING STRATEGY: *Heat conduction*

1. Identify the direction of heat flow in the problem; in Eq. (16–1), L is always measured along this direction, and A is always an area perpendicular to this direction. Sometimes a box or other container of irregular shape but uniform wall thickness can be approximated as a flat slab of the same thickness and total wall area.

2. In some problems the heat flows through two different materials in succession. The temperature at the interface between the two materials is then intermediate between T_1 and T_2; represent it by a symbol such as T_3. The temperature differences for the two materials are then $(T_2 - T_3)$ and $(T_3 - T_1)$. In steady-state heat flow, heat cannot accumulate within either material, so the total rate of heat flow H is equal to the heat flow in each material (the same in both materials). This is like an electric circuit with the elements connected in series.

3. If there are two *parallel* heat-flow paths, so that some heat flows through each, then the total H is the sum of the quantities H_1 and H_2 for the separate paths. In this case the temperature difference is the same for the two materials, but L, A, and k may be different for the two paths.

4. As always, it is essential to use a consistent set of units. If you use a value of k expressed in $J \cdot s^{-1} \cdot m^{-1} \cdot (C°)^{-1}$, don't use distances in cm, heat in calories, or T in °F!

EXAMPLE 16–1 A styrofoam box used to keep drinks cold at a picnic has total wall area (including the lid) of $0.8 \ m^2$ and wall thickness 2.0 cm. It is filled with ice and cans of orange soda at 0°C. What is the rate of heat flow into the box if the outside temperature is 30°C? How much ice melts in one day?

How fast does the ice melt in a styrofoam cooler?

SOLUTION We assume that the total heat flow is approximately the same as it would be through a flat slab of area $0.8 \ m^2$ and thickness 2.0 cm = 0.02 m. We find k from Table 16–1. The rate of heat flow, from Eq. (16–1), is

$$H = kA\frac{T_2 - T_1}{L} = (0.01 \text{ J·m}^{-1}\text{·s}^{-1}\text{·(C°)}^{-1})(0.8 \text{ m}^2)\frac{30 \text{ C°}}{0.02 \text{ m}} = 12 \text{ J·s}^{-1}.$$

There are 86,400 s in one day, so the total heat flow in one day is

$$(12 \text{ J·s}^{-1})(86,400 \text{ s}) = 1.04 \times 10^6 \text{ J}.$$

The heat of fusion of ice is 334 J·g^{-1}, so the quantity of ice melted by this quantity of heat is

$$m = \frac{Q}{L_F} = \frac{1.04 \times 10^6 \text{ J}}{334 \text{ J·g}^{-1}} = 3110 \text{ g}, \qquad \text{or about 3.1 kg.}$$

Heat flow in two bars in series

EXAMPLE 16–2 A steel bar 10 cm long is welded end to end to a copper bar 20 cm long. Each bar has a square cross section, 2 cm on a side. The free end of the steel bar is in contact with steam at 100°C, and the free end of the copper bar is in contact with ice at 0°C. Find the temperature at the junction of the two bars and the total rate of heat flow.

SOLUTION The key to the solution is the fact that the rates of heat flow in the two bars must be equal; otherwise some sections would have more heat flowing in than out, or the reverse, and steady-state conditions could not exist. Let T be the unknown junction temperature; we use Eq. (16–1) for each bar and equate the two expressions:

$$\frac{k_s A(100°\text{C} - T)}{L_s} = \frac{k_c A(T - 0°\text{C})}{L_c}.$$

The areas A are equal and may be divided out. Substituting numerical values, we find

$$\frac{(50.2 \text{ J·s}^{-1}\text{·m}^{-1}\text{·(C°)}^{-1})(100°\text{C} - T)}{0.1 \text{ m}} = \frac{(385 \text{ J·s}^{-1}\text{·m}^{-1}\text{·(C°)}^{-1})(T - 0°\text{C})}{0.2 \text{ m}}.$$

Rearranging and solving for T, we find

$$T = 20.7°\text{C}.$$

Even though the steel bar is shorter, the temperature drop across it is much greater than across the copper bar because steel is a much poorer conductor.

We now obtain the total heat current by substituting this value for T back into either of the above expressions:

$$H = \frac{(50.2 \text{ J·s}^{-1}\text{·m}^{-1}\text{·(C°)}^{-1})(0.02 \text{ m})^2(100°\text{C} - 20.7°\text{C})}{0.1 \text{ m}} = 15.9 \text{ J·s}^{-1} = 15.9 \text{ W},$$

or

$$H = \frac{(385 \text{ J·s}^{-1}\text{·m}^{-1}\text{·(C°)}^{-1})(0.02 \text{ m})^2(20.7°\text{C} - 0°\text{C})}{0.2 \text{ m}} = 15.9 \text{ J·s}^{-1} = 15.9 \text{ W}.$$

Heat flow in two bars in parallel

EXAMPLE 16–3 In Example 16–2, suppose the two bars are separated; one end of each bar is placed in contact with steam at 100°C, and the other end of each bar contacts ice at 0°C. What is the *total* rate of heat flow in the two bars?

SOLUTION In this case, the bars are in parallel rather than in series. The total heat flow is the sum of the flows in the two bars, and for each bar $T_2 - T_1 = 100°C - 0°C = 100$ C°. Thus

$$H = \frac{(50.2 \text{ J·s}^{-1}\text{·m}^{-1}\text{·(C°)}^{-1})(0.02 \text{ m})^2(100 \text{ C°})}{0.1 \text{ m}}$$

$$+ \frac{(385 \text{ J·s}^{-1}\text{·m}^{-1}\text{·(C°)}^{-1})(0.02 \text{ m})^2(100 \text{ C°})}{0.2 \text{ m}}$$

$$= 20.1 \text{ J·s}^{-1} + 77.0 \text{ J·s}^{-1} = 97.1 \text{ J·s}^{-1} = 97.1 \text{ W}.$$

The heat flow in the copper bar is much greater than in the steel bar, even though the copper bar is longer, because its thermal conductivity is much larger. The total heat flow is much larger than in Example 16–2 because the full 100 C° temperature difference appears across each bar.

16–2 CONVECTION

In heat conduction, energy is transferred by molecular vibrations and electron motion, but there is no overall bulk motion of the material. In contrast, **convection** is transfer of heat by actual motion of material from one region of space to another. The hot-air furnace, the hot-water heating system, and the flow of blood in the body are examples. If the material is forced to move by a blower or pump, the process is called *forced convection;* if the material flows owing to differences in density caused by thermal expansion, the process is called *natural* or *free convection.*

A simple example of natural convection is shown in Fig. 16–3. In (a) the water is at the same temperature throughout, so the levels are the same in the two sides of the U-tube. In (b) we heat the right side; the water expands, becoming less dense, and a taller column is needed to balance the pressure of the left-hand side. When we open the valve, water flows from the top of the warm column into the cooler column. If we supply heat continuously to the hot side and remove it from the cold side, this circulation continues indefinitely. The net result is a continuous heat transfer from the hot to the cold side. In hot-water heating systems, the cold side corresponds to the radiators and the hot side to the furnace.

Convective heat transfer is a complex process, and there is no simple equation to describe it, as there is for conduction. When a surface at one temperature is in contact with a fluid at a different temperature, the heat transfer depends on the shape and orientation of the surface, the properties of the fluid, and the nature of the fluid flow. An approximate rule of thumb often used for practical calculations is to represent the heat current H by means of the equation

$$H = hA \, \Delta T, \tag{16–5}$$

where A is the surface area, ΔT is the temperature difference between the surface and the main body of the fluid, and h is a quantity called the **convection coefficient.** Unlike the situation with conduction, h is not constant but depends on ΔT. In practice, values of h are determined by experiment and published as tables, which a physicist or engineer may then use for calculations

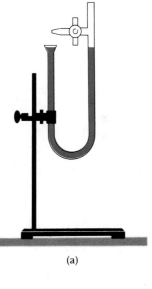

(a)

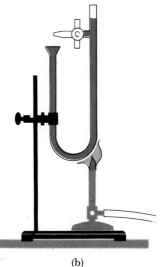

(b)

16–3 Free convection is caused by density differences due to thermal expansion. In (b), the level in the right side is higher than in the left side because the liquid on the right is warmer and therefore less dense.

TABLE 16–2   Coefficients of Natural Convection in Air at Atmospheric Pressure

Equipment	Convection Coefficient h, $J \cdot s^{-1} \cdot m^{-2} \cdot (C°)^{-1}$
Horizontal plate, facing upward	$2.49 \, (\Delta T)^{1/4}$
Horizontal plate, facing downward	$1.31 \, (\Delta T)^{1/4}$
Vertical plate	$1.77 \, (\Delta T)^{1/4}$
Horizontal or vertical pipe $\left(\dfrac{\text{diameter}}{D}\right)$	$1.32 \left(\dfrac{\Delta T}{D}\right)^{1/4}$

in specific problems. A few values are given in Table 16–2. In the last line, D is measured in meters.

A common situation is natural convection from a wall or a pipe that is at a constant temperature and is surrounded by air that is cooler by an amount ΔT. The convection coefficients applicable in this situation are given in Table 16–2. For a horizontal plate, if the air is *hotter* than the plate, the first two values of h should be interchanged.

Calculating heat transfer through a glass window: conduction and convection

EXAMPLE 16–5   The air in a room is at a temperature of 25°C and the outside air is at −15°C. How much heat is transferred per unit area of a glass windowpane of thermal conductivity $0.80 \, J \cdot m^{-1} \cdot s^{-1} \cdot (C°)^{-1}$ and thickness 2 mm?

SOLUTION   To assume that the inner surface of the glass is at 25°C and the outer surface is at −15°C is erroneous, as you can verify by touching the inner surface of a glass windowpane on a cold day. We must expect a much *smaller* temperature difference across the windowpane, so that in the steady state the rates of heat transfer (1) by convection in the room, (2) by conduction through the glass, and (3) by convection in the outside air are all equal.

As a first approximation in the solution of this problem, let us assume that the window is at a uniform temperature T. If $T = 5°C$, then the temperature difference between the inside air and the glass is the same as that between the glass and the outside air, or 20 C°. Hence the convection coefficient in both cases is

$$h = 1.77(20)^{1/4} J \cdot s^{-1} \cdot m^{-2} \cdot (C°)^{-1}$$
$$= 3.74 \, J \cdot s^{-1} \cdot m^{-2} \cdot (C°)^{-1},$$

and, from Eq. (16–5), the heat transferred per unit area is

$$\frac{H}{A} = (3.74 \, J \cdot s^{-1} \cdot m^{-2} \cdot (C°)^{-1})(20 \, C°)$$
$$= 74.8 \, J \cdot s^{-1} \cdot m^{-2}.$$

The glass, however, is *not* at uniform temperature; there must be a temperature difference ΔT across the glass sufficient to provide heat conduction at the rate of $74.9 \, J \cdot s^{-1} \cdot m^{-2}$. Using the conduction equation, Eq. (16–1), we obtain

$$\Delta T = \frac{LH}{kA} = \frac{(0.002 \text{ m})(74.8 \, J \cdot s^{-1} \cdot m^{-2})}{0.80 \, J \cdot m^{-1} \cdot s^{-1} \cdot (C°)^{-1}} = 0.19 \, C°.$$

Thus the inner surface is at about 5.09°C and the outer surface at 4.91°C.

16–3 RADIATION

Heat transfer by **radiation** is associated with energy carried by electromagnetic radiation, such as visible light, infrared, and ultraviolet radiation. Every body at a finite temperature (above absolute zero) emits electromagnetic radiation, with associated energy called **radiant energy.** This radiation contains a mixture of different wavelengths. At ordinary temperatures, say 20°C, nearly all the energy is carried by infrared waves having wavelengths longer than those of visible light. As the temperature rises, the wavelengths shift to shorter values. At 800°C a body emits enough visible radiation to be self-luminous and appears "red-hot," although even at this temperature most of the energy is carried by infrared waves. At 3000°C, the temperature of an incandescent lamp filament, the radiation contains enough visible light so the body appears "white-hot."

The total rate of radiation of energy from a surface increases very rapidly with temperature, in proportion to the *fourth power* of the absolute (Kelvin) temperature. For example, a copper block at a temperature of 100°C (373 K) radiates about 0.03 J·s^{-1} or 0.03 W from each square centimeter of its surface. At a temperature of 500°C (773 K), it radiates about 0.54 W from each square centimeter, and at 1000°C (1273 K), it radiates 4 W per square centimeter. This rate is 130 times as great as the rate at a temperature of 100°C.

Radiation is the mechanism by which the sun's energy reaches the earth. Near the earth, the rate of energy transfer due to solar radiation is about 1400 W for each square meter of surface area. Not all of this energy reaches the earth's surface; even on a clear day, about one fourth of this energy is absorbed by the earth's atmosphere.

Experiments show that the rate of radiation of energy from a surface is proportional not only to the fourth power of the absolute temperature but also to the surface area A. In addition, it depends on the nature of the surface; this dependence is described by a quantity e called **emissivity,** which is a dimensionless number between 0 and 1. Thus the heat current H due to radiation from a surface area A with emissivity e at absolute temperature T can be expressed as

$$H = Ae\sigma T^4, \tag{16–6}$$

where σ is a fundamental physical constant called the **Stefan-Boltzmann constant.** This relation, called the **Stefan-Boltzmann law,** was deduced by Josef Stefan (1835–1893) on the basis of experimental measurements made by John Tyndall (1820–1893) and was later derived from theoretical considerations by Ludwig Boltzmann (1844–1906).

In Eq. (16–6), H has units of power (energy per unit time). Thus in SI units σ has the units W·m^{-2}·K^{-4}. The numerical value is found experimentally to be

$$\sigma = 5.6699 \times 10^{-8} \text{ W·m}^{-2}\text{·K}^{-4}.$$

Emissivity is usually larger for dark, rough surfaces than for light, smooth ones. The emissivity of a smooth copper surface is about 0.3.

> Radiation: energy transfer by electromagnetic waves

> Emission of radiation by a hot body increases very rapidly with temperature.

> Some surfaces are better emitters of radiation than others at the same temperature.

EXAMPLE 16–6 A thin square steel plate, 10 cm on a side, is heated in a blacksmith's forge to a temperature of 800°C. If the emissivity is unity, what is the total rate of radiation of energy?

SOLUTION The total surface area, including both sides, is $2(0.01 \text{ m})^2 = 0.02 \text{ m}^2$. The temperature used in Eq. (16–6) must be *absolute* temperature; $800°C = 1073 \text{ K}$. Then Eq. (16–6) gives

$$H = (0.02 \text{ m}^2)(1)(5.67 \times 10^{-8} \text{ W·m}^{-2}\text{·K}^{-4})(1073 \text{ K})^4 = 1503 \text{ W}.$$

If the plate were heated by an electric heater instead, an electric power input of 1503 W would be required to maintain it at constant temperature.

Why doesn't a radiating body radiate away all its energy?

If the surfaces of all bodies continuously emit radiant energy according to Eq. (16–6), then why do they not eventually radiate away *all* their energy and cool down to a temperature of absolute zero? The answer is that they *would* do so if energy were not supplied to them in some way. In the case of an electric-heater element or the filament of an electric lamp, energy is supplied electrically to make up for the energy radiated. As soon as this energy supply is cut off, these bodies do, in fact, cool down very quickly to the temperature of their surroundings.

But why do they not continue to radiate and cool still more? The reason is that the surroundings are *also* radiating. Some of this radiated energy is intercepted and absorbed. The rate at which a body *radiates* energy is determined by the temperature of the *body*, but the rate at which it *absorbs* energy by radiation depends on the temperature of its *surroundings*. When a body is hotter than its surroundings, the rate of emission is greater than the rate of absorption; there is a net loss of energy, and the body cools down, unless its temperature is maintained by some other means. When a body is colder than its surroundings, the rate of absorption is greater than the rate of emission, and the body's temperature rises. At thermal equilibrium, the two rates must be equal.

Emission and absorption of radiation: a two-way process

In thermal equilibrium, the rate of emission equals the rate of absorption.

Thus when a body at temperature T is surrounded by walls also at temperature T, maintenance of thermal equilibrium requires that the body's rate of *absorption* of radiant energy from the walls be equal to the rate of radiation from its surface, which is $H = Ae\sigma T^4$. Hence, for such a body at a temperature T_1, surrounded by walls at a temperature T_2, the *net* rate of loss (or gain) of energy per unit area by radiation is

$$H_{net} = Ae\sigma T_1{}^4 - Ae\sigma T_2{}^4 = Ae\sigma(T_1{}^4 - T_2{}^4). \tag{16–7}$$

EXAMPLE 16–7 Assuming the total surface of the human body is 1.2 m^2 and the surface temperature is $30°C = 303 \text{ K}$, find the total rate of radiation of energy from the body. If the surroundings are at a temperature of $20°C$, what is the net rate of heat loss from the body by radiation?

SOLUTION For infrared radiation, the emissivity of the body is very close to unity, irrespective of skin pigmentation. The rate of energy transfer per unit area is given by Eq. (16–6). Taking $e = 1$, we find

$$
\begin{aligned}
H &= Ae\sigma T^4 \\
&= (1.2 \text{ m}^2)(1)(5.67 \times 10^{-8} \text{ W·m}^{-2}\text{·K}^{-4})(303 \text{ K})^4 \\
&= 574 \text{ W}.
\end{aligned}
$$

This loss is partially balanced by *absorption* of radiation, which depends on the temperature of the surroundings. The *net* rate of radiative energy transfer is given by Eq. (16–7):

$$H = Ae\sigma(T_1{}^4 - T_2{}^4)$$
$$= (1.2 \text{ m}^2)(1)(5.67 \times 10^{-8} \text{ W·m}^{-2}\text{·K}^{-4})[(303 \text{ K})^4 - (293 \text{ K})^4]$$
$$= 72 \text{ W}.$$

From the above discussion, we can see that a body that is a good absorber is also a good emitter. An ideal radiator, with an emissivity of unity, is also an ideal absorber, absorbing *all* the radiation that strikes it. Such an ideal surface is called an ideal black body, or simply a **blackbody.** Conversely, an ideal *reflector*, which absorbs *no* radiation at all, is also a very ineffective radiator. This is the reason for the silver coatings on vacuum ("thermos") bottles. A vacuum bottle is constructed with double glass walls; the air is pumped out of the spaces between the walls, so that heat transfer by conduction and convection is practically eliminated. The silver coating on the walls reflects most of the radiation from the contents back into the container, and the wall itself is a very poor emitter. The Dewar flask, used to store liquefied gases, is exactly the same in principle.

The blackbody: an ideal absorber and radiator

Heat transfer by radiation is more important than one might have guessed. A premature baby in an incubator can be dangerously cooled by radiation, if the walls of the incubator happen to be cold, even when the air in the incubator is warm. Some incubators regulate the air temperature by measuring the baby's skin temperature.

16–4 SOLAR ENERGY AND RESOURCE CONSERVATION

The principles of heat transfer discussed in this chapter have many very practical applications in a civilization such as ours with growing energy consumption and dwindling energy resources. A substantial fraction of all energy consumption in the United States is for heating and cooling of homes and other buildings, to maintain comfortable temperature and humidity inside when the outside temperature is much hotter or much colder.

For space heating, the objective is to prevent as much heat flow as possible from inside to outside; heating units replace the inevitable loss of heat. Walls insulated with material of low thermal conductivity, storm windows, and multiple-layer glass windows all help to reduce heat loss. It has been estimated that if all buildings used such materials, the total energy needed for space heating would be reduced by at least one-third.

Heat transfer considerations in heating and cooling buildings

Air conditioning in summer poses the reverse problem; heat flows from outside to inside, and energy must be expended in refrigerating units to remove it. Again, appropriate insulation can decrease this energy cost considerably.

Direct conversion of solar energy is a promising development in energy technology. In a typical household heating system, large black plates facing the sun are backed with pipes through which water circulates. The black surface absorbs most of the sun's radiation; the heat is transferred by conduction to the water, and then by forced convection to radiators inside the house. Heat loss by convection of air near the solar collecting panels is reduced by covering

Solar energy in household heating systems

them with glass, with a thin air space. An insulated heat reservoir provides for storage of collected energy for use at night and on cloudy days. Many larger-scale solar-energy conversion systems are also currently under study; a few examples are discussed in Section 19–10.

Solar energy has many environmental advantages.

From an environmental standpoint, solar energy has multiple advantages over the use of fossil fuels (burning of coal or oil) or nuclear power. Fossil fuels are being used up; solar power continues indefinitely. Obtaining fossil fuels may involve strip mining, with its associated destruction of landscape and elimination of other useful land functions, such as farming or timber. Off-shore oil drilling is a source of ocean water pollution. Air pollution from combustion products is a familiar problem, as is acid rain, directly attributable in many cases to coal smoke. The long-range effects on our climate of excess atmospheric carbon dioxide produced by combustion are unknown, but some scientists believe they may be serious or even catastrophic. With nuclear power there are radiation hazards and the problem of disposal of radioactive waste material. Solar power avoids all these problems.

Intelligent consideration of the effects of solar radiation in the design of buildings can also reduce energy consumption in heating and cooling. An example is the use of movable shades on the outsides of buildings, permitting the sun to enter windows in winter but keeping it out in summer. Architects have unfortunately tended in the past to ignore such solutions, but awareness of the need to design buildings with energy costs in mind is gradually growing.

SUMMARY

KEY TERMS

conduction

thermal conductivity

heat current

temperature gradient

thermal resistance

convection

convection coefficient

radiation

radiant energy

emissivity

Stefan-Boltzmann constant

Stefan-Boltzmann law

blackbody

The three mechanisms of heat transfer are conduction, convection, and radiation. Conduction is transfer of energy of molecular motion within materials without bulk motion of the materials. Convection involves mass motion from one region to another, and radiation is energy transfer through electromagnetic waves.

The heat current H for conduction depends on the area A through which the heat flows, the length L of the heat path, the temperature difference $(T_2 - T_1)$, and the thermal conductivity k of the material, according to

$$H = \frac{kA(T_2 - T_1)}{L}. \tag{16-1}$$

Good thermal conductors have large values of k, and poor conductors have small k's. The temperature change per unit length along the heat-flow path is called the temperature gradient.

Heat transfer by convection is a complex process; an approximate relation used for practical calculations of heat current H is

$$H = hA \, \Delta T, \tag{16-5}$$

where A is the surface area, ΔT is the temperature difference between the surface and the main body of fluid, and h is the convection coefficient, an empirical number that depends on the geometry and on ΔT.

The heat current H due to radiation is given by

$$H = Ae\sigma T^4, \tag{16-6}$$

where A is the surface area, e the emissivity of the surface (a pure number between 0 and 1), T the absolute temperature, and σ a fundamental constant called the Stefan-Boltzmann constant. When a body at temperature T_1 is surrounded by material at temperature T_2, the net heat current from the body to its surroundings is

$$H_{\text{net}} = Ae\sigma(T_1{}^4 - T_2{}^4). \qquad (16\text{–}7)$$

QUESTIONS

16–1 Why does a marble floor feel cooler to the feet than a carpet at the same temperature?

16–2 A beaker of boiling water can be picked up with bare fingers without burning if it is grasped only at the thin turned-out rim at the top. Why?

16–3 Old-time kitchen lore suggests that things cook better (evenly and without burning) in heavy cast-iron pots. What desirable characteristics would such pots have?

16–4 If you have wet hands and pick up a piece of metal that is below freezing, you may stick to it. This doesn't happen with wood. Why not?

16–5 A cold block of metal feels colder than a block of wood at the same temperature, but a hot block of metal feels hotter than a block of wood at the same temperature. Is there any temperature at which they feel equally hot or cold?

16–6 A person pours a cup of hot coffee, intending to drink it five minutes later. To keep it as hot as possible, should he put cream in it now or wait until just before he drinks it?

16–7 In late afternoon when the sun is about to set, swarms of insects such as mosquitoes are sometimes seen in vertical "plumes" above a tree or above the hood of a car with a hot engine. Why?

16–8 Mountaineers caught in a storm sometimes survive by digging a cave in snow or ice. How in the world can you keep warm in an ice cave?

16–9 Old-time pioneers sometimes kept themselves warm on cold winter nights by heating bricks in the fire, wrapping them in thick layers of cloth, and taking them to bed. Discuss the roles of conduction, convection, and radiation in this technique.

16–10 It is well known that a potato bakes faster if a large nail is stuck through it. Why? Is aluminum better than steel? There is also a gadget on the market to hasten roasting of meat, consisting of a hollow metal tube containing a wick and some water; this is claimed to be much better than a solid metal rod. How does this work?

16–11 The temperature in outer space, far from any solid body, is believed to be about 3 K. If you leave a spaceship and go for a space walk, do you get cold very quickly?

16–12 Aluminum foil used for food cooking and storage sometimes has one shiny surface and one dull surface. When food is wrapped for baking, should the shiny side be in or out? Which side should be out when it is to be frozen?

16–13 Some cooks claim that the bottom crust of a pie gets browner if the pie pan is glass then if it is metal. Why should this be?

16–14 Glider pilots in the midwest know that thermal updrafts are likely to occur above freshly plowed fields. Why?

16–15 We're lucky the earth isn't in thermal equilibrium with the sun. But why isn't it?

16–16 On a chilly fall morning, the grass is covered with frost but the concrete sidewalk isn't. Why?

16–17 Why can a microwave oven cook massive objects such as potatoes and beef roasts much faster than a conventional oven? What disadvantages are associated with the reasons for this greater speed?

16–18 Some folks claim that ice cubes freeze faster if the trays are filled with hot water because hot water cools off faster than cold water. What do you think?

EXERCISES

Section 16–1 Conduction

16–1 A slab of a thermal insulator is 100 cm^2 in cross section and 2 cm thick. Its thermal conductivity is $2.4 \times 10^{-4}\,\text{cal·s}^{-1}\text{·cm}^{-1}\text{·}(\text{C}°)^{-1}$. If the temperature difference between opposite faces is 100°C, how much heat flows through the slab in one day?

16–2 The oven in a kitchen range has a total wall area of 1.5 m^2 and is insulated with a layer of fiberglass 5 cm thick.

The inside is to be kept at a temperature of 200°C, and the outside is at room temperature, 20°C.

a) What is the heat current through the insulation, assuming it may be treated as a flat slab of area 1.5 m^2?

b) What electric-power input to the heating element is required to maintain this temperature?

16–3 A carpenter builds an outer house wall with a layer of wood 3 cm thick on the outside, and a layer of styrofoam insulation 3 cm thick at the inside surface of the wall. If the wood has $k = 9.5 \times 10^{-5}$ cal·s^{-1}·cm^{-1}·(C°)$^{-1}$, the interior temperature is 20°C, and the exterior temperature is -10°C,

a) what is the temperature at the plane where the wood meets the styrofoam?

b) what is the rate of heat flow, per square meter, through this wall?

16–4 One end of an insulated metal rod is maintained at 100°C and the other one is placed in an ice-water mixture. The bar has length 50 cm and cross-sectional area 0.8 cm^2. It is observed that the heat transported by the rod melts 2 g of ice in 5 min. Calculate the thermal conductivity k of the metal. Express your answer in J·s^{-1}·m^{-1}·(C°)$^{-1}$.

16–5 The ceiling of a room has an area of 200 ft^2. The ceiling contains insulation such that its R-value is 30. If the room is maintained at 72°F and the attic has a temperature of 105°F, how many Btu of heat flow through the ceiling into the room in 6 hr? Also express your answer in joules.

16–6 A long rod, insulated to prevent heat loss, has one end immersed in boiling water (at atmospheric pressure) and the other end in an ice-water mixture. The rod consists of 1.00 m of copper (one end in steam) and a length L_2 of steel (one end in ice). Both rods are of cross-sectional area 5 cm^2. The temperature of the copper-steel junction is 60°C, after a steady state has been set up.

a) How much heat per second flows from the steam bath to the ice-water mixture?

b) How long is L_2?

16–7 Suppose that the rod in Fig. 16–1 is of copper, of length 10 cm, and of cross-sectional area 1 cm^2. Let $T_2 = 100$°C and $T_1 = 0$°C.

a) What is the final steady-state temperature gradient along the rod?

b) What is the heat current in the rod, in the final steady state?

c) What is the final steady-state temperature at a point in the rod 2 cm from its left end?

16–8 A boiler with a steel bottom 1.5 cm thick rests on a hot stove. The area of the bottom of the boiler is 0.15 m^2 The water inside the boiler is at 100°C, and 0.75 kg are evaporated every 5 min. Find the temperature of the lower surface of the boiler, which is in contact with the stove.

Section 16–2 Convection

16–9

a) What would be the difference in height between the columns in the U-tube in Fig. 16–3 if the liquid is water and the left arm is 1 m high at 4°C while the other is at 75°C? (Recall that the density of water at various temperatures is given in Table 14–3.)

b) What is the difference between the pressures at the foot of two columns of water, each 10 m high, if the temperature of one is 4°C and that of the other is 75°C?

16–10 A flat plate is maintained at constant temperature of 100°C, and the air on both sides is at atmospheric pressure and at 20°C. How much heat is lost by natural convection from 1 m^2 of the plate (both sides) in 1 hr if

a) the plate is vertical?

b) the plate is horizontal?

16–11 A vertical steam pipe of outside diameter 7.5 cm and height 4 m has its outer surface at the constant temperature of 95°C. The surrounding air is at atmospheric pressure and at 20°C. How much heat is delivered to the air by natural convection in 1 hr?

Section 16–3 Radiation

16–12 What is the rate of energy radiation per unit area of a blackbody at a temperature of

a) 300 K?   b) 3000 K?

16–13 In Example 16–7, what is the net rate of heat loss by radiation if the temperature of the surroundings is 0°C?

16–14 The emissivity of tungsten is approximately 0.35. A tungsten sphere 1 cm in radius is suspended within a large evacuated enclosure whose walls are at 300 K. What power input is required to maintain the sphere at a temperature of 3000 K if heat conduction along the supports is neglected?

16–15 The operating temperature of a tungsten filament in an incandescent lamp is 2450 K, and its emissivity is 0.30. Find the surface area of the filament of a 25-W lamp.

Section 16–4 Solar Energy and Resource Conservation

16–16 A well-insulated house of moderate size in a temperate climate may require a maximum heat input rate of the order of 20 kW. If this heat is to be supplied by a solar collector with an average (night and day) energy input of 300 W·m^{-2} and a collection efficiency of 60%, what area of solar collector is required?

16–17 In Exercise 16–16, suppose a heat reservoir is required that can store enough energy for a week of cloudy days, by means of a large tank of water that is heated to 80°C by solar energy and cooled to 40°C as it circulates through the house. What volume of water is required?

16–18 A solar water heater for domestic hot-water supply uses solar collecting panels with collection efficiency of 50% in a location where the average solar-energy input is 200 W·m^{-2}. If the water comes into the house at 15°C and is to be heated to 65°C, what volume of water can be heated per hour if the collector area is 20 m^2?

16–19 The average rate of energy consumption of all forms in the United States is of the order of 3×10^{11} W. Suppose this energy were to be supplied by solar collectors in the desert. Assume an average solar-energy input of 500 W·m^{-2} for 12 hr each day, and a collection efficiency of 50%. What total collector area would be required? Express your result in square kilometers and square miles.

PROBLEMS

16–20 One experimental method of measuring the thermal conductivity of an insulating material is to construct a box of the material and measure the power input to an electric heater, inside the box, that maintains the interior at a measured temperature above the outside surface. Suppose that in such an apparatus a power input of 120 W is required to keep the interior surface of the box 65 C° (about 120 F°) above the temperature of the outer surface. The total area of the box is 2.32 m², and the wall thickness is 3.8 cm. Find the thermal conductivity of the material, in SI units.

16–21 A carpenter builds a solid wood door with dimensions 2.0 m × 0.8 m × 4 cm. Its thermal conductivity is

$$k = 0.04 \text{ J·s}^{-1}\text{·m}^{-1}\text{·(C°)}^{-1}.$$

The inside air temperature is 20°C, and the outside air temperature is −10°C.

a) What is the rate of heat flow through the door, assuming the surface temperatures are those of the surrounding air?

b) By what factor is the heat flow increased if a window 0.5 m square is inserted, assuming the glass is 0.5 cm thick and the surface temperatures are again those of the surrounding air?

16–22

a) A wood ceiling with thermal resistance R_1 is covered with a layer of insulation with thermal resistance R_2. Prove that the effective thermal resistance of the combination is given by $R = R_1 + R_2$.

b) Two metal bars of the same length L and cross-sectional area A are laid side by side, in good thermal contact, as shown in Fig. 16–4. If the bars have thermal conductivities k_1 and k_2, respectively, calculate the total heat current carried by the bars and from this deduce the effective thermal conductivity of the two combined bars. Does your result make sense in the special case where $k_1 = k_2$?

FIGURE 16–4

16–23 An engineer designs a camping icebox having a wall area of 2 m² and a thickness of 5 cm, using insulating material having a thermal conductivity of 0.05 J·s⁻¹·m⁻¹·(C°)⁻¹. The outside temperature is 20°C, and the inside of the box is to be maintained at 5°C by ice. The melted ice drains from the box at a temperature of 5°C. If ice costs 25 cents per kilogram, what will it cost to run the icebox for 1 hr?

16–24 A layer of ice 10 cm thick has formed on the surface of a lake, where the temperature is −10°C. The water under the slab is at 0°C. At what rate does the thickness of the slab increase if the heat of fusion of water freezing on the underside of the slab is conducted through the slab? Express your answer in units of cm·hr⁻¹.

16–25 Three rods of equal length and cross section are placed end to end, one copper, one steel, one aluminum, in that order. The free end of the copper rod is in contact with ice water at 0°C, and the free end of the aluminum rod is in contact with steam at 100°C. If no heat transfer occurs at the sides, find the temperature at each junction. (*Note:* The concept of thermal resistance R can be used here to reduce the algebra. See Problem 16–22.)

16–26 Rods of copper, brass, and steel are welded together to form a Y-shaped figure. The cross-sectional area of each rod is 2 cm². The free end of the copper rod is maintained at 100°C, and the free ends of the brass and steel rods at 0°C. Assume there is no heat loss from the surfaces of the rods. The lengths of the rods are: copper, 46 cm; brass, 13 cm; steel, 12 cm.

a) What is the temperature of the junction point?

b) What is the heat current in each of the three rods?

16–27 A compound bar 2 m long is constructed of a solid steel core 1 cm in diameter surrounded by a copper casing whose outside diameter is 2 cm. The outer surface of the bar is thermally insulated, and one end is maintained at 100°C, the other at 0°C.

a) Find the total heat current in the bar.

b) What fraction of this total is carried by each material?

16–28 A rod is initially at a uniform temperature of 0°C throughout. One end is kept at 0°C, and the other is brought into contact with a steam bath at 100°C. The surface of the rod is insulated so that heat can flow only lengthwise along the rod. The cross-sectional area of the rod is 2 cm², its length is 100 cm, its thermal conductivity is 335 J·s⁻¹·m⁻¹·(C°)⁻¹, its density is 1.0×10^4 kg·m⁻³, and its specific heat capacity is 419 J·kg⁻¹·(C°)⁻¹. Consider a short cylindrical element of the rod 1 cm in length.

a) If the temperature gradient at the cooler end of this element is 1000C°·m⁻¹, how many joules of heat energy flow across this end per second?

b) If the average temperature of the element is increasing at the rate of 2 C°·s⁻¹, what is the temperature gradient at the other end of the element?

16–29 An electrician installs an electric transformer in a cylindrical tank 0.60 m in diameter and 1 m high, with a flat top and bottom. If the tank transfers heat to the air only by natural convection, and electrical losses are to be dissipated at the rate of 1 kW, how many degrees will the tank surface rise above room temperature?

16–30 The rate at which radiant energy reaches the earth from the sun is about 1.4 kW·m⁻². The distance from earth to sun is about 1.5×10^{11} m, and the radius of the sun is about 7.0×10^8 m.

a) What is the rate of radiation of energy, per unit area, from the sun's surface?

b) If the sun radiates as an ideal blackbody, what is the temperature of its surface?

16–31 A physicist uses a cylindrical metal can 0.10 m high and 0.05 m in diameter to store liquid helium at 4 K, where its heat of vaporization is 2.0×10^4 J·kg^{-1}. Completely surrounding the helium can are walls maintained at the temperature of liquid nitrogen, 80 K, the intervening space being evacuated. How much helium is lost per hour? Assume the emissivity of the helium can to be 0.2.

CHALLENGE PROBLEMS

16–32 A solid cylindrical copper rod 0.10 m long has one end maintained at a temperature of 20.00 K. The other end is blackened and exposed to thermal radiation from a body at 300 K, no energy being lost or gained elsewhere. When equilibrium is reached, what is the temperature of the blackened end? (*Hint:* Since copper is a very good conductor of heat at low temperature, $k = 1670$ J·s^{-1}·m^{-1}·(C°)$^{-1}$, the temperature of the blackened end is slightly greater than 20 K.)

16–33

a) A spherical shell has inner and outer radii a and b, respectively, and the temperatures at the inner and outer surfaces are T_2 and T_1. The thermal conductivity of the material of which the shell is made is k. Derive an equation for the total heat current through the shell.

b) For the shell in part (a), derive an equation for the temperature variation within the shell. That is, calculate T as a function of r, the distance from the center of the shell.

c) A hollow cylinder has length L, inner radius a, and outer radius b, and the temperatures at the inner and outer surfaces are T_2 and T_1. (The cylinder could represent an insulated hot-water pipe, for example.) The thermal conductivity of the material of which the cylinder is made is k. Derive an equation for the total heat current through the walls of the cylinder.

d) For the cylinder of part (c), derive an equation for the temperature variation inside the cylinder walls.

e) For the spherical shell of part (a) and the hollow cylinder of part (c), show that the equation for the total heat current in each case reduces to Eq. (16–1) for linear heat flow when the shell or cylinder is very thin.

16–34 A steam pipe of radius 2 cm, carrying steam at 120°C, is surrounded by a cylindrical jacket of cork with inner and outer radii 2 cm and 4 cm, and this in turn is surrounded by a cylindrical jacket of styrofoam having inner and outer radii 4 cm and 6 cm. The outer surface of the styrofoam is in contact with air at 20°C.

a) What is the temperature at a radius of 4 cm, where the two insulating layers meet?

b) What is the total rate of transfer of heat out of a 2 m length of pipe?

(*Hint:* Use the expression derived in part (c) of Problem 16–33.)

16–35 Suppose that both ends of the rod in Fig. 16–1 are kept at a temperature of 0° C, and that the initial temperature distribution along the rod is given by $T = 100°C \sin \pi x/L$, where T is in °C. Let the rod be copper, of length $L = 0.10$ m and of cross section 1 cm^2.

a) Show the initial temperature distribution in a diagram.

b) What is the final temperature distribution after a very long time has elapsed?

c) Sketch curves that you think would represent the temperature distribution at intermediate times.

d) What is the initial temperature gradient at the ends of the rod?

e) What is the initial heat current from the ends of the rod into the bodies making contact with its ends?

f) What is the initial heat current at the center of the rod? Explain. What is the heat current at this point at any later time?

g) What is the value of $k/\rho c$ for copper, and in what unit is it expressed? Here k is the thermal conductivity, ρ is the density, and c is the specific heat capacity. (This quantity is called the *thermal diffusivity*.)

h) What is the initial time rate of change of temperature at the center of the rod?

i) How long a time would be required for the rod to reach its final temperature, if the temperature continued to decrease at this rate? (This time can be described as the *relaxation time* of the rod.)

j) From the graphs in part (c), would you expect the rate of change of temperature at the midpoint to remain constant, to increase, or to decrease?

k) What is the initial rate of change of temperature at a point in the rod 2.5 cm from its left end?

17

THERMAL PROPERTIES
OF MATTER

IN PREVIOUS CHAPTERS WE HAVE DISCUSSED SEVERAL PROPERTIES OF materials, including elasticity, thermal expansion, specific heat capacity, and thermal conductivity. All these properties describe how a material behaves when some aspect of its environment, such as temperature or pressure, changes. In describing the quantity and condition of the material itself, we have used several physical quantities, including total mass m, number of moles n, pressure p, volume V, and temperature T. We now want to consider a more general formulation of the relationships among these quantities, using the concept of the *equation of state* of a material. We also look in greater detail at the conditions that determine the *phase* of a material (e.g., solid, liquid, or gas) and the conditions under which *phase transitions* occur. This discussion provides some of the language needed for our study of *thermodynamics*, the relation of heat to other forms of energy, in the following chapters.

17–1 EQUATIONS OF STATE

Physical quantities such as pressure, volume, temperature, and amount of substance describe the conditions in which a particular specimen of material exists. They determine the *state* of the material, and they are called *state variables* or **state coordinates.** For some physical phenomena we need additional state variables, such as magnetization or electric polarization, but those named above are sufficient for our present discussion.

State variables are interrelated; ordinarily we cannot change one without also causing a change in one or more of the others. For example, when we heat a gas such as steam in a closed container of constant volume, the pressure *must* increase as the temperature increases. Indeed, if the temperature becomes high enough, the pressure may cause the container, such as a boiler, to explode. The volume V of a substance is determined by its pressure p, temperature T, and amount of substance, m or n.

In some cases the relationship among p, V, and T is simple enough that it can be expressed in the form of an equation, which we call the **equation of**

State coordinates: describing the conditions in which a material exists

The equation of state is an expression of the interrelation of the state coordinates.

389

state. We can construct a simple though approximate equation of state for a solid material. Recall that the temperature coefficient of volume expansion β is the fractional volume change per unit temperature change, and that the isothermal compressibility k is the fractional volume change per unit pressure change. Thus if a specimen of material has volume V_0 when the pressure is p_0 and the temperature T_0, the volume V at pressure p and temperature T is given approximately by

$$V = V_0[1 + \beta(T - T_0) - k(p - p_0)],$$

provided the temperature and pressure changes are not too great.

Another simple equation of state is that of an *ideal gas*, to be discussed in Section 17–2. In many cases, however, the relationship among p, V, and T is so complex that it can be described only by use of graphs or tables of numerical values. Even then, the relation among the variables still exists, and we call it an equation of state even when it is too complex to be expressed in actual equation form.

For the present we consider only **equilibrium states,** that is, states in which the system is in mechanical and thermal equilibrium. For such states the temperature and pressure are uniform throughout the system. We note that when a system changes from one state to another, nonequilibrium states must occur during the transition. For example, when a material expands or is compressed, there must be mass in motion; this change requires acceleration and nonuniform pressure. After the system has returned to mechanical equilibrium, the pressure is again the same throughout the system. Similarly, when heat conduction occurs within a system, different regions must have different temperatures. Only when thermal equilibrium is once again attained is the temperature again the same throughout.

An equation of state has meaning only for systems that are in mechanical and thermal equilibrium.

17–2 IDEAL GASES

The quantity of material in a system can be described by either the mass or the number of moles.

Gases at low pressures have particularly simple equations of state. We can study the behavior of a gas by means of a cylinder with a movable piston, equipped with a pressure gauge and a thermometer, as shown in Fig. 17–1.

17–1 A hypothetical setup for studying the behavior of gases. The pressure, volume, temperature, and number of moles of gas can be controlled and measured.

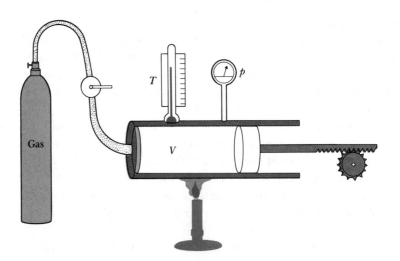

We can vary the pressure, volume, and temperature, and pump any desired mass of any gas into the cylinder, in order to explore the relationships between pressure, volume, temperature, and quantity of substance. It is often convenient to measure the amount of gas in terms of the number of *moles n*, rather than the mass m. Since the molecular mass M is the mass per mole, the total mass m is given by

$$m = nM. \tag{17-1}$$

That is, the total mass m of material is the mass of one mole M multiplied by the number of moles n.

From measurements of the pressure, volume, temperature, and number of moles of various gases, several conclusions emerge. First, the volume V is proportional to the number of moles n. If we double the number of moles, keeping pressure and temperature constant, the volume doubles. Second, the volume varies *inversely* with pressure p; if we double the pressure, holding the temperature T and quantity of material n constant, the gas is compressed to one-half its initial volume. Thus

$$pV = \text{constant} \qquad (T \text{ constant}), \tag{17-2}$$

or, if p_1 and V_1 are the pressure and volume in the initial state and p_2 and V_2 those of the final state, then

$$p_1V_1 = p_2V_2 \qquad (T \text{ constant}). \tag{17-3}$$

This relation is called Boyle's law, after Robert Boyle (1627–1691), a contemporary of Newton.

Third, the pressure is proportional to the absolute temperature. If we double the absolute temperature, keeping the volume and quantity of material constant, the pressure doubles. Thus

$$p = (\text{constant})T \qquad (V \text{ constant}), \tag{17-4}$$

or, if p_1 and T_1 are the pressure and absolute temperature in the initial state and p_2 and T_2 those of the final state, then

$$\frac{p_1}{T_1} = \frac{p_2}{T_2}. \tag{17-5}$$

This is called Charles's law, after Jacques Charles (1746–1823).

These three relationships can be combined neatly into a single equation of state:

$$pV = nRT. \tag{17-6}$$

The constant of proportionality R might be expected to have different values for different gases, but instead it turns out to have the same value for *all* gases, at least at sufficiently high temperature and low pressure. This quantity is called the **ideal gas constant.** The numerical value of R depends, of course, on the units in which p, V, n, and T are expressed. In SI units, where the unit of p is the Pa or N·m^{-2} and the unit of V is m^3, the numerical value of R is found to be

$$R = 8.314 \ (\text{N·m}^{-2})\text{·m}^3\text{·mol}^{-1}\text{·K}^{-1} = 8.314 \ \text{J·mol}^{-1}\text{·K}^{-1}.$$

A launch balloon used to lift a three-ton optical telescope. This balloon is used to lift the main balloon to an altitude of about 3000 m, where it inflates. The inflated main balloon system is 200 m tall, holds 150,000 m³ of hydrogen gas, and lifts the telescope to a final altitude of 24,000 m. The behavior of the hydrogen gas in the balloon is very close to that of an ideal gas. (Courtesy of NASA.)

Charles's law: At constant volume, pressure is directly proportional to absolute temperature.

The ideal-gas equation: a simple equation of state for an idealized system

The ideal-gas constant is the same number for all gases.

In cgs units, where the unit of p is $dyn \cdot cm^{-2}$ and the unit of V is cm^3,

$$R = 8.314 \times 10^7 \, (dyn \cdot cm^{-2}) \cdot cm^3 \cdot mol^{-1} \cdot K^{-1}$$
$$= 8.314 \times 10^7 \, erg \cdot mol^{-1} \cdot K^{-1}.$$

Several different systems of units can be used with the ideal-gas equation.

Note that the units of pressure times volume are the same as units of energy, so in *all* systems of units, R has units of energy per mole, per unit of absolute temperature. In terms of calories,

$$R = 1.99 \, cal \cdot mol^{-1} \cdot K^{-1}.$$

In chemical calculations, volumes are commonly expressed in liters (L), pressures in atmospheres, and temperatures in kelvins. In this system,

$$R = 0.08207 \, L \cdot atm \cdot mol^{-1} \cdot K^{-1}.$$

Real gases behave approximately like the ideal-gas model at sufficiently low pressures and high temperatures.

We now define an **ideal gas** as one for which Eq. (17–6) holds precisely for *all* pressures and temperatures. As the term suggests, an ideal gas is an idealized model, which represents the behavior of gases very well in some circumstances, less well in others. Generally, gas behavior approximates the ideal-gas model most closely at very low pressures, when the gas molecules are far apart. However, the deviations are not very great at moderate pressures (e.g., 1 atm) and at temperatures not too near those at which the gas liquefies.

For a *fixed mass* (or fixed number of moles) of an ideal gas, the product nR is constant, and hence pV/T is constant also. Thus if the subscripts 1 and 2 refer to two states of the same mass of a gas, but at different pressures, volumes, and temperatures,

$$\frac{p_1 V_1}{T_1} = \frac{p_2 V_2}{T_2} \tag{17–7}$$
$$= constant.$$

If the temperatures T_1 and T_2 are the same, then

$$p_1 V_1 = p_2 V_2 \tag{17–8}$$
$$= constant,$$

which is again Boyle's law.

The ideal-gas model plays an important role in the temperature scale defined by the gas thermometer.

Now you may recall that in Chapter 14 our most satisfactory definition of temperature was stated in terms of the constant-volume gas thermometer, for which pressure was proportional to absolute temperature. Hence the proportionality of pressure to absolute temperature, as expressed by Charles's law, Eqs. (17–4) and (17–5), is not at all surprising; indeed, it is inevitable!

You may have the feeling that you have been swindled into accepting as natural law something that is actually only a matter of definition. To this charge we plead guilty, but with mitigating circumstances. Toward the end of Chapter 19 we will arrive at a definition of temperature that really *is* independent of the properties of any particular material (including ideal gases). The temperature scale based on the gas thermometer will then emerge as a very close approximation of this truly absolute scale. Until then, consider Charles's law and the ideal-gas equation of state as based on this genuinely material-independent temperature scale, even though we have not actually defined it yet!

PROBLEM-SOLVING STRATEGY: Ideal gases

1. In some problems you will be concerned with only one state of the system; some of the quantities in Eq. (17–6) will be known, some unknown. Make a list of what you know and what you have to find: $p = 1.0 \times 10^6$ Pa, $V = 4$ m^3, $T = ?$, $n = 2$ mol. Or something like that.

2. In other problems there will be two different states of the same quantity of gas. Decide which is state 1, which state 2, and make a list of the quantities for each: p_1, p_2, V_1, V_2, T_1, T_2, and so on. You may be able to use Eq. (17–7) if all but one of the quantities in that equation are known; otherwise you may first have to use Eq. (17–6), for example, if p_1, V_1, and n are given.

3. As always, be sure to use a consistent set of units. First decide which value of the gas constant R you are going to use, and then convert the units of the other quantities accordingly. You may have to convert atmospheres to pascals or liters to cubic meters. Sometimes the problem statement will make one system of units clearly more convenient than others. Decide on your system, and stick to it.

4. Don't forget that T must always be an *absolute* temperature. If you are given temperatures in °C, be sure to add 273 to convert to Kelvin.

5. You may sometimes have to convert between mass m and number of moles n. The relationship is $m = Mn$, where M is the molecular mass. Here's a tricky point: If you replace n in Eq. (17–6) by (m/M), you *must* use the same mass units for m and M. So if M is in grams per mole (the usual units for molecular mass), then m must also be in grams. If you want to use m in kg, then you must convert M to kg·mol^{-1}. For example, the molecular mass of oxygen is 32 g·mol^{-1} or 32×10^{-3} kg·mol^{-1}. Be careful!

EXAMPLE 17–1 The condition called **standard temperature and pressure (STP)** for a gas is defined to be a temperature of 0°C = 273 K and a pressure of 1 atm = 1.013×10^5 Pa. Find the volume of one mole of any gas at STP.

Standard temperature and pressure: a convenient reference condition

SOLUTION From Eq. (17–6), using R in J·mol^{-1}·K^{-1},

$$V = \frac{nRT}{p} = \frac{(1 \text{ mol})(8.314 \text{ J·mol}^{-1}\text{·K}^{-1})(273 \text{ K})}{1.01 \times 10^5 \text{ Pa}}$$

$$= 0.0224 \text{ m}^3$$
$$= 22{,}400 \text{ cm}^3$$
$$= 22.4 \text{ L}.$$

Alternatively, using R in L·atm·mol^{-1}·K^{-1},

$$V = \frac{nRT}{p} = \frac{(1 \text{ mol})(0.0821 \text{ L·atm·mol}^{-1}\text{·K}^{-1})(273 \text{ K})}{1.00 \text{ atm}}$$

$$= 22.4 \text{ L}.$$

EXAMPLE 17–2 A tank attached to an air compressor contains 20.0 L of air at a temperature of 30°C and gauge pressure of 4.00×10^5 Pa (about 60 lb·in^{-2}). What is the mass of air, and what volume would it occupy at normal atmospheric pressure and 0°C? Air is a mixture of gases, consisting of about 78% nitrogen and 21% oxygen, with small percentages of other gases. The *average* molecular mass is $M = 28.8$ g·mol^{-1}.

SOLUTION Gauge pressure is the amount in excess of atmospheric pressure (1.01×10^5 Pa), so the total pressure is

$$p = 4.00 \times 10^5 \text{ Pa} + 1.01 \times 10^5 \text{ Pa} = 5.01 \times 10^5 \text{ Pa}.$$

The temperatures must be expressed in kelvins: 30°C = 303 K. We express the volume in cubic meters: 20 L = 20 × 10⁻³ m³. To find the mass of gas, we first use Eq. (17–6) to find the number of moles, and then use Eq. (17–1):

$$n = \frac{pV}{RT} = \frac{(5.01 \times 10^5 \text{ Pa})(20 \times 10^{-3} \text{ m}^3)}{(8.314 \text{ J·mol}^{-1}\text{·K}^{-1})(303 \text{ K})} = 3.98 \text{ mol};$$

$$m = nM = (3.98 \text{ mol})(28.8 \text{ g·mol}^{-1}) = 115 \text{ g} = 0.115 \text{ kg}.$$

The volume at 1 atm and 0°C (273 K) is

$$V = \frac{nRT}{p} = \frac{(3.98 \text{ mol})(8.314 \text{ J·mol}^{-1}\text{·K}^{-1})(273 \text{ K})}{1.01 \times 10^5 \text{ Pa}}$$

$$= 89.4 \times 10^{-3} \text{ m}^3 = 89.4 \text{ L}.$$

How does atmospheric pressure vary with elevation?

EXAMPLE 17–3 Find the variation of atmospheric pressure with elevation in the earth's atmosphere, assuming the temperature to be uniform throughout (which is not actually the case).

SOLUTION We begin with the hydrostatic relation of Eq. (13–3), $dp/dy = -\rho g$. The density ρ is not constant but varies with pressure. Replacing n in the ideal-gas equation by m/M, we obtain

$$pV = \frac{m}{M}RT \quad \text{and} \quad \rho = \frac{m}{V} = \frac{pM}{RT}.$$

We substitute this expression for ρ into Eq. (13–3), separate variables, and integrate, letting p_1 be the pressure at elevation y_1 and p_2 that at y_2:

$$\frac{dp}{dy} = -\frac{pMg}{RT}, \qquad \int_{p_1}^{p_2} \frac{dp}{p} = -\frac{Mg}{RT}\int_{y_1}^{y_2} dy, \qquad \ln\frac{p_2}{p_1} = -\frac{Mg}{RT}(y_2 - y_1).$$

If the pressure at $y = 0$ is p_0, then the pressure p at any y is given by

$$p = p_0 e^{-Mgy/RT} \tag{17–9}$$

At the summit of Mount Everest, where $y = 8882$ m,

$$\frac{Mgy}{RT} = \frac{(28.8 \times 10^{-3}\text{kg·mol}^{-1})(9.8 \text{ m·s}^{-2})(8882 \text{ m})}{(8.314 \text{ J·mol}^{-1}\text{·K}^{-1})(273 \text{ K})} = 1.10;$$

$$p = (1.01 \times 10^5 \text{ Pa})e^{-1.10} = 0.336 \times 10^5 \text{ Pa} = 0.333 \text{ atm}.$$

We have used a constant temperature of 0°C = 273 K, which is not completely realistic, but this example shows why mountaineers have so much difficulty breathing on Mount Everest; air pressure is only one-third its sea-level value. For similar reasons, jet airplanes, which typically fly at altitudes of 8000 to 10,000 m, must have pressurized cabins for passenger comfort and health.

17–3 *pV*-DIAGRAMS

Representing the behavior of a material by use of pressure–volume graphs

For a given quantity of a material, the equation of state is a relation among the three state coordinates p, V, and T. We could in principle represent this relationship graphically as a *surface* in a three-dimensional space with coordinates p, V, and T. This representation sometimes helps in grasping the overall behavior of the material, but usually ordinary two-dimensional graphs are more convenient. One of the most useful of these is a set of graphs of pressure as a function of volume for each of several constant temperatures. Such a diagram is called a *pV*-diagram. Each curve, representing a particular constant temperature, is called an **isotherm,** or a *pV-isotherm.*

A *pV*-diagram for an ideal gas is shown in Fig. 17–2. Each curve is an isotherm, showing the relation of pressure to volume at a certain constant temperature. We can think of this as a graphical representation of Boyle's law, Eq. (17–2), for each of several temperatures, and it is also a graphical representation of the ideal-gas equation of state. From these curves we can read off the volume V corresponding to any given pressure p and temperature T.

Figure 17–3 shows a *pV*-diagram for a material that *does not* obey the ideal-gas equation. At a high enough temperature, say T_4, the behavior is approximately that of an ideal gas, but at lower temperatures the behavior is rather different. At temperatures below a critical temperature T_c, the isotherms develop flat portions. In these portions, as the material is compressed at constant temperature, the pressure does not increase at all. As the volume decreases and the end of the flat portion of an isotherm is reached, the pressure increases very rapidly with further decrease of volume.

Visual inspection of the material shows that in the flat portion of the isotherm the material is *condensing* from the vapor to the liquid phase; this flat portion represents a condition of phase equilibrium. As compression continues, more and more material goes from vapor to liquid, but the pressure does not change. During this process, heat must be removed in order to keep the temperature constant; this is the heat of vaporization, discussed in Section 15–6.

Thus when a real (nonideal) gas is compressed at a constant temperature T_2, it remains in the gaseous phase until a in the diagram is reached. At point a it begins to liquefy, and with further decrease in volume more and more material liquefies, with both pressure and temperature remaining constant. At point b all the material is in the liquid state; any further compression is accompanied by a rapid rise of pressure, because in general liquids are much less compressible than gases. At a lower constant temperature T_1 similar behavior occurs, but the onset of condensation occurs at lower pressure and greater volume than at the constant temperature T_2.

The shaded area in Fig. 17–3 is the region of liquid-vapor phase equilibrium for this material. At temperatures greater than T_c, no phase transition occurs as the material is compressed. The temperature T_c is called the *critical temperature* for this material. We will explore its significance in greater detail in the next section.

The next two chapters provide many more opportunities to use *pV*-diagrams in connection with energy calculations. As we will see, the area under a *pV*-curve, whether or not it is an isotherm, represents the *work* done by or on the system during a volume change. This work, in turn, is directly related to heat transfer and changes in the *internal energy* of the system, which will be defined in Chapter 18.

17–4 PHASE DIAGRAMS, TRIPLE POINT, CRITICAL POINT

We have discussed *pV*-diagrams and their usefulness in describing the behavior of materials, including the equation of state. Another useful diagram, especially in reference to phase transitions, is a graph with axes p and T, called a **phase diagram.**

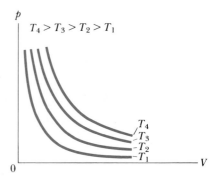

17–2 A *pV*-diagram for an ideal gas. Each curve represents the relation between pressure and volume at the indicated constant temperature. This is a graphical representation of Boyle's law; for each curve the product *pV* is constant; the value of that constant is different from different temperatures, increasing with increasing temperature. These curves are called *pV-isotherms* for the substance.

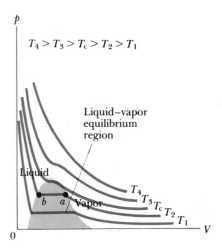

17–3 A *pV*-diagram for a nonideal gas, showing *pV*-isotherms for temperatures above and below the critical temperature T_c. The liquid–vapor equilibrium region is shown as a shaded area. At still lower temperature the material might undergo phase transitions from liquid to solid or from gas to solid; these are not shown on this diagram.

Each phase of a material can exist only in certain regions of temperature and pressure.

The phase diagram shows what phase exists for any given temperature and pressure.

In the discussion of phases of matter in Section 15–5 we found that each phase is stable only in certain ranges of temperature and pressure. A transition from one phase to another takes place ordinarily under conditions of **phase equilibrium** between the two phases, and for a given pressure this occurs at only one specific temperature. A phase diagram shows what phase occurs for each possible combination of temperature and pressure. A typical example is shown in Fig. 17–4. At each point on the diagram, only a single phase can exist, except for points on the lines, where two phases can coexist in phase equilibrium. Thus any point on a line represents a set of conditions under which phase changes can occur.

Curves on the phase diagram show conditions of phase equilibrium.

If we increase the temperature of a substance at constant pressure, it passes through a sequence of states represented by points on the broken horizontal line labeled (a). The melting and boiling temperatures at this pressure are the temperatures at which the broken line crosses the fusion and vaporization curves, respectively. Similarly, if we compress a material by increasing the pressure while we hold the temperature constant, as represented by the vertical broken line labeled (b), the material passes from vapor to liquid and then to solid at the points where the broken line crosses the vaporization curve and fusion curve, respectively.

When the pressure is low enough, a material may be transformed by constant-pressure heating from solid directly to vapor, as represented by line (c). This process, called *sublimation,* occurs where the line crosses the sublimation curve. At atmospheric pressure, solid carbon dioxide undergoes sublimation; there is no liquid phase at this pressure. Similarly, when a substance with initial temperature lower than the temperature at the point labeled "triple point" is compressed by increasing its pressure, it goes directly from the vapor to the solid phase, as shown by line (d).

Figure 17–4 is characteristic of materials that expand on melting. A few materials *contract* on melting; the most familiar examples are water and the metal antimony. For such a substance the fusion curve always slopes upward to the left, rather than to the right. The melting temperature decreases when the pressure increases, so these solid materials can be melted by an increase of pressure. This is part of the reason ice cubes stick together.

At the triple point, gas, liquid, and solid phases can all exist together.

The intersection point of the equilibrium curves in Fig. 17–4 is called the **triple point.** For any substance there is only one pressure and temperature at which *all three* phases can coexist, namely, the triple-point pressure and tem-

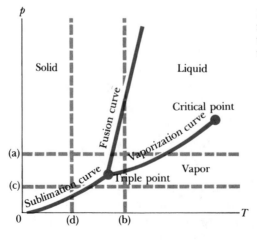

17–4 A pT phase diagram, showing regions of temperature and pressure at which the various phases exist and where phase changes occur.

TABLE 17–1 Triple-Point Data

Substance	Temperature, K	Pressure, Pa
Hydrogen	13.84	0.0704×10^5
Deuterium	18.63	0.171×10^5
Neon	24.57	0.432×10^5
Nitrogen	63.18	0.125×10^5
Oxygen	54.36	0.00152×10^5
Ammonia	195.40	0.0607×10^5
Carbon dioxide	216.55	5.17×10^5
Sulfur dioxide	197.68	0.00167×10^5
Water	273.16	0.00610×10^5

perature. Triple-point data for a few substances are given in Table 17–1. In Section 14–4 we saw the role of the triple-point temperature of water in defining a temperature scale.

Figure 17–3 shows that the liquid and gas (or vapor) phases can exist together only when the temperature and pressure are less than those at the point lying at the top of the tongue-shaped area labeled "liquid–vapor equilibrium region." This point also corresponds to the endpoint at the top of the vaporization curve in Fig. 17–4. It is called the **critical point,** and the corresponding values of p and T are called the critical pressure and temperature. A gas at a temperature *above* the critical temperature does not separate into two phases when compressed isothermally. Its properties change gradually and continuously from those we ordinarily associate with a gas (low density, large compressibility) to those of a liquid (high density, small compressibility), *without a phase transition.*

If this description stretches credibility, here is another point of view. When we look at liquid–vapor phase transitions at successively higher points on the vaporization curve, we find that as we approach the critical point, the changes in density, bulk modulus, index of refraction, viscosity, and all other physical properties become smaller and smaller. At the critical point they all become zero, and at this point the distinction between liquid and vapor disappears. The heat of vaporization also grows smaller and smaller as we approach the critical point, and it too becomes zero at the critical point.

Table 17–2 lists critical constants for a few substances. The very low critical temperatures of hydrogen and helium make it evident why these gases defied attempts to liquefy them for many years.

The information presented by the pT phase diagram is also contained in the pV-diagram. In Fig. 17–3, for example, we can read off for each temperature the pressure at which liquid and vapor are in equilibrium. This set of pairs of values of p and T can then be used to plot the vaporization curve of Fig. 17–4, and we can also obtain the critical-point temperature and pressure.

Many substances can exist in more than one solid phase. A familiar example is carbon, which exists as noncrystalline lampblack and crystalline graphite and diamond. Transitions from one solid phase to another occur at definite temperatures and pressures, just as with phase changes from liquid to solid, and so on. Water is one such substance; at least eight types of ice, differing in their crystal structure and physical properties, have been observed at very high pressures.

Above the critical point, the distinction between liquid and vapor disappears.

How can a liquid turn into a vapor without a phase transition?

Some materials have more than one solid phase.

TABLE 17–2   Critical Constants

Substance	Critical Temperature, K	Critical Pressure, Pa	Critical Volume, m³·mol⁻¹	Critical Density, kg·m⁻³
Helium (4)	5.3	2.29×10^5	57.8×10^{-6}	69.3
Helium (3)	3.34	1.17×10^5	72.6×10^{-6}	41.3
Hydrogen (normal)	33.3	13.0×10^5	65.0×10^{-6}	31.0
Deuterium (normal)	38.4	16.6×10^5	60.3×10^{-6}	66.3
Nitrogen	126.2	33.9×10^5	90.1×10^{-6}	311
Oxygen	154.8	50.8×10^5	78×10^{-6}	410
Ammonia	405.5	112.8×10^5	72.5×10^{-6}	235
Freon 12	384.7	40.1×10^5	218×10^{-6}	555
Carbon dioxide	304.2	73.9×10^5	94.0×10^{-6}	468
Sulfur dioxide	430.7	78.8×10^5	122×10^{-6}	524
Water	647.4	221.2×10^5	56×10^{-6}	320
Carbon disulfide	552.0	79.0×10^5	170×10^{-6}	440

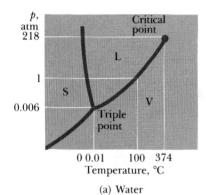

(a) Water

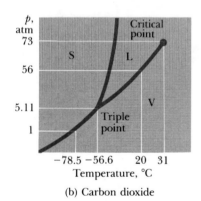

(b) Carbon dioxide

17–5 Pressure–temperature diagrams (not to a uniform scale).

A *pVT*-surface gives a bird's-eye view of the behavior of a material.

Examples of processes represented on a *pVT*-surface

As examples of phase diagrams, consider the *pT*-diagrams for water and carbon dioxide shown in Fig. 17–5. In (a), a horizontal line at a pressure of 1 atm intersects the freezing-point curve at 0°C and the boiling-point curve at 100°C. The boiling point increases with increasing pressure up to the critical temperature of 374°C. Solid, liquid, and vapor can remain in equilibrium only at the triple point, where the vapor pressure is 0.006 atm and the temperature is 0.01°C.

For carbon dioxide, the triple-point temperature is −56.6°C, and the corresponding pressure is 5.11 atm. Hence at atmospheric pressure CO_2 can exist only as a solid or vapor. When heat is supplied to solid CO_2, open to the atmosphere, it changes directly to vapor without passing through the liquid phase, whence the name "dry ice." Liquid CO_2 can exist only at a pressure greater than 5.11 atm. The steel tanks in which CO_2 is commonly stored contain liquid and vapor (both saturated). The pressure in these tanks is the vapor pressure of CO_2 at the temperature of the tank. If this temperature is 20°C, the vapor pressure is 56 atm or about 830 lb·in⁻².

We remarked at the beginning of Section 17–3 that the equation of state of any material can be represented graphically as a surface in a three-dimensional space with coordinates *p*, *V*, and *T*. Such a surface is seldom useful in representing detailed quantitative information, but it can enhance our general understanding of the behavior of materials at various temperatures and pressure. Figure 17–6 shows a typical *pVT*-surface. The light lines represent *pV*-isotherms, and we see that their projections on the *pV*-plane yield a diagram similar to Fig. 17–3. The *pV*-isotherms represent contour lines on the *pVT*-surface, just as contour lines on a topographic map represent the elevation (the third dimension) at each point. The projections of the edges of the surface onto the *pT*-plane yield the *pT*-phase diagram of Fig. 17–5b.

Three lines on the surface show specific processes. Line *abcdef* shows constant-volume heating, with melting at *bc* and vaporization at *de*. Note the volume changes that occur as *T* increases along this line. Line *ghjklm* is an isothermal compression; *hj* shows liquefaction and *kl* solidification. Between these, sections *gh* and *jk* represent constant-temperature compression with

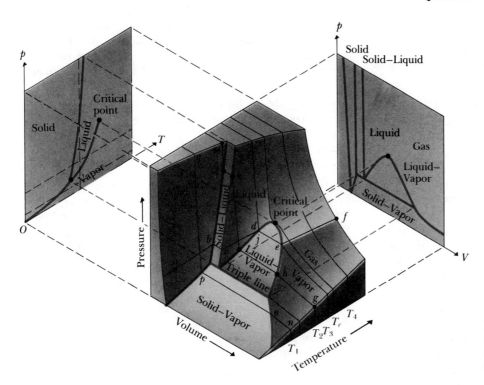

17-6 pVT-surface for a substance that expands on melting. Projections of the surface on the pT- and pV-planes are also shown.

increase in pressure; the pressure increases are much greater in the liquid region jk and the solid region lm than in the vapor region gh. Finally, $nopq$ represents isothermal solidification directly from the vapor phase. These three lines on the pVT-surface are worth careful study.

SUMMARY

The pressure, volume, and temperature of a given quantity of a substance are called state coordinates. They cannot all vary independently but are related by an equation of state. The equation of state pertains only to equilibrium states, where p and T are uniform throughout the system.

The ideal gas equation of state is

$$pV = nRT, \qquad (17\text{--}6)$$

where T is always absolute temperature, n is the number of moles, and R is a fundamental constant called the ideal-gas constant.

A pV-diagram is a set of graphs of pressure as a function of volume for several constant temperatures. The curves are called isotherms. A pT phase diagram shows what phase is stable for any given values of p and T and the conditions for which phase equilibrium can occur. A pVT-surface is a geometric representation of the equation of state of a substance.

KEY TERMS

state coordinates

equation of state

equilibrium states

ideal gas constant

ideal gas

standard temperature and pressure (STP)

pV-diagram

isotherm

phase diagram

phase equilibrium

triple point

critical point

QUESTIONS

17-1 If the ideal-gas model were perfectly valid at all temperatures, what would be the volume of a gas when the temperature is absolute zero?

17-2 In the ideal-gas equation of state, could Celsius temperature be used instead of Kelvin if an appropriate numerical value of the gas constant R is used?

17–3 Comment on this statement:

Equal masses of two different gases, placed in containers of equal volume at equal temperature, must exert equal pressures.

17–4 When a car is driven some distance, the air pressure in the tires increases. Why? Should you let some air out to reduce the pressure?

17–5 Section 17–1 states that pressure, volume, and temperature cannot change individually without one affecting the others. Yet when a liquid evaporates, its volume changes, even though its pressure and temperature are constant. Is this inconsistent?

17–6 At low pressure or high temperature, gas behavior deviates from the ideal-gas model. One reason is the finite size of gas molecules. From these deviations, how could we estimate what fraction of the total volume in a gas is occupied by the molecules themselves?

17–7 Radiator caps in automobile engines will release coolant when the pressure reaches a certain value, typically 15 psi or so. Why is it desirable to have the coolant under pressure? Why not just seal the system completely?

17–8 People speak of being angry enough to make their blood boil. Could your blood boil if you went to a high enough altitude? Is the pressure of your blood the same as atmospheric pressure? Does this matter?

17–9 Can it be too cold for good ice skating? What about skiing? Does a thin layer of ice under the skis melt?

17–10 Why does the air cool off more on a clear night than on a cloudy night? Desert areas often have exceptionally large day–night temperature variations. Why?

17–11 On a chilly evening you can "see your breath." Can you really? What are you really seeing? Does this phenomenon depend on the temperature of the air, the humidity, or both?

17–12 Why do frozen water pipes burst? Would a mercury thermometer break if the temperature went below the freezing temperature of mercury?

17–13 Why does moisture condense on the insides of windows on cold days? If it is cold enough outside, frost will form, but only when the outside temperature is far below freezing. Why is this so? Why are the bottoms frostier than the tops? How do storm windows change the situation?

17–14 "It's not the heat; it's the humidity!" Why are you more uncomfortable on a hot day when the humidity is high than when it is low?

17–15 Why does a shower room get steamy or foggy when the hot water runs for a long time?

17–16 The critical volumes of several substances are listed in Table 17–2, and with a few exceptions they are all in the range of 50 to 100×10^{-6} m^3·mol^{-1}. Why is there not more variation?

17–17 Why do flames from a fire always go up rather than down?

17–18 Unwrapped food placed in a freezer experiences dehydration, known as "freezer burn." Why? The same process, carried out intentionally, is called "freeze-drying." For freeze-drying, the food is usually frozen first and then placed in a partial vacuum. Why? What advantages might freeze-drying have, compared to ordinary drying?

17–19 An old or burned-out incandescent light bulb usually has a dark gray area on part of the inside of the bulb. What is this? Why is it not a uniform deposit all over the inside of the bulb?

17–20 Why do glasses containing cold liquids sweat in warm weather? Glasses containing alcoholic drinks sometimes form frost on the outside. How can this happen?

17–21 Why does a person perspire more when working hard than when resting?

17–22 Hiking down a steep trail ought to be much easier than hiking up, yet people perspire a lot when descending a steep trail. Why?

17–23 In a hot desert climate a person can survive a lot longer without food than without water. Why?

EXERCISES

Section 17–2 Ideal Gases

17–1 A cylindrical tank has a tight-fitting piston that allows the volume of the tank to be changed. The tank originally contains 0.12 m^3 of air at a pressure of 1 atm. The piston is slowly pushed in until the volume of the gas is decreased to 0.04 m^3. If the temperature remains constant, what is the final value of the pressure?

17–2 A 2-L tank contains air at 1 atm and 20°C. The tank is sealed and heated until the pressure is 3 atm.

a) What is the temperature now, in degrees Celsius? Assume that the volume of the tank is constant.

b) If the temperature is kept at the value found in (a) and the gas is permitted to expand, what is the volume when the pressure again becomes 1 atm?

17–3 A 20-L tank contains 0.20 kg of helium at 27°C.

a) What is the number of moles of helium?

b) What is the pressure in pascals? In atmospheres?

17–4 A room with dimensions 3 m × 4 m × 5 m is filled with pure oxygen at 20°C and 1 atm.

a) What is the number of moles of oxygen?

b) What is the mass of oxygen, in kilograms?

17–5 A tank contains 0.5 m^3 of nitrogen at 27°C and 1.5×10^5 Pa (absolute pressure). What is the pressure if the volume is increased to 5.0 m^3 and the temperature is increased to 327°C?

17–6 A welder using a tank of volume 0.10 m^3 fills it with oxygen at a gauge pressure of 4.0×10^5 Pa and temperature of 47°C. Later it is found that, because of a leak, the gauge pressure has dropped to 3.0×10^5 Pa and the temperature has decreased to 27°C. Find

a) the initial mass of oxygen;

b) the mass that has leaked out.

17–7 At the beginning of the compression stroke, a cylinder of a diesel engine contains 800 cm³ of air at atmospheric pressure and a temperature of 27°C. At the end of the stroke, the air has been compressed to a volume of 50 cm³ and the gauge pressure has increased to 40×10^5 Pa. Compute the temperature.

17–8 A diver observes a bubble of air rising from the bottom of a lake, where the pressure is 3 atm, to the surface, where the pressure is 1 atm. The temperature at the bottom is 7°C, and the temperature at the surface is 27°C. What is the ratio of the volume of the bubble as it reaches the surface to its volume at the bottom?

17–9

a) Derive from the equation of state of an ideal gas an equation for the density of an ideal gas in terms of pressure, temperature, and appropriate constants.

b) What is the density of air, in kg·m⁻³, at 1 atm and 20°C? The average molecular mass of air is 28.8 g·mol⁻¹.

17–10 What is the atmospheric pressure at an altitude of 5000 m, if the air temperature is a uniform 0°C? This is roughly the maximum altitude usually attained by aircraft with nonpressurized cabins.

Section 17–4 Phase Diagrams, Triple Point, Critical Point

17–11 A physicist places a piece of ice at 0°C alongside a beaker of water at 0°C in a glass vessel, from which all the air has been removed. If the ice, water, and vessel are all maintained at a temperature of 0°C by a suitable thermostat, describe the final equilibrium state inside the vessel.

17–12 Calculate the volume of 1 mol of liquid water at a temperature of 20°C (use Table 14–3), and compare this to the volume occupied by 1 mol of water at the critical point.

17–13 Solid nitrogen is slowly heated from a very low temperature.

a) What minimum external pressure p_1 must be applied to the solid if a melting-phase transition is to be observed? Describe the sequence of phase transitions that occur if the applied pressure p is such that $p < p_1$.

b) Above a certain maximum pressure p_2 no liquid-to-vapor (boiling) transition is observed. What is this pressure? Describe the sequence of phase transitions that occur if $p_1 < p < p_2$.

PROBLEMS

17–14 Partial vacuums where the residual gas pressure is about 10^{-10} atm are not difficult to obtain with modern technology. Calculate the mass of nitrogen (N_2) present in a volume of 2000 cm³ at this pressure. Assume that the temperature of the gas is 27°C.

17–15 An automobile tire has a volume of about 0.015 m³ on a cold day when the temperature of the air in the tire is 0°C. Under these conditions the gauge pressure is measured to be 2 atm (about 30 lb·in⁻²). After the car is driven on the highway for 30 min, the temperature of the air in the tires has risen to 47°C and the volume to 0.016 m³. What then is the gauge pressure?

17–16 A glassblower makes a barometer from a tube 0.90 m long and of cross section 1.5 cm². Mercury stands in this tube to a height of 0.75 m. The room temperature is 27°C. A small amount of nitrogen is introduced into the evacuated space above the mercury, and the column drops to a height of 0.70 m. How many grams of nitrogen were introduced?

17–17 A bicyclist uses a tire pump whose cylinder is initially full of air at an absolute pressure of 1.01×10^5 Pa. The length of stroke of the pump (the length of the cylinder) is 45.7 cm. At what part of the stroke (i.e., what is the length of air column) does air begin to enter a tire in which the gauge pressure is 2.76×10^5 Pa? Assume that the temperature remains constant during the compression.

17–18 The submarine *Squalus* sank at a point where the depth of water was 73 m. The temperature at the surface is 27°C, and at the bottom it is 7°C. The density of sea water is 1030 kg·m⁻³.

a) If a diving bell in the form of a circular cylinder 2.4 m high, open at the bottom and closed at the top, is lowered to this depth, to what height will water rise within it when it reaches the bottom?

b) At what gauge pressure must compressed air be supplied to the bell while on the bottom to expel all the water from it?

17–19 A flask of volume 2 L (2×10^{-3} m³), provided with a stopcock, contains oxygen at 300 K and atmospheric pressure. The system is heated to a temperature of 400 K, with the stopcock open to the atmosphere. The stopcock is then closed and the flask cooled to its original temperature.

a) What is the final pressure of the oxygen in the flask?

b) How many grams of oxygen remain in the flask?

17–20 A hot-air balloon makes use of the fact that hot air at atmospheric pressure is less dense than cooler air at the same pressure; the buoyant force is calculated as discussed in Chapter 13. If the volume of the balloon is 500 m³ and the surrounding air is at 0°C, what must be the temperature of the air in the balloon in order for it to lift a total load of 200 kg (in addition to the mass of the hot air)? The density of air at 0°C and atmospheric pressure is 1.29 kg·m⁻³.

17–21 A balloon whose volume is 500 m³ is to be filled with hydrogen at atmospheric pressure.

a) If the hydrogen is stored in cylinders of volume 2.5 m³ at an absolute pressure of 15×10^5 Pa, how many cylinders are required? Assume that the temperature of the hydrogen remains constant.

b) What is the total weight (in addition to the weight of the gas) that can be supported by the balloon, in air at standard conditions?

c) What weight could be supported if the balloon were filled with helium instead of hydrogen?

17–22 A vertical cylindrical tank 1 m high has its top end closed by a tightly fitting frictionless piston of negligible

weight. The air inside the cylinder is at an absolute pressure of 1 atm. The piston is depressed by pouring mercury on it slowly. How far will the piston descend before mercury spills over the top of the cylinder? The temperature of the air is maintained constant.

CHALLENGE PROBLEMS

17–23 A large tank of water has a hose connected to it, as shown in Fig. 17–7. The tank is sealed at the top and has compressed air between the water surface and the top. When the water height h_2 has the value 3.0 m, the absolute pressure p_1 is 2.00×10^5 Pa. Assume that the air above the water expands isothermally (i.e., at constant T). Take the atmospheric pressure to be 1.00×10^5 Pa.

a) What is the velocity of flow out of the hose when $h_2 = 3.0$ m?

b) As water flows out of the tank, h_2 decreases. Calculate the velocity of flow for $h_2 = 2.5$ and 2.0 m.

c) At what value of h_2 will the flow stop?

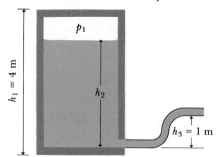

FIGURE 17–7

17–24 In the lower part of the atmosphere (the troposphere) the temperature is not uniform but decreases with increasing elevation.

a) Show that if the temperature variation is approximated by the linear relation

$$T = T_0 - \alpha y,$$

where T_0 is the temperature at the earth's surface and T is the temperature at height y, the pressure is given by

$$\ln\left(\frac{p}{p_0}\right) = \frac{Mg}{R\alpha} \ln\left(\frac{T_0 - \alpha y}{T_0}\right),$$

where p_0 is the pressure at the earth's surface and M is the molecular mass for air. The coefficient α is called the "lapse rate of temperature." Although it varies with atmospheric conditions, an average value is about 0.6 C°/100 m.

b) Show that the above result reduces to Eq. (17–9) in the limit that $\alpha \to 0$.

c) With $\alpha = 0.6$ C°/100 m, calculate p for $y = 8882$ m and compare your answer to the result of Example 17–3. Take $T_0 = 273$ K and $p_0 = 1$ atm, as in the example.

17–25 In pV-diagrams, the slope dp/dV along an isotherm is never positive. (You should be able to explain why.) As

shown in Section 17–3, and indicated in Figs. 17–3 and 17–6, regions where $dp/dV = 0$ represent equilibrium between two phases; volume can change with no change in pressure. We can use this fact to determine critical constants if we have an equation of state. With $p = p(V, T, n)$, if $T > T_c$, then $p(V)$ has no maximum along an isotherm, whereas if $T < T_c$, then $p(V)$ has a maximum. This leads to the following condition for determining the critical point:

$$\frac{dp}{dV} = 0 \quad \text{and} \quad \frac{d^2p}{dV^2} = 0 \quad \text{at the critical point.}$$

An equation of state that has isotherms resembling those in Fig. 17–3 is the van der Waals equation:

$$\left(p + \frac{an^2}{V^2}\right)(V - nb) = nRT,$$

where a and b are constants.

a) Solve the van der Waals equation for p. That is, find $p(V, T, n)$. Find dp/dV and d^2p/dV^2. Set these equal to zero, and thereby obtain two equations for V, T, and n.

b) Simultaneous solution of the two equations obtained in (a) gives the critical values T_c and $(V/n)_c$. Find these constants in terms of a and b. (*Hint:* Divide one equation by the other to eliminate T.)

c) Substitute these values into the equation of state to find p_c, the critical pressure.

d) Find the ratio

$$\frac{RT_c}{p_c(V/n)_c}.$$

This should not contain either a or b.

e) Compute the ratio in part (d) for the gases H_2, N_2, O_2, and H_2O by using the critical-point data of Table 17–2. Compare the results to the prediction you made based on the van der Waals equation.

f) The experimentally determined values of a and b for some gases are given below. Use your results of parts (b) and (c) to predict T_c, p_c, and $(V/n)_c$ for these gases. Compare these values to those in Table 17–2.

Gas	$a(\text{m}^6 \cdot \text{Pa} \cdot \text{mol}^{-2})$	$b(\text{m}^3 \cdot \text{mol}^{-1})$
H_2	0.0247	2.66×10^{-5}
N_2	0.141	3.91×10^{-5}
O_2	0.138	3.18×10^{-5}
H_2O	0.553	3.05×10^{-5}

18

THE FIRST LAW OF THERMODYNAMICS

THERMODYNAMICS IS THE STUDY OF ENERGY RELATIONSHIPS INVOLVING heat, mechanical work, and other aspects of energy and energy transfer. The first law of thermodynamics extends the principle of conservation of energy to include not only the forms of mechanical energy encountered in Chapter 7 but also energy associated with heat. Conservation of energy plays a vital role in every area of physical science, so the first law is widely applicable. Energy relationships are of central importance in the study of energy-conversion devices such as engines, batteries, and refrigerators, and also in the functioning of living organisms. To state these relationships precisely, we introduce the concept of a *thermodynamic system* and discuss *heat* and *work* as two means of transferring energy into or out of such a system. Emerging from this discussion and from a generalized view of the principle of conservation of energy is the concept of *internal energy* of a system.

18–1 ENERGY, HEAT, AND WORK

We have studied the transfer of energy through mechanical work (Chapter 7) and through heat transfer (Chapters 15 and 16); now we are ready to combine and generalize these principles. Most of our analysis will refer to a *system*, which will usually be a specified quantity of material, perhaps in the form of a device or an organism. A **thermodynamic system** is a system that can interact with its surroundings in at least two ways, one of which must be heat transfer. A familiar example of a thermodynamic system is a quantity of a gas confined in a cylinder with a piston. Energy can be added to the system by conduction of heat, and the system can also do *work*, since the gas exerts a force on the piston and can move it through a displacement.

We have already discussed the concept of a system. In Chapters 4, 5, and 9 we made extensive use of free-body diagrams to help identify the forces acting on a particular mechanical system so as to include only these forces in our analysis. Similarly, in Chapter 8 we studied the principle of conservation of momentum for an isolated mechanical system. In both examples, it is essential

Energy exchange in a thermodynamic system by means of heat and work

The concept of a system occurs throughout physics; a thermodynamic system is a specific example.

to define clearly at the outset precisely what is and is not included in the system. The same is true of thermodynamic systems; it is essential in solving problems to identify the system under discussion and to describe unambiguously the energy transfers into and out of that system.

The concept of a system can be broadened to include a variable quantity of matter; in that case the transport of *mass* into or out of a system provides an additional mechanism for energy exchange with the surroundings. This is a complication we will not consider here; all the systems discussed in this chapter have a constant amount of matter.

As already mentioned, thermodynamics has its roots in practical problems. For example, a steam engine or steam turbine uses the heat of combustion of coal or other fuel to perform mechanical work such as driving an electric generator, pulling a train, or performing some other useful function. The gasoline engine in an automobile and the jet engines in an airplane have similar functions. Muscle tissue in living organisms metabolizes chemical energy in food and performs mechanical work on the surroundings of the organism.

In all these situations, we will describe the energy relations in terms of the quantity of heat Q added *to* the system and the work W done *by* the system. Both Q and W may be positive or negative. A positive value of Q represents heat flow *into* the system, with a corresponding input of energy to it; negative Q represents heat flow *out of* the system. A positive value of W represents work done *by* the system against its surroundings, such as an expanding gas, and hence corresponds to energy leaving the system. Negative W, such as compression of a gas in which work is done *on it* by its surroundings, represents energy entering the system. These conventions will be clarified with many examples in this chapter and the next.

Heat can be understood on the basis of *microscopic* mechanical energy, that is, the kinetic and potential energies of individual molecules in a material; it is also possible to develop the principles of thermodynamics from a microscopic viewpoint. In this chapter we deliberately avoid that development in order to emphasize that the central principles and concepts of thermodynamics can be treated in a wholly *macroscopic* way, without reference to microscopic models. Indeed, part of the great power and generality of thermodynamics springs from the fact that it *does not* depend on details of the structure of matter. In Chapter 20, however, we will return to microscopic considerations and will examine their relation to the principles of thermodynamics.

18–2 WORK DONE DURING VOLUME CHANGES

A gas in a cylinder with a movable piston is a simple example of a thermodynamic system. In the next several sections this system will be used as an example to explore several kinds of processes involving energy considerations. First we consider the *work* done by the system during a volume change. When a gas expands, it pushes out on its boundary surfaces as they move outward; hence an expanding gas always does positive work. The same is true of any solid or fluid material confined under pressure. Figure 18–1 shows a solid or fluid in a cylinder with a movable piston. Suppose that the cylinder has a cross-sectional area A and that the pressure exerted by the system at the piston face is p. The force F exerted by the system is therefore $F = pA$. If the piston moves out a

Energy conversions involving heat and work have many practical applications.

Sign rules are an essential part of energy bookkeeping.

The relation of heat to molecular mechanical energy: not essential in thermodynamics, but useful later for additional insight

When a system under pressure expands, it does work against its surroundings.

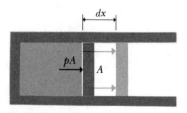

18–1 Force exerted *by* a system during a small expansion.

The catastrophic eruption of Mt. St. Helens on May 18, 1980, a dramatic illustration of the enormous pressure developed in a confined hot gas. As the gas expands, it does enough work on the overlying rock and soil to lift roughly a cubic mile of material several thousand feet into the air. (Courtesy of USGS.)

small distance dx, the work dW done by this force is equal to

$$dW = pA\,dx.$$

But

$$A\,dx = dV,$$

where dV is the change of volume of the system. Therefore

$$dW = p\,dV, \tag{18–1}$$

and in a finite change of volume from V_1 to V_2,

$$W = \int_{V_1}^{V_2} p\,dV. \tag{18–2}$$

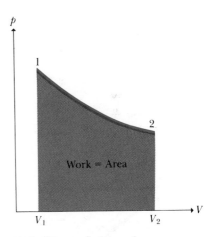

18–2 The work done when a substance changes its volume and pressure is the area under the curve on a pV-diagram.

In general, the pressure of the system may vary during the volume change, and the integral can be evaluated only when the pressure is known as a function of volume. It is customary to represent this relationship graphically by plotting p as a function of V, as in Fig. 18–2. Then Eq. (18–2) may be interpreted graphically as the *area* under the curve between the limits V_1 and V_2.

According to the convention stated above, work is *positive* when a system *expands*. Thus, if a system expands from 1 to 2 in Fig. 18–2, the area is regarded as positive. A *compression* from 2 to 1 gives rise to a *negative* area; when a system is compressed, its volume decreases and it does *negative* work on its surroundings.

An expanding system does positive work; a contracting system does negative work.

If the pressure remains constant while the volume changes, then the work is

$$W = p(V_2 - V_1) \qquad \text{(constant pressure only).} \tag{18–3}$$

EXAMPLE 18–1   An ideal gas at constant temperature T undergoes an *isothermal expansion* in which its volume changes from V_1 to V_2. How much work does the gas do?

SOLUTION   From Eq. (18–2),

$$W = \int_{V_1}^{V_2} p\, dV.$$

For an ideal gas,

$$p = \frac{nRT}{V}.$$

Since n, R, and T are constant,

$$W = nRT \int_{V_1}^{V_2} \frac{dV}{V} = nRT \ln \frac{V_2}{V_1}. \tag{18–4}$$

In an expansion, $V_2 > V_1$ and W is positive. At constant T,

$$p_1 V_1 = p_2 V_2, \qquad \text{or} \qquad \frac{V_2}{V_1} = \frac{p_1}{p_2},$$

so the isothermal work may be expressed also in the form

$$W = nRT \ln \frac{p_1}{p_2}. \tag{18–5}$$

When a system goes from an initial state to a final state, there are many different possible sequences of intermediate states.

On the pV-diagram in Fig. 18–3, an initial state 1 (characterized by pressure p_1 and volume V_1) and a final state 2 (characterized by pressure p_2 and volume V_2) are represented by the two points 1 and 2. To progress from the initial state to the final state, the system must go through a series of intermediate states, and infinitely many such series are possible. For example, the pressure might be kept constant at p_1 while the system expands to volume V_2 (point 3 on the diagram), and then the pressure might be reduced to p_2 with the volume remaining constant at V_2 (to point 2 on the diagram). The work during this process is the area under the line $1 \to 3$; no work is done during the constant-volume process $3 \to 2$. Or the system might traverse the path $1 \to 4 \to 2$, in which case the work is the area under the line $4 \to 2$. The wiggly line and the smooth curve from 1 to 2 are two other possibilities, and the work is different for each one.

We conclude that *the work depends not only on the initial and final states but also on the intermediate states, that is, on the path.* For this reason, we cannot speak of the amount of work contained in a system in a given state; such a concept has no meaning. In this respect, the character of work is very different from that of the state coordinates p, V, and T.

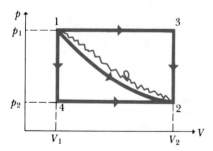

18–3   Work depends on the path.

18–3 HEAT TRANSFER DURING VOLUME CHANGES

As explained in Chapter 15, heat is energy transferred into or out of a system as a result of a temperature difference between the system and its surroundings. The performance of work, discussed in the preceding section, and the transfer of heat are means of energy transfer, that is, processes by which the energy of a system may increase or decrease.

We have seen in Section 18–2 that in a change of state of a thermodynamic system, the work done by the system depends not only on the initial and final states, but also on the *path* taken, that is, the series of intermediate states. This is also true of the *heat* added to the system in such a process. The following example illustrates this point.

In Fig. 18–4a, a quantity of gas is contained in a cylinder with a piston, with initial volume of two liters; it is kept at a constant temperature of 300 K by an electric stove. We then let the gas expand slowly; heat flows from the stove to the gas, keeping the temperature of the gas at 300 K. Suppose that the gas expands in this slow, controlled, isothermal manner until its volume becomes, say, seven liters. A finite amount of heat is absorbed by the gas during this process.

Figure 18–4b shows a container surrounded by insulating walls and divided into two compartments (the lower one having volume two liters, the upper one five liters) by a thin, breakable partition. In the lower compartment, the same gas has the same initial volume and temperature as in (a). Now we break the partition; the gas in (b) undergoes a rapid, uncontrolled expansion, with no heat passing through the insulating walls. The final volume is seven liters, as in (a). The rapid, uncontrolled expansion of a gas into a vacuum is called a **free expansion;** we will discuss it further in Section 18–6. Experiments have shown that in a free expansion of an ideal gas, no temperature change takes place; therefore the final state of the gas is the same as the final state in (a). But the states traversed by the gas (the pressures and volumes) while proceeding from state 1 to state 2 are entirely different, so that (a) and (b) represent *two different paths* connecting the *same states* 1 and 2. For path (b), *no* heat is transferred and no work is done. Like work, *heat depends not only on the initial and final states but also on the path.*

Thus it would *not* be correct to say that a body *contains* a certain amount of heat. Suppose we assigned an arbitrary value to "the heat in a body" in some standard reference state. The "heat in the body" in some other state would then equal the "heat" in the reference state plus the heat added when the body is carried to the second state. But the heat added depends on the path by which the body goes from one state to the other. There are infinitely many paths that might be followed, so there are also infinitely many values that might equally well be assigned to the "heat in the body" in the second state. Because there is no consistent way to define "heat in a body," this is not a useful concept. It *does* make sense, however, to speak of the amount of *internal energy* in a body, and this important concept is our next topic.

Heat depends on the path.

There is no such thing as the amount of heat in a body.

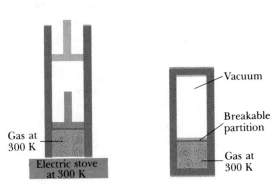

(a)   (b)

Gas at 300 K

Electric stove at 300 K

Vacuum

Breakable partition

Gas at 300 K

18–4 (a) Slow, controlled isothermal expansion of a gas from an initial state 1 to a final state 2. (b) Rapid, uncontrolled expansion of the same gas starting at the same state 1 and ending at the same state 2.

18–4 INTERNAL ENERGY AND THE FIRST LAW

The concept of internal energy of a thermodynamic system can be viewed in various ways, some simple, some subtle. To begin with the simple, we note that matter consists of atoms and molecules; these, in turn, are made up of particles having kinetic and potential energies. We tentatively define the **internal energy** of a system as the sum of all the kinetic and potential energies of its constituent particles. We use the symbol U for internal energy. During a change of state of the system, the internal energy may change from an initial value U_1 to a final value U_2; in that case we denote the change in internal energy in the usual way as $\Delta U = U_2 - U_1$.

Internal energy is associated with molecular kinetic and potential energies.

We also know that heat transfer is really energy transfer. Thus when a quantity of heat Q is added to a system, the result should be to increase the internal energy by an amount equal to Q; that is, $\Delta U = Q$. Also, when a system does work W by expanding against its surroundings, this represents energy leaving the system. When W is positive, ΔU is negative, and conversely, so $\Delta U = -W$. Finally, if both heat transfer and work occur, the total change in internal energy is

The internal energy of a system changes when heat is added and when the system does work.

$$U_2 - U_1 = \Delta U = Q - W. \tag{18–6}$$

This equation may be rearranged to the form

$$Q = \Delta U + W, \tag{18–7}$$

which can be read as follows: When heat Q is added to a system, some of this added energy remains within the system, increasing its internal energy by an amount ΔU; the remainder leaves the system again as the system does work W against its surroundings. Of course, as explained above, W and Q are algebraic quantities that may be either positive or negative, and correspondingly we expect ΔU to be positive for some processes and negative for others.

Equation (18–6) or (18–7) is the **first law of thermodynamics.** In the context of the above discussion, the law simply represents conservation of energy, generalized to include energy transfer through heat as well as mechanical work. It seems eminently reasonable that conservation of energy, which we encountered in Chapter 7 in a purely mechanical context, should have this more general validity; nevertheless, it is worth a somewhat closer look.

The first law of thermodynamics: energy conservation in a thermodynamic system

First of all, our definition of internal energy in terms of microscopic kinetic and potential energies is not entirely convincing. Actually calculating this total energy for any real system would be hopelessly complicated. Thus our definition is not an *operational* definition because it does not describe how to obtain the defined quantity from physical quantities that can be measured directly. We need to backtrack a little in defining internal energy, in order to show clearly the empirical and experimental basis of the first law of thermodynamics.

An operational definition of internal energy

Starting over, we *define* the internal energy change ΔU in any change of a system by means of Eq. (18–6). This *is* an operational definition, because Q and W can be measured. It does not define U itself, only ΔU. This is not a serious shortcoming; we can define the internal energy of a system to have a specified value in a certain standard state, and then use Eq. (18–6) to define the internal energy in any other state. This procedure is analogous to the definition of potential energy in Chapter 6, where we arbitrarily defined the potential energy of a mechanical system to be zero at a certain position.

This new definition trades one difficulty for another. If we define ΔU via Eq. (18–6), then we must ask the following question: If the system goes from some initial state to some final state by two different paths, is ΔU necessarily the same for the two paths? We have already seen that in such a case Q and W are in general both path-dependent. If ΔU, which is just $Q - W$, is also path-dependent, then ΔU is ambiguous, and the concept of internal energy in a system is subject to the same criticism as the erroneous concept of quantity of heat in a system, as discussed at the end of Section 18–3.

At this point the only recourse is to experiment. We study the properties of a variety of materials; in particular, we measure Q and W for various changes of state and various paths, in order to learn whether ΔU is or is not path-dependent. The results of many such investigations are clear and unambiguous: Within experimental error, ΔU is found to be path-independent. The change in internal energy in any thermodynamic process is found to depend only on the initial and final states and *not* on the path leading from one to the other.

Experiment, then, is the ultimate justification for believing that in any state, a thermodynamic system has a unique internal energy that depends only on the state the system is in. An equivalent statement is that the internal energy U of a system is a function of the state coordinates p, V, and T (or actually of any two of these, since the three variables are related by the equation of state).

Internal energy depends only on the state of the system: the real significance of the first law.

Let us now return to the first law of thermodynamics. It is perfectly correct to say that the first law, given by Eq. (18–6) or (18–7), is a statement of conservation of energy for thermodynamic processes. But we must also recognize that part of the content of the first law is the experimental verification that internal energy is a property of the state of a system; in changes of state, the change in internal energy is path-independent.

All this discussion may seem a little abstract to a reader who is satisfied to regard internal energy as microscopic mechanical energy. There is nothing wrong with that view. But in the interest of precise *operational* definitions, internal energy, like heat, can and must be defined in a way that is independent of the detailed microscopic structure of the material.

If a system is carried through a process that eventually returns it to its initial state (a *cyclic* process), then

In any cyclic process, the internal energy change is zero.

$$U_2 = U_1 \quad \text{and} \quad Q = W.$$

Thus, although net work W may be done by the system in the process, energy has not been created, since an equal amount of energy must have flowed into the system as heat Q.

An *isolated* system is one that does no external work and into which there is no heat flow. Then, for any process taking place in such a system,

$$W = Q = 0$$

and

$$U_2 - U_1 = 0 \quad \text{and} \quad \Delta U = 0.$$

That is, *the internal energy of an isolated system remains constant*. This is the most general statement of the principle of conservation of energy. The internal energy of an *isolated* system cannot be changed by any process (mechanical, electrical, chemical, nuclear, or biological) taking place *within* the system. The energy of a system can be changed only by a flow of heat across its boundary,

The internal energy of an isolated system cannot change.

or by the performance of work. If either of these takes place, the system is no longer isolated. The increase in energy of the system is then equal to the energy flowing in as heat, minus the energy flowing out as work.

PROBLEM-SOLVING STRATEGY: *First law of thermodynamics*

1. The internal energy change ΔU in any thermodynamic process or series of processes is independent of the path, whether or not the substance is an ideal gas. This is of utmost importance in the problems in this chapter and the next. Sometimes you will be given enough information about one path between given initial and final states to calculate ΔU for that path. Then you can use the fact that ΔU is the same for every other path between the same two states to relate the various energy quantities for other paths.

2. As usual, consistent units are essential. If p is in Pa and V in m^3, then W is in joules. Otherwise you may want to convert the pressure and volume units into Pa and m^3. If a heat capacity is given in terms of calories,

usually the simplest procedure is to convert it to joules. Be especially careful with moles. When using $n = m/M$ to convert between mass and number of moles, remember that if m is in kilograms, M must be in kilograms per mole. The usual units for M are *grams* per mole, so be careful.

3. When a process consists of several distinct steps, it is often helpful to make a chart showing Q, W, and ΔU for each step. Put these quantities for each step on a different line, and arrange them so the Q's, W's, and ΔU's form columns. Then you can apply the first law to each line; in addition, you can add each column and apply the first law to the sums. Do you see why?

An example of the usefulness of internal energy and the first law

EXAMPLE 18–2   A thermodynamic process is shown in the pV-diagram of Fig. 18–5. In process ab, 600 J of heat are added to the system, and in process bd 200 J of heat are added. Find (a) the internal energy change in process ab; (b) the internal energy change in process abd; and (c) the total heat added in process acd.

SOLUTION

a)   No volume change occurs in process ab, so $W = 0$ and $\Delta U = Q = 600$ J.

b)   Process bd occurs at constant pressure, so the work done by the system during this expansion is

$$W = p(V_2 - V_1) = (8 \times 10^4 \text{ Pa})(5 \times 10^{-3} \text{ m}^3 - 2 \times 10^{-3} \text{ m}^3)$$
$$= 240 \text{ J}.$$

Thus the total work for abd is $W = 240$ J, and the total heat is $Q = 800$ J. Equation (18–6) gives

$$\Delta U = Q - W = 800 \text{ J} - 240 \text{ J} = 560 \text{ J}.$$

c)   Because ΔU is independent of path, the internal energy change is the same for path acd as for abd, that is, 560 J. The total work for the path acd is

$$W = p(V_2 - V_1)$$
$$= (3 \times 10^4 \text{ Pa})(5 \times 10^{-3} \text{ m}^3 - 2 \times 10^{-3} \text{ m}^3)$$
$$= 90 \text{ J}.$$

Equation (18–7) then gives

$$Q = \Delta U + W = 560 \text{ J} + 90 \text{ J} = 650 \text{ J}.$$

We see that although ΔU is the same for abd and acd, W and Q are quite different for the two processes.

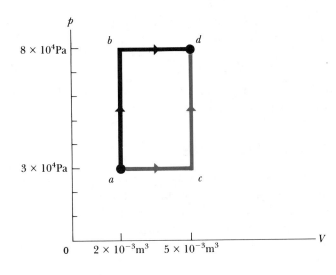

18–5 pV-diagram for Example 18–2, showing the various thermodynamic processes.

EXAMPLE 18–3   One gram of water (1 cm^3) becomes 1671 cm^3 of steam when boiled at a constant pressure of 1 atm. The heat of vaporization at this pressure is 2256 J·g^{-1}. Compute (a) the work done by the water when it vaporizes and (b) its increase in internal energy.

The first law applied to boiling: Where does the heat of vaporization go?

SOLUTION

a)   For a constant-pressure process we may use Eq. (18–3) to compute the work done by the water:

$$W = p(V_2 - V_1)$$
$$= (1.013 \times 10^5 \text{ Pa})(1671 \times 10^{-6} \text{ m}^3 - 1 \times 10^{-6} \text{ m}^3)$$
$$= 169 \text{ J}.$$

b)   The heat added is the heat of vaporization:

$$Q = mL_V$$
$$= (1 \text{ g})(2256 \text{ J·g}^{-1}) = 2256 \text{ J}.$$

From the first law of thermodynamics, Eq. (18–6), the change in internal energy is

$$\Delta U = Q - W = 2256 \text{ J} - 169 \text{ J} = 2087 \text{ J}.$$

That is, when one gram of water vaporizes, 2256 J of heat are added. Of this added energy, 2087 J remain in the system as an increase in internal energy, and the remaining 169 J leave the system again as it does work against the surroundings while expanding from liquid to vapor.

In the examples above we have applied the first law of thermodynamics to processes in which the initial and final states differ in pressure, volume, and temperature by a finite amount. Later we will often want to consider an *infinitesimal* change of state in which a small amount of heat dQ is transferred into the system, the system does a small amount of work dW, and its internal energy changes by an amount dU. For such a process, we state the first law as

The first law for an infinitesimal process

$$dU = dQ - dW. \qquad (18\text{–}8)$$

Also, in the systems we have been considering, the work dW is given by $dW = p\, dV$, and we can write the first law as

$$dU = dQ - p\, dV. \tag{18–9}$$

Equations (18–8) and (18–9) are often called the *differential form* of the first law of thermodynamics; we will use them several times later in this chapter.

18–5 THERMODYNAMIC PROCESSES

Here are four specific kinds of thermodynamic processes that occur often enough in practical problems to be worth special mention. Each one permits a particular simplification in applying the first law of thermodynamics, at least in some cases. Their characteristics can be summarized briefly as "no heat transfer," "constant volume," "constant pressure," and "constant temperature," respectively.

In an adiabatic process, no heat is transferred.

1. **Adiabatic process** An **adiabatic process** is one in which there is no heat transfer either into or out of a system; in other words, $Q = 0$. We can prevent heat flow either by surrounding the system with thermally insulating material or by carrying out the process so quickly that there is not enough time for appreciable heat flow. From the first law, we find for every adiabatic process

$$U_2 - U_1 = \Delta U = -W \qquad \text{(adiabatic process).} \tag{18–10}$$

If a system expands under adiabatic conditions, W is positive, ΔU is negative, and the internal energy decreases; when a system is compressed adiabatically, W is negative and U increases. An increase of internal energy is usually, although not always, accompanied by a rise in temperature.

The compression stroke in an internal-combustion engine is an example of a process that is approximately adiabatic. The temperature rises as the air-fuel mixture in the cylinder is compressed. Similarly, the expansion of the burned fuel during the power stroke is an approximately adiabatic expansion with a drop in temperature.

In an isochoric process, no work is done.

2. **Isochoric process** **Isochoric** means simply *constant-volume.* When the volume of a thermodynamic system is constant, it does no work on its surroundings. Thus $W = 0$, and

$$U_2 - U_1 = \Delta U = Q \qquad \text{(isochoric process).} \tag{18–11}$$

In an isochoric process, all heat added remains in the system as an increase in internal energy. Heating a gas in a closed constant-volume container is an example of an isochoric process.

In an isobaric process, it is easy to calculate the work.

3. **Isobaric process** **Isobaric** means *constant-pressure.* In general, *none* of the three quantities ΔU, Q, and W in the first law is zero, but calculating W is easy, as we have already seen in Section 18–2. For an isobaric process the integral in Eq. (18–2) becomes simply

$$W = p(V_2 - V_1) \qquad \text{(isobaric process).} \tag{18–12}$$

Example 18–3 (Section 18–4) is an example of an isobaric process.

4. **Isothermal process   Isothermal** means *constant-temperature*. For a process to be isothermal, the system must remain in thermal equilibrium; this means that the pressure and volume changes must occur slowly enough to maintain thermal equilibrium. In general, *none* of the quantities ΔU, Q, and W is zero. Example 18–1 (Section 18–2) is an example of an isothermal process.

Isothermal means constant-temperature.

There are some special cases, considered later in this chapter, in which the internal energy of a system depends *only* in its temperature, not on its pressure or volume. The most familiar system having this property is an ideal gas. For such a system, if the temperature is constant, the internal energy is also constant; in this case, $\Delta U = 0$ and $Q = W$. That is, any energy entering the system as heat Q must leave it again as work W done by the system. This does *not* hold for systems other than ideal gases, because in general the internal energy of a system depends on pressure as well as temperature, and it may vary even when T is constant.

18–6 INTERNAL ENERGY OF AN IDEAL GAS

We mentioned in Section 18–5 that for an ideal gas the internal energy U depends only on temperature, not on pressure or volume. To explore this special property in more detail, imagine a thermally insulated container with rigid walls, divided into two compartments by a partition, as in Fig. 18–4b. Suppose there is a gas in one compartment and that the other contains only vacuum. When the partition is removed or broken, the gas expands to fill both parts of the container. This process is called a *free expansion;* the gas does no work on its surroundings because the total volume of the container does not change. No heat is transferred because of the thermal insulation. From the first law, since both Q and W are zero, it follows that *the internal energy remains unchanged during a free expansion.* This is true of any substance, whether it is an ideal gas or not.

In a free expansion the internal energy is constant.

Whether or not the *temperature* of a gas changes during a free expansion is an important question, for it provides information concerning the dependence of internal energy on other quantities. If the temperature should change while the internal energy stays the same, we would have to conclude that the internal energy depends on both the temperature and the volume, or both the temperature and the pressure, but certainly not on the temperature alone. If, on the other hand, T remains unchanged during a free expansion in which we know U remains unchanged, then the only admissible conclusion is that *U is a function of T only,* not of p or V.

Experiments have shown that *in a free expansion an ideal gas undergoes no temperature change.* That is, when the behavior of a gas is described well by the ideal-gas equation of state, it is found that in a free expansion its temperature does not change. Thus *the internal energy of an ideal gas depends only on its temperature, not on its pressure or volume.* We may regard this property as part of the ideal-gas model, in addition to the ideal-gas equation of state. This property will be applied several times in the following sections.

The internal energy of an ideal gas does not depend on its pressure or volume.

For real gases that do not precisely obey the ideal-gas equation, experiments have shown that there is some temperature change during free expansions. For such gases, the internal energy does depend to some extent on the pressure as well as the temperature.

18–7 HEAT CAPACITIES OF AN IDEAL GAS

The temperature of a substance may be changed under a variety of conditions. The volume may be kept constant, or the pressure may be kept constant, or both may be allowed to vary in some definite way. In general, the amount of heat per mole needed to cause a unit rise in temperature is different for each type of process. In other words, a substance has *many different* molar heat capacities, depending on the conditions under which it is heated or cooled.

The basis of this variation is the first law of thermodynamics. In a constant-volume temperature change, no work is done, and the change in internal energy equals the heat added. In a constant-pressure temperature change, however, the volume *must* increase—otherwise the pressure could not remain constant—and as the material expands it does work. Thus for a given temperature change, the heat input for a constant-pressure process must be *greater* than for a constant-volume process because, in the former, additional energy must be supplied to account for the work done in the expansion. If the volume were to *decrease* during heating, the heat input would be *less* than in the constant-volume case.

The discussion of heat capacities in Chapter 15 ignored this distinction. Indeed, for solid and liquid materials the difference is not usually of great practical importance. The coefficients of volume expansion of such materials are quite small, and ordinarily the work done against the surroundings is so small compared to the total heat added that it may be ignored. The distinction *is* important in theoretical studies of heat capacities.

For gases the situation is quite different, and the dependence of heat capacity on the details of the process is significant. For air, for example, the heat capacity under constant-pressure conditions is 40 percent greater than under constant-volume conditions. Fortunately, for ideal gases we can derive a simple relation between two heat capacities of special importance.

From a practical standpoint, the two conditions of greatest interest are raising the temperature of a gas at *constant volume*, as in a container whose volume may be considered constant, and at *constant pressure*, where the gas expands just enough so the pressure remains constant. The two corresponding molar heat capacities are called the **molar heat capacity at constant volume,** denoted as C_v, and the **molar heat capacity at constant pressure,** denoted as C_p. The discussion above shows that ordinarily C_p should be larger than C_v. We will now derive the relation between these two quantities for an ideal gas.

First consider the constant-volume process. We place n moles of an ideal gas at temperature T in a rigid container and bring it in contact with a body at a slightly higher temperature $T + dT$. A quantity of heat dQ flows into the gas, and by definition of the molar heat capacity at constant volume, C_v,

$$dQ = nC_v \, dT \qquad \text{(constant volume)}. \tag{18–13}$$

The pressure of the gas increases during this process, but no *work* is done because the volume is constant. Hence, $W = 0$; from the first law in differential form, Eq. (18–8),

$$dU = dQ - dW = dQ,$$

we also have

$$dU = nC_v \, dT. \tag{18–14}$$

Now consider a constant-pressure process. We enclose the gas in a cylinder with a piston that moves so as to maintain constant pressure. Again the system is brought in contact with a body at a temperature $T + dT$. As heat flows into the gas, it expands at constant pressure and does work. By definition of the molar heat capacity at constant pressure, C_p, the heat dQ flowing into the gas is

$$dQ = nC_p \, dT \qquad \text{(constant pressure)}. \qquad (18\text{--}15)$$

The work dW done by the gas is

$$dW = p \, dV.$$

This may also be expressed in terms of the temperature change dT by use of the ideal-gas equation of state $pV = nRT$. Since p is constant, we have

$$p \, dV = nR \, dT,$$

so

$$dW = nR \, dT. \qquad (18\text{--}16)$$

Then, from the differential form of the first law of thermodynamics,

$$dQ = dU + dW,$$

or, using Eqs. (18–15) and (18–16),

$$nC_p \, dT = dU + nR \, dT. \qquad (18\text{--}17)$$

Now, here comes the crux of the calculation. The internal energy change dU is again given by Eq. (18–14), that is, $dU = nC_v \, dT$, *even though now the volume is not constant!* Why is this so? Recall the discussion of Section 18–6: One of the special properties of an ideal gas is that its internal energy depends *only* on temperature. Thus the *change* in internal energy in any process must be determined only by the temperature change. That is, if Eq. (18–14) is valid for an ideal gas for one particular kind of process, it must be valid for an ideal gas for *every* kind of process.

To complete our derivation, we replace dU in Eq. (18–17) by $nC_v \, dT$ and divide each term by the common factor $n \, dT$ to obtain

$$nC_p \, dT = nC_v \, dT + nR \, dT,$$

and

$$C_p = C_v + R. \qquad (18\text{--}18)$$

As predicted, the molar heat capacity of an ideal gas at constant pressure is *greater* than the molar heat capacity at constant volume; the difference is the gas constant R. Of course, R must be expressed in the same units as C_p and C_v, such as $J \cdot mol^{-1} \cdot K^{-1}$.

Although Eq. (18–18) was derived for an ideal gas, it is very nearly true for many real gases at moderate pressures, where their behavior approaches that of an ideal gas. Measured values of C_p and C_v are given in Table 18–1 for some real gases at low pressures; the difference in most cases is approximately $8.31 \, J \cdot mol^{-1} \cdot K^{-1}$.

Internal energy change for an ideal gas depends only on temperature change, even when the volume is not constant.

TABLE 18–1   Molar Heat Capacities of Gases at Low Pressure

Type of Gas	Gas	C_p, J·mol^{-1}·K^{-1}	C_v, J·mol^{-1}·K^{-1}	$C_p - C_v$	$\gamma = \dfrac{C_p}{C_v}$
Monatomic	He	20.78	12.47	8.31	1.67
	A	20.78	12.47	8.31	1.67
Diatomic	H_2	28.74	20.42	8.32	1.41
	N_2	29.07	20.76	8.31	1.40
	O_2	29.41	21.10	8.31	1.40
	CO	29.16	20.85	8.31	1.40
Polyatomic	CO_2	36.94	28.46	8.48	1.30
	SO_2	40.37	31.39	8.98	1.29
	H_2S	34.60	25.95	8.65	1.33

The last column of Table 18–1 lists the values of the ratio C_p/C_v, denoted by the Greek letter γ (gamma):

$$\gamma = \frac{C_p}{C_v}. \qquad (18-19)$$

The ratio of two heat capacities: a quantity that will come in handy later

The quantity γ plays an important role in adiabatic processes for an ideal gas, which we will study in the next section. We see that γ is 1.67 for monatomic gases and about 1.40 for diatomic gases. There is no simple regularity for polyatomic gases. Table 18–1 also shows that the molar heat capacity of a gas seems to be related to its molecular structure; all the *monatomic* gases have approximately equal values of C_v, and all the *diatomic* gases have about equal values. This is not an accident; in Chapter 20 we will undertake a more detailed theoretical study of heat capacities of gases.

Solids and liquids also expand with increasing temperature, if free to do so, and hence perform work. The coefficients of volume expansion of solids and liquids are, however, so much smaller than those of gases that the work is small. For solids and liquids, C_p is usually greater than C_v, but the difference is small and is not expressible as simply as for a gas. Because of the large stresses set up when solids or liquids are heated and *not* allowed to expand, most heating processes involving them take place at constant pressure. Hence C_p (rather than C_v) is the quantity usually measured for a solid or liquid.

The relation of heat capacities as seen on a pV-diagram

Finally, we emphasize once more that for an ideal gas, the internal energy change in *any* process is given by Eq. (18–14).

$$dU = nC_v\, dT,$$

whether the volume is constant or not. This relation holds for other substances *only* when the volume is constant.

This discussion may be represented graphically by using isotherms on a pV-diagram. Figure 18–6 shows two isotherms for an ideal gas, one representing possible states of the system at temperature T, the other at $T + dT$. Since the internal energy of an ideal gas depends only on the temperature, the internal energy is constant if the temperature is constant; the isotherms are also curves of *constant internal energy*. The internal energy therefore has a constant value U at every point on the isotherm at temperature T, and a constant value $U + dU$ at every point on the isotherm at $T + dT$. It follows that the *change* in internal energy, dU, is the same in all processes in which the gas

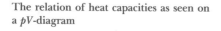

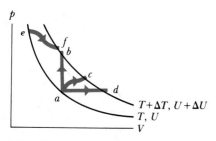

18–6 The change in internal energy of an ideal gas is the same in all processes between two given temperatures.

is taken from *any* point on one isotherm to *any* point on the other. The dU is the same for all the processes *ab*, *ac*, *ad*, and *ef* in Fig. 18–6. In particular, *ab* represents constant-volume heating from temperature T to $T + dT$, and *ad* represents constant-pressure heating through the same temperature interval. No work is done in process *ab*; the work in process *ad* is the area under the curve.

18–8 ADIABATIC PROCESSES FOR AN IDEAL GAS

An adiabatic process, defined in Section 18–5, is a process in which no heat transfer takes place between a system and its surroundings. A completely adiabatic process is an idealization, but a process may be approximately adiabatic if the system is well insulated or if the process takes place so rapidly that there is no time for appreciable heat flow to occur. The compression stroke in a gasoline or diesel engine is thus approximately adiabatic.

> In an adiabatic process, there is no heat transfer into or out of the system.

An **adiabatic process for an ideal gas** is shown on the *pV*-diagram of Fig. 18–7. As the gas expands from volume V_a to V_b, its temperature drops because of the decrease in internal energy. If the point representing the initial state lies on an isotherm at temperature $T + \Delta T$, then the point for the final state is on a different isotherm at a lower temperature T. Thus an adiabatic curve or *adiabat* at any point is always steeper than the isotherm passing through the same point. For an adiabatic *compression* from V_b to V_a, the situation is reversed and the temperature rises.

> Adiabatic compression causes a rise in temperature. How much?

The air in the output pipe of an air compressor, as used in gasoline stations and paint-spraying equipment, is always warmer than that entering the compressor, a result of the approximately adiabatic compression. When air is compressed in the cylinders of a diesel engine during the compression stroke, it becomes so hot that fuel injected into the cylinders during the power stroke ignites spontaneously.

It is useful to have quantitative relations between volume and temperature, or volume and pressure, for an adiabatic process. Consider first an infinitesimal change of state in which the temperature changes by dT and the volume by dV. Equation (18–14) gives the internal energy change for *any* process for an ideal gas, adiabatic or not, so we have $dU = nC_v \, dT$. Also, the work done by the gas during the process is given by $dW = p \, dV$. Then, since $dU = -dW$ for an adiabatic process, we have

$$nC_v \, dT = -p \, dV. \qquad (18\text{–}20)$$

To obtain a relation containing only the variables T and V, we may eliminate p by using the ideal-gas equation in the form $p = nRT/V$. Substituting this in Eq. (18–20) and rearranging, we obtain

$$nC_v \, dT = -\frac{nRT}{V} \, dV,$$

$$\frac{dT}{T} + \frac{R}{C_v} \frac{dV}{V} = 0,$$

or, since $R/C_v = (C_p - C_v)/C_v = C_p/C_v - 1 = \gamma - 1$,

$$\frac{dT}{T} + (\gamma - 1)\frac{dV}{V} = 0. \qquad (18\text{–}21)$$

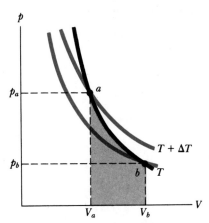

18–7 *pV*-diagram of an adiabatic process for an ideal gas. As the gas expands from V_a to V_b, its temperature drops from $T + \Delta T$ to T, corresponding to the decrease in internal energy due to the work done by the gas (indicated by shaded area). For an ideal gas, when an isotherm and an adiabat pass through the same point, the adiabat is always steeper.

This gives the relation between dT and dV for an infinitesimal adiabatic process. For a finite change, we integrate both sides of Eq. (18–21), obtaining

$$\ln T + (\gamma - 1) \ln V = \text{constant},$$
$$\ln T + \ln V^{\gamma-1} = \text{constant},$$
$$\ln [TV^{\gamma-1}] = \text{constant},$$

and finally

$$TV^{\gamma-1} = \text{constant}. \qquad (18\text{–}22)$$

Thus for an initial state (T_1, V_1) and a final state (T_2, V_2),

$$T_1 V_1^{\gamma-1} = T_2 V_2^{\gamma-1}. \qquad (18\text{–}23)$$

Temperature–volume relation for an adiabatic process

In applications of Eqs. (18–22) and (18–23) it is important to remember that the T's must be *absolute* temperatures (Kelvin or Rankine).

We can also convert Eq. (18–22) into a relation between pressure and volume by eliminating T, using the ideal-gas equation in the form $T = pV/nR$. We obtain

$$\left(\frac{pV}{nR}\right) V^{\gamma-1} = \text{constant},$$

or, since n and R are constants,

$$pV^{\gamma} = \text{constant}. \qquad (18\text{–}24)$$

Thus for an initial state (p_1, V_1) and a final state (p_2, V_2),

$$p_1 V_1^{\gamma} = p_2 V_2^{\gamma}. \qquad (18\text{–}25)$$

Why don't diesel engines need spark plugs? What ignites the fuel?

EXAMPLE 18–4   The compression ratio of a certain diesel engine is 15. This means that air in the cylinders is compressed to $\frac{1}{15}$ of its initial volume. If the initial pressure is 1.0×10^5 Pa and the initial temperature is 27°C (= 300 K), find the final pressure and temperature after compression. Air is mostly a mixture of oxygen and nitrogen, and $\gamma = 1.40$.

SOLUTION   From Eq. (18–23),

$$T_2 = T_1 \left(\frac{V_1}{V_2}\right)^{\gamma-1} = (300 \text{ K})(15)^{0.40} = 886 \text{ K} = 613°C.$$

From Eq. (18–25),

$$p_2 = p_1 \left(\frac{V_1}{V_2}\right)^{\gamma} = (1.0 \times 10^5 \text{ Pa})(15)^{1.40}$$
$$= 44.3 \times 10^5 \text{ Pa} \cong 44 \text{ atm}.$$

If the compression had been isothermal, the final pressure would have been 15 atm, but since the temperature also increases, the final pressure is much greater. The high temperature attained during compression causes the fuel to ignite spontaneously without the need for spark plugs, when it is injected into the cylinders at the end of the compression stroke.

How much work does a gas do during an adiabatic expansion?

Because $W = -dU$ for an adiabatic process, it is easy to calculate the work done by an ideal gas during an adiabatic process. If the initial and final temperatures are known, we have simply

$$W = nC_v(T_1 - T_2). \tag{18-26}$$

Or, if the pressure or volumes are known, we may use $pV = n RT$ to obtain

$$W = \frac{C_v}{R}(p_1V_1 - p_2V_2) = \frac{1}{\gamma - 1}(p_1V_1 - p_2V_2). \tag{18-27}$$

Note that if the process is an expansion, the temperature drops, T_1 is greater than T_2, p_1V_1 is greater than p_2V_2, and the work is *positive*, as we should expect. If it is a compression, the work is negative.

EXAMPLE 18–5 In Example 18–4, how much work does the gas do during the compression if the initial volume of the cylinder is $1.0 \text{ L} = 1.0 \times 10^{-3} \text{ m}^3$?

SOLUTION We may determine the number of moles and then use Eq. (18–26), or we may use Eq. (18–27). In the first case we have

Work in a diesel-engine cylinder

$$n = \frac{p_1V_1}{RT_1} = \frac{(1.0 \times 10^5 \text{ Pa})(1.0 \times 10^{-3} \text{ m}^3)}{(8.314 \text{ J·mol}^{-1}\text{·K}^{-1})(300 \text{ K})} = 0.040 \text{ mol}.$$

C_v for air is $20.8 \text{ J·mol}^{-1}\text{·K}^{-1}$; so

$$\begin{aligned} W &= nC_v(T_1 - T_2) \\ &= (0.040 \text{ mol})(20.8 \text{ J·mol·K}^{-1})(300 \text{ K} - 886 \text{ K}) \\ &= -488 \text{ J}. \end{aligned}$$

In the second case,

$$\begin{aligned} W &= \frac{1}{1.40 - 1}\Big[(1.0 \times 10^5 \text{ Pa})(1.0 \times 10^{-3} \text{ m}^3) \\ &\qquad\qquad - (44.3 \times 10^5 \text{ Pa})\Big(\frac{1.0}{15} \times 10^{-3} \text{ m}^3\Big)\Big] \\ &= -488 \text{ J}. \end{aligned}$$

The work is negative because the gas is compressed.

In the analysis above we have used the ideal-gas equation of state, which holds only when the state of the gas changes slowly enough so that at each step the pressure and temperature are *uniform* throughout the gas. Thus the validity of our results is limited to situations where the process is rapid enough to prevent appreciable heat exchange with the surroundings, yet slow enough so that the system does not depart very much from thermal and mechanical equilibrium.

SUMMARY

A thermodynamic system is a system that can exchange energy with its surroundings by heat transfer, through mechanical work, and in some cases by other mechanisms. When a system at pressure p expands from volume V_1 to V_2, it does an amount of work W given by

$$W = \int_{V_1}^{V_2} p \, dV. \tag{18-2}$$

KEY TERMS
thermodynamic system
free expansion
internal energy
first law of thermodynamics
adiabatic process
isochoric process

If the pressure p is constant during the expansion, the work W is given by

$$W = p(V_2 - V_1). \qquad (18\text{--}3)$$

In a thermodynamic process leading from an initial state to a final state, the heat added to the system and the work done by the system depend not only on the initial and final states but also on the path, that is, the series of intermediate states through which the system passes during the process.

In a thermodynamic process where heat Q is added to a system while it does work W, the internal energy U changes by an amount

$$\Delta U = U_2 - U_1 = Q - W. \qquad (18\text{--}6)$$

In an infinitesimal process,

$$dU = dQ - dW = dQ - p\,dV. \qquad (18\text{--}8)$$

The internal energy of any thermodynamic system depends only on its state, not on the processes by which it reached that state. The change in internal energy in any process depends only on the initial and final states, not on the path (the intermediate states through which the system passes). The internal energy of an isolated system is constant.

Four kinds of thermodynamic processes that occur are:

Adiabatic process: A process where there is no heat transfer in or out of a system. For any adiabatic process, $Q = 0$.

Isochoric process: A process that takes place at constant volume. For any isochoric process, $W = 0$.

Isobaric process: A process that takes place at constant pressure. For any isobaric process, $W = p(V_2 - V_1)$.

Isothermal process: A process that takes place at constant temperature.

The internal energy of an ideal gas depends only on its temperature, not on its pressure or volume. For other substances, the internal energy in general depends on both pressure and temperature.

The molar heat capacities of an ideal gas at constant volume (C_v) and at constant pressure (C_p) are related by

$$C_p = C_v + R, \qquad (18\text{--}18)$$

where R is the ideal gas constant. The ratio of C_p to C_v is denoted by γ:

$$\gamma = \frac{C_p}{C_v}. \qquad (18\text{--}19)$$

For an adiabatic process for an ideal gas, the quantities $TV^{\gamma-1}$ and pV^{γ} are constant. For an initial state (p_1, V_1, T_1) and a final state (p_2, V_2, T_2), the corresponding relations are

$$T_1 V_1^{\gamma-1} = T_2 V_2^{\gamma-1} \qquad (18\text{--}23)$$

and

$$p_1 V_1^{\gamma} = p_2 V_2^{\gamma}. \qquad (18\text{--}25)$$

QUESTIONS

18–1 It is not correct to say that a body contains a certain amount of heat, yet a body can transfer heat to another body. How can a body give away something it does not have in the first place?

18–2 Discuss the application of the first law of thermodynamics to a mountaineer who eats food, gets warm and sweats a lot during the climb, and does a lot of mechanical work in raising himself to the summit. What about the descent? One also gets warm during the descent. Is the source of this energy the same as during the ascent?

18–3 How can you cool a room (i.e., take heat out of it) by adding energy to it in the form of electric energy supplied to an air conditioner?

18–4 If you are told the initial and final states of a system and the associated change in internal energy, can you determine whether the internal energy change was due to work or to heat transfer?

18–5 Household refrigerators always have arrays or coils of tubing on the outside, usually at the back or bottom. When the refrigerator is running, the tubing becomes quite hot. Where does the heat come from?

18–6 There are a few materials that contract when heated, such as water between 0°C and 4°C. Would you expect C_p for such a material to be greater or less than C_v?

18–7 When ice melts (decreasing its volume), is the internal energy change greater or less than the heat added?

18–8 When one drives a car in cool, foggy weather, ice sometimes forms in the throat of the carburetor, even though the outside air temperature is above freezing. Why?

18–9 On a warm summer day a large cylinder of compressed gas (propane or butane) was used to supply several large gas burners at a cookout. After a while, frost formed on the outside of the tank. Why?

18–10 Air escaping from an air hose at a gas station always feels cold. Why?

18–11 The prevailing winds blow across the central valley of California and up the western slopes of the Sierra Nevada. They cool as they reach the slopes, and the precipitation there is much greater than in the valley. But what makes them cool?

18–12 Applying the same considerations as in Question 18–11, can you explain why Death Valley, on the opposite side of the Sierra from the central valley, is so hot and dry?

18–13 In the situation of Question 18–11, during certain seasons the wind blows in the opposite direction. Although the mountains are cool, the wind in the valley (called the "Santa Ana," after the notorious Mexican general) is always very hot. What heats it? A similar phenomenon in the Alps is called the "Foehn"; by local legend it is blamed for irrational behavior in man and beast, and has even been used as a defense in murder trials.

18–14 When a gas expands adiabatically, it does work on its surroundings. But if there is no heat input to the gas, where does the energy come from?

18–15 In a constant-volume process, $dU = nC_v \, dT$, but in a constant-pressure process it is *not* true that $dU = nC_p \, dT$. Why not?

18–16 Since C_v is defined with specific reference to a constant-volume process, how can it be correct that for an ideal gas $dU = nC_v \, dT$ even when the volume is not constant?

EXERCISES

Section 18–2 Work Done during Volume Changes

18–1 Two moles of oxygen are in a container with rigid walls. The gas is heated until the pressure doubles. Neglect the thermal expansion of the container. Calculate the work done by the gas.

18–2 A gas under a constant pressure of 3×10^5 Pa and an initial volume of 0.05 m³ is cooled until its volume becomes 0.04 m³. Calculate the work done by the gas.

18–3 Three moles of an ideal gas are heated at constant pressure from $T = 27°C$ to $127°C$. Calculate the work done by the gas.

18–4 One mole of an ideal gas has an initial temperature of 27°C. While the temperature is kept constant, the volume is decreased until the pressure doubles. Calculate the work done by the gas.

Section 18–4 Internal Energy and the First Law

18–5 A student performs a combustion experiment by burning a mixture of fuel and oxygen in a constant-volume "bomb" surrounded by a water bath. During the experiment the temperature of the water is observed to rise. Regarding the mixture of fuel and oxygen as the system,

a) has heat been transferred?

b) has work been done?

c) what is the sign of ΔU?

18–6 A liquid is irregularly stirred in a well-insulated container and thereby undergoes a rise in temperature. Regarding the liquid as the system,

a) has heat been transferred?

b) has work been done?

c) what is the sign of ΔU?

18–7 A gas in a cylinder expands from a volume of 0.4 m³ to 0.6 m³. Heat is added just rapidly enough to keep the pressure constant at 2.0×10^5 Pa during the expansion. The total heat added is 1.2×10^5 J.

a) Find the work done by the gas.

b) Find the change in internal energy of the gas.

c) Does it matter whether or not the gas is ideal?

18–8 A gas in a cylinder is cooled and compressed at a constant pressure of 2.0×10^5 Pa, from 1.2 m³ to 0.8 m³. A quantity of heat of magnitude 2.4×10^5 J is removed from the gas.

a) Find the work done by the gas.

b) Find the change in internal energy of the gas.

c) Does it matter whether or not the gas is ideal?

18–9 When a system is taken from state a to state b, in Fig. 18–8, along the path acb, 80 J of heat flow into the system, and 30 J of work are done.

a) How much heat flows into the system along path adb if the work is 10 J?

b) When the system is returned from b to a along the curved path, the magnitude of the work is 20 J. Does the system absorb or liberate heat, and how much?

c) If $U_a = 0$ and $U_d = 40$ J, find the heat absorbed in the processes ad and db.

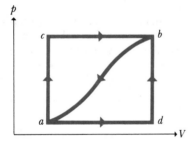

FIGURE 18–8

18–10 When water is boiled under a pressure of 2 atm, the heat of vaporization is 2.20×10^6 J·kg⁻¹ and the boiling point is 120°C. At this pressure, 1 kg of water has a volume of 10^{-3} m³, and 1 kg of steam a volume of 0.824 m³.

a) Compute the work done when 1 kg of steam is formed at this temperature.

b) Compute the increase in internal energy.

Section 18–6 Internal Energy of an Ideal Gas

Section 18–7 Heat Capacities of an Ideal Gas

18–11 Consider the isothermal compression of 0.10 mol of an ideal gas at $T = 0$°C. The initial pressure is 1 atm and the final volume is $\frac{1}{8}$ the initial volume.

a) Determine the work required.

b) What is the change in internal energy?

c) Does the gas exchange heat with its surroundings? If so, how much and in what direction?

18–12 A cylinder contains 1 mol of oxygen gas at a temperature of 27°C. The cylinder is provided with a frictionless piston, which maintains a constant pressure of 1 atm on the gas. The gas is heated until its temperature increases to 127°C.

a) Draw a diagram representing the process in the pV-plane.

b) How much work is done by the gas in this process?

c) On what is this work done?

d) What is the change in internal energy of the gas?

e) How much heat was supplied to the gas?

f) How much work would have been done if the pressure had been 0.5 atm?

18–13 A certain ideal gas has $\gamma = 1.33$. Determine the molar heat capacities at constant volume and at constant pressure.

Section 18–8 Adiabatic Processes for an Ideal Gas

18–14 An ideal gas initially at 10 atm and 300 K is permitted to expand adiabatically until its volume doubles. Find the final pressure and temperature if the gas is

a) monatomic;

b) diatomic.

18–15 A gasoline engine takes in air at 20°C and 1 atm, and compresses it adiabatically to $\frac{1}{10}$ the original volume. Find the final temperature and pressure.

18–16 An engineer is designing an engine that runs on compressed air. Air enters the engine at a pressure of 2×10^6 Pa and leaves at a pressure of 2×10^5 Pa. What must be the temperature of the compressed air so that there is no possibility of frost forming in the exhaust ports of the engine? Assume the expansion to be adiabatic. (*Note.* Frost frequently forms in the exhaust ports of an air-driven engine. This happens when the moist air is cooled below 0°C by the expansion that takes place in the engine.)

18–17 During an adiabatic expansion, the temperature of 0.1 mol of oxygen drops from 30°C to 10°C.

a) How much work does the gas do?

b) How much heat is added?

PROBLEMS

18–18 Nitrogen gas in an expandable container is raised from 0°C to 50°C, with the pressure held constant at 4×10^5 Pa. The total heat added is 3.0×10^4 J.

a) Find the number of moles of gas.

b) Find the change in internal energy of the gas.

c) Find the work done by the gas.

d) How much heat would be needed to cause the same temperature change if the volume were constant?

18–19 In a certain process, 2.11×10^5 J of heat are supplied to a system and at the same time the system expands against a constant external pressure of 6.89×10^5 Pa. The internal energy of the system is the same at the beginning and end of the process. Find the increase in volume of the system.

18–20 A chemical engineer studying the properties of glycerin uses a steel cylinder of cross-sectional area 0.01 m² that contains 1×10^{-2} m³ of glycerin. The cylinder is equipped with a tightly fitting piston that supports a load of 3×10^4 N. The temperature of the system is increased from 20°C to 70°C. The coefficient of volume expansion of glycerin is given in Table 14–2. Neglect the expansion of the steel cylinder. Find

a) the increase in volume of the glycerin;

b) the mechanical work of the 3×10^4 N force;

c) the amount of heat added to the glycerin [c_p for glycerin is 243×10^3 J·kg⁻¹·(C°)⁻¹];

d) the change in internal energy of the glycerin.

18–21 A pump compressing air from atmospheric pressure into a very large tank at 4.13×10^5 Pa gauge pressure has a cylinder 0.254 m long.

a) At what position in the stroke will air begin to enter the tank? Assume the compression to be adiabatic. (You are being asked to calculate the distance the piston has moved in the cylinder.)

b) If the air is taken into the pump at 27°C, what is the temperature of the compressed air?

18–22 Initially at a temperature of 60°C, 0.283 m³ of air expands at a constant gauge pressure of 1.38×10^5 Pa to a volume of 1.42 m³, and then expands further adiabatically to a final volume of 2.27 m³ and a final gauge pressure of 2.07×10^4 Pa. Sketch the process in the pV-plane and compute the total work done by the air. (Take atmospheric pressure to be 1.01×10^5 Pa.)

18–23 An ideal gas expands slowly to twice its original volume, doing 500 J of work in the process. Find the heat added and the change in internal energy if the process is

a) isothermal;   b) adiabatic.

18–24 In a cylinder, 2.0 mol of an ideal monatomic gas initially at 1.0×10^6 Pa and 300 K expands until its volume doubles. Compute the work done if the expansion is

a) isothermal;   b) adiabatic;   c) isobaric.

d) Show each process on a pV-diagram.

e) In which case is the magnitude of the heat transfer greatest? Least?

f) In which case is the magnitude of the internal-energy change greatest? Least?

18–25 Two moles of helium are initially at a temperature of 27°C and occupy a volume of 0.020 m³. The helium first expands at constant pressure until the volume has doubled, and then adiabatically until the temperature returns to its initial value.

a) Draw a diagram of the process in the pV-plane.

b) What is the total heat supplied in the process?

c) What is the total change in the internal energy of the helium?

d) What is the total work done by the helium?

e) What is the final volume?

18–26 A cylinder with a piston contains 0.5 mol of oxygen at 2×10^5 Pa and 300 K. The gas first expands at constant pressure to twice its original volume; it is then compressed isothermally back to its original volume, and finally it is cooled at constant volume to its original pressure.

a) Show the series of processes on a pV-diagram.

b) Compute the temperature during the isothermal compression.

c) Compute the maximum pressure.

18–27 Use the conditions and processes of Problem 18–26 to compute

a) the work done, the heat added, and the internal-energy change during the initial expansion;

b) the work done, the heat added, and the internal-energy change during the final cooling;

c) the internal-energy change during the isothermal compression.

CHALLENGE PROBLEM

18–28 The van der Waals equation of state, an approximate representation of the behavior of gases at high pressure, is

$$\left(p + \frac{an^2}{V^2}\right)(V - nb) = nRT,$$

where a and b are constants having different values for different gases. (In the special case of $a = b = 0$, this is the ideal-gas equation.) Calculate the work done by a van der Waals gas in an isothermal expansion from V_1 to V_2.

19

THE SECOND LAW OF THERMODYNAMICS

THUS FAR OUR DISCUSSION OF THERMODYNAMICS HAS CENTERED ON THE first law, which expresses conservation of energy in thermodynamic processes. But there is a whole family of questions that the first law cannot answer, having to do with the *directions* of thermodynamic processes. A study of the inherently one-way nature of such processes as the flow of heat from hotter to colder regions and the irreversible conversion of work into heat by friction leads to the *second law of thermodynamics*. This law places fundamental limitations on the efficiency with which an engine can convert heat into useful mechanical work; it also places limitations on the minimum energy input for a refrigerator. Hence the second law is directly relevant for many extremely important practical problems. We can also use the second law to define a temperature scale that is independent of the properties of any specific material. Finally, we restate the second law in terms of the concept of *entropy*, a quantitative measure of the degree of disorder or randomness of a system.

19–1 DIRECTIONS OF THERMODYNAMIC PROCESSES

Some energy-conversion processes go more easily in one direction than in the other.

Many thermodynamic processes proceed naturally in one direction but not the opposite. For example, heat always flows from a hot body to a cooler body, never the reverse. If heat were to flow from cooler to hotter regions, the first law would still be obeyed, but such flow does not occur in nature. Or suppose all the air in a box were to rush to one side, leaving vacuum in the other side, the reverse of the free expansion described in Section 18–3. This does not occur in nature either, although the first law does not forbid it. It is easy to convert mechanical energy completely into heat; we do this every time we use a car's brakes to stop it. It is not easy to convert heat into mechanical energy, and no one has ever built a machine that converts heat *completely* into mechanical energy. The world is full of would-be inventors who would like to cool the air a little, extracting heat from it, and convert that heat to mechanical energy to propel an airplane. No one has ever succeeded with such a scheme.

424

What these examples have in common is a preferred *direction*. In each case a process proceeds spontaneously in one direction but not in the other. This fact suggests that there must be some physical law that determines what the preferred direction is for a given process. As we will see in the following sections, that law is the *second law of thermodynamics*.

Closely related to the concept of directionality is another concept, **reversibility.** We call a process *reversible* if it involves a system that is always very close to being in thermodynamic equilibrium within itself and with its surroundings. When this is the case, any change of state that takes place can be reversed (i.e., made to go the other way) by making only an infinitesimal change in the conditions of the system. For example, heat flow between two bodies whose temperatures differ only infinitesimally can be reversed by making only a very small change in one temperature or the other. A gas expanding slowly and adiabatically can be compressed slowly by an infinitesimal increase in pressure.

A reversible process can be made to change direction by an infinitesimal change in conditions.

Reversible processes are thus **equilibrium processes.** By contrast, the processes cited above—heat flow with finite temperature difference, free expansion of a gas, and conversion of work to heat by friction—are all **irreversible processes;** no small change in conditions could make any of them go the other way. They are also all *nonequilibrium* processes.

An irreversible process is a nonequilibrium process.

You may notice an apparent contradiction in this discussion. If a system is *really* in thermodynamic equilibrium, how can any change of state at all take place? How can heat flow into or out of a system if it is at uniform temperature throughout? How can it start to move in order to expand and do work against its surroundings if it is really in mechanical equilibrium?

The answer to all these questions is that an equilibrium process is an idealization; strictly speaking, such a process can never be precisely attained in the real world. But by carrying out a process *slowly* enough, we can make the temperature gradients and the pressure differences in the substance as small as we like, and thus come as close as we like to the goal of keeping the system in equilibrium states. The term *quasi-equilibrium process* is sometimes used for such a process. We will not use that term; instead, we suggest you keep this discussion in mind when you read the words *equilibrium process*.

Finally, we will find that there is a relation between the direction of a process and the *disorder* or *randomness* of the resulting state. For example, imagine a tedious sorting job, such as alphabetizing a thousand book titles written on file cards. Throw the alphabetized stack of cards into the air. Do they come down in alphabetical order? No; the tendency is for them to come down in a random or disordered state. In the free-expansion example, the air is more disordered after it has expanded into the entire box than when it was confined to one side, because the molecules are scattered over more space.

Things always tend to get more mixed up.

Similarly, macroscopic kinetic energy is energy associated with organized, coordinated motions of many molecules. Energy associated with heat is random, disordered molecular motion, as we will see in Chapter 20. Hence conversion of mechanical energy into heat involves an increase of randomness or disorder.

In the following section we will explore the implications of the second law of thermodynamics by considering two specific classes of devices, namely, *heat engines,* which are partly successful in converting heat into work, and *refrigerators,* which are partly successful in transporting heat from cooler to hotter bodies.

19–2 HEAT ENGINES

The essence of a technological society is its ability to use sources of energy other than muscle power. Sometimes, as with water power, mechanical energy is directly available. Most of our energy, however, comes from the burning of fossil fuels (coal, oil, and gas) and from nuclear reactions; both of these supply energy that is transferred as *heat*. Some heat can be used directly for heating buildings, for cooking, and for chemical and metallurgical processing; but to operate a machine or propel a vehicle, we need *mechanical* energy.

How do we turn heat into mechanical work?

Thus a problem of the utmost practical importance is how to take heat from a source and convert as much of it as possible into mechanical energy or work. This is exactly what happens in gasoline engines in automobiles, jet engines in airplanes, steam turbines in electric power plants, and many other systems. Closely related processes occur in the animal kingdom, where food energy is "burned" (i.e., carbohydrates combine with oxygen to yield water, carbon dioxide, and energy) and partly converted to mechanical energy as the animal's muscles do work on their surroundings.

Any device for transforming heat into work or mechanical energy is called a **heat engine.** Typically, a certain quantity of matter in the engine undergoes addition and subtraction of heat, expansion and compression, and sometimes change of phase. We call this matter the **working substance** of the engine. In internal-combustion engines the working substance is a mixture of air and burnt fuel; in a steam turbine it is water.

In a cyclic process, a substance goes through the same series of states over and over.

For the sake of simplicity, we will discuss an engine in which the working substance undergoes a **cyclic process,** that is, a sequence of processes that eventually leaves the substance in the same state in which it started. In a steam turbine the water actually is recycled and used over and over. Internal-combustion engines do not use the same air over and over, but we can still analyze them in terms of cyclic processes that approximate the actual operation.

All the heat engines mentioned absorb heat from a source at a relatively high temperature, perform some mechanical work and discard some heat at a lower temperature. As far as the engine is concerned, the discarded heat is wasted. In internal-combustion engines the waste heat is in the hot exhaust gases and the cooling system; in a steam turbine it is heat that must be taken from the used steam to condense and recycle it.

The total internal-energy change in any cyclic process is zero.

When a system is carried through a cyclic process, its initial and final internal energies are equal. From the first law, for any number of complete cycles,

$$U_2 - U_1 = 0 = Q - W,$$
$$Q = W.$$

That is, the *net* heat flowing into the engine in a cyclic process equals the net work done by the engine.

Heat reservoirs: constant-temperature environments where a system can gain or lose heat

In the analysis of heat engines, it is useful to think of two bodies with which the working substance of the engine can interact. One of these, called the *hot reservoir,* can give the working substance large amounts of heat without appreciably changing its own temperature. The other body, called the *cold reservoir,* can absorb large amounts of discarded heat from the engine, also without appreciable change in its temperature. Thus in a steam-turbine system the flames and hot gases in the boiler are the hot reservoir, and the cold water and air used to condense and cool the used steam are the cold reservoir.

Heat transferred from the hot and cold reservoirs will be denoted by Q_H and Q_C, respectively. Each Q is considered positive when heat is transferred *from* a reservoir *to* the working substance, negative when the reverse. Thus in a heat engine, Q_H is positive but Q_C is negative, representing heat *leaving* the working substance. In the following discussion this sign convention may sometimes seem awkward; the alternatives are not any better, however, and it is best to keep the sign convention we have already learned for quantity of heat added to a body.

We can represent the energy transformations in a heat engine by the *flow diagram* of Fig. 19–1. The engine itself is represented by the circle. The heat Q_H supplied to the engine by the hot reservoir is proportional to the cross section of the incoming "pipeline" at the top of the diagram. The cross section of the outgoing pipeline at the bottom is proportional to the magnitude $|Q_C|$ of the heat rejected in the exhaust. The branch line to the right represents the portion of the heat supplied that the engine converts to mechanical work, W.

As mentioned above, we are talking about engines that repeat the same cycle over and over. In this case Q_H and Q_C represent the quantities of heat absorbed and rejected by the engine *during one cycle*. The *net* heat absorbed per cycle is

$$Q = Q_H + Q_C = Q_H - |Q_C|, \qquad (19\text{–}1)$$

where Q_C is a negative number. The useful output of the engine is the net work W done by the working substance; from the first law,

$$W = Q = Q_H + Q_C. \qquad (19\text{–}2)$$

Ideally we would like to convert *all* the heat Q_H into work; in that case we would have $Q_H = W$ and $Q_C = 0$. Experience shows that this is impossible; some heat is always wasted, and Q_C is never zero. We define the **thermal efficiency** of an engine, denoted by e, as the quotient

$$e = \frac{W}{Q_H}. \qquad (19\text{–}3)$$

Thus e represents the fraction of Q_H that *is* converted to work.

To put it another way, e is what you get, divided by what you pay for. This value is always less than unity, an all-too-familiar experience! In terms of the flow diagram of Fig. 19–1, the most efficient engine is the one for which the branch pipeline representing the work output is as *large* as possible and the exhaust pipeline representing the heat thrown away is as small as possible.

Using Eq. (19–2) and keeping in mind the signs of Q_H and Q_C, we can write the following equivalent expressions:

$$\begin{aligned} e = \frac{W}{Q_H} &= \frac{Q_H + Q_C}{Q_H} \\ &= 1 + \frac{Q_C}{Q_H} \\ &= 1 - \left| \frac{Q_C}{Q_H} \right|. \qquad (19\text{–}4) \end{aligned}$$

Note that e is a quotient of two energy quantities and thus is a pure number, without units. Of course, W and Q_H must always be expressed in the same units.

Be careful about algebraic signs of heat quantities.

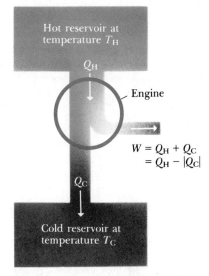

19–1 Schematic flow diagram of a heat engine.

Thermal efficiency: What fraction of the heat you pay for can you convert to useful work?

PROBLEM-SOLVING STRATEGY: *Heat engines*

We recommend you reread the strategy in Section 18–4; these suggestions are equally useful throughout the present chapter. The following points may need additional emphasis.

1. Be very careful with the sign conventions for W and the various Q's; W is positive when the system expands and does work, negative when it is compressed. Each Q is positive if it represents heat entering the working substance of the engine or other system, negative when heat leaves the system. When in doubt, use the first law if possible, to check consistency.

2. Some problems deal with power rather than energy quantities. Keep in mind that power is work per unit time ($P = W/t$,) and that rate of heat transfer (heat current) H is heat transfer per unit time ($H = Q/t$). Sometimes it helps to ask: "What is W or Q in one second (or one hour)?"

Thermal efficiency of a gasoline engine: How much heat from burning gasoline is wasted?

EXAMPLE 19–1 A gasoline engine takes in 2500 J of heat and delivers 500 J of mechanical work per cycle. The heat is obtained by burning gasoline having a heat of combustion of $L_C = 5.0 \times 10^4 \text{ J} \cdot \text{g}^{-1}$.

a) What is the thermal efficiency of this engine?

b) How much heat is discarded in each cycle?

c) How much gasoline is burned in each cycle?

d) If the engine goes through 100 cycles per second, what is its power output in watts? In horsepower?

e) How much gasoline is burned per second? per hour?

SOLUTION We have $Q_H = 2500$ J and $W = 500$ J.

a) From Eq. (19–3) the thermal efficiency is

$$e = \frac{W}{Q_H} = \frac{500 \text{ J}}{2500 \text{ J}} = 0.20 = 20\%.$$

b) From Eq. (19–2)

$$W = Q_H + Q_C,$$
$$500 \text{ J} = 2500 \text{ J} + Q_C, \qquad Q_C = -2000 \text{ J}.$$

That is, 2000 J of heat leave the engine during each cycle.

c) If m is the mass of gasoline burned, then

$$Q_H = mL_C,$$
$$2500 \text{ J} = m(5.0 \times 10^4 \text{ J} \cdot \text{g}^{-1}),$$
$$m = 0.05 \text{ g}.$$

d) The rate of doing work P is the work per cycle multiplied by the number of cycles per second:

$$P = (500 \text{ J})(100 \text{ s}^{-1}) = 50,000 \text{ W} = 50 \text{ kW}$$

$$= (50,000 \text{ W})\left(\frac{1 \text{ hp}}{746 \text{ W}}\right)$$

$$= 67 \text{ hp}.$$

e) The mass of gasoline burned per second is the mass per cycle multiplied by the number of cycles per second:

$$(0.05 \text{ g})(100 \text{ s}^{-1}) = 5 \text{ g·s}^{-1}.$$

The mass burned per hour is

$$(5 \text{ g·s}^{-1})\left(\frac{3600 \text{ s}}{1 \text{ hr}}\right) = 18,000 \text{ g·hr}^{-1} = 18 \text{ kg·hr}^{-1}.$$

The density of gasoline is about 0.70 g·cm^{-3}, so this is about $25,700 \text{ cm}^3$, 25.7 L, or 6.8 gal of gasoline per hour.

19–3 INTERNAL-COMBUSTION ENGINES

For our first example of a heat engine and a calculation of thermal efficiency, let us consider the common gasoline engine, as found in automobiles and many other types of machinery. The sequence of processes in the cycle is shown in Fig. 19–2. First a mixture of air and gasoline vapor flows into a cylinder through an open intake valve while the piston descends, increasing the volume of the cylinder from a minimum of V (when the piston is all the way up) to a maximum of rV (when it is all the way down). The ratio r is called the **compression ratio,** and for present-day automobile engines it is typically about 8. At the end of this *intake stroke* the intake valve closes, and the mixture is compressed to volume V during the *compression stroke*. The mixture is then ignited by the spark plug, and the heated gas expands back to volume rV, pushing on the piston and doing work; this is the *power stroke*. Finally, the exhaust valve opens and the combustion products are pushed out (during the *exhaust stroke*) to prepare the cylinder for the next intake stroke.

Figure 19–3 is a pV-diagram showing an idealized model of the corresponding thermodynamic processes. This model is called the **Otto cycle**. At point a the gasoline-air mixture has entered the cylinder. The mixture is com-

How does a gasoline engine work? The intake, compression, power, and exhaust strokes.

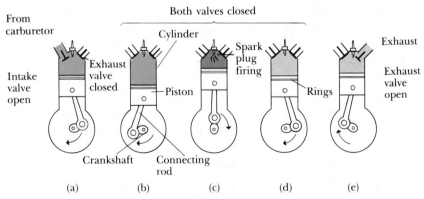

19–2 Cycle of a four-stroke internal-combustion engine. (a) Intake stroke: piston moves down, causing a partial vacuum in cylinder; gasoline and air are mixed in carburetor and flow through open intake valve into cylinder. (b) Compression stroke: intake valve closes and mixture is compressed as piston moves up. (c) Ignition: spark plug ignites mixture. (d) Power stroke: hot burned mixture pushes piston down, doing work. (e) Exhaust stroke: exhaust valve opens and piston moves up, pushing burned mixture out of cylinder. Engine is now ready for next intake stroke, and the cycle repeats.

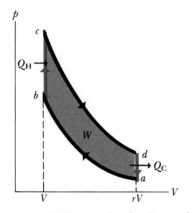

19–3 pV-diagram for the Otto cycle, an idealized model of the thermodynamic processes in a gasoline engine.

The thermal efficiency of an idealized gasoline engine depends on its compression ratio.

pressed adiabatically along line ab and is then ignited. Heat Q_H is added to the system by the burning gasoline (line bc), and the power stroke is the adiabatic expansion cd. The gas is cooled to the temperature of the outside air (da); during this process heat Q_C is exchanged. In practice, of course, this same air does not enter the engine again but, since an equivalent amount does enter, we may consider the process to be cyclic.

It is not hard to calculate the efficiency of this idealized cycle. Processes bc and da are constant-volume, so the heats Q_H and Q_C are related simply to the temperatures:

$$Q_H = nC_v(T_c - T_b),$$
$$Q_C = nC_v(T_a - T_d).$$

The thermal efficiency is given by Eq. (19–4); inserting the above expressions and canceling out the common factor nC_v, we obtain

$$e = \frac{T_c - T_b + T_a - T_d}{T_c - T_b}. \tag{19–5}$$

This equation may be further simplified by using the temperature–volume relation for adiabatic processes, Eq. (18–23). For the two adiabatic processes ab and cd, we find

$$T_a(rV)^{\gamma-1} = T_b V^{\gamma-1},$$
$$T_d(rV)^{\gamma-1} = T_c V^{\gamma-1}.$$

We divide out the common factor $V^{\gamma-1}$ and substitute the resulting expressions for T_b and T_c back into Eq. (19–5). The result is

$$e = \frac{T_d r^{\gamma-1} - T_a r^{\gamma-1} + T_a - T_d}{T_d r^{\gamma-1} - T_a r^{\gamma-1}} = \frac{(T_d - T_a)(r^{\gamma-1} - 1)}{(T_d - T_a)r^{\gamma-1}}.$$

Finally, dividing out the common factor $(T_d - T_a)$ yields the simple result

$$e = 1 - \frac{1}{r^{\gamma-1}}. \tag{19–6}$$

The thermal efficiency given by Eq. (19–6) is always less than unity, even for this idealized model. Using $r = 8$ and $\gamma = 1.4$ (the value for air), we find $e = 0.56$, or 56%. The efficiency can be increased by increasing r. However, this also increases the temperature at the end of the adiabatic compression of the air-fuel mixture. If the temperature is too high, the mixture explodes spontaneously and prematurely, instead of burning evenly after the spark plug ignites it. The maximum practical compression ratio for ordinary gasoline is about 10. Higher ratios can be used with more exotic fuels.

The cycle just described is of course a highly idealized model. It assumes that the mixture behaves as an ideal gas; it neglects friction, turbulence, loss of heat to cylinder walls, and many other effects that combine to reduce the efficiency of a real engine. Another source of inefficiency is incomplete combustion. A mixture of gasoline vapor with just enough air for complete combustion of the hydrocarbons to H_2O and CO_2 does not ignite readily. Reliable ignition requires a mixture "richer" in gasoline; the resulting incomplete combustion leads to CO and unburned hydrocarbons in the exhaust. The heat obtained from the gasoline is then less than the total heat of combustion; the difference is wasted, and the exhaust products contribute to air pollution. One attack on this problem is the stratified-charge engine, in which the concentration of gasoline vapor near the spark plug is greater than in the remainder of the combustion chamber. Efficiencies of real gasoline engines are typically around 20%.

The operation of the diesel engine is similar to that of the gasoline engine; the principal difference is that there is no fuel in the cylinder during compression. At the beginning of the power stroke, fuel is injected into the cylinder just rapidly enough to keep the pressure approximately constant during the first part of the power stroke. The fuel ignites spontaneously because of the high temperature developed during the adiabatic compression; no spark plugs are needed.

The idealized **Diesel cycle** is shown in Fig. 19–4. Starting at point a, air is compressed adiabatically to point b, heated at constant pressure to point c, expanded adiabatically to point d, and cooled at constant volume to point a.

Since there is no fuel in the cylinder of a Diesel engine during the compression stroke, preignition cannot occur, and the compression ratio r may be much higher than for a gasoline engine. Values of 15 to 20 are typical; with these values and $\gamma = 1.4$, the efficiency of the idealized Diesel cycle is about 0.65 to 0.70. As with the Otto cycle, the efficiency of any actual engine is substantially less than this. Although usually more efficient, Diesel engines are heavier (per unit power output) and often harder to start than gasoline engines. They need no carburetor or ignition system, but the fuel-injection system requires expensive high-precision machining.

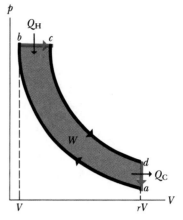

19–4 A pV-diagram of the Diesel cycle.

Why do diesel engines have higher compression ratios than gasoline engines?

19–4 REFRIGERATORS

We can think of a **refrigerator** as a heat engine operating in reverse. A heat engine takes heat from a hot place and gives off heat to a colder place. A refrigerator does the opposite; it takes heat from a cold place (the inside of the refrigerator) and gives it off to a warmer place (usually the air in the room where the refrigerator is located). A heat engine has a net *output* of mechanical work; the refrigerator requires a net *input* of mechanical work. Thus, with the symbols used in Section 19–1, Q_C is positive for a refrigerator, but both W and Q_H are negative.

A flow diagram for a refrigerator is shown in Fig. 19–5. From the first law for a cyclic process,

$$Q_H + Q_C - W = 0, \quad \text{or} \quad -Q_H = Q_C - W,$$

or, since both Q_H and W are negative,

$$|Q_H| = Q_C + |W|. \tag{19–7}$$

Thus, as the diagram shows, the heat Q_H leaving the working substance of the engine and given to the hot reservoir is always *greater* than the heat Q_C taken from the cold reservoir.

From an economic point of view, the best refrigeration cycle is one that removes the greatest amount of heat Q_C from the refrigerator for the least expenditure of mechanical work, W. We therefore define the **performance coefficient** (rather than the efficiency) of a refrigerator as the ratio $K = -Q_C/W$. Since $W = Q_H + Q_C$,

$$\text{Performance coefficient} = K = -\frac{Q_C}{W} = -\frac{Q_C}{Q_H + Q_C}. \tag{19–8}$$

A refrigerator transfers heat from a cool place to a warmer place.

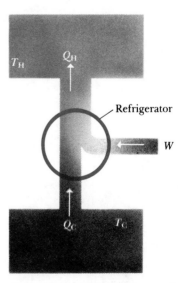

19–5 Schematic flow diagram of a refrigerator.

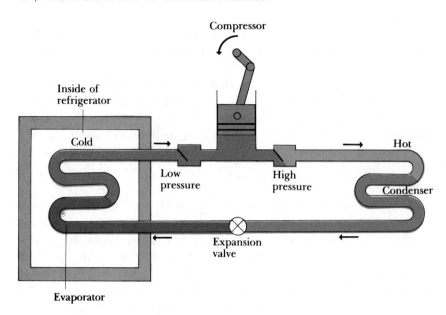

19–6 Principle of the mechanical refrigeration cycle.

The nuts and bolts of refrigerator operation

What's the difference between a refrigerator and an air conditioner?

As always, we assume that Q_H, Q_C, and W are all measured in the same energy units; K is then a dimensionless number.

The principles of the common refrigeration cycle are illustrated schematically in Fig. 19–6. The fluid "circuit" contains a refrigerant fluid (the working substance), typically CCl_2F_2 or another member of the "Freon" family. The left side is at low temperature and low pressure, the right side at high temperature and high pressure; ordinarily both sides contain liquid and vapor in phase equilibrium. The compressor takes in fluid, compresses it adiabatically, and delivers it to the condenser coil at high pressure. The fluid temperature is then higher than that of the air surrounding the condenser, so the refrigerant gives off heat (Q_H) and partially condenses to liquid. When the expansion valve opens, fluid expands adiabatically into the evaporator. As it does so, it cools considerably, enough so that the fluid in the evaporator coil is colder than its surroundings. It absorbs heat (Q_C) from its surroundings, cooling them and partially vaporizing. The fluid then enters the compressor to begin another cycle. The compressor, usually driven by an electric motor, requires energy input and does work $|W|$ on the working substance during each cycle.

An air conditioner operates on exactly the same principle. In this case the refrigerator box is a room or an entire building. The evaporator coils are inside, the condenser is outside, and fans are used to circulate air through these. In large installations the condenser coils are often cooled by water.

For air conditioners the quantity of greatest practical importance is the *rate* of heat removal (i.e., the heat current H) from the region being cooled. If heat Q_C is removed in time t, then $H = Q_C/t$. Similarly, the work W is usually expressed in terms of the *power* input $P = W/t$ to the compressor. The performance coefficient can then be expressed as

$$K = \frac{Q_C}{|W|} = \frac{Ht}{Pt} = \frac{H}{P}.$$

Typical room air conditioners used in homes have heat removal rates H of 5000 to 10,000 Btu·hr^{-1}, or about 1500 to 3000 W, and require electric power input of about 600 to 1500 W. Typical performance coefficients are roughly 2 to 3, with somewhat larger values for larger-capacity units.

Unfortunately, K is usually expressed commercially in mixed units, with H in Btu per hour and P in watts. In these units H/P is called the **energy efficiency rating** (EER); for room air conditioners it typically has a numerical value of 7 to 10. The units, customarily omitted, are $Btu \cdot hr^{-1} \cdot W^{-1}$.

A variation on this theme is the **heat pump,** used to heat buildings by cooling the outside air. It functions like a refrigerator turned inside out: The evaporator coils are outside, where they take heat from cold air, and the condenser coils are inside, where they give off heat to the warmer air. With proper design the heat Q_H delivered to the inside per cycle can be considerably greater than the work W required to get it there.

If *no* work were needed to operate a refrigerator, the performance coefficient (heat removed divided by work done) would be infinite. Performance coefficients of actual refrigerators are typically in the range from 2 to 6. Experience shows that some work is *always* needed to transfer heat from a colder to a hotter body. Heat flows spontaneously from a hotter to a colder body, and to reverse this flow requires the addition of work from the outside. It is impossible to make a refrigerator that transports heat from a cold body to a hotter body without the addition of work. Such a mythical device is called a *workless refrigerator.*

The heat pump: how to heat your house by cooling the outside. Are you getting something for nothing?

A refrigerator can't function without some mechanical work input.

19–5 THE SECOND LAW OF THERMODYNAMICS

We have discussed the fact that no one has ever been able to build a heat engine that converts heat completely to work, that is, an engine with 100% thermal efficiency. Experimental evidence suggests strongly that it is *impossible* to build such an engine. This impossibility forms the basis of one form of the **second law of thermodynamics,** as follows:

> It is impossible for any system to undergo a process in which it absorbs heat from a reservoir at a single temperature and converts it completely into mechanical work, while ending in the same state in which it began.

In other words, it is impossible *in principle* for any heat engine to have a thermal efficiency of 100%.

The basis of the second law of thermodynamics lies in the difference between the nature of internal energy and that of macroscopic mechanical energy. In a moving body the molecules have random motion, but superimposed on this is a coordinated motion of every molecule in the direction of the velocity of the body. The kinetic energy associated with this coordinated macroscopic motion is what we call the kinetic energy of the moving body. The kinetic and potential energies associated with the *random* motion constitute the internal energy.

When a moving body makes an inelastic collision or comes to rest as a result of friction, the organized part of the motion is converted to random motion. Since we cannot control the motions of individual molecules, we cannot convert this random motion completely back to organized motion. We can, however, convert *part* of it, and this is what a heat engine does.

If the second law were *not* true, it would be possible to power an automobile or to run a power plant by extracting heat from the surrounding air. Neither of these impossibilities violates the *first* law of thermodynamics. The second law, therefore, is not a deduction from the first but stands by itself as a separate law of nature, referring to an aspect of nature different from that

No engine can be 100% efficient. It's the (second) law!

The second law of thermodynamics limits energy-conversion processes.

19–7 Energy-flow diagrams for equivalent forms of the second law. (a) A workless refrigerator (left), if it existed, could be used in combination with an ordinary heat engine (right) to form a composite device that functions as an engine with 100% efficiency, converting heat Q_H completely to work. (b) An engine with 100% efficiency (left), if it existed, could be used in combination with an ordinary refrigerator (right) to form a workless refrigerator, transferring heat Q_C from the cold reservoir to the hot with no net input of work. Thus if either of these is impossible, the other must be also.

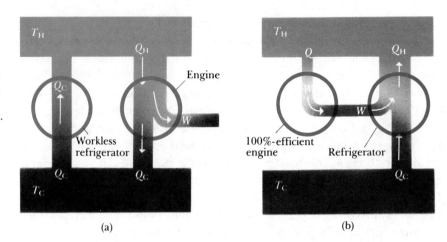

The "engine" and "refrigerator" statements of the second law are logically equivalent.

considered in the first law. The first law denies the possibility of creating or destroying energy; the second law limits the ways energy may be used and converted.

The analysis of refrigerators forms the basis for an alternative statement of the second law of thermodynamics. Heat flows spontaneously from hotter to colder bodies, never the reverse. A refrigerator does take heat from a colder to a hotter body, but its operation depends on input of mechanical energy or work. Generalizing this observation, we state:

> It is impossible for any process to have as its sole result the transfer of heat from a cooler to a hotter body.

This statement may not seem to be very closely related to the previous statement about heat engines. In fact, though, the two statements are completely equivalent. For example, if we could build a workless refrigerator, violating the second or "refrigerator" statement of the second law, we could use it in conjunction with a heat engine, pumping the heat rejected by the engine back to the hot reservoir to be reused. The composite machine (Fig. 19–7a) would violate the first or "engine" statement of the second law.

Alternatively, if we could make an engine with 100% thermal efficiency, in violation of the first statement, we could run it by using heat from the hot reservoir, and use the work output to drive a refrigerator that pumps heat from the cold reservoir to the hot (Fig. 19–7b). This would violate the "refrigerator" statement. Thus any device that violates one form of the second law can also be used to violate the other form. We conclude that if violations of the first form are impossible, so are violations of the second.

We have mentioned that heat flow across a finite temperature gradient and conversion of mechanical energy to heat, as in friction or turbulent fluid flow, are irreversible processes. Other examples can be cited. Gases always seep through an opening spontaneously from a region of high pressure to a region of low pressure; gases and liquids left by themselves always tend to mix, not to unmix. The second law of thermodynamics is an expression of the inherent one-way aspect of these and many other irreversible processes that take place naturally in only one direction. An irreversible process is always a *nonequilibrium* process. Irreversible heat flow accompanies departures from thermal equilibrium, free expansion of a gas involves states that are not in mechanical equilibrium, and so on. In each case the process tends to move the system toward an equilibrium state.

19–6 THE CARNOT CYCLE

According to the second law, no heat engine can have 100% efficiency. But what is the theoretical *maximum* possible efficiency of an engine, given two heat reservoirs at temperatures T_H and T_C? This question was answered in 1824 by the French engineer Sadi Carnot, who developed a hypothetical, idealized heat engine that has the maximum possible efficiency consistent with the second law. The cycle of this engine is called the **Carnot cycle.**

To understand the rationale of the Carnot cycle, we return to the matter of reversibility, discussed at the end of Section 19–5. Conversion of work to heat is an irreversible process; the purpose of a heat engine is a *partial* reversal of this process, the conversion of heat to work with as great efficiency as possible. For maximum heat-engine efficiency, therefore, we must *avoid* all irreversible processes.

Heat flow through a finite temperature drop is an irreversible process. Therefore, during heat transfer in the Carnot cycle there must be *no* finite temperature difference. When the engine takes heat from the hot reservoir at T_H, the engine itself must also be at T_H; otherwise irreversible heat flow would occur. Similarly, when heat is rejected to the cold reservoir at T_C, the engine itself must be at T_C. That is, every process that involves heat transfer must be *isothermal* at either T_H or T_C. Conversely, there must be *no* heat transfer between the engine and either reservoir in any process where the temperature of the engine changes; such heat transfer could not be reversible. In short, *every process* in our idealized cycle must be either *isothermal* or *adiabatic*. In addition, not only thermal but mechanical equilibrium must be maintained at all times, so that each process is completely reversible.

The Carnot cycle consists of two isothermal and two adiabatic processes. A Carnot cycle using an ideal gas as the working substance is shown on a pV-diagram in Fig. 19–8. It comprises the following steps:

1. The gas expands isothermally at temperature T_H, absorbing heat Q_H (*ab*).

2. It expands adiabatically until its temperature drops to T_C (*bc*).

3. It is compressed isothermally at T_C, rejecting heat Q_C (*cd*).

4. It is compressed adiabatically back to its initial state at temperature T_H (*da*).

When the working substance in a Carnot engine is an ideal gas, it is easy to calculate the thermal efficiency *e*. To carry out this calculation we will first find the ratio Q_C/Q_H of the quantities of heat transferred in the two isothermal processes, and then use Eq. (19–4) to find *e*. For an ideal gas the internal energy depends only on temperature and is thus constant in any isothermal process. For example, Q_H is equal to the work W_{ab} done by the gas during its isothermal expansion at temperature T_H. This work in turn is given by Eq. (18–4), and we find

$$Q_H = nRT_H \ln \frac{V_b}{V_a}. \tag{19–9}$$

The Carnot cycle: the best possible thermal efficiency for a heat engine

Complete reversibility in the Carnot cycle requires that every process be either isothermal or adiabatic.

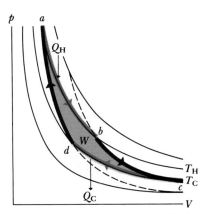

19–8 The Carnot cycle for an ideal gas. Color lines are isothermals; black lines are adiabatics.

Similarly,

$$Q_C = nRT_C \ln \frac{V_d}{V_c} = -nRT_C \ln \frac{V_c}{V_d}. \qquad (19\text{--}10)$$

This quantity is negative because V_d is less than V_c. The ratio of the two quantities of heat is thus

$$\frac{Q_C}{Q_H} = -\frac{T_C}{T_H} \frac{\ln (V_c/V_d)}{\ln (V_b/V_a)}. \qquad (19\text{--}11)$$

This equation can be simplified further by use of the temperature–volume relation for an adiabatic process, Eq. (18–23). We find for the two adiabatic processes:

$$T_H V_b{}^{\gamma-1} = T_C V_c{}^{\gamma-1}$$

and

$$T_H V_a{}^{\gamma-1} = T_C V_d{}^{\gamma-1}.$$

Dividing the first of these equations by the second, we find

$$\frac{V_b{}^{\gamma-1}}{V_a{}^{\gamma-1}} = \frac{V_c{}^{\gamma-1}}{V_d{}^{\gamma-1}} \quad \text{and} \quad \frac{V_b}{V_a} = \frac{V_c}{V_d}.$$

The thermal efficiency of a Carnot engine depends only on the temperatures of the two heat reservoirs.

Thus the two logarithms in Eq. (19–11) are equal, and that equation reduces to

$$\frac{Q_C}{Q_H} = -\frac{T_C}{T_H}. \qquad (19\text{--}12)$$

Then from Eq. (19–3) the efficiency of the Carnot engine is

$$e = 1 - \frac{T_C}{T_H} = \frac{T_H - T_C}{T_H}. \qquad (19\text{--}13)$$

This surprisingly simple result says that the efficiency of a Carnot engine depends only on the temperatures of the two heat reservoirs. The efficiency is large when the temperature *difference* is large, and it is very small when the temperatures are nearly equal. The efficiency can never be exactly unity, unless $T_C = 0$, and as we will see, this too is impossible.

EXAMPLE 19–2   A Carnot engine takes 2000 J of heat from a reservoir at 500 K, does some work, and discards some heat to a reservoir at 350 K. How much work does it do, how much heat is discarded, and what is the efficiency?

SOLUTION   From Eq. (19–12),

$$Q_C = -Q_H \frac{T_C}{T_H} = -(2000 \text{ J})\frac{350 \text{ K}}{500 \text{ K}} = -1400 \text{ J}.$$

Then from the first law,

$$W = Q_H + Q_C = 2000 \text{ J} + (-1400 \text{ J})$$
$$= 600 \text{ J}.$$

From Eq. (19–13) the efficiency is

$$e = 1 - \frac{350 \text{ K}}{500 \text{ K}} = 0.30 = 30\%.$$

Alternatively, from the basic definition of efficiency,

$$e = \frac{W}{Q_H} = \frac{600 \text{ J}}{2000 \text{ J}} = 0.30 = 30\%.$$

EXAMPLE 19–3 Suppose 0.2 mol of an ideal diatomic gas ($\gamma = 1.40$) undergoes a Carnot cycle with temperatures $T_H = 400$ K and $T_C = 300$ K. The initial pressure is $p_a = 10.0 \times 10^5$ Pa, and during the isothermal expansion at temperature T_H the volume doubles.

A detailed example of a specific Carnot cycle using an ideal gas

a) Find the pressure and volume at each of points a, b, c, and d in Fig. 19–8.

b) Find Q, W, and ΔU for each step in the cycle and for the entire cycle.

c) Determine the efficiency directly from the results of (b) and compare with the result from Eq. (19–13).

SOLUTION

a) First,

$$V_a = \frac{nRT_H}{p_a} = \frac{(0.2 \text{ mol})(8.314 \text{ J·mol}^{-1}\text{·K}^{-1})(400 \text{ K})}{10.0 \times 10^5 \text{ Pa}}$$

$$= 6.65 \times 10^{-4} \text{ m}^3,$$

$$V_b = 2V_a = 2(6.65 \times 10^{-4} \text{ m}^3) = 13.3 \times 10^{-4} \text{ m}^3.$$

For the isothermal expansion $a \rightarrow b$, $p_a V_a = p_b V_b$, so

$$p_b = \frac{p_a V_a}{V_b} = 5.0 \times 10^5 \text{ Pa}.$$

For the adiabatic expansion $b \rightarrow c$, $T_H V_b{}^{\gamma-1} = T_C V_c{}^{\gamma-1}$, so

$$V_c = V_b \left(\frac{T_H}{T_C}\right)^{1/(\gamma-1)} = (13.3 \times 10^{-4} \text{ m}^3)\left(\frac{4}{3}\right)^{2.5} = 27.3 \times 10^{-4} \text{ m}^3.$$

So

$$p_c = \frac{nRT_C}{V_c} = \frac{(0.2 \text{ mol})(8.314 \text{ J·mol}^{-1}\text{·K}^{-1})(300 \text{ K})}{27.3 \times 10^{-4} \text{ m}^3}$$

$$= 1.83 \times 10^5 \text{ Pa}.$$

For the adiabatic compression $d \rightarrow a$, $T_C V_d{}^{\gamma-1} = T_H V_a{}^{\gamma-1}$, and

$$V_d = V_a \left(\frac{T_H}{T_C}\right)^{1/\gamma-1} = (6.65 \times 10^{-4} \text{ m}^3)\left(\frac{4}{3}\right)^{2.5} = 13.65 \times 10^{-4} \text{ m}^3;$$

$$p_d = \frac{nRT_C}{V_d} = \frac{(0.2 \text{ mol})(8.314 \text{ J·mol}^{-1}\text{·K}^{-1})(300 \text{ K})}{13.65 \times 10^{-4} \text{ m}^3}$$

$$= 3.65 \times 10^5 \text{ Pa}.$$

b) For the isothermal expansion $a \rightarrow b$, $\Delta U = 0$ and, from Eq. (18–4),

$$W = Q_H = nRT_H \ln \frac{V_b}{V_a}$$

$$= (0.2 \text{ mol})(8.314 \text{ J·mol}^{-1}\text{·K}^{-1})(400 \text{ K}) \ln 2 = 461 \text{ J}.$$

For the adiabatic expansion $b \rightarrow c$, $Q = 0$ and, from Eq. (18–26),

$$W = -\Delta U = nC_v(T_H - T_C)$$

$$= (0.2 \text{ mol})(20.78 \text{ J·mol}^{-1}\text{·K}^{-1})(400 \text{ K} - 300 \text{ K}) = 416 \text{ J}.$$

For the isothermal compression $c \rightarrow d$, $\Delta U = 0$, and

$$W = Q_C = nRT_C \ln \frac{V_d}{V_c}$$

$$= (0.2 \text{ mol})(8.314 \text{ J·mol}^{-1}\text{·K}^{-1})(300 \text{ K}) \ln \frac{13.65 \times 10^{-4} \text{ m}^3}{27.3 \times 10^{-4} \text{ m}^3}$$

$$= -346 \text{ J}.$$

For the adiabatic compression $d \rightarrow a$, $Q = 0$; and

$$W = -\Delta U = nC_v(T_C - T_H)$$

$$= (0.2 \text{ mol})(20.78 \text{ J·mol}^{-1}\text{·K}^{-1})(300 \text{ K} - 400 \text{ K})$$

$$= -416 \text{ J}$$

Alternatively Eq. (18–27) may be used for either adiabatic process. For $d \rightarrow a$,

$$W = \frac{1}{\gamma - 1}(p_d V_d - p_a V_a)$$

$$= (2.5)[(3.65 \times 10^5 \text{ Pa})(13.65 \times 10^{-4} \text{ m}^3)$$
$$- (10.0 \times 10^5 \text{ Pa})(6.65 \times 10^{-4} \text{ m}^3)]$$

$$= -416 \text{ J}.$$

The results may be tabulated as follows:

Process	Q	W	ΔU
$a \rightarrow b$	461 J	461 J	0
$b \rightarrow c$	0	416 J	−416 J
$c \rightarrow d$	−346 J	−346 J	0
$d \rightarrow a$	0	−416 J	416 J
Total	115 J	115 J	0

Note that for the entire cycle, $Q = W$ and $\Delta U = 0$. Also note that the quantities of work in the two adiabatic processes are negatives of each other; it is easy to prove from the analysis leading to Eq. (19–13) that this must always be the case.

c)   From the table, the total work is 115 J and $Q_H = 461$ J. Thus

$$e = \frac{W}{Q_H} = \frac{115 \text{ J}}{461 \text{ J}} = 0.250.$$

From Eq. (19–13)

$$e = \frac{T_H - T_C}{T_H} = \frac{400 \text{ K} - 300 \text{ K}}{400 \text{ K}} = 0.250.$$

A Carnot engine run backward is a Carnot refrigerator.

Because each step in the Carnot cycle is reversible, the *entire cycle* may be reversed, thereby converting the engine into a refrigerator. The performance coefficient of the Carnot refrigerator is obtained by combining Eqs. (19–8) and (19–12), first rewriting Eq. (19–8) as

$$K = -\left(\frac{Q_H + Q_C}{Q_C}\right)^{-1}.$$

We invite you to fill in the details; the result is

$$K = \frac{T_C}{T_H - T_C}. \tag{19-14}$$

When the temperature difference is small, K is much larger than unity; in this case a lot of heat can be "pumped" from the lower to the higher temperature with only a little expenditure of work. But the greater the temperature difference, the smaller is K and the more work is required to transfer a given quantity of heat.

EXAMPLE 19–4   If the cycle described in Example 19–3 is run backward as a refrigerator, the performance coefficient is given by Eq. (19–8):

$$K = -\frac{Q_C}{W} = -\frac{(-346 \text{ J})}{115 \text{ J}} = 3.00.$$

Because the cycle is a Carnot cycle, we may also use Eq. (19–14):

$$K = \frac{T_C}{T_H - T_C} = \frac{300 \text{ K}}{400 \text{ K} - 300 \text{ K}} = 3.00.$$

For a Carnot cycle, e and K depend only on the temperatures, and it is not necessary to calculate Q and W. For cycles containing irreversible processes, however, more detailed calculations are necessary.

It is easy to prove that *no engine can be more efficient than a Carnot engine operating between the same two temperatures.* The key to the proof is the observation that because each step in the Carnot cycle is reversible, the *entire cycle* may be reversed. Run backward, the engine becomes a refrigerator. If any engine is more efficient than a Carnot engine, its work output may be used to drive a Carnot refrigerator and pump the rejected heat back to the reservoir, thus violating the engine statement of the second law. Hence the above statement is still another equivalent statement of the second law of thermodynamics. It also follows directly that *all Carnot engines operating between the same two temperatures have the same efficiency, irrespective of the nature of the working substance.* Thus, although Eqs. (19–13) and (19–14) were derived for a Carnot engine using an ideal gas as its working substance, they are in fact valid for *any* Carnot engine, no matter what its working substance.

No engine can be more efficient than a Carnot engine, for given temperatures.

For given temperatures, all Carnot engines have the same efficiency.

Equation (19–13) points the way to the conditions that a real engine, such as a steam turbine, must fulfill to approach as closely as possible its maximum attainable efficiency. These conditions are that the intake temperature T_H must be made as high as possible and the exhaust temperature T_C as low as possible.

The exhaust temperature cannot be lower than the lowest temperature available for cooling the exhaust. This is usually the temperature of the air, or perhaps of river water if available at the plant. The only recourse then is to raise the boiler temperature T_H. Because the vapor pressure of all liquids increases rapidly with increasing temperature, a limit is set by the mechanical strength of the boiler. At 500°C the vapor pressure of water is about 240 ×

10^5 Pa, 235 atm, or 3450 lb·in^{-2}, which is about the maximum practical pressure in large present-day steam boilers.

The unavoidable exhaust heat loss in electric-power plants creates a serious environmental problem. When a lake or river is used for cooling, the temperature of the body of water may be raised several degrees. Such a temperature change has a severely disruptive effect on the overall ecological balance, inasmuch as relatively small temperature changes can have significant effects on metabolic rates in plants and animals. Since **thermal pollution,** as this effect is called, is an inevitable consequence of the second law of thermodynamics, careful planning is essential to minimize the ecological impact of new power plants.

19–7 ENTROPY

The second law of thermodynamics, as stated above, is rather different in form from other familiar physical laws; it is not a quantitative relationship but rather a statement of *impossibility*. We can also express the second law in quantitative form, using the concept of **entropy,** the subject of this section.

In this chapter we have seen several examples of processes that proceed naturally in the direction of increasing disorder, although we have not yet given a quantitative definition of disorder. Irreversible heat flow increases disorder because initially the molecules are sorted into hotter and cooler regions; this sorting is lost when the system comes to thermal equilibrium. Adding heat to a body increases its disorder because it increases the randomness of molecular motion. Free expansion of a gas increases its disorder because the molecules have greater randomness of position after the expansion than before.

Entropy provides a quantitative measure of disorder. Consider first the entropy change of a substance undergoing a reversible addition of heat. We use the symbol S for the entropy of the system, and ΔS for the change in entropy during any process. Adding heat increases molecular motion and thus disorder, but the effect is greater if the substance is cold (with little molecular motion) at the beginning than if it is already hot (with a lot of molecular motion). A suitable definition of entropy change in such a process is

$$\Delta S = \frac{Q}{T} \qquad \text{(reversible isothermal process),} \qquad (19\text{–}15)$$

where Q is the heat added and T is the *absolute* temperature. This definition holds only for reversible isothermal equilibrium processes.

We can generalize the definition of entropy change to include *any* reversible process leading from one state to another, whether it is isothermal or not. We represent the process as a series of infinitesimal reversible steps. During a typical step an infinitesimal quantity of heat dQ is added to the system at absolute temperature T. Then we sum (integrate) the quotients dQ/T for the entire process; that is,

$$\Delta S = \int_1^2 \frac{dQ}{T} \qquad \text{(reversible process).} \qquad (19\text{–}16)$$

The limits 1 and 2 refer to the initial and final states.

Because entropy is a measure of the disorder of a system in any specific state, we expect it to depend only on the current state of the system, not on its

The waste heat from power plants can have a major impact on the ecology.

Entropy: how to measure mixed-up-ness

Ink mixing with water. There is greater disorder and more entropy after mixing than before; the water and ink mix spontaneously but will never unmix spontaneously. (Photo by Chip Clark.)

The entropy of a system depends only on its state.

past history. It can be proved, by use of the second law of thermodynamics, that when a system proceeds from an initial state with entropy S_1 to a final state with entropy S_2, the change in entropy $\Delta S = S_2 - S_1$ defined by Eq. (19–16) does not depend on the path leading from the initial to the final state but is the same for *all possible* processes leading from state 1 to state 2. Thus the entropy of a system must also have a definite value for any given state of the system. We recall that *internal energy,* introduced in Chapter 18, also has this property, although entropy and internal energy are very different quantities. The unit of entropy is $1\ \text{J·K}^{-1}$, $1\ \text{cal·K}^{-1}$, $1\ \text{Btu·(R°)}^{-1}$, and so on.

The fact that entropy is a function only of the state of a system enables us to compute entropy changes in nonequilibrium processes, where Eq. (19–16) is not applicable. We can simply invent a path connecting the given initial and final states that *does* consist entirely of reversible equilibrium processes and compute the total entropy change for that path. It is not the actual path, but the entropy change must be the same as for the actual path.

Entropy changes are independent of the path.

As with internal energy, this discussion does not define entropy itself but only the change in entropy in any given process. To complete the definition we may arbitrarily assign a value to the entropy of a system in a specified reference state and then calculate the entropy of any other state with reference to this value.

EXAMPLE 19–5   One kilogram of ice at 0°C is melted and converted to water at 0°C. Compute its change in entropy.

SOLUTION   The temperature is constant at 273 K, and

What is the entropy change of melting ice?

$$\Delta S = S_2 - S_1 = \frac{Q}{T}.$$

But Q is simply the total heat of fusion that must be supplied to melt the ice, or 334×10^3 J. Hence

$$S_2 - S_1 = \frac{334 \times 10^3\ \text{J}}{273\ \text{K}} = 1223\ \text{J·K}^{-1},$$

and the increase in entropy of the system is $1223\ \text{J·K}^{-1}$. In any *isothermal* reversible process, the entropy change equals the heat added divided by the absolute temperature.

EXAMPLE 19–6   One kilogram of water at 0°C is heated to 100°C. Compute its change in entropy.

SOLUTION   The temperature is not constant; to carry out the integral in Eq. (19–16) we replace dQ by $mc\ dT$, obtaining

$$\Delta S = S_2 - S_1 = \int_{T_1}^{T_2} mc\ \frac{dT}{T} = mc \ln \frac{T_2}{T_1}$$

$$= (1000\ \text{g})(4.19\ \text{J·g}^{-1}\text{·K}^{-1}) \ln \frac{373\ \text{K}}{273\ \text{K}}$$

$$= 1308\ \text{J·K}^{-1}.$$

EXAMPLE 19–7   A gas expands adiabatically and reversibly. What is its change in entropy?

SOLUTION In an adiabatic process, no heat enters or leaves the system. Hence $Q = 0$ and there is *no* change in entropy. Every *reversible* adiabatic process is a constant-entropy process.

EXAMPLE 19–8 A thermally insulated box is divided by a partition into two compartments, each having volume V. Initially one compartment contains n moles of an ideal gas at temperature T, and the other is evacuated. We then break the partition, and the gas expands to fill both compartments. What is its entropy change?

When a gas undergoes a free expansion, its entropy increases.

SOLUTION For this process $Q = 0$, $W = 0$, $\Delta U = 0$, and therefore (since it is an ideal gas) $\Delta T = 0$. We might think that the entropy change is zero because there is no heat exchange. But Eq. (19–16) is valid only for *reversible* processes; this free expansion is *not* reversible, and there *is* an entropy change. To calculate ΔS we can use the fact that the entropy change depends only on the initial and final states. We can devise a *reversible* process having the same endpoints, use Eq. (19–16) to calculate its entropy change, and thus determine the entropy change in the original process. The appropriate reversible process in this case is an isothermal expansion from V to $2V$ at temperature T. The gas does work during this expansion, so heat must be supplied in order to keep the internal energy constant. The total heat equals the total work, which is given by Eq. (18–4):

$$W = Q = nRT \ln 2.$$

Thus the entropy change is

$$\Delta S = \frac{Q}{T} = nR \ln 2,$$

which is also the entropy change for the free expansion. For one mole,

$$\Delta S = (1 \text{ mol})(8.314 \text{ J} \cdot \text{mol}^{-1} \cdot \text{K}^{-1})(0.693) = 5.76 \text{ J} \cdot \text{K}^{-1}.$$

EXAMPLE 19–9 For the Carnot engine in Example 19–2 (Section 19–6), find the total entropy change in the engine during one cycle.

Entropy changes in a Carnot engine

SOLUTION During the isothermal expansion at 500 K the engine takes in 2000 J, and its entropy change is

$$\Delta S = \frac{Q}{T} = \frac{2000 \text{ J}}{500 \text{ K}} = 4.0 \text{ J} \cdot \text{K}^{-1}.$$

During the isothermal compression at 350 K the engine gives off 1400 J of heat, and its entropy change is

$$\Delta S = \frac{-1400 \text{ J}}{350 \text{ K}} = -4.0 \text{ J} \cdot \text{K}^{-1}.$$

Thus the total entropy change is $4.0 \text{ J} \cdot \text{K}^{-1} - 4.0 \text{ J} \cdot \text{K}^{-1} = 0$. This is to be expected, of course, since the final state is the same as the initial state. The total entropy change of the two heat reservoirs is also zero. This cycle contains no irreversible processes, and the total entropy change is zero.

There is no such thing as conservation of entropy.

Unlike energy, entropy is *not* a conserved quantity. In fact, the reverse is true; the entropy of an isolated system *can* change but, as we will see, it can never decrease. An entropy increase occurs in every natural process, if all systems taking part in the process are included. In an idealized, completely reversible process involving only equilibrium states, no entropy change occurs, but all natural (i.e., irreversible) processes take place with an increase in entropy.

EXAMPLE 19–10 Suppose 1 kg of water at 100°C is placed in thermal contact with 1 kg of water at 0°C. What is the total change in entropy?

SOLUTION This process involves irreversible heat flow. We assume the specific heat capacity of water is constant in this temperature range. Then the first 4186 J of heat transferred cool the hot water to 99°C and warm the cold water from 0°C to 1°C. The net change of entropy is approximately

$$\Delta S = -\frac{4190 \text{ J}}{373 \text{ K}} + \frac{4190 \text{ J}}{273 \text{ K}} = 4.1 \text{ J} \cdot \text{K}^{-1}.$$

Mixing hot and cold water increases the entropy.

Further increases in entropy occur as the system approaches thermal equilibrium at 50°C. The *total* increase in entropy can be calculated by the same method as in Example 19–6. The entropy change of the hot water is

$$\Delta S = (1 \text{ kg})(4190 \text{ J} \cdot \text{kg}^{-1} \cdot \text{K}^{-1}) \int_{373 \text{ K}}^{323 \text{ K}} \frac{dT}{T}$$

$$= (4190 \text{ J} \cdot \text{K}^{-1}) \ln \frac{323 \text{ K}}{373 \text{ K}} = -602 \text{ J} \cdot \text{K}^{-1};$$

the entropy change of the cold water is

$$\Delta S = (4190 \text{ J} \cdot \text{K}^{-1}) \ln \frac{323 \text{ K}}{273 \text{ K}} = +704 \text{ J} \cdot \text{K}^{-1};$$

and the *total* entropy change of the system is

$$\Delta S = +704 \text{ J} \cdot \text{K}^{-1} - 602 \text{ J} \cdot \text{K}^{-1} = +102 \text{ J} \cdot \text{K}^{-1}.$$

Thus an irreversible heat flow in an isolated system is accompanied by an increase in entropy. The same end state could have been achieved by simply mixing the two quantities of water. This too is an irreversible process, and because entropy depends only on the state of the system, the total entropy change would be the same.

These examples of the mixing of substances at different temperatures, or the flow of heat from a higher to a lower temperature, are characteristic of *all* natural (i.e., irreversible) processes. When all the entropy changes in the process are included, the increases in entropy are always greater than the decreases. In the special case of a *reversible* process, the increases and decreases are equal. Hence we can state the general principle that *when all systems taking part in a process are included, the entropy either remains constant or increases.* In other words, *no process is possible in which the total entropy decreases,* when all systems taking part in the process are included. This is an alternative statement of the second law of thermodynamics, in terms of entropy. This statement is equivalent to the "engine" and "refrigerator" statements discussed earlier.

The increase of entropy that accompanies every natural (irreversible) process measures the increase of disorder or randomness in the universe in

Entropy always increases in an irreversible process.

that process. Consider again the example of mixing hot and cold water. We *might* have used the hot and cold water as the high- and low-temperature reservoirs of a heat engine. In the course of removing heat from the hot water and giving heat to the cold water, we could have obtained some mechanical work. But once the hot and cold water have been mixed and have come to a uniform temperature, this opportunity of converting heat to mechanical work is lost, and it is lost irretrievably. The lukewarm water will never *unmix* itself and separate into hotter and colder portions. No decrease in *energy* occurs when the hot and cold water are mixed; what has been lost in the mixing process is not *energy*, but *opportunity*—the opportunity to convert part of the heat from the hot water into mechanical work. Hence when entropy increases, energy becomes less *available*, and the universe becomes more random or "run down."

19–8 THE KELVIN TEMPERATURE SCALE

The Kelvin temperature scale does not depend on the properties of any particular material; it is truly absolute.

The Carnot cycle can be used to define a temperature scale that is completely independent of the properties of any particular material. As we have seen, the efficiency of a Carnot engine operating between reservoirs at two given temperatures is independent of the nature of the working substance and is a function only of the temperatures. If we consider a number of Carnot engines using different working substances and absorbing and rejecting heat to the same two reservoirs, the thermal efficiency is the same for all:

$$e = \frac{Q_H + Q_C}{Q_H} = 1 + \frac{Q_C}{Q_H}$$
$$= \text{constant.}$$

Hence the ratio Q_H/Q_C is the same for all Carnot engines operating between two given temperatures T_H and T_C. Kelvin proposed that the ratio of the reservoir temperatures be *defined* as equal to this constant ratio of the magnitudes of the quantities of heat absorbed and rejected or, since Q_C is a negative quantity, as equal to the negative of the ratio Q_H/Q_C. Thus

$$\frac{T_H}{T_C} = \frac{|Q_H|}{|Q_C|} = -\frac{Q_H}{Q_C}. \tag{19–17}$$

Equation (19–17) appears identical to Eq. (19–12), but there is a subtle and crucial difference. The temperatures in Eq. (19–12) are based on an ideal-gas thermometer, as defined in Sections 14–2 and 14–4, while Eq. (19–17) defines a temperature scale, based on the Carnot cycle and the second law of thermodynamics, that is independent of the behavior of any particular substance. Thus the **Kelvin temperature scale** is truly *absolute*. To complete the definition of the Kelvin scale we proceed, as in Chapter 14, to assign the arbitrary value of 273.16 K to the temperature of the triple point of water. When a substance is taken around a Carnot cycle, the ratio of the heats absorbed and rejected, $|Q_H|/|Q_C|$, is equal to the ratio of the temperatures of the reservoirs *as expressed on the gas scale*, defined in Chapter 14. Since, in both scales, the triple point of water is chosen to be 273.16 K, it follows that *the Kelvin and the ideal gas scales are identical.*

Absolute zero is as cold as you can get.

The zero point on the Kelvin scale is called **absolute zero.** There are theoretical reasons for believing that absolute zero cannot be attained experimentally, although temperatures as low as 10^{-6} K have been achieved. The more closely we approach absolute zero, the more difficult it is to get closer. Absolute zero can also be interpreted on a molecular level, although this must be done with some care. Because of quantum effects, it is *not* correct to say that at $T = 0$ all molecular motion ceases. Rather, at absolute zero the system has its *minimum* possible total energy (kinetic plus potential), although this minimum amount is in general not zero.

19–9 ENERGY CONVERSION

The laws of thermodynamics place very general limitations on conversion of energy from one form to another. In this day of increasing energy demand and diminishing resources, these matters are of the utmost practical importance. We conclude this chapter with a brief discussion of a few energy-conversion systems, present and proposed.

Over half the electric power generated in the United States is obtained from coal-fired steam-turbine generating plants. Modern boilers can transfer about 80% to 90% of the heat of combustion of coal into steam. The theoretical thermal efficiency of the turbine, given by Eq. (19–13), is usually limited to about 0.55, and the actual efficiency is typically 90% of the theoretical value, or about 0.50. The efficiency of large electrical generators in converting mechanical power to electrical is very large, typically 99%. Thus the overall thermal efficiency of such a plant is roughly (0.85)(0.50)(0.99), or about 40%.

In 1970 a generator with a capacity of about one gigawatt ($=1000$ MW $=10^9$ W) went into operation at the Tennessee Valley Authority's Paradise power plant. The steam is heated to 540°C (1003°F) at a pressure of 248 atm (2.51×10^7 Pa or 3650 lb·in^{-2}). The plant consumes 10,500 tons of coal per day, and the overall thermal efficiency is 39.3%.

Nuclear-power plants have the same theoretical efficiency limit as coal-fired plants. Because at present is is not practical to run nuclear reactors at as

Efficiencies of modern electric-power plants

Interior of the Paradise power plant, showing turbine housings and steam ducts. For scale, note the men in the right foreground.

Exterior of the Paradise power plant. The cooling towers on the left cool and condense the exhaust steam from the turbines so it can be recycled into the boilers. Waste heat is given off to the air flowing through the towers. (Photographs courtesy of Tennessee Valley Authority.)

(a)

(b)

19–9 (a) A solar-energy installation in the Mojave Desert, near Barstow, California. An array of mirrors concentrates the sun's energy on a boiler (central tower) to generate steam for turbines. The mirrors move continuously, controlled by a computer, to track the sun's motion. (b) An array of photovoltaic cells for direct conversion of sun power to electric power. Such an array supplies the electrical needs of the Headquarters and Visitors Center of Natural Bridges National Monument, Utah. (Photos by Dan McCoy/Rainbow.)

high temperatures and pressures as coal boilers, the theoretical thermal efficiency is usually lower. The overall thermal efficiency of a nuclear plant is typically 30%.

In both coal-fired and nuclear plants, the energy not converted to electrical energy is wasted and must be disposed of. A common practice is to locate such a plant near a lake or river and use the water for disposal of excess heat. This can raise the water temperature several degrees, often with serious ecological consequences.

Solar energy is an inviting possibility. The power in the sun's radiation (before it passes through the earth's atmosphere) is about 1.4 kW per square meter. A maximum of about 1.0 kW·m^{-2} reaches the surface of the earth on a clear day, and the time average, over a 24-hour period, is about 0.2 kW·m^{-2}. This radiation can be collected and focused with mirrors and used to generate steam for a heat engine, as in Fig. 19–9a. A different scheme is to use large banks of photocells for direct conversion of solar energy to electricity. Such a process is not a heat engine in the usual sense and is not limited by the Carnot efficiency. There are other fundamental limitations on photocell efficiency, but 50% seems attainable in multilayer semiconductor photocells. The energy of wind, which is fundamentally solar in origin, can be gathered and converted by "forests" of windmills, as shown in Fig. 19–10.

An indirect scheme for collection and conversion of solar energy would use the temperature gradient in the ocean. In the Caribbean, for example, the water temperature near the surface is about 25°C, while at a depth of a few hundred meters it may be 10°C. Although the second law of thermodynamics forbids taking heat from the ocean and converting it completely into work, nothing forbids running a heat engine between these two temperatures. The thermodynamic efficiency would be very low, but with such a vast reservoir of energy available, this would not be a serious problem.

This is only a small sample of present-day activity in energy-conversion research. Many other processes are under discussion or development, and the principles of thermodynamics outlined in this chapter are of central importance in all of them.

19–10 An array of windmills to collect and convert wind energy. The propellerlike blades turn electric generators, converting the kinetic energy of moving air into electrical energy. (Photo by Kurt Rogers, *San Francisco Examiner*.)

SUMMARY

A cyclic process is one in which the initial and final states are the same; the total internal-energy change is always zero.

A heat engine takes heat Q_H from a source, converts part of it into work W, and discards the remainder Q_C at a lower temperature. The thermal efficiency e of a heat engine is defined as

$$e = \frac{W}{Q_H} = \frac{Q_H + Q_C}{Q_H} = 1 + \frac{Q_C}{Q_H} = 1 - \left| \frac{Q_C}{Q_H} \right|. \qquad (19\text{--}4)$$

A gasoline engine operating on the Otto cycle with compression ratio r has a theoretical maximum thermal efficiency e given by

$$e = 1 - \frac{1}{r^{\gamma-1}}. \qquad (19\text{--}6)$$

A refrigerator takes heat Q_C from a cold place, has a work input W, and discards heat Q_H at a warmer place. The performance coefficient K is defined as

$$\text{Performance coefficient} = K = -\frac{Q_C}{W} = -\frac{Q_C}{Q_H + Q_C}. \qquad (19\text{--}8)$$

A reversible process is one whose direction can be reversed by an infinitesimal change in the conditions of the process. A reversible process is also an equilibrium process.

The second law of thermodynamics describes the directionality of natural thermodynamic processes. It can be stated in several equivalent forms; the two simplest are (1) the impossibility of a cyclic process in which heat is converted completely to work; (2) the impossibility of a cyclic process in which heat is transferred from a cold place to a hotter place with no input of mechanical work.

The Carnot cycle is a theoretical heat-engine cycle operating between two heat reservoirs at temperatures T_H and T_C and using only reversible processes. Its thermal efficiency is given by

$$e = 1 - \frac{T_C}{T_H} = \frac{T_H - T_C}{T_H}. \qquad (19\text{--}13)$$

No engine operating between the same two temperatures can be more efficient than a Carnot engine, and all Carnot engines operating between the same two temperatures have the same efficiency.

A Carnot engine run backward makes a Carnot refrigerator. Its performance coefficient is given by

$$K = \frac{T_C}{T_H - T_C}. \qquad (19\text{--}14)$$

No refrigerator operating between the same two temperatures can have a larger performance coefficient than a Carnot refrigerator, and all Carnot refrigerators operating between the same two temperatures have the same performance coefficient.

Entropy is a quantitative measure of the disorder of a system. The entropy change in any reversible thermodynamic process is defined as

$$\Delta S = \int_1^2 \frac{dQ}{T} \qquad \text{(reversible process)}, \qquad (19\text{--}16)$$

where T must always be the absolute temperature. Entropy depends only on

KEY TERMS

reversibility
equilibrium processes
irreversible processes
heat engine
working substance
cyclic process
thermal efficiency
compression ratio
Otto cycle
Diesel cycle
refrigerator
performance coefficient
energy efficiency rating
heat pump
second law of thermodynamics
Carnot cycle
thermal pollution
entropy
Kelvin temperature scale
absolute zero

the state of the system, and the change in entropy between given initial and final states is the same for all processes leading from one to the other. This fact can be used to find the entropy change in an irreversible process, where Eq. (19–16) is not applicable.

The second law of thermodynamics can be restated in terms of entropy. The entropy of an isolated system may increase but can never decrease. When a system interacts with its surroundings, the total entropy change of system and surroundings can never decrease. If the interaction involves only reversible processes, the total entropy is constant; if there is any irreversible process, the total entropy increases.

The Kelvin temperature scale is based on the efficiency of the Carnot cycle and is independent of the properties of any specific material. The zero point on the Kelvin scale is called absolute zero.

QUESTIONS

19–1 Suppose you want to increase the efficiency of a heat engine. Would it be better to increase T_H or to decrease T_C by an equal amount?

19–2 If an energy-conversion process involves two steps, each with its own efficiency, such as heat to work and work to electric energy, is the efficiency of the composite process equal to the product of the two efficiencies, or the sum, or the difference, or what?

19–3 In some climates it is practical to heat a house by using a heat pump, which acts as an air conditioner in reverse, cooling the outside air and heating the inside air. Can the heat delivered to the house ever exceed the electric-energy input to the pump?

19–4 What irreversible processes occur in a gasoline engine?

19–5 A housewife tries to cool her kitchen on a hot day by leaving the refrigerator door open. What happens? Would the result be different if an old-fashioned ice box were used?

19–6 Is it a violation of the second law to convert mechanical energy completely into heat? To convert heat completely into work?

19–7 A growing plant creates a highly complex and organized structure out of simple materials, such as air, water, and trace minerals. Does this violate the second law of thermodynamics? What is the plant's ultimate source of energy?

19–8 An electric motor has its shaft coupled to that of a generator. The motor drives the generator, and the current from the generator is used to run the motor. The excess current is used to power a television set. What is wrong with this scheme?

19–9 Think of some reversible and some irreversible processes in purely mechanical systems, such as blocks sliding on planes, springs, pulleys, and strings.

19–10 Why must a room air conditioner be placed in a window? Why can't it just be set on the floor and plugged in?

19–11 Discuss the following examples of increasing disorder or randomness: mixing of hot and cold water; free expansion of a gas; irreversible heat flow; and development of heat through mechanical friction. Are entropy increases involved in all these?

19–12 When the sun shines on a glass-roofed greenhouse, the temperature becomes higher inside than outside. Does this phenomenon violate the second law?

19–13 When a wet cloth is hung up in a hot wind in the desert, it is cooled by evaporation to a temperature that may be 20 C° or so below that of the air. Discuss this process in light of the second law.

19–14 Are the earth and sun in thermal equilibrium? Are there entropy changes associated with the transmission of energy from sun to earth? Does radiation differ from other modes of heat transfer with respect to entropy changes?

19–15 Discuss the entropy changes involved in the preparation and consumption of a hot-fudge sundae.

EXERCISES

Section 19–2 Heat Engines

19–1 A large diesel engine takes in 8000 J of heat and delivers 3000 J of work per cycle. The heat is obtained by burning diesel fuel with a heat of combustion of 5.0×10^4 J·g^{-1}.

a) What is the thermal efficiency?

b) How much heat is discarded in each cycle?

c) What mass of fuel is burned in each cycle?

d) If the engine goes through 50 cycles per second, what is its power output in watts? In horsepower?

19–2 A gasoline engine has a power output of 20 kW (about 27 hp). Its thermal efficiency is 20%.

a) How much heat must be supplied to the engine per second?

b) How much heat is discarded by the engine per second?

19–3 A nuclear-power plant has a mechanical-power output (used to drive an electric generator) of 200 MW. Its rate of heat input from the nuclear reactor is 800 MW.

a) What is the thermal efficiency of the system?

b) At what rate is heat discarded by the system?

19–4 A coal-fired steam-turbine power plant has a mechanical-power output of 500 MW and a thermal efficiency of 40%.

a) At what rate must heat be supplied by burning coal?

b) If the heat of combustion of coal is $2.5 \times 10^4 \, \text{J} \cdot \text{g}^{-1}$, what mass of coal is burned per second? Per day?

c) At what rate is heat discarded by the system?

d) If the discarded heat is given to water in a river, and its temperature rises by 5 C°, what volume of water is needed per second?

e) In part (d), if the river is 100 m wide and 5 m deep, what must be the minimum flow velocity of the water?

19–5 A heat engine takes 0.1 mol of an ideal gas around the cycle shown in the pV-diagram of Fig. 19–11. Process 1–2 is at constant volume, process 2–3 is adiabatic, and process 3–1 is at a constant pressure of 1 atm. The value of γ for this gas is $\frac{5}{3}$.

a) Find the pressure and volume at points 1, 2, and 3.

b) Find the net work done by the gas in the cycle.

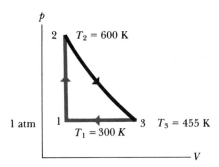

FIGURE 19–11

Section 19–3 Internal-Combustion Engines

19–6 For a gas with $\gamma = 1.4$, what compression ratio r must an Otto cycle have to achieve an ideal efficiency of 70%?

19–7 For an Otto cycle with $\gamma = 1.4$ and $r = 8$, the temperature of the gasoline-air mixture when it enters the cylinder is 22°C (point a of Fig. 19–3). What is the temperature at the end of the compression stroke (point b)?

Section 19–4 Refrigerators

19–8 A window air-conditioner unit absorbs 4000 J of heat per minute from the room being cooled and in the same time period deposits 12,000 J of heat to the outside air.

a) What is the power consumption of the unit, in watts?

b) What is the performance coefficient of the unit?

19–9 A freezer has a performance coefficient $K = 5$. The freezer is to convert 2 kg of water at $T = 20°C$ into 2 kg of ice at $T = -10°C$ in one hour.

a) What amount of heat must be removed from the water at 20°C to convert it into ice at $-10°C$?

b) How much electrical energy will be consumed by the freezer?

c) How much waste heat will be rejected to the room in which the freezer sits?

Section 19–6 the Carnot Cycle

19–10 Show that the efficiency e of a Carnot engine and the performance coefficient K of a Carnot refrigerator are related by $K = (1 - e)/e$.

19–11 A Carnot engine whose high-temperature reservoir is at 400 K takes in 420 J of heat at this temperature in each cycle and gives up 335 J to the low-temperature reservoir.

a) What is the temperature of the low-temperature reservoir?

b) What is the thermal efficiency of the cycle?

19–12 A Carnot engine is operated between two heat reservoirs at temperatures of 400 K and 300 K.

a) If the engine receives 5000 J of heat energy from the reservoir at 400 K in each cycle, how many joules per cycle does it reject to the reservoir at 300 K?

b) If the engine is operated in reverse, as a refrigerator, and receives 5000 J from the reservoir at 300 K, how many joules does it deliver to the reservoir at 400 K?

c) How many joules of mechanical work are required to operate the refrigerator in part (b)?

19–13 An ice-making machine operates in a Carnot cycle; it takes heat from water at 0°C and rejects heat to a room at 27°C. Suppose that 50 kg of water at 0°C are converted into ice at 0°C.

a) How much heat is rejected to the room?

b) How much energy must be supplied to the refrigerator?

Section 19–7 Entropy

19–14 A sophomore with nothing better to do adds heat to 0.5 kg of ice at 0°C until it is all melted.

a) What is the change in entropy of the water?

b) If the source of heat is a very massive body at a temperature of 20°C, what is the change in entropy of this body?

c) What is the total change in entropy of the water and the heat source?

19–15 Calculate the entropy change that occurs when 1 kg of water at 20°C is mixed with 2 kg of water at 80°C.

19–16 A block of aluminum of mass 1 kg, initially at 100°C, is dropped into 1 kg of water initially at 0°C.

a) What is the final temperature?

b) What is the total change in entropy of the system?

19–17 Two moles of an ideal gas undergo a reversible isothermal expansion from 0.02 m³ to 0.04 m³ at a temperature of 300 K. What is the change in entropy of the gas?

Section 19–9 Energy Conversion

19–18 An engine is to be built to extract power from the temperature gradient of the ocean. If the surface and deep-water temperatures are 25°C and 10°C, respectively, what is the maximum theoretical efficiency of such an engine?

19–19 A solar-power plant is to be built with a power output capacity of 1000 MW. What land area must the solar energy collectors occupy if they are

a) photocells with 60% efficiency?

b) mirrors that generate steam for a turbine-generator unit with overall efficiency of 30%?

Take the average power in the sun's radiation to be 200 W·m^{-2} at the earth's surface. Express your answers in square kilometers and square miles.

PROBLEMS

19–20 A cylinder contains oxygen at a pressure of 2 atm. The volume is 3 L and the temperature is 300 K. The oxygen is carried through the following processes:

1. Heated at constant pressure to 500 K.
2. Cooled at constant volume to 250 K.
3. Cooled at constant pressure to 150 K.
4. Heated at constant volume to 300 K.

a) Show these four processes in a pV-diagram, giving the numerical values of p and V at the end of each process.

b) Calculate the net work done by the oxygen.

c) What is the efficiency of this device as a heat engine?

19–21 What is the thermal efficiency of an engine that operates by taking n moles of an ideal gas through the following cycle? Let $C_v = 12$ J·mol^{-1}·K^{-1}.

1. Start with n moles at P_0, V_0, T_0.
2. Change to $2P_0, V_0$, at constant volume.
3. Change to $2P_0, 2V_0$, at constant pressure.
4. Change to $P_0, 2V_0$, at constant volume.
5. Change to P_0, V_0, at constant pressure.

19–22 A Carnot engine operates between two heat reservoirs at temperatures T_H and T_C. An inventor proposes to increase the efficiency by running one engine between T_H and an intermediate temperature T', and a second engine between T' and T_C, using the heat expelled by the first engine. Compute the efficiency of this composite system and compare it to that of the original engine.

19–23 An 0.08 kg cube of ice at an initial temperature of −15°C is placed in 0.50 kg of water at $T = 60$°C in an insulated container of negligible mass. Calculate the entropy change of the system.

19–24 A physics student performing a heat-conduction experiment immerses one end of a copper rod in boiling water at 100°C, the other end in an ice-water mixture at 0°C. The sides of the rod are insulated. During a certain time interval, 0.5 kg of ice melts. Find

a) the entropy change of the boiling water;

b) the entropy change of the ice-water mixture;

c) the entropy change of the copper rod;

d) the total entropy change of the entire system.

CHALLENGE PROBLEMS

19–25 Consider a Diesel cycle that starts (point a in Fig. 19–4) with 2.0 L of air at a temperature of 300 K and a pressure of 1.0×10^5 Pa. If the temperature at point c is $T_c = 1200$ K, derive an expression for the efficiency of the cycle in terms of the compression ratio r. What is the efficiency when $r = 20$?

19–26

a) Draw a graph of a Carnot cycle, plotting Kelvin temperature vertically and entropy horizontally (a temperature–entropy or TS diagram).

b) Show that the area under any curve in a temperature–entropy diagram represents the heat absorbed by the system.

c) Derive from your diagram the expression for the thermal efficiency of a Carnot cycle.

20

MOLECULAR PROPERTIES OF MATTER

WE HAVE STUDIED SEVERAL PROPERTIES OF MATTER IN BULK, INCLUDING elasticity, density, surface tension, equations of state, heat capacities, phase changes, internal energy, entropy, and others. We have discussed qualitatively the relationships of these properties to molecular structure, but have deliberately avoided any detailed discussion of these relationships. There is a good reason for this; all of the above macroscopic (bulk) properties can be used in practical calculations without any detailed understanding of their microscopic (molecular) basis. Even more important, quantities such as heat, internal energy, and entropy *must* be defined in a way that does not depend on the details of any microscopic picture. Indeed, much of the power and usefulness of thermodynamics lies in its generality and its lack of dependence on microscopic models.

Nevertheless, we can gain a lot of additional insight into the behavior of matter by looking at the relation of bulk behavior to microscopic structure. We will study a few examples of molecular models that enable us actually to *predict* some properties of matter. We begin with a general discussion of the molecular structure of matter. Then we develop the kinetic-molecular model of a gas, which helps us understand the equation of state and heat capacities of gases on the basis of a molecular model. Finally, we look briefly at the heat capacities of solids and their relation to molecular structure.

Understanding bulk properties of matter on the basis of molecular structure

20–1 MOLECULAR STRUCTURE OF MATTER

An abundance of physical and chemical evidence has shown conclusively that matter in all phases is made up of particles called **molecules.** For any specific chemical compound, the molecules are all identical. The smallest molecules are of the order of 10^{-10} m in size; the largest are at least 10,000 times this large. In some materials the molecules are dissociated into electrically charged substructures called *ions.*

In liquids and solids molecules are held together by intermolecular forces that are *electrical* in nature, arising from interactions of the electrically charged fundamental particles that make up the molecules. *Gravitational* forces be-

Forces between atoms and molecules are basically electrical in nature.

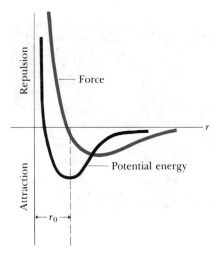

20–1 The force between two molecules (color curve) changes from an attraction when the separation is large to a repulsion when the separation is small. The potential energy (black curve) is minimum at r_0, where the force is zero

Forces between molecules are attractive at some distances and repulsive at others.

Molecules are always in motion, and the amount of motion increases with temperature.

20–2 Crystal of necrosis virus protein. The actual size of the entire crystal is about two-thousandths of a millimeter (0.002 mm). (Courtesy of Ralph W. G. Wyckoff. Reprinted with the permission of Educational Services, Inc., from *Physics*, D. C. Heath and Co., Boston, 1960.)

tween molecules are so weak compared with electrical forces that they are completely negligible.

The interaction of two point electric charges is described by a force (repulsive for like charges, attractive for unlike charges) whose magnitude is proportional to $1/r^2$, where r is the distance between the points. This relationship is called *Coulomb's law;* we will study it in detail in Chapter 24. Forces between *molecules,* however, do not follow a simple inverse-square law. Molecules are not point charges but complex structures containing both positive and negative charges, and their interactions are correspondingly complex. When molecules are far apart, as in a gas, the intermolecular force is very small and usually attractive. As a gas is compressed and its molecules are brought closer together, the force increases. In liquids, relatively large pressures are required to compress the substance appreciably. We conclude that at separations between molecules that are only slightly *less* than their normal spacing, the force becomes *repulsive* and relatively large.

Thus the intermolecular force must vary with the distance r between molecules somewhat as shown in Fig. 20–1. At large distances the force is small and attractive. As the molecules come closer together, the force of attraction becomes larger, passes through a maximum, and then decreases to zero at an equilibrium separation r_0. When the distance is less than r_0, the force becomes repulsive and increases quite rapidly. Figure 20–1 also shows the potential energy as a function of r. This function has a *minimum* at r_0, where the force is zero. Such a potential-energy function is often called a **potential well.** The force F and potential energy U are related by $F = -dU/dr$.

In view of these attractive intermolecular forces, why do all molecules not eventually coalesce into matter in the liquid or solid phase? The answer is that molecules are always in *motion*. There is kinetic energy associated with this motion, and it usually increases with temperature. At very low temperatures the average kinetic energy of a molecule may be much *less* than the maximum magnitude of potential energy, which is the "depth" of the potential well in Fig. 20–1. The molecules then condense into the liquid or solid phase with average intermolecular spacing of about r_0. But at higher temperatures the average kinetic energy becomes larger than the depth of the potential well; molecules can then escape the intermolecular force and become free to move independently, as in the gaseous phase of matter.

In *solids*, molecules execute vibratory motion about more-or-less fixed centers. The potential well is usually approximately parabolic in shape near its minimum, and the motion is then approximately simple harmonic. The amplitudes of the vibratory motions are relatively small, and the fixed centers form a space lattice, corresponding to the repeated spatial patterns and symmetry of crystals. A photograph of the individual, very large molecules of necrosis virus protein is shown in Fig. 20–2. It was taken with an electron microscope at a magnification of about 80,000. The molecules look like neatly stacked oranges. Each molecule is about 1.2×10^{-8} m in diameter.

In a *liquid* the intermolecular distances are usually only slightly greater than in the solid phase of the same substance. The molecules have vibratory motions of greater energy about centers that are free to move, but they remain at approximately the same distances from each other. Liquids show a certain regularity of structure only in the immediate neighborhood of a few molecules. This is called **short-range order,** in contrast to the **long-range order** of a solid crystal.

The molecules of a gas have greater average kinetic energy than those of liquids and solids. The molecules are usually widely separated and have only very small attractive forces. A molecule of a gas therefore moves in a straight line until it collides either with another molecule or with a wall of the container. In molecular terms, an *ideal gas* is a gas whose molecules exert *no* forces of attraction on each other. The mathematical analysis of a collection of such idealized molecules in random linear motion between collisions is called the *kinetic-molecular theory of gases*. The analysis of an ideal gas given in this chapter is a simple example of kinetic theory.

Most common substances exist in the solid phase at low temperatures. When the temperature is raised beyond a definite value, the liquid phase results, and when the temperature of the liquid is raised further, the substance exists in the gaseous phase. That is, from a large-scale, or *macroscopic,* point of view, the transition from solid to liquid to gas is in the direction of increasing temperature. From a *molecular* point of view, this transition is in the direction of increasing molecular kinetic energy. Thus temperature and molecular kinetic energy are closely related.

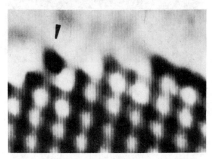

High-resolution electron micrographs showing the surface of a small crystal of gold. Most of the atoms are bound in a crystal lattice, but a few columns of gold atoms (shown by an arrow) are "hopping" with respect to the crystal lattice. (Courtesy of J.-O. Bovin, R. Wallenberg (Univ. of Lund, Sweden) and D. J. Smith (Arizona State Univ.).)

20–2 AVOGADRO'S NUMBER

In 1811 the chemist John Dalton suggested that the molecules of a given chemical substance are all alike. This hypothesis has since been confirmed by an overwhelming variety of evidence; it explains why, in chemical reactions, elements and compounds combine with each other in definite proportions by mass. It also leads directly to the concepts of *molecular mass* and *number of moles.* Specifically, one **mole** of any pure chemical element or compound contains a definite number of molecules, the same number for all elements and compounds. The official SI definition of the mole is as follows:

> The mole is the amount of substance that contains as many elementary entities as there are atoms in 0.012 kilogram of carbon 12.

In our discussion the "elementary entities" referred to are atoms or molecules.

The number of atoms or molecules in a mole is called **Avogadro's number,** denoted by N_A. (Avogadro was a contemporary of Dalton.) The most precise measurements of N_A have been obtained by using x-rays to measure the distance between layers of molecules in a crystal. The numerical value of N_A is now known with an uncertainty of less than six parts per million; to four significant figures, it is

$$N_A = 6.022 \times 10^{23} \text{ molecules·mol}^{-1}.$$

The **molecular mass** M of a compound is the mass of one mole; the mass m of a single molecule is thus given by

$$M = N_A m. \tag{20–1}$$

When the molecule consists of a single atom, the term *atomic mass* is often used.

Once Avogadro's number has been determined, it can be used to compute the mass of a molecule. For example, the mass of 1 mol of atomic hydrogen (i.e., the atomic mass) is 1.008 g. The mass of 1 mol of diatomic hydrogen molecules is 2.016 g. Since, by definition, the number of atoms or molecules in

A mole is a definite number of atoms or molecules.

Avogadro's number is the number of atoms or molecules in a mole.

Finding the mass of a single molecule from the molecular mass and Avogadro's number

20–3 How much is one mole? The photograph shows one mole each of sucrose (ordinary sugar, rear pile), iodine (metallic-looking chips), water, mercury, iron (cube), acetylsalicylic acid (aspirin tablets), and sodium chloride (ordinary salt, front pile). (Photo by Chip Clark.)

The physical properties of materials are determined by their molecular structure.

1 mol is Avogadro's number, it follows that the mass of a single atom of hydrogen is

$$m_H = \frac{1.008 \text{ g·mol}^{-1}}{6.022 \times 10^{23} \text{ molecules·mol}^{-1}} = 1.674 \times 10^{-24} \text{ g·molecule}^{-1}.$$

For an oxygen molecule of molecular mass 32 g·mol^{-1},

$$m_{O_2} = \frac{32 \text{ g·mol}^{-1}}{6.022 \times 10^{23} \text{ molecules·mol}^{-1}} = 53.14 \times 10^{-24} \text{ g·molecule}^{-1}.$$

Figure 20–3 shows one mole of each of several familiar materials.

20–3 MOLECULAR BASIS OF PROPERTIES OF MATTER

The usual goal of any molecular theory of matter is to understand the *macroscopic* properties of matter in terms of the properties, behavior, and interactions of the molecules of which it is composed. Such theories are of tremendous practical importance; once this understanding has been gained, it becomes possible actually to *design* materials with specific desired properties. Thus this kind of molecular analysis has led to the development of high-strength steels, glasses with special optical properties for use in optical instruments, semiconductor materials for solid-state electronic devices, and countless other materials essential to contemporary technology.

In the following sections we will consider a few simple examples of molecular theories of matter. For example, we can represent a monatomic gas with a model consisting of a large number of particles described completely by their

Installing heat-shield tiles on the Space Shuttle. The material must have exceptional strength and heat resistance to protect the space vehicle during re-entry into the earth's atmosphere. (Courtesy of Rockwell International.)

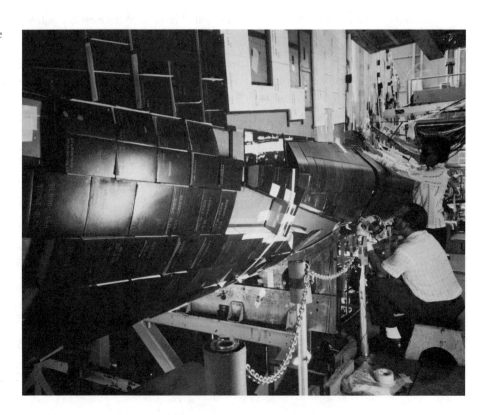

mass and velocities. We can then derive from this model the ideal-gas equation of state of a gas at low pressure and show that C_v for the gas should equal $12.5 \, \text{J} \cdot \text{mol}^{-1} \cdot \text{K}^{-1}$ (see Table 18–1). If we add the hypothesis that the "particles" are not simply points but have a finite size, then the general features of the viscosity, thermal conductivity, and coefficient of diffusion of a gas can be understood, as well as the fact that polyatomic gases have larger values of C_v than monatomic gases. By assuming that there are interaction forces between the particles, we can refine the equation of state to bring it into better agreement with the behavior of a real gas, and we can begin to understand the phenomena of liquefaction and solidification at low temperatures.

The electrical and magnetic properties of matter, and the emission and absorption of light by matter, call for a molecular model in which the molecules themselves are aggregates of subatomic particles. Some of these particles are electrically charged, and the forces between molecules originate in these electric charges. This development of molecular theory will take us into the area of atomic and molecular structure in Chapter 43. In the next section we return to the starting point and see what properties of a gas at low pressure can be explained by the simplest possible molecular model, an aggregate of particles having mass and velocity.

20–4 KINETIC-MOLECULAR THEORY OF AN IDEAL GAS

We are now ready to develop in detail the relation between the kinetic-molecular model of an ideal gas and the ideal-gas equation of state, discussed in Section 17–2. Consider a container of volume V, containing N identical molecules each of mass m. The molecules are in constant motion; each molecule collides from time to time with a wall of the container. During such collisions, the molecules exert forces on the walls, and this is the origin of the macroscopic *pressure* the gas exerts on the container walls. We assume that each collision of a molecule with a wall of the container is perfectly elastic, as shown in Fig. 20–4. In each collision the component of velocity parallel to the wall is unchanged and the component perpendicular to the wall is reversed.

Our program will be to determine for a given wall area A the number of collisions per unit time, the associated momentum change, and the force needed for the momentum change. Then we can obtain an expression for the pressure, which is force per unit area. Let v_x be the *magnitude* of the x-component of velocity of a molecule. At first we will assume that all molecules have the same v_x. This assumption, though unrealistic, helps clarify the basic ideas; and we will show soon that it is not really necessary.

In each collision the change in the x-component of momentum is $2mv_x$. To find the number of collisions with a given wall area A during a time interval Δt, we note that, in order to experience a collision during Δt, a molecule must be within a distance $v_x \, \Delta t$ from the wall at the beginning of Δt, as shown in Fig. 20–5, and must be headed toward the wall. Thus to collide with A during Δt, a molecule must, at the beginning of Δt, be within a cylinder of base area A and length $v_x \, \Delta t$. The volume of such a cylinder is $Av_x \, \Delta t$.

Assuming the number of molecules per unit volume (N/V) is uniform, the *number* of molecules in this cylinder is $(N/V) \, (Av_x \, \Delta t)$. But, on average, half of these molecules are moving *away from* the wall. Thus the number of collisions

The pressure of a gas results from collisions of molecules with the walls of the container.

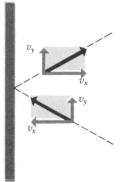

20–4 Elastic collision of a molecule with container wall. The component v_y parallel to the wall does not change; the component v_x perpendicular to the wall reverses direction. The speed v does not change.

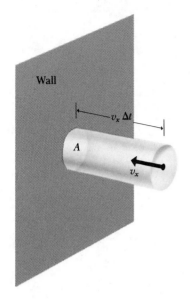

20–5 A molecule moving toward the wall with speed v_x collides with the area A during the time interval Δt only if it is within a distance $v_x \Delta t$ of the wall at the beginning of the interval. All such molecules are contained within a volume $A v_x \Delta t$.

with A during Δt is

$$\frac{1}{2}\left(\frac{N}{V}\right)(A v_x \Delta t). \tag{20–2}$$

The total momentum change ΔP_x due to all these collisions is $2m v_x$ times the *number* of collisions. (We are using capital P for momentum, and we will use small p for pressure; be careful!)

$$\Delta P_x = \frac{1}{2}\left(\frac{N}{V}\right)(A v_x \Delta t)(2m v_x) = \frac{N A m v_x^2 \Delta t}{V}, \tag{20–3}$$

and the *rate* of change of momentum is

$$\frac{\Delta P_x}{\Delta t} = \frac{N A m v_x^2}{V}. \tag{20–4}$$

According to Newton's second law, this rate of change of momentum equals the average force exerted by the wall area A on the molecules. From Newton's *third* law, this is the negative of the force exerted *on* the wall *by* the molecules. Finally, pressure p is force per unit area, and we obtain

$$p = \frac{F}{A} = \frac{N m v_x^2}{V}. \tag{20–5}$$

Now in fact v_x is *not* the same for all molecules. But we could have sorted the molecules into groups having the same v_x within each group and added up the resulting contributions to the pressure. The net effect is simply to replace v_x^2 in Eq. (20–5) by the *average* value of v_x^2, which we denote by $(v_x^2)_{av}$. Furthermore, $(v_x^2)_{av}$ is related simply to the *speeds* of the molecules. The speed v (magnitude of velocity) of any molecule is related to the velocity components v_x, v_y, and v_z by

$$v^2 = v_x^2 + v_y^2 + v_z^2.$$

We can average this relation over all molecules:

$$(v^2)_{av} = (v_x^2)_{av} + (v_y^2)_{av} + (v_z^2)_{av}.$$

But since the x-, y-, and z-directions are all equivalent,

$$(v_x^2)_{av} = (v_y^2)_{av} = (v_z^2)_{av}.$$

Hence

$$(v_x^2)_{av} = \frac{1}{3}(v^2)_{av},$$

and Eq. (20–5) becomes

$$pV = \frac{1}{3}Nm(v^2)_{av} = \frac{2}{3}N\left[\frac{1}{2}m(v^2)_{av}\right]. \tag{20–6}$$

The pressure of a gas is proportional to the average kinetic energy of its molecules.

But $\frac{1}{2}m(v^2)_{av}$ is the average kinetic energy of a single molecule, and the product of this energy and the total number of molecules N equals the total random kinetic energy, or internal energy U. Hence the product pV equals two-thirds of the internal energy:

$$pV = \frac{2}{3}U. \tag{20–7}$$

By experiment, at low pressures, the equation of state of a gas is

$$pV = nRT.$$

The theoretical and experimental laws will therefore be in complete agreement if we set

$$U = \frac{3}{2}nRT. \tag{20–8}$$

The total kinetic energy of gas molecules is proportional to the temperature.

The average kinetic energy of a single molecule is then

$$\frac{U}{N} = \frac{1}{2}m(v^2)_{av} = \frac{3nRT}{2N}.$$

But the number of moles, n, equals the total number of molecules, N, divided by Avogadro's number N_A, the number of molecules per mole:

$$n = \frac{N}{N_A}, \qquad \frac{n}{N} = \frac{1}{N_A}.$$

Hence

$$\frac{1}{2}m(v^2)_{av} = \frac{3}{2}\frac{R}{N_A}T. \tag{20–9}$$

The ratio R/N_A occurs frequently in molecular theory. It is called the **Boltzmann constant, k**:

The Boltzmann constant is a gas constant on a "per molecule" basis.

$$k = \frac{R}{N_A} = \frac{8.31 \text{ J·mol}^{-1}\text{·K}^{-1}}{6.02 \times 10^{23} \text{ molecules·mol}^{-1}}$$
$$= 1.38 \times 10^{-23} \text{ J·molecule}^{-1}\text{·K}^{-1}.$$

Since R and N_A are fundamental physical constants, the same is true of k. Then

$$\frac{1}{2}m(v^2)_{av} = \frac{3}{2}kT. \tag{20–10}$$

Thus the average kinetic energy *per molecule* depends only on the temperature, not on the pressure, volume, or molecular species. An equivalent statement can also be obtained from Eq. (20–9) by using the relation $M = N_A m$:

The rms speed of gas molecules depends on the molecular mass and on temperature.

$$N_A\left[\frac{1}{2}m(v^2)_{av}\right] = \frac{1}{2}M(v^2)_{av} = \frac{3}{2}RT. \tag{20–11}$$

That is, the kinetic energy of a *mole* of molecules depends only on T.

From Eqs. (20–10) and (20–11) we can obtain expressions for the square root of $(v^2)_{av}$, called the *root-mean-square speed* v_{rms}:

$$v_{rms} = \sqrt{(v^2)_{av}} = \sqrt{\frac{3kT}{m}} = \sqrt{\frac{3RT}{M}}. \tag{20–12}$$

Finally, it is sometimes convenient to rewrite the ideal-gas equation on a molecular basis. Since $N = N_A n$ and $R = N_A k$, an alternative form of the ideal-gas equation is

$$pV = NkT. \tag{20–13}$$

Thus k may be regarded as a gas constant on a "per molecule" basis instead of the usual "per mole" basis for R.

PROBLEM-SOLVING STRATEGY: *Kinetic-molecular theory*

1. As always, using a consistent set of units is essential. We list below several places where caution is needed.

2. The most common units for molecular mass M are grams per mole; the molecular mass of oxygen is 32 g·mol^{-1}, for example. These units are often omitted in tables. In equations with SI units, such as Eq. (20–12), M *must* be converted to kilograms per mole by dividing by 10^3. Thus in SI units $M = 32 \times 10^{-3}$ kg·mol^{-1} for oxygen.

3. Are you working on a "per molecule" basis or a "per mole" basis? Remember that m is the mass of a molecule, and M is the mass of a mole. Similarly, N is the number of molecules, and n is the number of moles; k is the gas constant per molecule, and R is the gas constant per mole. Although N, the number of molecules, is in one sense a dimensionless number, you can do a complete unit check if you think of N as having the unit "molecules"; m has units "mass per molecule," and k has units "joules per molecule per kelvin."

4. Remember that T is always *absolute* temperature. In the problems in this chapter, it is always in kelvins.

EXAMPLE 20–1   What is the average kinetic energy of a molecule of a gas at a temperature of 300 K?

SOLUTION   From Eq. (20–10),

$$\frac{1}{2}m(v^2)_{\text{av}} = \frac{3}{2}kT = \left(\frac{3}{2}\right)(1.38 \times 10^{-23} \text{ J·K}^{-1})(300 \text{ K})$$

$$= 6.21 \times 10^{-21} \text{ J}.$$

EXAMPLE 20–2   What is the total random kinetic energy of the molecules in one mole of a gas at a temperature of 300 K?

SOLUTION   From Eq. (20–8),

$$U = \frac{3}{2}nRT = \frac{3}{2}(1 \text{ mol})(8.314 \text{ J·mol}^{-1}\text{·K}^{-1})(300 \text{ K})$$

$$= 3741 \text{ J} = 894 \text{ cal}.$$

EXAMPLE 20–3   What is the root-mean-square speed of a hydrogen molecule at 300 K?

Finding the rms speed of hydrogen molecules

SOLUTION   The mass of a hydrogen molecule (see Section 20–2) is

$$m_{\text{H}_2} = (2)(1.674 \times 10^{-27} \text{ kg}) = 3.348 \times 10^{-27} \text{ kg}.$$

Hence, from Eq. (20–12),

$$v_{\text{rms}} = \sqrt{\frac{3kT}{m}} = \sqrt{\frac{3(1.38 \times 10^{-23} \text{ J·K}^{-1})(300 \text{ K})}{3.348 \times 10^{-27} \text{ kg}}}$$

$$= 1927 \text{ m·s}^{-1}.$$

Alternatively,

$$v_{\text{rms}} = \sqrt{\frac{3RT}{M}} = \sqrt{\frac{3(8.314 \text{ J·mol}^{-1}\text{·K}^{-1})(300 \text{ K})}{2(1.008 \times 10^{-3} \text{ kg·mol}^{-1})}}$$

$$= 1927 \text{ m·s}^{-1}.$$

Note that in Eq. (20–12), when we use the value of R in SI units, M must be expressed in *kilograms* per mole, not grams per mole. In this example, $M = 2.016 \times 10^{-3}$ kg·mol^{-1}, not 2.016 g·mol^{-1}.

EXAMPLE 20–4 Five gas molecules chosen at random are found to have speeds of 500, 600, 700, 800, and 900 m·s^{-1}. Find the rms speed. Is it the same as the *average* speed?

An exercise in calculating rms speed.

SOLUTION The average value of v^2 for the five molecules is

$$(v^2)_{av} = \frac{(500 \text{ m·s}^{-1})^2 + (600 \text{ m·s}^{-1})^2 + (700 \text{ m·s}^{-1})^2 + (800 \text{ m·s}^{-1})^2 + (900 \text{ m·s}^{-1})^2}{5}$$

$$= 510{,}000 \text{ m}^2\text{·s}^{-2},$$

and v_{rms} is the square root of this value:

$$v_{rms} = 714 \text{ m·s}^{-1}.$$

The *average* speed v_{av} is given by

$$v_{av} = \frac{500 \text{ m·s}^{-1} + 600 \text{ m·s}^{-1} + 700 \text{ m·s}^{-1} + 800 \text{ m·s}^{-1} + 900 \text{ m·s}^{-1}}{5}$$

$$= 700 \text{ m·s}^{-1}.$$

Clearly, v_{rms} and v_{av} are not, in general, the same.

EXAMPLE 20–5 Find the number of molecules in one cubic meter of air at atmospheric pressure and 0°C.

SOLUTION From Eq. (20–13),

$$N = \frac{pV}{kT} = \frac{(1.013 \times 10^5 \text{ Pa})(1 \text{ m}^3)}{(1.38 \times 10^{-23} \text{ J·K}^{-1})(273 \text{ K})} = 2.69 \times 10^{25}.$$

When a gas expands against a moving piston, it does work. This work is accompanied by a decrease in the random kinetic energy of the gas molecules. Conversely, when work is done on a gas during compression, the random kinetic energy of its molecules increases. But if the collisions with the walls are perfectly elastic, as we have assumed, how can a molecule gain or lose energy in a collision with a piston? To understand this phenomenon, we must consider the collision of a molecule with a *moving* wall.

When a molecule collides with a *stationary* wall, it exerts a momentary force on the wall but does no work, because the wall does not move. But if the wall is in motion, work *is* done during the collision. Thus if the wall in Fig. 20-4 is moving to the left, work is done on it by the molecules that strike it, and their speeds (and kinetic energies) after colliding are smaller than they were before the collision. The collision is still completely elastic because the work done on the moving piston is just equal to the decrease in the kinetic energy of the molecules. Similarly, if the piston is moving toward the right, the kinetic energy of a colliding molecule *increases,* by an amount equal to the work done on the molecule.

When a molecule collides with a moving wall, it does work on the wall.

The assumption that individual molecules undergo elastic collisions with the container wall is not strictly correct. More detailed investigation has shown that in most cases, molecules actually adhere to the wall for a short time, and then leave again with speeds characteristic of the temperature *of the wall.* The gas and the wall are ordinarily in thermal equilibrium, however, and the validity of our conclusions is not altered by this discovery.

20–5 MOLAR HEAT CAPACITY OF A GAS

The analysis in Section 20–4 can be extended to include a discussion of the molar heat capacities of an ideal gas. We begin with the molar heat capacity at constant volume, C_v. When we add heat to a gas under constant-volume conditions, the gas does no work. According to the first law, then, *all* the added energy goes to increase the internal energy. That is, $\Delta U = Q - W$, but $W = 0$; so $\Delta U = Q$. From the *molecular* viewpoint, the internal energy is the sum of the kinetic and potential energies of the molecules. If we know how this total internal energy depends on temperature, then from its rate of change with temperature we can make a theoretical prediction of the molar heat capacity of the system.

Predicting the heat capacity of a gas by using the kinetic-molecular model

The simplest system is a monatomic ideal gas. We represent each molecule (a single atom) as a point particle. There is no potential energy of interaction of the molecules, and the only energy is the translational kinetic energy of random motion of the molecules. For n molecules of gas at absolute temperature T, this energy is given by Eq. (20-8):

$$\text{Kinetic energy} = U = \frac{3}{2}nRT.$$

If the temperature increases by ΔT, the kinetic energy increases by

$$\Delta U = \frac{3}{2}nR\,\Delta T.$$

In a process in which the temperature increases by ΔT at constant volume, the energy flowing into a system is, by definition of C_v,

$$\Delta Q = nC_v\,\Delta T = \Delta U.$$

Hence

$$C_v = \frac{3}{2}R. \qquad (20\text{–}14)$$

In SI units,

$$C_v = \frac{3}{2}(8.314\ \text{J}\cdot\text{mol}^{-1}\cdot\text{K}^{-1}) = 12.47\ \text{J}\cdot\text{mol}^{-1}\cdot\text{K}^{-1}.$$

The experimental values of C_v listed in Table 18–1 for monatomic gases are, in fact, almost exactly equal to $\frac{3}{2}R$. This agreement is a striking confirmation of the basic correctness of the kinetic-molecular model of a gas.

Since $C_p = C_v + R$, it follows that the theoretical ratio of molar heat capacities for a monatomic ideal gas is

$$\frac{C_p}{C_v} = \gamma = \frac{\frac{3}{2}R + R}{\frac{3}{2}R} = \frac{5}{3} = 1.67.$$

This is also in good agreement with the experimental values in Table 18–1.

Gases with more than one atom in a molecule have additional energy associated with rotational and vibrational motion

In this analysis we have treated each molecule as a point particle. For a gas whose molecules have two or more atoms each—that is, *polyatomic* gases—the problem is more complicated. For example, a diatomic molecule can be thought of in terms of *two* point masses, with an interaction force of the kind shown in Fig. 20–1. We can picture such a molecule as a kind of elastic dumbbell. It can have additional kinetic energy associated with *rotation* about an axis through its center of mass, and the atoms may have a back-and-forth *vibrating* motion along the line joining them, with associated additional kinetic and potential energies.

The *temperature* of a gas is determined by the average random *translational* kinetic energy $\frac{1}{2}mv^2$ of its molecules. When heat flows into a *monatomic* gas, at constant volume, all of this energy goes into an increase in random *translational* molecular kinetic energy, as shown by the agreement between the measured values of C_v and the values computed from the increase in translational kinetic energy. But when heat flows into a *diatomic* or *polyatomic* gas, part of the energy goes into increasing rotational and vibrational motion. Hence for a given temperature change, a greater total amount of energy is required in order to increase the rotational and vibrational as well as translational energies. Thus polyatomic gases have larger molar heat capacities than monatomic gases. The experimental values in Table 18–1 show this effect.

To progress further in our theoretical understanding of heat capacities, we need some principle that will tell us *how much* energy is associated with each additional kind of motion of a complex molecule, compared to the translational kinetic energy we have already discussed. The necessary principle goes by the fancy name of **principle of equipartition of energy.** In the early days of kinetic-molecular theory, this principle was treated as an axiom. Today it can be derived from sophisticated statistical-mechanics considerations; but that derivation is beyond our scope, and we too will treat it as an axiom.

> Equipartition of energy: Each kind of motion has its energy quota.

According to the principle of equipartition of energy, each velocity component (either linear or angular) has, on the average, an associated kinetic energy per molecule of $\frac{1}{2}kT$. The number of velocity components needed to describe the motion of a molecule completely is called the number of **degrees of freedom.** For a monatomic gas, the number is three. For a diatomic molecule, there are two possible axes of rotation, perpendicular to each other and to the molecule's axis. (Rotation about the molecule's own axis is not counted because, in ordinary collisions, there is no way for this rotational motion to change.) Thus if we assign five degrees of freedom to a diatomic molecule, the average total kinetic energy per molecule is $5kT/2$ instead of $3kT/2$. The total internal energy of n moles is $U = 5nRT/2$, and the molar heat capacity (at constant volume) is

> Degrees of freedom: How many kinds of motion can a molecule have?

$$C_v = \frac{5}{2}R. \qquad (20\text{--}15)$$

In SI units,

$$C_v = \frac{5}{2}(8.314 \text{ J·mol}^{-1}\text{·K}^{-1}) = 20.78 \text{ J·mol}^{-1}\text{·K}^{-1}.$$

Reference to Table 18–1 shows that this value is in approximate agreement with measured values for diatomic gases. The corresponding value of γ is

$$\gamma = \frac{C_p}{C_v} = \frac{\frac{5}{2}R + R}{\frac{5}{2}R} = \frac{7}{5} = 1.40,$$

which is again in reasonable agreement with experimental values.

Additional contributions to the heat capacities of gases can arise from *vibrational* motion. Molecules are never perfectly rigid; the molecular bonds can stretch and bend, permitting internal vibrations to occur. There are additional degrees of freedom and energies associated with vibrational motion. For

> Vibrational motion of molecules: Does it contribute to heat capacities, or not, and why?

most diatomic gases, however, vibrational motion does *not* contribute appreciably to heat capacity, for reasons that involve concepts of quantum mechanics. Briefly, the energy of a vibrational motion can change only in finite steps. If the energy change of the first step is much larger than the energy possessed by most molecules, then nearly all the molecules will remain in the minimum-energy state of motion. In this case changing the temperature does not change their average vibrational energy appreciably, and the vibrational degrees of freedom are said to be "frozen out." In more complex molecules, the gaps between permitted energy levels are sometimes much smaller, and then vibration *does* contribute to heat capacity. In Table 18–1 the larger values of C_v for some polyatomic molecules show the contributions of vibrational energy. In addition, a molecule with three or more atoms not in a straight line has three, not two, rotational degrees of freedom.

Energy-level spacing determines whether vibrational motion contributes to heat capacities.

20–6 DISTRIBUTION OF MOLECULAR SPEEDS

Measuring the speeds of molecules in a gas. What is the distribution of speeds?

As mentioned in Section 20–4, the molecules in a gas do not all have the same speed. Direct measurements of the distribution of molecular speeds can be made; one experimental scheme is shown in Fig. 20–6. A substance is vaporized in a hot oven; molecules of the vapor escape through an aperture in the oven wall and into a vacuum chamber. A series of slits blocks all molecules except those in a narrow beam; the beam is aimed at a pair of rotating disks. A molecule passing through the slit in the first disk arrives at the second disk just as *its* slit is lined up with the beam only if the molecule has a certain speed. The setup thus functions as a speed selector that allows only molecules with a certain narrow range of speeds to pass. This speed can be varied by changing the disk speed, and we can measure how many molecules have each of various speeds.

The distribution function: describing how many molecules have how much speed

The results of such measurements can be represented graphically as shown in Fig. 20–7. We define a function $f(v)$ called a *distribution function*, such that if N molecules are observed, the number dN having speeds in any range between v and $v + dv$ is given by

$$dN = Nf(v)\, dv. \qquad (20\text{–}16)$$

The figure shows distribution functions for several different temperatures; at each temperature the height of the curve for any value of v is proportional to the number of molecules with speeds near v. The peak of the curve represents the *most probable speed* for the corresponding temperature. As the temperature increases, the peak shifts to higher and higher speeds, corresponding to the increase in average molecular kinetic energy with temperature. Figure 20–7 also shows that the area under a curve between any two values of v represents

20–6 Apparatus for producing a molecular beam and observing the distribution of molecular speeds in the beam.

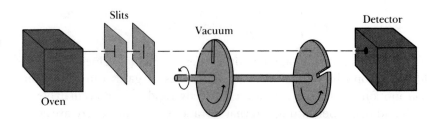

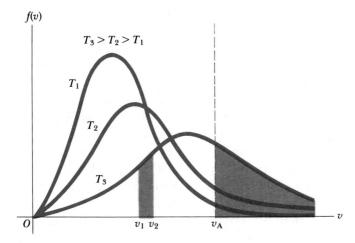

f(v)

$T_3 > T_2 > T_1$

T_1

T_2

T_3

O v_1 v_2 v_A v

20-7 Maxwell-Boltzmann distribution curves for various temperatures. As the temperature increases, the curve becomes flatter, and its maximum shifts to higher temperature. At temperature T_3, the number of molecules having speeds in the range v_1 to v_2 and the number having speeds greater than v_A are shown by the shaded areas under the T_3 curve.

the number of molecules having speeds in that range. The area corresponding to the range v_1 to v_2 and the area for all speeds greater than v_A are shown for the distribution for temperature T_3.

The function $f(v)$ describing the actual distribution of molecular speeds is called the **Maxwell–Boltzmann distribution.** It can be derived from statistical-mechanics considerations, but that derivation is beyond our scope. Here is the result:

$$f(v) = 4\pi \left(\frac{m}{2\pi kT}\right)^{3/2} v^2 e^{-mv^2/2kT}. \qquad (20\text{--}17)$$

The Maxwell–Boltzmann distribution describes both speed distribution and energy distribution in a gas.

We can also express this function in terms of the translational kinetic energy of a molecule, which we denote by ϵ (the Greek letter epsilon). That is, $\epsilon = \frac{1}{2}mv^2$. We invite you to verify that when this value is substituted into Eq. (20–17), the result is

$$f(v) = \frac{8\pi}{m} \left(\frac{m}{2\pi kT}\right)^{3/2} \epsilon e^{-\epsilon/kT}. \qquad (20\text{--}18)$$

In this form the Maxwell–Boltzmann distribution function shows that the exponent is $-\epsilon/kT$, and that the shape of the curve is determined by the relative magnitude of ϵ and kT at any point. In particular, we invite you to prove that the *peak* of each curve occurs when $\epsilon = kT$, corresponding to a speed v given by

$$v = \sqrt{\frac{2kT}{m}}. \qquad (20\text{--}19)$$

This is therefore the *most probable* speed. Note that it differs from the rms speed given by Eq. (20–12) by a simple numerical factor $\left(\frac{2}{3}\right)^{1/2}$.

Every molecule must have *some* speed, so the integral of $f(v)$ over all v must have the value unity. Also, the integral of the quantity $v^2 f(v)$ over all v must equal the average value of v^2; its square root is the rms speed, Eq. (20–12). To verify these two statements we have to evaluate some fairly complicated integrals, and the easiest procedure is to look them up in a table of definite integrals. We leave these calculations for problems.

The distribution of molecular speeds in liquids is similar, although not identical, to that for gases. We can understand the vapor pressure of a liquid and the phenomenon of boiling on this basis. Suppose a molecule must have a

Gas–liquid phase equilibrium: when the gas-to-liquid and liquid-to-gas molecular traffic just balance

speed at least as great as v_A in Fig. 20–7 to escape from the surface of a liquid into the adjacent vapor. The number of such molecules, represented by the area under each curve to the right of v_A, increases rapidly with temperature. Thus the rate at which molecules can escape is strongly temperature-dependent. This process is balanced by another one in which molecules in the vapor phase collide inelastically with the surface and are trapped back into the liquid phase. The number of molecules suffering this fate, per unit time, is proportional to the pressure in the vapor phase. Phase equilibrium between liquid and vapor occurs when these two competing processes proceed at exactly the same rate. Hence if we have an accurate collection of curves such as Fig. 20–7 for a variety of temperatures, we can make a theoretical prediction of the vapor pressure of a substance as a function of temperature.

Rates of chemical reactions are often strongly temperature-dependent, and the Maxwell–Boltzmann distribution contains the reason for this dependence. When two reacting molecules collide, the reaction can occur only when the molecules are close enough for the electric-charge distributions of their electrons to interact strongly. This requires a minimum energy, called the *activation energy,* and thus a certain minimum speed. We will call this speed v_A, although it is not necessarily equal to the v_A in the vapor-pressure discussion above. We have noted that the number of molecules whose speeds exceed some value v_A increases rapidly with temperature. Thus we expect the rate of any reaction that depends on an activation energy to increase rapidly with temperature. Similarly, many plant-growth processes have strongly temperature-dependent rates.

20–7 CRYSTALS

Many materials can exist in a variety of solid forms; a familiar example is the element *carbon.* The black soot deposited on a kettle by a smoky campfire is nearly pure carbon. The "lead" in a pencil is not lead at all, but chiefly a

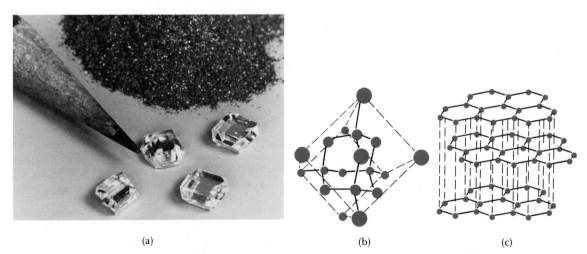

(a)   (b)   (c)

20–8 (a) Two different crystalline forms of carbon: graphite and diamond. (Courtesy of General Electric.) (b) Crystal structure of diamond; each atom is located at the center of a regular tetrahedron, with four equidistant nearest-neighbor atoms at the four corners. The diamonds shown are synthetic, made by subjecting graphite to extreme temperature and pressure; they are much more costly than natural diamonds of comparable size (about one carat). (c) Crystal structure of graphite, showing two-dimensional hexagonal arrays with strong bonds between atoms in each plane and much weaker bonds between planes.

20–9 Close packing of the sodium ions (black spheres) and chlorine ions (white spheres) in a sodium chloride crystal. (Courtesy of Alan Holden, reprinted with permission of Educational Services, Inc., from *Crystals and Crystal growing*, Doubleday and Co., New York, 1960.)

different form of carbon called graphite. Diamond is a third form of solid carbon.

The remarkably dissimilar mechanical, thermal, electrical, and optical properties of these three forms of carbon can be understood in terms of the different arrangement of the carbon atoms in the solid structures. In soot the atoms have no regular arrangement and are said to form an **amorphous solid.** A **crystalline solid,** as we discussed in Section 20–1, is characterized by an orderly geometric arrangement of atoms or molecules forming a recurrent pattern, called a crystal lattice, that extends over many molecules. Graphite and diamond are both crystals, but the orderly arrangement of atoms is different in the two crystal lattices, as shown in Fig. 20–8. Crystallography, the study of the spatial arrangements of atoms (or molecules or ions) in the various types of crystal lattices, is an important and interesting branch of solid-state physics. We will be concerned here with only a few basic ideas that are needed to understand some of the mechanical and thermal properties of single crystals or solids composed of an aggregate of crystals, that is, *polycrystalline solids.*

> In a crystalline solid the atoms or molecules form a regular, recurrent pattern.

The particles constituting a crystal lattice are *closely packed,* as seen in Fig. 20–2 and as suggested by the model in Fig. 20–9, where the sodium and chloride ions of a sodium chloride crystal are represented as spheres touching one another. Actually, each ion consists of a nucleus surrounded by electrons in motion, so that the boundary of an ion must be thought of as the average positions of the outermost electrons. It follows that the outer electrons of one particle in a crystal lattice come close to and even at times interpenetrate the outer electrons of a neighboring particle. In drawings we often exaggerate the distances between neighboring particles, as in the *face-centered cubic* lattice structure shown in Fig. 20–10.

If the chemical composition and density of a very small volume element of a substance are measured at many different places and are found to be the same at all points, the substance is said to be *homogeneous.* If the physical properties determining the transport of heat, electricity, light, and the like, are *the same in all directions,* the substance is said to be *isotropic.* Cubic crystals are

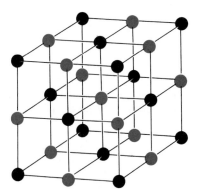

20–10 Symbolic representation of a sodium chloride crystal, with exaggerated distances between ions. Sodium ions are shown as black spheres, chlorine ions as colored spheres.

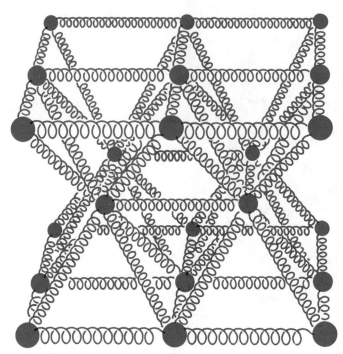

20-11 The forces between neighboring particles in a crystal may be visualized by imagining every particle to be connected to its neighbors by springs. In the case of a cubic crystal, all springs are assumed to have the same spring constant. Anisotropy is associated with differing spring constants in different directions.

both homogeneous and isotropic, whereas all other crystals, although homogeneous, are not isotropic—they are *anisotropic*. Anisotropic crystals have different properties for different directions in the crystal. Such a crystal may have a larger Young's modulus for stretching in one direction than for a perpendicular direction, it may conduct electricity better in some directions than in others; and so on.

The anisotropy of noncubic crystals can be understood in terms of the forces between neighboring particles in a lattice. A convenient way to visualize these forces, as shown in Fig. 20–11, is to use a model that represents a crystal lattice as a large number of spheres, each connected to its neighbors by springs. We imagine the force constants of parallel springs to be equal, but springs pointing in different directions may have different force constants. If we imagine each of the spheres in Fig. 20–11 as vibrating about its equilibrium position, we obtain a rough picture of the dynamic character of a crystal lattice.

The elastic properties of crystals can be partially understood from a diagram such as Fig. 20–11. Young's modulus is simply a measure of the stiffness of the springs in the direction in which the crystal is pulled or pushed. The shear modulus depends on the springs that are stretched and those that are compressed when the crystal is twisted. One of the first tests of the correctness of a lattice structure representing a particular material is to compare a calculated value of an elastic modulus with a measured value.

In Chapter 43 we return to a more detailed study of the electrical and optical properties of crystals. Meanwhile, in the next section we take a brief look at heat capacities of crystals.

The elastic properties of crystals are determined by the behavior of the intermolecular forces.

20–8 HEAT CAPACITY OF CRYSTALS

In Section 20–5 we studied the heat capacities of ideal gases on the basis of a kinetic-molecular model. We can carry out a similar analysis for a crystalline solid. Consider a crystal consisting of N identical atoms. Each atom is bound to an equilibrium position by forces that may be pictured in terms of springs, in the same spirit as Fig. 20–11. Thus each atom can vibrate about its equilibrium position; each atom has three degrees of freedom, corresponding to the three components of velocity needed to describe its vibrational motion. According to the equipartition principle, introduced in Section 20–5, each atom should have an average kinetic energy of $\frac{1}{2}kT$ for each of its three degrees of freedom (corresponding to the three components of velocity). In addition, each atom has on the average some *potential* energy associated with the elastic deformation. Now, for a simple harmonic oscillator (discussed in Chapter 11), it is not hard to show that the average kinetic energy of an atom is *equal* to its average potential energy. In our crystal lattice model, each atom is essentially a three-dimensional harmonic oscillator; and it can be shown that the equality of average kinetic and potential energies also holds here, provided the "spring" forces are proportional to the displacement from the equilibrium position.

Thus we expect each atom to have an average kinetic energy $\frac{3}{2}kT$ and an average potential energy $\frac{3}{2}kT$, or an average total energy $3kT$. The total energy of the whole crystal, which may be expressed in either molecular or molar terms, is

$$U = 3NkT = 3nRT. \tag{20–20}$$

From this relation we conclude that the molar heat capacity of a crystal of N identical atoms (n moles) should be

$$C = 3R. \tag{20–21}$$

In SI units,

$$c = (3)(8{,}314 \text{ J·mol}^{-1}\text{·K}^{-1}) = 24.9 \text{ J·mol}^{-1}\text{·K}^{-1}.$$

But this is just the **law of Dulong and Petit,** which we discussed in Section 15–4. There we regarded it as an *empirical* rule, but now we have *derived* it from kinetic theory. The agreement is only approximate, to be sure, but considering the very simple nature of the model, it is quite significant.

We have ignored the distinction between C_v and C_p. As we remarked in Section 15–4, the difference is usually insignificant for solid materials, unlike gases, where it is always significant.

At low temperatures, the heat capacities of most solids *decrease* with decreasing temperature. As with vibrational energy of molecules, this phenomenon requires quantum-mechanical concepts for its understanding. The vibrating atoms in the crystal lattice can give or lose energy only in certain finite increments. At very low temperatures, the quantity kT is much *smaller* than the smallest energy increment the vibrating atoms can accommodate. As a result, at low T most of the atoms remain in their lowest energy states because the next higher energy level is out of reach. Thus at low temperatures the average vibrational energy per vibrator is *less* than kT, and correspondingly the heat capacity per molecule is *less* than k. But in the other limit, where kT is *large* compared to the minimum energy increment, the classical equipar-

The heat capacity of a crystal is determined by the kinetic and potential energies associated with molecular vibrations.

The law of Dulong and Petit can be derived from a molecular picture of a solid.

Temperature variation of heat capacities of solids: an interesting test of quantum theory

tition theorem holds, and the total heat capacity is $3k$ per molecule, or $3R$ per mole, as the Dulong and Petit relation predicts. This behavior is illustrated by Fig. 15–5. Quantitative understanding of the temperature variation of heat capacities was one of the triumphs of quantum mechanics during its initial development, early in the twentieth century.

SUMMARY

Intermolecular forces are small and usually attractive when the molecules are far apart, as in a gas. At shorter distances the force becomes stronger, then weaker again, then zero, and finally becomes repulsive at sufficiently short distances. Crystalline solids have long-range order, a regular periodic structure extending over many molecules. Liquids and amorphous solids have only short-range order.

Avogadro's number N_A is the number of molecules in a mole. Molecular mass M is the mass of a mole of substance. These are related to the mass m of a single molecule by

$$M = N_A m. \tag{20–1}$$

The kinetic-molecular model of an ideal gas can be used to derive the ideal-gas equation of state ($pV = nRT$) and molar heat capacities of some gases. Several other useful relations also follow from this analysis. The total kinetic energy of translational motion of n moles of gas at temperature T is

$$U = \frac{3}{2}nRT. \tag{20–8}$$

The Boltzmann constant k may be thought of as the gas constant on a "per molecule" basis; $k = R/N_A$. The ideal-gas equation of state may be written in the alternative form

$$pV = NkT, \tag{20–13}$$

and Eq. (20–8) can also be expressed as

$$U = \frac{3}{2}NkT, \tag{20–10}$$

where N is the number of molecules, related to the number of moles n by $N = N_A n$.

The rms speed of molecules in an ideal gas is given by

$$v_{\text{rms}} = \sqrt{(v^2)_{\text{av}}} = \sqrt{\frac{3kT}{m}} = \sqrt{\frac{3RT}{M}}. \tag{20–12}$$

The molar heat capacity at constant volume for an ideal monatomic gas is given by

$$C_v = \frac{3}{2}R. \tag{20–14}$$

Diatomic and polyatomic gases have additional energy associated with rotational and sometimes vibrational motion. The principle of equipartition of energy states that on average there is a kinetic energy $\frac{1}{2}kT$ per molecule for each degree of freedom (component of velocity, either linear or angular).

The Maxwell–Boltzmann distribution describes the speed distribution of molecules in a gas. The height of the curve for any value of v is determined by the ratio of the corresponding kinetic energy of translational motion of a molecule, ϵ, to the quantity kT. This distribution function can be used to determine the fraction of all molecules that have speed greater than a certain value; this information can be used in the study of evaporation and chemical reactions, where a certain minimum molecular kinetic energy is required.

The equipartition principle can be used to calculate the molar heat capacity of a crystal; each molecule has, on average, $\frac{3}{2}kT$ of kinetic energy and an additional $\frac{3}{2}kT$ of potential energy. The resulting molar heat capacity is $C = 3R$, or about 25 J·mol^{-1}·K^{-1}. Thus the law of Dulong and Petit can be derived from the kinetic theory of gases.

QUESTIONS

20–1 Which has more atoms, a kilogram of hydrogen or a kilogram of lead? Which has more mass?

20–2 Chlorine is a mixture of two isotopes, one having molecular mass 35 g·mol^{-1}, the other 37 g·mol^{-1}. Which molecules move faster, on average?

20–3 The proportion of various gases in the earth's atmosphere changes somewhat with altitude. Would you expect the proportion of oxygen at high altitude to be greater or less than at sea level?

20–4 *Comment on this statement:* When two gases are mixed, if they are to be in thermal equilibrium they must have the same average molecular speed.

20–5 In deriving the ideal-gas equation from the kinetic-molecular model, we ignored potential energy due to the earth's gravity. Is this omission justified?

20–6 In deriving the ideal-gas equation, we assumed the number of molecules to be very large, so that we could compute the average force due to many collisions; however, the ideal-gas equation holds accurately only at low pressures, where the molecules are few and far between. Is this inconsistent?

20–7 Some elements of solid crystalline form have molar heat capacities *larger* than $3R$. What effects could account for this?

20–8 Considering molecular speeds, is v_{rms} equal to the *most probable* speed, as indicated by the peak on one of the curves in Fig. 20–7?

20–9 A gas storage tank has a small leak. The pressure in the tank drops more quickly if the gas is hydrogen or helium than if it is oxygen. Why?

20–10 Consider two specimens of gas at the same temperature, both having the same total mass but different molecular masses. Which has the greater internal energy? Does your answer depend on the molecular structure of the gases?

20–11 A process called *gaseous diffusion* is sometimes used to separate isotopes of uranium, i.e., atoms of the element having different masses, such as ^{235}U and ^{238}U. Can you speculate on how this might work?

20–12 When food is preserved by freeze-drying, it is first quick-frozen, then placed in a vacuum chamber and irradiated with infrared radiation. What is the purpose of the vacuum? The radiation?

EXERCISES

Molecular data

N_A = Avogadro's number
    = 6.02×10^{23} molecules·mol^{-1}.
Mass of a hydrogen atom = 1.67×10^{-27} kg.
Mass of a nitrogen molecule = $28 \times 1.66 \times 10^{-27}$ kg.
Mass of an oxygen molecule = $32 \times 1.66 \times 10^{-27}$ kg.

Section 20–2 Avogadro's Number

20–1 How many moles are there in a glass of water (0.2 kg)? How many molecules?

20–2 Consider an ideal gas at 0°C and 1 atm pressure. Imagine each molecule to be, on average, at the center of a small cube.

a) What is the length of an edge of this cube?

b) How does this distance compare with the diameter of a molecule?

20–3 Consider 1 mol of liquid water.

a) What volume is occupied by this amount of water?

b) Imagine each molecule to be, on average, at the center of a small cube. What is the length of an edge of this cube?

c) How does this distance compare with the diameter of a molecule?

20–4 What is the length of the side of a cube, in a gas at standard conditions, that contains a number of molecules equal to the population of the United States (about 200 million)?

Section 20–4 Kinetic-Molecular Theory of an Ideal Gas

20–5 At what temperature is the rms speed of oxygen molecules equal to the rms speed of hydrogen molecules at 0°C?

20–6 A flask contains a mixture of mercury vapor, neon, and helium. Compare

a) the average kinetic energies of the three types of atoms;

b) the root-mean-square speeds.

The molecular masses are: helium, 4 g·mol^{-1}; neon, 20 g·mol^{-1}; mercury 201 g·mol^{-1}

20–7 Isotopes of uranium are sometimes separated by gaseous diffusion, using the fact that the rms speeds of the molecules in vapor are slightly different, and hence the vapors diffuse at slightly different rates. If the atomic masses for ^{235}U and ^{238}U are 0.235 kg·mol^{-1} and 0.238 kg·mol^{-1}, respectively, what is the ratio of the rms speed of the ^{235}U atoms in the vapor to that of the ^{238}U atoms, assuming the temperature is uniform?

20–8 Smoke particles in the air typically have masses of the order of 10^{-16} kg. The Brownian motion of these particles resulting from collisions with air molecules can be observed with a microscope.

a) Find the root-mean-square speed of Brownian motion for such a particle in air at 300 K.

b) Would the speed be different if the particle were in hydrogen gas at the same temperature? Explain.

20–9

a) What is the average translational kinetic energy of a molecule of oxygen at a temperature of 300 K?

b) What is the average value of the square of its speed?

c) What is the root-mean-square speed?

d) What is the momentum of an oxygen molecule traveling at this speed?

e) Suppose a molecule traveling at this speed bounces back and forth between opposite sides of a cubical vessel 0.10 m on a side. What is the average force it exerts on

the walls of the container? (Assume that the molecule's velocity is perpendicular to the two sides that it strikes.)

f) What is the average force per unit area?

g) How many molecules traveling at this speed are necessary to produce an average pressure of 1 atm?

h) Compute the number of oxygen molecules actually contained in a vessel of this size, at 300 K and atmospheric pressure.

i) Your answer for (h) should be three times as large as the answer for (g). Where does this discrepancy arise?

Section 20–5 Molar Heat Capacity of a Gas

20–10 Compute the specific heat capacity at constant volume of hydrogen gas, and compare with the specific heat capacity of water.

20–11 Calculate the molar heat capacity at constant volume of water vapor, assuming the triatomic molecule has three translational and three rotational degrees of freedom and that vibrational motion does not contribute. The actual specific heat capacity of water vapor at low pressures is about 2000 J·kg^{-1}·K^{-1}. Compare this with your calculation, and comment on the actual role of vibrational motion.

Section 20–6 Distribution of Molecular Speeds

20–12 Derive Eq. (20–18) from Eq. (20–17).

20–13 Prove that $f(v)$ as given by Eq. (20–18) is maximum for $\epsilon = kT$.

Section 20–7 Crystals

20–14 The density of crystalline sodium chloride is 2160 kg·m^{-3}, and its crystal structure is shown in Fig. 20–10. Find the spacing of adjacent atoms in the crystal lattice. The atomic mass of sodium is 23.0 g·mol^{-1} and that of chlorine is 35.5 g·mol^{-1}.

PROBLEMS

20–15 The lowest pressures readily attainable in the laboratory are of the order of 10^{-13} atm. At this pressure and ordinary temperature (say $T = 300$ K), how many molecules are present in a volume of 1 cm^3?

20–16 Experiment shows that the size of an oxygen molecule is of the order 2×10^{-10} m. Make a rough estimate of the pressure at which the finite volume of the molecules should cause noticeable deviations from ideal-gas behavior at ordinary temperatures ($T = 300$ K).

20–17 The speed of propagation of a sound wave in air at 27°C is about 350 m·s^{-1}. Compare this with

a) v_{rms} for nitrogen molecules;

b) the root-mean-square value of v_x at this temperature.

(If sound propagation were an isothermal process, which it

ordinarily is not, the speed of sound would be equal to $(v_x)_{rms}$. See Section 21–6.)

20–18

a) Compute the increase in gravitational potential energy of an oxygen molecule for an increase in elevation of 1 m near the earth's surface.

b) At what temperature is this equal to the average kinetic energy of oxygen molecules?

20–19

a) For what mass of molecule or particle is v_{rms} equal to 1 m·s^{-1} at a temperature of 300 K?

b) If the particle is an ice crystal, how many molecules does it contain?

c) Estimate the size of the particle. Would it be visible to the naked eye? Through a microscope?

20–20 The surface of the sun has a temperature of about 6000 K and consists largely of hydrogen atoms. (In fact, most of the atoms will be ionized, so the hydrogen "atom" is actually a proton.)

a) Find the rms speed and average kinetic energy of a hydrogen atom at this temperature.

b) Show that the escape velocity for a particle to leave the gravitational influence of the sun is given by $(2GM/R)^{1/2}$, where M is the sun's mass, R its radius, and G the gravitational constant (Example 7–10).

c) Can appreciable quantities of hydrogen escape from the sun's gravitational field?

20–21

a) Show that a projectile of mass m can "escape" from the earth's gravitational field if it is launched vertically upward with a kinetic energy greater than mgR, where g is the acceleration due to gravity at the earth's surface, and R is the earth's radius. (See the preceding problem.)

b) At what temperature would an average oxygen molecule have this much energy? An average hydrogen molecule?

20–22 For each of the triatomic gases in Table 18–1, compute the value of C_v on the assumption that there is no vibrational energy. Compare with the measured values in the table, and compute the fraction of the total heat capacity due to vibration for each of the three gases. (*Note.* CO_2 is linear; SO_2 and H_2S are not.)

CHALLENGE PROBLEMS

20–23 A commonly used potential-energy function for the interaction of two molecules (Fig. 20–1) is the Lennard–Jones 6–12 potential:

$$U(r) = U_0\left[\left(\frac{\sigma}{r}\right)^{12} - \left(\frac{\sigma}{r}\right)^6\right].$$

a) What is the force $F(r)$ that corresponds to this potential? Sketch $U(r)$ and $F(r)$.

b) In terms of σ and U_0, what are the values of r_1 defined by $U(r_1) = 0$ and of r_2 defined by $F(r_2) = 0$? Also, what is the ratio r_1/r_2? Show the location of r_1 and r_2 on your sketch of $U(r)$.

c) If the molecules are located a distance r_2 apart, as calculated in (b), how much work must be done to pull them apart such that $r \to \infty$?

20–24

a) Show that

$$\int_0^\infty f(v)\, dv = 1,$$

where $f(v)$ is the Maxwell–Boltzmann distribution of Eq. (20–17).

b) Calculate

$$\int_0^\infty v^2 f(v)\, dv$$

and compare this result to $(v^2)_{av}$ as given by Eq. (20–12). (*Hint:* You may use the tabulated integral

$$\int_0^\infty x^{2n} e^{-ax}\, dx = \frac{1\cdot3\cdot5\cdots(2n-1)}{2^{n+1}a^n}\sqrt{\frac{\pi}{a}},$$

where n is an integer and a is a positive constant.)

PART FOUR

WAVES

PERSPECTIVE

In the past nine chapters we studied various aspects of the behavior of matter. We began with mechanical properties of deformable materials, including *elastic* deformations and the concepts of *stress* and *strain,* and with several aspects of the equilibrium and flow of *fluids.* We then expanded our study of properties of matter to include *thermal phenomena* and *thermal properties* of matter, including the concepts of *temperature* and *heat,* thermal expansion, and mechanisms for heat transfer. We generalized our description of the mechanical and thermal behavior of materials with the concept of the *equation of state* of a substance.

But the heart of this section of our book is the subject of *thermodynamics.* We studied two very general and very powerful principles. The first law of thermodynamics extends the principle of conservation of energy to include internal energy of matter, and the second law states in very general terms a principle governing the direction in which a thermodynamic process tends to proceed. In formulating these principles we took a strongly empirical and inductive approach, relying heavily on experimental observations of a wide variety of physical phenomena. Both these laws are new generalizations; neither can be derived from principles we have encountered previously. The fact that the principles of thermodynamics can be applied to such a wide variety of physical systems makes them among the most powerful and useful of all the laws of physics. Finally, we have seen that for the simple case of an ideal gas, some of the mechanical and thermal properties of a material can be understood on the basis of its microscopic or molecular structure. With respect to the foundations of thermodynamics, consideration of microscopic structure of a material is not a substitute for the careful observation of and induction from macroscopic behavior, but it does give us additional insight into the nature of such concepts as pressure, heat, internal energy, and entropy.

Next we turn to a discussion of *wave phenomena.* A wave is a disturbance from an equilibrium state of a system that can travel or propagate from one region of space to another. Waves and wave phenomena occur in all areas of physics, including acoustics (a branch of mechanics), electromagnetism, optics, and quantum mechanics. To introduce wave concepts in as simple and familiar a context as possible, we consider various wave phenomena in *mechanical* systems, including waves traveling on stretched strings or ropes and compression waves in solids and fluids. We learn the appropriate mathematical language for describing waves, and we study the special features of *periodic* waves, in which each particle of a material undergoes a periodic motion during wave propagation. Then we examine several applications of these concepts to vibrating systems and acoustical phenomena. Although we concentrate here on mechanical waves, the broad variety of wave phenomena found in all branches of physics makes the wave concept one of the important unifying threads that run through the entire fabric of physics, giving it strength and cohesiveness.

21
MECHANICAL WAVES

A **WAVE** IS ANY DISTURBANCE FROM AN EQUILIBRIUM CONDITION THAT travels or *propagates* with time from one region of space to another. Examples of wave phenomena are everywhere around us; sound, light, ocean waves, radio and television transmission, and earthquakes are all wave phenomena. One of the most familiar kinds of wave is shown in Fig. 21–1. Wave phenomena occur in all branches of physical and biological science, and the concept of waves is one of the most important unifying threads running through the entire fabric of the natural sciences.

In this chapter and the next two we consider **mechanical waves,** which always travel within some material substance called the **medium** for the wave. Some waves are **periodic;** in these the particles of the medium undergo periodic motions during wave propagation. If the periodic motions are *sinusoidal,* the result is a **sinusoidal wave,** a type of periodic wave of special importance. We can also develop relations between the speed of propagation of a wave and the mechanical properties of the medium. The concepts of this chapter form an important part of the foundation for the study of other kinds of wave phenomena, including electromagnetic waves, later in this book.

21–1 A small ball dropped vertically into water produces a wave pattern that moves radially outward from the source of the wave. The wave crests and troughs are concentric circles. (Photo by Fundamental Photographs, New York.)

21–1 MECHANICAL WAVE PHENOMENA

In this chapter we are concerned with *mechanical* waves. Each type of mechanical wave is associated with some material or substance called the *medium* for that type. As the wave travels through the medium, the particles that make up the medium undergo displacements of various kinds, depending on the nature of the wave.

Here are a few examples of mechanical waves. In Fig. 21–2a the medium is a long spring or "Slinky," or even just a wire or cord under tension. If we give the left end a small sideways shake or wriggle, the wriggle travels down the length of the spring, with successive sections of spring going through the same sideways motion that was given at the end, but at successively later times. Because the displacements of the medium are perpendicular (transverse) to

In a transverse wave, the motions of the particles of the medium are perpendicular to the direction of wave travel.

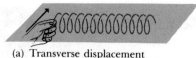

(a) Transverse displacement

(b) Longitudinal displacement

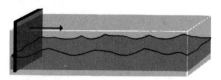

(c) Longitudinal and transverse
    displacement

21–2 Propagation of disturbances.

Many mechanical waves travel with a
definite speed, called the wave speed.

the direction of travel of the wave along the medium, this is called a **transverse wave.**

In Fig. 21–2b the medium is a liquid or a gas in a tube with a rigid wall at the right end and a movable piston at the left end. If the piston is given a back-and-forth motion, again a displacement travels down the length of the medium. This time the motions of the particles of the medium are back and forth along the same direction as the travel of the wave, so this type of wave is called a **longitudinal wave.**

In Fig. 21–2c the medium is water in a trough such as an irrigation ditch or canal. When the flat board at the left end is moved back and forth, a disturbance travels down the length of the trough; careful observation shows that in this case the displacements of the water have both longitudinal and transverse components.

These examples have several things in common. First, the medium itself does not travel through space; its individual particles undergo back-and-forth motions around their equilibrium positions. What *does* travel is the wave disturbance. Second, to set any of these systems in motion requires energy input in the form of mechanical work, and the wave motion has the effect of transporting this energy from one region of space to another. Third, the wave motion can be used to transmit information, a function that is familiar with sound waves. Thus waves transport energy, but not matter, from one region to another. Finally, observation shows that in each case the disturbance travels with a definite speed through the medium. This speed is called the **wave speed,** and it is determined by the mechanical properties of the medium. We will use the symbol c to denote wave speed.

Not all waves are mechanical in nature. Another broad class is *electromagnetic* waves, including light, radio waves, infrared and ultraviolet radiation, x-rays, and gamma rays. There is *no* medium for electromagnetic waves; they can travel through empty space. Still another class of wave phenomena is the wavelike behavior of fundamental particles in some situations; this behavior forms part of the foundation of quantum mechanics, the basic theory used for the analysis of atomic and molecular structure. We will return to electromagnetic waves in Chapter 35 and to the wave nature of particles in Chapter 42. Meanwhile, we can learn the essential language of waves in the context of mechanical waves.

21–2 PERIODIC WAVES

To introduce basic concepts as simply as possible, we concentrate our discussion first on one specific kind of mechanical wave, namely, transverse waves on a stretched string or rope. We tie one end of a long, flexible rope to a stationary object and hold the other end, stretching the rope tight. We then give this end some transverse (sideways) motion. If we give it a single "flip" or "wiggle," the result is a single *wave pulse*, which travels down the length of the string.

A more interesting situation develops when we give the free end of the rope a repetitive or *periodic* motion. (We studied periodic motion in Chapter 11, and we suggest you review that discussion now.) In particular, suppose we move it back and forth with *simple harmonic motion* of amplitude A, frequency f, and period τ, where as usual $f = 1/\tau$.

In a periodic wave, each particle of the
medium undergoes periodic motion with
a definite frequency and period.

A *continuous succession* of transverse sinusoidal waves then advances along the string. The shape of a portion of the string near the end, at intervals of $\frac{1}{8}$ period, is shown in Fig. 21–3 for a total time of one period. The waveform advances steadily toward the right, as indicated by the short arrow pointing to one particular wave crest, while any one point on the string (see the black dot) oscillates back and forth about its equilibrium position with simple harmonic motion. Be careful to distinguish between the motion of a *waveform*, which moves with constant speed *along* the string, and the motion of a *particle of the string*, which moves with simple harmonic motion *transverse* to the string.

The distance between two successive maxima (or between any two successive points at the same position in the repeating wave shape) is the **wavelength** of the wave, denoted by λ. Since the waveform, traveling with constant speed c, advances a distance of one wavelength in a time interval of one period, it follows that $c = \lambda/\tau$, or since $f = 1/\tau$,

$$c = \lambda f. \tag{21–1}$$

That is, *the speed of propagation equals the product of frequency and wavelength.*

To understand the mechanics of a *longitudinal* wave, consider a long tube filled with a compressible fluid, with a plunger at the left end, as shown in Fig. 21–4. Suppose we move the plunger with simple harmonic motion along a line parallel to the direction of the tube. During a part of each oscillation, a region whose pressure is greater than the equilibrium pressure is formed. Such a region is called a *condensation* and is represented in the figure by a darkly shaded area. Following the production of a condensation, a region forms where the pressure is lower than the equilibrium value. This region is called a *rarefaction* and is represented by a lightly shaded area in the figure. The condensations and rarefactions move to the right with constant speed c, as indicated by successive positions of the small vertical arrow. The speed of

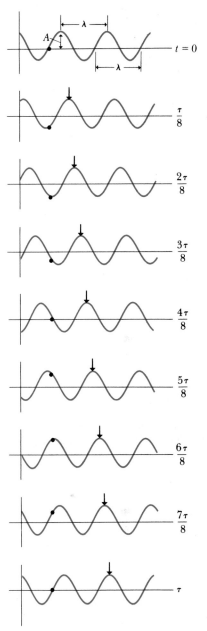

21–3 A sinusoidal transverse wave traveling toward the right. The shape of the string is shown at intervals of one-eighth of a period. The vertical scale is exaggerated.

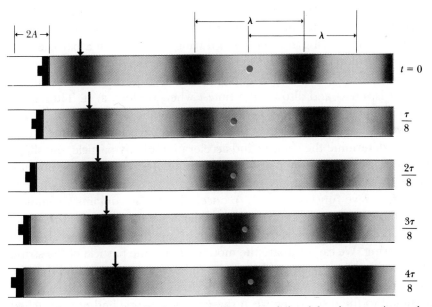

21–4 A sinusoidal longitudinal wave traveling toward the right, shown at intervals of one-eighth of a period.

longitudinal waves in air (i.e., sound) at 20°C is 344 m·s^{-1}, or 1130 ft·s^{-1}. The motion of a single particle of the medium, shown by a colored dot, is simple harmonic, parallel to the direction of propagation.

The wavelength is the distance between two successive condensations or two successive rarefactions. The same fundamental equation, $c = f\lambda$, holds in this example, as in all types of periodic or repetitive waves.

EXAMPLE 21–1   What is the wavelength of a sound wave having a frequency of 262 Hz (the approximate frequency of the note "middle C" on the piano)?

SOLUTION   At 20°C the speed of sound in air is 344 m·s^{-1}, as mentioned above. From Eq. (21–1),

$$\lambda = \frac{c}{f} = \frac{344 \text{ m·s}^{-1}}{262 \text{ s}^{-1}} = 1.31 \text{ m}.$$

The "high C" sung by coloratura sopranos is two octaves above middle C. The corresponding frequency is four times as large, $f = 4(262 \text{ Hz}) = 1048 \text{ Hz}$, and the wavelength is one-fourth as large, $\lambda = (1.31 \text{ m})/4 = 0.328 \text{ m}$.

What is the wavelength of "middle C"? Of "high C"?

21–3   MATHEMATICAL DESCRIPTION OF A WAVE

The wave function is a mathematical description of the displacements of points in the medium at various times.

For complete analysis of wave motion we need a mathematical language that provides a detailed description of the motion of the medium during wave propagation. A central element of this language is the concept of **wave function,** which is a function that describes the position of an arbitrary particle in the medium at any time. In this discussion we will concentrate primarily on sinusoidal waves, in which each particle of the medium undergoes simple harmonic motion about its equilibrium position.

As a concrete example, we look first at waves on a stretched string. If we ignore the sag of the string due to gravity, the equilibrium position of the string is along a straight line. We take this to be the *x*-axis of a coordinate system. Waves on a string are *transverse;* during wave motion a particle with equilibrium position *x* is displaced some distance *y* in the direction perpendicular to the *x*-axis. The value of *y* depends on which particle we are talking about (that is, on *x*) and also on the time *t* when we look at it. Thus *y* is a *function* of *x* and *t*; $y = f(x, t)$. If we know this function for a particular wave motion, we can use it to predict the position of any particle at any time. From this we can determine the velocity and acceleration of any particle, the shape of the string, its slope at any point, and anything else related to the position and motion of the string at any time.

In a sinusoidal wave, all points in the medium move with the same frequency but with phase differences.

Thus the wave function $y = f(x, t)$, once it is known, contains a complete description of the motion. Let us now consider wave functions for sinusoidal waves. Suppose a wave travels from left to right (the direction of increasing *x*) along the string. We can compare the motion of any one particle of the string with the motion of a second particle to the right of the first. We find that the second particle has the same motion as the first, but after a time lag that is proportional to the distance between the particles. If one end of a stretched string oscillates with simple harmonic motion, all other points oscillate with simple harmonic motion of the same amplitude and frequency. The *phase* of the motion, however, is different for different points. This means that the cyclic motions of various points are out of step with each other by various fractions of a cycle. For example, if one point has its maximum positive displacement at the same time another has its maximum negative displacement, the two are a half-cycle out of phase. In Section 11–2 we described phase relationships in terms of a *phase angle* θ_0. A phase angle of π (180°) corresponds to 1/2 cycle, $\pi/2$ to 1/4 cycle, and so on.

Suppose the displacement of a particle at the left end (at $x = 0$), where the motion originates, is given by

$$y = A \sin \omega t = A \sin 2\pi f t. \tag{21-2}$$

The time required for the wave disturbance to travel from $x = 0$ to some point x to the right of the origin is given by x/c, where c is the wave speed. The motion of point x at time t is the same as the motion of point $x = 0$ at the earlier time $(t - x/c)$. Thus the displacement of point x at time t is obtained simply by replacing t in Eq. (21–2) by $(t - x/c)$, and we find

$$y(x, t) = A \sin \omega \left(t - \frac{x}{c} \right)$$

$$= A \sin 2\pi f \left(t - \frac{x}{c} \right). \tag{21-3}$$

Wave functions for a sinusoidal wave can be written in several alternative forms.

The notation $y(x, t)$ is a reminder that the displacement y is a function of both the location x of the point and the time t.

Equation (21–3) can be rewritten in several alternative forms, conveying the same information in different ways. In terms of the period τ and wavelength λ, we find, using Eq. (21–1),

$$y(x, t) = A \sin 2\pi \left(\frac{t}{\tau} - \frac{x}{\lambda} \right). \tag{21-4}$$

Another convenient form is obtained by defining a quantity k, called the **propagation constant** or the **wave number**:

$$k = \frac{2\pi}{\lambda}. \tag{21-5}$$

In terms of k and the angular frequency ω, the wavelength-frequency relation $c = \lambda f$ becomes

The simplest form of the sinusoidal wave function, in terms of angular frequency and wave number

$$\omega = ck, \tag{21-6}$$

and we can rewrite Eq. (21–4) as

$$y(x, t) = A \sin (\omega t - kx). \tag{21-7}$$

Which of these various forms we use is a matter of convenience in a specific problem; fundamentally they all say the same thing. We should mention in passing that some authors define the wave number as $1/\lambda$ rather than $2\pi/\lambda$. In that case, our k is still called the propagation constant.

At any given *time t*, Eq. (21–3), (21–4), or (21–7) gives the displacement y of a particle from its equilibrium position, as a function of the *coordinate x* of the particle. If the wave is a transverse wave in a string, the equation repre-

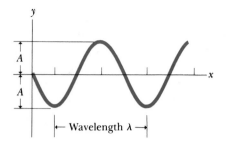

21–5 Waveform at $t = 0$.

sents the *shape* of the string at that instant, as if we have taken a photograph of the string. Thus at time $t = 0$,

$$y = A \sin (-kx) = -A \sin kx = -A \sin 2\pi \frac{x}{\lambda}.$$

This curve is plotted in Fig. 21–5.

At any given *coordinate* x, Eq. (21–3), (21–4), or (21–7) gives the displacement y of the particle at that coordinate, as a function of *time*. That is, it describes the motion of that particle. Thus at the position $x = 0$,

$$y = A \sin \omega t = A \sin 2\pi \frac{t}{\tau}.$$

This curve is plotted in Fig. 21–6.

Wave functions for a wave traveling in the negative direction

The above formulas may be used to represent a wave traveling in the *negative x*-direction by making a simple modification. In this case the displacement of point x at time t is the same as the motion of point $x = 0$ at the *later* time $(t + x/c)$. In Eq. (21–2) we must therefore replace t by $(t + x/c)$. Thus, for a wave traveling in the negative x-direction,

$$y = A \sin 2\pi f\left(t + \frac{x}{c}\right) = A \sin 2\pi \left(\frac{t}{\tau} + \frac{x}{\lambda}\right)$$

$$= A \sin (\omega t + kx). \tag{21–8}$$

We must be careful to distinguish between the *speed of propagation* c of the waveform and the *particle speed* v of a particle of the medium in which the wave is traveling. The wave speed c is given by

$$c = \lambda f = \frac{\omega}{k}. \tag{21–9}$$

The wave function contains a wealth of information about the wave motion it describes.

The particle speed v for any point in a transverse wave (that is, at a fixed value of x) is obtained by taking the derivative of y with respect to t, holding x constant. Such a derivative is called a *partial derivative*, written $\partial y / \partial t$. Thus for a sinusoidal wave given by

$$y = A \sin (\omega t - kx), \tag{21–10}$$

we have

$$v = \frac{\partial y}{\partial t} = \omega A \cos (\omega t - kx). \tag{21–11}$$

The *acceleration* of the particle is the *second* partial derivative:

$$a = \frac{\partial^2 y}{\partial t^2} = -\omega^2 A \sin (\omega t - kx). \tag{21–12}$$

We may also compute partial derivatives with respect to x, holding t constant. The first derivative $\partial y / \partial x$ is the *slope* of the string at any point. The second partial derivative with respect to x is

$$\frac{\partial^2 y}{\partial x^2} = -k^2 A \sin (\omega t - kx). \tag{21–13}$$

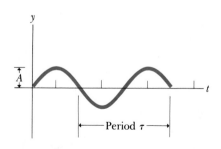

21–6 Vibration at $x = 0$.

It follows from Eqs. (21–12) and (21–13) that

$$\frac{\partial^2 y / \partial t^2}{\partial^2 y / \partial x^2} = \frac{\omega^2}{k^2} = c^2,$$

The wave equation: Every wave function for a physically possible wave must satisfy this equation.

since $c = \omega/k$. The *partial differential equation*

$$\frac{\partial^2 y}{\partial t^2} = c^2 \frac{\partial^2 y}{\partial x^2} \qquad (21\text{--}14)$$

is one of the most important equations in all physics. It is called the **wave equation,** and whenever it occurs, we can conclude immediately that the disturbance described by the function y propagates as a traveling wave along the x-axis with a wave speed c.

Although we have introduced the concept of wave function with reference to *transverse* waves on a string, the concept is equally useful with *longitudinal* waves. The quantity y still measures the displacement of a particle of the medium from its equilibrium position; the difference is that for a longitudinal wave this displacement is *parallel* to the x-axis instead of perpendicular to it. Thus the distance y is measured parallel to the x-axis, and x and y no longer represent distances measured in perpendicular directions, as in the usual xy-coordinate system. If this sounds confusing, don't panic! It will become clear as we study longitudinal waves later in this chapter and in the next.

21–4 SPEED OF A TRANSVERSE WAVE

How is the speed of propagation c of a transverse wave on a string related to the *mechanical* properties of the system? The relevant physical quantities are the *tension* in the string and its *mass per unit length*. Intuition suggests that increasing the tension should increase the speed, while increasing the mass should decrease the speed. We now develop this relationship by two different methods. The first is simple and considers a specific type of waveform; the second is more general but also more formal. Choose whichever you like better!

The speed of a wave on a string is determined by the tension and the mass per unit length.

For our first development, we consider a perfectly flexible string, as shown in Fig. 21–7, having linear mass density (mass per unit length) μ and stretched with a tension S. In Fig. 21–7a the string is at rest. At time $t = 0$, a constant transverse force F is applied at the left end of the string. We might expect that the end would move with constant acceleration; this would certainly occur if the force were applied to a point mass. But here the effect of the force is to set successively more and more mass in motion. The wave travels with constant speed, so the division point between moving and nonmoving portions also travels with definite speed. Hence the total mass in motion is proportional to the time the force has been acting and thus to the impulse of the force. This, in turn, is equal to the total momentum mv of the moving part of the string. The total momentum thus must increase proportionately with time, so the change of momentum must be associated entirely with the increasing amount of mass in motion, not with the increasing velocity of an individual mass element. Force is rate of change of momentum mv, and mv changes because m changes, not v. Hence the end of the string moves upward with constant *velocity v*.

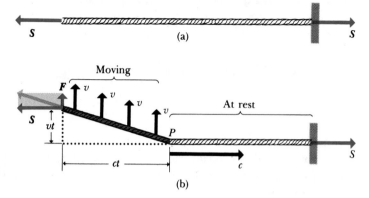

21–7 Propagation of a transverse disturbance in a string.

Figure 21–7b shows the shape of the string after a time t has elapsed. All particles of the string to the left of the point P are moving upward with speed v, and all particles to the right of P are still at rest. The boundary point P between the moving and the stationary portions is traveling to the right with the *speed of propagation c*. The left end of the string has moved up a distance vt, and the boundary point P has advanced a distance ct.

The tension at the left end of the string is the vector sum of the forces S and F. Because no motion occurs in a direction along the length of the string, there is no unbalanced horizontal force, so S, the magnitude of the horizontal component, does not change when the string is displaced. As a result of the increased tension, the string clearly stretches somewhat. It can be shown that, for small displacements, the amount of stretch is approximately proportional to the increase in tension, as we would expect from Hooke's law.

We can obtain an expression for the speed of propagation c by applying the impulse–momentum relation to the portion of the string in motion at time t; that is, the darkly shaded portion in Fig. 21–7. We set the transverse *impulse* (transverse force × time) equal to the change of transverse *momentum* of the moving portion (mass × transverse velocity). The impulse of the transverse force F in time t is Ft. By similar triangles,

$$\frac{F}{S} = \frac{vt}{ct}, \qquad F = S\frac{v}{c}.$$

Hence

$$\text{Transverse impulse} = S\frac{v}{c}t.$$

The transverse momentum of the string is equal to the transverse impulse of the force acting at the end.

The mass of the moving portion of the string is the product of the mass per unit length μ and the length ct. (In its displaced position, the section of string is stretched somewhat, making its mass per unit length somewhat *less* than μ and its length somewhat *greater* than ct. But the mass of the section is still μct, the same as in the undisplaced position.) Hence

$$\text{Transverse momentum} = \mu ctv.$$

Note again that the momentum increases with time *not* because mass is moving faster, as was usually the case in Chapter 8, but because *more mass* is brought into motion. Nevertheless, the impulse of the force F is still equal to

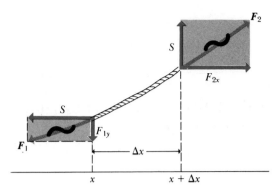

21–8 Free-body diagram for a segment of string whose length in its equilibrium position is Δx. The force at each end of the string is tangent to the string at the point of application; each force is represented in terms of its x- and y-components.

the total change in momentum of the system. Applying this relation, we obtain

$$S\frac{v}{c}t = \mu c t v,$$

and therefore

$$c = \sqrt{\frac{S}{\mu}} \qquad \text{(transverse wave).} \qquad (21\text{–}15)$$

Hence the speed of propagation of a transverse pulse in a string depends only on the tension (a force) and the mass per unit length. Although this calculation of wave speed considered only a very special kind of pulse, a little thought shows that *any* shape of wave disturbance can be considered as a series of pulses with different rates of transverse displacement. Thus, although derived for a special case, Eq. (21–15) is valid for *any* transverse wave motion on a string, including, in particular, the sinusoidal and other periodic waves discussed in Sections 21–2 and 21–3.

Here is an alternative derivation of Eq. (21–15). We apply Newton's second law, $F = ma$, to a small segment of string whose length in the equilibrium position is Δx, as shown in Fig. 21–8. The mass of the segment is $m = \mu\Delta x$; the forces at the ends are represented in terms of their x- and y-components. The x-components have equal magnitude S and add to zero, since the motion is transverse and there is no component of acceleration in the x-direction. To obtain F_{1y} and F_{2y}, we note that the ratio F_{1y}/S is equal in magnitude to the *slope* of the string at point x, and F_{2y}/S is equal to the slope at point $(x + \Delta x)$. Taking proper account of signs, we find

A derivation of the wave equation for a string, starting with Newton's second law

$$\frac{F_{1y}}{S} = -\left(\frac{\partial y}{\partial x}\right)_x, \qquad \frac{F_{2y}}{S} = \left(\frac{\partial y}{\partial x}\right)_{x+\Delta x}. \qquad (21\text{–}16)$$

The notation reminds us that the derivatives are evaluated at points x and $(x + \Delta x)$, respectively. Thus the net y-component of force is

$$F_{1y} + F_{2y} = S\left[\left(\frac{\partial y}{\partial x}\right)_{x+\Delta x} - \left(\frac{\partial y}{\partial x}\right)_x\right]. \qquad (21\text{–}17)$$

We now equate this to the mass ($\mu\Delta x$) times the y-component of acceleration $\partial^2 y/\partial t^2$ to obtain

$$S\left[\left(\frac{\partial y}{\partial x}\right)_{x+\Delta x} - \left(\frac{\partial y}{\partial x}\right)_x\right] = \mu\Delta x\,\frac{\partial^2 y}{\partial t^2}, \qquad (21\text{–}18)$$

or, dividing by $S\Delta x$,

$$\frac{\left(\frac{\partial y}{\partial x}\right)_{x+\Delta x} - \left(\frac{\partial y}{\partial x}\right)_{x}}{\Delta x} = \frac{\mu}{S}\frac{\partial^2 y}{\partial t^2}. \tag{21-19}$$

The speed of a wave on a string, derived from the wave equation

We now take the limit as $\Delta x \to 0$. In this limit, the left side of Eq. (21–19) becomes simply the *second* (partial) derivative of y with respect to x:

$$\frac{\partial^2 y}{\partial x^2} = \frac{\mu}{S}\frac{\partial^2 y}{\partial t^2}. \tag{21-20}$$

Now, finally, comes the punch line of our story. Equation (21–20) has exactly the same form as the general differential equation we derived at the end of Section 21–3, Eq. (21–14). Both that equation and Eq. (21–20), just derived, describe the very same wave motion, so they must be identical. Comparing the two equations, we see that for this to be so, we must have

$$c = \sqrt{\frac{S}{\mu}}, \tag{21-21}$$

which is the same expression as Eq. (21–15). We also note that this result confirms our prediction that c should increase when S increases but decrease when μ increases.

PROBLEM-SOLVING STRATEGY: *Mechanical waves*

1. It is helpful to distinguish between *kinematic* problems and *dynamic* problems. In kinematics problems we are concerned only with *describing* motion; the relevant quantities are wave speed, wavelength (or wave number), frequency (or angular frequency), amplitude, and the position, velocity, and acceleration of individual particles. In dynamics problems, concepts such as force and mass enter; the relation of wave speed to the mechanical properties of a system is an example.

2. If λ is given, it is easy to find k, and vice versa. If f is given, it is easy to find τ, and vice versa. If any two of the quantities c, λ, and f are known, it is easy to find the third. In some problems that is all you need. To determine the wave function completely, you need to know A and any two of c, λ, and f. Once this information is known, you can use it in Eq. (21–3),

(21–4), or (21–7) to get the specific wave function for the problem at hand. Once you have that, you can find values of y, v, a, and the slope of the string at any point (values of x) and at any time, by substituting into the wave function.

3. The wave speed c, the tension S, and the mass per unit length μ are related by Eq. (21–15). If you know any two of these, it is easy to find the third. Look to see what you are given in the problem statement. Later we will find similar relations for speeds of longitudinal waves.

4. The vertical component of force at any point in the string can be obtained from Eq. (21–16). This is useful in connection with energy transfer along the string, as we will see later.

Transverse waves on a clothesline: physics in everyday life

EXAMPLE 21–2   A clothesline has a linear mass density of 0.25 kg·m^{-1} and is stretched with a tension of 25 N. One end is given a sinusoidal motion with frequency 5 Hz and amplitude 0.01 m. At time $t = 0$ the end has zero displacement and is moving in the $+y$-direction.

a) Find the wave speed, amplitude, angular frequency, period, wavelength, and wave number.

b) Write a wave function describing the wave.

c) Find the position of the point at $x = 0.25$ m at time $t = 0.1$ s.

d) Find the transverse velocity at the point $x = 0.25$ m at time $t = 0.1$ s.

e) Find the slope of the string at the point $x = 0.25$ m at time $t = 0.1$ s.

SOLUTION

a) From Eq. (21–15), the wave speed is

$$c = \sqrt{\frac{S}{\mu}} = \sqrt{\frac{25 \text{ N}}{0.25 \text{ kg·m}^{-1}}} = 10 \text{ m·s}^{-1}.$$

The amplitude A is just the amplitude of the motion of the endpoint, $A = 0.01$ m. The angular frequency is

$$\omega = 2\pi f = 2\pi(5 \text{ s}^{-1}) = 31.4 \text{ s}^{-1}.$$

The period is $\tau = 1/f = 0.2$ s. The wavelength is obtained from Eq. (21–1):

$$\lambda = \frac{c}{f} = \frac{10 \text{ m·s}^{-1}}{5 \text{ s}^{-1}} = 2 \text{ m}.$$

The wave number is obtained from Eq. (21–5) or (21–6):

$$k = \frac{2\pi}{\lambda} = \frac{2\pi}{2 \text{ m}} = 3.14 \text{ m}^{-1},$$

or

$$k = \frac{\omega}{c} = \frac{31.4 \text{ s}^{-1}}{10 \text{ m·s}^{-1}} = 3.14 \text{ m}^{-1}.$$

b) The wave function, if we use the form of Eq. (21–4), is

$$y = (0.01 \text{ m}) \sin 2\pi\left(\frac{t}{0.2 \text{ s}} - \frac{x}{2 \text{ m}}\right)$$
$$= (0.01 \text{ m}) \sin [(31.4 \text{ s}^{-1})t - (3.14 \text{ m}^{-1})x].$$

This equation can also be obtained from Eq. (21–7), if we use the values of ω and k obtained above.

c) The displacement of the point $x = 0.25$ m at time $t = 0.1$ s is given by substituting these values into either of the above wave equations:

$$y = (0.01 \text{ m}) \sin 2\pi\left(\frac{0.1 \text{ s}}{0.2 \text{ s}} - \frac{0.25 \text{ m}}{2 \text{ m}}\right)$$
$$= (0.01 \text{ m}) \sin 2\pi(0.375) = 0.00707 \text{ m}.$$

d) The transverse velocity at any time is given by

$$v = \frac{\partial y}{\partial t} = \omega A \cos (\omega t - kx).$$

In the present problem,

$$v = (31.4 \text{ s}^{-1})(0.01 \text{ m}) \cos [(31.4 \text{ s}^{-1})(0.1 \text{ s}) - (3.14 \text{ m}^{-1})(0.25 \text{ m})]$$
$$= -0.22 \text{ m·s}^{-1}.$$

e) The slope at any point at any time is given by

$$\frac{\partial y}{\partial x} = -kA \cos (\omega t - kx).$$

In the present problem,

$$\frac{\partial y}{\partial x} = -(3.14 \text{ m}^{-1})(0.01 \text{ m}) \cos \left[(31.4 \text{ s}^{-1})(0.1 \text{ s}) - (3.14 \text{ m}^{-1})(0.25 \text{ m})\right]$$

$$= 0.022.$$

For a wave traveling in the $+x$-direction, the transverse velocity is positive at every point where the slope is negative, and negative where the slope is positive. Can you prove this?

Polarization: an important property of all transverse waves, including light

An important property of transverse waves is **polarization.** When we produce a transverse wave on a string, we have a choice between moving the end up and down or sideways; in either case the wave displacements are perpendicular or *transverse* to the length of the string. If the end moves up and down, the motion of the entire string is confined to a vertical plane; if the end moves sideways, the wave moves in a horizontal plane. In either case the wave is said to be *linearly polarized* because the individual particles move back and forth in straight lines perpendicular to the string.

The motion may also be more complex, containing both vertical and horizontal components. Combining two sinusoidal motions of equal amplitude, with a quarter-cycle phase difference, results in a wave in which each particle moves in a circular path perpendicular to the string. Such a wave is said to be *circularly polarized*.

A polarizing filter sorts out waves with differing states of polarization.

We can make a device to separate the various components of motion. We cut a thin slot in a flat board, thread the string through it, and orient the plane of the board perpendicular to the string. Then any transverse motion parallel to the slot passes through unimpeded, while any motion perpendicular to the slot is blocked. Such a device is called a *polarizing filter*. Analogous optical devices for polarized light form the basis of some kinds of sunglasses and also of polarizing filters used in photography. We will discuss polarization of light in Chapter 36.

21–5 SPEED OF A LONGITUDINAL WAVE

Propagation speeds of longitudinal as well as transverse waves are determined by the mechanical properties of the medium, and we can derive relations for longitudinal waves that are analogous to Eq. (21–15) for transverse waves on a string. Here is an example of such a derivation for longitudinal waves in a fluid in a tube. Our analysis will be quite similar to the derivation of Eq. (21–15), and we invite you to compare the two developments.

Figure 21–9 shows a fluid (either liquid or gas) with density ρ in a tube with cross-sectional area A. In the equilibrium state the fluid is under a uniform pressure p. In Fig. 21–9a the fluid is at rest. At time $t = 0$ we start the piston at the left end moving toward the right with constant speed v. This initiates a wave motion that travels to the right along the length of the tube, in which successive sections of fluid begin to move and become compressed at successively later times.

Figure 21–9b shows the fluid after a time t has elapsed. All portions of fluid to the left of point P are moving with speed v, and all portions to the right of P are still at rest. The boundary between the moving and stationary

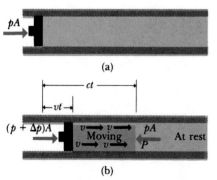

21–9 Propagation of a longitudinal disturbance in a fluid confined in a tube.

portions travels to the right with a speed equal to the speed of propagation c. At time t the piston has moved a distance vt, and the boundary has advanced a distance ct. As with a transverse disturbance in a string, we can compute the speed of propagation from the impulse–momentum theorem.

The quantity of fluid set in motion in time t is the amount that originally occupied a volume of length ct and of cross-sectional area A. The mass of this fluid is therefore ρctA, and the longitudinal momentum it has acquired is

$$\text{Longitudinal momentum} = \rho ctAv.$$

We next compute the increase of pressure, Δp, in the moving fluid. The original volume of the moving fluid, Act, has decreased by an amount Avt. From the definition of bulk modulus B (see Chapter 12),

$$B = \frac{\text{Change in pressure}}{\text{Fractional change in volume}} = \frac{\Delta p}{Avt/Act}.$$

Therefore

$$\Delta p = B\frac{v}{c}.$$

The pressure in the moving fluid is $p + \Delta p$, and the force exerted on it by the piston is $(p + \Delta p)A$. The net force on the moving fluid (see Fig. 21–9b) is ΔpA, and the longitudinal impulse is

$$\text{Longitudinal impulse} = \Delta pAt = B\frac{v}{c}At.$$

Applying the impulse–momentum theorem, we find

$$B\frac{v}{c}At = \rho ctAv\,;$$

hence

$$c = \sqrt{\frac{B}{\rho}} \qquad \text{(longitudinal wave)}. \qquad (21\text{–}22)$$

The speed of propagation of a longitudinal pulse in a fluid therefore depends only on the bulk modulus and the density of the medium. The form of this relation is similar to that of Eq. (21–15); in both cases the numerator is a quantity characterizing the strength of the restoring force, and the denominator is a quantity describing the inertial properties of the medium.

When a longitudinal wave propagates in a *solid* bar, the situation is somewhat different from that of a fluid confined in a tube of constant cross section, because the bar expands slightly sidewise when it is compressed longitudinally. It can be shown, by the same type of reasoning as that just given, that the speed of a longitudinal pulse in the bar is given by

$$c = \sqrt{\frac{Y}{\rho}} \qquad \text{(longitudinal wave)}, \qquad (21\text{–}23)$$

where Y is Young's modulus, defined in Chapter 12.

As with the calculation for a transverse wave on a string, Eqs. (21–22) and (21–23) are valid for all wave motions, not just the special case discussed here. In particular, they are valid for sinusoidal and other periodic waves.

The longitudinal momentum of the medium equals the longitudinal impulse of the force acting at the end.

The speed of a longitudinal wave in a solid depends on the density and Young's modulus.

The wave equation for longitudinal waves in a fluid: an alternative approach to finding the wave speed

21–10 Diagram for illustrating longitudinal traveling waves.

What does the speed of sound in water have to do with dolphin navigation?

We can also derive Eq. (21–22) by using a method analogous to the second derivation in Section 21–4 but applying $F = ma$ to a thin slice of fluid and deriving a partial differential equation. In this case the wave function $y(x, t)$ represents the *longitudinal* displacement of a point in the fluid. We omit the details; the result is

$$\frac{\partial^2 y}{\partial x^2} = \frac{\rho}{B} \frac{\partial^2 y}{\partial t^2}. \tag{21–24}$$

For this equation to be consistent with Eq. (21–14), c *must* be given by Eq. (21–22). A similar derivation for elastic solids leads to Eq. (21–23).

Longitudinal waves, unlike transverse waves, *do not* exhibit polarization. This concept has no meaning for a longitudinal wave.

Since the *shape* of a fluid does not change when a longitudinal wave passes through it, visualizing the relation between particle motion and wave motion is not as easy as for transverse waves in a string. Figure 21–10 may help in correlating these motions. To use this figure, cut a slit about $\frac{1}{16}$ in. wide and 3 in. long in a 3-in. × 5-in. card (or fasten two cards edge-to-edge with a $\frac{1}{16}$-in. gap). Place the card over the figure with the slit along the top of the diagram, and move the card downward with constant velocity. The portions of the sine curves that are visible through the slit correspond to a row of particles along which a longitudinal, sinusoidal wave is traveling. Each particle undergoes simple harmonic motion about its equilibrium position, with a phase that increases continuously along the slit, while the regions of maximum compression and expansion move from left to right with constant speed. Moving the card upward simulates a wave traveling from right to left.

EXAMPLE 21–3   Determine the speed of sound waves in water, and find the wavelength of a wave having a frequency of 262 Hz.

SOLUTION   We use Eq. (21–22) to find the wave speed. From Table 12–2, we find that the compressibility of water, which is the reciprocal of the bulk modulus, is $k = 45.8 \times 10^{-11}$ Pa^{-1}. Thus $B = (1/45.8) \times 10^{11}$ Pa. The density of water is $\rho = 1.00 \times 10^3$ kg·m^{-3}. We obtain

$$c = \sqrt{\frac{B}{\rho}} = \sqrt{\frac{(1/45.8) \times 10^{11} \text{ Pa}}{1.00 \times 10^3 \text{ kg·m}^{-3}}} = 1478 \text{ m·s}^{-1}.$$

This is over four times the speed of sound in air at ordinary temperatures. The wavelength is given by

$$\lambda = \frac{c}{f} = \frac{1478 \text{ m·s}^{-1}}{262 \text{ s}^{-1}} = 5.64 \text{ m}.$$

A wave of this frequency in air has a wavelength of 1.31 m, as obtained in Example 21–1 (Section 21–2).

Dolphins emit high-frequency sound waves (typically 100,000 Hz) and use the echoes for guidance and for hunting. The corresponding wavelength is 1.48 cm. With this high-frequency "sonar" system, they can sense objects of about the size of the wavelength, but not much smaller.

EXAMPLE 21–4   What is the speed of longitudinal sound waves in a steel rod?

SOLUTION We use Eq. (21–23). From Table 12–1, $Y = 2.0 \times 10^{11}$ Pa, and from Table 13–1, $\rho = 7.8 \times 10^3$ kg·m^{-3}. From these data,

$$c = \sqrt{\frac{Y}{\rho}} = \sqrt{\frac{2.0 \times 10^{11} \text{ Pa}}{7.8 \times 10^3 \text{ kg·m}^{-3}}} = 5064 \text{ m·s}^{-1}.$$

21–6 SOUND WAVES IN GASES

In Section 21–5 we derived an expression for the speed of sound in a fluid in a pipe, in terms of its density ρ and bulk modulus B. We learned in Section 18–8 that when a gas is compressed adiabatically, its temperature rises; when it expands adiabatically, its temperature drops. Does this happen when a wave travels through a gas, or does enough heat conduction occur between adjacent layers of gas to maintain a nearly constant temperature throughout? This is a crucial question, because it determines what we use for the bulk modulus B in Eq. (21–22). The bulk modulus is defined in general as in Eq. (12–7): $B = -V \, dp/dV$. If the temperature is constant, then according to Boyle's law the product pV is constant, and we can use this to evaluate dp/dV. But if the process is adiabatic, then Eq. (18–24) states that pV^γ is constant, and we get a different result for B.

The compressions and expansions of a gas during sound wave propagation are adiabatic processes.

Experiments show that for ordinary sound frequencies, say 20 to 20,000 Hz, the thermal conductivity of gases is so small that the propagation of sound is, in fact, very nearly *adiabatic*. Thus we must use the **adiabatic bulk modulus** B_{ad}, derived from the assumption

$$pV^\gamma = \text{constant.} \tag{21–25}$$

We take the derivative of Eq. (21–25) with respect to V:

$$\frac{dp}{dV}V^\gamma + \gamma pV^{\gamma-1} = 0.$$

Dividing by $V^{\gamma-1}$ and rearranging, we find that the adiabatic bulk modulus for an ideal gas is simply

$$-V\frac{dp}{dV} = B_{\text{ad}} = \gamma p. \tag{21–26}$$

For an isothermal process, however, $pV = \text{constant}$, and the isothermal bulk modulus is

$$B_{\text{iso}} = p. \tag{21–27}$$

In each case the bulk modulus (characterizing the material's resistance to being compressed) is proportional to the pressure, but the adiabatic modulus is *larger* than the isothermal by a factor γ.

Sound speed in a gas can be expressed in terms of the pressure, density, and γ.

Combining Eqs. (21–22) and (21–26), we obtain

$$c = \sqrt{\frac{\gamma p}{\rho}} \quad \text{(ideal gas).} \tag{21–28}$$

We can obtain an alternative form by using the relation

Sound speed in a gas depends on molecular mass, temperature, and γ.

$$\frac{p}{\rho} = \frac{RT}{M},$$

where R is the gas constant, M is the molecular mass, and T is the absolute temperature. Therefore

$$c = \sqrt{\frac{\gamma R T}{M}} \quad \text{(ideal gas)}. \qquad (21\text{–}29)$$

For a given gas, γ, R, and M are constants, so the speed of propagation is proportional to the square root of the absolute temperature. It is interesting to compare this result with Eq. (20–12), which gives the rms speed of molecules in an ideal gas. The two expressions are identical except for the numerical factor of 3 in one and γ in the other.

Computing the speed of sound in air

EXAMPLE 21–5 Compute the speed of longitudinal waves in air at an absolute temperature of 300 K.

SOLUTION The mean molecular mass of air is

$$28.8 \text{ g·mol}^{-1} = 28.8 \times 10^{-3} \text{ kg·mol}^{-1}.$$

Also, $\gamma = 1.40$ for air, and $R = 8.314 \text{ J·mol}^{-1}\text{·K}^{-1}$. At $T = 300$ K we obtain

$$c = \sqrt{\frac{(1.40)(8.314 \text{ J·mol}^{-1}\text{·K}^{-1})(300 \text{ K})}{28.8 \times 10^{-3} \text{ kg·mol}^{-1}}}$$

$$= 348 \text{ m·s}^{-1} = 1142 \text{ ft·s}^{-1} = 779 \text{ mi·hr}^{-1}.$$

This result agrees with the measured speed at this temperature to within 0.3%.

Sound consists of longitudinal waves. The ear is sensitive to a range of sound frequencies from about 20 Hz to about 20,000 Hz. From the relation $c = f\lambda$, the corresponding wavelength range is from about 17 m, corresponding to a 20-Hz note, to about 1.7 cm, corresponding to 20,000 Hz.

Bat and dolphin navigation: Why do they have to use such high frequencies?

Like the dolphin, the bat uses high-frequency sound waves for navigation. A typical frequency is 100,000 Hz; the corresponding wavelength in air is about 3.5 mm, small enough to permit detection of flying insects useful as food.

In this discussion we have ignored the *molecular* nature of a gas and have treated it as a continuous medium. Actually, we know that a gas is composed of molecules in random motion, separated by distances that are large compared with their diameters. The vibrations that constitute a wave in a gas are superposed on the random thermal motion. At atmospheric pressure, a molecule travels an average distance of about 10^{-5} cm between collisions, while the displacement amplitude of a faint sound may be only a few ten-thousandths of this amount. An element of gas in which a sound wave is traveling can be compared to a swarm of bees, where the swarm as a whole can be seen to oscillate slightly while individual insects move about through the swarm, apparently at random.

21–7 ENERGY IN WAVE MOTION

Every wave motion has energy associated with it. To initiate a wave motion, we exert a force on a portion of the wave medium; the point where the force is applied moves, so we do *work* on the system. A wave can transport energy from one region of space to another. For example, transmission of energy by electromagnetic waves is familiar, and the destructive power of ocean surf is a convincing demonstration of energy transported by water waves.

As an example of energy considerations in wave motion, we return to the first example of the chapter, transverse waves on a string. To understand how energy is transferred from one portion of string to another, picture a wave traveling from left to right (the positive x-direction) on the string. Considering a particular point on the string, note that according to Eq. (21–16), the portion of string to the left of this point exerts a transverse force F_y on the portion to the right, given by

$$F_y = -S \frac{\partial y}{\partial x}. \qquad (21\text{–}30)$$

Waves on a string transport energy because of the work done by one section on another by the transverse force.

If the instantaneous vertical velocity of the point is v, then the left side of the string does *work* on the right side. The corresponding *power* P (rate of doing work) is given by

$$P = F_y v = -S \left(\frac{\partial y}{\partial x} \right) \left(\frac{\partial y}{\partial t} \right). \qquad (21\text{–}31)$$

To be more specific, suppose the wave is a sinusoidal wave:

$$y(x, t) = A \sin(\omega t - kx). \qquad (21\text{–}32)$$

Then

$$\frac{\partial y}{\partial x} = -kA \cos(\omega t - kx),$$

$$\frac{\partial y}{\partial t} = \omega A \cos(\omega t - kx),$$

$$P = Sk\omega A^2 \cos^2(\omega t - kx). \qquad (21\text{–}33)$$

By using the relations $\omega = ck$ and $c^2 = S/\mu$, we can also express this in the alternative form

$$P = \sqrt{\mu S} \omega^2 A^2 \cos^2(\omega t - kx). \qquad (21\text{–}34)$$

Equation (21–34) gives the *instantaneous* rate of energy transmission along the string; it depends on both x and t. We can obtain the *average* power by using the familiar fact that the *average* value of the $\cos^2$ function, averaged over any whole number of cycles, is $\frac{1}{2}$. Hence the average power is

$$P_{\text{av}} = \tfrac{1}{2} \sqrt{\mu S} \omega^2 A^2. \qquad (21\text{–}35)$$

It is interesting to note that the rate of energy transfer is proportional to the *square* of the amplitude and is also proportional to the square of the frequency.

Analogous relationships can be worked out for longitudinal waves. We will not go into detail; the results are most easily stated in terms of the average power *per unit cross-sectional area* in the wave motion. This power is called the *intensity*, denoted by I. For fluids in a pipe it is given by

The intensity of a wave in a solid or fluid is the average power transmitted per unit area.

$$I = \tfrac{1}{2} \sqrt{\rho B} \omega^2 A^2,$$

and for a solid rod by

$$I = \tfrac{1}{2} \sqrt{\rho Y} \omega^2 A^2.$$

Again the power is proportional to A^2 and to ω^2.

SUMMARY

A wave is any disturbance from an equilibrium condition that propagates from one region to another. A mechanical wave always travels within some material called the medium. In a periodic wave, the motion of each point of the medium is cyclic or periodic; if the motion is sinusoidal, the wave is called a sinusoidal wave. The frequency f of a periodic wave is the number of repetitions per unit time, and the period τ is the time for one cycle. The wavelength λ is the distance over which the spatial pattern repeats. The speed of propagation c is the speed with which the wave disturbance travels. For any periodic wave these quantities are related by

$$c = \lambda f. \tag{21-1}$$

A wave function describes the displacements of individual particles in the medium. It is a function of the undisplaced position x and time t. The wave function for a sinusoidal wave traveling in the $+x$-direction can be written as

$$y(x, t) = A \sin \omega \left(t - \frac{x}{c} \right) = A \sin 2\pi f \left(t - \frac{x}{c} \right), \tag{21-3}$$

or

$$y(x, t) = A \sin 2\pi \left(\frac{t}{\tau} - \frac{x}{\lambda} \right), \tag{21-4}$$

or, using the wave number k, defined as $k = 2\pi/\lambda$, and the angular frequency ω defined as $\omega = 2\pi f$,

$$y(x, t) = A \sin (\omega t - kx). \tag{21-7}$$

In all three forms A is the amplitude, which is the maximum displacement of a particle from its equilibrium position.

The wave function must obey a partial differential equation called the wave equation:

$$\frac{\partial^2 y}{\partial t^2} = c^2 \frac{\partial^2 y}{\partial x^2}. \tag{21-14}$$

The speed of a transverse wave on a string having tension S and mass per unit length μ is given by

$$c = \sqrt{\frac{S}{\mu}} \quad \text{(transverse wave).} \tag{21-15}$$

Transverse waves have the property of polarization; longitudinal waves do not.

The speed of a longitudinal wave in a fluid having bulk modulus B and density ρ is given by

$$c = \sqrt{\frac{B}{\rho}} \quad \text{(longitudinal wave).} \tag{21-22}$$

The speed of a longitudinal wave in a solid rod having Young's modulus Y and density ρ is given by

$$c = \sqrt{\frac{Y}{\rho}} \quad \text{(longitudinal wave).} \tag{21-23}$$

Sound propagation is ordinarily an adiabatic process; the adiabatic bulk modulus for an ideal gas is given by

$$B_{ad} = \gamma p. \qquad\qquad (21\text{--}26)$$

The speed of sound in an ideal gas is given by

$$c = \sqrt{\frac{\gamma RT}{M}} \qquad \text{(ideal gas).} \qquad\qquad (21\text{--}29)$$

Wave motion conveys energy from one region to another. For a transverse wave on a stretched string, a portion of string exerts a transverse force on an adjacent portion while it undergoes a displacement; hence one section does work on another, and energy is transferred along the length of the string.

QUESTIONS

21–1 The term *wave* has a variety of meanings in ordinary language. Think of several, and discuss their relation to the precise physical sense in which the term is used in this chapter.

21–2 What kinds of energy are associated with waves on a stretched string? How could such energy be detected experimentally?

21–3 Is it possible to have a longitudinal wave on a stretched string? A transverse wave on a steel rod?

21–4 The speed of sound waves in air depends on temperature, but the speed of light waves does not. Why?

21–5 For the wave motions discussed in this chapter, does the speed of propagation depend on the amplitude?

21–6 Children make toy telephones by sticking each end of a long string through a hole in the bottom of a paper cup and knotting it so it will not pull out. When the string is pulled taut, sound can be transmitted from one cup to the other. How does this work? Why is the transmitted sound louder than the sound traveling through air for the same distance?

21–7 An echo is sound reflected from a distant object, such as a wall or a cliff. Explain how you can determine how far away the object is by timing the echo.

21–8 Why do you see lightning before you hear the thunder? A familiar rule of thumb is to start counting slowly, once per second, when you see the lightning; when you hear thunder, divide the number you have reached by 5 to obtain your distance (in miles) from the lightning. Why does this work? Or does it?

21–9 When ocean waves approach a beach, the crests are always nearly parallel to the shore, despite the fact that they must come from various directions. Why?

21–10 The amplitudes of ocean waves increase as they approach the shore, and often the crests bend over and "break." Why?

21–11 When a rock is thrown into a pond and the resulting ripples spread in ever-widening circles, the amplitude decreases with increasing distance from the center. Why?

21–12 In speaker systems designed for high-fidelity music reproduction, the "tweeters" that reproduce the high frequencies are always much smaller than the "woofers" used for low frequencies. Why?

21–13 When sound travels from air into water, does the frequency of the wave change? The wavelength? The speed?

21–14 Which of the quantities describing a sinusoidal wave is most closely related to musical pitch? To loudness?

21–15 Musical notes produced by different instruments (such as a flute and an oboe) may have the same pitch and loudness and yet sound different. What is the difference, in physical terms?

EXERCISES

Section 21–2 Periodic Waves

21–1 The speed of sound in air at 20°C is 344 m·s^{-1}.

a) What is the wavelength of a sound wave of frequency 32 Hz, the lowest pedal note on medium-sized pipe organs?

b) What is the frequency of a wave having a wavelength of 1.22 m (4 ft), corresponding approximately to the note D above middle C on the piano?

21–2 The speed of radio waves in vacuum (equal to the speed of light) is 3.00×10^8 m·s^{-1}. Find the wavelength for

a) an AM radio station with frequency 1000 kHz;

b) an FM radio station with frequency 100 MHz.

21–3 Provided the amplitude is sufficiently great, the human ear can respond to longitudinal waves over a range of frequencies from about 20 Hz to about 20,000 Hz. Compute the wavelengths corresponding to these frequencies

a) for waves in air ($c = 345$ m·s^{-1});

b) for waves in water ($c = 1480$ m·s^{-1}).

21–4 The sound waves from a loudspeaker spread out nearly uniformly in all directions when their wavelengths are large compared with the diameter of the speaker. When the wavelength is small compared with the diameter of the speaker, much of the sound energy is concentrated forward. For a speaker of diameter 25 cm, compute the frequency for which the wavelength of the sound waves, in air for which $c = 345$ m·s^{-1}, is

a) 10 times the diameter of the speaker;

b) equal to the diameter of the speaker;

c) $\frac{1}{10}$ the diameter of the speaker.

Section 21–3 Mathematical Description of a Wave

21–5 Show that Eq. (21–4) may be written

$$y = -A \sin \frac{2\pi}{\lambda}(x - ct).$$

21–6 The equation of a certain traveling transverse wave is

$$y = 2 \sin 2\pi\left(\frac{t}{0.01} - \frac{x}{30}\right),$$

where x and y are in centimeters and t is in seconds. What are the wave's

a) amplitude,

b) wavelength,

c) frequency,

d) speed of propagation?

21–7 A traveling transverse wave on a string is represented by the equation in Exercise 21–5. Let $A = 8$ cm, $\lambda = 16$ cm, and $c = 2$ cm·s^{-1}.

a) At time $t = 0$, compute the transverse displacement y at 2-cm intervals of x (that is, at $x = 0$, $x = 2$ cm, $x = 4$ cm, etc.) from $x = 0$ to $x = 32$ cm. Show the results in a graph. This is the shape of the string at time $t = 0$.

b) Repeat the calculations, for the same values of x, at times $t = 1$ s, $t = 2$ s, $t = 3$ s, and $t = 4$ s. Show on the same graph the shape of the string at these instants. In what direction is the wave traveling?

Section 21–4 Speed of a Transverse Wave

21–8 A steel wire 6 m long has a mass of 0.060 kg and is stretched with a tension of 1000 N. What is the speed of propagation of a transverse wave in the wire?

21–9 One end of a horizontal string is attached to a prong of an electrically driven tuning fork whose frequency of vibration is 240 Hz. The other end passes over a pulley and supports a mass of 5 kg. The linear mass density of the string is 0.02 kg·m^{-1}.

a) What is the speed of a transverse wave in the string?

b) What is the wavelength?

21–10 The equation of a transverse traveling wave on a string is

$$y = 2 \cos [\pi(0.5x - 200t)],$$

where x and y are in cm and t is in seconds.

a) Find the amplitude, wavelength, frequency, period, and velocity of propagation.

b) Sketch the shape of the string at the following values of t: 0; 0.0025 s; and 0.005 s.

c) If the mass per unit length of the string is 0.5 kg·m^{-1}, find the tension.

Section 21–5 Speed of a Longitudinal Wave

Section 21–6 Sound Waves in Gases

21–11 A steel pipe 100 m long is struck at one end. A person at the other end hears two sounds as a result of two longitudinal waves, one traveling in the metal pipe and the other traveling in the air. What is the time interval between the two sounds? Take Young's modulus of steel to be 2×10^{11} Pa, the density of steel to be 7800 kg·m^{-3}, and the speed of sound in air to be 345 m·s^{-1}.

21–12 At a temperature of 27°C, what is the speed of longitudinal waves in

a) argon?

b) hydrogen?

c) Compare your answers to (a) and (b) with the speed in air at the same temperature.

21–13 What is the difference between the speeds of longitudinal waves in air at −3°C and at 57°C?

21–14 Use the definition $B = -V(dp/dV)$ and the relation between p and V for an isothermal process to derive Eq. (21–27).

21–15

a) If the propagation of sound waves in gases were characterized by isothermal rather than adiabatic expansions and compressions, and assuming that the gas behaves as an ideal gas, show that the speed of sound would be given by $(RT/M)^{1/2}$.

b) What would be the speed of sound in air at 27°C in this case?

c) Under what circumstances might the wave propagation be expected to be isothermal?

Section 21–7 Energy in Wave Motion

21–16

a) A string of mass 4 g and length 2 m is stretched with a tension of 30 N. Waves of frequency $f = 60$ Hz and amplitude 8 cm are traveling along the string. Calculate the average power carried by these waves.

b) What happens to the average power if the amplitude of the waves is doubled?

21–17 Show that Eq. (21–35) can also be written as $P_{av} = \frac{1}{2}Sk\omega A^2$, where k is the wave number of the wave.

PROBLEMS

21–18 A student who could not get tickets to a Red Sox–Yankees baseball game is listening to the radio broadcast of the game in her dorm room while doing her physics homework. In the bottom of the fourth inning, however, a thunderstorm approaching from the west makes its presence known in three ways: (1) The student sees a lightning flash (and hears the electromagnetic pulse on her radio receiver); (2) 2.4 s later, she hears the thunder over the radio; (3) 4.0 s after the lightning flash, the thunder rattles her window. By a previous careful measurement, she knows that she is 1.12 km due north of the broadcast booth at the ballpark, and that the speed of sound is 350 m·s^{-1}. Where did the lightning flash occur, relative to the ballpark?

21–19 A transverse sine wave of amplitude 0.10 m and wavelength 2 m travels from left to right along a long horizontal stretched string with a speed of 1 m·s^{-1}. Take the origin at the left end of the undisturbed string. At time $t = 0$, the left end of the string is at the origin and is moving downward.

a) What is the frequency of the wave?

b) What is the angular frequency?

c) What is the propagation constant?

d) What is the equation of the wave?

e) What is the equation of motion of the left end of the string?

f) What is the equation of motion of a particle 1.5 m to the right of the origin?

g) What is the maximum magnitude of transverse velocity of any particle of the string?

h) Find the transverse displacement and the transverse velocity of a particle 1.5 m to the right of the origin, at time $t = 3.25$ s.

i) Make a sketch of the shape of the string, for a length of 4 m, at time $t = 3.25$ s.

21–20 Show that $y(x, t) = A \cos(\omega t + kx)$ satisfies the wave equation, Eq. (21–14).

21–21 One end of a rubber tube 20 m long, of total mass 1 kg, is fastened to a fixed support. A cord attached to the other end passes over a pulley and supports a body of mass 10 kg. The tube is struck a transverse blow at one end. Find the time required for the pulse to reach the other end.

21–22 One end of a stretched rope is given a periodic transverse motion with a frequency of 10 Hz. The rope is 50 m long, has a total mass of 0.5 kg, and is stretched with a tension of 400 N.

a) Find the wave speed and the wavelength.

b) If the tension is doubled, how must the frequency be changed to maintain the same wavelength?

21–23 What must be the stress (F/A) in a stretched wire of a material whose Young's modulus is Y, for the speed of longitudinal waves to equal ten times the speed of transverse waves?

21–24 A metal wire, of density 5×10^3 kg·m^{-3}, with a Young's modulus equal to 2.0×10^{11} Pa, is stretched between rigid supports. At one temperature the speed of a transverse wave is found to be 200 m·s^{-1}. When the temperature is raised 25 C°, the speed decreases to 160 m·s^{-1}. Determine the coefficient of linear expansion.

21–25 What is the ratio of the speed of sound in a diatomic gas to the rms speed of gas molecules at the same temperature?

21–26

a) By how many meters per second, at a temperature of 27°C, does the speed of sound in air increase per Celsius degree rise in temperature?

b) Is the change of speed for a 1 C° change in temperature the same at all temperatures?

CHALLENGE PROBLEMS

21–27 Derive Eq. (21–24) by applying $F = ma$ to a thin slice of fluid in which a longitudinal wave is propagating.

21–28 For longitudinal waves in a solid, prove that the power per unit cross-sectional area in the wave motion (the *intensity*) is given by $\frac{1}{2}\omega^2 A^2 \sqrt{\rho Y}$, where ρ and Y are the density and Young's modulus for the material, and ω and A are the angular frequency and amplitude of the waves.

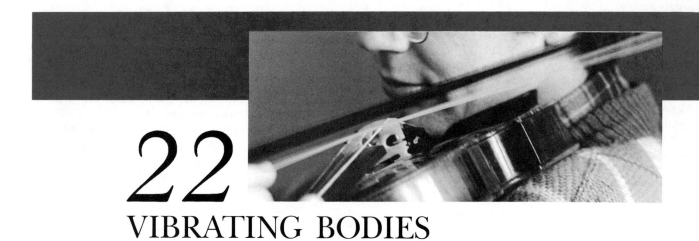

22

VIBRATING BODIES

IN CHAPTER 21 WE STUDIED THE PROPAGATION OF MECHANICAL WAVES IN media without ends or boundaries; we were not concerned with what happens when a wave arrives at an end or boundary of the medium in which it propagates. But in many wave phenomena such boundaries do play a significant role. A familiar example is the echo that occurs when a sound wave reflects from a rigid wall. Such reflections lead to overlap, or *superposition*, of two waves—the initial and reflected waves—in the same region of the medium. When there are two or more boundary points or surfaces, repeated reflections can occur. In such cases it turns out that sinusoidal wave motion is possible only for certain special values of the frequency of the wave, determined by the dimensions and mechanical properties of the medium. These special frequencies and their associated wave patterns are called *normal modes*. Many familiar phenomena are associated with normal modes. The pitches of most musical instruments are determined by normal-mode frequencies, and many other mechanical vibrations involve normal-mode motion. This concept will also reappear later in some unexpected places, such as the energy levels of atoms.

22–1 BOUNDARY CONDITIONS FOR A STRING

When a transverse wave reaches the end
of a string, it is reflected.

As a simple example of reflections and the role of boundaries, let us return to transverse waves on a stretched string. Consider what happens when a wave pulse or a continuous succession of waves (such as a sinusoidal wave) arrives at the end of the string. If the end is fastened to a rigid support, it must remain at rest. The arriving wave exerts a force on the support; the reaction to this force, exerted *by* the support *on* the string, "kicks back" on the string and sets up a *reflected* pulse or wave traveling in the reverse direction.

At the opposite extreme from a rigidly fixed end is one that is perfectly free to move in the direction transverse to the length of the string. For example, the string might be tied to a light ring that slides on a smooth rod perpendicular to the length of the string. The ring and rod maintain the tension of the string but exert no transverse force. At the free end the arriving pulse or wave train causes the end to "overshoot," and again a reflected wave is set up.

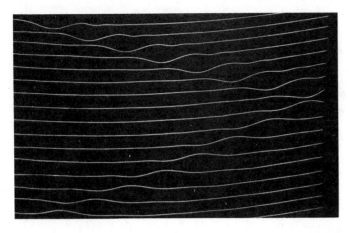

22–1 A pulse starts at the left in the top image, travels to the right, and is reflected from the fixed end of the string at the right.

The conditions imposed on the motion of the end of the string, such as attachment to a rigid support or the complete absence of transverse force, are called **boundary conditions.**

The multiflash photograph of Fig. 22–1 shows the reflection of a pulse at a fixed end of a string. (The camera was tipped vertically while the photographs were taken so that successive images lie one under the other. The "string" is a rubber tube, which sags somewhat.) The pulse is reflected with both its displacement and its direction of propagation reversed. When reflection takes place at a *free* end, the direction of propagation is reversed but the direction of the displacement is unchanged.

A useful way to think of the process of reflection is to imagine that the string extends indefinitely beyond its actual end. Consider the actual pulse to continue on into the imaginary portion as though the support were not there, while at the same time a "virtual" pulse, which has been traveling in the opposite direction in the imaginary portion, moves out into the real string and forms the reflected pulse. The nature of the reflected pulse depends on whether the end is fixed or free. The two cases are shown in Fig. 22–2.

The actual displacement of the string at a point where the actual and virtual pulses cross each other is the *algebraic sum* of the displacements in the

The total displacement of the string is the superposition of the original wave and the reflected wave.

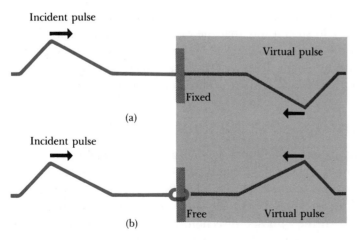

22–2 Description of the reflection of a pulse (a) at a fixed end of a string and (b) at a free end, in terms of an imaginary "virtual" pulse.

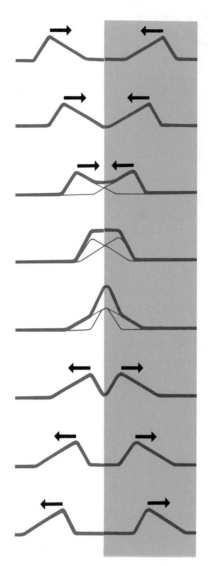

22–3 Reflection at a free end.

22–4 Reflection at a fixed end.

individual pulses. Figures 22–3 and 22–4 show the shape of the region near the end of the string for both types of reflected pulses, Fig. 22–3 for a free end and Fig. 22–4 for a fixed end. In the latter case, the incident and reflected pulses must combine in such a way that the displacement of the end of the string is always zero. This combining of the displacements of the separate pulses at each point to obtain the actual displacement is an example of the *principle of superposition,* which plays a central role in most of this chapter.

22–2 SUPERPOSITION AND STANDING WAVES

Consider a continuous succession of waves, such as a sinusoidal wave, arriving at a boundary point on a string. The boundary point may be a fixed end or an end that it is free to move transversely, as in Fig. 22–2. As we discussed in the preceding section, a corresponding succession of *reflected* waves originates at this end and travels in the opposite direction. The resulting motion of the string is determined by an extremely important principle called the **principle**

The principle of superposition: What happens when several waves are present in a medium at the same time?

of superposition. This principle states that the actual displacement of any point on the string, at any time, is obtained by adding the displacement that point would have if only the first wave were present, and the displacement it would have with only the second wave.

Mathematically speaking, the principle of superposition states that the wave function describing the resulting motion in the above situation, where a wave combines with a reflected wave, is obtained by adding the two wave functions for the two separate waves. This additive property of wave functions depends, in turn, on the form of the wave equation, Eq. (21–14), which every physically possible wave function must satisfy. Specifically, the wave equation is *linear;* that is, it contains derivatives of the function y only to the first power. As a result, if each of any two functions $y_1(x, t)$ and $y_2(x, t)$ satisfies the wave equation separately, their *sum* $y_1 + y_2$ automatically satisfies the wave equation also and hence is a physically possible motion. In view of this linearity of the wave equation and the corresponding linear-combination property of its solutions, the principle is also called the *principle of linear superposition.*

The principle of superposition is of central importance in all types of wave motion. It applies not only to waves on a string, but also to sound waves, electromagnetic waves (such as light), and all other wave phenomena in which the wave equation is linear. The general term **interference** is used to describe phenomena that result from two or more waves passing through the same region at the same time.

When a sinusoidal wave is reflected by a fixed point on a string, the appearance of the resulting motion gives no evidence of two waves traveling in opposite directions. If the frequency is sufficiently great so that the eye cannot follow the motion, the string appears subdivided into a number of segments, as in the time-exposure photograph of Fig. 22–5a. A multiflash photograph of the same string, in Fig. 22–5b, indicates a few of the instantaneous shapes of the string. At every instant its shape is a sine curve, but in a traveling wave the amplitude remains constant as the wave progresses, while here the waveform remains fixed in position (longitudinally) as the amplitude fluctuates. Certain points, known as the **nodes,** remain always at rest. Midway between these points, at the *loops* or **antinodes,** the fluctuations are maximum. The vibration as a whole is called a **standing wave.**

To understand the formation of a standing wave, consider the separate graphs of the waveform at four instants one-tenth of a period apart, shown in Fig. 22–6. The gray curves represent a wave traveling to the right. The light-color curves represent a wave of the same propagation speed, wavelength, and amplitude traveling to the left. The dark-color curves represent the resultant waveform, obtained by applying the principle of superposition, that is, by adding displacements. At those places marked N at the bottom of Fig. 22–6, the resultant displacements are *always* zero. These are the *nodes.* Midway between the nodes, the vibrations have the *largest* amplitude. These are the *antinodes.* It is evident from the figure that

$$\left\{ \begin{array}{c} \text{Distance between adjacent nodes} \\ \text{or} \\ \text{Distance between adjacent antinodes} \end{array} \right\} = \frac{\lambda}{2}.$$

The wave function for the standing wave of Fig. 22–6 may be obtained by adding the displacements of two waves of equal amplitude, period, and wave-

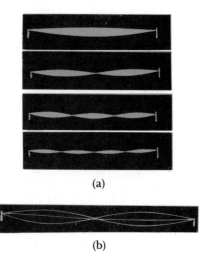

22–5 (a) Standing waves in a stretched string (time exposure). (b) Multiflash photograph of a standing wave, with nodes at the center and at the ends.

A node is a point that never moves during wave motion.

An antinode is a point with maximum amplitude of motion, between two nodes.

A standing wave does not appear to be moving in either the positive or negative direction.

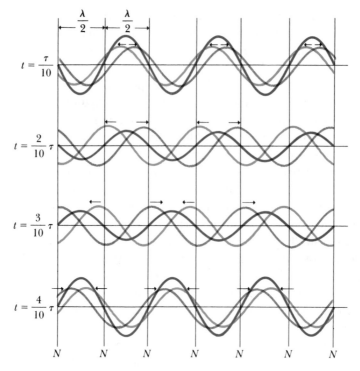

22–6 The formation of a standing wave. A wave traveling to the right (gray lines) combines with a wave traveling to the left (light-color lines) to form a standing wave (dark-color lines). The thin horizontal line in each part shows the equilibrium position of the string.

length, traveling in opposite directions. Thus if

$$y_1 = A \sin (\omega t - kx) \qquad \text{(positive x-direction)},$$
$$y_2 = -A \sin (\omega t + kx) \qquad \text{(negative x-direction)},$$

then

$$y_1 + y_2 = A[\sin (\omega t - kx) - \sin (\omega t + kx)].$$

Introducing the expressions for the sine of the sum and difference of two angles and combining terms, we obtain

A standing wave formed by superposition of two sinusoidal waves moving in opposite directions

$$y_1 + y_2 = -[2A \cos \omega t] \sin kx. \qquad (22\text{–}1)$$

The shape of the string at each instant is, therefore, a sine curve whose amplitude (the expression in brackets) varies with time. The appearance is not that of a traveling wave shape, but of a sinusoidal wave shape in one position with an amplitude that grows larger and smaller with time. Each point in the string still undergoes simple harmonic motion, but all points move *in phase* (or 180° out of phase). This is in contrast to the phase differences between motions of adjacent points that are seen with a wave traveling in one direction.

The nodes are spaced a half-wavelength apart.

The positions of the nodes can be obtained from Eq. (22–1). Wherever $\sin kx = 0$, the displacement is always zero. This occurs when $kx = 0, \pi, 2\pi, 3\pi, \ldots$, or

$$x = 0, \pi/k, 2\pi/k, 3\pi/k, \ldots$$
$$= 0, \lambda/2, \lambda, 3\lambda/2, \ldots. \qquad (22\text{–}2)$$

22–3 NORMAL MODES OF A STRING

We have mentioned the reflection or echo of a sound wave from a rigid wall, and the analogous reflection of a transverse wave on a string from a rigidly held end. Now suppose we have two parallel walls. If a sharp sound pulse, such as a hand clap, originates at a point between the walls, the result is a series of regularly spaced echoes caused by repeated back-and-forth reflection between the walls. In room acoustics this phenomenon is called "flutter echo"; it is the bane of acoustical engineers.

The analogous situation with transverse waves on a string is a string with some definite length L, rigidly held at *both* ends. If a sinusoidal wave is produced on such a string, the wave is reflected and re-reflected. Because the string is held at both ends, both ends must be nodes. Since adjacent nodes are one-half wavelength apart, the length of the string may be

$$\frac{\lambda}{2}, \quad \frac{2\lambda}{2}, \quad \frac{3\lambda}{2},$$

or, in general, *any* integer number of half-wavelengths. To put it differently, if we consider a particular string of length L, standing waves may be set up in the string by vibrations of a number of different frequencies—namely, those that give rise to waves of wavelengths

$$\lambda = 2L, \frac{2L}{2}, \frac{2L}{3}, \cdots = \frac{2L}{n} \qquad (n = 1, 2, 3, \ldots). \qquad (22\text{–}3)$$

The wave speed c is the same for all frequencies. From the relation $f = c/\lambda$, the possible frequencies are

$$f = \frac{c}{2L}, \frac{2c}{2L}, \frac{3c}{2L}, \cdots = \frac{nc}{2L} \qquad (n = 1, 2, 3, \ldots). \qquad (22\text{–}4)$$

The lowest frequency, $c/2L$, is called the **fundamental frequency** f_1. All the other frequencies are integer multiples of f_1, such as $2f_1, 3f_1, 4f_1$, and so on. Thus we can rewrite Eq. (22–4) as

$$f_n = n\frac{c}{2L} = nf_1 \qquad (n = 1, 2, 3, \ldots). \qquad (22\text{–}5)$$

This series of frequencies, all integer multiples of the fundamental, is called a **harmonic series.** Musicians sometimes call f_2, f_3, and so on **overtones;** f_2 is the second harmonic or the first overtone, f_3 is the third harmonic or the second overtone, and so on.

These results may also be obtained directly from Eq. (22–1). The boundary conditions require that $y_1 + y_2 = 0$ at the ends of the string, that is, at $x = 0$ and $x = L$. Since the sine of zero is zero, the first condition is satisfied automatically. The second condition requires that $\sin kL = 0$, and this is true only when k has certain special values. The sine of an angle is zero only when the angle is zero or an integer multiple of π (180°). Thus we must have

$$kL = n\pi \qquad (n = 1, 2, 3, \ldots).$$

We do not include the possibility $n = 0$ because that gives $k = 0$, that is, a wave with zero displacement *everywhere* (a possible case, to be sure, but not a very interesting one!).

Replacing k above by $2\pi/\lambda$, we obtain

$$\frac{2\pi L}{\lambda} = n\pi \quad \text{or} \quad \lambda = \frac{2L}{n}, \qquad (22\text{–}6)$$

in agreement with Eq. (22–3).

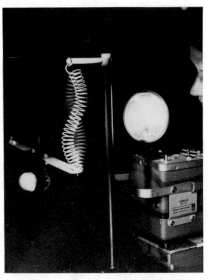

A stroboscopic photograph showing a transverse standing wave on a spring. One end of the spring is held stationary while the other end is vibrated with small amplitude by a motor. The maximum amplitude of the wave is much greater than the amplitude at the driven end, and both ends are approximately node points. (Dr. Harold Edgerton, M.I.T., Cambridge, Massachusetts.)

The frequencies are all integer multiples of the lowest frequency.

What is the difference between harmonics and overtones?

In a normal-mode motion, every particle vibrates sinusoidally with the same frequency. Each mode has its own vibration pattern.

Interior of a concert grand piano, showing the full-length bass strings and the progressively shorter strings in the treble range. (Courtesy of Steinway and Sons.)

Each of the frequencies given by Eq. (22–4) corresponds to a possible **normal mode** of motion, that is, a motion in which each particle of the string moves sinusoidally, all with the same frequency. As this analysis shows, there are an infinite number of normal modes, each with its characteristic frequency. This situation is in striking contrast with the simple harmonic oscillator system, consisting of a single mass and a spring. The harmonic oscillator has only one normal mode and one characteristic frequency, while the vibrating string has an infinite number.

If a string is initially displaced so that its shape is the same as *any one* of the possible harmonics, it will vibrate, when released, at the frequency of that particular harmonic. But when a piano string is struck or a guitar string is plucked, not only the fundamental but many of the overtones are present in the resulting vibration. This motion is therefore a combination or *superposition* of normal modes. Several frequencies and motions are present simultaneously, and the displacement of any point on the string is the sum (or superposition) of displacements associated with the individual modes. Indeed, every possible motion of the string can be represented as some superposition of normal-mode motions.

The fundamental frequency of the vibrating string is $f_1 = c/2L$, where, from Eq. (21–15), $c = \sqrt{S/\mu}$. It follows that

$$f_1 = \frac{1}{2L}\sqrt{\frac{S}{\mu}}. \tag{22–7}$$

Stringed instruments provide many examples of the implications of this equation. For example, all such instruments are "tuned" by varying the tension S. An increase of tension increases the frequency or pitch, and vice versa. The inverse dependence of frequency on length L is illustrated by the long strings of the bass section of the piano or the bass viol compared with the shorter strings on the piano treble or the violin. In playing the violin or guitar, the usual means of varying the pitch is to press the strings against the fingerboard with the fingers to change the length of the vibrating portion of the string. One reason for winding the bass strings of a piano with wire is to increase the mass per unit length μ, so as to obtain the desired low frequency without resorting to a string that is inconveniently long.

PROBLEM-SOLVING STRATEGY: *Standing waves*

1. As with the problems in Chapter 21, it is useful to distinguish between the purely kinematic quantities, such as wave speed c, wavelength λ, and frequency f, and the dynamic quantities involving the properties of the medium, including S, μ, and (in the next section) B and ρ. The latter determine the wave speed c. Try to determine, in the problem at hand, whether the properties of the medium are involved or whether the problem is only kinematic in nature.

2. In visualizing nodes and antinodes in standing waves, it is always helpful to draw diagrams. For a string you can draw the shape at one instant and label the nodes N and antinodes A. For longitudinal waves (discussed in the next section), it is not so easy to draw the shape, but you can still label the nodes and antinodes. Remember that the distance between a node and the adjacent antinode is *always* $\lambda/4$, and that the distance between two adjacent nodes or two adjacent antinodes is always $\lambda/2$.

22–4 LONGITUDINAL STANDING WAVES

When longitudinal waves propagate in a fluid in a tube of finite length, the waves are reflected from the ends in the same way that transverse waves on a string are reflected at its ends. The superposition of the waves traveling in opposite directions again forms a standing wave.

When reflection takes place at a *closed* end, the displacement of the particles there must always be zero. This situation is analogous to a fixed end of a string; in both cases there is no displacement at the end, and the end is a *node*. For clarity, in the following discussion we call a closed end of a tube or pipe a *displacement node*. If the end of the tube is open, the nature of the reflection is more complex and depends on whether the tube is wide or narrow compared with the wavelength. If the tube is narrow compared with the wavelength, which is the case in most musical instruments, the open end is a *displacement antinode,* for reasons we will discuss below. (A free end of a stretched string, as discussed in Section 22–1, is also a displacement antinode.) Thus longitudinal waves in a column of fluid are reflected at the closed and open ends of a tube in the same way that transverse waves in a string are reflected at fixed and free ends, respectively.

We can demonstrate longitudinal standing waves in a column of gas, and also measure the wave speed, by using the apparatus called Kundt's tube, shown in Fig. 22–7. A glass tube a meter or so long is closed at one end and has a flexible diaphragm at the other end that can transmit vibrations. We use a sound source, which might be a small loudspeaker driven by an audio oscillator and amplifier, to vibrate the diaphragm sinusoidally with a variable frequency. A small amount of light powder or cork dust is distributed uniformly along the bottom side of the tube. As we vary the frequency of the sound, we pass through frequencies where the amplitude of the standing waves becomes large enough for the moving gas to sweep the cork dust along the tube at all points where the gas is in motion. The powder therefore collects at the displacement nodes, which can be seen and measured easily.

In a standing wave, the distance between two adjacent nodes is one-half (*not* one!) wavelength. Thus we may measure the wavelength by measuring the distances $\lambda/2$ between adjacent clumps of powder. We read the frequency f from the oscillator dial, and we can calculate the speed c of the waves from the usual relation

$$c = \lambda f.$$

At a displacement node, the pressure variations above and below the average have their *maximum* value, while at a displacement antinode the pressure does not vary. To understand this, note that two small masses of gas on opposite sides of a displacement *node* vibrate in *opposite phase*. When the masses of

Reflections of longitudinal waves take place at a closed or open end of a pipe.

Adjacent nodes are a half-wavelength apart; so are adjacent antinodes.

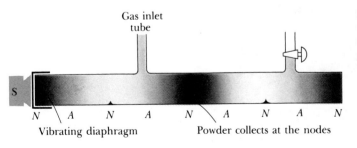

Gas inlet tube

$N \quad A \quad N \quad A \quad N \quad A \quad N \quad A \quad N$

Vibrating diaphragm   Powder collects at the nodes

22–7 Kundt's tube for determining the velocity of sound in a gas. The shading represents the density of the gas molecules at an instant when the pressure at the displacement nodes is a maximum or a minimum.

22–8 Diagram for illustrating longitudinal standing waves.

gas approach each other, the gas between them is compressed and the pressure rises; when they recede from each other, the pressure drops. But two small masses of gas on opposite sides of a displacement *antinode* vibrate *in phase,* and so cause *no* pressure variations at the antinode.

We can describe this relationship in terms of **pressure nodes,** which are the points where the pressure does not vary, and **pressure antinodes,** which are the points where its variation is greatest. *A pressure node is always a displacement antinode, and a pressure antinode is always a displacement node.* An *open* end of a thin tube or pipe is a pressure node because such an end is open to the atmosphere and is thus at constant pressure. But for this reason, an open end is always a displacement *antinode.*

We can visualize longitudinal standing waves with the help of Fig. 22–8, which is analogous to Fig. 21–10 for longitudinal traveling waves. Again use a 3-in. × 5-in. card with a slit $\frac{1}{16}$ in. wide and 3 in. long, or two cards fastened edge-to-edge with a $\frac{1}{16}$-in. gap. Place the card over the diagram with the slit horizontal and move it vertically with constant velocity. The portions of the sine curves that appear in the slit correspond to the oscillations of the particles in a longitudinal standing wave. Each particle moves with simple harmonic motion about its equilibrium positions. The particles at the nodes do not move, and the nodes are regions of maximum compression and expansion. Midway between the nodes are the antinodes, regions of maximum displacement but zero compression and expansion.

22–5 NORMAL MODES OF ORGAN PIPES

The behavior of an open organ pipe is similar to that of a string held at both ends.

Organ pipes provide us with a good example of standing waves and normal modes in vibrating air columns. Air is supplied by a blower, at a gauge pressure typically of the order of 10^3 Pa or 10^{-2} atm, to the left ends of the pipes in Figs. 22–9 and 22–10. A stream of air emerges from the narrow opening formed by the vertical surface and is directed against the right edge of the opening in the top surface of the pipe, called the *mouth* of the pipe. The column of air in the pipe is set into vibration; just as with the stretched string, there is a series of possible normal modes. For a pipe open at both ends, the fundamental frequency f_1 corresponds to a displacement antinode (pressure node) at each end and a displacement node in the middle, as shown at the top of Fig. 22–9. The other two parts of this figure show the second and third harmonics; their vibration patterns have two and three displacement nodes, respectively, and their frequencies are $f_2 = 2f_1$ and $f_3 = 3f_1$, respectively. *In an*

22–9 Modes of vibration of an open organ pipe.

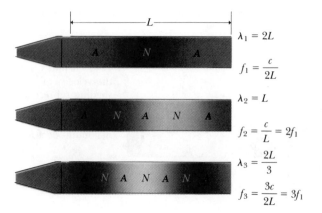

$$\lambda_1 = 2L$$

$$f_1 = \frac{c}{2L}$$

$$\lambda_2 = L$$

$$f_2 = \frac{c}{L} = 2f_1$$

$$\lambda_3 = \frac{2L}{3}$$

$$f_3 = \frac{3c}{2L} = 3f_1$$

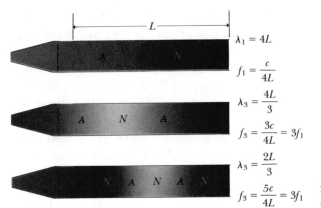

$$\lambda_1 = 4L$$

$$f_1 = \frac{c}{4L}$$

$$\lambda_3 = \frac{4L}{3}$$

$$f_3 = \frac{3c}{4L} = 3f_1$$

$$\lambda_3 = \frac{2L}{3}$$

$$f_3 = \frac{5c}{4L} = 3f_1$$

22–10 Modes of vibration of a stopped organ pipe.

open pipe the fundamental frequency is $c/2L$, and all harmonics in the series are possible.

The properties of a *stopped* pipe (open at one end, closed at the other) are shown in Fig. 22–10. The left (open) end is a displacement antinode (pressure node), but the right (closed) end is a displacement node (pressure antinode). We see that in the lowest-frequency mode the length of the pipe is a quarter-wavelength ($L = \lambda/4$). The fundamental frequency is $f_1 = c/4L$, which is one-half that of an open pipe of the same length. In musical language, the *pitch* of a closed pipe is one octave lower (a factor of two in frequency) than that of an open pipe of the same length. From the other parts of Fig. 22–10 we see that the second, fourth, and all *even* harmonics are missing. *In a stopped pipe the fundamental frequency is $f_1 = c/4L$, and only the odd harmonics in the series ($3f_1$, $5f_1$, . . .) are possible.*

In an organ pipe in actual use, several modes are almost always present at once. This situation is analogous to that of a string that is struck or plucked, producing several modes at the same time. The motion in each case is then a *superposition* of various modes. The extent to which modes higher than the fundamental are present depends on the cross-section of the pipe, the proportion of length to width, the shape of the mouth, and other more subtle factors. The harmonic content of the tone is an important factor in determining the tone quality, or timbre. A very narrow pipe produces a tone rich in higher harmonics, which the ear perceives as thin and "stringy"; a fatter pipe produces principally the fundamental mode, perceived as a softer, more flutelike tone.

Although this discussion has centered on organ pipes, it is also relevant for other wind instruments. The flute and the recorder are directly analogous. The most significant difference is that those instruments have holes along the pipe; opening and closing these with the fingers changes the effective length L of the air column and thus changes the pitch. Any individual organ pipe, by comparison, plays only a single note. The flute and recorder function as *open* pipes, while the clarinet acts as a *closed* pipe (closed at the reed end, open at the bell).

Console and a few of the thousands of pipes in the organ of the Shrine of the Immaculate Conception, Washington, DC. Each pipe produces a single note. The largest pipe is 38 feet long overall and weighs 1050 lb. (Courtesy of M. P. Möller, Inc.)

Several wind instruments are similar to organ pipes in their operation.

22–6 INTERFERENCE OF WAVES

Wave phenomena that occur when two or more waves overlap in the same region of space are grouped under the heading *interference*. As we have seen, standing waves are a simple example of an interference effect. Here are two

What happens when two waves overlap in the same region of space? They can either reinforce or cancel each other.

Source
S

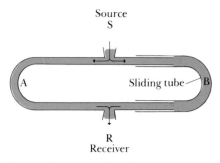

22–11 Apparatus for demonstrating interference of longitudinal waves.

more examples. Consider the trombonelike apparatus of Fig. 22–11. The source S emits a sinusoidal sound wave that enters the tube and divides into two waves. One wave follows the path SAR, which has constant length, and the other follows path SBR, whose length can be changed by moving the sliding tube. The two waves recombine and emerge at R, the receiver or detector.

Suppose the wavelength of the sound is 1.0 m. If the paths SAR and SBR have the same length, the waves take equal times to reach R and arrive there *in phase*. The resulting displacement has an amplitude equal to the sum of the two individual amplitudes. This addition of amplitudes is called *reinforcement* or *constructive interference*.

Now we slide the tube B out a distance of 0.25 m, making the path SBR 0.5 m longer than SAR. Then the right-hand wave travels a distance $\lambda/2$ farther than the left-hand wave. When the two waves arrive at R, their displacements differ in phase by a half-cycle and thus add to zero. This is called *cancellation* or *destructive interference,* and results in a sharp decrease in sound level at R.

If we now pull the tube B out another 0.25 m, so that the *path difference, SBR* minus *SAR,* is 1.0 m (1 wavelength), the two vibrations at R again reinforce each other. Thus

$$\left\{\begin{array}{l}\text{Reinforcement takes place}\\ \text{when the path difference}\end{array}\right\} = 0,\ \lambda,\ 2\lambda,\ \text{etc.}$$

$$\left\{\begin{array}{l}\text{Cancellation takes place}\\ \text{when the path difference}\end{array}\right\} = \frac{\lambda}{2},\ \frac{3\lambda}{2},\ \frac{5\lambda}{2},\ \text{etc.}$$

We can demonstrate a similar phenomenon with two loudspeakers driven by the same amplifier, as shown in Fig. 22–12. Suppose both speakers emit a pure sinusoidal sound wave of constant frequency. When we place a microphone at point A in the figure, equidistant from the speakers, it receives a strong acoustic signal, corresponding to the fact that the two waves emitted from the speakers in phase also arrive at point A in phase and add to each other. But at point B the signal is much *weaker* than when only one speaker is present, corresponding to the fact that the wave from one speaker travels a half-wavelength farther than that from the other. The two waves arrive a half-cycle out of phase and cancel each other out almost completely. An experiment closely analogous to this one, but using light waves, provided the first conclusive evidence of the wave nature of light. This experiment is discussed in detail in Chapter 39.

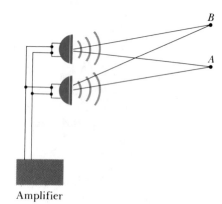

Amplifier

22–12 Two speakers driven by the same amplifier. The waves emitted by the speakers are in phase; they arrive at point A in phase because the two path lengths are the same. They arrive at point B a half-cycle out of phase because the path lengths differ by $\lambda/2$.

22–7 RESONANCE

In this chapter we have discussed several examples of mechanical systems having normal modes of oscillation. In each mode, every particle of the system oscillates with simple harmonic motion of the same frequency as the frequency of this mode. The systems we have discussed have an infinite series of normal modes, but the basic concept is closely related to the simple harmonic oscillator, discussed in Chapter 11, which has only a single normal mode.

Now suppose we apply a periodically varying force to a system having normal modes. The system vibrates with a frequency equal to that of the *force.* This motion is called a *forced oscillation.* In general, the amplitude of this motion is relatively small, but if the frequency of the force is close to one of the normal-mode frequencies, the amplitude can become quite large. We dis-

Resonance: When an applied force varies with the same frequency as a normal-mode frequency, the amplitude of oscillation builds up.

cussed forced oscillations of the harmonic oscillator in Section 11–8, and we suggest you review that discussion now.

If the frequency of the force were precisely *equal* to a normal-mode frequency, and if there were no friction or other energy-dissipating mechanism, then the force would continue to add energy to the system, and the amplitude would increase indefinitely. In any real system there is always some dissipation of energy, or damping, as we noted in Section 11–7. Even so, the "response" of the system (that is, the amplitude of the forced oscillation) is greatest when the force frequency is equal to one of the normal-mode frequencies. This behavior is called **resonance.**

Pushing a child's swing is a common example of mechanical resonance, as mentioned in Section 11–8. The swing is a pendulum; it has only a single natural frequency, determined by its length. If we give the swing a series of regularly spaced pushes, with a frequency equal to that of the swing, the motion may be made quite large. If the frequency of the pushes differs from the natural frequency of the swing, or if the pushes occur at irregular intervals, the swing will execute hardly any vibration at all.

Unlike a simple pendulum, which has only one natural frequency, a stretched string (and other systems discussed in this chapter) has an infinite number of natural frequencies. Suppose that one end of a stretched string is held stationary while the other is moved back and forth in a transverse direction. The amplitude at the driven end is fixed by the driving mechanism. Standing waves are set up in the string, whatever the value of the frequency f. If the frequency is *not* equal to one of the natural frequencies of the string, the amplitude at the antinodes is fairly small.

A pendulum has only one normal mode; a stretched string has an infinite number.

However, if the frequency is equal to *any one* of the natural frequencies, the string is in resonance and the amplitude at the antinodes is very much *larger* than that at the driven end. In other words, although the driven end is not a node, it lies much closer to a node than to an antinode when the string is in resonance. In Fig. 22–5a, the right end of the string was held stationary and the left end was forced to oscillate vertically with small amplitude. The photographs show the standing waves of relatively large amplitude that resulted when the frequency of oscillation of the left end was equal to the fundamental frequency or to one of the first three overtones.

A steel bridge or, for that matter, any elastic structure has normal modes and can vibrate with certain natural frequencies. If the regular footsteps of a marching band have a frequency equal to one of the natural frequencies of a bridge that the band is crossing, a vibration of dangerously large amplitude may result. Therefore, when crossing a bridge, a marching group should always "break step."

How to make a bridge collapse by marching a band across it. We don't recommend it.

We can demonstrate the phenomenon of resonance with two identical tuning forks. We place them some distance apart and strike one to set it into vibration. The resulting sound waves set the other fork into vibration, and its vibration can be heard if the first fork is suddenly damped. If we place a small piece of wax or modeling clay on one of the forks, the frequency of that fork will be altered enough to destroy the resonance.

A similar phenomenon can be demonstrated with a piano. Depress the damper pedal (the right-hand pedal) so the dampers are lifted and the strings are free to vibrate, and then sing a steady tone into the piano. When the singing stops, the piano seems to continue to sing the same note. The sound waves from your voice excite vibrations in the strings that have natural fre-

How to sing to a piano and have it sing back to you

quencies close to the frequencies (fundamental and harmonic) present in the note sung initially.

Resonance is a very important concept, not only in mechanical systems but in all areas of physics. Later we will see examples of resonance in electric circuits (Chapter 34) and in atomic systems (Chapter 42).

SUMMARY

A wave that reaches a boundary of the medium in which it propagates is reflected. The total wave displacement at any point where the initial and reflected waves overlap is the sum of the displacements of the individual waves; this statement is the principle of superposition.

When a wave is reflected from a fixed or free end of a stretched string, the incident and reflected waves combine to form a standing wave, which does not appear to travel in either direction. Its pattern contains nodes and antinodes; adjacent nodes are spaced a distance $\lambda/2$ apart, as are adjacent antinodes.

When both ends of a string having length L are held, standing waves can occur only when L is an integer multiple of $\lambda/2$; the corresponding possible frequencies are given by

$$f_n = n\frac{c}{2L} \qquad (n = 1, 2, 3, \ldots). \qquad (22\text{--}5)$$

Each frequency, and its associated vibration pattern, is called a normal mode. The lowest frequency f_1 is called the fundamental frequency. In terms of the mechanical properties S and μ of the string, the fundamental frequency is given by

$$f_1 = \frac{1}{2L}\sqrt{\frac{S}{\mu}}. \qquad (22\text{--}7)$$

Standing waves also occur in wave motion in pipes or tubes. A closed end is a displacement node and a pressure antinode; an open end is a displacement antinode and a pressure node.

For a pipe open at both ends, the normal-mode frequencies are given by

$$f_n = n\frac{c}{2L} \qquad (n = 1, 2, 3, \ldots).$$

For a pipe open at one end and closed at the other, the normal-mode frequencies are

$$f_n = n\frac{c}{4L} \qquad (n = 1, 3, 5, \ldots).$$

When two or more waves overlap in the same region of space, the resulting effects are called interference. The resulting amplitude can be either larger or smaller than the amplitude of each individual wave, depending on whether the waves are in phase or out of phase. If the waves are in phase, the result is called reinforcement or constructive interference; if they are out of phase, it is called cancellation or destructive interference.

When a periodically varying force is applied to a system having normal modes of vibration, the system vibrates with the same frequency as that of the force; this is called a forced oscillation. If the force frequency is equal or close to one of the normal-mode frequencies, the amplitude of the resulting forced oscillation can become very large; this phenomenon is called resonance.

QUESTIONS

22–1 When you inhale helium, your voice becomes high and squeaky. Why? (Don't try this with hydrogen; you might explode like the Hindenburg.) What happens when you inhale carbon dioxide?

22–2 A musical interval of an octave corresponds to a factor of two in frequency. By what factor must the tension in a piano or violin string be increased to raise its pitch one octave?

22–3 The pitch (or frequency) of an organ pipe changes with temperature. Does it increase or decrease with increasing temperature? Why?

22–4 By touching a string lightly at its center while bowing, a violinist can produce a note exactly one octave above the note to which the string is tuned, i.e., a note with exactly twice the frequency. Why is this possible?

22–5 In most modern wind instruments the pitch is changed by changing the length of the vibrating air column with keys or valves. The bugle, however, has no valves or keys, yet it can play many notes. How? Are there restrictions on what notes it can play?

22–6 Kettledrums (tympani) have a pitch determined by the frequencies of the normal modes. How can a kettledrum be tuned?

22–7 Energy can be transferred along a string by wave motion. However, in a standing wave no energy can ever be transferred past a node. Why?

22–8 Can a standing wave be produced on a string by superposing two waves with the same frequency but different amplitudes, traveling in opposite directions?

22–9 In discussing standing longitudinal waves in an organ pipe, we spoke of pressure nodes and antinodes. What physical quantity is analogous to pressure for standing transverse waves on a stretched string?

22–10 A glass of water sits on a kitchen counter just above a dishwasher that is running. The surface of the water has a set of concentric stationary circular ripples. What is happening?

22–11 Consider the peculiar sound of a large bell. Do you think the overtones are harmonic?

22–12 What is the difference between music and noise?

22–13 Is the mass per unit length the same for all strings on a piano? On a guitar? Why?

22–14 When a heavy truck drives up a steep hill, the windows in nearby houses sometimes vibrate. Why?

22–15 A popular children's pastime is lining up a row of dominoes, standing on edge, so that when the first one is pushed over, the whole row falls, one after another. Is this a wave motion? In what respects is it similar to waves on a stretched string? In what respects is it different?

22–16 A series of traffic lights along a street is timed so that a car going at the right speed can hit a succession of green lights and not have to stop. If several cars are on the same street, does this situation have wavelike properties?

22–17 Some opera singers are reputed to be able to break a glass by singing the appropriate note. What physical phenomenon could account for this?

22–18 A pipe organ in a church is tuned when the temperature is 20°C. On a cold winter day when the temperature in the room is 10°C, the organ still sounds "in tune," despite the phenomenon mentioned in Question 22–3. Why?

EXERCISES

Unless otherwise indicated, assume the speed of sound in air to be $c = 345 \text{ m·s}^{-1}$.

Section 22–2 Superposition and Standing Waves

22–1 Let

$$y_1(x, t) = A_1 \sin (\omega_1 t - k_1 x)$$

and

$$y_2(x, t) = A_2 \sin (\omega_2 t - k_2 x)$$

be two solutions to the wave equation, Eq. (21–14), for the *same c*. Show that $y(x, t) = y_1(x, t) + y_2(x, t)$ is also a solution to the wave equation.

22–2 Give the details of the derivation of Eq. (22–1) from $y_1 + y_2 = A[\sin (\omega t - kx) - \sin (\omega t + kx)]$.

22–3 Prove by direct substitution that

$$y = -[2A \cos \omega t] \sin kx$$

is a solution of the wave equation, for $c = \omega/k$.

Section 22–3 Normal Modes of a String

22–4 A piano tuner stretches a steel piano "string" 0.50 m long, of mass 5 g, with a tension of 400 N.

a) What is the frequency of its fundamental mode of vibration?

b) What is the number of the highest overtone that could be heard by a person capable of hearing frequencies up to 10,000 Hz?

22–5 A physics student observes that a stretched string vibrates with a frequency of 30 Hz in its fundamental mode when the supports are 0.60 m apart. The amplitude at the antinode is 3 cm. The string has a mass of 0.030 kg.

a) What is the speed of propagation of a transverse wave in the string?

b) Compute the tension in the string.

22-6

a) A string is vibrating in its fundamental mode. The waves have velocity c, frequency f, amplitude A, and wavelength λ. Calculate the maximum velocity and acceleration of points located at
   i) $x = \lambda/2$,   ii) $x = \lambda/4$,   iii) $x = \lambda/8$
   from the left-hand end of the string.

b) At each of the points in (a), what is the amplitude of the motion?

c) At each of the points in (a), how much time does it take the string to go from its largest upward displacement to its largest downward displacement?

Section 22-4 Longitudinal Standing Waves

Section 22-5 Normal Modes of Organ Pipes

22-7 The longest pipe in most medium-sized pipe organs is 4.88 m (16 ft) long. What is the frequency of the note corresponding to the fundamental mode if the pipe is

a) open at both ends?

b) open at one end, closed at the other?

22-8 Find the fundamental frequency and the first four overtones of a 15-cm pipe

a) if the pipe is open at both ends;

b) if the pipe is closed at one end.

c) How many overtones can be heard by a person with normal hearing for each of the above cases? (A person with normal hearing can hear frequencies in the range 20 Hz to 20,000 Hz.)

22-9 A certain pipe produces a frequency of 440 Hz in air. If the pipe is filled with helium at the same temperature, what frequency does it produce?

22-10 A standing wave of frequency 1100 Hz in a column of methane (CH_4) at 20°C produces nodes that are 0.20 m apart. What is the ratio of the heat capacity at constant pressure to that at constant volume?

22-11 A student in a physics lab sets up standing waves in a Kundt's tube (Fig. 22-7) by the longitudinal vibration of an iron rod 1 m long, clamped at the center. If the frequency of vibration of the iron rod is 2480 Hz and the powder heaps within the tube are 6.9 cm apart, what is the speed of the waves

a) in the iron rod?   b) in the gas?

PROBLEMS

22-12 A steel wire of length $L = 1.00$ m and density $\rho = 8000$ kg·m^{-3} is stretched tightly between two rigid supports. When the wire vibrates in its fundamental mode, the frequency is $f = 200$ Hz.

a) What is the speed of transverse waves on this wire?

b) What is the longitudinal stress (force per unit area) in the wire?

c) If the maximum acceleration at the midpoint of the wire is 800 m·s^{-2}, what is the amplitude of vibration at the midpoint?

22-13 A cellist tunes the A-string of her instrument to a fundamental frequency of 220 Hz. The vibrating portion of the string is 0.68 m long and has a mass of 1.29 g.

a) With what tension must it be stretched?

b) What percentage increase in tension is needed to increase the frequency from 220 Hz to 233 Hz, corresponding to a rise in pitch from A to A-sharp?

22-14 A solid aluminum sculpture is hung from a steel wire. The fundamental frequency for transverse standing waves on the wire is 300 Hz. The sculpture is then immersed in water so that one-half its volume is submerged. What is the new fundamental frequency?

22-15 A long tube contains air at a pressure of 1 atm and temperature of 77°C. The tube is open at one end and closed at the other by a movable piston. A tuning fork near the open end is vibrating with a frequency of 500 Hz. Resonance is produced when the piston is at distances 18.0, 55.5, and 93.0 cm from the open end.

a) From these measurements, what is the speed of sound in air at 77°C?

b) From the above result, what is the ratio of the specific heat capacities (γ) for air?

22-16 The atomic mass of iodine is 127 g·mol^{-1}. A standing wave in iodine vapor at 400 K produces nodes that are 6.77 cm apart when the frequency is 1000 Hz. Is iodine vapor monatomic or diatomic?

22-17 The frequency of middle C is 262 Hz.

a) If an organ pipe is open at both ends, what length must it have to produce this note at 20°C?

b) At what temperature will the frequency be 6% higher, corresponding to a rise in pitch from C to C-sharp?

22-18 The steel B string of an acoustic guitar is 63.5 cm long and has a diameter of 0.406 mm (16 gauge).

a) Under what tension must the string be placed to give a frequency for transverse waves of 247.5 Hz? Use 7.8 g·cm^{-3} as the density of steel, and assume that the string vibrates in its fundamental mode.

b) If the tension is changed by a small amount ΔS, the frequency changes by Δf. Show that

$$\frac{\Delta f}{f} = \frac{1}{2}\frac{\Delta S}{S}.$$

c) If the string is tuned indoors as in part (a), where the temperature is 25°C, and the guitar is then taken to an outdoor stage where $T = 15$°C, the frequency will change, with unpleasant results. Find Δf if the Young's modulus Y is 2.0×10^{11} Pa and the coefficient of linear expansion α is 1.2×10^{-5}(C °)$^{-1}$. Will the pitch be raised or lowered?

CHALLENGE PROBLEMS

22–19 Two identical loudspeakers are located at points A and B that are 2 m apart. The loudspeakers are driven by the same amplifier and produce sound waves of frequency 440 Hz. Take the speed of sound in air to be 340 m·s^{-1}. A small microphone is moved out from point B along a line perpendicular to the line connecting A and B (line BC in Fig. 22–13). At what distances from B will there be *destructive* interference?

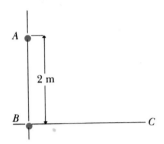

FIGURE 22–13

22–20 A simple example of resonance in a mechanical system is as follows. A mass m is attached to one end of a massless spring of force constant k and unstretched length l_0. The other end of the spring is free to turn about a nail driven into a frictionless horizontal surface (Fig. 22–14). The mass is then made to revolve in a circle with an angular frequency of revolution ω'. (Refer to Challenge Problem 6–44.)

a) Calculate the length l of the spring as a function of ω'.

b) What happens to the result in (a) when ω' approaches the natural frequency $\omega = \sqrt{k/m}$ of the mass-spring system. (If your result bothers you, remember that precisely massless springs and frictionless surfaces do not exist but can only be approximated. Also, Hooke's law is itself only an approximation to the way real springs behave; the greater the elongation of the spring, the greater the deviation from Hooke's law.)

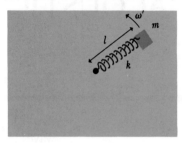

FIGURE 22–14

23
ACOUSTIC PHENOMENA

IN THIS CHAPTER WE ARE CONCERNED PRIMARILY WITH LONGITUDINAL waves in air, commonly called **sound.** The human ear is sensitive to waves in the frequency range from about 20 to 20,000 Hz, although the term *sound* is sometimes used for similar waves with frequencies outside the range of human hearing. Properties of sound waves include frequency, amplitude, and intensity. We will study the relations among displacement, pressure variation, and intensity, and the connections between these quantities and human sound perception. The interference of two sound waves differing in frequency by a few hertz causes phenomena called *beats.* When the source of sound or the observer, or both, are in motion relative to the air, frequency shifts known as the *Doppler effect* are observed. We will examine both of these phenomena.

23–1 SOUND WAVES

The simplest sound waves are sinusoidal waves with definite frequency, amplitude, and wavelength. When such a wave arrives at the ear, the air particles at the eardrum vibrate with definite frequency and amplitude. This vibration may also be described in terms of the variation of *air pressure* at the same point. The pressure fluctuates above and below atmospheric pressure with a sinusoidal variation having the same frequency as the motions of the air particles.

A sinusoidal sound wave in an elastic medium is described by a wave function of the form introduced in Section 21–3, Eq. (21–7):

$$y = A \sin (\omega t - kx), \tag{23–1}$$

where y is the displacement from equilibrium of a point in the medium, and the amplitude A is, as usual, the *maximum* displacement from equilibrium.

From a practical standpoint it is nearly always easier to measure the *pressure* variations in a sound wave than to measure the displacements, so it is worthwhile to develop a relation between the two. Let p be the instantaneous pressure fluctuation at any point; that is, the amount by which the pressure *differs* from normal atmospheric pressure. If the displacements of two neighboring points x and $x + \Delta x$ are the same, the air between these points is neither compressed nor expanded, there is no volume change, and consequently

What is sound?

It is easier to describe sound waves in terms of pressure variations than in terms of displacements.

In a sound wave, the pressure varies slightly above and below normal atmospheric pressure.

$p = 0$. Only when y varies from one point to a neighboring point is there a change of volume and therefore of pressure.

The fractional volume change $\Delta V/V$ in a volume element near point x turns out to be given simply by $\partial y/\partial x$, which is the rate of change of y with x as we go from one point to a neighboring point. To see why this is so, consider an imaginary cylinder of air, as in Fig. 23–1, with cross-sectional area A and axis along the direction of propagation. The color cylinder shows the undisplaced position, and the dashed lines show the displaced position. When no sound disturbance is present, the cylinder's length is Δx and its volume is $V = A\,\Delta x$. When a wave is present, the end of the cylinder initially at x is displaced a distance $y_1 = y(x, t)$, and the end initially at $x + \Delta x$ is displaced a distance $y_2 = y(x + \Delta x, t)$. The change in volume ΔV of this element is

$$A(y_2 - y_1) = A[y(x + \Delta x, t) - y(x, t)],$$

and in the limit as $\Delta x \to 0$, the fractional change in volume $\Delta V/V$ is given by

$$\frac{\Delta V}{V} = \frac{y(x + \Delta x, t) - y(x, t)}{\Delta x} = \frac{\partial y}{\partial x}. \qquad (23\text{–}2)$$

Now from the definition of the bulk modulus B, Eq. (12–7), $p = -B\,\Delta V/V$, and we find

$$p = -B\left(\frac{\partial y}{\partial x}\right). \qquad (23\text{–}3)$$

The negative sign arises because, when $\partial y/\partial x$ is positive, the displacement is greater at $x + \Delta x$ than at x, corresponding to an increase in volume and a *decrease* in pressure. For the sinusoidal wave of Eq. (23–1) we find

$$p = BkA \cos(\omega t - kx). \qquad (23\text{–}4)$$

This expression shows that the quantity BkA represents the maximum pressure variation. This maximum is called the **pressure amplitude** and is denoted by p_{max}. Thus

$$p_{max} = BkA. \qquad (23\text{–}5)$$

The pressure amplitude is directly proportional to the displacement amplitude A, as might be expected, and it also depends on wavelength. Waves of shorter wavelength (larger k) have greater pressure variations for a given amplitude because the maxima and minima are squeezed closer together.

> The pressure variation is proportional to the rate of change of displacement with position.

> The pressure amplitude is proportional to the displacement amplitude.

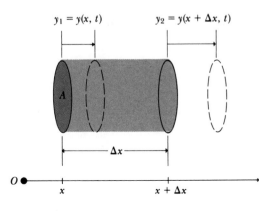

23–1 A cylindrical volume of gas with cross-sectional area A. The length in the undisplaced position is Δx. During wave propagation along the axis, the left end is displaced to the right a distance y_1, and the right end is displaced a different distance y_2. The resulting change in volume is $A(y_2 - y_1)$.

EXAMPLE 23–1 Measurements of sound waves show that the maximum pressure variations in the loudest sounds that the ear can tolerate without pain are of the order of 30 Pa (above and below atmospheric pressure, which is about 100,000 Pa). Find the corresponding maximum displacement if the frequency is 1000 Hz and $c = 350$ m·s^{-1}.

SOLUTION We have $\omega = (2\pi)(1000$ Hz$) = 6283$ s^{-1} and

$$k = \frac{\omega}{c} = \frac{6283 \text{ s}^{-1}}{350 \text{ m·s}^{-1}} = 18.0 \text{ m}^{-1}.$$

The adiabatic bulk modulus for air is

$$B = \gamma p = (1.4)(1.01 \times 10^5 \text{ Pa}) = 1.42 \times 10^5 \text{ Pa}.$$

The displacement amplitude of even the loudest sound is very small.

From Eq. (23–5) we find

$$A = \frac{p_{max}}{Bk} = \frac{(30 \text{ Pa})}{(1.42 \times 10^5 \text{ Pa})(18.0 \text{ m}^{-1})}$$

$$= 1.18 \times 10^{-5} \text{ m} = 0.0118 \text{ mm}.$$

Thus the displacement amplitude of even the loudest sound is *extremely* small. The maximum pressure variation in the *faintest* audible sound of frequency 1000 Hz is only about 3×10^{-5} Pa. The corresponding displacement amplitude is about 10^{-9} cm. For comparison, the wavelength of yellow light is 6×10^{-5} cm, and the diameter of a molecule is about 10^{-8} cm. The ear is an extremely sensitive organ!

23–2 INTENSITY

Intensity: a concept to describe the energy transported by a sound wave

An essential aspect of wave propagation of all sorts is transfer of energy. A familiar example is the energy supply of the earth, which reaches us from the sun via electromagnetic waves. The **intensity** I of a traveling wave is defined as *the time average rate at which energy is transported by the wave, per unit area*, across a surface perpendicular to the direction of propagation. More briefly, the intensity is the average *power* transported per unit area.

We have seen that the power developed by a force equals the product of force and velocity. Hence the power *per unit area* in a sound wave equals the product of the excess pressure (force per unit area), given by Eq. (23–4), and the *particle* velocity v, obtained by taking the time derivative of Eq. (23–1). We find

$$v = \omega A \cos(\omega t - kx), \tag{23–6}$$

$$pv = \omega BkA^2 \cos^2(\omega t - kx). \tag{23–7}$$

The intensity is, by definition, the average value of this quantity. The average value of the function $\cos^2 z$ is 1/2, so we find

$$I = \frac{1}{2} \omega BkA^2. \tag{23–8}$$

By using the relations $\omega = ck$ and $c^2 = B/\rho$, we can transform this into the form

$$I = \frac{1}{2}\sqrt{\rho B}\,\omega^2 A^2,$$

which we cited at the end of Section 21–8.

It is usually more convenient to express I in terms of the pressure amplitude $p_{\max}$. Using Eq. (23–5) and the relation $\omega = ck$, we find

$$I = \frac{\omega p_{\max}^2}{2Bk} = \frac{c p_{\max}^2}{2B}. \tag{23–9}$$

By using the wave speed relation $c^2 = B/\rho$, we can also write this in the alternative forms

$$I = \frac{p_{\max}^2}{2\rho c} = \frac{p_{\max}^2}{2\sqrt{\rho B}}. \tag{23–10}$$

The intensity of a sound wave of the largest amplitude tolerable to the human ear (about $p_{\max} = 30$ Pa) is

$$
\begin{aligned}
I &= \frac{(30\ \text{Pa})^2}{2(1.22\ \text{kg} \cdot \text{m}^{-3})(346\ \text{m} \cdot \text{s}^{-1})} \\
&= 1.07\ \text{J} \cdot \text{s}^{-1} \cdot \text{m}^{-2} = 1.07\ \text{W} \cdot \text{m}^{-2} \\
&= 1.07 \times 10^{-4}\ \text{W} \cdot \text{cm}^{-2}.
\end{aligned}
$$

The unit $1\ \text{W} \cdot \text{cm}^{-2}$ is a mixed unit, neither cgs nor mks. We mention it here because it is unfortunately in general use among acousticians.

The pressure amplitude of the *faintest* sound wave that can be heard is about 3×10^{-5} Pa, and the corresponding intensity is about $10^{-12}\ \text{W} \cdot \text{m}^{-2}$ or $10^{-16}\ \text{W} \cdot \text{cm}^{-2}$.

The *total* power carried across a surface by a sound wave equals the product of the intensity at the surface and the surface area, if the intensity over the surface is uniform. The average total sound power emitted by a person speaking in a conversational tone is about 10^{-5} W, while a loud shout corresponds to about 3×10^{-2} W. If all the residents of New York City were to talk at the same time, the total sound power would be about 100 W, equivalent to the electric-power requirement of a medium-sized light bulb. Yet the power required to fill a large auditorium with loud sound is considerable. Suppose the sound intensity over the surface of a hemisphere 20 m in radius is $1\ \text{W} \cdot \text{m}^{-2}$. The area of the surface is about $2500\ \text{m}^2$. Hence the acoustic power output of a speaker at the center of the sphere would have to be

$$(1\ \text{W} \cdot \text{m}^{-2})(2500\ \text{m}^2) = 2500\ \text{W},$$

or 2.5 kW. The electric-power input to the speaker would need to be considerably larger, since the efficiency of such devices is not very high.

Because of the extremely large range of intensities over which the ear is sensitive, a *logarithmic* rather than an arithmetic intensity scale is convenient. The **intensity level** β of a sound wave is defined by the equation

$$\beta = 10 \log \frac{I}{I_0}, \tag{23–11}$$

where I_0 is an arbitrary reference intensity, taken as $10^{-12}\ \text{W} \cdot \text{m}^{-2}$. This value corresponds roughly to the faintest sound that can be heard. Intensity levels are expressed in **decibels,** abbreviated dB. A decibel is 1/10 of a bel, a unit named for Alexander Graham Bell. The bel is inconveniently large for most purposes, and the decibel is the usual unit of sound intensity level.

If the intensity of a sound wave equals I_0 or $10^{-12}\ \text{W} \cdot \text{m}^{-2}$, its intensity level is 0 dB. The maximum intensity that the ear can tolerate without pain is

It doesn't take very much energy to make a very loud sound.

The decibel: a logarithmic scale that describes a very wide range of sound intensities

TABLE 23–1   Noise Levels Due To Various Sources (Representative Values)

Source or Description of Noise	Noise Level, dB	Intensity, $W \cdot m^{-2}$
Threshold of pain	120	1
Riveter	95	3.2×10^{-3}
Elevated train	90	10^{-3}
Busy street traffic	70	10^{-5}
Ordinary conversation	65	3.2×10^{-6}
Quiet automobile	50	10^{-7}
Quiet radio in home	40	10^{-8}
Average whisper	20	10^{-10}
Rustle of leaves	10	10^{-11}
Threshold of hearing	0	10^{-12}

The relation between sound intensity and human hearing: What is the smallest intensity that can be heard? The loudest that can be tolerated?

Is sound still sound if you can't hear it?

about 1 $W \cdot m^{-2}$, which corresponds to an intensity level of 120 dB. Table 23–1 gives the intensity levels in decibels of several familiar noises. It is taken from a survey made by the New York City Noise Abatement Commission.

Within the range of audibility, the sensitivity of the ear varies with frequency. The **threshold of audibility** at any frequency is the minimum intensity of sound at that frequency that can be detected. For a young adult with normal hearing, the threshold of audibility at 1000 Hz is about 0 dB; at 200 and 15,000 Hz it is about 20 dB; and at 50 and 18,000 Hz it is about 50 dB. Thus the ear's sensitivity drops off at the low and high ends of the frequency scale. Frequencies above 20,000 Hz (20 kHz) are not audible to humans at *any* intensity, and such frequencies are referred to as **ultrasonic.**

PROBLEM-SOLVING STRATEGY

1. Quite a few quantities are involved in characterizing the amplitude and intensity of a sound wave, and it's easy to get lost in the maze of relationships. It helps to put them in categories: The amplitude is described by A or p_{max}; the frequency by f, ω, k, or λ. These quantities are related through the wave speed c, which in turn is determined by the properties of the medium, B and ρ. Take a hard look at the problem at hand; identify which of these quantities are given and which you have to find; then start looking for relationships that take you where you want to go.

2. In using Eq. (23–11) for the sound intensity level, remember that I and I_0 must be in the same units, usually $W \cdot m^{-2}$. If they aren't, convert!

Frequency ratios are the basis of consonance and dissonance in music.

The **pitch** of a musical tone is determined by its frequency, and musical intervals can be defined in terms of frequency ratios. For example, middle C on the piano has a frequency of 262 Hz, and the C an octave higher has a frequency of 524 Hz, a factor of two larger. An octave is always a pair of tones having a frequency ratio of 2:1, and a perfect fifth corresponds to a ratio of 3:2. A major triad consists of three tones with frequency ratios of 4:5:6. These and similar relationships can be used to explore the physical basis of consonance and dissonance of combinations of musical tones.

23–3 BEATS

In Section 22–6 we discussed *interference* effects that occur when two different waves overlap in the same region of space. In that discussion we considered two waves with the same frequency. Now let us examine an interference effect resulting from the overlap of two waves with equal amplitude but slightly different frequency. This occurs, for example, when two tuning forks with slightly different frequencies are sounded together, or when two organ pipes that are supposed to have exactly the same frequency are slightly "out of tune."

Consider a particular point in space where the two waves overlap. The displacements of the individual waves at this point are plotted as functions of time in Fig. 23–2a. If the total length of the time axis represents about one second, the frequencies are 16 Hz and 18 Hz. Applying the principle of superposition, we add the two displacements at each instant of time to find the total displacement at that time, obtaining the graph of Fig. 23–2b. At certain times the two waves are in phase; their maxima coincide and their amplitudes add. But as time goes on, they become more and more out of phase because of their slightly different frequencies. Eventually a maximum of one wave coincides with a maximum in the opposite direction for the other wave. The two waves then cancel each other, and the total amplitude is zero.

Thus the amplitude of the composite wave varies from a maximum value to zero and back, as shown in Fig. 23–2b. The appearance is that of a single sinusoidal wave with a varying amplitude. In this example, the amplitude goes through two maxima and two minima in one second, so the frequency of this amplitude variation is 2 Hz. The amplitude variation causes variations of loudness called **beats,** and the frequency with which the amplitude varies is called the **beat frequency.** In this example the beat frequency is the *difference* of the two frequencies. If the beat frequency is a few hertz, it is perceived as a waver or pulsation in the tone.

We can prove that the beat frequency is *always* the difference of the two frequencies f_1 and f_2. Suppose f_1 is larger than f_2; the corresponding periods are τ_1 and τ_2, with $\tau_1 < \tau_2$. If the two waves start out in phase at time $t = 0$, they will again be in phase at a time T such that the first wave has gone through exactly one more cycle than the second. Let n be the number of cycles of the first wave in time T; then the number of cycles of the second wave in the same time is $(n - 1)$, and we have the relations

$$T = n\tau_1 = (n - 1)\tau_2.$$

We solve the second equation for n and substitute the result back into the first

Beats: superposing two waves with slightly different frequencies

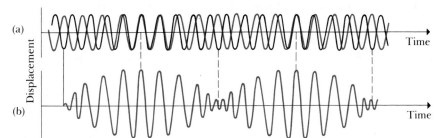

23–2 Beats are fluctuations in amplitude produced by two sound waves of slightly different frequency. (a) Individual waves. (b) Pattern formed by superposition of the two waves.

equation, obtaining

$$T = \frac{\tau_1 \tau_2}{\tau_2 - \tau_1}.$$

Now T is just the *period* of the beat, and its reciprocal is the beat *frequency*, $f_{beat} = 1/T$. Thus we find

$$f_{beat} = \frac{\tau_2 - \tau_1}{\tau_1 \tau_2} = \frac{1}{\tau_1} - \frac{1}{\tau_2},$$

and finally

$$f_{beat} = f_1 - f_2. \tag{23-12}$$

When a musical instrument sounds wobbly and out of tune, you may be hearing beats.

Beats between two tones can be detected by the ear up to a beat frequency of 6 or 7 Hz. Two piano strings or two organ pipes differing in frequency by 2 or 3 Hz sound wavery and "out of tune," although some organ stops contain two sets of pipes deliberately tuned to beat frequencies of about 1 Hz to 2 Hz for a gently undulating effect. Listening for beats is an important technique in tuning all musical instruments.

At higher-frequency differences, individual beats can no longer be distinguished. The sensation then merges into one of *consonance* or *dissonance*, depending on the frequency ratio of the two tones. In some cases the ear perceives a tone having a pitch corresponding to the difference of frequency of the two tones. Such a tone is called a *difference* tone.

23-4 THE DOPPLER EFFECT

Why does the pitch of a tone sound different when you are moving toward or away from the source?

When a source of sound or a listener, or both, are in motion relative to the air, the pitch of the sound, as heard by the listener, is in general not the same as when source and listener are at rest. This phenomenon is called the **Doppler effect.** A common example is the sudden drop in pitch of the sound from an automobile horn as one meets and passes a car proceeding in the opposite direction.

Let v_L and v_S represent the velocities of a listener and a source, relative to the air. We will consider only the special case in which the velocities lie along the line joining listener and source. Since these velocities may be in the same or opposite directions, and the listener may be either ahead of or behind the source, we need a sign convention. We will take the positive directions of v_L and v_S as that *from* the position of the listener, toward the position of the source. The speed of propagation of sound waves, c, will always be considered positive.

Consider first a listener L moving with velocity v_L toward a stationary source S, as in Fig. 23-3. The source emits a wave with frequency f_S and wavelength $\lambda = c/f_S$. The figure shows several wave crests, separated by equal distances λ. The waves approaching the moving listener have a speed of propagation *relative to the listener* of $(c + v_L)$. Thus the frequency f_L with which the listener encounters wave crests—that is, the frequency heard—is

$$f_L = \frac{c + v_L}{\lambda} = \frac{c + v_L}{c/f_S}, \tag{23-13}$$

or

$$f_L = f_S \left(\frac{c + v_L}{c} \right) = f_S \left(1 + \frac{v_L}{c} \right). \tag{23-14}$$

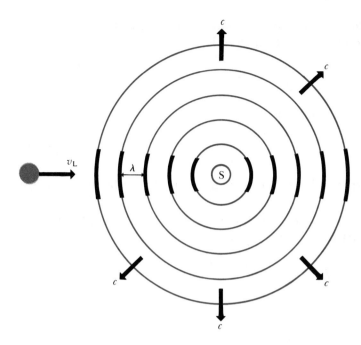

23–3 A listener moving toward a stationary source hears a frequency higher than the source frequency because the relative velocity of listener and wave is greater than c.

Hence an observer moving toward a source hears a larger frequency and higher pitch than a stationary observer. Similarly, a listener moving away from the source ($v_L < 0$) hears a lower pitch.

Now suppose the source is also moving with velocity v_S, as in Fig. 23–4. The wave speed relative to the air is still c; the speed of propagation of a wave is not altered by the motion of the source but is a property of the wave medium alone. But the wavelength is no longer given by c/f_S. The time for emission of one cycle of the wave is the period $\tau = 1/f_S$. During this time the wave travels a distance $c\tau = c/f_S$, and the source moves a distance $v_S\tau = v_S/f_S$. The wavelength is the distance between successive wave crests, and this is deter-

When the source moves relative to the medium, the wavelength changes.

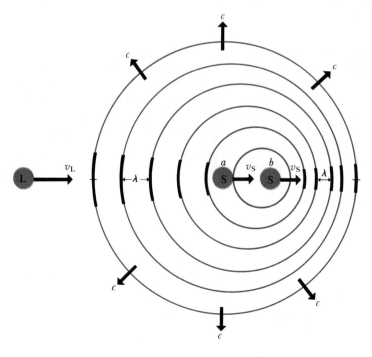

23–4 Wave surfaces emitted by a moving source are crowded together in front of the source and stretched out behind it.

mined by the *relative* displacement of source and wave. In the region to the right of the source in Fig. 23–4, the wavelength is

$$\lambda = \frac{c}{f_S} - \frac{v_S}{f_S} = \frac{c - v_S}{f_S}, \qquad (23\text{--}15)$$

and in the region to the left it is

$$\lambda = \frac{c + v_S}{f_S}. \qquad (23\text{--}16)$$

The waves are, respectively, compressed and stretched out by the motion of the source.

Both the source and the listener may be moving with respect to the medium.

The frequency measured by the listener is now given by substituting Eq. (23–16) into the first form of Eq. (23–13):

$$f_L = \left(\frac{c + v_L}{\lambda}\right) = \frac{c + v_L}{(c + v_S)/f_S},$$

or

$$\frac{f_L}{c + v_L} = \frac{f_S}{c + v_S}, \qquad (23\text{--}17)$$

which expresses the frequency f_L heard by the listener in terms of the frequency f_S of the source.

This general relation includes all possibilities for collinear motion of source and listener relative to the medium. If the listener is at rest in the medium, the corresponding velocity is zero, and of course when both source and listener are at rest or have the same velocity relative to the medium, then $f_L = f_S$. Whenever source or listener is moving in the opposite direction to what we have designated as positive, the corresponding velocity to be used in Eq. (23–17) is negative. The examples illustrate these sign conventions.

PROBLEM-SOLVING STRATEGY: Doppler effect

1. Establish a coordinate system, with the positive direction from the listener toward the source, and make sure you know the signs of all the relevant velocities, using the sign convention described above.

2. Use consistent notation to identify the various quantities: subscript S for source, L for listener.

3. When a wave is reflected from a surface, either stationary or moving, the analysis can be carried out in two stages. First find the frequency with which the wave crests arrive at the surface; this is f_L. Then think of the surface as a new source, emitting waves with this same frequency f_L, which becomes the frequency of the new source. Finally, determine what frequency is heard by a listener detecting this new wave.

Some examples of frequency shifts caused by the Doppler effect

EXAMPLE 23–2 Let $f_S = 300$ Hz and $c = 300$ m·s^{-1}. The wavelength of the waves emitted by a stationary source is then $c/f_S = 1.00$ m.

a) What are the wavelengths ahead of and behind the moving source in Fig. 23–4 if its velocity is 30 m·s^{-1}?

In front of the source,

$$\lambda = \frac{c - v_S}{f_S} = \frac{300 \text{ m·s}^{-1} - 30 \text{ m·s}^{-1}}{300 \text{ Hz}} = 0.90 \text{ m}.$$

Behind the source,

$$\lambda = \frac{c + v_S}{f_S} = \frac{300 \text{ m·s}^{-1} + 30 \text{ m·s}^{-1}}{300 \text{ Hz}} = 1.10 \text{ m}.$$

b) If the listener L in Fig. 23–4 is at rest and the source is moving away from L at 30 m·s^{-1}, what is the frequency as heard by the listener?

Since

$$v_L = 0 \quad \text{and} \quad v_S = 30 \text{ m·s}^{-1},$$

we have

$$f_L = f_S \frac{c}{c + v_S}$$

$$= 300 \text{ Hz} \left(\frac{300 \text{ m·s}^{-1}}{300 \text{ m·s}^{-1} + 30 \text{ m·s}^{-1}} \right) = 273 \text{ Hz}.$$

c) If the source in Fig. 23–4 is at rest and the listener is moving toward the left at 30 m·s^{-1}, what is the frequency as heard by the listener?

The positive direction (from listener to source) is still from left to right, so

$$v_L = -30 \text{ m·s}^{-1}, \qquad v_S = 0,$$

$$f_L = f_S \frac{c + v_L}{c} = 300 \text{ Hz} \left(\frac{300 \text{ m·s}^{-1} - 30 \text{ m·s}^{-1}}{300 \text{ m·s}^{-1}} \right) = 270 \text{ Hz}.$$

Thus, while the frequency f_L heard by the listener is less than the frequency f_S both when the source moves away from the listener and when the listener moves away from the source, the decrease in frequency is not the same for the same speed of recession.

In the preceding equations, the velocities v_L, v_S, and c are all *relative to the air*, or more generally, to the medium in which the waves are traveling. The Doppler effect exists also for electromagnetic waves in empty space, such as light waves or radio waves. In this case there is no "medium" relative to which a velocity can be defined, and we can speak only of the *relative* velocity v *of source and receiver*.

For light waves there is no material medium, but there is still a Doppler effect.

To derive the expression for the Doppler frequency shift for light requires the use of relativistic kinematic relations. These will be derived in Chapter 40, but meanwhile we quote the result without derivation. The wave speed c is the speed of light and is the same for both source and listener. In the frame of reference in which the listener is at rest, the source is moving away from the listener with velocity v. (If, instead, the source is *approaching* the listener, v is negative.) The source frequency is again f_S. The frequency f_L measured by the listener (i.e., the frequency of arrival of the waves at L) is then given by

Working out the Doppler effect for light requires methods of the special theory of relativity.

$$f_L = \left(\sqrt{\frac{c - v}{c + v}} \right) f_S. \qquad (23–18)$$

When v is positive, the source moves *away* from the listener and f_L is always *less* than f_S; when v is negative, the source moves *toward* the listener and f_L is *greater* than f_S. Thus the qualitative effect is the same as for sound, although the quantitative relationship is different.

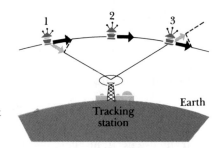

23–5 Change of velocity component along the line of sight of a satellite passing a tracking station.

The Doppler effect provides a convenient means of tracking a satellite that is emitting a radio signal of constant frequency f_S. The frequency f_L of the signal received on the earth decreases as the satellite is passing, since the velocity component *toward* the tracking station decreases from position 1 to position 2 in Fig. 23–5 and then points *away* from the earth from 2 to 3. If the received signal is combined with a constant signal generated in the receiver to give *beats*, then the beat frequency may be such as to produce an audible note whose pitch changes as the satellite passes overhead.

A similar technique is used by law-enforcement officers to measure automobile speeds. An electromagnetic wave is emitted by a source (sometimes called a "radar gun"), typically attached to a police car, at the side of the road. The wave is reflected from a moving car, which thus acts as a moving source; the reflected wave is Doppler-shifted in frequency. Measuring the frequency shift by using beats, as with satellite tracking, permits simple measurement of the speed.

The Doppler effect for *light* is important in astronomy. Analysis of light emitted by elements in distant stars shows shifts in wavelength compared to light from the same elements on earth. These shifts can be interpreted as Doppler shifts due to motion of the stars. The shift is nearly always toward the longer wavelength, or red end of the spectrum, and is therefore called the *red shift*. Such observations have provided practically all the evidence for the "exploding universe" cosmological theories, which represent the universe as having evolved from a great explosion several billion years ago in a relatively small region of space.

How do policemen use the Doppler effect to discourage drivers from exceeding the speed limit?

Astronomers use the Doppler effect to observe the motions of distant celestial objects.

SUMMARY

KEY TERMS

sound

pressure amplitude

intensity

intensity level

decibels

threshold of audibility

ultrasonic

pitch

beats

beat frequency

Doppler effect

Sound consists of longitudinal waves in air. A sinusoidal sound wave is characterized by its frequency f, wavelength λ, and amplitude A. The amplitude is also related to the pressure amplitude p_{max} by

$$p_{max} = BkA, \qquad (23–5)$$

where B is the bulk modulus and k is the wave number.

The intensity I of a wave is the time average rate at which energy is transported by the wave, per unit area. The intensity of a sound wave can be expressed in terms of the amplitude A as

$$I = \frac{1}{2}\omega BkA^2, \qquad (23–8)$$

or in terms of the pressure amplitude as

$$I = \frac{cp_{max}^2}{2B} = \frac{p_{max}^2}{2\rho c}. \qquad (23–9)$$

The intensity level β of a sound wave is defined as

$$\beta = 10 \log \frac{I}{I_0}, \qquad (23\text{--}11)$$

where I_0 is an arbitrary intensity defined to be $10^{-12}\ \text{W·m}^{-2}$. Intensity levels are expressed in decibels (dB).

The threshold of audibility at any frequency is the minimum intensity of sound that can be heard. The threshold of pain is the intensity at which the perception changes from hearing to feeling or pain.

Beats are heard when two tones with slightly different frequencies are sounded together. The beat frequency is the difference of the two frequencies.

The Doppler effect is the frequency shift that occurs when there is relative motion of a source of sound and a listener. The source and listener frequencies f_S and f_L and their velocities v_S and v_L are related by

$$\frac{f_L}{c + v_L} = \frac{f_S}{c + v_S}. \qquad (23\text{--}17)$$

QUESTIONS

23–1 Ultrasonic cleaners use ultrasonic waves of high intensity in water or a solvent to clean dirt from dishes, engine parts, and so on. How do they work? What advantages and disadvantages does this process have, compared with other cleaning methods?

23–2 Lane dividers on highways sometimes have regularly spaced ridges or ripples. When the tires of a moving car roll along such a divider, a musical note is produced. Why? Could this phenomenon be used to measure the car's speed? How?

23–3 Two tuning forks have identical frequencies, but one is stationary while the other is mounted on a rotating record turntable. What does a listener hear?

23–4 The organist in a cathedral plays a loud chord and then releases it. The sound persists for a few seconds and gradually dies away. Why does it persist? What happens to the energy when it dies away?

23–5 Why do foghorns always have very low pitches?

23–6 Some stereo amplifiers intended for home use have a maximum power output of 600 W or more. What would happen if you had 600 W of actual sound power in a moderate-sized room?

23–7 Why does your voice sound different over the telephone than in person?

23–8 Why does a guitar have a sharper, more metallic sound when played with a hard pick than when plucked with the bare fingertips?

23–9 When a record is being played with the volume control turned down all the way, a faint sound can be heard coming directly from the stylus. It seems to be all treble and no bass. How is the sound produced, and why are the frequencies so unbalanced?

23–10 The tone quality of a violin is different when the bow is near the bridge (the ends of the strings) than when it is nearer the centers of the strings. Why?

23–11 An engineer who likes to express everything in technical terms remarked that the price of gasoline rose by 2 dB in 1979. What do you think he meant?

23–12 When you are shouting at someone a fair distance away, it is easier for him to hear you if the wind is blowing from you to him, than if it is in the opposite direction. What is the physical basis for this difference?

23–13 Two notes an octave apart on a piano have a frequency ratio of 2:1. When one is slightly out of tune, beats are heard. How are they produced?

23–14 A large church has part of the organ in front and part in back. When both divisions are playing at once and a person walks rapidly down the center aisle, the two divisions sound out of tune. Why?

23–15 Can you think of circumstances in which a Doppler effect would be observed for surface waves in water? For elastic waves propagating in a body of water?

23–16 How does a person perceive the direction from which a sound comes? Is our directional perception more acute for some frequency ranges than for others? Why?

23–17 A sound source and a listener are both at rest on the earth, but a strong wind is blowing. Is there a Doppler effect?

EXERCISES

Unless indicated otherwise, assume the speed of sound in air to be $c = 345$ m·s^{-1}.

Section 23–1 Sound Waves

Section 23–2 Intensity

23–1 Consider a sound wave in air that has displacement amplitude 0.01 mm. Calculate the pressure amplitude for frequencies of

a) 500 Hz;

b) 20,000 Hz (not audible).

In each case compare the results to the pain threshold given in Example 23–1.

23–2

a) If the pressure amplitude in a sound wave is tripled. by what factor is the intensity of the wave increased?

b) By what factor must the pressure amplitude of a sound wave be increased in order to increase the intensity by a factor of 16?

23–3

a) Two sound waves of the same frequency, one in air and one in water, are equal in intensity. What is the ratio of the pressure amplitude of the wave in water to that of the wave in air?

b) If the pressure amplitudes of the waves are equal, what is the ratio of their intensities?

23–4 Derive Eq. (23–9) from the preceding equations.

23–5 A sound wave in air has frequency 400 Hz, wave velocity 345 m·s^{-1}, and displacement amplitude 0.005 mm. Calculate the intensity (in W·m^{-2}) and intensity level (in decibels) for this sound wave.

23–6

a) Relative to the arbitrary reference intensity of 10^{-12} W·m^{-2}, what is the intensity level in decibels of a sound wave whose intensity is 10^{-6} W·m^{-2}?

b) What is the intensity level of a sound wave in air whose pressure amplitude is 0.2 Pa?

23–7

a) Show that if β_1 and β_2 are the intensity levels in decibels of sounds of intensities I_1 and I_2, respectively, the difference in intensity levels of the sounds is

$$\beta_2 - \beta_1 = 10 \log \frac{I_2}{I_1}.$$

b) Show that if $(p_{max})_1$ and $(p_{max})_2$ are the pressure amplitudes of two sound waves, the difference in intensity levels of the waves is

$$\beta_2 - \beta_1 = 20 \log \frac{(p_{max})_2}{(p_{max})_1}.$$

c) Show that if the reference level of intensity is $I_0 = 10^{-12}$ W·m^{-2}, the intensity level of a sound of

intensity I (in W·m^{-2}) is

$$\beta = 120 + 10 \log I.$$

23–8 The intensity due to a number of independent sound sources is the sum of the individual intensities. How many decibels greater is the intensity level when all five quintuplets cry simultaneously than when a single one cries? How many more crying babies would be required to produce a further increase in the intensity level of an equal number of decibels?

Section 23–3 Beats

23–9 A trumpet player is tuning his instrument by playing an A note simultaneously with the first-chair trumpeter, who has perfect pitch. The first-chair player's note is exactly 440 Hz, and 3.6 beats per second are heard. What is the frequency of the other player's note?

23–10 Two identical piano strings, when stretched with the same tension, have a fundamental frequency of 440 Hz. By what fractional amount must the tension in one string be increased so that four beats per second will occur when both strings vibrate simultaneously?

Section 23–4 The Doppler Effect

23–11 A railroad train is traveling at 30 m·s^{-1} in still air. The frequency of the note emitted by the locomotive whistle is 500 Hz. What is the wavelength of the sound waves

a) in front of the locomotive?

b) behind the locomotive?

What would be the frequency of the sound heard by a stationary listener

c) in front of the locomotive?

d) behind the locomotive?

What frequency would be heard by a passenger on a train moving in the direction opposite to the first at 15 m·s^{-1} and

e) approaching the first?

f) receding from the first?

g) How would each of the preceding answers be altered if a wind speed of 10 m·s^{-1} were blowing in the same direction as that in which the first locomotive was traveling?

23–12 Two whistles, A and B, each have a frequency of 500 Hz. A is stationary and B is moving toward the right (away from A) at a speed of 50 m·s^{-1}. An observer is between the two whistles, moving toward the right with a speed of 25 m·s^{-1}.

a) What is the frequency from A as heard by the observer?

b) What is the frequency from B as heard by the observer?

c) What is the beat frequency heard by the observer?

PROBLEMS

23–13 A certain sound source radiates uniformly in all directions in air. At a distance of 5 m the sound level is 80 db. The frequency is 440 Hz.

a) What is the pressure amplitude at this distance?

b) What is the displacement amplitude?

c) At what distance is the sound level 60 db?

23–14 A window whose area is 1 m^2 opens on a street where the street noises result in an intensity level, at the window, of 60 db. How much "acoustic power" enters the window via the sound waves?

23–15 A very noisy chain saw operated by a tree surgeon emits a total sound power of 10 W uniformly in all directions. At what distance from the source is the sound level

a) 100 db?

b) 60 db?

23–16 Two loudspeakers, A and B, radiate sound uniformly in all directions. The output of acoustic power from A is 8×10^{-4} W, and from B it is 13.5×10^{-4} W. Both loudspeakers are vibrating in phase at a frequency of 172.5 Hz.

a) Determine the difference in phase of the two signals at a point C along the line joining A and B, 3 m from B and 4 m from A.

b) Determine the intensity at C from speaker A if speaker B is turned off, and the intensity at C from speaker B if speaker A is turned off.

c) With both speakers on, what are the intensity and intensity level at C?

23–17 The frequency ratio of a half-tone interval on the equally tempered scale is 1.059. Find the speed of an automobile passing a listener at rest in still air, if the pitch of the car's horn drops a half-tone between the times when the car is coming directly toward him and when it is moving directly away from him.

23–18

a) Show that Eq. (23–18) can be written

$$f_L = f_S\left(1 - \frac{v}{c}\right)^{1/2}\left(1 + \frac{v}{c}\right)^{-1/2}.$$

b) Use the binomial theorem to show that if $v \ll c$, this

expression is approximately equal to

$$f_L = f_S\left(1 - \frac{v}{c}\right).$$

c) An earth satellite emits a radio signal of frequency 10^8 Hz. An observer on the ground detects beats between the received signal and a local signal also of frequency 10^8 Hz. At a particular moment, the beat frequency is 2400 Hz. What is the component of the satellite's velocity directed toward the earth at this moment?

23–19 A sound wave of frequency f_0 and wavelength λ_0 travels horizontally toward the right. It strikes and is reflected from a large, rigid, vertical plane surface, perpendicular to the direction of propagation of the wave and moving toward the left with a speed v.

a) How many positive wave crests strike the surface in a time interval t?

b) At the end of this time interval, how far to the left of the surface is the wave that was reflected at the beginning of the time interval?

c) What is the wavelength of the reflected waves, in terms of λ_0?

d) What is the frequency, in terms of f_0?

e) A listener is at rest at the left of the moving surface. How many beats per second does she hear as a result of the combined effect of the incident and reflected waves?

23–20 A man stands at rest in front of a large, smooth wall. Directly in front of him, between him and the wall, he holds a vibrating tuning fork of frequency f_0. He now runs toward the wall with a speed v. How many beats per second will he hear between the sound waves reaching him directly from the fork and those reaching him after being reflected from the wall?

23–21 The sound source of a ship's sonar system operates at a frequency of 50,000 Hz. The velocity of sound in water can be taken as 1450 m·s^{-1}.

a) What is the wavelength of the waves emitted by the source?

b) What is the difference in frequency between the directly radiated waves and the waves reflected from a whale traveling directly away from the ship at 6.95 m·s^{-1}?

CHALLENGE PROBLEMS

23–22 This problem asks you to perform an alternative derivation of the phenomenon of beats.

a) Consider two sound waves of the same amplitude but different frequencies f_1 and f_2. The displacements of the air as a function of time produced by the individual waves are $y_1 = A \sin(2\pi f_1 t)$ and $y_2 = A \sin(2\pi f_2 t)$. (For convenience take $x = 0$ in Eq. [23–1].) Apply the principle of superposition and the appropriate trigonometric identities to show that the resultant

displacement produced by the two simultaneous waves is

$$y = \left[2A\cos 2\pi\left(\frac{f_1 - f_2}{2}\right)t\right]\sin 2\pi\left(\frac{f_1 + f_2}{2}\right)t.$$

b) From the result in (a), show that for f_1 and f_2 that are not too different, one will hear beats with a frequency of occurrence given by Eq. (23–12). What happens when $|f_1 - f_2|$ is large?

23–23 A source of sound waves, S, emitting waves of frequency f_0, is traveling toward the right in still air with a speed v_1. At the right of the source is a large, smooth, reflecting surface moving toward the left with a speed v_2. The velocity of the sound waves is c.

a) How far does an emitted wave travel in time t?

b) What is the wavelength of the emitted waves in front of (i.e., at the right of) the source?

c) How many waves strike the reflecting surface in time t?

d) What is the speed of the reflected waves?

e) What is the wavelength of the reflected waves?

f) What is the frequency of the reflected waves, as heard by a stationary listener?

g) Calculate a numerical value for the frequency in part (f), for $c = 345 \text{ m·s}^{-1}$, $v_1 = 30 \text{ m·s}^{-1}$, $v_2 = 60 \text{ m·s}^{-1}$, and $f_0 = 1000$ Hz.

PART FIVE

ELECTROSTATICS AND
ELECTRIC CURRENTS

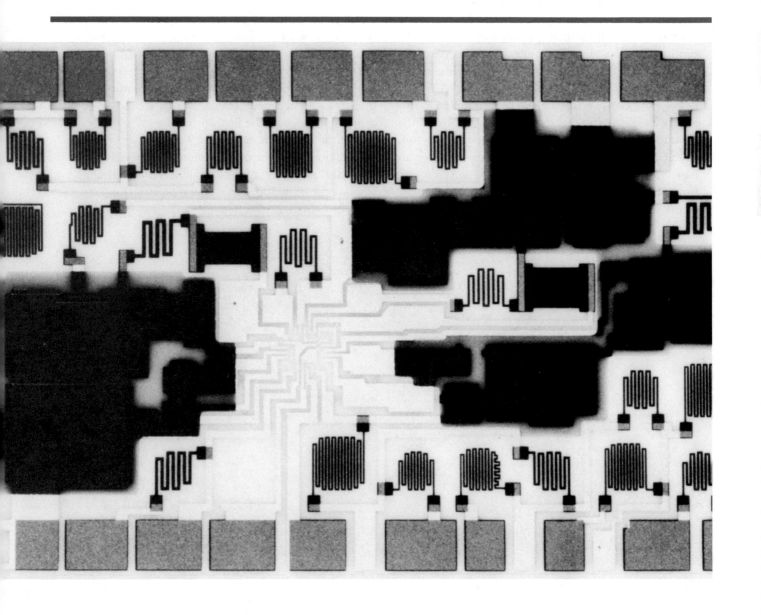

PERSPECTIVE

We are now at the halfway point in our study of physics. In the past three chapters we have been concerned with wave phenomena. We developed the most important wave concepts in the familiar context of mechanical waves, but we also tried to indicate their relevance in many other areas of physics and their role as a unifying concept throughout all areas of physics. In particular, we will explore in the last half of the book the indispensible role that wave phenomena play in optics; many optical phenomena simply cannot be understood without wave concepts. Equally important, as we will see later, wave concepts have proved to be an important gateway into the twentieth-century concepts of quantum mechanics, so vital to the understanding of atomic, molecular, and nuclear structure.

The next twelve chapters take us into the area of electricity and magnetism, a broad area of physics that includes a rich variety of fascinating physical phenomena and the fundamental principles on which they are based. We first introduce the concept of *electric charge;* we describe interactions among charges with the help of the concept of *electric field*. A study of the potential energies associated with these interactions leads to the concept of *electric potential*. Together, these concepts enable us to understand a wide range of phenomena in electrostatics, or "static electricity," as it is often called, and also to explore the subject of *capacitance,* so vital to contemporary electronics. Next we examine the *flow* of electric charge through materials we call *conductors*. A flow of charge is called an *electric current,* and this discussion leads us into a detailed study of electric *circuits*. We study the circuit concepts of resistance, voltage, current, and electromotive force and develop principles that enable us to analyze and understand a variety of practical devices and systems.

Electric circuits are at the heart of electric-power distribution systems, household wiring, the electrical systems of automobiles, and all of contemporary electronics, including radio, television, and computer circuitry and much of the instrumentation of present-day research in physics. Thus our study of electrostatics and of electric circuits has not only a great deal of inherent interest but also a broad range of practical applications to contemporary technology.

24
COULOMB'S LAW

INTERACTIONS BETWEEN ELECTRICALLY CHARGED BODIES ARE ONE OF THE
four fundamental classes of interactions found in nature, as discussed at the
beginning of Chapter 5. In this chapter we study the basic principles underlying the interactions of electric charges at rest, that is, *electrostatic* phenomena.
The basic force law for interaction of electric charges at rest is called *Coulomb's
law;* a variety of applications of this law are worked out in this chapter.

24–1 ELECTRIC CHARGE

The ancient Greeks discovered as early as 600 B.C. that when amber is rubbed
with wool, it becomes able to attract other objects. Today we say that the amber
has acquired an **electric charge,** or is *electrified.* Indeed, these terms are derived from the Greek word *elektron,* meaning "amber." A person may become
electrified by scuffing shoes across a nylon carpet; a comb is electrified by
passing it through dry hair; and so on.

Objects can be electrified by rubbing
them together.

For demonstrations of electric-charge interactions we often use plastic
rods and fur. We can electrify a plastic rod by rubbing it with fur, and then
touch the rod to two small, light balls of cork or pith, suspended by thin silk or
nylon threads. We find that the rod then *repels* the balls and that they also
repel each other.

We get the same results by rubbing a glass rod with silk. However, when a
pith ball that has been in contact with electrified plastic is placed near one that
has been in contact with electrified glass, the pith balls *attract* each other. We
conclude that there are two kinds of electric charge, the kind on the plastic rod
rubbed with fur and the kind on the glass rod rubbed with silk. Benjamin
Franklin suggested calling these *negative* and *positive,* respectively, and these
names are still used. These experiments lead to the fundamental conclusion
that *like charges repel each other, and unlike charges attract each other.*

There are two kinds of electric charge:
Like charges attract, and unlike charges
repel.

Two bodies may also interact by *magnetic* interaction; the most familiar
example is the attraction of iron or steel objects to a permanent magnet. Magnetic forces were once believed to be a fundamentally different type of interaction, but it is now known that magnetic interactions are actually the

interactions of electrically charged particles in motion. An electromagnet shows *magnetic* interactions when an *electric* current passes through its coils. We will study magnetic interactions in detail in later chapters, but for now we concentrate on *charges at rest*, that is, on **electrostatics.**

Here is another fundamental experiment. We rub a plastic rod with fur and then touch it to a suspended pith ball. Both the rod and the ball then have negative charge. If we now bring the *fur* near the pith ball, the ball is *attracted*, showing that the fur is *positively* charged. Thus when plastic is rubbed with fur, *opposite* charges appear on the two materials. The same thing happens with glass and silk. The implication is that in these experiments, electric charge is not *created* but is *transferred* from one body to another. It is now known that the plastic rod acquires extra electrons, which have negative charge. These elecrons are taken from the fur, which thus acquires a net positive charge.

Electrifying a body involves transfer of charge from one body to another, not creation of charge.

24–2 ATOMIC STRUCTURE

The interactions responsible for the structure of atoms and molecules (and hence of all ordinary matter) are primarily *electrical* interactions between electrically charged particles. The fundamental building blocks of ordinary matter are three particles: the negatively charged **electron,** the positively charged **proton,** and the electrically neutral **neutron.** The negative charge of the electron has the same magnitude as the positive charge of the proton. Thus in one sense the charge of a proton or an electron is the fundamental natural unit of charge. In the presently accepted theory of fundamental particles, however, the proton and neutron are not truly fundamental but are combinations of other entities called *quarks*, which have charges of $\pm\frac{1}{3}$ and $\pm\frac{2}{3}$ times the electron charge. In this theory the electron *is* a truly fundamental particle. We return to the quark model of fundamental particles in Chapter 44.

Fundamental particles have electric charges.

These three particles are arranged in the same general way in all atoms. The protons and neutrons always form a closely packed, dense cluster called the **nucleus,** which thus has a positive charge. The particles in the nucleus are held together by strong interactions, mentioned in Section 5–1. The diameter of the nucleus, if we think of it as roughly spherical, is of the order of 10^{-14} m. Outside the nucleus, at distances of the order of 10^{-10} m from it, are the electrons. In a neutral atom the number of electrons equals the number of protons in the nucleus. Thus the net electric charge of such an atom (the algebraic sum of all the charges) is zero. If one or more electrons are removed, the remaining positively charged structure is called a **positive ion; a negative ion** is an atom that has *gained* one or more electrons. This gaining or losing of electrons is called *ionization*.

The nucleus is very much smaller than the overall size of the atom.

In a neutral atom, the negative charge of the electrons exactly balances the positive charge of the nucleus.

In an atomic model proposed by the Danish physicist Niels Bohr in 1913, the electrons were pictured as whirling about the nucleus in circular or elliptical orbits. We now know that the electrons are more accurately represented as spread-out distributions of electric charge, governed by the principles of quantum mechanics, which we will discuss in Chapter 42. Nevertheless, the Bohr model is still useful for visualizing the structure of an atom. The diameters of the electron charge distributions, which the Bohr model pictures as circular orbits, determine the overall size of the atom as a whole. The diame-

A crude but useful picture of an atom: electrons in orbits around the nucleus

ters of these orbits are of the order of 10^{-10} m, or about ten thousand times as great as the diameter of the nucleus. A Bohr atom is analogous to a miniature solar system, with electrical forces taking the place of gravitational forces. The massive, positively charged central nucleus corresponds to the sun, while the electrons, moving around the nucleus under the electrical force of its attraction, correspond to the planets moving around the sun under the influence of its gravitational attraction.

The masses of the proton and neutron are nearly equal, and the mass of the proton is about 1836 times that of the electron. Nearly all the mass of an atom, therefore, is concentrated in its nucleus. Since one mole of monatomic hydrogen consists of 6.022×10^{23} particles (Avogadro's number) and its mass is 1.008 g, the mass of a single hydrogen atom is

Most of the mass of an atom is concentrated in its nucleus.

$$\frac{1.008 \text{ g}}{6.022 \times 10^{23}} = 1.674 \times 10^{-24} \text{ g} = 1.674 \times 10^{-27} \text{ kg}.$$

The nucleus of a hydrogen atom is a single proton, and around it moves a single electron. Hence, of the total mass of the hydrogen atom, 1/1837 part is the mass of the electron, and the remainder is the mass of the proton. To four significant figures,

$$\text{Mass of electron} = 9.110 \times 10^{-31} \text{ kg},$$

$$\text{Mass of proton} = 1.673 \times 10^{-27} \text{ kg},$$

$$\text{Mass of neutron} = 1.675 \times 10^{-27} \text{ kg}.$$

After hydrogen, the atom with the next simplest structure is helium. Its nucleus consists of two protons and two neutrons, and it has two electrons outside the nucleus. When these two electrons are removed, the doubly charged helium ion (which is the helium nucleus itself) is often called an *alpha particle*, or *α*-particle. The next element, lithium, has three protons in its nucleus and thus has a nuclear charge of three units. In the un-ionized state the lithium atom has three extranuclear electrons. Each element has a different number of protons in its nucleus and therefore a different positive nuclear charge. In the *periodic table of elements* at the end of this book, each element occupies a box with an associated number called the **atomic number.**

The atomic number is the number of nuclear protons; in the un-ionized state, this is equal to the number of extranuclear electrons.

Every material body contains a tremendous number of charged particles: positively charged protons in the nuclei of its atoms and negatively charged electrons outside the nuclei. When the total number of protons equals the total number of electrons, the body as a whole is electrically neutral. To give a body an excess negative charge, we may either *add negative* charges to a neutral body or *remove positive* charges from the body. Similarly, we can create an excess positive charge by either *adding positive* charge or *removing negative* charge. In most cases it is negative charge (electrons) that is added or removed, and a "positively charged body" is one that has lost some of its normal complement of electrons. When we speak of the charge of a body, we always mean its *net* charge. The net charge is always a very small fraction of the total positive or negative charge in the body.

Electrifying or charging a body usually involves adding or subtracting electrons.

The total electric charge in the universe is constant: Charge cannot be created or destroyed.

Implicit in the preceding statements is the principle of **conservation of charge.** This principle states that the algebraic sum of all the electric charges in any closed system is constant. Charge can be transferred from one body to another, but it cannot be created or destroyed. Conservation of charge is believed to be a *universal* conservation law; there is no experimental evidence for any violation of this principle.

Electrical interactions play a central and dominant role in most aspects of the structure of matter. The forces that hold atoms together in a molecule or in a solid crystal lattice, the adhesive force of glue, the forces associated with surface tension—all these are electrical in nature, arising from the electrical forces between the charged particles in the interacting atoms.

Electrical interactions are responsible for the structure of atoms, molecules, and condensed matter.

Nuclei are held together by strong interactions, despite the repulsions of their protons.

Electrical interactions alone are *not* sufficient to understand the structure of atomic *nuclei,* however. A nucleus consists of protons, which repel each other, and neutrons, which have no electric charge. For nuclei to be stable, there must be additional forces, attractive in nature, that hold the nucleus together despite the electrical repulsion. This additional interaction is called the *nuclear force;* it is an example of the *strong interaction* mentioned in Section 5–1. The nuclear force has a short range, of the order of nuclear dimensions, and its effects do not extend far beyond the nucleus. In analyzing the structure of atoms we can ignore the nuclear force, considering the nucleus as a rigid structure, and concentrate on the electrical interactions.

When the structure of the nucleus itself becomes the subject of our study, we do need to consider the strong interactions. Many phenomena associated with the stability or instability of nuclei arise from the competition between the repulsive electrical forces and the attractive nuclear forces. We will study these matters in greater detail in Chapter 44.

24–3 ELECTRICAL CONDUCTORS AND INSULATORS

Some materials permit electric charge to move from one region of the material to another, while others do not. For example, suppose we touch one end of a copper wire to an electrified plastic rod and the other end to a metal ball that is initially uncharged, as in Fig. 24–1. By studying the subsequent interaction of the ball with other charged objects, we discover that it has become charged. The copper wire is called a **conductor** of electricity. If we now repeat the experiment but use a rubber band or nylon thread in place of the wire, we find that *no* charge is transferred to the ball. These materials are called **insulators.** Conductors permit the passage of charge through them; insulators do not.

Electric charge can move through conductors but not through insulators.

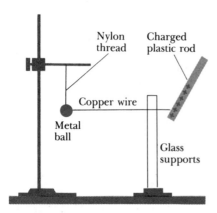

24–1 Copper is a conductor of electricity. Charge can move from the plastic rod through the wire to the metal ball.

In general, *metals* are good conductors, and most *nonmetals* are insulators. The positive valence of metallic elements and the fact that they form positive ions in solutions show that the atoms of a metal can easily lose one or more of their outer electrons. Within a metal, such as a copper wire, a few outer electrons become detached from each atom. These electrons can then move freely throughout the material, in much the same way that the molecules of a gas move freely through the spaces between grains of sand in a sand-filled container. In fact, these free electrons are sometimes described as an "electron gas." The positive nuclei and the other electrons remain fixed in their positions within the material. In an insulator, however, there are no, or at most very few, free electrons.

24–4 CHARGING BY INDUCTION

When we charge a pith ball by touching it with a plastic rod that has been rubbed with fur, some of the extra electrons on the plastic are transferred to the ball, leaving the plastic with a smaller negative charge. However, we can use a different technique in which the plastic rod can give another body a charge of *opposite* sign without losing any of its own charge. This process, called charging by **induction,** is shown in Fig. 24–2.

In Fig. 24–2a two neutral metal spheres are in contact, each supported on an insulating stand. When we bring a negatively charged rod near one of the spheres, without actually touching it, as in (b), the free electrons in the metal spheres are repelled by the rod and shift slightly toward the right, away from the rod. The electrons cannot escape from the spheres because the supporting stands and the surrounding air are insulators. Thus excess negative charge must accumulate at the right surface of the right sphere, and a deficiency of negative charge (i.e., a net positive charge) arises at the left surface of the left sphere. These excess charges are called **induced charges.**

Of course, not *all* the free electrons are forced to the surface of the right sphere. As soon as any induced charge develops, it also exerts forces on the other free electrons. This force is toward the left; it consists of a repulsion by the induced negative charge on the right and an attraction toward the induced positive charge on the left. Thus the system reaches an equilibrium state. At each point, the force toward the right on an electron, due to the charged rod, is just balanced by the force toward the left on the electron, due to the induced charge. The induced charge remains on the surfaces of the spheres as long as the rod is held nearby. When the rod is removed, the free electrons shift back to the left and the original neutral condition is restored.

Now suppose we separate the spheres slightly, as shown in (c), while the plastic rod is near. If we now remove the rod, as in (d), we are left with two oppositely charged metal spheres whose charges attract each other. When the two spheres are separated by a great distance, as in (e), each of the two charges becomes uniformly distributed over its sphere. We note that the charge on the negatively charged rod has not changed during this process.

Figure 24–3 shows a variation of the technique of charging by induction. In this figure a single metal sphere (on an insulating stand) is charged by induction. The symbol labeled "ground" in part (c) simply means that the sphere is connected to the earth (a conductor). The earth takes the place of the second sphere in Fig. 24–2. In step (c), electrons are repelled to ground either through a conducting wire or along the moist skin of a person who touches the sphere. The earth thus acquires a negative charge equal to the induced positive charge remaining on the sphere.

The process taking place in Figs. 24–2 and 24–3 could be explained equally well if the mobile charges in the spheres were *positive* or, in fact, if *both* positive and negative charges were mobile. Although we now know that in a metallic conductor it is actually the *negative* charges that move, it is

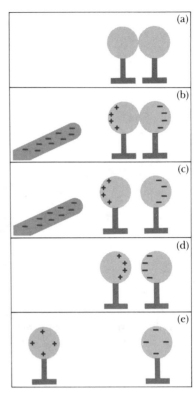

24–2 Two metal spheres are oppositely charged by induction.

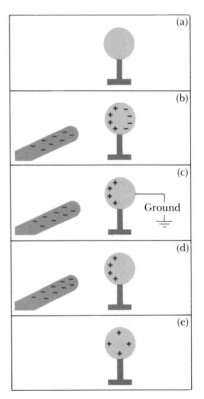

24–3 Charging a single metal sphere by induction.

often convenient to describe a process *as if* the positive charges moved. In ionic solutions, both positive ions and negative ions participate in the conduction process.

24–5 COULOMB'S LAW

Coulomb's law: a quantitative description of interactions between electric charges

We describe the electrical interaction between two charged particles in terms of the *forces* they exert on each other. A quantitative investigation of this interaction was carried out by Augustin de Coulomb (1736–1806) in 1784. For his force measurements he used a torsion balance similar to that used 13 years later by Cavendish to study the (much weaker) gravitational interaction, as we discussed in Section 6–3.

Coulomb studied the force of attraction or repulsion between two "point charges"; that is, charged bodies whose sizes are small compared with the distance between them. He found that the force grows weaker with increasing separation between the bodies. When the distance doubles, the force decreases to $\frac{1}{4}$ its initial value. That is, it varies inversely with the square of the distance. If r is the distance between the particles, then the force is proportional to $1/r^2$.

How to divide an electric charge into two equal parts

The force also depends on the quantity of charge on each body, which we will denote by q or Q. To explore this dependence, Coulomb devised a method for dividing a charge into two equal parts. He reasoned that if he brought a charged spherical conductor into contact with an identical conductor that was initially *uncharged*, then by symmetry the charge would be shared equally between the two conductors. Thus he could obtain one-half, one-quarter, and so on, of any given charge. He found that the force that each of two point charges q_1 and q_2 exerts on the other is proportional to each charge and is thus proportional to the *product* $q_1 q_2$ of the two charges.

The magnitude F of the force between two point charges can be expressed as

$$F = k \frac{|q_1 q_2|}{r^2}, \tag{24–1}$$

where k is a proportionality constant. The numerical value of k depends on the units in which F, q_1, q_2, and r are expressed. Equation (24–1) is the mathematical statement of what is known today as **Coulomb's law:**

> The force of attraction or repulsion between two point charges is directly proportional to the product of the charges and inversely proportional to the square of the distance between them.

The *direction* of the force on each particle is always along the line joining the two particles, pulling each particle toward the other in the case of attractive forces on unlike charges, and pushing them apart in the case of repulsive forces on like charges.

The exponent 2 in Coulomb's law appears to be exactly 2.

The proportionality of the electrical force to $1/r^2$ has been verified with great precision. There is no reason to suspect, for example, that the electrical force might vary as $1/r^{2.0001}$.

The charges q_1 and q_2 can be either positive or negative quantities, corresponding to the existence of two kinds of charge, but Eq. (24–1) gives the magnitude of the interaction force in all cases. When the charges are of like sign, the forces are repulsive; when they are unlike, they are attractive. In either case the forces obey Newton's third law; the force that q_1 exerts on q_2

is the negative of the force that q_2 exerts on q_1. The "absolute value" bars in Eq. (24–1) are needed because F, the magnitude of a vector quantity, is by definition always positive, while the product q_1q_2 is negative whenever the two charges have opposite signs.

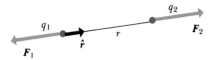

24–4 The unit vector $\hat{r}$, pointing from q_1 toward q_2, is used to describe the directions of the forces $\boldsymbol{F}_1$ and $\boldsymbol{F}_2$ on the charges.

The form of Eq. (24–1) is the same as that of the law of gravitation, discussed in Section 6–3, but electrical and gravitational interactions are two distinct classes of phenomena. Electrical interactions depend on electric charges and can be either attractive or repulsive, while gravitational interactions depend on mass and are always attractive.

We can incorporate both the magnitude and direction of the interaction force between two charged particles into a single vector equation. To do this we use a unit vector $\hat{r}$ lying along the line joining q_1 and q_2, in the direction from q_1 to q_2, as shown in Fig. 24–4. The force $\boldsymbol{F}_2$ on q_2 is then given by

Unit vectors can be used to describe the magnitude and direction of an electrical force in a single equation.

$$\boldsymbol{F}_2 = k\frac{q_1q_2}{r^2}\hat{r}, \qquad (24\text{–}2)$$

and the force $\boldsymbol{F}_1$ on q_1 is given by

$$\boldsymbol{F}_1 = -k\frac{q_1q_2}{r^2}\hat{r}. \qquad (24\text{–}3)$$

When two charges exert forces simultaneously on a third charge, the total force experienced by that charge is found to be the *vector sum* of the forces that the two charges would exert individually. This important property, called the **principle of superposition,** also holds for any number of charges. It permits the application of Coulomb's law to arrays of charge of any degree of complexity, although the computational problems can be very great. Examples 24–1 and 24–4 (Section 24–6) show applications of the superposition principle.

Superposition: how to calculate the total force exerted by several charges

If matter is present in the space between the charges, the *net force* acting on each charge is altered because charges are induced in the molecules of the intervening material. We will describe this effect later. As a practical matter, Coulomb's law can be used as stated for point charges in air, since even at atmospheric pressure the effect of the air changes the electrical force from its value in vacuum by only about one part in two thousand.

In the chapters of this book dealing with electrical phenomena, we will use SI units exclusively. The SI electrical units include most of the common electrical units such as the volt, the ampere, the ohm, and the watt. The cgs system is also used, more so in scientific work than in commerce and industry. However, there is *no* British system of electrical units. This is one of many reasons for abandoning the British system and adopting metric units universally.

The coulomb: the SI unit of electric charge

To the three basic SI units (the meter, the kilogram, and the second) we now add a fourth: the unit of electric charge. This unit is called one **coulomb** (1 C). The electrical constant k in Eq. (24–1) is, in this system,

$$k = 8.98755 \times 10^9 \text{ N·m}^2\text{·C}^{-2}.$$

This value of k can be regarded as an operational definition of the coulomb, since in principle we can measure the interaction force between any two charges and use Coulomb's law to determine the charge. As a practical matter, it is better to define the coulomb instead in terms of a unit of electric current (charge per unit time), the *ampere*. We will return to this definition in Chapter 31.

The value of the constant k in Coulomb's law is not as arbitrary as it seems.

This value of k may seem arbitrary and strange, but it really is not. Later, when we study electromagnetic radiation, we will show that k is closely related to the *speed of light*,

$$c = 2.998 \times 10^8 \text{ m·s}^{-1}.$$

Specifically, the numerical value of k is given by

$$k = 10^{-7} c^2.$$

A different (and more usual) form for the constant in Coulomb's law

In the cgs system of electrical units (not used in this book) the constant k is defined to be unity, without units. This defines a unit of electric charge called the *statcoulomb* or the *esu* (electrostatic unit). The conversion factor is

$$1 \text{ C} = 2.998 \times 10^9 \text{ esu}.$$

In SI units the constant k in Eq. (24–1) is usually written not as k but as $1/4\pi\epsilon_0$, where ϵ_0 is another constant. This appears to complicate matters, but it actually simplifies many formulas to be encountered later. Thus Coulomb's law is usually written as

$$F = \frac{1}{4\pi\epsilon_0} \frac{|q_1 q_2|}{r^2}. \tag{24-4}$$

In the vector form of Eqs. (24–2) and (24–3) this becomes

$$\boldsymbol{F} = \frac{1}{4\pi\epsilon_0} \frac{q_1 q_2}{r^2} \hat{r}. \tag{24-5}$$

In Eqs. (24–4) and (24–5),

$$\frac{1}{4\pi\epsilon_0} = 8.98755 \times 10^9 \text{ N·m}^2\text{·C}^{-2}$$

and

$$\epsilon_0 = 8.854188 \times 10^{-12} \text{ C}^2\text{·N}^{-1}\text{·m}^{-2}.$$

In examples and problems we often use the approximate value

$$\frac{1}{4\pi\epsilon_0} = 9.0 \times 10^9 \text{ N·m}^2\text{·C}^{-2},$$

which is within about 0.1% of the correct value.

The charge of the electron: a fundamental unit of electric charge

As we mentioned in Section 24–2, the most fundamental unit of charge is the magnitude of the charge of an electron or a proton. This quantity is denoted by e; the most precise measurements to date yield the value

$$e = 1.602192 \times 10^{-19} \text{ C} \cong 1.60 \times 10^{-19} \text{ C}.$$

One coulomb therefore represents the negative of the total charge carried by about 6×10^{18} electrons. For comparison, the population of the earth is estimated to be about 5×10^9 persons, while a cube of copper 1 cm on a side contains about 2.4×10^{24} electrons.

In most contexts, the coulomb is a very large unit of charge.

In electrostatics problems, charges as large as one coulomb are unusual. Two charges of magnitude 1 C, a distance 1 m apart, would exert forces of magnitude 9×10^9 N on each other! A more typical range of magnitude is 10^{-9} to 10^{-6} C. The microcoulomb (1 μC $= 10^{-6}$ C) is often used as a practical unit of charge.

24–6 APPLICATIONS OF COULOMB'S LAW

In this section we present several examples of problems using Coulomb's law. There are no new principles here, but the problem-solving methods are helpful preparation for the next several chapters.

PROBLEM-SOLVING STRATEGY: *Coulomb's law*

1. As always, consistent units are essential. With the value of $k = 1/4\pi\epsilon_0$ given above, distances *must* be in meters and charge in coulombs; the force is then in newtons. If you are given distances in centimeters, inches, or furlongs, don't forget to convert!

2. When a charge has forces acting on it due to two or more other charges, the total force is the *vector sum* of the individual forces. Don't forget what you learned about vector addition at the beginning of the book. Components in an xy-coordinate system, with or without unit vectors, are often helpful. If you are in doubt, review the material on vector addition in Chapter 1. And be sure to use correct vector notation; if a symbol represents a vector quantity, underline it or put an arrow over it. If you get sloppy with your notation, you will also get sloppy with your thinking. It is absolutely essential to distinguish between vector quantities and scalar quantities, and to treat vectors as vectors.

3. Some of the examples and problems in this and later chapters involve a continuous distribution of charge along a line or over a surface. In these cases the vector sum described in (2) becomes a vector integral, usually carried out by use of components. We divide the total charge distribution into infinitesimal pieces, use Coulomb's law for each piece, and then integrate to carry out the vector sum. Study Example 24–5 in this section carefully.

EXAMPLE 24–1 Two charges are located on the positive x-axis of a coordinate system, as shown in Fig. 24–5. Charge $q_1 = 2 \times 10^{-9}$ C is 2 cm from the origin, and charge $q_2 = -3 \times 10^{-9}$ C is 4 cm from the origin. What is the total force exerted by these two charges on a charge $q_3 = 5 \times 10^{-9}$ C located at the origin?

Using the superposition principle to find the total force on a charge due to two others

SOLUTION The total force on q_3 is the vector sum of the forces due to q_1 and q_2 individually. Converting distance to meters, we use Eq. (24–4) to find the magnitude F_1 of the force on q_3 due to q_1:

$$F_1 = \frac{(9.0 \times 10^9 \text{ N·m}^2\text{·C}^{-2})(2 \times 10^{-9} \text{ C})(5 \times 10^{-9} \text{ C})}{(0.02 \text{ m})^2} = 2.25 \times 10^{-4} \text{ N}.$$

This force has a negative x-component because q_3 is repelled (i.e., pushed in the negative x-direction) by q_1, which has the same sign. Similarly, the force due to q_2 is found to have magnitude

$$F_2 = \frac{(9.0 \times 10^9 \text{ N·m}^2\text{·C}^{-2})(3 \times 10^{-9} \text{ C})(5 \times 10^{-9} \text{ C})}{(0.04 \text{ m})^2} = 0.84 \times 10^{-4} \text{ N}.$$

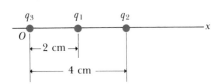

FIGURE 24–5

This force has a positive x-component because q_3 is attracted (i.e., pulled in the positive x-direction) by the opposite charge q_2. The sum of the x-components is

$$\Sigma F_x = -2.25 \times 10^{-4} \text{ N} + 0.84 \times 10^{-4} \text{ N} = -1.41 \times 10^{-4} \text{ N}.$$

There are no y- or z-components. Thus the total force on q_3 is directed to the left, with magnitude 1.41×10^{-4} N.

EXAMPLE 24–2 An α-particle is a nucleus of a helium atom. It has a mass m of 6.64×10^{-27} kg and a charge q of $+2e$ or 3.2×10^{-19} C. Compare the force of

Comparing the strengths of electrical and gravitational forces the electrostatic repulsion between two α-particles with the force of gravitational attraction between them.

SOLUTION The electrostatic force F_e is

$$F_e = \frac{1}{4\pi\epsilon_0}\frac{q^2}{r^2},$$

and the gravitational force F_g is

$$F_g = G\frac{m^2}{r^2}.$$

The ratio of the electrostatic to the gravitational force is

$$\frac{F_e}{F_g} = \frac{1}{4\pi\epsilon_0 G}\frac{q^2}{m^2} = \frac{(9.0\times10^9\ \text{N·m}^2\text{·C}^{-2})}{(6.67\times10^{-11}\ \text{N·m}^2\text{·kg}^{-2})}\frac{(3.2\times10^{-19}\ \text{C})^2}{(6.64\times10^{-27}\ \text{kg})^2}$$
$$= 3.1\times10^{35}.$$

Thus the gravitational force is negligible compared to the electrostatic force. This is always the case for interactions of atomic and subatomic particles, while for objects the size of the earth the positive and negative charges are nearly equal, and the net *electrical* interactions are usually much smaller than the gravitational.

The Bohr model: a simple dynamic model of the structure of the hydrogen atom

EXAMPLE 24–3 The Bohr model of the hydrogen atom (mentioned in Section 24–2 and discussed in detail in Chapter 42) consists of a single electron having electric charge $-e$, revolving in a circular orbit about a single proton of charge $+e$. The electrostatic force of attraction between electron and proton provides the centripetal force that retains the electron in its orbit. Hence if v is the orbital speed, Newton's second law gives

$$\frac{1}{4\pi\epsilon_0}\frac{e^2}{r^2} = m\left(\frac{v^2}{r}\right). \tag{24–6}$$

This motion is completely analogous to the motion of a satellite around the earth, discussed in Section 6–5.

In the Bohr model, the angular momentum must be an integer multiple of some minimum value.

In Bohr's theory, the electron may revolve only in some one of a number of specified orbits. The orbit of smallest radius is that for which the angular momentum L of the electron is $h/2\pi$, where h is a universal constant called *Planck's constant*, equal to 6.625×10^{-34} J·s. Then,

$$L = mvr = \frac{h}{2\pi}. \tag{24–7}$$

When v is eliminated between the preceding equations, we find

$$r = \frac{\epsilon_0 h^2}{\pi m e^2},$$

and when numerical values are inserted, we find, for the radius of the *first Bohr orbit*,

$$r = 5.29\times10^{-11}\ \text{m} = 0.529\times10^{-8}\ \text{cm}.$$

This result corresponds roughly with other estimates of the "size" of a hydrogen atom obtained from deviations from ideal-gas behavior, the density of hydrogen in the liquid and solid states, and other observations.

EXAMPLE 24-4 In Fig. 24–6, two equal positive charges $q = 2.0 \times 10^{-6}$ C interact with a third charge $Q = 4.0 \times 10^{-6}$ C. Find the magnitude and direction of the total (resultant) force on Q.

SOLUTION The key word is *total;* we must compute the force each charge exerts on Q, and then obtain the *vector sum* of the forces. This is most easily accomplished by using components. The figure shows the force on Q due to the upper charge q. From Coulomb's law,

$$F = (9.0 \times 10^9 \text{ N·C}^{-2}\text{·m}^{-2})\frac{(4.0 \times 10^{-6} \text{ C})(2.0 \times 10^{-6} \text{ C})}{(0.5 \text{ m})^2}$$

$$= 0.29 \text{ N}.$$

The components of this force are given by

$$F_x = F \cos \theta = (0.29 \text{ N})\left(\frac{0.4 \text{ m}}{0.5 \text{ m}}\right) = 0.23 \text{ N},$$

$$F_y = -F \sin \theta = -(0.29 \text{ N})\left(\frac{0.3 \text{ m}}{0.5 \text{ m}}\right) = -0.17 \text{ N}.$$

The lower charge q exerts a force of the same magnitude, but in a different direction. From symmetry we see that its x-component is the same as that due to the upper charge, but its y-component is opposite. Hence

$$\sum F_x = 2(0.23 \text{ N}) = 0.46 \text{ N},$$

$$\sum F_y = 0,$$

or

$$\sum \mathbf{F} = (0.46 \text{ N}) \, \mathbf{i}.$$

The total force on Q is horizontal, with magnitude 0.46 N. How would this solution differ if the lower charge were *negative*?

EXAMPLE 24-5 An electric charge Q is distributed uniformly along a line of length $2a$, lying along the y-axis, as shown in Fig. 24–7. A point charge q lies on the x-axis, at a distance x from the origin. Find the total force Q exerts on q.

Using the superposition principle with vector addition to find the total force on a charge

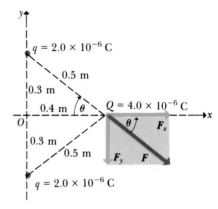

24-6 $\mathbf{F}$ is the force on Q due to the upper charge q.

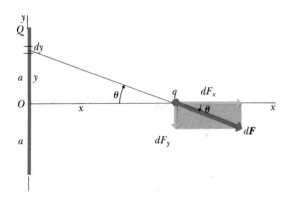

24-7 A line of charge with length $2a$ and total charge Q exerts an electrostatic force on point charge q.

Using the superposition principle for a continuous distribution of charge along a line

SOLUTION We divide the line charge into infinitesimal segments; let the length of a typical segment at height y be dy, as shown. To find the charge dQ in this segment, note that if the charge is distributed uniformly, the ratio of dQ to the total charge Q is equal to the ratio of dy to the total length $2a$. Thus

$$dQ/Q = dy/2a$$

and

$$dQ = Q\,dy/2a.$$

The distance r from this segment to q is $(x^2 + y^2)^{1/2}$, so the magnitude of force dF on q due to this segment is

$$dF = \frac{qQ}{4\pi\epsilon_0}\frac{dy}{2a(x^2 + y^2)}.$$

We now represent this force in terms of its x- and y-components, using the same procedure as in Example 24–4:

$$dF_x = dF\cos\theta,$$
$$dF_y = -dF\sin\theta.$$

We note that

$$\sin\theta = \frac{y}{\sqrt{x^2 + y^2}}, \qquad \cos\theta = \frac{x}{\sqrt{x^2 + y^2}}.$$

To find the components of the total force, we integrate the forces due to the separate charge elements.

Combining this with the above expression for dF, we find

$$dF_x = \frac{qQ}{4\pi\epsilon_0}\frac{x\,dy}{2a(x^2 + y^2)^{3/2}},$$
$$dF_y = -\frac{qQ}{4\pi\epsilon_0}\frac{y\,dy}{2a(x^2 + y^2)^{3/2}}.$$

To find the total force components F_x and F_y, we integrate these expressions, noting that to include all of Q we must integrate from $y = -a$ to $y = +a$. We invite you to work out the details of the integration; an integral table is helpful. The final results are

$$F_x = \int_{-a}^{a}\frac{qQ}{4\pi\epsilon_0}\frac{x\,dy}{2a(x^2 + y^2)^{3/2}} = \frac{qQ}{4\pi\epsilon_0}\frac{1}{x\sqrt{x^2 + a^2}},$$
$$F_y = -\int_{-a}^{a}\frac{qQ}{4\pi\epsilon_0}\frac{y\,dy}{2a(x^2 + y^2)^{3/2}} = 0,$$

or, in vector form,

$$\mathbf{F} = \frac{qQ}{4\pi\epsilon_0}\frac{1}{x\sqrt{x^2 + a^2}}\,\mathbf{i}.$$

We could have guessed from the symmetry of the situation that F_y would be zero; the upper half of Q pushes downward on q, but this is balanced by an equal upward push due to the lower half of Q (assuming both Q and q are positive). We also note that when x is much larger than a, we can neglect a in the denominator, and our result becomes approximately

$$\mathbf{F} \cong \frac{qQ}{4\pi\epsilon_0}\frac{1}{x^2}\,\mathbf{i},$$

which simply means that if the charge q is far away from the line charge, compared to its size, it looks like a point. When q and Q are both positive or both negative, the force is in the *positive x*-direction, corresponding to a repulsive interaction. If they have opposite signs, F_x is negative, corresponding to an attraction.

SUMMARY

The fundamental entity in electrostatics is electric charge. Charge can be transferred from one body to another by rubbing or other means, but it cannot be created or destroyed. There are two kinds of charge, positive and negative. Like charges repel each other; unlike charges attract.

The basic constituents of atoms are protons, neutrons, and electrons. The protons and neutrons are bound together by the nuclear force in a small, dense nucleus, with the electrons surrounding it at distances much greater than the nuclear size. The electrical interactions between the electrons and the positively charged nucleus are responsible for the structure of atoms, molecules, and solids.

Materials that permit electric charge to move within them are called conductors; those that do not are called insulators. Most metals are good conductors; most nonmetals are insulators. The presence of an electrically charged body near a conductor causes a redistribution of charge on the conductor, known as induced charge.

The basic law of interaction for point electric charges q_1 and q_2 separated by a distance r is Coulomb's law. The magnitude of the force is given by

$$F = k\frac{|q_1 q_2|}{r^2}. \tag{24-1}$$

The force on each charge is directed along the line joining the two charges; if q_1 and q_2 have the same sign, the forces are repulsive; if opposite signs, attractive. If $\hat{r}$ is a unit vector in the direction from q_1 to q_2, then the force $\boldsymbol{F}_2$ on q_2 is given by

$$\boldsymbol{F}_2 = k\frac{q_1 q_2}{r^2}\hat{r}, \tag{24-2}$$

and the force $\boldsymbol{F}_1$ on q_1 is given by

$$\boldsymbol{F}_1 = -k\frac{q_1 q_2}{r^2}\hat{r}. \tag{24-3}$$

This pair of forces is an action–reaction pair and obeys Newton's third law.

The principle of superposition states that when two or more charges each exert a force on a certain charge, the total force on that charge is the vector sum of the forces exerted by the individual charges.

In SI units the unit of electric charge is the coulomb, abbreviated C. The numerical value of the constant k in Coulomb's law is

$$k = 8.98755 \times 10^9 \text{ N·m}^2 \text{·C}^{-2}.$$

This constant is usually expressed in terms of another constant ϵ_0:

$$k = \frac{1}{4\pi\epsilon_0} = 8.98755 \times 10^9 \text{ N·m}^2 \text{·C}^{-2},$$

KEY TERMS

electric charge
electrostatics
electron
proton
neutron
nucleus
positive ion
negative ion
atomic number
conservation of charge
conductor
insulators
induction
induced charges
Coulomb's law
principle of superposition
coulomb

and

$$\epsilon_0 = 8.854188 \times 10^{-12} \, C^2 \cdot N^{-1} \cdot m^{-2}.$$

The usual methods of vector addition, including use of components, are used in calculating vector sums for electrical forces due to two or more charges. When charge is distributed over a line or surface or through a volume, the vector sums become integrals.

QUESTIONS

24-1 Plastic food wrap can be used to cover a container by simply stretching the material across the top and pressing the overhanging material against the sides. What makes it stick? Does it stick to itself with equal tenacity? Why? Does it matter whether the container is metallic?

24-2 Bits of paper are attracted to an electrified comb or rod even though they have no net charge. How is this possible?

24-3 How do we know that the magnitudes of electron and proton charge are *exactly* equal? With what precision is this really known?

24-4 When you walk across a nylon rug and then touch a large metal object, you may get a spark and a shock. Why does this tend to happen more in the winter than the summer? Why do you not get a spark when you touch a *small* metal object?

24-5 The free electrons in a metal have mass and therefore weight and are gravitationally attracted toward the earth. Why, then, do they not all settle to the bottom of the conductor, as sediment settles to the bottom of a river?

24-6 Simple electrostatics experiments, such as picking up bits of paper with an electrified comb, never work as well on rainy days as on dry days. Why?

24-7 High-speed printing presses sometimes use gas flames to reduce electric charge buildup on the paper passing through the press. Why does this help? (An added benefit is rapid drying of the ink.)

24-8 What similarities do electrical forces have to gravitational forces? What are the most significant differences?

24-9 Given two identical metal objects mounted on insulating stands, describe a procedure for placing charges of equal magnitude and opposite sign on the two objects.

24-10 How do we know that protons have positive charge and electrons negative charge, rather than the reverse?

24-11 Gasoline transport trucks sometimes have chains that hang down and drag on the ground at the rear end. What are these for?

24-12 When a nylon sleeping bag is dragged across a rubberized-cloth air mattress in a dark tent, small sparks are sometimes seen. What causes them?

24-13 Atomic nuclei are made of protons and neutrons. This fact by itself shows that there must be another kind of interaction in addition to the electrical forces. Explain.

24-14 When transparent plastic tape is pulled off a roll and one tries to position it precisely on a piece of paper, it often jumps over and sticks where it is not wanted. Why does it do this?

24-15 When a thunderstorm is approaching, sailors at sea sometimes observe a phenomenon called "St. Elmo's fire," a bluish flickering light at the tips of masts and along wet rigging. What causes this?

EXERCISES

Section 24-2 Atomic Structure

24-1 What is the total positive charge, in coulombs, of all the protons in 1 mole of hydrogen atoms?

24-2 What is the total negative charge, in coulombs, of all the electrons in 20 g of aluminum?

Section 24-5 Coulomb's Law

Section 24-6 Applications of Coulomb's Law

24-3 A negative charge -0.50×10^{-6} C exerts a repulsive force of magnitude 0.20 N on an unknown charge 0.20 m away. What is the unknown charge (magnitude and sign)?

24-4 At what distance would the repulsive force between two electrons have a magnitude of 1 N? Between two protons?

24-5 Two small plastic balls are given positive electric charges. When they are 5 cm apart, the repulsive forces between them have magnitude 0.10 N. What is the charge on each ball

a) if the two charges are equal?

b) if one ball has twice the charge of the other?

24-6 How many excess electrons must be placed on each of two small spheres spaced 3 cm apart if the spheres are to have equal charge and if the force of repulsion between them is to be 10^{-19} N?

24–7 If 6.02×10^{23} atoms of monatomic hydrogen have a mass of 1 g, how far would the electron of a hydrogen atom have to be removed from the nucleus for the force of attraction to equal the weight of the atom?

24–8 If all the positive charges in 1 mole of hydrogen atoms were lumped into a single charge, and all the negative charges into a single charge, what force would the two lumped charges exert on each other at a distance of

a) 1 m

b) 10^7 m (comparable to the diameter of the earth)?

24–9 Two copper spheres, each having mass 1 kg, are separated by 1 m.

a) How many electrons does each sphere contain?

b) How many electrons would have to be removed from one sphere and added to the other to cause an attractive force of 10^4 N (roughly 1 ton)?

c) What fraction of all the electrons on a sphere does this represent?

24–10 The dimensions of atomic nuclei are of the order of 10^{-14} m. Suppose that two α-particles are separated by this distance.

a) What is the force exerted on each α-particle by the other?

b) What is the acceleration of each?

(See Example 24–2 for numerical data.)

24–11 Use the Bohr model to calculate the speed of the electron in a hydrogen atom when the electron is in the orbit of smallest radius.

24–12 Two point charges are located in the xy-plane as follows: A charge 2.0×10^{-9} C is at the point ($x = 0$, $y = 4$ cm), and a charge -3.0×10^{-9} C is at the point ($x = 3$ cm, $y = 4$ cm).

a) If a third charge of 4.0×10^{-9} C is placed at the origin, find the x- and y-components of the total force on this third charge.

b) Find the magnitude and direction of the total force on the charge at the origin in (a).

24–13 Two positive point charges, each of magnitude q, are located on the y-axis at points $y = +a$ and $y = -a$. A third positive charge of the same magnitude is located at some point on the x-axis.

a) What is the force exerted on the third charge when it is at the origin?

b) What are the magnitude and direction of the force on the third charge when its coordinate is x?

c) Sketch a graph of the force on the third charge as a function of x, for values of x between $+4a$ and $-4a$. Plot forces to the right upward, forces to the left downward.

d) How does the force on the third charge vary with x when x is very large?

24–14 A negative point charge of magnitude q is located on the y-axis at the point $y = +a$, and a positive charge of the same magnitude is located at $y = -a$. A third charge that is positive and of the same magnitude q is located at some point on the x-axis.

a) What are the magnitude and direction of the force exerted on the third charge when it is at the origin?

b) What is the force on the third charge when its coordinate is x?

c) Sketch a graph of the force on the third charge as a function of x, for values of x between $+4a$ and $-4a$.

24–15 The pair of equal and opposite charges in Exercise 24–14 is called an *electric dipole*.

a) Show that when the x-coordinate of the third charge in Exercise 24–14 is large compared with the distance a, the force on it is inversely proportional to the *cube* of its distance from the midpoint of the dipole.

b) Show that if the third charge is located on the y-axis, at a y-coordinate that is large compared with the distance a, the forces on it are also inversely proportional to the cube of its distance from the midpoint of the dipole.

24–16 A positive electric charge Q is distributed uniformly along the positive x-axis from $x = 0$ to $x = a$. A positive point charge q is located on the x-axis at $x = a + r$, so it is a distance r to the right of the end of Q. Calculate the magnitude and direction of the force that the charge distribution Q exerts on q.

24–17 Positive charge Q is distributed uniformly along the positive y-axis between $y = 0$ and $y = a$. A negative point charge $-q$ lies on the positive x-axis, a distance x from the origin. Calculate the x- and y-components of the force that the charge distribution Q exerts on q.

PROBLEMS

24–18 Each of two small spheres is positively charged, the combined charge totaling 4×10^{-8} C. What is the charge on each sphere if they are repelled with a force of 27×10^{-5} N when placed 0.1 m apart?

24–19 Two small balls, each of mass 10 g, are attached to silk threads 1 m long and hung from a common point. When the balls are given equal quantities of negative charge, each thread makes an angle of 20° with the vertical.

a) Draw a diagram showing all the forces on each ball.

b) Find the magnitude of the charge on each ball.

c) The two threads are now shortened to length $l = 0.5$ m, while the charge on the balls is held fixed. What will be the new angle θ that the threads each make with the vertical? (*Hint:* This part of the problem can be solved numerically by using trial values for θ and adjusting the values of θ until a self-consistent answer is obtained.)

24–20 One gram of monatomic hydrogen contains 6.02×10^{23} atoms, each consisting of an electron with charge -1.60×10^{-19} C and a proton with charge $+1.60 \times 10^{-19}$ C.

a) Suppose all these electrons could be located at the north pole of the earth and all the protons at the south pole. What would be the total force of attraction exerted on each group of charges by the other? The diameter of the earth is 12,800 km.

b) What would be the magnitude and direction of the force exerted by the charges in part (a) on a third positive charge, equal in magnitude to the total charge at one of the poles and located at a point on the equator? Draw a diagram.

24–21 According to the Bohr theory of the hydrogen atom (see Example 24–3), the electron orbits the hydrogen nucleus (a proton) in a circular path of radius 0.529×10^{-10} m when in the orbit of smallest radius. Consider an atom whose constituents are an electron and a positron (positronium). The positron is the antiparticle of the electron and has the same mass as the electron and a charge that is the same in magnitude but opposite in sign. If such a system conforms to the Bohr theory, the orbital angular momentum of the electron-positron system, for the orbit of smallest radius, must equal $h/2\pi$. Assuming that the electron and positron each follow circular orbits about the center of mass of the system, determine

a) the distance between the particles;

b) the magnitude of the velocity of each particle with respect to the center of mass.

24–22 A charge -3×10^{-9} C is placed at the origin of an xy-coordinate system, and a charge 2×10^{-9} C is placed on the positive y-axis, at $y = 4$ cm.

a) If a third charge 4×10^{-9} C is now placed at the point $(x = 3$ cm, $y = 4$ cm), find the components of the total force exerted on this charge by the other two.

b) Find the magnitude and direction of this force.

24–23 Point charges of 2×10^{-9} C are situated at each of three corners of a square whose side is 0.20 m. What would be the magnitude and direction of the resultant force on a point charge of -1×10^{-9} C if it were placed

a) at the center of the square?

b) at the vacant corner of the square?

24–24 A small ball having a positive charge q_1 hangs by an insulating thread. A second ball with a negative charge $q_2 = -q_1$ is kept at a horizontal distance a to the right of the first. (The distance a is large compared with the diameter of the ball.)

a) Show in a diagram all the forces on the hanging ball in its final equilibrium position.

b) You are given a third ball having a positive charge $q_3 = 2q_1$. Find at least two points at which this ball can be placed so that the first ball will hang vertically.

24–25 Positive charge $+Q$ is distributed uniformly along the $+x$-axis from $x = 0$ to $x = a$. Negative charge $-Q$ is distributed uniformly along the $-x$-axis from $x = 0$ to $x = -a$.

a) A positive point charge q lies on the positive y-axis, a distance y from the origin. Find the magnitude and direction of the force that the charge distribution exerts on q. Show that this force is proportional to y^{-3} as y becomes large.

b) Suppose instead that the positive point charge q lies on the positive x-axis, a distance $x > a$ from the origin. Find the magnitude and direction of the force that the charge distribution exerts on q. Show that this force is proportional to x^{-3} as x becomes large.

CHALLENGE PROBLEMS

24–26 Two identical 5-g pith balls are each attached to insulating threads of length $l = 0.5$ m and hung from a common point. One ball is given charge q_1 and the other a different charge q_2, which causes the balls to separate such that each thread makes an angle of 30° with the vertical. A small wire is then connected between the pith balls, allowing charge to be transferred from one ball to the other until the two balls have equal charges. The wire is removed, and the threads from which the balls are hanging now each make an angle of 40° with the vertical.

a) Determine the electrostatic force F between the balls before equalization of the charges.

b) Determine the product q_1q_2 of the charges before equalization of the charges.

c) Determine the electrostatic force F' between the balls after equalization of the charges.

d) Determine the original charges q_1 and q_2. (*Hint:* The total charge on the pair of balls is conserved.)

24—27 Three charges are placed as shown in Fig. 24–8. It is known that the magnitude of q_1 is 4×10^{-6} C, but its sign and the value of the charge q_2 are not known. The charge q_3 equals $+1 \times 10^{-6}$ C, and the resultant force F on q_3 is measured to be entirely in the negative x-direction.

a) Considering the different possible signs of q_1 and q_2, there are four possible force diagrams representing the

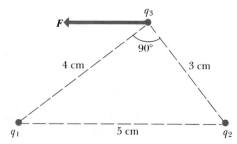

FIGURE 24–8

forces F_1 and F_2 that q_1 and q_2 exert on q_3. Sketch these four possible force configurations.

b) Using the sketches from (a) and the fact that the net force on q_3 has no y-component and a negative x-component, deduce the signs of the charges q_1 and q_2.

c) Calculate the magnitude of q_2.

d) Determine F.

24–28 Positive charge Q is distributed uniformly over the surface of a thin spherical shell of radius R. Calculate the force (magnitude and direction) that Q exerts on a positive point charge q located

a) a distance $r > R$ from the center of the shell (outside the shell);

b) a distance $r < R$ from the center of the shell (inside the shell).

(*Hint:* Divide the shell into infinitesimally thin coaxial rings, calculate the force due to each ring, and integrate to find the total force.)

25

THE ELECTRIC FIELD

THE ELECTRICAL INTERACTION BETWEEN CHARGED PARTICLES CAN BE reformulated by using the concept of *electric field*. We think of an electric charge as creating an electric field in the region of space surrounding it; that field in turn exerts a force on any other charge in that region. In this chapter we study methods for calculating the electric fields caused by various arrangements of charge. When a charge distribution has a high degree of symmetry, determining its electric field can be simplified by means of a principle called *Gauss's law.* This law is also useful in exploring several general properties of electric fields.

25–1 ELECTRIC FIELD AND ELECTRICAL FORCES

To introduce the concept of electric field, let us consider the mutual repulsion of two positively charged bodies A and B, as shown in Fig. 25–1a. In particular, consider the force on B, labeled **F** in the figure. This is an "action-at-a-distance" force; it can act across empty space and does not need any matter in the intervening space to transmit the force.

Now let us think of body A as having the effect of modifying some properties of the space in its vicinity. We remove body B and label its former position as point P (Fig. 25–1b). We say that the charged body A produces or causes an **electric field** at point P (and at all other points in the vicinity). Then when body B is placed at point P and experiences the force **F,** we take the point of view that the force is exerted on B *by the field at P,* rather than directly by A. Since B would experience a force at any point in space around A, the electric field exists at all points in the region around A. (We could equally well consider that body B sets up an electric field, and that the force on body A is exerted by the field due to B. The electric field due to B is of course not equal to that due to A.)

The concept of electric field is directly analogous to the concept of gravitational field, introduced in Section 6–4. We suggest you review that section as an aid to understanding what follows.

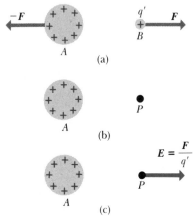

25–1 A charged body creates an electric field in the space around it.

The experimental test for the existence of an electric field at any point is simply to place a charged body, called a **test charge,** at the point. If the test charge experiences a force of electrical origin, then an electric field exists at that point.

A test charge can be used to detect and explore an electric field.

Since force is a vector quantity, electric field is a *vector quantity.* We define the *electric field* E at a point as the quotient obtained when the force F, acting on a positive test charge, is divided by the magnitude q' of the test charge (Fig. 25–1c). Thus

Electric field is force per unit charge.
Electric field is a vector quantity.

$$E = \frac{F}{q'}$$

or

$$F = q'E. \tag{25–1}$$

The direction of E is the direction of F. The force on a *negative* charge, such as an electron, has a direction *opposite* to that of the electric field.

The electric field is sometimes called *electric intensity* or *electric field intensity.* In SI units, where the unit of force is 1 N and the unit of charge is 1 C, the unit of electric field magnitude is one newton per coulomb ($1 \text{ N} \cdot \text{C}^{-1}$). Electric field may also be expressed in other equivalent units that will be introduced later.

The force experienced by the test charge q' varies from point to point, and so the electric field is also different at different points. Thus E is not a single vector quantity but an infinite set of vector quantities, one associated with each point in space. The electric field is an example of a **vector field.** Another example of a vector field is the description of motion of a flowing fluid. Different points in the fluid have different velocities, so the velocity is a vector field. If we use a rectangular coordinate system, then each component of E is a function of the coordinates (x, y, z) of a point in space. Vector fields are an important part of the mathematical language used in many areas of physics, particularly in electricity and magnetism.

One difficulty with our definition of electric field is that in Fig. 25–1 the force exerted by the test charge q' may change the charge distribution A, especially if the body is a conductor on which charge is free to move. The electric field around A when q' is present is not the same as when q' is absent. When q' is very small, however, the redistribution of charge on body A is also very small; thus the difficulty can be avoided by refining the definition of electric field to be *the limiting value of the force per unit charge on a test charge q' at the point, as the charge q' approaches zero:*

Does a test charge disturb the electric field by its presence?

$$E = \lim_{q' \to 0} \frac{F}{q'}.$$

If an electric field exists within a *conductor,* a force is exerted on every charge in the conductor. The motion of the free charges brought about by this force is called a *current.* Conversely, if there is *no* current in a conductor, and hence no motion of its free charges, *the electric field in the conductor must be zero.*

When all charges are at rest, the electric field inside a conductor is always zero.

In most instances, the magnitude and direction of an electric field vary from point to point. If the magnitude and direction are constant throughout a certain region, we say that the field is *uniform* in this region.

PROBLEM-SOLVING STRATEGY: *Electric field forces*

1. To determine the electric field at a point, we use Coulomb's law to find the total force F on a test charge q' placed at the point. Then we divide F by q' to obtain E. If we assume q' is positive, F and E have the same direction, and the magnitude of E is the magnitude of F divided by q'.

2. To analyze the motion of a charged particle in an electric field, we need to use Newton's second law,

$F = ma$, with F given in this case by $F = qE$. If the field is uniform, the acceleration is constant; we find its components and then use the tools we developed back in Chapter 3. It wouldn't hurt to review that chapter now.

EXAMPLE 25–1   What is the electric field 30 cm from a charge $q = 4 \times 10^{-9}$ C?

SOLUTION   From Coulomb's law, the force on a test charge q' 30 cm from q has magnitude

$$F = \frac{(9 \times 10^9 \ \text{N·m}^2\text{·C}^{-2})(4 \times 10^{-9} \ \text{C})(q')}{(0.3 \ \text{m})^2}$$
$$= (400 \ \text{N·C}^{-1})(q').$$

Then from Eq. (25–1), the magnitude of E is

$$E = \frac{F}{q'} = 400 \ \text{N·C}^{-1}.$$

The *direction* of E at this point is along the line joining q and q', away from q.

EXAMPLE 25–2   When the terminals of a 100-V battery are connected to two large parallel horizontal plates 1 cm apart, the electric field E in the region between the plates is very nearly uniform, with magnitude $E = 10^4 \ \text{N·C}^{-1}$. Suppose the direction of E is vertically upward. Compute the force on an electron in this field and compare it with the weight of the electron.

SOLUTION   We need the following data, found in Appendix F:

$$\text{Electron charge } e = 1.60 \times 10^{-19} \ \text{C},$$

$$\text{Electron mass } m = 9.11 \times 10^{-31} \ \text{kg}.$$

From Eq. (25–1),

$$F_{\text{elec}} = eE = (1.60 \times 10^{-19} \ \text{C})(10^4 \ \text{N·C}^{-1}) = 1.60 \times 10^{-15} \ \text{N};$$

$$F_{\text{grav}} = mg = (9.11 \times 10^{-31} \ \text{kg})(9.8 \ \text{m·s}^{-2}) = 8.93 \times 10^{-30} \ \text{N}.$$

Electrical forces are often much larger than the weights of the particles.

The ratio of the electrical to the gravitational force is therefore

$$\frac{F_{\text{elec}}}{F_{\text{grav}}} = \frac{1.60 \times 10^{-15} \ \text{N}}{8.93 \times 10^{-30} \ \text{N}} = 1.8 \times 10^{14}.$$

The gravitational force is negligibly small compared to the electrical force.

EXAMPLE 25–3 If the electron of Example 25–2 is released from rest at the upper plate, what speed does it acquire while traveling 1 cm? What is then its kinetic energy? How much time is required for it to travel this distance?

SOLUTION The force is constant, so the electron moves with constant acceleration a given by

$$a = \frac{F}{m} = \frac{eE}{m} = \frac{1.60 \times 10^{-15} \text{ N}}{9.11 \times 10^{-31} \text{ kg}}$$
$$= 1.76 \times 10^{15} \text{ m·s}^{-2}.$$

From Eq. (2–13), its speed after traveling 1 cm, or 10^{-2} m, is

$$v = \sqrt{2ax} = 5.93 \times 10^6 \text{ m·s}^{-1}.$$

Its kinetic energy is

$$\frac{1}{2}mv^2 = 1.60 \times 10^{-17} \text{ J}.$$

The time required is

$$t = \frac{v}{a} = 3.37 \times 10^{-9} \text{ s}.$$

EXAMPLE 25–4 If the electron of Example 25–2 is projected into the electric field with an initial horizontal velocity v_0, as in Fig. 25–2, find the equation of its trajectory.

SOLUTION The direction of the field is upward in Fig. 25–2, so the force on the (negatively charged) electron is downward. We take the positive x-direction to be the direction of the initial velocity. The x-acceleration is zero; the y-acceleration is $-(eE/m)$. Hence after a time t,

$$x = v_0 t,$$
$$y = \frac{1}{2}a_y t^2 = -\frac{1}{2}\left(\frac{eE}{m}\right)t^2.$$

Elimination of t gives

$$y = -\frac{1}{2}\left(\frac{eE}{mv_0{}^2}\right)x^2,$$

which is the equation of a parabola. The motion is the same as that of a body projected horizontally in the earth's gravitational field, as we discussed in Section 3–4. The deflection of electrons by an electric field is used to control the direction of an electron beam in many electronic devices, such as the TV picture tube and cathode-ray oscilloscope, which will be discussed in Section 26–8.

EXAMPLE 25–5 Figure 25–3 shows a structure incorporating two point charges, $+q$ and $-q$, having equal magnitude but opposite sign, separated by a constant distance l. We call such a pair of charges an **electric dipole.** The electric-charge distributions of many molecules, such as the water molecule, can be represented approximately as electric dipoles. In Fig. 25–3 the dipole is in a uniform electric field E. The direction of E makes an angle θ with the line joining the two charges, called the *dipole axis.* A force F_1, of magnitude qE, in the direction of the field, is exerted on the positive charge; and a force F_2, of the same magnitude but

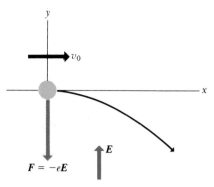

25–2 Trajectory of an electron in an electric field.

Electron trajectories in an electric field are very similar to ballistic trajectories.

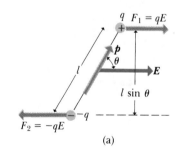

(a)

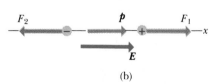

(b)

25–3 (a) The torque on the dipole is $\Gamma = pE \sin \theta$. (b) The dipole is in equilibrium in a uniform field when p and E are parallel.

in the opposite direction, is exerted on the negative charge. The resultant force on the dipole is zero, but since the two forces do not have the same line of action, they constitute a *couple* (see Section 10–4). The torque produced by a couple is

$$\Gamma = (qE)(l \sin \theta),$$

since $l \sin \theta$ is the perpendicular distance between the action lines of the forces.

Dipole in a uniform field: torque but no
net force

The product ql of the charge q and the distance l is called the **electric dipole moment** and is denoted by p:

$$p = ql.$$

The torque exerted by the couple is therefore

$$\Gamma = pE \sin \theta. \qquad (25\text{--}2)$$

We can also define a *vector dipole moment* $\boldsymbol{p}$, a vector of magnitude p lying along the dipole axis and pointing from negative toward positive charge. With this definition and the concept of vector torque introduced in Section 9–8, the vector torque $\boldsymbol{\Gamma}$ on the dipole is given by

$$\boldsymbol{\Gamma} = \boldsymbol{p} \times \boldsymbol{E}. \qquad (25\text{--}3)$$

This torque tends to rotate the dipole to a position in which $\boldsymbol{p}$ is parallel to $\boldsymbol{E}$, as in Fig. 25-3b. In this position $\boldsymbol{\Gamma} = \boldsymbol{0}$, and if $\boldsymbol{E}$ is uniform, the dipole is in equilibrium.

If the electric field is *not* uniform, then the dipole experiences not only a torque but also a net force, because the two charges are located at points of slightly different electric field and the forces on them do not precisely cancel. In this case, the net force can be expressed in terms of derivatives of the components of $\boldsymbol{E}$ with respect to the coordinates.

25–2 ELECTRIC-FIELD CALCULATIONS

How to calculate electric fields from
Coulomb's law

We can calculate the electric field at any point if we know the magnitudes and positions of all the charges that contribute to the field at that point. First we determine the electric field $\boldsymbol{E}$ at a point P caused by a single point charge q at a distance r from P. To do this we imagine a test charge q' at P. According to Coulomb's law, the force F on the test charge has magnitude

$$F = \frac{1}{4\pi\epsilon_0} \frac{qq'}{r^2},$$

so the electric field at P has magnitude

$$E = \frac{F}{q'} = \frac{1}{4\pi\epsilon_0} \frac{q}{r^2}.$$

The location of charge q is often called the *source point* and P the *field point*. We can define a unit vector $\hat{\boldsymbol{r}}$ pointing in the direction from the source point to the field point; that is, from q to point P. In terms of $\hat{\boldsymbol{r}}$, the field $\boldsymbol{E}$ at P due to charge q is given by

$$\boldsymbol{E} = \frac{1}{4\pi\epsilon_0} \frac{q}{r^2} \hat{\boldsymbol{r}}. \qquad (25\text{--}4)$$

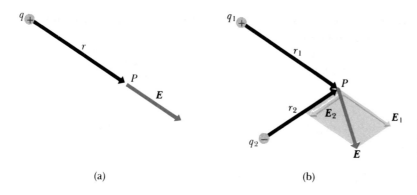

25–4 (a) The electric field **E** is in the same direction as the vector **r** when q is positive. (b) The resultant electric field at point P is the vector sum of $\boldsymbol{E}_1$ and $\boldsymbol{E}_2$.

When q is positive, the field is away from q, in the direction of $\hat{\boldsymbol{r}}$; when it is negative, the field is toward q, opposite in direction to $\hat{\boldsymbol{r}}$.

Now suppose the field at point P is caused by several point charges q_1, q_2, q_3, . . . , at distances r_1, r_2, r_3, . . . , from P, with associated unit vectors $\hat{\boldsymbol{r}}_1$ pointing from q_1 to P, and so on. Each unit vector points from its associated source point to the field point P. Then from the principle of superposition introduced in Section 24–5, it follows that the *total* electric field at P is the *vector sum* of the fields $\boldsymbol{E}_1$, $\boldsymbol{E}_2$, and so on, of the individual charges. That is,

Superposition: combining fields caused by several point charges

$$\boldsymbol{E} = \boldsymbol{E}_1 + \boldsymbol{E}_2 + \boldsymbol{E}_3 + \cdots$$

$$= \frac{1}{4\pi\epsilon_0}\frac{q_1}{r_1{}^2}\hat{\boldsymbol{r}}_1 + \frac{1}{4\pi\epsilon_0}\frac{q_2}{r_2{}^2}\hat{\boldsymbol{r}}_2 + \cdots. \qquad (25\text{–}5)$$

An example with two charges, one positive and one negative, is shown in Fig. 25–4.

PROBLEM-SOLVING STRATEGY: Electric-field calculations

The strategy outlined in Section 24–6 is directly relevant here, too. We suggest you review it now. Briefly, the key points are:

1. Be sure to use a consistent set of units.

2. Use proper vector notation; distinguish clearly between vectors and components of vectors. Indicate your coordinate axes clearly on your diagram, and be certain the components are consistent with your choice of axes. Use the methods you learned in Chapter 1 for finding vector sums.

3. In some problems you will have a continuous distribution of charge along a line, over a surface, or through a volume. You must then define a small element of charge that can be considered as a point, find its electric field, and then integrate over the whole charge distribution to find the total electric field. Usually it is easiest to do the integration for each component of **E** separately.

EXAMPLE 25–6 Point charges q_1 and q_2 of $+12 \times 10^{-9}$ C and -12×10^{-9} C, respectively, are placed 0.1 m apart, as in Fig. 25–5. Compute the electric fields due to these charges at points a, b, and c.

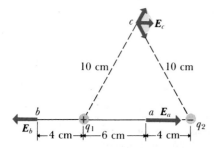

25–5 Electric field at three points, a, b, and c, in the field set up by charges q_1 and q_2.

A simple electric-field calculation: two point charges

SOLUTION At point a, the vector due to the positive charge q_1 is directed toward the right, and its magnitude is

$$E_1 = (9.0 \times 10^9 \text{ N·m}^2\text{·C}^{-2}) \frac{(12 \times 10^{-9} \text{ C})}{(0.06 \text{ m})^2}$$

$$= 3.00 \times 10^4 \text{ N·C}^{-1}$$

The vector due to the negative charge q_2 is also directed toward the right; its magnitude is

$$E_2 = (9.0 \times 10^9 \text{ N·m}^2\text{·C}^{-2}) \frac{(12 \times 10^{-9} \text{ C})}{(0.04 \text{ m})^2}$$

$$= 6.75 \times 10^4 \text{ N·C}^{-1}.$$

Hence, at point a,

$$E_a = (3.00 + 6.75) \times 10^4 \text{ N·C}^{-1}$$

$$= 9.75 \times 10^4 \text{ N·C}^{-1}, \qquad \text{toward the right.}$$

At point b, the vector due to q_1 is directed toward the left, with magnitude

$$E_1 = (9.0 \times 10^9 \text{ N·m}^2\text{·C}^{-2}) \frac{(12 \times 10^{-9} \text{ C})}{(0.04 \text{ m})^2}$$

$$= 6.75 \times 10^4 \text{ N·C}^{-1}.$$

The vector due to q_2 is directed toward the right, with magnitude

$$E_2 = (9.0 \times 10^9 \text{ N·m}^2\text{·C}^{-2}) \frac{(12 \times 10^{-9} \text{ C})}{(0.14 \text{ m})^2}$$

$$= 0.55 \times 10^4 \text{ N·C}^{-1}.$$

Hence, at point b

$$E_b = (6.75 - 0.55) \times 10^4 \text{ N·C}^{-1}$$

$$= 6.20 \times 10^4 \text{ N·C}^{-1}, \qquad \text{toward the left.}$$

At point c, the magnitude of each vector is

$$E = (9.0 \times 10^9 \text{ N·m}^2\text{·C}^{-2}) \frac{(12 \times 10^{-9} \text{ C})}{(0.1 \text{ m})^2}$$

$$= 1.08 \times 10^4 \text{ N·C}^{-1}.$$

The directions of these vectors are shown in the figure; their resultant is easily seen to be

$$E_c = 1.08 \times 10^4 \text{ N·C}^{-1}, \qquad \text{toward the right.}$$

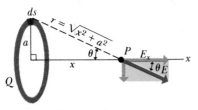

25–6 Electric field due to ring of charge.

EXAMPLE 25–7 A ring-shaped conductor of radius a carries a total charge Q. Find the electric field at a point P, a distance x from the center, along the line perpendicular to the plane of the ring, through its center.

SOLUTION The situation is shown in Fig. 25–6. Note the close resemblance of this problem to the gravitational-field calculation in Example 6–10 (Section 6–4). We represent the ring as made up of small segments ds, as shown. The charge of the segment ds is dQ. At the field point P the element of charge dQ causes an electric-field contribution dE having magnitude dE given by

$$dE = \frac{1}{4\pi\epsilon_0}\frac{dQ}{x^2 + a^2}.$$

The component dE_x of this field along the x-axis is

$$dE_x = dE\cos\theta = \frac{1}{4\pi\epsilon_0}\frac{dQ}{x^2 + a^2}\frac{x}{\sqrt{x^2 + a^2}}$$

$$= \frac{1}{4\pi\epsilon_0}\frac{x\,dQ}{(x^2 + a^2)^{3/2}}.$$

To find the total x-component of the field, E_x, we integrate this expression:

$$E_x = \int \frac{1}{4\pi\epsilon_0}\frac{x\,dQ}{(x^2 + a^2)^{3/2}}.$$

On the right side, x does not vary as we move from point to point on the ring. Thus everything except dQ may be taken outside the integral. The integral of dQ is simply the total charge Q, and we finally obtain

$$E_x = \frac{1}{4\pi\epsilon_0}\frac{Qx}{(x^2 + a^2)^{3/2}}. \tag{25–6}$$

In principle, this calculation should also be performed for the components perpendicular to the x-axis, but we can see from symmetry that these sum to zero.

Equation (25–6) shows that at the center of the ring ($x = 0$), the total field is zero, as might be expected; charges on opposite sides pull in opposite directions on a test charge at that point, and their fields cancel. When x is much larger than a, Eq. (25–6) becomes approximately

At the center of the ring, the field is zero.

$$E_x = \frac{1}{4\pi\epsilon_0}\frac{Q}{x^2},$$

corresponding to the fact that at distances much greater than the dimensions of the ring it appears as a point charge.

EXAMPLE 25–8 An electric charge Q is distributed uniformly along a line with length $2a$, lying along the y-axis, as shown in Fig. 25–7. Find the electric field at point P on the x-axis, at a distance x from the origin.

SOLUTION We encountered the same situation in Example 24–5 (Section 24–6). Looking back at that solution for the force F, we see that the electric field E at P is just equal to F/q:

$$E = \frac{1}{4\pi\epsilon_0}\frac{Q}{x\sqrt{x^2 + a^2}}\,i. \tag{25–7}$$

Again, when x is much larger than a, this becomes approximately

$$E = \frac{1}{4\pi\epsilon_0}\frac{Q}{x^2}\,i,$$

that is, the field of a point charge Q at a distance x.

We get an added dividend from Eq. (25–7) if we express it in terms of the charge per unit length $Q/2a$. We call this quantity the *linear charge density* and

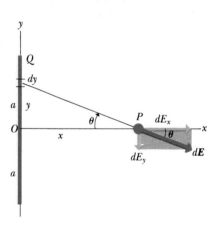

25–7 A line of charge with length $2a$ and total charge Q creates an electric field E at point P.

denote it by λ. In this case, $\lambda = Q/2a$. Substituting $Q = 2a\lambda$ into Eq. (25–7) and simplifying, we obtain

$$E = \frac{1}{2\pi\epsilon_0} \frac{\lambda}{x\sqrt{\dfrac{x^2}{a^2} + 1}} i. \tag{25–8}$$

Now what happens if we make the line of charge longer and longer, adding charge in proportion to the total length so that λ, the charge per unit length, remains constant? What is E at a distance x from a *very* long line of charge?

To answer the question we take the *limit* of Eq. (25–8) as a becomes very large. In this limit, the term x^2/a^2 in the denominator becomes much smaller than unity and can be discarded. What remains is

$$E = \frac{\lambda}{2\pi\epsilon_0 x} i. \tag{25–9}$$

From the symmetry of the situation, the field magnitude depends only on the distance of point P from the line of charge, so we can say that at any point P at a distance r from the line, the field has magnitude

$$E = \frac{\lambda}{2\pi\epsilon_0 r}. \tag{25–10}$$

Its direction is radially outward from the line. Thus the electric field due to an infinitely long line of charge is proportional to $1/r$ rather than to $1/r^2$ as for a point charge.

Electric field caused by charge distributed over a plane

EXAMPLE 25–9 In Fig. 25–8, electric charge is distributed uniformly over the entire *xy*-plane. We call the charge per unit area the *surface charge density* and denote it by σ. What is the electric field at point P, at a distance a above the plane?

SOLUTION Assume for the moment that σ is positive. We divide the charge into narrow strips of width dx, parallel to the *y*-axis. Each strip can be considered a *line* charge, and we can use the result of the preceding example.

25–8 Electric field created by a uniform infinite plane of charge.

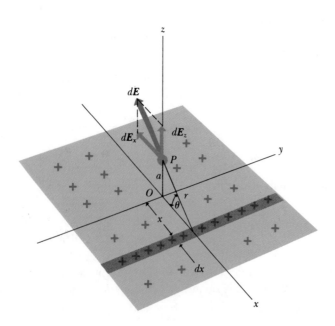

The area of a portion of a strip of length L is $L\,dx$, and the charge dq on the strip is

$$dq = \sigma L\,dx.$$

The charge per unit length, λ, is therefore

$$\lambda = \frac{dq}{L} = \sigma\,dx.$$

From Eq. (25–10), the strip sets up at point P a field $d\mathbf{E}$, lying in the xz-plane, of magnitude

$$dE = \frac{\sigma}{2\pi\epsilon_0}\frac{dx}{r}.$$

The field can be resolved into components dE_x and dE_z. By symmetry, the components dE_x will sum to zero when the entire sheet of charge is considered. (Be sure that you understand why.) The resultant field at P is therefore in the z-direction, perpendicular to the sheet of charge. From the diagram,

$$dE_z = dE \sin\theta.$$

The total z-component of $\mathbf{E}$ is given by

$$E = \int dE_z = \frac{\sigma}{2\pi\epsilon_0}\int_{-\infty}^{+\infty}\frac{\sin\theta\,dx}{r}.$$

But

$$\sin\theta = \frac{a}{r},$$

$$r^2 = a^2 + x^2,$$

and therefore

$$E = \frac{\sigma a}{2\pi\epsilon_0}\int_{-\infty}^{+\infty}\frac{dx}{a^2 + x^2} = \frac{\sigma a}{2\pi\epsilon_0}\left[\frac{1}{a}\arctan^{-1}\frac{x}{a}\right]_{-\infty}^{+\infty},$$

$$E = \frac{\sigma}{2\epsilon_0}. \tag{25–11}$$

Our final result does not contain the distance a from the plane. This correct but rather surprising result means that the electric field produced by an infinite-plane sheet of charge is *independent of the distance from the charge.* Thus the field is *uniform;* its direction is everywhere perpendicular to the sheet, away from it.

The field is independent of the distance from the plane.

If P is below the plane instead of above it, the result is the same except that the direction of $\mathbf{E}$ is vertically downward instead of upward. Also, if σ is a negative quantity, the directions of the fields both above and below the plane are toward the plane rather than away from it.

As we have seen, calculating the $\mathbf{E}$ fields of given charge distributions can be quite laborious. For problems having a lot of symmetry, such as the examples above, there is an alternative method using a principle called *Gauss's law;* we return to this important principle at the end of the chapter.

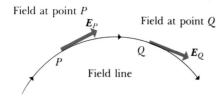

25–9 The direction of the electric field at any point is tangent to the field line through that point.

Field lines: a graphical representation of electric fields

Field lines begin and end at charges.

Field lines never cross.

25–3 FIELD LINES

Field lines are useful as an aid to visualizing electric (and also magnetic) fields. A **field line** is an imaginary line drawn through a region of space so that at every point it is tangent to the direction of the electric-field vector at that point. This idea is shown in Fig. 25–9. We will show in Section 25–4 that at every point the *number* of lines per unit cross-sectional area, perpendicular to E, is proportional to the magnitude of E at that point. The concept of field lines was introduced by Michael Faraday (1791–1867). He called them "lines of force," but the term *field lines* is preferable.

Thus field lines show the direction of E at each point, and their spacing gives a general indication of the magnitude of E at each point. Where the lines are bunched closely together, E is strong; where they are farther apart, E is weaker.

Figure 25–10 shows some of the field lines in two planes containing (a) a single positive charge; (b) two equal charges, one positive and one negative (an electric dipole); and (c) two equal positive charges. The direction of the resultant field at every point in each diagram is along the tangent to the field line passing through the point. Arrowheads on the field lines indicate the direction in which the tangent is to be drawn.

No field lines begin or end in the space surrounding a charge. Every field line in an *electrostatic* field is a continuous line terminated by a positive charge at one end and a negative charge at the other. Although we sometimes speak of an "isolated" charge and draw its field as in Fig. 25–10a, this simply means that the charges on which the lines end are at large distances from the charge under consideration. For example, if the charged body in Fig. 25–10a is a small sphere suspended by a thread from the laboratory ceiling, the negative charges on which its field lines end would be found on the walls, floor, or ceiling, as well as on other objects in the laboratory.

At any one point, the electric field can have but one direction. Hence only one field line can pass through each point of the field. In other words, *field lines never intersect*.

If a field line were drawn through every point of an electric field, all the space and the entire surface of a diagram would be filled with lines, and no individual line could be distinguished. By limiting the number of field lines, we can use them to indicate the *magnitude* of a field as well as its *direction*. In a region where the field magnitude is large, such as that between the positive

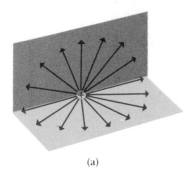

(a)

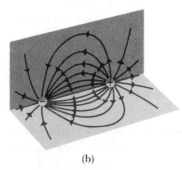

(b)

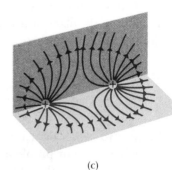

(c)

25–10 The mapping of an electric field with the aid of field lines.

Electric field produced by two point charges. The pattern is formed by grass seeds floating in a liquid above the charges. (*PSSC Physics*, 2d. ed., 1965; D. C. Heath and Company, Inc., with Education Development Center, Inc., Newton, Mass.)

and negative charges of Fig. 25–10b, the field lines are closely spaced. In a region where the field magnitude is small, such as that between the two positive charges of Fig. 25–10c, the lines are widely separated. In a *uniform* field, the field lines are straight, parallel, and uniformly spaced.

25–4 GAUSS'S LAW

Gauss's law provides an alternative formulation of the relationship between electric charge and electric field. It is logically equivalent to Coulomb's law, but in some situations it is more convenient to use. It was formulated by Karl Friedrich Gauss (1777–1855), one of the greatest mathematical geniuses of all time. Many areas of mathematics, from number theory and geometry to the theory of differential equations, bear the mark of his influence, and he made equally significant contributions to theoretical physics. In this section we will derive Gauss's law, and in the following section we will use it to find the electric fields caused by several kinds of charge distributions of practical importance.

Gauss's law: an alternative formulation of the relation between charge and electric field

The content of Gauss's law is suggested by consideration of field lines, discussed in Section 25–3. The field of an isolated positive point charge q is represented by lines radiating out in all directions. Suppose we imagine this charge to be surrounded by a spherical surface of radius R, with the charge at its center. The area of this imaginary surface is $4\pi R^2$, so if the total number of field lines emanating from q is N, then the number of lines *per unit surface area* on the spherical surface is $N/4\pi R^2$. Now imagine a second sphere concentric with the first, but with radius $2R$. Its area is $4\pi(2R)^2 = 16\pi R^2$, and the number of lines per unit area on this sphere in $N/16\pi R^2$, one-fourth that of the first sphere. This corresponds to the fact that at distance $2R$ the field has only one-fourth the magnitude it has at distance R. It also verifies our statement in Section 25–3 that the number of lines per unit area is proportional to the magnitude of the field.

For a point charge, how many field lines pass through a given area?

The fact that the *total* number of lines at distance $2R$ is the same as at R can be expressed another way. The field is inversely proportional to R^2, but

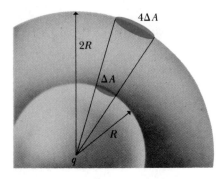

25–11 Projection of an element of area ΔA on a sphere of radius R, onto a sphere of radius $2R$. The projection multiplies each linear dimension by two, so the area element on the larger sphere is $4\,\Delta A$.

the *area* of the sphere is proportional to R^2. Thus the *product* of the two is independent of R. For a sphere of arbitrary radius r, the magnitude of E on the surface is

$$E = \frac{1}{4\pi\epsilon_0}\frac{q}{r^2},$$

the surface area is

$$A = 4\pi r^2,$$

and the product of the two is

$$EA = \frac{q}{\epsilon_0}. \tag{25–12}$$

This product is independent of r and depends *only* on the charge q. As we will see, this result is of crucial importance in the following development.

What is true of the entire sphere is also true of any portion of its surface. In the construction of Fig. 25–11, an area ΔA is outlined on a sphere of radius R and then projected onto the sphere of radius $2R$ by drawing lines from the center through points on the boundary of ΔA. The area projected on the larger sphere is clearly $4\,\Delta A$; thus again the product $E\,\Delta A$ is independent of the radius of the sphere.

This projection technique shows how this discussion may be extended to nonspherical surfaces. Instead of a second sphere, let us surround the sphere of radius R by a surface of irregular shape, as in Fig. 25–12a. Consider a small element of area ΔA; note that this area is *larger* than the corresponding element on a spherical surface at the same distance from q. If a line normal (perpendicular) to the irregular surface makes an angle θ with a radial line from q, two sides of the area projected onto the spherical surface are foreshortened by a factor $\cos\theta$, as shown in Fig. 25–12b. Thus the quantity corresponding to $E\,\Delta A$ for the spherical surface is $E\,\Delta A\cos\theta$ for the irregular surface.

Now we may divide the entire irregular surface into small elements ΔA, compute the quantity $E\,\Delta A\cos\theta$ for each, and sum the results. Each of these areas projects onto a corresponding element of area on the sphere, so summing the quantities $E\,\Delta A\cos\theta$ over the irregular surface must yield the same

25–12 (a) The outward normal makes an angle θ with the direction of E. (b) The projection of the area element ΔA onto the spherical surface is $\Delta A\cos\theta$.

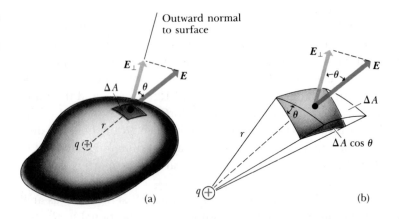

result as summing the quantities $E \, \Delta A$ over the sphere. But we have already performed that calculation; the result, given by Eq. (25–12), depends only on the charge q. Hence, for the irregular surface, the result is

$$\sum E \, \Delta A \cos \theta = \frac{q}{\epsilon_0}. \qquad (25\text{–}13)$$

This equation holds for *any* shape of surface, provided only that it is a *closed* surface enclosing the charge q. Correspondingly, for a closed surface enclosing *no* charge,

$$\sum E \, \Delta A \cos \theta = 0.$$

This is a mathematical statement of the fact that when a region contains no charge, any field lines (caused by charges outside the region) that enter on one side must leave again at some other point on the boundary surface of the region. *Field lines can begin or end inside a region of space only when there is charge in that region.*

Because the electric field varies from point to point on the irregular surface, Eq. (25–13) is strictly true only in the limit when the area elements become very small. In this limit, the sum becomes an integral called the *surface integral* of $E \cos \theta$, written

$$\oint E \cos \theta \, dA = \frac{q}{\epsilon_0}. \qquad (25\text{–}14)$$

The circle on the integral sign reminds us that the integral is always taken over a *closed* surface enclosing the charge q.

Since $E \cos \theta$ is the component of $\boldsymbol{E}$ perpendicular to the surface at each point, we may use the more compact notation $E_\perp = E \cos \theta$, and write

$$\oint E_\perp \, dA = \frac{q}{\epsilon_0}. \qquad (25\text{–}15)$$

If the point charge in Fig. 25–12 is negative, the $\boldsymbol{E}$ field is directed radially *inward;* the angle θ is then greater than $90°$, its cosine is negative, and the integral in Eq. (25–15) is negative. But since q is also negative, Eq. (25–15) still holds.

Now suppose the point charge lies *outside* a closed surface; construct your own diagram. The electric field is then inward at some points of the surface and outward at others. It is not difficult to show that the positive and negative contributions to the integral over the surface exactly cancel, and the sum is zero. But the charge *inside* the closed surface is also zero, so again Eq. (25–15) is obeyed.

Although thus far we have considered only a single point charge, it is easy to generalize our results to *any* charge distribution. The total electric field $\boldsymbol{E}$ at a point on the closed surface is the vector sum of the fields produced by the individual charges, and the quantity $E \, dA \cos \theta$ is therefore the sum of the contributions from these charges. Since Eq. (25–15) holds for each point charge, a corresponding relation holds for the *total* $\boldsymbol{E}$ field and the *total* charge enclosed by the surface. That is,

$$\oint E_\perp \, dA = \frac{1}{\epsilon_0} \sum q, \qquad (25\text{–}16)$$

The total number of field lines coming out of a surface is proportional to the charge enclosed by the surface.

When no charge is enclosed, the numbers of field lines entering and leaving the enclosed volume are equal.

where $E_\perp$ is now the normal component of the total electric field and Σq represents the *algebraic* sum of all charges enclosed by the surface.

Equation (25–16) is the mathematical statement of Gauss's law. It states that when we multiply each element of area of a closed surface by the normal component of the total electric field $\boldsymbol{E}$ at the element, and sum over the entire surface, the result is a constant times the total charge inside the surface.

The notation can be simplified by use of the *vector area d$\boldsymbol{A}$,* defined as a vector whose magnitude equals dA and whose direction is that of the *outward normal* at dA. The product $E_\perp\, dA = E \cos \theta\, dA$ can then be written as the *scalar product* or *dot product* of the vectors $\boldsymbol{E}$ and $d\boldsymbol{A}$:

$$E_\perp\, dA = \boldsymbol{E} \cdot d\boldsymbol{A}.$$

Denoting the total charge enclosed by $Q = \Sigma q$, we may write Gauss's law more compactly as

$$\oint \boldsymbol{E} \cdot d\boldsymbol{A} = \frac{Q}{\epsilon_0}. \qquad (25\text{–}17)$$

The quantity $E_\perp\, dA = E \cos \theta\, dA$ is also called the **electric flux** through the area dA. Electric flux may be denoted by Ψ, and an element of flux corresponding to a small area dA by $d\Psi$. Then

$$d\Psi = E_\perp\, dA = E \cos \theta\, dA = \boldsymbol{E} \cdot d\boldsymbol{A}. \qquad (25\text{–}18)$$

The total flux Ψ through a finite surface area is

$$\Psi = \int E_\perp\, dA = \int E \cos \theta\, dA = \int \boldsymbol{E} \cdot d\boldsymbol{A}. \qquad (25\text{–}19)$$

Thus Eq. (25–17) states that the total electric flux out of a closed surface is proportional to the total charge Q enclosed:

$$\Psi = \frac{Q}{\epsilon_0}. \qquad (25\text{–}20)$$

As we have mentioned, the flux of $\boldsymbol{E}$ through a surface, as well as Gauss's law, can be interpreted graphically in terms of field lines. The number of field lines per unit area, perpendicular to the direction of the lines, is proportional to $\boldsymbol{E}$. Thus the surface integral of $E_\perp$ over a closed surface (equal to the flux Ψ through that surface) is proportional to the total number of lines crossing the surface in the outward direction, minus the number crossing in the inward direction. Furthermore, the net charge (algebraic sum of all charges) enclosed by the surface is proportional to this number.

Here is an example. Figure 25–13 shows the field produced by two equal and opposite point charges (an electric dipole). Surface A encloses the positive charge only, and 18 lines cross it in an outward direction. Surface B encloses the negative charge only; it is also crossed by 18 lines, but in an inward direction. Surface C encloses *both* charges. It is intersected by lines at 16 points; at 8 intersections the lines are outward, and at 8 they are inward. The *net* number of lines crossing in an outward direction is zero, and the net charge inside the surface is also zero. Surface D is intersected at 6 points; at 3 intersections the lines are outward, and at the other 3 they are inward. The net number of lines crossing in an outward direction and the enclosed charge are both zero.

Actually evaluating the integral in Eq. (25–17) may appear impossible. In quite a few cases, however, we can exploit the symmetry of a charge distribu-

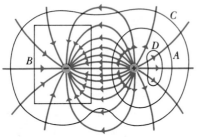

25–13 The net number of field lines leaving a closed surface is proportional to the total charge enclosed.

tion to accomplish the calculation quite simply. In these cases Gauss's law enables us to calculate the electric field of a charge distribution much more simply than with the methods used in Section 25–2. Several examples are worked out in the next section.

25–5 APPLICATIONS OF GAUSS'S LAW

In this section we present several examples of applications of Gauss's law. They are of two kinds: The first two are derivations of general properties of the electric field and the location of charges on conductors; the remainder are derivations of electric-field configurations caused by various specific charge distributions. In some of these examples we find that we can obtain much more simply some of the results derived in Section 25–2. The following problem-solving strategy is useful for both categories.

Gauss's law greatly simplifies some field calculations.

PROBLEM-SOLVING STRATEGY: Gauss's law

1. The first step is to select the surface (which we will often call a *Gaussian surface*) that you are going to use with Gauss's law. If you are trying to learn something about the field at a particular point, then obviously that point must lie on your Gaussian surface.

2. The Gaussian surface does not need to be a real physical surface, such as the surface of a solid body. In many applications of Gauss's law we use an imaginary geometric surface that may be in empty space, embedded in a solid body, or partly both.

3. The Gaussian surface must have enough *symmetry* to make it possible to evaluate the surface integral in Gauss's law. If the problem itself has cylindrical or spherical symmetry, the Gaussian surface will usually be cylindrical or spherical, respectively.

4. Often the surface can be thought of as several separate areas, such as the sides and ends of a cylinder. The integral of $E_\perp$ over the whole surface is always

equal to the sum of the integrals over the separate areas. Some of these integrals may be zero, as in points 6 and 7 below.

5. If E is perpendicular to a surface A at every point, and if it also has the same *magnitude* at every point on the surface, then $E_\perp = E = $ constant, and the integral $\oint E \cdot dA$ over that surface is simply EA.

6. If E is *parallel* to a surface at every point, then $E_\perp = 0$, and the integral is zero.

7. If $E = 0$ at every point on a surface, the integral is zero.

8. Finally, in the integral $\oint E \cdot dA$, E is always the *total* electric field at each point on the surface. In general this field is caused partly by charges within the volume and partly by charges outside. Even when there is *no* charge within the volume, the field at points on the surface is not necessarily zero. In that case, however, the integral over the surface is always zero.

EXAMPLE 25–10 *Location of excess charge on a conductor.* We have remarked that the electric field E is zero at all points in the interior of a conducting material when the charges in the conductor are at rest. If E were *not* zero, the charges would move and we would not have an electrostatic situation. We may construct a Gaussian surface inside the conductor, such as surface A in Fig. 25–14a. Because $E = 0$ everywhere on this surface, Eq. (25–17) requires that the net charge inside the surface be zero. Now imagine shrinking the surface down like a collapsing balloon, until it encloses a region so small that we may consider it a point; then the charge at that point must be zero. We can do this anywhere inside the conductor, so *there can be no net charge at any point within the conductor.* Thus any excess charge on the conductor must be located only on its surface, as shown.

The charge on a conductor must be located entirely on its surface.

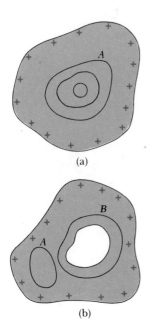

(a)

(b)

25–14 Excess charge on a conductor resides entirely on its outer surface.

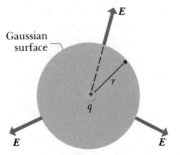

Gaussian surface

25–15 Gaussian surface used to derive Coulomb's law from Gauss's law.

Field of a charged conducting sphere: Outside the sphere, the field is the same as for a point charge.

If there is a cavity inside the conductor, as in Fig. 25–14b, the situation is somewhat different. We return to this case in Section 25–6.

EXAMPLE 25–11 *Coulomb's law.* We have considered Coulomb's law as the fundamental equation of electrostatics and have derived Gauss's law from it. An alternative procedure is to consider Gauss's law as a fundamental experiment relation. We can then derive Coulomb's law from Gauss's law by obtaining the expression for the electric field E due to a point charge.

Consider the electric field of a single positive point charge q, shown in Fig. 25–15. By symmetry, the field is everywhere radial; there is no reason why it should deviate to one side of a radial direction rather than to another. Also, its magnitude is the same at all points at the same distance r from the charge; any point at this distance is like any other. Hence, if we select as a Gaussian surface a spherical surface of radius r, $E_\perp = E =$ constant at all points of the surface. Then

$$\oint E_\perp \, dA = E_\perp \oint dA = EA = 4\pi r^2 E.$$

From Gauss's law

$$4\pi r^2 E = \frac{q}{\epsilon_0} \quad \text{and} \quad E = \frac{1}{4\pi\epsilon_0} \frac{q}{r^2}.$$

The force on a point charge q' at a distance r from the charge q is then

$$F = q'E = \frac{1}{4\pi\epsilon_0} \frac{qq'}{r^2},$$

which is Coulomb's law,

EXAMPLE 25–12 *Field of a charged conducting sphere.* Let us place a charge q on a solid conducting sphere. From Example 25–10 we know that all the charge is on the surface of the sphere, and from symmetry we know that it is distributed *uniformly* over the surface. In principle, we can find the field at any point outside the sphere by summing the contributions from elements of charge on the surface, but it is much easier to use Gauss's law.

As with the point charge, we conclude from spherical symmetry that the field is radial everywhere. Furthermore, the magnitude of the field is again uniform over a spherical surface of radius r, concentric with the conductor. Thus if we construct a Gaussian surface of radius r, where r is greater than the radius R of the sphere, and if q is the total charge on the sphere,

$$4\pi r^2 E = \frac{q}{\epsilon_0},$$

$$E = \frac{1}{4\pi\epsilon_0} \frac{q}{r^2}. \tag{25–21}$$

The field *outside* the sphere is, therefore, the same as though the entire charge was concentrated at a point at its center. Just outside the surface of the sphere, where $r = R$,

$$E = \frac{1}{4\pi\epsilon_0} \frac{q}{R^2}.$$

Inside the sphere, as in the interior of any conductor where no charge is in motion, the field is zero. Thus when r is less than R, $E = 0$.

Reasoning very similar to this can be used to prove the assertion in Chapter 6 that the *gravitational* field of any spherically symmetric mass distribution is the same, at any point outside the distribution, as though all the mass were concentrated at the center. This argument provides the justification for treating spherical bodies as points when calculating gravitational interactions.

Gauss's law can be applied to gravitational fields.

The same argument may be applied to a conducting, hollow, spherical *shell* (a spherical conductor with a concentric spherical hole in the center), if there is no charge in the hole. This time we take a spherical Gaussian surface of radius r less than the radius of the hole. If there *is* a field inside the hole, it must be spherically symmetric (radial) as before, so again $E = q/4\pi\epsilon_0 r^2$. But this time $q = 0$, so E must also be zero.

Here is a challenge: Can you use this same technique to find the electric field in the interspace between a charged sphere and a concentric, hollow conducting sphere surrounding it?

EXAMPLE 25–13 *Field of a line charge and of a charged cylindrical conductor.* Consider next the electric field caused by a long, thin, uniformly charged wire. We solved this problem in Section 25–2 (Example 25–8) by a straightforward application of Coulomb's law, requiring a somewhat involved integration. Gauss's law makes it possible to find the field by an almost trivial calculation.

Gauss's law applied to charge distributed along a line

If the wire is very long and we are not too near either end, then, by symmetry, the field lines outside the wire are *radial* and lie in planes perpendicular to the wire. Also, the field has the same magnitude at all points at the same radial distance from the wire. This suggests that we use as a Gaussian surface a *cylinder* of arbitrary radius r and arbitrary length l, with its ends perpendicular to the wire, as in Fig. 25–16. If λ is the charge per unit length on the wire, the charge within the Gaussian surface is λl. Since E is at right angles to the wire, the component of E normal to the end faces is zero. Thus the end faces make no contribution to the integral in Gauss's law. At all points of the curved surface, $E_\perp = E =$ constant, and since the area of this surface is $2\pi rl$, we have

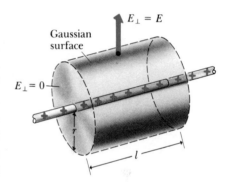

25–16 Cylindrical Gaussian surface for calculating the electric field due to a long charged wire.

$$(E)(2\pi rl) = \frac{\lambda l}{\epsilon_0},$$

$$(25\text{--}22)$$

$$E = \frac{1}{2\pi\epsilon_0}\frac{\lambda}{r}.$$

Note that although the *entire* charge on the wire contributes to the field E, only that portion of the total charge lying within the Gaussian surface is used when we apply Gauss's law. This may be puzzling at first; it appears as though we had somehow obtained the right answer by ignoring a part of the charge, and that the field of a *short* wire of length l would be the same as that of a very long wire. The existence of the entire charge on the wire *is*, however, taken into account when we consider the *symmetry* of this problem. If the wire had been short, we could not conclude by symmetry that the field at an end of the cylinder would equal that at the center, or that the field lines would everywhere be perpendicular to the wire. So the entire charge on the wire actually *is* taken into account, but in an indirect way.

Here is another challenge: (1) Show that the field outside a long, uniformly charged cylinder is the same, at points outside the cylinder, as though all the charge were concentrated on a line along its axis. (2) Calculate the electric field in the interspace between a charged cylinder and a coaxial hollow conducting cylinder surrounding it.

A charged conducting cylinder: a field problem made much simpler by use of Gauss's law

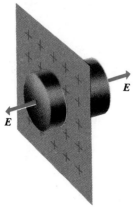

25–17 Gaussian surface in the form of a cylinder for finding the field of an infinite plane sheet of charge.

Field calculations with infinite planes of charge: Gauss's law makes it easy.

EXAMPLE 25–14 *Field of an infinite-plane sheet of charge.* To solve this problem, we construct the Gaussian surface shown by the shaded area in Fig. 25–17, consisting of a cylinder whose ends have an area A and whose walls are perpendicular to the sheet of charge. By symmetry, since the sheet is infinite, the electric field E has the same magnitude E on both sides of the surface, is uniform, and is directed normally away from the sheet of charge. No field lines cross the *side* walls of the cylinder; that is, the component of E normal to these walls is zero. At the ends of the cylinder the normal component of E is equal to E. The integral $\oint E_\perp \, dA$, calculated over the entire surface of the cylinder, therefore reduces to $2EA$. If σ is the charge per unit area in the plane sheet, the net charge within the Gaussian surface is σA. Hence

$$2EA = \frac{\sigma A}{\epsilon_0}, \qquad E = \frac{\sigma}{2\epsilon_0}. \tag{25–23}$$

This is the same result obtained in Section 25–2 (Example 25–9) by a much more complicated method. The magnitude of the field is *independent* of the distance from the sheet and does *not* decrease inversely with the square of the distance. The field lines are straight, parallel, and uniformly spaced. This is because the sheet was assumed infinitely large.

Of course, nothing in nature can really be infinitely large; this infinitely large plane sheet is an idealization. Our result is still useful, however; the real meaning of Eq. (25–23) is that it is a very good approximation to the behavior of the electric field caused by a large but finite sheet of charge, at points that are not near its edge and are close to the plane compared to its dimensions. In such cases the field is very nearly uniform and perpendicular to the plane.

EXAMPLE 25–15 *Field of an infinite-plane charged conducting plate.* When a flat metal plate is given a net charge, this charge distributes itself over the entire outer surface of the plate. If the plate is of uniform thickness and is infinitely large (or if we are not too near the edges of a finite plate), the charge per unit area is uniform and is the same on both surfaces. Hence the field of such a charged plate arises from the superposition of the fields of *two* sheets of charge, one on each surface of the plate. By symmetry, the field is perpendicular to the plate; it is directed away from it (if the plate has a positive charge); and it is uniform. The magnitude of the electric field at any point can be found from Gauss's law or by using the results already derived for a sheet of charge.

Figure 25–18 shows a portion of a large charged conducting plate. Let σ represent the charge per unit area in the sheet of charge on *each* surface. (I.e., the total charge per unit area on both surfaces together is 2σ.) At point a, outside the plate at the left, the component of electric field E_1, due to the sheet of charge on the left face of the plate, is directed toward the left, and its magnitude is $\sigma/2\epsilon_0$. The component E_2 due to the sheet of charge on the right face of the plate is also toward the left, and its magnitude is also $\sigma/2\epsilon_0$. The magnitude of the *resultant* E is therefore

$$E = E_1 + E_2 = \frac{\sigma}{2\epsilon_0} + \frac{\sigma}{2\epsilon_0} = \frac{\sigma}{\epsilon_0}. \tag{25–24}$$

At point b, inside the plate, the two components of electric field are in opposite directions and their resultant is zero, as it must be in any conductor where the charges are at rest. At point c, the components again add, and the magnitude of the resultant is σ/ϵ_0, directed toward the right.

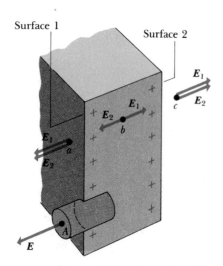

25–18 Electric field inside and outside a charged conducting plate.

To derive these results directly from Gauss's law, consider the cylinder shown in Fig. 25–18. Its end faces have area A; one end lies inside and one outside the plate. The field inside the conductor is zero. The field outside, by symmetry, is perpendicular to the plate, so the normal component of E is zero over the walls of the cylinder and is equal to E over the outside end face. Hence, from Gauss's law,

$$EA = \frac{\sigma A}{\epsilon_0}, \qquad E = \frac{\sigma}{\epsilon_0}. \tag{25–25}$$

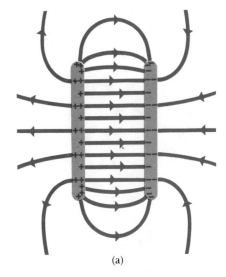

(a)

EXAMPLE 25–16 *Field between oppositely charged parallel conducting plates.* When two plane parallel conducting plates, having the size and spacing shown in Fig. 25–19, are given equal and opposite charges, the field between and around them is approximately as shown in Fig. 25–19a. Most of the charge accumulates at the opposing faces of the plates, and the field is nearly uniform in the space between them. A small quantity of charge resides on the outer surfaces of the plates, and some spreading or "fringing" of the field occurs at the edges of the plates.

As the plates are made larger and the distance between them smaller, the fringing becomes relatively less. Often the fringing is entirely negligible. We will usually assume that the field between two oppositely charged plates is uniform, as in Fig. 25–19b, and that the charges are distributed uniformly over the opposing surfaces.

The electric field at any point can be considered as the resultant of that due to two sheets of charge of opposite sign, or it may be found from Gauss's law. Thus at points a and c in Fig. 25–19b, the components E_1 and E_2 are each of magnitude $\sigma/2\epsilon_0$ but are oppositely directed, so their resultant is zero. At any point b between the plates, the components are in the same direction and their resultant is

$$E = \sigma/\epsilon_0.$$

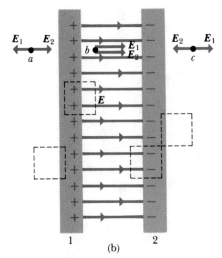

25–19 Electric field between oppositely charged parallel plates.

We leave it to you to show that the same results follow from applying Gauss's law to the surface shown by broken lines.

EXAMPLE 25–17 Electric charge is distributed uniformly throughout the volume of a sphere of radius R; the total charge is Q. Find the electric-field magnitude at a point P a distance r from the center of the sphere, where $r < R$.

SOLUTION We choose as our Gaussian surface a sphere of radius r, concentric with the charge distribution. The charge per unit volume, which we may call the volume charge density ρ, is given by

$$\rho = \frac{Q}{4\pi R^3/3}.$$

Charge distributed through a spherical volume: What is the electric field inside the charge distribution?

The volume V' enclosed by the Gaussian surface is $\frac{4}{3}\pi r^3$, so the total charge q enclosed by that surface is

$$q = \rho V' = \frac{Q}{4\pi R^3/3}\left(\frac{4}{3}\pi r^3\right) = Q\frac{r^3}{R^3}.$$

By symmetry, the electric-field magnitude has the same value E at every point on the Gaussian surface, and its direction at every point is radially outward. The total area of the surface is $4\pi r^2$, so the value of the integral in Gauss's law is simply $4\pi r^2 E$. We equate this to q/ϵ_0, with q given by the equation above. We find

$$4\pi r^2 E = \frac{Q r^3}{\epsilon_0 R^3},$$

or

$$E = \frac{1}{4\pi\epsilon_0} \frac{Qr}{R^3}.$$

The field magnitude is proportional to the distance of the field point from the center of the sphere. At the center ($r = 0$), $E = 0$, as we should expect from symmetry. At the surface of the sphere ($r = R$), the field magnitude is

$$E = \frac{1}{4\pi\epsilon_0} \frac{Q}{R^2}.$$

That is, at the surface the field has the same magnitude as though all the charge were concentrated at the center. As we have learned, this is also the case at any field point a distance greater than R from the center.

25–6 CHARGES ON CONDUCTORS

In any electrostatic situation (i.e., when all charges are at rest) *the electric field at every point within a conductor is zero.* We have also shown in Section 25–5 (Example 25–10) that when excess charge is placed on a solid conductor, *the charge is located entirely on the surface of the conductor*, in Fig. 25–20a. We assume the conductor is mounted on an insulating stand after it is given an initial charge.

What if there is a cavity inside the conductor, as in Fig. 25–20b? If there is no charge in the cavity, we can use a Gaussian surface such as B to show that the net charge on the surface *of the cavity* must be zero, because $\mathbf{E} = \mathbf{0}$ everywhere on the Gaussian surface. In fact, we can prove that in this situation no charge is present *anywhere* on the cavity surface, but detailed proof of that statement will be postponed until Chapter 26.

Next, suppose a conductor is placed inside the cavity, carrying a charge q, as in Fig. 25–20c. Again $\mathbf{E} = \mathbf{0}$ everywhere on surface B, so according to Gauss's law the net charge inside this surface must be zero. Thus there must be a total charge $-q$ on the cavity surface. Since the *total* charge on the conductor cannot change, a charge $+q$ must appear on its outer surface. If the outer surface originally had a charge q', the total charge after the charge q is inserted into the cavity must be $q + q'$.

Now let us consider the experiment shown in Fig. 25–21. We place a conducting container, such as a tin can, on an insulating stand. The container is initially uncharged. Then we hang a charged metal ball from an insulating thread, lower it into the can, and put the lid on, as in Fig. 25–21b. As we have just discussed, charges are induced on the walls of the container, as shown. But now we let the ball touch the inner wall, as in Fig. 25–21c. The surface of the ball becomes, in effect, part of the cavity surface. The situation is now the same as Fig. 25–20b, and according to Gauss's law the net charge on this surface must be zero. Finally, we pull the ball out; we find that it has indeed lost all its charge.

This experimental result confirms the validity of Gauss's law, and therefore of Coulomb's law. So what is the point? Simply this: Coulomb's experimental methods, using a torsion balance and dividing charges, were not very precise. It is difficult to confirm with great precision the $1/r^2$ dependence of the electrostatic force by direct measurement. Here, by contrast, is an experi-

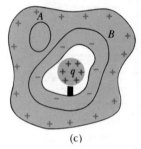

(a)

(b)

(c)

25–20 (a) Charge on a solid conductor resides entirely on its outer surface. (b) If there is no charge inside the cavity, the net charge on the surface of the cavity is zero. (c) If there is a charge q inside the cavity, the total charge on the cavity surface is $-q$.

A very sensitive test of the $1/r^2$ factor in Coulomb's law

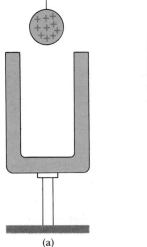
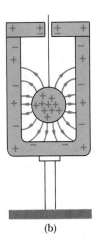
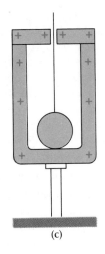

(a) (b) (c)

25-21 (a) A charged conducting ball suspended by an insulating thread outside a conducting container on an insulating stand. (b) The ball is lowered into the container and the lid put on. Charges are induced on the walls of the container. (c) When the ball is touched to the inner surface of the container, all its charge is transferred to the container and appears on its outer surface.

ment that tests the validity of Gauss's law and therefore of Coulomb's law with potentially much greater precision.

A contemporary version of this experiment is shown in Fig. 25–22. The details of the black box labeled "power supply" are not important; its function is to place charge on the outer sphere and remove it, on demand. The inner box with a dial is a sensitive electrometer, an instrument that can detect minute motion of charge between the outer and inner spheres. If Gauss's law is correct, there can never be any charge on the inner surface of the outer sphere, so there should be no flow of charge through the electrometer while the outer sphere is being charged and discharged. The fact that no flow is actually observed is a very sensitive confirmation of Gauss's law and therefore of Coulomb's law. The precision of the experiment is limited mainly by the sensitivity of the electrometer, which can be astonishing. The most recent (1971) experiments of this sort have shown that the exponent 2 in the $1/r^2$ expression in Coulomb's law does not differ from precisely 2 by more than 10^{-15}. So there is no reason to suspect that it is anything other than exactly 2.

This discussion also forms the basis for **electrostatic shielding,** shown in Fig. 25–23. Suppose we have a very sensitive electronic instrument that must be protected from stray electric fields that might give erroneous measurements. We surround the instrument with a conducting box, or we line the

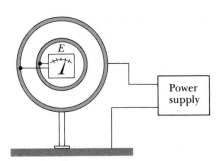

25-22 The outer surface can be alternately charged and discharged by the power supply to which it is connected. If there is any flow of charge between the inner and outer surfaces, it is detected by the electrometer inside the inner surface.

Electrostatic shielding: how to protect sensitive instruments from disruptive electric fields

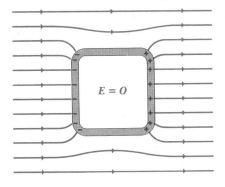

$E = 0$

25-23 A conducting box in a uniform electric field. The field pushes electrons toward the left, leaving a net negative charge on the left side and a net positive charge on the right. The total electric field at every point inside the box is zero; the shapes of the exterior field lines near the box are somewhat changed.

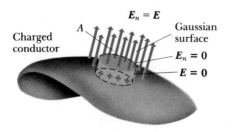

25–24 The field just outside a charged conductor is perpendicular to the surface and is equal to σ/ϵ_0.

walls, floor, and ceiling of the room with a conducting material such as sheet copper. The external electric field redistributes the free electrons in the conductor, leaving a net positive charge on the outer surface in some regions and a net negative charge in others, as shown in Fig. 25–23. This charge distribution causes an additional electric field such that the *total* field at every point inside the box is zero, as Gauss's law says it must be. The charge distribution on the box also alters the shapes of the field lines near the box, as the figure shows.

Finally, we note that there is a direct relation between the E field at any point just outside a conductor and the surface charge density σ on the conductor at that point. In general, σ varies from point to point on the surface. We will show in Chapter 26 that at any such point the direction of E is always perpendicular to the surface.

To find a relation between σ at any point on the surface and E at that point, we construct a Gaussian surface in the form of a small cylinder, as in Fig. 25–24. One end face, with area A, lies within the conductor, and the other lies just outside. The charge within the Gaussian surface is σA. The electric field is zero at all points within the conductor. Outside the conductor, the normal component of E is zero at the side walls of the cylinder (since E is normal to the conductor), while over the end face the normal component is equal to E. Hence from Gauss's law,

$$EA = \frac{\sigma A}{\epsilon_0}, \qquad E = \frac{\sigma}{\epsilon_0}. \qquad (25\text{--}26)$$

This equation agrees with the results already obtained for spherical, cylindrical, and plane surfaces. Just outside the surface of a sphere of radius R carrying a charge q, for example, the electric field is

$$E = \frac{1}{4\pi\epsilon_0}\frac{q}{R^2}.$$

But the surface density of charge on the sphere is $q/4\pi R^2$, so $E = \sigma/\epsilon_0$.

We also showed that the field outside an infinite charged conducting plate equals σ/ϵ_0. In this case, the field is the same at *all* distances from the plate, but in all other cases it decreases with increasing distance from the surface.

SUMMARY

Electric field is the force per unit charge exerted on a test charge at any point, provided the test charge is small enough so that it does not disturb the charges that cause the field. Electric field is a vector quantity; it is an example of a

The electric field just outside a conducting surface is proportional to the surface charge density.

KEY TERMS
electric field
test charge

vector field, that is, a vector quantity associated with each point in a region of space (different vectors at different points).

The electric field of a point charge can be calculated from Coulomb's law; the electric field of any combination of charges is the vector sum of the fields caused by the individual charges. This principle is called the principle of superposition.

An electric dipole is a pair of electric charges of equal magnitude q but opposite sign, separated by a distance l. The electric dipole moment p is defined to be $p = ql$. The vector dipole moment is the vector p having this magnitude and a direction from negative toward positive charge. An electric dipole in an electric field experiences a torque Γ given by

$$\Gamma = pE \sin \theta, \qquad (25\text{--}2)$$

where θ is the angle between the directions of p and E. The vector torque Γ is given by

$$\Gamma = p \times E. \qquad (25\text{--}3)$$

Calculating the electric field caused by a continuous distribution of charge is carried out by dividing the distribution into small elements, calculating the field caused by each element, and then carrying out the vector sum, usually by integrating. It is often simplest to compute one component of E at a time. The concepts of linear charge density λ (charge per unit length), surface charge density σ (charge per unit area), and volume charge density ρ (charge per unit volume) are often helpful in characterizing charge distributions.

Field lines are a useful graphical representation of electric fields. A field line at any point in space is tangent to the direction of E at that point, and the number of lines per unit area (perpendicular to their direction) is proportional to the magnitude of E at the point. Electrostatic field lines always begin and end on charges, and they never cross.

Gauss's law is logically equivalent to Coulomb's law, but its use simplifies some problems considerably. It states that the surface integral of the component of E normal to the surface, over any closed surface, equals a constant times the total charge Q enclosed by the surface; symbolically,

$$\oint E \cdot dA = \frac{Q}{\epsilon_0}. \qquad (25\text{--}17)$$

Electric flux Ψ is defined as

$$\Psi = \int E \cdot dA. \qquad (25\text{--}19)$$

If the field is uniform and perpendicular to the area, then this equation becomes simply

$$\Psi = EA.$$

When excess charge is placed on a conductor, it resides entirely on the surface. In any electrostatic situation, $E = 0$ everywhere in the interior of a conductor.

Table 25–1 lists electric fields caused by several simple charge distributions.

vector field
electric dipole
electric dipole moment
field line
Gauss's law
electric flux
electrostatic shielding

TABLE 25–1 Electric Fields around Simple Charge Distributions

Charge distribution responsible for the electric field	Arbitrary point in the electric field	Electric field at this point
Single point charge q	Distance r from q	$E = \dfrac{1}{4\pi\epsilon_0}\dfrac{q\hat{r}}{r^2}$
Several point charges, q_1, $q_2, \ldots$	Distance r_1 from q_1, r_2 from $q_2, \ldots$	$E = \dfrac{1}{4\pi\epsilon_0}\left(\dfrac{q_1\hat{r}_1}{r_1^2} + \dfrac{q_2\hat{r}_2}{r_2^2} + \cdots\right)$
Charge q uniformly distributed on the surface of a solid conducting sphere of radius R	(a) Outside, $r \geq R$ (b) Inside, $r < R$	(a) $E = \dfrac{1}{4\pi\epsilon_0}\dfrac{q\hat{r}}{r^2}$ (b) $E = 0$
Long cylinder of radius R, with charge per unit length λ	(a) Outside, $r \geq R$ (b) Inside, $r < R$	(a) $E = \dfrac{1}{2\pi\epsilon_0}\dfrac{\lambda}{r}$ (b) $E = 0$
Two oppositely charged conducting plates with charge per unit area σ	Any point between plates	$E = \dfrac{\sigma}{\epsilon_0}$
Any charged conductor	Just outside the surface	$E = \dfrac{\sigma}{\epsilon_0}$

QUESTIONS

25–1 It was shown in the text that the electric field inside a spherical hole in a conductor is zero. Is this also true for a cubical hole? Can the same argument be used?

25–2 Coulomb's law and Newton's law of gravitation have the same *form*. Can Gauss's law be applied to gravitational fields as well as electric fields? Is so, what modifications are needed?

25–3 By considering how the field lines must look near a conducting surface, can you see why the charge density and electric-field magnitude at the surface of an irregularly shaped solid conductor must be greatest in regions where the surface curves most sharply, and least in flat regions?

25–4 The electric field and the velocity field in a moving fluid are two examples of vector fields. Think of several other examples. There are also *scalar* fields, which associate a single number with each point in space. Temperature is an example; think of several others.

25–5 Consider the electric field caused by two point charges separated by some distance. Suppose there is a point where the field is zero; what does this tell you about the *signs* of the charges?

25–6 Does an electric charge experience a force due to the field that the charge itself produces?

25–7 A particle having electric charge and mass moves in an electric field. If it starts from rest, does it always move along the field line that passes through its starting point? Explain.

25–8 If the exponent 2 in the r^2 of Coulomb's law were 3 instead, would Gauss's law still be valid?

25–9 A student claimed that an appropriate unit for electric field magnitude is $1\ \mathrm{J \cdot C^{-1} \cdot m^{-1}}$. Is this correct?

25–10 A certain region of space bounded by an imaginary closed surface contains no charge. Is the electric field always zero everywhere on the surface? If not, under what circumstances is it zero on the surface?

25–11 A student claimed that the electric field produced by a dipole is represented by field lines that cross each other. Is this correct? Is there any simple rule governing field lines for a superposition of fields due to point charges, if the field lines due to the separate charges are known? Do field lines *ever* cross?

25–12 Nineteenth-century physicists liked to give everything mechanical attributes. Faraday and his contemporaries thought of field lines as elastic strings that repelled each other and arranged themselves in equilibrium under the action of their elastic tension and mutual repulsion. Try this picture on several examples and decide whether it makes any sense. (These mechanical properties are of course now known to be completely fictitious.)

25–13 Are Coulomb's law and Gauss's law *completely* equivalent? Are there any situations in electrostatics where one is valid and the other is not?

25–14 The text states that, in an electrostatic field, every field line must start on a positive charge and terminate on a negative charge. But suppose the field is that of a single positive point charge. Then what?

25–15 Is the total (net) electric charge in the universe positive, negative, or zero?

25–16 A lightning rod is a pointed copper rod mounted on top of a building and welded to a heavy copper cable running down into the ground. Lightning rods are used in prairie country to protect houses and barns from lightning; the lightning current runs through the copper rather than through the barn. Why? Why should the end of the rod be pointed? (The answer to Question 25–3 may be helpful.)

25–17 When a high-voltage power line falls on your car, you are safe as long as you stay in the car, but when you step out you may be electrocuted. Why? What about a car as a safe place in a thunderstorm?

EXERCISES

Section 25–1 Electric Field and Electrical Forces

25–1 Find the magnitude and direction of the electric field at a point 0.5 m directly above a particle having an electric charge of $+2.0 \times 10^{-6}$ C.

25–2 The electric field caused by a certain point charge has a magnitude of 5.0×10^3 N·C^{-1} at a distance of 0.10 m from the charge. What is the magnitude of the charge?

25–3 At what distance from a particle of charge 5.0×10^{-9} C does the electric field of that charge have a magnitude of 2.00 N·C^{-1}?

25–4

a) What is the electric field of a gold nucleus, at a distance of 10^{-14} m from the nucleus?

b) What is the electric field of a proton, at a distance of 5.28×10^{-11} m from the proton?

25–5 A small object carrying a charge of -5×10^{-9} C experiences a downward force of 20×10^{-9} N when placed at a certain point in an electric field. What is the electric field at the point, in magnitude and direction?

25–6 What must be the charge (sign and magnitude) on a particle of mass 2 g for it to remain stationary in the laboratory when placed in a downward-directed electric field of 5000 N·C^{-1}?

25–7 What is the magnitude of an electric field in which the force on an electron is equal in magnitude to the weight of the electron?

25–8 A uniform electric field exists in the region between two oppositely charged plane parallel plates. An electron is released from rest at the surface of the negatively charged plate and strikes the surface of the opposite plate, 2 cm distant from the first, in a time interval of 1.5×10^{-8} s.

a) Find the electric field.

b) Find the velocity of the electron when it strikes the second plate.

25–9 An electron is projected with an initial velocity $v_0 = 1.0 \times 10^7$ m·s^{-1} into the uniform field between the parallel plates in Fig. 25–25. The direction of the field is vertically downward, and the field is zero except in the space between the two plates. The electron enters the field at a point

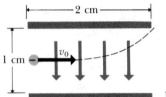

FIGURE 25–25

midway between the plates. If the electron just misses the upper plate as it emerges from the field, find the magnitude of the electric field.

Section 25–2 Electric-Field Calculations

25–10 Two particles having charges $q_1 = 1.00 \times 10^{-9}$ C and $q_2 = 2.00 \times 10^{-9}$ C are separated by a distance of 2.00 m. At what point is the total electric field due to the two charges equal to zero?

25–11 In a rectangular coordinate system a charge of 25×10^{-9} C is placed at the origin of coordinates, and a charge of -25×10^{-9} C is placed at the point $x = 6$ m, $y = 0$. What are the magnitude and direction of the electric field at

a) $x = 3$ m, $y = 0$?

b) $x = 3$ m, $y = 4$ m?

25–12 In a rectangular coordinate system, two positive point charges of 10^{-8} C each are fixed at the points $x = +0.1$ m, $y = 0$, and $x = -0.1$ m, $y = 0$. Find the magnitude and direction of the electric field at the following points:

a) the origin;   b) $x = 0.2$ m, $y = 0$;

c) $x = 0.1$ m, $y = 0.15$ m;   d) $x = 0$, $y = 0.1$ m.

25–13 Same as Exercise 25–12, except that the point charge at $x = +0.1$ m, $y = 0$ is positive and the other is negative.

25–14 A long, straight wire has charge per unit length 3.0×10^{-10} C·m^{-1}. At what distance from the wire is the electric field equal to 0.8 N·C^{-1}?

25–15 What must be the charge per unit area, in C·m^{-2}, of an infinite-plane sheet of charge if the electric field produced by the sheet of charge is to be 2.0 N·C^{-1}?

Section 25–4 Gauss's Law

25–16 The electric field **E** in Fig. 25–26 is everywhere parallel to the *x*-axis. The field has the same magnitude at all points in any given plane perpendicular to the *x*-axis (parallel to the *yz*-plane), but the magnitude is different for various planes. That is, E_x depends on *x* but not on *y* and *z*, and E_y and E_z are zero. At points *in* the *yz*-plane, $E_x = 400$ N·C^{-1}. (The volume shown could be a section of a large insulating slab 1 m thick, with its faces parallel to the *yz*-plane and with a uniform volume charge distribution embedded in it.)

a) What is the value of $\int E_\perp dA$ over surface I in the diagram?

b) What is the value of the surface integral of **E** over surface II?

c) There is a positive charge of 26.6×10^{-9} C within the volume. What are the magnitude and direction of **E** at the face opposite I?

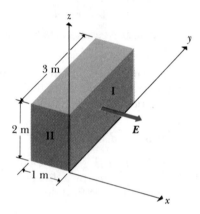

FIGURE 25–26

Section 25–5 Applications of Gauss's Law

Section 25–6 Charges on Conductors

25–17 How many excess electrons must be added to an isolated spherical conductor 0.10 m in diameter to produce a field of 1300 N·C^{-1} just outside the surface?

25–18 The earth has a net electric charge that causes a field at points near its surface of the order of 100 N·C^{-1}. If the earth is regarded as a conducting sphere of radius 6.38×10^6 m, what is the magnitude of its charge?

25–19 Prove that the electric field outside an infinitely long cylindrical conductor with a uniform surface charge is the same as if all the charge were on the axis.

25–20 Apply Gauss's law to the dotted Gaussian surfaces in Fig. 25–19b to calculate the electric field between and outside the plates.

25–21 The electric field in the region between a pair of oppositely charged plane parallel conducting plates, each 100 cm^2 in area, is 10^4 N·C^{-1}. What is the charge on each plate? Neglect edge effects.

25–22 A conducting sphere carrying charge *q* has radius *a*. It is inside a concentric hollow conducting sphere of inner radius *b* and outer radius *c*. The hollow sphere has no net charge. Calculate the electric field for

a) $r < a$,

b) $a < r < b$,

c) $b < r < c$,

d) $r > c$.

e) What is the charge on the inner surface of the hollow sphere?

f) On the outer surface?

25–23 A small conducting sphere of radius *a*, mounted on an insulating handle and having a positive charge *q*, is inserted through a hole in the walls of a hollow conducting sphere of inner radius *b* and outer radius *c*. The hollow sphere is supported on an insulating stand and is initially uncharged, and the small sphere is placed at the center of the hollow sphere. Neglect any effect of the hole.

a) Show that the electric field at a point in the region between the spheres, at a distance *r* from the center, is equal to

$$E = \frac{1}{4\pi\epsilon_0} \frac{q}{r^2}.$$

b) What is the electric field at a point outside the hollow sphere?

c) Sketch a graph of the magnitude of **E** as a function of *r*, from $r = 0$ to $r = 2c$.

d) Represent the charge of the small sphere by four + signs. Sketch the field lines of the system, within a spherical volume of radius $2c$.

PROBLEMS

25–24 A small sphere whose mass is 0.1 g carries a charge of 3×10^{-10} C and is attached to one end of a silk fiber 5 cm long. The other end of the fiber is attached to a large vertical conducting plate, which has a surface charge of 25×10^{-6} C·m^{-2} on each side. Find the angle the fiber makes with the vertical plate.

25–25 An electron is projected into a uniform electric field of 5000 N·C^{-1}. The direction of the field is vertically upward. The initial velocity of the electron is 1.0×10^7 m·s^{-1}, at an angle of 30° above the horizontal.

a) Find the maximum distance the electron rises vertically above its initial elevation.

b) After what horizontal distance does the electron return to its original elevation?

c) Sketch the trajectory of the electron.

25–26 A charge of 16×10^{-9} C is fixed at the origin of coordinates; a second charge of unknown magnitude is at $x = 3$ m, $y = 0$; and a third charge of 12×10^{-9} C is at $x = 6$ m, $y = 0$. What are the sign and magnitude of the unknown charge if the resultant field at $x = 8$ m, $y = 0$ is 20.25 N·C^{-1} directed to the right?

25–27 Two charges are placed as shown in Fig. 25–27. It is known that the magnitude of q_1 is 5×10^{-6} C, but its sign and the value of the other charge, q_2, are not known. The resultant electric field $\boldsymbol{E}$ at point P is measured to be entirely in the negative y-direction.

a) Considering the different possible signs of q_1 and q_2, there are four possible diagrams that could represent the electric fields $\boldsymbol{E}_1$ and $\boldsymbol{E}_2$ produced by q_1 and q_2. Sketch the four possible electric field configurations.

b) Using the sketches from (a) and the fact that the net electric field at P has no x-component and a negative y-component, deduce the signs of q_1 and q_2.

c) Determine the magnitude of the resultant field $\boldsymbol{E}$.

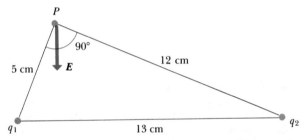

FIGURE 25–27

25–28 Electric charge is uniformly distributed around a semicircle of radius a, with total charge Q. What is the electric field at the center of curvature?

25–29 Negative electric charge is distributed uniformly around a quarter of a circle of radius a, with total charge $-Q$ (Fig. 25–28). What are the x- and y-components of the resultant electric field at the center of curvature (point P)?

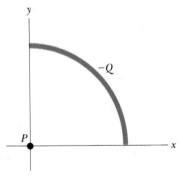

FIGURE 25–28

25–30 Electric charge is distributed uniformly along each of the sides of a square. Two adjacent sides have positive charge, with total charge $+Q$ on each.

a) If the other two sides have negative charge with total charge $-Q$ each (see Fig. 25–29), what are the x- and

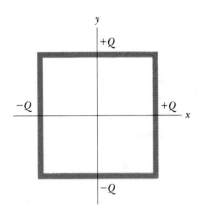

FIGURE 25–29

y-components of the resultant electric field at the center of the square? Each side of the square has length a.

b) Repeat the calculation of (a) if all four sides have positive charge $+Q$.

25–31 A uniform electric field, $\boldsymbol{E}_1 = -9 \times 10^4$ N·C$^{-1}\boldsymbol{i}$, is directed out of one face of a parallelopiped, and another uniform electric field, $\boldsymbol{E}_2 = -11 \times 10^4$ N·C$^{-1}\boldsymbol{i}$, is directed into the opposite face, as shown in Fig. 25–30. Assuming that there are no other electric-field lines crossing the surfaces of the parallelopiped, determine the net charge contained within.

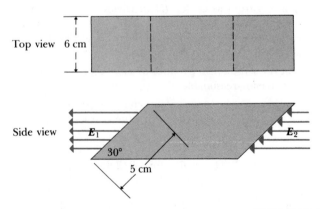

FIGURE 25–30

25–32 A conducting spherical shell of inner radius a and outer radius b has a positive point charge Q located at its center. The total charge on the shell is $-3Q$, and it is insulated from its surroundings.

a) Derive expressions for the electric-field magnitude in terms of the distance r from the center, for the regions $r < a$, $a < r < b$, and $r > b$.

b) What is the surface charge density on the inner surface of the conducting shell?

c) What is the surface charge density on the outer surface of the conducting shell?

d) Draw a sketch showing electric-field lines and the location of all charges.

e) Draw a graph of E as a function of r.

25–33 A long coaxial cable consists of an inner cylindrical conductor of radius a and an outer coaxial cylinder of inner radius b and outer radius c. The outer cylinder is mounted on insulating supports and has no net charge. The inner cylinder has a uniform positive charge λ per unit length. Calculate the electric field

a) at any point between the cylinders;

b) at any external point.

c) Sketch a graph of the magnitude of E as a function of the distance r from the axis of the cable, from $r = 0$ to $r = 2c$.

d) Find the charge per unit length on the inner surface of the outer cylinder, and on the outer surface.

25–34 Suppose that positive charge is uniformly distributed through a very long cylindrical volume of radius R, the charge per unit volume being ρ.

a) Derive the expression for the electric field inside the volume at a distance r from the axis of the cylinder, in terms of the charge density ρ.

b) What is the electric field at a point outside the volume, in terms of the charge per unit length λ in the cylinder?

c) Compare the answers to (a) and (b) when $r = R$.

d) Sketch a graph of the magnitude of E as a function of r, from $r = 0$ to $r = 3R$.

25–35 A nonuniform but spherically symmetric distribution of charge has a charge density ρ given as follows:

$$\rho = \rho_0(1 - r/R) \quad \text{for} \quad r \leq R,$$
$$\rho = 0 \quad \text{for} \quad r \geq R,$$

where $\rho_0 = 3Q/(\pi R^3)$ is a constant.

a) Show that the total charge contained in the charge distribution is Q.

b) Show that for the region defined by $r \geq R$ the electric field is identical to that produced by a point charge Q.

c) Obtain an expression for the electric field in the region $r \leq R$.

d) Compare your results in (b) and (c) for $r = R$.

CHALLENGE PROBLEMS

25–36 Electric charge is distributed uniformly over a disk of radius a, with total charge Q.

a) Find the electric field at a point on the axis of the disk, a distance x from its center. (*Hint:* Divide the disk into concentric rings; use the result of Example 25–7 to find the field due to each ring; and integrate to find the total field.)

b) What does your result reduce to as x becomes large? Is this behavior reasonable?

25–37 A region in space contains charge distributed spherically such that the volume charge density ρ is given by:

$$\rho = A \quad \text{for} \quad r \leq R/2,$$
$$\rho = 2A(1 - r/R) \quad \text{for} \quad R/2 \leq r \leq R,$$
$$\rho = 0 \quad \text{for} \quad r \geq R.$$

The total charge Q is 1×10^{-17} C, the radius R of the spherical charge distribution is 2×10^{-14} m, and A is a constant having units of C·m^{-3}.

a) Determine A in terms of Q and R, and also determine its numerical value.

b) Using Gauss's law, derive an expression for the electric field E as a function of the distance r from the center of the distribution. Do this separately for all three regions. Be sure to check that your results agree on the boundaries of the regions.

c) What fraction of the total charge is contained within the region where $r \leq R/2$?

d) If an electron of charge $q' = -e$ is oscillating back and forth about $r = 0$ (the center of the distribution) with an amplitude of $R/2$, show that the motion is simple harmonic. (*Hint:* Review the definition of simple

harmonic motion, as defined by Eq. (11–1). If it can be shown that the net force on the electron is of this form, then it follows that the motion is simple harmonic. Conversely, if the net force on the electron does not follow this form, the motion is not simple harmonic.)

e) For the motion in (d), what is the period? (Calculate a numerical value.)

f) If the amplitude of the motion described in (e) were greater than $R/2$, explain why the motion would no longer be simple harmonic.

25–38 A region in space contains charge distributed spherically such that the volume charge density ρ is given by:

$$\rho = 3Ar/(2R) \quad \text{for} \quad r \leq R/2,$$
$$\rho = A[1 - (r/R)^2] \quad \text{for} \quad R/2 \leq r \leq R,$$
$$\rho = 0 \quad \text{for} \quad r \geq R.$$

The total charge is Q, the radius R of the spherical charge distribution is 5×10^{-10} m, and $A = 6 \times 10^9$ C·m^{-3} and is constant.

a) Determine Q in terms of A and R, and also determine its numerical value.

b) Using Gauss's law, derive an expression for the electric field E as a function of the distance r from the center of the distribution. Do this separately for all three regions.

c) What fraction of the total charge is contained within the region where $R/2 \leq r \leq R$?

d) What is the magnitude of the electric field at $r = R/2$?

e) If an electron of charge $q' = -e$ is released from rest at any point in any of the three regions, the resulting motion will be oscillatory, although not simple harmonic. Why? (See Challenge Problem 25–37).

26
ELECTRICAL POTENTIAL

THE CONCEPT OF *POTENTIAL* IS DIRECTLY TIED TO POTENTIAL ENERGIES associated with electrical interactions. We studied the concepts of *work* and *energy* in Chapter 7; they provide an important method of analysis for many problems in mechanics. Energy plays an equally fundamental role in electricity and magnetism. In this chapter we apply work and energy considerations to the electric field. When a charged particle moves in an electric field, the electric-field force does *work* on the particle. This work can always be expressed in terms of a potential energy, which in turn is associated with a new concept called *electrical potential,* or simply *potential.* We will discuss several examples of calculations of potential and their applications to the motions of charged particles.

26–1 ELECTRICAL POTENTIAL ENERGY

When a charged particle moves in a region of space where an electric field is present, the field exerts a force on the particle and thus does *work* on it. Figure 26–1 shows a simple example; a pair of charged parallel metal plates sets up a uniform electric field of magnitude E in the region between them, and the resulting force on a test charge q' has magnitude $F = q'E$. When the charge moves from point a to point b, the work done by this force on the test charge is

$$W_{a \to b} = q'Ed. \tag{26–1}$$

This work can be represented by a **potential-energy** function, just as for gravitational potential energy, discussed in Section 7–5. If we take the potential energy to be zero at point b, then at point a it has the value $q'Ed$, and we may say

$$W_{a \to b} = -\Delta U = U_a - U_b. \tag{26–2}$$

More generally, the potential energy at any point a distance y above the bottom plate is given by

$$U(y) = q'Ey, \tag{26–3}$$

When an electric-field force acts on a moving charge, it does work.

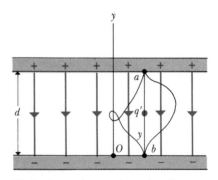

26–1 A test charge q' moving from a to b experiences a force of magnitude $q'E$; the work done by this force is $W_{a \to b} = q'Ed$, and is independent of the particle's path.

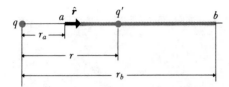

26–2 Charge q' moves along a straight line extending radially from charge q. As it moves from a to b, the distance varies from r_a to r_b.

The electric field is a conservative force field.

Work done on a point charge in the field of another point charge

and when the test charge moves from height y_1 to height y_2, the work done on the charge by the field is given by

$$W_{1 \to 2} = U(y_1) - U(y_2) = q'Ey_1 - q'Ey_2. \qquad (26\text{--}4)$$

When y_1 is greater than y_2, U decreases and the field does positive work; when y_1 is less than y_2, U increases and the field does negative work. This whole situation is directly analogous to the motion of a particle in a uniform gravitational field, with its associated work and potential energy.

Note that the work done by the electric-field force depends only on the change in the coordinate y and is independent of the *path* of the particle. Hence this force is an example of what we called in Section 7–7 a **conservative force field;** we suggest you review that discussion now. In fact, we will show very soon that *every* electric field produced by charges at rest is a conservative force field and thus can be associated with a potential-energy function.

Consider next the work done on a test charge q' moving in the electric field caused by a single stationary point charge q. Such a force field is clearly *not* uniform, and we have to integrate to calculate the work done on q'. We first calculate the work done on q' during a displacement along the *radial* line from point a to point b in Fig. 26–2. We introduce a unit vector $\hat{r}$ pointing from a toward b; the component of displacement in this direction is just dr, and the component of force on q' in this direction is

$$F = \frac{1}{4\pi\epsilon_0} \frac{qq'}{r^2}. \qquad (26\text{--}5)$$

The work done by this force on q' as it moves from r_a to r_b is given by

$$W_{a \to b} = \int_{r_a}^{r_b} F \, dr = \int_{r_a}^{r_b} \frac{1}{4\pi\epsilon_0} \frac{qq'}{r^2} \, dr$$

$$= \frac{qq'}{4\pi\epsilon_0} \left(\frac{1}{r_a} - \frac{1}{r_b} \right). \qquad (26\text{--}6)$$

Thus the work for this particular path depends only on the endpoints. We have not yet proved that the work is the same for *all possible* paths from a to b. To do this, we consider a more general displacement, in which a and b *do not* lie on the same radial line, as in Fig. 26–3. To find the work done by the field in the displacement from point a to point b we must now use the more general definition of work given by Eq. (7–8):

$$W = \int_a^b F \cos\theta \, dl = \int_a^b \boldsymbol{F} \cdot d\boldsymbol{l}. \qquad (26\text{--}7)$$

But from the figure, $\cos\theta \, dl = dr$. That is, the work done during a small displacement dl depends only on the change dr in the distance r between the charges, which is the *radial component* of the displacement. In any displacement in which r does not change, no work is done because $\boldsymbol{F}$ and $d\boldsymbol{l}$ are perpendicular, $\cos\theta = 0$, and $\boldsymbol{F} \cdot d\boldsymbol{l} = 0$.

Thus Eq. (26–6) gives the work done by the field even for this more general displacement. This result shows that the work done on q' by the $\boldsymbol{E}$ field depends only on the initial and final distances r_a and r_b and not on the path connecting these points. The work is the same for *any* path between these points or between any pair of points at distances r_a and r_b from the charge q.

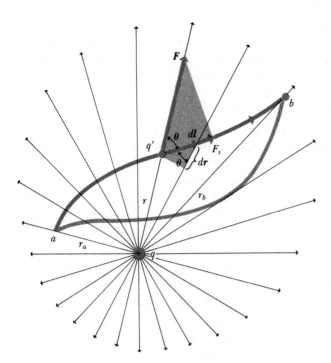

26–3 The work done by the electric-field force on charge q' depends only on the distances r_a and r_b.

Also, if the test charge returns from b to a along any path, the work done by the $\boldsymbol{E}$ field is the *negative* of that done in the displacement from a to b. So the total work done during a *loop* displacement returning to the starting point is always zero.

Comparing Eqs. (26–2) and (26–6), we see that the term $qq'/4\pi\epsilon_0 r_a$ is the potential energy U_a when q' is at point a, at distance r_a from q, and $qq'/4\pi\epsilon_0 r_b$ is the potential energy U_b when it is at point b, at distance r_b from q. Thus the potential energy U of the test charge q' at *any* distance r from charge q is given by

$$U = \frac{1}{4\pi\epsilon_0}\frac{qq'}{r}. \tag{26–8}$$

Potential energy of a point charge in the field of another point charge

If the field in which charge q' moves is not that of a single point charge but is due to a more general charge distribution, we can always divide the charge distribution into small elements and treat each element as a point charge. Then, since the total field is the *vector sum* of the fields due to the individual elements, and since the total work on q' is the sum of the contributions from the individual charge elements, we may conclude that *every electric field due to a static charge distribution is a conservative force field.* Furthermore, the potential energy of a test charge q' at point a in Fig. 26–4, due to a collection of charges q_1, q_2, q_3, and so on, at distances r_1, r_2, r_3, and so on, from the test charge q', is given by

Potential energy of a charge due to several point charges is the sum of the potential energies due to the individual charges.

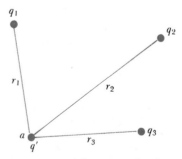

$$U = \frac{q'}{4\pi\epsilon_0}\left(\frac{q_1}{r_1} + \frac{q_2}{r_2} + \frac{q_3}{r_3} + \cdots\right)$$

$$= \frac{q'}{4\pi\epsilon_0}\sum\frac{q_i}{r_i}. \tag{26–9}$$

26–4 Potential energy of a charge q' at point a depends on charges q_1, q_2, and q_3, and on their distances r_1, r_2, and r_3 from point a.

At a second point b, the potential energy of q' is given by the same expression except that r_1, r_2, . . . now represent the distances from the respective charges to point b. The work done by the electric force in moving the test charge from a to b along any path is equal to the difference $U_a - U_b$ between its potential energies at a and at b.

The discussion of potential energy in Chapter 7 pointed out that it is always possible to add an arbitrary constant C to a potential-energy function U, since adding the same constant C to both U_a and U_b in Eq. (26–2) does not change the physically significant quantity, namely, the difference $U_a - U_b$. Thus we are always free to choose the constant C so that U is zero at some convenient reference position. In Chapter 7 the potential energy of a body in a uniform gravitational field was taken to be zero at some convenient reference level, often the surface of the earth. When the body is above this reference level, its potential energy is positive; when below, negative.

Potential energy of two point charges can be defined to be zero when they are infinitely far apart.

In Eqs. (26–8) and (26–9), the reference position for electrical potential energy (the position at which $U = 0$) has been chosen implicitly to be at a point where all the distances r_1, r_2, . . . are *infinite*. That is, the potential energy of the test charge is zero when it is very far removed from all the charges setting up the field. This reference level is the most convenient one for many electrostatic problems. When we consider electrical circuits, other reference levels are often more convenient.

From the preceding definitions, *the potential energy of a test charge at any point in an electric field is equal to the work done by the electric-field force when the test charge moves from the point in question to a zero reference level, often taken at infinity.*

EXAMPLE 26–1   A small plastic ball having mass 2.0 g and electric charge $+0.10$ μC moves in the vicinity of a stationary metal ball with a charge of $+2.0$ μC. When the plastic ball is 0.1 m from the metal one, it is moving directly away from it with a speed of 5.0 m·s^{-1}.

(a)   What is its speed when the two balls are 0.2 m apart?

(b)   How would the situation change if the plastic ball had a charge of -0.1 μC?

SOLUTION

A simple example of electrical potential energy

(a)   Because of the conservative nature of the force, total mechanical energy (kinetic plus potential) is conserved. Recalling the problem-solving strategy of Section 7–5, we list the initial and final kinetic and potential energies, K_1, K_2, U_1, and U_2, respectively:

$$K_1 = \tfrac{1}{2}mv_1{}^2 = \tfrac{1}{2}(2.0 \times 10^{-3} \text{ kg})(5.0 \text{ m·s}^{-1})^2 = 0.025 \text{ J};$$

$$K_2 = \tfrac{1}{2}mv_2{}^2 = \tfrac{1}{2}(2.0 \times 10^{-3} \text{ kg})v_2{}^2;$$

$$U_1 = \frac{1}{4\pi\epsilon_0} \frac{q_1 q_2}{r_1}$$

$$= (8.99 \times 10^9 \text{ N·m}^2\text{·C}^{-2})(0.10 \times 10^{-6} \text{ C})(2.0 \times 10^{-6} \text{ C})/(0.1 \text{ m})$$

$$= 0.0180 \text{ J};$$

$$U_2 = (8.99 \times 10^9 \text{ N·m}^2\text{·C}^{-2})(0.10 \times 10^{-6} \text{ C})(2.0 \times 10^{-6} \text{ C})/(0.2 \text{ m})$$

$$= 0.0090 \text{ J}.$$

From conservation of energy, we have

$$K_1 + U_1 = K_2 + U_2,$$
$$0.025 \text{ J} + 0.0180 \text{ J} = (1.0 \times 10^{-3} \text{ kg})v_2{}^2 + 0.0090 \text{ J},$$

and finally

$$v_2 = 5.83 \text{ m·s}^{-1}.$$

The force is repulsive, and the plastic ball speeds up as it moves away from the stationary charge.

(b) If the moving charge is negative, the force on it is attractive rather than repulsive, and we expect it to slow down rather than speed up. The only difference in the calculations above is that both potential-energy quantities are negative; the conservation-of-energy equation becomes

$$0.025 \text{ J} - 0.0180 \text{ J} = (1.0 \times 10^{-3} \text{ kg})v_2{}^2 - 0.0090 \text{ J},$$

and the final result is

$$v_2 = 4.00 \text{ m·s}^{-1}.$$

We conclude this section with a comment about our treatment of electrical potential energy. We have spoken consistently about the work done *by the electric-field force* on the charged particle moving in the field. In a displacement of the particle from point a to point b, this work, which we denote as $W_{a\rightarrow b}$, is always given by

Relation between work and potential energy for electrical forces

$$W_{a\rightarrow b} = U_a - U_b.$$

When U_a is greater than U_b, this work is positive, and the potential energy decreases during the displacement. In this case, the field does positive work on the particle as the particle "falls" from a point of higher potential energy to a point of lower potential energy. This viewpoint is consistent with the approach of Chapter 7, where we always spoke of the work done on a particle by a gravitational field or by an elastic force.

An alternative viewpoint used in some books is that in order to "raise" a particle from a point b where the potential energy has the value U_b to a point a where it has a greater value U_a, it would be necessary to apply an additional force that *opposes* the electric-field force and does positive work. The potential-energy difference $U_a - U_b$ would then be defined as the work done *by that additional force* during a displacement from b to a. This viewpoint is not wrong, but in our view it may sometimes be confusing. We prefer always to deal with the work done *by the electric-field force*, without introducing any hypothetical additional forces that may or may not be present in a particular problem. We continue to use this viewpoint in the next section, which introduces the concept of *electrical potential*. If you use other books for reference, beware of the possible confusion about these two alternative viewpoints.

Alternative definitions of electrical potential energy: Don't get confused!

26–2 POTENTIAL

Potential is potential energy per unit charge.

In the preceding section we discussed the potential energy U of a charged particle. Now we introduce the more general concept of *potential energy per unit charge*. This quantity is called **potential**. The potential at any point of an electrostatic field is defined as *the potential energy per unit charge* for a test charge q' at that point. Potential is represented by the letter V:

$$V = \frac{U}{q'}, \quad \text{or} \quad U = q'V. \tag{26–10}$$

The volt: a unit of potential

Potential energy and charge are both scalars, so potential is a scalar quantity. Its basic SI unit is one *joule per coulomb* (1 J·C^{-1}). For brevity, a potential of 1 J·C^{-1} is called one **volt** (1 V), named in honor of the Italian scientist Alessandro Volta (1745–1827):

$$1 \text{ V} = 1 \text{ J·C}^{-1}.$$

Commonly used multiples and submultiples of the volt are the kilovolt ($1 \text{ kV} = 10^3 \text{ V}$), the megavolt ($1 \text{ MV} = 10^6 \text{ V}$), the gigavolt ($1 \text{ GV} = 10^9 \text{ V}$), the millivolt ($1 \text{ mV} = 10^{-3} \text{ V}$), and the microvolt ($1 \text{ }\mu\text{V} = 10^{-6} \text{ V}$).

To put Eq. (26–2) on a "work-per-unit-charge" basis, we divide both sides by q', obtaining

$$\frac{W_{a \to b}}{q'} = \frac{U_a}{q'} - \frac{U_b}{q'} = V_a - V_b, \tag{26–11}$$

where $V_a = U_a/q'$ is the potential energy per unit charge at point a, and similarly for V_b. The terms V_a and V_b are called the *potential at point a* and *potential at point b*, respectively. From Eq. (26–9), the potential V at a point due to an arbitrary collection of point charges is given by

$$V = \frac{U}{q'} = \frac{1}{4\pi\epsilon_0} \sum \frac{q_i}{r_i}. \tag{26–12}$$

Potential difference: potential of one point with respect to another

The difference $V_a - V_b$ is called the *potential of a with respect to b*, and is often abbreviated V_{ab}. Note that potential, like electric field, is independent of the test charge q' used to define it.

The potential due to a collection of point charges is usually obtained most easily with Eq. (26–12), but in some problems where the E field is known, it is easier to work directly with the field. The force F on the test charge q' can be written as $F = q'E$. When this relation is applied in Eq. (26–7) and the result combined with Eq. (26–11), we obtain the useful result

$$V_{ab} = V_a - V_b = \int_a^b \boldsymbol{E} \cdot d\boldsymbol{l} = \int_a^b E \cos\theta \, dl. \tag{26–13}$$

Potential difference can be represented as a line integral of electric field.

Either integral on the right side is called the **line integral** of E. It represents the conceptual process of dividing a path into small elements dl, multiplying each magnitude dl by the component of E parallel to dl at that point, and summing the results for the entire path. We have already encountered the concept of line integral in our general definition of work in Section 7–3.

Equation (26–13) states that when a positive test charge moves from a region of high potential to one of lower potential (that is, $V_b < V_a$), the electric field does positive work on it. Thus a positive charge tends to "fall" from a high-potential region to a lower-potential one. The opposite is true for a negative charge.

An instrument that measures the difference of potential between two points is called a *voltmeter*. The principle of the common type of moving-coil voltmeter will be described later. There are also much more sensitive potential-measuring devices that utilize electronic amplification. Devices that can measure a potential difference of 1 μV are not unusual.

A voltmeter measures potential difference (voltage).

EXAMPLE 26–2   A particle having a charge $q = 3 \times 10^{-9}$ C moves from point a to point b along a straight line, a total distance $d = 0.5$ m. The electric field is uniform along this line, in the direction from a to b, with magnitude $E = 200$ N·C^{-1}. Determine (a) the force on q; (b) the work done on it by the field; and (c) the potential difference $V_a - V_b$.

SOLUTION

a)   The force is in the same direction as the electric field, and its magnitude is given by

$$F = qE = (3 \times 10^{-9} \text{ C})(200 \text{ N·C}^{-1}) = 600 \times 10^{-9} \text{ N}.$$

b)   The work done by this force is

$$W = Fd = (600 \times 10^{-9} \text{ N})(0.5 \text{ m}) = 300 \times 10^{-9} \text{ J}.$$

c)   The potential difference is the work per unit charge, which is

$$V_a - V_b = \frac{W}{q} = \frac{300 \times 10^{-9} \text{ J}}{3 \times 10^{-9} \text{ C}} = 100 \text{ J·C}^{-1} = 100 \text{ V}.$$

Alternatively, since E is force per unit charge, the work per unit charge is obtained by multiplying E by the distance d:

$$V_a - V_b = Ed = (200 \text{ N·C}^{-1})(0.5 \text{ m}) = 100 \text{ J·C}^{-1} = 100 \text{ V}.$$

EXAMPLE 26–3   Point charges of $+12 \times 10^{-9}$ C and -12×10^{-9} C are placed 10 cm apart, as in Fig. 26–5. Compute the potentials at points a, b, and c.

A simple potential calculation with point charges

SOLUTION   We must evaluate the *algebraic* sum

$$\frac{1}{4\pi\epsilon_0} \sum \frac{q_i}{r_i}$$

at each point. At point a, the potential due to the positive charge is

Potential is a scalar quantity; potentials combine algebraically, not vectorially.

$$(9.0 \times 10^9 \text{ N·m}^2\text{·C}^{-2}) \frac{12 \times 10^{-9} \text{ C}}{0.06 \text{ m}} = 1800 \text{ N·m·C}^{-1}$$

$$= 1800 \text{ J·C}^{-1} = 1800 \text{ V},$$

and the potential due to the negative charge is

$$(9.0 \times 10^9 \text{ N·m}^2\text{·C}^{-2}) \frac{-12 \times 10^{-9} \text{ C}}{0.04 \text{ m}} = -2700 \text{ J·C}^{-1} = -2700 \text{ V}.$$

Hence

$$V_a = 1800 \text{ V} - 2700 \text{ V} = -900 \text{ V}$$

$$= -900 \text{ J·C}^{-1}.$$

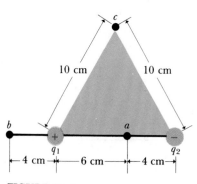

FIGURE 26–5

At point b, the potential due to the positive charge is $+2700$ V, and that due to the negative charge is -770 V. Hence

$$V_b = 2700 \text{ V} - 770 \text{ V} = 1930 \text{ V}$$
$$= 1930 \text{ J·C}^{-1}.$$

At point c the potential is

$$V_c = 1080 \text{ V} - 1080 \text{ V} = 0.$$

EXAMPLE 26–4 Refer to Fig. 26–5. Compute the potential energy of a point charge of $+4 \times 10^{-9}$ C if placed at points a, b, and c.

SOLUTION First, for any point charge q,

$$U = qV.$$

Hence at point a,

$$U = qV_a = (4 \times 10^{-9} \text{ C})(-900 \text{ J·C}^{-1})$$
$$= -36 \times 10^{-7} \text{ J}.$$

At point b,

$$U = qV_b$$
$$= (4 \times 10^{-9} \text{ C})(1930 \text{ J·C}^{-1})$$
$$= 77 \times 10^{-7} \text{ J}.$$

At point c,

$$U = qV_c = 0.$$

(All are relative to a point at infinity.)

EXAMPLE 26–5 In Fig. 26–6, a particle having mass $m = 5$ g and charge $q' = 2 \times 10^{-9}$ C starts from rest at point a and moves in a straight line to point b. What is its speed v at point b?

Conservation of energy applied to motion of a charged particle

SOLUTION Conservation of energy gives

$$K_a + U_a = K_b + U_b.$$

For this situation, $K_a = 0$ and $K_b = \frac{1}{2}mv^2$. The potential energies are given in terms of the potentials by Eq. (26–10): $U_a = q'V_a$ and $U_b = q'V_b$. Thus the energy equation becomes

$$0 + q'V_a = \frac{1}{2}mv^2 + q'V_b.$$

When solved for v, this yields

$$v = \sqrt{\frac{2q'(V_a - V_b)}{m}}.$$

We obtain the potentials just as we did in the preceding examples:

$$V_a = (9.0 \times 10^9 \text{ N·m}^2\text{·C}^{-2}) \left(\frac{3 \times 10^{-9} \text{ C}}{0.01 \text{ m}} + \frac{-3 \times 10^{-9} \text{ C}}{0.02 \text{ m}} \right) = 1350 \text{ V},$$

$$V_b = (9.0 \times 10^9 \text{ N·m}^2\text{·C}^{-2}) \left(\frac{3 \times 10^{-9} \text{ C}}{0.02 \text{ m}} + \frac{-3 \times 10^{-9} \text{ C}}{0.01 \text{ m}} \right) = -1350 \text{ V}.$$

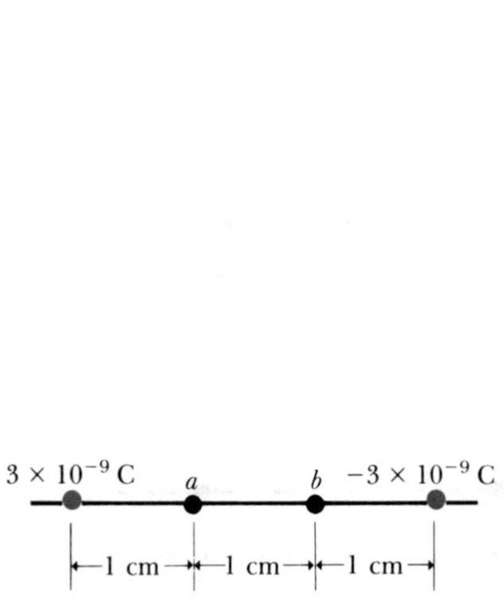

3×10^{-9} C a b -3×10^{-9} C

$\longmapsto$ 1 cm $\longmapsto$ 1 cm $\longmapsto$ 1 cm $\longmapsto$

FIGURE 26–6

Finally,

$$v = \sqrt{\frac{2(2 \times 10^{-9}\text{ C})(2700\text{ V})}{5 \times 10^{-3}\text{ kg}}} = 4.65 \times 10^{-2}\text{ m·s}^{-1} = 4.65\text{ cm·s}^{-1}.$$

We can check consistency of units by noting that $1\text{ V} = 1\text{ J·C}^{-1}$, and thus the numerator under the radical has units of J or $\text{kg·m}^2\text{·s}^{-2}$.

26–3 CALCULATION OF POTENTIALS

In principle we can always calculate the potential at a point by applying Eq. (26–12). We can then use the results of such calculations to find the potential *difference* between two points. An alternative approach, especially convenient when the electric field can be obtained easily, is to use Eq. (26–13) to calculate the needed potential difference from the known electric field. In some problems a combination of these two approaches is best. The following examples illustrate these comments.

Several different methods can be used to calculate potentials.

PROBLEM-SOLVING STRATEGY: *Potential calculations*

1. When you are dealing with a given continuous charge distribution, devise a way to divide it into infinitesimal elements identified by coordinates and coordinate differentials. For example, for a line charge along the *y*-axis, choose a segment of length dy. Next, determine how much charge dQ is contained in that element. If the total length is L and the charge is distributed uniformly, then $dy/L = dQ/Q$, where Q is the total charge. Then use Eq. (26–12), expressing the sum as an integral. Carry out the integration, using appropriate limits to include the entire charge distribution. Be careful about which geometric quantities vary in the integral and which are constant.

2. If the electric field is given or can be obtained with little effort, it is usually easier not to follow the above procedure but instead to use Eq. (26–13). When appropriate, make use of your freedom to define V to be zero at any convenient place. For point charges this will usually be at infinity, but for other distributions of charge (especially those that extend to infinity themselves) it is often necessary to set $V = 0$ at some finite distance from the charge distribution, say at point b. Then the potential at any other point, say a, can be found from Eq. (26–13), with $V_b = 0$.

3. Remember that potential is a *scalar* quantity, not a *vector* quantity. Don't get carried away and try to use components. A scalar quantity does not have components, and to try to use components would be wrong.

EXAMPLE 26–6 *Charged spherical conductor.* Consider a solid conducting sphere of radius R, with total charge q. From Gauss's law, as discussed in Chapter 25, we conclude that at all points *outside* the sphere, the field is the same as that of a point charge q at the center of the sphere. *Inside* the sphere the field is zero everywhere; otherwise charge would move within the sphere.

As mentioned above, it is convenient to take the reference level of potential (the point where $V = 0$) at a very large distance from all charges. Since the field at any point r greater than R is the same as for a point charge, the work done by the **E** field on a test charge moving from a finite value of r to infinity is also the same as for a point charge. Thus the potential V at a radial distance r, relative to a point at

Potential of a charged conducting sphere: Outside the sphere, the potential is the same as for a point charge.

infinity, is

$$V = \frac{1}{4\pi\epsilon_0} \frac{q}{r}. \qquad (26\text{--}14)$$

The potential is positive if q is positive, negative if q is negative.

Equation (26–14) applies to the field of a charged spherical conductor only when r is greater than or equal to the radius R of the sphere. The potential at the surface is

$$V = \frac{1}{4\pi\epsilon_0} \frac{q}{R}. \qquad (26\text{--}15)$$

Inside the sphere the field is zero everywhere, and no work is done on a test charge displaced from any point to any other point in this region. Thus the potential is the same at all points inside the sphere and is equal to its value $q/4\pi\epsilon_0 R$ at the surface. The field and potential are shown as functions of r in Fig. 26–7. The electric field $\boldsymbol{E}$ at the surface has magnitude

> **Inside the sphere, the potential is constant and the field is zero.**

$$E = \frac{1}{4\pi\epsilon_0} \frac{q}{R^2}. \qquad (26\text{--}16)$$

> **Air becomes conductive when very strong electric fields are present.**

Here is a simple practical application of this result. The maximum potential to which a conductor in air can be raised is limited by the fact that air molecules become ionized, and hence the air becomes a conductor, at an electric-field magnitude of about 0.8×10^6 N·C^{-1}. Comparing Eqs. (26–15) and (26–16), we note that at the surface of a conducting sphere the field and potential are related by $V = RE$. Thus if E_m represents electric-field magnitude at which air becomes conductive (known as the *dielectric strength* of air), the maximum potential V_m to which a spherical conductor can be raised is

$$V_m = RE_m.$$

For a sphere 1 cm in radius, in air,

$$V_m = (10^{-2}\ \text{m})(0.8 \times 10^6\ \text{N·C}^{-1}) = 8000\ \text{V},$$

and no amount of "charging" could raise the potential of a sphere of this size, in air, higher than about 8000 V. This fact necessitates the use of large spherical

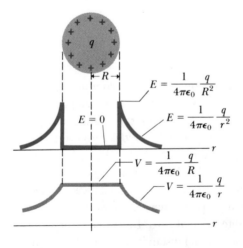

26–7 Electric-field magnitude E and potential V at points inside and outside a charged spherical conductor.

terminals on high-voltage machines. If we make $R = 2$ m, then

$$V_m = (2 \text{ m})(0.8 \times 10^6 \text{ N·C}^{-1}) = 1.6 \times 10^6 \text{ V} = 1.6 \text{ MV}.$$

At the other extreme is the effect produced by a surface of very *small* radius of curvature, such as a sharp point. Since the maximum potential is proportional to the radius, even relatively small potentials applied to sharp points in air will produce sufficiently high fields just outside the point to result in ionization of the surrounding air.

EXAMPLE 26–7 *Parallel plates.* We discussed the electric field between two parallel plates at the beginning of Section 26–1 and in Fig. 26–1. To obtain the potential difference between points y and b, also called the potential of y with respect to b, we note that the force on a test charge q' is $q'E$. The work done on the charge by the field during a displacement from y to b is

Finding the potential between two parallel plates.

$$W_{y \to b} = q'Ey. \qquad (26\text{--}17)$$

Hence the potential difference, or work per unit charge, is

$$V_y - V_b = Ey. \qquad (26\text{--}18)$$

The potential decreases linearly with y as we move from the upper to the lower plate. At point a, where $y = d$ and $V_y = V_a$,

$$V_a - V_b = Ed,$$

and

$$E = \frac{V_a - V_b}{d} = \frac{V_{ab}}{d}. \qquad (26\text{--}19)$$

That is, *the electric field equals the potential difference between the plates divided by the distance between them.* This equation is a more useful expression for E than is Eq. (25–11), because the potential difference V_{ab} can be measured easily with a voltmeter. There are no instruments that read surface charge density directly.

Equation (26–19) also shows that the unit of electric field can be expressed as one *volt per meter* (1 V·m^{-1}), as well as 1 N·C^{-1}. In practice, the volt per meter is the most commonly used unit of E.

EXAMPLE 26–8 *Line charge and charged conducting cylinder.* We found in Example 25–13 (Section 25–5) that the field at a distance r from a long straight-line charge or a long charged conducting cylinder is given by

Finding the potential due to a line charge distribution.

$$E = \frac{1}{2\pi\epsilon_0} \frac{\lambda}{r},$$

where λ is the charge per unit length.

The potential of any point a with respect to any other point b, at radial distances r_a and r_b from the line of charge, is

$$V_a - V_b = \frac{\lambda}{2\pi\epsilon_0} \int_{r_a}^{r_b} \frac{dr}{r} = \frac{\lambda}{2\pi\epsilon_0} \ln \frac{r_b}{r_a}. \qquad (26\text{--}20)$$

If we take point b at infinity and set $V_b = 0$, we find for the potential V_a,

$$V_a = \frac{\lambda}{2\pi\epsilon_0} \ln \frac{\infty}{r_a} = \infty.$$

This does not make sense, and we are forced to conclude that we cannot use a reference point at infinity for this field! We can, however, set $V = 0$ at some arbitrary radius r_0. Then at any radius r,

$$V = \frac{\lambda}{2\pi\epsilon_0} \ln \frac{r_0}{r}. \tag{26–21}$$

Equations (26–20) and (26–21) give the potential in the field of a cylinder only for values of r equal to or greater than the radius R of the cylinder. If r_0 is taken as the cylinder radius R, so that the potential at the surface of the cylinder is zero, the potential at any external point, relative to that of the cylinder, is

$$V = \frac{\lambda}{2\pi\epsilon_0} \ln \frac{R}{r}, \tag{26–22}$$

where r is the distance from the axis of the cylinder.

EXAMPLE 26–9 Electric charge is distributed uniformly around a thin ring of radius a, with total charge Q, as shown in Fig. 26–8. Find the potential at a point along the line perpendicular to the plane of the ring, through its center, at a point P a distance x from the center of the ring.

SOLUTION This situation is the same as in Section 25–2 (Example 25–7). Referring back to that example, we note that the entire charge is at a distance $r = (x^2 + a^2)^{1/2}$ from point P. We conclude immediately that the potential at point P, which is a function of x, is given by

$$V(x) = \frac{1}{4\pi\epsilon_0} \frac{Q}{\sqrt{x^2 + a^2}}. \tag{26–23}$$

Potential is a *scalar* quantity; there is no need to consider components of vectors in this calculation, as we had to do in finding the electric field at P. Hence the potential calculation in this case is much simpler than the field calculation. Keep this comment in mind; we will return to it later in this chapter. Also note that when x is much larger than a, Eq. (26–23) becomes approximately equal to

$$V(x) = \frac{1}{4\pi\epsilon_0} \frac{Q}{x},$$

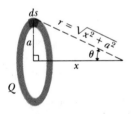

26–8 Potential at point P due to a ring of charge.

corresponding to the potential of a point charge Q at distance x; when we are far away from a ring, its charge looks like a point charge.

EXAMPLE 26–10 Electric charge is distributed uniformly along a thin rod of length $2a$, with total charge Q. Find the potential at a point P along the perpendicular bisector of the rod, at a distance x from its center.

SOLUTION This situation is the same as in Example 24–5 (Section 24–6) and Example 25–8 (Section 25–2). In Fig 25–7 the element of charge dQ corresponding to an element dy on the rod is again given by $dQ = (Q/2a)dy$. The distance from dQ to P is $(x^2 + y^2)^{1/2}$, and the contribution it makes to the potential at P is

$$dV = \frac{1}{4\pi\epsilon_0} \frac{Q}{2a} \frac{dy}{\sqrt{x^2 + y^2}}.$$

The potential at P is obtained by integrating this over the length of the rod, from

$y = -a$ to $y = a$. The integral may be found in a table of integrals; the result is

$$V(x) = \frac{1}{4\pi\epsilon_0} \frac{Q}{2a} \int_{-a}^{a} \frac{dy}{\sqrt{x^2 + y^2}} = \frac{1}{4\pi\epsilon_0} \frac{Q}{2a} \ln \frac{\sqrt{a^2 + x^2} + a}{\sqrt{a^2 + x^2} - a}. \quad (26\text{–}24)$$

Again note that this problem is simpler than the calculation of E at point P because potential is a scalar quantity and no vector calculations are involved.

26–4 EQUIPOTENTIAL SURFACES

The potential at various points in an electric field may be represented graphically by **equipotential surfaces.** An equipotential surface is a surface for which the potential has the same value at all points on the surface. An equipotential surface may be constructed through every point of an electric field, but it is usual to show only a few equipotentials in a diagram.

Equipotential surface: a surface where the potential is constant

Since the potential energy of a charged body is the same at all points of a given equipotential surface, it follows that no work is done by the E field when a charged body moves over such a surface. Hence the equipotential surface through any point must be at right angles to the direction of the field at that point. If this were not so, the field would have a component tangent to the surface, and work would be done by the electric-field force when a charge moved in the direction of this component. Thus field lines and equipotential surfaces are mutually perpendicular. In general, the field lines of a field are curves, and the equipotentials are curved surfaces. For the special case of a uniform field, where the field lines are straight and parallel, the equipotentials are parallel planes perpendicular to the field lines.

Figure 26–9 shows several arrangements of charges. The field lines are represented by colored lines, and cross sections of the equipotential surfaces are shown as black lines. The actual field is, of course, three-dimensional. At each crossing of an equipotential and a field line, the two are perpendicular.

When charges at rest reside on the surface of a conductor, the electric field just outside the conductor must be perpendicular to the surface at every point. That is, at the surface there can never be a component of E *parallel* to the surface. If such a component did exist, it would do work on a charge moving close to and parallel to the surface. But then the charge could go through the surface and return to its starting point by a path close to the surface but inside the conductor. No work would be done in the return trip because E is zero everywhere inside the conductor. Thus when it returned to the starting point, the total work would not be zero; this statement contradicts the general principle that an electrostatic field is always conservative. We conclude that a parallel component of E at the surface is an impossibility.

The electric field just outside a conducting surface is always perpendicular to the surface.

It also follows that when a test charge is moved from point to point along the surface of a conductor, the electric field does no work on the charge. Hence when all charges are at rest, *a conducting surface is always an equipotential surface.* These principles are shown in Fig. 26–10.

Finally, we can now prove a theorem that we quoted without proof in Section 25–6. The theorem is as follows: In an electrostatic situation, if a conductor contains a cavity, and if no charge is present inside the cavity, then there can be no charge *anywhere* on the surface of the cavity. To prove this theorem, we first prove that *every point in the cavity is at the same potential.* In Fig.

When all charges are at rest, a conducting surface is always an equipotential surface.

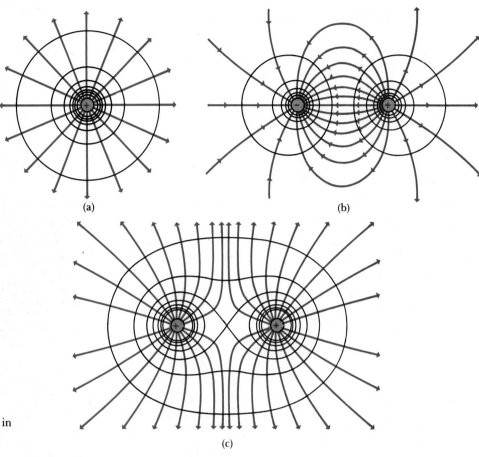

(a)   (b)

(c)

26-9 Equipotential surfaces (black lines) and field lines (colored lines) in the neighborhood of point charges.

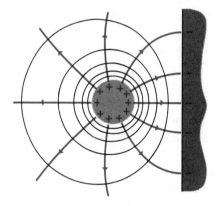

26-10 Field lines always meet charged conducting surfaces at right angles.

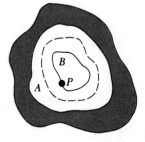

26-11 A cavity in a conductor. If the cavity contains no charge, every point in the cavity is at the same potential, the electric field is zero everywhere, and there is no charge on the surface of the cavity.

26-11, the surface A of the cavity is an equipotential surface, as we have just proved. Suppose point P in the cavity is at a different potential; then we can construct a different equipotential surface B including point P.

Now consider a Gaussian surface, shown as a broken line, between the two equipotential surfaces. Because of the relation between E and the equipotentials, we know that the field at every point between the equipotentials is from A toward B, or else at every point it is from B toward A, depending on which equipotential surface is at higher potential. In either case, the flux through this Gaussian surface is certainly not zero. But then Gauss's law says that the charge enclosed by the Gaussian surface cannot be zero. This conclusion contradicts our initial assumption that there is *no* charge in the cavity. So *P cannot* be at a different potential from the cavity wall.

We have proved that the entire region of the cavity is at the same potential. For this to be true, however, the electric field inside the cavity must be zero everywhere. Finally, because the electric field at any point on the surface of a conductor is proportional to the surface charge density σ at that point, according to Eq. (25–26), we conclude that *the surface charge density on the wall of the cavity is zero at every point.* This chain of reasoning may seem tortuous, but it is worth careful study.

26–5 POTENTIAL GRADIENT

Electric field and potential are closely related. Equation (26–13) expresses one aspect of that relationship; if we know E at various points, we can use Eq. (26–13) to calculate potential differences. Conversely, if we know the potential V at various points, we ought to be able to use it to determine E. This converse problem is the subject of this section. Regarding V as a function of the coordinates (x, y, z) of a point in space, we will show that the components of E are directly related to the *derivatives* of V with respect to x, y, and z.

In Eq. (26–13), $V_{ab} = V_a - V_b$ is the potential of a with respect to b, that is, the change of potential when a point moves from b to a. This may be written as

$$V_{ab} = \int_b^a dV = -\int_a^b dV,$$

where dV is the infinitesimal change of potential accompanying a small displacement dl. As usual, $E \cos \theta$ is the component of E parallel to dl. We may call this $E_\parallel$, then, using the above expression for V_{ab}, we may rewrite Eq. (26–13) as

$$-\int_a^b dV = \int_a^b E_\parallel \, dl.$$

Since these two integrals are equal for any pair of limits a and b, the *integrands* must be equal. Hence for any *infinitesimal* displacement dl,

$$-dV = E_\parallel \, dl,$$

or

$$E_\parallel = -\frac{dV}{dl}. \tag{26–25}$$

The derivative dV/dl is the rate of change of V for a displacement in the direction of dl. In particular, if dl is parallel to the x-axis, then the component of E parallel to dl is just the x-component of E; that is, E_x. Thus $E_x = -dV/dx$. Because V is, in general, also a function of y and z, we use partial derivative notation; a derivative in which only x varies is written $\partial V/\partial x$. The y- and z-components of E are related to the corresponding derivatives of V in the same way, so we have

$$E_x = -\frac{\partial V}{\partial x}, \qquad E_y = -\frac{\partial V}{\partial y}, \qquad E_z = -\frac{\partial V}{\partial z}. \tag{26–26}$$

Potential gradient: how to find the field when the potential is known

Derivatives of V in various directions

In terms of unit vectors, E may be written as

$$E = -\left(i\frac{\partial V}{\partial x} + j\frac{\partial V}{\partial y} + k\frac{\partial V}{\partial z}\right)$$

$$= -\left(i\frac{\partial}{\partial x} + j\frac{\partial}{\partial y} + k\frac{\partial}{\partial z}\right)V. \tag{26-27}$$

In vector notation the operation

$$i\frac{\partial}{\partial x} + j\frac{\partial}{\partial y} + k\frac{\partial}{\partial z}$$

The gradient: a vector differential operation

is called the **gradient** and is denoted by the symbol ∇, pronounced "del." Thus in vector notation, Eqs. (26–26) are summarized compactly as

$$E = -\nabla V. \tag{26-28}$$

This is read "E is the negative of the gradient of V," or "E equals negative del V."

Note again that the unit of electric field can be expressed in either of these equivalent forms:

$$1 \text{ V·m}^{-1} = 1 \text{ N·C}^{-1}.$$

EXAMPLE 26–11 We have shown that the potential at a radial distance r from a point charge q is

$$V = \frac{1}{4\pi\epsilon_0}\frac{q}{r}.$$

By symmetry, the electric field is in the radial direction, so

$$E = E_r = -\frac{dV}{dr} = -\frac{d}{dr}\left(\frac{1}{4\pi\epsilon_0}\frac{q}{r}\right) = \frac{1}{4\pi\epsilon_0}\frac{q}{r^2},$$

in agreement with Coulomb's law.

Some examples of the relation of potential to field

EXAMPLE 26–12 In Example 26–8 we found that the potential outside a charged conducting cylinder of radius R and charge per unit length λ is

$$V = \frac{\lambda}{2\pi\epsilon_0}\ln\frac{R}{r} = \frac{\lambda}{2\pi\epsilon_0}(\ln R - \ln r).$$

The electric field is radial, and its magnitude is given by

$$E = -\frac{dV}{dr} = \frac{\lambda}{2\pi\epsilon_0 r},$$

in agreement with our previous result.

EXAMPLE 26–13 In Example 26–9 we found that for a ring of charge with radius a and total charge Q, the potential at a point P, a distance x from the center of the ring on a line through the center perpendicular to the plane of the ring, is

$$V(x) = \frac{1}{4\pi\epsilon_0}\frac{Q}{\sqrt{x^2 + a^2}}.$$

From Eq. (26–26),

$$E_x = -\frac{\partial V}{\partial x} = \frac{1}{4\pi\epsilon_0} \frac{Qx}{(x^2 + a^2)^{3/2}}.$$

This expression agrees with the result we obtained in Section 25–2 (Example 25–7).

In this example, V is not a function of y, but it would *not* be correct to conclude that $\partial V/\partial y = 0$ and $E_y = 0$. The reason is that our expression for V is valid only for points on the x-axis. If we had the complete form of V, valid for all points in space, then we could use it to find $E_y = -\partial V/\partial y$ at any point, and so on.

EXAMPLE 26–14   A charge Q is uniformly distributed along a rod of length $2a$. In Example 26–10 (Section 26–3) we derived an expression for the potential at a point on the perpendicular bisector of the rod, at a distance x from its center:

$$V(x) = \frac{1}{4\pi\epsilon_0} \frac{Q}{2a} \ln \frac{\sqrt{a^2 + x^2} + a}{\sqrt{a^2 + x^2} - a}.$$

The negative of the derivative of this expression with respect to x is E_x. We challenge you to carry out the differentiation and show that the result is the same expression we found in Examples 24–5 (Section 24–6) and 25–8 (Section 25–2) by direct integration.

Several of these examples illustrate an important point. Often we can compute the electric field caused by a charge distribution in two ways; either directly, as in Section 25–2, or by first calculating the potential and then taking its gradient to find the field. The second method is often easier because potential is a *scalar* quantity, requiring at worst the integration of a scalar function, while the electric field is a *vector* quantity, requiring computation of components for each element of charge and a separate integration for each component. Thus, quite apart from its fundamental significance, potential offers a very useful computational technique in field calculations.

> Potential is a scalar, but electric field is a vector; that's why it's often easier to calculate potential.

Another important point is the relation of the *direction* of $\boldsymbol{E}$ to the behavior of V. Specifically, $\boldsymbol{E}$ always points in the direction in which V *decreases* most rapidly. In all these examples, the potential decreases as we move away from the charge distribution (assuming the charge is positive), and the electric field points away from it. If the charge is negative, the potential increases algebraically (becoming less negative) as we move away from the charge, and $\boldsymbol{E}$ points *toward* the charge. The situation is completely analogous to gravitational potential energy, which we studied in Section 7–5. Near the surface of the earth, for example, the gravitational field points down, the direction of most rapid decrease of gravitational potential energy.

26–6 THE MILLIKAN OIL-DROP EXPERIMENT

We can now describe one of the classic physics experiments of all time, the **Millikan oil-drop experiment.** In a brilliant series of investigations carried out at the University of Chicago in the period 1909–1913, Robert Andrews Millikan not only demonstrated conclusively the discrete nature of electric charge but actually measured the charge of an individual electron.

> How can you measure the charge of a single electron?

Millikan's apparatus is shown schematically in Fig. 26–12a. Two parallel horizontal metal plates, A and B, are insulated from each other and separated by a few millimeters. Oil is sprayed in fine droplets from an atomizer above the upper plate, and a few droplets are allowed to fall through a small hole in this plate. A beam of light is directed horizontally between the plates, and a telescope is set up with its axis at right angles to the light beam. The oil drops, illuminated by the light beam and viewed through the telescope, appear like tiny bright stars, falling slowly with a terminal velocity determined by their weight and by the viscous air-resistance force opposing their motion.

Some of the oil droplets are electrically charged because of frictional effects. The drops can also be charged if the air in the apparatus is ionized by x-rays or a bit of radioactive material. Some of the electrons or ions then collide with the drops and stick to them. The drops are usually negatively charged, but occasionally one with a positive charge is found.

In principle, the simplest method for measuring the charge on a drop is as follows. Suppose a drop has a negative charge and the plates are maintained at a potential difference such that a downward electric field of magnitude E ($= V_{AB}/l$) is set up between them. The forces on the drop are then its weight mg and the upward force qE. By adjusting the field E, we can make qE equal to mg, so that the drop remains at rest, as indicated in Fig. 26–10b. Under these circumstances,

$$q = \frac{mg}{E}.$$

The mass of the drop equals the product of its density ρ and its volume, $4\pi r^3/3$, and $E = V_{AB}/l$, so

$$q = \frac{4\pi}{3}\frac{\rho r^3 g l}{V_{AB}}. \tag{26–29}$$

All the quantities on the right can be measured easily, with the exception of the drop radius r, which is of the order of 10^{-5} cm and is much too small to be measured directly. It can be calculated, however, by cutting off the electric field and measuring the terminal velocity v_T of the drop as it falls through a known distance d defined by reference lines in the eyepiece of the telescope.

At the terminal velocity, the weight mg is just balanced by the viscous force f. The viscous force on a sphere of radius r, moving with a velocity v through a fluid of viscosity η, is given by Stokes's law, discussed in Section 13–8, Eq. (13–29):

$$f = 6\pi\eta rv.$$

If Stokes's law applies, and the drop is falling with its terminal velocity v_T, then

$$mg = f,$$

$$\tfrac{4}{3}\pi r^3 \rho g = 6\pi\eta rv_T,$$

and

$$r = 3\sqrt{\eta v_T/2\rho g}.$$

When this expression for r is inserted into Eq. (26–29), we find

$$q = 18\pi\frac{l}{V_{AB}}\sqrt{\frac{\eta^3 v_T^3}{2\rho g}}, \tag{26–30}$$

which expresses the charge q in terms of measurable quantities.

Studying the motion of electrically charged oil drops

Free fall of oil drops in air: The viscous force just balances the weight.

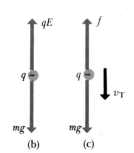

26–12 (a) Schematic diagram of Millikan apparatus. (b) Forces on a drop at rest. (c) Forces on a drop falling with its terminal velocity v_T.

In actual practice this procedure is modified somewhat. To correct for the buoyant force of the air through which the drop falls, the density ρ of the oil should be replaced by $(\rho - \rho_g)$, where ρ_g is the density of air. A correction to Stokes's law is also required, because air is not a continuous fluid but a collection of molecules separated by distances that are of the same order of magnitude as the dimensions of the drops.

Correcting for buoyancy effects of the drop in air

Millikan and his coworkers measured the charges of thousands of drops and found that, within the limits of their experimental error, every drop had a charge equal to some small integer multiple of a basic charge e. That is, drops were observed with charges of e, $2e$, $3e$, and so on, but never with such values as $0.76e$ or $2.49e$. The evidence is conclusive that electric charge is not something that can be divided indefinitely, but that it exists in nature only in units of magnitude e. When a drop is observed with charge e, we conclude it has acquired one extra electron; if its charge is $2e$, it has two extra electrons, and so on. As we stated in Section 24–5, the best experimental value of the charge e is

The charge of the electron: a fundamental constant of nature

$$e = 1.602192 \times 10^{-19}\,\text{C} \cong 1.60 \times 10^{-19}\,\text{C}.$$

The quark theory of fundamental particle structure, discussed in Section 44–11, incorporates particles called *quarks* having fractional charges $\pm e/3$ and $\pm 2e/3$. Physicists have conflicting views as to whether properties (such as charge) of an individual quark should be directly observable.

26–7 THE ELECTRONVOLT

The change in electrical potential energy of a particle having a charge q, when it moves from a point where the potential is V_a to a point where the potential is V_b, is

$$\Delta U = q(V_b - V_a) = qV_{ba}$$

In particular, if the charge q equals the electron charge $e = 1.60 \times 10^{-19}$ C, and the potential difference $V_{ba} = 1$ V, the change in energy is

The electronvolt: an electron moving through a potential difference of one volt

$$\Delta U = (1.60 \times 10^{-19}\,\text{C})(1\,\text{V}) = 1.60 \times 10^{-19}\,\text{J}.$$

This quantity of energy is called one **electronvolt** (1 eV):

$$1\,\text{eV} = 1.60 \times 10^{-19}\,\text{J}.$$

Other commonly used units are

$$1\,\text{keV} = 10^3\,\text{eV},$$
$$1\,\text{MeV} = 10^6\,\text{eV},$$
$$1\,\text{GeV} = 10^9\,\text{eV},$$
$$1\,\text{meV} = 10^{-3}\,\text{eV}.$$

The electronvolt is a convenient energy unit when we are dealing with the motions of electrons and ions in electric fields. For a particle having charge e, the change in potential *energy* between two points of the path of the particle, when expressed in electronvolts, is *numerically* equal to the potential difference between the points, in volts. If the charge is some multiple of e, say Ne, the change in potential energy in electronvolts is, numerically, N times the potential difference in volts. For example, if a particle having a charge $2e$ moves

The electronvolt is a convenient unit of energy in fundamental-particle calculations.

between two points for which the potential difference is 1000 V, the change in its potential energy is

$$\Delta U = qV_{ba} = (2)(1.60 \times 10^{-19} \text{ C})(10^3 \text{ V})$$
$$= 3.2 \times 10^{-16} \text{ J}$$
$$= 2000 \text{ eV}.$$

Although the electronvolt was defined above in terms of *potential* energy, energy of *any* form, such as the kinetic energy of a moving particle, can be expressed in terms of the electronvolt. Thus when we speak of a "one-million-volt electron," we mean an electron having a kinetic energy of one million electronvolts (1 MeV).

One of the principles of the special theory of relativity, to be developed in Chapter 40, is that the mass m of a particle is equivalent to a quantity of energy mc^2, where c is the speed of light. The rest mass of an electron is 9.11×10^{-31} kg, and the energy equivalent to this is

$$E_0 = mc^2 = (9.11 \times 10^{-31} \text{ kg})(3.00 \times 10^8 \text{ m·s}^{-1})^2$$
$$= 81.9 \times 10^{-15} \text{ J}.$$

Rest mass in relativity theory, expressed in electronvolts

Since 1 eV = 1.60×10^{-19} J, this is equivalent to

$$E_0 = 511,000 \text{ eV} = 0.511 \text{ MeV}.$$

When the kinetic energy of a particle becomes comparable in magnitude to its rest energy, Newton's laws of motion are not strictly valid and must be replaced by the more general relations of relativistic mechanics, developed in detail in Chapter 40. For example, an electron accelerated through a potential difference of 500 kV acquires a kinetic energy of 500 keV, approximately *equal* to its rest energy. A correct analysis of the motion of the particle requires the use of relativistic mechanics.

26–8 THE CATHODE-RAY TUBE

The cathode-ray tube: the heart and soul of oscilloscopes and television receivers

Figure 26–13 is a schematic diagram of the elements of a **cathode-ray tube.** Such tubes are found in oscilloscopes and computer terminal displays, and the principle of the TV picture tube is similar. A cathode-ray tube uses an electron beam that, before its basic nature was well understood, was called a *cathode-ray* beam. Cathode rays are now known to be electrons.

26–13 Basic elements of a cathode-ray tube.

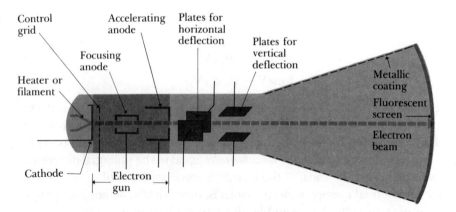

The interior of the tube is highly evacuated. Ordinarily the pressure is of the order of 0.01 Pa (10^{-7} atm); at any greater pressure, collisions of electrons with air molecules would scatter the electron beam excessively. The *cathode* at the left end is raised to a high temperature by the *heater*, and electrons evaporate from the surface of the cathode. The *accelerating anode*, which has a small hole at its center, is maintained at a high positive potential V_1 relative to the cathode, so that there is an electric field, directed from right to left, between the accelerating anode and cathode. This field is confined to the cathode–anode region, and electrons passing through the hole in the anode travel with constant horizontal velocity from the anode to the *fluorescent screen*. The area of impact of the electrons on the screen glows brightly.

The function of the *control grid* is to regulate the number of electrons that reach the anode (and hence the brightness of the spot on the screen). The *focusing anode* ensures that electrons leaving the cathode in slightly different directions are focused down to a narrow beam and all arrive at the same spot on the screen. These two electrodes need not be considered in the following analysis. The complete assembly of cathode, control grid, focusing anode, and accelerating electrode is referred to as an *electron gun*.

Creating an electron beam in a vacuum tube

The accelerated electrons pass between two pairs of *deflecting plates*. An electric field between the first pair of plates deflects the electrons horizontally, and an electric field between the second pair of plates deflects them vertically. If no such fields are present, the electrons travel in a straight line from the hole in the accelerating anode to the *fluorescent screen* and produce a bright spot on the screen where they strike it.

Deflecting the electron beam

To analyze the electron motion, we first calculate the speed v given to the electrons by the electron gun, just as in Example 26–5 (Section 26–2). We assume the electrons leave the cathode with zero initial speed. The electrons actually have some initial speed when they evaporate from the cathode, but it is very small compared to the final speed v and can be neglected. We find

$$v = \sqrt{\frac{2eV_1}{m}}. \qquad (26\text{–}31)$$

As a numerical example, if $V_1 = 2000$ V,

$$v = \sqrt{\frac{2(1.6 \times 10^{-19} \text{ C})(2 \times 10^3 \text{ V})}{9.11 \times 10^{-31} \text{ kg}}}$$
$$= 2.65 \times 10^7 \text{ m·s}^{-1}.$$

Note that the kinetic energy of an electron at the anode depends only on the *potential difference* between anode and cathode, and not at all on the details of the fields within the electron gun or on the shape of the electron trajectory within the gun.

The speeds of electrons in the beam are determined by the accelerating voltage.

If there is no electric field between the plates for horizontal deflection, the electrons enter the region between the other plates with a speed equal to v and represented by v_x in Fig. 26–14. If there is a potential difference V_2 between the plates, and the upper plate is positive, a downward electric field of magnitude $E = V_2/l$ is set up between the plates. A constant upward force eE then acts on the electrons, and their upward acceleration is

$$a_y = \frac{eE}{m} = \frac{eV_2}{ml}. \qquad (26\text{–}32)$$

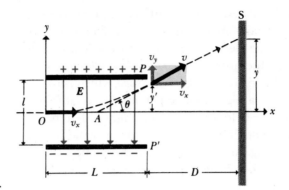

26–14 Electrostatic deflection of an electron beam in a cathode-ray tube.

The *horizontal* component of velocity v_x remains constant, so the time required for the electrons to travel the length L of the plates is

$$t = \frac{L}{v_x}. \qquad (26\text{--}33)$$

In this time, they acquire an upward velocity component given by

$$v_y = a_y t. \qquad (26\text{--}34)$$

Calculating the trajectories of deflected electrons

Combining this equation with Eqs. (26–32) and (26–33), we find

$$v_y = \left(\frac{eV_2}{ml}\right)\frac{L}{v_x}. \qquad (26\text{--}35)$$

When the electrons emerge from the deflecting field, their velocity v makes an angle θ with the x-axis, where

$$\tan\theta = \frac{v_y}{v_x}. \qquad (26\text{--}36)$$

From this point on, the electrons travel in a straight line to the screen. It is not difficult to show that this straight line, if projected backward, intersects the x-axis at a point A that is midway between the end of the plates. Then if y is the vertical component of the point of impact with screen S,

$$\tan\theta = \frac{y}{D + (L/2)}. \qquad (26\text{--}37)$$

Combining this with Eq. (26–36), we find

$$\frac{y}{D + (L/2)} = \frac{v_y}{v_x},$$

or, using Eq. (26–35),

$$y = \left(D + \frac{L}{2}\right)\frac{eV_2 L}{ml v_x{}^2}. \qquad (26\text{--}38)$$

Finally, v_x is given by Eq. (26–31); using this in Eq. (26–38), we obtain

$$y = \left[\frac{L}{2l}\left(D + \frac{L}{2}\right)\right]\frac{V_2}{V_1}. \qquad (26\text{--}39)$$

The term in brackets is a purely geometrical factor. If the accelerating voltage V_1 is constant, *the deflection y is proportional to the deflecting voltage V_2.*

If a field is set up between the *horizontal* deflecting plates, the beam is also deflected in the horizontal direction, perpendicular to the plane of Fig. 26–14. The coordinates of the luminous spot on the screen are then proportional, respectively, to the horizontal and vertical deflecting voltages. This is the principle of the *cathode-ray oscilloscope*. If the horizontal deflection voltage sweeps the beam from left to right at a uniform rate, the beam traces out a graph of the vertical voltage as a function of time. Oscilloscopes are extremely useful laboratory instruments in many areas of pure and applied science.

The oscilloscope: a very useful laboratory instrument

The picture tube in a television set is similar, but the beam is deflected by magnetic fields (to be discussed in later chapters) rather than electric fields. The electron beam traces out the area of the picture 30 times per second in an array of 525 horizontal lines, as the intensity of the beam is varied to make bright and dark areas on the screen. The accelerating voltage in TV picture tubes (V_1 in the preceding discussion) is typically about 20 kV. Computer terminal displays operate on the same principle, using a magnetically deflected electron beam to trace out images on a fluorescent screen. In this context the device is called a CRT (cathode-ray tube) display or a VDT (video display terminal).

Television picture tubes: a familiar example of cathode-ray tubes

SUMMARY

The electric field caused by any collection of charges at rest is always a conservative force field. Thus the work W done by the electric-field force on a charged particle moving in a field can always be represented by a potential-energy function U:

$$W_{a\to b} = U_a - U_b. \tag{26–2}$$

The potential energy of a charge q' in the electric field of a collection of charges q_i is given by

$$U = \frac{q'}{4\pi\epsilon_0} \sum \frac{q_i}{r_i}, \tag{26–9}$$

where r_i is the distance from q_i to q'. This expression assumes that $U = 0$ at a point an infinite distance from all the charges.

Potential is potential energy per unit charge; it is denoted by V. The potential at any point due to a collection of charges q_i is given by

$$V = \frac{U}{q'} = \frac{1}{4\pi\epsilon_0} \sum \frac{q_i}{r_i}. \tag{26–12}$$

The potential difference between two points a and b, also called the potential of a with respect to b, is given by the line integral of $\boldsymbol{E}$:

$$V_{ab} = V_a - V_b = \int_a^b \boldsymbol{E} \cdot d\boldsymbol{l} = \int_a^b E \cos\theta\, dl. \tag{26–13}$$

Potentials can be calculated either by evaluating the sum in Eq. (26–12) or by first finding $\boldsymbol{E}$ and then using Eq. (26–13).

KEY TERMS

potential energy
conservative force field
potential
volt
line integral
equipotential surfaces
gradient
Millikan oil-drop experiment
electronvolt
cathode-ray tube

An equipotential surface is a surface such that the potential has the same value at every point on the surface. When a field line crosses an equipotential surface, the two are perpendicular at the point of intersection; that is, the tangent to the field line is perpendicular to the tangent plane to the equipotential surface at the point of intersection.

When all charges are at rest, the surface of a conductor is always an equipotential surface, and all points in the interior of a conductor are at the same potential. When a cavity within a conductor contains no charge, the entire cavity is an equipotential region, and there is no surface charge anywhere on the surface of the cavity.

If the potential is known as a function of the spatial coordinates x, y, and z, the electric field E at any point is given by

$$E = -\left(i\frac{\partial V}{\partial x} + j\frac{\partial V}{\partial y} + k\frac{\partial V}{\partial z}\right) = -\left(i\frac{\partial}{\partial x} + j\frac{\partial}{\partial y} + k\frac{\partial}{\partial z}\right)V, \quad (26\text{--}27)$$

which is the negative of the gradient of V:

$$E = -\nabla V. \quad (26\text{--}28)$$

Two equivalent sets of units for electric-field magnitude are the volt per meter ($V\cdot m^{-1}$) and the newton per coulomb ($N\cdot C^{-1}$). One volt is one joule per coulomb ($1\ V = 1\ J\cdot C^{-1}$).

The Millikan oil-drop experiment measured the electric charge of individual electrons by measuring the motion of electrically charged oil drops in an electric field. The size of the drop is determined by measuring its speed of free fall under gravity and the viscous force of air resistance given by Stokes's law.

The electronvolt, abbreviated eV, is the energy corresponding to a particle with a charge equal to that of the electron moving through a potential difference of one volt. The conversion factor is

$$1\ eV = 1.60 \times 10^{-19}\ J.$$

The cathode-ray tube uses an electron beam created by a set of electrodes, collectively called an electron gun, and deflected by two sets of deflecting plates. The beam strikes a fluorescent screen and forms an image on it. Such tubes are used in television sets, oscilloscopes, and video display terminals (VDT) for computers.

QUESTIONS

26–1 Are there cases in electrostatics where a conducting surface is *not* an equipotential surface? If so, give an example.

26–2 If the electrical potential at a single point is known, can the electric field at that point be determined?

26–3 If two points are at the same potential, is the electric field necessarily zero everywhere between them?

26–4 If the electric field is zero throughout a certain region of space, is the potential also zero in this region? If not, what *can* be said about the potential?

26–5 A student said: "Since electrical potential is always proportional to potential energy, why bother with the concept of potential at all?" How would you respond?

26–6 Is potential gradient a scalar quantity or a vector quantity?

26–7 A conducting sphere is to be charged by bringing in positive charge, a little at a time, until the total charge is Q. The total work required for this process is alleged to be proportional to Q^2. Is this correct? Why or why not?

26–8 The potential (relative to a point at infinity) midway between two charges of equal magnitude and opposite sign is zero. Can you think of a way to bring a test charge from infinity to this midpoint in such a way that no work is done in any part of the displacement?

26–9 A high-voltage dc power line falls on a car, so the entire metal body of the car is at a potential of 10,000 V with respect to the ground. What happens to the occupants

a) when they are sitting in the car?

b) when they step out of the car?

26–10 In electronics it is customary to define the potential of ground (thinking of the earth as a large conductor) as zero. Is this consistent with the fact that the earth has a net electric charge that is not zero? (Cf. Problem 25–18.)

26–11 A positive point charge is placed near a very large conducting plane. A professor of physics asserted that the

field caused by this configuration is the same as would be obtained by removing the plane and placing a negative point charge of equal magnitude in the mirror-image position behind the initial position of the plane. Is this correct?

26–12 It is easy to produce a potential of several thousand volts on your body by scuffing your shoes across a nylon carpet; yet contact with a power line of comparable voltage would probably be fatal. What is the difference?

EXERCISES

Section 26–1 Electrical Potential Energy.

26–1 A point charge with charge $Q = +4.0 \ \mu C$ is held fixed at the origin.

a) A second point charge $q = -0.5 \ \mu C$ and mass 3×10^{-4} kg is placed on the x-axis, 0.8 m from the origin. What is the electrical potential energy of the pair of charges?

b) The second point charge is released at rest. What is its speed when it is 0.4 m from the origin?

26–2 Three equal point charges of 3×10^{-7} C are placed at the corners of an equilateral triangle whose side is 1 m. What is the potential energy of the system? Take as zero potential the energy of the three charges when they are infinitely far apart.

Section 26–2 Potential

26–3 There is a uniform electric field of magnitude E and directed in the positive x-direction. Consider point a at $x = 0.8$ m and point b at $x = 1.2$ m. The potential difference between these two points is 600 V.

a) Which point, a or b, is at the higher potential?

b) Calculate the magnitude E of the electric field.

c) A negative point charge of magnitude $q = -0.2 \ \mu C$ is moved from b to a. Calculate the work done on the point charge by the electric field.

26–4 The potential at a certain distance from a point charge is 600 V, and the electric field is 200 N·C^{-1}.

a) What is the distance to the point charge?

b) What is the magnitude of the charge?

26–5 Two point charges $q_1 = +40 \times 10^{-9}$ C and $q_2 = -30 \times 10^{-9}$ C are 10 cm apart. Point A is midway between them; point B is 8 cm from q_1 and 6 cm from q_2. Find

a) the potential at point A;

b) the potential at point B;

c) the work that must be done to carry a charge of 25×10^{-9} C from point B to point A.

26–6 Two point charges whose magnitudes are $+2.0 \times 10^{-10}$ C and -1.2×10^{-10} C are separated by a distance of 5 cm. An electron is released from rest between the two charges, 1 cm from the negative charge, and moves along the line connecting the two charges. What is its velocity when it is 1 cm from the positive charge?

26–7 Two positive point charges, each of magnitude q, are fixed on the y-axis at the points $y = +a$ and $y = -a$.

a) Draw a diagram showing the positions of the charges.

b) What is the potential V_0 at the origin?

c) Show that the potential at any point on the x-axis is

$$V = \frac{1}{4\pi\epsilon_0} \frac{2q}{\sqrt{a^2 + x^2}}.$$

d) Sketch a graph of the potential on the x-axis as a function of x over the range from $x = +4a$ to $x = -4a$.

e) At what value of x is the potential one-half that at the origin?

26–8 A positive charge $+q$ is located at the point $(x = -a, y = -a)$, and an equal negative charge $-q$ is located at the point $(x = +a, y = -a)$.

a) Draw a diagram showing the positions of the charges.

b) What is the potential at the origin?

c) What is the expression for the potential at a point on the x-axis, as a function of x?

d) Sketch a graph of the potential as a function of x, in the range from $x = +4a$ to $x = -4a$. Plot positive potentials upward, negative potentials downward.

Section 26–3 Calculation of Potentials

26–9 Two large parallel metal plates carry opposite charges. They are separated by 0.1 m, and the potential difference between them is 500 V.

a) What is the magnitude of the electric field, if it is uniform, in the region between the plates?

b) Compute the work done by this field on a charge of 2.0×10^{-9} C as it moves from the higher-potential plate to the lower.

c) Compare the result of (b) to the change of potential energy of the same charge, computed from the electrical potential.

26–10 Two large parallel metal sheets carrying equal and opposite electric charges are separated by a distance of 0.05 m. The electric field between them is approximately uniform and has magnitude 600 N·C^{-1}.

a) What is the potential difference between the plates?

b) Which plate is at higher potential?

26–11 A potential difference of 2000 V is established between parallel plates in air. If the air becomes electrically conducting when the electric field exceeds 0.8×10^6 N·C^{-1}, what is the minimum separation of the plates?

26–12 Some cell walls in the human body have a double layer of surface charge, with a layer of negative charge inside and a layer of positive charge of equal magnitude on the outside. If the surface charge densities are $\pm 0.5 \times 10^{-3}$ C·m^{-2} and the cell wall is 5×10^{-9} m thick, find

a) the electric-field magnitude in the wall, between the two charge layers;

b) the potential difference between inside and outside the cell. Which is at higher potential?

26–13 A total electric charge of 4×10^{-9} C is distributed uniformly over the surface of a sphere of radius 0.20 m. If the potential is zero at a point at infinity, what is the value of the potential

a) at a point on the surface of the sphere?

b) at a point inside the sphere, 0.1 m from the center?

26–14 A particle of charge $+3 \times 10^{-9}$ C is in a uniform electric field directed to the left. It is released from rest and moves a distance of 5 cm, after which its kinetic energy is found to be $+4.5 \times 10^{-5}$ J.

a) What work was done by the electrical force?

b) What is the magnitude of the electric field?

c) What is the potential of the starting point with respect to the endpoint?

26–15 A charge of 2.5×10^{-8} C is placed in an upwardly directed uniform electric field having magnitude 5×10^4 N·C^{-1}. What work is done by the electrical force when the charge is moved

a) 0.45 m to the right?

b) 0.80 m downward?

c) 2.60 m at an angle of 45° upward from the horizontal?

26–16 A simple type of vacuum tube known as a *diode* consists essentially of two electrodes within a highly evacuated enclosure. One electrode, the *cathode,* is maintained at a high temperature and emits electrons from its surface. A potential difference of a few hundred volts is maintained between the cathode and the other electrode, known as the *anode,* with the anode at the higher potential. Suppose that a diode consists of a cylindrical cathode of radius 0.05 cm, mounted coaxially within a cylindrical anode 0.45 cm in radius. The potential of the anode is 300 V higher than that of the cathode. An electron leaves the surface of the cathode with zero initial speed. Find its speed when it strikes the anode.

26–17 Refer to Example 25–17.

a) From the expression for E obtained in Example 25–17, find the expression for the potential V as a function of r, both inside and outside the sphere, relative to a point at infinity.

b) Sketch graphs of V and E as functions of r from $r = 0$ to $r = 3\,R$, and compare with Fig. 26–7.

Section 26–5 Potential Gradient

26–18 A metal sphere of radius r_a is supported on an insulating stand at the center of a hollow metal sphere of inner radius r_b. There is a charge $+q$ on the inner sphere and a charge $-q$ on the outer.

a) Calculate the potential $V(r)$ for

   i) $r < r_a$,   ii) $r_a < r < r_b$,   iii) $r > r_b$.

Use Eq. (26–14) and the fact that the net potential is the sum of the potentials due to the individual spheres.

b) Show that the potential of the inner sphere with respect to the outer is

$$V_{ab} = \frac{q}{4\pi\epsilon_0}\left(\frac{1}{r_a} - \frac{1}{r_b}\right).$$

c) Use Eq. (26–26) and the results from (a) to show that the electric field at any point between the spheres has the magnitude

$$E = \frac{V_{ab}}{(1/r_a - 1/r_b)}\cdot\frac{1}{r^2}.$$

d) Find the electric field at a point outside the larger sphere, at a distance r from the center, where $r > r_b$.

e) Suppose the charge on the outer sphere is not $-q$ but a negative charge of different magnitude, say $-Q$. Show that the answers for (b) and (c) are the same as before, but the answer for (d) is different.

26–19 Evaluate $-dV/dx$ for $V(x)$ given in Example 26–14. Show that this value gives an E_x that agrees with that calculated in Example 25–8.

Section 26–6 The Millikan Oil-Drop Experiment

26–20 In an apparatus for measuring the electronic charge e by Millikan's method, an electric field of 6.34×10^4 V·m^{-1} is required to maintain a certain charged oil drop at rest. If the plates are 1.5 cm apart, what potential difference between them is required?

26–21 An oil droplet of mass 3×10^{-14} kg and of radius 2×10^{-6} m carries ten excess electrons. What is its terminal velocity

a) when falling in a region in which there is no electric field?

b) when falling in an electric field of magnitude 3×10^5 N·C^{-1} directed downward?

(The viscosity of air is 180×10^{-7} N·s·m^{-2}. Neglect the buoyant force of the air.)

Section 26–7 The Electronvolt

26–22 Use the relation $E_0 = mc^2$ to find the energy equivalent of the rest mass of the proton; express your result in MeV.

26–23 Find the potential energy of the interaction of two protons at a distance of 1×10^{-15} m, typical of the dimensions of atomic nuclei. Express your result in MeV.

26–24

a) Prove that when a particle is accelerated from rest in an electric field, its final velocity is proportional to the square root of the potential difference through which it is accelerated.

b) What is the final velocity of an electron accelerated through a potential difference of 1136 V if it has an initial velocity of $1.0 \times 10^7 \, \text{m·s}^{-1}$?

26–25

a) What is the maximum potential difference through which an electron can be accelerated, from rest, if its kinetic energy is not to exceed 1% of the rest energy?

b) What is the speed of such an electron, expressed as a fraction of the speed of light, c?

c) Make the same calculations for a *proton*.

Section 26–8 The Cathode-Ray Tube

26–26 The electric field in the region between the deflecting plates of a certain cathode-ray oscilloscope is $30,000 \, \text{N·C}^{-1}$.

a) What force is on an electron in this region?

b) What is the acceleration of an electron when acted on by this force?

26–27 In Fig. 26–15, an electron is projected along the axis midway between the plates of a cathode-ray tube with an initial velocity of $2 \times 10^7 \, \text{m·s}^{-1}$. The uniform electric field between the plates has a magnitude of $20,000 \, \text{N·C}^{-1}$ and is upward.

a) How far below the axis has the electron moved when it reaches the end of the plates?

b) At what angle with the axis is it moving as it leaves the plates?

c) How far below the axis will it strike the fluorescent screen S?

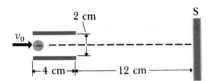

FIGURE 26–15

PROBLEMS

26–28 A small sphere of mass 0.2 g hangs by a thread between two parallel vertical plates 5 cm apart. The charge on the sphere is 6×10^{-9} C. What potential difference between the plates will cause the thread to assume an angle of 30° with the vertical?

26–29 In Exercise 26–14, suppose that another force in addition to the electrical force acts on the particle, so that when it is released from rest, it moves to the right. After it has moved 5 cm, the additional force has done 9×10^{-5} J of work and the particle has 4.5×10^{-5} J of kinetic energy.

a) What work was done by the electrical force?

b) What is the magnitude of the electric field?

c) What is the potential of the starting point with respect to the endpoint?

26–30 Consider the same distribution of charges as in Exercise 26–7.

a) Find the potential at a point on the y-axis, a distance y from the origin. Use your result to sketch a graph of the potential on the y-axis as a function of y, over the range from $y = +4a$ to $y = -4a$.

b) Discuss the physical meaning of the graph at the points $+a$ and $-a$.

c) At what point or points on the y-axis is the potential equal to its value at the origin?

d) At what points on the y-axis is the potential equal to half its value at the origin?

26–31 Consider the same distribution of charges as in Exercise 26–7.

a) Suppose a positively charged particle of charge q' and mass m is placed precisely at the origin and released from rest. What happens?

b) What will happen if the charge in part (a) is constrained not to move in the x-direction but is displaced slightly along the $+y$-axis and then released?

c) What will happen if it is displaced slightly in the direction of the $+x$-axis and then released?

26–32 Again consider the charge distribution of Exercise 26–7. Suppose a positively charged particle of charge q' and mass m is displaced slightly from the origin in the direction of the x-axis.

a) What is its speed at infinity?

b) Sketch a graph of the velocity of the particle as a function of x.

c) If the particle is projected toward the left along the x-axis from a point at a large distance to the right of the origin, with a velocity half that acquired in part (a), at what distance from the origin will it come to rest?

d) If a negatively charged particle, of charge $-q'$, were released from rest on the x-axis at a very large distance to the left of the origin, what would be its velocity as it passed the origin?

26–33 A potential difference of 1600 V is established between two parallel plates 4 cm apart. An electron is released from the negative plate at the same instant that a proton is released from the positive plate.

a) How far from the positive plate will they pass each other?

b) How do their velocities compare when they strike the opposite plates?

c) How do their kinetic energies compare when they strike the opposite plates?

26–34 In the Bohr model of the hydrogen atom, a single electron revolves around a single proton in a circle of radius R. Assume that the proton remains at rest.

a) By equating the electrical force to the electron mass times its acceleration, derive an expression for the electron's speed.

b) Obtain an expression for the electron's kinetic energy, and show that its magnitude is just half that of the electrical potential energy.

c) Obtain an expression for the total energy, and evaluate it by using $R = 5.29 \times 10^{-11}$ m.

Give your numerical result in joules and in electronvolts.

26–35 Refer to Problem 24–12.

a) Calcualte the potential at the origin, and at the point (3 cm, 0) due to the first two point charges.

b) Calculate the work the electric field would do on the third charge if it moved from the origin to the point (3 cm, 0).

26–36 Refer to Problem 24–22.

a) Calculate the potential at the point (3 cm, 0) and at the point (3 cm, 4 cm) due to the first two charges.

b) If the third charge moves from the point (3 cm, 0) to the point (3 cm, 4 cm), calculate the work done on it by the field of the first two charges. Comment on the *sign* of this work. Is your result reasonable?

26–37 A vacuum tube diode was described in Exercise 26–16. Because of the accumulation of charge near the cathode, the electrical potential between the electrodes is not a linear function of the position, even for planar geometry, but is given by

$$V = Cx^{4/3},$$

where, for given operating conditions, C is a constant, characteristic of a particular diode and operating conditions, and x is the distance from the cathode (negative electrode). If the distance between the cathode and anode (positive electrode) is 8 mm and the potential difference between electrodes is 160 V,

a) determine the value of C;

b) obtain a formula for the electric field between the electrodes as a function of x;

c) determine the force on an electron when the electron is halfway between the electrodes.

26–38 A long metal cylinder of radius r_a is supported on an insulating stand on the axis of a long, hollow metal cylinder of inner radius r_b. The positive charge per unit length on the inner cylinder is λ, and there is an equal negative charge per unit length on the outer cylinder.

a) Calculate the potential $V(r)$ for
   i) $r < r_a$,
   ii) $r_a < r < r_b$,
   iii) $r > r_b$.

Use the results of Example 26–8 and the fact that the net potential is the sum of the potentials due to the individual conductors. It is useful to take $V = 0$ at the same r_0 for both conductors.

b) Show that the potential of the inner cylinder with respect to the outer is

$$V_{ab} = \frac{\lambda}{2\pi\epsilon_0} \ln \frac{r_b}{r_a}.$$

c) Use Eq. (26–26) and the result from (a) to show that the electric field at any point between the cylinders has magnitude

$$E = \frac{V_{ab}}{\ln (r_b/r_a)} \cdot \frac{1}{r}.$$

d) What is the potential difference between the two cylinders if the outer cylinder has no net charge?

26–39 Refer to Problem 25–34.

a) From the expression for E obtained in Problem 25–34, find the expressions for the potential V as a function of r, both inside and outside the cylinder. Let $V = 0$ at the surface of the cylinder. In each case express your result in terms of the charge per unit length λ of the charge distribution.

b) Sketch graphs of V and E as functions of r, from $r = 0$ to $r = 3R$.

26–40 Four lines of charge are arranged to form a square with sides of length a. Calculate the potential at the center of the square for a zero of potential at infinity if

a) two opposite sides are positively charged with charge $+Q$ each and the other two sides are negatively charged with charge $-Q$ each;

b) if each side has positive charge $+Q$.

(*Hint:* Use the result of Example 26–10.)

26–41 Electric charge is distributed uniformly along a thin rod of length a, with total charge Q. Take the zero of potential to be at infinity. Find the potential at the following points (see Fig. 26–16):

a) point P, a distance x to the right of the rod;

b) point R, a distance y above the right-hand end of the rod.

c) In (a) and (b), what does your result reduce to as x or y becomes large?

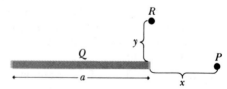

FIGURE 26–16

26–42 Electric charge is distributed uniformly along a semicircle of radius a. Calculate the potential at the center of curvature if the potential is assumed to be zero at infinity.

26–43 A charged oil drop, in a Millikan oil-drop apparatus, is observed to fall a distance of 1 mm in a time of 27.4 s, in the absence of any external field. The same drop can be held stationary in a field of 2.37×10^4 N·C^{-1}. How many excess electrons has the drop acquired? The viscosity of air is 180×10^{-7} N·s·m^{-2}. The density of the oil is 824 kg·m^{-3}, and the density of air is 1.29 kg·m^{-3}.

26–44 An alpha particle with kinetic energy 10 MeV makes a head-on collision with a gold nucleus at rest. What is the distance of closest approach of the two particles? (Assume that the gold nucleus remains stationary and that it may be treated as a point charge.)

26–45 Consider the charge distribution of Problem 25–35. Use the electric field calculated in that problem to do the following:

a) Show that for $r \geq R$ the potential is identical to that produced by a point charge Q. (Take the potential to be zero at infinity.)

b) Obtain an expression for the electrical potential valid in the region $r \leq R$.

CHALLENGE PROBLEMS

26–46 An electron with kinetic energy 100 MeV collides head-on with a gold nucleus at rest. Assume that the gold nucleus can be treated as a uniform distribution of charge through a sphere of radius 7×10^{-15} m and that the electron can penetrate into the nucleus, which remains at rest. What is the electron's kinetic energy when it reaches the center of the nucleus?

26–47 Two point charges are moving toward the right along the x-axis. Point charge 1 has charge $q_1 = 2$ μC, mass $m_1 = 6 \times 10^{-5}$ kg, and a velocity of magnitude v_1. Point charge 2 is to the right of q_1 and has charge $q_2 = -5$ μC, mass $m_2 = 3 \times 10^{-5}$ kg, and a velocity of magnitude v_2. At a particular instant the charges are separated by a distance of 9 mm and have velocities $v_1 = 700$ m·s^{-1} and $v_2 = 1200$ m·s^{-1}. The only forces on the particles are the forces they exert on each other.

a) Determine the velocity v_{cm} of the center of mass of the system.

b) The "relative energy" E_r of the system is defined as the total energy minus the kinetic energy contributed by the motion of the center of mass. That is,

$$E_r = E - \tfrac{1}{2}(m_1 + m_2)v_{cm}^2,$$

where $E = \tfrac{1}{2}m_1 v_1^2 + \tfrac{1}{2}m_2 v_2^2 + q_1 q_2/4\pi\epsilon_0 r$ is the total energy of the system and r is the distance between the charges. Show that $E_r = \tfrac{1}{2}\mu v^2 + q_1 q_2/4\pi\epsilon_0 r$, where μ is called the *reduced mass* of the system and is equal to $m_1 m_2/(m_1 + m_2)$, and $v = v_2 - v_1$ is the relative velocity of the moving particles.

c) For the numerical values given above, calculate the numerical value of E_r.

d) The value of E_r can be used to determine whether or not the particles will "escape" from one another. For the conditions given above, will the particles escape from one another? Explain.

e) If the particles do escape, what will be their final relative velocity? That is, what will their relative velocity be when $r \rightarrow \infty$? If the particles do not escape, what will be their distance of maximum separation? That is, what will be the value of r when $v = 0$?

26–48 The electrical potential V in a region of space is given by

$$V = ax^2 + ay^2 - 2az^2,$$

where a is a constant.

a) Derive an expression for the electric field $\boldsymbol{E}$, valid at all points in space.

b) The work done by the field when a 1 μC test charge moves from the point $(x, y, z) = (0, 0, 0.1$ m) to the origin is measured to be -5.0×10^{-5} J. Determine a.

c) Determine the electric field at the point $(0, 0, 0.1$ m).

d) Show that in every plane parallel to the xy-plane, the equipotential lines are circles.

e) What is the radius of the equipotential line corresponding to $V = 5000$ V and $z = \sqrt{2}$ m?

26–49 In a certain region a charge distribution exists that is spherical but nonuniform. That is, the volume charge density $\rho(r)$ depends on the distance r from the center of the distribution but not on the spherical polar angles θ and ϕ. The electric potential $V(r)$ due to this charge distribution is given by

$$V(r) = \left[\frac{\rho_0 a^2}{18\epsilon_0}\right][1 - 3(r/a)^2 + 2(r/a)^3] \quad \text{for} \quad r \leq a,$$
$$V = 0 \quad \text{for} \quad r \geq a$$

where ρ_0 is a constant having units of C·m^{-3}, and a is a constant having units of meters.

a) Derive expressions for the electric field for the regions $r \leq a$ and $r \geq a$. (*Hint:* Use the gradient operator for spherical polar coordinates,

$$\nabla = \boldsymbol{i}_r \frac{\partial}{\partial r} + \boldsymbol{i}_\theta \frac{1}{r}\frac{\partial}{\partial \theta} + \boldsymbol{i}_\phi \frac{1}{r \sin\theta}\frac{\partial}{\partial \phi},$$

where $\boldsymbol{i}_r$, $\boldsymbol{i}_\theta$, and $\boldsymbol{i}_\phi$ are unit vectors in the r-, θ-, and ϕ-directions.)

b) Derive an expression for $\rho(r)$ in each of the two regions $r \leq a$ and $r \geq a$. (*Hint:* Use Gauss's law for two spherical shells, one of radius r and the other of radius $r + dr$. Then use the fact that $dq = 4\pi r^2 \rho \, dr$, where dq is the charge contained in the infinitesimal spherical shell of thickness dr.)

c) Show that the net charge contained in the volume of a sphere of radius greater than or equal to a is zero. (*Hint:* Integrate the expressions derived in (b) for ρ over a spherical volume of radius greater than or equal to a.)

27
CAPACITANCE AND DIELECTRICS

Capacitors play a vital role in modern electronics.

A CAPACITOR IS A DEVICE CONSISTING OF TWO CONDUCTORS SEPARATED by vacuum or an insulating material. Capacitors are used in a wide variety of electric circuits and are a vital part of modern electronics. When charges of equal magnitude and opposite sign are placed on the conductors of a capacitor, an electric field is established in the region between them, with a corresponding potential difference between the conductors. The relations among charge, field, and potential can be analyzed by using the results of the two preceding chapters. For a given capacitor, the ratio of charge to potential difference is a constant, called the *capacitance*. Placing charges on the conductors requires an input of energy; this energy is stored in the capacitor and can be regarded as associated with the electric field in the space between conductors. When this space contains an insulating material (a dielectric) rather than vacuum, the capacitance is increased. This change can be understood on the basis of redistribution of charge within the material. We have previously studied mechanical and thermal properties of materials; electrical properties are another important class of properties of materials.

27-1 CAPACITORS

A capacitor: two charged conductors with a potential difference

Any two conductors separated by an insulator form a **capacitor.** In most cases of practical interest, the conductors usually have charges of equal magnitude and opposite sign, so that the *net* charge on the capacitor as a whole is zero. The electric field in the region between the conductors is proportional to the magnitude Q of charge on each conductor, and it follows that the *potential difference* V_{ab} between the conductors is also proportional to Q. If we double the magnitude of charge on each conductor, the charge density at each point doubles, the electric field at each point doubles, the potential difference between conductors doubles, and the ratio of charge to potential difference does not change.

When we speak of a capacitor as having charge Q, we mean that the conductor at higher potential has a charge Q and the conductor at lower potential has a charge $-Q$ (assuming Q is a positive quantity). This interpretation should be kept in mind in the following discussion and examples.

The **capacitance** C of a capacitor is defined as the ratio of the magnitude of the charge Q on *either* conductor to the magnitude of the potential difference V_{ab} between the conductors:

$$C = \frac{Q}{V_{ab}}. \qquad (27{-}1)$$

Capacitance: how much charge for a given voltage?

It follows from this definition that the unit of capacitance is one *coulomb per volt* ($C{\cdot}V^{-1}$). A capacitance of one coulomb per volt is called one **farad** (1 F), in honor of Michael Faraday. A capacitor is represented by the symbol

The farad: one coulomb per volt, the SI unit of capacitance

$$\dashv\vdash$$

Capacitors have numerous practical uses, and contemporary electronics could not exist without them. They are an essential element in tuning circuits in radio transmitters and receivers, in circuits that smooth and regulate the output of electronic power supplies, in engine ignition systems, and in other areas. The study of capacitors helps us develop insight into the behavior of electric fields and their interactions with matter. So there are many reasons to study capacitors!

Another term for capacitor is *condenser*. This term is seldom used except in reference to engine ignition systems, but you may find it occasionally in older literature.

27–2 THE PARALLEL-PLATE CAPACITOR

The most common type of capacitor consists of two conducting plates *parallel* to each other and separated by a distance that is small compared with the linear dimensions of the plates (see Fig. 27–1). Practically the entire field of such a capacitor is localized in the region between the plates, as shown. Some "fringing" of the field occurs at the edges of the plates, but if the plates are sufficiently close, the fringing may be neglected. The field between the plates is then uniform, and the charges on the plates are uniformly distributed over their opposing surfaces. This arrangement is known as a **parallel-plate capacitor.**

The electric-field magnitude between a pair of closely spaced parallel plates in vacuum, as discussed in Section 25–4, is

$$E = \frac{\sigma}{\epsilon_0} = \frac{Q}{\epsilon_0 A},$$

where σ is the magnitude of surface charge density on either plate, A is the area of each plate, and Q is the magnitude of total charge on each plate. Since the electric field (potential gradient) between the plates is uniform, the potential difference ("voltage") between the plates is

$$V_{ab} = Ed = \frac{1}{\epsilon_0}\frac{Qd}{A},$$

The capacitance of a parallel-plate capacitor is determined by its dimensions.

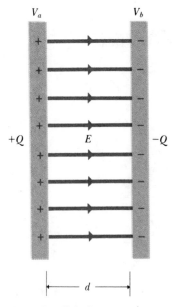

27–1 Parallel-plate capacitor.

where d is the separation of the plates. Hence the capacitance of a parallel-plate capacitor in vacuum is

$$C = \frac{Q}{V_{ab}} = \epsilon_0 \frac{A}{d}. \qquad (27\text{--}2)$$

Since ϵ_0, A, and d are constants for a given capacitor, the capacitance is a constant, independent of the charge on the capacitor, and is directly proportional to the area of the plates and inversely proportional to their separation.

The SI unit of capacitance is the *farad*, abbreviated F. In Eq. (27–2), if A is in square meters and d in meters, C is in farads. The units of ϵ_0 are $(C^2 \cdot N^{-1} \cdot m^{-2})$, so we see that

$$1\ F = 1\ C^2 \cdot N^{-1} \cdot m^{-1}.$$

A one-farad capacitor is huge!

As an example, let us compute the area of the plates of a 1-F parallel-plate capacitor if the separation of the plates is 1 mm and the plates are in vacuum:

$$C = \epsilon_0 \frac{A}{d},$$

$$A = \frac{Cd}{\epsilon_0} = \frac{(1\ F)(10^{-3}\ m)}{8.85 \times 10^{-12}\ C^2 \cdot N^{-1} \cdot m^{-2}}$$

$$= 1.13 \times 10^8\ m^2.$$

This area corresponds to a square whose side measures 10,600 m, or 34,800 ft, or about $6\frac{1}{2}$ miles!

Smaller and more practical units of capacitance

The farad is thus an extremely large unit of capacitance; units of more convenient size are the *microfarad* $(1\ \mu F = 10^{-6}\ F)$ and the *picofarad* $(1\ pF = 10^{-12}\ F)$. For example, a common radio contains in its power supply several capacitors whose capacitances are of the order of 10 or more microfarads, while the capacitances of the tuning capacitors are of the order of 10 to 100 picofarads.

Some numerical calculations for a parallel-plate capacitor

EXAMPLE 27–1 The plates of a parallel-plate capacitor are 5 mm apart and 2 m² in area. The plates are in vacuum. A potential difference of 10,000 V is applied across the capacitor. Compute (a) the capacitance, (b) the charge on each plate, and (c) the magnitude of electric field in the space between them.

SOLUTION

a) From Eq. (27–2),

$$C = \epsilon_0 \frac{A}{d} = \frac{(8.85 \times 10^{-12}\ C^2 \cdot N^{-1} \cdot m^{-2})(2\ m^2)}{5 \times 10^{-3}\ m}$$

$$= 3.54 \times 10^{-9}\ F = 0.00354\ \mu F.$$

b) The charge on the capacitor is

$$Q = CV_{ab} = (3.54 \times 10^{-9}\ C \cdot V^{-1})(10^4 V)$$

$$= 3.54 \times 10^{-5}\ C = 35.4\ \mu C.$$

That is, the plate at higher potential has charge $+35.4\ \mu C$, and the other plate has charge $-35.4\ \mu C$.

c) The electric field is

$$E = \frac{\sigma}{\epsilon_0} = \frac{Q}{\epsilon_0 A} = \frac{3.54 \times 10^{-5} \text{ C}}{(8.85 \times 10^{-12} \text{ C}^2 \cdot \text{N}^{-1} \cdot \text{m}^{-2})(2 \text{ m}^2)}$$

$$= 20 \times 10^5 \text{ N} \cdot \text{C}^{-1};$$

or, since the electric field equals the potential gradient,

$$E = \frac{V_{ab}}{d} = \frac{10^4 \text{ V}}{5 \times 10^{-3} \text{ m}} = 20 \times 10^5 \text{ V} \cdot \text{m}^{-1}.$$

The newton per coulomb and the volt per meter are equivalent units.

27–3 CAPACITORS IN SERIES AND PARALLEL

The arrangement shown in Fig. 27–2a is called a **series** connection. Two capacitors are connected in series between points a and b, and a constant potential difference V_{ab} is maintained. The capacitors are both initially uncharged. In this connection, both capacitors always have the same charge Q. The lower plate of C_1 and the upper plate of C_2 cannot have charges different from those on the remaining two plates. If they did, the net charge on each capacitor would not be zero, and the resulting electric field in the conductor connecting the two capacitors would cause a current to flow until the total charge on each capacitor is zero. Hence, *in a series connection the magnitude of charge on all plates is the same.*

Referring again to Fig. 27–2a, we have

$$V_{ac} \equiv V_1 = \frac{Q}{C_1}, \qquad V_{cb} \equiv V_2 = \frac{Q}{C_2},$$

$$V_{ab} \equiv V = V_1 + V_2 = Q\left(\frac{1}{C_1} + \frac{1}{C_2}\right),$$

and

$$\frac{V}{Q} = \frac{1}{C_1} + \frac{1}{C_2}. \tag{27–3}$$

The **equivalent capacitance** C of the series combination is defined as the capacitance of a *single* capacitor for which the charge Q is the same as for the combination, when the potential difference V is the same. For such a capacitor, shown in Fig. 27–2b,

$$Q = CV, \qquad \frac{V}{Q} = \frac{1}{C}. \tag{27–4}$$

Hence, from Eqs. (27–3) and (27–4),

$$\frac{1}{C} = \frac{1}{C_1} + \frac{1}{C_2}.$$

Similarly, for any number of capacitors in series,

$$\frac{1}{C} = \frac{1}{C_1} + \frac{1}{C_2} + \frac{1}{C_3} + \cdots. \tag{27–5}$$

Capacitors connected together: What is the equivalent capacitance?

Capacitors in series all have the same charge.

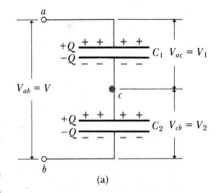

(a)

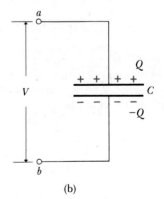

(b)

27–2 (a) Two capacitors in series, and (b) their equivalent.

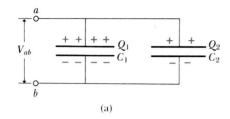

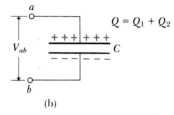

27–3 (a) Two capacitors in parallel, and (b) their equivalent.

Capacitors in parallel add directly.

The reciprocal of the equivalent capacitance of a series combination equals the sum of the reciprocals of the individual capacitances.

The arrangement shown in Fig. 27–3a is called a **parallel** connection. Two capacitors are connected in parallel between points a and b. In this case, the upper plates of the two capacitors are connected to form an equipotential surface, and the lower plates form another. The potential difference is the same for both capacitors and is equal to $V_{ab} = V$. The charges Q_1 and Q_2, not necessarily equal, are

$$Q_1 = C_1V, \qquad Q_2 = C_2V.$$

The *total* charge Q supplied by the source was

$$Q = Q_1 + Q_2 = V(C_1 + C_2),$$

and

$$\frac{Q}{V} = C_1 + C_2. \tag{27–6}$$

The *equivalent* capacitance C of the parallel combination is defined as that of a single capacitor, shown in Fig. 27–3b, for which the total charge is the same as in part (a). For this capacitor,

$$\frac{Q}{V} = C,$$

and hence

$$C = C_1 + C_2.$$

In the same way, for any number of capacitors in parallel

$$C = C_1 + C_2 + C_3 + \cdots. \tag{27–7}$$

The equivalent capacitance of a parallel combination equals the *sum* of the individual capacitances.

PROBLEM-SOLVING STRATEGY: Equivalent capacitance

1. Keep in mind that when we say a capacitor has a charge Q, we always mean that one plate has a charge Q and the other plate has a charge $-Q$.

2. When two capacitors are connected in series, as in Fig. 27–2a, they always have the same charge, provided they were initially uncharged before they were connected. The potential differences are *not* equal, unless the capacitances are equal. In this case the total potential difference across the combination is the sum of the individual potential differences.

3. When two capacitors have their plates connected as in Fig. 27–3a, so they are in parallel, the potential difference V is always the same for both. The charges on the two are *not* equal unless the capacitances are equal, but the total charge on the combination is the sum of the individual charges.

EXAMPLE 27–2 In Figs. 27–2 and 27–3, let $C_1 = 6\ \mu\text{F}$, $C_2 = 3\ \mu\text{F}$, and $V_{ab} = 18$ V. Find the equivalent capacitance, and find the charge and potential difference for each capacitor when the two capacitors are connected in (a) series, (b) parallel.

SOLUTION

a) The equivalent capacitance of the series combination in Fig. 27–2a is given by

$$\frac{1}{C} = \frac{1}{6\ \mu\text{F}} + \frac{1}{3\ \mu\text{F}}, \qquad C = 2\ \mu\text{F}.$$

The charge Q is

$$Q = CV = (2\ \mu\text{F})(18\ \text{V}) = 36\ \mu\text{C}.$$

The potential differences across the capacitors are

$$V_{ac} \equiv V_1 = \frac{Q}{C_1} = \frac{36\ \mu\text{C}}{6\ \mu\text{F}} = 6\ \text{V}, \qquad V_{cb} \equiv V_2 = \frac{Q}{C_2} = \frac{36\ \mu\text{C}}{3\ \mu\text{F}} = 12\ \text{V}.$$

The *larger* potential difference appears across the *smaller* capacitor.

b) The equivalent capacitance of the parallel combination in Fig. 27–3a is

$$C = C_1 + C_2 = 9\ \mu\text{F}.$$

The charges Q_1 and Q_2 are

$$Q_1 = C_1 V = (6\ \mu\text{F})(18\ \text{V}) = 108\ \mu\text{C},$$

$$Q_2 = C_2 V = (3\ \mu\text{F})(18\ \text{V}) = 54\ \mu\text{C}.$$

The potential difference across each capacitor is 18 V.

An example of capacitors in series and parallel

27–4 ENERGY OF A CHARGED CAPACITOR

To charge a capacitor we must transfer charge from the plate at lower potential to the plate at higher potential. This process requires work; energy is added to the capacitor and stored as *potential energy*. The positive and negative charges that are separated but attract each other are analogous to a stretched spring or a body lifted in the earth's gravitational field. The potential energy corresponds to the work done by the electrical forces when the capacitor becomes discharged, just as the spring or the earth's gravity does work when the system returns from its displaced position to the reference position.

Charging a capacitor requires the addition of energy.

To calculate the potential energy U of a charged capacitor, we calculate the work W required to charge it. The final charge Q and the final potential difference V are related by

$$Q = CV.$$

At a stage of the charging process at which the magnitude of the net charge on either plate is q, the potential difference v between the plates is $v = q/C$. The work dW required to transfer the next charge dq is

$$dW = v\, dq = \frac{q\, dq}{C}.$$

The total work W needed to increase the charge from zero to a final value Q is

$$W = \int dW = \frac{1}{C} \int_0^Q q\, dq = \frac{Q^2}{2C}.$$

The energy of a charged capacitor is proportional to the square of the charge.

If we define the potential energy of an *uncharged* capacitor to be zero, then W is equal to the potential energy U of the charged capacitor. The final potential difference V between the plates is $V = Q/C$. Thus we can write

$$U = W = \frac{Q^2}{2C} = \frac{1}{2}CV^2 = \frac{1}{2}QV. \qquad (27\text{–}8)$$

When Q is expressed in coulombs and V in volts (joules per coulomb), W is expressed in joules.

The last form of Eq. (27–8) also shows that the total work is equal to the *average* potential $V/2$ during the charging process, multiplied by the total charge Q transferred.

A charged capacitor is the electrical analog of a stretched spring, whose elastic potential energy equals $\frac{1}{2}kx^2$. The charge Q is analogous to the elongation x, and the *reciprocal* of the capacitance, $1/C$, is analogous to the force constant k. The energy supplied to a capacitor in the charging process is stored by the capacitor and is released when the capacitor discharges.

It is often useful to consider the energy stored in a capacitor as localized in the *electric field* between the capacitor plates. The capacitance of a parallel-plate capacitor in vacuum, from Eq. (27–2), is

$$C = \epsilon_0 \frac{A}{d}.$$

The electric field fills the space between the plates, of volume Ad, and its magnitude E is

$$E = \frac{V}{d}.$$

The energy per unit volume, or the **energy density,** denoted by u, is

$$u = \text{Energy density} = \frac{\frac{1}{2}CV^2}{Ad}.$$

Making use of the preceding equations, we can express this relation as

$$u = \tfrac{1}{2}\epsilon_0 E^2. \qquad (27\text{–}9)$$

Although we have derived Eq. (27–9) only for one specific situation, it turns out to be valid in general. The energy per unit volume associated with *any* electric-field configuration is given by Eq. (27–9). We will need this result in Chapter 35 in connection with the energy transported by electromagnetic waves.

EXAMPLE 27–3 In Fig. 27–4 we charge a capacitor C_1 by connecting it to a source of potential difference V_0 (not shown in the figure). Let $C_1 = 8\ \mu\text{F}$ and $V_0 = 120$ V. The charge Q_0 is

$$Q_0 = C_1 V_0 = 960\ \mu\text{C},$$

and the energy of the capacitor is

$$\tfrac{1}{2}Q_0 V_0 = \tfrac{1}{2}(960 \times 10^{-6}\ \text{C})(120\ \text{V}) = 0.0576\ \text{J}.$$

After we close the switch S, the positive charge Q_0 is distributed over the upper plates of both capacitors, and the negative charge $-Q_0$ is distributed over

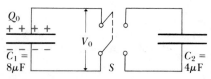

27–4 When the switch S is closed, the charged capacitor C_1 is connected to an uncharged capacitor C_2.

the lower plates of both. Let Q_1 and Q_2 represent the magnitudes of the final charges on the respective capacitors. Then

$$Q_1 + Q_2 = Q_0.$$

When the motion of charges has ceased, both upper plates are at the same potential; they are connected by a wire and so form a single equipotential surface. Both lower plates are at the same potential, different from that of the upper plates. The final potential difference between the plates, V, is therefore the same for both capacitors, and

$$Q_1 = C_1 V, \qquad Q_2 = C_2 V.$$

When we combine this with the preceding equation, we find

$$V = \frac{Q_0}{C_1 + C_2} = \frac{960\ \mu\text{C}}{12\ \mu\text{F}} = 80\ \text{V},$$
$$Q_1 = 640\ \mu\text{C}, \qquad Q_2 = 320\ \mu\text{C}.$$

The final energy of the system is

$$\tfrac{1}{2}Q_1 V + \tfrac{1}{2}Q_2 V = \tfrac{1}{2}Q_0 V = \tfrac{1}{2}(960 \times 10^{-6}\ \text{C})(80\ \text{V}) = 0.0384\ \text{J}.$$

This result is less than the original energy of 0.0576 J; the difference has been converted to energy of some other form. The conductors become a little warmer, and some energy is radiated as electromagnetic waves.

This process is exactly analogous to an inelastic collision of a moving car with a stationary car. In the electrical case, the charge $Q = CV$ is conserved. In the mechanical case, the momentum $p = mv$ is conserved. The electrical energy $\tfrac{1}{2}CV^2$ is *not* conserved, and the mechanical energy $\tfrac{1}{2}mv^2$ is *not* conserved.

27–5 EFFECT OF A DIELECTRIC

Most capacitors have a solid, nonconducting material or **dielectric** between their plates. A common type of capacitor incorporates strips of metal foil, forming the plates, separated by strips of wax-impregnated paper or plastic sheet such as Mylar, which serves as the dielectric. A sandwich of these materials is rolled up, forming a compact unit that can provide a capacitance of several microfarads in a relatively compact package.

What happens when an insulating material is placed between the conductors?

Electrolytic capacitors have as their dielectric an extremely thin layer of nonconducting oxide between a metal plate and a conducting solution. Because of the thinness of the dielectric, electrolytic capacitors of relatively small dimensions may have a capacitance of the order of 100 or 1000 μF. (From Eq. [27–2], the capacitance is inversely proportional to the distance d between the plates.)

Placing a solid dielectric between the plates of a capacitor serves three functions. First, it solves the mechanical problem of maintaining two large metal sheets at an extremely small separation but without actual contact.

The dielectric in a capacitor serves several purposes.

Second, any dielectric material, when subjected to a sufficiently large electric field, experiences *dielectric breakdown*, which is a partial ionization that permits conduction through a material that is supposed to insulate. Many insulating materials can tolerate stronger electric fields without breakdown than air can tolerate.

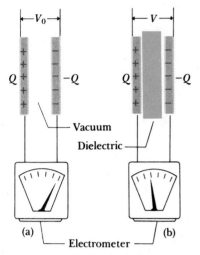

27–5 Effect of a dielectric between the plates of a parallel-plate capacitor. The electrometer measures potential difference. (a) With a given charge, the potential difference is V_0. (b) With the same charge, the potential difference V is smaller than V_0.

The dielectric constant: characterizing the effect of a dielectric on the capacitance and the field

Third, the capacitance of a capacitor of given dimensions is *larger* when a dielectric material is placed between the plates than when the plates are separated only by air or vacuum. This effect can be demonstrated with the aid of a sensitive electrometer, a device that can measure the potential difference between two conductors without permitting any charge to flow from one to the other. Figure 27–5a shows a charged capacitor, with magnitude of charge Q on each plate and potential difference V_0. When a sheet of dielectric, such as glass, paraffin, or polystyrene, is inserted between the plates, the potential difference is found to *decrease* to a smaller value V. When the dielectric is removed, the potential difference returns to its original value, showing that the original charges on the plates have not been affected by inserting the dielectric.

The original capacitance of the capacitor, C_0, was

$$C_0 = \frac{Q}{V_0}.$$

Since Q does not change and V is observed to be less than V_0, it follows that the capacitance C with the dielectric present is *greater* than C_0. The ratio of C to C_0 is called the **dielectric constant** of the material, K:

$$K = \frac{C}{C_0} \qquad (27\text{–}10)$$

Since C is always greater than C_0, the dielectric constants of all dielectrics are greater than unity. Some representative values of K are given in Table 27–1. For vacuum, of course, $K = 1$ by definition, and K for air is so nearly equal to 1 that for most purposes an air capacitor is equivalent to one in vacuum; the original measurement of V_0 in Fig. 27–5a could have been made with the plates in air instead of in vacuum.

With vacuum (or air) between the plates, the electric field E_0 in the region between the plates of a parallel-plate capacitor is

$$E_0 = \frac{V_0}{d} = \frac{\sigma}{\epsilon_0}.$$

The observed reduction in potential difference, when a dielectric is inserted between the plates, implies a reduction in the electric field, which in turn implies a reduction in the charge per unit area. Since no charge has leaked off the plates, such a reduction could be caused only by induced

TABLE 27–1 **Dielectric Constant K at 20°C**

Material	K	Material	K
Vacuum	1	Strontium titanate	310
Glass	5–10	Titanium dioxide (rutile)	173($\perp$),
Mica	3–6		86($\parallel$)
Mylar	3.1	Water	80.4
Neoprene	6.70	Glycerin	42.5
Plexiglas	3.40	Liquid ammonia (−78°C)	25
Polyethylene	2.25	Benzene	2.284
Polyvinyl chloride	3.18	Air (1 atm)	1.00059
Teflon	2.1	Air (100 atm)	1.0548
Germanium	16		

charges of opposite sign appearing on the two surfaces of the *dielectric.* That is, the dielectric surface adjacent to the positive plate must have an *induced negative charge,* and the surface adjacent to the negative plate must have an *induced positive charge of equal magnitude,* as shown in Fig. 27–6. These induced surface charges are a result of redistribution of charge within the dielectric material, a phenomenon called **polarization.**

If σ_i is the magnitude of the induced charge per unit area on the surfaces of the dielectric, then the *net* surface charge on each side of the capacitor that contributes to the electric field within the dielectric has magnitude $(\sigma - \sigma_i)$, and the electric field in the dielectric is

$$E = \frac{V}{d} = \frac{\sigma - \sigma_i}{\epsilon_0}. \qquad (27\text{--}11)$$

However,

$$K = \frac{C}{C_0} = \frac{Q/V}{Q/V_0} = \frac{V_0}{V} = \frac{E_0}{E} = \frac{\sigma}{\sigma - \sigma_i}; \qquad (27\text{--}12)$$

therefore

$$\sigma - \sigma_i = \frac{\sigma}{K}. \qquad (27\text{--}13)$$

Substituting Eq. (27–13) into Eq. (27–11), we obtain

$$E = \frac{\sigma}{K\epsilon_0}. \qquad (27\text{--}14)$$

The product $K\epsilon_0$ is called the **permittivity** of the dielectric and is represented by ϵ:

$$\epsilon = K\epsilon_0. \qquad (27\text{--}15)$$

The electric field within the dielectric may therefore be written

$$E = \frac{\sigma}{\epsilon}. \qquad (27\text{--}16)$$

Also,

$$C = KC_0 = K\epsilon_0 \frac{A}{d},$$

hence the capacitance of a parallel-plate capacitor with a dielectric between its plates is

$$C = \epsilon \frac{A}{d}. \qquad (27\text{--}17)$$

In empty space, where $K = 1$, $\epsilon = \epsilon_0$, and therefore ϵ_0 may be described as the "permittivity of empty space" or the "permittivity of vacuum." Since K is a pure number, the units of ϵ and ϵ_0 are evidently the same, $C^2 \cdot N^{-1} \cdot m^{-2}$.

The derivation of Eq. (27–9) for the energy density in an electric field can be repeated for a dielectric, by using the relations presented above. The result is that the energy density u for the electric field in a dielectric is given by

$$u = \tfrac{1}{2}K\epsilon_0 E^2 = \tfrac{1}{2}\epsilon E^2. \qquad (27\text{--}18)$$

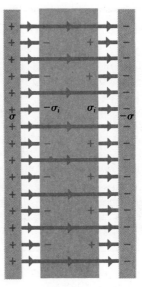

27–6 Induced charges on the faces of a dielectric in an external field.

Permittivity: an alternative description of the effect of a dielectric

Energy density in a dielectric

PROBLEM-SOLVING STRATEGY: *Dielectrics*

1. As usual, be careful with units. Distances must be in meters; remember that a microfarad is 10^{-6} farads, and so on. Don't confuse the numerical value of ϵ_0 with that of $1/4\pi\epsilon_0$. Several equivalent sets of units for electric-field magnitude exist including $N \cdot C^{-1}$ and $V \cdot m^{-1}$. Always check for consistency of units; it's a bit more of a nuisance with electrical quantities than it was in mechanics, but checking pays off!

2. In problems such as the following examples, it is easy to get lost in a blizzard of formulas. Ask yourself at each step what kind of quantity each symbol

represents. For example, distinguish clearly between charges and charge densities, and between electric fields and potentials. In checking numerical values, remember that the capacitance with a dielectric present is always greater than without, and that the induced surface charge density σ_i on the dielectric is always less than the free charge density σ on the capacitor plates. With a given charge on a capacitor, the electric field and potential difference are less with a dielectric present than without it.

A detailed analysis of the effect of a dielectric on the field, potential, and surface charges

EXAMPLE 27–4   The parallel plates in Fig. 27–6 have an area of 2000 cm^2 $(2 \times 10^{-1} \text{ m}^2)$ and are 1 cm (10^{-2} m) apart. The original potential difference between them, V_0, is 3000 V, and it decreases to 1000 V when a sheet of dielectric is inserted between the plates. Compute (a) the original capacitance C_0; (b) the charge Q on each plate; (c) the capacitance C after insertion of the dielectric; (d) the dielectric constant K of the dielectric; (e) the permittivity ϵ of the dielectric; (f) the induced charge Q_i on each face of the dielectric; (g) the original electric field E_0 between the plates; and (h) the electric field E after insertion of the dielectric.

SOLUTION

a) $C_0 = \epsilon_0 \dfrac{A}{d} = (8.85 \times 10^{-12} \text{ C}^2 \cdot \text{N}^{-1} \cdot \text{m}^{-2}) \dfrac{2 \times 10^{-1} \text{ m}^2}{10^{-2} \text{ m}}$

   $= 17.7 \times 10^{-11} \text{ F} = 177 \text{ pF}.$

b) $Q = C_0 V_0 = (17.7 \times 10^{-11} \text{ F})(3 \times 10^3 \text{ V}) = 53.1 \times 10^{-8} \text{ C}.$

c) $C = \dfrac{Q}{V} = \dfrac{53.1 \times 10^{-8} \text{ C}}{10^3 \text{ V}} = 53.1 \times 10^{-11} \text{ F} = 531 \text{ pF}.$

d) $K = \dfrac{C}{C_0} = \dfrac{53.1 \times 10^{-11} \text{ F}}{17.7 \times 10^{-11} \text{ F}} = 3.$

The dielectric constant could also be found from Eq. (27–12),

$$K = \frac{V_0}{V} = \frac{3000 \text{ V}}{1000 \text{ V}} = 3.$$

e) $\epsilon = K\epsilon_0 = (3)(8.85 \times 10^{-12} \text{ C}^2 \cdot \text{N}^{-1} \cdot \text{m}^{-2})$
    $= 26.6 \times 10^{-12} \text{ C}^2 \cdot \text{N}^{-1} \cdot \text{m}^{-2}.$

f) $Q_i = A\sigma_i, \qquad Q = A\sigma,$

   $\sigma - \sigma_i = \dfrac{\sigma}{K}, \qquad \sigma_i = \sigma\left(1 - \dfrac{1}{K}\right),$

   $Q_i = Q\left(1 - \dfrac{1}{K}\right) = (53.1 \times 10^{-8} \text{ C})\left(1 - \dfrac{1}{3}\right)$

   $= 35.4 \times 10^{-8} \text{ C}.$

g) $E_0 = \dfrac{V_0}{d} = \dfrac{3000 \text{ V}}{10^{-2} \text{ m}} = 3 \times 10^5 \text{ V}\cdot\text{m}^{-1}.$

h) $E = \dfrac{V}{d} = \dfrac{1000 \text{ V}}{10^{-2} \text{ m}} = 1 \times 10^5 \text{ V}\cdot\text{m}^{-1};$

or

$$E = \frac{\sigma}{\epsilon} = \frac{Q}{A\epsilon} = \frac{53.1 \times 10^{-8} \text{ C}}{(2 \times 10^{-1} \text{ m}^2)(26.6 \times 10^{-12} \text{ C}^2\cdot\text{N}^{-1}\cdot\text{m}^{-2})}$$

$$= 1 \times 10^5 \text{ V}\cdot\text{m}^{-1};$$

or

$$E = \frac{\sigma - \sigma_i}{\epsilon_0} = \frac{Q - Q_i}{A\epsilon_0}$$

$$= \frac{(53.1 - 35.4) \times 10^{-8} \text{ C}}{(2 \times 10^{-1} \text{ m}^2)(8.85 \times 10^{-12} \text{ C}^2\cdot\text{N}^{-1}\cdot\text{m}^{-2})}$$

$$= 1 \times 10^5 \text{ V}\cdot\text{m}^{-1};$$

or, from Eq. (27–12),

$$E = \frac{E_0}{K} = \frac{3 \times 10^5 \text{ V}\cdot\text{m}^{-1}}{3} = 1 \times 10^5 \text{ V}\cdot\text{m}^{-1}.$$

As we mentioned earlier, when any dielectric material is subjected to a sufficiently strong electric field, it becomes a conductor, a phenomenon known as dielectric breakdown. The onset of conduction, associated with cumulative ionization of molecules of the material, is often quite sudden and may be characterized by spark or arc discharges. When a capacitor is subjected to excessive voltage, an arc may form through a layer of dielectric, burning or melting a hole in it. This hole permits the two metal foils to come in contact, creating a short circuit and rendering the device permanently useless as a capacitor.

Dielectric breakdown: when an insulator quits insulating

The maximum electric field a material can withstand without the occurrence of breakdown is called its **dielectric strength.** This quantity is affected significantly by impurities in the material, small irregularities in the metal electrodes, and other factors difficult to control. For this reason we can give only approximate figures for dielectric strengths. The dielectric strength of dry air is about $0.8 \times 10^6 \text{ V}\cdot\text{m}^{-1}$. Typical values for plastic and ceramic materials commonly used to insulate capacitors and current-carrying wires are of the order of $10^7 \text{ V}\cdot\text{m}^{-1}$. For example, a layer of such a material, 10^{-4} m in thickness, could withstand a maximum voltage of 1000 V, since $1000 \text{ V}/10^{-4} \text{ m} = 10^7 \text{ V}\cdot\text{m}^{-1}$.

27–6 MOLECULAR MODEL OF INDUCED CHARGE

We now discuss briefly how surface charges on a dielectric, described in the preceding section, can come about. If the material were a *conductor*, the answer would be simple: Conductors contain charge that is free to move, and in the presence of an electric field some of the charge redistributes itself on the surface so that there is no electric field inside the conductor. But dielectrics have no charges that are free to move, so how can a surface charge occur?

The molecular basis of induced surface charge on dielectrics

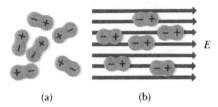

27–7 Behavior of polar molecules (a) in the absence and (b) in the presence of an electric field.

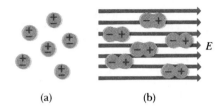

27–8 Behavior of nonpolar molecules (a) in the absence and (b) in the presence of an electric field.

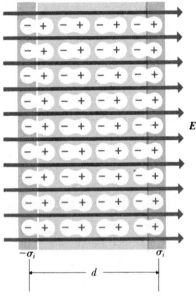

27–9 Polarization of a dielectric in an electric field gives rise to thin layers of bound charges on the surfaces.

To understand this phenomenon, we have to look at rearrangement of charge at the *molecular* level. First, some molecules, such as H_2O and N_2O, have equal amounts of positive and negative charge, but a lopsided distribution, with excess positive charge concentrated on one side of the molecule and negative on the other. Such a molecule is a little electric dipole, as defined in Example 25–5 (Section 25–1). It has an electric dipole moment and is called a *polar molecule*. When no electric field is present, the dipole moments of the molecules are randomly oriented. When polar molecules are placed in an electric field, however, they tend to orient themselves as in Fig. 27–7, as a result of the torques caused by the electric-field forces.

Even a molecule that is *not* polar acquires a dipole moment when placed in an electric field because the field forces cause some redistribution of charge within the molecule, as shown in Fig. 27–8. Such dipoles are called *induced* dipoles. With either polar or nonpolar molecules, an external field causes the formation of a layer of charge on each surface of the dielectric material, as shown in Fig. 27–9. These layers are the surface charges described in Section 27–5; their surface charge density is denoted by σ_i. The charges are not free to move indefinitely, as they would be in a conductor, because each charge is bound to a molecule. They are in fact called **bound charges** to distinguish them from the **free charges** that are added to and removed from the conducting capacitor plates. In the interior of the material the net charge per unit volume remains zero. We say that in this state the material is *polarized*.

The four parts of Fig. 27–10 show the behavior of a slab of dielectric when it is inserted in the field between a pair of oppositely charged capacitor plates. Part (a) shows the original field. Part (b) is the situation after the dielectric has

27–10 (a) Electric field between two charged plates. (b) Introduction of a dielectric. (c) Induced surface charges and their field. (d) Resultant field when a dielectric is between charged plates.

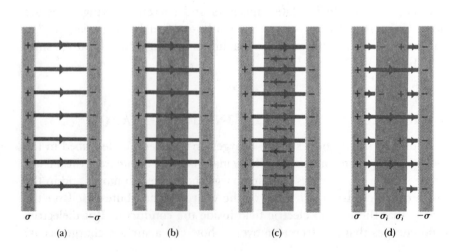

been inserted but before any rearrangement of charges has occurred. Part (c) shows by thinner lines the additional field set up in the dielectric by its induced surface charges. This field is *opposite* to the original field but not enough to cancel it completely, since the charges in the dielectric are not free to move indefinitely. The field in the dielectric is therefore decreased in magnitude. The resultant field is shown in Fig. 27–10d. Some of the field lines leaving the positive plate penetrate the dielectric; others terminate on the induced charges on the faces of the dielectric.

The charges induced on the surface of a dielectric in an external field afford an explanation of the attraction of an *uncharged* object such as a pith ball or bit of paper by a charged rod of rubber or glass. Figure 27–11 shows an uncharged dielectric sphere B in the radial field of a positive charge A. The induced positive charges on B experience a force toward the right, while the force on the negative charges is toward the left. Since the negative charges are closer to A and therefore in a stronger field than are the positive, the force toward the left exceeds that toward the right, and B, although its net charge is zero, experiences a resultant force toward A. The sign of A's charge does not affect the conclusion. Furthermore, the effect is not limited to dielectrics; a conducting sphere would be similarly attracted.

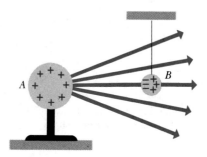

27–11 An uncharged dielectric sphere B in the radial field of a positive charge A.

How can an electric field exert a force on an uncharged object?

27–7 GAUSS'S LAW IN DIELECTRICS

With a slight extension of the analysis of Section 27–5, we can reformulate Gauss's law in a form that is particularly useful for dielectrics. Figure 27–12 shows the left capacitor plate and left surface of the dielectric in Fig. 27–6. Let us apply Gauss's law to the surface shown by the broken line; the surface area of each side is A. The left surface is embedded in the conductor forming the left capacitor plate, and so E everywhere on that surface is zero. The right surface is embedded in the dielectric, where the electric field has magnitude E. The total charge enclosed, including both the free charge on the capacitor plate and the bound charge on the dielectric surface, is $(\sigma - \sigma_i)A$, so Gauss's law gives us

$$EA = (\sigma - \sigma_i)A/\epsilon_0.$$

This is not very illuminating, as it stands, because it relates two unknown quantities, E in the dielectric and the induced surface charge density σ_i. But now we can use Eq. (27–13), developed for this same situation, to rewrite this equation as

$$EA = \sigma A/K\epsilon_0 \quad or \quad KEA = \sigma A/\epsilon_0. \qquad (27\text{–}19)$$

More generally, we can rewrite Gauss's law as

$$\int K\mathbf{E} \cdot d\mathbf{A} = Q/\epsilon_0. \qquad (27\text{–}20)$$

where Q is the total *free* charge (not bound charge) enclosed by the Gaussian surface. The significance of these results is that the right sides contain only the *free* charge, not the bound charge.

In fact, although we have not proved it, Eq. (27–20) remains valid when different parts of the Gaussian surface are embedded in dielectrics having different values of K.

Gauss's law: Both free and bound charge must be included.

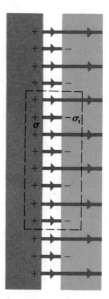

27–12 Gauss's law in a dielectric.

Another useful form of this relation, employing the notation $\epsilon = K\epsilon_0$ introduced in Section 27–5, is

$$\oint \epsilon E \cdot dA = Q. \tag{27–21}$$

This equation will be useful to our analysis of electromagnetic waves in matter in Chapter 35. For historical reasons of dubious validity, the quantity ϵE is given the name **electric displacement** and is denoted by **D**. So still another form of Gauss's law in the presence of dielectric materials is

$$\oint D \cdot dA = Q, \tag{27–22}$$

where again Q represents only the *free* charge enclosed by the Gaussian surface, not the polarization charge.

SUMMARY

A capacitor is any pair of conductors separated by an insulating material. When the capacitor is charged, with charges of equal magnitude Q and opposite sign on the two conductors, the potential difference V_{ab} between them is proportional to Q. The capacitance C is defined as

$$C = \frac{Q}{V_{ab}}. \tag{27–1}$$

A parallel-plate capacitor is made with two plates of area A, a distance d apart. If they are separated by vacuum, the capacitance is

$$C = \frac{Q}{V_{ab}} = \epsilon_0 \frac{A}{d}. \tag{27–2}$$

The SI unit of capacitance is the farad, abbreviated F. One farad is one coulomb per volt, and also $1 \text{ C}^2 \cdot \text{N}^{-1} \cdot \text{m}^{-1}$. This is a very large unit; the microfarad ($1 \ \mu\text{F} = 10^{-6}$ F) and the picofarad (1 pF $= 10^{-12}$ F) are more commonly used.

When capacitors having capacitances C_1, C_2, C_3, . . . are connected in series, the equivalent capacitance C is given by

$$\frac{1}{C} = \frac{1}{C_1} + \frac{1}{C_2} + \frac{1}{C_3} + \cdots. \tag{27–5}$$

When they are connected in parallel, the equivalent capacitance is

$$C = C_1 + C_2 + C_3 + \cdots. \tag{27–7}$$

The energy required to charge a capacitor C to a potential difference V and a charge Q is equal to the energy stored in the capacitor and is given by

$$U = \frac{Q^2}{2C} = \frac{1}{2}CV^2 = \frac{1}{2}QV. \tag{27–8}$$

This energy can be thought of as residing in the electric field between the conductors, with an energy density (energy per unit volume) u given by

$$u = \tfrac{1}{2}\epsilon_0 E^2. \tag{27–9}$$

When the space between the conductors is filled with a dielectric material, the capacitance increases by a factor K, called the dielectric constant of the material. Surface charges are induced on the surface of the dielectric; these have the effect of decreasing the electric field and potential difference between conductors by a factor K. The surface charge results from a microscopic rearrangement of charge in the dielectric, called polarization.

Under sufficiently strong fields, dielectrics become conductors; this phenomenon is called dielectric breakdown. The maximum field that a material can withstand without breakdown is called its dielectric strength.

The microscopic basis of polarization of dielectrics is the reorientation of polar molecules in an applied $\boldsymbol{E}$ field, or the creation of induced dipole moments in nonpolar materials.

The energy density u in an electric field in a dielectric is given by

$$u = \tfrac{1}{2}K\epsilon_0 E^2 = \tfrac{1}{2}\epsilon E^2. \tag{27-18}$$

Gauss's law can be reformulated for dielectrics, as follows:

$$\oint \epsilon \boldsymbol{E} \cdot d\boldsymbol{A} = \oint \boldsymbol{D} \cdot d\boldsymbol{A} = Q, \tag{27-21}$$

where Q is only the free charge (not bound charge or polarization charge) enclosed by the Gaussian surface.

QUESTIONS

27-1 A student claimed that two capacitors connected in parallel always have an effective capacitance *greater* than that of either capacitor, but when they are in series the effective capacitance is always *less* than that of either capacitor. Confirm or refute these claims.

27-2 Could one define a capacitance for a single conductor? What would be a reasonable definition?

27-3 Suppose the two plates of a capacitor have different areas. When the capacitor is charged by connecting it to a battery, do the charges on the two plates have equal magnitude, or may they be different?

27-4 Can you think of a situation in which the two plates of a capacitor *do not* have equal magnitudes of charge?

27-5 A capacitor is charged by being connected to a battery, and is then disconnected from the charging agency. The plates are then pulled apart a little. How does the electric field change? The potential difference? The total energy?

27-6 According to the text, one can consider the energy in a charged capacitor to be located in the field between the plates. But suppose there is vacuum between the plates; can there be energy in vacuum?

27-7 The charged plates of a capacitor attract each other. To pull the plates farther apart therefore requires work by some external force. What becomes of the energy added by this work?

27-8 A solid slab of metal is placed between the plates of a capacitor without touching either plate. Does the capacitance increase, decrease, or remain the same?

27-9 The two plates of a capacitor are given charges $\pm Q$, and then they are immersed in a tank of oil. Does the electric field between them increase, decrease, or remain the same? How may this field be measured?

27-10 Is dielectric strength the same thing as dielectric constant?

27-11 Liquid dielectrics having polar molecules (such as water) always have dielectric constants that decrease with increasing temperature. Why?

27-12 A capacitor made of aluminum foil strips separated by Mylar film was subjected to excessive voltage, and the resulting dielectric breakdown melted holes in the Mylar. After this the capacitance was found to be about the same as before, but the breakdown voltage was much less. Why?

27-13 Two capacitors have equal capacitance, but one has a higher maximum voltage rating than the other. Which one is likely to be bulkier? Why?

27-14 A capacitor is made by rolling a sandwich of aluminum foil and Mylar, as described in Section 27-5. A student claimed that the capacitance when the sandwich is rolled up is twice the value when it is flat. Discuss this allegation.

EXERCISES

Section 27–2 The Parallel-Plate Capacitor

27–1 A parallel-plate air capacitor has a capacitance of 500 pF and a charge of magnitude 0.2 μC on each plate. The plates are 0.2 mm apart.

a) What is the potential difference between the plates?

b) What is the area of each plate?

c) What is the electric-field magnitude between plates?

d) What is the surface charge density on each plate?

27–2 The plates of a parallel-plate capacitor are 4 mm apart and each carries a charge of 8×10^{-8} C. The plates are in vacuum. The electric field between the plates has magnitude 40×10^5 V·m^{-1}.

a) What is the potential difference between the plates?

b) What is the area of each plate?

c) What is the capacitance?

27–3 A capacitor has a capacitance of 8.5 μF. How much charge must be removed to lower the potential difference of its plates by 50 V?

Section 27–3 Capacitors in Series and Parallel

27–4 In Fig. 27–2a let $C_1 = 4$ μF, $C_2 = 6\mu$F, and $V_{ab} = 36$ V. Calculate

a) the charge on each capacitor;

b) the potential difference across each capacitor.

27–5 In Fig. 27–3a let $C_1 = 4$ μF, $C_2 = 6$ μF, and $V_{ab} = 36$ V. Calculate

a) the charge on each capacitor;

b) the potential difference across each capacitor.

27–6 In the circuit shown in Fig. 27–13, $C_1 = 2$ μF, $C_2 = 4$ μF, and $C_3 = 6$ μF. The applied potential is $V_{ab} = 24$ V. Calculate

a) the charge on each capacitor;

b) the potential difference across each capacitor;

c) the potential difference between points a and d.

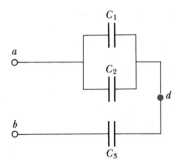

FIGURE 27–13

27–7 In Fig. 27–14 each capacitor has $C = 2$ μF and $V_{ab} = 48$ V. Calculate

a) the charge on each capacitor;

b) the potential difference across each capacitor;

c) the potential difference between points a and d.

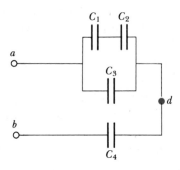

FIGURE 27–14

Section 27–4 Energy of a Charged Capacitor

27–8 An air capacitor is made from two flat parallel plates 0.5 mm apart. The magnitude of charge on each plate is 0.01 μC when the potential difference is 200 V.

a) What is the capacitance?

b) What is the area of each plate?

c) What maximum voltage can be applied without dielectric breakdown? (Dielectric breakdown for air occurs at an electric-field strength 8.0×10^5 V·m^{-1}.)

d) When the charge is 0.01 μC, what total energy is stored?

27–9 A parallel-plate air capacitor has a capacitance of 0.001 μF.

a) What potential difference is required for a charge of 0.5 μC on each plate?

b) In (a), what is the total stored energy?

c) If the plates are 1.0 mm apart, what is the area of each plate?

d) What potential difference is required for dielectric breakdown? (See Exercise 27–8c.)

27–10 An air capacitor consisting of two closely spaced parallel plates has a capacitance of 1000 pF. The charge on each plate is 1 μC.

a) What is the potential difference between the plates?

b) If the charge is kept constant, what will be the potential difference between the plates when the separation is doubled?

c) How much work is required to double the separation?

27–11 A 8-μF parallel-plate capacitor has a plate separation of 4 mm and is charged to a potential difference of 500 V. Calculate the energy density in the region between the plates, in units of J·m^{-3}.

27–12 A 20-μF capacitor is charged to a potential difference of 1000 V. The terminals of the charged capacitor are then connected to those of an uncharged 5-μF capacitor. Compute

a) the original charge of the system;

b) the final potential difference across each capacitor;

c) the final energy of the system;

d) the decrease in energy when the capacitors are connected.

27–13 A parallel-plate capacitor with plate area A and separation x is charged to a charge of magnitude q on each plate.

a) What is the total energy stored in the capacitor?

b) The plates are pulled apart an additional distance dx; now what is the total energy?

c) If F is the force with which the plates attract each other, then the difference in the two energies above must equal the work $dW = F\,dx$ done in pulling the plates apart. Show that $F = q^2/2\epsilon_0 A$.

d) Explain why F is not equal to qE, where E is the electric field between the plates.

Section 27–5 Effect of a Dielectric

27–14 Show that Eq. (27–18) holds for a parallel-plate capacitor with a dielectric material between the plates; use a derivation analogous to that used for Eq. (27–9). Explain why, even though a dielectric *decreases E,* the energy density *increases.*

27–15 A parallel-plate capacitor is to be constructed by using, as a dielectric, rubber with a dielectric constant of 3 and a dielectric strength of $2 \times 10^7 \text{V·m}^{-1}$. The capacitor is to have a capacitance of 0.15 μF and must be able to withstand a maximum potential difference of 6000 V. What is the minimum area the plates of the capacitor can have?

27–16 The paper dielectric in a paper-and-foil capacitor is 0.005 cm thick. Its dielectric constant is 2.5, and its dielectric strength is $50 \times 10^6 \text{ V·m}^{-1}$.

a) What area of paper and tinfoil is required for a 0.1-μF capacitor?

b) If the electric intensity in the paper is not to exceed one-half the dielectric strength, what is the maximum potential difference that can be applied across the capacitor?

27–17 Two parallel plates have equal and opposite charges. When the space between the plates is evacuated, the electric field is $2 \times 10^5 \text{ V·m}^{-1}$. When the space is filled with dielectric, the electric field is $1.2 \times 10^5 \text{ V·m}^{-1}$.

a) What is the charge density on the surface of the dielectric?

b) What is the dielectric constant?

27–18 Two oppositely charged conducting plates, with numerically equal quantities of charge per unit area, are separated by a dielectric 5 mm thick, of dielectric constant 3. The resultant electric field in the dielectric is $1 \times 10^6 \text{ V·m}^{-1}$. Compute

a) the charge per unit area on the conducting plate;

b) the charge per unit area on the surfaces of the dielectric.

27–19 Two parallel plates of area 100 cm^2 are given equal and opposite charges of 1.0×10^{-7} C. The space between the plates is filled with a dielectric material, and the electric field within the dielectric is $3.3 \times 10^5 \text{ V·m}^{-1}$.

a) What is the dielectric constant of the dielectric?

b) What is the total induced charge on either face of the dielectric?

PROBLEMS

27–20 A parallel-plate air capacitor is made from two plates 0.2 m square, spaced 1 cm apart. It is connected to a 50-V battery.

a) What is the capacitance?

b) What is the charge on each plate?

c) What is the electric field between the plates?

d) What is the energy stored in the capacitor?

e) If the battery is disconnected and the plates are pulled apart to a separation of 2 cm, what are the answers to parts (a), (b), (c), and (d)?

27–21 In Problem 27–20, suppose the battery remains connected while the plates are pulled apart. What are the answers to parts (a), (b), (c), and (d) after the plates have been pulled apart?

27–22 Several 0.5-μF capacitors are available. The voltage across each is not to exceed 400 V. A capacitor of capacitance 0.5 μF must be connected across a potential difference of 600 V.

a) Show in a diagram how an equivalent capacitor having the desired properties can be obtained.

b) No dielectric is a perfect insulator, of infinite resistance. Suppose that the dielectric in one of the capacitors in your diagram is a moderately good conductor. What will happen?

27–23 In Fig. 27–15, each capacitance C_3 is 3 μF and each capacitance C_2 is 2 μF.

a) Compute the equivalent capacitance of the network between points a and b.

b) Compute the charge on each of three capacitors nearest a and b when $V_{ab} = 900$ V.

c) With 900 V across a and b, compute V_{cd}.

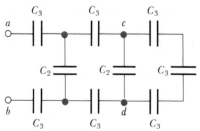

FIGURE 27–15

27–24 A 1-μF capacitor and a 2-μF capacitor are connected in series across a 1200-V supply line.

a) Find the charge on each capacitor and the voltage across each.

b) The charged capacitors are disconnected from the line and from each other and then reconnected, with terminals of like sign together. Find the final charge on each and the voltage across each.

27–25 A 1-μF capacitor and a 2-μF capacitor are connected in parallel across a 1200-V supply line.

a) Find the charge on each capacitor and the voltage across each.

b) The charged capacitors are disconnected from the line and from each other and then reconnected with terminals of unlike sign together. Find the final charge on each and the voltage across each.

27–26 In Fig. 27–3a, let $C_1 = 6\ \mu$F, $C_2 = 3\ \mu$F, and $V_{ab} = 18$ V. Suppose that the charged capacitors are disconnected from the source and from each other and then reconnected, with plates of *opposite* sign together. By how much does the energy of the system decrease?

27–27 Three capacitors having capacitances of 8, 8, and 4 μF are connected in series across a 12-V line.

a) What is the charge on the 4-μF capacitor?

b) What is the total energy of all three capacitors?

c) The capacitors are disconnected from the line and reconnected in parallel, with the positively charged plates connected together. What is the voltage across the parallel combination?

d) What is the total energy now stored in the capacitors?

27–28 The capacitors in Fig. 27–16 are initially uncharged and are connected as in the diagram with switch S open. The applied potential difference is $V_{ab} = +200$ V.

a) What is the potential difference V_{cd}?

b) What is the potential difference V_{ad} after switch S is closed?

c) How much charge flowed through the switch when it was closed?

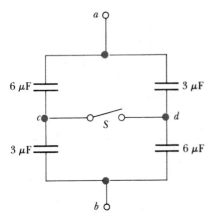

FIGURE 27–16

27–29 A capacitor consists of two parallel plates of area 25 cm^2 separated by a distance of 0.2 cm. The material between the plates has a dielectric constant of 5. The plates of the capacitor are connected to a 300-V battery.

a) What is the capacitance of the capacitor?

b) What is the charge on either plate?

c) What is the energy in the charged capacitor?

d) What is the energy density in the dielectric?

27–30 An air capacitor is made from two flat plates of area A separated by a distance d. Then a metal slab with thickness a (less than d) and the same shape and size as the plates is inserted between them, parallel to the plates and not touching either plate.

a) What is the capacitance of this arrangement?

b) Express the capacitance as a multiple of the capacitance when the metal slab is not present.

27–31 A spherical capacitor consists of an inner metal sphere of radius r_a supported on an insulating stand at the center of a hollow metal sphere of inner radius r_b; there is a charge $+Q$ on the inner sphere and a charge $-Q$ on the outer. (See Exercise 26–18.)

a) What is the potential difference V_{ab} between the spheres?

b) Prove that the capacitance is

$$C = 4\pi\epsilon_0\,\frac{r_a r_b}{r_b - r_a}.$$

c) If $r_b - r_a = d$, show that the equation obtained in (b) reduces to Eq. (27–2) when $d \ll r_a$, with A being the surface area of each sphere.

27–32 A coaxial cable consists of an inner solid cylindrical conductor of radius r_a supported by insulating disks on the axis of a conducting tube of inner radius r_b. The two cylinders are oppositely charged with a charge λ per unit length. (See Problem 26–38.)

a) What is the potential difference between the two cylinders?

b) Prove that the capacitance of a length L of the cable is

$$C = \frac{2\pi\epsilon_0 L}{\ln(r_b/r_a)}.$$

Neglect any effect of the supporting disks.

c) If $r_b - r_a = d$, show that the equation obtained in (b) reduces to Eq. (27–2) when $d \ll r_a$, with A being the surface area of each cylinder.

27–33 A parallel-plate capacitor has the space between the plates filled with two slabs of dielectric, one with constant K_1 and one with constant K_2. Each slab has thickness $d/2$, where d is the plate separation. Show that the capacitance is

$$C = \frac{2\epsilon_0 A}{d}\left(\frac{K_1 K_2}{K_1 + K_2}\right).$$

CHALLENGE PROBLEMS

27-34 A fuel gauge uses a capacitor to determine the height of the fuel in a tank. The effective dielectric constant K_{eff} changes from a value of 1 when the tank is empty to a value of K, the dielectric constant of the fuel, when the tank is full. The appropriate electronic circuitry can determine the effective dielectronic constant of the combined air and fuel between the capacitor plates. Each of the two rectangular plates has a width w (not shown) and a length l. (See Fig. 27–17.) The height of the fuel between the plates is h. Neglect any fringing effects.

a) Derive an expression for K_{eff} as a function of h.

b) What would be the effective dielectronic constant for a tank one-quarter full, one-half full, and three-quarters full if the fuel were gasoline ($K = 1.95$)?

c) Repeat part (b) for methanol ($K = 33$).

d) For which fuel would this fuel gauge be more practical?

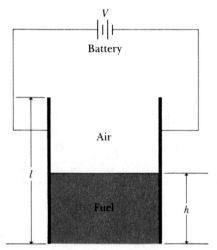

FIGURE 27–17

27-35 Three square metal plates A, B, and C, each 10 cm on a side and 3 mm thick, are arranged as in Fig. 27–18. The plates are separated by sheets of paper 0.5 mm thick and of dielectric constant 5. The outer plates are connected together and to point b. The inner plate is connected to point a.

a) Copy the diagram and show by + and − signs the charge distribution on the plates when point a is maintained at a positive potential relative to point b.

b) What is the capacitance between points a and b?

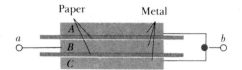

FIGURE 27–18

27-36 A parallel-plate capacitor consists of two horizontal conducting plates of equal area A. The bottom plate is resting on a fixed support, and the top plate is suspended by four springs of spring constant k, positioned at each of the four corners of the top plate. (See Fig. 27–19.) The plates, when uncharged, are separated by a distance z_0. A battery connected to the plates produces a potential difference V between them. This causes the plate separation to decrease to z. Neglect any fringing effects.

a) Show that the electrostatic force between the charged plates has a magnitude $\epsilon_0 A V^2 / 2z^2$. (*Note.* See Exercise 27–13.)

b) Obtain an expression that relates the plate separation z to the potential difference V between the plates. The resulting equation will be cubic in z.

c) Given the values $A = 0.25$ m^2, $z_0 = 1$ mm, $k = 25$ N·m^{-1}, and $V = 100$ V, find the two values of z for which the top plate will be in equilibrium. (*Hint:* You should solve this last part by using an iterative technique. Rearrange the equation to isolate z on the left side of the equation. Insert a trial value of z on the right side of the equation and calculate a new value of z. Insert the new value of z on the right side of the equation and calculate a new value again. Continue this process until the old and new values of z agree to within three significant figures. The third root of the cubic equation has a nonphysical negative value.)

d) For each of the two values of z found in (c), is the equilibrium stable or unstable? For stable equilibrium a small displacement of the object in question will give rise to a net force tending to return the object to the equilibrium position. For unstable equilibrium a small displacement gives rise to a net force that takes the object farther away from equilibrium.

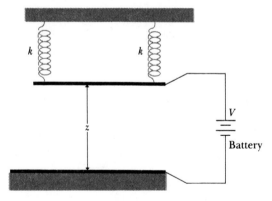

FIGURE 27–19

27-37 It is not always possible to combine capacitors in a simple series or parallel relationship. Consider the capacitors C_x, C_y, and C_z in the network depicted in Fig. 27–20a. Such a configuration of capacitors, referred to as a delta network, cannot be transformed into a single equivalent capacitor because three terminals a, b, and c exist in that network. It can be shown that as far as any effect on the external circuit is concerned, a delta network can be transformed into what is called a Y network. For example, the delta network of Fig. 27–20a can be replaced by the Y network of Fig. 27–20b.

a) Show that the transformation equations that give C_1, C_2, and C_3 in terms of C_x, C_y, and C_z are

$$C_1 = [C_xC_y + C_yC_z + C_zC_x]/C_x,$$
$$C_2 = [C_xC_y + C_yC_z + C_zC_x]/C_y,$$
$$C_3 = [C_xC_y + C_yC_z + C_zC_x]/C_z.$$

(*Hint:* The potential difference V_{ac} must be the same in both circuits, as must also be V_{bc}. Furthermore, the charge q_1 that flows from point a along the wire as indicated must be the same in both circuits, as must q_2. Obtain a relationship for V_{ac} as a function of q_1 and q_2 [and the capacitances] for each network and a separate relationship for V_{bc} as a function of the charges for each network. The coefficients of corresponding charges in corresponding equations must be the same for both networks.)

b) Determine the equivalent capacitance of the network of capacitors, between the terminals at the left-hand end of the network, for the network shown in Fig. 27–20c. (*Hint:* Use the delta-Y transformation derived in [a]. Use points a, b, and c to form the delta, and transform the delta into a Y. The capacitors can then be easily combined by using the relationships for series and parallel combinations of capacitors.)

c) Determine the charges of the 72-μF, 27-μF, 18-μF, 6-μF, 28-μF, and 21-μF capacitors and the potential differences across those capacitors.

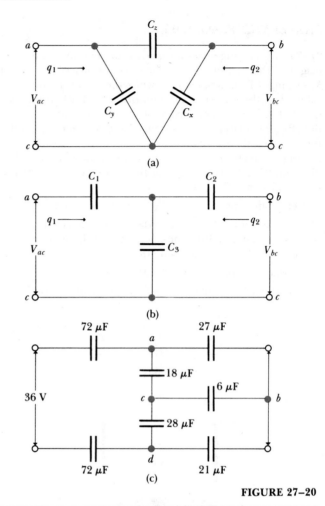

FIGURE 27–20

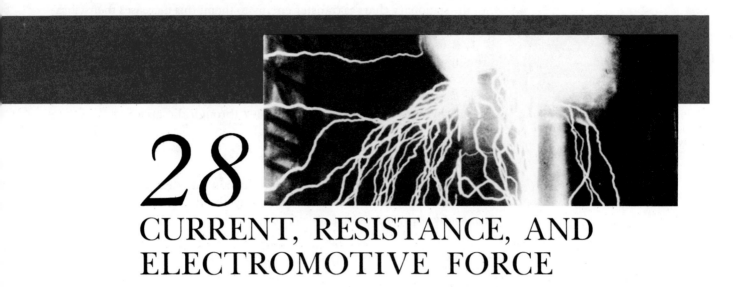

28

CURRENT, RESISTANCE, AND ELECTROMOTIVE FORCE

IN THIS CHAPTER WE STUDY THE BASIC PRINCIPLES OF ELECTRIC CURRENT (flow of charge within conducting materials) and electric circuits (continuous closed paths for electric currents). These principles form the basis of electric-circuit analysis, and as such underlie the whole fabric of our modern electronic age. We characterize the relation of current to potential difference or "voltage" by means of the concepts of resistivity and resistance and a relation called Ohm's law. Both resistivity and resistance are significantly temperature dependent. The basic principles of circuit analysis are contained in two principles called Kirchhoff's laws. Energy and power considerations are of primary importance in many electric circuits, and we study the energy relations in circuits. Finally, we discuss briefly a simple microscopic model of electrical conduction in metals.

28–1 CURRENT

Any motion of charge from one region of a conductor to another constitutes **current.** When an isolated conductor is placed in an electrostatic field, the mobile charges in the conductor rearrange themselves so as to make the entire interior of the conductor a field-free region and its surface an equipotential surface. The motion of charges during this rearrangement is a temporary, or *transient,* current. To maintain a *continuous* current, we must maintain a steady force on the mobile charge in the conductor, either with an electrostatic field or by other means to be described later in this chapter. For the present we assume that an electric field E is present within the conductor, so that a particle of charge q experiences a force $F = qE$.

The motion of a charged particle in a conductor is very different from the continuously accelerated motion that results when a constant field is applied to a charge in vacuum. In a conductor, a charge bumps along randomly, accelerating until it collides with a stationary particle. The charged particle thus gives up some of its kinetic energy, accelerates again until it bumps into something else, and so on. Thus there is a lot of back-and-forth motion, with a gradual *drift* in the direction of the electric-field force. The inelastic collisions

Current is charge in motion.

Motion of charge in a conductor: random motion plus a steady drift

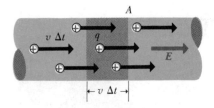

28–1 All the particles, and only those particles, within the shaded cylinder will cross its base in time Δt.

One ampere is one coulomb per second: the SI unit of current.

How is current related to drift velocity and density of charge in the conductor?

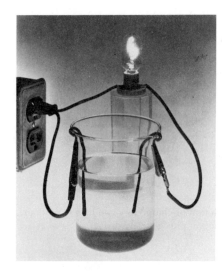

A simple laboratory demonstration of electrolytic conductivity. The sodium chloride solution in the beaker is part of the circuit, and the bulb glows brightly. (Photo by Chip Clark.)

with the stationary charges transfer energy to them; this increases their vibrational energy and hence the temperature of the conductor.

It is useful to consider the motion of charge in a conductor, across or through some imaginary area; the area might be, for example, a cross section of a wire used as a conductor. We define the current through an area as *the net charge flowing through the area per unit time*. Thus if a net charge ΔQ flows through an area in a time Δt, the current I through the area is

$$I = \frac{\Delta Q}{\Delta t}.$$

The rate of flow of charge may vary with time; in that case we generalize the definition of current in a natural way, using the derivative. We define the *instantaneous current I* as

$$I = \frac{dQ}{dt}. \tag{28–1}$$

Current is a *scalar* quantity.

The SI unit of current, one *coulomb per second*, is called one **ampere** ($1\ \text{A} = 1\ \text{C}\cdot\text{s}^{-1}$), in honor of the French scientist André Marie Ampère (1775–1836). Small currents are more conveniently expressed in *milliamperes* ($1\ \text{mA} = 10^{-3}\ \text{A}$) or in *microamperes* ($1\ \mu\text{A} = 10^{-6}\ \text{A}$).

The current through an area can be expressed in terms of the drift velocity v of the moving charges. Consider a portion of a conductor of cross-sectional area A through which there is an electric field E directed from left to right. Suppose first that the conductor contains free *positively* charged particles; these move in the same direction as the field. A few positive particles are shown in Fig. 28–1. Suppose there are n such particles per unit volume, all moving with a drift velocity v. In a time Δt, each particle advances a distance $v\ \Delta t$. Hence all the particles within the shaded cylinder of length $v\ \Delta t$, and only those particles, flow through the end of the cylinder during time Δt. The volume of the cylinder is $Av\ \Delta t$, and the number of particles within it is $nAv\ \Delta t$. If each particle has a charge q, the charge ΔQ flowing through the end of the cylinder in time Δt is

$$\Delta Q = nqvA\ \Delta t.$$

The current carried by the positively charged particles is therefore

$$I = \frac{\Delta Q}{\Delta t} = nqvA. \tag{28–2}$$

If the moving charges are negative rather than positive, the electric-field force is opposite to E, and the drift velocity is right to left, opposite to the direction shown in Fig. 28–1. But the current is still left to right; the reason is that negative charge moving right to left increases the positive charge at the right of the section, just as positive charge moving from left to right does. Thus the motion of *both* kinds of charge has the same effect, namely, to increase the positive charge at the right of the section. In both cases particles flowing out through an end of the cylindrical section are continuously replaced by particles flowing *in* through the opposite end.

In general, a conductor may contain several different kinds of charged particles having charges $q_1, q_2, \ldots$; densities $n_1, n_2, \ldots$; and drift velocities v_1,

$v_2, \ldots$. The total current is then

$$I = A(n_1 q_1 v_1 + n_2 q_2 v_2 + \cdots). \qquad (28\text{–}3)$$

In metals, the moving charges are always (negative) electrons, while in an ionized gas (plasma) both electrons and positively charged ions are moving. In a semiconductor material such as germanium or silicon, conduction is partly by electrons and partly by motion of *vacancies,* also known as *holes;* these are sites of missing electrons and act like positive charges. Conduction by motion of holes in semiconductors is discussed in Section 43–8.

Current in conductors can be due to either positive or negative charge motion, or both.

The current *per unit cross-sectional area* is called the **current density** J. For each kind of charged particle, $J = I/A = nqv$, and in general,

Current density: current per unit area

$$J = \frac{I}{A} = n_1 q_1 v_1 + n_2 q_2 v_2 + \cdots. \qquad (28\text{–}4)$$

We can also define a vector current density $\boldsymbol{J}$ that includes the directions of the drift velocities:

$$\boldsymbol{J} = n_1 q_1 \boldsymbol{v}_1 + n_2 q_2 \boldsymbol{v}_2 + \cdots. \qquad (28\text{–}5)$$

The direction of the drift velocity $\boldsymbol{v}$ of a positive charge is the same as that of the electric field $\boldsymbol{E}$, and the direction of the velocity of a negative charge is opposite to $\boldsymbol{E}$. But because the charge q is negative, each of the vectors $nq\boldsymbol{v}$ is in the same direction as $\boldsymbol{E}$, and hence the *vector current density $\boldsymbol{J}$ always has the same direction as the field $\boldsymbol{E}$.* Even in a metallic conductor, where the moving charges are negative electrons only and move in the *opposite* direction to $\boldsymbol{E}$, the *vector* current density $\boldsymbol{J}$ is in the *same* direction as $\boldsymbol{E}$.

How is current density related to the electric field that causes the charge motion?

Thus the effect of a current is the same, whether it consists of positive charges moving in the direction of $\boldsymbol{E}$, of negative charges moving in the opposite direction, or a combination of the two. When the current consists of a flow of positive charges, the direction of the current is the same as that of the motion of the charges; when it is a flow of negative charges, the direction of the current is *opposite* to that of the motion of the charges. These statements are consistent with Eq. (28–5): When q is negative, $\boldsymbol{J}$ and $\boldsymbol{v}$ have opposite directions. In describing circuit behavior it is customary to describe currents as though they consisted entirely of positive charge flow, even in cases where the actual current is known to be due to electrons. We will follow this convention in the following sections. In Chapter 30, when we study the effect of a *magnetic* field on a moving charge, we will consider a phenomenon called the Hall effect, in which the sign of the moving charges *is* important.

When there is a steady current in a closed loop (a "complete circuit"), the total charge in every segment of the conductor is constant. The rate of flow of charge *out* at one end of a segment equals the rate of flow of charge *in* at the other end of the segment. In other words, *the current is the same at all cross sections.* Current is *not* something that squirts out of the positive terminal of a battery and is consumed or used up by the time it reaches the negative terminal. These statements are direct consequences of the principle of conservation of charge, introduced in Section 24–2.

Conservation of charge: The current is the same at all points in a conducting loop.

EXAMPLE 28–1   A copper conductor of square cross section 1 mm on a side carries a constant current of 20 A. The density of free electrons is 8×10^{28} electrons per cubic meter. Find the current density and the drift velocity.

An example of charge motion in a conductor

SOLUTION The current density in the wire is

$$J = \frac{I}{A} = 20 \times 10^6 \text{ A}\cdot\text{m}^{-2}.$$

From Eq. (28–4),

$$v = \frac{J}{nq} = \frac{(20 \times 10^6 \text{ A}\cdot\text{m}^{-2})}{(8 \times 10^{28} \text{ m}^{-3})(1.6 \times 10^{-19} \text{ C})}$$
$$= 1.6 \times 10^{-3} \text{ m}\cdot\text{s}^{-1},$$

or about $1.6 \text{ mm}\cdot\text{s}^{-1}$. At this speed an electron would require 625 s or about 10 min to travel the length of a wire 1 m long.

28–2 RESISTIVITY

Resistivity: the ratio of electric field to current density

The current density J in a conductor depends on the electric field E and on the nature of the conductor. In general, the dependence of J on E can be quite complex. For some materials, especially the metals, however, it can be represented quite well by a direct proportionality. For such materials the ratio of E to J is *constant*.

We define the **resistivity** ρ of a particular materials as the ratio of electric field to current density:

$$\rho = \frac{E}{J}. \tag{28–6}$$

That is, the resistivity is the *electric field per unit current density*. The greater the resistivity, the greater the field needed to establish a given current density, or the smaller the current density caused by a given field. Representative values of resistivity are given in Table 28–1. The unit $\Omega\cdot\text{m}$ (ohm·meter) will be explained in the following section. A "perfect" conductor would have zero resistivity, and a "perfect" insulator would have infinite resistivity. Metals and alloys have the lowest resistivities and are the best conductors. The resistivities of insulators exceed those of the metals by a factor of the order of 10^{22}

Resistivities of conductors and insulators cover an enormous range of magnitudes.

Comparison with Table 16–1 shows that *thermal* insulators have thermal conductivities that differ from those of good thermal conductors by factors of only 10^3. By the use of electrical insulators, *electric* currents can be confined to well-defined paths in good electrical conductors, while it is impossible to con-

TABLE 28–1 Resistivities at Room Temperature

Substance		ρ, $\Omega\cdot$m	Substance		ρ, $\Omega\cdot$m
Conductors			Semiconductors		
Metals	Silver	1.47×10^{-8}	Pure	Carbon	3.5×10^{-5}
	Copper	1.72×10^{-8}		Germanium	0.60
	Gold	2.44×10^{-8}		Silicon	2300
	Aluminum	2.63×10^{-8}	Insulators		
	Tungsten	5.51×10^{-8}		Amber	5×10^{14}
	Steel	20×10^{-8}		Glass	10^{10}–10^{14}
	Lead	22×10^{-8}		Lucite	$>10^{13}$
	Mercury	95×10^{-8}		Mica	10^{11}–10^{15}
Alloys	Manganin	44×10^{-8}		Quartz (fused)	75×10^{16}
	Constantan	49×10^{-8}		Sulfur	10^{15}
	Nichrome	100×10^{-8}		Teflon	$>10^{13}$
				Wood	10^{8}–10^{11}

TABLE 28–2 Temperature Coefficients of Resistivity (Approximate Values Near Room Temperature)

Material	α, $C^{\circ-1}$	Material	α, $C^{\circ-1}$
Aluminum	0.0039	Lead	0.0043
Brass	0.0020	Manganin (Cu 84, Mn 12, Ni 4)	0.000000
Carbon	−0.0005		
Constantan (Cu 60, Ni 40)	+0.000002	Mercury	0.00088
Copper (commercial annealed)	0.00393	Nichrome	0.0004
		Silver	0.0038
Iron	0.0050	Tungsten	0.0045

fine *heat* currents to a comparable extent. Just as good electrical conductors, such as the metals, are also good conductors of heat, poor electrical conductors, such as ceramic and plastic materials, are also poor thermal conductors. The free electrons in a metal that carry charge in electrical conductions also play an important role in the conduction of heat; hence we expect a correlation between electrical and thermal conductivity.

The *semiconductors* form a class of materials intermediate between the metals and the insulators. They are important not primarily because of their resistivities, but because of the way in which these are affected by temperature and by small amounts of impurities.

The discovery that J is proportional to E for a metallic conductor at constant temperature was made by G. S. Ohm (1789–1854) and is called **Ohm's law.** A material obeying Ohm's law is called an *ohmic* conductor, or a *linear* conductor. If Ohm's law is *not* obeyed, the conductor is called *nonlinear*. Ohm's law, like the ideal-gas equation, Hooke's law, and many other relations describing the properties of materials, is an *idealized model* that describes the behavior of certain materials reasonably well but is by no means a general description of all matter.

The resistivity of all *metallic* conductors increases with increasing temperature, as shown in Fig. 28–2a. Over a moderate temperature range, the resistivity of a metal can be represented approximately by the equation

$$\rho_T = \rho_0[1 + \alpha(T - T_0)], \tag{28–7}$$

where ρ_0 is the resistivity at a reference temperature T_0 (often taken as 0°C or 20°C) and ρ_T is the resistivity at temperature T. The factor α is called the **temperature coefficient of resistivity.** Some representative values are given in Table 28–2. The resistivity of carbon (a nonmetal) *decreases* with increasing temperature, and its temperature coefficient of resistivity is negative. The resistivity of the alloy manganin is practically independent of temperature.

A number of materials have been found to exhibit the property of *superconductivity.* As the temperature is decreased, the resistivity at first decreases regularly, like that of any metal. At a certain transition temperature, usually in the range of 0.1 K to 20 K, a phase transition occurs, and the resistivity suddenly drops to zero, as shown in Fig. 28–2b. A current once established in a superconducting ring will continue indefinitely without the presence of any driving field. Superconductivity is discussed in greater detail in Section 43–10.

The resistivity of a *semiconductor* decreases rapidly with increasing temperature, as shown in Fig. 28–2c. A tiny bead of semiconducting material, called a *thermistor,* can be used as the temperature-sensing element in a sensitive electronic thermometer. That is, its resistivity is used as a thermometric property.

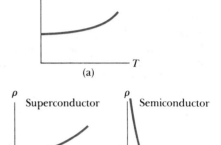

28–2 Variation of resistivity with temperature for three conductors: (a) an ordinary metal; (b) a superconducting metal, alloy, or compound; and (c) a semiconductor.

For most materials, resistivity increases with temperature.

Superconductivity: At very low temperatures, some materials have zero resistivity.

28–3 RESISTANCE

The current density J, at a point within a conductor where the electric field is E, is given by Eq. (28–6):

$$E = \rho J.$$

It is often difficult to measure E and J directly, so it is useful to state this relation in a form involving readily measured quantities, such as total current and potential difference. To do this, consider a conductor with uniform cross-sectional area A and length l, as shown in Fig. 28–3. Assuming a constant current density over a cross section, and a uniform electric field along the length of the conductor, the total current I is given by

$$I = JA,$$

and the potential difference V between the ends of the conductor is

$$V = El. \tag{28–8}$$

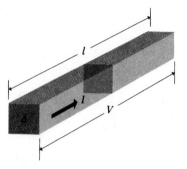

28–3 A conductor of uniform cross section. The current density is uniform over any cross section, and the electric field is constant along the length.

Solving these equations for J and E, respectively, and substituting the results in Eq. (28–6), we obtain

$$\frac{V}{I} = \frac{\rho l}{A}. \tag{28–9}$$

Thus the total current is proportional to the potential difference.

The quantity $\rho l/A$ for a particular specimen of material is called its **resistance** R:

$$R = \frac{\rho l}{A}. \tag{28–10}$$

Equation (28–9) then becomes

$$V = IR. \tag{28–11}$$

This relation is also called *Ohm's law;* in this form it refers to a specific piece of material, not to a general property of the material as with Eq. (28–6).

Equation (28–10) shows that the resistance of a wire or other conductor of uniform cross section is directly proportional to its length and inversely proportional to its cross-sectional area. It is of course also proportional to the resistivity of the material of which the conductor is made.

The SI unit of resistance is one *volt per ampere* (1 V·A^{-1}). This unit is given the name 1 **ohm** (1 Ω). The unit of resistivity is therefore one *ohm·meter* (1 Ω·m). Large resistances are conveniently expressed in *kilohms* (1 k$\Omega = 10^3$ Ω) or *megohms* (1 M$\Omega = 10^6$ Ω). Resistivities are also expressed in a variety of hybrid units, most common of which is the ohm·centimeter (1 Ω·cm $= 10^{-2}$ Ω·m).

Because the resistance of any sample of material is proportional to its resistivity, which varies with temperature, resistance also varies with temperature. For temperature ranges that are not too great, this variation may be represented approximately as a linear relation analogous to Eq. (28–7):

$$R_T = R_0[1 + \alpha(T - T_0)]. \tag{28–12}$$

Here R_T is the resistance at temperature T, and R_0 is the resistance at temperature T_0, often taken to be 20°C or 0°C. Within the limits of validity of Eq. (28–12), the *change* in resistance resulting from a temperature change $T - T_0$ is given by $R_0\alpha(T - T_0)$, where α is the temperature coefficient of resistivity, given for several common materials in Table 28–2.

EXAMPLE 28–2 In Example 28–1 (Section 28–1), find (a) the electric field; (b) the potential difference between two points 100 m apart; and (c) the resistance of a 100-m length of this copper conductor.

SOLUTION

(a) From Eq. (28–6), the electric field is given by

$$E = \rho J = (1.72 \times 10^{-8}\ \Omega\text{·m})(20 \times 10^6\ \text{A·m}^{-2})$$
$$= 0.344\ \text{V·m}^{-1}.$$

(b) The potential difference is given by

An example of a resistance calculation

$$V = El = (0.344\ \text{V·m}^{-1})(100\ \text{m})$$
$$= 34.4\ \text{V}.$$

(c) The resistance of a piece of this wire 100 m in length is

$$R = \frac{V}{I} = \frac{34.4\ \text{V}}{20\ \text{A}}$$
$$= 1.72\ \Omega.$$

This result can also be obtained directly from Eq. (28–10):

$$R = \frac{\rho l}{A} = \frac{(1.72 \times 10^{-8}\ \Omega\text{·m})(100\ \text{m})}{(1 \times 10^{-3}\ \text{m})^2}$$
$$= 1.72\ \Omega.$$

EXAMPLE 28–3 In Example 28–2, suppose the resistance is 1.72 Ω at a temperature of 20°C. Find the reistance at 0°C and at 100°C.

SOLUTION We use Eq. (28–12). In this instance $T_0 = 20°C$ and $R_0 = 1.72\ \Omega$. From Table 28–2, the temperature coefficient of resistivity of copper is $\alpha = 0.00393$ $(\text{C}°)^{-1}$. Thus at $T = 0°C$,

An example of the temperature variation of resistance

$$R = (1.72\ \Omega)[1 + (0.00393\text{C}^{°-1})(0°C - 20°C)]$$
$$= 1.58\ \Omega,$$

and at $T = 100°C$,

$$R = (1.72\ \Omega)[1 + (0.00393\text{C}^{°-1})(100°C - 20°C)]$$
$$= 2.26\ \Omega.$$

EXAMPLE 28–4 The space between two metallic coaxial cylinders of radii a and b is filled with a material of resistivity ρ. What is the resistance between the cylinders?

Radial current flow: an interesting variation

SOLUTION Equation (28–10) cannot be used directly because the cross section through which the charge travels varies from $2\pi a l$ at the inner cylinder to $2\pi b l$ at the outer cylinder. Instead, we consider a thin cylindrical shell of inner radius r and thickness dr. The area A is then $2\pi rl$, and the length of the current path through the shell is dr. Thus the resistance dR of the shell is

$$dR = \frac{\rho \, dr}{2\pi rl}$$

The current must pass successively through each one of these infinitesimal coaxial shells, so the total resistance between the cylinders is the *sum* of the resistances of the shells:

$$R = \frac{\rho}{2\pi l} \int_a^b \frac{dr}{r} = \frac{\rho}{2\pi l} \ln \frac{b}{a}.$$

In this last step we have used the fact that the total resistance of several resistors in series is the sum of their individual resistances. We will derive this relation in detail in Section 29–1.

28–4 ELECTROMOTIVE FORCE AND CIRCUITS

In order for a steady current to exist in a conducting path, that path must form a closed loop, or **complete circuit.** Otherwise charge would accumulate at the ends of the conductor, the resulting electric field would change with time, and the current could not be constant.

Electromotive force: Charge can't flow downhill forever.

However, such a path cannot consist entirely of resistance. Current in a resistor requires an electric field and an associated potential difference. The field always does *positive* work on the charge, which always moves in the direction of *decreasing* potential. But after a complete trip around the loop, the charge returns to its starting point, and the potential there must be the same as when it left that point. This is impossible if the trip around the loop involves only *decreases* in potential.

We can compare this situation with that of a decorative water fountain. Water emerges from openings at the top, cascades down over terraces and spouts, and eventually reaches the basin in the bottom. It collects there and runs into a pump that lifts it back to the top for another trip. Without the pump, the water would not be able to circulate continuously.

Thus in the electric circuit there must be some part of the loop where a charge travels "uphill," from lower to higher potential, despite the fact that the electrostatic force is trying to push it from higher to lower potential. The influence that makes charge move from lower to higher potential is called **electromotive force.** Every complete circuit having a steady current must include some device that provides electromotive force, usually abbreviated emf and pronounced "ee-em-eff."

Batteries, electric generators, solar cells, thermocouples, and fuel cells are all examples of sources of emf. Any such device has the ability to convert energy of some form (mechanical, chemical, thermal, and so on) into electrical energy and transfer it into the circuit where it is connected. An ideal source of emf would maintain a constant potential difference between its terminals, independent of the current through it. As we will see, such an ideal source is a mythical beast, like the unicorn, the frictionless plane, and the ideal gas. But it is a useful idealized model, and we will discuss later how real-life sources of emf differ in their behavior from this model.

Lightning striking the Eiffel Tower in Paris. The steel of the tower forms a good conducting path to ground. (Courtesy of the Lockyer Collection.)

Figure 28–4 is a schematic diagram of a source of emf that maintains a potential difference between conductors a and b, called the *terminals* of the device. Terminal a, marked +, is maintained at *higher* potential than terminal b, marked −. Associated with this potential difference is an electric field E in the region around the terminals, both inside and outside the source. The electric field inside the device is directed from a to b, as shown. A charge q within the source experiences an electrical force $\boldsymbol{F}_e = q\boldsymbol{E}$ caused by this field. But the source is itself a conductor, and if this were the *only* force on the free charges, then positive charge would move through the source from a toward b, and negative charge would move from b toward a. The excess charges on the conductors would decrease, and the potential difference would eventually decrease to zero.

But this is *not* the way batteries and generators actually behave; they maintain a potential difference even when there is a steady current through them from b to a. Thus in a source some additional influence must be present that tends to push positive charges from lower to higher potential, *opposite* to the direction of the electric-field force $\boldsymbol{F}_e$. The origin of this additional influence and its associated energy depends on the nature of the source. In a generator it consists of an additional electrical force that results from redistribution of charge in a conductor; this redistribution is caused by magnetic-field forces on moving charges. In a battery or fuel cell it is associated with diffusion processes caused by varying electrolyte concentrations associated with chemical reactions and their energies. In an electrostatic machine such as a Van de Graaff or Wimshurst generator, mechanical force is applied by a moving belt or wheel.

In all these cases we can represent this additional influence in terms of an equivalent non-electrostatic force $\boldsymbol{F}_n$ that is present in addition to the electrostatic force $\boldsymbol{F}_e$. In some cases, such as the Van de Graaff generator, $\boldsymbol{F}_n$ is an actual mechanical force. In other cases it is a simplified representation of a complex set of phenomena.

The potential V_{ab} of point a with respect to point b is defined, as always, as the work per unit charge performed by the electrostatic force $\boldsymbol{F}_e$ on a charge moving from a to b. Similarly, the emf of the source, denoted by $\mathcal{E}$, is the energy per unit charge supplied by the additional influence mentioned above during the "uphill" displacement from b to a. (It is the work per unit charge done by the non-electrostatic force $\boldsymbol{F}_n$.) For a source on open circuit, the potential difference V_{ab} is equal to the electromotive force $\mathcal{E}$:

$$V_{ab} = \mathcal{E} \qquad \text{(source on open circuit).} \qquad (28\text{–}13)$$

The SI unit of emf is the same as that of potential or potential difference, namely, $1\ \mathrm{J \cdot C^{-1}}$ or 1 V. There is a subtle but important distinction between emf and potential difference. Electromotive force refers to the energy supplied (per unit charge) by the non-electrostatic influences within a source (the work done by the non-electrostatic force). Potential difference is associated with electrostatic fields caused by distributions of electric charge. The emf may often be taken to be constant, independent of current, for a given source, while V_{ab} typically depends on current. In the following discussion we usually assume the emf of a source to be constant.

Now suppose that the terminals of a source are connected by a wire, as shown schematically in Fig. 28–5, forming a *complete circuit*. The driving force on the free charges *in the wire* is due solely to the electrostatic field $\boldsymbol{E}_e$ set up by

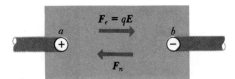

28–4 Schematic diagram of a source of emf in an "open-circuit" situation. The electric-field force $\boldsymbol{F}_e = q\boldsymbol{E}$ and the nonelectrostatic force $\boldsymbol{F}_n$ on a charge q are shown. The work done by $\boldsymbol{F}_n$ on a charge q moving from b to a is equal to $q\,\mathcal{E}$, where $\mathcal{E}$ is the electromotive force. In the open-circuit situation, $\boldsymbol{F}_e$ and $\boldsymbol{F}_n$ have equal magnitude.

Representing the effect of a source of emf in terms of an equivalent non-electrostatic field

When there is no current, the terminal potential difference of a source equals its emf.

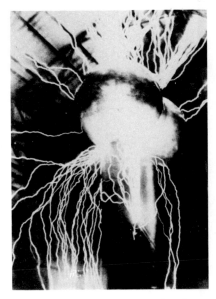

A Van de Graaff generator. Electric charge is carried to the dome at the top by a moving belt inside the vertical column. Potential differences as large as 10^6 can be developed. (American Institute of Physics.)

because in going in the opposite direction to the current we are going "uphill" from lower to higher potential.

For the circuit in Fig. 28–7, if we start at point b and go clockwise, the resulting equation is

$$I(4\ \Omega) + I(2\ \Omega) - 12\ V = 0,$$

which is the same as the previous equation except for an overall factor of -1, which does not change the value of I. But if we assume that I is clockwise, then starting at b and going counterclockwise around the circle yields the equation

$$12\ V + I(2\ \Omega) + I(4\ \Omega) = 0, \qquad I = -2\ A.$$

Here the negative sign on the result shows that our initial assumption about the current direction was wrong; the actual direction is counterclockwise.

PROBLEM-SOLVING STRATEGY: Electric circuits

1. Make an assumption about the direction of current in the circuit, and mark it clearly on your diagram. It doesn't matter whether the assumption is right or wrong. If it's wrong, your solution for I will give you a negative number, which means the actual current direction is opposite to your assumption. But you *must* pick a direction at the start and use it consistently when you apply Kirchhoff's loop rule.

2. Decide in which direction you will go around the loop when you add up the potential differences in applying Kirchhoff's loop rule. Again, it doesn't matter which way you go, but you *must* keep going the same way until you are back where you started.

3. Remember the sign rules: An emf is counted as positive if you go through the source from $-$ to $+$,

negative if from $+$ to $-$. The change in potential going through a resistor (IR) is negative if you go through in the same direction as the assumed current, and positive if in the opposite direction.

4. This same bookkeeping system may be used to find the potential difference between any two points a and b in a circuit. To find $V_{ab} = V_a - V_b$ (the potential of a with respect to b), start at b and add the potential changes you encounter in going from b to a. Use the same sign rules you used in step 3. An emf is positive when you go from $-$ to $+$, negative otherwise. An IR term is positive when you go "uphill," against the assumed current, and negative when "downhill," in the same direction as the current. Then the sum of these changes is V_{ab}.

An example of the use of Kirchhoff's loop rule

EXAMPLE 28–8   The circuit shown in Fig. 28–9 contains two batteries, each having an emf and an internal resistance, and two resistors. Find the current in the circuit and the potential difference V_{ab}.

SOLUTION   We assume a direction for the current, as shown. Then, starting at a and going counterclockwise, we add potential increases and decreases and equate the sum to zero. The resulting equation is

$$-I(4\ \Omega) - 4\ V - I(7\ \Omega) + 12\ V - I(2\ \Omega) - I(3\ \Omega) = 0.$$

Collecting terms containing I and solving for I, we find

$$8\ V = I(16\ \Omega) \quad \text{and} \quad I = 0.5\ A.$$

The result for I is positive, showing that our assumed current direction is correct. For exercise, try assuming the opposite direction for I; you should then get $I = -0.5\ A$, indicating that the actual current is opposite to this assumption.

To find V_{ab}, the potential at a with respect to b, we start at b and go toward a, adding potential changes. There are two possible paths from b to a; taking the lower one first, we find

$$V_{ab} = (0.5 \text{ A})(7 \text{ }\Omega) + 4 \text{ V} + (0.5 \text{ A})(4 \text{ }\Omega) = 9.5 \text{ V}.$$

Point a is at 9.5 V higher potential than b. All the terms in this sum are positive because each represents an *increase* in potential as we go from b toward a. If instead we use the upper path, the resulting equation is

$$V_{ab} = 12 \text{ V} - (0.5 \text{ A})(2 \text{ }\Omega) - (0.5 \text{ A})(3 \text{ }\Omega) = 9.5 \text{ V}.$$

Here the *IR* terms are negative because our path goes in the direction of the current, with potential decreases through the resistors. The result is the same as for the lower path, as it must be in order for the total potential change around the complete loop to be zero. In each case, potential rises are taken as positive, and drops as negative.

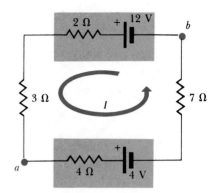

FIGURE 28–9

28–5 CURRENT–VOLTAGE RELATIONS

The amount of current in a device such as a resistor depends on the potential difference (voltage) between its terminals. For a device obeying Ohm's law, the current is *directly proportional* to voltage, as shown in Eq. (28–11). But in many devices current depends on voltage in a nonproportional way, and the current resulting from a given potential difference may depend on the *polarity* of the potential difference. This is the case with *diodes*, devices constructed deliberately to conduct much better in one direction than in the other.

Current–voltage relations: Some devices don't obey Ohm's law.

It is often convenient to represent the current–voltage relation as a graph, and Fig. 28–10 shows several examples. Part (a) shows the behavior of a resistor that obeys Ohm's law, for which the graph is a straight line. Part (b) shows the relation for a vacuum diode, a vacuum tube used to convert high-voltage alternating current to direct current. For positive potentials of anode with respect to cathode, I is approximately proportional to $V^{3/2}$; for negative potentials, the current is several orders of magnitude smaller and for most purposes may be assumed to be zero. Semiconductor diode behavior (c) is somewhat different but still strongly asymmetric, acting as a one-way valve in a circuit. Diodes are used to convert alternating current to direct current and to perform a wide variety of logic functions in computer circuitry. We will study the microscopic basis of diode behavior in later chapters.

A diode conducts better in one direction than in the other.

The current–voltage relation is temperature dependent for nearly all materials. At low temperatures the curve in Fig. 28–10c rises more steeply for

Current–voltage relations can be represented graphically.

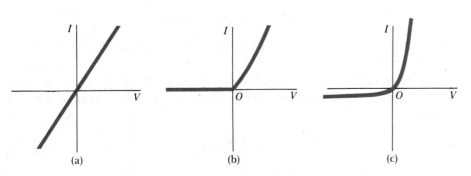

28–10 Current–voltage relations for (a) a resistor obeying Ohm's law; (b) a vacuum diode; (c) a semiconductor diode.

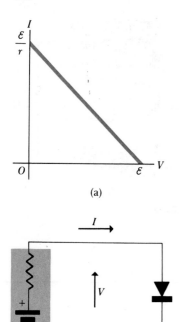

(a)

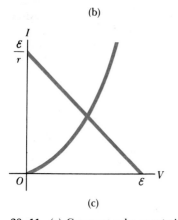

(b)

(c)

28–11 (a) Current–voltage relation for a source with emf $\mathcal{E}$ and internal resistance r; (b) a circuit containing a source and a nonlinear element; (c) simultaneous solution of I–V equations for this circuit.

A current transfers energy from one part of a circuit to another.

positive V than at higher temperatures, and at successively higher temperatures the asymmetry in the curve becomes less and less pronounced.

The current–voltage relation for a *source* may also be represented graphically. For a source represented by Eq. (28–16), that is,

$$V = \mathcal{E} - Ir,$$

the graph appears as in Fig. 28–11a. The intercept on the V-axis, corresponding to the open-circuit condition ($I = 0$), is at $V = \mathcal{E}$, and the intercept on the I-axis, corresponding to a short-circuit situation ($V = 0$), is at $I = \mathcal{E}/r$.

This graph may be used to find the current in a circuit containing a nonlinear device, as in Fig. 28–11b. The current–voltage relation is shown in Fig. 28–11c, and Eq. (28–16) is also plotted on this graph. Each curve represents a current–voltage relation that must be satisfied, so the intersection represents the only possible values of V and I. This amounts to a graphical solution of two simultaneous equations for V and I, one of which is nonlinear.

Finally, we remark that Eq. (28–16) is not always an adequate representation of the behavior of a source. What we have described as an internal resistance may actually be a more complex voltage–current relation. Nevertheless, the concept of internal resistance frequently provides an adequate description of batteries, generators, and other energy converters. The difference between a fresh flashlight battery and an old one is not in the emf, which decreases only slightly with use, but principally in the internal resistance, which may increase from a few ohms when fresh to as much as 1000 Ω or more after long use. Similarly, the current a car battery can deliver to the starter motor on a cold morning is less than when the battery is warm, not because the emf is appreciably less but because the internal resistance is temperature-dependent, increasing with decreasing temperature. Residents of northern Wisconsin have been known to soak their car batteries in warm water to provide greater starting power on very cold mornings!

28–6 ENERGY AND POWER IN ELECTRIC CIRCUITS

We are now ready for a detailed study of energy and power relations in electric circuits. The rectangle in Fig. 28–12 represents a portion of a circuit having current I and potential difference $V_a - V_b = V_{ab}$ between the two conductors leading to and from this point of the circuit. The detailed nature of this circuit element does not matter. As charge passes through the circuit element, the electric field does work on the charge. In a time interval Δt, an amount of charge $\Delta Q = I \, \Delta t$ passes through, and the work ΔW done by the electric field is given by the product of the potential difference (work per unit charge) and the quantity of charge:

$$\Delta W = V_{ab} \, \Delta Q = V_{ab} I \, \Delta t.$$

By means of this work the electric field transfers energy into this portion of the circuit.

The *rate* of the energy transfer is *power*, denoted by P. Dividing the above relation by Δt, we obtain the rate at which energy enters this part of the circuit:

$$P = \frac{\Delta W}{\Delta t} = V_{ab} I. \tag{28–20}$$

28–12 The power input P to the portion of the circuit between a and b is $P = V_{ab}I$.

It may happen that the potential at b is higher than that at a; in this case, V_{ab} is negative. The charge then *gains* potential energy (at the expense of some other form of energy), and there is a corresponding transfer of electrical energy *out of* this portion of the circuit.

Equation (28–20) is the general expression for the magnitude of the electric-power input to (or the power output from) any portion of an electric circuit. The unit of V_{ab} is one volt, or one joule per coulomb, and the unit of I is one ampere, or one coulomb per second. The SI unit of power is therefore

> A resistor converts electrical energy to heat.

$$(1 \text{ J} \cdot \text{C}^{-1})(1 \text{ C} \cdot \text{s}^{-1}) = 1 \text{ J} \cdot \text{s}^{-1} = 1 \text{ W} = 1 \text{ watt.}$$

We now consider some special cases.

1. *Pure resistance.* If the portion of the circuit in Fig. 28–12 is a pure resistance, the potential difference is given by $V_{ab} = IR$ and

$$P = V_{ab}I = I^2R = \frac{V^2}{R}. \qquad (28\text{–}21)$$

The potential at a must be higher than at b, so there is a power *input* to the resistor. The circulating charges give up energy to the atoms of the resistor when they collide with them, and the temperature of the resistor increases unless there is a flow of heat out of it. We say that energy is *dissipated* in the resistor at a rate I^2R.

> Power dissipation in a resistor is proportional to the square of the current.

Because of this heat, every resistor has a maximum power rating, which is the maximum power that can be dissipated without overheating the device. When this rating is exceeded, the resistance may change unpredictably; in extreme cases, the resistor may melt or even explode. In practical applications, the power rating of a resistor is often as important a characteristic as its resistance value.

2. *Power output of a source.* The upper rectangle in Fig. 28–13 represents a source having emf $\mathcal{E}$ and internal resistance r, connected by ideal (resistanceless) conductors to an external circuit represented by the lower rectangle; the precise nature of the external circuit does not matter. We assume only

> Power relations for a source delivering energy to a circuit

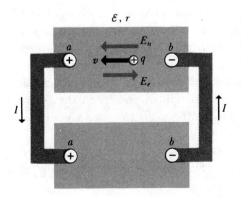

28–13 The rate of conversion of nonelectrical to electrical energy in the source equals $\mathcal{E}I$. The rate of energy dissipation in the source is I^2r. The difference $\mathcal{E}I - I^2r$ is the power output of the source.

that a current I is present in the circuit in the direction shown, from a to b in the external circuit and from b to a within the source. Point a is at higher potential than point b; $V_a > V_b$. The external circuit then corresponds to the rectangle in Fig. 28–12, and the power *input* to it is

$$P = V_{ab}I.$$

If a and b are considered to be the terminals of the source, then, as we have shown,

$$V_{ab} = \mathcal{E} - Ir,$$

and hence

$$P = V_{ab}I = \mathcal{E}I - I^2r. \tag{28–22}$$

What do the terms $\mathcal{E}I$ and I^2r mean? The emf $\mathcal{E}$ has been defined as the work per unit charge performed on the charges by the non-electrostatic force as the charges are pushed "uphill" from b to a in the source. If a charge ΔQ flows in time Δt, this force does work (adds energy) $\Delta W = \mathcal{E} \, \Delta Q$, and its *rate* of doing work, or power, is

$$\frac{\Delta W}{\Delta t} = \mathcal{E} \, \frac{\Delta Q}{\Delta t} = \mathcal{E}I. \tag{28–23}$$

Hence the product $\mathcal{E}I$ is the rate at which work is done on the circulating charges by the agency that causes the non-electrostatic force in the source.

The term I^2r is the rate at which energy is *dissipated* in the internal resistance of the source, and the difference $\mathcal{E}I - I^2r$ is the rate at which energy is delivered by the source to the remainder of the circuit. In other words, the power P in Eq. (28–22) represents the *net power output* of the source, or the power delivered to the remainder of the circuit.

3. *Power input to a source.* Suppose that the lower rectangle in Fig. 28–13 is itself a source, with an emf *larger* than that of the upper source and with its emf opposite to that of the upper source. The current I in the circuit is then *opposite* to that shown in Fig. 28–13. That is, the lower source pushes current backward through the upper source. The power output of the lower source is

$$P = V_{ab}I.$$

Considering a and b as the terminals of the upper source, we have

$$V_{ab} = \mathcal{E} + Ir,$$

and

$$P = V_{ab}I = \mathcal{E}I + I^2r. \tag{28–24}$$

We can interpret the terms $\mathcal{E}I$ and I^2r as follows. The charges in the upper source now move from left to right, in a direction opposite the field $\mathbf{E}_n$, and the work done by the non-electrostatic force $\mathbf{F}_n$ is *negative*. That is, work is done *on* the agent maintaining the non-electrostatic field. The product $\mathcal{E}I$ equals the rate at which work is done on (energy transferred to) this agent, and the term I^2r again equals the rate of dissipation of energy in the internal resistance of the source. The sum $\mathcal{E}I + I^2r$ is therefore the total *power input* to the upper source.

Thus the upper source has become a sink, *absorbing* electrical energy from the circuit and converting it to non-electrical energy. This is exactly what happens when a rechargeable battery (a storage battery) is connected to a charger. The charger supplies electrical energy to the battery; part of it is stored in the battery as chemical energy, to be reconverted later, and the remainder is dissipated as heat in the battery's internal resistance.

PROBLEM-SOLVING STRATEGY: Power and energy in circuits

1. A set of sign rules is useful in calculating the energy balance in a circuit. A source of emf puts positive power into a circuit when the current I runs through it from $-$ to $+$; the energy is converted from chemical energy in a battery, from mechanical energy in a generator, or whatever. But when current passes through a source in the direction from $+$ to $-$, the source is taking power out of the circuit (as in charging a storage battery, when electrical energy is converted back to chemical energy). In that case we count $\mathcal{E}I$ as negative. The internal resistance r of a source always removes energy from the circuit, converting it into heat at a rate I^2r, irrespective of the direction of the current. This represents a *negative* power input to the circuit. Similarly, a resistor R always removes energy from the circuit, at a rate given by $VI = I^2R = V^2/R$.

2. The energy balance can be expressed in one of two forms: either "net power input = net power output" or "the algebraic sum of the power inputs to the circuit is zero."

3. When every term in Kirchhoff's loop equation for a circuit is multiplied through by the current I in the circuit, the result is always an equation relating the power inputs and outputs in the circuit.

EXAMPLE 28–9 The rate of energy conversion in the source in Fig. 28–7 is

$$\mathcal{E}I = (12 \text{ V})(2 \text{ A}) = 24 \text{ W}.$$

The rate of dissipation of energy in the source is

$$I^2r = (2 \text{ A})^2(2 \text{ }\Omega) = 8 \text{ W}.$$

The power *output* of the source is the difference between these, or 16 W. The power output is also given by

$$IV_{ab} = (2 \text{ A})(8 \text{ V}) = 16 \text{ W}.$$

The power input to the resistor is

$$V_{a'b'}I = (2 \text{ A})(8 \text{ V}) = 16 \text{ W}.$$

This value equals the rate of dissipation of energy in the resistor:

$$I^2R = (2 \text{ A})^2(4 \text{ }\Omega) = 16 \text{ W}.$$

In terms of the sign convention suggested above, we may also write

$$\mathcal{E}I - I^2r - I^2R = 0,$$

or

$$(12 \text{ V})(2 \text{ A}) - (2 \text{ A})^2(2 \text{ }\Omega) - (2 \text{ A})^2(4 \text{ }\Omega) = 0.$$

Finally, Kirchhoff's loop equation for this circuit is

$$\mathcal{E} - Ir - IR = 0.$$

When we multiply this equation through by I, we again get the power-balance equation given above.

Some examples of energy and power relations in simple circuits

EXAMPLE 28–10 The rate of energy conversion in the source in Fig. 28–8 is

$$\mathcal{E}I = (12 \text{ V})(6 \text{ A}) = 72 \text{ W}.$$

The rate of dissipation of energy in the source is

$$I^2r = (6 \text{ A})^2(2 \text{ }\Omega) = 72 \text{ W}.$$

The power *output* of the source (also given by $V_{ab}I$) equals zero. *All* of the converted energy is dissipated within the source.

EXAMPLE 28–11 The rate of energy conversion in the upper source in Fig. 28–9 is

$$\mathcal{E}I = (12 \text{ V})(0.5 \text{ A}) = 6 \text{ W}.$$

The rate of dissipation of energy in this source is

$$I^2r = (0.5 \text{ A})^2(2 \text{ }\Omega) = 0.5 \text{ W}.$$

The power output of the upper source is 6 W − 0.5 W = 5.5 W.

The rates of dissipation of energy in the 3-Ω and 7-Ω resistors are, respectively,

$$(0.5 \text{ A})^2(3 \text{ }\Omega) = 0.75 \text{ W},$$

$$(0.5 \text{ A})^2(7 \text{ }\Omega) = 1.75 \text{ W}.$$

The rate of energy conversion in the lower source is

$$\mathcal{E}I = (4 \text{ V})(0.5 \text{ A}) = 2 \text{ W},$$

and the rate of energy dissipation in this source is

$$I^2r = (0.5 \text{ A})^2(4 \text{ }\Omega) = 1 \text{ W}.$$

The power *input* to the lower source is 2 W + 1 W = 3 W. Thus, of the 5.5-W output of the upper source, 2.5 W are dissipated in the two resistors and the remaining 3 W are partly converted, partly dissipated in the lower source. Conversion of energy in the sources is often partly *reversible,* as in a storage battery where electrical energy is converted to chemical energy for later retrieval as electrical energy. Energy dissipated in resistors is converted *irreversibly* to heat.

Finally, we write a power-balance equation, starting at point *b* and going counterclockwise:

$$(12 \text{ V})(0.5 \text{ A}) - (0.5 \text{ A})^2(2 \text{ }\Omega) - (0.5 \text{ A})^2(3 \text{ }\Omega)$$
$$- (0.5 \text{ A})^2(4 \text{ }\Omega) - (4 \text{ V})(0.5 \text{ A}) - (0.5 \text{ A})^2(7 \text{ }\Omega) = 0.$$

We invite you to add this up and verify that the power is accounted for.

28–7 THEORY OF METALLIC CONDUCTION

Microscopic picture of conduction in metals: electrons in random motion

We can gain additional insight into the phenomenon of conduction by examining the microscopic mechanisms of conductivity. Here we consider only a crude and primitive model that treats the electrons as classical particles and ignores their inherently quantum-mechanical behavior in solids. Thus this model is not entirely correct conceptually; yet it is useful in helping to develop an intuitive idea of the microscopic basis of conduction.

In the simplest microscopic model of metallic conduction, each atom in the crystal lattice is assumed to give up one or more of its outer electrons. These electrons are then free to move through the crystal lattice, colliding at intervals with the stationary positive ions. The motion of the electrons is like that of the molecules of gas in a container, and they are often referred to as an "electron gas." In the absence of an electric field, the electrons move in straight lines between collisions; but if an electric field is present, the paths are slightly curved, as in Fig. 28–14, which represents schematically a few free paths of an electron in an electric field directed from right to left. We assume that at each collision, the electron loses any energy it may have acquired from the field and makes a fresh start. The energy given up in these collisions increases the thermal energy of vibration of the positive ions.

A force $F = eE$ is exerted on each electron by the field and produces an acceleration a in the direction of the force given by

$$a = \frac{F}{m} = \frac{eE}{m},$$

where m is the electron mass. Let u represent the average *random* speed of an electron, and let λ be the *mean free path* (the average distance traveled between collisions). The average time t between collisions, called the *mean free time*, is

$$t = \frac{\lambda}{u}.$$

In this time, the electron acquires a final velocity component v_f in the direction of the force, given by

$$v_f = at = \frac{eE}{m}\frac{\lambda}{u}.$$

Its *average* velocity v in the direction of the force, which is superposed on its random velocity and which we interpret as the *drift* velocity, is one-half the final velocity, so

$$v = \frac{1}{2}v_f = \frac{1}{2}\frac{e\lambda}{mu}E.$$

The drift velocity is therefore proportional to the electric field E.

The current density is

$$J = nev = \frac{ne^2\lambda}{2mu}E,$$

and the resistivity is

$$\rho = \frac{E}{J} = \frac{2mu}{ne^2\lambda}. \qquad (28\text{–}25)$$

This is the theoretical expression for the resistivity, and it is in *qualitative* agreement with experiment. At a given temperature, the quantities $m, u, n, e,$ and λ are constant. The resistivity is then constant, and Ohm's law is obeyed. When the temperature is increased, the random speed u increases, so the theory predicts that the resistivity of a metal increases with increasing temperature.

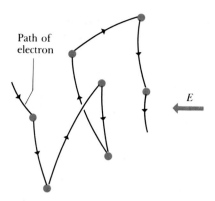

Path of electron

E

FIGURE 28–14

How far does an electron travel between collisions? How much time does it take?

The drift velocity is proportional to the electric field.

What is the difference between a
conductor and a semiconductor?

In a semiconductor, the number of charge carriers per unit volume, n, increases rapidly with increasing temperature. The increase in n far outweighs any increase in u, and the resistivity decreases. At low temperatures, n is very small and the resistivity becomes so large that the material can be considered an insulator.

The modern theory of superconductivity predicts that, in effect, at temperatures below the critical temperature, the electrons move freely throughout the lattice. The mean free path λ then becomes very large and the resistivity very small. We discuss superconductivity in greater detail in Section 43–10.

SUMMARY

Current is the amount of charge flowing through a boundary surface, per unit time. The SI unit of current is the ampere, equal to one coulomb per second $(1 \text{ A} = 1 \text{ C·s}^{-1})$. In terms of the charges q_i and drift velocities v_i of the charge carriers in a material, the current I is given by

$$I = A(n_1 q_1 v_1 + n_2 q_2 v_2 + \cdots). \qquad (28\text{–}3)$$

Current density J is current per unit cross-sectional area; $J = I/A$. In terms of the above quantities, current density is given by

$$J = n_1 q_1 v_1 + n_2 q_2 v_2 + \cdots. \qquad (28\text{–}4)$$

The corresponding vector current density $\boldsymbol{J}$ is given by

$$\boldsymbol{J} = n_1 q_1 \boldsymbol{v_1} + n_2 q_2 \boldsymbol{v_2} + \cdots. \qquad (28\text{–}5)$$

Current is described in terms of a flow of positive charge, even when the actual charge carriers are negative or of both signs.

Ohm's law, obeyed approximately by many materials, states that current density is proportional to electric-field magnitude. For such materials, resistivity is defined as electric-field magnitude per unit current density. The SI unit of resistivity is the ohm-meter $(1 \; \Omega\text{·m})$. Good conductors have small resistivity; good insulators have large resistivity. Resistivity usually increases with temperature; for small temperature changes this variation can be represented approximately by

$$\rho_T = \rho_0[1 + \alpha(T - T_0)], \qquad (28\text{–}7)$$

where α is the temperature coefficient of resistivity.

For materials obeying Ohm's law, the potential difference V across a particular sample of material is proportional to the current I through the material:

$$V = IR, \qquad (28\text{–}11)$$

where R is the resistance of the sample. In terms of the resistivity ρ, length l, and cross-sectional area A of the sample, the resistance is given by

$$R = \frac{\rho l}{A}. \qquad (28\text{–}10)$$

The SI unit of resistance is the ohm $(1 \; \Omega = 1 \text{ V·A}^{-1})$.

A complete circuit is a conductor in the form of a loop, providing a continuous current-carrying path. A complete circuit carrying a steady current must contain a source of electromotive force (emf), which is an influence within a source that makes charge move from a region of lower potential (the negative [−] terminal) to a region of higher potential (the positive [+] terminal) within the device. The SI unit of electromotive force is the volt (1 V). An ideal source of emf maintains a constant potential difference, independent of current through the device, but every real source of emf has some internal resistance r. The terminal potential difference then depends on current:

$$V_{ab} = \mathcal{E} - Ir. \tag{28–16}$$

Kirchhoff's loop rule states that the algebraic sum of potential differences around any closed circuit loop is zero. This rule follows from the conservative nature of the electrostatic field.

Many devices do not obey Ohm's law but show a more complicated voltage–current relationship; often this must be represented graphically.

A circuit element with a potential difference V and a current I puts energy into a circuit if the current direction is from lower to higher potential in the device, and takes energy out if the current is opposite. The power P (rate of energy transfer) is given by $P = VI$. A resistor R always takes energy out of a circuit, converting it to heat at a rate given by

$$P = V_{ab}I = I^2R = \frac{V^2}{R}. \tag{28–21}$$

When current in a source is from − to +, the source puts energy into the circuit at a rate given by

$$P = V_{ab}I = \mathcal{E}I - I^2r. \tag{28–22}$$

In this equation $\mathcal{E}I$ is the power corresponding to the emf, and I^2r is the power dissipated in the internal resistance.

The microscopic basis of conduction in metals is the motion of electrons that move freely through the crystal lattice sites, bumping into ion cores in the lattice. In a crude classical model of this motion, the resistivity of the material can be related to the electron mass, charge, speed of random motion, density, and mean free path between collisions.

QUESTIONS

28–1 A rule of thumb used to determine the internal resistance of a source is that it is the open-circuit voltage divided by the short-circuit current. Is this correct?

28–2 The energy that can be extracted from a storage battery is always less than the energy that goes into it while it is being charged. Why?

28–3 In circuit analysis one often assumes that a wire connecting two circuit elements has no potential difference between its ends; yet there must be an electric field within the wire to make the charges move, and so there must be a potential difference. How do you resolve this discrepancy?

28–4 Long-distance electric-power transmission lines always operate at very high voltage, sometimes as much as 750 kV. What are the advantages of such high voltages? The disadvantages?

28–5 Ordinary household electric lines usually operate at 120 V. Why is this a desirable voltage, rather than a value considerably larger or smaller? What about cars, which usually have 12-V electrical systems?

28–6 What is the difference between an emf and a potential difference?

28–7 Electric power for household and commercial use always uses *alternating current*, which reverses direction 120 times each second. A student claimed that the power conveyed by such a current would have to average out to zero, since it is going one way half the time and the other way the other half. What is your response?

28–8 As discussed in the text, the drift velocity of electrons in a good conductor is very slow. Why then does the light come on so quickly when the switch is turned on?

28–9 A fuse is a device designed to break a circuit, usually by melting, when the current exceeds a certain value. What characteristics should the material of the fuse have?

28–10 What considerations determine the maximum current-carrying capacity of household wiring?

28–11 The text states that good thermal conductors are also good electrical conductors. If so, why don't the cords used to connect toasters, irons, and similar heat-producing appliances get hot by conduction of heat from the heating element?

28–12 Eight flashlight batteries in series have an emf of about 12 V, similar to that of a car battery. Could they be used to start a car with a dead battery?

28–13 High-voltage power supplies are sometimes designed intentionally to have rather large internal resistance, as a safety precaution. Why is such a power supply with a large internal resistance safer than one with the same voltage but lower internal resistance?

28–14 How would you expect the resistivity of a good insulator such as glass or polystyrene to vary with temperature? Why?

28–15 A would-be inventor proposed to increase the power supplied by a battery to a light bulb by using thick wire near the battery and thinner wire near the bulb. In the thin wire, the electrons from the thick wire would become more densely packed, more electrons per second would reach the bulb, and the energy received by the bulb would be greater than that emitted by the battery. What do you think of this scheme?

EXERCISES

Section 28–1 Current

28–1 A silver wire 1 mm in diameter transfers a charge of 90 C in 1 hr and 15 min. Silver contains 5.8×10^{28} free electrons per m^3.

a) What is the current in the wire?

b) What is the drift velocity of the electrons in the wire?

28–2 When a sufficiently high potential difference is applied between two electrodes in a gas, the gas ionizes; electrons move toward the positive electrode, and positive ions move toward the negative electrode.

a) What is the current in a hydrogen discharge if, in each second, 4×10^{18} electrons and 1.5×10^{18} protons move in opposite directions past a cross section of the tube?

b) What is the direction of the current?

28–3 The current in a wire varies with time according to the relation

$$I = 4 \text{ A} + (2 \text{ A·s}^{-2})t^2.$$

a) How many coulombs pass a cross section of the wire in the time interval between $t = 5$ s and $t = 10$ s?

b) What constant current would transport the same charge in the same time interval?

Section 28–2 Resistivity

Section 28–3 Resistance

28–4 A copper wire has a square cross section 2.0 mm on a side. It is 4 m long and carries a current of 10 A. The density of free electrons is $8 \times 10^{28} \text{ m}^{-3}$.

a) What is the current density in the wire?

b) What is the electric field?

c) How much time is required for an electron to travel the length of the wire?

28–5 In household wiring, a copper wire commonly known as 12-gauge is often used. Its diameter is 2.05 mm. Find the resistance of a 50-m length of this wire.

28–6 What length of copper wire 1.0 mm in diameter would have a resistance of 1.00 Ω?

28–7 An aluminum bar 2.5 m long has a rectangular cross section 1 cm by 5 cm.

a) What is its resistance?

b) What would be the length of a copper wire 1.5 mm in diameter having the same resistance?

28–8 What diameter must an aluminum wire have if its resistance is to be the same as that of an equal length of copper wire of diameter 2.0 mm?

28–9

a) What is the resistance of a Nichrome wire at 0°C, if its resistance is 100.00 Ω at 12°C?

b) What is the resistance of a carbon rod at 30°C, if its resistance is 0.0150 Ω at 0°C?

28–10 A certain resistor has a resistance of 150.4 Ω at 20°C and a resistance of 162.4 Ω at 40°C. What is its temperature coefficient of resistivity?

28–11 A carbon resistor is to be used as a thermometer. On a winter day when the temperature is 0°C, its resistance is 217.3 Ω. What is the temperature on a hot summer day when the resistance is 214.2 Ω?

28–12 Refer to Example 28–4. Let the resistivity of the material between the cylinders be 10 Ω ·m and let $r_a = 10$ cm, $r_b = 20$ cm, and $l = 5$ cm.

a) Find the resistance between the cylinders.

b) Find the current between the cylinders if $V_{ab} = 8$ V.

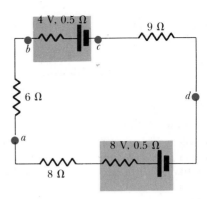

FIGURE 28–16

Section 28–4 Electromotive Force and Circuits

28–13 When switch S is open, the voltmeter V, connected across the terminals of the dry cell in Fig. 28–15, reads 1.52 V. When the switch is closed, the voltmeter reading drops to 1.37 V and the ammeter A reads 1.5 A. Find the emf and internal resistance of the cell. Neglect meter corrections.

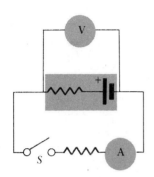

FIGURE 28–15

28–14 A closed circuit consists of a 12-V battery, a 3.7-Ω resistor, and a switch. The internal resistance of the battery is 0.3 Ω. The switch is opened. What would a high-resistance voltmeter read when placed

a) across the terminals of the battery?

b) across the resistor?

c) across the switch?

d) Repeat (a), (b), and (c) for the case when the switch is closed.

28–15 The internal resistance of a dry cell increases gradually with age, even though the cell is not used. The emf, however, remains fairly constant at about 1.5 V. Dry cells may be tested for age at the time of purchase by connecting an ammeter directly across the terminals of the cell and reading the current. The resistance of the ammeter is so small that the cell is practically short-circuited.

a) The short-circuit current of a fresh No. 6 dry cell (1.5-V emf) is about 30 A. Approximately what is the internal resistance?

b) What is the internal resistance if the short-circuit current is only 10 A?

c) The short-circuit current of a 6-V storage battery may be as great as 1000 A. What is its internal resistance?

28–16

a) What is the potential difference V_{ad} in the circuit of Fig. 28–16?

b) What is the terminal voltage of the 4-V battery?

c) A battery of emf 17 V and internal resistance 1 Ω is inserted in the circuit at d, with its positive terminal connected to the positive terminal of the 8-V battery. What is now the difference of potential V_{bc} between the terminals of the 4-V battery?

Section 28–5 Current–Voltage Relations

28–17 The following measurements of current and potential difference were made on a resistor constructed of Nichrome wire:

I, A	V_{ab}, V
0.5	2.18
1.0	4.36
2.0	8.72
4.0	17.44

a) Make a graph of V_{ab} as a function of I.

b) Does Nichrome obey Ohm's law?

c) What is the resistance of the resistor, in ohms?

28–18 The following measurements were made on a Thyrite resistor:

I, A	V_{ab}, V
0.5	4.76
1.0	5.81
2.0	7.05
4.0	8.56

a) Make a graph of V_{ab} as a function of I. Does Thyrite have a constant resistance?

b) Construct a graph of the resistance $R = V_{ab}/I$.

Section 28–6 Energy and Power in Electric Circuits

28–19 Consider a resistor of length L, uniform cross section A, and uniform resistivity ρ that is carrying a current of uniform current density J. Use Eq. (28–21) to find the electric power dissipated per unit volume, p. Express your result in terms of

a) E and J,

b) J and ρ,

c) E and ρ.

28–20 A resistor develops heat at the rate of 360 W when the potential difference across its ends is 180 V. What is its resistance?

28–21 A "660-W" electric heater is designed to operate from 120-V lines.

a) What is its resistance?

b) What current does it draw?

c) If the line voltage drops to 110 V, what power does the heater take, in watts? (Assume the resistance to be constant. Actually, it will change because of the change in temperature.)

28–22 A typical small flashlight contains two batteries, each having an emf of 1.5 V, connected in series with a bulb having resistance 15 Ω.

a) If the internal resistance of the batteries is negligible, what power is delivered to the bulb?

b) If the batteries last for five hours, what is the total energy delivered to the bulb?

c) The resistance of real batteries increases as they run down. If the initial internal resistance is negligible, what is the internal resistance of each battery when the power to the bulb has decreased to half its initial value?

28–23 The capacity of a storage battery, such as those used in automobile electrical systems, is rated in ampere-hours (A·hr). A 50-A·hr battery can supply a current of 50 A for one hour, or 25 A for two hours, and so on.

a) What total energy is stored in a 12-V, 50-A·hr battery if its internal resistance is negligible?

b) What volume (in liters) of gasoline has a total heat of combustion equal to the energy obtained in (a)? (See Section 15–5, near the end. Take the density of gasoline to be 900 kg·m^{-3}.)

c) If a windmill-powered generator has an average electric-power output of 300 W, how much time will be required for it to charge the battery fully?

28–24 In the circuit in Fig. 28–17, find

a) the rate of conversion of internal (chemical) energy to electrical energy within the battery;

b) the rate of dissipation of electrical energy in the battery;

c) the rate of dissipation of electrical energy in the external resistor.

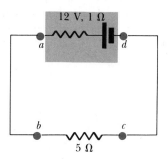

FIGURE 28–17

28–25 A person with body resistance between hands of 10 kΩ accidentally grasps the terminals of a 20-kV power supply.

a) If the internal resistance of the power supply is 1000 Ω, what is the current through the person's body?

b) What is the power dissipated in the body?

c) If the power supply is to be made safe by increasing its internal resistance, what should the internal resistance be for the maximum current in the above situation to be 0.001 A or less?

28–26 The average bulk resistivity of the human body (apart from surface resistance of the skin) is about 5 Ω·m. The conducting path between the hands can be represented approximately as a cylinder 1.6 m long and 0.1 m in diameter. The skin resistance can be made negligible by soaking the hands in salt water (or sea water).

a) What is the resistance between the hands if the skin resistance is negligible?

b) What potential difference between the hands is needed for a lethal shock current of 100 mA?

c) With the current in (b), what power is dissipated in the body?

d) Does the result of (b) increase your respect for electrical-shock hazards?

PROBLEMS

28–27 In the Bohr model of the hydrogen atom, the electron makes about 6×10^{15} rev·s^{-1} around the nucleus. What is the average current at a point on the orbit of the electron?

28–28 A vacuum diode (see Exercise 26–16) can be approximated by a plane cathode and a plane anode, parallel to each other and 5 mm apart. The area of both cathode and anode is 2 cm^2. In the region between cathode and anode the current is carried solely by electrons. If the electron current is 50 mA, and the electrons strike the anode surface with a speed of 1.2×10^7 m·s^{-1}, find the number of electrons per cubic millimeter in the space just outside the surface of the anode.

28–29 A certain electrical conductor has a square cross section 2.0 mm on a side and is 12 m long. The resistance between its ends is 0.072 Ω.

a) What is the resistivity of the material?

b) If the electric-field magnitude in the conductor is 0.12 V·m^{-1}, what is the total current?

c) If the material has 8.0×10^{28} free electrons per cubic meter, find the average drift velocity under conditions of part (b).

28–30 The current in a wire varies with time according to the relation

$$I = (20 \text{ A}) \sin (377 \text{ s}^{-1}t).$$

a) How many coulombs pass a cross section of the wire in the time interval between $t = 0$ and $t = \frac{1}{120}$ s?

b) In the interval between $t = 0$ and $t = \frac{1}{60}$ s?

c) What constant current would transport the same charge in each of the intervals above?

28–31 Two parallel plates of a capacitor have equal and opposite charges Q. The dielectric has a dielectric constant K and a resistivity ρ. Show that the "leakage" current carried by the dielectric is given by the relationship $i = Q/K\epsilon_0\rho$.

28–32 The region between two concentric conducting spheres of radii r_a and r_b is filled with a conducting material of resistivity ρ.

a) Show that the resistance between the spheres is given by

$$R = \frac{\rho}{4\pi}\left(\frac{1}{r_a} - \frac{1}{r_b}\right).$$

b) Derive an expression for the current density as a function of radius, if the potential difference between the spheres is V_{ab}.

c) Show that the result in (a) reduces to Eq. (28–10) when the separation $l = r_b - r_a$ between the spheres is small.

28–33 A piece of wire has a resistance R. It is cut into three pieces of equal length, and the pieces are twisted together in parallel. What is the resistance of the resulting wire?

28–34 A toaster using a Nichrome heating element operates on 120 V. When it is switched on at 0°C, it carries an initial current of 1.5 A. A few seconds later the current reaches the steady value of 1.33 A. What is the final temperature of the element? The average value of the temperature coefficient of Nichrome over the temperature range is $0.00045(\text{C}°)^{-1}$.

28–35 The potential difference across the terminals of a battery is 8.5 V when there is a current of 3 A in the battery from the negative to the positive terminal. When the current is 2 A in the reverse direction, the potential difference becomes 11 V.

a) What is the internal resistance of the battery?

b) What is the emf of the battery?

28–36 The open-circuit terminal voltage of a source is 10 V and its short-circuit current is 4.0 A.

a) What will be the current when the source is connected to a linear resistor of resistance 2 Ω?

b) What will be the current in the Thyrite resistor of Exercise 28–18 when connected across the terminals of this source?

c) What is the terminal voltage at this current?

28–37 A certain 12-V storage battery has a capacity of 60 A·hr. (See Exercise 28–23.) Its internal resistance is 0.2 Ω. The battery is charged by passing a 15-A current through it for 4 hr.

a) What is the terminal voltage during charging?

b) What total electrical energy is supplied to the battery during charging?

c) What electrical energy is dissipated in the internal resistance during charging?

The battery is now completely discharged through a resistor, again with a constant current of 15 A.

d) What is the external circuit resistance?

e) What total electrical energy is supplied to the external resistor?

f) What total electrical energy is dissipated in the internal resistance?

g) Why are the answers to (b) and (e) not equal?

28–38 Repeat Problem 28–37 with charge and discharge currents of 30 A. The charging and discharging times will now be 2 hr rather than 4 hr. What differences in performance do you see?

28–39 In the circuit of Fig. 28–18, find

a) the current through the 8-Ω resistor;

b) the total rate of dissipation of electrical energy (conversion to heat) in the 8-Ω resistor and in the internal resistance of the batteries.

c) In one of the batteries chemical energy is being converted into electrical energy. In which one is this happening, and at what rate?

d) In one of the batteries electrical energy is being converted into chemical energy. In which one is this happening, and at what rate?

e) Show that the overall rate of production of electrical energy equals the overall rate of consumption of electrical energy in the circuit.

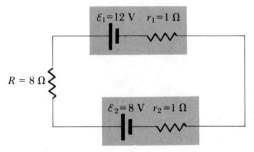

FIGURE 28–18

CHALLENGE PROBLEMS

28–40 A source whose emf is $\mathcal{E}$ and whose internal resistance is r is connected to an external circuit.

a) Show that the power output of the source is maximum when the current in the circuit is one-half the short-circuit current of the source.

b) If the external circuit consists of a resistance R, show that the power output is maximum when $R = r$, and that the maximum power is $\mathcal{E}^2/4r$.

28–41 The definition of α, the temperature coefficient of resistivity, is given by

$$\alpha = \left(\frac{1}{\rho}\right)\frac{d\rho}{dT},$$

where ρ is the resistivity at the temperature T. Eq. (28–7) then follows if α is assumed constant and much smaller than $(T - T_0)^{-1}$. (See Challenge Problem 14–36 for the analogous case of the coefficient of linear expansion.)

a) If α is not constant but is given by $\alpha = -n/T$, where T is the Kelvin temperature and n is a constant, show that the resistivity ρ is given by $\rho = a/T^n$ where a is a constant.

b) From Fig. 28–2c, it is seen that such a relation might be used as a rough approximation for a semiconductor. Using the values of ρ for carbon from Table 28–1 and of α from Table 28–2, determine a and n. (In Table 28–1 assume that "room temperature" means 293 K.)

c) Using your result from part (b), determine the resistivity of carbon at the freezing and normal boiling points of water. (Remember to express T in kelvins.)

28–42 A semiconductor diode is a nonlinear device whose current–voltage relation is described by

$$I = I_0[\exp(eV/kT) - 1],$$

where I and V are, respectively, the current through and the voltage across the diode; I_0 is a constant characteristic of the device; e is the electron charge; k is Boltzmann's constant; and T is the Kelvin temperature. Such a diode is connected in series with a 1-Ω resistor and a 2-V battery. The polarity of the battery is such that the current through the diode is in the forward direction. (See Fig. 28–19.)

a) Obtain an equation for V. Note that V cannot be solved for analytically.

b) Since V cannot be solved for analytically, the value of V must be obtained by using a numerical method, such as an iterative technique. Using $I_0 = 0.5$ A and $T = 293$ K, obtain a solution (accurate to three significant figures) for the voltage drop V across the diode and the current I through it. (For a brief discussion of how to do an iterative calculation, see Challenge Problem 27–36, part [c]. This procedure can also be very effectively done on a computer or programmable calculator.)

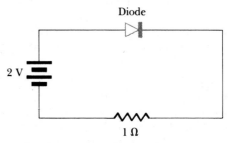

FIGURE 28–19

29

DIRECT-CURRENT CIRCUITS

IN THE LAST CHAPTER WE STUDIED SOME OF THE BASIC PRINCIPLES UNDER-lying the behavior of electric currents in simple circuits. But many circuits used in the real world contain several sources, resistors, and other circuit elements such as capacitors and motors, interconnected in a *network*. In this chapter we study general methods for analyzing networks, including finding unknown voltages, currents, and properties of circuit elements.

When several resistors are connected in series or in parallel, they can always be represented as a single equivalent resistor. To analyze more general networks we need two rules called *Kirchhoff's rules*. One is basically the principle of conservation of charge applied to a junction; the other is based on the sum of potential differences around a closed loop. We also discuss instruments for measuring various electrical quantities. Finally, we examine the concept of *displacement current* in the charging of capacitors.

29–1 RESISTORS IN SERIES AND PARALLEL

Figure 29–1 shows four different ways in which three resistors having resistances R_1, R_2, and R_3 might be connected between points a and b. In (a), the resistors provide only a single path between these points. Any number of circuit elements such as resistors, cells, and motors are in **series** with one another between two points if they are connected as in (a) so as to provide only a single current path between the points. The *current* is the same in each element.

The resistors in Fig. 29–1b are said to be in **parallel** between points a and b. Each resistor provides an alternative path between the points, and any number of circuit elements similarly connected are in parallel with one another. The *potential difference* is the same across each element.

In Fig. 29–1c, resistors R_2 and R_3 are in parallel with each other, and this combination is in series with the resistor R_1. In Fig. 29–1d, R_2 and R_3 are in series, and this combination is in parallel with R_1.

It is always possible to find a single resistor that could replace a combination of resistors and leave unaltered both the potential difference between the

Combining resistors in a circuit: What is the equivalent resistance of a combination?

What do we mean by equivalent resistance?

653

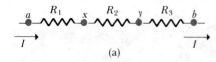

(a)

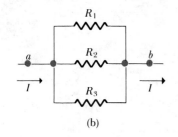

(b)

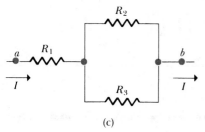

(c)

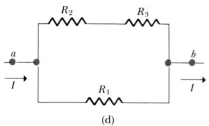

(d)

29–1 Four different ways of connecting three resistors.

For resistors in series, the currents are all the same.

Resistances in series add directly.

For resistors in parallel, the voltages are all the same.

Resistances in parallel add reciprocally.

terminals of the combination and the current in the rest of the circuit. The resistance of this single resistor is called the **equivalent resistance** of the combination. If any one of the networks in Fig. 29–1 were replaced by its equivalent resistance R, we could write

$$V_{ab} = IR \quad \text{or} \quad R = \frac{V_{ab}}{I},$$

where V_{ab} is the potential difference between terminals a and b of the network and I is the current at point a or b. To compute an equivalent resistance, we assume a potential difference V_{ab} across the actual network, compute the corresponding current I, and take the ratio of one to the other.

If the resistors are in *series*, as in Fig. 29–1a, the current I must be the same in all of them. Hence

$$V_{ax} = IR_1, \qquad V_{xy} = IR_2, \qquad V_{yb} = IR_3,$$

and

$$V_{ab} = V_{ax} + V_{xy} + V_{yb} = I(R_1 + R_2 + R_3), \quad \frac{V_{ab}}{I} = R_1 + R_2 + R_3.$$

But V_{ab}/I is, by definition, the equivalent resistance R. Therefore

$$R = R_1 + R_2 + R_3. \qquad (29–1)$$

The equivalent resistance of *any number* of resistors in series equals the sum of their individual resistances.

If the resistors are in *parallel*, as in Fig. 29–1b, the potential difference between the terminals of each must be the same and equal to V_{ab}. If the currents in each are denoted by I_1, I_2, and I_3, respectively,

$$I_1 = \frac{V_{ab}}{R_1}, \qquad I_2 = \frac{V_{ab}}{R_2}, \qquad I_3 = \frac{V_{ab}}{R_3}.$$

Charge is delivered to point a by the current I and removed from a by the currents I_1, I_2, and I_3. Since charge is not accumulating at a, it follows that

$$I = I_1 + I_2 + I_3 = V_{ab}\left(\frac{1}{R_1} + \frac{1}{R_2} + \frac{1}{R_3}\right), \quad \text{or} \quad \frac{I}{V_{ab}} = \frac{1}{R_1} + \frac{1}{R_2} + \frac{1}{R_3}.$$

But

$$\frac{I}{V_{ab}} = \frac{1}{R},$$

so

$$\frac{1}{R} = \frac{1}{R_1} + \frac{1}{R_2} + \frac{1}{R_3}. \qquad (29–2)$$

For *any number* of resistors in parallel, the *reciprocal* of the equivalent resistance equals the *sum of the reciprocals* of their individual resistances.

For the special case of *two* resistors in parallel,

$$\frac{1}{R} = \frac{1}{R_1} + \frac{1}{R_2} = \frac{R_2 + R_1}{R_1 R_2}$$

and

$$R = \frac{R_1 R_2}{R_1 + R_2}.$$

Also, since $V_{ab} = I_1R_1 = I_2R_2$,

$$\frac{I_1}{I_2} = \frac{R_2}{R_1}, \tag{29–3}$$

and the currents carried by two resistors in parallel are *inversely proportional* to their resistances.

For two resistors in parallel, the current in each is inversely proportional to its resistance.

The equivalent resistances of the networks in Figs. 29–1c and 29–1d could be found by the same general method, but it is simpler to consider them as combinations of series and parallel arrangements. Thus in (c) the combination of R_2 and R_3 in parallel is first replaced by its equivalent resistance, which then forms a simple series combination with R_1. In (d) the combination of R_2 and R_3 in series forms a simple parallel combination with R_1.

PROBLEM-SOLVING STRATEGY: *Resistors in series and parallel*

1. It helps to remember that when two or more resistors are connected in series, the total potential difference across the combination is the sum of the individual potential differences. When they are connected in parallel, the potential difference is the same for every resistor, and the total potential difference across the combination is equal to that for each individual resistor.

2. Also keep in mind the analogous statements for current. When two or more resistors are connected in series, the current is the same through every resistor, and this is also equal to the current through the combination. When resistors are connected in parallel, the total current through the combination is equal to the sum of currents through the individual resistors.

EXAMPLE 29–1 Compute the equivalent resistance of the network in Fig. 29–2, and find the current in each resistor.

Examples of combining resistors in series and parallel

SOLUTION Successive stages in the reduction to a single equivalent resistance are shown in parts (b) and (c). From Eq. (29–2), the 6-Ω and the 3-Ω resistors in parallel in part (a) are equivalent to the single 2-Ω resistor in part (b), and the series combination of this with the 4-Ω resistor results in the single equivalent 6-Ω resistor in part (c).

In the simple series circuit of (c), the current is 3 A, and hence the current in the 4-Ω and 2-Ω resistors in part (b) is also 3 A. The potential difference V_{cb} is therefore 6 V, and since it must be 6 V in part (a) as well, the currents in the 6-Ω and 3-Ω resistors in part (a) are 1 A and 2 A, respectively.

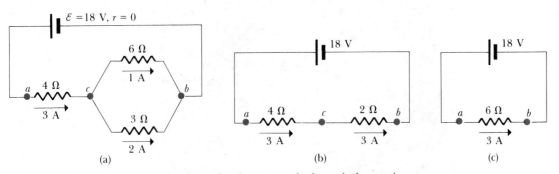

(a)   (b)   (c)

29–2 Steps in reducing a combination of resistors to a single equivalent resistor.

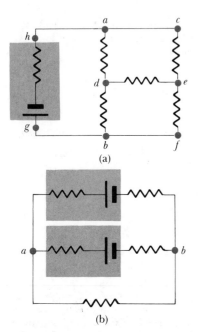

(a)

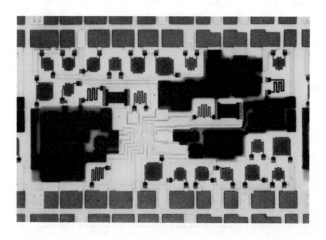

(b)

29–3 Two networks that cannot be reduced to simple series–parallel combinations of resistors.

A printed circuit using deposited thin films for conducting paths. This circuit cannot be reduced to simple series and parallel combinations, but it can be analyzed using Kirchhoff's rules. (Courtesy of AT&T Bell Laboratories.)

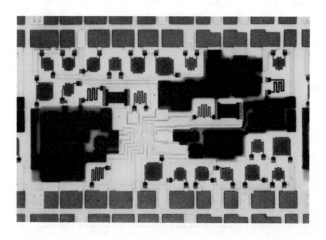

29–2 KIRCHHOFF'S RULES

Not all networks can be reduced to simple series–parallel combinations. An example is a resistance network with a cross connection, as in Fig. 29–3a. A circuit like that in Fig. 29–3b, which contains sources in parallel paths, is another example. No new *principles* are required to compute the currents in these networks, but a number of techniques are available that help us handle such problems systematically. We will describe one of these, first developed by Gustav Robert Kirchhoff (1824–1887).

We first define two terms. A **branch point** in a network is a point where three or more conductors are joined. A **loop** is any closed conducting path. In Fig. 29–3a, for example, points *a*, *b*, *d*, and *e* are branch points but *c* and *f* are not. The circuit in Fig. 29–3b has only two branch points, *a* and *b*. Some possible loops in Fig. 29–3a are the closed paths *aceda*, *defbd*, *hadbgh*, and *hadefbgh*.

Kirchhoff's rules consist of the following two statements. **Point rule:** *The algebraic sum of the currents toward any branch point is zero:*

$$\sum I = 0. \tag{29–4}$$

Loop rule: *The algebraic sum of the potential differences in any loop*, including those associated with emf's and those of resistive elements, *must equal zero.*

The point rule is an application of the principle of *conservation of electric charge*. Since no charge can accumulate at a branch point, the total current entering the point must equal the total current leaving, or (considering those entering as positive and those leaving as negative) the algebraic sum of currents into a point must be zero.

We have already seen the loop rule in Section 28–4. It is an expression of an *energy* relationship; as a charge goes around a loop and returns to its starting point, the algebraic sum of the changes in potential must be zero. Rises in potential are associated with sources of emf, and drops are associated with resistors and other circuit elements. No matter what the detailed nature of the circuit elements is, or whether Ohm's law is obeyed, the algebraic sum of potential differences around every closed loop must be zero.

These basic rules are all we need to solve a wide variety of network problems. Usually some of the emf's, currents, and resistances are known, and others are unknown. We must always obtain from Kirchhoff's rules a number of equations equal to the number of unknowns, so we can solve the equations simultaneously. Often the hardest part of the solution is not in understanding the basic principles but in keeping track of algebraic signs!

PROBLEM-SOLVING STRATEGY: *Kirchhoff's Rules*

1. Draw a *large* circuit diagram so you have plenty of room for labels. Label all quantities, known and unknown, including an assumed sense of direction for each unknown current and emf. Often you will not know in advance the actual direction of an unknown current or emf, but this doesn't matter. Carry out your solution, using the assumed direction. If the actual direction of a particular quantity is opposite to your assumption, the result will come out with a negative sign. Hence if you use Kirchhoff's rules correctly, they give you the directions as well as magnitudes of unknown currents and emf's. We have made this point before, in Section 28–4; you may want to review that discussion now. We will illustrate it further in the following examples.

2. Usually, when you label currents it is best to use the point rule immediately to express the currents in terms of as few quantities as possible. For example, Fig. 29–4a shows a circuit correctly labeled; Fig. 29–4b shows the same circuit, relabeled by applying the point rule to point a to eliminate I_3.

3. Choose any closed loop in the network, and designate a direction (clockwise or counterclockwise) to traverse the loop in applying the loop rule.

4. Go around the loop in the designated direction, adding potential differences as you cross them. An emf is counted as positive when it is traversed from $-$ to $+$, and negative when from $+$ to $-$. An IR product is negative if your path passes through the resistor in the *same* direction as the assumed current, positive if in the opposite direction. "Uphill" potential changes are always positive, "downhill" changes negative.

5. Equate the sum in step 4 to zero.

6. If necessary, choose another loop to get a different relation among the unknowns, and continue until you

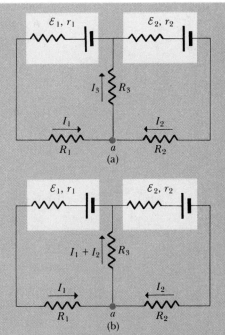

29–4 Application of the point rule to point a reduces the number of unknown currents from three to two.

have as many equations as unknowns, or until every circuit element has been included in at least one of the chosen loops.

7. Finally, solve the equations simultaneously to determine the unknowns. This step is algebra, not physics, but it can be fairly complex. Be careful with algebraic manipulations; one sign error is fatal to the entire solution.

EXAMPLE 29–2   In the circuit shown in Fig. 29–5, find the unknown current I, resistance R, and emf $\mathcal{E}$.

Some examples of circuit problems using Kirchhoff's rules

SOLUTION   Application of the point rule to point a yields the relation

$$I + 1\ \text{A} - 6\ \text{A} = 0,$$
$$I = 5\ \text{A}.$$

To determine R we apply the loop rule to the loop labeled (1), obtaining

$$18\ \text{V} - (5\ \text{A})R + (1\ \text{A})(2\ \Omega) = 0,$$
$$R = 4\ \Omega.$$

The term for resistance R is negative because our loop traverses that element in the same direction as the current and hence finds a potential *drop*, while the term

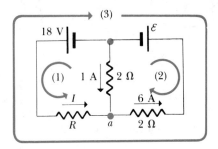

29–5   Circuit for Example 29–2.

for the 2-Ω resistor is positive because in traversing it in the direction opposite to the current we find a potential *rise*. If we had chosen to traverse loop (1) in the opposite direction, every term would have had the opposite sign, and the result for R would have been the same.

To determine $\mathcal{E}$, we apply the loop rule to loop (2):

$$\mathcal{E} + (6\ \text{A})(2\ \Omega) + (1\ \text{A})(2\ \Omega) = 0,$$
$$\mathcal{E} = -14\ \text{V}.$$

This result shows that the actual polarity of this emf is opposite to that assumed, and that the positive terminal of this source is really on the left side. Alternatively, we could use loop (3), obtaining the equation

$$\mathcal{E} + (6\ \text{A})(2\ \Omega) + (5\ \text{A})(4\ \Omega) - 18\ \text{V} = 0,$$

from which again $\mathcal{E} = -14\ \text{V}$.

EXAMPLE 29–3   In Fig. 29–6, find the current in each resistor and the equivalent resistance of the network.

SOLUTION   As pointed out at the beginning of this section, it is not possible to represent this network in terms of series and parallel combinations. There are five different currents to determine, but by applying the point rule to junctions a and b we represent them in terms of three unknown currents, as indicated in the figure. The current in the battery is $(I_1 + I_2)$.

We now apply the loop rule to the three loops shown, obtaining the following three equations:

$$13\ \text{V} - I_1(1\ \Omega) - (I_1 - I_3)(1\ \Omega) = 0; \qquad (1)$$
$$-I_2(1\ \Omega) - (I_2 + I_3)(2\ \Omega) + 13\ \text{V} = 0; \qquad (2)$$
$$-I_1(1\ \Omega) - I_3(1\ \Omega) + I_2(1\ \Omega) = 0. \qquad (3)$$

This is a set of three simultaneous equations for the three unknown currents. They may be solved by various methods; one straightforward procedure is to solve the third for I_2, obtaining $I_2 = I_1 + I_3$, and then substitute this expression into the first two equations to eliminate I_2. When this is done we obtain the two equations

$$13\ \text{V} = I_1(2\ \Omega) - I_3(1\ \Omega), \qquad (1')$$
$$13\ \text{V} = I_1(3\ \Omega) - I_3(5\ \Omega), \qquad (2')$$

Now I_3 may be eliminated by multiplying the first of these by 5 and adding the two equations, obtaining

$$78\ \text{V} = I_1(13\ \Omega),$$

from which $I_1 = 6$ A. This result may be substituted back into $(1')$ to obtain $I_3 = -1$ A, and finally from (3) we find $I_2 = 5$ A. Note that the direction of I_3 is opposite that of the initial assumption.

The total current through the network is $I_1 + I_2 = 11$ A, and the potential drop across it is equal to the battery emf, namely, 13 V. Thus the equivalent resistance of the network is

$$R = \frac{13\ \text{V}}{11\ \text{A}} = 1.18\ \Omega.$$

Sometimes you have to solve several simultaneous equations.

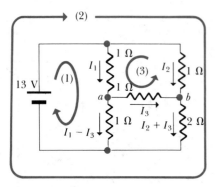

29–6 Circuit for Examples 29–3 and 29–4.

Once the currents in a circuit are determined, the potential difference between any two points a and b can be found by the procedure discussed in Section 28–4, just preceding Example 28–8. To find $V_{ab} = V_a - V_b$ (the potential at a with respect to b), we start at b and add the potential differences as we go from b to a. An emf is considered positive when we go from $-$ to $+$, negative otherwise. An IR term is positive when we go "uphill," against the current direction, negative when in the same direction as the current.

How to find the potential difference between two points when the currents are known

EXAMPLE 29–4 In the circuit of Example 29–3 (Fig. 29–6) find the potential difference V_{ab}.

SOLUTION Starting at point b, we follow a path to point a, adding potential rises and drops as we go. The simplest path is through the center 1-Ω resistor. We have found $I_3 = -1$ A, showing that the actual current direction in this branch is from right to left. Thus as we go from b to a there is a drop of potential of magnitude $IR = (1\text{ A})(1\text{ }\Omega) = 1$ V, and $V_{ab} = -1$ V. Alternatively, we may go around the lower loop. We then have

$$I_2 + I_3 = 5\text{ A} + (-1\text{ A}) = 4\text{ A},$$
$$I_1 - I_3 = 6\text{ A} - (-1\text{ A}) = 7\text{ A},$$

and

$$V_{ab} = -(4\text{ A})(2\text{ }\Omega) + (7\text{ A})(1\text{ }\Omega) = -1\text{ V}.$$

We suggest you try some other paths from b to a to verify that they also give this result.

29–3 ELECTRICAL INSTRUMENTS

Many familiar instruments for measuring potential difference (voltage), current, or resistance use a device called a **d'Arsonval galvanometer.** A pivoted coil of fine wire is placed in the magnetic field of a permanent magnet, as shown in Fig. 29–7. When there is a current in the coil, the magnetic field exerts on the coil a *torque* that is proportional to the current. (This magnetic interaction is discussed in detail in Chapter 30.) The torque is opposed by a spring, similar to the hairspring on the balance wheel of a watch, which exerts a restoring torque proportional to the angular displacement.

The d'Arsonval galvanometer: a common current-measuring instrument

Thus the angular deflection of the pointer attached to the pivoted coil is directly proportional to the coil current, and the device can be calibrated to measure current. The maximum deflection for which the meter is designed, typically 90° to 120°, is called *full-scale deflection*. The current required to produce full-scale deflection (typically of the order of 10 μA to 10 mA) and the resistance of the coil (typically of the order of 10 to 1000 Ω) are the essential electrical characteristics of the meter.

The meter deflection is proportional to the *current* in the coil, but if the coil obeys Ohm's law, the current is proportional to the *potential difference* between the terminals of the coil. Thus the deflection is also proportional to this potential difference. For example, consider a meter whose coil has a resistance of 20 Ω, and which deflects full scale with a current of 1 mA in its coil. The

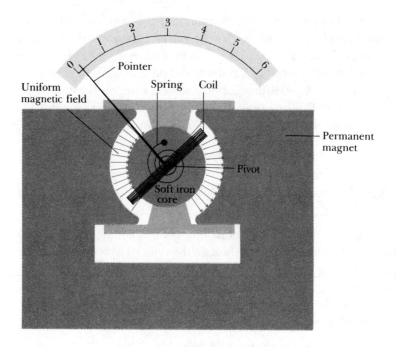

29–7 A d'Arsonval meter movement, showing pivoted coil with attached pointer, permanent magnet supplying uniform magnetic field, and spring to provide restoring torque, which opposes magnetic-field torque.

corresponding potential difference is

$$V_{ab} = IR = (10^{-3} \text{ A})(20 \ \Omega) = 0.020 \text{ V} = 20 \text{ mV}.$$

A current-measuring instrument is usually called an **ammeter** (or milliammeter, microammeter, etc., depending on the range). Such a device always measures the current passing through it. A meter such as the one just described can be adapted to measure currents larger than its full-scale reading by connecting a resistor in parallel with it, as in Fig. 29–8a, so that some of the current bypasses the meter. The parallel resistor is called a **shunt resistor** or simply a *shunt*, symbol R_{sh}.

For example, suppose we need an ammeter with a range of 0 A to 10 A, based on the 1-mA meter described above. We must choose a shunt such that the total current I through both meter and shunt is 10 A when the current through the meter itself is 1 mA = 0.001 A. Thus the current in the shunt is 9.999 A at full-scale deflection. The potential difference across the shunt is the same as that across the meter, namely 0.020 V, so the shunt resistance must be (from Ohm's law)

Using a d'Arsonval meter to measure larger currents: using a shunt to bypass some of the current

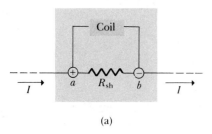

(a)

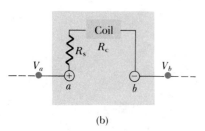

(b)

29–8 (a) Internal connections of an ammeter. (b) Internal connections of a voltmeter.

$$R_{sh} = \frac{0.020 \text{ V}}{9.999 \text{ A}}$$
$$= 0.00200 \ \Omega.$$

The equivalent resistance R of the instrument is

$$\frac{1}{R} = \frac{1}{R_c} + \frac{1}{R_{sh}},$$

and

$$R = 0.00200 \ \Omega.$$

Thus we have a low-resistance instrument with the desired range of 0 to 10 A. Of course, if the current I is *less* than 10 A, the coil current and the deflection are correspondingly less.

An *ideal* ammeter would have zero resistance, so that including it in a branch of a circuit would not affect the current in that branch. Real ammeters always have some finite resistance, but it is always desirable for an ammeter to have the smallest practical resistance.

An ideal ammeter has zero internal resistance.

This same 1-mA meter may also be used to measure potential difference or *voltage;* any voltage-measuring device is usually called a **voltmeter** (or millivoltmeter, etc., depending on range). A voltmeter always measures the potential difference between two points, and its terminals must be connected to these points. Our 1-mA meter may be used as a voltmeter, but the maximum voltage it can measure is 20 mV. The range may be extended by connecting a resistor R_s in series with the meter, as in Fig. 29–8b, so that only some fraction of the total potential difference appears across the meter itself, and the remainder across R_s.

Using a d'Arsonval meter to measure voltage (potential difference)

For example, suppose we need a voltmeter with a maximum range of 10 V. Then when the voltage across the meter is 20 mV = 0.020 V, the voltage across the series resistor R_s must be 10 V − 0.020 V, or 9.98 V. The current through the meter at full-scale deflection is still 1 mA or 0.001 A, so from Ohm's law the value of R_s must be

$$R_s = \frac{9.98 \text{ V}}{0.001 \text{ A}} = 9980 \ \Omega.$$

The equivalent resistance of the device is then

$$R = R_c + R_s = 10{,}000 \ \Omega.$$

An ideal voltmeter would have infinite resistance, so that connecting it between two points in a circuit would not alter any of the currents. Real voltmeters always have finite resistance; to be useful, however, a voltmeter must have large enough resistance so that connecting it in a circuit does not change the other currents appreciably.

An ideal voltmeter has infinite internal resistance.

A voltmeter and an ammeter can be used together to measure *resistance* and *power*. The resistance of a resistor equals the potential difference V_{ab} between its terminals, divided by the current I:

Measuring resistance with voltmeter and ammeter: Finite internal resistances of the meters cause errors that can be corrected.

$$R = \frac{V_{ab}}{I},$$

and the power input to any portion of a circuit equals the product of the potential difference across this portion and the current:

$$P = V_{ab}I.$$

The most straightforward method of measuring R or P is therefore to measure V_{ab} and I simultaneously.

In Fig. 29–9a, ammeter A reads correctly the current I in the resistor R. Voltmeter V, however, reads the *sum* of the potential difference V_{ab} across the resistor and the potential difference V_{bc} across the ammeter.

If we transfer the voltmeter terminal from c to b, as in Fig. 29–9b, the voltmeter reads correctly the potential difference V_{ab}, but the ammeter now reads the *sum* of the current I in the resistor and the current I_V in the voltmeter. Thus, whichever connection is used, we must correct the reading of one instrument or the other to obtain the true values of V_{ab} or I (unless, of course, the corrections are small enough to be neglected).

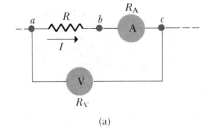

(a)

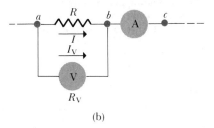

(b)

29–9 Ammeter–voltmeter method for measuring resistance or power.

EXAMPLE 29–5   Suppose we want to measure an unknown resistance R by
using the circuit of Fig. 29–9a. The meter resistances are $R_V = 10,000\ \Omega$ and
$R_A = 2.0\ \Omega$. If the voltmeter reads 12.0 V and the ammeter reads 0.10 A, what is
the true resistance?

SOLUTION   If the meters were ideal (i.e., $R_V = \infty$ and $R_A = 0$), the resistance would
be simply $R = V/I = (12.0\ \text{V})/(0.10\ \text{A}) = 120\ \Omega$. But the voltmeter reading includes the potential V_{bc} across the ammeter as well as that (V_{ab}) across the resistor.
We have $V_{bc} = IR_A = (0.10\ \text{A})(2.0\ \Omega) = 0.2\ \text{V}$, so the actual potential drop V_{ab}
across the resistor is 12.0 V − 0.2 V = 11.8 V, and the resistance is

$$R = \frac{V_{ab}}{I} = \frac{11.8\ \text{V}}{0.10\ \text{A}} = 118\ \Omega.$$

EXAMPLE 29–6   Suppose the meters of Example 29–5 are connected to a different resistor, in the circuit shown in Fig. 29–9b, and the readings noted above
are obtained. What is the true resistance?

SOLUTION   In this case the voltmeter measures the potential across the resistor
correctly; the difficulty is that the ammeter measures the voltmeter current I_V as
well as the current I in the resistor. We have $I_V = V/R_V = (12.0\ \text{V})/(10,000\ \Omega) =$
1.2 mA; so the actual current I in the resistor is $I = 0.10\ \text{A} - 0.0012\ \text{A} =$
0.0988 A. Thus the resistance is

$$R = \frac{V_{ab}}{I} = \frac{12.0\ \text{V}}{0.0988\ \text{A}} = 121.5\ \Omega.$$

An alternative method for measuring resistance is to use a d'Arsonval
meter in an arrangement called an **ohmmeter.** It consists of a meter, a resistor,
and a source (often a flashlight cell) connected in series, as in Fig. 29–10. The
resistance R to be measured is connected between terminals x and y.

The series resistance R_s is chosen so that when terminals x and y are short-
circuited (that is, when $R = 0$), the meter deflects full scale. When the circuit
between x and y is open (that is, when $R = \infty$), the galvanometer shows no
deflection. For a value of R between zero and infinity, the meter deflects to
some intermediate point depending on the value of R, and hence the meter
scale can be calibrated to read the resistance R. Larger currents correspond to
smaller resistances, so this scale reads backward compared to the current scale.

In situations where high precision is required, instruments containing
d'Arsonval meters have been supplanted by electronic instruments with direct
digital readouts. These are more precise, stable, and mechanically rugged
than d'Arsonval meters, but they are also considerably more expensive. Digital
voltmeters can be made with extremely high internal resistance, of the order
of 100 MΩ.

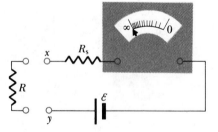

29–10 Ohmmeter circuit. The
backward scale on the meter is
calibrated to read resistance directly.

29–4 THE *R–C* SERIES CIRCUIT

Thus far in our discussion of circuits, we have assumed that the emf's and
resistances are constant, so that all potentials and currents are constant, independent of time. Figure 29–11 shows a simple example of a circuit in which

the current and voltages are *not* constant. The capacitor is initially uncharged; at some initial time $t = 0$ we close the switch, completing the circuit and permitting current around the loop to begin charging the capacitor. The current begins at the same instant in every part of the circuit, and at each instant the current is the same in every part.

Because the capacitor is initially uncharged, the initial potential difference across it is zero. The entire battery voltage appears across the resistor, causing an initial current $I = V/R$. As the capacitor charges, its voltage increases, and the potential difference across the resistor decreases, corresponding to a decrease in current. After a long time the capacitor becomes fully charged and the entire battery voltage V appears across the capacitor. There is then no potential difference across the resistor, and the current is zero.

We let q represent the charge on the capacitor and i the current in the circuit at some time t after the switch has been closed. It is customary in circuit analysis to use small letters for quantities that vary with time, and capital letters for constant quantities; we follow that convention here. We note that i and q are related by $i = dq/dt$; the instantaneous potential differences v_{ac} and v_{cb} are

$$v_{ac} = iR, \qquad v_{cb} = \frac{q}{C}. \tag{29–5}$$

Therefore

$$V_{ab} = V = v_{ac} + v_{cb} = iR + \frac{q}{C}, \tag{29–6}$$

where V is the terminal voltage of the battery, assumed to be constant. (That is, we neglect the internal resistance of the battery; V is then equal to its emf.) Solving this equation for i, we find

$$i = \frac{V}{R} - \frac{q}{RC}. \tag{29–7}$$

At time $t = 0$, when the switch is first closed, the capacitor is uncharged, so $q = 0$. The *initial* current, which we call I_0, is, from Eq. (29–7), given by $I_0 = V/R$. This would be the (constant) current if the capacitor were not in the circuit.

As the charge q increases, the term q/RC becomes larger, and the current decreases and eventually becomes zero. When $i = 0$,

$$\frac{V}{R} = \frac{q}{RC}, \qquad q = CV = Q_f,$$

where Q_f is the final charge on the capacitor. The behavior of the current and capacitor charge as functions of time are shown in Fig. 29–12. At the instant the switch is closed ($t = 0$), the current jumps from zero to its initial value $I_0 = V/R$, and then gradually decreases to zero. The capacitor charge starts at zero and gradually approaches the final value $Q_f = CV$.

We can derive expressions for the charge q and current i as functions of time. To do this, we replace i in Eq. (29–7) with dq/dt, obtaining

$$\frac{dq}{dt} = \frac{V}{R} - \frac{q}{RC}, \tag{29–8}$$

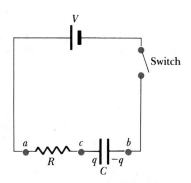

29–11 A capacitor C charged by a circuit containing a battery V and a resistor R.

The initial voltage across the capacitor is zero.

The final current in the circuit is zero.

Applying Kirchhoff's loop law to the capacitor-charging circuit

A differential equation for capacitor charge as a function of time

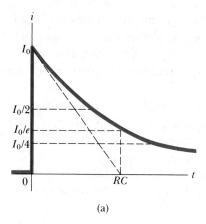

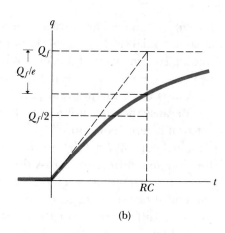

29–12 Current and capacitor charge as functions of time for the circuit of Fig. 29–11.

(a)   (b)

which may be rearranged as

$$\frac{dq}{VC - q} = \frac{dt}{RC}.$$

Integrating both sides, we obtain

$$-\ln(VC - q) = t/RC + \text{constant}.$$

To evaluate the constant, note that at $t = 0$, $q = 0$, so

$$-\ln(VC - 0) = 0 + \text{constant}.$$

Rearranging again, we obtain

The current is an exponentially decreasing function of time.

$$\ln(VC - q) - \ln VC = \ln\frac{VC - q}{VC} = -\frac{t}{RC},$$

$$1 - \frac{q}{VC} = e^{-t/RC}, \tag{29–9}$$

$$q = VC(1 - e^{-t/RC}) = Q_f(1 - e^{-t/RC}).$$

The time derivative of this expression is the current:

$$i = \frac{dq}{dt} = \frac{V}{R}e^{-t/RC} = I_0 e^{-t/RC}. \tag{29–10}$$

The charge and current are therefore both *exponential* functions of time. Figure 29–12a is a graph of Eq. (29–10), and Fig. 29–12b is a graph of Eq. (29–9). At a time $t = RC$, the current has decreased to $1/e$ (about 0.368) of its initial value and the charge has increased to *within* $1/e$ of its final value. The product RC is called the **time constant,** or the **relaxation time,** of the circuit, denoted by τ:

The time constant is the time needed for the current to drop to $1/e$ of its initial value.

$$\tau = RC. \tag{29–11}$$

EXAMPLE 29–7 A resistor of resistance $R = 10\ \text{M}\Omega$ is connected in series with a capacitor of capacitance $1\ \mu\text{F}$. The time constant is

$$\tau = RC = (10 \times 10^6\ \Omega)(10^{-6}\ \text{F}) = 10\ \text{s}.$$

But if $R = 10\ \Omega$, the time constant is only 10×10^{-6} s, or $10\ \mu$s.

Suppose now that after the capacitor in Fig. 29–11 has acquired some charge Q_0, we remove the battery from the circuit and connect points a and b directly together. We reset our stopwatch so the connection is made at time $t = 0$. The capacitor then *discharges* through the resistor, and its charge eventually decreases to zero.

Again let i and q represent the time-varying current and charge at some instant after the switch is thrown. Since V_{ab} is now zero, we have, from Eq. (29–6),

$$0 = v_{ac} + v_{cb}.$$

The direction of the current in the resistor is now from c to a, so $v_{ca} = -v_{ac} = iR$, and

$$i = \frac{q}{RC}. \tag{29–12}$$

When $t = 0$, $q = Q_0$ and the initial current I_0 is

$$I_0 = \frac{Q_0}{RC} = \frac{V_0}{R},$$

where V_0 is the initial potential difference across the capacitor. As the capacitor discharges, both q and i decrease.

The same procedures as above can be followed to obtain $i(t)$ and $q(t)$. If we replace i in Eq. (29–12) by $-dq/dt$ (the charge q is now *decreasing*), we get

$$\frac{dq}{dt} = -\frac{q}{RC}. \tag{29–13}$$

Integration of this equation gives $q(t)$, and by differentiation we find $i(t)$.

Alternatively, differentiation of Eq. (29–12) gives

$$\frac{di}{dt} = -\frac{i}{RC}, \tag{29–14}$$

from which we can get $i(t)$ and, by a second integration, get $q(t)$. We leave it to you to show that

$$i = I_0 e^{-t/RC}, \tag{29–15}$$

$$q = Q_0 e^{-t/RC}. \tag{29–16}$$

Both the current and the charge decrease exponentially with time. Comparing these results with Eqs. (29–9) and (29–10), we note that the expressions for the current are identical. The capacitor charge approaches zero asymptotically in Eq. (29–16), while the *difference* between q and Q_f approaches zero asymptotically in Eq. (29–9).

Energy considerations offer us additional insight into the behavior of an RC circuit. The instantaneous rate at which the battery delivers energy to the circuit is $P = Vi$. The instantaneous rate at which energy is dissipated in the resistor is i^2R, and the rate at which energy is stored in the capacitor is $iv_{cb} = qi/C$. Multiplying Eq. (29–6) by i, we find

$$Vi = i^2R + iq/C. \tag{29–17}$$

This expression means that of the power Vi supplied by the battery, part (i^2R) is dissipated in the resistor and part (iq/C) is stored in the capacitor.

A capacitor discharges through a resistor: The behavior is similar to the charging process.

The charge and current are exponentially decreasing functions of time.

The time constant is the same for charging and discharging.

Energy relations for a charging or
discharging capacitor

The *total* energy supplied by the battery during charging of the capacitor equals the battery potential difference V multiplied by the total charge Q_f, or $Q_f V$. The total energy stored in the capacitor, from Eq. (27–8), is $Q_f V/2$. Thus of the energy supplied by the battery, *exactly half* is stored in the capacitor, and the other half is dissipated in the resistor. It is a little surprising that this half-and-half division of energy does not depend on C, R, or V. This result can also be verified in detail by taking the integral over time of each of the power quantities mentioned above; we leave this calculation for your amusement.

29–5 DISPLACEMENT CURRENT

For a *conducting* circuit the total current *into* any given portion must equal the current *out of* that portion. This statement forms the basis of Kirchhoff's point rule, discussed in Section 29–2. However, this rule is *not* obeyed for a capacitor that is being charged. In Fig. 29–13, a conduction current goes *into* the left plate, but no conduction current comes *out of* this plate; similarly, there is conduction current out of the right plate, but none into it.

James Clerk Maxwell (1831–1879) showed that it is possible to generalize the definition of current so that we can still say that the current out of each plate is equal to the current into it. As the capacitor charges, the conduction current increases the charge on each plate, and this, in turn, increases the electric field between the plates. The *rate* of increase of field is proportional to the conduction current. Maxwell's idea was to associate an equivalent current density with this rate of increase of field.

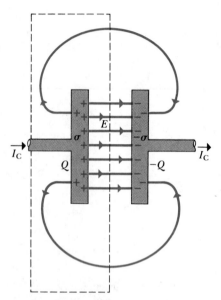

29–13 The conduction current into the left plate of the capacitor equals the displacement current between the plates.

Consider a parallel-plate capacitor with uniform charge density σ on the plates. The field magnitude E between the plates is then given by

$$E = \frac{\sigma}{\epsilon_0} = \frac{Q}{\epsilon_0 A}. \tag{29–18}$$

In a time interval dt, the charge Q increases by $dQ = I_C\, dt$; the corresponding change in E is

$$dE = \frac{dQ}{\epsilon_0 A} = \frac{I_C\, dt}{\epsilon_0 A},$$

Maxwell's breakthrough: associating a
changing electric field with an effective
current density

and the *rate* of change of E is

$$\frac{dE}{dt} = \frac{I_C}{\epsilon_0 A}. \tag{29–19}$$

We now define an *equivalent current density* J_D between the plates as

$$J_D = \epsilon_0 \frac{dE}{dt}. \tag{29–20}$$

Then the total equivalent current between the plates, which we may call I_D, is

$$I_D = J_D A = \epsilon_0 \left(\frac{dE}{dt} \right) A = I_C. \tag{29–21}$$

When both conduction and displacement
currents are included, Kirchhoff's point
rule is obeyed for one plate of a
capacitor.

Thus, if we include this equivalent current as well as the conduction current, the current *into* the region bounded by the broken line in Fig. 29–13 equals the current *out of* this region. The subscript D is chosen because Maxwell gave this equivalent current the name **displacement current.**

Displacement current may also be expressed in terms of *electric flux*, defined in Section 25–4, Eqs. (25–18) and (25–19). For a uniform field E perpendicular to an area A, $\Psi = EA$, and Eq. (29–21) may be rewritten as

$$I_D = \epsilon_0 \frac{d\Psi}{dt}. \tag{29–22}$$

If the space between the capacitor plates contains a dielectric, Eq. (29–18) must be replaced by the more general relation derived in Section 27–5:

$$E = \frac{\sigma}{\epsilon} = \frac{Q}{\epsilon A} = \frac{Q}{K\epsilon_0 A}. \tag{29–23}$$

The corresponding modification of the definition of displacement current density is

$$J_D = \epsilon\left(\frac{dE}{dt}\right) = K\epsilon_0\left(\frac{dE}{dt}\right). \tag{29–24}$$

Maxwell's generalized view of current and current density may appear to be merely an artifice introduced to preserve Kirchhoff's current rule even in cases where charge is accumulating in a certain region of space, such as a capacitor plate. However, it is much more than this. In Chapter 31 we will study the role of current as a source of *magnetic* field, and we will see that a *displacement* current sets up a magnetic field in exactly the same way as an ordinary *conduction* current. Thus displacement current, far from being an artifice, is a fundamental fact of nature, and Maxwell's discovery of it was the bold step of an extraordinary genius. As we will see, the concept of displacement current provided the necessary basis for the understanding of electromagnetic waves in the last third of the nineteenth century.

Displacement current isn't just a gimmick; it is a basic fact of nature.

29–6 POWER DISTRIBUTION SYSTEMS

We conclude this chapter with a brief discussion of practical household and automotive electric-power distribution systems. Automobiles use direct-current (dc) systems, while nearly all household, commercial, and industrial systems use alternating current (ac). Most of the same basic wiring concepts apply to both. Alternating-current circuits are discussed in greater detail in Chapter 34.

Parallel circuits are the usual thing in household wiring.

The various lamps, motors, and other appliances to be operated are always connected in *parallel* to the power source—the wires from the power company in the case of houses, the battery and alternator for a car. The basic idea of house wiring is shown in Fig. 29–14. One side of the "line," as the pair of conductors is called, is always connected to "ground." For houses this is an actual electrode driven into the earth (usually a good conductor) and also connected to the household water pipes. Electricians speak of the "hot" side and the "ground" side of the line.

The voltage–current–power relations are the same as those we saw in Section 28–5. Household voltage is nominally 120 V in the United States and Canada, often 240 V in Europe. (These are actually the root–mean–square voltages, as we will discuss in Section 34–4). The current in a 100-W light

Voltage, current, and resistance for an ordinary light bulb

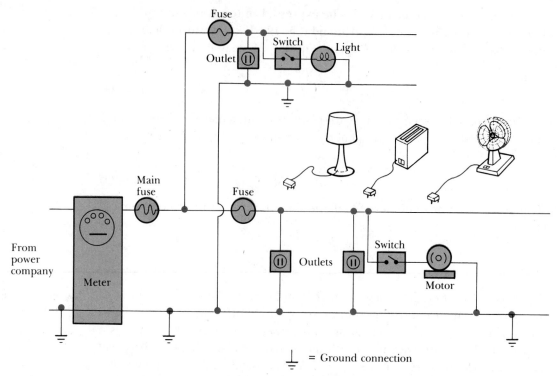

29–14 Schematic diagram of a house wiring system. Only two branch circuits are shown; an actual system might have 4 to 30 branch circuits. The conventional symbol for ground is shown. Lamps and appliances may be plugged into the outlets. The grounding conductors, which normally carry no current, are not shown.

bulb, for example, can be determined from Eq. (28–21):

$$I = \frac{P}{V} = \frac{100 \text{ W}}{120 \text{ V}} = 0.833 \text{ A}.$$

The resistance of this bulb at operating temperature is given by

$$R = \frac{V}{I} = \frac{120 \text{ V}}{0.833 \text{ A}} = 144 \text{ }\Omega,$$

or

$$R = \frac{V^2}{P} = \frac{(120 \text{ V})^2}{100 \text{ W}} = 144 \text{ }\Omega.$$

Similarly, a 1500-W waffle iron draws a current of (1500 W)/(120 V) = 12.5 A and has a resistance, at operating temperature, of 9.6 Ω.

The maximum current available from an individual circuit is determined by the resistance of the wires; the power loss in the wires from I^2R heat causes them to become hot, and in extreme cases this heat can cause a fire or melt the wires. Ordinary lighting and outlet wiring in houses is usually 12-gauge wire, which has a diameter of 2.05 mm and can carry a maximum current of 20 A without overheating. Larger sizes such as 8-gauge (3.26 mm) or 6-gauge (4.11 mm) are used for high-current appliances such as ranges and clothes dryers, and 2-gauge (6.54 mm) or larger is used for the main entrance lines.

Protection against overloading and overheating of circuits is provided by fuses or circuit breakers. A *fuse* contains a link of lead–tin alloy with a very low

The safe current-carrying capacity of a wire is limited by resistive heating; bigger wires can carry more current.

Fuses and circuit breakers: protecting wiring against dangerous overloads

melting temperature; the link melts and breaks the circuit when its rated current is exceeded. A *circuit breaker* is an electromechanical device that performs the same function, using an electromagnet or a bimetallic strip to "trip" the breaker and interrupt the circuit when the current exceeds a specified value. Circuit breakers have the advantage that they can be reset after they are tripped; a blown fuse must be replaced, but fuses are somewhat more reliable in operation than circuit breakers.

You may have had the experience of blowing a fuse by plugging too many high-current appliances into the same outlet or into two outlets fed by the same circuit. *Do not* replace the fuse with one of larger rating; to do so risks overheating the wires and starting a fire. The only safe solution is to distribute the appliances among several circuits. Modern kitchens often have three or four separate 20-A circuits.

Accidental contact between the hot and ground sides of the line causes a **short circuit.** Such a situation, which can be caused by faulty insulation or by any of a variety of mechanical malfunctions, provides a very low-resistance current path, permitting a very large current that would quickly melt the wires and ignite their insulation if the current were not interrupted by a fuse or circuit breaker. The opposite situation, a broken wire that interrupts the current path, creates an **open circuit.** An open circuit can be hazardous because of the sparking that can occur at the point of intermittent contact.

Dangers of short circuits and open circuits

In approved wiring practice, a fuse or breaker is placed *only* in the hot side of the line, never in the ground side. Otherwise, if a short circuit should develop because of faulty insulation or other malfunction, the ground-side fuse could blow, leaving the hot side alive and posing a shock hazard to a person who touched the live conductor and a grounded object such as a water pipe. For similar reasons, the wall switch for a light fixture is always in the hot side of the line, never the ground side.

Further protection against shock hazard is provided by a third conductor included in all present-day wiring. This conductor (corresponding to the long, round prong of the three-prong connector plug on an appliance or power tool) normally carries no current, but it connects the metal case or frame of the device to ground. If a conductor in the hot side of the line accidentally contacts the frame or case, the grounding conductor provides a current path and the fuse blows. Without the ground wire, the frame could become "live," that is, at a potential 120 V above ground; then touching it and a water pipe (or even a damp basement floor) at the same time would offer a shock hazard. In some special situations, especially outlets located outdoors or near a sink or other water pipes, a special kind of circuit breaker called a *ground-fault interrupter* is used. This device senses the difference in current between the hot and neutral conductors (which is normally zero) and trips when this difference exceeds some very small value, typically 10 mA.

Grounding conductors: What is the third wire for?

All the discussion above can be applied directly to automobile wiring. The voltage is 12 V; the power is supplied by the battery and by the alternator, which charges the battery when the engine is running. The ground side of the circuit is connected to the body and frame of the vehicle. There is no separate grounding conductor, and indeed the frame itself sometimes provides the ground side of the circuit. The fuse or circuit breaker arrangement is the same in principle as in household wiring. Because of the lower voltage, more current is required for the same power; a 100-W headlight bulb requires a current of $(100 \text{ W})/(12 \text{ V}) = 8.3 \text{ A}$.

Automotive wiring: same principles, lower voltage

29–15 Simplified schematic diagram of a 120–240-V house wiring system. Only one circuit on each side is shown; in actual systems there would be several 120-V circuits on each side of the neutral line. Grounding wires are not shown.

Color codes for wiring: not always consistent

Power-company employees read a "light meter." This meter records the total input of electrical energy into the facility being supplied; the meter indicates the energy input directly in kilowatt-hours. (Courtesy of Pacific Gas and Electric.)

The cost of electrical energy

KEY TERMS

series

parallel

equivalent resistance

branch point

loop

Kirchhoff's point rule

Kirchhoff's loop rule

To help prevent wiring errors, household wiring uses a standardized color code in which the hot side of a line has black or red insulation, the ground side has white insulation, and the grounding conductor is bare or has green insulation. In electronic devices and equipment, however, the ground side of the line is usually black. Beware!

Most household wiring systems actually use a slight elaboration of the system described above. The power company provides *three* conductors. One is grounded or "neutral"; the other two are both at 120 V with respect to the neutral, but with opposite polarity, giving a voltage between them of 240 V. Thus 120-V lamps and appliances can be connected between neutral and either hot conductor, and high-power devices requiring 240 V are connected between the two hot lines, as shown in Fig. 29–15. Ranges and dryers usually are designed for 240-V power input.

Although we have spoken of *power* in this discussion, what we buy from the power company is *energy*. Power is energy transferred per unit time, so energy is power multiplied by time. The usual unit of energy sold by the power company is the kilowatt-hour (1 kWh):

$$1 \text{ kWh} = (10^3 \text{ W})(3600 \text{ s}) = 3.6 \times 10^6 \text{ W·s} = 3.6 \times 10^6 \text{ J}.$$

One kilowatt-hour typically costs 5 to 10 cents, depending on location and quantity of energy purchased. To operate a 1500-W (1.5-kW) waffle iron for one hour requires 1.5 kWh of energy and costs 7.5 to 15 cents. The cost of operating any lamp or appliance for a specified time can be calculated in the same way if the power rating is known.

SUMMARY

When several resistors $R_1, R_2, R_3, \ldots,$ are connected in series, the equivalent resistance R is the sum of the individual resistances;

$$R = R_1 + R_2 + R_3 + \cdots. \tag{29–1}$$

When several resistors are connected in parallel, the equivalent resistance R is given by

$$\frac{1}{R} = \frac{1}{R_1} + \frac{1}{R_2} + \frac{1}{R_3}. \tag{29–2}$$

Kirchhoff's point rule states that the algebraic sum of the currents into any branch point must be zero. This rule follows from conservation of electric charge. Kirchhoff's loop rule states that the algebraic sum of potential differences around any loop must be zero. This rule follows from conservation of energy and the conservative nature of electrostatic fields.

Many simple electrical measuring instruments incorporate a d'Arsonval galvanometer. The deflection is proportional to the current in the coil. For a larger current range, a shunt resistor is added, so some of the current bypasses the meter. Such an instrument is called an ammeter. If the coil and any additional series resistance included obey Ohm's law, the meter can also be calibrated to read potential difference, or voltage. The instrument is then called a voltmeter.

When a capacitor is charged by a battery in series with a resistor, the current and capacitor charge are not constant. The charge approaches its final value asymptotically, and the current approaches zero asymptotically. The charge and current in the circuit are given by exponential functions. The time τ at which the charge has approached within $1/e$ of its final value is called the time constant or the relaxation time, given by $\tau = RC$. Discharge of a capacitor through a resistor is also characterized by exponential variation of the charge and current; the time constant is the same for charging and discharging.

A single plate of a charging or discharging capacitor does not obey Kirchhoff's point rule because there is current in the conductor attached to the plate but none in the gap between plates. Maxwell generalized the definition of current by postulating an equivalent current density J_D between the plates, given by

$$J_D = \epsilon_0 \frac{dE}{dt}. \qquad (29\text{--}20)$$

With this definition, the conduction current in the conductor is equal to the displacement current in the gap. When a dielectric is present between the plates, displacement current density is defined as

$$J_D = \epsilon \left(\frac{dE}{dt} \right) = K\epsilon_0 \left(\frac{dE}{dt} \right). \qquad (29\text{--}34)$$

Displacement current acts as a source of magnetic field, just as conduction current does. Thus it is not an artifice but a real fact of nature.

In household wiring systems, the various electrical devices are connected in parallel across the power line, which consists of a pair of conductors, one "hot" and the other "ground." The current capacity of a circuit is determined by the size of the wires and the maximum temperature they can tolerate. Protection against excessive current and the resulting fire hazard is provided by fuses or circuit breakers.

d'Arsonval galvanometer
ammeter
shunt resistor
voltmeter
ohmmeter
time constant (relaxation time)
displacement current
short circuit
open circuit

QUESTIONS

29–1 Can the potential difference between terminals of a battery ever be opposite in direction to the emf?

29–2 Why do the lights on a car become dimmer when the starter is operated?

29–3 What determines the maximum current that can be carried safely by household wiring? (Typical limits are 15 A for 14-gauge wire, 20 A for 12-gauge, and so on.)

29–4 Lights in a house often dim momentarily when a

motor, such as a washing machine or a power saw, is turned on. Why does this happen?

29–5 Compare the formulas for resistors in series and parallel with those for capacitors in series and parallel. What similarities and differences do you see? Sometimes in circuit analysis one uses the quantity *conductance*, denoted as *g* and defined as the reciprocal of resistance: $g = 1/R$. What is the corresponding comparison for conductance and capacitance?

29–6 Is it possible to connect resistors together in a way that cannot be reduced to some combination of series and parallel combinations? If so, give examples; if not, state why not.

29–7 In a two-cell flashlight, the batteries are usually connected in series. Why not connect them in parallel?

29–8 Some Christmas-tree lights have the property that, when one bulb burns out, all the lights go out, while with others only the burned-out bulb goes out. Discuss this difference in terms of series and parallel circuits.

29–9 What possible advantage could there be in connecting several identical batteries in parallel?

29–10 Two 120-V light bulbs, one 25-W and one 200-W, were connected in series across a 240-V line. It seemed like a good idea at the time, but one bulb burned out almost instantaneously. Which one burned out, and why?

29–11 When the direction of current in a battery reverses, does the direction of its emf also reverse?

29–12 Under what conditions would the terminal voltage of a battery be zero?

29–13 What sort of meter should be used to test the condition of a dry cell (such as a flashlight battery) having constant emf but internal resistance that increases with age and use?

29–14 For very large resistances, it is easy to construct *RC* circuits having time constants of several seconds or minutes. How might this fact be used to measure very large resistances, too large to measure by more conventional means?

EXERCISES

Section 29–1 Resistors in Series and Parallel

29–1 A 60-Ω resistor and a 90-Ω resistor are connected in parallel, and the combination is connected across a 120-V dc line.

a) What is the resistance of the parallel combination?

b) What is the total current through the parallel combination?

c) What is the current through each resistor?

29–2 Three resistors having resistances of 1 Ω, 2 Ω, and 3 Ω are connected in series to a 12-V battery having negligible internal resistance.

a) Find the equivalent resistance of the combination.

b) Find the current in each resistor.

c) Find the total current through the battery.

d) Find the voltage across each resistor.

e) Find the power dissipated in each resistor.

29–3 In Exercise 29–2, the same three resistors are connected in parallel to the same battery. Answer the same questions for this situation.

29–4 A 25-W, 120-V light bulb and a 100-W, 120-V light bulb are connected in series across a 240-V line. Assume that the resistance of each bulb does not vary with current.

a) Find the current through the bulbs.

b) Find the power dissipated in each bulb.

c) One bulb burns out very quickly. Which one? Why?

29–5

a) The power rating of a 10,000-Ω resistor is 2 W. (The power rating is the maximum power the resistor can safely dissipate without too great a rise in temperature.) What is the maximum allowable potential difference across the terminals of the resistor?

b) A 20,000-Ω resistor is to be connected across a potential difference of 300 V. What power rating is required?

c) It is desired to connect an equivalent resistance of 1000 Ω across a potential difference of 200 V. A number of 10-W, 1000-Ω resistors are available. How should they be connected?

Section 29–2 Kirchhoff's Rules

29–6 Find the emf's $\mathcal{E}_1$ and $\mathcal{E}_2$ in the circuit of Fig. 29–16, and the potential difference between points *a* and *b*.

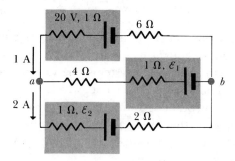

FIGURE 29–16

29–7 In the circuit shown in Fig. 29–17, find

a) the current in resistor *R*;

b) the resistance *R*;

c) the unknown emf $\mathcal{E}$.

d) If the circuit is broken at point *x*, what is the current in the 28-V battery?

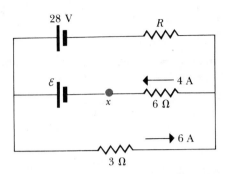

FIGURE 29–17

29–8 In the circuit shown in Fig. 29–18, find

a) the current in each branch;

b) the potential difference V_{ab}.

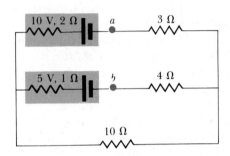

FIGURE 29–18

29–9

a) Find the potential difference between points a and b in Fig. 29–19.

b) If a and b are connected, find the current in the 12-V cell.

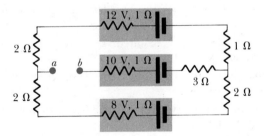

FIGURE 29–19

Section 29–3 Electrical Instruments

29–10 The resistance of a galvanometer coil is 50 Ω, and the current required for full-scale deflection is 500 μA.

a) Show in a diagram how to convert the galvanometer to an ammeter reading 5 A full scale, and compute the shunt resistance.

b) Show how to convert the galvanometer to a voltmeter reading 150 V full scale, and compute the series resistance.

29–11 The resistance of the coil of a pivoted-coil galvanometer is 10 Ω, and a current of 0.02 A causes it to deflect full scale. We want to convert this galvanometer to an ammeter reading 10 A full scale. The only shunt available has a resistance of 0.04 Ω. What resistance R must be connected in series with the coil? (See Fig. 29–20.)

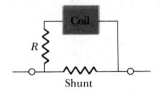

FIGURE 29–20

29–12 The resistance of the moving coil of the galvanometer G in Fig. 29–21 is 25 Ω; the meter deflects full scale with a current of 0.010 A. Find the magnitudes of the resistances R_1, R_2, and R_3 required to convert the galvanometer to a multirange ammeter deflecting full scale with currents of 10 A, 1 A, and 0.1 A.

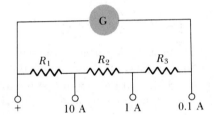

FIGURE 29–21

29–13 Figure 29–22 shows the internal wiring of a "three-scale" voltmeter whose binding posts are marked +, 3 V, 15 V, 150 V. The resistance of the moving coil, R_G, is 15 Ω, and a current of 1 mA in the coil causes it to deflect full scale. Find the resistances R_1, R_2, R_3, and the overall resistance of the meter on each of its ranges.

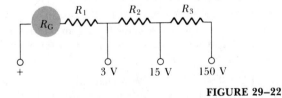

FIGURE 29–22

29–14 A 100-V battery has an internal resistance of $r = 5\ \Omega$.

a) What is the reading of a voltmeter having a resistance of $R_V = 500\ \Omega$ when placed across the terminals of the battery?

b) What maximum value may the ratio r/R_V have if the error in the reading of the emf of a battery is not to exceed 5%?

29–15 Two 150-V voltmeters, one with resistance 15,000 Ω and the other with resistance 150,000 Ω, are connected in series across a 120-V dc line. Find the reading of each voltmeter.

29–16 In the ohmmeter in Fig. 29–23, M is a 1-mA meter having a resistance of 100 Ω. The battery B has an emf of 3 V and negligible internal resistance. R is so chosen that, when terminals a and b are shorted ($R_x = 0$), the meter reads full scale. When a and b are open ($R_x = \infty$), the meter reads zero.

a) What should be the value of the resistor R?

b) What current would indicate a resistance R_x of 600 Ω?

c) What resistances correspond to meter deflections of 1/4, 1/2, and 3/4 full scale, if the deflection is proportional to the current through the galvanometer?

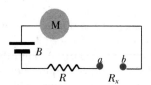

FIGURE 29–23

Section 29–4 The R–C Series Circuit

29–17 A 10-μF capacitor is connected through a 1-MΩ resistor to a constant potential difference of 100 V.

a) Compute the charge on the capacitor at the following times after the connections are made: 0, 5 s, 10 s, 20 s, 100 s.

b) Compute the charging current at the same instants.

c) How much time would be required for the capacitor to acquire its final charge if the charging current remained constant at its initial value?

d) Find the time required for the charge to increase from zero to 5×10^{-4} C.

e) Construct graphs of the results of parts (a) and (b) for a time interval of 20 s.

29–18 A capacitor is charged to a potential of 10 V and is then connected to a voltmeter having an internal resistance of 1.0 MΩ. After a time of 5 s, the voltmeter reads 5 V. What is the capacitance?

29–19 A 0.05-μF capacitor is charged to a potential of 200 V and is then permitted to discharge through a 10-MΩ resistor. How much time is required for the charge to decrease to

a) $1/e$

b) $1/e^2$

of its initial value?

29–20

a) The differential equation for the instantaneous charge q of a capacitor a moment after its terminals have been disconnected from a source and connected to a resistance R is given by Eq. (29–13), namely,

$$\frac{dq}{dt} = -\frac{q}{RC}.$$

Show that

$$q = Q_0 e^{-t/RC}.$$

b) The current in the circuit of part (a) is given by Eq. (29–14), namely,

$$\frac{di}{dt} = -\frac{i}{RC}.$$

Show that

$$i = I_0 e^{-t/RC}.$$

Section 29–5 Displacement Current

29–21 In Fig. 29–13 the capacitor plates have area 4 cm^2 and separation 3 mm. The plates are in vacuum. The charging current I_C has a *constant* value of 2 mA. At $t = 0$ the charge on the plates is zero.

a) Calculate the charge on the plates, the electric field between the plates, and the potential difference between the plates when $t = 5.0 \times 10^{-6}$ s.

b) Calculate dE/dt, the time rate of change of the electric field between the plates. Does dE/dt vary with time?

c) Calculate the displacement current density J_D between the plates, and from this the total displacement current I_D. How do I_C and I_D compare?

29–22 Suppose that the parallel plates in Fig. 29–13 have an area of 2 cm^2 and are separated by a sheet of dielectric 1 mm thick, of dielectric constant 3. (Neglect edge effects.) At a certain instant, the potential difference between the plates is 100 V and the current I_C equals 2 mA.

a) What is the charge Q on each plate at this instant?

b) What is the rate of change of charge on the plates?

c) What is the displacement current in the dielectric?

Section 29–6 Power Distribution Systems

29–23 A 100-W driveway light is left on night and day for a year.

a) What total energy is required? Express your results in joules and kilowatt-hours.

b) What does this energy cost if the power company's rate is 10 cents per kWh?

29–24 An electric dryer is rated at 5.0 kW when connected to a 240-V line.

a) What is the current in the dryer? Is 12-gauge wire large enough to supply this current?

b) What is the resistance of the dryer's heating element?

c) At 10 cents per kWh, how much does it cost to operate the dryer for 1 hr?

29–25 A 1500-W toaster, a 1200-W electric frypan, and a 100-W lamp are plugged into the same outlet in a 20-A, 120-V circuit.

a) What current is drawn by each device?

b) Will this combination blow a fuse?

29–26 How many 100-W light bulbs can be connected to a 20-A, 120-V circuit without tripping the circuit breaker?

PROBLEMS

29–27 Prove that when two resistors are connected in parallel, the equivalent resistance of the combination is always smaller than that of either resistor.

29–28

a) A resistance R_2 is connected in parallel with a resistance R_1. What resistance R_3 must be connected in series with the combination of R_1 and R_2 so that the equivalent resistance is equal to the resistance R_1? Draw a diagram.

b) A resistance R_2 is connected in series with a resistance R_1. What resistance R_3 must be connected in parallel with the combination of R_1 and R_2 so that the equivalent resistance is equal to R_1? Draw a diagram.

29–29 A 1000-Ω 2-W resistor is needed, but only several 1000-Ω 1-W resistors are available.

a) How can the required resistance and power rating be obtained by a combination of the available units?

b) What power is then dissipated in each resistor?

29–30

a) Calculate the equivalent resistance of the circuit of Fig. 29–24 between x and y.

b) What is the potential difference between x and a if the current in the 8-Ω resistor is 0.5 A?

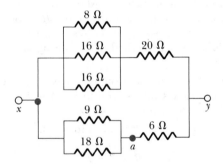

FIGURE 29–24

29–31 Each of three resistors in Fig. 29–25 has a resistance of 2 Ω and can dissipate a maximum of 18 W without becoming excessively heated. What is the maximum power the circuit can dissipate?

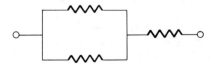

FIGURE 29–25

29–32 Three equal resistors are connected in series. When a certain potential difference is applied across the combination, the total power dissipated is 10 W. What power would be dissipated if the three resistors were connected in parallel across the same potential difference?

29–33 Calculate the three currents indicated in the circuit diagram of Fig. 29–26.

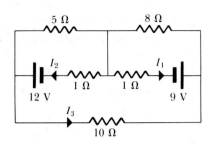

FIGURE 29–26

29–34 Find the current in each branch of the circuit shown in Fig. 29–27.

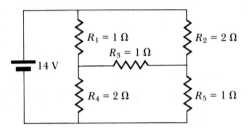

FIGURE 29–27

29–35
Note. Figure 29–28a employs a convention often used in circuit diagrams. The battery (or other power supply) is not shown explicitly. It is understood that the point at the top, labeled "36 V," is connected to the positive terminal of a 36-V battery having negligible internal resistance, and the "ground" symbol at the bottom is connected to its negative terminal. The circuit is completed through the battery, even though it is not shown on the diagram.

a) In Fig. 29–28a, what is the potential difference V_{ab} when the switch S is open?

b) What is the current through switch S when it is closed?

c) In Fig. 29–28b, what is the potential difference V_{ab} when the switch is open?

d) What is the current through switch S when it is closed?

e) What is the equivalent resistance in Fig. 29–28b when switch S is open?

f) When switch S is closed?

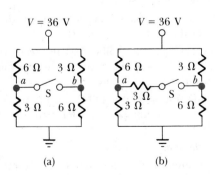

FIGURE 29–28

29–36 (See note with Problem 29–35.)

a) What is the potential of point a with respect to point b in Fig. 29–29 when switch S is open?

b) Which point, a or b, is at the higher potential?

c) What is the final potential of point b when switch S is closed?

d) How much does the charge on each capacitor change when S is closed?

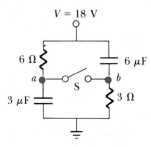

FIGURE 29–29

29–37 (See note with Problem 29–35.)

a) What is the potential of point a with respect to point b in Fig. 29–30 when switch S is open?

b) Which point, a or b, is at the higher potential?

c) What is the final potential of point b when switch S is closed?

d) How much charge flows through switch S when it is closed?

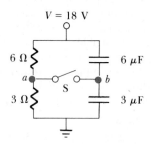

FIGURE 29–30

29–38 A certain galvanometer has a resistance of 200 Ω and deflects full scale with a current of 1 mA in its coil. We want to replace this with a second galvanometer whose resistance is 50 Ω and which deflects full scale with a current of 50 μA in its coil. Devise a circuit incorporating the second galvanometer such that

a) the equivalent resistance of the circuit equals the resistance of the first galvanometer;

b) the second galvanometer will deflect full scale when the current into and out of the circuit equals the full-scale current of the first galvanometer.

29–39 A 600-Ω resistor and a 400-Ω resistor are connected in series across a 90-V line. A voltmeter across the 600-Ω resistor reads 45 V.

a) Find the voltmeter resistance.

b) Find the reading of the same voltmeter if connected across the 400-Ω resistor.

29–40 Point a in Fig. 29–31 is maintained at a constant potential of 300 V above ground. (See note with Problem 29–35.)

a) What is the reading of a voltmeter of the proper range and of resistance 3×10^4 Ω when connected between point b and ground?

b) What would be the reading of a voltmeter of resistance 3×10^6 Ω?

c) What would be the reading of a voltmeter of infinite resistance?

FIGURE 29–31

29–41 A 150-V voltmeter has resistance of 20,000 Ω. When connected in series with a large resistance R across a 110-V line, the meter reads 5 V. Find the resistance R. (This problem illustrates one method of measuring large resistances.)

29–42 Let V and I represent the readings of the voltmeter and ammeter, respectively, shown in Fig. 29–9, and R_V and R_A their equivalent resistances.

a) When the circuit is connected as in Fig. 29–9a, show that

$$R = \frac{V}{I} - R_A.$$

b) When the connections are as in Fig. 29–9b, show that

$$R = \frac{V}{I - (V/R_V)}.$$

c) Show that the power delivered to the resistor in part (a) is $IV - I^2 R_A$, and in part (b) $IV - (V^2/R_V)$.

29–43 In Fig. 29–32, a resistor of resistance 75 Ω is connected between points a and b. The resistance of the galvanometer G is 90 Ω. What should be the resistance between b and the sliding contact c if the galvanometer current I_G is to be ⅓ of the current I?

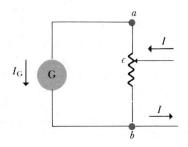

FIGURE 29–32

29-44 The current shown in Fig. 29-33, called a *Wheatstone bridge*, is used to determine the value of an unknown resistor X by comparison with three resistors M, N, and P whose resistance can be varied. For each setting, the resistance of each resistor is precisely known. With switches K_1 and K_2 closed, these resistors are varied until the current in the galvanometer G is zero; the bridge is then said to be *balanced*.

a) Show that under this condition the unknown resistance is given by $X = MP/N$. (This method permits very high precision in comparing resistors.)

b) If the galvanometer G shows zero deflection when $M = 1000$ Ω, $N = 10.00$ Ω, and $P = 27.49$ Ω, what is the unknown resistance X?

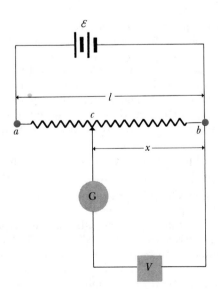

FIGURE 29-33

29-45 The circuit shown in Fig. 29-34 is called a *potentiometer*. It permits measurements of potential difference without drawing current from the circuit being measured, and hence acts as an infinite-resistance voltmeter. The resistor between a and b is a uniform wire of length l, with a sliding contact c at a distance x from b. An unknown potential difference V is measured by sliding the contact until the galvanometer G reads zero.

a) Show that under this condition the unknown potential difference is given by $V = (x/l)\,\mathcal{E}$.

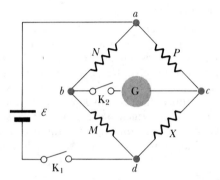

FIGURE 29-34

b) Why is the internal resistance of the galvanometer not important?

Now suppose $\mathcal{E} = 12.00$ V and $l = 1.000$ m. The galvanometer G reads zero when $x = 0.793$ m.

c) What is the potential difference V?

d) Suppose V is the emf of a battery. Can its internal resistance be determined by this method?

29-46 The current in a discharging capacitor is given by Eq. (29-15).

a) Using Eq. (29-15), derive an expression for the instantaneous power $P = i^2R$ dissipated in the resistor.

b) Integrate the expression for P to find the total energy dissipated in the resistor, and show that this value is equal to the total energy initially stored in the capacitor.

29-47 The current in a charging capacitor is given by Eq. (29-10).

a) The instantaneous power supplied by the battery is Vi. Integrate this expression to find the total energy supplied by the battery.

b) The instantaneous power dissipated in the resistor is i^2R. Integrate this expression to find the total energy dissipated in the resistor.

c) Find the final energy stored in the capacitor, and show that this value equals the total energy supplied by the battery, less the energy dissipated in the resistor, as obtained in parts (a) and (b).

d) What fraction of the energy supplied by the battery is stored in the capacitor? How does this fraction depend on R?

29-48 Two capacitors are charged in series by a 12-V battery that has an internal resistance of 1 Ω. There is a 5-Ω resistor in series between the capacitors (Fig. 29-35).

a) What is the time constant of the charging circuit?

b) After the switch has been closed for the time determined in (a), what is the voltage across the 6-μF capacitor?

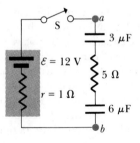

FIGURE 29-35

29-49 In a certain copper conductor ($\rho = 1.72 \times 10^{-8}$ Ω·m) carrying a current, the electric field varies sinusoidally with time according to $E = E_0 \sin \omega t$, where $E_0 = 0.1$ V·m^{-1}, $\omega = 2\pi f$, and the frequency is $f = 60$ Hz.

a) Find the magnitude of the maximum conduction current density in the wire.

b) Assuming $\epsilon = \epsilon_0$, find the maximum displacement current density in the wire, and compare it with the result of part (a).

29–50

a) Repeat the calculations of Problem 29–49 for a rod of pure silicon having $\rho = 2300 \ \Omega\cdot m$.

b) At what frequency f would the maximum conduction and displacement current densities become equal, if $\epsilon = \epsilon_0$ (which is not actually the case)?

c) At the frequency determined in part (b), what is the relative *phase* of the conduction and displacement currents?

CHALLENGE PROBLEMS

29–51 Prove that the resistance of the infinite network shown in Fig. 29–36 is equal to $(1 + \sqrt{3})r$.

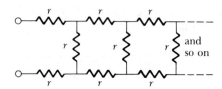

FIGURE 29–36

29–52 Suppose a resistor R lies along each edge of a cube (12 resistors in all) with connections at the corners. Find the equivalent resistance between two diagonally opposite corners of the cube.

29–53 A parallel-plate capacitor is made from two plates, each having area A, spaced a distance d apart. The space between plates is filled with a material having dielectric constant K. The material is not a perfect insulator but has resistivity ρ. The capacitor is initially charged with charge of magnitude Q_0 on each plate, which gradually discharges by conduction through the dielectric. Show that at any instant the displacement current density in the dielectric is equal in magnitude to the conduction current density but is opposite in direction, so that the *total* current density is zero at every instant.

29–54 The capacitance of a capacitor can be affected by dielectric material that, although not inside the capacitor, is near enough to the capacitor to be polarized by the fringing electric field that exists near a charged capacitor. This effect is usually of the order of picofarads (pF), but it can be used with appropriate electronic circuitry to detect a change in the dielectric material surrounding the capacitor. Such a dielectric material might be the human body, and the effect described above might be used in the design of a burglar alarm.

  Consider the simplified circuit shown in Fig. 29–37. The voltage source has emf $\mathcal{E}_0 = 1000$ V, and the capacitor has capacitance $C = 10$ pF. The electronic circuitry for detecting the current, represented as an ammeter in the diagram, has negligible resistance and is capable of detecting a current that persists at a level of at least 1 μA for at least 200 μs after the capacitance has changed abruptly from C to C'. The burglar alarm is to be designed to be activated if the capacitance changes by 10%.

a) Determine the charge on the 10-pF capacitor when it is fully charged.

b) If the capacitor is fully charged before the intruder is detected, and assuming that the time taken for the capacitance to change by 10% is small enough to be neglected, derive an equation that expresses the current through the resistor R as a function of the time t since the capacitance has changed.

c) Determine the range of values of the resistance R that will meet the design specifications of the burglar alarm. What happens if R is too small? Too large? (*Hint*: You will not be able to solve this part analytically but must use numerical methods. R can be expressed as a logarithmic function of R plus known quantities. Use a trial value of R and calculate from the expression a new value. Continue to do this until the input and output values of R agree to within three significant figures.)

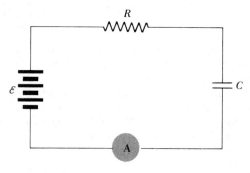

FIGURE 29–37

29–55 Thevenin's theorem, as it applies to a dc circuit, can be stated as follows: *Any linear, two-terminal network branch can be replaced by an equivalent source of emf in series with an equivalent resistance.* This theorem can be used to simplify the analysis of a circuit if we want to modify one part of the circuit and then see what effect that modification will have on that part of the circuit without having to use Kirchhoff's rules to reanalyze the entire circuit each time. Consider the circuit shown in Fig. 29–38a.

a) Using Kirchhoff's rules, determine the branch currents in each of the three branches.

b) According to Thevenin's theorem, the part of the circuit to the left of points a and b can be replaced by an equivalent emf and, in series with that emf, by an equivalent resistance, as depicted in Fig. 29–38b. Find the equivalent emf by considering that circuit as an open circuit between a and b. That is, remove the branch containing the 60-Ω resistor and the 150-V source and do not replace the branch with anything. The voltage drop between a and b is the emf of the equivalent voltage source.

c) Find the equivalent resistance that must be in series with the source found in part (b). Do this once again by removing the 60-Ω resistor and the 150-V source, but this time provide a short circuit between a and b. The current through that short circuit must be the equivalent emf divided by the equivalent resistance.

d) Suppose we want to replace the 150-V source with another source such that the current through the 60-Ω resistor will be 1 A. Using the Thevenin-equivalent circuit elements found in parts (b) and (c), determine the emf of the source that should replace the 150-V source.

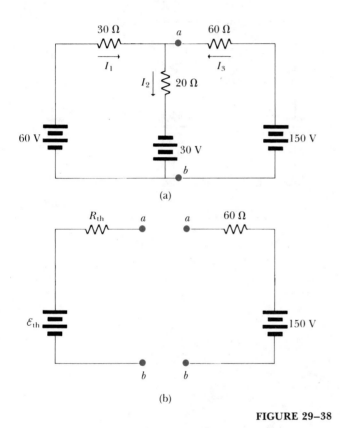

FIGURE 29–38

29–56 Norton's theorem, as it applies to a dc circuit, can be stated as follows: *Any linear, two-terminal network branch can be replaced by an equivalent current source in parallel with an equivalent shunt resistance.* As with Thevenin's theorem, Norton's theorem can be used to simplify the analysis of a circuit if we want to modify one part of the circuit and then see what effect that modification will have on that part of the circuit without having to use Kirchhoff's rules to reanalyze the entire circuit each time. Consider the circuit shown in Fig. 29–39a.

a) Using Kirchhoff's rules, determine the branch currents in each of the three branches.

b) According to Norton's theorem, the part of the circuit to the left of points a and b can be replaced by an equivalent current source and, in parallel with that current source, by an equivalent resistance, as depicted in

Fig. 29–39b. Find the equivalent current source by providing a short circuit between a and b. That is, remove the branch containing the 60-Ω resistor and the 30-V source and replace it with a short circuit. The current that passes through the short circuit between a and b is the current produced by the equivalent current source.

c) Find the equivalent resistance that must be in parallel with the source found in part (b). Do this by once again removing the 60-Ω resistor and the 30-V source, but this time consider that circuit as an open circuit between a and b. That is, remove the branch containing the 60-Ω resistor and the 150-V source and do not replace it with anything. The voltage drop across the open circuit must be the product of the current of the equivalent current source and the resistance of the equivalent resistance.

d) If a 20-Ω resistor is connected between a and b in the circuit of Fig. 29–39a, what will be the current through the 60-Ω resistor? Use the Norton-equivalent-circuit elements found in parts (b) and (c) to answer this part.

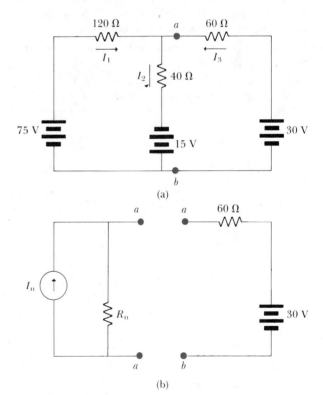

FIGURE 29–39

29–57 According to the theorem of superposition, the response (current) in a circuit is proportional to the stimulus (voltage) that causes it. This is true even if there are multiple sources in a circuit. The theorem can be used to analyze a circuit without resorting to Kirchhoff's rules by considering the currents in the circuit to be the superposition of currents caused by each source independently. In this way, the circuit can be analyzed by computing equivalent resistances rather than by using the (sometimes) more cumbersome method of Kirchhoff's rules.

Furthermore, by using the superposition theorem, it is possible to examine how the modification of a source in one part of the circuit will affect the currents in all parts of the circuit without having to use Kirchhoff's rules to recalculate all the currents.

Consider the circuit shown in Fig. 29–40. If the circuit were redrawn with the 55-V and 57-V sources replaced by short circuits, the circuit could be analyzed by the method of equivalent resistances without resorting to Kirchhoff's rules, and the current in each branch could be found in a simple manner. Similarly, by redrawing the circuit with the 92-V and the 55-V sources replaced by short circuits, the circuit could again be analyzed simply. Finally, by replacing the 92-V and the 57-V sources with a short circuit, the circuit could also be analyzed simply. By superposing the respective currents found in each of the branches using the three simplified circuits, the actual current in each branch can be found.

a) Using Kirchhoff's rules, find the branch currents in the 140-Ω, 210-Ω, and 35-Ω resistors.

b) Using a circuit similar to that in Fig. 29–40, but with the 55-V and 57-V sources replaced by a short circuit, determine the current in each resistance.

c) Repeat part (b) by replacing the 92-V and 55-V sources by short circuits, leaving the 57-V source intact.

d) Repeat part (b) by replacing the 92-V and 57-V sources by short circuits, leaving the 55-V source intact.

e) Verify the superposition theorem by taking the currents calculated in parts (b), (c), and (d) and comparing them with the currents calculated in part (a).

f) If the 57-V source is replaced by a 100-V source, what will be the new currents in all branches of the circuit? (*Hint:* Applying the superposition theorem, recalculate the partial currents calculated in part [c] using the fact that those currents are proportional to the source that is being replaced. Then superpose the new partial currents with those found in parts [b] and [d].)

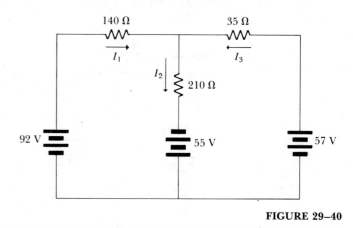

FIGURE 29–40

ELECTRODYNAMICS

PERSPECTIVE

Electromagnetism is the study of the interaction of electric charges. We began our study with *electrostatics,* the study of charges *at rest.* In this case the forces are *conservative* forces; we described them and their associated potential energies using the concepts of *electric field* and *electric potential.* We analyzed several fundamental experiments and practical devices that use electrostatic principles. We also analyzed the behavior of *capacitors* and some aspects of the electrical properties of materials.

Next we studied electric *circuits;* we described the conducting properties of materials and their temperature variation, and we introduced the concepts of *resistance* and *electromotive force.* Circuits are the heart of all contemporary electronics and electric-power distribution systems. Although we confined our study mostly to *constant* currents, many of the same principles are applicable in circuits with time-varying currents.

Now we are ready to broaden our study to include electric charges in motion. The forces now depend on the *velocities* of the charged particles. These interactions are best described in terms of *magnetic field.* Our analysis takes the same two-part path as for electric field. First, moving charge (such as a current in a conductor) acts as a *source* of magnetic field, establishing a magnetic field in the surrounding region. Second, a charged particle moving in a magnetic field experiences a *force* that depends on the magnitude of the field, the particle's speed, and its charge.

We begin by taking a magnetic field as a given quantity and describing the force it exerts on a moving charge. This force provides the basis for a broad variety of phenomena, including electric motors, the deflections of electron beams in TV picture tubes, and experiments with fundamental particles. Next we discuss in detail how magnetic fields are *produced* by currents and moving charges. This discussion enables us to understand electromagnets, motors, and many other practical devices. We also look at some magnetic properties of materials, including *magnetization,* which alters the magnetic field inside a material.

An essential aspect of electromagnetic interactions is *electromagnetic induction,* in which a time-varying magnetic field acts as a source of electric field. An electric current can be induced in a circuit by a changing magnetic field, whether caused by a moving permanent magnet, motion of the circuit, or changing current in a nearby electromagnet. To analyze induction phenomena we introduce the concept of *magnetic flux,* closely related to magnetic field. Induction is the basis for inductors and transformers, two indispensible elements in modern electronics and power-distribution systems. We study the role of these and other circuit elements in alternating-current (ac) circuits.

Finally, a changing *electric* field acts as a source of *magnetic* field. All the fundamental relations of electrodynamics can be summarized in four compact but general equations called *Maxwell's equations.* These equations, comparable in power and elegance to Newton's laws in mechanics, show once again the beauty and generality of physics. The culminating achievement of Maxwell's equations is the prediction of the existence of *electromagnetic waves:* time-varying electric and magnetic fields that *propagate* from one region to another.

Thus Maxwell's equations show that light is electromagnetic radiation. Electromagnetic waves also include radio and TV transmission, infrared, ultraviolet, x-rays, and gamma rays, and they play a significant role in nearly every area of contemporary science and technology.

30

MAGNETIC FIELD AND MAGNETIC FORCES

THE MOST FAMILIAR ASPECTS OF MAGNETISM ARE THOSE ASSOCIATED WITH permanent magnets, which attract unmagnetized iron objects and can also attract or repel other magnets. The fundamental nature of magnetism, however, is to be found in interactions involving electric charges in motion. A magnetic field is established by a permanent magnet, by an electric current in a conductor, or by moving charges. This magnetic field, in turn, exerts forces on moving charges and current-carrying conductors. In this chapter we study these magnetic forces, and in Chapter 31 we examine the ways in which magnetic fields are produced by moving charges. Magnetic forces are an essential aspect of the interactions of electrically charged particles. Electric motors, TV picture tubes, high-energy particle accelerators, magnetrons in microwave ovens, and many other devices depend in part on magnetic forces for their operation.

30–1 MAGNETISM

Magnetic phenomena were first observed at least 2000 years ago, in fragments of magnetized iron ore found near the ancient city of Magnesia. It was found that when an iron rod was brought in contact with a natural magnet, the rod also became a magnet. Such a rod, when suspended by a string from its center, tended to line itself up in a north–south direction, like a compass needle; magnets have been used for navigation at least since the eleventh century.

Before the relation of magnetic interactions to moving charges was understood, the interactions of bar magnets and compass needles were described in terms of **magnetic poles.** The end of a compass needle that points north is called a *north pole* or *N-pole,* and the other end is a *south* or *S-pole.* Two opposite poles attract each other, and two like poles repel each other. The concept of magnetic poles is of limited usefulness and is somewhat misleading because a single isolated magnetic pole has never been found; poles always appear in pairs. If a bar magnet is broken in two, each broken end becomes a pole. The existence of an isolated magnetic pole, or *magnetic monopole,* would have

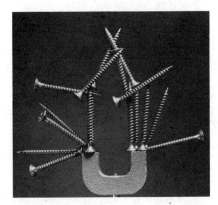

Screws made of steel (a ferromagnetic material) are attracted to a permanent magnet. Screws made of brass or aluminum (nonferromagnetic materials) would not be attracted. (Photo by Chip Clark.)

(a) A compass needle (a permanent magnet) deflects in the magnetic field of another permanent magnet. Magnetization is the result of motion of electrons in the magnetized material. (b) A compass needle deflects in the magnetic field produced by the current in the conductor. The field exerts torques on the circulating electric charges in the needle associated with its magnetization. (Photos by Chip Clark.)

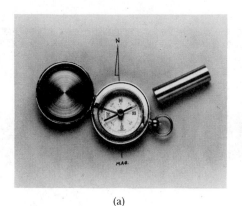

(a)

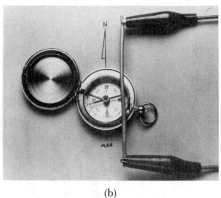

(b)

A compass responds to the earth's magnetic field.

sweeping implications for theoretical physics; extensive searches for magnetic monopoles have been carried out, so far without success.

A compass needle points north because the earth is a magnet; its north geographic pole is a magnetic *south* pole. The earth's magnetic axis is not quite parallel to its geographic axis (the axis of rotation), so a compass reading deviates somewhat from geographic north; this deviation, which varies with location, is called *magnetic declination*. A sketch of the earth's magnetic field is shown in Fig. 30–1. The lines show the direction a compass would point at each location. These lines are actually magnetic-field lines, to be discussed in Section 30–3. The direction of the field at any point can be defined as the direction of the force the field would exert on a magnetic north pole. In Section 30–2 we will introduce a more fundamental way to define magnetic field.

Magnetic effects of an electric current

In 1819 the Danish scientist Hans Christian Oersted discovered that a compass needle was deflected by a current-carrying wire. A few years later Michael Faraday in England and Joseph Henry in the United States discovered that moving a magnet near a conducting loop can cause a current in the loop, and that a changing current in one conducting loop can cause a current in another separate loop. These observations were the first evidence of the relationship of magnetism to moving charges.

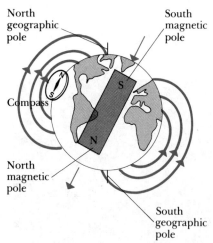

North geographic pole

South magnetic pole

Compass

North magnetic pole

South geographic pole

30–1 A sketch of the earth's magnetic field. A compass placed at any point in this field would point in the direction of the field line at that point. Representing the earth's field as that of a tilted bar magnet is only a crude approximation of the actual fairly complex field configuration.

30–2 MAGNETIC FIELD

In describing the interaction between two charges *at rest*, we introduced the concept of *electric field*, and we described the interaction in two stages:

1. One charge sets up or creates an electric field E in the space surrounding it.
2. The electric field E exerts a force $F = qE$ on a charge q placed in the field.

We will follow the same pattern in describing the interactions of charges in motion:

1. A *moving* charge or a current sets up or creates a **magnetic field** in the space surrounding it.
2. The magnetic field exerts a *force* on a *moving* charge or a current in the field.

Like electric field, magnetic field is a *vector field,* that is, a vector quantity associated with each point in space. We use the symbol **B** for magnetic field.

In this chapter we consider the *second* aspect of the interaction: Given the presence of a magnetic field, what force does it exert on a moving charge or a current? In Chapter 31 we return to the problem of how magnetic fields are *created* by moving charges and currents.

Some properties of the magnetic force on a moving charge are analogous to corresponding properties of the electric-field force. The magnitude of the magnetic force is proportional to the charge. If a 1-μC charge and a 2-μC charge move through a given magnetic field with the same velocity, the force on the 2-μC charge is twice as great as that on the 1-μC charge. The force is also proportional to the magnitude or "strength" of the field; if a given charge moves with the same velocity in two magnetic fields, one having twice the magnitude (and the same direction) as the other, the charge experiences twice as great a force in the larger field as in the smaller.

*The magnetic force also depends on the particle's motion; this is quite different from the electric-field force, which is the same whether the charge is moving or not. The *magnetic* force is found to have a magnitude that is directly proportional to the particle's speed. A particle at rest experiences no magnetic force at all. The *direction* of the force is determined by the directions of the magnetic field **B** and the velocity **v** in an interesting way. The force **F** *does not* have the same direction as **B,** but instead is always *perpendicular* to both **B** and **v.** The magnitude F of the force is found to be proportional to the component of **v** perpendicular to the field; when that component is zero (that is, when **v** and **B** are parallel or antiparallel) there is *no* force!

These characteristics of the magnetic force can be summarized, with reference to Fig. 30–2, as follows: The direction of **F** is perpendicular to the plane containing **v** and **B**; its magnitude is given by

$$F = qv_\perp B = qvB \sin \phi, \tag{30–1}$$

where q is the charge and ϕ is the angle between the vectors **v** and **B,** as shown in the figure.

This description does not specify the direction of **F** completely; there are always two directions perpendicular to the plane of **v** and **B** but opposite to each other. To complete the description we use the same right-hand-thread rule used to define the vector product in Section 1–9 and for angular mechanical quantities in Section 9–14. Imagine turning **v** into **B,** through the smaller of the two possible angles; the direction of **F** is the direction in which a right-hand-thread screw would advance if turned the same way. Alternatively, wrap the fingers of your right hand around the line perpendicular to the plane of **v** and **B** so that they curl around with this sense of rotation from **v** to **B**; your thumb then points in the direction of **F.**

Thus the force on a charge q moving with velocity **v** in a magnetic field **B** is given, both in magnitude and in direction, by

$$\mathbf{F} = q\mathbf{v} \times \mathbf{B}. \tag{30–2}$$

This is the first of several applications of the vector product we will encounter in our study of magnetic-field relationships.

Spatial relations for magnetic forces can be troublesome; the following review of the definition of the vector product may help. First, always draw the

The magnetic force on a charged particle depends on its velocity.

The magnetic force on a charged particle is always perpendicular to its velocity.

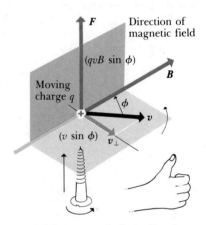

30–2 The magnetic force **F** acting on a charge q moving with velocity **v** is perpendicular to both the magnetic field **B** and to **v.**

A right-hand rule for magnetic force on a moving charge

The magnetic force on a moving charge can be expressed neatly as a vector product.

vectors v and B from a common point, that is, with their tails together. Second, visualize and if possible draw the *plane* in which these two vectors lie. The force vector always lies along a line perpendicular to this plane. To determine its direction along this line, imagine rotating v into B, as in Fig. 30–2. The direction of F is the direction of advance of a right-hand-thread screw rotated in this direction, or the direction of advance of the extended thumb of your right hand if your fingers are curled around the perpendicular line with this direction of rotation.

Finally, we note that Eq. (30–1) can be interpreted in a different but equivalent way. Recalling that ϕ is the angle between the direction of vectors v and B, we may interpret ($B \sin \phi$) as the component of B perpendicular to v, that is, $B_\perp$. With this notation, the force expression becomes

$$F = qvB_\perp. \tag{30–3}$$

Although equivalent to Eq. (30–1), this form is sometimes more convenient, especially in problems involving currents rather than individual particles. We will discuss forces on currents in conductors later in this chapter.

The *units* of B can be deduced from Eq. (30–1); they must be the same as the units of F/qv. Therefore the SI unit of B is $1 \text{ N}\cdot\text{s}\cdot\text{C}^{-1}\cdot\text{m}^{-1}$, or, since one ampere is one coulomb per second ($1 \text{ A} = 1 \text{ C}\cdot\text{s}^{-1}$), $1 \text{ N}\cdot\text{A}^{-1}\cdot\text{m}^{-1}$. This unit is called one **tesla** (1 T), in honor of Nikolai Tesla, the prominent nineteenth-century Russian scientist and inventor. The cgs unit of B, the **gauss** ($1 \text{ G} = 10^{-4} \text{ T}$) is also in common use. Instruments for measuring magnetic field are often called gaussmeters. To summarize,

$$1 \text{ T} = 1 \text{ N}\cdot\text{A}^{-1}\cdot\text{m}^{-1} = 10^4 \text{ G}.$$

In this discussion we assumed that q is a *positive* charge. If q is negative, the direction of F is opposite to that shown in Fig. 30–2 and given by the right-hand rule. Thus if two charges of equal magnitude and opposite sign move in the same B field with the same velocity, the forces on the two charges have equal magnitude and opposite direction.

To explore an unknown magnetic field, we can measure the magnitude and direction of the force on a *moving* test charge and use the relationship $F = qv \times B$. The cathode-ray tube, discussed in Section 26–8, is a convenient device for making such measurements. The electron gun shoots out a narrow beam of electrons at a known speed. If there is no deflecting force on the beam, it strikes the center of the screen.

In the presence of a magnetic field, the electron beam is in general deflected. If the beam is parallel or antiparallel to the field, however, then $v \times B = 0$; then there is no force and no deflection. If we find that the electron beam is undeflected when its direction is parallel to the z-axis, as in Fig. 30–3, the B-vector must point either up or down.

When we turn the tube 90°, so that its axis is along the x-axis in Fig. 30–3, the beam is deflected in a direction showing a force perpendicular to the plane of B and v. We can perform additional experiments in which the angle between B and v is between zero and 90° to confirm Eq. (30–1) and the accompanying discussion. We note that the electron has a negative charge, and the force in Fig. 30–3 is opposite in direction to the force on a positive charge.

When a charged particle moves through a region of space where *both* electric and magnetic fields are present, both fields exert forces on the parti-

An alternative way to describe the magnitude of the magnetic force on a moving charge

The tesla: the SI unit of magnetic field

Using a cathode-ray tube to explore a magnetic field

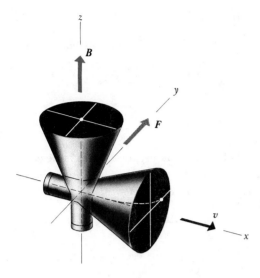

30–3 The electron beam of the cathode-ray tube is undeflected when the beam is parallel to the z-axis. The **B**-vector then points either up or down. When the tube axis is parallel to the x-axis, the beam is deflected in the positive y-direction. Then the **B**-vector points upward, and the force **F** on the electrons points along the positive y-axis, opposite to the rule of Fig. 30–2 because q is negative.

cle, and the total force is the vector sum of the electric-field and magnetic-field forces:

$$F = q(E + v \times B). \qquad (30\text{–}4)$$

PROBLEM-SOLVING STRATEGY: *Magnetic-field forces*

1. The biggest difficulty is in relating the directions of the vector quantities. In evaluating $v \times B$, draw the two vectors with their tails together so you can visualize and draw the plane in which they lie. This also helps to identify the angle ϕ between the two vectors and to avoid getting its complement, or some other erroneous angle. Then remember that **F** is always perpendicular to this plane. The direction is determined by the right-hand rule; keep referring to Fig. 30–2 and back to Fig. 1–13 until you're sure you have this down cold.

2. Whenever possible, do the problem two ways. Do it directly from the geometric definition of the vector product. Then find the components of the vectors in some convenient axis system and calculate the vector product from the components. Check that both methods give the same result.

EXAMPLE 30–1 A proton beam moves through a region of space where there is a uniform magnetic field, of magnitude 2.0 T, directed along the positive z-axis, as in Fig. 30–4. The protons have velocity of magnitude 3×10^5 m·s^{-1}, in the xz-plane at an angle of 30° to the positive z-axis. Find the force on a proton ($q = 1.6 \times 10^{-19}$ C).

SOLUTION The right-hand rule shows that the direction of the force is along the negative y-axis. The magnitude of the force, from Eq. (30–1), is

$$
\begin{aligned}
F &= qvB \sin \phi \\
&= (1.6 \times 10^{-19} \text{ C})(3 \times 10^5 \text{ m·s}^{-1})(2.0 \text{ T})(\sin 30°) \\
&= 4.8 \times 10^{-14} \text{ N}.
\end{aligned}
$$

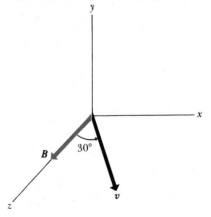

30–4 Directions of v and B for Example 30–1.

Alternatively, in vector language, with Eq. (30–2),

$$v = (3 \times 10^5 \text{ m·s}^{-1})(\sin 30°)i + (3 \times 10^5 \text{ m·s}^{-1})(\cos 30°)k,$$
$$B = (2.0 \text{ T})k,$$
$$F = qv \times B = (1.6 \times 10^{-19} \text{ C})(3 \times 10^5 \text{ m·s}^{-1})(2.0 \text{ T})$$
$$\cdot (\sin 30° i + \cos 30° k) \times k$$
$$= (-4.8 \times 10^{-14} \text{ N})j,$$

since $i \times k = -j$ and $k \times k = 0$.

If the beam consists of *electrons* rather than protons, the charge is negative ($q = -1.6 \times 10^{-19}$ C), and the direction of the force is reversed. The solution proceeds just as before; the force is now directed along the *positive* y-axis: $F = +(4.8 \times 10^{-14} \text{ N})j$.

30–3 MAGNETIC-FIELD LINES AND MAGNETIC FLUX

Magnetic-field lines: a useful graphical representation of magnetic fields

A magnetic field can be represented by lines, just as we represented an electric field by lines in Section 25–3. We draw the lines so that the line through any point is tangent to the magnetic-field vector B at that point, and so that the number of lines per unit area (perpendicular to the lines at a given point) is proportional to the magnitude of the field at that point. We call these lines **magnetic-field lines.** They are sometimes called magnetic lines of force, but this term is unfortunate because, unlike electric-field lines, they *do not* point in the direction of the force on a charge.

The field lines for a uniform field are equally spaced parallel straight lines.

In a uniform magnetic field, where the B vector has the same magnitude and direction at every point in a region, the field lines are straight and parallel. If the poles of an electromagnet are large, flat, and close together, the magnetic field in the region between them is very nearly uniform. Magnetic-field lines produced by a few examples of magnetic-field sources are shown in Figs. 30–5 and 30–6. The magnetic field of the earth is shown in Fig. 30–1; the field lines resemble those of a bar magnet with its axis tilted somewhat with respect to the earth's axis of rotation. The earth's field changes with time, and there is geologic evidence that it has actually changed direction many times in the last 100 million years.

Magnetic flux: the magnetic analog of electric flux

We define the **magnetic flux** Φ through a surface in direct analogy with the electric-field flux used with Gauss's law in Section 25–4. We can divide any surface into elements of area dA, as shown in Fig. 30–7. For each element, we determine the components of B normal and tangent to the surface at the position of that element, as shown. In general, these components will vary from point to point on the surface. From the figure, $B_\perp = B \cos \theta$. We define the magnetic flux $d\Phi$ through this area as

$$d\Phi = B_\perp \, dA = B \cos \theta \, dA = \boldsymbol{B} \cdot d\boldsymbol{A}. \tag{30–5}$$

The *total* magnetic flux through the surface is the sum of the contributions from the individual area elements, given by

$$\Phi = \int B_\perp \, dA = \int \boldsymbol{B} \cdot d\boldsymbol{A}. \tag{30–6}$$

In the special case where B is uniform over a plane surface with total area A,

$$\Phi = B_\perp A = BA \cos \theta. \tag{30–7}$$

If **B** happens to be perpendicular to the surface, $\cos \theta = 1$, and this expression reduces to $\Phi = BA$. The chief utility of the concept of magnetic flux is in our study of electromagnetic induction in Chapter 32.

A sign convention for magnetic flux

The definition of magnetic flux given above has an ambiguity of sign associated with the direction of the vector area element $d\mathbf{A}$. In Gauss's law we

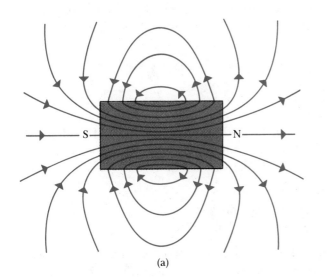

(a)

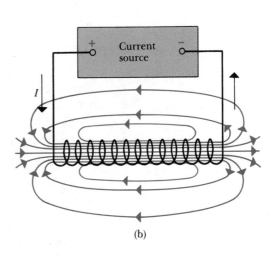

(b)

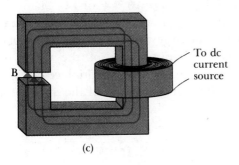

(c)

30–5 Magnetic-field lines produced by (a) a bar-shaped permanent magnet, (b) a coil of wire wound on a cylindrical form, that is, a solenoid, and (c) a laboratory electromagnet with an iron core. In (c) the magnetic field is nearly uniform in the gap in the core. In all three cases the field lines are continuous curves closing on themselves; there are no endpoints.

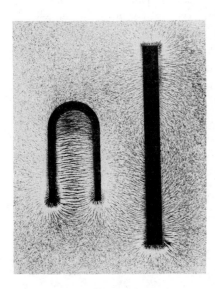

30–6 Magnetic fields can be visualized by use of iron filings, which orient themselves parallel to the field direction. This photograph shows the concentration of the fields near the poles. (Photo by Fundamental Photographs, New York.)

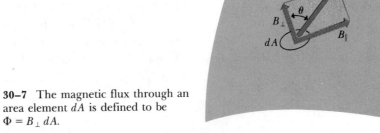

30–7 The magnetic flux through an area element dA is defined to be $\Phi = B_\perp \, dA$.

always took $d\mathbf{A}$ as pointing *out of* a closed surface. Some applications of *magnetic* flux involve an *open* surface with an edge. In these cases we define the direction of $d\mathbf{A}$ as follows: Walk around the edge, keeping the surface on your left. The direction of $d\mathbf{A}$ is from your feet toward your head.

The SI unit of magnetic field $\mathbf{B}$ is one tesla = one newton per ampere meter; hence the unit of magnetic flux is one *newton meter per ampere* (1 N·m·A^{-1}). In honor of Wilhelm Weber (1804–1890), 1 N·m·A^{-1} is called one **weber** (1 Wb). In the cgs system, the unit of magnetic flux is one **maxwell.**

If the element of area $d\mathbf{A}$ in Eq. (30–5) is at right angles to the field lines, $B_\perp = B$; in this case,

$$B = \frac{d\Phi}{dA}. \qquad (30\text{--}8)$$

That is, the magnetic field equals the *flux per unit area* across an area at right angles to the magnetic field. Magnetic field $\mathbf{B}$ is sometimes called **flux density.** Since the unit of flux is one weber, the unit of field, one tesla, is equal to one *weber per square meter* (1 Wb·m^{-2}). Similarly, in cgs units, one gauss equals one maxwell per square centimeter:

$$1 \text{ T} = 1 \text{ Wb·m}^{-2},$$
$$1 \text{ G} = 1 \text{ maxwell·cm}^{-2}.$$

We can picture the total flux through a surface as proportional to the number of field lines crossing the surface, and the field (the flux density) as the number of lines *per unit area*.

The magnetic field of the earth is of the order of 10^{-4} T, or 1 G. Magnetic fields of the order of 10 T(10^5 G) occur in the interior of atoms and are important in the analysis of atomic spectra. The largest values of steady magnetic field that have been achieved in the laboratory are of the order of 30 T = 300,000 G. Some pulsed-current electromagnets can produce fields of the order of 120 T = 1.2×10^6 G for short time intervals of the order of a millisecond.

30–4 MOTION OF CHARGED PARTICLES IN A MAGNETIC FIELD

Here is a simple example of the motion of a charged particle in a magnetic field. In Fig. 30–8 a particle with positive charge q is at point O, moving with velocity v in a uniform magnetic field $\mathbf{B}$ directed into the plane of the figure. An upward force $\mathbf{F} = q\mathbf{v} \times \mathbf{B}$, of magnitude qvB, acts on the particle at this point. The force is always perpendicular to v, so it cannot change the *magnitude* of the velocity, but only its direction. Thus the magnitudes of both $\mathbf{F}$ and v are constant. At points such as P and Q the directions of force and velocity have changed as shown; the magnitude of the force is constant, since the magnitudes of q, v, and $\mathbf{B}$ are constant. The particle therefore moves under the influence of a force whose *magnitude* is constant but whose *direction* is always at right angles to the velocity of the particle. The orbit of the particle is therefore a *circle* described with constant tangential speed v. Since the centripetal accleration is v^2/R, as shown in Section 3–5, we have, from Newton's second law,

$$F = qvB = m\left(\frac{v^2}{R}\right),$$

The weber: the SI unit of magnetic flux

Magnetic-field magnitude is magnetic flux per unit area.

Some magnitudes of real-life magnetic fields

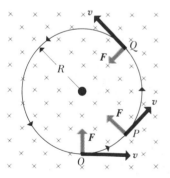

30–8 The orbit of a charged particle in a uniform magnetic field is a circle when the initial velocity is perpendicular to the field. The crosses represent a uniform magnetic field directed *away from* the reader.

Circular motion of a charged particle in a uniform magnetic field

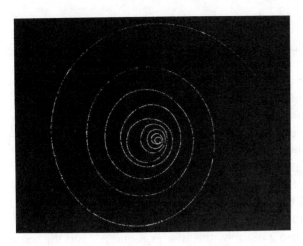

30–9 A track in a liquid-hydrogen bubble chamber made by an electron with initial kinetic energy of about 27 MeV. The magnetic field strength is 1.8 T. As the particle loses energy, the radius of curvature decreases; the maximum radius is about 5 cm. (Courtesy of Lawrence Berkeley Laboratory, University of California.)

where m is the mass of the particle. The radius of the circular orbit is

$$R = \frac{mv}{Bq}. \tag{30–9}$$

If the direction of the initial velocity is *not* perpendicular to the field, the velocity component parallel to the field remains constant and the particle moves in a helix. In this case, v in Eq. (30–9) is the component of velocity perpendicular to the field.

Note that the radius is proportional to the *momentum* of the particle, mv. Note also that the magnetic force acting on a charged particle can never do *work*, because the force is always at *right angles* to the motion. The only effect of a magnetic force is to change the *direction* of motion, never to increase or decrease the *magnitude* of the velocity. Thus, *motion of a charged particle under the action of a magnetic field alone is always motion with constant speed.*

Figure 30–9 is a photograph of the track made in a liquid-hydrogen bubble chamber by a high-energy electron moving in a magnetic field perpendicular to the plane of the paper. As the particle loses energy (and speed), the radius of curvature decreases according to Eq. (30–9). Figure 30–10 shows electron–positron pair production, also in a bubble chamber. Similar experiments using a cloud chamber provided the first experimental evidence in 1932 for the existence of the *positron*, or positive electron.

30–10 Electron–positron pair production in a liquid-hydrogen bubble chamber. A high-energy gamma ray coming in from above scatters on an atomic electron. The paths of the recoil electron and of the electron–positron pair can be seen; the directions of curvature in the magnetic field show the signs of the charges. (Courtesy of Lawrence Berkeley Laboratory, University of California.)

PROBLEM-SOLVING STRATEGY: Motion in magnetic fields

1. In analyzing the motion of a charged particle in electric and magnetic fields, you are combining the use of Newton's laws of motion with what you have learned about electric and magnetic forces. So you are still dealing with $\mathbf{F} = m\mathbf{a}$, with $\mathbf{F}$ given by $q(\mathbf{E} + \mathbf{v} \times \mathbf{B})$. Many of the problems are similar to the trajectory problems we encountered in Sections 3–4, 3–5, and 6–1, and it wouldn't do any harm to review those sections.

2. Often the use of components is the most efficient approach. Set up a coordinate system, and then express all the vector quantities (including $\mathbf{E}$, $\mathbf{B}$, $\mathbf{v}$, $\mathbf{F}$, and $\mathbf{a}$) in terms of their components in this system. Then use $\mathbf{F} = m\mathbf{a}$ in component form: $F_x = ma_x$, and so forth. This approach is particularly useful when you have both electric and magnetic fields present.

3. The next two sections, although not explicitly labeled as examples, are in fact applications of the principles introduced in this chapter and this strategy. Study them carefully!

30–5 THOMSON'S MEASUREMENT OF *e/m*

One of the landmark experiments in modern physics at the turn of the century was the measurement of the charge-to-mass ratio of an electron, *e/m*. This quantity was first measured by Sir J. J. Thomson in 1897 at the Cavendish Laboratory in Cambridge, England. The discovery that this ratio is *constant* provided the best experimental evidence available at that time of the *existence* of electrons, particles with definite mass and charge. Thomson's term for these particles was "cathode corpuscles." His experiment offers an important and interesting example of magnetic-field forces.

Thomson's apparatus (Fig. 30–11) is very similar in principle to the cathode-ray tube discussed in Section 26–8. It consists of a highly evacuated glass tube into which several metal electrodes are sealed. Electrons from the hot cathode C are formed into a beam by the anodes A and A′, and the beam passes into the region between the two plates P and P′. After passing between the plates, the electrons strike the end of the tube, where they cause fluorescent material at S to glow. The speed of the electrons depends on the accelerating potential *V*; the kinetic energy $\frac{1}{2}mv^2$ equals the loss of potential energy *eV* (where *e* is the magnitude of the electron charge):

$$\tfrac{1}{2}mv^2 = eV, \quad \text{or} \quad v = \sqrt{\frac{2eV}{m}}. \tag{30–10}$$

If a potential difference is established between the two deflecting plates P and P′, as shown, the resulting downward electric field deflects the negatively charged electrons *upward*. Alternatively, we may impose a *magnetic* field directed into the plane of the figure, as shown by the × × ×'s; this field results in a *downward* deflection of the beam. (Can you verify this direction?) Finally, if *both* **E** and **B** fields are applied simultaneously, we can adjust their relative magnitudes so that the two forces cancel and the beam is undeflected. The condition that must be satisfied, obtained by equating the two force magnitudes, is

$$eE = evB, \quad \text{or} \quad v = \frac{E}{B}. \tag{30–11}$$

Finally, we may combine this with Eq. (30–10) to eliminate *v* and obtain an expression for the charge-to-mass ratio *e/m* in terms of the other quantities:

$$\frac{E}{B} = \sqrt{\frac{2eV}{m}}, \qquad \frac{e}{m} = \frac{E^2}{2B^2V}. \tag{30–12}$$

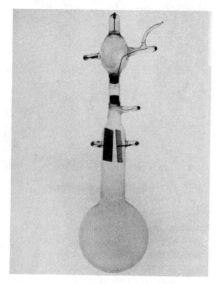

Electron-beam apparatus used by J. J. Thomson (1856–1940) for his crucial *e/m* experiment in 1897. A sketch of the apparatus is shown in Fig. 30–11. (Copyright Cavendish Laboratory, University of Cambridge.)

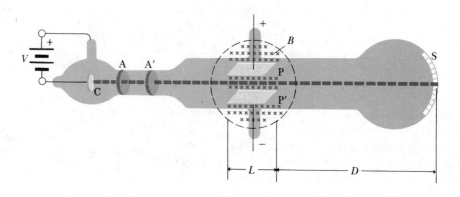

30–11 Thomson's apparatus for measuring the ratio *e/m* for cathode rays.

All the quantities on the right side can be measured, so e/m can be determined. Note that it is *not* possible to measure e or m separately by this method, but only their ratio.

Thomson measured e/m for his "cathode corpuscles" and found a unique value for this quantity that was independent of the cathode material and the residual gas in the tube. This independence indicated that cathode corpuscles are a common constituent of all matter. The most precise value of e/m available at present is $(1.758803 \pm 0.000003) \times 10^{11}$ C·kg^{-1}. Thus Thomson is credited with discovery of the first subatomic particle, the electron. He also found that the speed of the electrons in the beam was about one-tenth the speed of light, much larger than any previously measured material-particle speed.

Fifteen years after Thomson's experiments, Millikan succeeded in measuring the charge of the electron with his famous oil-drop experiment, described in Section 26–6. The magnitude of the electron charge e is 1.6022×10^{-19} C. Thus the *mass* of the electron can be obtained:

$$m = \frac{1.6022 \times 10^{-19} \text{ C}}{1.7588 \times 10^{11} \text{ C·kg}^{-1}} = 9.110 \times 10^{-31} \text{ kg}.$$

The charge and mass of the electron: two fundamental constants of nature

30–6 ISOTOPES AND MASS SPECTROSCOPY

Thomson devised a method similar to the above e/m measurement, for measuring the charge-to-mass ratio for positive ions. In Thomson's day it was difficult to produce a beam of positive ions all having the same speed. Because the e/m electron experiment depends on the particles having a common speed (in order for the electric- and magnetic-field forces to balance), this method is not directly applicable for a beam of particles having various velocities. Thomson's idea was to make the electric and magnetic fields *parallel,* so that the deflections due to these fields are in perpendicular directions. The net deflection can then never be zero, but it turns out that the relation between the x- and y-deflections for a beam permits determination of the charge-to-mass ratio of the particles.

Using magnetic-field forces to measure molecular masses

Thomson assumed that each positive ion had a charge equal in magnitude to that of the electron because each ion was an atom that had lost one electron. He could then identify particular values of q/m with particular ions. Positive ions move more slowly than electrons and have lower values of q/m because they are much more massive. The *largest* q/m for positive particles is that for the *lightest* element, hydrogen. From the value of q/m it was found that the mass of the *hydrogen ion* or *proton* is 1836.13 ± 0.01 times the mass of an electron. This showed for the first time that electrons contribute only a small fraction of the mass of material objects.

The most striking result of these experiments was that certain chemically pure gases had *more than one* value of q/m. Most notable was the case of neon, which has atomic mass 20.2 g·mol^{-1}. Thomson obtained *two* values of q/m, corresponding to 20 and 22 g·mol^{-1}, and after trying and discarding various explanations he concluded that there must be two kinds of neon atoms with different masses.

Some elements have more than one molecular mass: the discovery of isotopes.

Soon afterward, Francis Aston, a student of Thomson, succeeded in actually separating these two atomic species. Aston permitted the gas to diffuse repeatedly through a porous plug between two containers. He made use of

Varieties of mass spectrometers, and their uses to identify isotopes of many elements

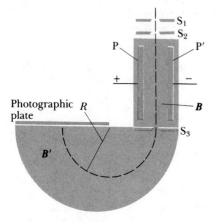

30–12 Bainbridge's mass spectrometer, utilizing a velocity selector.

Velocity selectors: obtaining a beam of electrons that all have the same speed

The atomic mass unit: a convenient unit for masses of atoms and molecules

the fact that at a given temperature T, each atom has an average kinetic energy equal to $\frac{3}{2}kT$, independent of the atom's mass, as discussed in Section 20–4. If the average of $\frac{1}{2}mv^2$ is the same for all atoms, then the more massive atoms must have, on the average, somewhat smaller speeds. Because of this speed difference, the gas emerging from the plug had a slightly greater concentration of the less massive atoms than the gas entering the plug. Thus Aston demonstrated directly the existence of two species of neon atoms with different masses.

Since these experiments in the early twentieth century, many other elements have been shown to have several kinds of atoms, identical in their chemical behavior but differing in mass. Such forms of an element are called **isotopes.** We will see in Chapter 44 that the mass differences are due to differing numbers of neutrons in the *nuclei* of the atoms.

A detailed search for the isotopes of all the elements required precise experimental technique. Aston built the first of many instruments called **mass spectrometers** in 1919; his instrument had a precision of one part in 10,000. A variation built by Bainbridge incorporates a velocity selector to produce a beam of ions all with the same speed. In Fig. 30–12, a source of ions (not shown) is situated above S_1. The ions under study pass through slits S_1 and S_2 and move down into the electric field between the two plates P and P'. In this region there is also a magnetic field **B**, perpendicular to the page. Thus the ions enter a region of crossed electric and magnetic fields like those used in the Thomson e/m experiment. Only those ions whose speed is equal to E/B pass straight through this region; ions with other speeds are deflected and blocked by slit S_3. Thus all ions emerging from S_3 have the same velocity. The region of crossed fields is called a *velocity selector.*

Below S_3 the ions enter a region where there is another magnetic field **B'**, perpendicular to the page, but no electric field. Here the ions move in circular paths of radius R. From Eq. (30–9), we find that

$$m = \frac{qB'R}{v}. \qquad (30\text{–}13)$$

Usually the charges on all the ions are the same, so the mass of each ion is proportional to the radius of its path. Ions of different isotopes converge at different points on the photographic plate. The relative abundance of the isotopes is measured from the densities of the photographic images they produce.

In present-day mass spectrometers the photographic plate is replaced by a more sophisticated particle detector. Figure 30–13 shows a modern mass spectrometer and a typical isotope analysis. For each isotope, the number given represents the **mass number,** equal to the total number of protons and neutrons in the nucleus of the atom. Different isotopes of an element have the same number of protons but differing numbers of neutrons in the nucleus.

Masses of atoms are often expressed in **atomic mass units.** By definition, one atomic mass unit (1 u) is $\frac{1}{12}$ the mass of one atom of the most abundant isotope of carbon, ^{12}C. Since the mass of an atom in grams is equal to its atomic mass divided by Avogadro's number, it follows that

$$1\,\text{u} = \frac{(1/12)(12\ \text{g·mol}^{-1})}{6.022 \times 10^{23}\ \text{mol}^{-1}}$$

$$= 1.661 \times 10^{-24}\,\text{g} = 1.661 \times 10^{-27}\,\text{kg}.$$

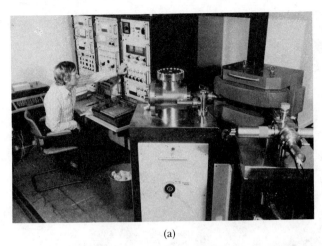

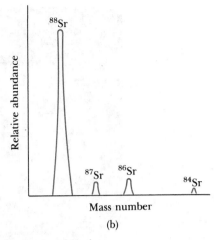

(a)   (b)

30–13 (a) A modern solid-source mass spectrometer. The specimen is placed in the cylindrical chamber in the center; the ion beam is deflected by the large magnet toward the collector and detector (right front). The instrument is controlled and the data processed by the equipment in front of the operator. (Courtesy of Institute of Geological Studies.) (b) A typical isotope analysis, showing several isotopes of strontium.

30–7 MAGNETIC FORCE ON A CONDUCTOR

When a current-carrying conductor lies in a magnetic field, the field exerts magnetic forces on the moving charges within the conductor. These forces are transmitted to the material of the conductor, and the conductor as a whole experiences a force distributed along its length. The electric motor and the moving-coil galvanometer both depend for their operation on the magnetic forces on conductors carrying currents.

Figure 30–14 represents a segment of a conducting wire, with length l and cross-sectional area A, in which the current density J is from left to right. The wire is in a magnetic field B, perpendicular to the plane of the diagram, and directed *into* the plane. For generality, we assume the wire contains moving charges of both signs. In metals the moving charges are always negatively charged electrons, but in semiconductors charges of both signs may be present. A positive charge q_1 within the wire, moving with its drift velocity v_1, is acted on by an upward force F_1 given by Eq. (30–2), $F_1 = q_1(v_1 \times B)$. As the figure shows, the direction of this force is upward, and since in this case v_1 and B are perpendicular, the magnitude of the force is $F_1 = q_1 v_1 B$. Similarly, a negative charge q_2, with drift velocity v_2 in a direction opposite to that of J, experiences a force $F_2 = q_2 v_2 \times B$. Because q_1 and q_2 have opposite signs and v_1 and v_2 have opposite directions, F_2 has the *same* direction as F_1, as shown.

Magnetic force on a moving charge in a conductor

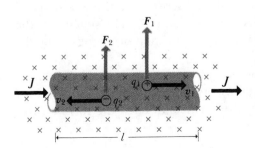

30–14 Forces on the moving charges in a current-carrying conductor. The forces on both positive and negative charges are in the same direction.

An example of magnetic levitation in railroads. This German system using attractive magnetic forces with conventional electromagnetic technology can achieve speeds of 400 km/hr (250 mi/hr). (Courtesy of Transit America, Inc.)

The *total* force on all the moving charges in a length l of conductor can be expressed in terms of the current, using the considerations of Section 28–1, Eqs. (28–3), (28–4), and (28–5). Let n_1 and n_2 represent the numbers of positive and negative charges, respectively, per unit volume. The numbers of charges in the portion having length l are then n_1Al and n_2Al. The total force F on all charges (and hence the total force on the wire) has magnitude

$$F = (n_1Al)(q_1v_1B) + (n_2Al)(q_2v_2B)$$
$$= (n_1q_1v_1 + n_2q_2v_2)AlB.$$

The magnetic force on the moving charges in a conductor can be expressed in terms of the current.

But $n_1q_1v_1 + n_2q_2v_2$ (or more generally, Σnqv) equals the current density J, and the product JA equals the current I, so finally

$$F = IlB. \tag{30–14}$$

If the B field is not perpendicular to the wire but makes an angle ϕ with it, the situation is just like that discussed in Section 30–2 for a single charge. The component of B parallel to the wire (and thus to the drift velocities of the charges) exerts no force; the component perpendicular to the wire is given by $B_\perp = B \sin \phi$. Thus for this more general case

$$F = IlB_\perp = IlB \sin \phi. \tag{30–15}$$

Expressing the magnetic force on a conductor using the vector product

To find the direction of the force on a current-carrying conductor placed in a magnetic field, we may use the same right-hand rule that we used for a moving positive charge. Rotate a right-hand-thread screw from the direction of I toward B; the direction of advance is the direction of F. The situation is the same as that shown in Fig. 30–2, but with the direction of v replaced by the direction of I.

Thus this magnetic force, like the force on a single moving charge, may be expressed as a vector product. We represent the section of wire by a vector l along the wire and in the direction of the current. Then the force on this section is given by

$$F = Il \times B. \tag{30–16}$$

EXAMPLE 30–2   A straight horizontal wire carries a current of 50 A from west to east in a region between the poles of a large electromagnet, which provides a horizontal magnetic field toward the northeast (i.e., 45° north of east) with magnitude 1.2 T. Find the magnitude and direction of the force on a 1-m section of wire.

SOLUTION   The angle ϕ between the directions of current and field is 45°. From Eq. (31–2) we obtain

$$F = (50 \text{ A})(1 \text{ m})(1.2 \text{ T})(\sin 45°) = 42.4 \text{ N}.$$

An example of the magnetic-field force on a conductor

Consistency of units may be checked by observing, as noted in Section 30–2, that $1 \text{ T} = 1 \text{ N} \cdot \text{A}^{-1} \cdot \text{m}^{-1}$. The direction of the force is perpendicular to the plane of the current and the field, both of which lie in the horizontal plane. Thus the force must be along the vertical; the right-hand rule shows that it is vertically upward.

Alternatively, we can use a coordinate system with the x-axis pointing east, the y-axis north, and the z-axis up. Then we have

$$\boldsymbol{l} = (1 \text{ m})\boldsymbol{i},$$

$$\boldsymbol{B} = (1.2 \text{ T})(\cos 45° \, \boldsymbol{i} + \sin 45° \, \boldsymbol{j}),$$

$$\boldsymbol{F} = I\boldsymbol{l} \times \boldsymbol{B}$$
$$= (50 \text{ A})(1 \text{ m})(1.2 \text{ T})\boldsymbol{i} \times (\cos 45° \, \boldsymbol{i} + \sin 45° \, \boldsymbol{j})$$
$$= (42.4 \text{ N})\boldsymbol{k}.$$

30–8 FORCE AND TORQUE ON A CURRENT LOOP

We can represent any current-carrying conductor in terms of straight-line segments; curved portions can be represented by a large number of very small segments. Thus we can use Eq. (30–15) or (30–16) to calculate the magnetic-field force on *any* conductor. In general, this requires integration; but when the magnetic field is uniform, the calculation is quite simple. As an example, we will consider a rectangular current loop in a uniform field. We will find that the total *force* on the loop is zero but that there is a *torque* that tends to turn the loop to a particular orientation with respect to the field.

Figure 30–15 shows a rectangular loop of wire having sides of lengths a and b. The normal to the plane of the loop makes an angle α with the direction of a uniform magnetic field, and the loop carries a current I. (Provision must be made for leading the current into and out of the loop, or for inserting a source of emf. This is omitted from the diagram, for simplicity.)

The total magnetic-field force on a closed current loop in a uniform field is zero.

The force $\boldsymbol{F}$ on the right side of the loop (length a), is in the direction of the x-axis, toward the right, as shown. On this side $\boldsymbol{B}$ is perpendicular to the current direction, and the total force on this side (the sum of the forces distributed along the side) has magnitude

$$F = IaB.$$

A force of the same magnitude but in the opposite direction acts on the opposite side, as shown in the figure.

The forces on the sides of length b, represented by the vectors $\boldsymbol{F'}$ and $-\boldsymbol{F'}$, have magnitude

Torque on a current loop in a magnetic field

$$IbB \sin (90° - \alpha), \quad \text{or} \quad IbB \cos \alpha.$$

The lines of action of both lie along the y-axis.

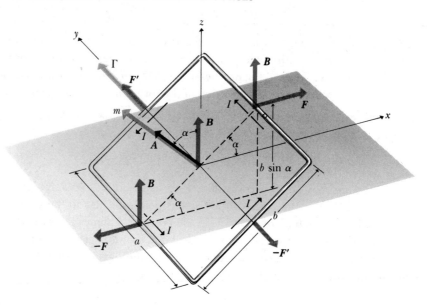

30–15 Forces on the sides of a current-carrying loop in a magnetic field. The resultant of the set of forces is a couple of moment $\Gamma = IAB \sin \alpha$.

The total force on the loop is zero, since the forces on opposite sides cancel out in pairs. The two forces F' and $-F'$ lie along the same lines and so have no torque with respect to any axis, but the two forces F and $-F$ constitute a *couple,* as defined in Section 10–4. The torque of a couple has the same value with respect to any point, and is given by the magnitude of either force multiplied by the distance between the lines of action of the two forces. From Fig. 30–15, this distance is $b \sin \alpha$, so the torque is

$$\Gamma = (IBa)(b \sin \alpha). \qquad (30\text{--}17)$$

The torque is greatest when $\alpha = 90°$ (i.e., when the plane of the coil is parallel to the field), and it is zero when α is zero or $180°$ and the plane of the coil is perpendicular to the field. (One of the latter positions is a position of *stable* equilibrium, the other of *unstable* equilibrium; which is which?)

Since ab is the area A of the coil, Eq. (30–17) may also be written

$$\Gamma = IBA \sin \alpha. \qquad (30\text{--}18)$$

Magnetic moment: a convenient way to represent the interaction of a current loop with a magnetic field

The product IA is called the **magnetic moment** m of the loop:

$$m = IA.$$

This is analogous to the electric dipole moment introduced in Section 25–1. The torque on a magnetic moment associated with a current loop can be expressed simply as

$$\Gamma = mB \sin \alpha. \qquad (30\text{--}19)$$

Because of the directional relations indicated, the torque Γ tends to rotate the loop in the direction of *decreasing* α; that is, toward its equilibrium position, in which it lies in the xy-plane, perpendicular to the direction of the field B.

Using the vector product to describe the torque on a current loop in a magnetic field

We can express the interaction between a conducting loop and a magnetic field more compactly in terms of vector torque, introduced in Section 9–8. We first define a vector area A (as in Section 25–4) having magnitude A and direction perpendicular to its plane; the sense is determined by the right-hand rule applied to the direction of circulation of current around the loop, as shown in Fig. 30–15. Then, as Eq. (30–18) and the figure show, the vector torque $\boldsymbol{\Gamma}$ is given by

$$\boldsymbol{\Gamma} = I\boldsymbol{A} \times \boldsymbol{B}. \qquad (30\text{–}20)$$

We may also define a vector magnetic moment

$$\boldsymbol{m} = I\boldsymbol{A}; \qquad (30\text{–}21)$$

in terms of $\boldsymbol{m}$ we can express the torque simply as

$$\boldsymbol{\Gamma} = \boldsymbol{m} \times \boldsymbol{B}. \qquad (30\text{–}22)$$

In the equilibrium position of the loop, its vector magnetic moment $\boldsymbol{m}$ is parallel to the field $\boldsymbol{B}$. The torque $\boldsymbol{\Gamma}$ tends to rotate the loop toward this position. The torque is greatest when $\boldsymbol{m}$ and $\boldsymbol{B}$ are perpendicular, and zero when they are parallel or antiparallel. Equation (30–22) is the analog of Eq. (25–3) (Section 25–1) for the torque on an *electric* dipole in an electric field.

When a magnetic dipole changes its orientation in a magnetic field, the field does work on it, given by $\int \Gamma\, d\theta$, and there is a corresponding potential energy. As this discussion suggests, the potential energy is least when $\boldsymbol{m}$ and $\boldsymbol{B}$ are parallel and greatest when they are antiparallel. If the potential energy U is taken as zero when $\alpha = \pi/2$, then the potential energy $U(\alpha)$ at any other position is given by

$$U(\alpha) = \int_{\alpha}^{\pi/2} \Gamma\, d\theta.$$

We can express the torque Γ in terms of the variable angle θ by use of Eq. (30–19):

$$\Gamma = -mB \sin \theta.$$

The extra minus sign is included here because the torque is in the direction of *decreasing* θ. Combining the relations given above and evaluating the integral, we find

$$\begin{aligned}
U(\alpha) &= -mB \int_{\alpha}^{\pi/2} \sin \theta\, d\theta \\
&= -mB \cos \theta,
\end{aligned}$$

or

$$U = -\boldsymbol{m} \cdot \boldsymbol{B}. \qquad (30\text{–}23)$$

Although we have derived Eqs. (30–18), (30–22), and (30–23) for a rectangular current loop, it is not hard to show that all these relations are valid for a plane loop of any shape at all. A *circular* loop, for example, may be approximated as closely as we wish by a very large number of rectangular loops. If these loops all carry equal currents in the same sense, then the forces on the sides of two loops adjacent to each other cancel out, and the only forces that do not cancel are around the boundary. Pursuing this line of reasoning, we can prove that all the relations above are valid not only for a circular loop but for *a plane loop of any shape*. In particular, for a circular loop of radius R,

$$\Gamma = \pi IBR^2 \sin \alpha \qquad (30\text{–}24)$$

and

$$\boldsymbol{\Gamma} = \pi IR^2 \boldsymbol{n} \times \boldsymbol{B}, \qquad (30\text{–}25)$$

A solenoid can be represented as many circular current loops side-by-side.

where **n** is a unit vector perpendicular to the plane of the loop as determined by the sense of circulation of current.

An arrangement of particular interest is the **solenoid,** which is a helical winding of wire, such as a coil wound on a circular cylinder. If the windings are closely spaced, the solenoid can be approximated by a number of circular loops lying in planes at right angles to its long axis. The total torque on a solenoid in a magnetic field is simply the sum of the torques on the individual turns. Hence, for a solenoid of N turns in a uniform field B,

$$\Gamma = NIAB \sin \alpha, \tag{30–26}$$

where α is the angle between the axis of the solenoid and the direction of the field. The torque is maximum when the magnetic field is parallel to the planes of the individual turns or perpendicular to the long axis of the solenoid. The effect of this torque, if the solenoid is free to turn, is to rotate it into a position in which each turn is perpendicular to the field and the axis of the solenoid is parallel to the field.

A solenoid in a magnetic field acts a lot like a bar magnet.

The behavior of a solenoid in a magnetic field resembles that of a bar magnet or compass needle; both the solenoid and the magnet, if free to turn, orient themselves with their axes parallel to a magnetic field. The torque on a solenoid could be incorporated into an alternative *definition* of magnetic field. The behavior of a bar magnet or compass needle is sometimes described in terms of magnetic forces on "poles" at its ends; for the solenoid, no such concept is needed. In fact, the moving electrons in a bar of magnetized iron play the same role as the current in the windings of a solenoid, and the observed torque arises from the same cause in both instances. We will return to this subject later.

An example of a current loop interacting with a magnetic field

EXAMPLE 30–3 A circular coil of wire 0.05 m in radius, having 30 turns, lies in a horizontal plane, as shown in Fig. 30–16. It carries a current of 5 A, in a counterclockwise sense when viewed from above. The coil is in a magnetic field directed toward the right, with magnitude 1.2 T. Find the magnetic moment and the torque on the coil.

SOLUTION The area of the coil is

$$A = \pi r^2 = \pi(0.05 \text{ m})^2 = 7.85 \times 10^{-3} \text{ m}^2.$$

The magnetic moment of each turn of the coil is

$$m = IA = (5 \text{ A})(7.85 \times 10^{-3} \text{ m}^2) = 3.93 \times 10^{-2} \text{ A·m}^2,$$

and the total magnetic moment of all 30 turns is

$$m = (30)(3.93 \times 10^{-2} \text{ A·m}^2) = 1.18 \text{ A·m}^2.$$

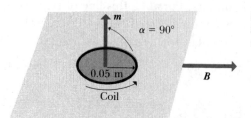

30–16 A circular coil of wire and its associated magnetic moment, in a magnetic field (Example 30–3).

The angle α between the direction of B and the normal to the plane of the coil is 90°, and from Eq. (30–18) the torque on each turn of the coil is

$$\Gamma = IBA \sin \alpha = (5 \text{ A})(1.2 \text{ T})(7.85 \times 10^{-3} \text{ m}^2)(\sin 90°)$$
$$= 0.047 \text{ N·m},$$

and the total torque on the coil is

$$\Gamma = (30)(0.047 \text{ N·m}) = 1.41 \text{ N·m}.$$

Alternatively, from Eq. (30–19),

$$\Gamma = mB \sin \alpha = (1.18 \text{ A·m}^2)(1.2 \text{ T})(\sin 90°)$$
$$= 1.41 \text{ N·m}.$$

The direction of the torque is such as to tend to push the right side of the coil down and the left side up and to rotate it into a position where the normal to its plane is parallel to B.

EXAMPLE 30–4 If the coil in Example 30–3 rotates from its initial position to a position where its magnetic moment is parallel to B, what is the change in potential energy?

An example of energy relations for a current loop in a magnetic field

SOLUTION From Eq. (30–23), the initial potential energy U_1 is

$$U_1 = -(1.18 \text{ A·m}^2)(1.2 \text{ T})(\cos 90°) = 0,$$

and the final potential energy U_2 is

$$U_2 = -(1.18 \text{ A·m}^2)(1.2 \text{ T})(\cos 0°) = -1.41 \text{ J}.$$

Thus the change in potential energy is -1.41 J.

EXAMPLE 30–5 What vertical forces applied to the left and right edges of the coil of Fig. 30–16 would be required to hold it in equilibrium in its initial position?

SOLUTION An upward force of magnitude F at the right side and a downward force of equal magnitude on the left side would have a total torque of

$$\Gamma = (2)(0.05 \text{ m})F.$$

This must be equal to the magnitude of the magnetic-field torque of 1.41 N·m, and we find that the required forces have magnitude 14.1 N.

The d'Arsonval galvanometer, which we described in Section 29–3, makes use of a magnetic-field torque on a coil carrying a current. As Fig. 29–7 shows, the magnetic field is not uniform but is *radial*, so the side thrusts on the coil are always perpendicular to its plane. Thus the magnetic-field torque is directly proportional to the current, no matter what the orientation of the coil. A restoring torque proportional to the angular displacement of the coil is provided by two hairsprings, which also serve as current leads to the coil. When current is supplied to the coil, it rotates, along with its attached pointer, until the restoring spring torque just balances the magnetic-field torque. Thus the pointer deflection is proportional to the current. Such instruments can measure currents as small as 10^{-7} A, and modified versions currents as small as 10^{-10} A. For extreme sensitivity and reliability, however, d'Arsonval instruments have been mostly replaced by digital electronic instruments.

The d'Arsonval galvanometer: an application of magnetic torque on a current loop

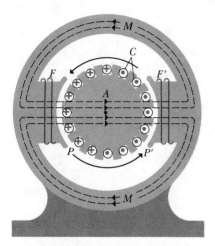

30–17 Schematic diagram of a dc motor. The armature or rotor A rotates on a shaft through its center, perpendicular to the plane of the figure. The conductors on the rotor are shown in cross section; those with dots at their centers carry current out of the plane; those with crosses, into the plane.

The rotor and field windings can be connected in various ways.

Voltage–current relations for a direct-current motor

An example of a series dc motor

30–9 THE DIRECT-CURRENT MOTOR

No one needs to be reminded of the importance of electric motors in contemporary society. Their operation depends on magnetic-field forces on current-carrying conductors. As an example, let us consider the operation of one type of direct-current motor, shown schematically in Fig. 30–17. The center part A is the *armature*, or *rotor;* it is a cylinder of soft steel mounted on a shaft so that it can rotate about its axis (perpendicular to the plane of the figure).

Embedded in slots in the rotor surface (parallel to its axis) are insulated copper conductors C. Current is led into and out of these conductors through graphite brushes making contact with a cylinder of the shaft called the *commutator* (not shown in Fig. 30–17). The commutator is an automatic switching arrangement that maintains the currents in the conductors in the directions shown in the figure, whatever the position of the rotor. The current in the field coils F and F' sets up a magnetic field in the motor frame M and in the gap between the pole pieces P and P' and the rotor. Some of the magnetic field lines are shown as broken lines. With the directions of field and rotor currents shown, the side thrust on each conductor in the rotor is such as to produce a *counterclockwise* torque on the rotor.

If the rotor and the field windings are connected in series, we have a *series* motor; if they are connected in parallel, we have a *shunt* motor. In some motors the field windings are in two parts, one in series with the rotor and the other in parallel with it; such a motor is called *compound.*

A motor converts electrical energy to mechanical energy or work, and thus requires electrical energy input. If the potential difference between its terminals is V_{ab} and the current is I, then the power input is $P = V_{ab}I$. Even if the motor coils have negligible resistance, there must be a potential difference between the terminals if P is to be different from zero. This potential difference results principally from magnetic forces exerted on the charges in the conductors of the rotor as they rotate through the magnetic field. These forces are an example of nonelectrostatic forces; the resulting electromotive force $\mathcal{E}$ is called an *induced* emf or sometimes a *back* emf, referring to the fact that its sense is opposite to that of the current. Induced emf's resulting from motion of conductors in magnetic fields will be considered in greater detail in Chapter 32.

In a series motor having internal resistance r, V_{ab} is greater than $\mathcal{E}$, and the difference is the drop Ir across the internal resistance. Thus, for either a motor or a battery being charged,

$$V_{ab} = \mathcal{E} + Ir. \tag{30–27}$$

Because of the nature of the magnetic-field force, $\mathcal{E}$ is *not* constant but depends on the speed of rotation of the rotor.

The behavior of a dc motor is analogous to that of a battery being charged, as discussed in Section 28–4. In that case, electrical energy is converted to chemical rather than mechanical energy.

EXAMPLE 30–6   A dc motor with its rotor and field coils connected in series has an internal resistance of 2.0 Ω. When running at full load on a 120-V line, it draws a current of 4.0 A.

a) What is the emf in the rotor?

$$V_{ab} = \mathcal{E} + Ir,$$

$$120 \text{ V} = \mathcal{E} + (4.0 \text{ A})(2.0 \text{ }\Omega),$$

$$\mathcal{E} = 112 \text{ V}.$$

b) What is the power delivered to the motor?

$$P = V_{ab}I = (120 \text{ V})(4.0 \text{ A}) = 480 \text{ W}.$$

c) What is the rate of dissipation of energy in the resistance of the motor?

$$P = I^2 r = (4.0 \text{ A})^2 (2.0 \text{ }\Omega) = 32 \text{ W}.$$

d) What is the mechanical power developed?

The mechanical power output is the electrical power input, minus the rate of dissipation of energy in the motor's resistance:

$$P = 480 \text{ W} - 32 \text{ W} = 448 \text{ W}.$$

30–10 THE HALL EFFECT

The reality of the forces acting on the moving charges in a conductor in a magnetic field is strikingly demonstrated by the **Hall effect.** The conductor in Fig. 30–18 is in the form of a flat strip. The charges within it are driven toward the upper edge of the strip by the magnetic force qvB exerted on them. Here v is the drift velocity of the moving charges in the material.

If the charge carriers are electrons, as in Fig. 30–18a, an excess negative charge accumulates at the upper edge of the strip, leaving an excess positive charge at its lower edge. This accumulation continues to a point where the resulting transverse electrostatic field E_e causes a force of magnitude qE_e that is equal and opposite to the magnetic field force of magnitude qvB. Then there is no longer any net force to deflect the moving charges sideways. Associated with this electric field is a potential difference between opposite edges

The Hall effect: sideways potential difference in a conductor, caused by a magnetic field

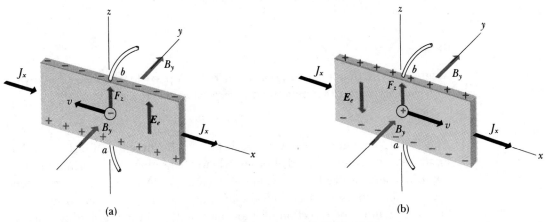

(a)　　　　　　　　　　　　　(b)

30–18 Forces on charge carriers in a conductor in a magnetic field. (a) Negative current carriers (electrons) are pushed toward top of slab, leading to charge distribution as shown. Point a is at higher potential than point b. (b) Positive current carriers; polarity of potential difference is opposite to that of (a).

of the strip. This potential difference, which can be measured with a potentiometer, is called the *Hall voltage*, or the *Hall emf*. Experiments show that for metals, the upper edge of the strip in Fig. 30–18a *does* become negatively charged, showing that the charge carriers in a metal are indeed negative electrons.

However, if the charge carriers are *positive*, as in Fig. 30–18b, then *positive* charge accumulates at the upper edge, and the potential difference is *opposite* to that resulting from the deflection of negative charges. Soon after the discovery of the Hall effect, in 1879, it was observed that some materials, notably the *semiconductors*, exhibit a Hall emf opposite to that of the metals, *as if* their charge carriers were *positively* charged. We now know that these materials conduct by a process known as *hole conduction*. Within such a material there are locations, called *holes*, that would normally be occupied by an electron but are actually empty, and a *missing negative* charge is equivalent to a *positive* charge. When an electron moves in one direction to fill a hole, it leaves another hole behind it, and the result is that the hole (equivalent to a positive charge) migrates in the direction *opposite* to that of the electron.

In terms of the coordinate axes in Fig. 30–18a, the electrostatic field E_e is in the z-direction, and we write it as E_z. The magnetic field is in the y-direction, and we write it as B_y. The magnetic-field force (in the $-z$-direction) is qvB_y. The current density J_x is in the x-direction. In the final steady state, when the forces qE_z and qvB_y are equal,

$$E_z = vB_y.$$

The current density J_x is

$$J_x = nqv.$$

When v is eliminated, we have

$$nq = \frac{J_x B_y}{E_z}. \tag{30–28}$$

Thus, from measurements of J_x, B_y, and E_z, we can compute the product nq. In both metals and semiconductors, q is equal in magnitude to the electron charge, so the Hall effect permits a direct measurement of n, the density of current-carrying charges in the material.

SUMMARY

Magnetic field, denoted by B, is a vector field. A particle with charge q moving with velocity v in a magnetic field B experiences a force F given by

$$F = qv \times B. \tag{30–2}$$

The SI unit of magnetic field is the tesla (1 T). The direction of B is defined by Eq. (30–2); it can be defined alternatively in terms of the equilibrium orientation of a compass needle or a solenoid permitted to rotate freely. A magnetic field can be represented graphically by magnetic-field lines; at each point a line is tangent to the direction of B at that point, and the number of lines per unit area perpendicular to B is proportional to the magnitude of B.

Magnetic flux through an area is defined as

$$\Phi = \int B_\perp \, dA = \int B \cdot dA. \tag{30–6}$$

The SI unit of magnetic flux is the weber (1 Wb). Magnetic field is also called magnetic flux density, and $1 \, T = 1 \, Wb \cdot m^{-2}$.

Because the magnetic-field force is always perpendicular to v, it does no work on the particle and cannot change the magnitude of v. Thus a particle moving under the action of a magnetic field alone moves with constant speed.

J. J. Thomson used crossed electric and magnetic fields to measure the charge-to-mass ratio (e/m) for electrons. The electric- and magnetic-field forces exactly cancel when $v = E/B$.

Different atoms of an element having different atomic masses are called isotopes; molecular masses are measured with a mass spectrometer. For each isotope, the mass number is the total number of protons and neutrons in the nucleus. Masses of atoms are often expressed in atomic mass units. One atomic mass unit (1 u) is defined to be $1/12$ the mass of one atom of ^{12}C.

The force F on a segment l of a conductor carrying current I in a magnetic field B is given by

$$F = Il \times B. \qquad (30\text{--}16)$$

A current loop of vector area A carrying current I in a uniform magnetic field B experiences no net force but a torque Γ given by

$$\Gamma = IA \times B. \qquad (30\text{--}20)$$

The magnetic moment m of the loop is defined as

$$m = IA. \qquad (30\text{--}21)$$

The torque can also be expressed as

$$\Gamma = m \times B. \qquad (30\text{--}22)$$

The potential energy U of a magnetic moment m in a magnetic field B is given by

$$U = -m \cdot B. \qquad (30\text{--}23)$$

The magnetic moment of a loop depends only on the current and the area, and is independent of the shape of the loop.

In a d'Arsonval galvanometer, the magnetic forces on the conducting loop are perpendicular to its plane. The magnetic-field torque is proportional to the current in the coil; the spring restoring torque is proportional to angular displacement, so the pointer deflection is proportional to current.

In a dc motor the magnetic field set up by the field windings exerts tangential forces on the currents in the rotor. In a series motor the rotor and field coils are in series; in a parallel motor they are in parallel. Motion of the rotor through the magnetic field causes an induced emf. For a series motor the terminal voltage is the sum of the induced emf and the drop Ir across the internal resistance.

The Hall effect is a potential difference perpendicular to the direction of current in a conductor, when the conductor is placed in a magnetic field. The Hall potential is determined by the requirement that the associated electric field must just balance the magnetic-field force on a moving charge. Hall-effect measurements can be used to determine the density n of charge carriers and their sign.

QUESTIONS

30–1 Does the earth's magnetic field have a significant effect on the electron beam in a TV picture tube?

30–2 If an electron beam in a cathode-ray tube travels in a straight line, can you be sure there is no magnetic field present?

30–3 If the magnetic-field force does no work on a charged particle, how can it have any effect on the particle's motion? Are there other examples of forces that do no work but have a significant effect on a particle's motion?

30–4 A permanent magnet can be used to pick up a string of nails, tacks, or paper clips, even though these are not magnets by themselves. How can this be?

30–5 How could a compass be used for a *quantitative* determination of magnitude and direction of magnetic field at a point?

30–6 The direction in which a compass points (magnetic north) is not in general exactly the same as the direction toward the north pole (true north). The difference is called *magnetic declination;* it varies from point to point on the earth and also varies with time. What are some possible explanations for magnetic declination?

30–7 Can a charged particle move through a magnetic field without experiencing any force? How?

30–8 Could the electron beam in an oscilloscope tube (cathode-ray tube) be used as a compass? How? What advantages and disadvantages would it have, compared with a conventional compass?

30–9 Does a magnetic field exert forces on the electrons within atoms? What observable effect might such interaction have on the behavior of the atom?

30–10 How might a loop of wire carrying a current be used as a compass? Could such a compass distinguish between north and south?

30–11 How could the direction of a magnetic field be determined by making only *qualitative* observations of the magnetic force on a straight wire carrying a current?

30–12 Do the currents in the electrical system of a car have a significant effect on a compass placed in the car?

30–13 A student claimed that if lightning strikes a metal flagpole, the force exerted by the earth's magnetic field on the current in the pole can be large enough to bend it. Typical lightning currents are of the order of 10^4 to 10^5 A; is the student's opinion justified?

30–14 A student tried to make an electromagnetic compass by suspending a coil of wire from a thread (with the plane of the coil vertical) and passing a current through it. He expected the coil to align itself perpendicular to the horizontal component of the earth's magnetic field; but instead, the coil went into what appeared to be angular simple harmonic motion (cf. Section 11–4), turning back and forth past the expected direction. What was happening? Was the motion truly simple harmonic?

30–15 When the polarity of the voltage applied to a dc motor is reversed, the direction of rotation *does not* reverse. Why not? How *could* the direction of rotation be reversed?

30–16 If an emf is produced in a dc motor, would it be possible to use the motor somehow as a *generator* or *source,* taking power out of it instead of putting power into it? How might this be done?

30–17 Hall-effect voltages are much *larger* for relatively poor conductors such as germanium than for good conductors such as copper, for comparable currents, fields, and dimensions. Why?

30–18 Is it possible that, in a Hall-effect experiment, *no* transverse potential difference will be observed? Under what circumstances might this happen?

EXERCISES

Section 30–2 Magnetic Field

30–1 In a magnetic field directed vertically upward, a particle initially moving north is deflected toward the west. What is the sign of the charge of the particle?

30–2 A particle with mass 0.001 kg and charge 1.2×10^{-8} C has at a given instant a velocity $v = (2 \times 10^5 \text{ m·s}^{-1})j$. What are the magnitude and direction of the force exerted on the particle by a uniform magnetic field $B = -(0.8 \text{ T})i$?

30–3 A particle with charge -2.5×10^{-8} C is moving with instantaneous velocity

$$v = (-3 \times 10^4 \text{ m·s}^{-1})i + (5 \times 10^4 \text{ m·s}^{-1})j.$$

What is the force exerted on this particle by a magnetic field

a) $B = (2 \text{ T})i;$ b) $B = (2 \text{ T})k?$

30–4 A particle having a mass of 0.5 g carries a charge of 2.5×10^{-8} C. The particle is given an initial horizontal velocity of 6×10^4 m·s^{-1}. What are the magnitude and direction of the magnetic field needed to keep the particle moving in a horizontal direction?

30–5 Each of the lettered circles at the corners of the cube in Fig. 30–19 represents a positive charge q moving with a velocity of magnitude v in the direction indicated. The region in the figure is a uniform magnetic field B, parallel to the x-axis and directed toward the right. Copy the figure, find the magnitude and direction of the force on each charge, and show the force in your diagram.

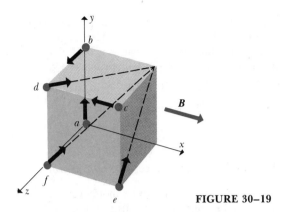

FIGURE 30–19

Section 30–3 Magnetic-Field Lines and Magnetic Flux

30–6 The magnetic field **B** in a certain region has magnitude 2 T, and its direction is that of the positive x-axis in Fig. 30–20.

a) What is the magnetic flux across the surface abcd in the figure?

b) What is the magnetic flux across the surface befc?

c) What is the magnetic flux across the surface aefd?

d) What is the net flux through all five surfaces that enclose the shaded volume?

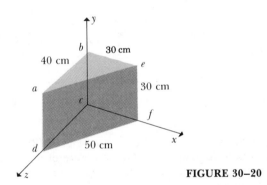

FIGURE 30–20

Section 30–4 Motion of Charged Particles in a Magnetic Field

30–7 An electron at point A in Fig. 30–21 has a speed v_0 of 1.0×10^7 m·s^{-1}. Find

a) the magnitude and direction of the magnetic field that will cause the electron to follow the semicircular path from A to B;

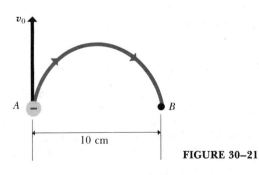

FIGURE 30–21

b) the time required for the electron to move from A to B.

30–8 In Exercise 30–7, suppose the particle is a proton rather than an electron. Answer the same questions as in that problem.

30–9 A deuteron, an isotope of hydrogen whose mass is very nearly 2 u, travels in a circular path of radius 0.40 m in a magnetic field of magnitude 1.5 T.

a) Find the speed of the deuteron.

b) Find the time required for it to make one-half a revolution.

c) Through what potential difference would the deuteron have to be accelerated to acquire this speed?

30–10 A singly charged ^{7}Li ion has a mass of 1.16×10^{-26} kg. It is accelerated through a potential difference of 500 V and then enters a magnetic field of 0.4 T, moving perpendicular to the field. What is the radius of its path in the magnetic field?

30–11 In a TV picture tube, an electron in the beam is accelerated by a potential difference of 20,000 V. Then it passes through a region of transverse magnetic field, where it moves in a circular arc with radius 0.12 m. What is the magnitude of the field?

30–12 An electron moves in a circular path of radius 1.2 cm perpendicular to a uniform magnetic field. The speed of the electron is 1×10^6 m·s^{-1}. What is the total magnetic flux encircled by the orbit?

Section 30–5 Thomson's Measurement of e/m

Section 30–6 Isotopes and Mass Spectroscopy

30–13

a) What is the velocity of a beam of electrons when the simultaneous influence of an electric field of 34×10^4 V·m^{-1} and a magnetic field of 2×10^{-2} T, both fields being normal to the beam and to each other, produces no deflection of the electrons?

b) Show in a diagram the relative orientation of the vectors **v**, **E**, and **B**.

c) What is the radius of the electron orbit when the electric field is removed?

30–14 In the Bainbridge mass spectrometer (Fig. 30–12), suppose the magnetic field B in the velocity selector is 1.0 T, and ions having a speed of 4.0×10^6 m·s^{-1} pass through undeflected.

a) What should be the electric field between the plates P and P'?

b) If the separation of the plates is 0.5 cm, what is the potential difference between plates?

30–15 The electric field between the plates of the velocity selector in a Bainbridge mass spectrometer is 1.20×10^5 V·m^{-1}, and the magnetic field in both regions is 0.6 T. A stream of singly charged neon moves in a circular path of 0.728-m radius in the magnetic field. Determine the mass number of the neon isotope.

Section 30–7 Magnetic Force on a Conductor

30–16 An electromagnet produces a magnetic field of 1.2 T in a cylindrical region of radius 5 cm between its poles. A wire carrying a current of 20 A passes through this region, intersecting the axis of the cylinder and perpendicular to it. What force is exerted on the wire?

30–17 A horizontal rod 0.2 m long is mounted on a balance and carries a current. In the vicinity of the rod is a uniform horizontal magnetic field of magnitude 0.05 T, perpendicular to the rod. The magnetic force on the rod is measured by the balance and is found to be 0.24 N. What is the current?

30–18 In Exercise 30–17, suppose the magnetic field is horizontal but makes an angle of 30° with the rod. What is the current in the rod?

30–19 A wire along the x-axis carries a current of 5 A in the positive direction. Calculate the force (expressed in terms of unit vectors) on a 1 cm section of the wire exerted by the following magnetic fields:

a) $\boldsymbol{B} = -(0.6\ \text{T})\boldsymbol{j}$ b) $\boldsymbol{B} = +(0.5\ \text{T})\boldsymbol{k}$ c) $\boldsymbol{B} = -(0.3\ \text{T})\boldsymbol{i}$

d) $\boldsymbol{B} = +(0.2\ \text{T})\boldsymbol{i} - (0.3\ \text{T})\boldsymbol{k}$ e) $\boldsymbol{B} = +(0.9\ \text{T})\boldsymbol{j} - (0.4\ \text{T})\boldsymbol{k}$

Section 30–8 Force and Torque on a Current Loop

30–20 What is the maximum torque on a rectangular coil 5 cm × 12 cm and of 600 turns when carrying a current of 1×10^{-5} A in a uniform field of magnitude 0.1 T?

30–21 A circular coil of wire 8 cm in diameter has 12 turns and carries a current of 5 A. The coil is in a region where the magnetic field is 0.6 T.

a) What is the maximum torque on the coil?

b) In what position would the torque be one-half as great as in (a)?

30–22 The plane of a rectangular loop of wire 5 cm × 8 cm is parallel to a magnetic field of magnitude 0.15 T.

a) The loop carries a current of 10 A. What torque acts on it?

b) What is the magnetic moment of the loop?

c) What is the maximum torque that can be obtained with the same total length of wire carrying the same current in this magnetic field?

30–23 A circular coil of area A and N turns is free to rotate about a diameter that coincides with the x-axis. Current I is circulating in the coil. There is a uniform magnetic field $\boldsymbol{B}$ in the positive y-direction. Calculate the magnitude and direction of the torque $\boldsymbol{\Gamma}$ and the value of the potential energy U, as given in Eq. (30–23), when the coil is oriented as shown in parts (a)–(d) of Fig. 30–22.

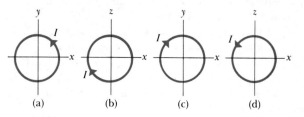

(a)   (b)   (c)   (d)

FIGURE 30–22

30–24 The coil of a pivoted-coil galvanometer has 50 turns and encloses an area of 6 cm². The magnetic field in the region in which the coil swings is 0.01 T and is radial. The torsional constant of the hairsprings is

$$k' = 1.0 \times 10^{-8}\ \text{N·m·(degree)}^{-1}.$$

Find the angular deflection of the coil for a current of 1 mA.

Section 30–9 The Direct-Current Motor

30–25 In a shunt-wound dc motor (Fig. 30–23), the resistance R_f of the field coils is 150 Ω, and the resistance R_r of the rotor is 2 Ω. When a difference of potential of 120 V is applied to the brushes, and the motor is running at full speed delivering mechanical power, the current supplied to it is 4.5 A.

a) What is the current in the field coils?

b) What is the current in the rotor?

c) What is the induced emf developed by the motor?

d) How much mechanical power is developed by this motor?

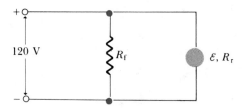

FIGURE 30–23

30–26 A shunt-wound dc motor (Fig. 30–23) operates from a 120-V dc power line. The resistance of the field windings, R_f, is 240 Ω. The resistance of the rotor, R_r, is 3 Ω. When the motor is running, the rotor develops an emf $\mathcal{E}$. The motor draws a current of 4.5 A from the line. Compute

a) the field current,

b) the rotor current,

c) the emf $\mathcal{E}$,

d) the rate of development of heat in the field windings,

e) the rate of development of heat in the rotor,

f) the power input to the motor,

g) the efficiency of the motor,

if friction losses amount to 50 W.

Section 30–10 The Hall Effect

30–27 Figure 30–24 shows a portion of a silver ribbon with $z_1 = 2$ cm and $y_1 = 1$ mm, carrying a current of 200 A in the positive x-direction. The ribbon lies in a uniform magnetic field, in the y-direction, of magnitude 1.5 T. If there are 7.4×10^{28} free electrons per m³, find

a) the drift velocity of the electrons in the x-direction;

b) the magnitude and direction of the electric field in the z-direction due to the Hall effect;

c) the Hall emf.

30–28 Let Fig. 30–24 represent a strip of copper of the same dimensions as those of the silver ribbon of the preceding exercise. When the magnetic field is 5 T and the current is 100 A, the Hall emf is found to be 45.4 μV. What is the density of free electrons in the copper?

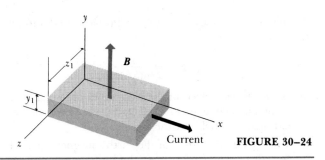

FIGURE 30–24

PROBLEMS

30–29 A particle with initial velocity $v_0 = (4 \times 10^3 \text{m·s}^{-1})i$ enters a region of uniform electric and magnetic fields. The magnetic field in the region is $B = -(0.3 \text{ T})j$. Calculate the magnitude and direction of the electric field in the region if the particle is to pass through undeflected, for a particle of charge

a) $+0.4 \times 10^{-8}$ C   b) -0.4×10^{-8} C

Neglect the weight of the particle.

30–30 Estimate the effect of the earth's magnetic field on the electron beam in a TV picture tube. Suppose the accelerating voltage is 20,000 V; calculate the approximate deflection of the beam over a distance of 0.4 m from the electron gun to the screen, under the action of a transverse field of magnitude 5.0×10^{-5} T (comparable to the magnitude of the earth's field), assuming there are no other deflecting fields. Is this deflection significant?

30–31 A particle carries a charge of 4×10^{-9} C. When it moves with a velocity v_1 of 3×10^4 m·s^{-1} at 45° above the x-axis in the xy-plane, a uniform magnetic field exerts a force F_1 along the negative z-axis. When the particle moves with a velocity of v_2 of 2×10^4 m·s^{-1} along the z-axis, there is a force F_2 of 4×10^{-5} N exerted on it along the x-axis. What are the magnitude and direction of the magnetic field? (See Fig. 30–25.)

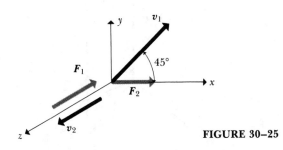

FIGURE 30–25

30–32 A particle having a charge $q = 2 \times 10^{-6}$ C is traveling with a velocity $v = (1 \times 10^3 \text{ m·s}^{-1})j$. The particle experiences a force $F = 2 \times 10^{-4}(3i - 4k)$N due to a magnetic field B.

a) Determine F, the magnitude of F.

b) Determine B_x, B_y, and B_z, or at least as many of the three components as possible, from the information given.

c) If it is given in addition that the magnitude of the magnetic field is 0.5 T, determine the remaining components of B.

30–33 An electron and an alpha particle (a doubly ionized helium atom) both move in circular paths in a magnetic field with the same tangential speed. Compare the number of revolutions they make per second. The mass of the alpha particle is 6.64×10^{-27} kg.

30–34 A particle of mass m and charge q moves with velocity v in a magnetic field B. The velocity of the particle is perpendicular to the field, and the particle moves in a circle whose radius is given by Eq. (30–9).

a) Calculate the period of the motion, the time it takes for the particle to complete one revolution.

b) From your answer in (a) obtain the frequency $f = 1/\tau$ of the circular motion. This frequency is called the cyclotron frequency.

30–35 Suppose the electric field between the plates P and P' in Fig. 30–12 is 1.5×10^4 V·m^{-1}, and the magnetic field is 0.5 T. If the source contains the three isotopes of magnesium, ^{24}Mg, ^{25}Mg, and ^{26}Mg, and the ions are singly charged, find the distance between the lines formed by the three isotopes on the photographic plate. Assume the atomic masses of the isotopes are equal to their mass numbers.

30–36 An electron is moving in a circular path of radius $r = 4$ cm in the space between two concentric cylinders. The inner cylinder is a positively charged wire of radius $a = 1$ mm and the outer cylinder is a negatively charged cylinder of radius $b = 5$ cm. The potential difference between the inner and outer cylinders is $V_{ab} = 100$ V, with the wire being at the higher potential. See Fig. 30–26. The electric field E in the region between the cylinders is

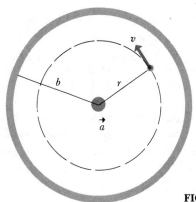

FIGURE 30–26

radially outward and was shown in Problem 26–38 to have the magnitude

$$E = \frac{V_{ab}}{r \ln(b/a)}.$$

a) Determine the speed of the electron in order for it to maintain its circular orbit. Neglect both the gravitational and magnetic fields of the earth.

b) Now include the effect of the earth's magnetic field. If the axis of symmetry of the cylinders is positioned parallel to the magnetic field of the earth, at what speed must the electron move in order to maintain the same circular orbit? Assume that the magnetic field of the earth has magnitude 1.0×10^{-4} T, and that its direction is out of the plane of the paper in Fig. 30–26.

c) Redo the calculation of part (b) for the case in which the magnetic field is in the opposite direction to that of part (b).

30–37 A particle of positive charge q and mass $m = 1 \times 10^{-15}$ kg is traveling through a region containing a uniform magnetic field $\boldsymbol{B} = -(0.1 \text{ T})\boldsymbol{k}$. At a particular instant of time, the velocity of the particle is given by

$$\boldsymbol{v} = (4\boldsymbol{i} - 3\boldsymbol{j} + 12\boldsymbol{k}) \times 10^6 \text{ m·s}^{-1}$$

and the force $\boldsymbol{F}$ on the particle has a magnitude of 2 N.

a) Determine the charge q.

b) Determine the acceleration $\boldsymbol{a}$ of the particle.

c) Explain why the path of the particle is a helix and determine the radius of curvature R of the circular component of the helical path.

d) Determine the cyclotron frequency of the particle. (See Problem 30–34.)

e) Although helical motion is not periodic in the full sense of the word, it is periodic with respect to the motion projected onto the xy-plane. If the coordinates of the particle at $t = 0$ are $(x, y, z) = (R, 0, 0)$, determine the coordinates of the particle at a time $t = 2\tau$, where τ is the "period" of the motion in the xy-plane.

30–38 A wire 0.5 m long lies along the y-axis and carries a current of 10 A in the $+y$-direction. The magnetic field is uniform and has components

$$B_x = 0.3 \text{ T}, \qquad B_y = -1.2 \text{ T}, \qquad \text{and } B_z = 0.5 \text{ T}.$$

a) Find the components of force on the wire.

b) What is the magnitude of the total force on the wire?

30–39 The cube in Fig. 30–27, 0.5 m on a side, is in a uniform magnetic field of 0.6 T, parallel to the x-axis. The wire $abcdef$ carries a current of 4 A in the direction indicated.

a) Determine the magnitude and direction of the force acting on each of the segments ab, bc, cd, de, and ef.

b) What are the magnitude and direction of the total force on the wire?

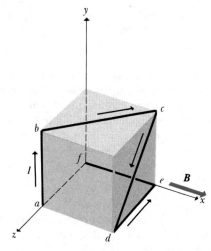

FIGURE 30–27

30–40 The rectangular loop in Fig. 30–28 is pivoted about the y-axis and carries a current of 10 A in the direction indicated.

a) If the loop is in a uniform magnetic field of magnitude 0.2 T, parallel to the x-axis, find the torque required to hold the loop in the position shown.

b) Same as (a), except that the field is parallel to the z-axis.

c) For each of the above magnetic fields, what torque would be required if the loop were pivoted about an axis through its center, parallel to the y-axis?

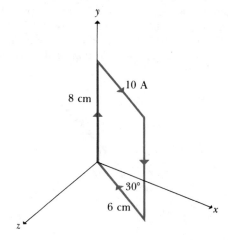

FIGURE 30–28

30–41 The rectangular loop of wire in Fig. 30–29 has a mass of 0.1 g per centimeter of length and is pivoted about side ab as a frictionless axis. The current in the wire is 10 A in the direction shown. Find the magnitude and sense of the magnetic field, parallel to the y-axis, that will cause the loop to swing up until its plane makes an angle of 30° with the yz-plane.

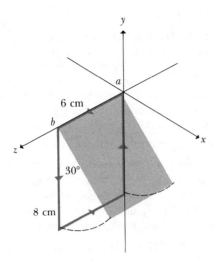

FIGURE 30–29

30–42 The neutron is a particle with zero charge but with a nonzero magnetic moment of magnitude $m = 9.66 \times 10^{-27}$ J·T^{-1}. If the neutron is considered to be a fundamental entity with no internal structure, the two properties listed above seem to be contradictory. According to present theory in particle physics, a neutron is composed of three more fundamental particles called "quarks." (See the discussion of quarks in Section 44–10.) In this model the neutron consists of an "up quark" having a charge of $+2e/3$ and two "down quarks" each having a charge of $-e/3$. The combination of the three quarks produces a net charge of $2e/3 - e/3 - e/3 = 0$, as required, and if the quarks are in motion they could also produce a nonzero magnetic moment. As a very simple model, suppose the up quark is moving in a counterclockwise circular path and the down quarks are moving in a clockwise path, all of radius r and all with the same speed v. (See Fig. 30–30).

a) Obtain an expression for the current due to the circulation of the up quark.

b) Obtain an expression for the magnitude m_u of the magnetic moment due to the circulating up quark.

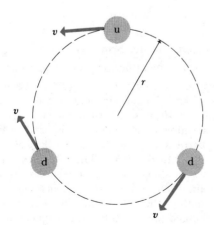

FIGURE 30–30

c) Obtain an expression for the magnitude of the magnetic moment of the three-quark system. (Be careful to use the correct magnetic moment directions.)

d) With what speed v must the quarks move if this model is to reproduce the magnetic moment of the neutron? For the radius of the orbits use $r = 1.2 \times 10^{-15}$ m, the radius of the neutron.

30–43 A wire of length 35 cm and mass $m = 9.79 \times 10^{-5}$ kg is bent into the shape of an inverted U such that the horizontal part is of length $l = 25$ cm. The bent ends of the wire, which are each 5 cm long, are completely immersed in two pools of mercury, and the entire structure is in a region containing a magnetic field of magnitude 0.01 T and a direction into the page. (See Fig. 30–31.) An electrical connection from the mercury pools is made through wires that are connected to a 1.5-V battery and a switch S. Switch S is closed, and the wire jumps 1.00 m into the air, measured from its initial position.

a) Determine the speed v of the wire as it leaves the mercury.

b) Assuming that the current I through the wire was constant from the time the switch was closed until the wire left the mercury, determine I.

c) Neglecting the resistance of the mercury and the circuit wires, determine the resistance of the moving wire.

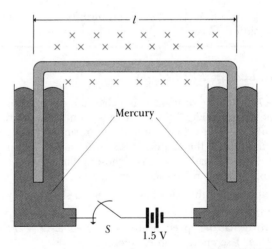

FIGURE 30–31

30–44 We can derive Eq. (30–25) explicitly without great difficulty. Consider a wire ring in the xy-plane, with its center at the origin. The ring carries a counterclockwise current I (Fig. 30–32). Let the magnetic field $\boldsymbol{B}$ be in the x-direction, $\boldsymbol{B} = B_x \boldsymbol{i}$. (The result is easily extended to $\boldsymbol{B}$ in an arbitrary direction.)

a) In Fig. 30–32 show that the element

$$d\boldsymbol{l} = R\, d\theta(-\boldsymbol{i} \sin \theta + \boldsymbol{j} \cos \theta),$$

and find $d\boldsymbol{F} = I\, d\boldsymbol{l} \times \boldsymbol{B}$.

b) Integrate $d\boldsymbol{F}$ around the loop to show that the net force is zero.

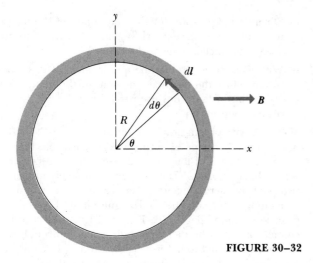

FIGURE 30–32

c) From part (a), find $d\boldsymbol{\Gamma} = \boldsymbol{r} \times d\boldsymbol{F}$, where

$$\boldsymbol{r} = R(\boldsymbol{i} \cos \theta + \boldsymbol{j} \sin \theta).$$

(Note that $d\boldsymbol{l}$ is perpendicular to $\boldsymbol{r}$.)

d) Integrate $d\boldsymbol{\Gamma}$ over the loop to find the total torque $\boldsymbol{\Gamma}$ on the loop. Show that the result can be written as

$$\boldsymbol{\Gamma} = I A \boldsymbol{n} \times \boldsymbol{B}.$$

(*Note*. $\int \cos^2 x \, dx = \frac{1}{2}x + \frac{1}{4} \sin 2x$, $\int \sin^2 x \, dx = \frac{1}{2}x - \frac{1}{4} \sin 2x$, and $\int \sin x \cos x \, dx = \frac{1}{2} \sin^2 x$.)

30–45 A circular loop of wire of area 10 cm² carries a current of 10 A. The loop lies in the *xy*-plane. As viewed in toward the origin along the *z*-axis, the current is circulating counterclockwise. The torque produced by an external magnetic field $\boldsymbol{B}$ is given by $\boldsymbol{\Gamma} = (-6 \times 10^{-3}\boldsymbol{i} + 8 \times 10^{-3}\boldsymbol{j})$N·m, and for this orientation of the loop the magnetic potential energy $U = -\boldsymbol{m} \cdot \boldsymbol{B}$ is negative. The magnitude of the magnetic field is 2.6 T.

a) Determine the magnetic moment of the current loop.

b) Determine the components B_x, B_y, and B_z of $\boldsymbol{B}$.

30–46 A circular ring of area 10 cm² and negligible mass is carrying a current of 50 A. The ring is free to rotate about a diameter. The ring, initially at rest, is immersed in a region of magnetic field where $\boldsymbol{B}$ is given by

$$\boldsymbol{B} = (3\boldsymbol{i} - 4\boldsymbol{j} - 12\boldsymbol{k}) \times 10^{-2} \text{ T}.$$

The ring is positioned initially such that its magnetic moment $\boldsymbol{m}_i$ is given by

$$\boldsymbol{m}_i = m(-0.6\boldsymbol{i} + 0.8\boldsymbol{j}),$$

where m is the (positive) magnitude of the magnetic moment. The ring is released and turns through an angle of 90°, at which point its magnetic moment is given by

$$\boldsymbol{m}_f = -m\boldsymbol{k}.$$

a) Determine the decrease in potential energy of the ring.

b) If the moment of inertia of the ring about a diameter is 1.70×10^{-6} kg·m², determine the angular velocity of the ring as it passes through the second position.

CHALLENGE PROBLEMS

30–47 A particle of mass m and charge $+q$ starts from rest at the origin in Fig. 30–33. There is a uniform electric field $\boldsymbol{E}$ in the positive *y*-direction and a uniform magnetic field $\boldsymbol{B}$ directed toward the reader. It is shown in more advanced books that the path is a *cycloid* whose radius of curvature at the top points is twice the *y*-coordinate at that level.

a) Explain why the path has this general shape and why it is repetitive.

b) Prove that the speed at any point is equal to $\sqrt{2qEy/m}$. (*Hint:* Use energy conservation.)

c) Applying Newton's second law at the top point and taking as given that the radius of curvature here equals $2y$, prove that the speed at this point is $2E/B$.

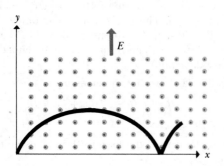

FIGURE 30–33

30–48 Two positive ions having the same charge q but different masses, m_1 and m_2, are accelerated horizontally from rest through a potential difference V. They then enter a region where there is a uniform magnetic field $\boldsymbol{B}$ normal to the plane of the trajectory.

a) Show that if the beam entered the magnetic field along the *x*-axis, the value of the *y*-coordinate for each ion at any time t is approximately

$$y = Bx^2 \left(\frac{q}{8mV}\right)^{1/2},$$

provided y remains much smaller than x.

b) Can this arrangement be used for isotope separation?

30–49 A particle of charge $q = 4 \times 10^{-6}$ C and mass $m = 1 \times 10^{-11}$ kg is initially traveling in the *y*-direction with a speed $v_0 = 2 \times 10^5$ m·s⁻¹. It then enters a region containing a uniform magnetic field, directed away from the reader and perpendicular to the page in Fig. 30–34. The magnitude of the field is 0.5 T. The region extends a distance of 25 cm along the initial direction of travel; 75 cm from the point of entry into the magnetic field region is a wall. The length of the field-free region is thus 50 cm. When the charged particle enters the magnetic field, it will follow a curved path whose radius of curvature is R. It then leaves the magnetic field after a time t_1, having been

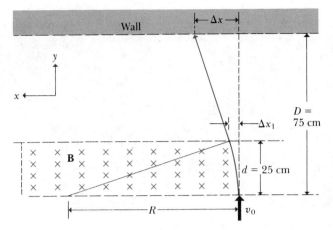

FIGURE 30-34

deflected a distance Δx_1. The particle then travels in the field-free region and strikes the wall after undergoing a total deflection Δx.

a) Determine the radius R of the curved part of the path.

b) Determine t_1, the time the particle spends in the magnetic field.

c) Determine Δx_1, the horizontal deflection at the point of exit from the field.

d) Determine Δx, the total horizontal deflection.

30-50 Magnetic forces acting on conducting fluids provide a convenient means of pumping these fluids. This problem deals with such an electromagnetic pump. A horizontal tube of rectangular cross section (height h, width w) is placed at right angles to a uniform magnetic field of magnitude B, so that a length l is in the field. (See Fig. 30-35.) The tube is filled with liquid sodium and an electric current of density J is maintained in the third mutually perpendicular direction.

a) Show that the difference of pressure between a point in the liquid on a vertical plane through ab (Fig. 30-35) and a point in the liquid on another vertical plane through cd, under conditions in which the liquid is prevented from flowing, is $\Delta p = JlB$.

b) What current density would be needed to provide a pressure difference of 1 atm between these two points if $B = 1$ T and $l = 0.1$ m?

30-51 A wire of length 60 cm, mass $m = 1.27 \times 10^{-5}$ kg, and resistance $R = 2.01$ Ω is bent into the shape of an inverted U such that the horizontal part is of length $l = 40$ cm. The bent ends of the wire, which are each 10 cm long, are completely immersed in two pools of mercury, and the entire structure is in a region containing a magnetic field of magnitude $B = 4$ T and directed into the page in Fig. 30-36. (See Problem 30-43.) An electrical connection from the mercury pools is made through wires connected to a capacitor whose capacitance C is 20 μ F and then to a switch S. With a potential difference of 1000 V across the capacitor, switch S is closed and the resulting upward magnetic force causes the wire to jump into the air.

a) Derive an expression which relates the speed with which the wire leaves the mercury pool to the charge Δq that passes through the wire. You may assume that the magnetic force on the wire is much greater than the weight.

b) Show that

$$md = lBq_0(\Delta t - RC\,[1 - e^{-\Delta t/RC}]),$$

where q_0 is the initial charge on the capacitor, d is the length of the bend in the wire at each end and is 10 cm, and Δt is the time it takes for the wire to leave the pool of mercury, that is, the time to travel the first 10 cm. It should be assumed that all resistances can be neglected except that of the moving wire.

c) Solve for a numerical value for Δt. (*Hint:* You will not be able to solve for Δt analytically but must use a numerical method such as iteration. Rearrange the equation with $\Delta t/(RC)$ by itself on the left side. Use a trial guess for $\Delta t/(RC)$ (make an astute guess) in the right-hand side of the equation. Compute a revised value for $\Delta t/(RC)$ and use this new value to recompute still another value. Continue to do this until a self-consistent value of $\Delta t/(RC)$, to within three significant figures, is found.)

d) Determine the total charge Δq that passed through the wire.

e) Determine the speed v of the wire as it leaves the mercury.

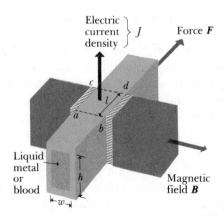

FIGURE 30-35

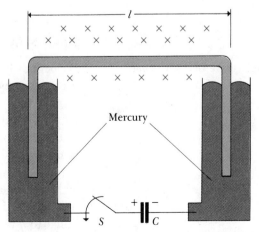

FIGURE 30-36

31

SOURCES OF MAGNETIC FIELD

IN CHAPTER 30 WE STUDIED ONE ASPECT OF THE MAGNETIC INTERACTION of moving charges: the *forces* exerted on moving charges and on currents in conductors when a magnetic field is present. We did not worry about how the magnetic field got there, but simply took its existence as a given fact. Now we are ready to return to the other aspect of this interaction, the principles that describe how magnetic fields are *produced* by moving charges and by currents.

The most basic relationship is the magnetic field produced by a single moving point charge. This relationship can be used to derive an equation for the field produced by a small segment of a current-carrying conductor. From this equation, called the law of Biot and Savart, we derive relationships for the magnetic fields produced by several specific shapes of conductor, including long, straight wires; circular loops; and solenoids. Ampere's law, the magnetic analog of Gauss's law in electrostatics, forms a useful alternative formulation of the relation of magnetic fields to their sources. Finally, we study the role of displacement current, which we encountered in Section 29–5, as a source of magnetic field. This relationship will be one of the key elements in our study of electromagnetic waves in Chapter 35.

31–1 MAGNETIC FIELD OF A MOVING CHARGE

A moving charge creates a magnetic field.

We begin with the basics, the magnetic field of a single moving point charge q. We call the location of the charge the **source point** and the point P where we want to find the field the **field point.** In our study of *electric* fields in Chapter 25, we found that the E field of a point charge q, at a field point located a distance r from the charge, is proportional to q and to $1/r^2$, and that its direction (for positive q) is along the line from source point to field point. The corresponding relationship for the *magnetic* field B of a point charge q moving with velocity v has some similarities to this relationship and some interesting differences. First, experiments show that the magnitude B is again proportional to q and to $1/r^2$. But the *direction* of B is *not* along the line from the source point to the field point; instead it is perpendicular to the plane containing this line and the particle's velocity vector *v,* as shown in Fig. 31–1. Furthermore, the field magnitude is proportional to the sine of the angle θ between these two directions. Thus the magnitude of the magnetic field at point P is given by

$$B = k' \frac{qv \sin \theta}{r^2}, \qquad (31\text{–}1)$$

where k' is a proportionality constant.

We can incorporate both the magnitude and direction of **B** into a single vector equation using the vector product. First we introduce a unit vector $\hat{r} = r/r$, pointing in the direction from charge q to point P, that is, from the source point to the field point. Then the **B** field of a moving point charge is

$$\boldsymbol{B} = k' \frac{q\boldsymbol{v} \times \hat{r}}{r^2}. \qquad (31\text{–}2)$$

Figure 31–1 shows the magnetic field **B** at several points in the vicinity of the charge. At all points along a line through the charge parallel to the velocity *v,* the field is zero, because $\sin \theta = 0$ at all such points. At any distance r from q, **B** has its greatest magnitude at points lying in the plane through the charge perpendicular to *v,* because at all such points $\theta = 90°$ and $\sin \theta = 1$. The charge also produces an *electric* field in its vicinity; the electric-field vectors are not shown in the figure.

Recall that the field lines for the *electric* field of a point charge radiate outward from the charge. The *magnetic*-field lines are completely different in character; review of the above discussion shows that for a point charge moving with velocity *v,* the magnetic-field lines are *circles* with centers along the line of *v,* lying in planes perpendicular to this line. The directions of these lines are given by a right-hand rule: Grasp the velocity vector *v* with your right hand, so that your right thumb points in the direction of *v;* your fingers then curl around the line of *v* in the same sense as the magnetic-field lines.

As we discussed in Section 30–2, the units of B are

$$1\ \text{T} = 1\ \text{N·s·C}^{-1}\text{·m}^{-1} = 1\ \text{N·A}^{-1}\text{·m}^{-1}.$$

Using this with Eq. (31–1) or (31–2) and solving for k', we find that the units

The magnetic field of a moving charge shows inverse-square behavior.

B is perpendicular to both *v* and *r.*

The tesla: the SI unit of magnetic field

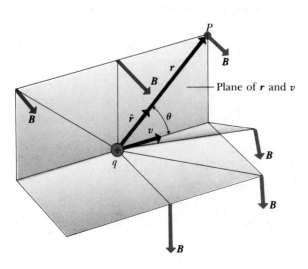

31–1 Magnetic field vectors due to a moving positive point charge q.

of the constant k' are

$$1 \text{ N·s}^2\text{·C}^{-2} = 1 \text{ N·A}^{-2} = 1 \text{ Wb·A}^{-1}\text{·m}^{-1}$$
$$= 1 \text{ T·A}^{-1}\text{·m}.$$

An arbitrary constant in magnetic-field relations

In SI units, the numerical value of k' is arbitrarily assigned to be exactly 10^{-7}. Thus

$$k' = 10^{-7} \text{ N·s}^2\text{·C}^{-2} = 10^{-7} \text{ N·A}^{-2}$$
$$= 10^{-7} \text{ Wb·A}^{-1}\text{·m}^{-1} = 10^{-7} \text{ T·A}^{-1}\text{·m} \quad \text{(exactly)}.$$

In electrostatics we found it convenient to express electric-field relations not in terms of the constant k in Eq. (24–1) but in terms of ϵ_0, where $k = 1/4\pi\epsilon_0$. Similarly, in magnetic-field relations it is convenient to introduce a constant μ_0, defined by the relation

$$k' = \frac{\mu_0}{4\pi}.$$

Thus

$$\mu_0 = 4\pi \times 10^{-7} \text{ Wb·A}^{-1}\text{·m}^{-1} = 4\pi \times 10^{-7} \text{ T·A}^{-1}\text{·m}.$$

Magnetic field of a moving charge can be expressed as a vector product.

In terms of μ_0, Eqs. (31–1) and (31–2) become

$$B = \frac{\mu_0}{4\pi} \frac{qv \sin\theta}{r^2}, \tag{31–3}$$

$$\boldsymbol{B} = \frac{\mu_0}{4\pi} \frac{q\boldsymbol{v} \times \hat{\boldsymbol{r}}}{r^2}. \tag{31–4}$$

Interaction force between two moving charges

We can now write the expression for the magnetic *force* between two point charges, both of which are in motion relative to an observer. A charge q', moving with velocity v' in a magnetic field $\boldsymbol{B}$, experiences a force

$$\boldsymbol{F} = q'\boldsymbol{v}' \times \boldsymbol{B},$$

and if the field $\boldsymbol{B}$ is set up by a charge q moving with velocity v, then

$$\boldsymbol{F} = \frac{\mu_0}{4\pi} \frac{qq'\boldsymbol{v}' \times (\boldsymbol{v} \times \hat{\boldsymbol{r}})}{r^2}. \tag{31–5}$$

This equation is analogous to Coulomb's law for the *electrical* force between the charges. In Eq. (31–5), $\boldsymbol{F}$ is the force exerted on q' by q, and $\hat{\boldsymbol{r}}$ is in the direction from q toward q'.

Recall from Section 24–5 that the electrical constant k has the value $8.98755 \times 10^9 \text{ N·m}^2\text{·C}^{-2}$. The ratio k/k' is therefore

$$\frac{k}{k'} = \frac{8.98755 \times 10^9 \text{ N·m}^2\text{·C}^{-2}}{10^{-7} \text{ N·s}^2\text{·C}^{-2}}$$
$$= 8.98755 \times 10^{16} \text{ m}^2\text{·s}^{-2},$$

which is equal to the square of the speed of light, c. We invite you to verify that the corresponding relation between ϵ_0 and μ_0 is

$$\epsilon_0\mu_0 = \frac{1}{c^2}. \tag{31–6}$$

Thus there is a close relationship between these constants and the nature of light. We will return to this matter in Chapter 35.

The arbitrary constant isn't really so arbitrary.

What if we have several point charges? For electric fields, we could take the vector sum of the E fields of the individual charges, and experiments show that this also works for magnetic fields. That is, the magnetic field also obeys the **superposition principle:** *The total magnetic field caused by several moving charges is the vector sum of the fields caused by the individual charges.* We can use this fact to calculate the field caused by a current in a circuit, in terms of the fields caused by individual segments of conductor. We develop this relationship in the next section.

Superposition principle for magnetic fields: finding the total field caused by several moving charges

31–2 MAGNETIC FIELD OF A CURRENT ELEMENT

The magnetic field produced at any field point P by the current in a conductor is the vector sum of the fields due to all the moving charges in the conductor; this statement is the *principle of superposition.* We can think of the conductor as divided into short segments of length dl; a typical segment is shown in Fig. 31–2. We can use the relations of Section 31–1 to find the magnetic field caused by the current in this segment.

The volume of the segment is $A\,dl$, where A is the cross-sectional area. Suppose we have n charges q per unit volume; then the total moving charge dQ in the segment is

The moving charge in an element of a current-carrying conductor

$$dQ = nqA\,dl.$$

The moving charges are therefore equivalent to a single charge dQ, traveling with a velocity equal to the *drift* velocity v. (Fields due to the *random* velocities of the carriers will, on average, cancel out at every point.) From Eq. (31–3), the magnitude of the resulting field dB at any point is

$$dB = \frac{\mu_0}{4\pi}\frac{dQ\,v\,\sin\theta}{r^2} = \frac{\mu_0}{4\pi}\frac{nqvA\,dl\,\sin\theta}{r^2}.$$

But $(nqvA)$ equals the current I in the element, so

The law of Biot and Savart: magnetic field caused by an element of a current-carrying conductor

$$dB = \frac{\mu_0}{4\pi}\frac{I\,dl\,\sin\theta}{r^2}, \tag{31–7}$$

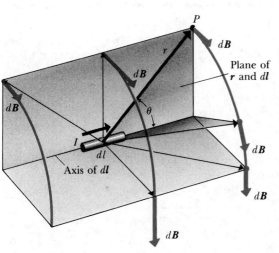

31–2 Magnetic-field vectors due to a current element.

or in vector form,

$$dB = \frac{\mu_0}{4\pi} \frac{I\, dl \times \hat{r}}{r^2}, \qquad (31\text{--}8)$$

where dl is a vector of length dl, in the same direction as the current in the conductor, and $\hat{r}$ is a unit vector in the direction from the current element to point P. As shown in Fig. 31–2, the field vectors dB are exactly like those set up by a positive charge dQ moving in the direction of the drift velocity v.

Integrating to find the total magnetic field caused by a current-carrying conductor

Equations (31–7) and (31–8) are called the **law of Biot and Savart.** To find the total magnetic field B any point in space due to the current in a complete circuit, we have to integrate one of these expressions; symbolically,

$$B = \frac{\mu_0}{4\pi} \int \frac{I\, dl \times \hat{r}}{r^2}. \qquad (31\text{--}9)$$

We will learn how to do this in specific cases in the next several sections.

The magnetic-field lines due to an element of a conductor are concentric circles.

The field lines of the magnetic field produced by a current element $I\,dl$ look just like those of a single moving point charge; they are circles in planes perpendicular to dl, centered on the line of dl. Their directions are given by the same right-hand rule we introduced for point charges in Section 31–1. A few field lines are shown in Fig. 31–2.

An individual element of a conductor cannot be isolated.

To be completely honest, we must point out that it is impossible to verify Eq. (31–8) directly because we can never experiment with an isolated segment of a current-carrying circuit. The only thing we can measure experimentally is the *total* B for a complete circuit. But we can still verify Eq. (31–8) indirectly by calculating B for various current configurations and comparing the results with experimental measurements.

If matter is present in the space around a conductor, the field at a field point P in its vicinity will have an additional contribution resulting from the *magnetization* of the material. We return to this point in Section 31–8. Unless the material is iron or some other ferromagnetic material, however, the additional field is so small that it is usually negligible.

The next three sections include several applications of the law of Biot and Savart to particular conductor shapes. The following comments will help you understand these examples and also help you with additional problems.

PROBLEM-SOLVING STRATEGY: Magnetic-field calculations

1. Be careful about the directions of vector quantities. The current element dl always points in the direction of the current. The unit vector $\hat{r}$ is always directed *from* the current element *toward* the point P at which the field is to be determined, that is, from the source point toward the field point.

2. In some problems the dB's at point P have the same direction for all the current elements. In that case, the magnitude of the total B field is the sum of the magnitudes of the dB's. But often the dB's have different directions for different current elements; then you have to set up a coordinate system and represent each dB in terms of its components. The integral for

the total B is then expressed in terms of an integral for each component. Sometimes you can see from the symmetry of the situation that one component integrates to zero. Always be on the lookout for ways to use symmetry to simplify the problem.

3. Look for ways to use the superposition principle. If you know the fields produced by certain simple conductor shapes, and if you encounter a complex shape that can be represented as a combination of simple shapes, then it is easy to use superposition to find the field of the complex shape. Examples: a rectangular loop, or a semicircle with straight-line segments on both sides.

31–3 MAGNETIC FIELD OF A STRAIGHT CONDUCTOR

Our first application of the law of Biot and Savart is finding the magnetic field of a conductor of length $2a$ carrying a current I, at a point on its perpendicular bisector, at a distance x from the conductor. The situation is shown in Fig. 31–3; the geometry is reminiscent of the electric-field problem of Example 25–8 (Section 25–2), where we found the electric field caused by a line of charge of length $2a$. Although the *shapes* of the sources are the same in these two problems, the behavior of the magnetic field is completely different from that of the electric field.

We begin by identifying an element of the conductor, having length $dl = dy$. To use Eq. (31–7), we note that $r = \sqrt{x^2 + y^2}$ and $\sin \theta = \sin(\pi - \theta) = x/\sqrt{x^2 + y^2}$. The *direction* of $d\boldsymbol{B}$ is perpendicular to the plane of the figure, into the plane, and in this case the directions of the $d\boldsymbol{B}$'s from all elements of the conductor are the same. Thus in integrating over the conductor, we can just add their magnitudes.

Putting the pieces together, we find that the magnitude B of the total $\boldsymbol{B}$ is

$$B = \frac{\mu_0 I}{4\pi} \int_{-a}^{a} \frac{x\, dy}{(x^2 + y^2)^{3/2}}. \tag{31–10}$$

We can integrate this by trigonometric substitution or by using an integral table. We challenge you to fill in the details and show that the final result is

$$B = \frac{\mu_0 I}{4\pi} \frac{2a}{x\sqrt{x^2 + a^2}}. \tag{31–11}$$

An alternative route to Eq. (31–11) uses Eqs. (31–8) and (31–9) with the appropriate vectors. We have $d\boldsymbol{l} = dy\,\boldsymbol{j}$, where $\boldsymbol{j}$ is the usual unit vector in the y-direction. Also,

$$\hat{\boldsymbol{r}} = \frac{\boldsymbol{r}}{r} = \frac{x\boldsymbol{i} - y\boldsymbol{j}}{\sqrt{x^2 + y^2}}. \tag{31–12}$$

Putting the pieces together, we find

$$
\begin{aligned}
\boldsymbol{B} &= \frac{\mu_0 I}{4\pi} \int_{-a}^{a} \frac{(dy\boldsymbol{j}) \times (x\boldsymbol{i} - y\boldsymbol{j})}{(x^2 + y^2)^{3/2}} \\
&= \frac{\mu_0 I}{4\pi} \int_{-a}^{a} \frac{(-\boldsymbol{k})x\, dy}{(x^2 + y^2)^{3/2}} \\
&= \frac{\mu_0 I}{4\pi} \frac{2a(-\boldsymbol{k})}{x\sqrt{x^2 + a^2}}.
\end{aligned}
\tag{31–13}
$$

In a right-handed axis system, the z-axis points out of the plane of Fig. 31–3. Thus the negative sign in Eq. (31–13) shows that the direction of $\boldsymbol{B}$ at point P is *into* the plane of the figure. At points in the xy-plane to the left of the y-axis, of course, the direction of $\boldsymbol{B}$ is opposite, *out of* the plane.

When the length ($2a$) of the conductor is very large compared to its distance x from point P, we can consider it to be infinitely long. When a is much larger than x, $\sqrt{x^2 + a^2}$ is approximately equal to a, and so in the limit as $a \to \infty$, Eq. (31–11) becomes

$$B = \frac{\mu_0 I}{2\pi x}. \tag{31–14}$$

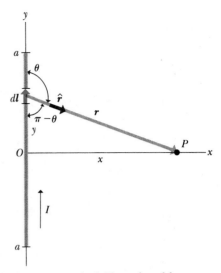

31–3 Magnetic field produced by a straight conductor of length $2a$. The $\boldsymbol{B}$ field at point P is directed into the plane of the figure.

Finding the magnetic field of a long, straight conductor by integrating over its length

Using unit vectors to evaluate vector products in magnetic-field calculations

For a long, straight wire, B is inversely proportional to the distance from the wire.

Because the physical situation has axial symmetry about the y-axis, B has the same *magnitude* at all points on a circle centered on the conductor and lying in a plane perpendicular to the conductor, and its *direction* is everywhere tangent to such a circle. Thus, at all points on a circle of radius r around the conductor, the magnitude B is given by

$$B = \frac{\mu_0 I}{2\pi r} \qquad \text{(long, straight wire).} \qquad (31\text{–}15)$$

EXAMPLE 31–1 A long, straight conductor carries a current of 100 A. At what distance from the conductor is the magnetic field caused by the current equal in magnitude to the earth's magnetic field in Pittsburgh (about 0.5×10^{-4} T)?

SOLUTION We use Eq. (31–15). Everything except r is known, so we solve for r and insert the appropriate numbers

$$r = \frac{\mu_0 I}{2\pi B} = \frac{(4\pi \times 10^{-7} \text{ T·m·A}^{-1})(100 \text{ A})}{(2\pi)(0.5 \times 10^{-4} \text{ T})} = 0.4 \text{ m.}$$

At smaller distances, the field becomes stronger; for example, when $r = 0.2$ m, $B = 1.0 \times 10^{-4}$ T, and so on.

Magnetic-field lines for a long, straight wire are circles centered on the wire.

Part of the magnetic field around a long, straight conductor is shown in Fig. 31–4. The shape of the magnetic-field lines in this situation is completely different from that of the electric-field lines in the analogous electrical situation. Electric-field lines radiate outward from the charges that are their sources (or inward for negative charges). By contrast, magnetic-field lines *encircle* the current that acts as their source. Electric-field lines begin and end at charges, while experiments have shown that magnetic-field lines *never* have endpoints, irrespective of the shape of the conductor that sets up the field. If lines *did* begin or end at a point, this point would correspond to a "magnetic charge," or a single magnetic pole. There is no experimental evidence that such entities exist.

31–4 Magnetic field around a long, straight conductor. The field lines are circles, with directions determined by the right-hand rule.

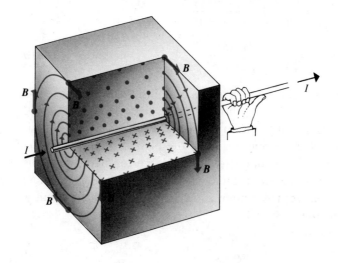

Thus if we construct an imaginary closed surface in a magnetic field, no field line can start or end inside this surface. The number of lines emerging from the surface must equal the number entering it. We have shown that the number of lines crossing a surface is proportional to the flux Φ across the surface. Hence in a magnetic field, the flux across any *closed* surface is zero, or

The total magnetic flux out of a closed surface is zero.

$$\oint B_\perp \, dA = \oint \boldsymbol{B} \cdot d\boldsymbol{A} = 0. \qquad (31\text{–}16)$$

This equation should be compared with Gauss's law for electrostatic fields, in which the surface integral of $\boldsymbol{E}$ over a closed surface equals $1/\epsilon_0$ times the enclosed charge. Thus Eq. (31–16) is an alternative statement of the fact that there is no such thing as "magnetic charge" to act as a source of $\boldsymbol{B}$. The sources of $\boldsymbol{B}$ are moving electric charges, as outlined above.

31–4 FORCE BETWEEN PARALLEL CONDUCTORS

We are now ready to examine the interaction force between two long current-carrying conductors. This problem comes up in a variety of practical problems, and it also has fundamental significance in connection with the definition of the ampere. Figure 31–5 shows segments of two long, straight, parallel conductors separated by a distance r and carrying currents I and I', respectively, in the same direction. Since each conductor lies in the magnetic field set up by the other, each experiences a force. The diagram shows some of the field lines set up by the current in the lower conductor.

From Eq. (31–15), the magnitude of the $\boldsymbol{B}$-vector at the upper conductor is

$$B = \frac{\mu_0 I}{2\pi r}.$$

From Eq. (30–14), the force on a length l of the upper conductor is

$$F = I'lB = \frac{\mu_0 II'l}{2\pi r},$$

and the force *per unit length* is therefore

$$\frac{F}{l} = \frac{\mu_0 II'}{2\pi r}. \qquad (31\text{–}17)$$

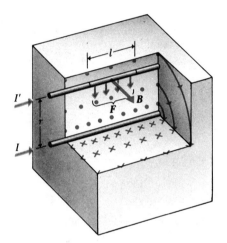

31–5 Parallel conductors carrying currents in the same direction attract each other.

Two parallel currents attract each other.

The right-hand rule shows that the direction of the force on the upper conductor is downward. An equal and opposite force per unit length acts on the lower conductor, as may be seen by considering the field set up by the upper conductor. Hence the conductors *attract* each other.

If the direction of either current is reversed, the forces reverse also. Parallel conductors carrying currents in *opposite* directions *repel* each other.

The fact that two straight, parallel conductors exert forces of attraction or repulsion on one another is the basis of the official SI definition of the ampere. The ampere is defined as follows:

Defining the ampere in terms of forces between parallel conductors

> One **ampere** is that unvarying current which, if present in each of two parallel conductors of infinite length and one meter apart in empty space, causes each conductor to experience a force of exactly 2×10^{-7} newton per meter of length.

It follows from this definition and Eq. (31–17) that, by definition, the constant μ_0 is exactly $4\pi \times 10^{-7}$ N·A^{-2}, as we stated in Section 31–1.

From the definition above, the ampere can be established, in principle, with the help of a meter stick and a spring balance. For high-precision standardization of the ampere, coils of wire are used instead of straight wires, and their separation is made only a few centimeters. The complete instrument, which is capable of measuring currents with a high degree of precision, is called a *current balance*.

Note that this definition of the ampere is an *operational definition;* it gives us an actual experimental procedure for measuring current and defining the unit of current. With this definition, we can now return to electrostatics and define the *coulomb* as *the quantity of charge that in one second crosses a section of a circuit in which there is a constant current of one ampere.* Finally, we have to regard the constant $k = 1/4\pi\epsilon_0$ in Coulomb's law as an empirical constant, determined by experiment. As we mentioned in Section 31–1, however, ϵ_0 and μ_0 are related to the speed of light. This fact provides the most precise determination of ϵ_0. We return to this relationship in Chapter 35.

Mutual forces of attraction exist not only between *wires* carrying currents in the same direction, but also between the longitudinal elements of a single current-carrying conductor. If the conductor is a liquid or an ionized gas (a plasma), these forces result in a constriction of the conductor, as if its surface were acted on by an external, inward pressure. The constriction of the conductor is called the *pinch effect.* The high temperature produced by the pinch effect in a plasma has been used in one technique to bring about nuclear fusion.

31–5 MAGNETIC FIELD OF A CIRCULAR LOOP

In many practical devices where a current is used to establish a magnetic field, such as an electromagnet or transformer, the wire carrying the current is wound into a *coil,* often consisting of many circular loops. Thus it is useful to derive an expression for the magnetic field produced by a single circular conducting loop carrying a current. Figure 31–6 shows a circular conductor of radius a, carrying a current I. The current is led into and out of the loop

31–6 Magnetic field of a circular loop. The segment dl causes the field $d\mathbf{B}$, lying in the xy-plane. Other dl's have different components perpendicular to the x-axis; these add to zero, while the x-components combine to give the total $\mathbf{B}$ field at point P.

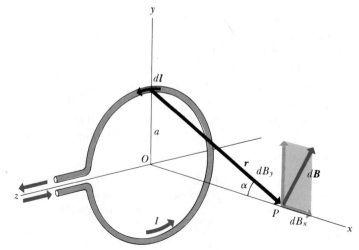

through two long, straight wires side by side; the currents in these straight wires are in opposite directions, and their magnetic fields cancel each other.

We can use the law of Biot and Savart, Eq. (31–7) or (31–8), to find the magnetic field at a point P along a line through the center of the loop perpendicular to its plane, at a distance x from the center. As the figure shows, dl and r are always perpendicular, and the direction of the field dB caused by element dl lies in the xy-plane. Also, $r = \sqrt{x^2 + y^2}$. Thus for the element dl we find

$$dB = \frac{\mu_0 I}{4\pi} \frac{dl}{x^2 + a^2}. \tag{31–18}$$

In terms of components,

$$dB_x = dB \sin \alpha = \frac{\mu_0 I}{4\pi} \frac{dl}{x^2 + a^2} \cdot \frac{a}{(x^2 + a^2)^{1/2}}, \tag{31–19}$$

$$dB_y = dB \cos \alpha = \frac{\mu_0 I}{4\pi} \frac{dl}{x^2 + a^2} \cdot \frac{x}{(x^2 + a^2)^{1/2}}. \tag{31–20}$$

Because of symmetry about the x-axis, the components of B perpendicular to this axis must sum to zero. To see why this must be, consider an element dl on the opposite side of the loop, having direction opposite to that of the dl shown. This element gives a contribution to the field having the same x-component as given by Eq. (31–19) but an *opposite* y-component. Similarly, the components of B perpendicular to the x-axis set up by pairs of diametrically opposite elements cancel, leaving only the x-components. To obtain the total x-component, we integrate Eq. (31–19). Everything is constant in this integration except dl, and the integral of dl for the entire loop is simply the circumference of the circle, $2\pi a$. Thus we obtain

Using symmetry to prove that the component perpendicular to the axis is zero

$$B_x = \int \frac{\mu_0 I}{4\pi} \cdot \frac{a \, dl}{(x^2 + a^2)^{3/2}} = \frac{\mu_0 I a}{4\pi (x^2 + a^2)^{3/2}} \int dl,$$

or, since $\int dl = 2\pi a$,

$$B_x = \frac{\mu_0 I a^2}{2(x^2 + a^2)^{3/2}} \qquad \text{(circular loop).} \tag{31–21}$$

We can use unit vectors to obtain Eqs. (31–19) and (31–20) somewhat more compactly. We write $dl = dl \, k$, $r = xi - aj$, and

Using unit vectors in the circular-loop calculation

$$\frac{dl \times \hat{r}}{r^2} = \frac{dl \times r}{r^3} = \frac{(dl \, k) \times (xi - aj)}{(x^2 + a^2)^{3/2}}$$

$$= \frac{x \, dl \, j + a \, dl \, i}{(x^2 + a^2)^{3/2}}.$$

As before, the components perpendicular to the x-axis sum to zero, and only the term containing i survives. Using the above expression in Eq. (31–9) and again using the fact that $\int dl = 2\pi a$, we obtain

$$B = \int \frac{\mu_0 I}{4\pi} \frac{a \, dl \, i}{(x^2 + a^2)^{3/2}} = \frac{\mu_0 I a^2 i}{2(x^2 + a^2)^{3/2}}, \tag{31–22}$$

which is equivalent to Eq. (31–21). At the center of the loop, $x = 0$, and Eq. (31–21) reduces to

$$B_x = \frac{\mu_0 I}{2a} \qquad \text{(center of circular loop).} \tag{31–23}$$

If the coil has N turns, the magnetic field is multiplied by N.

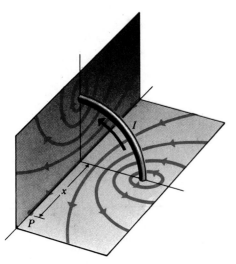

31–7 Field lines surrounding a circular loop.

Line integral of magnetic field around a closed path

The integration path need not be formed by a physical object.

If, instead of a single loop as in Fig. 31–6, we have a coil of N closely spaced loops, all having the same radius, each loop contributes equally to the field, and Eq. (31–23) becomes

$$B_x = \frac{\mu_0 N I}{2a} \qquad (N \text{ circular loops}). \qquad (31\text{–}24)$$

Similarly, we can adapt Eqs. (31–21) and (31–22) to the case of N loops by inserting a factor N.

Some of the magnetic-field lines surrounding a circular loop and lying in planes through the axis are shown in Fig. 31–7. Again we see that the field lines encircle the conductor, and that their directions are given by the right-hand rule, as for a long, straight conductor. The field lines for the circular loop are *not* circles, but they are closed curves that link the conductor.

31–6 AMPERE'S LAW

Ampere's law provides an alternative formulation of the relation between a magnetic field and its sources. In some problems it is more convenient than the law of Biot and Savart. The role of Ampere's law for magnetic fields is analogous to that of Gauss's law for electric fields, which we studied in Chapter 25. Recall that Gauss's law relates the integral of the normal component of electric field over a closed surface to the total electric charge enclosed by that surface. Ampere's law relates the *tangential* component of *magnetic* field at points on a closed *curve* to the current passing through the *area* bounded by that curve.

Ampere's law is formulated in terms of the line integral of B around a closed path, denoted by

$$\oint \boldsymbol{B} \cdot d\boldsymbol{l}.$$

This is the same sort of integral we used to define work in Chapter 7 and electric potential in Chapter 26. We divide the path into infinitesimal segments $d\boldsymbol{l}$ and calculate the scalar product of $\boldsymbol{B}$ and $d\boldsymbol{l}$ for each segment. In general $\boldsymbol{B}$ varies from point to point, and we must use the $\boldsymbol{B}$ at the location of each $d\boldsymbol{l}$. An alternative notation is $\oint B_\parallel \, dl$, where $B_\parallel$ is the component of $\boldsymbol{B}$ parallel to $d\boldsymbol{l}$ at each point. The circle on the integral sign indicates that this integral is always computed for a *closed* path whose beginning and end points are the same. As with Gauss's law, the path used for Ampere's law need not be the outline of any actual physical object; usually it is a purely geometric curve that we construct in order to apply Ampere's law to a specific situation.

To introduce the basic idea, we consider a long, straight conductor carrying a current I. Imagine a circle of radius r in a plane perpendicular to the conductor and centered on it. From Eq. (31–15), the magnetic field has magnitude $\mu_0 I / 2\pi r$ at every point on the circle, and it is tangent to the circle at each point. Thus for every point on the circle $B_\parallel = B = \mu_0 I / 2\pi r$. The line integral of $\boldsymbol{B}$ around the circle is

$$\oint \boldsymbol{B} \cdot d\boldsymbol{l} = \oint \frac{\mu_0 I}{2\pi r} dl = \frac{\mu_0 I}{2\pi r} \oint dl,$$

or, since $\oint dl$ is just the circumference ($2\pi r$) of the circle,

$$\oint \boldsymbol{B} \cdot d\boldsymbol{l} = \mu_0 I. \qquad (31\text{–}25)$$

That is, the line integral of B is equal to μ_0 multiplied by the current passing through the area bounded by the circle.

We can also derive this result for a more general integration path, such as the one in Fig. 31–8. At the position of the line element dl, the angle between dl and B is θ, and

$$B \cdot dl = B \, dl \cos \theta.$$

From the figure, $dl \cos \theta = r \, d\phi$, where $d\phi$ is the angle subtended by dl at the position of the conductor. Thus

$$\oint B \cdot dl = \oint \left(\frac{\mu_0 I}{2\pi r}\right)(r \, d\phi) = \frac{\mu_0 I}{2\pi} \oint d\phi.$$

But $\oint d\phi$ is just the total angle swept out by the radial line from the conductor to dl during a complete trip around the path, that is, 2π. Thus

$$\oint B \cdot dl = \mu_0 I. \qquad (31\text{--}26)$$

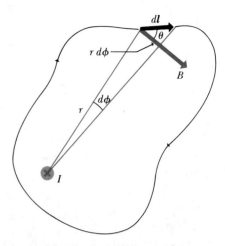

31–8 Line integral of B around a closed path linking a long, straight conductor, carrying current I *into* the plane of the figure. The conductor is seen end-on.

The line integral does not depend on the shape of the path or on the position of the wire within it. If the current in the wire is opposite to that shown, the integral has the opposite sign.

Now suppose several long, straight conductors pass through the surface bounded by the integration path. The total magnetic field at any point on the path is the vector sum of the fields produced by the individual conductors. Thus the line integral of the total B equals μ_0 times the *algebraic sum* of the currents, with currents going into the page in Fig. 31–8 considered as positive and those coming out of it considered negative. In general, the positive direction for current is determined by a right-hand rule. Considering a line perpendicular to the surface, we wrap the fingers of the right hand around this line so our fingers curl around in the same direction we plan to go around the integration path; then the thumb indicates the positive current direction.

If the integration path does not encircle the wire (or if a wire lies outside the path), the line integral of the B field of that wire is zero, because then the angle ϕ does not sweep through 2π as it did before; instead, its net change is zero. Hence if there are other conductors present that do not pass through a given path, they may contribute to the value of B at every point, but the *line integrals* of their fields are zero.

It follows that if we interpret I in Eq. (31–26) to mean the *algebraic sum* of the currents through an area bounded by a closed path, this equation implies all of the above statements. Thus the general form of Ampere's law is

> The line integral of the magnetic field B around any closed path is equal to μ_0 times the net current through any area bounded by the path.

Although we have derived Ampere's law only for the special case of the field of a number of long, straight, parallel conductors, it is true for conductors and paths of any shape. The general derivation is no different in principle from what we have presented, but it is more complicated geometrically.

Finally, we emphasize that the statement $\oint B \cdot dl = 0$ *does not* necessarily mean that $B = 0$ everywhere along the path, but only that the total current through an area bounded by the path is zero.

Ampere's law: The line integral of B around a closed path is proportional to the current through the area enclosed by the path.

Sign and direction conventions used with Ampere's law

31-7 APPLICATIONS OF AMPERE'S LAW

Ampere's law is particularly helpful when we can make use of the symmetry of a situation to evaluate the line integral of B. Following are several examples. Note that the problem-solving strategy we suggest is directly analogous to the strategy proposed for applications of Gauss's law in Section 25–5; we invite you to review that strategy now and compare the two methods.

PROBLEM-SOLVING STRATEGY: *Ampere's law*

1. The first step is to select the integration path you will use with Ampere's law. If you want to learn something about the magnetic field at a certain point, then the path must pass through that point.

2. The integration path doesn't have to be any actual physical boundary. Usually it is a purely geometric curve; it may be in empty space, embedded in a solid body, or some of each.

3. The integration path has to have enough *symmetry* to make evaluation of the integral possible. If the problem itself has cylindrical symmetry, the integration path will usually be a circle coaxial with the cylinder axis.

4. If B is tangent to the integration path and has the same magnitude B at every point, then its line integral equals B multiplied by the circumference of the path.

5. If B is everywhere perpendicular to the path, for all or some portion of the path, that portion of the path makes no contribution to the integral.

6. In the integral $\oint B \cdot dl$, B is always the *total* magnetic field at each point on the path. In general this field is caused partly by currents enclosed by the path and partly by currents outside. Even when *no* net current is enclosed by the path, the field at points on the path need not be zero. In that case, however, $\oint B \cdot dl$ is always zero.

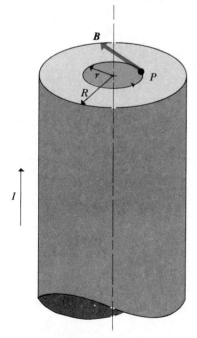

31-9 The current through the color area is $(r^2/R^2)I$. To find the magnetic field at point P we apply Ampere's law to the circle around the color area.

EXAMPLE 31–2 *Field of a long, straight conductor:* Because of the way we derived Ampere's law in Section 31–6, we could reverse the process and use it to derive Eq. (31–15) for a long, straight conductor. We take as our integration path a circle of radius r centered on the conductor and in a plane perpendicular to it. Then in Eq. (31–26), $\oint B \cdot dl = 2\pi r B$, and Eq. (31–15) follows immediately.

EXAMPLE 31–3 *Field inside a long, cylindrical conductor.* To find the magnetic field inside a cylindrical conductor of radius R, at a point a distance r from the axis, as in Fig. 31–9, we consider a circle of radius r. By symmetry, B has the same magnitude at every point on this circle and is tangent to it. Thus the line integral is simply $B(2\pi r)$. To find the current enclosed by the path, note that the current *density* J (current per unit area) is $J = I/\pi R^2$, so the current through the path is $I_r = J(\pi r^2) = Ir^2/R^2$. Finally, Ampere's law gives

$$2\pi r B = \mu_0 Ir^2/R^2,$$

or

$$B = \frac{\mu_0 I}{2\pi} \frac{r}{R^2}. \tag{31-27}$$

Note that at $r = R$, that is, at the surface of the conductor, Eq. (31–15) (derived for $r > R$) and Eq. (31–27) (derived for $r < R$) agree.

EXAMPLE 31–4 *Field of a solenoid:* A *solenoid* consists of a helical winding of wire around the surface of a cylindrical form, usually circular in cross section. Ordinarily the turns are so closely spaced that each one is very nearly a circular loop. There may be several layers of windings. For simplicity, we have shown a solenoid in Fig. 31–10 as having only a few turns. All turns carry the same current I, and the total $\boldsymbol{B}$ field at every point is the vector sum of the fields caused by the individual turns. The figure shows field lines in the xy- and yz-planes. Exact calculations show that for a long, closely wound solenoid, half the lines passing through a cross section at the center emerge from the ends, and half "leak out" through the windings between center and end.

A solenoid is a coil wound on a cylindrical form.

If the length of the solenoid is large compared with its cross-sectional diameter, the *internal* field near its center is very nearly uniform and parallel to the axis, and the *external* field near the center is very small. The internal field at or near the center can then be found by use of Ampere's law.

What is the magnetic field inside a solenoid?

We choose as our integration path the broken-line rectangle *abcd* in Fig. 31–10. Side *ab*, of length l, is parallel to the axis of the solenoid. Sides *bc* and *da* are taken to be very long, so that side *cd* is far from the solenoid, and the field at this side is negligibly small.

By symmetry, the $\boldsymbol{B}$ field along side *ab* is parallel to this side and is constant, so for this side, $B_\parallel = B$ and

$$\int \boldsymbol{B} \cdot d\boldsymbol{l} = Bl.$$

Along sides *bc* and *da*, $B_\parallel = 0$ since B is perpendicular to these sides; and along side *cd*, $B_\parallel = 0$ also since $B = 0$. The sum around the entire closed path therefore reduces to Bl.

Let n be the number of turns *per unit length* in the windings. The number of turns in length l is then nl. Each of these turns passes once through the rectangle *abcd* and carries a current I, where I is the current in the windings. The total current through the rectangle is then nlI and, from Ampere's law,

$$Bl = \mu_0 nlI,$$

$$B = \mu_0 nI \qquad \text{(solenoid).} \qquad (31\text{–}28)$$

Since side *ab* need not lie on the axis of the solenoid, the field is uniform over the entire cross section.

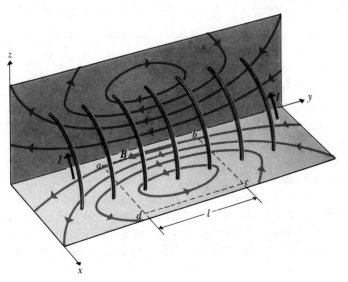

31–10 Magnetic-field lines surrounding a solenoid. The dashed rectangle *abcd* is used to compute the flux density $\boldsymbol{B}$ in the solenoid from Ampere's law.

Toroidal solenoid: a solenoid that meets itself coming back

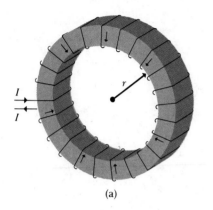

(a)

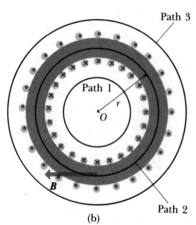

(b)

31–11 (a) A toroidal winding. (b) Closed paths (black circles) used to compute the magnetic field **B** set up by a current in a toroidal winding. The field is very nearly zero at all points except those within the space enclosed by the windings.

What if the space around a conductor contains a material?

Magnetized materials make additional contributions to magnetic field.

EXAMPLE 31–5 *Field of a toroidal solenoid:* Figure 31–11a shows a **toroidal solenoid** wound with wire carrying a current I. The black lines in Fig. 31–11b are paths to which we wish to apply Ampere's law. Consider first path 1. By symmetry, if there is any field at all in this region, it will be *tangent* to the path at all points, and $\oint \mathbf{B} \cdot d\mathbf{l}$ will equal the product of B and the circumference $l = 2\pi r$ of the path. The current through the path, however, is zero; hence, from Ampere's law, the field **B** must be zero.

Similarly, if there is any field at path 3, it will also be tangent to the path at all points. Each turn of the winding passes *twice* through the area bounded by this path, carrying equal currents in opposite directions. The *net* current through this area is therefore zero, and hence $B = 0$ at all points of the path. *The field of the toroidal solenoid is therefore confined wholly to the space enclosed by the windings.* The toroid may be thought of as a solenoid that has been bent into a circle.

Finally, consider path 2, a circle of radius r. Again by symmetry, the **B** field is tangent to the path, and $\oint \mathbf{B} \cdot d\mathbf{l}$ equals $2\pi r B$. Each turn of the winding passes *once* through the area bounded by path 2, and the total current through the area is NI, where N is the *total* number of turns in the winding. Then, from Ampere's law,

$$2\pi r B = \mu_0 N I,$$

and

$$B = \frac{\mu_0 N I}{2\pi r} \qquad \text{(toroidal solenoid).} \qquad (31\text{–}29)$$

The magnetic field is *not* uniform over a cross section of the core, because the path length l is larger at the outer side of the section than at the inner side. However, if the radial thickness of the core is small compared with the toroid radius r, the field varies only slightly across a section. In that case, considering that $2\pi r$ is the circumferential length of the toroid and that $N/2\pi r$ is the number of turns per unit length n, the field may be written as

$$B = \mu_0 n I,$$

just as at the center of the long, *straight* solenoid.

The questions derived above for the field in a closely wound straight or toroidal solenoid are strictly correct only for windings in *vacuum*. For most practical purposes, however, they can be used for windings in air or on a core of any nonferromagnetic material. We will show in the next section how they are modified if the core is of iron.

31–8 MAGNETIC MATERIALS

In the preceding discussion of magnetic fields caused by currents, we assumed that the space surrounding the conductors contained only vacuum. If matter is present in the surrounding space, the magnetic field is changed. The atoms that make up all matter contain electrons in motion, and these electrons form microscopic current loops capable of producing magnetic fields of their own. In many materials these currents are randomly oriented so as to cause no net magnetic field. But in some materials the presence of an externally caused field can cause the loops to become oriented preferentially with the field, so that their magnetic fields *add* to the external field. The material is said to become *magnetized*.

A material showing this behavior is said to be **paramagnetic.** The result is that the magnetic field at any point in such a material is greater by a factor K_m then it would be if the material were not present. The value of K_m is different for different materials; it is called the *relative permeability* of the material. Values of K_m for common paramagnetic materials are typically 1.0001 to 1.0020.

All the equations in this chapter that relate magnetic fields to their sources can be adapted to the situation where the conductor is embedded in a paramagnetic material by simply replacing μ_0 everywhere by $K_m\mu_0$. This product is usually denoted as μ; it is called the **permeability** of the material:

$$\mu = K_m\mu_0. \qquad (31\text{–}30)$$

The amount by which the relative permeability differs from unity is called the **magnetic susceptibility,** symbol χ_m:

$$\chi_m = K_m - 1. \qquad (31\text{–}31)$$

Values of magnetic susceptibility for several materials are given in Table 31–1.

In some materials the total field due to the electrons in each atom sums to zero when there is no external field, and these materials have *no* net atomic current loops. But even in these materials magnetic effects are present because an external field alters the electron motions to cause additional current loops. In this case the additional field caused by these current loops is always *opposite* in direction to the external field. Such materials are said to be **diamagnetic;** they always have relative permeabilities very slightly less than unity, typically of the order of 0.99990 to 0.99999. The susceptibility of a diamagnetic material is negative, as Table 31–1 shows.

There is a third class of materials, called **ferromagnetic** materials, in which the atomic current loops tend to line up parallel to each other even when no external field is present. This cooperative phenomenon leads to a relative permeability that is *much larger* than unity, typically of the order of 1000 to 10,000. Iron, cobalt, nickel, and many alloys containing these elements are ferromagnetic.

Ferromagnetic materials differ in their behavior from paramagnetic and diamagnetic materials in two other important ways. First, the additional field caused by the microscopic current loops is not directly proportional to the external field except at relatively small field magnitudes. Thus for a ferromagnetic material, K_m is not constant. As the external field increases, a point is reached where nearly all the microscopic current loops have their axes parallel to the external field. This condition is called *saturation magnetization;* after it is reached, a further increase in the external field causes no increase in magnetization or in the additional field caused by the material.

Second, some ferromagnetic materials retain their magnetization even when there is no externally caused field at all. These materials can thus become *permanent magnets.* Many kinds of steel and many alloys, such as Alnico, are commonly used for permanent magnets. The magnetic field in such a material, when it is magnetized to near saturation, is typically of the order of 1 T.

More generally, for many ferromagnetic materials the relation of magnetization to magnetic field is different when the field is decreasing from when it is increasing. Thus the magnetization for a given field depends somewhat on the past history of the material. This phenomenon is called **hysteresis.** One consequence is the magnetization that remains when the field is reduced to

A paramagnetic material makes the field a little stronger.

A diamagnetic material makes the field a little weaker.

A ferromagnetic material makes the field much stronger.

Magnetization of ferromagnetic materials is nonlinear.

TABLE 31–1 Magnetic Susceptibilities of Paramagnetic and Diamagnetic Materials, at $T = 20°C$

Materials	$\chi_m = K_m - 1$
Paramagnetic	
Iron ammonium alum	66×10^{-5}
Uranium	40×10^{-5}
Platinum	26×10^{-5}
Aluminum	2.2×10^{-5}
Sodium	0.72×10^{-5}
Oxygen gas	0.19×10^{-5}
Diamagnetic	
Bismuth	-16.6×10^{-5}
Mercury	-2.9×10^{-5}
Silver	-2.6×10^{-5}
Carbon (diamond)	-2.1×10^{-5}
Lead	-1.8×10^{-5}
Rock salt	-1.4×10^{-5}
Copper	-1.0×10^{-5}

Hysteresis: The material may stay magnetized when the external magnetic field is removed.

zero, as discussed above. Magnetizing and demagnetizing a material that has hysteresis involves the dissipation of energy, and the temperature of such a material increases in this process.

Ferromagnetic materials are widely used in electromagnets, transformer cores, and motors and generators, where it is desirable to have as large a magnetic field as possible for a given current. In these applications it is usually desirable for the material *not* to have permanent magnetization; that is, for it to have as little hysteresis as possible. Soft iron is often used; it has high permeability without appreciable hysteresis or permanent magnetization.

Electrical machines often require high permeability and low hysteresis.

31–9 MAGNETIC FIELD AND DISPLACEMENT CURRENT

Displacement current: an additional source of magnetic field

When we discussed Ampere's law in Section 31–6, we assumed that the current I was a *conduction* current. However, experimental evidence shows that in situations that include time-varying electric fields, *displacement current I_D*, as defined in Section 29–5, must also be included in Ampere's law. That is, in using Ampere's law in problems involving the relation of magnetic field to its sources (currents), we must treat conduction and displacement currents on an equal footing; both act as sources of magnetic field. Thus the general form of Ampere's law is

$$\oint \boldsymbol{B} \cdot d\boldsymbol{l} = \mu_0 (I_C + I_D). \tag{31–32}$$

Generalizing Ampere's law to include displacement current

Here is an example of the role of displacement current in Ampere's law in a magnetic-field calculation. Figure 31–12 is a cross-sectional view of two circular conducting plates of radius R, separated by a narrow gap, in vacuum. They are, in other words, a parallel-plate capacitor. If the gap is small compared to R, the $\boldsymbol{E}$ field caused by the charge distributions on the plates is confined almost entirely to the region between the plates and is uniform, as shown. Suppose we want to calculate the $\boldsymbol{B}$ field at some point in the midplane, shown in the figure as a thin line. In principle this field can be obtained by applying the law of Biot and Savart to the conduction currents in the plates and the wires leading to them. Because of the radial currents in the circular plates, this is a fairly involved calculation. Instead, we may take advantage of the axial symmetry of the situation and use Ampere's law.

Using displacement current to find the magnetic field between the plates of a charging capacitor.

There is no *conduction* current in the region between plates, but there is *displacement current*. From the definition of displacement current in Section 29–5, the total displacement current out of the left plate is equal to the conduction current into the plate. Also, by symmetry, the $\boldsymbol{B}$ field lines are circles with centers on the axis. Thus, at any point on a circle perpendicular to the axis and passing through points such as a and b, the $\boldsymbol{B}$-vector has the same magnitude and is tangent to the circle. At point a, the $\boldsymbol{B}$-vector points out of the page, and at point b it points into the page, as indicated by the symbols $\odot$ and $\otimes$.

The displacement current density J_D is given by Eq. (29–20):

$$J_D = \epsilon_0 \frac{dE}{dt}. \tag{31–33}$$

Displacement-current density is proportional to rate of change of electric field.

At any instant E is uniform over the region between the plates, so J_D is also uniform. The *total* displacement current I_D between the plates is just J_D times the area πR^2 of the plates. As we remarked above, this must also equal the

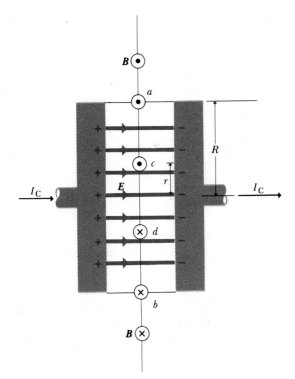

31–12 A capacitor being charged by a current I_C has a displacement current I_D between its plates equal to I_C, with displacement-current density $J_D = \epsilon_0 \, dE/dt$.

total *conduction* current I_C. Thus

$$I_D = I_C = \pi R^2 J_D, \tag{31–34}$$

$$J_D = \frac{I_C}{\pi R^2}. \tag{31–35}$$

Now we may find the magnetic field at a point in the region between plates, at a distance r from the axis. We apply Ampere's law to a circle of radius r passing through the point; such a circle passes through points c and d in Fig. 31–12. The total current enclosed by the circle is J_D times its area, or $(I_C/\pi R^2)(\pi r^2)$. The integral $\oint \mathbf{B} \cdot d\mathbf{l}$ in Ampere's law is just B times the circumference $2\pi r$ of the circle, and Ampere's law becomes

$$\oint \mathbf{B} \cdot d\mathbf{l} = 2\pi rB = \mu_0 \left(\frac{r^2}{R^2}\right) I_C,$$

or

$$B = \frac{\mu_0}{2\pi}\left(\frac{r}{R^2}\right) I_C. \tag{31–36}$$

This result shows that in the region between the plates $\mathbf{B}$ is zero at the axis, increasing linearly with distance from the axis. A similar calculation shows that *outside* the region between the plates, $\mathbf{B}$ is the same as though the wire were continuous and the plates not present at all.

This calculation bears a strong resemblance to the one in Example 31–3 (Section 31–7), where we used Ampere's law to find the magnetic field inside a cylindrical conductor. In a certain sense we can think of the space between the conducting plates as a cylindrical conductor. Of course, there is no actual charge in motion to constitute a conduction current, but there is a displacement current. Thus it should not be too surprising that Eq. (31–36) is in fact identical to Eq. (31–27).

Ampere's law with displacement current expressed in terms of electric flux

The role of displacement current as a source of magnetic field can be expressed in a different way by using the concept of *electric flux* Ψ, which we introduced in Sections 25–4 and 29–5. As stated in Eq. (29–22), the displacement current may be expressed as

$$I_D = \epsilon_0 \frac{d\Psi}{dt}.$$

Thus in empty space, where there is no conduction current ($I_C = 0$), Ampere's law, Eq. (31–32), may be rewritten

$$\oint \boldsymbol{B} \cdot d\boldsymbol{l} = \mu_0 \epsilon_0 \frac{d\Psi}{dt}. \tag{31–37}$$

We will see in Chapter 32 that this equation bears a close resemblance to another relation, *Faraday's law,* which describes the *electric* field induced by a changing *magnetic* field.

The magnetic field caused by displacement current plays a vital role in electromagnetic-wave propagation.

The fact that *displacement* current acts as a source of magnetic field plays an essential role in the understanding of electromagnetic *waves.* A changing *electric* field in a region of space induces a *magnetic* field in neighboring regions, even when no conduction current and no matter are present. This relationship, first proposed by Maxwell in 1865, provides the key to a theoretical understanding of electromagnetic radiation and of light as a particular example of this radiation. We return to this topic in Chapter 35.

SUMMARY

The magnetic field $\boldsymbol{B}$ produced by a charge q moving with velocity v is described by

$$B = \frac{\mu_0}{4\pi} \frac{qv \sin\theta}{r^2}, \tag{31–3}$$

$$\boldsymbol{B} = \frac{\mu_0}{4\pi} \frac{qv \times \hat{\boldsymbol{r}}}{r^2}, \tag{31–4}$$

where $\boldsymbol{r}$ is the vector from the source point (the location of q) to the field point P, and $\hat{\boldsymbol{r}}$ is a unit vector in the direction of $\boldsymbol{r}$. Magnetic fields produced by moving charges obey the superposition principle; the total $\boldsymbol{B}$ produced by several moving charges is the vector sum of the fields produced by the individual charges.

The magnetic interaction force $\boldsymbol{F}$ exerted on a charge q' moving with velocity v' by a charge q moving with velocity v is given by

$$\boldsymbol{F} = \frac{\mu_0}{4\pi} \frac{qq'v' \times (v \times \hat{\boldsymbol{r}})}{r^2}, \tag{31–5}$$

where $\boldsymbol{r}$ points from q toward q'.

An element $d\boldsymbol{l}$ of a conductor carrying a current I produces a magnetic field $d\boldsymbol{B}$ described by

$$dB = \frac{\mu_0}{4\pi} \frac{I\,dl \sin\theta}{r^2}, \tag{31–7}$$

or in vector form,

$$d\boldsymbol{B} = \frac{\mu_0}{4\pi} \frac{I\,d\boldsymbol{l} \times \hat{\boldsymbol{r}}}{r^2}. \tag{31–8}$$

These relations are called the law of Biot and Savart. The total field of a finite conductor is obtained by integrating these equations over the length of the conductor; this usually requires vector calculations, often best carried out by use of components.

Magnetic-field lines never have endpoints, and the surface integral of **B** over any closed surface is always zero. This is the magnetic analog of Gauss's law and shows that there are no isolated magnetic poles.

The interaction force, per unit length, between two long, parallel conductors carrying currents I and I' is

$$\frac{F}{l} = \frac{\mu_0 I I'}{2\pi r}. \tag{31-17}$$

This relation is the basis for the operational definition of the ampere.

The magnetic field **B** at a distance r from a long, straight wire carrying a current I has magnitude

$$B = \frac{\mu_0 I}{2\pi r} \qquad \text{(long, straight wire).} \tag{31-15}$$

The magnetic-field lines for a long, straight wire are circles coaxial with the wire, with directions given by the right-hand rule.

The magnetic field produced by a circular conducting loop of radius a, carrying current I, at a distance x from its center, along its axis, is given by

$$B_x = \frac{\mu_0 I a^2}{2(x^2 + a^2)^{3/2}} \qquad \text{(circular loop).} \tag{31-21}$$

At the center of the loop, where $x = 0$, this equation reduces to

$$B_x = \frac{\mu_0 I}{2a} \qquad \text{(center of circular loop).} \tag{31-23}$$

If there are N loops, this expression is multiplied by N.

Ampere's law states that the line integral of **B** around any closed path equals μ_0 times the net current through the area enclosed by the path. The positive sense of current is determined by a right-hand rule.

The magnetic field in the interior of a long, cylindrical conductor of radius R carrying current I is given by

$$B = \frac{\mu_0 I}{2\pi} \frac{r}{R^2}. \tag{31-27}$$

The magnetic field at the center of a long solenoid with n turns per unit length, carrying current I, is

$$B = \mu_0 n I. \tag{31-28}$$

The magnetic field inside a toroidal solenoid with N turns, carrying current I, at a distance r from its axis, is

$$B = \frac{\mu_0 N I}{2\pi r} \qquad \text{(toroidal solenoid).} \tag{31-29}$$

When magnetic materials are present, the magnetization of the material causes an additional contribution to **B.** For paramagnetic and diamagnetic materials, all the relations between magnetic field and its sources are still correct if we replace μ_0 by $\mu = K_m\mu_0$, where μ is the permeability of the material and K_m is its relative permeability. The magnetic susceptibility χ_m is defined $\chi_m = K_m - 1$. Magnetic susceptibilities for paramagnetic materials are small positive quantities; those for diamagnetic materials are small negative quantities. For ferromagnetic materials K_m is much larger than unity and is not constant. Some ferromagnetic materials also can retain magnetization even after the external magnetic field is removed.

Displacement current acts as a source of magnetic field in exactly the same way as conduction current. Ampere's law is generalized to include displacement current, as follows:

$$\oint \boldsymbol{B} \cdot dl = \mu_0(I_C + I_D). \tag{31–32}$$

The concept of displacement current as a source of magnetic field is essential to the understanding of electromagnetic waves.

QUESTIONS

31–1 Streams of charged particles emitted from the sun during unusual sunspot activity create a disturbance in the earth's magnetic field. How does this happen?

31–2 A topic of current interest in physics research is the search (thus far unsuccessful) for an isolated magnetic pole, or magnetic *monopole*. If such an entity were found, how could it be recognized? What would its properties be?

31–3 What are the relative advantages and disadvantages of Ampere's law and the law of Biot and Savart for practical calculations of magnetic fields?

31–4 A student proposed to obtain an isolated magnetic pole by taking a bar magnet (N pole at one end, S at the other) and breaking it in half. Would this work?

31–5 Pairs of conductors carrying current into and out of the power-supply components of electronic equipment are sometimes twisted together to reduce magnetic-field effects. Why does this help?

31–6 The text discusses the magnetic field of an infinitely long, straight conductor carrying a current. Of course, there is no such thing as an infinitely long *anything*. How do you decide whether a particular wire is long enough to be considered infinite?

31–7 Suppose one has three long, parallel wires, arranged so that in cross section they are at the corners of an equilateral triangle. Is there any way to arrange the currents so that all three wires attract each other? So that all three wires repel each other?

31–8 Two parallel conductors carrying current in the same direction attract each other. If they are permitted to move toward each other, the forces of attraction do work. Where does the energy come from? Does this contradict the assertion in Chapter 30 that magnetic forces on moving charges do no work?

31–9 Considering the magnetic field of a circular loop of wire, would you expect the field to be greatest at the center,

or is it greater at some points in the plane of the loop but off-center?

31–10 Two concentric circular loops of wire, of different diameter, carry currents in the same direction. Describe the nature of the forces exerted on the inner loop.

31–11 A current was sent through a helical coil spring. The spring appeared to contract, as though it had been compressed. Why?

31–12 Using the fact that magnetic-field lines never have a beginning or an end, explain why it is reasonable for the field of a toroidal solenoid to be confined entirely to its interior, while a straight solenoid *must* have some field outside it.

31–13 Can one have a displacement current as well as a conduction current within a conductor?

31–14 A character in a popular comic strip has at various times proposed the possibility of "harnessing the earth's magnetic field" as a nearly inexhaustible source of energy. Comment on this concept.

31–15 Why should the permeability of a paramagnetic material be expected to decrease with increasing temperature?

31–16 In the discussion of magnetic forces on current loops in Section 30–8, it was found that no net force is exerted on a complete loop in a uniform magnetic field, only a torque. Yet magnetized materials, which contain atomic current loops, certainly *do* experience net forces in magnetic fields. How is this discrepancy resolved?

31–17 What features of atomic structure determine whether an element is diamagnetic or paramagnetic?

31–18 The magnetic susceptibility of paramagnetic materials is quite strongly temperature dependent, while that of diamagnetic materials is nearly independent of temperature. Why the difference?

EXERCISES

$$1 \text{ T} = 1 \text{ Wb·m}^{-2} = 1 \text{ N·A}^{-1}\text{·m}^{-1},$$
$$\mu_0 = 4\pi \times 10^{-7} \text{ N·A}^{-2} = 4\pi \times 10^{-7} \text{ T·A}^{-1}\text{·m}$$
$$= 4\pi \times 10^{-7} \text{ Wb·A}^{-1}\text{·m}^{-1}.$$

Section 31–1 Magnetic Field of a Moving Charge

31–1 A positive point charge of magnitude $q = 5\ \mu\text{C}$ has velocity $\boldsymbol{v} = (8 \times 10^6\text{m·s}^{-1})\boldsymbol{i}$. At the instant when the point charge is at the origin, what is the magnetic field vector $\boldsymbol{B}$ it produces at the following points:

a) $x = 0.5$ m, $y = 0$, $z = 0$?

b) $x = 0$, $y = -0.5$ m, $z = 0$?

c) $x = 0$, $y = 0$, $z = +0.5$ m?

d) $x = 0$, $y = -0.5$ m, $z = +0.5$ m?

31–2 Two positive point charges q and q' are moving relative to an observer as shown in Fig. 31–13.

a) What is the direction of the force that q' exerts on q?

b) What is the direction of the force that q exerts on q'?

c) If $v = v' = 5 \times 10^6\text{m·s}^{-1}$, what is the ratio of the magnitude of the magnetic force to that of the Coulomb force between the two charges?

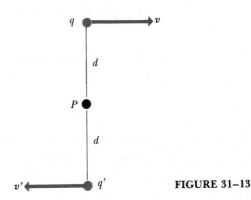

FIGURE 31–13

31–3 A pair of point charges, where $q = +5\mu\text{C}$ and $q' = -3\mu\text{C}$, are moving as shown in Fig. 31–14. At this instant, what are the magnitude and direction of the magnetic field produced at the origin (point P)? Take $v = v' = 4 \times 10^5\text{m·s}^{-1}$.

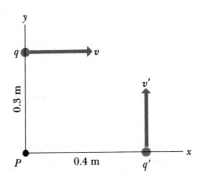

FIGURE 31–14

Section 31–2 Magnetic Field of a Current Element

31–4 A long, straight wire, carrying a current of 200 A, runs through a cubical wooden box, entering and leaving through holes in the centers of opposite faces, as in Fig. 31–15. The length of each side of the box is 20 cm. Consider an element of the wire 1 cm long at the center of the box. Compute the magnitude of the magnetic field ΔB produced by this element at points a, b, c, d, and e in Fig. 31–15. Points a, c, and d are at the centers of the faces of the cube; point b is at the midpoint of one edge; and point e is at a corner. Copy the figure and show by vectors the directions and relative magnitudes of the field vectors. (*Note.* Assume that Δl is small compared to the distances from the current element to the points where $\boldsymbol{B}$ is to be calculated.)

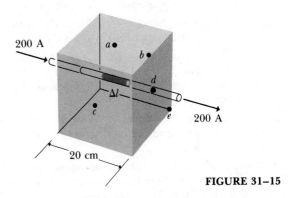

FIGURE 31–15

31–5 Calculate the magnitude and direction of the magnetic field at point P due to the current in the semicircular section of wire shown in Fig. 31–16. (Does the current in the long, straight section of the wire produce any field at P?)

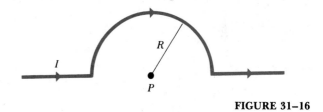

FIGURE 31–16

31–6 Calculate the magnetic field at point P of Fig. 31–17 in terms of R, I_1, and I_2. What does your expression give when $I_1 = I_2$?

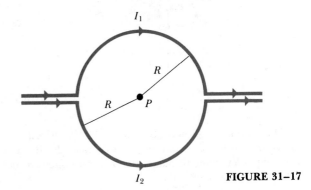

FIGURE 31–17

Section 31–3 Magnetic Field of a Straight Conductor

31–7 Two hikers are reading a compass under an overhead transmission line that is 5.0 m above the ground and carries a current of 400 A in a direction from south to north.

a) Find the magnitude and direction of the magnetic field at a point on the ground directly under the conductor.

b) One hiker suggests they walk on another 50 m to avoid inaccurate compass readings caused by the current. Considering that the magnitude of the earth's field is of the order of 0.5×10^{-4} T, is the current really a problem?

31–8 A magnetic field of magnitude 5.0×10^{-4} T is to be produced at a distance of 0.05 m from a long, straight wire.

a) What current is required to produce this field?

b) With the current found in (a), what is the magnitude of the field at a distance of 0.10 m from the wire? At 0.20 m?

31–9 A long, straight telephone cable contains six wires, each carrying a current of 0.5 A. The distances between wires can be neglected.

a) If the currents in all six wires are in the same direction, what is the magnitude of the magnetic field 0.10 m from the cable?

b) If four wires carry currents in one direction and the other two in the opposite direction, what is the field magnitude 0.10 m from the cable?

31–10 A long, straight wire carries a current of 10 A directed along the y-axis, as shown in Fig. 31–18. A uniform magnetic field $\boldsymbol{B}_0$ of magnitude 1×10^{-6} T is directed parallel to the x-axis. What is the resultant field at the following points:

a) $x = 0$, $z = 2$ m?

b) $x = 2$ m, $z = 0$?

c) $x = 0$, $z = -0.5$ m?

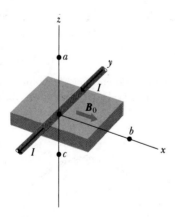

FIGURE 31–18

31–11 Two long, straight, horizontal parallel wires, one above the other, are separated by a distance $2a$. If the wires carry equal currents of magnitude I in opposite directions,

what is the field magnitude in the plane of the wires at a point

a) midway between them?

b) at a distance a above the upper wire?

If the wires carry equal currents in the same direction, what is the field magnitude in the plane of the wires at a point

c) midway between them?

d) at a distance a above the upper wire?

Section 31–4 Force between Parallel Conductors

31–12 Two long, parallel wires are separated by a distance of 0.4 m, as shown in Fig. 31–19. The currents I_1 and I_2 have the directions shown.

a) Calculate the magnitude and direction of the force exerted by each wire on a 1-m length of the other.

b) If each current is doubled, so that I_1 becomes 10 A and I_2 becomes 4 A, what is now the magnitude of the force each wire exerts on a 1-m length of the other?

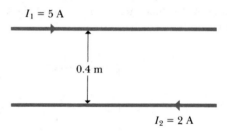

FIGURE 31–19

31–13 Two long, parallel wires are separated by a distance of 0.1 m. The force per unit length that each wire exerts on the other is 3.0×10^{-5} N·m^{-1}, and the wires repel each other. If the current in one wire is 2 A,

a) what is the current in the second wire?

b) are the two currents in the same or in opposite directions?

31–14 Three parallel wires each carry current I in the directions shown in Fig. 31–20. If the separation between adjacent wires is d, calculate the magnitude and direction of the resultant magnetic force per unit length on each wire.

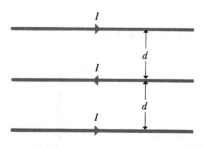

FIGURE 31–20

31–15 A long, horizontal wire *AB* rests on the surface of a table. (See Fig. 31–21.) Another wire *CD* vertically above the first is 1.0 m long and free to slide up and down on the two vertical metal guides *C* and *D*. The two wires are connected through the sliding contacts and carry a current of 50 A. The mass per unit length of the wire *CD* is 0.005 kg·m^{-1}. To what equilibrium height *h* will the wire *CD* rise, assuming the magnetic force on it to be due wholly to the current in the wire *AB*?

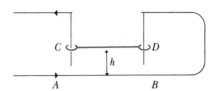

FIGURE 31–21

Section 31–5 Magnetic Field of a Circular Loop

31–16 Refer to Fig. 31–6. Sketch a graph of the magnitude of the **B** field on the axis of the loop, from $x = -3a$ to $x = +3a$.

31–17 A circular coil of radius 5 cm has 200 turns and carries a current 0.2 A. What is the magnetic field

a) at the center of the coil?

b) at a point on the axis of the coil 10 cm from its center?

Section 31–7 Applications of Ampere's Law

31–18 A solenoid of length 20 cm and radius 3 cm is closely wound with 200 turns of wire. The current in the winding is 5 A. Compute the magnetic field at a point near the center of the solenoid.

31–19 A wooden ring of mean diameter 0.10 m is wound with a closely spaced toroidal winding of 500 turns. Compute the field at a point on the mean circumference of the ring when the current in the windings is 0.3 A.

31–20 A solenoid is to be designed to produce a magnetic field of 0.1 T at its center. The radius is to be 5 cm and the length 50 cm, and the available wire can carry a maximum current of 10 A.

a) What is the minimum number of turns per unit length the solenoid must have?

b) What total length of wire is required?

31–21 A coaxial cable consists of a solid conductor of radius R_1, supported by insulating disks on the axis of a tube of inner radius R_2 and outer radius R_3. If the central conductor and tube carry equal currents *I* in opposite directions, find the magnetic field

a) at points outside the central solid conductor but inside the tube;

b) at points outside the tube.

31–22 Repeat Exercise 31–21 for the case where the current in the central solid conductor is I_1 and in the tube I_2. Also take these currents to be in the same direction rather than in opposite directions.

Section 31–8 Magnetic Materials

31–23 Experimental measurements of the magnetic susceptibility of iron ammonium alum are given in Table 31–2. Make a graph of $1/\chi$ against Kelvin temperature; what conclusion can you draw?

TABLE 31–2

T,°C	χ
−258.15	129×10^{-4}
−173	19.4×10^{-4}
−73	9.7×10^{-4}
27	6.5×10^{-4}

31–24 A toroidal solenoid having 500 turns of wire and a mean circumferential length of 0.50 m carries a current of 0.3 A. The relative permeability of the core is 600.

a) What is the magnetic field in the core?

b) What part of the magnetic field is due to atomic currents?

31–25 A toroidal solenoid with 400 turns is wound on a ring having a cross section of 2 cm^2 and a mean circumference of 30 cm. Find the current in the winding that is required to set up a magnetic field of 0.1 T in the ring

a) if the ring is of annealed iron ($K_m = 1400$);

b) if the ring is of silicon steel ($K_m = 5200$).

Section 31–9 Magnetic Field and Displacement Current

31–26 A copper wire with circular cross section of area 4 mm^2 carries a current of 50 A. The resistivity of the material is 2.0×10^{-8} Ω·m.

a) What is the electric field in the material?

b) If the current is changing at the rate of 5000 A·s^{-1}, at what rate is the electric field in the material changing?

c) What is the displacement-current density in the material in (b)?

d) If the current is changing as in (b), what is the magnetic field 5 cm from the wire? Note that both the conduction current and the displacement current must be included in the calculation of *B*. Is the contribution from the displacement current significant?

PROBLEMS

31–27 A long, straight wire carries a current of 1.5 A. An electron travels with a speed of $5 \times 10^4 \, \mathrm{m \cdot s^{-1}}$ parallel to the wire, 0.10 m from it, and in the same direction as the current. What are the magnitude and direction of the force that the magnetic field of the current exerts on the moving electron?

31–28 Figure 31–22 is an end view of two long, parallel wires perpendicular to the *xy*-plane, each carrying a current *I* but in opposite directions.

a) Copy the diagram, and show by vectors the **B** field of each wire and the resultant **B** field at point *P*.

b) Derive the expression for the magnitude of **B** at any point on the *x*-axis, in terms of the coordinate *x* of the point.

c) Construct a graph of the magnitude of **B** at points on the *x*-axis.

d) At what value of *x* is **B** a maximum?

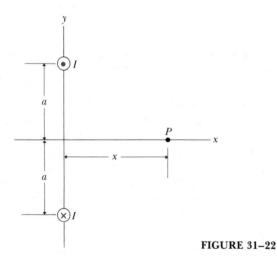

FIGURE 31–22

31–29 Same as Problem 31–28, except that the current in both wires is *away* from the reader.

31–30 In Fig. 31–22, suppose a third long, straight wire, parallel to the other two, passes through point *P*, and that each wire carries a current *I* = 20 A. Let *a* = 0.30 m and *x* = 0.40 m. Find the magnitude and direction of the force per unit length on the third wire

a) if the current in it is away from the reader;

b) if current is toward the reader

31–31 Two long, straight, parallel wires are 1.0 m apart, as in Fig. 31–23. The upper wire carries a current I_1 of 6 A into the plane of the paper.

a) What must be the magnitude and direction of the current I_2 for the resultant field at point *P* to be zero?

b) What is then the resultant field at *Q*?

c) At *S*?

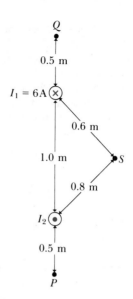

FIGURE 31–23

31–32 Two long, parallel wires are hung by cords 4 cm long from a common axis, as shown in Fig. 31–24. The wires have a mass per unit length of $0.050 \, \mathrm{kg \cdot m^{-1}}$ and carry the same current in opposite directions. What is the current if the cords hang at an angle of 6° with the vertical?

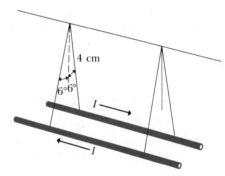

FIGURE 31–24

31–33 The long, straight wire *AB* in Fig. 31–25 carries a current of 20 A. The rectangular loop whose long edges are parallel to the wire carries a current of 10 A. Find the magnitude and direction of the resultant force exerted on the loop by the magnetic field of the wire.

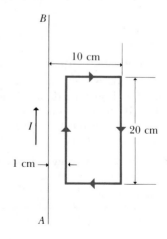

FIGURE 31–25

31–34 The long, straight wire AB in Fig. 31–26 carries a constant current I.

a) What is the magnetic field at the shaded area a perpendicular distance x from the wire?

b) What is the magnetic flux $d\Phi$ through the shaded area?

c) What is the total flux through the rectangle $CDEF$?

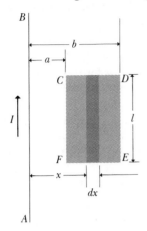

FIGURE 31–26

31–35 In Fig. 31–27, calculate the magnitude and direction of the magnetic field produced at point P by current I in the rectangular wire loop. (Point P is at the center of the rectangle.) (*Hint:* The gap on the left-hand side where the wires enter and leave the rectangle is so small that this side of the rectangle can be taken to be a continuous wire of length b.)

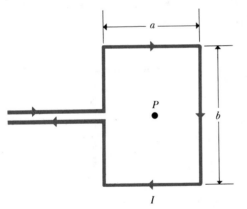

FIGURE 31–27

31–36 The wire semicircles in Fig. 31–28 have radii a and b. Calculate the resultant magnetic field (magnitude and direction) at point P.

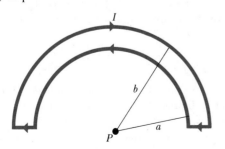

FIGURE 31–28

31–37 Figure 31–29 is a sectional view of two circular coils of radius a, each wound with N turns of wire carrying a current I, circulating in the same direction in both coils. The coils are separated by a distance a equal to their radii.

a) Derive the expression for the magnetic field at a point on the axis a distance x to the right of point P that is midway between the coils.

b) From (a) obtain an expression for the magnetic field at point P.

c) Calculate the magnitude of $\mathbf{B}$ at P if $N = 100$ turns, $I = 5$ A, and $a = 0.30$ m.

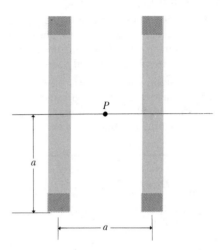

FIGURE 31–29

31–38 A conductor is made in the form of a hollow cylinder with inner and outer radii a and b, respectively. It carries a current I, uniformly distributed over the cross section. Derive expressions for the magnetic field in the regions

a) $r < a$,

b) $a < r < b$,

c) $r > b$.

31–39 The current in the windings on a toroid is 2.0 A. There are 400 turns, and the mean circumferential length is 0.40 m. The magnetic field is found to be 1.0 T. Calculate

a) the relative permeability;

b) the magnetic susceptibility

31–40 Consider the coaxial cable of Exercise 31–21. Assume the currents to be uniformly distributed over the cross sections of the solid conductor and the tube. Calculate the magnetic field at

a) $r < R_1$ (inside the central conductor; does your answer give the expected result for $r = R_1$?);

b) $R_2 < r < R_3$ (inside the outer tube; does your answer give the expected results for $r = R_3$ and for $r = R_2$?).

31–41 A long, straight wire with circular cross section of radius R carries current I. Assume that the current density is not constant across the cross section of the wire but rather varies as $J = \alpha r$, where α is a constant.

a) By the requirement that J integrated over the cross section of the wire gives the total current I, calculate the constant α in terms of I and R.

b) Use Ampere's law to calculate the magnetic field $B(r)$ for
   i) $r \leq R$;   ii) $r \geq R$.

31–42 A capacitor has two parallel plates of area A separated by a distance d. The capacitor is given an initial charge Q but, because the damp air between the plates is slightly conductive, the charge slowly leaks through the air. The charge initially changes at a rate dQ/dt.

a) In terms of dQ/dt, what is the initial rate of change of the electric field between the plates?

b) Show that the displacement-current density has the same magnitude as the conduction-current density, but the opposite direction. Hence show that the magnetic field in the air between the plates is exactly zero at all times.

31–43 A long, straight, solid cylinder, oriented with its axis in the z-direction, carries a current whose current density is J. The current density, although symmetrical about the cylinder axis, is not uniform but varies according to the relation

$$J = \frac{2I_0}{\pi a^2}[1 - (r/a)^2]k \quad \text{for} \quad r \leq a,$$
$$J = 0 \qquad\qquad\qquad \text{for} \quad r \geq a,$$

where a is the radius of the cylinder, r is the radial distance from the cylinder axis, and I_0 is a constant having units of amperes.

a) Show that I_0 is the total current passing through the entire cross section of the wire.

b) Using Ampere's law, derive an expression for the magnetic field B in the region $r \geq a$.

c) Obtain an expression for the current I contained in a circular cross section of radius $r \leq a$ and centered at the cylinder axis.

d) Using Ampere's law, derive an expression for the magnetic field B in the region $r \leq a$.

31–44 Integrate B_x as given in Eq. (31–21) from $-\infty$ to $+\infty$. That is, calculate

$$\int_{-\infty}^{+\infty} B_x \, dx.$$

Explain the significance of your result.

CHALLENGE PROBLEMS

31–45 A thin disk of dielectric material of radius a has a total charge $+Q$ distributed uniformly over its surface. It rotates n times per second about an axis perpendicular to the surface of the disk and passing through its center. Find the magnetic field at the center of the disk. (*Hint:* Divide the disk into concentric rings of infinitesimal width.)

31–46 Wire in the shape of a semicircle of radius a is oriented in the zy- plane, with its center of curvature at the origin. (See Fig. 31–30.) If the current in the wire is I, calculate the magnetic-field components produced at point P, a distance x out along the x-axis. (*Note:* Do not forget the contribution from the straight wire at the bottom of the

semicircle that runs from $z = -a$ to $z = +a$. The fields of the two antiparallel currents at $z > a$ cancel.)

31–47 The wire in Fig. 31–31 is infinitely long and carries current I. Calculate the magnitude and direction of the magnetic field this current produces at point P.

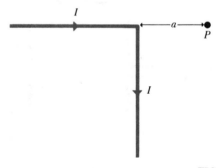

FIGURE 31–31

31–48 A long, straight, solid cylinder, oriented with its axis in the z-direction, carries a current whose current density is J. The current density, although symmetrical about the cylinder axis, is not uniform but varies according to the relation

$$J = (b/r)e^{(r-a)/\delta}k \quad \text{for} \quad r \leq a,$$
$$J = 0 \qquad\qquad\quad \text{for} \quad r \geq a,$$

where a is the radius of the cylinder and is equal to 10 cm, r is the radial distance from the cylinder axis, b is a constant equal to 400 A·m^{-1}, and δ is a constant equal to 5 cm.

a) Let I_0 be the total current passing through the entire cross section of the wire. Obtain an expression for I_0 in

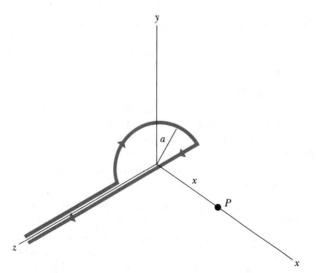

FIGURE 31–30

terms of b, δ, and a. Evaluate your expression to obtain a numerical value for I_0.

b) Using Ampere's law derive an expression for the magnetic field B in the region $r \geq a$. Express your answer in terms of I_0 rather than b.

c) Obtain an expression for the current I contained in a circular cross section of radius $r \leq a$ and centered at the cylinder axis. Express your answer in terms of I_0 rather than b.

d) Using Ampere's law, derive an expression for the magnetic field B in the region $r \leq a$.

e) Evaluate the magnitude of the magnetic field at $r = \delta$, a, and $2a$.

31-49 Two long, straight conducting wires of linear mass density $\lambda = 1.75 \times 10^{-3}\,\text{kg}\cdot\text{m}^{-1}$ are suspended from cords so that they are each horizontal, parallel to each other, and a distance $d = 2$ cm apart. The back ends of the wires are connected to one another by a slack low-resistance connecting wire. The positive side of a 1.00-μF capacitor that has been charged by a 1000-V source is connected to the front end of one of the wires, and the negative side of the capacitor is connected to the front end of the other wire. See Fig. 31-32. When the connection is made, the wires are pushed aside by the repulsive force between the wires, and each wire has an initial horizontal velocity v_0. It can be assumed that the time for the capacitor to discharge is negligible compared to the time it takes for any appreciable displacement in the position of the wires to occur.

a) Show that the initial velocity v_0 of either wire is given by

$$v_0 = \frac{\mu_0 Q_0^{\,2}}{4\pi\lambda R C d},$$

where Q_0 is the initial charge on the capacitor of capacitance C, and R is the total resistance of the circuit.

b) Determine v_0 numerically if $R = 0.01\ \Omega$.

c) To what height h will each wire rise as a result of the circuit connection?

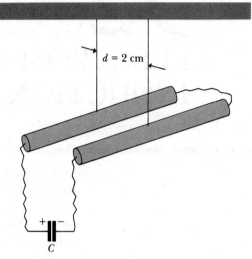

FIGURE 31-32

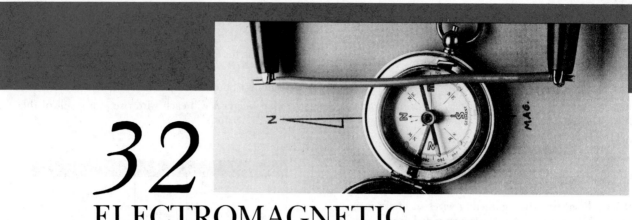

32
ELECTROMAGNETIC
INDUCTION

WE RETURN IN THIS CHAPTER TO THE SUBJECT OF ELECTROMOTIVE FORCE, a topic we studied in Chapter 28 in connection with current, energy, and power in electric circuits. Here our concern is electromotive force of *magnetic* origin. The present large-scale production, distribution, and use of electrical energy all depend directly on the principles underlying magnetically induced emf's; none of these would be possible if we had to depend on chemical sources of emf, such as batteries. Magnetically induced emf's are essential to the understanding of energy-conversion devices such as electric motors, generators, and transformers.

We begin by studying the emf developed in a conductor moving in a magnetic field. Then we develop the relationship between emf and changing magnetic flux in a stationary conducting loop; this relationship is called Faraday's law. We also discuss Lenz's law, which helps us determine the directions of induced emf's and currents. Finally, we discuss the summing up of all the essential principles of electromagnetism in four equations called Maxwell's equations. These pave the way to the analysis of electromagnetic waves in Chapter 35.

32–1 INDUCTION PHENOMENA

When a conductor moves in a magnetic field, an emf is induced.

Our study of magnetically induced electromotive force begins with a look at several pioneering experiments carried out in the 1830s by Michael Faraday in England and Joseph Henry (first director of the Smithsonian Institution) in the United States. Figure 32–1a shows a coil of wire connected to a current-measuring device, such as a d'Arsonval galvanometer of the sort described in Section 29–3. Near the coil is a bar magnet. When the magnet is stationary, the galvanometer shows no current flow, as we would expect with no source of emf in the circuit. But if we move the magnet either toward or away from the coil, the meter shows current in the circuit *while the magnet is moving*. If we hold the magnet stationary and move the coil and galvanometer, the same thing happens. So something about the *changing* magnetic field through the coil is

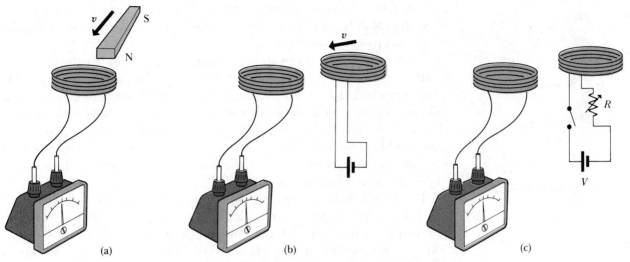

32-1 (a) A magnet moving near a coil of wire connected to a galvanometer induces a current in the coil. (b) A second coil carrying a current moves toward the coil connected to the galvanometer, inducing a current in it. (c) A changing current in the second coil induces a current in the first coil.

causing a current in the circuit. We call this an **induced current,** and the corresponding emf that has to be present to cause this current is called an **induced emf.**

The same effect occurs when the source of magnetic field is current in another circuit, as in Fig. 32–1b. We find that when the second coil is stationary, there is no current in the first coil. But when the second coil is moved toward or away from the first, or if the first coil is moved toward or away from the second, there is current in the first coil during the motion. With these experiments, as well as those with the bar magnet, the essential feature seems to be the *relative* motion of the two parts of the setup.

Finally, using the two-coil setup in Fig. 32–1c, we keep both coils stationary and vary the current in the second coil, either by opening and closing the switch or by changing the resistance R in its circuit. We find that as we open or close the switch, there is a momentary current pulse in the first circuit. When we vary the current in the second coil, there is an induced current in the first circuit *while the current in the second circuit is changing.*

What do all these experiments have in common? The answer, which we will explore in detail in this chapter, is that *changing magnetic flux* in the galvanometer circuit, from whatever cause, induces a current in that circuit. This statement forms the basis of Faraday's law of induction, the main subject of this chapter.

> A changing magnetic flux through a closed circuit induces an emf and a current in the circuit.

32-2 MOTIONAL ELECTROMOTIVE FORCE

As the first step in our detailed study of magnetically induced electromotive force, we consider a conductor moving in a magnetic field. Figure 32–2 shows a conductor of length l in a uniform magnetic field perpendicular to the plane of the figure, directed *into* the page. We give the conductor a constant velocity

> Magnetic forces on charges in a moving conductor

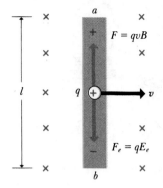

32–2 Conducting rod in uniform magnetic field.

Finding the emf due to motion of a straight conductor in a uniform magnetic field

Motional emf in a circuit with resistance is analogous to a battery with emf and internal resistance.

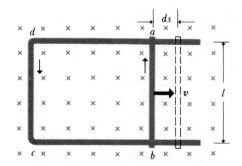

32–3 Current produced by the motion of a conductor in a magnetic field.

v to the right; a charged particle q in the conductor then experiences a force F equal to $qv \times B$. If q is positive, the direction of this force is upward along the conductor, from b toward a.

Under the action of this force the free charges in the conductor move, creating an excess of positive charge at end a and of negative charge at end b. This charge buildup in turn creates an electric field E in the direction from a to b, which exerts a downward force qE on the charges. The accumulation of charge at the ends of the conductor continues until E is large enough for the downward electric-field force qE to cancel exactly the upward magnetic-field force $qv \times B$. In this condition, the charges are in equilibrium, with point a at higher potential than point b.

Let us determine the magnitude of this potential difference. Because v and B are perpendicular, the magnitude of the magnetic-field force is simply qvB, and as we have pointed out, this is equal in the equilibrium state to qE. Thus the electric-field magnitude is $E = vB$, and the potential difference is

$$V_{ab} = El = vBl. \qquad (32-1)$$

Now suppose the moving conductor slides along a stationary U-shaped conductor, as in Fig. 32–3. No magnetic force acts on the charges in the stationary conductor, but it lies in the electrostatic field caused by the charge accumulations at a and b. Under the action of this field, a current is established in the counterclockwise sense around this complete circuit. The moving conductor has become a source of electromotive force; charge moves within it from lower to higher potential, and in the remainder of the circuit charge moves from higher to lower potential. We call this a **motional electromotive force,** denoted by $\mathcal{E}$; we can write

$$\mathcal{E} = vBl. \qquad (32-2)$$

If the resistance of the sliding bar is negligible, then $\mathcal{E}$ is also equal to the potential difference V_{ab}. For a real conductor with finite resistance r, there is a potential drop Ir, and the potential difference between the ends of the bar is *less* than $\mathcal{E}$ by this amount:

$$V_{ab} = \mathcal{E} - Ir. \qquad (32-3)$$

Note that this equation is exactly the same form of relationship that we used for the emf, terminal potential difference, and current in a battery, in Section 28–4, Eq. (28–16), although the origins of the two emf's are completely different.

If the velocity v of the conductor is not perpendicular to the field B but makes an angle ϕ with it, then we replace v in Eq. (32–2) by $v \sin \phi$, and the induced emf is

$$\mathcal{E} = v \sin \phi \, Bl. \qquad (32-4)$$

More generally, we can define the electromotive force $\mathcal{E}$ as the line integral of the associated non-electrostatic force, divided by the charge q to put it on a "force-per-unit-charge" basis. This corresponds directly to our definition of potential difference as the line integral of E, the electrostatic force per unit charge. In the present situation the non-electrostatic force per unit charge is

$v \times B$, and we define the emf as

$$\mathcal{E} = \int_b^a v \times B \cdot dl. \qquad (32\text{–}5)$$

In Fig. 33–2, v and B are perpendicular; both are constant over the length of the bar, and their vector product is parallel to dl. Thus in this case Eq. (32–5) reduces to vBl, in agreement with Eq. (32–2).

As we have mentioned, the emf associated with the moving conductor in Fig. 32–3 is analogous to that of a battery with its positive terminal at a and negative terminal at b. In each case a non-electrostatic force acts in the direction from b to a. When the device is connected to an external circuit, the direction of current is from a to b in the external circuit and from b to a in the device. Electromotive force is not a vector quantity, but the quantity defined by Eq. (32–5) may be positive or negative, depending on the directions of v and B.

Determining the sign of the induced emf in a moving conductor

If v is expressed in m·s^{-1}, B in T, and l in meters, $\mathcal{E}$ is in joules per coulomb or volts; we invite you to verify this statement.

***

EXAMPLE 32–1   Let the length l in Fig. 32–3 be 0.1 m; the velocity v, 0.1 m·s^{-1}; the resistance of the loop, 0.01 Ω; and B, 1 T. Find $\mathcal{E}$, the induced current, the force acting on the loop, and the mechanical power needed to keep the loop moving.

SOLUTION The emf $\mathcal{E}$ is

$$\begin{aligned} \mathcal{E} &= vBl = (0.1 \text{ m·s}^{-1})(1 \text{ T})(0.1 \text{ m}) \\ &= 0.01 \text{ V.} \end{aligned}$$

The current in the loop is

$$I = \frac{\mathcal{E}}{R} = \frac{0.01 \text{ V}}{0.01 \text{ }\Omega} = 1 \text{ A.}$$

Because of this current, a force F acts on the loop, in the direction *opposite* to its motion, equal to

$$\begin{aligned} F = IBl &= (1 \text{ A})(1 \text{ T})(0.1 \text{ m}) \\ &= 0.1 \text{ N.} \end{aligned}$$

To make the loop move with constant velocity, despite this resisting force, requires an equal and opposite additional force. The rate at which this additional force does work is the *power* needed to move the loop,

Energy conversion associated with induced current in a loop

$$P = Fv = (0.1 \text{ N})(0.1 \text{ m·s}^{-1}) = 0.01 \text{ W.}$$

The product $\mathcal{E}I$ is

$$\mathcal{E}I = (0.01 \text{ V})(1 \text{ A}) = 0.01 \text{ W.}$$

Thus the rate of energy conversion, $\mathcal{E}I$, equals the mechanical power input, Fv, to the system.

An emf is induced in a loop that rotates in a magnetic field.

EXAMPLE 32–2   The rectangular loop in Fig. 32–4, of length a and width b, is rotating with uniform angular velocity ω about the y-axis. The entire loop lies in a uniform, constant $\boldsymbol{B}$ field, parallel to the z-axis. We wish to calculate the induced emf in the loop, from Eq. (32–2).

SOLUTION   The velocity $\boldsymbol{v}$ of either side of the loop of length a has magnitude

$$v = \omega\left(\frac{b}{2}\right).$$

The motional emf in each side of length a is

$$\mathcal{E} = vB\sin\theta\,a = \tfrac{1}{2}\omega Bab\sin\theta.$$

These two emf's add, so the total emf due to these two sides is

$$\mathcal{E} = \omega Bab\sin\theta.$$

The magnetic forces on the other two sides of the loop are transverse to these sides and so contribute nothing to the emf.

As an alternative approach, we may use Eq. (32–5). The direction of $\boldsymbol{v}\times\boldsymbol{B}$ in each side of length a is shown in the diagram. Its magnitude is

$$|\boldsymbol{v}\times\boldsymbol{B}| = vB\sin\theta = \omega\frac{b}{2}B\sin\theta.$$

The motional forces in the other two sides of the loop are transverse to these sides and contribute nothing to the emf. The line integral of $\boldsymbol{v}\times\boldsymbol{B}$ around the loop reduces to $2|\boldsymbol{v}\times\boldsymbol{B}|a$, so

$$\mathcal{E} = \int \boldsymbol{v}\times\boldsymbol{B}\cdot d\boldsymbol{l} = 2\left(\omega\frac{b}{2}B\sin\theta\right)a$$
$$= \omega Bab\sin\theta.$$

The product ab equals the area A of the loop, and if the loop lies in the xy-plane at $t = 0$, then $\theta = \omega t$. Hence

$$\mathcal{E} = \omega AB\sin\omega t. \tag{32–6}$$

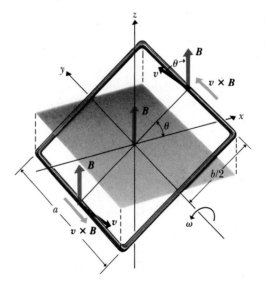

32–4 Rectangular loop rotating with constant angular velocity in a uniform magnetic field.

The emf therefore varies sinusoidally with time. The *maximum* emf $\mathcal{E}_m$, which occurs when $\sin \omega t = 1$, is

$$\mathcal{E}_m = \omega AB,$$

so we can write Eq. (32–6) as

$$\mathcal{E} = \mathcal{E}_m \sin \omega t. \tag{32–7}$$

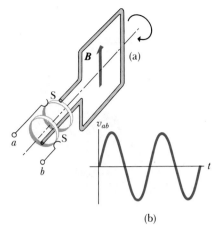

32–5 (a) Schematic diagram of an alternator, using a conducting loop rotating in a magnetic field. Connections to the external circuit are made by means of the slip rings S. (b) The resulting emf at terminals *ab*.

The rotating loop is the prototype of one type of alternating-current generator, or *alternator*; it develops a sinusoidally varying emf. The emf is maximum (in absolute value) when $\theta = 90°$ or $270°$ and the long sides are moving at right angles to the field. The emf is zero when $\theta = 0$ or $180°$ and the long sides are moving parallel or antiparallel to the field. We will show in the next section that the emf depends only on the *area A* of the loop and not on its shape.

The rotating loop in Fig. 32–4 can serve as the source in an external circuit by use of *slip rings* S, S, which rotate with the loop, as shown in Fig. 32–5a. Stationary brushes sliding on the rings are connected to the output terminals *a* and *b*. The instantaneous terminal voltage v_{ab}, on open circuit, equals the instantaneous emf. Figure 32–5b is a graph of v_{ab} as a function of time.

A similar scheme may be used to obtain an emf that, though varying with time, always has the same sign. We connect the loop to a split ring, or *commutator*, as in Fig. 32–6a. At angular positions where the emf reverses, the connections to the external circuit are interchanged. The resulting emf is shown in Fig. 32–6b. This device is the prototype of a dc generator. Commercial dc generators have a large number of coils and commutator segments; their terminal voltage is not only unidirectional but also practically constant.

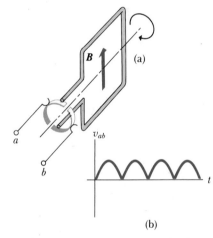

32–6 (a) Schematic diagram of a dc generator, using a split-ring commutator. (b) The resulting induced emf at terminals *ab*.

A rotating disk as a dc generator: principle of the Faraday disk dynamo

---

EXAMPLE 32–3 A disk of radius R, shown in Fig. 32–7, lies in the *xy*-plane and rotates with uniform angular velocity ω about the *z*-axis. The disk is in a uniform, constant B field parallel to the *z*-axis. Consider a short portion of a narrow radial segment of the disk, of length dr. Its velocity is $v = \omega r$, and since v is at right angles to B, the magnitude of $v \times B$ in the segment is

$$|v \times B| = vB = \omega rB.$$

The direction of $v \times B$ is radially outward. The motional emf due to this segment is

$$d\mathcal{E} = v \times B \cdot dl = vB\, dr = \omega rB\, dr.$$

The total emf between center and rim is

$$\mathcal{E} = \int_0^R v \times B \cdot dl = \omega B \int_0^R r\, dr = \tfrac{1}{2}\omega BR^2. \tag{32–8}$$

All the radial segments of the disk are in *parallel*, so the emf between center and rim is the same as that in any radial segment. Thus the entire disk is a *source*,

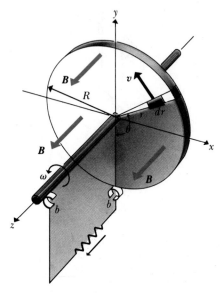

32–7 A Faraday disk dynamo. The emf is induced along radial lines of the rotating disk and is connected to an external circuit through the sliding contacts *bb*.

Relating induced emf to changing magnetic flux in a circuit

and the emf between center and rim equals $\omega BR^2/2$. The source can be included in a closed circuit by completing the circuit through sliding contacts of brushes *b*, *b*.

The emf in such a disk was studied by Faraday; the device is a called a *Faraday disk dynamo*, or a *homopolar generator*.

We pointed out in Section 30–4 that the magnetic-field force on a moving charged particle never does any *work* on the particle because ***F*** and ***v*** are always perpendicular. Yet in defining motional emf, we have spoken about the work done by the equivalent non-electrostatic force $q\mathbf{v} \times \mathbf{B}$. How is this apparent paradox resolved?

The resolution is somewhat subtle, and we cannot discuss it in detail here. Briefly, the vertical motion of the charges in the moving conductor in Fig. 32–3 causes a transverse (horizontal) magnetic-field force and thus a transverse displacement of charge, corresponding to the Hall effect discussed in Section 30–10. Hence there is a transverse *electrostatic* field in the conductor, from left to right in the example of Fig. 32–3. As the conductor moves to the right, it is this electrostatic force that actually does work on the moving charges. Detailed analysis shows that this work is the same *as though* it had been done by the vertical non-electrostatic or motional force $q\mathbf{v} \times \mathbf{B}$ during the vertical displacement of the moving charges.

32–3 FARADAY'S LAW

We analyzed the induced emf in the circuit of Fig. 32–3 by computing the magnetic-field forces on the mobile charges in the moving conductor. We can also consider this situation from another viewpoint based on the changing *magnetic flux* through the circuit. When the conductor moves toward the right a distance *ds*, the *area* enclosed by the circuit *abcd* increases by

$$dA = l\,ds,$$

and the change in magnetic *flux* through the circuit is

$$d\Phi = B\,dA = Bl\,ds.$$

The *rate of change* of flux is therefore

$$\frac{d\Phi}{dt} = \left(\frac{ds}{dt}\right)Bl = vBl. \tag{32–9}$$

But the product *vBl* equals the induced emf $\mathcal{E}$, so this equation states that *the induced emf in a circuit is numerically equal to the rate of change of the magnetic flux through it.*

To state this relationship in equation form, we need some sign rules. When we defined magnetic flux in Section 30–3 as the integral $\int \mathbf{B} \cdot d\mathbf{A},$ we pointed out the ambiguity in defining the direction of *dA* for an open surface. (We resolved that ambiguity for a *closed* surface with Gauss's law by making *dA* point always *out of* the surface.) For an open surface with a boundary curve, we can arbitrarily pick the direction for *d**A***. Then if the magnetic-field lines pass through the surface in the same direction as *d**A***, we count the corresponding flux as positive (because $\boldsymbol{B} \cdot d\boldsymbol{A}$ is positive); when they pass through in the direction opposite to $d\boldsymbol{A}$, the flux is negative.

That is half of our sign convention; the other half has to do with the sign of $\mathcal{E}$. We define a positive sense of circulation around the boundary curve by curling the fingers of the right hand around the $d\boldsymbol{A}$ vector with the thumb in the direction of $d\boldsymbol{A}$. The direction the fingers curl then defines the positive sense of going around the boundary. If an emf causes a current in this direction in a conductor lying along the boundary, it is positive; if in the opposite direction, negative.

Comparing these definitions with Fig. 32–3, suppose we choose $d\boldsymbol{A}$ to point out the plane, toward the reader. Then with the given direction of $\boldsymbol{B}$, the *flux* is negative. As the bar moves, the flux becomes larger in magnitude, and hence more negative, and $d\Phi/dt$ is a *negative* quantity. But as our discussion has shown, the induced current is counterclockwise, and according to the above definition, the associated $\mathcal{E}$ is *positive*. So, at least in this situation, $\mathcal{E}$ and $d\Phi/dt$ have opposite signs, and we write

$$\mathcal{E} = -\frac{d\Phi}{dt}. \tag{32–10}$$

This sign rule will become clearer with continued study and additional examples. We must be very precise about signs right from the beginning, in order to relate this equation to other phenomena that we will study later.

Equation (32–10) is called **Faraday's law,** or *Faraday's law of induction*. It may appear to be just an alternative form of Eq. (32–2) for the emf in a moving conductor. It turns out, though, that it has much broader significance than we might expect from our derivation. It is applicable to *any* circuit through which there is a varying flux, even when no part of the circuit is moving and thus no emf appears that we can attribute directly to a $v \times \boldsymbol{B}$ force.

Suppose, for example, that two loops of wire are located as in Fig. 32–8. A current in circuit 1 sets up a magnetic field whose magnitude at every point is proportional to this current. Part of this flux passes through circuit 2, and if the current in circuit 1 increases or decreases, the flux through circuit 2 will also vary. Circuit 2 is not moving in a magnetic field, so no "motional" emf is induced in it. However, a change does occur in the flux through the circuit, and it is found experimentally that an emf appears in circuit 2, given by $\mathcal{E} = -d\Phi/dt$. In such a situation no one portion of circuit 2 can be considered the source of emf; the *entire circuit* constitutes the source.

Here is another example. Suppose we set up a magnetic field within the toroidal winding of Fig. 32–9, link the toroid with a conducting ring, and vary the current in the winding of the toroid. We have shown that the flux lines set up by a steady current in a toroidal winding are confined to the space enclosed by the winding; not only is the ring not *moving* in a magnetic field, but if the current were steady, the ring would not even be *in* a magnetic field! Field lines do pass through the area bounded by the ring, however, and the flux changes as the current in the windings changes. Equation (32–10) predicts an induced emf in the ring, and we find by experiment that the emf actually exists. In case you have not recognized it, the apparatus in Fig. 32–9 is a *transformer* with a one-turn secondary; the phenomenon we are now discussing is the basis of operation for every transformer.

A slightly complicated sign convention for changing flux and emf

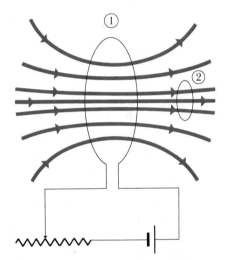

32–8 As the current in circuit 1 is varied, the magnetic flux through circuit 2 changes.

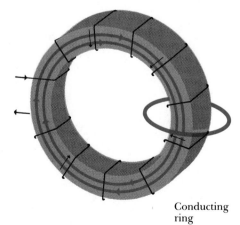

Conducting ring

32–9 An emf is induced in the ring when the flux in the toroid varies.

Faraday's law works both for moving conductors and for stationary conductors in changing magnetic fields.

To sum up, then, an emf is induced in a circuit whenever the magnetic flux through the circuit varies with time. The flux may change in various ways: For example, (1) a conductor may move through a stationary magnetic field, as in Figs. 32–2 and 32–3, or (2) the magnetic field through a stationary conducting loop may change with time, as in Figs. 32–8 and 32–9. For case (1), the emf may be computed *either* from

$$\mathcal{E} = vBl$$

or from

$$\mathcal{E} = -\frac{d\Phi}{dt}.$$

For case (2), the emf may be computed *only* from

$$\mathcal{E} = -\frac{d\Phi}{dt}.$$

If a coil has N turns, the emf is multiplied by N.

If we have a coil of N turns and the flux varies at the same rate through each turn, the induced emf's in the turns are in *series* and must be added; the total emf is

$$\mathcal{E} = -N\frac{d\Phi}{dt}. \tag{32–11}$$

PROBLEM-SOLVING STRATEGY: *Faraday's law*

1. First, be sure you understand what is making the flux change. Is the conductor moving? Is the magnetic field changing? Or both?

2. Define a positive direction for your area, and use the sign rule introduced above to write an expression for $d\Phi/dt$ that is consistent with that rule. If your conductor has N turns in a coil, don't forget to multiply by N.

3. Use Faraday's law to get the emf. Interpret the sign of your result with reference to the sign rules to determine the direction of the induced current. If the circuit resistance is known, the current can be calculated.

EXAMPLE 32–4   A certain coil of wire consists of 500 circular loops of radius 4 cm. It is placed between the poles of a large electromagnet, where the magnetic field is uniform, perpendicular to the plane of the coil, and increasing at a rate of 0.2 T·s^{-1}. What is the magnitude of the resulting induced emf?

Emf induced in a coil between the poles of an electromagnet

SOLUTION   The flux Φ at any time is given by $\Phi = BA$, and the rate of change of flux by $d\Phi/dt = (dB/dt)A$. In our problem, $A = \pi(0.04 \text{ m})^2 = 0.00503 \text{ m}^2$ and

$$\frac{d\Phi}{dt} = (0.2 \text{ T·s}^{-1})(0.00503 \text{ m}^2)$$

$$= 0.00101 \text{ T·m}^2\text{·s}^{-1} = 0.00101 \text{ Wb·s}^{-1}.$$

From Eq. (32–11), the magnitude of the induced emf is

$$\mathcal{E} = \left| N\frac{d\Phi}{dt} \right| = (500)(0.00101 \text{ Wb·s}^{-1}) = 0.503 \text{ V}.$$

If the coil is tilted so that a line perpendicular to its plane makes an angle of 30° with **B**, then only the component $B \cos 30°$ contributes to the flux through the coil. In that case the induced emf has magnitude $\mathcal{E} = (0.503 \text{ V})(\cos 30°) = 0.435 \text{ V}$.

EXAMPLE 32–5   Consider again the rotating rectangular loop in Fig. 32–4, discussed in Example 32–2 (Section 32–1). This time we compute the emf from Eq. (32–10). The flux through the loop equals its area A multiplied by the component of B perpendicular to the area; that is, $B \cos \theta$. This is also equal to the magnitude B multiplied by the projected area, shaded in Fig. 32–4, on a plane (the *xy*-plane) perpendicular to **B**:

$$\Phi = \boldsymbol{B} \cdot \boldsymbol{A} = BA \cos \theta = BA \cos \omega t.$$

Then

Emf induced in a rotating loop (again)

$$\mathcal{E} = -\frac{d\Phi}{dt} = \omega BA \sin \omega t,$$

in agreement with Eq. (32–6).

Note that the maximum value of $\mathcal{E}$ occurs when $\theta = 90°$ and the flux through the loop is zero, and that $\mathcal{E} = 0$ when $\theta = 0$ and the flux is maximum. The emf depends not on the flux through the loop, but on its *rate of change*.

EXAMPLE 32–6   Consider again the rotating disk in Fig. 32–7, discussed in Example 32–3 (Section 32–2). To compute the emf from Faraday's law, Eq. (32–10), consider the circuit to be the periphery of the shaded areas in Fig. 32–7. The rectangular portion in the *yz*-plane is fixed. The area of the shaded section in the *xy*-plane is $\frac{1}{2}R^2\theta$, and the flux through it is

Faraday's law applied to Faraday's disk: an alternative derivation of the emf

$$\Phi = \tfrac{1}{2}BR^2\theta.$$

As the disk rotates, the shaded area increases. In a time dt, the angle θ increases by $d\theta = \omega dt$. The flux increases by

$$d\Phi = \tfrac{1}{2}BR^2 \, d\theta = \tfrac{1}{2}BR^2\omega \, dt,$$

and the induced emf has magnitude

$$\mathcal{E} = \left| \frac{d\Phi}{dt} \right| = \tfrac{1}{2}BR^2\omega,$$

in agreement with the previous result.

EXAMPLE 32–7   *The search coil.* A useful experimental method for measuring magnetic-field strength uses a small, closely wound coil of N turns called a *search coil*. Suppose first that the plane of the windings of the search coil is initially perpendicular to a magnetic field **B**. If the area enclosed by the coil is A, the flux Φ through it is $\Phi = BA$. Now if the coil is quickly given a quarter-turn about one of its diameters so that its plane becomes parallel to the field, or if it is quickly snatched from its position to another where the field is known to be zero, the flux through it decreases rapidly from BA to zero. During the time that the flux is decreasing, an emf of short duration is induced in the coil, and a momentary induced current occurs in the external circuit to which the coil is connected. As we will see, the total flux change is proportional to the total *charge* that flows around the circuit.

The search coil: a way to measure magnetic fields

The current at any instant is

$$i = \frac{\mathcal{E}}{R},$$

where R is the combined resistance of external circuit and search coil, $\mathcal{E}$ is the instantaneous induced emf, and i is the instantaneous current. From Faraday's law, disregarding the negative sign,

$$\mathcal{E} = N\frac{d\Phi}{dt}, \qquad i = \frac{N}{R}\frac{d\Phi}{dt}.$$

The total charge Q that flows through the circuit is given by

$$Q = \int_0^t i\, dt = \frac{N}{R}\int d\Phi = \frac{N\Phi}{R}.$$

The field is proportional to the total charge that flows in the circuit.

Thus Φ and B are given by

$$\Phi = \frac{RQ}{N} \quad \text{and} \quad B = \frac{\Phi}{A} = \frac{RQ}{NA}. \tag{32-12}$$

It is not difficult to construct an instrument that measures the total charge passing through it. If the external circuit contains such an instrument, Q may be measured directly. We may then use Eq. (32-12) to compute B.

Strictly speaking, this method gives only the *average* field over the area of the coil. But if the area is small, this value is very nearly equal to the field at the center of the coil.

32–4 INDUCED ELECTRIC FIELDS

Faraday's law applied to changing fields and stationary conductors

The examples of induced emf analyzed in Section 32–3 were of two types, involving either a conductor moving in a magnetic field or changing flux through a stationary conductor. Another example of the second type is shown in Fig. 32–10, where a solenoid is encircled by a conducting loop with a galvanometer in it. A current I in the winding of the solenoid sets up a magnetic field B along the solenoid axis, as shown, and there is a magnetic flux $\Phi = BA$ through the surface bounded by the loop. When the current I changes, causing the flux to change, the galvanometer indicates a current in the wire *during the time that the flux is changing*. The associated emf is equal in magnitude to the *time rate of change of flux* through the surface bounded by the wire. Furthermore, the directions of the induced emf and current are given by the same sign rule described in Section 32–3. If we take the surface element dA to point in the direction of B, and if I is increasing, then $d\Phi/dt$ is positive. The induced current is in the counterclockwise sense around the loop, as shown in Fig. 32–10. Thus we have

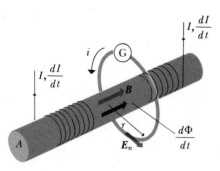

32–10 The windings of a long solenoid carry a current I that is increasing at a rate dI/dt. The magnetic flux in the solenoid is increasing at a rate $d\Phi/dt$, and this changing flux passes through a wire loop of arbitrary size and shape. An emf $\mathcal{E}$ is induced in the loop, given by $\mathcal{E} = d\Phi/dt$.

$$\mathcal{E} = -\frac{d\Phi}{dt}. \tag{32-13}$$

Equation (32–13) is again Faraday's law. We emphasize once more that it is valid for flux changes caused by motion of a conductor in a magnetic field and by magnetic-field changes in the area bounded by a stationary conductor.

EXAMPLE 32–8 Suppose the long solenoid in Fig. 32–10 is wound with 1000 turns per meter, and the current in its windings is increasing at the rate of $100 \, \text{A} \cdot \text{s}^{-1}$. The cross-sectional area of the solenoid is $4 \, \text{cm}^2 = 4 \times 10^{-4} \, \text{m}^2$.

The magnetic-field magnitude inside the solenoid, at points not too near its ends, is given by Eq. (31–28): $B = \mu_0 nI$. The flux Φ in the solenoid is

$$\Phi = BA = \mu_0 nIA.$$

The rate of change of flux is

$$\begin{aligned}
\frac{d\Phi}{dt} &= \mu_0 nA \frac{dI}{dt} \\
&= (4\pi \times 10^{-7} \, \text{Wb} \cdot \text{A}^{-1} \cdot \text{m}^{-1})(1000 \, \text{turns} \cdot \text{m}^{-1}) \\
&\quad \times (4 \times 10^{-4} \, \text{m}^2)(100 \, \text{A} \cdot \text{s}^{-1}) \\
&= 16\pi \times 10^{-6} \, \text{Wb} \cdot \text{s}^{-1}.
\end{aligned}$$

The magnitude of the induced emf is

$$|\mathcal{E}| = \frac{d\Phi}{dt} = 16\pi \times 10^{-6} \, \text{V} = 16\pi \, \mu\text{V}.$$

For conductors moving in magnetic fields, we derived Faraday's law for induced emf from the line integral of $v \times B$, the magnetic force per unit charge. But when the conductors are not moving, we must conclude that the induced current in the loop is caused by an **induced electric field.** This field is not caused by a distribution of electric charge but rather is associated with the changing magnetic field. To emphasize this difference, we call it a **non-electrostatic field,** and denote it by E_n. Then the induced emf in this case is the line integral of E_n around the loop, and we can restate Faraday's law as

An induced electric field is associated with an induced emf.

$$\oint E_n \cdot dl = -\frac{d\Phi}{dt}. \tag{32–14}$$

As an example, suppose the loop in Fig. 32–10 is a circle of radius r. Because of axial symmetry, the non-electrostatic field E_n has the same magnitude at every point on the circle and is tangent to it at each point. The line integral in Eq. (32–14) becomes simply the magnitude E_n times the circumference $2\pi r$ of the circle. Thus at a distance r from the axis, the magnitude of the induced electric field is given by

$$E_n = \frac{1}{2\pi r} \frac{d\Phi}{dt}. \tag{32–15}$$

The direction of E_n is as shown in the figure. We know that E_n has to have this direction because the integral of $E_n \cdot dl$ has to be negative when $d\Phi/dt$ is positive.

In summary, Eq. (32–10) is valid for two rather different situations. In one, an emf is induced by the magnetic forces on charges in a conductor moving through a magnetic field. In the other, a time-varying magnetic field induces an electric field of nonelectrostatic nature in a stationary conductor and hence induces an emf. The E_n field in the latter case differs from an *electrostatic* field in an important way. Its line integral around a closed path is

The analogy between magnetically induced electric fields and magnetic fields caused by displacement current

not zero, so it is not a *conservative* field. In contrast, an electrostatic field is *always* conservative, as discussed in Section 26–1. Despite this difference, however, the effect of an electric field in exerting a force $F = qE$ on a charged particle is the same, whether the field is produced by a charge distribution or by a changing magnetic field.

Electric field produced by a changing magnetic field is analogous to the role of *displacement current,* discussed in Section 29–5, where a changing *electric* field acts as a source of *magnetic* field. These relations thus exhibit a symmetry in the behavior of the two fields. We will return to this relationship in Section 32–7, and again in Chapter 35 in connection with electromagnetic waves.

32–5 LENZ'S LAW

Lenz's law is a convenient alternative method for determining the sign or direction of an induced emf or current, or the direction of the associated non-electrostatic field. It always gives the same results as the sign rules we have introduced for $\mathcal{E}$ and $d\Phi/dt$ in connection with Faraday's law, but it is often easier to use. It also helps us gain intuitive understanding of various induction phenomena. H. F. E. Lenz (1804–1864) was a German scientist who, without knowing about the work of Faraday and Henry, duplicated many of their discoveries at about the same time. **Lenz's law** states:

> The direction of an induced current is such as to oppose the cause producing it.

The "cause" of the current may be the motion of a conductor in a magnetic field or the change of flux through a stationary circuit. In the first case, the direction of the induced current in the moving conductor is such that the direction of the sideways force exerted on the conductor by the magnetic field is opposite in direction to its motion. The motion of the conductor is therefore "opposed."

In the second case, the induced current sets up a magnetic field of its own that, within the area bounded by the circuit, is (a) *opposite* to the original field if this is *increasing,* but (b) is in the *same* direction as the original field if the latter is *decreasing.* Thus it is the *change in flux* through the circuit (not the flux itself) that is "opposed" by the induced current.

In order to have an induced current, we must have a closed circuit. If a conductor does not form a closed circuit, then we mentally complete the circuit between the ends of the conductor and use Lenz's law to determine the direction of the current. The polarity of the ends of the open-circuit conductor may then be deduced.

Lenz's law: an alternative way to determine directions of induced emf's and currents

The effect always opposes its cause.

EXAMPLE 32–9   In Fig. 32–3, when the conductor moves to the right, a counterclockwise current is induced in the loop. The force exerted by the field on the moving conductor as a result of this current is to the left, *opposing* the conductor's motion.

EXAMPLE 32–10   In Fig. 32–4, the direction of the induced current in the loop is the same as that of the non-electrostatic field E_n. The magnetic field exerts forces on the loop as a result of this current; the force on the right side of the loop (of length *a*) is in the +*x*-direction, and that on the left side is in the −*x*-direction. The resulting torque is opposite in direction to *ω* and thus opposes the motion.

EXAMPLE 32–11   In Fig. 32–10, when the solenoid current is increasing, the induced current in the loop is counterclockwise. The additional field caused by this current, at points inside the loop, is opposite in direction to that of the solenoid; hence the induced current tends to oppose the increase in flux through the loop by causing flux in the opposite sense.

Lenz's law is also directly related to energy conservation. For instance, in Example 32–9 above, the induced current in the loop dissipates energy at the rate I^2R, and this energy must be supplied by the force that makes the conductor move despite the magnetic force opposing its motion. The work done by this applied force, in fact, must equal the energy dissipated in the circuit resistance. If the induced current were to have the opposite direction, the resulting force on the moving conductor would make it move faster and faster, violating energy conservation.

Lenz's law is a consequence of energy conservation.

The content of Lenz's law is contained in the sign rules introduced for use with Faraday's law. Let us review those briefly. We designate a positive direction for the surface through which we calculate the flux; this is the direction of the area element *dA* in the flux integral. This direction defines positive and negative flux and its rate of change. So, for example, if **B** has the same direction as *dA* and is increasing in magnitude, or if the area is increasing, then $d\Phi/dt$ is positive. If **B** has the same direction as *dA* but is decreasing in magnitude, or if the area is decreasing, $d\Phi/dt$ is negative. We invite you to think of other possibilities. Finally, in evaluating $\oint E_n \cdot dl$, we apply the right-hand rule to *dA*. For example, if the area element lies in the plane of this page and we define *dA* to point toward you, then we go counterclockwise around the circuit in evaluating the line integral of E_n.

Relation of Lenz's law to sign conventions in Faraday's law

EXAMPLE 32–12   In Fig. 32–3 let us take *dA* as parallel to **B**, away from the reader. Then Φ and $d\Phi/dt$ are both positive. Here $v \times B$ plays the role of E_n. The integral $\oint E_n \cdot dl$ must be taken clockwise around the loop. E_n is different from zero only along the line *ab*, where its direction is upward. Thus

Some exercise in using Lenz's law

$$\oint E_n \cdot dl = -E_n l.$$

This quantity is negative because E_n and *dl* have opposite directions, and it is consistent with Eq. (32–10), in which both sides are positive. Alternatively, we could have taken *dA* as toward the reader. In that case Φ and $d\Phi/dt$ are both negative, the loop is traversed in the counterclockwise sense, and $\oint E_n \cdot dl$ is positive. Either way, consistency with Eq. (32–10) requires that the direction of E_n must be from *b* to *a*, which is the direction of current in the moving conductor.

EXAMPLE 32–13   In Fig. 32–10, suppose the current in the solenoid has the direction shown and is decreasing. What is the direction of the induced current in the loop?

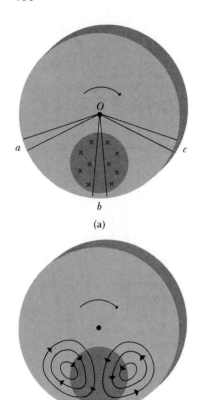

(a)

(b)

32–11 Eddy currents in a rotating disk.

Eddy currents dissipate energy: eddy-current brakes.

Eddy currents in transformer cores are a nuisance: How can we minimize them?

SOLUTION Take the area elements $d\boldsymbol{A}$ as pointing to the right, parallel to $\boldsymbol{B}$. Then the flux is positive but decreasing, and $d\Phi/dt$ is negative. Hence, according to Eq. (32–10), when we calculate $\mathcal{E} = \oint\boldsymbol{E}_n \cdot d\boldsymbol{l}$ by traversing the loop clockwise (looking in the direction of $\boldsymbol{B}$) the result must be positive. Hence $\boldsymbol{E}_n$ points clockwise around the loop, and the induced current must have this direction. Lenz's law gives the same result.

32–6 EDDY CURRENTS

In the examples of induction phenomena considered thus far, the induced currents resulting from induced emf's have been confined to well-defined paths in wires and other components forming a *circuit*. However, many pieces of electrical equipment contain masses of metal moving in magnetic fields or located in changing magnetic fields. In such situations it is possible to have induced currents that circulate throughout the volume of the material; because of their circulating nature, we call them **eddy currents.**

As an example, consider a disk that rotates in a magnetic field perpendicular to the plane of the disk but confined to a limited portion of the disk's area, as shown in Fig. 32–11. Element Ob is moving across the field and has an emf induced in it. Elements Oa and Oc are not in the field but, in common with all other elements located outside the field, provide return conducting paths along which charges displaced along Ob can return from b to O. A general eddy-current circulation is therefore set up in the disk, somewhat as sketched in Fig. 32–11b.

The currents in the neighborhood of radius Ob experience a side thrust that *opposes* the motion of the disk, while the return currents, since they lie outside the field, do not experience such a thrust. The interaction between the eddy currents and the field therefore results in a braking action on the disk. This apparatus finds some technical applications and is known as an eddy-current brake.

As a second example of eddy currents, consider the core of an alternating-current transformer, shown in Fig. 32–12a. The alternating current in the primary winding P sets up an alternating flux within the core, and an induced emf develops in the secondary winding S because of the continual change in flux through it. The iron core, however, is also a conductor, and any section such as AA can be thought of as a number of closed conducting circuits, one within the other (Fig. 32–12b). The flux through each of these circuits is

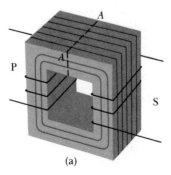

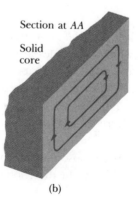

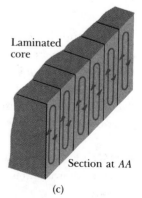

32–12 Reduction of eddy currents by use of a laminated core.

continually changing, so eddy currents circulate in the entire volume of the core, the lines of flow lying in planes perpendicular to the flux. These eddy currents are very undesirable both because of the energy they dissipate and because of the flux they themselves set up.

In all actual transformers, the eddy currents are greatly reduced by the use of a *laminated* core, that is, one built up of thin sheets, or laminae. The electrical resistance between the surfaces of the laminations (due either to a natural coating of oxide or to an insulating varnish) effectively confines the eddy currents to individual laminae (Fig. 32–12c). The flux through each loop and the resulting emf are small, and the currents and their heating effects are minimized.

In small transformers where eddy-current losses must be kept to an absolute minimum, the cores are sometimes made of *ferrites*, which are complex oxides of iron and other metals. These materials are ferromagnetic but have relatively high resistivity.

Ferrites: nonconducting ferromagnetic materials

32–7 MAXWELL'S EQUATIONS

We are now in a position to wrap up in a wonderfully neat package all the relationships between electric and magnetic fields and their sources that we have studied in the past several chapters. The package consists of four equations, called **Maxwell's equations.** You may remember Maxwell as the discoverer of the concept of displacement current, which we studied in Sections 29–5 and 31–9.

Two of Maxwell's equations involve integrals of E and B over closed surfaces. The first is simply Gauss's law, stating that the surface integral of $E_\perp$ over any closed surface equals $1/\epsilon_0$ times the total charge Q enclosed within the surface:

Gauss's law is one of Maxwell's equations.

$$\oint E \cdot dA = \frac{Q}{\epsilon_0}. \tag{32–16}$$

The second is the analogous relation for *magnetic* fields, stating that the surface integral of $B_\perp$ over any closed surface is always zero:

$$\oint B \cdot dA = 0. \tag{32–17}$$

This statement means, among other things, that there is no such thing as magnetic charge as a source of magnetic field.

There are no isolated magnetic poles: the basis of another of Maxwell's equations.

The third equation is Ampere's law, including displacement current, and the fourth is Faraday's law. Ampere's law, as stated in Section 31–9, states that both conduction current I_C and displacement current $\epsilon_0 \, d\Psi/dt$, where Ψ is electric flux, act as sources of magnetic field:

Ampere's law, as generalized by Maxwell to include displacement current as a source of magnetic field

$$\oint B \cdot dl = \mu_0 \left(I_C + \epsilon_0 \frac{d\Psi}{dt} \right). \tag{32–18}$$

Faraday's law, which we have studied in this chapter, states that a changing magnetic field or magnetic flux induces an electric field:

Faraday's law: the fourth Maxwell equation

$$\oint E \cdot dl = -\frac{d\Phi}{dt}. \tag{32–19}$$

In our discussions we have been careful to describe this induced electric field as a non-electrostatic field and to denote it as E_n. But if an *electrostatic* field, produced by electric charges, is also present in the same region, it is always *conservative,* and so its line integral around any closed path is always zero. Hence in Eq. (32–19) *E* is the *total* electric field, including both electrostatic and non-electrostatic contributions.

Comparing Eqs. (32–18) and (32–19), we see a remarkable symmetry. Equation (32–18) says that a changing electric field or electric flux creates a magnetic field, and Eq. (32–19) says that a changing magnetic field or magnetic flux creates an electric field. Indeed, in empty space, where there is no conduction current and $I_C = 0$, the two equations have the same form, apart from a numerical constant and a negative sign, with the roles of *E* and *B* reversed in the two equations. This same symmetry is also evident in the first two equations; in empty space, where there is no charge, the equations are identical in form, one containing *E,* the other *B.*

But the most remarkable thing about the third and fourth equations is this: A time-varying field of *either* kind induces a field of the other kind in neighboring regions of space. As Maxwell recognized, these relationships predict the existence of electromagnetic disturbances consisting of time-varying electric and magnetic fields that travel or *propagate* from one region of space to another, even if no matter is present in the intervening space. Such disturbances are called *electromagnetic waves,* and we now know that they provide the physical basis for light, radio and television waves, infrared, ultraviolet, x-rays, and the rest of the electromagnetic spectrum. We will return to this vitally important topic for further study in Chapter 35.

We can rewrite Eqs. (32–18) and (32–19) in a different but equivalent form by introducing the definitions of electric and magnetic flux, $\Psi = \int E \cdot dA$ and $\Phi = \int B \cdot dA,$ respectively. In empty space, where $I_C = 0$, we obtain

$$\oint B \cdot dl = \mu_0 \epsilon_0 \frac{d}{dt} \oint E \cdot dA, \tag{32–20}$$

$$\oint E \cdot dl = -\frac{d}{dt} \oint B \cdot dA. \tag{32–21}$$

Again we notice the symmetry between the roles of the *E* and *B* fields in these expressions.

Although it may not be obvious, *all* the basic relations between fields and their sources are contained in Maxwell's equations. We can derive Coulomb's law from Gauss's law, we can derive the law of Biot and Savart from Ampere's law, and so on. When we add the equation that gives the force on a charged particle in *E* and *B* fields, namely,

$$F = q(E + v \times B),$$

we have all the fundamental relations of electromagnetism!

The fact that electromagnetism can be wrapped up so neatly and elegantly is a very satisfying discovery. In conciseness and generality, Maxwell's equations are comparable to Newton's law of motion and to the laws of thermodynamics. Indeed, that is really what science is all about—learning how to express very broad and general physical laws in a concise and compact form. Maxwell's synthesis of electromagnetism stands as a towering intellectual achievement, comparable to the Newtonian synthesis described at the end of Section 6–5 and to the development of relativity, quantum mechanics, and the understanding of DNA in our own century. All beautiful, and all monuments to the achievements of which the human intellect is capable!

SUMMARY

When a conducting loop moves in a magnetic field or is stationary in a region of changing magnetic field, there is an induced current with a corresponding induced emf, whether the magnetic field is caused by a permanent magnet or by another current loop. The induced current depends on the rate of change of magnetic flux through the loop.

When a conductor of length l moves with velocity v in a uniform magnetic field B, the induced emf is given by

$$\mathcal{E} = vBl. \tag{32-2}$$

More generally, the induced emf in a conductor moving in a B field is

$$\mathcal{E} = \int_b^a v \times B \cdot dl. \tag{32-5}$$

When a conducting loop of area A rotates with angular velocity ω in a uniform magnetic field of magnitude B, the induced emf is

$$\mathcal{E} = \omega AB \sin \omega t. \tag{32-6}$$

When a disk of radius R rotates with angular velocity ω in a magnetic field of magnitude B, with the disk axis parallel to B, the induced emf is

$$\mathcal{E} = \frac{1}{2} \omega BR^2. \tag{32-8}$$

Faraday's law of induction states that the induced emf in a conducting loop through which the magnetic flux Φ is changing is given by

$$\mathcal{E} = -\frac{d\Phi}{dt}. \tag{32-10}$$

This equation is valid whether the changing flux is caused by motion of the coil or by time variation of the magnetic field. If the conductor is a coil of N turns, the emf is multiplied by N. When a stationary conductor is in a changing magnetic field, the induced emf is associated with a non-electrostatic field E_n, such that

$$\oint E_n \cdot dl = -\frac{d\Phi}{dt}. \tag{32-14}$$

Lenz's law, a useful rule to find the directions of induced currents and emf's, states that the induced current or emf always acts to oppose or cancel the change that caused it. Lenz's law can be derived from Faraday's law but is sometimes easier to use.

When a bulk piece of conducting material, such as a metal, is in a changing magnetic field or moves through a field, currents called eddy currents are induced in the volume of the material.

The relationships between electric and magnetic fields and their sources can be stated compactly in four equations called Maxwell's equations. They are

$$\oint E \cdot dA = \frac{Q}{\epsilon_0}, \tag{32-16}$$

$$\oint B \cdot dA = 0, \tag{32-17}$$

$$\oint \boldsymbol{B} \cdot d\boldsymbol{l} = \mu_0 \left(I_C + \epsilon_0 \frac{d\Psi}{dt} \right), \qquad (32\text{--}18)$$

$$\oint \boldsymbol{E} \cdot d\boldsymbol{l} = -\frac{d\Phi}{dt}. \qquad (32\text{--}19)$$

The first equation is Gauss's law of electrostatics; the second states the absence of magnetic charge; the third is Ampere's law, generalized by Maxwell to include displacement current; and the fourth is Faraday's law. Together they form a complete basis for the relation of $\boldsymbol{E}$ and $\boldsymbol{B}$ fields to their sources. These equations enabled Maxwell to predict the existence of electromagnetic waves, an outstanding achievement in the development of physical science.

QUESTIONS

32–1 In most parts of the northern hemisphere the earth's magnetic field has a vertical component directed *into* the earth. An airplane flying east generates an emf between its wingtips. Which wingtip acquires an excess of electrons and which a deficiency?

32–2 A sheet of copper is placed between the poles of an electromagnet, so the magnetic field is perpendicular to the sheet. When it is pulled out, a considerable force is required, and the force required increases with speed. What is happening?

32–3 In Fig. 32–4, if the angular velocity ω of the loop is doubled, then the frequency with which the induced current changes direction doubles, and the maximum emf also doubles. Why? Does the torque required to turn the loop change?

32–4 If we compare the conventional dc generator (Fig. 32–6) and the Faraday disk dynamo (Fig. 32–7), what are some advantages and disadvantages of each?

32–5 Some alternating-current generators use a rotating permanent magnet and stationary coils. What advantages does this scheme have? What disadvantages?

32–6 When a conductor moves through a magnetic field, the magnetic forces on the charges in the conductor cause an emf. But if this phenomenon is viewed in a frame of reference moving with the conductor, there is no motion, yet there is still an emf. How is this paradox resolved?

32–7 Two circular loops lie adjacent to each other. One is connected to a source that supplies an increasing current; the other is a simple closed ring. Is the induced current in the ring in the same direction as that in the ring connected to the source, or opposite? What if the current in the first ring is decreasing?

32–8 A farmer claimed that the high-voltage transmission lines running parallel to his fence induced dangerously large voltages on the fence. Is this within the realm of possibility?

32–9 Small one-cylinder gasoline engines sometimes use a device called a *magneto* to supply current to the spark plug. A permanent magnet is attached to the flywheel, and a stationary coil is mounted adjacent to it. What happens when the magnet passes the coil?

32–10 A current-carrying conductor passes through the center of a metal ring, perpendicular to its plane. If the current in the conductor increases, is a current induced in the ring?

32–11 A student asserted that if a permanent magnet is dropped down a vertical copper pipe, it eventually reaches a terminal velocity, even if there is no air resistance. Why should this be? Or should it?

EXERCISES

Section 32–2 Motional Electromotive Force

32–1 In Fig. 32–3 a rod of length $l = 0.40$ m moves in a magnetic field of magnitude $B = 1.2$ T. The emf induced in the moving rod is found to be 2.40 V.

a) What is the speed of the rod?

b) If the total circuit resistance is 1.2 Ω, what is the induced current?

c) What force (magnitude and direction) does the field exert on the rod as a result of this current?

32–2 In Fig. 32–2 a rod of length $l = 0.25$ m moves with constant speed of 6.0 m·s^{-1} in the direction shown. The induced emf is found to be 3.0 V.

a) What is the magnitude of the magnetic field?

b) Which point is at higher potential, a or b?

32–3 In Fig. 32–13 a rod of length $l = 0.15$ m moves in a magnetic field $\boldsymbol{B}$ directed into the plane of the figure. If $B = 0.5$ T and the rod moves with velocity $v = 4$ m·s^{-1} in the direction shown,

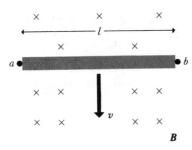

FIGURE 32-13

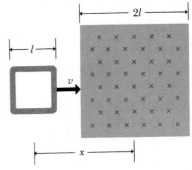

FIGURE 32-15

a) What is the motional emf induced in the rod?

b) What is the potential difference between the ends of the rod?

c) Which point, a or b, is at higher potential?

32-4 A conducting rod AB in Fig. 32-14 makes contact with metal rails CA and DB. The apparatus is in a uniform magnetic field 0.5 T, perpendicular to the plane of the diagram.

a) Find the magnitude and direction of the emf induced in the rod when it is moving toward the right with a speed 4 m·s^{-1}.

b) If the resistance of circuit $ABDC$ is 0.2 Ω (assumed constant), find the force required to maintain the rod in motion. Neglect friction.

c) Compare the rate at which mechanical work is done by the force (Fv) with the rate of development of heat in the circuit (i^2R).

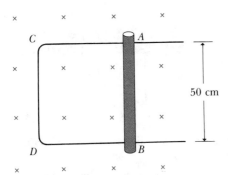

FIGURE 32-14

32-5 A square loop of wire with resistance R is moved at constant velocity v across a uniform magnetic field confined to a square region whose sides are twice the length of those of the square loop. (See Fig. 32-15.)

a) Sketch a graph of the external force F needed to move the loop at constant velocity, as a function of the distance x, from $x = -2l$ to $x = +2l$. (The distance x is negative when the center of the loop is to the left of the center of the magnetic-field region. Take positive force to be to the right.)

b) Sketch a graph of the induced current in the loop as a function of x. Take clockwise currents to be positive.

Section 32-3 Faraday's Law

32-6 A coil of 1000 turns enclosing an area of 20 cm² is rotated from a position where its plane is perpendicular to the earth's magnetic field to one where its plane is parallel to the field, in 0.02 s. What average emf is induced if the earth's magnetic field is 6×10^{-5} T?

32-7 A closely wound rectangular coil of 50 turns has dimensions of 12 cm × 25 cm. The plane of the coil is rotated from a position where it makes an angle of 45° with a magnetic field 2 T to a position perpendicular to the field in time $t = 0.1$ s. What is the average emf induced in the coil?

32-8 A flat, square coil of ten turns has sides of length 0.12 m. The coil rotates in a magnetic field of 0.025 T.

a) What is the angular velocity of the coil if the maximum emf produced is 20 mV?

b) What is the average emf at this velocity?

32-9 A Faraday disk dynamo is to be used to supply current to a large electromagnet that requires 20,000 A at 1.0 V. The disk is to be 0.6 m in radius, and it turns in a magnetic field of 1.2 T supplied by a smaller electromagnet.

a) How many revolutions per second must the disk turn?

b) What torque is required to turn the disk, assuming that all the mechanical energy is dissipated as heat in the large electromagnet?

32-10 The cross-sectional area of a closely wound search coil having 20 turns is 1.5 cm² and its resistance is 4 Ω. The coil is connected through leads of negligible resistance to a charge-measuring instrument having internal resistance 16 Ω. Find the quantity of charge displaced when the coil is pulled quickly out of a region where $B = 1.8$ T to a point where the magnetic field is zero. The plane of the coil, when in the field, makes an angle of 90° with the magnetic field.

32-11 A closely wound search coil has an area of 4 cm², 160 turns, and a resistance of 50 Ω. It is connected to a charge-measuring instrument whose resistance is 30 Ω. When the coil is rotated quickly from a position parallel to a uniform magnetic field to one perpendicular to the field, the instrument indicates a charge of 4×10^{-5} C. What is the magnitude of the field?

32–12 A coil 4 cm in radius, containing 500 turns, is placed in a magnetic field that varies with time according to $B = 0.01t + (2 \times 10^{-4})t^3$, where B is in teslas and t is in seconds. The coil is connected to a 500-Ω resistor, and its plane is perpendicular to the magnetic field.

a) Find the induced emf in the coil as a function of time.

b) What is the current in the resistor at time $t = 10$ s?

Section 32–4 Induced Electric Fields

32–13 A long, straight solenoid of cross-sectional area 6 cm^2 is wound with ten turns of wire per centimeter, and the windings carry a current of 0.25 A. A secondary winding of two turns encircles the solenoid. When the primary circuit is opened, the magnetic field of the solenoid becomes zero in 0.05 s. What is the average induced emf in the secondary?

32–14 The magnetic field within a long, straight solenoid of circular cross section and radius R is increasing at a rate of dB/dt.

a) What is the rate of change of flux through a circle of radius r_1 inside the solenoid, normal to the axis of the solenoid, and with center on the solenoid axis?

b) Find the induced electric field E_n inside the solenoid, at a distance r_1 from its axis. Show the direction of this field in a diagram.

c) What is the induced electric field *outside* the solenoid, at a distance r_2 from the axis?

d) Sketch a graph of the magnitude of E_n as a function of the distance r from the axis, from $r = 0$ to $r = 2R$.

e) What is the induced emf in a circular turn of radius $R/2$?

f) Of radius R?

g) Of radius $2R$?

32–15 The magnetic field B at all points within the colored circle of Fig. 32–16 has an initial magnitude of 0.5 T. It is directed into the plane of the diagram and is decreasing at the rate of -0.1 T·s^{-1}.

a) What is the shape of the field lines of the induced E_n field in Fig. 32–16, within the colored circle?

b) What are the magnitude and direction of this field at any point of the circular conducting ring of radius 0.10 m, and what is the emf in the ring?

c) What is the current in the ring if its resistance is 2 Ω?

d) What is the potential difference between points a and b of the ring?

Section 32–5 Lenz's Law

32–16 A circular loop of wire is in a region of spatially uniform magnetic field, as shown in Fig. 32–16. The magnetic field is directed into the plane of the figure. Calculate the direction (clockwise or counterclockwise) of the induced current in the loop when

a) B is increasing,

b) B is decreasing,

c) B is constant with value B_0.

32–17 A cardboard tube is wound with two windings of insulated wire, as in Fig. 32–17. Terminals a and b of winding A may be connected to a battery through a reversing switch. State whether the induced current in the resistor R is from left to right, or from right to left, in the following circumstances:

a) The current in winding A is from a to b and is increasing;

b) the current is from b to a and is decreasing;

c) the current is from b to a and is increasing.

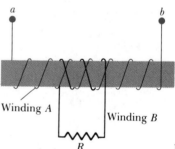

Winding A   Winding B

R

FIGURE 32–17

32–18 Using Lenz's law, determine the direction of the current in resistor ab of Fig. 32–18 when

a) switch S is opened after having been closed for several minutes;

b) coil B is brought closer to coil A, with the switch closed;

c) the resistance of R is decreased while the switch remains closed.

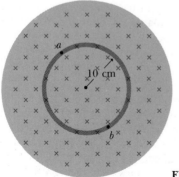

FIGURE 32–16

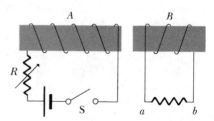

FIGURE 32–18

PROBLEMS

32–19 The long rectangular loop in Fig. 32–19, of width l, mass m, and resistance R, starts from rest in the position shown and is acted on by a constant force F. At all points in the colored area there is a uniform magnetic field B perpendicular to the plane of the diagram.

a) Sketch a graph of the velocity of the loop as a function of time.

b) Find the terminal velocity.

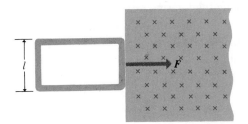

FIGURE 32–19

32–20 The cube in Fig. 32–20, 1 m on a side, is in a uniform magnetic field of 0.2 T, directed along the positive y-axis. Wires A, C, and D move in the directions indicated, each with a speed of 0.5 m·s^{-1}.

a) What is the motional emf along the length of each wire?

b) What is the potential difference between the ends of each wire?

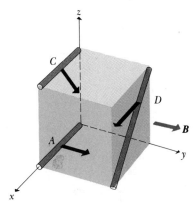

FIGURE 32–20

32–21 A slender rod 1 m long rotates about an axis through one end and perpendicular to the rod, with an angular velocity of 2 rev·s^{-1}. The plane of rotation of the rod is perpendicular to a uniform magnetic field of 0.5 T.

a) What is the induced emf in the rod?

b) What is the potential difference between its ends?

32–22 The rectangular loop in Fig. 32–21, of area A and resistance R, rotates at uniform angular velocity ω about the y-axis. The loop lies in a uniform magnetic field B in the direction of the x-axis. Sketch the following graphs:

a) the flux Φ through the loop as a function of time (let $t = 0$ in the position shown in Fig. 32–21);

b) the rate of change of flux $d\Phi/dt$;

c) the induced emf in the loop;

d) the torque Γ needed to keep the loop rotating at constant angular velocity;

e) the induced emf if the angular velocity is doubled.

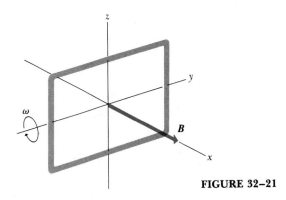

FIGURE 32–21

32–23 In Problem 32–22 and Fig. 32–21, let $A = 400 \text{ cm}^2$, $R = 2 \, \Omega$, $\omega = 10 \text{ rad·s}^{-1}$, $B = 0.5 \text{ T}$. Find

a) the maximum flux through the loop,

b) the maximum induced emf,

c) the maximum torque.

d) Show that the work of the external torque in one revolution is equal to the energy dissipated in the loop during one revolution.

32–24 Suppose the loop in Fig. 32–21 is

a) rotated about the z-axis;

b) rotated about the x-axis;

c) rotated about an edge parallel to the y-axis.

What is the maximum induced emf in each case, if the numerical parameters are as given in Problem 32–23?

32–25 A flexible circular loop 0.10 m in diameter lies in a magnetic field of magnitude 1.2 T, directed into the plane of the diagram in Fig. 32–22. The loop is pulled at the points indicated by the arrows, forming a loop of zero area in 0.2 s.

a) Find the average induced emf in the circuit.

b) What is the direction of the current in R?

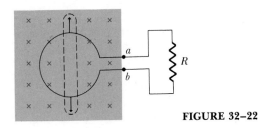

FIGURE 32–22

32–26 The current in the wire AB of Fig. 32–23 is upward and increasing steadily at a rate di/dt.

a) At an instant when the current is i, what are the magnitude and direction of the field B at a distance r from the wire?

b) What is the flux $d\Phi$ through the narrow shaded strip?

c) What is the total flux through the loop?

d) What is the induced emf in the loop?

e) Evaluate the numerical value of the induced emf if $a = 0.10$ m, $b = 0.30$ m, $l = 0.20$ m, and $di/dt = 2$ A·s^{-1}.

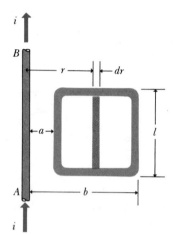

FIGURE 32–23

32–27 The magnetic field B at all points within a circular region of radius R is uniform in space and directed into the plane of Fig. 32–24. If the magnetic field is increasing at a rate dB/dt, what are the magnitude and direction of the force on a stationary point charge of positive charge q located at points a, b, and c? Point a is a distance r above the center of the region, point b is a distance r to the right of the center, and point c is at the center of the region.

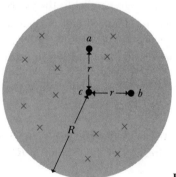

FIGURE 32–24

32–28 A circular conducting ring of radius $r_0 = 0.5$ m is oriented in the xy-plane and is in a region of magnetic field B, where

$$B = B_0[1 - 3(t/t_0)^2 + 2(t/t_0)^3]k.$$

Assume $t_0 = 0.01$ s and is constant, r is the distance of an arbitrary point from the center of the ring, t is the time,

and k is a unit vector in the positive z-direction. Also, $B_0 = 8 \times 10^{-2}$ T and is constant. At points a and b (see Fig. 32–25) there is a small gap in the ring with wires leading to an external circuit of resistance $R = 12\ \Omega$. There is no magnetic field at the location of the external circuits.

a) Derive an expression, as a function of time, for the total magnetic flux Φ enclosed by the ring.

b) Determine the emf induced in the ring at time $t = 0.005$ s. What is the polarity?

c) Because of the internal resistance of the ring, the current through R at the time given in part (b) is only 0.3 A. Determine the internal resistance of the ring.

d) Determine the emf in the ring at a time $t = 0.0121$ s. What is the polarity?

e) Determine the time at which the current through R reverses its direction.

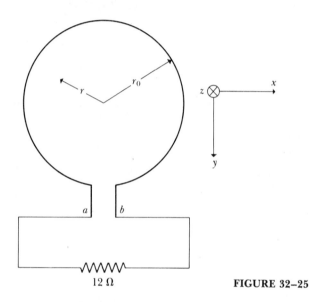

FIGURE 32–25

32–29 A search coil used to measure magnetic fields is to be made with a radius of 2 cm. It is to be designed so that flipping it 180° in a field of 0.1 T causes a total charge of 1×10^{-4} C to flow in a charge-measuring instrument when the total circuit resistance is 50 Ω. How many turns should the coil have?

32–30 A solenoid 0.50 m long and 0.08 m in diameter is wound with 500 turns. A closely wound coil of 20 turns of insulated wire surrounds the solenoid at its midpoint, and the terminals of the coil are connected to a charge-measuring instrument. The total circuit resistance is 25 Ω.

a) Find the quantity of charge displaced through the instrument when the current in the solenoid is quickly decreased from 3 A to 1 A.

b) Draw a sketch of the apparatus, showing clearly the directions of the windings of the solenoid and coil, and of the current in the solenoid. What is the direction of the current in the coil when the solenoid current is decreasing?

32–31 The long, straight wire in Fig. 32–26a carries constant current I. A metal bar of length l is moving at constant velocity v, as shown in the figure. Point a is a distance d from the wire.

a) Calculate the emf induced in the rod.

b) Which point, a or b, is at higher potential?

c) If the bar is replaced by a rectangular wire loop of resistance R, as shown in Fig. 32–26b, what is the magnitude of the current induced in the loop?

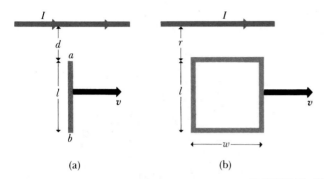

(a)   (b)

FIGURE 32–26

32–32 A rectangular loop of width a and slide wire of mass m are as shown in Fig. 32–27. A uniform magnetic field $\boldsymbol{B}$ is directed perpendicular to the plane of the loop, into the plane of the figure. The slide wire is given an initial velocity of v_0 and then released. Assume that there is no friction between the slide wire and the loop and that the resistance of the loop is negligible compared to the resistance R of the slide wire.

a) Obtain an expression for F, the magnitude of the force exerted on the wire while it is moving at speed v.

b) Show that the distance x which the wire moves before coming to rest is given by $x = mv_0 R/(a^2 B^2)$.

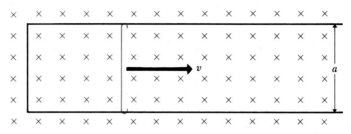

FIGURE 32–27

32–33 A circular ring of radius $a = 20$ cm is rotated about its vertical diameter with a constant angular velocity $\omega = 1000$ rad·s^{-1}. At time $t = 0$ the plane of the ring is perpendicular to a magnetic field that has a fixed direction into the page and whose magnitude varies with time according to the relation

$$B = B_0 e^{-t/\tau},$$

where t is the time, $B_0 = 0.1$ T and $\tau = 0.02$ s. The resistance R of the ring is 0.01 Ω.

a) Obtain an expression for the magnetic flux Φ enclosed by the ring as an explicit function of time.

b) Using Faraday's law, derive an expression for the induced emf $\mathcal{E}$ as an explicit function of time.

c) Determine the current I at $t = 0$.

d) Determine the first time t_0 at which $\mathcal{E} = 0$. (Be careful; the answer is not π/ω.)

e) Determine the first time t_m at which $\mathcal{E}$ becomes a maximum and determine its maximum value.

CHALLENGE PROBLEMS

32–34 A metal bar of length l, mass m, and resistance R is placed on frictionless metal rails inclined at an angle ϕ above the horizontal. The rails have negligible resistance. There is a uniform magnetic field of magnitude B directed downward in Fig. 32–28. The bar is released from rest and slides down the rails.

a) Is the direction of the current induced in the bar from a to b, or from b to a?

b) What is the terminal velocity of the bar?

c) What is the induced current in the bar when the terminal velocity has been reached?

d) After the terminal velocity has been reached, at what rate is electrical energy being converted to heat in the resistance of the bar?

e) After the terminal velocity has been reached, at what rate is work being done on the bar by gravity? Compare your answer to that in (d).

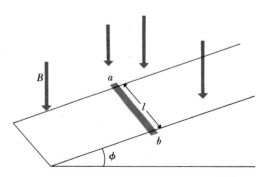

FIGURE 32–28

32–35 A thin rod of length $l = 1$ m and mass $m = 0.1$ kg is free to rotate in a horizontal plane about one of its ends. The other end rests on a frictionless circular rail whose resistance per unit length λ is 2×10^{-2} Ω·m^{-1}. There is a narrow gap in the rail at point a; this gap has infinite resistance. A flexible wire of negligible resistance connects the end of the rod at the pivot and point a of the rail. The rod has negligible resistance. The rod and rail are in the region of a uniform magnetic field whose magnitude is 0.3 T and whose direction is perpendicular to the plane of rotation, into the page. (See Fig. 32–29.) The angular position θ of the rod is measured counterclockwise from a line drawn from the pivot to point a. The rod is initially set into motion such that when the rod makes an angle of 90°, the middle of the rod is moving with an initial velocity $v_0 = 20$ m·s^{-1}.

a) Obtain an expression for the current I induced in the rod when its midpoint is moving with a velocity v and its angular position is θ.

b) Obtain an expression for the rate P at which electrical energy is converted into thermal energy when the angular position of the rod is θ.

c) Show that the velocity v of the midpoint of the rod is related to the angular position θ of the rod by the relation

$$v = v_0 - \left[\frac{3B^2l^2}{8m\lambda}\right] \ln (\theta/\theta_0),$$

where θ_0 is the initial angular position of the rod. [Hint: Equate the answer from part [b] to the time rate of decrease of the kinetic energy of the rod.]

d) Calculate θ_f, the angular position of the rod after it comes to rest. (Obtain a numerical value.)

e) Repeat the problem, but this time for a circular rail that is continuous, with no gap. When you redo part (c), you should obtain the relation

$$v = v_0 - \left[\frac{3B^2l^2}{8m\lambda}\right] \ln \left[\frac{\theta(2\pi - \theta_0)}{\theta_0(2\pi - \theta)}\right].$$

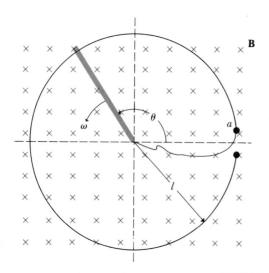

FIGURE 32–29

32–36 A square conducting loop, 20 cm on a side, is placed in the same magnetic field as in Exercise 32–15. (See Fig. 32–30.)

a) Copy Fig. 32–30, and show by vectors the directions and relative magnitudes of the induced electric field $\boldsymbol{E}_n$ at points a, b, and c.

b) Prove that the component of $\boldsymbol{E}_n$ along the loop has the same value at every point of the loop and is equal to that of the ring of Fig. 32–16 (Exercise 32–15).

c) What is the current induced in the loop if its resistance is 2 Ω?

d) What is the potential difference between points a and b?

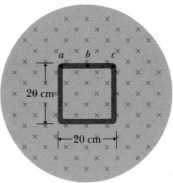

FIGURE 32–30

32–37 A uniform square conducting loop, 20 cm on a side, is placed in the same magnetic field as in Exercise 32–15, with side ac along a diameter and with point b at the center of the field. (See Fig. 32–31.)

a) Copy Fig. 32–31, and show by vectors the directions and relative magnitudes of the induced electric field $\boldsymbol{E}_n$ at the lettered points.

b) What is the induced emf in side ac?

c) What is the induced emf in the loop?

d) What is the current in the loop if its resistance is 2 Ω?

e) What is the potential difference between points a and c? Which is at higher potential?

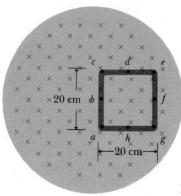

FIGURE 32–31

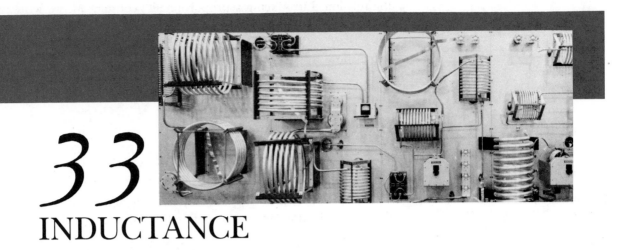

33
INDUCTANCE

THE PHENOMENA OF INDUCTION AND INDUCED EMF, WHICH WE STUDIED IN Chapter 32, have immediate practical applications in a variety of electric-circuit devices, including transformers and inductors. When two coils are adjacent, a changing current in one induces an emf in the other; this is the operating principle of the *transformer,* and the coupling between the coils is characterized by their *mutual inductance.* A changing current in a single coil also causes an induced emf in that same coil, and the relationship of current to emf is described by the *self-inductance* of the coil. A study of energy relationships in inductors leads to the concept of energy stored in magnetic fields. We study several simple circuits containing inductors, including one that can undergo electrical oscillations analogous to those of a mechanical harmonic oscillator. This circuit analysis is fundamental to our study of alternating-current circuits in Chapter 34.

33–1 MUTUAL INDUCTANCE

An emf is induced in a stationary circuit whenever the magnetic flux through the circuit varies with time. If the variation in flux is brought about by a varying current in a second circuit, it is convenient to express the induced emf in terms of the varying *current,* rather than in terms of the varying *flux.* We will use the symbol i to represent the instantaneous value of a varying current.

Figure 33–1 is a cross-sectional view of two closely wound coils of wire. A current i_1 in coil 1 sets up a magnetic field as indicated by the color lines, and some of these lines pass through coil 2. Let the resulting flux through coil 2 be Φ_2. The magnetic field is proportional to i_1, so Φ_2 is also proportional to i_1. When i_1 changes, Φ_2 changes; this changing flux induces an emf $\mathcal{E}_2$ in coil 2, given by

$$\mathcal{E}_2 = N_2 \frac{d\Phi_2}{dt}. \tag{33–1}$$

The proportionality of Φ_2 and i_1 could be represented in the form $\Phi_2 =$ (constant) i_1, but it is more convenient to include the number of turns N_2 in

A changing current in one coil induces an emf in a neighboring coil.

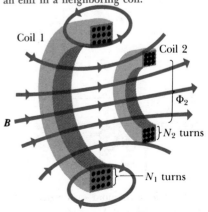

33–1 A portion of the flux set up by a current in coil 1 links with coil 2.

767

the relation. Introducing a proportionality constant M, we write

$$N_2\Phi_2 = Mi_1. \tag{33–2}$$

From this,

$$N_2\frac{d\Phi_2}{dt} = M\frac{di_1}{dt},$$

and we can rewrite Eq. (33–1) as

$$\mathcal{E}_2 = M\frac{di_1}{dt}. \tag{33–3}$$

Mutual inductance: the relation between changing current and induced emf

The constant M, which depends only on the geometry of the two coils, is called their **mutual inductance.** It is defined by Eq. (33–2), which may be written

$$M = \frac{N_2\Phi_2}{i_1}. \tag{33–4}$$

The entire discussion can be repeated for the case where a changing current i_2 in coil 2 causes a changing flux Φ_1 and hence an emf $\mathcal{E}_1$ in coil 1. We might expect that the constant M would be different in this case, since the two coils are not, in general, symmetric. It turns out, however, that M is always the same in this case as in the case considered above, so *a single value of the mutual inductance characterizes completely the induced-emf interaction of two coils.*

The henry: the SI unit of inductance

The SI unit of mutual inductance, from Eq. (33–4), is one *weber per ampere.* An equivalent unit, obtained by reference to Eq. (33–3), is one *volt-second per ampere.* These two equivalent units are called one **henry** (1 H), in honor of Joseph Henry (1797–1878), one of the discoverers of electromagnetic induction. Thus the unit of mutual inductance is

$$1\ \text{H} = 1\ \text{Wb·A}^{-1} = 1\ \text{V·s·A}^{-1}.$$

A simple example of a mutual-inductance calculation

EXAMPLE 33–1   A long solenoid of length l and cross-sectional area A is closely wound with N_1 turns of wire. A small coil of N_2 turns surrounds it at its center, as in Fig. 33–2. A current i_1 in the solenoid sets up a *B* field at its center; from Eq. (31–28), the magnitude of *B* is

$$B = \mu_0 n i_1 = \frac{\mu_0 N_1 i_1}{l}.$$

The flux through the central section equals BA, and since all of this flux links with the small coil, the mutual inductance is

$$M = \frac{N_2\Phi_2}{i_1} = \frac{N_2}{i_1}\left(\frac{\mu_0 N_1 i_1}{l}\right)A = \frac{\mu_0 A N_1 N_2}{l}.$$

If $l = 0.50$ m, $A = 10$ cm^2 $= 10^{-3}$ m^2, $N_1 = 1000$ turns, $N_2 = 10$ turns,

$$M = \frac{(4\pi \times 10^{-7}\ \text{Wb·A}^{-1}\text{·m}^{-1})(10^{-3}\text{m}^2)(1000)(10)}{0.5\ \text{m}}$$

$$= 25.1 \times 10^{-6}\ \text{Wb·A}^{-1}$$

$$= 25.1 \times 10^{-6}\ \text{H} = 25.1\ \mu\text{H}.$$

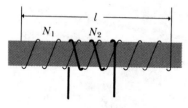

33–2 A long solenoid with cross-sectional area A and N_1 turns, surrounded by a small coil with N_2 turns.

33–2 SELF-INDUCTANCE

In discussing mutual inductance, we assumed that one circuit acted as the source of magnetic field and that the emf under consideration was induced in a separate independent circuit linking some of the flux created by the first circuit. But whenever a current is present in any circuit, this current sets up a magnetic field that links with the *same* circuit and varies when the current varies. Hence any circuit with a varying current has an induced emf in it resulting from the variation in *its own* magnetic field. Such an emf is called a **self-induced electromotive force (emf)**.

As an example, suppose a circuit contains a coil with N turns of wire, as in Fig. 33–3, and that a flux Φ passes through each turn. In analogy to Eq. (33–4), we define the **self-inductance** L of the circuit, or simply its **inductance**:

$$L = \frac{N\Phi}{i}. \qquad (33\text{–}5)$$

This relation can be written as

$$N\Phi = Li.$$

If Φ and i change with time, then

$$N\frac{d\Phi}{dt} = L\frac{di}{dt},$$

and, since the self-induced emf $\mathcal{E}$ has magnitude

$$\mathcal{E} = N\frac{d\Phi}{dt},$$

it follows that

$$\mathcal{E} = L\frac{di}{dt}. \qquad (33\text{–}6)$$

The self-inductance of a circuit is therefore *the self-induced emf per unit rate of change of current.* The SI unit of self-inductance is one henry.

A circuit, or part of a circuit, that has inductance is called an **inductor.** An inductor is represented by the symbol

The direction of a self-induced (non-electrostatic) field can be found from Lenz's law. We consider the "cause" of this field, and hence of the emf associated with it, to be the *changing current* in the conductor. If the current is increasing, the direction of the induced field is *opposite* to that of the current. If the current is *decreasing,* the induced field is in the *same* direction as the current. Thus it is the *change* in current, not the current itself, that is "opposed" by the induced field.

EXAMPLE 33–2 An air-core toroidal solenoid of cross-sectional area A and mean radius r is closely wound with N turns of wire. We neglect the variation of B across the section, assuming its average value to be very nearly equal to the value at the center of the cross section, as given by Eq. (31–29). Then the flux in the

A changing current in a coil induces an emf in that coil: pulling yourself up by your bootstraps.

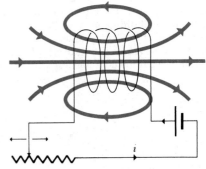

33–3 A flux Φ linking a coil of N turns. When the current in the circuit changes, the flux changes also, and a self-induced emf appears in the circuit.

Self-inductance: the ratio of induced emf to rate of change of current

The sign of the induced emf in an inductor

Self-inductance in a toroidal solenoid

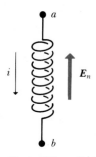

33–4 When di/dt is positive, the non-electrostatic induced field $\mathbf{E}_n$ is in the direction shown, and the emf is treated as a potential difference with V_{ab} positive; that is, a is at higher potential than b. When di/dt is negative, $\mathbf{E}_n$ has the opposite direction, and V_{ab} is negative.

toroid is

$$\Phi = BA = \frac{\mu_0 N i A}{2\pi r}.$$

Since all of the flux links with each turn, the self-inductance is

$$L = \frac{N\Phi}{i} = \frac{\mu_0 N^2 A}{2\pi r}.$$

Thus if $N = 100$ turns, $A = 10 \text{ cm}^2 = 10^{-3}\text{m}^2$, $r = 0.10$ m,

$$L = \frac{(4\pi \times 10^{-7}\text{ Wb·A}^{-1}\text{·m}^{-1})(100)^2(10^{-3}\text{m}^2)}{2\pi(0.10\text{ m})}$$

$$= 20 \times 10^{-6}\text{ H} = 20\ \mu\text{H}.$$

EXAMPLE 33–3 If the current in the coil above increases uniformly from zero to 1 A in 0.1 s, find the magnitude and direction of the self-induced emf.

SOLUTION

$$\mathcal{E} = L\frac{di}{dt} = (20 \times 10^{-6}\text{ H})\frac{1\text{ A}}{0.1\text{ s}}$$

$$= 2.0 \times 10^{-4}\text{ V}.$$

The current is increasing, so according to Lenz's law the direction of the emf is opposite to that of the current. In Fig. 33–4, suppose the inductor terminals are a and b and there is an *increasing* current from a to b in the inductor. Then the induced field $\mathbf{E}_n$ and the emf are in the direction from b to a, like a battery with a as the $(+)$ terminal and b as the $(-)$ terminal.

Sign rules for using Kirchhoff's rules with inductors

When Kirchhoff's loop rule (Section 29–2) is used with circuits containing inductors, this self-induced emf is treated as though it were a potential difference, with a at higher potential than b. Thus if we define the positive direction of the current i as from a to b in the inductor, then

$$V_{ab} = L\frac{di}{dt}. \tag{33–7}$$

The self-inductance of a circuit depends on its size, shape, number of turns, and so on. It also depends on the magnetic properties of the material enclosed by the circuit.

PROBLEM-SOLVING STRATEGY: *Self-inductance*

1. In this chapter we view the inductor primarily as a *circuit* device, so circuit analysis is of primary importance. In general all the voltages, currents, and capacitor charges are now functions of time, not constants as they have been in most of our previous analyses. But Kirchhoff's rules, which we studied in Chapter 29, are still valid. When the voltages and currents vary with time, Kirchhoff's loop rule is a relationship that holds at each instant of time, and it often gives us *differential equations* to solve, rather than just algebraic equations. This means that the solutions will depend on initial conditions as well as on the constants that describe the circuits.

2. As in all circuit analysis, getting the signs right is often more challenging than understanding the principles. We suggest you review the strategy in Section 29–2 as preparation for study of the circuits in Sections 33–4 through 33–6. In addition, give close attention to the sign rule described with Eq. (33–7). Then look for the applications of Kirchhoff's loop rule in these discussions.

In the examples above we calculated the magnetic field by assuming that the conductor was surrounded by vacuum. If matter is present, then when we calculate the field we must replace the constant μ_0, the permeability of vacuum, by the permeability of the material, $\mu = K_m\mu_0$, as discussed in Section 31–8. If the material is diamagnetic or paramagnetic, this makes very little difference. If the material is *ferromagnetic*, however, the difference is of crucial importance. An inductor wound on a soft iron core having $K_m = 5000$ has an inductance approximately 5000 times as great as the same coil with an air core. Iron-core inductors are very widely used in a variety of electronic and electric-power applications. An added complication is that with ferromagnetic materials, the magnetization is not always a linear function of magnetizing current, especially as saturation is approached. As a result, the inductance can depend on current in a fairly complicated way. In our discussion we will ignore this complication and assume the inductance to be constant. This is a reasonable assumption even for a ferromagnetic material if the magnetization remains well below the saturation level.

A magnetic material multiplies the inductance by K_m.

33–3 ENERGY IN AN INDUCTOR

A changing current in an inductor causes an emf; while the current is changing, the source supplying the current must maintain a potential difference between its terminals and therefore must supply energy to the inductor. Let us calculate the total energy input needed to establish a final current I in an inductor of inductance L if the initial current is zero.

If the current at some instant is i and is changing at the rate di/dt, the induced emf at that instant is $\mathcal{E} = L\,di/dt$, and the instantaneous power P supplied by the current source is

To establish a current in an inductor, the circuit must supply energy to it.

$$P = \mathcal{E}i = Li\frac{di}{dt}.$$

The energy dW supplied in time dt is $P\,dt$, or

$$dW = Li\,di,$$

and the total energy supplied while the current increases from zero to a final value I is

$$W = L\int_0^I i\,di = \frac{1}{2}LI^2. \tag{33–8}$$

After the current has reached its final steady value, $di/dt = 0$ and the power input is zero. The energy that has been supplied to the inductor is needed to establish the magnetic field in and around the inductor. We can think of this as a *potential energy* associated with the current. We define this energy to be zero when there is no current; then when the current is I, the potential energy has the value $\frac{1}{2}LI^2$. In the discussion of *mechanical* potential energy in Chapter 7, we defined the potential energy of a system in a certain position as the work done by the forces of the system when it returns from this position to the position where the potential energy is defined to be zero. Our inductor's energy is consistent with this viewpoint; when the inductor current decreases from an initial value I to the reference value $I = 0$, the inductor does an amount of electrical work on the external circuit equal to $\frac{1}{2}LI^2$. If the current in the circuit is interrupted suddenly by opening a switch, the energy may be dissipated in an arc across the switch contacts.

The energy stored in an inductor is proportional to the square of the current.

Stored energy in an inductor is associated with the magnetic field produced by the current.

We can also consider the energy to be associated with the magnetic field itself, and we can develop a relation analogous to the one obtained for electric-field energy in Section 27–4, Eqs. (27–9) and (27–18). As we did then, we consider here only one simple case, the toroidal solenoid; this system has the advantage that its magnetic field is confined completely to a finite region of space in its interior. As in Example 33–2 (Section 33–2), we assume the cross-sectional area A is small enough that we can consider the magnetic field to be uniform over the area. The volume in the toroid is then approximately equal to the circumferential length $l = 2\pi r$ multiplied by the area A. From the preceding example, the self-inductance of the toroidal solenoid is

$$L = \frac{\mu_0 N^2 A}{l}.$$

When the current in the windings is I, the energy stored in the toroid is

$$W = \frac{1}{2}LI^2 = \frac{1}{2}\left(\frac{\mu_0 N^2 A}{l}\right)I^2.$$

We can think of this energy as localized in the volume enclosed by the windings, equal to lA. The energy *per unit volume*, or **energy density** u, is then

$$u = \frac{W}{lA} = \frac{1}{2}\mu_0\left(\frac{N^2 I^2}{l^2}\right).$$

We can express this relation in terms of the magnetic field B inside the toroid. From Eq. (31–29),

$$B = \frac{\mu_0 NI}{2\pi r} = \frac{\mu_0 NI}{l}. \tag{33–9}$$

Energy density in a magnetic field

Squaring this and rearranging, we find

$$\frac{N^2 I^2}{l^2} = \frac{B^2}{\mu_0{}^2}.$$

When we substitute this expression into the above equation, we finally find

$$u = \frac{B^2}{2\mu_0}. \tag{33–10}$$

This is the analog of the expression for the energy per unit volume in the electric field of an air capacitor, $\frac{1}{2}\epsilon_0 E^2$, derived in Section 27–4. We will need these expressions for electric- and magnetic-field energy when we study energy associated with electromagnetic waves in Chapter 35.

When a magnetic material is present, the energy density is smaller by a factor of K_m.

When the material inside the toroid is not vacuum but a material having magnetic permeability $\mu = K_m \mu_0$, we have to replace μ_0 by μ in Eq. (31–29). The energy per unit volume in the magnetic field is then

$$u = \frac{B^2}{2\mu}. \tag{33–11}$$

33–4 THE *R–L* CIRCUIT

Time-varying current in a circuit containing a resistor and an inductor

In our discussion of inductors thus far we have neglected the *resistance* of the windings. But every inductor found in the real world must have some resistance, unless its windings are superconducting. We can represent a real-life inductor as an ideal zero-resistance inductor with self-inductance L, in series with a resistance R, as shown in Fig. 33–5. This diagram can also represent a resistor in series with an inductor; in that case R is the *total* resistance of the combination.

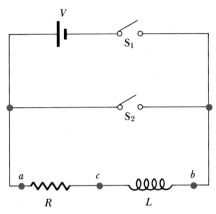

33–5 An *R–L* series circuit.

The circuit of Fig. 33–5 shows some interesting behavior. By means of switch S_1, the *R–L* combination can be connected to a source with constant terminal voltage V, or it may be "shorted" by closing switch S_2. Suppose both switches are initially open, and then at some initial time $t = 0$ switch S_1 is closed. Because of the self-induced emf in the inductor, the current does not immediately rise to its final value at the instant S_1 is closed. Instead, it grows at a rate that depends on the values of R and L in the circuit.

Let i represent the instantaneous current at some time t after the switch is closed; then di/dt is its rate of increase at that time. The potential difference across the inductor is

$$v_{cb} = L\frac{di}{dt},$$

and the potential difference across the resistor is

$$v_{ac} = iR.$$

Since $V = v_{ac} + v_{cb}$, it follows that

$$V = L\frac{di}{dt} + iR. \tag{33–12}$$

The rate of increase of current is therefore

$$\frac{di}{dt} = \frac{V - iR}{L} = \frac{V}{L} - \frac{R}{L}i. \tag{33–13}$$

Applying Kirchhoff's loop rule to the circuit

At the instant the circuit is first closed, $i = 0$ and the current starts to grow at the rate

$$\left(\frac{di}{dt}\right)_{\text{initial}} = \frac{V}{L}.$$

The greater the self-inductance L, the more slowly the current increases.

As the current increases, the term Ri/L increases also, and hence the *rate* of increase of current becomes smaller and smaller, as Eq. (33–13) shows. When the current reaches its final *steady-state* value I, its rate of increase is zero. Then

$$0 = \frac{V}{L} - \left(\frac{R}{L}\right)I$$

and

$$I = \frac{V}{R}.$$

That is, the *final* current does not depend on the self-inductance; it is the same as it would be in a pure resistance R connected to a source having emf V.

To obtain an expression for the current as a function of time, we proceed just as we did for the problem of the charging capacitor in Section 29–4. We

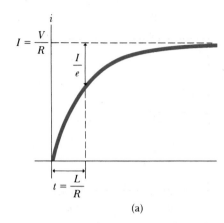

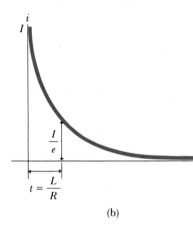

33-6 (a) Growth of current in a circuit containing inductance and resistance. (b) Decay of current in a circuit containing inductance and resistance.

When the inductance is large, the current in the circuit grows slowly.

Energy relations in an R–L circuit

first rearrange Eq. (33–13):

$$\frac{di}{(V/R) - i} = \frac{R}{L}dt.$$

Integrating both sides, we find

$$-\ln\left(\frac{V}{R} - i\right) = \left(\frac{R}{L}\right)t + \text{constant}.$$

The integration constant is evaluated by noting that the initial current is zero, so $i = 0$ at time $t = 0$:

$$\text{constant} = -\ln\frac{V}{R}.$$

Rearranging again, we obtain

$$\ln\left(\frac{V}{R} - i\right) - \ln\frac{V}{R} = \ln\left(1 - \frac{Ri}{V}\right) = -\left(\frac{R}{L}\right)t, \qquad (33\text{–}14)$$

$$i = \frac{V}{R}(1 - e^{-(R/L)t}).$$

From this,

$$\frac{di}{dt} = \frac{V}{L}e^{-(R/L)t}. \qquad (33\text{–}15)$$

We note that at time $t = 0$, $i = 0$ and $di/dt = V/L$, and that as $t \to \infty$, $i \to V/R$ and $di/dt \to 0$, as predicted above.

Figure 33–6a is a graph of Eq. (33–14) and shows the variation of current with time. The instantaneous current i first rises rapidly, then increases more slowly and approaches asymptotically the final value $I = V/R$. At a time τ equal to L/R, the current has risen to $(1 - 1/e)$ or about 0.63 of its final value. This time is called the **time constant** or the *decay constant* for the circuit:

$$\tau = \frac{L}{R}. \qquad (33\text{–}16)$$

For a given value of R, the time constant increases when the inductance L increases. Thus, although the graph of i versus t has the same general shape whatever the inductance, the current rises rapidly to its final value if L is small and slowly if L is large. For example, if $R = 100\ \Omega$ and $L = 10$ H,

$$\frac{L}{R} = \frac{10\text{ H}}{100\Omega} = 0.1\text{ s},$$

and the current increases to about 63% of its final value in 0.1 s. But if $L = 0.01$ H,

$$\frac{L}{R} = \frac{0.01\text{ H}}{100\Omega} = 10^{-4}\text{ s},$$

and only 10^{-4} s is required for the current to increase to 63% of its final value.

Energy considerations offer us additional insight into the behavior of an R–L circuit. The instantaneous rate at which the source delivers energy to the circuit is $P = Vi$. The instantaneous rate at which energy is dissipated in the resistor is i^2R, and the rate at which energy is stored in the inductor is $iv_{cb} = Li\,di/dt$. Multiplying Eq. (33–12) by i, we find

$$Vi = Li\frac{di}{dt} + i^2R. \qquad (33\text{--}17)$$

This means that part of the power Vi supplied by the source is dissipated (i^2R) in the resistor, and part is stored ($Li\,di/dt$) in the inductor. This discussion is completely analogous to our power analysis for a charging capacitor at the end of Section 29–4.

Now suppose switch S_1 in the circuit of Fig. 33–5 has been closed for a long time, so that the final current $I = V/R$ has been reached. Redefining our initial time, we close switch S_2 at time $t = 0$, short-circuiting the battery. (We can then open S_1 to save the battery from ruin.) The current through R and L does not instantaneously go to zero but decays smoothly, as shown in Fig. 33–6b. We challenge you to follow the same pattern as in the analysis above to show that the current varies with time according to

When the battery is short-circuited, the current doesn't suddenly drop to zero.

$$i = Ie^{-Rt/L}. \qquad (33\text{--}18)$$

The time constant, $\tau = L/R$, is the time for the current to decrease to $1/e$, or about 37% of its original value. The energy needed to maintain the current during this decay is provided by the energy stored in the magnetic field of the inductor. The detailed energy analysis is simpler this time. In place of Eq. (33–12) we have

$$0 = L\frac{di}{dt} + iR.$$

Multiplying through by i, we find

$$0 = Li\frac{di}{dt} + i^2R. \qquad (33\text{--}19)$$

In this case, of course, $Li\,di/dt$ is negative, and this equation shows that the rate of *decrease* of energy stored in the inductor is equal to the rate of dissipation of energy, i^2R, in the resistor.

Energy relations for decaying current in an *R–L* circuit

33–5 THE *L–C* CIRCUIT

We studied the behavior of an *R–C* circuit in Section 29–4 and that of an *R–L* circuit in Section 33–4. In both cases the behavior is characterized by an exponential approach to some steady-state situation. When a circuit contains *both* inductance L and capacitance C, entirely new modes of behavior appear, characterized by *oscillating* current and charge.

Consider first the *L–C* circuit in Fig. 33–7, containing an ideal resistanceless inductor and a capacitor. We charge the capacitor to a potential difference V_m, as shown in Fig. 33–7a, and then close the switch. The capacitor immediately starts to discharge through the inductor. Figure 33–7b shows the situation later when the capacitor has completely discharged and the potential difference between its terminals (and those of the inductor) has decreased to zero. During this discharge, the current in the inductor has established a magnetic field in the space around it. This magnetic field now decreases, inducing an emf in the inductor in the same direction as the cur-

The *L–C* circuit: an electrical system that vibrates

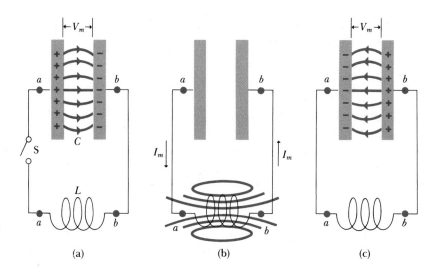

33-7 Energy transfer between electric and magnetic fields in an oscillating $L–C$ circuit.

(a)   (b)   (c)

rent. The current therefore persists, although with decreasing magnitude, until the magnetic field has disappeared and the capacitor has been charged in the *opposite* sense to its initial polarity, as in Fig. 33–7c. The process now repeats itself in the reverse direction. In the absence of energy losses, the charges on the capacitor surge back and forth indefinitely. This process is called an **electrical oscillation.**

From the energy standpoint, the oscillations of an electrical circuit consist of a transfer of energy back and forth from the electric field of the capacitor to the magnetic field of the inductor. The *total* energy associated with the circuit remains constant. This is analogous to the transfer of energy in an oscillating mechanical system from kinetic to potential, and vice versa, with the total energy remaining constant.

The frequency of the electrical oscillations of a circuit containing inductance and capacitance only (a so-called $L–C$ circuit) may be calculated in exactly the same way as the frequency of oscillation of a body attached to a spring, discussed in Chapter 11. We suggest you review that discussion before going on. In the mechanical problem, a body of mass m is attached to a spring of force constant k. Suppose we displace the body a distance A from its equilibrium position and release it from rest at time $t = 0$. Then, as shown in the left column of Table 33–1, the kinetic energy of the system at any later time is

An electrical version of the harmonic oscillator

TABLE 33–1 Oscillation of a Mass on a Spring Compared with the Electrical Oscillation in an $L–C$ Circuit

Mass on a Spring	Circuit Containing Inductance and Capacitance
Kinetic energy $= \frac{1}{2}mv^2$	Magnetic energy $= \frac{1}{2}Li^2$
Potential energy $= \frac{1}{2}kx^2$	Electrical energy $\dfrac{q^2}{2C}$
$\frac{1}{2}mv^2 + \frac{1}{2}kx^2 = \frac{1}{2}kA^2$	$\dfrac{1}{2}Li^2 + \dfrac{q^2}{2C} = \dfrac{Q^2}{2C}$
$v = \pm\sqrt{k/m}\ \sqrt{A^2-x^2}$	$i = \pm\sqrt{1/LC}\ \sqrt{Q^2-q^2}$
$v = \dfrac{dx}{dt}$	$i = \dfrac{dq}{dt}$
$x = A\cos\sqrt{k/m}\ t = A\cos\omega t$	$q = Q\cos\sqrt{1/LC}\ t = Q\cos\omega t$
$v = -\omega A\sin\omega t = -v_{max}\sin\omega t$	$i = -\omega Q\sin\omega t = -I\sin\omega t$

$\frac{1}{2}mv^2$, and its elastic potential energy is $\frac{1}{2}kx^2$. Because the system is conservative, the sum of these values equals the initial energy of the system, $\frac{1}{2}kA^2$. The velocity v at any position is obtained just as in Section 11–2, Eq. (11–4):

$$v = \pm\sqrt{\frac{k}{m}}\ \sqrt{A^2-x^2}. \qquad (33\text{–}20)$$

The velocity v equals dx/dt, and we found in Section 11–2 that the coordinate x as a function of t is

$$x = A\cos\left(\sqrt{\frac{k}{m}}\right)t = A\cos\omega t, \qquad (33\text{–}21)$$

where the angular frequency ω is

$$\omega = \sqrt{\frac{k}{m}}.$$

We recall also that the ordinary frequency f, the number of cycles per unit time, is given by $f = \omega/2\pi$.

In the electrical problem, also a conservative system, a capacitor of capacitance C is given an initial charge Q and, at time $t = 0$, is connected to the terminals of an inductor of self-inductance L. The magnetic energy of the inductor at any later time is analogous to the kinetic energy of the vibrating body and is given by $\frac{1}{2}Li^2$. The electrical energy of the capacitor corresponds to the elastic potential energy of the spring and is given by $q^2/2C$, where q is the charge on the capacitor. The sum of these equals the initial energy of the system, $Q^2/2C$. That is,

$$\frac{1}{2}Li^2 + \frac{q^2}{2C} = \frac{Q^2}{2C}.$$

Energy relations in the L–C circuit

Solving for i, we find that when the charge on the capacitor is q, the current i is

$$i = \pm\sqrt{\frac{1}{LC}}\ \sqrt{Q^2 - q^2}. \qquad (33\text{–}22)$$

Comparing this expression with Eq. (33–20), we see that the current $i = dq/dt$ varies with time in the same way as the velocity $v = dx/dt$ in the mechanical problem. Continuing the analogy, we conclude that q is given as a function of time by

$$q = Q\cos\left(\sqrt{\frac{1}{LC}}\right)t = Q\cos\omega t. \qquad (33\text{–}23)$$

The angular frequency ω of the electrical oscillations is therefore

$$\omega = \sqrt{\frac{1}{LC}}. \qquad (33\text{–}24)$$

This frequency is called the *natural frequency* of the *L–C* circuit. As Table 33–1 shows, it is analogous to the equation $\omega = \sqrt{k/m}$ for the angular frequency of a harmonic oscillator.

How to find the frequency of oscillation of an L–C circuit

These results may also be derived directly from an analysis of the *L–C* circuit. At each instant the capacitor voltage $v_{ab} = q/C$ must equal that of the inductor, $v_{ab} = L\,di/dt$. Also, because of the choice of positive direction for i, we have $i = -dq/dt$. Combining these relations, we find

$$\frac{d^2q}{dt^2} + \frac{1}{LC} q = 0, \tag{33–25}$$

which is analogous to Eq. (11–15) for the harmonic oscillator. The solutions of this differential equation, functions whose second derivative is equal to $-1/LC$ times the original function, are

$$q = Q \cos \omega t, \tag{33–26a}$$

$$q = Q \sin \omega t, \tag{33–26b}$$

$$q = Q \cos (\omega t + \phi), \tag{33–26c}$$

where $\omega = 1/\sqrt{LC}$, and Q and ϕ are constants. Just as with the harmonic oscillator, the choice of one of these functions is determined by the initial conditions. If at time $t = 0$, the capacitor has maximum charge and $i = 0$, as in the discussion above, then we use Eq. (33–26a). If at $t = 0$, $q = 0$ but i is different from zero, we use Eq. (33–26b). And if both q and i are different from zero at time $t = 0$, the more general form, Eq. (33–26c), must be used.

The striking parallel between the mechanical and electrical systems shown in Table 33–1 is only one of many such examples in physics. So close is the parallel between electrical and mechanical (and acoustical) systems that it is possible to solve complicated mechanical and acoustical problems by setting up analogous electrical circuits and measuring the currents and voltages that correspond to the mechanical and acoustical quantities to be determined. This is the basic principle of one kind of *analog computer*. We see that in our comparison between the harmonic oscillator and the L–C circuit, m corresponds to L, k to $(1/C)$, x to q, and v to i. This analogy can be extended to *damped oscillations*, which we consider in the next section.

Analogs between mechanical and electrical systems are used in analog computers.

33–6 THE L–R–C CIRCUIT

In our discussion of the L–C circuit we did not include any *resistance*. This omission is an idealization, of course; every real inductor has resistance in its windings, and there may also be resistance in the connecting wires. The effect of resistance is to dissipate the electromagnetic energy in the circuit and convert it to heat; thus resistance in an electric circuit plays a role analogous to that of friction in a mechanical system.

The L–R–C circuit: the electrical analog of damped harmonic motion

To study this situation in greater detail, we consider an inductor with inductance L and a resistor of resistance R connected in series across the terminals of a charged capacitor. As before, the capacitor starts to discharge as soon as the circuit is completed, but because of i^2R losses in the resistor, the energy of the inductor when the capacitor is completely discharged is *less* than the original energy of the capacitor. In the same way, the energy of the capacitor when the magnetic field has collapsed is still smaller, and so on.

The oscillations die out because the resistor dissipates energy.

If the resistance R is relatively small, the circuit still oscillates, but with **damped harmonic motion,** as shown in Fig. 33–8a. If we increase R, the oscillations die out more rapidly. When R reaches a certain value, the circuit no longer oscillates, and we say that it is **critically damped,** as in Fig. 33–8b. For still larger values of R the circuit is **overdamped,** as in Fig. 33–8c. You

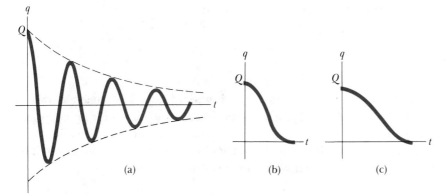

(a)   (b)   (c)

33-8 Graphs of q versus t in an *L–R–C* circuit. (a) Small damping. (b) Critically damped. (c) Overdamped.

may recall that we used these same terms in Section 11–8 for the analogous mechanical situation, the damped harmonic oscillator.

To analyze this behavior in detail, consider the circuit shown in Fig. 33–9. Let us find the current i and the capacitor charge q as functions of time; the analysis is sketched below. First we close the switch in the upward position for a long enough time so that the capacitor acquires its final charge $Q = CV$ and any initial oscillations have died out. Then at time $t = 0$ we flip the switch to the downward position. To find how q and i vary with time, we apply Kirchhoff's loop rule. Starting at point a and going around the loop in the direction *acdba*, we obtain the equation

$$iR + L\frac{di}{dt} + \frac{q}{C} = 0.$$

Replacing i with dq/dt and rearranging, we obtain

$$\frac{d^2q}{dt^2} + \frac{R}{L}\frac{dq}{dt} + \frac{1}{LC}q = 0. \qquad (33\text{–}27)$$

Note that when $R = 0$, this equation reduces to Eq. (33–25).

Solutions of Eq. (33–27) can be obtained by general methods of differential equations. The form of this solution depends on whether R is large or small. When R is less than $2(L/C)^{1/2}$, the solution has the form

$$q = VCe^{-(R/2L)t} \cos\left(\sqrt{\frac{1}{LC} - \left(\frac{R}{2L}\right)^2}\right)t. \qquad (33\text{–}28)$$

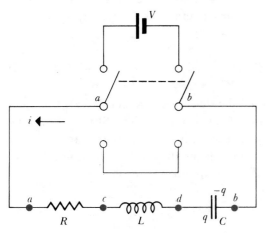

33-9 An *L–R–C* series circuit.

When R is greater than $2(L/C)^{1/2}$, the solution is

$$q = e^{-(R/2L)t}\left[Ae^{\left(\sqrt{\left(\frac{R}{2L}\right)^2 - \frac{1}{LC}}\right)t} + Be^{\left(-\sqrt{\left(\frac{R}{2L}\right)^2 - \frac{1}{LC}}\right)t}\right], \qquad (33\text{--}29)$$

where A and B are constants determined by V, R, L, and C.

The frequency of the damped oscillator is smaller than it would be with no resistance.

Equation (33–28) corresponds to the **underdamped** behavior shown in Fig. 33–8a; the function represents a sinusoidal oscillation with an exponentially decaying amplitude. Note that the angular frequency of the oscillation is no longer $1/(LC)^{1/2}$ but is *less* than this because of the term containing R. The frequency ω' of the damped oscillations is thus given by

$$\omega' = \sqrt{\frac{1}{LC} - \frac{R^2}{4L^2}}. \qquad (33\text{--}30)$$

If the resistance is too large, there is no oscillation at all.

As R increases, ω' becomes smaller and smaller. When $R^2 = 4L/C$, the quantity under the radical becomes zero and the case of *critical damping* has been reached (Fig. 33–8b). For still larger values of R, the behavior is no longer oscillatory but is described as the sum of two exponential functions, as in Fig. 33–8c; the circuit is then *overdamped*.

We emphasize once more that this behavior is completely analogous to that of the damped harmonic oscillator studied in Section 11–8. We invite you to verify, for example, that if you start with Eq. (11–44) and substitute L for m, $1/C$ for k, and R for b, the result is Eq. (33–30). Similarly, the crossover point between underdamping and overdamping occurs at $b^2 = 4km$ for the mechanical system and at $R^2 = 4L/C$ for the electrical one. Can you find still other aspects of this analogy?

There are ways to overcome energy losses due to resistance, and to produce sustained oscillations.

It is possible, with appropriate electronic circuitry, to feed energy *into* an L–R–C circuit at the same rate as the rate of dissipation by i^2R losses. It is as though we had inserted a *negative resistance* into the circuit to make the *total* circuit resistance exactly zero. The circuit then oscillates with sustained oscillations of constant amplitude, just as the idealized L–C circuit with no resistance does.

Additional interesting aspects of the behavior of this circuit emerge when a sinusoidally varying emf is included in the circuit. This leads us to the study of alternating-current (ac) circuit analysis, the principal topic of Chapter 34.

SUMMARY

KEY TERMS
mutual inductance
henry
self-induced electromotive force (emf)
inductance (self-inductance)
inductor
energy density
time constant
electrical oscillation
damped harmonic motion
critically damped
overdamped
underdamped

A changing current i_1 in one circuit induces an emf $\mathcal{E}$ in a second circuit if some of the magnetic flux created by the current links the second circuit. The induced emf is given by

$$\mathcal{E}_2 = M\frac{di_1}{dt}, \qquad (33\text{--}3)$$

where M is a constant called the mutual inductance. It depends on the geometry of the two coils and on the material between them. The SI unit of mutual inductance, the weber per ampere, is called the henry, abbreviated H.

A changing current i in any circuit induces an emf $\mathcal{E}$ in that same circuit, called a self-induced emf. The self-induced emf is given by

$$\mathcal{E} = L\frac{di}{dt}, \qquad (33\text{--}6)$$

where L is a constant depending on the geometry of the circuit and the material surrounding it, called the self-inductance, or simply inductance. A circuit device, usually including a coil of wire, intended to have a substantial inductance is called an inductor.

An inductor with inductance L carrying current I has energy $\frac{1}{2}LI^2$. This energy was fed into the inductor when the current was established, and it is fed back into the circuit when the current decreases to zero again. This energy is associated with the magnetic field of the inductor, with an energy density u (energy per unit volume) given by

$$u = \frac{B^2}{2\mu_0} \tag{33–10}$$

if the field is in vacuum, or

$$u = \frac{B^2}{2\mu} \tag{33–11}$$

if it is in a material having magnetic permeability μ.

In a circuit containing a resistor R, an inductor L, and a source of emf V, the growth and decay of current are exponential, with a characteristic time τ called the time constant, given by

$$\tau = \frac{L}{R}. \tag{33–16}$$

This is the time required for the current to approach within a fraction $1/e$ of its final value.

A circuit containing an inductance L and a capacitance C undergoes electrical oscillations, with angular frequency ω given by

$$\omega = \sqrt{\frac{1}{LC}}. \tag{33–24}$$

Such a circuit is analogous to a mechanical harmonic oscillator, with the mass m analogous to inductance L, the force constant k to the reciprocal of capacitance $1/C$, the displacement x to charge q, and the velocity v to current i.

A series circuit containing inductance, resistance, and capacitance undergoes damped oscillations for sufficiently small resistance. As R increases, the damping increases; at a certain value of R the behavior becomes overdamped and no longer oscillates. The crossover between underdamping and overdamping occurs when

$$R^2 = 4L/C,$$

and the frequency ω' of damped oscillations when R is less than this critical value is

$$\omega' = \sqrt{\frac{1}{LC} - \frac{R^2}{4L^2}}. \tag{33–30}$$

There is a direct analogy between every aspect of the behavior of the L–R–C circuit and the mechanical damped harmonic oscillator. This analogy and similar ones are widely used in analog computers.

QUESTIONS

33–1 A resistor is to be made by winding a wire around a cylindrical form. In order to make the inductance as small as possible, it is proposed that we wind half the wire in one direction and the other half in the opposite direction. Would this achieve the desired result? Why or why not?

33–2 In Fig. 33–1, if coil 2 is turned 90° so that its axis is vertical, does the mutual inductance increase or decrease?

33–3 The toroidal solenoid is one of the few configurations for which it is easy to calculate self-inductance. What features of the toroidal solenoid give it this simplicity?

33–4 Two identical closely wound circular coils, each having self-inductance L, are placed side by side, close together. If they are connected in series, what is the self-inductance of the combination? What if they are connected in parallel? Can they be connected so that the total inductance is zero?

33–5 If two inductors are separated enough so that practically no flux from either links the coils of the other, show that the equivalent inductance of two inductors in series or parallel is obtained by the same rules for combining resistance.

33–6 Two closely wound circular coils have the same number of turns, but one has twice the radius of the other. How are the self-inductances of the two coils related?

33–7 One of the great problems in the field of energy resources and utilization is the difficulty of storing electrical energy in large quantities economically. Discuss the possibility of storing large amounts of energy by means of currents in large inductors.

33–8 In what regions in a toroidal solenoid is the energy density greatest? Least?

33–9 Suppose there is a steady current in an inductor. If one attempts to reduce the current to zero instantaneously by opening a switch, a big fat arc appears at the switch contacts. Why? What happens to the induced emf in this situation? Is it physically possible to stop the current instantaneously?

33–10 In the $R–L$ circuit of Fig. 33–5, is the current in the resistor always the same as that in the inductor? How do you know?

33–11 In the $R–L$ circuit of Fig. 33–5, when switch S_1 is closed, the potential V_{ab} changes suddenly and discontinuously, but the current does not. Why can the voltage change suddenly but not the current?

33–12 In the $R–L–C$ circuit, what criteria could be used to decide whether the system is overdamped or underdamped? For example, could one compare the maximum energy stored during one cycle to the energy dissipated during one cycle?

EXERCISES

Section 33–1 Mutual Inductance

33–1 A solenoid of length 10 cm and radius 2 cm is wound uniformly with 1000 turns. A second coil of 50 turns is wound around the solenoid at its center. What is the mutual inductance of the two coils?

33–2 A toroidal solenoid (cf. Section 31–7) has a radius of 10 cm and a cross-sectional area of 5 cm^2, and is wound uniformly with 1000 turns. A second coil with 500 turns is wound uniformly on top of the first. What is the mutual inductance?

33–3 Two coils have mutual inductance $M = 0.01$ H. The current i_1 in the first coil increases at a uniform rate of 0.05 A·s^{-1}.

a) What is the induced emf in the second coil? Is it constant?

b) Suppose that the current described is in the second coil rather than the first; what is the induced emf in the first coil?

Section 33–2 Self-Inductance

33–4

a) Show that the two expressions for self-inductance, namely,

$$\frac{N\Phi}{i} \quad \text{and} \quad \frac{\mathcal{E}}{di/dt},$$

have the same units.

b) Show that L/R and RC both have units of time.

c) Show that 1 Wb·s^{-1} equals 1 V.

33–5 An inductor of inductance 5 H carries a current that decreases at a uniform rate, $di/dt = -0.02$ A·s^{-1}. Find the self-induced emf; what is its polarity?

33–6 An inductor with $L = 40$ H carries a current i that varies with time according to $i = (0.1 \text{ A}) \sin ([120\pi \text{ s}^{-1}]t)$. Find an expression for the induced emf. What is the phase of $\mathcal{E}$ relative to i?

33–7

a) Find the self-inductance of the toroidal solenoid in Exercise 33–2 if only the 1000-turn coil is used.

b) If both coils are used, connected in series with each other, what is the self-inductance of the combination?

33–8 A solenoid has length 15 cm, radius 4 cm, and 2000 turns. It is filled with a core of relative permeability 600. Calculate the self-inductance of the solenoid.

Section 33–3 Energy in an Inductor

33–9 A toroidal solenoid has a mean radius of 0.12 m and a cross-sectional area of 20×10^{-4} m^2. It is found that when the current is 20 A, the energy stored is 0.1 J. How many turns does the winding have?

33–10 An inductor used in a dc power supply has an inductance of 20 H and a resistance of 200 Ω and carries a current of 0.1 A.

a) What is the energy stored in the magnetic field?

b) At what rate is energy dissipated in the resistor?

33–11 Derive in detail Eq. 33–11 for the energy density in a toroidal solenoid filled with a magnetic material.

33–12 A magnetic field of magnitude $B = 0.4$ T is uniform across a volume of 0.02 m^3. Calculate the total magnetic energy in the volume if

a) the volume is free space;

b) the volume is filled with material of relative permeability 600.

Section 33–4 The R–L Circuit

33–13 An inductor of inductance 3 H and resistance 6 Ω is connected to the terminals of a battery of emf 12 V and of negligible internal resistance. Find

a) the initial rate of increase of current in the circuit,

b) the rate of increase of current at the instant when the current is 1 A,

c) the current 0.2 s after the circuit is closed,

d) the final steady-state current.

33–14 The resistance of a 10-H inductor is 200 Ω. The inductor is suddenly connected across a potential difference of 10 V.

a) What is the final steady current in the inductor?

b) What is the initial rate of increase of current?

c) At what rate is the current increasing when its value is one-half the final current?

d) At what time after the circuit is closed does the current equal 99% of its final value?

e) Compute the current at the following times after the circuit is closed: 0, 0.025 s, 0.05 s, 0.075 s, 0.10 s, and 0.20 s. Show the results in a graph.

33–15 Write an equation corresponding to Eq. (33–12) for the current in Fig. 33–5 just after switch S_2 is closed and S_1 is opened, if the initial current is I. Solve the resulting differential equation and verify Eq. (33–18).

33–16 In Fig. 33–5 let $V = 200$ V, $R = 500$ Ω, and $L = 0.1$ H. With switch S_2 open, switch S_1 is closed and left

until a constant current is established. Then S_2 is closed and S_1 opened, so the battery is taken out of the circuit.

a) What is the initial current in the resistor?

b) What is the current in the resistor at $t = 0.2 \times 10^{-3}$ s?

c) What is the potential difference between points b and c at $t = 0.2 \times 10^{-3}$ s? Which point is at higher potential?

d) How long does it take the current to decrease to half its initial value?

Section 33–5 The L–C Circuit

33–17 An inductor having $L = 40$ mH is to be combined with a capacitor to make an L–C circuit with natural frequency 2×10^6 Hz. What value of capacitance should be used?

33–18 The maximum capacitance of a variable air capacitor is 35 pF.

a) What should be the self-inductance of a coil connected to this capacitor if the natural frequency of the L–C circuit is to be 550×10^3 Hz, corresponding to one end of the AM radio broadcast band, when the capacitor is set to its maximum capacitance?

b) The frequency at the other end of the broadcast band is 1550×10^3 Hz. What must be the minimum capacitance of the capacitor if the natural frequency is to be adjustable over the range of the broadcast band?

33–19 Show that differential Eq. (33–25) is satisfied by the function $q = Q \cos \omega t$, with ω given by $1/(LC)^{1/2}$.

Section 33–6 The L–R–C Circuit

33–20 Show that the quantity $(L/C)^{1/2}$ has units of resistance (ohms).

33–21 An L–R–C circuit has $L = 0.5$ H, $C = 0.1 \times 10^{-3}$ F, and resistance R.

a) What is the natural angular frequency of the circuit when $R = 0$?

b) What value must R have to give a 10% decrease in natural frequency compared to the value calculated in part (a)?

PROBLEMS

33–22 A solenoid has length l_1, radius r_1, and number of turns N_1. A second, smaller solenoid of length l_2, radius r_2, and number of turns N_2 is placed at the center of the first solenoid, such that their axes coincide.

a) What is the mutual inductance of the pair of solenoids?

b) If the current in the larger solenoid is increasing at the rate di_1/dt, what is the emf induced in the small solenoid?

c) If the current in the smaller solenoid is increasing at the

rate di_2/dt, what is the emf induced in the larger solenoid?

33–23 A coil of wire, initially carrying no current, has a current of 50 A established in 10 s. The current increases at a steady (constant) rate during this time, and the charging current induces an emf of 25 V in the wire.

a) Determine the self-inductance of the coil

b) Determine the total magnetic flux through the coil when the current is 50 A.

c) If the resistance of the coil is 25 Ω, determine the ratio of the rate at which energy is being stored in the magnetic field to the rate at which energy is being dissipated by the resistance at the instant when the current is 50 A.

33–24 A toroidal solenoid has two coils with N_1 and N_2 turns, respectively; it has radius r and cross-sectional area A.

a) Derive an expression for the self-inductance L_1 when only the first coil is used, and that for L_2 when only the second coil is used.

b) Derive an expression for the mutual inductance of the two coils.

c) Show that $M^2 = L_1L_2$. This result is valid whenever all the flux linked by one coil is also linked by the other.

33–25 The current in a resistanceless inductor is caused to vary with time as in the graph of Fig. 33–10.

a) Sketch the pattern that would be observed on the screen of an oscilloscope connected to the terminals of the inductor. (The oscilloscope spot sweeps horizontally across the screen at a constant speed, and its vertical deflection is proportional to the potential difference between the inductor terminals.)

b) Explain why the inductor can be described as a "differentiating circuit."

FIGURE 33–10

33–26 A coaxial cable consists of a small solid conductor of radius r_a supported by insulating disks on the axis of a thin-walled tube of inner radius r_b. Show that the self-inductance of a length l of the cable is

$$L = l\frac{\mu_0}{2\pi}\ln\left[\frac{r_b}{r_a}\right].$$

Assume the inner and outer conductors carry equal currents in opposite directions. (*Hint:* Use Ampere's law to find the magnetic field at any point in the space between the conductors. Write the expression for the flux $d\Phi$ through a narrow strip of length l parallel to the axis, of width dr, at a distance r from the axis of the cable and lying in a plane containing the axis. Integrate to find the total flux linking a current i in the central conductor.)

33–27 Uniform electric and magnetic fields E and B occupy the same region of free space. If $E = 400$ V·m^{-1}, what is B if the energy densities in the electric and magnetic fields are equal?

33–28 The 1000-turn toroidal solenoid described in Exercise 33–2 carries a current of 5.0 A.

a) What is the energy density in the magnetic field?

b) What is the total magnetic-field energy? Find the answer by using Eq. (33–8) and also by multiplying the energy density from (a) by the volume of the toroid, which is $2\pi rA$; compare the two results.

33–29 An inductor having inductance L and resistance R carries a current I. Show that the time constant is equal to *twice the ratio* of the energy stored in the magnetic field to the rate of dissipation of energy in the resistance.

33–30 Refer to Exercise 33–13.

a) What is the power input to the inductor from the battery, as a function of time, if the circuit is completed at $t = 0$?

b) What is the rate of dissipation of energy in the resistance of the inductor, as a function of time?

c) What is the rate at which the energy of the magnetic field in the inductor is increasing, as a function of time?

d) Compare the results of (a), (b), and (c).

33–31 Refer to Exercise 33–13.

a) How much energy is stored in the magnetic field of the inductor one time constant after the battery has been connected? Compute this value both by integrating the expression in Problem 33–30(c) and by using Eq. (33–8), and compare the results.

b) Integrate the expression obtained in Problem 33–30(a) to find the *total* energy supplied by the battery during the time interval considered in (a).

c) Integrate the expression obtained in Problem 33–30(b) to find the *total* energy dissipated in the resistance of the inductor during the same time period.

d) Compare the results obtained in (a), (b), and (c).

33–32 Refer to Exercise 33–16.

a) What is the total energy initially stored in the inductor?

b) At $t = 0.2 \times 10^{-3}$ s, at what rate is the energy stored in the inductor decreasing?

c) At $t = 0.2 \times 10^{-3}$ s, at what rate is electrical energy being converted into heat in the resistor?

d) Obtain an expression for the rate at which electrical energy is being converted into heat in the resistor, as a function of time. Integrate this expression from $t = 0$ to $t = \infty$ to obtain the total electrical energy dissipated in the resistor. Compare your result to that of part (a).

33–33 An inductor is made with two coils wound close together on a form, so all the flux linking one coil also links the other. The number of turns is the same in each. If the inductance of one coil is L, what is the inductance when the two coils are connected

a) in series?

b) in parallel?

In each case the current travels in the same sense around each coil.

c) If an L–C circuit with this inductor has natural frequency ω using one coil, what is the natural frequency when the two coils are used in series?

33–34 An L–C circuit consists of an inductor of $L = 0.8$ H and a capacitor of $C = 0.4 \times 10^{-3}$ F. The initial charge on the capacitor is $5\ \mu$C.

a) What is the maximum voltage across the capacitor?

b) What is the maximum current in the inductor?

c) What is the maximum energy stored in the inductor?

d) When the current in the inductor has half its maximum value, what is the charge on the capacitor and what is the energy stored in the inductor?

33–35 The equation preceding Eq. (33–27) may be converted into an energy relation. Multiply both sides of this equation by $i = dq/dt$. The first term then becomes i^2R. Show that the second can be written $d(\tfrac{1}{2}Li^2)/dt$, and the third can be written $d(q^2/2C)/dt$. What does the resulting equation say about energy conservation in the circuit?

33–36 An inductor of resistance R and self-inductance L is connected in series with a noninductive resistor of resistance R_0 to a constant potential difference V (Fig. 33–11).

a) Find the expression for the potential difference v_{cb} across the inductor at any time t after switch S_1 is closed.

b) Let $V = 20$ V, $R_0 = 50\ \Omega$, $R = 150\ \Omega$, $L = 5$ H. Compute v_{ac} and v_{cb} for $t = 0$, 0.5τ, τ, 1.5τ, and 2.5τ, where τ is the time constant of the circuit. Also calculate v_{ac} and v_{cb} for $t \to \infty$. Sketch graphs of v_{ac} and v_{cb} versus time from time zero to infinity.

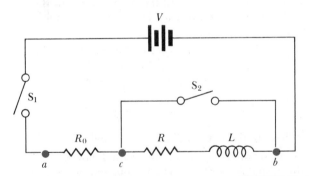

FIGURE 33–11

33–37 After the current in the circuit of Fig. 33–11 has reached its final steady value, the switch S_2 is closed, thus short-circuiting the inductor. (Switch S_1 remains closed.)

a) Derive an expression for the currents through R, R_0, and S_2 as functions of time.

b) What will be the magnitude and direction of the current in S_2, 0.01 s after S_2 is closed? Use the numerical values given in Problem 33–36.

33–38 A popular demonstration of self-inductance often used in a physics course employs a circuit such as the one

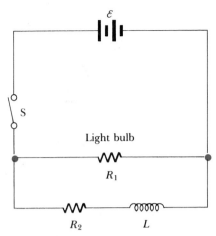

FIGURE 33–12

shown in Fig. 33–12. Switch S is closed and the light bulb, represented by resistance R_1, is seen to just barely glow. After a period of time switch S is opened and the bulb is then seen to light up brightly for a short time. This effect is easy to understand if one thinks of an inductor as a device that imparts an "inertia" to the current, preventing a discontinuous change in the current through it.

a) Derive, as explicit functions of time, expressions for i_1 and i_2, the currents through the light bulb and inductor, after switch S is closed.

b) After a long time, steady-state conditions can be assumed. Obtain expressions for the steady-state currents in the bulb and the inductor.

c) Switch S is now opened. Obtain an expression for the current through the inductor and light bulb as an explicit function of time.

d) You have been asked to design demonstration apparatus using the circuit shown in Fig. 33–12 with a 50-H inductor and a 60-W light bulb. You are to connect a resistor in series with the inductor, and R_2 represents the sum of that resistance plus the internal resistance of the inductor. When switch S is opened, a transient current is to be set up that starts at 0.9 A and is not to fall below 0.3 A until after 0.2 s. For simplicity assume the resistance of the light bulb is constant and equals the resistance the bulb must have to dissipate 60 W at 120 V. Determine R_2 and $\mathcal{E}$ for the given design considerations.

e) With the numerical values determined in (d), what is the current through the light bulb just before the switch is opened? Does the result confirm the qualitative description of what one observes in the demonstration?

CHALLENGE PROBLEMS

33–39 A certain toroidal solenoid has a rectangular cross section, as shown in Fig. 33–13. It has N uniformly spaced turns, with air inside. The magnetic field at a point inside the toroid is given by Eq. (31–29). *Do not* assume the field to be uniform over the cross section.

a) Show that the magnetic flux through a cross section of

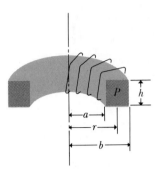

FIGURE 33–13

the toroid is

$$\Phi = \frac{\mu_0 N I h}{2\pi} \ln\left[\frac{b}{a}\right].$$

b) Show that the self-inductance of the toroidal solenoid is given by

$$L = \frac{\mu_0 N^2 h}{2\pi} \ln\left[\frac{b}{a}\right].$$

c) The fraction b/a may be written as

$$\frac{b}{a} = \frac{a + b - a}{a} = 1 + \frac{b - a}{a}.$$

The power series expansion for $\ln(1 + x)$ is $\ln(1 + x) = x + x^2/2 + \cdots$. Hence show that when $b - a$ is much less than a, the self-inductance is approximately equal to

$$L = \frac{\mu_0 N^2 h(b - a)}{2\pi a}.$$

Compare this result with that obtained in Example 33–2.

33–40 For the toroidal solenoid described in Challenge Problem 33–39 calculate the total energy when the current is I. *Do not* assume the field to be uniform over the cross section. Obtain the total energy in two ways:

a) Use Eq. (33–8) with the expression for L obtained in Problem 33–39.

b) Integrate Eq. (33–10) through the volume of the toroid. Compare the results of parts (a) and (b).

33–41 Consider the circuit shown in Fig. 33–14. The values of the circuit elements are as follows: $\mathcal{E} = 50$ V, $L = 10$ H, $C = 20$ μF, $R_1 = 25$ Ω, and $R_2 = 5000$ Ω. Switch S is closed at time $t = 0$, causing a current i_1 through the inductive branch and a current i_2 through the capacitive branch. The initial charge on the capacitor is zero, and the charge at time t is q_2.

a) Derive expressions for i_1, i_2, and q_2 as explicit functions of time.

b) What is the initial current through the inductive branch? What is the initial current through the capacitive branch?

c) What will be the currents through the inductive and capacitive branches a long time after the switch has been closed? How long is a "long time"? Explain.

d) At what time t_1 will the currents i_1 and i_2 be equal? (*Hint:* Consider using series expansions for the exponentials.)

e) For the conditions given in part (d), determine i_1.

f) The total current through the battery is $i = i_1 + i_2$. At what time t_2 will i equal one-half its final value? (*Hint:* As in (d), the numerical work is greatly simplified if one makes suitable approximations.)

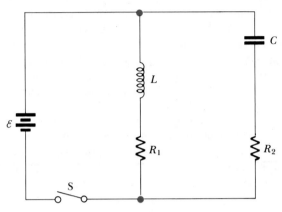

FIGURE 33–14

33–42 Consider the circuit shown in Fig. 33–15. The circuit elements are as follows: $\mathcal{E} = 50$ V, $L = 20$ H, $C = 6.25$ μF, and $R = 1000$ Ω. At time $t = 0$, switch S is closed. The current through the inductance is i_1, the current through the capacitor branch is i_2, and the charge on the capacitor is q_2.

a) Using Kirchhoff's rules, verify the circuit equations

$$R(i_1 + i_2) + L\left(\frac{di_1}{dt}\right) = \mathcal{E},$$

$$R(i_1 + i_2) + \frac{q_2}{C} = \mathcal{E}.$$

b) What are the initial values of i_1, i_2, and q_2?

c) Show by direct substitution that the following solutions for i_1 and q_2 satisfy the circuit equations from part (a). Also show that they satisfy the initial conditions.

$$i_1 = (\mathcal{E}/R)(1 - [(2\omega RC)^{-1}\sin(\omega t) + \cos(\omega t)]e^{-\beta t}),$$

$$q_2 = \left(\frac{\mathcal{E}}{\omega R}\right)e^{-\beta t}\sin(\omega t),$$

where $\beta = (2RC)^{-1}$ and $\omega = [(LC)^{-1} - (2RC)^{-2}]^{\frac{1}{2}}$.

d) Determine the time t_1 at which i_2 first becomes zero.

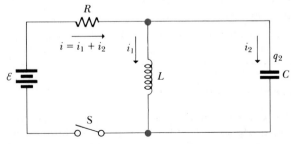

FIGURE 33–15

33–43 A tank containing a liquid has turns of wire wrapped around it, causing it to act as an inductor. The tank can be used to measure the liquid content by using its inductance to determine the height of the liquid. The inductance of the tank changes from a value of L_0, corresponding to a relative permeability of 1, when the tank is empty to a value of L_f, corresponding to a relative permeability of K_m (the relative permeability of the liquid) when the tank is full. The appropriate electronic circuitry can determine the inductance to five significant figures, and hence the effective relative permeability of the combined air and liquid within the rectangular cavity of the tank. Each of the four sides of the tank has a width W and a height D. (See Fig. 33–16.) The height of the liquid in the tank is d. Neglect any fringing effects, and also assume that the relative permeability of the tank material can be neglected.

a) Derive an expression for d as a function of L, the inductance corresponding to a certain fluid height, and L_0 and L_f.

b) What would be the inductance (to five significant figures) for a tank one-quarter full, one-half full, and three-quarters full, if the tank contains liquid oxygen? Take $L_0 = 1.25$ H. The magnetic susceptibility of liquid oxygen is $\chi_m = 1.52 \times 10^{-3}$.

c) Repeat part (b) for mercury. The magnetic susceptibility of mercury is given in Table 31–1.

d) For which material would this gauge be more practical?

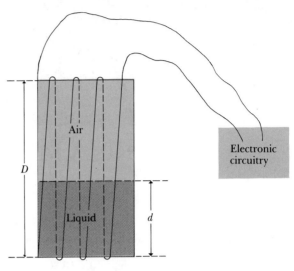

FIGURE 33–16

34
ALTERNATING CURRENTS

IN OUR STUDY OF CIRCUITS IN CHAPTERS 28 AND 29, WE CONCENTRATED primarily on *direct-current* circuits, in which all the currents, voltages, and emf's are *constant,* that is, not functions of time. However, many electric circuits of practical importance, including most household and industrial power-distribution systems, use **alternating current,** in which the voltages and currents vary with time, often in a *sinusoidal* manner. Alternating currents are of utmost importance in technology and industry. Transmission of power over long distances is much more economical with alternating than with direct current. Circuits used in modern communication equipment, including radio and television, make extensive use of alternating current. In this chapter we study the behavior of circuits with sinusoidally varying voltages and currents. Many of the same principles used in Chapters 28, 29, and 33 are applicable here as well, and we also introduce several new concepts related to the circuit behavior of inductors and capacitors.

34–1 AC SOURCES AND PHASORS

An alternator is a source of sinusoidally varying voltage and current.

We have already studied several sources of alternating emf or voltage. A coil of wire, rotating with constant angular velocity in a magnetic field, develops a sinusoidal alternating emf, as discussed in Section 32–2. This simple device is the prototype of the commercial alternating-current generator, or **alternator.** An *L–C* circuit, as discussed in Section 33–5, oscillates sinusoidally; and with the proper circuitry it provides an alternating potential difference having a frequency, depending on its purpose, that may range from a few hertz to many millions of hertz.

We first consider several circuits connected to an alternator or oscillator that maintains between its terminals a sinusoidal alternating potential difference

$$v = V \cos \omega t. \tag{34–1}$$

Sinusoidally varying voltages and currents are described in terms of their amplitude, frequency, and phase.

Here V is the maximum potential difference, or the **voltage amplitude;** v is the *instantaneous* potential difference; and ω is the *angular frequency,* equal to 2π times the frequency f. In the United States, commercial electric-power distribution systems always use a frequency of $f = 60$ Hz, corresponding to $\omega = 377$ s^{-1}. For brevity, the alternator or oscillator will be referred to as an **ac source.** The circuit-diagram symbol for an ac source is $\otimes$.

Analysis of alternating-current circuits is facilitated by use of rotating vector diagrams similar to those used in the study of harmonic motion in Section 11–3. In such diagrams the instantaneous value of a quantity that varies sinusoidally with time is represented by the *projection,* onto a horizontal axis, of a vector whose length corresponds to the amplitude of the quantity; the vector rotates counterclockwise with constant angular velocity ω. These rotating vectors are called **phasors,** and diagrams containing them are called **phasor diagrams.** In Section 11–3 we used a single phasor to represent the position and motion of a point mass undergoing simple harmonic motion. In this chapter we will find phasors convenient for *adding* sinusoidal voltages and currents; with them we can apply the method of vector addition to combine sinusoidal quantities with phase differences. We will find a similar use for phasors in Chapter 39, in our study of interference phenomena in optics.

Phasor diagrams: a handy way to handle phase relationships in ac circuits

34–2 RESISTANCE, INDUCTANCE, AND CAPACITANCE

The simplest problem in ac-circuit analysis consists of a resistor of resistance R connected between the terminals of an ac source, as in Fig. 34–1a. Suppose the instantaneous potential of point a with respect to point b is given by $v = V \cos \omega t$. (We could equally well have chosen the sine function instead of the cosine.) The instantaneous current i in the resistor is

In a resistor, the voltage and current are in phase.

$$i = \frac{v}{R} = \frac{V}{R} \cos \omega t.$$

The maximum current I, or the **current amplitude,** is

$$I = \frac{V}{R}, \tag{34–2}$$

and we can therefore write

$$i = I \cos \omega t. \tag{34–3}$$

The current and voltage are both proportional to cos ωt, so the current is *in phase* with the voltage. The current and voltage amplitudes, from Eq. (34–2), are related in the same way as in a dc circuit.

Figure 34–1b shows graphs of i and v as functions of time. The fact that the curve representing current has the greater amplitude in the diagram is of no significance, because the choice of vertical scales for i and v is arbitrary. The corresponding phasor diagram is given in Fig. 34–1c. Because i and v are in phase and have the same frequency, the current and voltage phasors rotate together. Their projections on the horizontal axis represent the instantaneous current and voltage, respectively.

Next, suppose that a capacitor of capacitance C is connected across the source, as in Fig. 34–2a. The instantaneous charge q on the capacitor is

$$q = Cv = CV \cos \omega t. \tag{34–4}$$

In this case the instantaneous current is equal to the *rate of change* of the capacitor charge and so is proportional to the rate of change of voltage. Tak-

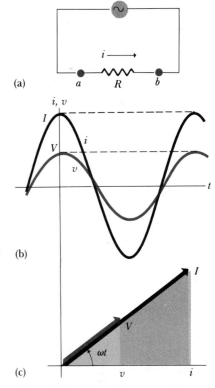

34–1 (a) Resistance R connected across an ac source. (b) Graphs of instantaneous voltage and current. (c) Phasor diagram; current and voltage are in phase.

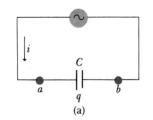

(a)

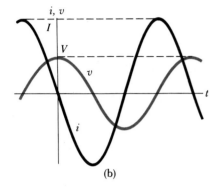

(b)

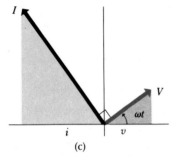

(c)

34-2 (a) Capacitor C connected across an ac source. (b) Graphs of instantaneous voltage and current. (c) Phasor diagram; voltage *lags* current by 90°.

Capacitive reactance: the ratio of voltage amplitude to current amplitude for a capacitor

The ohm: the unit of reactance as well as resistance

ing the derivative of Eq. (34–4), we find

$$i = \frac{dq}{dt} = -\omega CV \sin \omega t. \qquad (34\text{–}5)$$

Thus if the voltage is represented by a *cosine* function, the current is represented by a *negative sine* function, as shown in Fig. 34–2b. The current is *not* in phase with the voltage; the current is greatest at times when the v curve is rising or falling most steeply, and is zero at times when the v curve "levels off"; that is, when it reaches its maximum and minimum values.

The voltage and current are "out of step" or *out of phase* by a quarter-cycle, with the current a quarter-cycle ahead. The peaks of current occur a quarter-cycle *before* the corresponding voltage peaks. This result is also represented by the diagram of Fig. 34–2c, which shows a current phasor ahead of the voltage phasor by a quarter-cycle, or 90°. This *phase difference* between current and voltage can also be obtained by rewriting Eq. (34–5), using the trigonometric identity $\cos (A + 90°) = -\sin A$:

$$i = \omega CV \cos (\omega t + 90°). \qquad (34\text{–}6)$$

We can think of the expression for i in a capacitor as a cosine function with a "head start" of 90° compared with that for the voltage. We say that the current *leads* the voltage by 90° or that the voltage *lags* the current by 90°. Alternatively, if i is given by $i = \omega CV \cos \omega t$, then $v = V \cos (\omega t - 90°)$.

For uniformity of language in later discussions, we will usually describe the phase of the voltage relative to the current. Thus if the current i in a circuit is given by

$$i = I \cos \omega t$$

and the voltage v between two points is

$$v = V \cos (\omega t + \phi),$$

we call ϕ the **phase angle,** understanding that it gives the phase of the voltage relative to the current. It is always an angle between −90° and +90°. Hence for the capacitor we have $\phi = -90°$.

Equations (34–5) and (34–6) show that the maximum current I is given by

$$I = \omega CV. \qquad (34\text{–}7)$$

This expression can be put in the same form as that for the maximum current in a resistor $(I = V/R)$ if we write Eq. (34–7) as

$$I = \frac{V}{1/\omega C}$$

and define a quantity X_C, called the **capacitive reactance** of the capacitor, as

$$X_C = \frac{1}{\omega C}. \qquad (34\text{–}8)$$

Then

$$I = \frac{V}{X_C}. \qquad (34\text{–}9)$$

From Eq. (34–9) the SI unit of capacitive reactance is one *volt per ampere* (1 V·A^{-1}), or one *ohm* $(1 \ \Omega)$.

The reactance of a capacitor is inversely proportional both to the capacitance C and to the angular frequency ω; the greater the capacitance and the higher the frequency, the *smaller* the reactance X_C.

Capacitive reactance is inversely proportional to frequency.

EXAMPLE 34–1   At an angular frequency of 1000 rad·s^{-1}, the reactance of a 1-μF capacitor is

$$X_C = \frac{1}{\omega C} = \frac{1}{(10^3 \text{ rad·s}^{-1})(10^{-6} \text{ F})} = 1000 \ \Omega.$$

At a frequency of 10,000 rad·s^{-1}, the reactance of the same capacitor is only 100 Ω, and at a frequency of 100 rad·s^{-1}, it is 10,000 Ω.

Finally, suppose we connect a pure inductor having a self-inductance L and zero resistance to an ac source, as in Fig. 34–3. The potential v of point a with respect to point b is given by $v = L\,di/dt$, and we have

$$L\frac{di}{dt} = V \cos \omega t \qquad (34\text{--}10)$$

and

$$di = \frac{V}{L} \cos \omega t \, dt.$$

Integrating both sides gives

$$i = \frac{V}{\omega L} \sin \omega t + \text{constant}.$$

If $i = 0$ at time $t = 0$, then constant $= 0$ and

$$i = \frac{V}{\omega L} \sin \omega t. \qquad (34\text{--}11)$$

Again, the voltage and current are a quarter-cycle out of phase, but this time the voltage *leads* the current by 90°. This may also be seen by rewriting Eq. (34–11) using the trigonometric identity $\cos (A - 90°) = \sin A$:

$$i = \frac{V}{\omega L} \cos (\omega t - 90°). \qquad (34\text{--}12)$$

This result and the phasor diagram of Fig. 34–3c show that the voltage can be viewed as a cosine function with a "head start" of 90°. Alternatively, if i is given by

$$i = \frac{V}{\omega L} \cos \omega t,$$

then the voltage v is

$$v = V \cos (\omega t + 90°).$$

The phase angle ϕ of the voltage with respect to the current is $\phi = +90°$.

In Fig. 34–3b the points of maximum voltage correspond on the graphs to points of maximum steepness of the current curve, and the points of zero

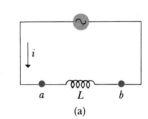

(a)

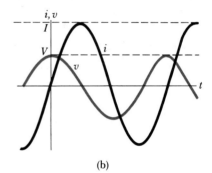

(b)

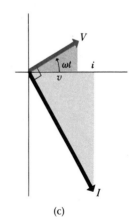

(c)

34–3 (a) Inductance L connected across an ac source. (b) Graphs of instantaneous voltage and current. (c) Phasor diagram; voltage *leads* current by 90°.

The phase angle describes the phase of the voltage relative to the current in any circuit.

voltage correspond to the points where the current curve levels off at its maximum and minimum values.

From Eq. (34–11) or (34–12) the maximum current I is

$$I = \frac{V}{\omega L},\tag{34–13}$$

so we can also write

$$i = I \sin \omega t = I \cos (\omega t - 90°).\tag{34–14}$$

Inductive reactance: the ratio of voltage amplitude to current amplitude for an inductor

We define the **inductive reactance** X_L of an inductor as

$$X_L = \omega L,\tag{34–15}$$

and we can write Eq. (34–13) in the same form as for a resistor:

$$I = \frac{V}{X_L}.\tag{34–16}$$

The SI unit of inductive reactance is also one *ohm*.

Inductive reactance is directly proportional to frequency.

The reactance of an inductor is directly proportional both to its inductance L and to the angular frequency ω; the greater the inductance and the higher the frequency, the *larger* the reactance.

EXAMPLE 34–2 At an angular frequency of 1000 rad·s^{-1}, the reactance of a 1-H inductor is

$$X_L = \omega L = (10^3 \text{ rad·s}^{-1})(1 \text{ H}) = 1000 \text{ }\Omega.$$

At a frequency of 10,000 rad·s^{-1} the reactance of the same inductor is 10,000 Ω, while at a frequency of 100 rad·s^{-1} it is only 100 Ω.

The graphs in Fig. 34–4 summarize the variations with frequency of the resistance of a resistor and the reactances of an inductor and of a capacitor. As the frequency increases, the reactance of the inductor approaches infinity, and that of the capacitor approaches zero. As the frequency decreases, the inductive reactance approaches zero, and the capacitive reactance approaches infinity. The limiting case of zero frequency corresponds to a dc circuit.

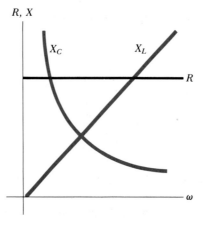

34–4 Graphs of R, X_L, and X_C as functions of frequency.

PROBLEM-SOLVING STRATEGY: *Alternating-current circuits*

1. In ac-circuit problems it is nearly always easiest to work with angular frequency ω. But in problems you may be given the ordinary frequency f, expressed in Hz. Don't forget to make the conversion, using $\omega = 2\pi f$.

2. Keep in mind a few basic facts about phase relationships. For a resistor, voltage and current are always *in phase*, and the two corresponding phasors in a phasor diagram always have the same direction. For a capacitor, the voltage always *lags* the current by 90° (i.e., $\phi = -90°$), and the voltage phasor is always turned 90° clockwise from the current phasor. For an inductor, the voltage always *leads* the current by 90° (i.e., $\phi = +90°$), and the voltage phasor is always turned 90° counterclockwise from the current phasor.

3. Remember that Kirchhoff's rules are just as applicable for ac circuits as for dc. All the voltages and currents are sinusoidal functions of time instead of being constant, but Kirchhoff's rules hold at each instant. Thus in a series circuit, the instantaneous current is the same in each circuit element; in a parallel circuit, the instantaneous potential difference is the same for each circuit element.

4. Reactance and impedance (coming in the next section) are a lot like resistance; each represents the ratio of voltage amplitude V to current amplitude I in a circuit element or combination of elements. But keep in mind that phase relations are always lurking in the shadows; resistance, reactance, and impedance have to be combined by *vector* addition of the corresponding phasors. When you have several circuit elements in series, for example, it isn't correct simply to *add* all the numerical values of resistance and reactance; that would ignore the phase relations.

34-3 THE *L-R-C* SERIES CIRCUIT

Many ac circuits of great usefulness include resistance, inductive reactance, and capacitive reactance. A simple series circuit is shown in Fig. 34-5a. In analyzing this and similar circuits, we will use a phasor diagram that includes the voltage and current phasors for each of the components. Here, because of Kirchhoff's loop rule, the instantaneous *total* voltage across all three components is equal to the source voltage at that instant, and its phasor is the *vector sum* of the phasors for the individual voltages. The complete phasor diagram for this circuit is shown in Fig. 34-5b. This may appear complex, but it is not as bad as it looks; we will explain it step by step.

The instantaneous current i has the same value at all points of the circuit. Thus a *single phasor I*, of length proportional to the current amplitude, represents the current in each circuit element.

Let us use the symbols v_R, v_L, and v_C for the instantaneous voltages across R, L, and C, and V_R, V_L, and V_C for their maximum values. We denote the instantaneous and maximum source voltages by v and V. Then $v = v_{ab}$, $v_R = v_{ac}$, $v_L = v_{cd}$, and $v_C = v_{db}$.

We have shown that the potential difference between the terminals of a resistor is *in phase* with the current in the resistor, and that its maximum value V_R is

$$V_R = IR.$$

Thus the phasor V_R in Fig. 34-5b, in phase with the current vector, represents the voltage across the resistor. Its projection on the horizontal axis, at any instant, gives the instantaneous potential difference v_R.

The voltage across an inductor *leads* the current by 90°. The voltage amplitude is

$$V_L = IX_L.$$

Using phasor diagrams to analyze voltage and current relations in an *L-R-C* circuit

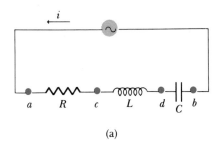

(a)

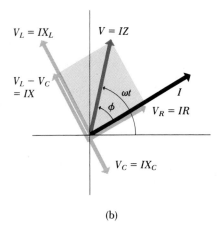

(b)

34-5 (a) A series *L-R-C* circuit. (b) Phasor diagram.

The phasor V_L in Fig. 34–5b represents the voltage across the inductor, and its projection at any instant onto the horizontal axis equals v_L.

The voltage across a capacitor *lags* the current by 90°. The voltage amplitude is

$$V_C = IX_C.$$

The phasor V_C in Fig. 34–5b represents the voltage across the capacitor, and its projection at any instant onto the horizontal axis equals v_C.

The instantaneous potential difference v between terminals a and b equals at every instant the (algebraic) sum of the potential differences v_R, v_L, and v_C; that is, it equals the sum of the projections of the phasors V_R, V_L, and V_C. But the *projection* of the *vector sum* of these phasors is equal to the *sum* of their *projections*, so this vector sum V must be the phasor that represents the source voltage. To form the vector sum, we first subtract the phasor V_C from the phasor V_L (since these always lie along a straight line), giving the phasor $V_L - V_C$. Since this is at right angles to the phasor V_R, the magnitude of the phasor V is

$$V = \sqrt{V_R{}^2 + (V_L - V_C)^2} = \sqrt{(IR)^2 + (IX_L - IX_C)^2}$$

$$= I\sqrt{R^2 + (X_L - X_C)^2}.$$

The quantity $X_L - X_C$ is called the **reactance** of the circuit, denoted by X:

$$X = X_L - X_C. \qquad (34\text{–}17)$$

Impedance: the ratio of voltage amplitude to current amplitude for the entire circuit

Finally, we define the **impedance** Z of the circuit as

$$Z = \sqrt{R^2 + (X_L - X_C)^2} = \sqrt{R^2 + X^2}, \qquad (34\text{–}18)$$

so we can write

$$V = IZ \quad \text{or} \quad I = \frac{V}{Z}. \qquad (34\text{–}19)$$

Thus the *form* of the equation relating current and voltage *amplitudes* is the same as that for a dc circuit; impedance Z plays the same role as resistance R in a dc circuit. Note, however, that the impedance is actually a function of R, L, and C, as well as of the frequency ω. The complete expression for Z, for a series circuit, is

$$Z = \sqrt{R^2 + X^2}$$

$$= \sqrt{R^2 + (X_L - X_C)^2}$$

$$= \sqrt{R^2 + [\omega L - (1/\omega C)]^2}. \qquad (34\text{–}20)$$

The ohm is the unit of resistance, reactance, and impedance.

The unit of impedance, from Eq. (34–19) or (34–20), is one *volt per ampere* (1 V·A^{-1}) or one *ohm*.

The expressions for the impedance Z of (a) an R–L series circuit, (b) an R–C series circuit, and (c) an L–C series circuit can be obtained from Eq. (34–20) by letting (a) $X_C = 0$, (b) $X_L = 0$, and (c) $R = 0$. Can you verify this statement?

Equation (34–20) gives the impedance Z only for a *series* L–R–C circuit. But the impedance of *any* network can be *defined* by Eq. (34–19) as the ratio of the voltage amplitude to the current amplitude.

The angle ϕ shown in Fig. 34–5b is the phase angle of the source voltage v with respect to the current i. From the diagram,

$$\tan \phi = \frac{V_L - V_C}{V_R} = \frac{I(X_L - X_C)}{IR} = \frac{X}{R}. \qquad (34–21)$$

How do we find the phase of voltage with respect to current for an *L–R–C* circuit?

Hence if the current is represented by a cosine function,

$$i = I \cos \omega t,$$

the source voltage *leads* the current by an angle ϕ between 0 and 90°, and its equation is

$$v = V \cos (\omega t + \phi).$$

Figure 34–5b has been constructed for a circuit in which $X_L > X_C$. If $X_L < X_C$, vector V lies on the opposite side of the current vector I and the voltage *lags* the current. In this case $X = X_L - X_C$ is a *negative* quantity, $\tan \phi$ is negative, and ϕ is a negative angle between 0 and $-90°$.

To summarize, we can say that the *instantaneous* potential differences in an ac-series circuit add *algebraically*, just as in a dc circuit, while the voltage *amplitudes* add *vectorially*.

Combining voltages in the series *L–R–C* circuit requires vector addition.

EXAMPLE 34–3 In the series circuit of Fig. 34–5, let $R = 300\ \Omega$, $L = 0.9$ H, $C = 2.0\ \mu$F, and $\omega = 1000$ rad·s^{-1}. Then

$$X_L = \omega L = 900\ \Omega,$$

$$X_C = \frac{1}{\omega C} = 500\ \Omega.$$

The reactance X of the circuit is

$$X = X_L - X_C = 400\ \Omega,$$

and the impedance Z is

$$Z = \sqrt{R^2 + X^2} = 500\ \Omega.$$

If the circuit is connected across an ac source of voltage amplitude 50 V, the current amplitude is

A numerical example of an *L–R–C* circuit

$$I = \frac{V}{Z} = 0.10\ A.$$

The phase angle ϕ is

$$\phi = \arctan \frac{X}{R} = 53°.$$

The voltage amplitude across the resistor is

$$V_R = IR = 30\ V.$$

The voltage amplitudes across the inductor and capacitor are, respectively,

$$V_L = IX_L = 90\ V, \qquad V_C = IX_C = 50\ V.$$

Transients: additional voltages and
currents that die out with time

The analysis above describes the *steady-state* condition of a circuit, the condition that prevails after the circuit has been connected to the source for a long time. When the source is first connected, there may be additional voltages and currents, called *transients*, whose nature depends on the time in the cycle when the circuit is initially completed. A detailed analysis of transients is beyond our scope. They always die out after a sufficiently long time and therefore do not affect the steady-state behavior of the circuit.

34–4 AVERAGE AND ROOT–MEAN–SQUARE VALUES

In this section we consider various ways of *measuring* ac quantities. The *instantaneous* potential difference between any two points in an ac circuit can be measured by connecting a calibrated oscilloscope between the points. The instantaneous current can be measured by connecting an oscilloscope across a resistor in the circuit. But the usual d'Arsonval galvanometer, described in Section 29–3, has too large a moment of inertia to follow the instantaneous values of an alternating current, except possibly at extremely low frequencies. At ordinary frequencies, such as 60-Hz household power, it averages out the sinusoidally varying torque on its coil; the meter deflection is thus proportional to the *average* current.

How do we calculate the average value of a quantity that varies with time?

The **average value** f_{av} of any quantity $f(t)$ that varies with time, over a time interval from t_1 to t_2, is defined as

$$f_{av} = \frac{1}{t_2 - t_1} \int_{t_1}^{t_2} f(t)\, dt. \qquad (34\text{–}22)$$

This equation may be interpreted graphically. The integral represents the *area* bounded by the curve of $f(t)$ and by the two vertical lines t_1 and t_2. The product $f_{av}(t_2 - t_1)$ is the area of a rectangle of height f_{av} and base $(t_2 - t_1)$. Thus f_{av} is the height of a rectangle having the same area as the area under the curve between the same two values of t.

Let us apply this definition to a sinusoidally varying quantity—for example, a current given by

$$i = I \sin \omega t.$$

The period τ of this sinusoidal current is given by $\tau = 1/f = 2\pi/\omega$. From Eq. (34–22) the average value of i for a *half-period* or half-cycle, from $t = 0$ to $t = \pi/\omega$, is

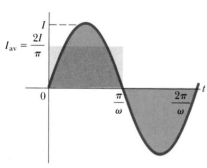

34–6 The average value of a sinusoidal current over a half-cycle is $2I/\pi$. The average over a complete cycle is zero.

$$I_{av} = \frac{\omega}{\pi} \int_0^{\pi/\omega} I \sin \omega t\, dt = \frac{2}{\pi} I. \qquad (34\text{–}23)$$

That is, the average current is $2/\pi$ (about $\frac{2}{3}$) times the maximum current, and the area under the rectangle in Fig. 34–6 equals the area under one loop of the sine curve.

The average current for a *complete cycle* (or any number of complete cycles) is

The average value of any sinusoidally varying quantity is zero.

$$I_{av} = \frac{\omega}{2\pi} \int_0^{2\pi/\omega} I \sin \omega t\, dt = 0,$$

as we should expect, since the *positive* area of the loop between 0 and π/ω is equal to the *negative* area of the loop between π/ω and $2\pi/\omega$. Hence if a sinusoidal current is sent through a d'Arsonval meter, the meter reads zero!

Such a meter can be used in an ac circuit, however, if it is connected in the *full-wave rectifier circuit* shown in Fig. 34–7a. As explained in Section 28–5, an ideal rectifier, or diode, offers zero resistance to current in the forward direction and infinite resistance to current in the opposite direction. Study of Fig. 34–7a shows that the current through the galvanometer is always upward, for either polarity of the source. Thus for a sinusoidal source, the current through the meter has the waveform shown as a solid curve in Fig. 34–7b. Although pulsating, it is always in the same direction, and its average value is *not* zero. Then, when provided with the necessary series resistance or shunt, as for a dc voltmeter or ammeter, the meter can serve as an ac voltmeter or ammeter.

The average value of the rectified current is the same as the average current in a half-cycle in Fig. 34–6, or $2/\pi$ times the maximum current I. Hence if the meter deflects full scale with a steady current I_0, it also deflects full scale when the *average value* of the rectified current, $2I/\pi$, is equal to I_0. The current amplitude I, when the meter deflects full scale, is then

$$I = \frac{\pi I_0}{2}.$$

For example, if $I_0 = 1$ A, $I = 1.57$ A.

Most ac meters are calibrated to read not the maximum value of the current or voltage, but the **root–mean–square (rms) value,** that is, the *square root* of the *average* value of the *square* of the current or voltage. Because i^2 is always positive, the average of i^2 is never zero, even when the average of i itself is zero. Also, rms values of voltage and current are useful in relationships for *power* in ac circuits; we return to this point in Section 34–5.

Figure 34–8 shows graphs of a sinusoidally varying current and of its square. If $i = I \sin \omega t$, then

$$i^2 = I^2 \sin^2 \omega t = I^2[\tfrac{1}{2}(1 - \cos 2\omega t)]$$
$$= \tfrac{1}{2}I^2 - \tfrac{1}{2}I^2 \cos 2\omega t.$$

The average value of i^2, or the *mean square current*, is equal to the constant term $\tfrac{1}{2}I^2$, since the average value of $\cos 2\omega t$, over any number of complete cycles, is zero:

$$(i^2)_{\mathrm{av}} = \frac{I^2}{2}.$$

The root–mean–square current is the square root of this quantity, or

$$I_{\mathrm{rms}} = \sqrt{(i^2)_{\mathrm{av}}} = \frac{I}{\sqrt{2}}. \tag{34–24}$$

In the same way, the root–mean–square value of a sinusoidal voltage is

$$V_{\mathrm{rms}} = \frac{V}{\sqrt{2}}. \tag{34–25}$$

Voltages and currents in power-distribution systems are always referred to in terms of their rms values. Thus when we speak of our household power supply as "120-volt ac," we mean that the rms voltage is 120 V. The voltage

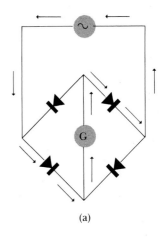
(a)

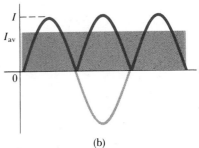
(b)

34–7 (a) A full-wave rectifier circuit. (b) Graph of a full-wave rectified current and its average value.

The root–mean–square value of a sinusoidally varying quantity is proportional to the amplitude.

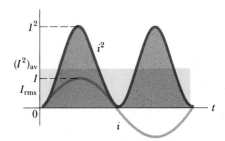

34–8 The average value of the square of a sinusoidally varying current, over any number of half-cycles, is $I^2/2$. The root–mean–square value is $I/\sqrt{2}$.

Voltages and currents in ac circuits are usually stated in terms of rms values, not amplitudes.

amplitude is

$$V = \sqrt{2}\,V_{\text{rms}} = 170 \text{ V}.$$

All the voltage-current relations developed in preceding sections have been stated in terms of amplitudes (maximum values) of voltage and current, but they remain valid when rms values are used; in each case the difference is simply a multiplicative factor of $1/\sqrt{2}$.

34–5 POWER IN AC CIRCUITS

How to calculate the electrical power input to a circuit element

Because electric power plays a central role in our systems for distributing, converting, and using energy, it is a concept of utmost practical importance. When a source with an instantaneous potential difference v supplies an instantaneous current i to an ac circuit, the instantaneous power p it supplies is given by

$$p = vi.$$

First we consider power relations for individual circuit elements.

If the circuit consists of a pure resistance R, as in Fig. 34–1, i and v are *in phase*. The graph representing p is obtained by multiplying together at every instant the ordinates of the graphs of v and i in Fig. 34–1b, and it is shown in Fig. 34–9a. The product vi is positive when v and i are both positive or both negative. That is, energy is supplied *to* the resistor at every instant, for either direction of instantaneous current, although the *rate* at which it is supplied is not constant.

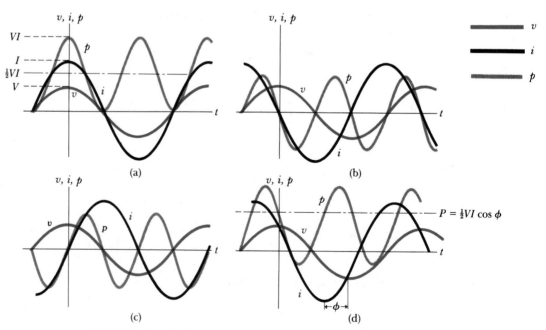

34–9 Graphs of voltage, current, and power as functions of time, for various circuits. (a) Instantaneous power input to a resistor. The average power is $\frac{1}{2}VI$. (b) Instantaneous power input to a capacitor. The average power is zero. (c) Instantaneous power input to a pure inductor. The average power is zero. (d) Instantaneous power input to an arbitrary ac circuit. The average power is $\frac{1}{2}VI\cos\phi = V_{\text{rms}}I_{\text{rms}}\cos\phi$.

The power curve is symmetrical about a value equal to one-half its maximum ordinate VI, so the *average power P* is

$$P = \tfrac{1}{2}VI. \qquad (34\text{--}26)$$

The average power can also be written

$$P = \frac{V}{\sqrt{2}}\frac{I}{\sqrt{2}} = V_{rms}I_{rms}. \qquad (34\text{--}27)$$

Furthermore, since $V_{rms} = I_{rms}R$, we have

$$P = I_{rms}{}^2R. \qquad (34\text{--}28)$$

The average power input to a resistor depends on the rms current and the resistance.

Note that Eqs. (34–27) and (34–28) have the same form as those for a dc circuit.

Suppose next that the circuit consists of a capacitor, as in Fig. 34–2. The current and voltage are then 90° out of phase. When the curves of v and i are multiplied together (the product vi is *negative* when v and i have *opposite* signs), we get the power curve in Fig. 34–9b, which is symmetrical about the horizontal axis. The average power is therefore *zero!*

The average power input to a pure inductor or capacitor is zero.

To see why this is so, recall that positive power means that energy is supplied *to* a device, and negative power means that energy is supplied *by* a device. The process we are considering is merely the charge and discharge of a capacitor. During the intervals when p is positive, energy is supplied to charge the capacitor; when p is negative, the capacitor is discharging and returning energy to the source.

Figure 34–9c is the power curve for a pure inductor. As with a capacitor, the current and voltage are out of phase by 90°, and the average power is zero. Energy is supplied to establish a magnetic field around the inductor and is returned to the source when the field collapses.

In a circuit containing any combination of resistors, capacitors, and inductors, the current and voltage differ in phase by an angle ϕ, and

$$p = [V \cos (\omega t + \phi)][I \cos \omega t]. \qquad (34\text{--}29)$$

The power input to a circuit element or a circuit depends on the phase of the voltage with respect to the current.

The instantaneous power curve has the form shown in Fig. 34–9d. The area under the positive loops is greater than that under the negative loops, and the net average power is positive.

The preceding analyses have shown that when v and i are *in phase,* the average power equals $\tfrac{1}{2}VI$; when v and i are 90° *out of phase,* the average power is zero. In the general case, when v and i differ by an angle ϕ, the average power equals $\tfrac{1}{2}V$, multiplied by $I \cos \phi$, the component of I that is *in phase* with V. That is,

$$P = \tfrac{1}{2}VI \cos \phi = V_{rms}I_{rms} \cos \phi. \qquad (34\text{--}30)$$

This equation is the general expression for the power input to *any* ac circuit. The factor $\cos \phi$ is called the **power factor** of the circuit. For a pure resistance, $\phi = 0$, $\cos \phi = 1$, and $P = V_{rms}I_{rms}$. For a pure (resistanceless) capacitor or inductor, $\phi = \pm 90°$, $\cos \phi = 0$, and $P = 0$.

A low power factor is often a liability.

A low power factor (large angle of lag or lead) is usually undesirable in power circuits because, for a given potential difference, a large current is needed to supply a given amount of power, resulting in large heat losses in the transmission lines. Many types of ac machinery draw a lagging current; the

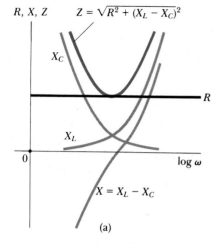

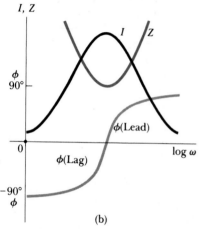

34–10 (a) Reactance, resistance, and impedance as functions of frequency (logarithmic frequency scale). (b) Impedance, current, and phase angle as functions of frequency (logarithmic frequency scale).

At resonance, the inductor and capacitor voltages cancel exactly.

At resonance, the voltages across individual circuit elements can be larger than the source voltage.

The phase of voltage relative to current changes sign as the frequency goes through resonance.

power factor can be corrected by connecting a capacitor in parallel with the load. The leading current drawn by the capacitor compensates for the lagging current in the other branch of the circuit. The capacitor itself takes no net power from the line.

34–6 SERIES RESONANCE

The impedance of an L–R–C series circuit depends on the frequency, since inductive reactance is directly proportional to frequency and capacitive reactance is inversely proportional to frequency. Equation (34–20) shows explicitly the frequency dependence of the impedance of such a circuit. Figure 34–10 shows graphs of R, X_L, X_C, and Z as functions of ω. We have used a logarithmic frequency scale because of the wide range of frequencies covered. Note that at one particular frequency, X_L and X_C are numerically equal and $X = X_L - X_C$ is zero. Hence at this frequency the impedance Z has its *minimum* value, equal simply to the resistance R.

If we connect an ac source having constant voltage amplitude but variable frequency across the circuit, the current amplitude I varies with frequency as shown in Fig. 34–10b; it has its *maximum* value at the frequency where the impedance Z is *minimum*. This peaking of the current amplitude at a certain frequency is called **resonance.** It is easy to find the frequency ω_0 at which the resonance peak occurs; when the inductive and capacitive reactances are equal, we have

$$X_L = X_C, \qquad \omega_0 L = \frac{1}{\omega_0 C}, \qquad \omega_0 = \frac{1}{\sqrt{LC}}. \qquad (34\text{--}31)$$

Note that this frequency is equal to the natural frequency of oscillation of an L–C circuit, derived in Section 33–5, Eq. (33–24).

To understand the resonance peak in current amplitude more fully, consider the voltages in the circuit of Fig. 34–5. The current at any instant is the same in L and C. As we have learned, the voltage across an inductor always *leads* the current by 90°, or a quarter-cycle, and the voltage across a capacitor always *lags* the current by 90°. Thus, comparing the phase of the instantaneous voltage v_{cd} across L and the voltage v_{db} across C, we find that these two voltages always differ in phase by 180°, or a half-cycle. Therefore they have opposite signs at each instant. If, in addition, the *amplitudes* of the two voltages are equal, then they add to zero at each instant, and the *total* voltage v_{cb} across the L–C combination is exactly zero! As we found above, this phenomenon occurs only at one particular frequency, which we call the **resonant frequency.** Depending on the numerical values of R, L, and C, the voltages across L and C individually can be larger than that across R, so at frequencies sufficiently close to the resonant frequency, the voltages across L and C individually can be much *larger* than the source voltage!

The *phase* of the voltage in this circuit, relative to the current, also varies with frequency in an interesting way. At frequencies below resonance, X_C is greater than X_L; the phase relation is dominated by the capacitive reactance. The voltage *lags* the current, and the phase angle ϕ is between zero and −90°. At frequencies above resonance, X_L is greater than X_C, and the inductive reactance dominates. The voltage *leads* the current, and the phase angle is between zero and +90°. This variation of ϕ with frequency is shown in Fig. 36–10b. We can obtain an expression for ϕ as a function of ω from the phasor diagram in Fig. 34–5b and from Eq. (34–21):

$$\phi = \arctan \frac{X}{R} = \arctan \frac{\omega L - (1/\omega C)}{R}. \qquad (34-32)$$

If the inductance L or the capacitance C of a circuit can be varied, the resonant frequency can also be varied. This is the procedure by which a radio or television receiving set may be "tuned" to receive the signal from the desired station. In the early days of radio this was accomplished by use of capacitors with movable metal plates whose overlap could be varied to change C. Nowadays it is more common to use a variable inductor with a ferrite core that slides in and out of a coil to vary L.

EXAMPLE 34–4 The series circuit in Fig. 34–11 is connected to the terminals of an ac source whose frequency is variable but whose rms terminal voltage is constant and equal to 100 V. Find (a) the resonant frequency, (b) the inductive and capacitive reactances and the impedance at the resonant frequency, (c) the rms current at resonance, and (d) the rms potential difference across each circuit element at resonance.

A numerical example of series resonance in an ac circuit

SOLUTION a) The resonant frequency is

$$\omega_0 = \frac{1}{\sqrt{LC}} = \frac{1}{\sqrt{(2\ \text{H})(0.5 \times 10^{-6}\ \text{F})}} = 1000\ \text{rad·s}^{-1}.$$

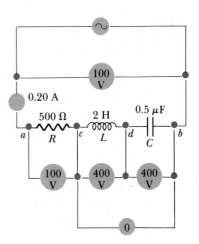

34–11 Series resonant circuit.

b) At this frequency,

$$X_L = (1000\ \text{rad·s}^{-1})(2\ \text{H}) = 2000\ \Omega,$$

$$X_C = \frac{1}{(1000\ \text{rad·s}^{-1})(0.5 \times 10^{-6}\ \text{F})} = 2000\ \Omega,$$

$$X = X_L - X_C = 0.$$

From Eq. (34–20), the impedance Z at resonance is equal to the resistance: $Z = R = 500\ \Omega$.

c) At resonance, the rms current is

$$I = \frac{V}{Z} = \frac{V}{R} = 0.20\ \text{A}.$$

d) The rms potential difference across the resistor is

$$V_R = IR = 100\ \text{V}.$$

The rms potential differences across the inductor and capacitor are, respectively,

$$V_L = IX_L = (0.20\ \text{A})(2000\ \Omega) = 400\ \text{V},$$

$$V_C = IX_C = (0.20\ \text{A})(2000\ \Omega) = 400\ \text{V}.$$

The rms potential difference across the inductor-capacitor combination (V_{cb}) is

$$V = IX = I(X_L - X_C) = 0.$$

This is true because the instantaneous potential differences across the inductor and the capacitor have equal amplitudes but are 180° out of phase, and so add to zero at each instant. Note also that at resonance, V_R is equal to the source voltage V, while V_L and V_C are both considerably *larger* than V.

The current in a series L–R–C circuit reaches its maximum value at the resonant frequency. Figure 34–12 shows a graph of current as a function of frequency for the circuit in Fig. 34–11, discussed in the example above. This curve is often called a *response curve* or a *resonance curve*. As expected, the curve has a peak at $\omega = 1000\ s^{-1}$, the resonant frequency.

What happens when we change R? The figure also shows graphs of I as a function of ω for $R = 200\ \Omega$ and for $R = 2000\ \Omega$. Note that the smaller R is, the taller and sharper the peak is, and the more rapidly it drops off as the frequency ω of the source moves away from the resonant frequency ω_0. Looking back at the expression for the impedance Z of the circuit, Eq. (34–20), we can see the reason for this behavior. When ω is far away from ω_0, either greater or less, Z is dominated by one of the reactance terms; ωL becomes large at high frequencies, $1/\omega C$ at low frequencies. This behavior does not change when we change R, nor does the value of ω_0 change. But at resonance, where the two reactances cancel, $Z = R$; the height of the peak is determined by R, and the behavior near resonance is also dominated by this value. Thus a small R gives a sharply peaked response curve, and a large R gives a broad, flat curve.

This distinction is crucially important in the design of radio and television receiving circuits. The sharply peaked curve is what makes it possible to discriminate between two stations broadcasting on adjacent frequency bands. But if the peak is *too* sharp, some of the information in the received signal is lost. Finally, note that the shape of the resonance curve is related to the overdamped and underdamped oscillations studied in Section 33–6. Reviewing that discussion, we see that a sharply peaked resonance curve corresponds to a small value of R and a lightly damped oscillating system; a broad, flat curve goes with a large value of R and a heavily damped system.

Resonance phenomena occur in all areas of physics; we have already seen one example in the forced oscillation of the harmonic oscillator, studied in Section 11–8. In that case the amplitude of a mechanical oscillation peaked at a driving-force frequency close to the natural frequency of the system, and the

The resonance curve is tall and sharp when R is small, smaller and broader when R is larger.

Practical applications of resonance in tuning radio and TV receivers

Resonance behavior occurs with both electrical and mechanical oscillating systems.

34–12 Graph of rms current I as a function of frequency ω (the dark color curve) for the circuit of Example 34–4. The other curves show the relationship for different values of the circuit resistance R, 2000 Ω (black curve) and 200 Ω (light color curve).

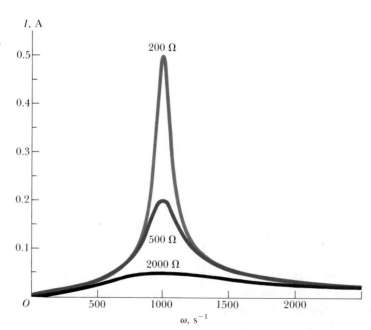

behavior of the *L–R–C* circuit is analogous to this. We suggest you review that discussion now, looking for the analogies. Other important examples of resonance occur in acoustics, in atomic and nuclear physics, and in the study of fundamental particles (high-energy physics).

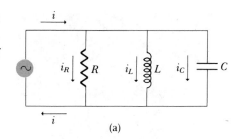

34–7 PARALLEL RESONANCE

A different kind of resonance occurs when *L*, *R*, and *C* are connected in *parallel*, as shown in Fig. 34–13a. We can analyze this circuit by using the same procedure as for the series circuit. In this case the instantaneous potential difference *v* is the same for all elements and is equal to the source voltage. Figure 34–13b shows a phasor diagram; the single phasor *V* represents this common voltage. There are three separate currents, one in each branch, and the three corresponding current phasors are also shown. The phasor I_R, with amplitude V/R and in phase with *V*, represents the current in the resistor. Phasor I_L, with amplitude V/X_L and lagging *V* by 90°, represents the current in the inductor. Phasor I_C, with amplitude V/X_C and leading *V* by 90°, represents the current in the capacitor.

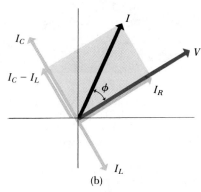

The instantaneous current *i*, by Kirchhoff's point rule, equals the (algebraic) sum of the instantaneous currents i_R, i_L, and i_C and is represented by the phasor *I*, the vector sum of phasors I_R, I_L, and I_C. Angle ϕ is the phase angle of current with respect to source voltage (the negative of the phase angle of voltage with respect to current).

34–13 (a) A parallel *L–R–C* circuit. (b) Phasor diagram showing current phasors for the three branches. The single voltage phasor *V* represents the voltage across all three branches.

From Fig. 34–13,

$$I = \sqrt{I_R{}^2 + (I_C - I_L)^2} = \sqrt{\left(\frac{V}{R}\right)^2 + \left(\omega C V - \frac{V}{\omega L}\right)^2}$$

$$= V \sqrt{\frac{1}{R^2} + \left(\omega C - \frac{1}{\omega L}\right)^2}. \tag{34–33}$$

The maximum current *I* is frequency-dependent, as expected. It is *minimum* when the second factor in the radical is zero; this occurs when the two reactances have equal magnitudes, at the resonant frequency ω_0 given by Eq. (34–31).

Comparing this equation with Eq. (34–19), we see that the *impedance Z* of this parallel combination is given by

Impedance of a parallel *L–R–C* circuit

$$\frac{1}{Z} = \sqrt{\frac{1}{R^2} + \left(\omega C - \frac{1}{\omega L}\right)^2}. \tag{34–34}$$

At resonance $1/Z$ is minimum, so *Z* itself has its *maximum* value at $\omega = \omega_0 = (1/LC)^{1/2}$.

Thus at resonance the total current in the parallel *L–R–C* circuit is *minimum*, in contrast to the *L–R–C* series circuit, which has *maximum* current at resonance. This distinction can be understood by noting that, in the parallel circuit, the currents in *L* and *C* are *always* exactly a half-cycle out of phase. When they also have equal *magnitudes*, they cancel each other completely, and the total current is simply the current through *R*. Indeed, when $\omega C = 1/\omega L$, Eq. (34–33) becomes simply $I = V/R$. This does *not* mean that there is *no* current in *L* or *C* at resonance, but only that the two currents cancel. If *R* is large, the equivalent impedance of the circuit near resonance is much *larger* that the individual reactances X_L and X_C.

At resonance, the inductor and capacitor currents in a parallel circuit cancel exactly.

Transformers are used in power-distribution systems to "step down" the voltage from the large values (typically 500 kV) used for long-distance transmission to values used for local distribution and consumption (typically 120 V to 2400 V). (Courtesy of Pacific Gas and Electric.)

Transformers: the key to efficient long-distance power transmission

Small but inevitable energy losses in a transformer

An idealized transformer has no energy losses at all.

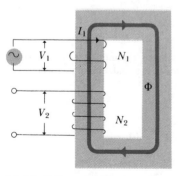

34–14 Schematic diagram of a transformer with secondary open.

34–8 TRANSFORMERS

One of the great advantages of ac over dc for electric-power distribution is that it is much easier to step voltage levels up and down with ac than with dc. For long-distance power transmission it is desirable to use as high a voltage and as small a current as possible; this reduces I^2R heating losses in the transmission lines, and smaller wires can be used, saving on material costs. Present-day transmission lines routinely operate at rms voltages of the order of 500 kV. Yet safety considerations and insulation requirements dictate relatively low voltages in generating equipment and in household and industrial power distribution. The standard voltage for household wiring is 120 V in the United States and Canada, 240 V in most of Western Europe. The necessary voltage conversion is accomplished by use of **transformers.**

In principle, the transformer consists of two coils, electrically insulated from each other but wound on the same core. Transformers used in power-distribution systems have soft iron cores. An alternating current in one winding sets up an alternating flux in the core, and the induced electric field produced by this varying flux induces an emf in the other winding, as we learned in Sections 32–3 and 33–1. Energy is thus transferred from one winding to another via the core flux and its associated induced electric field. The winding to which power is supplied is called the **primary;** the winding from which power is delivered is called the **secondary.** The circuit symbol for an iron core transformer is

The power output of a transformer is necessarily less than the power input because of unavoidable losses. These losses include resistance (I^2R) losses in the primary and secondary windings, and losses in the core due to hysteresis (Section 31–8) and eddy currents (Section 32–6). Hysteresis losses are minimized by the use of soft iron having a narrow hysteresis loop; and eddy currents are minimized by laminating the core. In spite of these losses, transformer efficiencies are usually well over 90%; in large installations they may reach 99%.

We will consider only an idealized transformer in which there are *no* losses and in which *all* of the flux is confined to the iron core, so that the same flux links both primary and secondary. The transformer is shown schematically in Fig. 34–14. A primary winding of N_1 turns and a secondary winding of N_2 turns both encircle the core in the same sense. An ac source with voltage amplitude V_1 is connected to the primary; we first assume the secondary to be open, so that there is no secondary current. The primary winding then functions merely as an inductor.

The primary current, which is small, lags the primary voltage by 90° and is called the *magnetizing* current. Because the primary voltage is 90° out of phase with the current, the power input to the transformer is zero. The core flux is in phase with the primary current. Since the same flux links both primary and secondary, the induced emf *per turn* is the same in each. The ratio of primary to secondary induced emf is therefore equal to the ratio of primary to secondary turns, or

$$\frac{\mathcal{E}_2}{\mathcal{E}_1} = \frac{N_2}{N_1}.$$

Since the windings are assumed to have zero resistance, the induced emf's $\mathcal{E}_1$ and $\mathcal{E}_2$ are numerically equal to the corresponding terminal voltages V_1 and V_2, and

$$\frac{V_2}{V_1} = \frac{N_2}{N_1}. \tag{34–35}$$

A transformer functions as a voltage-conversion device.

Hence by properly choosing the turn ratio N_2/N_1, we may obtain any desired secondary voltage from a given primary voltage. If $V_2 > V_1$, we have a *step-up* transformer; if $V_2 < V_1$, we have a *step-down* transformer.

Consider next the effect of closing the secondary circuit. The secondary current I_2 and its phase angle ϕ_2 will, of course, depend on the nature of the secondary circuit. As soon as the latter is closed, some power must be delivered by the secondary (except when $\phi_2 = 90°$); and, from energy considerations, an equal amount of power must be supplied to the primary. The process by which the transformer is able to draw the necessary power is as follows. When the secondary circuit is open, the core flux is produced by the primary current only. But when the secondary circuit is closed, *both* primary and secondary currents set up flux in the core. The secondary current, by Lenz's law, tends to weaken the core flux and therefore to decrease the induced emf in the primary. But (in the absence of losses) the induced emf in the primary must equal the primary terminal voltage, which is assumed to have constant amplitude. The primary *current* therefore increases until the core flux is restored to its original no-load magnitude.

Faraday's law in a transformer core: flux relations under various operating conditions

If the secondary circuit is completed by a resistance R, $I_2 = V_2/R$. From energy considerations, the power delivered to the primary equals that taken out of the secondary (neglecting losses), so

$$V_1 I_1 = V_2 I_2. \tag{34–36}$$

Combining this equation with the above expression for I_2 and Eq. (34–35) to eliminate V_2 and I_2, we find

$$I_1 = \frac{V_1}{(N_1/N_2)^2 R}. \tag{34–37}$$

Hence, when the secondary circuit is completed through a resistance R, the result is the same as if the *source* had been connected directly to a resistance equal to R multiplied by the reciprocal of the *square* of the turns ratio. In other words, the transformer "transforms" not only voltages and currents but resistances (more generally, impedances) as well. It can be shown that maximum power is supplied by a source to a resistor when its resistance equals the internal resistance of the source. (See Challenge Problem 28–40.) The same principle applies in both dc and ac circuits. When a high-impedance ac source must be connected to a low-impedance circuit, as when an audio amplifier is connected to a loudspeaker, the impedance of the source can be *matched* to that of the circuit by use of a transformer having the correct turns ratio.

A transformer changes the effective impedance of a circuit.

SUMMARY

An alternator or ac source produces an emf that varies sinusoidally with time. Voltages and currents that vary sinusoidally with time can be represented by vectors called phasors. A phasor rotates counterclockwise with constant angular velocity ω equal to the angular frequency of the sinusoidal quantity, and its projection on the horizontal axis at any instant represents the instantaneous value of the quantity.

If the current i in an ac circuit is given by

$$i = I \cos \omega t$$

and the voltage v between two points by

$$v = V \cos (\omega t + \phi),$$

then ϕ is called the phase angle of the voltage relative to the current.

The voltage across a resistor R is in phase with the current, and the amplitudes are related by

$$V = IR. \tag{34–2}$$

The voltage across a capacitor C lags the current by 90°; the amplitudes are related by

$$V = IX_C, \tag{34–9}$$

where $X_C = 1/\omega C$ is the capacitive reactance of the capacitor.

The voltage across an inductor L leads the current by 90°; the amplitudes are related by

$$V = IX_L, \tag{34–16}$$

where $X_L = \omega L$ is the inductive reactance of the inductor.

In an $L–R–C$ series circuit, the voltage and current amplitudes are related by

$$V = IZ, \tag{34–19}$$

where Z is the impedance of the circuit, given by

$$Z = \sqrt{R^2 + X^2}$$

$$= \sqrt{R^2 + (X_L - X_C)^2}$$

$$= \sqrt{R^2 + [\omega L - (1/\omega C)]^2}. \tag{34–20}$$

The phase angle ϕ of the voltage relative to the current is given by

$$\tan \phi = \frac{V_L - V_C}{V_R} = \frac{I(X_L - X_C)}{IR} = \frac{X}{R}. \tag{34–21}$$

The SI unit of capacitive or inductive reactance or impedance is the ohm.

The average value of a sinusoidal current with amplitude I over a half-cycle is given by $I_{av} = 2I/\pi$. This is also equal to the rectified average for a full cycle. The root–mean–square (rms) value of a sinusoidally varying quantity is $1/\sqrt{2}$ times the amplitude; thus $I_{rms} = I/\sqrt{2}$ and $V_{rms} = V/\sqrt{2}$.

The average power input P to an ac circuit is given by

$$P = V_{rms}I_{rms} \cos \phi, \tag{34–30}$$

where ϕ is the phase angle of voltage with respect to current. The quantity $\cos\phi$ is called the power factor.

The current in an L–R–C series circuit becomes maximum, and the impedance minimum, at a frequency $\omega_0 = 1/(LC)^{1/2}$ called the resonant frequency. This phenomenon is called resonance. At resonance the voltage and current are in phase, and the impedance Z is equal to the resistance R.

The current in an L–R–C parallel circuit becomes minimum, and the impedance maximum, at this same resonant frequency ω_0. The impedance Z of this circuit at any frequency is given by

$$\frac{1}{Z} = \sqrt{\frac{1}{R^2} + \left(\omega C - \frac{1}{\omega L}\right)^2}. \tag{34–34}$$

At resonance, $Z = R$.

A transformer is used to transform the voltage and current levels in an ac circuit. If the primary winding has N_1 turns and the secondary N_2, the two voltages are related by

$$\frac{V_2}{V_1} = \frac{N_2}{N_1}. \tag{34–35}$$

If there are no energy losses in the transformer, then the primary and secondary voltages and currents are related by

$$V_1 I_1 = V_2 I_2. \tag{34–36}$$

QUESTIONS

34–1 Some electric-power systems formerly used 25-Hz alternating current instead of the 60 Hz that is now standard. The lights flickered noticeably. Why is this not a problem with 60 Hz?

34–2 Power-distribution systems in airplanes sometimes use 400-Hz ac. What advantages and disadvantages does this have compared to the standard 60 Hz?

34–3 Fluorescent lights often use an inductor, called a "ballast," to limit the current through the tubes. Why is it better to use an inductor than a resistor for this purpose?

34–4 At high frequencies a capacitor becomes a short circuit. Discuss.

34–5 At high frequencies an inductor becomes an open circuit. Discuss.

34–6 Household electric power in most of Western Europe is 240 V, rather than the 120 V that is standard in the United States and Canada. What advantages and disadvantages does each system have?

34–7 The current in an ac-power line changes direction 120 times per second, and its average value is zero. So how is it possible for power to be transmitted in such a system?

34–8 Electric-power connecting cords, such as lamp cords, always have two conductors that carry equal currents in opposite directions. How might one determine, by using measurements only at the midpoints along the lengths of the wires, the direction of power transmission in the cord?

34–9 Are the equations for the average and rms values of current, Eqs. (34–23) and (34–24), correct when the variation with time is not sinusoidal? Explain.

34–10 Electric-power companies like to have their power factors (cf. Section 34–5) as close to unity as possible. Why?

34–11 Some electrical appliances operate equally well on ac or dc, while others work only on ac or only on dc. Give examples of each, and explain the differences.

34–12 When a series-resonant circuit is connected across a 120-V ac line, the voltage rating of the capacitor may be exceeded even if it is rated at 200 or 400 V. How can this be?

34–13 In a parallel-resonant circuit connected across a 120-V line, it is possible for the maximum current rating in the inductor to be exceeded even if the total current through the circuit is very small. How can this be?

34–14 Can a transformer be used with dc? What happens if a transformer designed for 120-V ac is connected to a 120-V dc line?

34–15 During the last quarter of the nineteenth century there was great and acrimonious controversy over whether ac or dc should be used for power transmission. Edison favored dc, George Westinghouse ac. What arguments might each proponent have used to promote his scheme?

EXERCISES

$$\text{Angular frequency} = 2\pi \text{ (frequency)},$$
$$\omega(\text{rad·s}^{-1}) = 2\pi f(\text{Hz}).$$

Section 34–2 Resistance, Inductance, and Capacitance

34–1

a) What is the reactance of a 1-H inductor at a frequency of 60 Hz?

b) What is the inductance of an inductor whose reactance is 1 Ω at 60 Hz?

c) What is the reactance of a 1-μF capacitor at a frequency of 60 Hz?

d) What is the capacitance of a capacitor whose resistance is 1 Ω at 60 Hz?

34–2

a) Compute the reactance of a 10-H inductor at frequencies of 60 Hz and 600 Hz.

b) Compute the reactance of a 10-μF capacitor at the same frequencies.

c) At what frequency is the reactance of a 10-H inductor equal to that of a 10-μF capacitor?

34–3 A 1-μF capacitor is connected across an ac source whose voltage amplitude is kept constant at 50 V but whose frequency can be varied. Find the current amplitude when the angular frequency is

a) 100 rad·s^{-1},   b) 1000 rad·s^{-1},   c) 10,000 rad·s^{-1}.

34–4 The voltage amplitude of an ac source is 50 V, and its angular frequency is 1000 rad·s^{-1}. Find the current amplitude if the capacitance of a capacitor connected across the source is

a) 0.01 μF,   b) 1.0 μF,   c) 100 μF.

d) Construct a log-log plot of current amplitude versus capacitance.

34–5 An inductor of self-inductance 10 H and of negligible resistance is connected across the source in Exercise 34–3. Find the current amplitude when the angular frequency is

a) 100 rad·s^{-1},   b) 1000 rad·s^{-1},   c) 10,000 rad·s^{-1}.

d) Construct a log-log plot of current amplitude versus frequency.

34–6 Find the current amplitude if the self-inductance of a resistanceless inductor connected across the source of Exercise 34–4 is

a) 0.01 H,   b) 1.0 H,   c) 100 H.

d) Construct a log-log plot of current amplitude versus self-inductance.

Section 34–3 The L–R–C Series Circuit

34–7 The expression for the impedance Z of an R–L series circuit can be obtained from Eq. (34–20) by setting $X_C = 0$, which corresponds to $C = \infty$, whereas for an R–C series circuit one obtains the impedance Z from Eq. (34–20) by setting $L = 0$. Explain.

34–8 In an L–R–C series circuit, the source has a constant voltage amplitude of 50 V and a frequency of 1000 rad·s^{-1}. $R = 300$ Ω, $L = 0.9$ H, and $C = 2.0$ μF. Suppose a series circuit contains only the resistor and the inductor in series.

a) What is the impedance of the circuit?

b) What is the current amplitude?

c) What are the voltage amplitudes across the resistor and across the inductor?

d) What is the phase angle ϕ of the source voltage with respect to the current? Does the source voltage lag or lead the current?

e) Construct the phasor diagram.

34–9 Same as Exercise 34–8, except that the circuit consists of only the resistor and the capacitor in series. For part (c) calculate the voltage amplitudes across the resistor and across the capacitor.

34–10 Same as Exercise 34–8, except that the circuit consists of only the capacitor and the inductor in series. For part (c), calculate the voltage amplitudes across the capacitor and across the inductor.

34–11 A 400-Ω resistor is in series with a 0.1-H inductor and a 0.5-μF capacitor. Compute the impedance of the circuit and draw the phasor diagram

a) at a frequency of 500 Hz;

b) at a frequency of 1000 Hz.

Compute, in each case, the phase angle of the source voltage with respect to the current, and state whether the source voltage lags or leads the current.

34–12

a) Compute the impedance of an L–R–C series circuit at angular frequencies of 1000, 750, and 500 rad·s^{-1}. Take $R = 300$ Ω, $L = 0.9$ H, and $C = 2.0$ μF.

b) Describe how the current amplitude varies as the frequency of the source is slowly reduced from 1000 rad·s^{-1} to 500 rad·s^{-1}.

c) What is the phase angle of the source voltage with respect to the current when $\omega = 1000$ rad·s^{-1}? Construct the phasor diagram when $\omega = 1000$ rad·s^{-1}.

d) Repeat part (c) for $\omega = 500$ rad·s^{-1}.

Section 34–5 Power in AC Circuits

34–13 The circuit in Exercise 34–11 carries an rms current of 0.25 A with a frequency of 100 Hz.

a) What average power is delivered by the source?

b) What average power is consumed in the resistor?

c) In the capacitor?

d) In the inductor?

e) Compare your answers in (a) to the sum of (b), (c), and (d).

34–14 A series circuit has a resistance of 75 Ω and an impedance of 150 Ω. What power is consumed in the circuit when a voltage of 120 V (rms) is impressed across it?

Section 34–6 Series Resonance

34–15 Consider the *L–R–C* series circuit of Exercise 34–12. The voltage amplitude of the source is 50 V.

a) At what angular frequency is the circuit in resonance?

b) What is the power factor at resonance? Sketch the phasor diagram at the resonant frequency.

c) What is the reading of each voltmeter in Fig. 34–15 when the source frequency equals the resonant frequency? The voltmeters are calibrated to read rms voltages.

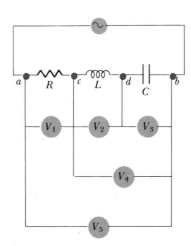

FIGURE 34–15

d) What would be the resonant angular frequency if the resistance were reduced to 100 Ω?

e) What would then be the rms current at resonance?

34–16 In an *L–R–C* series circuit, $R = 250$ Ω, $L = 0.5$ H, and $C = 0.02$ μF.

a) What is the resonant angular frequency of the circuit?

b) The capacitor can withstand a peak voltage of 350 V. If the voltage source operates at the resonant frequency, what maximum voltage amplitude can it have if the maximum capacitor voltage is not exceeded?

Section 34–7 Parallel Resonance

34–17 For the circuit of Fig. 34–13a, let $V = 120$ V, $R = 200$ Ω, $L = 0.5$ H, and $C = 0.2$ μF.

a) What is the resonant angular frequency of the circuit?

b) Sketch the phasor diagram at the resonant frequency.

c) At the resonant frequency, what is the current through the source?

d) At the resonant frequency, what is the current through the resistor?

34–18 Consider the circuit of Fig. 34–13a, with the same numerical values as in Exercise 34–17. At resonance, what is

a) the average rate at which electrical energy is being delivered by the source?

b) the average rate at which electrical energy is being dissipated in the resistor? Compare to the result of (a).

c) Is the current through the inductor, and hence the energy stored in its magnetic field, zero at all times? If not, how can the result obtained in (b) be explained?

d) Calculate the maximum energy stored in the inductor.

e) Calculate the maximum energy stored in the capacitor.

Section 34–8 Transformers

34–19 A transformer connected to a 120-V ac line is to supply 12 V to a low-voltage lighting system for a model-railroad village. The total equivalent resistance of the system is 2 Ω.

a) What should be the ratio of primary to secondary turns of the transformer?

b) What current must the secondary supply?

c) What power is delivered to the load?

d) What resistance connected directly across the 120-V line would draw the same power as the transformer? Show that this quantity is equal to 2 Ω times the square of the ratio of primary to secondary turns.

34–20 A step-up transformer connected to a 120-V ac line is to supply 18,000 V for a neon sign. To reduce shock hazard, a fuse is to be inserted in the primary circuit; the fuse is to blow when the current in the secondary circuit exceeds 10 mA.

a) What is the ratio of secondary to primary turns of the transformer?

b) What power must be supplied to the transformer when the secondary current is 10 mA?

c) What current rating should the fuse in the primary circuit have?

34–21 The internal resistance of an ac source is 10,000 Ω.

a) What should be the ratio of primary to secondary turns of a transformer to match the source to a load of resistance of 10 Ω? ("Matching" means that the effective load resistance equals the internal resistance of the source. Refer to Exercise 34–19, part d.)

b) If the voltage amplitude of the source is 100 V, what is the voltage amplitude in the secondary circuit, under open-circuit conditions?

PROBLEMS

34–22 At a certain frequency ω_1, the reactance of a certain capacitor equals that of a certain inductor.

a) If the frequency is changed to $\omega_2 = 2\omega_1$, what is the ratio of the reactance of the inductor to that of the capacitor, and which reactance is larger?

b) If the frequency is changed to $\omega_3 = \omega_1/3$, what is the ratio of the reactance of the inductor to that of the capacitor, and which reactance is larger?

34–23 A coil has a resistance of 20 Ω. At a frequency of 100 Hz, the voltage across the coil leads the current in it by 30°. Determine the inductance of the coil.

34–24 Five infinite-impedance voltmeters, calibrated to read rms values, are connected as shown in Fig. 34–15. Take R, L, C, and V as given in Exercise 34–8. What is the reading of each voltmeter

a) if $\omega = 500 \text{ rad·s}^{-1}$?

b) if $\omega = 1000 \text{ rad·s}^{-1}$?

34–25 An inductor having a reactance of 25 Ω and a resistance R gives off heat at the rate of 10 J·s^{-1} when it carries a current of 0.5 A (rms). What is the impedance of the inductor?

34–26 A circuit draws 330 W from a 110-V (rms), 60-Hz ac line. The power factor is 0.6, and the current lags the source voltage.

a) What is the net resistance R of the circuit?

b) Find the capacitance of the series capacitor that will result in a power factor of unity when it is added to the original circuit.

c) What power will then be drawn from the supply line?

34–27 A series circuit has an impedance of 50 Ω and a power factor of 0.6 at 60 Hz, the voltage lagging the current.

a) Should an inductor or a capacitor be placed in series with the circuit to raise its power factor?

b) What size element will raise the power factor to unity?

34–28 A resistor of 500 Ω and a capacitor of 2 μF are connected in parallel to an ac generator supplying a constant voltage amplitude of 282 V and having an angular frequency of 377 rad·s^{-1}. Find

a) the current amplitude in the resistor,

b) the current amplitude in the capacitor,

c) the phase angle (sketch the phasor diagram for the current phasors),

d) the line-current amplitude.

34–29 A 100-Ω resistor, a 0.1-μF capacitor, and a 0.1-H inductor are connected in parallel to a voltage source with an amplitude of 100 V.

a) What is the resonant frequency? The resonant angular frequency?

b) What is the maximum total current through the parallel combination at the resonant frequency?

c) What is the maximum current in the resistor at resonance?

d) What is the maximum current in the inductor at resonance?

e) What is the maximum current in the branch containing the capacitor at resonance?

f) What is the maximum energy stored in the inductor at resonance? In the capacitor?

34–30 The same three components as in Problem 34–29 are connected in *series* to a voltage source with an amplitude of 100 V.

a) What is the resonant frequency? The resonant angular frequency?

b) What is the maximum current in the resistor at resonance?

c) What is the maximum voltage across the capacitor at resonance?

d) What is the maximum voltage across the inductor at resonance?

e) What is the maximum energy stored in the capacitor at resonance? In the inductor?

34–31 The current in a certain circuit varies with time as shown in Fig. 34–16. Find the average current and the rms current, in terms of I_0.

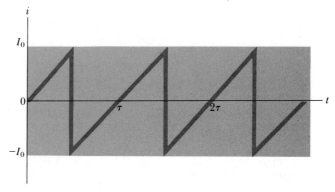

FIGURE 34–16

34–32 Consider a series circuit with a source of terminal voltage amplitude $V = 120$ V and frequency ω. The inductor inductance is $L = 1.2$ H, the capacitor capacitance is $C = 0.8$ μF, and the resistor resistance is $R = 400$ Ω.

a) What is the resonant angular frequency ω_0 of the circuit?

b) What is the current through the source at resonance?

c) For what two values of the angular frequency, ω_1 and ω_2, will the current be half the resonant value? The quantity $|\omega_1 - \omega_2|$ defines the *width* of the resonance.

d) Calculate the resonance width for $R = 4$ Ω, 40 Ω, and 400 Ω.

CHALLENGE PROBLEMS

34–33

a) At what angular frequency is the voltage amplitude across the *resistor* in an L–R–C series circuit at maximum value?

b) At what angular frequency is the voltage amplitude across the *inductor* at maximum value?

c) At what angular frequency is the voltage amplitude across the *capacitor* at maximum value?

34–34 The voltage drop across a circuit element in an ac circuit is not necessarily in phase with the current through that circuit element; hence the voltage amplitudes across the circuit elements in a branch in an ac circuit do not add algebraically. A method commonly employed to simplify the analysis of an ac circuit driven by a sinusoidal source is to represent the impedance Z as a complex number rather than as a real number. The resistance R is taken to be the real part of the impedance, and the reactance $X = X_L - X_C$ is taken to be the imaginary part. Thus for a branch containing a resistor, inductor, and capacitor in series, the complex impedance would be given by $Z_{cpx} = R + iX$, where $i^2 = -1$. If the voltage amplitude across the branch is V_{cpx} and the current amplitude is I_{cpx}, these quantities are related by $I_{cpx} = V_{cpx}/Z_{cpx}$. The actual current amplitude would be computed by taking the absolute value of the complex representation of the current amplitude; that is, $I = (I^*_{cpx}I_{cpx})^{1/2}$. The phase angle ϕ of the current with respect to the source voltage is given by

$$\tan(\phi) = Im(I_{cpx})/Re(I_{cpx}).$$

The voltage amplitudes V_{Rcpx}, V_{Lcpx}, and V_{Ccpx} across the resistance, inductance, and capacitance, respectively, are found by multiplying I_{cpx} by R, iX_L, or $-iX_C$, respectively. If we use the complex representation for the voltage amplitudes, the voltage drop across a branch can be found by algebraic addition of the voltage drops across each circuit element; that is,

$$V_{cpx} = V_{Rcpx} + V_{Lcpx} + V_{Ccpx}.$$

When we want to find the actual value of any current amplitude or voltage amplitude, the absolute value of the appropriate quantity is then used.

Consider the series L–R–C circuit shown in Fig. 34–17. The angular frequency ω is 1000 s^{-1}, and the voltage amplitude of the source is 100 V. The values of the circuit elements are as shown.

Use the phasor-diagram techniques presented in Section 34–3 to solve for

a) the current amplitude;

b) the phase angle ϕ of the current with respect to the source voltage. (Note that this angle is the negative of the phase angle defined in Fig. 34–5.)

Now apply the procedure outlined above to the same circuit.

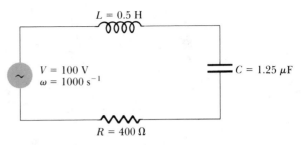

FIGURE 34–17

c) Determine the impedance of the circuit, Z_{cpx}, as represented by a complex number. Take the absolute value to obtain Z, the actual impedance of the circuit.

d) Take the voltage amplitude of the source, V_{cpx}, to be real, and find the complex current amplitude I_{cpx}. Find the actual current amplitude by taking the absolute value of I_{cpx}.

e) Find the phase angle ϕ of the current with respect to the source voltage by using the real and imaginary parts of I_{cpx}, as explained above.

f) Find the complex representations of the voltages across the resistance, the inductance, and the capacitance.

g) Add the answers found in part (f) and verify that the sum of these complex numbers is real and equal to 100 V, the voltage of the source.

34–35 The use of complex numbers to represent impedance, voltage amplitudes, and current amplitudes in an ac circuit is not restricted to series circuits but can be applied to parallel circuits as well. In the case of parallel circuits, the (complex) impedances are combined by using the same reciprocal combination rule used in a dc circuit for finding the equivalent resistance of resistors in parallel.

Consider the circuit shown in Fig. 34–18. The voltage amplitude of the source is 100 V, and its angular frequency is $\omega = 1000 \text{ s}^{-1}$. The values of the passive circuit elements are as shown.

a) Determine the (complex) impedance of the branch containing the 0.3-H inductance and label it Z_{1cpx}.

b) Determine the (complex) impedance of the branch containing the 2.5-μF capacitor and label it Z_{2cpx}.

c) Determine the (complex) equivalent impedance Z_{cpx} of the circuit. Do this by using the relation $1/Z_{cpx} = 1/Z_{1cpx} + 1/Z_{2cpx}$.

d) Determine the current I_{cpx} that passes through the source by using the relation $I_{cpx} = V_{cpx}/Z_{cpx}$, where $V_{cpx} = 100 \text{ V}$.

e) Determine the actual circuit current amplitude I and the phase angle ϕ of the current through the source relative to the source voltage.

f) Determine the (complex) current I_{1cpx} through the branch containing the 0.3-H inductor.

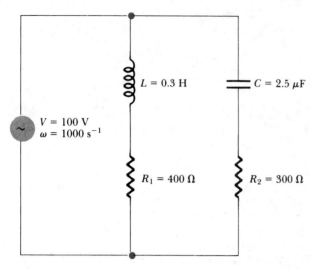

$L = 0.3$ H

$C = 2.5$ μF

$V = 100$ V
$\omega = 1000$ s^{-1}

$R_1 = 400$ Ω

$R_2 = 300$ Ω

g) Determine the actual current amplitude I_1 through this branch, as well as the phase angle ϕ_1 of the branch current with respect to the circuit voltage.

h) Determine the complex current I_{2cpx} through the branch containing the capacitor.

i) Determine the actual current I_2 through this branch, as well as the phase angle ϕ_2 of the branch current with respect to the circuit voltage.

j) Verify that $I_{cpx} = I_{1cpx} + I_{2cpx}$.

FIGURE 34–18

35

ELECTROMAGNETIC WAVES

AN ELECTROMAGNETIC WAVE CONSISTS OF TIME-VARYING ELECTRIC AND magnetic fields that travel or propagate through space with a definite speed. The theoretical basis for electromagnetic waves, discovered by Maxwell in 1865, rests on the ability of a time-varying *electric* field and its associated displacement current to act as a source of *magnetic* field, and on the ability of a time-varying *magnetic* field to act as a source of *electric* field, according to Faraday's law. Electromagnetic waves carry energy and momentum and show the property of polarization. The simplest electromagnetic waves are sinusoidal waves, in which electric and magnetic fields vary sinusoidally with time. The spectrum of electromagnetic waves covers an extremely broad range of frequency and wavelength, and these waves play a central role in a wide variety of physical phenomena and life processes. In particular, light consists of electromagnetic waves. This chapter forms a bridge between electromagnetism and optics; we can understand such basic optical concepts as reflection and refraction on the basis of the electromagnetic nature of light.

35–1 INTRODUCTION

In the last several chapters we studied various aspects of electric and magnetic fields, falling in two general categories. The first category includes fields that do not vary with time. The electrostatic field of a distribution of charges at rest and the magnetic field of a steady current in a conductor are examples of fields that do not vary with time at any individual point, although they may vary from point to point in space. For these situations we could treat the electric and magnetic fields independently, without worrying very much about interactions between them.

The second category includes fields that *do* vary with time, and in all such cases we found that it is *not* possible to treat the fields independently. Faraday's law tells us that a time-varying magnetic field acts as a source of electric field. This field is manifested in the induced emf's in inductances and transformers. Similarly, in developing the general formulation of Ampere's law (Section 31–9), which is valid for charging capacitors and similar situations as

A bootstrap effect: Each field acts as a source of the other field.

well as for ordinary conductors, we found that a changing electric field acts as a source of magnetic field. This mutual interaction between the two fields is summarized neatly in Maxwell's equations, which we studied in Section 32–7.

Thus, when *either* an electric or a magnetic field is changing with time, a field of the other kind is induced in adjacent regions of space. We are led (as Maxwell was) to consider the possibility of an electromagnetic disturbance, consisting of time-varying electric and magnetic fields, that can propagate through space from one region to another even when no matter is present in the intervening region. Such a disturbance, if it exists, will have the properties of a *wave*, and an appropriate term is **electromagnetic wave.** Such waves do exist, of course, and it is a familiar fact that radio and television transmission, light, x-rays, and many other phenomena are examples of electromagnetic radiation. In this chapter we will show how the existence of such waves is related to the principles of electromagnetism we have studied thus far, and we will examine the properties of these waves.

As so often happens in the development of science, the theoretical understanding of electromagnetic waves originally took a considerably more devious path than the one just outlined. In the early days of electromagnetic theory (the early nineteenth century), two different units of electric charge were used, one for electrostatics, the other for magnetic phenomena involving currents. In the particular system of units in common use at the time, these two units of charge had different physical dimensions; their *ratio* turned out to have units of velocity. This in itself was not so astounding, but experimental measurements revealed that the ratio had a numerical value precisely equal to the speed of light, 3.00×10^8 m·s^{-1}. At the time, physicists regarded this as an extraordinary coincidence and had no idea how to explain it.

During his search for understanding of this result, Maxwell proved in 1865 that an electromagnetic disturbance should propagate in free space with a speed equal to that of light, and hence that light waves were very likely to be electromagnetic in nature. At the same time he discovered that the basic principles of electromagnetism could be expressed in terms of the four equations we now call **Maxwell's equations.** To review our discussion of Section 32–7, these four relationships are (1) Gauss's law, (2) the absence of magnetic charge, (3) Ampere's law, including displacement current, and (4) Faraday's law. Maxwell's equations, in the integral form used in this text, are

$$\oint \boldsymbol{E} \cdot d\boldsymbol{A} = \frac{Q}{\epsilon_0}, \qquad \text{(Gauss's law)} \qquad (25\text{--}17)$$

$$\oint \boldsymbol{B} \cdot d\boldsymbol{A} = 0, \qquad \text{(no magnetic charge)} \qquad (31\text{--}16)$$

$$\oint \boldsymbol{B} \cdot d\boldsymbol{l} = \mu_0 \left(I_{\mathrm{C}} + \epsilon_0 \frac{d\Psi}{dt} \right), \qquad \text{(Ampere's law)} \qquad (31\text{--}32)$$

$$\oint \boldsymbol{E} \cdot d\boldsymbol{l} = -\frac{d\Phi}{dt}. \qquad \text{(Faraday's law)} \qquad (32\text{--}14)$$

These equations apply to electric and magnetic fields *in vacuum.* If a material is present, the permittivity ϵ_0 and permeability μ_0 of free space are replaced by the permittivity ϵ and permeability μ of the material. We have already remarked that Maxwell's synthesis of electromagnetism in these four equations is one of the great milestones of theoretical physics, comparable to the Newtonian synthesis 150 years earlier.

In 1887 Heinrich Hertz actually *produced* electromagnetic waves, with the aid of oscillating circuits, and received and detected these waves with other circuits tuned to the same frequency. Hertz then produced electromagnetic standing waves and measured the distance between adjacent nodes (one-half wavelength), to determine the wavelength. Knowing the resonant frequency of his circuits, he then found the speed of the waves from the fundamental wave equation $c = \lambda f$ and verified Maxwell's theoretical value directly. The SI unit of frequency, one cycle per second, is named the *hertz* in honor of Hertz.

The possible use of electromagnetic waves for purposes of long-distance communication does not seem to have occurred to Hertz. It remained for the enthusiasm and energy of Marconi and others to make radio communication a familiar household phenomenon.

35–2 SPEED OF AN ELECTROMAGNETIC WAVE

In developing the relation of electromagnetic-wave propagation to familiar electromagnetic principles, we begin with a particularly simple and somewhat artificial example of an electromagnetic wave. We first postulate a particular configuration of fields, and then test whether it is consistent with the principles mentioned above, particularly Faraday's law and Ampere's law including displacement current.

Using an x–y–z-coordinate system, as shown in Fig. 35–1, we imagine that all space is divided into two regions by a plane perpendicular to the x-axis (parallel to the yz-plane). At all points to the left of this plane, there is a uniform electric field E in the $+y$-direction and a uniform magnetic field B in the $+z$-direction, as shown. Furthermore, we suppose that the boundary surface, which may also be called the *wave front*, moves to the right with a constant speed c, as yet unknown. Thus the E and B fields travel to the right into previously field-free regions, with a definite speed. The situation, in short, describes a rudimentary electromagnetic wave, provided that it is consistent with the laws of electromagnetism.

We will not worry about how such a field configuration might actually be produced. Instead, we simply ask whether it is consistent with the laws of electrodynamics, that is, with Maxwell's equations. First we apply Faraday's law to a rectangle in the xy-plane, as in Fig. 35–2. The rectangle is located so that at some instant the wave front has progressed partway through it, as shown. In a time dt, the boundary surface moves a distance $c\,dt$ to the right, sweeping out an area $ac\,dt$ of the rectangle. In this time the magnetic flux increases by $B(ac\,dt)$. We take the vector area element to point in the $+z$-direction, so this is a positive quantity. The rate of change of magnetic flux is given by

$$\frac{d\Phi}{dt} = Bac. \qquad (35\text{–}1)$$

According to Faraday's law, this expression must equal the negative of the line integral $\oint E \cdot dl$, evaluated counterclockwise around the boundary of the area. The field at the right end is zero, and the top and bottom sides do not contribute because there the component of E along dl is zero. Thus only the left side contributes to the integral. On that side, E and dl have opposite directions, so the value of the integral is $-Ea$. When we divide out a, Faraday's law gives

$$E = cB. \qquad (35\text{–}2)$$

Hertz produced electromagnetic waves in his laboratory: a dramatic confirmation of Maxwell's theoretical predictions.

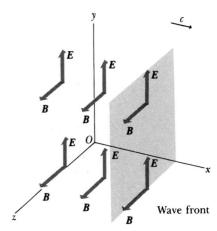

35–1 An electromagnetic wave front. The E and B fields are uniform over the region to the left of the plane, but are zero everywhere to the right of it. The plane representing the wave front moves to the right with speed c.

Checking for consistency with Faraday's law

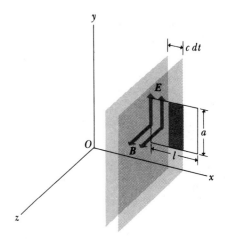

35–2 In time dt the wave front moves to the right a distance $c\,dt$. The magnetic flux through the rectangle in the xy-plane increases by an amount $d\Phi$ equal to the flux through the shaded rectangle of area $ac\,dt$; that is, $d\Phi = Bac\,dt$.

Thus the wave we have postulated is consistent with Faraday's law only if E, B, and c are related as in Eq. (35–2).

Next we apply Ampere's law to a rectangle in the xz-plane, as shown in Fig. 35–3. There is no conduction current, so $I_C = 0$. In terms of electric flux Ψ, Ampere's law with displacement current but no conduction current is

$$\oint B \cdot dl = \mu_0 \epsilon_0 \frac{d\Psi}{dt}. \qquad (35\text{–}3)$$

Checking for consistency with Ampere's law (with displacement current)

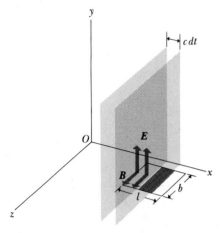

35–3 In time dt the electric flux through the rectangle in the xz-plane increases by an amount equal to E times the area $bc\,dt$ of the shaded rectangle; that is, $d\Psi = Ebc\,dt$. Thus $d\Psi/dt = Ebc$.

In Fig. 35–3 we take the vector area to point in the $+y$-direction, so Ψ is positive and increasing. The line integral is then evaluated counterclockwise around the boundary of the area. The change in electric flux $d\Psi$ in time dt is the area $bc\,dt$ swept out by the wave front, multiplied by E. In evaluating the line integral of B, we note that B is zero on the right end of the rectangle and is perpendicular to the boundary on the front and back sides. Thus only the left end, where B and dl are parallel, contributes to the integral, and we find $\oint B \cdot dl = Bb$. Combining these results with Eq. (35–3) and dividing out the common factor b, we obtain

$$B = \epsilon_0 \mu_0 cE. \qquad (35\text{–}4)$$

Thus Ampere's law is obeyed only if B, c, and E are related as in Eq. (35–4).

Since *both* Ampere's law and Faraday's law must be obeyed at the same time, Eqs. (35–2) and (35–4) must both be satisfied. This can happen only when $\epsilon_0 \mu_0 c = 1/c$, or

$$c = \frac{1}{\sqrt{\epsilon_0 \mu_0}}. \qquad (35\text{–}5)$$

Inserting the numerical values of these quantities, we find

$$c = \frac{1}{\sqrt{(8.85 \times 10^{-12})(4\pi \times 10^{-7})}} = 3.00 \times 10^8 \text{ m·s}^{-1}.$$

The postulated field configuration *is* consistent with the laws of electrodynamics, provided the wave front moves with the speed given above, which of course is recognized as the speed of light.

We have chosen a particularly simple wave for study in order to avoid undue mathematical complexity, but this special case nevertheless illustrates several important features of *all* electromagnetic waves:

Some significant properties of electromagnetic waves

1. The wave is **transverse;** both E and B are perpendicular to the direction of propagation of the wave and to each other.

2. There is a definite ratio between the magnitudes of E and B.

3. The wave travels in vacuum with a definite and unchanging speed.

Constructing any electromagnetic wave as a superposition from our primitive example

It is not difficult to generalize this discussion to a more realistic situation. Suppose we have several wave fronts in the form of parallel planes perpendicular to the x-axis and all moving to the right with speed c. And suppose that, within a single region between two planes, the E and B fields are the same at all points in the region, but that they differ from region to region. An extension of the development above shows that such a situation is also consistent with Ampere's and Faraday's laws, provided the wave fronts all move with the speed c given by Eq. (35–5). From this picture it is only a short additional step to a wave picture in which the E and B fields at any instant vary smoothly rather than in steps, as we move along the x-axis, and the entire field pattern moves to the right with speed c. In Section 35–5 we consider waves in which the dependence of E and B on position and time is *sinusoidal;* first, however, we consider the *energy* associated with an electromagnetic wave.

35–3 ENERGY IN ELECTROMAGNETIC WAVES

It is a familiar fact that there is energy associated with electromagnetic waves. Two simple examples are the energy in the sun's radiation and cooking with microwave ovens. To derive detailed relationships for the energy in an electromagnetic wave, we begin with the expressions derived in Sections 27–4 and 33–3 for the **energy densities** associated with electric and magnetic fields; we suggest you review those derivations now. Specifically, Eqs. (27–9) and (33–10) show that the total energy density u in a region of space where E and B fields are present is given by

The total energy density is the sum of the energy densities associated with the electric and magnetic fields.

$$u = \frac{1}{2}\epsilon_0 E^2 + \frac{1}{2\mu_0}B^2. \qquad (35\text{–}6)$$

We found in Section 35–2 that E and B in an electromagnetic wave are not independent but are related by

$$B = \frac{E}{c} = \sqrt{\epsilon_0\mu_0}E. \qquad (35\text{–}7)$$

Thus the energy density may also be expressed as

$$\begin{aligned} u &= \frac{1}{2}\epsilon_0 E^2 + \frac{1}{2\mu_0}\left(\sqrt{\epsilon_0\mu_0}E\right)^2 \\ &= \epsilon_0 E^2, \qquad (35\text{–}8) \end{aligned}$$

which shows that, in a wave, the energy density associated with the E field is equal to that of the B field.

Because the E and B fields in the simple wave considered above advance with time into regions where originally no fields were present, it is clear that the wave transports energy from one region to another. We can describe this energy transfer in terms of energy transferred *per unit time, per unit cross-sectional area* for an area perpendicular to the direction of wave travel. This quantity will be denoted by S. This is analogous to the concept of current density, which is the charge per unit time transferred across unit area perpendicular to the direction of flow.

To see how the energy flow is related to the fields, consider a stationary plane, perpendicular to the x-axis, that coincides with the wave front at a certain time. In a time dt after this, the wave front moves a distance $c\,dt$ to the right. Considering an area A on the stationary plane, we note that the energy in the space to the right of this area must have passed through it to reach its new location. The volume dV of the relevant region is the base area A times the length $c\,dt$, and the total energy dU in this region is the energy density u times this volume:

$$dU = \epsilon_0 E^2 A c\,dt. \qquad (35\text{–}9)$$

Since this much energy passed through area A in time dt, the energy flow S

The comet Mrkos, photographed August 22 through 27, 1957. The tail of the comet is pushed away from the sun and split into two distinct parts by radiation pressure from the sun's electromagnetic radiation and by the "solar wind," a stream of particles emitted by the sun. (Courtesy of Hale Observatories.)

Energy flow can be expressed as power per unit area.

per unit time, per unit area, is

$$S = \frac{1}{A}\frac{dU}{dt} = \epsilon_0 c E^2.$$

Using Eq. (35–7), we obtain the alternative forms

$$S = \frac{\epsilon_0}{\sqrt{\epsilon_0\mu_0}}E^2 = \sqrt{\frac{\epsilon_0}{\mu_0}}E^2 = \frac{EB}{\mu_0}. \qquad (35\text{–}10)$$

The unit of S is energy per unit time, per unit area. The SI unit of S is $1\ \text{J·s}^{-1}\text{·m}^{-2}$ or $1\ \text{W·m}^{-2}$.

We can define a *vector* quantity that describes both the magnitude and the direction of the energy-flow rate:

$$S = \frac{1}{\mu_0}E \times B. \qquad (35\text{–}11)$$

The Poynting vector: both magnitude and direction of energy flow

The average magnitude of the Poynting vector is called intensity.

S is called the **Poynting vector;** its magnitude is given by Eq. (35–10), and its direction is the direction of propagation of the wave. The magnitude EB/μ_0 *gives the flow of energy through a cross section perpendicular to the propagation direction, per unit area and per unit time.* The total energy flow per unit time (power, P) through any surface is given by the integral

$$P = \int S \cdot dA$$

over the surface.

Intensity is average power per unit area.

The electric and magnetic fields at any point in a wave vary with time, so the Poynting vector at any point is also a function of time. The *average* value of the magnitude of the Poynting vector at a point is called the **intensity** of the radiation at that point. As mentioned above, the SI unit of intensity is the watt per square meter (W·m^{-2}). In Section 35–5 we consider the relation between the intensity of a sinusoidal wave and the amplitudes of the sinusoidally varying electric and magnetic fields.

EXAMPLE 35–1   For the wave described in Section 35–2, suppose

$$E = 100\ \text{V·m}^{-1} = 100\ \text{N·C}^{-1}.$$

Find the value of B, the energy density, and the rate of energy flow per unit area S.

SOLUTION   From Eq. (35–7),

$$B = \frac{E}{c} = \frac{100\ \text{V·m}^{-1}}{3.0 \times 10^8\ \text{m·s}^{-1}} = 3.33 \times 10^{-7}\ \text{T}.$$

From Eq. (35–8),

$$\begin{aligned}
u &= \epsilon_0 E^2 = (8.85 \times 10^{-12}\ \text{C}^2\text{·N}^{-1}\text{·m}^{-2})(100\ \text{N·C}^{-1})^2 \\
&= 8.85 \times 10^{-8}\ \text{N·m}^{-2} = 8.85 \times 10^{-8}\ \text{J·m}^{-3};
\end{aligned}$$

$$\begin{aligned}
S &= EB/\mu_0 = (100\ \text{V·m}^{-1})(3.33 \times 10^{-7}\ \text{T})/(4\pi \times 10^{-7}\ \text{Wb·A}^{-1}\text{·m}^{-1}) \\
&= 26.5\ \text{V·A·m}^{-2} = 26.5\ \text{W·m}^{-2}.
\end{aligned}$$

Alternatively,

$$S = \epsilon_0 c E^2$$
$$= (8.85 \times 10^{-12}\ \text{C}^2 \cdot \text{N}^{-1} \cdot \text{m}^{-2})(3.0 \times 10^8\ \text{m} \cdot \text{s}^{-1})(100\ \text{N} \cdot \text{C}^{-1})^2$$
$$= 26.5\ \text{W} \cdot \text{m}^{-2}.$$

The idea that energy can travel through empty space without the aid of any matter in motion may seem strange, yet this is the very mechanism by which energy reaches us from the sun. The conclusion that electromagnetic waves transport energy is as inescapable as the conclusion that energy is required to establish electric and magnetic fields. It is also possible to show that electromagnetic waves carry *momentum p*, with a corresponding momentum density (momentum p per unit volume V) of magnitude

Electromagnetic waves transport momentum as well as energy.

$$\frac{p}{V} = \frac{EB}{\mu_0 c^2} = \frac{S}{c^2}. \qquad (35\text{–}12)$$

This momentum is a property of the field alone and is not associated with moving mass. There is also a corresponding momentum-flow rate; just as the energy density u corresponds to S, the rate of energy flow per unit area, the momentum density given by Eq. (35–12) corresponds to the momentum-flow rate per unit area

$$\frac{1}{A}\frac{dp}{dt} = \frac{S}{c} = \frac{EB}{\mu_0 c}, \qquad (35\text{–}13)$$

which represents the momentum transferred per unit surface area, per unit time.

This momentum is responsible for the phenomenon of **radiation pressure.** When an electromagnetic wave is completely absorbed by a surface perpendicular to the propagation direction, the time rate of change of momentum equals the force on the surface. Thus the force per unit area, or pressure, is equal to S/c. If the wave is totally reflected, the momentum change is twice as great, and the pressure is $2S/c$. For example, the value of S for direct sunlight is about $1.4\ \text{kW} \cdot \text{m}^{-2}$, and the corresponding pressure on a completely absorbing surface is

Radiation pressure: a manifestation of momentum in electromagnetic waves

$$\frac{S}{c} = \frac{1.4 \times 10^3\ \text{W} \cdot \text{m}^{-2}}{3.0 \times 10^8\ \text{m} \cdot \text{s}^{-1}} = 4.7 \times 10^{-6}\ \text{Pa}$$
$$= 4.7 \times 10^{-6}\ \text{N} \cdot \text{m}^{-2}.$$

The pressure on a totally reflecting surface is $9.4 \times 10^{-6}\ \text{N} \cdot \text{m}^{-2}$.

Radiation pressure is important in the structure of stars. Gravitational attractions tend to shrink a star, but this tendency is balanced by radiation pressure in maintaining the size of the star through most stages of its evolution. The pressure of the sun's radiation is responsible for pushing the tail of a comet away from the sun.

The usefulness of the Poynting vector is not limited to wave situations; it also describes energy flow in situations where the fields are constant.

The Poynting vector is meaningful even when there is no wave motion.

EXAMPLE 35–2   A long cylindrical conductor of radius R and length l carries a steady current I, and the potential difference between its ends is V. Find the magnitude and direction of the Poynting vector at a point on the surface, and relate this to the energy input to the conductor.

SOLUTION   The electric field is uniform inside the conductor, with magnitude $E = V/l$ and direction parallel to the cylinder's axis. The magnitude of the magnetic field at a point on the surface is given by Eq. (31–14), with $x = R$: $B = \mu_0 I / 2\pi R$. Its direction is tangent to the surface, perpendicular to the axis and to E. The magnitude of the Poynting vector is

$$S = \frac{EB}{\mu_0} = \frac{VI}{2\pi Rl}.$$

Equation (35–11) and the right-hand rule show that at each point on the surface, the direction of S is radially inward, toward the axis. Now VI is just the total *power* input P to the conductor, and $2\pi Rl$ is its surface area. Thus these quantities are related by

$$P = SA.$$

Although this situation is not a wave phenomenon, it does illustrate the usefulness of the Poynting vector in computing energy flow associated with electric and magnetic fields.

35–4 ELECTROMAGNETIC WAVES IN MATTER

Electromagnetic waves can travel through dielectrics.

Thus far we have talked only about electromagnetic waves in vacuum, but it is easy to extend our analysis to include electromagnetic waves in dielectrics. The wave speed is not the same as in vacuum, so we denote it by v instead of c. Faraday's law is unaltered, but Eq. (35–2) is replaced by $E = vB$. In Ampere's law the displacement-current density is given not by $\epsilon_0\, dE/dt$ but by $\epsilon\, dE/dt = K\epsilon_0\, dE/dt$. In addition, the constant μ_0 in Ampere's law must be replaced by $\mu = K_m\mu_0$. Thus Eq. (35–4) is replaced by

$$B = \mu\epsilon vE,$$

and the wave speed is given by

$$v = \frac{1}{\sqrt{\epsilon\mu}} = \frac{1}{\sqrt{KK_m}}\,\frac{1}{\sqrt{\epsilon_0\mu_0}}. \tag{35–14}$$

For many dielectrics the relative permeability K_m is very nearly equal to unity; in such cases we say

$$v \simeq \frac{1}{\sqrt{K}}\,\frac{1}{\sqrt{\epsilon_0\mu_0}} = \frac{c}{\sqrt{K}}.$$

What is the speed of electromagnetic waves in a dielectric?

The speed depends on the electric and magnetic properties of the material.

Because K is always greater than unity, the speed v of electromagnetic waves in a dielectric is always *less* than the speed c in vacuum, by a factor of $1/\sqrt{K}$. The ratio of the speed c in vacuum to the speed v in a material is known in optics as the **index of refraction** n of the material. For most dielectrics, where $K_m \simeq 1$, n is given by

$$\frac{c}{v} = n = \sqrt{KK_m} \simeq \sqrt{K}. \tag{35–15}$$

We also have to make corresponding modifications in the expressions for the energy density and the Poynting vector. The energy density is now given by

$$u = \frac{1}{2}\epsilon E^2 + \frac{1}{2}\frac{1}{\mu}B^2 = \epsilon E^2. \qquad (35\text{--}16)$$

The energy densities in the *E* and *B* fields are still equal. The Poynting vector is now

$$S = \frac{1}{\mu}E \times B. \qquad (35\text{--}17)$$

The waves described above cannot propagate any appreciable distance in a *conducting* material because the *E* field leads to currents that provide a mechanism for dissipating the energy of the wave. For an ideal conductor with zero resistivity, *E* must be zero everywhere inside the material. When an electromagnetic wave strikes such a material, it is totally reflected. Real conductors with finite resistivity permit some penetration of the wave into the material, with partial reflection. A polished metal surface is usually a good *reflector* of electromagnetic waves, but metals are not *transparent* to radiation.

Why can't electromagnetic waves travel through a conductor?

35–5 SINUSOIDAL WAVES

Sinusoidal electromagnetic waves are directly analogous to sinusoidal transverse mechancal waves on a stretched string. We studied these in Chapter 21; we suggest you review that discussion now, especially Section 21–3. In a sinusoidal electromagnetic wave, the *E* and *B* fields at any point in space are sinusoidal functions of time, and at any instant of time the spatial variation of the fields is also sinusoidal.

The simplest sinusoidal electromagnetic waves share with the waves described in Section 35–2 the property that at any instant the fields are uniform over any plane perpendicular to the direction of propagation. Such a wave is called a **plane wave.** The entire pattern travels in the direction of propagation with speed *c*. The directions of *E* and *B* are perpendicular to the direction of propagation (and to each other), so the wave is *transverse*.

Plane waves travel in one direction and are particularly simple to describe.

The frequency *f*, the wavelength λ, and the speed of propagation *c* are related by the equation applicable to all periodic-wave motion, namely, $c = f\lambda$. If the frequency *f* is the power-line frequency of 60 Hz, the wavelength is

$$\lambda = \frac{c}{f} = \frac{3 \times 10^8 \text{ m·s}^{-1}}{60 \text{ Hz}} = 5 \times 10^6 \text{ m} = 5000 \text{ km},$$

which is of the order of the earth's radius! Hence at this frequency even a distance of many miles includes only a small fraction of a wavelength. But if the frequency is 10^8 Hz (100 MHz), typical of commercial FM radio stations, the wavelength is

$$\lambda = \frac{3 \times 10^8 \text{ m·s}^{-1}}{10^8 \text{ Hz}} = 3 \text{ m},$$

and a moderate distance can include many complete waves.

Figure 35–4 shows a sinusoidal electromagnetic wave traveling in the +*x*-direction. The *E* and *B* vectors are shown only for a few points on the *x*-axis. If

An example of a plane sinusoidal electromagnetic wave

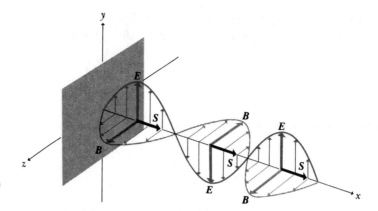

35–4 *E*, *B*, and *S* vectors in a sinusoidal electromagnetic wave traveling in the positive *x*-direction. The position of the wave at time $t = 0$ is shown.

The *E* and *B* fields are perpendicular to each other and to the direction of propagation.

we construct a plane perpendicular to the *x*-axis at a particular point and at a particular instant, the fields have the same values at all points in that plane. The values are, of course, different on different planes. In those planes in which the *E* vector is in the +*y*-direction, *B* is in the +*z*-direction; where *E* is in the −*y*-direction, *B* is in the −*z*-direction. In both cases the direction of the Poynting vector, given by Eq. (35–11), is along the +*x*-direction.

We can describe electromagnetic waves by means of *wave functions*, just as we did in Section 21–3 for waves on a string. One form of the equation of a transverse wave traveling to the right along a stretched string is Eq. (21–7):

$$y = A \sin (\omega t - kx),$$

where *y* is the transverse displacement from its equilibrium position, at time *t*, of a point of the string whose coordinate is *x*. The quantity *A* is the maximum displacement, or *amplitude*, of the wave; ω is its *angular frequency*, equal to 2π times the frequency *f*; and *k* is the *propagation constant*, equal to $2\pi/\lambda$, where λ is the wavelength.

Wave functions for a plane sinusoidal electromagnetic wave

Let *E* and *B* represent the instantaneous values, and E_{max} and B_{max} the maximum values, or *amplitudes*, of the electric and magnetic fields in Fig. 35–4. The equations of the traveling electromagnetic wave are then

$$E = E_{max} \sin (\omega t - kx), \qquad B = B_{max} \sin (\omega t - kx). \qquad (35–18)$$

The sine curves in Fig. 35–4 represent instantaneous values of *E* and *B*, as functions of *x*, at time $t = 0$. The wave travels to the right with speed *c*.

The instantaneous value *S* of the magnitude of the Poynting vector is

$$S = \frac{EB}{\mu_0} = \frac{E_{max}B_{max}}{\mu_0} \sin^2 (\omega t - kx)$$

$$= \frac{E_{max}B_{max}}{2\mu_0}[1 - \cos 2(\omega t - kx)].$$

The time-average value of $\cos 2(\omega t - kx)$ is zero, so the average value S_{av} of the Poynting vector magnitude is

$$S_{av} = \frac{E_{max}B_{max}}{2\mu_0}.$$

This is the *average power* transmitted per unit area; as noted in Section 35–3, it is called the *intensity* of the radiation, denoted by *I*. By using the relations

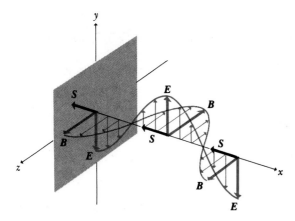

35-5 Electric and magnetic fields of a sinusoidal wave traveling in the negative x-direction. The position of the wave at time $t = 0$ is shown.

$E_{max} = B_{max}c$ and $\epsilon_0\mu_0 = 1/c^2$, we can express the intensity in several equivalent forms:

$$I = S_{av} = \frac{E_{max}B_{max}}{2\mu_0} = \frac{E_{max}{}^2}{2\mu_0 c} = \frac{1}{2}\sqrt{\frac{\epsilon_0}{\mu_0}}E_{max}{}^2 = \frac{1}{2}\epsilon_0 c E_{max}{}^2. \quad (35-19)$$

We invite you to verify that these expressions are all equivalent.

We can also obtain various expressions for the intensity of a wave in a material. In Eq. (35-19) we simply replace ϵ_0 by ϵ, μ_0 by μ, and c by v.

Figure 35-5 represents schematically the electric and magnetic fields of a wave traveling in the *negative* x-direction. At points where **E** is in the positive y-direction, **B** is in the *negative* z-direction, and where **E** is in the negative y-direction, **B** is in the *positive* z-direction. The Poynting vector is in the negative x-direction at all points. (Compare this with Fig. 35-4, which shows a wave traveling in the *positive* x-direction.) The equations of the wave are

$$E = -E_{max}\sin(\omega t + kx),$$
$$B = B_{max}\sin(\omega t + kx). \quad (35-20)$$

Finding the intensity of a sinusoidal electromagnetic wave

EXAMPLE 35-3 A radio station radiates a sinusoidal wave with an average total power of 50 kW. Assuming it radiates equally in all directions (which is unlikely in real-world situations), find the amplitudes E_{max} and B_{max} at a distance of 100 km from the antenna.

A radio transmitter: an example of the relation of power to amplitude in a wave

SOLUTION First we find the magnitude S of the Poynting vector. We surround the antenna with an imaginary sphere of radius 100 km $= 1.00 \times 10^5$ m. This sphere has area

$$A = 4\pi R^2 = 4\pi(1.00 \times 10^5 \text{ m})^2 = 12.6 \times 10^{10} \text{ m}^2.$$

All the power radiated passes through this surface, so the power per unit area is

$$S = \frac{P}{A} = \frac{P}{4\pi R^2} = \frac{5.00 \times 10^4 \text{ W}}{12.6 \times 10^{10} \text{ m}^2} = 3.98 \times 10^{-7} \text{ W} \cdot \text{m}^{-2}.$$

But from Eq. (35-19),

$$S_{av} = \frac{E_{max}{}^2}{2\mu_0 c} \quad \text{and} \quad E_{max} = \sqrt{2\mu_0 c S_{av}},$$

so

$$E_{max} = \sqrt{2(4\pi \times 10^{-7} \text{ Wb·A}^{-1}\text{·m}^{-1})(3 \times 10^8 \text{ m·s}^{-1})(3.98 \times 10^{-7} \text{ W·m}^{-2})}$$
$$= 1.73 \times 10^{-2} \text{ V·m}^{-1},$$

$$B_{max} = E_{max}/c = 5.77 \times 10^{-11} \text{ T}.$$

Note that E_{max} is of an order of magnitude comparable to many laboratory phenomena, but that B_{max} is extremely small compared to the **B** fields we have seen in previous chapters.

What is polarization?

Electromagnetic waves exhibit *polarization*. This concept was introduced at the end of Section 21–4 in the context of transverse waves on a string, and we suggest you review that discussion now. In the present context, note that for a wave traveling in the *x*-direction the choice of the *y*-direction for **E** was arbitrary. We could just as well have specified the *z*-axis for **E**; then when **E** is in the +*z*-direction, **B** is in the −*y*-direction, and so on.

A wave in which **E** always lies along a certain axis is said to be *linearly polarized* along that axis. More generally, we can think of *any* wave traveling in the *x*-direction as a superposition of waves linearly polarized in the *y*- and *z*-directions. A superposition of two linearly polarized waves with the same frequency and amplitude but a 90° phase difference yields a wave that is *circularly polarized*. We will study polarization phenomena in greater detail, with special emphasis on polarization of light, in Chapter 36.

35–6 STANDING WAVES

Electromagnetic waves can be reflected by a conducting surface.

Electromagnetic waves can be *reflected;* a conducting surface can serve as a reflector. The superposition principle holds for electromagnetic waves just as for all electric and magnetic fields, and the superposition of an incident wave and a reflected wave can form a **standing wave.** The situation is analogous to standing waves on a stretched string; we studied these in Section 22–2, and you should review that discussion now.

Suppose a sheet of an ideal conductor, having zero resistivity, is placed in the *yz*-plane of Fig. 35–5, and that the wave shown, traveling in the negative *x*-direction, is incident on it. The essential characteristic of an ideal conductor is that no electric field can ever exist within it; any attempt to establish a field is immediately canceled by rearrangement of the mobile charges in the conductor. Thus **E** must always be zero everywhere in this plane, and the **E** field of the incident wave induces sinusoidal currents in the conductor so that **E** is zero inside it.

Superposition of an incident and a reflected wave forms a standing wave.

These induced currents produce a reflected wave, traveling out from the plane, to the right. From the superposition principle, the total **E** field at any point to the right of the plane is the vector sum of the **E** fields of the incident and reflected waves; the same is true for the total **B** field.

Suppose the incident wave is described by the wave functions of Eqs. (35–20) and the reflected wave by the wave functions of Eqs. (35–18). (Compare these with Eqs. [21–7] and [21–8] for transverse waves on a string.) From the superposition principle, the total fields at any point are given by

$$E = E_{max}[-\sin(\omega t + kx) + \sin(\omega t - kx)],$$

$$B = B_{max}[\sin(\omega t + kx) + \sin(\omega t - kx)].$$

These expressions may be expanded and simplified by using the identity

Wave functions for a standing wave

$$\sin (A \pm B) = \sin A \cos B \pm \cos A \sin B.$$

The results are

$$E = -2E_{max} \cos \omega t \sin kx, \qquad (35\text{–}21a)$$

$$B = 2B_{max} \sin \omega t \cos kx. \qquad (35\text{–}21b)$$

The first of these is analogous to Eq. (22–1) for a stretched string. We see that at $x = 0$, E is *always* zero; this condition is required by the nature of the ideal conductor, which plays the same role as a fixed point at the end of the string. Furthermore, E is zero at all times in those planes for which $\sin kx = 0$; that is, $kx = 0, \pi, 2\pi, \ldots$, or

$$x = 0, \quad \frac{\lambda}{2}, \quad \lambda, \quad \frac{3\lambda}{2}, \quad \ldots .$$

These are called the **nodal planes** of the **E** field.

The total magnetic is zero at all times in those planes for which $\cos kx = 0$, or at which

The nodal planes of the **B** field do not coincide with those of the **E** field.

$$x = \frac{\lambda}{4}, \quad 3\frac{\lambda}{4}, \quad 5\frac{\lambda}{4}, \quad \ldots .$$

These are the nodal planes of the **B** field. The magnetic field is *not* zero at the conducting surface ($x = 0$), and there is no reason it should be. The nodal planes of one field are midway between those of the other, and the nodal planes of either field are separated by one-half wavelength. Figure 35–6 shows a standing-wave pattern at one instant of time.

The total electric field is a *cosine* function of t, and the total magnetic field is a *sine* function of t. The fields are therefore 90° out of phase. At times when $\cos \omega t = 0$, the electric field is zero *everywhere* and the magnetic field is maximum. When $\sin \omega t = 0$, the magnetic field is zero *everywhere* and the electric field is maximum.

Pursuing the stretched-string analogy, we may now insert a second conduction plane, parallel to the first and a distance L from it, along the $+x$-axis. This is analogous to the stretched string held at the points $x = 0$ and $x = L$. A standing wave can exist only when L is an integer multiple of $\lambda/2$. Hence the

When there are two reflecting surfaces, normal modes result.

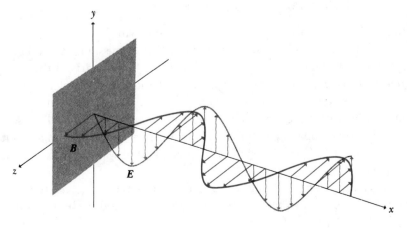

35–6 **E** and **B** vectors in a standing wave. The pattern does not move along the x-axis, but the **E** and **B** vectors grow and diminish with time at each point. At each point **E** is maximum when **B** is minimum, and conversely. The position of the wave at time $t = 0$ is shown.

possible wavelengths are

$$\lambda_n = \frac{2L}{n}, \qquad n = 1, 2, 3, \ldots, \tag{35-22}$$

and the corresponding frequencies are

$$f_n = \frac{c}{\lambda_n} = n\frac{c}{2L}, \qquad n = 1, 2, 3, \ldots. \tag{35-23}$$

Thus there is a set of *normal modes,* each with a characteristic frequency, wave shape, and node pattern. Measurement of the node positions makes it possible to measure the wavelength. If the frequency is known, the wave speed can be determined. This technique was, in fact, used by Hertz in his pioneering investigations of electromagnetic waves.

Reflection and transmission at a dielectric interface

Conducting surfaces are not the only reflectors of electromagnetic waves; reflections also occur at an interface between two insulating materials having different dielectric or magnetic properties. The mechanical analog is a junction of two strings with equal tension but different linear mass density. In general, a wave incident on such a boundary surface is partly transmitted into the second material and partly reflected back into the first. The partial transmission and reflection of light at a glass surface is a familiar phenomenon; light is transmitted through a sheet of glass, but its surfaces also reflect light.

PROBLEM-SOLVING STRATEGY: *Electromagnetic waves*

1. In the problems of this chapter, the most important advice we can give is to make sure you know what the basic relationships are, such as the relation of **E** to **B,** how the wave speed is determined, the transverse nature of the waves, and so on.

2. In the discussion of sinusoidal waves in Sections 35-5 and 35-6 you will need to use the language of sinusoidal waves you learned in Chapters 21 and 22. Don't hesitate to go back and review that material, including the problem-solving strategies suggested in those chapters. In particular, keep in mind the basic relationships for periodic waves, $c = \lambda f$ and $\omega = ck$. Be careful to distinguish between the ordinary frequency f, usually expressed in hertz, and the angular frequency $\omega = 2\pi f$, expressed in s^{-1}. Also remember that the wave number k is $k = 2\pi/\lambda$.

3. In the discussion of standing waves, make sure what you mean by nodes and antinodes, and which field you are talking about. Nodes of **E** coincide with antinodes of **B,** and conversely. Compare this situation to the distinction between pressure nodes and displacement nodes in Section 22-4.

35-7 THE ELECTROMAGNETIC SPECTRUM

Electromagnetic waves occur in nature with an enormous range of frequencies and wavelengths.

Electromagnetic waves cover an extremely broad spectrum of wavelength and frequency, as shown in Fig. 35-7. Radio and TV transmission, visible light, infrared and ultraviolet radiation, and gamma rays all form parts of the **electromagnetic spectrum.** It is well established that light consists of electromagnetic waves; it is a small part of a very broad class of electromagnetic radiations, all having the general characteristic described in preceding sections, including the propagation speed (in vacuum) $c = 3.00 \times 10^8$ m·s^{-1} but differing in frequency f and wavelength λ. The general wave relation $c = \lambda f$ holds for each.

Wavelengths of light are less than a thousandth of a millimeter.

The wavelengths of visible light (i.e., of electromagnetic waves that are perceived by the sense of sight) can be measured by methods to be explained

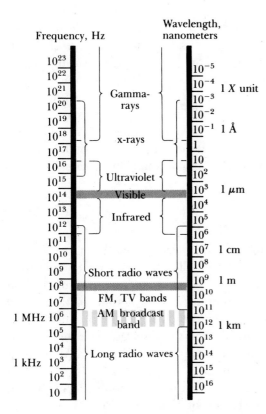

35–7 A chart of the electromagnetic spectrum.

in Chapter 39; they are in the range 4 to 7×10^{-7} m (400 to 700 nm). The corresponding range of frequencies is about 7.5 to 4.3×10^{14} Hz.

Because of the very small magnitudes of light wavelengths, it is convenient to measure them in small units of length. Three such units are commonly used: the micrometer (1 μm), the nanometer (1 nm) (both accented on the *first* syllable), and the angstrom (1 Å):

$$1 \; \mu m \; = \; 10^{-6} \, m = 10^{-4} \, cm,$$

$$1 \; nm \; = \; 10^{-9} \, m = 10^{-7} \, cm,$$

$$1 \; \text{Å} \; = \; 10^{-10} \, m = 10^{-8} \, cm.$$

In older literature, the micrometer is sometimes called the *micron*, and the nanometer is sometimes called the *millimicron;* these terms are now obsolete. Most workers in the fields of optical-instrument design, color, and physiological optics express wavelengths in *nanometers*. For example, the wavelength of the yellow light from a sodium-vapor lamp is 589 nm, but many spectroscopists would identify this same wavelength as 5890 Å.

Different parts of the visible spectrum evoke the sensations of different colors. Wavelengths for colors in the visible spectrum are (very approximately) as follows:

The color of light depends on its wavelength or frequency.

400 to 440 nm	Violet
440 to 480 nm	Blue
480 to 530 nm	Green
530 to 590 nm	Yellow
590 to 630 nm	Orange
630 to 700 nm	Red

Monochromatic (single-frequency) light: a useful idealization

By the use of special sources or special filters, it is possible to select a narrow band of wavelengths, with a range of, say, from 1 to 10 nm. Such light is approximately *monochromatic* (single-color) light. Light consisting of only one wavelength is an idealization that is impossible to attain in practice but is useful in theoretical calculations. When the expression "monochromatic light of wavelength 550 nm" is used in theoretical discussions, it refers to one wavelength, but in descriptions of laboratory experiments it means a small band of wavelengths *around* 550 nm. One distinguishing characteristic of light from a *laser* is that it is much more nearly monochromatic than light obtainable in any other way.

35–8 RADIATION FROM AN ANTENNA

Radiation from an antenna is not usually a plane wave.

The waves we have been discussing are called *plane waves;* this name refers to the fact that if we construct any plane perpendicular to the direction of propagation of the wave, then at any instant the **E** and **B** fields are uniform over that plane. Although plane waves are the simplest of all electromagnetic waves to describe and analyze, they are by no means the simplest to produce experimentally. Any charge or current distribution that oscillates sinusoidally with time produces sinusoidal electromagnetic waves, but in general there is no reason to expect them to be plane waves.

The simplest example of an oscillating charge distribution is an **oscillating dipole,** which is a pair of electric charges of equal magnitude and opposite sign, with the charge magnitude varying sinusoidally with time. Such an oscillating dipole can be constructed in various ways, but we need not be concerned with the details.

The radiation from an oscillating dipole is *not* a plane wave, but it travels out in all directions from the source. The wave fronts are not planes but expanding concentric spheres centered at the source. At points far from the source, the **E** and **B** fields are perpendicular to the direction from the source and to each other; in this sense the wave is still transverse. The value of S drops off as the square of the distance from the source. The intensity (the average value of S) depends on the direction from the source; it is greatest in directions perpendicular to the dipole axis, and $S = 0$ in directions parallel to the axis.

The radiation pattern from a dipole source is shown schematically in Fig. 35–8. The figure shows a cross section of the radiation pattern at one instant. The oscillating dipole **P** is located at the centers of the spheres. At all points in the plane of the figure, the **E** field lies in the plane and the **B** field is perpendicular to it. The **E** field is shown by colored arrows, and the direction of **B** by crosses (where it points into the plane) and circles with dots (where it points out of the plane). We invite you to verify that the direction of the Poynting vector **S** is radially outward from the source at every point.

Electromagnetic waves can be *reflected* by conducting surfaces. When the surface is large compared to the wavelength of the radiation, the reflection behaves like reflection of light rays from a mirror, which we will study in Chapter 37. Large parabolic mirrors several meters in diameter are used as both transmitting and receiving antennas for microwave communications signals; typical wavelengths are a few centimeters. The transmitting reflector produces a wave that radiates in a narrow, well-defined beam; the receiving reflector gathers wave energy over its whole area and reflects it to the focus of

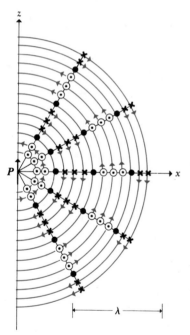

35–8 Cross section in the *xz*-plane of radiation from an oscillating electric dipole **P**. The wave fronts are expanding concentric spheres centered at **P**. The **E** field at every point lies in the plane; the colored lines show the directions of **E** at various points. The **B** field at every point is perpendicular to the plane. At points with circles, **B** comes out of the plane, and at points with crosses it is into the plane. The direction of the Poynting vector **S** is radially outward at every point.

35-9 A microwave-receiving antenna 64 m in diameter at Goldstone Tracking Station, California, the prime station of a worldwide network of stations used by NASA to monitor unmanned interplanetary spacecraft.

the parabola, where a detecting device is placed. Figure 35-9 shows an antenna installation using a large parabolic reflector.

SUMMARY

Maxwell's equations, which incorporate all the basic relationships of electric and magnetic fields and their sources (charges and currents), predict the existence of electromagnetic disturbances that can propagate through empty space and travel with a speed equal to the measured value of the speed of light. The simplest such wave is a plane wave, in which E and B are uniform over any plane perpendicular to the propagation direction, such that E and B are zero everywhere to the left of a certain plane and have constant values everywhere to the right. For such a wave disturbance to be consistent with Faraday's law, the two field magnitudes must be related by

$$E = cB, \tag{35-2}$$

and for consistency with Ampere's law they must be related by

$$B = \epsilon_0 \mu_0 cE, \tag{35-4}$$

where c is the propagation speed. For both of these requirements to be satisfied, c must be given by

$$c = \frac{1}{\sqrt{\epsilon_0 \mu_0}} = 3.00 \times 10^8 \text{ m·s}^{-1}. \tag{35-5}$$

Electromagnetic waves are transverse; the E and B fields are perpendicular to the direction of propagation and to each other. There is a definite ratio between E and B in a wave, and the waves travel in vacuum with a definite and unchanging speed c, given by Eq. (35-5).

The energy density in an electromagnetic wave can be expressed as

$$u = \frac{1}{2}\epsilon_0 E^2 + \frac{1}{2\mu_0}B^2 = \epsilon_0 E^2. \tag{35-8}$$

KEY TERMS

electromagnetic wave
Maxwell's equations
transverse wave
energy densities
Poynting vector
intensity
radiation pressure
index of refraction
plane wave
standing wave
nodal planes
electromagnetic spectrum
oscillating dipole

The energy-flow rate (power per unit area) is given by the Poynting vector

$$S = \frac{1}{\mu_0} E \times B. \tag{35-11}$$

The time-average value of the magnitude EB/μ_0 of the Poynting vector is called the intensity of the wave. These waves also carry momentum; the momentum per unit volume has magnitude

$$\frac{EB}{\mu_0 c^2} = \frac{S}{c^2}, \tag{35-12}$$

and the rate of transfer of momentum per unit cross-sectional area is

$$\frac{1}{A}\frac{dp}{dt} = \frac{S}{c} = \frac{EB}{\mu_0 c}. \tag{35-13}$$

When an electromagnetic wave travels through a dielectric, the wave speed v is given by

$$v = \frac{1}{\sqrt{\epsilon\mu}} = \frac{1}{\sqrt{KK_m}}\frac{1}{\sqrt{\epsilon_0\mu_0}}. \tag{35-14}$$

If the relative permeability K_m can be taken as unity, then

$$v \simeq \frac{1}{\sqrt{K}}\frac{1}{\sqrt{\epsilon_0\mu_0}} = \frac{c}{\sqrt{K}}.$$

In this case the index of refraction $n = c/v$ is given by

$$\frac{c}{v} = n = \sqrt{KK_m} \simeq \sqrt{K}. \tag{35-15}$$

For a sinusoidal electromagnetic wave traveling in the $+x$-direction, both E and B are sinusoidal functions of the quantity $(\omega t - kx)$, and at each point the sinusoidal variations of E and B are in phase. For a wave in the $-x$-direction, E and B are sinusoidal functions of $(\omega t + kx)$.

If a reflecting surface is placed at $x = 0$, it becomes a nodal plane for the E field; that is, E is zero everywhere in that plane. A wave traveling in the $-x$-direction is reflected, and the incident and reflected waves form a standing wave. There are nodal planes for E at $kx = 0$, π, 2π, and so on, and nodal planes for B at $kx = \pi/2$, $3\pi/2$, $5\pi/2$, and so on. At each point the sinusoidal variations of E and B are 90° out of phase; the nodes of B coincide with the antinodes of E, and conversely.

The electromagnetic spectrum covers a range of frequencies from at least 1 to 10^{24} Hz and a correspondingly broad range of wavelengths. Visible light is a very small part of this spectrum, with wavelengths of 4 to 7×10^{-7} m or 400 to 700 nm.

An oscillating dipole produces a wave that radiates out in all directions. At points far from the source, E and B are perpendicular to each other and to the radial direction, so this wave is transverse. The intensity depends on direction; it is zero along the dipole axis and greatest in the plane perpendicular to that axis.

QUESTIONS

35–1 In Ampere's law, is it possible to have both a conduction current and a displacement current at the same time? Is it possible for the effects of the two kinds of current to cancel each other exactly, so that *no* magnetic field is produced?

35–2 By measuring the electric and magnetic fields at a point in space where there is an electromagnetic wave, can one determine the direction from which the wave came?

35–3 Sometimes neon signs located near a powerful radio station are seen to glow faintly at night, even though they are not turned on. What is happening?

35–4 Light can be *polarized;* is this a property of all electromagnetic waves, or is it unique to light? What about polarization of sound waves? What fundamental distinction in wave properties is involved?

35–5 How does a microwave oven work? Why does it heat materials that are conductors of electricity, including most foods, but not insulators, such as glass or ceramic dishes? Why do the directions forbid putting anything metallic in the oven?

35–6 Electromagnetic waves can travel through vacuum, where there is no matter. We usually think of vacuum as empty space, but is it *really* empty if electric and magnetic fields are present? What *is* vacuum, anyway?

35–7 Give several examples of electromagnetic waves that are encountered in everyday life. How are they all alike? How do they differ?

35–8 We are surrounded by electromagnetic waves emitted by many radio and television stations. How is a radio or television receiver able to select a single station among all this mishmash of waves? What happens inside a radio receiver when the dial is turned to change stations?

35–9 The metal conducting rods on a television antenna are always in a horizontal plane. Would they work as well if they were vertical?

35–10 If a light beam carries momentum, should a person holding a flashlight feel a recoil analogous to the recoil of a rifle when it is fired? Why is this recoil not actually observed?

35–11 The nineteenth-century inventor Nikolai Tesla proposed to transmit large quantities of electrical energy across space by using electromagnetic waves instead of conventional transmission lines. What advantages would this scheme have? What disadvantages?

35–12 Does an electromagnetic *standing wave* have energy? Momentum? What distinction can be drawn between a standing wave and a propagating wave on this basis?

35–13 If light is an electromagnetic wave, what is its frequency? Is this a proper question to ask?

35–14 When an electromagnetic wave is reflected from a moving reflector, the frequency of the reflected wave is different from that of the initial wave. Explain physically how this can happen. (Some radar systems used for highway-speed control operate on this principle.)

35–15 The ionosphere is a layer of ionized air 100 km or so above the earth's surface. It acts as a reflector of radio waves of frequency less than about 30 MHz, but not of higher frequency. How does this reflection occur? Why does it work better for lower frequencies than for higher?

EXERCISES

Section 35–2 Speed of an Electromagnetic Wave

35–1 In a TV picture, ghost images are formed when the signal from the transmitter travels directly to the receiver and also indirectly after reflection from a building or other large metallic mass. In a 25-in. set the ghost is about 1 cm to the right of the principal image if the reflected signal arrives 1 μs after the principal signal. In this case, what is the difference in path length for the two signals?

35–2 The maximum electric field in the vicinity of a certain radio transmitter is 1.0×10^{-3} V·m^{-1}. What is the maximum magnitude of the *B* field? How does this compare in magnitude with the earth's field?

35–3 A certain radio station broadcasts at a frequency of 1020 kHz. At a point some distance from the transmitter, the maximum magnetic field of the electromagnetic wave it emits is found to be 1.6×10^{-11} T.

a) What is the wavelength of the wave?

b) What is the maximum electric field?

Section 35–3 Energy in Electromagnetic Waves

Section 35–5 Sinusoidal Waves

35–4 Consider each of the electric- and magnetic-field orientations given below. In each case what is the direction of propagation of the wave?

a) $E = Ei$, $B = Bj$;

b) $E = -Ej$, $B = Bi$;

c) $E = Ek$, $B = -Bi$;

d) $E = Ej$, $B = -Bk$.

35–5 Verify that all the expressions in Eq. (35–19) are equivalent.

35–6 A certain plane electromagnetic wave emitted by a microwave antenna has a wavelength of 3.0 cm and a maximum magnitude of electric field of 2.0×10^{-2} V·m^{-1}.

a) What is the frequency of the wave?

b) What is the maximum magnetic field?

c) What is the intensity (average power per unit area) of the wave, if the wave is sinusoidal?

35–7 At a distance of 50 km from a radio station antenna, the electric-field amplitude is measured to be $E_{max} = 2 \times 10^{-2}$ V·m^{-1}.

a) What is the magnetic-field amplitude B_{max} at this same point?

b) Assuming that the antenna radiates equally in all directions (which is probably not the case), what is the total power output of the station?

c) At what distance from the antenna would $E_{max} = 1 \times 10^{-2}$ V·m^{-1}, half the above value?

35–8 Assume that 10% of the power input to a 100-W lamp is radiated uniformly as light of wavelength 500 nm (1 nm = 10^{-9} m). At a distance of 2 m from the source, the electric and magnetic fields vary sinusoidally according to the equations $E = E_{max} \sin(\omega t + \phi)$ and $B = B_{max} \sin(\omega t + \phi)$. Calculate E_{max} and B_{max} for the 500-nm light.

35–9 If the intensity of direct sunlight is 1.4 kW·m^{-2}, find

a) the momentum density (momentum per unit volume);

b) the momentum flow rate (momentum carried through a surface area A in unit time) in the sunlight. (*Note.* This equals the radiation pressure.)

35–10 The intensity of a bright light source is 900 W·m^{-2}. Find the radiation pressure (in pascals) on

a) a totally absorbing surface;

b) a totally reflecting surface.

Also express your results in atmospheres.

Section 35–4 Electromagnetic Waves in Matter

35–11 An electromagnetic wave propagates in a ferrite material having $K = 10$ and $K_m = 1000$. Find

a) the speed of propagation;

b) the wavelength of a wave having a frequency of 100 MHz.

Section 35–6 Standing Waves

35–12 For the standing wave given by Eqs. (35–21):

a) Plot the energy density as a function of x, $0 < x < \pi/k$, for the times $t = 0$, $\pi/4\omega$, $\pi/2\omega$, $3\pi/4\omega$, and π/ω.

b) Find the direction of S in the regions $0 < x < \pi/2k$ and $\pi/2k < x < \pi/k$ at the times $t = \pi/4\omega$ and $3\pi/4\omega$.

c) Use your results in (b) to explain the plots obtained in (a).

Section 35–7 The Electromagnetic Spectrum

35–13 What is the wavelength in meters, microns, nanometers, and angstrom units of

a) soft x-rays of frequency 2×10^{17} Hz?

b) green light of frequency 5.6×10^{14} Hz?

PROBLEMS

35–14 The energy flow to the earth associated with sunlight is about 1.4 kW·m^{-2}.

a) Find the maximum values of E and B for a wave of this intensity.

b) The distance from the earth to the sun is about 1.5×10^{11} m. Find the total power radiated by the sun.

35–15 For a sinusoidal electromagnetic wave in vacuum, such as that described by Eqs. (35–18), show that the average density of energy in the electric field is the same as that in the magnetic field.

35–16 A plane sinusoidal electromagnetic wave has a wavelength of 3.0 cm and an E-field amplitude of 30 V·m^{-1}.

a) What is the frequency?

b) What is the B-field amplitude?

c) What is the intensity?

d) What average force does this radiation exert on a totally absorbing surface of area 0.5 m^2 perpendicular to the direction of propagation?

35–17 A very long solenoid of n turns per unit length and radius a carries a current i that is increasing at a constant rate di/dt.

a) Calculate the induced electric field at a point inside the solenoid at a distance r from the solenoid axis.

b) Compute the magnitude and direction of the Poynting vector at this point.

35–18 A capacitor consists of two circular plates of radius r separated by a distance l. Neglecting fringing, show that while the capacitor is being charged, the rate at which energy flows into the space between the plates is equal to the rate at which the electrostatic energy stored in the capacitor increases. (*Hint:* Integrate the Poynting vector over the surface of the space between the plates.)

35–19 A circular loop of wire can be used as a radio frequency antenna. If a 0.5-m diameter antenna is located 100 m from a 10-MHz source with a total power of 1 MW, what is the maximum emf induced in the loop? (Assume that the plane of the antenna loop is perpendicular to the direction of the magnetic field of the radiation, and that the source radiates uniformly in all directions.)

35–20 It has been proposed that to aid in meeting the energy needs of the United States, solar-power-generating satellites could be placed in earth orbit, and the power they generate could be beamed down to earth as microwave radiation. For a microwave beam with a cross-sectional area of 50 m^2 and a total power of 1 kW at the earth's surface, what is the magnitude of the electric field of the beam on the earth's surface?

35–21 The nineteenth-century inventor Nikolai Tesla proposed to transmit electric power via electromagnetic waves. Suppose power is to be transmitted in a beam of cross-sectional area 100 m^2; what E and B strengths are required to transmit an amount of power comparable to that handled by modern transmisson lines (of the order of 500 kV and 1000 A)?

35–22 A space-walking astronaut has run out of fuel for her jet-pack and is floating 20 m from the space shuttle with zero relative velocity. The astronaut and all her equipment have a total mass of 200 kg. If she uses her 100-W flashlight as a light rocket, how long will it take her to reach the shuttle?

CHALLENGE PROBLEMS

35–23 The concept of solar sailing has appeared in science fiction and proposals to NASA. A solar sailcraft uses a large low-mass sail and the energy and momentum of sunlight for propulsion. The total power output of the sun is 4×10^{26} W.

a) Should the sail be absorbing or reflective? Why?

b) How large a sail is necessary to propel a 10^5 kg spacecraft against the gravitational force of the sun? (Express your result in square miles.)

c) Explain why your answer to (b) is independent of the distance from the sun.

35–24 Electromagnetic radiation is emitted by accelerating charges. The rate of energy emission from an accelerating charge that has charge q and acceleration a is given by

$$\frac{dE}{dt} = \frac{q^2 a^2}{6\pi\epsilon_0 c^3},$$

where c is the speed of light.

a) Verify that this equation is dimensionally correct.

b) If a proton with a kinetic energy of 5 MeV is traveling in a particle accelerator in a circular orbit of radius 1 m, what fraction of its energy does it radiate per second?

c) Consider an electron orbiting with the same velocity and radius. What fraction of its energy does it radiate per second?

35–25 The electron in a hydrogen atom can be considered to be in a circular orbit with a radius of 0.053 nm and an energy of 13.6 eV. If the electron behaved classically, how much energy would it radiate per second? (See Challenge Problem 35–24). What does this tell you about the use of classical physics in describing the atom?

PART SEVEN

OPTICS

PERSPECTIVE

We have now completed our study of the principles of electromagnetism. We learned that the concept of magnetic field enables us to describe the interactions of moving charges and currents, including the action of these as *sources* of magnetic field and the *forces* that the field exerts on moving charges and currents. We also studied the interplay between electric and magnetic fields; a time-varying field of one kind acts as a source of the other kind of field in neighboring regions of space. This two-way interaction, which can take place in empty space even when there is no electric charge present, forms the basis of electromagnetic waves.

Our study of electromagnetism has had both fundamental and practical significance. On the practical side, we have learned how magnetic forces are used to deflect electron beams in TV picture tubes, how electric motors and generators work, and the operation of ac circuits, so vital in communications, electric-power distribution, and every area of contemporary electronics. From a fundamental point of view, we have seen how Maxwell's equations wrap up all of electromagnetism in a neat, concise package comparable to Newton's laws and the laws of thermodynamics. As we have remarked before, this is what science is all about—learning to generalize from experimental evidence and to express broad and general physical laws concisely and compactly. Maxwell's synthesis of electromagnetism stands as a towering intellectual achievement, comparable to the Newtonian synthesis in the preceding century, and to the development of relativity, quantum mechanics, and the understanding of DNA in our own time. All beautiful, and all monuments to the achievements of the human intellect!

Now we proceed to the subject of *optics*, originally the study of light but now broadened to include other electromagnetic radiation as well. Light is known to be electromagnetic radiation; it forms a small part of the electromagnetic spectrum, with frequencies ranging from less than 1 Hz to at least 10^{25} Hz. The wave model of light and other electromagnetic radiation provides the basis for understanding a variety of optical phenomena, including polarization, interference, and diffraction.

We begin with the special case where the wave picture can be simplified further by representing light in terms of *rays*. We study the reflection of rays and their bending or *refraction* when they pass from one material to another. The ray representation forms the basis of *geometrical optics* and is the model used to analyze mirrors, lenses, and common optical instruments. Geometrical optics is more limited in its scope than the general wave picture, but it is much simpler. We use it to study the optical behavior of several practical devices, including cameras, projectors, optical systems, the human eye, and various kinds of microscopes and telescopes.

Next we proceed to several optical phenomena that require the more general wave description of light for their understanding. Interference and diffraction phenomena show departures of light from the straight-line propagation that would occur if the ray picture were exactly correct. The corresponding branch of optics is called *physical optics,* to distinguish it from the more restricted geometrical optics. Interference effects enable us to measure wavelengths of light, design nonreflective coatings for lenses, measure atomic spacings in crystal lattices, and understand the fundamental limitations on the resolution of optical instruments. Finally, we study the principles of holography, one of the most exciting and useful developments in modern optics and a striking application of the wave nature of light.

36

THE NATURE AND PROPAGATION OF LIGHT

WE BEGIN THIS CHAPTER WITH A GENERAL DISCUSSION OF THE NATURE OF light and of light sources. Light is electromagnetic radiation and is thus a *wave* phenomenon. Many aspects of the propagation of light can be described by a *ray* model; rays travel in straight lines in homogeneous materials but can be reflected and also bent, or *refracted,* at interfaces between materials. In this chapter we study several basic phenomena associated with reflection and refraction of light. Like all transverse waves, light waves are *polarized;* we examine several aspects of polarization phenomena.

36–1 NATURE OF LIGHT

Until the time of Newton (1642–1727), most scientists (including Newton) thought that light consisted of streams of some sort of particles (which they called *corpuscles*) emanating from light sources. But at about this time, Huygens and others suggested that light might be a *wave* phenomenon. Indeed, diffraction effects that are now recognized as wave phenomena were observed by Grimaldi as early as 1665, but their significance was not understood then. By the early nineteenth century, evidence that light is a wave phenomenon grew more persuasive. The experiments of Fresnel, Thomas Young, and others revealed many phenomena that can be understood on the basis of a wave picture but not on the corpuscular model. We will study these phenomena in detail in Chapter 39.

What is light, anyway?

The next great step was taken in 1873 by Maxwell, who predicted the existence of electromagnetic waves and calculated their speed of propagation, as we learned in Chapter 35. In 1887 Hertz succeeded in producing short-wavelength electromagnetic waves and showing that they had all the properties of light waves. They could be reflected, refracted, focused by a lens, polarized, and so on. Thus the evidence grew more and more conclusive that light is indeed an electromagnetic-wave phenomenon.

Maxwell established that light is electromagnetic radiation with very short wavelengths.

Successful as the wave picture of light is, it is not the whole story. Several phenomena associated with emission and absorption of light reveal a particle

Particle aspects of light: twentieth-century concepts

aspect, in the sense that the energy carried by light waves is packaged in discrete bundles called *photons* or *quanta*. We will explore some of these phenomena in Chapter 41. These apparently contradictory wave and particle properties have been reconciled only since 1930 with the development of quantum electrodynamics, a comprehensive theory that includes *both* wave and particle properties. *Propagation* of light is best described by a wave model, but emission and absorption phenomena require a particle approach.

36–2 SOURCES OF LIGHT

Emission of electromagnetic radiation by hot bodies

The fundamental sources of all electromagnetic radiation are electric charges in motion. All bodies emit electromagnetic radiation as a result of thermal motion of their molecules; this radiation, called *thermal radiation*, is a mixture of different wavelengths. At a temperature of 300°C the most intense of these waves has a wavelength of 5000×10^{-9} m or 5000 nm, which is in the *infrared* region. At a temperature of 800°C a body emits enough visible radiant energy to be self-luminous and appears "red hot," although most of the energy emitted is still carried by infrared waves. At 3000°C, about the temperature of an incandescent lamp filament, the radiation contains enough of the visible wavelengths, between 400 nm and 700 nm, that the body appears "white hot." In modern incandescent lamps, the filament is a coil of fine tungsten wire. An inert gas such as argon is introduced to reduce evaporation of the filament. Incandescent lamps vary in size from one no larger than a grain of wheat to one with a power input of 5000 W, used for illuminating airfields.

Carbon-arc lamps are among the brightest light sources.

One of the brightest sources of light is the *carbon arc*. Two carbon rods, typically 10 cm to 20 cm long and 1 cm in diameter, are connected to a 120-V or 240-V dc source. They are touched together momentarily and then pulled apart a few millimeters. An arc forms, and the resulting intense electron bombardment of the positive rod causes an extremely hot crater to form at its end. This crater, whose temperature is typically 4000°C, is the source of light. Carbon-arc lights are used in most theater motion-picture projectors and in large searchlights and lighthouses.

Light produced by electrical discharge in a gas or vapor

Some light sources use an arc discharge through a conducting metal vapor, such as mercury or sodium. The vapor is contained in a sealed bulb with two electrodes, which are connected to a power source. Argon is sometimes added to permit a glow discharge that helps vaporize and ionize the metal. The bluish light of mercury-arc lamps and the bright orange-yellow of sodium-vapor lamps are familiar in highway and other outdoor lighting.

How do fluorescent lamps work?

An important variation of the mercury-arc lamp is the *fluorescent* lamp, consisting of a glass tube containing argon and mercury vapor, with tungsten electrodes. When an electric discharge takes place in the mercury-argon mixture, the emitted radiation is mostly in the ultraviolet region. The ultraviolet radiation is absorbed in a thin layer of material, called a *phosphor*, which is the white coating on the interior walls of the glass tube. The phosphor has the property of *fluorescence*, which means that it emits visible light when illuminated by ultraviolet radiation. Various phosphors can be used to obtain various colors of light. Fluorescent lamps have much higher efficiency of conversion of electrical energy to visible light than do incandescent lamps.

What is so special about laser light? A partial answer.

A special light source that has attained prominence in the last twenty years is the *laser*. It can produce a very narrow beam of enormously intense radiation. High-intensity lasers have been used to cut through steel, fuse high-melting-point materials, and bring about many other effects that are important in physics, chemistry, biology, and engineering. An equally significant characteristic of laser light is that it is much more nearly *monochromatic,* or single-frequency, than any other light source. We will study the operation of one type of laser in Chapter 41.

36–3 THE SPEED OF LIGHT

The speed of light in empty space is a fundamental constant of nature. The first successful measurement of the speed of light was made in 1676 by the Danish astronomer Olaf Roemer, who measured the period of revolution of one of Jupiter's satellites. The period (about 42 hr) appeared longer when Earth was moving in its orbit *away from* Jupiter than when it was moving *toward* Jupiter. Roemer correctly concluded that the difference was due to the displacement of the earth during one revolution of the satellite. When Earth is moving away from Jupiter, light leaving the satellite at the end of a particular revolution has to travel farther than light leaving at the beginning of the same revolution. In that case the apparent time for one revolution, measured on Earth, is a few seconds longer than the actual time. Six months later, when Earth is moving toward Jupiter, the situation is reversed, and the apparent period is a few seconds shorter than the actual value.

Measuring the speed of light three centuries ago

The first successful determination of the speed of light from purely *terrestrial* measurements was made by the French scientist Fizeau in 1849. A schematic diagram of his apparatus is shown in Fig. 36–1. The light source S directs light toward a toothed wheel T that can be rotated at high speed. G is an inclined plate of clear glass. When the wheel is stationary, light passes through one of the openings between the teeth. The light is reflected from M, retraces its path, and is in part reflected from the glass plate G into the eye of an observer at E. The role of the lenses is to concentrate the light on the toothed wheel and on mirror M.

Measuring the speed of light on Earth: a series of increasingly precise refinements

When the wheel T is rotating, the light from S is "chopped up" into a succession of wave pulses of limited length. A pulse travels through an opening between two teeth of the wheel, is reflected by the mirror M, and returns to the wheel. At a certain speed of rotation, the wheel turns just enough during this time so the return path is blocked by a tooth of the wheel, and *no* reflected light is seen by the observer at E. From a knowledge of (a) the angular velocity and radius of the wheel, (b) the distance between openings, and

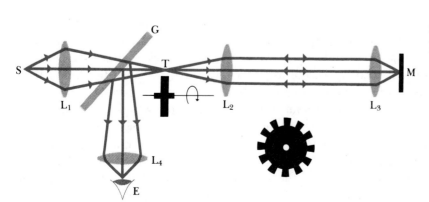

36–1 Fizeau's toothed-wheel method for measuring the speed of light. S is a light source; L_1, L_2, L_3, and L_4 are lenses. T is the toothed wheel, M is a mirror, and G is a glass plate.

(c) the distance from wheel to mirror, the speed of light may be computed. Fizeau obtained the value 3.15×10^8 m·s^{-1}.

Fizeau's apparatus was modified by Foucault, who replaced the toothed wheel with a rotating mirror. The most precise measurements by the Foucault method were made by the American physicist Albert A. Michelson (1852–1931). His first experiments were performed in 1878; the last, underway at the time of his death, were completed in 1935 by Pease and Pearson.

Does it make sense to *define* the speed of light to have a certain value?

From analysis of all measurements up to 1983, the most probable value for the speed of light is

$$c = 2.99792458 \times 10^8 \text{ m·s}^{-1}.$$

This number is based on the definition of the meter in terms of the krypton wavelength and the definition of the second in terms of the cesium clock, as described in Section 1–2. The definition of the second is precise to within one part in 10 trillion (10^{13}), whereas the definition of the meter is much less precise, about four parts in a billion (10^9). Thus it has become advantageous to redefine the meter in terms of the unit of time by *defining* the speed of light to have a specific value and then defining the meter in terms of the distance traveled by light in one second.

Accordingly, in November 1983 the General Conference of Weights and Measures *defined* the speed of light in vacuum to be precisely 299,792,458 m·s^{-1} and defined one meter to be the distance traveled by light in a time of 1/299,792,458 s, with the second defined by the cesium clock.

The speed of light is closely related to the constant in Coulomb's law.

As we found in Chapter 35, the speed of any electromagnetic wave in empty space is given by

$$c = \frac{1}{\sqrt{\epsilon_0 \mu_0}}.$$

In SI units, μ_0 is assigned the value of precisely $4\pi \times 10^{-7}$ N·s^2·C^{-2}. Thus the preceding equation provides a very precise means of evaluating the electrical constant ϵ_0:

$$\epsilon_0 = \frac{1}{\mu_0 c^2} = 8.85418782 \times 10^{-12} \text{ C}^2\text{·N}^{-1}\text{·m}^{-2}.$$

This value has much greater precision than could be obtained from direct measurements of forces between electric charges.

36–4 WAVES, WAVE FRONTS, AND RAYS

Describing waves in terms of wave fronts and rays: some basic language

The concept of **wave front** provides a convenient language for describing the propagation of any kind of wave. We define a wave front as *the locus of all points at which the phase of vibration of a physical quantity associated with the wave is the same*. A familiar example is the crest of a water wave; when we drop a pebble in a calm pool, the expanding circles formed by the wave crests are wave fronts. When sound waves spread out in all directions from a pointlike source, any spherical surface concentric with the source is a wave front. The surfaces over which the pressure is maximum and those over which it is minimum form sets of expanding spheres as the wave travels outward from the source. The *phase* of the pressure variation is the same at all points on one of the spherical surfaces. In diagrams of wave motion we usually draw only a few wave fronts, often those that correspond to the maxima and minima of the disturbance, such as the crests and troughs of a water wave. For a sinusoidal wave, wave fronts corresponding to maximum displacements in opposite directions are separated from each other by one-half wavelength. Two consecutive wave fronts corresponding to maximum displacement in the same direction are separated by one wavelength.

For a light wave (or any other electromagnetic wave), the quantity that corresponds to the pressure in a sound wave is the electric or magnetic field. Often it is not necessary to indicate in a diagram either the magnitude or the direction of the field; instead we simply show the *shapes* of the wave fronts or their intersections with some reference plane. For example, the electromagnetic waves radiated by a small light source may be represented by *spherical* surfaces concentric with the source or, as in Fig. 36–2a, by the intersections of these surfaces with the plane of the diagram. At a sufficiently great distance from the source, where the radii of the spheres have become very large, the spherical surfaces can be considered planes and we have a *plane* wave, as in Fig. 36–2b.

For some phenomena, it is convenient to represent a light wave by **rays** rather than by wave fronts. The corresponding branch of optics is called *geometrical optics*. Indeed, rays were used to describe light long before its wave nature was firmly established and, in a particle theory of light, rays are merely the paths of the particles. From the wave viewpoint, *a ray is an imaginary line drawn in the direction in which the wave is traveling*. Thus in Fig. 36–2a the rays are the radii of the spherical wave fronts, and in Fig. 36–2b they are straight lines perpendicular to the wave fronts. In fact, in every case where waves travel in a homogeneous isotropic material, the rays are straight lines normal to the wave fronts. At a boundary surface between two materials, such as the surface between a glass plate and the air outside it, the direction of a ray changes, but the portions in the air and in the glass are straight lines.

Although the ray picture provides an adequate description of many reflection and refraction phenomena found in mirrors and lenses, several other optical phenomena, such as polarization and diffraction, require a more detailed wave theory for their understanding.

36–5 REFLECTION AND REFRACTION

In many familiar optical phenomena, a wave strikes an interface between two optical materials, such as air and glass or water and glass. When the interface is smooth (i.e., when its irregularities are small compared with the wavelength), the wave is in general partly reflected and partly transmitted into the second material, as shown in Fig. 36–3a. For example, when you look into a store window from the street, you see a reflection of the street scene, but a person in the store can look *through* the window at the same scene.

The segments of plane waves shown in Fig. 36–3a can be represented by bundles of rays forming *beams* of light, as in Fig. 36–3b. For simplicity in discussing the various angles, we often consider only one ray in each beam, as in Fig. 36–3c. Representing these waves in terms of rays is the basis of the branch of optics called **geometrical optics.** This chapter and the next two are concerned primarily with geometrical optics and with optical phenomena that can be understood on the basis of rays, without explicit use of the wave nature of light.

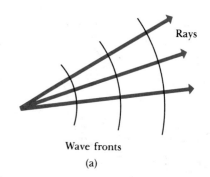

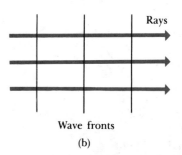

36–2 Wave fronts and rays. (a) When the wave fronts are spherical, the rays radiate out from the center of the spheres. (b) When the wave fronts are planes, the rays are parallel.

Some optical phenomena can't be analyzed by using rays.

Geometrical optics: a very simple model based on reflection and refraction of rays

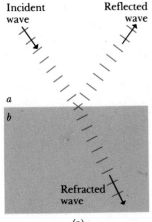

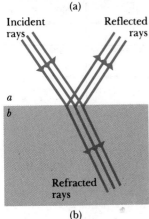

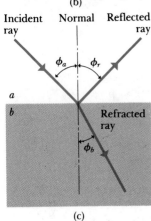

36–3 (a) A plane wave is in part reflected and in part refracted at the boundary between two media. (b) The waves in (a) are represented by rays. (c) For simplicity, only one example of incident, reflected, and refracted rays is drawn.

Index of refraction: describing the refractive properties of an optical material

At an interface between two optical materials, the directions of the incident, reflected, and refracted (transmitted) beams of light are described in terms of the angles they make with the *normal* to the surface at the point of incidence. For this purpose we need indicate only one ray, as in Fig. 36–3c, although a single ray of light is a geometrical abstraction. Experimental studies of the directions of the incident, reflected, and refracted beams lead to the following results

1. *The incident, reflected, and refracted rays, and the normal to the surface, all lie in the same plane.* Hence, if the incident ray is in the plane of the diagram, and the boundary surface between the two materials is perpendicular to this plane, the reflected and refracted rays are in the plane of the diagram.

2. *The angle of reflection ϕ_r is equal to the angle of incidence ϕ_a for all wavelengths and any pair of substances.* Thus

$$\phi_r = \phi_a. \tag{36–1}$$

The experimental result that $\phi_r = \phi_a$, and that the incident and reflected rays and the normal all lie in the same plane, is known as the **law of reflection.**

3. For monochromatic light and for a given pair of substances, a and b, on opposite sides of the surface of separation, *the ratio of the sine of the angle ϕ_a (between the ray in substance a and the normal) and the sine of the angle ϕ_b (between the ray in substance b and the normal) is constant.* Thus

$$\frac{\sin \phi_a}{\sin \phi_b} = \text{constant.} \tag{36–2}$$

This experimental result, together with the fact that the incident and refracted rays and the normal to the surface all lie in the same plane, is known as the **law of refraction** or **Snell's law,** after Willebrord Snell (1591–1626). There is some doubt that Snell actually discovered it.

The laws of reflection and refraction relate the *directions* of the rays; they say nothing about the *intensities* of the reflected and refracted rays. These depend on the angle of incidence; for the present we simply state that the fraction reflected is smallest at *normal* incidence, where it is a few percent, and that it increases with increasing angle of incidence to 100% at grazing incidence, when $\phi_a = 90°$.

When a ray of light approaches the interface from *below* in Fig. 36–3, there are again reflected and refracted rays; these, together with the incident ray and the normal, all lie in the same plane. The laws of reflection and refraction apply, whether the incident ray is in material a or b. The passage of a refracted ray, transmitted from one material to another, is reversible. It follows the same path in going from b to a as when going from a to b.

Let us now consider a beam of monochromatic light traveling *in vacuum*, making an angle of incidence ϕ_0 with the normal to the surface of a substance a, and let ϕ_a be the angle of refraction in the substance. The constant in Snell's law is then called the **index of refraction** of substance a and is denoted by n_a:

$$\frac{\sin \phi_0}{\sin \phi_a} = n_a. \tag{36–3}$$

This definition of index of refraction may seem unrelated to the definition we encountered in Section 35–4. In fact, though, they are equivalent. We will return to this point in Section 36–12.

The index of refraction (also called *refractive index*) depends not only on the substance but also on the wavelength of the light. If no wavelength is stated, the index is often assumed to be that corresponding to the yellow light from a sodium lamp, of wavelength 589 nm. This wavelength is near the middle of the visible spectrum, and sodium lamps are simple, inexpensive, nearly monochromatic, and easy to use.

The index of refraction of most common glasses used in optical instruments lies between 1.46 and 1.96. A few substances have indexes larger than this value; diamond is one, with an index of 2.42, and rutile (crystalline titanium dioxide) is another, with an index of 2.62. Indexes of refraction for several solids and liquids are given in Table 36–1. The index of refraction of *air* at standard conditions is about 1.0003; for most purposes the *index of refraction of air can be assumed to be unity.* The index of refraction of a gas increases uniformly as the density of the gas increases.

For reasons to be discussed in Section 36–12, all materials have a refractive index greater than unity; for example, note the values in Table 36–1. Thus for a ray passing from vacuum into a material, the angle of refraction ϕ_a is always *less than* the angle of incidence ϕ_0. That is, the ray is bent *toward* the normal. When light travels in the opposite direction, from a material into vacuum, the reverse is true and the ray is bent *away from* the normal.

We can express the constant in Eq. (36–2) in terms of the indexes of refraction of the two materials. To develop the relation, we consider two parallel-sided plates of substances a and b placed parallel to each other with space between them, as in Fig. 36–4a. We assume that the surrounding medium is vacuum, although the behavior would differ only very slightly if it were air. A ray of monochromatic light starts at the lower left with an angle of incidence ϕ_0. The angle between the ray and the normal in substance a is ϕ_a, and the light emerges from substance a at an angle ϕ_0 equal to its incident angle. The light ray therefore enters plate b with an angle of incidence ϕ_0, makes an angle ϕ_b in substance b, and emerges again at an angle ϕ_0. Exactly the same path would be traversed if the same light ray were to start at the upper right and enter substance b at an angle ϕ_0. Moreover, *the angles are independent of the thickness of the space between the two plates* and are the same when the space shrinks to nothing, as in Fig. 36–4b.

Applying Snell's law to the refractions at the surface between vacuum and substance a, and at the surface between vacuum and substance b, we have

$$\frac{\sin \phi_0}{\sin \phi_a} = n_a,$$

$$\frac{\sin \phi_0}{\sin \phi_b} = n_b.$$

Dividing the second equation by the first, we obtain

$$\frac{\sin \phi_a}{\sin \phi_b} = \frac{n_b}{n_a}, \tag{36–4}$$

which shows that *the constant in Snell's law for the refraction between substances a and b is the ratio of the indices of refraction.* From Eq. (36–4) we see that the simplest and most symmetrical way of writing Snell's law for any two substances a and b, and for any direction, is

$$n_a \sin \phi_a = n_b \sin \phi_b. \tag{36–5}$$

TABLE 36–1 Index of Refraction for Yellow Sodium Light (λ = 589 nm)

Substance	Index of Refraction
Solids	
Ice (H_2O)	1.309
Fluorite (CaF_2)	1.434
Polystyrene	1.49
Rock salt (NaCl)	1.544
Quartz (SiO_2)	1.544
Zircon ($ZrO_2 \cdot SiO_2$)	1.923
Diamond (C)	2.417
Fabulite ($SrTiO_3$)	2.409
Rutile (TiO_2)	2.62
Glasses (typical values)	
Crown	1.52
Light flint	1.58
Medium flint	1.62
Dense flint	1.66
Lanthanum flint	1.80
Liquids at 20°C	
Methyl alcohol (CH_3OH)	1.329
Water (H_2O)	1.333
Ethyl alcohol (C_2H_5OH)	1.36
Carbon tetrachloride (CCl_4)	1.460
Turpentine	1.472
Glycerine	1.473
Benzene	1.501
Carbon disulfide (CS_2)	1.628

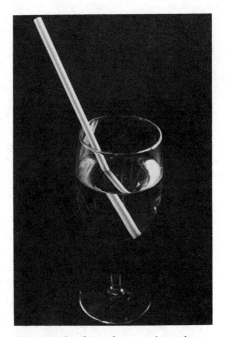

Photograph of a soda straw in a glass of water, showing refraction at the top and side surfaces of the water. (Nancy Rodger, Exploratorium.)

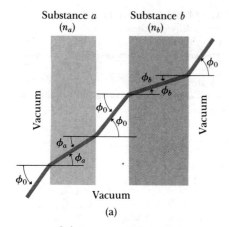

Substance a   Substance b
(n_a)   (n_b)

Vacuum

Vacuum

Vacuum

(a)

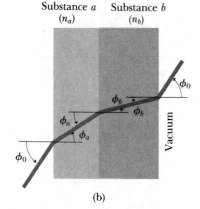

Substance a   Substance b
(n_a)   (n_b)

Vacuum

(b)

36–4 The transmission of light through parallel plates of different substances. The incident and emerging rays are parallel, regardless of direction and regardless of the thickness of the space between adjacent slabs.

The frequency of light doesn't change when it passes from one material to another.

The wavelength is smaller in a material than for the same light in vacuum.

EXAMPLE 36–1 In Fig. 36–3, material a is water and material b is a glass with index of refraction 1.50. If the incident ray makes an angle of 60° with the normal, find the directions of the reflected and refracted rays.

SOLUTION According to Eq. (36–1), the angle the reflected ray makes with the normal is the same as that of the incident ray. Hence $\phi_r = \phi_a = 60°$. To find the direction of the refracted ray we use Eq. (36–5), with $n_a = 1.33$, $n_b = 1.50$, and $\phi_a = 60°$. We find

$$n_a \sin \phi_a = n_b \sin \phi_b,$$

$$(1.33)(\sin 60°) = (1.50)(\sin \phi_b),$$

$$\phi_b = 50.2°.$$

The second material has a larger refractive index than the first, and the refracted ray is bent toward the normal. As Eq. (36–5) shows, this is always the case when the second index is larger than the first. In the opposite case, where the second index is *smaller* than the first, the ray is always bent *away from* the normal.

The index of refraction of a material is closely related to the speed of light in the material. We will develop this relationship in detail in Section 36–12, but we state the principal results here for reference. Light always travels *more slowly* in a material than in vacuum, and the ratio of the two speeds is equal to the index of refraction. Thus the speed of light v in a material having index of refraction n is given by

$$v = \frac{c}{n}, \quad \text{or} \quad n = \frac{c}{v}. \tag{36–6}$$

When light passes from one material to another, its frequency f does not change, for the following reason. When light interacts with matter, the electrons in the material absorb energy from the light and undergo vibration motion with the same frequency as the light. This motion causes reradiation of the energy *with the same frequency*. Thus, since $v = \lambda f$, when v is less than the wave speed c in vacuum, λ is also correspondingly reduced. Thus the wavelength λ of light in a material is less than the wavelength λ_0 of the same light in vacuum, by a factor n:

$$\lambda = \frac{\lambda_0}{n}. \tag{36–7}$$

Two final comments about reflection and refraction need to be made. First, reflection occurs at a highly polished surface of an *opaque* material such as a metal. There is *no* refracted ray, but the reflected ray still behaves according to Eq. (36–1). Second, if the reflecting surface of either a transparent or an opaque material is rough, with irregularities on a scale comparable to or larger than the wavelength of light, reflection occurs not in a single direction but in all directions; such reflection is called *diffuse* reflection. Conversely, reflections in a single direction from smooth surfaces are called *regular* reflections or *specular* reflections.

PROBLEM-SOLVING STRATEGY: *Reflection and refraction*

1. In geometrical optics problems involving rays and angles, *always* start by drawing a big, neat diagram. Use a ruler and protractor. Label all known angles and indexes of refraction.

2. Don't forget that by convention we always measure the angles of incidence, reflection, and refraction from the *normal* to the surface where the reflection and refraction occur, never from the surface itself.

3. You will often have to use some simple geometry or trigonometry in working out angular relations. The sum of the interior angles in a triangle is 180°, and so on. Often it helps to think through the problem, asking yourself, "What do I need to know in order to find this angle?" or "What other angles or other quantities can I compute using the information given in the problem?"

36-6 TOTAL INTERNAL REFLECTION

Figure 36-5a shows a number of rays diverging from a point source P in material a having index of refraction n_a. The rays strike the surface of a second material b having index n_b, where $n_a > n_b$. From Snell's law,

$$\sin \phi_b = \frac{n_a}{n_b} \sin \phi_a.$$

Because n_a/n_b is greater than unity, $\sin \phi_b$ is larger than $\sin \phi_a$. Thus there must be some value of ϕ_a *less than* 90° for which $\sin \phi_b = 90°$. This is illustrated by ray 3 in the diagram, which emerges just grazing the surface at an angle of refraction of 90°. The angle of incidence for which the refracted ray emerges tangent to the surface is called the **critical angle** and is denoted by ϕ_{crit} in the diagram. If the angle of incidence is *greater than* the critical angle, the sine of the angle of refraction, as computed by Snell's law, would have to be greater

> Total internal reflection: light can be trapped inside a material (but not outside).

> Total internal reflection occurs when the angle of incidence exceeds the critical value.

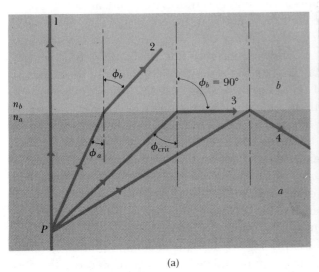

(a)

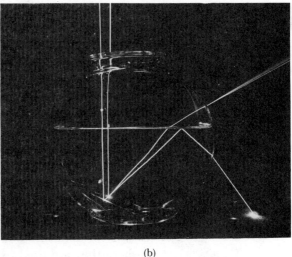

(b)

36-5 (a) Total internal reflection. The angle of incidence ϕ_a for which the angle of refraction is 90°, is called the critical angle. (b) Rays of laser light enter the water in the fishbowl from above; they are reflected at the bottom by mirrors tilted at slightly different angles, and one ray undergoes total internal reflection at the air–water interface. (The Exploratorium.)

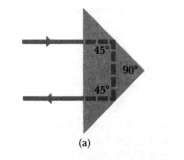

(a)

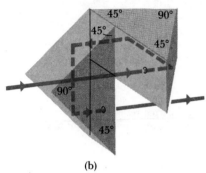

(b)

36–6 (a) A Porro prism. (b) A combination of two Porro prisms.

Total internal reflection is used in binocular prisms.

Fiber optics: a dramatic example of total internal reflection

Fiber optics for communication: the wave of the future

36–7 A light ray "trapped" by internal reflections.

than unity. This is impossible; beyond the critical angle, the ray *cannot pass* into the upper material but is completely reflected internally at the boundary surface. **Total internal reflection** occurs only when a ray is incident on the surface of a material whose index of refraction is *smaller than* that of the material in which the ray is traveling.

The critical angle for two given materials may be found by setting $\phi_b = 90°$ or $\sin \phi_b = 1$ in Snell's law. We then have

$$\sin \phi_{\text{crit}} = \frac{n_b}{n_a}. \tag{36–8}$$

The critical angle of a glass–air surface, taking 1.50 as a typical index of refraction of glass, is

$$\sin \phi_{\text{crit}} = \frac{1}{1.50} = 0.67, \qquad \phi_{\text{crit}} = 42°.$$

The fact that this angle is slightly less than 45° makes it possible to use a prism with angles of 45°, 45°, and 90° as a totally reflecting surface. The advantages of totally reflecting prisms over metallic surfaces as reflectors are, first, that the light is *totally* reflected, while no metallic surface reflects 100% of the light incident on it, and second, the reflecting properties are permanent and not affected by tarnishing. Offsetting these is the fact that some loss of light occurs by reflection at the surfaces where light enters and leaves the prism, although coating the surfaces with so-called nonreflecting films can reduce this loss considerably.

A 45°–45°–90° prism, used as in Fig. 36–6a, is called a *Porro* prism. Light enters and leaves at right angles to the hypotenuse and is totally reflected at each of the shorter faces. The total deviation (change of direction of the rays) is 180°. Two Porro prisms are sometimes combined, as in Fig. 36–6b, an arrangement often found in binoculars.

If a beam of light enters at one end of a transparent rod, as in Fig. 36–7, the light is totally reflected internally and is "trapped" within the rod even if it is curved, provided the curvature is not too great. Such a rod is sometimes referred to as a *light pipe*. A bundle of fine glass fibers behaves in the same way and has the advantage of being flexible. A bundle may consist of thousands of individual fibers, each of the order of 0.002 mm to 0.01 mm in diameter. If the fibers are assembled in the bundle so that the relative positions of the ends are the same (or mirror images) at both ends, the bundle can transmit an image, as shown in Fig. 36–8.

Fiber-optic devices have found a wide range of medical applications in instruments called *endoscopes*, which can be inserted directly into the bronchial tubes, the bladder, the colon, and so on, for direct visual examination. A bundle of fibers can be enclosed in a hypodermic needle for study of tissues and blood vessels far beneath the skin.

Fiber optics are now also finding applications in communication systems, where they are used to transmit a modulated laser beam. Because the frequency of the modulated beam is very much higher than those used in wire or radio communication, an enormous amount of information can be transmitted through one fiber-optic cable. For example, the Carnegie-Mellon University computer system, consisting of several thousand microcomputers networked with mainframe computers, will be linked partly by fiber-optic ca-

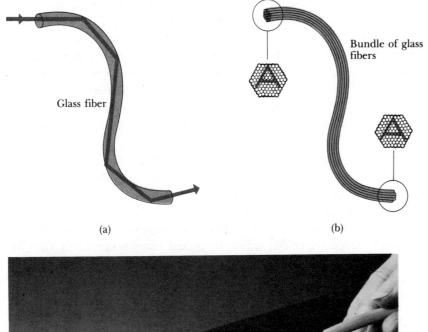

Bundle of glass fibers

(a)   (b)

(c)

36-8 (a) Total internal reflection in a single fiber. (b) Image transmission by a bundle of fibers. (c) A fiber-optic cable used to transmit a modulated laser beam for communication purposes compared to a larger copper cable that has equal information-transmitting capacity. (Courtesy of Corning Glass Works.)

bles as shown in Fig. 36-8c. In the future, most telephone systems are likely to be connected by fiber optics.

36-7 DISPERSION

Most light beams are a superposition of waves with wavelengths extending throughout the visible spectrum. The speed of light *in vacuum* is the same for all wavelengths, but the speed in a material substance is different for different wavelengths. Hence the index of refraction of a material depends on wavelength. Any wave medium in which the speed of a wave varies with wavelength is said to show **dispersion.** Figure 36-9 shows the variation of index of refraction with wavelength for a few common optical materials. The value of n usually *decreases* with increasing wavelength and so *increases* with increasing frequency. Thus light of longer wavelength usually has greater speed in a material than does light of shorter wavelength.

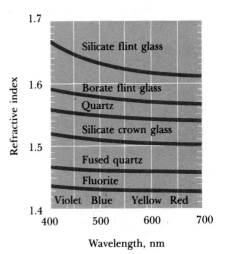

36-9 Variation of index of refraction with wavelength.

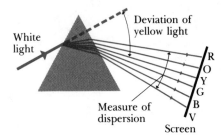

36–10 Dispersion by a prism. The band of colors on the screen is called a spectrum.

What makes a diamond sparkle? Do fake diamonds sparkle as much?

Figure 36–10 shows a ray of white light (a superposition of all visible wavelengths) incident on a prism. The deviation (change of direction) produced by the prism increases with increasing index of refraction and decreasing wavelength. Violet light is deviated most and red least, with other colors in intermediate positions. When it comes out of the prism, the light is spread out into a fan-shaped beam, as shown. The light is said to be *dispersed* into a spectrum. The amount of dispersion depends on the *difference* between the refractive index for violet light and that for red light. From Fig. 36–9 we can see that for a substance such as fluorite, whose refractive index for yellow light is small, the difference between the indexes for red and violet is also small. But, for silicate flint glass, both the index for yellow light and the difference between extreme indexes are large. In other words, for most transparent materials, the greater the deviation, the greater the dispersion.

The brilliance of diamond is due in part to its large dispersion and in part to its unusually large refractive index. In recent years synthetic crystals of titanium dioxide and of strontium titanate, with about eight times the dispersion of diamond, have been produced.

36–8 POLARIZATION

Polarization: a property of all transverse waves, electromagnetic and otherwise

Polarization phenomena occur with transverse waves. Our principal concern in this chapter is with electromagnetic waves, especially light, but to introduce basic concepts we first consider the mechanical example of transverse waves on a string, as discussed in Chapter 21. For a string whose equilibrium position is along the x-axis, the displacements may be along the y-direction, as in Fig. 36–11a. In this case the string always lies in the xy-plane. But the displacements might instead be along the z-axis, as in Fig. 36–11b, so that the string lies in the xz-plane.

36–11 (a) Transverse wave on a string, polarized in the y-direction. (b) Wave polarized in the z-direction. (c) Barrier with a frictionless vertical slot passes components polarized in the y-direction but blocks those polarized in the z-direction, acting as a polarizing filter.

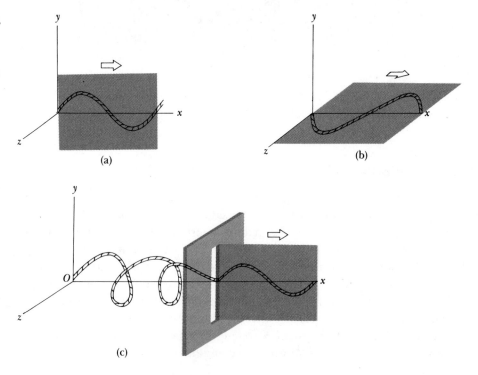

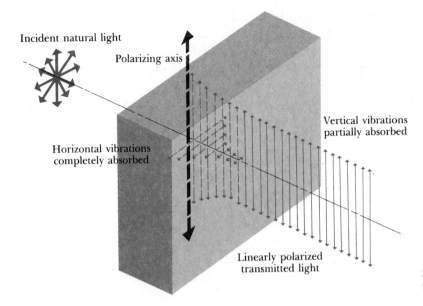

Incident natural light

Polarizing axis

Vertical vibrations
partially absorbed

Horizontal vibrations
completely absorbed

Linearly polarized
transmitted light

36–12 Linearly polarized light
transmitted by a polarizing filter.

A wave having only y-displacements in the discussion above is said to be **linearly polarized** in the y-direction, and the one with only z-displacements is linearly polarized in the z-direction. It is easy in principle to construct a mechanical *filter* that permits only waves with a certain polarization direction to pass. An example is shown in Fig. 36–11c; the string can slide vertically in the slot without friction, but no horizontal motion is possible. This filter passes waves polarized in the y-direction but blocks those polarized in the z-direction.

Linear polarization: when the disturbance is always along one direction

This same language can be applied to light and other electromagnetic waves, which also exhibit polarization. As we learned in Chapter 35, an electromagnetic wave consists of fluctuating electric and magnetic fields, perpendicular to each other and to the direction of propagation. By convention, the direction of polarization is taken to be that of the *electric*-field vector, not the magnetic field, because most mechanisms for detecting electromagnetic waves employ principally the electric-field forces on electrons in materials. That is, the most common manifestations of electromagnetic radiation are due chiefly to the electric-field force, not the magnetic-field force.

Polarizing filters, or **polarizers,** can be made for electromagnetic waves; the details of construction depend on the wavelength. For microwaves having a wavelength of a few centimeters, a grid of closely spaced, parallel conducting wires insulated from each other will pass waves whose E fields are perpendicular to the wires but not those with E fields parallel to the wires. For light, the most common polarizing filter is a material known by the trade name Polaroid, widely used for sunglasses and polarizing filters for camera lenses. This material works on the principle of preferential absorption, passing waves polarized parallel to a certain axis in the material (called the **polarizing axis**) with 80% or more transmission, but offering only 1% or less transmission to waves with polarization perpendicular to this axis. The action of such a polarizing filter is shown schematically in Fig. 36–12.

A polarizing filter transmits only waves polarized along a certain direction.

Light emerging from a polarizing filter is always linearly polarized.

Waves emitted by a radio transmitter are usually linearly polarized; a vertical rod antenna of the type widely used for CB radios emits waves that in a horizontal plane around the antenna are polarized in the vertical direction

Ordinary light is a random mixture of all states of polarization.

(parallel to the antenna). Light from ordinary sources is *not* polarized, for a slightly subtle reason. The "antennas" that radiate light waves are the molecules of which the light sources are composed. The electrically charged particles in the molecules acquire energy in some way and radiate this energy as electromagnetic waves of short wavelength. The waves from any one molecule may be linearly polarized, like those from a radio antenna; but since any actual light source contains a tremendous number of molecules, oriented at random, the light emitted is a random mixture of waves linearly polarized in all possible transverse directions.

Behavior of an ideal polarizing filter (or polarizer).

36–9 POLARIZING FILTERS

An ideal polarizer has the property that it passes 100% of the incident light polarized in the direction of the filter's polarizing axis but blocks completely all light polarized perpendicular to this axis. Such a device is an unattainable idealization, but the concept is useful in clarifying the basic ideas. Some real polarizers approximate this ideal behavior very closely. In Fig. 36–13, unpolarized light is incident on a polarizer in the form of a flat plate. The polarizing axis is represented by the broken line. As the polarizer is rotated about an axis parallel to the incident ray, the intensity of the transmitted light (measured by the photocell) does not change. (Recall from Sections 35–3 and 35–5 that intensity, the average magnitude of the Poynting vector, is the power per unit area transmitted by the light.) The polarizer transmits the components of the

36–13 The intensity of the transmitted linearly polarized light, measured by the photocell, is the same for all orientations of the polarizing filter.

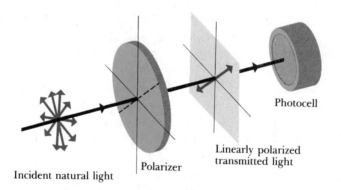

Incident natural light — Polarizer — Linearly polarized transmitted light — Photocell

36–14 The analyzer transmits only the component parallel to its transmission direction or polarizing axis.

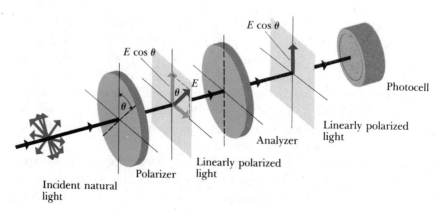

Incident natural light — Polarizer — Linearly polarized light — $E \cos \theta$ — E — Analyzer — $E \cos \theta$ — Linearly polarized light — Photocell

incident waves in which the **E** vector is parallel to the transmission direction of the polarizer, and by symmetry the components are equal for all angles perpendicular to the beam direction.

The intensity of the light transmitted through the polarizer can be measured with the photocell in Fig. 36–13; it is found to be exactly one-half that of the incident light. We can understand why this should be, as follows: The incident light can always be resolved into components polarized parallel to the polarizer axis and components polarized perpendicular to it. Because the incident light is a random mixture of all states of polarization, these two components are, on the average, equal. Thus (in the ideal polarizer) exactly half of the incident intensity, the part corresponding to the component parallel to the polarizer axis, is transmitted.

Suppose now that we insert a second polarizer between the first polarizer and the photocell, as in Fig. 36–14. Let the transmission direction of the second polarizer, or *analyzer*, be vertical, and let that of the first polarizer make an angle θ with the vertical. The linearly polarized light transmitted by the polarizer may be resolved into two components as shown, one parallel and the other perpendicular to the transmission direction of the analyzer. Only the parallel component, of amplitude $E \cos \theta$, is transmitted by the analyzer. The transmitted intensity is maximum when $\theta = 0$ and is zero when $\theta = 90°$, or when polarizer and analyzer are *crossed*.

To find the transmitted intensity at intermediate angles, we recall the discussion of energy in electromagnetic waves in Sections 35–3 and 35–5. In particular, Eq. (35–18) shows that the intensity is proportional to the *square* of the amplitude; thus we have

$$I = I_{max} \cos^2 \theta, \qquad (36–9)$$

where I_{max} is the maximum intensity of light transmitted (at $\theta = 0$) and I is the amount transmitted at angle θ. This relation, discovered experimentally by Etienne Louis Malus in 1809, is called **Malus' law.**

The angle θ is the angle between the transmission directions of polarizer and analyzer. If either the analyzer or the polarizer is rotated, the amplitude of the transmitted beam varies with the angle between them according to Eq. (36–19).

When unpolarized light strikes a polarizing filter, the transmitted light has half the incident intensity.

Using vector components to analyze transmission of polarized light by polarizing filters

---

PROBLEM-SOLVING STRATEGY: *Linear polarization*

1. Remember that in light or any electromagnetic wave the **E** field is perpendicular to the propagation direction and is the direction of polarization (or opposite to that direction). The polarization direction can be thought of as a two-headed arrow. When working with polarizing filters, you are really dealing with components of **E** parallel and perpendicular to the polarizing axis. Everything you know about components of vectors is applicable here.

2. The intensity (average power per unit area) of a wave is proportional to the *square* of its amplitude, as shown by Eq. (35–18). If you find that two waves differ in amplitude by a certain factor, their intensities differ by the square of that factor.

3. In Section 36–11 we will encounter problems involving superposition of two linearly polarized waves with perpendicular directions of polarization. Their relative phase is crucial; if they are in phase, the resultant is again linearly polarized, but if they are not, the resultant is circularly or elliptically polarized. In such cases, pay close attention to phase relationships.

EXAMPLE 36–2   In Fig. 36–14, the incident unpolarized light has intensity I_0. Find the intensity transmitted by the first polarizer and by the second, if the angle θ is 30°.

SOLUTION   As explained above, the intensity after the first filter is $I_0/2$. According to Eq. (36–9), the second filter reduces the intensity by a factor of $\cos^2 30° = \frac{3}{4}$. Thus the intensity transmitted by the second polarizer is $(I_0/2)(\frac{3}{4}) = 3I_0/8$.

Reflected light is preferentially polarized.

Unpolarized light can be partially polarized by *reflection*. When unpolarized light strikes a reflecting surface between two optical materials, preferential reflection occurs for those waves in which the electric-field vector is perpendicular to the plane of incidence (the plane containing the incident ray and the normal to the surface). At one particular angle of incidence, called the **polarizing angle** ϕ_p, no light whatever is reflected except that in which the electric vector is perpendicular to the plane of incidence. This case is shown in Fig. 36–15.

When light is incident at the polarizing angle, *none* of the component parallel to the plane of incidence is reflected; this component is 100% transmitted in the *refracted* beam. For the component perpendicular to the plane of incidence, the fraction reflected depends on the index of the reflecting material. About 15% is reflected if the reflecting surface is glass. Hence the *reflected* light is weak and *completely* polarized. The *refracted* light is a mixture of the parallel component, all of which is refracted, and the remaining 85% of the perpendicular component. It is therefore strong but only *partially* polarized. At angles of incidence other than the polarizing angle, some of the component parallel to the plane of incidence is reflected, so that, except at the polarizing angle, the reflected light is not completely linearly polarized.

When the angle of incidence is Brewster's angle, the reflected light is completely polarized.

In 1812 Sir David Brewster noticed that when the angle of incidence is equal to the polarizing angle ϕ_p, the reflected ray and the refracted ray are

36–15 When light is incident at the polarizing angle, the reflected light is linearly polarized.

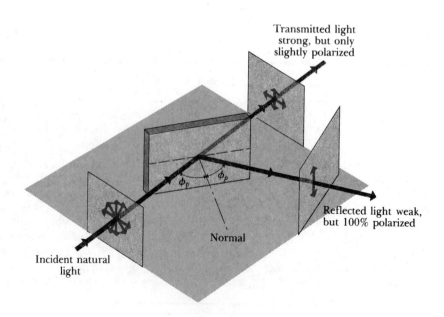

Transmitted light strong, but only slightly polarized

Reflected light weak, but 100% polarized

Normal

Incident natural light

perpendicular to each other, as shown in Fig. 36–16. When this is the case, the angle of refraction ϕ' becomes the complement of ϕ_p, so that $\sin \phi' = \cos \phi_p$. Since, according to the law of refraction,

$$n \sin \phi_p = n' \sin \phi', \tag{36–10}$$

we find $n \sin \phi_p = n' \cos \phi_p$ and

$$\tan \phi_p = \frac{n'}{n}, \tag{36–11}$$

a relation known as **Brewster's law.**

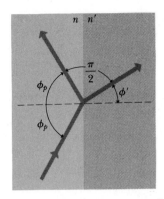

36–16 At the polarizing angle the reflected and transmitted rays are perpendicular to each other.

Some crystalline materials exhibit different refractive indexes for different directions of polarization. A common example is calcite; when a calcite crystal is oriented appropriately in a beam of unpolarized light, its refractive index (for $\lambda = 589$ nm) is 1.658 for one direction of polarization and 1.486 for the perpendicular direction. Such crystals are said to be *doubly refracting,* or **birefringent.** They can be used in arrangements or prisms that separate spatially the two polarized components of an unpolarized beam and reflect one component out of the beam direction. These polarizing filters are more efficient than Polaroid filters and also much more expensive.

Birefringence: different indexes of refraction for different polarizations

Some doubly refracting crystals exhibit **dichroism,** a selective absorption in which one of the polarized components is *absorbed* much more strongly than the other. If the crystal is cut of the proper thickness, one component is practically extinguished by absorption, while the other is transmitted in appreciable amount, as indicated in Fig. 36–12. Tourmaline is one example of such a dichroic crystal.

Dichroism: preferential absorption of some polarization states

An early form of Polaroid, invented by Edwin H. Land in 1928, consisted of a thin layer of tiny needlelike dichroic crystals of herapathite (iodoquinine sulfate), in parallel orientation, embedded in a plastic matrix and enclosed for protection between two transparent plates. A more recent modification, developed by Land in 1938 and known as an H-sheet, is a *molecular* polarizer. It consists of long polymeric molecules of polyvinyl alcohol (PVA) that have been given a preferred direction by stretching and have been stained with an ink containing iodine that causes the sheet to exhibit dichroism. The PVA sheet is laminated to a support sheet of cellulose acetate butyrate.

Dichroic materials can make polarizing filters.

Polaroid sheet is widely used in sunglasses, where, from the standpoint of its polarizing properties, it plays the role of the analyzer in Fig. 36–14. We have seen that when unpolarized light is reflected, there is a preferential reflection for light polarized perpendicular to the plane of incidence. When sunlight is reflected from a horizontal surface, the plane of incidence is vertical. Hence the reflected light contains a preponderance of light polarized in the horizontal direction. When such reflection occurs at smooth asphalt road surfaces or the surfaces of a lake, it causes unwanted "glare," and vision is improved by eliminating it. The transmission direction of the Polaroid sheet in the sunglasses is vertical, so none of the horizontally polarized light is transmitted to the eyes.

What are the advantages of polarizing sunglasses?

Apart from this polarizing feature, these glasses serve the same purpose as any dark glasses, absorbing 50% of the incident light. In an unpolarized beam, half the light can be considered as polarized horizontally and half vertically, and only the vertically polarized light is transmitted. The sensitivity of the eye is independent of the state of polarization of the light.

36–10 SCATTERING OF LIGHT

The sky is blue. Sunsets are red. Skylight is partially polarized, as can readily be verified by looking at the sky directly overhead through a polarizing filter. It turns out that one phenomenon is responsible for all three of these effects.

In Fig. 36–17, sunlight (unpolarized) comes from the left along the z-axis and passes over an observer looking vertically upward along the y-axis. Molecules of the earth's atmosphere are located at point O. The electric field in the beam of sunlight sets the electric charges in the molecules in vibration. Since light is a transverse wave, the direction of the electric field in any component of the sunlight lies in the xy-plane, and the motion of the charges takes place in this plane. There is no field, and hence no vibration, in the direction of the x- and y-axes.

A component of the incident light at an angle θ with the x-axis sets the electric charges in the molecules vibrating in the same direction, as indicated by the heavy line through point O. We can resolve this vibration into two components, one along the x-axis and the other along the y-axis. Each component in the incident light produces the equivalent of two molecular "antennas," oscillating with the same frequency as the incident light, and lying along the x- and y-axes.

We mentioned in Section 35–8 that an antenna does not radiate in the direction of its own length. Hence the antenna along the y-axis does not send any light to the observer directly below it. It does, of course, send out light in other directions. The only light reaching the observer comes from the component of vibration along the x-axis, and, as is the case with the waves from any antenna, this light is linearly polarized, with the electric field parallel to the antenna. The vectors on the y-axis below point O show the direction of polarization of the light reaching the observer.

The process just described is called **scattering.** The energy of the scattered light is removed from the original beam, which becomes weakened in the process. Detailed analysis of the scattering processes shows that the intensity

Scattering of light leads to preferential polarization.

Blue light is scattered more than red.

Why is the sky blue?
Why are sunsets red?

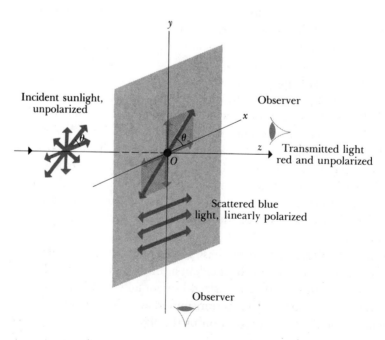

36–17 Scattered light is linearly polarized.

of the scattered light increases with increasing frequency; blue light is scattered more than red, with the result that the hue of the scattered light is blue.

Because skylight is partially polarized, polarizers are useful in photography. The sky can be darkened in a photograph by appropriate orientation of the polarizer axis. The effect of atmospheric haze can be reduced in exactly the same way, and unwanted reflections can be controlled just as with polarizing sunglasses, discussed in Section 36–9.

Toward evening, when sunlight has to travel a large distance through the earth's atmosphere to reach a point over or nearly over an observer, a large proportion of the blue light in sunlight is removed from it by scattering. White light minus blue light is yellow or red in hue. Thus when sunlight, with the blue component removed, is incident on a cloud, the light reflected from the cloud to the observer has the yellow or red hue so commonly seen at sunset. If the earth had no atmosphere, we would receive *no* skylight at the earth's surface, and the sky would appear as black in the daytime as it does at night. To an astronaut in a spaceship or on the moon, the sky appears black, not blue.

In outer space, the sky is black, not blue.

36–11 CIRCULAR AND ELLIPTICAL POLARIZATION

Up to this point we have discussed polarization phenomena in terms of *linearly polarized* light. Light (and all other electromagnetic radiation) may also have *circular* or *elliptical* polarization. To introduce these new concepts, we return again to mechanical waves on a stretched string. In Fig. 36–11, suppose the two linearly polarized waves in parts (a) and (b) are in phase and have equal amplitude. When they are *superposed,* each point in the string has simultaneously y- and z-displacements of equal magnitude, and a little thought shows that the resultant wave lies in a plane oriented at 45° to the x- and z-axes (i.e., in a plane making a 45° angle with the xy- and xz-planes). The amplitude of the resultant wave is larger by a factor of $\sqrt{2}$ than that of either component wave, and the resultant wave is again linearly polarized.

Superposing two linearly polarized waves with phase differences

But now suppose one of the equal-amplitude component waves differs in phase by a quarter-cycle from the other. Then the resultant motion of each point corresponds to a superposition of two simple harmonic motions at right angles with a quarter-cycle phase difference. The motion is then no longer confined to a single plane, and it can be shown that each point on the rope moves in a *circle* in a plane parallel to the yz-plane. Successive points on the rope have successive phase differences, and the overall motion of the string then has the appearance of a rotating helix. This particular superposition of two linearly polarized waves is called **circular polarization.** By convention, the wave is said to be *right circularly polarized* when, as in the present instance, the sense of motion of a particle of the string, to an observer looking *backward* along the direction of propagation, is *clockwise.* The wave is *left circularly polarized* when it appears counterclockwise to that observer. Left circular polarization would be the result if the phase difference between y- and z-components were opposite to that in our example.

In circularly polarized light, the E and B vectors rotate around the direction of propagation.

If the phase difference between the two component waves is something other than a quarter-cycle, or if the two component waves have different amplitudes, then each point on the string traces out not a circle but an *ellipse.* The resulting wave is said to be **elliptically polarized.**

For electromagnetic waves of radio frequencies, circular or elliptical polarization can be produced by using two antennas at right angles, fed from the same transmitter but with a phase-shifting network that introduces the appropriate phase difference. For light, the phase shift can be introduced by use of a birefringent material. If two waves with perpendicular directions of polarization are in phase as they enter the material, they travel with different speeds because the refractive index is different for the two waves. In general, they are no longer in phase when they emerge from the crystal. If the thickness of the crystal is such as to introduce a quarter-cycle phase difference, then the crystal converts linearly polarized light to circularly polarized light. Such a crystal is called a **quarter-wave plate.** We challenge you to show that such a plate also converts circularly polarized light to linearly polarized light!

When a polarizer and an analyzer are mounted in the "crossed" position, that is, with their transmission directions at right angles to each other, no light is transmitted through the combination. But if a doubly refracting crystal is inserted between polarizer and analyzer, the light transmitted through the crystal is, in general, elliptically polarized, and some light will be transmitted by the analyzer. Thus the field of view, dark in the absence of the crystal, becomes light when the crystal is inserted.

Some substances, such as glass and various plastics, though not normally doubly refracting, become so when subjected to mechanical stress. This is the basis of the science of **photoelasticity.** Stresses in opaque engineering materials such as girders, boiler plates, and gear teeth can be analyzed by constructing a transparent model of the object, usually of a plastic, subjecting it to stress and examining it between a polarizer and an analyzer in the crossed position. Very complicated stress distributions, such as those around a hole or gear tooth, that are practically impossible to analyze mathematically, may thus be studied by optical methods. Figure 36–18 shows two photographs of photoelastic models under stress.

Liquids are not normally doubly refracting, but some become so when an electric field is established within them. This phenomenon is known as the *Kerr effect.* The existence of the Kerr effect makes it possible to construct an electrically controlled "light valve." A cell with transparent walls contains the liquid between a pair of parallel plates. The cell is inserted between crossed Polaroid disks. Light is transmitted when an electric field is set up between the plates and is cut off when the field is removed.

When a beam of linearly polarized light is sent through certain types of crystals and certain liquids, the direction of polarization of the emerging linearly polarized light is found to be different from the original direction. This phenomenon is called *rotation of the direction of polarization,* and substance that exhibit this effect are said to be *optically active.* Those that rotate the direction of polarization to the right, looking along the advancing beam, are called *dextrorotatory,* or right-handed; those that rotate it to the left are *levorotatory,* or left-handed.

Optical activity may be due to an asymmetry of the molecules of a substance, or it may be a property of a crystal as a whole. For example, solutions of cane sugar are dextrorotatory, indicating that the optical activity is a property of the sugar molecule. The molecules of the sugars dextrose and levulose are mirror images, and their optical activities are opposite. The rotation of the direction of polarization by a sugar solution is used commercially as a method

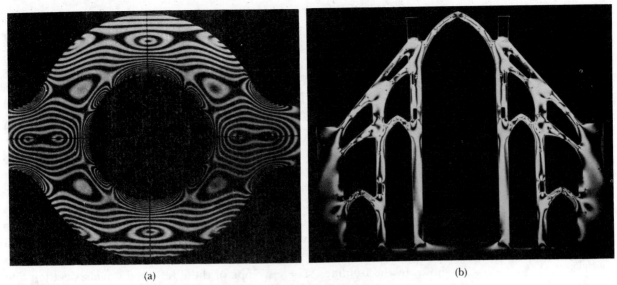

(a)   (b)

36–18 (a) Photoelastic stress analysis of a plastic model of a machine part. (Courtesy of Dr. W. M. Murray, Massachusetts Institute of Technology.) (b) Stress analysis of a model of a cross section of a Gothic cathedral. The masonry construction used for this kind of building had great strength in compression but very little in tension. Inadequate buttressing and high winds sometimes caused tensile stresses in normally compressed structural elements, leading to some spectacular collapses. (Sepp Seitz/Woodfin Camp.)

of determining the proportion of cane sugar in a given sample. Crystalline quartz is also optically active; some natural crystals are right-handed and others left-handed. Here the optical activity is a consequence of the crystalline structure, since the activity disappears when the quartz is melted and allowed to resolidify into a glassy, noncrystalline state called fused quartz.

36–12 HUYGENS' PRINCIPLE

The principles of reflection and refraction of light rays that we introduced in Section 36–5 were discovered experimentally long before the wave nature of light was firmly established. These principles may, however, be derived from *wave* considerations and thus shown to be consistent with the wave nature of light. To establish this connection we use a principle called **Huygens' principle.** This principle, stated originally by Christian Huygens in 1678, is a geometrical method for finding, from the known shape of a wave front at some instant, the shape of the wave front at some later time. Huygens assumed that *every point of a wave front may be considered the source of secondary wavelets that spread out in all directions with a speed equal to the speed of propagation of the wave.* The new wave front is then found by constructing a surface *tangent* to the secondary wavelets or, as it is called, the *envelope* of the wavelets.

Huygens' principle is illustrated in Fig. 36–19. The original wave front AA' is traveling as indicated by the small arrows. We wish to find the shape of the wave front after a time interval t. Let v represent the speed of propagation. We construct several circles (traces of spherical wavelets) of radius $r = vt$, with centers along AA'. The trace of the envelope of these wavelets, which is the new wave front, is the curve BB'. The speed v is assumed to be the same at all points and in all directions.

Huygens' principle: an important concept in wave propagation

36–19 Geometric construction illustrating Huygens' principle.

To derive the law of reflection from Huygens' principle, we consider a plane wave approaching a plane reflecting surface. In Fig. 36–20a, the lines AA', BB', and CC' represent successive positions of a wave front approaching the surface MM'. The actual planes are perpendicular to the plane of the figure. Point A on the wave front AA' has just arrived at the reflecting surface. The position of the wave front after a time interval t may be found by applying Huygens' principle. With points on AA' as centers, we draw several secondary wavelets of radius vt, where v is the speed of propagation of the wave. Those wavelets originating near the upper end of AA' spread out unhindered, and their envelope gives that portion of the new wave surface OB'. The wavelets originating near the lower end of AA', however, strike the reflecting surface. If the surface had not been there, they would have reached the positions shown by the broken circular arcs. The effect of the reflecting surface is to *reverse the direction* of travel of those wavelets that strike it, so that part of a wavelet that would have penetrated the surface actually lies to the left of it, as shown by the full lines. The envelope of these reflected wavelets is then that portion of the wave front OB. The trace of the entire wave front at this instant is the broken line BOB'. A similar construction gives the line CNC' for the wave front after another time interval t.

The angle ϕ between the incident *wave front* and the *surface* is the same as that between the incident *ray* and the *normal* to the surface and is therefore the angle of incidence. Similarly, ϕ_r is the angle of reflection. To find the relation between these angles, we consider Fig. 36–20b. From O, we draw $OP = vt$, perpendicular to AA'. Now OB, by construction, is tangent to a circle of radius vt with center at A. Hence, if we draw AQ from A to the point of tangency, the triangles APO and OQA are equal (right triangles with the side AO in common and with $AQ = OP$). The angle ϕ therefore equals the angle ϕ_r, and we have the law of reflection.

We can derive the law of *refraction* by a similar procedure. In Fig. 36–21a, we consider a wave front, represented by line AA', for which point A has just arrived at the boundary surface SS' between two transparent materials a and b, with indexes of refraction n_a and n_b. (The *reflected* waves are not shown in the figure; they proceed exactly as in Fig. 36–20.) We can apply Huygens' principle to find the position of the refracted wave fronts after a time t.

With points on AA' as centers, we draw a number of secondary wavelets. Those originating near the upper end of AA' travel with speed v_a and, after a time interval t, are spherical surfaces of radius $v_a t$. The wavelet originating at point A, however, is traveling in the second material b with speed v_b and at time t is a spherical surface of radius $v_b t$. The envelope of the wavelets from the original wave front is the plane whose trace is the broken line BOB'. A similar construction leads to the trace CPC' after a second interval t.

The angles ϕ_a and ϕ_b between the surface and the incident and refracted wave fronts are, respectively, the angle of incidence and the angle of refraction. To find the relation between these angles, refer to Fig. 36–21b. Draw $OQ = v_a t$, perpendicular to AQ, and draw $AB = v_b t$, perpendicular to BO. From the right triangle AOQ,

$$\sin \phi_a = \frac{v_a t}{AO},$$

and from the right triangle AOB,

$$\sin \phi_b = \frac{v_b t}{AO}.$$

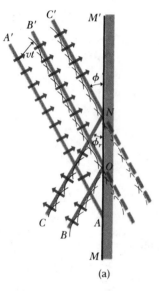

(a)

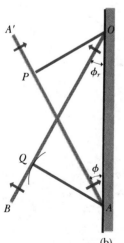

(b)

36–20 (a) Successive positions of a plane wave AA' as it is reflected from a plane surface. (b) A portion of (a).

Hence

$$\frac{\sin \phi_a}{\sin \phi_b} = \frac{v_a}{v_b}. \tag{36-12}$$

Since v_a/v_b is a constant, Eq. (36–12) expresses Snell's law, and we have derived Snell's law from a wave theory! This result emphasizes that the general wave theory gives the same results as the more specialized ray picture in cases where the ray picture is applicable.

The most general form of Snell's law is given by Eq. (36–4), namely,

$$\frac{\sin \phi_a}{\sin \phi_b} = \frac{n_b}{n_a}.$$

Comparing this with Eq. (36–12), we see that

$$\frac{v_a}{v_b} = \frac{n_b}{n_a},$$

and

$$n_a v_a = n_b v_b.$$

When either material is vacuum, $n = 1$ and the speed is c. Hence

$$n_a = \frac{c}{v_a}, \qquad n_b = \frac{c}{v_b}, \tag{36-13}$$

showing that *the index of refraction of any material is the ratio of the speed of light in vacuum to the speed in the material.* We stated Eq. (36–13) without proof at the end of Section 36–5. The speed of light in a material is always *less* than in vacuum; hence for any material n is always greater than unity.

In Fig. 36–21, if t is chosen to be the period τ of the wave, the spacing is $v\tau$, which is the wavelength λ. The figure shows that when v_b is less than v_a, the wavelength in the second material is smaller than in the first. When a light wave proceeds from one material to another, where the speed is different, the wavelength changes *but not the frequency*, as we discussed in Section 36–5. Since $v_a = f\lambda_a$ and $v_b = f\lambda_b$,

$$\frac{\lambda_a}{v_a} = \frac{\lambda_b}{v_b}$$

and

$$\lambda_a \frac{c}{v_a} = \lambda_b \frac{c}{v_b}.$$

Therefore,

$$\lambda_a n_a = \lambda_b n_b.$$

If either material is vacuum, the index is 1 and the wavelength in vacuum is represented by λ_0. Hence

$$\lambda_a = \frac{\lambda_0}{n_a}, \qquad \lambda_b = \frac{\lambda_0}{n_b}, \tag{36-14}$$

showing that *the wavelength in any material is the wavelength in vacuum divided by the index of refraction of the medium.* We stated Eq. (36–14) without proof in Section 36–5.

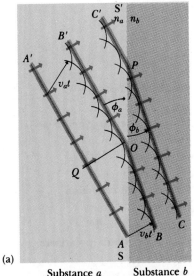

(a)

Substance *a*   Substance *b*

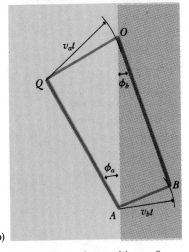

(b)

36–21 (a) Successive positions of a plane wave front AA' as it is refracted by a plane surface. (b) A portion of (a). The case $v_b < v_a$ is shown.

The wavelength is smaller in a material than for the same wave in vacuum.

SUMMARY

Light is electromagnetic radiation with wavelengths in the range 400 to 700 nm. It is emitted by moving electric charges within atoms and molecules that have been given excess energy by heating or by electrical discharge. Commonly used light sources include incandescent lamps, carbon arcs, sodium- and mercury-arc lamps, and fluorescent lamps. The light emitted by a laser is much more nearly monochromatic (single-frequency) than that of any other source.

The speed of light is a fundamental physical constant. It is defined to have a certain precise value; this, along with the unit of time defined by an atomic clock, defines the unit of length.

A wave front is a line or surface of constant phase; wave fronts move with a speed equal to the propagation speed of the wave. The propagation of light can also be represented by rays, which are lines along the direction of propagation, perpendicular to the wave fronts. Representation of light by rays is the basis of geometrical optics.

The law of reflection and the law of refraction (Snell's law) govern the behavior of light rays at an interface between two optical materials. The incident, reflected, and refracted rays and the normal to the plane of interface all lie in a single plane. The law of reflection states that the angles of incidence and reflection are equal; the law of refraction states that the ratio of the sines of the angles of incidence and refraction is constant for a given pair of materials. Angles of incidence, reflection, and refraction are always measured from the normal to the surface.

When a ray is incident in vacuum on material a, with incident and refracted angles ϕ_0 and ϕ_a, respectively, the index of refraction n_a of the material is defined as

$$\frac{\sin \phi_0}{\sin \phi_a} = n_a. \tag{36-3}$$

The general statement of the law of refraction, in terms of indexes of refraction, is

$$n_a \sin \phi_a = n_b \sin \phi_b. \tag{36-5}$$

The speed of light is less in a material than in vacuum by a factor n, and the wavelength in a material is smaller than in vacuum by the same factor n.

When a ray travels from a material of greater index of refraction toward one of smaller index, total internal reflection occurs when the angle of incidence exceeds a critical value ϕ_{crit} given by

$$\sin \phi_{crit} = \frac{n_b}{n_a}. \tag{36-8}$$

The index of refraction of a material depends on the wavelength of light, usually decreasing with increasing wavelength. This is called dispersion.

Electromagnetic waves, like all transverse waves, exhibit polarization. The direction of polarization of a linearly polarized wave is defined as the direction of the E field. A polarizing filter passes radiation that is linearly polarized in the direction of its polarizing axis, and blocks radiation polarized perpendicularly to that axis. When linearly polarized light is incident on a polarizing filter with its axis at an angle θ to the direction of polarization, the transmitted intensity I is given by

$$I = I_{max} \cos^2 \theta, \qquad (36\text{--}9)$$

where I_{max} is the intensity when $\theta = 0$. This relation is called Malus' law.

When unpolarized light strikes an interface between two materials, the reflected light is completely polarized perpendicular to the plane of incidence if the angle of incidence ϕ_p is given by

$$\tan \phi_p = \frac{n'}{n}. \qquad (36\text{--}11)$$

This relation is called Brewster's law.

Materials having different indexes of refraction for two perpendicular directions of polarization are said to be birefringent. Those that show preferential absorption for one polarization direction are dichroic; they are used in polarizing filters for light. Some materials become birefringent under mechanical stress; these form the basis of photoelastic stress analysis.

Light can be scattered by air molecules. The scattered light is preferentially polarized. When two linearly polarized waves with a phase difference are superposed, the result is circularly or elliptically polarized light. In this case the E vector is not confined to a plane containing the direction of propagation but describes a circle or ellipse in the planes perpendicular to the direction of propagation.

Huygens' principle states that if the position of a wave front at one instant is known, the position of the front at a later time can be constructed by imagining the front as a source of secondary wavelets. Huygens' principle can be used to derive the laws of reflection and refraction.

QUESTIONS

36–1 During a thunderstorm one always sees the flash of lightning before hearing the accompanying thunder. Discuss this in terms of the various wave speeds. Can this phenomenon be used to determine how far away the storm is?

36–2 When hot air rises around a radiator or from a heating duct, objects behind it appear to shimmer or waver. What is happening?

36–3 Light requires about 8 min to travel from the sun to the earth. Is it delayed appreciably by the earth's atmosphere?

36–4 Sometimes when looking at a window one sees two reflected images, slightly displaced from each other. What causes this?

36–5 An object submerged in water appears to be closer to the surface than it actually is. Why? Swimming pools are always deeper than they look; is this the same phenomenon?

36–6 A ray of light in air strikes a glass surface. Is there a range of angles for which total reflection occurs?

36–7 As shown in Table 36–1, diamond has a much larger refractive index than glass. Is there a larger or smaller range of angles for which total internal reflection occurs for

diamond, than for glass? Does this have anything to do with the fact that a real diamond has more sparkle than a glass imitation?

36–8 Sunlight or starlight passing through the earth's atmosphere is always bent toward the vertical. Why? Does this mean that a star is not really where it appears to be?

36–9 The sun or moon usually appears flattened just before it sets. Is this related to refraction in the earth's atmosphere, mentioned in Question 36–8?

36–10 A student claimed that, because of atmospheric refraction (cf. Question 36–8), the sun can be seen after it has set, and that the day is therefore longer than it would be if the earth had no atmosphere. First, what does he mean by saying the sun can be seen after it has set? Second, comment on the validity of his conclusion.

36–11 It has been proposed that automobile windshields and headlights should have polarizing filters to reduce the glare of oncoming lights during night driving. Would this work? How should the polarizing axes be arranged? What advantages would this scheme have? What disadvantages?

36–12 A salesperson at a bargain counter claims that a certain pair of sunglasses has Polaroid filters; you suspect they are just tinted plastic. How could you find out for sure?

36–13 When unpolarized light is incident on two crossed polarizers, no light is transmitted. A student asserted that if a third polarizer is inserted between the other two, some transmission may occur. Does this make sense? How can adding a third filter *increase* transmission?

36–14 How could you determine the direction of the polarizing axis of a single polarizer?

36–15 In three-dimensional movies, two images are projected on the screen, and the viewers wear special glasses to sort them out. How does this work?

36–16 In Fig. 36–17, since the light scattered out of the incident beam is polarized, why is the transmitted beam not also partially polarized?

36–17 Light from blue sky is strongly polarized because of the nature of the scattering process described in Section 36–10. But light scattered from white clouds is usually *not* polarized. Why not?

36–18 When a sheet of plastic food wrap is placed between two crossed polarizers, no light is transmitted. When the sheet is stretched in one direction, some light passes through. What is happening?

36–19 Television transmission usually uses plane-polarized waves. It has been proposed to use circularly polarized waves to improve reception. Why? (*Hint:* See Problem 36–24.)

36–20 Can sound waves be reflected? Refracted? Give examples. Does Huygens' principle apply to sound waves?

36–21 Why should the wavelength of light change, but not its frequency, in passing from one material to another?

36–22 When light is incident on an interface between two materials, the angle of the refracted ray depends on the wavelength, but the angle of the reflected ray does not. Why should this be?

EXERCISES

Section 36–3 The Speed of Light

36–1 The orbital period of the moon of Jupiter studied by Roemer is 42 hr.

a) Using data from Appendix F, how far along its orbit around the sun does the earth travel in 42 hr?

b) How much time does it take light to travel the distance calculated in (a)?

36–2 Fizeau's measurements of the speed of light were continued by Cornu, using Fizeau's apparatus but with the distance from the toothed wheel to the mirror increased to 22.9 km. One of the toothed wheels used was 40 mm in diameter and had 180 teeth. Find the angular velocity at which it should rotate so that light transmitted through one opening will return through the next.

Section 36–5 Reflection and Refraction

36–3 A ray of light is incident on a plane surface separating two transparent substances of refractive indices 1.60 and 1.40. The angle of incidence is 30°, and the ray originates in the medium of higher index. Compute the angle of refraction.

36–4 A parallel-sided plate of glass having a refractive index of 1.60 is held on the surface of water in a tank. A ray coming from above makes an angle of incidence of 45° with the top surface of the glass.

a) What angle does the ray make with the normal in the water?

b) How does this angle vary with the refractive index of the glass?

36–5 A parallel beam of light makes an angle of 30° with the surface of a glass plate having a refractive index of 1.50.

a) What is the angle between the refracted beam and the surface of the glass?

b) What should be the angle of incidence ϕ with this plate for the angle of refraction to be $\phi/2$, where both angles are measured relative to the normal?

36–6 The density of the earth's atmosphere increases as the surface of the earth is approached. This increase in density is accompanied by a corresponding increase in refractive index.

a) Draw a diagram showing how the light from a star or planet is bent as it goes through the atmosphere. Indicate the apparent position of the light source.

b) Explain how one can see the sun after it has set.

c) Explain why the setting sun appears flattened.

36–7 The speed of light of wavelength 656 nm in heavy flint glass is 1.60×10^8 m·s^{-1}. What is the index of refraction of this glass?

36–8 Light of a certain frequency has a wavelength in water of 442 nm. What is the wavelength of this light when it passes into carbon disulfide?

36–9

a) What is the speed of light of wavelength 500 nm (in vacuum) in glass whose index of refraction at this wavelength is 1.50?

b) What is the wavelength of these waves in the glass?

36–10 Prove that a ray of light reflected from a plane mirror rotates through an angle of 2θ when the mirror rotates through an angle θ about an axis perpendicular to the plane of incidence.

36–11 A parallel beam of light is incident on a prism, as shown in Fig. 36–22. Part of the light is reflected from one face and part from another. Show that the angle θ between the two reflected beams is twice the angle A between the two reflecting surfaces.

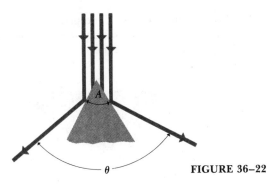

FIGURE 36–22

Section 36–6 Total Internal Reflection

36–12 A ray of light in glass of index of refraction 1.50 is incident on an interface with air. What is the *largest* angle the ray can make with the normal and not be totally reflected back into the glass?

36–13 The speed of a sound wave is 330 m·s⁻¹ in air and 1320 m·s⁻¹ in water.

a) What is the critical angle for a sound wave incident on the surface between air and water?

b) Which medium has the higher "index of refraction" for sound?

36–14 A point source of light is 20 cm below the surface of a body of water. Find the diameter of the largest circle at the surface through which light can emerge from the water.

Section 36–9 Polarizing Filters

36–15 Unpolarized light of intensity I_0 is incident on a polarizing filter, and the emerging light strikes a second polarizing filter with its axis 45° to that of the first. Determine

a) the intensity of the emerging beam;

b) its state of polarization.

36–16 A polarizer and an analyzer are oriented so that the maximum amount of light is transmitted. To what fraction of its maximum value is the intensity of the transmitted light reduced when the analyzer is rotated through

a) 30°?

b) 45°?

c) 60°?

36–17 Three polarizing filters are stacked, with the polarizing axes of the second and third at 45° and 90°, respectively, with that of the first.

a) If unpolarized light of intensity I_0 is incident on the stack, find the intensity and state of polarization of light emerging from each filter.

b) If the second filter is removed, how does the situation change?

36–18 Light traveling in air strikes a glass plate at an angle of incidence of 60°; part of the beam is reflected and part

refracted. It is observed that the reflected and refracted portions make an angle of 90° with each other. What is the index of refraction of the glass?

36–19 The critical angle of light in a certain substance is 45°. What is the polarizing angle?

36–20

a) At what angle above the horizontal must the sun be for sunlight reflected from the surface of a calm body of water to be completely polarized?

b) What is the plane of the **E** vector in the reflected light?

36–21 A parallel beam of "natural" light is incident at an angle of 58° (with respect to the normal) on a plane glass surface. The reflected beam is completely linearly polarized.

a) What is the refractive index of the glass?

b) What is the angle of refraction of the transmitted beam?

Section 36–10 Scattering of Light

36–22 A beam of light, after passing through the Polaroid disk P_1 in Fig. 36–23, traverses a cell containing a scattering medium. The cell is observed at right angles through another Polaroid disk P_2. Originally the disks are oriented so that the brightness of the field as seen by the observer is a maximum.

a) Disk P_2 is now rotated through 90°. Is extinction produced?

b) Disk P_1 is now rotated through 90°. Is the field bright or dark?

c) Disk P_2 is then restored to its original position. Is the field bright or dark?

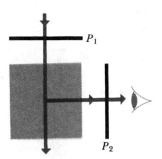

FIGURE 36–23

Section 36–11 Circular and Elliptical Polarization

36–23 What is the state of polarization of the light transmitted by a quarter-wave plate when the electric vector of the incident linearly polarized light makes an angle of 30° with the optic axis?

36–24 A beam of right circularly polarized light is reflected at normal incidence from a reflecting surface. Is the reflected beam right or left circularly polarized? Explain.

PROBLEMS

36–25 An inside corner of a cube is lined with mirrors. A ray of light is reflected successively from each of three mutually perpendicular mirrors; show that its final direction is always exactly opposite to its initial direction. This principle is used in tail-light lenses and reflecting highway signs.

36–26 High frequency (f = 1 to 5 MHz) soundwaves, called ultrasound, are now being used by physicians to image internal organs. The speed of these ultrasound waves is 1480 m·s^{-1} in muscle and 350 m·s^{-1} in air.

a) At what angle would an ultrasound beam enter the heart if it left the lungs at an angle of 8° from the normal to the heart wall?

b) What is the critical angle for sound waves in air incident on muscle?

36–27 Old photographic plates were made of glass with a light-sensitive emulsion on the front surface. This emulsion was somewhat transparent. When a bright point source is focused on the front of the plate, the developed photograph will show a halo around the image of the spot. If the glass plate is 3 mm thick and the halos have a radius of 2.2 mm, what is the index of refraction of the glass? (*Hint:* Consider total internal reflection of the light from the focused spot at the back surface of the glass.)

36–28 A glass plate 3 mm thick, of index of refraction 1.50, is placed between a point source of light of wavelength 600 nm (in vacuum) and a screen. The distance from source to screen is 3 cm. How many waves are there between source and screen?

36–29 The prism of Fig. 36–24 has a refractive index of 1.414, and the angles A are 30°. Two light rays m and n are parallel as they enter the prism. What is the angle between them after they emerge?

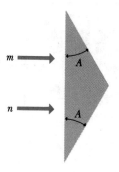

FIGURE 36–24

36–30 A 45°–45°–90° prism is immersed in water. A ray of light is incident normally on one of its shorter faces. What is the minimum index of refraction the prism must have if this ray is to be totally reflected within the glass, at the long face of the prism?

36–31 Light is incident normally on the short face of a 30°–60°–90° prism, as in Fig. 36–25. A drop of liquid is placed on the hypotenuse of the prism. If the index of the prism is 1.50, find the maximum index the liquid may have if the light is to be totally reflected.

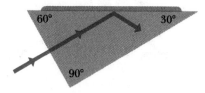

FIGURE 36–25

36–32 The glass vessel shown in Fig. 36–26a contains a large number of small, irregular pieces of glass and a liquid. The dispersion curves of the glass and of the liquid are shown in Fig. 36–26b. Explain the behavior of a parallel beam of white light as it traverses the vessel. (This is known as a *Christiansen filter*.)

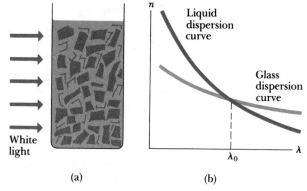

(a)   (b)

FIGURE 36–26

36–33 Three polarizing filters are stacked, with the polarizing axes of the second and third at angles θ and 90°, respectively, with that of the first. Unpolarized light of intensity I_0 is incident on the stack.

a) Derive an expression for the intensity of light transmitted through the stack, as a function of I_0 and θ.

b) For what value of θ does the maximum transmission occur?

36–34 It is desired to rotate the direction of polarization of linearly polarized light 90°, by using two Polaroid filters. Explain how this can be done, and find the final intensity in terms of the incident intensity.

36–35 A light source is made of an unpolarized component of intensity I_0 and a polarized component of intensity I_p. The plane of polarization of the polarized component is oriented at an angle of θ with respect to the vertical. The following data give the intensity measured through a polarizer with an orientation of ϕ with respect to the vertical.

$\phi(°)$	$I_{\text{total}}(\text{W·m}^{-2})$
0	18.4
10	21.4
20	23.7
30	24.8
40	24.8
50	23.7
60	21.4
70	18.4
80	15.0
90	11.6
100	8.6
110	6.3
120	5.2
130	5.2
140	6.3
150	8.6
160	11.6
170	15.0
180	18.4

Rotation(°)		Concentration (g/100 ml)
l-leucine	l-glutamic acid	
− 0.11	0.124	1.0
− 0.22	0.248	2.0
− 0.55	0.620	5.0
− 1.10	1.24	10.0
− 2.20	2.48	20.0
− 5.50	6.20	50.0
−11.0	12.4	100.0

36–38 In Fig. 36–27, A and C are Polaroid sheets whose transmission directions are as indicated. B is a sheet of doubly refractive material whose optic axis is vertical. All three sheets are parallel. Unpolarized light enters from the left. Discuss the state of polarization of the light at points 2, 3, and 4.

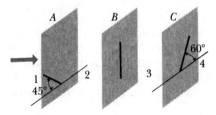

FIGURE 36–27

a) What is the orientation of the polarized component? (That is, what is the angle θ?)

b) What are the values of I_0 and I_p?

36–36 The refractive index of a certain flint glass is 1.65. For what incident angle is light reflected from the surface of this glass completely polarized if the glass is immersed in

a) air?

b) water?

36–37 Many biologically active molecules are also optically active. Plane polarized light on traversing a solution of these compounds has its plane of polarization rotated. Some compounds rotate the polarization clockwise, while others rotate the polarization counterclockwise. The amount of rotation depends on the amount of material in the light path. The following data give the amount of rotation through two amino acids over a path length of 100 cm. From these data find the relationship between the concentration and polarization rotation. Find both the functional form valid for both materials and the value of any compound-specific parameters.

36–39 A certain birefringent material has indexes of refraction n_1 and n_2 for the two perpendicular components of linearly polarized light passing through it. The corresponding wavelengths are $\lambda_1 = \lambda_0/n_1$ and $\lambda_2 = \lambda_0/n_2$, where λ_0 is the wavelength in vacuum.

a) If the crystal is to function as a quarter-wave plate, the number of wavelengths of each component within the material must differ by $\frac{1}{4}$. Thus show that the minimum thickness for a quarter-wave plate is

$$d = \frac{\lambda_0}{4(n_1 - n_2)}.$$

b) Find the minimum thickness of a quarter-wave plate made of calcite, if the indexes of refraction are 1.658 and 1.486 and the wavelength is $\lambda_0 = 589$ nm.

CHALLENGE PROBLEMS

36–40 A ray of light traveling with speed c leaves point 1 of Fig. 36–28 and is reflected to point 2. The ray strikes the reflecting surface a horizontal distance x from point 1.

a) Show that time t required for the light to travel from 1 to 2 is

$$t = \frac{\sqrt{y_1{}^2 + x^2} + \sqrt{y_2{}^2 + (l - x)^2}}{c}.$$

b) Take the derivative of t with respect to x. Set the

derivative equal to zero to show that this time reaches its *minimum* value when $\theta_1 = \theta_2$, which is the law of reflection and corresponds to the actual path taken by the light. This is an example of Fermat's *principle of least time*, which states that among all possible paths between two points, the one actually taken by a ray of light is that for which the time of travel is a *minimum*. (In fact, there are some cases where the time is a maximum rather than a minimum.)

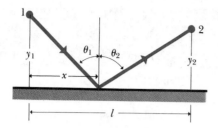

FIGURE 36–28

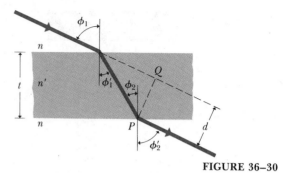

FIGURE 36–30

36–41 A ray of light goes from point A in a medium where the velocity of light is v_1 to point B in a medium where the velocity is v_2, as in Fig. 36–29. The ray strikes the interface a horizontal distance x to the right of point A.

a) Show that the time required for the light to go from A to B is

$$t = \frac{\sqrt{h_1^2 + x^2}}{v_1} + \frac{\sqrt{h_2^2 + (l - x)^2}}{v_2}.$$

b) Take the derivative of t with respect to x. Set this derivative equal to zero to show that this time reaches its *minimum* value when $n_1 \sin \theta_1 = n_2 \sin \theta_2$. This is Snell's law, and corresponds to the actual path taken by the light. This problem is a second example of Fermat's principle of least time, discussed in Challenge Problem 36–40.

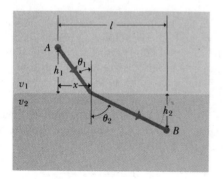

FIGURE 36–29

36–42 Light is incident at an angle ϕ_1 (as in Fig. 36–30) on the upper surface of a transparent plate, the surfaces of the plate being plane and parallel to each other.

a) Prove that $\phi_1 = \phi_2'$.

b) Show that this relation is true for any number of different parallel plates.

c) Prove that the lateral displacement d of the emergent beam is given by the relation

$$d = t \frac{\sin (\phi_1 - \phi_1')}{\cos \phi_1'},$$

where t is the thickness of the plate.

d) A ray of light is incident at an angle of 60° on one surface of a glass plate 2 cm thick, of index 1.50. The

medium on either side of the plate is air. Find the lateral displacement between the incident and emergent rays.

36–43 Light passes symmetrically through a prism having apex angle A, as shown in Fig. 36–31.

a) Show that the angle of deviation δ (the angle between the initial and final directions of the ray) is given by

$$\sin \frac{A + \delta}{2} = n \sin \frac{A}{2}.$$

b) Use the result of (a) to find the angle of deviation for a ray of light passing symmetrically through a prism having three equal angles ($A = 60°$) and $n = 1.50$.

c) A certain glass has a refractive index of 1.50 for red light (700 nm) and 1.52 for violet light (400 nm). If both colors pass through symmetrically, as described in (a), and if $A = 60°$, find the difference between the angles of deviation for the two colors.

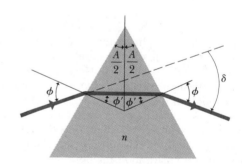

FIGURE 36–31

36–44 Consider two vibrations, one along the x-axis,

$$x = a \sin (\omega t - \alpha),$$

and the other along the y-axis, of equal amplitude and frequency, but differing in phase,

$$y = a \sin (\omega t - \beta).$$

Let us write them as follows:

$$\frac{x}{a} = \sin \omega t \cos \alpha - \cos \omega t \sin \alpha, \tag{1}$$

$$\frac{y}{a} = \sin \omega t \cos \beta - \cos \omega t \sin \beta. \tag{2}$$

a) Multiply Eq. (1) by $\sin \beta$ and Eq. (2) by $\sin \alpha$ and then subtract the resulting equations.

b) Multiply Eq. (1) by $\cos \beta$ and Eq. (2) by $\cos \alpha$ and then subtract the resulting equations.

c) Square and add the results of (a) and (b).

d) Derive the equation $x^2 + y^2 - 2xy \cos \delta = a^2 \sin^2 \delta$, where $\delta = \alpha - \beta$.

e) Use the above result to justify each of the diagrams in Fig. 36–32. In the figure the angle given is the phase difference between two simple harmonic motions, one horizontal (along the x-axis) and the other vertical (along the y-axis), and of the same frequency and amplitude. The figure thus shows the resultant motion from the superposition of the two perpendicular harmonic motions.

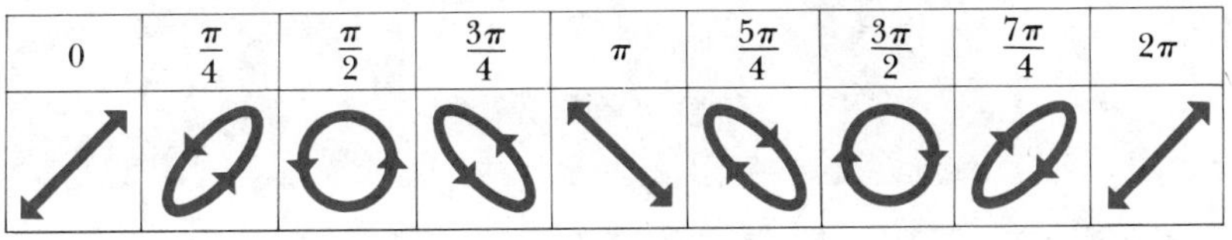

0	$\dfrac{\pi}{4}$	$\dfrac{\pi}{2}$	$\dfrac{3\pi}{4}$	π	$\dfrac{5\pi}{4}$	$\dfrac{3\pi}{2}$	$\dfrac{7\pi}{4}$	2π

FIGURE 36–32

37

IMAGES FORMED BY A SINGLE SURFACE

IN THE PRECEDING CHAPTER WE DISCUSSED THE PRINCIPLES THAT GOVERN reflection and refraction of a ray of light at a reflecting surface or an interface between two materials. We now consider the behavior of several rays that diverge from a common point and strike a reflecting or refracting surface. A central idea in this discussion is the concept of *image*. After reflection or refraction, the rays emerge with directions characteristic of having passed through some other common point called the *image point*. This discussion lays the foundation for analysis of many familiar optical instruments, including camera lenses, magnifiers, the human eye, microscopes, and telescopes. These instruments will be discussed in Chapter 38.

37–1 REFLECTION AT A PLANE SURFACE

Consider the situation shown in Fig. 37–1. Rays diverge from point P, called an **object point,** and are reflected or refracted (or both) at an interface between two transparent materials. The direction of each *reflected* ray is given by the law of reflection, and that of each transmitted or *refracted* ray by Snell's law. Here and in the following discussion we denote the two refractive indexes as n and n', without using the more elaborate subscript notation of Chapter 36.

Let us concentrate first on the *reflected* rays. Reflection can occur at an interface between two transparent materials or at a highly polished surface of an opaque material such as a metal, in which case the surface is usually called a *mirror*. We will use the term *mirror* to include all these possibilities.

As mentioned above, the key concept is that of **image.** After reflection the rays appear to have diverged from some common point, which we call the *image point*. In some cases, the rays coming from the surface after reflection really *do* meet at a common point and then diverge again after passing it; such a point is called a **real-image** point. In other cases the rays diverge *as though* they had passed through such a point, which is then called a **virtual-image** point. In many situations the image point exists only in an approximate sense, that is, when certain approximations are used in the calculations. This chapter and the next are devoted primarily to a study of the formation and properties of images.

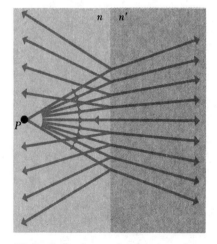

37–1 Reflection and refraction of rays at a plane interface between two transparent materials.

What's the difference between a real image and a virtual image?

Figure 37–2a shows two rays diverging from a point P at a distance s to the left of a plane mirror. The ray PV, incident normally on the mirror, returns along its original path. The ray PB, making an angle u with PV, strikes the mirror at an angle of incidence $\phi = u$ and is reflected at an angle $r = \phi = u$. When extended backward, the reflected ray intersects the normal to the surface at point P'. The angle u' is equal to r and hence to u.

Figure 37–2b shows several rays diverging from P. The construction of Fig. 37–2a can be repeated for each of them, and we see that the directions of the outgoing rays are the same as though they had originated at point P', which is therefore the *image* of P. The rays do not, of course, actually pass through this point; in fact, if the mirror is opaque, there is no light at all on the right side. Thus P' is a *virtual* image. Nevertheless, P' is a very real point in the sense that it describes the final directions of all the rays that originally diverged from P.

From the symmetry of the figure, we see that P' lies on a line perpendicular to the mirror passing through P, and that P and P' are equidistant from the mirror, on opposite sides. Thus *for a plane mirror, the image of an object point lies on the extension of the normal from the object point to the mirror, and the object and image points are equidistant from the mirror.*

Before proceeding further, we pause here to introduce some conventions that anticipate later situations where the object and image may be on either side of a reflecting or refracting surface. We adopt the following:

1. When the object is on the same side of the reflecting surface as the incoming light, the object distance s is positive; otherwise it is negative.

2. When the image is on the same side of the reflecting surface as the outgoing light, the image distance s is positive; otherwise it is negative.

For a mirror, the incoming and outgoing sides are always the same; in Fig. 37–2 they are both the left side. The rules above have been stated in this form so they may be applied also to *refracting* surfaces, where light comes in one side and goes out the opposite side.

In Fig. 37–2 the object distance s is *positive* because the object point P is on the incoming side—that is, the left side—of the reflecting surface. The image distance s' is *negative* because the image point P' is *not* on the outgoing side of the surface. Thus object and image distances are related simply by

$$s = -s'. \tag{37–1}$$

Next we consider an object of finite size, parallel to the mirror, represented by the arrow PQ in Fig. 37–3. Two of the rays from Q are shown, and *all* rays from Q diverge from its image Q' after reflection. Other points of the object PQ have image points between P' and Q'. As the figure shows, the object PQ and image $P'Q'$ have the same size and orientation: $y = y'$.

The ratio of image to object size, y'/y, is called the **lateral magnification** m; that is,

$$m = \frac{y'}{y}. \tag{37–2}$$

Thus we have found that *the lateral magnification for a plane mirror is unity.* When you look at yourself in a plane mirror, you do not look any larger or smaller than you really are.

Using the law of reflection to locate a virtual image

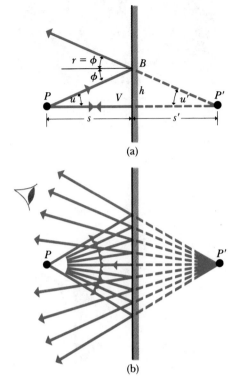

(a)

(b)

37–2 After reflection at a plane surface, all rays originally diverging from the object point P now diverge from the point P', although they do not *originate* at P'. Point P' is called the *virtual image* of point P. The eye sees some of the outgoing rays and perceives them as having come from point P'.

Lateral magnification is the ratio of image height to object height.

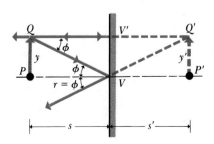

37–3 Construction for determining the height of an image formed by reflection at a plane surface.

The image formed by a plane mirror is reversed; the image of a right hand is a left hand, and so on. (Photo by Chip Clark.)

In general, if an object transverse to the optic axis is represented by an arrow, its image may point in the same direction as the object or in the opposite direction. When, as in Fig. 37–3, the directions are the same, the image is called **erect;** if they are opposite, the image is **inverted.** The image formed by a plane mirror is always erect.

The three-dimensional virtual image of a three-dimensional object, formed by a plane mirror, is shown in Fig. 37–4. The images $P'Q'$ and $P'S'$ are parallel to their objects, but $P'R'$ is reversed relative to PR. The image of a three-dimensional object formed by a plane mirror is the same size as the object in both its longitudinal and transverse dimensions. The image and object are not identical in all respects, however, but are related in the same way as are a right hand and a left hand. To verify this statement, point your thumbs along PR and $P'R'$, your forefingers along PQ and $P'Q'$, and your middle fingers along PS and $P'S'$. When an object and its image are related in this way, the image is said to be **reversed.** When the transverse dimensions of object and image are in the same direction, the image is erect. Thus a plane mirror forms an erect but reversed image.

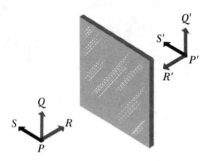

37–4 The image formed by a plane mirror is virtual, erect, and reversed, and is the same size as the object.

Some terms to describe spherical reflecting surfaces

37–2 REFLECTION AT A SPHERICAL SURFACE

Next we consider the formation of an image of a *spherical* mirror. Figure 37–5a shows a spherical mirror of radius of curvature R, with its concave side facing the incident light. The **center of curvature** of the surface is at C. Point P is an object point; for the moment, assume that P is farther from V than is the center of curvature. The ray PV, passing through C, strikes the mirror normally and is reflected back on itself. Point V is called the **vertex** of the mirror, and the line PCV is the **optic axis.**

Ray PB, at an angle u with the axis, strikes the mirror at B, where the angle of incidence is ϕ and the angle of reflection is $r = \phi$. The reflected ray intersects the axis at point P'. We will show that *all* rays from P intersect the axis at the *same* point P', as in Fig. 37–5b, no matter what u is, provided that u is a *small* angle. Point P' is therefore the *image* of object point P. The object distance, measured from the vertex V, is s, and the image distance is s'. The object point P is on the same side as the incident light, so the object distance s is positive. The image point P' is on the same side as the reflected light, so the image distance s' is also positive.

Unlike the reflected rays in Fig. 37–2, the reflected rays in Fig. 37–5b actually intersect at point P' and then diverge from P' *as if* they had originated at this point. The image P' is called *real*, and a real image corresponds to a *positive* image distance.

Making use of the fact that an exterior angle of a triangle equals the sum of the two opposite interior angles, and considering the triangles PBC and $P'BC$ in Fig. 37–5a, we have

$$\theta = u + \phi, \qquad u' = \theta + \phi.$$

Eliminating ϕ between these equations gives

$$u + u' = 2\theta. \tag{37–3}$$

We may now introduce a sign convention for radii of curvature of spherical surfaces:

> When the center of curvature C is on the same side as the outgoing (reflected) light, the radius of curvature is positive; otherwise it is negative.

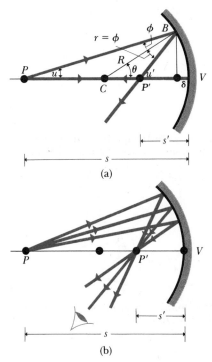

In Fig. 37–5, R is positive because the center of curvature C is on the same side of the mirror as the reflected light. This is always the case for reflection from the concave side of a surface; for a convex surface, the center of curvature is on the opposite end from the reflected light, and R is negative.

We may now compute the image distance s'. Let h represent the height of point B above the axis, and δ the short distance from V to the foot of this vertical line. We now write expressions for the tangents of u, u', and θ, remembering that s, s', and R are all positive quantities:

$$\tan u = \frac{h}{s - \delta}, \qquad \tan u' = \frac{h}{s' - \delta}, \qquad \tan \theta = \frac{h}{R - \delta}.$$

These trigonometric equations cannot be solved as simply as the corresponding algebraic equations for a plane mirror. *If the angle u is small,* however, the angles u' and θ will be small also. Since the tangent of a small angle is nearly equal to the angle itself (in radians), we can replace $\tan u'$ by u', and so on, in the equations above. Also if u is small, the distance δ can be neglected compared with s', s, and R. Hence, approximately, for small angles,

$$u = \frac{h}{s}, \qquad u' = \frac{h}{s'}, \qquad \theta = \frac{h}{R}.$$

Substituting in Eq. (37–3) and canceling h, we obtain

$$\frac{1}{s} + \frac{1}{s'} = \frac{2}{R} \qquad\qquad (37\text{–}4)$$

37–5 (a) Construction for finding the position of the image P' of a point object P, formed by a concave spherical mirror. (b) If the angle u is small, *all* rays from P intersect at P'. The eye sees some of the outgoing rays and perceives them as having come from P'.

as a general relation among the three quantities s, s', and R. Since this equation does not contain the angle u, *all* rays from P making sufficiently small angles with the axis will, after reflection, intersect at P'. Such rays, nearly parallel to the axis, are called *paraxial rays.*

We must understand that Eq. (37–4), as well as many similar relations to be derived later in this chapter and the next, is the result of a calculation containing approximations and is valid only for paraxial rays. (The term **paraxial approximation** is often used for the approximations just described.) As the angle increases, the point P' moves closer to the vertex; a spherical mirror, unlike a plane mirror, does not form precisely a point image of a point object. This property of a spherical mirror is called **spherical aberration.**

If $R = \infty$, the mirror becomes *plane,* and Eq. (37–4) reduces to Eq. (37–1), previously derived for this special case.

Now suppose we have an object of finite size, represented by the arrow PQ in Fig. 37–6, perpendicular to the axis PV. The image of P formed by paraxial rays is at P'. Since the object distance for point Q is very nearly equal to that for the point P, the image of $P'Q'$ is nearly straight and is perpendicular to the axis. Note that object and image have different sizes, y and y', respectively, and that they have opposite orientation. We have defined the *lateral magnification m*

A fundamental relation for spherical mirrors

What is the paraxial approximation?

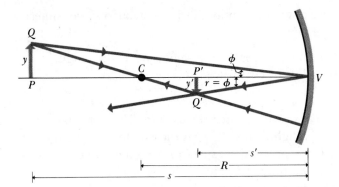

37–6 Construction for determining the position, orientation, and height of an image formed by a concave spherical mirror.

as the ratio of image to object size:

$$m = \frac{y'}{y}.$$

How to find the lateral magnification of a spherical mirror

Because triangles PVQ and $P'VQ'$ in Fig. 37–6 are *similar*, we also have the relation $y/s = -y'/s'$. The negative sign is needed because object and image are on opposite sides of the optic axis; if y is positive, y' is negative. Hence

$$m = \frac{y'}{y} = -\frac{s'}{s}. \tag{37–5}$$

A negative value of m indicates that the image is *inverted* relative to the object, as the figure shows. In cases to be considered later, where m may be either positive or negative, a positive value always corresponds to an erect image, a negative value to an inverted one. For a *plane* mirror, $s = -s'$, and hence $y' = y$, as we have already shown.

When the image is smaller than the object, the magnification is less than unity.

Although the ratio of image size to object size is called the *magnification*, the image formed by a mirror or lens may be *smaller* than the object. The magnification is then less than unity. The image formed by an astronomical telescope mirror or by a camera lens is much smaller than the object. It is interesting to note that for three-dimensional objects the ratio of image-to-object distances measured *along* the optic axis is different from the ratio of *lateral* distances (which we have called the lateral magnification). In particular, if m is a small fraction, the three-dimensional image of a three-dimensional object is reduced *longitudinally* much more than it is reduced *transversely*. Figure 37–7 illustrates this effect. Also, the image formed by a spherical mirror, like that of a plane mirror, is always reversed.

37–7 Schematic diagram of an object and its real, inverted, reduced image formed by a concave mirror.

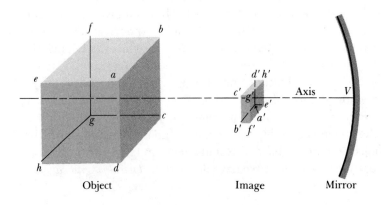

EXAMPLE 37–1 A concave mirror forms an image, on a wall 3 m from the mirror, of the filament of a headlight lamp 10 cm in front of the mirror. (a) What is the radius of curvature of the mirror? (b) What is the height of the image if the height of the object is 5 mm?

SOLUTION

a) Both object distance and image distance are positive, so from Eq. (37–4),

$$s = 10 \text{ cm}, \qquad s' = 300 \text{ cm},$$

$$\frac{1}{10 \text{ cm}} + \frac{1}{300 \text{ cm}} = \frac{2}{R},$$

$$R = 19.4 \text{ cm}.$$

The mirror in a headlight lamp forms a real image of the lamp filament.

Since the radius is positive, a concave mirror is required.

b) From Eq. (37–5),

$$m = \frac{y'}{y} = -\frac{s'}{s} = -\frac{300 \text{ cm}}{10 \text{ cm}} = -30.$$

The image is therefore inverted (m is negative) and is 30 times the height of the object, or $(30)(5 \text{ mm}) = 150 \text{ mm}$.

In Fig. 37–8a the *convex* side of a spherical mirror faces the incident light, so R is negative. Ray PB is reflected with the angle of reflection r equal to the angle of incidence ϕ, and the reflected ray, projected backward, intersects the axis at P'. As in the case of a concave mirror, *all* rays from P will, after reflection, diverge from the same point P', provided that the angle u is small; so P' is the image of P. The object distance s is positive, the image distance s' is negative, and the radius of curvature R is negative.

Convex mirrors: The same formulas can be used, but the results are different.

Figure 37–8b shows two rays diverging from the head of the arrow PQ, and the virtual image $P'Q'$ of this arrow. We leave it to you to show, by the same procedure used for a concave mirror, that

$$\frac{1}{s} + \frac{1}{s'} = \frac{2}{R},$$

(a)   (b)

37–8 Construction for finding (a) the position and (b) the magnification of the image formed by a convex mirror.

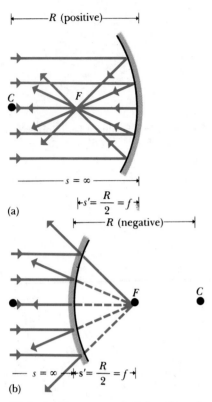

37–9 Incident rays parallel to the axis (a) converge to the focus F of a concave mirror, (b) diverge as though coming from the focus F of a convex mirror.

and that the lateral magnification is

$$m = \frac{y'}{y} = -\frac{s'}{s}.$$

These expressions are exactly the same as those for a concave mirror, as they must be when a consistent sign convention is adopted.

37–3 FOCUS AND FOCAL LENGTH

When an object point is at a very large distance from a spherical mirror, all rays from that point that strike the mirror are (in the paraxial approximation) parallel to one another. The object distance is $s = \infty$ and, from Eq. (37–4),

$$\frac{1}{\infty} + \frac{1}{s'} = \frac{2}{R}, \qquad s' = \frac{R}{2}.$$

When R is positive (concave mirror), the situation is as shown in Fig. 37–9a. A beam of incident parallel rays converges after reflection at a point F a distance $R/2$ from the vertex of the mirror. Point F is called the *focal point* or simply the **focus,** and its distance from the vertex, denoted by f, is called the **focal length.**

When R is negative (convex mirror), as in Fig. 37–9b, the image point is behind the mirror and f is negative. In that case the outgoing rays do not converge at a point but instead diverge as though they had come from the point F behind the mirror. In this case F is called a *virtual focus.*

The entire discussion may be reversed, as shown in Fig. 37–10. When the image distance s' is very large, the outgoing rays are parallel to the optic axis. The object distance s is then given by

$$\frac{1}{s} + \frac{1}{\infty} = \frac{2}{R}, \qquad s = \frac{R}{2}.$$

In Fig. 37–10a, rays coming in toward the mirror pass through the focus and diverge from it, and after reflection they are parallel to the optic axis. In Fig. 37–10b, the incoming rays are converging as though they would meet at the virtual focus F, and they are reflected parallel to the optic axis.

Thus for both concave and convex mirrors, the focal length f is related to the radius of curvature R by

$$f = \frac{R}{2}. \tag{37–6}$$

How is the focal length of a mirror related to its radius of curvature?

37–10 Rays from a point object at the focus of a spherical mirror are parallel to the axis after reflection. The object in part (b) is virtual.

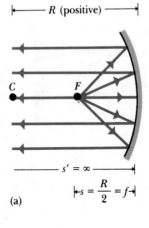

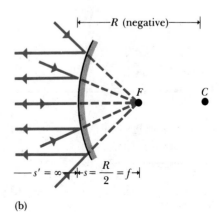

For a concave mirror both f and R are positive, and for a convex mirror both are negative.

The relation between object and image distances for a mirror, Eq. (37–4), may now be written as

$$\frac{1}{s} + \frac{1}{s'} = \frac{1}{f}. \tag{37–7}$$

37–4 GRAPHICAL METHODS

The position and size of the image formed by a mirror may be found by a simple graphical method. This method consists of finding the point of intersection, after reflection from the mirror, of a few particular rays diverging from some point of the object *not* on the mirror axis, such as point Q in Fig. 37–11. Then (neglecting aberrations) *all* rays from this point that strike the mirror will intersect at the same point. Four rays that we can always draw easily are shown in Fig. 37–11. These are called **principal rays.**

Principal rays: four rays you can always draw without guessing

1. *A ray parallel to the axis*, after reflection, passes through the focus of a concave mirror or appears to come from the focus of a convex mirror.

2. *A ray through (or proceeding toward) the focus* is reflected parallel to the axis.

3. *A ray along the radius* through C (extended if necessary) intersects the surface normally and is reflected back along its original path.

4. *A ray to the vertex* is reflected and forms equal angles with the optic axis.

Once we have found the position of the image point by means of the intersection of any two of these principal rays (1, 2, 3, 4), we can draw the path of any other ray from the same object point.

37–11 Rays used in the graphical method of locating an image.

Several examples of principal-ray diagrams for spherical mirrors, along with the corresponding calculations

EXAMPLE 37–2   A concave mirror has a radius of curvature of magnitude 20 cm. Find graphically the image of an object in the form of an arrow perpendicular to the axis of the mirror at each of the following object distances: 30 cm, 20 cm, 10 cm, and 5 cm. Check the construction by computing the size and magnification of the image.

SOLUTION The graphical constructions are shown in the four parts of Fig. 37–12. Study each of these diagrams carefully, comparing each numbered ray with the description above. Several points are worth noting. First, in (b) the object and image distances are equal; ray 3 cannot be drawn in this case because a ray from Q through the center of curvature C does not strike the mirror. For the same reason, ray 2 cannot be drawn in (c); in this case the outgoing rays are parallel, corresponding to an infinite image distance. In (d) the outgoing rays have no real intersection point; they must be extended backward to find the point from which they appear to diverge, that is, from the *virtual-image point Q'*.

Measurement of the figures, with appropriate scaling, gives the following approximate image distances: (a) 15 cm, (b) 20 cm, (c) ∞ or −∞, (d) −10 cm. To *compute* these distances, we first note that $f = R/2 = 10$ cm; then we use Eq. (37–7):

a)
$$\frac{1}{30 \text{ cm}} + \frac{1}{s'} = \frac{1}{10 \text{ cm}}, \qquad s' = 15 \text{ cm},$$

b)
$$\frac{1}{20 \text{ cm}} + \frac{1}{s'} = \frac{1}{10 \text{ cm}}, \qquad s' = 20 \text{ cm},$$

37–12 Image of an object at various distances from a concave mirror, showing principal-ray construction.

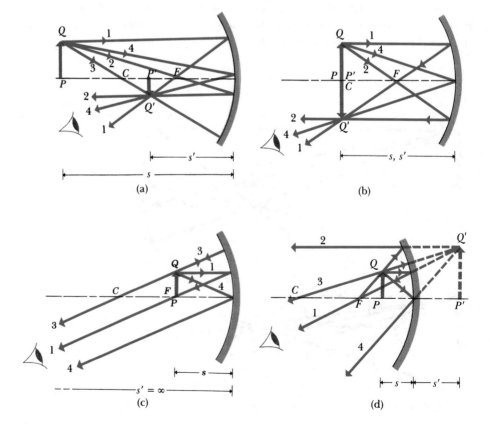

(a)   (b)

(c)   (d)

c)
$$\frac{1}{10 \text{ cm}} + \frac{1}{s'} = \frac{1}{10 \text{ cm}}, \qquad s' = \infty \text{ (or } -\infty),$$

d)
$$\frac{1}{5 \text{ cm}} + \frac{1}{s'} = \frac{1}{10 \text{ cm}}, \qquad s' = -10 \text{ cm}.$$

The lateral magnifications measured from the figures are approximately (a) $-\frac{1}{2}$, (b) -1, (c) ∞ or $-\infty$, and (d) $+2$. *Computing* the magnifications from Eq. (37–4), we find

An image may be at infinity; if so, it is infinitely large.

a)
$$m = -(15 \text{ cm})/(30 \text{ cm}) = -\tfrac{1}{2},$$

b)
$$m = -(20 \text{ cm})/(20 \text{ cm}) = -1,$$

c)
$$m = -(\pm\infty \text{ cm})/(10 \text{ cm}) = \mp\infty,$$

d)
$$m = -(-10 \text{ cm})/(5 \text{ cm}) = +2.$$

In (a) and (b) the image is inverted; in (d) it is erect.

PROBLEM-SOLVING STRATEGY: *Image formation by mirrors*

1. The principal-ray diagram is to geometrical optics what the free-body diagram is to mechanics! When the problem concerns image formation by a mirror, *always* draw a principal-ray diagram first. The same advice should be applied to lenses in the next chapter. It is usually best to orient your diagrams consistently, with the incoming rays traveling from left to right. Don't draw a million other rays at random; stick with the principal rays, the ones you know something about. If your principal rays don't converge at a real-image point, you may have to extend them backward to locate a virtual-image point. We recommend drawing the extensions with broken lines. Another useful aid is to color-code your principal rays, using red for (1), green for (2), black for (3), and blue for (4), in the list above; or something like that.

2. Pay careful attention to signs on object and image distances, radii of curvature, and object and image heights. Make certain you understand that the same set of sign rules works for all four cases in this chapter: reflection and refraction from plane and spherical surfaces. A negative sign on one of the quantities mentioned above *always* has significance; apply the equations carefully and with consistent use of the sign rules, and they will give you correct results!

3. Later in the chapter we will get into refraction as well as reflection. Remember that when a ray passes from a material of smaller index of refraction to one of larger index, it is always bent *toward* the normal; when going from larger to smaller, it is bent *away from* the normal.

37–5 REFRACTION AT A PLANE SURFACE

Suppose we want to find the image of a point object formed by rays *refracted* at a plane or spherical surface. The method is essentially the same as for reflection; the only difference is that Snell's law replaces the law of reflection. We let n represent the refractive index of the material on the "incoming" side of the surface and n' that of the material on the "outgoing" side, and we use the same convention of signs as in reflection.

Consider first a plane surface, shown in Fig. 37–13, and assume $n' > n$. This is not a necessary restriction, but the picture looks a little different if $n' < n$. A ray from the object point P toward the vertex V is incident normally and passes into the second material without deviation. A ray making an angle u with the axis is incident at B with an angle of incidence $\phi = u$. We find the

Applying the law of refraction to a ray bent at a plane interface between two materials

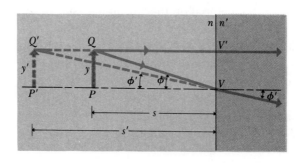

37–13 Construction for finding the position of the image P' of a point object P, formed by a refraction at a plane surface.

angle of refraction, ϕ', from Snell's law:

$$n \sin \phi = n' \sin \phi'.$$

The two rays both appear to come from the image point P' after refraction. From the triangles PVB and $P'VB$,

$$\tan \phi = \frac{h}{s}, \qquad \tan \phi' = \frac{h}{-s'}. \tag{37–8}$$

We must write $-s'$, since the image point is on the side *opposite* that of the refracted (outgoing) light.

If the angle u is small, the angles ϕ, u', and ϕ' are small also and therefore approximately

Again, small-angle approximations simplify things greatly.

$$\tan \phi = \sin \phi, \qquad \tan \phi' = \sin \phi'.$$

Then Snell's law can be written as

$$n \tan \phi = n' \tan \phi',$$

and from Eq. (37–8), after canceling h, we have

Image position and size for a plane refracting surface

$$\frac{n}{s} = -\frac{n'}{s'},$$

or

$$\frac{s'}{s} = -\frac{n'}{n}. \tag{37–9}$$

This is an *approximate* relation, *valid for paraxial rays only.* That is, a plane refracting surface does *not* image all rays from a point object at the same image point, but only those rays that are nearly perpendicular to the refracting surface.

Consider next the image of a finite object, as in Fig. 37–14. The two rays diverging from point Q appear to diverge from its image Q' after refraction,

37–14 Construction for determining the height of an image formed by refraction at a plane surface.

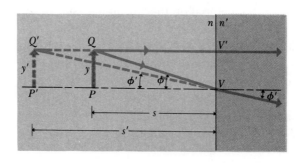

and $P'Q'$ is the image of the object PQ. As the figure shows, the object and the image are the same size, and so the lateral magnification is unity:

$$m = \frac{y'}{y} = 1. \qquad (37\text{--}10)$$

The image *distance* is greater than the object distance, but image and object are the same height.

Here is a familiar example of refraction at a plane surface. When you look vertically downward into the quiet water of a pond or swimming pool, the apparent depth is less than the actual depth. Figure 37–15 illustrates this situation. Two rays are shown diverging from a point Q at a distance s below the surface. Here, n' (air) is less than n (water), and the ray through V is deviated *away from* the normal. The rays after refraction appear to diverge from Q', and the arrow PQ, to an observer looking vertically downward, appears lifted to the position $P'Q'$. From Eq. (37–9),

$$s' = -\frac{n'}{n}s = -\frac{1.00}{1.33}s = -0.75\,s.$$

The apparent depth s' is therefore only three-fourths of the actual depth s. The same phenomenon accounts for the apparent sharp bend in an oar when a portion of it extends below a water surface. The submerged portion appears lifted above its actual position.

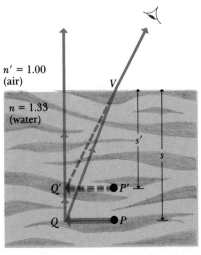

37–15 Arrow $P'Q'$ is the image of the underwater object PQ. The angles of the rays with the vertical are exaggerated for clarity.

Water is deeper than it appears to be.

37–6 REFRACTION AT A SPHERICAL SURFACE

Finally, we consider refraction at a spherical surface. In Fig. 37–16, P is an object point at a distance s to the left of a spherical surface of radius R, with center of curvature C. The refractive indexes at the left and right of the surface are n and n', respectively. Ray PV, incident normally at V, passes into the second material without deviation. Ray PB, making an angle u with the axis, is incident at an angle ϕ with the normal and is refracted at an angle ϕ'. These rays intersect at P' at a distance s' to the right of the vertex. The figure is drawn for the case $n' > n$.

We can prove that if the angle u is small, *all* rays from P intersect at the same point P', so P' is the *real image* of P. The object and image distances are both positive. The radius of curvature is positive also, because the center of curvature is on the side of the outgoing or refracted light.

Applying the law of refraction to a ray bent at a spherical surface

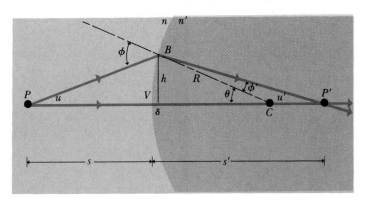

37–16 Construction for finding the position of the image P' of a point object P, formed by refraction at a spherical surface.

When we considered rays from P that are *reflected* at the surface, as in Fig. 37–8, the radius of curvature was negative. Thus the radius of curvature of a given surface has one sign for reflected light and the opposite sign for refracted light. This apparent inconsistency is resolved by noting that in both cases R is positive when the center of curvature is on the "outgoing" side of the reflecting or refracting surface and negative when it is not on the "outgoing" side.

From the triangles PBC and $P'BC$, we have

$$\phi = \theta + u, \qquad \theta = u' + \phi'. \tag{37–11}$$

From Snell's law,

$$n \sin \phi = n' \sin \phi'.$$

Small-angle approximations, one more time

Also, the tangents of u, u', and θ are

$$\tan u = \frac{h}{s + \delta}, \qquad \tan u' = \frac{h}{s' - \delta}, \qquad \tan \theta = \frac{h}{R - \delta}.$$

For paraxial rays, we may approximate both the sine and tangent of an angle by the angle itself and neglect the small distance δ. Snell's law then becomes

$$n\phi = n'\phi',$$

and, combining with the first of Eqs. (37–11), we obtain

$$\phi' = \frac{n}{n'}(u + \theta).$$

The basic formula for image formation by a spherical refracting surface

Substituting this in the second of Eqs. (37–11) gives

$$nu + n'u' = (n' - n)\theta.$$

Using the small-angle approximations for the tangents of u, u', and θ, and canceling h, we obtain

$$\frac{n}{s} + \frac{n'}{s'} = \frac{n' - n}{R}. \tag{37–12}$$

This equation does not contain the angle u, so the image distance is the same for all paraxial rays from P.

If the surface is plane, $R = \infty$ and this equation reduces to Eq. (37–9), already derived for the special case of a plane surface.

The magnification is found from the construction in Fig. 37–17. We draw two rays from point Q, one through the center of curvature C and the other incident at the vertex V. From the triangles PVQ and $P'Q'V$,

$$\tan \phi = \frac{y}{s}, \qquad \tan \phi' = \frac{-y'}{s'},$$

and from Snell's law

$$n \sin \phi = n' \sin \phi'.$$

Magnification of a spherical refracting surface

For small angles,

$$\tan \phi = \sin \phi, \qquad \tan \phi' = \sin \phi',$$

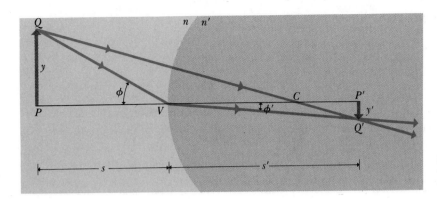

37–17 Construction for determining the height of an image formed by refraction at a spherical surface.

and hence

$$\frac{ny}{s} = -\frac{n'y'}{s'},$$

or

$$m = \frac{y'}{y} = -\frac{ns'}{n's}. \qquad (37\text{-}13)$$

This is the same relation previously derived for a *plane* surface. In that case, from Eq. (37–9), $ns' = -n's$, and $m = 1$.

EXAMPLE 37–3 One end of a cylindrical glass rod (Fig. 37–18) is ground to a hemispherical surface of radius $R = 20$ mm. Find the image distance of a point object on the axis of the rod, 80 mm to the left of the vertex. The rod is in air.

SOLUTION We are given

$$n = 1, \quad n' = 1.5,$$
$$R = +20 \text{ mm}, \quad s = +80 \text{ mm}.$$

From Eq. (37–12),

$$\frac{1}{80 \text{ mm}} + \frac{1.5}{s'} = \frac{1.5 - 1}{+20 \text{ mm}},$$
$$s' = +120 \text{ mm}.$$

The image is therefore formed at the right of the vertex (s' is positive) and at a distance of 120 mm from it.

Suppose that the object is an arrow 1 mm high, perpendicular to the axis. Then, from Eq. (37–13),

$$m = -\frac{ns'}{n's} = -\frac{(1)(120 \text{ mm})}{(1.5)(80 \text{ mm})} = -1.$$

That is, the image is the same height as the object but is inverted.

EXAMPLE 37–4 The rod in Example 37–3 is immersed in water of index 1.33; the other quantities have the same values as before. Find the image distance (Fig. 37–19).

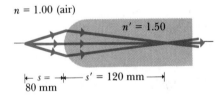

FIGURE 37–18

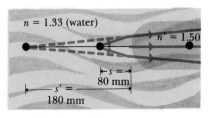

FIGURE 37–19

SOLUTION

$$\frac{1.33}{80 \text{ mm}} + \frac{1.5}{s'} = \frac{1.5 - 1.33}{+20 \text{ mm}},$$

$$s' = -185 \text{ mm}.$$

The fact that s' is negative means that the rays, after refraction by the surface, are not converging but *appear* to diverge from a point 180 mm to the *left* of the vertex. We have met a similar case before in the refraction of spherical waves by a plane surface and have called the point a *virtual image*. In this example, then, the surface forms a virtual image 180 mm to the left of the vertex.

Equations (37–12) and (37–13) can be applied to both convex and concave refracting surfaces when the appropriate sign convention is used, and they apply whether n' is greater or less than n. You should construct diagrams like Figs. 37–16 and 37–17, when R is negative and $n' < n$, and use them to derive Eqs. (37–12) and (37–13) for these cases.

SUMMARY

KEY TERMS

object point

image

real image

virtual image

lateral magnification

erect image

inverted image

reversed image

center of curvature

vertex

optic axis

paraxial approximation

focus

focal length

principal rays

When rays diverge from an object point P and are reflected or refracted, their directions after reflection or refraction are as though they had diverged from a point P' called the image point. If they really do converge at P' and diverge again beyond it, P' is a real image of P; if they only appear to have come from P', then it is a virtual image. Images can be either erect or inverted. An image formed by a plane or spherical mirror is always reversed; the image of a right hand is a left hand.

The lateral magnification in any reflecting or refracting situation is defined as the ratio of image height to object height:

$$m = \frac{y'}{y}. \tag{37–2}$$

When m is positive, the image is erect; when negative, inverted.

Relations derived in this chapter for the object and image positions in reflection or refraction with a spherical surface are valid only for rays nearly parallel to the optic axis; these are called paraxial rays, and the corresponding approximation is the paraxial approximation. The failure of nonparaxial rays to converge precisely at an image point is called spherical aberration.

The focus of a mirror is the point at which parallel rays converge after reflection from a concave mirror, or the point from which they appear to diverge after reflection from a convex mirror. Conversely, rays diverging from the focus of a concave mirror are parallel after reflection, and rays converging toward the focus of a convex mirror are parallel after reflection. The distance from the focus to the vertex is called the focal length, denoted as f.

All the formulas for object distance s and image distance s' for plane and spherical mirrors and refracting surfaces are summarized in Table 37–1. The equation for a plane surface can be obtained from the corresponding equation for a spherical surface by setting $R = \infty$.

TABLE 37–1

	Plane Mirror	Spherical Mirror	Plane Refracting Surface	Spherical Refracting Surface
Object and image distances	$\dfrac{1}{s} + \dfrac{1}{s'} = 0$	$\dfrac{1}{s} + \dfrac{1}{s'} = \dfrac{2}{R} = \dfrac{1}{f}$	$\dfrac{n}{s} + \dfrac{n'}{s'} = 0$	$\dfrac{n}{s} + \dfrac{n'}{s'} = \dfrac{n'-n}{R}$
Lateral magnification	$m = -\dfrac{s'}{s} = 1$	$m = -\dfrac{s'}{s}$	$m = -\dfrac{ns'}{n's} = 1$	$m = -\dfrac{ns'}{n's}$

The following set of sign rules can be applied uniformly to plane and spherical reflecting and refracting surfaces:

s is positive when the object is on the incoming side of surface, negative otherwise.

s' is positive when the image is on the outgoing side of surface, negative otherwise.

R is positive when the center of curvature is on the outgoing side of surface, negative otherwise.

m is positive when the image is erect, negative when inverted.

QUESTIONS

37–1 Can a person see a real image by looking backward along the direction from which the rays come? A virtual image? Can you tell by looking whether an image is real or virtual? How *can* the two be distinguished?

37–2 Why does a plane mirror reverse left and right but not top and bottom?

37–3 For a spherical mirror, if $s = f$, then $s' = \infty$, and the lateral magnification m is infinite. Does this make sense? If so, what does it mean?

37–4 According to the discussion of the preceding chapter, light rays are reversible. Are the formulas in Table 37–1 still valid if object and image are interchanged? What does reversibility imply with respect to the *forms* of the various formulas?

37–5 If a spherical mirror is immersed in water, does its focal length change?

37–6 For what range of object positions does a concave spherical mirror form a real image? What about a convex spherical mirror?

37–7 If a piece of photographic film is placed at the location of a real image, the film will record the image. Can this be done with a virtual image? How might one record a virtual image?

37–8 When a room has mirrors on two opposite walls, an infinite series of reflections can be seen. Discuss this phenomenon in terms of images. Why do the distant images appear darker?

37–9 When observing fish in an aquarium filled with water, one can see clearly only when looking nearly perpendicularly to the glass wall; objects viewed at an oblique angle always appear blurred. Why? Do the fish have the same problem when looking at you?

37–10 Can an image formed by one reflecting or refracting surface serve as an object for a second reflection or refraction? Does it matter whether the first image is real or virtual?

37–11 A concave mirror (sometimes surrounded by lights) is often used as an aid for applying cosmetics to the face. Why is such a mirror always concave rather than convex? What considerations determine its radius of curvature?

37–12 A student claimed that one can start a fire on a sunny day by use of the sun's rays and a concave mirror. How is this done? Is the concept of image relevant? Could one do the same thing with a convex mirror?

37–13 A person looks at his reflection in the concave side of a shiny spoon. Is it right side up or inverted? What if he looks in the convex side?

37–14 In Example 37–2 (Section 37–4), there appears to be an ambiguity for the case $s = 10$ cm, as to whether s' is ∞ or $-\infty$, and as to whether the image is erect or inverted. How is this resolved? Or is it?

37–15 "See yourself as others see you." Can you do this with an ordinary plane mirror? If not, how *can* you do it?

37–16 The shadow formed under a tree by sunlight partially blocked by leaves ordinarily is sprinkled with small circles of light and irregular patterns. During a solar eclipse, however, the shadow is sprinkled instead with a pattern of overlapping crescents that copy the shape of the occluded sun. Why?

EXERCISES

Section 37–1 Reflection at a Plane Surface

37–1 A candle 6 cm tall is 80 cm to the left of a plane mirror. Where is the image formed by the mirror, and what is the height of this image?

37–2 The image of a tree just covers the length of a 5-cm plane mirror when the mirror is held 30 cm from the eye. The tree is 100 m from the mirror. What is its height?

Section 37–2 Reflection at a Spherical Surface

Section 37–3 Focus and Focal Length

Section 37–4 Graphical Methods

37–3 The diameter of the moon is 3480 km, and its distance from the earth is 386,000 km. Find the diameter of the image of the moon formed by a spherical concave telescope mirror of focal length 4 m.

37–4 A spherical concave shaving mirror has a radius of curvature of 30 cm.

a) What is the magnification when the face is 10 cm from the vertex of the mirror?

b) Where is the image?

37–5 An object 2 cm high is placed 12 cm away from a concave spherical mirror having radius of curvature of 20 cm.

a) Draw a principal-ray diagram showing formation of the image.

b) Determine the position, size, orientation, and nature of the image.

37–6 A concave mirror has a radius of curvature of 20 cm.

a) What is its focal length?

b) If the mirror is immersed in water (refractive index 1.33), what is its focal length?

37–7 Prove that the image formed of a real object by a convex mirror is always virtual, no matter what the object position.

37–8 Repeat Exercise 37–5 for the case where the mirror is convex.

37–9 An object 1.5 cm tall is placed 8 cm away from the vertex of a convex spherical mirror whose radius of curvature has magnitude 20 cm.

a) Draw a principal-ray diagram showing formation of the image.

b) Determine the position, size, orientation, and nature of the image.

Section 37–5 Refraction at a Plane Surface

37–10 A ray of light in air makes an angle of incidence of 45° at the surface of a sheet of ice. The ray is refracted within the ice at an angle of 30°.

a) What is the critical angle for the ice?

b) A speck of dirt is embedded 3 cm below the surface of the ice. What is its apparent depth when viewed at normal incidence?

37–11 A skin diver is 2 m below the surface of a lake. A bird flies overhead 3 m above the surface of the lake. When the bird is directly overhead, how far above the diver does it appear to be?

37–12 A tank whose bottom is a mirror is filled with water to a depth of 20 cm. A small object hangs motionless 8 cm under the surface of the water. What is the apparent depth of its *image* when viewed at normal incidence?

Section 37–6 Refraction at a Spherical Surface

37–13 Equations (37–12) and (37–13) were derived in the text for the case where R is positive and $n < n'$. (See Figs. 37–16 and 37–17.)

a) Carry through the derivation of these two equations for the case where $R > 0$ and $n > n'$.

b) Carry through the derivation for $R < 0$ and $n < n'$.

37–14 The end of a long glass rod 8 cm in diameter has a convex hemispherical surface 4 cm in radius. The refractive index of the glass is 1.50. Determine the position of the image if an object is placed on the axis of the rod at the following distances to the left of its end:

a) infinitely far,   b) 16 cm,

c) 4 cm.

37–15 The rod of Exercise 37–14 is immersed in a liquid. An object 60 cm from the end of the rod and on its axis is imaged at a point 100 cm inside the rod. What is the refractive index of the liquid?

37–16 The left end of a long glass rod 10 cm in diameter, of index 1.50, is ground and polished to a convex hemispherical surface of radius 5 cm. An object in the form of an arrow 1 mm long, at right angles to the axis of the rod, is located on the axis 20 cm to the left of the vertex of the convex surface. Find the position and magnification of the image of the arrow formed by paraxial rays incident on the convex surface.

37–17 Repeat Exercise 37–16 for the case where the end of the rod is ground to a *concave* hemispherical surface of radius 5 cm.

37–18 A small tropical fish is at the center of a spherical fish bowl 30 cm in diameter. Find its apparent position and magnification to an observer outside the bowl. The effect of the thin walls of the bowl may be neglected.

PROBLEMS

37–19 What is the size of the smallest vertical plane mirror in which an observer standing erect can see his full-length image?

37–20 An object is placed between two mirrors arranged at right angles to each other.

a) Locate all the images of the object.

b) Draw the paths of rays from the object to the eye of an observer.

37–21 A concave mirror is to form an image of the filament of a headlight lamp on a screen 4 m from the mirror. The filament is 5 mm high, and the image is to be 40 cm high.

a) How far in front of the vertex of the mirror should the filament be placed?

b) What should be the radius of curvature of the mirror?

37–22 An object is 16 cm from the center of a silvered spherical glass Christmas tree ornament 8 cm in diameter. What are the position and magnification of its image?

37–23 If the light striking a convex mirror does not diverge from an object point but instead converges toward a point at a (negative) distance s to the right of the mirror, this point is called a *virtual object*.

a) For a convex mirror having radius of curvature 10 cm, for what range of virtual-object positions is a real image formed?

b) What is the orientation of this real image?

c) Draw a principal-ray diagram showing formation of such an image.

37–24 A microscope is focused on the upper surface of a glass plate. A second plate is then placed over the first. In order to focus on the bottom surface of the second plate, the microscope must be raised 1 mm. In order to focus on the upper surface, it must be raised 2 mm *farther*. Find the index of refraction of the second plate. (This problem illustrates one method of measuring index of refraction.)

37–25 What should be the index of refraction of a transparent sphere so that paraxial rays from an infinitely distant object will focus at the vertex of the surface opposite the point of incidence?

37–26 A transparent rod 40 cm long is cut flat at one end and rounded to a hemispherical surface of 12-cm radius at the other end. A small object is embedded within the rod along its axis and halfway between its ends. When viewed from the flat end of the rod, the apparent depth of the object is 12.5 cm. What is its apparent depth when viewed from the curved end?

37–27 A solid glass hemisphere having a radius of 10 cm and a refractive index of 1.50 is placed with its flat face downward on a table. A parallel beam of light of circular cross section 1 cm in diameter travels directly downward and enters the hemisphere along its diameter.

a) What is the diameter of the circle of light formed on the table?

b) How does your result depend on the radius of the hemisphere?

37–28 In Fig. 37–20a the first focal length f is the value of s corresponding to $s' = \infty$; in Fig. 37–20b the second focal length f' is the value of s' when $s = \infty$.

a) Prove that $n/n' = f/f'$.

b) Prove that the general relation between object and image distance is

$$\frac{f}{s} + \frac{f'}{s'} = 1.$$

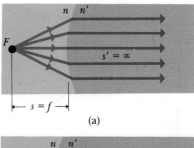

(a)

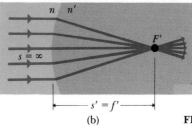

(b)   **FIGURE 37–20**

37–29 The longitudinal magnification is defined as $m' = ds'/ds$. It relates the longitudinal dimension of a small object to the longitudinal dimension of its image.

a) For a spherical mirror show that $m' = -m^2$. What is the significance of the fact that m' is *always* negative?

b) A wire frame in the form of a small cube 1 cm on a side is placed with its center on the axis of a concave mirror of radius of curvature 30 cm. The sides of the cube are all either parallel or perpendicular to the axis. The cube face toward the mirror is 60 cm to the left of the mirror vertex. Find

i) the location of the image of this face and of the opposite face of the cube;

ii) the lateral and longitudinal magnifications;

iii) the dimensions of each of the six cube faces of the image.

37–30 Refer to Problem 37–29. Show that the longitudinal magnification m' for refraction at a spherical surface is given by

$$m' = -\frac{n'}{n} m^2.$$

CHALLENGE PROBLEMS

37–31 You are sitting in your parked car and notice a jogger approach in the convex side mirror, which has a radius of curvature 2 m. If the jogger is running at a speed of 4 m·s^{-1}, how fast does the runner appear to be moving when she is

a) 10 m away?   b) 2 m away?

37–32 Two mirrors are placed together as shown in Fig. 37–21.

a) Show that a source in front of these mirrors and its two images lie on a circle.

b) Find the center of the circle.

c) Where should an observer stand to be able to see both images?

FIGURE 37–21

37–33 Spherical aberration is a blurring of the image formed by a mirror or lens because parallel rays striking the mirror far from the optic axis are focused at a different point than are rays near the axis. This problem is usually minimized by using only the center of a spherical mirror.

a) Show that for a spherical concave mirror the focus moves toward the mirror as the parallel rays move toward the outer edge of the mirror. (*Hint:* Derive an analytic expression for the distance from the vertex to the focus of the ray for a particular parallel ray, in terms of the radius of curvature R of the mirror and the angle θ. θ is the angle between the incident ray and the line connecting the center of curvature of the mirror and the point where the ray strikes the mirror.)

b) What value of θ produces a 2% change in the location of the focus, compared to the $\theta \approx 0$ value?

37–34 Parabolic mirrors are preferred in critical applications because they do not show any spherical aberration. (See Challenge Problem 37–33.) Show that for a mirror with a concave parabolic surface ($y = ax^2$), parallel rays are focused at $f = \frac{1}{4}a$ no matter how far from the axis they strike the mirror.

38

LENSES AND OPTICAL INSTRUMENTS

THE MOST FAMILIAR AND SIGNIFICANT SIMPLE OPTICAL DEVICE (EXCEPT for the plane mirror) is the *lens.* In this chapter we introduce the concepts of focus and focal length for a lens and then study the formation of images by lenses; we also develop graphical methods for analyzing image formation. Lenses are central to the operation of many familiar optical devices, including the human eye, magnifiers, projectors, and cameras. Other devices, such as telescopes and compound microscopes, use combinations of lenses or of lenses and mirrors. Here, as in the preceding chapter, the concept of *image* provides the key to analyzing and understanding these optical devices. We again base our analysis on the ray model of light, an adequate representation of many aspects of lens and mirror behavior.

38–1 THE THIN LENS

A lens is an optical system that includes two refracting surfaces. For now we concentrate primarily on lenses with two *spherical* surfaces sufficiently close together that the distance between them (the thickness of the lens) can be neglected. Such a device is called a **thin lens.** The behavior of a lens can be analyzed in detail by repeated application of the results of Section 37–6 for refraction by a single spherical surface. We postpone this derivation until Section 38–4 in order to move immediately into a discussion of the properties of thin lenses.

A lens of the type shown in Fig. 38–1 has the property that a beam of parallel rays converges, after passing through the lens, at a point F', as in Fig. 38–1a. Similarly, rays passing through point F emerge from the lens as a beam of parallel rays, as in Fig. 38–1b. The points F and F' are called the first and second *focal points,* or *foci* (plural form of **focus**), and the distance f is called the **focal length.** The concepts of focus and focal lengths are directly analogous to the same concepts used for spherical mirrors in Section 37–3. The two focal lengths in Fig. 38–1, both labeled f, are always equal for a thin lens, even when the two sides have different curvatures. We will derive this somewhat surprising result in Section 38–4. The central horizontal line is called the

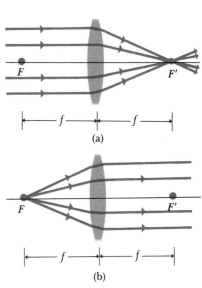

38–1 Focal points of a thin lens.

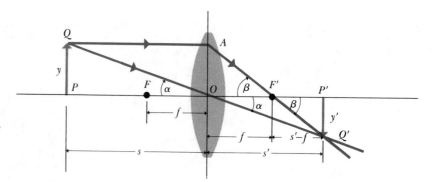

38–2 Construction used to find image position for a thin lens. The ray QAQ' is shown as bent at the midplane of the lens rather than at the two surfaces, to emphasize that the thickness of the lens is assumed to be very small.

Using the properties of the foci to derive the equation relating object and image positions for a thin lens

optic axis, as with spherical mirrors. The centers of curvature of the two spherical surfaces lie on (and in fact define) the optic axis.

The focal length f depends on the refractive index of the lens material (and on that of the surrounding matter if it is not vacuum) and on the radii of curvature of the spherical surfaces. We will derive this relationship, called the *lensmaker's equation,* in Section 38–4.

A thin lens forms an image of an object of finite size. Figure 38–2 shows how the position of the image may be calculated. Let s and s' be the object and image distances, respectively, and let y and y' be the object and image heights. This is the same notation we used in Chapter 37. Ray QA, parallel to the optic axis before refraction, passes through the second focus F' after refraction. Ray QOQ' passes undeflected straight through the center of the lens, because at the center the two surfaces are parallel and (we have assumed) very close together.

The angles labeled α in Fig. 38–2 are vertical angles and are equal. Therefore the two right triangles PQO and $P'Q'O$ are *similar,* and the ratios of corresponding sides are equal. Thus

$$\frac{y}{s} = -\frac{y'}{s'}, \quad \text{or} \quad \frac{y'}{y} = -\frac{s'}{s}. \tag{38–1}$$

(The negative sign arises because y' is negative, according to the convention used in Chapter 37.) Also, the two angles labeled β are equal, and the two right triangles OAF' and $P'Q'F'$ are similar, so

$$\frac{y}{f} = -\frac{y'}{s'-f}, \quad \text{or} \quad \frac{y'}{y} = -\frac{s'-f}{f}. \tag{38–2}$$

The basic object–image relation for a thin lens

We now equate Eqs. (38–1) and (38–2), divide by s', and rearrange to obtain

$$\frac{1}{s} + \frac{1}{s'} = \frac{1}{f}. \tag{38–3}$$

Magnification of a thin lens: a simple relationship

This analysis also gives us immediately the magnification $m = y'/y$ of the system; from Eq. (38–1),

$$m = -\frac{s'}{s}. \tag{38–4}$$

The negative sign tells us that when s and s' are both positive, the image is *inverted,* and y and y' have opposite signs.

Equations (38–3) and (38–4) are the basic equations for thin lenses; their *form* is exactly the same as the corresponding equations for spherical mirrors, Eqs. (37–5) and (37–7). As we will see, the same sign conventions we used for spherical mirrors are also applicable to lenses.

The three-dimensional image of a three-dimensional object, formed by a lens, is shown in Fig. 38–3. Since point R is nearer the lens than point P, its image, from Eq. (38–3), is farther from the lens than is point P', and the image $P'R'$ points in the same direction as the object PR. Arrows $P'S'$ and $P'Q'$ are reversed in space, relative to PS and PQ. Although we speak of the image as "inverted," only its transverse dimensions are reversed.

We invite you to compare Fig. 38–3 with Fig. 37–4, showing the image formed by a plane mirror. Note that the image formed by a lens, although inverted, is *not* reversed. That is, if the object is a left hand, its image is also a left hand. This fact may be verified by pointing the left thumb along PR, the left forefinger along PQ, and the left middle finger along PS. A rotation of 180° about the thumb as an axis then brings the fingers into coincidence with $P'Q'$ and $P'S'$. In other words, *inversion* of an image is equivalent to a rotation of 180° about the lens axis.

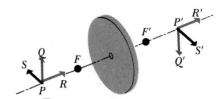

38–3 A lens forms a three-dimensional image of a three-dimensional object.

A lens doesn't turn a right hand into a left hand.

38–2 DIVERGING LENSES

A bundle of parallel rays incident on the lens shown in Fig. 38–1 *converges* to a real image after passing through the lens. Thus the lens is called a **converging lens.** Because its focal length is a positive quantity, it is also called a *positive lens.*

A bundle of parallel rays incident on the lens shown in Fig. 38–4 *diverges* after refraction, and so the lens is called a **diverging lens.** Because its focal length is a negative quantity, it is also called a *negative lens.* The foci of a negative lens are reversed relative to those of a positive lens. The second focus F' of a negative lens is the point from which rays, originally parallel to the axis, *appear to diverge* after refraction, as in Fig. 38–4a. Incident rays converging toward the first focus F, as in Fig. 38–4b, emerge from the lens parallel to its axis.

Equations (38–3) and (38–4) apply both to negative and to positive lenses. Various types of lenses, both converging and diverging, are shown in Fig. 38–5. In Section 38–4 we will see that any lens thicker in the center than at the edges is a converging lens with positive f, and any lens thicker at the edges than in the center is a diverging lens with negative f.

What's the difference between a converging lens and a diverging lens?

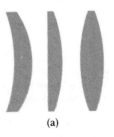

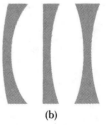

(a)

(b)

38–5 (a) Meniscus, plano-convex, and double-convex converging lenses. (b) Meniscus, plano-concave, and double concave diverging lenses. A converging lens is always thicker at its center than at its edges, and the reverse is true for all diverging lenses.

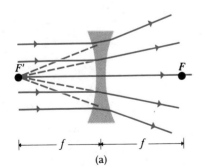

(a)

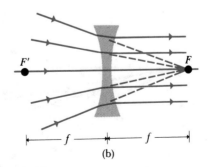

(b)

38–4 Focal points and focal length of a diverging lens.

38–3 GRAPHICAL METHODS

We can determine the position and size of an image formed by a thin lens with a graphical method very similar to the one used in Section 37–4 for spherical mirrors. Again we draw a few special rays, called **principal rays,** diverging from a point of the object that is *not* on the optic axis. The intersection of these rays, after passing through the lens, determines the position and size of the image. In using this graphical method, we will consider the entire deviation of a ray as occurring at the midplane of the lens, as shown in Fig. 38–6; this is consistent with the assumption that the distance between the lens surfaces is negligible.

The three principal rays whose paths can be traced easily are shown in Fig. 38–6:

1. *A ray parallel to the axis,* after refraction by the lens, passes through the second focal point of a converging lens or appears to come from the second focal point of a diverging lens.

2. *A ray through the center of the lens* is not appreciably deviated, since the two lens surfaces through which the central ray passes are very nearly parallel and close together if the lens is thin.

3. *A ray through (or proceeding toward) the first focal point* emerges parallel to the axis.

Once the position of the image point has been found by means of the intersection of any two rays 1, 2, and 3, the path of any other ray from the same point, such as ray 4 in Fig. 38–6, may be drawn. A few examples of this procedure are given in Fig. 38–7. The unnumbered rays in Fig. 38–7 are not principal rays.

We urge you to study each of these diagrams (drawn for several different object distances) very carefully, comparing each numbered ray with the above description. Several points are worth noting. In (d) the object is at the focus; ray 3 cannot be drawn because it does not pass through the lens. In (e) the

38–6 Principal-ray diagram, showing graphical method of locating an image. (a) A converging lens; (b) a diverging lens.

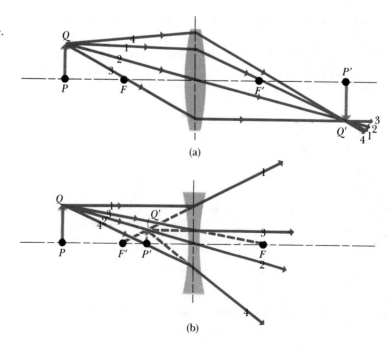

(a)

(b)

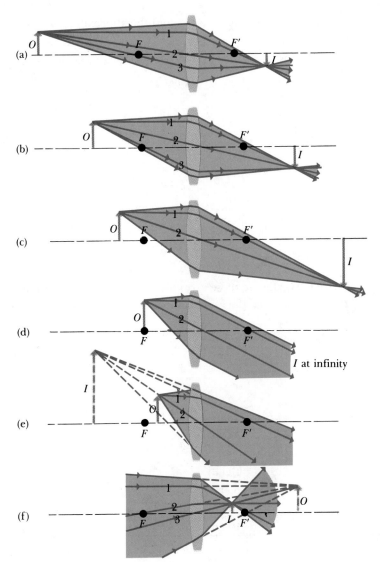

38–7 Formation of an image by a thin lens. The principal rays are labeled.

object distance is less than the focal length. The outgoing rays are divergent, and the *virtual image* is located by extending the outgoing rays backward. In this case the image distance s' is negative. Part (f) corresponds to a *virtual object*. The incoming rays are not diverging from a real-object point but are converging as though they would meet at the virtual-object point O on the right side. The object distance s is negative in this case. The image is real, and the image distance s' is positive and less than f.

EXAMPLE 38–1 A converging lens has a focal length of 20 cm. Find graphically the image location for an object at each of the following distances from the lens: 50 cm, 20 cm, 15 cm, and −40 cm. Determine the magnification in each case. Check your results by calculating the image position and magnification from Eqs. (38–3) and (38–4).

SOLUTION The appropriate principal-ray diagrams are shown in Fig. 38–7, parts (a), (d), (e), and (f). The approximate image distances, from measurements of

Several principal-ray diagrams for a
converging lens, along with the
calculated image positions and
magnifications

these diagrams, are 35 cm, ∞, −40 cm, and 15 cm, and the approximate magnifications are $-\frac{2}{3}$, ∞, +3, and $+\frac{1}{3}$.

Calculating the image positions from Eq. (38–3), we find

$$\frac{1}{s} + \frac{1}{s'} = \frac{1}{f},$$

a) $\qquad \dfrac{1}{50 \text{ cm}} + \dfrac{1}{s'} = \dfrac{1}{20 \text{ cm}}, \qquad s' = 33.3 \text{ cm},$

d) $\qquad \dfrac{1}{20 \text{ cm}} + \dfrac{1}{s'} = \dfrac{1}{20 \text{ cm}}, \qquad s' = \infty,$

e) $\qquad \dfrac{1}{15 \text{ cm}} + \dfrac{1}{s'} = \dfrac{1}{20 \text{ cm}}, \qquad s' = -60 \text{ cm},$

f) $\qquad \dfrac{1}{-40 \text{ cm}} + \dfrac{1}{s'} = \dfrac{1}{20 \text{ cm}}, \qquad s' = 13.3 \text{ cm}.$

The graphical results are fairly close to these except for (e), where the precision of
the diagram is limited by the fact that the rays extended backward have nearly the
same direction.

From Eq. (38–4), the magnifications are

$$m = -\frac{s'}{s},$$

a) $\qquad m = -\dfrac{33.3 \text{ cm}}{30 \text{ cm}} = -\dfrac{2}{3},$

d) $\qquad m = -\dfrac{\infty}{20 \text{ cm}} = -\infty,$

e) $\qquad m = -\dfrac{-60 \text{ cm}}{15 \text{ cm}} = +4,$

f) $\qquad m = -\dfrac{13.3 \text{ cm}}{-40 \text{ cm}} = +\dfrac{1}{3}.$

PROBLEM-SOLVING STRATEGY: *Image formation by a thin lens*

1. The strategy outlined at the end of Section 37–4 is equally applicable here, and we suggest you review it now. Always begin with a principal-ray diagram; orient your diagrams consistently so that light travels from left to right. For a lens there are only three principal rays, compared to four for a mirror. Don't just sketch these diagrams; draw the rays with a ruler, measuring the distances carefully. Draw them so they bend at the midplane of the lens, as shown in Fig. 38–6. Be sure to draw *all three* principal rays. The intersection of any two determines the image, but if the third doesn't pass through the same intersection point, you know you have made a mistake. Some redundancy can be a virtue! When there is a virtual image, you will have to extend the outgoing rays backward; in that case the image lies on the incoming side of the lens. Don't let that alarm you.

2. The same set of sign rules used in Chapter 37 is still applicable for thin lenses, and we will extend it in the next section to include radii of curvature of lens surfaces. Be extremely careful to get your signs right and to interpret the signs of results correctly.

3. Always determine the image position and size *both* graphically and by calculating. This procedure gives an extremely useful consistency check.

4. An additional point we will encounter in the next section is that the *image* from one lens or mirror may serve as the *object* for another. In that case, be careful in finding the object and image *distances* for this intermediate image; be sure you include the correct distance between the two elements (lenses and/or mirrors).

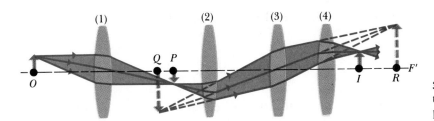

38–8 The object for each lens, after the first, is the image formed by the preceding lens.

38–4 IMAGES AS OBJECTS

An image formed by one lens or refracting surface can serve as the object for a second lens or refracting surface. Many optical systems, such as camera lenses, microscopes, and telescopes, use more than one lens. In each case the image formed by any one lens serves as the object for the next lens. Figure 38–8 shows various possibilities. Lens 1 forms a real image at P of a real object at O. This real image serves as a real object for lens 2. The virtual image at Q formed by lens 2 is a real object for lens 3. If lens 4 were not present, lens 3 would form a real image at R. Although this image is never formed, it serves as a **virtual object** for lens 4, which forms a real image at I.

What is a virtual object?

These considerations can also be applied to images formed by individual refracting surfaces. In particular, we can derive the thin-lens equation, Eq. (38–3), by repeated application of the single-surface equation, Eq. (37–12). For this derivation, consider the situation of Fig. 38–9. The figure shows several rays diverging from point Q of an object PQ. The first surface forms a virtual image of Q at Q'. This virtual image serves as a real object for the second surface of the lens, which forms a real image of Q' at Q''. Distance s_1 is the object distance for the first surface; s_1' is the corresponding image distance. The object distance for the second surface is s_2, equal to the sum of s_1' and the lens thickness t, and s_2' is the image distance for the second surface.

If the lens is thin enough so that its thickness t is negligible in comparison with the distances s_1, s_1', s_2, and s_2', we may assume that s_1' equals $-s_2$ and measure object and image distances from either vertex of the lens. We will also assume that the medium on both sides of the lens is air, with index of refraction 1.00. For the first refraction, Eq. (37–12) becomes

$$\frac{1}{s_1} + \frac{n}{s_1'} = \frac{n-1}{R_1}.$$

Refraction at the second surface yields the equation

$$\frac{n}{s_2} + \frac{1}{s_2'} = \frac{1-n}{R_2}.$$

The object–image relation for a lens, in terms of the properties of the lens

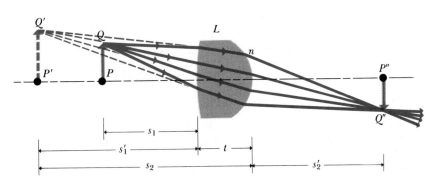

38–9 The image formed by the first surface of a lens serves as the object for the second surface.

Adding these two equations, and remembering that the lens is so thin that $s_2 = -s_1'$, we find

$$\frac{1}{s_1} + \frac{1}{s_2'} = (n - 1)\left(\frac{1}{R_1} - \frac{1}{R_2}\right).$$

Since s_1 is the object distance for the thin lens and s_2' is the image distance, the subscripts may be omitted, and we finally obtain

$$\frac{1}{s} + \frac{1}{s'} = (n - 1)\left(\frac{1}{R_1} - \frac{1}{R_2}\right). \tag{38–5}$$

The usual sign conventions apply to this equation. In Fig. 38–10, s, s', and R_1 are positive quantities, but R_2 is negative.

Comparing Eq. (38–5) with our previous form of the thin-lens equation, Eq. (38–3), we see that the focal length is given by

$$\frac{1}{f} = (n - 1)\left(\frac{1}{R_1} - \frac{1}{R_2}\right), \tag{38–6}$$

which is known as the **lensmaker's equation.** Thus in this derivation we obtain the thin-lens equation and also find how the focal length is related to the refractive index and the radii of curvature of the surfaces.

We can also obtain the focal length directly from Eq. (38–5) by recalling that the focus is the image of an infinitely distant object; when $s = \infty$, $s' = f$. Inserting these values into Eq. (38–5) yields Eq. (38–6) immediately.

How to calculate the focal length of a lens from its shape and material

---

EXAMPLE 38–2 In Fig. 38–10, let the absolute magnitudes of the radii of curvature of the lens surfaces be 20 cm and 5 cm, respectively. Since the center of curvature of the first surface is on the side of the outgoing light, $R_1 = +20$ cm, and since that of the second surface is not, $R_2 = -5$ cm. Let $n = 1.50$. Then

$$\frac{1}{f} = (1.50 - 1)\left(\frac{1}{20 \text{ cm}} - \frac{1}{-5 \text{ cm}}\right),$$

$$f = +8 \text{ cm}.$$

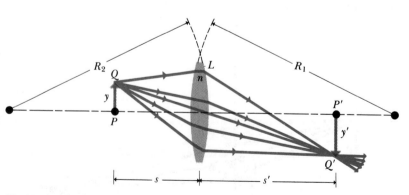

38–10 A thin lens. The radius of curvature of the first surface is R_1, and for the second surface R_2. In this case R_1 is positive but R_2 is negative. The focal length f is positive, and the lens is a converging lens.

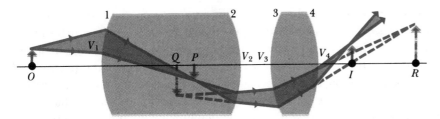

38–11 The object for each surface, after the first, is the image formed by the preceding surface.

Equation (38–6) can be generalized to the situation where the lens is immersed in a material having a refractive index greater than unity. We invite you to work out the lensmaker's equation for this more general situation.

We can extend the same method of analysis to systems consisting of several *thick* lenses; an example is shown in Fig. 38–11. The arrow at point O represents a small object at right angles to the optic axis. A narrow cone of rays diverging from the head of the arrow is traced through the system. Surface 1 forms a real image of the arrow at point P. Distance OV_1 is the object distance for the first surface, and distance V_1P is the image distance. Both distances are positive.

The image at P, formed by surface 1, serves as the object for surface 2. The object distance is PV_2 and is positive, since the direction from P to V_2 is the same as that of the incident light. The second surface forms a virtual image at point Q. The image distance is V_2Q and is negative because the direction from V_2 to Q is opposite to that of the refracted light.

The image at Q, formed by surface 2, serves as the object for surface 3. The object distance is QV_3 and is positive. The image at Q, although virtual, constitutes a *real object* so far as surface 3 is concerned. The rays incident on surface 3 are rendered converging and, except for the interposition of surface 4, would converge to a real image at point R. Even though this image is never formed, distance V_3R is the image distance for surface 3 and is positive.

The rays incident on surfaces 1, 2, and 3 have all been diverging, and the object distance has been the distance from the surface to the point from which the rays were actually or apparently diverging. The rays incident on surface 4, however, are *converging,* and there is no point at the left of the vertex from which they diverge or appear to diverge. The *image at R, toward which the rays are converging*, is the object for surface 4, and since the direction from R to V_4 is opposite to that of the incident light, the object distance RV_4 is negative. The image at R is called a *virtual object* for surface 4. In general, whenever a *converging* cone of rays is incident on (and interrupted by) a surface, the point toward which the rays are converging serves as the object, the object distance is negative, and the point is called a virtual object.

Finally, surface 4 forms a real image at I; the image distance is V_4I and is positive.

In our analysis of compound microscopes and telescopes in Sections 38–11 and 38–12, we will see examples where the image formed by a lens or mirror serves as the object for another lens.

An example of formation of successive images: The image from each surface is the object for the next.

38–5 LENS ABERRATIONS

The relatively simple equations we have derived for object and image distances, focal lengths, radii of curvature, and so on, have been based on the *paraxial approximation;* all rays were assumed to be *paraxial,* that is, to make

Lens aberrations: What happens when the paraxial approximation isn't good enough?

small angles with the axis. In general, however, a lens must form images not only of points on its axis but also of points that lie off the axis. Furthermore, because of the finite size of the lens, the cone of rays that forms the image of any point is of finite size. Nonparaxial rays proceeding from a given object point *do not*, in general, all intersect at precisely the same point after refraction by a lens, and the image formed by these rays is never a perfectly sharp one. Furthermore, the focal length of a lens depends on its index of refraction, which varies with wavelength. Therefore, if the light proceeding from an object is a mixture of wavelengths, different wavelengths are imaged at different points.

The departures of an actual image from the predictions of simple theory are called **aberrations.** Those caused by the variation of index with wavelength are the **chromatic aberrations.** Others, which would arise even if the light were monochromatic, are **monochromatic aberrations.** Lens aberrations are not caused by faulty construction of the lens, such as the failure of its surfaces to conform to a truly spherical shape, but are inevitable consequences of the laws of refraction at spherical surfaces.

Monochromatic aberrations are all related to the limitations of the paraxial approximation. *Spherical aberration* is the failure of rays from a point object on the optic axis to converge to a point image. Instead, the rays converge within a circle of minimum radius, called *the circle of least confusion,* and then diverge again, as shown in Fig. 38–12. The corresponding effect for points off the axis produces images that are comet-shaped figures rather than circles; this effect is called *coma.*

Astigmatism is the imaging of a point off the axis as two perpendicular *lines.* In this aberration the rays from a point object converge, at a certain distance from the lens, to a line in the plane defined by the optic axis and the object point. At a somewhat different distance from the lens, they converge to a second line *perpendicular* to this plane. The circle of least confusion appears between these two positions at a location that depends on the object point's distance from the axis as well as its distance from the lens. As a result, object points lying in a plane are, in general, imaged not in a plane but in some curved surface; this effect is called *curvature of field.* Finally, the image of a straight line that does not pass through the optic axis may be curved. As a result, the image of a square with the axis through its center may resemble a barrel (sides bent outward) or a pincushion (sides bent inward). This effect, called *distortion,* is not related to lack of sharpness of the image but results from a change in lateral magnification with distance from the axis.

Chromatic aberrations result directly from the variation of index of refraction with wavelength. When an object is illuminated with white light containing a mixture of wavelengths, different wavelengths are imaged at different points. The magnification of a lens also varies with wavelength; this effect is

Lens aberrations show themselves in several interesting ways.

Chromatic aberrations: What happens when the index of refraction varies with wavelength?

38–12 Spherical aberration. The circle of least confusion is shown by $C-C$.

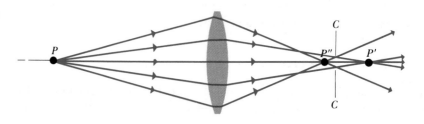

responsible for the rainbow-fringed images seen with inexpensive binoculars or telescopes.

It is impossible to eliminate all these aberrations from a single lens, but in a compound lens of several elements, the aberrations of one element may partially cancel those of another element. Design of such lenses is an extremely complex problem, which has been aided greatly in recent years by the use of computers. It is still impossible to eliminate all aberrations, but it *is* possible to decide which ones are most troublesome for a particular application and to design accordingly.

38–6 THE EYE

The essential parts of the human eye, considered as an optical system, are shown in Fig. 38–13. The eye is nearly spherical in shape and about 2.5 cm in diameter. The front portion is somewhat more sharply curved and is covered by a tough, transparent membrane C, called the *cornea*. The region behind the cornea contains a liquid A called the *aqueous humor*. Next comes the *crystalline lens L*, which is a capsule containing a fibrous jelly, hard at the center and progressively softer at the outer portions. The crystalline lens is held in place by ligaments that attach it to the ciliary muscle M. Behind the lens, the eye is filled with a thin, watery jelly V, called the *vitreous humor*. The indices of refraction of both the aqueous humor and the vitreous humor are nearly equal to that of water, about 1.336. The crystalline lens, while not homogeneous, has an average index of 1.437. This is not very different from the indices of the aqueous and vitreous humors; most of the refraction of light entering the eye occurs at the cornea.

Refraction at the cornea and the surfaces of the lens produces a *real image* of the object being viewed; the image is formed on the light-sensitive *retina R*, lining the rear inner surface of the eye. The *rods* and *cones* in the retina act like an array of miniature photocells; they sense the image and transmit it via the *optic nerve O* to the brain. Vision is most acute in a small central region called the *fovea centralis Y*, about 0.25 mm in diameter.

In front of the lens is the *iris*, containing an aperture of variable diameter called the *pupil P*, which opens and closes to adapt to changing light intensity. The receptors of the retina also have intensity-adaptation mechanisms.

For an object to be seen sharply, the image must be formed exactly at the location of the retina. The lens-to-retina distance, corresponding to *s'*, does not change, but the eye accommodates to different object distances *s* by changing the focal length of its lens. When the ciliary muscle surrounding the lens contracts, the lens bulges and the radii of curvature of its surfaces decrease, decreasing the focal length. For the normal eye, an object at infinity is sharply focused when the ciliary muscle is relaxed, and maximum tension in this muscle provides the appropriate focal length for an object at a distance of about 25 cm. This process is called *accommodation*.

The extremes of the range over which distinct vision is possible are known as the *far point* and the *near point* of the eye. The far point of a normal eye is at infinity. The position of the near point depends on the amount by which the curvature of the crystalline lens may be increased. The range of accommodation gradually diminishes with age as the crystalline lens loses its flexibility. For this reason, the near point gradually recedes as one grows older. This recession of the near point is called *presbyopia*. Following is a table of the

The eye is an optical instrument: using geometrical optics to analyze its image formation.

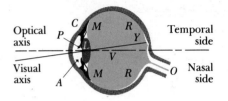

38–13 The eye.

How can the eye focus on objects at varying distances?

Why can't older people focus on close objects?

approximate average position of the near point at various ages:

Age (years)	Near Point (cm)
10	7
20	10
30	14
40	22
50	40
60	200

For example, an average person 50 years of age cannot focus on an object closer than about 40 cm.

38–7 DEFECTS OF VISION

Several common defects of vision result from incorrect distance relations between the parts of the optical system of the eye. A normal eye forms an image on the retina of an object at infinity when the eye is relaxed, as in Fig. 38–14a. In the *myopic* (nearsighted) eye, the eyeball is too long from front to back in comparison with the radius of curvature of the cornea, and rays from an object at infinity are focused in front of the retina. The most distant object for which an image can be formed on the retina is then nearer than infinity. In the *hyperopic* (farsighted) eye, the eyeball is too short, and the image of an infinitely distant object is behind the retina. The myopic eye produces too much convergence in a parallel bundle of rays for an image to be formed on the retina; the hyperopic eye, not enough convergence.

Astigmatism refers to a defect in which the surface of the cornea is not spherical but is more sharply curved in one plane than another. Astigmatism makes it impossible, for example, to focus clearly on the horizontal and vertical bars of a window at the same time.

These defects can be corrected by the use of corrective lenses ("glasses" or contact lenses). The near point of either a presbyopic or a hyperopic eye is farther from the eye than normal. To see clearly an object at normal reading distance (usually assumed to be 25 cm), we must place in front of the eye a lens of such focal length that it forms a virtual image of the object at or beyond the near point. Thus the function of the lens is not to make the object appear larger, but in effect to move the object farther away from the eye to a point where a sharp retinal image can be formed.

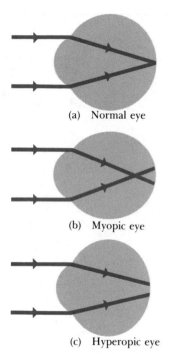

(a)   Normal eye

(b)   Myopic eye

(c)   Hyperopic eye

38–14 Refractive errors for myopic (nearsighted) and hyperopic (farsighted) eye viewing a very distant object.

Using lenses to correct for farsightedness

EXAMPLE 38–3   The near point of a certain hyperopic eye is 100 cm in front of the eye. What lens should be used to see clearly an object 25 cm in front of the eye?

SOLUTION   We want the lens to form an image of the object at a location corresponding to the near point of the eye, 100 cm from it. Thus we have

$$s = +25 \text{ cm}, \qquad s' = -100 \text{ cm},$$

$$\frac{1}{f} = \frac{1}{s} + \frac{1}{s'} = \frac{1}{+25 \text{ cm}} + \frac{1}{-100 \text{ cm}},$$

$$f = +33 \text{ cm}.$$

That is, a converging lens of focal length 33 cm is required.

The far point of a *myopic* eye is nearer than infinity. To see clearly objects beyond the far point, a lens must be used that will form an image of such objects not farther from the eye than the far point.

EXAMPLE 38–4 The far point of a certain myopic eye is 1 m in front of the eye. What lens should be used to see clearly an object at infinity?

SOLUTION Assume the image to be formed at the far point. Then

$$s = \infty, \qquad s' = -100 \text{ cm},$$

$$\frac{1}{f} = \frac{1}{s} + \frac{1}{s'} = \frac{1}{\infty} + \frac{1}{-100 \text{ cm}},$$

$$f = -100 \text{ cm}.$$

A *diverging* lens of focal length 100 cm is required.

Astigmatism is corrected by means of a *cylindrical* lens. The curvature of the cornea in a horizontal plane may have the proper value such that rays from infinity are focused on the retina. In the vertical plane, however, the curvature may not be sufficient to form a sharp retinal image. When a cylindrical lens with its axis horizontal is placed before the eye, the rays in a horizontal plane are unaffected, while the additional convergence of the rays in a vertical plane now causes these to be sharply imaged on the retina.

Optometrists describe the converging or diverging effect of lenses in terms not of the focal length, but of its *reciprocal*. The reciprocal of the focal length of a lens is called its *power*. If the focal length is in meters, the power is in *diopters*. Thus the power of a positive lens whose focal length is 1 m is 1 diopter; if the focal length is negative, the power is negative also. In the two examples above, the required powers are +3.0 diopter and −1.0 diopter, respectively.

Using cylindrical surfaces to correct astigmatism

A diopter is not an ancient flying dinosaur.

38–8 THE MAGNIFIER

The apparent size of an object is determined by the size of its retinal image. If the eye is unaided, this size depends on the *angle* subtended by the object at the eye, called its *angular size*. To examine a small object in detail, we bring it close to the eye in order to make the subtended angle and the retinal image as large as possible. Since the eye cannot focus sharply on objects closer than the near point, a particular object subtends the maximum possible viewing angle at an unaided eye when placed at this point. (We assume here that the near point is 25 cm from the eye.)

When a converging lens is placed in front of the eye, it forms a virtual image farther from the eye than the object. Thus the object may be moved closer to the eye, and the angular size of the image may be substantially larger than the angular size of the object at 25 cm without the lens. A lens used in this way is called a *magnifying glass*, or simply a *magnifier*. Viewing of the virtual image is most comfortable when it is placed at infinity, and in the following discussion we assume this is done.

The principle of the magnifier is shown in Fig. 38–15. In (a) the object is at the near point, where it subtends an angle u at the eye. In (b) a magnifier in

The magnifying glass: getting the image in the right place

Angular magnification: making things look bigger than they are

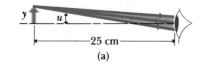

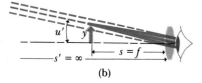

38–15 A simple magnifier.

front of the eye forms an image at infinity, and the angle subtended at the magnifier is u'. The **angular magnification** M (not to be confused with the *lateral magnification m*) is defined as the ratio of the angle u' to the angle u. We can find the value of M as follows. From Fig. 38–15, u and u' are given (in radians) by

$$u = \frac{y}{25 \text{ cm}} \qquad \text{(approximately)},$$

$$u' = \frac{y}{f} \qquad \text{(approximately)}.$$

Hence

$$M = \frac{u'}{u} = \frac{y/f}{y/25} = \frac{25}{f} \qquad (f \text{ in centimeters}). \qquad (38\text{--}7)$$

It appears at first that we can make the angular magnification as large·as we like by decreasing the focal length f. In fact, the aberrations of a simple double-convex lens set a limit to M of about 2X or 3X. If these aberrations are corrected, the magnification may be carried as high as 20X. It is not ordinarily practical to correct aberrations adequately for greater magnification; instead, greater magnification is attained by use of a compound microscope, discussed in Section 38–11.

38–9 THE CAMERA

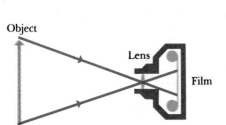

38–16 Essential elements of a camera.

Controlling the amount of light that strikes the film

The essential elements of a camera are a lens equipped with a shutter, a light-tight enclosure, and a light-sensitive film to record an image. An example is shown in Fig. 38–16. The lens forms a real image, on the film, of the object being photographed, just as the lens of the human eye forms a real image on the retina, the eye's "film." The lens may be moved closer to or farther from the film to provide proper image distances for various object distances. All but the most inexpensive lenses have several elements to permit partial correction of various aberrations. A typical example is the Zeiss "Tessar" design, shown in Fig. 38–17.

In order for the image to be recorded properly on the film, the total light energy per unit area reaching the film (the "exposure") must fall within certain limits; this is controlled by the *shutter* and the *lens aperture*. The shutter controls the time during which light enters the lens, typically adjustable in steps corresponding to factors of about two, from 1 s to $\frac{1}{1000}$ s or thereabouts. The light-gathering capacity of the lens is proportional to its effective area; this may be varied by means of an adjustable aperture, or *diaphragm,* which is a nearly circular hole of variable diameter. The aperture size is usually described in terms of its "f-number," which is the number by which the focal length of the lens must be divided to obtain the diameter of the aperture. Hence a lens having $f = 50$ mm and an aperture diameter of 25 mm is said to have an aperture of $f/2$, or an f-number of 2.

Because the light-gathering capacity of a lens is proportional to its area and thus to the *square* of its diameter, changing the diameter by a factor of $\sqrt{2}$ corresponds to a factor of two in exposure. Thus adjustable apertures usually have scales labeled with successive numbers related by factors of $\sqrt{2}$, such as

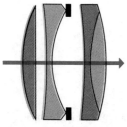

38–17 Zeiss "Tessar" lens design.

$$f/2, \quad f/2.8, \quad f/4, \quad f/5.6, \quad f/8, \quad f/11, \quad f/16,$$

and so on. The larger numbers represent smaller apertures and exposures, and each step corresponds to a factor of two in exposure.

The choice of focal length for a camera lens depends on the film size and the desired angle of view, or *field.* For the popular 35-mm cameras, with image size of 24 × 36 mm, the normal lens is usually about 50 mm in focal length and permits an angle of about 45°. A longer focal-length lens, used with the same film size, provides a smaller angle of view and a larger image of part of the object, compared with a normal lens. This gives the impression that the camera is closer than it really is; such a lens is called a *telephoto* lens. At the other extreme, a lens of shorter focal length, such as 35 mm or 28 mm, permits a wider angle of view and is called a *wide-angle* lens.

The focal length of the lens determines the angle of view.

The optical system for a television camera is the same in principle as for an ordinary camera. The film is replaced by an electronic system that scans the image with a series of 525 parallel lines. The image brightness at points along these lines is translated into electrical impulses that can be broadcast by using electromagnetic waves with frequencies of the order of 100 to 400 MHz. The entire picture is scanned 30 times each second, so 30 × 525, or 15,750, lines are scanned each second. Some TV receivers emit a faint high-pitched sound at this frequency.

The optical system of a television camera

38–10 THE PROJECTOR

A projector for slides or motion pictures operates very much like a camera in reverse. The essential elements are shown in Fig. 38–18. Light from the source (an incandescent lamp bulb or, in large motion-picture projectors, a carbon-arc lamp) shines through the film, and the projection lens forms a real, enlarged, inverted image of the film on the projection screen. Additional lenses called *condenser* lenses are placed between lamp and film. Their function is to direct the light from the source so that most of it enters the projection lens after passing through the film. A concave mirror behind the lamp also helps direct the light. The condenser lenses must be large enough to cover the entire area of the film. The image on the screen is always inverted; this is why slides have to be put into a projector upside down.

A projector is like a camera in reverse: making an enlarged real image on a projection screen.

The position and size of the image projected on the screen are determined by the position and focal length of the projection lens.

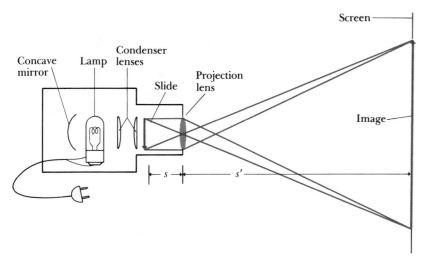

38–18 A slide projector. The concave mirror and condenser lenses gather and direct the light from the lamp so it will enter the projection lens after passing through the slide. The cooling fan, usually needed to take away excessive heat from the lamp, is not shown.

EXAMPLE 38–5 An ordinary 35-mm color slide has a picture area 24 × 36 mm. What focal-length projection lens would be needed to project an image 1.2 m × 1.8 m on a screen 5 m from the lens?

SOLUTION We need a lateral magnification (apart from sign) of 1.2 m/24 mm = 50. Thus, from Eq. (38–4), the ratio s'/s must also be 50. We are given $s' = 5$ m, so $s = 5$ m/50 = 0.1 m. Then from Eq. (38–3),

$$\frac{1}{f} = \frac{1}{0.1 \text{ m}} + \frac{1}{5 \text{ m}},$$

$$f = 0.098 \text{ m} = 98 \text{ mm}.$$

A commercially available lens would probably have $f = 100$ mm; this is a popular focal length for home slide projectors.

38–11 THE COMPOUND MICROSCOPE

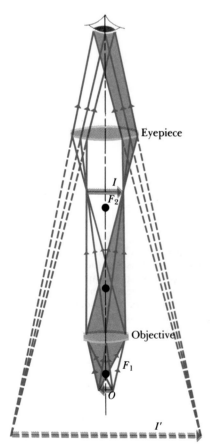

The compound microscope uses two lens systems; the second looks at the image formed by the first.

38–19 The optical system of a microscope.

When an angular magnification larger than that attainable with a simple magnifier is desired, it is necessary to use a *compound microscope*, usually called merely a *microscope*. The essential elements of a microscope are illustrated in Fig. 38–19. The object O to be examined is placed just beyond the first focal point F_1 of the **objective** lens, which forms a real and enlarged image I. This image lies just within the first focal point F_2 of the **eyepiece,** which forms a final virtual image of I at I'. As we stated earlier, the position of I' may be anywhere between the near and far points of the eye. Although both the objective and eyepiece of an actual microscope are highly corrected compound lenses, they are shown as simple thin lenses for simplicity.

Since the objective lens forms an enlarged real image that is viewed through the eyepiece, the overall angular magnification M of the compound microscope is the product of the *lateral* magnification m_1 of the objective and the *angular* magnification M_2 of the eyepiece. The former is given by

$$m_1 = -\frac{s_1'}{s_1},$$

where s_1 and s_1' are the object and image distances, respectively, for the objective lens. Ordinarily the object is very close to the focus, resulting in an image whose distance from the objective is much larger than its focal length f_1. Thus s_1 is approximately equal to f_1, and $m_1 = -s_1'/f_1$, approximately. The angular magnification of the eyepiece, from Eq. (38–7), is $M_2 = (25 \text{ cm})/f_2$, where f_2 is the focal length of the eyepiece, considered as a simple lens. Hence the overall magnification M of the compound microscope is, apart from a negative sign, which is customarily ignored,

$$M = m_1 M_2 = \frac{(25 \text{ cm}) s_1'}{f_1 f_2}, \tag{38–8}$$

where s_1', f_1, and f_2 are measured in centimeters. Microscope manufacturers customarily specify the values of m_1 and M_2 for microscope components rather than the focal lengths of the objective and eyepiece.

38-12 TELESCOPES

The optical system of a refracting telescope is similar to that of a compound microscope. In both instruments the image formed by an objective is viewed through an eyepiece. The difference is that the telescope is used to examine large objects at large distances and the microscope is used to examine small objects close at hand.

The *astronomical* telescope is illustrated in Fig. 38-20. The objective lens forms a real, reduced image I of the object, and a virtual image of I is formed by the eyepiece. As with the microscope, the image I' may be formed anywhere between the near and far points of the eye. In practice, the objects examined by a telescope are at such large distances from the instrument that the image I is formed very nearly at the second focal point of the objective. Furthermore, if the image I' is at infinity, the image I is at the first focal point of the eyepiece. The distance between objective and eyepiece, or the length of the telescope, is therefore the *sum* of the focal lengths of objective and eyepiece, $f_1 + f_2$.

The astronomical telescope uses two lens systems; the second looks at the image formed by the first.

The angular magnification M of a telescope is defined as the ratio of the angle subtended at the eye by the final image I', to the angle subtended at the (unaided) eye by the object. This ratio may be expressed in terms of the focal lengths of objective and eyepiece as follows. In Fig. 38-20 the ray passing through F_1, the first focal point of the objective, and through F_2', the second focal point of the eyepiece, has been emphasized. The object (not shown) subtends an angle u at the objective and would subtend essentially the same angle at the unaided eye. Also, since the observer's eye is placed just to the right of the focal point F_2', the angle subtended at the eye by the final image is very nearly equal to the angle u'. Because bd is parallel to the optic axis, the distances ab and cd are equal to each other and to the height y' of the image I. Since u and u' are small, they may be approximated by their tangents. From the right triangles F_1ab and $F_2'cd$,

The overall magnification of a telescope.

$$u = \frac{-y'}{f_1}, \qquad u' = \frac{y'}{f_2}.$$

Hence

$$M = \frac{u'}{u} = -\frac{y'/f_2}{y'/f_1} = -\frac{f_1}{f_2}. \tag{38-9}$$

The angular magnification M of a telescope is therefore equal to the ratio of the focal length of the objective to that of the eyepiece. The negative sign denotes an inverted image.

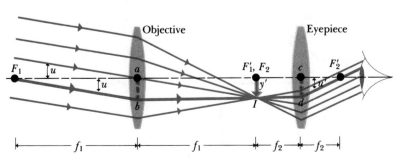

38-20 Optical system of a telescope; final image at infinity.

38–21 The prism binocular. (Courtesy of Bushnell Division of Bausch & Lomb Optical Co.)

Using prisms to turn an inverted image right side up

An inverted image is not a disadvantage if the instrument is to be used for astronomical observations, but it is desirable for a terrestrial telescope to form an erect image. This is accomplished in the *prism binocular* by a pair of 45°– 45°–90° totally reflecting prisms inserted between objective and eyepiece, as shown in Fig. 38–21. The image is inverted by the four reflections from the inclined faces of the prisms. It is customary to stamp on a flat metal surface of a binocular two numbers separated by a multiplication sign, such as 7×50. The first number is the magnification and the second is the diameter of the objective lenses in millimeters, which determines the brightness of the image.

The reflecting telescope: a mirror and a lens

In the *reflecting telescope,* the objective lens is replaced by a concave mirror, as shown in Fig. 38–22. In large telescopes this scheme has many advantages, both theoretical and practical. The mirror is intrinsically free of chromatic aberrations, and spherical aberrations are much easier to correct than with a lens. The material need not be transparent, and the reflector can be made more rigid than a lens, which has to be supported only at its edges. The largest reflecting telescope in the world has a mirror over 5 m in diameter.

Huge mirrors are a lot easier to make than huge lenses.

Because the image is formed in a region traversed by incoming rays, this image can be observed directly with an eyepiece only by blocking off part of the incoming beam; this is practical only for the largest telescopes. Alternative schemes use a mirror to reflect the image out the side or through a hole in the mirror, as shown in Figs. 38–22b and 38–22c. This scheme is also used in some long-focal-length telephoto lenses for cameras. In the context of photography, such a system is called a *catadioptric lens,* which is a fancy name for an

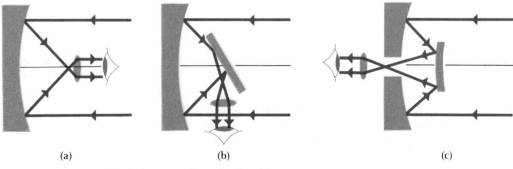

(a)   (b)   (c)

38–22 Optical systems for reflecting telescopes.

38–23 The reflector of the 200-inch Hale telescope at Palomar Observatory, being cleaned in preparation for the application of a new aluminum reflective surface. (Palomar Observatory Photograph.)

optical system containing both reflecting and refracting elements. The reflector for a large astronomical telescope is shown in Fig. 38–23.

SUMMARY

A thin lens has two focal points, or foci, at equal distances from the lens; this distance is called the focal length, denoted by f. An incoming parallel light beam converges to the focus for a converging lens (positive f) or diverges as though the rays had come from the focus for a diverging lens (negative f). Rays diverging from the focus of a converging lens emerge parallel, and rays directed toward the focus of a diverging lens (on the far side) emerge parallel, after refraction.

For an object at distance s from a thin lens, the lens forms an image at distance s'. The two distances are related by

$$\frac{1}{s} + \frac{1}{s'} = \frac{1}{f}, \tag{38–3}$$

and the magnification is given by

$$m = -\frac{s'}{s}. \tag{38–4}$$

The distances and focal length follow the same sign rules as in Chapter 37: s is positive if the object is on the incoming side, negative otherwise. The image distance s' is positive when the image is on the outgoing side, negative otherwise. The focal length f is positive for a converging lens, negative for a diverging lens.

A principal-ray diagram for a thin lens contains three rays that can be easily drawn; two are related to the foci, and the third passes through the center undeflected.

The focal length f of a thin lens is related to its index of refraction n and the radii of curvature R_1 and R_2 of its surfaces by the lensmaker's equation:

$$\frac{1}{f} = (n - 1)\left(\frac{1}{R_1} - \frac{1}{R_2}\right). \tag{38–6}$$

KEY TERMS
thin lens
focus
focal length
optic axis
converging lens
diverging lens
principal rays
virtual object
lensmaker's equation
aberrations
chromatic aberrations
monochromatic aberrations
angular magnification
objective
eyepiece

A radius is positive if its center of curvature is on the outgoing side, negative otherwise. The same method of analysis used to derive this result can also be used for thick lenses, by successive application of the single-surface equation, Eq. (37–12).

Lens aberrations describe the failure of a lens to form a perfectly sharp image of an object. Monochromatic aberrations occur because of limitations of the paraxial approximation; chromatic aberrations result from the dependence of index of refraction on wavelength.

In the eye, a real image is formed on the retina and transmitted to the optic nerve. Adjustment for various object distances is made by the ciliary muscle, which squeezes the lens, making it bulge and decreasing its focal length. A nearsighted eye is too long for its lens, a farsighted eye too short. Focal lengths of corrective lenses are described in terms of their power; the power of a lens, in diopters, is the reciprocal of the focal length, in meters.

The simple magnifier creates a virtual image whose angular size is larger than that of the object itself at a distance of 25 cm, the nominal closest distance for comfortable viewing. The angular magnification is the ratio of the angular size of the virtual image to that of the object, at this distance.

The camera forms a real, inverted, reduced image (on film) of the object being photographed. The amount of light striking the film is controlled by the shutter speed and the aperture. The projector is essentially a camera in reverse; a lens forms an enlarged, real, inverted image on a screen of the slide or motion-picture film. Other lenses, called condenser lenses, concentrate the light from the lamp on the slide or film.

In the compound microscope, the objective lens forms a first image in the barrel of the instrument, and the eyepiece forms a final virtual image at infinity of this first image. The telescope operates on the same principle, but the object is far away. In a reflecting telescope the objective lens is replaced by a concave mirror, which eliminates chromatic aberrations.

QUESTIONS

38–1 Sometimes a wine glass filled with white wine forms an image of an overhead light on a white tablecloth. Would the same image be formed with an empty glass? With a glass of gin? Gasoline?

38–2 How could one very quickly make an approximate measurement of the focal length of a converging lens? Could the same method be applied if we wished to use a diverging lens?

38–3 If you look closely at a shiny Christmas tree ball, you can see nearly the entire room. Does the room appear right side up or upside down? Discuss your observations in terms of images.

38–4 A student asserted that any lens with spherical surfaces has a positive focal length if it is thicker at the center than at the edge, and negative if it is thicker at the edge. Do you agree?

38–5 The focal length of a simple lens depends on the color (wavelength) of light passing through it. Why? Is it possible for a lens to have a positive focal length for some colors and negative for others?

38–6 The human eye is often compared to a camera. In what ways is it similar to a camera? In what ways does it differ?

38–7 How could one make a lens for sound waves?

38–8 A student proposed to use a plastic bag full of air, immersed in water, as an underwater lens. Is this possible? If the lens is to be a converging lens, what shape should the air pocket have?

38–9 When a converging lens is immersed in water, does its focal length increase or decrease, compared with the value in air?

38–10 You are marooned on a desert island and want to use your eyeglasses to start a fire. Can this be done if you are nearsighted? If you are farsighted?

38–11 While lost in the mountains, a person who was nearsighted in one eye and farsighted in the other made a crude emergency telescope from the two lenses of his eyeglasses. How did he do this?

38–12 In using a magnifying glass, is the magnification greater when the glass is close to the object or when it is close to the eye?

38–13 When a slide projector is turned on without a slide in it, and the focus adjustment is moved far enough in one direction, a gigantic image of the light-bulb filament can be seen on the screen. Explain how this happens.

38–14 A spherical air bubble in water can function as a lens. Is it a converging or diverging lens? How is its focal length related to its radius?

38–15 As discussed in the text, some binoculars use prisms to invert the final image. Why are prisms better than ordinary mirrors for this purpose?

38–16 There have been reports of round fishbowls starting fires by focusing the sun's rays coming in a window. Is this possible?

38–17 How does a person judge distance? Can a person with vision in only one eye judge distance? What is meant by "binocular vision"?

38–18 Zoom lenses are widely used in television cameras and conventional photography. Such a lens has, effectively, a variable focal length; changes in focal length are accomplished by moving some lens elements relative to others. Try to devise a scheme to accomplish this effect.

38–19 Why can't you see clearly when your head is under water? Could you wear glasses so you *could* see under water? Would the lenses be converging or diverging?

EXERCISES

Section 38–1 The Thin Lens

Section 38–2 Diverging Lenses

Section 38–3 Graphical Methods

38–1 A converging lens has a focal length of 10 cm. For object distances of 30 cm, 20 cm, 15 cm, and 5 cm, determine

a) image position,

b) magnification,

c) whether the image is real or virtual,

d) whether the image is erect or inverted.

Be sure to draw a principal-ray diagram in each case.

38–2 A converging lens of focal length 10 cm forms a real image 1 cm high, 12 cm to the right of the lens. Determine the position and size of the object. Is the image erect or inverted? Draw a principal-ray diagram for this situation.

38–3 Repeat Exercise 38–1 for the case where the lens is diverging, with focal length −10 cm.

38–4 An object is 10 cm from a lens that forms an image 15 cm from the lens, on the side opposite to the object.

a) What is the focal length of the lens? Is the lens converging or diverging?

b) If the object is 1 cm high, how high is the image? Is it erect or inverted?

c) Draw a principal-ray diagram.

38–5 A lens forms an image of an object 20 cm from it. The image is 4 cm from the lens on the same side as the object.

a) What is the focal length of the lens? Is the lens converging or diverging?

b) If the object is 2 cm high, how high is the image? Is it erect or inverted?

c) Draw a principal-ray diagram.

38–6 Prove that the image of a real object formed by a diverging lens is *always* virtual.

Section 38–4 Images as Objects

38–7 A diverging meniscus lens (see Fig. 38–5b) of refractive index 1.48 has spherical surfaces whose radii are 4.0 cm and 2.5 cm. What would be the position of the image if an object were placed 15 cm in front of the lens?

38–8 Sketch the various possible thin lenses obtainable by combining two surfaces whose radii of curvature are, in absolute magnitude, 10 cm and 20 cm. Which are converging and which are diverging? Find the focal length of each if made of glass of index 1.50.

38–9 A layer of benzene ($n = 1.50$) 2 cm deep floats on water ($n = 1.33$) that is 4 cm deep. What is the apparent distance from the upper benzene surface to the bottom of the water layer, when viewed at normal incidence?

38–10 A transparent rod 40 cm long and of refractive index 1.50 is cut flat at one end and rounded to a hemispherical surface of 12-cm radius at the other end. An object is placed on the axis of the rod 10 cm from the hemispherical end.

a) What is the position of the final image?

b) What is its magnification?

38–11 Both ends of a glass rod 10 cm in diameter, of index 1.50, are ground and polished to convex hemispherical surfaces of radius 5 cm at the left end and radius 10 cm at the right end. The length of the rod between vertexes is 60 cm. An arrow 1 mm long, at right angles to the axis and 20 cm to the left of the first vertex, constitutes the object for the first surface.

a) What constitutes the object for the second surface?

b) What is the object distance for the second surface?

c) Is this object real or virtual?

d) What is the position of the image formed by the second surface?

38–12 The same rod as in Exercise 38–11 is now shortened to a distance of 10 cm between its vertices, the curvatures of its ends remaining the same.

a) What is the object distance for the second surface?

b) Is the object real or virtual?

c) What is the position of the image formed by the second surface?

d) Is the image real or virtual and is it erect or inverted, with respect to the original object?

e) What is the height of the final image?

38–13 A narrow beam of parallel rays enters a solid glass sphere in a radial direction. At what point outside the sphere are these rays brought to a focus? The radius of the sphere is 3 cm, and its index is 1.50.

38–14 The radii of curvature of the surfaces of a thin lens are $R_1 = +10$ cm and $R_2 = +30$ cm. The index is 1.50.

a) Compute the position and size of the image of an object in the form of an arrow 1 cm high, perpendicular to the lens axis, 40 cm to the left of the lens.

b) A second similar lens is placed 160 cm to the right of the first. Find the position and size of the final image.

c) Same as (b), except the second lens is 40 cm to the right of the first.

d) Same as (c), except the second lens, of focal length −40 cm, is diverging.

38–15 Three thin lenses, each of focal length 20 cm, are aligned on a common axis, and adjacent lenses are separated by 30 cm. Find the position of the image of a small object on the axis, 60 cm to the left of the first lens.

38–16 Two thin lenses with focal length of magnitude 10 cm, the first converging and the second diverging, are placed 5 cm apart. An object 2 mm tall is placed 20 cm in front of the first (converging) lens.

a) How far from this lens will the final image be formed?

b) Is the final image real or virtual?

c) What is the height of the final image?

38–17 An eyepiece consists of two similar positive thin lenses having focal lengths of 6 cm, separated by a distance of 3 cm. Where are the focal points of the eyepiece?

Section 38–6 The Eye

Section 38–7 Defects of Vision

38–18 What is the power of the lens required

a) by a hyperopic eye whose near point is at 125 cm?

b) by a myopic eye whose far point is at 50 cm?

38–19

a) Where is the near point of an eye for which a lens of power +2 diopters is prescribed?

b) Where is the far point of an eye for which a lens of power −0.5 diopter is prescribed for distant vision?

38–20 In a simplified model of the human eye, the aqueous and vitreous humors and the lens all have a refractive index of 1.40, and all the refraction occurs at the cornea, about 2.50 cm from the retina. What should be the radius of curvature of the cornea in order to focus an infinitely distant object on the retina?

38–21 For the model of the eye described in Exercise 38–20, what should be the radius of curvature of the cornea in order to focus an object 25 cm from the cornea?

Section 38–8 The Magnifier

38–22 A thin lens of focal length 10 cm is used as a simple magnifier.

a) What angular magnification is obtainable with the lens?

b) When an object is examined through the lens, how close can it be brought to the eye? Assume that the image viewed by the eye is at infinity.

38–23 The focal length of a simple magnifier is 10 cm.

a) How far in front of the magnifier should an object be placed if the image is formed at the observer's near point, 25 cm in front of her eye?

b) If the object is 1 mm high, what is the height of its image formed by the magnifier?

Assume the magnifier to be a thin lens placed very close to the eye.

Section 38–9 The Camera

38–24 During a lunar eclipse, a picture of the moon (diameter 3.48×10^6 m, distance from earth 3.8×10^8 m) is taken with a camera whose lens has a focal length of 50 mm. What is the diameter of the image on the film?

38–25 The picture size on ordinary 35-mm camera film is 24×36 mm. Focal lengths of lenses available for 35-mm cameras typically include 28, 35, 50 (the "standard" lens), 85, 100, 135, 200, and 300 mm, among others. Which of these lenses should be used to photograph the following objects, assuming the object is to fill most of the picture area?

a) A cathedral 100 m high and 150 m long, at a distance of 150 m.

b) An eagle with a wingspan of 2.0 m, at a distance of 15 m.

38–26 Camera A, having an $f/8$ lens 2.5 cm in diameter, photographs an object with the correct exposure of $\frac{1}{100}$ s. What exposure should camera B use in photographing the same object if it has an $f/4$ lens 5 cm in diameter?

38–27 The focal length of an $f/2.8$ camera lens is 8 cm.

a) What is the aperature diameter of the lens?

b) If the correct exposure of a certain scene is $\frac{1}{200}$ s at $f/2.8$, what would be the correct exposure at $f/5.6$?

Section 38–10 The Projector

38–28 The dimensions of the picture on a 35-mm color slide are 24×36 mm. An image of the slide is projected on a screen 10 m from the projector lens. The focal length of the projector lens is 12 cm.

a) How far should the slide be from the lens?

b) What will be the dimensions of the image on the screen?

38–29 In Example 38–5, for a 98-mm focal-length lens, it was found that the slide needs to be placed 10 cm in front of the lens to focus the image on a screen 5 m from the lens. Now assume instead that a 100-mm-focal-length lens is used.

a) If the screen remains 5 m from the lens, how far should the slide be in front of the lens?

b) Could the distance between the slide and lens be kept at 10 cm and the screen moved to achieve focus? Is so, how far and in what direction would it have to be moved?

Section 38–11 The Compound Microscope

38–30 A certain microscope is provided with objectives of focal lengths 16 mm, 4 mm, and 1.9 mm, and with eyepieces of angular magnification 5X and 10X. What is

a) the largest overall magnification obtainable?

b) the least?

Each objective forms an image 160 mm beyond its second focal point.

38–31 The focal length of the eyepiece of a certain microscope is 2.5 cm. The focal length of the objective is 16 mm. The distance between objective and eyepiece is 22.1 cm. The final image formed by the eyepiece is at infinity. Treat all lenses as thin.

a) What should be the distance from the objective to the object viewed?

b) What is the linear magnification produced by the objective?

c) What is the overall magnification of the microscope?

38–32 The image formed by a microscope objective of focal length 4 mm is 180 mm from its second focal point. The eyepiece has a focal length of 31.25 mm.

a) What is the magnification of the microscope?

b) The unaided eye can distinguish two points as separate if they are about 0.1 mm apart. What is the minimum separation that can be resolved with this microscope?

Section 38–12 Telescopes

38–33 A crude telescope is constructed of two eyeglass lenses of focal lengths 100 cm and 20 cm, respectively.

a) Find its angular magnification.

b) Find the height of the image formed by the objective of a building 80 m high at a distance of 2 km.

38–34 The eyepiece of a telescope has a focal length of 10 cm. The distance between objective and eyepiece is 2.1 m. What is the angular magnification of the telescope?

38–35 The moon subtends an angle at the earth of approximately $\frac{1}{2}°$. What is the diameter of the image of the moon produced by the objective of the Lick Observatory telescope, a refractor with a focal length of 18 m?

38–36 A reflecting telescope is made of a mirror of radius of curvature 0.50 m and an eyepiece of focal length 1.0 cm.

a) What is the angular magnification?

b) What should be the position of the eyepiece if both the object and final image are at infinity?

38–37 The new space telescope uses an optical system similar to that shown in Fig. 38–24 (Cassegrain system). The image of a far distant object is focused on the detector through a hole in the large (primary) mirror. The primary mirror has a focal length of 2.5 m, the distance between the vertexes of the two mirrors is 1.5 m, and the distance from the vertex of the primary mirror to the detector is 0.25 m. Should the smaller mirror (the secondary) be concave or convex, and what is its radius of curvature?

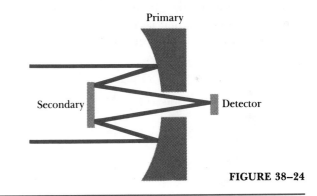

FIGURE 38–24

PROBLEMS

38–38 An object is placed 18 cm from a screen.

a) At what points between object and screen may a converging lens of 4-cm focal length be placed to obtain an image on the screen?

b) What is the magnification of the image for these positions of the lens?

38–39 An object is imaged by a lens on a screen placed 12 cm from the lens. When the lens is moved 2 cm farther from the object, the screen must be moved 2 cm closer to the object to refocus it. Determine the focal length of the lens.

38–40

a) Prove that when two thin lenses of focal lengths f_1 and f_2 are placed *in contact*, the focal length f of the combination is given by the relation

$$\frac{1}{f} = \frac{1}{f_1} + \frac{1}{f_2}.$$

b) A converging meniscus lens (Fig. 38–5a) has an index of refraction of 1.50, and the radii of its surfaces are 5 cm and 10 cm. The concave surface is placed upward and filled with water. What is the focal length of the water–glass combination?

38–41 When an object is placed at the proper distance in front of a converging lens, the image falls on a screen 20 cm from the lens. A diverging lens is now placed halfway between the converging lens and the screen, and it is found that the screen must be moved 20 cm farther away from the converging lens to obtain a sharp image. What is the focal length of the diverging lens?

38–42 A lens operates via Snell's law, bending light rays at each surface an amount determined by the index of the lens and the index of the medium in which the lens is located.

a) Equation (38–6) assumes that the lens is surrounded by air. Consider instead a thin lens immersed in a liquid of refractive index n_{liq}. Derive the equation analogous to Eq. (38–6) for the focal length f' of the lens.

b) A thin lens of index n has a focal length in vacuum of f. Use the result of (a) to show that when this lens is immersed in a liquid of index n_{liq}, it will have a new focal length of

$$f' = \left[\frac{n_{liq}(n - 1)}{n - n_{liq}}\right]f.$$

38–43 A convex mirror and a concave mirror are placed on the same optic axis separated by a distance $L = 0.8$ m. The radius of curvature of each mirror has a magnitude of 0.4 m. A light source is located a distance x from the concave mirror, as shown in Fig. 38–25.

a) What distance x will result in the rays from the source returning to the source after reflecting first from the convex mirror and then from the concave mirror?

b) Repeat part (a), but now let the rays reflect first from the concave mirror and then from the convex one.

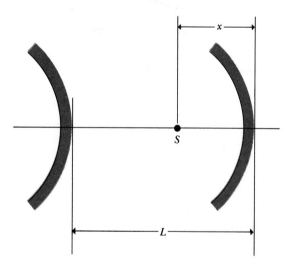

FIGURE 38–25

38–44 In the arrangement shown in Fig. 38–26, the candle is at the center of curvature of the concave mirror, whose focal length is 10 cm. The converging lens has a focal length of 35 cm and is 85 cm to the right of the candle. The candle is viewed by looking through the lens from the right.

a) Draw a principal-ray diagram that locates the final image.

b) Where is the final image?

c) Is the final image real or virtual?

d) Is the final image erect or inverted? (*Hint:* Are one or two images formed?)

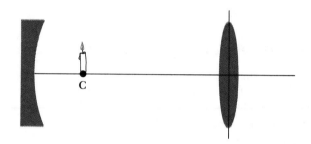

FIGURE 38–26

38–45 Your camera has a 50-mm-focal-length lens and a viewfinder that is 20 mm (high) by 30 mm (long). In taking a picture of a 3-m-long automobile, you find that the image of the auto fills only two-thirds of the viewfinder.

a) How far are you from the auto?

b) How close should you stand if you want to fill the viewfinder frame with the auto's image?

38–46 A glass rod of refractive index 1.50 is ground and polished at both ends to hemispherical surfaces of 5-cm radius. When an object is placed on the axis of the rod, 20 cm from one end, the final image is formed 40 cm from the opposite end. What is the length of the rod?

38–47 A thin-walled glass sphere of radius R is filled with water. An object is placed a distance $3R$ from the surface of the sphere. Determine the position of the final image. The effect of the glass wall may be neglected.

38–48 A thick-walled wine goblet can be considered a glass sphere of radius 5 cm with a spherical cavity of radius 4 cm. The index of refraction of the goblet glass is 1.50.

a) A beam of parallel light rays enters the empty goblet along a radius. Where, if anywhere, will an image be formed?

b) The goblet is filled with white wine ($n = 1.37$). Where is the image now formed?

38–49 A glass plate 2 cm thick, of index 1.50, having plane parallel faces, is held with its faces horizontal and its lower face 8 cm above a printed page. Find the position of the image of the page formed by rays making a small angle with the normal to the plate.

38–50 Rays from a lens are converging toward a point image P, as in Fig. 38–27. What thickness t of glass of index 1.50 must be interposed, as in the figure, for the image to be formed at P'?

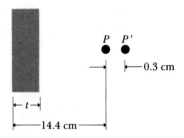

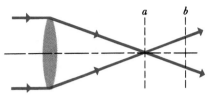

FIGURE 38–28

FIGURE 38–27

38–51 The *resolution* of a camera lens can be defined as the maximum number of lines per millimeter in the image that can barely be distinguished as separate lines. A certain lens has a focal length of 50 mm and resolution of 100 lines·mm^{-1}. What is the minimum separation of two lines in an object 100 m away if they are to be visible in the image as separate lines?

38–52

a) Show that when two thin lenses are placed in contact, the *power* of the combination in diopters, as defined in Section 38–7, is the sum of the powers of the separate lenses. Is this relation valid even when one lens has positive power and the other negative?

b) Two thin lenses of 25-cm and 40-cm focal lengths are in contact. What is the power of the combination?

38–53 A certain very nearsighted person cannot focus anything farther than 10 cm from the eye. If the radius of curvature of the cornea is 0.70 cm when the eye is focusing on an object 10 cm from it, and the indexes of refraction are as described in Exercise 38–20, what is the cornea-to-retina distance? What does this tell you about the shape of the nearsighted eye?

38–54 A camera lens is focused on a distant point source of light; the image is formed on a screen at *a* (Fig. 38–28). When the screen is moved backward a distance of 2 cm to *b*,

the circle of light on the screen has a diameter of 4 mm. What is the f/number of the lens?

38–55 A microscope with an objective of focal length 9 mm and an eyepiece of focal length 5 cm is used to project an image on a screen 1 m from the eyepiece. Let the image distance of the objective be 18 cm.

a) What is the lateral magnification of the image?

b) What is the distance between the objective and the eyepiece?

38–56 A certain reflecting telescope, constructed as in Fig. 38–22a, has a mirror 10 cm in diameter with radius of curvature 1.0 m and an eyepiece of focal length 1.0 cm. If the angular magnification has a magnitude of 48 and the object is at infinity, find the position of the lens and the position and nature of the final image. (*Note:* $|M|$ is *not* equal to $|f_1/f_2|$, so the image formed by the eyepiece is *not* at infinity.)

38–57 Figure 38–29 is a diagram of a *Galilean telescope,* or *opera glass,* with both object and final image at infinity. The image *I* serves as a virtual object for the eyepiece. The final image is virtual and erect.

a) Prove that the angular magnification is $M = -f_1/f_2$.

b) A Galilean telescope is to be constructed with the same objective lens as in Exercise 38–33. What focal length should the eyepiece have if this telescope is to have the same magnification as the one in Exercise 38–33?

c) Compare the lengths of the telescopes.

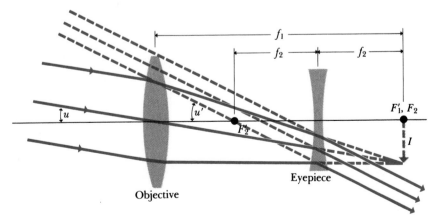

FIGURE 38–29

CHALLENGE PROBLEMS

38–58 A 20-cm long pencil is placed at a 45° angle, with its center 15 cm above the optic axis and 45 cm from a 25-cm-focal-length lens, as shown in Fig. 38–30.

a) Where is the image of the pencil? (Give the location of the images of the points A, B, and C on the object, which are located at the eraser, point, and center of the pencil, respectively.)

b) What is the length of the image?

c) How is the image oriented?

Neglect aberration effects.

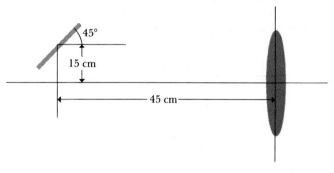

FIGURE 38–30

38–59

a) For a lens of focal length f, find the smallest distance possible between the object and its real image.

b) Sketch a graph of the distance between the object and real image as a function of the distance of the object from the lens. Does your sketch agree with the result you found in (a)?

38–60 A solid glass sphere of radius R and index 1.50 is silvered over one hemisphere, as in Fig. 38–31. A small object is located on the axis of the sphere at a distance $2R$ from the pole of the unsilvered hemisphere. Find the position of the final image after all refractions and reflections have taken place.

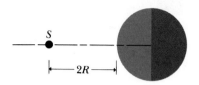

FIGURE 38–31

38–61 A symmetric double-convex thin lens made of glass of index 1.50 has a focal length in air of 30 cm. The lens is sealed into an opening in one end of a tank filled with water (index = 1.33). At the end of the tank opposite the lens is a plane mirror 80 cm from the lens.

a) Find the position of the image formed by the lens–water–mirror system, of a small object outside the tank on the lens axis and 90 cm to the left of the lens.

b) Is the image real or virtual?

c) Is it erect or inverted?

d) If the object has a height of 4 mm, what is the height of the image?

38–62 A person with normal vision cannot focus his or her eyes underwater.

a) Why not?

b) With the simplified model of the eye described in Exercise 38–20, what corrective lens (specified by focal length, as measured in air) would be needed to enable a person underwater to focus an infinitely distant object? (Be careful—the focal length of a lens underwater is not the same as in air! See Problem 38–42. Assume the corrective lens has a refractive index in air of 1.60.)

39

INTERFERENCE AND DIFFRACTION

IN OUR ANALYSIS OF IMAGE FORMATION BY LENSES AND MIRRORS, WE HAVE represented light as *rays* that travel in straight lines in a homogeneous material and that are bent according to simple laws at a reflecting or refracting surface. This simple model forms the basis of *geometrical optics;* as we have seen, this model is an adequate basis for analyzing a wide variety of phenomena involving lenses and mirrors.

In this chapter we will study the phenomena of *interference* and *diffraction.* These are inherently *wave* phenomena, and the principles of geometrical optics are *not* an adequate basis for this study. Instead, we must return to the more general view that light is a *wave motion.* When several waves overlap at a point, their total effect depends on the *phases* of the waves as well as their amplitudes. This part of the subject is called *physical optics.* When light passes through apertures or around obstacles, patterns are formed that could not be predicted on the basis of a ray model, but instead depend directly on the wave nature of light. In several simple situations we can predict the characteristics of these patterns in detail. We will look at several practical applications of physical optics, including diffraction gratings, x-ray diffraction, and holography.

Light doesn't always behave like rays: Interference and diffraction phenomena show the wave nature of light.

39–1 INTERFERENCE AND COHERENT SOURCES

In our discussions of mechanical waves in Chapter 21 and electromagnetic waves in Chapter 35, we often considered *sinusoidal* waves having a single frequency and a single wavelength. In the context of optics, such a wave is called a **monochromatic** (single-color) **wave.** Common sources of light, such as an incandescent light bulb or a flame, *do not* emit monochromatic light but rather emit a continuous distribution of wavelengths. A strictly monochromatic light wave is an unattainable idealization.

Monochromatic light can be *approximated* in the laboratory. For example, we can pass continuous-spectrum light through a filter that blocks all but a narrow range of wavelengths. Gas-discharge lamps, such as the mercury-arc lamp, emit line spectra in which the light consists of a discrete set of colors,

Monochromatic light has only one wavelength and frequency: a useful though unattainable idealization.

Laser light is very nearly monochromatic.

each having a narrow band of wavelengths called a *spectrum line*. For example, the bright green line in the mercury spectrum has an average wavelength of 546.1 nm, with a spread of wavelength of the order of ±0.001 nm, depending on the pressure and temperature of the mercury vapor in the lamp. By far the most nearly monochromatic source available at present is the *laser,* to be discussed in Chapter 41. The familiar helium-neon laser, inexpensive and readily available, emits visible light at 632.8 nm with a line width (wavelength range) of the order of ±0.000001 nm, or about one part in 10^9. In the following discussion of interference and diffraction phenomena, we will nearly always assume that we are dealing with monochromatic waves.

The term **interference** refers to any situation in which two or more waves overlap in space. We introduced this term in Section 22–2 in connection with standing waves on a stretched string, formed by the superposition of two sinusoidal waves traveling in opposite directions. In such cases the total displacement at any point at any instant of time is governed by the **principle of linear superposition,** also introduced in Section 22–2. This principle, the most important in all of physical optics, states that *when two or more waves overlap, the resultant displacement at any point and at any instant may be found by adding the instantaneous displacements that would be produced at the point by the individual waves if each were present alone.* The term *displacement,* as used here, is a general one. If we are considering surface ripples on a liquid, the displacement means the actual displacement of the surface above or below its normal level. If the waves are sound waves, the term refers to the excess or deficiency of pressure. If the waves are electromagnetic, the displacement means the magnitude of the electric or magnetic field. When light of extremely high intensity passes through matter, the principle of linear superposition is *not* precisely obeyed, and the resulting phenomena are classified under the heading *nonlinear optics.* (These effects are beyond the scope of this book.)

To introduce the essential ideas of interference, consider first the problem of two identical sources of monochromatic waves, S_1 and S_2, separated in space by a certain distance. The two sources are permanently *in phase,* so that at every point in space there is a definite and unchanging phase relation between waves from the two sources. They might be, for example, two agitators in a ripple tank, two loudspeakers driven by the same amplifier, two radio antennas powered by the same transmitter, or two small apertures in an opaque screen, illuminated by the same monochromatic light source.

We position the sources S_1 and S_2 along the y-axis, equidistant from the origin, as shown in Fig. 39–1. Let P_0 be any point on the x-axis. From symmetry, the two distances S_1P_0 and S_2P_0 are equal; waves from the two sources thus require equal times to travel to P_0, and having left S_1 and S_2 in phase, they arrive at P_0 in phase. The total amplitude at P_0 is thus twice the amplitude of each individual wave.

Next we consider a point P_1, located so that its distance from S_2 is exactly one wavelength greater than its distance from S_1. That is,

$$S_2P_1 - S_1P_1 = \lambda.$$

Then any given wave crest from S_1 arrives at P_1 exactly one cycle earlier than the crest emitted at the same time from S_2, and again the two waves arrive *in phase.* Similarly, waves arrive in phase at all points P_2 for which the path difference is *two* wavelengths $(S_2P_2 - S_1P_2 = 2\lambda)$, or indeed for *any* positive or negative integer number of wavelengths.

Superposition: the underlying principle in all interference and diffraction calculations

For two-source interference, the sources must have a definite phase relationship.

The two waves arriving at a point are not in phase because they travel different distances from the sources.

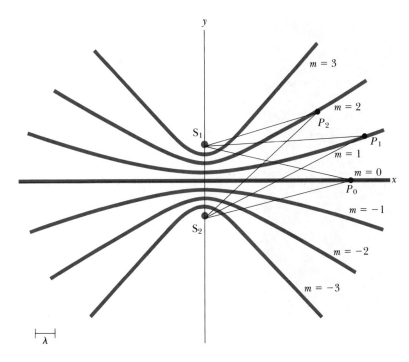

39–1 Curves of maximum intensity in the interference pattern of two monochromatic point sources. In this example the distance between sources is four times the wavelength.

The addition of amplitudes that results when waves from two or more sources arrive at a point *in phase* is often called **constructive interference** or *reinforcement,* and our discussion shows that constructive interference occurs whenever the path difference for the two sources is an integral multiple of the wavelength:

$$S_2P - S_1P = m\lambda \qquad (m = 0, \pm 1, \pm 2, \pm 3, \ldots). \qquad (39\text{--}1)$$

In our example, the points satisfying this condition lie on the set of curves shown in Fig. 39–1.

Intermediate between these lines is a set of other lines for which the path difference for the two sources is a *half-integral* number of wavelengths. Waves from the two sources arrive at a point on one of these lines exactly a half-cycle out of phase, and the resultant amplitude is the *difference* between the two individual amplitudes. If the amplitudes are equal, which is approximately true when the distance from either source to P is much greater than the distance between sources, then the total amplitude at such a point is zero! This condition is called **destructive interference** or *cancellation.* In our example, the condition for destructive interference is

$$S_2P - S_1P = (m + \tfrac{1}{2})\lambda \qquad (m = 0, \pm 1, \pm 2, \pm 3, \ldots). \qquad (39\text{--}2)$$

An example of this interference pattern is the familiar ripple-tank pattern shown in Fig. 39–2. The two wave sources are two agitators driven by the same vibrating mechanism. The regions of both maximum and zero amplitude are clearly visible. The superposed color lines, corresponding to those in Fig. 39–1, show the lines of maximum amplitude.

A ripple tank is an inherently two-dimensional situation, but in some cases, such as two loudspeakers or two radio-transmitter antennas, the pattern is three-dimensional. We may think of rotating the color curves of Fig. 39–1

When two waves arrive in phase, they add.

When two waves arrive a half-cycle out of phase, they subtract.

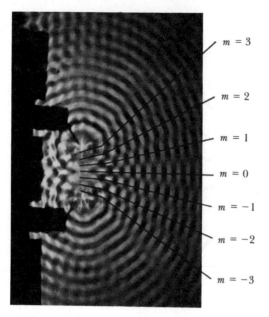

$m = 3$

$m = 2$

$m = 1$

$m = 0$

$m = -1$

$m = -2$

$m = -3$

39–2 Photograph of an interference pattern produced by water waves in a shallow ripple tank. The two wave sources are small balls moved up and down by the same vibrating mechanism. As waves move outward from the sources, they overlap and produce an interference pattern. The lines of maximum amplitude in the pattern are shown by the superposed color lines. (*PSSC Physics*, second edition, 1965; D. C. Heath and Co., with Educational Development Center, Inc., Newton, Mass.)

A definite phase relation between the sources is essential.

Coherent sources have a definite phase relationship.

about the horizontal line through the center; the resulting surfaces are the points where maximum constructive interference occurs.

In this discussion, the constant *phase* relationship between the sources is an essential requirement. If the relative phase of the sources changes, the positions of the maxima and minima in the resulting interference pattern also change. When the radiation is light, two sources can have a definite and constant phase relation *only when they both emit light coming from a single primary source.* There is no practical way to achieve such a relationship with two separate sources. The reason is a fundamental one associated with the mechanisms of light emission. In ordinary light sources, atoms of the material of the source are given excess energy by thermal agitation or by impact with accelerated electrons. An atom thus "excited" begins to radiate energy and continues until it has lost all the energy it can, typically in a time of the order of 10^{-8} s. A source ordinarily contains a very large number of atoms, and they radiate in an unsynchronized and random phase relationship. Thus emission from two such sources has a rapidly varying phase relation. The result is an interference pattern that constantly changes in a random manner, and ordinary observation does not show a visible interference pattern at all.

However, if the light from a single source is split so that parts of it emerge from two or more regions of space, forming two or more *secondary sources*, any random phase change in the source affects these secondary sources equally and does not change their *relative* phase. Two such sources derived from a single source and having a definite phase relation are said to be **coherent.**

The distinguishing feature of light from a *laser* is that the emission of light from many atoms is *synchronized* in frequency and phase, by mechanisms to be discussed in Chapter 41. As a result, the random phase changes mentioned above occur *much* less frequently. Definite phase relations are preserved over correspondingly much greater lengths in the beam. Accordingly, laser light is said to be much more *coherent* than ordinary light.

39-2 TWO-SOURCE INTERFERENCE

One of the earliest experimental demonstrations of the fact that light can produce interference effects was performed in 1800 by the English scientist Thomas Young. This was a crucial experiment because it added significantly to the evidence for the wave nature of light. His experiment involved the interference of light from two sources, as described in the preceding section.

A landmark experiment to demonstrate the wave nature of light and measure its wavelengths

Young's apparatus is shown schematically in Fig. 39-3a. Monochromatic light emerging from a narrow slit S_0 falls on a screen having two other narrow slits S_1 and S_2, 0.1 mm or so wide and 1.0 mm or less apart. According to Huygens' principle (Section 36-12), cylindrical wavelets spread out from slit S_0 and reach slits S_1 and S_2 in phase because they travel equal distances from S_0. A succession of Huygens wavelets then emerges from each slit; the two sets leave in phase, and therefore the two slits act as two *coherent* sources. But the waves do not necessarily arrive at point P in phase because of the path difference $(r_1 - r_2)$ for the two waves.

Coherent sources can be obtained by dividing the light from a single source.

In the following analysis we will assume that the distance R from the slits to the screen is so large compared to the slit spacing d that the lines from the slits to P are very nearly parallel, as shown in Fig. 39-3c. This is not an essential assumption, but it simplifies the following analysis considerably and is usually valid for the experiments we will discuss. In this case the path difference is given by

$$r_1 - r_2 = d \sin \theta. \tag{39-3}$$

According to the discussion in the preceding section, constructive interference (reinforcement) occurs at point P when the path difference $d \sin \theta$ is some integral number of wavelengths, say $m\lambda$ ($m = 0, \pm 1, \pm 2, \pm 3$, etc.). Thus,

$$d \sin \theta = m\lambda \quad \text{or} \quad \sin \theta = \frac{m\lambda}{d}. \tag{39-4}$$

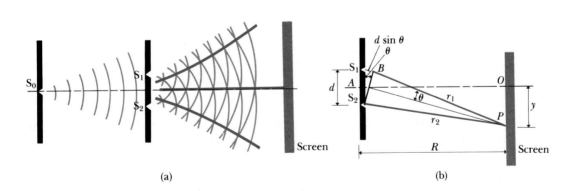

(a)

(b)

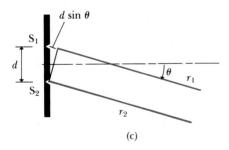

(c)

39-3 (a) Interference of light waves passing through two slits.
(b) Geometrical analysis of Young's experiment. (c) Approximate geometry when R is much larger than d.

Zeroth fringe

39–4 Interference fringes produced by Young's double-slit experiment.

Similarly, complete cancellation or destructive interference occurs when the path difference is a half-integral number of wavelengths, $(m + \frac{1}{2})\lambda$, where again $m = 0, \pm 1, \pm 2$, and so on:

$$d \sin \theta = \left(m + \frac{1}{2}\right)\lambda \quad \text{or} \quad \sin \theta = \frac{(m + \frac{1}{2})\lambda}{d}. \tag{39–5}$$

An interference pattern is a series of alternating bright and dark bands or fringes.

Thus the pattern on the screen of Fig. 39–3b is a succession of bright and dark bands; an actual photograph of such a pattern is shown in Fig. 39–4.

Let us derive an expression for the positions of the centers of the bright bands. Let y_m be the distance of the center of the mth bright band from the center of the central band at $\theta = 0$. The corresponding value of θ is θ_m, and

$$y_m = R \tan \theta_m.$$

In optical interference experiments, the y_m's are nearly always much smaller than R. In that case $\tan \theta_m$ is very nearly equal to $\sin \theta_m$, and

$$y_m = R \sin \theta_m.$$

Combining this with Eq. (39–4), we find

$$y_m = R\frac{m\lambda}{d} \tag{39–6}$$

and

$$\lambda = \frac{y_m d}{mR}. \tag{39–7}$$

A simple and direct way to measure the wavelength of light

We can measure R and d, as well as the positions y_m of the bright fringes. Thus this experiment provides a direct measurement of the wavelength λ.

EXAMPLE 39–1 In a two-slit interference experiment with two slits 0.2 mm apart, and a screen at a distance of 1 m, the third bright fringe is found to be displaced 7.5 mm from the central fringe. Find the wavelength of the light used.

SOLUTION Let λ be the unknown wavelength. Then

$$\lambda = \frac{y_m d}{mR} = \frac{(0.75 \text{ cm})(0.02 \text{ cm})}{(3)(100 \text{ cm})} = 5 \times 10^{-5} \text{ cm}$$
$$= 500 \times 10^{-9} \text{ m} = 500 \text{ nm}.$$

Although we have described this experiment as Young performed it with visible light, a completely analogous experiment could be done with any other electromagnetic waves, such as radio waves.

EXAMPLE 39–2 A radio station operating at a frequency of 1500 kHz = 1.5×10^6 Hz has two identical vertical dipole antennas spaced 400 m apart. Where are the intensity maxima in the resulting radiation pattern?

Interference with radio antennas: how to make a directional antenna

SOLUTION The wavelength is $\lambda = c/f = 200$ m. The directions of the intensity *maxima* are those for which the path difference is zero or an integer number of wavelengths, as given by Eq. (39–4). Inserting the numerical values, we find

$$\sin \theta = \frac{m\lambda}{d} = \frac{m}{2}, \qquad \theta = 0, \pm 30°, \pm 90°.$$

In this example, values of m greater than 2 give values of $\sin \theta$ greater than unity and thus have no meaning; there is *no* direction for which the path difference is three or more wavelengths. Similarly, the directions having zero intensity (complete destructive interference) are given by Eq. (39–5):

$$\sin \theta = \frac{(m + \frac{1}{2})\lambda}{d} = \frac{m + \frac{1}{2}}{2}, \qquad \theta = \pm 14.5°, \pm 48.6°.$$

In this case values of m greater than 1 have no meaning, for the reason just mentioned.

39–3 INTENSITY DISTRIBUTION IN INTERFERENCE PATTERNS

In Section 39–2 we computed the directions of maximum and minimum intensity in the two-source interference pattern. We may also find the intensity at *any* point in the pattern. To do this we have to combine the two sinusoidally varying fields at a point P in the radiation pattern caused by the two sources, taking proper account of their phase difference. The intensity is then proportional to the square of the resultant electric-field amplitude, as we learned in Chapter 35.

Calculating the details of the intensity distribution in an interference pattern

Here is our program. The phase difference δ between the two sinusoidally varying fields at point P depends on the difference in path length. We have to relate δ to the location of P, and then we have to learn how to *add* the two sinusoidal functions with a phase difference of δ. From the amplitude E_P of the resultant sinusoidal wave at point P, we compute the **intensity** I. To obtain I we recall from Section 35–3 that I is equal to the average magnitude of the Poynting vector, S_{av}. For a sinusoidal wave with E-field amplitude E_P, this is given by Eq. (35–18) with E_{max} replaced by E_P. By using the relation $\epsilon_0 \mu_0 = c^2$, we can express this in the following equivalent forms:

$$S_{av} = I = \frac{1}{2} \frac{E_P{}^2}{\mu_0 c} = \frac{1}{2} \sqrt{\frac{\epsilon_0}{\mu_0}} E_P{}^2 = \frac{1}{2} \epsilon_0 c E_P{}^2. \qquad (39\text{–}8)$$

We will assume the two sinusoidal functions have equal amplitude E and that the **E** fields lie along the same line; this assumes the sources are identical and neglects the slight amplitude difference caused by the unequal path lengths. If the sources are in phase, then the waves that arrive at P differ in phase by an amount proportional to the difference in their path lengths, $(r_1 - r_2)$. If the phase angle between these arriving waves is δ, then the two electric fields superposed at P might be

$$E_1 = E \cos \omega t,$$

$$E_2 = E \cos (\omega t + \delta).$$

To find δ, we note that if the path difference is one wavelength, then $\delta = 2\pi$ or $360°$. If it is equal to $\lambda/2$, the phase difference δ is π, and so on. That is, the ratio of δ to 2π is equal to the ratio of $(r_1 - r_2)$ to λ:

$$\frac{\delta}{2\pi} = \frac{r_1 - r_2}{\lambda}.$$

Thus a path difference $(r_1 - r_2)$ causes a phase difference δ given by

$$\delta = \frac{2\pi}{\lambda}(r_1 - r_2) = k(r_1 - r_2), \tag{39–9}$$

where $k = 2\pi/\lambda$ is the *wave number* introduced in Section 21–3.

If the material in the space between the sources and P is anything other than vacuum, the wavelength *in the material* must be used in Eq. (39–9). If the material has refractive index n, then

$$\lambda = \frac{\lambda_0}{n} \quad \text{and} \quad k = nk_0, \tag{39–10}$$

where λ_0 and k_0 are the values of λ and k, respectively, in vacuum.

To add the two sinusoidal functions with a phase difference, we can use the *phasor* representation that we have previously used for simple harmonic motion (Section 11–3) and for voltages and currents in ac circuits (Section 34–1). We suggest you review these sections so that phasors are fresh in mind. Each sinusoidal function is represented by a rotating vector (phasor) whose projection on the horizontal axis at any instant represents the instantaneous value of the sinusoidal function. In Fig. 39–5, E_1 is the phasor representing the wave from source S_1, and E_2 is the phasor for S_2. E_2 is *ahead* of E_1 in phase by an angle δ, as shown in the diagram. The amplitude of the resultant sinusoidal wave at P is the magnitude of the phasor labeled E_P in the diagram, the *vector sum* of E_1 and E_2. To find E_P, we use the law of cosines:

$$E_P{}^2 = E_1{}^2 + E_2{}^2 - 2E_1E_2 \cos (\pi - \delta).$$

If the amplitudes of the two waves are equal, then $E_1 = E_2 = E$, and

$$E_P{}^2 = E^2 + E^2 - 2E^2 \cos (\pi - \delta)$$

$$= 2E^2 + 2E^2 \cos \delta = 2E^2(1 + \cos \delta)$$

$$= 4E^2 \cos^2 (\delta/2). \tag{39–11}$$

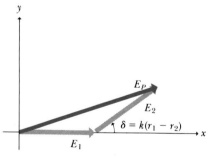

39–5 Phasor diagram of the variations of the electric field at a point P where two waves are superposed.

When the two waves are in phase, $\delta = 0$ and $E_p = 2E$. When they are exactly a half-cycle out of phase, $\delta = \pi$ ($180°$), $\delta/2 = \pi/2$, $\cos^2 (\delta/2) = 0$, and $E = 0$.

Thus the superposition of two sinusoidal waves with the same amplitude and frequency but in a phase difference yields a sinusoidal wave with amplitude between zero and twice the individual amplitudes, depending on the phase difference.

Next we relate δ to the geometry of the situation. The path difference is given by Eq. (39–3):

$$r_1 - r_2 = d \sin \theta.$$

Combining this with Eq. (39–9), we find

$$\delta = k(r_1 - r_2) = kd \sin \theta = \frac{2\pi d}{\lambda} \sin \theta. \qquad (39\text{–}12)$$

When we substitute this into Eq. (39–11), we find

$$E_P^2 = 4E^2 \cos^2 \left(\frac{1}{2} kd \sin \theta \right) = 4E^2 \cos^2 \left(\frac{\pi d}{\lambda} \sin \theta \right). \qquad (39\text{–}13)$$

Finally, the intensity I is given by Eq. (39–8):

$$I = \frac{1}{2} \epsilon_0 c E_P^2$$

$$= 2\epsilon_0 c E^2 \cos^2 \left(\frac{1}{2} kd \sin \theta \right)$$

$$= 2\epsilon_0 c E^2 \cos^2 \left(\frac{\pi d}{\lambda} \sin \theta \right). \qquad (39\text{–}14)$$

The intensity in a two-source interference pattern depends on the direction from the source.

We can simplify this result by expressing it in terms of the intensity I_0 at points where $\delta = 0$, such as $\theta = 0$. From Eq. (39–14),

$$I_0 = 2\epsilon_0 c E^2,$$

and finally

$$I = I_0 \cos^2 \left(\frac{1}{2} kd \sin \theta \right) \qquad (39\text{–}15)$$

$$= I_0 \cos^2 \left(\frac{\pi d}{\lambda} \sin \theta \right).$$

Note that at points of maximum intensity (maximum constructive interference), the intensity is *four times* (not twice) as great as it would be from either source alone. Of course, there are other points where it is *less* than the intensity from either source by itself.

In Section 39–2 we noted that in experiments with light we can describe the positions of the bright fringes with the coordinate y, as in Eq. (39–6), and that ordinarily $y \ll R$. In that case $\sin \theta$ is approximately equal to y/R, and we obtain the even simpler expressions

$$I = I_0 \cos^2 \left(\frac{kdy}{2R} \right) = I_0 \cos^2 \left(\frac{\pi dy}{\lambda R} \right). \qquad (39\text{–}16)$$

We can use these to calculate the intensity of light at any point and to compare the results with the photographically recorded pattern of Fig. 39–4.

EXAMPLE 39–3 In Fig. 39–1, suppose the two sources are identical radio antennas 10 m apart, radiating waves in all directions with a frequency $f = 30$ MHz. If the intensity in the $+x$-direction (corresponding to $\theta = 0$ in Fig. 39–2) is $I_0 = 0.02$ W·m^{-2}, what is the intensity in the direction $\theta = 45°$? In what direction is the intensity zero?

SOLUTION We want to use Eq. (39–15); the approximate relation of Eq. (39–16) cannot be used in this case because θ is not small. First we must find the wavelength, using the relation $c = \lambda f$. We find

$$\lambda = \frac{c}{f} = \frac{3.0 \times 10^8 \text{ m·s}^{-1}}{30 \times 10^6 \text{ s}^{-1}} = 10 \text{ m}.$$

The spacing between sources is $d = 10$ m, and Eq. (39–15) becomes

$$I = (0.02 \text{ W·m}^{-2}) \cos^2\left[\frac{\pi(10 \text{ m})}{(10 \text{ m})}\sin \theta\right]$$
$$= (0.02 \text{ W·m}^{-2}) \cos^2 (\pi \sin \theta).$$

When $\theta = 45°$,

$$I = (0.02 \text{ W·m}^{-2}) \cos^2 (\pi \sin 45°) = 0.0073 \text{ W·m}^{-2}.$$

This is about 37% of the intensity at $\theta = 0$.

The intensity is zero when $\cos (\pi \sin \theta) = 0$; this occurs when $\pi \sin \theta = \pi/2$, $\sin \theta = \frac{1}{2}$, and $\theta = 30°$. By symmetry, the intensity is also zero when $\theta = 150°, 210°$, and $330°$ (or $-150°, -30°$).

This is not just a hypothetical problem. It is often desirable to beam most of the radiated energy from a radio transmitter in particular directions rather than uniformly in all directions. Pairs or rows of antennas are often used to produce the desired radiation pattern.

39–4 INTERFERENCE IN THIN FILMS

The bright bands of color that you often see when light is reflected from a soap bubble or from a thin layer of oil floating on water are the result of an interference effect. Light waves are reflected from opposite surfaces of the thin films, and constructive interference between the two reflected waves occurs in different places for different wavelengths. The situation is represented schematically in Fig. 39–6. The line ab is a ray in a beam of light shining on the upper surface of a thin film. Part of the light is reflected at the first surface (ray bc), and part is transmitted (bd). At the second surface another partial reflection occurs, and some of the reflected light emerges (ray ef). The rays bc and ef come together at a point on the retina of the eye. Depending on the phase relationship, they may interfere constructively or destructively. Because different colors have different wavelengths, the interference may be constructive for some colors and destructive for others; hence the appearance of colored rings or fringes.

To keep things as simple as possible, let us consider interference of *monochromatic* light reflected from two nearly parallel surfaces. Figure 39–7 shows two plates of glass separated by a thin wedge of air; we want to consider interference between the two light waves reflected from the surfaces adjacent

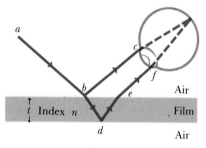

39–6 Interference between rays reflected from the upper and lower surfaces of a thin film.

to the air wedge, as shown. The situation is the same as in Fig. 39–6 except that the film thickness is not uniform. The refractions at the upper surface of the top plate in Fig. 39–7 do not change the situation in any essential way. If the observer is at a great distance compared to the other dimensions of the experiment and the rays are nearly perpendicular to the reflecting surfaces, then the path difference between the two waves (corresponding to $r_1 - r_2$ in the preceding section) is just twice the thickness d of the air wedge at each point. At points for which this path difference is an integer number of wavelengths, we expect to see constructive interference and a bright area, and where it is a half-integer number of wavelengths, destructive interference and a dark area. Along the line where the plates are in contact, there is *no* path difference and we expect a bright area.

It is easy enough to carry out this experiment; the bright and dark fringes appear as expected, but they are interchanged! Along the line of contact a *dark* fringe, not a bright one, is found. What happened? The inescapable conclusion is that one of the waves has undergone a half-cycle phase shift during its reflection, so that the two reflected waves are a half-cycle out of phase even though they have the same path length.

Further experiments show that a half-cycle phase change occurs whenever the material in which the wave is initially traveling before reflection has a *smaller* refractive index than the second material forming the interface. But when the first material has a *greater* refractive index than the second, such as a wave in glass reflected internally at a glass–air interface, *no* phase change occurs. Thus in Fig. 39–6 the wave reflected at point b undergoes the half-cycle phase shift, while the wave reflected at d does not. Similarly, in Fig. 39–7 the wave reflected from the upper surface of the lower plate has a half-cycle phase shift, while the other wave has none.

The path difference for the two waves is twice the thickness of the film.

39–7 Interference between two light waves reflected from the two sides of an air wedge separating two glass plates. The path difference is *2d*.

In some cases reflection causes an additional half-cycle phase shift.

EXAMPLE 39–4 Suppose the two glass plates in Fig. 39–7 are two microscope slides 10 cm long. At one end they are in contact, and at the other end they are separated by a thin piece of tissue paper 0.02 mm thick. What is the spacing of the resulting interference fringes? Is the fringe adjacent to the line of contact bright or dark? Assume monochromatic light with $\lambda = 500$ nm.

A simple example of interference from an air space between two glass plates

SOLUTION To answer the second question first, the fringe at the line of contact is dark because the wave reflected from the lower surface of the air wedge has undergone a half-cycle phase shift, while that from the upper surface has not. For this reason, the condition for *destructive* interference (a dark fringe) is that the path difference $(2d)$ be an integer number of wavelengths:

$$2d = m\lambda \qquad (m = 0, 1, 2, 3, \ldots). \qquad (39\text{–}17)$$

From similar triangles in Fig. 39–7, d is proportional to the distance x from the line of contact:

$$\frac{d}{x} = \frac{h}{l}.$$

Combining this with Eq. (39–17), we find

$$\frac{2xh}{l} = m\lambda,$$

or

$$x = m\frac{l\lambda}{2h} = m\frac{(0.1\text{ m})(500 \times 10^{-9}\text{ m})}{(2)(0.02 \times 10^{-3}\text{ m})} = m(1.25\text{ mm}).$$

If the film has an index of refraction different from unity, the wavelength in the material is less than in vacuum.

Thus successive dark fringes, corresponding to successive integer values of m, are spaced 1.25 mm apart.

In this example, if the space between plates contains water ($n = 1.33$) instead of air, the phase changes are the same but the wavelength is $\lambda = \lambda_0/n = 376$ nm, and the fringe spacing is 0.94 mm. But now suppose the top plate is a glass with $n = 1.4$, the wedge is filled with a silicone grease having $n = 1.5$, and the bottom plate has $n = 1.6$. In this case there are half-cycle phase shifts at *both* surfaces bounding the wedge, and the line of contact corresponds to a *bright* fringe, not a dark one. The fringe spacing is again obtained by using the wavelength in the wedge (i.e., in the silicone grease), $\lambda = 500\text{ nm}/1.5 = 333$ nm; we invite you to show that the fringe spacing is 0.83 mm.

39–8 Air film between a convex and a plane surface.

Colored interference fringes formed by reflection of white light.

If we illuminate the arrangement in Fig. 39–7 first with blue, then with red light, the spacing of the red fringes is greater than that of the blue, as we should expect from the greater wavelength of the red light. The fringes produced by intermediate wavelengths occupy intermediate positions. If it is illuminated by white light, the color at any point is that due to the mixture of those colors that may be reflected at that point, while the colors for which the thickness is such as to result in destructive interference are absent. Just those colors that are absent in the reflected light, however, are found to predominate in the transmitted light. At any point, the color of the wedge as viewed by reflected light is *complementary* to its color as seen by transmitted light! Roughly speaking, the complement to any color is obtained by removing that color from white light. For example, the complement of blue is yellow, and the complement of green is magenta.

If the convex surface of a lens is placed in contact with a plane glass plate, as in Fig. 39–8, a thin film of air is formed between the two surfaces. The resulting circular interference fringes are shown in Fig. 39–9; they were studied by Newton and are called *Newton's rings*. When viewed by reflected light, the center of the pattern is black. Can you use the discussion of this section to show why this should be expected?

The surface of an optical part that is being ground to some desired curvature may be compared with that of another surface, known to be correct, by bringing the two in contact and observing the interference fringes. Figure 39–10 is a photograph made during the grinding of a telescope objective lens. The larger-diameter, thicker lower disk is the master, and the smaller upper disk is the lens under test. The "contour lines" are Newton's interference fringes; each one indicates an additional departure of the specimen from the master of $\frac{1}{2}$ wavelength of light. At ten lines from the center spot the distance between the two surfaces is five wavelengths, or about 0.0001 inch. This specimen is very poor; high-quality lenses are routinely ground with a precision of less than one wavelength!

Thin-film interference occurs in nonreflective coatings for lenses. A thin layer or film of hard transparent material with an index of refraction smaller

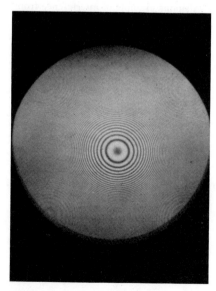

39–9 Newton's rings formed by interference in the air film between a convex and a plane surface. (Courtesy of Bausch & Lomb Optical Co.)

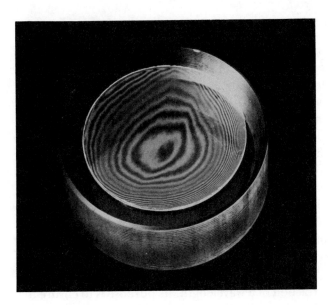

39–10 The surface of a telescope objective under inspection during manufacture. (Courtesy of Bausch & Lomb Optical Co.)

than that of the glass is deposited on the surface of the glass, as in Fig. 39–11. If the coating has the proper index of refraction, equal quantities of light will be reflected from its outer surface and from the boundary surface between it and the glass. Furthermore, since in both reflections the light is reflected from a medium of greater index than that in which it is traveling, the same phase change occurs in each reflection. It follows that if the film thickness is $\frac{1}{4}$ wavelength *in the film* (normal incidence is assumed), the light reflected from the first surface will be 180° out of phase with that reflected from the second, and complete destructive interference will result.

Of course, the thickness can be $\frac{1}{4}$ wavelength for only one particular wavelength. This is usually chosen in the yellow-green portion of the spectrum (about 550 mm), where the eye is most sensitive. Some reflection then takes place at both longer and shorter wavelengths, and the reflected light has a purple hue. The overall reflection from a lens or prism surface can be reduced in this way from 4% to 5% to a fraction of 1%. The treatment is extremely effective in eliminating stray reflected light and increasing the contrast in an image formed by highly corrected lenses having a large number of air–glass surfaces. A commonly used coating material is magnesium fluoride, MgF_2, with an index of 1.38. With this coating, the wavelength of green light in the coating is

$$\lambda = \frac{\lambda_0}{n} = \frac{550 \times 10^{-9} \text{ m}}{1.38} = 4 \times 10^{-5} \text{ cm},$$

and the thickness of a "nonreflecting" film of MgF_2 is 10^{-5} cm.

If a material whose index of refraction is *greater* than that of glass is deposited on glass to a thickness of $\frac{1}{4}$ wavelength, then the reflectivity is *increased*. For example, a coating of index 2.5 will allow 38% of the incident energy to be reflected, instead of the usual 4% when there is no coating. By use of multilayer coatings, it is possible to achieve reflectivity for a particular wavelength of almost 100%. These coatings are used for "one-way" windows and reflective sunglasses.

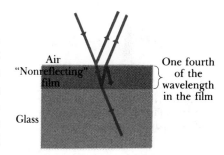

39–11 Destructive interference results when the film thickness is one-quarter of the wavelength in the film.

A film can't be perfectly nonreflecting for all wavelengths; some compromises are necessary.

39–5 THE MICHELSON INTERFEROMETER

The **Michelson interferometer** played an interesting role in the history of science during the latter part of the nineteenth century and has had an equally important role in establishing high-precision standards of the unit of length. In contrast to the Young two-slit experiment, which uses light from two very narrow sources, the Michelson interferometer uses light from a broad, spread-out source. Figure 39–12 is a diagram of its principal features. The figure shows the path of one ray from a point A of an extended but monochromatic source. This ray strikes a glass plate C, the right side of which has a thin coating of silver. Part of the light is reflected from the silvered surface at point P to the mirror M_2 and back through C to the observer's eye. The remainder of the light passes through the silvered surface and the compensator plate D and is reflected from mirror M_1. It then returns through D and is reflected from the silvered surface of C to the observer. The compensator plate D is cut from the same piece of glass as plate C, so that its thickness will not differ from that of C by more than a fraction of a wavelength. Its purpose is to ensure that rays 1 and 2 pass through the same thickness of glass. Plate C is called a *beam splitter*.

The whole apparatus is mounted on a very rigid frame, and a fine, very accurate screw thread is used to move the mirror M_2. A common commercial model of the interferometer is shown in Fig. 39–13. The light source is placed to the left, and the observer is directly in front of the handle that turns the screw.

If the distances L_1 and L_2 in Fig. 39–12 are exactly equal, and the mirrors M_1 and M_2 are exactly at right angles, the virtual image of M_1 formed by reflection at the silvered surface of plate C coincides with mirror M_2. If L_1 and L_2 are not exactly equal, the image of M_1 is displaced slightly from M_2; and if the angle between the mirrors is not exactly a right angle, the image of M_1 makes a slight angle with M_2. Then the mirror M_2 and the virtual image of M_1 play the same roles as the two surfaces of a thin film, discussed in Section 39–3, and the same sort of interference fringes result from the light reflected from these surfaces.

Suppose that the extended source in Fig. 39–12 is monochromatic with wavelength λ, and that the angle between mirror M_2 and the virtual image of

39–12 The Michelson interferometer.

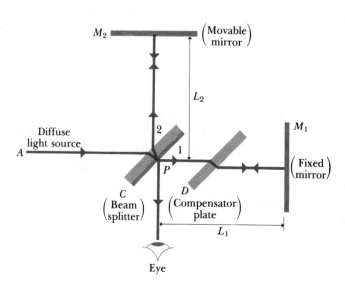

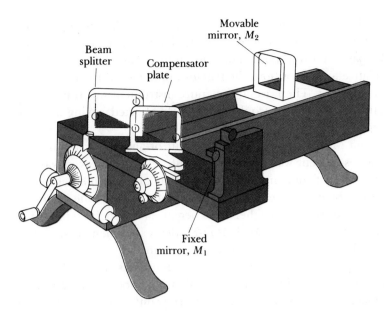

39–13 A common type of Michelson interferometer.

M_1 is such that five or six vertical fringes are present in the field of view. If the mirror M_2 is now moved slowly either backward or forward by a distance $\lambda/2$, the effective film thickness changes by λ and each of the fringes moves either to the right or to the left through a distance equal to the spacing of the fringes. If the fringes are observed through a telescope whose eyepiece is equipped with a cross hair, and m fringes cross the cross hair when the mirror is moved a distance x, then

$$x = m\frac{\lambda}{2} \quad \text{or} \quad \lambda = \frac{2x}{m}. \tag{39–18}$$

If m is several thousand, the distance x is large enough so that it can be measured with good precision, and hence a precise value of the wavelength λ can be obtained.

Until recently the meter was defined as a length equal to a specified number of wavelengths of the orange-red light of krypton-86. Before this standard could be established, it was necessary to measure as accurately as possible the number of these wavelengths in the *former* standard meter, defined as the distance between two scratches on a bar of platinum-iridium. The measurement was made with a modified Michelson interferometer many times and under very carefully controlled conditions. The number of wavelengths in a distance equal to the old standard meter was found to be 1,650,763.73 wavelengths. The meter was then defined as *exactly* this number of wavelengths. As we mentioned in Section 1–2, this definition has recently been superseded by a new length standard based on the unit of *time*.

Another application of the Michelson interferometer with considerable historical interest is the Michelson-Morley experiment. To understand the purpose of this experiment, we must recall that before the electromagnetic theory of light and Einstein's special theory of relativity became established, physicists believed that the propagation of light waves occurred in a medium called the **ether,** which was believed to permeate all space. In 1887 Michelson and Morley used the Michelson interferometer in an attempt to detect the motion of the earth through the ether. Suppose the interferometer in Fig.

The role of interferometry in establishing a standard of length

The Michelson-Morley experiment: a death-blow to the ether theory of light

39–12 were moving from left to right relative to the ether. According to nineteenth-century theory, this would lead to changes in the speed of light in the portions of the path shown as horizontal lines in the figure. There would be fringe shifts relative to the positions the fringes would have if the instrument were at rest in the ether. Then, when the entire instrument was rotated 90°, the other portions of the paths would be similarly affected, giving a fringe shift in the opposite direction.

Michelson and Morley expected a fringe shift of about four-tenths of a fringe when the instrument was rotated. The shift actually observed was less than a hundredth of a fringe and, within the limits of experimental uncertainty, appeared to be exactly zero. Despite its orbital velocity, the earth appeared to be at rest relative to the ether. This negative result baffled physicists of the time, and to this day the Michelson-Morley experiment is the most significant "negative-result" experiment ever performed.

Understanding of this result had to wait for Einstein's special theory of relativity, published in 1905. Einstein realized that the velocity of a light wave has the same magnitude c relative to *all* reference frames, whatever their velocity may be relative to each other. The presumed ether then plays no role, and the very concept of an ether has been given up. The theory of relativity is a well-established cornerstone of modern physics, and we will study it in detail in Chapter 40.

Evidence for the correctness of the special theory of relativity

39–6 FRESNEL DIFFRACTION

According to *geometrical* optics, if an opaque object is placed between a point light source and a screen, as in Fig. 39–14, the edges of the object cast a sharp shadow on the screen. No light reaches the screen at points within the geometrical shadow, while outside the shadow the screen is uniformly illuminated. But as we have seen, geometrical optics is an idealized model of the behavior of light, and there are many situations where the ray model of light is inadequate. Another important class of such phenomena is grouped under the heading **diffraction.**

The photograph in Fig. 39–15 was made by placing a razor blade halfway between a pinhole illuminated by monochromatic light and a photographic film, so that the film made a record of the shadow cast by the blade. Figure 39–16 is an enlargement of a region near the shadow of an edge of the blade. The boundary of the *geometrical* shadow is indicated by the short arrows. The geometrical shadow is bordered by alternating bright and dark bands. Some light has "bent" around the edge into the shadow region, although this is not visible in the figure. Note also that the first bright band, just outside the geometrical shadow, is actually *brighter* than in the region of uniform illumination to the extreme left. This simple experimental setup serves to give some idea of the true complexity of what is often considered the most elementary of optical phenomena, the shadow cast by a small source of light.

Diffraction patterns such as that in Fig. 39–15 are not commonly observed in everyday life because most ordinary light sources are not point sources of monochromatic light. If the shadow of a razor blade is cast by a frosted-bulb incandescent lamp, for example, the light from every point of the surface of the lamp forms its own diffraction pattern, but these overlap to such an extent that no individual pattern can be observed.

Diffraction phenomena: When is a shadow not really a shadow?

39–14 Geometrical shadow of a straight edge.

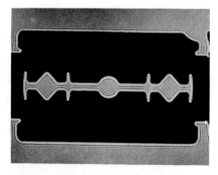

39–15 Shadow of a razor blade.

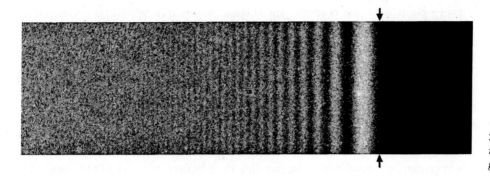

39–16 Shadow of a straight edge. The arrows show the position of the *geometric* shadow.

The term *diffraction* is applied to problems involving *the resultant effect produced by a limited portion of a wave front.* Since in most diffraction problems some light is found within the region of geometrical shadow, diffraction is sometimes defined as "the bending of light around an obstacle." It should be emphasized, however, that the process by which diffraction effects are produced is going on continuously in the propagation of *every* wave. Only if part of the wave is cut off by some obstacle do we observe diffraction effects. But every optical instrument uses only a limited portion of a wave; for example, a telescope uses only that portion of a wave admitted by the objective lens. Thus diffraction plays a role in practically all optical phenomena.

Figure 39–17 shows a diffraction pattern formed by a steel ball about 3 mm in diameter. Note the rings in the pattern, both outside and inside the geometrical shadow area, and the bright spot at the very center of the shadow. The existence of this spot was predicted in 1818 by the French mathematician Poisson. Ironically, Poisson himself was not a believer in the wave theory of light, and he published this *apparently* absurd result as a final ridicule of the wave theory. But almost immediately, the bright spot was observed experimentally by Arago. It had, in fact, been observed much earlier, in 1723 by Maraldi, but its significance was not recognized then.

Do you believe that the shadow of a circular object can have a bright spot at its center?

The essential features observed in diffraction effects can be predicted with the help of Huygens' principle: Every point of a wave surface can be considered the source of a secondary wavelet that spreads out in all directions. At

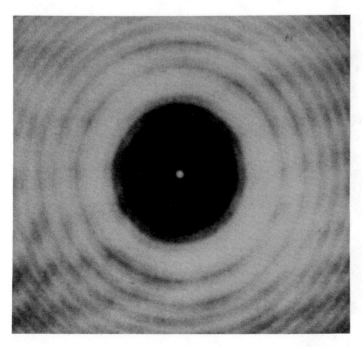

39–17 Fresnel diffraction pattern formed by a steel ball 3 mm in diameter. The Poisson bright spot is seen at the center of the shadow area. (Courtesy Prof. E. Hecht.)

every point we must combine the displacements that would be produced by the secondary wavelets, taking into account their amplitudes and relative phases. The mathematical operations are often quite complicated.

In Fig. 39–14 both the point source and the screen are at finite distances from the obstacle forming the diffraction pattern. This situation is described as **Fresnel diffraction** (after Augustin Jean Fresnel, 1788–1827), and the resulting pattern on the screen is called a *Fresnel diffraction pattern.* If the source, obstacle, and screen are far enough away so that the lines from the source to the obstacle and from the obstacle to a point in the pattern formed on the screen can be considered to be parallel, the phenomenon is called **Fraunhofer diffraction** (after Joseph von Fraunhofer, 1787–1826). The latter situation is simpler to treat in detail, and our analysis will be restricted to Fraunhofer diffraction.

39–7 FRAUNHOFER DIFFRACTION FROM A SINGLE SLIT

Suppose a monochromatic plane wave (in the ray picture, a beam of parallel rays) is incident on an opaque plate having a narrow slit. According to geometrical optics, the transmitted beam should have the same cross section as the slit, and a screen in the path of the beam would be illuminated uniformly over an area of the same size and shape as the slit, as in Fig. 39–18a. What is *actually* observed is the pattern shown in Fig. 39–18b. The beam spreads out horizontally after passing through the slit, and the diffraction pattern consists of a central bright band, which may be much wider than the slit width, bordered by alternating dark bands and bright bands of decreasing intensity. You can easily observe a diffraction pattern of this sort by looking at a point source such as a distant street light through a narrow slit formed between two fingers in front of your eye. The retina of your eye then corresponds to the screen.

Figure 39–19 shows a side view of the setup of Fig. 39–18, with the slit turned to a horizontal position. That is, in Fig. 39–19 we are looking along the

39–18 (a) Geometrical "shadow" of a slit. (b) Diffraction pattern of a slit. The slit width has been greatly exaggerated.

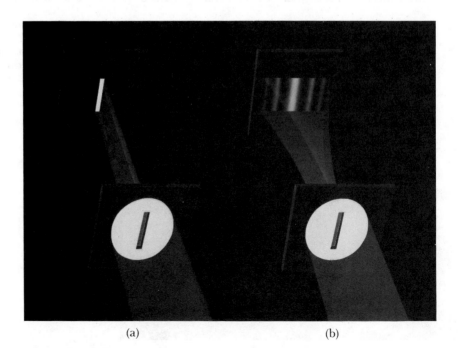

(a)   (b)

length of the slit. According to Huygens' principle, each element of area of the slit opening can be considered a source of secondary wavelets. In particular, we may imagine elements of area formed by subdividing the slit into narrow strips parallel to the long edges, perpendicular to the page in Fig. 39–19a. From each strip, cylindrical secondary wavelets spread out, as shown in cross section.

In Fig. 39–19b a screen is placed to the right of the slit. The resultant intensity at a point P is calculated by adding the contributions from the individual wavelets, taking proper account of their various phases and amplitudes. The problem becomes much simpler if the screen is far away, so the rays from the slit to the screen are parallel, as in Fig. 39–19c, or when a lens is placed as in Fig. 39–19d, in which case the rays to the lens are parallel and the lens forms in its focal plane a reduced *image* of the pattern that would be formed on an infinitely distant screen without the lens. You might ask whether the lens introduces additional phase shifts that are different for different parts of the wave front; it can be shown quite generally that the lens *does not* cause any additional phase shifts.

The situation of Fig. 39–19b is Fresnel diffraction; those in Figs. 39–19c and 39–19d, where the outgoing rays can be considered parallel, are Fraunhofer diffraction. Some aspects of Fraunhofer diffraction from a single slit can be deduced easily. First consider two narrow strips, one just below the top edge of the slit and one at its center, as in Fig. 39–20. The difference in path length to point P is $(a/2) \sin \theta$, where a is the slit width. Suppose this path difference happens to be equal to $\lambda/2$; then light from these two strips arrives at point P with a half-cycle phase difference, and cancellation occurs. Similarly, light from two strips just below these two will also arrive a half-cycle out of phase; and, in fact, light from *every* strip in the top half cancels out that from a corresponding strip in the bottom half, resulting in complete cancellation and giving a dark fringe in the interference pattern. Thus a dark fringe occurs whenever

$$\frac{a}{2} \sin \theta = \pm \frac{\lambda}{2} \quad \text{or} \quad \sin \theta = \pm \frac{\lambda}{a}. \tag{39–19}$$

We may also divide the screen into quarters, sixths, and so on, and use the argument above to show that a dark fringe occurs whenever $\sin \theta = \pm 2\lambda/a$,

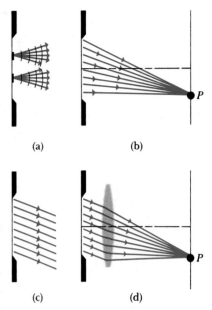

(a)   **(b)**

(c)   **(d)**

39–19 Diffraction by a single slit. Relation of Fresnel diffraction to Fraunhofer diffraction by a single slit.

Using the superposition principle to find the dark areas in the diffraction pattern

39–20 The wavefront is divided into a large number of narrow strips.

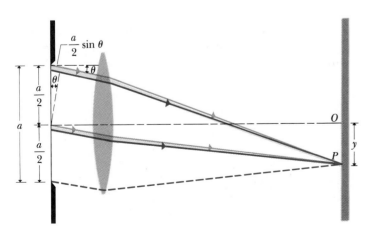

$\pm 3\lambda/a$, and so on. Hence the condition for a *dark* fringe is

$$\sin \theta = \frac{n\lambda}{a} \qquad (n = \pm 1, \pm 2, \pm 3, \ldots). \qquad (39\text{–}20)$$

An important difference between the single-slit pattern and the two-slit interference pattern

For example, if the slit width is equal to ten wavelengths, dark fringes occur at $\sin \theta = \pm \frac{1}{10}, \pm \frac{2}{10}, \pm \frac{3}{10}, \ldots$. Between the dark fringes are bright fringes. We also note that $\sin \theta = 0$ is a *bright* band, since then light from the entire slit arrives at P in phase. The central bright fringe is therefore twice as wide as the others, as Fig. 39–18 shows.

We can derive an expression for the intensity distribution for the single-slit pattern by the same method we used to obtain Eq. (39–15) for the two-slit pattern. We again imagine a plane wavefront at the slit subdivided into a large number of strips, each of which sends out rays in *all* directions toward the lens, as shown in Fig. 39–20. If we choose an arbitrary point P on a screen in the focal plane of the lens, only those rays making an angle θ with the axis will arrive at P. Point O on the screen is the special point at which all rays making the angle $\theta = 0$ arrive. Figure 39–21a is a phasor diagram for the situation, showing that when the slit is subdivided into 14 sections and each section emits a Huygens wavelet at the angle $\theta = 0$, all wavelets arrive in phase. The resultant amplitude at O is denoted by S.

Using phasor addition to find the intensity distribution in a single-slit diffraction pattern

With the same subdivision of the wavefront into 14 strips, the wavelets that make the angle θ and arrive at P have a slight phase difference between succeeding wavelets, and the corresponding phasor diagram is shown in Fig. 39–21b. The sum S is now the perimeter of a portion of a many-sided polygon and E_P, the amplitude of the resultant electric field at P, is the *chord*. The angle δ is the total phase difference between the wave from the bottom strip of Fig. 39–20 and the wave from the top strip.

In the limit, as the number of strips into which the slit is subdivided is increased indefinitely, the phasor diagram becomes an *arc of a circle*, as shown in Fig. 39–21c, with the arc length S equal to the length S in Fig. 39–21a. We can find the center C of this arc by constructing perpendiculars at A and B. From the definition of radian measure of angles, the radius of the arc is S/δ, and the resultant amplitude E_P (distance AB) is $2(S/\delta) \sin (\delta/2)$. We then have

$$E_P = S\frac{\sin \delta/2}{\delta/2},$$

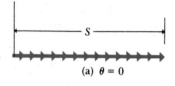

(a) $\theta = 0$

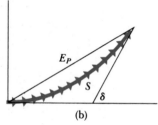

(b)

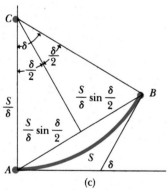

(c)

39–21 (a) Phasor diagram when all elementary electric fields are in phase ($\theta = 0$, $\delta = 0$). (b) Phasor diagram when each elementary electric field differs in phase slightly from the preceding one. (c) Limit reached by the phasor diagram when the slit is subdivided infinitely.

where δ is the phase difference between the two wavelets at the extreme top and bottom edges of the slit.

From Eq. (39–9), the phase difference is $2\pi/\lambda$ times the path difference. From Fig. 39–20, the path difference between the ray from the top of the slit and that from the bottom is $a \sin \theta$. Therefore

$$\delta = \frac{2\pi}{\lambda} a \sin \theta, \qquad (39\text{–}21)$$

and

$$E_P = S\frac{\sin[\pi a(\sin \theta)/\lambda]}{\pi a(\sin \theta)/\lambda}.$$

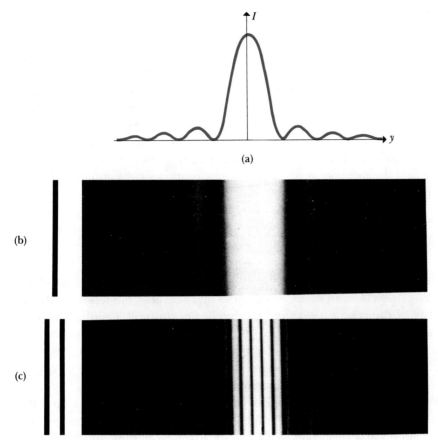

39–22 (a) Intensity distribution. (b) Photograph of the Fraunhofer diffraction pattern of a single slit. (c) Fraunhofer diffraction pattern of a double slit.

Since the intensity I is proportional to the *square* of the amplitude,

$$I = I_0\left(\frac{\sin\delta/2}{\delta/2}\right)^2 = I_0\left\{\frac{\sin[\pi a(\sin\theta)/\lambda]}{\pi a(\sin\theta)/\lambda}\right\}^2, \qquad (39\text{--}22)$$

where I_0 is the intensity at O in Fig. 39–20 where $\theta = 0$.

Equation (39–22) is plotted in Fig. 39–22a, and a photograph of an actual diffraction pattern is shown directly under it in Fig. 39–22b. Note that the intensity of the central maximum is much greater than at any of the others, and that the peak intensities drop off rapidly as we go away from the center of the pattern.

We can calculate the positions of the peaks and the peak intensities from Eq. (39–22). The central peak occurs when $\delta = 0$. Substituting this value into Eq. (39–22) gives an indeterminate form $(0/0)$, but by taking the limit as $\delta \to 0$, we find that when $\delta = 0$, $I = I_0$. The side maxima occur approximately where $\sin(\delta/2) = \pm 1$, or

$$\frac{\delta}{2} = \frac{3\pi}{2}, \frac{5\pi}{2}, \cdots,$$

or in general

$$\frac{\delta}{2} = \left(m + \frac{1}{2}\right)\pi \qquad (m = 1, 2, 3, \ldots), \qquad (39\text{--}23)$$

or, using Eq. (39–21),

$$\sin \theta = \left(m + \frac{1}{2}\right)\frac{\lambda}{a}. \tag{39–24}$$

Note that these maximum intensities occur between the minima that we found earlier, as given by Eq. (39–20). Note also that there is *no* side maximum at $\delta/2 = \pi/2$. The function of δ in Eq. (39–22) does not have a maximum at this value of δ but drops off steadily from $\delta = 0$ to $\delta = 2\pi$. That is, the values $\delta = \pm\pi$ actually lie within the central maximum.

The intensities of the side maxima are much less than for the central maximum.

To find the intensity at the *m*th side maximum, we substitute Eq. (39–23) back into Eq. (39–22) to obtain

$$I = \frac{I_0}{(m + \frac{1}{2})^2\pi^2}. \tag{39–25}$$

Putting in $m = 1$, $m = 2$, . . . , we find the series of intensities $0.0450I_0$, $0.0162I_0$, $0.0083I_0$, and so on. So even the first side peak has less than 5% the intensity of the central peak, and the intensities drop off rapidly. This is in contrast to the two-slit pattern we studied in Section 39–3, where the side maxima were approximately as intense as the central maximum. Also, the central maximum in the single-slit pattern is twice as wide as the others, an effect not seen in the two-slit pattern.

With light, the wavelength λ is ordinarily much smaller than the slit width a, and as a result the values of θ in Eq. (39–22) are so small that the approximation $\sin \theta = \theta$ is very good. With this approximation, the position θ_1 of the first minimum beside the central maximum, corresponding to $\delta/2 = \pi$, is, from Eq. (39–21),

$$\theta_1 = \frac{\lambda}{a}. \tag{39–26}$$

When a is of the order of a centimeter or more, θ_1 is so small that we can consider practically all the light to be concentrated at the geometrical focus.

Superposition of interference and diffraction patterns resulting from two slits of finite width

The photograph in Fig. 39–22c shows the Fraunhofer diffraction pattern of *two* slits, each with the same width a as for Fig. 39–22b but separated by a distance $d = 4a$. The two-slit interference pattern, represented by the narrow fringes, is modified by the diffraction curve of part (a) because of the finite widths of the slits. Thus the first minimum in the single-slit pattern in Fig. 39–21c completely extinguishes the fourth bright fringe of the two-slit pattern.

39–8 THE DIFFRACTION GRATING

The diffraction grating: many equally spaced sources

Suppose that instead of a single slit or two slits side by side, we have a very large number of parallel slits, all with the same width and spaced equal distances apart. Such an arrangement is called a **diffraction grating;** the first one was constructed by Fraunhofer with fine wires. Gratings are now made by using a diamond point to scratch a large number of equally spaced grooves on a glass or metal surface, or by photographic reduction of a black-and-white pattern drawn with a pen.

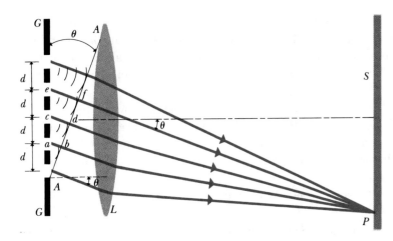

39–23 The plane diffraction grating.

In Fig. 39–23, GG represents the grating; the slits are perpendicular to the plane of the page. Only five slits are shown in the diagram, but an actual grating may contain several thousand, with spacing d of the order of 0.002 mm. A plane monochromatic wave is incident normally on the grating from the left side. The problem of finding the intensity pattern in the light transmitted by the grating then combines the principles of interference and diffraction. Each slit creates a diffraction pattern, and these then interfere with each other to produce the final pattern. The lens is included so we can view the pattern on a screen at a finite distance from the grating and still meet the conditions for Fraunhofer diffraction, that is, parallel rays emerging from the grating.

Let us assume that the slits are very narrow, so the diffracted beam from each spreads out over a wide enough angle for it to interfere with all the other diffracted beams. Consider first the light proceeding from elements of infinitesimal width at the lower edges of each opening and traveling in a direction making an angle θ with that of the incident beam, as in Fig. 39–23. A lens at the right of the grating forms in its focal plane a diffraction pattern similar to that which would appear on a screen at infinity.

Suppose the angle θ in Fig. 39–23 is taken so that the distance ab equals λ, the wavelength of the incident light. Then $cd = 2\lambda$, $ef = 3\lambda$, and so on. The waves from all these elements, since they are in phase at the plane of the grating, are also in phase along the plane AA and therefore reach the point P in phase. The same holds true for any set of elements in corresponding positions in the various slits.

If the angle θ is increased slightly, the waves from the various grating slits no longer arrive at AA in phase, and even an extremely small change in angle results in almost complete destructive interference among them, provided there are a large number of slits in the grating. Roughly speaking, for each slit in the grating we can find another slit whose wave arrives a half-cycle out of phase with the first one and cancels it. Hence the maximum at the angle θ is an extremely sharp one, differing from the rather broad maxima that result from interference or diffraction effects with a small number of openings.

As the angle θ is increased still further, a position is eventually reached in which the distance ab in Fig. 39–23 becomes equal to 2λ. Then cd equals 4λ, cf equals 6λ, and so on. The waves at AA are again all in phase; the path differ-

A crucial difference between the grating pattern and a two-slit pattern

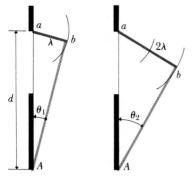

39-24 First-order maximum when $ab = \lambda$; second-order maximum when $ab = 2\lambda$.

Use of diffraction gratings in spectrometry

ence between adjacent waves is now 2λ, and another maximum results. Still others appear when $ab = 3\lambda$, 4λ, ... We also find maxima at corresponding angles on the opposite side of the grating normal, as well as along the normal itself, since in the latter position the phase difference between waves reaching AA is zero.

From Fig. 39–24 we can find the angles of deviation at which the maxima occur. Consider the right triangle Aba. Let d be the distance between successive grating elements, called the *grating spacing*. The necessary condition for a maximum is that $ab = m\lambda$, where $m = 0$, ± 1, ± 2, ± 3, and so on. It follows that

$$\sin \theta = m\frac{\lambda}{d} \qquad (m = 0, \pm 1, \pm 2, \ldots) \qquad (39\text{-}27)$$

is the necessary condition for a maximum. The angle θ is also the angle by which the rays corresponding to the maxima have been *deviated* from the direction of the incident light.

In practice, the parallel beam of rays incident on the grating is usually produced by a lens with a narrow illuminated slit at its first focal point. These are not shown in Fig. 39–23. Each of the maxima in the diffraction pattern formed by the grating is then a sharp image of the slit. If the slit is illuminated by light consisting of a mixture of two or more wavelengths, the grating forms two or more series of images of the slit in different positions; each wavelength in the original light gives rise to a set of slit images deviated by the appropriate angles. If the slit is illuminated with white light, the diffraction grating forms a continuous group of images side by side. That is, white light is dispersed into continuous spectra. In contrast with the single spectrum produced by a prism, a grating forms several spectra on either side of the normal. Those that correspond to $m = \pm 1$ in Eq. (39–27) are called *first order;* those that correspond to $m = \pm 2$ are called *second order;* and so on. Since for $m = 0$ the deviation is zero, all colors combine to produce a white image of the slit in the direction of the incident beam.

As Eq. (39–27) shows, the sines of the deviation angles of the maxima are proportional to the ratio λ/d. Thus for substantial deviation to occur, the grating spacing d should be of the same order of magnitude as the wavelength λ. Gratings for use in or near the visible spectrum are ruled with from about 500 to 1500 lines per millimeter.

The diffraction grating is widely used in spectrometry, instead of a prism, as a means of dispersing a light beam into spectra. If the grating spacing is known, then from a measurement of the angle of deviation of any wavelength, the value of this wavelength may be computed. This is not true for a prism; the angles of deviation are not related in any simple way to the wavelengths but depend on the characteristics of the prism material. Since the index of refraction of optical glass varies more rapidly at the violet end than at the red end of the spectrum, the spectrum formed by a prism is always spread out more at the violet end than it is at the red. Also, while a prism deviates red light the least and violet the most, the reverse is true of a grating.

Two examples of diffraction-grating calculations

EXAMPLE 39–5 The wavelengths of the visible spectrum are approximately 400 nm to 700 nm. Find the angular breadth of the first-order visible spectrum produced by a plane grating having 6000 lines per centimeter, when white light is incident normally on the grating.

SOLUTION The grating spacing d is

$$d = \frac{1}{6000 \text{ lines·cm}^{-1}} = 1.67 \times 10^{-6} \text{ m}.$$

The angular deviation of the violet is

$$\sin \theta = \frac{400 \times 10^{-9} \text{ m}}{1.67 \times 10^{-6} \text{ m}} = 0.240,$$

$$\theta = 13.9°.$$

The angular deviation of the red is

$$\sin \theta = \frac{700 \times 10^{-9} \text{ m}}{1.67 \times 10^{-6} \text{ m}} = 0.420,$$

$$\theta = 24.8°.$$

Hence the first-order visible spectrum includes an angle of

$$24.8° - 13.9° = 10.9°.$$

EXAMPLE 39–6 Show that the violet of the third-order spectrum overlaps the red of the second-order spectrum.

SOLUTION The angular deviation of the third-order violet is

$$\sin \theta = \frac{(3)(400 \times 10^{-9} \text{ m})}{d}$$

and of the second-order red it is

$$\sin \theta = \frac{(2)(700 \times 10^{-9} \text{ m})}{d}.$$

Since the first angle is smaller than the second, whatever the grating spacing, the third order will *always* overlap the second.

PROBLEM-SOLVING STRATEGY: *Diffraction*

Here are several general comments about the kinds of problems you will encounter in this chapter. You may want to refer back to earlier sections and look for applications of these ideas.

1. Always draw a diagram; label all the distances and angles clearly. Several of the equations derived in this chapter contain an angle θ. Make absolutely sure you know where θ is on your diagram.

2. The equations for an intensity *maximum* in the two-source or grating pattern, Eqs. (39–4) and (39–27), look just like the equation for an intensity *minimum* in the single-slit pattern, Eq. (39–20). Be careful not to get these confused, and be sure you understand why the same equation gives intensity maxima in some situations and minima in others.

3. In x-ray diffraction, discussed in the next section, the angle θ is defined differently from the angles of incidence and reflection in geometrical optics. In this case, θ is the angle of the incoming beam relative to the crystal planes, not their normals. Be careful! Also notice that there are *two* conditions for an intensity maximum; the incident and scattering angles are equal, and the Bragg condition, Eq. (39–28), must be satisfied.

4. One more time we caution you about units. Light wavelengths are usually in nanometers (1 nm = 10^{-9} m), but if you refer to other books you may also find Ångstrom units; 1 Å = 10^{-10} m = $\frac{1}{10}$ nm. Dimensions of slits and apertures may be expressed in millimeters, and lens-to-screen distances in centimeters or meters. Lens focal lengths are traditionally expressed in millimeters. In coping with all this, be very careful not to drop any powers of ten!

39–9 X-RAY DIFFRACTION

X-rays have much shorter wavelengths than visible light.

X-rays were discovered by Röntgen in 1895, and early experiments suggested that they were electromagnetic waves with wavelengths of the order of 10^{-10} m. It was also strongly suspected at this time that in a crystalline solid the atoms are arranged in a lattice in a regular repeating pattern, with spacing between adjacent atoms also of the order of 10^{-10} m. Putting these two ideas together, Max von Laue (1879–1960) suggested in 1913 that a crystal might serve as a kind of three-dimensional diffraction grating for x-rays. That is, a beam of x-rays might be scattered by the individual atoms in a crystal, and the scattered waves might interfere just as the individual waves from a diffraction grating interfere.

Using a crystal as a diffraction grating for x-rays.

The first **x-ray diffraction** experiments were performed by Friederich and Knipping, and interference effects *were* observed. These experiments thus verified in a single stroke the hypothesis that x-rays *are* waves, or at least have wavelike properties, and that the atoms in a crystal *are* arranged in a regular pattern. Since that time, the phenomenon of x-ray diffraction by a crystal has proved an invaluable research tool, both as a way to measure x-ray wavelengths and as a method of studying the structure of crystals. Figure 39–25 is a diagram of the structure of a familiar crystal, sodium chloride.

To introduce the basic idea in a simple context, we consider first a two-dimensional scattering situation, as shown in Fig. 39–26a, where a plane wave is incident on a square array of scattering centers. The situation might be a ripple tank with an array of small posts, or 3-cm microwaves with an array of small conducting spheres, or x-rays with an array of atoms. In the case of electromagnetic waves, the wave induces an oscillating electric dipole moment in each scatterer, and each one emits a scattered wave. The total interference pattern is the superposition of all these scattered waves. To compute its nature we have to consider the total path differences for the various scattered waves, including the distances both from source to scatterer and from scatterer to observer.

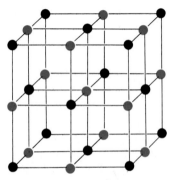

39–25 Model of arrangement of ions in a crystal of NaCl. Black circles, Na; color circles, Cl. The spacing of adjacent atom is 0.282 nm.

A crystal is analogous to a three-dimensional diffraction grating for x-rays.

As Fig. 39–26a shows, the path length from source to observer is the same for all the scatterers in a single row if the two angles θ are equal, as shown. Scattered radiation from adjacent rows is *also* in phase if the path difference is an integer number of wavelengths. Figure 39–26b shows that the path difference for adjacent rows is $2d \sin \theta$. Thus the conditions for radiation from the *entire array* to reach the observer in phase are that (1) the angle of incidence must equal the angle of scattering, and (2) the path difference for adjacent

39–26 Scattering of radiation from a square array. Interference from successive scatterers in a row is constructive when the angles of incidence and reflection are equal. Interference from adjacent rows is also constructive when Eq. (39–28) is satisfied.

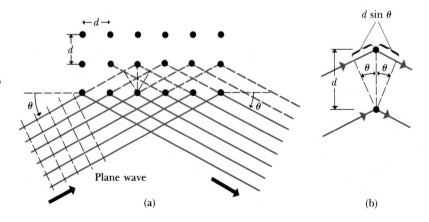

rows must equal $n\lambda$, where n is an integer. We can express the second condition as

$$2d \sin \theta = n\lambda \qquad (n = 1, 2, 3, \ldots). \qquad (39\text{–}28)$$

At points where this condition is satisfied, a strong maximum in the interference pattern is observed; otherwise no strong maximum is seen because there is no point where the radiation from all scatterers arrives in phase. We can describe this interference in terms of *reflections* of the wave from the horizontal rows of scatterers in Fig. 39–26a. Strong interference occurs at angles such that the incident and scattered angles are equal and Eq. (39–28) is satisfied.

We can extend this discussion to a three-dimensional array. Instead of *rows*, we consider *planes* of scatterers. Figure 39–27 shows that we can construct several different sets of parallel planes that pass through all the scatterers. Waves from all the scatterers in a given plane interfere constructively if the angles of incidence and scattering are equal. There is also constructive interference between planes when Eq. (39–28) is satisfied, where d is now the distance between adjacent planes. Because there are many different sets of parallel planes, there are also many values of d and many sets of angles corresponding to constructive interference for the whole crystal lattice. This phenomenon is called **Bragg reflection,** and Eq. (39–28) is called the **Bragg condition,** in honor of Sir William Bragg and his son Laurence Bragg, two pioneers in x-ray analysis. We must not let the term *reflection* obscure the fact that we are dealing with an *interference* effect.

Figure 39–28 is a photograph made by directing a narrow beam of x-rays at a thin section of a quartz crystal and allowing the scattered beam to strike a photographic film. As we predicted, nearly complete cancellation occurs for all but certain very specific directions, where constructive interference occurs and forms bright spots. Such a pattern is usually called an x-ray *diffraction* pattern, although as we have seen there is no fundamental distinction between the phenomena we call *interference* and those we call *diffraction*. Such patterns are also called Laue patterns.

If the crystal lattice spacing is known, we can determine the wavelength from the diffraction pattern, just as we determined wavelengths of visible light from measurements on diffraction patterns from slits or gratings. For example, we can determine the crystal lattice spacing for sodium chloride from its density and Avogadro's number. Conversely, once we know the x-ray wavelength, we can use x-ray diffraction to explore the structure and lattice spacing of crystals of unknown structure.

Indeed, x-ray diffraction has been by far the most important experimental tool in the investigation of crystal structure of solids. Atomic spacings in crystals can be measured precisely, and the details of the lattice arrangement of complex crystals can be determined. More recently, x-ray diffraction has played an important role in studies of the structures of liquids and of organic molecules. But the basic principles, superposition and interference, are exactly the same as for all the other phenomena we have studied in this chapter.

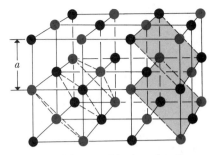

39–27 Cubic crystal lattice, showing two different families of crystal planes. The spacing of the planes on the left is $a/\sqrt{3}$; that of the planes on the right is $a/\sqrt{2}$. There are also three sets of planes parallel to the cube faces, with spacing a.

X-ray diffraction can be described in terms of reflection from crystal planes. Waves reflected from adjacent planes interfere with each other.

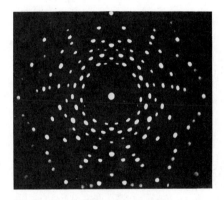

39–28 Laue diffraction pattern formed by directing a beam of x-rays at a thin section of quartz crystal. (Courtesy of Dr. B. E. Warren.)

X-ray diffraction: an important tool for investigating crystal structure

39–10 CIRCULAR APERTURES AND RESOLVING POWER

We have studied in detail the diffraction patterns formed by long, thin slits or arrays of slits. We also know from our introductory discussion that an aperture of *any* shape forms a diffraction pattern. The pattern formed by a *circular*

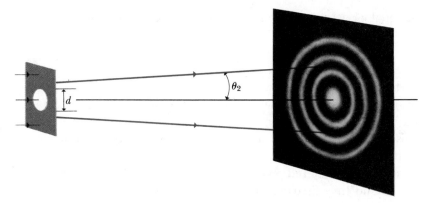

39–29 Diffraction pattern formed by a circular aperture of diameter d, consisting of a central bright spot and alternating dark and bright rings. The angular size θ_2 of the second dark ring is shown.

aperture is of special interest because of its role in limiting the resolving power of optical instruments. In principle we could compute the intensity at any point P in the diffraction pattern by dividing the area of the aperture into small elements, finding the resulting wave amplitude and phase at P, and then integrating over the aperture area to find the resultant amplitude and intensity at that point. In practice, the integration cannot be carried out in terms of elementary functions and has to be done by numerical approximation. We will simply describe the pattern and quote some relevant numbers.

The diffraction pattern formed by a circular aperture consists of a central bright spot surrounded by a series of bright and dark rings. Considering the axial symmetry of the situation, this is not surprising. In Fig. 39–29 we can characterize the size of the pattern in terms of the angle θ, representing the angular size of each ring. If the diameter of the aperture is d and the wavelength is λ, the angular size θ_1 of the first dark ring is found to be given by

$$\sin \theta_1 = 1.22 \frac{\lambda}{d}; \tag{39–29}$$

that of the second dark ring is given by

$$\sin \theta_2 = 2.23 \frac{\lambda}{d}; \tag{39–30}$$

the third by

$$\sin \theta_3 = 3.24 \frac{\lambda}{d}; \tag{39–31}$$

and so on. Between the dark rings are bright rings with angular size given by

$$\sin \theta = 1.63 \frac{\lambda}{d}, \quad 2.68 \frac{\lambda}{d}, \quad 3.70 \frac{\lambda}{d}, \tag{39–32}$$

and so on. The central bright spot is called the **Airy disk,** in honor of Sir George Airy (1801–1892), Astronomer Royal of England, who first derived the complete expression for the intensity in the pattern.

The *intensities* of the bright rings drop off very quickly; the peak intensity in the center of the first ring is only 1.7% of the value at the center of the Airy disk, and at the center of the second ring it is only 0.4%. Thus nearly all the radiant power (about 85%) is concentrated in the Airy disk. The angular size of the Airy disk can be taken to be that of the first dark ring, given by Eq.

(39–29). Figure 39–30 is a photograph of a diffraction pattern from a circular aperture 1.0 mm in diameter.

All of this has far-reaching implications for image formation of lenses and mirrors. In our study of optical instruments in Chapter 38, we assumed that a lens of focal length f focuses a parallel beam (plane wave) to a *point* at a distance f from the lens. This assumption ignored diffraction effects. We now see that what we get is not a point but the diffraction pattern just described. To put it another way, if we have two point objects, their images are not two points but two diffraction patterns. If the objects are close together, their diffraction patterns overlap; if they are close enough, their patterns overlap almost completely and cannot be distinguished. The effect is shown in Fig. 39–31, which shows the patterns for four (very small) "point" objects. In (a) the two images on the right have merged together; in (b), where we use a

39–30 Diffraction pattern formed by a circular aperture 1.0 mm in diameter.

39–31 Diffraction patterns of four "point" sources, with a circular opening in front of the lens. In (a) the opening is so small that the patterns at the right are just resolved, by Rayleigh's criterion. Increasing the aperture decreases the size of the diffraction patterns, as in (b) and (c).

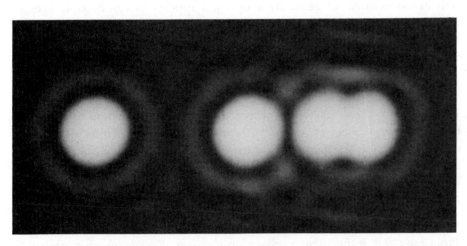

(a)

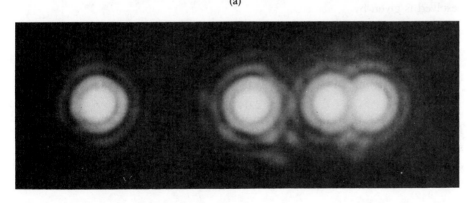

(b)

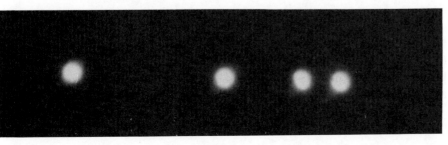

(c)

larger aperture diameter d with resulting smaller angular size of the Airy disk, the two right images are barely resolved. In (c), with a still larger aperture, they are well resolved.

A widely used criterion for resolution of two point objects, proposed by Lord Rayleigh and called **Rayleigh's criterion,** is that the points are just barely resolved (i.e., distinguishable) if the center of one diffraction pattern coincides with the first minimum of the other. In that case the angular separation of the image centers is given by Eq. (39–29). Because the angular separation of the *objects* is the same as that of the *images*, this means that two point objects are barely resolved, according to Rayleigh's criterion, if their angular separation is given by Eq. (39–29).

The minimum separation of two points that can just be resolved by an optical instrument is called the **limit of resolution** of the instrument. The smaller the limit of resolution, the greater the **resolving power** of the instrument. Diffraction sets the ultimate limits on resolution of lenses. If we look only at the formulas of geometrical optics, it appears we can make images as large as we like. While this is true, we always reach a point eventually where the image becomes larger but does not gain in detail.

Diffraction limits the sharpness of photographic images. Sharpness decreases when the lens is stopped down.

EXAMPLE 39–7   A camera lens of focal length $f = 50$ mm and maximum aperture $f/2$ forms an image of an object 10 m away. (a) If the resolution is limited by diffraction, what is the minimum distance between two points on the object that are barely resolved, and what is the corresponding distance between image points? (b) How does the situation change if the lens is "stopped down" to $f/16$? Assume $\lambda = 500$ nm in both cases.

SOLUTION   (a) The aperture diameter is $d = (50 \text{ mm})/2 = 25 \text{ mm} = 25 \times 10^{-3}$ m. From Eq. (39–29) the angular separation θ of two object points that are barely resolved is given by

$$\sin \theta \cong \theta = 1.22 \frac{\lambda}{d}$$
$$= 1.22 \frac{500 \times 10^{-9} \text{ m}}{25 \times 10^{-3} \text{ m}}$$
$$= 2.44 \times 10^{-5}.$$

Let y be the separation of the object points and y' the separation of the corresponding image points. We know from our thin-lens analysis in Section 38–1 that, apart from sign, $y/s = y'/s'$. Thus the angular separations of the object points and the corresponding image points are both equal to θ. Thus

$$\frac{y}{10 \text{ m}} = 2.44 \times 10^{-5}, \quad y = 2.44 \times 10^{-4} \text{ m} = 0.244 \text{ mm};$$
$$\frac{y'}{25 \text{ mm}} = 2.44 \times 10^{-5}, \quad y' = 6.10 \times 10^{-4} \text{ mm} = 0.00061 \text{ mm}.$$

(b) The aperture is now $(50 \text{ mm})/16$, or one-eighth as large as before. The angular separation is eight times as great, and the values of y and y' are also eight times as great as before:

$$y = 1.95 \text{ mm}, \quad y' = 0.00488 \text{ mm}.$$

Because setups to test the resolution of lenses often use series of parallel lines with varying spacing, the resolution in the image is often described in "lines per millimeter." Our lens would be described as having a resolution of about 1600 lines per mm when "wide open" and about 200 lines per mm when stopped down to $f/16$. Only the best-quality camera lenses approach this resolution. Photographers who always use the smallest possible aperture for maximum depth of focus and (presumably) maximum sharpness should be aware that diffraction effects become more significant at small apertures. Thus there is an optimization problem, balancing one cause of fuzzy images against another.

Another lesson to be learned is that resolution improves with shorter wavelengths. Ultraviolet microscopes have higher resolution than visible-light microscopes. In electron microscopes, which we will study in Chapter 42, the resolution is limited by the wavelengths associated with wavelike behavior of electrons. As we will see, electron wavelengths can be made 100,000 times smaller than wavelengths of visible light, with a corresponding gain in resolution. Finally, we note that part of the motivation for building very large reflecting telescopes is to increase the aperture diameter and thus minimize diffraction effects. The other reason, of course, is to provide greater light-gathering area for viewing very faint stars.

Why can electron microscopes have greater magnification than optical microscopes?

39–11 HOLOGRAPHY

Holography is a technique for recording and reproducing an image of an object without the use of lenses. Unlike the two-dimensional images recorded by an ordinary photograph or television system, a holographic image is truly three-dimensional. Such an image can be viewed from different directions to reveal different sides, and from various distances to reveal changing perspective.

Holography: using interference to create three-dimensional images

The basic procedure for making a hologram is very simple in principle. A possible arrangement is shown in Fig. 39–32a. We illuminate the object to be holographed with monochromatic light, and we place a photographic film so

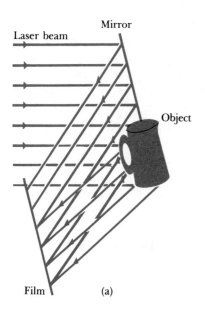

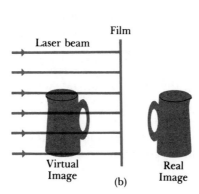

39–32 (a) The hologram is the record on film of the interference pattern formed with light directly from the source and light scattered from the object. (b) Images are formed when light is projected through the hologram.

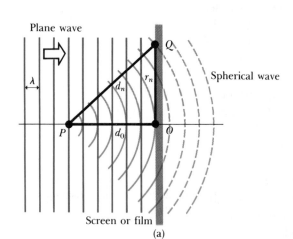

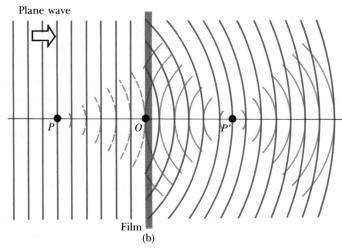

39–33 (a) Constructive interference of the plane and spherical waves occurs in the plane of the film at every point Q for which the distance d_n from P is greater than the distance d_0 from P to O by an integer number of wavelengths $n\lambda$. For the point shown, $n = 2$. (b) When a plane wave strikes the developed film, the diffracted wave consists of a wave converging to P' and then diverging again, and a diverging wave that appears to originate at P. These waves form the real and virtual images, respectively.

that it is struck by scattered light from the object and also by direct light from the source. In practice, the source must be a laser, for reasons to be discussed later. Interference between the direct and scattered light leads to the formation and recording of a complex interference pattern on the film.

To form the images, we simply project laser light through the developed film, as shown in Fig. 39–32b. Two images are formed, a virtual image on the side of the film nearer the source, and a real image on the opposite side.

A complete analysis of holography is beyond our scope, but we can gain some insight into the process by examining how a single point is holographed and imaged. Consider the interference pattern formed on a photographic film by the superposition of an incident plane wave and a spherical wave, as shown in Fig. 39–33a. The spherical wave originates at a point source P a distance d_0 from the film; P may in fact be a small object that scatters part of the incident plane wave. In any event, we assume that the two waves are monochromatic and coherent, and that the phase relation is such that constructive interference occurs at point O on the diagram. Then constructive interference will *also* occur at any point Q on the film that is farther from P than O is, by an integer number of wavelengths. That is, if $d_n - d_0 = n\lambda$, where n is an integer, then constructive interference occurs. The points where this condition is satisfied form circles centered at O, with radii r_n given by

$$d_n - d_0 = \sqrt{d_0{}^2 + r_n{}^2} - d_0 = n\lambda \qquad (n = 1, 2, 3, \ldots).\qquad (39\text{–}33)$$

Solving this equation for $r_n{}^2$, we find

$$r_n{}^2 = \lambda(2nd_0 + n^2\lambda).$$

How to make a hologram

Ordinarily d_0 is very much larger than λ, so we neglect the second term in parentheses, obtaining

$$r_n = \sqrt{2n\lambda d_0}.\qquad (39\text{–}34)$$

Since n must be an integer, the interference pattern consists of a series of concentric bright circular fringes, with the radii of the brightest regions given by Eq. (39–34). Between these bright fringes are darker fringes.

Now we develop the film and make a transparent positive print, so the bright-fringe areas have the greatest transparency on the film. It is then illuminated with monochromatic plane-wave light of the same wavelength as that used initially. In Fig. 39–33b, consider a point P' at a distance d_0 along the axis from the film. The centers of successive bright fringes differ in their distances from P' by an integer number of wavelengths, and therefore a strong *maximum* in the diffracted wave occurs at P'. That is, light converges to P' and then diverges from it on the opposite side, and P' is therefore a *real image* of point P.

This is not the entire diffracted wave, however; there is also a diverging spherical wave, which would represent a continuation of the wave originally emanating from P if the film had not been present. Thus the *total* diffracted wave is a superposition of a converging spherical wave forming a real image at P' and a diverging spherical wave shaped as though it had originated at P, forming a virtual image at P.

Because of the principle of linear superposition, what is true for the imaging of a single point is also true for the imaging of any number of points. The film records the superposed interference pattern from the various points, and when light is projected through the film the various image points are reproduced simultaneously. Thus the images of an extended object can be recorded and reproduced just as for a single point object.

In making a hologram, several practical problems must be overcome. First, the light used must be *coherent* over distances that are large compared to the dimensions of the object and its distance from the film. Ordinary light sources *do not* satisfy this requirement, for reasons discussed in Section 39–1, and laser light is essential. Second, extreme mechanical stability is needed. If any relative motion of source, object, or film occurs during exposure, even by as much as a wavelength, the interference pattern on the film is blurred enough to prevent satisfactory image formation. These obstacles are not insurmountable, however, and holography promises to become increasingly important in research, entertainment, and a wide variety of technological applications.

Making a hologram isn't as easy as it sounds.

SUMMARY

Monochromatic light is light having a single definite frequency. Coherence is a definite, unchanging phase relationship between two waves. When two coherent sources emit monochromatic light, their waves overlap, causing an interference pattern. The principle of linear superposition states that the total wave disturbance at a point at any instant is the sum of the disturbances from the separate waves. When the sources are in phase, constructive interference at a point occurs when the difference in path length from the two sources is zero or an integer number of wavelengths; destructive interference occurs when the path difference is a half-integer number of wavelengths. When the line from the sources to a point P makes an angle θ with the line perpendicular to the line of the sources, and when the distance between sources is d, the

KEY TERMS
monochromatic wave
interference
principle of linear superposition
constructive interference
destructive interference
coherent light
intensity
Michelson interferometer
ether
diffraction

condition for constructive interference is

$$d \sin \theta = m\lambda \qquad (m = 0, \pm1, \pm2, \pm3, \ldots), \qquad (39\text{--}4)$$

and the condition for destructive interference is

$$d \sin \theta = (m + \tfrac{1}{2})\lambda \qquad (m = 0, \pm1, \pm2, \pm3, \ldots). \qquad (39\text{--}5)$$

When θ is very small, the position y_m of the mth bright fringe is given by

$$\lambda = \frac{y_m d}{mR}. \qquad (39\text{--}7)$$

When two sources emit waves in phase, the phase difference δ of the waves arriving at point P is related to the difference in path length $(r_1 - r_2)$ by

$$\delta = \frac{2\pi}{\lambda}(r_1 - r_2) = k(r_1 - r_2). \qquad (39\text{--}9)$$

When two sinusoidal waves of amplitude E and phase difference δ are superposed, the resultant amplitude E_P is

$$E_P{}^2 = 4E^2 \cos^2(\delta/2), \qquad (39\text{--}11)$$

and the intensity I is given by

$$I = 2\epsilon_0 c E^2 \cos^2\left(\frac{\pi d}{\lambda}\sin\theta\right). \qquad (39\text{--}14)$$

When light is reflected from both sides of a thin film of thickness d, constructive interference between the reflected waves occurs when

$$2d = m\lambda \qquad (m = 0, 1, 2, 3, \ldots) \qquad (39\text{--}17)$$

unless a half-cycle phase shift occurs at one surface; then this is the condition for destructive interference. A half-cycle phase shift occurs during reflection whenever the index of refraction in the second material is greater than that in the first.

The Michelson interferometer uses an extended monochromatic source and can be used for high-precision measurements of wavelengths. Its original purpose was to detect motion of the earth relative to a hypothetical ether, the supposed medium for electromagnetic waves. The concept of ether has been abandoned; the speed of light is the same relative to all observers. This is part of the foundation of the special theory of relativity.

Diffraction occurs when light passes through an aperture or around an edge. When source and observer are so far away from the obstructing surface that the outgoing rays can be considered parallel, it is called Fraunhofer diffraction, and when source or observer is at a finite distance it is Fresnel diffraction. For a single narrow slit of width a, the condition for destructive interference at a point P at an angle θ from the perpendicular to the surface of the slit is

$$a \sin \theta = n\lambda \qquad (n = 0, \pm1, \pm2, \pm3, \ldots). \qquad (39\text{--}20)$$

The complete expression for the intensity I at any angle θ, in terms of the intensity I_0 at $\theta = 0$, is

$$I = I_0\left(\frac{\sin \delta/2}{\delta/2}\right)^2 = I_0\left\{\frac{\sin[\pi a(\sin \theta)/\lambda]}{\pi a(\sin \theta)/\lambda}\right\}^2. \qquad (39\text{--}22)$$

A diffraction grating consists of a large number of thin parallel slits, spaced a distance d apart. The condition for maximum intensity in the interference pattern is

$$d \sin \theta = m\lambda \qquad (m = 0, \pm 1, \pm 2, \pm 3, \ldots). \qquad (39\text{--}27)$$

This is the same condition for the two-source pattern, but for the grating the maxima are very sharp and narrow.

A crystal serves as a three-dimensional diffraction grating for waves having wavelengths of the order of magnitude of the lattice spacing; namely, x-rays. For a set of crystal planes spaced a distance d apart, constructive interference occurs when the angles of incidence and scattering are equal and when

$$2d \sin \theta = n\lambda \qquad (n = 0, 1, 2, 3, \ldots). \qquad (39\text{--}28)$$

This is called the Bragg condition.

The diffraction pattern from a circular aperture of diameter d consists of a central bright spot, called the Airy disk, and a series of concentric dark and bright rings. The angular size θ_1 of the first dark ring, which is also the angular size of the Airy disk, is given by

$$\sin \theta_1 = 1.22\lambda/d. \qquad (39\text{--}29)$$

Diffraction sets the ultimate limit on resolution (image sharpness) of optical instruments. According to Rayleigh's criterion, two object points are just barely resolved when their angular separation θ is given by Eq. (39–29).

A hologram is photographic record of an interference pattern formed by light scattered from an object and light coming directly from the source. It can be used to form three-dimensional images of the object.

QUESTIONS

39–1 Could an experiment similar to Young's two-slit experiment be performed with sound? How might this be carried out? Does it matter that sound waves are longitudinal and electromagnetic waves transverse?

39–2 At points of constructive interference between waves of equal amplitude, the intensity is four times that of either individual wave. Does this violate energy conservation? If not, why not?

39–3 In using the superposition principle to calculate intensities in interference and diffraction patterns, could one add the intensities of the waves instead of their amplitudes? What is the difference?

39–4 A two-slit interference experiment is set up and the fringes displayed on a screen. Then the whole apparatus is immersed in the nearest swimming pool. How does the fringe pattern change?

39–5 Would the headlights of a distant car form a two-source interference pattern? If so, how might it be observed? If not, why not?

39–6 A student asserted that it is impossible to observe interference fringes in a two-source experiment if the distance between sources is less than half the wavelength of the wave. Do you agree? Explain.

39–7 An amateur scientist proposed to record a two-source interference pattern by using only one source, placing it first in position S_1 in Fig. 39–1 and turning it on for a certain time, then placing it at S_2 and turning it on for an equal time. Does this work?

39–8 When a thin oil film spreads out on a puddle of water, the thinnest part of the film looks lightest in the resulting interference pattern. What does this tell you about the relative magnitudes of the refractive indexes of oil and water?

39–9 A glass windowpane with a thin film of water on it reflects less than when it is perfectly dry. Why?

39–10 In high-quality camera lenses, the resolution in the image is determined by diffraction effects. Is the resolution best when the lens is "wide open" or when it is "stopped down" to a smaller aperture? How does this behavior compare with the effect of aperture size on the depth of focus, that is, on the limit of resolution due to imprecise focusing?

39–11 If a two-slit interference experiment were done with white light, what would be seen?

39–12 Why is a diffraction grating better than a two-slit setup for measuring wavelengths of light?

39–13 Would the interference and diffraction effects described in this chapter still be seen if light were a longitudinal wave instead of transverse?

39–14 One sometimes sees rows of evenly spaced radio antenna towers. A student remarked that these act like diffraction gratings. What did he mean? Why would one *want* them to act like a diffraction grating?

39–15 Could x-ray diffraction effects with crystals be observed by using visible light instead of x-rays? Why or why not?

39–16 Does a microscope have better resolution with red light or blue light?

39–17 How could an interference experiment, such as one using a Michelson interferometer or fringes caused by a thin air space between glass plates, be used to measure the refractive index of air?

EXERCISES

Section 39–2 Two-Source Interference

39–1 Two slits are spaced 0.3 mm apart and are placed 50 cm from a screen. What is the distance between the second and third dark lines of the interference pattern when the slits are illuminated with light of 600-nm wavelength?

39–2 Young's experiment is performed with sodium light ($\lambda = 589$ nm). Fringes are measured carefully on a screen 100 cm away from the double slit, and the center of the twentieth fringe is found to be 11.78 mm from the center of the zeroth fringe. What is the separation of the two slits?

39–3 Light from a mercury-arc lamp is passed through a filter that blocks everything except for one spectrum line in the green region of the spectrum. It then falls on two slits separated by 0.6 mm. In the resulting interference pattern on a screen 2.5 m away, adjacent bright fringes are separated by 2.27 mm. What is the wavelength?

Section 39–3 Intensity Distribution in Interference Patterns

39–4 An AM radio station has a frequency of 1000 kHz; it uses two identical antennas at the same elevation, 150 m apart.

a) In what direction is the intensity maximum, considering points in a horizontal plane?

b) Calling the maximum intensity in (a) I_0, determine in terms of I_0 the intensity in directions making angles of 30°, 45°, 60°, and 90° to the direction of maximum intensity.

39–5 Consider a two-slit interference pattern, for which the intensity distribution is given by Eq. (39–15), Let θ_m be the angular position of the mth bright fringe, where the intensity is I_0. Assume that θ_m is small, so that $\sin \theta_m \approx \theta_m$. Let θ_m^+ and θ_m^- be the two angles on either side of θ_m for which $I = \frac{1}{2}I_0$. The quantity $\Delta\theta_m = |\theta_m^+ - \theta_m^-|$ is the half-width of the mth fringe. Calculate $\Delta\theta_m$. How does $\Delta\theta_m$ depend on m?

Section 39–4 Interference in Thin Films

39–6 Light of wavelength 500 nm is incident perpendicularly from air on a film 1×10^{-4} cm thick and of refractive index 1.375. Part of the light is reflected from the first surface of the film, and part enters the film and is reflected back at the second surface.

a) How many waves are contained along the path of this second part of the light in the film?

b) What is the phase difference between these two parts of the light as they leave the film?

39–7 In Example 39–4 suppose the top plate is glass with $n = 1.4$, the wedge is filled with silicone grease having $n = 1.5$, and the bottom plate is glass with $n = 1.6$. Calculate the spacing between the dark fringes.

39–8 A sheet of glass 10 cm long is placed in contact with a second sheet and is held at a small angle with it by a metal strip 0.1 mm thick placed under one end. The glass is illuminated from above with light of 546-nm wavelength. How many interference fringes are observed per centimeter in the reflected light?

39–9 Two rectangular pieces of plane glass are laid one upon the other on a table. A thin strip of paper is placed between them at one edge so that a very thin wedge of air is formed. The plates are illuminated by a beam of sodium light at normal incidence ($\lambda = 589$ nm). Interference fringes are formed, with ten fringes per centimeter length of wedge measured normal to the edges in contact. Find the angle of the wedge.

39–10

a) Is a thin film of quartz suitable as a nonreflecting coating for fabulite? (See Table 36–1.)

b) If so, what is the minimum thickness of the film required?

39–11

a) What is the thinnest film of a 1.40 refractive index coating on glass ($n = 1.50$) for which destructive interference of the violet component (400 nm) of an incident white light beam in air can take place by reflection?

b) What then is the residual color of the beam?

Section 39–5 The Michelson Interferometer

39–12 How far must the mirror M_2 (Fig. 39–12) of the Michelson interferometer be moved so that 3000 fringes of krypton-86 light ($\lambda = 606$ nm) will move across a line in the field of view?

39–13 A Mach-Zehnder interferometer is shown in Fig. 39–34. The mirrors at B and C are totally reflecting, while the mirrors at A and D are half-silvered, so they reflect $\frac{1}{2}$ the light and transmit $\frac{1}{2}$ the light. Light polarized in the plane of the mirrors is used so that the light waves will undergo a 90° phase shift at each mirror. Show that if the optical paths are such that the output at F is zero (total destructive interference), then the output at E will show total constructive interference.

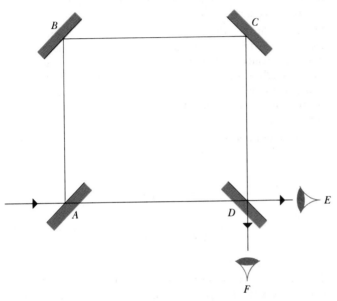

FIGURE 39–34

Section 39–7 Fraunhofer Diffraction from a Single Slit

39–14 Parallel rays of green mercury light of wavelength 546 nm pass through a slit of width 0.437 mm covering a lens of focal length 40 cm. In the focal plane of the lens, what is the distance from the central maximum to the first minimum?

39–15 Monochromatic light from a distant source is incident on a slit 0.8 mm wide. On a screen 3.0 m away, the distance from the central maximum of the diffraction pattern to the first minimum is measured to be 2 mm. Calculate the wavelength of the light.

39–16 Light of wavelength 589 nm from a distant source is incident on a slit 1.0 mm wide, and the resulting diffraction pattern is observed on a screen 2.0 m away. What is the distance between the two dark fringes on either side of the central bright fringe?

Section 39–8 The Diffraction Grating

39–17 Plane monochromatic waves of wavelength 600 nm are incident normally on a plane transmission grating having 500 lines·mm^{-1}. Find the angles of deviation in the first, second, and third orders.

39–18

a) What is the wavelength of light that is deviated in the first order through an angle of 20° by a transmission grating having 6000 lines·cm^{-1}?

b) What is the second-order deviation of this wavelength? Assume normal incidence.

39–19 A plane transmission grating is ruled with 4000 lines·cm^{-1}. Compute the angular separation in degrees between the α and δ lines of atomic hydrogen in the second-order spectrum. The wavelengths of these lines are, respectively, 656 nm and 410 nm. Assume normal incidence.

Section 39–10 Circular Apertures and Resolving Power

39–20 In Fig. 39–35, two point sources of light, a and b, at a distance of 50 m from lens L and 6 mm apart, produce images at c that are just resolved by Rayleigh's criterion. The focal length of the lens is 20 cm. What is the diameter of the diffraction circles at c?

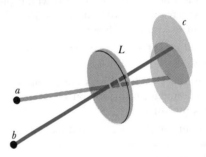

FIGURE 39–35

39–21 A telescope is used to observe two distant point sources 1 m apart. ($\lambda = 500$ nm.) The objective of the telescope is covered with a slit of width 1 mm. What is the maximum distance in meters at which the two sources may be distinguished?

39–22 A converging lens 8 cm in diameter has a focal length of 40 cm. If the resolution is diffraction limited, how far away can an object be if points on it 4 mm apart are to be resolved (by Rayleigh's criterion)? (Use $\lambda = 550$ nm.)

Section 39–11 Holography

39–23 If a hologram is made by using 600-nm light and then viewed with 500-nm light, how will the images look compared to those observed with 600-nm light?

39–24 A hologram is made by using 600-nm light and is then viewed with continuous-spectrum white light from an incandescent bulb. What will be seen?

39–25 Ordinary photographic film reverses black and white, in the sense that the most brightly illuminated areas become blackest upon development (hence the term *negative*). Suppose a hologram negative is viewed directly, without making a positive transparency. How will the resulting images differ from those obtained with the positive hologram?

PROBLEMS

39–26 Parallel light rays of wavelength $\lambda = 600$ nm fall on a single slit. On a screen 3 m away, the distance from the center of the central maximum to the center of the next maximum is 4 mm. What is the width of the slit?

39–27 In Exercise 39–1, suppose the entire apparatus is immersed in water. Then what is the distance between the second and third dark lines?

39–28 An FM radio station has a frequency of 100 MHz and uses two identical antennas mounted at the same elevation, 12 m apart. The resulting radiation pattern has a maximum intensity along a horizontal line perpendicular to the line joining the antennas.

a) At what other angles (measured from the line of maximum intensity) is the intensity maximum?

b) At what angles is it zero?

39–29 A glass plate 0.40 μm thick is illuminated by a beam of white light normal to the plate. The index of refraction of the glass is 1.50. What wavelengths within the limits of the visible spectrum ($\lambda = 400$ nm to $\lambda = 700$ nm) will be intensified in the reflected beam?

39–30 An oil tanker spills a large amount of oil ($n = 1.4$) into the sea.

a) If you are overhead and look down onto the oil spill, what predominant color do you see at a point where the oil is 410 nm thick?

b) If you swam under the slick and looked up at the same place in the slick as in (a), what color would predominate in the transmitted light?

39–31 The radius of curvature of the convex surface of a plano-convex lens is 1.20 m. The lens is placed convex side down on a plane glass plate and illuminated from above with red light of wavelength 650 nm. Find the diameter of the third bright ring in the interference pattern.

39–32 Newton's rings can be seen when a plano-convex lens is placed on a flat glass surface (Problem 39–31). If the lens has an index of refraction of $n = 1.50$ and the glass plate an index of $n = 1.80$, the rings are seen with a spacing of 0.50 mm. If a liquid of index 1.65 is added to the space between the lens and the plate, what will be the new spacing between the interference rings?

39–33 In a Young's two-slit experiment, a piece of glass with an index of refraction n and a thickness L is placed in front of the upper slit.

a) Describe qualitatively what happens to the interference pattern.

b) Derive an expression for the intensity I of the light at points on a screen as a function of n, L, and θ. Here θ is

the usual angle measured from the center of the two slits. That is, determine the equation analogous to Eq. (39–15).

c) From your result in (b) derive an expression for the values of θ that locate the maxima in the interference pattern. That is, derive an equation analogous to Eq. (39–4).

39–34 In Exercise 39–16, suppose the entire apparatus is immersed in water. Then what is the distance between the two dark fringes?

39–35 A slit of width a was placed in front of a lens of focal length 0.80 m. The slit was illuminated by parallel light of wavelength 600 nm, and the diffraction pattern of Fig. 39–22b was formed on a screen in the second focal plane of the lens. If the photograph of Fig. 39–22b represents an enlargement to twice the actual size, what was the slit width?

39–36 The intensity of light in the Fraunhofer diffraction pattern of a single slit is

$$I = I_0(\sin \beta/\beta)^2,$$

where

$$\beta = (\pi a \sin \theta)/\lambda.$$

Show that the equation for the values of β at which I is a maximum is $\tan \beta = \beta$. How can you solve such an equation graphically?

39–37 What is the longest wavelength that can be observed in the fourth order for a transmission grating having 5000 lines per centimeter? Assume normal incidence.

39–38 A Michelson interferometer can be used to measure the index of refraction of gases by placing an initially evacuated tube in one arm of the interferometer. The gas is then slowly added to the tube, and the number of fringes that cross the telescope cross hairs are counted. If the length of the tube is 4 cm and the light source is a sodium lamp (589 nm), what is the index of refraction of the gas if 35 fringes are seen to pass the view of the telescope? (*Note*. For gases it is convenient to give the value of $n - 1$ rather than n itself, as the index differs little from unity.)

39–39 An astronaut in the space shuttle can just resolve two point sources on earth that are 20 m apart. Assume that the resolution is diffraction limited and use Rayleigh's criterion. What is his altitude above the earth? Treat his eye as a circular aperture of diameter 4.0 mm (the diameter of his pupil), and take 550 nm as the wavelength of the light.

CHALLENGE PROBLEMS

39–40 During the Battle of Britain (WW II), it was found that aircraft flying at certain low altitudes over the English Channel could not receive radio signals from transmission towers located on the cliffs of Dover, 200 m above the

water. At other altitudes the signals received were very strong. For an airplane flying 10 km from England and a radio signal with wavelength 4 m, at what altitudes above the water will the received signal be strongest? (Consider only altitudes much less than 10 km. Interference occurs between radio waves that travel directly to the plane and those that first reflect off the water.)

39–41 Consider a single-slit diffraction pattern. The center of the central maximum, where the intensity is I_0, is located at $\theta = 0$. Let θ_+ and θ_- be the two angles on either side of $\theta = 0$ for which $I = \frac{1}{2}I_0$. $\Delta\theta = |\theta_+ - \theta_-|$ is called the half-width of the central diffraction maximum. Solve for $\Delta\theta$, when the ratio between the slit width a and wavelength λ is

a) $a/\lambda = 2$,   b) $a/\lambda = 5$,

c) $a/\lambda = 10$.

(*Hint:* Your equation for θ_+ or θ_- cannot be solved analytically. You must use iteration—see Challenge Problem 27–35—or solve it graphically.)

39–42 Figure 39–36 shows an interferometer known as *Fresnel's biprism*. The magnitude of the prism angle A is extremely small.

a) If S_0 is a very narrow source slit, show that the separation of the two virtual coherent sources S_1 and S_2 is given by $d = 2aA(n - 1)$, where n is the index of refraction of the material of the prism.

b) Calculate the spacing of the fringes of green light of wavelength 500 nm on a screen 2 m from the biprism. Take $a = 0.20$ m, $A = 0.005$ rad, and $n = 1.5$.

39–43 The index of refraction of a glass rod is 1.48 at $T = 20°C$ and varies linearly with temperature, with a coefficient of $2.0 \times 10^{-5}(C°)^{-1}$. The coefficient of linear expansion of the glass is $5.0 \times 10^{-6}(C°)^{-1}$. At 20°C the length of the rod is 2.00 cm. Derive a relationship for the speed with which fringes will cross the field of view of a Michelson interferometer that has this glass rod in one arm, if the rod is being heated at a rate of 5C°·min^{-1}. The light source has wavelength $\lambda = 589$ nm, and the rod initially is at $T = 20°C$.

39–44 The yellow sodium D lines are a doublet with wavelengths of 589.0 and 589.6 nm and equal intensities.

a) How many lines/cm are required for a diffraction grating to resolve these two lines in the first-order spectrum? Assume the grating is placed 0.5 m from a screen and that the source image is 0.1 mm wide, so that resolution of the lines means their centers are 0.1 mm apart on the screen.

b) The sodium D lines are used as a source in a Michelson interferometer. As the mirror is moved, it is noted that in addition to moving across the field of view, the interference fringes periodically appear and disappear. Why? How far is the mirror moved between disappearances of the fringes?

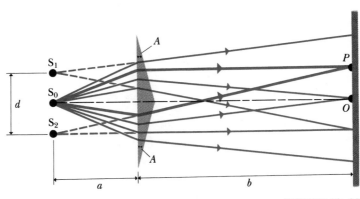

FIGURE 39–36

PART EIGHT

MODERN PHYSICS

PERSPECTIVE

In the past several chapters we have seen how the subject of optics grows out of electromagnetic theory. Maxwell's equations predict the existence of electromagnetic waves. Many optical phenomena, including polarization, interference, and diffraction, can be understood on the basis of an electromagnetic wave model of light. The simpler ray model of geometrical optics can be used to analyze mirrors, lenses, and optical instruments.

We have now arrived at a very significant milepost in our study of physics. We have stressed the roles of Newton's laws, the laws of thermodynamics, and Maxwell's equations in summarizing the most important laws of physics. All these principles were formulated in their present form by the year 1900. It would have been tempting to speculate then that the laws of physics were essentially complete, that all that remained was to work out the details of some complicated problems using well-established principles. Nothing would have been further from the truth! The revolution in scientific thought that has occurred since 1900 is at least as significant as the sum of all developments up to that year. Part of this revolution was the theory of relativity. Starting with the innocent-sounding premise that the laws of physics should be the same in all inertial frames of reference, Einstein showed that consistency with this premise required sweeping revisions of our concepts of space and time.

Optics, far from being a closed, nineteenth-century subject, has formed a significant gateway into twentieth-century physics. Ray and wave optics are an adequate description of the *propagation* of light. But for many other phenomena, especially the *emission* and *absorption* of light, classical optics is *not* adequate. A variety of experimental evidence shows that the energy in light is emitted and absorbed only in discrete packages called *quanta* or *photons*. These packages, in turn, are related to the existence of characteristic spectra of elements and hence to the internal structure of atoms.

Understanding of atomic structure and the emission and absorption of light requires a new theory called *quantum mechanics* that involves fundamental changes in the language we use to describe mechanical systems. We abandon the description of a particle as a geometric point and regard it instead as an inherently spread-out entity that can have wave-like properties. Thus both particles (such as electrons) and radiation (photons) exhibit both particle and wave aspects. Like relativity, quantum mechanics requires sweeping revisions of our fundamental notions of space, time, and the description of motion. Quantum mechanics also provides the foundation for understanding many macroscopic (bulk) properties of matter on the basis of its microscopic (molecular) structure. Thus quantum mechanics is the starting point for modern materials science.

Another area of study is the atomic nucleus. Nuclear structure involves a kind of interaction distinct from gravitational and electromagnetic interactions. Some nuclei are unstable, leading to radioactivity. Some nuclear reactions, such as fission and fusion, have practical applications as sources of energy and as weapons.

We end the final chapter with a brief overview of high-energy physics, one of the present-day frontiers of physics. Many new particles have been discovered during the past forty years; the familiar particles, such as protons and neutrons, are not the most fundamental level of the structure of matter. There are still many unanswered questions in this vital area of present-day research, and our understanding of the physical world is far from complete!

40

RELATIVISTIC MECHANICS

IN EARLIER CHAPTERS OF THIS BOOK, ESPECIALLY THOSE ON MECHANICS, we have stressed the importance of inertial frames of reference. Newton's laws of motion are valid only in inertial frames, but they are valid in *all* inertial frames. Any frame moving with constant velocity with respect to an inertial frame is itself an inertial frame, and all such frames are equivalent with respect to expressing the basic principles of mechanics. *The laws of mechanics are the same in every inertial frame of reference.*

In 1905 Einstein proposed that this principle be extended to include *all* the basic laws of physics. This innocent-sounding proposition, often called the *principle of relativity,* has far-reaching and startling consequences. We will find, for example, that if the principle of conservation of momentum is to be valid in all inertial systems, the definition of momentum for particles moving at speeds comparable to the speed of light must be modified, and kinetic energy must also be redefined. Equally fundamental are the modifications needed in the *kinematic* aspects of motion. The resulting generalizations of the laws of mechanics are part of the **special theory of relativity,** the subject of this chapter. In studying this material you must be prepared to confront some ideas that at first sight will seem too strange to be believed, and you must learn to mistrust intuition when dealing with phenomena far removed from everyday experience!

Our discussion will center mostly on *mechanical* concepts, but the theory of relativity has significant consequences in *all* areas of physics, including thermodynamics, electromagnetism, optics, atomic and nuclear physics, and high-energy physics.

40–1 INVARIANCE OF PHYSICAL LAWS

Einstein's **principle of relativity** states that *the laws of physics are the same in every inertial frame of reference.* A familiar example of this principle in electromagnetism is the electromotive force induced in a coil of wire as a result of motion of a nearby permanent magnet. In the frame of reference where the *coil* is sta-

Describing physical phenomena in various frames of reference

tionary, the moving magnet causes a change of magnetic flux through the coil and hence an induced emf. In a different frame, where the *magnet* is stationary, the motion of the coil through a magnetic field causes magnetic-field forces on the mobile charges in the conductor, inducing an emf. According to the principle of relativity, both points of view have equal validity and both must predict the same result for the induced emf. As we have seen in Chapter 32, Faraday's law of electromagnetic induction can be applied to either description, so it does indeed satisfy this requirement.

Of equal significance is the prediction of the speed of light and other electromagnetic radiation, emerging from the development in Chapter 35. According to the principle of relativity, the speed of propagation of this radiation must be independent of the frame of reference and must be the same for all inertial frames.

Does light need a mechanical medium in which to travel?

Indeed, the speed of light plays a special role in the theory of relativity. According to the principle of relativity, light travels in vacuum with speed c, independent of the motion of the source. As we mentioned in Section 1–2, the numerical value of c is *defined* to be exactly 299,792,458 m·s^{-1}, and this number is used to define the unit of length in terms of the unit of time. The approximate value $c = 3.00 \times 10^8$ m·s^{-1} is within one part in 1000 of the exact value, and we will often use it when we do not need greater precision.

The ether is a mythical beast, like the unicorn and phlogiston.

At one time it was thought that light traveled through a hypothetical medium called the *ether,* just as sound waves travel through air. In that case, its speed would depend on the motion of the observer relative to the ether and hence would be different in different directions. During the late nineteenth and early twentieth centuries, intensive efforts were made to find experimental evidence for the existence of the ether. The Michelson-Morley experiment, described in Section 39–5, was an effort to detect motion of the earth relative to the ether. This and all similar experiments yielded consistently negative results, and the ether concept has been discarded. Thus this experimental evidence confirms the prediction of the principle of relativity that the speed of light is the same in all frames of reference. With this result in mind, let us suppose the speed of light is measured by two observers, one at rest with respect to the light source, the other moving away from it. Both are in inertial frames of reference, and according to Einstein's principle of relativity, the laws of physics—in particular the speed of light—must be the same in both frames.

Some strange conclusions about relative velocities

If this does not impress you, consider the following situation. A spaceship moving away from Earth at 1000 m·s^{-1} fires a missile with a speed of 2000 m·s^{-1} in a direction directly away from Earth. What is the missile's speed relative to Earth? Simple, you say. An elementary problem in relative velocity. The correct answer is 3000 m·s^{-1}. But now suppose there is a searchlight in the spaceship, pointing in the same direction that the missile was fired. An observer on the spaceship measures the speed of light emitted by the searchlight and obtains the value c. But, according to our previous discussion, the motion of the light after it has left the source cannot depend on the motion of the source. So when an observer on Earth measures the speed of this same light, he must also obtain the value c. This contradicts our elementary notion of relative velocities and may not appear to agree with common sense. But we have to recognize that "common sense" is intuition based on everyday experience, and this does not usually include measurements of the speed of light. Sometimes we must be prepared to accept results that seem not to make sense when they involve realms far removed from everyday observation.

Thus the principle of relativity, supported by experimental evidence, requires that the speed of light (in vacuum) is independent of the motion of the source and is the same in all frames of reference. To explore the consequences of this statement, consider first the *Newtonian* relationship between two inertial frames, labeled S and S' in Fig. 40–1. To keep things as simple as possible, we have not shown the z-axis on the figure. Let the x-axes of the two frames lie along the same line, but let the origin O' of S' move relative to the origin O of S with constant velocity u along the common x-axis. If the two origins coincide at time $t = 0$, then their separation at a later time t is ut.

We can describe a point P by coordinates (x, y, z) in S or by coordinates (x', y', z') in S'. Reference to the figure shows that these are related by

$$x = x' + ut, \qquad y = y', \qquad z = z'. \tag{40–1}$$

These equations are called the **Galilean coordinate transformation.**

If point P moves in the x-direction, its velocity v relative to S is given by $v = \Delta x / \Delta t$, and its velocity v' relative to S' is $v' = \Delta x'/\Delta t$. Intuitively it is clear that these are related by

$$v = v' + u. \tag{40–2}$$

This relation may also be obtained formally from Eqs. (40–1). Suppose the particle is at a point described by coordinate x_1 or x_1' at time t_1, and at x_2 or x_2' at time t_2. Then $\Delta t = t_2 - t_1$, and, from Eq. (40–1),

$$\Delta x = x_2 - x_1 = (x_2' - x_1') + u(t_2 - t_1)$$
$$= \Delta x' + u\, \Delta t,$$
$$\frac{\Delta x}{\Delta t} = \frac{\Delta x'}{\Delta t} + u,$$

and, in the limit as $\Delta t \to 0$,

$$v = v' - u,$$

in agreement with Eq. (40–2).

A fundamental problem now appears. Applied to the speed of light, Eq. (40–2) says $c = c' + u$. Einstein's principle of relativity, supported by experimental observations, says $c = c'$. This is a genuine inconsistency, not an illusion, and it demands resolution. If we accept the principle of relativity, we are forced to conclude that Eqs. (40–1) and (40–2), intuitively appealing as they are, *cannot* be correct but need to be modified to bring them into harmony with this principle.

The resolution involves modifying our basic kinematic concepts. The first modification involves an assumption so fundamental that it might seem unnecessary, namely, the assumption that the same *time scale* is used in frames S and S'. This may be stated formally by adding to Eqs. (40–1) a fourth equation:

$$t = t'.$$

Obvious though this assumption may seem, it is not correct when the relative speed u of the two frames of reference is comparable to the speed of light. The difficulty lies in the concept of *simultaneity*, which we examine next.

Transforming coordinates from one system to another

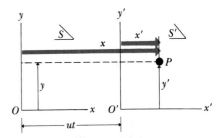

40–1 The position of point P can be described by the coordinates x and y in frame of reference S, or by x' and y' in S'. S' moves relative to S with constant velocity u along the common x–x' axis. The two origins O and O' coincide at time $t = t' = 0$.

Addition of velocities, according to the Galilean coordinate transformation

The assumption that all frames of reference have the same time scale needs to be reexamined.

40–2 RELATIVE NATURE OF SIMULTANEITY

To measure time intervals we need to use the concept of simultaneity.

Measuring times and time intervals involves the concept of **simultaneity.** When a person says he awoke at seven o'clock, he means that two *events* (his awakening and the arrival of the hour hand of his clock at the number seven) occurred *simultaneously*. The fundamental problem in measuring time intervals is that, in general, two events that appear simultaneous in one frame of reference *do not* appear simultaneous in a second frame that is moving relative to the first, even if both are inertial frames.

The following thought experiment, devised by Einstein, illustrates this point. Consider a long train moving with uniform velocity, as shown in Fig. 40–2a. Two lightning bolts strike the train, one at each end. Each bolt leaves a mark on the train and one on the ground at the same instant. The points on the ground are labeled A and B in the figure, and the corresponding points on the train are A' and B'. An observer on the ground is located at O, midway between A and B; another observer is at O', moving with the train and midway between A' and B'. Both observers observe the lightning bolts by means of the light signals they emit.

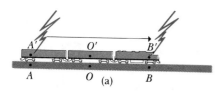

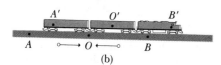

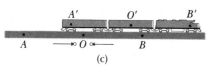

40–2 (a) To the stationary observer at point O, two lightning bolts appear to strike simultaneously. (b) The moving observer at point O' sees the light from the front of the train first and thinks that the bolt at the front struck first. (c) The two light pulses arrive at O simultaneously.

Simultaneity is not an absolute concept; events that seem simultaneous to one observer do not to another.

Suppose the two light signals reach the observer at O simultaneously. He concludes that the two events took place at A and B simultaneously. But the observer at O' is moving with the train, and the light pulse from B' reaches him before the light pulse from A' does. He concludes that the lightning bolt at the front of the train happened *earlier* than the one at the rear. Thus the two events appear simultaneous to one observer but not to the other. *Whether two events at different space points are simultaneous depends on the state of motion of the observer.* It follows that *the time interval between two events at different space points is, in general, different for two observers in relative motion.*

You may want to argue that in this example the lightning bolts really *are* simultaneous, and that if the observer at O' could communicate with the distant points without time delay, he would realize this. But that would be erroneous; the finite speed of information transmission is not the real issue. If O' is really midway between A' and B', then, in his frame of reference, the time for a signal to travel from A' to O' is the same as from B' to O'. Two signals arrive simultaneously at O' only if they were emitted simultaneously at A' and B'. In this example they *do not* arrive simultaneously at O', and so O' must conclude that the events at A' and B' were *not* simultaneous.

Furthermore, there is no basis for saying either that O is right and O' is wrong, or the reverse, since, according to the principle of relativity, no inertial frame of reference is preferred over any other in the formulation of physical laws. Each observer is correct *in his own frame of reference*. In other words, simultaneity is not an absolute concept. Whether two events are simultaneous depends on the frame of reference, and the time interval between two events also depends on the frame of reference.

40–3 RELATIVITY OF TIME

Comparing a time interval between two events in two coordinate systems

To derive a quantitative relation between time intervals in different coordinate systems, we consider another thought experiment. As before, a frame of reference S' moves along the common x–x' axis with constant speed u relative to a frame S. For reasons that will become clear later, we assume that u is always less than the speed of light c. An observer O' in S' directs a source of light at a mirror a distance d away, as shown in Fig. 40–3a, and measures the time interval $\Delta t'$ for light to make the "round trip" to the mirror and back. The total distance is $2d$, so the time interval is

$$\Delta t' = \frac{2d}{c}. \qquad (40\text{--}3)$$

As measured in frame S, the time for the round trip is a different interval Δt. During this time the source moves relative to S a distance $u\,\Delta t$, and the total round-trip distance is not just $2d$ but is $2l$, where

$$l = \sqrt{d^2 + \left(\frac{u\,\Delta t}{2}\right)^2}.$$

In writing this expression, we have used the fact that the distance d looks the same to both observers. This can (and indeed must) be justified by other thought experiments, but we will not go into this matter now. The speed of light is the same for both observers, so the relation in S analogous to Eq. (40–3) is

$$\Delta t = \frac{2l}{c} = \frac{2}{c}\sqrt{d^2 + \left(\frac{u\,\Delta t}{2}\right)^2}. \qquad (40\text{--}4)$$

To obtain a relation between Δt and $\Delta t'$ that does not contain d, we solve Eq. (40–3) for d and substitute the result into Eq. (40–4), obtaining

$$\Delta t = \frac{2}{c}\sqrt{\left(\frac{c\,\Delta t'}{2}\right)^2 + \left(\frac{u\,\Delta t}{2}\right)^2}.$$

This may now be squared and solved for Δt; the result is

$$\Delta t = \frac{\Delta t'}{\sqrt{1 - u^2/c^2}}. \qquad (40\text{--}5)$$

We may generalize this important result: If two events (in our case, the departure and arrival of the light signal at O') occur at the same space point in a frame of reference S' and are separated in time by an interval $\Delta t'$, then the time interval Δt between these two events as observed in S is given by Eq. (40–5). Because the denominator is always smaller than unity, Δt is always *larger* than $\Delta t'$. Thus when the rate of a clock at rest in S' is measured by an observer in S, the rate measured in S is *slower* than the rate observed in S'. This effect is called **time dilation.** We note that Eq. (40–5) makes sense only when $u < c$; otherwise the denominator is imaginary.

It is important to note that the observer in S measuring the time interval Δt cannot do so with a single clock. In Fig. 40–3 the points of departure and return of the light pulse are different space points in S, although they are the same point in S'. If S tries to use a single clock, the finite time of communication between two points will cloud the issue. To avoid this, S may use two assistants with two clocks at the two relevant points. There is no difficulty in synchronizing two clocks in the same frame of reference; one procedure is to send a light pulse simultaneously to two clocks from a point midway between them, with the two assistants setting their clocks to a predetermined time when the pulses arrive. In thought experiments it is often helpful to imagine a large number of observers with synchronized clocks distributed conveniently in a single frame of reference. Only when a clock is moving relative to a given frame of reference do ambiguities of synchronization or simultaneity arise.

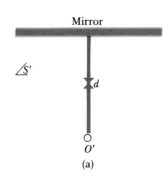

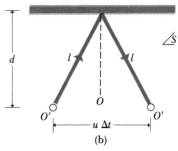

40–3 (a) Light pulse emitted from source at O' and reflected back along the same line, as observed in S'. (b) Path of the same light pulse, as observed in S. The positions of O' at the times of departure and return of the pulse are shown. The speed of the pulse is the same in S as in S', but the path is longer in S.

The relation between time intervals in two coordinate systems: Moving clocks run slower.

For some measurements we have to imagine many synchronized clocks distributed around a particular frame of reference.

EXAMPLE 40–1   A spaceship flies past earth with a speed of $0.99c$ (about 2.97×10^8 m·s^{-1}). A high-intensity signal light (perhaps a pulsed laser) blinks on and off, each pulse lasting 2×10^{-6} s as measured on the spaceship. At a certain instant the ship appears to an observer on earth to be directly overhead at an altitude of 1000 km and to be traveling perpendicular to the line of sight. What is the duration of each light pulse, as measured by this observer, and how far does the ship travel relative to the earth during each pulse?

SOLUTION   The observer does not see the pulse at the instant it is emitted, because the light signal requires a time equal to $(1000 \times 10^3 \text{ m})/(3 \times 10^8 \text{ m·s}^{-1})$, or $(\frac{1}{300})$ s, to travel from the ship to earth. But if the distance from the spaceship to the observer is essentially constant during the emission of a pulse, the time delays at the beginning and end of the pulse are equal, and the time *interval* is not affected.

Let S be the earth's frame of reference, S' that of the spaceship. Then, in the notation of Eq. (40–5), $\Delta t' = 2 \times 10^{-6}$ s. This interval refers to two events occurring at the same point relative to S', namely, the starting and stopping of the pulse. The corresponding interval in S is given by Eq. (40–5):

$$\Delta t = \frac{\Delta t'}{\sqrt{1 - u^2/c^2}} = \frac{2 \times 10^{-6} \text{ s}}{\sqrt{1 - (0.99)^2}} = 14.2 \times 10^{-6} \text{ s}.$$

Thus the time dilation in S is about a factor of seven. The distance d traveled in S during this interval is

$$d = u \, \Delta t = (0.99)(3 \times 10^8 \text{ m·s}^{-1})(14.2 \times 10^{-6} \text{ s})$$

$$= 4210 \text{ m} = 4.21 \text{ km}.$$

If the spaceship is traveling directly *toward* the observer, the time interval cannot be measured directly by a single observer because the time delay is not the same at the beginning and end of the pulse. One possible scheme, at least in principle, is to use *two* observers at rest in S, with synchronized clocks, one at the position of the ship when the pulse starts, the other at its position at the end of the pulse. These observers will again measure a time interval in S of 14.2×10^{-6} s.

Time-dilation effects are not observed in everyday life because the speeds of all practical modes of transportation are much less than the speed of light. For example, for a jet airplane flying at 270 m·s^{-1} (about 600 mi·hr^{-1}),

$$\frac{u^2}{c^2} = \left(\frac{270 \text{ m·s}^{-1}}{3 \times 10^8 \text{ m·s}^{-1}} \right)^2 = 8.1 \times 10^{-13},$$

and the time-dilation factor in Eq. (40–5) is approximately $1 + (4 \times 10^{-13})$. Thus to observe time dilation in this situation requires a clock with a precision of the order of one part in 10^{13}. However, as noted in Section 1–2, atomic clocks capable of this precision have recently been developed, and in the past few years experiments with such clocks in jet airplanes have verified Eq. (40–5) directly.

From the derivation of Eq. (40–5) and the spaceship example, we can see that a time interval between two events occurring at *the same point* in a given frame of reference is a more fundamental quantity than an interval between events at different points. The term *proper time* is used to denote an interval between two events occurring at the same space point. Hence Eq. (40–5) may be used *only* when $\Delta t'$ is a proper time interval in S'; in that case Δt is *not* a proper time interval in S. If, instead, Δt is proper in S, then Δt and $\Delta t'$ must be interchanged in Eq. (40–5).

When the relative velocity u of S and S' is very small, the factor $(1 - u^2/c^2)$ is very nearly equal to unity, and Eq. (40–5) approaches the Newtonian relation $\Delta t = \Delta t'$ (i.e., the same time scale for all frames of reference).

Equation (40–5) suggests an apparent paradox called the **twin paradox.** Consider identical-twin astronauts named Eartha and Astro. Eartha remains on earth while Astro takes off on a high-speed trip through the galaxy. Because of time dilation, Eartha sees Astro's heartbeat and all other life processes proceeding more slowly than his own. Therefore Eartha thinks Astro ages more slowly, so when Astro returns to earth he is younger than Eartha.

Now here is the paradox: Since all inertial frames are equivalent, cannot Astro make exactly the same arguments, to conclude that Eartha is in fact the younger? Thus each twin would think the other is younger, which is a paradox.

The resolution of this paradox is the recognition that the twins are *not* identical in all respects. If Eartha remains in an inertial frame at all times, Astro must at times have an acceleration with respect to inertial frames in order to turn around and come back. Eartha remains always at rest in the same inertial frame; Astro does not. There is a real physical difference between the circumstances of the twins. Careful analysis shows that Eartha is correct: When Astro returns, he *is* younger than Eartha.

> The twin paradox: How can one twin grow old faster than the other?

> The twin paradox isn't actually a paradox if you understand relativity.

40–4 RELATIVITY OF LENGTH

Just as the time interval between two events depends on the observer's frame of reference, the *distance* between two points may also depend on the observer's frame of reference. To measure a distance one must, in principle, observe the positions of two points, such as the two ends of a ruler, simultaneously; but what is simultaneous in one reference frame is not simultaneous in another.

To develop a relation between lengths in various coordinate systems, we consider another thought experiment. We attach a source of light pulses to one end of a ruler and a mirror to the other end, as shown in Fig. 40–4. Let the ruler be at rest in reference frame S' and its length in this frame be l'. Then the time $\Delta t'$ required for a light pulse to make the round trip from source to mirror and back is given by

$$\Delta t' = \frac{2l'}{c}. \tag{40–6}$$

This is a proper time interval, since departure and return occur at the same point in S'.

In S the ruler is moving with speed u, and it is displaced during this travel of the light pulse. Let the length of the ruler in S be l, and let the time of travel from source to mirror, as measured in S, be Δt_1. During this interval the ruler, with source and mirror attached, moves a distance $u\,\Delta t_1$, and the total length of path d from source to mirror is not l but

$$d = l + u\,\Delta t_1. \tag{40–7}$$

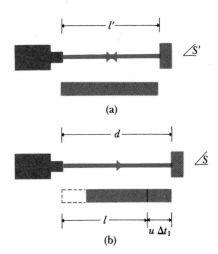

40–4 (a) A light pulse is emitted from a source at one end of a ruler, reflected from a mirror at the opposite end, and returned to the source position. (b) Motion of the light pulse as seen by an observer in S. The distance traveled from source to mirror is greater than the length l measured in S, by the amount $u\,\Delta t_1$, as shown.

But since the pulse travels with speed c, it is also true that

$$d = c\,\Delta t_1. \qquad (40\text{--}8)$$

Combining Eqs. (40–7) and (40–8) to eliminate d, we find

$$c\,\Delta t_1 = l + u\,\Delta t_1,$$

or

$$\Delta t_1 = \frac{l}{c - u}. \qquad (40\text{--}9)$$

In the same way we can show that the time Δt_2 for the return trip from mirror to source is

$$\Delta t_2 = \frac{l}{c + u}. \qquad (40\text{--}10)$$

The *total* time $\Delta t = \Delta t_1 + \Delta t_2$ for the round trip, as measured in S, is

$$\Delta t = \frac{l}{c - u} + \frac{l}{c + u} = \frac{2l}{c(1 - u^2/c^2)}. \qquad (40\text{--}11)$$

We also know that Δt and $\Delta t'$ are related by Eq. (40–5), since $\Delta t'$ is proper in S'. Thus Eq. (40–6) becomes

$$\Delta t\sqrt{1 - \frac{u^2}{c^2}} = \frac{2l'}{c}. \qquad (40\text{--}12)$$

Moving objects appear shorter.

Finally, combining this expression with Eq. (40–11) to eliminate Δt, and simplifying, we obtain

$$l = l'\sqrt{1 - \frac{u^2}{c^2}}. \qquad (40\text{--}13)$$

Thus the length l measured in S, in which the ruler is moving, is *shorter* than the length l' in S', where it is at rest. A length measured in the rest frame of the body is called a proper length; thus l' above is a proper length in S', and the length measured in any other frame is less than l'. This effect is called **length contraction.**

EXAMPLE 40–2 In Example 40–1 (Section 40–3), what distance does the spaceship travel during emission of a pulse, as measured in its rest frame?

SOLUTION The question is somewhat ambiguous since, of course, in its own frame of reference the ship is at rest. But suppose it leaves markers in space (such as small smoke bombs) that remain at rest relative to the earth, at the instants when the pulse starts and stops. Then an observer on the ship measures the distance between these markers, with the aid of observers behind the ship but moving with it, each with a clock synchronized with that of the ship. The distance d between the markers is a proper length in the earth's frame S. In the spaceship's frame S', the distance d' is contracted by the factor given in Eq. (40–13):

$$d' = d\sqrt{1 - \frac{u^2}{c^2}} = (4210\text{ m})\sqrt{1 - (0.99)^2}$$

$$= 594\text{ m}.$$

(Note that because d, not d', is a proper length, we must reverse the roles of l and l'.) An observer in the spaceship can calculate its speed relative to earth from this set of data:

$$u = \frac{d'}{\Delta t'} = \frac{594 \text{ m}}{2 \times 10^{-6} \text{ s}} = 2.97 \times 10^8 \text{ m·s}^{-1},$$

which agrees with the initial data.

When u is very small compared to c, the contraction factor in Eq. (40–13) approaches unity, and in the limit of small speeds we recover the Newtonian relation $l = l'$. This and the corresponding result for time dilation shows that Eqs. (40–1) retain their validity in the limit of speeds much smaller than c; only at speeds comparable to c are modifications needed.

We have derived Eq. (40–13) for lengths measured in the direction *parallel* to the relative motion of the two frames of reference. Lengths measured *perpendicular* to the direction of motion are *not* contracted. To prove this, consider two identical rulers; one ruler lies along the y-axis with one end at the origin O of frame of reference S, and the other lies along the y'-axis with one end at the origin O' of S'. At the instant the two origins coincide, observers in the two frames of reference S and S' observe the positions of the upper ends of the rulers. If the observer in S thinks his ruler is shorter, the observer in S' must think his ruler is longer. But this would mean that there is some distinction between the two frames of reference, which violates our basic premise that all inertial frames of reference are equivalent. Thus both observers must conclude that the two rulers have the same length, despite the fact that to each observer, one is stationary and one is moving. Hence there is no length contraction perpendicular to the direction of relative motion of the coordinate systems.

Lengths perpendicular to the observer's motion do not contract.

40-5 THE LORENTZ TRANSFORMATION

The Galilean coordinate transformation, given by Eqs. (40–1) and the relation $t = t'$, is valid only in the limit when u is much smaller than c. We are now ready to derive a more general transformation that is not subject to this limitation. The more general relations are called the **Lorentz transformation.** In the limit of very small u, they reduce to the Galilean transformation, but they may also be used when u is comparable to c.

The Lorentz transformation: a coordinate transformation consistent with requirements of the principle of relativity

The basic problem is this: When an event occurs at point (x, y, z) at time t, as observed in a frame of reference S, what are the coordinates (x', y', z') and time t' of the event as observed in a second frame S' moving relative to S with constant velocity u along the x-direction?

To derive the transformation equations, we refer to Fig. 40–5, which is the same as Fig. 40–1. As before, we assume that the origins coincide at the initial time $t = t' = 0$. Then in S the distance from O to O' at time t is still ut. The coordinate x' is a proper length in S', so in S it appears contracted by the factor given in Eq. (40–13). Thus the distance x from O to P in S is given not simply by $x = ut + x'$ as in the Galilean transformation, but by

$$x = ut + x'\sqrt{1 - \frac{u^2}{c^2}}. \qquad (40\text{--}14)$$

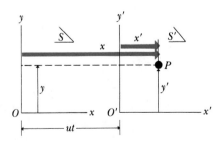

40–5 The distance x' is a proper length in S'. In S it appears contracted by the factor $\sqrt{1 - u^2/c^2}$. The distance between O and O', as seen in S, is ut, and x is a proper length in S. Thus $x = ut + x'\sqrt{1 - u^2/c^2}$.

Solving this equation for x', we obtain

$$x' = \frac{x - ut}{\sqrt{1 - u^2/c^2}}. \tag{40–15}$$

This equation is half of the Lorentz transformation; the other half is the equation giving t' in terms of x and t. To obtain it we note that the principle of relativity requires that the *form* of the transformation from S to S' be identical to that from S' to S, the only difference being a change in the sign of the relative velocity u. Thus, from Eq. (40–14), it must be true that

$$x' = -ut' + x\sqrt{1 - \frac{u^2}{c^2}}. \tag{40–16}$$

We may now equate Eqs. (40–15) and (40–16) to eliminate x' from this expression, obtaining the desired relation for t' in terms of x and t. We will leave the algebraic details for you to work out; the result is

$$t' = \frac{t - ux/c^2}{\sqrt{1 - u^2/c^2}}. \tag{40–17}$$

As we discussed previously, lengths perpendicular to the direction of relative motion are not affected by the motion, so $y' = y$ and $z' = z$.

Collecting all the transformation equations, we have

$$x' = \frac{x - ut}{\sqrt{1 - u^2/c^2}}, \qquad y' = y, \qquad z' = z, \qquad t' = \frac{t - ux/c^2}{\sqrt{1 - u^2/c^2}}. \tag{40–18}$$

These are the *Lorentz transformation* equations, the relativistic generalization of the Galilean transformation, Eqs. (40–1). When u is much smaller than c, the two transformations become identical, and $t = t'$. In both cases we assume that the relative motion of the two coordinate systems is along the common xx'-axis.

Using the Lorentz transformation to derive a relativistically correct velocity transformation

Next we consider the relativistic generalization of the Galilean velocity-transformation equations, Eq. (40–2). As we have noted, the Galilean relation is valid only in the limit when u is very small. The relativistic expression can easily be obtained from the Lorentz transformation. In the following discussion, we continue to use u as the velocity of frame of reference S' relative to S, and we use v and v' for the x-components of velocity of a particle as measured in S and S', respectively. Suppose a particle observed in S' is at point x_1' at time t_1' and point x_2' at time t_2'. Then its x-component of velocity v' is given by

$$v' = \frac{x_2' - x_1'}{t_2' - t_1'} = \frac{\Delta x'}{\Delta t'}. \tag{40–19}$$

To obtain the x-component of velocity in S, we use Eqs. (40–18) to translate this expression into terms of the corresponding positions x_1 and x_2 and times t_1 and t_2 observed in S. We find

$$x_2' - x_1' = \frac{x_2 - x_1 - u(t_2 - t_1)}{\sqrt{1 - u^2/c^2}} = \frac{\Delta x - u\,\Delta t}{\sqrt{1 - u^2/c^2}},$$

$$t_2' - t_1' = \frac{t_2 - t_1 - u(x_2 - x_1)/c^2}{\sqrt{1 - u^2/c^2}} = \frac{\Delta t - u\,\Delta x/c^2}{\sqrt{1 - u^2/c^2}}.$$

Using these results in Eq. (40–19), we find

$$v' = \frac{\Delta x - u\,\Delta t}{\Delta t - u\,\Delta x/c^2}$$

$$= \frac{(\Delta x/\Delta t) - u}{1 - (u/c^2)(\Delta x/\Delta t)}.$$

In the limit $\Delta t \to 0$, $\Delta x/\Delta t$ is just the x-component of velocity v measured in S, so we finally obtain

$$v' = \frac{v - u}{1 - uv/c^2}. \tag{40–20}$$

Note that when u and v are much smaller than c, the denominator approaches unity, and we obtain the nonrelativistic result $v' = v - u$. The opposite extreme is the case $v = c$; then we find

Anything moving with speed c relative to one observer also moves with speed c relative to any other observer.

$$v' = \frac{c - u}{1 - uc/c^2} = c.$$

That is, anything moving with speed c relative to S also has speed c relative to S', despite the relative motion of the two frames. This result demonstrates the consistency of Eq. (40–20) with our initial assumption that the speed of light is the same in all frames of reference.

We can also rearrange Eq. (40–20) to give v in terms of v'. We leave the algebraic details as a problem; the result is

$$v = \frac{v' + u}{1 + uv'/c^2}. \tag{40–21}$$

EXAMPLE 40–3 A spaceship moving away from earth with a speed $0.9c$ fires a missile in the same direction as its motion, with a speed $0.9c$ relative to the spaceship. What is the missile's speed relative to the earth?

SOLUTION Let the earth's frame of reference be S, the spaceship S'. Then $v' = 0.9c$ and $u = 0.9c$. The nonrelativistic velocity addition formula would give a velocity relative to earth of $1.8c$. The correct relativistic result, obtained from Eq. (40–21), is

$$v = \frac{0.9c + 0.9c}{1 + (0.9c)(0.9c)/c^2}$$

$$= 0.994c.$$

When u is less than c, a body moving with a speed less than c in one frame of reference also has a speed less than c in *every other* frame of reference. This is one reason for thinking that no material body can travel with a speed greater than that of light, relative to *any* frame of reference. The relativistic generalizations of energy and momentum, to be considered in the following sections, give further support to this hypothesis.

Nothing can travel faster than light.

1. Try hard to understand the concepts of proper time and proper length. A time interval between two events that happen at the same point in a particular frame of reference is a proper time in that frame, and the time interval between the same two events is longer in any other frame. A length measured between two points at rest in a frame of reference is a proper length in that frame, and the distance between the points is shorter in any other frame.

2. You can get a lot of mileage out of the Lorentz transformation equations. It helps to make a list of what you know and don't know. Do you know the coordinates in one frame? The time of an event in one frame? What are your knowns and unknowns?

3. In velocity-transformation problems, if you have two observers measuring the motion of a body, decide which you want to call S and which S', identify the velocities v and v' clearly, and make sure you know the velocity u of S' relative to S. Use either form of the velocity-transformation equation, Eq. (40–20) or (40–21), whichever is more convenient.

4. Don't be discouraged if some of your results don't seem to make sense or if they disagree with your intuition. Reliable intuition about relativity takes time to develop. Keep trying to develop intuitive understanding; it will come.

40–6 MOMENTUM

To save conservation of momentum we have to revise our definition of momentum.

We have discussed the fact that Newton's laws of motion are *invariant* under the Galilean coordinate transformation, but that to satisfy the principle of relativity, this transformation must be replaced by the more general Lorentz transformation. This requires corresponding generalizations in the laws of motion and the definitions of momentum and energy.

The principle of conservation of momentum states that *when two bodies collide, the total momentum is constant,* provided there is no interaction except that of the two bodies with each other. However, when we consider a collision in one coordinate system S, in which momentum is conserved, and then use the Lorentz transformation to obtain the velocities in a second system S', we find that if the Newtonian definition of momentum ($\boldsymbol{p} = m\boldsymbol{v}$) is used, momentum is *not* conserved in the second system. According to the basic principle of relativity, if momentum conservation is a valid physical law, it must hold *exactly* in all inertial frames of reference and for all particle speeds, whether or not they are small compared to c. Thus if the Lorentz transformation is correct, we must revise the *definition* of momentum.

Deriving the correct relativistic generalization is beyond our scope, and we simply quote the result. The **relativistic momentum $\boldsymbol{p}$** of a particle of mass m moving with velocity $\boldsymbol{v}$ is given by

$$p = \frac{mv}{\sqrt{1 - v^2/c^2}}. \tag{40–22}$$

Note that, as usual when the particle speed v is much less than c, this is approximately equal to the Newtonian expression $\boldsymbol{p} = m\boldsymbol{v},$ but that in general the momentum is greater in magnitude than mv.

In these momentum expressions, m is a *constant* characterizing the inertial properties of a particle. Because $\boldsymbol{p} = m\boldsymbol{v}$ is still valid in the limit of very small velocities, we see that m must be the same quantity we used (and learned to measure) in our study of Newtonian mechanics early in the book. In relativistic mechanics it is often called the **rest mass** of a particle.

In Newtonian mechanics, the second law of motion can be stated in the form

What is the relativistic relation between force and motion?

$$F = \frac{d\boldsymbol{p}}{dt}. \tag{40–23}$$

That is, force equals time rate of change of momentum. Experiments show that this result is still valid in relativistic mechanics, provided we use the relativistic momentum given by Eq. (40–22). That is, the relativistically correct generalization of Newton's second law is found to be

$$\boldsymbol{F} = \frac{d}{dt}\frac{m\boldsymbol{v}}{\sqrt{1 - v^2/c^2}}. \tag{40–24}$$

This relation has the consequence that constant force no longer causes constant acceleration. For example, when force and velocity are both along a single line, causing straight-line motion, it is not difficult to show that Eq. (40–24) becomes

$$F = \frac{m}{(1 - v^2/c^2)^{3/2}}\frac{dv}{dt},$$

or

$$\frac{dv}{dt} = a = \frac{F}{m}(1 - v^2/c^2)^{3/2}. \tag{40–25}$$

Why can't we accelerate an object up to the speed of light?

We leave the derivation of Eq. (40–25) as a problem. This equation shows that as a particle's speed increases, the acceleration caused by a given force continuously *decreases*. As the speed approaches c, the acceleration approaches zero, no matter how great a force is applied. Thus it is impossible to accelerate a particle from a state of rest to a speed equal to or greater than c, and the speed of light is sometimes referred to as "the ultimate speed."

Equation (40–22) is sometimes interpreted to mean that a rapidly moving particle undergoes an increase in mass. If the mass at zero velocity (the rest mass) is denoted by m, then the "relativistic mass" m_{rel} is given by

$$m_{\text{rel}} = \frac{m}{\sqrt{1 - v^2/c^2}}.$$

Relativistic mass: a concept with uses and pitfalls

Indeed, when we consider the motion of a system of particles (such as gas molecules in a moving container), the total mass of the system is the sum of the relativistic masses of the particles rather than the sum of their rest masses.

The concept of relativistic mass also has its pitfalls, however. It is *not* correct to say that the relativistic generalization of Newton's second law is $\boldsymbol{F} = m_{\text{rel}}\boldsymbol{a}$, and it is *not* correct that the relativistic kinetic energy of a particle is $K = \frac{1}{2}m_{\text{rel}}v^2$. So this concept must be approached with great caution. For the present discussion it is better to regard Eq. (40–22) as a generalized definition of momentum and to retain the meaning of m as an inertial quantity characteristic of each particle and independent of its state of motion.

40–7 WORK AND ENERGY

When we developed the relationship between work and kinetic energy in Chapter 7, we used Newton's laws of motion. As we have just seen, Newton's laws have to be generalized to bring them into harmony with the principle of

Deriving a relativistically correct work–energy relation

relativity; hence it is not surprising that we also need to generalize the definition of kinetic energy and the work–energy relation.

Because a constant force on a body does not cause a constant acceleration (except when the speed of the body is very small), even the simplest dynamics problems require the use of calculus. But we can derive fairly easily the relativistic generalization of the work–energy principle. We begin with the definition of work: $W = \int F\,dx$. According to Eq. (40–23), $F = dp/dt$. Combining these, we write

$$W_{12} = \int_{x_1}^{x_2} F\,dx = \int_{x_1}^{x_2} \frac{dp}{dt}\,dx. \qquad (40\text{–}26)$$

In order to derive the generalized expression for kinetic energy K as a function of speed v, we would like to convert this to an integral on v. To do so, we first note that dp, dx, and dv are the infinitesimal increments in p, x, and v, respectively, in the time interval dt. Thus it is permissible to interchange dp and dx in the right side of Eq. (40–26) and rewrite it as

$$\int_{x_1}^{x_2} F\,dx = \int_{x_1}^{x_2} \frac{dp}{dt}\,dx = \int_{x_1}^{x_2} \frac{dx}{dt}\,dp. \qquad (40\text{–}27)$$

Now $dx/dt = v$, and dp may be expressed in terms of dv as $dp = (dp/dv)dv$. Thus Eq. (40–27) becomes

$$\int_{x_1}^{x_2} F\,dx = \int_{v_1}^{v_2} v\frac{dp}{dv}\,dv. \qquad (40\text{–}28)$$

Note that the limits on the integral on the right are the speeds v_1 and v_2 at positions x_1 and x_2, respectively. To find dp/dv in terms of v, we take the derivative of Eq. (40–22). We will let you work out the details; the result is

$$\frac{dp}{dv} = \frac{m}{(1 - v^2/c^2)^{3/2}}.$$

We substitute this into Eq. (40–28) to obtain, finally, an expression that we can integrate easily

$$W_{12} = \int_{x_1}^{x_2} F\,dx = \int_{v_1}^{v_2} \frac{mv\,dv}{(1 - v^2/c^2)^{3/2}}$$

$$= \frac{mc^2}{\sqrt{1 - v_2^2/c^2}} - \frac{mc^2}{\sqrt{1 - v_1^2/c^2}}, \qquad (40\text{–}29)$$

where v_1 and v_2 are the initial and final speeds of the particle, respectively.

This result suggests defining kinetic energy as

$$\frac{mc^2}{\sqrt{1 - v^2/c^2}}. \qquad (40\text{–}30)$$

The relativistic expression for kinetic energy

But this expression is not zero when $v = 0$; instead it becomes equal to mc^2. Thus the correct relativistic generalization of kinetic energy K is

$$K = \frac{mc^2}{\sqrt{1 - v^2/c^2}} - mc^2. \qquad (40\text{–}31)$$

This expression, if correct, must reduce to the Newtonian expression $K = \frac{1}{2}mv^2$ when v is much smaller than c. It is not obvious that this is the case; to

A nuclear-fission explosion, a dramatic and frightening example of conversion of nuclear mass into energy. (Courtesy of U.S. Dept. of Energy.)

demonstrate that it is so, we can expand the radical by using the binomial theorem:

$$\left(1 - \frac{v^2}{c^2}\right)^{-1/2} = 1 + \frac{1}{2}\frac{v^2}{c^2} + \frac{3}{8}\frac{v^4}{c^4} + \frac{5}{16}\frac{v^6}{c^6} + \cdots.$$

Combining this with Eq. (40–31), we find

At slow speeds the relativistic kinetic energy expression reduces to the familiar Newtonian relation.

$$K = mc^2\left(1 + \frac{1}{2}\frac{v^2}{c^2} + \frac{3}{8}\frac{v^4}{c^4} + \cdots\right) - mc^2$$

$$= \frac{1}{2}mv^2 + \frac{3}{8}m\frac{v^4}{c^2} + \cdots. \qquad (40\text{–}32)$$

In each expression the dots stand for omitted terms. When v is much smaller than c, all the terms in the series except the first are negligibly small, and we obtain the classical $\frac{1}{2}mv^2$.

But what is the significance of the term mc^2 that had to be subtracted in Eq. (40–31)? Although Eq. (40–30) does not give the *kinetic energy* of the particle, perhaps it represents some kind of *total* energy, including both the kinetic energy and an additional energy mc^2, which the particle possesses even when it is not moving. Calling this total energy E and using Eq. (40–31), we find

$$E = K + mc^2 = \frac{mc^2}{\sqrt{1 - v^2/c^2}}. \qquad (40\text{–}33)$$

The energy mc^2 associated with mass rather than motion may be called the **rest energy** of the particle. This speculation does not prove that the concept of rest energy is meaningful, but it points the way toward further investigation.

Rest energy: energy associated with mass rather than motion

There is in fact direct experimental evidence of the existence of rest energy. The simplest example is the decay of the π° meson, an unstable particle that "decays"; in the decay process, the particle disappears and electromagnetic radiation appears. When the particle is at rest (and therefore has no

Conversion of mass into energy: direct confirmation of the validity of the concept of rest mass

kinetic energy) before its decay, the total energy of the radiation produced is found to be exactly equal to mc^2. There are many other examples of fundamental particle transformations in which the total mass of the system changes; in every case, a corresponding energy change occurs, consistent with the assumption of a rest energy mc^2 associated with a rest mass m.

Although the principles of conservation of mass and of energy originally were developed quite independently, the theory of relativity shows that they are but two special cases of a single broader conservation principle, the *principle of conservation of mass and energy*. In some physical phenomena, neither mass nor energy is separately conserved, but the changes in these quantities are governed by a more general conservation principle: When a change in rest mass m occurs in an isolated system, an opposite change mc^2 in the total energy of other types must accompany this change.

The conversion of mass into energy is the fundamental principle involved in the generation of power through nuclear reactions, a subject we will discuss in Chapter 44. When a uranium nucleus undergoes fission in a nuclear reactor, the total mass of the resulting fragments is less than that of the parent nucleus, and the total kinetic energy of the fragments is equal to this mass deficit times c^2. This kinetic energy can be used to produce steam to operate turbines for electric-power generators or in a variety of other ways.

We can relate the total energy E (kinetic plus rest) of a particle directly to its momentum by combining Eqs. (40–22) and (40–33) to eliminate the particle's velocity. The simplest procedure is to rewrite these equations in the following forms:

$$\left(\frac{E}{mc^2}\right)^2 = \frac{1}{1 - v^2/c^2}; \qquad \left(\frac{p}{mc}\right)^2 = \frac{v^2/c^2}{1 - v^2/c^2}.$$

Subtracting the second of these equations from the first and rearranging, we find

$$E^2 = (mc^2)^2 + (pc)^2. \qquad (40\text{--}34)$$

Again we see that for a particle at rest ($p = 0$), $E = mc^2$.

Equation (40–34) also suggests that a particle may have energy and momentum even when it has no rest mass. In such a case, $m = 0$ and

$$E = pc. \qquad (40\text{--}35)$$

In fact, massless particles do exist. The most familiar examples are photons, the quanta of electromagnetic radiation. The existence of these particles is well established, and we will study them in greater detail in later chapters. They always travel with the speed of light; they are emitted and absorbed during changes of state of atomic systems, accompanied by corresponding changes in the energy and momentum of these systems.

40–8 INVARIANCE

Invariance: things that don't change
when you go from one coordinate system
to another

We have used the term **invariance** to describe the idea that, according to the principle of relativity, all the fundamental laws of physics must keep the same form, that is, must be *invariant,* when we shift our description from one inertial frame of reference to another. This word is also used in a more specific sense to describe *physical quantities* that have the same value in all inertial frames of reference. The speed of light is one obvious example, but there are others that are important.

For example, we can describe the position (x, y, z) and time t of an event in a frame of reference S, and the position (x', y', z') and time t' of that same event in a frame S' moving with constant velocity relative to S. In general, of course, x is not equal to x', nor y to y', nor t to t'. But we can prove that the quantity $(x^2 + y^2 + z^2 - c^2t^2)$ *does* have the same value in both systems, no matter what the event. That is, it is always true that the position and time in S are related to the position and time in S' by

$$x^2 + y^2 + z^2 - c^2t^2 = x'^2 + y'^2 + z'^2 - c^2t'^2. \qquad (40\text{–}36)$$

We leave the detailed proof of this as a problem; just substitute Eqs. (40–18) into the right side of Eq. (40–36) and simplify the result to show that it equals the left side.

We say that the quantity $(x^2 + y^2 + z^2 - c^2t^2)$ is *invariant* under the Lorentz transformation. This invariance is directly related to the invariance of the speed of light. To see this, suppose a light pulse starts from the origin O of frame S at time $t = 0$; the square of its distance from the origin at any later time t is $(x^2 + y^2 + z^2)$ and must equal $(ct)^2$. But when the same light pulse is observed in frame S', an exactly similar equation must hold for the coordinates and time in S'. Thus Eq. (40–36) is not so surprising.

We can obtain another useful invariant quantity from Eq. (40–34). The rest mass m of a particle is an inherent property of the particle, independent of the frame of reference. Thus the energy–momentum relation of Eq. (40–34) tells us that the quantity $(E^2 - p^2c^2)$ is *invariant*. That is, if a particle has energy E and momentum of magnitude p, computed from the velocity v measured in frame of reference S, and energy E' and momentum of magnitude p' computed from the velocity v' in S', then

Invariance of a quantity involving energy and momentum

$$E^2 - p^2c^2 = E'^2 - p'^2c^2. \qquad (40\text{–}37)$$

It is not very hard to show that this relation is also valid for a collection of particles. That is, if as usual we define the total energy (rest plus kinetic) of a set of particles as the sum of the energies of the individual particles, and the total momentum as the vector sum of momenta of particles, then Eq. (40–37) also holds if we interpret E as the *total* energy and p as the magnitude of the *total* momentum. These relations are useful in analyzing relativistic collision problems.

40–9 THE DOPPLER EFFECT

An additional important consequence of relativistic kinematics is the Doppler effect for electromagnetic radiation. In our previous discussion of the Doppler effect in Section 23–5, we quoted without proof the formula, Eq. (23–18), for the frequency shift that results from motion of a source relative to an observer. We can now derive that result.

The Doppler effect: The frequency seems different when the source and observer are in relative motion.

Suppose a light source at rest in a frame of reference S' emits light pulses with frequency f', as measured by an observer in that frame. The period, or time between pulses, is $\tau' = 1/f'$ as seen in S'. To an observer in frame S, the source is moving with speed u; the frequency of the pulses as seen by this observer is f, and the time between successive pulses measured by this observer

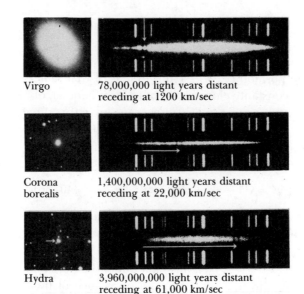

Virgo — 78,000,000 light years distant receding at 1200 km/sec

Corona borealis — 1,400,000,000 light years distant receding at 22,000 km/sec

Hydra — 3,960,000,000 light years distant receding at 61,000 km/sec

The Doppler effect for light. Each photograph shows two dark bands characteristic of the absorption spectrum of calcium in the spectra of three galaxies. The recession of the sources from Earth causes Doppler "redshifts" to smaller frequencies and longer wavelengths. For a nearby galaxy in Virgo the redshift is small, but in the Hydra galaxy, nearly 50 times as far away, the shift is much greater. (Palomar Observatory Photograph.)

is $\tau = 1/f$. Because τ' is a proper time interval in S', τ and τ' are related by

$$\tau = \frac{\tau'}{\sqrt{1 - u^2/c^2}}. \tag{40-38}$$

During the interval τ between pulses, as seen in S, the source moves to the right a distance of $u\tau$ and the first pulse moves to the left a distance $c\tau$. Hence the distance between successive pulses approaching the observer in S is $(c + u)\tau$. But this is also equal to the wavelength λ, which in turn is equal to c/f. Thus we have

$$\lambda = (c + u)\tau = \frac{c}{f}. \tag{40-39}$$

We now use Eq. (40–38) to express τ in terms of τ' and then use the relation $\tau' = 1/f'$ to obtain

$$(c + u)\frac{\tau'}{\sqrt{1 - u^2/c^2}} = \frac{1}{f'} \frac{c + u}{\sqrt{1 - u^2/c^2}} = \frac{c}{f}. \tag{40-40}$$

Finally, we rearrange this to obtain

$$f = \sqrt{\frac{c - u}{c + u}} f'. \tag{40-41}$$

This relation has the same form as Eq. (23–18), with minor differences in notation.

The relativistic Doppler effect is different from the effect with sound.

Unlike the situation with sound, with light there is no distinction between motion of source and motion of observer; only the *relative* velocity of the two is significant. Equation (40–41) shows that when the source is moving away from the observer, the observed frequency is less than the frequency measured in the rest frame of the source. If the source is moving *toward* the observer, we change the sign of u in Eq. (40–41) to find the apparent *increase* in frequency. The last three paragraphs of Section 23–5 discuss several practical applications of the Doppler effect with light and other electromagnetic radiation; we suggest you review those paragraphs now.

40–10 RELATIVITY AND NEWTONIAN MECHANICS

The sweeping changes required by the principle of relativity go to the very roots of Newtonian mechanics, including the concepts of length and time, the equations of motion, and the conservation principles. Thus it may appear that the foundations on which Newton's mechanics are built have been destroyed. While this is true in one sense, it is essential to keep in mind that the Newtonian formulation still retains its validity whenever speeds are small compared with the speed of light. In such cases time dilation, length contraction, and the modifications of the laws of motion are so small that they are unobservable. In fact, every one of the principles of Newtonian mechanics survives as a special case of the more general relativistic formulation.

Is there anything left of Newtonian mechanics?

Newtonian mechanics is not *wrong;* it is *incomplete.* It is a limiting case of relativistic mechanics. It is approximately correct when all speeds are small compared to c; and in the limit when all speeds approach zero, it becomes exactly correct.

Newtonian mechanics is a special case of relativistic mechanics, when all the speeds are much smaller than c.

Thus relativity does not completely destroy Newtonian mechanics but *generalizes* it. After all, Newton's laws rest on a very solid base of experimental evidence, and it would be strange indeed to advance a new theory inconsistent with this evidence. So it always is with the development of physical theory. Whenever a new theory is in partial conflict with an older, established theory, it nevertheless must yield the same predictions as the old in areas where the old theory is supported by experimental evidence. Every new physical theory must pass this test, called the **correspondence principle,** which has come to be regarded as a fundamental procedural rule in all physical theory. There are many problems for which Newtonian mechanics is clearly inadequate, including all situations where particle speeds are comparable to that of light or where direct conversion of mass to energy occurs. But there is still a large area, including nearly all the behavior of macroscopic bodies in mechanical systems, in which Newtonian mechanics is still perfectly adequate.

At this point it is legitimate to ask whether the relativistic mechanics just discussed is the final word on this subject or whether *further* generalizations are possible or necessary. For example, inertial frames of reference have occupied a privileged position in all our discussions thus far. Should the principle of relativity be extended to noninertial frames as well?

Generalizations to noninertial frames of reference

Here is an example that illustrates some implications of this question. A student decides to go over Niagara Falls while enclosed in a large wooden box. During his free fall he can, in principle, perform experiments inside the box. An object released inside the box does not fall to the floor because both the box and the object are in free fall with a downward acceleration of 9.8 m·s^{-2}. But an alternative interpretation, from this man's point of view, is that the force of gravity has suddenly been turned off. Provided he remains in the box and it remains in free fall, he cannot tell whether he is indeed in free fall or whether the force of gravity has vanished. A similar problem appears in a space station in orbit around the earth. Objects in the spaceship appear weightless, but without going outside the ship there is no way to determine whether gravity has disappeared or the spaceship is in an accelerated (i.e., noninertial) frame of reference.

These considerations form the basis of Einstein's **general theory of relativity.** If we cannot distinguish experimentally between a gravitational field at a particular location and an accelerated reference system, then there cannot be any real distinction between the two. Pursuing this concept, we may try to

The general theory of relativity: What's the difference between gravitation and an accelerated frame of reference?

represent *any* gravitational field in terms of special characteristics of the coordinate system. This turns out to require even more sweeping revisions of our space–time concepts than did the special theory of relativity, and we find that, in general, the geometric properties of space are noneuclidean.

The basic ideas of the general theory of relativity are now well established, but some of the details remain speculative in nature. Its chief application is in cosmological investigations of the structure of the universe, the formation and evolution of stars, and related matters. It is not believed to have any relevance for atomic or nuclear phenomena or macroscopic mechanical problems of less than astronomical dimensions.

SUMMARY

The principle of relativity states that all the fundamental laws of physics have the same form in all inertial frames of reference. In particular, the speed of light is the same in all inertial frames and is independent of the motion of the source. This is inconsistent with the Galilean coordinate transformation, which has to be modified. Simultaneity is not an absolute concept; two events that appear simultaneous in one frame of reference in general do not appear simultaneous in a second frame moving relative to the first.

If the time interval between two events occurring at the same space point in a frame S' is $\Delta t'$, and if S' moves with constant velocity u relative to S, the time interval Δt between the events, as observed in S, is given by

$$\Delta t = \frac{\Delta t'}{\sqrt{1 - u^2/c^2}}. \qquad (40\text{--}5)$$

This effect is called time dilation, and $\Delta t'$ is called a proper time interval in S'.

If the distance between two points at rest in a frame S' is l', and if S' moves with constant velocity u relative to S, the distance l between the points as measured in S is

$$l = l'\sqrt{1 - \frac{u^2}{c^2}}. \qquad (40\text{--}13)$$

This effect is called length contraction, and l' is called a proper length in S'.

The Lorentz transformation relates the coordinates and time of an event in one inertial coordinate system to the coordinates and time of the same event as observed in a second inertial coordinate system moving with constant velocity u relative to the first. The transformation equations are

$$x' = \frac{x - ut}{\sqrt{1 - u^2/c^2}},$$
$$y' = y,$$
$$z' = z,$$
$$t' = \frac{t - ux/c^2}{\sqrt{1 - u^2/c^2}}. \qquad (40\text{--}18)$$

If a particle's velocity in S' is v', and S' moves relative to S with velocity u, then the particle's velocity v in S is given by

$$v = \frac{v' + u}{1 + uv'/c^2}. \qquad (40\text{--}21)$$

In order for momentum conservation in collisions to hold in all coordinate systems, the definition of momentum must be generalized. For a particle of mass m moving with velocity v, the momentum p is defined as

$$p = \frac{mv}{\sqrt{1 - v^2/c^2}}. \tag{40-22}$$

Generalizing the work–energy relation requires that we generalize the definition of kinetic energy K. For a particle of mass m moving with speed v,

$$K = \frac{mc^2}{\sqrt{1 - v^2/c^2}} - mc^2. \tag{40-31}$$

This form suggests assigning a rest energy mc^2 to a particle, so that the total energy E, kinetic energy plus rest energy, is

$$E = K + mc^2 = \frac{mc^2}{\sqrt{1 - v^2/c^2}}. \tag{40-33}$$

The total energy E and magnitude of momentum p for a particle of rest mass m are related by

$$E^2 = (mc^2)^2 + (pc)^2. \tag{40-34}$$

The quantity

$$x^2 + y^2 + z^2 - c^2t^2$$

is invariant; for a given event it has the same value in all inertial frames. Similarly, the quantity

$$E^2 - p^2c^2$$

is invariant.

The special theory of relativity is a generalization of Newtonian mechanics. All the principles of Newtonian mechanics are present as limiting cases when all the speeds are small compared to c. Further generalization to include accelerated frames of reference and their relation to gravitational fields leads to the general theory of relativity.

QUESTIONS

40-1 What do you think would be different in everyday life if the speed of light were 10 m·s^{-1} instead of its actual value?

40-2 The average life span in the United States is about 70 years. Does this mean that it is impossible for an average person to travel a distance greater than 70 light-years away from the earth? (A light-year is the distance light travels in a year.)

40-3 What are the fundamental distinctions between an inertial frame of reference and a noninertial frame?

40-4 A physicist claimed that it is impossible to define what is meant by a rigid body in a relativistically correct way. Why?

40-5 Two events occur at the same space point in a particular frame of reference and appear simultaneous in that frame. Is it possible that they may not appear simultaneous in another frame?

40-6 Does the fact that simultaneity is not an absolute concept also destroy the concept of *causality*? If event A is to *cause* event B, A must occur first. Is is possible that in some frames A may appear to cause B, and in others B may appear to cause A?

40-7 A social scientist who has done distinguished work in fields far removed from physics has written a book purporting to refute the special theory of relativity. He begins with a premise that might be paraphrased as follows: "Either two events occur at the same time, or they don't; that's just common sense." How would you respond to this in the light of our discussion of the relative nature of simultaneity?

40–8 When an object travels across an observer's field of view at a relativistic speed, it appears not only foreshortened but also slightly rotated, with the side toward the observer shifted in the direction of motion relative to the side away from him. How does this come about?

40–9 According to the twin paradox mentioned in Section 40–3, if one twin stays on earth while the other takes off in a spaceship at relativistic speed and then returns, one will be older than the other. Can you think of a practical experiment, perhaps using two very precise atomic clocks, that would test this conclusion?

40–10 When a monochromatic light source moves toward an observer, its wavelength appears to be shorter than the value measured when the source is at rest. Does this contradict the hypothesis that the speed of light is the same for all observers? What about the apparent frequency of light from a moving source?

40–11 A student asserted that a massive particle must always have a speed less than that of light, while a massless particle must always travel at exactly the speed of light. Is he correct? If so, how do massless particles such as photons and neutrinos acquire this speed? Can they not start from rest and accelerate?

40–12 The theory of relativity sets an upper limit on the speed a particle can have. Are there also limits on its energy and momentum?

40–13 In principle, does a hot gas have more mass than the same gas when it is cold? Explain. In practice, would this be a measurable effect?

40–14 Why do you think the development of Newtonian mechanics preceded the more refined relativistic mechanics by so many years?

EXERCISES

Section 40–2 Relative Nature of Simultaneity

40–1 For the two trains discussed in Section 40–2, suppose the two lightning bolts appear simultaneous to an observer on the train. Show that they *do not* appear simultaneous to an observer on the ground. Which appears to come first?

Section 40–3 Relativity of Time

40–2 The π^+ meson, an unstable particle, lives on average about 2.6×10^{-8} s (measured in its own frame of reference) before decaying.

a) If such a particle is moving with respect to the laboratory with a speed of $0.8c$, what lifetime is measured in the laboratory?

b) What distance, measured in the laboratory, does the particle move before decaying?

40–3 The μ^+ meson (or positive muon) is an unstable particle with a lifetime of about 2.3×10^{-6} s (measured in the rest frame of the muon).

a) If the muon is made to travel at very high velocity relative to a laboratory, its lifetime is measured to be 1.6×10^{-5} s. Calculate the speed of the muon, expressed as a fraction of c.

b) What distance, measured in the laboratory, does the particle travel during its lifetime?

40–4 Two atomic clocks are carefully synchronized. One remains in New York while the other is loaded on a supersonic airplane that travels at an average speed of 400 m·s^{-1} and then returns to New York. When the plane returns, the elapsed time on the clock that stayed behind is 5 hr. By how much will the reading of the two clocks differ, and which clock will show the smaller elapsed time? (*Hint*: Use the fact that $v \ll c$ to simplify $\sqrt{1 - v^2/c^2}$ by making a binomial expansion.)

Section 40–4 Relativity of Length

40–5 In the year 2010 a spacecraft flies over Moon Station III at a speed of $0.8c$. A scientist on the moon measures the length of the moving spacecraft to be 200 m. The spacecraft later lands on the moon, and the same scientist measures the length of the now stationary spacecraft. What value does she get?

40–6 Two events are observed in a frame of reference S to occur at the same space point, the second occurring 2 s after the first. In a second frame S' moving relative to S, the second event is observed to occur 3 s after the first. What is the difference between the positions of the two events as measured in S'?

Section 40–5 The Lorentz Transformation

40–7 Show in detail the derivation of Eq. (40–17) from Eqs. (40–15) and (40–16).

40–8 Solve Eqs. (40–18) to obtain x and t in terms of x' and t', and show that the resulting transformation has the same form as the original one except for a change of sign for u.

40–9 Show the details of the derivation of Eq. (40–21) from Eq. (40–20).

40–10 Two particles emerge from a high-energy accelerator in opposite directions, each with a speed $0.6c$ as measured in the laboratory. What is the relative velocity of the particles?

Section 40–6 Momentum

40–11 At what speed is the momentum of a particle twice as great as the result obtained from the nonrelativistic expression mv?

40–12

a) At what speed does the momentum of a particle differ by 1% from the value obtained by using the nonrelativistic expression mv?

b) Is the correct relativistic value greater or less than that obtained from the nonrelativistic expression?

40–13 A particle moves along a straight line under the action of a force lying along the same line. Show that the acceleration $a = dv/dt$ of the particle is given by Eq. (40–25).

Section 40–7 Work and Energy

40–14 For a particle moving in the x-direction, use Eq. (40–22) to show that

$$\frac{dp}{dv} = \frac{m}{(1 - v^2/c^2)^{3/2}}.$$

40–15 What is the speed of a particle whose kinetic energy is equal to

a) its rest energy?   b) ten times its rest energy?

40–16

a) How much work must be done to accelerate a particle of mass m from rest to a speed of $0.1c$?

b) From a speed of $0.9c$ to a speed of $0.99c$?

40–17 Compute the kinetic energy of an electron (mass 9.11×10^{-31} kg) by using both the nonrelativistic and relativistic expressions, and compute the ratio of the two results, for speeds of

a) 1.0×10^8 m·s^{-1},   b) 2.8×10^8 m·s^{-1}.

40–18 In *positron annihilation,* an electron and a positron (a positively charged electron) collide and disappear, producing electromagnetic radiation. If each particle has a mass of 9.11×10^{-31} kg and they are at rest just before the annihilation, find the total energy of the radiation.

40–19 The total consumption of electrical energy per year in the United States is of the order of 10^{19} J. If matter could be converted completely into energy, how many kilograms

of matter would have to be converted to produce this much energy?

40–20 In a hypothetical nuclear-fusion reactor, two deuterium nuclei combine or "fuse" to form one helium nucleus. The mass of a deuterium nucleus, expressed in atomic mass units (u), is 2.0136 u; that of a helium nucleus is 4.0015 u. (1 u = 1.66×10^{-27} kg.)

a) How much energy is released when 1 kg of deuterium undergoes fusion?

b) The annual consumption of electrical energy in the United States is of the order of 10^{19} J. How much deuterium must react to produce this much energy?

40–21 Starting from Eq. (40–34), show that in the classical limit ($pc \ll mc^2$) the energy approaches the classical kinetic energy plus the rest mass energy.

40–22 Calculate, relativistically, the amount of work in MeV that must be done

a) to bring an electron from rest to a velocity of $0.4c$;

b) to increase its velocity from $0.4c$ to $0.8c$.

c) What is the ratio of the kinetic energy of the electron at the velocity of $0.8c$ to that of $0.4c$ when computed (1) from relativistic values and (2) from classical values?

Section 40–8 Invariance

40–23 Use the Lorentz transformation, Eq. (40–18), to prove Eq. (40–36).

Section 40–9 The Doppler Effect

40–24 In terms of c, what relative velocity u between source and observer produces

a) a 1% decrease in frequency?

b) a doubling of the frequency of the observed light?

40–25 How fast must you be approaching a red traffic light ($\lambda = 675$ nm) for it to appear green ($\lambda = 525$ nm)? (If you used this as an excuse for not getting a ticket for running a red light, do you think you might get a speeding ticket instead?)

PROBLEMS

40–26 A cube of metal with sides of length a and rest mass m sits at rest in a frame S_1. In S_1, therefore, the cube has density $\rho_1 = m/a^3$. Frame S_2 moves along the x-axis with a velocity v. To an observer in frame S_2, what is the density of the metal cube?

40–27 The Stanford Linear Accelerator Center (SLAC) uses a 3-km long tube in accelerating subatomic particles. The π^+ meson has a half-life of 2.6×10^{-8} s.

a) How fast must a π^+ meson travel if it is not to decay before it reaches the end of the tube? (Since v will be very close to c, write $v = [1 - \Delta]c$ and give your answer in terms of Δ rather than v.)

b) With a rest mass of 139.6 MeV, what is the π^+ meson's energy at the velocity calculated in (a)?

40–28 A photon of energy E is emitted by an atom of mass m, which recoils in the opposite direction.

a) Assuming that the atom can be treated nonrelativistically, compute the recoil velocity of the atom.

b) From the result of (a), show that the recoil velocity is much smaller than c whenever E is much smaller than the rest energy mc^2 of the atom.

40–29 A radioactive isotope of cobalt, ^{60}Co, emits an electromagnetic photon (γ ray) of energy 1.33 MeV. The cobalt nucleus contains 27 protons and 33 neutrons, each with a mass of about 1.66×10^{-27} kg.

a) If the nucleus is at rest before emission, what is the speed afterward?

b) Is it necessary to use the relativistic generalization of momentum?

40–30 In Problem 40–29, suppose the cobalt atom is in a metallic crystal containing 0.01 mol of cobalt (about 6.02×10^{21} atoms) and that the entire crystal recoils as a unit, rather than just the single nucleus. Find the recoil velocity. (This recoil of the entire crystal, rather than a single nucleus, is called the *Mössbauer effect*, in honor of its discoverer, who first observed it in 1958.)

40–31 A particle is said to be in the *extreme relativistic range* when its kinetic energy is much larger than its rest energy.

a) What is the speed of a particle (expressed as a fraction of c) such that the total energy is ten times the rest energy?

b) For such a particle, what percent error in the energy–momentum relation of Eq. (40–34) results if the term $(mc^2)^2$ is neglected? That is, what is the percentage difference between the left and right sides of Eq. (40–34) if $(mc^2)^2$ is neglected for a particle with the speed calculated in (a)?

40–32 A nuclear bomb containing 20 kg of plutonium explodes. The rest mass of the products of the explosion is less than the original rest mass by one part in 10^4.

a) How much energy is released in the explosion?

b) If the explosion takes place in $1\,\mu$s, what is the average power developed by the bomb?

c) How much water could the released energy lift to a height of 1 km?

40–33 An electron in a certain x-ray tube is accelerated from rest through a potential difference of 180,000 V in going from the cathode to the anode. When it arrives at the anode, what is

a) its kinetic energy in eV?

b) its total energy?

c) its velocity?

d) What is the velocity of the electron, calculated classically?

40–34 A certain electron accelerator accelerates electrons through a potential difference of 6.5×10^7 V, so that their kinetic energy is 6.5×10^7 eV.

a) What is the ratio of the speed v of an electron having this energy to the speed of light c?

b) What would the speed be if computed from the principles of classical mechanics?

40–35 The Soviet physicist P. A. Cerenkov discovered that a charged particle traveling in a solid with a speed exceeding the speed of light in that material would radiate electromagnetic radiation. What is the minimum kinetic energy an electron must have while traveling inside a slab of crown glass ($n = 1.52$) in order to create Cerenkov radiation?

40–36 Construct a right triangle in which one of the angles is α, where $\sin \alpha = v/c$. (v is the speed of a particle, c the speed of light.) If the base of the triangle (the side adjacent to α) is the rest energy mc^2, show that

a) the hypotenuse is the total energy;

b) the side opposite α is c times the relativistic momentum.

c) Describe a simple graphical procedure for finding the kinetic energy K.

40–37 One of the wavelengths of light emitted by hydrogen atoms under normal laboratory conditions is $\lambda = 121.6$ nm. In the light emitted from a distant interstellar region, this same spectral line is observed to be Doppler-shifted to $\lambda = 430.4$ nm. How fast are the emitting atoms moving relative to the earth, and are they approaching the earth or receding from it?

40–38 A highway patrolman measures the speed of cars approaching him with a device that sends out electromagnetic waves of frequency f_0 and then measures the shift in frequency Δf of the waves reflected from the moving car. What fractional frequency shift $\Delta f/f_0$ is produced by a car speeding at 80 mph (35.8 m·s^{-1})? (See Problem 23–19.)

40–39 A particle moves along the x-axis with speed v. A force in the $+y$-direction, having magnitude F, is applied. Show that the magnitude of the acceleration initially is

$$a = \frac{F}{m}(1 - v^2/c^2)^{1/2}.$$

This result and that of Exercise 40–13 were sometimes interpreted in the early days of relativity as meaning that a particle has a "longitudinal mass" given by $m/(1 - v^2/c^2)^{3/2}$ and a "transverse mass" given by $m/(1 - v^2/c^2)^{1/2}$. These terms are no longer in common use.

40–40 An astronaut, her jet pack, and her spacesuit have a combined rest mass of 100 kg and a velocity of 1×10^4 m·s^{-1}.

a) What is the difference between her Newtonian energy and her correct relativistic kinetic energy? Note that you cannot simply calculate both values and subtract, as the precision of most calculators is insufficient.

b) What fraction of the Newtonian energy is this difference?

40–41 A particle of mass m accelerated by a constant force F will, according to Newtonian mechanics, continue to accelerate without bound. That is, as $t \to \infty$, $v \to \infty$. Show that according to relativistic mechanics the particle's speed approaches c as $t \to \infty$ (*Note.* A standard integral is $\int [1 - x^2]^{-3/2}\, dx = x/\sqrt{1 - x^2}$.)

CHALLENGE PROBLEMS

40–42 The classical velocity of a particle of rest mass m and charge q in a circular orbit of radius R in a magnetic field is

$$v = (qBR)/m.$$

Find the corresponding velocity for a particle orbiting at relativistic speeds, and show that your result reduces to the classical result when v/c is small. (*Hint*: Use Eq. [40–24]. The relation between velocity and acceleration in circular motion, and between the velocity and the magnetic force, is the same relativistically as nonrelativistically.)

40–43 In high-energy-particle physics, new particles are created by colliding fast-moving projectile particles with stationary particles. Some of the kinetic energy of the incident particle is used to create mass of the new particle. A proton–proton collision can result in the creation of 2 pions (π^- and π^+):

$$p + p \rightarrow p + p + \pi^- + \pi^+.$$

a) Calculate the threshold kinetic energy of the incident proton that will allow this reaction to occur if the second proton is initially at rest. The rest mass of each π is 139.6 MeV. (*Hint*: Working in the center-of-mass frame is useful here. See Challenge Problem 8–57. But now the Lorentz transformation must be used to relate the velocities in the laboratory and the center-of-mass frames.)

b) How does this calculated threshold kinetic energy compare with the total rest mass energy of the created particles?

40–44 The French physicist Fizeau was the first to measure accurately the speed of light. (See Section 36–3.) He also found experimentally that the speed relative to the lab frame of light traveling in a tank of water that is itself moving at a speed V relative to the lab frame is

$$v = c/n + kV.$$

Fizeau called k the dragging coefficient and obtained an experimental value of $k = 0.44$. What is the value of k you would calculate from relativistic transformations?

40–45 Two events observed in a frame of reference S have positions and times given by (x_1, t_1) and (x_2, t_2), respectively.

a) Show that in a frame S' moving along the x-axis just fast enough so the two events occur at the same point in S', the time interval $\Delta t'$ between the two events is given by

$$\Delta t' = \sqrt{(\Delta t)^2 - \left(\frac{\Delta x}{c}\right)^2},$$

where $\Delta x = x_2 - x_1$, and $\Delta t = t_2 - t_1$. Hence show that, if $\Delta x \geq c \, \Delta t$, there is *no* frame S' in which the two events occur at the same point. The interval $\Delta t'$ is sometimes called the *proper-time interval* for the event. Is this term appropriate?

b) Show that if $\Delta x > c \, \Delta t$, there is a frame of reference S' in which the two events occur *simultaneously*. Find the distance between the two events in S'. This distance is sometimes called a *proper length*. Is this term appropriate?

c) Two events are observed in a frame of reference S' to occur simultaneously at points separated by a distance of 1 m. In a second frame S moving relative to S' along the line joining the two points in S', the two events appear to be separated by 2 m. What is the time interval between the events, as measured in S? (*Hint*: Apply the result obtained in [b].)

41

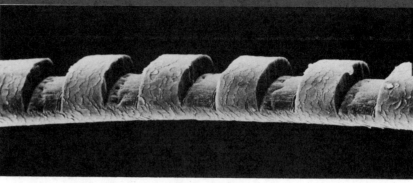

PHOTONS, ELECTRONS, AND ATOMS

IN THE PAST SEVERAL CHAPTERS WE HAVE STUDIED VARIOUS ASPECTS OF the propagation of light on the basis of an electromagnetic wave theory. The work of Maxwell, Hertz, and others has established firmly the electromagnetic nature of light. We understand interference, diffraction, and polarization on the basis of a *wave* model; when interference effects can be neglected, the further simplifications of *ray* optics permit us to analyze the behavior of lenses and mirrors. Together, these phenomena form the subject of *classical optics*.

Many other phenomena, however, show a different aspect of the nature of light, in which it seems to behave as a stream of *particles*. Among these are the photoelectric effect, the liberation of electrons from a surface by absorption of light energy; line spectra of elements; and the production and scattering of x-rays. All these point to a model in which the energy of light is carried in packages of a definite size, called *quanta* or *photons*. The energy of a single photon is proportional to the frequency of the radiation, and photons also carry momentum. Understanding the role of photons in emission and absorption phenomena requires some radical changes in our view of the nature of radiation and of matter itself. The new theory that emerges is called *quantum mechanics*. It provides the key to understanding many aspects of the structure and behavior of atoms and molecules that cannot be understood on the basis of the older "classical" theories of mechanics and electromagnetism.

41–1 EMISSION AND ABSORPTION OF LIGHT

The electromagnetic wave model of light provides an adequate basis for understanding many aspects of the propagation of light, including interference, diffraction, and polarization phenomena. It is considerably *less* successful with respect to the emission and absorption of light and other electromagnetic radiation from matter. In Section 35–1 we discussed the electromagnetic waves of Hertz; they were produced by oscillations with frequencies of the order of 10^8 Hz in a resonant $L–C$ circuit similar to those studied in Chapter 33. Frequencies of visible light are much larger, of the order of 10^{15} Hz, far higher than the highest frequencies attainable with conventional electronic equipment.

In the mid-nineteenth century theorists speculated that visible light might be produced by motion of electric charges within individual atoms rather than in macroscopic circuits. In fact, in 1862 Faraday placed a light source in a strong magnetic field in an attempt to determine whether the emitted radiation was changed by the effect of the field on the source. He was not able to detect any change, but when his experiments were repeated thirty years later by Zeeman with greatly improved equipment, changes *were* observed.

Particularly puzzling was the existence of *line spectra*. We have seen how a prism or grating spectrometer functions to disperse a beam of light into a *spectrum*. If the light source is an incandescent solid or liquid, the spectrum is *continuous;* that is, light of all wavelengths is present. But if the source is a gas through which an electrical discharge is passing, or a flame in which a volatile salt is present, the spectrum is of an entirely different character. Instead of a continuous band of color, only a few colors appear, in the form of isolated parallel lines. (Each "line" is an image of the spectrograph slit, deviated through an angle that depends on the frequency of the light forming the image.) A spectrum of this sort is termed a **line spectrum.**

The wavelengths of spectral lines are characteristic of the element emitting the light. That is, hydrogen always gives a set of lines in the same pattern, sodium produces another set, iron still another, and so on. The line structure of the spectrum extends into both the ultraviolet and infrared regions, where photographic or other means are required for its detection.

The existence of a characteristic spectrum for each element, discovered early in the nineteenth century, suggested a direct relation between the characteristics and internal structure of an atom and its spectrum. Attempts to understand this relation on the basis of Newtonian mechanics and classical electricity and magnetism were not successful, however.

There were other mysteries associated with the absorption of light. The *photoelectric effect,* discovered by Hertz in 1887 during his investigations of electromagnetic wave propagation, is the liberation of electrons from the surface of a conductor when light strikes the surface. When light is absorbed by the surface, energy is transferred to electrons near the surface, and some of the electrons acquire enough energy to surmount the potential-energy barrier at the surface and escape from the material into space. More detailed investigation revealed some puzzling features that could *not* be understood on the basis of classical optics. We will discuss these in the next section.

Still another area of unsolved problems centered around the production and scattering of *x-rays*, electromagnetic radiation with wavelengths shorter than those of visible light by a factor of the order of 10^4 and with correspondingly greater frequencies. These rays were produced in high-voltage glow discharge tubes, but the details of this process eluded understanding. Even worse, when these rays collided with matter, the scattered rays sometimes had a longer wavelength than the original ray. This is like directing a beam of blue light at a mirror and having it reflect as red!

All these phenomena, and several others, pointed forcefully to the conclusion that classical optics, successful though it was in explaining ray optics, interference, and polarization, nevertheless had its limitations. Understanding the phenomena cited above would require at least some generalization of classical theory. In fact, it has required something much more radical than that.

Several mysteries about production of light

Each element has a characteristic line spectrum. How does it come about?

The photoelectric effect: a puzzling aspect of the interaction of light and matter

All these phenomena are concerned with the *quantum* theory of radiation, which includes the assumption that despite the *wave* nature of electromagnetic radiation, it has some properties akin to those of *particles*. In particular, the *energy* conveyed by an electromagnetic wave is always carried in units whose magnitude is proportional to the frequency of the wave. These units of energy are called photons or quanta.

Thus electromagnetic radiation emerges as an entity with a dual nature, having both wave and particle aspects. The remainder of this chapter will be devoted to the applications of this duality to some of the phenomena mentioned above, and to a study of this seemingly (but not actually) inconsistent nature of electromagnetic radiation.

Is light a wave or a particle? Yes!

41–2 THE PHOTOELECTRIC EFFECT

The **photoelectric effect** is the liberation of electrons from the surface of a conductor when light strikes the surface. The electrons absorb energy from the incident radiation and are thus able to overcome the potential-energy barrier that normally confines them in the material.

Two mechanisms for electrons to be liberated from a surface

The photoelectric effect was first observed in 1887 by Hertz, quite by accident. He noticed that a spark would jump more readily between two spheres when their surfaces were illuminated by the light from another spark. Light seemed to facilitate the escape of charges from the surfaces. This idea in itself was not revolutionary. The existence of the surface potential-energy barrier was already known. In 1883 Edison had discovered **thermionic emission,** in which the escape energy is supplied by heating the material to a very high temperature, liberating electrons by a process analogous to boiling a liquid.

The photoelectric effect was investigated in detail by Hallwachs and Lenard, with quite unexpected results. We will describe their work in terms of more contemporary equipment.

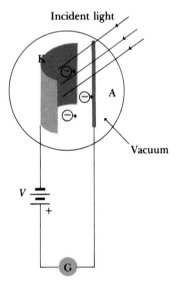

41–1 Schematic diagram of a photocell circuit.

A modern phototube is shown schematically in Fig. 41–1. A beam of light, indicated by the arrows, falls on a photosensitive surface K called the *cathode*. The battery or other source of potential difference creates an electric field in the direction from A (called the *collector* or *anode*) toward K, and electrons emitted from K are pushed by this field to the anode A. Anode and cathode are enclosed in a container that can be pumped down to a good vacuum; a residual pressure of 0.01 Pa (10^{-7} atm) or less is needed to avoid collisions of electrons with gas molecules. The photoelectric current is measured by the galvanometer G.

Electrons are emitted only when the light frequency is less than some critical value.

It is found that with a given material as emitter, no photoelectrons at all are emitted unless the wavelength of the light is *shorter* than some critical value. The corresponding *minimum* frequency is called the **threshold frequency** of the particular surface. The threshold frequency for most metals is in the ultraviolet region (corresponding to wavelengths of 200 to 300 nm), but for potassium and cesium oxide it lies in the visible spectrum (400 to 700 nm).

In energy relations involving electrons, it is often convenient to use the electronvolt. We introduced this unit in Section 26–7 during our study of electrical potential, and we suggest you review that section now. The volt, equal to one joule per coulomb, is a unit of *electrical potential*, but the electronvolt (1 eV) is a unit of *energy*. If we express a potential difference V in volts and the electron charge e in coulombs ($e = 1.602 \times 10^{-19}$ C), then the corresponding potential energy eV is in joules. But if V is in volts and $e = 1$ electron charge, then the product eV is in electronvolts. Thus if $V = 5$ V, then $eV = 5$ eV. From the definition of the electronvolt, we see that the conversion factor is

$$1 \text{ eV} = 1.602 \times 10^{-19} \text{ J}.$$

Some electrons are emitted from the cathode with substantial initial speeds; this is shown by the fact that, even with *no* emf in the external circuit, a few electrons reach the collector, causing a small current in the external circuit. Indeed, even when the polarity of the potential difference V is reversed and the associated electric-field force on the electrons is back toward the cathode, some electrons still reach the anode. Only when the reversed potential V is made large enough so that the potential energy eV is greater than the maximum kinetic energy $\frac{1}{2}mv_{\text{max}}^2$ with which the electrons leave the cathode does the electron flow stop completely. The critical reversed potential is called the *stopping potential,* denoted by V_0, and it provides a direct measurement of the maximum kinetic energy with which electrons leave the cathode, through the relation

$$\frac{1}{2}mv_{\text{max}}^2 = eV_0. \tag{41–1}$$

Surprisingly, it turns out that the maximum kinetic energy of an emitted electron *does not* depend on the *intensity* of the incident light, but *does* depend on its *wavelength* or *frequency.* When the light intensity increases, the photoelectric current also increases, but only because the *number* of emitted electrons increases, not the energy of an individual electron. This surprising result is hard to understand on the basis of classical theory, which would predict that greater intensity should yield some electrons with higher and higher energies.

Each electron absorbs the same amount of energy from the radiation.

The correct explanation of the photoelectric effect was given by Einstein in 1905. Extending a proposal made two years earlier by Planck, Einstein postulated that a beam of light consisted of small bundles of energy called **quanta** or **photons.** The energy E of a photon is proportional to its frequency f or is equal to its frequency multiplied by a constant. That is,

$$E = hf, \tag{41–2}$$

Photons: bundles of energy in electromagnetic radiation

where h is a universal constant, called **Planck's constant,** whose numerical value is 6.626×10^{-34} J·s. When a photon collides with an electron at or just within the surface of a metal, the photon may transfer its energy to the electron. This transfer is an "all-or-nothing" process, the electron getting all the photon's energy or none at all. The photon then simply drops out of existence. The energy acquired by the electron may enable it to surmount the surface potential-energy barrier and escape from the surface of the metal.

Planck's constant: a new fundamental constant of nature

The height of the surface potential-energy barrier is called the **work function,** denoted by ϕ. That is, in leaving the surface of the metal, the electron loses an amount of kinetic energy ϕ. Some electrons may lose more than this if they start at some distance below the metal surface, but the *maximum* energy with which an electron can emerge is the energy gained from a photon minus the work function. Hence the maximum kinetic energy of the photoelectrons ejected by light of frequency f is

How much energy is needed to pull an electron away from a surface?

$$\frac{1}{2}mv_{\text{max}}^2 = hf - \phi. \tag{41–3}$$

Combining this with Eq. (41–1) leads to the relation

$$eV_0 = hf - \phi. \qquad (41\text{–}4)$$

Thus by measuring the stopping potential V_0 required for each of several values of frequency f for a given cathode material, we can determine both the work function ϕ for the material and the value of the quantity h/e. This experiment provides a direct measurement of the value of Planck's constant, in addition to a direct confirmation of Einstein's interpretation of photoelectric emission. Typical work functions are of the order of 1 to 5 eV.

A simple experiment to measure Planck's constant and work functions

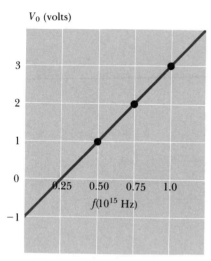

41–2 Stopping potential as a function of frequency. For a different cathode material having a different work function, the line would be displaced up or down but would have the same slope.

EXAMPLE 41–1 For a certain cathode material used in a photoelectric-effect experiment, a stopping potential of 3.0 V was required for light of wavelength 300 nm, 2.0 V for 400 nm, and 1.0 V for 600 nm. Determine the work function for this material and the value of Planck's constant.

SOLUTION According to Eq. (41–4), a graph of V_0 as a function of f should be a straight line. We rewrite this equation as

$$V_0 = \frac{h}{e}f - \frac{\phi}{e}.$$

In this form we see that the *slope* of the line is h/e and the *intercept* on the vertical axis (corresponding to $f = 0$) is at $-\phi/e$. The frequencies, obtained from $f = c/\lambda$ and $c = 3.00 \times 10^8 \text{ m}\cdot\text{s}^{-1}$, are 1.0, 0.75, and $0.5 \times 10^{15} \text{ s}^{-1}$, respectively. The graph is shown in Fig. 41–2. From it we find

$$-\frac{\phi}{e} = -1.0 \text{ V}, \qquad \phi = 1.0 \text{ eV} = 1.60 \times 10^{-19} \text{ J},$$

and

$$\frac{h}{e} = \frac{1.0 \text{ V}}{0.25 \times 10^{15} \text{ s}^{-1}} = 4.0 \times 10^{-15} \text{ J}\cdot\text{C}^{-1}\cdot\text{s},$$

$$h = (4.0 \times 10^{-15} \text{ J}\cdot\text{C}^{-1}\cdot\text{s})(1.60 \times 10^{-19} \text{ C})$$

$$= 6.4 \times 10^{-34} \text{ J}\cdot\text{s}.$$

This experimental value differs by about 3.4% from the currently accepted value of $6.626 \times 10^{-34} \text{ J}\cdot\text{s}$.

For a photon, energy, wavelength, frequency, and momentum are all directly related.

The particlelike nature of electromagnetic radiation, used in our analysis of the photoelectric effect, has been established beyond any reasonable doubt. Figure 41–3 shows a very direct illustration of the particle aspects of light. A photon of electromagnetic radiation of frequency f and corresponding wavelength $\lambda = c/f$ has energy E given by

$$E = hf = \frac{hc}{\lambda}. \qquad (41\text{–}5)$$

Furthermore, according to relativity theory, every particle having energy must also have momentum, even if it has no rest mass. Photons have zero rest mass; according to Eq. (40–35), the momentum p of a photon of energy E has

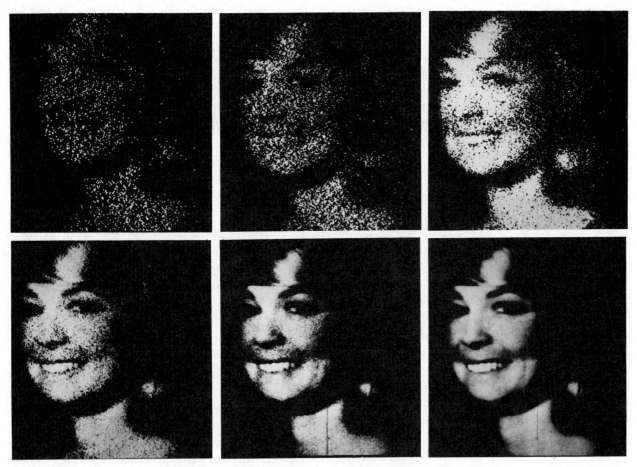

41–3 These photographs are images made by small numbers of photons, using electronic image amplification. In the upper pictures each spot corresponds to one photon; under extremely faint light (few photons) the pattern seems almost random, but it emerges more distinctly as the light level increases (lower pictures). (Used with permission of RCA Corporation.)

magnitude p given by

$$E = pc. \qquad (41\text{–}6)$$

Thus the wavelength of a photon and its momentum are related simply by

$$p = \frac{h}{\lambda}. \qquad (41\text{–}7)$$

We will use these relations frequently in the remainder of this chapter. We also note that Eq. (41–6) is consistent with the energy–momentum relation for electromagnetic waves that we derived in Section 35–3; this further confirms our view of the photon as a particle with zero rest mass.

Photons have zero rest mass and travel always with speed c.

41–3 LINE SPECTRA

The quantum hypothesis, used in the preceding section for the analysis of the photoelectric effect, also plays an important role in understanding atomic spectra, particularly the line spectra described in Section 41–1. We might

Trying to make sense out of the line spectrum of hydrogen

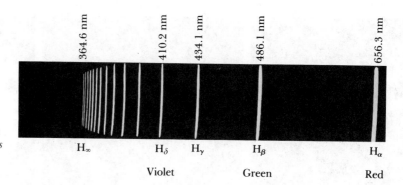

41–4 The Balmer series of atomic hydrogen. (Reproduced by permission from *Atomic Spectra and Atomic Structures* by Gerhard Herzberg. Copyright 1937 by Prentice-Hall, Inc.)

expect that the spectrum frequencies of the light emitted by a particular element would be arranged in some regular way. For instance, a radiating atom might be analogous to a vibrating string, emitting a fundamental frequency and its harmonics. At first sight, there does not seem to be any semblance of order or regularity in the lines of a typical spectrum. For many years unsuccessful attempts were made to correlate the observed frequencies with those of a fundamental and its overtones. Finally, in 1885 Johann Jakob Balmer (1825–1898) found by trial and error a simple formula that gives the frequencies of a group of lines emitted by atomic hydrogen. Since the spectrum of this element is relatively simple, we will consider it in more detail.

A numerical rule for the hydrogen-spectrum wavelengths

Under the proper conditions of excitation, atomic hydrogen may be made to emit the sequence of lines shown in Fig. 41–4. This sequence is called the **Balmer series.** A pattern is evident in this spectrum; the lines become closer and closer together as the end of the series is approached. The line of longest wavelength or lowest frequency, in the red, is known as H_α; the next line, in the blue-green, as H_β; the third is H_γ; and so on. Balmer found that the wavelengths of these lines were given accurately by the simple formula

$$\frac{1}{\lambda} = R\left(\frac{1}{2^2} - \frac{1}{n^2}\right), \tag{41-8}$$

where λ is the wavelength, R is a constant called the **Rydberg constant,** and n may have the integer values 3, 4, 5, and so on. If λ is in meters,

$$R = 1.097 \times 10^7 \text{ m}^{-1}.$$

Letting $n = 3$ in Eq. (41–8), we obtain the wavelength of the H_α-line:

$$\frac{1}{\lambda} = 1.097 \times 10^7 \text{ m}^{-1}\left(\frac{1}{4} - \frac{1}{9}\right)$$
$$= 1.524 \times 10^6 \text{ m}^{-1},$$

and

$$\lambda = 656.2 \text{ nm}.$$

For $n = 4$, we obtain the wavelength of the H_β-line, and so on. For $n = \infty$, we obtain the limit of the series, at $\lambda = 364.6$ nm. This is the *shortest* wavelength in the series.

Hydrogen has several spectral series in the ultraviolet, visible, and infrared regions of the spectrum.

Other series spectra for hydrogen have since been discovered. These are known, after their discoverers, as the Lyman, Paschen, Brackett, and Pfund series. The formulas for these follow.

Lyman series:

$$\frac{1}{\lambda} = R\left(\frac{1}{1^2} - \frac{1}{n^2}\right), \qquad n = 2, 3, \ldots,$$

Paschen series:

$$\frac{1}{\lambda} = R\left(\frac{1}{3^2} - \frac{1}{n^2}\right), \qquad n = 4, 5, \ldots,$$

Brackett series:

$$\frac{1}{\lambda} = R\left(\frac{1}{4^2} - \frac{1}{n^2}\right), \qquad n = 5, 6, \ldots,$$

Pfund series:

$$\frac{1}{\lambda} = R\left(\frac{1}{5^2} - \frac{1}{n^2}\right), \qquad n = 6, 7, \ldots.$$

The Lyman series is in the ultraviolet region, and the Paschen, Brackett, and Pfund series are in the infrared. The Balmer series evidently fits into the scheme between the Lyman and Paschen series.

The Balmer formula, Eq. (41–8), may also be written in terms of the frequency of the light, by use of the relation

$$c = f\lambda, \quad \text{or} \quad \frac{1}{\lambda} = \frac{f}{c}.$$

Thus Eq. (41–8) becomes

Expressing frequencies of spectrum lines as differences of two terms

$$f = Rc\left(\frac{1}{2^2} - \frac{1}{n^2}\right), \tag{41–9}$$

or

$$f = \frac{Rc}{2^2} - \frac{Rc}{n^2}. \tag{41–10}$$

Each of the fractions on the right-hand side of Eq. (41–10) is called a **term,** and the frequency of every line in the series is given by the difference between two terms.

Only a few elements (hydrogen, singly ionized helium, doubly ionized lithium) have spectra that can be represented by a simple formula of the Balmer type. Nevertheless, it is possible to separate the more complicated spectra of other elements into series, and to express the frequency of each line in the series as the difference of two terms. The first term is constant for any one series, while the various values of the second term can be labeled by values of an integer index n analogous to the n appearing in Eq. (41–10). In a few simple cases the numerical values of the terms can be *calculated* from theoretical considerations, as we will see in Chapter 42. For complex atoms, however, the term values must be determined experimentally by analysis of spectra, often an extremely complex problem.

41–4 ENERGY LEVELS

Energy levels in atoms: the connection between spectra and photons

Every element has a characteristic line spectrum that must be related to the characteristics and structure of the *atoms* of that element. Part of the key to understanding the relation of atomic structure to atomic spectra was supplied in 1913 by the Danish physicist Niels Bohr, who applied to spectra the same concept of light quanta or *photons* that Einstein had used earlier in analyzing the photoelectric effect.

Bohr's hypothesis was as follows: Each atom, as a result of its internal structure (and, presumably, internal motion), can have a variable amount of *internal energy.* But the energy of an atom cannot change by any arbitrary amount; rather, each atom has a series of discrete **energy levels.** An atom can have an amount of internal energy corresponding to any one of these levels, but it cannot have an energy *intermediate* between two levels. All atoms of a given element have the same set of energy levels, but atoms of different elements have different sets.

An atom radiates only while going from one energy level to another.

While an atom is in one of these states corresponding to a definite energy, it does not radiate. However, an atom can make a transition from one energy level to a lower level by emitting a photon whose energy is equal to the energy difference between the initial and final states. If E_i is the initial energy of the atom, before such a transition, and E_f is its final energy, after the transition, then, since the photon's energy is hf, we have

$$hf = E_i - E_f. \qquad (41–11)$$

For example, a photon of orange light of wavelength 600 nm has a frequency f given by

$$f = \frac{c}{\lambda} = \frac{3.00 \times 10^8 \text{ m·s}^{-1}}{600 \times 10^{-9} \text{ m}} = 5.00 \times 10^{14} \text{ Hz}$$
$$= 5.00 \times 10^{14} \text{ s}^{-1}.$$

The corresponding photon energy is

$$E = hf = (6.63 \times 10^{-34} \text{ J·s})(5.00 \times 10^{14} \text{ s}^{-1})$$
$$= 3.31 \times 10^{-19} \text{ J} = 2.07 \text{ eV}.$$

Energy levels provide a simple explanation for the Balmer formula and others.

Thus this photon must be emitted in a transition between two states of the atom differing in energy by 2.07 eV.

The Bohr hypothesis, if correct, would shed new light on the analysis of spectra on the basis of *terms,* as described in Section 41–3. For example, Eq. (41–10) gives the frequencies of the Balmer series in the hydrogen spectrum. Multiplied by Planck's constant h, this becomes

$$hf = \frac{Rch}{2^2} - \frac{Rch}{n^2}. \qquad (41–12)$$

A formula for the energy levels of the hydrogen atom

If we now compare Eqs. (41–11) and (41–12), identifying $-Rch/n^2$ with the initial energy of the atom E_i and $-Rch/2^2$ with its final energy E_f, before and after a transition in which a photon of energy $hf = E_i - E_f$ is emitted, then Eq. (41–12) takes on the same form as Eq. (41–11). More generally, if we assume that the possible energy levels for the hydrogen atom are given by

$$E_n = -\frac{Rch}{n^2}, \qquad n = 1, 2, 3, \ldots, \qquad (41–13)$$

then *all* the series spectra of hydrogen can be understood on the basis of transitions from one energy level to another. For the Lyman series the final state is always $n = 1$; for the Paschen series it is $n = 3$; and so on. Similarly, complex spectra of other elements, represented by terms, are understood on the basis that each term corresponds to an energy level; and a frequency, represented as a difference of two terms, corresponds to a transition between the two corresponding energy levels.

In 1914 a series of experiments by Franck and Hertz provided more direct experimental evidence for energy levels in atoms. In studying the motion of electrons through mercury vapor, under the action of an electric field, they found that a spectrum line at 254 nm was emitted by the vapor when the electron kinetic energy was greater than 4.9 eV, but not when it was less. This strongly suggests the existence of an energy level 4.9 eV above the ground state. A mercury atom is excited to this level by collision with an electron and subsequently decays to the ground state by emitting a photon. The energy of the photon, according to Eq. (41–2), should be

The Franck-Hertz experiment: independent confirmation of the existence of energy levels in atoms

$$E = hf = \frac{hc}{\lambda} = \frac{(6.63 \times 10^{-34}\ \text{J·s})(3.00 \times 10^8\ \text{m·s}^{-1})}{254 \times 10^{-9}\ \text{m}}$$

$$= 7.82 \times 10^{-19}\ \text{J} = \frac{7.82 \times 10^{-19}\ \text{J}}{1.60 \times 10^{-19}\ \text{J·eV}^{-1}}$$

$$= 4.9\ \text{eV},$$

in excellent agreement with the measured electron energy.

Although the Bohr hypothesis permits partial understanding of line spectra on the basis of energy levels of atoms, it is not yet complete because it provides no basis for *predicting* what the energy levels for any particular kind of atom should be. We will return to this problem in Chapter 42, where we will develop the principles of quantum mechanics needed for the understanding of the structure and energy levels of atoms.

PROBLEM-SOLVING STRATEGY: *Photons and energy levels*

1. Remember that with photons, as with any other periodic wave, the wavelength λ and frequency f are related by $f = c/\lambda$. The energy E of a photon can be expressed as hf or hc/λ, whichever is more convenient for the problem at hand. Be careful with units; if E is in joules, h must be in (J·s), λ in meters, and f in (seconds)$^{-1}$ or hertz. The magnitudes are in such an unfamiliar realm that common sense may not help if you goof by a factor of 10^{10}, so be careful with powers of ten.

2. It is often convenient to measure energy in electronvolts. The conversion $1\ \text{eV} = 1.602 \times 10^{-19}\ \text{J}$ comes in handy. When energies are in eV, you may want to express h in eV·s; in those units, $h = 4.136 \times 10^{-15}\ \text{eV·s}$. We invite you to verify this value.

3. Keep in mind that an electron moving through a potential difference of one volt gains or loses an amount of energy equal to one electronvolt. You will use the electronvolt a lot in this chapter and the next three, so it's worthwhile to become familiar with it now.

41–5 ATOMIC SPECTRA

As we have seen, the key to understanding atomic spectra is the concept of atomic *energy levels*. Every spectrum line corresponds to a specific transition between two energy levels of an atom, and the corresponding frequency is

Atomic spectra are related directly to atomic-energy levels.

given in each case by Eq. (41–11). Thus the fundamental problem for the spectroscopist is to determine the energy levels of an atom from the measured values of the wavelengths of the spectral lines that are emitted when the atom goes from one energy level to another. In the case of complicated spectra emitted by the heavier atoms, this is a complex task requiring tremendous ingenuity. Nevertheless, almost all atomic spectra have been analyzed, and the resulting energy levels have been tabulated with the aid of diagrams similar to the one shown for sodium in Fig. 41–5.

Every atom has a lowest energy level, representing the *minimum* energy the atom can have. This lowest energy level is called the **ground state,** and all higher levels are called **excited states.** As we have seen, a photon corresponding to a particular spectral line is emitted when an atom makes a transition from an excited state to a lower state. The only means discussed so far for raising the atom from the ground state to an excited state has been with the aid of an electric discharge. Let us consider now another method, involving *absorption* of radiant energy.

From Fig. 41–5 we can see that a sodium atom emits characteristic yellow light of wavelengths 589.0 and 589.6 nm when it makes transitions from the

An atom can gain or lose energy only in units determined by its energy levels.

41–5 Energy levels of the sodium atom. Numbers on the lines between levels are wavelengths. The column labels, such as $^2S_{1/2}$, refer to the quantum states of the electron, to be discussed in Chapter 43.

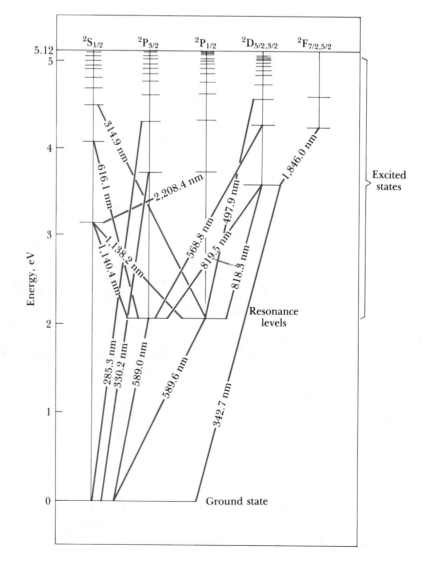

41–6 Absorption spectrum of sodium. This is an image on photographic film. Black and white are reversed; the bright regions of the spectrum show as black, and the dark absorption lines show as light lines.

two closely spaced levels marked *resonance levels* to the ground state. Suppose a sodium atom in the ground state were to *absorb* a quantum of radiant energy of wavelength 589.0 or 589.6 nm. It would then undergo a transition in the opposite direction and be raised to one of the resonance levels. After a short time the atom returns to the ground state, emitting a photon. The average time spent in the excited state is called the *lifetime* of the state; for the resonance levels of the sodium atom, the lifetime is about 1.6×10^{-8} s.

This emission process is called **resonance radiation;** we can demonstrate it as follows. We pump the air out of a glass bulb and then introduce a small amount of pure metallic sodium. Then we concentrate a strong beam of the yellow light from a sodium-vapor lamp on the bulb. When we warm the bulb to evaporate some of the sodium, some of the atoms in the vapor absorb the 589-nm photons from the beam, then reemit them in all directions. This happens throughout the whole bulb, which glows with the yellow light characteristic of sodium.

A sodium atom in the ground state may absorb radiant energy of wavelengths other than the yellow resonance lines. All wavelengths corresponding to spectral lines *emitted* when the sodium atom returns to its ground state may also be *absorbed* by a sodium atom in the ground state. If, therefore, the continuous-spectrum light from a carbon arc is sent through an absorption tube containing sodium vapor and is then examined with a spectroscope, a series of dark lines appears, corresponding to the wavelengths absorbed, as shown in Fig. 41–6. This is called an **absorption spectrum.**

The sun's spectrum is an absorption spectrum. The main body of the sun emits a continuous spectrum, whereas the cooler vapors in the sun's atmosphere emit line spectra corresponding to all the elements present. When the intense light from the main body of the sun passes through the cooler vapor, the lines of these elements are absorbed. The light *emitted* by the cooler vapors is so small compared with the unabsorbed continuous spectrum that the continuous spectrum appears to be crossed by many faint *dark* lines. These were first observed by Fraunhofer and are called *Fraunhofer lines.* They may be observed with any student spectroscope pointed toward any part of the sky. Figure 41–7 shows Fraunhofer lines in a portion of the sun's spectrum.

Absorption spectra: An atom in its ground state can absorb only certain wavelengths, depending on its energy levels.

41–7 A portion of the solar spectrum, between 390 nm and 460 nm, showing the Fraunhofer lines corresponding to the absorption spectrum. (Courtesy of Mt. Wilson & Las Campanas Observatories, Carnegie Institute of Washington.)

41–6 THE LASER

Stimulated emission: The presence of photons stimulates emission of other photons of the same energy.

If the energy difference between the ground state and the first excited state of an atom is E, the atom is capable of absorbing a photon whose frequency f is given by the Planck equation $E = hf$. The *absorption* of a photon by a normal atom A is depicted schematically in Fig. 41–8a. After absorbing the photon, the atom becomes an excited atom A*. A short time later, *spontaneous emission* takes place, and the excited atom returns to the ground state by emitting a photon of the same frequency as the one originally absorbed. The direction and phase of this photon are random, as shown in Fig. 41–8b. There is also a third process, first proposed by Einstein, called **stimulated emission,** shown schematically in Fig. 41–8c. Stimulated emission takes place when an incident photon encounters an excited atom and forces it to emit *another* photon with the same frequency, the same direction, the same phase, and the same polarization as the incident photon. The two photons thus have a definite phase relation and go off together as *coherent* radiation.

Suppose we have a large number of identical atoms in a gas or vapor, in a container with transparent walls, as in Fig. 41–8a. At moderate temperatures, if no radiation is incident on the container, most of the atoms are in the ground state, and only a few are in excited states. If there is an energy level E above the ground state, the ratio of the number of atoms n_E in this state (the *population* of the state) to the number of atoms n_0 in the ground state is very small.

Population inversion: how to get more atoms into higher energy states, with a top-heavy distribution

Now suppose we send through the container a beam of radiation with frequency f corresponding to the energy difference E. Some of the atoms absorb photons of energy E and are raised to the excited state, and the population ratio n_E/n_0 increases. Because n_0 is originally so much larger than n_E, an enormously intense beam of light would be required to increase n_E to a value comparable to n_0. Therefore the rate at which energy is absorbed from the beam by the n_0 ground-state atoms far outweighs the rate at which energy is added to the beam by stimulated emission from the relatively rare (n_E) excited atoms.

But now suppose we can create a situation in which n_E is substantially increased compared to the normal equilibrium value; this condition is called **population inversion.** In this case the rate of energy radiation by stimulated emission may actually *exceed* the rate of absorption. The system then acts as a

41–8 Three interaction processes between an atom and radiation. The heavier waves on the right side of (c) show the presence of additional photons from stimulated emission.

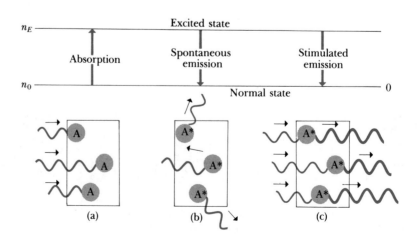

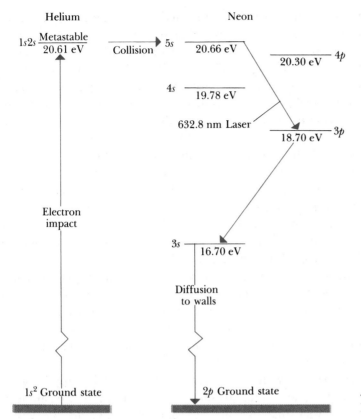

41-9 Energy-level diagram for helium-neon laser.

source of radiation with photon energy *E*. Furthermore, since the photons are the result of stimulated emission, they all have the same frequency, phase, polarization, and direction. The resulting radiation is therefore very much more *coherent* than light from ordinary sources, where the emissions of individual atoms are *not* coordinated.

The necessary population inversion can be achieved in a variety of ways. As an example, consider the helium-neon laser, a simple, inexpensive laser available in many undergraduate laboratories. A mixture of helium and neon, each typically at a pressure of the order of 10^2 Pa (or 10^{-3} atm), is sealed in a glass enclosure provided with two electrodes. When a sufficiently high voltage is applied, a glow discharge occurs. Collisions between ionized atoms and electrons carrying the discharge current excite atoms to various energy states.

Figure 41-9 shows an energy-level diagram for the system. The notation used to label the various energy levels, such as $1s2s$ or $5s$, refers to the energy states of the electrons in the atoms and will be discussed in Section 43-1. A helium atom excited to the $1s2s$ state cannot return to the ground state by emitting a 20.61-eV photon, as might be expected. The reason, which we cannot discuss in detail here, is related to restrictions imposed by conservation of angular momentum. Such a state, in which single-photon radiative decay is impossible, is called a **metastable state.**

However, excited helium atoms *can* lose energy by energy-exchange collisions with neon atoms initially in the ground state. A $1s2s$ helium atom, with its internal energy of 20.61 eV and a little additional kinetic energy, can collide with a neon atom in the ground state, exciting the neon atom to the $5s$ excited

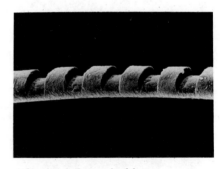

A human hair notched by an ultraviolet laser. The ultraviolet photons have enough energy to break molecular bonds; the fragments are swept away by remaining photon energy, causing very clean cuts with very little heating. Such lasers can be used for integrated-circuit fabrication and corneal surgery. (Courtesy of IBM Corp.)

Metastable states: atoms getting stuck in states from which radiative decay is impossible

Induced emission produces a beam with many photons of the same energy, all in phase.

state at 20.66 eV and leaving the helium atom in the $1s^2$ ground state. Thus we have the necessary mechanism for a population inversion in neon, greatly enhancing the population in the $5s$ state. Stimulated emission from this state then results in the emission of highly coherent light at 632.8 nm, as shown on the diagram. In practice the beam is sent back and forth through the gas many times by a pair of parallel mirrors, so as to stimulate emission from as many excited atoms as possible. One of the mirrors is partially transparent, so a portion of the beam emerges as an external beam.

The net effect of all these processes taking place in a laser tube is a beam of radiation that is (1) very intense, (2) almost perfectly parallel, (3) almost monochromatic, and (4) spatially *coherent* at all points within a given cross section. To understand this fourth characteristic, recall the simple double-slit interference experiment. A mercury arc placed directly behind the double slit would not give rise to interference fringes because the light issuing from the two slits would come from different points of the arc and would not retain a constant phase relationship. When we use common laboratory arc-lamp sources, we must use the light from a very small portion of the source to illuminate the double slit. The slightly diverging beam from a laser, however, may be allowed to fall directly on a double slit (or other interferometer) because the light rays from any two points of a cross section are in phase and are said to exhibit "spatial coherence."

Lasers have many practical applications, some of them spectacular.

In recent years lasers have found a wide variety of practical applications. The high intensity of a laser beam makes it a convenient drill. For example, a laser beam can drill a very small hole in a diamond for use as a die in drawing very small-diameter wire. Because the photons in a laser beam are strongly correlated in their directions, a laser beam can travel long distances without appreciable spreading. This makes it a very useful tool for surveyors, especially in situations where great precision is required, such as a long tunnel drilled from both ends.

41–7 CONTINUOUS SPECTRA

Hot condensed matter emits a continuous spectrum, not a line spectrum.

Our discussion of the particle aspect of electromagnetic radiation has centered so far on its role in the understanding of line spectra emitted by elements in the gaseous state, where each atom behaves as an isolated system and interactions between atoms are negligible. Very hot materials in the solid and liquid states also emit radiation, but with a **continuous spectrum,** that is, continuous distribution of wavelengths rather than a line spectrum. By 1900 this radiation had been studied extensively, and several characteristics had been established.

The total rate of energy radiation depends on the fourth power of the absolute temperature.

First, the total rate of radiation of energy, per unit surface area, is proportional to the fourth power of the absolute temperature. We have seen this relationship before in Section 16–3, in our study of heat transfer mechanisms, and we suggest you review that section. The total heat current H (power, or energy per unit time) radiated from an ideal radiator of area A at absolute temperature T is given by the **Stefan-Boltzmann law:**

$$H = A\sigma T^4, \tag{41–14}$$

where σ is a fundamental physical constant called the **Stefan-Boltzmann constant.** In SI units,

$$\sigma = 5.6699 \times 10^{-8} \text{ W·m}^{-2}\text{·K}^{-4}. \tag{41–15}$$

An alternative statement is that the *power emitted per unit area* for an ideal radiator is equal to σT^4.

Second, the power is not uniformly distributed over all wavelengths, but it can always be described by a power distribution function $F(\lambda)$ having the property that $F(\lambda)\,d\lambda$ is the power per unit area radiated with wavelengths in the interval $d\lambda$. The *integral* of $F(\lambda)$ over all wavelengths, represented by the area under the curve when $F(\lambda)$ is plotted as a function of λ, is the *total* power per unit area. The distribution functions for three different temperatures are shown in Fig. 41–10. Each has a peak wavelength λ_m, where the power is most concentrated. Experiment shows that λ_m is inversely proportional to T, and that in fact

$$\lambda_m T = \text{constant} = 2.90 \times 10^{-3}\ \text{m·K}. \qquad (41\text{–}16)$$

This rule is called the **Wien displacement law.** As the temperature rises, the peak of $F(\lambda)$ becomes higher and shifts to shorter wavelengths. This corresponds to the familiar fact that a body that glows yellow is hotter and brighter than one that glows red; yellow has a shorter wavelength than red. Finally, experiments show that the *shape* of the distribution function is the same for all temperatures; we can make a curve for one temperature fit any other temperature by simply changing the scales on the graph.

In the last decade of the nineteenth century, many attempts were made to *derive* these empirical results from basic principles. It was known by then that light is electromagnetic radiation, and it seemed natural to assume that light is produced by vibration of elementary charges in a material. Attempts were made to derive a function for the curves of Fig. 41–10 by treating the electrons in a material as harmonic oscillators with energies governed by the principle of equipartition of energy. We studied this principle in Section 20–5 in connection with heat capacities of gases. Finally, in 1900 Max Planck succeeded in deriving a function, now called the **Planck radiation law,** that agreed with experimentally obtained power-distribution curves. To do this he added to the classical equipartition theorem the additional assumption that a harmonic oscillator with frequency f can gain or lose energy only in discrete steps of magnitude hf, where h is the same constant that now bears his name.

Ironically, Planck himself originally regarded this *quantum hypothesis* as a calculational trick rather than a fundamental principle. But as we have seen, evidence for the quantum aspects of light accumulated, and by 1920 there was no longer any doubt about the validity of the concept. Indeed, the concept of discrete energy levels of microscopic systems really originated with Planck, not Bohr, and we have departed from the historical order of things by discussing atomic spectra before continuous spectra.

Planck's derivation of the power-distribution function is rather involved, and we quote his result here without derivation:

$$F(\lambda) = \frac{2\pi hc^2}{\lambda^5}\ \frac{1}{e^{hc/\lambda kT}-1}. \qquad (41\text{–}17)$$

where h is Planck's constant, c is the speed of light, k is Boltzmann's constant, T is the *absolute* temperature, and λ is the wavelength. This function turns out to agree well with experimentally determined power-distribution curves such as those in Fig. 41–10. It also contains the Wien displacement law and the Stefan-Boltzmann law as consequences.

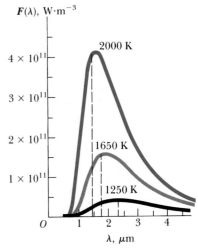

41–10 Power distribution function $F(\lambda)$ for continuous-spectrum radiation from an ideal radiator. The power per unit area in a wavelength interval $d\lambda$ at wavelength λ is $F(\lambda)\,d\lambda$. The total area under the curve at any temperature represents the total radiated power per unit area. Curves are plotted for three different temperatures, with darker color corresponding to higher temperature. The vertical broken lines show the value of λ_m in Eq. (41–18) for each temperature. As the temperature increases, the peak grows larger and shifts to shorter wavelengths; the total area under the curve is proportional to T^4.

The quantum hypothesis: calculational trick or fundamental truth?

The Wien displacement law and the Stefan-Boltzmann law can be derived from the Planck radiation formula.

To derive the Wien law we take the derivative of Eq. (41–17) and set it equal to zero to find the value of λ at which $F(\lambda)$ is maximum. We leave you the details as a problem; the result is

$$\lambda_{\mathrm{m}} = \frac{hc}{4.965kT}. \tag{41–18}$$

To obtain this result, you have to solve the equation

$$5 - x = 5e^{-x}. \tag{41–19}$$

The root of this equation is (approximately) 4.965. We invite you to evaluate the constant $hc/4.965k$ and show that it agrees with the empirical value given in Eq. (41–16).

We can obtain the Stefan-Boltzmann law by integrating Eq. (41–17) over all λ to find the *total* radiated power per unit area. This calculation involves looking up an integral in a table. Again we leave you the details as a problem; the result is

$$\int_0^\infty F(\lambda)\, d\lambda = \frac{2\pi^5 k^4}{15c^2 h^3}T^4 = \sigma T^4. \tag{41–20}$$

This is the Stefan-Boltzmann law, and it also shows that the constant σ in that law can be expressed in terms of a rather unlikely looking combination of other fundamental constants:

$$\sigma = \frac{2\pi^5 k^4}{15c^2 h^3}. \tag{41–21}$$

We invite you to plug in the values of k, c, and h, and verify that you get the value of σ quoted above.

The general form of Eq. (41–18) can be predicted from kinetic theory. If photons are emitted by microscopic oscillating charges, and if their energies are typically of the order of kT, as suggested by the equipartition theorem, then for a typical photon we would expect

$$E = \frac{hc}{\lambda} \quad \text{and} \quad \lambda = \frac{hc}{kT}. \tag{41–22}$$

Indeed, a photon with wavelength given by Eq. (41–18) has an energy of $4.965kT$.

41–8 X-RAY PRODUCTION AND SCATTERING

Production of x-rays: additional evidence for the validity of the photon concept

The production and scattering of **x-rays** gives additional support to the quantum view of electromagnetic radiation. X-rays are produced when rapidly moving electrons that have been accelerated through a potential difference of the order of 10^3 to 10^6 V strike a metal target. They were first produced by Wilhelm Röntgen (1845–1923) in 1895 with an apparatus similar in principle to that shown in Fig. 41–11a. Electrons are "boiled off" from the heated cathode by thermionic emission and are accelerated toward the anode (the target) by a large potential difference V. Most of the air is pumped out of the bulb so that electrons can travel from cathode to anode with only a small probability of collision with air molecules. The residual gas pressure is of the

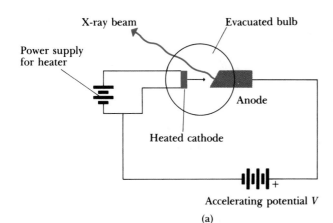

X-ray beam

Evacuated bulb

Power supply for heater

Anode

Heated cathode

Accelerating potential V

(a)

(b)

41–11 (a) Apparatus used to produce x-rays. Electrons are emitted thermionically from the heated cathode and are accelerated toward the anode; when they strike it, x-rays are produced. (b) A Coolidge-type x-ray tube.

order of 0.01 Pa or 10^{-7} atm. When V is a few thousand volts or more, a very penetrating radiation is emitted from the anode surface. A common x-ray tube of the type invented by Coolidge is shown in Fig. 41–11b.

Because of their origin, it is clear that x-rays are electromagnetic waves; like light, they are governed by quantum relations in their interaction with matter. Thus we can talk about x-ray photons or quanta, and the energy of an x-ray photon is related to its frequency and wavelength just as for photons of light, $E = hf = hc/\lambda$. Typical x-ray wavelengths are 0.001 to 1 nm (10^{-12} to 10^{-9} m). X-ray wavelengths can be measured quite precisely by crystal diffraction techniques, such as those we studied in Section 39–9.

Two distinct processes are involved in x-ray emission. Some electrons are stopped by the target (anode), and all their kinetic energy is converted directly to an x-ray photon. Other electrons transfer their energy partly or completely to atoms in the target. These atoms are left in excited states, and when they decay back to the ground state, they emit x-ray photons with energies characteristic of the element in the target.

X-ray photons have much higher energy than visible-light photons.

EXAMPLE 41–2 Electrons are accelerated by a potential difference of 10 kV. If an electron produces a photon on impact with the target, what is the wavelength of the resulting x-rays?

SOLUTIONS   The photon energy $hf = hc/\lambda$ is equal to the kinetic energy eV of the electron just before impact; hence

$$eV = hf = \frac{hc}{\lambda}, \tag{41-23}$$

and

$$\lambda = \frac{hc}{eV} = \frac{(6.626 \times 10^{-34}\ \mathrm{J \cdot s})(3.00 \times 10^{8}\ \mathrm{m \cdot s^{-1}})}{(1.602 \times 10^{-19}\ \mathrm{C})(1.00 \times 10^{4}\ \mathrm{V})}$$

$$= 1.24 \times 10^{-10}\ \mathrm{m}$$

$$= 0.124\ \mathrm{nm}.$$

As we have mentioned, x-ray wavelengths can be measured directly by crystal diffraction, so this relation between accelerating potential and wavelength can be confirmed directly.

X-ray energy levels in atoms: the origin of x-ray line spectra

The atomic energy levels associated with excitation by x-rays are rather different in character from those associated with visible spectra. To understand them, we need some understanding of the arrangement of electrons in complex atoms, a topic we will study in detail in Chapter 43. For the present, we simply state that in a many-electron atom, the electrons are always arranged in concentric *shells* at increasing distances from the nucleus. These shells are labeled K, L, M, N, and so on. The K shell is closest to the nucleus, the L shell next, and so on. For any given atom in the ground state there is a definite number of electrons in each shell. Each shell has a maximum number of electrons it can accommodate, and we may speak of *filled shells* and *partially filled shells*. The K shell can contain at most 2 electrons. The next, the L shell, can contain 8. The third, the M shell, has a capacity for 18 electrons, while the N shell can hold 32. The sodium atom, for example, which contains 11 electrons, has 2 in the K shell, 8 in the L shell, and a single electron in the M shell. Molybdenum, with 42 electrons, has 2 in the K shell, 8 in the L shell, 18 in the M shell, 13 in the N shell, and 1 in the O shell.

Transitions involving the inner electrons in a many-electron atom produce x-ray spectra.

The *outer* electrons of an atom are responsible for the optical spectra of the elements. Relatively small amounts of energy (typically a few electronvolts) are needed to promote these electrons to excited states, and when they return to their normal states, wavelengths in or near the visible region are emitted. The inner electrons of a complex atom are closer to the nucleus and are more tightly bound; much more energy is required to displace them from their normal levels. As a result, we would expect a photon of much larger energy, and hence much higher frequency, to be emitted when the atom returns to its normal state after the displacement of an inner electron. The displacement of the *inner* electrons gives rise to the emission of x-rays.

What are x-ray energy levels?

If they have enough energy, some of the electrons accelerated in an x-ray tube will dislodge one of the inner electrons of a target atom, say one of the K electrons. This leaves a vacant space in the K shell, which is immediately filled by an electron from one of the outer shells, such as the L, M, N, O, . . . shell. The readjustment of the electrons is accompanied by a decrease in the energy of the atom, and an x-ray photon is emitted with energy just equal to this

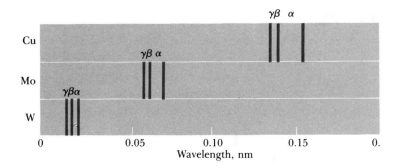

41–12 Wavelengths of the K_α-, K_β-, and K_γ-lines of copper, molybdenum, and tungsten.

decrease. Since the energy change is perfectly definite for atoms of a given element, the emitted x-rays should have definite frequencies. In other words, the **x-ray spectrum** should be a *line spectrum*. If the outermost electrons are in the N shell, there should be just three lines in the series, corresponding to the three possibilities that the vacant space may have been filled by an L, M, or N electron.

This is precisely what is observed. Figure 41–12 illustrates the so-called K series of the elements copper, molybdenum, and tungsten. Each series consists of three lines, known as the K_α-, K_β-, and K_γ-lines. The K_α-line is produced by the transition of an L electron to the vacated space in the K shell, the K_β-line by an M electron, and the K_γ-line by an N electron.

In addition to the K series, there are other series of x-ray lines, called the L, M, and N series, produced by the ejection of electrons from the L, M, and N shells rather than the K shell. Electrons in these outer shells are farther away from the nucleus and are not held as tightly as those in the K shell, so their removal requires less energy. The corresponding x-ray photons emitted when these vacancies are again filled have lower energy and longer wavelength.

Along with this x-ray *line* spectrum is a *continuous* spectrum of x-ray radiation, resulting from the stopping of electrons without exciting x-ray energy levels in the target. This spectrum extends indefinitely at the long-wavelength (low-energy) end but cuts off sharply at the short-wavelength end at a wavelength corresponding to a photon energy just equal to the electron energy before impact, as in Example 41–2.

This energy relationship is exactly the same as for the photoelectric effect (Section 41–2) except for the omission of the work-function term; this is negligible here because the x-ray energies are much larger than the work function. Indeed, x-ray emission can be thought of as an *inverse photoelectric effect.* In photoelectric emission the energy of a photon is transformed into kinetic energy of an electron; in x-ray production the kinetic energy of an electron is transformed into energy of a photon.

A phenomenon called **Compton scattering,** first observed in 1924 by A. H. Compton, provides additional direct confirmation of the quantum nature of electromagnetic radiation. When x-rays impinge on matter, some of the radiation is *scattered,* just as visible light falling on a rough surface undergoes diffuse reflection. Observations show that some of the scattered radiation has smaller frequency and longer wavelength than the incident radiation, and that the change in wavelength depends on the angle through which the radiation is scattered. Specifically, if the scattered radiation emerges at an angle ϕ

Continuous x-ray spectra: What determines the short-wavelength limit?

X-ray scattering: How can the wavelength change?

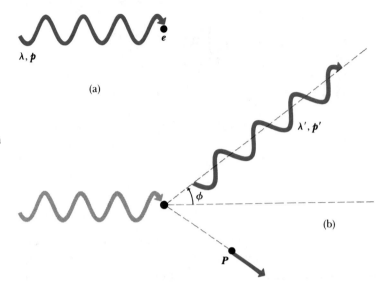

41–13 Schematic diagram of Compton scattering showing (1) electron initially at rest with incident photon of wavelength λ and momentum p; (b) scattered photon with longer wavelength λ' and momentum p' and recoiling electron with momentum P. The direction of the scattered photon makes an angle ϕ with that of the incident photon, and the angle between p and p' is also ϕ.

with respect to the incident direction, as shown in Fig. 41–13, and if λ and λ' are the wavelengths of the incident and scattered radiation, respectively, we find that

$$\lambda' - \lambda = \frac{h}{mc}(1 - \cos\phi), \qquad (41\text{--}24)$$

where m is the electron mass.

Compton scattering cannot be understood on the basis of classical electromagnetic theory. On the basis of classical principles, the scattering mechanism is induced motion of electrons in the material, caused by the incident radiation. This motion must have the same frequency as that of the incident wave, and so the scattered wave radiated by the oscillating charges should have the same frequency. The frequency cannot be *shifted* by this mechanism.

A simple analysis of Compton scattering in terms of a collision of particles

In contrast, the quantum theory provides a beautifully simple explanation. We imagine the scattering process as a collision of two *particles*, the incident photon and an electron initially at rest, as in Fig. 41–13. The photon gives up some of its energy and momentum to the electron, which recoils as a result of this impact. The final scattered photon has less energy, smaller frequency, and longer wavelength than the initial one.

We can derive Eq. (41–24) from the principles of conservation of energy and momentum. We outline the derivation below and invite you to fill in the details. The electron energy may be in the relativistic range, so we have to use the relativistic energy–momentum relations, Eqs. (40–34) and (40–35). The initial photon has momentum p and energy pc; the final photon has momentum p' and energy $p'c$. The electron is initially at rest, so its initial momentum is zero and its initial energy is mc^2. The final electron momentum is P, and the final electron energy E is given by $E^2 = (mc^2)^2 + (Pc)^2$. Then energy conservation gives us the relation

$$pc + mc^2 = p'c + E,$$

or

$$(pc - p'c + mc^2)^2 = E^2 = (mc^2)^2 + (Pc)^2. \qquad (41\text{--}25)$$

We may eliminate the electron momentum P from this equation by using momentum conservation:

$$p = p' + P,$$

or

$$p - p' = P. \tag{41–26}$$

We take the scalar product of this quantity with itself, noting that $p \cdot p' = pp' \cos \phi$, and obtain

$$P^2 = p^2 + p'^2 - 2pp' \cos \phi. \tag{41–27}$$

This expression for P^2 may now be substituted into Eq. (41–25) and the left side multiplied out. A common factor c^2 is divided out; several terms cancel, and when the resulting equation is divided through by (pp'), the result is

$$\frac{mc}{p'} - \frac{mc}{p} = 1 - \cos \phi. \tag{41–28}$$

Finally, we substitute $p = h/\lambda$ and $p' = h/\lambda'$ and rearrange again to obtain Eq. (41–24).

X-rays have many practical applications in medicine and industry. Because they can penetrate several centimeters of solid matter, they can be used to visualize the interiors of materials opaque to ordinary light, such as broken bones or defects in structural steel. The object to be visualized is placed between an x-ray source and a large sheet of photographic film; the darkening of the film is proportional to the radiation exposure. A crack or air bubble allows greater transmission and shows as a dark area. Bones appear lighter than the surrounding flesh because they contain greater proportions of elements with high atomic number (and greater absorption) than flesh, where the light elements carbon, hydrogen, and oxygen predominate. This technique is not very effective in discriminating slightly different absorption characteristics, as found with many kinds of tumors.

Using x-rays to see into things and through things

In the past decade, several vastly improved x-ray techniques have been developed. One widely used system is *computerized axial tomography;* the corresponding instrument is called a CAT-scanner. The x-ray source produces a thin fan-shaped beam that is detected on the opposite side of the subject by an array of several hundred detectors in a line. Each detector measures absorption along a thin line through the subject. The entire apparatus is rotated around the subject in the plane of the beam during a few seconds. The changing reactions of the detectors are recorded digitally; a computer processes this information and reconstructs a picture of density over an entire cross section of the subject. Density differences as small as 1% can be detected with CAT-scans, and tumors and other anomalies much too small to be seen with older x-ray techniques can be detected.

More sophisticated x-ray imaging techniques

X-rays cause damage to living tissues. As x-ray photons are absorbed in tissues, they break molecular bonds and create highly reactive free radicals (such as neutral H and OH), which in turn can disturb the molecular structure of proteins and especially genetic material. Young and rapidly growing cells are particularly susceptible; hence x-rays are useful for selective destruction of

Biological effects of x-rays: some good, some bad

cancer cells. Conversely, however, a cell may be damaged by radiation but survive, continue dividing, and produce generations of defective cells; hence x-rays can *cause* cancer. Even when the organism itself shows no apparent damage, excessive radiation exposure can cause changes in the reproductive system that will affect the organism's offspring. The use of x-rays in medical diagnosis has become an area of great concern in recent years; a careful assessment of the balance between risks and benefits of radiation exposure is essential in each individual case.

SUMMARY

The wave and ray pictures of electromagnetic radiation provide an adequate model to analyze the propagation of light. However, several phenomena involving emission and absorption point to a particle aspect of its behavior, in which the energy comes in quanta or photons. The energy E of a photon is proportional to its frequency f: $E = hf$, where h is Planck's constant, a fundamental constant of nature.

In the photoelectric effect, an electron is liberated from a conducting surface by absorption of a photon. The energy required to surmount the potential-energy barrier at the surface and escape is called the work function. The kinetic energy of the liberated electrons can be determined by measuring their stopping potential. Thus both the work function and Planck's constant can be determined.

Line spectra of atoms are associated with the existence of discrete energy levels. An atom emits a photon of a certain energy when it makes a transition between two energy levels differing in energy by that amount. The energy levels of the hydrogen atom are given by

$$E_n = -\frac{Rch}{n^2} \qquad (n = 1, 2, 3, \ldots). \qquad (41-13)$$

The lowest-energy state of an atom is called the ground state, and all higher-energy states are called excited states. An absorption spectrum results when continuous-spectrum radiation is partially absorbed while exciting atoms from the ground state.

The laser operates on the principle of stimulated emission, where an atom is stimulated to emit a photon by the presence of other photons of the same energy. For a laser to operate, there must be a means for creating a population inversion in which there are excited metastable states that cannot decay directly to the ground state.

Continuous spectra are emitted by very hot solid and liquid materials. The total rate of radiation is proportional to T^4, where T is the absolute temperature, and is given by

$$H = A\sigma T^4. \qquad (41-14)$$

This is the Stefan-Boltzmann radiation law. The distribution of the radiated power among various wavelengths is described by a distribution function $F(\lambda)$. At any temperature T, the wavelength λ_m at which this function has a maximum is given by

$$\lambda_m T = \text{constant} = 2.90 \times 10^{-3} \text{ m·K}. \qquad (41-16)$$

This relation is called the Wien displacement law. This and the Stefan-Boltzmann law can both be derived from the Planck radiation law. In deriving this law, Planck treated the electrons in a material as harmonic oscillators, with the added assumption that each electron could gain or lose energy only in increments hf, where f is the oscillator frequency. Historically, this was the first use of the quantum hypothesis. The Planck radiation law is

$$F(\lambda) = \frac{2\pi hc^2}{\lambda^5} \frac{1}{e^{hc/\lambda kT} - 1}. \qquad (41\text{--}17)$$

X-rays are produced when rapidly moving electrons strike a target. The kinetic energy of an electron is converted directly to energy of a photon, or it excites an atom to an x-ray energy level; when the atom decays back to the ground state, it emits an x-ray photon. The first process gives a continuous spectrum of x-rays, with maximum energy (and minimum wavelength) determined by the electron's initial kinetic energy; the second process yields an x-ray line spectrum that depends on the element of the target.

In Compton scattering, an x-ray photon is scattered by an electron; the electron recoils, absorbing some of the photon's energy and momentum, and the scattered photon has lower energy and longer wavelength than the original one. This process can be analyzed as a collision between two particles; for a photon scattered through an angle ϕ, the increase in wavelength is given by

$$\lambda' - \lambda = \frac{h}{mc}(1 - \cos \phi). \qquad (41\text{--}24)$$

QUESTIONS

41–1 In analyzing the photoelectric effect, how can we be sure that each electron absorbs only *one* photon?

41–2 In what ways do photons resemble other particles such as electrons? In what ways do they differ? Do they have mass? Electric charge? Can they be accelerated? What mechanical properties do they have?

41–3 Considering a two-slit interference experiment, if the photons are not synchronized with each other (i.e., are not coherent) and if half go through each slit, how can they possibly interfere with each other? Is there any way out of this paradox?

41–4 Can you devise an experiment to measure the work function of a material?

41–5 How might the energy levels of an atom be measured directly, that is, without recourse to analysis of spectra?

41–6 Would you expect quantum effects to be generally more important at the low-frequency end of the electromagnetic spectrum (radio waves) or at the high-frequency end (x-rays and gamma rays)? Why?

41–7 Most black-and-white photographic film (with the exception of some special-purpose films) is less sensitive at the far red end of the visible spectrum than at the blue end and has almost no sensitivity to infrared. How can these properties be understood on the basis of photons?

41–8 Human skin is relatively insensitive to visible light, but ultraviolet radiation can be quite destructive. Does this have anything to do with photon energies?

41–9 Does the concept of photon energy shed any light (no pun intended) on the question of why x-rays are so much more penetrating than visible light?

41–10 The phosphorescent materials that coat the inside of a fluorescent lamp tube convert ultraviolet radiation (from the mercury-vapor discharge inside the tube) to visible light. Could one also make a phosphor that converts visible light to ultraviolet?

41–11 As a body is heated to very high temperature and becomes self-luminous, the apparent color of the emitted radiation shifts from red to yellow and finally to blue as the temperature increases. Why the color shift?

41–12 Elements in the gaseous state emit line spectra with well-defined wavelengths; but hot solid bodies usually emit a continuous spectrum, that is, a continuous smear of wavelengths. Can you account for this difference?

41–13 Could Compton scattering occur with protons as well as electrons? Suppose, for example, one directed a beam of x-rays at a liquid-hydrogen target. What similarities and differences in behavior would be expected?

EXERCISES

$$e = 1.602 \times 10^{-19} \text{ C}$$
$$m = 9.110 \times 10^{-31} \text{ kg}$$
$$h = 6.626 \times 10^{-34} \text{ J·s}$$
$$N_A = 6.022 \times 10^{23} \text{ atoms·mol}^{-1}$$

Energy equivalent of 1 u = 931.5 MeV

$$\frac{e}{m} = 1.758 \times 10^{11} \text{ C·kg}^{-1}$$
$$k = 1.381 \times 10^{-23} \text{ J·K}^{-1}$$
$$1 \text{ eV} = 1.602 \times 10^{-19} \text{ J}$$
$$1 \text{ u} = 1.661 \times 10^{-27} \text{ kg}$$
$$\epsilon_0 = 8.854 \times 10^{-12} \text{ C}^2\text{·N·}^{-1}\text{·m}^{-2}$$

Section 41–2 The Photoelectric Effect

41–1 A nucleus in an excited state emits a γ-ray photon of energy 1 MeV.

a) What is the photon frequency?

b) What is the photon wavelength?

c) How does the wavelength compare with typical nuclear radii (of the order of 10^{-15} m)?

41–2 A sodium-vapor lamp emits light of wavelength 589 nm. If the total power of the emitted light is 6 W, how many photons are emitted per second?

41–3 A photon of orange light has a wavelength of 600 nm. Find the frequency, momentum, and energy of the photon; express the energy both in joules and in electronvolts.

41–4 A laser used to weld detached retinas emits light of wavelength 633 nm with a power of 0.5 W, in pulses 20 ms in duration.

a) How much energy is in each pulse, in joules? In electronvolts?

b) What is the energy of one photon, in joules? In electronvolts?

c) How many photons are in each pulse?

41–5 A radio station broadcasts at a frequency of 100 MHz, with a total power output of 50 kW.

a) What is the energy of the emitted photons, in joules? In electronvolts?

b) How many photons are emitted per second?

41–6 In the photoelectric effect, what is the relation between the threshold frequency f_0 and the work function ϕ?

41–7 A photoelectric surface has a work function of 4.00 eV. What is the maximum speed of the photoelectrons emitted by light of frequency 3×10^{15} Hz?

41–8 The photoelectric threshold wavelength of tungsten is 2.73×10^{-5} cm. Calculate the maximum kinetic energy of the electrons ejected from a tungsten surface by ultraviolet radiation of wavelength 1.80×10^{-5} cm. (Express the answer in electronvolts.)

41–9 When ultraviolet light of wavelength 2.54×10^{-5} cm from a mercury arc falls on a clean copper surface, the retarding potential necessary to stop emission of photoelectrons is 0.59 V. What is the photoelectric threshold wavelength for copper?

41–10 The photoelectric work function of potassium is 2.0 eV. If light having a wavelength of 360 nm falls on potassium, find

a) the stopping potential;

b) the kinetic energy in electronvolts of the most-energetic electrons ejected;

c) the speeds of these electrons.

Section 41–3 Line Spectra

Section 41–4 Energy Levels

Section 41–5 Atomic Spectra

41–11 Calculate (a) the frequency and (b) the wavelength of the H_β-line of the Balmer series for hydrogen. This line is emitted in the transition from $n = 4$ to $n = 2$.

41–12 Find the longest and shortest wavelengths in the Lyman, Balmer, and Paschen series for hydrogen. In what region of the electromagnetic spectrum does each series lie?

41–13 The silicon-silicon single bond that forms the basis of the (mythical) silicon-based creature the Horta has a bond strength of 3.2 eV. What wavelength photon would you need in a (mythical) phasor disintegration gun to destroy the Horta?

41–14 The energy-level scheme for the mythical one-electron element Searsium is shown in Fig. 41–14. The potential energy of an electron is taken to be zero at an infinite distance from the nucleus.

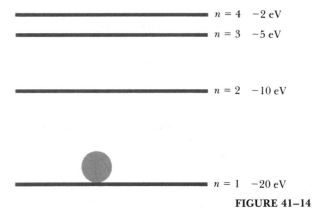

$n = 4 \quad -2 \text{ eV}$

$n = 3 \quad -5 \text{ eV}$

$n = 2 \quad -10 \text{ eV}$

$n = 1 \quad -20 \text{ eV}$

FIGURE 41–14

a) How much energy (in electronvolts) does it take to ionize an electron from the ground state?

b) A 15-eV photon is absorbed by a Searsium atom. When the atom returns to its ground state, what possible energies can the emitted photons have?

c) What will happen if a photon with an energy of 8 eV strikes a Searsium atom? Why?

d) If photons emitted from Searsium transitions $n = 4$ to $n = 2$ and from $n = 2$ to $n = 1$ will eject photoelectrons from an unknown metal, but the photon emitted from the transition $n = 3$ to $n = 2$ will not, what are the limits (maximum and minimum possible values) of the work function of the metal?

Section 41–6 The Laser

41–15 In Fig. 41–9, compute the energy difference for the $5s$–$3p$ transition in neon; express your result in electronvolts and in joules. Compute the wavelength of a photon having this energy, and compare your result with the observed wavelength of the laser light.

41–16 In the helium-neon laser, what wavelength corresponds to the $3p$–$3s$ transition in neon? Why is this not observed in the beam with the same intensity as the 632.8-nm laser line?

41–17 How many photons per second are emitted by a 0.5-mW He-Ne laser that has a wavelength of 633 nm?

Section 41–7 Continuous Spectra

41–18 What is λ_m, the wavelength at the peak of the Planck distribution, and the corresponding frequency f_m, at the following Kelvin temperatures:

a) 3 K?   b) 300 K?   c) 3000 K?

41–19

a) Show that the maximum in the Planck distribution, Eq. (41–17), occurs at a wavelength λ_m given by $\lambda_m = hc/4.965kT$ (Eq. [41–18]). As discussed in the text, 4.965 is the root of Eq. (41–19).

b) Evaluate the constants in the expression derived in (a) to show that $\lambda_m T$ has the numerical value given in the Wien displacement law, Eq. (41–16).

Section 41–8 X-Ray Production and Scattering

41–20

a) What is the minimum potential difference between the filament and the target of an x-ray tube if the tube is to produce x-rays of wavelength 0.05 nm?)

b) What is the shortest wavelength produced in an x-ray tube operated at 2×10^6 V?

41–21 Complete the derivation of the Compton-scattering formula, Eq. (41–24), following the outline given in Eqs. (41–25) through (41–28).

41–22 X-rays are produced in a tube operating at 50.0 kV. After emerging from the tube, some x-rays strike a target and are Compton-scattered through an angle of 60°.

a) What is the original x-ray wavelength?

b) What is the wavelength of the scattered x-rays?

c) What is the energy of the scattered x-rays (in electronvolts)?

41–23 X-rays with initial wavelength 0.5×10^{-10} m undergo Compton scattering. For what scattering angle is the wavelength of the scattered x-rays greater by 1% than that of the incident x-rays?

PROBLEMS

41–24

a) If the average wavelength emitted by a 100-W light bulb is 600 nm, and 10% of the input power is emitted as visible light, approximately how many visible light photons are emitted per second?

b) At what distance would this correspond to 100 photons per square centimeter, per second, if the light is emitted uniformly in all directions?

41–25 The light-sensitive compound on most photographic films is silver bromide, AgBr. A film is "exposed" when the light energy absorbed dissociates this molecule into its atoms. (The actual process is more complex, but the quantitative result does not differ greatly.) The energy of dissociation of AgBr is 1.00×10^5 J·mol^{-1}. Find

a) the energy in electronvolts,

b) the wavelength,

c) the frequency of the photon that is just able to dissociate a molecule of silver bromide.

d) What is the energy in electronvolts of a quantum of radiation having a frequency of 100 MHz?

e) Explain the fact that light from a firefly can expose a photographic film, whereas the radiation from a TV station transmitting 50,000 W at 100 MHz cannot.

f) Will photographic films stored in a light-tight container be ruined (exposed) by the radio waves passing through them? Explain.

41–26 The directions of emission of photons from a source of radiation are random. According to the wave theory, intensity of radiation from a point source varies inversely as the square of the distance from the source. Show that the number of photons from a point source passing out through a unit area is also given by an inverse-square law.

41–27 What will be the change in the stopping potential for photoelectrons emitted from a surface if the wavelength of the incident light is reduced from 400 nm to 360 nm?

41–28 The photoelectric work functions for particular samples of certain metals are as follows: cesium, 2.00 eV; copper, 4.00 eV; potassium, 2.25 eV; and zinc, 3.60 eV.

a) What is the threshold wavelength for each metal?

b) Which of these metals could not emit photoelectrons when irradiated with visible light?

41–29 When a certain photoelectric surface is illuminated with light of different wavelengths, the stopping potentials in the table below are observed.

Wavelength (nm)	Stopping Potential (V)
366	1.48
405	1.15
436	0.93
492	0.62
546	0.36
579	0.24

Plot the stopping potential as ordinate against the frequency of the light as abscissa. Determine

a) the threshold frequency,

b) the threshold wavelength,

c) the photoelectric work function of the material (in electronvolts),

d) the value of Planck's constant h (assuming the value of e is known).

41–30 An unknown element is found to have an absorption spectrum with lines at 4, 7, and 9 eV, and its ionization potential is 10 eV.

a) Draw an energy-level diagram for this element.

b) If a 9-eV photon is absorbed, what energies can the subsequently emitted photons have?

41–31

a) What is the least amount of energy in electronvolts that must be given to a hydrogen atom initially in its ground state so that it can emit the H_β-line (see Exercise 41–11) in the Balmer series?

b) How many different possibilities of spectral line emissions are there for this atom, when the electron starts in the $n = 4$ level and eventually ends up in the ground state? Calculate the wavelength of the emitted photon in each case.

41–32 If hydrogen were monatomic, at what kinetic temperature would the average translational kinetic energy be equal to the energy required to raise a hydrogen atom from the ground state to the $n = 2$ excited state?

41–33 If electrons in a metal had the same energy distribution as molecules in a gas at the same temperature (which is not actually the case), at what temperature would the average electron kinetic energy equal 1 eV, typical of work functions of metals?

41–34 An x-ray tube is operating at 150,000 V and 10 mA.

a) If only 1% of the electric power supplied is converted into x-rays, at what rate is the target being heated in joules per second?

b) If the target has a mass of 0.300 kg and a specific heat of 147 $J \cdot kg^{-1} \cdot C^{\circ -1}$, at what average rate would its temperature rise if there were no thermal losses?

c) What must be the physical properties of a practical target material? What would be some suitable target elements?

41–35

a) Calculate the maximum increase in x-ray wavelength that can occur during Compton scattering.

b) What is the energy (in electronvolts) of the smallest-energy x-ray photon for which Compton scattering could result in doubling the original wavelength?

41–36 A photon of wavelength 0.1200 nm is Compton-scattered through an angle of 180°.

a) What is the wavelength of the scattered photon?

b) How much energy is given to the electron?

c) What is the recoil speed of the electron? Is it necessary to use the relativistic kinetic-energy relationship?

41–37 A photon with $\lambda = 0.100$ nm collides with an electron at rest. After the collision the photon's wavelength is 0.110 nm.

a) What is the kinetic energy of the electron after the collision?

b) If the electron is suddenly stopped (for example, in a solid target), it emits a photon. What is the wavelength of this photon?

41–38 An electron with an energy of 1000 eV suddenly emits a photon and then continues on with an energy of 600 eV. This photon is then totally absorbed by another electron. How much momentum is transferred to the second electron by the photon?

CHALLENGE PROBLEMS

41–39

a) Write the Planck distribution law in terms of the frequency f rather than the wavelength λ, to obtain $F(f)$.

b) Show that

$$\int_0^\infty F(\lambda)\, d\lambda = \frac{2\pi^5 k^4}{15c^2 h^3} T^4,$$

where $F(\lambda)$ is the Planck distribution formula of Eq. (41–

17). *Hint:* Change the integration variable from λ to f. You will need to use the following tabulated integral:

$$\int_0^\infty \frac{x^3}{e^{\alpha x} - 1}\, dx = \frac{1}{240}\left(\frac{2\pi}{\alpha}\right)^4.$$

c) The result of (b) is H/A and has the form of the Stefan-Boltzmann law, $H/A = \sigma T^4$ (Eq. 41–14). Evaluate the constants in (b) to show that σ has the value given in Eq. (41–15).

42

QUANTUM MECHANICS

WE HAVE SEEN IN THE PRECEDING CHAPTER THAT SOME ASPECTS OF emission and absorption of light, including atomic spectra, can be understood on the basis of the photon concept, together with the concept of discrete energy levels in atoms. But a complete theory should also offer some means of *predicting,* on theoretical grounds, the values of these energy levels for any particular atom. We begin this chapter with a discussion of a partially successful attempt to do so for the hydrogen atom. This attempt, the Bohr model of the hydrogen atom, cannot be generalized to atoms having more than one electron, and more drastic departures from nineteenth-century ideas are needed. In particular, we discuss the extension of the wave-particle duality, firmly established for electromagnetic radiation, to include particles as well as radiation. The entities (such as electrons) that we are accustomed to calling *particles* may in some situations exhibit *wavelike* behavior.

The new theory we will study in this chapter requires fundamental changes in the language we use to describe the state of a mechanical system. A particle can no longer be described as a single point moving in space but is an inherently spread-out entity. As the particle moves, the spread-out character has some of the properties of a *wave;* for example, particles can undergo *diffraction.* This new theory is called *quantum mechanics;* in it we find the key to understanding the structure of atoms and molecules, including their spectra, chemical behavior, and many other properties. Quantum mechanics has the happy effect of restoring unity to our description of both particles and radiation, and wave concepts are central to the entire theory.

42–1 THE BOHR ATOM

At the same time (1915) that Bohr advanced his hypothesis about the relation of spectrum-line frequencies to energy levels of atoms, he also proposed a mechanical model of the simplest atom, hydrogen. By combining some classical mechanics with a postulate that had no basis in previous physics, he was able to calculate the energy levels of hydrogen and obtain agreement with

The Bohr model: predicting the energy levels of the hydrogen atom

values determined from spectra. His description is called the **Bohr model** of the hydrogen atom.

Bohr's was not by any means the first attempt to understand the internal structure of atoms. Starting in 1906, Rutherford and his coworkers had performed experiments on the scattering of alpha particles (helium nuclei emitted from radioactive elements) by thin metallic foils. These experiments, which we will discuss in Chapter 44, showed that each atom contains a dense, compact nucleus whose size (of the order of 10^{-15} m) is very much smaller than the overall size of the atom (of the order of 10^{-10} m). The nucleus is surrounded by a swarm of electrons.

A dynamic picture of the atom: electrons in orbits

To account for the fact that the negatively charged electrons remain at relatively large distances from the positively charged nucleus despite the electrostatic attraction the nucleus exerts on the electrons, Rutherford postulated that the electrons *revolve* about the nucleus in orbits, more or less as the planets in the solar system revolve around the sun, but with the electrical attraction providing the necessary centripetal force.

Why doesn't an orbiting electron radiate energy continuously?

This assumption, however, has an unfortunate consequence. A body moving in a circle accelerates continuously toward the center of the circle. According to classical electromagnetic theory, an accelerating electron radiates energy. The total energy of the electrons would therefore decrease continuously, their orbits would become smaller and smaller, and eventually they would spiral into the nucleus and come to rest. Furthermore, according to classical theory the *frequency* of the electromagnetic waves emitted by a revolving electron should equal the frequency of revolution. As the electrons radiated energy, their angular velocities would change continuously and they would emit a *continuous* spectrum (a mixture of all frequencies), in contradiction to the *line* spectrum actually observed.

Electrons in stable nonradiating orbits

Faced with the dilemma that electromagnetic theory predicted an unstable atom emitting radiant energy of all frequencies, while observation showed stable atoms emitting only a few frequencies, Bohr concluded that, in spite of the success of electromagnetic theory in explaining large-scale phenomena, it could not be applied to processes on an atomic scale. He therefore postulated that an electron in an atom can revolve in certain **stable orbits,** each having a definite associated energy, *without* emitting radiation, contrary to the predictions of classical electromagnetic theory. According to Bohr, an atom radiates only when it makes a transition from one of these stable orbits to another, at the same time emitting (or absorbing) a photon of appropriate energy and frequency, given by Eq. (41–11).

The permitted orbits have definite values of angular momentum.

To determine the radii of the "permitted" orbits, Bohr introduced what must be regarded in hindsight as a brilliant intuitive guess. He noted that the *units* of Planck's constant h, usually written as J·s, are the same as the units of angular momentum, usually written as kg·m^2·s^{-1}, and he postulated that only those orbits are permitted for which the angular momentum is an integral multiple of $h/2\pi$. Recall from Section 9–12 that the angular momentum of a particle of mass m, moving with tangential speed v in a circle of radius r, is mvr. Hence the condition above may be stated as

$$mvr = n\frac{h}{2\pi},$$

where $n = 1, 2, 3$, and so on. Each value of n corresponds to a permitted value of the orbit radius, which we denote from now on by r_n, and a corresponding speed v_n. With this notation, the equation above becomes

$$mv_n r_n = n\frac{h}{2\pi}. \tag{42–1}$$

We now incorporate this condition into the analysis of the hydrogen atom. This atom consists of a single electron of charge $-e$, revolving about a single proton of charge $+e$. The proton is nearly 2000 times as massive as the electron, and we will assume the proton does not move. The electrostatic force of attraction between the charges,

$$F = \frac{1}{4\pi\epsilon_0}\frac{e^2}{r_n{}^2},$$

provides the centripetal force and, from Newton's second law,

$$\frac{1}{4\pi\epsilon_0}\frac{e^2}{r_n{}^2} = \frac{mv_n{}^2}{r_n}. \tag{42–2}$$

When Eqs. (42–1) and (42–2) are solved simultaneously for r_n and v_n, we obtain

$$r_n = \epsilon_0\frac{n^2h^2}{\pi me^2}, \tag{42–3}$$

$$v_n = \frac{1}{\epsilon_0}\frac{e^2}{2nh}. \tag{42–4}$$

Let

$$\epsilon_0\frac{h^2}{\pi me^2} = r_1. \tag{42–5}$$

Then Eq. (42–3) becomes

$$r_n = n^2 r_1,$$

and the permitted, nonradiating orbits have radii r_1, $4r_1$, $9r_1$, and so on. The appropriate value of n is called the **quantum number** of the orbit.

The numerical values of the quantities on the left side of Eq. (42–5) are

$$\epsilon_0 = 8.854 \times 10^{-12}\ \mathrm{C^2 \cdot N^{-1} \cdot m^{-2}},$$

$$h = 6.626 \times 10^{-34}\ \mathrm{J \cdot s},$$

$$m = 9.110 \times 10^{-31}\ \mathrm{kg},$$

$$e = 1.602 \times 10^{-19}\ \mathrm{C}.$$

Hence r_1, the radius of the first Bohr orbit, is

$$r_1 = \frac{(8.854 \times 10^{-12}\ \mathrm{C^2 \cdot N^{-1} \cdot m^{-2}})(6.626 \times 10^{-34}\ \mathrm{J \cdot s})^2}{(3.14)(9.110 \times 10^{-31}\ \mathrm{kg})(1.602 \times 10^{-19}\ \mathrm{C})^2}$$

$$= 0.53 \times 10^{-10}\ \mathrm{m} = 0.53 \times 10^{-8}\ \mathrm{cm}.$$

This is in good agreement with atomic diameters as estimated by other methods, namely, about 10^{-8} cm.

The kinetic energy of the electron in any orbit is

$$K_n = \frac{1}{2}mv_n{}^2 = \frac{1}{\epsilon_0{}^2}\frac{me^4}{8n^2h^2},$$

Calculating the energy corresponding to a particular electron orbit

The permitted states are labeled by a value of an integer called a quantum number.

The energy of each state depends on the value of its quantum number in a simple way.

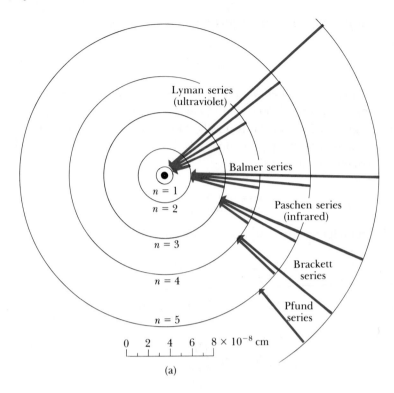

Lyman series
(ultraviolet)

Balmer series

$n = 1$

$n = 2$

Paschen series
(infrared)

$n = 3$

$n = 4$

Brackett
series

$n = 5$

Pfund
series

0   2   4   6 8×10^{-8} cm

(a)

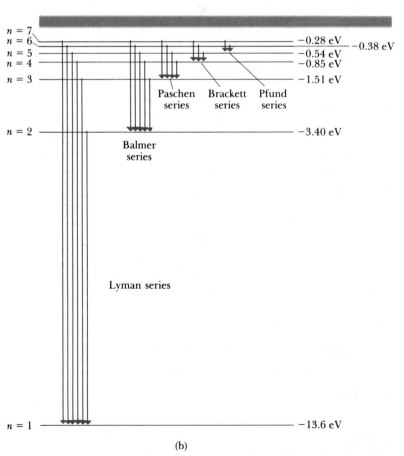

$n = 7$
$n = 6$ -0.28 eV
$n = 5$ -0.38 eV
$n = 4$ -0.54 eV
                     -0.85 eV
$n = 3$ -1.51 eV

Paschen   Brackett   Pfund
series   series   series

$n = 2$ -3.40 eV

Balmer
series

Lyman series

$n = 1$ -13.6 eV

(b)

42–1 (a) "Permitted" orbits of an electron in the Bohr model of a hydrogen atom. The transitions responsible for some of the lines of the various series are indicated by arrows. (b) Energy-level diagram, showing transitions corresponding to the various series.

and the potential energy is

$$U_n = -\frac{1}{4\pi\epsilon_0}\frac{e^2}{r_n} = -\frac{1}{\epsilon_0^2}\frac{me^4}{4n^2h^2},$$

The total energy, E_n, is therefore

$$E_n = K_n + U_n = -\frac{1}{\epsilon_0^2}\frac{me^4}{8n^2h^2}. \tag{42–6}$$

The total energy has a negative sign because the reference level of potential energy is taken to be zero with the electron at an infinite distance from the nucleus. Since we are interested only in energy *differences,* this is not of importance. The energy levels can be displayed graphically as in Fig. 42–1b.

Thus the possible states of the atoms are labeled by values of the integer n, which we call the *quantum number* for the system. For each value of n there are corresponding values of orbit radius, angular momentum, and total energy. The energy of the atom is least when $n = 1$, for then E_n has its largest negative value. This is the **ground state** of the atom; in this state the electron is in the smallest orbit, with radius r_1. For $n = 2, 3, \ldots$, the absolute value of E_n is smaller and the energy is progressively larger (less negative). The orbit radius increases as n^2, according to Eq. (42–3).

As a result of collisions with rapidly moving electrons in an electrical discharge, or by other means, the atom may temporarily acquire enough energy to raise the electron to some higher energy and larger orbit. The atom is then said to be in an **excited state.** This state is unstable; the electron eventually falls back to a state of lower energy, emitting a photon in the process.

Let n_1 be the quantum number of some excited state and n_2 be the quantum number of the lower state to which the electron returns after the emission process. Then E_i, the initial energy, is

$$E_i = -\frac{1}{\epsilon_0^2}\frac{me^4}{8n_1^2h^2},$$

and E_f, the final energy, is

$$E_f = -\frac{1}{\epsilon_0^2}\frac{me^4}{8n_2^2h^2}.$$

The decrease in energy, $E_i - E_f$, which we set equal to the energy hf of the emitted photon, is

$$E_i - E_f = hf = -\frac{1}{\epsilon_0^2}\frac{me^4}{8n_1^2h^2} + \frac{1}{\epsilon_0^2}\frac{me^4}{8n_2^2h^2},$$

or

$$f = \frac{1}{\epsilon_0^2}\frac{me^4}{8h^3}\left(\frac{1}{n_2^2} - \frac{1}{n_1^2}\right). \tag{42–7}$$

This equation has the same form as the Balmer formula, Eq. (41–9), for the frequencies in the hydrogen spectrum if we place

$$\frac{1}{\epsilon_0^2}\frac{me^4}{8h^3} = Rc, \tag{42–8}$$

and let $n_2 = 1$ for the Lyman series, $n_2 = 2$ for the Balmer series, and so on.

The ground state has the smallest energy and the smallest quantum number.

Photon energies are differences between the initial and final energy states of an atom.

The Lyman series is therefore the group of lines emitted by electrons returning from some excited state to the ground state. The Balmer series is the group of lines emitted by electrons returning from some higher excited state, but stopping in the *second orbit* ($n = 2$) instead of falling directly to the ground state. That is, an electron returning from the third orbit ($n = 3$) to the second orbit ($n = 2$) emits the H_α-line. One returning from the fourth orbit ($n = 4$) to the second ($n = 2$) emits the H_β-line, and so on. These transitions are shown in Fig. 42–1.

Every quantity in Eq. (42–8) may be determined quite independently of the Bohr theory. Apart from this theory, we have no reason to expect these quantities to be related in this particular way. The quantities m and e, for instance, are found from experiments on free electrons; h may be found from the photoelectric effect; R is determined by measurements of wavelengths; and c is the speed of light. But if we substitute the values of these quantities, obtained by such diverse means, into Eq. (42–8), we find it *does* hold exactly, within the limits of experimental error, providing direct confirmation of Bohr's theory.

The ionization energy of the hydrogen atom (the energy required to remove the electron completely) can also be predicted from the Bohr theory. Ionization corresponds to a transition from the ground state ($n = 1$) to an infinitely large orbit radius ($n = \infty$). The predicted energy is 13.6 eV, and again this is in excellent agreement with the experimentally measured value.

The Bohr model can be extended easily to other one-electron atoms, such as the singly ionized helium atom, the doubly ionized lithium atom, and so on. If the nuclear charge is Ze (where Z is the atomic number) instead of just e, the effect in the analysis above is to replace e^2 everywhere by Ze^2. In particular, the orbits, described by Eq. (42–3), become smaller by a factor of Z, and the energy levels, given by Eq. (42–6), are all multiplied by Z^2. We invite you to verify these statements.

Although the Bohr model was successful in predicting the energy levels of the hydrogen atom, it raised as many questions as it answered. It combined elements of classical physics with new postulates that were inconsistent with classical ideas. It provided no insight into what happens *during* a transition from one orbit to another, and the stability of certain orbits was achieved at the expense of discarding the only picture available at the time of the electromagnetic mechanism for the atom to radiate energy. There was no clear justification for restricting the angular momentum to multiples of $h/2\pi$, except that it led to the right answer. Furthermore, attempts to extend the model to atoms with two or more electrons were not successful. We will see in the next section that an even more radical departure from classical concepts was required before the understanding of atomic structure could progress further.

42–2 WAVE NATURE OF PARTICLES

The next major advance in understanding atomic structure came in 1923, about ten years after the Bohr theory. This was a suggestion by de Broglie that since light is dualistic in nature, behaving in some situations like waves and in others like particles, the same might be true of matter. That is, electrons and protons, which until that time had been thought to be purely particlelike, might in some circumstances behave like *waves*. Specifically, de Broglie postulated that a free electron of mass m, moving with speed v, should have a wavelength λ related to its momentum $p = mv$ in exactly the same way as the wavelength and momentum of a photon are related, as expressed by Eq. (41–7), $\lambda = h/p$. Thus the **de Broglie wavelength** of an electron is given by

$$\lambda = \frac{h}{mv}, \qquad (42\text{–}9)$$

where h is the same Planck's constant that appears in the frequency–energy relation for photons.

This wave hypothesis, unorthodox though it seemed at the time, almost immediately received direct experimental confirmation. We described in Section 39–9 how the layers of atoms in a crystal can serve as a diffraction grating for x-rays. An x-ray beam is strongly reflected when it strikes a crystal at such an angle that the waves scattered from the atomic layers combine to reinforce one another. The essential point here is that the existence of these strong reflections is evidence of the *wave* nature of x-rays.

In 1927 Davisson and Germer, working in the Bell Telephone Laboratories, were studying the surface of a crystal of nickel by directing a beam of *electrons* at the surface and observing the electrons reflected at various angles. It might be expected that even the smoothest surface attainable would still look rough to an electron, and that the electron beam would therefore be diffusely reflected. But the **Davisson-Germer experiment** showed that the electrons were reflected in almost the same way that x-rays would be reflected from the same crystal; that is, they were being *diffracted*. The de Broglie wavelengths of the electrons in the beam were computed from their known speed, with the help of Eq. (42–9); and the angles at which strong reflection took place were found to be the same as those at which x-rays of the same wavelength would be reflected. The discovery of **electron diffraction** gave strong support to de Broglie's hypothesis.

> A diffraction experiment with electrons: direct confirmation of the wave nature of particles

This wave hypothesis clearly required sweeping revisions of our fundamental concepts regarding the description of matter. What we are accustomed to calling a *particle* actually behaves like a particle only if we do not look too closely. In general, a particle has to be described as a spread-out entity that is not entirely localized in space; and at least in some cases this spreading out appears as a periodic pattern suggesting *wavelike* properties. The wave and particle aspects are not inconsistent; the particle model is an *approximation* of a more general wave picture. We are reminded of the ray picture of geometrical optics, a special case of the more general wave picture of physical optics. Indeed, there is a very close analogy between optics and the description of the motion of particles.

Within a few years after 1923, the wave hypothesis of de Broglie was developed by Heisenberg, Schrödinger, Dirac, Born, and many others, into a complete theory called **quantum mechanics.** In the following sections we will sketch the main lines of thought in a nonmathematical way and describe some of the experimental evidence for the wave nature of material particles. We will show how the quantum numbers that were introduced in such an artificial way by Bohr now enter naturally into the theory of atomic structure.

One of the essential features of quantum mechanics is that a particle is no longer described as located at a single point, but is described instead in terms of a *function* that has various values at various points in space. The spatial

> Quantum mechanics: some drastic revisions in our ideas of how to describe position and motion of a particle

distribution describing a *free* electron may have a recurring pattern characteristic of a wave that propagates through space. Electrons *in atoms* can be visualized as diffuse clouds surrounding the nucleus. The idea that the electrons in an atom move in definite orbits such as those in Fig. 42–1 has been abandoned. The orbits themselves, however, were never an essential part of Bohr's theory, since the quantities that determine the frequencies of the emitted photons are the *energies* corresponding to the orbits. The new theory still assigns definite energy states to an atom. In the hydrogen atom the energies are the same as those given by Bohr's theory; in more complicated atoms, where the Bohr theory does not work, the quantum mechanical picture is in excellent agreement with observation.

To show how quantization arises in atomic structure, we can use an analogy with the classical mechanical problem of a vibrating string held at its ends. We worked out the normal modes of this system in Section 22–3, and we suggest that you review that discussion. When the string vibrates, the ends must be nodes, but nodes may occur at other points also. The general requirement is that the length of the string shall equal some *integral* number of half-wavelengths.

In a similar way, the principles of quantum mechanics lead to a differential equation (Schrödinger's equation) that must be satisfied by an electron in an atom, subject also to certain boundary conditions. Let us think of an electron as a wave extending in a circle around the nucleus. In order for the wave to "come out even" and join onto itself smoothly, the circumference of this circle must include some *integral number* of wavelengths, as suggested by Fig. 42–2. The wavelength of a particle of mass m, moving with speed v, is given, according to wave mechanics, by Eq. (42–9), $\lambda = h/mv$. Then if r is the radius and $2\pi r$ the circumference of the circle occupied by the wave, we must have $2\pi r = n\lambda$, where $n = 1, 2, 3$, and so on. Since $\lambda = h/mv$, this equation becomes

$$2\pi r = n\frac{h}{mv}, \qquad mvr = n\frac{h}{2\pi}. \qquad (42\text{–}10)$$

But mvr is the angular momentum of the electron, so we see that the wave-mechanical picture leads naturally to Bohr's postulate that the angular momentum equals some integral multiple of $h/2\pi$.

To be sure, the idea of wrapping a wave around in a circular orbit is a rather vague notion. But the agreement of Eq. (42–10) with Bohr's hypothesis is much too remarkable to be a coincidence; and it strongly suggests that the wave properties of electrons do indeed have something to do with atomic structure.

A wave picture of atomic structure: standing waves in an atom

The Bohr quantum condition as a consequence of the de Broglie wave hypothesis

42–2 Diagrams showing the idea of wrapping a standing wave around a circular orbit. For the wave to join onto itself smoothly, the circumference of the orbit must be an integral number n of wavelengths. Examples are shown for $n = 2, 3$, and 4.

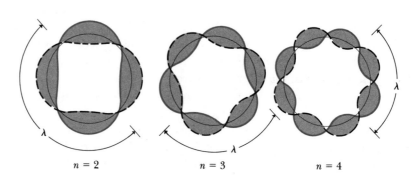

$n = 2$ $n = 3$ $n = 4$

PROBLEM-SOLVING STRATEGY: *Atomic physics*

1. In atomic physics, the orders of magnitude of physical quantities are so unfamiliar that often common sense isn't much help in judging the reasonableness of a result. It helps to remind yourself of some typical magnitudes of various quantities:

Size of an atom, 10^{-10} m,

Mass of an atom, 10^{-26} kg,

Mass of an electron, 10^{-30} kg,

Energy of an atomic state, 1 to 10 eV or 10^{-18} J (but some interaction energies are much *smaller* than this),

Speed of an electron in the Bohr atom, 10^6 m·s^{-1},

Electric charge, 10^{-19} C,

kT at room temperature, $\frac{1}{40}$ eV.

You may want to add items to the list. This will also help you in Chapter 44, where we have to deal with magnitudes characteristic of nuclear rather than atomic structure, often different by factors of 10^6 or so. In working out problems, be very careful to handle powers of ten properly. A gross error may not be obvious.

2. As in the last chapter, energies may be expressed either in joules or in electronvolts. Be sure you use consistent units. Lengths, such as wavelengths, are always in meters if you use the other quantities consistently in SI units, such as $h = 6.626 \times 10^{-34}$ J·s. If you want nanometers or something else, don't forget to convert.

3. Aside from these calculational details, the main challenges of this chapter are conceptual, not computational. Try to keep an open mind when you encounter new and sometimes jarring ideas. Eventually you will come to appreciate the fact that a photon *can* have both wavelike and particlelike properties. Don't get discouraged; intuitive understanding of quantum mechanics takes some time to develop. Keep trying!

EXAMPLE 42–1 Find the speed and the kinetic energy of a neutron ($m = 1.675 \times 10^{-27}$ kg) having a de Broglie wavelength of 0.1 nm, typical of atomic spacing in crystals. Compare the energy with the average kinetic energy of a gas molecule at room temperature ($T = 20°C$).

An example of a wavelength calculation for a particle

SOLUTION From Eq. (42–9),

$$v = \frac{h}{\lambda m} = \frac{6.626 \times 10^{-34} \text{ J·s}}{(0.1 \times 10^{-9} \text{ m})(1.675 \times 10^{-27} \text{ kg})}$$

$$= 3.96 \times 10^3 \text{ m·s}^{-1};$$

$$K = \tfrac{1}{2}mv^2 = \tfrac{1}{2}(1.675 \times 10^{-27} \text{ kg})(3.96 \times 10^3 \text{ m·s}^{-1})^2$$

$$= 1.31 \times 10^{-20} \text{ J} = 0.0818 \text{ eV}.$$

The average translational kinetic energy of a molecule of an ideal gas is given by Eq. (20–10):

$$K = \tfrac{3}{2}kT = (\tfrac{3}{2})(1.38 \times 10^{-23} \text{ J·K}^{-1})(293 \text{ K})$$

$$= 6.06 \times 10^{-21} \text{ J} = 0.0378 \text{ eV}.$$

Thus the two energies are of comparable magnitude, and indeed a neutron having an energy in this range is called a *thermal neutron*. Diffraction of thermal neutrons can be used to study crystal and molecular structure in the same way as x-ray diffraction; neutron diffraction has proved especially useful in the study of large organic molecules.

42–3 THE ELECTRON MICROSCOPE

Using an electron beam to form greatly enlarged images of microscopic objects

An electron beam can be used to form an image of an object in exactly the same way as a light beam. A ray of light is bent by reflection or refraction, and an electron trajectory is bent by an electric or magnetic field. Rays of light diverging from a point on an object can be brought to convergence by a converging lens, and electrons diverging from a small region can be brought to convergence by an electrostatic or magnetic lens. Figure 42–3 (parts a and b) shows the behavior of a simple type of electrostatic lens, and Figure 42–3c shows the analogous optical system. In each case the image can be made larger than the object by appropriate design; hence both devices can act as magnifiers.

The analogy between light rays and electrons goes deeper. The ray model of geometrical optics is an approximate representation of the more general wave picture, and geometrical optics is valid whenever interference and diffraction effects can be neglected. Similarly, we have seen in Section 42–2 that the model of an electron as a point particle following a line trajectory is an approximate description of the actual behavior of the electron, valid when effects associated with the wave nature of electrons can be neglected.

Resolution is better with short-wavelength electrons than with much longer wavelength visible light.

Herein lies the value of the **electron microscope.** The *resolution* of an optical microscope is limited by diffraction effects, as discussed in Section 39–10. Using a wavelength of 500 nm, typical of visible light, no optical microscope can resolve objects smaller than a few hundred nanometers, no matter how carefully its lenses are made. The resolution of an electron microscope is similarly limited by the wavelengths of the electrons, but these may be many thousands of times *smaller* than wavelengths of visible light. Hence the useful

42–3 (a) An electrostatic lens. The two cylinders are at different electrical potentials V_a and V_b; there is an electric field in the region between them, and the corresponding equipotential lines are shown in black. (b) Cross-sectional view from the side. The electron trajectories are shown in color; electrons diverging from point A are brought to a focus at point B. The behavior of magnetic lenses is similar. (c) Optical analog of the electrostatic lens in part (a).

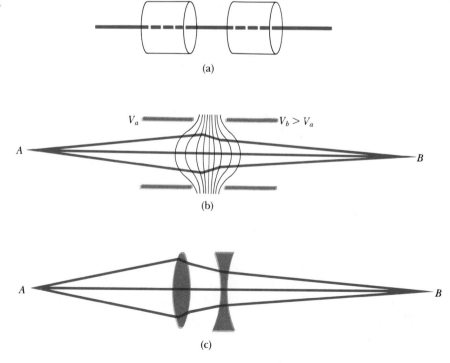

magnification of an electron microscope can be thousands of times as great as that of an optical microscope.

It is important to understand that the ability of the electron microscope to form an image *does not* depend on the wave properties of electrons. Their trajectories can be computed by treating them as charged particles under the action of electric- and magnetic-field forces. In the matter of *resolution,* however, their wave properties do become significant.

EXAMPLE 42–2 An electron beam is formed by a setup similar to that of the cathode-ray tube, discussed in Section 26–8. If the accelerating voltage is 10 kV, what is the wavelength of the electrons?

SOLUTION The wavelength is determined by Eq. (42–9). To find the speed we use conservation of energy; the kinetic energy $\frac{1}{2}mv^2$ of an electron equals the loss of potential energy eV. Thus

$$\frac{1}{2}mv^2 = eV, \qquad v = \sqrt{\frac{2eV}{m}}.$$

Inserting this result into Eq. (42–9), we find

$$\lambda = \frac{h}{m}\sqrt{\frac{m}{2eV}} = \frac{h}{\sqrt{2meV}}$$

$$= \frac{6.63 \times 10^{-34}\ \text{J·s}}{\sqrt{2(9.11 \times 10^{-31}\ \text{kg})(1.60 \times 10^{-19}\ \text{C})(10^4\ \text{V})}} \qquad (42\text{–}11)$$

$$= 1.23 \times 10^{-11}\ \text{m}$$

$$= 0.0123\ \text{nm}.$$

This value is smaller than typical wavelengths of visible light (around 500 nm) by a factor of about 40,000.

Most electron microscopes use magnetic rather than electrostatic lenses, for practical reasons. A common setup includes three lenses in a compound-microscope arrangement, as shown in Fig. 42–4. Electrons are emitted from a hot cathode and accelerated by a potential difference, typically 10 to 100 kV. The electrons pass through a condenser lens and are formed into a parallel beam before passing through the specimen or object to be viewed. The objective lens then forms an intermediate image of this object, and the projection lens produces a final real image of that image. These lenses play the roles of the objective and eyepiece lenses, respectively, of a compound optical microscope. The final image is recorded on photographic film or projected on a fluorescent screen for viewing or photographing. The entire apparatus, including the specimen, must be enclosed in a vacuum container, just as with the cathode-ray tube; otherwise electrons would collide with air molecules, muddling up the image. The specimen to be viewed is very thin, typically 10 nm to 100 nm, so the electrons are not slowed appreciably as they pass through.

We might think that with electrons of wavelength 0.01 nm, as in the example above, the resolution would be 0.01 nm or less. In fact, it is seldom better

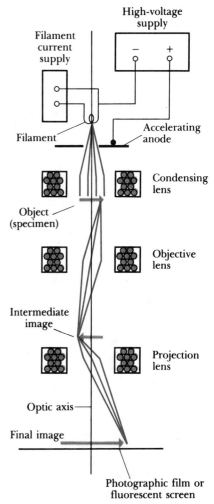

42–4 An electron microscope. The magnetic lenses, consisting of coils of wire carrying currents, are shown in cross section. The condensing lens forms a parallel beam of electrons that strikes the object. The objective lens forms an intermediate image that serves as the object for the final image formed by the projection lens. All images are *real*. The final image is projected on photographic film or a fluorescent screen. The magnification of each lens may be of the order of 100X, and the overall magnification of the order of 10,000X. For greater magnification an additional intermediate lens may be used. The color lines are not continuous electron trajectories through the instrument but are drawn from lens to lens to show the formation of images. The angles of the electron paths with the optic axis are greatly exaggerated; in actual instruments these angles are usually less than 0.01 rad or 0.5°. The entire apparatus is enclosed in a vacuum chamber, not shown in the diagram.

than 0.5 nm, for several reasons. Large-aperture magnetic lenses have aberrations analogous to those of optical lenses, as discussed in Section 38–5. In addition, the focal length of a magnetic lens depends on the current in the coil, which must be controlled precisely, and on the electron speed, which is never a single precise value; the latter effect is the equivalent of chromatic aberration.

An improved electron microscope concept for studying surface details

A useful variation is the *scanning electron microscope*. The electron beam is focused to a very fine line and is swept across the specimen just as the electron beam in a TV picture tube traces out the picture. As the beam scans the specimen, electrons are knocked off; these are collected by a collecting anode kept at a potential a few hundred volts positive with respect to the specimen. The current in the collecting anode is amplified and used to modulate the electron beam in a cathode-ray tube, which is swept in synchronism with the microscope beam. Thus the cathode-ray tube traces out a greatly magnified image of the specimen. This scheme has the advantages that the beam need not pass through the specimen and that knock-off electron production depends on the angle at which the beam strikes the surface. Thus scanning electron micrographs have a much greater three-dimensional appearance than conventional ones. The resolution is not as great, typically of the order of 10 nm, but still much greater than the optical microscope. A scanning electron microscope is shown in Fig. 42–5, and a photograph made with such an instrument is shown in Fig. 42–6.

42–5 A scanning electron microscope, showing a vacuum chamber in the background and cathode-ray tube monitors in the foreground. (Courtesy of General Electric.)

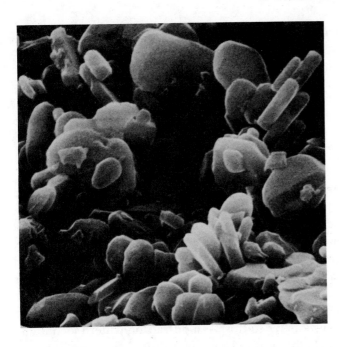

42–6 Photograph of powdered aluminum oxide. The overall magnification is about 4500X. (Manfred Kage-Peter Arnold, Inc.)

42–4 PROBABILITY AND UNCERTAINTY

We have learned that the entities we customarily call *particles* can, in some circumstances, behave like *waves*. Figure 42–7 shows a *wave packet* or *wave pulse* that has both wave and particle properties. The regular spacing λ_{av} between successive maxima is characteristic of a wave, but there is also a particlelike localization in space. To be sure, the wave pulse is not localized at a single point, but in any experiment that detects only dimensions much larger than Δx, the wave pulse appears to be a localized particle. Although it would be simplistic to say that Fig. 42–7 is a picture of an electron, the figure does suggest that wave and particle properties are not necessarily incompatible.

It would *not* be correct to regard the wave of Fig. 42–7 as having only a single wavelength. A sinusoidal wave with a definite wavelength has no beginning and no end; to make a wave *pulse* we must superpose many sinusoidal waves having various wavelengths. Thus Fig. 42–7 is a wave having such a distribution of wavelengths, and λ_{av} is an average value. This distribution has additional important implications, which we will explore at the end of this section.

The discovery of the dual wave-particle nature of matter has forced a drastic revision of the language used to describe the behavior of a particle. In classical Newtonian mechanics, we think of a particle as an idealized geometrical point that, at any instant of time, has a perfectly definite location in space and is moving with a definite velocity. As we will see, such a specific description is, in general, not possible. On a sufficiently small scale, there are fundamental limitations on the precision with which the position and velocity of a particle can be described; and some aspects of a particle's behavior can be stated only in terms of *probabilities*.

To illustrate the nature of the problem, let us consider again the single-slit diffraction experiment described in Section 39–7. We learned there that most (85%) of the intensity in the diffraction pattern is concentrated in the central maximum; the angular size of this maximum is determined by the positions of

Can something be a particle and a wave at the same time?

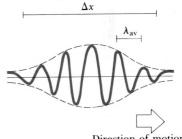

Direction of motion

42–7 A wave pulse or packet. There is an average wavelength λ_{av}, the distance between adjacent peaks, but the wave is localized at any instant in a region with length of the order of Δx. The broken lines are called the envelope of the pulse; all the peaks lie on the envelope curves, which approach zero at both ends of the pulse.

Particle and wave aspects of a single-slit diffraction experiment

the first intensity *minimum* on either side of the central maximum. Using Eq. (39–20), with $n = 1$, we find that the angle θ in Fig. 39–20 between the central peak and the minimum on either side is given by $\sin\theta = \lambda/a$, where a is the slit width. If λ is much smaller than a, then θ is very small, $\sin\theta$ is very nearly equal to θ, and the relation may be simplified further to

$$\theta = \frac{\lambda}{a}. \qquad (41\text{–}12)$$

Now we perform the same experiment again but use a beam of *electrons* instead of a beam of monochromatic light. The apparatus must be evacuated to avoid collisions of electrons with air molecules; and there are other experimental details that need not concern us. The electron beam can be produced with a setup similar in principle to the electron gun in a cathode-ray tube, which produces a narrow beam of electrons all having the same direction and speed, and therefore also the same wavelength. Such an experiment is shown schematically in Fig. 42–8.

The result of this experiment, as again recorded on photographic film or by means of more sophisticated detectors, is a diffraction pattern identical to that shown in Fig. 39–22b; this provides additional direct evidence of the wave nature of electrons. Most electrons strike the film in the vicinity of the central maximum, but a few strike farther from the center, near the edges of that maximum and also in the subsidiary maxima on both sides. Thus, if we believe that electrons are waves, the wave behavior in this experiment presents no surprises.

Interpreted in terms of *particles,* however, this experiment poses very serious problems. First, although the electrons all have the same initial state of motion, they do not all follow the same path. In fact, *the trajectory of an individual electron cannot be predicted from knowledge of its initial state.* The best we can do is to say that most of the electrons go to a certain region, fewer go to other regions, and so on; alternatively, we can describe the *probability* for an individual electron to strike each of various areas on the film. This fundamental indeterminacy has no counterpart in Newtonian mechanics, where the motion of a particle or a system is always predictable if the initial position and motion are known with sufficient precision.

Second, there are fundamental *uncertainties* in both position and momentum of an individual particle, and these two uncertainties are related inseparably. To illustrate this point, we note that, in Fig. 42–8, an electron striking the

We cannot predict precisely the trajectory of an individual particle.

The diffraction pattern is a probability distribution.

42–8   An electron diffraction experiment. The graph at the right shows the degree of blackening of the film, which in any region is proportional to the number of electrons striking that region. The components of momentum of an electron striking the outer fringe of the central maximum are shown.

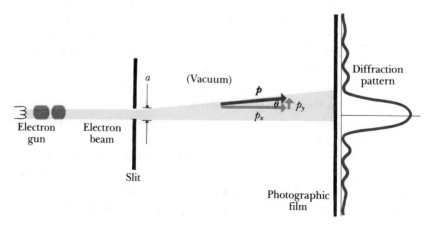

film at the outer edge of the central maximum, at angle θ, must have a component of momentum p_y in the y-direction, as well as a component p_x in the x-direction, despite the fact that initially the beam was directed along the x-axis. From the geometry of the situation, the two components are related by $p_y/p_x = \tan \theta$; and if θ is small, we may approximate $\tan \theta = \theta$, obtaining

$$p_y = p_x \theta. \tag{42-13}$$

Neglecting any electrons striking the film outside the central maximum (that is, at angles greater than λ/a), we see that the y-component of momentum may be as large as

$$p_y = p_x \left(\frac{\lambda}{a} \right). \tag{42-14}$$

The uncertainties in the transverse position and momentum have a reciprocal relationship.

Hence the *uncertainty* Δp_y in the y-component of momentum is at least as great as $p_x \lambda/a$:

$$\Delta p_y \geq p_x \frac{\lambda}{a}.$$

Thus the narrower the slit width a, the broader the diffraction pattern and the greater the uncertainty in the y-component of momentum.

Now the electron wavelength λ is related to the momentum $p_x = mv_x$ by the de Broglie relation, Eq. (42-9), which may be rewritten as $\lambda = h/p_x$. Using this result in the equation above and simplifying, we find

The more accurately the position is known, the less accurately the momentum can be known.

$$\Delta p_y \geq p_x \left(\frac{h}{ap_x} \right) = \frac{h}{a},$$

or

$$\Delta p_y a \geq h. \tag{42-15}$$

To interpret this result, we note that the slit width a represents the uncertainty in *position* of an electron as it passes through the slit; we do not know through which particular part of the slit each particle passes. Thus the y-components of *both* position and momentum have uncertainties, and the two uncertainties are related by Eq. (42-15). We can reduce the *momentum* uncertainty only by increasing the slit width, which increases the *position* uncertainty; and conversely, when we decrease the position uncertainty by narrowing the slit, the diffraction pattern broadens and the corresponding momentum uncertainty increases.

All of this may be bitter medicine for a reader steeped in the tradition of nearly three centuries of Newtonian mechanics; but the weight of experimental evidence leaves us no alternative. To those who protest that the lack of a definite position and momentum is contrary to common sense, we reply that what we call common sense is based on familiarity gained through experience, and that our usual experience includes very little contact with the microscopic behavior of particles. Thus we must sometimes be prepared to accept conclusions that seem contrary to intuition when we are dealing with areas far removed from everyday experience.

Some rather jarring conclusions about what we can and cannot know about particle motion

In more general discussions of uncertainty relations, it is customary to describe the uncertainty of a quantity in terms of the statistical concept of *standard deviation*, a measure of the spread or dispersion of a set of numbers around their average value. If a coordinate x has an uncertainty Δx, defined in this way, and if the corresponding momentum component p_x has an uncertainty Δp_x, then the two uncertainties are found to be related in general by the inequality

$$\Delta x \, \Delta p_x \geq \frac{h}{2\pi}. \qquad (42\text{–}16)$$

Equation (42–16) is one form of the **Heisenberg uncertainty principle;** it states that, in general, neither the momentum nor the position of a particle can be predicted with arbitrarily great precision, as classical physics would predict. Instead, the two quantities play complementary roles, as described above. One might protest that greater precision could be attained by using more sophisticated particle detectors in various areas of the slit or by other means; but this turns out to be impossible. To detect a particle the detector must *interact* with it, and this interaction unavoidably changes the state of motion of the particle. A more detailed analysis of such hypothetical experiments shows that the uncertainties we have described are fundamental and intrinsic; they cannot be circumvented even in principle by any experimental technique, no matter how sophisticated.

Additional insight into the uncertainty principle is provided by the wave packet shown in Fig. 42–7. We can construct such a packet by superposing several sinusoidal waves with various wavelengths. Each individual component wave has no beginning or end but extends indefinitely in both directions. We choose the wavelengths and the amplitude of each component wave so that constructive interference occurs only in a small region of width Δx, as shown in the figure, and the interference is destructive everywhere else. It turns out that a rather broad wave packet can be obtained by superposing waves with a relatively small range of wavelengths; but to make a narrow packet requires a wider range of wavelengths. Yet a wide range of wavelengths also means a correspondingly wide range of values of momentum, because of the de Broglie relation. So again we see that a small uncertainty in position must be accompanied by a large uncertainty in momentum, and conversely.

In addition, the *energy* of a system, as well as its position and momentum, always has uncertainty. The uncertainty ΔE is found to depend on the time interval Δt during which the system remains in the given state. The relation is

$$\Delta E \, \Delta t \geq \frac{h}{2\pi}. \qquad (42\text{–}17)$$

Thus a system that remains in a certain state for a long time can have a very well defined energy, but if it remains in that state for only a short time, the uncertainty in energy must be correspondingly greater.

EXAMPLE 42–3   A sodium atom in one of the "resonance levels" shown in Fig. 41–5 remains in that state for an average time of 1.6×10^{-8} s before making a transition to the ground state by emitting a photon of wavelength 589 nm and energy 2.109 eV. What is the uncertainty in energy of the resonance level?

SOLUTION   From Eq. (42–17),

$$\Delta E = \frac{h}{2\pi \, \Delta t} = \frac{(6.626 \times 10^{-34} \text{ J·s})}{(2\pi)(1.6 \times 10^{-8} \text{ s})}$$

$$= 6.59 \times 10^{-27} \text{ J} = 4.11 \times 10^{-8} \text{ eV}.$$

The atom remains an indefinitely long time in the ground state, so there is *no* uncertainty there; the uncertainty of the resonance level energy and of the corresponding photon energy amounts to about two parts in 10^8. This irreducible uncertainty is called the *natural line width* of this particular spectrum line. Ordinarily the natural line width is much smaller than line broadening from other causes, such as collisions among atoms.

Let us now take a brief look at a quantum interpretation of a *two-slit* interference pattern. We studied these patterns in detail for light in Sections 39–2 and 39–3. In terms of photons, the intensity-distribution pattern must correspond to the numbers of photons striking various regions of the screen where the pattern is formed. In fact, it is possible to detect individual photons with a device called a *photomultiplier*. Using the setup shown in Fig. 42–9, we can place the photomultiplier at various positions for equal time intervals, count photons at each position, and plot out the intensity distribution. We find that *on the average*, the distribution of photons agrees with our predictions from Section 39–3.

If we now reduce the light intensity to a point where only a few photons per second pass through the slits, there is no way to predict where an individual photon will go. Thus the interference pattern has to be viewed as a *statistical distribution* of photons. It tells us how many will go in each direction, or the *probability* for an individual photon to go in each of various directions, but *not* where any particular photon will go.

It is tempting to think that each individual photon must pass through one or the other of the slits. But if this were the case, we would be able to record the interference pattern on film by opening one slit for a time, then closing it and opening the other. As we know, this does not work; to form an interference pattern, the waves from the two slits have to be *coherent*. To resolve this apparent paradox, we are forced to conclude that *every* photon goes partly through *both* slits, and that *each photon interferes only with itself*. If this sounds like nonsense, remember that the photon is *not* a single point in space; it is a conceptual framework to describe the quantization of energy in electromagnetic waves. There is nothing conceptually wrong with saying that every photon passes through both slits, and indeed this is the *only* consistent viewpoint!

Particle and wave aspects of a two-slit interference experiment

Is it possible for a photon to pass through both slits?

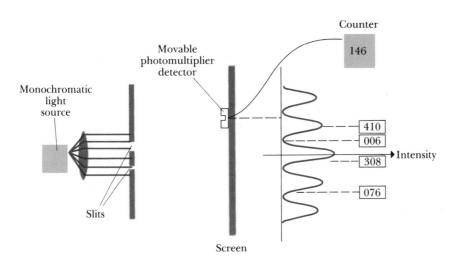

42–9 Two-slit interference pattern observed with photomultipliers. The curve shows the intensity distribution predicted by the wave picture, and the photon distribution is shown by the numbers of photons counted at various positions.

Two-slit interference with electrons: Every particle must pass through both slits.

Finally, what happens when we do a two-slit interference experiment with electrons? Exactly the same thing as with photons! Again we can do a particle-counting experiment to trace out the interference pattern, as we did with photons, but we *cannot* predict where in the pattern an individual electron will land. Furthermore, again because of the coherence requirement, we have to assume that every electron goes through both slits! If the idea of a photon passing through both slits made you uncomfortable, you'll really hate this. But since the electron is not an entity described by a single geometric point, there is nothing conceptually wrong with having it pass through both slits. Like photons, each electron interferes only with itself.

42–5 WAVE FUNCTIONS

Wave functions: the fundamental kinematic language of quantum mechanics

As we have seen, the dual wave-particle nature of electrons and other fundamental particles requires us to generalize the kinematic language we use to describe the position and motion of a particle. The classical notion of a particle as a point having at each instant a definite position in space (described by three coordinates) and a definite velocity (described by three components) must be replaced by more general language.

In making the needed generalizations, we are guided by the language of classical wave motion. When studying transverse waves on a string in Chapter 21, we described the motion of the string by specifying the position of each point in the string at each instant of time. This was done by means of a **wave function,** introduced in Section 21–3. If y represents the displacement from equilibrium at time t of a point on the string whose equilibrium position is a distance x from the origin, then the function $y = f(x, t)$ or, more briefly, $y(x, t)$ represents the displacement of point x at time t. If we know the wave function for a given motion, we know everything there is to know about the motion; from the function we can determine the shape of the string at any time, the slope at each point, the velocity and acceleration of each point, and any other needed information.

Similarly, in Chapter 23 we discussed a sound wave propagating in the x-direction. Letting p represent the variation in air pressure from its equilibrium value at any point, we write $p(x, t)$ as the pressure variation at any point x at any time t; again this is a *wave function*. If the wave is three-dimensional, we can describe p at a space point with coordinates (x, y, z) at any time t by means of a wave function $p(x, y, z, t)$, which contains all the space coordinates and time. The same pattern reappears in the description of *electromagnetic* waves in Section 35–5, where we use two wave functions to describe the electric and magnetic fields at any point in space, at any time.

Thus it is natural to use a wave function as the central element of our generalized language for describing particles. The symbol usually used for this wave function is Ψ, and it is, in general, a function of all the space coordinates and time. Just as the wave function $y(x, t)$ for mechanical waves on a string provides a complete description of the motion, the wave function $\Psi(x, y, z, t)$ for a particle contains all the information that can be known about the particle.

The wave function describes the distribution of the particle in space.

Two questions immediately arise. First, what is the *meaning* of the wave function Ψ for a particle? Second, how is Ψ determined for any given physical situation?

With reference to the first question, the wave function describes the distribution of the particle in space. It is related to the *probability* of finding the particle in each of various regions; the particle is most likely to be found in regions where Ψ is large, and so on. If the particle has charge, the wave function can be used to find the *charge density* at any point in space. In addition, from Ψ we can calculate the *average* position of the particle; its average velocity; and dynamic quantities such as momentum, energy, and angular momentum. The required techniques are far beyond the scope of this discussion, but they are well established and well supported by experimental results.

The answer to the second question is that the wave function must be one of a set of solutions of a certain differential equation called the **Schrödinger equation,** developed by Erwin Schrödinger in 1925. In principle we can set up a Schrödinger equation for any given physical situation, such as the electron in a hydrogen atom; the functions that are solutions of this equation represent various possible physical states of the system. Furthermore, it turns out for some systems that acceptable solutions exist only when some physical quantity, such as the energy of the system, has certain special values. Thus the solutions of the Schrödinger equation are also associated with *energy levels.* This discovery is of the utmost importance; before the development of the Schrödinger equation, there was no way to predict energy levels from any fundamental theory, except for the very limited success of the Bohr model for hydrogen.

Soon after it was developed, the Schrödinger equation was applied to the problem of the hydrogen atom. The predicted energy levels for the simplest model turned out to be identical to those from the Bohr model, Eq. (42–6), and thus agreed with experimental values from spectrum analysis. The energy levels are labeled with the quantum number n.

In addition, the solutions have *quantized* values of angular momentum; that is, only certain discrete values of the magnitude of angular momentum and its components are possible. Recall that quantization of angular momentum was put into the Bohr model as an *ad hoc* assumption with no fundamental justification; with the Schrödinger equation it appears automatically! Specifically, it is found that the magnitude L of the angular momentum of an electron in the hydrogen atom in a state with energy E_n and quantum number n must be given by

$$L = \sqrt{l(l+1)}\left(\frac{h}{2\pi}\right), \qquad (42\text{--}18)$$

where l is zero or a positive integer no larger than $n - 1$. The *component* of L in a given direction, say the z-component L_z, can have only the set of values

$$L_z = m\frac{h}{2\pi}, \qquad (42\text{--}19)$$

where m can be zero or a positive or negative integer up to but no larger than l.

The quantity $h/2\pi$ appears so often in quantum mechanics that it is given a special symbol, $\hbar$. That is,

$$\hbar = \frac{h}{2\pi} = 1.054 \times 10^{-34}\ \text{J·s}.$$

In terms of $\hbar$, the two preceding equations become

$$L = \sqrt{l(l+1)}\hbar \qquad (l = 0, 1, 2, \ldots, n-1) \qquad (42\text{--}20)$$

The wave functions must be solutions of a certain differential equation.

The energy levels come out of the Schrödinger equation automatically.

The quantization of angular momentum comes out automatically.

and

$$L_z = m\hbar \qquad (m = 0, \pm 1, \pm 2, \ldots, \pm l). \qquad (42\text{--}21)$$

Uncertainty in the direction of the angular momentum vector

Note that the component L_z can never be quite as large as L. For example, when $l = 4$ and $m = 4$, we find

$$L = \sqrt{4(4 + 1)}\,\hbar = 4.47\hbar,$$
$$L_z = 4\hbar.$$

This inequality arises from the uncertainty principle, which makes it impossible to predict the *direction* of the angular momentum vector with complete certainty. Thus the component of L in a given direction can never be quite as large as the magnitude L, except when $l = 0$ and both L and L_z are zero. Unlike the Bohr model, the Schrödinger equation gives values for the magnitude L of angular momentum that are *not* integer multiples of $\hbar$.

Describing the quantum state of the hydrogen atom by use of quantum numbers

Another interesting feature of Eqs. (42–20) and (42–21) is that there are states for which the angular momentum is *zero*. This is a result that has no classical analog; in the Bohr model the electron always moved in an orbit and thus had nonzero angular momentum. But in the new mechanics we find states having zero angular momentum.

The possible wave functions for the hydrogen atom can be labeled according to the values of the three integers n, l, and m, called, respectively, the **principal quantum number,** the **angular momentum quantum number,** and the **magnetic quantum number.** For each energy level E_n, there are several distinct states having the same energy but different values of l and m, the only exception being the ground state $n = 1$, for which only $l = 0, m = 0$ is possible. Examination of the spatial extent of the wave functions shows that, in each case, the wave function is confined primarily to a region of space around the nucleus having a radius of the same order of magnitude as the corresponding Bohr radius. For increasing values of n, the electron is, on the average, farther away from the nucleus and hence has less negative potential energy and becomes less tightly bound.

The new mechanics described above, which we call quantum mechanics, is much more complex, both conceptually and mathematically, than Newtonian mechanics. However, quantum mechanics enables us to understand physical phenomena and to analyze physical problems for which classical mechanics is completely powerless. Added complexity is the price we pay for this greatly expanded understanding.

42–6 THE ZEEMAN EFFECT

Interaction of the atom with a magnetic field causes shifts in energy levels and spectrum wavelengths.

In Section 41–1 we mentioned Faraday's suggestion that shifts of the spectrum wavelengths in a line spectrum might occur when the source is placed in a magnetic field. Faraday's spectroscopic techniques in 1862 were not refined enough to observe such shifts, but the Dutch physicist Pieter Zeeman, using improved instruments, was able in 1896 to detect small shifts, an effect now called the *Zeeman effect.*

We first analyze the interaction of atoms with a magnetic field using the Bohr model. This interaction can be represented by use of the concept of *magnetic moment,* which we studied in Section 30–8. To avoid confusion in notation, we denote magnetic moment by $\boldsymbol{\mu}$, the electron mass by m_e, and the magnetic quantum number by m. The interaction energy U of a magnetic moment $\boldsymbol{\mu}$ in a magnetic field $\boldsymbol{B}$ was derived in Section 30–8, Eq. (30–23). It is given by

$$U = -\boldsymbol{\mu} \cdot \boldsymbol{B}. \qquad (42\text{–}22)$$

Let us consider the electron in the first Bohr orbit ($n = 1$). The orbiting charge is equivalent to a current loop of radius r and area πr^2. The average charge per unit time passing a point of the orbit is the average current I, and this is given by e/τ, where τ is the time for one revolution: $\tau = 2\pi r/v$. Thus $I = ev/2\pi r$. The magnetic moment μ is given by

$$\mu = IA = \left(\frac{ev}{2\pi r}\right)\pi r^2 = \frac{evr}{2}. \qquad (42\text{–}23)$$

But according to the Bohr theory, the angular momentum $m_e vr$ in the $n = 1$ state is equal to $\hbar$, so

$$vr = \frac{\hbar}{m_e}$$

and

$$\mu = \frac{e}{2m_e}\hbar = \frac{1}{2}(1.758 \times 10^{11}\ \mathrm{C\cdot kg^{-1}})(1.054 \times 10^{-34}\ \mathrm{J\cdot s})$$
$$= 9.27 \times 10^{-24}\ \mathrm{A\cdot m^2}.$$

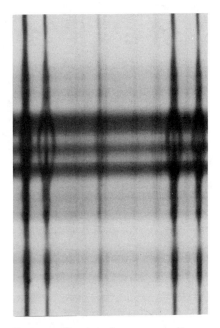

Zeeman effect in solar spectrum lines. The lines emitted by gaseous iron near a sunspot are broadened and split into doublets and triplets by the very strong magnetic field created by the sunspot. (Courtesy of Kitt Peak Observatory.)

EXAMPLE 42–4 Find the interaction potential energy when the hydrogen atom described above is placed in a magnetic field of 2 T.

SOLUTION According to Eq. (42–22), the interaction energy U when $\boldsymbol{B}$ and $\boldsymbol{\mu}$ are parallel is

$$U = -\mu B = -(9.27 \times 10^{-24}\ \mathrm{A\cdot m^2})(2\ \mathrm{T})$$
$$= -1.85 \times 10^{-23}\ \mathrm{J}$$
$$= -1.16 \times 10^{-4}\ \mathrm{eV}.$$

An example showing the order of magnitude of the interaction energy

When $\boldsymbol{\mu}$ and $\boldsymbol{B}$ are antiparallel, the energy is $+1.16 \times 10^{-4}\ \mathrm{eV}$. We note that these energies are *smaller* than the energy levels of the atom, by a factor of the order of 10^{-4}.

In the $n = 1$ state of the Bohr atom, the magnetic moment μ is equal to $e/2m_e$ times the angular momentum L. The ratio of magnetic moment to angular momentum is called the **gyromagnetic ratio,** and it can be shown that it has the value $e/2m_e$ for *every* Bohr orbit. It can also be shown that electrons described by the Schrödinger equation have this same ratio of μ to L.

We can generalize this discussion to states described by Schrödinger wave functions. Suppose the magnetic field $\boldsymbol{B}$ is directed along the $+z$-axis; then the interaction energy of the atom's magnetic moment with the field is given by

The ratio of magnetic moment to angular momentum is the same in the Schrödinger equation as for the Bohr model.

$$U = -\boldsymbol{\mu} \cdot \boldsymbol{B} = -\mu_z B. \qquad (42\text{–}24)$$

From the discussion above,

$$\mu_z = \frac{e}{2m_e}L_z. \qquad (42\text{–}25)$$

But $L_z = m\hbar$, with $m = 0, \pm 1, \pm 2, \ldots, \pm l$, so

$$\mu_z = \frac{e}{2m_e} L_z = m\frac{e\hbar}{2m_e}. \qquad (42\text{-}26)$$

Finally, the interaction energy of Eq. (42-24) becomes

$$U = -\mu_z B = -m\frac{e\hbar B}{2m_e} \qquad (m = 0, \pm 1, \pm 2, \ldots, \pm l). \qquad (42\text{-}27)$$

An example of line shift caused by interaction of the atom with a magnetic field

The effect of the magnetic field is to shift the energy level by an amount U, given by Eq. (42-27), which depends on the orientation of the angular momentum, that is, on the value of the magnetic quantum number m. An energy level having a certain value of the quantum number l is thus split into $2l + 1$ sublevels, with adjacent sublevels differing in energy by $e\hbar B/2m_e$. Spectrum lines corresponding to transitions involving this energy level are correspondingly split and appear as a series of closely spaced spectrum lines replacing a single line. This effect, called the **Zeeman effect,** is a very direct experimental confirmation of the quantization of angular momentum.

EXAMPLE 42–5 An atom in a state having $l = 1$ emits a photon with wavelength 600 nm as it decays to its ground state (with $l = 0$). If the atom is placed in a magnetic field of magnitude $B = 1.0$ T, determine the shifts in the energy levels and in the wavelengths.

SOLUTION The energy of a 600-nm photon, as obtained in Section 41–4, is 3.31×10^{-19} J or 2.07 eV. The ground-state level has $l = 0$ and is not split by the field. The splitting of levels in the $l = 1$ state is given by Eq. (42-27):

$$\begin{aligned} U &= -m\frac{e\hbar B}{2m_e} \\ &= -m\frac{(1.60 \times 10^{-19} \text{ C})(1.054 \times 10^{-34} \text{ J·s})(1.00 \text{ T})}{2(9.11 \times 10^{-31} \text{ kg})} \\ &= -m(9.26 \times 10^{-24} \text{ J}) = -m(5.79 \times 10^{-5} \text{ eV}). \end{aligned}$$

When $l = 1$, the possible values of m are -1, 0, and $+1$, and the three resulting levels are split by equal intervals of 5.79×10^{-5} eV. This is a small fraction of the photon energy: $(5.79 \times 10^{-5} \text{ eV})/(2.07 \text{ eV}) = 2.80 \times 10^{-5}$. Thus we expect the corresponding *wavelength* shifts to be approximately $(2.80 \times 10^{-5})(600 \text{ nm}) = 0.017$ nm. The original 600-nm line is split into a triplet with wavelengths 599.983, 600.000, and 600.017 nm. This splitting is well within the limit of resolution of modern spectrometers.

42–7 ELECTRON SPIN

Electron spin: additional angular momentum not due to orbital motion

For the analysis of more complex atoms, and even for some details of the spectrum of hydrogen, we need one additional concept, **electron spin.** To illustrate this concept, we consider the motion of the earth around the sun. The earth travels in a nearly circular orbit and at the same time *rotates* on its axis. Each motion has its associated angular momentum, called the *orbital* and *spin* angular momentum, respectively, and the total angular momentum of the system is the sum of the two. If we were to model the earth as a single point, no spin angular momentum would be possible; but a more refined model, with an earth of finite size, includes the possibility of spin angular momentum.

This discussion can be translated to the language of the Bohr model. Suppose the electron is not a point charge moving in an orbit, but a small spinning sphere in orbit. Then the electron has not only orbital angular momentum but also additional angular momentum associated with the spin motion. Because the sphere carries an electric charge, the spinning motion leads to current loops and to a magnetic moment, as we discussed in Section 30–8. If this magnetic moment really exists, then it interacts with a magnetic field and there is an associated interaction energy. These effects are in addition to the interaction of the orbital magnetic moment with a magnetic field, which we discussed in Section 42–6. Thus there should be additional small shifts in the energy levels of the atom and in the wavelengths of the associated spectrum lines.

Such shifts *are* indeed observed in precise spectroscopic analysis; this and other experimental evidence have shown conclusively that the electron *does* have angular momentum and a magnetic moment that are not related to the orbital motion but are intrinsic to the particle itself. Like orbital angular momentum, spin angular momentum (usually denoted by S) is found to be *quantized*. Denoting the z-component of S by S_z, we find that the only possible values are

Spin angular momentum is quantized.

$$S_z = \pm \frac{1}{2}\hbar. \tag{42–28}$$

This relation is reminiscent of Eq. (42–21) for the z-component of orbital angular momentum, but the component is one-half of $\hbar$ instead of an *integral* multiple.

In quantum mechanics, where the Bohr orbits are superseded by wave functions, it is not really possible to *picture* electron spin. If the wave functions are visualized as clouds surrounding the nucleus, then we can imagine many tiny arrows distributed throughout the cloud, all pointing in the same direction, either all $+z$ or all $-z$. Of course, this picture should not be taken too seriously; there is no hope of actually seeing an atom's structure because it is thousands of times smaller than wavelengths of light and because interactions with light photons would seriously disturb the very structure we are trying to observe.

In any event, the concept of electron spin is well established by a variety of experimental evidence. To label completely the state of the electron in a hydrogen atom, we now need a fourth quantum number s to specify the electron spin orientation. If s can take the values $+1$ or -1, then the z-component of spin angular momentum is given by

$$S_z = \tfrac{1}{2}s\hbar \qquad (s = \pm 1). \tag{42–29}$$

The corresponding component of magnetic moment, which we again denote by μ_z, turns out to be related to S_z by

The gyromagnetic ratio for electron spin is twice as large as for orbital angular momentum.

$$\mu_z = \frac{e}{m_e} S_z, \tag{42–30}$$

where e and m_e are again the charge and mass of the electron, respectively. Note that for electron spin the gyromagnetic ratio is just *twice* the value for *orbital* angular momentum and magnetic moment. When the atom is placed in a magnetic field, the interaction of the electron spin magnetic moment with the field causes further splittings in energy levels and in the corresponding spectrum lines.

Somewhat more subtle is the level splitting caused by the electron spin magnetic moment even when there is *no* external field. In the Bohr model, an observer moving with the electron sees the positively charged nucleus moving around him; this moving charge causes a magnetic field at the location of the electron, and the resulting interaction energy with the spin magnetic moment causes a twofold splitting of this level, corresponding to the two possible orientations of electron spin.

Magnetic interactions without an external field

Although the Bohr model is now known to be inadequate, a similar result can be derived from a more complete quantum-mechanical treatment based on the Schrödinger equation. The effect is called **spin-orbit coupling;** it is responsible for the small energy difference between the two closely spaced "resonance levels" of sodium shown in Fig. 41–5 and for the corresponding familiar doublet (589.0, 589.6 nm) in the spectrum of sodium.

Interactions with nuclear magnetic moments: additional, very small shifts in energy levels

The various line splittings resulting from these magnetic interactions are collectively called *fine structure.* There are additional, much smaller splittings resulting from the fact that the *nucleus* of the atom also has a magnetic moment and interacts with the orbital and spin angular momentum of the electrons. These effects are called *hyperfine structure.*

SUMMARY

KEY TERMS
Bohr model
stable orbits
quantum number
ground state
excited state
de Broglie wavelength
Davisson-Germer experiment
electron diffraction
quantum mechanics
electron microscope
Heisenberg uncertainty principle
wave function
Schrödinger equation
principal quantum number
angular momentum quantum number
magnetic quantum number
gyromagnetic ratio
Zeeman effect
electron spin
spin-orbit coupling

In the Bohr model of the hydrogen atom, the electron travels in a circular orbit around the nucleus, with the centripetal force supplied by the $1/r^2$ electrical attraction. Bohr further assumed that the angular momentum must be an integral multiple of $h/2\pi$. The resulting orbit radii r_n are given by

$$r_n = \epsilon_0 \frac{n^2 h^2}{\pi m e^2}, \tag{42–3}$$

and the energies are

$$E_n = K_n + U_n = -\frac{1}{\epsilon_0^2} \frac{m e^4}{8 n^2 h^2}. \tag{42–6}$$

These equations agree with deductions from the Balmer formula and the quantum nature of light.

Electrons show wavelike properties in some situations; the first experimental observation of electron diffraction was the Davisson-Germer experiment. The wavelength λ of a particle with mass m and speed v is given by

$$\lambda = \frac{h}{mv}. \tag{42–9}$$

The electron microscope uses an electron beam to form an enlarged image of an object. The wave nature of electrons is not an essential requirement for image formation, but the resolution is limited by diffraction effects. Electrons can have much shorter wavelengths than visible light, so the electron microscope has much greater resolution.

Analysis of a single-slit diffraction experiment with electrons shows that there are always uncertainties in position and momentum, related by the Heisenberg uncertainty principle:

$$\Delta x \, \Delta p_x \geq \frac{h}{2\pi}. \qquad (42\text{--}16)$$

There is a corresponding uncertainty relation for energy: If a system exists in a certain energy state for a time Δt, the uncertainty ΔE in its energy is given by

$$\Delta E \, \Delta t \geq \frac{h}{2\pi}. \qquad (42\text{--}17)$$

Analysis of interference and diffraction experiments for both photons and electrons leads us to a statistical interpretation of the result. The motion of an individual electron or photon cannot be predicted.

The new kinematic language needed to describe the position of a particle includes a wave function, which must be a solution of the Schrödinger equation. Solution of this equation for the hydrogen atom yields the same energy levels as the Bohr model and in addition shows that angular momentum and its component in a given direction are quantized, according to

$$L = \sqrt{l(l+1)}\hbar \qquad (l = 0, 1, 2, \ldots, n-1), \qquad (42\text{--}20)$$

and

$$L_z = m\hbar \qquad (m = 0, \pm 1, \pm 2, \ldots, \pm l). \qquad (42\text{--}21)$$

Three quantum numbers are used to identify the spatial states of the hydrogen atom; they are the principal, angular momentum, and magnetic quantum numbers, n, l, and m.

The Zeeman effect is the shifting of energy levels caused by interactions of an atom's magnetic moments with a magnetic field. The interaction energy associated with the orbital magnetic moment is

$$U = -\mu_z B = -m\frac{e\hbar B}{2m_e} \qquad (m = 0, \pm 1, \pm 2, \ldots, \pm l). \qquad (42\text{--}27)$$

The electron also has an inherent or "spin" angular momentum S that can have a component $S_z = \pm\frac{1}{2}\hbar$ in a given direction; the associated magnetic moment is given by

$$\mu_z = \frac{e}{m_e}S_z. \qquad (42\text{--}30)$$

The gyromagnetic ratio is the ratio of magnetic momentum to angular momentum. For orbital angular momentum it is $e/2m_e$; for spin angular momentum it is e/m_e.

QUESTIONS

42–1 In analyzing the absorption spectrum of hydrogen at room temperature, one finds absorption lines corresponding to wavelengths in the Lyman series but not to those in the Balmer series. Why not?

42–2 A singly ionized helium atom has one of its two electrons removed, and the energy levels of the remaining electron are closely related to those of the hydrogen atom. The nuclear charge for helium is $+2e$ instead of just $+e$;

exactly how are the energy levels related to those of hydrogen? How is the size of the ion in the ground state related to that of the hydrogen atom?

42–3 Consider the line spectrum emitted from a gas discharge tube such as a neon sign or a sodium-vapor or mercury-vapor lamp. It is found that when the pressure of the vapor is increased, the spectrum lines spread out, that is, are less sharp and less monochromatic. Why?

42–4 Suppose a two-slit interference experiment is carried out by using an electron beam. Would the same interference pattern result if one slit at a time is uncovered instead of both at once? If not, why not? Does not each electron go through one slit or the other? Or does every electron go through both slits? Does the latter possibility make sense?

42–5 Is the wave nature of electrons significant in the function of a television picture tube? For example, do diffraction effects limit the sharpness of the picture?

42–6 A proton and an electron have the same speed. Which has longer wavelength?

42–7 A proton and an electron have the same kinetic energy. Which has longer wavelength?

42–8 Does the uncertainty principle have anything to do with marksmanship? That is, is the accuracy with which a bullet can be aimed at a target limited by the uncertainty principle?

42–9 Is the Bohr model of the hydrogen atom consistent with the uncertainty principle?

42–10 If the energy of a system can have uncertainty, as stated by Eq. (42–17), does this mean that the principle of conservation of energy is no longer valid?

42–11 If quantum mechanics replaces the language of Newtonian mechanics, why do we not have to use wave functions to describe the motion of macroscopic objects such as baseballs and cars?

42–12 Why is analysis of the helium atom much more complex than that of the hydrogen atom, either in a Bohr type of model or using the Schrödinger equation?

42–13 Do gravitational forces play a significant role in atomic structure?

EXERCISES

$$e = 1.602 \times 10^{-19} \text{ C}$$
$$m = 9.110 \times 10^{-31} \text{ kg}$$
$$h = 6.626 \times 10^{-34} \text{ J·s}$$
$$N_A = 6.022 \times 10^{23} \text{ atoms·mol}^{-1}$$

Energy equivalent of 1 u = 931.5 MeV

$$\frac{e}{m} = 1.758 \times 10^{11} \text{ C·kg}^{-1}$$
$$k = 1.381 \times 10^{-23} \text{ J·K}^{-1}$$
$$1 \text{ eV} = 1.602 \times 10^{-19} \text{ J}$$
$$1 \text{ u} = 1.661 \times 10^{-27} \text{ kg}$$
$$\epsilon_0 = 8.854 \times 10^{-12} \text{ C}^2\text{·N}^{-1}\text{·m}^{-2}$$

Section 42–1 The Bohr Atom

42–1

a) Calculate the Bohr-model speed of the electron in a hydrogen atom in the $n = 1$, 2, and 3 states.

b) Calculate the orbital period in each of these states.

c) The average lifetime of the first excited state of a hydrogen atom is 10^{-8} s. How many orbits does an electron in an excited atom complete before returning to the ground state?

42–2 According to the Bohr model, the Rydberg constant R is equal to $me^4/8\epsilon_0^2 h^3 c$.

a) Calculate R in m^{-1} and compare with the experimental value.

b) Calculate the energy (in electronvolts) of a photon whose wavelength equals R^{-1}. (This quantity is known as the Rydberg energy.)

42–3 For a hydrogen atom in the ground state, determine in electronvolts:

a) the kinetic energy of the electron;

b) its potential energy;

c) its total energy;

d) the energy required to remove the electron completely.

e) What wavelength would a photon with the energy calculated in (d) have? In what region of the electromagnetic spectrum does it lie?

42–4 A hydrogen atom initially in the ground state absorbs a photon, which excites it to the $n = 4$ state. Determine the wavelength and frequency of the photon.

42–5 A singly ionized helium ion (a helium atom with one electron removed) behaves very much like a hydrogen atom, except that the nuclear charge is twice as great.

a) How do the energy levels differ in magnitude from those of the hydrogen atom?

b) Which spectral series for He$^+$ have lines in the visible spectrum? (Refer to Exercise 41–12.)

c) For a given value of n how does the radius of an orbit in He$^+$ relate to that for H?

Section 42–2 Wave Nature of Particles

Section 42–3 The Electron Microscope

42–6

a) An electron moves with a speed of 3×10^6 m·s^{-1}. What is its de Broglie wavelength?

b) A proton moves with the same speed. Determine its de Broglie wavelength.

42–7 Approximately what range of photon energies (in electronvolts) corresponds to the visible spectrum? Approximately what range of wavelengths would electrons in this energy range have?

42–8 For crystal diffraction experiments, wavelengths of the order of 0.1 nm are often appropriate. Find the energy, in electronvolts, for a particle with this wavelength if the particle is

a) a photon;   b) an electron.

42–9

a) What is the de Broglie wavelength of an electron accelerated through 500 V?

b) What is the de Broglie wavelength of a proton accelerated through the same potential difference?

42–10

a) What is the de Broglie wavelength of an electron that has been accelerated through a potential difference of 200 V?

b) Would this electron exhibit particlelike or wavelike characteristics on meeting an obstacle or opening 1 mm in diameter?

Section 42–4 Probability and Uncertainty

42–11

a) Suppose that the uncertainty in position of a particle is on the order of its de Broglie wavelength. Show that in this case the uncertainty in its momentum is on the order of its momentum.

b) Suppose the uncertainty in position of an electron is equal to the radius of the $n = 1$ Bohr orbit, about 0.5×10^{-10} m. Estimate the uncertainty in its momentum, and compare this with the magnitude of the momentum of the electron in the $n = 1$ Bohr orbit.

42–12 A certain atom has an energy level 2.0 eV above the ground state. When excited to this state, it remains on the average 2.0×10^{-6} s before emitting a photon and returning to the ground state.

a) What is the energy of the photon? Its wavelength?

b) What is the smallest possible uncertainty in energy of the photon?

c) Show that $|\Delta E / E| = |\Delta\lambda / \lambda|$ when $|\Delta\lambda / \lambda|$ is small. Use this to calculate the magnitude of the smallest possible uncertainty in the wavelength of the photon.

42–13 Suppose an unstable particle produced in a high-energy collision has a mass three times that of the proton and an uncertainty in mass that is 1% of the particle's mass. Assuming mass and energy are related by $E = mc^2$, estimate the lifetime of the particle.

Section 42–5 Wave Functions

42–14 Make a chart showing all the possible sets of quantum numbers l and m for the states of the electron in the hydrogen atom when $n = 3$. How many combinations are there?

42–15 Consider states with $l = 3$.

a) In units of $\hbar$, what is the largest possible value of L_z?

b) In units of $\hbar$, what is the value of L? Which is larger, L or the maximum possible L_z?

c) Assume a model where L is described as a classical vector. For each allowed value of L_z, what angle does the vector L make with the $+z$-axis?

Section 42–6 The Zeeman Effect

42–16 Consider an atom in an $l = 2$ state. In the absence of an external magnetic field, the states with different m have (approximately) the same energy. (One says that the states are *degenerate*.)

a) If the effect of electron spin can be ignored (which is not actually the case), calculate the splitting in eV of the m-levels when the atom is put in a 0.6-T magnetic field.

b) Which m-level will have the lowest energy?

c) Draw an energy-level diagram that shows the $l = 2$ levels with and without the external magnetic field.

Section 42–7 Electron Spin

42–17 If you treat an electron as a classical spherical particle with a radius of 1×10^{-17} m, what is the angular velocity necessary to produce a spin angular momentum of $\hbar$?

42–18 A hydrogen atom in the $n = 1$ state is placed in a magnetic field of magnitude 1.2 T. Find the interaction energy (in electronvolts) of the atom with the field due to the electron spin.

PROBLEMS

42–19 The negative μ-meson (or muon) has a charge equal to that of an electron but a mass about 207 times as great. Consider a hydrogenlike atom consisting of a proton and a muon. (For simplicity, assume that the muon orbits around the proton, which is stationary. In reality both revolve about the center of mass of the system.)

a) What is the ground-state energy (in electronvolts)?

b) What is the radius of the $n = 1$ Bohr orbit?

c) What is the wavelength of the radiation emitted in the transition from the $n = 2$ state to the $n = 1$ state?

42–20 Consider a hydrogenlike atom of nuclear charge Z. (Refer to Exercise 42–5.)

a) For what value of Z (rounded to the nearest integer value) is the Bohr speed of the electron in the ground state equal to 5% of the speed of light?

b) For what value of Z is the ionization energy of the ground state equal to 1% of the rest mass energy of the electron?

42–21 Refer to Example 42–4. Suppose a hydrogen atom makes a transition from the $n = 3$ state to the $n = 2$ state (the Balmer H_α line at 656.3 nm) while in a magnetic field of magnitude 2 T. If the magnetic moment of the atom is parallel to the field in both the initial and final states,

a) by how much is each energy level shifted from the zero-field value?

b) by how much is the wavelength of the spectrum line shifted?

42–22 A 10-kg satellite circles the earth once every 2 hr in an orbit having a radius of 8060 km.

a) Assuming that Bohr's angular-momentum postulate applies to satellites just as it does to an electron in the hydrogen atom, find the quantum number of the orbit of the satellite.

b) Show from Bohr's first postulate and Newton's law of gravitation that the radius of an earth-satellite orbit is directly proportional to the square of the quantum number, $r = kn^2$, where k is the constant of proportionality.

c) Using the result from part (b), find the distance between the orbit of the satellite in this problem and its next "allowed" orbit. (Calculate a numerical value.)

d) Comment on the possibility of observing the separation of the two adjacent orbits.

e) Do quantized and classical orbits correspond for this satellite? Which is the "correct" method for calculating the orbits?

42–23 The radii of atomic nuclei are of the order of 10^{-15} m.

a) Estimate the minimum uncertainty in the momentum of an electron if it is confined within a nucleus.

b) Take this uncertainty in momentum to be an estimate of the magnitude of the momentum. (Refer to Exercise 42–11.) Use the relativistic expression of Eq. (40–34) to obtain an estimate of the energy of an electron confined within a nucleus.

c) Compare the energy calculated in (b) to the Coulomb potential energy of a proton and an electron separated by 1×10^{-15} m. On the basis of your result, could there be electrons within the nucleus?

42–24 The average kinetic energy of a thermal neutron is $\frac{3}{2}kT$. What is the de Broglie wavelength associated with the neutrons in thermal equilibrium with matter at 300 K? (The mass of a neutron is approximately 1 u.)

42–25 In a certain television picture tube, the accelerating voltage is 20,000 V and the electron beam passes through an aperture 0.5 mm in diameter to a screen 0.3 m away.

a) What is the uncertainty in position of the point where the electrons strike the screen?

b) Does this uncertainty affect the clarity of the picture significantly? (Use nonrelativistic expressions for the motion of the electrons. This is fairly accurate and is certainly adequate for obtaining an estimate of uncertainty-principle effects.)

42–26 The π^0 meson is an unstable particle produced in high-energy particle collisions. Its mass is about 264 times that of the electron, and it exists for an average lifetime of 0.8×10^{-16} s before decaying into two gamma-ray photons. Assuming that the mass and energy of the particle are related by the Einstein relation $E = mc^2$, find the uncertainty in the mass of the particle and express it as a fraction of the mass.

42–27 Show that the total number of hydrogen-atom states (including different spin states) for a given value of the principal quantum number n is $2n^2$. (*Hint:* The sum of the first N numbers, $1 + 2 + 3 + \cdots + N$ is given by $N[N + 1]/2$.)

42–28 When a photon is emitted by an atom, the atom must recoil to conserve momentum. This means that the photon and the recoiling atom share the transition energy. For a hydrogen atom, calculate the correction due to recoil to the wavelength of the photon emitted when an electron in the $n = 5$ state returns to the ground state.

CHALLENGE PROBLEMS

42–29 You are entered in a contest to drop a marble of mass 40 g from the roof of a building onto a small target 30 m below. From uncertainty considerations, what is the typical distance by which you will miss the target, given that you aim with the highest possible precision? (*Hint:* The uncertainty Δx_f in the x-coordinate of the marble when it reaches the ground comes in part from the uncertainty Δx_i in the x-coordinate initially and in part from the initial uncertainty in v_x. The latter gives rise to an uncertainty in the horizontal motion of the marble as it falls. Δx_i and Δv_x are related by the uncertainty principle. A small Δx_i gives rise to a large Δv_x, and vice versa. Find the Δx_i that gives the smallest total uncertainty in x at the ground.)

42–30 The wave nature of particles results in the quantum mechanical situation that a particle confined in a box can assume only wavelengths that result in standing waves in the box, with nodes at the box walls.

a) Show that an electron confined in a one-dimensional box of length L will have energy levels given by

$$E_n = \frac{n^2 h^2}{8mL^2}.$$

(*Hint:* Recall that the relation between the de Broglie wavelength and the speed of a particle is $mv = h/\lambda$. The energy of the particle is $\frac{1}{2}mv^2$.)

b) If a hydrogen atom were modeled as a one-dimensional box with length equal to the Bohr radius, what would be the energy (in eV) of the ground state of the electron?

42–31 Consider a particle of mass m moving in a potential $U = \frac{1}{2}kx^2$, as in a mass-spring system. The total energy of the particle is $E = p^2/2m + \frac{1}{2}kx^2$. Assume that p and x are approximately related by the Heisenberg uncertainty principle, $px \sim \hbar$.

a) Calculate the minimum possible value of the energy E, and the value of x that gives this minimum E. This lowest possible energy, which is not zero, is called the *zero-point* energy.

b) For the x calculated in (a), what is the ratio of the kinetic to the potential energy of the particle?

42–32

a) Show that the frequency of revolution of an electron in its circular orbit in the Bohr model of the hydrogen atom is $f = me^4/4\epsilon_0^2 n^3 h^3$.

b) Show that when n is very large, the frequency of revolution equals the radiated frequency calculated from Eq. (42–7) for a transition from

$$n_1 = n + 1 \quad \text{to} \quad n_2 = n.$$

(This problem illustrates Bohr's *correspondence principle*, which is often used as a check on quantum calculations. When n is small, quantum physics gives results that are very different from those of classical physics. When n is large, the differences are not significant, and the two methods then "correspond." In fact, when Bohr first tackled the hydrogen atom problem, he sought to determine f as a function of n such that it would correspond to classical physics for large n.)

43

ATOMS, MOLECULES, AND SOLIDS

THE PRINCIPLES OF QUANTUM MECHANICS THAT WE STUDIED IN CHAPTER 42 provide the basis for understanding a wide variety of phenomena associated with the structure of atoms, molecules, and solid materials. One additional principle is needed, the *exclusion principle*, which states that two electrons may not occupy the same quantum-mechanical state. With this principle, the most important features of the structure and chemical behavior of multielectron atoms can be derived, including the periodic table of the elements.

The nature of *molecular bonds*, by which two or more atoms combine in a stable structure, can be understood on the basis of the electron configurations of the atoms. We will see that transitions among vibrational and rotational energy states of molecules give rise to *molecular spectra*. Interatomic forces are also responsible for the large-scale binding of atoms into solid structures. Several different types of interatomic interactions are possible, and many properties of solid materials can be understood at least qualitatively on the basis of the types of bonding that are present. *Semiconductors*, a particular class of solid materials, are discussed in some detail because of their inherent interest and their great practical importance in present-day technology.

43–1 THE EXCLUSION PRINCIPLE

How do we deal with the interactions among electrons in a multielectron atom?

The hydrogen atom is the simplest of all atoms, containing one electron and one proton. Analysis of atoms with more than one electron increases in complexity very rapidly; each electron interacts not only with the positively charged nucleus but also with all the other electrons. The motion of the electrons is governed by the Schrödinger equation, mentioned in Section 42–5, but the mathematical problem of finding appropriate solutions of this equation is so complex that it has not been accomplished exactly even for the helium atom, with two electrons.

Various approximation schemes can be used to apply the Schrödinger equation to many-electron atoms. The simplest and most drastic scheme is simply to ignore the interactions between electrons and to regard each electron as influenced only by the electric field of the nucleus, considered to be a point charge. A less drastic and more useful approximation is to think of all the electrons together as making up a charge cloud that is, on the average, *spherically symmetric*. We can then think of each individual electron as moving in the total electric field due to the nucleus and this averaged-out electron cloud. This model is called the **central-field approximation;** it provides a useful starting point for the understanding of atomic structure.

An additional principle is also needed, the *exclusion principle*. To understand the need for this principle, we consider the lowest-energy state or *ground state* of a many-electron atom. The central-field model suggests that each electron has a lowest-energy state (roughly corresponding to the $n = 1$ state for the hydrogen atom). We might expect that, in the ground state of a complex atom, all the electrons should be in this lowest state. If this is the case, then when we examine the behavior of atoms with increasing numbers of electrons, we should find gradual changes in physical and chemical properties of elements as the number of electrons in the atoms increases.

A variety of evidence shows conclusively that this is *not* what happens at all. For example, the elements fluorine, neon, and sodium have, respectively, 9, 10, and 11 electrons per atom. Fluorine is a *halogen* and tends strongly to form compounds in which each fluorine atom acquires an extra electron. Sodium, an *alkali metal*, forms compounds in which it *loses* an electron; and neon is an *inert gas*, forming no compounds at all. This and many other observations show that in the ground state of a complex atom the electrons *cannot* all be in the lowest-energy state.

The key to this puzzle, discovered by the Swiss physicist Wolfgang Pauli in 1925, is called the **Pauli exclusion principle,** or the *exclusion principle*. Briefly, it states that *no two electrons can occupy the same quantum-mechanical state*. Since different states correspond to different spatial distributions, including different distances from the nucleus, this means that in a complex atom there is not enough room for all the electrons in the states nearest the nucleus; some are forced into states farther away, having higher energies.

To apply the Pauli principle to atomic structure, we first review some results quoted in Sections 42–5 and 42–7. The quantum-mechanical state of the electron in the hydrogen atom is identified by the four quantum numbers n, l, m, and s, which determine the energy, angular momentum, and components of orbital and spin angular momentum in a particular direction. It turns out that we can still use this scheme when the electron moves not in the electric field of a point charge, as in the hydrogen atom, but in the electric field of *any spherically symmetric charge distribution*, as in the central-field approximation. One important difference is that the energy of a state is no longer given by Eq. (42–6); in general the energy corresponding to a given set of quantum numbers depends on *both* n and l, usually increasing with increasing l for a given value of n.

We can now make a list of all the possible sets of quantum numbers and thus of the possible states of electrons in an atom. Such a list is given in Table 43–1, which also indicates two alternative notations. It is customary to designate the value of l by a letter, according to this scheme:

$l = 0$: s state
$l = 1$: p state
$l = 2$: d state
$l = 3$: f state
$l = 4$: g state

A brief look at electron arrangement in multielectron atoms

A zoning ordinance for electrons in atomic systems: Let's not all try to crowd into the same space.

Describing one-electron states by use of four quantum numbers

TABLE 43-1

n	l	m	Spectroscopic Notation	Maximum Number of Electrons	Shell
1	0	0	$1s$	2	K
2	0	0	$2s$	2 } 8	L
2	1	−1		
2	1	0	$2p$	
2	1	1		6
3	0	0	$3s$	2	
3	1	−1			
3	1	0	$3p$	6	
3	1	1			
3	2	−2			18	M
3	2	−1			
3	2	0	$3d$	10	
3	2	1			
3	2	2			
4	0	0	$4s$	2	
4	1	−1			
4	1	0	$4p$	6 } 32	N
4	1	1			
	etc.				

The electrons arrange themselves in concentric shells around the nucleus.

The origins of these letters are rooted in the early days of spectroscopy and need not concern us. A state for which $n = 2$ and $l = 1$ is called a $2p$ state, and so on, as shown in Table 43-1. This table also shows the relation between values of n and the x-ray levels ($K, L, M, \ldots$) described in Section 41-8. The $n = 1$ levels are designated as K; $n = 2$ as L; and so on. Because the average electron distance from the nucleus increases with n, each value of n corresponds roughly to a region of space around the nucleus in the form of a spherical **shell.** Hence we speak of the L shell as the region occupied by the electrons in the $n = 2$ states, and so on. States with the same n but different l are said to form *subshells*, such as the $3p$ subshell.

The exclusion principle can be stated in terms of quantum numbers.

We are now ready for a more precise statement of the exclusion principle: *In any atom, only one electron can occupy any given quantum state.* That is, no two electrons in an atom can have the same values of all four quantum numbers. Since each quantum state corresponds to a certain distribution of the electron "cloud" in space, the principle says, in effect: "Not more than two electrons (with opposite values of the spin quantum number s) can occupy the same region of space." We should not take this statement too seriously, since the clouds and associated wave functions describing electron distributions do not have definite sharp boundaries; but the exclusion principle limits the degree of overlap of electron wave functions that is permitted. The maximum number of electrons in each shell and subshell is shown in Table 43-1.

The exclusion principle plays an essential role in understanding the structure of complex atoms. In the next section we will see how the periodic table of the elements can be understood on the basis of this principle.

43-2 ATOMIC STRUCTURE

The number of electrons in an atom in its normal state is called the **atomic number,** denoted by Z. The nucleus contains Z protons and some number of neutrons. The proton and electron charges have the same magnitude but opposite sign, so in the normal atom the net electrical charge is zero. Because the electrons are attracted to the nucleus, we expect the quantum states corresponding to regions near the nucleus to have the lowest energies. We may imagine starting with a bare nucleus with Z protons, and adding electrons one by one until the normal complement of Z electrons for a neutral atom is reached. We expect the lowest-energy states, ordinarily those with the smallest values of n and l, to fill first, and we use successively higher states until all electrons are accommodated.

The chemical properties of an atom are determined principally by interactions involving the outermost electrons, so it is of particular interest to find out how these electrons are arranged. For example, when an atom has one electron considerably farther from the nucleus (on the average) than the others, this electron is relatively loosely bound. The atom tends to lose this electron and form what chemists call an *electrovalent* or *ionic* bond, with valence +1. This behavior is characteristic of the alkali metals lithium, sodium, potassium, and so on.

We now proceed to describe ground-state electron configurations for the first few atoms (in order of increasing Z). For hydrogen, the ground state is $1s$; the single electron is in the state $n = 1$, $l = 0$, $m = 0$, and $s = \pm 1$. In the helium atom ($Z = 2$), *both* electrons are in the $1s$ states, with opposite spins; this state is denoted as $1s^2$. For helium, the K shell is completely filled and all others are empty.

Lithium ($Z = 3$) has three electrons; in the ground state two are in the $1s$ state, and one in a $2s$ state. We denote this state as $1s^2 2s$. On the average, the $2s$ electron is considerably farther from the nucleus than the $1s$ electrons, as shown schematically in Fig. 43–1. Hence, according to Gauss's law, the *net* charge influencing the $2s$ electron is $+e$, rather than $+3e$ as it would be without the $1s$ electrons present. Thus the $2s$ electron is loosely bound, as the chemical behavior of lithium suggests. An alkali metal, lithium forms ionic compounds in which each atom loses an electron. In such compounds, lithium has a valence of +1.

Next is beryllium ($Z = 4$); its ground-state configuration is $1s^2 2s^2$, with two electrons in the L shell. Beryllium is the first of the *alkaline-earth* elements, forming ionic compounds in which they have a valence of +2.

Table 43–2 shows the ground-state electron configurations of the first thirty elements. The L shell can hold a total of eight electrons; we invite you to verify this from the rules in Section 43–1. At $Z = 10$ the K and L shells are filled and there are no electrons in the M shell. We expect this to be a particularly stable configuration, with little tendency to gain or lose electrons, and in fact this element is neon, a "noble gas" with no known compounds. The next element after neon is sodium ($Z = 11$), with filled K and L shells and one electron in the M shell. Thus its "filled-shell-plus-one-electron" structure resembles that of lithium; both are alkali metals. The element *before* neon is fluorine, with $Z = 9$. It has a vacancy in the L shell and might be expected to have an affinity for an electron, normally forming ionic compounds in which

Electron structure of complex atoms: filling up the lowest-energy states first

The chemical behavior of elements is determined mostly by the arrangement of outer electrons in their atoms.

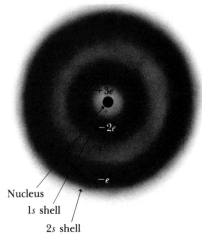

Nucleus
1s shell
2s shell

43–1 Schematic representation of charge distribution in a lithium atom. The nucleus has a charge $3e$; the two $1s$ electrons are closer to the nucleus than the $2s$ electron, which moves in a field approximately equal to that of a point charge of $3e - 2e$, or simply e.

When there are only one or two electrons in the outermost shell, they are only loosely bound.

All the noble gases have filled-shell electron configurations.

All the halogens have one vacancy in the outermost shell.

TABLE 43–2   Ground-state Electron Configurations

Element	Symbol	Atomic Number (Z)	Electron Configuration
Hydrogen	H	1	$1s$
Helium	He	2	$1s^2$
Lithium	Li	3	$1s^2 2s$
Beryllium	Be	4	$1s^2 2s^2$
Boron	B	5	$1s^2 2s^2 2p$
Carbon	C	6	$1s^2 2s^2 2p^2$
Nitrogen	N	7	$1s^2 2s^2 2p^3$
Oxygen	O	8	$1s^2 2s^2 2p^4$
Fluorine	F	9	$1s^2 2s^2 2p^5$
Neon	Ne	10	$1s^2 2s^2 2p^6$
Sodium	Na	11	$1s^2 2s^2 2p^6 3s$
Magnesium	Mg	12	$1s^2 2s^2 2p^6 3s^2$
Aluminum	Al	13	$1s^2 2s^2 2p^6 3s^2 3p$
Silicon	Si	14	$1s^2 2s^2 2p^6 3s^2 3p^2$
Phosphorus	P	15	$1s^2 2s^2 2p^6 3s^2 3p^3$
Sulfur	S	16	$1s^2 2s^2 2p^6 3s^2 3p^4$
Chlorine	Cl	17	$1s^2 2s^2 2p^6 3s^2 3p^5$
Argon	Ar	18	$1s^2 2s^2 2p^6 3s^2 3p^6$
Potassium	K	19	$1s^2 2s^2 2p^6 3s^2 3p^6 4s$
Calcium	Ca	20	$1s^2 2s^2 2p^6 3s^2 3p^6 4s^2$
Scandium	Sc	21	$1s^2 2s^2 2p^6 3s^2 3p^6 3d 4s^2$
Titanium	Ti	22	$1s^2 2s^2 2p^6 3s^2 3p^6 3d^2 4s^2$
Vanadium	V	23	$1s^2 2s^2 2p^6 3s^2 3p^6 3d^3 4s^2$
Chromium	Cr	24	$1s^2 2s^2 2p^6 3s^2 3p^6 3d^5 4s$
Manganese	Mn	25	$1s^2 2s^2 2p^6 3s^2 3p^6 3d^5 4s^2$
Iron	Fe	26	$1s^2 2s^2 2p^6 3s^2 3p^6 3d^6 4s^2$
Cobalt	Co	27	$1s^2 2s^2 2p^6 3s^2 3p^6 3d^7 4s^2$
Nickel	Ni	28	$1s^2 2s^2 2p^6 3s^2 3p^6 3d^8 4s^2$
Copper	Cu	29	$1s^2 2s^2 2p^6 3s^2 3p^6 3d^{10} 4s$
Zinc	Zn	30	$1s^2 2s^2 2p^6 3s^2 3p^6 3d^{10} 4s^2$

it has a valence of -1. This behavior is characteristic of the *halogens* (fluorine, chlorine, bromine, iodine, astatine), all of which have "filled-shell-minus-one" configurations.

The periodic table of the elements is a result of the numbers of electron states and the exclusion principle.

By similar analysis, we can understand all the regularities in chemical behavior exhibited by the **periodic table of the elements,** on the basis of electron configurations. A slight complication occurs with the M and N shells because the $3d$ and $4s$ subshells ($n = 3$, $l = 2$, and $n = 4$, $l = 0$, respectively) overlap in energy. Thus argon ($Z = 18$) has all the $1s$, $2s$, $2p$, $3s$, and $3p$ states filled, but in potassium ($Z = 19$) the additional electron goes into a $4s$ level rather than a $3d$ level. The next several elements have one or two electrons in the $4s$ states and increasing numbers in the $3d$ states. These elements are all metals with rather similar chemical and physical properties; they form the first *transition series*, starting with scandium ($Z = 21$) and ending with zinc ($Z = 30$), for which the $3d$ and $4s$ levels are filled.

Each vertical column in the periodic table has a characteristic outer-electron configuration.

Thus the similarity of elements in each *group* (vertical column) of the periodic table reflects corresponding similarity in outer-electron configuration. All the noble gases (helium, neon, argon, krypton, xenon, and radon) have filled-shell configurations. All the alkali metals (lithium, sodium, potassium, rubidium, cesium, and francium) have "filled-shell-plus-one" configurations. All the alkaline-earth metals (beryllium, magnesium, calcium, strontium, barium, and radium) have "filled-shell-plus-two" configurations, and all the halogens (fluorine, chlorine, bromine, iodine, and astatine) have "filled-shell-minus-one" structures.

This whole theory can be refined to account for differences between elements in a group and to account for various aspects of chemical behavior. Indeed, some physicists like to claim that all of chemistry is contained in the Schrödinger equation! This statement is a bit extreme, but we can see that even this qualitative discussion of atomic structure takes us a considerable distance toward understanding the atomic basis of many chemical phenomena. Next we look in more detail at the nature of the chemical bond.

PROBLEM-SOLVING STRATEGY: Atomic structure

1. Some numerology has to be mastered in counting the energy levels for electrons in the central-field approximation. There are four quantum numbers: n, l, m, and s, all integers; n is always positive, l can be zero or positive, and the other two can be positive or negative. Be sure you know how to count the number of levels in each shell and subshell; study Tables 43–1 and 43–2 carefully.

2. As in Chapter 42, familiarizing yourself with some numerical magnitudes is useful. Here are two examples to work out: The electrical potential energy of a proton and an electron 0.1 nm apart (typical of atomic dimensions) is 14.4 eV or 2.31×10^{-18} J. The moment of inertia of two protons 0.1 nm apart (about an axis through the center of mass, perpendicular to the line joining them) is 0.836×10^{-47} kg·m². Think of other examples like this and work them out, to help you know what kinds of magnitudes to expect in atomic physics.

3. As in Chapter 42, you will need to use both electronvolts and joules. The conversion 1 eV = 1.602×10^{-19} J and Planck's constant in eV, $h = 4.136 \times 10^{-15}$ eV·s, are handy. Nanometers are convenient for atomic and molecular dimensions, but don't forget to convert to meters in calculations.

43–3 DIATOMIC MOLECULES

As we learned in Section 43–2, the study of electron configurations in atoms provides valuable insight into the nature of the chemical bond; that is, the interactions that hold atoms together to form stable structures such as molecules and crystalline solids. There are several types of chemical bonds; the simplest to understand is the **ionic bond,** also called the *electrovalent* or *heteropolar* bond. The most familiar example is sodium chloride (NaCl), in which the sodium atom gives its one $3s$ electron to the chlorine atom, filling the vacancy in the $3p$ subshell of chlorine. Energy is required to make this transfer if the atoms are far apart, but the result is two ions, one positively charged and one negatively charged, that attract each other. As they come together, their potential energy decreases, so that the final bound state of Na^+Cl^- has lower total energy than the state in which the two atoms are separated and neutral.

Removing the $3s$ electron from the sodium atom requires 5.1 eV of energy; this is called the **ionization energy** or *ionization potential* of sodium. Chlorine has an **electron affinity** of 3.8 eV. That is, the neutral chlorine atom can attract an extra electron; once this electron takes its place in the $3p$ level, 3.8 eV of energy are required to remove it. Thus, creating the separate Na^+ and Cl^- ions requires a net expenditure of only 5.1 − 3.8 eV = 1.3 eV. The negative potential energy associated with the mutual attraction of the ions is determined by the closeness to which they can approach each other; this in turn is determined by the electrical interactions and by the exclusion principle,

What holds atoms together in a molecule: the types of chemical bond

How much energy is required to gain or lose an electron?

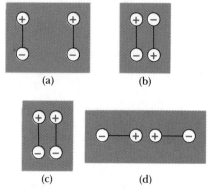

43–2 When two dipoles (a) are brought together, the interaction may be (b) attractive, or (c and d) repulsive.

A bond formed by sharing a pair of electrons between two atoms

The electrons in a covalent bond must have opposite spins.

An atom can form several covalent bonds with other atoms.

which forbids extensive overlap of the electron clouds of the two atoms. The minimum potential energy is -5.5 eV at a separation of 0.24 nm. This is more than enough to repay the initial investment in creating the ions. At lesser distances, the interaction becomes repulsive. Thus the net energy involved in creating the ions and letting them come together to the equilibrium separation of 0.24 nm is $(-5.5 + 1.3)$ or -4.2 eV, which is the binding energy of the molecule. To put it another way, 4.2 eV of energy are needed to dissociate the molecule into separate atoms.

Ionic bonds can involve more than one electron per atom. The alkaline-earth elements form ionic compounds in which each atom loses *two* electrons; an example is $Mg^{2+}Cl_2^-$. Loss of more than two electrons is relatively rare; instead a *different kind of bond* comes into operation.

The **covalent** or *homopolar* **bond** is characterized by a more nearly symmetric participation of the two atoms, as contrasted with the complete asymmetry involved in the electron-transfer process of the ionic bond. The simplest example of a covalent bond is the hydrogen molecule, a structure containing two protons and two electrons. As a preliminary to understanding this bond, consider first the interaction of two electric dipoles, as in Fig. 43–2. In (a) the dipoles are far apart; in (b) the like charges are farther apart than the unlike charges, and there is a net *attractive* force. In (c) and (d) the interaction is repulsive.

In a molecule the charges are not at rest, but Fig. 43–2b at least makes it seem plausible that if the electrons are localized primarily in the region between the protons, the electrons exert an attractive force on each proton that may more than counteract the repulsive interactions of the protons on each other and the electrons on each other. That is, in the covalent bond in the hydrogen molecule, the attractive interaction is supplied by a pair of electrons, one contributed by each atom, whose charge clouds are concentrated primarily in the region between the two atoms. Hence this bond may also be thought of as a shared-electron or electron-pair bond.

According to the exclusion principle, two electrons can occupy the same region of space only when they have opposite spin orientations. When the spins are parallel, the state that would be most favorable from energy considerations is forbidden by the exclusion principle, and the lowest-energy state permitted is one in which the electron clouds are concentrated *outside* the central region between atoms. The nuclei then repel each other, and the interaction is repulsive rather than attractive. Thus opposite spins are an essential requirement for an electron-pair bond, and no more than two electrons can participate in such a bond.

This is not to say, however, that an atom cannot have several electron-pair bonds. On the contrary, an atom having several electrons in its outermost shell can form covalent bonds with several other atoms. The bonding of carbon and hydrogen atoms, of central importance in organic chemistry, is an example. In the *methane* molecule (CH_4) the carbon atom is at the center of a regular tetrahedron, with a hydrogen atom at each corner. The carbon atom has four electrons in its L shell, and one of these electrons forms a covalent bond with each of the four hydrogen atoms, as shown in Fig. 43–3. Similar patterns occur in more complex organic molecules.

Ionic and covalent bonds represent two extremes in the nature of molecular bonds, but there is no sharp division between the two types. In many situations there is a *partial* transfer of one or more electrons (corresponding to a greater or smaller distortion of the electron wave functions) from one atom to another. As a result, many molecules having dissimilar atoms have electric dipole moments, that is, a preponderance of positive charge at one end and of negative charge at the other. Such molecules are said to be *polar*. Water molecules have large dipole moments that are responsible for the exceptionally large dielectric constant of liquid water.

The bonds discussed so far typically have energies in the range of 1 eV to 5 eV. They are called *strong* bonds to distinguish them from several types of much *weaker* bonds having energies typically of the order of 0.1 eV or less. One of these, the **van der Waals bond,** is an interaction between the electric dipole moments of two atoms or molecules. The existence of dipole moments in molecules was mentioned above. Even when an atom or molecule has no permanent dipole moment, fluctuating charge densities in the interior of the structure can lead to fluctuating dipole moments; these in turn can induce dipole moments in neighboring structures. The resulting dipole–dipole interaction can be attractive and can lead to weak bonding of atoms or molecules. For example, the low-temperature liquefaction and solidification of such molecules as H_2, O_2, and N_2 and of the noble gases is due to interaction of the induced-dipole type. Not much energy of thermal agitation is needed to break these weak bonds, so these substances exist in the liquid and solid states only at very low temperatures.

A second type of important weak bond is the **hydrogen bond,** which is analogous to the covalent bond. In the latter an electron pair serves to bind two positively charged structures together. In the hydrogen bond a hydrogen atom acts as the glue to bond two negatively charged structures. Again the bond energy is only about 0.1 eV, but nevertheless the hydrogen bond plays an essential role in many organic molecules, including, for example, cross-link bonding between the two strands of the famous double-helix DNA molecule.

All these bond types play roles in the structure of *solids* as well as of molecules. Indeed, a solid is in many respects a gigantic molecule. Still another type of bonding, the *metallic bond*, comes into play in the structure of metallic solids. We return to this subject in Section 43–5.

43–3 Methane (CH_4) molecule, showing four covalent bonds. The electron cloud between the central carbon atom and each of the four hydrogen nuclei represents the two electrons of a covalent bond. The hydrogen nuclei are at the corners of a regular tetrahedron.

The hydrogen bond: another type of shared-particle bond

43–4 MOLECULAR SPECTRA

The energy levels of atoms are associated with the kinetic energy of electron motion relative to the nucleus and with the potential energy of interaction of the electrons with the nucleus and with each other. Energy levels of *molecules* have additional features that result from motion of the *nuclei* of the atoms relative to each other. Associated with these energy levels are characteristic **molecular spectra.**

To keep things as simple as possible, we consider only *diatomic* molecules. Viewed as a rigid dumbbell, a diatomic molecule can *rotate* about an axis through its center of mass. According to classical mechanics, the kinetic energy K of a rigid body with moment of inertia I rotating with angular velocity ω can be expressed as $K = \frac{1}{2}I\omega^2$. This is Eq. (9–13); you may want to review its derivation in Section 9–5. The kinetic energy can also be expressed in terms of the magnitude L of angular momentum, which is given by $L = I\omega$. Combining this with the preceding expression, we can express the kinetic energy alternatively as $L^2/2I$. If the molecule rotates as a rigid body, there is no potential

The rotational kinetic energy of a diatomic molecule has energy levels.

energy; the kinetic energy is then equal to the total energy E:

$$E = \frac{L^2}{2I}. \qquad (43-1)$$

Determining the rotational energy levels
from quantization of angular momentum

In a quantum-mechanical discussion of this molecular rotation, it is reasonable to assume that the angular momentum is quantized in the same way as for electrons in an atom, as given by Eq. (42–20):

$$L^2 = l(l + 1)\hbar^2 \qquad (l = 0, 1, 2, 3, \ldots). \qquad (43-2)$$

Combining Eqs. (43–1) and (43–2), we obtain the **rotational energy levels:**

$$E = l(l + 1)\left(\frac{\hbar^2}{2I}\right) \qquad (l = 0, 1, 2, 3, \ldots). \qquad (43-3)$$

A more detailed analysis, using the Schrödinger equation, confirms that the angular momentum is given by Eq. (43–2) and the energy levels by Eq. (43–3).

Considering the magnitudes involved, we note that for an oxygen molecule, $I = 5 \times 10^{-46}$ kg·m². Thus the constant $\hbar^2/2I$ in Eq. (43–3) is approximately equal to

$$\frac{\hbar^2}{2I} = \frac{(1.05 \times 10^{-34} \text{ J·s})^2}{2(5 \times 10^{-46} \text{ kg·m}^2)} = 1.10 \times 10^{-23} \text{ J} = 0.69 \times 10^{-4} \text{ eV}.$$

Rotational energy-level spacing is much smaller than typical atomic energy levels.

This energy is much *smaller* than typical atomic energy levels (of the order of a few eV) associated with optical spectra. The energies of photons emitted or absorbed in transitions among rotational levels are correspondingly small, and they fall in the far infrared region of the spectrum.

The rigid-dumbbell model of a diatomic molecule suggests that the distance between the two nuclei is constant. In fact, a more realistic model would represent the connection as a *spring* rather than a rigid rod. The atoms can undergo *vibrational* motion relative to the center of mass, and there are additional kinetic and potential energies associated with this motion. Application of the Schrödinger equation shows that the corresponding **vibrational energy levels** are given by

$$E = \left(n + \frac{1}{2}\right)hf \qquad (n = 0, 1, 2, 3, \ldots), \qquad (43-4)$$

where f is the frequency of vibration. For typical diatomic molecules this turns out to be of the order of 10^{13} Hz; thus the constant hf in Eq. (43–4) is of the order of

$$hf = (6.6 \times 10^{-34} \text{ J·s})(10^{13} \text{ s}^{-1}) = 6.6 \times 10^{-21} \text{ J}$$
$$= 0.041 \text{ eV}.$$

Hence the vibrational energies, while still much smaller than those of atomic spectra, are typically considerably *larger* than the rotational energies.

An energy-level diagram for a diatomic molecule has the general appearance of Fig. 43–4. For each value of n, there are many values of l, leading to a series of closely spaced levels. Transitions between different pairs of n values give different series of spectrum lines, and the resulting spectrum has the appearance of a series of *bands*. Each band corresponds to a particular vibrational transition, and each individual line in a band, to a particular rotational transition. A typical **band spectrum** is shown in Fig. 43–5.

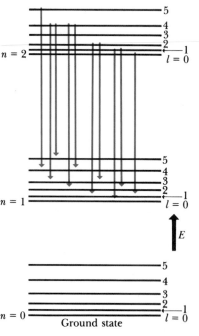

43–4 Energy-level diagram for vibrational and rotational energy levels of a diatomic molecule. For each vibrational level (n), there is a series of more closely spaced rotational levels (l). Several transitions corresponding to a single band in a band spectrum are shown.

43-5 Typical band spectrum (Courtesy of R. C. Herman.)

The same considerations can be applied to more complex molecules. A molecule with three or more atoms has several different kinds or *modes* of vibratory motion, each with its own set of energy levels, related to the frequency by Eq. (43–4). The resulting energy-level scheme and associated spectra can be quite complex, but the general considerations discussed above still apply. In nearly all cases the associated radiation lies in the infrared region of the electromagnetic spectrum. Analysis of molecular spectra has proved to be an extremely valuable analytical tool, providing a great deal of information about the strength and rigidity of molecular bonds and the structure of complex molecules.

Molecular spectroscopy: an important tool for investigating molecular structure

43-5 STRUCTURE OF SOLIDS

At ordinary temperatures and pressures, most materials are in the *solid state*. This is a condensed state of matter in which the interactions among the atoms or molecules are strong enough to give the material a definite volume and shape that change relatively little with stress. The distances between adjacent atoms in a solid are of the same order of magnitude as the diameters of the electron clouds of the atoms.

A solid may be **amorphous** or **crystalline.** We have mentioned crystalline solids briefly in Section 20–7, and we suggest you review that discussion. A crystalline solid is characterized by *long-range order,* which is a recurring pattern in the arrangement of atoms. This pattern is called the *crystal structure* or the *lattice structure.* Amorphous solids have more in common with liquids than with crystalline solids; liquids also have short-range order but not long-range order.

Some solid materials have a definite crystal structure; some don't.

The forces responsible for the regular arrangement of atoms in a crystal are, in some cases, the same as those involved in molecular bonds. Corresponding to the classes of strong molecular bonds are ionic and covalent crystals. The alkali halides, of which ordinary salt (NaCl) is the most common variety, are the most familiar **ionic crystals.** The positive sodium ions and the negative chlorine ions occupy alternate positions in a cubic crystal lattice, as in Fig. 43–6. The forces are the familiar Coulomb's-law forces between charged particles; these forces are not directional, and the particular arrangement in which the material crystallizes is determined by the relative size of the two ions.

Crystals can be classified in terms of the kinds of bonds that hold them together.

The simplest example of a **covalent crystal** is the *diamond structure,* a structure found in the diamond form of carbon and also in silicon, germanium, and tin. All these elements are in Group IV of the periodic table, with four electrons in the outermost shell. Each atom in this structure is situated at the center of a regular tetrahedron, with four nearest-neighbor atoms at the corners, and it forms a covalent bond with each of these atoms. These bonds are strongly directional because of the asymmetric electron distribution, leading to the tetrahedral structure.

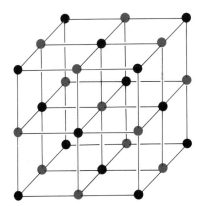

43-6 Symbolic representation of a sodium chloride crystal, with exaggerated distances between ions.

Metallic crystals: electron sharing on a vast scale

A third crystal type, which is less directly related to the chemical bond than are ionic or covalent crystals, is the **metallic crystal.** In this structure the outermost electrons are not localized at individual atomic lattice sites but are detached from their parent atoms and free to move through the crystal. The corresponding charge clouds (and their associated wave functions) extend over many atoms. Thus we can picture a metallic crystal roughly as an array of positive ions (atoms from which one or more electrons have been removed) immersed in a sea of electrons whose attraction for the positive ions holds the crystal together. This "sea" has many of the properties of a gas, and indeed we speak of the *electron-gas model* of metallic solids.

Close packing: How many atoms can you stuff into a given volume?

In a metallic crystal the situation is as though the atoms would like to form shared-electron bonds but do not have enough valence electrons. Instead, electrons are shared among *many* atoms. This bonding is not strongly directional in nature, and the shape of the crystal lattice is determined primarily by considerations of **close packing,** that is, the maximum number of atoms that can fit into a given volume. The two most common metallic crystal lattices, the face-centered cubic and the hexagonal close-packed, are shown in Fig. 43–7. In each of these lattices, each atom has 12 nearest neighbors.

As we mentioned in Section 43–3, van der Waals interactions and hydrogen bonding also play a role in the structure of some solids. In polyethylene and similar polymers, covalent bonding of atoms forms long-chain molecules, and hydrogen bonding forms cross links between adjacent chains. In solid water, both van der Waals forces and hydrogen bonds are significant, and together they determine the crystal structure of ice. Many other examples might be cited.

Crystals can have several kinds of imperfections in the lattice.

The discussion in this section has centered around *perfect crystals,* crystals in which the crystal lattice extends uninterrupted throughout the entire material. Real crystals show a variety of departures from this idealized structure. Materials are often *polycrystalline,* composed of many small single crystals bonded together at *grain boundaries.* Within a single crystal, *interstitial* atoms may occur in places where they do not belong, and there may be *vacancies,* lattice sites that should be occupied by an atom but are not. An imperfection of particular interest in semiconductors, to be discussed in Section 43–7, is the

Models illustrating the six basic crystallographic types: orthorhombic, hexagonal, monoclinic, tetragonal, triclinic, and isometric. (Photo by Chip Clark.)

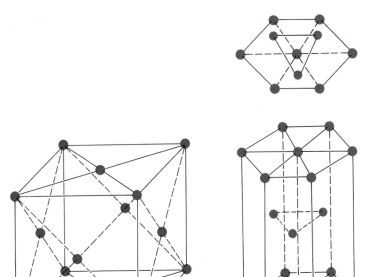

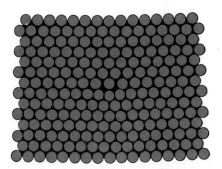

43–7 Close-packed crystal lattice structures. (a) Face-centered cubic; (b) hexagonal close-packed.

(a)   (b)

impurity atom, a foreign atom (e.g., arsenic in a silicon crystal) occupying a regular lattice site.

A more complex kind of imperfection is the *dislocation,* shown schematically in Fig. 43–8, in which one plane of atoms slips relative to another. The mechanical properties of metallic crystals are influenced strongly by the presence of dislocations. The ductility and malleability of some metals depend on the presence of dislocations that move through the lattice during plastic deformations.

43–8 A dislocation. The concept is seen most easily by viewing the figure from various directions at a grazing angle with the page.

43–6 PROPERTIES OF SOLIDS

We can understand many *macroscopic* properties of solids, including mechanical, thermal, electrical, magnetic, and optical properties, by considering their relation to the *microscopic* structure of the material. Various aspects of the relation of structure to properties are part of the vigorous program of research in the physics of solids being carried on throughout the world. Although we cannot discuss these topics in detail in a book such as this, a few examples will indicate the kinds of insights to be gained through study of the microscopic structure of solids.

Understanding thermal and electrical properties on the basis of the microscopic structure of materials

We have already discussed, in Section 20–8, the subject of specific heat capacities of crystals. We applied the same principle of equipartition of energy as in the kinetic theory of gases. A simple analysis permits understanding of the empirical rule of Dulong and Petit on the basis of a microscopic model. This analysis has its limitations, to be sure; it does not include the energy of electron motion, which in metals makes a small additional contribution to specific heat; nor does it predict the temperature dependence of specific heats resulting from the *quantization* of the lattice-vibration energy discussed in Section 20–8. But these additional refinements *can* be included in the model to permit more detailed comparison of observed macroscopic properties with theoretical predictions.

The electrical resistivity of a material depends on the mobility or lack of mobility of electrons in the material. In a *metallic* crystal, the valence electrons are not bound to individual lattice sites but are free to move through the crystal. Thus we expect metals to be good conductors, and they usually are. In a *covalent* crystal, the valence electrons are tied up in the bonds responsible for the crystal structure and are therefore *not* free to move. Because no mobile charges are available for conduction, we expect such materials to be insulators. Similarly, an ionic crystal such as NaCl has no charges that are free to move, and solid NaCl is an insulator. When salt is melted, however, the ions are no longer locked to their individual lattice sites but are free to move, and *molten* NaCl is a good conductor.

There are, of course, no perfect conductors or insulators (except for superconductors at very low temperatures), although the resistivity of good insulators is greater than that of good conductors by an enormous factor, of the order of at least 10^{15}. This great difference is one of the factors that make extremely precise electrical measurements possible. In addition, the resistivity of all materials depends on temperature; in general, the large resistivity of insulators *decreases* with temperature, but that of good conductors usually *increases* at increased temperatures.

Why do some materials conduct better at higher temperatures, and others not so well?

Two competing effects are responsible for this difference. In metals the *number* of electrons available for conduction is nearly independent of temperature, and the resistivity is determined by the frequency of collisions between electrons and the lattice. Roughly speaking, lattice vibrations increase with increased temperature, and the ion cores present a larger target area for collisions with electrons. Hence resistivities of metals usually increase with temperature. In insulators, the small amount of conduction that does take place is due to electrons that have gained enough energy from thermal motion of the lattice to break away from their "home" atoms and wander through the lattice. The number of electrons able to acquire the needed energy is very strongly temperature dependent; a twofold increase for a 10 C° temperature rise is typical. There is also increased scattering at higher temperatures, as with metals, but the increased number of carriers is a far larger effect. Thus insulators invariably become better conductors at higher temperatures.

Why are good electrical conductors also good thermal conductors?

A similar analysis can be made for *thermal* conductivity, which involves transport of microscopic mechanical energy rather than electric charge. The wave motion associated with lattice vibrations is one mechanism for energy transfer, and in metals the mobile electrons also carry kinetic energy from one region to another. This effect is much larger than that of the lattice vibrations, and so metals are usually much better thermal conductors than are electrical insulators, which have at most very few free electrons available to transport energy.

Why are some materials transparent and others opaque?

Optical properties are also related directly to microscopic structure. Good electrical conductors *cannot* be transparent to electromagnetic waves, for the electric fields of the waves induce currents in the material. These induced currents dissipate the wave energy into heat as the electrons collide with the atoms in the lattice. All transparent solid materials are very good insulators. Metals are good *reflectors* of radiation, however, and again the reason is the presence of free electrons at the surface of the material, which can move in response to the incident wave and generate a reflected wave. Reflection from the polished surface of an insulator is a somewhat more subtle phenomenon, dependent on polarization of the material; but again, it is possible to relate macroscopic properties to microscopic structure.

43–7 BAND THEORY OF SOLIDS

The concept of **energy bands** in solids offers us additional insight into some of the properties of solids discussed in Section 43–6. It also allows more detailed theoretical analysis of these properties on the basis of electron energy levels and wave functions.

To introduce the idea, suppose we have a large number N of identical atoms, far enough apart so their interactions are negligible. Then every atom has the same energy-level diagram, showing the possible quantum states for the electrons and their associated energy levels. We can draw an energy-level diagram for the entire system; it looks just like the diagram for a single atom, except that the exclusion principle permits each state to be occupied by N electrons instead of just one.

Now we begin to push the atoms closer together. Because of the electrical interactions and the exclusion principle, the wave functions begin to distort, especially those of the outer, or *valence*, electrons. The energy levels also shift somewhat; some move upward, some downward, depending on the environment of each individual atom. Each valence-electron state for the system, formerly a sharp energy level that could accommodate N electrons, becomes spread out into a *band* containing N closely spaced levels, as shown in Fig. 43–9. Since N is ordinarily very large, of the order of Avogadro's number (10^{23}), we can think of the levels as forming a continuous distribution of energies within a band. Between adjacent energy bands are gaps or forbidden regions where there are *no* possible energy levels.

The inner electrons in an atom are affected much less by nearby atoms than the valence electrons, and their energy levels remain relatively sharp. Ordinarily the behavior of an atom in a solid structure can be modeled as that of a rigid, unchanging *ion core* consisting of the positively charged nucleus and the inner electrons, surrounded by the valence electrons, whose states and energies *are* altered significantly by interactions with neighboring atoms.

Now let us see what this has to do with electrical conductivity. In insulators and semiconductors the valence electrons completely fill the highest occupied band, called the **valence band;** the next higher band, called the **conduction band,** is completely empty. The gap separating the two may be of the order of 1 to 5 eV. This situation is shown in Fig. 43–10b. The electrons in the valence band are not free to move in response to an applied electric field; to move, an electron would have to go to a different quantum-mechanical state with slightly different energy, but all the neighboring states are already occupied. The only way an electron can move is to jump into the conduction band, which would require an additional energy of a few electronvolts. Ordinarily that much energy is not available; for electric fields of reasonable magnitude the potential difference between two points spaced one atom apart is very much *less* than 1 eV. There is no way such a field can do enough work on an electron to raise it across the gap into the conduction band. The situation is like a completely filled parking lot; none of the cars can move because there is no place to go. If a car could jump over the others, it could move!

In solid materials, electron energy levels form bands of allowed energies, separated by forbidden bands.

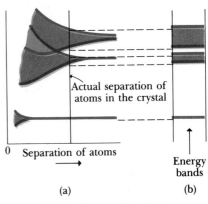

43–9 Origin of energy bands in a solid. (a) As the atoms are pushed together, the energy levels spread into bands. The vertical line shows the actual atomic spacing in the crystal lattice. (b) Symbolic representation of energy bands.

A completely filled band is like a grid-locked traffic jam; nobody can move.

43–10 Three types of energy-band structure. (a) A conductor; there is a partially filled band, and electrons in this band are free to move when an electric field is applied. (b) An insulator; a completely full band is separated by a gap of several electronvolts from a completely empty band, and electrons in the full band cannot move. At finite temperatures a few electrons can reach the upper "conduction band." (c) A semiconductor; a completely filled band is separated by a small gap of 1 eV or less from an empty band; at finite temperatures substantial numbers of electrons can reach the upper "conduction band," where they are free to move.

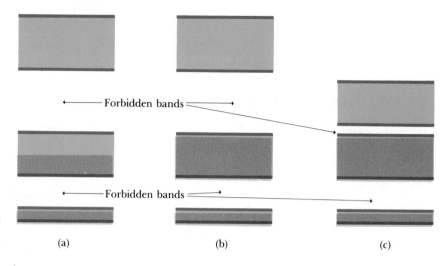

Some electrons can jump into the conduction band, where they are free to move.

Materials with partially filled bands are always conductors.

Dielectric breakdown: when an insulator becomes a conductor

At any finite temperature, however, the crystal lattice has some vibrational motion, and there is some probability that an electron can gain enough energy from *thermal* motion to jump to the conduction band. Once in the conduction band, an electron is free to move in response to an applied electric field because plenty of nearby empty states are available. At any finite temperature a few electrons are present in the conduction band, so no material is a perfect insulator. Furthermore, as the temperature increases, the population in the conduction band increases very rapidly. An increase of a factor of two for a temperature rise of 10 C° is typical.

With metals, the situation is different because the valence band is only partially filled. The metal sodium is a simple example. The energy-level diagram for sodium in Fig. 41–5 shows that for an isolated atom, the $3p$ "resonance-level" states for the valence electron are about 2.1 eV above the $3s$ ground state. (The wavelength of the familiar yellow sodium doublet corresponds to this energy difference.) But in the crystal lattice of solid sodium, the $3s$ and $3p$ *bands* spread out enough so that they actually overlap, forming a single band that is only $\frac{1}{4}$ filled. The situation is similar to the one shown in Fig. 43–10a. Electrons in states near the top of the filled portion of the band have many adjacent unoccupied states available, and they can easily gain or lose small amounts of energy in response to an applied electric field. Thus these electrons are mobile and can contribute to electrical and thermal conductivity. Partly filled bands are always characteristic of the metallic bond found in most metallic crystals. In the ionic NaCl crystal, however, there is no overlapping of bands; the valence band is completely filled, and solid sodium chloride is an insulator.

The band picture also adds insight to the phenomenon of *dielectric breakdown*, in which insulators subjected to a large enough electric field become conductors. If the electric field in a material is so large that there is a potential difference of a few volts over a distance comparable to atomic sizes (i.e., a field of the order of 10^{10} V·m^{-1}), then the field can do enough work on a valence electron to boost it over the forbidden region and into the conduction band. Real insulators usually have dielectric strengths much less than this because of structural imperfections that provide some energy states in the forbidden region.

The concept of energy bands is very useful in understanding the properties of semiconductors, which we will study in the next section.

43–8 SEMICONDUCTORS

As the name implies, a **semiconductor** has an electrical resistivity intermediate between those of good conductors and of good insulators. This is only one aspect of the behavior of this important class of materials, so vital to present-day electronics. To keep things as simple as possible, we will discuss mostly the elements germanium and silicon. These are the simplest semiconductors, but they illustrate the most important concepts.

Both silicon and germanium have four electrons in the outermost electron subshell, and both crystallize in the diamond structure described in Section 43–5. Each atom lies at the center of a regular tetrahedron, with four nearest neighbors at the corners and a covalent bond with each. Thus all the valence electrons are involved in the bonding, and the materials should be insulators. An unusually small amount of energy, however, is needed to break one of the bonds and set an electron free to roam around the lattice: 1.1 eV for silicon and only 0.7 eV for germanium. This corresponds to the energy gap between the valence and conduction bands, shown in Fig. 43–10c. So even at room temperature, a substantial number of electrons are dissociated from their parent atoms, and this number increases rapidly with temperature.

Furthermore, when an electron is removed from a covalent bond, it leaves a vacancy where there would ordinarily be an electron. An electron in a neighboring atom can drop into this vacancy, leaving the neighbor with the vacancy. In this way the vacancy, usually called a **hole,** can travel through the lattice and serve as an additional current carrier. A hole behaves like a positively charged particle. In a pure semiconductor, holes and electrons are always present in equal numbers; the resulting conductivity is called *intrinsic conductivity* to distinguish it from conductivity due to impurities, to be discussed later.

The parking-lot analogy mentioned in Section 43–7 is useful in clarifying the mechanism of conduction in a semiconductor. A crystal with no bonds broken is like a completely filled floor of a parking garage. No cars (electrons) can move because there is nowhere for them to go. But if one car is removed to the empty floor above, it can move freely, and the vacancy it leaves also permits cars to move on the nearly filled floor. This motion is most easily described in terms of *motion of the vacant space* from which the car has been removed. This corresponds to a vacancy or hole in the normally filled valence band.

Now suppose we mix into melted germanium a small amount of arsenic, a Group V element having *five* valence electrons. When one of these electrons is removed, the remaining electron structure is essentially that of germanium; the only difference is that it is scaled down in size by the insignificant factor $\frac{32}{33}$ because the arsenic nucleus contains a charge of $+33e$ rather than $+32e$. Thus an arsenic atom can take the place of a germanium atom in the lattice. Four of its five valence electrons form the necessary covalent bonds with the nearest neighbors; the fifth is very loosely bound, with a binding energy of only about 0.01 eV. This corresponds in the band picture to an energy level 0.01 eV below the bottom of the conduction band. Even at ordinary temperature this electron can very easily escape and wander about the lattice. The corresponding positive charge is associated with the nuclear charge and is *not* free to move, in contrast to the situation with electrons and holes in pure germanium.

Because at ordinary temperatures only a very small fraction of the valence electrons in germanium are able to escape their sites and participate in conduction, a concentration of arsenic atoms as small as one part in 10^{10} can

A crystal puller used to produce exceptionally pure single crystals of germanium. A seed crystal is pulled out of the melt in a quartz crucible heated by radiofrequency induction in an atmosphere of hydrogen. Slices of the crystal are used in making highly sensitive particle detectors. (Courtesy of Lawrence Berkeley Laboratory, University of California.)

Vacancies (holes) can act as current carriers.

The conductivity of a semiconductor can be changed drastically by trace impurities.

Impurity atoms can contribute extra electrons or steal electrons from the atoms of the host material.

increase the conductivity so drastically that conduction due to impurities becomes by far the dominant mechanism. In such a case the conductivity is due almost entirely to *negative* charge motion; the material is called an **n-type semiconductor** and is said to have *n*-type impurities.

Adding atoms of an element in Group III, with only *three* valence electrons, has an analogous effect. An example is gallium; placed in the germanium lattice, the gallium atom would like to form four covalent bonds, but it has only three outer electrons. It can, however, steal an electron from a neighboring germanium atom to complete the bonding. This leaves the neighboring atom with a *hole,* or missing electron, and this hole can then move through the lattice just as with intrinsic conductivity. In this case the corresponding negative charge is associated with the deficiency of positive charge of the gallium nucleus ($+31e$ instead of $+32e$), so it is *not* free to move. This situation is characteristic of **p-type semiconductors,** materials with *p*-type impurities. The two types of impurities, *n* and *p*, are also called *donors* and *acceptors*, respectively, and the deliberate addition of these impurity elements is called *doping*.

The assertion that in *n* and *p* semiconductors the current *is* actually carried by electrons and holes, respectively, can be verified by using the Hall effect (Section 30–10). The direction of the Hall emf is opposite in the two cases, and measurements of the Hall effect in various semiconductor materials confirm our analysis of the conduction mechanisms.

43–9 SEMICONDUCTOR DEVICES

Semiconductor devices are the heart and soul of modern electronics.

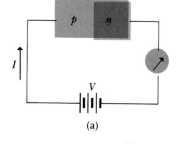

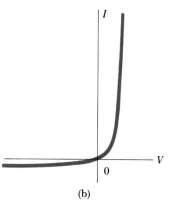

43–11 (a) A semiconductor *p-n* junction in a circuit; (b) graph showing the asymmetric voltage–current relationship given by Eq. (43–5).

Semiconductor devices are of tremendous importance in contemporary electronics. Although radio and television transmitting and receiving equipment originally relied on vacuum tubes, these have been almost completely replaced in the last two decades by transistors, diodes, integrated circuits, and other semiconductor devices. The only surviving vacuum tubes in radio and TV equipment are the picture tube in a TV receiver and the imaging tubes in TV cameras. Equally significant are large-scale integrated circuits that incorporate the equivalent of many thousands of transistors, capacitors, resistors, and diodes on a silicon chip less than 1 cm square. Such chips are at the heart of every pocket calculator, personal computer, and large mainframe computer.

The usefulness of semiconductor materials in these devices results directly from the fact that the conductivity of the material can be controlled within wide limits and changed from one region of a device to another by changing impurity concentrations, deposition of metallic and oxide layers, and etching of patterns to form current paths. The simplest example of controlling conductivity is the **p-n junction,** which is a crystal of germanium or silicon with *p*-type impurities in one region and *n*-type impurities in another. The two regions are separated by a boundary region called the *junction*. Originally such junctions were produced by growing a crystal, usually by pulling a seed crystal very slowly away from the surface of a melted semiconductor. If the concentration of impurities in the melt is changed as the crystal is grown, the result is a crystal with two or more regions of varying conductivity and conductivity type. There are now much better ways to fabricate *p-n* junctions, but we do not need to go into the details.

When a *p-n* junction is connected in an external circuit as shown in Fig. 43–11a, and the potential *V* across the device is varied, the behavior of the current is as shown in Fig. 43–11b. The device conducts much more readily in the direction $p \rightarrow n$ than in the reverse, in striking contrast to the behavior of materials that obey Ohm's law. In the language of electronics, such a one-way device is called a **diode.**

We can understand the behavior of a p-n junction diode at least roughly on the basis of the conductivity mechanisms in the two regions. When the p region is at higher potential than the n, holes in the p region flow into the n region and electrons in the n region move into the p region, so both contribute substantially to current. When the polarity is reversed, the resulting electric fields tend to push electrons from p to n and holes from n to p. But there are very few mobile electrons in the p region, only those associated with intrinsic conductivity and some that diffuse over from the n region. Similarly, there are very few holes in the n region. As a result, the current is much smaller than with the opposite polarity.

A more detailed analysis of this process, taking into account the effects of drift under the applied field and the diffusion that takes place even in the absence of a field, shows that the voltage–current relationship in a p-n junction diode is given by

$$I = I_0[\exp(eV/kT) - 1], \tag{43–5}$$

where I_0 is a constant characteristic of the device, e is the electron charge, k is Boltzmann's constant, and T is absolute temperature.

A **transistor** includes two p-n junctions in a "sandwich" configuration, which may be either p-n-p or n-p-n. A p-n-p transistor is shown in Fig. 43–12. The three regions are usually called the emitter, base, and collector, as shown. In the absence of current in the left loop of the circuit, there is only a very small current through the resistor R because the voltage across the base–collector junction is in the "reverse" direction, that is, the direction of small current flow, as in a simple p-n junction. But when a voltage is applied between emitter and base, as shown, the holes traveling from emitter to base can travel *through* the base to the second junction, where they come under the influence of the collector-to-base potential difference and thus flow on through the resistor.

Hence the current in the collector circuit is *controlled* by the current in the emitter circuit. Furthermore, since V_c may be considerably larger than V_e, the power dissipated in R may be much larger than that supplied to the emitter circuit by the battery V_e. Thus the device functions as a power amplifier. If the potential drop across R is greater than V_e, it may also be a voltage amplifier.

In this configuration the *base* is the common element between the "input" and "output" sides of the circuit. Another widely used arrangement is the *common-emitter* circuit, shown in Fig. 43–13. In this circuit the current in the collector side of the circuit is much larger than in the base side, and the result is current amplification.

Transistors have taken over most of the functions for which vacuum tubes were formerly used following the invention of the triode by de Forest in 1907. They offer many advantages over vacuum tubes, including mechanical ruggedness (since no vacuum container or fragile electrodes are involved), small size, long life (because they operate at relatively low temperatures and deteriorate very little with age), and efficiency (because no power is wasted in heating a cathode for thermionic emission). Invented in 1948, semiconductor devices have completely revolutionized the electronics industry, including applica-

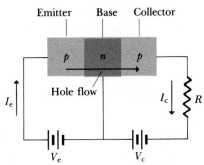

43–12 Schematic diagram of a p-n-p transistor and circuit. When $V_e = 0$, the current in the collector circuit is very small. When a potential V_e is applied between emitter and base, holes travel from emitter to base, as shown; when V_c is sufficiently large, most of them continue into the collector. The collector current I_c is controlled by the emitter current I_e.

Two p-n junctions form a transistor.

A transistor can be a voltage amplifier or a current amplifier.

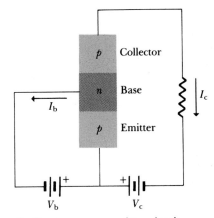

43–13 A common-emitter circuit. When $V_b = 0$, I_c is very small and most of the voltage V_c appears across the base–collector junction. As V_b increases, the base–collector potential decreases and more holes can diffuse into the collector; thus I_c increases. Ordinarily I_c is much larger than I_b.

A silicon wafer containing many identical individual chips. These will be separated and encapsulated in various kinds of envelopes as shown, depending on number of connections, heat-dissipation requirements, and other considerations. (Courtesy of Philips Electronics Co.)

Photocells and light sources using semiconductor devices

tions in communications, computer systems, control systems, and many other areas.

A further refinement in semiconductor technology is the **integrated circuit.** By successively depositing layers of material and etching patterns to define current paths, we can combine the functions of several transistors, capacitors, and resistors on a single square of semiconductor material that may be only a few millimeters on a side. A further elaboration of this idea leads to the *large-scale integrated circuit* and *very-large-scale integration* (VLSI). Starting on a base consisting of a silicon chip, various layers are built up, including evaporated-metal layers for conducting paths and silicon-dioxide layers for insulators and for dielectric layers in capacitors. Appropriate patterns are etched into each layer by use of photosensitive etch-resistant materials and optically reduced patterns. This makes it possible to build up a circuit containing the functional equivalent of many thousands of transistors, diodes, resistors, and capacitors on a single chip. These so-called MOS (metal-oxide-semiconductor) chips are the heart of pocket calculators and microprocessors, as well as of many large-scale computers. An example is shown in Fig. 43–14.

Semiconductors have many other practical applications. A thin slab of pure silicon or germanium can serve as a *photocell.* When the material is irradiated with light whose photons have at least as much energy as the gap between the valence and conduction bands, an electron in the valence band can absorb a photon and jump to the conduction band, where it contributes to the conductivity. Thus the conductivity increases when the material is exposed to light, and this change can be used to control the current in a circuit containing the device. Detectors for charged particles operate on the same principle. A high-energy charged particle passing through the semiconductor material interacts with the electrons, and as it loses energy some of the electrons are excited from the valence to the conduction band. The conductivity increases momentarily, causing a pulse of current in the external circuit. Solid-state detectors are widely used in nuclear and high-energy physics research.

The inverse of the semiconductor photocell is the light-emitting diode (LED), which acts as a source of light. When a diode such as that shown in Fig. 43–11a is given a large *forward* voltage, many holes are injected across the junction into the *n* region and electrons into the *p* region. When these minor-

43–14 A large-scale integrated circuit. This dime-sized chip is a 32-bit microprocessor chip containing the equivalent of 150,000 transistors. (Courtesy of AT&T Bell Laboratories.)

ity carriers recombine with majority carriers in the respective regions, they lose energy, which in some cases is radiated as photons of visible light. Light-emitting diodes are widely used for digital displays in clocks and electronic equipment.

43–10 SUPERCONDUCTIVITY

At very low temperatures (below about 20 K), some materials show a complete disappearance of all electrical resistance. Such materials are called **superconductors.** Superconductivity was discovered in 1911 by the Dutch physicist H. Kamerlingh-Onnes during experiments on the temperature dependence of resistivity of materials. This area of research served at the time as a testing ground for competing theories of the behavior of electrons in solids. Three years earlier, Kamerlingh-Onnes had succeeded in liquefying helium for the first time, thereby attaining lower temperatures than those available by any other means. Helium boils at 4.2 K at a pressure of 1 atm, and at lower temperatures under reduced pressure. Using liquid helium cooling, he found that when very pure solid mercury was cooled to 4.16 K, its resistivity suddenly dropped to zero.

Subsequently many other superconducting metals and alloys have been found. Each one has a characteristic superconducting transition temperature called its *critical temperature.* The highest critical temperature found so far is 23 K for a niobium-germanium alloy, and several niobium alloys have critical temperatures in the range 15 to 20 K.

Below the superconducting transition temperature, the resistivity of a material is believed to be *exactly* zero. Experimental upper limits are of the order of 10^{-25} $\Omega \cdot$m, compared with typical values of 10^{-8} $\Omega \cdot$m for good conductors such as silver and copper at ordinary temperatures. When a current is magnetically induced in a superconducting ring, the current continues without measurable decrease for many months!

A magnetic field can never exist inside a superconducting material. Any attempt to establish such a field results in induced eddy currents that exactly cancel the applied field everywhere inside the material. Related to this behavior is the fact that an applied magnetic field lowers the critical temperature, as shown in Fig. 43–15; a sufficiently strong field eliminates the superconducting transition completely. This transition can be regarded as a *phase transition* (although there is no associated heat of transition), and Fig. 43–15 is a *phase diagram.*

Although discovered in 1911, superconductivity was not well understood on a theoretical basis until 1957, when Bardeen, Cooper, and Schrieffer published the theory (now called the BCS theory) that was to earn them the Nobel Prize in 1972. It is difficult to describe the BCS theory in terms of elementary concepts. It is based on the existence of an *energy gap* in what would ordinarily be a continuum of electron energy states in a partly filled band. This gap is created through the collective motions of pairs of electrons with opposite spins and momenta. Once excited to energy states above the gap, electrons cannot decay back to their "normal" states below it by usual means, such as loss of energy by collisions with the ion cores in the crystal lattice. Thus these electrons become free to move through the lattice without any scattering by the ion cores.

Strange things happen at very low temperatures.

A phase transition to a state with zero resistance

A superconducting magnet used with a 12-foot-diameter hydrogen bubble chamber at Argonne National Laboratory. Its special niobium-titanium coils become superconducting at a temperature of about 4.5 K, and their electrical resistance becomes zero. The magnet weighs about 110 tons and produces a magnetic field of 1.8 T (18,000 gauss). A conventional electromagnet with similar characteristics would consume about 10 megawatts of electrical power. (Courtesy of Argonne National Laboratory.)

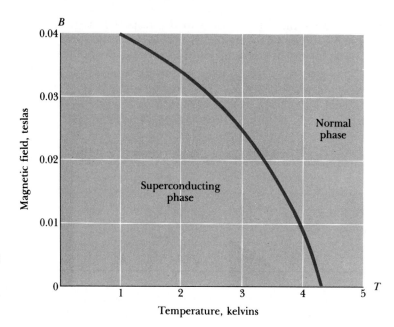

43-15 Phase diagram for pure mercury, showing the superconducting transition critical temperature and its dependence on magnetic field. Other superconducting materials have similar curves but with different scales.

Practical applications of superconductivity: powerful magnets, efficient power transmission, levitating trains

Many important and exciting applications of superconductors are under development. Superconducting electromagnets have been used in research laboratories for several years. Once a current is established in the coil of such a magnet, no additional power input is required because there is no resistive energy loss. The coils can also be made more compact because there is no need to provide channels for the circulation of cooling fluids. Hence superconducting magnets can attain very large fields much more easily and economically than conventional magnets; fields of the order of 10 T are fairly routine. These considerations also make superconductors attractive for long-distance electric-power transmission, an active area of development.

One of the most glamorous applications of superconductors is in the field of magnetic levitation. Imagine a superconducting ring mounted on a railroad car that runs on a magnetized rail. The current induced in the ring leads to a repulsive interaction with the rail, and levitation is possible. Magnetically levitated trains have been running on an experimental basis in Japan for several years, and similar development projects are also under way in Germany.

SUMMARY

KEY TERMS

central-field approximation

Pauli exclusion principle

shell

atomic number

periodic table of the elements

ionic bond

ionization energy

electron affinity

covalent bond

van der Waals bond

In the central-field approximation, each electron in an atom moves under the action of a spherically symmetric electric field caused by the nucleus and all the other electrons. In this approximation, the quantum state of each electron is identified by the four quantum numbers n, l, m, and s. The Pauli exclusion principle states that no two electrons can occupy the same quantum-mechanical state, that is, have the same four quantum numbers. The general nature of the structure of complex atoms, including the main features of the periodic table of the elements, can be understood on the basis of filling up successive energy levels as the atomic number Z increases.

In an ionic bond, one atom gives one or more electrons completely to another; the resulting positive and negative ions attract each other. In a covalent bond, one or more pairs of electrons occupy the space between atoms and cause an attractive interaction. To satisfy the exclusion principle, the two electrons in a pair must have opposite spins. The van der Waals bond is a much weaker dipole–dipole interaction, and the hydrogen bond uses the positive charge of the hydrogen nucleus to form an attractive interaction.

Diatomic molecules have sets of energy levels associated with vibrational and rotational motion; the rotational levels are usually much more closely spaced than the vibrational ones. Transitions among these levels result in molecular spectra. The energies are typically in the range of 0.0001 to 0.1 eV, and the corresponding photons are in the far infrared region of the spectrum.

Solids may be amorphous (having no long-range pattern) or crystalline (having a definite pattern extending over many atoms or molecules). The ionic, covalent, van der Waals, and hydrogen bonds play the same role in binding atoms or molecules into a solid structure as in molecules. An additional type of solid is the metallic solid, in which one or more electrons are detached from each atom and can wander freely through the lattice. Electrical and thermal conductivity of the various types of solids can be understood on the basis of the presence or absence of mobile electrons. The electron states can be further understood on the basis of energy bands, which are ranges of allowed electron energies, separated by ranges of forbidden energies.

Semiconductors have electrical conductivity intermediate between those of good conductors and good insulators. Their conductivity can be changed drastically by addition of minute quantities of impurities, which can lead to conduction predominantly by electrons (n-type) or by holes (p-type). Diodes and transistors contain p-n junctions, regions of transition between p-type and n-type material. A large number of junctions, capacitors, resistors, and conductors can be built up on a silicon chip by successive deposition of layers and etching.

Superconductivity is the complete disappearance of a material's electrical resistance at very low temperature. The transition to the superconducting state is a phase transition, and the temperature at which it occurs depends on the magnetic field. There can be no magnetic field inside a superconducting material.

hydrogen bond
molecular spectra
rotational energy levels
vibrational energy levels
band spectrum
amorphous solid
crystalline solid
ionic crystal
covalent crystal
metallic crystal
close packing
energy bands
valence band
conduction band
semiconductor
hole
*n***-type semiconductor**
*p***-type semiconductor**
p-n **junction**
diode
transistor
integrated circuit
superconductors

QUESTIONS

43–1 In the ground state of the helium atom, the electrons must have opposite spins. Why?

43–2 The central-field approximation is more accurate for alkali metals than for transition metals (Group IV of the periodic table). Why?

43–3 The outermost electron in the potassium atom is in a $4s$ state. What does this tell you about the relative positions of the $3d$ and $4s$ states for this atom?

43–4 A student asserted that any filled shell (i.e., all the levels for a given n occupied by electrons) must have zero total angular momentum and hence must be spherically symmetric. Do you believe this? What about a filled *subshell* (all values of m for given values of n and l)?

43–5 What factors determine whether a material is a conductor of electricity or an insulator?

43–6 The nucleus of a gold atom contains 79 protons. How would you expect the energy required to remove a $1s$

electron completely from a gold atom to compare with the energy required to remove the electron from a hydrogen atom? In what region of the electromagnetic spectrum would a photon of the appropriate energy lie?

43–7 Elements can be identified by their visible spectra. Could analogous techniques be used to identify compounds from their molecular spectra? In what region of the electromagnetic spectrum would the appropriate radiation lie?

43–8 The ionization energies of the alkali metals (i.e., the energy required to remove an outer electron) are in the range from 4 eV to 6 eV, while those of the inert gases are in the range from 15 eV to 25 eV. Why the difference?

43–9 The energy required to remove the $3s$ electron from a sodium atom in its ground state is about 5 eV. Would you expect the energy required to remove an additional electron to be about the same, or more or less? Why?

43–10 Electrical conductivities of most metals decrease gradually with increasing temperature, but the intrinsic conductivity of semiconductors always *increases* rapidly with increasing temperature. Why the difference?

43–11 What are some advantages of transistors compared to vacuum tubes, for electronic devices such as amplifiers? What are some disadvantages? Are there any situations in which vacuum tubes *cannot* be replaced by solid-state devices?

EXERCISES

Section 43–1 The Exclusion Principle

Section 43–2 Atomic Structure

43–1

a) List the different possible combinations of quantum numbers l and m for the $n = 5$ shell.

b) How many electrons can be placed in the $n = 5$ shell?

43–2 For germanium ($Z = 32$) make a list of the number of electrons in each state ($1s$, $2s$, $2p$, etc.).

43–3 Work the two examples described in Problem-Solving Strategy step 2 in Section 43–2. That is,

a) show that the magnitude of the electrical potential energy of a proton and an electron 0.1 nm apart is 14.4 eV, or 2.31×10^{-18} J;

b) show that the moment of inertia of two protons 0.1 nm apart about an axis through the center of mass and perpendicular to the line joining them is 0.836×10^{-47} kg·m².

Section 43–3 Diatomic Molecules

Section 43–4 Molecular Spectra

43–4

a) Calculate the electrical potential energy for a K^+ and a Br^- ion separated by a distance of 0.29 nm, the equilibrium separation in the KBr molecule.

b) The ionization energy of the potassium atom is 4.3 eV. Atomic bromine has an electron affinity of 3.5 eV. Use these data and the results of (a) to *estimate* the binding energy of the KBr molecule. Do you expect the actual binding energy to be larger or smaller than your estimate?

43–5 Show that the frequencies in a pure rotation spectrum (no change in vibrational level) are all integer multiples of the quantity $\hbar/2\pi I$. (Assume that the rotational quantum number l changes by ± 1 in the rotational transitions.)

43–6 If the distance between atoms in a diatomic oxygen molecule is 0.20 nm, calculate the moment of inertia about an axis through the center of mass perpendicular to the line joining the atoms.

43–7 The distance between atoms in a hydrogen molecule is 0.074 nm.

a) Calculate the moment of inertia of a hydrogen molecule about an axis perpendicular to the line joining the nuclei, at its center.

b) Find the energies (in electronvolts) of the $l = 0$, $l = 1$, and $l = 2$ rotational states.

c) Find the wavelength and frequency of the photon emitted in the transition from $l = 2$ to $l = 0$.

43–8 The vibrational frequency of the hydrogen molecule is 1.29×10^{14} Hz.

a) What is the spacing of adjacent vibrational energy levels, in electronvolts?

b) What is the wavelength of radiation emitted in the transition from the $n = 2$ to $n = 1$ vibrational state of the hydrogen molecule?

c) From what initial values of n do transitions to the ground state of vibrational motion yield radiation in the visible spectrum (400 to 700 nm)?

Section 43–5 Structure of Solids

Section 43–6 Properties of Solids

43–9 The spacing of adjacent atoms in a sodium-chloride crystal is 0.282 nm. Calculate the density of sodium chloride.

43–10 Potassium bromide, KBr, has a density of 2.75×10^3 kg·m⁻³, and the same crystal structure as NaCl.

a) Calculate the average spacing between adjacent atoms in a KBr crystal.

b) How does the value calculated in (a) compare with the spacing in NaCl? Is the relation between the two values qualitatively what you would expect?

Section 43–8 Semiconductors

Section 43–9 Semiconductor Devices

43–11

a) Suppose a piece of very pure germanium is to be used as a light detector by observing the increase in conductivity resulting from generation of electron-hole pairs by absorption of photons. If each pair requires 0.7 eV of energy, what is the maximum wavelength that can be detected? In what portion of the spectrum does it lie?

b) What are the answers to (a) if the material is silicon, with an energy requirement of 1.1 eV per pair, corresponding to the gap between valence and conduction bands in silicon?

PROBLEMS

43–12 For magnesium, the first ionization potential is 7.6 eV; the second (additional energy required to remove a second electron) is almost twice this, 15 eV; and the third ionization potential is much larger, about 80 eV. How can these numbers be understood?

43–13 The vibration frequency for the molecule HCl is 8.6×10^{13} Hz.

a) If this molecule can be regarded as a simple harmonic oscillator with the Cl atom stationary (because it is much more massive than the H atom), what is the effective "spring constant" k corresponding to the interatomic force?

b) What is the spacing between adjacent vibrational energy levels, in joules? In electronvolts?

c) What is the wavelength of a photon emitted in a transition between two adjacent vibrational levels? In what region of the spectrum does it lie?

43–14 In Problem 43–13, suppose the hydrogen atom is replaced by an atom of deuterium, an isotope of hydrogen with nuclear mass 2 u. The effective spring constant k is determined by the electron configuration, so it is the same as for the normal HCl molecule.

a) What is the vibrational frequency of this molecule?

b) What is the wavelength of light emitted in the transition $n = 2$ to $n = 1$? In what region of the spectrum does it lie?

43–15

a) For the sodium chloride molecule (NaCl) discussed at the beginning of Section 43–3, what is the maximum separation of the ions for stability if they may be regarded as point charges? That is, what is the largest separation for which the energy of $Na^+ + Cl^-$, calculated in this model, is lower than the energy of the two separate atoms Na and Cl?

b) Calculate this distance for the potassium bromide (KBr) molecule (Exercise 43–4).

43–16 The rotational spectrum of HCl contains the following wavelengths:

$$60.4 \ \mu m$$
$$69.0 \ \mu m$$
$$80.4 \ \mu m$$
$$96.4 \ \mu m$$
$$120.4 \ \mu m$$

Find the moment of inertia of the HCl molecule.

43–17 The dissociation energy of the hydrogen molecule (i.e., the energy required to separate the two atoms) is 4.72 eV. At what temperature is the average kinetic energy of a molecule equal to this energy?

CHALLENGE PROBLEMS

43–18

a) If we consider the hydrogen molecule (H_2) to be a simple harmonic oscillator with an equilibrium spacing of $r_0 = 0.074$ nm, estimate the vibrational energy-level spacing for H_2. (*Hint:* Estimate the force constant k by equating the net Coulomb repulsion of the protons, if the atoms move slightly closer together than r_0, with the "spring" force. That is, assume that the chemical binding force remains approximately constant as r is decreased slightly from r_0.)

b) Use the results of part (a) to calculate the vibrational energy-level spacing for the deuterium molecule, D_2. As in Exercises 43–13 and 43–14, assume that the spring constant is the same for these two molecules.

(In each case above, the spring constant is related to the angular frequency ω by $\omega = \sqrt{k/\mu}$, where

$\mu = m_A m_B / [m_A + m_B]$ is the *reduced mass* for the diatomic molecule whose atoms have masses m_A and m_B.)

43–19

a) Use the result of Problem 43–16 to calculate the equilibrium separation of the atoms in HCl. Assume a mass of the chlorine atom of 35 u.

b) Given that l changes by ± 1 in rotational transitions, what is the value of l for the upper level of the transition that gives rise to each of the wavelengths listed in Problem 43–16?

c) What will be the longest wavelength line in the rotational spectrum of HCl?

d) Calculate the wavelengths of the emitted light for the corresponding transitions in the DCl molecule. Assume that the equilibrium separation between the atoms is the same as for HCl.

44

NUCLEAR AND HIGH-ENERGY PHYSICS

EVERY ATOM CONTAINS AT ITS CENTER AN EXTREMELY DENSE, POSITIVELY charged *nucleus,* much smaller than the overall size of the atom but nevertheless containing most of its total mass. In this chapter we will review the experimental evidence for the existence of nuclei and then describe their most important properties in terms of the protons and neutrons of which they are composed. The stability or instability of a particular nuclear structure is determined by the competition between the attractive nuclear force among the protons and neutrons and the repulsive electrical interactions among the protons. Unstable nuclei *decay,* or transform themselves spontaneously into other structures, by a variety of decay processes. Structure-altering nuclear reactions can also be induced by bombardment of a nucleus with particles or other nuclei. Two classes of reactions of special interest are *fission* and *fusion.*

Protons and neutrons are not among the most fundamental of particles; they are believed to be composed of more basic entities called *quarks,* which also play a central role in understanding the relationships among the four types of interactions: strong, electromagnetic, weak, and gravitational. Research into the nature and interactions of fundamental particles has required the construction of very large experimental facilities such as particle accelerators, as scientists continue to probe more and more deeply into this most basic aspect of the nature of our physical universe.

44–1 THE NUCLEAR ATOM

In 1900 nobody knew what the inside of an atom was like.

The first experimental evidence for the existence of nuclei in atoms was provided by the **Rutherford scattering** experiments, carried out in 1910–1911 by Sir Ernest Rutherford and two of his students, Hans Geiger and Ernest Marsden, at Cambridge, England. The electron had been discovered 13 years earlier, in 1897, by Sir J. J. Thomson, and Millikan had measured its charge in 1910. It was also known by this time that all atoms except hydrogen contain more than one electron. Thomson had proposed a model of the atom that included a relatively large (of the order of 10^{-10} m) sphere of positive charge, with the electrons embedded in it like chocolate chips in a cookie.

Rutherford's experiment consisted of projecting other charged particles at the atoms under study. Observations of the ways the projected particles were deflected, or *scattered,* provided information about the internal structure and charge distribution of the atoms. The particle accelerators now in common use in high-energy physics laboratories had not yet been invented, and Rutherford's projectiles were alpha particles emitted from naturally radioactive elements. We now know that alpha particles are identical with the nuclei of helium atoms, two protons and two neutrons bound together. They are ejected from unstable nuclei with speeds of the order of 10^7 m·s^{-1}, and they can travel several centimeters through air, or 0.1 mm or so through solid matter, before they are brought to rest by collisions.

Exploring atomic structure by shooting atomic bullets into atoms

Rutherford's experimental setup is shown schematically in Fig. 44–1. A radioactive material at the left emits alpha particles. Thick lead screens stop all particles except those in a narrow *beam* defined by small holes. The beam then passes through a thin gold, silver, or copper foil and strikes a plate coated with zinc sulfide. A momentary flash, or *scintillation,* can be seen on the screen whenever it is struck by an alpha particle. (This is the same phenomenon that makes your TV screen glow when the electron beam strikes it.) The numbers of particles deflected through various angles can therefore be determined.

Primitive but crucial experiments in nuclear structure

According to the Thomson model, the atoms of a solid are packed together like marbles in a box. An alpha particle can pass through a thin sheet of metal foil; thus if the Thomson model is correct, the alpha particle must actually penetrate the spheres of positive charge. Assuming for the moment that this happens, we can compute the deflection it would undergo. The Thomson atom is electrically neutral, so outside the atom no force would be exerted on the alpha particle. Within the atom, the electrical force would be due in part to the electrons and in part to the sphere of positive charge. The mass of an alpha particle, however, is about 7400 times that of an electron; momentum considerations show that the alpha particle can be scattered only a negligible amount by its interaction with the electrons. It is like driving a car through a hailstorm; the hailstones do not deflect the car much. Only interactions with the *positive* charge, which makes up most of the mass of the atom, can deflect the alpha particle appreciably.

Assuming the positive charge is distributed through the whole atom, as in the Thomson model, we can calculate the maximum deflection of an alpha

44–1 The scattering of alpha particles by a thin metal foil.

44–2 (a) Alpha particle scattered through a small angle by the Thomson atom. (b) Alpha particle scattered through a large angle by the Rutherford nuclear atom.

The nucleus is extremely small and extremely dense.

particle. It turns out that the interaction potential energy is much *smaller* than the kinetic energy of the alpha particles, and that the maximum deflection to be expected is only a few degrees, as Fig. 44–2a suggests.

The results were very different from this and were totally unexpected. Some alpha particles were scattered by nearly 180°, that is, almost straight backward. Rutherford wrote later:

> It was quite the most incredible event that ever happened to me in my life. It was almost as incredible as if you had fired a 15-inch shell at a piece of tissue paper and it came back and hit you.

Back to the drawing board! Suppose the positive charge, instead of being distributed through a sphere of atomic dimensions (of the order of 10^{-10} m), is all concentrated in a much *smaller* volume. Rutherford called this concentration of charge the **nucleus.** Then it would act like a point charge down to much smaller distances. The maximum repulsive force on the alpha particle would be much larger, and large-angle scattering would be possible, as in Fig. 44–2b. Rutherford again computed the numbers of particles expected to be scattered through various angles. Within the precision of his experiments, the computed and measured results agreed, down to distances of the order of 10^{-14} m. Thus his experiments confirmed the concept of the nucleus of the atom as a very small, very dense structure containing all the positive charge and most of the mass of the atom. They also showed that the size of the nucleus cannot be larger than 10^{-14} m.

44–2 PROPERTIES OF NUCLEI

As we have just seen, the most obvious feature of the atomic nucleus is its size, 20,000 to 200,000 times smaller than that of the atom itself. Since Rutherford's initial experiments, many additional scattering experiments have been performed, using high-energy protons, electrons, and neutrons as well as alpha particles. Although the "surface" of a nucleus is not a sharp boundary, these experiments have determined an approximate radius for each nucleus. The radius is found to depend on the mass, which in turn depends on the total number A of neutrons and protons in the nucleus, usually called the **mass number.** The radii of most nuclei are represented fairly well by the empirical equation

$$r = r_0 A^{1/3}, \qquad (44-1)$$

where r_0 is an empirical constant equal to 1.2×10^{-15} m and is the same for all nuclei.

All nuclei have more or less the same density.

Since the volume of a sphere is proportional to r^3, Eq. (44–1) shows that the *volume* of a nucleus is proportional to A (i.e., to the total mass), and therefore that the mass per unit volume (proportional to A/r^3) is the same for all nuclei. That is, *all nuclei have approximately the same density.* This fact is of crucial importance in understanding nuclear structure.

Two additional important properties of nuclei are angular momentum and magnetic moment. The particles in the nucleus are in motion, just as the electrons in an atom are in motion. Associated with this motion is angular momentum, and, because circulating charge constitutes a current, there is also magnetic moment. Experimental evidence for the existence of nuclear angular momentum (often called **nuclear spin**) and nuclear magnetic moment came originally from spectroscopy. Some spectrum lines are found to be split into series of very closely spaced lines, called *hyperfine structure,* which can be understood on the basis of interactions between electrons and the nuclear magnetic moment. Detailed analysis indicates that the nuclear angular momentum is *quantized,* just as it is for electrons and for molecular rotation. The component of angular momentum in a specified axis direction is a multiple of $\hbar$, but some nuclei have *integral* multiples (as with orbital angular momentum of electrons) and some *half-integral* multiples of $\hbar$ (as with electron spin). Nuclear spin can be understood in detail by analysis of the motions of the particles making up the nucleus.

> Nuclei have spin angular momentum and magnetic moment.

Although it was once believed that nuclei were made of protons and electrons, the discovery of the neutron (discussed in Section 44–9) and many other experiments have established that the basic building blocks of the nucleus are the **proton** and the **neutron.** The total number of protons, equal in a neutral atom to the number of electrons, is the **atomic number** Z. The total number of **nucleons** (protons and neutrons) is called the mass number A. The number of neutrons, denoted by N, is called the **neutron number.** For any nucleus, these three quantities are related by

> Nuclei are made of protons and neutrons.

$$A = Z + N. \tag{44-2}$$

Table 44–1 lists values of A, Z, and N for several nuclei. As the table shows, some nuclei have the same Z but different N. Since the electron structure of an atom, which determines its chemical properties, depends on the charge of the nucleus, these are nuclei of the same element, but they have different masses and can be distinguished in precise experiments. Nuclei of a given element having different mass numbers are called **isotopes** of the element, and a single nuclear species (unique values of both Z and N) is called a **nuclide.** We studied the experimental investigation of *isotopes* through mass spectroscopy in Section 30–6, and we suggest you review that section.

Table 44–1 also shows the usual notation for individual nuclides; we write the symbol of the element, with a presubscript equal to Z and a presuperscript equal to the mass number A. This is redundant, since the element is determined by the atomic number, but the notation is a useful aid to memory.

> Notation for describing the constituents of nuclei

The total mass of a nucleus is always *less* than the total mass of its constituent parts because of the mass equivalent ($E = mc^2$) of the (negative) potential energy associated with the attractive forces that hold the nucleus together. This mass difference is sometimes called the **mass defect.** In fact, the best way to determine the total potential energy, or **binding energy,** of a nucleus is to compare its mass with the total mass of its constituents. The proton and neutron masses are

$$m_p = 1.6726 \times 10^{-27} \text{ kg,}$$

$$m_n = 1.6750 \times 10^{-27} \text{ kg.}$$

TABLE 44–1   Compositions of Some Common Nuclei

Nucleus	Mass Number (Total Number of Nuclear Particles), A	Atomic Number (Number of Protons), Z	Neutron Number, $N = A - Z$
$^{1}_{1}H$	1	1	0
$^{2}_{1}D$	2	1	1
$^{4}_{2}He$	4	2	2
$^{6}_{3}Li$	6	3	3
$^{7}_{3}Li$	7	3	4
$^{9}_{4}Be$	9	4	5
$^{10}_{5}B$	10	5	5
$^{11}_{5}B$	11	5	6
$^{12}_{6}C$	12	6	6
$^{13}_{6}C$	13	6	7
$^{14}_{7}N$	14	7	7
$^{16}_{8}O$	16	8	8
$^{23}_{11}Na$	23	11	12
$^{65}_{29}Cu$	65	29	36
$^{200}_{80}Hg$	200	80	120
$^{235}_{92}U$	235	92	143
$^{238}_{92}U$	238	92	146

Since these values are nearly equal, it is not surprising that many nuclear masses are approximately integer multiples of the proton or neutron mass.

This observation suggests that we define a new mass unit equal to the proton or neutron mass. Instead, for reasons of precision of measurement, it has been found more convenient to define a new unit, called the **atomic mass unit** (u), as $\frac{1}{12}$ the mass of the neutral carbon atom having mass number $A = 12$. It is found that

The atomic mass unit: a convenient unit for nuclear masses

$$1 \text{ u} = 1.660566 \times 10^{-27} \text{ kg}.$$

In atomic units, the masses of the proton, neutron, and electron are found to be

$$m_p = 1.007276 \text{ u},$$

$$m_n = 1.008665 \text{ u},$$

$$m_e = 0.000549 \text{ u}.$$

The masses of some common atoms, including their electrons, are shown in Table 44–2. The masses of the bare nuclei are obtained by subtracting Z times the electron mass. The energy equivalent of 1 u, which we need for calculations of mass defect and binding energy, is found from the relation $E = mc^2$:

The energy equivalent of the atomic mass unit

$$E = (1.660566 \times 10^{-27} \text{ kg})(2.998 \times 10^8 \text{ m·s}^{-1})^2$$

$$= 1.492 \times 10^{-10} \text{ J}$$

$$= 931.5 \text{ MeV}.$$

TABLE 44–2 Atomic Masses of Light Elements

Element	Atomic Number, Z	Neutron Number, N	Atomic Mass, u	Mass Number, A
Hydrogen H	1	0	1.00783	1
Deuterium H	1	1	2.01410	2
Helium He	2	1	3.01603	3
Helium He	2	2	4.00260	4
Lithium Li	3	3	6.01512	6
Lithium Li	3	4	7.01600	7
Beryllium Be	4	5	9.01218	9
Boron B	5	5	10.01294	10
Boron B	5	6	11.00931	11
Carbon C	6	6	12.00000	12
Carbon C	6	7	13.00336	13
Nitrogen N	7	7	14.00307	14
Nitrogen N	7	8	15.00011	15
Oxygen O	8	8	15.99491	16
Oxygen O	8	9	16.99913	17
Oxygen O	8	10	17.99916	18

Source: Encyclopedia of Physics, Lerner and Trigg, eds., (Reading, Mass.: Addison-Wesley, 1981).

EXAMPLE 44–1 Find the mass defect, the total binding energy, and the binding energy per nucleon for the common isotope of carbon, ^{12}C.

SOLUTION The mass of the neutral carbon atom, including the nucleus and the six electrons, is, according to Table 44–2, 12.00000 u. The mass of the bare nucleus is obtained by subtracting the mass of the six electrons:

The mass of a nucleus is always less than the total mass of its constituents.

$$m = 12.00000 \text{ u} - (6)(0.000549 \text{ u})$$

$$= 11.996706 \text{ u}.$$

The total mass of the six protons and six neutrons in the nucleus is

$$(6)(1.007276 \text{ u}) + (6)(1.008665 \text{ u}) = 12.095646 \text{ u}.$$

The mass defect is therefore

The mass defect: mass equivalent of the binding energy

$$12.095646 \text{ u} - 11.996706 \text{ u} = 0.09894 \text{ u}.$$

The energy equivalent of this mass is

$$(0.09894 \text{ u})(931.5 \text{ MeV·u}^{-1}) = 92.16 \text{ MeV}.$$

Thus the total binding energy for the 12 nucleons is 92.16 MeV. To pull the carbon nucleus completely apart into 12 separate nucleons would require a minimum of 92.16 MeV. The binding energy *per nucleon* is 1/12 of this amount, or 7.68 MeV per nucleon. Nearly all stable nuclei, from the lightest to the most massive, have binding energies in the range of 6 to 9 MeV per nucleon. Figure 44–3 is a graph of binding energy per nucleon, as a function of the mass number A.

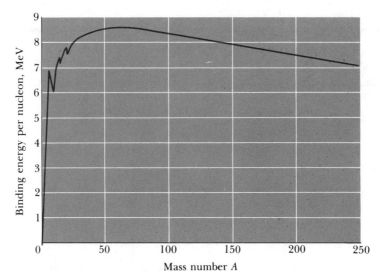

44–3 Binding energy per nucleon as a function of mass number A (the total number of nucleons). The curve reaches a peak of about 8.7 MeV at about $A = 60$, corresponding to the element iron. The spike at $A = 4$ shows the unusual stability of the α-particle structure.

Nuclear energy levels are a million times as large as atomic energy levels.

Because the nucleus is a composite structure, it can have internal motion, with corresponding energy levels. Thus each nucleus has a set of allowed energy levels, corresponding to a *ground state* (state of lowest energy) and several *excited states*. Because of the strength of nuclear interactions, excitation energies of nuclei are typically of the order of 1 MeV, compared with a few eV for atomic energy levels. In ordinary physical and chemical transformations, the nucleus always remains in its ground state. When a nucleus is placed in an excited state, either by bombardment by high-energy particles or by a radioactive transformation, it can decay to the ground state by emission of one or more photons, called in this case **gamma rays,** or *gamma-ray photons*.

Characteristics of the nuclear force: What holds nucleons together?

The forces that hold protons and neutrons together in the nucleus, despite the electrical repulsion of the protons, are not of any familiar sort but are unique to the nucleus. Some aspects of the **nuclear force** are still incompletely understood, but we can describe several qualitative features. First, it does not depend on charge; neutrons as well as protons must be bound, and the binding is found to be the same for both. Second, it must be of short range; otherwise the nucleus would pull in additional protons and neutrons. But within its range, the nuclear force must be stronger than electrical forces; otherwise the nucleus could never be stable. Third, the nearly constant density of nuclear matter and the nearly constant binding energy per nucleon indicate that a given nucleon cannot interact simultaneously with *all* the other nucleons in a nucleus, but only with those few in its immediate vicinity. This feature is again in contrast to the behavior of electrical forces, in which *every* proton in the nucleus repels every other one. This limitation on the maximum number of nucleons with which a nucleon can interact is called *saturation;* it is analogous in some respects to covalent bonding in solids. Finally, the nuclear force favors binding of *pairs* of protons or neutrons with opposite spins, and of *pairs of pairs,* a pair of protons and a pair of neutrons, with each pair having total spin zero. For example, the alpha particle is an exceptionally stable nuclear structure.

These qualitative features of the nuclear force are helpful in understanding the various kinds of nuclear instability, discussed in the following sections.

PROBLEM-SOLVING STRATEGY: *Nuclear structure*

1. As in Chapters 42 and 43, some familiarity with numerical magnitudes is helpful. The scale of things in nuclear structures is very different from that in atomic structures. The size of a nucleus is of the order of 10^{-15} m; the potential energy of interaction of two protons at this distance is 2.31×10^{-13} J or 1.44 MeV. Hence characteristic nuclear energies are of the order of a few MeV, rather than a few eV as with atoms. Protons and neutrons are about 1840 times as massive as electrons. The binding energy per nucleon is roughly 1% of the rest energy of a nucleon; compare this with the ionization energy of the hydrogen atom, which is only 0.003% of its rest energy. Angular momentum is of the same order of magnitude in both atoms and nuclei because it is determined by the value of Planck's constant. But magnetic moments of nuclei are typically much *smaller* than those of electrons in atoms because the nuclear gyromagnetic ratio (the ratio of magnetic moment to angular momentum) is $e/2m_p$ instead of $e/2m_e$, smaller than for orbital electron motion by a factor of the order of 2000. Check out all these numbers, and try to think of other magnitudes to check.

2. In energy calculations involving the mass defect, binding energies, and so on, the mass tables nearly always list the masses of neutral atoms, including their full complements of electrons. To get the mass of a bare nucleus, you have to subtract the masses of these electrons. In principle, you should also take into account the binding energies of the electrons; we won't worry about that in this book, but in very precise work it does have to be considered. Binding-energy calculations often involve subtracting two quantities that are nearly equal; to get enough precision in the difference you often have to carry five or six significant figures, if that many are available. If not, you may have to be content with an approximate result.

44–3 NUCLEAR STABILITY

Of about 1500 different nuclides now known, only about one-fifth are stable. The others are **radioactive;** this means they are unstable structures that decay to form other nuclides by emitting particles and electromagnetic radiation. The time scale of these decay processes ranges from a small fraction of a microsecond to billions of years. The stable nuclides are shown by dots on the graph in Fig. 44–4, where the neutron number N and charge number (or atomic number) Z are plotted. Such a chart is called a *Segre chart,* after its inventor.

The mass number A is equal to the sum $N + Z$, so a curve of constant A is a straight line perpendicular to the line $N = Z$. In general, lines of constant A pass through only one or two stable nuclides; that is, there are usually only one or two stable nuclides with a given mass number. The lines at $A = 20$, $A = 40$, $A = 60$, and $A = 80$ are examples. In four cases, these lines pass through *three* stable nuclides, namely, at $A = 96$, 124, 130, and 136. Only four stable nuclides have both odd Z and odd N:

$$^{2}_{1}\text{H}, \quad ^{6}_{3}\text{Li}, \quad ^{10}_{5}\text{B}, \quad ^{14}_{7}\text{N};$$

these are called *odd-odd nuclides.* Also, there is *no* stable nuclide with $A = 5$ or $A = 8$.

The points representing stable nuclides define a rather narrow stability region. For low mass numbers, $N/Z = 1$. This ratio increases and becomes about 1.6 at large mass numbers. Points to the right of the stability region represent nuclides that have too many protons, or not enough neutrons, to be stable. To the left of the stability region are the points representing nuclei with

Why are some nuclear structures stable and others unstable?

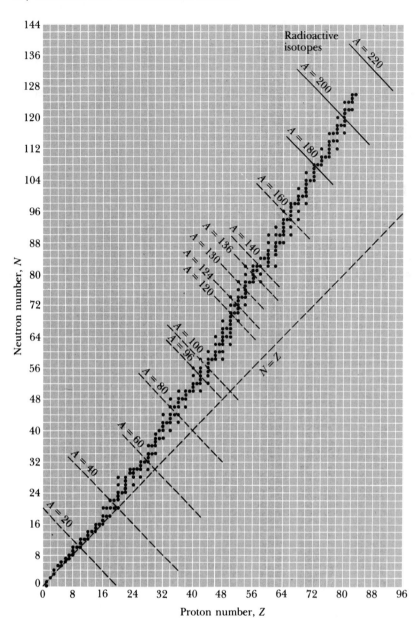

44–4 Segre chart, showing neutron number and proton number for stable nuclides.

A nucleus can be too big to be stable.

too many neutrons or not enough protons. The graph also shows that there are maximum values of A and Z; no nuclide with A greater than 209 or Z greater than 83 is stable.

The stability of nuclei can be understood qualitatively on the basis of the nature of the nuclear force and the competition between the attractive nuclear force and the repulsive electrical force. As we mentioned in Section 44–2, the nuclear force favors *pairs* of nucleons and *pairs of pairs*. In the absence of electrical interactions, the most stable nuclei would be those having equal numbers of neutrons and protons, $N = Z$. The electrical repulsion shifts the balance to favor greater numbers of neutrons, but a nucleus with *too many* neutrons is unstable because not enough of them are paired with protons. A nucleus with too many *protons* has too much repulsive electrical interaction, compared with the attractive nuclear interaction, to be stable.

The neutron–proton balance has to be right for the nucleus to be stable.

As the number of nucleons increases, the total energy of electrical interaction increases faster than that of the nuclear interaction. To understand this behavior, recall the discussion of electrostatic energy in Section 27–4. The energy of a capacitor with a charge Q is proportional to Q^2. It can be shown that to bring a total charge Q to form a uniform volume charge distribution inside a sphere of radius a requires a total energy $3Q^2/20\pi\epsilon_0 a$. Thus the (positive) electric potential energy in the nucleus increases approximately as Z^2, while the (negative) nuclear potential energy increases approximately as A, with corrections for pairing effects. At large A, the electric energy *per nucleon* grows faster than the nuclear energy per nucleon, until the point is reached where stability is impossible. Thus the competition of electric and nuclear forces accounts for the fact that the neutron–proton ratio in stable nuclei increases with Z, and also for the fact that there are maximum values of A and Z for stability. Unstable nuclei respond to these conditions in various ways; in the next several sections we discuss various types of decay of unstable nuclei.

44–4 RADIOACTIVE TRANSFORMATIONS

The study of naturally occurring radioactivity began in 1896, only one year after Röntgen discovered x-rays, when Becquerel found that uranium salts emit a radiation that seemed similar to x-rays. Intensive investigation in the following two decades by Marie and Pierre Curie, Rutherford, and many others revealed that the emissions consist of positively and negatively charged and neutral particles, which were christened **alpha, beta,** and **gamma particles.** It was later established that alpha particles are helium nuclei (two protons and two neutrons bound together), betas are high-energy electrons, and gammas are high-energy electromagnetic-wave photons. The Curies discovered the elements radium and polonium, and showed that the emission of radiation from radium, per unit mass, is of the order of a million times more intense than that from uranium.

Three kinds of radiation from unstable nuclei

Not all three particles are emitted at once by all radioactive elements. Some radioactive elements emit alphas, some emit betas, and gammas sometimes accompany one and sometimes the other. No change in physical or chemical environment, such as chemical reactions or heating or cooling, affects the rate of decay. As soon as the existence of the nucleus was established by Rutherford, it was suspected that radioactivity was a nuclear process. Emission of a charged particle from a nucleus leaves behind a nucleus with a different charge, which therefore must belong to a different chemical element. Thus radioactivity has to involve the transformation of one element into another, which is called *transmutation of elements.*

The first measurement of the charge of the alpha particle was made with a *Geiger counter,* a device that detects high-energy charged particles by the ionization they cause in a gas enclosed in a tube. The gas becomes temporarily conductive when a high-energy charged particle passes through. Rutherford and Geiger counted the number of alpha particles emitted from a radium source in a known time interval. They then allowed the alpha particles from the same source to fall upon a conducting plate and measured its rate of increase of charge. The charge on an alpha particle was found to be equal, within experimental error, to twice the charge on an electron, but opposite in sign. The *mass* of the alpha particle was determined by measuring the ratio of charge to mass by the electric and magnetic deflection method described in Section 30–5. When the result of this measurement was combined with the

A detector for charged particles emitted by nuclei

charge of an alpha particle, the mass was found to be 6.62×10^{-27} kg, almost exactly four times the mass of a hydrogen atom.

Since a helium atom has a mass four times that of a hydrogen atom and, stripped of its two electrons (as a bare nucleus), has a charge equal in magnitude and opposite in sign to two electrons, it seemed certain that alpha particles are helium nuclei. To make the identification certain, however, Rutherford and Royds collected alpha particles in a glass discharge tube over a period of about six days and then established an electric discharge in the tube. In the spectrum of the emitted light they identified the characteristic helium spectrum and established without doubt that alpha particles *are* helium nuclei.

The speed of an alpha particle can be determined from the curvature of its path in a transverse magnetic field. The alpha particles emitted by radium, $^{226}_{88}$Ra, have speeds of about 1.5×10^7 m·s^{-1}. The corresponding kinetic energy is

$$K = \frac{1}{2}(6.62 \times 10^{-27} \text{ kg})(1.5 \times 10^7 \text{ m·s}^{-1})^2$$

$$= 7.4 \times 10^{-13} \text{ J} = 4.6 \times 10^6 \text{ eV}$$

$$= 4.6 \text{ MeV}.$$

This speed, although large, is only 5% of the speed of light, so the nonrelativistic kinetic-energy expression may be used. The energy is *larger* than typical energies of atomic electrons by a factor of the order of a million. Because of these large energies, alpha particles are capable of traveling several centimeters in air, or a few tenths or hundredths of a millimeter through solids, before they are brought to rest by collisions.

Beta particles are *negatively* charged and are therefore deflected in an electric or magnetic field. Deflection experiments similar to those described in Section 30–5 prove conclusively that beta particles have the same charge and mass as electrons. They are emitted with tremendous speeds, up to 0.9995 that of light. Thus relativistic relations must be used in analyzing their motion. Unlike alpha particles, which are emitted from a given nucleus with one speed or a few definite speeds, beta particles are emitted with various speeds within a range from zero up to a maximum that depends on the emitting nucleus.

In order to satisfy conservation of energy and momentum in beta emission, it is necessary to assume that the emission of a beta particle is accompanied by the emission of another particle with no charge. This particle, called a **neutrino,** has zero rest mass and zero charge and therefore produces very little measurable effect, even in traversing the densest matter. Nevertheless, Reines and Cowan were able in 1953 to detect its existence in a series of extraordinary experiments. Subsequent investigation has shown that in fact there are at least three varieties of neutrinos, the one associated with beta decay and two others associated with the decay of unstable particles, the μ mesons and the τ particles.

Gamma rays are not deflected by a magnetic field, so they cannot be charged particles. However, they are diffracted at the surface of a crystal in a manner similar to x-rays, but with extremely small angles of diffraction. Diffraction experiments of this sort led to the conclusion that gamma rays are actually electromagnetic waves of extremely short wavelength, of the order of $\frac{1}{100}$ that of x-rays, with correspondingly higher-energy photons.

The gamma-ray spectrum of any individual element is a *line spectrum;* this fact suggests that gamma emission is analogous to emission of spectrum lines from atoms. That is, a gamma-ray photon is emitted when a nucleus proceeds

Identifying the alpha particle as a helium nucleus: an interesting detective mystery

Beta particles are electrons moving very fast.

The neutrino: a particle predicted from conservation principles long before it was actually observed

Gamma rays are emitted during transitions in nuclear energy levels.

from a state of higher energy to one of lower energy. For example, alpha particles emitted from radium have a kinetic energy of either 4.879 MeV or 4.695 MeV. When a radium nucleus emits an alpha particle with the smaller energy, the resulting nucleus (which corresponds to the element *radon*) has *more* energy than if the higher-energy alpha particle had been emitted. Hence the radon nucleus is left in an excited state. It can then undergo a transition from this state to its ground state, emitting a gamma-ray photon of energy

$$(4.879 - 4.695) \text{ MeV} = 0.184 \text{ MeV}.$$

The *measured* energy of the gamma-ray photon is 0.189 MeV, in excellent agreement.

When a radioactive nucleus decays by alpha or beta emission, the resulting nucleus may also be unstable, and there may be a series of successive decays until a stable configuration is reached. The most abundant radioactive nucleus found on earth is that of uranium $^{238}_{92}\text{U}$, which undergoes a series of 14 decays, including eight alpha emissions and six beta emissions, terminating at the stable isotope of lead, $^{206}_{82}\text{Pb}$.

In alpha decay, the neutron number N and the charge number Z each decrease by two, and the mass number A decreases by four, corresponding to the values $N = 2$, $Z = 2$, $A = 4$ for the alpha particle. The situation in beta decay is less obvious; how can a nucleus composed of protons and neutrons emit an *electron*? The answer is that, in beta decay, a neutron in the nucleus is transformed into a proton, an electron, and a neutrino. We will study such transformations of fundamental particles in Section 44–10. The effect is to *increase* the charge number Z by one, decrease the neutron number N by one, and leave the mass number A unchanged. Finally, gamma emission leaves all three numbers unchanged. Both alpha and beta emissions are often accompanied by gamma emission.

The number of radioactive nuclei in any sample of radioactive material decreases continuously as some of the nuclei disintegrate. The *rate* at which the number decreases, however, varies widely for different kinds of nuclei. Let N or $N(t)$ represent the number of radioactive nuclei in a sample at time t, and let dN be the number that undergo transformations in a short time interval dt. Since every transformation results in a *decrease* in the number N, the corresponding change in N is $-dN$ and the rate of change of N is $-dN/dt$. The larger the number of nuclei in the sample, the larger the number that will undergo transformations, so the rate of change of N is proportional to N, or is equal to a constant λ multiplied by N. Therefore

$$\frac{dN}{dt} = -\lambda N. \tag{44–3}$$

The constant λ is called the **decay constant,** and it has different values for different nuclides. Clearly, a large value of λ corresponds to rapid decay, and conversely. If $N_0 = N(0)$ is the number of nuclei at time $t = 0$, then the solution of this differential equation is an exponential function:

$$N(t) = N_0 e^{-\lambda t}. \tag{44–4}$$

A graph of this function is shown in Fig. 44–5.

The **half-life** $t_{1/2}$ of a radioactive sample is defined as the time at which the number of radioactive nuclei has decreased to one-half the number at

Sometimes several successive decays occur before a stable configuration is reached.

The number of radioactive nuclei in a specimen decreases exponentially with time.

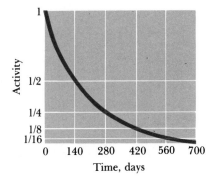

44–5 Decay curve for the radioactive element polonium. Polonium has a half-life of 140 days.

$t = 0$. At this time,

$$e^{-\lambda t_{1/2}} = \frac{N}{N_0} = \frac{1}{2}.$$

Taking natural logarithms of both sides and solving for $t_{1/2}$, we find

$$\lambda t_{1/2} = \ln 2,$$

$$t_{1/2} = \frac{\ln 2}{\lambda} = \frac{0.693}{\lambda}. \qquad (44\text{--}5)$$

Half of the original nuclei in a radioactive sample decay in a time interval $t_{1/2}$, half of those remaining at this time decay in a second interval $t_{1/2}$, and so on; thus the number remaining after successive intervals of $t_{1/2}$ is $N_0/2$, $N_0/4$, $N_0/8$, and so on.

The **mean lifetime** (or average lifetime) of a nucleus or of an unstable particle is related to the half-life $t_{1/2}$ as follows:

$$t_{\text{mean}} = \frac{1}{\lambda} = \frac{t_{1/2}}{\ln 2} = \frac{t_{1/2}}{0.693}. \qquad (44\text{--}6)$$

In particle physics, the life of an unstable particle is usually described in terms of the mean lifetime rather than the half-life.

The *activity* of a sample is defined to be the number of disintegrations per unit time. A commonly used unit is the *curie,* abbreviated Ci, defined to be 3.70×10^{10} decays per second. This is approximately equal to the activity of one gram of radium. Since the number of disintegrations is proportional to the number of radioactive nuclei in the sample, the activity decreases exponentially with time in the same way as the number N. Figure 44–5 is a graph of the activity of polonium, $^{210}_{84}\text{Po}$, which has a half-life of 140 days.

The SI unit of activity is the *becquerel,* abbreviated Bq. One becquerel is one disintegration per second. Thus 1 Bq = 1 s^{-1}, and 1 Ci = 3.70×10^{10} Bq.

In studying sequences of radioactive decays, the following questions are relevant:

1. What is the parent nucleus?
2. What particle is emitted from this nucleus?
3. What is the resulting nucleus (often called the *daughter nucleus*)?
4. What is the half-life of the parent nucleus?
5. Is the daughter nucleus radioactive, and if so, what are the answers to questions 2, 3, and 4 for this nucleus?

Extensive investigations of radioactive decays have been carried out in the last 75 years, and these questions have been answered for many nuclides. The results for any particular parent nuclide are most conveniently presented on a Segre chart such as that shown in Fig. 44–6. The neutron number N is plotted along the vertical axis and the atomic number (or charge number) Z on the horizontal axis. Unit increase of Z with unit decrease of N indicates beta emission; decrease of two in both N and Z indicates alpha emission. The half-lives are given either in years (y), days (d), hours (h), minutes (m), or seconds (s).

Figure 44–6 shows the uranium decay series, which begins with the common uranium isotope ^{238}U and ends with an isotope of lead, ^{206}Pb. Each arrow represents a decay in which an alpha or beta particle is emitted. The

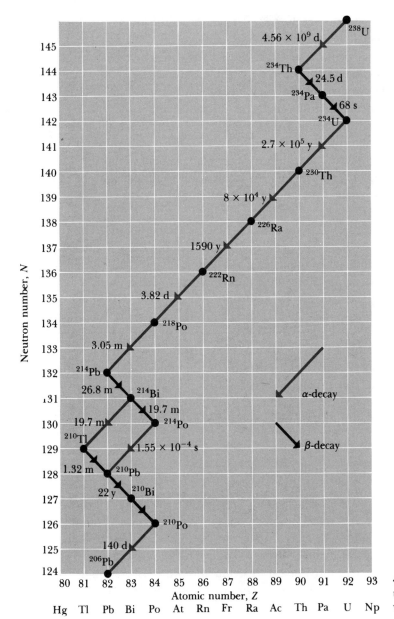

44–6 Segre chart showing the uranium ^{238}U decay series, terminating with the stable nuclide $^{206}_{82}$Pb.

decays can also be represented in equation form; the first two decays in the series are written as

$$^{238}\text{U} \longrightarrow {}^{234}\text{Th} + \alpha,$$

$$^{234}\text{Th} \longrightarrow {}^{234}\text{Pa} + \beta,$$

or, even more briefly, as

$$^{238}\text{U} \xrightarrow{\ \alpha\ } {}^{234}\text{Th},$$

$$^{234}\text{Th} \xrightarrow{\ \beta\ } {}^{234}\text{Pa}.$$

In some cases there are alternate modes of decay of a particular nucleus.

In the second decay a gamma emission follows the beta emission. The reason is that the beta decay leaves the daughter nucleus in an excited state, from which it decays to the ground state by emitting a photon.

An interesting feature of the ^{238}U decay series is the branching that occurs at ^{214}Bi. This nuclide decays to ^{210}Pb by emission of an alpha and a beta, which can occur in either order. We also note that the series includes unstable isotopes of several elements that also have stable isotopes, including Tl, Pb, and Bi. The unstable isotopes of these elements that occur in the ^{238}U series all have too many neutrons to be stable, as discussed in Section 44–3.

Three other decay series are known; two of these occur in nature, one starting with the uncommon isotope ^{235}U, the other with thorium (^{232}Th). The last series starts with neptunium (^{237}Np), an element not found in nature but produced in nuclear reactors. In each case the series continues until a stable nucleus is reached; for these series the final members are ^{207}Pb, ^{208}Pb, and ^{209}Bi, respectively.

Using radioisotopes for archaeological and geological dating

An interesting application of radioactivity is the dating of archaeological and geological specimens by measuring the concentration of radioactive isotopes. The most familiar example is carbon dating. The unstable isotope ^{14}C is produced by nuclear reactions in the atmosphere, caused by cosmic-ray bombardment, and there is a small proportion of ^{14}C in the CO_2 in the atmosphere. Plants, which obtain their carbon from this source, contain the same proportion of ^{14}C as the atmosphere. When a plant dies, it stops taking in carbon, and the ^{14}C it has already taken in decays, with a half-life of 5568 years. Thus by measuring the proportion of ^{14}C in the remains, we can determine how long ago the organism died. Similar techniques are used with other isotopes for dating geologic specimens. A difficulty with carbon dating is that the ^{14}C concentration in the atmosphere changes with time, over long intervals.

44–5 NUCLEAR REACTIONS

In Section 44–4 we studied the decay of unstable nuclei. The processes we described there were natural, spontaneous emission of an alpha or beta particle, sometimes followed by gamma emission. Nothing was done to initiate this emission, and nothing could be done to control it. The first step in artificially starting and controlling a nuclear reaction was Rutherford's suggestion in 1919 that a massive particle with sufficient kinetic energy might be able to penetrate a nucleus. The result would be either a new nucleus with greater atomic number and mass number or a disintegration of the original nucleus. Rutherford bombarded nitrogen with alpha particles and obtained an oxygen nucleus and a proton, according to the reaction

$$\ce{^{4}_{2}He + ^{14}_{7}N \longrightarrow ^{17}_{8}O + ^{1}_{1}H}. \tag{44–7}$$

Rearranging nuclear structures by bombarding a nucleus with high-energy particles

Note that the sum of the initial atomic numbers is equal to the sum of the final atomic numbers, a condition imposed by conservation of charge. The sum of the initial mass numbers is also equal to the sum of the final mass numbers, but the initial rest mass is *not* equal to the final rest mass. Such a process is called a **nuclear reaction,** a rearrangement of nuclear components that results from bombardment by a particle rather than a spontaneous natural process. This experiment was the first nuclear reaction produced under laboratory conditions.

The difference between the rest masses before and after the reaction corresponds to the **reaction energy,** according to the mass–energy relation $E = mc^2$. If the sum of the final rest masses *exceeds* the sum of the initial rest masses, energy is *absorbed* in the reaction. If the final sum is *less than* the initial sum, energy is released in the form of kinetic energy of the final particles. (1 u is equivalent to 931.5 MeV.)

For example, in the nuclear reaction represented by Eq. (44–7), the rest masses of the various particles, in u, are found from Table 44–2 to be

$$\begin{array}{ll} {}^4_2\text{He} = 4.00260 \text{ u} & {}^{17}_8\text{O} = 16.99913 \text{ u} \\ \underline{{}^{14}_7\text{N} = 14.00307 \text{ u}} & \underline{{}^1_1\text{H} = 1.00783 \text{ u}} \\ \phantom{{}^{14}_7\text{N} =} 18.00567 \text{ u} & \phantom{{}^1_1\text{H} =} 18.00696 \text{ u} \end{array}$$

(These values include nine electron masses in each case.) The total rest mass of the final products exceeds that of the initial particles by 0.00129 u, which is equivalent to 1.20 MeV. This amount of energy is *absorbed* in the reaction. If the initial particles do not have at least this much kinetic energy, the reaction cannot take place.

Here is another example. When lithium is bombarded by a proton, two alpha particles are produced:

$$ {}^1_1\text{H} + {}^7_3\text{Li} \longrightarrow {}^4_2\text{He} + {}^4_2\text{He}. \tag{44–8}$$

The sum of the final masses is *smaller* than the sum of the initial values, as shown by the following data:

$$\begin{array}{ll} {}^1_1\text{H} = 1.00783 \text{ u} & {}^4_2\text{He} = 4.00260 \text{ u} \\ \underline{{}^7_3\text{Li} = 7.01600 \text{ u}} & \underline{{}^4_2\text{He} = 4.00260 \text{ u}} \\ \phantom{{}^7_3\text{Li} =} 8.02383 \text{ u} & \phantom{{}^4_2\text{He} =} 8.00520 \text{ u} \end{array}$$

(Four electron masses are included on each side.) Since the decrease in mass is 0.01863 u, 17.4 MeV of energy are liberated and appear as kinetic energy of the two separating alpha particles. We can verify this computation by observing the distance the alpha particles travel in air at atmospheric pressure before being brought to rest by collisions with molecules. We can then compare this value with the range of alpha particles of known energy. The range is found to be 8.31 cm. This corresponds to an alpha particle kinetic energy of 8.7 MeV. The energy of the two alphas together is therefore $2 \times 8.7 = 17.4$ MeV, the same value obtained from the mass decrease.

For positively charged particles such as the proton and the alpha particle to penetrate nuclei of the other atoms, they must have enough initial kinetic energy to overcome the potential-energy barrier caused by the repulsive electrostatic forces. For example, in the reaction of Eq. (44–8), if the lithium nucleus has a radius of the order of 2.3×10^{-15} m, as suggested by Eq. (44–1), the repulsive potential energy of the proton (charge $+e$) and the lithium nucleus (charge $+3e$) at this distance is

$$ U = \frac{1}{4\pi\epsilon_0}\frac{(e)(3e)}{r} = \frac{3(9.0 \times 10^9 \text{ N·m}^2\text{·C}^{-2})(1.6 \times 10^{-19} \text{ C})^2}{2.3 \times 10^{-15} \text{ m}} $$
$$ = 3.01 \times 10^{-13} \text{ J} = 1.88 \times 10^6 \text{ eV} = 1.88 \text{ MeV}. $$

Even though energy is liberated in this reaction, the proton must have a minimum, or **threshold,** energy of about 1.9 MeV for the reaction to occur.

Absorption of *neutrons* by nuclei forms an important class of nuclear reactions. Heavy nuclei bombarded by neutrons in a nuclear reactor can undergo a series of neutron absorptions alternating with beta decays, in which the mass

Several new elements have been produced in nuclear physics laboratories; they are all unstable.

number A increases by as much as 25. The *transuranic elements*, elements having Z larger than 92, are produced in this way. These elements do not occur in nature. Seventeen transuranic elements, having Z up to 109 and A up to about 265, have been identified.

The analytical technique of *neutron activation analysis* uses similar reactions. When stable nuclei are bombarded by neutrons, some absorb neutrons and then undergo beta decay. The energies of the beta and gamma emissions depend on the parent nucleus and hence provide a means of identifying it. The presence of elements in quantities far too small for conventional chemical analysis can be detected in this way.

44–6 NUCLEAR FISSION

A nucleus can sometimes split into two massive fragments.

Up to this point, all the nuclear reactions we have considered involve the ejection of relatively light particles, such as alpha particles, beta particles, protons, or neutrons. This is not always the case, as Hahn and Strassman discovered in Germany in 1939. These scientists bombarded uranium ($Z = 92$) with neutrons, and after a careful chemical analysis they discovered barium ($Z = 56$) and krypton ($Z = 36$) among the products. Cloud chamber photographs showed the two heavy particles traveling in opposite directions with tremendous speed. In this process, the uranium nucleus is said to undergo **fission,** and the two pieces resulting from this split are called **fission fragments.** Measurement showed that an enormous amount of energy, 200 MeV, is released when the uranium nucleus splits up in this way. The rest mass of a uranium atom is greater than the sum of the rest masses of the fission products. The energy released during fission emerges as kinetic energy of the fission fragments. Uranium fission is usually accompanied by the release of a few free neutrons. Both of the two most abundant isotopes of uranium, ^{238}U and ^{235}U, may be split by neutron bombardment.

Energy is liberated when fission occurs.

Fission can be triggered by neutron absorption.

When uranium undergoes fission, barium and krypton are not the only products. Over 100 different isotopes of more than 20 different elements have been detected among the fission products. All of these atoms are in the middle of the periodic table, however, with atomic numbers ranging from 34 to 58. Because the neutron–proton ratio needed for stability in this range is much *smaller* than that of the original uranium nucleus, the fission fragments always have too many neutrons for stability. A few free neutrons are liberated during fission, and the fission fragments undergo a series of beta decays (each of which increases Z by one and decreases N by one) until a stable nucleus is reached. During decay of the fission fragments, an average of 15 MeV of additional energy is liberated.

Chain reactions: Neutrons emitted during fission trigger additional fissions, ad infinitum.

Discovery of the facts that 200 MeV of energy are released when a uranium nucleus undergoes fission triggered by neutron bombardment, and that other neutrons are liberated from the uranium nucleus during fission, suggested the possibility of a **chain reaction;** that is, a self-sustaining series of events that, once started, continues until much of the uranium in a given sample is used up (provided the sample stays together). In the case of a uranium chain reaction, a neutron causes one uranium atom to undergo fission, during which a large amount of energy and several neutrons are emitted. These neutrons then cause fission in neighboring uranium nuclei, which also give out energy and more neutrons. The chain reaction may be made to proceed slowly and in a controlled manner; the device for accomplishing this effect is called a **nuclear reactor.** If the chain reaction is fast and uncontrolled, the device is a bomb.

The probability of neutron absorption by a nucleus is much larger for low-energy (less than 1 eV) neutrons than for the higher-energy neutrons liberated during fission. In a nuclear reactor, the fissionable isotope is contained in *fuel elements;* neutrons emitted during fission are slowed down by collisions with nuclei in the surrounding material, so that they can cause further fissions. The material where the neutrons are slowed down is called the *moderator;* in reactors in nuclear-power plants it is often water. On the average each fission produces about 2.5 free neutrons, so 40% of the neutrons are needed to sustain a chain reaction. The *rate* of the reaction is controlled by inserting or withdrawing *control rods* made of elements (often cadmium) whose nuclei *absorb* neutrons without undergoing any additional reaction.

What does the moderator in a nuclear reactor do?

The most familiar application of nuclear reactors is for the generation of electric power. To illustrate some of the numbers involved, consider a hypothetical nuclear-power plant with a generating capacity of 1000 MW; this figure is typical of large plants currently being built. As noted above, the fission energy appears as kinetic energy of the fission fragments, and its immediate result is to heat the fuel elements and the surrounding water. This heat generates steam to drive turbines, which in turn drive the electrical generators. The turbines, being heat engines, are subject to the efficiency limitations imposed by the second law of thermodynamics, as we discussed in Chapter 19. In modern nuclear plants the overall efficiency is about one-third, so 3000 MW of thermal power from the fission reaction are needed to generate 1000 MW of electrical power.

How much uranium does a nuclear power plant use?

It is easy to calculate how much uranium must undergo fission per unit time to provide 3000 MW of thermal power. Each second, we need 3000 MJ

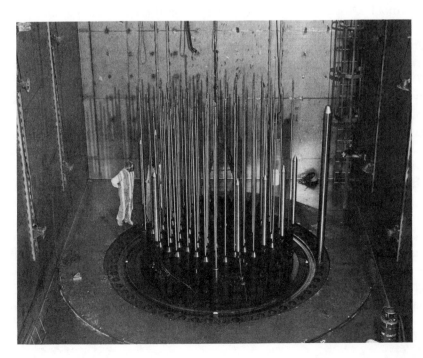

Control rods in a nuclear-fission reactor. The rods are made of a material that absorbs neutrons without undergoing any further nuclear reaction. Inserting or withdrawing the rods from the interior of the reactor controls the rate of the nuclear-fission reaction. (Courtesy of Southern California Edison.)

or 3000×10^6 J. Each fission provides 200 MeV, which is

$$200 \text{ MeV} = (200 \text{ MeV})(1.6 \times 10^{-13} \text{ J} \cdot \text{MeV}^{-1}) = 3.2 \times 10^{-11} \text{ J}.$$

Thus the number of fissions needed per second is

$$\frac{3000 \times 10^6 \text{ J}}{3.2 \times 10^{-11} \text{ J}} = 0.94 \times 10^{20}.$$

Each uranium atom has a mass of about $(235)(1.67 \times 10^{-27} \text{ kg}) = 3.9 \times 10^{-25}$ kg, so the mass of uranium needed per second is

$$(0.94 \times 10^{20})(3.9 \times 10^{-25} \text{ kg}) = 3.7 \times 10^{-5} \text{ kg} = 37 \text{ mg}.$$

In one day (86,400 seconds), the consumption of uranium is

$$(3.7 \times 10^{-5} \text{ kg} \cdot \text{s}^{-1})(86,400 \text{ s} \cdot \text{d}^{-1}) = 3.2 \text{ kg} \cdot \text{d}^{-1}.$$

For comparison, note that the 1000-MW coal-fired power plant described in Section 19–10 burns 10,600 tons (about 10^7 kg) of coal per day! Fission of one uranium nucleus liberates 200 MeV of energy, while combustion of one carbon atom yields about 2 eV.

Nuclear fission reactors have several other practical uses. Among these are the production of artificial radioactive isotopes for medical and other research; production of high-intensity neutron beams for research in nuclear structure; and production of fissionable transuranic elements such as plutonium from the common isotope ^{238}U. The last is the function of *breeder reactors*.

Additional energy is evolved by beta decay of the radioactive fission fragments.

As noted above, about 15 MeV of the energy liberated as a result of fission of a ^{235}U nucleus comes from the subsequent beta decay of the fission fragments rather than from the kinetic energy of the fragments themselves. This fact poses a serious problem with respect to control and safety of reactors. Even after the chain reaction has been completely stopped by insertion of control rods into the core, heat continues to be evolved by the beta decays, which cannot be stopped. For a 1000-MW reactor, this heat power amounts to about 200 MW, which, in the event of total loss of cooling water, is more than enough to cause a catastrophic "meltdown" of the reactor core and possible penetration of the containment vessel. The difficulty in achieving a "cold shutdown" following an accident at the Three Mile Island nuclear power plant in Pennsylvania in March 1979 was a result of the continued evolution of heat due to beta decays.

How big can a nucleus be?

Fission appears to set an upper limit on the production of transuranic nuclei, discussed in Section 44–5. When a nucleus with $Z = 109$ is bombarded with neutrons, fission occurs essentially instantaneously; no $Z = 110$ nucleus is formed even for a short time. There are theoretical reasons to expect that nuclei in the vicinity of $Z = 114$, $N = 184$, might be stable with respect to spontaneous fission. These numbers correspond to *filled shells* in the nuclear energy-level structure, analogous to the filled shells of electrons in the noble gases, as discussed in Section 43–2. Such nuclei, called *superheavy nuclei*, would still be unstable with respect to alpha emission, but they might live long enough to be identified. Attempts to produce superheavy nuclei in the laboratory have not been successful; whether they exist in nature is still an open question.

44–7 NUCLEAR FUSION

In any nuclear reaction where the total rest mass of the products is less than the original rest mass, energy is liberated. The fission of uranium, which we have just described, is an example of one type of energy-liberating reaction. Another type involves the *combination* of two light nuclei to form a nucleus that is more complex but whose rest mass is less than the sum of the rest masses of the original nuclei. These are called **fusion reactions.** Here are three examples of energy-liberating fusion reactions:

When two light nuclei combine, energy is evolved.

$$_1^1H + _1^1H \longrightarrow _1^2H + \beta^+ + \nu,$$
$$_1^2H + _1^1H \longrightarrow _2^3He + \gamma,$$
$$_2^3He + _2^3He \longrightarrow _2^4He + _1^1H + _1^1H.$$

In the first reaction, two protons combine to form a deuteron and a β^+ or *positron* (a positively charged electron, to be discussed in Section 44–9). In the second, a proton and a deuteron unite to form the light isotope of helium. For the third reaction to occur, the first two reactions must occur twice, in which case two nuclei of light helium unite to form ordinary helium. These fusion reactions, known as the *proton-proton* chain, are believed to take place in the interior of the sun and other stars.

The positrons produced during the first step of the proton-proton chain collide with electrons; mutual annihilation takes place, and their energy is converted into gamma radiation. The net effect of the chain, therefore, is the combination of four hydrogen nuclei into a helium nucleus and gamma radiation. The net amount of energy released may be calculated from the mass balance as follows:

Mass of four hydrogen atoms (including electrons)	= 4.03132 u
Mass of one helium plus two additional electrons	= 4.00370 u
Difference in mass	= 0.02762 u
	= 25.7 MeV

In the case of the sun, 1 g of its mass contains about 2×10^{23} protons. Hence, if all these protons were fused into helium, the energy released would be about 57,000 kWh. If the sun were to continue to radiate at its present rate, it would take about 30 billion years to exhaust its supply of protons.

Fusion reactions in the sun are the source of its energy and ours.

For fusion of two nuclei to occur, they must come together to within the range of the nuclear force, typically of the order of 2×10^{-15} m. To do this, they must overcome the electrical repulsion of their positive charges. For two protons at this distance, the corresponding potential energy is of the order of 1.1×10^{-13} J or 0.7 MeV; this represents the initial *kinetic* energy the fusion nuclei must have.

Such energies are available at extremely high temperatures. According to Section 20–4, the average translational kinetic energy of a gas molecule at temperature T is $\frac{3}{2}kT$, where k is Boltzmann's constant. For this value to be equal to 1.1×10^{-13} J, the temperature must be of the order of 5×10^9 K. Not all the nuclei have to have this energy, but this calculation shows that the temperature must be of the order of millions of kelvins if any appreciable fraction of the nuclei are to have enough kinetic energy to surmount the electrical repulsion and achieve fusion.

Photograph showing fusion fuel at Lawrence Livermore Laboratories, given a fan-shaped appearance by magnetic mirrors. The goal of the magnetic fusion energy program is the development of a nuclear-fusion reactor to generate electricity. (Courtesy of Lawrence Livermore Laboratory.)

44–7 The Novette laser system at Lawrence Livermore National Laboratory, used for fusion research. This system went into operation in January 1983; it can deliver a power of 50×10^{12} W for a period of the order of 1 ns. (Courtesy Lawrence Livermore National Laboratory.)

Such temperatures occur in stars as a result of gravitational contraction and its associated liberation of gravitational potential energy. When the temperature gets high enough, fusion reactions occur, more energy is liberated, and the pressure of the resulting radiation prevents further contraction. Only after most of the hydrogen has been converted into helium do further contraction and an accompanying increase of temperature result. Conditions are then suitable for the formation of heavier elements.

Intensive efforts are under way in many laboratories to achieve controlled fusion reactions, which potentially represent an enormous energy resource. In one kind of experiment, a plasma is heated to extremely high temperature by an electrical discharge, while being contained by appropriately shaped magnetic fields. In another experiment, pellets of the material to be fused are heated by a high-intensity laser beam. One current laser experiment setup is shown in Fig. 44–7. Some of the reactions being studied are the following:

$$
\begin{aligned}
{}^{2}_{1}\text{H} + {}^{2}_{1}\text{H} &\longrightarrow {}^{3}_{1}\text{H} + {}^{1}_{1}\text{H} + 4 \text{ MeV}, & (1)\\
{}^{3}_{1}\text{H} + {}^{2}_{1}\text{H} &\longrightarrow {}^{4}_{2}\text{He} + {}^{1}_{0}\text{n} + 17.6 \text{ MeV}, & (2)\\
{}^{2}_{1}\text{H} + {}^{2}_{1}\text{H} &\longrightarrow {}^{3}_{2}\text{He} + {}^{1}_{0}\text{n} + 3.3 \text{ MeV}, & (3)\\
{}^{3}_{2}\text{He} + {}^{2}_{1}\text{H} &\longrightarrow {}^{4}_{2}\text{He} + {}^{1}_{1}\text{H} + 18.3 \text{ MeV}. & (4)
\end{aligned}
$$

In the first reaction, two deuterons combine to form tritium and a proton. In the second, the tritium nucleus combines with another deuteron to form helium and a neutron. The result of both these reactions together is the conversion of three deuterons into a helium-4 nucleus, a proton, and a neutron, with the liberation of 21.6 MeV of energy. Reactions (3) and (4) together achieve the same conversion. In a plasma containing deuterium, the two pairs of reactions occur with roughly equal probability. As yet no one has succeeded in producing these reactions under controlled conditions in such a way as to yield a net surplus of usable energy, but the practical problems do not appear to be insurmountable.

44–8 PARTICLE ACCELERATORS

Many important experiments in nuclear and high-energy physics during the last 60 years or so have made use of beams of charged particles, such as protons or electrons, that have been accelerated to high speeds. Any device that uses electric and magnetic fields to guide and accelerate a beam of charged particles is called a **particle accelerator.** In a sense, the cathode-ray tubes of Thomson and his contemporaries were the first accelerators. In more recent times accelerators have grown enormously in size, complexity, and energy range.

The **cyclotron,** developed in 1931 by Lawrence and Livingston at the University of California, is important historically because it was the first accelerator to use a magnetic field to guide particles in a nearly circular path so that they could be accelerated repeatedly by an electric field in a cyclic process.

In the cyclotron, shown schematically in Fig. 44–8, particles with mass m and charge q move inside a vacuum chamber in a uniform magnetic field B perpendicular to the plane of their trajectories. We learned in Section 30–4 that in such a situation, a particle with speed v moves in a circular path with a radius of curvature r given by

$$r = \frac{mv}{qB}. \tag{44–9}$$

The angular velocity ω of the particles is

$$\omega = \frac{v}{r} = \frac{qB}{m}. \tag{44–10}$$

Note that ω is independent of r.

Now we apply an alternating potential difference between the two hollow electrodes D_1 and D_2, which are called *dees*. If this potential difference has the same frequency as the particles' circular motions, it gives them a push twice each revolution, as they pass the gaps between the dees, boosting them into paths with larger radius and greater kinetic energy. The maximum radius is determined by the radius R of the electromagnet poles. We can find the corresponding maximum energy by solving Eq. (44–9) for v and substituting the

Particle accelerators: more and more powerful tools for research in fundamental-particle interactions

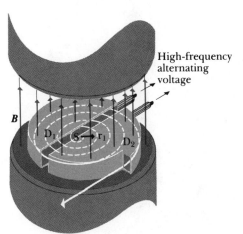

44–8 Schematic diagram of a cyclotron.

result into the relation $K = \frac{1}{2}mv^2$. The result is

$$K_{max} = \frac{1}{2}mv^2 = \frac{q^2B^2R^2}{2m}. \qquad (44-11)$$

If the particles are protons,

$$\frac{q}{m} = \frac{1.60 \times 10^{-19} \text{ C}}{1.67 \times 10^{-27} \text{ kg}} = 9.58 \times 10^7 \text{ C·kg}^{-1}.$$

Limitations on the maximum energy of particles in a cyclotron

Taking values typical of early cyclotrons, we assume $B = 1.5$ T and $R = 0.5$ m. Then from Eq. (44–11), the maximum kinetic energy is

$$K = \frac{1}{2}(1.67 \times 10^{-27} \text{ kg})(9.58 \times 10^7 \text{ C·kg}^{-1})^2(1.5 \text{ T})^2(0.5 \text{ m})^2$$

$$= 4.31 \times 10^{-12} \text{ J} = 2.69 \times 10^7 \text{ eV} = 26.7 \text{ MeV}.$$

This energy, considerably larger than the average binding energy per nucleon, is enough to cause a variety of interesting nuclear reactions.

The energy attainable with the cyclotron is limited by relativistic effects. For Eq. (44–10) to be relativistically correct, m should be replaced by $m/(1 - v^2/c^2)^{1/2}$. As the particles speed up, their angular velocity *decreases;* if the decrease is appreciable, the particle motion is no longer in the correct phase relative to the alternating dee voltage. In a variation called the **synchrocyclotron,** the particles are accelerated in bursts, and for each burst the frequency of the alternating voltage is decreased at just the right rate to maintain the correct phase relation with the particles' motion. A practical limitation of the cyclotron is the expense of building very large electromagnets. The largest synchrocyclotron ever built has a vacuum chamber about 8 m in diameter and accelerates protons to energies of about 600 MeV. The synchrocyclotron, incidentally, provides a very direct confirmation of relativistic mechanics.

The synchrotron: a giant doughnut for accelerating particles to extremely high energies

To attain higher energies, another type of machine, called the **synchrotron,** is more practical. In a synchrotron, the vacuum chamber in which the particles move is in the form of a thin doughnut, called the *accelerating ring.* The particles are forced to move within this chamber by a series of magnets

Interior view of part of the Stanford Linear Accelerator (SLAC). (Courtesy of Stanford Linear Accelerator, Stanford University.)

44–9 An aerial view of the 800-GeV accelerator at the Fermi National Accelerator Laboratory, Batavia, Illinois. (Courtesy of Fermi National Accelerator Laboratory.)

placed around it. As the particles speed up, the magnetic field is increased so that the particles retrace the same trajectory over and over. The synchrotron located at the Fermi National Accelerator Laboratory (Fermilab) in Batavia, Illinois, can accelerate protons to an energy of 800 GeV (800×10^9 eV), and modifications are under way that will permit a maximum energy of 1000 GeV. The accelerating ring is 2 km in diameter, and the accelerator and associated facilities cost about $400 million to build. In each machine cycle, of a few seconds' duration, it accelerates approximately 10^{13} protons. An aerial view of the Fermilab accelerator is shown in Fig. 44–9.

As higher and higher energies are sought in the attempt to investigate new phenomena in particle interactions, a new problem emerges. In an experiment where a beam of high-energy particles collides with a stationary target, not all the kinetic energy of the incident particles is available to form new particle states. Because momentum must be conserved, the particles emerging from the collision must have some motion and thus some kinetic energy. The energy E_a available for creating new particles or particle configurations is the *difference* between initial and final kinetic energies.

In the extreme-relativistic range, where the kinetic energies of the particles are large compared to their rest energies, this is a very severe limitation. When beam and target particles have equal mass, as with protons bombarding a hydrogen target, it can be shown from relativistic mechanics that the available energy E_a is related to the total energy E of the bombarding particle and to its mass m by

$$E_a = \sqrt{2mc^2E}. \qquad (44\text{–}12)$$

For example, for the proton, $mc^2 = 931$ MeV $= 0.931$ GeV. If $E = 800$ GeV, as for the Fermilab accelerator, then

$$E_a = \sqrt{2(0.931 \text{ GeV})(800 \text{ GeV})} = 38.6 \text{ GeV},$$

and if $E = 1000$ GeV, $E_a = 43.1$ GeV. Thus, increasing the beam energy by 200 GeV increases the available energy by only 4.5 GeV.

This limitation may be circumvented in part by *colliding-beam* experiments, in which there is no stationary target but in which beams of particles and their antiparticles (such as electrons and positrons, or protons and antiprotons)

Available energy: a law of diminishing returns in accelerator energies

Colliding-beam experiments: making all the beam energy available

circulate in opposite directions in arrangements called *storage rings*. In regions where the rings intersect, the beams are focused sharply onto one another, and collisions can occur. Because the total momentum in such two-particle collisions is zero, the available energy E_a is the total kinetic energy of the two particles.

An example is the storage-ring facility at the Stanford Linear Accelerator Center (SLAC), where electron and positron beams collide with total available energy of up to 36 GeV. Other storage-ring facilities are located at DESY (German Electron Synchrotron) in Hamburg, West Germany (E_a up to 38 GeV total in electron-positron collisions), and at the Cornell Electron Storage Ring Facility (CESR), with 16 GeV maximum total available energy. At the CERN (European Council for Nuclear Research) laboratory in Geneva, Switzerland, construction has begun for a large electron-positron storage ring that will transport beams of particles with energies of 50 GeV or more, for a total available energy of at least 100 GeV. This facility is expected to have usable beams in 1989.

44–9 FUNDAMENTAL PARTICLES

The physics of fundamental particles has been a recognized field of research for only the past 50 years. The electron and proton were known by the turn of the century, but the existence of the neutron was not established definitely until 1930; its discovery is an interesting story and a useful illustration of nuclear reactions.

Discovery of the neutron: another interesting detective mystery

In 1930, two German physicists, Bothe and Becker, observed that when beryllium, boron, or lithium was bombarded by fast alpha particles, the bombarded material emitted something, either particles or electromagnetic waves, of much greater penetrating power than the original alpha particles. Further experiments in 1932 by I. Curie and Joliot in Paris confirmed these results, but all attempts to explain them in terms of gamma rays were unsuccessful. Chadwick in England repeated the experiments and found that they could be satisfactorily interpreted on the assumption that *uncharged* particles of mass approximately equal to that of the proton were emitted from the nuclei of the bombarded material. He called the particles *neutrons*. The emission of a neutron from a beryllium nucleus takes place according to the reaction

$$^4_2\text{He} + {}^9_4\text{Be} \longrightarrow {}^{12}_6\text{C} + {}^1_0\text{n}, \qquad (44\text{–}13)$$

where ^1_0n is the symbol for a neutron.

Because neutrons have no charge, they produce no ionization in their passage through gases. They are not deflected by the electric field around a nucleus and can be stopped only by colliding with a nucleus in a direct hit, in which case they may either undergo an elastic impact or penetrate the nucleus. We showed in Section 8–5 that if a particle collides elastically with a stationary particle of the same mass, the first particle stops and the second moves off with the same speed as the first. Since the proton and neutron masses are almost the same, fast neutrons can be stopped during collisions with hydrogen atoms in hydrogenous materials such as water or paraffin. A common laboratory method of obtaining slow neutrons is to surround the fast-neutron source with water or blocks of paraffin.

Detecting neutrons by the reactions they cause

Once the neutrons are moving slowly, they may be detected by means of the alpha particles they eject from the nucleus of a boron atom, according to the reaction

$$\frac{1}{0}n + \frac{10}{5}B \longrightarrow \frac{7}{3}Li + \frac{4}{2}He. \tag{44-14}$$

The ejected alpha particle then produces ionization that may be detected in a Geiger counter or other particle detector.

The discovery of the neutron gave the first real clue to the structure of the nucleus. Before 1930, it had been thought that the total mass of a nucleus was due to protons only. We now know that a nucleus consists of both protons and neutrons (except hydrogen, whose nucleus consists of a lone proton) and that (1) the mass number A equals the total number of nuclear particles and (2) the atomic number Z equals the number of protons.

In the early days of particle physics, the *cloud chamber* and the *bubble chamber* were used to visualize paths of charged particles. In the cloud chamber, supercooled vapor condenses around a line of ions created by passage of a charged particle; the result is a visible track. In the bubble chamber, superheated liquid boils locally around a similar line of ions, creating a visible track of tiny bubbles. By placing either instrument in a magnetic field and measuring the radius of curvature of a particle trajectory, we can determine the momentum of the particle.

The positive electron, or **positron,** was first observed during an investigation of cosmic rays by Carl D. Anderson in 1932, in a historic cloud-chamber photograph reproduced in Fig. 44–10. The photograph was made with the cloud chamber in a magnetic field perpendicular to the plane of the paper. A lead plate crosses the chamber, and evidently the particle has passed through it. Since the curvature of the track is greater above the plate than below it, the velocity is less above than below; the inference is that the particle was moving upward, since it could not have *gained* energy going through the lead.

The *density* of droplets along the path is the same as would be expected if the particle were an electron. But the direction of the magnetic field and the direction of motion are consistent only with a particle of *positive* charge. Hence Anderson concluded that the track had been made by a positive electron, or *positron*. The mass of the positron is equal to that of an ordinary (negative) electron, and its charge is equal in magnitude but of opposite sign to that of the electron. Pairs of particles related to each other in this way are said to be *antiparticles* of each other.

Positrons have only a transitory existence; they do not form a part of ordinary matter. They are produced in high-energy collisions of charged particles or gamma rays with matter in a process called *pair production*, in which an ordinary electron and a positron are produced simultaneously. Electric charge is conserved in this process, but enough energy must be available to account for the energy equivalent of the rest masses of the two particles, about 0.5 MeV each. The inverse process, e^+e^- *annihilation*, occurs when a positron and an electron collide. Both particles disappear, and two or three gamma-ray photons appear, with total energy $2mc^2$, where m is the electron rest mass. Decay into a *single* gamma is impossible because such a process cannot possibly conserve both energy and momentum.

Positrons also occur in the decay of some unstable nuclei. Recall that nuclei having too many neutrons for stability often emit a beta particle (electron), decreasing N by one and increasing Z by one. Similarly, a nucleus having *too few neutrons* for stability may respond by converting a proton to a neutron, emitting a positron, increasing N by one and decreasing Z by one. Such nu-

Historical instruments for visualizing paths of high-energy particles

Discovery of the positron: still another detective thriller

Electrons and positrons can be created or destroyed only in pairs.

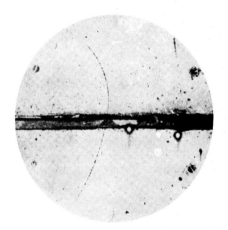

44–10 Track of a positive electron traversing a lead plate 6 mm thick. (Photograph by C. D. Anderson.)

clides do not occur in nature, but they can be produced artificially by neutron bombardment of stable nuclides in nuclear reactors. An example is the unstable nuclide $^{22}_{11}$Na, which has one less neutron that the stable $^{23}_{11}$Na. It emits a positron, leaving the stable nuclide $^{22}_{10}$Ne.

More mysteries: prediction of a new particle and discovering a different one

In 1935 the Japanese physicist Hideki Yukawa inferred, from theoretical considerations, the existence of a particle having a mass intermediate between that of the electron and the proton. A particle of intermediate mass, but *not* identical with that predicted by Yukawa, was discovered one year later by Anderson and Neddermeyer as a component of cosmic radiation. This particle is now known as a μ *meson* (or *muon*). The μ^- has charge equal to that of the electron, and its antiparticle the μ^+ has a positive charge of equal magnitude. The two particles have equal mass, about 207 times the electron mass. The muons are unstable; each decays into an electron of the same sign, plus two neutrinos, with a lifetime of about 2.2×10^{-6} s.

Pions: Yukawa's predicted nuclear glue

Yukawa first proposed the mesons as a basis for transmitting nuclear forces. He suggested that nucleons could interact by emitting and absorbing unstable particles, just as two basketball players interact by tossing the ball back and forth, or by snatching it away from each other. It was established soon after the discovery of the muons that they could not be Yukawa's particles because their interactions with nuclei were far too weak. But in 1947 *another* family of mesons was discovered, called π *mesons* or *pions*. There are three types, positive, negative, and neutral. The charged pions have masses of about 273 times the electron mass and decay into muons with the same sign, plus a neutrino, with a lifetime of about 2.6×10^{-8} s. The neutral pion has a smaller mass, about 264 electron masses, and decays, with a very short lifetime of about 0.8×10^{-16} s, into two gamma-ray photons. The pions interact strongly with nuclei, and they *are* the particles predicted by Yukawa.

In the years since 1947, *high-energy physics* has emerged as a distinct branch of physics. These years have witnessed the attainment of higher and higher energies in particle accelerators, the discovery of many new particles, and intensive efforts to understand the properties of these new particles and their interactions.

Particle detectors have become very sophisticated.

Along with higher energies and the creation of new particles has come the need for new and more sophisticated detectors. Since energy and momentum must be conserved in any reaction or decay process, most large present-day detectors are designed to handle the problems of mass, charge, and momentum identification. Early detectors such as scintillation counters, proportional tubes, and cloud and bubble chambers have been replaced by vertex detectors, electromagnetic calorimeters, wire proportional chambers, large solenoidal magnets, and various kinds of Cerenkov counters. Modern electronics, in the form of on-line computers and microprocessors, has followed, along with increased intensity of accelerator beams. Present-day electronic detectors can comfortably distinguish individual interactions (events) that occur one microsecond apart.

44–10 HIGH-ENERGY PHYSICS

All fundamental particles can be created and destroyed in appropriate circumstances.

It was recognized, even in the early years of high-energy physics, that fundamental particles are not *permanent* entities but can be created or destroyed in interactions with other particles. The earliest such interactions to be observed were creation and destruction of electron-positron pairs. Such pairs are *created* in collisions of high-energy cosmic-ray particles with stationary targets; when an electron and a positron collide, both *disappear* and two or three gamma-ray photons are created to carry away the energy. This transitory nature of the fundamental particles may seem disturbing, but in one sense it is a welcome development. We have seen that photons and electrons (and indeed all particles) share the dual wave-particle nature discussed in Section 42–2, and photons are known to be created and destroyed (or emitted and absorbed) in atomic transitions. Thus it seems natural that other particles can also be created and destroyed.

For example, it was speculated as early as 1932 that there might be an *antiproton,* bearing the same relation to the ordinary proton as the positron does to the electron; that is, a particle with the same mass as the proton but negatively charged. Finally, in 1955 proton-antiproton pairs were created by impact on a stationary target of a beam of protons with kinetic energy 6 GeV (6×10^9 eV) from the Bevatron at the University of California at Berkeley.

Discovery of the antiproton: This time they knew what they were looking for.

In the years after 1960, as higher-energy accelerators and more sophisticated detectors were developed, a veritable blizzard of new unstable particles was identified. To describe them, we have to create a small blizzard of new terms. Initially the particles were classified according to *mass.* The particles having the smallest masses (electrons, muons, and their associated neutrinos) are called **leptons.** (The recently discovered τ particles are also classified as leptons, even though their masses are greater than those of nucleons; we will return to this point later.) Particles with masses between those of muons and nucleons are called **mesons.** Particles that resemble nucleons but are more massive are called **hyperons,** and nucleons and hyperons collectively are called **baryons.** Particles are further classified according to electric charge, spin, and two additional quantum numbers, *isospin* (the number that determines the number of different charges a particular type of particle can have) and *strangeness* (a number needed to account for the production and decay modes of certain particles). A partial list of some known particles is shown in Table 44–3. All particles having zero or integer spin (including photons and π and K mesons) are called **bosons,** and all particles having half-integer spin (including all leptons and baryons) are called **fermions.** These terms, derived from the names Bose and Fermi, refer to the different statistical energy-distribution functions describing the behaviors of the two classes of particles.

Several new terms used to classify particles

During this period it became clear that particles could also be classified in terms of the types of *interactions* in which they participate and in terms of the conservation laws associated with these interactions. In Section 5–1 we spoke briefly about kinds of interactions. There appear to be four classes of interactions; in order of decreasing strength, they are:

Classifying particles in terms of their interactions

1. Strong interactions,
2. Electromagnetic interactions,
3. Weak interactions,
4. Gravitational interactions.

Particles that experience strong interactions are called **hadrons;** these include all the mesons and baryons in Table 44–3. The strong interactions are responsible for the nuclear force and also for the creation of pions, heavy mesons, and hyperons in high-energy collisions. Electrons, muons, and neutrinos have *no* strong interactions.

TABLE 44–3   Some Known Particles and Their Properties

Particle	Mass (MeV/c^2)	Charge	Spin	Isopin	Strangeness	Mean Lifetime (s)	Typical Decay Modes	Quark Content
e^-	0.511	−1	$\frac{1}{2}$	—	0	stable	—	—
ν_e	0 ($<5 \times 10^{-5}$)	0	$\frac{1}{2}$	—	0	stable	—	—
μ^-	105.7	−1	$\frac{1}{2}$	—	0	2.2×10^{-6}	$e^- \bar{\nu}_e \nu_\mu$	—
ν_μ	0 (<0.52)	0	$\frac{1}{2}$	—	0	stable	—	—
τ^-	1784	−1	$\frac{1}{2}$	—	0	5×10^{-13}	$\mu^- \bar{\nu}_\mu \nu_\tau$	—
ν_τ	0 (<250)	0	$\frac{1}{2}$	—	0	stable	—	—
π^0	135.0	0	0	1	0	0.83×10^{-16}	$\gamma\gamma$	$u\bar{u}, d\bar{d}$
π^+	139.6	+1	0	1	0	2.6×10^{-8}	$\mu^+ \nu_\mu$	$u\bar{d}$
π^-	139.6	−1	0	1	0	2.6×10^{-8}	$\mu^- \bar{\nu}_\mu$	$\bar{u}d$
K^+	493.7	+1	0	$\frac{1}{2}$	+1	1.24×10^{-8}	$\mu^+ \nu_\mu$	$u\bar{s}$
K^-	492.67	−1	0	$\frac{1}{2}$	−1	1.24×10^{-8}	$\mu^- \bar{\nu}_\mu$	$\bar{u}s$
η^0	548.8	0	0	0	0	$\sim 10^{-18}$	$\gamma\gamma$	$u\bar{u}, d\bar{d}, s\bar{s}$
p	938.3	+1	$\frac{1}{2}$	$\frac{1}{2}$	0	stable	—	uud
n	939.6	0	$\frac{1}{2}$	$\frac{1}{2}$	0	917	$pe^- \bar{\nu}_e$	udd
Λ	1115	0	$\frac{1}{2}$	0	−1	2.63×10^{-10}	$p\pi^-$ or $n\pi^0$	uds
Σ^+	1189	+1	$\frac{1}{2}$	1	−1	0.80×10^{-10}	$p\pi^0$ or $n\pi^+$	uus
Δ^{++}	1232	+2	$\frac{3}{2}$	$\frac{3}{2}$	0	$\sim 10^{-23}$	$p\pi^+$	uuu
Ξ^-	1321	−1	$\frac{1}{2}$	$\frac{1}{2}$	−2	1.64×10^{-10}	$\Lambda\pi^-$	dss
Ω^-	1672	−1	$\frac{3}{2}$	0	−3	0.82×10^{-10}	ΛK^-	sss
Λ_c^+	2273	1	$\frac{1}{2}$	0	0	$\sim 7 \times 10^{-13}$	$\Lambda\pi\pi\pi$	udc

The *electromagnetic* interactions are those associated directly with electric charge. As noted previously, the electromagnetic interaction between two protons is weaker at distances of the order of nuclear dimensions than the strong interaction, but the electromagnetic interaction has longer range. Neutral particles other than photons have no electromagnetic interactions, with the exception of effects due to the magnetic moments of neutral baryons. These magnetic moments are believed to be associated with the emission and absorption of charged pions and heavy mesons.

The *weak* interaction is responsible for beta decay, such as the conversion of a neutron into a proton, an electron, and a neutrino. It is also responsible for the decay of many unstable particles (pions into muons, muons into electrons, Λ particles into protons, and so on). The *gravitational* interaction, although of central importance for the large-scale structure of celestial bodies, is not believed to be of significance in the analysis of fundamental-particle interactions. For example, the gravitational attraction of two electrons is smaller than their electrical repulsion by a factor of about 2.4×10^{-43}.

Several conservation laws are believed to be obeyed by *all* of the interactions mentioned above. These include the laws growing out of classical physics: energy, momentum, angular momentum, and electrical charge. In addition, several new quantities having no classical analog have been introduced to help characterize the properties of particles. These include *baryon number* (the number of baryons involved in an interaction, minus the number of antibaryons), *isospin* (used also to describe the charge independence of nuclear forces), *parity* (the comparative behavior of two systems that are mirror images of each other), *strangeness* (a quantum number used to classify particle

Classifying interactions in terms of the conservation principles that are obeyed or not obeyed

production and decay reactions), and *lepton number* (the number of leptons involved in an interaction, minus the number of antileptons). Baryon number is conserved in *all* interactions; isospin is conserved in strong interactions but not in electromagnetic or weak interactions. Parity and strangeness are conserved in strong and electromagnetic interactions but not in weak interactions. Lepton number is thought to be conserved in all interactions. Thus the new conservation laws are not absolute but instead serve as a means for *classifying* interactions.

The large number of supposedly fundamental particles discovered since 1960 (well over a hundred) suggests strongly that these particles *do not* represent the most fundamental level of the structure of matter, but that there is at least one additional level of structure. There is now fairly general agreement among physicists concerning the nature of this level; the theory is based on a proposal made initially in 1964 by Gell-Mann and his collaborators. We cannot discuss this theory in detail, but the following is a very brief sketch of some of its features.

Leptons are indeed fundamental particles. In addition to the electrons and muons and their associated neutrinos, a third massive lepton called the *tau* (τ), having spin $\frac{1}{2}$ and mass 1784 MeV, was discovered in 1974. The τ neutrino has thus far escaped detection, but experiments designed to establish its existence are under way. In Table 44–3 upper limits on the three neutrino masses are given. Although zero-mass neutrinos are postulated in most theories, a small mass could be accommodated. If neutrinos do have mass, oscillations in which one type of neutrino changes into another type are possible, and these oscillations allow for experimental detection of finite neutrino mass. Experiments designed to detect neutrino oscillations are under way, but no positive results have yet been obtained. Meanwhile, neutrino research continues; an example is shown in Fig. 44–11.

It now appears that hadrons are *not* fundamental particles but are composite structures whose constituents are spin-$\frac{1}{2}$ fermions called **quarks.** In fact, all known hadrons can be constructed as follows: baryons are composed of three quarks (qqq), and mesons are composed of quark-antiquark pairs ($q\bar{q}$). No other combinations seem to be necessary. This scheme requires that quarks have properties not previously allowed for fundamental particles. For example, quarks have electric charge of magnitude $\frac{1}{3}$ and $\frac{2}{3}$ of the electron charge, which was previously thought to be the fundamental unit of electric charge. Quarks have a strong affinity for each other through a new kind of charge known as "color" charge. Thus color charge is responsible for strong interactions, and the force is known as the color force. The color force is mediated (transmitted) by exchange of color *gluons*, massless spin-one bosons that play the role in strong interactions that the pions played in the Yukawa theory of nuclear force and the photon plays in a quantum theory of electromagnetic interactions.

The weak and gravitational forces are also mediated by exchange of particles. In these cases the particles exchanged are the weak bosons ($W^{\pm}$ and Z^{0}) and the graviton, respectively. The theory of strong forces is known as **quantum chromodynamics** (QCD). Most QCD theories require that phenomena associated with the creation of quark-antiquark pairs make it impossible to observe a free, isolated quark. The binding energy between quarks is thought to be so strong that any stray quark or antiquark in matter will always be reabsorbed, and only baryons or mesons can emerge.

Quarks: A more fundamental level in the structure of matter.

44–11 Brookhaven National Laboratory's solar neutrino experiment, located 4900 feet underground in a gold mine in South Dakota in order to shield out cosmic rays and all other particles except neutrinos. The tank contains 100,000 gallons of perchloroethylene. Neutrinos from the interior of the sun are captured by ^{37}Cl nuclei, which then beta-decay into ^{37}A. The argon is then trapped and measured.

A unified theory of the interactions of all the strongly interacting particles (hadrons)

TABLE 44–4   Properties of Quarks

Symbol	Q/e	Spin	Baryon Number	Strange-ness	Charm	Bottom-ness	Top-ness
u	$\frac{2}{3}$	$\frac{1}{2}$	$\frac{1}{3}$	0	0	0	0
d	$-\frac{1}{3}$	$\frac{1}{2}$	$\frac{1}{3}$	0	0	0	0
s	$-\frac{1}{3}$	$\frac{1}{2}$	$\frac{1}{3}$	-1	0	0	0
c	$\frac{2}{3}$	$\frac{1}{2}$	$\frac{1}{3}$	0	$+1$	0	0
b	$-\frac{1}{3}$	$\frac{1}{2}$	$\frac{1}{3}$	0	0	$+1$	0
t	$\frac{2}{3}$	$\frac{1}{2}$	$\frac{1}{3}$	0	0	0	$+1$

These strange new quark properties have challenged many experimentalists. To date, isolated free quarks have not been observed; the observation of fractional electric charge has been claimed by a few experimenters and disputed by most others. Many indirect observations, however, lead us to believe that the quark structure of hadrons is correct and that quantum chromodynamics may aid in the understanding of the strong force.

Early quark theory suggested the existence of three types (flavors) of quarks; these were labeled *u* (up), *d* (down), and *s* (strange). (See Table 44–4.) Protons, neutrons, π and K mesons, and so on, can all be constructed from these three quarks. For convenience, we describe the charge Q of a particle as a multiple of the magnitude e of the electron charge. For example, a proton has $Q/e = +1$, baryon number $(B) = +1$, strangeness $(S) = 0$. The proton quark content is *uud* if the *u* quark has $Q/e = \frac{2}{3}$ and $B = \frac{1}{3}$ and the *d* quark has $Q/e = -\frac{1}{3}$ and $B = \frac{1}{3}$. The neutron would then have quark content *udd*, the π^+ meson *ud̄*, and the K^+ meson *us̄*. Antiparticles can easily be accommodated: $\bar{p} = \overline{uud}$, $\pi^- = \bar{u}d$, and so on. Particles can then be arranged according to quark content, and families of particles can be classified according to intrinsic orbital angular momentum, spin, and parity. Thus a state *qq̄*, for example, could represent particles in different families depending on the spin configuration of the quarks and the orbital angular momentum.

Because of the Pauli exclusion principle (Section 42–1), quarks are required to have a property that distinguishes one quark from another of the same flavor. This new property was labeled *color;* each quark flavor has three colors. This "color charge" is then responsible for the strong force between quarks. Figure 44–12 shows how we can picture the decay process $K^0 \rightarrow \pi^- + e^+ + \nu_e$, according to QCD theory.

For symmetry and other compelling reasons, theorists later predicted the existence of a fourth quark flavor. This quark was labeled *c* (charm); it was required to have $Q/e = \frac{2}{3}$, $B = \frac{1}{3}$, $S = 0$, and a new quantum number $C = +1$. Charm was discovered in 1974 at both the SLAC and Brookhaven accelerator laboratories by the observation of a meson of mass 3100 MeV. This meson, named ψ at SLAC and J at Brookhaven, was found to have several decay modes, decaying into e^+e^-, $\mu^+\mu^-$, or into hadrons. The mean lifetime was found to be $\sim 10^{-20}$ s. This is consistent with J/ψ being the ground state of a bound $c\bar{c}$ system, much like the way in which the hydrogen atom is a bound p-e system. Immediately after this finding, excited $c\bar{c}$ states or energy levels were observed; and finally (a few years later) isolated mesons having the charm quantum number were also observed. These mesons, D^0 ($c\bar{u}$) and D^+ ($c\bar{d}$), and their excited states are now firmly established, and a charmed baryon, Λ_c^+, has been observed.

More quarks: Six flavors and three colors

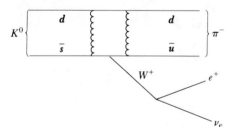

44–12 Diagram showing decay of the K^0 meson. The time sequence is shown from left to right. The initial K^0 meson consists of *d* and *s̄* quarks, bound together by gluon exchange (wiggly line). The quark *s̄* emits the weak boson W^+ and becomes quark *ū*. This is bound by gluon exchange to the *d* quark, forming the final π^- meson, while the W^+ decays into an electron and a neutrino.

In 1977 a meson of mass 9460 MeV, called upsilon (Y) was discovered at Brookhaven. Because it had properties similar to J/ψ, it was conjectured that the meson was really the bound system of a new quark b and its antiquark $\overline{b}$. Excited states of Y were soon observed, and the B^+ ($b\overline{u}$) and B^0 ($b\overline{d}$) mesons are now well established also.

Thus five flavors of quarks (u, d, s, c, b) are thought to exist along with six flavors of leptons ($e, \mu, \tau, \nu_e, \nu_\mu$, and ν_τ). If we assume that quarks and leptons are the basic particles of matter, we can explain many of the strong and weak interactions of hadrons and mesons and the weak interactions of leptons. But it is easy to conjecture that nature is symmetric in its building blocks, and so a sixth quark is thought to exist. This quark, labeled t (top), should have $Q/e = \frac{2}{3}$, $B = \frac{1}{3}$, and a new quantum number, $T = 1$. No compelling experimental evidence requiring the existence of the t quark has yet been found. Table 44–4 lists some properties of the six quarks.

Particle theorists have long tried to combine all four forces of nature into a single unified theory, but with little success. Recently, the weak and electromagnetic forces were successfully unified by Weinberg, Salam, and Glashow. Thus there is a fundamental force in nature, the electro-weak force. The electro-weak theory was successfully verified in 1983 with the discovery of the weak force propagators, the Z^0 and $W^\pm$ bosons, by two experimental groups working at the $\overline{p}p$ collider at CERN, in Geneva. Not only were these particles observed, but their masses are as predicted in the electro-weak theory. It is possible that quantum electrodynamics, the theory of strong interaction, and the electro-weak theory, when correctly unified, will give a valid theory of strong, weak, and electromagnetic forces. At present the photon, lepton, quarks, Z^0, and $W^\pm$ particles have been established and provide the basis for attempts to construct this "grand unified" theory. Although such schemes are still speculative in nature, the entire area is a very active field of present-day theoretical and experimental research.

Quarks and leptons are the fundamental building blocks of matter. At least we think so.

Attempts to combine all four kinds of interactions into one comprehensive theory: the ultimate dream of the particle theorist.

SUMMARY

In Rutherford scattering, alpha particles collide with atoms; their scattering pattern reveals the charge distribution in the atom. Most of the mass and all the positive charge are concentrated in the nucleus, which is much smaller than the overall size of the atom.

All nuclei have nearly the same density, and the radius r of a nucleus is given approximately by

$$r = r_0 A^{1/3}. \tag{44–1}$$

where $r_0 = 1.2 \times 10^{-15}$ m and A is the mass number, equal to the total number of protons and neutrons. The number of protons Z is the atomic number, and the number of neutrons is N. Thus

$$A = Z + N. \tag{44–2}$$

Nuclei of a given element all have the same Z; those with different values of N are called isotopes, and a single species (single values of Z and N) is called a nuclide.

The mass of a nucleus is less than the total mass of its constituents because of the mass equivalent of the binding energy. Binding energy per nucleon is typically about 8 MeV. Nuclear masses are conveniently measured in terms of

KEY TERMS
Rutherford scattering
nucleus
mass number
nuclear spin
proton
neutron
atomic number
nucleons
neutron number
isotopes
nuclide
mass defect
binding energy
atomic mass unit
gamma rays
nuclear force
radioactive
alpha particle

the atomic mass unit (u), equal to $\frac{1}{12}$ the mass of the neutral carbon atom with $A = 12$.

Nuclei have energy levels; each has a ground state and excited states. Gamma-ray photons are emitted and absorbed during transitions between states; excitation energies are typically of the order of 1 MeV. Nuclei are bound together by the nuclear force, which is short range, saturating, and favoring pairs of particles. Within its range, it is stronger than the electrical repulsion of the protons, and the stability or lack of stability of a nucleus is determined by the competition between the attractive nuclear forces and the repulsive electrical forces.

Unstable nuclei undergo spontaneous radioactive decay, emitting alpha or beta particles, sometimes followed by one or more gamma photons. The alpha particle is identical to the helium nucleus: two protons and two neutrons bound together. The beta particle is a high-energy electron. In alpha decay, N and Z both decrease by two; in beta decay, N decreases by one and Z increases by one.

If N_0 is the number of radioactive nuclei present at time $t = 0$, the number $N(t)$ at a later time t is given by

$$N(t) = N_0 e^{-\lambda t}, \qquad (44-4)$$

where λ is the decay constant, different for different nuclides. The half-life $t_{1/2}$ is the time for the number of nuclei to decrease to half the original number. These quantities and the mean lifetime t_{mean} are related by

$$t_{1/2} = \frac{\ln 2}{\lambda} \qquad (44-5)$$

and

$$t_{\text{mean}} = \frac{1}{\lambda}. \qquad (44-6)$$

When a naturally radioactive nucleus decays, a series of successive decays occurs until a stable configuration is reached. No nucleus with Z greater than 83 or A greater than 209 is stable.

A nuclear reaction is a rearrangement of the constituents of a nucleus, caused by bombardment by a particle or a photon. When the bombarding particle is positively charged, there is always a threshold energy because of electrostatic repulsion. The final kinetic energy of all parts of the system may be either greater or less than the initial energy, depending on the total mass change.

In nuclear fission, a heavy nucleus such as uranium or thorium splits into two fission fragments of nearly equal size. Fission can occur spontaneously or can be triggered by absorption of a neutron. The fission fragments have substantial kinetic energy, 200 MeV for uranium. A few free neutrons are released during fission, making a self-sustaining fission chain reaction possible.

In nuclear fusion, two light nuclei fuse into a single nucleus, with the liberation of energy.

Particle accelerators produce beams of high-energy particles for research in nuclear and high-energy physics. In cyclotrons and synchrotrons, particles are accelerated in circles under the action of electric and magnetic fields. In recent installations, storage rings permit the collision of two beams (particles and their antiparticles) traveling in opposite directions, which makes the total energy of the particles available to cause reactions.

Fundamental particles are classified as leptons, mesons, and baryons. Particles having strong interactions are called hadrons. Several new quantities, including isospin and strangeness, are used to characterize particle interactions. Interactions are classified as strong, electromagnetic, weak, or gravitational, although attempts are under way to unify these classes. Hadrons are not fundamental particles but are composed of six kinds of quarks, held together with an interaction mediated by gluons. The leptons (electrons, muons, and tau particles) are themselves fundamental particles.

QUESTIONS

44–1 How can you be sure that nuclei are not made of protons and electrons, rather than of protons and neutrons?

44–2 In calculations of nuclear binding energies such as those in the examples of Sections 44–2 and 44–5, should the binding energies of the *electrons* in the atoms be included?

44–3 If different isotopes of the same element have the same chemical behavior, how can they be separated?

44–4 In beta decay a neutron becomes a proton, an electron, and a neutrino. This decay also occurs with free neutrons, with a half-life of about 15 min. Could a free *proton* undergo a similar decay?

44–5 Since lead is a stable element, why doesn't the ^{238}U decay series shown in Fig. 44–6 stop at lead, ^{214}Pb?

44–6 In the ^{238}U decay chain shown in Fig. 44–6, some nuclides in the chain are found much more abundantly in nature than others, despite the fact that every ^{238}U nucleus goes through every step in the chain before finally becoming ^{206}Pb. Why are the abundances of the intermediate nuclides not all the same?

44–7 Radium has a half-life of about 1600 years. If the universe was formed five billion or more years ago, why is there any radium left now?

44–8 Why is the decay of an unstable nucleus unaffected by the *chemical* situation of the atom, such as the nature of the molecule in which it is bound, and so on?

44–9 Fusion reactions, which liberate energy, occur only with light nuclei, and fission reactions only with heavy nuclei. A student asserted that this shows that the binding energy per nucleon increases with A at small A but decreases at large A and hence must have a maximum somewhere in between. Do you agree?

44–10 Nuclear power plants use nuclear fission reactions to generate steam to run steam-turbine generators. How does the nuclear reaction produce heat?

44–11 There are cases where a nucleus having too few neutrons for stability can capture one of the electrons in the K shell of the atom. What is the effect of this process on N, A, and Z? Is this the same effect as that of β^+ emission? Might there be situations where K capture is energetically possible while β^+ emission is not? Explain.

44–12 Is it possible that some parts of the universe contain antimatter whose atoms have nuclei made of antiprotons and antineutrons, surrounded by positrons? How could we detect this condition without actually going there? What problems might arise if we actually *did* go there?

44–13 When x-rays are used to diagnose stomach disorders such as ulcers, the patient first drinks a thick mixture of (insoluble) barium sulfate and water. What does this do? What is the significance of the choice of barium for this purpose?

44–14 Why are so many health hazards associated with fission fragments that are produced during fission of heavy nuclei?

EXERCISES

Section 44–1 The Nuclear Atom

44–1 A beam of alpha particles is incident on gold nuclei. A particular alpha particle comes in "head-on" and stops 1×10^{-14} m away from the center of a gold nucleus. Assume that the gold nucleus remains at rest.

a) Calculate the electrostatic potential energy of the alpha particle when it has stopped. Express your result in joules and in MeV.

b) What initial kinetic energy did the alpha particle have? Express in joules and in MeV.

c) What was the initial velocity of the alpha particle?

44–2 A 4.7-MeV alpha particle from a radium ^{226}Ra decay makes a head-on collision with a gold nucleus.

a) What is the distance of closest approach of the alpha particle to the center of the nucleus? Assume that the gold nucleus remains at rest.

b) What is the force on the alpha particle at the instant when it is at the distance of closest approach?

Section 44–2 Properties of Nuclei

44–3 How many protons and neutrons are there in a nucleus of

a) hydrogen, ^{1}H?

b) iron, ^{56}Fe?

c) gold, ^{197}Au?

44–4 Consider the three nuclei of Exercise 44–3.

a) Estimate the radius of each nucleus.

b) Estimate the surface area of each.

c) Estimate the volume of each.

d) Determine the mass density (in kg·m^{-3}) for each.

e) Determine the particle density (in particles·m^{-3}) for each.

44–5 Using the data in Table 44–2, calculate the binding energy of the deuterium nucleus. (Express your result in MeV.)

44–6 Calculate

a) the mass defect,

b) the binding energy,

c) the binding energy per nucleon, for the common isotope of oxygen, ^{16}O.

Section 44–3 Nuclear Stability
Section 44–4 Radioactive Transformations

44–7 Tritium is an unstable isotope of hydrogen, $^3_1$H; its mass, including one electron, is 3.01647 u.

a) Show that it must be unstable with respect to beta decay because $^3_2$He plus an emitted electron has less total mass.

b) Determine the total kinetic energy of the decay products, taking care to account for the electron masses correctly.

44–8 Radium (^{226}Ra) undergoes alpha emission, leading to radon (^{222}Rn). The masses, including all electrons in each atom, are 226.0254 u and 222.0163 u, respectively. Find the maximum kinetic energy that the emitted alpha particle can have.

44–9 The common isotope of uranium, ^{238}U, has a half-life of 4.50×10^9 years, decaying by alpha emission.

a) What is the decay constant?

b) What mass of uranium would be required for an activity of one curie?

c) How many alpha particles are emitted per second by 1 g of uranium?

44–10 An unstable isotope of cobalt, ^{60}Co, has one more neutron in its nucleus than the stable ^{59}Co and is a beta emitter with a half-life of 5.3 years. This isotope is widely used in medicine. A certain radiation source in a hospital contains 0.01 g of ^{60}Co.

a) What is the decay constant for this isotope?

b) How many atoms are in the source?

c) How many decays occur per second?

d) What is the activity of the source, in curies? How does this compare with the activity of an equal mass of radium?

44–11 The unstable isotope ^{40}K is used for dating rock samples. Its half-life is 2.4×10^8 years.

a) How many decays occur per second in a sample containing 2×10^{-6} g of ^{40}K?

b) What is the activity of the sample, in curies?

Section 44–5 Nuclear Reactions
Section 44–6 Nuclear Fission
Section 44–7 Nuclear Fusion

44–12 In the fission of one ^{238}U nucleus, 200 MeV of energy are released. Express this energy in joules per mole and compare with typical heats of combustion, which are on the order of 1.0×10^5 J·mol^{-1}.

44–13 Consider the nuclear reaction

$$^2_1\text{H} + {}^9_4\text{Be} \longrightarrow {}^7_3\text{Li} + {}^4_2\text{He}.$$

a) How much energy is liberated?

b) Estimate the threshold energy for this reaction.

44–14 Consider the fusion reaction

$$^2\text{H} + {}^2\text{H} \longrightarrow {}^4\text{He} + \text{energy}.$$

a) Compute the energy liberated in this reaction, in MeV and in joules.

b) Compute the energy *per mole* of deuterium, remembering that the gas is diatomic, and compare with the heat of combustion of hydrogen, about 2.9×10^5 J·mol^{-1}.

44–15. Consider the nuclear reaction

$$^2_1\text{H} + {}^{14}_7\text{N} \longrightarrow {}^6_3\text{Li} + {}^{10}_5\text{B}.$$

Is energy absorbed or liberated, and how much?

Section 44–8 Particle Accelerators

44–16 The magnetic field in a cyclotron that is accelerating protons is 0.8 T.

a) How many times per second should the potential across the dees reverse?

b) The maximum radius of the cyclotron is 0.25 m. What is the maximum velocity of the proton?

c) Through what potential difference would the proton have to be accelerated to give it the maximum cyclotron velocity?

d) What is the energy of the protons when they emerge? Express your result in joules and in MeV.

44–17 Deuterons in a cyclotron describe a circle of radius 32.0 cm just before emerging from the dees. The frequency of the applied alternating voltage is 10 MHz. Find

a) the magnetic field,

b) the energy and speed of the deuterons upon emergence.

Section 44-9 Fundamental Particles
Section 44-10 High-Energy Physics

44-18 If two gamma-ray photons are produced in e^+e^- annihilation, find the energy, frequency, and wavelength of each photon.

44-19 A neutral pion at rest decays into two gamma-ray photons. Find the energy, frequency, and wavelength of each photon.

44-20 Beams of π^- mesons are being used experimentally in the treatment of cancer. What is the minimum total energy a pion can release in a tumor?

44-21 Determine the electric charge, baryon number, strangeness, and charm quantum numbers for the following quark combinations:

a) uus, b) $c\bar{s}$, c) $\overline{ddu}$, d) $c\bar{b}$.

PROBLEMS

44-22 Compute the approximate density of nuclear matter, and compare your result with typical densities of ordinary matter.

44-23 The starship *Enterprise* is powered by the controlled combination of matter and antimatter. If the entire 100 kg antimatter fuel supply of the *Enterprise* were to combine with matter, how much energy would be released?

44-24 The results of activity measurements on a radioactive sample are given below.

a) Find the half-life.

b) How many radioactive nuclei were present in the sample at $t = 0$?

c) How many were present after 7 hr?

Time (hr)	Counts·s^{-1}
0.	20,000
0.5	14,800
1.0	11,000
1.5	8,130
2.0	6,020
2.5	4,460
3.0	3,300
4.0	1,810
5.0	1,000
6.0	550
7.0	300

44-25 A carbon specimen found in a cave believed to have been inhabited by cavemen contained ⅛ as much ^{14}C as an equal amount of carbon in living matter. Find the approximate age of the specimen.

44-26

a) What is the binding energy of the least strongly bound proton in $^{12}_{6}$C?

b) The least strongly bound neutron in $^{13}_{6}$C?

44-27 A free neutron at rest decays into a proton, an electron, and a neutrino, with a half-life of about 15 min. Calculate the total kinetic energy of the decay products.

44-28 A K^+ meson at rest decays into two π mesons.

a) What are the allowed combinations of π^0, π^+, and π^- as decay products?

b) Find the total kinetic energy of the π mesons.

44-29 A Λ hyperon at rest decays into a proton and a π^-.

a) Find the total kinetic energy of the decay products.

b) What fraction of the energy is carried off by each particle? (For simplicity, use nonrelativistic momentum and kinetic-energy expressions.)

44-30 The measured energy width of the ϕ meson is 4 MeV and its mass is 1020 MeV/c^2. Using the uncertainty principle, Eq. (42-17), estimate the lifetime of the ϕ meson.

44-31 A ϕ meson (Problem 44-30) at rest decays via $\phi \rightarrow K^+K^-$.

a) Find the kinetic energy of the K^+ meson.

b) Suggest a reason why the decay $\phi \rightarrow K^+K^-\pi^0$ has not been observed.

c) Suggest reasons why the decays $\phi \rightarrow K^+\pi^-$ and $\phi \rightarrow K^+\mu^-$ have not been observed.

CHALLENGE PROBLEMS

44-32 The results of activity measurements on a radioactive sample that is a mixture of radioactive elements is shown in the following table.

a) How many different nuclides are present in the mixture?

b) What are their half-lives?

c) How many nuclei of each type are initially present in the sample?

d) How many of each type are present at $t = 6$ hr?

Time (hr)	Counts·s^{-1}	Time (hr)	Counts·s^{-1}
0.	7500	5.0	414
0.5	4120	6.0	288
1.0	2570	7.0	201
1.5	1790	8.0	140
2.0	1350	9.0	98
2.5	1070	10.0	68
3.0	872	12.0	33
4.0	596		

44–33 ^{128}I is created by the irradiation of ^{127}I with neutrons. The half-life of ^{128}I is 25 min. A neutron beam is used that creates 1×10^6 ^{128}I nuclei per second.

a) Sketch a graph of the number of ^{128}I nuclei present as a function of time.

b) What is the activity of the sample 1, 10, 25, 50, 75, and 180 min after irradiation is begun?

c) What is the maximum number of ^{128}I atoms that can be created in the sample, after being irradiated for a long time? (This steady-state situation is called *saturation*.)

d) What therefore is the maximum activity that can be produced?

44–34 Radioisotopes are used in a variety of manufacturing and testing techniques. Wear measurements can be made by the following method. An automobile engine is produced using piston rings with a total mass of 100 g, which includes 10μ Ci of ^{59}Fe whose half-life is 45 days. The engine is test run for 1000 hr, after which the oil is drained and its activity measured. If the activity of the engine oil is 50 decays·s^{-1}, how much mass was worn from the piston rings per hour of operation?

APPENDIX A

THE INTERNATIONAL SYSTEM OF UNITS

The Système International d'Unités, abbreviated SI, is the system developed by the General Conference on Weights and Measures and adopted by nearly all the industrial nations of the world. It is based on the mksa (meter-kilogram-second-ampere) system. The following material is adapted from NBS Special Publication 330 (1981 edition) of the National Bureau of Standards.

Quantity	Name of unit	Symbol	
SI Base Units			
length	meter	m	
mass	kilogram	kg	
time	second	s	
electric current	ampere	A	
thermodynamic temperature	kelvin	K	
luminous intensity	candela	cd	
amount of substance	mole	mol	
SI Derived Units			**Equivalent Units**
area	square meter	m^2	
volume	cubic meter	m^3	
frequency	hertz	Hz	s^{-1}
mass density (density)	kilogram per cubic meter	$kg \cdot m^{-3}$	
speed, velocity	meter per second	$m \cdot s^{-1}$	
angular velocity	radian per second	$rad \cdot s^{-1}$	
acceleration	meter per second squared	$m \cdot s^{-2}$	
angular acceleration	radian per second squared	$rad \cdot s^{-2}$	
force	newton	N	$kg \cdot m \cdot s^{-2}$
pressure (mechanical stress)	pascal	Pa	$N \cdot m^{-2}$
kinematic viscosity	square meter per second	$m^2 \cdot s^{-1}$	
dynamic viscosity	newton-second per square meter	$N \cdot s \cdot m^{-2}$	
work, energy, quantity of heat	joule	J	$N \cdot m$
power	watt	W	$J \cdot s^{-1}$
quantity of electricity	coulomb	C	$A \cdot s$
potential difference, electromotive force	volt	V	$W \cdot A^{-1}, J \cdot C^{-1}$
electric field strength	volt per meter	$V \cdot m^{-1}$	$N \cdot C^{-1}$
electric resistance	ohm	Ω	$V \cdot A^{-1}$
capacitance	farad	F	$A \cdot s \cdot V^{-1}$

Quantity	Name of unit	Symbol	Equivalent Units
magnetic flux	weber	Wb	$V \cdot s$
inductance	henry	H	$V \cdot s \cdot A^{-1}$
magnetic flux density	tesla	T	$Wb \cdot m^{-2}$
magnetic field strength	ampere per meter	$A \cdot m^{-1}$	
magnetomotive force	ampere	A	
luminous flux	lumen	lm	$cd \cdot sr$
luminance	candela per square meter	$cd \cdot m^{-2}$	
illuminance	lux	lx	$lm \cdot m^{-2}$
wave number	1 per meter	m^{-1}	
entropy	joule per kelvin	$J \cdot K^{-1}$	
specific heat capacity	joule per kilogram kelvin	$J \cdot kg^{-1} \cdot K^{-1}$	
thermal conductivity	watt per meter kelvin	$W \cdot m^{-1} \cdot K^{-1}$	
radiant intensity	watt per steradian	$W \cdot sr^{-1}$	
activity (of a radioactive source)	becquerel	Bq	s^{-1}
radiation dose	gray	Gy	$J \cdot kg^{-1}$
radiation dose equivalent	sievert	Sv	$J \cdot kg^{-1}$
SI Supplementary Units			
plane angle	radian	rad	
solid angle	steradian	sr	

DEFINITIONS OF SI UNITS

meter (m) The *meter* is the length equal to the distance traveled by light, in vacuum, in a time of 1/299,792,458 second.

kilogram (kg) The *kilogram* is the unit of mass; it is equal to the mass of the international prototype of the kilogram. (The international prototype of the kilogram is a particular cylinder of platinum-iridium alloy that is preserved in a vault at Sèvres, France, by the International Bureau of Weights and Measures.)

second (s) The *second* is the duration of 9,192,631,770 periods of the radiation corresponding to the transition between the two hyperfine levels of the ground state of the cesium-133 atom.

ampere (A) The *ampere* is that constant current that, if maintained in two straight parallel conductors of infinite length, of negligible circular cross section, and placed 1 meter apart in vacuum, would produce between these conductors a force equal to 2×10^{-7} newton per meter of length.

kelvin (K) The *kelvin*, unit of thermodynamic temperature, is the fraction 1/273.16 of the thermodynamic temperature of the triple point of water.

ohm (Ω) The *ohm* is the electric resistance between two points of a conductor when a constant difference of potential of 1 volt, applied between these two points, produces in this conductor a current of 1 ampere, this conductor not being the source of any electromotive force.

coulomb (C) The *coulomb* is the quantity of electricity transported in 1 second by a current of 1 ampere.

candela (cd) The *candela* is the luminous intensity, in a given direction, of a source that emits monochromatic radiation of frequency 540×10^{12} hertz and that has a radiant intensity in that direction of 1/683 watt per steradian.

mole (mol) The *mole* is the amount of substance of a system that contains as many elementary entities as there are carbon atoms in 0.012 kg of carbon 12. The elementary entities must be specified and may be atoms, molecules, ions, electrons, other particles, or specified groups of such particles.

newton (N) The *newton* is that force that gives to a mass of 1 kilogram an acceleration of 1 meter per second per second.

joule (J) The *joule* is the work done when the point of application of 1 newton is displaced a distance of 1 meter in the direction of the force.

watt (W) The *watt* is the power that gives rise to the production of energy at the rate of 1 joule per second.

volt (V) The *volt* is the difference of electric potential between two points of a conducting wire carrying a constant current of 1 ampere, when the power dissipated between these points is equal to 1 watt.

weber (Wb) The *weber* is the magnetic flux that, linking a circuit of one turn, produces in it an electromotive force of 1 volt as it is reduced to zero at a uniform rate in 1 second.

lumen (lm) The *lumen* is the luminous flux emitted in a solid angle of 1 steradian by a uniform point source having an intensity of 1 candela.

farad (F)   The *farad* is the capacitance of a capacitor between the plates of which there appears a difference of potential of 1 volt when it is charged by a quantity of electricity equal to 1 coulomb.

henry (H)   The *henry* is the inductance of a closed circuit in which an electromotive force of 1 volt is produced when the electric current in the circuit varies uniformly at a rate of 1 ampere per second.

radian (rad)   The *radian* is the plane angle between two radii of a circle that cut off on the circumference an arc equal in length to the radius.

steradian (sr)   The *steradian* is the solid angle that, having its vertex in the center of a sphere, cuts off an area of the surface of the sphere equal to that of a square with sides of length to the radius of the sphere.

SI Prefixes   The names of multiples and submultiples of SI units may be formed by application of the prefixes listed in Table 1–1, page 6.

APPENDIX B
USEFUL MATHEMATICAL RELATIONS

ALGEBRA

$$a^{-x} = \frac{1}{a^x} \qquad a^{(x+y)} = a^x a^y \qquad a^{(x-y)} = \frac{a^x}{a^y}$$

Logarithms: If $\log a = x$, then $a = 10^x$. $\log a + \log b = \log (ab)$ $\log a - \log b = \log (a/b)$ $\log (a^n) = n \log a$

If $\ln a = x$, then $a = e^x$. $\ln a + \ln b = \ln (ab)$ $\ln a - \ln b = \ln (a/b)$ $\ln (a^n) = n \ln a$

Quadratic formula: If $ax^2 + bx + c = 0$, $x = \dfrac{-b \pm \sqrt{b^2 - 4ac}}{2a}$.

BINOMIAL THEOREM

$$(a + b)^n = a^n + na^{n-1}b + \frac{n(n-1)a^{n-2}b^2}{2!} + \frac{n(n-1)(n-2)a^{n-3}b^3}{3!} + \cdots$$

TRIGONOMETRY

In the right triangle ABC, $x^2 + y^2 = r^2$.

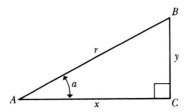

Definitions of the trigonometric functions: $\sin a = y/r$ $\cos a = x/r$ $\tan a = y/x$

Identities: $\sin^2 a + \cos^2 a = 1$ $\tan a = \dfrac{\sin a}{\cos a}$

$\sin 2a = 2 \sin a \cos a$ $\cos 2a = \cos^2 a - \sin^2 a = 2 \cos^2 a - 1$

$\sin \tfrac{1}{2}a = \sqrt{\dfrac{1 - \cos a}{2}}$ $\cos \tfrac{1}{2}a = \sqrt{\dfrac{1 + \cos a}{2}}$

$\sin (-a) = -\sin a$ $\sin (a \pm b) = \sin a \cos b \pm \cos a \sin b$

$\cos (-a) = \cos a$ $\cos (a \pm b) = \cos a \cos b \mp \sin a \sin b$

$\sin (a \pm \pi/2) = \pm\cos a$ $\sin a + \sin b = 2 \sin \tfrac{1}{2}(a + b) \cos \tfrac{1}{2}(a - b)$

$\cos (a \pm \pi/2) = \mp\sin a$ $\cos a + \cos b = 2 \cos \tfrac{1}{2}(a + b) \cos \tfrac{1}{2}(a - b)$

GEOMETRY

Circumference of circle of radius r: $C = 2\pi r$

Area of circle of radius r: $A = \pi r^2$

Volume of sphere of radius r: $V = 4\pi r^3/3$

Surface area of sphere of radius r: $A = 4\pi r^2$

Volume of cylinder of radius r and height h: $V = \pi r^2 h$

CALCULUS

Derivatives:

$$\frac{d}{dx}x^n = nx^{n-1}$$

$$\frac{d}{dx}\sin ax = a\cos ax$$

$$\frac{d}{dx}\cos ax = -a\sin ax$$

$$\frac{d}{dx}e^{ax} = ae^{ax}$$

$$\frac{d}{dx}\ln ax = \frac{1}{x}$$

Integrals:

$$\int x^n\,dx = \frac{x^{n+1}}{n+1}$$

$$\int \frac{dx}{x} = \ln x$$

$$\int \sin ax\,dx = -\frac{1}{a}\cos ax$$

$$\int \cos ax\,dx = \frac{1}{a}\sin ax$$

$$\int e^{ax}\,dx = \frac{1}{a}e^{ax}$$

$$\int \frac{dx}{\sqrt{a^2 - x^2}} = \arcsin\frac{x}{a}$$

$$\int \frac{dx}{\sqrt{x^2 + a^2}} = \ln(x + \sqrt{x^2 + a^2})$$

$$\int \frac{dx}{x^2 + a^2} = \frac{1}{a}\arctan\frac{x}{a}$$

$$\int \frac{dx}{(x^2 + a^2)^{3/2}} = \frac{1}{a^2}\frac{x}{\sqrt{x^2 + a^2}}$$

$$\int \frac{x\,dx}{(x^2 + a^2)^{3/2}} = -\frac{1}{\sqrt{x^2 + a^2}}$$

Power series (convergent for range of x shown):

$$\sin x = x - \frac{x^3}{3!} + \frac{x^5}{5!} - \frac{x^7}{7!} + \cdots \quad (\text{all } x)$$

$$\cos x = 1 - \frac{x^2}{2!} + \frac{x^4}{4!} - \frac{x^6}{6!} + \cdots \quad (\text{all } x)$$

$$\tan x = x + \frac{x^3}{3} + \frac{2x^5}{15} + \frac{17x^7}{315} + \cdots \quad (|x| < \pi/2)$$

$$e^x = 1 + x + \frac{x^2}{2!} + \frac{x^3}{3!} + \cdots \quad (\text{all } x)$$

$$\ln(1 + x) = x - \frac{x^2}{2} + \frac{x^3}{3} - \frac{x^4}{4} + \cdots \quad (|x| < 1)$$

APPENDIX C

THE GREEK ALPHABET

Name	Capital	Lowercase	Name	Capital	Lowercase
Alpha	A	α	Nu	N	ν
Beta	B	β	Xi	Ξ	ξ
Gamma	Γ	γ	Omicron	O	o
Delta	Δ	δ	Pi	Π	π
Epsilon	E	ϵ	Rho	P	ρ
Zeta	Z	ζ	Sigma	Σ	σ
Eta	H	η	Tau	T	τ
Theta	Θ	θ	Upsilon	Y	υ
Iota	I	ι	Phi	Φ	ϕ
Kappa	K	κ	Chi	X	χ
Lambda	Λ	λ	Psi	Ψ	ψ
Mu	M	μ	Omega	Ω	ω

APPENDIX D

PERIODIC TABLE OF THE ELEMENTS

Period	IA	IIA	IIIB	IVB	VB	VIB	VIIB	VIIIB	VIIIB	IB	IIB	IIIA	IVA	VA	VIA	VIIA	Noble gases
1	1 **H** 1.008																2 **He** 4.003
2	3 **Li** 6.941	4 **Be** 9.012										5 **B** 10.811	6 **C** 12.011	7 **N** 14.007	8 **O** 15.999	9 **F** 18.998	10 **Ne** 20.179
3	11 **Na** 22.990	12 **Mg** 24.305										13 **Al** 26.982	14 **Si** 28.086	15 **P** 30.974	16 **S** 32.064	17 **Cl** 35.453	18 **Ar** 39.948
4	19 **K** 39.098	20 **Ca** 40.08	21 **Sc** 44.956	22 **Ti** 47.90	23 **V** 50.942	24 **Cr** 51.996	25 **Mn** 54.938	26 **Fe** 55.847	27 **Co** 58.933	28 **Ni** 58.70	29 **Cu** 63.546	30 **Zn** 65.38	31 **Ga** 69.72	32 **Ge** 72.59	33 **As** 74.922	34 **Se** 78.96	35 **Br** 79.904	36 **Kr** 83.80
5	37 **Rb** 85.468	38 **Sr** 87.62	39 **Y** 88.906	40 **Zr** 91.22	41 **Nb** 92.906	42 **Mo** 95.94	43 **Tc** (99)	44 **Ru** 101.07	45 **Rh** 102.905	46 **Pd** 106.4	47 **Ag** 107.868	48 **Cd** 112.41	49 **In** 114.82	50 **Sn** 118.69	51 **Sb** 121.75	52 **Te** 127.60	53 **I** 126.905	54 **Xe** 131.30
6	55 **Cs** 132.905	56 **Ba** 137.33	57 **La** 138.905	72 **Hf** 178.49	73 **Ta** 180.948	74 **W** 183.85	75 **Re** 186.2	76 **Os** 190.2	77 **Ir** 192.22	78 **Pt** 195.09	79 **Au** 196.966	80 **Hg** 200.59	81 **Tl** 204.37	82 **Pb** 207.19	83 **Bi** 208.2	84 **Po** (210)	85 **At** (210)	86 **Rn** (222)
7	87 **Fr** (223)	88 **Ra** (226)	89 **Ac** (227)	104 **Rf(?)** (261)	105 **Ha(?)** (262)	106 (257)	107 (260)											

58 **Ce** 140.12	59 **Pr** 140.907	60 **Nd** 144.24	61 **Pm** (145)	62 **Sm** 150.35	63 **Eu** 151.96	64 **Gd** 157.25	65 **Tb** 158.925	66 **Dy** 162.50	67 **Ho** 164.930	68 **Er** 167.26	69 **Tm** 168.934	70 **Yb** 173.04	71 **Lu** 174.96
90 **Th** (232)	91 **Pa** (231)	92 **U** (238)	93 **Np** (239)	94 **Pu** (239)	95 **Am** (240)	96 **Cm** (242)	97 **Bk** (245)	98 **Cf** (246)	99 **Es** (247)	100 **Fm** (249)	101 **Md** (256)	102 **No** (254)	103 **Lr** (257)

For each element the average atomic mass of the mixture of isotopes occurring in nature is shown. For elements having no stable isotope, the approximate atomic mass of the most common isotope is shown in parentheses.

APPENDIX E
UNIT CONVERSION FACTORS

LENGTH

1 m = 100 cm = 1000 mm = 10^6 μm = 10^9 nm
1 km = 1000 m = 0.6214 mi
1 m = 3.281 ft = 39.37 in.
1 cm = 0.3937 in.
1 in. = 2.540 cm
1 ft = 30.48 cm
1 yd = 91.44 cm
1 mi = 5280 ft = 1.609 km
1 Å = 10^{-10} m = 10^{-8} cm = 10^{-1} nm
1 nautical mile = 6080 ft
1 light year = 9.461×10^{15} m

AREA

1 cm^2 = 0.155 in^2
1 m^2 = 10^4 cm^2 = 10.76 ft^2
1 in^2 = 6.452 cm^2
1 ft^2 = 144 in^2 = 0.0929 m^2

VOLUME

1 liter = 1000 cm^3 = $10^{-3}m^3$ = 0.03531 ft^3 = 61.02 in^3
1 ft^3 = 0.02832 m^3 = 28.32 liters = 7.477 gallons
1 gallon = 3.788 liters

TIME

1 min = 60 s
1 hr = 3600 s
1 da = 86,400 s
1 yr = 365.24 da = 3.156×10^7 s

ANGLE

1 rad = 57.30° = 180°/π
1° = 0.01745 rad = π/180 rad
1 revolution = 360° = 2π rad
1 rev·min^{-1} (rpm) = 0.1047 rad·s^{-1}

SPEED

1 m·s^{-1} = 3.281 ft·s^{-1}
1 ft·s^{-1} = 0.3048 m·s^{-1}
1 mi·min^{-1} = 60 mi·hr^{-1} = 88 ft·s^{-1}
1 km·hr^{-1} = 0.2778 m·s^{-1} = 0.6214 mi·hr^{-1}
1 mi·hr^{-1} = 1.466 ft·s^{-1} = 0.4470 m·s^{-1} = 1.609 km·hr^{-1}
1 furlong·$fortnight^{-1}$ = 1.662×10^{-4} m·s^{-1}

ACCELERATION

1 m·s^{-2} = 100 cm·s^{-2} = 3.281 ft·s^{-2}
1 cm·s^{-2} = 0.01 m·s^{-2} = 0.03281 ft·s^{-2}
1 ft·s^{-2} = 0.3048 m·s^{-2} = 30.48 cm·s^{-2}
1 mi·hr^{-1}·s^{-1} = 1.467 ft·s^{-2}

MASS

1 kg = 10^3 g = 0.0685 slug
1 g = 6.85×10^{-5} slug
1 slug = 14.59 kg
1 u = 1.661×10^{-27} kg
1 kg has a weight of 2.205 lb when g = 9.80 m·s^{-2}

FORCE

1 N = 10^5 dyn = 0.2248 lb
1 lb = 4.448 N = 4.448×10^5 dyn

PRESSURE

1 Pa = 1 N·m^{-2} = 1.451×10^{-4} lb·in^{-2} = 0.209 lb·ft^{-2}
1 bar = 10^5 Pa
1 lb·in^{-2} = 6891 Pa
1 lb·ft^{-2} = 47.85 Pa
1 atm = 1.013×10^5 Pa = 1.013 bar
        = 14.7 lb·in^{-2} = 2117 lb·ft^{-2}
1 mm Hg = 1 torr = 133.3 Pa

ENERGY

1 J = 10^7 ergs = 0.239 cal
1 cal = 4.186 J (based on 15° calorie)
1 ft·lb = 1.356 J
1 Btu = 1055 J = 252 cal = 778 ft·lb
1 eV = 1.602×10^{-19} J
1 kWh = 3.600×10^6 J

MASS–ENERGY EQUIVALENCE

1 kg ↔ 8.988×10^{16} J
1 u ↔ 931.5 MeV
1 eV ↔ 1.073×10^{-9} u

POWER

1 W = 1 J·s^{-1}
1 hp = 746 W = 550 ft·lb·s^{-1}
1 Btu·hr^{-1} = 0.293 W

APPENDIX F
NUMERICAL CONSTANTS

FUNDAMENTAL PHYSICAL CONSTANTS

Name	Symbol	Value
Speed of light	c	2.9979×10^8 m·s^{-1}
Charge of electron	e	1.602×10^{-19} C
Gravitational constant	G	6.673×10^{-11} N·m^2·kg^{-2}
Planck's constant	h	6.626×10^{-34} J·s
Boltzmann's constant	k	1.381×10^{-23} J·K^{-1}
Avogadro's number	N_0	6.022×10^{23} molecules·mol^{-1}
Gas constant	R	8.314 J·mol^{-1}·K^{-1}
Mass of electron	m_e	9.110×10^{-31} kg
Mass of neutron	m_n	1.675×10^{-27} kg
Mass of proton	m_p	1.673×10^{-27} kg
Permittivity of free space	ϵ_0	8.854×10^{-12} C^2·N^{-1}·m^{-2}
	$1/4\pi\epsilon_0$	8.987×10^9 N·m^2·C^{-2}
Permeability of free space	μ_0	$4\pi \times 10^{-7}$ Wb·A^{-1}·m^{-1}

OTHER USEFUL CONSTANTS

Name	Symbol	Value
Mechanical equivalent of heat		4.186 J·cal^{-1} (15° calorie)
Standard atmospheric pressure	1 atm	1.013×10^5 Pa
Absolute zero	0 K	-273.15°C
Electronvolt	1 eV	1.602×10^{-19} J
Atomic mass unit	1 u	1.661×10^{-27} kg
Electron rest energy	mc^2	0.511 MeV
Energy equivalent of 1 u	Mc^2	931.5 MeV
Volume of ideal gas (0°C and 1 atm)	V	22.4 liter·mol^{-1}
Acceleration due to gravity (sea level, at equator)	g	9.78049 m·s^{-2}

ASTRONOMICAL DATA

Body	Mass, kg	Radius, m	Orbit radius, m	Orbit period
Sun	1.99×10^{30}	6.95×10^8	—	—
Moon	7.36×10^{22}	1.74×10^6	0.38×10^9	27.3 d
Mercury	3.28×10^{23}	2.57×10^6	5.8×10^{10}	88.0 d
Venus	4.82×10^{24}	6.31×10^6	1.08×10^{11}	224.7 d
Earth	5.98×10^{24}	6.38×10^6	1.49×10^{11}	365.3 d
Mars	6.34×10^{23}	3.43×10^6	2.28×10^{11}	687.0 d
Jupiter	1.88×10^{27}	7.18×10^7	7.78×10^{11}	11.86 y
Saturn	5.63×10^{26}	6.03×10^7	1.43×10^{12}	29.46 y
Uranus	8.61×10^{25}	2.67×10^7	2.87×10^{12}	84.02 y
Neptune	9.99×10^{25}	2.48×10^7	4.49×10^{12}	164.8 y
Pluto	5×10^{23}	4×10^5	5.90×10^{12}	247.7 y

ANSWERS TO ODD-NUMBERED PROBLEMS

CHAPTER 1

1–1 1.61 km

1–3 a) 0.403 mi·s^{-1} b) 648 m·s^{-1}

1–5 120 in^3

1–7 40.0 mi·gal^{-1}

1–9 a) 7.04×10^{-10} s   b) 5.11×10^{12} cycles·hr^{-1}
c) 4.48×10^{26} cycles   d) 4.60×10^4 s

1–11 0.0038

1–13 5.50×10^3 kg·m^{-3}

1–15 Ten thousand

1–17 No; no

1–19 10^5

1–21 $\$1 \times 10^{15}$

1–23 7000

1–25 a) 5.0 cm, $-53.1°$ b) 13.0 cm, $-112.6°$
c) 3.6 km, $123.7°$

1–27 5.0 km, $36.9°$ E of S

1–29 a) 15.0 m, $53.3°$ b) 25.9 m, $27.7°$ below $-x$-axis

1–31 a) $N_x = -1.0$ cm, $N_y = +1.0$ cm   b) 1.41 cm, $135°$

1–33 a) $A = 3.61$, $B = 2.24$ b) $3\boldsymbol{i} + \boldsymbol{j}$
c) 3.16, $18.4°$ above $+x$-axis   d) $\boldsymbol{i} + 5\boldsymbol{j}$
e) 5.10, $78.7°$ above $+x$-axis

1–35 $\boldsymbol{i} \cdot \boldsymbol{i} = \boldsymbol{j} \cdot \boldsymbol{j} = \boldsymbol{k} \cdot \boldsymbol{k} = 1$
$\boldsymbol{i} \cdot \boldsymbol{j} = \boldsymbol{i} \cdot \boldsymbol{k} = 0$
$\boldsymbol{j} \cdot \boldsymbol{i} = \boldsymbol{j} \cdot \boldsymbol{k} = 0$
$\boldsymbol{k} \cdot \boldsymbol{i} = \boldsymbol{k} \cdot \boldsymbol{j} = 0$

1–37 $130°$

1–39 $\boldsymbol{i} \times \boldsymbol{i} = \boldsymbol{j} \times \boldsymbol{j} = \boldsymbol{k} \times \boldsymbol{k} = 0$
$\boldsymbol{i} \times \boldsymbol{j} = \boldsymbol{k}$ $\boldsymbol{j} \times \boldsymbol{i} = -\boldsymbol{k}$ $\boldsymbol{k} \times \boldsymbol{i} = \boldsymbol{j}$
$\boldsymbol{i} \times \boldsymbol{k} = -\boldsymbol{j}$ $\boldsymbol{j} \times \boldsymbol{k} = \boldsymbol{i}$ $\boldsymbol{k} \times \boldsymbol{j} = -\boldsymbol{i}$

1–41 a) $\boldsymbol{A} \cdot \boldsymbol{B} = 0$, $\boldsymbol{A} \times \boldsymbol{B} = 29\,\boldsymbol{k}$ b) $\boldsymbol{A} \cdot \boldsymbol{B} = 58$,
$\boldsymbol{A} \times \boldsymbol{B} = 0$

1–45 70.0 m, $26.3°$ E of N

1–47 $A = \sqrt{(x_2 - x_1)^2 + (y_2 - y_1)^2}$;
$\theta = \arctan\,(y_2 - y_1/x_2 - x_1)$

1–49 b) $\theta = 120°$ or $240°$ c) $\sqrt{A^2 + B^2 - 2AB \cos \theta}$
d) $\theta = 60°$ or $300°$

1–51 14

1–53 $90°$

1–57 b) $\boldsymbol{C} \cdot (\boldsymbol{A} \times \boldsymbol{B})$

CHAPTER 2

2–1 a) 15.0 mi·hr^{-1} b) 22.0 ft·s^{-1}

2–3 5.0 cm·s^{-1}; -4.0 cm·s^{-1}

2–5 60 cm·s^{-1}

2–7 a) 0; 1.0 m·s^{-2}; 1.5 m·s^{-2}; 2.5 m·s^{-2}; 2.5 m·s^{-2};
2.5 m·s^{-2}; 1.0 m·s^{-2}; 0. No. Yes, from 6 s to 12 s.
b) 2.5 m·s^{-2}; 1.3 m·s^{-2}; 0

2–11 a) 3.95 ft·s^{-2} b) 620 ft

2–13 a) 5.0 m·s^{-1} b) 1.67 m·s^{-2}

2–15 a) 0, 6.25 m·s^{-2}, -11.2 m·s^{-2} b) 100 m, 230 m, 320 m

2–17 a) 30.5 ft   b) 260 ft

2–19 a) 6000 m·s^{-1} b) 0.99 c) 1120 min

2–21 a) $x = 0.20t^3 - 0.010t^4$; $v = 0.60t^2 - 0.040t^3$
b) 20 m·s^{-1}

2–23 a) 19.8 m·s^{-1} b) 4.04 s

2–25 a) 29.6 m·s^{-1} b) 39.6 m   c) 17.2 m·s^{-1}
d) 50.0 m·s^{-2} e) 2.01 s   f) 29.7 m·s^{-1}

2–27 a) 48 ft·s^{-1} b) 36 ft   c) 0 d) 32 ft·s^{-2}, downward   e) 80 ft·s^{-1}

2–29 a) 815 ft·s^{-2} b) 25.5 times longer   c) 1320 ft
d) $a = 20.7\,g$; not consistent if you assume constant acceleration

2–31 a) 50 s   b) 150 s

2–33 a) $B + 3Ct^2$ b) $6Ct$

2–35 a) 9.0 m·s^{-1}; 11.0 m·s^{-1}; 13.0 m·s^{-1}
b) 1.0 m·s^{-2} c) 8.0 m·s^{-1} d) 8.0 s
e) 8.5 m   f) 1.06 s   g) 9.06 m·s^{-1}

2–37 a) 100 m   b) 20 m·s^{-1}

2–39 $x = (2 \text{ m·s}^{-2})t^2 + (0.75 \text{ m·s}^{-4})t^4$,
$a = 4 \text{ m·s}^{-2} + (9 \text{ m·s}^{-4})t^2$

2–41 a) $A\omega \cos \omega t$ b) $-A\omega^2 \sin \omega t$
c) $v = \pm\omega\sqrt{A^2 - x^2}$; max at $x = 0$, min at $x = \pm A$
d) $-\omega^2 x$; max at $x = \pm A$, min at $x = 0$ e) A

f) $\pm\omega A$ g) $\pm\omega^2 A$

2–43 a) 11.3 m   b) 0.51 s   c) $v = 9.97\ m\cdot s^{-1}$,
$a = -9.8\ \text{m}\cdot\text{s}^{-2}$

2–45 a) 4.0 m·s^{-2} b) 6.0 m·s^{-1} c) 4.5 m

2–47 2.54 s after first is dropped

2–49 a) 17.3 s   b) 421 m   c) 28.6 m·s^{-1}

2–53 a) 20 s, 120 m   b) 4.0 m·s^{-1} d) 40 s,
8.0 m·s^{-1} e) No   f) 5.66 m·s^{-1}; 28.3 s, 160 m

2–55 a) 25   b) 91

2–57 a) A b) A/B

CHAPTER 3

3–1 a) $v = -4i + 6tj$; $a = 6j$ b) $v = 18.4$ m·s^{-1},
102.5°; $a = 6.0$ m·s^{-2}, 90°

3–3 a) $t = 0$: $x = 0$, $y = 0$
$t = 1$ s: $x = 20$ m, $y = -4.9$ m
$t = 2$ s: $x = 40$ m, $y = -19.6$ m
$t = 3$ s: $x = 60$ m, $y = -44.1$ m
$t = 4$ s: $x = 80$ m, $y = -78.4$ m
b) $v = 20$ m·s^{-1}i, $a = -9.8$ m·s^{-2}j
c) $v_x = 20.0$ m·s^{-1}, $v_y = -19.6$ m·s^{-1}; yes

3–5 a) 1.22 m   b) 2.00 m   c) $v_x = 4.00$ m·s^{-1},
$v_y = 4.90$ m·s^{-1}

3–7 a) 0.446 ft   b) 4.00 ft

3–9 a) 52.1 m   b) 3.26 s   c) 0.91 s and 5.61 s
d) $t = 0.91$ s: $v_x = 24.1$ m·s^{-1}, $v_y = 23.0$ m·s^{-1}
$t = 5.61$ s: $v_x = 24.1$ m·s^{-1}, $v_y = -23.0$ m·s^{-1}
e) 40.0 m·s^{-1}, 53° below horizontal

3–11 a) 66.5 m   b) 67.7 m·s^{-1} c) 410 m

3–13 0.0337 m·s^{-2}

3–15 a) 6.75 m·s^{-2}, upward   b) 8.38 s

3–17 a) 19.3° W of N   b) 274 km·hr^{-1}

3–19 a) 3.61 m·s^{-1}; 33.7° N of E   b) 333 s
c) 666 m

3–21 a) 11.7 m   b) 8.25 m·s^{-1}, 76.0° below x-axis
c) 4.0 m·s^{-2}, $-y$-direction   d) $t = 0$ e) $t = 0$ s,
$x = 0$, $y = 19$ m; $t = 3.00$ s, $x = 6.0$ m, $y = 1.0$ m
f) 6.08 m at $t = 3.00$ s

3–23 a) $r(t) = [2\ \text{m·s}^{-1}t - 1\ \text{m·s}^{-3}t^3]i + [2.5\ \text{m·s}^{-2}t^2]j$;
$a(t) = -6\ \text{m·s}^{-3}ti + 5\ \text{m·s}^{-2}j$ b) 5.0 m

3–25 a) 140 ft·s^{-1} b) 138 ft

3–27 a) 227 m·s^{-1} b) 908 m   c) $v_x = 182$ m·s^{-1},
$v_y = -185$ m·s^{-1}

3–33 a) 100 km·hr^{-1}, 36.9° W of S   b) 30.0° N of W

3–35 18.7 km·hr^{-1}, 31.0° E of N

3–37 17.8 m·s^{-1}

3–39 a) $[2v_0^2 \cos^2(\phi + \theta)/g \cos \theta][\tan(\phi + \theta) - \tan \theta]$
b) $\pi/4 - \theta/2$

3–41 $\Delta t = 0.5$ s: $a = 11.70$ m·s^{-2}, 54.2° from direction
of v_1
$\Delta t = 0.1$ s: $a = 12.47$ m·s^{-2}, 82.8°
$\Delta t = 0.05$ s: $a = 12.49$ m·s^{-2}, 86.4°
Exact result is $a = 12.50$ m·s^{-2}, 90°

CHAPTER 4

4–1 $F_x = 34.6$ N (to right); $F_y = 20.0$ N (down)

4–3 109 lb, 18.9°

4–5 a) 0.102 kg   b) 78.4 N

4–7 $m = 5.31$ kg; $W = 19.6$ N

4–9 a) Earth's gravity (downward)   b) Gravitational
force on earth, by bottle

4–11 a) 5.0 kg   b) 200 m

4–13 a) $x = 0.025$ m, $v = 0.010$ m·s^{-1} b) $x = 0.200$ m,
$v = 0.020$ m·s^{-1}

4–15 6 Bmt

4–17 14.1 N, 135° counterclockwise from F_1

4–19 46.6 N, 60° clockwise from F_2

4–21 a) 2.2 m·s^{-2} b) 10.3 m·s^{-2}

4–23 a) $F_x = -60$ N·s^{-1}t, $F_y = 64$ N   b) 191 N, 160°
counterclockwise from $+x$-axis

4–25 $r(t) = [(k_1/2m)t^2 + (k_2k_3/120m^2)t^5]i + [(k_3/6m)t^3]j$,
$v(t) = [(k_1/m)t + (k_2k_3/24m^2)t^4]i + [(k_3/2m)t^2]j$

CHAPTER 5

5–3 a) 3.84 lb   b) 2.34 s

5–5 a) 46.5 m   b) 15.1 m·s^{-1} (33.8 mi·hr^{-1})

5–7 Low pressure: 0.0210; high pressure: 0.00449

5–9 a) 10 N   b) 20 N

5–11 a) 1540 N   b) 1.15°

5–13 a) $F_1 = 14.1$ N, $F_2 = 14.1$ N   b) 14.1 N

5–15 a) $w \sin \theta$ b) $2w \sin \theta$

5–17 a) $\mu_k(w_A + w_B)$ b) $\mu_k w_A$

5–19 a) Held back   b) 480 N

5–21 a) $\mu_k w/(\cos \theta - \mu_k \sin \theta)$ b) 1 tan θ

5–25 5.20 m·s^{-2}

5–27 a) 16,200 lb   b) 12,150 lb   c) 8100 lb
d) 4050 lb   e) Same magnitude as above, but
opposite direction

5–29 a) 4.90 m·s^{-2} b) 3.20 m·s^{-2}

5–31 a) 3.27 m·s^{-2}, upward   b) 2.45 m·s^{-2}, downward
c) Yes (free-fall)

5–33 a) 0.98 m·s^{-2} b) 35.3 N

5–35 a) 3.92 m   b) 2.35 N before, 1.96 N after
c) 4.70 N before

5–37 $T = w$ each chain; $F = w/2$

5–39 $w \tan \theta$

5–41 $\mu_s/(\mu_s + 1)$

5–43 a) 20 N   b) 30 N

5–45 a) 10.8 N   b) 6.47 N

5–47 0.251

5–49 a) 100 N, toward front of truck   b) 78.4 N,
toward rear of truck

5–51 a) Yes   b) 1.58 m·s^{-2}, toward front of truck

5–53 a) 10.7 kg   b) 23.6 N, 83.2 N

5–55 $a_1 = 2m_2g/(4m_1 + m_2)$, $a_2 = m_2g/(4m_1 + m_2)$

5–57 a) 2.70 m·s^{-2} b) 112.5 N   c) 87.5 N

5–59 2.43 m above the floor

5–61 $a = g/\mu_s$

5–63 a) $\mu_k w/(\cos \phi + \mu_k \sin \phi)$ b) $\phi = 0°$, $P = 160$ N;
$\phi = 10°$, $P = 152$ N; $\phi = 20°$, $P = 149$ N; $\phi = 30°$,
$P = 150$ N; $\phi = 40°$, $P = 156$ N; $\phi = 50°$,
$P = 169$ N; $\phi = 60°$, $P = 189$ N; $\phi = 70°$, $P =$
223 N; $\phi = 80°$, $P = 282$ N; $\phi = 90°$, $P = 400$ N
c) $\phi = \tan^{-1}(\mu_k)$; 21.8°

5–65 101 m

5–67 c) $v = 50$ m·s^{-1}, $m = 0.25$ kg, $T = 0.01$ s gives
$F_0 = 940$ N

CHAPTER 6

6–1 22.4 m·s^{-1}

6–3 $7.97°$

6–5 1.34 rev·min^{-1}

6–7 3.13 m·s^{-1}

6–9   a) 472 m   b) 6270 N

6–11 2.16

6–13 6.18×10^{24} kg

6–15 3.71×10^{-11} m·s^{-2}

6–17 $g_x = -10$ m·s^{-2}, $g_y = +40$ m·s^{-2}

6–19 7.35×10^3 m·s^{-1}

6–21   a) 97.1 min   b) 8.14 m·s^{-2}

6–23   a) 50 N   b) 46 N

6–25   a) 15.5 m·s^{-1} b) 4.95 m·s^{-1}

6–27   b) 0.331 c) No

6–29   a) $51.6°$ b) No   c) Bead rides at the bottom of the hoop ($\theta = 0°$)

6–31   a) $0.403w$ b) 4.49 s   c) $2.00w$ d) Hits ground 10 m from center of wheel

6–33 2.58×10^8 m

6–35   a) $g_x = g_y = 4.55 \times 10^{-10}$ m·s^{-2}
b) 6.44×10^{-12} N, $45°$ above x-axis

6–37   a) $m = 2.88 \times 10^{15}$ kg, $g = 0.00768$ m·s^{-2}
b) 6.20 m·s^{-1}; yes

6–39 1.41 hr

6–41 $(2GmM/a^2)[1 - x/\sqrt{x^2 + a^2}$, attractive and along line connecting point mass and center of the disk

6–43 $\tau_{\min} = \{[(\tan\theta - \mu_s)/(1 + \mu_s \tan\theta)]4\pi^2 h \tan\theta/g\}^{1/2}$;
$\tau_{\max} = \{[(\tan\theta + \mu_s)/(1 - \mu_s \tan\theta)]4\pi^2 h \tan\theta/g\}^{1/2}$

CHAPTER 7

7–1   a) 4.0 J   b) -0.8 J

7–3   a) 270 ft·lb   b) -165 ft·lb   c) 0
d) 105 ft·lb

7–5   a) 25 N; 50 N   b) 1.25 J; 5.00 J

7–7   a) -6 N   b) -48 N   c) -22.5 J

7–9 250 J

7–11 14.1 m·s^{-1}

7–13   a) $v_0^2/2\mu_k g$ b) 315 ft

7–15   a) 1.35×10^4 m·s^{-1} b) 5.45×10^{-8} m

7–17 2.98×10^6 J

7–19   a) 261 N   b) 784 J

7–21   a) 96.7 ft·s^{-1} b) 96.7 ft·s^{-1}

7–23 -0.230 J

7–25   a) 2400 J   b) -470 J   c) 1415 J   d) 515 J

7–27 7.30 m·s^{-1}; just barely

7–29   a) 60 J; 15 J   b) 14.4 J

7–31 0.100 m

7–33 2.04 m

7–35 0.204

7–37 0.927 hp

7–39   a) 8.23×10^3 J   b) 412 W

7–41 1200 m^3·s^{-1}

7–43 3000 N

7–45   a) 814 W (1.09 hp)   b) 565 W (0.757 hp)
c) 83.6 W (0.112 hp)

7–47 5.42 m·s^{-1}

7–49 $48.2°$

7–51   a) 4.43 m·s^{-1} b) 14.7 N

7–53   a) 0.272 b) -3.6 J

7–55   a) 7.84×10^4 J   b) 1.60×10^5 J
c) 3.97×10^3 W

7–57   a) $nRT \ln(V_2/V_1)$ b) $k(V_2^{1-\gamma} - V_1^{1-\gamma})/(1 - \gamma)$

7–59   a) 24 J   b) 0   c) 16 J

7–61 $h/R_E = 0.01$; 63.8 km

7–63 $v = \sqrt{2gR_E[1 - R_E/(R_E + h)]}$

7–65   a) $40x^2 + 5x^3$ b) 5.87 m·s^{-1}

7–67 334 J

7–69   a) 3.87 m·s^{-1} b) 0.10 m

7–71   a) $U(x_0) = 0$ b) $v(x) = \pm\sqrt{(2\alpha/mx_0 x)(1 - x_0/x)}$
c) $x = 2x_0$; $v_{\max} = \pm\sqrt{\alpha/2mx_0^2}$ d) 0
e) $v(x) = \pm\sqrt{(2\alpha/9mx^2 x_0^2)(9xx_0 - 2x^2 - 9x_0^2)}$
f) released at x_0: $x_{\min} = x_0$, $x_{\max} \to \infty$
released at x_1: $x_{\min} = 3x_0/2$, $x_{\max} = 3x_0$

CHAPTER 8

8–1   a) 2.00×10^5 kg·ms^{-1} b) 40.0 m·s^{-1}
c) 28.3 m·s^{-1}

8–3   a) 18.5 m·s^{-1}, to the right   b) 11.5 m·s^{-1}, to the left

8–5 2500 N; no

8–7   a) $At_2 + Bt_2^3/3$ b) $(t_2/m)(A + t_2^2/3)$

8–9   a) 0.20 m·s^{-1} b) 4780 J

8–11   a) $v_A = 22.0$ m·s^{-1}, $v_B = 15.5$ m·s^{-1} b) 19.6%

8–13   a) 0.667 m·s^{-1} b) 1.33×10^4 J   c) 1.0 m·s^{-1}

8–15 24.0 m·s^{-1}, $33.7°$ S of E

8–17 595 m·s^{-1}

8–21 $v_A = 26.0$ m·s^{-1}; $v_B = 15.0$ m·s^{-1} at $60°$ from initial direction of A and $90°$ from final direction of A

8–23   a) 1.0 m·s^{-1} b) 0.75 J

8–25 0.898 m·s^{-1}

8–27 4.62×10^6 m from center of earth

8–29   a) 250 N   b) Yes

8–31   a) $5.6 \times 10^{-53}\%$ b) 28.7%

8–33   b) 3.0×10^{-3} s   c) 0.60 N·s   d) 2.0×10^{-3} kg

8–35 $v = 0.417$ m·s$^{-1}\boldsymbol{i} - 0.625$ m·s$^{-1}\boldsymbol{j}$

8–37   Station wagon: 8.90 m·s^{-1}; truck: 18.4 m·s^{-1}

8–39   a) 0.163 b) 240 J   c) 0.320 J

8–41 0.300 m

8–43 20

8–45   a) 12.5 m·s^{-1}, $36.9°$ from initial direction of bullet and hence $126.9°$ from final direction of bullet
b) No

8–47   a) 8.95% b) 32.6%

8–49 1.30×10^6 m·s^{-1} and 5.54×10^5 m·s^{-1};
$v_r = 1.53 v_{Ba}$

8–51   b) $\frac{1}{2} MV^2$

8–53   a) $1.39v_{Kr}$ b) $1.18v_r$ c) $2.57v_r$
d) 3.12 km·s^{-1}

8–55   a) $F = -\lambda yg$, where y is the length of rope hanging over edge; $W = \lambda gl^2/18$ b) same W as in (a)

8–57   a) $v_{A1}m_A/(m_A + m_B)$ b) yes
c) $u_{A1} = v_{A1}m_B/(m_A + m_B)$, $u_{B1} = -v_{A1}m_A/(m_A + m_B)$,
$P = 0$ d) $p_{B2} = -p_{B1}$, $p_{A2} = -p_{A1}$; $u_{A2} = -u_{A1}$,
$u_{B2} = -u_{B1}$ e) $v_{A2} = -(2/3)$ m·s^{-1},
$v_{B2} = +(4/3)$ m·s^{-1}

8–59   b) 0.312 s^{-1} e) 8.55 g   f) 166 m·s^{-2}

CHAPTER 9

9–1 a) 1.5 rad   b) 1.57 rad; $90°$ c) 1.2 m

9–3 a) $2.0 \text{ rad·s}^{-1} + (0.15 \text{ rad·s}^{-3})t^2$ b) 2.0 rad·s^{-1}
    c) 5.75 rad·s^{-1}; 3.25 rad·s^{-1}

9–5 $2b + 6ct$

9–7 a) -2.00 rev·s^{-2}; 58.3 rev   b) 3.33 s

9–9 7.5 s

9–11 a) 5.89 m·s^{-1} b) 229 rev·min^{-1}

9–13 a) 50.0 m·s^{-2} b) 5.0 m·s^{-1}; 50.0 m·s^{-2}

9–15 a) 0.15 m·s^{-2}; 0; 0.15 m·s^{-2} b) 0.15 m·s^{-2};
    0.628 m·s^{-2}; 0.646 m·s^{-2} c) 0.15 m·s^{-2};
    1.26 m·s^{-2}; 1.27 m·s^{-2}

9–17 a) 1.50 kg·m^2 b) 0.75 kg·m^2

9–19 0.138 kg·m^2

9–21 0.80 kg·m^2

9–23 a) 0.18 kg·m^2 b) 0.32 kg·m^2 c) 0.50 kg·m^2

9–25 $2Ma^2/3$

9–27 a) 80.0 ft·lb, counterclockwise   b) 69.3 ft·lb,
    counterclockwise   c) 40.0 ft·lb, counterclockwise
    d) 34.6 ft·lb clockwise   e) 0   f) 0

9–29 $(-0.10 \text{ N·m})\boldsymbol{k}$

9–31 0.589

9–33 a) 65.3 N   b) 16.2 m·s^{-1} c) 2.48 s
    d) 261 N

9–35 a) 19.6 N·m   b) 3.08×10^3 J   c) 3.08×10^3 J

9–37 a) 5.97×10^3 N·m   b) 2.39×10^4 N
    c) 62.8 m·s^{-1}

9–39 a) 5.88 N   b) 0.452 s   c) 27.7 rad·s^{-1}

9–41 $6.06 \times 10^{-8} \text{ kg·m}^2\text{·s}^{-1}$

9–43 $116 \text{ kg·m}^2\text{·s}^{-1}$

9–45 a) $43.0 \text{ kg·m}^2\text{·s}^{-1}$ b) 7.96 rad·s^{-1} c) Before:
    108 J; after: 171 J. Increase in K equals work done
    by man.

9–47 a) 12 rad·s^{-1} b) 0.027 J   c) 0.027 J

9–49 218 rev·s^{-1}

9–51 0.149 rad·s^{-1}

9–53 a) 99.1 min   b) 1.65×10^5 N·m

9–55 $0.371 MR^2$

9–57 a) 247 rev·min^{-1} b) 800 W

9–59 a) 5.42 rad·s^{-1} b) 5.42 m·s^{-1}

9–61 $v = \sqrt{2gd(m_B - \mu_k m_A)/(m_A + m_B + I/R^2)}$

9–63 a) 17.4 kg·m^2 b) -1.82 N·m   c) 91.6 rev

9–65 0.597 s

9–67 a) 10 N at block on table; 39 N at hanging block
    b) 0.145 kg·m^2

9–69 a) 1.97 m·s^{-2} b) 9.85 N

9–71 $a = F/(2M)$; $\mathcal{F} = F/2$

9–73 0.299 m

9–75 3000 J

9–77 $2.94 \text{ rad·s}^{-1}(28.1 \text{ rev·min}^{-1})$

9–79 $3MR^2/5$ (larger than $MR^2/2$)

9–81 a) $mv_1^2 r_1^2/r^3$ b) $1/2 \, mv_1^2[(r_1/r_2)^2 - 1]$
    c) $\Delta K = 1/2 \, mv_1^2[(r_1/r_2)^2 - 1]$; same

9–83 a) $a = \mu_k g$, $\alpha = -2\mu_k g/R$ b) $R^2 \omega_0^2/18 \mu_k g$
    c) $-\frac{1}{6} MR^2 \omega_0^2$

CHAPTER 10

10–1 Between the balls, 0.1 m from the 3 kg ball

10–3 $x = 0.0667$ m, $y = 0.0667$ m

10–5 1000 N; 0.8 m from the end where the 600 N force
    is applied

10–7 b) 2.50 m beyond pont B c) 2.88 m

10–9 a) $T_1 = 1.73w$, $T_2 = w$; $F = 2w$ at $30°$ above the
    horizontal
    b) $T_1 = 2.73w$, $T_2 = w$; $F = 3.35w$ at $45°$ above the
    horizontal

10–11 Top hinge: $F_{\text{horiz}} = -66.7$ N, $F_{\text{vert}} = 100$ N;
    Bottom hinge: $F_{\text{horiz}} = 66.7$ N, $F_{\text{vert}} = 100$ N

10–13 a) 2.71 N   b) 2.71 N   c) 2.71 N

10–15 a) 2.31 m   b) 213 N

10–17 $F_{\text{horiz}} = 60$ N, $F_{\text{vert}} = 80$ N

10–19 a) 60 N   b) $53.1°$

10–21 a) 900 N   b) 727 N

10–23 a) 214 N   b) 186 N   c) 293 N

10–25 b) Yes; slides at $21.8°$, tips at $26.6°$ c) Tips first;
    tips at $26.6°$, slides at $31.0°$

10–27 a) A: 100 N; B: 700 N   b) 2 m

10–29 a) 390 N   b) 11,700 N   c) Slide:
    $0.3w/[\sin \theta - 0.3 \cos \theta]$; tip:
    $0.9w/[1.8 \sin \theta + 0.1 \cos \theta]$; $\theta = 39.8°$

10–31 $X = 0$, $Y = 0.424a$

CHAPTER 11

11–1 $\omega = 31.4 \text{ rad·s}^{-1}$; $\tau = 0.2$ s

11–3 $\tau = 1.26$ s, $f = 0.796$ Hz, $\omega = 5.00 \text{ rad·s}^{-1}$

11–5 $\tau = 0.333$ s, $\omega = 18.8 \text{ rad·s}^{-1}$, $m = 0.563$ kg

11–7 a) 94.7 m·s^{-2}, 3.77 m·s^{-1} b) -56.8 m·s^{-2},
    $\pm 3.02 \text{ m·s}^{-1}$ c) 0.0369 s

11–9 $A = 1.22$ m, $\theta_0 = 1.41$ rad, $E = 74.0$ J,
    $x(t) = (1.22 \text{ m}) \cos[(5 \text{ s}^{-1})t + 1.41]$

11–11 a) 98.0 N·m^{-1} b) 0.898 s   c) 1.27 s

11–13 0.0620 m

11–15 a) 39.2 J, 0, 0, 39.2 J   b) 9.8 J, 9.8 J, 19.6 J,
    39.2 J   c) 0, 0, 39.2 J, 39.2 J

11–17 0.190 m

11–19 a) $1.80 \times 10^{-7} \text{ kg·m}^2$ b) $2.84 \times 10^{-5} \text{ N·m·rad}^{-1}$

11–21 0.248 m

11–23 a) $7 \, mL^2/48$ b) $2\pi\sqrt{7L/12g}$

11–25 a) 4.24 Hz   b) 21.9 kg·s^{-1}

11–27 a) $7.11 \times 10^3 \text{ m·s}^{-2}$ b) 3.55×10^3 N
    c) 18.8 m·s^{-1}

11–29 a) 2.09 s   b) 0.0916 m   c) 0.0918

11–31 1.67 s

11–33 a) $\pm 0.314 \text{ m·s}^{-1}$ b) 0.493 m·s^{-2}, downward
    c) 0.333 s   d) 0.993 m

11–35 0.719 m

11–37 a) $n(t) = m[g - (2\pi f)^2 A \cos(2\pi f t + \theta_0)]$
    b) $(2\pi f_b)^2 A$

11–39 a) $r = 0$, $\tau \to \infty$; $r = R/4$, $\tau = 1.04$ s; $r = R/2$,
    $\tau = 0.815$ s;
    $r = 3R/4$, $\tau = 0.827$ s; $r = R$, $\tau = 0.851$ s;
    $r = 3R/2$, $\tau = 0.941$ s
    c) $r/R = 0.707$; $\tau = 0.827$ s

11–41 a) 3.97 m   b) Stick of length 0.5 m, pivoted
    0.525 cm above its center

11–43 a) $l_1 = 0.35$ m, $l_2 = 0.25$ m   b) 0.993 s

11–45 a) $-2\pi\sqrt{L/g}(\Delta g/2g)$ b) $-\frac{1}{2}\Delta g/g$
    c) 9.7977 m·s^{-2}

11–47 a) $\sqrt{(1 + 3d^2/l^2)/(1 + 2d/l)}$, where $l = 1$ m is length of meterstick   b) $2l/3$

CHAPTER 12

12–1 4.80×10^{11} Pa
12–3 1.0×10^8 N·m^{-2}; 5.0×10^{-4}; 2.5×10^{-3} m; $-9.5 \times 10^{-3}\%$
12–5 a) Upper wire: 1.84×10^{-3}; lower wire: 1.23×10^{-3} b) Upper wire: 9.20×10^{-4} m; lower wire: 6.15×10^{-4} m
12–7 1.38 m
12–9 $B = 4.0 \times 10^9$ Pa; $k = 2.5 \times 10^{-10}$ Pa^{-1}
12–11 4.71×10^4 N
12–13 13.1 m·s^{-2}
12–15 1.66×10^{-3} m
12–17 a) 0.700 m to right of A
b) 0.600 m to right of A
12–19 1.22×10^6 Pa
12–21 a) 1.82 m   b) Steel: 3.0×10^8 Pa; copper: 1.5×10^8 Pa   c) Steel: 1.5×10^{-3}; copper: 1.36×10^{-3}
12–23 a) $F \cos^2 \theta/A$ b) $F \sin 2\theta/2A$ c) $0°$ d) $45°$
12–25 $(A^2 x - k_0 V_0 F)/F V_0$
12–27 a) 1.23×10^{-3} m   b) 20.0 N   c) 0.0245 J
d) 0.0050 J   e) 0.0295 J   f) 0.0295 J

CHAPTER 13

13–1 a) 889 kg·m^{-3} b) Yes
13–3 a) 6.06×10^6 Pa   b) 1.07×10^5 N
13–5 a) 882 Pa   b) 3.33×10^3 Pa
13–7 a) 1.077×10^5 Pa   b) 1.037×10^5 Pa
c) 1.037×10^5 Pa   d) 6.66×10^3 Pa
e) 5.0 cm of Hg   f) 68 cm of water
13–9 2.04×10^{-4} m^3; 6.00×10^3 kg·m^{-3}
13–11 1.00 m^3
13–13 a) 118 Pa   b) 784 Pa   c) 0.680 kg
13–15 4.00 Pa
13–17 a) 6.37 m·s^{-1} b) 0.259 m
13–19 a) 88.9 m·s^{-1} b) 8.14×10^3 N
13–21 5.35×10^{-4} m^2
13–23 1.96×10^5 Pa
13–25 665 N, downward
13–27 a) 2.25 m·s^{-1} b) 0
13–29 a) 0.0533 m^3·s^{-1} b) 1.28×10^4 Pa
c) 0.114 m^3·s^{-1}
13–31 167 m·s^{-1}
13–33 a) 22,400   b) Turbulent
13–35 1.96×10^4 N·m
13–37 a) 1.10×10^8 Pa   b) 1.085 kg·m^{-3}; 5.3% increase
13–39 a) 8.33×10^3 m^3 b) 9260 kg
13–41 8.29×10^6 kg; yes
13–43 7.77×10^{-5} m^3
13–45 0.10087 kg
13–47 25.9 N
13–49 7.02 N·m
13–51 0.10 m
13–53 0.0268 m^3·min^{-1}

13–55 107 m·s^{-1}
13–57 $3h_1$
13–59 a) $r = r_0 \sqrt{v_0}/(v_0^2 + 2gy)^{1/4}$, where r_0 and v_0 are the radius and velocity at the opening and y is the distance the water has fallen   b) 0.765 m
13–61 a) 3.26 mm·s^{-1} b) 0.541 m·s^{-1}
13–63 a) 5.6×10^{-5} m^3·s^{-1} b) 0.56 m·s^{-1}; 1.12 m·s^{-1}; 2.80 m·s^{-1} c) 0.384 m; 0.336 m   d) 0.180 m
e) 0.718 m   f) 1.12 m·s^{-1}, 2.24 m·s^{-1}, 5.60 m·s^{-1} g) Yes ($N_R = 355$)
h) No ($N_R = 22,000$)
13–65 a) 88.2 N
13–67 a) la/g b) $\omega^2 l^2/2g$ c) Yes; yes
13–69 a) $2\gamma \cos \theta/\rho x g$ b) 2.97 cm
13–71 a) $\sqrt{2gh}$ b) $p_1/\rho g - h$, where p_1 is atmospheric pressure

CHAPTER 14

14–1 a) Yes (104°F)   b) 37°   c) -40
14–3 90.18 K = 162.3°R
14–5 -286°C
14–7 79.97 m
14–9 1.5×10^{-5}(C°)$^{-1}$
14–11 2.509 cm
14–13 53.3°C
14–15 5.62×10^{-3} m^3
14–17 a) 1.0×10^{-5}(C°)$^{-1}$ b) 1×10^9 Pa
14–19 7.14×10^7 Pa
14–21 a) 1.44 cm^3 b) -3.89 kg·m^{-3}
14–23 a) 75.6°C   b) -63.2°C
14–25 20 cm; 10 cm
14–27 2.8×10^{-5}(C°)$^{-1}$
14–29 4.8×10^8 Pa
14–31 5.06×10^7 Pa
14–33 a) -0.012% b) 5.18 s   c) 1.93°C
14–35 b) 1.009×10^{-4}(C°)$^{-1}$

CHAPTER 15

15–1 1.88×10^4 J
15–3 a) 8380 J   b) 324°C   c) 42.3°C
15–5 1.05×10^6 J
15–7 3.01×10^3 J·kg^{-1}·(C°)$^{-1}$
15–9 3029 J, 724 cal, 2.87 Btu
15–11 20%
15–13 357 m·s^{-1}
15–15 3.53×10^3 W, 1.20×10^4 Btu·hr^{-1}
15–17 25.0°C
15–19 1.83 kg
15–21 108 grams
15–23 0.102 kg
15–25 170 m
15–27 a) 273 J   b) 3.41 J·mol^{-1}·K^{-1}
c) 10.9 J·mol^{-1}·K^{-1}
15–29 a) 1.91×10^{11} J   b) 12.8 m on a side
15–31 a) 45.6 m^3 b) 8.75 m^3
15–33 2.99 grams
15–35 40.0°C
15–37 3.69×10^8 J·m^{-3}

CHAPTER 16

16–1 4.34×10^5 J
16–3 a) -4.8°C b) $1.65\ \text{cal}\cdot\text{s}^{-1}\cdot\text{m}^{-2}$
16–5 1320 Btu = 1.39×10^6 J
16–7 a) $-1000\ \text{C}°\cdot\text{m}^{-1}$ b) $38.5\ \text{J}\cdot\text{s}^{-1}$ c) 80 C°
16–9 a) 2.58 cm b) 2.46×10^3 Pa
16–11 1.89×10^6 J
16–13 196 W
16–15 0.408 cm^2
16–17 72.3 m^3
16–19 $2.40 \times 10^3\ \text{km}^2 = 927\ \text{mi}^2$
16–21 a) $48\ \text{J}\cdot\text{s}^{-1}$ b) 25.8
16–23 7.61¢
16–25 Copper-steel junction: 9.48°C; steel-aluminum junction: 82.2°C
16–27 a) $4.73\ \text{J}\cdot\text{s}^{-1}$ b) Copper: 95.8%; steel: 4.2%
16–29 84.5°C
16–31 $1.64\ \text{grams}\cdot\text{hr}^{-1}$
16–33 a) $4\pi k(T_2 - T_1)ab/(b - a)$
b) $T(r) = T_2 - (T_2 - T_1)(b/r)[(r - a)/(b - a)]$
c) $2\pi kL(T_2 - T_1)/\ln (b/a)$
d) $T(r) = T_2 - (T_2 - T_1)[\ln (r/a)/\ln (b/a)]$
16–35 b) T = 0°C throughout d) $31.4°\text{C}\cdot\text{cm}^{-1}$
e) $121\ \text{J}\cdot\text{s}^{-1}$ f) 0 g) $1.11 \times 10^{-4}\ \text{m}^2\cdot\text{s}^{-1}$
h) $-10.9°\text{C}\cdot\text{s}^{-1}$ i) 9.13 s j) Decrease
k) $-7.74°\text{C}\cdot\text{s}^{-1}$

CHAPTER 17

17–1 3.0 atm
17–3 a) 50 moles b) 6.24×10^6 Pa = 61.6 atm
17–5 3.0×10^4 Pa
17–7 486°C
17–9 a) pM/RT b) $1.20\ \text{kg}\cdot\text{m}^{-3}$
17–11 Solid + vapor in equilibrium (no liquid)
17–13 a) 0.125×10^5 Pa; solid → vapor
b) 33.9×10^5 Pa; solid → liquid, liquid → vapor
17–15 2.30 atm ($33.8\ \text{lb}\cdot\text{in}^{-2}$)
17–17 12.2 cm, so piston has traveled 33.5 cm
17–19 a) 7.60×10^4 Pa b) 1.95 grams
17–21 a) 13.5 cylinders b) 5880 N c) 5440 N
17–23 a) $15.5\ \text{m}\cdot\text{s}^{-1}$
b) $h_2 = 2.5$ m: $9.80\ \text{m}\cdot\text{s}^{-1}$; $h_2 = 2.0$ m: $4.43\ \text{m}\cdot\text{s}^{-1}$
c) 1.82 m
17–25 a) $n\,RT/(V-nb)^2 = 2an^2/V^3$; $2n\,RT/(V - nb)^3 = 6an^2/V^4$ b) $(V/n)_c = 3b$; $T_c = 8a/27bR$
c) $a/27b^2$ d) 8/3 e) H_2: 3.28; N_2: 3.44; O_2: 3.25; H_2O: 4.35
f) H_2: $T_c = 33.1$ K, $p_c = 12.9 \times 10^5$ Pa, $(V/n)_c = 79.8 \times 10^{-6}\ \text{m}^3\cdot\text{mol}^{-1}$
N_2: $T_c = 128$ K, $p_c = 34.1 \times 10^5$ Pa, $(V/n)_c = 117 \times 10^{-6}\ \text{m}^3\cdot\text{mol}^{-1}$
O_2: $T_c = 155$ K, $p_c = 50.5 \times 10^5$ Pa, $(V/n)_c = 95.4 \times 10^{-6}\ \text{m}^3\cdot\text{mol}^{-1}$
H_2O: $T_c = 646$ K, $p_c = 220 \times 10^5$ Pa, $(V/n)_c = 91.5 \times 10^{-6}\ \text{m}^3\cdot\text{mol}^{-1}$

CHAPTER 18

18–1 0
18–3 2494 J
18–5 a) Yes b) No c) Negative
18–7 a) 4.0×10^4 J b) 8.0×10^4 J c) No
18–9 a) 60 J b) 70 J liberated c) *ad*: 50 J; *db*: 10 J
18–11 a) 365 J b) 0 c) Yes, 365 J liberated
18–13 $C_v = 25.2\ \text{J}\cdot\text{mol}^{-1}\cdot\text{K}^{-1}$; $C_p = 33.5\ \text{J}\cdot\text{mol}^{-1}\cdot\text{K}^{-1}$
18–15 736 K, 25.1 atm
18–17 a) 42.2 J b) 0
18–19 0.306 m^3
18–21 a) 0.174 m b) 204°C
18–23 a) 500 J; 0 b) 0; -500 J
18–25 b) 1.25×10^4 J c) 0 d) 1.25×10^4 J
e) 0.112 m^3
18–27 a) 1247 J; 4412 J; 3165 J b) 0; -3165 J; -3165 J c) 0

CHAPTER 19

19–1 a) 37.5% b) 5000 J c) 0.16 grams
d) 1.50×10^5 W = 201 hp
19–3 a) 25% b) $600 \times 10^6\ \text{J}\cdot\text{s}^{-1}$
19–5 a) $p_1 = 1.0$ atm, $V_1 = 2.46 \times 10^{-3}\ \text{m}^3$;
$p_2 = 2.0$ atm, $V_2 = 2.46 \times 10^{-3}\ \text{m}^3$
$p_3 = 1.0$ atm, $V_3 = 3.73 \times 10^{-3}\ \text{m}^3$
b) 52 J
19–7 678 K
19–9 a) 8.76×10^5 J b) 1.75×10^5 J
c) 1.05×10^6 J
19–11 a) 319 K b) 20.2%
19–13 a) 1.84×10^7 J b) 1.65×10^6 J
19–15 $47.4\ \text{J}\cdot\text{K}^{-1}$
19–17 $11.5\ \text{J}\cdot\text{K}^{-1}$
19–19 a) $8.33\ \text{km}^2 = 3.22\ \text{mi}^2$ b) $16.7\ \text{km}^2 = 6.44\ \text{mi}^2$
19–21 15.8%
19–23 $21\ \text{J}\cdot\text{K}^{-1}$
19–25 $e = 1 - (6.96r^{-0.56} - 1)/1.4(4 - r^{0.4})$; 68.7%

CHAPTER 20

20–1 11.1 mol; 6.69×10^{24} molecules
20–3 a) 18 cm^3 b) 3.10×10^{-10} m c) Comparable
20–5 4095°C
20–7 1.006
20–9 a) 6.21×10^{-21} J b) $2.34 \times 10^5\ \text{m}^2\cdot\text{s}^{-2}$
c) $484\ \text{m}\cdot\text{s}^{-1}$ d) $2.57 \times 10^{-23}\ \text{kg}\cdot\text{m}\cdot\text{s}^{-1}$
e) 1.24×10^{-19} N f) 1.24×10^{-17} Pa
g) 8.15×10^{21} molecules
h) 2.44×10^{22} molecules i) $(v_x^2)_{av} = (v^2)_{av}/3$
20–11 $24.9\ \text{J}\cdot\text{mol}^{-1}\cdot\text{K}^{-1}$; vibration is significant
20–15 2.45×10^6 molecules
20–17 a) $v_{rms} = 517\ \text{m}\cdot\text{s}^{-1}$ b) $v_{x,rms} = 298\ \text{m}\cdot\text{s}^{-1}$
29–19 a) 1.24×10^{-20} kg b) 4.15×10^5 molecules
c) Sphere of radius 1.5×10^{-6} cm; no; no
20–21 b) 1.61×10^5 K; 1.00×10^4 K
20–23 a) $F(r) = (U_0/\sigma)[12(\sigma/r)^{13} - 6(\sigma/r)^7]$ b) $r_1 = \sigma$, $r_2 = 1.122\sigma$, $r_1/r_2 = 0.820$ c) $U_0/4$

CHAPTER 21

21–1 a) 10.8 m b) 282 Hz
21–3 a) 17.2 m; 0.0172 m b) 74.0 m; 0.0740 m
21–7 b) $+x$-direction

21–9 a) 49.5 m·s^{-1} b) 0.206 m

21–11 0.270 s

21–13 35 m·s^{-1} greater at 57°C

21–15 b) 294 m·s^{-1} c) Large thermal conductivity, small λ

21–19 a) 0.50 Hz   b) 3.14 rad·s^{-1} c) 3.14 m^{-1}
d) $y(x, t) = -(0.10) \sin (\pi(t - x))$, x and y in meters, t in seconds   e) $y(t) = -(0.10) \sin (\pi t)$, y in meters, t in seconds   f) $y(t) = -(0.10) \cos (\pi t)$, y in meters, t in seconds   g) 0.314 m·s^{-1}
h) +0.0707 m; −0.222 m·s^{-1}

21–21 0.452 s

21–23 $Y/100$

21–25 $c = 0.683 v_{\text{rms}}$

CHAPTER 22

22–5 a) 36.0 m·s^{-1} b) 64.8 N

22–7 a) 35.3 Hz   b) 17.7 Hz

22–9 1290 Hz

22–11 a) 4960 m·s^{-1} b) 342 m·s^{-1}

22–13 a) 170 N   b) 12.2%

22–15 a) 375 m·s^{-1} b) 1.39

22–17 a) 0.657 m   b) 56°C

22–19 4.98 m, 1.15 m, 0.069 m

CHAPTER 23

23–1 a) 12.8 Pa (below pain threshold)   b) 514 Pa (well above pain threshold)

23–3 a) 58.7   b) 2.90×10^{-4}

23–5 0.0325 W·m^{-2}; 105 dB

23–9 443.6 Hz or 436.4 Hz

23–11 a) 0.63 m   b) 0.75 m   c) 548 Hz
d) 460 Hz   e) 571 Hz   f) 440 Hz
g) a: 0.65 m; b: 0.73 m; c: 546 Hz; d: 459 Hz; e: 569 Hz; f: 438 Hz

23–13 a) 0.286 Pa   b) 2.53×10^{-7} m   c) 50 m

23–15 a) 8.92 m   b) 892 m

23–17 9.89 m·s^{-1}

23–19 a) $t(c + v)/\lambda_0$ b) $(c - v)t$
c) $\lambda_0[(c - v)/(c + v)]$ d) $f_0[(c + v)/(c - v)]$
e) $2v f_0/(c - v)$

23–21 a) 0.029 m   b) 477 Hz

23–23 a) ct b) $(c - v_1)/f_0$ c) $f_0 t(c + v_2)/(c - v_1)$
d) c e) $f_0(c - v_2)(c - v_1)/(c + v_2)$
f) $[c(c + v_2)/(c - v_2)(c - v_1)]f_0$ g) 1560 Hz

CHAPTER 24

24–1 9.65×10^4 C

24–3 -1.78×10^{-6} C

24–5 a) 1.67×10^{-7} C, 1.67×10^{-7} C
b) 1.18×10^{-7} C, 2.36×10^{-7} C

24–7 0.119 m

24–9 a) 2.75×10^{26} b) 6.58×10^{15}
c) 2.39×10^{-9}%

24–11 2.19×10^6 m·s^{-1}

24–13 a) 0   b) $(1/4\pi\epsilon_0)2xq^2/(a^2 + x^2)^{3/2}$, +x-direction   d) $(1/4\pi\epsilon_0)2q^2/x^2$

24–15 a) $F = aq^2/2\pi\epsilon_0 x^3$ b) $F = aq^2/\pi\epsilon_0 y^3$

24–17 $F_x = -Q/(4\pi\epsilon_0 x \sqrt{a^2 + x^2})$;
$F_y = (Q/4\pi\epsilon_0 a)[1/x - 1/\sqrt{a^2 + x^2}]$

24–19 b) 1.36×10^{-6} C   c) 31.7°

24–21 a) 1.06×10^{-10} m   b) 1.09×10^6 m·s^{-1}

24–23 a) 9.00×10^{-7} N, away from the vacant corner
b) 8.61×10^{-7} N, toward the opposite corner

24–25 a) $(Q/2\pi\epsilon_0 a)[1/y - 1/\sqrt{a^2 + y^2}]$ in the −x-direction; when $y \rightarrow \infty$, $F \rightarrow Qa/4\pi\epsilon_0 y^3$
b) $(Q/4\pi\epsilon_0 a)[1/(x - a) + 1/(x + a) - 2/x]$ in +x-direction; when $x \rightarrow \infty$, $F \rightarrow Qa/2\pi\epsilon_0 x^3$

24–27 b) $q_1 < 0$, $q_2 > 0$ c) 1.69×10^{-6} C   d) 28.1 N

CHAPTER 25

25–1 7.20×10^4 N·C^{-1}, upward

25–3 4.74 m

25–5 4.0 N·C^{-1}, upward

25–7 5.57×10^{-11} N·C^{-1}

25–9 1.42×10^4 N·C^{-1}

25–11 a) 50.0 N·C^{-1}, +x-direction   b) 10.8 N·C^{-1},+x-direction

25–13 a) 1.80×10^4 N·C^{-1}, −x-direction
b) 8.00×10^3 N·C^{-1}, +x-direction
c) 3.34×10^3 N·C^{-1}, 69.9° above −x-axis
d) 6.36×10^3 N·C^{-1}, −x-direction

25–15 3.54×10^{-11} C·m^{-2}

25–17 2.26×10^9 electrons

25–21 8.85×10^{-10} C

25–23 b) $q/4\pi\epsilon_0 r^2$

25–25 a) 1.42 cm   b) 9.85 cm

25–27 b) $q_1 < 0$, $q_2 < 0$ c) 1.95×10^7 N·C^{-1}

25–29 $E_x = E_y = Q/2\pi^2\epsilon_0 a^2$

25–31 -2.66×10^{-10} C

25–33 a) $\lambda/2\pi\epsilon_0 r$ b) $\lambda/2\pi\epsilon_0 r$ d) $-\lambda$ inner, $+\lambda$ outer

25–35 c) $(4 - 3r/R)Qr/4\pi\epsilon_0 R^3$ d) $Q/4\pi\epsilon_0 R^2$, both expressions

25–37 a) $8Q/5\pi R^3$; 6.37×10^{23} C·m^{-3}
b) $r \leq R/2$: $8Qr/15\pi\epsilon_0 R^3$;
$R/2 \leq r \leq R$: $(4 - 3r/R)4Qr/15\pi\epsilon_0 R^3$
$-Q/60\pi\epsilon_0 r^2$; $r \geq R$: $Q/4\pi\epsilon_0 r^2$ c) 4/15
e) 9.68×10^{-23} s

CHAPTER 26

26–1 a) −0.0225 J   b) 12.2 m·s^{-1}

26–3 a) point a b) 1500 V·m^{-1} c) 1.20×10^{-4} J

26–5 a) 1.80×10^3 V   b) 0   c) 4.50×10^{-5} J

26–7 b) $q/2\pi\epsilon_0 a$ e) $\pm\sqrt{3}a$

26–9 a) 5000 V·m^{-1} b) 1.0×10^{-6} J
c) 1.0×10^{-6} J

26–11 2.5 mm

26–13 a) 180 V   b) 180 V

26–15 a) 0   b) -1.00×10^{-3} J   c) 2.30×10^{-3} J

26–17 a) Inside: $q(3 - r^2/R^2)/8\pi\epsilon_0 R = (\rho/2\epsilon_0)(R^2 - r^2/3)$
outside: $q/4\pi\epsilon_0 r = \rho R^3/3\epsilon_0 r$

26–21 a) 4.33×10^{-4} m·s^{-1} (downward)
b) -2.75×10^{-4} m·s^{-1} (upward)

26–23 1.44 MeV

26–25 a) 5.11×10^3 V   b) $0.141c$
c) 9.39×10^6 V; $0.141c$

26–27 a) 0.703 cm   b) 19.4°   c) 4.92 cm
26–29 a) -4.5×10^{-5} J   b) 3.0×10^5 V·m^{-1}
c) -1.5×10^4 V
26–31 a) Remains at rest   b) Oscillates about the origin, along the y-axis   c) Accelerates away from the origin along the x-axis
26–33 a) 2.18×10^{-5} m   b) Electron's is 42.8 times the proton's   c) Equal
26–35 a) -90 V; -315 V   b) $+9.0 \times 10^{-7}$ J
26–37 a) 1.00×10^5 V·m$^{-4/3}$
b) $-4/3\ Cx^{1/3}$
c) 3.39×10^{-15} N
26–39 a) Inside: $\lambda(R^2 - r^2)/4\pi\epsilon_0 R^2$
outside: $-\lambda \ln(r/R)/2\pi\epsilon_0$
26–41 a) $(Q/4\pi\epsilon_0 a) \ln[(x + a)/x]$
b) $(Q/4\pi\epsilon_0 a) \ln[a + \sqrt{a^2 + y^2})/y]$
c) Point P: $Q/4\pi\epsilon_0 x$;   point Q: $Q/4\pi\epsilon_0 y$
26–43 2
26–45 b) $Q(2R^2 - 2r^2 + r^3/R)/4\pi\epsilon_0 R^3$
26–47 a) 867 m·s^{-1} c) -7.50 J   d) Will not escape because $E_r < 0$ e) 12.0 mm
26–49 a) $r \leq a$: $E_r = (\rho_0 r/3\epsilon_0)(1 - r/a)$, $E_\theta = 0$, $E_\phi = 0$
$r \geq a$: $E = 0$
b) $r \leq a$: $\rho(r) = (\rho_0/3)(3 - 4r/a)$
$r \geq a$: $\rho(r) = 0$

CHAPTER 27

27–1 a) 400 V   b) 113 cm^2 c) 2.00×10^6 V·m^{-1}
d) 1.77×10^{-5} C·m^{-2}
27–3 4.25×10^{-4} C
27–5 a) $Q_1 = 1.44 \times 10^{-4}$ C, $Q_2 = 2.16 \times 10^{-4}$ C
b) $V_1 = V_2 = 36.0$ V
27–7 a) $Q_1 = 1.92 \times 10^{-5}$ C, $Q_2 = 1.92 \times 10^{-5}$ C, $Q_3 = 3.84 \times 10^{-5}$ C, $Q_4 = 5.76 \times 10^{-5}$ C
b) $V_1 = 9.6$ V, $V_2 = 9.6$ V, $V_3 = 19.2$ V, $V_4 = 28.8$ V
c) 19.2 V
27–9 a) 500 V   b) 1.25×10^{-4} J   c) $0.113\,\text{m}^2$
d) 800 V
27–11 0.0692 V·m^{-3}
27–13 a) $xq^2/2\epsilon_0 A$ b) $(x + dx)q^2/2\epsilon_0 A$ d) E is due to superposition of fields of each plate; must calculate force of one plate due to electric field from other plate alone
27–15 1.69 m^2
27–17 a) 0.71×10^{-6} C·m^{-2} b) 1.67
27–19 a) 3.42   b) 7.08×10^{-8} C
27–21 a) 1.77×10^{-11} F   b) 8.85×10^{-10} C
c) 2500 V·m^{-1} d) 2.21×10^{-8} J
27–23 a) 1 μF   b) $Q_2 = 6.0 \times 10^{-4}$ C, $Q_3 = 9.0 \times 10^{-4}$ C   c) 100 V
27–25 a) 1 μF: $q = 1.2 \times 10^{-3}$ C, $V = 1200$ V
2 μF: $q = 2.4 \times 10^{-3}$ C, $V = 1200$ V
b) 1 μF: $q = 4.0 \times 10^{-4}$ C, $V = 400$ V
2 μF: $q = 8.0 \times 10^{-4}$ C, $V = 400$ V
27–27 a) 2.4×10^{-5} b) 1.44×10^{-4} J   c) 3.6 V
d) 1.30×10^{-4} J
27–29 a) 5.53×10^{-11} F   b) 1.66×10^{-8} C
c) 2.49×10^{-6} J   d) 0.498 J·m^{-3}
27–31 a) $(Q/4\pi\epsilon_0)(1/r_a - 1/r_b)$
27–35 b) 1.77×10^{-9} F

27–37 b) 14 μF   c) 72 μF: 504 μC, 7 V;   27 μF: 270 μC, 10 V;
18 μF: 234 μC, 13 V;   6 μF: 18 μC, 3 V;
28 μF: 252 μC, 9 V;   21 μF: 252 μC, 12 V

CHAPTER 28

28–1 a) 0.020 A   b) 2.74×10^{-6} m·s^{-1}
28–3 a) 603 C   b) 121 A
28–5 0.261 Ω
28–7 a) 1.32×10^{-4} Ω b) 1.35 cm
28–9 a) 99.5 Ω b) 0.0148 Ω
28–11 28.5°C
28–13 $\mathcal{E} = 1.52$ V, $r = 0.10$ Ω
28–15 a) 0.050 Ω b) 0.15 Ω c) 0.0060 Ω
28–17 b) Yes   c) 4.36 Ω
28–19 a) EJ b) $J^2\rho$ c) E^2/ρ
28–21 a) 21.8 Ω b) 5.50 A   c) 555 W
28–23 a) 2.16×10^6 J   b) 0.0522 l   c) 2.0 hr
28–25 a) 1.82 A   b) 3.31×10^4 W   c) 20×10^6 Ω
28–27 9.6×10^{-4} A
28–29 a) 2.40×10^{-8} Ω·m   b) 20.0 A
c) 3.90×10^{-4} m·s^{-1}
28–33 $R/9$
28–35 a) 0.50 Ω b) 10.0 V
28–37 a) 15.0 V   b) 3.24×10^6 J   c) 6.48×10^5 J
d) 0.60 Ω e) 1.94×10^6 J   f) 6.48×10^5 J
g) Energy dissipated in internal resistance of battery
28–39 a) 0.40 A   b) 1.6 W   c) In 12-V battery; 4.8 W   d) In 8-V battery; 3.2 W
28–41 b) $a = 8.04 \times 10^{-5}$ Ω·m·K$^{0.146}$; $n = 0.146$
c) Freezing point: 3.54×10^{-5} Ω·m; boiling point: 3.38×10^{-5} Ω·m

CHAPTER 29

29–1 a) 36.0 Ω b) 3.33 A   c) 2.00 A through 60 Ω, 1.33 A through 90 Ω
29–3 a) 0.545 Ω b) 1 Ω: 12 A; 2 Ω: 6 A; 3 Ω: 4 A   c) 22 A   d) 12 V for each   e) 1 Ω: 144 W; 2 Ω: 72 W; 3 Ω: 48 W
29–5 a) 141 V   b) 4.5 W   c) Two in series, connected in parallel with two others in series
29–7 a) 2.0 A   b) 5.0 Ω c) 42.0 V   d) 3.5 A
29–9 a) 0.222 V   b) 0.464 A
29–11 9.96 Ω
29–13 $R_1 = 2.99 \times 10^3$ Ω, $R_2 = 1.20 \times 10^4$ Ω, $R_3 = 1.35 \times 10^5$ Ω
3 V: $R = 3.0 \times 10^3$ Ω;
15 V: $R = 1.50 \times 10^4$ Ω;
150 V: $R = 1.50 \times 10^5$ Ω
29–15 15,000 Ω: 10.9 V; 150,000 Ω: 109 V
29–17 a) 0, 3.93×10^{-4} C, 6.32×10^{-4} C, 8.65×10^{-4} C, 1.00×10^{-3} C (or, more precisely, 0.99995×10^{-3} C)   b) 1.00×10^{-4} A, 6.07×10^{-5} A, 3.68×10^{-5} A; 1.35×10^{-5} A, 4.54×10^{-9} A   c) 10.0 s   d) 6.93 s
29–19 a) 0.5 s   b) 1.0 s
29–21 a) 1.00×10^{-8} C; 2.82×10^6 V·m^{-1}; 8.47×10^3 V
b) 5.65×10^{11} V·m^{-1}·s^{-1} no

c) $J_D = 5.00$ A·m^{-2}; $I_D = 2.00 \times 10^{-3}$ A; equal

29–23 a) 3.15×10^9 J $= 876$ kW h   b) $87.60

29–25 a) Toaster: 12.5 A; frypan: 10.0 A; lamp: 0.833 A
b) Yes ($I = 23.3$ A)

29–29 a) Two in series in parallel with two in series
b) 0.5 W

29–31 27 W

29–33 $I_1 = 0.848$ A, $I_2 = 2.14$ A, $I_3 = 0.171$ A

29–35 a) -12 V   b) 3 A   c) -12 V   d) 12/7 A,
from b to a e) 4.5 Ω f) 4.2 Ω

29–37 a) -6 V   b) b c) $+6$ V   d) 54 μ C, from b
to a

29–39 a) 1200 Ω b) 30 V

29–41 4.2×10^5 Ω

29–43 55 Ω

29–45 b) No current through galvanometer   c) 9.52 V
d) No

29–47 a) CV^2 b) $\frac{1}{2}CV^2$ c) $\frac{1}{2}CV^2$ d) 50%,
independent of R

29–49 a) 5.81×10^6 A·m^{-2} b) 3.34×10^{-10} A·m^{-2}

29–55 a) $I_1 = 1$ A, $I_2 = 3$ A, $I_3 = 2$ A   b) 6 V
c) 12 Ω d) 78 V

29–57 a) $I_1 = 0.300$ A, $I_2 = 0.500$ A, $I_3 = 0.200$ A
b) $I_1 = 0.541$ A, $I_2 = 0.0773$ A, $I_3 = -0.464$ A
c) $I_1 = -0.287$ A, $I_2 = 0.192$ A, $I_3 = 0.479$ A
d) $I_1 = 0.0462$ A, $I_2 = 0.231$ A, $I_3 = 0.185$ A
f) $I_1 = 0.0833$ A, $I_2 = 0.644$ A, $I_3 = 0.561$ A

CHAPTER 30

30–1 Negative

30–3 a) $F = +(2.5 \times 10^{-3}$ N)k
b) $F = -(2.5 \times 10^{-3}$ N)$i - (1.5 \times 10^{-3}$ N)j

30–5 a: $-qv\,Bk$; b: $qv\,Bj$; c: 0; d: $-(qv\,B/\sqrt{2})j$;
e: $-(qv\,B/\sqrt{2})(j + k)$; f: $-(qv\,B/\sqrt{3})(j + k)$

30–7 a) 1.14×10^{-3} T, into page   b) 1.57×10^{-8} s

30–9 a) 2.89×10^7 m·s^{-1} b) 4.34×10^{-8} s
c) 8.68×10^6 V

30–11 3.97×10^{-3} T

30–13 a) 1.70×10^7 m·s^{-1} c) 4.83×10^{-3} m

30–15 21

30–17 24 A

30–19 a) -0.030 Nk b) -0.025 Nj c) 0
d) 0.015 Nj e) 0.020 Nj + 0.045 Nk

30–21 a) 0.181 N·m   b) Angle of 30° between B and
normal to the coil

30–23 a) $\Gamma = NIAB$, $-x$-direction $U = 0$
b) $\Gamma = 0$; $U = -NIAB$
c) $\Gamma = NIAB$, $+x$-direction; $U = 0$
d) $\Gamma = 0$; $U = +NIAB$

30–25 a) 0.80 A   b) 3.7 A   c) 113 V   d) 417 W

30–27 a) 8.44×10^{-4} m·s^{-1} b) 1.27×10^{-3} V·m^{-1},
$+z$-direction   c) 2.53×10^{-5} V

30–29 a) 1.2×10^3 V·m^{-1}, $+z$-direction
b) 1.2×10^3 V·m^{-1}, $+z$-direction

30–31 0.5 T, $-y$-direction

30–33 Number for alpha particle $= 3.64 \times 10^3$ times
number for electron

30–35 1.25 mm

30–37 a) 4.0×10^{-6} C   b) $(1.2i + 1.6j) \times 10^{15}$ m·s^{-2}
c) 1.25 cm   d) 6.37×10^7 Hz
e) $(R, 0, 37.7$ cm)

30–39 a) ab: -1.2 Nk; bc: -1.2 Nj; cd: $+1.2$ N$(j + k)$;
de: -1.2 Nj; ef: 0   b) -1.2 Nj

30–41 0.0132 T, in $+y$-direction

30–43 a) 4.32 m·s^{-1} b) 7.69 A   c) 0.195 Ω

30–45 a) 1.0×10^{-2} J·T^{-1}, $+z$-direction   b) $B_x = 0.8$ T,
$B_y = 0.6$ T, $B_z = 2.4$ T

30–49 a) 1.0 m   b) 1.26×10^{-6} s   c) 0.0318 m
d) 0.161 m

30–51 a) $v = lB\Delta q/m$ c) 7.34×10^{-5} s   d) 0.0168 C
e) 2.11×10^3 m·s^{-1}

CHAPTER 31

31–1 a) 0   b) -1.60×10^{-5} Tk c) -1.60×10^{-5} Tj
d) -5.66×10^{-6} $T(j + k)$

31–3 2.97×10^{-6} T, into the page

31–5 $\mu_0 I/4R$, into page (no)

31–7 a) 1.6×10^{-5} T, west   b) Yes

31–9 a) 6.0×10^{-6} T   b) 2.0×10^{-6} T

31–11 a) $\mu_0 I/\pi a$ b) $\mu_0 I/3\pi a$ c) 0   d) $2\mu_0 I/3\pi a$

31–13 a) 7.5 A   b) Opposite

31–15 1.02 cm

31–17 a) 5.03×10^{-4} T   b) 4.50×10^{-5} T

31–19 6.0×10^{-4} T

31–21 a) $\mu_0 I/2\pi r$ b) 0

31–23 χ varies inversely with Kelvin temperature

31–25 a) 0.0426 A   b) 0.0115 A

31–27 2.40×10^{-20} N, away from the wire

31–29 b) $\mu_0 Ix/\pi(x^2 + a^2)$ d) $x = \pm a$

31–31 a) 2 A, out of page   b) 2.13×10^{-6} T, to the
right   c) 2.06×10^{-6} T, 39.0° below horizontal
and to the left

31–33 7.2×10^{-4} N, toward the wire

31–35 $(2\mu_0 I/\pi ab)\sqrt{a^2 + b^2}$, into page

31–37 a) $1/2\,N\mu_0 Ia^2(1/[a^2 + (x + a/2)^2]^{3/2}$
$+1/[a^2 + (x - a/2)^2]^{3/2})$
b) $8\,N\mu_0 I/\sqrt{125}a$ c) 1.50×10^{-3} T

31–39 a) 398   b) 397

31–41 a) $3I/2\pi R^3$ b) (i) $\mu_0 Ir^2/2\pi R^3$; (ii) $\mu_0 I/2\pi r$

31–43 b) $\mu_0 I_0/2\pi r$ c) $I_0 r^2(2 - r^2/a^2)/a^2$
d) $\mu_0 I_0 r(2 - r^2/a^2)/2\pi a^2$

31–45 $Qn\mu_0/a$

31–47 $\mu_0 I/4\pi a$, out of page

31–49 b) 0.286 m·s^{-1} c) 4.16×10^{-3} m

CHAPTER 32

32–1 a) 5.0 m·s^{-1} b) 2.0 A   c) 0.96 N, to the left

32–3 a) 0.30 V   b) 0.30 V   c) b

32–7 8.79 V

32–9 a) 0.737 rev·s^{-1} b) 4.32×10^3 N·m

32–11 5.0×10^{-2} T

32–13 7.54×10^{-6} V

32–15 a) Clockwise concentric circles
b) 5.0×10^{-3} V·m^{-1}, tangent to the ring and
clockwise; 3.14×10^{-3} V   c) 1.57×10^{-3} A
d) 0

32–17 a) Right to left   b) Right to left   c) Left to right

32–19 b) RF/B^2l^2

32–21 a) 3.14 V   b) 3.14 V

32–23 a) 2.00×10^{-2} W b   b) 0.200 V
c) 2.00×10^{-3} N·m

32–25 a) 0.0471 V   b) a to b

32–27 Point a: $(qr/2)\,dB/dt$, to the left in the plane of the figure
Point b: $(qr/2)\,dB/dt$, upward in the plane of the figure
Point c: 0

32–29 20

32–31 a) $(\mu_0 Iv/2\pi)\ln(1 + l/d)$ b) a c) 0

32–33 a) $B_0\pi a^2 e^{-t/\tau}\cos\omega t$
b) $B_0\pi a^2 e^{-t/\tau}[(1/\tau)\cos\omega t + \omega\sin\omega t]$ c) 62.8 A
d) 3.09×10^{-3} s   e) 1.47×10^{-3} s; 11.7 V

32–35 a) $Bv/\lambda\theta$ b) $B^2v^2l/\lambda\theta$ d) 294°
e) a: $(Bv/\lambda)(2\pi/\theta[2\pi - \theta])$; b: $B^2v^2l2\pi/\lambda\theta(2\pi - \theta)$;
d: 188°

32–37 b) 0   c) 4.0×10^{-3} V   d) 2.0×10^{-3} A
e) 1.0×10^{-3} V; a

CHAPTER 33

33–1 7.90×10^{-4} H

33–3 a) 5.0×10^{-4} V; yes   b) 5.0×10^{-4} V

33–5 0.10 V, in the direction of the current

33–7 a) 1.00×10^{-3} H   b) 2.25×10^{-3} H if both coils are wound in same sense, 2.5×10^{-4} H if one coil is wound in opposite sense to other

33–9 387

33–13 a) 4.00 A·s^{-1} b) 2.00 A·s^{-1} c) 0.659 A
d) 2.00 A

33–17 0.158 pF

33–21 a) 141 rad·s^{-1} b) 61.6 Ω

33–23 a) 5.0 H   b) 250 Wb   c) 0.020

33–25 b) $\mathcal{E}_L$ is proportional to di/dt

33–27 1.33×10^{-6} T

33–31 a) 2.40 J   b) 4.41 J   c) 2.02 J
d) (a) + (c) = (b)

33–33 a) $4L$ b) $2L$ c) $\omega/2$

33–35 Rate of decrease of energy in capacitor equals rate of storage in inductor plus rate of dissipation in resistor.

33–37 a) $i_R = (0.10\text{ A})e^{-30t}$; $i_{R_0} = 0.40$ A;
$i_{s_2} = 0.40\text{ A} - (0.10\text{ A})e^{-30t}$
b) 0.326 A, to the right

33–41 a) $i_1(t) = (\mathcal{E}/R_1)(1 - e^{-R_1t/L})$; $i_2(t) = (\mathcal{E}/R_2)e^{-t/R_2C}$;
$q_2(t) = \mathcal{E}C(1 - e^{-t/R_2C})$ b) $i_1(0) = 0$; $i_2(0) =$
0.010 A   c) $i_1(\infty) = 2.0$ A; $i_2(\infty) = 0$; several time constants   d) 1.97×10^{-3} s   e) 9.81 m A
f) 0.277 s

33–43 a) $D(L - L_0)/(L_f - L_0)$
b) One-quarter: 1.2505 H; one-half: 1.2510 H; three-quarters: 1.2514 H
c) One-quarter: 1.2500 H; one-half: 1.2500 H; three-quarters: 1.2500 H
d) Completely ineffective for mercury, marginal for oxygen

CHAPTER 34

34–1 a) 377 Ω   b) 2.65×10^{-3} H   c) 2.65×10^3 Ω
d) 2.65×10^{-3} F

34–3 a) 5.0×10^{-3} A   b) 5.0×10^{-2} A   c) 0.50 A

34–5 a) 5.0×10^{-2} A   b) 5.0×10^{-3} A
c) 5.0×10^{-4} A

34–7 $V_C = Q/C$, whereas $\mathcal{E}_L = L\,di/dt$

34–9 a) 583 Ω   b) 0.0857 A   c) 25.7 V, 42.9 V
d) $-59.1°$; lags

34–11 a) 514 Ω: $-38.9°$; lags   b) 506 Ω; $+37.8°$; leads

34–13 a) 25.0 W   b) 25.0 W   c) 0   d) 0
e) (a) = (b) + (c) + (d)

34–15 a) 745 rad·s^{-1} b) 1   c) $V_1 = 35.4$ V,
$V_2 = 79.1$ V, $V_3 = 79.1$ V, $V_4 = 0$, $V_5 = 35.4$ V
d) 745 rad·s^{-1} e) 0.354 A

34–17 a) 3160 rad·s^{-1} c) 0.60 A   d) 0.60 A

34–19 a) 10   b) 6.0 A   c) 72.0 W   d) 200 Ω

34–21 a) 31.6   b) 3.16 V

34–23 0.0183 H

34–25 47.2 Ω

34–27 a) Inductor   b) 0.106 H

34–29 a) 1.59×10^3 Hz; 1.00×10^4 rad·s^{-1} b) 1.00 A
c) 1.00 A   d) 0.10 A   e) 0.10 A
f) 5.00×10^{-4} J; 5.00×10^{-4} J

34–31 0; $I_0/\sqrt{3}$

34–33 a) $[LC]^{-1/2}$ b) $[LC - R^2C^2/2]^{-1/2}$
c) $[1/LC - R^2/2L^2]^{1/2}$

34–35 a) 400 Ω + (300 Ω)i b) 300 Ω − (400 Ω)i
c) 350 Ω − (50 Ω)i d) 0.280 A + (0.040 A)i
e) 0.283 A; 8.13°   f) 0.160 A − (0.120 A)i
g) 0.200 A; $-36.9°$ h) 0.120 A + (0.160 A)i
i) 0.200 A; 53.1

CHAPTER 35

35–1 300 m

35–3 a) 294 m   b) 4.80×10^{-3} V·m^{-1}

35–7 a) 6.67×10^{-11} T   b) 1.67×10^4 W
c) 100 km

35–9 a) 1.56×10^{-14} kg·m^{-2}·s^{-1} b) 4.67×10^{-6} Pa

35–11 a) 3.0×10^6 m·s^{-1} b) 0.030 m

35–13 a) 1.50×10^{-9} m = 1.50×10^{-3} microns =
1.50 nm = 15.0 Å   b) 5.35×10^{-7} m =
0.535 microns = 535 nm = 5350 Å

35–17 a) $(1/2)r\mu_0 n\,di/dt$, in tangential direction
b) $(1/2)\mu_0 n^2 ri\,di/dt$, radially inward

35–19 3.19 V

35–21 6.14×10^4 V·m^{-1}, 2.05×10^{-4} T

35–23 a) Reflective; gives twice the radiation pressure
b) 24.1 mi^2

35–25 2.90×10^{11} eV·s^{-1}; doesn't apply

CHAPTER 36

36–1 a) 4.48×10^9 m   b) 15.0 s

36–3 34.8°

36–5 a) 54.7°   b) 82.8°

36–7 1.87

36–9 a) 2.00×10^8 m·s^{-1} b) 333 nm

36–13 a) 14.5°   b) Air

36–15 a) $I_0/4$ b) Linearly polarized, parallel to axis of second polarizer

36–17 a) Transmitted intensities are $I_0/2$, $I_0/4$, $I_0/8$. In each case the light is linearly polarized, along the axis of the polarizer.
b) $I_0/2$ for first filter, 0 for the next.

36–19 54.7°

36–21 a) 1.60   b) 32.0°

36–23 Elliptical

36–27 1.69

36–29 30°

36–31 1.30

36–33 a) $\frac{1}{4}I_0 (\sin 2\theta)^2$ b) 45°

36–35 a) 35.0°   b) 10.1 W·m^{-2}, 19.9 W·m^{-2}

36–37 l-leucine: $\theta = -0.110\,c$, where c is the concentration in grams/100 ml
l-glutamic: $\theta = 0.124\,c$

36–39 b) 856 nm

36–43 b) 37.2°   c) 1.75°

CHAPTER 37

37–1 80 cm to right of mirror; 6 cm

37–3 3.61 cm

37–5 b) 60 cm in front of mirror, 10 cm, inverted, real

37–9 b) 4.44 cm behind mirror, 0.833 cm, erect, virtual

37–11 6.0 cm

37–15 1.35

37–17 10 cm to left of vertex, +1/3

37–19 Half the observer's height

37–21 a) 5.0 cm   b) 9.88 cm

37–23 a) $-5\,\text{cm} < s < 0$ b) Erect

37–25 2.00

37–27 a) 0.667 cm   b) Independent of the radius

37–29 a) As object point moves closer to mirror, image moves away   b) (i) 20.00 cm, 19.89 cm; (ii) $-1/3$, $-1/9$; (iii) transverse dimension is 0.33 cm, longitudinal dimension is 0.11 cm

37–31 a) $-0.0331\ \text{m·s}^{-1}$ b) $-0.444\ \text{m·s}^{-1}$

37–33 b) 11.4°

CHAPTER 38

38–1 $s = 30$ cm: a) 15 cm   b) $-1/2$ c) real   d) inverted
$s = 20$ cm: a) 20 cm   b) -1 c) real   d) inverted
$s = 15$ cm: a) 30 cm   b) -2 c) real   d) inverted
$s = 5$ cm: a) -10 cm   b) $+2$ c) virtual   d) erect

38–3 $s = 30$ cm: a) -7.50 cm   b) $+1/4$ c) virtual   d) erect
$s = 20$ cm: a) -6.67 cm   b) $+1/3$ c) virtual   d) erect
$s = 15$ cm: a) -6.00 cm   b) $+2/5$ c) virtual   d) erect
$s = 5$ cm: a) -3.33 cm   b) $+2/3$ c) virtual   d) erect

38–5 a) -5.0 cm; diverging   b) 0.40 cm; erect

38–7 7.21 cm in front of lens

38–9 4.34 cm

38–11 a) Image formed by first surface   b) +30 cm   c) Real   d) At ∞

38–13 4.5 cm from center of sphere

38–15 60 cm to right of third lens

38–17 2 cm to left of first lens and 2 cm to right of second lens

38–19 a) 50 cm   b) 200 cm

38–21 0.667 cm

38–23 a) 7.14 cm   b) 3.5 mm

38–25 a) 35 mm   b) 200 mm

38–27 a) 2.86 cm   b) $1/50\ s$

38–29 a) 10.2 cm   b) No

38–31 a) 1.74 cm   b) -11.3 c) -113

38–33 a) -5.0 b) 4.0 cm

38–35 15.7 cm

38–37 Convex, 4.66 m

38–39 4.0 cm

38–41 -15 cm

38–43 a) 0.254 m   b) 0.254 m

38–45 a) 7.55 m   b) 5.05 m

38–47 $4R$ from center of sphere

38–49 0.67 cm above page

38–51 2.0 cm

38–53 2.97 cm; elongated

38–55 a) 361   b) 23.3 cm

38–57 b) -20 cm   c) Galilean telescope is 80 cm long, telescope in Exercise 38-33 is 120 cm long.

38–59 a) $4f$

38–61 a) 103 cm from lens   b) Virtual   c) Inverted   d) 13.2 mm

CHAPTER 39

39–1 1.00 mm

39–3 545 nm

39–5 $\lambda/2d$; independent of m

39–7 0.833 mm

39–9 2.94×10^{-4} rad $= 0.0168°$

39–11 a) 7.14×10^{-8} m   b) Red

39–15 533 nm

39–17 17.5°, 36.9°, 64.2°

39–19 12.5°

39–21 2000 m

39–23 Smaller by a factor of 5/6

39–25 No difference

39–27 0.750 mm

39–29 480 nm

39–31 2.79 mm

39–33 a) Shifted downward on screen
b) $I(\theta) = I_0 \cos^2[(\pi/\lambda)\{d \sin \theta + L(1 - 1/n)\}]$
c) $\sin \theta = [m\lambda - L(1 - 1/n)n)]/d$

39–35 0.10 mm

39–37 500 nm

39–39 119 km

39–41 a) 25.6°   b) 10.2°   c) 5.08°

39–43 3.80 fringes·min^{-1}

CHAPTER 40

40–1 Bolt at A

40–3 a) $0.990c$ b) 4.75×10^3 m

40–5   333 m

40–11 2.60×10^8 m·s^{-1}

40–15   a) $0.866c$ b) $0.996c$

40–17   a) Nonrelativistic: 4.56×10^{-15} J; relativistic: 4.97×10^{-15} J; ratio = 0.916
b) Nonrelativistic: 3.57×10^{-14} J; relativistic: 1.47×10^{-13} J; ratio = 0.243

40–19   111 kg

40–25 7.38×10^7 m·s^{-1}

40–27   a) 3.38×10^{-6} b) 5.37×10^4 MeV

40–29   a) 7.14×10^3 m·s^{-1} b) No

40–31   a) $0.995c$ b) 1.0%

40–33   a) 1.80×10^5 eV   b) 6.91×10^5 eV
c) 2.02×10^8 m·s^{-1} d) 2.52×10^8 m·s^{-1}

40–35   168 keV

40–37 2.55×10^8 m·s^{-1}; receding

40–43   a) 600 MeV   b) 2.15 times larger

40–45   c) 5.78×10^{-9} s

CHAPTER 41

41–1   a) 2.42×10^{20} Hz   b) 1.24×10^{-12} m
c) 1000 times as large

41–3 5.00×10^{14} Hz; 1.10×10^{-27} kg·m·s^{-1};
3.31×10^{-19} J = 2.07 eV

41–5   a) 6.63×10^{-26} J = 4.14×10^{-7} eV
b) 7.55×10^{29} photons·s^{-1}

41–7 1.72×10^6 m·s^{-1}

41–9   289 nm

41–11   a) 6.17×10^{14} Hz   b) 486 nm

41–13   387 nm

41–15   1.96 eV = 3.14×10^{-19} J;   633 nm

41–17 1.59×10^{15} photons·s^{-1}

41–23   37.4°

41–25   a) 1.04 eV   b) 1197 nm   c) 2.51×10^{14} Hz
d) 4.14×10^{-7} eV   f) No

41–27   0.344 V greater

41–29   a) 4.59×10^{14} Hz   b) 652 nm   c) 1.89 eV
d) 6.59×10^{-34} J·s

41–31   a) 12.8 eV   b) 6; $n = 4 \rightarrow 3$, 1875 nm;
$n = 4 \rightarrow 2$, 486 nm; $n = 4 \rightarrow 1$, 97.2 nm; $n = 3 \rightarrow 2$,
656 nm; $n = 3 \rightarrow 1$, 103 nm; $n = 2 \rightarrow 1$, 122 nm

41–33   7730 K

41–35   a) 4.85×10^{-3} nm   b) 2.56×10^5 eV

41–37   a) 1.13×10^3 eV   b) 11 Å

CHAPTER 42

42–1   a) 2.19×10^6 m·s^{-1}, 1.09×10^6 m·s^{-1},
7.29×10^5 m·s^{-1}
b) 1.52×10^{-16} s, 1.22×10^{-15} s, 4.10×10^{-15} s
c) 8.22×10^6

42–3   a) 13.6 eV   b) -27.2 eV   c) -13.6 eV
d) 13.6 eV   e) 91.2 nm; ultraviolet

42–5   a) 4 times larger   b) Paschen, Brackett, Pfund
c) Smaller by a factor of 2

42–7   1.77 eV to 3.10 eV; 0.696 nm to 0.922 nm

42–9   a) 5.48×10^{-2} nm   b) 1.28×10^{-3} nm

42–11   b) 2×10^{-24} kg·m·s^{-1}; 2×10^{-24} kg·m·s^{-1}

42–13 2.34×10^{-23} s

42–15   a) $3\hbar$ b) $3.46\hbar$; L is larger   c) $m = 3$, 30.0°;
$m = 2$, 54.7°; $m = 1$, 73.2°; $m = 0$, 90.0°; $m = -1$,
106.8°; $m = -2$, 125.3°; $m = -3$, 150.0°

42–17 2.89×10^{30} rad·s^{-1}

42–19   a) -2820 eV   b) 2.56×10^{-13} m   c) 0.586 nm

42–21   a) $n = 3$, -3.48×10^{-4} eV; $n = 2$, -2.32×10^{-4} eV
b) Increased by 0.0403 nm

42–23   a) 1.05×10^{-19} kg·m·s^{-1} b) 197 MeV
c) -1.44 MeV; no

42–25   a) 1.0×10^{-8} m   b) No

42–29 1.61×10^{-16} m

42–31   a) $\hbar\omega$ b) 1

CHAPTER 43

43–1   a) $l = 0$, $m = 0$; $l = 1$, $m = 0$, ± 1;
$l = 2$, $m = 0$, ± 1, ± 2; $l = 3$, $m = 0$, ± 1, ± 2, ± 3;
$l = 4$, $m = 0$, ± 1, ± 2, ± 3, ± 4 b) 50

43–7   a) 4.58×10^{-48} kg·m^2 b) 0; 0.0151 eV;
0.0454 eV   c) 27.3 μm, 1.10×10^{13} Hz

43–9 2.16×10^3 kg·m^{-3}

43–11   a) 1770 nm;   infrared   b) 1126 nm;
infrared

43–13   489 N·m^{-1} b) 5.70×10^{-20} J; 0.356 eV
c) 3490 nm; infrared

43–15   a) 1.11 nm   b) 1.80 nm

43–17 3.65×10^4 K

43–19   a) 0.129 nm   b) 8, 7, 6, 5, 4   c) 484 μm
d) 118 μm, 134 μm, 157 μm, 188 μm, 234 μm

CHAPTER 44

44–1   a) 3.64×10^{-12} J, 22.8 MeV   b) 3.64×10^{-12} J,
22.8 MeV   c) 3.31×10^7 m·s^{-1}

44–3   a) 1 proton, 1 neutron   b) 28 protons, 28
neutrons   c) 79 protons, 118 neutrons

44–5   2.23 MeV

44–7   b) 0.410 MeV

44–9   a) 4.88×10^{-18} s^{-1} b) 3.00×10^3 kg
c) 1.24×10^4 alpha-particles

44–11   a) 2.76 decays·s^{-1} b) 7.45×10^{-11} Ci

44–13   a) 7.15 MeV   b) 2.31 MeV

44–15   10.1 MeV is absorbed

44–17   a) 1.30 T   b) 4.19 MeV, 2.01×10^7 m·s^{-1}

44–19   67.5 MeV, 1.63×10^{22} Hz, 1.84×10^{-5} nm

44–21   a) $Q = +e$, $B = 1$, $S = -1$, $C = 0$
b) $Q = +e$, $B = 0$, $S = 1$, $C = 1$
c) $Q = 0$, $B = -1$, $S = 0$, $C = 0$
d) $Q = +e$, $B = 0$, $S = 0$, $C = 1$

44–23 1.80×10^{19} J

44–25 1.67×10^4 yr

44–27   0.782 MeV

44–29   a) 37 MeV   b) Proton: 13%, π^-: 87%

44–31   a) 16.3 MeV   b) Not energetically allowed
c) Violate conservation of strangeness

44–33   a) $N = (\alpha/\lambda)(1 - e^{-\lambda t})$ b) 2.73×10^4 counts·s^{-1},
2.42×10^5 counts·s^{-1}, 5.00×10^5 counts·s^{-1},
7.50×10^5 counts·s^{-1}, 8.75×10^5 counts·s^{-1},
9.93×10^5 counts·s^{-1} c) 2.16×10^9 atoms
d) 1.0×10^6 counts·s^{-1}

INDEX